To convert from	To	Multiply by
Mass Flow Rate		
lb_m/h	kg/s	1.26×10^{-4}
lb_m/s	kg/s	0.4536
Mass Flux, Mass Velocity		
$lb_m/h \cdot ft^2$	$kg/s \cdot m^2$	1.356×10^{-3}
Power		
$ft \cdot lb_f/h$	$W = J/s = N \cdot m/s$	3.766×10^{-4}
$ft \cdot lb_f/s$	$W = J/s = N \cdot m/s$	1.356
hp	$W = J/s = N \cdot m/s$	745.7
Btu (IT)/h	$W = J/s = N \cdot m/s$	0.2931
Pressure		
lb_f/ft^2	$Pa = N/m^2$	47.88
lb_f/in^2	$Pa = N/m^2$	6895
atm	$Pa = N/m^2$	1.013×10^5
bar	$Pa = N/m^2$	1×10^5
torr = mmHg	$Pa = N/m^2$	133.3
in. Hg	$Pa = N/m^2$	3386
in. H_2O	$Pa = N/m^2$	249.1
Specific Heat		
Btu (IT)/$lb_m \cdot °F$	$J/kg \cdot K = N \cdot m/kg \cdot K$	4187
cal/g $\cdot °C$	$J/kg \cdot K = N \cdot m/kg \cdot K$	4187
Surface Tension		
lb_f/ft	N/m	14.59
dyne/cm	N/m	0.001
erg/cm^2	N/m	0.001
Thermal Conductivity		
Btu (I/T) $\cdot ft/h \cdot ft^2 \cdot °F$	$W/m \cdot K = J/s \cdot m \cdot K$	1.731
cal (IT) $\cdot cm/s \cdot cm^2 \cdot °C$	$W/m \cdot K = J/s \cdot m \cdot K$	418.7
Velocity		
ft/h	m/s	8.467×10^{-5}
ft/s	m/s	0.3048
Viscosity		
$lb_m/ft \cdot s$	$kg/m \cdot s$	1.488
$lb_m/ft \cdot h$	$kg/m \cdot s$	4.134×10^{-4}
cP	$kg/m \cdot s$	0.001
Volume		
ft^3	m^3	0.02832
L	m^3	1×10^{-3}
gal (US)	m^3	3.785×10^{-3}

SEPARATION PROCESS PRINCIPLES
Chemical and Biochemical Operations

THIRD EDITION

J. D. Seader

Department of Chemical Engineering
University of Utah

Ernest J. Henley

Department of Chemical Engineering
University of Houston

D. Keith Roper

Ralph E. Martin Department of Chemical Engineering
University of Arkansas

WILEY

John Wiley & Sons, Inc.

Vice President and Executive Publisher: Don Fowley
Acquisitions Editor: Jennifer Welter
Developmental Editor: Debra Matteson
Editorial Assistant: Alexandra Spicehandler
Marketing Manager: Christopher Ruel
Senior Production Manager: Janis Soo
Assistant Production Editor: Annabelle Ang-Bok
Designer: RDC Publishing Group Sdn Bhd

This book was set in 10/12 Times Roman by Thomson Digital and printed and bound by Courier Westford. The cover was printed by Courier Westford.

This book is printed on acid free paper.

Founded in 1807, John Wiley & Sons, Inc. has been a valued source of knowledge and understanding for more than 200 years, helping people around the world meet their needs and fulfill their aspirations. Our company is built on a foundation of principles that include responsibility to the communities we serve and where we live and work. In 2008, we launched a Corporate Citizenship Initiative, a global effort to address the environmental, social, economic, and ethical challenges we face in our business. Among the issues we are addressing are carbon impact, paper specifications and procurement, ethical conduct within our business and among our vendors, and community and charitable support. For more information, please visit our website: www.wiley.com/go/citizenship.

Library of Congress Cataloging-in-Publication Data

Seader, J. D.
    Separation process principles : chemical and biochemical operations / J. D. Seader, Ernest J. Henley, D. Keith Roper.—3rd ed.
        p. cm.
    Includes bibliographical references and index.
    ISBN 978-0-470-48183-7 (hardback)
  1. Separation (Technology)–Textbooks. I. Henley, Ernest J. II. Roper, D. Keith. III. Title.
    TP156.S45S364 2010
    660^1.2842—dc22   2010028565

Printed in the United States of America

10 9 8 7 6 5 4 3 2 1

About the Authors

J. D. Seader is Professor Emeritus of Chemical Engineering at the University of Utah. He received B.S. and M.S. degrees from the University of California at Berkeley and a Ph.D. from the University of Wisconsin-Madison. From 1952 to 1959, he worked for Chevron Research, where he designed petroleum and petrochemical processes, and supervised engineering research, including the development of one of the first process simulation programs and the first widely used vapor-liquid equilibrium correlation. From 1959 to 1965, he supervised rocket engine research for the Rocketdyne Division of North American Aviation on all of the engines that took man to the moon. He served as a Professor of Chemical Engineering at the University of Utah for 37 years. He has authored or coauthored 112 technical articles, 9 books, and 4 patents, and also coauthored the section on distillation in the 6th and 7th editions of *Perry's Chemical Engineers' Handbook*. He was a founding member and trustee of CACHE for 33 years, serving as Executive Officer from 1980 to 1984. From 1975 to 1978, he served as Chairman of the Chemical Engineering Department at the University of Utah. For 12 years he served as an Associate Editor of the journal, *Industrial and Engineering Chemistry Research*. He served as a Director of AIChE from 1983 to 1985. In 1983, he presented the 35th Annual Institute Lecture of AIChE; in 1988 he received the Computing in Chemical Engineering Award of the CAST Division of AIChE; in 2004 he received the CACHE Award for Excellence in Chemical Engineering Education from the ASEE; and in 2004 he was a co-recipient, with Professor Warren D. Seider, of the Warren K. Lewis Award for Chemical Engineering Education of the AIChE. In 2008, as part of the AIChE Centennial Celebration, he was named one of 30 authors of groundbreaking chemical engineering books.

Ernest J. Henley is Professor of Chemical Engineering at the University of Houston. He received his B.S. degree from the University of Delaware and his Dr. Eng. Sci. from Columbia University, where he served as a professor from 1953 to 1959. He also has held professorships at the Stevens Institute of Technology, the University of Brazil, Stanford University, Cambridge University, and the City University of New York. He has authored or coauthored 72 technical articles and 12 books, the most recent one being *Probabilistic Risk Management for Scientists and Engineers*. For 17 years, he was a trustee of CACHE, serving as President from 1975 to 1976 and directing the efforts that produced the seven-volume *Computer Programs for Chemical Engineering Education* and the five-volume *AIChE Modular Instruction*. An active consultant, he holds nine patents, and served on the Board of Directors of Maxxim Medical, Inc., Procedyne, Inc., Lasermedics, Inc., and Nanodyne, Inc. In 1998 he received the McGraw-Hill Company Award for "Outstanding Personal Achievement in Chemical Engineering," and in 2002, he received the CACHE Award of the ASEE for "recognition of his contribution to the use of computers in chemical engineering education." He is President of the Henley Foundation.

D. Keith Roper is the Charles W. Oxford Professor of Emerging Technologies in the Ralph E. Martin Department of Chemical Engineering and the Assistant Director of the Microelectronics-Photonics Graduate Program at the University of Arkansas. He received a B.S. degree (magna cum laude) from Brigham Young University in 1989 and a Ph.D. from the University of Wisconsin-Madison in 1994. From 1994 to 2001, he conducted research and development on recombinant proteins, microbial and viral vaccines, and DNA plasmid and viral gene vectors at Merck & Co. He developed processes for cell culture, fermentation, biorecovery, and analysis of polysaccharide, protein, DNA, and adenoviral-vectored antigens at Merck & Co. (West Point, PA); extraction of photodynamic cancer therapeutics at Frontier Scientific, Inc. (Logan, UT); and virus-binding methods for Millipore Corp (Billerica, MA). He holds adjunct appointments in Chemical Engineering and Materials Science and Engineering at the University of Utah. He has authored or coauthored more than 30 technical articles, one U.S. patent, and six U.S. patent applications. He was instrumental in developing one viral and three bacterial vaccine products, six process documents, and multiple bioprocess equipment designs. He holds memberships in Tau Beta Pi, ACS, ASEE, AIChE, and AVS. His current area of interest is interactions between electromagnetism and matter that produce surface waves for sensing, spectroscopy, microscopy, and imaging of chemical, biological, and physical systems at nano scales. These surface waves generate important resonant phenomena in biosensing, diagnostics and therapeutics, as well as in designs for alternative energy, optoelectronics, and micro-electromechanical systems.

Preface to the Third Edition

Separation Process Principles was first published in 1998 to provide a comprehensive treatment of the major separation operations in the chemical industry. Both equilibrium-stage and mass-transfer models were covered. Included also were chapters on thermodynamic and mass-transfer theory for separation operations. In the second edition, published in 2006, the separation operations of ultrafiltration, microfiltration, leaching, crystallization, desublimation, evaporation, drying of solids, and simulated moving beds for adsorption were added. This third edition recognizes the growing interest of chemical engineers in the biochemical industry, and is renamed *Separation Process Principles—Chemical and Biochemical Operations*.

In 2009, the National Research Council (NRC), at the request of the National Institutes of Health (NIH), National Science Foundation (NSF), and the Department of Energy (DOE), released a report calling on the United States to launch a new multiagency, multiyear, multidisciplinary initiative to capitalize on the extraordinary advances being made in the biological fields that could significantly help solve world problems in the energy, environmental, and health areas. To help provide instruction in the important bioseparations area, we have added a third author, D. Keith Roper, who has extensive industrial and academic experience in this area.

NEW TO THIS EDITION

Bioseparations are corollaries to many chemical engineering separations. Accordingly, the material on bioseparations has been added as new sections or chapters as follows:

- Chapter 1: An introduction to bioseparations, including a description of a typical bioseparation process to illustrate its unique features.
- Chapter 2: Thermodynamic activity of biological species in aqueous solutions, including discussions of pH, ionization, ionic strength, buffers, biocolloids, hydrophobic interactions, and biomolecular reactions.
- Chapter 3: Molecular mass transfer in terms of driving forces in addition to concentration that are important in bioseparations, particularly for charged biological components. These driving forces are based on the Maxwell-Stefan equations.
- Chapter 8: Extraction of bioproducts, including solvent selection for organic-aqueous extraction, aqueous two-phase extraction, and bioextractions, particularly in Karr columns and Podbielniak centrifuges.

- Chapter 14: Microfiltration is now included in Section 3 on transport, while ultrafiltration is covered in a new section on membranes in bioprocessing.
- Chapter 15: A revision of previous Sections 15.3 and 15.4 into three sections, with emphasis in new Sections 15.3 and 15.6 on bioseparations involving adsorption and chromatography. A new section on electrophoresis for separating charged particles such as nucleic acids and proteins is added.
- Chapter 17: Bioproduct crystallization.
- Chapter 18: Drying of bioproducts.
- Chapter 19: Mechanical Phase Separations. Because of the importance of phase separations in chemical and biochemical processes, we have also added this new chapter on mechanical phase separations covering settling, filtration, and centrifugation, including mechanical separations in biotechnology and cell lysis.

Other features new to this edition are:

- Study questions at the end of each chapter to help the reader determine if important points of the chapter are understood.
- Boxes around important fundamental equations.
- Shading of examples so they can be easily found.
- Answers to selected exercises at the back of the book.
- Increased clarity of exposition: This third edition has been completely rewritten to enhance clarity. Sixty pages were eliminated from the second edition to make room for biomaterial and updates.
- More examples, exercises, and references: The second edition contained 214 examples, 649 homework exercises, and 839 references. This third edition contains 272 examples, 719 homework exercises, and more than 1,100 references.

SOFTWARE

Throughout the book, reference is made to a number of software products. The solution to many of the examples is facilitated by the use of spreadsheets with a Solver tool, Mathematica, MathCad, or Polymath. It is particularly important that students be able to use such programs for solving nonlinear equations. They are all described at websites on the Internet. Frequent reference is also made to the use of process simulators, such as

ASPEN PLUS, ASPEN HYSYS.Plant, BATCHPLUS, CHEMCAD, PRO/II, SUPERPRO DESIGNER, and UNI-SIM. Not only are these simulators useful for designing separation equipment, but they also provide extensive physical property databases, with methods for computing thermodynamic properties of mixtures. Hopefully, those studying separations have access to such programs. Tutorials on the use of ASPEN PLUS and ASPEN HYSYS. Plant for making separation and thermodynamic-property calculations are provided in the Wiley multimedia guide, "Using Process Simulators in Chemical Engineering, 3rd Edition" by D. R. Lewin (see www.wiley.com/college/lewin).

TOPICAL ORGANIZATION

This edition is divided into five parts. Part 1 consists of five chapters that present fundamental concepts applicable to all subsequent chapters. Chapter 1 introduces operations used to separate chemical and biochemical mixtures in industrial applications. Chapter 2 reviews organic and aqueous solution thermodynamics as applied to separation problems. Chapter 3 covers basic principles of diffusion and mass transfer for rate-based models. Use of phase equilibrium and mass-balance equations for single equilibrium-stage models is presented in Chapter 4, while Chapter 5 treats cascades of equilibrium stages and hybrid separation systems.

The next three parts of the book are organized according to separation method. Part 2, consisting of Chapters 6 to 13, describes separations achieved by phase addition or creation. Chapters 6 through 8 cover absorption and stripping of dilute solutions, binary distillation, and ternary liquid–liquid extraction, with emphasis on graphical methods. Chapters 9 to 11 present computer-based methods widely used in process simulation programs for multicomponent, equilibrium-based models of vapor–liquid and liquid–liquid separations. Chapter 12 treats multicomponent, rate-based models, while Chapter 13 focuses on binary and multicomponent batch distillation.

Part 3, consisting of Chapters 14 and 15, treats separations using barriers and solid agents. These have found increasing applications in industrial and laboratory operations, and are particularly important in bioseparations. Chapter 14 covers rate-based models for membrane separations, while Chapter 15 describes equilibrium-based and rate-based models of adsorption, ion exchange, and chromatography, which use solid or solid-like sorbents, and electrophoresis.

Separations involving a solid phase that undergoes a change in chemical composition are covered in Part 4, which consists of Chapters 16 to 18. Chapter 16 treats selective leaching of material from a solid into a liquid solvent. Crystallization from a liquid and desublimation from a vapor are discussed in Chapter 17, which also includes evaporation. Chapter 18 is concerned with the drying of solids and includes a section on psychrometry.

Part 5 consists of Chapter 19, which covers the mechanical separation of phases for chemical and biochemical processes by settling, filtration, centrifugation, and cell lysis.

Chapters 6, 7, 8, 14, 15, 16, 17, 18, and 19 begin with a detailed description of an industrial application to familiarize the student with industrial equipment and practices. Where appropriate, theory is accompanied by appropriate historical content. These descriptions need not be presented in class, but may be read by students for orientation. In some cases, they are best understood after the chapter is completed.

HELPFUL WEBSITES

Throughout the book, websites that present useful, supplemental material are cited. Students and instructors are encouraged to use search engines, such as Google or Bing, to locate additional information on old or new developments. Consider two examples: (1) McCabe–Thiele diagrams, which were presented 80 years ago and are covered in Chapter 7; (2) bioseparations. A Bing search on the former lists more than 1,000 websites, and a Bing search on the latter lists 40,000 English websites.

Some of the terms used in the bioseparation sections of the book may not be familiar. When this is the case, a Google search may find a definition of the term. Alternatively, the "Glossary of Science Terms" on this book's website or the "Glossary of Biological Terms" at the website: www.phschool.com/science/biology_place/glossary/a.html may be consulted.

Other websites that have proven useful to our students include:

(1) www.chemspy.com—Finds terms, definitions, synonyms, acronyms, and abbreviations; and provides links to tutorials and the latest news in biotechnology, the chemical industry, chemistry, and the oil and gas industry. It also assists in finding safety information, scientific publications, and worldwide patents.

(2) webbook.nist.gov/chemistry—Contains thermochemical data for more than 7,000 compounds and thermophysical data for 75 fluids.

(3) www. ddbst.com—Provides information on the comprehensive Dortmund Data Bank (DDB) of thermodynamic properties.

(4) www.chemistry.about.com/od/chemicalengineerin1/index.htm—Includes articles and links to many websites concerning topics in chemical engineering.

(5) www.matche.com—Provides capital cost data for many types of chemical processing

(6) www.howstuffworks.com—Provides sources of easy-to-understand explanations of how thousands of things work.

RESOURCES FOR INSTRUCTORS

Resources for instructors may be found at the website: *www. wiley.com/college/seader*. Included are:

(1) Solutions Manual, prepared by the authors, giving detailed solutions to all homework exercises in a tutorial format.

(2) Errata to all printings of the book

(3) A copy of a Preliminary Examination used by one of the authors to test the preparedness of students for a course in separations, equilibrium-stage operations, and mass transfer. This closed-book, 50-minute examination, which has been given on the second day of the course, consists of 10 problems on topics studied by students in prerequisite courses on fundamental principles of chemical engineering. Students must retake the examination until all 10 problems are solved correctly.

(4) Image gallery of figures and tables in jpeg format, appropriate for inclusion in lecture slides.

These resources are password-protected, and are available only to instructors who adopt the text. Visit the instructor section of the book website at www.wiley.com/college/seader to register for a password.

RESOURCES FOR STUDENTS

Resources for students are also available at the website: *www.wiley.com/college/seader*. Included are:

(1) A discussion of problem-solving techniques

(2) Suggestions for completing homework exercises

(3) Glossary of Science Terms

(4) Errata to various printings of the book

SUGGESTED COURSE OUTLINES

We feel that our depth of coverage is one of the most important assets of this book. It permits instructors to design a course that matches their interests and convictions as to what is timely and important. At the same time, the student is provided with a resource on separation operations not covered in the course, but which may be of value to the student later. Undergraduate instruction on separation processes is generally incorporated in the chemical engineering curriculum following courses on fundamental principles of thermodynamics, fluid mechanics, and heat transfer. These courses are prerequisites for this book. Courses that cover separation processes may be titled: Separations or Unit Operations, Equilibrium-Stage Operations, Mass Transfer and Rate-Based Operations, or Bioseparations.

This book contains sufficient material to be used in courses described by any of the above four titles. The Chapters to be covered depend on the number of semester credit hours. It should be noted that Chapters 1, 2, 3, 8, 14, 15, 17, 18, and 19 contain substantial material relevant to bioseparations, mainly in later sections of each chapter. Instructors who choose not to cover bioseparations may omit those sections. However, they are encouraged to at least assign their students Section 1.9, which provides a basic awareness of biochemical separation processes and how they differ from chemical separation processes. Suggested chapters for several treatments of separation processes at the undergraduate level are:

SEPARATIONS OR UNIT OPERATIONS:

3 Credit Hours: Chapters 1, 3, 4, 5, 6, 7, 8, (14, 15, or 17)

4 Credit Hours: Chapters 1, 3, 4, 5, 6, 7, 8, 9, 14, 15, 17

5 Credit Hours: Chapters 1, 3, 4, 5, 6, 7, 8, 9, 10, 13, 14, 15, 16, 17, 18, 19

EQUILIBRIUM-STAGE OPERATIONS:

3 Credit Hours: Chapters 1, 4, 5, 6, 7, 8, 9, 10

4 Credit Hours: Chapters 1, 4, 5, 6, 7, 8, 9, 10, 11, 13

MASS TRANSFER AND RATE-BASED OPERATIONS:

3 Credit Hours: Chapters 1, 3, 6, 7, 8, 12, 14, 15

4 Credit Hours: Chapters 1, 3, 6, 7, 8, 12, 14, 15, 16, 17, 18

BIOSEPARATIONS:

3 Credit Hours: Chapter 1, Sections 1.9, 2.9, Chapters 3, 4, Chapter 8 including Section 8.6, Chapters 14, 15, 17, 18, 19

Note that Chapter 2 is not included in any of the above course outlines because solution thermodynamics is a prerequisite for all separation courses. In particular, students who have studied thermodynamics from "Chemical, Biochemical, and Engineering Thermodynamics" by S.I. Sandler, "Physical and Chemical Equilibrium for Chemical Engineers" by N. de Nevers, or "Engineering and Chemical Thermodynamics" by M.D. Koretsky will be well prepared for a course in separations. An exception is Section 2.9 for a course in Bioseparations. Chapter 2 does serve as a review of the important aspects of solution thermodynamics and has proved to be a valuable and popular reference in previous editions of this book.

Students who have completed a course of study in mass transfer using "Transport Phenomena" by R.B. Bird, W.E. Stewart, and E.N. Lightfoot will not need Chapter 3. Students who have studied from "Fundamentals of Momentum, Heat, and Mass Transfer" by J.R. Welty, C.E. Wicks, R.E. Wilson, and G.L. Rorrer will not need Chapter 3, except for Section 3.8 if driving forces for mass transfer other than concentration need to be studied. Like Chapter 2, Chapter 3 can serve as a valuable reference.

Although Chapter 4 is included in some of the outlines, much of the material may be omitted if single equilibrium-stage calculations are adequately covered in sophomore courses on mass and energy balances, using books like "Elementary Principles of Chemical Processes" by R.M. Felder and R.W. Rousseau or "Basic Principles and Calculations in Chemical Engineering" by D.M. Himmelblau and J.B. Riggs.

Considerable material is presented in Chapters 6, 7, and 8 on well-established graphical methods for equilibrium-stage calculations. Instructors who are well familiar with process simulators may wish to pass quickly through these chapters and emphasize the algorithmic methods used in process simulators, as discussed in Chapters 9 to 13. However, as reported by P.M. Mathias in the December 2009 issue of *Chemical Engineering Progress*, the visual approach of graphical methods continues to provide the best teaching tool for developing insight and understanding of equilibrium-stage operations.

As a further guide, particularly for those instructors teaching an undergraduate course on separations for the first time or using this book for the first time, we have designated in the Table of Contents, with the following symbols, whether a section (§) in a chapter is:

* Important for a basic understanding of separations and therefore recommended for presentation in class, unless already covered in a previous course.

O Optional because the material is descriptive, is covered in a previous course, or can be read outside of class with little or no discussion in class.

• Advanced material, which may not be suitable for an undergraduate course unless students are familiar with a process simulator and have access to it.

B A topic in bioseparations.

A number of chapters in this book are also suitable for a graduate course in separations. The following is a suggested course outline for a graduate course:

GRADUATE COURSE ON SEPARATIONS

2–3 Credit Hours: Chapters 10, 11, 12, 13, 14, 15, 17

ACKNOWLEDGMENTS

The following instructors provided valuable comments and suggestions in the preparation of the first two editions of this book:

Richard G. Akins, Kansas State University
Paul Bienkowski, University of Tennessee
C. P. Chen, University of Alabama in Huntsville
William A. Heenan, Texas A&M University–Kingsville
Richard L. Long, New Mexico State University
Jerry Meldon, Tufts University
William L. Conger, Virginia Polytechnic Institute and State University
Kenneth Cox, Rice University
R. Bruce Eldridge, University of Texas at Austin
Rafiqul Gani, Institut for Kemiteknik
Ram B. Gupta, Auburn University
Shamsuddin Ilias, North Carolina A&T State University
Kenneth R. Jolls, Iowa State University of Science and Technology
Alan M. Lane, University of Alabama
John Oscarson, Brigham Young University
Timothy D. Placek, Tufts University
Randel M. Price, Christian Brothers University
Michael E. Prudich, Ohio University
Daniel E. Rosner, Yale University
Ralph Schefflan, Stevens Institute of Technology
Ross Taylor, Clarkson University
Vincent Van Brunt, University of South Carolina

The preparation of this third edition was greatly aided by the following group of reviewers, who provided many excellent suggestions for improving added material, particularly that on bioseparations. We are very grateful to the following Professors:

Robert Beitle, University of Arkansas
Rafael Chavez-Contreras, University of Wisconsin-Madison
Theresa Good, University of Maryland, Baltimore County
Ram B. Gupta, Auburn University
Brian G. Lefebvre, Rowan University
Joerg Lahann, University of Michigan
Sankar Nair, Georgia Institute of Technology
Amyn S. Teja, Georgia Institute of Technology
W. Vincent Wilding, Brigham Young University

Paul Barringer of Barringer Associates provided valuable guidance for Chapter 19. Lauren Read of the University of Utah provided valuable perspectives on some of the new material from a student's perspective.

J. D. Seader
Ernest J. Henley
D. Keith Roper

Brief Contents

Contents

---

* Suitable for an UG course
° Optional
• Advanced
^{B} Bioseparations

Nomenclature

All symbols are defined in the text when they are first used. Symbols that appear infrequently are not listed here.

Latin Capital and Lowercase Letters

A	area; absorption factor $= L/KV$; Hamaker constant
A_M	membrane surface area
a	activity; interfacial area per unit volume; molecular radius
a_v	surface area per unit volume
B	bottoms flow rate
B^0	rate of nucleation per unit volume of solution
b	molar availability function $= h - T_0 s$; component flow rate in bottoms
C	general composition variable such as concentration, mass fraction, mole fraction, or volume fraction; number of components; rate of production of crystals
C_D	drag coefficient
C_F	entrainment flooding factor
C_P	specific heat at constant pressure
$C_{P_V}^o$	ideal-gas heat capacity at constant pressure
c	molar concentration; speed of light
$c*$	liquid concentration in equilibrium with gas at its bulk partial pressure
c'	concentration in liquid adjacent to a membrane surface
$\bar{c}_b$	volume averaged stationary phase solute concentration in (15-149)
c_d	diluent volume per solvent volume in (17-89)
c_f	bulk fluid phase solute concentration in (15-48)
c_m	metastable limiting solubility of crystals
c_o	speed of light in a vacuum
c_p	solute concentration on solid pore surfaces of stationary phase in (15-48)
c_s	humid heat; normal solubility of crystals; solute concentration on solid pore surfaces of stationary phase in (15-48); solute saturation concentration on the solubility curve in (17-82)
c^s	concentration of crystallization-promoting additive in (17-101)
c_t	total molar concentration
Δc_{limit}	limiting supersaturation
$D, \mathcal{D}$	diffusivity; distillate flow rate; diameter

D_{ij}'	multicomponent mass diffusivity
D_B	bubble diameter
D_E	eddy-diffusion coefficient
D_{eff}	effective diffusivity
D_i	impeller diameter
D_{ij}	mutual diffusion coefficient of i in j
D_K	Knudsen diffusivity
D_L	longitudinal eddy diffusivity
$\bar{D}_N$	arithmetic-mean diameter
D_P	particle diameter
$\bar{D}_p$	average of apertures of two successive screen sizes
D_S	surface diffusivity
D_s	shear-induced hydrodynamic diffusivity in (14-124)
$\bar{D}_S$	surface (Sauter) mean diameter
D_T	tower or vessel diameter
$\bar{D}_V$	volume-mean diameter
$\bar{D}_W$	mass-mean diameter
d	component flow rate in distillate
d_e	equivalent drop diameter; pore diameter
d_H	hydraulic diameter $= 4r_H$
d_i	driving force for molecular mass transfer
d_m	molecule diameter
d_p	droplet or particle diameter; pore diameter
d_{vs}	Sauter mean diameter
E	activation energy; extraction factor; amount or flow rate of extract; turbulent-diffusion coefficient; voltage; evaporation rate; convective axial-dispersion coefficient
E^0	standard electrical potential
E_b	radiant energy emitted by a blackbody
E_{MD}	fractional Murphree dispersed-phase efficiency
E_{MV}	fractional Murphree vapor efficiency
E_{OV}	fractional Murphree vapor-point efficiency
E_o	fractional overall stage (tray) efficiency
ΔE^{vap}	molar internal energy of vaporization
e	entrainment rate; charge on an electron
$F, \mathfrak{I}$	Faraday's contant $= 96{,}490$ coulomb/g-equivalent; feed flow rate; force
F_d	drag force
f	pure-component fugacity; Fanning friction factor; function; component flow rate in feed

G	Gibbs free energy; mass velocity; rate of growth of crystal size
g	molar Gibbs free energy; acceleration due to gravity
g_c	universal constant $= 32.174 \ \text{lb}_m \cdot \text{ft/lb}_f \cdot \text{s}^2$
H	Henry's law constant; height or length; enthalpy; height of theoretical chromatographic plate
ΔH_{ads}	heat of adsorption
ΔH_{cond}	heat of condensation
ΔH_{crys}	heat of crystallization
ΔH_{dil}	heat of dilution
$\Delta H_{\text{sol}}^{\text{sat}}$	integral heat of solution at saturation
$\Delta H_{\text{sol}}^{\infty}$	heat of solution at infinite dilution
ΔH^{vap}	molar enthalpy of vaporization
H_G	height of a transfer unit for the gas phase $= l_T/N_G$
H_L	height of a transfer unit for the liquid phase $= l_T/N_L$
H_{OG}	height of an overall transfer unit based on the gas phase $= l_T/N_{OG}$
H_{OL}	height of an overall transfer unit based on the liquid phase $= l_T/N_{OL}$
$\mathcal{H}$	humidity
$\mathcal{H}'_m$	molal humidity
$\mathcal{H}_P$	percentage humidity
$\mathcal{H}_R$	relative humidity
$\mathcal{H}_S$	saturation humidity
$\mathcal{H}_W$	saturation humidity at temperature T_w
HETP	height equivalent to a theoretical plate
HETS	height equivalent to a theoretical stage (same as HETP)
HTU	height of a transfer unit
h	plate height/particle diameter in Figure 15.20
I	electrical current; ionic strength
i	current density
J_i	molar flux of i by ordinary molecular diffusion relative to the molar-average velocity of the mixture
j_D	Chilton–Colburn j-factor for mass transfer $\equiv N_{\text{St}_M}(N_{\text{Sc}})^{2/3}$
j_H	Chilton–Colburn j-factor for heat transfer $\equiv N_{\text{St}}(N_{\text{Pr}})^{2/3}$
j_M	Chilton–Colburn j-factor for momentum transfer $\equiv f/2$
j_i	mass flux of i by ordinary molecular diffusion relative to the mass-average velocity of the mixture
K	equilibrium ratio for vapor–liquid equilibria; overall mass-transfer coefficient
K_a	acid ionization constant
K_D	equilibrium ratio for liquid–liquid equilibria; distribution or partition ratio; equilibrium dissociation constant for biochemical receptor-ligand binding
K'_D	equilibrium ratio in mole- or mass-ratio compositions for liquid–liquid equilibria; equilibrium dissociation constant
K_e	equilibrium constant
K_G	overall mass-transfer coefficient based on the gas phase with a partial-pressure driving force
K_L	overall mass-transfer coefficient based on the liquid phase with a concentration-driving force
K_w	water dissociation constant
K_X	overall mass-transfer coefficient based on the liquid phase with a mole ratio driving force
K_x	overall mass-transfer coefficient based on the liquid phase with a mole fraction driving force
K_Y	overall mass-transfer coefficient based on the gas phase with a mole ratio driving force
K_y	overall mass-transfer coefficient based on the gas phase with a mole-fraction driving force
K_r	restrictive factor for diffusion in a pore
k	thermal conductivity; mass-transfer coefficient in the absence of the bulk-flow effect
k'	mass-transfer coefficient that takes into account the bulk-flow effect
k_A	forward (association) rate coefficient
k_B	Boltzmann constant
k_c	mass-transfer coefficient based on a concentration, c, driving force
$k_{c,tot}$	overall mass-transfer coefficient in linear driving approximation in (15-58)
k_D	reverse (dissociation) rate coefficient
k_i	mass-transfer coefficient for integration into crystal lattice
$k_{i,j}$	mass transport coefficient between species i and j
k_p	mass-transfer coefficient for the gas phase based on a partial pressure, p, driving force
k_T	thermal diffusion factor
k_x	mass-transfer coefficient for the liquid phase based on a mole-fraction driving force
k_y	mass-transfer coefficient for the gas phase based on a mole-fraction driving force
$\bar{L}$	liquid molar flow rate in stripping section
L	liquid; length; height; liquid flow rate; crystal size; biochemical ligand
L'	solute-free liquid molar flow rate; liquid molar flow rate in an intermediate section of a column
L_B	length of adsorption bed
L_e	entry length
L_p	hydraulic membrane permeability
L_{pd}	predominant crystal size
L_S	liquid molar flow rate of sidestream

LES	length of equilibrium (spent) section of adsorption bed	N_{Re}	Reynolds number $= du\rho/\mu =$ inertial force/ viscous force ($d =$ characteristic length)
LUB	length of unused bed in adsorption	N_{Sc}	Schmidt number $= \mu/\rho D =$ momentum diffusivity/mass diffusivity
l_M	membrane thickness	N_{Sh}	Sherwood number $= dk_c/D =$ concentration gradient at wall or interface/concentration gra-
l_T	packed height		dient across fluid ($d =$ characteristic length)
M	molecular weight	N_{St}	Stanton number for heat transfer $= h/GC_P$
M_i	moles of i in batch still	N_{St_M}	Stanton number for mass transfer $= k_c\rho/G$
M_T	mass of crystals per unit volume of magma	NTU	number of transfer units
M_t	total mass	N_t	number of equilibrium (theoretical) stages
m	slope of equilibrium curve; mass flow rate; mass; molality	N_{We}	Weber number $=$ inertial force/surface force
		$\mathcal{N}$	number of moles
m_c	mass of crystals per unit volume of mother liquor; mass in filter cake	n	molar flow rate; moles; crystal population density distribution function in (17-90)
$\bar{m}_i$	molality of i in solution	P	pressure; power; electrical power
m_s	mass of solid on a dry basis; solids flow rate	P_c	critical pressure
m_v	mass evaporated; rate of evaporation	P_i	molecular volume of component i/molecular volume of solvent
MTZ	length of mass-transfer zone in adsorption bed	P_M	permeability
N	number of phases; number of moles; molar flux $= n/A$; number of equilibrium (theoreti-	$\bar{P}_M$	permeance
	cal, perfect) stages; rate of rotation; number of	P_r	reduced pressure, P/P_c
	transfer units; number of crystals/unit volume	P^s	vapor pressure
	in (17-82)	p	partial pressure
N_A	Avogadro's number $= 6.022 \times 10^{23}$ molecules/mol	$p*$	partial pressure in equilibrium with liquid at its bulk concentration
N_a	number of actual trays	pH	$= -\log(a_{H+})$
N_{Bi}	Biot number for heat transfer	pI	isoelectric point (pH at which net charge is
N_{Bi_M}	Biot number for mass transfer		zero)
N_D	number of degrees of freedom	pK_a	$= -\log(K_a)$
N_{Eo}	Eotvos number	Q	rate of heat transfer; volume of liquid;
N_{Fo}	Fourier number for heat transfer $= \alpha t/a^2 =$ dimensionless time		volumetric flow rate
		Q_C	rate of heat transfer from condenser
N_{Fo_M}	Fourier number for mass transfer $= Dt/a^2 =$ dimensionless time	Q_L	volumetric liquid flow rate
		Q_{ML}	volumetric flow rate of mother liquor
N_{Fr}	Froude number $=$ inertial force/gravitational force	Q_R	rate of heat transfer to reboiler
N_G	number of gas-phase transfer units	q	heat flux; loading or concentration of adsorb-
N_L	number of liquid-phase transfer units		ate on adsorbent; feed condition in distillation
N_{Le}	Lewis number $= N_{Sc}/N_{Pr}$		defined as the ratio of increase in liquid molar
N_{Lu}	Luikov number $= 1/N_{Le}$		flow rate across feed stage to molar feed rate;
N_{min}	mininum number of stages for specified split		charge
N_{Nu}	Nusselt number $= dh/k =$ temperature gradi-	R	universal gas constant; raffinate flow rate;
	ent at wall or interface/temperature gradient		resolution; characteristic membrane resist-
	across fluid ($d =$ characteristic length)		ance; membrane rejection coefficient,
N_{OG}	number of overall gas-phase transfer units		retention coefficient, or solute reflection
N_{OL}	number of overall liquid-phase transfer units		coefficient; chromatographic resolution
N_{Pe}	Peclet number for heat transfer $= N_{Re}N_{Pr} =$ convective transport to molecular transfer	R_i	membrane rejection factor for solute i
		R_{min}	minimum reflux ratio for specified split
N_{Pe_M}	Peclet number for mass transfer $= N_{Re}N_{Sc} =$ convective transport to molecular transfer	R_p	particle radius
		r	radius; ratio of permeate to feed pressure for a
N_{Po}	Power number		membrane; distance in direction of diffusion;
N_{Pr}	Prandtl number $= C_P\mu/k =$ momentum diffusivity/thermal diffusivity		reaction rate; molar rate of mass transfer per

	unit volume of packed bed; separation distance between atoms, colloids, etc.
r_c	radius at reaction interface
r_H	hydraulic radius = flow cross section/wetted perimeter
S	entropy; solubility; cross-sectional area for flow; solvent flow rate; mass of adsorbent; stripping factor = KV/L; surface area; Svedberg unit, a unit of centrifugation; solute sieving coefficient in (14-109); Siemen (a unit of measured conductivity equal to a reciprocal ohm)
S_o	partial solubility
S_T	total solubility
s	molar entropy; relative supersaturation; sedimentation coefficient; square root of chromatographic variance in (15-56)
s_p	particle external surface area
T	temperature
T_c	critical temperature
T_0	datum temperature for enthalpy; reference temperature; infinite source or sink temperature
T_r	reduced temperature = T/T_c
T_s	source or sink temperature
T_v	moisture-evaporation temperature
t	time; residence time
$\bar{t}$	average residence time
t_{res}	residence time
U	overall heat-transfer coefficient; liquid side-stream molar flow rate; internal energy; fluid mass flowrate in steady counterflow in (15-71)
u	velocity; interstitial velocity
$\bar{u}$	bulk-average velocity; flow-average velocity
u_L	superficial liquid velocity
u_{mf}	minimum fluidization velocity
u_s	superficial velocity after (15-149)
u_t	average axial feed velocity in (14-122)
V	vapor; volume; vapor flow rate
V'	vapor molar flow rate in an intermediate section of a column; solute-free molar vapor rate
V_B	boilup ratio
V_V	volume of a vessel
$\bar{V}$	vapor molar flow rate in stripping section
$\bar{V}_i$	partial molar volume of species i
$\hat{V}_i$	partial specific volume of species i
V_{max}	maximum cumulative volumetric capacity of a dead-end filter
v	molar volume; velocity; component flow rate in vapor
$\bar{v}$	average molecule velocity
v_i	species velocity relative to stationary coordinates

v_{i_D}	species diffusion velocity relative to the molar-average velocity of the mixture
v_c	critical molar volume
v_H	humid volume
v_M	molar-average velocity of a mixture
v_r	reduced molar volume, v/v_c
v_0	superficial velocity
W	rate of work; moles of liquid in a batch still; moisture content on a wet basis; vapor sidestream molar flow rate; mass of dry filter cake/filter area
W_D	potential energy of interaction due to London dispersion forces
W_{min}	minimum work of separation
WES	weight of equilibrium (spent) section of adsorption bed
WUB	weight of unused adsorption bed
W_s	rate of shaft work
w	mass fraction
X	mole or mass ratio; mass ratio of soluble material to solvent in underflow; moisture content on a dry basis
X^*	equilibrium-moisture content on a dry basis
X_B	bound-moisture content on a dry basis
X_c	critical free-moisture content on a dry basis
X_T	total-moisture content on a dry basis
X_i	mass of solute per volume of solid
x	mole fraction in liquid phase; mass fraction in raffinate; mass fraction in underflow; mass fraction of particles; ion concentration
x'	normalized mole fraction = $x_i / \sum_{j=1}^{N} x_j$
Y	mole or mass ratio; mass ratio of soluble material to solvent in overflow
y	mole fraction in vapor phase; mass fraction in extract; mass fraction in overflow
Z	compressibility factor = Pv/RT; height
z	mole fraction in any phase; overall mole fraction in combined phases; distance; overall mole fraction in feed; charge; ionic charge

Greek Letters

α	thermal diffusivity, $k/\rho C_P$; relative volatility; average specific filter cake resistance; solute partition factor between bulk fluid and stationary phases in (15-51)
α^*	ideal separation factor for a membrane
α_{ij}	relative volatility of component i with respect to component j for vapor–liquid equilibria; parameter in NRTL equation
α_T	thermal diffusion factor
β_{ij}	relative selectivity of component i with respect to component j for liquid–liquid

	equilibria; phenomenological coefficients in the Maxwell–Stefan equations
Γ	concentration-polarization factor; counterflow solute extraction ratio between solid and fluid phases in (15-70)
γ	specific heat ratio; activity coefficient; shear rate
γ_w	fluid shear at membrane surface in (14-123)
Δ	change (final – initial)
δ	solubility parameter
δ_{ij}	Kronecker delta
$\delta_{i,j}$	fractional difference in migration velocities between species i and j in (15-60)
$\delta_{i,m}$	friction between species i and its surroundings (matrix)
ϵ	dielectric constant; permittivity
ϵ_b	bed porosity (external void fraction)
ϵ_D	eddy diffusivity for diffusion (mass transfer)
ϵ_H	eddy diffusivity for heat transfer
ϵ_M	eddy diffusivity for momentum transfer
ϵ_p	particle porosity (internal void fraction)
ϵ_p^*	inclusion porosity for a particular solute
ζ	zeta potential
ζ_{ij}	frictional coefficient between species i and j
η	fractional efficiency in (14-130)
κ	Debye–Hückel constant; $1/\kappa$ = Debye length
λ	mV/L; radiation wavelength
λ_+, λ_-	limiting ionic conductances of cation and anion, respectively
λ_{ij}	energy of interaction in Wilson equation
μ	chemical potential or partial molar Gibbs free energy; viscosity
μ_o	magnetic constant
ν	momentum diffusivity (kinematic viscosity), μ/ρ; wave frequency; stoichiometric coefficient; electromagnetic frequency
π	osmotic pressure
ρ	mass density
ρ_b	bulk density
ρ_p	particle density
σ	surface tension; interfacial tension; Stefan–Boltzmann constant = $5.671 \times 10^{-8} \, \text{W/m}^2 \cdot \text{K}^4$
σ_T	Soret coefficient
σ_I	interfacial tension
τ	tortuosity; shear stress
τ_w	shear stress at wall
Φ	volume fraction; statistical cumulative distribution function in (15-73)
φ	electrostatic potential
ϕ	pure-species fugacity coefficient; volume fraction
ϕ_s	particle sphericity

Ψ	electrostatic potential
Ψ_E	interaction energy
ψ	sphericity
ω	acentric factor; mass fraction; angular velocity; fraction of solute in moving fluid phase in adsorptive beds

Subscripts

A	solute
a, ads	adsorption
avg	average
B	bottoms
b	bulk conditions; buoyancy
bubble	bubble-point condition
C	condenser; carrier; continuous phase
c	critical; convection; constant-rate period; cake
cum	cumulative
D	distillate, dispersed phase; displacement
d, db	dry bulb
des	desorption
dew	dew-point condition
ds	dry solid
E	enriching (absorption) section
e	effective; element
eff	effective
F	feed
f	flooding; feed; falling-rate period
G	gas phase
GM	geometric mean of two values, A and B = square root of A times B
g	gravity; gel
gi	gas in
go	gas out
H, h	heat transfer
I, I	interface condition
i	particular species or component
in	entering
irr	irreversible
j	stage number; particular species or component
k	particular separator; key component
L	liquid phase; leaching stage
LM	log mean of two values, A and B = (A – B)/ln (A/B)
LP	low pressure
M	mass transfer; mixing-point condition; mixture
m	mixture; maximum; membrane; filter medium
max	maximum
min	minimum
N	stage
n	stage

O	overall
$o, 0$	reference condition; initial condition
out	leaving
OV	overhead vapor
P	permeate
R	reboiler; rectification section; retentate
r	reduced; reference component; radiation
res	residence time
S	solid; stripping section; sidestream; solvent; stage; salt
SC	steady counterflow
s	source or sink; surface condition; solute; saturation
T	total
t	turbulent contribution
V	vapor
w	wet solid–gas interface
w, wb	wet bulb
ws	wet solid
X	exhausting (stripping) section
x, y, z	directions
0	surroundings; initial
∞	infinite dilution; pinch-point zone

Superscripts

a	α-amino base
c	α-carboxylic acid
E	excess; extract phase
F	feed
floc	flocculation
ID	ideal mixture
(k)	iteration index
LF	liquid feed
o	pure species; standard state; reference condition
p	particular phase
R	raffinate phase
s	saturation condition
VF	vapor feed
$^-$	partial quantity; average value
∞	infinite dilution
(1), (2)	denotes which liquid phase
I, II	denotes which liquid phase
$*$	at equilibrium

Abbreviations and Acronyms

AFM	atomic force microscopy
Angstrom	1×10^{-10} m
ARD	asymmetric rotating-disk contactor
ATPE	aqueous two-phase extraction
atm	atmosphere
avg	average
B	bioproduct
BET	Brunauer–Emmett–Teller
BOH	undissociated weak base
BP	bubble-point method
BSA	bovine serum albumin
B–W–R	Benedict–Webb–Rubin equation of state
bar	0.9869 atmosphere or 100 kPa
barrer	membrane permeability unit, 1 barrer = 10^{-10} cm^3 (STP)-cm/(cm^2-s cm Hg)
bbl	barrel
Btu	British thermal unit
C	coulomb
C_i	paraffin with i carbon atoms
$C_i^=$	olefin with i carbon atoms
CBER	Center for Biologics Evaluation and Research
CF	concentration factor
CFR	Code of Federal Regulations
cGMP	current good manufacturing practices
CHO	Chinese hamster ovary (cells)
CMC	critical micelle concentration
CP	concentration polarization
CPF	constant-pattern front
C–S	Chao–Seader equation
CSD	crystal-size distribution
°C	degrees Celsius, K-273.2
cal	calorie
cfh	cubic feet per hour
cfm	cubic feet per minute
cfs	cubic feet per second
cm	centimeter
cmHg	pressure in centimeters head of mercury
cP	centipoise
cw	cooling water
Da	daltons (unit of molecular weight)
DCE	dichloroethylene
DEAE	diethylaminoethyl
DEF	dead-end filtration
DLVO	theory of Derajaguin, Landau, Vervey, and Overbeek
DNA	deoxyribonucleic acid
dsDNA	double-stranded DNA
rDNA	recombinant DNA
DOP	diisoctyl phthalate
ED	electrodialysis
EMD	equimolar counter-diffusion
EOS	equation of state
EPA	Environmental Protection Agency
ESA	energy-separating agent

ESS	error sum of squares	lb_m	pound-mass
EDTA	ethylenediaminetetraacetic acid	lbmol	pound-mole
eq	equivalents	ln	logarithm to the base e
°F	degrees Fahrenheit, °R- 459.7	log	logarithm to the base 10
FDA	Food and Drug Administration	M	molar
FUG	Fenske–Underwood–Gilliland	MF	microfiltration
ft	feet	MIBK	methyl isobutyl ketone
GLC-EOS	group-contribution equation of state	MSMPR	mixed-suspension, mixed-product-removal
GLP	good laboratory practices	MSC	molecular-sieve carbon
GP	gas permeation	MSA	mass-separating agent
g	gram	MTZ	mass-transfer zone
gmol	gram-mole	MW	molecular weight; megawatts
gpd	gallons per day	MWCO	molecular-weight cut-off
gph	gallons per hour	m	meter
gpm	gallons per minute	meq	milliequivalents
gps	gallons per second	mg	milligram
H	high boiler	min	minute
HA	undissociated (neutral) species of a weak acid	mm	millimeter
HCP	host-cell proteins	mmHg	pressure in mm head of mercury
HEPA	high-efficiency particulate air	mmol	millimole (0.001 mole)
HHK	heavier than heavy key component	mol	gram-mole
HIV	Human Immunodeficiency Virus	mole	gram-mole
HK	heavy-key component	N	newton; normal
HPTFF	high-performance TFF	NADH	reduced form of nicotinamide adenine dinucleotide
hp	horsepower		
h	hour	NF	nanofiltration
I	intermediate boiler	NLE	nonlinear equation
IMAC	immobilized metal affinity chromatography	NMR	nuclear magnetic resonance
IND	investigational new drug	NRTL	nonrandom, two-liquid theory
in	inches	nbp	normal boiling point
J	Joule	ODE	ordinary differential equation
K	degrees Kelvin	PBS	phosphate-buffered saline
kg	kilogram	PCR	polymerase chain reaction
kmol	kilogram-mole	PEG	polyethylene glycol
L	liter; low boiler	PEO	polyethylene oxide
LES	length of an ideal equilibrium adsorption section	PES	polyethersulfones
		PDE	partial differential equation
LHS	left-hand side of an equation	POD	Podbielniak extractor
LK	light-key component	P–R	Peng–Robinson equation of state
LLE	liquid–liquid equilibrium	PSA	pressure-swing adsorption
LLK	lighter than light key component	PTFE	poly(tetrafluoroethylene)
L–K–P	Lee–Kessler–Plöcker equation of state	PVDF	poly(vinylidene difluoride)
LM	log mean	ppm	parts per million (usually by weight for liquids and by volume or moles for gases)
LMH	liters per square meter per hour		
LRV	log reduction value (in microbial concentration)	psi	pounds force per square inch
		psia	pounds force per square inch absolute
LUB	length of unused sorptive bed	PV	pervaporation
LW	lost work	PVA	polyvinylalcohol
lb	pound	QCMD	quartz crystal microbalance/dissipation
lb_f	pound-force	R	amino acid side chain; biochemical receptor

RDC	rotating-disk contactor
RHS	right-hand side of an equation
R–K	Redlich–Kwong equation of state
R–K–S	Redlich–Kwong–Soave equation of state (same as S–R–K)
RNA	ribonucleic acid
RO	reverse osmosis
RTL	raining-bucket contactor
°R	degrees Rankine
rph	revolutions per hour
rpm	revolutions per minute
rps	revolutions per second
SC	simultaneous-correction method
SDS	sodium docecylsulfate
SEC	size exclusion chromatography
SF	supercritical fluid
SFE	supercritical-fluid extraction
SG	silica gel
S.G.	specific gravity
SOP	standard operating procedure
SPM	stroke speed per minute; scanning probe microscopy
SPR	surface plasmon resonance
SR	stiffness ratio; sum-rates method
S–R–K	Soave–Redlich–Kwong equation of state
STP	standard conditions of temperature and pressure (usually 1 atm and either 0°C or 60°F)
s	second
scf	standard cubic feet
scfd	standard cubic feet per day
scfh	standard cubic feet per hour
scfm	standard cubic feet per minute
stm	steam
TBP	tributyl phosphate
TFF	tangential-flow filtration
TIRF	total internal reflectance fluorescence
TLL	tie-line length
TMP	transmembrane pressure drop
TOMAC	trioctylmethylammonium chloride
TOPO	trioctylphosphine oxide

Tris	tris(hydroxymethyl) amino-methane
TSA	temperature-swing adsorption
UF	ultrafiltration
UMD	unimolecular diffusion
UNIFAC	Functional Group Activity Coefficients
UNIQUAC	universal quasichemical theory
USP	United States Pharmacopeia
UV	ultraviolet
vdW	van der Waals
VF	virus filtration
VOC	volatile organic compound
VPE	vibrating-plate extractor
vs	versus
VSA	vacuum-swing adsorption
WFI	water for injection
WHO	World Health Organization
wt	weight
X	organic solvent extractant
y	year
yr	year
μm	micron = micrometer

Mathematical Symbols

d	differential
∇	del operator
e, exp	exponential function
erf$\{x\}$	error function of $x = \frac{1}{\sqrt{\pi}} \int_0^x \exp(-\eta^2) d\eta$
erfc$\{x\}$	complementary error function of $x = 1 - \mathrm{erf}(x)$
f	function
i	imaginary part of a complex value
ln	natural logarithm
log	logarithm to the base 10
∂	partial differential
$\{\ \}$	delimiters for a function
$\|\ \|$	delimiters for absolute value
Σ	sum
π	product; pi $\cong$ 3.1416

Dimensions and Units

Chemical engineers must be proficient in the use of three systems of units: (1) the International System of Units, SI System (Systeme Internationale d'Unites), which was established in 1960 by the 11th General Conference on Weights and Measures and has been widely adopted; (2) the AE (American Engineering) System, which is based largely upon an English system of units adopted when the Magna Carta was signed in 1215 and is a preferred system in the United States; and (3) the CGS (centimeter-gram-second) System, which was devised in 1790 by the National Assembly of France, and served as the basis for the development of the SI System. A useful index to units and systems of units is given on the website: http://www.sizes.com/units/index.htm

Engineers must deal with dimensions and units to express the dimensions in terms of numerical values. Thus, for 10 gallons of gasoline, the dimension is volume, the unit is gallons, and the value is 10. As detailed in NIST (National Institute of Standards and Technology) Special Publication 811 (2009 edition), which is available at the website: http://www.nist.gov/physlab/pubs/sp811/index.cfm, units are *base* or *derived*.

BASE UNITS

The base units are those that are independent, cannot be subdivided, and are accurately defined. The base units are for dimensions of length, mass, time, temperature, molar amount, electrical current, and luminous intensity, all of which can be measured independently. Derived units are expressed in terms of base units or other derived units and include dimensions of volume, velocity, density, force, and energy. In this book we deal with the first five of the base dimensions. For these, the base units are:

Base	SI Unit	AE Unit	CGS Unit
Length	meter, m	foot, ft	centimeter, cm
Mass	kilogram, kg	pound, lb_m	gram, g
Time	second, s	hour, h	second, s
Temperature	kelvin, K	Fahrenheit, °F	Celsius, °C
Molar amount	gram-mole, mol	pound-mole, lbmol	gram-mole, mol

ATOM AND MOLECULE UNITS

atomic weight = atomic mass unit = the mass of one atom

molecular weight (MW) = molecular mass (M) = formula weight* = formula mass* = the sum of the atomic weights of all atoms in a molecule (*also applies to ions)

1 atomic mass unit (amu or u) = 1 universal mass unit = 1 dalton (Da) = 1/12 of the mass of one atom of carbon-12 = the mass of one proton or one neutron

The units of MW are amu, u, Da, g/mol, kg/kmol, or lb/lbmol (the last three are most convenient when MW appears in a formula).

The number of molecules or ions in one mole = Avogadro's number = 6.022×10^{23}.

DERIVED UNITS

Many derived dimensions and units are used in chemical engineering. Several are listed in the following table:

Derived Dimension	SI Unit	AE Unit	CGS Unit
Area = Length^2	m^2	ft^2	cm^2
Volume = Length^3	m^3	ft^3	cm^3
Mass flow rate = Mass/Time	kg/s	lb$_m$/h	g/s
Molar flow rate = Molar amount/Time	mol/s	lbmol/h	mol/s
Velocity = Length/Time	m/s	ft/h	cm/s
Acceleration = Velocity/Time	m/s^2	ft/h^2	cm/s^2
Force = Mass · Acceleration	newton, N = 1 kg · m/s^2	lb$_f$	dyne = 1 g · cm/s^2
Pressure = Force/Area	pascal, Pa = 1 N/m^2 = 1 kg/m · s^2	lb$_f$/in.^{2}	atm
Energy = Force · Length	joule, J = 1 N · m = 1 kg · m^2/s^2	ft · lb$_f$, Btu	erg = 1 dyne · cm = 1 g · cm^2/s^2, cal
Power = Energy/Time = Work/Time	watt, W = 1 J/s = 1 N · m/s 1 kg · m^2/s^3	hp	erg/s
Density = Mass/Volume	kg/m^3	lb$_m$/ft^3	g/cm^3

OTHER UNITS ACCEPTABLE FOR USE WITH THE SI SYSTEM

A major advantage of the SI System is the consistency of the derived units with the base units. However, some acceptable deviations from this consistency and some other acceptable base units are given in the following table:

Dimension	Base or Derived	Acceptable SI Unit
Time	s	minute (min), hour (h), day (d), year (y)
Volume	m^3	liter (L) = 10^{-3} m^3
Mass	kg	metric ton or tonne (t) = 10^3 kg
Pressure	Pa	bar = 10^5 Pa

PREFIXES

Also acceptable for use with the SI System are decimal multiples and submultiples of SI units formed by prefixes. The following table lists the more commonly used prefixes:

Prefix	Factor	Symbol
tera	10^{12}	T
giga	10^9	G
mega	10^6	M
kilo	10^3	k

deci	10^{-1}	d
centi	10^{-2}	c
milli	10^{-3}	m
micro	10^{-6}	μ
nano	10^{-9}	n
pico	10^{-12}	p

USING THE AE SYSTEM OF UNITS

The AE System is more difficult to use than the SI System because of the units for force, energy, and power. In the AE System, the force unit is the pound-force, lb_f, which is defined to be numerically equal to the pound-mass, lb_m, at sea-level of the earth. Accordingly, Newton's second law of motion is written,

$$F = m\frac{g}{g_c}$$

where F = force in lb_f, m = mass in lb_m, g = acceleration due to gravity in ft/s^2, and, to complete the definition, the constant $g_c = 32.174 \, lb_m \cdot ft/lb_f \cdot s^2$, where 32.174 ft/s^2 is the acceleration due to gravity at sea-level of the earth. The constant g_c is not used with the SI System or the CGS System because the former does not define a kg_f and the CGS System does not use a g_f.

Thus, when using AE units in an equation that includes force and mass, incorporate g_c to adjust the units.

EXAMPLE

A 5.000-pound-mass weight, m, is held at a height, h, of 4.000 feet above sea-level. Calculate its potential energy above sea-level, P.E. = mgh, using each of the three systems of units. Factors for converting units are given on the inside front cover of this book.

SI System:

$$m = 5.000 \, lb_m = 5.000(0.4536) = 2.268 \, kg$$
$$g = 9.807 \, m/s^2$$
$$h = 4.000 \, ft = 4.000(0.3048) = 1.219 \, m$$
$$P.E. = 2.268(9.807)(1.219) = 27.11 \, kg \cdot m^2/s^2 = 27.11 \, J$$

CGS System:

$$m = 5.000 \, lb_m = 5.000(453.6) = 2268 \, g$$
$$g = 980.7 \, cm/s^2$$
$$h = 4.000 \, ft = 4.000(30.48) = 121.9 \, cm$$
$$P.E. = 2268(980.7)(121.9) = 2.711 \times 10^8 \, g \cdot cm^2/s^2$$
$$= 2.711 \times 10^8 \, erg$$

AE System:

$$m = 5.000 \, lb_m$$
$$g = 32.174 \, ft/s^2$$
$$h = 4.000 \, ft$$
$$P.E. = 5.000(32.174)(4.000) = 643.5 \, lb_m \cdot ft^2/s^2$$

However, the accepted unit of energy for the AE System is $ft \cdot lb_f$, which is obtained by dividing by g_c. Therefore, P.E. = 643.5/32.174 = 20.00 ft lb_f.

Another difficulty with the AE System is the differentiation between energy as work and energy as heat. As seen in the above table of derived units, the work unit is $ft \cdot lb_f$, while the heat unit is Btu. A similar situation exists in the CGS System with corresponding units of erg and calorie (cal). In older textbooks, the conversion factor between work and heat is often incorporated into an equation with the symbol J, called Joule's constant or the mechanical equivalent of heat, where

$$J = 778.2 \, ft \, lb_f/Btu = 4.184 \times 10^7 erg/cal$$

Thus, in the previous example, the heat equivalents are

AE System:

$$20.00/778.2 = 0.02570 \, Btu$$

CGS System:

$$2.711 \times 10^8 / 4.184 \times 10^7 = 6.479 \text{ cal}$$

In the SI System, the prefix M, mega, stands for million. However, in the natural gas and petroleum industries of the United States, when using the AE System, M stands for thousand and MM stands for million. Thus, MBtu stands for thousands of Btu, while MM Btu stands for millions of Btu.

It should be noted that the common pressure and power units in use for the AE System are not consistent with the base units. Thus, for pressure, pounds per square inch, psi or $lb_f/in.^2$, is used rather than lb_f/ft^2. For power, hp is used instead of $ft \cdot lb_f/h$, where the conversion factor is

$$1 \text{ hp} = 1.980 \times 10^6 \text{ft} \cdot lb_f/h$$

CONVERSION FACTORS

Physical constants may be found on the inside back cover of this book. Conversion factors are given on the inside front cover. These factors permit direct conversion of AE and CGS values to SI values. The following is an example of such a conversion, together with the reverse conversion.

EXAMPLE

1. Convert 50 psia ($lb_f/in.^2$ absolute) to kPa:
   The conversion factor for $lb_f/in.^2$ to Pa is 6,895, which results in

$$50(6,895) = 345,000 \text{ Pa or } 345 \text{ kPa}$$

2. Convert 250 kPa to atm:
   250 kPa = 250,000 Pa. The conversion factor for atm to Pa is 1.013×10^5. Therefore, dividing by the conversion factor,

$$250,000/1.013 \times 10^5 = 2.47 \text{ atm}$$

Three of the units [gallons (gal), calories (cal), and British thermal unit (Btu)] in the list of conversion factors have two or more definitions. The gallons unit cited here is the U.S. gallon, which is 83.3% of the Imperial gallon. The cal and Btu units used here are international (IT). Also in common use are the thermochemical cal and Btu, which are 99.964% of the international cal and Btu.

FORMAT FOR EXERCISES IN THIS BOOK

In numerical exercises throughout this book, the system of units to be used to solve the problem is stated. Then when given values are substituted into equations, units are not appended to the values. Instead, the conversion of a given value to units in the above tables of base and derived units is done prior to substitution into the equation or carried out directly in the equation, as in the following example.

EXAMPLE

Using conversion factors on the inside front cover of this book, calculate a Reynolds number, $N_{Re} = Dv\rho/\mu$, given $D = 4.0$ ft, $v = 4.5$ ft/s, $\rho = 60$ lb_m/ft^3, and $\mu = 2.0$ cP (i.e., centipoise).

Using the SI System (kg-m-s),

$$N_{Re} = \frac{Dv\rho}{\mu} = \frac{[(4.00)(0.3048)][(4.5)(0.3048)][(60)(16.02)]}{[(2.0)(0.0001)]} = 804,000$$

Using the CGS System (g-cm-s),

$$N_{Re} = \frac{Dv\rho}{\mu} = \frac{[(4.00)(30.48)][(4.5)(30.48)][(60)(0.01602)]}{[(0.02)]} = 804,000$$

Using the AE System (lb_m-ft-h) and converting the viscosity 0.02 cP to lb_m/ft-h,

$$N_{Re} = \frac{Dv\rho}{\mu} = \frac{(4.00)[(4.5)(3600)](60)}{[(0.02)(241.9)]} = 804,000$$

Part One

Fundamental Concepts

Chapters 1–5 present concepts that describe methods for the separation of chemical mixtures by industrial processes, including bioprocesses. Five basic separation techniques are enumerated. The equipment used and the ways of making mass balances and specifying component recovery and product purity are also illustrated.

Separations are limited by thermodynamic equilibrium, while equipment design depends on the rate of mass transfer. Chapter 2 reviews thermodynamic principles and Chapter 3 discusses component mass transfer under stagnant, laminar-flow, and turbulent-flow conditions. Analogies to conductive and convective heat transfer are presented.

Single-stage contacts for equilibrium-limited multiphase separations are treated in Chapters 4 and 5, as are the enhancements afforded by cascades and multistage arrangements. Chapter 5 also shows how degrees-of-freedom analysis is used to set design parameters for equipment. This type of analysis is used in process simulators such as ASPEN PLUS, CHEMCAD, HYSYS, and SuperPro Designer.

Chapter 1

Separation Processes

§1.0 INSTRUCTIONAL OBJECTIVES

After completing this chapter, you should be able to:

- Explain the role of separation operations in the chemical and biochemical industries.
- Explain what constitutes the separation of a mixture and how each of the five basic separation techniques works.
- Calculate component material balances around a separation operation based on specifications of component recovery (split ratios or split fractions) and/or product purity.
- Use the concept of key components and separation factor to measure separation between two key components.
- Understand the concept of sequencing of separation operations, particularly distillation.
- Explain the major differences between chemical and biochemical separation processes.
- Make a selection of separation operations based on factors involving feed and product property differences and characteristics of separation operations.

Separation processes developed by early civilizations include (1) extraction of metals from ores, perfumes from flowers, dyes from plants, and potash from the ashes of burnt plants; (2) evaporation of sea water to obtain salt; (3) refining of rock asphalt; and (4) distilling of liquors. In addition, the human body could not function if it had no kidney—an organ containing membranes that separates water and waste products of metabolism from blood.

Chemists use chromatography, an *analytical separation method*, to determine compositions of complex mixtures, and *preparative separation techniques* to recover chemicals. Chemical engineers design industrial facilities that employ separation methods that may differ considerably from those of laboratory techniques. In the laboratory, chemists separate light-hydrocarbon mixtures by chromatography, while a manufacturing plant will use distillation to separate the same mixture.

This book develops methods for the design of large-scale separation operations, which chemical engineers apply to produce chemical and biochemical products economically. Included are distillation, absorption, liquid–liquid extraction, leaching, drying, and crystallization, as well as newer methods such as adsorption, chromatography, and membrane separation.

Engineers also design small-scale industrial separation systems for manufacture of specialty chemicals by batch processing, recovery of biological solutes, crystal growth of semiconductors, recovery of chemicals from wastes, and development of products such as lung oxygenators and the artificial kidney. The design principles for these smaller-scale operations are also covered in this book. Both large- and small-scale industrial operations are illustrated in examples and homework exercises.

§1.1 INDUSTRIAL CHEMICAL PROCESSES

The chemical and biochemical industries manufacture products that differ in composition from feeds, which are (1) naturally occurring living or nonliving materials, (2) chemical intermediates, (3) chemicals of commerce, or (4) waste products. Especially common are oil refineries (Figure 1.1), which produce a variety of products [1]. The products from, say, 150,000 bbl/day of crude oil depend on the source of the crude and the refinery processes, which include distillation to separate crude into boiling-point fractions or cuts, alkylation to combine small molecules into larger molecules, catalytic reforming to change the structure of hydrocarbon molecules, catalytic cracking to break apart large molecules, hydrocracking to break apart even larger molecules, and processes to convert crude-oil residue to coke and lighter fractions.

A chemical or biochemical plant is operated in a *batchwise*, *continuous*, or *semicontinuous* manner. The operations may be *key operations* unique to chemical engineering because they involve changes in chemical composition, or *auxiliary operations*, which are necessary to the success of the key operations but may be designed by mechanical engineers because the operations do not involve changes in chemical composition. The key operations are (1) chemical reactions and (2) separation of chemical mixtures. The auxiliary operations include phase separation, heat addition or removal (heat exchangers), shaft work (pumps or compressors), mixing or dividing of streams, solids agglomeration, size reduction of

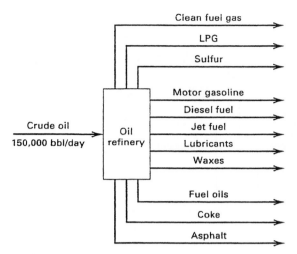

Figure 1.1 Refinery for converting crude oil into a variety of marketable products.

solids, and separation of solids by size. Most of the equipment in biochemical or chemical plants is there to purify raw material, intermediates, and products by the separation techniques discussed in this book.

Block-flow diagrams are used to represent processes. They indicate, by square or rectangular blocks, chemical reaction and separation steps and, by connecting lines, the process streams. More detail is shown in *process-flow diagrams*, which also include auxiliary operations and utilize symbols that depict the type of equipment employed. A block-flow diagram for manufacturing hydrogen chloride gas from chlorine and hydrogen [2] is shown in Figure 1.2. Central to the process is a reactor, where the gas-phase combustion reaction, $H_2 + Cl_2 \rightarrow 2HCl$, occurs. The auxiliary equipment required consists of pumps, compressors, and a heat exchanger to cool the product. No separation operations are necessary because of the complete conversion of chlorine. A slight excess of hydrogen is used, and the product, 99% HCl and small amounts of H_2, N_2, H_2O, CO, and CO_2, requires no purification. Such simple processes that do not require separation operations are very rare, and most chemical and biochemical processes are dominated by separations equipment.

Many industrial chemical processes involve at least one chemical reactor, accompanied by one or more separation trains [3]. An example is the continuous hydration of

ethylene to ethyl alcohol [4]. Central to the process is a reactor packed with catalyst particles, operating at 572 K and 6.72 MPa, in which the reaction, $C_2H_4 + H_2O \rightarrow C_2H_5OH$, occurs. Due to equilibrium limitations, conversion of ethylene is only 5% per pass through the reactor. However, by recovering unreacted ethylene and recycling it to the reactor, near-complete conversion of ethylene feed is achieved.

Recycling is a common element of chemical and biochemical processes. If pure ethylene were available as a feedstock and no side reactions occurred, the simple process in Figure 1.3 could be realized. This process uses a reactor, a partial condenser for ethylene recovery, and distillation to produce aqueous ethyl alcohol of near-azeotropic composition (93 wt%). Unfortunately, impurities in the ethylene feed—and side reactions involving ethylene and feed impurities such as propylene to produce diethyl ether, isopropyl alcohol, acetaldehyde, and other chemicals—combine to increase the complexity of the process, as shown in Figure 1.4. After the hydration reaction, a partial condenser and high-pressure water absorber recover ethylene for recycling. The pressure of the liquid from the bottom of the absorber is reduced, causing partial vaporization. Vapor is then separated from the remaining liquid in the low-pressure flash drum, whose vapor is scrubbed with water to remove alcohol from the vent gas. Crude ethanol containing diethyl ether and acetaldehyde is distilled in the crude-distillation column and catalytically hydrogenated to convert the acetaldehyde to ethanol. Diethyl ether is removed in the light-ends tower and scrubbed with water. The final product is prepared by distillation in the final purification tower, where 93 wt% aqueous ethanol product is withdrawn several trays below the top tray, light ends are concentrated in the so-called pasteurization-tray section above the product-withdrawal tray and recycled to the catalytic-hydrogenation reactor, and wastewater is removed with the bottoms. Besides the equipment shown, additional equipment may be necessary to concentrate the ethylene feed and remove impurities that poison the catalyst. In the development of a new process, experience shows that more separation steps than originally anticipated are usually needed. Ethanol is also produced in biochemical fermentation processes that start with plant matter such as barley, corn, sugar cane, wheat, and wood.

Sometimes a separation operation, such as absorption of SO_2 by limestone slurry, is accompanied by a chemical reaction that facilitates the separation. Reactive distillation is discussed in Chapter 11.

More than 95% of industrial chemical separation operations involve feed mixtures of organic chemicals from coal, natural gas, and petroleum, or effluents from chemical reactors processing these raw materials. However, concern has been expressed in recent years because these fossil feedstocks are not renewable, do not allow sustainable development, and result in emission of atmospheric pollutants such as particulate matter and volatile organic compounds (VOCs). Many of the same organic chemicals can be extracted from renewable biomass, which is synthesized biochemically by cells in agricultural or fermentation reactions and recovered by bioseparations. Biomass components include carbohydrates, oils,

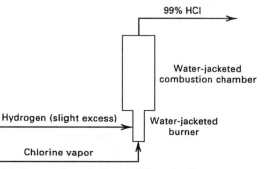

Figure 1.2 Process for anhydrous HCl production.

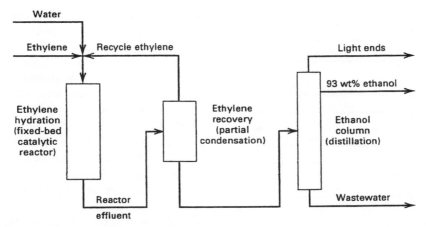

Figure 1.3 Hypothetical process for hydration of ethylene to ethanol.

and proteins, with carbohydrates considered to be the predominant raw materials for future biorefineries, which may replace coal and petroleum refineries if economics prove favorable [18, 19, 20].

Biochemical processes differ significantly from chemical processes. Reactors for the latter normally operate at elevated temperatures and pressures using metallic or chemical catalysts, while reactors for the former typically operate in aqueous solutions at or near the normal, healthy, nonpathologic (i.e., physiologic) state of an organism or bioproduct. Typical physiologic values for the human organism are 37°C, 1 atm, pH of 7.4 (that of arterial blood plasma), general salt content of 137 mM/L of NaCl, 10 mM/L of phosphate, and 2.7 mM/L of KCl. Physiologic conditions vary with the organism, biological component, and/or environment of interest.

Bioreactors make use of catalytic enzymes (products of *in vivo* polypeptide synthesis), and require residence times of hours and days to produce particle-laden aqueous broths that are dilute in bioproducts that usually require an average of six separation steps, using less-mature technology, to produce the final products.

Bioproducts from fermentation reactors may be inside the microorganism (*intracellular*), or in the fermentation broth (*extracellular*). Of major importance is the extracellular case, which can be used to illustrate the difference between chemical separation processes of the type shown in Figures 1.3 and 1.4, which use the more-mature technology of earlier chapters in Part 2 of this book, and bioseparations, which often use the less-mature technology presented in Parts 3, 4, and 5.

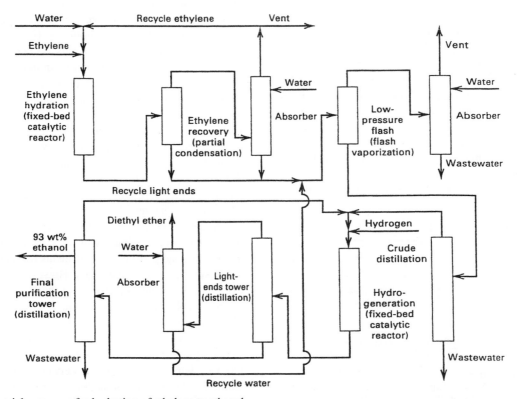

Figure 1.4 Industrial processes for hydration of ethylene to ethanol.

Consider the manufacture of citric acid. Although it can be extracted from lemons and limes, it can also be produced in much larger quantities by submerged, batch *aerobic fermentation* of starch. As in most bioprocesses, a sequence of reactions is required to go from raw material to bioproduct, each reaction catalyzed by an *enzyme* produced in a living cell from its DNA and RNA. In the case of citric acid, the cell is a strain of *Aspergillus niger*, a *eukaryotic fungus*. The first step in the reaction is the hydrolysis of starch at 28°C and 1 atm in an aqueous media to a *substrate* of dextrin using the enzyme α-amylase, in the absence of the fungus. A small quantity of viable fungus cells, called an *inoculum*, is then added to the reactor. As the cells grow and divide, dextrin diffuses from the aqueous media surrounding the cells and crosses the fungus cell wall into the cell cytoplasm. Here a series of interrelated biochemical reactions that comprise a *metabolic pathway* transforms the dextrin into citric acid. Each reaction is catalyzed by a particular enzyme produced within the cell. The first step converts dextrin to glucose using the enzyme, glucoamylase. A series of other enzyme-catalyzed reactions follow, with the final product being citric acid, which, in a process called secretion, moves from the cytoplasm, across the cell wall, and into the aqueous broth media to become an extracellular bioproduct. The total residence time in the fermentation reactor is 6–7 days. The reactor effluent is processed in a series of continuous steps that include vacuum filtration, ultrafiltration, ion exchange, adsorption, crystallization, and drying.

Chemical engineers also design products. One product that involves the separation of chemicals is the espresso coffee machine, which leaches oil from the coffee bean, leaving behind the ingredients responsible for acidity and bitterness. The machine accomplishes this by conducting the leaching operation rapidly in 20–30 seconds with water at high temperature and pressure. The resulting cup of espresso has (1) a topping of creamy foam that traps the extracted chemicals, (2) a fullness of body due to emulsification, and (3) a richness of aroma. Typically, 25% of the coffee bean is extracted, and the espresso contains less caffeine than filtered coffee. Cussler and Moggridge [17] and Seider, Seader, Lewin, and Widagdo [7] discuss other examples of products designed by chemical engineers.

§1.2 BASIC SEPARATION TECHNIQUES

The creation of a mixture of chemical species from the separate species is a spontaneous process that requires no energy input. The inverse process, separation of a chemical mixture into pure components, is not a spontaneous process and thus requires energy. A mixture to be separated may be single or multiphase. If it is multiphase, it is usually advantageous to first separate the phases.

A general separation schematic is shown in Figure 1.5 as a box wherein species and phase separation occur, with arrows to designate feed and product movement. The feed and products may be vapor, liquid, or solid; one or more separation operations may be taking place; and the products differ in composition and may differ in phase. In each separation

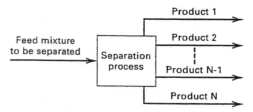

Figure 1.5 General separation process.

operation, the mixture components are induced to move into different, separable spatial locations or phases by any one or more of the five basic separation methods shown in Figure 1.6. However, in most instances, the separation is not perfect, and if the feed contains more than two species, two or more separation operations may be required.

The most common separation technique, shown in Figure 1.6a, *creates* a second phase, immiscible with the feed phase, by energy (heat and/or shaft-work) transfer or by pressure reduction. Common operations of this type are distillation, which involves the transfer of species between vapor and liquid phases, exploiting differences in volatility (e.g., vapor pressure or boiling point) among the species; and crystallization, which exploits differences in melting point. A second technique, shown in Figure 1.6b, *adds* another fluid phase, which selectively absorbs, extracts, or strips certain species from the feed. The most common operations of this type are liquid–liquid extraction, where the feed is liquid and a second, immiscible liquid phase is added; and absorption, where the feed is vapor, and a liquid of low volatility is added. In both cases, species solubilities are significantly different in the added phase. Less common, but of growing importance, is the use of a barrier (shown in Figure 1.6c), usually a polymer membrane, which involves a gas or liquid feed and exploits differences in species permeabilities through the barrier. Also of growing importance are techniques that involve contacting a vapor or liquid feed with a solid agent, as shown in Figure 1.6d. Most commonly, the agent consists of particles that are porous to achieve a high surface area, and differences in species adsorbability are exploited. Finally, external fields (centrifugal, thermal, electrical, flow, etc.), shown in Figure 1.6e, are applied in specialized cases to liquid or gas feeds, with electrophoresis being especially useful for separating proteins by exploiting differences in electric charge and diffusivity.

For the techniques of Figure 1.6, the size of the equipment is determined by rates of mass transfer of each species from one phase or location to another, relative to mass transfer of all species. The driving force and direction of mass transfer is governed by the departure from thermodynamic equilibrium, which involves volatilities, solubilities, etc. Applications of thermodynamics and mass-transfer theory to industrial separations are treated in Chapters 2 and 3. Fluid mechanics and heat transfer play important roles in separation operations, and applicable principles are included in appropriate chapters of this book.

The extent of separation possible depends on the exploitation of differences in molecular, thermodynamic, and transport properties of the species. Properties of importance are:

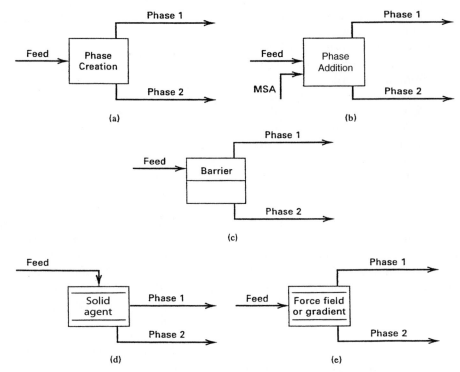

Figure 1.6 Basic separation techniques: (a) separation by phase creation; (b) separation by phase addition; (c) separation by barrier; (d) separation by solid agent; (e) separation by force field or gradient.

1. Molecular properties

Molecular weight	Polarizability
van der Waals volume	Dielectric constant
van der Waals area	Electric charge
Molecular shape (acentric factor)	Radius of gyration
Dipole moment	

2. Thermodynamic and transport properties

Vapor pressure	Adsorptivity
Solubility	Diffusivity

Values of these properties appear in handbooks, reference books, and journals. Many can be estimated using process simulation programs. When property values are not available, they must be estimated or determined experimentally if a successful application of the separation operation is to be achieved.

EXAMPLE 1.1   Feasibility of a separation method.

For each of the following binary mixtures, a separation operation is suggested. Explain why the operation will or will not be successful.

(a) Separation of air into oxygen-rich and nitrogen-rich products by distillation.

(b) Separation of *m*-xylene from *p*-xylene by distillation.

(c) Separation of benzene and cyclohexane by distillation.

(d) Separation of isopropyl alcohol and water by distillation.

(e) Separation of penicillin from water in a fermentation broth by evaporation of the water.

Solution

(a) The normal boiling points of O_2 ($-183°C$) and N_2 ($-195.8°C$) are sufficiently different that they can be separated by distillation, but elevated pressure and cryogenic temperatures are required. At moderate to low production rates, they are usually separated at lower cost by either adsorption or gas permeation through a membrane.

(b) The close normal boiling points of *m*-xylene (139.3°C) and *p*-xylene (138.5°C) make separation by distillation impractical. However, their widely different melting points of $-47.4°C$ for *m*-xylene and 13.2°C for *p*-xylene make crystallization the separation method of choice.

(c) The normal boiling points of benzene (80.1°C) and cyclohexane (80.7°C) preclude a practical separation by distillation. Their melting points are also close, at 5.5°C for benzene and 6.5°C for cyclohexane, making crystallization also impractical. The method of choice is to use distillation in the presence of phenol (normal boiling point of 181.4°C), which reduces the volatility of benzene, allowing nearly pure cyclohexane to be obtained. The other product, a mixture of benzene and phenol, is readily separated in a subsequent distillation operation.

(d) The normal boiling points of isopropyl alcohol (82.3°C) and water (100.0°C) seem to indicate that they could be separated by distillation. However, they cannot be separated in this manner because they form a minimum-boiling azeotrope at 80.4°C and 1 atm of 31.7 mol% water and 68.3 mol% isopropanol. A feasible separation method is to distill the mixture in the presence of benzene, using a two-operation process. The first step produces almost pure isopropyl alcohol and a heterogeneous azeotrope of the three components. The azeotrope is separated into two phases, with the benzene-rich phase recycled to the first step and the water-rich phase sent to a second step, where almost pure water is produced by distillation, with the other product recycled to the first step.

(e) Penicillin has a melting point of 97°C, but decomposes before reaching the normal boiling point. Thus, it would seem that it could be isolated from water by evaporation of the water. However, penicillin and most other antibiotics are heat-sensitive, so a near-ambient temperature must be maintained. Thus, water evaporation would have to take place at impractical, high-vacuum conditions. A practical separation method is liquid–liquid extraction of the penicillin with *n*-butyl acetate or *n*-amyl acetate.

§1.3 SEPARATIONS BY PHASE ADDITION OR CREATION

If the feed is a single-phase solution, a second separable phase must be developed before separation of the species can be achieved. The second phase is created by an *energy-separating agent* (ESA) and/or added as a *mass-separating agent* (MSA). An ESA involves heat transfer or transfer of shaft work to or from the mixture. An example of shaft work is the creation of vapor from a liquid phase by reducing the pressure. An MSA may be partially immiscible with one or more mixture components and frequently is the constituent of highest concentration in the added phase. Alternatively, the MSA may be miscible with a liquid feed mixture, but may selectively alter partitioning of species between liquid and vapor phases. This facilitates a separation when used in conjunction with an ESA, as in extractive distillation.

Disadvantages of using an MSA are (1) need for an additional separator to recover the MSA for recycle, (2) need for MSA makeup, (3) possible MSA product contamination, and (4) more difficult design procedures.

When immiscible fluid phases are contacted, intimate mixing is used to enhance mass-transfer rates so that the maximum degree-of-partitioning of species can be approached rapidly. After phase contact, the phases are separated by employing gravity and/or an enhanced technique such as centrifugal force. Table 1.1 includes the more common separation operations based on interphase mass transfer between two phases, one of which is created by an ESA or added as an MSA. Design procedures have become routine for the operations prefixed by an asterisk (∗) in the first column. Such procedures are incorporated as mathematical models into process simulators.

When the feed mixture includes species that differ widely in volatility, expressed as vapor–liquid equilibrium ratios (*K*-values)—*partial condensation* or *partial vaporization*—Operation (1) in Table 1.1 may be adequate to achieve the desired separation. Two phases are created when a vapor feed is partially condensed by removing heat, and a liquid feed is partially vaporized by adding heat. Alternatively, partial vaporization can be initiated by *flash vaporization*, Operation (2), by reducing the feed pressure with a valve or turbine. In both operations, after partitioning of species has occurred by interphase mass transfer, the resulting vapor phase is enriched with respect to the species that are more easily vaporized, while the liquid phase is enriched with respect to the less-volatile species. The two phases are then separated by gravity.

Often, the degree of separation achieved by a single contact of two phases is inadequate because the volatility differences among species are not sufficiently large. In that case, distillation, Operation (3) in Table 1.1 and the most widely utilized industrial separation method, should be considered. Distillation involves multiple contacts between countercurrently flowing liquid and vapor phases. Each contact, called a *stage*, consists of mixing the phases to promote rapid partitioning of species by mass transfer, followed by phase separation. The contacts are often made on horizontal trays arranged in a column, as shown in the symbol for distillation in Table 1.1. Vapor, flowing up the column, is increasingly enriched with respect to the more-volatile species, and liquid flowing down the column is increasingly enriched with respect to the less-volatile species. Feed to the column enters on a tray somewhere between the top and bottom trays. The portion of the column above the feed entry is the *enriching* or *rectification section*, and that portion below is the *stripping section*. Vapor feed starts up the column; feed liquid starts down. Liquid is required for making contacts with vapor above the feed tray, and vapor is required for making contacts with liquid below the feed tray. Commonly, at the top of the column, vapor is condensed to provide down-flowing liquid called *reflux*. Similarly, liquid at the bottom of the column passes through a reboiler, where it is heated to provide up-flowing vapor called *boilup*.

When the volatility difference between two species to be separated is so small as to necessitate more than about 100 trays, *extractive distillation*, Operation (4), is considered. Here, a miscible MSA, acting as a solvent, increases the volatility difference among species in the feed, thereby reducing the number of trays. Generally, the MSA is the least volatile species and is introduced near the top of the column. Reflux to the top tray minimizes MSA content in the top product. A subsequent operation, usually distillation, is used to recover the MSA for recycling.

If it is difficult to condense the vapor leaving the top of a distillation column, a liquid MSA called an *absorbent* may be fed to the top tray in place of reflux. The resulting operation is called *reboiled absorption*, (5). If the feed is vapor and the stripping section of the column is not needed, the operation is referred to as *absorption*, (6). Absorbers generally do not require an ESA and are frequently conducted at ambient temperature and elevated pressure. Species in the feed vapor dissolve in the absorbent to extents depending on their solubilities.

The inverse of absorption is *stripping*, Operation (7) in Table 1.1, where liquid mixtures are separated, at elevated temperature and ambient pressure, by contacting the feed with a vapor stripping agent. This MSA eliminates the need to reboil the liquid at the bottom of the column, which may be important if the liquid is not thermally stable. If trays are needed above the feed tray to achieve the separation, a *refluxed stripper*, (8), may be employed. If the bottoms

Table 1.1 Separation Operations Based on Phase Creation or Addition

Separation Operation	Symbol[a]	Initial or Feed Phase	Created or Added Phase	Separating Agent(s)	Industrial Example[b]
Partial condensation or vaporization* (1)		Vapor and/or liquid	Liquid or vapor	Heat transfer (ESA)	Recovery of H_2 and N_2 from ammonia by partial condensation and high-pressure phase separation
Flash vaporization* (2)		Liquid	Vapor	Pressure reduction	Recovery of water from sea water
Distillation* (3)		Vapor and/or liquid	Vapor and liquid	Heat transfer (ESA) and sometimes work transfer	Purification of styrene
Extractive distillation* (4)		Vapor and/or liquid	Vapor and liquid	Liquid solvent (MSA) and heat transfer (ESA)	Separation of acetone and methanol
Reboiled absorption* (5)		Vapor and/or liquid	Vapor and liquid	Liquid absorbent (MSA) and heat transfer (ESA)	Removal of ethane and lower molecular weight hydrocarbons for LPG production
Absorption* (6)		Vapor	Liquid	Liquid absorbent (MSA)	Separation of carbon dioxide from combustion products by absorption with aqueous solutions of an ethanolamine
Stripping* (7)		Liquid	Vapor	Stripping vapor (MSA)	Stream stripping of naphtha, kerosene, and gas oil side cuts from crude distillation unit to remove light ends
Refluxed stripping (steam distillation)* (8)		Vapor and/or liquid	Vapor and liquid	Stripping vapor (MSA) and heat transfer (ESA)	Separation of products from delayed coking

8

Separation operation	Initial or feed phase	Developed or added phase	Separating agent(s)	Industrial example
Reboiled stripping* (9)	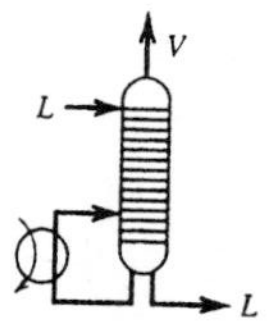 Liquid	Vapor	Heat transfer (ESA)	Recovery of amine absorbent
Azeotropic distillation* (10)	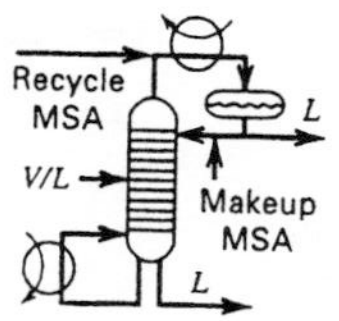 Vapor and/or liquid	Vapor and liquid	Liquid entrainer (MSA) and heat transfer (ESA)	Separation of acetic acid from water using n-butyl acetate as an entrainer to form an azeotrope with water
Liquid–liquid extraction* (11)	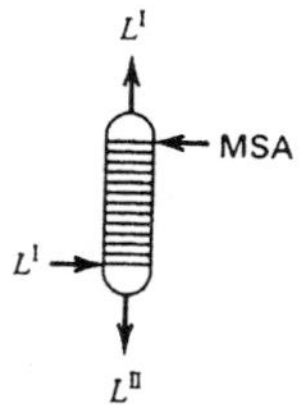 Liquid	Liquid	Liquid solvent (MSA)	Recovery of penicillin from aqueous fermentation medium by methyl isobutyl ketone. Recovery of aromatics
Liquid–liquid extraction (two-solvent)* (12)	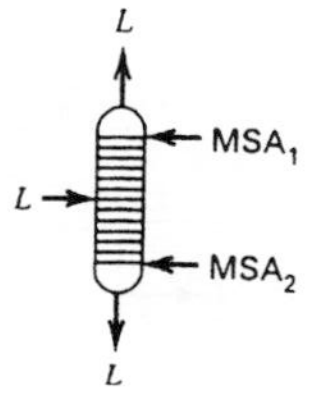 Liquid	Liquid	Two liquid solvents (MSA_1 and MSA_2)	Use of propane and cresylic acid as solvents to separate paraffins from aromatics and naphthenes
Drying (13)	 Liquid and often solid	Vapor	Gas (MSA) and/or heat transfer (ESA)	Removal of water from polyvinylchloride with hot air in a fluid-bed dryer
Evaporation (14)	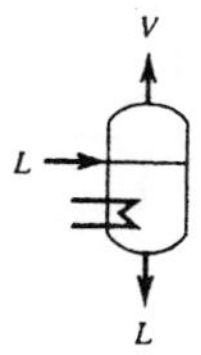 Liquid	Vapor	Heat transfer (ESA)	Evaporation of water from a solution of urea and water
Crystallization (15)	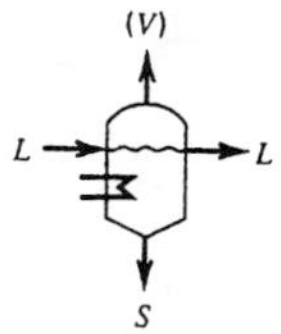 Liquid	Solid (and vapor)	Heat transfer (ESA)	Recovery of a protease inhibitor from an organic solvent. Crystallization of p-xylene from a mixture with m-xylene

(Continued)

9

Table 1.1 (Continued)

Separation Operation	Symbol[a]	Initial or Feed Phase	Created or Added Phase	Separating Agent(s)	Industrial Example[b]
Desublimation (16)		Vapor	Solid	Heat transfer (ESA)	Recovery of phthalic anhydride from non-condensible gas
Leaching (liquid–solid extraction) (17)		Solid	Liquid	Liquid solvent	Extraction of sucrose from sugar beets with hot water
Foam fractionation (18)		Liquid	Gas	Gas bubbles (MSA)	Recovery of detergents from waste solutions

*Design procedures are fairly well accepted.

[a]Trays are shown for columns, but alternatively packing can be used. Multiple feeds and side streams are often used and may be added to the symbol.

[b]Details of examples may be found in *Kirk-Othmer Encyclopedia of Chemical Technology*, 5th ed., John Wiley & Sons, New York (2004–2007).

product from a stripper is thermally stable, it may be reboiled without using an MSA. In that case, the column is a *reboiled stripper*, (9). Additional separation operations may be required to recover MSAs for recycling.

Formation of minimum-boiling azeotropes makes *azeotropic distillation* (10) possible. In the example cited in Table 1.1, the MSA, *n*-butyl acetate, which forms a two-liquid (heterogeneous), minimum-boiling azeotrope with water, is used as an *entrainer* in the separation of acetic acid from water. The azeotrope is taken overhead, condensed, and separated into acetate and water layers. The MSA is recirculated, and the distillate water layer and bottoms acetic acid are the products.

Liquid–liquid extraction, (11) and (12), with one or two solvents, can be used when distillation is impractical, especially when the mixture to be separated is temperature-sensitive. A solvent selectively dissolves only one or a fraction of the components in the feed. In a two-solvent extraction, each has its specific selectivity for the feed components. Several countercurrently arranged stages may be necessary. As with extractive distillation, additional operations are required to recover solvent from the streams leaving the extraction operation. Extraction is widely used for recovery of bioproducts from fermentation broths. If the extraction temperature and pressure are only slightly above the critical point of the solvent, the operation is termed *supercritical-fluid extraction*. In this region, solute solubility in the supercritical fluid can change drastically with small changes in temperature and pressure. Following extraction, the pressure of the solvent-rich product is reduced to release the solvent, which is recycled. For the processing of foodstuffs, the supercritical fluid is an inert substance, with CO_2 preferred because it does not contaminate the product.

Since many chemicals are processed wet but sold as dry solids, a common manufacturing step is *drying*, Operation (13). Although the only requirement is that the vapor pressure of the liquid to be evaporated from the solid be higher than its partial pressure in the gas stream, dryer design and operation represents a complex problem. In addition to the effects of such external conditions as temperature, humidity, air flow, and degree of solid subdivision on drying rate, the effects of internal diffusion conditions, capillary flow, equilibrium moisture content, and heat sensitivity must be considered. Because solid, liquid, and vapor phases coexist in drying, equipment-design procedures are difficult to devise and equipment size may be controlled by heat transfer. A typical dryer design procedure is for the process engineer to send a representative feed sample to one or two reliable dryer manufacturers for pilot-plant tests and to purchase equipment that produces a dried product at the lowest cost. Commercial dryers are discussed in [5] and Chapter 18.

Evaporation, Operation (14), is defined as the transfer of volatile components of a liquid into a gas by heat transfer. Applications include humidification, air conditioning, and concentration of aqueous solutions.

Crystallization, (15), is carried out in some organic, and in almost all inorganic, chemical plants where the desired product is a finely divided solid. Crystallization is a purification step, so the conditions must be such that impurities do not precipitate with the product. In *solution crystallization*, the mixture, which includes a solvent, is cooled and/or the solvent is evaporated. In *melt crystallization*, two or more soluble species are separated by partial freezing. A versatile melt-crystallization technique is *zone melting* or *refining*, which relies on selective distribution of impurities between a liquid and a solid phase. It involves moving a molten zone slowly through an ingot by moving the heater or drawing the ingot past the heater. Single crystals of very high-purity silicon are produced by this method.

Sublimation is the transfer of a species from the solid to the gaseous state without formation of an intermediate liquid phase. Examples are separation of sulfur from impurities, purification of benzoic acid, and freeze-drying of foods. The reverse process, *desublimation*, (16), is practiced in the recovery of phthalic anhydride from gaseous reactor effluent. A common application of sublimation is the use of dry ice as a refrigerant for storing ice cream, vegetables, and other perishables. The sublimed gas, unlike water, does not puddle.

Liquid–solid extraction, *leaching*, (17), is used in the metallurgical, natural product, and food industries. To promote rapid solute diffusion out of the solid and into the liquid solvent, particle size of the solid is usually reduced.

The major difference between solid–liquid and liquid–liquid systems is the difficulty of transporting the solid (often as slurry or a wet cake) from stage to stage. In the pharmaceutical, food, and natural product industries, countercurrent solid transport is provided by complicated mechanical devices.

In adsorptive-bubble separation methods, surface-active material collects at solution interfaces. If the (very thin) surface layer is collected, partial solute removal from the solution is achieved. In ore flotation processes, solid particles migrate through a liquid and attach to rising gas bubbles, thus floating out of solution. In *foam fractionation*, (18), a natural or chelate-induced surface activity causes a solute to migrate to rising bubbles and is thus removed as foam.

The equipment symbols shown in Table 1.1 correspond to the simplest configuration for each operation. More complex versions are frequently desirable. For example, a more complex version of the reboiled absorber, Operation (5) in Table 1.1, is shown in Figure 1.7. It has two feeds, an intercooler, a side stream, and both an interreboiler and a bottoms reboiler. Design procedures must handle such complex equipment. Also, it is possible to conduct chemical reactions simultaneously with separation operations. Siirola [6] describes the evolution of a commercial process for producing methyl acetate by esterification. The process is conducted in a single column in an integrated process that involves three reaction zones and three separation zones.

§1.4 SEPARATIONS BY BARRIERS

Use of microporous and nonporous membranes as semipermeable barriers for selective separations is gaining adherents. Membranes are fabricated mainly from natural fibers and synthetic polymers, but also from ceramics and metals. Membranes are fabricated into flat sheets, tubes, hollow fibers, or spiral-wound sheets, and incorporated into commercial

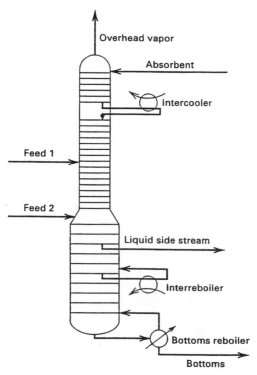

Figure 1.7 Complex reboiled absorber.

modules or cartridges. For microporous membranes, separation is effected by rate of species diffusion through the pores; for nonporous membranes, separation is controlled by differences in solubility in the membrane and rate of species diffusion. The most complex and selective membranes are found in the trillions of cells in the human body.

Table 1.2 lists membrane-separation operations. *Osmosis*, Operation (1), involves transfer, by a concentration gradient, of a solvent through a membrane into a mixture of solute and solvent. The membrane is almost impermeable to the solute. In *reverse osmosis*, (2), transport of solvent in the opposite direction is effected by imposing a pressure, higher than the osmotic pressure, on the feed side. Using a nonporous membrane, reverse osmosis desalts brackish water commercially. *Dialysis*, (3), is the transport by a concentration gradient of small solute molecules, sometimes called crystalloids, through a porous membrane. The molecules unable to pass through the membrane are small, insoluble, nondiffusible particles.

Microporous membranes selectively allow small solute molecules and/or solvents to pass through the membrane, while preventing large dissolved molecules and suspended solids from passing through. *Microfiltration*, (4), refers to the retention of molecules from 0.02 to 10 μm. *Ultrafiltration*, (5), refers to the retention of molecules that range from 1 to

Table 1.2 Separation Operations Based on a Barrier

Separation Operation	Symbol[a]	Initial or Feed Phase	Separating Agent	Industrial Example[b]
Osmosis (1)		Liquid	Nonporous membrane	—
Reverse osmosis[*] (2)		Liquid	Nonporous membrane with pressure gradient	Desalinization of sea water
Dialysis[*] (3)		Liquid	Porous membrane with pressure gradient	Recovery of caustic from hemicellulose
Microfiltration[*] (4)		Liquid	Microporous membrane with pressure gradient	Removal of bacteria from drinking water
Ultrafiltration[*] (5)		Liquid	Microporous membrane with pressure gradient	Separation of whey from cheese
Pervaporation[*] (6)		Liquid	Nonporous membrane with pressure gradient	Separation of azeotropic mixtures
Gas permeation[*] (7)		Vapor	Nonporous membrane with pressure gradient	Hydrogen enrichment
Liquid membrane (8)		Vapor and/or liquid	Liquid membrane with pressure gradient	Removal of hydrogen sulfide

[*]Design procedures are fairly well accepted.

[a]Single units are shown. Multiple units can be cascaded.

[b]Details of examples may be found in *Kirk-Othmer Encyclopedia of Chemical Technology*, 5th ed., John Wiley & Sons, New York (2004–2007).

20 nm. To retain molecules down to 0.1 nm, nonporous membranes can be used in *hyperfiltration.*

To achieve high purities, reverse osmosis requires high pressures. Alternatively, *pervaporation*, (6), wherein the species transported through the nonporous membrane is evaporated, can be used. This method, which is used to separate azeotropic mixtures, uses much lower pressures than reverse osmosis, but the heat of vaporization must be supplied.

Separation of gases by selective *gas permeation*, (7), using a pressure driving force is a process that was first used in the 1940s with porous fluorocarbon barriers to separate $^{235}UF_6$ and $^{238}UF_6$. It required enormous amounts of electric power. Today, centrifuges are used to achieve enrichment more economically. Nonporous polymer membranes are employed to enrich mixtures containing H_2, recover hydrocarbons from gas streams, and produce O_2-enriched air.

Liquid membranes, (8), only a few molecules thick, can be formed from surfactant-containing mixtures at the interface between two fluid phases. With liquid membranes, aromatic/paraffinic hydrocarbons can be separated. Alternatively, a liquid membrane can be formed by imbibing the micropores with liquids doped with additives to facilitate transport of solutes such as CO_2 and H_2S.

§1.5 SEPARATIONS BY SOLID AGENTS

Separations that use solid agents are listed in Table 1.3. The solid, in the form of a granular material or packing, is the adsorbent itself, or it acts as an inert support for a thin layer of adsorbent by selective adsorption or chemical reaction with species in the feed. Adsorption is confined to the surface of the solid adsorbent, unlike absorption, which occurs throughout the absorbent. The active separating agent eventually becomes saturated with solute and must be regenerated or replaced. Such separations are often conducted batchwise or semicontinuously. However, equipment is available to simulate continuous operation.

Adsorption, Operation (1) in Table 1.3, is used to remove species in low concentrations and is followed by desorption to regenerate the adsorbents, which include activated carbon, aluminum oxide, silica gel, and synthetic sodium or calcium aluminosilicate zeolites (molecular sieves). The sieves are crystalline and have pore openings of fixed dimensions, making them very selective. Equipment consists of a cylindrical vessel packed with a bed of solid adsorbent particles through which the gas or liquid flows. Because regeneration is conducted periodically, two or more vessels are used, one desorbing while the other(s) adsorb(s), as indicated in Table 1.3. If the vessel is vertical, gas flow is best employed downward. With upward flow, jiggling can cause particle attrition, pressure-drop increase, and loss of material. However, for liquid mixtures, upward flow achieves better flow distribution. Regeneration occurs by one of four methods: (1) vaporization of the adsorbate with a hot purge gas (*thermal-swing adsorption*), (2) reduction of pressure to vaporize the adsorbate (*pressure-swing adsorption*), (3) inert purge stripping without change in temperature or pressure, and (4) displacement desorption by a fluid containing a more strongly adsorbed species.

Chromatography, Operation (2) in Table 1.3, separates gas or liquid mixtures by passing them through a packed bed. The bed may be solid particles (gas–solid chromatography)

Table 1.3 Separation Operations Based on a Solid Agent

Separation Operation	Symbol[a]	Initial or Feed Phase	Separating Agent	Industrial Example[b]
Adsorption* (1)		Vapor or liquid	Solid adsorbent	Purification of *p*-xylene
Chromatography* (2)		Vapor or liquid	Solid adsorbent or liquid adsorbent on a solid support	Separation and purification of proteins from complex mixtures. Separation of xylene isomers and ethylbenzene
Ion exchange* (3)		Liquid	Resin with ion-active sites	Demineralization of water

*Design procedures are fairly well accepted.

[a]Single units are shown. Multiple units can be cascaded.

[b]Details of examples may be found in *Kirk-Othmer Encyclopedia of Chemical Technology*, 5th ed., John Wiley & Sons, New York (2004–2007).

or a solid–inert support coated with a viscous liquid (gas–liquid chromatography). Because of selective adsorption on the solid surface, or absorption into liquid absorbents followed by desorption, components move through the bed at different rates, thus effecting the separation. In *affinity chromatography*, a macromolecule (a *ligate*) is selectively adsorbed by a *ligand* (e.g., an ammonia molecule in a coordination compound) covalently bonded to a solid-support particle. Ligand–ligate pairs include inhibitors–enzymes, antigens–antibodies, and antibodies–proteins. Chromatography is widely used in bioseparations.

Ion exchange, (3), resembles adsorption in that solid particles are used and regenerated. However, a chemical reaction is involved. In water softening, an organic or inorganic polymer in its sodium form removes calcium ions by a calcium–sodium exchange. After prolonged use, the (spent) polymer, saturated with calcium, is regenerated by contact with a concentrated salt solution.

§1.6 SEPARATIONS BY EXTERNAL FIELD OR GRADIENT

External fields can take advantage of differing degrees of response of molecules and ions to force fields. Table 1.4 lists common techniques and combinations.

Centrifugation, Operation (1) in Table 1.4, establishes a pressure field that separates fluid mixtures according to molecular weight. It is used to separate $^{235}UF_6$ from $^{238}UF_6$, and large polymer molecules according to molecular weight.

If a temperature gradient is applied to a homogeneous solution, concentration gradients are established, and *thermal diffusion*, (2), is induced. This process has been used to enhance separation of isotopes in permeation processes.

Water contains 0.000149 atom fraction of deuterium. When it is decomposed by *electrolysis*, (3), into H_2 and O_2, the deuterium concentration in the hydrogen is lower than it was in the water. Until 1953, this process was the only source of heavy water (D_2O), used to moderate the speed of nuclear reactions. In *electrodialysis*, (4), cation- and anion-permeable membranes carry a fixed charge, thus preventing migration of species of like charge. This phenomenon is applied in seawater desalination. A related process is *electrophoresis*, (5), which exploits the different migration velocities of charged colloidal or suspended species in an electric field. Positively charged species, such as dyes, hydroxide sols, and colloids, migrate to the cathode, while most small, suspended, negatively charged particles go to the anode. By changing from an acidic to a basic condition, migration direction can be changed, particularly for proteins. Electrophoresis is thus a versatile method for separating biochemicals.

Another separation technique for biochemicals and heterogeneous mixtures of micromolecular and colloidal materials is *field-flow fractionation*, (6). An electrical or magnetic field or thermal gradient is established perpendicular to a laminar-flow field. Components of the mixture travel in the flow direction at different velocities, so a separation is achieved. A related device is a small-particle collector where the particles are charged and then collected on oppositely charged metal plates.

§1.7 COMPONENT RECOVERIES AND PRODUCT PURITIES

If no chemical reaction occurs and the process operates in a continuous, steady-state fashion, then for each component i, in a mixture of C components, the molar (or mass) flow rate in the feed, $n_i^{(F)}$, equals the sum of the product molar (or mass) flow rates, $n_i^{(p)}$, for that component in the N product phases, p. Thus, referring to Figure 1.5,

$$n_i^{(F)} = \sum_{p=1}^{N} n_i^{(p)} = n_i^{(1)} + n_i^{(2)} + \cdots + n_i^{(N-1)} + n_i^{(N)} \quad (1\text{-}1)$$

To solve (1-1) for values of $n_i^{(p)}$ from specified values of $n_i^{(F)}$, an additional $N - 1$ independent expressions involving $n_i^{(p)}$ are required. This gives a total of NC equations in NC unknowns. If a single-phase feed containing C components is separated into N products, $C(N - 1)$ additional expressions are needed. If more than one stream is fed to the separation process, $n_i^{(F)}$ is the summation for all feeds.

§1.7.1 Split Fractions and Split Ratios

Chemical plants are designed and operated to meet specifications given as *component recoveries* and *product purities*. In Figure 1.8, the feed is the bottoms product from a reboiled absorber used to deethanize—i.e., remove ethane and lighter components from—a mixture of petroleum refinery gases and liquids. The separation process of choice, shown in Figure 1.8, is a sequence of three multistage distillation columns, where feed components are rank-listed by decreasing volatility, and hydrocarbons heavier (i.e., of greater

Table 1.4 Separation Operations by Applied Field or Gradient

Separation Operation	Initial or Feed Phase	Force Field or Gradient	Industrial Example[a]
Centrifugation (1)	Vapor or liquid	Centrifugal force field	Separation of uranium isotopes
Thermal diffusion (2)	Vapor or liquid	Thermal gradient	Separation of chlorine isotopes
Electrolysis (3)	Liquid	Electrical force field	Concentration of heavy water
Electrodialysis (4)	Liquid	Electrical force field and membrane	Desalinization of sea water
Electrophoresis (5)	Liquid	Electrical force field	Recovery of hemicelluloses
Field-flow fractionation (6)	Liquid	Laminar flow in force field	—

[a]Details of examples may be found in *Kirk-Othmer Encyclopedia of Chemical Technology*, 5th ed., John Wiley & Sons, New York (2004–2007).

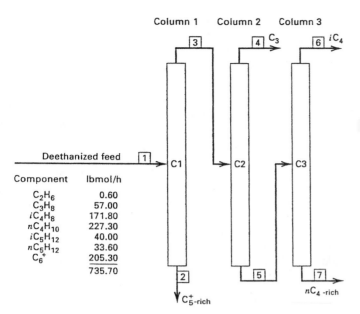

Figure 1.8 Hydrocarbon recovery process.

$$SR_{i,k} = \frac{n_{i,k}^{(1)}}{n_{i,k}^{(2)}} = \frac{SF_{i,k}}{\left(1 - SF_{i,k}\right)} \qquad (1\text{-}3)$$

where $n^{(2)}$ refers to a component flow rate in the second product.

If the process shown in Figure 1.8 operates with the material balance of Table 1.5, the computed split fractions and split ratios are given in Table 1.6. In Table 1.5, it is seen that only two of the products are relatively pure: C_3 overhead from Column C2 and iC_4 overhead from Column C3. Molar purity of C_3 in Column C2 overhead is (54.80/56.00), or 97.86%, while the iC_4 purity is (162.50/175.50), or 92.59% iC_4. The nC_4 bottoms from Column C3 has an nC_4 purity of (215.80/270.00), or 79.93%.

Each column is designed to make a split between two adjacent key components in the feed, whose components are ordered in decreasing volatility. As seen by the horizontal lines in Table 1.6, the key splits are nC_4H_{10}/iC_5H_{12}, C_3H_8/iC_4H_{10}, and iC_4H_{10}/nC_4H_{10} for Columns C1, C2, and C3, respectively. From Table 1.6, we see that splits are sharp (SF > 0.95 for the light key and SF < 0.05 for the heavy key), except for Column C1, where the heavy-key split (iC_5H_{12}) is not sharp and ultimately causes the nC_4-rich bottoms to be impure in nC_4, even though the key-component split in the third column is sharp.

In Table 1.6, for each column we see that SF and SR decrease as volatility decreases, and SF may be a better degree-of-separation indicator than SR because SF is bounded between 0 and 1, while SR can range from 0 to a large value.

Two other measures of success can be applied to each column or to the entire process. One is the *percent recovery* of a designated product. These values are listed in the last column of Table 1.6. The recoveries are high (>95%), except for the pentane isomers. Another measure is *product purity*. Purities were computed for all except the C_5^+-rich product, which is [(11.90 + 16.10 + 205.30)/234.10], or 99.66% pure with respect to pentanes and heavier products. Such a product is a *multicomponent product*, an example of which is gasoline.

Impurity and impurity levels are included in *specifications* for chemicals in commerce. The computed product purity of

molecular weight) than *n*-pentane, and in the hexane (C_6)-to-undecane (C_{11}) range, are lumped together in a C_6^+ fraction. The three distillation columns of Figure 1.8 separate the feed into four products: a C_5^+-rich bottoms, a C_3-rich distillate, an iC_4-rich distillate, and an nC_4-rich bottoms. For each column, feed components are partitioned between the overhead and the bottoms according to a *split fraction* or *split ratio* that depends on (1) the component thermodynamic and transport properties, (2) the number of stages, and (3) the vapor and liquid flows through the column. The *split fraction*, SF, for component *i* in separator *k* is the fraction found in the first product:

$$SF_{i,k} = \frac{n_{i,k}^{(1)}}{n_{i,k}^{(F)}} \qquad (1\text{-}2)$$

where $n^{(1)}$ and $n^{(F)}$ refer to component flow rates in the first product and feed. Alternatively, a *split ratio*, SR, between two products is

Table 1.5 Operating Material Balance for Hydrocarbon Recovery Process

	lbmol/h in Stream						
	1	2	3	4	5	6	7
Component	Feed to C1	C_5^+-rich	Feed to C2	C_3	Feed to C3	iC_4	nC_4-rich
C_2H_6	0.60	0.00	0.60	0.60	0.00	0.00	0.00
C_3H_8	57.00	0.00	57.00	54.80	2.20	2.20	0.00
iC_4H_{10}	171.80	0.10	171.70	0.60	171.10	162.50	8.60
nC_4H_{10}	227.30	0.70	226.60	0.00	226.60	10.80	215.80
iC_5H_{12}	40.00	11.90	28.10	0.00	28.10	0.00	28.10
nC_5H_{12}	33.60	16.10	17.50	0.00	17.50	0.00	17.50
C_6^+	205.30	205.30	0.00	0.00	0.00	0.00	0.00
Total	735.60	234.10	501.50	56.00	445.50	175.50	270.00

Table 1.6 Computed Split Fractions (SF) and Split Ratios (SR) for Hydrocarbon Recovery Process

Component	Column 1 SF	Column 1 SR	Column 2 SF	Column 2 SR	Column 3 SF	Column 3 SR	Overall Percent Recovery
C_2H_6	1.00	Large	1.00	Large	—	—	100
C_3H_8	1.00	Large	0.9614	24.91	1.00	Large	96.14
iC_4H_{10}	0.9994	1,717	0.0035	0.0035	0.9497	18.90	94.59
nC_4H_{10}	0.9969	323.7	0.00	0.00	0.0477	0.0501	94.94
iC_5H_{12}	0.7025	2.361	0.00	0.00	0.00	0.00	29.75
nC_5H_{12}	0.5208	1.087	0.00	0.00	0.00	0.00	47.92
C_6^+	0.00	Small	—	—	—	—	100

the three products for the process in Figure 1.8 is given in Table 1.7, where the values are compared to the specified maximum allowable percentages of impurities set by the govenment or trade associations. The C_5^+ fraction is not included because it is an intermediate. From Table 1.7, it is seen that two products easily meet specifications, while the iC_4 product barely meets its specification.

§1.7.2 Purity and Composition Designations

The product purities in Table 1.7 are given in mol%, a designation usually restricted to gas mixtures for which vol% is equivalent to mol%. Alternatively, *mole fractions* can be used. For liquids, purities are more often specified in wt% or *mass fraction* (ω). To meet environmental regulations, small amounts of impurities in gas, liquid, and solid streams are often specified in *parts of solute per million parts* (ppm) or *parts of solute per billion parts* (ppb), where if a gas, the parts are moles or volumes; if a liquid or solid, the parts are mass or weight. For aqueous solutions, especially those containing acids and bases, common designations for composition are *molarity* (M), or molar concentration in moles of solute per liter of solution (m/L); millimoles per liter (mM/L); *molality* (m) in moles of solute per kilogram of solvent; or *normality* (N) in number of equivalent weights of solute per liter of solution. *Concentrations* (c) in mixtures can be in units of moles or mass per volume (i.e., mol/L, g/L, lbmol/ft^3, and lb/ft^3). For some chemical products, an attribute such as color may be used in place of a purity in terms of composition. For

biochemical processes, a biological *activity* specification is added for bioproducts such as pharmaceuticals, as discussed in §1.9.

§1.7.3 Separation Sequences

The three-column recovery process shown in Figure 1.8 is only one of five alternative sequences of distillation operations that can separate the feed into the four products when each column has a single feed and produces an overhead product and a bottoms product. For example, consider a hydrocarbon feed that consists, in order of decreasing volatility, of propane (C_3), isobutane (iC_4), n-butane (nC_4), isopentane (iC_5), and n-pentane (nC_5). A sequence of distillation columns is to be used to separate the feed into three nearly pure products of C_3, iC_4, and nC_4; and one multicomponent product of iC_5 and nC_5. The five alternative sequences are shown in Figure 1.9.

If only two products are desired, only a single column is required. For three final products, there are two alternative sequences. As the number of final products increases, the number of alternative sequences grows rapidly, as shown in Table 1.8.

Methods for determining the optimal sequence from the possible alternatives are discussed by Seider et al. [7]. For initial screening, the following heuristics are useful and easy to apply, and do not require column design or cost estimation:

Table 1.7 Comparison of Calculated Product Purities with Specifications

Component	mol% in Product Propane Data	Propane Spec	Isobutane Data	Isobutane Spec	Normal Butane Data	Normal Butane Spec
C_2H_6	1.07	5 max	0	—	0	—
C_3H_8	97.86	93 min	1.25	3 max	0	1 max
iC_4H_{10}	1.07	2 min	92.60	92 min	⎰ 83.11	⎰ 80 min
nC_4H_{10}	0	—	6.15	7 max	⎱	⎱
C_5^+	0	—	0	—	16.89	20 max
Total	100.00		100.00		100.00	

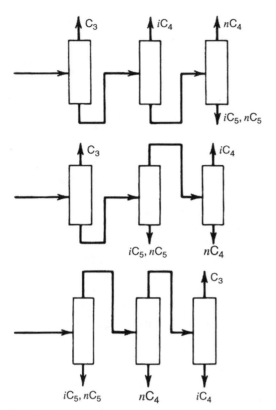

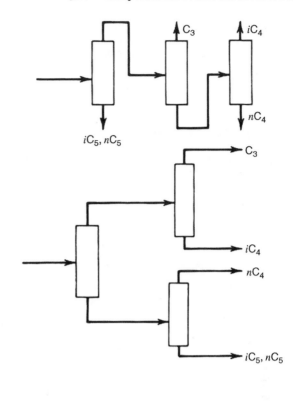

Figure 1.9 Distillation sequences to produce four products.

1. Remove unstable, corrosive, or chemically reactive components early in the sequence.

2. Remove final products one by one as overhead distillates.

3. Remove, early in the sequence, those components of greatest molar percentage in the feed.

4. Make the most difficult separations in the absence of the other components.

5. Leave for later in the sequence those separations that produce final products of the highest purities.

6. Select the sequence that favors near-equimolar amounts of overhead and bottoms in each column.

Unfortunately, these heuristics sometimes conflict with each other, and thus a clear choice is not always possible. Heuristic 1 should always be applied if applicable. The most common industrial sequence is that of Heuristic 2. When energy costs are high, Heuristic 6 is favored. When one of the separations, such as the separation of isomers, is particularly difficult, Heuristic 4 is usually

applied. Seider et al. [7] present more rigorous methods, which do require column design and costing to determine the optimal sequence. They also consider complex sequences that include separators of different types and complexities.

EXAMPLE 1.2   Selection of a separation sequence using heuristics.

A distillation sequence produces the same four final products from the same five components in Figure 1.9. The molar percentages in the feed are C_3 (5.0%), iC_4 (15%), nC_4 (25%), iC_5 (20%), and nC_5 (35%). The most difficult separation by far is that between the isomers, iC_4 and nC_4. Use the heuristics to determine the best sequence(s). All products are to be of high purity.

Solution

Heuristic 1 does not apply. Heuristic 2 favors taking C_3, iC_4, and nC_4 as overheads in Columns 1, 2, and 3, respectively, with the iC_5, nC_5 multicomponent product taken as the bottoms in Column 3, as in Sequence 1 in Figure 1.9. Heuristic 3 favors the removal of the iC_5, nC_5 multicomponent product (55% of the feed) in Column 1, as in Sequences 3 and 4. Heuristic 4 favors the separation of iC_4 from nC_4 in Column 3, as in Sequences 2 and 4. Heuristics 3 and 4 can be combined, with C_3 taken as overhead in Column 2 as in Sequence 4. Heuristic 5 does not apply. Heuristic 6 favors taking the multicomponent product as bottoms in Column 1 (45/55 mole split), nC_4 as bottoms in Column 2 (20/25 mole split), and C_3 as overhead, with iC_4 as bottoms in Column 3 as in Sequence 3. Thus, the heuristics lead to four possible sequences as being most favorable.

Table 1.8   Number of Alternative Distillation Sequences

Number of Final Products	Number of Columns	Number of Alternative Sequences
2	1	1
3	2	2
4	3	5
5	4	14
6	5	42

However, because of the large percentage of the iC_5/nC_5 multicomponent product in the feed, and the difficulty of the separation between iC_4 and nC_4, the best of the four favored sequences is Sequence 4, based on Heuristics 3 and 4.

§1.8 SEPARATION FACTOR

Some separation operations in Table 1.1 are incapable of making a sharp split between key components and can effect the desired recovery of only a single component. Examples are Operations 1, 2, 6, 7, 8, 9, 11, 13, 14, 15, 16, and 17. For these, either a single separation stage is utilized, as in Operations 1, 2, 13, 14, 15, 16, and 17, or the feed enters at one end (not near the middle) of a multistage separator, as in Operations 6, 7, 8, 9, and 11. The split ratio (SR), split fraction (SF), recovery, or purity that can be achieved for the single key component depends on a number of factors. For the simplest case of a single separation stage, these factors include: (1) the relative molar amounts of the two phases leaving the separator and (2) thermodynamic, mass transport, and other component properties. For multistage separators, additional factors are the number of stages and their configurations. The relationships involving these factors are unique to each type of separator, and are discussed in detail in Chapters 5 and 6.

If the feed enters near the middle of the column as in distillation (discussed in Chapter 7), it has both enriching and stripping sections, and it is often possible to achieve a sharp separation between two key components. The enriching section purifies the light key and the stripping section purifies the heavy key. Examples are Operations 3, 4, 5, 10, and 12 in Table 1.1. For these, a measure of the relative degree of separation between two key components, i and j, is the *separation factor* or *power*, SP, defined in terms of the component splits as measured by the compositions of the two products, (1) and (2):

$$\boxed{SP_{i,j} = \frac{C_i^{(1)}/C_i^{(2)}}{C_j^{(1)}/C_j^{(2)}}} \qquad (1\text{-}4)$$

where C is some measure of composition. SP is readily converted to the following forms in terms of split fractions or split ratios:

$$SP_{i,j} = \frac{SR_i}{SR_j} \qquad (1\text{-}5)$$

$$SP_{i,j} = \frac{SF_i/SF_j}{(1 - SF_i)/(1 - SF_j)} \qquad (1\text{-}6)$$

Achievable values of SP depend on the number of stages and the properties of components i and j. In general, components i and j and products 1 and 2 are selected so that $SP_{i,j} > 1.0$. Then, a large value corresponds to a relatively high degree of separation or separation factor, and a small value close to 1.0 corresponds to a low degree of separation factor. For example, if $SP = 10,000$ and $SR_i = 1/SR_j$, then, from (1-5), $SR_i = 100$ and $SR_j = 0.01$, corresponding to a sharp separation. However, if $SP = 9$ and $SR_i = 1/SR_j$, then $SR_j = 3$ and $SR_j = 1/3$, corresponding to a nonsharp separation.

Table 1.9 Key Component Separation Factors for Hydrocarbon Recovery Process

Key-Component Split	Column	Separation Factor, SP
nC_4H_{10}/iC_5H_{12}	C1	137.1
C_3H_{10}/iC_4H_{10}	C2	7103
iC_4H_{10}/nC_4H_{10}	C3	377.6

For the process of Figure 1.8, the values of SP in Table 1.9 are computed from Table 1.5 or 1.6 for the main split in each separator. The SP in Column C1 is small because the split for the heavy key, iC_5H_{12}, is not sharp. The largest SP occurs in Column C2, where the separation is relatively easy because of the large volatility difference. Much more difficult is the butane-isomer split in Column C3, where only a moderately sharp split is achieved.

Component flows and recoveries are easily calculated, while split ratios and purities are more difficult, as shown in the following example.

EXAMPLE 1.3 Using recovery and purity specifications.

A feed, F, of 100 kmol/h of air containing 21 mol% O_2 and 79 mol% N_2 is to be partially separated by a membrane unit according to each of four sets of specifications. Compute the amounts, in kmol/h, and compositions, in mol%, of the two products (retentate, R, and permeate, P). The membrane is more permeable to O_2.

Case 1: 50% recovery of O_2 to the permeate and 87.5% recovery of N_2 to the retentate.

Case 2: 50% recovery of O_2 to the permeate and 50 mol% purity of O_2 in the permeate.

Case 3: 85 mol% purity of N_2 in the retentate and 50 mol% purity of O_2 in the permeate.

Case 4: 85 mol% purity of N_2 in the retentate and a split ratio of O_2 in the permeate to the retentate equal to 1.1.

Solution

The feed is

$$n_{O_2}^{(F)} = 0.21(100) = 21 \text{ kmol/h}$$
$$n_{N_2}^{(F)} = 0.79(100) = 79 \text{ kmol/h}$$

Case 1: Because two recoveries are given:

$$n_{O_2}^{(P)} = 0.50(21) = 10.5 \text{ kmol/h}$$
$$n_{N_2}^{(R)} = 0.875(79) = 69.1 \text{ kmol/h}$$
$$n_{O_2}^{(R)} = 21 - 10.5 = 10.5 \text{ kmol/h}$$
$$n_{N_2}^{(P)} = 79 - 69.1 = 9.9 \text{ kmol/h}$$

Case 2: O_2 recovery is given; the product distribution is:

$$n_{O_2}^{(P)} = 0.50(21) = 10.5 \text{ kmol/h}$$
$$n_{O_2}^{(R)} = 21 - 10.5 = 10.5 \text{ kmol/h}$$

Using the fractional purity of O_2 in the permeate, the total permeate is

$$n^{(P)} = 10.5/0.5 = 21 \, \text{kmol/h}$$

By a total permeate material balance,

$$n_{N_2}^{(P)} = 21 - 10.5 = 10.5 \, \text{kmol/h}$$

By an overall N_2 material balance,

$$n_{N_2}^{(R)} = 79 - 10.5 = 68.5 \, \text{kmol/h}$$

Case 3: Two material-balance equations, one for each component, can be written.
For nitrogen, with a fractional purity of $1.00 - 0.50 = 0.50$ in the permeate,

$$n_{N_2} = 0.85 n^{(R)} + 0.50 n^{(P)} = 79 \, \text{kmol/h} \tag{1}$$

For oxygen, with a fractional purity of $1.00 - 0.85 = 0.15$ in the retentate,

$$n_{O_2} = 0.50 n^{(P)} + 0.15 n^{(R)} = 21 \, \text{kmol/h} \tag{2}$$

Solving (1) and (2) simultaneously for the total products gives

$$n^{(P)} = 17.1 \, \text{kmol/h}, \quad n^{(R)} = 82.9 \, \text{kmol/h}$$

Therefore, the component flow rates are

$$n_{N_2}^{(R)} = 0.85(82.9) = 70.5 \, \text{kmol/h}$$
$$n_{O_2}^{(R)} = 82.9 - 70.5 = 12.4 \, \text{kmol/h}$$
$$n_{O_2}^{(P)} = 0.50(17.1) = 8.6 \, \text{kmol/h}$$
$$n_{N_2}^{(P)} = 17.1 - 8.6 = 8.5 \, \text{kmol/h}$$

Case 4: First compute the O_2 flow rates using the split ratio and an overall O_2 material balance,

$$\frac{n_{O_2}^{(P)}}{n_{O_2}^{(R)}} = 1.1, \quad 21 = n_{O_2}^{(P)} + n_{O_2}^{(R)}$$

Solving these two equations simultaneously gives

$$n_{O_2}^{(R)} = 10 \, \text{kmol/h}, \quad n_{O_2}^{(P)} = 21 - 10 = 11 \, \text{kmol/h}$$

Since the retentate contains 85 mol% N_2 and, therefore, 15 mol% O_2, the flow rates for N_2 are

$$n_{N_2}^{(R)} = \frac{85}{15}(10) = 56.7 \, \text{kmol/h}$$
$$n_{N_2}^{(P)} = 79 - 56.7 = 22.3 \, \text{kmol/h}$$

§1.9 INTRODUCTION TO BIOSEPARATIONS

Bioproducts are products extracted from plants, animals, and microorganisms to sustain life and promote health, support agriculture and chemical enterprises, and diagnose and remedy disease. From the bread, beer, and wine produced by ancient civilizations using fermented yeast, the separation and purification of biological products (bioproducts) have grown in commercial significance to include process-scale recovery of antibiotics from mold, which began in the 1940s, and isolation of recombinant DNA and proteins from transformed bacteria in biotechnology protocols initiated in the 1970s. Bioproducts

used in pharmaceutical, agrichemical, and biotechnology market sectors—excluding commodity foods, beverages, and biofuels—accounted for an estimated $28.2 billion in sales in 2005, with an average annual growth rate of 12% that projects to $50 billion in sales by 2010.

§1.9.1 Bioproducts

To identify features that allow selection and specification of processes to separate bioproducts from other *biological species*[1] of a host cell, it is useful to classify biological species by their complexity and size as *small molecules*, *biopolymers*, and *cellular particulates* (as shown in Column 1 of Table 1.10), and to further categorize each type of species by name in Column 2, according to its biochemistry and function within a biological host in Column 3.

Small molecules include *primary metabolites*, which are synthesized during the primary phase of cell growth by sets of enzyme-catalyzed biochemical reactions referred to as *metabolic pathways*. Energy from organic nutrients fuels these pathways to support cell growth and relatively rapid reproduction. Primary metabolites include organic commodity chemicals, amino acids, mono- and disaccharides, and vitamins. *Secondary metabolites* are small molecules produced in a subsequent stationary phase, in which growth and reproduction slows or stops. Secondary metabolites include more complex molecules such as antibiotics, steroids, phytochemicals, and cytotoxins. Small molecules range in complexity and size from H_2 (2 daltons, Da), produced by cyanobacteria, to vitamin B-12 (1355 Da) or vancomycin antibiotic (1449 Da), whose synthesis originally occurred in bacteria.

Amino acid and monosaccharide metabolites are building blocks for higher-molecular-weight **biopolymers,** from which cells are constituted. Biopolymers provide mechanical strength, chemical inertness, and permeability; and store energy and information. They include *proteins, polysaccharides, nucleic acids,* and *lipids*.

Cellular particulates include cells and cell derivatives such as *extracts* and *hydrolysates* as well as subcellular components.

Proteins, the most abundant biopolymers in cells, are long, linear sequences of 20 naturally occurring amino acids, covalently linked end-to-end by peptide bonds, with molecular weights ranging from 10,000 Da to 100,000 Da. Their structure is often helical, with an overall shape ranging from globular to sheet-like, with loops and folds as determined largely by attraction between oppositely charged groups on the amino acid chain and by hydrogen bonding. Proteins participate in storage, transport, defense, regulation, inhibition, and catalysis. The first products of biotechnology were biocatalytic proteins that initiated or inhibited specific biological cascades [8]. These included hormones, thrombolytic agents, clotting factors, and

---

[1]The term "biological species" as used in this book is not to be confused with the word "species," a taxonomic unit used in biology for the classification of living and fossil organisms, which also includes genus, family, order, class, phylum, kingdom, and domain.

Table 1.10 Products of Bioseparations

Biological Species Classification	Types of Species	Examples
Small Molecules		
Primary metabolites	Gases, organic alcohols, ketones	H_2, CO_2, ethanol (biofuels, beverages), isopropanol, butanol (solvent), acetone
	Organic acids	Acetic acid (vinegar), lactic acid, propionic acid, citric acid, glutamic acid (MSG flavor)
	Amino acids	Lysine, phenylalanine, glycine
	Monosaccharides	Aldehydes: D-glucose, D-ribose; Ketones: D-fructose (in corn syrup)
	Disaccharides	Sucrose, lactose, maltose
	Vitamins	Fat soluble: A, E, and C (ascorbic acid); Water soluble: B, D, niacin, folic acid
Secondary metabolites	Antibiotics	Penicillin, streptomycin, gentamycin
	Steroids	Cholesterol, cortisone, estrogen derivatives
	Hormones	Insulin, human grown
	Phytochemicals	Resveratrol® (anti-aging agent)
	Cytotoxins	Taxol® (anti-cancer)
Biopolymers		
Proteins	Enzymes	Trypsin, ribonuclease, polymerase, cellulase, whey protein, soy protein, industrial enzymes (detergents)
	Hormones	Insulin, growth hormone, cytokines, erythropoietin
	Transport	Hemoglobin, β_1-lipoprotein
	Thrombolysis/clotting	Tissue plasminogen activator, Factor VIII
	Immune agents	α-interferon, interferon β-1a, hepatitis B vaccine
	Antibodies	Herceptin®, Rituxan®, Remicade®, Enbrel®
Polysaccharides		Dextrans (thickeners); alginate, gellan, pullulan (edible films); xanthan (food additive)
Nucleic acids		Gene vectors, antisense oligonucleotides, small interfering RNA, plasmids, ribozymes
Lipids		Glycerol (sweetener), prostaglandins
Virus		Retrovirus, adenovirus, adeno-associated virus (gene vectors), vaccines
Cellular Particulates		
Cells	Eubacteria	*Bacillus thuringensis* (insecticide)
	Eukaryotes	*Saccharomyces cerevisia* (baker's yeast), diatoms, single cell protein (SCP)
	Archae	Methanogens (waste treatment), acidophiles
Cell extracts and hydrolysates		Yeast extract, soy extract, animal tissue extract, soy hydrolysate, whey hydrolysate
Cell components		Inclusion bodies, ribosomes, liposomes, hormone granules

immune agents. Recently, bioproduction of monoclonal antibodies for pharmaceutical applications has grown in significance. Monoclonal antibodies are proteins that bind with high specificity and affinity to particles recognized as foreign to a host organism. Monoclonal antibodies have been introduced to treat breast cancer (Herceptin®), B-cell lymphoma (Rituxan®), and rheumatoid arthritis (Remicade® and Enbrel®).

Carbohydrates are mono- or polysaccharides with the general formula $(CH_2O)_n$, $n \geq 3$, photosynthesized from CO_2. They primarily store energy as cellulose and starch in plants, and as glycogen in animals. *Monosaccharides* ($3 \leq n \leq 9$) are aldehydes or ketones. Condensing two monosaccharides forms a *disaccharide*, like sucrose (α-D-glucose plus β-D-fructose), lactose (β-D-glucose plus β-D-galactose) from milk or whey, or maltose, which is hydrolyzed from germinating cereals like

barley. *Polysaccharides* form by condensing >2 monosaccharides. They include the starches amylase and amylopectin, which are partially hydrolyzed to yield glucose and dextrin, and cellulose, a long, unbranched D-glucose chain that resists enzymatic hydrolysis.

Nucleic acids are linear polymers of *nucleotides*, which are nitrogenous bases covalently bonded to a pentose sugar attached to one or more phosphate groups. They preserve the genetic inheritance of the cell and control its development, growth, and function by regulated translation of proteins. Linear deoxyribonucleic acid (DNA) is *transcribed* by polymerase into messenger ribonucleic acid (mRNA) during cell growth and metabolism. mRNA provides a template on which polypeptide sequences are formed (i.e., *translated*) by amino acids transported to the ribosome by transfer RNA (tRNA). Plasmids are circular, double-stranded DNA segments used to introduce genes into cells using recombinant DNA (rDNA), a process called genetic engineering.

Lipids are comprised primarily of *fatty acids*, which are straight-chain aliphatic hydrocarbons terminated by a hydrophilic carboxyl group with the formula $CH_3-(CH_2)_n-COOH$, where $12 \leq n \leq 20$ is typical. Lipids form membrane bilayers, provide reservoirs of fuel (e.g., fats), and mediate biological activity (e.g., phosopholipids, steroids). Fats are esters of fatty acids with glycerol, $C_3H_5(OH)_3$, a sweetener and preservative. Biodiesel produced by caustic transesterification of fats using caustic methanol or ethanol yields 1 kg of crude glycerol for every 9 kg of biodiesel. *Steroids* like cholesterol and cortisone are cyclical hydrocarbons that penetrate nonpolar cell membranes, bind to and modify intracellular proteins, and thus act as hormone regulators of mammalian development and metabolism.

Viruses are protein shells containing DNA or RNA genes that replicate inside a host such as a bacterium (e.g., bacteriophages) or a plant or mammalian cell. Viral *vectors* may be used to move genetic material into host cells in a process called *transfection*. Transfection is used in *gene therapy* to introduce nucleic acid that complements a mutated or inactive gene of a cell. Transfection also allows *heterologous* protein production (i.e., from one product to another) by a nonnative host cell via rDNA methods. Viruses may be inactivated for use as vaccines to stimulate a prophylactic humoral immune response.

Cellular particulates include cells themselves, crude cell *extracts* and cell *hydrolysates*, as well as subcellular components. Cells are mostly *aerobes* that require oxygen to grow and metabolize. *Anaerobes* are inhibited by oxygen, while *facultative anaerobes*, like yeast, can switch metabolic pathways to grow with or without O_2. As shown in Figure 1.10, *eukaryotic* cells have a nuclear membrane envelope around genetic material. Eukaryotes are single-celled organisms and multicelled systems consisting of fungi (yeasts and molds), algae, protists, animals, and plants. Their DNA is associated with small proteins to form chromosomes. Eukaryotic cells contain specialized *organelles* (i.e., membrane-enclosed domains). Plant cell walls consist of cellulose fibers embedded in pectin aggregates. Animal cells have only a sterol-containing cytoplasmic membrane, which makes them shear-sensitive and fragile. *Prokaryotic* cells, as shown in Figure 1.11, lack a nuclear membrane and organelles like mitochondria or endoplasmic reticulum. Prokaryotes are classified as *eubacteria* or *archae*. Eubacteria are single cells that double in size, mass, and number in 20 minutes to several hours. Most eubacteria are categorized as gram-negative or gram-positive using a dye method. *Gram-negative* bacteria have an outer membrane supported by *peptidoglycan* (i.e., cross-linked polysaccharides and amino acids) that is separated from an inner (*cytoplasmic*) membrane. *Gram-positive* bacteria lack an outer membrane (and more easily secrete protein) but have a rigid cell wall ($\sim$200 Å) of multiple peptidoglycan layers.

§1.9.2 Bioseparation Features

Several features are unique to removal of contaminant biological species and recovery of bioproducts. These features distinguish the specification and operation of bioseparation equipment and process trains from traditional chemical engineering unit operations. Criteria for selecting a bioseparation method are based on the ability of the method to differentiate the targeted bioproduct from contaminants based on physical property differences, as well as its capacity to accommodate the following six features of bioproducts.

Activity: Small primary and secondary metabolites are uniquely defined by a chemical composition and structure that are quantifiable by precise analytical methods (e.g., spectroscopic, physical, and chemical assays). In contrast, biopolymers and cellular particulates are valued for their *activity* in biological systems. Proteins, for example, act in enzyme catalysis and cell regulatory roles. Plasmid DNA or virus is valued as a vector (i.e., delivery vehicle of genetic information into target cells). Biological activity is a function of the assembly of the biopolymer, which results in a complex structure and surface functionality, as well as the presence of organic or inorganic *prosthetic groups*. A subset of particular structural features may be analyzable by spectroscopic, microscopic, or physicochemical assays. Surrogate in vitro assays that approximate biological conditions *in vivo* may provide a limited measure of activity. For biological products, whose origin and characteristics are complex, the manufacturing process itself defines the product.

Complexity: Raw feedstocks containing biological products are complex mixtures of cells and their constituent biomolecules as well as residual species from the cell's native environment. The latter may include *macro-* and *micronutrients* from media used to culture cells in vitro, woody material from harvested fauna, or tissues from mammalian extracts. Bioproducts themselves range from simple, for primary metabolites such as organic alcohols or acids, to complex, for infectious virus particles composed of polymeric proteins and nucleic acids. To recover a target species from a complex matrix of biological species usually requires a series of complementary separation operations that rely on differences in size, density, solubility, charge, hydrophobicity, diffusivity, or volatility to distinguish bioproducts from contaminating host components.

Lability: Susceptibility of biological species to phase change, temperature, solvents or exogenous chemicals, and mechanical shear is determined by bond energies, which maintain native configuration, reaction rates of enzymes and cofactors present in the feedstock, and biocolloid interactions. Small organic alcohols, ketones, and acids maintained by high-energy covalent bonds can resist substantial variations in thermodynamic state. But careful control of solution conditions (e.g., pH buffering, ionic strength, temperature) and suppression of enzymatic reactions (e.g., actions of proteases, nucleases, and lipases) are required to maintain biological activity of polypeptides, polynucleotides, and polysaccharides. Surfactants and organic solvents may

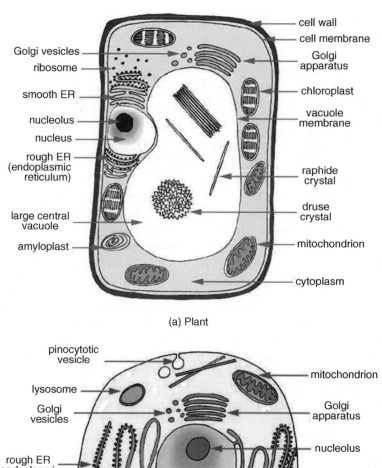

Golgi vesicles

ribosome

smooth ER

nucleolus

nucleus

rough ER
(endoplasmic
reticulum)

large central
vacuole

amyloplast

cell wall

cell membrane

Golgi
apparatus

chloroplast

vacuole
membrane

raphide
crystal

druse
crystal

mitochondrion

cytoplasm

(a) Plant

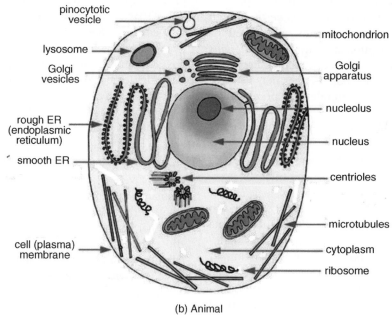

pinocytotic
vesicle

lysosome

Golgi
vesicles

rough ER
(endoplasmic
reticulum)

smooth ER

cell (plasma)
membrane

mitochondrion

Golgi
apparatus

nucleolus

nucleus

centrioles

microtubules

cytoplasm

ribosome

(b) Animal

Figure 1.10 Typical eukaryotic cells.

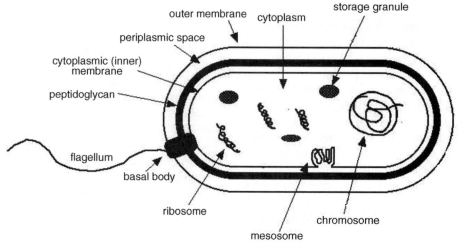

outer membrane   cytoplasm   storage granule

periplasmic space

cytoplasmic (inner)
membrane

peptidoglycan

flagellum

basal body

ribosome

mesosome

chromosome

Figure 1.11 Typical prokaryotic bacterial cell.

disrupt weaker hydrophobic bonds that maintain native configuration of proteins. Fluid–solid or gas–liquid interfaces that absorb dissolved biopolymers may unfold, inactivate, and aggregate biopolymers, particularly when mechanical shear is present.

Process scale: Small primary metabolites may be commodity chemicals with market demands of tons per year. Market requirements for larger biopolymers, proteins in particular, are typically 1 to 10 kg/yr in rDNA hosts. The hosts are usually Chinese Hamster Ovary (CHO) cells, *Escherichia coli* bacteria, and yeast. CHO cells are cultured in batch volumes of 8,000–25,000 liters and yield protein titers of about 1 to 3 g/L. Antibodies are required in quantities of approximately 1,000 kg/yr. Production in transgenic milk, which can yield up to 10 g/L, is being evaluated to satisfy higher demand for antibodies. The initially low concentration of bioproducts in aqueous fermentation and cell culture feeds, as illustrated in

Figure 1.12, results in excess water, which is removed early in the bioprocess train to reduce equipment size and improve process economics.

Purity: The mass of host-cell proteins (HCP), product variants, DNA, viruses, endotoxins, resin and membrane leachables, and small molecules is limited in biotechnology products for therapeutic and prophylactic application.

The Center for Biologics Evaluation and Research (CBER) of the Food and Drug Administration (FDA) approves HCP limits established by the manufacturer after review of process capability and safety testing in toxicology and clinical trials. The World Health Organization (WHO) sets DNA levels at ≤ 10 μg per dose. Less than one virus particle per 10^6 doses is allowed in rDNA-derived protein products. Sterility of final products is ensured by sterile filtration of the final product as well as by controlling microbial contaminant levels throughout the process.

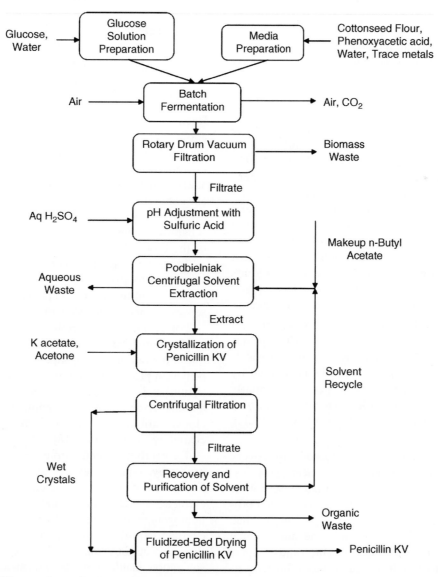

Figure 1.12 Block-Flow Diagram for Penicillin KV Process.

Approval and Manufacturing: The FDA ensures safety and efficacy of bioproducts used in human diagnostic, prophylactic, and therapeutic applications. They review clinical trial data as well as manufacturing process information, eventually approving approximately 1 in 10 candidates for introduction into the market as an *investigational new drug* (IND). Manufacture of drugs under current good manufacturing practices (cGMP) considers facility design and layout, equipment and procedures including operation, cleaning and sterilization documented by standard operating procedures (SOPs), analysis in labs that satisfy good laboratory practices (GLP), personnel training, control of raw materials and cultures, and handling of product. Drug manufacturing processes must be validated to assure that product reproducibly meets predetermined specifications and quality characteristics that ensure biological activity, purity, quality, and safety.

Bioseparation synthesis: Bioprocesses are required to economically and reliably recover purified bioproducts from chemical and biological species in complex cell matrices in quantities sufficient to meet market demands. Beginning with a raw cellular source: (1) cellular particulates are recovered or retained by sedimentation or filtration; (2) biopolymers are usually purified by filtration, adsorption, extraction, or precipitation; and (3) small biomolecules are often recovered by extraction. Economics, documentation, consideration of genetic engineering, and ordering of process steps are key features of bioseparation synthesis.

Bioprocess economics: Large-scale recovery operations must be efficient, since the cost of recovering biomolecules and treating aqueous, organic, and solid wastes can dominate total product manufacturing costs. Inefficient processes consume inordinate volumes of expensive solvent, which must be recovered and recycled, or disposed of. Costs resulting from solvent tankage and consumption during downstream recovery represent a significant fraction of biological-recovery costs. Development of a typical pharmaceutical bioproduct cost $400 million in 1996 and required 14 years—6.5 years from initial discovery through preclinical testing and another 7.5 years for clinical trials in human volunteers.

Bioprocess documentation: The reliability of process equipment must be well-documented to merit approval from governmental regulatory agencies. Such approval is important to meet cGMP quality standards and purity requirements for recovered biological agents, particularly those in prophylactic and therapeutic applications, which require approval by subdivisions of the FDA, including the CBER.

Genetic engineering: Conventional bioproduct-recovery processes can be enhanced via genetic engineering by fusing proteins to active species or intracellular insertion of active DNA to stimulate *in vivo* production of desired proteins. *Fusion proteins* consist of a target protein attached to an affinity peptide tag such as histidine hexamer, which binds transition metals (e.g., nickel, zinc, and copper) immobilized on sorptive or filtration surfaces. Incorporating purification considerations into early upstream cell culture manufacturing decisions can help streamline purification.

§1.9.3 Bioseparation Steps

A series of bioseparation steps are commonly required upstream of the bioreactor (e.g., filtration of incoming gases and culture media), after the bioreactor (i.e., *downstream* or *recovery processes*), and during (e.g., centrifugal removal of spent media) fermentation and cell culture operations. A general sequence of biorecovery steps is designed to remove solvent, insolubles (e.g., particle removal), unrelated soluble species, and similar species. A nondenaturing protein-recovery process, for example, consists of consecutive steps of *extraction, clarification, concentration, fractionation,* and *purification.* The performance of each purification step is characterized in terms of product *purity, activity,* and *recovery,* which are evaluated by:

$$\text{purity} = \frac{\text{bioproduct mass}}{\text{bioproduct mass} + \text{impurities mass}}$$

$$\text{activity} = \frac{\text{units of biological activity}}{\text{bioproduct mass}}$$

$$\text{yield} = \frac{\text{bioproduct mass recovered}}{\text{bioproduct mass in feed}}$$

Recovery yields of the final product can range from about 20% to 60–70% of the initial molecule present in the feed stream. Some clarification of raw fermentation or cell-culture feed streams prior is usually required to analyze their bioproduct content, which makes accurate assessment of recovery yields difficult. It is particularly important to preserve biological activity during the bioseparation steps by maintaining the structure or assembly of the bioproduct.

Table 1.11 classifies common bioseparation operations according to their *type, purpose,* and illustrative *species removed.* Subsequent chapters in this book discuss these bioseparation operations in detail.

Following this subsection, the production of penicillin KV is summarized to illustrate integration of several bioseparation operations into a sequence of steps. Modeling of the penicillin process as well as processes to produce citric acid, pyruvic acid, cysing, riboflavin, cyclodextrin, recombinant human serum albumin, recombinant human insulin, monoclonal antibodies, antitrypsin, and plasmid DNA are discussed by Heinzle et al. [18].

Extraction of cells from fermentation or cell culture broths by removing excess water occurs in a harvest step. Extraction of soluble biological species from these cellular extracts, which contain unexcreted product, occurs by homogenization, which renders the product soluble and accessible to solid–fluid and solute–solute separations. Lysis (breaking up) of whole cells by enzymatic degradation, ultrasonication, Gaulin-press homogenization, or milling releases and solubilizes intracellular enzymes.

Table 1.11 Synthesis of Bioseparation Sequences

Separation Operation	Purpose	Species Removed
Homogenization	**Extract** target from cells	
Cell disruption		
Fluid–Solid Separations	**Reduce volume**	Solvent
Flocculation	**Clarify** target species	Culture media
Precipitation/Centrifugation		Fermentation broth
Crystallization		Insolubles
Extraction		Host-cell debris
Filtration		Aggregates
Evaporation/Drying		
Solute–Solute Separations	**Fractionate** target species	Unrelated Solutes
Chromatography		Small metabolites
Extraction		Proteins
Crystallization		Lipids
Tangential-flow filtration		Nucleic acids
		Carbohydrates
	Purify target species	Related Solutes
		Truncated/misfolded
		Oligomers
Fluid–Solid Separations	**Formulation (Polishing)**	
Precipitation/Centrifugation	Preserve target species	Buffers
Crystallization	Prepare for injection	Solutions
Filtration		
Evaporation/Drying		

Clarification of solid cell debris, nucleic acids, and insoluble proteins by centrifugal precipitation or membrane filtration decreases fouling in later process steps. Selective precipitation is effected by adding salt, organic solvent, detergent, or polymers such as polyethyleneimine and polyethylene glycol to the buffered cell lysate. Size-selective membrane microfiltration may also be used to remove cell debris, colloidal or suspended solids, or virus particles from the clarified lysate. Ultrafiltration, tangential-flow filtration, hollow fibers, and asymmetrical membrane filtration are commonly used membrane-based configurations for clarification. Incompletely clarified lysate has been shown to foul dead-end stacked-membrane adsorbers, in concentrations as low as 5%.

Concentration reduces the volume of total material that must be processed, thereby improving process economics. Extraction of cells from media during harvest involves concentration, or solvent removal. Diafiltration of clarified extract into an appropriate buffer prepares the solution for concentration via filtration. Alternatively, the targeted product may be concentrated by batch adsorption onto a solid resin. The bioproduct of interest and contaminants with similar physical properties are removed by an eluting solvent. Microfiltration to clarify lysate and concentrate by adsorption has been performed simultaneously using a spiral-wound membrane adsorber.

Fractionation of the targeted product usually requires one or more complementary separation processes to distinguish between the product and the contaminants based on differences in their physicochemical features. As examples, filtration, batch adsorption, isoelectric focusing, and isotachophoresis are methods used to separate biological macromolecules based on differences in size, mass, isoelectric point, charge density, and hydrophobicity, respectively. Additional complementary separation steps are often necessary to fractionate the product from any number of similar contaminants. Due to its high specificity, adsorption using affinity, ion exchange, hydrophobic interaction, and reversed-phase chemistries is widely used to fractionate product mixtures.

Purification of the concentrated, fractionated product from closely related variants occurs by a high-resolution technique prior to final formulation and packaging of pharmaceutical bioproducts. Purification often requires differential absorption in an adsorptive column that contains a large number of theoretical stages or plates to attain the required purity. Batch electrophoresis achieves high protein resolution at laboratory scale, while production-scale, continuous apparatus for electrophoresis must be cooled to minimize ohmic heating of bioproducts. Crystallization is preferred, where possible, as a final purification step prior to formulation and packaging. Counterflow resolution of closely related species has also been used.

Formulation: The dosage form of a pharmaceutical bioproduct results from *formulating* the bioactive material by adding *excipients* such as stabilizers (e.g., reducing compounds, polymers), tablet solid diluents (e.g., gums, PEG, oils), liquid diluent (e.g., water for injection, WFI), or adjuvant (e.g., alum) to support activity, provide stability during storage, and provide deliverability at final concentration.

EXAMPLE 1.4   Using properties to select bioseparations

Proteins, nucleic acids, and viral gene vectors are current pharmaceutical products. Identify five physical and biochemical properties of these biological species by which they could be distinguished in a bioseparation. Identify a separation operation that could be used to selectively remove or retain each species from a mixture of the other two. Summarize important considerations that might constrain the bioseparation operating parameters.

Solution

Properties that could allow separation are density, size, solubility, charge, and hydrophobicity. Separation operations that could be considered are: CsCl gradient ultracentrifugation (UC), which selectively retains viral vectors from proteins and nucleic acids, based primarily on density; ultrafiltration (UF), which is commonly used to size-selectively remove nucleic acids enzymatically digested by nuclease (e.g., Benzonase™) from proteins and viral vectors; and ion-exchange adsorption (IEX), which is used to selectively remove proteins and/or nucleic acids from viral suspensions (or vice versa). Some important considerations in UC, UF, and IEX are maintaining temperature ($\sim 4°C$), water activity (i.e., ionic strength), pH to preserve virus stability, proper material selection to limit biological reactivity of equipment surfaces in order to prevent protein denaturation or virus disassembly, and aseptic operating procedures to prevent batch contamination.

Many industrial chemical separation processes introduced in earlier sections of this chapter have been adapted for use in bioproduct separation and purification. Specialized needs for adaptation arise from features unique to recovery of biological species. *Complex* biological feedstocks must often be processed at moderate temperatures, with low shear and minimal gas–liquid interface creation, in order to maintain *activity* of *labile* biological species. Steps in a recovery sequence to remove biological species from the feed milieu, whose complexity and size range broadly, are often determined by unique cell characteristics. For example, extracellular secretion of product may eliminate the need for cell disruption. High *activity* of biological species allows market demand to be satisfied at *process scales* suited to batch operation. Manufacturing process validation required to ensure *purity* of biopharmaceutical products necessitates batch, rather than continuous, operation. Batchwise expansion of seed inoculums in fermented or cultured cell hosts is also conducive to batch bioseparations in downstream processing. The value per gram of biopharmaceutical products relative to commodity chemicals is higher due to the higher specific activity of the former in vivo. This permits cost-effective use of high-resolution chromatography or other specialized operations that may be cost-prohibitive for commodity chemical processes.

§1.9.4 Bioprocess Example: Penicillin

In the chemical industry, a unit operation such as distillation or liquid–liquid extraction adds pennies to the sale price of an average product. For a 40 cents/lb commodity chemical, the component separation costs do not generally account for more than 10–15% of the manufacturing cost. An entirely different economic scenario exists in the bioproduct industry. For example, in the manufacture of tissue plasminogen activator (tPA), a blood clot dissolver, Datar et al., as discussed by Shuler and Kargi [9], enumerate 16 processing steps when the bacterium *E. coli* is the cell culture. Manufacturing costs for this process are $22,000/g, and it takes a $70.9 million investment to build a plant to produce 11 kg/yr of product. Purified product yields are only 2.8%. Drug prices must also include recovery of an average $400 million cost of development within the product's lifetime. Furthermore, product lifetimes are usually shorter than the nominal 20-year patent life of a new drug, since investigational new drug (IND) approval typically occurs years after patent approval. Although some therapeutic proteins sell for $100,000,000/kg, this is an extreme case; more efficient tPA processes using CHO cell cultures have separation costs averaging $10,000/g.

A more mature, larger-scale operation is the manufacture of penicillin, the first modern antibiotic for the treatment of bacterial infections caused by gram-positive organisms. Penicillin typically sells for about $15/kg and has a market of approximately $4.5 billion/yr. Here, the processing costs represent about 60% of the total manufacturing costs, which are still much higher than those in the chemical industry. Regulatory burden to manufacture and market a drug is significantly higher than it is for a petrochemical product. Figure 1.12 is a block-flow diagram for the manufacture of penicillin, a secondary metabolite secreted by the common mold *Penicillium notatum*, which, as Alexander Fleming serendipitously observed in September 1928 in St. Mary's Hospital in London, prevents growth of the bacterium *Staphylococcus aureus* on a nutrient surface [2]. Motivated to replace sulfa drugs in World War II, Howard Florey and colleagues cultured and extracted this delicate, fragile, and unstable product to demonstrate its effectiveness. Merck, Pfizer, Squibb, and USDA Northern Regional Research Laboratory in Peoria, Illinois, undertook purification of the product in broths with original titers of only 0.001 g/L using a submerged tank process. Vessel sizes grew to 10,000 gal to produce penicillin for 100,000 patients per year by the end of World War II. Ultraviolet irradiation of *Penicillium* spores has since produced mutants of the original strain, with increased penicillin yields up to 50 g/L.

The process shown in Figure 1.12 is one of many that produces 1,850,000 kg/yr of the potassium salt of penicillin V by the fermentation of phenoxyacetic acid (a side-chain precursor), aqueous glucose, and cottonseed flour in the presence of trace metals and air. The structures of phenoxyacetic acid and penicillin V are shown in Figure 1.13. *Upstream processing* includes preparation of culture medias. It is followed, after fermentation, by *downstream processing* consisting of a

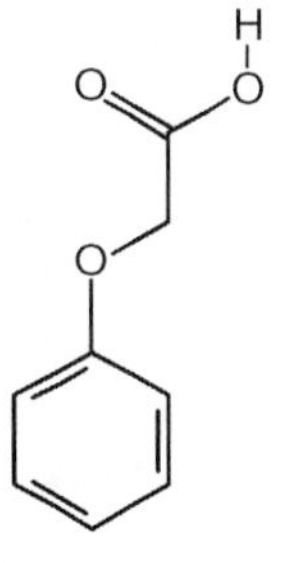

Phenoxyacetic acid

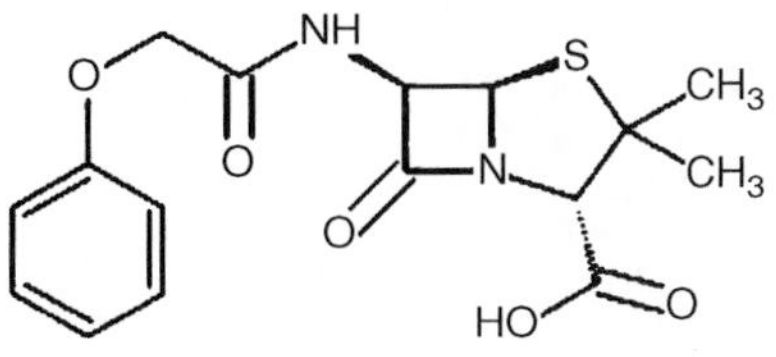

Penicillin V

Figure 1.13 Structure of penicillin V and its precursor.

number of bioseparation steps to recover and purify the penicillin and the *n*-butyl acetate, a recycled solvent used to extract the penicillin. A total of 20 steps are involved, some batch, some semicontinuous, and some continuous. Only the main steps are shown in Figure 1.12.

One upstream processing step, at 25°C, mixes the feed media, consisting of cottonseed flour, phenoxyacetic acid, water, and trace metals. The other step mixes glucose with water. These two feeds are sent to one of 10 batch fermentation vessels.

In the fermentation step, the penicillin product yield is 55 g/L, which is a dramatic improvement over the earliest yield of 0.001 g/L. Fermentation temperature is controlled at 28°C, vigorous aeration and agitation is required, and the fermentation reaction residence time is 142 hours (almost 6 days), which would be intolerable in the chemical industry. The fermentation effluent consists, by weight, of 79.3% water; 9.3% intracellular water; 1.4% unreacted cottonseed flour, phenoxyacetic acid, glucose, and trace metals; 4.5% mycelia (biomass); and only 5.5% penicillin V.

The spent mycelia (biomass), is removed using a rotary-drum vacuum filter, resulting in (following washing with water) a filter cake of 20 wt% solids and 3% of the penicillin entering the filter. Prior to the solvent-extraction step, the filtrate, which is now 6.1% penicillin V, is cooled to 2°C to minimize chemical or enzymatic product degradation, and its pH is adjusted to between 2.5 and 3.0 by adding a 10 wt% aqueous solution of sulfuric acid to enhance the penicillin partition coefficient for its removal by solvent extraction, which is extremely sensitive to pH. This important adjustment, which is common to many bioprocesses, is considered in detail in §2.9.1.

Solvent extraction is conducted in a Podbielniak centrifugal extractor (POD), which provides a very short residence time, further minimizing product degradation. The solvent is

n-butyl acetate, which selectively dissolves penicillin V. The extract is 38.7% penicillin V and 61.2% solvent, while the raffinate contains almost all of the other chemicals in the filtrate, including 99.99% of the water. Unfortunately, 1.6% of the penicillin is lost to the raffinate.

Following solvent extraction, potassium acetate and acetone are added to promote the crystallization of the potassium salt of penicillin V (penicillin KV). A basket centrifuge with water washing then produces a crystal cake containing only 5 wt% moisture. Approximately 4% of the penicillin is lost in the crystallization and centrifugal filtration steps. The crystals are dried to a moisture content of 0.05 wt% in a fluidized-bed dryer. Not shown in Figure 1.12 are subsequent finishing steps to produce, if desired, 250 and 500 mg tablets, which may contain small amounts of lactose, magnesium stearate, povidone, starch, stearic acid, and other inactive ingredients. The filtrate from the centrifugal filtration step contains 71 wt% solvent *n*-butyl acetate, which must be recovered for recycle to the solvent extraction step. This is accomplished in the separation and purification step, which may involve distillation, adsorption, three-liquid-phase extraction, and/or solvent sublation (an adsorption-bubble technique). The penicillin process produces a number of waste streams—e.g., wastewater containing *n*-butyl acetate—that require further processing, which is not shown in Figure 1.12.

Overall, the process has only a 20% yield of penicillin V from the starting ingredients. The annual production rate is achieved by processing 480 batches that produce 3,850 kg each of the potassium salt, with a batch processing time of 212 hours. It is no wonder that bioprocesses, which involve (1) research and development costs, (2) regulatory burden, (3) fermentation residence times of days, (4) reactor effluents that are dilute in the product, and (5) multiple downstream bioseparation steps, many of which are difficult, result in high-cost biopharmaceutical products.

Alternative process steps such as direct penicillin recovery from the fermentation broth by adsorption on activated carbon, or crystallization of penicillin from amyl or butyl acetate without a water extraction, are described in the literature. Industrial processes and process conditions are proprietary, so published flowsheets and descriptions are not complete or authoritative.

Commercially, penicillin V and another form, G, are also sold as intermediates or converted, using the enzyme penicillin acylase, to 6-APA (6-aminopenicillic acid), which is further processed to make semisynthetic penicillin derivatives. Another medicinal product for treating people who are allergic to penicillin is produced by subjecting the penicillin to the enzyme penicillinase.

§1.10 SELECTION OF FEASIBLE SEPARATIONS

Only an introduction to the separation-selection process is given here. A detailed treatment is given in Chapter 8 of Seider, Seader, Lewin, and Widagdo [7]. Key factors in the selection are listed in Table 1.12. These deal with feed and product conditions, property differences, characteristics of

Table 1.12 Factors That Influence the Selection of Feasible Separation Operations

A. Feed conditions
  1. Composition, particularly of species to be recovered or separated
  2. Flow rate
  3. Temperature
  4. Pressure
  5. Phase state (solid, liquid, or gas)

B. Product conditions
  1. Required purities
  2. Temperatures
  3. Pressures
  4. Phases

C. Property differences that may be exploited
  1. Molecular
  2. Thermodynamic
  3. Transport

D. Characteristics of separation operation
  1. Ease of scale-up
  2. Ease of staging
  3. Temperature, pressure, and phase-state requirements
  4. Physical size limitations
  5. Energy requirements

E. Economics
  1. Capital costs
  2. Operating costs

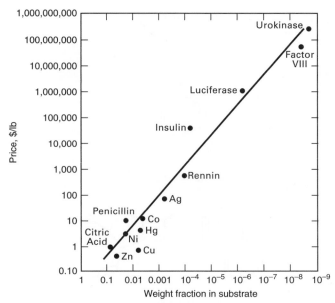

Figure 1.14 Effect of concentration of product in feed material on price [13].

the candidate separation operations, and economics. The most important feed conditions are composition and flow rate, because the other conditions (temperature, pressure, and phase) can be altered to fit a particular operation. However, feed vaporization, condensation of a vapor feed, or compression of a vapor feed can add significant energy costs to chemical processes. Some separations, such as those based on the use of barriers or solid agents, perform best on dilute feeds. The most important product conditions are purities because the other conditions listed can be altered by energy transfer after the separation is achieved.

Sherwood, Pigford, and Wilke [11], Dwyer [12], and Keller [13] have shown that the cost of recovering and purifying a chemical depends strongly on its concentration in the feed. Keller's correlation, Figure 1.14, shows that the more dilute the feed, the higher the product price. The five highest priced and most dilute in Figure 1.14 are all proteins.

When a very pure product is required, large differences in volatility or solubility or significant numbers of stages are needed for chemicals in commerce. For biochemicals, especially proteins, very expensive separation methods may be required. Accurate molecular and bulk thermodynamic and transport properties are also required. Data and estimation methods for the properties of chemicals in commerce are given by Poling, Prausnitz, and O'Connell [14], Daubert and Danner [15], and others.

A survey by Keller [13], Figure 1.15, shows that the degree to which a separation operation is technologically mature correlates with its commercial use. Operations based on

barriers are more expensive than operations based on the use of a solid agent or the creation or addition of a phase. All separation equipment is limited to a maximum size. For capacities requiring a larger size, parallel units must be provided. Except for size constraints or fabrication problems, capacity of a single unit can be doubled, for an additional investment cost of about 60%. If two parallel units are installed, the additional investment is 100%. Table 1.13 lists operations ranked according to ease of scale-up. Those ranked near the top are frequently designed without the need for pilot-plant or laboratory data provided that neither the process nor the final product is new and equipment is guaranteed by vendors. For new processes, it is never certain that product specifications will be met. If there is a potential impurity, possibility of corrosion, or other uncertainties such as

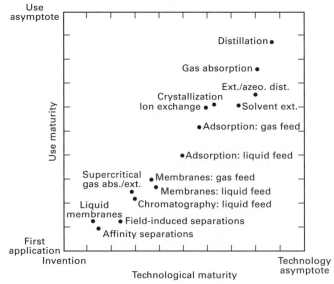

Figure 1.15 Technological and use maturities of separation processes [13].

Table 1.13 Ease of Scale-up of the Most Common Separation Operations

Operation in Decreasing Ease of Scale-up	Ease of Staging	Need for Parallel Units
Distillation	Easy	No need
Absorption	Easy	No need
Extractive and azeotropic distillation	Easy	No need
Liquid–liquid extraction	Easy	Sometimes
Membranes	Repressurization required between stages	Almost always
Adsorption	Easy	Only for regeneration cycle
Crystallization	Not easy	Sometimes
Drying	Not convenient	Sometimes

product degradation or undesirable agglomeration, a pilot-plant is necessary. Operations near the middle usually require laboratory data, while those near the bottom require pilot-plant tests.

Included in Table 1.13 is an indication of the ease of providing multiple stages and whether parallel units may be required. Maximum equipment size is determined by height limitations, and shipping constraints unless field fabrication is possible and economical. The selection of separation techniques for both homogeneous and heterogeneous phases, with many examples, is given by Woods [16]. Ultimately, the process having the lowest operating, maintenance, and capital costs is selected, provided it is controllable, safe, nonpolluting, and capable of producing products that meet specifications.

EXAMPLE 1.5   Feasible separation alternatives.

Propylene and propane are among the light hydrocarbons produced by cracking heavy petroleum fractions. Propane is valuable as a fuel and in liquefied natural gas (LPG), and as a feedstock for producing propylene and ethylene. Propylene is used to make acrylonitrile for synthetic rubber, isopropyl alcohol, cumene, propylene oxide, and polypropylene. Although propylene and propane have close boiling points, they are traditionally separated by distillation. From Figure 1.16, it is seen that a large number of stages is needed and that the reflux and boilup flows are large. Accordingly, attention has been given to replacement of distillation with a more economical and less energy-intensive process. Based on the factors in Table 1.12, the characteristics in Table 1.13, and the list of species properties given at the end of §1.2, propose alternatives to Figure 1.16.

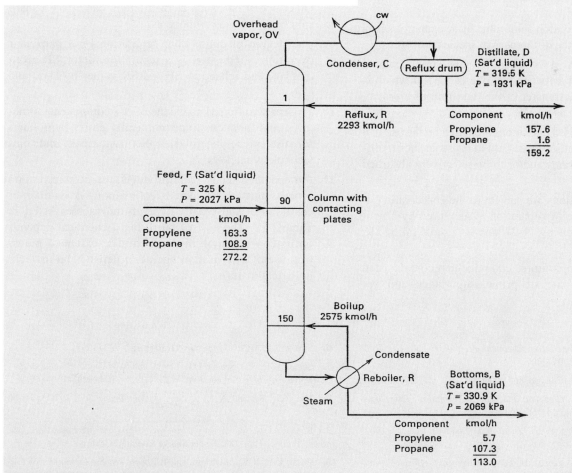

Figure 1.16 Distillation of a propylene–propane mixture.

Solution

First, verify that the component feed and product flows in Figure 1.16 satisfy (1-1), the conservation of mass. Table 1.14 compares properties taken mainly from Daubert and Danner [15]. The only listed property that might be exploited is the dipole moment. Because of the asymmetric location of the double bond in propylene, its dipole moment is significantly greater than that of propane, making propylene a weakly polar compound. Operations that can exploit this difference are:

1. Extractive distillation with a polar solvent such as furfural or an aliphatic nitrile that will reduce the volatility of propylene (Ref.: U.S. Patent 2,588,056, March 4, 1952).

2. Adsorption with silica gel or a zeolite that selectively adsorbs propylene [Ref.: *J.Am. Chem. Soc*, **72**, 1153–1157 (1950)].

3. Facilitated transport membranes using impregnated silver nitrate to carry propylene selectively through the membrane [Ref.: *Recent Developments in Separation Science*, Vol. IX, 173–195 (1986)].

Table 1.14 Comparison of Properties for Example 1.5

Property	Propylene	Propane
Molecular weight	42.081	44.096
van der Waals volume, $m^3/kmol$	0.03408	0.03757
van der Waals area, $m^2/kmol \times 10^{-8}$	5.060	5.590
Acentric factor	0.142	0.152
Dipole moment, debyes	0.4	0.0
Radius of gyration, $m \times 10^{10}$	2.254	2.431
Normal melting point, K	87.9	85.5
Normal boiling point, K	225.4	231.1
Critical temperature, K	364.8	369.8
Critical pressure, MPa	4.61	4.25

SUMMARY

1. Industrial chemical processes include equipment for separating chemicals in the process feed(s) and/or species produced in reactors within the process.

2. More than 25 different separation operations are commercially important.

3. The extent of separation achievable by a separation operation depends on the differences in species properties.

4. The more widely used separation operations involve transfer of species between two phases, one of which is created by energy transfer or the reduction of pressure, or by introduction as an MSA.

5. Less commonly used operations are based on the use of a barrier, solid agent, or force field to cause species to diffuse at different rates and/or to be selectively absorbed or adsorbed.

6. Separation operations are subject to the conservation of mass. The degree of separation is measured by a split fraction, SF, given by (1-2), and/or a split ratio, SR, given by (1-3).

7. For a train of separators, component recoveries and product purities are of prime importance and are related by material balances to individual SF and/or SR values.

8. Some operations, such as absorption, are capable of only a specified degree of separation for one component. Other operations, such as distillation, can effect a sharp split between two components.

9. The degree of component separation by a particular operation is indicated by a separation factor, SP, given by (1-4) and related to SF and SR values by (1-5) and (1-6).

10. Bioseparations use knowledge of cell components, structures, and functions to economically purify byproducts for use in agrichemical, pharmaceutical, and biotechnology markets.

11. For specified feed(s) and products, the best separation process must be selected from among a number of candidates. The choice depends on factors listed in Table 1.12. The cost of purifying a chemical depends on its concentration in the feed. The extent of industrial use of a separation operation depends on its technological maturity.

REFERENCES

1. *Kirk-Othmer Encyclopedia of Chemical Technology*, 5th ed., John Wiley & Sons, New York (2004–2007).

2. Maude, A.H., *Trans. AIChE*, 38, 865–882 (1942).

3. Considine, D.M., Ed., *Chemical and Process Technology Encyclopedia*, McGraw-Hill, New York, pp. 760–763 (1974).

4. Carle, T.C., and D.M. Stewart, *Chem. Ind.* (London), May 12, 1962, 830–839.

5. Perry, R.H., and D.W. Green, Eds., *Perry's Chemical Engineers' Handbook*, 8th ed., McGraw-Hill, New York, (2008).

6. Siirola, J.J., *AIChE Symp. Ser*, **91**(304), 222–233 (1995).

7. Seider, W.D., J.D. Seader, D.R. Lewin, and S. Widagdo, *Product & Process Design Principles*, 3rd ed., John Wiley & Sons, Hoboken, NJ (2009).

8. Van Reis, R., and A. L. Zydney, *J. Membrane Sci.*, **297**(1), 16–50 (2007).

9. Shuler, M. L., and F. Kargi, *Bioprocess Engineering Basic Concepts*, 2nd ed., Prentice Hall PTR, Upper Saddle River, NJ (2002).

10. Datar, R.V., R.V. Cartwright, and T. Rosen, Process Economics of Animal Cell and Bacterial Fermentation, *Bio/Technol.*, 11, 349–357 (1993).

11. Sherwood, T.K., R.L. Pigford, and C.R. Wilke, *Mass Transfer*, McGraw-Hill, New York (1975).

12. Dwyer, J.L., *Biotechnology*, **1**, 957 (Nov. 1984).

13. Keller, G.E., II, *AIChE Monogr. Ser*, **83**(17) (1987).

14. Poling, B.E., J.M. Prausnitz, and J.P. O'Connell, *The Properties of Gases and Liquids*, 5th ed., McGraw-Hill, New York (2001).

15. Daubert, T.E., and R.P Danner, *Physical and Thermodynamic Properties of Pure Chemicals—Data Compilation*, DIPPR, AIChE, Hemisphere, New York (1989).

16. Woods, D.R., *Process Design and Engineering Practice*, Prentice Hall, Englewood Cliffs, NJ (1995).

17. Cussler, E.L., and G.D. Moggridge, *Chemical Product Design*, Cambridge University Press, Cambridge, UK (2001).

18. Heinzle, E., A.P. Biwer, and C.L. Cooney, *Development of Sustainable Bioprocesses*, John Wiley & Sons, Ltd, England (2006).

19. Clark, J.H., and F.E.I. Deswarte, *Introduction to Chemicals from Biomass*, John Wiley & Sons, Ltd., West Sussex (2008).

20. Kamm, B., P.R. Gruber, and M. Kamm, Eds., *Biorefineries—Industrial Processes and Products, Volumes 1 and 2*, Wiley-VCH, Weinheim (2006).

STUDY QUESTIONS

1.1. What are the two key process operations in chemical engineering?

1.2. What are the main auxiliary process operations in chemical engineering?

1.3. What are the five basic separation techniques, and what do they all have in common?

1.4. Why is mass transfer a major factor in separation processes?

1.5. What limits the extent to which the separation of a mixture can be achieved?

1.6. What is the most common method used to separate two fluid phases?

1.7. What is the difference between an ESA and an MSA? Give three disadvantages of using an MSA.

1.8. What is the most widely used industrial separation operation?

1.9. What is the difference between adsorption and absorption?

1.10. The degree of separation in a separation operation is often specified in terms of component recoveries and/or product purities. How do these two differ?

1.11. What is a key component?

1.12. What is a multicomponent product?

1.13. What are the three types of bioproducts and how do they differ?

1.14. Identify the major objectives of the steps in a biopurification process.

1.15. Give examples of separation operations used for the steps in a bioprocess.

EXERCISES

Section 1.1

1.1. Fluorocarbons process.

Shreve's Chemical Process Industries, 5th edition, by George T. Austin (McGraw-Hill, New York, 1984), contains process descriptions, flow diagrams, and technical data for commercial processes. For each of the fluorocarbons processes on pages 353–355, draw a block-flow diagram of the reaction and separation steps and describe the process in terms of just those steps, giving attention to the chemicals formed in the reactor and separator.

Section 1.2

1.2. Mixing vs. separation.

Explain, using thermodynamic principles, why mixing pure chemicals to form a homogeneous mixture is a spontaneous process, while separation of that mixture into its pure species is not.

1.3. Separation of a mixture requires energy.

Explain, using the laws of thermodynamics, why the separation of a mixture into pure species or other mixtures of differing compositions requires energy to be transferred to the mixture or a degradation of its energy.

Section 1.3

1.4. Use of an ESA or an MSA.

Compare the advantages and disadvantages of making separations using an ESA versus using an MSA.

1.5. Producing ethers from olefins and alcohols.

Hydrocarbon Processing published a petroleum-refining handbook in November 1990, with process-flow diagrams and data for commercial processes. For the ethers process on page 128, list the separation operations of the type given in Table 1.1 and indicate what chemical(s) is (are) being separated.

1.6. Conversion of propylene to butene-2s.

Hydrocarbon Processing published a petrochemical handbook in March 1991, with process-flow diagrams and data for commercial processes. For the butene-2 process on page 144, list the separation operations of the type given in Table 1.1 and indicate what chemical(s) is (are) being separated.

Section 1.4

1.7. Use of osmosis.

Explain why osmosis is not an industrial separation operation.

1.8. Osmotic pressure for recovering water from sea water.

The osmotic pressure, π, of sea water is given by $\pi = RTc/M$, where c is the concentration of the dissolved salts (solutes) in g/cm^3 and M is the average molecular weight of the solutes as ions. If pure water is to be recovered from sea water at 298 K and containing 0.035 g of salts/cm^3 of sea water and $M = 31.5$, what is the minimum required pressure difference across the membrane in kPa?

1.9. Use of a liquid membrane.

A liquid membrane of aqueous ferrous ethylenediaminetetraacetic acid, maintained between two sets of microporous, hydrophobic, hollow fibers packed in a permeator cell, can selectively and continuously remove sulfur dioxide and nitrogen oxides from the flue gas of power plants. Prepare a drawing of a device to carry out such a separation. Show locations of inlet and outlet streams, the arrangement of the hollow fibers, and a method for handling the membrane liquid. Should the membrane liquid be left in the cell or circulated? Is a sweep fluid needed to remove the oxides?

Section 1.5

1.10. Differences in separation methods.

Explain the differences, if any, between adsorption and gas–solid chromatography.

1.11. Flow distribution in a separator.

In gas–liquid chromatography, is it essential that the gas flow through the packed tube be plug flow?

Section 1.6

1.12. Electrical charge for small particles.

In electrophoresis, explain why most small, suspended particles are negatively charged.

1.13. Flow field in field-flow fractionation.

In field-flow fractionation, could a turbulent-flow field be used? Why or why not?

Section 1.7

1.14. Material balance for a distillation sequence.

The feed to Column C3 in Figure 1.8 is given in Table 1.5. The separation is to be altered to produce a distillate of 95 mol% pure isobutane with a recovery (SF) in the distillate of 96%. Because of the sharp separation in Column C3 between iC_4 and nC_4, assume all propane goes to the distillate and all C_5's go to the bottoms.

(a) Compute the flow rates in lbmol/h of each component in each of the two products leaving Column C3.

(b) What is the percent purity of the n-butane bottoms product?

(c) If the isobutane purity in the distillate is fixed at 95%, what % recovery of isobutane in the distillate will maximize the % purity of normal butane in the bottoms product?

1.15. Material balance for a distillation sequence.

Five hundred kmol/h of liquid alcohols containing, by moles, 40% methanol (M), 35% ethanol (E), 15% isopropanol (IP), and 10% normal propanol (NP) are distilled in two distillation columns. The distillate from the first column is 98% pure M with a 96% recovery of M. The distillate from the second is 92% pure E with a 95% recovery of E from the process feed. Assume no propanols in the distillate from Column C1, no M in the bottoms from Column C2, and no NP in the distillate from Column C2.

(a) Assuming negligible propanols in the distillate from the first column, compute the flow rates in kmol/h of each component in each feed, distillate, and bottoms. Draw a labeled block-flow diagram. Include the material balances in a table like Table 1.5.

(b) Compute the mol% purity of the propanol mixture leaving as bottoms from the second column.

(c) If the recovery of ethanol is fixed at 95%, what is the maximum mol% purity of the ethanol in the distillate from the second column?

(d) If instead, the purity of the ethanol is fixed at 92%, what is the maximum recovery of ethanol (based on the process feed)?

1.16. Pervaporation to separate ethanol and benzene.

Ethanol and benzene are separated in a network of distillation and membrane separation steps. In one step, a near-azeotropic liquid mixture of 8,000 kg/h of 23 wt% ethanol in benzene is fed to a pervaporation membrane consisting of an ionomeric film of perfluorosulfonic polymer cast on a Teflon support. The membrane is selective for ethanol, so the vapor permeate contains 60 wt% ethanol, while the non-permeate liquid contains 90 wt% benzene.

(a) Draw a flow diagram of the pervaporation step using symbols from Table 1.2, and include all process information.

(b) Compute the component flow rates in kg/h in the feed stream and in the product streams, and enter these results into the diagram.

(c) What operation could be used to purify the vapor permeate?

Section 1.8

1.17. Recovery of hydrogen by gas permeation.

The Prism gas permeation process developed by the Monsanto Company is selective for hydrogen when using hollow-fiber membranes made of silicone-coated polysulphone. A gas at 16.7 MPa and 40°C, and containing in kmol/h: 42.4 H_2, 7.0 CH_4, and 0.5 N_2 is separated into a nonpermeate gas at 16.2 MPa and a permeate gas at 4.56 MPa.

(a) If the membrane is nonpermeable to nitrogen; the Prism membrane separation factor (SP), on a mole basis for hydrogen relative to methane, is 34.13; and the split fraction (SF) for hydrogen to the permeate gas is 0.6038, calculate the flow of each component and the total flow of non-permeate and permeate gas.

(b) Compute the mol% purity of the hydrogen in the permeate gas.

(c) Using a heat-capacity ratio, γ, of 1.4, estimate the outlet temperatures of the exiting streams, assuming the ideal gas law, reversible expansions, and no heat transfer between gas streams.

(d) Draw a process-flow diagram and include pressure, temperature, and component flow rates.

1.18. Nitrogen injection to recover natural gas.

Nitrogen is injected into oil wells to increase the recovery of crude oil (enhanced oil recovery). It mixes with the natural gas that is produced along with the oil. The nitrogen must then be separated from the natural gas. A total of 170,000 SCFH (based on 60°F and 14.7 psia) of natural gas containing 18% N_2, 75% CH_4, and 7% C_2H_6 at 100°F and 800 psia is to be processed so that it contains less than 3 mol% nitrogen in a two-step process: (1) membrane separation with a nonporous glassy polyimide membrane, followed by (2) pressure-swing adsorption using molecular sieves highly selective for $N_2(SP_{N_2,CH_4} = 16)$ and completely impermeable to ethane. The pressure-swing adsorption step selectively adsorbs methane, giving 97% pure methane in the adsorbate, with an 85% recovery of CH_4 fed to the adsorber. The non-permeate (retentate) gas from the membrane step and adsorbate from the pressure-swing adsorption step are combined to give a methane stream that contains 3.0 mol% N_2. The pressure drop across the membrane is 760 psi. The permeate at 20°F is compressed to 275 psia and cooled to 100°F before entering the adsorption step. The adsorbate, which exits the adsorber during regeneration at 100°F and 15 psia, is compressed to 800 psia and cooled to 100°F before being combined with non-permeate gas to give the final pipeline natural gas.

(a) Draw a flow diagram of the process using appropriate symbols. Include compressors and heat exchangers. Label the diagram with the data given and number all streams.

(b) Compute component flow rates of N_2, CH_4, and C_2H_6 in lbmol/h in all streams and create a material-balance table similar to Table 1.5.

1.19. Distillation sequences.

The feed stream in the table below is to be separated into four nearly pure products. None of the components is corrosive and, based on the boiling points, none of the three separations is difficult. As seen in Figure 1.9, five distillation sequences are possible. (a) Determine a suitable sequence of three columns using the heuristics of §1.7. (b) If a fifth component were added to give five products, Table 1.8 indicates that 14 alternative distillation sequences are possible. Draw, in a manner similar to Figure 1.9, all 14 of these sequences.

Component	Feed rate, kmol/h	Normal boiling point, K
Methane	19	112
Benzene	263	353
Toluene	85	384
Ethylbenzene	23	409

Section 1.9

1.20. Bioproduct separations.

Current and future pharmaceutical products of biotechnology include proteins, nucleic acids, and viral gene vectors. Example 1.4 identified five physical and biochemical features of these biological species by which they could be distinguished in a bioseparation; identified a bioseparation operation that could be used to selectively remove or retain each species from a mixture of the other two; and summarized important considerations in maintaining the activity of each species that would constrain the operating parameters of each bioseparation. Extend that example by listing the purity requirements for FDA approval of each of these three purified species as a parenteral product, which is one that is introduced into a human organism by intravenous, subcutaneous, intramuscular, or intramedullary injection.

1.21. Separation processes for bioproducts from *E. coli*.

Recombinant protein production from *E. coli* resulted in the first products from biotechnology. (a) List the primary structures and components of *E. coli* that must be removed from a fermentation broth to purify a heterologous protein product (one that differs from any protein normally found in the organism in question) expressed for pharmaceutical use. (b) Identify a sequence of steps to purify a conjugate heterologous protein (a compound comprised of a protein molecule and an attached non-protein prosthetic group such as a carbohydrate) that remained soluble in cell paste. (c) Identify a separation operation for each step in the process and list one alternative for each step. (d) Summarize important considerations in establishing operating procedures to preserve the activity of the protein. (e) Suppose net yield in each step in your process was 80%. Determine the overall yield of the process and the scale of operation required to produce 100 kg per year of the protein at a titer of 1 g/L.

1.22. Purification process for adeno-associated viral vector.

An AAV viral gene vector must be purified from an anchorage-dependent cell line. Repeat Exercise 1.21 to develop a purification process for this vector.

Section 1.10

1.23. Separation of a mixture of ethylbenzene and xylenes.

Mixtures of ethylbenzene (EB) and the three isomers (ortho, meta, and para) of xylene are available in petroleum refineries.

(a) Based on differences in boiling points, verify that the separation between *meta*-xylene (MX) and *para*-xylene (PX) by distillation is more difficult than the separations between EB and PX, and MX and *ortho*-xylene (OX).

(b) Prepare a list of properties for MX and PX similar to Table 1.14. Which property differences might be the best ones to exploit in order to separate a mixture of these two xylenes?

(c) Explain why melt crystallization and adsorption are used commercially to separate MX and PX.

1.24. Separation of ethyl alcohol and water.

When an ethanol–water mixture is distilled at ambient pressure, the products are a distillate of near-azeotropic composition (89.4 mol % ethanol) and a bottoms of nearly pure water. Based on differences in certain properties of ethanol and water, explain how the following operations might be able to recover pure ethanol from the distillate: (a) Extractive distillation. (b) Azeotropic distillation. (c) Liquid–liquid extraction. (d) Crystallization. (e) Pervaporation. (f) Adsorption.

1.25. Removal of ammonia from water.

A stream of 7,000 kmol/h of water and 3,000 parts per million (ppm) by weight of ammonia at 350 K and 1 bar is to be processed to remove 90% of the ammonia. What type of separation would you use? If it involves an MSA, propose one.

1.26. Separation by a distillation sequence.

A light-hydrocarbon feed stream contains 45.4 kmol/h of propane, 136.1 kmol/h of isobutane, 226.8 kmol/h of *n*-butane, 181.4 kmol/h of isopentane, and 317.4 kmol/h of *n*-pentane. This stream is to be separated by a sequence of three distillation columns into four products: (1) propane-rich, (2) isobutane-rich, (3) *n*-butane-rich, and (4) combined pentanes-rich. The first-column distillate is the propane-rich product; the distillate from Column 2 is the isobutane-rich product; the distillate from Column 3 is the *n*-butane-rich product; and the combined pentanes are the Column 3 bottoms. The recovery of the main component in each product is 98%. For example, 98% of the propane in the process feed stream appears in the propane-rich product.

(a) Draw a process-flow diagram similar to Figure 1.8.

(b) Complete a material balance for each column and summarize the results in a table similar to Table 1.5. To complete the balance, you must make assumptions about the flow rates of: (1) isobutane in the distillates for Columns 1 and 3 and (2) *n*-butane in the distillates for Columns 1 and 2, consistent with the specified recoveries. Assume that there is no propane in the distillate from Column 3 and no pentanes in the distillate from Column 2.

(c) Calculate the mol% purities of the products and summarize your results as in Table 1.7, but without the specifications.

1.27. Removing organic pollutants from wastewater.

The need to remove organic pollutants from wastewater is common to many industrial processes. Separation methods to be considered are: (1) adsorption, (2) distillation, (3) liquid–liquid extraction, (4) membrane separation, (5) stripping with air, and (6) stripping with steam. Discuss the advantages and disadvantages of each method. Consider the fate of the organic material.

1.28. Removal of VOCs from a waste gas stream.

Many waste gas streams contain volatile organic compounds (VOCs), which must be removed. Recovery of the VOCs may be accomplished by (1) absorption, (2) adsorption, (3) condensation, (4) freezing, (5) membrane separation, or (6) catalytic oxidation. Discuss the pros and cons of each method, paying particular attention to the fate of the VOC. For the case of a stream containing 3 mol% acetone in air, draw a flow diagram for a process based on absorption. Choose a reasonable absorbent and include in your process a means to recover the acetone and recycle the absorbent.

1.29. Separation of air into nitrogen and oxygen.

Describe three methods suitable for the separation of air into nitrogen and oxygen.

1.30. Separation of an azeotrope.

What methods can be used to separate azeotropic mixtures of water and an organic chemical such as ethanol?

1.31. Recovery of magnesium sulfate from an aqueous stream.

An aqueous stream contains 5% by weight magnesium sulfate. Devise a process, and a process-flow diagram, for the production of dry magnesium sulfate heptahydrate crystals from this stream.

1.32. Separation of a mixture of acetic acid and water.

Explain why the separation of a stream containing 10 wt% acetic acid in water might be more economical by liquid–liquid extraction with ethyl acetate than by distillation.

1.33. Separation of an aqueous solution of bioproducts.

Clostridium beijerinckii is a gram-positive, rod-shaped, motile bacterium. Its BA101 strain can ferment starch from corn to a mixture of acetone (A), *n*-butanol (B), and ethanol (E) at 37°C under anaerobic conditions, with a yield of more than 99%. Typically, the molar ratio of bioproducts is 3(A):6(B):1(E). When a semidefined fermentation medium containing glucose or maltodextrin supplemented with sodium acetate is used, production at a titer of up to 33 g of bioproducts per liter of water in the broth is possible. After removal of solid biomass from the broth by centrifugation, the remaining liquid is distilled in a sequence of distillation columns to recover: (1) acetone with a maximum of 10% water; (2) ethanol with a maximum of 10% water; (3) *n*-butanol (99.5% purity with a maximum of 0.5% water); and (4) water (W), which can be recycled to the fermentation reactor. If the four products distill according to their normal boiling points in °C of 56.5 (A), 117 (B), 78.4 (E), and 100 (W), devise a suitable distillation sequence using the heuristics of §1.7.3.

Chapter 2

Thermodynamics of Separation Operations

§2.0 INSTRUCTIONAL OBJECTIVES

After completing this chapter, you should be able to:

- Make energy, entropy, and availability balances around a separation process.
- Explain phase equilibria in terms of Gibbs free energy, chemical potential, fugacity, fugacity coefficient, activity, and activity coefficient.
- Understand the usefulness of equilibrium ratios (K-values and partition coefficients) for liquid and vapor phases.
- Derive K-value expressions in terms of fugacity coefficients and activity coefficients.
- Explain how computer programs use equations of state (e.g., Soave–Redlich–Kwong or Peng–Robinson) to compute thermodynamic properties of vapor and liquid mixtures, including K-values.
- Explain how computer programs use liquid-phase activity-coefficient correlations (e.g., Wilson, NRTL, UNIQUAC, or UNIFAC) to compute thermodynamic properties, including K-values.
- For a given weak acid or base (including amino acids), calculate pH, pK_a, degree of ionization, pI, and net charge.
- Identify a buffer suited to maintain activity of a biological species at a target pH and evaluate effects of temperature, ionic strength, solvent and static charge on pH, and effects of pH on solubility.
- Determine effects of electrolyte composition on electrostatic double-layer dimensions, energies of attraction, critical flocculation concentration, and structural stability of biocolloids.
- Characterize forces that govern ligand–receptor–binding interactions and evaluate dissociation constants from free energy changes or from batch solution or continuous sorption data.

Thermodynamic properties play a major role in separation operations with respect to energy requirements, phase equilibria, biological activity, and equipment sizing. This chapter develops equations for energy balances, for entropy and availability balances, and for determining densities and compositions for phases at equilibrium. The equations contain thermodynamic properties, including specific volume, enthalpy, entropy, availability, fugacities, and activities, all as functions of temperature, pressure, and composition. Both ideal and nonideal mixtures are discussed. Equations to determine ionization state, solubility, and interaction forces of biomolecular species are introduced. However, this chapter is not a substitute for any of the excellent textbooks on thermodynamics.

Experimental thermodynamic property data should be used, when available, to design and analyze the operation of separation equipment. When not available, properties can often be estimated with reasonable accuracy. Many of these estimation methods are discussed in this chapter. The most comprehensive source of thermodynamic properties for pure compounds and nonelectrolyte and electrolyte mixtures—including excess volume, excess enthalpy, activity coefficients at infinite dilution, azeotropes, and vapor–liquid,

liquid–liquid, and solid–liquid equilibrium—is the computerized Dortmund Data Bank (DDB) (www.ddbst.com), initiated by Gmehling and Onken in 1973. It is updated annually and is widely used by industry and academic institutions. In 2009, the DDB contained more than 3.9 million data points for 32,000 components from more than 64,000 references. Besides openly available data from journals, DDB contains a large percentage of data from non-English sources, chemical industry, and MS and PhD theses.

§2.1 ENERGY, ENTROPY, AND AVAILABILITY BALANCES

Industrial separation operations utilize large quantities of energy in the form of heat and/or shaft work. Distillation separations account for about 3% of the total U.S. energy consumption (Mix et al. [1]). The distillation of crude oil into its fractions is very energy-intensive, requiring about 40% of the total energy used in crude-oil refining. Thus, it is important to know the energy consumption in a separation process, and to what degree energy requirements can be reduced.

Consider the continuous, steady-state, flow system for the separation process in Figure 2.1. One or more feed streams flowing into the system are separated into two or more

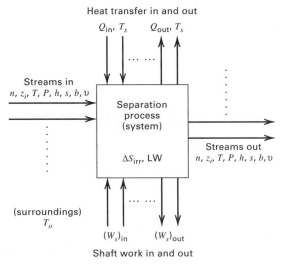

Figure 2.1 General separation system.

product streams. For each stream molar flow rates are denoted by n, the component mole fractions by z_i, the temperature by T, the pressure by P, the molar enthalpies and entropies by h and s, respectively, and the molar availabilities by b. If chemical reactions occur in the process, enthalpies and entropies are referred to the elements, as discussed by Felder and Rousseau [2]; otherwise they can be referred to the compounds. Flows of heat in or out are denoted by Q, and shaft work crossing the boundary of the system by W_s. At steady state, if kinetic, potential, and surface energy changes are neglected, the first law of thermodynamics states that the sum of energy flows into the system equals the sum of the energy flows leaving the system.

In terms of symbols, the energy balance is given by Eq. (1) in Table 2.1, where all flow-rate, heat-transfer, and shaft-work terms are positive. Molar enthalpies may be positive or negative, depending on the reference state.

The first law of thermodynamics provides no information on energy efficiency, but the second law of thermodynamics, given by Eq. (2) in Table 2.1, does.

In the entropy balance, the heat sources and sinks in Figure 2.1 are at absolute temperatures, T_s. For example, if condensing steam at 150°C supplies heat, Q, to the reboiler of a distillation column, $T_s = 150 + 273 = 423$ K. Unlike the energy balance, which states that energy is conserved, the entropy balance predicts the production of entropy, ΔS_{irr}, which is the irreversible increase in the entropy of the universe. This term, which must be positive, is a measure of the thermodynamic inefficiency. In the limit, as a reversible process is approached, ΔS_{irr} tends to zero. Unfortunately, ΔS_{irr} is difficult to apply because it does not have the units of energy/unit time (power).

A more useful measure of process inefficiency is lost work, LW. It is derived by combining Eqs. (1) and (2) to obtain a combined statement of the first and second laws, which is given as Eq. (3) in Table 2.1. To perform this derivation, it is first necessary to define an infinite source or sink available for heat transfer at the absolute temperature, $T_s = T_0$, of the surroundings. This temperature, typically 300 K, represents the largest source of coolant (heat sink) available. This might

Table 2.1 Universal Thermodynamic Laws for a Continuous, Steady-State, Flow System

Energy balance:

$$(1) \quad \sum_{\substack{\text{out of} \\ \text{system}}} (nh + Q + W_s) - \sum_{\substack{\text{in to} \\ \text{system}}} (nh + Q + W_s) = 0$$

Entropy balance:

$$(2) \quad \sum_{\substack{\text{out of} \\ \text{system}}} \left(ns + \frac{Q}{T_s} \right) - \sum_{\substack{\text{in to} \\ \text{system}}} \left(ns + \frac{Q}{T_s} \right) = \Delta S_{irr}$$

Availability balance:

$$(3) \quad \sum_{\substack{\text{in to} \\ \text{system}}} \left[nb + Q\left(1 - \frac{T_0}{T_s}\right) + W_s \right]$$
$$- \sum_{\substack{\text{out of} \\ \text{system}}} \left[nb + Q\left(1 - \frac{T_0}{T_s}\right) + W_s \right] = LW$$

Minimum work of separation:

$$(4) \quad W_{min} = \sum_{\substack{\text{out of} \\ \text{system}}} nb - \sum_{\substack{\text{in to} \\ \text{system}}} nb$$

Second-law efficiency:

$$(5) \quad \eta = \frac{W_{min}}{LW + W_{min}}$$

where $b = h - T_0 s$ = availability function
$LW = T_0 \Delta S_{irr}$ = lost work

be the average temperature of cooling water, air, or a nearby river, lake, or ocean. Heat transfer associated with this coolant and transferred from (or to) the process is Q_0. Thus, in both (1) and (2) in Table 2.1, the Q and Q/T_s terms include contributions from Q_0 and Q_0/T_0.

The derivation of (3) in Table 2.1 is made, as shown by de Nevers and Seader [3], by combining (1) and (2) to eliminate Q_0. It is referred to as an *availability* (or *exergy*) balance, where availability means "available for complete conversion to shaft work." The availability function, b, a thermodynamic property like h and s, is defined by

$$b = h - T_0 s \qquad (2\text{-}1)$$

and is a measure of the maximum amount of energy converted into shaft work if the stream is taken to the reference state. It is similar to Gibbs free energy, $g = h - Ts$, but differs in that the temperature, T_0, appears in the definition instead of T. Terms in (3) in Table 2.1 containing Q are multiplied by $(1 - T_0/T_s)$, which, as shown in Figure 2.2, is the reversible Carnot heat-engine cycle efficiency, representing the maximum amount of shaft work producible from Q at T_s, where the residual amount of energy $(Q - W_s)$ is transferred as heat to a sink at T_0. Shaft work, W_s, remains at its full value in (3). Thus, although Q and W_s have the same thermodynamic worth in (1) of Table 2.1, heat transfer has less worth in (3). Shaft work can be converted completely to heat, but heat cannot be converted completely to shaft work.

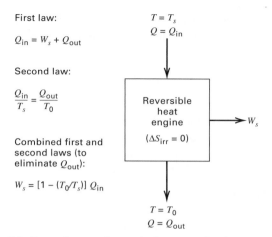

First law:

$$Q_{in} = W_s + Q_{out}$$

Second law:

$$\frac{Q_{in}}{T_s} = \frac{Q_{out}}{T_0}$$

Combined first and second laws (to eliminate Q_{out}):

$$W_s = [1 - (T_0/T_s)]\, Q_{in}$$

Figure 2.2 Carnot heat-engine cycle for converting heat to shaft work.

Availability, like entropy, is not conserved in a real, irreversible process. The total availability (i.e., ability to produce shaft work) into a system is always greater than the total availability leaving. Thus (3) in Table 2.1 is written with the "in to system" terms first. The difference is the *lost work*, LW, also called the loss of availability (or exergy), and is

$$LW = T_0 \Delta S_{irr} \qquad (2\text{-}2)$$

Lost work is always positive. The greater its value, the greater the energy inefficiency. In the lower limit, for a reversible process, it is zero. The lost work has the units of energy, thus making it easy to attach significance to its numerical value.

Lost work can be computed from Eq. (3) in Table 2.1. Its magnitude depends on process irreversibilities, which include fluid friction, heat transfer due to finite temperature-driving forces, mass transfer due to finite concentration- or activity-driving forces, chemical reactions proceeding at finite displacements from chemical equilibrium, mixing streams of differing temperature, pressure, and/or composition, etc. To reduce lost work, driving forces for momentum, heat, and mass transfer; and chemical reaction must be reduced. Practical limits to reduction exist because, as driving forces decrease, equipment sizes increase, tending to infinity as driving forces approach zero.

For a separation without chemical reaction, the summation of the stream availability functions leaving the process is usually greater than that for streams entering the process. In the limit for a reversible process (LW = 0), (3) of Table 2.1 reduces to (4), where W_{min} is the minimum shaft work for the separation and is equivalent to the difference in the heat-transfer and shaft-work terms in (3). This minimum work is a property independent of the nature (or path) of the separation. The work of separation for an irreversible process is greater than the minimum value from (4).

Equation (3) of Table 2.1 shows that as a process becomes more irreversible, and thus more energy-inefficient, the increasing LW causes the required work of separation to increase. Thus, the equivalent work of separation for an irreversible process is the sum of lost work and the minimum

work of separation. The *second-law efficiency*, therefore, is defined by (5) in Table 2.1.

EXAMPLE 2.1 Use of Thermodynamic Laws.

For the propylene–propane separation of Figure 1.16, using the following thermodynamic properties and the relations given in Table 2.1, compute in SI units: **(a)** the condenser duty, Q_C; **(b)** the reboiler duty, Q_R; **(c)** the irreversible entropy production, assuming 303 K for the condenser cooling-water sink and 378 K for the reboiler steam source; **(d)** the lost work, assuming $T_0 = 303$ K; **(e)** the minimum work of separation; and **(f)** the second-law efficiency.

Stream	Phase Condition	Enthalpy (h), kJ/kmol	Entropy (s), kJ/kmol-K
Feed (F)	Liquid	13,338	−4.1683
Overhead vapor (OV)	Vapor	24,400	24.2609
Distillate (D) and reflux (R)	Liquid	12,243	−13.8068
Bottoms (B)	Liquid	14,687	−2.3886

Solution

Let Q_C and Q_R cross the boundary of the system. The following calculations are made using the stream flow rates in Figure 1.16 and the properties above.

(a) From (1), Table 2.1, noting that the overhead-vapor molar flow rate is given by $n_{OV} = n_R + n_D$ and $h_R = h_D$, the condenser duty is

$$\begin{aligned} Q_C &= n_{OV}(h_{OV} - h_R) \\ &= (2,293 + 159.2)(24,400 - 12,243) \\ &= 29,811,000 \text{ kJ/h} \end{aligned}$$

(b) An energy balance around the reboiler cannot be made because data are not given for the boilup rate. From (1), Table 2.1, an energy balance around the column is used instead:

$$\begin{aligned} Q_R &= n_D h_D + n_B h_B + Q_C - n_F h_F \\ &= 159.2(12,243) + 113(14,687) \\ &\quad + 29,811,000 - 272.2(13,338) \\ &= 29,789,000 \text{ kJ/h} \end{aligned}$$

(c) Compute the production of entropy from an entropy balance around the entire distillation system. From Eq. (2), Table 2.1,

$$\begin{aligned} \Delta S_{irr} &= n_D s_D + n_B s_B + Q_C/T_C - n_F s_F - Q_R/T_R \\ &= 159.2(-13.8068) + 113(-2.3886) \\ &\quad + 29,811,000/303 - 272.2(-4.1683) \\ &\quad - 29,789,000/378 \\ &= 18,246 \text{ kJ/h-K} \end{aligned}$$

(d) Compute lost work from its definition at the bottom of Table 2.1:

$$\begin{aligned} LW &= T_0 \Delta S_{irr} \\ &= 303(18,246) = 5,529,000 \text{ kJ/h} \end{aligned}$$

Alternatively, compute lost work from an availability balance around the system. From (3), Table 2.1, where the availability function, b, is defined near the bottom of Table 2.1,

$$LW = n_F b_F + Q_R(1 - T_0/T_R)$$
$$-n_D b_D - n_B b_B - Q_C(1 - T_0/T_C)$$
$$= 272.2[13,338 - (303)(-4.1683)]$$
$$+ 29,789,000(1 - 303/378)$$
$$- 159.2[12,243 - (303)(-13.8068)]$$
$$- 113[14,687 - (303)(-2.3886)]$$
$$- 29,811,000(1 - 303/303)$$
$$= 5,529,000 \text{ kJ/h (same result)}$$

(e) Compute the minimum work of separation for the entire distillation system. From (4), Table 2.1,

$$W_{min} = n_D b_D + n_B b_B - n_F b_F$$
$$= 159.2[12,243 - (303)(-13.8068)]$$
$$+ 113[14,687 - (303)(-2.3886)]$$
$$- 272.2[13,338 - (303)(-4.1683)]$$
$$= 382,100 \text{ kJ/h}$$

(f) Compute the second-law efficiency for the entire distillation system. From (5), Table 2.1,

$$\eta = \frac{W_{min}}{LW + W_{min}}$$
$$= \frac{382,100}{5,529,000 + 382,100}$$
$$= 0.0646 \text{ or } 6.46\%$$

This low second-law efficiency is typical of a difficult distillation separation, which in this case requires 150 theoretical stages with a reflux ratio of almost 15 times the distillate rate.

§2.2 PHASE EQUILIBRIA

Many separations are determined by the extent to which the species are distributed among two or more phases at equilibrium at a specified T and P. The distribution is determined by application of the Gibbs free energy. For each phase in a multiphase, multicomponent system, the total Gibbs free energy is

$$G = G(T, P, N_1, N_2, \ldots, N_C)$$

where N_i = moles of species i. At equilibrium, the total G for all phases is a minimum, and methods for determining this are referred to as *free-energy minimization techniques*. Gibbs free energy is also the starting point for the derivation of commonly used equations for phase equilibria. From classical thermodynamics, the total differential of G is

$$dG = -S\,dT + V\,dP + \sum_{i=1}^{C} \mu_i dN_i \qquad (2\text{-}3)$$

where μ_i is the chemical potential or partial molar Gibbs free energy of species i. For a closed system consisting of two or more phases in equilibrium, where each phase is an open system capable of mass transfer with another phase,

$$dG_{system} = \sum_{p=1}^{N}\left[\sum_{i=1}^{C} \mu_i^{(p)} dN_i^{(p)}\right]_{P,T} \qquad (2\text{-}4)$$

where superscript (p) refers to each of N phases. Conservation of moles of species, in the absence of chemical reaction, requires that

$$dN_i^{(1)} = -\sum_{p=2}^{N} dN_i^{(p)} \qquad (2\text{-}5)$$

which, upon substitution into (2-4), gives

$$\sum_{p=2}^{N}\left[\sum_{i=1}^{C}\left(\mu_i^{(p)} - \mu_i^{(1)}\right) dN_i^{(p)}\right] = 0 \qquad (2\text{-}6)$$

With $dN_i^{(1)}$ eliminated in (2-6), each $dN_i^{(p)}$ term can be varied independently of any other $dN_i^{(p)}$ term. But this requires that each coefficient of $dN_i^{(p)}$ in (2-6) be zero. Therefore,

$$\mu_i^{(1)} = \mu_i^{(2)} = \mu_i^{(3)} = \cdots = \mu_i^{(N)} \qquad (2\text{-}7)$$

Thus, the chemical potential of a species in a multicomponent system is identical in all phases at physical equilibrium.

§2.2.1 Fugacities and Activity Coefficients

Chemical potential is not an absolute quantity, and the numerical values are difficult to relate to more easily understood physical quantities. Furthermore, the chemical potential approaches an infinite negative value as pressure approaches zero. Thus, the chemical potential is not favored for phase-equilibria calculations. Instead, *fugacity*, invented by G. N. Lewis in 1901, is employed as a surrogate.

The partial fugacity of species i in a mixture is like a pseudo-pressure, defined in terms of the chemical potential by

$$\bar{f}_i = C \exp\left(\frac{\mu_i}{RT}\right) \qquad (2\text{-}8)$$

where C is a temperature-dependent constant. Regardless of the value of C, it is shown by Prausnitz, Lichtenthaler, and de Azevedo [4] that (2-7) can be replaced with

$$\bar{f}_i^{(1)} = \bar{f}_i^{(2)} = \bar{f}_i^{(3)} = \cdots = \bar{f}_i^{(N)} \qquad (2\text{-}9)$$

Thus, at equilibrium, a given species has the same partial fugacity in each phase. This equality, together with equality of phase temperatures and pressures,

$$T^{(1)} = T^{(2)} = T^{(3)} = \cdots = T^{(N)} \qquad (2\text{-}10)$$

and

$$P^{(1)} = P^{(2)} = P^{(3)} = \cdots = P^{(N)} \qquad (2\text{-}11)$$

constitutes the conditions for phase equilibria. For a pure component, the partial fugacity, $\bar{f}_i$, becomes the pure-component fugacity, f_i. For a pure, ideal gas, fugacity equals the total pressure, and for a component in an ideal-gas mixture, the partial fugacity equals its partial pressure, $p_i = y_i P$. Because of the close relationship between fugacity and pressure, it is convenient to define their ratio for a pure substance as

$$\phi_i = \frac{f_i}{P} \qquad (2\text{-}12)$$

Table 2.2 Thermodynamic Quantities for Phase Equilibria

Thermodynamic Quantity	Definition	Physical Significance	Limiting Value for Ideal Gas and Ideal Solution
Chemical potential	$\mu_i \equiv \left(\dfrac{\partial G}{\partial N_i}\right)_{P,T,N_j}$	Partial molar free energy, $\bar{g}_i$	$\mu_i = \bar{g}_i$
Partial fugacity	$\bar{f}_i \equiv C \exp\left(\dfrac{\mu_i}{RT}\right)$	Thermodynamic pressure	$\bar{f}_{iV} = y_i P$ $\bar{f}_{iL} = x_i P_i^s$
Fugacity coefficient of a pure species	$\phi_i \equiv \dfrac{f_i}{P}$	Deviation to fugacity due to pressure	$\phi_{iV} = 1.0$ $\phi_{iL} = \dfrac{P_i^s}{P}$
Partial fugacity coefficient of a species in a mixture	$\bar{\phi}_{iV} \equiv \dfrac{\bar{f}_{iV}}{y_i P}$ $\bar{\phi}_{iL} \equiv \dfrac{\bar{f}_{iL}}{x_i P}$	Deviations to fugacity due to pressure and composition	$\bar{\phi}_{iV} = 1.0$ $\bar{\phi}_{iL} = \dfrac{P_i^s}{P}$
Activity	$a_i \equiv \dfrac{\bar{f}_i}{f_i^o}$	Relative thermodynamic pressure	$a_{iV} = y_i$ $a_{iL} = x_i$
Activity coefficient	$\gamma_{iV} \equiv \dfrac{a_{iV}}{y_i}$ $\gamma_{iL} \equiv \dfrac{a_{iL}}{x_i}$	Deviation to fugacity due to composition	$\gamma_{iV} = 1.0$ $\gamma_{iL} = 1.0$

where ϕ_i is the pure-species fugacity coefficient, which is 1.0 for an ideal gas. For a mixture, partial fugacity coefficients are

$$\bar{\phi}_{iV} \equiv \frac{\bar{f}_{iV}}{y_i P} \tag{2-13}$$

$$\bar{\phi}_{iL} \equiv \frac{\bar{f}_{iL}}{x_i P} \tag{2-14}$$

such that as ideal-gas behavior is approached, $\bar{\phi}_{iV} \rightarrow 1.0$ and $\bar{\phi}_{iL} \rightarrow P_i^s/P$, where $P_i^s = $ vapor pressure.

At a given temperature, the ratio of the partial fugacity of a component to its fugacity in a standard state is termed the *activity*. If the standard state is selected as the pure species at the same pressure and phase as the mixture, then

$$a_i \equiv \frac{\bar{f}_i}{f_i^o} \tag{2-15}$$

Since at phase equilibrium, the value of f_i^o is the same for each phase, substitution of (2-15) into (2-9) gives another alternative condition for phase equilibria,

$$a_i^{(1)} = a_i^{(2)} = a_i^{(3)} = \cdots = a_i^{(N)} \tag{2-16}$$

For an ideal solution, $a_{iV} = y_i$ and $a_{iL} = x_i$.

To represent departure of activities from mole fractions when solutions are nonideal, *activity coefficients* based on concentrations in mole fractions are defined by

$$\gamma_{iV} \equiv \frac{a_{iV}}{y_i} \tag{2-17}$$

$$\gamma_{iL} \equiv \frac{a_{iL}}{x_i} \tag{2-18}$$

For ideal solutions, $\gamma_{iV} = 1.0$ and $\gamma_{iL} = 1.0$.

For convenient reference, thermodynamic quantities useful in phase equilibria are summarized in Table 2.2.

§2.2.2 *K*-Values

A *phase-equilibrium ratio* is the ratio of mole fractions of a species in two phases at equilibrium. For vapor–liquid systems, the constant is referred to as the *K-value* or *vapor–liquid equilibrium ratio*:

$$K_i \equiv \frac{y_i}{x_i} \tag{2-19}$$

For the liquid–liquid case, the ratio is a *distribution* or *partition coefficient*, or *liquid–liquid equilibrium ratio*:

$$K_{D_i} \equiv \frac{x_i^{(1)}}{x_i^{(2)}} \tag{2-20}$$

For equilibrium-stage calculations, separation factors, like (1-4), are defined by forming ratios of equilibrium ratios. For the vapor–liquid case, *relative volatility* $\alpha_{i,j}$ between components i and j is given by

$$\alpha_{ij} \equiv \frac{K_i}{K_j} \tag{2-21}$$

Separations are easy for very large values of $\alpha_{i,j}$, but become impractical for values close to 1.00.

Similarly for the liquid–liquid case, the *relative selectivity* $\beta_{i,j}$ is

$$\beta_{ij} \equiv \frac{K_{D_i}}{K_{D_j}} \qquad (2\text{-}22)$$

Equilibrium ratios can contain the quantities in Table 2.2 in a variety of formulations. The ones of practical interest are formulated next.

For vapor–liquid equilibrium, (2-9) becomes, for each component,

$$\bar{f}_{iV} = \bar{f}_{iL}$$

To form an equilibrium ratio, partial fugacities are commonly replaced by expressions involving mole fractions. From the definitions in Table 2.2:

$$\bar{f}_{iL} = \gamma_{iL} x_i f_{iL}^o \qquad (2\text{-}23)$$

or $$\bar{f}_{iL} = \bar{\phi}_{iL} x_i P \qquad (2\text{-}24)$$

and $$\bar{f}_{iV} = \bar{\phi}_{iV} y_i P \qquad (2\text{-}25)$$

If (2-24) and (2-25) are used with (2-19), a so-called *equation-of-state form* of the K-value follows:

$$K_i = \frac{\bar{\phi}_{iL}}{\bar{\phi}_{iV}} \qquad (2\text{-}26)$$

Applications of (2-26) include the Starling modification of the Benedict, Webb, and Rubin (B–W–R–S) equation of state [5], the Soave modification of the Redlich–Kwong (S–R–K or R–K–S) equation of state [6], the Peng–Robinson (P–R) equation of state [7], and the Plöcker et al. modification of the Lee–Kesler (L–K–P) equation of state [8].

If (2-23) and (2-25) are used, a so-called *activity coefficient form* of the K-value is obtained:

$$K_i = \frac{\gamma_{iL} f_{iL}^o}{\bar{\phi}_{iV} P} = \frac{\gamma_{iL} \phi_{iL}}{\bar{\phi}_{iV}} \qquad (2\text{-}27)$$

Since 1960, (2-27) has received some attention, with applications to industrial systems presented by Chao and Seader (C–S) [9], with a modification by Grayson and Streed [10].

Table 2.3 is a summary of formulations for vapor–liquid equilibrium K-values. Included are the two rigorous expressions (2-26) and (2-27) from which the other approximate formulations are derived. The Raoult's law or ideal K-value is obtained from (2-27) by substituting, from Table 2.2, for an ideal gas and for ideal gas and liquid solutions, $\gamma_{iL} = 1.0$, $\phi_{iL} = P_i^s/P$ and $\bar{\phi}_{iV} = 1.0$. The modified Raoult's law relaxes the assumption of an ideal liquid by including the liquid-phase activity coefficient. The Poynting-correction form for moderate pressures is obtained by approximating the pure-component liquid fugacity coefficient in (2-27) by

$$\phi_{iL} = \phi_{iV}^s \frac{P_i^s}{P} \exp\left(\frac{1}{RT} \int_{P_i^s}^{P} \upsilon_{iL} dP\right) \qquad (2\text{-}28)$$

where the exponential term is the Poynting correction. If the liquid molar volume is reasonably constant over the pressure range, the integral in (2-28) becomes $\upsilon_{iL}\left(P - P_i^s\right)$.

For a light gas species, whose critical temperature is less than the system temperature, the Henry's law form for the K-value is convenient, provided H_i, the Henry's law coefficient, is available. This constant depends on composition, temperature, and pressure. Included in Table 2.3 are recommendations for the application of each of the vapor–liquid K-value expressions.

Table 2.3 Useful Expressions for Estimating K-Values for Vapor–Liquid Equilibria ($K_i = y_i/x_i$)

	Equation	Recommended Application
Rigorous forms:		
(1) Equation-of-state	$K_i = \dfrac{\bar{\phi}_{iL}}{\bar{\phi}_{iV}}$	Hydrocarbon and light gas mixtures from cryogenic temperatures to the critical region
(2) Activity coefficient	$K_i = \dfrac{\gamma_{iL}\phi_{iL}}{\bar{\phi}_{iV}}$	All mixtures from ambient to near-critical temperature
Approximate forms:		
(3) Raoult's law (ideal)	$K_i = \dfrac{P_i^s}{P}$	Ideal solutions at near-ambient pressure
(4) Modified Raoult's law	$K_i = \dfrac{\gamma_{iL} P_i^s}{P}$	Nonideal liquid solutions at near-ambient pressure
(5) Poynting correction	$K_i = \gamma_{iL}\phi_{iV}^s \left(\dfrac{P_i^s}{P}\right) \exp\left(\dfrac{1}{RT}\displaystyle\int_{P_i^s}^{P}\upsilon_{iL} dP\right)$	Nonideal liquid solutions at moderate pressure and below the critical temperature
(6) Henry's law	$K_i = \dfrac{H_i}{P}$	Low-to-moderate pressures for species at supercritical temperature

Regardless of which thermodynamic formulation is used for estimating K-values, their accuracy depends on the correlations used for the thermodynamic properties (vapor pressure, activity coefficient, and fugacity coefficients). For practical applications, the choice of K-value formulation is a compromise among accuracy, complexity, convenience, and past experience.

For liquid–liquid equilibria, (2-9) becomes

$$\bar{f}_{iL}^{(1)} = \bar{f}_{iL}^{(2)} \qquad (2\text{-}29)$$

where superscripts (1) and (2) refer to the immiscible liquids. A rigorous formulation for the distribution coefficient is obtained by combining (2-23) with (2-20) to obtain an expression involving only activity coefficients:

$$K_{D_i} = \frac{x_i^{(1)}}{x_i^{(2)}} = \frac{\gamma_{iL}^{(2)} f_{iL}^{o(2)}}{\gamma_{iL}^{(1)} f_{iL}^{o(1)}} = \frac{\gamma_{iL}^{(2)}}{\gamma_{iL}^{(1)}} \qquad (2\text{-}30)$$

For vapor–solid equilibria, if the solid phase consists of just one of the components of the vapor phase, combination of (2-9) and (2-25) gives

$$f_{iS} = \bar{\phi}_{iV} y_i P \qquad (2\text{-}31)$$

At low pressure, $\bar{\phi}_{iV} = 1.0$ and the fugacity of the solid is approximated by its vapor pressure. Thus for the vapor-phase mole fraction of the component forming the solid phase:

$$y_i = \frac{\left(P_i^s\right)_{\text{solid}}}{P} \qquad (2\text{-}32)$$

For liquid–solid equilibria, if the solid phase is a pure component, the combination of (2-9) and (2-23) gives

$$f_{iS} = \gamma_{iL} x_i f_{iL}^o \qquad (2\text{-}33)$$

At low pressure, fugacity of a solid is approximated by vapor pressure to give, for a component in the solid phase,

$$x_i = \frac{\left(P_i^s\right)_{\text{solid}}}{\gamma_{iL}\left(P_i^s\right)_{\text{liquid}}} \qquad (2\text{-}34)$$

EXAMPLE 2.2 K-Values from Raoult's and Henry's Laws.

Estimate the K-values and α of a vapor–liquid mixture of water (W) and methane (M) at $P = 2$ atm, $T = 20$ and $80°C$. What is the effect of T on the component distribution?

Solution

At these conditions, water exists mainly in the liquid phase and will follow Raoult's law, as given in Table 2.3. Because methane has a critical temperature of $-82.5°C$, well below the temperatures of interest, it will exist mainly in the vapor phase and follow Henry's law, in the form given in Table 2.3. From *Perry's Chemical*

Engineers' Handbook, 6th ed., pp. 3-237 and 3-103, the vapor pressure data for water and Henry's law coefficients for CH_4 are:

T, °C	P^s for H_2O, atm	H for CH_4, atm
20	0.02307	3.76×10^4
80	0.4673	6.82×10^4

K-values for water and methane are estimated from (3) and (6), respectively, in Table 2.3, using $P = 2$ atm, with the following results:

T, °C	K_{H_2O}	K_{CH_4}	$\alpha_{M,W}$
20	0.01154	18,800	1,629,000
80	0.2337	34,100	146,000

These K-values confirm the assumptions of the phase distribution of the two species. The K-values for H_2O are low, but increase with temperature. The K-values for methane are extremely high and do not change rapidly with temperature.

§2.3 IDEAL-GAS, IDEAL-LIQUID-SOLUTION MODEL

Classical thermodynamics provides a means for obtaining fluid properties in a consistent manner from P–υ–T relationships, which are *equation-of-state* models. The simplest model applies when both liquid and vapor phases are ideal solutions (all activity coefficients equal 1.0) and the vapor is an ideal gas. Then the thermodynamic properties can be computed from unary constants for each species using the equations given in Table 2.4. These ideal equations apply only at pressures up to about 50 psia (345 kPa), for components of similar molecular structure.

For the vapor, the molar volume, υ, and mass density, ρ, are computed from (1), the ideal-gas law in Table 2.4, which requires the mixture molecular weight, M, and the gas constant, R. It assumes that Dalton's law of additive partial pressures and Amagat's law of additive volumes apply.

The vapor enthalpy, h_V, is computed from (2) by integrating an equation in temperature for the zero-pressure heat capacity at constant pressure, $C_{P_V}^o$, starting from a reference (datum) temperature, T_0, to the temperature of interest, and then summing the resulting species vapor enthalpies on a mole-fraction basis. Typically, T_0 is taken as 0 K or 25°C. Pressure has no effect on the enthalpy of an ideal gas. A common empirical representation of the effect of temperature on the zero-pressure vapor heat capacity of a component is the fourth-degree polynomial:

$$C_{P_V}^o = \left[a_0 + a_1 T + a_2 T^2 + a_3 T^3 + a_4 T^4\right] R \qquad (2\text{-}35)$$

where the constants depend on the species. Values of the constants for hundreds of compounds, with T in K, are tabulated by Poling, Prausnitz, and O'Connell [11]. Because $C_P = dh/dT$, (2-35) can be integrated for each species to give the

Table 2.4 Thermodynamic Properties for Ideal Mixtures

Ideal gas and ideal-gas solution:

$$(1) \quad \upsilon_V = \frac{V}{\sum_{i=1}^{C} N_i} = \frac{M}{\rho_V} = \frac{RT}{P}, \quad M = \sum_{i=1}^{C} y_i M_i$$

$$(2) \quad h_V = \sum_{i=1}^{C} y_i \int_{T_0}^{T} (C_P^o)_{iV} \, dT = \sum_{i=1}^{C} y_i h_{iv}^o$$

$$(3) \quad s_V = \sum_{i=1}^{C} y_i \int_{T_0}^{T} \frac{(C_P^o)_{iV}}{T} \, dT - R \ln\left(\frac{P}{P_0}\right) - R \sum_{i=1}^{C} y_i \ln y_i$$

where the first term is s_V^o

Ideal-liquid solution:

$$(4) \quad \upsilon_L = \frac{V}{\sum_{i=1}^{C} N_i} = \frac{M}{\rho_L} = \sum_{i=1}^{C} x_i \upsilon_{iL}, \quad M = \sum_{i=1}^{C} x_i M_i$$

$$(5) \quad h_L = \sum_{i=1}^{C} x_i \left(h_{iV}^o - \Delta H_i^{\text{vap}} \right)$$

$$(6) \quad s_L = \sum_{i=1}^{C} x_i \left[\int_{T_0}^{T} \frac{(C_P^o)_{iV}}{T} \, dT - \frac{\Delta H_i^{\text{vap}}}{T} \right]$$
$$- R \ln\left(\frac{P}{P_0}\right) - R \sum_{i=1}^{C} x_i \ln x_i$$

Vapor–liquid equilibria:

$$(7) \quad K_i = \frac{P_i^s}{P}$$

Reference conditions (datum): h, ideal gas at T_0 and zero pressure; s, ideal gas at T_0 and 1 atm pressure.
Refer to elements if chemical reactions occur; otherwise refer to components.

ideal-gas species molar enthalpy:

$$h_V^o = \int_{T_0}^{T} C_{P_V}^o \, dT = \sum_{k=1}^{5} \frac{a_{k-1}\left(T^k - T_0^k\right) R}{k} \quad (2\text{-}36)$$

The vapor entropy is computed from (3) in Table 2.4 by integrating $C_{P_V}^o / T$ from T_0 to T for each species; summing on a mole-fraction basis; adding a term for the effect of pressure referenced to a datum pressure, P_0, which is generally taken to be 1 atm (101.3 kPa); and adding a term for the entropy change of mixing. Unlike the ideal vapor enthalpy, the ideal vapor entropy includes terms for the effects of pressure and mixing. The reference pressure is not zero, because the entropy is infinity at zero pressure. If (2-35) is used for the heat capacity,

$$\int_{T_0}^{T} \left(\frac{C_{P_V}^o}{T}\right) dT = \left[a_0 \ln\left(\frac{T}{T_0}\right) + \sum_{k=1}^{4} \frac{a_k\left(T^k - T_0^k\right)}{k} \right] R \quad (2\text{-}37)$$

The liquid molar volume and mass density are computed from the pure species using (4) in Table 2.4 and assuming additive volumes (not densities). The effect of temperature on pure-component liquid density from the freezing point to the critical region at saturation pressure is correlated well by the two-constant equation of Rackett [12]:

$$\rho_L = AB^{-(1-T/T_c)^{2/7}} \quad (2\text{-}38)$$

where values of the empirical constants A and B, and the critical temperature, T_c, are tabulated for approximately 700 organic compounds by Yaws et al. [13].

The vapor pressure of a liquid species is well represented over temperatures from below the normal boiling point to the critical region by an extended Antoine equation:

$$\ln P^s = k_1 + k_2/(k_3 + T) + k_4 T + k_5 \ln T + k_6 T^{k_7} \quad (2\text{-}39)$$

where the constants k_k depend on the species. Values of these constants for hundreds of compounds are built into the physical-property libraries of all process simulation programs. Constants for other vapor-pressure equations are tabulated by Poling et al. [11]. At low pressures, the enthalpy of vaporization is given in terms of vapor pressure by classical thermodynamics:

$$\Delta H^{\text{vap}} = RT^2 \left(\frac{d \ln P^s}{dT}\right) \quad (2\text{-}40)$$

If (2-39) is used for the vapor pressure, (2-40) becomes

$$\Delta H^{\text{vap}} = RT^2 \left[-\frac{k_2}{(k_3 + T)^2} + k_4 + \frac{k_5}{T} + k_7 k_6 T^{k_7-1} \right] \quad (2\text{-}41)$$

The enthalpy of an ideal-liquid mixture is obtained by subtracting the enthalpy of vaporization from the ideal vapor enthalpy for each species, as given by (2-36), and summing, as shown by (5) in Table 2.4. The entropy of the ideal-liquid mixture, given by (6), is obtained in a similar manner from the ideal-gas entropy by subtracting the molar entropy of vaporization, given by $\Delta H^{\text{vap}}/T$.

The final equation in Table 2.4 gives the expression for the ideal K-value, previously included in Table 2.3. It is the K-value based on Raoult's law, given as

$$p_i = x_i P_i^s \quad (2\text{-}42)$$

where the assumption of Dalton's law is also required:

$$p_i = y_i P \quad (2\text{-}43)$$

Combination of (2-42) and (2-43) gives the Raoult's law K-value:

$$\boxed{K_i \equiv \frac{y_i}{x_i} = \frac{P_i^s}{P}} \quad (2\text{-}44)$$

The extended Antoine equation, (2-39), can be used to estimate vapor pressure. The ideal K-value is independent of compositions, but exponentially dependent on temperature because of the vapor pressure, and inversely proportional to pressure. From (2-21), the relative volatility using (2-44) is pressure independent.

EXAMPLE 2.3 Thermodynamic Properties of an Ideal-Gas Mixture.

Styrene is manufactured by catalytic dehydrogenation of ethylbenzene, followed by vacuum distillation to separate styrene from unreacted ethylbenzene [14]. Typical conditions for the feed are 77.5°C (350.6 K) and 100 torr (13.33 kPa), with the following vapor and liquid flows at equilibrium:

Component	n, kmol/h	
	Vapor	Liquid
Ethylbenzene (EB)	76.51	27.31
Styrene (S)	61.12	29.03

Based on the property constants given, and assuming that the ideal-gas, ideal-liquid-solution model of Table 2.4 is suitable at this low pressure, estimate values of v_V, ρ_V, h_V, s_V, v_L, ρ_L, h_L, and s_L in SI units, and the K-values and relative volatility, α.

Property Constants for (2-35), (2-38), (2-39) (In all cases, T is in K)

	Ethylbenzene	Styrene
M, kg/kmol	106.168	104.152
$C_{P_V}^o$, J/kmol-K:		
$a_0 R$	−43,098.9	−28,248.3
$a_1 R$	707.151	615.878
$a_2 R$	−0.481063	−0.40231
$a_3 R$	1.30084×10^{-4}	9.93528×10^{-5}
$a_4 R$	0	0
P^s, Pa:		
k_1	86.5008	130.542
k_2	−7,440.61	−9,141.07
k_3	0	0
k_4	0.00623121	0.0143369
k_5	−9.87052	−17.0918
k_6	4.13065×10^{-18}	1.8375×10^{-18}
k_7	6	6
ρ_L, kg/m³:		
A	289.8	299.2
B	0.268	0.264
T_c, K	617.9	617.1
$R = 8.314$ kJ/kmol-K or kPa-m³/kmol-K = 8,314 J/kmol-K		

Solution

Phase mole-fraction compositions and average molecular weights:

From $y_i = (n_{iV})/n_V, x_i = (n_{iL})/n_L,$

	Ethylbenzene	Styrene
y	0.5559	0.4441
x	0.4848	0.5152

From (1), Table 2.4,

$$M_V = (0.5559)(106.168) + (0.4441)(104.152) = 105.27$$
$$M_L = (0.4848)(106.168) + (0.5152)(104.152) = 105.13$$

Vapor molar volume and density: From (1), Table 2.4,

$$v_V = \frac{RT}{P} = \frac{(8.314)(350.65)}{(13.332)} = 219.2 \text{ m}^3/\text{kmol}$$

$$\rho_V = \frac{M_V}{v_V} = \frac{105.27}{219.2} = 0.4802 \text{ kg/m}^3$$

Vapor molar enthalpy (datum = ideal gas at 298.15 K and 0 kPa):

From (2-36) for ethylbenzene,

$$h_{EB_V}^o = -43098.9(350.65 - 298.15)$$
$$+ \left(\frac{707.151}{2}\right)(350.65^2 - 298.15^2)$$
$$- \left(\frac{0.481063}{3}\right)(350.65^3 - 298.15^3)$$
$$+ \left(\frac{1.30084 \times 10^{-4}}{4}\right)(350.65^4 - 298.15^4)$$
$$= 7,351,900 \text{ J/kmol}$$

Similarly, $\qquad h_{S_V}^o = 6,957,700$ J/kmol

From (2), Table 2.4, for the mixture,

$$h_V = \sum y_i h_{iV}^o = (0.5559)(7,351,900)$$
$$+ (0.4441)(6,957,100) = 7,176,800 \text{ J/kmol}$$

Vapor molar entropy (datum = pure components as vapor at 298.15 K, 101.3 kPa):

From (2-37), for each component,

$$\int_{T_0}^{T} \left(\frac{C_{P_V}^o}{T}\right) dT = 22,662 \text{ J/kmol-K for ethylbenzene}$$
$$\text{and} \quad 21,450 \text{ J/kmol-K for styrene}$$

From (3), Table 2.4, for the mixture,

$$s_V = [(0.5559)(22,662.4) + (0.4441)(21,450.3)$$
$$- 8,314 \ln\left(\frac{13.332}{101.3}\right) - 8,314[(0.5559) \ln (0.5559)$$
$$+ (0.4441) \ln (0.4441)] = 44,695 \text{ J/kmol-K}$$

Note that the pressure effect and the mixing effect are significant.

Liquid molar volume and density:

From (2-38), for ethylbenzene,

$$\rho_{EB_L} = (289.8)(0.268)^{-(1-350.65/617.9)^{2/7}} = 816.9 \text{ kg/m}^3$$

$$v_{EB_L} = \frac{M_{EB}}{\rho_{EB_L}} = 0.1300 \text{ m}^3/\text{kmol}$$

Similarly, $\qquad \rho_{S_L} = 853.0$ kg/m³

$$v_{S_L} = 0.1221 \text{ m}^3/\text{kmol}$$

From (4), Table 2.4, for the mixture,

$$v_L = (0.4848)(0.1300) + (0.5152)(0.1221) = 0.1259 \text{ m}^3/\text{kmol}$$

$$\rho_L = \frac{M_L}{v_L} = \frac{105.13}{0.1259} = 835.0 \text{ kg/m}^3$$

Liquid molar enthalpy (datum = ideal gas at 298.15 K):

Use (5) in Table 2.4 for the mixture. For the enthalpy of vaporization of ethylbenzene, from (2-41),

$$\Delta H_{EB}^{vap} = 8{,}314(350.65)^2 \left[\frac{-(-7{,}440.61)}{(0+350.65)^2} + 0.00623121 \right.$$

$$\left. + \frac{-(9.87052)}{(350.65)} + 6\left(4.13065 \times 10^{-18}\right)(350.65)^5 \right]$$

$$= 39{,}589{,}800 \text{ J/kmol}$$

Similarly, $\qquad \Delta H_S^{vap} = 40{,}886{,}700$ J/kmol

Then, applying (5), Table 2.4, using $h_{EB_V}^o$ and $h_{S_V}^o$ from above,

$$h_L = [(0.4848)(7{,}351{,}900 - 39{,}589{,}800) + (0.5152)(6{,}957{,}700$$

$$- 40{,}886{,}700)] = -33{,}109{,}000 \text{ J/kmol}$$

Liquid molar entropy (datum = pure components as vapor at 298.15 K and 101.3 kPa):

From (6), Table 2.4 for the mixture, using values for $\int_{T_0}^{T} \left(C_{P_V}^o / T\right) dT$ and ΔH^{vap} of EB and S from above,

$$s_L = (0.4848)\left(22{,}662 - \frac{39{,}589{,}800}{350.65}\right)$$

$$+ (0.5152)\left(21{,}450 - \frac{40{,}886{,}700}{350.65}\right)$$

$$- 8{,}314 \ln\left(\frac{13.332}{101.3}\right)$$

$$- 8{,}314[0.4848 \ln(0.4848) + 0.5152 \ln(0.5152)]$$

$$= -70{,}150 \text{ J/kmol-K}$$

K-values: Because (7), Table 2.4, will be used to compute the K-values, first estimate the vapor pressures using (2-39). For ethylbenzene,

$$\ln P_{EB}^s = 86.5008 + \left(\frac{-7{,}440.61}{0+350.65}\right)$$

$$+ 0.00623121(350.65) + (-9.87052)\ln(350.65)$$

$$+ 4.13065 \times 10^{-18}(350.65)^6$$

$$= 9.63481$$

$$P_{EB}^s = \exp(9.63481) = 15{,}288 \text{ Pa} = 15.288 \text{ kPa}$$

Similarly, $\qquad P_S^s = 11.492$ kPa

From (7), Table 2.4,

$$K_{EB} = \frac{15.288}{13.332} = 1.147$$

$$K_S = \frac{11.492}{13.332} = 0.862$$

Relative volatility: From (2-21),

$$\alpha_{EB,S} = \frac{K_{EB}}{K_S} = \frac{1.147}{0.862} = 1.331$$

§2.4 GRAPHICAL CORRELATIONS OF THERMODYNAMIC PROPERTIES

Plots of thermodynamic properties are useful not only for the data they contain, but also for the pictoral representation, which permits the user to make general observations, establish correlations, and make extrapolations. All process simulators that contain modules that calculate thermodynamic properties also contain programs that allow the user to make plots of the computed variables. Handbooks and all thermodynamic textbooks contain generalized plots of thermodynamic properties as a function of temperature and pressure. A typical plot is Figure 2.3, which shows vapor pressures of common chemicals for temperatures from below the normal boiling point to the critical temperature where the vapor pressure curves terminate. These curves fit the extended Antoine equation (2-39) reasonably well and are useful in establishing the phase of a pure species and for estimating Raoult's law K-values.

Nomographs for determining effects of temperature and pressure on K-values of hydrocarbons and light gases are presented in Figures 2.4 and 2.5, which are taken from Hadden and Grayson [15]. In both charts, all K-values collapse to 1.0 at a pressure of 5,000 psia (34.5 MPa). This *convergence pressure* depends on the boiling range of the components in the mixture. In Figure 2.6 the components (N_2 to nC_{10}) cover a wide boiling-point range, resulting in a convergence pressure of close to 2,500 psia. For narrow-boiling mixtures such as ethane and propane, the convergence pressure is generally less than 1,000 psia. The K-value charts of Figures 2.4 and 2.5 apply to a convergence pressure of 5,000 psia. A procedure for correcting for the convergence pressure is given by Hadden and Grayson [15]. Use of the nomographs is illustrated in Exercise 2.4.

No simple charts are available for estimating liquid–liquid distribution coefficients because of the pronounced effect of composition. However, for ternary systems that are dilute in the solute and contain almost immiscible solvents, a tabulation of distribution (partition) coefficients for the solute is given by Robbins [16].

EXAMPLE 2.4   K-Values from a Nomograph.

Petroleum refining begins with distillation of crude oil into different boiling-range fractions. The fraction boiling from 0 to 100°C is light naphtha, a blending stock for gasoline. The fraction boiling from 100 to 200°C, the heavy naphtha, undergoes subsequent processing. One such process is steam cracking, which produces a gas containing ethylene, propylene, and other compounds including benzene and toluene. This gas is sent to a distillation train to separate the mixture into a dozen or more products. In the first column, hydrogen and methane are removed by distillation at 3.2 MPa (464 psia). At a tray in the column where the temperature is 40°F, use Figure 2.4 to estimate K-values for H_2, CH_4, C_2H_4, and C_3H_6.

Solution

The K-value of hydrogen depends on the other compounds in the mixture. Because benzene and toluene are present, locate a point A

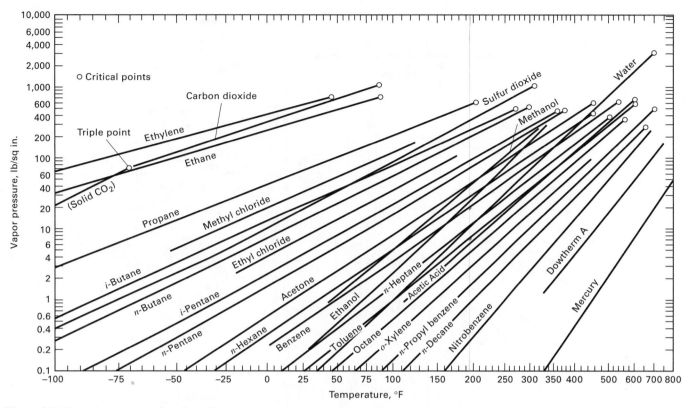

Figure 2.3 Vapor pressure as a function of temperature.

[Adapted from A.S. Faust, L.A. Wenzel, C.W. Clump, L. Maus, and L.B. Andersen, *Principles of Unit Operations*, John Wiley & Sons, New York (1960).]

midway between the points for "H_2 in benzene" and "H_2 in toluene." Next, locate a point B at 40°F and 464 psia on the *T-P* grid. Connect points A and B with a straight line and read $K = 100$ where the line intersects the *K* scale. With the same location for point B, read $K = 11$ for methane. For ethylene (ethene) and propylene (propene), the point A is located on the normal boiling-point scale; the same point is used for B. Resulting *K*-values are 1.5 and 0.32, respectively.

§2.5 NONIDEAL THERMODYNAMIC PROPERTY MODELS

Two types of models are used: (1) *P–υ–T* equation-of-state models and (2) activity coefficient or free-energy models. Their applicability depends on the nature of the components in the mixture and the reliability of the equation constants.

§2.5.1 *P–υ–T* Equation-of-State Models

A relationship between molar volume, temperature, and pressure is a *P–υ–T* equation of state. Numerous such equations have been proposed. The simplest is the ideal-gas law, which applies only at low pressures or high temperatures because it neglects the volume occupied by the molecules and the intermolecular forces. All other equations of state attempt to correct for these two deficiencies. The most widely used equations of state are listed in Table 2.5. These and other such equations are discussed by Poling et al. [11].

Not included in Table 2.5 is the van der Waals equation, $P = RT/(υ - b) - a/υ^2$, where a and b are species-dependent constants. The van der Waals equation was the first successful formulation of an equation of state for a nonideal gas. It is rarely used anymore because of its narrow range of application. However, its development suggested that all species have equal reduced molar volumes, $υ_r = υ/υ_c$, at the same reduced temperature, $T_r = T/T_c$, and reduced pressure, $P_r = P/P_c$. This finding, referred to as the *law of corresponding states*, was utilized to develop (2) in Table 2.5 and defines the *compressibility factor*, Z, which is a function of P_r, T_r, and the critical compressibility factor, Z_c, or the *acentric factor*, ω. This was introduced by Pitzer et al. [17] to account for differences in molecular shape and is determined from the vapor pressure curve by:

$$\omega = \left[-\log\left(\frac{P^s}{P_c}\right)_{T_r=0.7} \right] - 1.000 \qquad (2\text{-}45)$$

The value for ω is zero for symmetric molecules. Some typical values of ω are 0.264, 0.490, and 0.649 for toluene, *n*-decane, and ethyl alcohol, respectively, as taken from the extensive tabulation of Poling et al. [11].

In 1949, Redlich and Kwong [18] published an equation of state that, like the van der Waals equation, contains only two constants, both of which can be determined from T_c and P_c, by applying the critical conditions

$$\left(\frac{\partial P}{\partial υ}\right)_{T_c} = 0 \quad \text{and} \quad \left(\frac{\partial^2 P}{\partial υ^2}\right)_{T_c} = 0$$

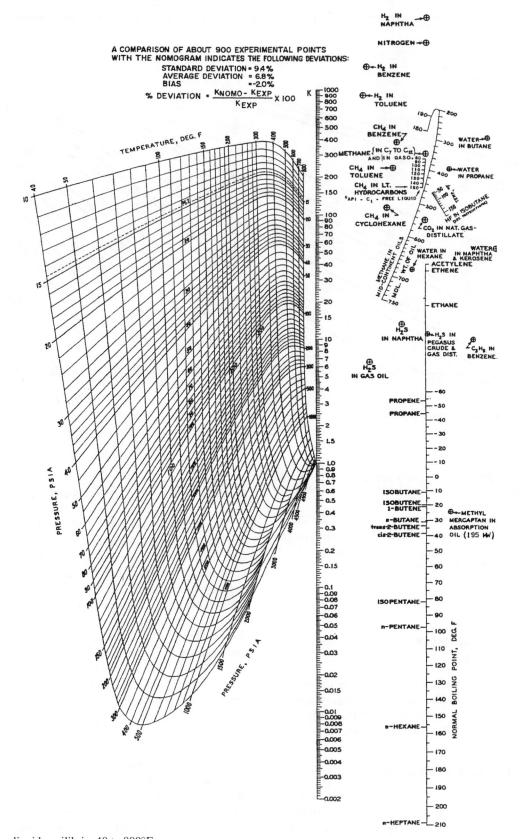

Figure 2.4 Vapor–liquid equilibria, 40 to 800°F.

[From S.T. Hadden and H.G. Grayson, *Hydrocarbon Proc. and Petrol. Refiner*, **40**, 207 (Sept. 1961), with permission.]

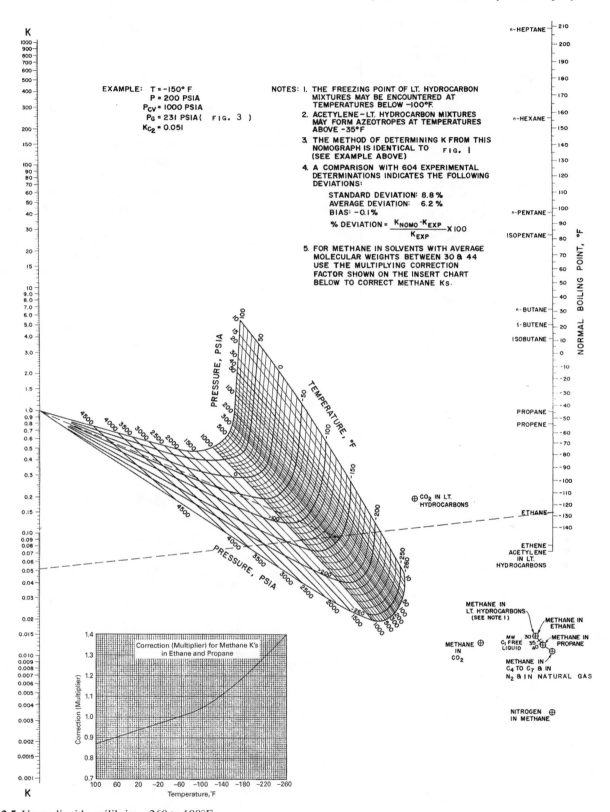

Figure 2.5 Vapor–liquid equilibria, −260 to 100°F.

[From S.T. Hadden and H.G. Grayson, *Hydrocarbon Proc. and Petrol. Refiner*, **40**, 207 (Sept. 1961), with permission.]

The R–K equation, given as (3) in Table 2.5, is an improvement over the van der Waals equation. Shah and Thodos [19] showed that the R–K equation, when applied to nonpolar compounds, has accuracy comparable with that of equations containing many more constants. Furthermore, the R–K equation can approximate the liquid-phase region.

If the R–K equation is expanded to obtain a common denominator, a cubic equation in υ results. Alternatively, (2)

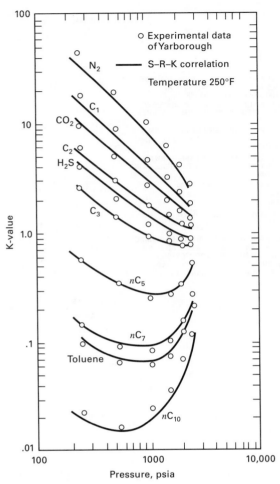

Figure 2.6 Comparison of experimental K-value data and S–R–K correlation.

and (3) in Table 2.5 can be combined to eliminate v to give the compressibility factor, Z, form of the equation:

$$Z^3 - Z^2 + (A - B - B^2)Z - AB = 0 \qquad (2\text{-}46)$$

where

$$A = \frac{aP}{R^2 T^2} \qquad (2\text{-}47)$$

$$B = \frac{bP}{RT} \qquad (2\text{-}48)$$

Equation (2-46), a cubic in Z, can be solved analytically for three roots (e.g., see *Perry's Handbook*, 8th ed., p. 3-10). At supercritical temperatures, where only one phase exists, one real root and a complex conjugate pair of roots are obtained. Below the critical temperature, where vapor and/or liquid phases can exist, three real roots are obtained, with the largest value of Z applying to the vapor and the smallest root corresponding to the liquid (Z_V and Z_L).

The intermediate value of Z is discarded.

To apply the R–K equation to mixtures, *mixing rules* are used to average the constants a and b for each component. The recommended rules for vapor mixtures of C components are

$$a = \sum_{i=1}^{C} \left[\sum_{j=1}^{C} y_i y_j (a_i a_j)^{0.5} \right] \qquad (2\text{-}49)$$

$$b = \sum_{i=1}^{C} y_i b_i \qquad (2\text{-}50)$$

EXAMPLE 2.5 Specific Volume of a Mixture from the R–K Equation.

Use the R–K equation to estimate the specific volume of a vapor mixture containing 26.92 wt% propane at 400°F (477.6 K) and a saturation pressure of 410.3 psia (2,829 kPa). Compare the results with the experimental data of Glanville et al. [20].

Table 2.5 Useful Equations of State

Name	Equation	Equation Constants and Functions
(1) Ideal-gas law	$P = \dfrac{RT}{v}$	None
(2) Generalized	$P = \dfrac{ZRT}{v}$	$Z = Z\{P_r, T_r, Z_c \text{ or } \omega\}$ as derived from data
(3) Redlich–Kwong (R–K)	$P = \dfrac{RT}{v-b} - \dfrac{a}{v^2 + bv}$	$b = 0.08664 RT_c/P_c$ $a = 0.42748 R^2 T_c^{2.5}/P_c T^{0.5}$
(4) Soave–Redlich–Kwong (S–R–K or R–K–S)	$P = \dfrac{RT}{v-b} - \dfrac{a}{v^2 + bv}$	$b = 0.08664 RT_c/P_c$ $a = 0.42748 R^2 T_c^2 [1 + f_\omega(1 - T_r^{0.5})]^2/P_c$ $f_\omega = 0.48 + 1.574\omega - 0.176\omega^2$
(5) Peng–Robinson (P–R)	$P = \dfrac{RT}{v-b} - \dfrac{a}{v^2 + 2bv - b^2}$	$b = 0.07780 RT_c/P_c$ $a = 0.45724 R^2 T_c^2 [1 + f_\omega(1 - T_r^{0.5})]^2/P_c$ $f_\omega = 0.37464 + 1.54226\omega - 0.26992\omega^2$

Solution

Let propane be denoted by P and benzene by B. The mole fractions are

$$y_P = \frac{0.2692/44.097}{(0.2692/44.097) + (0.7308/78.114)} = 0.3949$$

$$y_B = 1 - 0.3949 = 0.6051$$

The critical constants are given by Poling et al. [11]:

	Propane	Benzene
T_c, K	369.8	562.2
P_c, kPa	4,250	4,890

From Table 2.5, b and a for propane in the R–K equation, in SI units, are:

$$b_P = \frac{0.08664(8.3144)(369.8)}{4,250} = 0.06268 \text{ m}^3/\text{kmol}$$

$$a_P = \frac{0.42748(8.3144)^2(369.8)^{2.5}}{(4,250)(477.59)^{0.5}}$$

$$= 836.7 \text{ kPa-m}^6/\text{kmol}^2$$

Similarly, for benzene, $b_B = 0.08263$ m^3/kmol and $a_B = 2,072$ kPa-m^6/kmol^2.

From (2-50),

$$b = (0.3949)(0.06268) + (0.6051)(0.08263) = 0.07475 \text{ m}^3/\text{kmol}$$

From (2-49),

$$a = y_P^2 a_P + 2y_P y_B (a_P a_B)^{0.5} + y_B^2 a_B$$

$$= (0.3949)^2(836.7) + 2(0.3949)(0.6051)[(836.7)(2,072)]^{0.5}$$

$$+ (0.6051)^2(2,072) = 1,518 \text{ kPa-m}^6/\text{kmol}^2$$

From (2-47) and (2-48) using SI units,

$$A = \frac{(1,518)(2,829)}{(8.314)^2(477.59)^2} = 0.2724$$

$$B = \frac{(0.07475)(2,829)}{(8.314)(477.59)^2} = 0.05326$$

From (2-46), the cubic Z form of the R–K equation is obtained:

$$Z^3 - Z^2 + 0.2163Z - 0.01451 = 0$$

This equation gives one real root and a pair of complex roots:

$$Z = 0.7314, \quad 0.1314 + 0.04243i, \quad 0.1314 - 0.04243i$$

The one real root is assumed to be that for the vapor phase. From (2) of Table 2.5, the molar volume is

$$v = \frac{ZRT}{P} = \frac{(0.7314)(8.314)(477.59)}{2,829} = 1.027 \text{ m}^3/\text{kmol}$$

The average molecular weight of the mixture is 64.68 kg/kmol. The specific volume is

$$\frac{v}{M} = \frac{1.027}{64.68} = 0.01588 \text{ m}^3/\text{kg} = 0.2543 \text{ ft}^3/\text{lb}$$

Glanville et al. report experimental values of $Z = 0.7128$ and $v/M = 0.2478$ ft^3/lb, which are within 3% of the estimated values.

Following the work of Wilson [21], Soave [6] added a third parameter, the acentric factor, ω, to the R–K equation. The resulting Soave–Redlich–Kwong (S–R–K) or Redlich–Kwong–Soave (R–K–S) equation, given as (4) in Table 2.5, was accepted for application to mixtures of hydrocarbons because of its simplicity and accuracy. It makes the parameter a a function of ω and T, thus achieving a good fit to vapor pressure data and thereby improving the ability of the equation to predict liquid-phase properties.

Four years after the introduction of the S–R–K equation, Peng and Robinson [7] presented a modification of the R–K and S–R–K equations to achieve improved agreement in the critical region and for liquid molar volume. The Peng–Robinson (P–R) equation of state is (5) in Table 2.5. The S–R–K and P–R equations of state are widely applied in calculations for saturated vapors and liquids. For mixtures of hydrocarbons and/or light gases, the mixing rules are given by (2-49) and (2-50), except that (2-49) is often modified to include a binary interaction coefficient, k_{ij}:

$$a = \sum_{i=1}^{C}\left[\sum_{j=1}^{C} y_i y_j (a_i a_j)^{0.5}(1 - k_{ij})\right] \quad (2\text{-}51)$$

Values of k_{ij} from experimental data have been published for both the S–R–K and P–R equations, e.g., Knapp et al. [22]. Generally k_{ij} is zero for hydrocarbons paired with hydrogen or other hydrocarbons.

Although the S–R–K and P–R equations were not intended to be used for mixtures containing polar organic compounds, they are applied by employing large values of k_{ij} in the vicinity of 0.5, as back-calculated from data. However, a preferred procedure for polar organics is to use a mixing rule such as that of Wong and Sandler, which is discussed in Chapter 11 and which bridges the gap between a cubic equation of state and an activity-coefficient equation.

Another model for polar and nonpolar substances is the virial equation of state due to Thiesen [23] and Onnes [24]. A common representation is a power series in $1/v$ for Z:

$$Z = 1 + \frac{B}{v} + \frac{C}{v^2} + \cdots \quad (2\text{-}52)$$

A modification of the virial equation is the Starling form [5] of the Benedict–Webb–Rubin (B–W–R) equation for hydrocarbons and light gases. Walas [25] presents a discussion of B–W–R-type equations, which—because of the large number of terms and species constants (at least 8)—is not widely used except for pure substances at cryogenic temperatures. A more useful modification of the B–W–R equation is a generalized corresponding-states form developed by Lee and Kesler [26] with an extension to mixtures by Plöcker et al. [8]. All of the constants in the L–K–P equation are given in terms of the acentric factor and reduced temperature and pressure, as developed from $P–v–T$ data for three simple fluids ($\omega = 0$), methane, argon, and krypton, and a reference fluid ($\omega = 0.398$), n-octane. The equations, constants, and mixing rules are given by Walas [25]. The L–K–P equation describes vapor and liquid mixtures of hydrocarbons and/or light gases over wide ranges of T and P.

§2.5.2 Derived Thermodynamic Properties from P–υ–T Models

If a temperature-dependent, ideal-gas heat capacity or enthalpy equation such as (2-35) or (2-36) is available, along with an equation of state, all other vapor- and liquid-phase properties can be derived from the integral equations in Table 2.6. These equations, in the form of departure from the ideal-gas equations of Table 2.4, apply to vapor or liquid.

When the ideal-gas law, $P = RT/\upsilon$, is substituted into Eqs. (1) to (4) of Table 2.6, the results for the vapor are

$$(h - h_V^o) = 0, \quad \phi = 1$$

$$(s - s_V^o) = 0, \quad \phi = 1$$

When the R–K equation is substituted into the equations of Table 2.6, the results for the vapor phase are:

$$h_V = \sum_{i=1}^{C} \left(y_i h_{iV}^o \right) + RT \left[Z_V - 1 - \frac{3A}{2B} \ln \left(1 + \frac{B}{Z_V} \right) \right] \quad (2\text{-}53)$$

$$s_V = \sum_{i=1}^{C} \left(y_i s_{iV}^o \right) - R \ln \left(\frac{P}{P^o} \right) \quad (2\text{-}54)$$
$$- R \sum_{i=1}^{C} \left(y_i \ln y_i \right) + R \ln \left(Z_V - B \right)$$

$$\phi_V = \exp \left[Z_V - 1 - \ln \left(Z_V - B \right) - \frac{A}{B} \ln \left(1 + \frac{B}{Z_V} \right) \right] \quad (2\text{-}55)$$

Table 2.6 Integral Departure Equations of Thermodynamics

At a given temperature and composition, the following equations give the effect of pressure above that for an ideal gas.
Mixture enthalpy:

(1) $\quad \left(h - h_V^o \right) = P\upsilon - RT - \int_{\infty}^{\upsilon} \left[P - T \left(\frac{\partial P}{\partial T} \right)_{\upsilon} \right] d\upsilon$

Mixture entropy:

(2) $\quad \left(s - s_V^o \right) = \int_{\infty}^{\upsilon} \left(\frac{\partial P}{\partial T} \right)_{\upsilon} d\upsilon - \int_{\infty}^{\upsilon} \frac{R}{\upsilon} d\upsilon$

Pure-component fugacity coefficient:

(3) $\quad \phi_{iV} = \exp \left[\frac{1}{RT} \int_{0}^{P} \left(\upsilon - \frac{RT}{P} \right) dP \right]$
$$= \exp \left[\frac{1}{RT} \int_{\upsilon}^{\infty} \left(P - \frac{RT}{\upsilon} \right) d\upsilon - \ln Z + (Z - 1) \right]$$

Partial fugacity coefficient:

(4) $\quad \bar{\phi}_{iV} = \exp \left\{ \frac{1}{RT} \int_{V}^{\infty} \left[\left(\frac{\partial P}{\partial N_i} \right)_{T,V,N_j} - \frac{RT}{V} \right] dV - \ln Z \right\}$

where $V = \upsilon \sum_{i=1}^{C} N_i$

$$\bar{\phi}_{iV} = \exp \left[(Z_V - 1) \frac{B_i}{B} - \ln(Z_V - B) \right. \quad (2\text{-}56)$$
$$\left. - \frac{A}{B} \left(2 \sqrt{\frac{A_i}{A}} - \frac{B_i}{B} \right) \ln \left(1 + \frac{B}{Z_V} \right) \right]$$

The results for the liquid phase are identical if y_i and Z_V (but not h_{iV}^o) are replaced by x_i and Z_L, respectively. The liquid-phase forms of (2-53) and (2-54) account for the enthalpy and entropy of vaporization. This is because the R–K equation of state, as well as the S–R–K and P–R equations, are continuous functions through the vapor and liquid regions, as shown for enthalpy in Figure 2.7. Thus, the liquid enthalpy is determined by accounting for four effects, at a temperature below the critical. From (1), Table 2.6, and Figure 2.7:

$$h_L = h_V^o + P\upsilon - RT - \int_{\infty}^{\upsilon} \left[P - T \left(\frac{\partial P}{\partial T} \right)_{\upsilon} \right] d\upsilon$$
$$= \underbrace{h_V^o}_{}$$

(1) Vapor at zero pressure

$$+ \underbrace{(P\upsilon)_{V_s} - RT - \int_{\infty}^{\upsilon_{V_s}} \left[P - T \left(\frac{\partial P}{\partial T} \right)_{\upsilon} \right] d\upsilon}_{}$$

(2) Pressure correction for vapor to saturation pressure

$$- \underbrace{T \left(\frac{\partial P}{\partial T} \right)_s \left(\upsilon_{V_s} - \upsilon_{L_s} \right)}_{}$$

(3) Latent heat of vaporization

$$+ \underbrace{\left[(P\upsilon)_L - (P\upsilon)_{L_s} \right] - \int_{\upsilon_{L_s}}^{\upsilon_L} \left[P - T \left(\frac{\partial P}{\partial T} \right)_{\upsilon} \right] d\upsilon}_{}$$

(4) Correction to liquid for pressure in excess of saturation pressure

$$(2\text{-}57)$$

where the subscript s refers to the saturation pressure.

The fugacity coefficient, ϕ, of a pure species from the R–K equation, as given by (2-55), applies to the vapor for $P < P_i^s$. For $P > P_i^s$, ϕ is the liquid fugacity coefficient. Saturation pressure corresponds to the condition of $\phi_V = \phi_L$. Thus, at a temperature $T < T_c$, the vapor pressure, P^s, can be

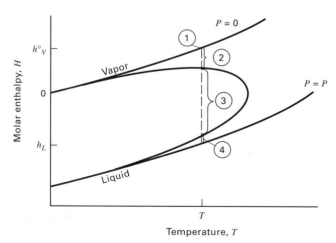

Figure 2.7 Contributions to enthalpy.

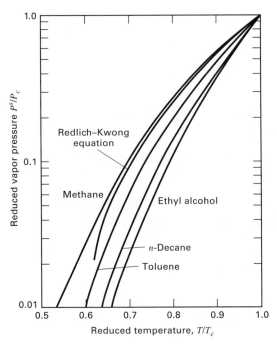

Figure 2.8 Reduced vapor pressure.

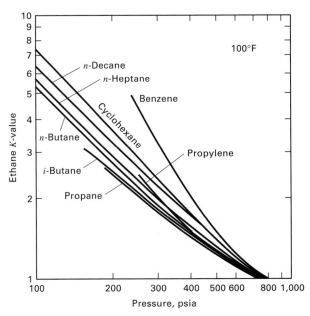

Curves represent experimental data of:
Kay et al. (Ohio State Univ.) Robinson et al. (Univ. Alberta)
Sage et al. (Calif. Inst. Tech.) Thodos (Northwestern)

Figure 2.9 K-values of ethane in binary hydrocarbon mixtures at 100°F.

estimated from the R–K equation of state by setting (2-55) for the vapor equal to (2-55) for the liquid and solving for P, which equals P^s.

The results of Edmister [27] are plotted in Figure 2.8. The R–K vapor-pressure curve does not satisfactorily represent data for a wide range of molecular shapes, as witnessed by the data for methane, toluene, n-decane, and ethyl alcohol. This failure represents a shortcoming of the R–K equation and is why Soave [6] modified the R–K equation by introducing the acentric factor, ω. Thus, while the critical constants T_c and P_c are insufficient to generalize thermodynamic behavior, a substantial improvement results by incorporating a third parameter that represents the generic differences in the reduced-vapor-pressure curves.

As seen in (2-56), partial fugacity coefficients depend on pure-species properties, A_i and B_i, and mixture properties, A and B. Once $\bar{\phi}_{iV}$ and $\bar{\phi}_{iL}$ are computed from (2-56), a K-value can be estimated from (2-26).

The most widely used $P–\upsilon–T$ equations of state are the S–R–K, P–R, and L–K–P. These are combined with the integral departure equations of Table 2.6 to obtain equations for estimating enthalpy, entropy, fugacity coefficients, partial fugacity coefficients, and K-values. The results of the integrations are complex and unsuitable for manual calculations. However, the calculations are readily made by computer programs incorporated into all process simulation programs.

Ideal K-values as determined from Eq. (7) in Table 2.4 depend only on temperature and pressure. Most frequently, they are suitable for mixtures of nonpolar compounds such as paraffins and olefins. Figure 2.9 shows experimental K-value curves for ethane in binary mixtures with other, less volatile hydrocarbons at 100°F (310.93 K), which is close to ethane's critical temperature of 305.6 K, for pressures from 100 psia (689.5 kPa) to *convergence pressures* between 720

and 780 psia (4.964 MPa to 5.378 MPa). At the convergence pressure, separation by distillation is impossible because K-values become 1.0. Figure 2.9 shows that at 100°F, ethane does not form ideal solutions with all the other components because the K-values depend on the other component, even for paraffin homologs. For example, at 300 psia, the K-value of ethane in benzene is 80% higher than the K-value of ethane in propane.

The ability of equations of state, such as S–R–K, P–R, and L–K–P equations, to predict the effects of composition, temperature, and pressure on K-values of mixtures of hydrocarbons and light gases is shown in Figure 2.6. The mixture contains 10 species ranging in volatility from nitrogen to n-decane. The experimental data points, covering almost a 10-fold range of pressure at 250°F, are those of Yarborough [28]. Agreement with the S–R–K equation is very good.

EXAMPLE 2.6   Effect of EOS on Calculations.

In the thermal hydrodealkylation of toluene to benzene ($C_7H_8 + H_2 \rightarrow C_6H_6 + CH_4$), excess hydrogen minimizes cracking of aromatics to light gases. In practice, conversion of toluene per pass through the reactor is only 70%. To separate and recycle hydrogen, hot reactor-effluent vapor of 5,597 kmol/h at 500 psia (3,448 kPa) and 275°F (408.2 K) is partially condensed to 120°F (322 K), with phases separated in a flash drum. If the composition of the reactor effluent is as given below and the flash pressure is 485 psia (3,344 kPa), calculate equilibrium compositions and flow rates of vapor and liquid leaving the drum and the amount of heat transferred, using a process simulation program for each of the equation-of-state models discussed above. Compare the results, including K-values and enthalpy and entropy changes.

Component	Mole Fraction
Hydrogen (H)	0.3177
Methane (M)	0.5894
Benzene (B)	0.0715
Toluene (T)	0.0214
	1.0000

Solution

The computations were made using the S–R–K, P–R, and L–K–P equations of state. The results at 120°F and 485 psia are:

	Equation of State		
	S–R–K	P–R	L–K–P
Vapor flows, kmol/h:			
Hydrogen	1,777.1	1,774.9	1,777.8
Methane	3,271.0	3,278.5	3,281.4
Benzene	55.1	61.9	56.0
Toluene	6.4	7.4	7.0
Total	5,109.6	5,122.7	5,122.2
Liquid flows, kmol/h:			
Hydrogen	1.0	3.3	0.4
Methane	27.9	20.4	17.5
Benzene	345.1	338.2	344.1
Toluene	113.4	112.4	112.8
Total	487.4	474.3	474.8
K-values:			
Hydrogen	164.95	50.50	466.45
Methane	11.19	14.88	17.40
Benzene	0.01524	0.01695	0.01507
Toluene	0.00537	0.00610	0.00575
Enthalpy change, GJ/h	35.267	34.592	35.173
Entropy change, MJ/h-K	−95.2559	−93.4262	−95.0287
Percent of benzene and toluene condensed	88.2	86.7	87.9

Because the reactor effluent is mostly hydrogen and methane, the effluent at 275°F and 500 psia, and the equilibrium vapor at 120°F and 485 psia, are nearly ideal gases (0.98 < Z < 1.00), despite the moderately high pressures. Thus, the enthalpy and entropy changes are dominated by vapor heat capacity and latent heat effects, which are independent of which equation of state is used. Consequently, the enthalpy and entropy changes differ by less than 2%.

Significant differences exist for the K-values of H_2 and CH_4. However, because the values are large, the effect on the amount of equilibrium vapor is small. Reasonable K-values for H_2 and CH_4, based on experimental data, are 100 and 13, respectively. K-values for benzene and toluene differ among the three equations of state by as much as 11% and 14%, respectively, which, however, causes less than a 2% difference in the percentage of benzene and toluene condensed. Raoult's law K-values for benzene and toluene are 0.01032 and 0.00350, which are considerably lower than the values computed from the three equations of state because deviations to fugacities due to pressure are important.

Note that the material balances are always precisely satisfied. Users of simulation programs should never take this as an indication that the results are correct but instead should always verify results in all possible ways.

§2.6 LIQUID ACTIVITY-COEFFICIENT MODELS

Predictions of liquid properties based on *Gibbs free-energy models* for predicting liquid-phase activity coefficients, and other excess functions such as volume and enthalpy of mixing, are developed in this section. Regular-solution theory, which describes mixtures of nonpolar compounds using only constants for the pure components, is presented first, followed by models useful for mixtures containing polar compounds, which require experimentally determined *binary interaction parameters*. If these are not available, group-contribution methods can be used to make estimates. All models can predict vapor–liquid equilibria; and some can estimate liquid–liquid and even solid–liquid and polymer–liquid equilibria.

For polar compounds, dependency of K-values on composition is due to nonideal behavior in the liquid phase. For hydrocarbons, Prausnitz, Edmister, and Chao [29] showed that the relatively simple *regular-solution theory* of Scatchard and Hildebrand [30] can be used to estimate deviations due to nonideal behavior. They expressed K-values in terms of (2-27), $K_i = \gamma_{iL}\phi_{iL}/\bar{\phi}_{iV}$. Chao and Seader [9] simplified and extended application of this equation to hydrocarbons and light gases in the form of a compact set of equations. These were widely used before the availability of the S–R–K and P–R equations.

For hydrocarbon mixtures, regular-solution theory is based on the premise that nonideality is due to differences in van der Waals forces of attraction among the molecules present. Regular solutions have an endothermic heat of mixing, and activity coefficients are greater than 1. These solutions are regular in that molecules are assumed to be randomly dispersed. Unequal attractive forces between like and unlike molecule pairs cause segregation of molecules. However, for regular solutions the species concentrations on a molecular level are identical to overall solution concentrations. Therefore, excess entropy due to segregation is zero and entropy of regular solutions is identical to that of ideal solutions, where the molecules are randomly dispersed.

§2.6.1 Activity Coefficients from Gibbs Free Energy

Activity-coefficient equations are often based on Gibbs free-energy models. The molar Gibbs free energy, g, is the sum of the molar free energy of an ideal solution and an excess molar free energy g^E for nonideal effects. For a liquid

$$g = \sum_{i=1}^{C} x_i g_i + RT \sum_{i=1}^{C} x_i \ln x_i + g^E$$
$$= \sum_{i=1}^{C} x_i \left(g_i + RT \ln x_i + \bar{g}_i^E \right)$$

(2-58)

where $g = h - Ts$ and excess molar free energy is the sum of the partial excess molar free energies, which are related to the liquid-phase activity coefficients by

$$\frac{\bar{g}_i^E}{RT} = \ln \gamma_i = \left[\frac{\partial (N_t g^E / RT)}{\partial N_i} \right]_{P,T,N_j}$$

$$= \frac{g^E}{RT} - \sum_k x_k \left[\frac{\partial (g^E / RT)}{\partial x_k} \right]_{P,T,x_r} \quad (2\text{-}59)$$

where $j \neq i, r \neq k, k \neq i$, and $r \neq i$.

The relationship between excess molar free energy and excess molar enthalpy and entropy is

$$g^E = h^E - Ts^E = \sum_{i=1}^C x_i \left(\bar{h}_i^E - T\bar{s}_i^E \right) \quad (2\text{-}60)$$

§2.6.2 Regular-Solution Model

For a regular liquid solution, the excess molar free energy is based on nonideality due to differences in molecular size and intermolecular forces. The former are expressed in terms of liquid molar volume, and the latter in terms of the enthalpy of vaporization. The resulting model is

$$g^E = \sum_{i=1}^C (x_i \upsilon_{iL}) \left[\frac{1}{2} \sum_{i=1}^C \sum_{j=1}^C \Phi_i \Phi_j (\delta_i - \delta_j)^2 \right] \quad (2\text{-}61)$$

where Φ is the volume fraction assuming additive molar volumes, given by

$$\Phi_i = \frac{x_i \upsilon_{iL}}{\sum_{j=1}^C x_j \upsilon_{jL}} = \frac{x_i \upsilon_{iL}}{\upsilon_L} \quad (2\text{-}62)$$

and δ is the solubility parameter, which is defined in terms of the volumetric internal energy of vaporization as

$$\delta_i = \left(\frac{\Delta E_i^{\text{vap}}}{\upsilon_{iL}} \right)^{1/2} \quad (2\text{-}63)$$

Combining (2-59) with (2-61) yields an expression for the activity coefficient in a regular solution:

$$\ln \gamma_{iL} = \frac{\upsilon_{iL} \left(\delta_i - \sum_{j=1}^C \Phi_j \delta_j \right)^2}{RT} \quad (2\text{-}64)$$

Because $\ln \gamma_{iL}$ varies almost inversely with absolute temperature, υ_{iL} and δ_j are taken as constants at a reference temperature, such as 25°C. Thus, the estimation of γ_L by regular-solution theory requires only the pure-species constants υ_L and δ. The latter parameter is often treated as an empirical constant determined by back-calculation from experimental data. For species with a critical temperature below 25°C, υ_L and δ at 25°C are hypothetical. However, they can be evaluated by back-calculation from data.

When molecular-size differences—as reflected by liquid molar volumes—are appreciable, the Flory–Huggins size correction given below can be added to the regular-solution free-energy contribution:

$$g^E = RT \sum_{i=1}^C x_i \ln \left(\frac{\Phi_i}{x_i} \right) \quad (2\text{-}65)$$

Substitution of (2-65) into (2-59) gives

$$\ln \gamma_{iL} = \ln \left(\frac{\upsilon_{iL}}{\upsilon_L} \right) + 1 - \left(\frac{\upsilon_{iL}}{\upsilon_L} \right) \quad (2\text{-}66)$$

Thus, the activity coefficient of a species in a regular solution, including the Flory–Huggins correction, is

$$\gamma_{iL} = \exp \left[\frac{\upsilon_{iL} \left(\delta_i - \sum_{j=1}^C \Phi_j \delta_j \right)^2}{RT} + \ln \left(\frac{\upsilon_{iL}}{\upsilon_L} \right) + 1 - \frac{\upsilon_{iL}}{\upsilon_L} \right] \quad (2\text{-}67)$$

EXAMPLE 2.7 Activity Coefficients from Regular-Solution Theory.

Yerazunis et al. [31] measured liquid-phase activity coefficients for the n-heptane/toluene system at 1 atm (101.3 kPa). Estimate activity coefficients using regular-solution theory both with and without the Flory–Huggins correction. Compare estimated values with experimental data.

Solution

Experimental liquid-phase compositions and temperatures for 7 of 19 points are as follows, where H denotes heptane and T denotes toluene:

T, °C	x_H	x_T
98.41	1.0000	0.0000
98.70	0.9154	0.0846
99.58	0.7479	0.2521
101.47	0.5096	0.4904
104.52	0.2681	0.7319
107.57	0.1087	0.8913
110.60	0.0000	1.0000

At 25°C, liquid molar volumes are $\upsilon_{H_L} = 147.5 \text{ cm}^3/\text{mol}$ and $\upsilon_{T_L} = 106.8 \text{ cm}^3/\text{mol}$. Solubility parameters are 7.43 and 8.914 $(\text{cal/cm}^3)^{1/2}$, respectively, for H and T. As an example, consider 104.52°C. From (2-62), volume fractions are

$$\Phi_H = \frac{0.2681(147.5)}{0.2681(147.5) + 0.7319(106.8)} = 0.3359$$

$$\Phi_T = 1 - \Phi_H = 1 - 0.3359 = 0.6641$$

Substitution of these values, together with the solubility parameters, into (2-64) gives

$$\gamma_H = \exp \left\{ \frac{147.5[7.430 - 0.3359(7.430) - 0.6641(8.914)]^2}{1.987(377.67)} \right\}$$

$$= 1.212$$

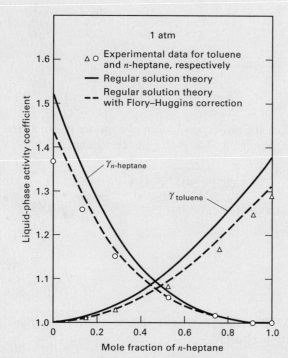

Figure 2.10 Liquid-phase activity coefficients for *n*-heptane/toluene system at 1 atm.

Values of γ_H and γ_T computed in this manner for all seven liquid-phase conditions are plotted in Figure 2.10.

Applying (2-67) at 104.52°C with the Flory–Huggins correction gives

$$\gamma_H = \exp\left[0.1923 + \ln\left(\frac{147.5}{117.73}\right) + 1 - \left(\frac{147.5}{117.73}\right)\right] = 1.179$$

Values of the computed γ_H and γ_T are included in Figure 2.10.

Deviations from experimental data are not greater than 12% for regular-solution theory and not greater than 6% with the Flory–Huggins correction. Unfortunately, good agreement is not always obtained for nonpolar hydrocarbon solutions, as shown, for example, by Hermsen and Prausnitz [32], who studied the cyclopentane/benzene system.

§2.6.3 Nonideal Liquid Solutions

With dissimilar polar species that can form or break hydrogen bonds, the ideal-liquid-solution assumption is invalid and regular-solution theory is also not applicable. Ewell, Harrison, and Berg [33] provide a classification based on the potential for association or solvation due to hydrogen-bond formation. If a molecule contains a hydrogen atom attached to a donor atom (O, N, F, and in certain cases C), the active hydrogen atom can form a bond with another molecule containing a donor atom. The classification in Table 2.7 permits qualitative estimates of deviations from Raoult's law for binary pairs when used in conjunction with Table 2.8. Positive deviations correspond to values of $\gamma_{iL} > 1$. Nonideality results in variations of γ_{iL} with composition, as shown in Figure 2.11 for several binary systems, where the Roman numerals refer to classification in Tables 2.7 and 2.8. Starting with Figure 2.11a, the following explanations for the nonidealities are offered: *n*-heptane (V) breaks ethanol (II) hydrogen bonds, causing strong positive deviations. In Figure 2.11b, similar, but less positive, deviations occur when acetone (III) is added to formamide (I). Hydrogen bonds are broken and formed with chloroform (IV) and methanol (II) in Figure 2.11c, resulting in an unusual deviation curve for chloroform that passes through a maximum. In Figure 2.11d, chloroform (IV) provides active hydrogen atoms that form hydrogen bonds with oxygen atoms of acetone (III), thus causing negative deviations. For water (I) and *n*-butanol (II) in Figure 2.11e, hydrogen bonds of both molecules are broken, and nonideality is sufficiently strong to cause formation of two immiscible liquid phases.

Nonideal-solution effects can be incorporated into K-value formulations by the use of the partial fugacity coefficient, $\bar{\Phi}_i$, in conjunction with an equation of state and adequate mixing rules. This method is most frequently used for handling nonidealities in the vapor phase. However, $\bar{\Phi}_{iV}$ reflects the combined effects of a nonideal gas and a nonideal gas solution. At low pressures, both effects are negligible. At moderate pressures, a vapor solution may still be ideal even

Table 2.7 Classification of Molecules Based on Potential for Forming Hydrogen Bonds

Class	Description	Example
I	Molecules capable of forming three-dimensional networks of strong H-bonds	Water, glycols, glycerol, amino alcohols, hydroxylamines, hydroxyacids, polyphenols, and amides
II	Other molecules containing both active hydrogen atoms and donor atoms (O, N, and F)	Alcohols, acids, phenols, primary and secondary amines, oximes, nitro and nitrile compounds with α-hydrogen atoms, ammonia, hydrazine, hydrogen fluoride, and hydrogen cyanide
III	Molecules containing donor atoms but no active hydrogen atoms	Ethers, ketones, aldehydes, esters, tertiary amines (including pyridine type), and nitro and nitrile compounds without α-hydrogen atoms
IV	Molecules containing active hydrogen atoms but no donor atoms that have two or three chlorine atoms on the same carbon as a hydrogen or one chlorine on the carbon atom and one or more chlorine atoms on adjacent carbon atoms	$CHCl_3$, CH_2Cl_2, CH_3CHCl_2, CH_2ClCH_2Cl, $CH_2ClCHClCH_2Cl$, and $CH_2ClCHCl_2$
V	All other molecules having neither active hydrogen atoms nor donor atoms	Hydrocarbons, carbon disulfide, sulfides, mercaptans, and halohydrocarbons not in class IV

Table 2.8 Molecule Interactions Causing Deviations from Raoult's Law

Type of Deviation	Classes	Effect on Hydrogen Bonding
Always negative	III + IV	H-bonds formed only
Quasi-ideal; always positive or ideal	III + III III + V IV + IV IV + V V + V	No H-bonds involved
Usually positive, but some negative	I + I I + II I + III II + II II + III	H-bonds broken and formed
Always positive	I + IV (frequently limited solubility) II + IV	H-bonds broken and formed, but dissociation of Class I or II is more important effect
Always positive	I + V II + V	H-bonds broken only

though the gas mixture does not follow the ideal-gas law. Nonidealities in the liquid phase, however, can be severe even at low pressures. When polar species are present, mixing rules can be modified to include binary interaction parameters, k_{ij}, as in (2-51). The other technique for handling solution nonidealities is to retain $\bar{\phi}_{iV}$ in the K-value formulation but replace $\bar{\phi}_{iL}$ by the product of γ_{iL} and ϕ_{iL}, where the former accounts for deviations from nonideal solutions. Equation (2-26) then becomes

$$K_i = \frac{\gamma_{iL}\phi_{iL}}{\bar{\phi}_{iV}} \qquad (2\text{-}68)$$

which was derived previously as (2-27). At low pressures, from Table 2.2, $\phi_{iL} = P_i^s/P$ and $\bar{\phi}_{iV} = 1.0$, so (2-68) reduces to a modified Raoult's law K-value, which differs from (2-44) in the added γ_{iL} term:

$$K_i = \frac{\gamma_{iL}P_i^s}{P} \qquad (2\text{-}69)$$

At moderate pressures, (5) of Table 2.3 is preferred.

Regular-solution theory is useful only for estimating values of γ_{iL} for mixtures of nonpolar species. Many semitheoretical equations exist for estimating activity coefficients of binary mixtures containing polar species. These contain binary interaction parameters back-calculated from experimental data. Six of the more useful equations are listed in Table 2.9 in binary-pair form. For a given activity-coefficient correlation, the equations of Table 2.10 can be used to determine excess volume, excess enthalpy, and excess entropy. However, unless the dependency on pressure is known, excess liquid volumes cannot be determined directly from (1) of Table 2.10. Fortunately, the contribution of excess volume to total volume is small for solutions of nonelectrolytes. For example, a 50 mol% solution of ethanol in n-heptane at 25°C is shown in Figure 2.11a to be a nonideal but miscible liquid mixture. From the data of Van Ness, Soczek, and Kochar [34], excess volume is only 0.465 cm^3/mol, compared to an estimated ideal-solution molar

volume of 106.3 cm^3/mol. By contrast, excess liquid enthalpy and excess liquid entropy may not be small. Once the partial molar excess functions for enthalpy and entropy are estimated for each species, the excess functions for the mixture are computed from the mole fraction sums.

§2.6.4 Margules Equations

Equations (1) and (2) in Table 2.9 date back to 1895, yet the two-constant form is still in use because of its simplicity. These equations result from power-series expansions for $\bar{g}_i^E$ and conversion to activity coefficients by (2-59). The one-constant form is equivalent to symmetrical activity-coefficient curves, which are rarely observed.

§2.6.5 van Laar Equation

Because of its flexibility, simplicity, and ability to fit many systems well, the van Laar equation is widely used. It was derived from the van der Waals equation of state, but the constants, shown as A_{12} and A_{21} in (3) of Table 2.9, are, in theory, constant only for a particular binary pair at a given temperature. In practice, the constants are best back-calculated from isobaric data covering a range of temperatures. The van Laar theory expresses the temperature dependence of A_{ij} as

$$A_{ij} = \frac{A'_{ij}}{RT} \qquad (2\text{-}70)$$

Regular-solution theory and the van Laar equation are equivalent for a binary solution if

$$A_{ij} = \frac{v_{iL}}{RT}\left(\delta_i - \delta_j\right)^2 \qquad (2\text{-}71)$$

The van Laar equation can fit activity coefficient–composition curves corresponding to both positive and negative deviations from Raoult's law, but cannot fit curves that exhibit minima or maxima such as those in Figure 2.11c.

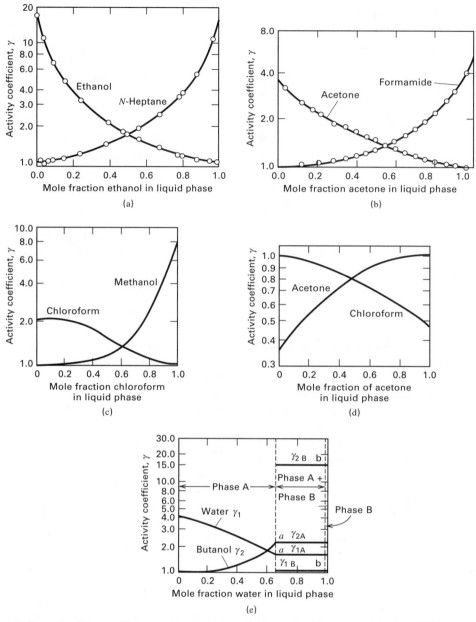

Figure 2.11 Typical variations of activity coefficients with composition in binary liquid systems: (a) ethanol(II)/n-heptane(V); (b) acetone (III)/formamide(I); (c) chloroform(IV)/methanol(II); (d) acetone(III)/chloroform(IV); (e) water(I)/n-butanol(II).

When data are isothermal or isobaric over only a narrow range of temperature, determination of van Laar constants is conducted in a straightforward manner. The most accurate procedure is a nonlinear regression to obtain the best fit to the data over the entire range of binary composition, subject to minimization of some objective function. A less accurate, but extremely rapid, manual-calculation procedure can be used when experimental data can be extrapolated to infinite-dilution conditions. Modern experimental techniques are available for accurately and rapidly determining activity coefficients at infinite dilution. Applying (3) of Table 2.9 to the conditions $x_i = 0$ and then $x_j = 0$,

$$A_{ij} = \ln \gamma_i^\infty, \quad x_i = 0$$

and

$$A_{ji} = \ln \gamma_j^\infty, \quad x_j = 0 \tag{2-72}$$

It is important that the van Laar equation predict azeotrope formation, where $x_i = y_i$ and $K_i = 1.0$. If activity coefficients are known or can be computed at the azeotropic composition say, from (2-69) ($\gamma_{iL} = P/P_i^s$, since $K_i = 1.0$), these coefficients can be used to determine the van Laar constants by solving (2-73) and (2-74), which describe activity-coefficient data at any single composition:

$$A_{12} = \ln \gamma_1 \left(1 + \frac{x_2 \ln \gamma_2}{x_1 \ln \gamma_1} \right)^2 \tag{2-73}$$

$$A_{21} = \ln \gamma_2 \left(1 + \frac{x_1 \ln \gamma_1}{x_2 \ln \gamma_2} \right)^2 \tag{2-74}$$

Table 2.9 Empirical and Semitheoretical Equations for Correlating Liquid-Phase Activity Coefficients of Binary Pairs

Name	Equation for Species 1	Equation for Species 2
(1) Margules	$\log \gamma_1 = A x_2^2$	$\log \gamma_2 = A x_1^2$
(2) Margules (two-constant)	$\log \gamma_1 = x_2^2 [\bar{A}_{12} + 2x_1(\bar{A}_{21} - \bar{A}_{12})]$	$\log \gamma_2 = x_1^2 [\bar{A}_{21} + 2x_2(\bar{A}_{12} - \bar{A}_{21})]$
(3) van Laar (two-constant)	$\ln \gamma_1 = \dfrac{A_{12}}{[1 + (x_1 A_{12})/(x_2 A_{21})]^2}$	$\ln \gamma_2 = \dfrac{A_{21}}{[1 + (x_2 A_{21})/(x_1 A_{12})]^2}$
(4) Wilson (two-constant)	$\ln \gamma_1 = -\ln(x_1 + \Lambda_{12} x_2)$ $\quad + x_2 \left(\dfrac{\Lambda_{12}}{x_1 + \Lambda_{12} x_2} - \dfrac{\Lambda_{21}}{x_2 + \Lambda_{21} x_1} \right)$	$\ln \gamma_2 = -\ln(x_2 + \Lambda_{21} x_1)$ $\quad - x_1 \left(\dfrac{\Lambda_{12}}{x_1 + \Lambda_{12} x_2} - \dfrac{\Lambda_{21}}{x_2 + \Lambda_{21} x_1} \right)$
(5) NRTL (three-constant)	$\ln \gamma_1 = \dfrac{x_2^2 \tau_{21} G_{21}^2}{(x_1 + x_2 G_{21})^2} + \dfrac{x_1^2 \tau_{12} G_{12}}{(x_2 + x_1 G_{12})^2}$ $G_{ij} = \exp(-\alpha_{ij} \tau_{ij})$	$\ln \gamma_2 = \dfrac{x_1^2 \tau_{12} G_{12}^2}{(x_2 + x_1 G_{12})^2} + \dfrac{x_2^2 \tau_{21} G_{21}}{(x_1 + x_2 G_{21})^2}$ $G_{ij} = \exp(-\alpha_{ij} \tau_{ij})$
(6) UNIQUAC (two-constant)	$\ln \gamma_1 = \ln \dfrac{\Psi_1}{x_1} + \dfrac{\bar{Z}}{2} q_1 \ln \dfrac{\theta_1}{\Psi_1}$ $\quad + \Psi_2 \left(l_1 - \dfrac{r_1}{r_2} l_2 \right) - q_1 \ln(\theta_1 + \theta_2 T_{21})$ $\quad + \theta_2 q_1 \left(\dfrac{T_{21}}{\theta_1 + \theta_2 T_{21}} - \dfrac{T_{12}}{\theta_2 + \theta_1 T_{12}} \right)$	$\ln \gamma_2 = \ln \dfrac{\Psi_2}{x_2} + \dfrac{\bar{Z}}{2} q_2 \ln \dfrac{\theta_2}{\Psi_2}$ $\quad + \Psi_1 \left(l_2 - \dfrac{r_2}{r_1} l_1 \right) - q_2 \ln(\theta_2 + \theta_1 T_{12})$ $\quad + \theta_1 q_2 \left(\dfrac{T_{12}}{\theta_2 + \theta_1 T_{12}} - \dfrac{T_{21}}{\theta_1 + \theta_2 T_{21}} \right)$

Mixtures of self-associated polar molecules (class II in Table 2.7) with nonpolar molecules (class V) can exhibit strong nonideality of the positive-deviation type shown in Figure 2.11a. Figure 2.12 shows experimental data of Sinor and Weber [35] for ethanol (1)/n-hexane (2), a system of this type, at 101.3 kPa. These data were correlated with the van Laar equation by Orye and Prausnitz [36] to give $A_{12} = 2.409$ and $A_{21} = 1.970$. From $x_1 = 0.1$ to 0.9, the data fit to the van Laar equation is good; in the dilute regions, however, deviations are quite severe and the predicted activity coefficients for ethanol are low. An even more serious problem with highly nonideal mixtures is that the van Laar equation may erroneously predict phase splitting (formation of two liquid phases) when values of activity coefficients exceed approximately 7.

Table 2.10 Partial Molar Excess Functions

Excess volume:

$$(1) \quad \left(\bar{\upsilon}_{iL} - \bar{\upsilon}_{iL}^{ID} \right) \equiv \bar{\upsilon}_{iL}^{E} = RT \left(\frac{\partial \ln \gamma_{iL}}{\partial P} \right)_{T,x}$$

Excess enthalpy:

$$(2) \quad \left(\bar{h}_{iL} - \bar{h}_{iL}^{ID} \right) \equiv \bar{h}_{iL}^{E} = -RT^2 \left(\frac{\partial \ln \gamma_{iL}}{\partial T} \right)_{P,x}$$

Excess entropy:

$$(3) \quad \left(\bar{s}_{iL} - \bar{s}_{iL}^{ID} \right) \equiv \bar{s}_{iL}^{E} = -R \left[T \left(\frac{\partial \ln \gamma_{iL}}{\partial T} \right)_{P,x} + \ln \gamma_{iL} \right]$$

ID = ideal mixture; E = excess because of nonideality.

§2.6.6 Local-Composition Concept and the Wilson Model

Following its publication in 1964, the Wilson equation [37], in Table 2.9 as (4), received wide attention because of its ability to fit strongly nonideal, but miscible systems. As shown in Figure 2.12, the Wilson equation, with binary

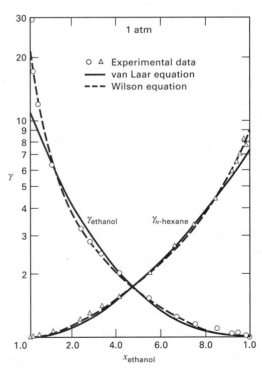

Figure 2.12 Activity coefficients for ethanol/n-hexane.
[Data from J.E. Sinor and J.H. Weber, *J. Chem. Eng. Data*, **5**, 243–247 (1960).]

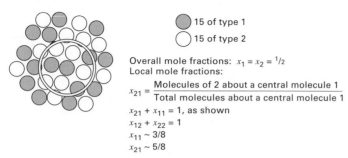

15 of type 1

15 of type 2

Overall mole fractions: $x_1 = x_2 = {}^1\!/_2$
Local mole fractions:

$$x_{21} = \frac{\text{Molecules of 2 about a central molecule 1}}{\text{Total molecules about a central molecule 1}}$$

$x_{21} + x_{11} = 1$, as shown
$x_{12} + x_{22} = 1$
$x_{11} \sim 3/8$
$x_{21} \sim 5/8$

Figure 2.13 The concept of local compositions.

[From P.M. Cukor and J.M. Prausnitz, *Int. Chem. Eng. Symp. Ser. No. 32*, **3**, 88 (1969).]

interaction parameters $\Lambda_{12} = 0.0952$ and $\Lambda_{21} = 0.2713$ from Orye and Prausnitz [36], fits experimental data well even in dilute regions where the variation of γ_1 becomes exponential. Corresponding infinite-dilution activity coefficients computed from the Wilson equation are $\gamma_1^\infty = 21.72$ and $\gamma_2^\infty = 9.104$.

The Wilson equation accounts for differences in both molecular size and intermolecular forces, consistent with the Flory–Huggins relation (2-65). Overall solution volume fractions ($\Phi_i = x_i \upsilon_{iL}/\upsilon_L$) are replaced by local-volume fractions, $\bar{\Phi}_i$, related to local-molecule segregations caused by differing energies of interaction between pairs of molecules. The concept of local compositions that differ from overall compositions is illustrated for an overall, equimolar, binary solution in Figure 2.13, from Cukor and Prausnitz [38]. About a central molecule of type 1, the local mole fraction of type 2 molecules is shown to be 5/8, while the overall composition is 1/2.

For local-volume fraction, Wilson proposed

$$\bar{\Phi}_i = \frac{\upsilon_{iL} x_i \exp(-\lambda_{ii}/RT)}{\sum\limits_{j=1}^{C} \upsilon_{jL} x_j \exp\left(-\lambda_{ij}/RT\right)} \qquad (2\text{-}75)$$

where energies of interaction $\lambda_{ij} = \lambda_{ji}$, but $\lambda_{ii} \neq \lambda_{jj}$. Following Orye and Prausnitz [36], substitution of the binary form of (2-75) into (2-65) and defining the binary interaction parameters as

$$\Lambda_{12} = \frac{\upsilon_{2L}}{\upsilon_{1L}} \exp\left[-\frac{(\lambda_{12} - \lambda_{11})}{RT}\right] \qquad (2\text{-}76)$$

$$\Lambda_{21} = \frac{\upsilon_{1L}}{\upsilon_{2L}} \exp\left[-\frac{(\lambda_{12} - \lambda_{22})}{RT}\right] \qquad (2\text{-}77)$$

leads to an equation for a binary system:

$$\frac{g^E}{RT} = -x_1 \ln\left(x_1 + \Lambda_{12} x_2\right) - x_2 \ln\left(x_2 + \Lambda_{21} x_1\right) \qquad (2\text{-}78)$$

The Wilson equation is effective for dilute compositions where entropy effects dominate over enthalpy effects. The Orye–Prausnitz form for activity coefficient, in Table 2.9, follows from combining (2-59) with (2-78). Values of $\Lambda_{ij} < 1$ correspond to positive deviations from Raoult's law, while values > 1 signify negative deviations. Ideal solutions result when $\Lambda_{ij} = 1$. Studies indicate that λ_{ii} and λ_{ij} are temperature-dependent. Values of $\upsilon_{iL}/\upsilon_{jL}$ depend on temperature

also, but the variation is small compared to the effect of temperature on the exponential terms in (2-76) and (2-77).

The Wilson equation is extended to multicomponent mixtures by neglecting ternary and higher interactions and assuming a pseudo-binary mixture. The following multicomponent Wilson equation involves only binary interaction constants:

$$\ln \gamma_k = 1 - \ln\left(\sum_{j=1}^{C} x_j \Lambda_{kj}\right) - \sum_{i=1}^{C}\left(\frac{x_i \Lambda_{ik}}{\sum\limits_{j=1}^{C} x_j \Lambda_{ij}}\right) \qquad (2\text{-}79)$$

where $\Lambda_{ii} = \Lambda_{jj} = \Lambda_{kk} = 1$.

For highly nonideal, but still miscible, mixtures, the Wilson equation is markedly superior to the Margules and van Laar equations. It is consistently superior for multicomponent solutions. The constants in the Wilson equation for many binary systems are tabulated in the DECHEMA collection of Gmehling and Onken [39] and the Dortmund Data Bank. Two limitations of the Wilson equation are its inability to predict immiscibility, as in Figure 2.11e, and maxima and minima in activity-coefficient–mole fraction relationships, as in Figure 2.11c.

When insufficient data are available to determine binary parameters from a best fit of activity coefficients, infinite-dilution or single-point values can be used. At infinite dilution, the Wilson equation in Table 2.9 becomes

$$\ln \gamma_1^\infty = 1 - \ln \Lambda_{12} - \Lambda_{21} \qquad (2\text{-}80)$$

$$\ln \gamma_2^\infty = 1 - \ln \Lambda_{21} - \Lambda_{12} \qquad (2\text{-}81)$$

If temperatures corresponding to γ_1^∞ and γ_2^∞ are not close or equal, (2-76) and (2-77) should be substituted into (2-80) and (2-81)—with values of $(\lambda_{12} - \lambda_{11})$ and $(\lambda_{12} - \lambda_{22})$ determined from estimates of pure-component liquid molar volumes—to estimate Λ_{12} and Λ_{21}.

When the data of Sinor and Weber [35] for *n*-hexane/ethanol, shown in Figure 2.12, are plotted as a y–x diagram in ethanol (Figure 2.14), the equilibrium curve crosses the 45° line at $x = 0.332$. The temperature corresponding to this composition is 58°C. Ethanol has a normal boiling point of 78.33°C, which is higher than the boiling point of 68.75°C

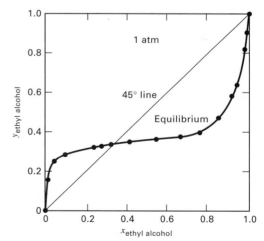

1 atm

45° line

Equilibrium

$y$ethyl alcohol

$x$ethyl alcohol

Figure 2.14 Equilibrium curve for *n*-hexane/ethanol.

for *n*-hexane. Nevertheless, ethanol is more volatile than *n*-hexane up to an ethanol mole fraction of $x = 0.322$, the minimum-boiling azeotrope. This occurs because of the close boiling points of the two species and the high activity coefficients for ethanol at low concentrations. At the azeotropic composition, $y_i = x_i$; therefore, $K_i = 1.0$. Applying (2-69) to both species,

$$\boxed{\gamma_1 P_1^s = \gamma_2 P_2^s} \tag{2-82}$$

If pure species 2 is more volatile $(P_2^s > P_1^s)$, the criteria for formation of a minimum-boiling azeotrope are

$$\gamma_1 \geq 1 \tag{2-83}$$

$$\gamma_2 \geq 1 \tag{2-84}$$

and

$$\frac{\gamma_1}{\gamma_2} < \frac{P_2^s}{P_1^s} \tag{2-85}$$

for x_1 less than the azeotropic composition. These criteria are most readily applied at $x_1 = 0$. For example, for the *n*-hexane (2)/ethanol (1) system at 1 atm when the liquid-phase mole fraction of ethanol approaches zero, the temperature approaches 68.75°C (155.75°F), the boiling point of pure *n*-hexane. At this temperature, $P_1^s = 10$ psia (68.9 kPa) and $P_2^s = 14.7$ psia (101.3 kPa). Also from Figure 2.12, $\gamma_1^\infty = 21.72$ when $\gamma_2 = 1.0$. Thus, $\gamma_1^\infty / \gamma_2 = 21.72$, but $P_2^s / P_1^s = 1.47$. Therefore, a minimum-boiling azeotrope will occur.

Maximum-boiling azeotropes are less common. They occur for close-boiling mixtures when negative deviations from Raoult's law arise, giving $\gamma_i < 1.0$. Criteria are derived in a manner similar to that for minimum-boiling azeotropes. At $x_1 = 1$, where species 2 is more volatile,

$$\gamma_1 = 1.0 \tag{2-86}$$

$$\gamma_2^\infty < 1.0 \tag{2-87}$$

and

$$\frac{\gamma_2^\infty}{\gamma_1} < \frac{P_1^s}{P_2^s} \tag{2-88}$$

For azeotropic binary systems, interaction parameters Λ_{12} and Λ_{21} can be determined by solving (4) of Table 2.9 at the azeotropic composition, as shown in the following example.

EXAMPLE 2.8 Wilson Constants from Azeotropic Data.

From measurements by Sinor and Weber [35] of the azeotropic condition for the ethanol (E)/*n*-hexane (H) system at 1 atm (101.3 kPa, 14.696 psia), calculate Λ_{12} and Λ_{21}.

Solution

The azeotrope occurs at $x_E = 0.332$, $x_H = 0.668$, and $T = 58°C$ (331.15 K). At 1 atm, (2-69) can be used to approximate K-values. Thus, at azeotropic conditions, $\gamma_i = P/P_i^s$. The vapor pressures at 58°C are $P_E^s = 6.26$ psia and $P_H^s = 10.28$ psia. Therefore,

$$\gamma_E = \frac{14.696}{6.26} = 2.348$$

$$\gamma_H = \frac{14.696}{10.28} = 1.430$$

Substituting these values together with the above corresponding values of x_i into the binary form of the Wilson equation in Table 2.9 gives

$$\ln 2.348 = -\ln(0.332 + 0.668\Lambda_{EH})$$
$$+0.668 \left(\frac{\Lambda_{EH}}{0.332 + 0.668\Lambda_{EH}} - \frac{\Lambda_{HE}}{0.332\Lambda_{HE} + 0.668} \right)$$

$$\ln 1.430 = -\ln(0.668 + 0.332\Lambda_{HE})$$
$$-0.332 \left(\frac{\Lambda_{EH}}{0.332 + 0.668\Lambda_{EH}} - \frac{\Lambda_{HE}}{0.332\Lambda_{HE} + 0.668} \right)$$

Solving these two nonlinear equations simultaneously, $\Lambda_{EH} = 0.041$ and $\Lambda_{HE} = 0.281$. From these constants, the activity-coefficient curves can be predicted if the temperature variations of Λ_{EH} and Λ_{HE} are ignored. The results are plotted in Figure 2.15. The fit of experimental data is good except, perhaps, for near-infinite-dilution conditions, where $\gamma_E^\infty = 49.82$ and $\gamma_H^\infty = 9.28$. The former is considerably greater than the value of 21.72 obtained by Orye and Prausnitz [36] from a fit of all data points. A comparison of Figures 2.12 and 2.15 shows that widely differing γ_E^∞ values have little effect on γ in the region $x_E = 0.15$ to 1.00, where the Wilson curves are almost identical. For accuracy over the entire composition range, data for at least three liquid compositions per binary are preferred.

The Wilson equation can be extended to liquid–liquid or vapor–liquid–liquid systems by multiplying the right-hand side of (2-78) by a third binary-pair constant evaluated from experimental data [37]. However, for multicomponent systems of three or more species, the third binary-pair constants must be the same for all binary pairs. Furthermore, as shown by Hiranuma [40], representation of ternary systems

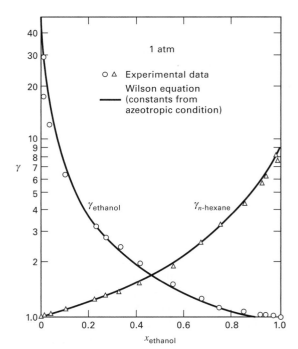

Figure 2.15 Liquid-phase activity coefficients for ethanol/*n*-hexane system.

involving only one partially miscible binary pair can be extremely sensitive to the third binary-pair Wilson constant. For these reasons, the Wilson equation is not favored for liquid–liquid systems.

§2.6.7 NRTL Model

The nonrandom, two-liquid (NRTL) equation developed by Renon and Prausnitz [41, 42], given in Table 2.9, represents an extension of Wilson's concept to multicomponent liquid–liquid, and vapor–liquid–liquid systems. It is widely used for liquid–liquid extraction. For multicomponent vapor–liquid systems, only binary-pair constants from binary-pair experimental data are required. For a multicomponent system, the NRTL expression is

$$\ln \gamma_i = \frac{\sum\limits_{j=1}^{C} \tau_{ji} G_{ji} x_j}{\sum\limits_{k=1}^{C} G_{ki} x_k} + \sum_{j=1}^{C} \left[\frac{x_j G_{ij}}{\sum\limits_{k=1}^{C} G_{kj} x_k} \left(\tau_{ij} - \frac{\sum\limits_{k=1}^{C} x_k \tau_{kj} G_{kj}}{\sum\limits_{k=1}^{C} G_{kj} x_k} \right) \right]$$

$$(2\text{-}89)$$

where

$$G_{ji} = \exp\left(-\alpha_{ji} \tau_{ji}\right) \qquad (2\text{-}90)$$

The coefficients τ are given by

$$\tau_{ij} = \frac{g_{ij} - g_{jj}}{RT} \qquad (2\text{-}91)$$

$$\tau_{ji} = \frac{g_{ji} - g_{ii}}{RT} \qquad (2\text{-}92)$$

where the double-subscripted g values are energies of interaction for molecule pairs. In the equations, $G_{ji} \neq G_{ij}$, $\tau_{ij} \neq \tau_{ji}$, $G_{ii} = G_{jj} = 1$, and $\tau_{ii} = \tau_{jj} = 0$. Often $(g_{ij} - g_{jj})$ and other constants are linear in temperature. For ideal solutions, $\tau_{ji} = 0$.

The parameter α_{ji} characterizes the tendency of species j and i to be distributed nonrandomly. When $\alpha_{ji} = 0$, local mole fractions equal overall solution mole fractions. Generally, α_{ji} is independent of temperature and depends on molecule properties similar to the classifications in Tables 2.7 and 2.8. Values of α_{ji} usually lie between 0.2 and 0.47. When $\alpha_{ji} > 0.426$, phase immiscibility is predicted. Although α_{ji} can be treated as an adjustable parameter determined from experimental binary-pair data, commonly α_{ji} is set according to the following rules, which are occasionally ambiguous:

1. $\alpha_{ji} = 0.20$ for hydrocarbons and polar, nonassociated species (e.g., n-heptane/acetone).

2. $\alpha_{ji} = 0.30$ for nonpolar compounds (e.g., benzene/ n-heptane), except fluorocarbons and paraffins; nonpolar and polar, nonassociated species (e.g., benzene/ acetone); polar species that exhibit negative deviations from Raoult's law (e.g., acetone/chloroform) and moderate positive deviations (e.g., ethanol/water); mixtures of water and polar nonassociated species (e.g., water/ acetone).

3. $\alpha_{ji} = 0.40$ for saturated hydrocarbons and homolog perfluorocarbons (e.g., n-hexane/perfluoro-n-hexane).

4. $\alpha_{ji} = 0.47$ for alcohols or other strongly self-associated species with nonpolar species (e.g., ethanol/benzene); carbon tetrachloride with either acetonitrile or nitromethane; water with either butyl glycol or pyridine.

§2.6.8 UNIQUAC Model

In an attempt to place calculations of activity coefficients on a more theoretical basis, Abrams and Prausnitz [43] used statistical mechanics to derive an expression for excess free energy. Their model, UNIQUAC (universal quasichemical), generalizes an analysis by Guggenheim and extends it to molecules that differ in size and shape. As in the Wilson and NRTL equations, local concentrations are used. However, rather than local volume fractions or local mole fractions, UNIQUAC uses local area fraction θ_{ij} as the primary concentration variable.

The local area fraction is determined by representing a molecule by a set of bonded segments. Each molecule is characterized by two structural parameters determined relative to a standard segment, taken as an equivalent sphere of a unit of a linear, infinite-length, polymethylene molecule. The two structural parameters are the relative number of segments per molecule, r (volume parameter), and the relative surface area, q (surface parameter). These parameters, computed from bond angles and bond distances, are given for many species by Abrams and Prausnitz [43–45] and Gmehling and Onken [39]. Values can also be estimated by the group-contribution method of Fredenslund et al. [46].

For a multicomponent liquid mixture, the UNIQUAC model gives the excess free energy as

$$\frac{g^E}{RT} = \sum_{i=1}^{C} x_i \ln\left(\frac{\Psi_i}{x_i}\right) + \frac{\bar{Z}}{2} \sum_{i=1}^{C} q_i x_i \ln\left(\frac{\theta_i}{\Psi_i}\right)$$
$$- \sum_{i=1}^{C} q_i x_i \ln\left(\sum_{j=1}^{C} \theta_i T_{ji}\right) \qquad (2\text{-}93)$$

The first two terms on the right-hand side account for *combinatorial* effects due to differences in size and shape; the last term provides a *residual* contribution due to differences in intermolecular forces, where

$$\Psi_i = \frac{x_i r_i}{\sum\limits_{i=1}^{C} x_i r_i} = \text{segment fraction} \qquad (2\text{-}94)$$

$$\theta = \frac{x_i q_i}{\sum\limits_{i=1}^{C} x_i q_i} = \text{area fraction} \qquad (2\text{-}95)$$

where $\bar{Z}$ = lattice coordination number set equal to 10, and

$$T_{ji} = \exp\left(\frac{u_{ji} - u_{ii}}{RT}\right) \qquad (2\text{-}96)$$

Equation (2-93) contains two adjustable parameters for each binary pair, $(u_{ji} - u_{ii})$ and $(u_{ij} - u_{jj})$. Abrams and Prausnitz show that $u_{ji} = u_{ij}$ and $T_{ii} = T_{jj} = 1$. In general, $(u_{ji} - u_{ii})$ and $(u_{ij} - u_{jj})$ are linear functions of absolute temperature.

If (2-59) is combined with (2-93), the activity coefficient for a species in a multicomponent mixture becomes:

$$\ln \gamma_i = \ln \gamma_i^C + \ln \gamma_i^R$$

$$= \underbrace{\ln \left(\Psi_i / x_i \right) + (\bar{Z}/2) q_i \ln \left(\theta_i / \Psi_i \right) + l_i - \left(\Psi_i / x_i \right) \sum_{j=1}^{C} x_j l_j}_{C, \text{ combinatorial}}$$

$$\underbrace{+ q_i \left[1 - \ln \left(\sum_{j=1}^{C} \theta_j T_{ji} \right) - \sum_{j=1}^{C} \left(\frac{\theta_j T_{ij}}{\sum_{k=1}^{C} \theta_k T_{kj}} \right) \right]}_{R, \text{ residual}}$$

(2-97)

where

$$l_j = \left(\frac{\bar{Z}}{2} \right) \left(r_j - a_j \right) - \left(r_j - 1 \right) \quad (2\text{-}98)$$

For a mixture of species 1 and 2, (2-97) reduces to (6) in Table 2.9 for $\bar{Z} = 10$.

§2.6.9 UNIFAC Model

Liquid-phase activity coefficients are required for design purposes even when experimental equilibria data are not available and the assumption of regular solutions is not valid because polar compounds are present. For such situations, Wilson and Deal [47], and then Derr and Deal [48], in the 1960s presented estimation methods based on functional groups instead of molecules. In a solution of toluene and acetone, the contributions might be 5 aromatic CH groups, 1 aromatic C group, and 1 CH_3 group from toluene; and 2 CH_3 groups plus 1 CO carbonyl group from acetone. Alternatively, larger groups might be employed to give 5 aromatic CH groups and 1 CCH_3 group from toluene; and 1 CH_3 group and 1 CH_3CO group from acetone. As larger functional groups are used, the accuracy increases, but the advantage of the group-contribution method decreases because more groups are required. In practice, about 50 functional groups represent thousands of multicomponent liquid mixtures.

For partial molar excess free energies, $\bar{g}_i^E$, and activity coefficients, size parameters for each functional group and interaction parameters for each pair are required. Size parameters can be calculated from theory. Interaction parameters are back-calculated from existing phase-equilibria data and used with the size parameters to predict properties of mixtures for which data are unavailable.

The UNIFAC (UNIQUAC Functional-group Activity Coefficients) group-contribution method—first presented by Fredenslund, Jones, and Prausnitz [49] and further developed by Fredenslund, Gmehling, and Rasmussen [50], Gmehling, Rasmussen, and Fredenslund [51], and Larsen, Rasmussen, and Fredenslund [52]—has advantages over other methods in that: (1) it is theoretically based; (2) the parameters are essentially independent of temperature; (3) size and binary interaction parameters are available for a range of functional

groups; (4) predictions can be made over a temperature range of 275–425 K and for pressures to a few atmospheres; and (5) extensive comparisons with experimental data are available. All components must be condensable at near-ambient conditions.

The UNIFAC method is based on the UNIQUAC equation (2-97), wherein the molecular volume and area parameters are replaced by

$$r_i = \sum_k v_k^{(i)} R_k \quad (2\text{-}99)$$

$$q_i = \sum_k v_k^{(i)} Q_k \quad (2\text{-}100)$$

where $v_k^{(i)}$ is the number of functional groups of type k in molecule i, and R_k and Q_k are the volume and area parameters, respectively, for the type-k functional group.

The residual term in (2-97), which is represented by $\ln \gamma_i^R$, is replaced by the expression

$$\ln \gamma_i^R = \underbrace{\sum_k v_k^{(i)} \left(\ln \Gamma_k - \ln \Gamma_k^{(i)} \right)}_{\text{all functional groups in mixture}} \quad (2\text{-}101)$$

where Γ_k is the residual activity coefficient of group k, and $\Gamma_k^{(i)}$ is the same quantity but in a reference mixture that contains only molecules of type i. The latter quantity is required so that $\gamma_i^R \rightarrow 1.0$ as $x_i \rightarrow 1.0$. Both Γ_k and $\Gamma_k^{(i)}$ have the same form as the residual term in (2-97). Thus,

$$\ln \Gamma_k = Q_k \left[1 - \ln \left(\sum_m \theta_m T_{mk} \right) - \sum_m \frac{\theta_m T_{mk}}{\sum_n \theta_n T_{nm}} \right] \quad (2\text{-}102)$$

where θ_m is the area fraction of group m, given by an equation similar to (2-95),

$$\theta_m = \frac{X_m Q_m}{\sum_n X_n Q_m} \quad (2\text{-}103)$$

where X_m is the mole fraction of group m in the solution,

$$X_m = \frac{\sum_j v_m^{(j)} x_j}{\sum_j \sum_n \left(v_n^{(j)} x_j \right)} \quad (2\text{-}104)$$

and T_{mk} is a group interaction parameter given by an equation similar to (2-96),

$$T_{mk} = \exp \left(-\frac{a_{mk}}{T} \right) \quad (2\text{-}105)$$

where $a_{mk} \neq a_{km}$. When $m = k$, then $a_{mk} = 0$ and $T_{mk} = 1.0$. For $\Gamma_k^{(i)}$, (2-102) also applies, where θ terms correspond to the pure component i. Although R_k and Q_k differ for each functional group, values of a_{mk} are equal for all subgroups within a main group. For example, main group CH_2 consists of subgroups CH_3, CH_2, CH, and C. Accordingly,

$$a_{CH_3,CHO} = a_{CH_2,CHO} = a_{CH,CHO} = a_{C,CHO}$$

Thus, the experimental data required to obtain values of a_{mk} and a_{km} and the size of the corresponding bank of data for these parameters are not as great as might be expected.

The group-contribution method was improved by the introduction of a modified UNIFAC method by Gmehling et al. [51], referred to as UNIFAC (Dortmund). For mixtures having a range of molecular sizes, they modified the combinatorial part of (2-97). For temperature dependence they replaced (2-105) with a three-coefficient equation. These changes permit reliable predictions of activity coefficients (including dilute solutions and multiple liquid phases), heats of mixing, and azeotropic compositions. Values of UNIFAC (Dortmund) parameters for 51 groups have been available in publications starting in 1993 with Gmehling, Li, and Schiller [53] and more recently with Wittig, Lohmann, and Gmehling [54], Gmehling et al. [92], and Jakob et al. [93].

§2.6.10 Liquid–Liquid Equilibria

When species are notably dissimilar and activity coefficients are large, two or more liquid phases may coexist. Consider the binary system methanol (1) and cyclohexane (2) at 25°C. From measurements of Takeuchi, Nitta, and Katayama [55], van Laar constants are $A_{12} = 2.61$ and $A_{21} = 2.34$, corresponding, respectively, to infinite-dilution activity coefficients of 13.6 and 10.4 from (2-72). Parameters A_{12} and A_{21} can be used to construct an equilibrium plot of y_1 against x_1 assuming 25°C. Combining (2-69), where $K_i = y_i/x_i$, with

$$P = \sum_{i=1}^{C} x_i \gamma_{iL} P_i^s \qquad (2\text{-}106)$$

gives the following for computing y_i from x_i:

$$y_1 = \frac{x_1 \gamma_1 P_1^s}{x_1 \gamma_1 P_1^s + x_2 \gamma_2 P_2^s} \qquad (2\text{-}107)$$

Vapor pressures at 25°C are $P_1^s = 2.452$ psia (16.9 kPa) and $P_2^s = 1.886$ psia (13.0 kPa). Activity coefficients can be computed from the van Laar equation in Table 2.9. The resulting equilibrium is shown in Figure 2.16, where over much of the liquid-phase region, three values of x_1 exist for a given y_1. This indicates phase instability with the formation of two

liquid phases. Single liquid phases can exist only for cyclohexane-rich mixtures of $x_1 = 0.8248$ to 1.0 and for methanol-rich mixtures of $x_1 = 0.0$ to 0.1291. Because a coexisting vapor phase exhibits one composition, two coexisting liquid phases prevail at opposite ends of the dashed line in Figure 2.16. The liquid phases represent solubility limits of methanol in cyclohexane and cyclohexane in methanol.

For two coexisting liquid phases, $\gamma_{iL}^{(1)} x_i^{(1)} = \gamma_{iL}^{(2)} x_i^{(2)}$ must hold. This permits determination of the two-phase region in Figure 2.16 from the van Laar or other suitable activity-coefficient equation for which the constants are known. Also shown in Figure 2.16 is an equilibrium curve for the binary mixture at 55°C, based on data of Strubl et al. [56]. At this higher temperature, methanol and cyclohexane are miscible. The data of Kiser, Johnson, and Shetlar [57] show that phase instability ceases to exist at 45.75°C, the critical solution temperature. Rigorous methods for determining phase instability and, thus, existence of two liquid phases are based on free-energy calculations, as discussed by Prausnitz et al. [4]. Most of the semitheoretical equations for the liquid-phase activity coefficient listed in Table 2.9 apply to liquid–liquid systems. The Wilson equation is a notable exception. The NRTL equation is the most widely used.

§2.7 DIFFICULT MIXTURES

The equation-of-state and activity-coefficient models in §2.5 and §2.6 are inadequate for estimating K-values of mixtures containing: (1) polar and supercritical (light-gas) components, (2) electrolytes, (3) polymers and solvents, and (4) biomacromolecules. For these difficult mixtures, special models are briefly described in the following subsections. Detailed discussions are given by Prausnitz, Lichtenthaler, and de Azevedo [4].

§2.7.1 Predictive Soave–Redlich–Kwong (PSRK) Model

Equation-of-state models are useful for mixtures of nonpolar and slightly polar compounds. Gibbs free-energy activity-coefficient models are suitable for liquid subcritical nonpolar and polar compounds. When a mixture contains both polar compounds and supercritical gases, neither method applies. To describe vapor–liquid equilibria for such mixtures, more theoretically based mixing rules for use with the S–R–K and P–R equations of state have been developed. To broaden the range of applications of these models, Holderbaum and Gmehling [58] formulated a group-contribution equation of state called the predictive Soave–Redlich–Kwong (PSRK) model, which combines the S–R–K equation of state with UNIFAC. To improve the ability of the S–R–K equation to predict vapor pressure of polar compounds, they make the pure-component parameter, a, in Table 2.5 temperature dependent. To handle mixtures of nonpolar, polar, and supercritical components, they use a mixing rule for a that includes the UNIFAC model for nonideal effects. Pure-component and group-interaction parameters for use in the PSRK model are provided by Fischer and Gmehling [59]. In

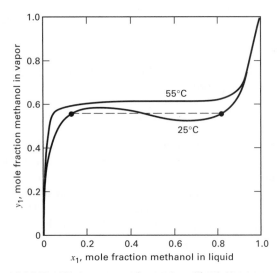

Figure 2.16 Equilibrium curves for methanol/cyclohexane.
[Data from K. Strubl, V. Svoboda, R. Holub, and J. Pick, *Collect. Czech. Chem. Commun.*, **35**, 3004–3019 (1970).]

particular, [58] and [59] provide parameters for nine light gases in addition to UNIFAC parameters for 50 groups.

§2.7.2 Electrolyte Solution Models

Solutions of electrolytes are common in the chemical and biochemical industries. For example, sour water, found in many petroleum plants, consists of water and five dissolved gases: CO, CO_2, CH_4, H_2S, and NH_3. Because of dissociation, the aqueous solution includes ionic as well as molecular species. For sour water, the ionic species include H^+, OH^-, HCO_3^-, $CO_3^=$, HS^-, $S^=$, NH_4^+, and NH_2COO^-, with the positive and negative ions subject to electroneutrality. For example, while the apparent concentration of NH_3 in the solution might be 2.46 moles per kg of water, the molality is 0.97 when dissociation is taken into account, with NH_4^+ having a molality of 1.49. All eight ionic species are nonvolatile, while all six molecular species are volatile to some extent. Calculations of vapor–liquid equilibrium for multicomponent electrolyte solutions must consider both chemical and physical equilibrium, both of which involve liquid-phase activity coefficients.

Models have been developed for predicting activity coefficients in multicomponent systems of electrolytes. Of particular note are those of Pitzer [60] and Chen and associates [61, 62, 63] , both of which are included in process simulation programs. Both models can handle dilute to concentrated solutions, but only the model of Chen and associates, which is a substantial modification of the NRTL model, can handle mixed-solvent systems.

§2.7.3 Polymer Solution Models

Polymer processing commonly involves solutions of solvent, monomer, and soluble polymer, thus requiring vapor–liquid and, sometimes, liquid–liquid phase-equilibria calculations, for which activity coefficients of all components are needed. In general, the polymer is nonvolatile, but the solvent and monomer are volatile. When the solution is dilute in the polymer, activity-coefficient methods of §2.6, such as the NRTL method, are suitable. Of more interest are mixtures with appreciable concentrations of polymer, for which the methods of §2.5 and §2.6 are inadequate, so, special-purpose models have been developed. One method, which is available in process simulation programs, is the modified NRTL model of Chen [64], which combines a modification of the Flory–Huggins equation (2-65) for widely differing molecular size with the NRTL concept of local composition. Because Chen represents the polymer with segments, solvent–solvent, solvent–segment, and segment–segment binary interaction parameters are required. These are available from the literature and may be assumed to be independent of temperature, polymer chain length, and polymer concentration.

Aqueous two-phase extraction (ATPE) is a nondenaturing and nondegrading method for recovering and separating large biomolecules such as cells, cell organelles, enzymes, lipids, proteins, and viruses from fermentation broths and solutions of lysed cells. An aqueous two-phase system (ATPS) consists

of water and two polymers [e.g., polyethylene glycol (PEG) and dextran] or one polymer (e.g., PEG) and a salt (e.g., K_2SO_4, Na_2SO_4, and KCl). At equilibrium, the aqueous top phase is enriched in PEG and depleted in dextran or salt, while the aqueous bottom phase is depleted in PEG and enriched in dextran or a salt. When an ATPS is dilute in biochemical solutes, they partition between the two aqueous phases, leaving the phase equilibria for the ATPS essentially unaltered. Therefore, the design of an ATPE separation requires a phase diagram for the ATPS and partition coefficients for the biochemical solutes. As discussed in §8.6, ternary phase diagrams, similar to Figure 8.44, for over 100 ATPSs have been published. Also, partition coefficients have been measured for a number of biomolecules in several ATPSs, e.g., Madeira et al. [89]. When the ternary phase diagram and/or the partition coefficients are not available, they may be estimated by methods developed by King et al. [90] and Haynes et al. [91].

§2.8 SELECTING AN APPROPRIATE MODEL

Design or analysis of a separation process requires a suitable thermodynamic model. This section presents recommendations for making at least a preliminary selection.

The procedure includes a few models not covered in this chapter but for which a literature reference is given. The procedure begins by characterizing the mixture by chemical types: Light gases (LG), Hydrocarbons (HC), Polar organic compounds (PC), and Aqueous solutions (A), with or without Electrolytes (E) or biomacromolecules (B).

If the mixture is (A) with no (PC), and if electrolytes are present, select the modified NRTL equation. Otherwise, select a special model, such as one for sour water (containing NH_3, H_2S, CO_2, etc.) or aqueous amine solutions.

If the mixture contains (HC), with or without (LG), for a wide boiling range, choose the corresponding-states method of Lee–Kesler–Plöcker [8, 65]. If the HC boiling range is not wide, selection depends on the pressure and temperature. The Peng–Robinson equation is suitable for all temperatures and pressures. For all pressures and noncryogenic temperatures, the Soave–Redlich–Kwong equation is applicable. For all temperatures, but not pressures in the critical region, the Benedict–Webb–Rubin–Starling [5, 66, 67] method is viable.

If the mixture contains (PC), selection depends on whether (LG) are present. If they are, the PSRK method is recommended. If not, then a liquid-phase activity-coefficient method should be chosen. If the binary interaction coefficients are not available, select the UNIFAC method, which should be considered as only a first approximation. If the binary interaction coefficients are available and splitting into two liquid phases will not occur, select the Wilson or NRTL equation. Otherwise, if phase splitting is probable, select the NRTL or UNIQUAC equation.

All process simulators have expert systems that chose what the program designers believe to be the optimal thermodynamic package for the chemical species involved. However, since temperature, composition, and pressure in the various processing units vary, care must be taken to use the expert system correctly.

§2.9 THERMODYNAMIC ACTIVITY OF BIOLOGICAL SPECIES

Effective bioseparations economically and reliably recover active biological product. Common petrochemical separations (e.g., distillation) occur by creating or adding a second phase, often a vapor. This is seldom possible with bioproduct separations, since most small molecules, polymers, and particulates of biological origin are unstable in the vapor phase. Thermodynamics of bioseparations thus focuses on molecular ionization states, interactions, and forces at physiologic conditions rather than the state of a continuous fluid phase. Biological activity is influenced by (1) *solution conditions* (e.g., pH buffering, ionization, solubility); (2) *biocolloid interactions* (e.g., van der Waals interactions, electrostatic forces, solvation forces, and hydrophobic effects); and (3) *biomolecule reactions* (e.g., actions of proteases, nucleases, lipases; effects of divalent cations, metals, chelating agents; rate/equilibrium of enzyme/substrate interactions and de-activation). Understanding these interrelated influences allows an engineer to (1) choose effectively between alternative bioseparation process options and (2) optimize operational parameters of a selected bioseparation operation to maintain activity of target biological species.

Solution conditions, biocolloid interactions, and biomolecule reactions affect separation of bioproducts by extraction (§8.6), membranes (§14.8), electrophoresis (§15.8), adsorption (§15.3), and crystallization (§17.11). In this section are important fundamental concepts upon which discussion in these later sections will be based. Biological suspensions contain a large number of complex biochemical species and are often incompletely specified. Therefore, the application of fundamental principles is balanced in practice with relevant experience from similar systems and careful attention to detail.

§2.9.1 Solution Conditions

Effects of temperature, ionic strength, solvent, and static charge on pH buffering impact biological stability as well as chromatographic adsorption and elution, membrane selectivity and fouling, and precipitation of biological molecules and entities.

pH buffers

Controlling pH by adding a well-suited buffer to absorb (or release) protons produced (or consumed) in biochemical reactions is important to maintain activity of biological products (e.g., to preserve catalytic activity). For example, reducing pH to <5.0 is sufficient to dissociate the rigid, 80-nm icosahedral protein coat of adenovirus, a nonenveloped, 165 MDalton double-stranded DNA (dsDNA) virus used as a viral vector for gene therapy [68]. Suitability is determined by several buffer attributes: (1) *acid-ionization* constant,

which can vary with temperature and ionic strength; (2) charge(s)—positive or negative; (3) interactions with solution components such as metal ions or chelating agents; (4) solubility and expense; and (5) interference with analytical methods.

Simple monovalent ($n = 1$) buffers in aqueous solution yield an uncharged form of weak acid, HA, after proton addition [69,70]:

$$H_2O + HA \rightleftarrows H_3O^+ + A^- \qquad (2\text{-}108)$$

or a charged form, HB^+, after proton addition to an uncharged weak base, B:

$$H_2O + HB^+ \rightleftarrows H_3O^+ + B \qquad (2\text{-}109)$$

Acid-ionization constant

In practice, buffers and biological species are commonly employed in complex solutions at dilutions sufficient to evaluate activity coefficients at unity and to neglect deviation in water concentration away from 55.5 mol/L, M, in calculating pH. Apparent acid-ionization constants (i.e., apparent equilibrium constants) are rarely distinguished from true equilibrium constants by using the Debye–Hückel limiting law or its extensions to calculate activity coefficients. Use of dilute buffers allows rearranging the expression for the equilibrium constant, K_c, for proton dissociation in order to write the acid ionization constant, K_a, for an uncharged weak acid as

$$\boxed{K_a = K_c[H_2O] = \frac{[H^+][A^-]}{[HA]}} \qquad (2\text{-}110)$$

Similarly, the K_a for an uncharged base is

$$\boxed{K_a = K_c[H_2O] = \frac{[H^+][B]}{[HB^+]}} \qquad (2\text{-}111)$$

Adding a small volume of simple, dilute, weak acid (acetic acid, ≤1 M) or weak base (Tris, ≤1 M) to a well-stirred protein solution, for example, allows its pH to be adjusted between 5 and 8 in the presence of buffering salts with minimal risk of inactivation. Acetic acid, CH_3COOH, is an important weak biochemical acid responsible for vinegar's pungent odor. It is excreted from bacteria such as *Acetobacter* and *Clostridium* which oxidize vinegar from ethanol and sugar substrates during food spoilage. Acetic acid has a measured $K_c = 3.19 \times 10^{-7}$. By comparison, the ion product for the dissociation of water, K_w, at 25°C is

$$K_w = [H^+][OH^-] = 1.0 \times 10^{-14} \qquad (2\text{-}112)$$

Determining pKₐ, pH, and ionization

Small values of K_c are typical, and so K_a is conveniently expressed in logarithmic form as

$$\boxed{pK_a = -\log(K_a)} \qquad (2\text{-}113)$$

Substituting (2-110) or (2-111) into (2-113) and using the definition of pH,

$$pH = -\log(H^+) \qquad (2\text{-}114)$$

yields, upon rearrangement, the *Henderson–Hasselbalch equation*

$$pH = pK_a + \log \frac{[\textit{basic form}]}{[\textit{acid form}]} \qquad (2\text{-}115)$$

which gives the pH of a solution containing both forms of a buffer. The pK of an acid or base is the pH at which it is half-dissociated, which corresponds to the *inflection point* in a *titration curve*.

Equation (2-115) allows the ratio of ionized basic form (A^-) to un-ionized acid form (HA) of a weak acid buffer to be determined using a measured pH and a known value of pK_a (and, similarly, for a weak base ionized acid —HB^+— and un-ionized basic—B—forms).

Determining pI *and net charge*

Amino acids each contain a primary –COOH that ionizes at pH $\sim$2 to 3 and a primary –NH$_2$ group that ionizes at pH $\sim$8 to 10. In some cases another ionizable group appears on a side chain. Amino acids exist mostly as *zwitterions*, which contain both acidic and basic groups (i.e., *amphoteric*) across a wide range of pH values that brackets a physiologic value of $\sim$7.4. Zwitterions buffer solutions at high pH by releasing H^+, as well as at low pH by accepting H^+. Table 2.11 lists pK values for α-carboxylic acid, α-amino base, and the ionizable side chain (if present) on representative amino acids.

Using the base-to-acid ratio in (2-115) for α-carboxylic acid, α-amino base, and side-chain groups, the net charge at a given pH and *isolectric point*, pI (i.e., pH at which net

Table 2.11 pK Values of Some Amino Acids, R-C(NH$_2$)(H)-COOH

Amino Acid	pK Values (25°C)		
	α-COOH group	α-NH$_3^+$ group	Side chain
Alanine	2.3	9.9	
Glycine	2.4	9.8	
Phenylalanine	1.8	9.1	
Serine	2.1	9.2	
Valine	2.3	9.6	
Aspartic acid	2.0	10.0	3.9
Glutamic acid	2.2	9.7	4.3
Histidine	1.8	9.2	6.0
Cysteine	1.8	10.8	8.3
Tyrosine	2.2	9.1	10.9
Lysine	2.2	9.2	10.8
Arginine	1.8	9.0	12.5

Source: Stryer [69].

charge is zero), for an amino acid can be determined. This is illustrated in Example 2.9 and Exercises 2.25 and 2.26.

EXAMPLE 2.9 Net Charge on an Amino Group.

The amino acid glycine has hydrogen as its side-chain (R) group. Values of p$K_a^c = 2.36$ and p$K_a^a = 9.56$ have been previously reported for glycine. Superscripts *c*, *a*, and *s* refer to α-carboxylic acid, α-amino base, and side chain, respectively. Using these values, determine the net charge on the α-amino group of glycine at a physiologic pH of 7.2 and a pH of 3 [70].

Solution

The ionization reactions for glycine are:

The fraction of α-carboxylic groups present as glycine zwitterions with net neutral charge is obtained by rearranging (2-115) to obtain the *deprotonation ratio*:

$$\frac{[A^-]}{[HA]} = 10^{\left(pH - pK_a^c\right)} \qquad (2\text{-}116)$$

The corresponding fraction of the protonated α-carboxylic group present as glycinium cation is then:

$$\frac{[HA]}{[HA] + [A^-]} = \frac{1}{1 + 10^{\left(pH - pK_a^c\right)}} \qquad (2\text{-}117)$$

This fraction is equivalent to the magnitude of the fractional charge of the amino group. To obtain the net charge, multiply the fraction by the unit charge of +1. At a physiologic pH of 7.2, this gives +1.4 $\times 10^{-5}$. In other words, there are $\sim$14 glycinium cations with protonated α-amino groups and a net molecular charge of +1 for every 10^6 glycine zwitterions in solution. At a pH of 3, the net charge increases to 1.9×10^{-1}—nearly 2 glycinium cations with protonated α-amino groups and net positive charge per 10 amphoteric glycines. An analogous development for net charge on α-carboxylic acid shows that the *isoelectric point*, pI, of an amino acid with a nonionizing side group is (see Exercise 2.25):

$$pH_{\text{zero. net. charge}} = pI = \frac{pK_a^c + pK_a^a}{2} \qquad (2\text{-}118)$$

Protein pI values are similar to those of the predominant constituent amino acid residue and are almost always <7.0 (anionic at physiologic pH). Exact calculation of protein pI using (2-118) is precluded by complex factors that influence ionizability. Ionization of one amino acid group in a polypeptide chain, for example, affects ionization of functional groups farther along the same chain. Tertiary structure, desolvation, and post-translational modifications also influence ionizability.

pK_a *criterion for buffer selection*

Equation (2-115) and corresponding acid/base titration curves show that the pH of solution changes less per proton absorbed (or released) as [basic form] approaches [acid form]. For this reason, it is preferable to (1) use a buffer whose pK_a is ± 0.5 unit of desired pH—on the (−) side if acidification is anticipated, or on the (+) side for expected basification; and (2) prepare the buffer with equal portions of acid and basic forms. A constant value of targeted pH is maintained by selecting from among buffers such as MES, 2-(N-morpholino) ethanesulfonic acid ($pK_a = 6.8$); inorganic orthophosphate ($pK_a = 7.2$); HEPES, N-hydroxyethylpiperazine-N'-2-ethanesulfonic acid ($pK_a = 6.8$); or Tris, tris (hydroxymethyl) amino-methane ($pK_a = 8.3$).

Ionic strength effects

Values of pK_a change more with buffer concentrations in solutions of multivalent buffers like phosphate or citrate than in simple monovalent buffers like acetate or Tris. The amount of change is indicated by the simplified *Debye–Hückel equation*

$$pK_a = pK_a^0 + \frac{0.5nI^{1/2}}{1 + 1.6I^{1/2}} \quad \text{(at 25°C)} \quad (2\text{-}119)$$

where I represents *ionic strength*, given by

$$\boxed{I = \frac{1}{2}\sum_i c_i(z_i)^2} \quad (2\text{-}120)$$

where c_i = concentration of ionic species i, which has charge z_i; pK_a^0 represents a value of pK_a in (2-119) extrapolated to $I = 0$; and $n = 2z - 1$ for a given charge (valence) z on the acid buffer form. Examples of ionic strength effects on buffers with representative values of pK_a^0, n, and z are illustrated in Table 2.12.

Note that the value of n in (2-119) is always odd and increases pK_a as ionic strength is increased only when the acid form of the buffer has valence $+1$ or higher.

Temperature effects

Increasing temperature, T, decreases the pK_a value of a buffer. Tris buffer provides an extreme example: its pK_a in solution at 37°C physiologic conditions is 1 pH unit lower than at 4°C (on ice). Heating increases a buffer's standard-state free energy, ΔG^o, in proportion to the value of pK_a:

$$\Delta G^o = \Delta H^o - T\Delta S^o = -RT \ln K_a = 2.3RT(pK_a) \quad (2\text{-}121)$$

Equation (2-121) shows that the standard-state enthalpy of dissociation, ΔH^o, likewise increases with pK_a, particularly

for small values of standard-state entropy of dissociation, ΔS^o. This decreases pK_a as T increases, viz.

$$\frac{-d\log K_a}{dT} = \frac{\Delta H^o}{2.3RT^2} \quad (2\text{-}122)$$

The value of ΔS^o depends mainly on the number of H_2O molecules that hydrate dissolved species, a process that produces order in a solution. Dissociation of amine buffers (e.g., Tris; $n = +1$), for example, yields no net formation of charged ions, so entropy increases slightly—$\Delta S^o = 5.7$ J/mol-K—and the free-energy change is absorbed by ΔH^o. Equation (2-122) thus shows that pK_a decreases substantially upon heating. On the other hand, physiologic buffers, acetic acid ($n = -1$) and sodium phosphate ($n = -3$), dissociate to form two new charged ions. This orders surrounding water molecules and significantly decreases entropy: $\Delta S^o = -90.5$ J/mol-K for acetic acid and $\Delta S^o = -122$ J/mol-K for phosphoric acid at pK_2. Therefore, ΔH^o changes less for acetic and phosphoric acid buffers, which exhibit relatively constant pK_a values upon heating. This preserves cell and organism viability.

Freezing

Lowering the temperature of biological samples in order to slow microbial growth and preserve enzymatic activity can significantly change local pH and solute concentrations during phase changes [71]. Free water freezes first, growing ice crystals that destroy membrane layers and organelles and locally concentrating electrolytes and biocolloids, which precipitates insoluble salts, changes pH, increases osmolarity, and depresses local freezing point. Proteins are left largely intact, so freeze-thaw cycles are used to lyse mammalian cells to release intracellular proteins and propagated virus. Any dissolved salt crystallizes as temperature approaches its eutectic point. Freezer temperatures between $-15°$ and $-25°$C allow protease and nuclease degradation at reduced rates. Consequently, flash freezing to $-80°$C and rapid thawing without local overheating preserves cell viability and enzymatic activity more effectively than freezing to about $-20°$C in a conventional freezer.

EXAMPLE 2.10 Preparation of Phosphate-Buffered Saline (PBS).

Phosphate-buffered saline (PBS) finds wide usage in biological processing, formulation, and research. Its pK_a and ionic strength mimic physiologic conditions and maintain pH across temperature changes expected in biological systems (4–37°C). PBS contains sodium chloride, sodium phosphate, and (in some formulations) potassium chloride and potassium phosphate at osmolarity (the concentration of osmotically active particles in solution) and ion concentrations that match those in the human body (isotonic). One common formulation is to prepare 10-liter stock of $10 \times$ PBS by dissolving 800 g NaCl, 20 g KCl, 144 g $Na_2HPO_4 \cdot 2H_2O$, and 24 g KH_2PO_4 in 8 L of distilled water, and topping up to 10 L. Estimate the pH of this solution in its concentrated form when stored at 4°C, after being diluted to $1 \times$ PBS at 4°C, and after being warmed to 37°C.

Table 2.12 Effect of Ionic Strength on pK_a of Some Characteristic Buffers

Common acid form	z	n	pK_a^0	pK_a $I = 0.01$	pK_a $I = 0.1$
Tris HCl, $Cl^- H^+NH_2C(CH_2OH)_3$	1	1	8.06	8.10	8.16
acetic acid, CH_3COOH	0	-1	4.76	4.72	4.66
monobasic sodium phosphate, NaH_2PO_4	-1	-3	7.2	7.08	6.89
disodium citrate 2-hydrate, $HOC(COOH)(CH_2COONa)_2 \cdot 2H_2O$	-2	-5	6.4	6.29	5.88

Solution

The ionic strength of the $10 \times$ PBS solution is

$$I = \frac{1}{2}\sum_i c_i(z_i)^2 = \frac{1}{2}\left[\begin{array}{l} 2\cdot\left(\dfrac{800\text{g}}{58.44\text{g/mol}\cdot 10\text{L}}\right)\cdot(\pm 1)^2 \\[2mm] +2\cdot\left(\dfrac{20\text{g}}{74.55\text{g/mol}\cdot 10\text{L}}\right)\cdot(\pm 1)^2 \end{array}\right] = 1.396$$

The effect of ionic strength on pK_a from Table 2.12 using the simplified Debye–Hückel equation (2-119) is

$$pK_a = 7.2 + \frac{0.5(-3)(1.396)^{1/2}}{1+1.6(1.396)^{1/.2}} = 6.59$$

Using $\Delta H^o = 4.2$ kJ/mol for monobasic phosphate, decreasing temperature increases pK_a according to

$$\begin{aligned} pK_a^{277\text{K}} &= pK_a^{298\text{K}} - \frac{4200\,\text{Jmol}^{-1}}{(2.3)8.314\,\text{Jmol}^{-1}\text{K}^{-1}\cdot(287.5\,\text{K})^2} \\ &\quad \times (277-298) \\ &= 6.59 + 0.06 = 6.65 \end{aligned}$$

Using MW for the mono- and dibasic phosphate salts of 178.0 g/mol and 136.1 g/mol, respectively, the pH of a solution with the given masses is, using (2-115),

$$pH = pK_a + \log\frac{\left[\dfrac{144}{178\cdot 10}\right]}{\left[\dfrac{24}{136.1\cdot 10}\right]} = 7.31$$

After $10\times$ dilution, $I = 0.1396$, $pK_a^{298\,K} = 6.85$, $pK_a^{277\,K} = 6.91$, and pH $= 7.57$.

After increasing T to a physiologic value of 310 K, $pK_a^{310\,K} = 6.82$, and pH $= 7.48$.

Upon dilution, the resultant $1\times$ PBS at physiologic conditions has a final concentration of 137 mM NaCl, 10 mM phosphate, and 2.7 mM KCl, at a pH near that of arterial blood plasma, pH $= 7.4$.

pH affects solubility

Water solubility of ionized species is high. An uncharged or undissociated weak acid, HA, exhibits partial solubility, S_o. In a solution saturated with undissolved HA, the partial solubility is

$$S_o = M_{\text{HA}} \tag{2-123}$$

where M_i represents molality (moles of solute per kg solvent) of component i. Total solubility, S_T, the concentration of both undissociated and ionized acid is given by

$$S_T = M_{\text{HA}} + M_{\text{A}^-} = S_o + M_{\text{A}^-} \tag{2-124}$$

Substituting the relations for M_{A^-} and M_{HA} given by (2-123) and (2-124), respectively, into the corresponding terms in the Henderson–Hasselbalch equation (2-115) and solving for S_T provides total solubility in terms of pH:

$$S_T(\text{pH}) = S_o\left(1 + 10^{-[pK_{HA}-pH]}\right) = S_o\left(1 + \frac{K_{HA}}{M_{\text{H}^+}}\right) \tag{2-125}$$

Equation (2-125) shows that solubility of a weak acid increases above the partial solubility of its undissociated form

to the extent that $K_{\text{HA}} > M_{\text{H}^+}$. Relative to a solubility value reported for pure water, $S_T^{pH=7}$, the pH effect on total solubility is given by

$$S_T(\text{pH}) = S_T^{pH=7}\frac{\left[1+10^{-(pK_{HA}-pH)}\right]}{\left[1+10^{-(pK_{HA}-7)}\right]} \tag{2-126}$$

Similarly, the total solubility of a partially soluble weak base that ionizes according to

$$\text{BOH} \rightleftharpoons \text{B}^+ + \text{OH}^- \tag{2-127}$$

with an ionization constant given by

$$K_{\text{BOH}} = \frac{[\text{B}^+][\text{OH}^-]}{[\text{BOH}]} \tag{2-128}$$

relative to solubility in water is

$$S_T(\text{pH}) = S_T^{pH=7}\frac{\left[1+10^{-(pK_{BOH}+pH-pK_W)}\right]}{\left[1+10^{-(pK_{BOH}+7-pK_W)}\right]} \tag{2-129}$$

It can be shown that the total solubility of a zwitterionic amino acid relative to water solubility is given by

$$S_T(\text{pH}) = S_T^{pH=7}\frac{\left[1+10^{-\left(pK_a^a-pH\right)}+10^{-\left(pH-pK_a^c\right)}\right]}{\left[1+10^{-\left(pK_a^a-7\right)}+10^{-\left(7-pK_a^c\right)}\right]} \tag{2-130}$$

EXAMPLE 2.11   Effect of pH on Solubility in Biological Systems.

As an example of solubility of weak organic acids, bases, and zwitterions in biological systems, prepare total solubility curves for the following species across a broad pH range ($\sim$pH $= 1$ to pH $= 11$).

1. Caprylic acid ($C_8H_{16}O_2$) is an oily, naturally occurring liquid in coconuts that has anti-fungal and anti-bacterial properties. Its water solubility is 0.068 g/100 g at 20°C with a value of $pK_a = 4.89$.

2. Thymidine (T, $C_5H_6N_2O_2$, 5-methyluracil) is a pyrimidine base that forms two hydrogen bonds with adenine (A) to stabilize dsDNA. Its water solubility is 4.0 g/kg at 25°C and its pK_{BOH} is 9.9 (neglect further dissociation that occurs at higher pH).

3. The hydrophobic amino acid valine ($C_5H_{11}NO_2$), which substitutes for hydrophilic glutamic acid in hemoglobin, causes misfolding, which results in sickle-cell anemia. Its water solubility is 8.85 g/100 mL at 25°C. Its ionization constants are $pK_a^c = 2.3$ and $pK_a^c = 9.6$.

Calculate the solubility of each component in the following bodily fluids: cell cytosol, pH $= 7.2$; saliva, pH $= 6.4$; urine, pH $= 5.8$.

Solution

The solubilities are calculated from (2-126) for the acid, (2-129) for the base, and (2-130) for the zwitterion. In (2-130), the pK for water is computed to be 14 from (2-129). Table 2.13 shows solubility values in each bodily fluid. Figure 2.17 shows extended, calculated solubility curves for each component. Caprylic acid solubility

Table 2.13 Solubility of Organic Acid, Base, and Amino Acid in Bodily Fluids

	Solubilities (g/kg)		
Fluid: pH	Caprylic acid	Thymidine	Valine
Cell cytosol: 7.2	1.07	4.00	88.6
Saliva: 6.4	0.175	4.01	88.3
Urine: 5.8	0.0478	4.07	88.3

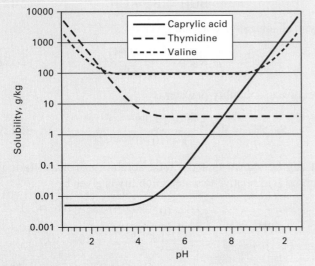

Figure 2.17 Solubility curves for caprylic (biological acid), thymidine (base), and valine (zwitterion) acids.

increases at pH $\geq$ pK_a. Thymidine solubility increases at pOH $= -\log[OH^-] = 14$ pH $\geq$ pK_{BOH}. Valine exhibits solubility increases at pH $\geq$ pK_a^a and pOH $\geq$ pK_a^c. Reduced water solubility at the pI of a biomolecule may be used for selective precipitation or for extraction into a less-polar phase. Effect of pH on bioproduct solubility is further illustrated in Examples 8.13 and 8.14 of §8.6.1. Pharmaceutical formulation for drug delivery requires good understanding of solubility.

Solvent effects

Adding miscible organic solvents increases pK_a values of acetate, phosphate, and other buffers with negative n values by reducing activity of water, which causes resistance to proton dissociation. For example, the pK_a of phosphate increases 1 pH unit upon increasing the volumetric content of ethanol from 0 to 50%. On the other hand, values of pK_a for buffers like Tris, which have positive n values, are only slightly affected.

Static charges (Donnan effects)

Static negative (positive) charges attract (repel) protons, producing an adjacent micro-environment with lower (higher) pH—the *Donnan effect*. For example, porous cation exchange media used for adsorptive biochromatography (see §15.3.3) has a pH in the matrix about 1 unit lower than that in the eluting buffer—a difference that increases with decreasing buffer ionic strength. An enzyme that remains stable in

eluent buffered at pH 5.5 may be denatured upon purification by cation exchange adsorption on media whose local pH is 4.5. On the other hand, since most enzymes are anionic (pI < 7.0) and remain stable at mildly alkaline conditions (pH = 8–10), fewer problems occur in anion exchange purification. Protein elution from ion-exchange media can produce sudden, large, local pH changes, particularly at low ionic strength, which decreases resolution of closely related species. On the other hand, some ion-exchange adsorbents like DEAE and some proteins may provide significant local buffering capacity. Donnan effects thus affect selection of operational pH ranges of ion-exchange adsorption separations.

Applications to Bioseparations

In *solvent selection* (§8.6.1, "Organic-Aqueous Extraction of Bioproducts"), it is shown how pH, I, T, and S_T affect the state of a bioproduct (e.g., pI) to influence its solvent partition coefficient, K_D (e.g., see Eq. 8-82) and guide selection of a suitable solvent for liquid–liquid extraction. The charge on bioproducts affects rejection and passage through ultrafiltration membranes (see §14.8.3) and determines adsorptive partitioning in ion-exchange chromatography (see §15.3.3). Interactions between pH and bioproduct isoelectric points are the basis for bioproduct separation using different modes for electrophoresis, distinguished in §15.8.2.

§2.9.2 Biocolloid Interactions

Biological polymers, macromolecular species, and cells exhibit many features of colloids—0.001 to ~1.0 μm particles that interact via long-range *hydrophobic, electrostatic*, and *van der Waals* forces, which arise from colloid size and accessible surface features [72, 73]. Such forces contrast with short-range *hydrogen bonding* and *dipole-dipole* interactions, which originate from electron orbitals of chemical functional groups on biocolloid surfaces and occur at separations approximately the length of a chemical bond. Separation distances and bond energies of interactions between biocolloids are summarized in Table 2.14. Colloid forces impact

Table 2.14 Bond Energies and Separation Distances of Biomolecular Interactions

Interaction	Equilibrium Separation (nm)	Bond Energy (kJ/mol)
Ionic bond	0.23	580–1050
Covalent bond		60–960
Metallic bond		105–355
Ion–dipole interaction	0.24	84
Dipole–dipole (hydrogen bond) interaction	0.28 (0.18–0.30)	5–30
Dipole-induced dipole interaction		<21
Hydrophobic	0.30	4.0
Dispersion forces	0.33	<42 (~0.25)

dissolution, aggregation, and other interactions between biological species such as cells, micelles, vesicles, and virus particles [74]. For example, Allison and Valentine [75, 76] reported binding of fowl plague and *vaccinia* virus to suspended HeLa cells at one-third the rate predicted by Fick's first and second laws of diffusion, due to electrostatic repulsion. The impact of colloid forces on biomolecular interaction affects properties in many bioseparations, including salting out effects in precipitation (see §17.11), particle aggregation in flocculation, solute mobility in electrophoresis (see §15.8.3), and charge dependence in phase partitioning (e.g., liquid–liquid extraction in §8.6 and liquid–solid adsorption in §15.3), and filtration in Chapter 19.

Consider the forces that contribute to the double-helix structure of DNA. *Covalent bonds* link adjacent nucleotides in each individual DNA strand. *Hydrogen bonds* between nucleotide bases (Watson–Crick base pairing) and *van der Waals' interactions* between stacked purine and pyrimidine bases hold the two complementary DNA strands in a helix together. Relatively *hydrophobic* nucleotide bases are buried in the helical interior, while charged phosphate groups and polar ribose sugars *solvate* dsDNA in aqueous solutions. *Electrostatic* interactions between adjacent negatively charged backbone phosphates are minimized by the extended backbone.

Cyclic manipulation of noncovalent force interactions by precise, consecutive, temperature adjustments—a technique called *polymerase chain reaction* (PCR)—allows a specific gene fragment (template) to be amplified for purposes such as gene sequencing or analysis of genetic mutations. In PCR, complementary DNA strands are thermally denatured (separated) at 94°C and then annealed to 21 base-pair complementary *primer* strands at ~55°C. Primers are subsequently elongated by polymerase at ~72°C using dissolved nucleotide base pairs. Repeating this cycle n times allows 2^n amplification of the original DNA template. The forces that bind DNA and influence other biomolecular interactions originate in colloidal interactions between suspended particles, small solutes, and solvent molecules.

DLVO theory

The theory of Derajaguin and Landau (of Russia) and Vervey and Overbeek (DLVO) describes attractive forces such as *van der Waals* (vdW) interactions and repulsive *double-layer* (*electrostatic*) forces between suspended colloids at approach distances >2 nm in the limit of low surface potentials [77], which occur when an elementary charge on the colloid surface has a potential energy $\ll k_BT$ (the thermal energy scale), where k_B is the Boltzmann constant. *Potential energy*, $\psi(r)$, between two like-charged colloids separated by distance r is the sum of attractive vdW and repulsive electrostatic forces. Coulomb's law holds that the force of electrical interaction between two charged particles varies directly with the product of their charges and inversely to the square of the distance between them, r^2. Adjacent colloids are repelled—thereby stabilizing colloid suspensions—by net respective surface charges (same sign) of 30–40 mV. These charges

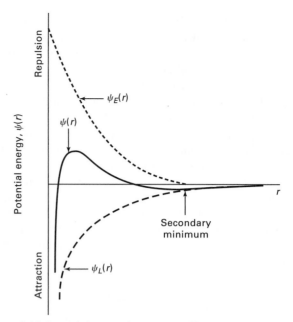

Figure 2.18 Potential energy between two like-charged particles.

arise from intrinsically high-area colloid surfaces that encourage (1) adsorption of ions; (2) molecular ionization on the surface, which leaves behind a surface charge; and/or (3) dissolution of ions from the solid into the suspending liquid [78].

The Gouy–Chapman theory postulates that an *electrostatic potential*, $\psi_E(r)$, forms at a surface with uniformly distributed charge and decays exponentially away with distance, r, due to thermal motion of oppositely charged ions in the adjacent solution

$$\Psi_E\{r\} = \Psi_{E,o}\exp(-\kappa r) \qquad (2\text{-}131)$$

where $\psi_{E,o}(r)$ is the potential at the surface and κ is the *Debye–Hückel constant*. An *attractive potential* $\psi_L(r)$ arises due to long-range induction, orientation, and dispersion forces (simple systems consider only attractive vdW forces). Evaluated together, as in Figure 2.18, these potentials yield an overall interaction energy curve—$\psi(r)$—that exhibits two minima: a shallow *secondary minimum* at $r \sim 4/\kappa$ in which colloids remain stably suspended (>5 nm apart at low I values); and a deep *primary minimum* associated with rapid coagulation. Coagulation results from increasing I such that free energy due to electrostatic interactions between particles decreases to $\leq k_BT$, corresponding to particle separations $\leq \kappa^{-1}$, the *Debye length*.

Electrostatic double-layer interactions

Colloid surface charges form an electrical *double layer* that surrounds ionic charged particles and exerts electrostatic forces on adjacent colloids when their double layers overlap [80]. To maintain electrical neutrality, a charged surface dissolved in a medium attracts hydrated ions of opposite charge (counterions), which become strongly bound, forming an inner *Stern* (or *Helmholtz*) *layer* of thickness, δ. A less-homogeneous, diffuse outer shell of moderately bound *Gouy* (or *Gouy–Chapman*) *second-layer* ions is comprised of hydrated ions of alternating charge. An external electric current will move the hydrated ions in the Gouy layer toward the electrode, forming a line of shear between fixed counterions and migrating ions. The measurable difference in electro-kinetic potential between the fixed boundary layer of charges on a surface and the shear line of mobile charges in the bulk of the solution is the *zeta potential* (or *Stern potential*), ζ. It indicates the fixed-charge density unsatisfied by counterions in the Stern layer and the corresponding distance into the solution required to satisfy it. That distance, the Debye length, κ^{-1}, or *double-layer thickness*, may be estimated with the Debye–Hückel approximation for modest surface potentials ($\Psi_o < 25$ mV) as

$$\frac{1}{\kappa} = \left(\frac{2000 e^2 N_A I}{\varepsilon k_B T} \right)^{-1/2} \tag{2-132}$$

in an electrolyte or colloid dispersion, where N_A is Avogadro's number. The ionic strength, I, is given by (2-120) for c_i (M_i) in mol/dm^3 (M)

$$I = \frac{1}{2} \sum_i z_i^2 M_i \tag{2-133}$$

The static permittivity for zero wavelength, ε, of the medium is given by

$$\varepsilon = \varepsilon_r \varepsilon_o = 78.54 \times 8.85 \times 10^{-12} \frac{C^2}{J \cdot m} \tag{2-134}$$

for water, using the relative static permittivity, ε_r, of H$_2$O at 25°C (commonly called the *dielectric constant*) and the electric constant, ε_o (i.e., vacuum permittivity, $\varepsilon_o = \mu_o^{-1} c_o^{-2}$, where μ_o is a magnetic constant and c_o is speed of light in a vacuum). The Boltzmann constant, k_B, is

$$k_B = 1.381 \times 10^{-23} J/K \cdot molecule \tag{2-135}$$

The charge on an electron, e, is

$$e = 1.60 \times 10^{-19} coulomb \tag{2-136}$$

The *dielectric constant* measures how much a material reduces the magnitude of an electric field. At 25°C for I in mol/dm^3, this gives [81]

$$\kappa^{-1} = 3.05 \times 10^{-10} I^{-1/2} \tag{2-137}$$

in meters.

In practice, it is easier to measure the streaming current potential by anchoring charged particles and moving the hydrated ions past them than it is to measure ζ. The Debye length decreases as ionic strength increases from ~100 Å (10^{-3} M); 30 Å (10^{-2} M) to 10 Å (10^{-1} M) for a 1-1 (monovalent) electrolyte solution. These values are small relative to particle diameter and decrease as zeta potential is lowered. Repulsive forces are <1 mN/m for spheres with moderate surface potentials ($\Psi_o < 100$ mV). Electrostatic repulsion can be felt at up to 60 nm in deionized water, but weakens

to ~1–3 nm at physiologic conditions normally found in biological fluids ($0.01 < c_i < 0.2$ M).

EXAMPLE 2.12 Colloidal Forces of Dissolved Biomolecules.

Colloidal forces on dissolved biomolecules affect their separability. Selection of appropriate molecular-weight cut-off values for an ultrafiltration bioseparation requires accurate estimation of bioproduct size (see §14.8.3). Determine the concentration of a 1:1 salt in moles/dm^3 below which the apparent radius of *bovine serum albumin* (BSA) is more than 10% larger than its actual radius of $a = 3.6$ nm.

Solution

For a 1:1 salt, ionic strength, I, in (2-133) reduces to c in mol/dm^3. The apparent radius ($a + \kappa^{-1}$) of a molecule is significantly different from its actual radius a if $\kappa^{-1}/a > 0.1$. Substituting this expression into (2-137) and solving for c in mol/dm^3 yields

$$c < \left(\frac{3.05 \times 10^{-9}}{a} \right)^2 \tag{2-138}$$

for a in meters. Therefore, the apparent radius of BSA is greater than its actual radius, $a = 3.6 \times 10^{-9}$ m, if salt concentration is less than 0.72 M.

Van der Waals forces

Instantaneous quantum fluctuations in charge distribution of one molecule distort the electron cloud in a neighboring atom or molecule to induce momentary polarization, and vice versa, resulting in short-range attraction between the transient dipole moments [82]. The cumulative effect is van der Waals (vdW), or *London dispersion* forces. Between atoms, these forces act over distances that are of the order of atomic dimensions, while between colloids, they are of the order of colloid dimensions. vdW forces are responsible for phase transitions (e.g., condensation of gas to liquid) and interfacial tension between adjacent phases, reaching up to 10 mN/m at ~1 nm separations. The potential energy of interaction, W_D, via dispersion forces between two identical atoms separated by distance r is

$$W_D = -\frac{\lambda}{r^6} \tag{2-139}$$

where λ is a constant parameter of interaction given by

$$\lambda = \frac{3}{4} \alpha^2 h v_o \tag{2-140}$$

for atom polarizability α and ionization potential $h v_o$. The characteristic energy for the quantized energy at frequency v_o constitutes the ionization potential. A single atom that closely approaches a macroscopic body of volume V produces an overall dispersion force felt by each atom that is estimated by integrating the energy of dispersion, W_D,

$$F_{12} = \int_V \frac{dW_D}{dr} dV \tag{2-141}$$

Table 2.15 Interaction Energies for Several Geometries

Geometry	Interaction Energy
Atom—flat body	$w = -\dfrac{\pi q \lambda}{6r^3}$
Two flat bodies	$w = -\dfrac{A}{12\pi r^2}$
Sphere—flat body	$w = -\dfrac{Aa}{6r}$
Two spheres	$w = -\dfrac{Aa_1 a_2}{6r(a_1 + a_2)}$

Source: A.A. Garcia et al. [73].

Potential energies of interaction, w, for more complex geometries may be obtained by extending this approach. Table 2.15 shows values of w for spheres of radius a_i, as well as other superatomic scale geometries. Negative interaction energies are associated with spontaneous processes.

Comparing (2-139) with the interaction energies in Table 2.15 shows that dispersion forces decay more slowly as separation increases for colloidal bodies than for atoms. At separations exceeding ~ 10 nm, vdW interaction potential decays faster than r^{-6} due to retardation by phase difference between instantaneous dipoles on opposing bodies. The *Hamaker constant*, A, for vdW interactions between body (1) and body (2) defined by

$$A_{1-2} = \pi^2 \lambda \rho_1 \rho_2 \tag{2-142}$$

indicates the interaction parameter, λ, and respective number of atoms per unit volume, ρ_i, (i.e., atomic densities) in each body, which determine the attractive force between them. The moderating influence of a fluid (1) intervening between two same bodies (2) may be determined using a pseudo-chemical model to modify the Hamaker constant

$$A_{2-1-2} = \left(A_{1-1}^{1/2} - A_{2-2}^{1/2}\right)^2 \tag{2-143}$$

Values of the Hamaker constant vary only slightly from one medium to another: 4.3×10^{-20} J for water, 7.8 to 9.8×10^{-20} J for polystyrene, and 8.6×10^{-20} J for natural rubber [83].

Colloid flocculation by electrolytes

Like-charged electrostatic double layers and lubrication forces inhibit close approach of repellant surfaces and association due to London dispersion or vdW forces. On the other hand, surfaces with oppositely charged ζ values are adherent. Addition of ions (electrolyte salts or hydronium ions) to a colloid suspension structures the waters, attenuates the zeta potential, reduces the Debye length, and allows closer approach of like-charged surfaces, while reducing adherence of oppositely charged double layers. Coulombic repulsive forces dominate until the distance separating two surfaces is reduced sufficiently (i.e., to $r \sim 4/\kappa$) for attractive forces to be asserted. DLVO theory predicts that the critical *flocculating electrolyte concentration*, c_i, for *symmetric* (e.g., 1:1, 2:2, 3:3, but not 1:2, etc.), indifferent (not chemically

adsorbed into the Stern layer) electrolytes is [84]

$$c_i^{floc} = \frac{9.85 \times 10^4 \varepsilon^3 k_B^5 T^5 \gamma^4}{N_A e^6 A^2 z_i^6} \tag{2-144}$$

where γ is a constant that approaches 1.0 at high potentials (> 240 mV) and $ze\psi_{E\delta}/4k_B T$ at low potentials ($\psi_{E\delta} \sim 75$ mV), and c_i^{floc} is the lower molar concentration of the electrolyte that induces particle coagulation. In water at 25°C, this relation becomes

$$c_i^{floc} = \frac{\left(3.38 \times 10^{-36} \text{J}^2\text{-mol-m}^{-3}\right)\gamma^4}{A^2 z_i^6} \tag{2-145}$$

with c_i^{floc} in mol/m^3 and the Hamaker constant A in J. Equation (2-145) predicts that the relative values for critical flocculating concentration of electrolytes such as K^+, Ca^{2+}, and Al^{3+} containing counterions with charge numbers $z = 1, 2,$ and 3 will be $1:2^{-6}:3^{-6}$ or 1000:15.6:1.37 at wall potentials > 240 mV, where $\gamma \sim 1$. This is the *Schulze–Hardy rule*. It is illustrated by the common practice of settling colloids during water treatment by adding alum, a double salt of aluminum and ammonium sulfates, to increase ionic strength.

EXAMPLE 2.13 **Temperature, Charge, and Colloid Effects on Flocculation by an Electrolyte.**

The first step in recovering bioproducts expressed in bacterial fermentation is often removal of aqueous culture broth to reduce process volume. Flocculation of bacteria by adding electrolyte enhances settling and broth removal. The ease of flocculation is a function of the electrolyte concentration. Determine the critical flocculation concentration, c_i^{floc} (in mol/dm^3), of an indifferent 1-1 electrolyte at 25°C and low potential. Calculate the effect on c_i^{floc} of (a) lowering temperature to 4°C; (b) changing to an indifferent 2-2 electrolyte at 4°C (maintaining low potential); (c) using electrolyte in part (b) to flocculate a viral colloid at 1/10 the concentration; and (d) flocculating at high potential.

Solution

At 75 mV,

$$\gamma = \frac{ze\Psi_{E\delta}}{4k_B T} = \frac{(1)(1.6 \times 10^{-19}\text{C})(0.075\text{V})}{4(1.38 \times 10^{-23}\text{J/K})(298\text{K})} = 0.7295$$

$$c_i^{floc} = \frac{(3.38 \times 10^{-36}\text{J}^2 \cdot \text{mol} \cdot \text{m}^{-3})(0.7295)^4}{(8 \times 10^{-20}\text{J})(1)^6}\left(\frac{\text{m}^3}{1000\,\text{dm}^3}\right) = 0.15\,\text{M}$$

(a) $c_i^{floc}(277°\text{C}) = c_i^{floc}(298°\text{C})\dfrac{277}{298} = 0.14\,\text{M}$

(b) $c_i^{floc}(277°\text{C}; 2-2) = c_i^{floc}(298°\text{C}; 1-1)\dfrac{277}{298}\dfrac{1}{(2)^2} = 0.035\,\text{M}$

(c) DLVO theory and (2-145) indicate that while critical electrolyte flocculation concentration is sensitive to temperature and electrolyte valence, it is independent of colloid particle size or concentration.

(d) Equation (2-145) indicates that at high potential, c_i^{floc} is proportional to T^6 and z^{-6} rather than to T and z^{-2}, respectively, at low potential.

Protein aggregation, crystallization, and adsorption have been shown to be controlled by long-range DLVO forces. However, nonclassical DLVO forces due to solvation, and hydrodynamic and steric interactions also influence protein interactions—generally at short range and with magnitudes up to the order of several $k_B T$—by altering biomolecular association rates or enhancing stability of protein complexes.

Solvation or hydration forces

At separation distances closer than 3 to 4 nm, continuum DLVO forces based on bulk properties of intervening solvent (e.g., density, dielectric constant, refractive index) give way to non-DLVO forces that account for structures formed by individual solvent molecules at solid interfaces based on their discrete size, shape, and chemistry. Normal to a surface and within several molecular diameters, solvent molecules form ordered layers. Attractive interactions between the surface and liquid molecules and the constraining effect of an approaching surface, which squeezes one layer after another out of the closing gap, cause *desolvation* ("lubrication") forces between the surfaces, F_{SOL} (known as *hydration, hydrodynamic,* or *drainage* forces when the solvent is water). These are decaying oscillatory functions of separation distance, D, for spherical molecules between two hard, smooth surfaces

$$F_{SOL}\{D\} = K \exp\left(-\frac{D}{l}\right) \qquad (2\text{-}146)$$

where $K > 0$ and $K < 0$ relate to hydrophilic repulsion and hydrophobic attraction forces, respectively, and l is the correlation length of orientational ordering of water molecules. Equation (2-146) explains short-range forces measured between neutral lipid bilayer membranes, DNA polyelectrolytes, and charged polysaccharides that are relatively insensitive to ionic strength. Polar solvent molecules like water that intervene between adjacent colloids form head-to-tail (positive-to-negative) conduit chains of partial charge interactions that can either attenuate forces between charged colloids or increase the effective distance of typically short-range ion–ion or acid–base interactions.

Molecular dipoles align in an orientation that opposes (and thereby diminishes) the originating electric field. Water dipoles also surround charged ions (e.g., electrolytic salt molecules) and displace ionic bonds, solvating the individual ions. Long-range (>10 nm) *hydrophobic* effects are also produced by water molecules.

Hydrophobic interactions

The free energy of attraction of water for itself due to hydrogen bonding is significant. *Hydrophobic* ("water-fearing") groups restrict the maximum number of energetically favorable hydrogen bonds (i.e., degrees of freedom) available to adjacent water molecules. Water thus forms ordered and interconnected tetrahedral hydrogen-bonded structures that exclude hydrophobic entities like hydrocarbons or surfaces to minimize the number of affected water molecules. These structures reduce *interfacial area* and *surface free energy* of hydrocarbon–water systems as they assemble, minimizing entropy (maximizing degrees of freedom) and maximizing enthalpy (from hydrogen bonding). This process drives attraction of nonpolar groups in aqueous solution via forces up to 100 mN/m at separations <3 nm, culminating in *phase separation*. Hydrophobic interactions allow formation of reverse micelles (§8.6.1) into which bioproducts partition for subsequent extraction. Other biological examples of hydrophobic interactions include aggregation of mycobacteria to form cords and clustering of hydrophobic protein patches due to side chains of phenylalanine (Phe), tyrosine (Tyr), tryptophan (Trp), leucine (Leu), isoleucine (Ile), methionine (Met), and valine (Val). Hydrophobic forces between macroscopic surfaces are one to two orders of magnitude greater than vdW attraction, decaying exponentially with a characteristic length of 1–2 nm in the range 0–10 nm and remaining significant at distances up to 80 nm. Hydrophobicity scales developed for amino acids are useful to explain protein partitioning in liquid–liquid extraction (see "Reverse Micelles" in §8.6.1 and "Salt Effect" in §8.6.2), retention order, and retention time in reversed-phase and hydrophobic-interaction chromatography.

Structuring water, kosmotropes and chaotropes

Hydrophobic forces are influenced by biomolecule structures and the nature of dissolved ionic and nonionic species. Hydrophobic interactions are increased by dispersion forces between planar surfaces (i.e., stacking) of aromatic groups in aromatic amino acids, purines or pyrimidines, chlorophyll, and haem, for example. Small or multiple-charged ionic *kosmotropes* ("order-maker," e.g., citrate, sulfate, phosphate, hydroxide, magnesium, and calcium) interact more strongly with water than water itself. This enhances formation of water structure, stabilizes biomolecule structures in solution, and promotes hydrophobic interactions like aggregation. Large, singly charged ions with low charge density, called *chaotropes* (e.g., guanidinium, tetramethylammonium, and thiocyanate), interact weakly with water and disrupt water structure formation, which solvates hydrophobic structures and unfolds amphipathic proteins, which have interior regions rich in nonpolar characteristics as well as polar functional groups on the exterior. Chaotropic ions like octylmethylammonium chloride enhance aqueous/organic extraction (§8.6.1) of polar zwitterions like amino acids.

Nonionic kosmotropes including zwitterions (e.g., proline, ectoine, glycine, betaine) and polyhydroxy compounds (e.g., trehalose) hydrogen-bond strongly to water (e.g., sugar hydration). This reduces availability of water freely diffusing around proteins, exchange rates of backbone amide protons, and hydration of larger surfaces exposed by denaturation. Dehydration reduces biomolecule flexibility, which promotes stability in solution and prevents thermal denaturation, but reduces enzymatic activity. Kosmotropic ions and nonionic polymers are used to form partially miscible aqueous phases that allow stable, two-phase aqueous extraction of biomolecules (§8.6.2). Nonionic chaotropes (e.g., urea) weaken hydrogen bonding, decrease the order of water, and increase its surface tension, which weakens macromolecular structure.

The Hofmeister series

The *Hofmeister series* classifies ions in order of their ability to change water structure. It was initially developed to rank cations and anions in terms of their ability to increase solvent surface tension and lower solubility of ("salt out") proteins [85]:

Anions: $F^- \approx SO_4^{2-} > HPO_4^{2-} > acetate > Cl^- > NO_3^- > Br^- > ClO_3^- > I^- > ClO_4^- > SCN^-$

Cations: $NH_4^+ > K^+ > Na^+ > Li^+ > Mg^{2+} > Ca^{2+} > guanidinium$

Early ions in the series strengthen hydrophobic interactions (e.g., ammonium sulfate is commonly used to precipitate proteins). Later ions in the series increase solubility of nonpolar molecules, "salting in" proteins at low concentrations: iodide and thiocyanate, which weaken hydrophobic interactions and unfold proteins, are strong *denaturants*. Effects of pH, temperature, T, and ionic strength, I, of an ionic salt on solubility, S, of a given solute may also be correlated using

$$\log S = \alpha - K_s I \qquad (2\text{-}147)$$

where K_s is a constant specific to a salt–solute pair and α is a function of pH and T for the solute. The relative positions of ions in the Hofmeister series change depending on variations in protein, pH, temperature, and counterion. Anions typically have a larger influence. The series explains ionic effects on 38 observed phenomena, including biomolecule denaturation, pH variations, promotion of hydrophobic association, and ability to precipitate protein mixtures. The latter two phenomena occur in roughly reverse order. Hofmeister ranking of an ionic salt influences its effect on solvent extraction of biomolecules, including formation of and partitioning into reverse micelles, and lowers partition coefficients, K_D, of *anionic* proteins between upper PEG-rich and lower dextran-rich partially miscible aqueous phases (§8.6.2, "Aqueous Two-Phase Extraction"). Table 2.16 orders ions in terms of their ability to stabilize protein structure and to accumulate

or exclude proteins from chaotropically disordered (e.g. low-density) water.

Steric forces

Either repulsive stabilizing or attractive coagulating forces may be produced by dissolved biopolymers in solution. The close approach of two surfaces at which polymers are anchored confines the thermal mobility of dangling chain molecules, resulting in a repulsive entropic force. Adding polysaccharides or proteins can sterically stabilize coagulative colloids with a force that depends on surface coverage, reversibility of anchoring, and solvent. One example is gelatin, an irreversibly hydrolyzed form of collagen, which constitutes ~50% of proteins in mammals and is commonly used as a gelling agent in foods, pharmaceuticals, and cosmetics. On the other hand, colloid flocculation may be induced by polymeric *nonionic or ionic surfactants*. Nonionic [nondissociating, e.g., polyethylene oxide (PEO), polyethylene glycol (PEG)] or ionic [e.g., polyacrylamide and sodium dodecylsulfate (SDS)] polymer surfactants may adsorb to and envelope adjacent repellant surfaces via hydrophobic interactions. This results in steric (or entropic) stabilization, which can match the potential energy barrier that prevents the approach of repellent colloids and induce flocculation. At the same time, attractive interactions like biomolecular interactions or affinity adsorption may be buffered or prevented by polymer surfactants. A surfactant may disrupt mutual hydrophobic interactions by masking a hydrophobic ligand, while the hydrophilic moiety of the surfactant interacts with water.

Surface force measurements of protein interactions

Particle detachment and peeling experiments, force-measuring spring or balance, and surface tension and contact angle measurements are used to gauge surface interaction; but these conventional methods do not provide forces as a function of distance. Scanning force probes [atomic force microscopy (AFM), scanning probe microscopy (SPM)] use improved piezoelectric crystals, transducers and stages, nanofabricated tips and microcantilevers, and photodiode-

Table 2.16 Ability of Cations and Anions to Stabilize Protein Structure

Effect of Ion on Proteins	Cations	Anions	Effect of Ion on Proteins
Protein stabilizing	$N(CH_3)_4^+$	Citrate $C_3H_4(OH)(COO)_3^{3-}$	Protein stabilizing
Weakly hydrated	NH_4^+	Sulfate SO_4^{2-}	Strongly hydrated
Accumulate in low-density water	Cs^+	Phosphate HPO_4^{2-}	Excluded from low-density water
	Rb^+	Acetate CH_3COO^-	
	K^+	F^-	
	Na^+	Cl^-	
	H^+	Br^-	
Protein destabilizing	Ca^{2+}	I^-	Protein destabilizing
Strongly hydrated	Mg^{2+}	NO_3^-	Weakly hydrated
Excluded from low-density water	Al^{3+}	ClO_4^-	Accumulate in low-density water

detectable laser light to measure the accuracy of calculated DLVO and non-DLVO potentials for biological systems.

Applications to bioseparations

In §8.6, it is shown how biocolloid interactions that influence solvation, hydrophobicity, water structure, and steric forces can enhance extraction of bioproducts via organic/aqueous and aqueous two-phase extraction. In §14.9.2, it is observed how biocolloid interactions affect membrane selectivity, sieving, and prediction of permeate flux. In §15.3.3, effects of biocolloid interactions on ion-exchange interactions are described. Bioproduct crystallization, §17.11, is also affected by biocolloid interactions.

§2.9.3 Biomolecule Reactions

Unique structural features of ligands or their functional groups allow specific, noncovalent interactions (e.g., ionic and hydrogen bonding, hydrophobic effects, vdW forces, and steric forces) with complementary structures of target biomolecules like receptor proteins that result in biochemical reactions, which sustain viability of cells. These interactions typically operate over short ($<$2 nm) intermolecular distances with binding energies associated with noncovalent bond formation, typically $>4k_BT$. Examples include: (1) immunologic recognition of a specific region (*epitope*) on a foreign substance (*antigen*) by protein immunoglobulins (*antibodies*); (2) regulation of gene expression by protein transcription factors that bind to control regions of DNA with high-affinity domains (*motifs*), such as zinc finger, leucine zipper, helix-turn-helix, or TATA box; (3) cell surface receptor (e.g., ion-channel, enzyme-linked, and g-protein linked receptor classes) binding to chemical substances (*ligands*) secreted by another cell in order to transduce cell-behavior-modifying signals; and (4) specific binding of a monosaccharide isomer by a carbohydrate-specific lectin—a protein with 2+ carbohydrate-binding sites.

Binding energies of noncovalent interactions require intermolecular proximities on the order of 0.1 nm and contact areas up to 10 nm^2 (usually ~1% of total solvent-accessible surface area) for these reactions to occur. Hydrogen bonds involving polar charged groups, for example, add 2.1 to 7.5 kJ/mol, or 12.6 to 25.1 kJ/mol when uncharged groups are involved. Hydrophobic bonds may generate 10 to 21 kJ/mol/cm^2 contact area.

Ligand–receptor binding cascade

Recognition and binding of the receptor by a ligand is initiated by electrostatic interactions. Solvent displacement and steric selection follow, after which charge and conformational rearrangement occur. Rehydration of the stabilized complex completes the process. From 20 to 1 nm, the approach and complementary pre-orientation of ligand and receptor maximize dominant coulombic attractive forces. Binding progresses from 10 to 1 nm as hydrogen-bonded

solvent (water) molecules are displaced from hydrated polar groups on hydrophilic exteriors of water-soluble biomolecules. Release of bound water decreases its fugacity and increases the solvent entropic effect associated with surface reduction, which contributes to binding energy. Solvent displacement can make sterically hindered ligands more accessible for interaction through short-range, dipole–dipole, or charge-transfer forces. Next, conformational adjustments in ligand and receptor produce a steric fit (i.e., "lock-and-key" interaction). Steric effects and redistribution of valence electrons in the ligand perturbed by solvent displacement produce conformational rearrangements that yield a stable complex. Rehydration completes the binding process. Methods to compute forces that control protein interactions account for (1) absorbed solvent molecules and ions; (2) non-uniform charge distributions; (3) irregular molecular surfaces that amplify potential profiles at dielectric interfaces; (4) shifts in ionization pK's due to desolvation and interactions with other charged groups; and (5) spatially varying dielectric permittivity across the hydration shell between the protein surface and bulk solvent. Interaction with a PEG-coupled ligand may be used to increase selective partitioning of a target biomolecule into an upper PEG-rich phase during aqueous two-phase extraction (§8.6.2).

Ionic interactions

Ionic interactions between a net charge on a ligand and counterions in the receptor have high dissociation energies, up to 10^3 kJ/mol. As an example, amino groups RNH_3^+, protonated at physiologic pH on lysine [R $=$ (CH$_2$)$_4$; pK $=$ 10.5], or arginine [R $=$ (CH$_2$)$_3$NHCNH; pK = 12.5] can interact with carboxyl groups, R'COO$^-$, on ionized aspartate (R' $=$ CH$_2$; pK = 3.9) or glutamate [R' $=$ (CH$_2$)$_2$; pK = 4.2] forms of aspartic acid and glutamic acid, respectively. Hard (soft) acids form faster and stronger ionic bonds with hard (soft) bases. Hard acids (e.g., H$^+$, Na$^+$, K$^+$, Ti^{4+}, Cr^{3+}) and bases (e.g., OH$^-$, Cl$^-$, NH$_3$, CH$_3$COO$^-$, CO$_3^{2-}$) are small, highly charged ions that lack sharable pairs of valence electrons (i.e., the nucleus strongly attracts valence electrons precluding their distortion or removal), resulting in high electronegativity and low polarizability. Soft acids (e.g., Pt^{4+}, Pd^{2+}, Ag$^+$, Au$^+$) and bases (e.g., RS$^-$, I$^-$, SCN$^-$, C$_6$H$_6$) are large, weakly charged ions that have sharable p or d valence electrons, producing low electronegativity and high polarizability. Examples of ionic stabilization of ligand–biomolecule or intrabiomolecule interactions are bonding between oppositely charged groups (i.e., salt bridges) and bonding between charged groups and transition-series metal cations. Example 8.15 in §8.6.1 demonstrates how desolvation via ion-pairing or acid–base pairing of organic extractants can enhance organic/aqueous extraction of bioproducts.

EXAMPLE 2.14   Selection of a Metal for Immobilized Metal Affinity Chromatography (IMAC).

Using the stability constant of metal-ion base complexes in Table 2.17, identify an appropriate metal to entrap in the solid phase via

Table 2.17 Stability Constants of Metal-Ion Base Complexes

		Thiourea (Log K)	Histidine (Log K)
Soft metal ions	Ag^+	$7.11 \pm 0.07^*$	–
	Cd^{2+}	$1.3 \pm 0.1^*$	5.74^*
Borderline metal ions	Cu^{2+}	$0.8^\#$	$10.16 \pm 0.06^*$
	Zn^{2+}	0.5^*	$6.51 \pm 0.06^*$

*in aqueous solution: $I = 0.1$, $25°C$

$^\#$in aqueous solution: $I = 1.0$, $25°C$

Source: A.A. Garcia et al. [73]

chelation to perform immobilized metal affinity chromatography (IMAC) of thiourea and histidine.

Solution

The R-group nitrogen in the amino acid histidine donates an electron pair to borderline soft-metal ions such as Cu^{2+}, Zn^{2+}, Ni^{2+}, or Co^{2+} to form a coordination covalent bond that can be displaced upon elution with imidazole. Data in the table suggest $Cu^{2+} > Zn^{2+}$ for histidine and $Ag^+ > Cd^{2+}$ at the conditions shown. rDNA techniques can be used to genetically modify target proteins to include multiples of this amino acid as a tag (His_6-tag) on a *fusion* protein to facilitate purification.

Amino acid–metal bonds

Nearly all amino acids exhibit affinity for divalent metal ions. Log K-values for interacting amino acids range from 1.3 to 2.4 for Mg^{2+} and Ca^{2+}, and from 2.0 to 10.2 for Mn^{2+}, Fe^{2+}, Co^{2+}, Ni^{2+}, Cu^{2+}, and Zn^{2+}. Interactions between amino acids and copper generally exhibit the highest log K-values. Histidine and cysteine exhibit the strongest metal affinities. Double-stranded nucleic acids contain accessible nitrogen (i.e., N7 atom in guanidine and adenine) and oxygen (i.e., O4 atom in thymine and uracil) sites that donate electrons to metal ions in biospecific interactions. Soft- and borderline-metal ions generally have stronger affinity for N7 and O4, while hard-metal ions have greater affinity for oxygen atoms in the phosphate groups of nucleic acids.

Hydrogen bonds

Hydrogen bonds between biomolecules result from electrostatic dipole–dipole interactions between a hydrogen atom covalently bonded to a small, highly electronegative *donor* atom (e.g., amide nitrogen, N) and an electronegative *acceptor* (e.g., backbone O or N), which contributes a lone pair of unshared electrons. Regular spacing of hydrogen-bonded amino acid groups between positions i and $i + 4$ forms an α-helix protein $2°$ structure. Hydrogen bonding between alternate residues on each of two participating amino acid strands forms a β-pleated sheet. Tertiary protein structures form in part through hydrogen bonding between R-groups. Hydrogen bonding between G–C and A–T base pairs forms the *antiparallel* double-helical DNA structure.

Affinity thermodynamics and equilibrium

From a thermodynamics perspective, the overall Gibbs free-energy change in forming the receptor-ligand complex, ΔG, consists of free-energy contributions due to water displacement from receptor and ligand ($-\Delta G_{R,L\_hydration}$), receptor-ligand interactions ($\Delta G_{RL\_interactions}$), and rehydration of the stabilized complex ($\Delta G_{RL\_hydration}$), viz.,

$$\Delta G = -\Delta G_{R,L-hydration} + \Delta G_{RL-interaction} + \Delta G_{RL-hydration}$$

(2-148)

The free-energy change contributed by receptor-ligand interactions results from an increase in receptor potential energy from a low-energy, unbound state to a high-energy, complex state, ΔU_{conf}, a change in potential energy due to receptor-ligand interactions in the complex, ΔU_{RL}; and an entropy change due to receptor-ligand interactions, viz.,

$$\Delta G_2 = \Delta U_{conf} + \Delta U_{RL} - T\Delta S \qquad (2-149)$$

This thermodynamic model is supplemented by computational chemistry calculations that quantify alterations in arrangements of chemical bonds using quantum mechanics and statistical physics. From an equilibrium perspective, bioaffinity interaction between a ligand, L, and a complementary receptor, R, may be expressed as

$$R + L \underset{k_D}{\overset{k_A}{\rightleftharpoons}} RL \qquad (2-150)$$

where k_A and k_D are forward (association) and reverse (dissociation) rate coefficients, respectively. Diffusion-controlled binding rates are typically less than 10^8/s, compared with diffusion-controlled collision rates of 10^{10}/s, because spatial localization reduces the probability of binding. Rate constants define the *equilibrium dissociation constant, K_D*,

$$K_D = \frac{k_D}{k_A} = \frac{[R][L]}{[RL]} = \frac{\left(z_R - [RL]/[R]_o\right)[L]}{[RL]/[R]_o} \qquad (2-151)$$

where bracketed terms denote concentrations (unity activity coefficients have been applied), subscript o represents an initial value, and z_R is the receptor valency, the number of ligand binding sites per receptor molecule. Reaction thermodynamics is related to equilibrium by

$$\Delta G = -RT \ln K_D^{-1} \qquad (2-152)$$

Table 2.18 shows values of K_D decrease from 10^{-3} M for enzyme-substrate interactions to 10^{-15} M for avidin-biotin complexation.

Scatchard plots

Equilibrium concentration values of L obtained from dialyzing a known initial mass of ligand against a number of solutions of known initial receptor concentration may be used to determine the dissociation constant and receptor valency by plotting $[RL]/[R]_o)/[L]$ versus $[RL]/[R]_o$ in a *Scatchard plot*, viz.,

$$\frac{[RL]/[R]_o}{[L]} = -\frac{[RL]/[R]_o}{K_D} + \frac{z_R}{K_D} \qquad (2-153)$$

Table 2.18 Dissociation Constants of Some Bioaffinity Interactions

Ligand–Receptor Pair	K_D (M)
Enzyme-substrate	$10^{-3} - 10^{-5}$
Lectin-monosaccharide	$10^{-3} - 10^{-5}$
Lectin-oligosaccharide	$10^{-5} - 10^{-7}$
Antibody-antigen	$10^{-7} - 10^{-11}$
DNA-protein	$10^{-8} - 10^{-9}$
Cell receptor—ligand	$10^{-9} - 10^{-12}$
Avidin/Streptavidin—biotin	$\sim 10^{-15}$

Inhomogeneous receptor populations with varying affinity values (i.e., K_D) will produce nonlinearity in the plot.

EXAMPLE 2.15   Scatchard Analysis of Ligand-Receptor Binding.

From the Scatchard plot for two antibodies (Ab1 and Ab2) interacting with an antigen in Figure 2.19, determine: (a) the respective dissociation constants; (b) the valency for each antibody; and (c) the homogeneity of population for each antibody.

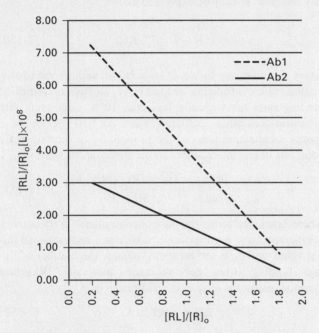

Figure 2.19 Scatchard plot for ligand–receptor interaction.

Solution

Equation (2-153) shows that the slope and intercept of each line correspond to K_D^{-1} and z_R/K_D, respectively. The dissociation constants are therefore $[(8-0)/(2-0)]^{-1} \times 10^{-8} = 0.25 \times 10^{-8}$ M and $[(3-0.33)/(1.8-0.2)]^{-1} \times 10^{-8} = 0.60 \times 10^{-8}$ M for antibody 1 and 2, respectively. These values lie within the range given in Table 2.18 for antibody–antigen interactions. From (2-153), the valency values are 2 for each antibody, meaning each antibody can bind 2 antigens. Homogeneity of the antibody and receptor preparations is indicated by the linear data in the Scatchard plot.

Bioaffinity interaction rate measurements

Quantifying interactions of macromolecules such as DNA, a protein or a virus is important to developing bioseparations, biocompatible materials, biosensors, medical devices, and pharmaceuticals. Intrinsic sorption (i.e., forward and reverse interaction) rates provide information about specificity, affinity, and kinetics [86, 87]. Intrinsic sorption rates of macromolecules including whole virus [88] can be measured by *surface plasmon resonance* (SPR) in label-free methods that are simpler, faster, and potentially more sensitive than *total internal reflectance fluorescence* (TIRF) or *nuclear magnetic resonance* (NMR) spectroscopy. Alternatives for quantitative sorption rate measurement include *quartz crystal microbalance/dissipation* (QCMD), calorimetry, ellipsometry, and voltammetry. SPR has the advantage of providing direct, unambiguous measurements to evaluate effects of binding site and concentration down to subnanomolar analyte levels without secondary reagents to enhance the signal. SPR measurements are ≥ 15 times faster and consume ≤ 100-fold less sample than curve-fitting chromatographic breakthrough profiles to determine protein sorption rates.

Applications to bioseparations

In "Extractant/diluent Systems" (§8.6.1), it is shown how desolvation via ion pairing or acid–base pairing can enhance organic/aqueous extraction of bioproducts. Ion pairing and acid–base pairing are also used to enhance selectivity of high-performance tangential flow filtration (see §14.9.2) and biochromatographic adsorption (§15.3.3).

SUMMARY

1. Separation processes are energy-intensive. Energy requirements are determined by applying the first law of thermodynamics. Estimates of irreversibility and minimum energy needs use the second law of thermodynamics with an entropy or availability balance.

2. Phase equilibrium is expressed in terms of vapor–liquid and liquid–liquid K-values, which are formulated in terms of fugacity and activity coefficients.

3. For separation systems involving an ideal-gas and an ideal-liquid solution, thermodynamic properties can be estimated from the ideal-gas law, a vapor heat-capacity equation, a vapor-pressure equation, and an equation for the liquid density.

4. Graphical representations of thermodynamic properties are widely available and useful for making manual calculations, and for visualizing effects of temperature and pressure.

5. For nonideal mixtures containing nonpolar components, P–v–T equation-of-state models such as S–R–K, P–R, and L–K–P can be used to estimate density, enthalpy, entropy, fugacity, and K-values.

6. For nonideal liquid solutions of nonpolar and/or polar components, free-energy models such as Margules, van Laar, Wilson, NRTL, UNIQUAC, and UNIFAC are used to estimate activity coefficients, volume and enthalpy of mixing, excess entropy of mixing, and K-values.

7. Special models are available for polymer solutions, electrolyte solutions, mixtures of polar and supercritical components, and biochemical systems.

8. Effects of solution conditions on solubility and recovery of active biological products can be quantified by evaluating the ionization of water and organic acids and bases as a function of temperature, ionic strength, solvent, and electrostatic interactions.

9. Evaluating effects of electrolyte and solvent composition on electrostatic double layers and forces due to vdW, hydrophobic, solvation, and steric interactions allows engineering of separation systems that control solubility and maintain structural stability of biocolloid suspensions.

10. Characterizing noncovalent interaction forces and free-energy changes by interpreting measurements using applicable theory allows quantitative evaluation of biospecific interactions in order to enhance biorecovery operations.

REFERENCES

1. Mix, T.W., J.S. Dweck, M. Weinberg, and R.C. Armstrong, *AIChE Symp. Ser., No. 192*, Vol. **76**, 15–23 (1980).

2. Felder, R.M., and R.W. Rousseau, *Elementary Principles of Chemical Processes, 3rd ed.*, John Wiley & Sons, New York (2000).

3. de Nevers, N., and J.D. Seader, *Latin Am. J. Heat and Mass Transfer*, **8**, 77–105 (1984).

4. Prausnitz, J.M., R.N. Lichtenthaler, and E.G. de Azevedo, *Molecular Thermodynamics of Fluid-Phase Equilibria, 3rd ed.*, Prentice-Hall, Upper Saddle River, NJ (1999).

5. Starling, K.E., *Fluid Thermodynamic Properties for Light Petroleum Systems*, Gulf Publishing, Houston, TX (1973).

6. Soave, G., *Chem. Eng. Sci.*, **27**, 1197–1203 (1972).

7. Peng, D.Y., and D.B. Robinson, *Ind. Eng. Chem. Fundam.*, **15**, 59–64 (1976).

8. Plöcker, U., H. Knapp, and J.M. Prausnitz, *Ind. Eng. Chem. Process Des. Dev.*, **17**, 324–332 (1978).

9. Chao, K.C., and J.D. Seader, *AIChE J.*, **7**, 598–605 (1961).

10. Grayson, H.G., and C.W Streed, Paper 20-P07, Sixth World Petroleum Conference, Frankfurt, June 1963.

11. Poling, B.E., J.M. Prausnitz, and J.P. O'Connell, *The Properties of Gases and Liquids, 5th ed.*, McGraw-Hill, New York (2001).

12. Rackett, H.G., *J. Chem. Eng. Data*, **15**, 514–517 (1970).

13. Yaws, C.L., H.-C. Yang, J.R. Hopper, and W.A. Cawley, *Hydrocarbon Processing*, **71**(1), 103–106 (1991).

14. Frank, J.C., G.R. Geyer, and H. Kehde, *Chem. Eng. Prog.*, **65**(2), 79–86 (1969).

15. Hadden, S.T., and H.G. Grayson, *Hydrocarbon Process., Petrol. Refiner*, **40**(9), 207–218 (1961).

16. Robbins, L.A., Section 15, "Liquid–liquid Extraction Operations and Equipment," in R.H. Perry, D. Green, and J.O. Maloney Eds., *Perry's Chemical Engineers' Handbook, 7th ed.*, McGraw-Hill, New York (1997).

17. Pitzer, K.S., D.Z. Lippman, R.F Curl, Jr., C.M. Huggins, and D.E. Petersen, *J. Am. Chem. Soc*, **77**, 3433–3440 (1955).

18. Redlich, O., and J.N.S. Kwong, *Chem. Rev.*, **44**, 233–244 (1949).

19. Shah, K.K., and G. Thodos, *Ind. Eng. Chem.*, **57**(3), 30–37 (1965).

20. Glanville, J.W., B.H. Sage, and W.N. Lacey, *Ind. Eng. Chem.* **42**, 508–513 (1950).

21. Wilson, G.M., *Adv. Cryogenic Eng.*, **11**, 392–400 (1966).

22. Knapp, H., R. Doring, L. Oellrich, U. Plöcker, and J.M. Prausnitz, *Vapor-liquid Equilibria for Mixtures of Low Boiling Substances*, Chem. Data. Ser., Vol. VI, DECHEMA (1982).

23. Thiesen, M., *Ann. Phys.*, **24**, 467–492 (1885).

24. Onnes, K., *Konink. Akad. Wetens*, p. 633 (1912).

25. Walas, S.M., *Phase Equilibria in Chemical Engineering*, Butterworth, Boston (1985).

26. Lee, B.I., and M.G. Kesler, *AIChE J.*, **21**, 510–527 (1975).

27. Edmister, W.C., *Hydrocarbon Processing*, **47**(9), 239–244 (1968).

28. Yarborough, L., *J. Chem. Eng. Data*, **17**, 129–133 (1972).

29. Prausnitz, J.M., W.C. Edmister, and K.C. Chao, *AIChE J.*, **6**, 214–219 (1960).

30. Hildebrand, J.H., J.M. Prausnitz, and R.L. Scott, *Regular and Related Solutions*, Van Nostrand Reinhold, New York (1970).

31. Yerazunis, S., J.D. Plowright, and F.M. Smola, *AIChE J.*, **10**, 660–665 (1964).

32. Hermsen, R.W., and J.M. Prausnitz, *Chem. Eng. Sci.*, **18**, 485–494 (1963).

33. Ewell, R.H., J.M. Harrison, and L. Berg, *Ind. Eng. Chem.*, **36**, 871–875 (1944).

34. Van Ness, H.C., C.A. Soczek, and N.K. Kochar, *J. Chem. Eng. Data*, **12**, 346–351 (1967).

35. Sinor, J.E., and J.H. Weber, *J. Chem. Eng. Data*, **5**, 243–247 (1960).

36. Orye, R.V., and J.M. Prausnitz, *Ind. Eng. Chem.*, **57**(5), 18–26 (1965).

37. Wilson, G.M., *J. Am. Chem. Soc* **86**, 127–130 (1964).

38. Cukor, P.M., and J.M. Prausnitz, *Inst. Chem. Eng. Symp. Ser. No. 32*, **3**, 88 (1969).

39. Gmehling, J., and U. Onken, *Vapor-liquid Equilibrium Data Collection*, DECHEMA Chem. Data Ser., 1–8 (1977–1984).

40. Hiranuma, M., *J. Chem. Eng. Japan*, **8**, 162–163 (1957).

41. Renon, H., and J.M. Prausnitz, *AIChE J.*, **14**, 135–144 (1968).

42. Renon, H., and J.M. Prausnitz, *Ind. Eng. Chem. Process Des. Dev.*, **8**, 413–419 (1969).

43. Abrams, D.S., and J.M. Prausnitz, *AIChE J.*, **21**, 116–128 (1975).

44. Abrams, D.S.,Ph.D. thesis in chemical engineering, University of California, Berkeley, 1974.

45. Prausnitz, J.M., T.F. Anderson, E.A. Grens, C.A. Eckert, R. Hsieh, and J.P. O'Connell, *Computer Calculations for Multicomponent Vapor-*

Liquid and Liquid-liquid Equilibria, Prentice-Hall, Englewood Cliffs, NJ (1980).

46. Fredenslund, A., J. Gmehling, M.L. Michelsen, P. Rasmussen, and J.M. Prausnitz, *Ind. Eng. Chem. Process Des. Dev.*, **16**, 450–462 (1977).

47. Wilson, G.M., and C.H. Deal, *Ind. Eng. Chem. Fundam.*, **1**, 20–23 (1962).

48. Derr, E.L., and C.H. Deal, *Inst. Chem. Eng. Symp. Ser. No. 32*, **3**, 40–51 (1969).

49. Fredenslund, A., R.L. Jones, and J.M. Prausnitz, *AIChE J.*, **21**, 1086–1099 (1975).

50. Fredenslund, A., J. Gmehling, and P. Rasmussen, *Vapor-liquid Equilibria Using UNIFAC, A. Group Contribution Method*, Elsevier, Amsterdam (1977).

51. Gmehling, J., P. Rasmussen, and A. Fredenslund, *Ind. Eng. Chem. Process Des. Dev.*, **21**, 118–127 (1982).

52. Larsen, B.L., P. Rasmussen, and A. Fredenslund, *Ind. Eng. Chem. Res.* **26**, 2274–2286 (1987).

53. Gmehling, J., J. Li, and M. Schiller, *Ind. Eng. Chem. Res.*, **32**, 178–193 (1993).

54. Wittig, R., J. Lohmann, and J. Gmehling, *Ind. Eng. Chem. Res.*, **42**, 183–188 (2003).

55. Takeuchi, S., T. Nitta, and T. Katayama, *J. Chem. Eng. Japan*, **8**, 248–250 (1975).

56. Strubl, K., V. Svoboda, R. Holub, and J. Pick, *Collect. Czech. Chem. Commun.*, **35**, 3004–3019 (1970).

57. Kiser, R.W., G.D. Johnson, and M.D. Shetlar, *J. Chem. Eng. Data*, **6**, 338–341 (1961).

58. Holderbaum, T., and J. Gmehling, *Fluid Phase Equilibria*, **70**, 251–265 (1991).

59. Fischer, K., and J. Gmehling, *Fluid Phase Equilibria*, **121**, 185–206 (1996).

60. Pitzer, K.S., *J. Phys. Chem.*, **77**, No. 2, 268–277 (1973).

61. Chen, C.-C., H.I. Britt, J.F. Boston, and L.B. Evans, *AIChE Journal*, **28**, 588–596 (1982).

62. Chen, C.-C., and L.B. Evans, *AIChE Journal*, **32**, 444–459 (1986).

63. Mock, B., L.B. Evans, and C.-C. Chen, *AIChE Journal*, **28**, 1655–1664 (1986).

64. Chen, C.-C., *Fluid Phase Equilibria*, **83**, 301–312 (1993).

65. Lee, B.I., and M.G. Kesler, *AIChE Journal*, **21**, 510–527 (1975).

66. Benedict, M., G.B. Webb, and L.C. Rubin, *Chem. Eng. Progress*, **47**(8), 419 (1951).

67. Benedict, M., G.B. Webb, and L.C. Rubin, *Chem. Eng. Progress*, **47**(9), 449 (1951).

68. Russell, W.C., *Journal of General Virology*, **81**, 2573–2604 (2000).

69. Stryer, L., *Biochemistry, 3rd ed.*, W.H. Freeman & Co., New York (1988).

70. Sandler, S.I., *Chemical, Biochemical and Engineering Thermodynamics*, John Wiley & Sons, Hoboken, NY (2006).

71. Scopes, R.K., *Protein Purification. Principles and Practice*, Springer-Verlag, New York (1987).

72. Meltzer, T.H., *Modus of filtration*, in *Adv. Biochem. Engin./Biotechnol.* Springer-Verlag, Heidelberg Vol. 98, pp. 27–71 (2006).

73. Garcia, A.A., M.R. Bonen, J. Ramirez-Vick, M. Sadaka, and A. Vuppu, *Bioseparation Process Science*, Blackwell Science, Malden, MA (1999).

74. Leckband, D., and S. Sivasankar, *Coll. Surf. B: Biointerfaces*, **14**, 83–97 (1999).

75. Valentine, R.C., and A.C. Allison, *Biochimica et Biophysica Acta*, **34**, 10–23 (1959).

76. Allison, A.C., and R.C. Valentine, *Biochem Biophys Acta*, **40**, 393–399 (1960).

77. Liang, Y., N. Hilal, P. Langston, and V. Sterov, *Adv. Coll. Int. Science*, 134–145, 151–166 (2007).

78. Pall, D.B., E.A. Kimbauer, and B.T. Allen, *Colloids Surf.*, **1**, 235–256 (1980).

79. Harrison, R.G., P. Todd, S.R. Rudge, and D.P. Petrides, *Bioseparations Science and Engineering*, Oxford University Press, New York (2003).

80. Hunter, R.J., *Foundations of Colloid Science*, Vol. I, Clarendon Press, Oxford (1986).

81. Hiemenz, P.C., *Principles of Colloid and Surface Chemistry, 2nd ed.*, Marcel Dekker, New York (1986).

82. Gabler, R., *Electrical Interactions in Molecular Biophysics*, Academic Press, New York (1978).

83. Hiemenz, P.C., and R. Rajagopalan, *Principles of Surface and Colloid Chemistry, 3rd ed.*, Dekker, New York (1997).

84. Shaw, D.J., *Introduction to Colloid and Surface Chemistry, 3rd ed.*, Butterworths, London (1980).

85. Hofmeister, F., *Arch. Exp. Pathol. Pharmacol*, **24**, 247–260 (1888).

86. Rich, R.L., and D.G. Myszka, *Journal of Molecular Recognition*, **16**(6), 351–382 (2003).

87. Rich, R.L., and D.G. Myszka, *Journal of Molecular Recognition*, **15**(6), 352–376 (2002).

88. Roper, D.K., and S. Nakra, *Anal. Biochem*, **348**, 75–83 (2006).

89. Madeira, P.P., J.A. Teixeira, E.A. Macedo, L.M. Mikheeva, and B.Y. Zaslavsky, *Fluid Phase Equilibria*, **267**, 150–157 (2008).

90. King, R.S., H.W. Blanch, and J.M. Prausnitz, *AIChE Journal*, **34**, 1585–1594 (1988).

91. Haynes, C.A., R.A. Beynon, R.S. King, H.W. Blanch, and J.M. Prausnitz, *J. Phys. Chem.*, **93**, 5612–5617 (1989).

92. Gmehling, J., R. Wittig, J. Lohmann, and R. Joh, *Ind. Eng. Chem. Res.*, **41**, 1678–1688 (2002).

93. Jakob, A., H. Grensemann, J. Lohmann, and J. Gmehling, *Ind. Eng. Chem. Res.*, **45**, 7924–7933 (2006).

STUDY QUESTIONS

2.1. In an energy balance, what are the two most common references (datums) used for enthalpy and entropy? Does one have an advantage over the other?

2.2. How does availability differ from Gibbs free energy?

2.3. Why is fugacity used in place of chemical potential to determine phase equilibria? Who invented fugacity?

2.4. How is the *K*-value for vapor–liquid equilibria defined?

2.5. How is the distribution coefficient for a liquid–liquid mixture defined?

2.6. What are the definitions of relative volatility and relative selectivity?

2.7. What are the two types of models used to estimate thermodynamic properties?

2.8. What is the limitation of the Redlich–Kwong equation of state? How did Wilson and Soave modify it to overcome the limitation?

2.9. What is unique about regular-solution theory compared to other activity-coefficient models for nonideal solutions? (This difference makes it much easier to use regular-solution theory when it is applicable.)

2.10. What are the six most widely used methods for estimating liquid-phase activity coefficients?

2.11. What very important concept did Wilson introduce in 1964?

2.12. What is a minimum-boiling azeotrope? What is a maximum-boiling azeotrope? Which type is by far the most common?

2.13. What is the critical solution temperature?

2.14. Why must electrolyte-solution activity-coefficient models consider both chemical and physical equilibrium?

2.15. Describe three effects of pH on ionization of a weak acid or base that impact biological stability of a protein.

2.16. Compare Tris and PBS as buffers in terms of temperature, ionic strength, and solvent effects.

2.17. What colloidal features do proteins and DNA exhibit?

2.18. What is the relation between the Debye length and the zeta potential?

2.19. Describe the role that the following colloidal forces play in biomolecular reactions: electrostatic, steric, solvent, hydrogen-bonding, ionic.

2.20. What is responsible for the large range in values of dissociation constants listed in Table 2.18?

EXERCISES

Section 2.1

2.1. Minimum work of separation.

A refinery stream is separated at 1,500 kPa into two products under the conditions shown below. Using the data given, compute the minimum work of separation, W_{min}, in kJ/h for $T_0 = 298.15$ K.

	kmol/h	
Component	Feed	Product 1
Ethane	30	30
Propane	200	192
n-butane	370	4
n-pentane	350	0
n-hexane	50	0

	Feed	Product 1	Product 2
Phase condition	Liquid	Vapor	Liquid
Temperature, K	364	313	394
Enthalpy, kJ/kmol	19,480	25,040	25,640
Entropy, kJ/kmol-K	36.64	33.13	54.84

2.2. Minimum work of separation.

In refineries, a mixture of paraffins and cycloparaffins is reformed in a catalytic reactor to produce blending stocks for gasoline and aromatic precursors for petrochemicals. A typical product from catalytic reforming is ethylbenzene with three xylene isomers. If this mixture is separated, these four chemicals can be processed to make styrene, phthalic anhydride, isophthalic acid, and terephthalic acid. Compute the minimum work of separation in Btu/h for $T_0 = 560°$R if the mixture below is separated at 20 psia into three products.

		Split Fraction (SF)		
Component	Feed, lbmol/h	Product 1	Product 2	Product 3
Ethylbenzene	150	0.96	0.04	0.000
p-xylene	190	0.005	0.99	0.005
m-xylene	430	0.004	0.99	0.006
o-xylene	230	0.00	0.015	0.985

	Feed	Product 1	Product 2	Product 3
Phase condition	Liquid	Liquid	Liquid	Liquid
Temperature, °F	305	299	304	314
Enthalpy, Btu/lbmol	29,290	29,750	29,550	28,320
Entropy, Btu/lbmol-°R	15.32	12.47	13.60	14.68

2.3. Second-law analysis of a distillation.

Column C3 in Figure 1.8 separates stream 5 into streams 6 and 7, according to the material balance in Table 1.5. The separation is carried out at 700 kPa in a distillation column with 70 plates and a condenser duty of 27,300,000 kJ/h. Using the following data and an infinite surroundings temperature T_0, of 298.15 K, compute: (a) the duty of the reboiler in kJ/h; (b) the irreversible production of entropy in kJ/h-K, assuming condenser cooling water at 25°C and reboiler steam at 100°C; (c) the lost work in kJ/h; (d) the minimum work of separation in kJ/h; and (e) the second-law efficiency.

Assume the shaft work of the reflux pump is negligible.

	Feed (Stream 5)	Distillate (Stream 6)	Bottoms (Stream 7)
Phase condition	Liquid	Liquid	Liquid
Temperature, K	348	323	343
Pressure, kPa	1,950	700	730
Enthalpy, kJ/~~mol~~ kmol	17,000	13,420	15,840
Entropy, kJ/kmol-K	25.05	5.87	21.22

2.4. Second-law analysis of membrane separation.

A spiral-wound, nonporous cellulose acetate membrane separator is used to separate a gas containing H_2, CH_4, and C_2H_6. The permeate is 95 mol% pure H_2 and contains no ethane. The relative split ratio (separation factor, SP) for H_2 relative to methane is 47. Using the following data and an infinite surroundings temperature of 80°F, compute the: (a) irreversible production of entropy in Btu/h-R; (b) lost work in Btu/h; and (c) minimum work of separation in Btu/h. Why is it negative? What other method(s) might be used to make the separation?

Stream flow rates and properties:

Feed flow rates, lbmol/h	
H_2	3,000
CH_4	884
C_2H_6	120

	Feed	Permeate	Retentate
Phase condition	Vapor	Vapor	Vapor
Temperature, °F	80	80	80
Pressure, psia	365	50	365
Enthalpy, Btu/lbmol	8,550	8,380	8,890
Entropy, Btu/lbmol-K	1.520	4.222	2.742

Section 2.2

2.5. Expressions for computing K-values.

Which of the following K-value expressions are rigorous? For the nonrigorous expressions, cite the assumptions.

(a) $K_i = \bar{\phi}_{iL}/\bar{\phi}_{iV}$

(b) $K_i = \phi_{iL}/\phi_{iV}$

(c) $K_i = \phi_{iL}$

(d) $K_i = \gamma_{iL}\phi_{iL}/\bar{\phi}_{iV}$

(e) $K_i = P_i^s/P$

(f) $K_i = \gamma_{iL}\phi_{iL}/\gamma_{iV}\phi_{iV}$

(g) $K_i = \gamma_{iL}P_i^s/P$

2.6. Comparison of experimental K-values to Raoult's law predictions.

Experimental measurements of Vaughan and Collins [*Ind. Eng. Chem.*, **34**, 885 (1942)] for the propane–isopentane system, at 167°F and 147 psia, show a propane liquid-phase mole fraction of 0.2900 in equilibrium with a vapor-phase mole fraction of 0.6650. Calculate:

(a) The K-values for C_3 and iC_5 from the experimental data.

(b) The K-values of C_3 and iC_5 from Raoult's law, assuming vapor pressures at 167°F of 409.6 and 58.6 psia, respectively.

Compare the results of (a) and (b). Assuming the experimental values are correct, how could better estimates of the K-values be

achieved? To respond to this question, compare the rigorous $K_i = \gamma_{iL}\phi_{iL}/\bar{\phi}_{iV}$ to the Raoult's law expression $K_i = P_i^s/P$.

2.7. Distribution coefficients from L/L data.

Mutual solubility data for the isooctane (1) furfural (2) system at 25°C [*Chem. Eng. Sci.*, **6**, 116 (1957)] are:

	Liquid Phase I	Liquid Phase II
x_1	0.0431	0.9461

Compute:

(a) The distribution (partition) coefficients for isooctane and furfural

(b) The selectivity for isooctane relative to that of furfural

(c) The activity coefficient of isooctane in phase 1 and an activity coefficient of furfural in phase 2, assuming $\gamma_2^{(1)}$ and $\gamma_1^{(2)} = 1.0$

2.8. Activity coefficients of solids dissolved in solvents.

In refineries, alkylbenzene and alkylnaphthalene streams result from catalytic cracking operations. They can be hydrodealkylated to yield valuable products such as benzene and naphthalene. At 25°C, solid naphthalene (normal melting point = 80.3°C) has the following solubilities in liquid solvents including benzene [*Naphthalene*, API Publication 707, Washington, DC (Oct. 1978)]:

Solvent	Mole Fraction Naphthalene
Benzene	0.2946
Cyclohexane	0.1487
Carbon tetrachloride	0.2591
n-hexane	0.1168
Water	0.18×10^{-5}

For each solvent, compute the activity coefficient of naphthalene in the solvent phase using the following equations (with T in K) for the vapor pressure in torr of solid and liquid naphthalene:

$$\ln P^s_{\text{solid}} = 26.708 - 8,712/T$$
$$\ln P^s_{\text{liquid}} = 16.1426 - 3992.01/(T - 71.29)$$

Section 2.3

2.9. Minimum isothermal work of separation.

An ideal-gas mixture of A and B undergoes an isothermal, isobaric separation at T_0, the infinite surroundings temperature. Starting with Eq. (4), Table 2.1, derive an equation for the minimum work of separation, $W_{\min}$, in terms of mole fractions of the feed and the two products. Use your equation to plot the dimensionless group, $W_{\min}/RT_0n_F$, as a function of mole fraction of A in the feed for:

(a) A perfect separation

(b) A separation with $SF_A = 0.98$, $SF_B = 0.02$

(c) A separation with $SR_A = 9.0$ and $SR_B = 1/9$

(d) A separation with $SF = 0.95$ for A and $SP_{A,B} = 361$

How sensitive is $W_{\min}$ to product purities? Does $W_{\min}$ depend on the separation operation used? Prove, by calculus, that the largest value of $W_{\min}$ occurs for a feed with equimolar quantities of A and B.

2.10. Relative volatility from Raoult's law.

The separation of isopentane from *n*-pentane by distillation is difficult (approximately 100 trays are required), but is commonly practiced in industry. Using the extended Antoine vapor pressure equation, (2-39), with the constants below and in conjunction with Raoult's law, calculate relative volatilities for the isopentane/n-pentane system and compare the values on a plot with the following experimental values [*J. Chem. Eng. Data*, **8**, 504 (1963)]:

Temperature, °F	$\alpha_{iC5,nC5}$
125	1.26
150	1.23
175	1.21
200	1.18
225	1.16
250	1.14

What do you conclude about the applicability of Raoult's law in this temperature range for this binary system? Vapor pressure constants for (2-39) with vapor pressure in kPa and T in K are

	iC_5	nC_5
k_1	13.6106	13.9778
k_2	−2,345.09	−2,554.60
k_3	−40.2128	−36.2529
k_4, k_5, k_6	0	0

2.11. Calculation of condenser duty.

Conditions at the top of a vacuum distillation column for the separation of ethylbenzene from styrene are given below, where the overhead vapor is condensed in an air-cooled condenser to give subcooled reflux and distillate. Using the property constants in Example 2.3, estimate the heat-transfer rate (duty) for the condenser in kJ/h, assuming an ideal gas and ideal-gas and liquid solutions. Are these valid assumptions?

	Overhead Vapor	Reflux	Distillate
Phase condition	Vapor	Liquid	Liquid
Temperature, K	331	325	325
Pressure, kPa	6.69	6.40	6.40
Component flow rates, kg/h:			
Ethylbenzene	77,500	66,960	10,540
Styrene	2,500	2,160	340

2.12. Calculation of mixture properties

Toluene is hydrodealkylated to benzene, with a conversion per pass through the reactor of 70%. The toluene must be recovered and recycled. Typical conditions for the feed to a commercial distillation unit are 100°F, 20 psia, 415 lbmol/h of benzene, and 131 lbmol/h of toluene. Using the property constants below, and assuming the ideal-gas, ideal-liquid-solution model of Table 2.4, prove that the mixture is a liquid and estimate v_L and ρ_L in American Engineering units.

Property constants for (2-39) and (2-38), with T in K, are:

	Benzene	Toluene
M, kg/kmol	78.114	92.141
P^s, torr:		
k_1	15.900	16.013
k_2	−2,788.51	−3,096.52
k_3	−52.36	−53.67
k_4, k_5, k_6	0	0
ρ_L, kg/m³:		
A	304.1	290.6
B	0.269	0.265
T_c	562.0	593.1

Section 2.4

2.13. Liquid density of a mixture.

Conditions for the bottoms at 229°F and 282 psia from a depropanizer distillation unit in a refinery are given below, including the pure-component liquid densities. Assuming an ideal-liquid solution (volume of mixing = 0), compute the liquid density in lb/ft³, lb/gal, lb/bbl (42 gal), and kg/m³.

Component	Flow rate, lbmol/h	Liquid density, g/cm³
Propane	2.2	0.20
Isobutane	171.1	0.40
n-butane	226.6	0.43
Isopentane	28.1	0.515
n-pentane	17.5	0.525

2.14. Condenser duty for two-liquid-phase distillate.

Isopropanol, with 13 wt% water, can be dehydrated to obtain almost pure isopropanol at a 90% recovery by azeotropic distillation with benzene. When condensed, the overhead vapor from the column forms two immiscible liquid phases. Use Table 2.4 with data in *Perry's Handbook* and the data below to compute the heat-transfer rate in Btu/h and kJ/h for the condenser.

	Overhead	Water-Rich Phase	Organic-Rich Phase
Phase	Vapor	Liquid	Liquid
Temperature, °C	76	40	40
Pressure, bar	1.4	1.4	1.4
Flow rate, kg/h:			
Isopropanol	6,800	5,870	930
Water	2,350	1,790	560
Benzene	24,600	30	24,570

2.15. Vapor tendency from K-values.

A vapor–liquid mixture at 250°F and 500 psia contains N_2, H_2S, CO_2, and all the normal paraffins from methane to heptane. Use Figure 2.4 to estimate the K-value of each component. Which components will be present to a greater extent in the equilibrium vapor?

2.16. Recovery of acetone from air by absorption.

Acetone can be recovered from air by absorption in water. The conditions for the streams entering and leaving are listed below. If the absorber operates adiabatically, obtain the temperature of the exiting liquid phase using a simulation program.

	Feed Gas	Absorbent	Gas Out	Liquid Out
Flow rate, lbmol/h:				
Air	687	0	687	0
Acetone	15	0	0.1	14.9
Water	0	1,733	22	1,711
Temperature, °F	78	90	80	—
Pressure, psia	15	15	14	15
Phase	Vapor	Liquid	Vapor	Liquid

Concern has been expressed about a possible feed-gas explosion hazard. The lower and upper flammability limits for acetone in air are 2.5 and 13 mol%, respectively. Is the mixture within the explosive range? If so, what can be done to remedy the situation?

Section 2.5

2.17. Volumetric flow rates for an adsorber.

Subquality natural gas contains an intolerable amount of N_2 impurity. Separation processes that can be used to remove N_2 include cryogenic distillation, membrane separation, and pressure-swing adsorption. For the last-named process, a set of typical feed and product conditions is given below. Assume a 90% removal of N_2 and a 97% methane natural-gas product. Using the R–K equation of state with the constants listed below, compute the flow rate in thousands of actual ft³/h for each of the three streams.

	N_2	CH_4
Feed flow rate, lbmol/h:	176	704
T_c, K	126.2	190.4
P_c, bar	33.9	46.0

Stream conditions are:

	Feed (Subquality Natural Gas)	Product (Natural Gas)	Waste Gas
Temperature, °F	70	100	70
Pressure, psia	800	790	280

2.18. Partial fugacity coefficients from R–K equation.

Use the R–K equation of state to estimate the partial fugacity coefficients of propane and benzene in the vapor mixture of Example 2.5.

2.19. K-values from the P–R and S–R–K equations.

Use a process simulation program to estimate the K-values, using the P–R and S–R–K equations of state, of an equimolar mixture of the two butane isomers and the four butene isomers at 220°F and 276.5 psia. Compare these values with the following experimental results [*J. Chem. Eng. Data*, **7**, 331 (1962)]:

Component	K-value
Isobutane	1.067
Isobutene	1.024
n-butane	0.922
1-butene	1.024
trans-2-butene	0.952
cis-2-butene	0.876

2.20. Cooling and partial condensation of a reactor effluent.

The disproportionation of toluene to benzene and xylenes is carried out in a catalytic reactor at 500 psia and 950°F. The reactor effluent is cooled in a series of heat exchangers for heat recovery until a temperature of 235°F is reached at a pressure of 490 psia. The effluent is then further cooled and partially condensed by the transfer of heat to cooling water in a final exchanger. The resulting two-phase equilibrium mixture at 100°F and 485 psia is then separated in a flash drum. For the reactor-effluent composition given below, use a process simulation program with the S–R–K and P–R equations of state to compute the component flow rates in lbmol/h in both the resulting vapor and liquid streams, the component K-values for the equilibrium mixture, and the rate of heat transfer to the cooling water. Compare the results.

Component	Reactor Effluent, lbmol/h
H_2	1,900
CH_4	215
C_2H_6	17
Benzene	577
Toluene	1,349
p-xylene	508

Section 2.6

2.21. Minimum work for separation of a nonideal liquid mixture.

For a process in which the feed and products are all nonideal solutions at the infinite surroundings temperature, T_0, Equation (4) of Table 2.1 for the minimum work of separation reduces to

$$\frac{W_{min}}{RT_0} = \sum_{out} n\left[\sum_i x_i \ln (\gamma_i x_i)\right] - \sum_{in} n\left[\sum_i x_i \ln (\gamma_i x_i)\right]$$

For the separation at ambient conditions (298 K, 101.3 kPa) of a 35 mol% mixture of acetone (1) in water (2) into 99 mol% acetone and 98 mol% water, calculate the minimum work in kJ/kmol of feed. Activity coefficients at ambient conditions are correlated by the van Laar equations with $A_{12} = 2.0$ and $A_{21} = 1.7$. What is the minimum work if acetone and water formed an ideal solution?

2.22. Relative volatility and activity coefficients of an azeotrope.

The sharp separation of benzene (B) and cyclohexane (CH) by distillation is impossible because of an azeotrope at 77.6°C, as shown by the data of K.C. Chao [PhD thesis, University of Wisconsin (1956)]. At 1 atm:

T, °C	x_B	y_B	γ_B	γ_{CH}
79.7	0.088	0.113	1.300	1.003
79.1	0.156	0.190	1.256	1.008
78.5	0.231	0.268	1.219	1.019
78.0	0.308	0.343	1.189	1.032
77.7	0.400	0.422	1.136	1.056
77.6	0.470	0.482	1.108	1.075
77.6	0.545	0.544	1.079	1.102
77.6	0.625	0.612	1.058	1.138
77.8	0.701	0.678	1.039	1.178
78.0	0.757	0.727	1.025	1.221
78.3	0.822	0.791	1.018	1.263
78.9	0.891	0.863	1.005	1.328
79.5	0.953	0.938	1.003	1.369

Vapor pressure is given by (2-39), where constants for benzene are in Exercise 2.12 and constants for cyclohexane are $k_1 = 15.7527$, $k_2 = -2766.63$, and $k_3 = -50.50$.

(a) Use the data to calculate and plot the relative volatility of benzene with respect to cyclohexane versus benzene composition in the liquid phase. What happens in the vicinity of the azeotrope?

(b) From the azeotropic composition for the benzene/cyclohexane system, calculate the van Laar constants, and then use the equation to compute the activity coefficients over the entire range of composition and compare them, in a plot like Figure 2.12, with the experimental data. How well does the van Laar equation fit the data?

2.23. Activity coefficients from the Wilson equation.

Benzene can break the ethanol/water azeotrope to produce nearly pure ethanol. Wilson constants for the ethanol (1)/benzene (2) system at 45°C are $\Lambda_{12} = 0.124$ and $\Lambda_{21} = 0.523$. Use these with the Wilson equation to predict liquid-phase activity coefficients over the composition range and compare them, in a plot like Figure 2.12, with the experimental results [*Austral. J. Chem.*, **7**, 264 (1954)]:

x_1	$\ln \gamma_1$	$\ln \gamma_2$
0.0374	2.0937	0.0220
0.0972	1.6153	0.0519
0.3141	0.7090	0.2599
0.5199	0.3136	0.5392
0.7087	0.1079	0.8645
0.9193	0.0002	1.3177
0.9591	−0.0077	1.3999

2.24. Activity coefficients over the composition range from infinite-dilution values.

For ethanol(1)-isooctane(2) mixtures at 50°C, the infinite-dilution, liquid-phase activity coefficients are $\gamma_1^\infty = 21.17$ and $\gamma_2^\infty = 9.84$.

(a) Calculate the constants A_{12} and A_{21} in the van Laar equations.

(b) Calculate the constants Λ_{12} and Λ_{21} in the Wilson equations.

(c) Using the constants from (a) and (b), calculate γ_1 and γ_2 over the composition range and plot the points as log γ versus x_1.

(d) How well do the van Laar and Wilson predictions agree with the azeotropic point $x_1 = 0.5941$, $\gamma_1 = 1.44$, and $\gamma_2 = 2.18$?

(e) Show that the van Laar equation erroneously predicts two liquid phases over a portion of the composition range by calculating and plotting a y–x diagram like Figure 2.16.

Section 2.9

2.25. Net charge and isoelectric point of an amino acid with an un-ionizable side group.

Consider the net charge and isoelectric point of an amino acid with an un-ionizable side group.

(a) Identify the amino acids that lack an ionizable R-group (Group I).

For an amino acid with a side (R-) chain that cannot ionize, derive a general expression in terms of measured pH and known pK_a values of α-carboxyl (pK_a^c) and α-amino (pK_a^a), respectively, for:

(b) the deprotonation ratio of the α-carboxyl group

(c) the fraction of un-ionized weak-acid α-carboxyl group in solution

(d) the positive charge of the amino acid group

(e) the fraction of ionized weak base α-amino group in solution

(f) the net charge of the amino acid

(g) the isoelectric point of the amino acid

(h) using the result in part (f), estimate the isoelectric point of glycine ($pK_a^c = 2.36$ and $pK_a^a = 9.56$). Compare this with the reported value of the pI.

2.26. Net charge and isoelectric point of an amino acid with an ionizable side group.

Consider the net charge and isoelectric point of an amino acid with ionizable side (R-) group.

(a) Identify the acidic amino acid(s) capable of having a negatively charged carboxyl side group.

(b) Identify the basic amino acid(s) capable of having a positively charged amino side group.

(c) For an amino acid with a side (R-) chain that can ionize to a negative charge, derive a general expression in terms of measured pH and known pK_a values of α-carboxyl (pK_a^c), α-amino (pK_a^a), and side group (pK_a^R), respectively, for the net charge of the amino acid.

(d) For an amino acid with a side (R-) chain that can ionize to a positive charge, derive a general expression in terms of measured pH and known pK_a values of α-carboxyl (pK_a^c), α-amino (pK_a^a), and side group (pK_a^R), respectively, for the net charge of the amino acid.

(e) Determine the isoelectric point of aspartic acid (the pH at which the net charge is zero) using the result in part (c) and pK values obtained from a reference book.

2.27. Effect of pH on solubility of caproic acid and tyrosine in water.

Prepare total solubility curves for the following species across a broad pH range ($\sim$pH 1 to pH 11).

(a) Caproic acid ($C_8H_{16}O_2$) is a colorless, oily, naturally occurring fatty acid in animal fats and oils. Its water solubility is 9.67 g/kg 25°C with a value of $pK_a = 4.85$.

(b) The least-soluble amino acid is tyrosine (Tyr, Y, $C_9H_{11}NO_3$), which occurs in proteins and is involved in signal transduction as well as photosynthesis, where it donates an electron to reduce oxidized chlorophyll. Its water solubility is 0.46 g/kg at 25°C, at which its $pK_a^c = 2.24$ and its $pK_a^a = 9.04$ (neglect deprotonation of phenolic OH-group, $pK_a = 10.10$).

Calculate the solubility of each component in the following bodily fluids:

Arterial blood plasma	pH = 7.4
Stomach contents	pH = 1.0 to 3.0

2.28. Total solubility of a zwitterionic amino acid.

Derive a general expression for the total solubility of a zwitterionic amino acid in terms of pH, pK_a^c, and pK_a^a from definitions of the respective acid dissociation constants, the expression for total solubility,

$$S_T = S_o + M_{-NH_3} + M_{-COO^-}$$

and the definition of solubility of the uncharged species,

$S_o = M_{uncharged\ species}.$

2.29. Thermodynamics of Warfarin binding to human plasma albumin.

Warfarin (coumadin) binds to human plasma albumin to prevent blood clotting in the reaction

$$W + A \rightleftharpoons WA$$

Measured thermodynamic values for this reaction at 25°C are $\Delta G = -30.8$ kJ/mol, $\Delta H = -13.1$ kJ/mol, and $\Delta C_p \sim 0$.

(a) Determine the entropy change for this reaction at 25°C.

(b) Determine the fraction of unbound albumin over a temperature range of 0 to 50°C for a solution initially containing warfarin and albumin at 0.1 mM.

Source: Sandler [70].

2.30. Affinity of drugs to a given receptor.

Different drug candidates are analyzed to determine their affinity to a given receptor. Measured equilibrium dissociation constants are listed in the following table. Rank-order the drug candidates from highest affinity to weakest affinity for the receptor.

Drug	K_D (M)
A	0.02×10^{-6}
B	7.01×10^{-6}
C	0.20×10^{-6}

2.31. Binding of hormone to two different receptors.

Examination of the binding of a particular hormone to two different receptors yields the data in the following table. What is the reverse (dissociation) rate coefficient, k_D, for the release of the hormone from the receptor?

Receptor	K_D(M)	k_A(M^{-1}s^{-1})
A	1.3×10^{-9}	2.0×10^7
B	2.6×10^{-6}	2.0×10^7

Chapter 3

Mass Transfer and Diffusion

§3.0 INSTRUCTIONAL OBJECTIVES

After completing this chapter, you should be able to:

- Explain the relationship between mass transfer and phase equilibrium, and why models for both are useful.
- Discuss mechanisms of mass transfer, including bulk flow.
- State Fick's law of diffusion for binary mixtures and discuss its analogy to Fourier's law of heat conduction.
- Estimate, in the absence of data, diffusivities for gas, liquid, and solid mixtures.
- Calculate multidimensional, unsteady-state molecular diffusion by analogy to heat conduction.
- Calculate rates of mass transfer by molecular diffusion in laminar flow for three common cases.
- Define a mass-transfer coefficient and explain its analogy to the heat-transfer coefficient.
- Use analogies, particularly those of Chilton and Colburn, and Churchill et al., to calculate rates of mass transfer in turbulent flow.
- Calculate rates of mass transfer across fluid–fluid interfaces using two-film theory and penetration theory.
- Relate molecular motion to potentials arising from chemical, pressure, thermal, gravitational, electrostatic, and friction forces.
- Compare the Maxwell–Stefan formulation with Fick's law for mass transfer.
- Use simplified forms of the Maxwell–Stefan relations to characterize mass transport due to chemical, pressure, thermal, centripetal, electrostatic, and friction forces.
- Use a linearized form of the Maxwell–Stefan relations to describe film mass transfer in stripping and membrane polarization.

*M*ass transfer is the net movement of a species in a mixture from one location to another. In separation operations, the transfer often takes place across an interface between phases. Absorption by a liquid of a solute from a carrier gas involves transfer of the solute through the gas to the gas–liquid interface, across the interface, and into the liquid. Mathematical models for this process—as well as others such as mass transfer of a species through a gas to the surface of a porous, adsorbent particle—are presented in this book.

Two mechanisms of mass transfer are: (1) *molecular diffusion* by random and spontaneous microscopic movement of molecules as a result of thermal motion; and (2) *eddy* (turbulent) *diffusion* by random, macroscopic fluid motion. Both molecular and eddy diffusion may involve the movement of different species in opposing directions. When a bulk flow occurs, the total rate of mass transfer of individual species is increased or decreased by this *bulk flow*, which is a third mechanism of mass transfer.

Molecular diffusion is extremely slow; eddy diffusion is orders of magnitude more rapid. Therefore, if industrial separation processes are to be conducted in equipment of reasonable size, the fluids must be agitated and interfacial areas maximized. For solids, the particle size is decreased to increase the area for mass transfer and decrease the distance for diffusion.

In multiphase systems the extent of the separation is limited by phase equilibrium because, with time, concentrations equilibrate by mass transfer. When mass transfer is rapid, equilibration takes seconds or minutes, and design of separation equipment is based on phase equilibrium, not mass transfer. For separations involving barriers such as membranes, mass-transfer rates govern equipment design.

Diffusion of species A with respect to B occurs because of driving forces, which include gradients of species concentration (ordinary diffusion), pressure, temperature (thermal diffusion), and external force fields that act unequally on different species. Pressure diffusion requires a large gradient, which is achieved for gas mixtures with a centrifuge. Thermal diffusion columns can be employed to separate mixtures by establishing a temperature gradient. More widely applied is forced diffusion of ions in an electrical field.

This chapter begins by describing only molecular diffusion driven by concentration gradients, which is the most common type of diffusion in chemical separation processes.

Emphasis is on binary systems, for which molecular-diffusion theory is relatively simple and applications are straightforward. The other types of diffusion are introduced in §3.8 because of their importance in bioseparations. Multicomponent ordinary diffusion is considered briefly in Chapter 12. It is a more appropriate topic for advanced study using texts such as Taylor and Krishna [1].

Molecular diffusion occurs in fluids that are stagnant, or in laminar or turbulent motion. Eddy diffusion occurs in fluids when turbulent motion exists. When both molecular diffusion and eddy diffusion occur, they are additive. When mass transfer occurs under bulk turbulent flow but across an interface or to a solid surface, flow is generally laminar or stagnant near the interface or solid surface. Thus, the eddy-diffusion mechanism is dampened or eliminated as the interface or solid surface is approached.

Mass transfer can result in a total net rate of bulk flow or flux in a direction relative to a fixed plane or stationary coordinate system. When a net flux occurs, it carries all species present. Thus, the molar flux of a species is the sum of all three mechanisms. If N_i is the molar flux of i with mole fraction x_i, and N is the total molar flux in moles per unit time per unit area in a direction perpendicular to a stationary plane across which mass transfer occurs, then

$$N_i = \text{molecular diffusion flux of } i$$
$$+ \text{ eddy diffusion flux of } i + x_i N \qquad (3\text{-}1)$$

where $x_i N$ *is* the bulk-flow flux. Each term in (3-1) is positive or negative depending on the direction of the flux relative to the direction selected as positive. When the molecular and eddy-diffusion fluxes are in one direction and N is in the opposite direction (even though a gradient of i exists), the net species mass-transfer flux, N_i, can be zero.

This chapter covers eight areas: (1) steady-state diffusion in stagnant media, (2) estimation of diffusion coefficients, (3) unsteady-state diffusion in stagnant media, (4) mass transfer in laminar flow, (5) mass transfer in turbulent flow, (6) mass transfer at fluid–fluid interfaces, (7) mass transfer across fluid–fluid interfaces, and (8) molecular mass transfer in terms of different driving forces in bioseparations.

§3.1 STEADY-STATE, ORDINARY MOLECULAR DIFFUSION

Imagine a cylindrical glass vessel partly filled with dyed water. Clear water is carefully added on top so that the dyed solution on the bottom is undisturbed. At first, a sharp boundary exists between layers, but as mass transfer of the dye occurs, the upper layer becomes colored and the layer below less colored. The upper layer is more colored near the original interface and less colored in the region near the top. During this color change, the motion of each dye molecule is random, undergoing collisions with water molecules and sometimes with dye molecules, moving first in one direction and then in another, with no one direction preferred. This type of motion is sometimes called a *random-walk process*, which yields a mean-square distance of travel in a time interval but not in a direction interval. At a given horizontal plane through the solution, it is not possible to determine whether, in a given time interval, a molecule will cross the plane or not. On the average, a fraction of all molecules in the solution below the plane cross over into the region above and the same fraction will cross over in the opposite direction. Therefore, if the concentration of dye in the lower region is greater than that in the upper region, a net rate of mass transfer of dye takes place from the lower to the upper region. Ultimately, a dynamic equilibrium is achieved and the dye concentration will be uniform throughout. Based on these observations, it is clear that:

1. Mass transfer by ordinary molecular diffusion in a binary mixture occurs because of a concentration gradient; that is, a species diffuses in the direction of decreasing concentration.

2. The mass-transfer rate is proportional to the area normal to the direction of mass transfer. Thus, the rate can be expressed as a flux.

3. Net transfer stops when concentrations are uniform.

§3.1.1 Fick's Law of Diffusion

The three observations above were quantified by Fick in 1855. He proposed an analogy to Fourier's 1822 first law of heat conduction,

$$q_z = -k \frac{dT}{dz} \qquad (3\text{-}2)$$

where q_z is the heat flux by conduction in the z-direction, k is the thermal conductivity, and dT/dz is the temperature gradient, which is negative in the direction of heat conduction. Fick's first law also features a proportionality between a flux and a gradient. For a mixture of A and B,

$$\boxed{J_{A_z} = -D_{AB} \frac{dc_A}{dz}} \qquad (3\text{-}3a)$$

and

$$\boxed{J_{B_z} = -D_{BA} \frac{dc_B}{dz}} \qquad (3\text{-}3b)$$

where J_{A_z} is the molar flux of A by ordinary molecular diffusion relative to the molar-average velocity of the mixture in the z-direction, D_{AB} is the *mutual diffusion coefficient* or *diffusivity* of A in B, c_A is the molar concentration of A, and dc_A/dz the concentration gradient of A, which is negative in the direction of diffusion. Similar definitions apply to (3-3b). The fluxes of A and B are in opposite directions. If the medium through which diffusion occurs is isotropic, then values of k and D_{AB} are independent of direction. Nonisotropic (anisotropic) materials include fibrous and composite solids as well as noncubic crystals.

Alternative driving forces and concentrations can be used in (3-3a) and (3-3b). An example is

$$J_A = -c D_{AB} \frac{dx_A}{dz} \qquad (3\text{-}4)$$

where the z subscript on J has been dropped, c = total molar concentration, and x_A = mole fraction of A.

Equation (3-4) can also be written in an equivalent mass form, where j_A is the mass flux of A relative to the mass-average velocity of the mixture in the positive z-direction, ρ is the mass density, and w_A is the mass fraction of A:

$$j_A = -\rho D_{AB} \frac{dw_A}{dz} \qquad (3\text{-}5)$$

§3.1.2 Species Velocities in Diffusion

If velocities are based on the molar flux, N, and the molar diffusion flux, J, then the molar average mixture velocity, υ_M, relative to stationary coordinates for the binary mixture, is

$$\upsilon_M = \frac{N}{c} = \frac{N_A + N_B}{c} \qquad (3\text{-}6)$$

Similarly, the velocity of species i in terms of N_i, relative to stationary coordinates, is:

$$\upsilon_i = \frac{N_i}{c_i} \qquad (3\text{-}7)$$

Combining (3-6) and (3-7) with $x_i = c_i/c$ gives

$$\upsilon_M = x_A \upsilon_A + x_B \upsilon_B \qquad (3\text{-}8)$$

Diffusion velocities, υ_{i_D}, defined in terms of J_i, are relative to molar-average velocity and are defined as the difference between the species velocity and the molar-average mixture velocity:

$$\upsilon_{i_D} = \frac{J_i}{c_i} = \upsilon_i - \upsilon_M \qquad (3\text{-}9)$$

When solving mass-transfer problems involving net mixture movement (bulk flow), fluxes and flow rates based on υ_M as the frame of reference are inconvenient to use. It is thus preferred to use mass-transfer fluxes referred to stationary coordinates. Thus, from (3-9), the total species velocity is

$$\upsilon_i = \upsilon_M + \upsilon_{i_D} \qquad (3\text{-}10)$$

Combining (3-7) and (3-10),

$$N_i = c_i \upsilon_M + c_i \upsilon_{i_D} \qquad (3\text{-}11)$$

Combining (3-11) with (3-4), (3-6), and (3-7),

$$N_A = \frac{n_A}{A} = x_A N - c D_{AB} \left(\frac{dx_A}{dz} \right) \qquad (3\text{-}12)$$

and

$$N_B = \frac{n_B}{A} = x_B N - c D_{BA} \left(\frac{dx_B}{dz} \right) \qquad (3\text{-}13)$$

In (3-12) and (3-13), n_i is the molar flow rate in moles per unit time, A is the mass-transfer area, the first right-hand side terms are the fluxes resulting from bulk flow, and the second terms are the diffusion fluxes. Two cases are important: (1) equimolar counterdiffusion (EMD); and (2) unimolecular diffusion (UMD).

§3.1.3 Equimolar Counterdiffusion (EMD)

In EMD, the molar fluxes in (3-12) and (3-13) are equal but opposite in direction, so

$$N = N_A + N_B = 0 \qquad (3\text{-}14)$$

Thus, from (3-12) and (3-13), the diffusion fluxes are also equal but opposite in direction:

$$J_A = -J_B \qquad (3\text{-}15)$$

This idealization is approached in distillation of binary mixtures, as discussed in Chapter 7. From (3-12) and (3-13), in the absence of bulk flow,

$$N_A = J_A = -c D_{AB} \left(\frac{dx_A}{dz} \right) \qquad (3\text{-}16)$$

and

$$N_B = J_B = -c D_{BA} \left(\frac{dx_B}{dz} \right) \qquad (3\text{-}17)$$

If the total concentration, pressure, and temperature are constant and the mole fractions are constant (but different) at two sides of a stagnant film between z_1 and z_2, then (3-16) and (3-17) can be integrated from z_1 to any z between z_1 and z_2 to give

$$J_A = \frac{c D_{AB}}{z - z_1} (x_{A_1} - x_A) \qquad (3\text{-}18)$$

and

$$J_B = \frac{c D_{BA}}{z - z_1} (x_{B_1} - x_B) \qquad (3\text{-}19)$$

At steady state, the mole fractions are linear in distance, as shown in Figure 3.1a. Furthermore, because total

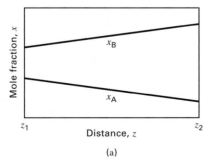

(a)

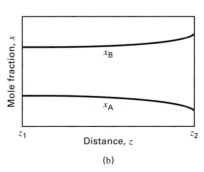

(b)

Figure 3.1 Concentration profiles for limiting cases of ordinary molecular diffusion in binary mixtures across a stagnant film: (a) equimolar counterdiffusion (EMD); (b) unimolecular diffusion (UMD).

concentration c is constant through the film, where

$$c = c_A + c_B \qquad (3\text{-}20)$$

by differentiation,

$$dc = 0 = dc_A + dc_B \qquad (3\text{-}21)$$

Thus,

$$dc_A = -dc_B \qquad (3\text{-}22)$$

From (3-3a), (3-3b), (3-15), and (3-22),

$$\frac{D_{AB}}{dz} = \frac{D_{BA}}{dz} \qquad (3\text{-}23)$$

Therefore, $D_{AB} = D_{BA}$. This equality of diffusion coefficients is always true in a binary system.

EXAMPLE 3.1 EMD in a Tube.

Two bulbs are connected by a straight tube, 0.001 m in diameter and 0.15 m in length. Initially the bulb at End 1 contains N_2 and the bulb at End 2 contains H_2. Pressure and temperature are constant at 25°C and 1 atm. At a time after diffusion starts, the nitrogen content of the gas at End 1 of the tube is 80 mol% and at End 2 is 25 mol%. If the binary diffusion coefficient is 0.784 cm²/s, determine:

(a) The rates and directions of mass transfer in mol/s

(b) The species velocities relative to stationary coordinates, in cm/s

Solution

(a) Because the gas system is closed and at constant pressure and temperature, no bulk flow occurs and mass transfer in the connecting tube is EMD.

The area for mass transfer through the tube, in cm², is $A = 3.14(0.1)^2/4 = 7.85 \times 10^{-3}$ cm². By the ideal gas law, the total gas concentration (molar density) is $c = \frac{P}{PT} = \frac{1}{(82.06)(298)} = 4.09 \times 10^{-5}$ mol/cm³. Take as the reference plane End 1 of the connecting tube. Applying (3-18) to N_2 over the tube length,

$$n_{N_2} = \frac{cD_{N_2,H_2}}{z_2 - z_1}\big[(x_{N_2})_1 - (x_{N_2})_2\big]A$$

$$= \frac{(4.09 \times 10^{-5})(0.784)(0.80 - 0.25)}{15}(7.85 \times 10^{-3})$$

$$= 9.23 \times 10^{-9} \text{ mol/s in the positive } z\text{-direction}$$

$$n_{H_2} = 9.23 \times 10^{-9} \text{ mol/s in the negative } z\text{-direction}$$

(b) For EMD, the molar-average velocity of the mixture, v_M, is 0. Therefore, from (3-9), species velocities are equal to species diffusion velocities. Thus,

$$v_{N_2} = (v_{N_2})_D = \frac{J_{N_2}}{c_{N_2}} = \frac{n_{N_2}}{Acx_{N_2}}$$

$$= \frac{9.23 \times 10^{-9}}{[(7.85 \times 10^{-3})(4.09 \times 10^{-5})x_{N_2}]}$$

$$= \frac{0.0287}{x_{N_2}} \text{ in the positive } z\text{-direction}$$

Similarly, $\quad v_{H_2} = \dfrac{0.0287}{x_{H_2}}$ in the negative z-direction

Thus, species velocities depend on mole fractions, as follows:

z, cm	x_{N_2}	x_{H_2}	v_{N_2}, cm/s	v_{H_2}, cm/s
0 (End 1)	0.800	0.200	0.0351	−0.1435
5	0.617	0.383	0.0465	−0.0749
10	0.433	0.567	0.0663	−0.0506
15 (End 2)	0.250	0.750	0.1148	−0.0383

Note that species velocities vary along the length of the tube, but at any location z, $v_M = 0$. For example, at $z = 10$ cm, from (3-8),

$$v_M = (0.433)(0.0663) + (0.567)(-0.0506) = 0$$

---

§3.1.4 Unimolecular Diffusion (UMD)

In UMD, mass transfer of component A occurs through stagnant B, resulting in a bulk flow. Thus,

$$N_B = 0 \qquad (3\text{-}24)$$

and

$$N = N_A \qquad (3\text{-}25)$$

Therefore, from (3-12),

$$N_A = x_A N_A - cD_{AB}\frac{dx_A}{dz} \qquad (3\text{-}26)$$

which can be rearranged to a Fick's-law form by solving for N_A,

$$\boxed{N_A = -\frac{cD_{AB}}{(1 - x_A)}\frac{dx_A}{dz} = -\frac{cD_{AB}}{x_B}\frac{dx_A}{dz}} \qquad (3\text{-}27)$$

The factor $(1 - x_A)$ accounts for the bulk-flow effect. For a mixture dilute in A, this effect is small. But in an equimolar mixture of A and B, $(1 - x_A) = 0.5$ and, because of bulk flow, the molar mass-transfer flux of A is twice the ordinary molecular-diffusion flux.

For the stagnant component, B, (3-13) becomes

$$0 = x_B N_A - cD_{BA}\frac{dx_B}{dz} \qquad (3\text{-}28)$$

or

$$x_B N_A = cD_{BA}\frac{dx_B}{dz} \qquad (3\text{-}29)$$

Thus, the bulk-flow flux of B is equal to but opposite its diffusion flux.

At quasi-steady-state conditions (i.e., no accumulation of species with time) and with constant molar density, (3-27) in integral form is:

$$\int_{z_1}^{z} dz = -\frac{cD_{AB}}{N_A}\int_{x_{A_1}}^{x_A}\frac{dx_A}{1 - x_A} \qquad (3\text{-}30)$$

which upon integration yields

$$N_A = \frac{cD_{AB}}{z - z_1}\ln\left(\frac{1 - x_A}{1 - x_{A_1}}\right) \qquad (3\text{-}31)$$

Thus, the mole-fraction variation as a function of z is

$$x_A = 1 - (1 - x_{A_1})\exp\left[\frac{N_A(z - z_1)}{cD_{AB}}\right] \qquad (3\text{-}32)$$

Figure 3.1b shows that the mole fractions are thus nonlinear in z.

A more useful form of (3-31) can be derived from the definition of the log mean. When $z = z_2$, (3-31) becomes

$$N_A = \frac{cD_{AB}}{z_2 - z_1} \ln\left(\frac{1 - x_{A_2}}{1 - x_{A_1}}\right) \qquad (3-33)$$

The log mean (LM) of $(1 - x_A)$ at the two ends of the stagnant layer is

$$(1 - x_A)_{LM} = \frac{(1 - x_{A_2}) - (1 - x_{A_1})}{\ln[(1 - x_{A_2})/(1 - x_{A_1})]}$$

$$= \frac{x_{A_1} - x_{A_2}}{\ln[(1 - x_{A_2})/(1 - x_{A_1})]} \qquad (3-34)$$

Combining (3-33) with (3-34) gives

$$N_A = \frac{cD_{AB}}{z_2 - z_1} \frac{(x_{A_1} - x_{A_2})}{(1 - x_A)_{LM}} = \frac{cD_{AB}}{(1 - x_A)_{LM}} \frac{(-\Delta x_A)}{\Delta z}$$

$$= \frac{cD_{AB}}{(x_B)_{LM}} \frac{(-\Delta x_A)}{\Delta z} \qquad (3-35)$$

EXAMPLE 3.2 Evaporation from an Open Beaker.

In Figure 3.2, an open beaker, 6 cm high, is filled with liquid benzene (A) at 25°C to within 0.5 cm of the top. Dry air (B) at 25°C and 1 atm is blown across the mouth of the beaker so that evaporated benzene is carried away by convection after it transfers through a stagnant air layer in the beaker. The vapor pressure of benzene at 25°C is 0.131 atm. Thus, as shown in Figure 3.2, the mole fraction of benzene in the air at the top of the beaker is zero and is determined by Raoult's law at the gas–liquid interface. The diffusion coefficient for benzene in air at 25°C and 1 atm is 0.0905 cm²/s. Compute the: (a) initial rate of evaporation of benzene as a molar flux in mol/cm²-s; (b) initial mole-fraction profiles in the stagnant air layer; (c) initial fractions of the mass-transfer fluxes due to molecular diffusion; (d) initial diffusion velocities, and the species velocities (relative to stationary coordinates) in the stagnant layer; (e) time for the benzene level in the beaker to drop 2 cm if the specific gravity of benzene is 0.874.

Neglect the accumulation of benzene and air in the stagnant layer with time as it increases in height (quasi-steady-state assumption).

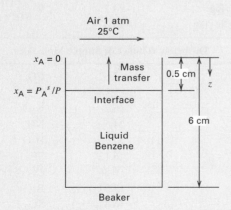

Figure 3.2 Evaporation of benzene from a beaker—Example 3.2.

Solution

The total vapor concentration by the ideal-gas law is:

$$c = \frac{P}{RT} = \frac{1}{(82.06)(298)} = 4.09 \times 10^{-5} \text{ mol/cm}^3$$

(a) With z equal to the distance down from the top of the beaker, let $z_1 = 0$ at the top of beaker and z_2 = the distance from the top of the beaker to gas–liquid interface. Then, initially, the stagnant gas layer is $z_2 - z_1 = \Delta z = 0.5$ cm. From Dalton's law, assuming equilibrium at the liquid benzene–air interface,

$$x_{A_1} = \frac{p_{A_1}}{P} = \frac{0.131}{1} = 0.131, \quad x_{A_2} = 0$$

$$(1 - x_A)_{LM} = \frac{0.131}{\ln[(1 - 0)/(1 - 0.131)]} = 0.933 = (x_B)_{LM}$$

From (3-35),

$$N_A = \frac{(4.09 \times 10^{-6})(0.0905)}{0.5}\left(\frac{0.131}{0.933}\right) = 1.04 \times 10^{-6} \text{ mol/cm}^2\text{-s}$$

(b) $$\frac{N_A(z - z_1)}{cD_{AB}} = \frac{(1.04 \times 10^{-6})(z - 0)}{(4.09 \times 10^{-5})(0.0905)} = 0.281 z$$

From (3-32),

$$x_A = 1 - 0.869 \exp(0.281 z) \qquad (1)$$

Using (1), the following results are obtained:

z, cm	x_A	x_B
0.0	0.1310	0.8690
0.1	0.1060	0.8940
0.2	0.0808	0.9192
0.3	0.0546	0.9454
0.4	0.0276	0.9724
0.5	0.0000	1.0000

These profiles are only slightly curved.

(c) Equations (3-27) and (3-29) yield the bulk-flow terms, $x_A N_A$ and $x_B N_A$, from which the molecular-diffusion terms are obtained.

	$x_i N$ Bulk-Flow Flux, mol/cm²-s × 10⁶		J_i Molecular-Diffusion Flux, mol/cm²-s × 10⁶	
z, cm	A	B	A	B
0.0	0.1360	0.9040	0.9040	−0.9040
0.1	0.1100	0.9300	0.9300	−0.9300
0.2	0.0840	0.9560	0.9560	−0.9560
0.3	0.0568	0.9832	0.9832	−0.9832
0.4	0.0287	1.0113	1.0113	−1.0113
0.5	0.0000	1.0400	1.0400	−1.0400

Note that the molecular-diffusion fluxes are equal but opposite and that the bulk-flow flux of B is equal but opposite to its molecular diffusion flux; thus N_B is zero, making B (air) stagnant.

(d) From (3-6),

$$v_M = \frac{N}{c} = \frac{N_A}{c} = \frac{1.04 \times 10^{-6}}{4.09 \times 10^{-5}} = 0.0254 \text{ cm/s} \quad (2)$$

From (3-9), the diffusion velocities are given by

$$v_{i_d} = \frac{J_i}{c_i} = \frac{J_i}{x_i c} \quad (3)$$

From (3-10), species velocities relative to stationary coordinates are

$$v_i = v_{i_d} + v_M \quad (4)$$

Using (2) to (4), there follows

	v_{i_d} Molecular-Diffusion Velocity, cm/s		J_i Species Velocity, cm/s	
z, cm	A	B	A	B
0.0	0.1687	−0.0254	0.1941	0
0.1	0.2145	−0.0254	0.2171	0
0.2	0.2893	−0.0254	0.3147	0
0.3	0.4403	−0.0254	0.4657	0
0.4	0.8959	−0.0254	0.9213	0
0.5	∞	−0.0254	∞	0

Note that v_A is zero everywhere, because its molecular-diffusion velocity is negated by the molar-mean velocity.

(e) The mass-transfer flux for benzene evaporation equals the rate of decrease in the moles of liquid benzene per unit cross section area of the beaker.

Using (3-35) with $\Delta z = z$,

$$N_A = \frac{cD_{AB}}{z} \frac{(-\Delta x_A)}{(1 - x_A)_{LM}} = \frac{\rho_L}{M_L} \frac{dz}{dt} \quad (5)$$

Separating variables and integrating,

$$\int_0^t dt = t = \frac{\rho_L(1 - x_A)_{LM}}{M_L c D_{AB}(-\Delta x_A)} \int_{z_1}^{z_2} z \, dz \quad (6)$$

where now z_1 = initial location of the interface and z_2 = location of the interface after it drops 2 cm.

The coefficient of the integral on the RHS of (6) is constant at

$$\frac{0.874(0.933)}{78.11(4.09 \times 10^{-5})(0.0905)(0.131)} = 21,530 \text{ s/cm}^2$$

$$\int_{z_1}^{z_2} z \, dz = \int_{0.5}^{2.5} z \, dz = 3 \text{ cm}^2$$

From (6), $t = 21,530(3) = 64,590$ s or 17.94 h, which is a long time because of the absence of turbulence.

§3.2 DIFFUSION COEFFICIENTS (DIFFUSIVITIES)

Diffusion coefficients (diffusivities) are defined for a binary mixture by (3-3) to (3-5). Measurement of diffusion coefficients involve a correction for bulk flow using (3-12) and (3-13), with the reference plane being such that there is no net molar bulk flow.

The binary diffusivities, D_{AB} and D_{BA}, are called mutual or binary diffusion coefficients. Other coefficients include D_{i_M}, the diffusivity of i in a multicomponent mixture; D_{ii}, the self-diffusion coefficient; and the tracer or interdiffusion coefficient.

In this chapter and throughout this book, the focus is on the mutual diffusion coefficient, which will be referred to as the diffusivity or diffusion coefficient.

§3.2.1 Diffusivity in Gas Mixtures

As discussed by Poling, Prausnitz, and O'Connell [2], equations are available for estimating the value of $D_{AB} = D_{BA}$ in gases at low to moderate pressures. The theoretical equations based on Boltzmann's kinetic theory of gases, the theorem of corresponding states, and a suitable intermolecular energy-potential function, as developed by Chapman and Enskog, predict D_{AB} to be inversely proportional to pressure, to increase significantly with temperature, and to be almost independent of composition. Of greater accuracy and ease of use is the empirical equation of Fuller, Schettler, and Giddings [3], which retains the form of the Chapman–Enskog theory but utilizes empirical constants derived from experimental data:

$$D_{AB} = D_{BA} = \frac{0.00143 T^{1.75}}{P M_{AB}^{1/2} [(\sum_V)_A^{1/3} + (\sum_V)_B^{1/3}]^2} \quad (3-36)$$

where D_{AB} is in cm^2/s, P is in atm, T is in K,

$$M_{AB} = \frac{2}{(1/M_A) + (1/M_B)} \quad (3-37)$$

and $\sum_V$ = summation of atomic and structural diffusion volumes from Table 3.1, which includes diffusion volumes of simple molecules.

Table 3.1 Diffusion Volumes from Fuller, Ensley, and Giddings [*J. Phys. Chem.*, **73**, 3679–3685 (1969)] for Estimating Binary Gas Diffusivities by the Method of Fuller et al. [3]

Atomic Diffusion Volumes and Structural Diffusion-Volume Increments			
C	15.9	F	14.7
H	2.31	Cl	21.0
O	6.11	Br	21.9
N	4.54	I	29.8
Aromatic ring	−18.3	S	22.9
Heterocyclic ring	−18.3		

Diffusion Volumes of Simple Molecules			
He	2.67	CO	18.0
Ne	5.98	CO_2	26.7
Ar	16.2	N_2O	35.9
Kr	24.5	NH_3	20.7
Xe	32.7	H_2O	13.1
H_2	6.12	SF_6	71.3
D_2	6.84	Cl_2	38.4
N_2	18.5	Br_2	69.0
O_2	16.3	SO_2	41.8
Air	19.7		

Table 3.2 Experimental Binary Diffusivities of Gas Pairs at 1 atm

Gas pair, A-B	Temperature, K	D_{AB}, cm^2/s
Air—carbon dioxide	317.2	0.177
Air—ethanol	313	0.145
Air—helium	317.2	0.765
Air—n-hexane	328	0.093
Air—water	313	0.288
Argon—ammonia	333	0.253
Argon—hydrogen	242.2	0.562
Argon—hydrogen	806	4.86
Argon—methane	298	0.202
Carbon dioxide—nitrogen	298	0.167
Carbon dioxide—oxygen	293.2	0.153
Carbon dioxide—water	307.2	0.198
Carbon monoxide—nitrogen	373	0.318
Helium—benzene	423	0.610
Helium—methane	298	0.675
Helium—methanol	423	1.032
Helium—water	307.1	0.902
Hydrogen—ammonia	298	0.783
Hydrogen—ammonia	533	2.149
Hydrogen—cyclohexane	288.6	0.319
Hydrogen—methane	288	0.694
Hydrogen—nitrogen	298	0.784
Nitrogen—benzene	311.3	0.102
Nitrogen—cyclohexane	288.6	0.0731
Nitrogen—sulfur dioxide	263	0.104
Nitrogen—water	352.1	0.256
Oxygen—benzene	311.3	0.101
Oxygen—carbon tetrachloride	296	0.0749
Oxygen—cyclohexane	288.6	0.0746
Oxygen—water	352.3	0.352

From Marrero, T. R., and E. A. Mason, *J. Phys. Chem. Ref. Data*, **1**, 3–118 (1972).

Experimental values of binary gas diffusivity at 1 atm and near-ambient temperature range from about 0.10 to 10.0 cm^2/s. Poling et al. [2] compared (3-36) to experimental data for 51 different binary gas mixtures at low pressures over a temperature range of 195–1,068 K. The average deviation was only 5.4%, with a maximum deviation of 25%.

Equation (3-36) indicates that D_{AB} is proportional to $T^{1.75}/P$, which can be used to adjust diffusivities for T and P. Representative experimental values of binary gas diffusivity are given in Table 3.2.

EXAMPLE 3.3 Estimation of a Gas Diffusivity.

Estimate the diffusion coefficient for oxygen (A)/benzene (B) at 38°C and 2 atm using the method of Fuller et al.

Solution

From (3-37), $M_{AB} = \dfrac{2}{(1/32) + (1/78.11)} = 45.4$

From Table 3.1, $(\sum_V)_A = 16.3$ and $(\sum_V)_B = 6(15.9) + 6(2.31) - 18.3 = 90.96$

From (3-36), at 2 atm and 311.2 K,

$$D_{AB} = D_{BA} = \frac{0.00143(311.2)^{1.75}}{(2)(45.4)^{1/2}[16.3^{1/3} + 90.96^{1/3}]^2} = 0.0495 \text{ cm}^2/\text{s}$$

At 1 atm, the predicted diffusivity is 0.0990 cm^2/s, which is about 2% below the value in Table 3.2. The value for 38°C can be corrected for temperature using (3-36) to give, at 200°C:

$$D_{AB} \text{ at } 200°\text{C and 1 atm} = 0.102\left(\frac{200 + 273.2}{38 + 273.2}\right)^{1.75}$$

$$= 0.212 \text{ cm}^2/\text{s}$$

For light gases, at pressures to about 10 atm, the pressure dependence on diffusivity is adequately predicted by the inverse relation in (3-36); that is, $PD_{AB} = $ a constant. At higher pressures, deviations are similar to the modification of the ideal-gas law by the compressibility factor based on the theorem of corresponding states. Takahashi [4] published a corresponding-states correlation, shown in Figure 3.3, patterned after a correlation by Slattery [5]. In the Takahashi plot, $D_{AB}P/(D_{AB}P)_{LP}$ is a function of reduced temperature and pressure, where $(D_{AB}P)_{LP}$ is at low pressure when (3-36) applies. Mixture critical temperature and pressure are molar-average values. Thus, a finite effect of composition is predicted at high pressure. The effect of high pressure on diffusivity is important in supercritical extraction, discussed in Chapter 11.

EXAMPLE 3.4 Estimation of a Gas Diffusivity at High Pressure.

Estimate the diffusion coefficient for a 25/75 molar mixture of argon and xenon at 200 atm and 378 K. At this temperature and 1 atm, the diffusion coefficient is 0.180 cm^2/s. Critical constants are:

	T_c, K	P_c, atm
Argon	151.0	48.0
Xenon	289.8	58.0

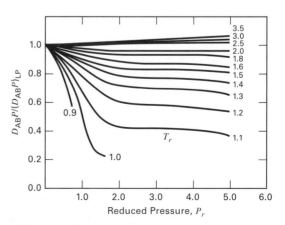

Figure 3.3 Takahashi [4] correlation for effect of high pressure on binary gas diffusivity.

Solution

Calculate reduced conditions:

$T_c = 0.25(151) + 0.75(289.8) = 255.1 \text{ K}$

$T_r = T/T_c = 378/255.1 = 1.48$

$P_c = 0.25(48) + 0.75(58) = 55.5$

$P_r = P/P_c = 200/55.5 = 3.6$

From Figure 3.3, $\dfrac{D_{AB}P}{(D_{AB}P)_{LP}} = 0.82$

$$D_{AB} = \frac{(D_{AB}P)_{LP}}{P}\left[\frac{D_{AB}P}{(D_{AB}P)_{LP}}\right] = \frac{(0.180)(1)}{200}(0.82)$$

$$= 7.38 \times 10^{-4} \text{cm/s}$$

§3.2.2 Diffusivity in Nonelectrolyte Liquid Mixtures

For liquids, diffusivities are difficult to estimate because of the lack of a rigorous model for the liquid state. An exception is a dilute solute (A) of large, rigid, spherical molecules diffusing through a solvent (B) of small molecules with no slip at the surface of the solute molecules. The resulting relation, based on the hydrodynamics of creeping flow to describe drag, is the Stokes–Einstein equation:

$$(D_{AB})_\infty = \frac{RT}{6\pi\mu_B R_A N_A} \tag{3-38}$$

where R_A is the solute-molecule radius and N_A is Avogadro's number. Equation (3-38) has long served as a starting point for more widely applicable empirical correlations for liquid diffusivity. Unfortunately, unlike for gas mixtures, where $D_{AB} = D_{BA}$, in liquid mixtures diffusivities can vary with composition, as shown in Example 3.7. The Stokes–Einstein equation is restricted to dilute binary mixtures of not more than 10% solutes.

An extension of (3-38) to more concentrated solutions for small solute molecules is the empirical Wilke–Chang [6] equation:

$$(D_{AB})_\infty = \frac{7.4 \times 10^{-8}(\phi_B M_B)^{1/2}T}{\mu_B \upsilon_A^{0.6}} \tag{3-39}$$

where the units are cm^2/s for D_{AB}; cP (centipoises) for the solvent viscosity, μ_B; K for T; and cm^3/mol for υ_A, the solute molar volume, at its normal boiling point. The parameter ϕ_B is a solvent association factor, which is 2.6 for water, 1.9 for methanol, 1.5 for ethanol, and 1.0 for unassociated solvents such as hydrocarbons. The effects of temperature and viscosity in (3-39) are taken identical to the prediction of the Stokes–Einstein equation, while the radius of the solute molecule is replaced by υ_A, which can be estimated by summing atomic contributions tabulated in Table 3.3. Some

Table 3.3 Molecular Volumes of Dissolved Light Gases and Atomic Contributions for Other Molecules at the Normal Boiling Point

	Atomic Volume $(\text{m}^3/\text{kmol}) \times 10^3$		Atomic Volume $(\text{m}^3/\text{kmol}) \times 10^3$
C	14.8	Ring	
H	3.7	Three-membered, as in	−6
O (except as below)	7.4	ethylene oxide	
Doubly bonded as carbonyl	7.4	Four-membered	−8.5
Coupled to two other elements:		Five-membered	−11.5
In aldehydes, ketones	7.4	Six-membered	−15
In methyl esters	9.1	Naphthalene ring	−30
In methyl ethers	9.9	Anthracene ring	−47.5
In ethyl esters	9.9		
In ethyl ethers	9.9		Molecular Volume $(\text{m}^3/\text{kmol}) \times 10^3$
In higher esters	11.0		
In higher ethers	11.0	Air	29.9
In acids (—OH)	12.0	O_2	25.6
Joined to S, P, N	8.3	N_2	31.2
N		Br_2	53.2
Doubly bonded	15.6	Cl_2	48.4
In primary amines	10.5	CO	30.7
In secondary amines	12.0	CO_2	34.0
Br	27.0	H_2	14.3
Cl in RCHClR′	24.6	H_2O	18.8
Cl in RCl (terminal)	21.6	H_2S	32.9
F	8.7	NH_3	25.8
I	37.0	NO	23.6
S	25.6	N_2O	36.4
P	27.0	SO_2	44.8

Source: G. Le Bas, *The Molecular Volumes of Liquid Chemical Compounds*, David McKay, New York (1915).

Table 3.4 Experimental Binary Liquid Diffusivities for Solutes, A, at Low Concentrations in Solvents, B

Solvent, B	Solute, A	Temperature, K	Diffusivity, D_{AB}, cm^2/s × 10^5
Water	Acetic acid	293	1.19
	Aniline	293	0.92
	Carbon dioxide	298	2.00
	Ethanol	288	1.00
	Methanol	288	1.26
Ethanol	Allyl alcohol	293	0.98
	Benzene	298	1.81
	Oxygen	303	2.64
	Pyridine	293	1.10
	Water	298	1.24
Benzene	Acetic acid	298	2.09
	Cyclohexane	298	2.09
	Ethanol	288	2.25
	n-heptane	298	2.10
	Toluene	298	1.85
n-hexane	Carbon tetrachloride	298	3.70
	Methyl ethyl ketone	303	3.74
	Propane	298	4.87
	Toluene	298	4.21
Acetone	Acetic acid	288	2.92
	Formic acid	298	3.77
	Nitrobenzene	293	2.94
	Water	298	4.56

From Poling et al. [2].

representative experimental values of solute diffusivity in dilute binary liquid solutions are given in Table 3.4.

EXAMPLE 3.5 Estimation of a Liquid Diffusivity.

Use the Wilke–Chang equation to estimate the diffusivity of aniline (A) in a 0.5 mol% aqueous solution at 20°C. The solubility of aniline in water is 4 g/100 g or 0.77 mol%. Compare the result to the experimental value in Table 3.4.

Solution

$$\mu_B = \mu_{H_2O} = 1.01 \text{ cP at } 20°C$$

υ_A = liquid molar volume of aniline at its normal boiling point of 457.6 K = 107 cm^3/mol

$\phi_B = 2.6$ for water, $M_B = 18$ for water, $T = 293$ K

From (3-39),

$$D_{AB} = \frac{(7.4 \times 10^{-8})[2.6(18)]^{0.5}(293)}{1.01(107)^{0.6}} = 0.89 \times 10^{-5} \text{cm}^2/\text{s}$$

This value is about 3% less than the experimental value of 0.92 × 10^{-5} cm^2/s for an infinitely dilute solution of aniline in water.

More recent liquid diffusivity correlations due to Hayduk and Minhas [7] give better agreement than the Wilke–Chang

equation with experimental values for nonaqueous solutions. For a dilute solution of one normal paraffin (C$_5$ to C$_{32}$) in another (C$_5$ to C$_{16}$),

$$(D_{AB})_{\infty} = 13.3 \times 10^{-8} \frac{T^{1.47} \mu_B^{\epsilon}}{\upsilon_A^{0.71}} \quad (3\text{-}40)$$

where

$$\epsilon = \frac{10.2}{\upsilon_A} - 0.791 \quad (3\text{-}41)$$

and the other variables have the same units as in (3-39). For nonaqueous solutions in general,

$$(D_{AB})_{\infty} = 1.55 \times 10^{-8} \frac{T^{1.29} (\mathscr{P}_B^{0.5} / \mathscr{P}_A^{0.42})}{\mu_B^{0.92} \upsilon_B^{0.23}} \quad (3\text{-}42)$$

where $\mathscr{P}$ is the parachor, which is defined as

$$\mathscr{P} = \upsilon \sigma^{1/4} \quad (3\text{-}43)$$

When units of liquid molar volume, υ, are cm^3/mol and surface tension, σ, are g/s^2 (dynes/cm), then the units of the parachor are cm^3-g$^{1/4}$/s$^{1/2}$-mol. Normally, at near-ambient conditions, $\mathscr{P}$ is treated as a constant, for which a tabulation is given in Table 3.5 from Quayle [8], who also provides in Table 3.6 a group-contribution method for estimating the parachor for compounds not listed.

The restrictions that apply to (3-42) are:

1. Solvent viscosity should not exceed 30 cP.

2. For organic acid solutes and solvents other than water, methanol, and butanols, the acid should be treated as a dimer by doubling the values of $\mathscr{P}_A$ and υ_A.

3. For a nonpolar solute in monohydroxy alcohols, values of υ_B and $\mathscr{P}_B$ should be multiplied by 8 μ_B, where viscosity is in centipoise.

Liquid diffusivities range from 10^{-6} to 10^{-4} cm^2/s for solutes of molecular weight up to about 200 and solvents with viscosity up to 10 cP. Thus, liquid diffusivities are five orders of magnitude smaller than diffusivities for gas mixtures at 1 atm. However, diffusion rates in liquids are not necessarily five orders of magnitude smaller than in gases because, as seen in (3-5), the product of concentration (molar density) and diffusivity determines the rate of diffusion for a given gradient in mole fraction. At 1 atm, the molar density of a liquid is three times that of a gas and, thus, the diffusion rate in liquids is only two orders of magnitude smaller than in gases at 1 atm.

EXAMPLE 3.6 Estimation of Solute Liquid Diffusivity.

Estimate the diffusivity of formic acid (A) in benzene (B) at 25°C and infinite dilution, using the appropriate correlation of Hayduk and Minhas.

Solution

Equation (3-42) applies, with $T = 298$ K

$\mathscr{P}_A = 93.7$ cm^3-g$^{1/4}$/s$^{1/2}$-mol $\mathscr{P}_B = 205.3$ cm^3-g$^{1/4}$/s$^{1/2}$-mol
$\mu_B = 0.6$ cP at 25°C $\upsilon_B = 96$ cm^3/mol at 80°C

Table 3.5 Parachors for Representative Compounds

	Parachor, cm^3-$g^{1/4}$/$s^{1/2}$-mol		Parachor, cm^3-$g^{1/4}$/$s^{1/2}$-mol		Parachor, cm^3-$g^{1/4}$/$s^{1/2}$-mol
Acetic acid	131.2	Chlorobenzene	244.5	Methyl amine	95.9
Acetone	161.5	Diphenyl	380.0	Methyl formate	138.6
Acetonitrile	122	Ethane	110.8	Naphthalene	312.5
Acetylene	88.6	Ethylene	99.5	n-octane	350.3
Aniline	234.4	Ethyl butyrate	295.1	1-pentene	218.2
Benzene	205.3	Ethyl ether	211.7	1-pentyne	207.0
Benzonitrile	258	Ethyl mercaptan	162.9	Phenol	221.3
n-butyric acid	209.1	Formic acid	93.7	n-propanol	165.4
Carbon disulfide	143.6	Isobutyl benzene	365.4	Toluene	245.5
Cyclohexane	239.3	Methanol	88.8	Triethyl amine	297.8

Source: Meissner, *Chem. Eng. Prog.*, **45**, 149–153 (1949).

However, for formic acid, $\mathscr{P}_A$ is doubled to 187.4. From (3-41),

$$(D_{AB})_\infty = 1.55 \times 10^{-8} \left[\frac{298^{1.29}(205.3^{0.5}/187.4^{0.42})}{0.6^{0.92}96^{0.23}} \right]$$

$$= 2.15 \times 10^{-5} cm^2/s$$

which is within 6% of the experimental value of 2.28×10^{-5} cm^2/s.

Table 3.6 Structural Contributions for Estimating the Parachor

Carbon–hydrogen:		R—[—CO—]—R' (ketone)	
C	9.0	R + R' = 2	51.3
H	15.5	R + R' = 3	49.0
CH$_3$	55.5	R + R' = 4	47.5
CH$_2$ in —(CH$_2$)$_n$		R + R' = 5	46.3
$n < 12$	40.0	R + R' = 6	45.3
$n > 12$	40.3	R + R' = 7	44.1
		—CHO	66
Alkyl groups			
1-Methylethyl	133.3	O (not noted above)	20
1-Methylpropyl	171.9	N (not noted above)	17.5
1-Methylbutyl	211.7	S	49.1
2-Methylpropyl	173.3	P	40.5
1-Ethylpropyl	209.5	F	26.1
1,1-Dimethylethyl	170.4	Cl	55.2
1,1-Dimethylpropyl	207.5	Br	68.0
1,2-Dimethylpropyl	207.9	I	90.3
1,1,2-Trimethylpropyl	243.5	Ethylenic bonds:	
C$_6$H$_5$	189.6	Terminal	19.1
		2,3-position	17.7
Special groups:		3,4-position	16.3
—COO—	63.8		
—COOH	73.8	Triple bond	40.6
—OH	29.8		
—NH$_2$	42.5	Ring closure:	
—O—	20.0	Three-membered	12
—NO$_2$	74	Four-membered	6.0
—NO$_3$ (nitrate)	93	Five-membered	3.0
—CO(NH$_2$)	91.7	Six-membered	0.8

Source: Quale [8].

The Stokes–Einstein and Wilke–Chang equations predict an inverse dependence of liquid diffusivity with viscosity, while the Hayduk–Minhas equations predict a somewhat smaller dependence. The consensus is that liquid diffusivity varies inversely with viscosity raised to an exponent closer to 0.5 than to 1.0. The Stokes–Einstein and Wilke–Chang equations also predict that $D_{AB}\mu_B/T$ is a constant over a narrow temperature range. Because μ_B decreases exponentially with temperature, D_{AB} is predicted to increase exponentially with temperature. Over a wide temperature range, it is preferable to express the effect of temperature on D_{AB} by an Arrhenius-type expression,

$$(D_{AB})_\infty = A \exp\left(\frac{-E}{RT}\right) \tag{3-44}$$

where, typically, the activation energy for liquid diffusion, E, is no greater than 6,000 cal/mol.

Equations (3-39), (3-40), and (3-42) apply only to solute A in a dilute solution of solvent B. Unlike binary gas mixtures in which the diffusivity is almost independent of composition, the effect of composition on liquid diffusivity is complex, sometimes showing strong positive or negative deviations from linearity with mole fraction.

Vignes [9] has shown that, except for strongly associated binary mixtures such as chloroform-acetone, which exhibit a rare negative deviation from Raoult's law, infinite-dilution binary diffusivities, $(D)_\infty$, can be combined with mixture activity-coefficient data or correlations thereof to predict liquid binary diffusion coefficients over the entire composition range. The Vignes equations are:

$$D_{AB} = (D_{AB})_\infty^{x_B}(D_{BA})_\infty^{x_A}\left(1 + \frac{\partial \ln \gamma_A}{\partial \ln x_A}\right)_{T,P} \tag{3-45}$$

$$D_{BA} = (D_{BA})_\infty^{x_A}(D_{AB})_\infty^{x_B}\left(1 + \frac{\partial \ln \gamma_B}{\partial \ln x_B}\right)_{T,P} \tag{3-46}$$

EXAMPLE 3.7 Effect of Composition on Liquid Diffusivities.

At 298 K and 1 atm, infinite-dilution diffusion coefficients for the methanol (A)–water (B) system are 1.5×10^{-5} cm^2/s and 1.75×10^{-5} cm^2/s for AB and BA, respectively.

Activity-coefficient data over a range of compositions as estimated by UNIFAC are:

x_A	γ_A	x_B	γ_B
0.5	1.116	0.5	1.201
0.6	1.066	0.4	1.269
0.7	1.034	0.3	1.343
0.8	1.014	0.2	1.424
1.0	1.000	0.0	1.605

Use the Vignes equations to estimate diffusion coefficients over a range of compositions.

Solution

Using a spreadsheet to compute the derivatives in (3-45) and (3-46), which are found to be essentially equal at any composition, and the diffusivities from the same equations, the following results are obtained with $D_{AB} = D_{BA}$ at each composition. The calculations show a minimum diffusivity at a methanol mole fraction of 0.30.

x_A	D_{AB}, cm^2/s	D_{BA}, cm^2/s
0.20	1.10×10^{-5}	1.10×10^{-5}
0.30	1.08×10^{-5}	1.08×10^{-5}
0.40	1.12×10^{-5}	1.12×10^{-5}
0.50	1.18×10^{-5}	1.18×10^{-5}
0.60	1.28×10^{-5}	1.28×10^{-5}
0.70	1.38×10^{-5}	1.38×10^{-5}
0.80	1.50×10^{-5}	1.50×10^{-5}

If the diffusivity is assumed to be linear with the mole fraction, the value at $x_A = 0.50$ is 1.625×10^{-5}, which is almost 40% higher than the predicted value of 1.18×10^{-5}.

§3.2.3 Diffusivities of Electrolytes

For an electrolyte solute, diffusion coefficients of dissolved salts, acids, or bases depend on the ions. However, in the absence of an electric potential, diffusion only of the electrolyte is of interest. The infinite-dilution diffusivity in cm^2/s of a salt in an aqueous solution can be estimated from the Nernst–Haskell equation:

$$(D_{AB})_\infty = \frac{RT[(1/n_+) + (1/n_-)]}{F^2[(1/\lambda_+) + (1/\lambda_-)]} \quad (3\text{-}47)$$

where n_+ and n_- = valences of the cation and anion; λ_+ and λ_- = limiting ionic conductances in (A/cm^2)(V/cm) (g-equiv/cm^3), with A in amps and V in volts; F = Faraday's constant = 96,500 coulombs/g-equiv; T = temperature, K; and R = gas constant = 8.314 J/mol-K.

Values of λ_+ and λ_- at 25°C are listed in Table 3.7. At other temperatures, these values are multiplied by $T/334 \mu_B$, where T and μ_B are in K and cP, respectively. As the concentration of the electrolyte increases, the diffusivity at first decreases 10% to 20% and then rises to values at a concentration of 2 N (normal) that approximate the infinite-

Table 3.7 Limiting Ionic Conductance in Water at 25°C, in (A/cm^2)(V/cm)(g-equiv/cm^3)

Anion	λ_-	Cation	λ_+
OH^-	197.6	H^+	349.8
Cl^-	76.3	Li^+	38.7
Br^-	78.3	Na^+	50.1
I^-	76.8	K^+	73.5
NO_3^-	71.4	NH_4^+	73.4
ClO_4^-	68.0	Ag^+	61.9
HCO_3^-	44.5	Tl^+	74.7
HCO_2^-	54.6	$(\frac{1}{2})Mg^{2+}$	53.1
$CH_3CO_2^-$	40.9	$(\frac{1}{2})Ca^{2+}$	59.5
$ClCH_2CO_2^-$	39.8	$(\frac{1}{2})Sr^{2+}$	50.5
$CNCH_2CO_2^-$	41.8	$(\frac{1}{2})Ba^{2+}$	63.6
$CH_3CH_2CO_2^-$	35.8	$(\frac{1}{2})Cu^{2+}$	54
$CH_3(CH_2)_2CO_2^-$	32.6	$(\frac{1}{2})Zn^{2+}$	53
$C_6H_5CO_2^-$	32.3	$(\frac{1}{3})La^{3+}$	69.5
$HC_2O_4^-$	40.2	$(\frac{1}{3})Co(NH_3)_6^{3+}$	102
$(\frac{1}{2})C_2O_4^{2-}$	74.2		
$(\frac{1}{2})SO_4^{2-}$	80		
$(\frac{1}{3})Fe(CN)_6^{3-}$	101		
$(\frac{1}{4})Fe(CN)_6^{4-}$	111		

Source: Poling, Prausnitz, and O'Connell [2].

dilution value. Some representative experimental values from Volume V of the International Critical Tables are given in Table 3.8.

Table 3.8 Experimental Diffusivities of Electrolytes in Aqueous Solutions

Solute	Concentration, mol/L	Temperature, °C	Diffusivity, D_{AB}, cm^2/s × 10^5
HCl	0.1	12	2.29
HNO$_3$	0.05	20	2.62
	0.25	20	2.59
H$_2$SO$_4$	0.25	20	1.63
KOH	0.01	18	2.20
	0.1	18	2.15
	1.8	18	2.19
NaOH	0.05	15	1.49
NaCl	0.4	18	1.17
	0.8	18	1.19
	2.0	18	1.23
KCl	0.4	18	1.46
	0.8	18	1.49
	2.0	18	1.58
MgSO$_4$	0.4	10	0.39
Ca(NO$_3$)$_2$	0.14	14	0.85

EXAMPLE 3.8 Diffusivity of an Electrolyte.

Estimate the diffusivity of KCl in a dilute solution of water at 18.5°C. Compare your result to the experimental value, 1.7×10^{-5} cm^2/s.

Solution

At 18.5°C, $T/334\ \mu_B = 291.7/[(334)(1.05)] = 0.832$. Using Table 3.7, at 25°C, the limiting ionic conductances are

$$\lambda_+ = 73.5(0.832) = 61.2 \quad \text{and} \quad \lambda_- = 76.3(0.832) = 63.5$$

From (3-47),

$$D_\infty = \frac{(8.314)(291.7)[(1/1) + (1/1)]}{96{,}500^2[(1/61.2) + (1/63.5)]} = 1.62 \times 10^{-5}\,\text{cm}^2/\text{s}$$

which is 95% of the experimental value.

§3.2.4 Diffusivity of Biological Solutes in Liquids

The Wilke–Chang equation (3-39) is used for solute molecules of liquid molar volumes up to 500 cm^3/mol, which corresponds to molecular weights to almost 600. In biological applications, diffusivities of soluble protein macromolecules having molecular weights greater than 1,000 are of interest. Molecules with molecular weights to 500,000 have diffusivities at 25°C that range from 1×10^{-6} to 1×10^{-9} cm^2/s, which is three orders of magnitude smaller than values of diffusivity for smaller molecules. Data for globular and fibrous protein macromolecules are tabulated by Sorber [10], with some of these diffusivities given in Table 3.9, which includes diffusivities of two viruses and a bacterium. In the absence of data, the equation of Geankoplis [11], patterned after the Stokes–Einstein equation, can be used to estimate protein diffusivities:

$$D_{AB} = \frac{9.4 \times 10^{-15} T}{\mu_B (M_A)^{1/3}} \tag{3-48}$$

where the units are those of (3-39).

Also of interest in biological applications are diffusivities of small, nonelectrolyte molecules in aqueous gels containing up to 10 wt% of molecules such as polysaccharides (agar), which have a tendency to swell. Diffusivities are given by Friedman and Kraemer [12]. In general, the diffusivities of small solute molecules in gels are not less than 50% of the values for the diffusivity of the solute in water.

§3.2.5 Diffusivity in Solids

Diffusion in solids takes place by mechanisms that depend on the diffusing atom, molecule, or ion; the nature of the solid structure, whether it be porous or nonporous, crystalline, or amorphous; and the type of solid material, whether it be metallic, ceramic, polymeric, biological, or cellular. Crystalline materials are further classified according to the type of bonding, as molecular, covalent, ionic, or metallic, with most inorganic solids being ionic. Ceramics can be ionic, covalent, or a combination of the two. Molecular solids have relatively weak forces of attraction among the atoms. In covalent solids, such as quartz silica, two atoms share two or more electrons equally. In ionic solids, such as inorganic salts, one atom loses one or more of its electrons by transfer to other atoms, thus forming ions. In metals, positively charged ions are bonded through a field of electrons that are free to move. Diffusion coefficients in solids cover a range of many orders of magnitude. Despite the complexity of diffusion in solids, Fick's first law can be used if a measured diffusivity is available. However, when the diffusing solute is a gas, its solubility in the solid must be known. If the gas dissociates upon dissolution, the concentration of the dissociated species must be used in Fick's law. The mechanisms of diffusion in solids are complex and difficult to quantify. In the next subsections, examples of diffusion in solids are given,

Table 3.9 Experimental Diffusivities of Large Biological Materials in Aqueous Solutions

	MW or Size	Configuration	T, °C	Diffusivity, D_{AB}, cm^2/s $\times 10^5$
Proteins:				
Alcohol dehydrogenase	79,070	globular	20	0.0623
Aprotinin	6,670	globular	20	0.129
Bovine serum albumin	67,500	globular	25	0.0681
Cytochrome C	11,990	globular	20	0.130
γ–Globulin, human	153,100	globular	20	0.0400
Hemoglobin	62,300	globular	20	0.069
Lysozyme	13,800	globular	20	0.113
Soybean protein	361,800	globular	20	0.0291
Trypsin	23,890	globular	20	0.093
Urease	482,700	globular	25	0.0401
Ribonuclase A	13,690	globular	20	0.107
Collagen	345,000	fibrous	20	0.0069
Fibrinogen, human	339,700	fibrous	20	0.0198
Lipoxidase	97,440	fibrous	20	0.0559
Viruses:				
Tobacco mosaic virus	40,600,000	rod-like	20	0.0046
T4 bacteriophage	90 nm × 200 nm	head and tail	22	0.0049
Bacteria:				
P. aeruginosa	~0.5 μm × 1.0 μm	rod-like, motile	ambient	2.1

together with measured diffusion coefficients that can be used with Fick's first law.

Porous solids

For porous solids, predictions of the diffusivity of gaseous and liquid solute species in the pores can be made. These methods are considered only briefly here, with details deferred to Chapters 14, 15, and 16, where applications are made to membrane separations, adsorption, and leaching. This type of diffusion is also of importance in the analysis and design of reactors using porous solid catalysts. Any of the following four mass-transfer mechanisms or combinations thereof may take place:

1. Molecular diffusion through pores, which present tortuous paths and hinder movement of molecules when their diameter is more than 10% of the pore

2. Knudsen diffusion, which involves collisions of diffusing gaseous molecules with the pore walls when pore diameter and pressure are such that molecular mean free path is large compared to pore diameter

3. Surface diffusion involving the jumping of molecules, adsorbed on the pore walls, from one adsorption site to another based on a surface concentration-driving force

4. Bulk flow through or into the pores

When diffusion occurs only in the fluid in the pores, it is common to use an effective diffusivity, D_{eff}, based on (1) the total cross-sectional area of the porous solid rather than the cross-sectional area of the pore and (2) a straight path, rather than the tortuous pore path. If pore diffusion occurs only by molecular diffusion, Fick's law (3-3) is used with the effective diffusivity replacing the ordinary diffusion coefficient, D_{AB}:

$$D_{eff} = \frac{D_{AB}\epsilon}{\tau} \qquad (3\text{-}49)$$

where ϵ is the fractional solid porosity (typically 0.5) and τ is the pore-path *tortuosity* (typically 2 to 3), which is the ratio of the pore length to the length if the pore were straight. The effective diffusivity is determined by experiment, or predicted from (3-49) based on measurement of the porosity and tortuosity and use of the predictive methods for molecular diffusivity. As an example of the former, Boucher, Brier, and Osburn [13] measured effective diffusivities for the leaching of processed soybean oil (viscosity = 20.1 cP at 120 °F) from 1/16-in.-thick porous clay plates with liquid tetrachloroethylene solvent. The rate of extraction was controlled by diffusion of the soybean oil in the clay plates. The measured D_{eff} was 1.0×10^{-6} cm²/s. Due to the effects of porosity and tortuosity, this value is one order of magnitude less than the molecular diffusivity, D_{AB}, of oil in the solvent.

Crystalline solids

Diffusion through nonporous crystalline solids depends markedly on the crystal lattice structure. As discussed in Chapter 17, only seven different crystal lattice structures exist. For a cubic lattice (simple, body-centered, and face-centered), the

diffusivity is equal in all directions (isotropic). In the six other lattice structures (including hexagonal and tetragonal), the diffusivity, as in wood, can be anisotropic. Many metals, including Ag, Al, Au, Cu, Ni, Pb, and Pt, crystallize into the face-centered cubic lattice structure. Others, including Be, Mg, Ti, and Zn, form anisotropic, hexagonal structures. The mechanisms of diffusion in crystalline solids include:

1. Direct exchange of lattice position, probably by a ring rotation involving three or more atoms or ions

2. Migration by small solutes through interlattice spaces called interstitial sites

3. Migration to a vacant site in the lattice

4. Migration along lattice imperfections (dislocations), or gain boundaries (crystal interfaces)

Diffusion coefficients associated with the first three mechanisms can vary widely and are almost always at least one order of magnitude smaller than diffusion coefficients in low-viscosity liquids. Diffusion by the fourth mechanism can be faster than by the other three. Experimental diffusivity values, taken mainly from Barrer [14], are given in Table 3.10. The diffusivities cover gaseous, ionic, and metallic solutes. The values cover an enormous 26-fold range. Temperature effects can be extremely large.

Metals

Important applications exist for diffusion of gases through metals. To diffuse through a metal, a gas must first dissolve in the metal. As discussed by Barrer [14], all light gases do

Table 3.10 Diffusivities of Solutes in Crystalline Metals and Salts

Metal/Salt	Solute	T, °C	D, cm²/s
Ag	Au	760	3.6×10^{-10}
	Sb	20	3.5×10^{-21}
	Sb	760	1.4×10^{-9}
Al	Fe	359	6.2×10^{-14}
	Zn	500	2×10^{-9}
	Ag	50	1.2×10^{-9}
Cu	Al	20	1.3×10^{-30}
	Al	850	2.2×10^{-9}
	Au	750	2.1×10^{-11}
Fe	H_2	10	1.66×10^{-9}
	H_2	100	1.24×10^{-7}
	C	800	1.5×10^{-8}
Ni	H_2	85	1.16×10^{-8}
	H_2	165	1.05×10^{-7}
	CO	950	4×10^{-8}
W	U	1727	1.3×10^{-11}
AgCl	Ag^+	150	2.5×10^{-14}
	Ag^+	350	7.1×10^{-8}
	Cl^-	350	3.2×10^{-16}
KBr	H_2	600	5.5×10^{-4}
	Br_2	600	2.64×10^{-4}

not dissolve in all metals. Hydrogen dissolves in Cu, Al, Ti, Ta, Cr, W, Fe, Ni, Pt, and Pd, but not in Au, Zn, Sb, and Rh. Nitrogen dissolves in Zr but not in Cu, Ag, or Au. The noble gases do not dissolve in common metals. When H_2, N_2, and O_2 dissolve in metals, they dissociate and may react to form hydrides, nitrides, and oxides. Molecules such as ammonia, carbon dioxide, carbon monoxide, and sulfur dioxide also dissociate. Example 3.9 illustrates how hydrogen gas can slowly leak through the wall of a small, thin pressure vessel.

EXAMPLE 3.9 Diffusion of Hydrogen in Steel.

Hydrogen at 200 psia and 300°C is stored in a 10-cm-diameter steel pressure vessel of wall thickness 0.125 inch. Solubility of hydrogen in steel, which is proportional to the square root of the hydrogen partial pressure, is 3.8×10^{-6} mol/cm^3 at 14.7 psia and 300°C. The diffusivity of hydrogen in steel at 300°C is 5×10^{-6} cm^2/s. If the inner surface of the vessel wall remains saturated at the hydrogen partial pressure and the hydrogen partial pressure at the outer surface is zero, estimate the time for the pressure in the vessel to decrease to 100 psia because of hydrogen loss.

Solution

Integrating Fick's first law, (3-3), where A is H_2 and B is the metal, assuming a linear concentration gradient, and equating the flux to the loss of hydrogen in the vessel,

$$-\frac{dn_A}{dt} = \frac{D_{AB} A \Delta c_A}{\Delta z} \quad (1)$$

Because $p_A = 0$ outside the vessel, $\Delta c_A = c_A =$ solubility of A at the inside wall surface in mol/cm^3 and $c_A = 3.8 \times 10^{-6} \left(\frac{p_A}{14.7}\right)^{0.5}$, where p_A is the pressure of A inside the vessel in psia. Let p_{A_o} and n_{A_o} be the initial pressure and moles of A in the vessel. Assuming the ideal-gas law and isothermal conditions,

$$n_A = n_{A_o} p_A / p_{A_o} \quad (2)$$

Differentiating (2) with respect to time,

$$\frac{dn_A}{dt} = \frac{n_{A_o}}{p_{A_o}} \frac{dp_A}{dt} \quad (3)$$

Combining (1) and (3),

$$\frac{dp_A}{dt} = -\frac{D_{AB} A (3.8 \times 10^{-6}) p_A^{0.5} p_{A_o}}{n_{A_o} \Delta z (14.7)^{0.5}} \quad (4)$$

Integrating and solving for t,

$$t = \frac{2 n_{A_o} \Delta z (14.7)^{0.5}}{3.8 \times 10^{-6} D_{AB} A p_{A_o}} (p_{A_o}^{0.5} - p_A^{0.5})$$

Assuming the ideal-gas law,

$$n_{A_o} = \frac{(200/14.7)[(3.14 \times 10^3)/6]}{82.05(300 + 273)} = 0.1515 \text{ mol}$$

The mean-spherical shell area for mass transfer, A, is

$$A = \frac{3.14}{2} \left[(10)^2 + (10.635)^2\right] = 336 \text{ cm}^2$$

The time for the pressure to drop to 100 psia is

$$t = \frac{2(0.1515)(0.125 \times 2.54)(14.7)^{0.5}}{3.8 \times 10^{-6}(5 \times 10^{-6})(336)(200)} (200^{0.5} - 100^{0.5})$$

$$= 1.2 \times 10^6 \text{ s or } 332 \text{ h}$$

Silica and glass

Another area of interest is diffusion of light gases through silica, whose two elements, Si and O, make up about 60% of the earth's crust. Solid silica exists in three crystalline forms (quartz, tridymite, and cristobalite) and in various amorphous forms, including fused quartz. Table 3.11 includes diffusivities, D, and solubilities as Henry's law constants, H, at 1 atm for helium and hydrogen in fused quartz as calculated from correlations of Swets, Lee, and Frank [15] and Lee [16]. The product of diffusivity and solubility is the permeability, P_M. Thus,

$$P_M = DH \quad (3-50)$$

Unlike metals, where hydrogen usually diffuses as the atom, hydrogen diffuses as a molecule in glass.

For hydrogen and helium, diffusivities increase rapidly with temperature. At ambient temperature they are three orders of magnitude smaller than they are in liquids. At high temperatures they approach those in liquids. Solubilities vary slowly with temperature. Hydrogen is orders of magnitude less soluble in glass than helium. Diffusivities for oxygen are included in Table 3.11 from studies by Williams [17] and Sucov [18]. At 1000°C, the two values differ widely because, as discussed by Kingery, Bowen, and Uhlmann [19], in the former case, transport occurs by molecular diffusion, while in the latter, transport is by slower network diffusion as oxygen jumps from one position in the network to another. The activation energy for the latter is much larger than that for the former (71,000 cal/mol versus 27,000 cal/mol). The choice of glass can be critical in vacuum operations because of this wide range of diffusivity.

Ceramics

Diffusion in ceramics has been the subject of numerous studies, many of which are summarized in Figure 3.4, which

Table 3.11 Diffusivities and Solubilities of Gases in Amorphous Silica at 1 atm

Gas	Temp, °C	Diffusivity, cm^2/s	Solubility mol/cm^3-atm
He	24	2.39×10^{-8}	1.04×10^{-7}
	300	2.26×10^{-6}	1.82×10^{-7}
	500	9.99×10^{-6}	9.9×10^{-8}
	1,000	5.42×10^{-5}	1.34×10^{-7}
H_2	300	6.11×10^{-8}	3.2×10^{-14}
	500	6.49×10^{-7}	2.48×10^{-13}
	1,000	9.26×10^{-6}	2.49×10^{-12}
O_2	1,000	6.25×10^{-9}	(molecular)
	1,000	9.43×10^{-15}	(network)

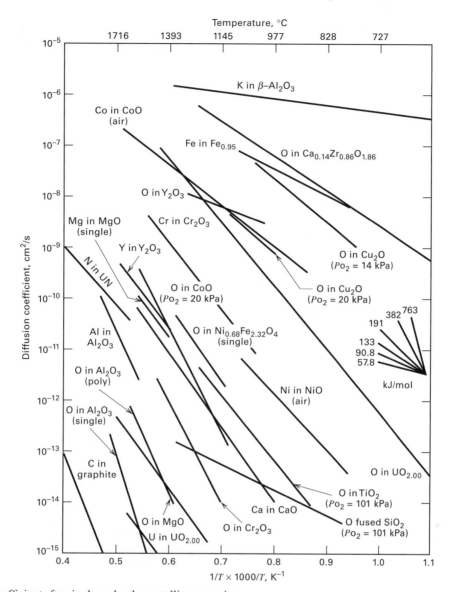

Figure 3.4 Diffusion coefficients for single and polycrystalline ceramics.

[From W.D. Kingery, H.K. Bowen, and D.R. Uhlmann, *Introduction to Ceramics*, 2nd ed., Wiley Interscience, New York (1976) with permission.]

is from Kingery et al. [19], where diffusivity is plotted as a function of the inverse of temperature in the high-temperature range. In this form, the slopes of the curves are proportional to the activation energy for diffusion, E, where

$$D = D_o \exp\left(-\frac{E}{RT}\right) \qquad (3\text{-}51)$$

An insert in Figure 3.4 relates the slopes of the curves to activation energy. The diffusivity curves cover a ninefold range from 10^{-6} to 10^{-15} cm^2/s, with the largest values corresponding to the diffusion of potassium in β-Al$_2$O$_3$ and one of the smallest for the diffusion of carbon in graphite. As discussed in detail by Kingery et al. [19], diffusion in crystalline oxides depends not only on temperature but also on whether the oxide is stoichiometric or not (e.g., FeO and Fe$_{0.95}$O) and on impurities. Diffusion through vacant sites of nonstoichiometric oxides is often classified as metal-deficient or oxygen-

deficient. Impurities can hinder diffusion by filling vacant lattice or interstitial sites.

Polymers

Diffusion through nonporous polymers is dependent on the type of polymer, whether it be crystalline or amorphous and, if the latter, glassy or rubbery. Commercial crystalline polymers are about 20% amorphous, and it is through these regions that diffusion occurs. As with the transport of gases through metals, transport through polymer membranes is characterized by the solution-diffusion mechanism of (3-50). Fick's first law, in the following integrated forms, is then applied to compute the mass-transfer flux.

Gas species:

$$N_i = \frac{H_i D_i}{z_2 - z_1}(p_{i_1} - p_{i_2}) = \frac{P_{M_i}}{z_2 - z_1}(p_{i_1} - p_{i_2}) \qquad (3\text{-}52)$$

where p_i is the partial pressure at a polymer surface.

Table 3.12 Diffusivities of Solutes in Rubbery Polymers

Polymer	Solute	Temperature, K	Diffusivity, cm²/s
Polyisobutylene	n-Butane	298	1.19×10^{-9}
	i-Butane	298	5.3×10^{-10}
	n-Pentane	298	1.08×10^{-9}
	n-Hexadecane	298	6.08×10^{-10}
Hevea rubber	n-Butane	303	2.3×10^{-7}
	i-Butane	303	1.52×10^{-7}
	n-Pentane	303	2.3×10^{-7}
	n-Hexadecane	298	7.66×10^{-8}
Polymethylacrylate	Ethyl alcohol	323	2.18×10^{-10}
Polyvinylacetate	n-Propyl alcohol	313	1.11×10^{-12}
	n-Propyl chloride	313	1.34×10^{-12}
	Ethyl chloride	343	2.01×10^{-9}
	Ethyl bromide	343	1.11×10^{-9}
Polydimethylsiloxane	n-Hexadecane	298	1.6×10^{-6}
1,4-Polybutadiene	n-Hexadecane	298	2.21×10^{-7}
Styrene-butadiene rubber	n-Hexadecane	298	2.66×10^{-8}

Liquid species:

$$N_i = \frac{K_i D_i}{z_2 - z_1}(c_{i_1} - c_{i_2}) \qquad (3\text{-}53)$$

where K_i, the equilibrium partition coefficient, is the ratio of the concentration in the polymer to the concentration, c_i, in the liquid at the polymer surface. The product $K_i D_i$ is the liquid permeability.

Diffusivities for light gases in four polymers, given in Table 14.6, range from 1.3×10^{-9} to 1.6×10^{-6} cm²/s, which is magnitudes less than for diffusion in a gas.

Diffusivity of liquids in rubbery polymers has been studied as a means of determining viscoelastic parameters. In Table 3.12, taken from Ferry [20], diffusivities are given for solutes in seven different rubber polymers at near-ambient conditions. The values cover a sixfold range, with the largest diffusivity being that for n-hexadecane in polydimethylsiloxane. The smallest diffusivities correspond to the case in which the temperature approaches the glass-transition temperature, where the polymer becomes glassy in structure. This more rigid structure hinders diffusion. As expected, smaller molecules have higher diffusivities. A study of n-hexadecane in styrene-butadiene copolymers at 25°C by Rhee and Ferry [21] shows a large effect on diffusivity of polymer fractional free volume.

Polymers that are 100% crystalline permit little or no diffusion of gases and liquids. The diffusivity of methane at 25°C in polyoxyethylene oxyisophthaloyl decreases from 0.30×10^{-9} to 0.13×10^{-9} cm²/s when the degree of crystallinity increases from 0 to 40% [22]. A measure of crystallinity is the polymer density. The diffusivity of methane at 25°C in polyethylene decreases from 0.193×10^{-6} to 0.057×10^{-6} cm²/s when specific gravity increases from 0.914 to 0.964 [22]. Plasticizers cause diffusivity to increase.

When polyvinylchloride is plasticized with 40% tricresyl triphosphate, the diffusivity of CO at 27°C increases from 0.23×10^{-8} to 2.9×10^{-8} cm²/s [22].

EXAMPLE 3.10 Diffusion of Hydrogen through a Membrane.

Hydrogen diffuses through a nonporous polyvinyltrimethylsilane membrane at 25°C. The pressures on the sides of the membrane are 3.5 MPa and 200 kPa. Diffusivity and solubility data are given in Table 14.9. If the hydrogen flux is to be 0.64 kmol/m²-h, how thick in micrometers (μm) should the membrane be?

Solution

Equation (3-52) applies. From Table 14.9,

$$D = 160 \times 10^{-11} \text{m}^2/\text{s}, \quad H = S = 0.54 \times 10^{-4} \text{mol/m}^3\text{-Pa}$$

From (3-50), $P_M = DH = (160 \times 10^{-11})(0.64 \times 10^{-4})$

$$= 86.4 \times 10^{-15} \text{mol/m-s-Pa}$$

$$p_1 = 3.5 \times 10^6 \text{ Pa}, \quad p_2 = 0.2 \times 10^6 \text{ Pa}$$

Membrane thickness $= z_2 - z_1 = \Delta z = P_M(p_1 - p_2)/N$

$$\Delta z = \frac{86.4 \times 10^{-15}(3.5 \times 10^6 - 0.2 \times 10^6)}{[0.64(1000)/3600]}$$

$$= 1.6 \times 10^{-6} \text{m} = 1.6 \text{ μm}$$

Membranes must be thin to achieve practical permeation rates.

Cellular solids and wood

A widely used cellular solid is wood, whose structure is discussed by Gibson and Ashby [23]. Chemically, wood consists of lignin, cellulose, hemicellulose, and minor amounts of organic chemicals and elements. The latter are extractable, and the former three, which are all polymers, give wood its structure. Green wood also contains up to 25 wt% moisture in the cell walls and cell cavities. Adsorption or desorption of moisture in wood causes anisotropic swelling and shrinkage.

Wood often consists of (1) highly elongated hexagonal or rectangular cells, called tracheids in softwood (coniferous species, e.g., spruce, pine, and fir) and fibers in hardwood (deciduous or broad-leaf species, e.g., oak, birch, and walnut); (2) radial arrays of rectangular-like cells, called rays; and (3) enlarged cells with large pore spaces and thin walls, called sap channels because they conduct fluids up the tree.

Many of the properties of wood are anisotropic. For example, stiffness and strength are 2 to 20 times greater in the axial direction of the tracheids or fibers than in the radial and tangential directions of the trunk. This anisotropy extends to permeability and diffusivity of wood penetrants, such as moisture and preservatives. According to Stamm [24], the permeability of wood to liquids in the axial direction can be up to 10 times greater than in the transverse direction.

Movement of liquids and gases through wood occurs during drying and treatment with preservatives, fire retardants, and other chemicals. It takes place by capillarity, pressure permeability, and diffusion. All three mechanisms of movement of gases and liquids in wood are considered by Stamm [24]. Only diffusion is discussed here.

The simplest form of diffusion is that of a water-soluble solute through wood saturated with water, so no dimensional changes occur. For the diffusion of urea, glycerine, and lactic acid into hardwood, Stamm [24] lists diffusivities in the axial direction that are 50% of ordinary liquid diffusivities. In the radial direction, diffusivities are 10% of the axial values. At 26.7°C, the diffusivity of zinc sulfate in water is 5×10^{-6} cm^2/s. If loblolly pine sapwood is impregnated with zinc sulfate in the radial direction, the diffusivity is 0.18×10^{-6} cm^2/s [24].

The diffusion of water in wood is complex. Water is held in the wood in different ways. It may be physically adsorbed on cell walls in monomolecular layers, condensed in preexisting or transient cell capillaries, or absorbed into cell walls to form a solid solution.

Because of the practical importance of lumber drying rates, most diffusion coefficients are measured under drying conditions in the radial direction across the fibers. Results depend on temperature and specific gravity. Typical results are given by Sherwood [25] and Stamm [24]. For example, for beech with a swollen specific gravity of 0.4, the diffusivity increases from a value of 1×10^{-6} cm^2/s at 10°C to 10×10^{-6} cm^2/s at 60°C.

§3.3 STEADY- AND UNSTEADY-STATE MASS TRANSFER THROUGH STATIONARY MEDIA

Mass transfer occurs in (1) stagnant or stationary media, (2) fluids in laminar flow, and (3) fluids in turbulent flow, each requiring a different calculation procedure. The first is presented in this section, the other two in subsequent sections.

Fourier's law is used to derive equations for the rate of heat transfer by conduction for steady-state and unsteady-state conditions in stationary media consisting of shapes such as slabs, cylinders, and spheres. Analogous equations can be derived for mass transfer using Fick's law.

In one dimension, the molar rate of mass transfer of A in a binary mixture is given by a modification of (3-12), which includes bulk flow and molecular diffusion:

$$n_A = x_A(n_A + n_B) - cD_{AB}A\left(\frac{dx_A}{dz}\right) \tag{3-54}$$

If A is undergoing mass transfer but B is stationary, $n_B = 0$. It is common to assume that c is a constant and x_A is small. The bulk-flow term is then eliminated and (3-54) becomes Fick's first law:

$$n_A = -cD_{AB}A\left(\frac{dx_A}{dz}\right) \tag{3-55}$$

Alternatively, (3-55) can be written in terms of a concentration gradient:

$$n_A = -D_{AB}A\left(\frac{dc_A}{dz}\right) \tag{3-56}$$

This equation is analogous to Fourier's law for the rate of heat conduction, Q:

$$Q = -kA\left(\frac{dT}{dz}\right) \tag{3-57}$$

§3.3.1 Steady-State Diffusion

For steady-state, one-dimensional diffusion with constant D_{AB}, (3-56) can be integrated for various geometries, the results being analogous to heat conduction.

1. Plane wall with a thickness, $z_2 - z_1$:

$$\boxed{n_A = D_{AB}A\left(\frac{c_{A_1} - c_{A_2}}{z_2 - z_2}\right)} \tag{3-58}$$

2. Hollow cylinder of inner radius r_1 and outer radius r_2, with diffusion in the radial direction outward:

$$n_A = 2\pi L \frac{D_{AB}(c_{A_1} - c_{A_2})}{\ln(r_2/r_1)} \tag{3-59}$$

or $\qquad n_A = D_{AB}A_{LM}\left(\dfrac{c_{A_1} - c_{A_2}}{r_2 - r_1}\right) \qquad$ (3-60)

where

A_{LM} = log mean of the areas $2\pi rL$ at r_1 and r_2
L = length of the hollow cylinder

3. Spherical shell of inner radius r_1 and outer radius r_2, with diffusion in the radial direction outward:

$$n_A = \frac{4\pi r_1 r_2 D_{AB}(c_{A_1} - c_{A_2})}{r_2 - r_1} \tag{3-61}$$

or

$$n_A = D_{AB} A_{GM} \left(\frac{c_{A_1} - c_{A_2}}{r_2 - r_1} \right) \tag{3-62}$$

where A_{GM} = geometric mean of the areas $4\pi r^2$.

When $r_1/r_2 < 2$, the arithmetic mean area is no more than 4% greater than the log mean area. When $r_1/r_2 < 1.33$, the arithmetic mean area is no more than 4% greater than the geometric mean area.

§3.3.2 Unsteady-State Diffusion

Consider one-dimensional molecular diffusion of species A in stationary B through a differential control volume with diffusion in the z-direction only, as shown in Figure 3.5. Assume constant diffusivity and negligible bulk flow. The molar flow rate of species A by diffusion in the z-direction is given by (3-56):

$$n_{A_z} = -D_{AB} A \left(\frac{\partial c_A}{\partial z} \right)_z \tag{3-63}$$

At the plane, $z = z + \Delta z$, the diffusion rate is

$$n_{A_{z+\Delta z}} = -D_{AB} A \left(\frac{\partial c_A}{\partial z} \right)_{z+\Delta z} \tag{3-64}$$

The accumulation of species A in the control volume is

$$A \frac{\partial c_A}{\partial t} \Delta z \tag{3-65}$$

Since rate in − rate out = accumulation,

$$-D_{AB} A \left(\frac{\partial c_A}{\partial z} \right)_z + D_{AB} A \left(\frac{\partial c_A}{\partial z} \right)_{z+\Delta z} = A \left(\frac{\partial c_A}{\partial t} \right) \Delta z \tag{3-66}$$

Rearranging and simplifying,

$$D_{AB} \left[\frac{(\partial c_A/\partial z)_{z+\Delta z} - (\partial c_A/\partial z)_z}{\Delta z} \right] = \frac{\partial c_A}{\partial t} \tag{3-67}$$

In the limit, as $\Delta z \to 0$,

$$\frac{\partial c_A}{\partial t} = D_{AB} \frac{\partial^2 c_A}{\partial z^2} \tag{3-68}$$

Equation (3-68) is Fick's second law for one-dimensional diffusion.

The more general form for three-dimensional rectangular coordinates is

$$\frac{\partial c_A}{\partial t} = D_{AB} \left(\frac{\partial^2 c_A}{\partial x^2} + \frac{\partial^2 c_A}{\partial y^2} + \frac{\partial^2 c_A}{\partial z^2} \right) \tag{3-69}$$

For one-dimensional diffusion in the radial direction only for cylindrical and spherical coordinates, Fick's second law becomes, respectively,

$$\frac{\partial c_A}{\partial t} = \frac{D_{AB}}{r} \frac{\partial}{\partial r} \left(r \frac{\partial c_A}{\partial r} \right) \tag{3-70}$$

and

$$\frac{\partial c_A}{\partial t} = \frac{D_{AB}}{r^2} \frac{\partial}{\partial r} \left(r^2 \frac{\partial c_A}{\partial r} \right) \tag{3-71}$$

Equations (3-68) to (3-71) are analogous to Fourier's second law of heat conduction, where c_A is replaced by temperature, T, and diffusivity, D_{AB}, by thermal diffusivity, $\alpha = k/\rho C_P$. The analogous three equations for heat conduction for constant, isotropic properties are, respectively:

$$\frac{\partial T}{\partial t} = \alpha \left(\frac{\partial^2 T}{\partial x^2} + \frac{\partial^2 T}{\partial y^2} + \frac{\partial^2 T}{\partial z^2} \right) \tag{3-72}$$

$$\frac{\partial T}{\partial t} = \frac{\alpha}{r} \frac{\partial}{\partial r} \left(r \frac{\partial T}{\partial r} \right) \tag{3-73}$$

$$\frac{\partial T}{\partial t} = \frac{\alpha}{r^2} \frac{\partial}{\partial r} \left(r^2 \frac{\partial T}{\partial r} \right) \tag{3-74}$$

Analytical solutions to these partial differential equations in either Fick's-law or Fourier's-law form are available for a variety of boundary conditions. They are derived and discussed by Carslaw and Jaeger [26] and Crank [27].

§3.3.3 Diffusion in a Semi-infinite Medium

Consider the semi-infinite medium shown in Figure 3.6, which extends in the z-direction from $z = 0$ to $z = \infty$. The x and y coordinates extend from $-\infty$ to $+\infty$ but are not of interest because diffusion is assumed to take place only in the z-direction. Thus, (3-68) applies to the region $z \geq 0$. At time $t \leq 0$, the concentration is c_{A_o} for $z \geq 0$. At $t = 0$, the surface of the semi-infinite medium at $z = 0$ is instantaneously brought to the concentration $c_{A_s} > c_{A_o}$ and held there for $t > 0$, causing diffusion into the medium to occur. Because the medium is infinite in the z-direction, diffusion cannot extend to $z = \infty$ and, therefore, as $z \to \infty$, $c_A = c_{A_o}$ for all $t \geq 0$. Because (3-68) and its one boundary (initial) condition in time and two boundary conditions in distance are linear in the dependent variable, c_A, an exact solution can be obtained by combination of variables [28] or the Laplace transform method [29]. The result, in terms of fractional concentration change, is

$$\theta = \frac{c_A - c_{A_o}}{c_{A_s} - c_{A_o}} = \text{erfc} \left(\frac{z}{2\sqrt{D_{AB} t}} \right) \tag{3-75}$$

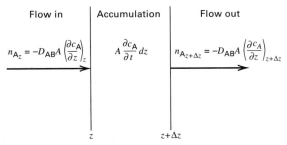

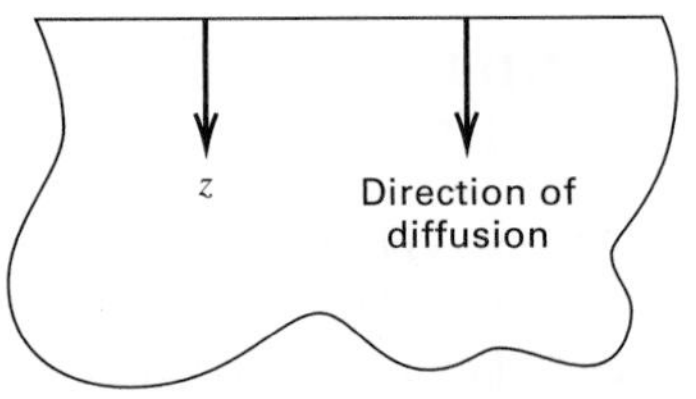

Figure 3.6 One-dimensional diffusion into a semi-infinite medium.

Flow in   Accumulation   Flow out

$$n_{A_z} = -D_{AB} A \left. \frac{\partial c_A}{\partial z} \right|_z \qquad A \frac{\partial c_A}{\partial t} dz \qquad n_{A_{z+\Delta z}} = -D_{AB} A \left. \frac{\partial c_A}{\partial z} \right|_{z+\Delta z}$$

$z \qquad\qquad z + \Delta z$

Figure 3.5 Unsteady-state diffusion through a volume $A\,dz$.

where the complementary error function, erfc, is related to the error function, erf, by

$$\text{erfc}(x) = 1 - \text{erf}(x) = 1 - \frac{2}{\sqrt{\pi}}\int_0^x e^{-\eta^2}d\eta \qquad (3\text{-}76)$$

The error function is included in most spreadsheet programs and handbooks, such as *Handbook of Mathematical Functions* [30]. The variation of erf(x) and erfc(x) is:

x	erf(x)	erfc(x)
0	0.0000	1.0000
0.5	0.5205	0.4795
1.0	0.8427	0.1573
1.5	0.9661	0.0339
2.0	0.9953	0.0047
∞	1.0000	0.0000

Equation (3-75) determines the concentration in the semi-infinite medium as a function of time and distance from the surface, assuming no bulk flow. It applies rigorously to diffusion in solids, and also to stagnant liquids and gases when the medium is dilute in the diffusing solute.

In (3-75), when $(z/2\sqrt{D_{AB}t}) = 2$, the complementary error function is only 0.0047, which represents less than a 1% change in the ratio of the concentration change at $z = z$ to the change at $z = 0$. It is common to call $z = 4\sqrt{D_{AB}t}$ the penetration depth, and to apply (3-75) to media of finite thickness as long as the thickness is greater than the penetration depth.

The instantaneous rate of mass transfer across the surface of the medium at $z = 0$ can be obtained by taking the derivative of (3-75) with respect to distance and substituting it into Fick's first law applied at the surface of the medium. Then, using the Leibnitz rule for differentiating the integral of (3-76), with $x = z/2\sqrt{D_{AB}t}$,

$$n_A = -D_{AB}A\left(\frac{\partial c_A}{\partial z}\right)_{z=0}$$

$$= D_{AB}A\left(\frac{c_{A_s} - c_{A_o}}{\sqrt{\pi D_{AB}t}}\right)\exp\left(-\frac{z^2}{4D_{AB}t}\right)\Bigg|_{z=0} \qquad (3\text{-}77)$$

Thus,

$$n_A\big|_{z=0} = \sqrt{\frac{D_{AB}}{\pi t}}A(c_{A_s} - c_{A_o}) \qquad (3\text{-}78)$$

The total number of moles of solute, $\mathcal{N}_A$, transferred into the semi-infinite medium is obtained by integrating (3-78) with respect to time:

$$\mathcal{N}_A = \int_o^t n_A\big|_{z=0}dt = \sqrt{\frac{D_{AB}}{\pi}}A(c_{A_s} - c_{A_o})\int_o^t\frac{dt}{\sqrt{t}}$$
$$= 2A(c_{A_s} - c_{A_o})\sqrt{\frac{D_{AB}t}{\pi}} \qquad (3\text{-}79)$$

EXAMPLE 3.11   Rates of Diffusion in Stagnant Media.

Determine how long it will take for the dimensionless concentration change, $\theta = (c_A - c_{A_o})/(c_{A_s} - c_{A_o})$, to reach 0.01 at a depth $z = 100$

cm in a semi-infinite medium. The medium is initially at a solute concentration c_{A_o}, after the surface concentration at $z = 0$ increases to c_{A_s}, for diffusivities representative of a solute diffusing through a stagnant gas, a stagnant liquid, and a solid.

Solution

For a gas, assume $D_{AB} = 0.1$ cm^2/s. From (3-75) and (3-76),

$$\theta = 0.01 = 1 - \text{erf}\left(\frac{z}{2\sqrt{D_{AB}t}}\right)$$

Therefore,

$$\text{erf}\left(\frac{z}{2\sqrt{D_{AB}t}}\right) = 0.99$$

From tables of the error function, $\left(\dfrac{z}{2\sqrt{D_{AB}t}}\right) = 1.8214$

Solving, $\quad t = \left[\dfrac{100}{1.8214(2)}\right]^2\dfrac{1}{0.10} = 7{,}540\,\text{s} = 2.09\,\text{h}$

In a similar manner, the times for typical gas, liquid, and solid media are found to be drastically different, as shown below.

Semi-infinite Medium	D_{AB}, cm^2/s	Time for $\theta = 0.01$ at 1 m
Gas	0.10	2.09 h
Liquid	1×10^{-5}	2.39 year
Solid	1×10^{-9}	239 centuries

The results show that molecular diffusion is very slow, especially in liquids and solids. In liquids and gases, the rate of mass transfer can be greatly increased by agitation to induce turbulent motion. For solids, it is best to reduce the size of the solid.

§3.3.4 Medium of Finite Thickness with Sealed Edges

Consider a rectangular, parallelepiped stagnant medium of thickness $2a$ in the z-direction, and either infinitely long dimensions in the y- and x-directions or finite lengths $2b$ and $2c$. Assume that in Figure 3.7a the edges parallel to the z-direction are sealed, so diffusion occurs only in the z-direction, and that initially, the concentration of the solute in the medium is uniform at c_{A_o}. At time $t = 0$, the two unsealed surfaces at $z = \pm a$ are brought to and held at concentration $c_{A_s} > c_{A_o}$. Because of symmetry, $\partial c_A/\partial z = 0$ at $z = 0$. Assume constant D_{AB}. Again (3-68) applies, and an exact solution can be obtained because both (3-68) and the boundary conditions are linear in c_A. The result from Carslaw and Jaeger [26], in terms of the fractional, unaccomplished concentration change, E, is

$$E = 1 - \theta = \frac{c_{A_s} - c_A}{c_{A_s} - c_{A_o}} = \frac{4}{\pi}\sum_{n=0}^{\infty}\frac{(-1)^n}{(2n+1)}$$

$$\times \exp\left[-D_{AB}(2n+1)^2\pi^2t/4a^2\right]\cos\frac{(2n+1)\pi z}{2a}$$

$$(3\text{-}80)$$

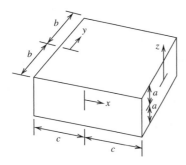

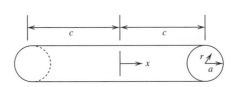

(a) Slab. Edges at $x = +c$ and $-c$ and at $y = +b$ and $-b$ are sealed.

(b) Cylinder. Two circular ends at $x = +c$ and $-c$ are sealed.

(c) Sphere

Figure 3.7 Unsteady-state diffusion in media of finite dimensions.

or, in terms of the complementary error function,

$$E = 1 - \theta = \frac{c_{A_s} - c_A}{c_{A_s} - c_{A_o}} = \sum_{n=0}^{\infty} (-1)^n$$
$$\times \left[\text{erfc} \frac{(2n+1)a - z}{2\sqrt{D_{AB}t}} + \text{erfc} \frac{(2n+1)a + z}{2\sqrt{D_{AB}t}} \right] \quad (3\text{-}81)$$

For large values of $D_{AB}t/a^2$, called the Fourier number for mass transfer, the infinite series solutions of (3-80) and (3-81) converge rapidly, but for small values (e.g., short times), they do not. However, in the latter case, the solution for the semi-infinite medium applies for $D_{AB}t/a^2 < \frac{1}{16}$. A plot of the solution is given in Figure 3.8.

The instantaneous rate of mass transfer across the surface of either unsealed face of the medium (i.e., at $z = \pm a$) is obtained by differentiating (3-80) with respect to z,

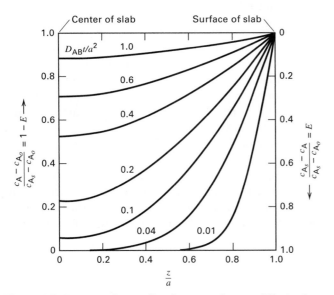

Figure 3.8 Concentration profiles for unsteady-state diffusion in a slab.

Adapted from H.S. Carslaw and J.C. Jaeger, *Conduction of Heat in Solids*, 2nd ed., Oxford University Press, London (1959).]

evaluating the result at $z = a$, and substituting into Fick's first law to give

$$n_A|_{z=a} = \frac{2D_{AB}(c_{A_s} - c_{A_o})A}{a} \times \sum_{n=0}^{\infty} \exp\left[-\frac{D_{AB}(2n+1)^2\pi^2 t}{4a^2} \right] \quad (3\text{-}82)$$

The total moles transferred across either unsealed face is determined by integrating (3-82) with respect to time:

$$\mathscr{N}_A = \int_o^t n_A|_{z=a} dt = \frac{8(c_{A_s} - c_{A_o})Aa}{\pi^2}$$
$$\times \sum_{n=0}^{\infty} \frac{1}{(2n+1)^2} \left\{ 1 - \exp\left[-\frac{D_{AB}(2n+1)^2\pi^2 t}{4a^2} \right] \right\} \quad (3\text{-}83)$$

For a slab, the average concentration of the solute $c_{A_{avg}}$, as a function of time, is

$$\frac{c_{A_s} - c_{A_{avg}}}{c_{A_s} - c_{A_o}} = \frac{\int_o^a (1-\theta)dz}{a} \quad (3\text{-}84)$$

Substitution of (3-80) into (3-84), followed by integration, gives

$$E_{avg_{slab}} = (1 - \theta_{ave})_{slab} = \frac{c_{A_s} - c_{A_{avg}}}{c_{A_s} - c_{A_o}}$$
$$= \frac{8}{\pi^2} \sum_{n=0}^{\infty} \frac{1}{(2n+1)^2} \exp\left[-\frac{D_{AB}(2n+1)^2\pi^2 t}{4a^2} \right] \quad (3\text{-}85)$$

This equation is plotted in Figure 3.9. The concentrations are in mass of solute per mass of dry solid or mass of solute/volume. This assumes that during diffusion, the solid does not shrink or expand; thus, the mass of dry solid per unit volume of wet solid remains constant. In drying it is common to express moisture content on a dry-solid basis.

When the edges of the slab in Figure 3.7a are not sealed, the method of Newman [31] can be used with (3-69) to determine concentration changes within the slab. In this method,

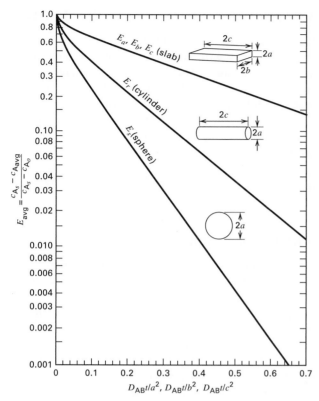

Figure 3.9 Average concentrations for unsteady-state diffusion.

[Adapted from R.E. Treybal, *Mass-Transfer Operations*, 3rd ed., McGraw-Hill, New York (1980).]

E or E_{avg} is given in terms of E values from the solution of (3-68) for each of the coordinate directions by

$$E = E_x E_y E_z \qquad (3-86)$$

Corresponding solutions for infinitely long, circular cylinders and spheres are available in Carslaw and Jaeger [26] and are plotted in Figures 3.9 to 3.11. For a short cylinder whose

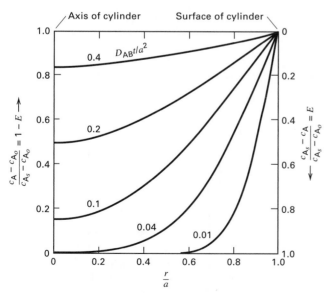

Figure 3.10 Concentration profiles for unsteady-state diffusion in a cylinder.

[Adapted from H.S. Carslaw and J.C. Jaeger, *Conduction of Heat in Solids*, 2nd ed., Oxford University Press, London (1959).]

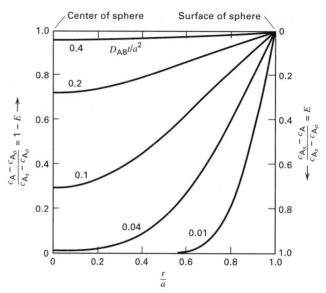

Figure 3.11 Concentration profiles for unsteady-state diffusion in a sphere.

[Adapted from H.S. Carslaw and J.C. Jaeger, *Conduction of Heat in Solids*, 2nd ed., Oxford University Press, London (1959).]

ends are not sealed, E or E_{avg} is given by the method of Newman as

$$E = E_r E_x \qquad (3-87)$$

For anisotropic materials, Fick's second law in the form of (3-69) does not hold. Although the general anisotropic case is exceedingly complex, as shown in the following example, its mathematical treatment is relatively simple when the principal axes of diffusivity coincide with the coordinate system.

EXAMPLE 3.12   Diffusion of Moisture from Lumber.

A board of lumber $5 \times 10 \times 20$ cm initially contains 20 wt% moisture. At time zero, all six faces are brought to an equilibrium moisture content of 2 wt%. Diffusivities for moisture at 25°C are 2×10^{-5} cm²/s in the axial (z) direction along the fibers and 4×10^{-6} cm²/s in the two directions perpendicular to the fibers. Calculate the time in hours for the average moisture content to drop to 5 wt% at 25°C. At that time, determine the moisture content at the center of the slab. All moisture contents are on a dry basis.

Solution

In this case, the solid is anisotropic, with $D_x = D_y = 4 \times 10^{-6}$ cm²/s and $D_z = 2 \times 10^{-5}$ cm²/s, where dimensions 2c, 2b, and 2a in the x-, y-, and z-directions are 5, 10, and 20 cm, respectively. Fick's second law for an isotropic medium, (3-69), must be rewritten as

$$\frac{\partial c_A}{\partial t} = D_x \left[\frac{\partial^2 c_A}{\partial x^2} + \frac{\partial^2 c_A}{\partial y^2} \right] + D_z \frac{\partial^2 c_A}{\partial z^2} \qquad (1)$$

To transform (1) into the form of (3-69) [26], let

$$x_1 = x\sqrt{\frac{D}{D_x}}, \quad y_1 = y\sqrt{\frac{D}{D_x}}, \quad z_1 = z\sqrt{\frac{D}{D_z}} \qquad (2)$$

where D is arbitrarily chosen. With these changes, (1) becomes

$$\frac{\partial c_A}{\partial t} = D\left(\frac{\partial^2 c_A}{\partial x_1^2} + \frac{\partial^2 c_A}{\partial y_1^2} + \frac{\partial^2 c_A}{\partial z_1^2}\right) \tag{3}$$

This is the same form as (3-69), and since the boundary conditions do not involve diffusivities, Newman's method applies, using Figure 3.9, where concentration c_A is replaced by weight-percent moisture on a dry basis. From (3-86) and (3-85),

$$E_{\text{ave}_{\text{slab}}} = E_{\text{avg}_x} E_{\text{avg}_y} E_{\text{avg}_z} = \frac{c_{A_{\text{ave}}} - c_{A_s}}{c_{A_o} - c_{A_s}} = \frac{5-2}{20-2} = 0.167$$

Let $D = 1 \times 10^{-5}$ cm^2/s.

z_1 **Direction (axial):**

$$a_1 = a\left(\frac{D}{D_z}\right)^{1/2} = \frac{20}{2}\left(\frac{1 \times 10^{-5}}{2 \times 10^{-5}}\right)^{1/2} = 7.07 \text{ cm}$$

$$\frac{Dt}{a_1^2} = \frac{1 \times 10^{-5}t}{7.07^2} = 2.0 \times 10^{-7}t, \text{ s}$$

y_1 **Direction:**

$$b_1 = b\left(\frac{D}{D_y}\right)^{1/2} = \frac{20}{2}\left(\frac{1 \times 10^{-5}}{4 \times 10^{-6}}\right)^{1/2} = 7.906 \text{ cm}$$

$$\frac{Dt}{b_1^2} = \frac{1 \times 10^{-5}t}{7.906^2} = 1.6 \times 10^{-7}t, \text{ s}$$

x_1 **Direction:**

$$c_1 = c\left(\frac{D}{D_x}\right)^{1/2} = \frac{5}{2}\left(\frac{1 \times 10^{-5}}{4 \times 10^{-6}}\right)^{1/2} = 3.953 \text{ cm}$$

$$\frac{Dt}{c_1^2} = \frac{1 \times 10^{-5}t}{3.953^2} = 6.4 \times 10^{-7}t, \text{ s}$$

Figure 3.9 is used iteratively with assumed values of time in seconds to obtain values of E_{avg} for each of the three coordinates until (3-86) equals 0.167.

t, h	t, s	$E_{\text{avg}_{z_1}}$	$E_{\text{avg}_{y_1}}$	$E_{\text{avg}_{x_1}}$	E_{avg}
100	360,000	0.70	0.73	0.46	0.235
120	432,000	0.67	0.70	0.41	0.193
135	486,000	0.65	0.68	0.37	0.164

Therefore, it takes approximately 136 h.
For 136 h = 490,000 s, Fourier numbers for mass transfer are

$$\frac{Dt}{a_1^2} = \frac{(1 \times 10^{-5})(490,000)}{7.07^2} = 0.0980$$

$$\frac{Dt}{b_1^2} = \frac{(1 \times 10^{-5})(490,000)}{7.906^2} = 0.0784$$

$$\frac{Dt}{c_1^2} = \frac{(1 \times 10^{-5})(490,000)}{3.953^2} = 0.3136$$

From Figure 3.8, at the center of the slab,

$$E_{\text{center}} = E_{z_1} E_{y_1} E_{x_1} = (0.945)(0.956)(0.605) = 0.547$$

$$= \frac{c_{A_s} - c_{A_{\text{center}}}}{c_{A_s} - c_{A_o}} = \frac{2 - c_{A_{\text{center}}}}{2 - 20} = 0.547$$

Solving, c_A at the center = 11.8 wt% moisture

§3.4 MASS TRANSFER IN LAMINAR FLOW

Many mass-transfer operations involve diffusion in fluids in laminar flow. As with convective heat-transfer in laminar flow, the calculation of such operations is amenable to well-defined theory. This is illustrated in this section by three common applications: (1) a fluid falling as a film down a wall; (2) a fluid flowing slowly along a horizontal, flat plate; and (3) a fluid flowing slowly through a circular tube, where mass transfer occurs, respectively, between a gas and the falling liquid film, from the surface of the flat plate into the flowing fluid, and from the inside surface of the tube into the flowing fluid.

§3.4.1 Falling Laminar, Liquid Film

Consider a thin liquid film containing A and nonvolatile B, falling in laminar flow at steady state down one side of a vertical surface and exposed to pure gas, A, which diffuses into the liquid, as shown in Figure 3.12. The surface is infinitely wide in the x-direction (normal to the page), flow is in the downward y-direction, and mass transfer of A is in the z-direction. Assume that the rate of mass transfer of A into the liquid film is so small that the liquid velocity in the z-direction, u_z, is zero. From fluid mechanics, in the absence of end effects the equation of motion for the liquid film in fully developed laminar flow in the y-direction is

$$\mu \frac{d^2 u_y}{dz^2} + \rho g = 0 \tag{3-88}$$

Usually, fully developed flow, where u_y is independent of the distance y, is established quickly. If δ is the film thickness and the boundary conditions are $u_y = 0$ at $z = \delta$ (no slip at the solid surface) and $du_y/dz = 0$ at $z = 0$ (no drag at the gas–liquid interface), (3-88) is readily integrated to give a parabolic velocity profile:

$$u_y = \frac{\rho g \delta^2}{2\mu}\left[1 - \left(\frac{z}{\delta}\right)^2\right] \tag{3-89}$$

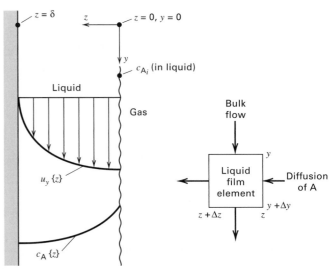

Figure 3.12 Mass transfer from a gas into a falling, laminar liquid film.

The maximum liquid velocity occurs at $z = 0$,

$$(u_y)_{max} = \frac{\rho g \delta^2}{2\mu} \tag{3-90}$$

The bulk-average velocity in the liquid film is

$$\bar{u}_y = \frac{\int_0^\delta u_y \, dz}{\delta} = \frac{\rho g \delta^2}{3\mu} \tag{3-91}$$

Thus, with no entrance effects, the film thickness for fully developed flow is independent of location y and is

$$\delta = \left(\frac{3\bar{u}_y \mu}{\rho g}\right)^{1/2} = \left(\frac{3\mu\Gamma}{\rho^2 g}\right)^{1/3} \tag{3-92}$$

where Γ = liquid film flow rate per unit width of film, W. For film flow, the Reynolds number, which is the ratio of the inertial force to the viscous force, is

$$N_{Re} = \frac{4r_H \bar{u}_y \rho}{\mu} = \frac{4\delta \bar{u}_y \rho}{\mu} = \frac{4\Gamma}{\mu} \tag{3-93}$$

where r_H = hydraulic radius = (flow cross section)/(wetted perimeter) = $(W\delta)/W = \delta$ and, by continuity, $\Gamma = \bar{u}_y \rho \delta$.

Grimley [32] found that for $N_{Re} < 8$ to 25, depending on surface tension and viscosity, flow in the film is laminar and the interface between the liquid film and gas is flat. The value of 25 is obtained with water. For 8 to $25 < N_{Re} < 1,200$, the flow is still laminar, but ripples may appear at the interface unless suppressed by the addition of wetting agents.

For a flat interface and a low rate of mass transfer of A, Eqs. (3-88) to (3-93) hold, and the film velocity profile is given by (3-89). Consider a mole balance on A for an incremental volume of liquid film of constant density, as shown in Figure 3.12. Neglect bulk flow in the z-direction and axial diffusion in the y-direction. Thus, mass transfer of A from the gas into the liquid occurs only by molecular diffusion in the z-direction. Then, at steady state, neglecting accumulation or depletion of A in the incremental volume (quasi-steady-state assumption),

$$-D_{AB}(\Delta y)(\Delta x)\left(\frac{\partial c_A}{\partial z}\right)_z + u_y c_A|_y (\Delta z)(\Delta x)$$
$$= -D_{AB}(\Delta y)(\Delta x)\left(\frac{\partial c_A}{\partial z}\right)_{z+\Delta z} + u_y c_A|_{y+\Delta y}(\Delta z)(\Delta x) \tag{3-94}$$

Rearranging and simplifying (3-94),

$$\left[\frac{u_y c_A|_{y+\Delta y} - u_y c_A|_y}{\Delta y}\right] = D_{AB}\left[\frac{(\partial c_A/\partial z)_{z+\Delta z} - (\partial c_A/\partial z)_z}{\Delta z}\right] \tag{3-95}$$

which, in the limit, as $\Delta z \to 0$ and $\Delta y \to 0$, becomes

$$u_y \frac{\partial c_A}{\partial y} = D_{AB} \frac{\partial^2 c_A}{\partial z^2} \tag{3-96}$$

Substituting the velocity profile of (3-89) into (3-96),

$$\frac{\rho g \delta^2}{2\mu}\left[1 - \left(\frac{z}{\delta}\right)^2\right]\frac{\partial c_A}{\partial y} = D_{AB}\frac{\partial^2 c_A}{\partial z^2} \tag{3-97}$$

This PDE was solved by Johnstone and Pigford [33] and Olbrich and Wild [34] for the following boundary conditions, where the initial concentration of A in the liquid film is c_{A_o}:

$$c_A = c_{A_i} \quad \text{at } z = 0 \quad \text{for } y > 0$$
$$c_A = c_{A_o} \quad \text{at } y = 0 \quad \text{for } 0 < z < \delta$$
$$\partial c_A/\partial z = 0 \quad \text{at } z = \delta \quad \text{for } 0 < y < L$$

where L = height of the vertical surface. The solution of Olbrich and Wild is in the form of an infinite series, giving c_A as a function of z and y. Of greater interest, however, is the average concentration of A in the film at the bottom of the wall, where $y = L$, which, by integration, is

$$\bar{c}_{A_y} = \frac{1}{\bar{u}_y \delta}\int_0^\delta u_y c_{A_y} \, dz \tag{3-98}$$

For the condition $y = L$, the result is

$$\frac{c_{A_i} - \bar{c}_{A_L}}{c_{A_i} - c_{A_o}} = 0.7857 e^{-5.1213\eta} + 0.09726 e^{-39.661\eta}$$
$$+ 0.036093 e^{-106.25\eta} + \cdots \tag{3-99}$$

where

$$\eta = \frac{2D_{AB}L}{3\delta^2 \bar{u}_y} = \frac{8/3}{N_{Re}N_{Sc}(\delta/L)} = \frac{8/3}{(\delta/L)N_{Pe_M}} \tag{3-100}$$

$$\boxed{N_{Sc} = \text{Schmidt number} = \frac{\mu}{\rho D_{AB}} \\ = \frac{\text{momentum diffusivity}, \mu/\rho}{\text{mass diffusivity}, D_{AB}}} \tag{3-101}$$

$$N_{Pe_M} = N_{Re}N_{Sc} = \text{Peclet number for mass transfer}$$
$$= \frac{4\delta \bar{u}_y}{D_{AB}} \tag{3-102}$$

The Schmidt number is analogous to the Prandtl number, used in heat transfer:

$$\boxed{N_{Pr} = \frac{C_P \mu}{k} = \frac{(\mu/\rho)}{(k/\rho C_P)} = \frac{\text{momentum diffusivity}}{\text{thermal diffusivity}}}$$

The Peclet number for mass transfer is analogous to the Peclet number for heat transfer:

$$N_{Pe_H} = N_{Re}N_{Pr} = \frac{4\delta \bar{u}_y C_P \rho}{k}$$

Both are ratios of convective to molecular transport.

The total rate of absorption of A from the gas into the liquid film for height L and width W is

$$n_A = \bar{u}_y \delta W(\bar{c}_{A_L} - c_{A_o}) \tag{3-103}$$

§3.4.2 Mass-Transfer Coefficients

Mass-transfer problems involving flowing fluids are often solved using *mass-transfer coefficients*, which are analogous to heat-transfer coefficients. For the latter, *Newton's law of cooling* defines a heat-transfer coefficient, h:

$$Q = hA\Delta T \tag{3-104}$$

where Q = rate of heat transfer, A = area for heat transfer (normal to the direction of heat transfer), and ΔT = temperature-driving force.

For mass transfer, a composition-driving force replaces ΔT. Because composition can be expressed in a number of ways, different mass-transfer coefficients result. If concentration is used, Δc_A is selected as the driving force and

$$\boxed{n_A = k_c A \Delta c_A} \tag{3-105}$$

which defines a mass-transfer coefficient, k_c, in mol/time-area-driving force, for a concentration driving force. Unfortunately, no name is in general use for (3-105).

For the falling laminar film, $\Delta c_A = c_{A_i} - \bar{c}_A$, where $\bar{c}_A$ is the bulk average concentration of A in the film, which varies with vertical location, y, because even though c_{A_i} is independent of y, the average film concentration of A increases with y. A theoretical expression for k_c in terms of diffusivity is formed by equating (3-105) to Fick's first law at the gas–liquid interface:

$$\boxed{k_c A (c_{A_i} - \bar{c}_A) = -D_{AB} A \left(\frac{\partial c_A}{\partial z} \right)_{z=0}} \tag{3-106}$$

Although this is the most widely used approach for defining a mass-transfer coefficient, for a falling film it fails because $(\partial c_A / \partial z)$ at $z = 0$ is not defined. Therefore, another approach is used. For an incremental height,

$$n_A = \bar{u}_y \delta W \, d\bar{c}_A = k_c (c_{A_\perp} - \bar{c}_A) W \, dy \tag{3-107}$$

This defines a local value of k_c, which varies with distance y because $\bar{c}_A$ varies with y. An average value of k_c, over height L, can be defined by separating variables and integrating (3-107):

$$
\begin{aligned}
k_{c_{avg}} &= \frac{\int_0^L k_c dy}{L} = \frac{\bar{u}_y \delta \int_{c_{A_o}}^{c_{A_L}} [d\bar{c}_A / (c_{A_i} - \bar{c}_A)]}{L} \\
&= \frac{\bar{u}_y \delta}{L} \ln \frac{c_{A_i} - c_{A_o}}{c_{A_i} - \bar{c}_{A_L}}
\end{aligned}
\tag{3-108}
$$

The argument of the natural logarithm in (3-108) is obtained from the reciprocal of (3-99). For values of η in (3-100) greater than 0.1, only the first term in (3-99) is significant (error is less than 0.5%). In that case,

$$k_{c_{avg}} = \frac{\bar{u}_y \delta}{L} \ln \frac{e^{5.1213\eta}}{0.7857} \tag{3-109}$$

Since $\ln e^x = x$,

$$k_{c_{avg}} = \frac{\bar{u}_y \delta}{L} (0.241 + 5.1213\eta) \tag{3-110}$$

In the limit for large η, using (3-100) and (3-102), (3-110) becomes

$$k_{c_{avg}} = 3.414 \frac{D_{AB}}{\delta} \tag{3-111}$$

As suggested by the Nusselt number, $N_{Nu} = h\delta/k$ for heat transfer, where δ is a characteristic length, a Sherwood number for mass transfer is defined for a falling film as

$$\boxed{N_{Sh_{avg}} = \frac{k_{c_{avg}} \delta}{D_{AB}}} \tag{3-112}$$

From (3-111), $N_{Sh_{avg}} = 3.414$, which is the smallest value the Sherwood number can have for a falling liquid film. The average mass-transfer flux of A is

$$N_{A_{avg}} = \frac{n_{A_{avg}}}{A} = k_{c_{avg}} (c_{A_i} - \bar{c}_A)_{mean} \tag{3-113}$$

For $\eta < 0.001$ in (3-100), when the liquid-film flow regime is still laminar without ripples, the time of contact of gas with liquid is short and mass transfer is confined to the vicinity of the interface. Thus, the film acts as if it were infinite in thickness. In this limiting case, the downward velocity of the liquid film in the region of mass transfer is $u_{y_{max}}$, and (3-96) becomes

$$u_{y_{max}} \frac{\partial c_A}{\partial y} = D_{AB} \frac{\partial^2 c_A}{\partial z^2} \tag{3-114}$$

Since from (3-90) and (3-91) $u_{y_{max}} = 3u_y/2$, (3-114) becomes

$$\frac{\partial c_A}{\partial y} = \left(\frac{2 D_{AB}}{3 \bar{u}_y} \right) \frac{\partial^2 c_A}{\partial z^2} \tag{3-115}$$

where the boundary conditions are

$$
\begin{aligned}
c_A &= c_{A_o} \text{ for } z > 0 \text{ and } y > 0 \\
c_A &= c_{A_i} \text{ for } z = 0 \text{ and } y > 0 \\
c_A &= c_{A_i} \text{ for large } z \text{ and } y > 0
\end{aligned}
$$

Equation (3-115) and the boundary conditions are equivalent to the case of the semi-infinite medium in Figure 3.6. By analogy to (3-68), (3-75), and (3-76), the solution is

$$E = 1 - \theta = \frac{c_{A_i} - c_A}{c_{A_i} - c_{A_o}} = \text{erf} \left(\frac{z}{2\sqrt{2 D_{AB} y / 3 \bar{u}_y}} \right) \tag{3-116}$$

Assuming that the driving force for mass transfer in the film is $c_{A_i} - c_{A_0}$, Fick's first law can be used at the gas–liquid interface to define a mass-transfer coefficient:

$$N_A = -D_{AB} \frac{\partial c_A}{\partial z} \bigg|_{z=0} = k_c (c_{A_i} - c_{A_o}) \tag{3-117}$$

To obtain the gradient of c_A at $z = 0$ from (3-116), note that the error function is defined as

$$\text{erf } z = \frac{2}{\sqrt{\pi}} \int_0^z e^{-t^2} dt \tag{3-118}$$

Combining (3-118) with (3-116) and applying the Leibnitz differentiation rule,

$$\frac{\partial c_A}{\partial z} \bigg|_{z=0} = -(c_{A_i} - c_{A_o}) \sqrt{\frac{3 \bar{u}_y}{2 \pi D_{AB} y}} \tag{3-119}$$

Substituting (3-119) into (3-117) and introducing the Peclet number for mass transfer from (3-102), the local mass-transfer coefficient as a function of distance down from the top of the wall is obtained:

$$k_c = \sqrt{\frac{3D_{AB}^2 N_{Pe_M}}{8\pi y \delta}} = \sqrt{\frac{3D_{AB}\Gamma}{2\pi y \delta \rho}} \quad (3\text{-}120)$$

The average value of k_c over the film height, L, is obtained by integrating (3-120) with respect to y, giving

$$k_{c_{avg}} = \sqrt{\frac{6D_{AB}\Gamma}{\pi \delta \rho L}} = \sqrt{\frac{3D_{AB}^2}{2\pi \delta L} N_{Pe_M}} \quad (3\text{-}121)$$

Combining (3-121) with (3-112) and (3-102),

$$N_{Sh_{avg}} = \sqrt{\frac{3\delta}{2\pi L} N_{Pe_M}} = \sqrt{\frac{4}{\pi \eta}} \quad (3\text{-}122)$$

where, by (3-108), the proper mean concentration driving force to use with $k_{c_{avg}}$ is the log mean. Thus,

$$\boxed{\begin{aligned} (c_{A_i} - \bar{c}_A)_{mean} &= (c_{A_i} - \bar{c}_A)_{LM} \\ &= \frac{(c_{A_i} - c_{A_o}) - (c_{A_i} - c_{A_L})}{\ln[(c_{A_i} - c_{A_o})/(c_{A_i} - \bar{c}_{A_L})]} \end{aligned}} \quad (3\text{-}123)$$

When ripples are present, values of $k_{c_{avg}}$ and $N_{Sh_{avg}}$ are considerably larger than predicted by the above equations.

The above development shows that asymptotic, closed-form solutions are obtained with relative ease for large and small values of η, as defined by (3-100). These limits, in terms of the average Sherwood number, are shown in Figure 3.13. The general solution for intermediate values of η is not available in closed form. Similar limiting solutions for large and small values of dimensionless groups have been obtained for a large variety of transport and kinetic phenomena (Churchill [35]). Often, the two limiting cases can be patched together to provide an estimate of the intermediate solution, if an intermediate value is available from experiment or the general numerical solution. The procedure is discussed by Churchill and Usagi [36]. The general solution of Emmert and Pigford [37] to the falling, laminar liquid film problem is included in Figure 3.13.

EXAMPLE 3.13 Absorption of CO_2 into a Falling Water Film.

Water (B) at 25°C, in contact with CO_2 (A) at 1 atm, flows as a film down a wall 1 m wide and 3 m high at a Reynolds number of 25. Estimate the rate of absorption of CO_2 into water in kmol/s:

$$D_{AB} = 1.96 \times 10^{-5} \text{cm}^2/\text{s}; \quad \rho = 1.0 \text{ g/cm}^3;$$
$$\mu_L = 0.89 \text{ cP} = 0.00089 \text{ kg/m-s}$$

Solubility of CO_2 at 1 atm and 25°C $= 3.4 \times 10^{-5}$ mol/cm^3.

Solution

From (3-93), $\quad \Gamma = \dfrac{N_{Re}\mu}{4} = \dfrac{25(0.89)(0.001)}{4} = 0.00556 \dfrac{\text{kg}}{\text{m-s}}$

From (3-101),

$$N_{Sc} = \frac{\mu}{\rho D_{AB}} = \frac{(0.89)(0.001)}{(1.0)(1,000)(1.96 \times 10^{-5})(10^{-4})} = 454$$

From (3-92),

$$\delta = \left[\frac{3(0.89)(0.001)(0.00556)}{1.0^2(1.000)^2(9.807)}\right]^{1/3} = 1.15 \times 10^{-4} \text{ m}$$

From (3-90) and (3-91), $\bar{u}_y = (2/3)u_{y_{max}}$. Therefore,

$$\bar{u}_y = \frac{2}{3}\left[\frac{(1.0)(1,000)(9.807)(1.15 \times 10^{-4})^2}{2(0.89)(0.001)}\right] = 0.0486 \text{ m/s}$$

From (3-100),

$$\eta = \frac{8/3}{(25)(454)[(1.15 \times 10^{-4})/3]} = 6.13$$

Therefore, (3-111) applies, giving

$$k_{c_{avg}} = \frac{3.41(1.96 \times 10^{-5})(10^{-4})}{1.15 \times 10^{-4}} = 5.81 \times 10^{-5} \text{ m/s}$$

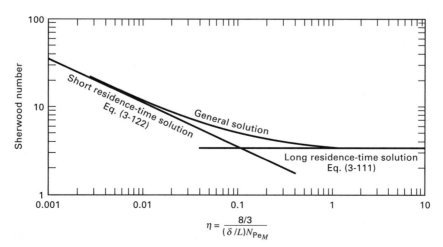

Figure 3.13 Limiting and general solutions for mass transfer to a falling, laminar liquid film.

To obtain the rate of absorption, $\bar{c}_{A_L}$ is determined. From (3-103) and (3-113),

$$n_A = \bar{u}_y \delta W (\bar{c}_{A_L} - c_{A_o}) = k_{c_{avg}} A \frac{(\bar{c}_{A_L} - c_{A_o})}{\ln[(c_{A_i} - c_{A_o})/(c_{A_i} - \bar{c}_{A_L})]}$$

Thus,

$$\ln\left(\frac{c_{A_i} - c_{A_o}}{c_{A_i} - \bar{c}_{A_L}}\right) = \frac{k_{c_{avg}} A}{\bar{u}_y \delta W}$$

Solving for $\bar{c}_{A_L}$,

$$\bar{c}_{A_L} = c_{A_i} - (c_{A_i} - c_{A_o}) \exp\left(-\frac{k_{c_{avg}} A}{\bar{u}_y \delta W}\right)$$

$$L = 3\,m, \quad W = 1\,m, \quad A = WL = (1)(3) = 3\,m^2$$

$$c_{A_o} = 0, \quad c_{A_i} = 3.4 \times 10^{-5}\,mol/cm^3 = 3.4 \times 10^{-2}\,kmol/m^3$$

$$\bar{c}_{A_L} = 3.4 \times 10^{-2}\left\{1 - \exp\left[-\frac{(5.81 \times 10^{-5})(3)}{(0.0486)(1.15 \times 10^{-4})(1)}\right]\right\}$$

$$= 3.4 \times 10^{-2}\,kmol/m^3$$

Thus, the exiting liquid film is saturated with CO_2, which implies equilibrium at the gas–liquid interface. From (3-103),

$$n_A = 0.0486(1.15 \times 10^{-4})(3.4 \times 10^{-2}) = 1.9 \times 10^{-7}\,kmol/s$$

§3.4.3 Molecular Diffusion to a Fluid Flowing Across a Flat Plate—The Boundary Layer Concept

Consider the flow of fluid (B) over a thin, horizontal, flat plate, as shown in Figure 3.14. Some possibilities for mass transfer of species A into B are: (1) the plate consists of material A, which is slightly soluble in B; (2) A is in the pores of an inert solid plate from which it evaporates or dissolves into B; and (3) the plate is a dense polymeric membrane through which A can diffuse and pass into fluid B. Let the fluid velocity upstream of the plate be uniform at a free-system velocity of u_o. As the fluid passes over the plate, the velocity u_x in the direction x of flow is reduced to zero at the wall, which establishes a velocity profile due to drag. At a certain distance z that is normal to and upward out from the plate surface, the fluid velocity is 99% of u_o. This distance, which increases with increasing distance x from the leading edge of the plate, is defined as the *velocity boundary-layer* thickness, δ. Essentially all flow retardation is assumed to occur in the boundary layer, as first suggested by Prandtl [38]. The buildup of this layer, the velocity profile, and the drag force

can be determined for laminar flow by solving the Navier–Stokes equations.

For a Newtonian fluid of constant density and viscosity, with no pressure gradients in the x- or y-directions, these equations for the boundary layer are

$$\frac{\partial u_x}{\partial x} + \frac{\partial u_z}{\partial z} = 0 \tag{3-124}$$

$$u_x \frac{\partial u_x}{\partial x} + u_z \frac{\partial u_x}{\partial z} = \frac{\mu}{\rho}\left(\frac{\partial^2 u_x}{\partial z^2}\right) \tag{3-125}$$

The boundary conditions are

$$u_x = u_o \text{ at } x = 0 \text{ for } z > 0, \quad u_x = 0 \text{ at } z = 0 \text{ for } x > 0$$
$$u_x = u_o \text{ at } z = \infty \text{ for } x > 0, \quad u_z = 0 \text{ at } z = 0 \text{ for } x > 0$$

A solution of (3-124) and (3-125) was first obtained by Blasius [39], as described by Schlichting [40]. The result in terms of a local friction factor, f_x; a local shear stress at the wall, τ_{w_x}; and a local drag coefficient at the wall, C_{D_x}, is

$$\frac{C_{D_x}}{2} = \frac{f_x}{2} = \frac{\tau_{w_x}}{\rho u_o^2} = \frac{0.322}{N_{Re_x}^{0.5}} \tag{3-126}$$

where

$$N_{Re_x} = \frac{x u_o \rho}{\mu} \tag{3-127}$$

The drag is greatest at the leading edge of the plate, where the Reynolds number is smallest. Values of the drag coefficient obtained by integrating (3-126) from $x = 0$ to L are

$$\frac{C_{D_{avg}}}{2} = \frac{f_{avg}}{2} = \frac{0.664}{(N_{Re_L})^{0.5}} \tag{3-128}$$

The thickness of the velocity boundary layer increases with distance along the plate:

$$\frac{\delta}{x} = \frac{4.96}{N_{Re_x}^{0.5}} \tag{3-129}$$

A reasonably accurate expression for a velocity profile was obtained by Pohlhausen [41], who assumed the empirical form of the velocity in the boundary layer to be $u_x = C_1 z + C_2 z^3$.

If the boundary conditions

$$u_x = 0 \text{ at } z = 0, \ u_x = u_o \text{ at } z = \delta, \ \partial u_x/\partial z = 0 \text{ at } z = \delta$$

are applied to evaluate C_1 and C_2, the result is

$$\frac{u_x}{u_o} = 1.5\left(\frac{z}{\delta}\right) - 0.5\left(\frac{z}{\delta}\right)^3 \tag{3-130}$$

This solution is valid only for a laminar boundary layer, which by experiment persists up to $N_{Re_x} = 5 \times 10^5$.

When mass transfer of A from the surface of the plate into the boundary layer occurs, a species continuity equation applies:

$$u_x \frac{\partial c_A}{\partial x} + u_z \frac{\partial c_A}{\partial z} = D_{AB}\left(\frac{\partial^2 c_A}{\partial x^2}\right) \tag{3-131}$$

If mass transfer begins at the leading edge of the plate and the concentration in the fluid at the solid–fluid interface is

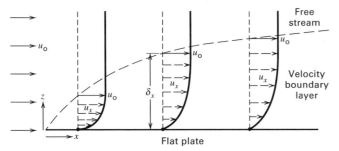

Figure 3.14 Laminar boundary layer for flow across a flat plate.

maintained constant, the additional boundary conditions are

$$c_A = c_{A_o} \text{ at } x = 0 \quad \text{for } z > 0,$$
$$c_A = c_{A_i} \text{ at } z = 0 \quad \text{for } x > 0,$$
$$\text{and } c_A = c_{A_o} \text{ at } z = \infty \text{ for } x > 0$$

If the rate of mass transfer is low, the velocity profiles are undisturbed. The analogous heat-transfer problem was first solved by Pohlhausen [42] for $N_{Pr} > 0.5$, as described by Schlichting [40]. The analogous result for mass transfer is

$$\frac{N_{Sh_x}}{N_{Re_x} N_{Sc}^{1/3}} = \frac{0.332}{N_{Re_x}^{0.5}} \qquad (3\text{-}132)$$

where

$$\boxed{N_{Sh_x} = \frac{x k_{c_x}}{D_{AB}}} \qquad (3\text{-}133)$$

and the driving force for mass transfer is $c_{A_i} - c_{A_o}$.

The concentration boundary layer, where essentially all of the resistance to mass transfer resides, is defined by

$$\frac{c_{A_i} - c_A}{c_{A_i} - c_{A_o}} = 0.99 \qquad (3\text{-}134)$$

and the ratio of the concentration boundary-layer thickness, δ_c, to the velocity boundary thickness, δ, is

$$\delta_c / \delta = 1/N_{Sc}^{1/3} \qquad (3\text{-}135)$$

Thus, for a liquid boundary layer where $N_{Sc} > 1$, the concentration boundary layer builds up more slowly than the velocity boundary layer. For a gas boundary layer where $N_{Sc} \approx 1$, the two boundary layers build up at about the same rate. By analogy to (3-130), the concentration profile is

$$\frac{c_{A_i} - c_A}{c_{A_i} - c_{A_o}} = 1.5 \left(\frac{z}{\delta_c} \right) - 0.5 \left(\frac{z}{\delta_c} \right)^3 \qquad (3\text{-}136)$$

Equation (3-132) gives the local Sherwood number. If this expression is integrated over the length of the plate, L, the average Sherwood number is found to be

$$N_{Sh_{avg}} = 0.664 \, N_{Re_L}^{1/2} N_{Sc}^{1/3} \qquad (3\text{-}137)$$

where

$$N_{Sh_{avg}} = \frac{L k_{c_{avg}}}{D_{AB}} \qquad (3\text{-}138)$$

EXAMPLE 3.14   Sublimation of Naphthalene from a Flat Plate.

Air at 100°C, 1 atm, and a free-stream velocity of 5 m/s flows over a 3-m-long, horizontal, thin, flat plate of naphthalene, causing it to sublime. Determine the: (a) length over which a laminar boundary layer persists, (b) rate of mass transfer over that length, and (c) thicknesses of the velocity and concentration boundary layers at the point of transition of the boundary layer to turbulent flow. The physical properties are: vapor pressure of naphthalene = 10 torr; viscosity of air = 0.0215 cP; molar density of air = 0.0327 kmol/m³; and diffusivity of naphthalene in air = 0.94×10^{-5} m²/s.

Solution

(a) $N_{Re_x} = 5 \times 10^5$ for transition to turbulent flow. From (3-127),

$$x = L = \frac{\mu N_{Re_x}}{u_o \rho} = \frac{[(0.0215)(0.001)](5 \times 10^5)}{(5)[(0.0327)(29)]} = 2.27 \text{ m}$$

at which transition to turbulent flow begins.

(b) $c_{A_o} = 0$, $c_{A_i} = \dfrac{10(0.0327)}{760} = 4.3 \times 10^{-4} \text{ kmol/m}^3$.

From (3-101),

$$N_{Sc} = \frac{\mu}{\rho D_{AB}} = \frac{[(0.0215)(0.001)]}{[(0.0327)(29)](0.94 \times 10^{-5})} = 2.41$$

From (3-137),

$$N_{Sh_{avg}} = 0.664(5 \times 10^5)^{1/2}(2.41)^{1/3} = 630$$

From (3-138),

$$k_{c_{avg}} = \frac{630(0.94 \times 10^{-5})}{2.27} = 2.61 \times 10^{-3} \text{ m/s}$$

For a width of 1 m, $A = 2.27$ m²,

$$n_A = k_{c_{avg}} A(c_{A_i} - c_{A_o}) = 2.61 \times 10^{-3}(2.27)(4.3 \times 10^{-4})$$
$$= 2.55 \times 10^{-6} \text{ kmol/s}$$

(c) From (3-129), at $x = L = 2.27$ m,

$$\delta = \frac{3.46(2.27)}{(5 \times 10^5)^{0.5}} = 0.0111 \text{ m}$$

From (3-135), $\delta_c = \dfrac{0.0111}{(2.41)^{1/3}} = 0.0083 \text{ m}$

§3.4.4 Molecular Diffusion from the Inside Surface of a Circular Tube to a Flowing Fluid—The Fully Developed Flow Concept

Figure 3.15 shows the development of a laminar velocity boundary layer when a fluid flows from a vessel into a straight, circular tube. At the entrance, a, the velocity profile is flat. A velocity boundary layer then begins to build up, as shown at b, c, and d in Figure 3.15. The central core outside the boundary layer has a flat velocity profile where the flow is accelerated over the entrance velocity. Finally, at plane e, the boundary layer fills the tube. Now the flow is fully developed. The distance from plane a to plane e is the entry region.

The entry length L_e is the distance from the entrance to the point at which the centerline velocity is 99% of fully developed flow. From Langhaar [43],

$$L_e / D = 0.0575 \, N_{Re} \qquad (3\text{-}139)$$

For fully developed laminar flow in a tube, by experiment the Reynolds number, $N_{Re} = D \bar{u}_x \rho / \mu$, where $\bar{u}_x$ is the flow-average velocity in the axial direction, x, and D is the inside diameter of the tube, must be less than 2,100. Then the equation of motion in the axial direction is

$$\frac{\mu}{r} \frac{\partial}{\partial r} \left(r \frac{\partial u_x}{\partial r} \right) - \frac{dP}{\partial x} = 0 \qquad (3\text{-}140)$$

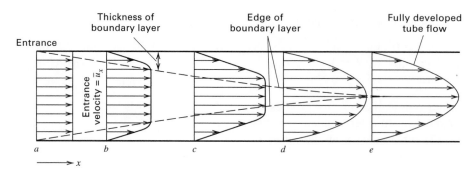

Figure 3.15 Buildup of a laminar velocity boundary layer for flow in a circular tube.

with boundary conditions:

$$r = 0 \text{ (axis of the tube)}, \quad \partial u_x / \partial r = 0$$

and $\qquad r = r_w \text{(tube wall)}, \quad u_x = 0$

Equation (3-140) was integrated by Hagen in 1839 and Poiseuille in 1841. The resulting equation for the velocity profile, in terms of the flow-average velocity, is

$$u_x = 2\bar{u}_x \left[1 - \left(\frac{r}{r_w} \right)^2 \right] \qquad (3\text{-}141)$$

or, in terms of the maximum velocity at the tube axis,

$$u_x = u_{x_{max}} \left[1 - \left(\frac{r}{r_w} \right)^2 \right] \qquad (3\text{-}142)$$

According to (3-142), the velocity profile is parabolic.

The shear stress, pressure drop, and Fanning friction factor are obtained from solutions to (3-140):

$$\tau_w = -\mu \left(\frac{\partial u_x}{\partial r} \right)\bigg|_{r=r_w} = \frac{4\mu\bar{u}_x}{r_w} \qquad (3\text{-}143)$$

$$-\frac{dP}{dx} = \frac{32\mu\bar{u}_x}{D^2} = \frac{2f\rho\bar{u}_x^2}{D} \qquad (3\text{-}144)$$

with $\qquad\qquad f = \frac{16}{N_{Re}} \qquad (3\text{-}145)$

At the upper limit of laminar flow, $N_{Re} = 2,100$, and $L_e/D = 121$, but at $N_{Re} = 100$, L_e/D is only 5.75. In the entry region, the friction factor is considerably higher than the fully developed flow value given by (3-145). At $x = 0$, f is infinity, but it decreases exponentially with x, approaching the fully developed flow value at L_e. For example, for $N_{Re} = 1,000$, (3-145) gives $f = 0.016$, with $L_e/D = 57.5$. From $x = 0$ to $x/D = 5.35$, the average friction factor from Langhaar is 0.0487, which is three times the fully developed value.

In 1885, Graetz [44] obtained a solution to the problem of convective heat transfer between the wall of a circular tube, at a constant temperature, and a fluid flowing through the tube in fully developed laminar flow. Assuming constant properties and negligible conduction in the axial direction,

the energy equation, after substituting (3-141) for u_x, is

$$2\bar{u}_x \left[1 - \left(\frac{r}{r_w} \right)^2 \right] \frac{\partial T}{\partial x} = \frac{k}{\rho C_p} \left[\frac{1}{r} \frac{\partial}{\partial r} \left(r \frac{\partial T}{\partial r} \right) \right] \qquad (3\text{-}146)$$

with boundary conditions:

$$x = 0 \text{ (where heat transfer begins)}, \quad T = T_0, \text{ for all } r$$
$$x > 0, \quad r = r_w, \ T = T_i \ \text{ and } \ x > 0, \ r = 0, \ \partial T / \partial r = 0$$

The analogous species continuity equation for mass transfer, neglecting bulk flow in the radial direction and axial diffusion, is

$$2\bar{u}_x \left[1 - \left(\frac{r}{r_w} \right)^2 \right] \frac{\partial c_A}{\partial x} = D_{AB} \left[\frac{1}{r} \frac{\partial}{\partial r} \left(r \frac{\partial c_A}{\partial r} \right) \right] \qquad (3\text{-}147)$$

with analogous boundary conditions.

The Graetz solution of (3-147) for the temperature or concentration profile is an infinite series that can be obtained from (3-146) by separation of variables using the method of Frobenius. A detailed solution is given by Sellars, Tribus, and Klein [45]. The concentration profile yields expressions for the mass-transfer coefficient and the Sherwood number. For large x, the concentration profile is fully developed and the local Sherwood number, N_{Sh_x}, approaches a limiting value of 3.656. When x is small, such that the concentration boundary layer is very thin and confined to a region where the fully developed velocity profile is linear, the local Sherwood number is obtained from the classic Leveque [46] solution, presented by Knudsen and Katz [47]:

$$N_{Sh_x} = \frac{k_{c_x}D}{D_{AB}} = 1.077 \left[\frac{N_{Pe_M}}{(x/D)} \right]^{1/3} \qquad (3\text{-}148)$$

where $\qquad\qquad N_{Pe_M} = \frac{D\bar{u}_x}{D_{AB}} \qquad (3\text{-}149)$

The limiting solutions, together with the general Graetz solution, are shown in Figure 3.16, where $N_{Sh_x} = 3.656$ is valid for $N_{Pe_M}/(x/D) < 4$ and (3-148) is valid for $N_{Pe_M}/(x/D) > 100$. These solutions can be patched together if a point from the general solution is available at the intersection in a manner like that discussed in §3.4.2.

Where mass transfer occurs, an average Sherwood number is derived by integrating the general expression for the local

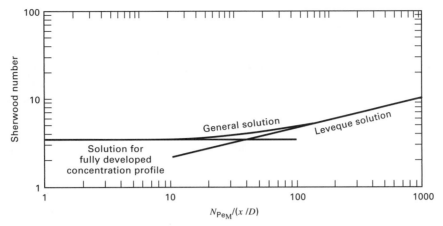

Figure 3.16 Limiting and general solutions for mass transfer to a fluid in laminar flow in a straight, circular tube.

Sherwood number. An empirical representation for that average, proposed by Hausen [48], is

$$N_{Sh_{avg}} = 3.66 + \frac{0.0668[N_{Pe_M}/(x/D)]}{1 + 0.04[N_{Pe_M}/(x/D)]^{2/3}} \qquad (3\text{-}150)$$

which is based on a log-mean concentration driving force.

EXAMPLE 3.15 Mass Transfer of Benzoic Acid into Water Flowing in Laminar Motion Through a Tube.

Linton and Sherwood [49] dissolved tubes of benzoic acid (A) into water (B) flowing in laminar flow through the tubes. Their data agreed with predictions based on the Graetz and Leveque equations. Consider a 5.23-cm-inside-diameter, 32-cm-long tube of benzoic acid, preceded by 400 cm of straight metal pipe wherein a fully developed velocity profile is established. Water enters at 25°C at a velocity corresponding to a Reynolds number of 100. Based on property data at 25°C, estimate the average concentration of benzoic acid leaving the tube before a significant increase in the inside diameter of the benzoic acid tube occurs because of dissolution. The properties are: solubility of benzoic acid in water = 0.0034 g/cm^3; viscosity of water = 0.89 cP = 0.0089 g/cm-s; and diffusivity of benzoic acid in water at infinite dilution = 9.18×10^{-6} cm^2/s.

Solution

$$N_{Sc} = \frac{0.0089}{(1.0)(9.18 \times 10^{-6})} = 970$$

$$N_{Re} = \frac{D\bar{u}_x \rho}{\mu} = 100$$

from which $\quad \bar{u}_x = \frac{(100)(0.0089)}{(5.23)(1.0)} = 0.170$ cm/s

From (3-149), $\quad N_{Pe_M} = \frac{(5.23)(0.170)}{9.18 \times 10^{-6}} = 9.69 \times 10^4$

$$\frac{x}{D} = \frac{32}{5.23} = 6.12$$

$$\frac{N_{Pe_M}}{(x/D)} = \frac{9.69 \times 10^4}{6.12} = 1.58 \times 10^4$$

From (3-150),

$$N_{Sh_{avg}} = 3.66 + \frac{0.0668(1.58 \times 10^4)}{1 + 0.04(1.58 \times 10^4)^{2/3}} = 44$$

$$k_{c_{avg}} = N_{Sh_{avg}}\left(\frac{D_{AB}}{D}\right) = 44\frac{(9.18 \times 10^{-6})}{5.23} = 7.7 \times 10^{-5} \text{cm/s}$$

Using a log mean driving force,

$$n_A = \bar{u}_x S(\bar{c}_{A_x} - c_{A_o}) = k_{c_{avg}} A \frac{[(c_{A_i} - c_{A_o}) - (c_{A_i} - \bar{c}_{A_x})]}{\ln[(c_{A_i} - c_{A_o})/(c_{A_i} - \bar{c}_{A_x})]}$$

where S is the cross-sectional area for flow. Simplifying,

$$\ln\left(\frac{c_{A_i} - c_{A_o}}{c_{A_i} - \bar{c}_{A_x}}\right) = \frac{k_{c_{avg}} A}{\bar{u}_x S}$$

$$c_{A_o} = 0 \quad \text{and} \quad c_{A_i} = 0.0034 \text{ g/cm}^3$$

$$S = \frac{\pi D^2}{4} = \frac{(3.14)(5.23)^2}{4} = 21.5 \text{ cm}^2 \quad \text{and}$$

$$A = \pi D x = (3.14)(5.23)(32) = 526 \text{ cm}^2$$

$$\ln\left(\frac{0.0034}{0.0034 - \bar{c}_{A_x}}\right) = \frac{(7.7 \times 10^{-5})(526)}{(0.170)(21.5)}$$

$$= 0.0111$$

$$\bar{c}_{A_x} = 0.0034 - \frac{0.0034}{e^{0.0111}} = 0.000038 \text{ g/cm}^3$$

Thus, the concentration of benzoic acid in the water leaving the cast tube is far from saturation.

§3.5 MASS TRANSFER IN TURBULENT FLOW

The two previous sections described mass transfer in stagnant media (§3.3) and laminar flow (§3.4), where in (3-1), only two mechanisms needed to be considered: molecular diffusion and bulk flow, with the latter often ignored. For both cases, rates of mass transfer can be calculated theoretically using Fick's law of diffusion. In the chemical industry, turbulent flow is more common because it includes eddy diffusion, which results in much higher heat and mass-transfer rates, and thus, requires smaller equipment. Lacking a fundamental theory for eddy diffusion, estimates of mass-transfer rates rely on empirical correlations developed from experimental

Table 3.13   Some Useful Dimensionless Groups

Name	Formula	Meaning	Analogy
Fluid Mechanics			
Drag Coefficient	$C_D = \dfrac{2F_D}{Au^2\rho}$	$\dfrac{\text{Drag force}}{\text{Projected area} \times \text{Velocity head}}$	
Fanning Friction Factor	$f = \dfrac{\Delta P}{L}\dfrac{D}{2\bar{u}^2\rho}$	$\dfrac{\text{Pipe wall shear stress}}{\text{Velocity head}}$	
Froude Number	$N_{Fr} = \dfrac{\bar{u}^2}{gL}$	$\dfrac{\text{Inertial force}}{\text{Gravitational force}}$	
Reynolds Number	$N_{Re} = \dfrac{L\bar{u}\rho}{\mu} = \dfrac{L\bar{u}}{v} = \dfrac{LG}{\mu}$	$\dfrac{\text{Inertial force}}{\text{Viscous force}}$	
Weber Number	$N_{We} = \dfrac{\bar{u}^2\rho L}{\sigma}$	$\dfrac{\text{Inertial force}}{\text{Surface-tension force}}$	
Heat Transfer			
j-Factor for Heat Transfer	$j_H = N_{St_H}(N_{Pr})^{2/3}$		j_M
Nusselt Number	$N_{Nu} = \dfrac{hL}{k}$	$\dfrac{\text{Convective heat transfer}}{\text{Conductive heat transfer}}$	N_{Sh}
Peclet Number for Heat Transfer	$N_{Pe_H} = N_{Re}N_{Pr} = \dfrac{L\bar{u}\rho C_p}{k}$	$\dfrac{\text{Bulk transfer of heat}}{\text{Conductive heat transfer}}$	N_{Pe_M}
Prandtl Number	$N_{Pr} = \dfrac{C_p\mu}{k} = \dfrac{v}{\alpha}$	$\dfrac{\text{Momentum diffusivity}}{\text{Thermal diffusivity}}$	N_{Sc}
Stanton Number for Heat Transfer	$N_{St_H} = \dfrac{N_{Nu}}{N_{Re}N_{Pr}} = \dfrac{h}{C_p G}$	$\dfrac{\text{Heat transfer}}{\text{Thermal capacity}}$	N_{St_M}
Mass Transfer			
j-Factor for Mass Transfer (analogous to the j-Factor for Heat Transfer)	$j_M = N_{St_M}(N_{Sc})^{2/3}$		j_H
Lewis Number	$N_{Le} = \dfrac{N_{Sc}}{N_{Pr}} = \dfrac{k}{\rho C_p D_{AB}} = \dfrac{\alpha}{D_{AB}}$	$\dfrac{\text{Thermal diffusivity}}{\text{Mass diffusivity}}$	
Peclet Number for Mass Transfer (analogous to the Peclet Number for Heat Transfer)	$N_{Pe_M} = N_{Re}N_{Sc} = \dfrac{L\bar{u}}{D_{AB}}$	$\dfrac{\text{Bulk transfer of mass}}{\text{Molecular diffusion}}$	N_{Pe_H}
Schmidt Number (analogous to the Prandtl Number)	$N_{Sc} = \dfrac{\mu}{\rho D_{AB}} = \dfrac{v}{D_{AB}}$	$\dfrac{\text{Momentum diffusivity}}{\text{Mass diffusivity}}$	N_{Pr}
Sherwood Number (analogous to the Nusselt Number)	$N_{Sh} = \dfrac{k_c L}{D_{AB}}$	$\dfrac{\text{Convective mass transfer}}{\text{Molecular diffusion}}$	N_{Nu}
Stanton Number for Mass Transfer (analogous to the Stanton Number for Heat Transfer)	$N_{St_M} = \dfrac{N_{Sh}}{N_{Re}N_{Sc}} = \dfrac{k_c}{\bar{u}\rho}$	$\dfrac{\text{Mass transfer}}{\text{Mass capacity}}$	N_{St_H}

$L =$ characteristic length, $G =$ mass velocity $= \bar{u}\rho$,   Subscripts: M = mass transfer   H = heat transfer

data. These correlations are comprised of the same dimensionless groups of §3.4 and use analogies with heat and momentum transfer. For reference as this section is presented, the most useful dimensionless groups for fluid mechanics, heat transfer, and mass transfer are listed in Table 3.13. Note that most of the dimensionless groups used in empirical equations for mass transfer are analogous to dimensionless groups used in heat transfer. The Reynolds number from fluid mechanics is used widely in empirical equations of heat and mass transfer.

As shown by a famous dye experiment conducted by Osborne Reynolds [50] in 1883, a fluid in laminar flow moves parallel to the solid boundaries in streamline patterns. Every fluid particle moves with the same velocity along a streamline, and there are no normal-velocity components. For a Newtonian fluid in laminar flow, momentum, heat, and mass transfer are by molecular transport, governed by Newton's law of viscosity, Fourier's law of heat conduction, and Fick's law of molecular diffusion, as described in the previous section.

In turbulent flow, where transport processes are orders of magnitude higher than in laminar flow, streamlines no longer exist, except near a wall, and eddies of fluid, which are large compared to the mean free path of the molecules in the fluid, mix with each other by moving from one region to another in fluctuating motion. This eddy mixing by velocity fluctuations occurs not only in the direction of flow but also in directions normal to flow, with the former referred to as axial transport but with the latter being of more interest. Momentum, heat, and mass transfer now occur by the two parallel mechanisms given in (3-1): (1) molecular diffusion, which is slow; and (2) turbulent or eddy diffusion, which is rapid except near a solid surface, where the flow velocity accompanying turbulence tends to zero. Superimposed on molecular and eddy diffusion is (3) mass transfer by bulk flow, which may or may not be significant.

In 1877, Boussinesq [51] modified Newton's law of viscosity to include a parallel eddy or turbulent viscosity, μ_t. Analogous expressions were developed for turbulent-flow heat and mass transfer. For flow in the x-direction and transport in the z-direction normal to flow, these expressions are written in flux form (in the absence of bulk flow in the z-direction) as:

$$\tau_{zx} = -(\mu + \mu_t)\frac{du_x}{dz} \qquad (3\text{-}151)$$

$$q_z = -(k + k_t)\frac{dT}{dz} \qquad (3\text{-}152)$$

$$N_{A_z} = -(D_{AB} + D_t)\frac{dc_A}{dz} \qquad (3\text{-}153)$$

where the double subscript zx on the shear stress, τ, stands for x-momentum in the z-direction. The molecular contributions, μ, k, and D_{AB}, are properties of the fluid and depend on chemical composition, temperature, and pressure; the turbulent contributions, μ_t, k_t, and D_t, depend on the mean fluid velocity in the flow direction and on position in the fluid with respect to the solid boundaries.

In 1925, Prandtl [52] developed an expression for μ_t in terms of an eddy mixing length, l, which is a function of position and is a measure of the average distance that an eddy travels before it loses its identity and mingles with other eddies. The mixing length is analogous to the mean free path of gas molecules, which is the average distance a molecule travels before it collides with another molecule. By analogy, the same mixing length is valid for turbulent-flow heat transfer and mass transfer. To use this analogy, (3-151) to (3-153) are rewritten in diffusivity form:

$$\frac{\tau_{zx}}{\rho} = -(v + \epsilon_M)\frac{du_x}{dz} \qquad (3\text{-}154)$$

$$\frac{q_z}{C_p\rho} = -(\alpha + \epsilon_H)\frac{dT}{dz} \qquad (3\text{-}155)$$

$$N_{A_z} = -(D_{AB} + \epsilon_D)\frac{dc_A}{dz} \qquad (3\text{-}156)$$

where ϵ_M, ϵ_H, and ϵ_D are momentum, heat, and mass eddy diffusivities, respectively; v is the momentum diffusivity (kinematic viscosity, μ/ρ); and α is the thermal diffusivity, $k/\rho C_P$. As an approximation, the three eddy diffusivities may be assumed equal. This is valid for ϵ_H and ϵ_D, but data indicate that $\epsilon_M/\epsilon_H = \epsilon_M/\epsilon_D$ is sometimes less than 1.0 and as low as 0.5 for turbulence in a free jet.

§3.5.1 Reynolds Analogy

If (3-154) to (3-156) are applied at a solid boundary, they can be used to determine transport fluxes based on transport coefficients, with driving forces from the wall (or interface), i, at $z = 0$, to the bulk fluid, designated with an overbar, $\bar{\ }$:

$$\frac{\tau_{zx}}{\bar{u}_x} = -(v + \epsilon_M)\frac{d(\rho u_x/\bar{u}_x)}{dz}\bigg|_{z=0} = \frac{f\rho}{2}\bar{u}_x \qquad (3\text{-}157)$$

$$q_z = -(\alpha + \epsilon_H)\frac{d(\rho C_P T)}{dz}\bigg|_{z=0} = h(T_i - \overline{T}) \qquad (3\text{-}158)$$

$$N_{A_z} = -(D_{AB} + \epsilon_D)\frac{dc_A}{dz}\bigg|_{z=0} = k_c(c_A - \bar{c}_A) \qquad (3\text{-}159)$$

To develop useful analogies, it is convenient to use dimensionless velocity, temperature, and solute concentration, defined by

$$\theta = \frac{u_x}{\bar{u}_x} = \frac{T_i - T}{T_i - \overline{T}} = \frac{c_{A_j} - c_A}{c_{A_i} - \bar{c}_A} \qquad (3\text{-}160)$$

If (3-160) is substituted into (3-157) to (3-159),

$$\begin{aligned}\frac{\partial\theta}{\partial z}\bigg|_{z=0} &= \frac{f\bar{u}_x}{2(v + \epsilon_M)} = \frac{h}{\rho C_P(\alpha + \epsilon_H)} \\ &= \frac{k_c}{(D_{AB} + \epsilon_D)}\end{aligned} \qquad (3\text{-}161)$$

which defines analogies among momentum, heat, and mass transfer. If the three eddy diffusivities are equal and molecular diffusivities are everywhere negligible or equal, i.e., $v = \alpha = D_{AB}$, (3-161) simplifies to

$$\frac{f}{2} = \frac{h}{\rho C_P\bar{u}_x} = \frac{k_c}{\bar{u}_x} \qquad (3\text{-}162)$$

Equation (3-162) defines the Stanton number for heat transfer listed in Table 3.13,

$$N_{St_H} = \frac{h}{\rho C_P\bar{u}_x} = \frac{h}{GC_P} \qquad (3\text{-}163)$$

where G = mass velocity = $\bar{u}_x\rho$. The Stanton number for mass transfer is

$$\boxed{N_{St_M} = \frac{k_c}{\bar{u}_x} = \frac{k_c\rho}{G}} \qquad (3\text{-}164)$$

Equation (3-162) is referred to as the *Reynolds analogy*. Its development is significant, but its application for the estimation of heat- and mass-transfer coefficients from measurements of the Fanning friction factor for turbulent flow is valid only when $N_{Pr} = v/\alpha = N_{Sc} = v/D_{AB} = 1$. Thus, the Reynolds analogy has limited practical value and is rarely used. Reynolds postulated its existence in 1874 [53] and derived it in 1883 [50].

§3.5.2 Chilton–Colburn Analogy

A widely used extension of the Reynolds analogy to Prandtl and Schmidt numbers other than 1 was devised in the 1930s by Colburn [54] for heat transfer and by Chilton and Colburn [55] for mass transfer. Using experimental data, they corrected the Reynolds analogy for differences in dimensionless velocity, temperature, and concentration distributions by incorporating the Prandtl number, N_{Pr}, and the Schmidt number, N_{Sc}, into (3-162) to define empirically the following three *j-factors* included in Table 3.13.

$$j_M \equiv \frac{f}{2} = j_H \equiv \frac{h}{G C_P} (N_{Pr})^{2/3}$$
$$= j_D \equiv \frac{k_c \rho}{G} (N_{Sc})^{2/3} \qquad (3\text{-}165)$$

Equation (3-165) is the *Chilton–Colburn analogy* or the Colburn analogy for estimating transport coefficients for turbulent flow. For $N_{Pr} = N_{Sc} = 1$, (3-165) equals (3-162).

From experimental studies, the *j*-factors depend on the geometric configuration and the Reynolds number, N_{Re}. Based on decades of experimental transport data, the following representative *j*-factor correlations for turbulent transport to or from smooth surfaces have evolved. Additional correlations are presented in later chapters. These correlations are reasonably accurate for N_{Pr} and N_{Sc} in the range 0.5 to 10.

1. Flow through a straight, circular tube of inside diameter D:

$$j_M = j_H = j_D = 0.023 (N_{Re})^{-0.2} \qquad (3\text{-}166)$$

for $10,000 < N_{Re} = DG/\mu < 1,000,000$

2. Average transport coefficients for flow across a flat plate of length L:

$$j_M = j_H = j_D = 0.037 (N_{Re})^{-0.2} \qquad (3\text{-}167)$$

for $5 \times 10^5 < N_{Re} = L u_o/\mu < 5 \times 10^8$

3. Average transport coefficients for flow normal to a long, circular cylinder of diameter D, where the drag coefficient includes both form drag and skin friction, but only the skin friction contribution applies to the analogy:

$$(j_M)_{\text{skin friction}} = j_H = j_D = 0.193 (N_{Re})^{-0.382} \qquad (3\text{-}168)$$

for $4,000 < N_{Re} < 40,000$

$$(j_M)_{\text{skin friction}} = j_H = j_D = 0.0266 (N_{Re})^{-0.195} \qquad (3\text{-}169)$$

for $40,000 < N_{Re} < 250,000$

$$\text{with} \qquad N_{Re} = \frac{DG}{\mu}$$

4. Average transport coefficients for flow past a single sphere of diameter D:

$$(j_M)_{\text{skin friction}} = j_H = j_D = 0.37 (N_{Re})^{-0.4}$$
$$\text{for} \quad 20 < N_{Re} = \frac{DG}{\mu} < 100,000 \qquad (3\text{-}170)$$

5. Average transport coefficients for flow through beds packed with spherical particles of uniform size D_P:

$$j_H = j_D = 1.17 (N_{Re})^{-0.415}$$
$$\text{for} \quad 10 < N_{Re} = \frac{D_P G}{\mu} < 2,500 \qquad (3\text{-}171)$$

The above correlations are plotted in Figure 3.17, where the curves are not widely separated but do not coincide because of necessary differences in Reynolds number definitions. When using the correlations in the presence of appreciable temperature and/or composition differences, Chilton and Colburn recommend that N_{Pr} and N_{Sc} be evaluated at the average conditions from the surface to the bulk stream.

§3.5.3 Other Analogies

New theories have led to improvements of the Reynolds analogy to give expressions for the Fanning friction factor and Stanton numbers for heat and mass transfer that are less empirical than the Chilton–Colburn analogy. The first major improvement was by Prandtl [56] in 1910, who divided the flow into two regions: (1) a thin laminar-flow *sublayer* of thickness δ next to the wall boundary, where only molecular transport occurs; and (2) a turbulent region dominated by eddy transport, with $\epsilon_M = \epsilon_H = \epsilon_D$.

Further improvements to the Reynolds analogy were made by von Karman, Martinelli, and Deissler, as discussed in

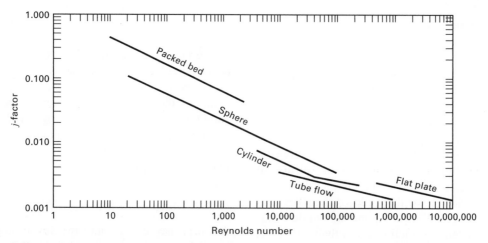

Figure 3.17 Chilton–Colburn *j*-factor correlations.

detail by Knudsen and Katz [47]. The first two investigators inserted a buffer zone between the laminar sublayer and turbulent core. Deissler gradually reduced the eddy diffusivities as the wall was approached. Other advances were made by van Driest [64], who used a modified form of the Prandtl mixing length; Reichardt [65], who eliminated the zone concept by allowing the eddy diffusivities to decrease continuously from a maximum to zero at the wall; and Friend and Metzner [57], who obtained improved accuracy at Prandtl and Schmidt numbers to 3,000. Their results for flow through a circular tube are

$$N_{St_H} = \frac{f/2}{1.20 + 11.8\sqrt{f/2}(N_{Pr} - 1)(N_{Pr})^{-1/3}} \tag{3-172}$$

$$N_{St_M} = \frac{f/2}{1.20 + 11.8\sqrt{f/2}(N_{Sc} - 1)(N_{Sc})^{-1/3}} \tag{3-173}$$

where the Fanning friction factor can be estimated over Reynolds numbers from 10,000 to 10,000,000 using the empirical correlation of Drew, Koo, and McAdams [66],

$$f = 0.00140 + 0.125(N_{Re})^{-0.32} \tag{3-174}$$

which fits the experimental data of Nikuradse [67] and is preferred over (3-165) with (3-166), which is valid only to N_{Re} = 1,000,000. For two- and three-dimensional turbulent-flow problems, some success has been achieved with the κ (kinetic energy of turbulence)–ϵ (rate of dissipation) model of Launder and Spalding [68], which is widely used in computational fluid dynamics (CFD) computer programs.

§3.5.4 Theoretical Analogy of Churchill and Zajic

An alternative to (3-154) to (3-156) for developing equations for turbulent flow is to start with *time-averaged* equations of Newton, Fourier, and Fick. For example, consider a form of Newton's law of viscosity for molecular and turbulent transport of momentum in parallel, where, in a turbulent-flow field in the axial x-direction, instantaneous velocity components u_x and u_z are

$$u_x = \bar{u}_x + u_x'$$
$$u_z = u_z'$$

The "overbarred" component is the time-averaged (mean) local velocity, and the primed component is the local fluctuating velocity that denotes instantaneous deviation from the local mean value. The mean velocity in the perpendicular z-direction is zero. The mean local velocity in the x-direction over a long period Θ of time θ is given by

$$\bar{u}_x = \frac{1}{\Theta}\int_0^\Theta u_x d\theta = \frac{1}{\Theta}\int_0^\Theta (\bar{u}_x + u_x')d\theta \tag{3-175}$$

Time-averaged fluctuating components u_x' and u_z' are zero.

The local instantaneous rate of momentum transfer by turbulence in the z-direction of x-direction turbulent momentum per unit area at constant density is

$$\rho u_z'(\bar{u}_x + u_x') \tag{3-176}$$

The time-average of this turbulent momentum transfer is equal to the turbulent component of the shear stress, τ_{zx_t},

$$\tau_{zx_t} = \frac{\rho}{\Theta}\int_0^\Theta u_z'(\bar{u}_x + u_x')d\theta$$
$$= \frac{\rho}{\Theta}\left[\int_0^\Theta u_z'(\bar{u}_x)d\theta + \int_0^\Theta u_z'(u_x')d\theta\right] \tag{3-177}$$

Because the time-average of the first term is zero, (3-177) reduces to

$$\tau_{zx_t} = \rho(\overline{u_z'u_x'}) \tag{3-178}$$

which is referred to as a Reynolds stress. Combining (3-178) with the molecular component of momentum transfer gives the following turbulent-flow form of Newton's law of viscosity, where the second term on the right-hand side accounts for turbulence,

$$\tau_{zx} = -\mu\frac{du_x}{dz} + \rho(\overline{u_z'u_x'}) \tag{3-179}$$

If (3-179) is compared to (3-151), it is seen that an alternative approach to turbulence is to develop a correlating equation for the Reynolds stress, $(\overline{u_z'u_x'})$ first introduced by Churchill and Chan [73], rather than an expression for turbulent viscosity μ_t. This stress is a complex function of position and rate of flow and has been correlated for fully developed turbulent flow in a straight, circular tube by Heng, Chan, and Churchill [69]. In generalized form, with tube radius a and $y = (a - z)$ representing the distance from the inside wall to the center of the tube, their equation is

$$(\overline{u_z'u_x'})^{++} = \left(\left[0.7\left(\frac{y^+}{10}\right)^3\right]^{-8/7} + \left|\exp\left\{\frac{-1}{0.436y^+}\right\}\right.\right.$$
$$\left.\left. - \frac{1}{0.436a^+}\left(1 + \frac{6.95y^+}{a^+}\right)\right|^{-8/7}\right)^{-7/8} \tag{3-180}$$

where
$$\begin{aligned}(\overline{u_z'u_x'})^{++} &= -\rho\overline{u_z'u_x'}/\tau \\ a^+ &= a(\tau_w\rho)^{1/2}/\mu \\ y^+ &= y(\tau_w\rho)^{1/2}/\mu\end{aligned}$$

Equation (3-180) is an accurate representation of turbulent flow because it is based on experimental data and numerical simulations described by Churchill and Zajic [70] and Churchill [71]. From (3-142) and (3-143), the shear stress at the wall, τ_w, is related to the Fanning friction factor by

$$f = \frac{2\tau_w}{\rho\bar{u}_x^2} \tag{3-181}$$

where $\bar{u}_x$ is the flow-average velocity in the axial direction. Combining (3-179) with (3-181) and performing the required integrations, both numerically and analytically, leads to the following implicit equation for the Fanning friction factor as a function of the Reynolds number, $N_{Re} = 2a\bar{u}_x\rho/\mu$:

Table 3.14 Comparison of Fanning Friction Factors for Fully Developed Turbulent Flow in a Smooth, Straight, Circular Tube

N_{Re}	f, Drew et al. (3-174)	f, Chilton–Colburn (3-166)	f, Churchill–Zajic (3-182)
10,000	0.007960	0.007291	0.008087
100,000	0.004540	0.004600	0.004559
1,000,000	0.002903	0.002902	0.002998
10,000,000	0.002119	0.001831	0.002119
100,000,000	0.001744	0.001155	0.001573

$$\left(\frac{2}{f}\right)^{1/2} = 3.2 - 227\frac{\left(\frac{2}{f}\right)^{1/2}}{\frac{N_{Re}}{2}} + 2500\left[\frac{\left(\frac{2}{f}\right)^{1/2}}{\frac{N_{Re}}{2}}\right]^2 \tag{3-182}$$

$$+ \frac{1}{0.436}\ln\left[\frac{\frac{N_{Re}}{2}}{\left(\frac{2}{f}\right)^{1/2}}\right]$$

This equation is in agreement with experimental data over a Reynolds number range of 4,000–3,000,000 and can be used up to a Reynolds number of 100,000,000. Table 3.14 presents a comparison of the Churchill–Zajic equation, (3-182), with (3-174) of Drew et al. and (3-166) of Chilton and Colburn. Equation (3-174) gives satisfactory agreement for Reynolds numbers from 10,000 to 10,000,000, while (3-166) is useful only for Reynolds numbers from 100,000 to 1,000,000.

Churchill and Zajic [70] show that if the equation for the conservation of energy is time-averaged, a turbulent-flow form of Fourier's law of conduction can be obtained with the fluctuation term $(\overline{u'_z T'})$. Similar time-averaging leads to a turbulent-flow form of Fick's law with $(\overline{u'_z c'_A})$. To extend (3-180) and (3-182) to obtain an expression for the Nusselt number for turbulent-flow convective heat transfer in a straight,

circular tube, Churchill and Zajic employ an analogy that is free of empiricism but not exact. The result for Prandtl numbers greater than 1 is

$$N_{Nu} = \frac{1}{\left(\frac{N_{Pr_t}}{N_{Pr}}\right)\frac{1}{N_{Nu_1}} + \left[1 - \left(\frac{N_{Pr_t}}{N_{Pr}}\right)^{2/3}\right]\frac{1}{N_{Nu_\infty}}} \tag{3-183}$$

where, from Yu, Ozoe, and Churchill [72],

$$N_{Pr_t} = \text{turbulent Prandtl number} = 0.85 + \frac{0.015}{N_{Pr}} \tag{3-184}$$

which replaces $(\overline{u'_z T'})$, as introduced by Churchill [74],

$$N_{Nu_1} = \text{Nusselt number for}(N_{Pr} = N_{Pr_t})$$

$$= \frac{N_{Re}\left(\frac{f}{2}\right)}{1 + 145\left(\frac{2}{f}\right)^{-5/4}} \tag{3-185}$$

$$N_{Nu_\infty} = \text{Nusselt number for}(N_{Pr} = \infty)$$

$$= 0.07343\left(\frac{N_{Pr}}{N_{Pr_t}}\right)^{1/3}N_{Re}\left(\frac{f}{2}\right)^{1/2} \tag{3-186}$$

The accuracy of (3-183) is due to (3-185) and (3-186), which are known from theoretical considerations. Although (3-184) is somewhat uncertain, its effect on (3-183) is negligible.

A comparison is made in Table 3.15 of the Churchill et al. correlation (3-183) with that of Friend and Metzner (3-172) and that of Chilton and Colburn (3-166), where, from Table 3.13, $N_{Nu} = N_{St}N_{Re}N_{Pr}$.

In Table 3.15, at a Prandtl number of 1, which is typical of low-viscosity liquids and close to that of most gases, the Chilton–Colburn correlation is within 10% of the Churchill–Zajic equation for Reynolds numbers up to 1,000,000. Beyond that, serious deviations occur (25% at $N_{Re} = 10,000,000$

Table 3.15 Comparison of Nusselt Numbers for Fully Developed Turbulent Flow in a Smooth, Straight, Circular Tube

Prandtl number, $N_{Pr} = 1$			
N_{Re}	N_{Nu}, Friend–Metzner (3-172)	N_{Nu}, Chilton–Colburn (3-166)	N_{Nu}, Churchill–Zajic (3-183)
10,000	33.2	36.5	37.8
100,000	189	230	232
1,000,000	1210	1450	1580
10,000,000	8830	9160	11400
100,000,000	72700	57800	86000

Prandt number, $N_{Pr} = 1000$			
N_{Re}	N_{Nu}, Friend–Metzner (3-172)	N_{Nu}, Chilton–Colburn (3-166)	N_{Nu}, Churchill–Zajic (3-183)
10,000	527	365	491
100,000	3960	2300	3680
1,000,000	31500	14500	29800
10,000,000	267800	91600	249000
100,000,000	2420000	578000	2140000

and almost 50% at $N_{Re} = 100,000,000$). Deviations of the Friend–Metzner correlation vary from 15% to 30% over the entire range of Reynolds numbers. At all Reynolds numbers, the Churchill–Zajic equation predicts higher Nusselt numbers and, therefore, higher heat-transfer coefficients.

At a Prandtl number of 1,000, which is typical of high-viscosity liquids, the Friend–Metzner correlation is in fairly close agreement with the Churchill–Zajic equation. The Chilton–Colburn correlation deviates over the entire range of Reynolds numbers, predicting values ranging from 74 to 27% of those from the Churchill–Zajic equation as the Reynolds number increases. The Chilton–Colburn correlation should not be used at high Prandtl numbers for heat transfer or at high Schmidt numbers for mass transfer.

The Churchill–Zajic equation for predicting the Nusselt number provides a power dependence on the Reynolds number. This is in contrast to the typically cited constant exponent of 0.8 for the Chilton–Colburn correlation. For the Churchill–Zajic equation, at $N_{Pr} = 1$, the exponent increases with Reynolds number from 0.79 to 0.88; at a Prandtl number of 1,000, the exponent increases from 0.87 to 0.93.

Extension of the Churchill–Zajic equation to low Prandtl numbers typical of molten metals, and to other geometries is discussed by Churchill [71], who also considers the effect of boundary conditions (e.g., constant wall temperature and uniform heat flux) at low-to-moderate Prandtl numbers.

For calculation of convective mass-transfer coefficients, k_c, for turbulent flow of gases and liquids in straight, smooth, circular tubes, it is recommended that the Churchill–Zajic equation be employed by applying the analogy between heat and mass transfer. Thus, as illustrated in the following example, in (3-183) to (3-186), using Table 3.13, the Sherwood number is substituted for the Nusselt number, and the Schmidt number is substituted for the Prandtl number.

EXAMPLE 3.16 Analogies for Turbulent Transport.

Linton and Sherwood [49] conducted experiments on the dissolving of tubes of cinnamic acid (A) into water (B) flowing turbulently through the tubes. In one run, with a 5.23-cm-i.d. tube, $N_{Re} = 35,800$, and $N_{Sc} = 1,450$, they measured a Stanton number for mass transfer, N_{St_M}, of 0.0000351. Compare this experimental value with predictions by the Reynolds, Chilton–Colburn, and Friend–Metzner analogies and the Churchill–Zajic equation.

Solution

From either (3-174) or (3-182), the Fanning friction factor is 0.00576.
Reynolds analogy. From (3-162), $N_{St_M} = f/2 = 0.00576/2 = 0.00288$, which, as expected, is in very poor agreement with the experimental value because the effect of the large Schmidt number is ignored.

Chilton–Colburn analogy. From (3-165),

$$N_{St_M} = \left(\frac{f}{2}\right)/(N_{Sc})^{2/3} = \left(\frac{0.00576}{2}\right)/(1450)^{2/3} = 0.0000225,$$

which is 64% of the experimental value.

Friend–Metzner analogy: From (3-173), $N_{St_M} = 0.0000350$, which is almost identical to the experimental value.

Churchill–Zajic equation. Using mass-transfer analogs,

(3-184) gives $N_{Sc_t} = 0.850$, (3-185) gives $N_{Sh_1} = 94$,

(3-186) gives $N_{Sh_\infty} = 1686$, and (3-183) gives $N_{Sh} = 1680$

From Table 3.13,

$$N_{St_M} = \frac{N_{Sh}}{N_{Re}N_{Sc}} = \frac{1680}{(35800)(1450)} = 0.0000324,$$

which is an acceptable 92% of the experimental value.

§3.6 MODELS FOR MASS TRANSFER IN FLUIDS WITH A FLUID–FLUID INTERFACE

The three previous sections considered mass transfer mainly between solids and fluids, where the interface was a smooth, solid surface. Applications occur in adsorption, drying, leaching, and membrane separations. Of importance in other separation operations is mass transfer across a fluid–fluid interface. Such interfaces exist in absorption, distillation, extraction, and stripping, where, in contrast to fluid–solid interfaces, turbulence may persist to the interface. The following theoretical models have been developed to describe such phenomena in fluids with a fluid-to-fluid interface. There are many equations in this section and the following section, but few applications. However, use of these equations to design equipment is found in many examples in: Chapter 6 on absorption and stripping; Chapter 7 on distillation; and Chapter 8 on liquid–liquid extraction.

§3.6.1 Film Theory

A model for turbulent mass transfer to or from a fluid-phase boundary was suggested in 1904 by Nernst [58], who postulated that the resistance to mass transfer in a given turbulent fluid phase is in a thin, relatively stagnant region at the interface, called a film. This is similar to the laminar sublayer that forms when a fluid flows in the turbulent regime parallel to a flat plate. It is shown schematically in Figure 3.18a for a gas–liquid interface, where the gas is component A, which diffuses into non-volatile liquid B. Thus, a process of absorption of A into liquid B takes place, without vaporization of B, and there is no resistance to mass transfer of A in the gas phase, because it is pure A. At the interface, phase equilibrium is assumed, so the concentration of A at the interface, c_{A_i}, is related to the partial pressure of A at the interface, p_A, by a solubility relation like Henry's law, $c_{A_i} = H_A p_A$. In the liquid film of thickness δ, molecular diffusion occurs with a driving force of $c_{A_i} - c_{A_b}$, where c_{A_b} is the bulk-average concentration of A in the liquid. Since the film is assumed to be very thin, all of the diffusing A is assumed to pass through the film and into the bulk liquid. Accordingly, integration of Fick's first law, (3-3a), gives

$$J_A = \frac{D_{AB}}{\delta}(c_{A_i} - c_{A_b}) = \frac{cD_{AB}}{\delta}(x_{A_i} - x_{A_b}) \quad (3\text{-}187)$$

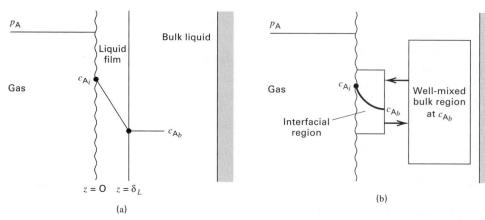

Figure 3.18 Theories for mass transfer from a fluid–fluid interface into a liquid: (a) film theory; (b) penetration and surface-renewal theories.

If the liquid phase is dilute in A, the bulk-flow effect can be neglected so that (3-187) applies to the total flux, and the concentration gradient is linear, as in Figure 3.18a.

$$N_A = \frac{D_{AB}}{\delta}(c_{A_i} - c_{A_b}) = \frac{cD_{AB}}{\delta}(x_{A_i} - c_{A_b}) \qquad (3\text{-}188)$$

If the bulk-flow effect is not negligible, then, from (3-31),

$$N_A = \frac{cD_{AB}}{\delta}\ln\left[\frac{1 - x_{A_b}}{1 - x_{A_i}}\right] = \frac{cD_{AB}}{\delta(1 - x_A)_{LM}}(x_{A_i} - x_{A_b})$$

$$(3\text{-}189)$$

where

$$(1 - x_A)_{LM} = \frac{x_{A_i} - x_{A_b}}{\ln[(1 - x_{A_b})/(1 - x_{A_i})]} = (x_B)_{LM} \quad (3\text{-}190)$$

In practice, the ratios D_{AB}/δ in (3-188) and $D_{AB}/[\delta\,(1 - x_A)_{LM}]$ in (3-189) are replaced by empirical mass-transfer coefficients k_c and k_c', respectively, because the film thickness, δ, which depends on the flow conditions, is unknown. The subscript, c, on the mass-transfer coefficient refers to a concentration driving force, and the prime superscript denotes that k_c includes both diffusion mechanisms and the bulk-flow effect.

The film theory, which is easy to understand and apply, is often criticized because it predicts that the rate of mass transfer is proportional to molecular diffusivity. This dependency is at odds with experimental data, which indicate a dependency of D^n, where n ranges from 0.5 to 0.75. However, if D_{AB}/δ is replaced with k_c, which is then estimated from the Chilton–Colburn analogy (3-165), k_c is proportional to $D_{AB}^{2/3}$, which is in better agreement with experimental data. In effect, δ is not a constant but depends on D_{AB} (or N_{Sc}). Regardless of whether the criticism is valid, the film theory continues to be widely used in design of mass-transfer separation equipment.

EXAMPLE 3.17 Mass-Transfer Flux in a Packed Absorption Tower.

SO_2 is absorbed from air into water in a packed absorption tower. At a location in the tower, the mass-transfer flux is 0.0270 kmol

SO_2/m^2-h, and the liquid-phase mole fractions are 0.0025 and 0.0003, respectively, at the two-phase interface and in the bulk liquid. If the diffusivity of SO_2 in water is 1.7×10^{-5} cm²/s, determine the mass-transfer coefficient, k_c, and the corresponding film thickness, neglecting the bulk flow effect.

Solution

$$N_{SO_2} = \frac{0.027(1,000)}{(3,600)(100)^2} = 7.5 \times 10^{-7}\ \frac{\text{mol}}{\text{cm}^2\text{-s}}$$

For dilute conditions, the concentration of water is

$$c = \frac{1}{18.02} = 5.55 \times 10^{-2}\ \text{mol/cm}^3$$

From (3-188),

$$k_c = \frac{D_{AB}}{\delta} = \frac{N_A}{c(x_{A_i} - x_{A_b})}$$

$$= \frac{7.5 \times 10^{-7}}{5.55 \times 10^{-2}(0.0025 - 0.0003)} = 6.14 \times 10^{-3}\ \text{cm/s}$$

Therefore, $\quad \delta = \dfrac{D_{AB}}{k_c} = \dfrac{1.7 \times 10^{-5}}{6.14 \times 10^{-3}} = 0.0028$ cm

which is small and typical of turbulent-flow processes.

§3.6.2 Penetration Theory

A more realistic mass-transfer model is provided by Higbie's penetration theory [59], shown schematically in Figure 3.18b. The stagnant-film concept is replaced by Boussinesq eddies that: (1) move from the bulk liquid to the interface; (2) stay at the interface for a short, fixed period of time during which they remain static, allowing molecular diffusion to take place in a direction normal to the interface; and (3) leave the interface to mix with the bulk stream. When an eddy moves to the interface, it replaces a static eddy. Thus, eddies are alternately static and moving. Turbulence extends to the interface.

In the penetration theory, unsteady-state diffusion takes place at the interface during the time the eddy is static. This process is governed by Fick's second law, (3-68), with boundary conditions

$$c_A = c_{A_b} \quad \text{at} \quad t = 0 \quad \text{for} \quad 0 \leq z \leq \infty;$$
$$c_A = c_{A_i} \quad \text{at} \quad z = 0 \quad \text{for} \quad t > 0; \quad \text{and}$$
$$c_A = c_{A_b} \quad \text{at} \quad z = \infty \quad \text{for} \quad t > 0$$

These are the same boundary conditions as in unsteady-state diffusion in a semi-infinite medium. The solution is a rearrangement of (3-75):

$$\frac{c_{A_i} - c_A}{c_{A_i} - c_{A_b}} = \text{erf}\left(\frac{z}{2\sqrt{D_{AB}t_c}}\right) \tag{3-191}$$

where t_c = "contact time" of the static eddy at the interface during one cycle. The corresponding average mass-transfer flux of A in the absence of bulk flow is given by the following form of (3-79):

$$N_A = 2\sqrt{\frac{D_{AB}}{\pi t_c}}(c_{A_i} - c_{A_b}) \tag{3-192}$$

or
$$N_A = k_c(c_{A_i} - c_{A_b}) \tag{3-193}$$

Thus, the penetration theory gives

$$k_c = 2\sqrt{\frac{D_{AB}}{\pi t_c}} \tag{3-194}$$

which predicts that k_c is proportional to the square root of the diffusivity, which is at the lower limit of experimental data.

Penetration theory is most useful for bubble, droplet, or random-packing interfaces. For bubbles, the contact time, t_c, of the liquid surrounding the bubble is approximated by the ratio of bubble diameter to its rise velocity. An air bubble of 0.4-cm diameter rises through water at a velocity of about 20 cm/s, making the estimated contact time $0.4/20 = 0.02$ s. For a liquid spray, where no circulation of liquid occurs inside the droplets, contact time is the total time it takes the droplets to fall through the gas. For a packed tower, where the liquid flows as a film over random packing, mixing is assumed to occur each time the liquid film passes from one piece of packing to another. Resulting contact times are about 1 s. In the absence of any estimate for contact time, the mass-transfer coefficient is sometimes correlated by an empirical expression consistent with the 0.5 exponent on D_{AB}, as in (3-194), with the contact time replaced by a function of geometry and the liquid velocity, density, and viscosity.

EXAMPLE 3.18   Contact Time for Penetration Theory.

For the conditions of Example 3.17, estimate the contact time for Higbie's penetration theory.

Solution

From Example 3.17, $k_c = 6.14 \times 10^{-3}$ cm/s and $D_{AB} = 1.7 \times 10^{-5}$ cm^2/s. From a rearrangement of (3-194),

$$t_c = \frac{4D_{AB}}{\pi k_c^2} = \frac{4(1.7 \times 10^{-5})}{3.14(6.14 \times 10^{-3})^2} = 0.57 \text{ s}$$

---

§3.6.3 Surface-Renewal Theory

The penetration theory is inadequate because the assumption of a constant contact time for all eddies that reach the surface is not reasonable, especially for stirred tanks, contactors with random packings, and bubble and spray columns where bubbles and droplets cover a range of sizes. In 1951, Danckwerts [60] suggested an improvement to the penetration theory that involves the replacement of constant eddy contact time with the assumption of a residence-time distribution, wherein the probability of an eddy at the surface being replaced by a fresh eddy is independent of the age of the surface eddy.

Following Levenspiel's [61] treatment of residence-time distribution, let $F(t)$ be the fraction of eddies with a contact time of less than t. For $t = 0$, $F\{t\} = 0$, and $F\{t\}$ approaches 1 as t goes to infinity. A plot of $F\{t\}$ versus t, as shown in Figure 3.19, is a residence-time or age distribution. If $F\{t\}$ is differentiated with respect to t,

$$\phi\{t\} = dF\{t\}/dt \tag{3-195}$$

where $\phi\{t\}dt$ = the probability that a given surface eddy will have a residence time t. The sum of probabilities is

$$\int_0^\infty \phi\{t\}dt = 1 \tag{3-196}$$

Typical plots of $F\{t\}$ and $\phi\{t\}$ are shown in Figure 3.19, where $\phi\{t\}$ is similar to a normal probability curve.

For steady-state flow into and out of a well-mixed vessel, Levenspiel shows that

$$F\{t\} = 1 - e^{-t/\bar{t}} \tag{3-197}$$

where $\bar{t}$ is the average residence time. This function forms the basis, in reaction engineering, of the ideal model of a continuous, stirred-tank reactor (CSTR). Danckwerts selected the same model for his surface-renewal theory, using the corresponding $\phi\{t\}$ function:

$$\phi\{t\} = se^{-st} \tag{3-198}$$

where
$$s = 1/\bar{t} \tag{3-199}$$

is the fractional rate of surface renewal. As shown in Example 3.19 below, plots of (3-197) and (3-198) are much different from those in Figure 3.19.

The instantaneous mass-transfer rate for an eddy of age t is given by (3-192) for penetration theory in flux form as

$$N_{A_t} = \sqrt{\frac{D_{AB}}{\pi t}}(c_{A_i} - c_{A_b}) \tag{3-200}$$

The integrated average rate is

$$(N_A)_{\text{avg}} = \int_0^\infty \phi\{t\}N_{A_t}dt \tag{3-201}$$

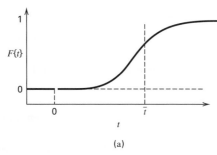

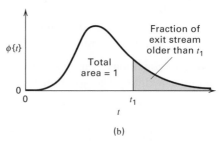

(a)

(b)

Figure 3.19 Residence-time distribution plots: (a) typical F curve; (b) typical age distribution.
[Adapted from O. Levenspiel, *Chemical Reaction Engineering*, 2nd ed., John Wiley & Sons, New York (1972).]

Combining (3-198), (3-200), and (3-201) and integrating:

$$(N_A)_{avg} = \sqrt{D_{AB}s}(c_{A_i} - c_{A_b}) \quad (3\text{-}202)$$

Thus,

$$k_c = \sqrt{D_{AB}s} \quad (3\text{-}203)$$

The surface-renewal theory predicts the same dependency of the mass-transfer coefficient on diffusivity as the penetration theory. Unfortunately, s, the fractional rate of surface renewal, is as elusive a parameter as the constant contact time, t_c.

EXAMPLE 3.19 Application of Surface-Renewal Theory.

For the conditions of Example 3.17, estimate the fractional rate of surface renewal, s, for Danckwert's theory and determine the residence time and probability distributions.

Solution

From Example 3.17,

$$k_c = 6.14 \times 10^{-3}\,\text{cm/s} \quad \text{and} \quad D_{AB} = 1.7 \times 10^{-5}\,\text{cm}^2/\text{s}$$

From (3-203),

$$s = \frac{k_c^2}{D_{AB}} = \frac{(6.14 \times 10^{-3})^2}{1.7 \times 10^{-5}} = 2.22\,\text{s}^{-1}$$

Thus, the average residence time of an eddy at the surface is $1/2.22 = 0.45$ s.

From (3-198),

$$\phi\{t\} = 2.22e^{-2.22t} \quad (1)$$

From (3-197), the residence-time distribution is

$$F(t) = 1 - e^{-t/0.45} \quad (2)$$

where t is in seconds. Equations (1) and (2) are shown in Figure 3.20. These curves differ from the curves of Figure 3.19.

---

§3.6.4 Film-Penetration Theory

Toor and Marchello [62] combined features of the film, penetration, and surface-renewal theories into a film-penetration model, which predicts that the mass-transfer coefficient, k_c, varies from $\sqrt{D_{AB}}$ to D_{AB}, with the resistance to mass transfer residing in a film of fixed thickness δ. Eddies move to and from the bulk fluid and this film. Age distributions for time spent in the film are of the Higbie or Danckwerts type. Fick's second law, (3-68), applies, but the boundary conditions are now

$$c_A = c_{A_b} \quad \text{at} \quad t = 0 \quad \text{for} \quad 0 \le z \le \infty;$$
$$c_A = c_{A_i} \quad \text{at} \quad z = 0 \quad \text{for} \quad t > 0; \quad \text{and}$$
$$c_A = c_{A_b} \quad \text{at} \quad z = \delta \quad \text{for} \quad t > 0$$

They obtained the following infinite series solutions using Laplace transforms. For small values of time, t,

$$N_{A_{avg}} = k_c(c_{A_i} - c_{A_b}) = (c_{A_i} - c_{A_b})(sD_{AB})^{1/2}$$
$$\times \left[1 + 2\sum_{n=1}^{\infty} \exp\left(-2n\delta\sqrt{\frac{s}{D_{AB}}}\right)\right] \quad (3\text{-}204)$$

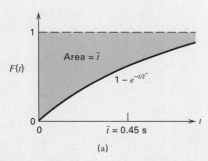

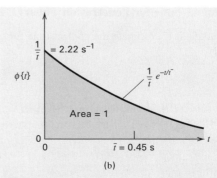

(a)

(b)

Figure 3.20 Age distribution curves for Example 3.19: (a) F curve; (b) $\phi\{t\}$ curve.

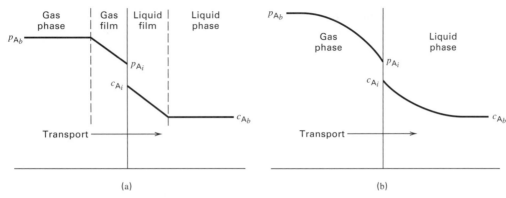

Figure 3.21 Concentration gradients for two-resistance theory: (a) film theory; (b) more realistic gradients.

converges rapidly. For large values of t, the following converges rapidly:

$$N_{A_{avg}} = k_c(c_{A_i} - c_{A_b}) = (c_{A_i} - c_{A_b})\left(\frac{D_{AB}}{\delta}\right)$$

$$\times \left[1 + 2\sum_{n=0}^{\infty} \frac{1}{1 + n^2\pi^2\frac{D_{AB}}{s\delta^2}}\right] \qquad (3\text{-}205)$$

In the limit for a high rate of surface renewal, $s\delta^2/D_{AB}$, (3-204) reduces to the surface-renewal theory, (3-202). For low rates of renewal, (3-205) reduces to the film theory, (3-188). In between, k_c is proportional to D_{AB}^n, where n is 0.5–1.0. Application of the film-penetration theory is difficult because of lack of data for δ and s, but the predicted effect of molecular diffusivity brackets experimental data.

§3.7 TWO-FILM THEORY AND OVERALL MASS-TRANSFER COEFFICIENTS

Gas–liquid and liquid–liquid separation processes involve two fluid phases in contact and require consideration of mass-transfer resistances in both phases. In 1923, Whitman [63] suggested an extension of the film theory to two films in series. Each film presents a resistance to mass transfer, but concentrations in the two fluids at the interface are assumed to be in phase equilibrium. That is, there is no additional interfacial resistance to mass transfer.

The assumption of phase equilibrium at the interface, while widely used, may not be valid when gradients of interfacial tension are established during mass transfer. These gradients give rise to interfacial turbulence, resulting, most often, in considerably increased mass-transfer coefficients. This phenomenon, the *Marangoni effect*, is discussed in detail by Bird, Stewart, and Lightfoot [28], who cite additional references. The effect occurs at vapor–liquid and liquid–liquid interfaces, with the latter having received the most attention. By adding surfactants, which concentrate at the interface, the Marangoni effect is reduced because of interface stabilization, even to the extent that an interfacial mass-transfer resistance (which causes the mass-transfer coefficient to be reduced) results. Unless otherwise indicated, the Marangoni effect will be ignored here, and phase equilibrium will always be assumed at the phase interface.

§3.7.1 Gas (Vapor)–Liquid Case

Consider steady-state mass transfer of A from a gas, across an interface, and into a liquid. It is postulated, as shown in Figure 3.21a, that a thin gas film exists on one side of the interface and a thin liquid film exists on the other side, with diffusion controlling in each film. However, this postulation is not necessary, because instead of writing

$$N_A = \frac{(D_{AB})_G}{\delta_G}(c_{A_b} - c_{A_i})_G = \frac{(D_{AB})_L}{\delta_L}(c_{A_i} - c_{A_b})_L \quad (3\text{-}206)$$

the rate of mass transfer can be expressed in terms of mass-transfer coefficients determined from any suitable theory, with the concentration gradients visualized more realistically as in Figure 3.21b. Any number of different mass-transfer coefficients and driving forces can be used. For the gas phase, under dilute or equimolar counterdiffusion (EMD) conditions, the mass-transfer rate in terms of partial pressures is:

$$\boxed{N_A = k_p(p_{A_b} - p_{A_i})} \qquad (3\text{-}207)$$

where k_p is a gas-phase mass-transfer coefficient based on a partial-pressure driving force.

For the liquid phase, with molar concentrations:

$$\boxed{N_A = k_c(c_{A_i} - c_{A_b})} \qquad (3\text{-}208)$$

At the interface, c_{A_i} and p_{A_i} are in equilibrium. Applying a version of Henry's law different from that in Table 2.3,[1]

$$c_{A_i} = H_A p_{A_i} \qquad (3\text{-}209)$$

Equations (3-207) to (3-209) are commonly used combinations for vapor–liquid mass transfer. Computations of mass-transfer rates are made from a knowledge of bulk concentrations c_{A_b} and p_{A_b}. To obtain an expression for N_A in terms of an overall driving force for mass transfer that includes both

---

[1]Different forms of Henry's law are found in the literature. They include

$$p_A = H_A x_A, \quad p_A = \frac{c_A}{H_A}, \quad \text{and} \quad y_A = H_A x_A$$

When a Henry's law constant, H_A, is given without citing the defining equation, the equation can be determined from the units of the constant. For example, if the constant has the units of atm or atm/mole fraction, Henry's law is given by $p_A = H_A x_A$. If the units are mol/L-mmHg, Henry's law is $p_A = c_A/H_A$.

fluid phases, (3-207) to (3-209) are combined to eliminate the interfacial concentrations, c_{A_i} and p_{A_i}. Solving (3-207) for p_{A_i}:

$$p_{A_i} = p_{A_b} - \frac{N_A}{k_p} \tag{3-210}$$

Solving (3-208) for c_{A_i}:

$$c_{A_i} = c_{A_b} + \frac{N_A}{k_c} \tag{3-211}$$

Combining (3-211) with (3-209) to eliminate c_{A_i} and combining the result with (3-210) to eliminate p_{A_i} gives

$$N_A = \frac{p_{A_b} H_A - c_{A_b}}{(H_A/k_p) + (1/k_c)} \tag{3-212}$$

Overall Mass-Transfer Coefficients. It is customary to define: (1) a fictitious liquid-phase concentration $c_A^* = p_{A_b} H_A$, which is a fictitious liquid concentration of A in equilibrium with the partial pressure of A in the bulk gas; and (2) an overall mass-transfer coefficient, K_L. Now (3-212) is

$$N_A = K_L(c_A^* - c_{A_b}) = \frac{(c_A^* - c_{A_b})}{(H_A/k_p) + (1/k_c)} \tag{3-213}$$

where K_L is the *overall mass-transfer coefficient* based on the liquid phase and defined by

$$\frac{1}{K_L} = \frac{H_A}{k_p} + \frac{1}{k_c} \tag{3-214}$$

The corresponding overall driving force for mass transfer is also based on the liquid phase, given by $(c_A^* - c_{A_b})$. The quantities H_A/k_p and $1/k_c$ are measures of gas and liquid mass-transfer resistances. When $1/k_c \gg H_A/k_p$, the resistance of the gas phase is negligible and the rate of mass transfer is controlled by the liquid phase, with (3-213) simplifying to

$$N_A = k_c(c_A^* - c_{A_b}) \tag{3-215}$$

so that $K_L \approx k_c$. Because resistance in the gas phase is negligible, the gas-phase driving force becomes $(p_{A_b} - p_{A_i}) \approx 0$, so $p_{A_b} \approx p_{A_i}$.

Alternatively, (3-207) to (3-209) combine to define an overall mass-transfer coefficient, K_G, based on the gas phase:

$$N_A = \frac{p_{A_b} - c_{A_b}/H_A}{(1/k_p) + (1/H_A k_c)} \tag{3-216}$$

In this case, it is customary to define: (1) a fictitious gas-phase partial pressure $p_A^* = c_{A_b}/H_A$, which is the partial pressure of A that would be in equilibrium with the concentration of A in the bulk liquid; and (2) an overall mass-transfer coefficient for the gas phase, K_G, based on a partial-pressure driving force. Thus, (3-216) becomes

$$N_A = K_G(p_{A_b} - p_A^*) = \frac{(p_{A_b} - p_A^*)}{(1/k_p) + (1/H_A k_c)} \tag{3-217}$$

where

$$\frac{1}{K_G} = \frac{1}{k_p} + \frac{1}{H_A k_c} \tag{3-218}$$

Now the resistances are $1/k_p$ and $1/H_A k_c$. If $1/k_p \gg 1/H_A k_c$,

$$N_A = k_p(p_{A_b} - p_A^*) \tag{3-219}$$

so $K_G \approx k_p$. Since the resistance in the liquid phase is then negligible, the liquid-phase driving force becomes $(c_{A_i} - c_{A_b}) \approx 0$, so $c_{A_i} \approx c_{A_b}$.

The choice between (3-213) or (3-217) is arbitrary, but is usually made on the basis of which phase has the largest mass-transfer resistance; if the liquid, use (3-213); if the gas, use (3-217); if neither is dominant, either equation is suitable.

Another common combination for vapor–liquid mass transfer uses mole-fraction driving forces, which define another set of mass-transfer coefficients k_y and k_x:

$$N_A = k_y(y_{A_b} - y_{A_i}) = k_x(x_{A_i} - x_{A_b}) \tag{3-220}$$

Now equilibrium at the interface can be expressed in terms of a K-value for vapor–liquid equilibrium, instead of as a Henry's law constant. Thus,

$$K_A = y_{A_i}/x_{A_i} \tag{3-221}$$

Combining (3-220) and (3-221) to eliminate y_{A_i} and x_{A_i},

$$N_A = \frac{y_{A_b} - x_{A_b}}{(1/K_A k_y) + (1/k_x)} \tag{3-222}$$

Alternatively, fictitious concentrations and overall mass-transfer coefficients can be used with mole-fraction driving forces. Thus, $x_A^* = y_{A_b}/K_A$ and $y_A^* = K_A x_{A_b}$. If the two values of K_A are equal,

$$N_A = K_x(x_A^* - x_{A_b}) = \frac{x_A^* - x_{A_b}}{(1/K_A k_y) + (1/k_x)} \tag{3-223}$$

and $\quad N_A = K_y(y_{A_b} - y_A^*) = \frac{y_{A_b} - y_A^*}{(1/k_y) + (K_A/k_x)} \tag{3-224}$

where K_x and K_y are overall mass-transfer coefficients based on mole-fraction driving forces with

$$\frac{1}{K_x} = \frac{1}{K_A k_y} + \frac{1}{k_x} \tag{3-225}$$

and

$$\frac{1}{K_y} = \frac{1}{k_y} + \frac{K_A}{k_x} \tag{3-226}$$

When using handbook or literature correlations to estimate mass-transfer coefficients, it is important to determine which coefficient (k_p, k_c, k_y, or k_x) is correlated, because often it is not stated. This can be done by checking the units or the form of the Sherwood or Stanton numbers. Coefficients correlated by the Chilton–Colburn analogy are k_c for either the liquid or the gas phase. The various coefficients are related by the following expressions, which are summarized in Table 3.16.

Table 3.16 Relationships among Mass-Transfer Coefficients

Equimolar Counterdiffusion (EMD):

Gases: $N_A = k_y \Delta y_A = k_c \Delta c_A = k_p \Delta p_A$

$k_y = k_c \dfrac{P}{RT} = k_p P$ if ideal gas

Liquids: $N_A = k_x \Delta x_A = k_c \Delta c_A$

$k_x = k_c c$, where c = total molar concentration (A + B)

Unimolecular Diffusion (UMD) with bulk flow:

Gases: Same equations as for EMD with k replaced

by $k' = \dfrac{k}{(y_B)_{LM}}$

Liquids: Same equations as for EMD with k

replaced by $k' = \dfrac{k}{(x_B)_{LM}}$

When working with concentration units, it is convenient to use:

$k_G(\Delta c_G) = k_c(\Delta c)$ for the gas phase

$k_L(\Delta c_L) = k_c(\Delta c)$ for the liquid phase

Liquid phase:

$$k_x = k_c c = k_c \left(\frac{\rho_L}{M}\right) \qquad (3\text{-}227)$$

Ideal-gas phase:

$$k_y = k_p P = (k_c)_g \frac{P}{RT} = (k_c)_g c = (k_c)_g \left(\frac{\rho_G}{M}\right) \qquad (3\text{-}228)$$

Typical units are

	SI	AE
k_c	m/s	ft/h
k_p	kmol/s-m²-kPa	lbmol/h-ft²-atm
k_y, k_x	kmol/s-m²	lbmol/h-ft²

When unimolecular diffusion (UMD) occurs under nondilute conditions, bulk flow must be included. For binary mixtures, this is done by defining modified mass-transfer coefficients, designated with a prime as follows:

For the liquid phase, using k_c or k_x,

$$k' = \frac{k}{(1 - x_A)_{LM}} = \frac{k}{(x_B)_{LM}} \qquad (3\text{-}229)$$

For the gas phase, using k_p, k_y, or k_c,

$$k' = \frac{k}{(1 - y_A)_{LM}} = \frac{k}{(y_B)_{LM}} \qquad (3\text{-}230)$$

Expressions for k' are convenient when the mass-transfer rate is controlled mainly by one of the two resistances. Literature mass-transfer coefficient data are generally correlated in terms of k rather than k'. Mass-transfer coefficients

estimated from the Chilton–Colburn analogy [e.g. equations (3-166) to (3-171)] are k_c, not k_c'.

§3.7.2 Liquid–Liquid Case

For mass transfer across two liquid phases, equilibrium is again assumed at the interface. Denoting the two phases by $L^{(1)}$ and $L^{(2)}$, (3-223) and (3-224) become

$$N_A = K_x^{(2)}(x_A^{(2)*} - x_{A_b}^{(2)}) = \frac{x_A^{(2)*} - x_{A_b}^{(2)}}{(1/K_{D_A}k_x^{(1)}) + (1/k_x^{(2)})} \qquad (3\text{-}231)$$

and

$$N_A = K_x^{(1)}(x_{A_b}^{(1)} - x_A^{(1)*}) = \frac{x_{A_b}^{(1)} - x_A^{(1)*}}{(1/k_x^{(1)}) + (K_{D_A}/k_x^{(2)})} \qquad (3\text{-}232)$$

where

$$K_{D_A} = \frac{x_{A_i}^{(1)}}{x_{A_i}^{(2)}} \qquad (3\text{-}233)$$

§3.7.3 Case of Large Driving Forces for Mass Transfer

Previously, phase equilibria ratios such as H_A, K_A, and K_{D_A} have been assumed constant across the two phases. When large driving forces exist, however, the ratios may not be constant. This commonly occurs when one or both phases are not dilute with respect to the solute, A, in which case, expressions for the mass-transfer flux must be revised. For mole-fraction driving forces, from (3-220) and (3-224),

$$N_A = k_y(y_{A_b} - y_{A_i}) = K_y(y_{A_b} - y_A^*) \qquad (3\text{-}234)$$

Thus,

$$\frac{1}{K_y} = \frac{y_{A_b} - y_A^*}{k_y(y_{A_b} - y_{A_i})} \qquad (3\text{-}235)$$

or

$$\frac{1}{K_y} = \frac{(y_{A_b} - y_{A_i}) + (y_{A_i} - y_A^*)}{k_y(y_{A_b} - y_{A_i})} = \frac{1}{k_y} + \frac{1}{k_y}\left(\frac{y_{A_i} - y_A^*}{y_{A_b} - y_{A_i}}\right) \qquad (3\text{-}236)$$

From (3-220),

$$\frac{k_x}{k_y} = \frac{(y_{A_b} - y_{A_i})}{(x_{A_i} - x_{A_b})} \qquad (3\text{-}237)$$

Combining (3-234) and (3-237),

$$\frac{1}{K_y} = \frac{1}{k_y} + \frac{1}{k_x}\left(\frac{y_{A_i} - y_A^*}{x_{A_i} - x_{A_b}}\right) \qquad (3\text{-}238)$$

Similarly

$$\frac{1}{K_x} = \frac{1}{k_x} + \frac{1}{k_y}\left(\frac{x_A^* - x_{A_i}}{y_{A_b} - y_{A_i}}\right) \qquad (3\text{-}239)$$

Figure 3.22 shows a curved equilibrium line with values of y_{A_b}, y_{A_i}, y_A^*, x_A^*, x_{A_i}, and x_{A_b}. Because the line is curved, the vapor–liquid equilibrium ratio, $K_A = y_A/x_A$, is not constant. As shown, the slope of the curve and thus, K_A, decrease with

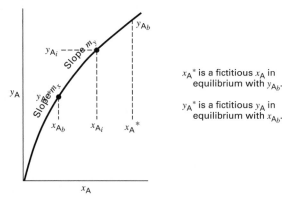

Figure 3.22 Curved equilibrium line.

x_A^* is a fictitious x_A in equilibrium with y_{A_b}.

y_A^* is a fictitious y_A in equilibrium with x_{A_b}.

increasing concentration of A. Denoting two slopes of the equilibrium curve by

$$m_x = \left(\frac{y_{A_i} - y_A^*}{x_{A_i} - x_{A_b}}\right) \quad (3\text{-}240)$$

and

$$m_y = \left(\frac{y_{A_b} - y_{A_i}}{x_A^* - x_{A_i}}\right) \quad (3\text{-}241)$$

then substituting (3-240) and (3-241) into (3-238) and (3-239), respectively,

$$\frac{1}{K_y} = \frac{1}{k_y} + \frac{m_x}{k_x} \quad (3\text{-}242)$$

and

$$\frac{1}{K_x} = \frac{1}{k_x} + \frac{1}{m_y k_y} \quad (3\text{-}243)$$

EXAMPLE 3.20 Absorption of SO_2 into Water.

Sulfur dioxide (A) is absorbed into water in a packed column, where bulk conditions are 50°C, 2 atm, $y_{A_b} = 0.085$, and $x_{A_b} = 0.001$. Equilibrium data for SO_2 between air and water at 50°C are

p_{SO_2}, atm	c_{SO_2}, lbmol/ft^3
0.0382	0.00193
0.0606	0.00290
0.1092	0.00483
0.1700	0.00676

Experimental values of the mass-transfer coefficients are:

Liquid phase: $k_c = 0.18$ m/h

Gas phase: $k_p = 0.040 \dfrac{\text{kmol}}{\text{h-m}^2\text{-kPa}}$

For mole-fraction driving forces, compute the mass-transfer flux: (a) assuming an average Henry's law constant and a negligible bulk-flow effect; (b) utilizing the actual curved equilibrium line and assuming a negligible bulk-flow effect; (c) utilizing the actual curved equilibrium line and taking into account the bulk-flow effect. In addition, (d) determine the magnitude of the two resistances and the values of the mole fractions at the interface that result from part (c).

Solution

Equilibrium data are converted to mole fractions by assuming Dalton's law, $y_A = p_A/P$, for the gas and $x_A = c_A/c$ for the liquid. The concentration of liquid is close to that of water, 3.43 lbmol/ft^3 or 55.0 kmol/m^3. Thus, the mole fractions at equilibrium are:

y_{SO_2}	x_{SO_2}
0.0191	0.000563
0.0303	0.000846
0.0546	0.001408
0.0850	0.001971

These data are fitted with average and maximum absolute deviations of 0.91% and 1.16%, respectively, by the equation

$$y_{SO_2} = 29.74 x_{SO_2} + 6{,}733 x_{SO_2}^2 \quad (1)$$

Differentiating, the slope of the equilibrium curve is

$$m = \frac{dy}{dx} = 29.74 + 13{,}466 x_{SO_2} \quad (2)$$

The given mass-transfer coefficients are converted to k_x and k_y by (3-227) and (3-228):

$$k_x = k_c c = 0.18(55.0) = 9.9 \frac{\text{kmol}}{\text{h-m}^2}$$

$$k_y = k_p P = 0.040(2)(101.3) = 8.1 \frac{\text{kmol}}{\text{h-m}^2}$$

(a) From (1) for $x_{A_b} = 0.001$, $y_A^* = 29.74(0.001) + 6{,}733(0.001)^2 = 0.0365$. From (1) for $y_{A_b} = 0.085$, solving the quadratic equation yields $x_A^* = 0.001975$.

The average slope in this range is

$$m = \frac{0.085 - 0.0365}{0.001975 - 0.001} = 49.7$$

Examination of (3-242) and (3-243) shows that the liquid-phase resistance is controlling because the term in k_x is much larger than the term in k_y. Therefore, from (3-243), using $m = m_x$,

$$\frac{1}{K_x} = \frac{1}{9.9} + \frac{1}{49.7(8.1)} = 0.1010 + 0.0025 = 0.1035$$

or $\quad K_x = 9.66 \dfrac{\text{kmol}}{\text{h-m}^2}$

From (3-223),

$$N_A = 9.66(0.001975 - 0.001) = 0.00942 \frac{\text{kmol}}{\text{h-m}^2}$$

(b) From part (a), the gas-phase resistance is almost negligible. Therefore, $y_{A_i} \approx y_{A_b}$ and $x_{A_i} \approx x_A^*$.

From (3-241), the slope m_y is taken at the point $y_{A_b} = 0.085$ and $x_A^* = 0.001975$ on the equilibrium line.

By (2), $m_y = 29.74 + 13{,}466(0.001975) = 56.3$. From (3-243),

$$K_x = \frac{1}{(1/9.9) + [1/(56.3)(8.1)]} = 9.69 \frac{\text{kmol}}{\text{h-m}^2}$$

giving $N_A = 0.00945$ kmol/h-m^2. This is a small change from part (a).

(c) Correcting for bulk flow, from the results of parts (a) and (b),

$$y_{A_b} = 0.085, y_{A_i} = 0.085, x_{A_i} = 0.1975, x_{A_b} = 0.001,$$
$$(y_B)_{LM} = 1.0 - 0.085 = 0.915, \text{ and } (x_B)_{LM} \approx 0.9986$$

From (3-229),

$$k'_x = \frac{9.9}{0.9986} = 9.9 \frac{kmol}{h\text{-}m^2} \text{ and } k'_y = \frac{8.1}{0.915} = 8.85 \frac{kmol}{h\text{-}m^2}$$

From (3-243),

$$K_x = \frac{1}{(1/9.9) + [1/56.3(8.85)]} = 9.71 \frac{kmol}{h\text{-}m^2}$$

From (3-223),

$$N_A = 9.7(0.001975 - 0.001) = 0.00947 \frac{kmol}{h\text{-}m^2}$$

which is only a very slight change from parts (a) and (b), where the bulk-flow effect was ignored. The effect is very small because it is important only in the gas, whereas the liquid resistance is controlling.

(d) The relative magnitude of the mass-transfer resistances is

$$\frac{1/m_y k'_y}{1/k'_x} = \frac{1/(56.3)(8.85)}{1/9.9} = 0.02$$

Thus, the gas-phase resistance is only 2% of the liquid-phase resistance. The interface vapor mole fraction can be obtained from (3-223), after accounting for the bulk-flow effect:

$$y_{A_i} = y_{A_b} - \frac{N_A}{k'_y} = 0.085 - \frac{0.00947}{8.85} = 0.084$$

Similarly, $\quad x_{A_i} = \dfrac{N_A}{k'_x} + x_{A_b} = \dfrac{0.00947}{9.9} + 0.001 = 0.00196$

§3.8 MOLECULAR MASS TRANSFER IN TERMS OF OTHER DRIVING FORCES

Thus far in this chapter, only a concentration driving force (in terms of concentrations, mole fractions, or partial pressures) has been considered, and only one or two species were transferred. Molecular mass transfer of a species such as a charged biological component may be driven by other forces besides its concentration gradient. These include gradients in *temperature*, which induces thermal diffusion via the Soret effect; *pressure*, which drives ultracentrifugation; *electrical potential*, which governs electrokinetic phenomena (dielectrophoresis and magnetophoresis) in ionic systems like permselective membranes; and *concentration gradients* of other species in systems containing three or more components. Three postulates of nonequilibrium thermodynamics may be used to relate such driving forces to frictional motion of a species in the Maxwell–Stefan equations [28, 75, 76]. Maxwell, and later Stefan, used kinetic theory in the mid- to late-19th century to determine diffusion rates based on momentum transfer between molecules. At the same time, Graham and Fick described ordinary diffusion based on binary mixture experiments. These three postulates and

applications to bioseparations are presented in this section. Application of the Maxwell–Stefan equations to rate-based models for multicomponent absorption, stripping, and distillation is developed in Chapter 12.

§3.8.1 The Three Postulates of Nonequilibrium Thermodynamics

This brief introduction summarizes a more detailed synopsis found in [28].

First postulate

The first (*quasi-equilibrium*) postulate states that equilibrium thermodynamic relations apply to systems not in equilibrium, provided departures from local equilibrium (gradients) are sufficiently small. This postulate and the second law of thermodynamics allow the diffusional driving force per unit volume of solution, represented by $cRT\mathbf{d}_i$ and which moves species i relative to a solution containing n components, to be written as

$$cRT\mathbf{d}_i \equiv c_i \nabla_{T,P}\mu_i + \left(c_i\overline{V}_i - \omega_i\right)\nabla P - \rho_i\left(\mathbf{g}_i - \sum_{k=1}^{n}\omega_k\mathbf{g}_k\right) \tag{3-244}$$

$$\sum_{i=1}^{n}\mathbf{d}_i = 0 \tag{3-245}$$

where $\mathbf{d}_i$ are driving forces for molecular mass transport, c_i is molar concentration, μ_i is chemical potential, ω_i is mass fraction, $\mathbf{g}_i$ are total body forces (e.g., gravitational or electrical potential) per unit mass, $\overline{V}_i$ is partial molar volume, and ρ_i is mass concentration, all of which are specific to species i. Each driving force is given by a negative spatial gradient in potential, which is the work required to move species i relative to the solution volume. In order from left to right, the three collections of terms on the RHS of (3-244) represent driving forces for *concentration diffusion*, *pressure diffusion*, and *forced diffusion*. The term $c_i\overline{V}_i$ in (3-244) corresponds to the volume fraction of species i, ϕ_i.

Second postulate

The second (*linearity*) postulate allows forces on species in (3-244) to be related to a vector mass flux, $\mathbf{j}_i$. It states that all fluxes in the system may be written as linear relations involving all the forces. For mass flux, the *thermal-diffusion* driving force, $-\beta_{i0}\nabla \ln T$, is added to the previous three forces to give

$$\mathbf{j}_i = -\beta_{i0}\nabla \ln T - \rho_i\sum_{j=1}^{n}\frac{\beta_{ij}}{\rho_i\rho_j}cRT\mathbf{d}_j \tag{3-246}$$

$$\beta_{ij} + \sum_{\substack{k=1\\k\neq j}}^{n}\beta_{ik} = 0 \tag{3-247}$$

where β_{i0} and β_{ij} are phenomenological coefficients (i.e., transport properties). The vector mass flux, $\mathbf{j}_i = \rho_i(\mathbf{v}_i - \mathbf{v})$, is the arithmetic average of velocities of all molecules of

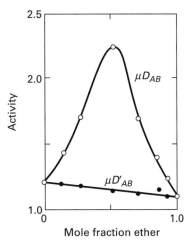

Figure 3.23 Effect of activity on the product of viscosity and diffusivity for liquid mixtures of chloroform and diethyl ether [R.E. Powell, W.E. Roseveare, and H. Eyring, *Ind. Eng. Chem.*, **33**, 430–435 (1941)].

species i in a tiny volume element ($\mathbf{v}_i$) relative to a mass-averaged value of the velocities of all such components in the mixture, $\mathbf{v} = \sum \omega_i \mathbf{v}_i$. It is related to molar flux, $\mathbf{J}_i$, in (3-251).

Third postulate

According to the third postulate, *Onsager's reciprocal relations*—developed using statistical mechanics and supported by data—the matrix of the β_{ij} coefficients in the flux-force relation (3-246) are symmetric ($\beta_{ij} = \beta_{ji}$) in the absence of magnetic fields. These coefficients may be rewritten as multicomponent mass diffusivities D'_{ij}

$$D'_{ij} = \frac{x_i x_j}{\beta_{ij}} cRT \left(= D'_{ji}\right) \qquad (3\text{-}248)$$

which exhibit less composition dependency than the transport properties, β_{ij}, and reduce to the more familiar binary diffusivity of Fick's Law, D_{AB}, for ideal binary solutions, as shown in Figure 3.23, and illustrated in Example 3.21.

§3.8.2 Maxwell–Stefan Equations

To show the effects of forces on molecular motion of species i, (3-248) is substituted into (3-246), which is solved for the driving forces, $\mathbf{d}_i$, and set equal to (3-244). Using $\mathbf{j}_i = \rho_i(\mathbf{v}_i - \mathbf{v})$, discussed above, a set of $n-1$ independent rate expressions, called the *Maxwell–Stefan equations*, is obtained:

$$
\begin{aligned}
\sum_{j=1}^{n} \frac{x_i x_j}{D'_{ij}} \left(\mathbf{v}_i - \mathbf{v}_j\right) &= \frac{1}{cRT} \left[c_i \nabla_{T,P} \mu_i + (\phi_i - \omega_i)\nabla P \right. \\
&\left. - \rho_i \left(\mathbf{g}_i - \sum_{k=1}^{n} \omega_k \mathbf{g}_k\right) \right] - \sum_{j=1}^{n} \frac{x_i x_j}{D'_{ij}} \left(\frac{\beta_{i0}}{\rho_i} - \frac{\beta_{j0}}{\rho_j}\right) \nabla \ln T
\end{aligned}
$$

$$(3\text{-}249)$$

The set of rate expressions given by (3-249) shows molecular mass transport of species i driven by gradients in

pressure, temperature, and concentration of species i for $j \neq i$ in systems containing three or more components, as well as driven by body forces that induce gradients in potential. The total driving force for species i due to potential gradients collected on the LHS of (3-249) is equal to the sum on the RHS of the cumulative *friction force* exerted on species i—$\zeta_{i,j}x_j$ $(x_i - x_j)$—by every species j in a mixture, where frictional coefficient ζ_{ij} is given by x_i/D'_{ij} in (3-249). The friction exerted by j on i is proportional to the mole fraction of j in the mixture and to the difference in average molecular velocity between species j and i.

Body forces in (3-249) may arise from gravitational acceleration, $\mathbf{g}$; electrostatic potential gradients, $\nabla \varphi$, or mechanically restraining matrices (e.g., permselective membranes and friction between species i and its surroundings), denoted by δ_{im}. These can be written as

$$\mathbf{g}_i = \mathbf{g} - \left(\frac{z_i \Im}{M_i}\right)\nabla\varphi + \delta_{im}\frac{1}{\rho_m}\nabla P \qquad (3\text{-}250)$$

where z_i is elementary charge and Faraday's constant, $\Im$, = 96,490 absolute coulombs per gram-equivalent.

Chemical versus physical potentials

Potential can be defined as the reversible work required to move an entity relative to other elements in its surroundings. The change in potential per unit distance provides the force that drives local velocity of a species relative to its environment in (3-249). For molecules, potential due to gravity in (3-250)—an external force that affects the whole system—is insignificant relative to chemical potential in (3-249), an internal force that results in motion within the system but not in the system as whole. Gravity produces a driving force downward at height z, resulting from a potential difference due to the work performed to attain the height, $-mg\Delta z$, divided by the height, Δz, which reduces to mg. For gold (a dense molecule), this driving force = $(0.197\,\text{kg/mol})$ $(10\,\text{m/s}^2) \cong 2\,\text{N/mol}$. Gravitational potential of gold across the distance of a centimeter is therefore 2×10^{-2} N/mol.

Chemical potential can be defined as the reversible work needed to separate one mole of species i from a large amount of a mixture. Its magnitude increases logarithmically with the species activity, or $\Delta\mu = -RT\Delta \ln(\gamma_i x_i)$. Gold, in an ideal solution ($\gamma_i = 1$) and for ambient conditions at $x_i = 1/e = 0.368$, experiences a driving force times distance due to a chemical potential of $-(8.314\,\text{J/mol–K})(298\,\text{K}) \times \ln$ $(0.37) = 2,460$ J/mol. The predominance of chemical potential leads to an approximate linear simplification of (3-249)—which neglects potentials due to pressure, temperature, and external body forces—that is applicable in many practical situations, as illustrated later in Example 3.25. Situations in which the other potentials are significant are also considered. For instance, Example 3.21 below shows that ultra-centrifugation provides a large centripetal ("center-seeking") force to induce molecular momentum, ∇P, sufficient to move species i in the positive direction, if its mass fraction is greater than its volume fraction (i.e., if component i is denser than its surroundings).

Driving forces for species velocities

Effects of driving forces on species velocity are illustrated in the following seven examples reduced from [28], [75], [76], and [77]. These examples show how to apply (3-249) and (3-250), together with species equations of continuity, the equation of motion, and accompanying auxiliary (bootstrap) relations such as (3-245) and (3-248). The auxiliary expressions are needed to provide the molecular velocity of the selected reference frame because the velocities in (3-249) are relative values. The first example considers concentration-driving forces in binary systems. Measured data for D'_{ij}, which requires simultaneous measurement of γ_i as a function of x_i, are rare. Instead, the multicomponent diffusivity values may be estimated from phenomenological Fick's law diffusivities.

EXAMPLE 3.21 Maxwell–Stefan Equations Related to Fick's Law.

Consider a binary system containing species A and B that is isotropic in all but concentration [75]. Show the correspondence between D_{AB} and D'_{AB} by relating (3-249) to the diffusive molar flux of species A relative to the molar-average velocity of a mixture, $\mathbf{J}_A$, which may be written in terms of the mass-average velocity, $\mathbf{j}_A$:

$$\mathbf{J}_A = -cD_{AB}\nabla x_A = c_A(\mathbf{v}_A - \mathbf{v}_M) = \mathbf{j}_A \frac{cx_A x_B}{\rho\omega_A\omega_B} \quad (3\text{-}251)$$

where $\mathbf{v}_M = x_A\mathbf{v}_A + x_B\mathbf{v}_B$ is the molar-average velocity of a mixture.

Solution

In a binary system, $x_B = 1 - x_A$, and the LHS of (3-249) may be written as

$$\frac{x_A x_B}{D'_{AB}}(\mathbf{v}_B - \mathbf{v}_A) = \frac{x_A}{D'_{AB}}(x_B\mathbf{v}_B + x_A\mathbf{v}_A - \mathbf{v}_A) = -\frac{x_A}{D'_{AB}}\frac{\mathbf{J}_A}{c_A}$$
$$(3\text{-}252)$$

which relates the friction force to the molar flux. Substituting $\nabla\mu_i = RT\nabla\ln(a_i)$ into the RHS of (3-249), setting it equal to (3-252), and rearranging, gives

$$\mathbf{J}_A = -cD'_{AB}\left\{ x_A\nabla\ln a_A + \frac{1}{cRT}[(\phi_A - \omega_A)\nabla P \right.$$
$$\left. -\rho\omega_A\omega_B(\mathbf{g}_A - \mathbf{g}_B)] + k_T\nabla\ln T \right\} \quad (3\text{-}253)$$

$$k_T = \frac{\beta_{A0}}{\rho D'_{AB}}\frac{x_A x_B}{\omega_A\omega_B} = \alpha_T x_A x_B = \sigma_T x_A x_B T \quad (3\text{-}254)$$

Equation (3-253) describes binary diffusion in gases or liquids. It is a specialized form of the *generalized Fick equations*. Equation (3-254) relates the *thermal diffusion ratio*, k_T, to the *thermal diffusion factor*, α_T, and the *Soret coefficient*, σ_T. For liquids, σ_T is preferred. For gases, α_T is nearly independent of composition.

Table 3.17 shows concentration- and temperature-dependent k_T values for several binary gas and liquid pairs. Species A moves to the colder region when the value of k_T is positive. This usually corresponds to species A having a larger molecular weight (M_A) or diameter. The sign of k_T may change with temperature.

In this example, the effects of pressure, thermal diffusion, and body force terms in (3-253) may be neglected. Then from the

Table 3.17 Experimental Thermal Diffusion Ratios for Low-Density Gas and Liquid Mixtures

Species A-B	$T(K)$	x_A	$k_T\{x_A,T\}$
Gas			
Ne-He	330	0.80	0.0531
		0.40	0.1004
N_2-H_2	264	0.706	0.0548
		0.225	0.0663
D_2-H_2	327	0.90	0.1045
		0.50	0.0432
		0.10	0.0166
Liquid			
$C_2H_2Cl_4$-n-C_6H_{14}	298	0.5	1.08
$C_2H_4Br_2$-$C_2H_4Cl_2$	298	0.5	0.225
$C_2H_2Cl_4$-CCl_4	298	0.5	0.060
CBr_4-CCl_4	298	0.09	0.129
CCl_4-CH_3OH	313	0.5	1.23
CH_3OH-H_2O	313	0.5	-0.137
Cyclo-C_6H_{12}-C_6H_6	313	0.5	0.100

Data from Bird et al. [28].

properties of logarithms,

$$x_A\nabla\ln a_A = \nabla x_A + x_A\nabla\ln\gamma_A = \nabla x_A\left(1 + \frac{\partial\ln\gamma_A}{\partial\ln x_A}\right) \quad (3\text{-}255)$$

By substituting (3-255) into (3-253) and comparing with (3-251), it is found that

$$D_{AB} = \left(1 + \frac{\partial\ln\gamma_A}{\partial\ln x_A}\right)D'_{AB} \quad (3\text{-}256)$$

The activity-based diffusion coefficient D'_{AB} is less concentration-dependent than D_{AB} but requires accurate activity data, so it is used less widely. Multicomponent mixtures of low-density gases have $\gamma_i = 1$ and $\mathbf{d}_i = \nabla x_i$ for concentration diffusion and $D_{AB} = D'_{AB}$ from kinetic theory.

EXAMPLE 3.22 Diffusion via a Thermal Gradient (thermal diffusion).

Consider two bulbs connected by a narrow, insulated tube that are filled with a binary mixture of ideal gases [28]. (Examples of binary mixtures are given in Table 3.17.) Maintaining the two bulbs at constant temperatures T_2 and T_1, respectively, typically enriches the larger species at the cold end for a positive value of k_T. Derive an expression for $(x_{A2} - x_{A1})$, the mole-fraction difference between the two bulbs, as a function of k_T, T_2, and T_1 at steady state, neglecting convection currents in the connecting tube.

Solution

There is no net motion of either component at steady state, so $\mathbf{J}_A = 0$. Use (3-253) for the ideal gases ($\gamma_A = 1$), setting the connecting tube on the z-axis, neglecting pressure and body forces, and applying the properties of logarithms to obtain

$$\frac{dx_A}{dz} = -\frac{k_T}{T}\frac{dT}{dz} \quad (3\text{-}257)$$

The integral of (3-257) may be evaluated by neglecting composition effects on k_T for small differences in mole fraction and using a value of k_T at a mean temperature, T_m, to yield

$$x_{A2} - x_{A1} = -k_T\{T_m\}\ln\frac{T_2}{T_1} \tag{3-258}$$

where the mean temperature at which to evaluate k_T is

$$T_m = \frac{T_1 T_2}{T_2 - T_1}\ln\frac{T_2}{T_1} \tag{3-259}$$

Substituting values of k_T from Table 3.17 into (3-258) suggests that a very large temperature gradient is required to obtain more than a small composition difference. During World War II, uranium isotopes were separated in cascades of Clausius–Dickel columns based on thermal diffusion between sets of vertical heated and cooled walls. The separation supplemented thermal diffusion with free convection to allow species A, enriched at the cooled wall, to descend and species B, enriched at the heated wall, to ascend. Energy expenditures were enormous.

EXAMPLE 3.23 Diffusion via a Pressure Gradient (pressure diffusion).

Components A and B in a small cylindrical tube of length L, held at radial position $R_o \gg L$ inside an ultracentrifuge, are rotated at constant angular velocity Ω [28]. The species experience a change in molecular momentum, ∇p, due to centripetal ("center-seeking") acceleration $\mathbf{g}_\Omega = \Omega^2 r$ given by the equation of motion,

$$\frac{dp}{dr} = \rho g_\Omega = \rho\Omega^2 r = \rho\frac{v_\theta^2}{r} \tag{3-260}$$

where $v_\theta = \delta_\theta \Omega r$ is the linear velocity. Derive expressions for (1) the migration velocity, v_{migr}, of dilute A in B (e.g., protein in H_2O) in terms of relative molecular weight, and for (2) the distribution of the two components at steady state in terms of their partial molar volumes, $\bar{V}_i$, $i = $ A, B, and the pressure gradient, neglecting changes in $\bar{V}_i$ and γ_i over the range of conditions in the centrifuge tube.

Solution

The radial motion of species A is obtained by substituting (3-255) into the radial component of the binary Maxwell–Stefan equation in (3-253) for an isothermal tube free of external body forces to give

$$\mathbf{J}_A = -cD'_{AB}\left[\left(1 + \frac{\partial\ln\gamma_A}{\partial\ln x_A}\right)\frac{dx_A}{dr} + \frac{1}{cRT}(\phi_A - \omega_A)\frac{dP}{dr}\right] \tag{3-261}$$

where the pressure gradient of the migration term in (3-261) remains relatively constant in the tube since $L \ll R_o$. Molecular-weight dependence in this term in the limit of a dilute solution of protein (A) in H_2O (B) arises in the volume and mass fractions, respectively,

$$\varphi_A = c_A\bar{V}_A = x_A c\bar{V}_A \approx x_A\frac{\bar{V}_A}{\bar{V}_B} = x_A\frac{M_A}{M_B}\frac{\hat{V}_A}{\hat{V}_B} \tag{3-262}$$

$$\omega_A = \frac{\rho_A}{\rho} = \frac{c_A M_A}{cM} = x_A\frac{M_A}{x_A M_A + x_B M_B} \approx x_A\frac{M_A}{M_B} \tag{3-263}$$

where $\hat{V}_i = \bar{V}_i/M_i$ is the partial specific volume of species i, which is 1 mL/g for H_2O and ∼0.75 mL/g for a globular protein (see Table 3.18). A pseudo-binary Fickian diffusivity given by (3-256) to be

Table 3.18 Protein Molecular Weights Determined by Ultracentrifugation

Protein	M	$s_{20,w}$ (S)	$\overline{V}_2$ (cm^{3}g^{-1})
Ribonuclease (bovine)	12,400	1.85	0.728
Lysozyme (chicken)	14,100	1.91	0.688
Serum albumin (bovine)	66,500	4.31	0.734
Hemoglobin	68,000	4.31	0.749
Tropomysin	93,000	2.6	0.71
Fibrinogen (human)	330,000	7.6	0.706
Myosin (rod)	570,000	6.43	0.728
Bushy stunt virus	10,700,000	132	0.74
Tobacco mosaic virus	40,000,000	192	0.73

Data from Cantor and Schimmel [78].

substituted into (3-261) may be estimated using Stokes law:

$$D_{AB} = \frac{\kappa T}{6\pi\mu_B R_A f_A} \tag{3-264}$$

where R_A is the radius of a sphere whose volume equals that of the protein, and protein nonsphericity is accounted for by a hydrodynamic shape factor, f_A. Substituting (3-256), (3-260), (3-262), and (3-263) into (3-261) gives

$$\mathbf{J}_A = -cD_{AB}\frac{dx_A}{dr} + c_A\left\{-\frac{D'_{AB}}{cRT}\left[\frac{M_A}{M_B}\left(\frac{\hat{V}_A}{\hat{V}_B} - 1\right)\right]\rho\Omega^2 r\right\} \tag{3-265}$$

where the term inside the curly brackets on the RHS of (3-265) corresponds to the *migration velocity*, v_{migr}, in the $+r$ direction driven by centripetal force in proportion to the relative molecular weight, M_A/M_B. The ratio of v_{migr} to centripetal force in (3-265) is the sedimentation coefficient, s, which is typically expressed in Svedberg (S) units (1 S = 10^{-13} sec), named after the inventor of the ultracentrifuge. Protein molecular-weight values obtained by photoelectric scanning detection of v_{migr} to determine s in pure water (w) at 20° (i.e., $s_{20,w}$) are summarized in Table 3.18. Equation (3-265) is the basis for analyzing transient behavior, steady polarization, and preparative application of ultracentrifugation.

Concentration and pressure gradients balance at steady state ($\mathbf{J}_A = 0$), and with constant $\bar{V}_i$ and γ_i in the tube and $x_A \sim x_B$ locally, writing (3-253) for species A gives

$$0 = \frac{dx_A}{dr} + \frac{M_A x_A}{RT}\left(\frac{\bar{V}_A}{M_A} - \frac{1}{\rho}\right)\frac{dp}{dr} \tag{3-266}$$

Multiplying (3-266) by $(\bar{V}_B/x_A)dr$, and substituting a constant centripetal force ($r \approx R_o$) from (3-260), gives

$$\bar{V}_B\frac{dx_A}{x_A} = \bar{V}_B\frac{g_\Omega}{RT}(\rho\bar{V}_A - M_A)dr \tag{3-267}$$

Writing an equation analogous to (3-267) for species B, and subtracting it from (3-267), gives

$$\bar{V}_B\frac{dx_A}{x_A} - \bar{V}_A\frac{dx_B}{x_B} = \frac{g_\Omega}{RT}(M_A\bar{V}_B - M_B\bar{V}_A)dr \tag{3-268}$$

Integrating (3-268) from $x_i\{r = 0\} = x_{i0}$ to $x_i\{r\}$ for $i = $ A,B, using $r = 0$ at the distal tube end, gives

$$\bar{V}_B\ln\frac{x_A}{x_{A0}} - \bar{V}_A\ln\frac{x_B}{x_{B0}} = \frac{g_\Omega}{RT}(M_B\bar{V}_A - M_A\bar{V}_B)r \tag{3-269}$$

Using the properties of logarithms and taking the exponential of both sides of (3-269) yields the *steady-state species distribution* in terms of the partial molar volumes:

$$\left(\frac{x_A}{x_{A0}}\right)^{\overline{V}_B}\left(\frac{x_{B0}}{x_B}\right)^{\overline{V}_A} = \exp\left[\frac{g_\Omega r}{RT}\left(M_B\overline{V}_A - M_A\overline{V}_B\right)\right] \quad (3\text{-}270)$$

The result in (3-269) is independent of transport coefficients and may thus be obtained in an alternative approach using equilibrium thermodynamics.

EXAMPLE 3.24 Diffusion in a Ternary System via Gradients in Concentration and Electrostatic Potential.

A 1-1 electrolyte M^+X^- (e.g., NaCl) diffuses in a constriction between two well-mixed reservoirs at different concentrations containing electrodes that exhibit a potential difference, $\Delta\varphi$, measured by a potentiometer under current-free conditions [75]. Derive an expression for salt flux in the system.

Solution

Any pressure difference between the two reservoirs is negligible relative to the reference pressure, $cRT \sim 1,350$ atm, at ambient conditions. Electroneutrality, in the absence of current flow through the potentiometer, requires that

$$x_{M^+} = x_{X^-} = x_S = 1 - x_W \quad (3\text{-}271)$$

$$N_{M^+} = N_{X^-} = N_S \quad (3\text{-}272)$$

Substituting (3-271) into (3-249) and rearranging yields the $n-1$ Maxwell–Stefan relations:

$$\frac{1}{cD'_{M^+W}}\left(x_W N_{M^+} - x_{M^+} N_W\right)$$

$$= -x_{M^+}\nabla_{T,P}a_{M^+} + \frac{\rho_{M^+}}{cRT}\left(\mathbf{g}_{M^+} - \sum_{k=1}^{n}\omega_k\mathbf{g}_k\right) \quad (3\text{-}273)$$

$$\frac{1}{cD'_{X^-W}}\left(x_W N_{X^-} - x_{X^-} N_W\right)$$

$$= -x_{X^-}\nabla_{T,P}a_{X^-} + \frac{\rho_{X^-}}{cRT}\left(\mathbf{g}_{X^-} - \sum_{k=1}^{n}\omega_k\mathbf{g}_k\right) \quad (3\text{-}274)$$

No ion-ion diffusivity appears because $\mathbf{v}_{M^+} - \mathbf{v}_{X^-} = 0$ in the absence of current. Substituting (3-250), (3-271), and (3-272) into (3-273) and (3-274) and rearranging yields

$$\frac{1}{cD'_{M^+W}}\left(x_W N_S - x_S N_W\right) = -\frac{\partial\ln a_{M^+}}{\partial\ln x_S}\nabla x_S - \frac{x_S}{RT}\Im\nabla\varphi \quad (3\text{-}275)$$

$$\frac{1}{cD'_{X^-W}}\left(x_W N_S - x_S N_W\right) = -\frac{\partial\ln a_{X^-}}{\partial\ln x_S}\nabla x_S + \frac{x_S}{RT}\Im\nabla\varphi \quad (3\text{-}276)$$

Adding (3-275) and (3-276) eliminates the electrostatic potential, to give

$$N_S = -\left(\frac{1}{cD'_{M^+W}} + \frac{1}{cD'_{X^-W}}\right)^{-1}\frac{\partial\ln\left(a_{M^+}a_{X^-}\right)}{\partial\ln x_S}\nabla x_S + x_S(N_S + N_W) \quad (3\text{-}277)$$

which has the form of Fick's law after a concentration-based diffusivity is defined:

$$D_{SW} = 2\left(\frac{D'_{M^+W}D'_{X^-W}}{D'_{M^+W} + D'_{X^-W}}\right)\left(1 + \frac{\partial\ln\gamma_S}{\partial\ln x_S}\right) \quad (3\text{-}278)$$

$$\gamma_S = \gamma_{M^+}\gamma_{X^-} \quad (3\text{-}279)$$

where γ_S is the mean activity coefficient given by $a_S = a_M + a_{X^-}$ $= x_s^2[(\gamma_M + g_{X^-})^{1/2}]^2 = x_s^2[(\gamma_S)^{1/2}]^2$. Equation (3-278) shows that while fast diffusion of small counterions creates a potential gradient that speeds large ions, the overall diffusivity of the salt pair is dominated by the slower ions (e.g., proteins).

EXAMPLE 3.25 Film Mass Transfer.

Species velocity in (3-249) is due to (1) bulk motion; (2) gradient of a potential $\Delta\psi_i = \psi_{i\delta} - \psi_{io}$ of species i across distance δ (which moves species i relative to the mixture); and (3) friction between species and surroundings [77]. Develop an approximate expression for film mass transfer using linearized potential gradients.

Solution

The driving force that results from the potential gradient, $-d\psi_i/dz$, is approximated by the difference in potential across a film of thickness δ, $-\Delta\psi_i/\delta$. Linearizing the chemical potential difference by

$$\Delta\mu_i = RT\Delta\ln(\gamma_i x_i) \approx RT\frac{x_{i\delta} - x_{io}}{(x_{i\delta} + x_{io})/2} = RT\frac{\Delta x_i}{\overline{x}_i} \quad (3\text{-}280)$$

provides a tractable approximation that has reasonable accuracy over a wide range of compositions [77].

Friction from hydrodynamic drag of fluid (1) of viscosity μ_1 on a spherical particle (2) of diameter d_2 is proportional to their relative difference in velocity, $\mathbf{v}$, viz.,

$$-\frac{d\mu_2}{dz} = 3N_A\pi\mu_1(\mathbf{v}_2 - \mathbf{v}_1)d_2 \quad (3\text{-}281)$$

where N_A (Avogadro's number) represents particles per mole. A large force is produced when the drag is summed over a mole of particles.

Rearranging (3-281) yields an expression for the Maxwell–Stefan diffusivity in terms of hydrodynamic drag:

$$-\frac{d}{dz}\left(\frac{\mu_2}{RT}\right) = \frac{\mathbf{v}_2 - \mathbf{v}_1}{D'_{12}} \quad (3\text{-}282)$$

$$D'_{12} = \frac{RT}{N_A 3\pi\eta_1 d_2} \quad (3\text{-}283)$$

Substituting (3-280) into (3-282) and rearranging, after linearizing the derivative across a film of thickness δ, yields the mass transport coefficient, k_{12},

$$\frac{\Delta x_2}{\overline{x}_2} = \frac{\overline{\mathbf{v}}_1 - \overline{\mathbf{v}}_2}{k_{12}} \quad (3\text{-}284)$$

$$k_{12} = \frac{D'_{12}}{\delta} \quad (3\text{-}285)$$

where k_{ij} is $\sim 10^{-1}$ m/s for gases and 10^{-4} m/s for liquids. These values decrease by approximately a factor of 10 for gases and liquids in porous media.

In a general case that includes any number of components, friction between components j and i per mole of i is proportional to the difference between the mean velocities of j and i, respectively.

Taking friction proportional to the local concentration of j decreases the composition dependence of k_{ij}, viz.,

$$\bar{x}_j \frac{\bar{\mathbf{v}}_j - \bar{\mathbf{v}}_i}{k_{ij}} \tag{3-286}$$

Because assigning a local concentration to a component like a solid membrane component, m, is difficult, a membrane coefficient, k_i, may be introduced instead:

$$\frac{\bar{x}_m}{k_{im}} = \frac{1}{k_i} \tag{3-287}$$

§3.8.3 Maxwell–Stefan Difference Equation

Linearization allows application of a difference form of the Maxwell–Stefan equation, which is obtained by setting the negative driving force on species i equal to the friction on species i, viz. [77]:

$$\frac{\Delta x_i}{\bar{x}_i} + \ldots = \sum_j \bar{x}_j \frac{\bar{\mathbf{v}}_j - \bar{\mathbf{v}}_i}{k_{ij}} \tag{3-288}$$

where the ellipsis . . . allows addition of relevant linearized potentials in addition to the chemical potential. The accuracy of (3-288) is adequate for many engineering calculations. This is illustrated by determining molar solute flux of dilute and nondilute solute during binary stripping, and by estimating concentration polarization and permeate flux in tangential-flow filtration.

Dilute stripping

Consider stripping a trace gas (1) ($\bar{x}_2 \approx 1$) from a liquid through a gas film into an ambient atmosphere. The atmosphere is taken at a reference velocity ($\mathbf{v}_2 = 0$). Application of (3-288) yields

$$k_{12} \frac{\Delta x_1}{\bar{x}_1} = -\bar{\mathbf{v}}_1 \tag{3-289}$$

or

$$N_1 = c\bar{\mathbf{v}}_1 \bar{x}_1 = -ck_{12}\Delta x_1 \tag{3-290}$$

The result in (3-290), obtained from the Maxwell–Stefan difference equation, is consistent with (3-35) for dilute ($x_2 \sim 1$) solutions, which was obtained from Fick's law.

Nondilute stripping

For this situation, $\bar{x}_1 = 0.5 = \bar{x}_2$, and drift occurs in the gas film. From (3-288),

$$k_{12} \frac{\Delta x_1}{\bar{x}_1} = -0.5 \cdot \bar{\mathbf{v}}_1 \tag{3-291}$$

for which

$$N_1 = c\bar{\mathbf{v}}_1 \bar{x}_1 = -2ck_{12}\Delta x_1 \tag{3-292}$$

The latter result is easily obtained using (3-288) without requiring a drift-correction, as Fick's law would have.

Concentration polarization in tangential flow filtration

Now consider the flux of water (2) through a semipermeable membrane that completely retains a dissolved salt (1) at dilute concentration ($x_2 \approx 1$), as discussed in [77]. Set the velocity of the salt equal to a stationary value in the frame of reference ($\mathbf{v}_1 = 0$). The average salt concentration in a film of thickness δ adjacent to the membrane is

$$\bar{x}_1 = x_{1o} + \frac{x_{1\delta} - x_{1o}}{2} = x_{1o} + \frac{\Delta x_1}{2} \tag{3-293}$$

Using (3-288) gives

$$k_{12} \frac{\Delta x_1}{\bar{x}_1} = \bar{\mathbf{v}}_2 \tag{3-294}$$

Combining (3-293) and (3-294) gives the increase in salt concentration in the film relative to its value in the bulk:

$$\frac{\Delta x_1}{x_{1o}} = \frac{2\dfrac{\bar{\mathbf{v}}_2}{k_{12}}}{2 - \dfrac{\bar{\mathbf{v}}_2}{k_{12}}} \tag{3-295}$$

EXAMPLE 3.26 Flux in Tangential-Flow Filtration.

Relate flux of permeate, j, in tangential-flow filtration to local wall concentration of a completely retained solute, i, using the Maxwell–Stefan difference equation.

Solution

Local permeate flux is given by $N_j = \bar{c}_j \bar{\mathbf{v}}_j$. An expression for local water velocity is obtained by solving (3-295) for $\bar{\mathbf{v}}_j$:

$$\bar{\mathbf{v}}_j = k_{ij} \frac{\Delta x_i}{\Delta x_i/2 + x_{i,b}} \approx k_{ij} \ln \frac{x_{i,w}}{x_{i,b}} \tag{3-296}$$

where subscripts b and w represent bulk feed and wall, respectively. In a film, $k_{ij} = D_{ij}/\delta$. Local permeate flux is then

$$N_j = c_j \frac{D_{ij}}{\delta} \ln \frac{x_{i,w}}{x_{i,b}} \tag{3-297}$$

The result is consistent with the classical stagnant-film model in (14-108), which was obtained using Fick's law.

EXAMPLE 3.27 Maxwell–Stefan Difference Equations Related to Fick's Law.

For a binary system containing species A and B, show how D_{AB} relates to D'_{AB} in the Maxwell–Stefan difference equation by relating (3-288) with the diffusive flux of species A relative to the molar-average velocity of a mixture in (3-3a),

$$J_{A_z} = -D_{AB} \frac{dc_A}{dz} = c_A(\mathbf{v}_A - \mathbf{v}_M) \tag{3-298}$$

where $\mathbf{v}_M = x_A\mathbf{v}_A + x_B\mathbf{v}_B$ is the molar-average velocity of a mixture.

Solution

For a binary system, (3-288) becomes

$$\frac{1}{x_A} \frac{dx_A}{dz} = \bar{x}_B \frac{\bar{\mathbf{v}}_B - \bar{\mathbf{v}}_A}{D'_{AB}} = \frac{\bar{x}_B \bar{\mathbf{v}}_B + \bar{x}_A \bar{\mathbf{v}}_A - \bar{\mathbf{v}}_A}{D'_{AB}} = -\frac{\bar{\mathbf{v}}_A - \mathbf{v}_M}{\mathbf{D}'_{AB}} \tag{3-299}$$

Comparing (3-298) and (3-299) shows that for the Maxwell–Stefan difference equation

$$D'_{AB} = D_{AB} \qquad (3\text{-}300)$$

The result in (3-300) is consistent with kinetic theory for multi-component mixtures of low-density gases, for which $\gamma_i = 1$ and $\mathbf{d}_i = \nabla x_i$ for concentration diffusion.

This abbreviated introduction to the Maxwell–Stefan relations has shown how this kinetic formulation yields diffusive flux of species proportional to its concentration gradient like Fick's law for binary mixtures, *and* provides a basis for examining molecular motion in separations based on additional driving forces such as temperature, pressure, and body forces.

For multicomponent mixtures that are typical of bioseparations, the relations also quantitatively identify how the flux of each species affects the transport of any one species. This approach yields concentration gradients of each species in terms of the fluxes of the other species, which often requires expensive computational inversion. Fick's law may be generalized to obtain single-species flux in terms of concentration gradients for all species, but the resulting Fickian multicomponent diffusion coefficients are conjugates of the binary diffusion coefficients. The linearized Maxwell–Stefan difference equation allows straightforward analysis of driving forces due to concentration-, pressure-, body force-, and temperature-driving forces in complex separations like bioproduct purification, with accuracy adequate for many applications.

SUMMARY

1. Mass transfer is the net movement of a species in a mixture from one region to a region of different concentration, often between two phases across an interface. Mass transfer occurs by molecular diffusion, eddy diffusion, and bulk flow. Molecular diffusion occurs by a number of different driving forces, including concentration (ordinary), pressure, temperature, and external force fields.

2. Fick's first law for steady-state diffusion states that the mass-transfer flux by ordinary molecular diffusion is equal to the product of the diffusion coefficient (diffusivity) and the concentration gradient.

3. Two limiting cases of mass transfer in a binary mixture are equimolar counterdiffusion (EMD) and unimolecular diffusion (UMD). The former is also a good approximation for distillation. The latter includes bulk-flow effects.

4. When data are unavailable, diffusivities (diffusion coefficients) in gases and liquids can be estimated. Diffusivities in solids, including porous solids, crystalline solids, metals, glass, ceramics, polymers, and cellular solids, are best measured. For some solids, e.g., wood, diffusivity is anisotropic.

5. Diffusivities vary by orders of magnitude. Typical values are 0.10, 1×10^{-5}, and 1×10^{-9} cm^2/s for ordinary molecular diffusion of solutes in a gas, liquid, and solid, respectively.

6. Fick's second law for unsteady-state diffusion is readily applied to semi-infinite and finite stagnant media, including anisotropic materials.

7. Molecular diffusion under laminar-flow conditions is determined from Fick's first and second laws, provided velocity profiles are available. Common cases include falling liquid-film flow, boundary-layer flow on a flat plate, and fully developed flow in a straight, circular tube. Results are often expressed in terms of a mass-transfer coefficient embedded in a dimensionless group called the Sherwood number. The mass-transfer flux is given by the product of the mass-transfer coefficient and a concentration-driving force.

8. Mass transfer in turbulent flow can be predicted by analogy to heat transfer. The Chilton–Colburn analogy utilizes empirical *j*-factor correlations with a Stanton number for mass transfer. A more accurate equation by Churchill and Zajic should be used for flow in tubes, particularly at high Reynolds numbers.

9. Models are available for mass transfer near a two-fluid interface. These include film theory, penetration theory, surface-renewal theory, and the film-penetration theory. These predict mass-transfer coefficients proportional to the diffusivity raised to an exponent that varies from 0.5 to 1.0. Most experimental data provide exponents ranging from 0.5 to 0.75.

10. Whitman's two-film theory is widely used to predict the mass-transfer flux from one fluid, across an interface, and into another fluid, assuming equilibrium at the interface. One resistance is often controlling. The theory defines an overall mass-transfer coefficient determined from the separate coefficients for each of the phases and the equilibrium relationship at the interface.

11. The Maxwell–Stefan relations express molecular motion of species in multicomponent mixtures in terms of potential gradients due to composition, pressure, temperature, and body forces such as gravitational, centripetal, and electrostatic forces. This formulation is useful to characterize driving forces in addition to chemical potential, that act on charged biomolecules in typical bioseparations.

REFERENCES

1. Taylor, R., and R. Krishna, *Multicomponent Mass Transfer*, John Wiley & Sons, New York (1993).

2. Poling, B.E., J.M. Prausnitz, and J.P. O'Connell, *The Properties of Liquids and Gases*, 5th ed., McGraw-Hill, New York (2001).

3. Fuller, E.N., P.D. Schettler, and J.C. Giddings, *Ind. Eng. Chem.*, **58**(5), 18–27 (1966).

4. Takahashi, S., *J. Chem. Eng. Jpn.*, **7**, 417–420 (1974).

5. Slattery, J.C., M.S. thesis, University of Wisconsin, Madison (1955).

6. Wilke, C.R., and P. Chang, *AIChE J.*, **1**, 264–270 (1955).

7. Hayduk, W., and B.S. Minhas, *Can. J. Chem. Eng.*, **60**, 295–299 (1982).

8. Quayle, O.R., *Chem. Rev.*, **53**, 439–589 (1953).

9. Vignes, A., *Ind. Eng. Chem. Fundam.*, **5**, 189–199 (1966).

10. Sorber, H.A., *Handbook of Biochemistry, Selected Data for Molecular Biology*, 2nd ed., Chemical Rubber Co., Cleveland, OH (1970).

11. Geankoplis, C.J., *Transport Processes and Separation Process Principles*, 4th ed., Prentice-Hall, Upper Saddle River, NJ (2003).

12. Friedman, L., and E.O. Kraemer, *J. Am. Chem. Soc.*, **52**, 1298–1314, (1930).

13. Boucher, D.F., J.C. Brier, and J.O. Osburn, *Trans. AIChE*, **38**, 967–993 (1942).

14. Barrer, R.M., *Diffusion in and through Solids*, Oxford University Press, London (1951).

15. Swets, D.E., R.W., Lee, and R.C Frank, *J. Chem. Phys.*, **34**, 17–22 (1961).

16. Lee, R.W., *J. Chem, Phys.*, **38**, 448–455 (1963).

17. Williams, E.L., *J. Am. Ceram. Soc.*, **48**, 190–194 (1965).

18. Sucov, E.W, *J. Am. Ceram. Soc.*, **46**, 14–20 (1963).

19. Kingery, W.D., H.K. Bowen, and D.R. Uhlmann, *Introduction to Ceramics*, 2nd ed., John Wiley & Sons, New York (1976).

20. Ferry, J.D., *Viscoelastic Properties of Polymers*, John Wiley & Sons, New York (1980).

21. Rhee, C.K., and J.D. Ferry, *J. Appl. Polym. Sci.*, **21**, 467–476 (1977).

22. Brandrup, J., and E.H. Immergut, Eds., *Polymer Handbook*, 3rd ed., John Wiley & Sons, New York (1989).

23. Gibson, L.J., and M.F. Ashby, *Cellular Solids, Structure and Properties*, Pergamon Press, Elmsford, NY (1988).

24. Stamm, A.J., *Wood and Cellulose Science*, Ronald Press, New York (1964).

25. Sherwood, T.K., *Ind. Eng. Chem.*, **21**, 12–16 (1929).

26. Carslaw, H.S., and J.C. Jaeger, *Heat Conduction in Solids*, 2nd ed., Oxford University Press, London (1959).

27. Crank, J., *The Mathematics of Diffusion*, Oxford University Press, London (1956).

28. Bird, R.B., W.E. Stewart, and E.N. Lightfoot, *Transport Phenomena*, 2nd ed., John Wiley & Sons, New York (2002).

29. Churchill, R.V., *Operational Mathematics*, 2nd ed., McGraw-Hill, New York (1958).

30. Abramowitz, M., and I. A. Stegun, Eds., *Handbook of Mathematical Functions*, National Bureau of Standards, Applied Mathematics Series 55, Washington, DC (1964).

31. Newman, A.B., *Trans. AIChE*, **27**, 310–333 (1931).

32. Grimley, S.S., *Trans. Inst. Chem. Eng.* (London), **23**, 228–235 (1948).

33. Johnstone, H.F., and R.L. Pigford, *Trans. AIChE*, **38**, 25–51 (1942).

34. Olbrich, W.E., and J.D. Wild, *Chem. Eng. Sci.*, **24**, 25–32 (1969).

35. Churchill, S.W., *The Interpretation and Use of Rate Data: The Rate Concept*, McGraw-Hill, New York (1974).

36. Churchill, S.W., and R. Usagi, *AIChE J.*, **18**, 1121–1128 (1972).

37. Emmert, R.E., and R.L. Pigford, *Chem. Eng. Prog.*, **50**, 87–93 (1954).

38. Prandtl, L., *Proc. 3rd Int. Math. Congress*, Heidelberg (1904); reprinted in *NACA Tech. Memo 452* (1928).

39. Blasius, H., *Z. Math Phys.*, **56**, 1–37 (1908) reprinted in *NACA Tech. Memo 1256* (1950).

40. Schlichting, H., *Boundary Layer Theory*, 4th ed., McGraw-Hill, New York (1960).

41. Pohlhausen, E., *Z. Angew. Math Mech.*, **1**, 252 (1921).

42. Pohlhausen, E., *Z. Angew. Math Mech.*, **1**, 115–121 (1921).

43. Langhaar, H.L., *Trans. ASME*, **64**, A–55 (1942).

44. Graetz, L., *Ann. d. Physik*, **25**, 337–357 (1885).

45. Sellars, J.R., M. Tribus, and J.S. Klein, *Trans. ASME*, **78**, 441–448 (1956).

46. Leveque, J., *Ann. Mines*, [12], **13**, 201, 305, 381 (1928).

47. Knudsen, J.G., and D.L. Katz, *Fluid Dynamics and Heat Transfer*, McGraw-Hill, New York (1958).

48. Hausen, H., *Verfahrenstechnik Beih. z. Ver. Deut. Ing.*, **4**, 91 (1943).

49. Linton, W.H., Jr., and T.K. Sherwood, *Chem. Eng. Prog.*, **46**, 258–264 (1950).

50. Reynolds, O., *Trans. Roy. Soc. (London)*, **174A**, 935–982 (1883).

51. Boussinesq, J., *Mem. Pre. Par. Div. Sav.*, XXIII, Paris (1877).

52. Prandtl, L., *Z. Angew, Math Mech.*, **5**, 136 (1925); reprinted in *NACA Tech. Memo 1231* (1949).

53. Reynolds, O., *Proc. Manchester Lit. Phil. Soc.*, **14**, 7 (1874).

54. Colburn, A.P., *Trans. AIChE*, **29**, 174–210 (1933).

55. Chilton, T.H., and A.P. Colburn, *Ind. Eng. Chem.*, **26**, 1183–1187 (1934).

56. Prandtl, L., *Physik. Z.*, **11**, 1072 (1910).

57. Friend, W.L., and A.B. Metzner, *AIChE J.*, **4**, 393–402 (1958).

58. Nernst, W., *Z. Phys. Chem.*, **47**, 52 (1904).

59. Higbie, R., *Trans. AIChE*, **31**, 365–389 (1935).

60. Danckwerts, P.V., *Ind. Eng. Chem.*, **43**, 1460–1467 (1951).

61. Levenspiel, O., *Chemical Reaction Engineering*, 3rd ed., John Wiley & Sons, New York (1999).

62. Toor, H.L., and J.M. Marchello, *AIChE J.*, **4**, 97–101 (1958).

63. Whitman, W.G., *Chem. Met. Eng.*, **29**, 146–148 (1923).

64. van Driest, E.R., *J. Aero Sci.*, 1007–1011, 1036 (1956).

65. Reichardt, H., Fundamentals of Turbulent Heat Transfer, NACA Report TM-1408 (1957).

66. Drew, T.B., E.C. Koo, and W.H. McAdams, *Trans. Am. Inst. Chem. Engrs.*, **28**, 56 (1933).

67. Nikuradse, J., *VDI-Forschungsheft*, p. 361 (1933).

68. Launder, B.E., and D.B. Spalding, *Lectures in Mathematical Models of Turbulence*, Academic Press, New York (1972).

69. Heng, L., C. Chan, and S.W. Churchill, *Chem. Eng. J.*, **71**, 163 (1998).

70. Churchill, S.W., and S.C. Zajic, *AIChE J.*, **48**, 927–940 (2002).

71. Churchill, S.W., "Turbulent Flow and Convection: The Prediction of Turbulent Flow and Convection in a Round Tube," in J.P. Hartnett and T.F. Irvine, Jr., Ser. Eds., *Advances in Heat Transfer*, Academic Press, New York, Vol. 34, pp. 255–361 (2001).

72. Yu, B., H. Ozoe, and S.W. Churchill, *Chem. Eng. Sci.*, **56**, 1781 (2001).

73. Churchill, S.W., and C. Chan, *Ind. Eng. Chem. Res.*, **34**, 1332 (1995).

74. Churchill, S.W., *AIChE J.*, **43**, 1125 (1997).

75. Lightfoot, E.N., *Transport Phenomena and Living Systems*, John Wiley & Sons, New York (1974).

76. Taylor, R., and R. Krishna, *Multicomponent Mass Transfer*, John Wiley & Sons, New York (1993).

77. Wesselingh, J.A., and R. Krishna, *Mass Transfer in Multicomponent Mixtures*, Delft University Press, Delft (2000).

78. Cantor, C.R., and P.R. Schimmel, *Biophysical Chemistry Part II. Techniques for the study of biological structure and function*, W.H. Freeman and Co., New York (1980).

STUDY QUESTIONS

3.1. What is meant by diffusion?

3.2. Molecular diffusion occurs by any of what four driving forces or potentials? Which one is the most common?

3.3. What is the bulk-flow effect in mass transfer?

3.4. How does Fick's law of diffusion compare to Fourier's law of heat conduction?

3.5. What is the difference between equimolar counterdiffusion (EMD) and unimolecular diffusion (UMD)?

3.6. What is the difference between a mutual diffusion coefficient and a self-diffusion coefficient?

3.7. At low pressures, what are the effects of temperature and pressure on the molecular diffusivity of a species in a binary gas mixture?

3.8. What is the order of magnitude of the molecular diffusivity in cm^2/s for a species in a liquid mixture? By how many orders of magnitude is diffusion in a liquid slower or faster than diffusion in a gas?

3.9. By what mechanisms does diffusion occur in porous solids?

3.10. What is the effective diffusivity?

3.11. Why is diffusion in crystalline solids much slower than diffusion in amorphous solids?

3.12. What is Fick's second law of diffusion? How does it compare to Fourier's second law of heat conduction?

3.13. Molecular diffusion in gases, liquids, and solids ranges from slow to extremely slow. What is the best way to increase the rate of mass transfer in fluids? What is the best way to increase the rate of mass transfer in solids?

3.14. What is the defining equation for a mass-transfer coefficient? How does it differ from Fick's law? How is it analogous to Newton's law of cooling?

3.15. For laminar flow, can expressions for the mass-transfer coefficient be determined from theory using Fick's law? If so, how?

3.16. What is the difference between Reynolds analogy and the Chilton–Colburn analogy? Which is more useful?

3.17. For mass transfer across a phase interface, what is the difference between the film, penetration, and surface-renewal theories, particularly with respect to the dependence on diffusivity?

3.18. What is the two-film theory of Whitman? Is equilibrium assumed to exist at the interface of two phases?

3.19. What advantages do the Maxwell–Stefan relations provide for multicomponent mixtures containing charged biomolecules, in comparison with Fick's law?

3.20. How do transport parameters and coefficients obtained from the Maxwell–Stefan relations compare with corresponding values resulting from Fick's law?

EXERCISES

Section 3.1

3.1. Evaporation of liquid from a beaker.

A beaker filled with an equimolar liquid mixture of ethyl alcohol and ethyl acetate evaporates at 0°C into still air at 101 kPa (1 atm). Assuming Raoult's law, what is the liquid composition when half the ethyl alcohol has evaporated, assuming each component evaporates independently? Also assume that the liquid is always well mixed. The following data are available:

	Vapor Pressure, kPa at 0°C	Diffusivity in Air m^2/s
Ethyl acetate (AC)	3.23	6.45×10^{-6}
Ethyl alcohol (AL)	1.62	9.29×10^{-6}

3.2. Evaporation of benzene from an open tank.

An open tank, 10 ft in diameter, containing benzene at 25°C is exposed to air. Above the liquid surface is a stagnant air film 0.2 in. thick. If the pressure is 1 atm and the air temperature is 25°C, what is the loss of benzene in lb/day? The specific gravity of benzene at 60°F is 0.877. The concentration of benzene outside the film is negligible. For benzene, the vapor pressure at 25°C is 100 torr, and the diffusivity in air is 0.08 cm^2/s.

3.3. Countercurrent diffusion across a vapor film.

An insulated glass tube and condenser are mounted on a reboiler containing benzene and toluene. The condenser returns liquid reflux down the wall of the tube. At one point in the tube, the temperature is 170°F, the vapor contains 30 mol% toluene, and the reflux contains 40 mol% toluene. The thickness of the stagnant vapor film is estimated to be 0.1 in. The molar latent heats of benzene and toluene are equal. Calculate the rate at which toluene and benzene are being interchanged by equimolar countercurrent diffusion at this point in the tube in lbmol/h-ft^2, assuming that the rate is controlled by mass transfer in the vapor phase.

Gas diffusivity of toluene in benzene = 0.2 ft^2/h. Pressure = 1 atm (in the tube). Vapor pressure of toluene at 170°F = 400 torr.

3.4. Rate of drop in water level during evaporation.

Air at 25°C and a dew-point temperature of 0°C flows past the open end of a vertical tube filled with water at 25°C. The tube has an inside diameter of 0.83 inch, and the liquid level is 0.5 inch below the top of the tube. The diffusivity of water in air at 25°C is 0.256 cm^2/s.

(a) How long will it take for the liquid level in the tube to drop 3 inches?

(b) Plot the tube liquid level as a function of time for this period.

3.5. Mixing of two gases by molecular diffusion.

Two bulbs are connected by a tube, 0.002 m in diameter and 0.20 m long. Bulb 1 contains argon, and bulb 2 contains xenon. The pressure and temperature are maintained at 1 atm and 105°C. The diffusivity is 0.180 cm²/s. At time $t = 0$, diffusion occurs between the two bulbs. How long will it take for the argon mole fraction at End 1 of the tube to be 0.75, and 0.20 at the other end? Determine at the later time the: (a) Rates and directions of mass transfer of argon and xenon; (b) Transport velocity of each species; (c) Molar-average velocity of the mixture.

Section 3.2

3.6. Measurement of diffusivity of toluene in air.

The diffusivity of toluene in air was determined experimentally by allowing liquid toluene to vaporize isothermally into air from a partially filled, 3-mm diameter, vertical tube. At a temperature of 39.4°C, it took 96×10^4 s for the level of the toluene to drop from 1.9 cm below the top of the open tube to a level of 7.9 cm below the top. The density of toluene is 0.852 g/cm³, and the vapor pressure is 57.3 torr at 39.4°C. The barometer reading was 1 atm. Calculate the diffusivity and compare it with the value predicted from (3-36). Neglect the counterdiffusion of air.

3.7. Countercurrent molecular diffusion of H_2 and N_2 in a tube.

An open tube, 1 mm in diameter and 6 in. long, has hydrogen blowing across one end and nitrogen across the other at 75°C.

(a) For equimolar counterdiffusion, what is the rate of transfer of hydrogen into nitrogen in mol/s? Estimate the diffusivity (3-36).

(b) For part (a), plot the mole fraction of hydrogen against distance from the end of the tube past which nitrogen is blown.

3.8. Molecular diffusion of HCl across an air film.

HCl gas diffuses through a film of air 0.1 in. thick at 20°C. The partial pressure of HCl on one side of the film is 0.08 atm and zero on the other. Estimate the rate of diffusion in mol HCl/s-cm², if the total pressure is (a) 10 atm, (b) 1 atm, (c) 0.1 atm. The diffusivity of HCl in air at 20°C and 1 atm is 0.145 cm²/s.

3.9. Estimation of gas diffusivity.

Estimate the diffusion coefficient for a binary gas mixture of nitrogen (A)/toluene (B) at 25°C and 3 atm using the method of Fuller et al.

3.10. Correction of gas diffusivity for high pressure.

For the mixture of Example 3.3, estimate the diffusion coefficient at 100 atm using the method of Takahashi.

3.11. Estimation of infinite-dilution liquid diffusivity.

Estimate the diffusivity of carbon tetrachloride at 25°C in a dilute solution of: (a) methanol, (b) ethanol, (c) benzene, and (d) n-hexane by the methods of Wilke–Chang and Hayduk–Minhas. Compare values with the following experimental observations:

Solvent	Experimental D_{AB}, cm²/s
Methanol	1.69×10^{-5} cm²/s at 15°C
Ethanol	1.50×10^{-5} cm²/s at 25°C
Benzene	1.92×10^{-5} cm²/s at 25°C
n-Hexane	3.70×10^{-5} cm²/s at 25°C

3.12. Estimation of infinite-dilution liquid diffusivity.

Estimate the liquid diffusivity of benzene (A) in formic acid (B) at 25°C and infinite dilution. Compare the estimated value to that of Example 3.6 for formic acid at infinite dilution in benzene.

3.13. Estimation of infinite-dilution liquid diffusivity in solvents.

Estimate the liquid diffusivity of acetic acid at 25°C in a dilute solution of: (a) benzene, (b) acetone, (c) ethyl acetate, and (d) water. Compare your values with the following data:

Solvent	Experimental D_{AB}, cm²/s
Benzene	2.09×10^{-5} cm²/s at 25°C
Acetone	2.92×10^{-5} cm²/s at 25°C
Ethyl acetate	2.18×10^{-5} cm²/s at 25°C
Water	1.19×10^{-5} cm²/s at 20°C

3.14. Vapor diffusion through an effective film thickness.

Water in an open dish exposed to dry air at 25°C vaporizes at a constant rate of 0.04 g/h-cm². If the water surface is at the wet-bulb temperature of 11.0°C, calculate the effective gas-film thickness (i.e., the thickness of a stagnant air film that would offer the same resistance to vapor diffusion as is actually encountered).

3.15. Diffusion of alcohol through water and N_2.

Isopropyl alcohol undergoes mass transfer at 35°C and 2 atm under dilute conditions through water, across a phase boundary, and then through nitrogen. Based on the data given below, estimate for isopropyl alcohol: (a) the diffusivity in water using the Wilke–Chang equation; (b) the diffusivity in nitrogen using the Fuller et al. equation; (c) the product, $D_{AB}\rho_M$, in water; and (d) the product, $D_{AB}\rho_M$, in air, where ρ_M is the mixture molar density.

Compare: (e) the diffusivities in parts (a) and (b); (f) the results from parts (c) and (d). (g) What do you conclude about molecular diffusion in the liquid phase versus the gaseous phase?

Data:	Component	T_c, °R	P_c, psia	Z_c	υ_L, cm³/mol
	Nitrogen	227.3	492.9	0.289	—
	Isopropyl alcohol	915	691	0.249	76.5

3.16. Estimation of liquid diffusivity over the entire composition range.

Experimental liquid-phase activity-coefficient data are given in Exercise 2.23 for ethanol-benzene at 45°C. Estimate and plot diffusion coefficients for both chemicals versus composition.

3.17. Estimation of the diffusivity of an electrolyte.

Estimate the diffusion coefficient of NaOH in a 1-M aqueous solution at 25°C.

3.18. Estimation of the diffusivity of an electrolyte.

Estimate the diffusion coefficient of NaCl in a 2-M aqueous solution at 18°C. The experimental value is 1.28×10^{-5} cm²/s.

3.19. Estimation of effective diffusivity in a porous solid.

Estimate the diffusivity of N_2 in H_2 in the pores of a catalyst at 300°C and 20 atm if the porosity is 0.45 and the tortuosity is 2.5. Assume ordinary molecular diffusion in the pores.

3.20. Diffusion of hydrogen through a steel wall.

Hydrogen at 150 psia and 80°F is stored in a spherical, steel pressure vessel of inside diameter 4 inches and a wall thickness of 0.125 inch. The solubility of hydrogen in steel is 0.094 lbmol/ft³, and the diffusivity of hydrogen in steel is 3.0×10^{-9} cm²/s. If the inner surface of the vessel remains saturated at the existing hydrogen pressure and the hydrogen partial pressure at the outer surface is assumed to be zero, estimate the: (a) initial rate of mass transfer of hydrogen through the wall; (b) initial rate of pressure decrease inside the vessel; and (c) time in hours for the pressure to decrease to 50 psia, assuming the temperature stays constant at 80°F.

3.21. Mass transfer of gases through a dense polymer membrane.

A polyisoprene membrane of 0.8-μm thickness is used to separate methane from H_2. Using data in Table 14.9 and the following partial pressures, estimate the mass-transfer fluxes.

	Partial Pressures, MPa	
	Membrane Side 1	Membrane Side 2
Methane	2.5	0.05
Hydrogen	2.0	0.20

Section 3.3

3.22. Diffusion of NaCl into stagnant water.

A 3-ft depth of stagnant water at 25°C lies on top of a 0.10-in. thickness of NaCl. At time < 0, the water is pure. At time = 0, the salt begins to dissolve and diffuse into the water. If the concentration of salt in the water at the solid–liquid interface is maintained at saturation (36 g NaCl/100 g H_2O) and the diffusivity of NaCl is 1.2×10^{-5} cm^2/s, independent of concentration, estimate, by assuming the water to act as a semi-infinite medium, the time and the concentration profile of salt in the water when: (a) 10% of the salt has dissolved; (b) 50% of the salt has dissolved; and (c) 90% of the salt has dissolved.

3.23. Diffusion of moisture into wood.

A slab of dry wood of 4-inch thickness and sealed edges is exposed to air of 40% relative humidity. Assuming that the two unsealed faces of the wood immediately jump to an equilibrium moisture content of 10 lb H_2O per 100 lb of dry wood, determine the time for the moisture to penetrate to the center of the slab (2 inches from each face). Assume a diffusivity of water of 8.3×10^{-6} cm^2/s.

3.24. Measurement of moisture diffusivity in a clay brick.

A wet, clay brick measuring $2 \times 4 \times 6$ inches has an initial uniform water content of 12 wt%. At time = 0, the brick is exposed on all sides to air such that the surface moisture content is maintained at 2 wt%. After 5 h, the average moisture content is 8 wt%. Estimate: (a) the diffusivity of water in the clay in cm^2/s; and (b) the additional time for the average moisture content to reach 4 wt%. All moisture contents are on a dry basis.

3.25. Diffusion of moisture from a ball of clay.

A spherical ball of clay, 2 inches in diameter, has an initial moisture content of 10 wt%. The diffusivity of water in the clay is 5×10^{-6} cm^2/s. At time $t = 0$, the clay surface is brought into contact with air, and the moisture content at the surface is maintained at 3 wt%. Estimate the time for the average sphere moisture content to drop to 5 wt%. All moisture contents are on a dry basis.

Section 3.4

3.26. Diffusion of oxygen in a laminar-flowing film of water.

Estimate the rate of absorption of oxygen at 10 atm and 25°C into water flowing as a film down a vertical wall 1 m high and 6 cm in width at a Reynolds number of 50 without surface ripples.

Diffusivity of oxygen in water is 2.5×10^{-5} cm^2/s and the mole fraction of oxygen in water at saturation is 2.3×10^{-4}.

3.27. Diffusion of carbon dioxide in a laminar-flowing film of water.

For Example 3.13, determine at what height the average concentration of CO_2 would correspond to 50% saturation.

3.28. Evaporation of water from a film on a flat plate into flowing air.

Air at 1 atm flows at 2 m/s across the surface of a 2-inch-long surface that is covered with a thin film of water. If the air and water are at 25°C and the diffusivity of water in air is 0.25 cm^2/s, estimate the water mass flux for the evaporation of water at the middle of the surface, assuming laminar boundary-layer flow. Is this assumption reasonable?

3.29. Diffusion of a thin plate of naphthalene into flowing air.

Air at 1 atm and 100°C flows across a thin, flat plate of subliming naphthalene that is 1 m long. The Reynolds number at the trailing edge of the plate is at the upper limit for a laminar boundary layer. Estimate: (a) the average rate of sublimation in kmol/s-m^2; and (b) the local rate of sublimation 0.5 m from the leading edge. Physical properties are given in Example 3.14.

3.30. Sublimation of a circular naphthalene tube into flowing air.

Air at 1 atm and 100°C flows through a straight, 5-cm i.d. tube, cast from naphthalene, at a Reynolds number of 1,500. Air entering the tube has an established laminar-flow velocity profile. Properties are given in Example 3.14. If pressure drop is negligible, calculate the length of tube needed for the average mole fraction of naphthalene in the exiting air to be 0.005.

3.31. Evaporation of a spherical water drop into still, dry air.

A spherical water drop is suspended from a fine thread in still, dry air. Show: (a) that the Sherwood number for mass transfer from the surface of the drop into the surroundings has a value of 2, if the characteristic length is the diameter of the drop. If the initial drop diameter is 1 mm, the air temperature is 38°C, the drop temperature is 14.4°C, and the pressure is 1 atm, calculate the: (b) initial mass of the drop in grams; (c) initial rate of evaporation in grams per second; (d) time in seconds for the drop diameter to be 0.2 mm; and (e) initial rate of heat transfer to the drop. If the Nusselt number is also 2, is the rate of heat transfer sufficient to supply the required heat of vaporization and sensible heat? If not, what will happen?

Section 3.5

3.32. Dissolution of a tube of benzoic acid into flowing water.

Water at 25°C flows turbulently at 5 ft/s through a straight, cylindrical tube cast from benzoic acid, of 2-inch i.d. If the tube is 10 ft long, and fully developed, turbulent flow is assumed, estimate the average concentration of acid in the water leaving the tube. Physical properties are in Example 3.15.

3.33. Sublimation of a naphthalene cylinder to air flowing normal to it.

Air at 1 atm flows at a Reynolds number of 50,000 normal to a long, circular, 1-in.-diameter cylinder made of naphthalene. Using the physical properties of Example 3.14 for a temperature of 100°C, calculate the average sublimation flux in kmol/s-m^2.

3.34. Sublimation of a naphthalene sphere to air flowing past it.

For the conditions of Exercise 3.33, calculate the initial average rate of sublimation in kmol/s-m^2 for a spherical particle of 1-inch initial diameter. Compare this result to that for a bed packed with naphthalene spheres with a void fraction of 0.5.

Section 3.6

3.35. Stripping of CO_2 from water by air in a wetted-wall tube.

Carbon dioxide is stripped from water by air in a wetted-wall tube. At a location where pressure is 10 atm and temperature 25°C, the flux of CO_2 is 1.62 lbmol/h-ft^2. The partial pressure of CO_2 is 8.2 atm at the interface and 0.1 atm in the bulk gas. The diffusivity of CO_2 in air at these conditions is 1.6×10^{-2} cm^2/s. Assuming turbulent flow, calculate by film theory the mass-transfer coefficient k_c for the gas phase and the film thickness.

3.36. Absorption of CO_2 into water in a packed column.

Water is used to remove CO_2 from air by absorption in a column packed with Pall rings. At a region of the column where the partial pressure of CO_2 at the interface is 150 psia and the concentration in the bulk liquid is negligible, the absorption rate is 0.017 lbmol/h-ft^2. The CO_2 diffusivity in water is 2.0×10^{-5} cm^2/s. Henry's law for CO_2 is $p = Hx$, where $H = 9,000$ psia. Calculate the: (a) liquid-phase mass-transfer coefficient and film thickness; (b) contact time for the penetration theory; and (c) average eddy residence time and the probability distribution for the surface-renewal theory.

3.37. Determination of diffusivity of H_2S in water.

Determine the diffusivity of H_2S in water, using penetration theory, from the data below for absorption of H_2S into a laminar jet of water at 20°C. Jet diameter = 1 cm, jet length = 7 cm, and solubility of H_2S in water = 100 mol/m^3. Assume the contact time is the time of exposure of the jet. The average rate of absorption varies with jet flow rate:

Jet Flow Rate, cm^3/s	Rate of Absorption, mol/s $\times 10^6$
0.143	1.5
0.568	3.0
1.278	4.25
2.372	6.15
3.571	7.20
5.142	8.75

Section 3.7

3.38. Vaporization of water into air in a wetted-wall column.

In a test on the vaporization of H_2O into air in a wetted-wall column, the following data were obtained: tube diameter = 1.46 cm; wetted-tube length = 82.7 cm; air rate to tube at 24°C and 1 atm = 720 cm^3/s; inlet and outlet water temperatures are 25.15°C and 25.35°C, respectively; partial pressure of water in inlet air is 6.27 torr and in outlet air is 20.1 torr. The diffusivity of water vapor in air is 0.22 cm^2/s at 0°C and 1 atm. The mass velocity of air is taken relative to the pipe wall. Calculate: (a) rate of mass transfer of water into the air; and (b) K_G for the wetted-wall column.

3.39. Absorption of NH_3 from air into aq. H_2SO_4 in a wetted-wall column.

The following data were obtained by Chamber and Sherwood [*Ind. Eng. Chem.*, **29**, 1415 (1937)] on the absorption of ammonia from an ammonia-air mixture by a strong acid in a wetted-wall column 0.575 inch in diameter and 32.5 inches long:

Inlet acid (2-N H_2SO_4) temperature, °F	76
Outlet acid temperature, °F	81
Inlet air temperature, °F	77
Outlet air temperature, °F	84
Total pressure, atm	1.00
Partial pressure NH_3 in inlet gas, atm	0.0807
Partial pressure NH_3 in outlet gas, atm	0.0205
Air rate, lbmol/h	0.260

The operation was countercurrent, the gas entering at the bottom of the vertical tower and the acid passing down in a thin film on the vertical, cylindrical inner wall. The change in acid strength was negligible, and the vapor pressure of ammonia over the liquid is negligible because of the use of a strong acid for absorption. Calculate the mass-transfer coefficient, k_p, from the data.

3.40. Overall mass-transfer coefficient for a packed cooling tower.

A cooling-tower packing was tested in a small column. At two points in the column, 0.7 ft apart, the data below apply. Calculate the overall volumetric mass-transfer coefficient $K_y a$ that can be used to design a large, packed-bed cooling tower, where a is the mass-transfer area, A, per unit volume, V, of tower.

	Bottom	Top
Water temperature, °F	120	126
Water vapor pressure, psia	1.69	1.995
Mole fraction H_2O in air	0.001609	0.0882
Total pressure, psia	14.1	14.3
Air rate, lbmol/h	0.401	0.401
Column cross-sectional area, ft^2	0.5	0.5
Water rate, lbmol/h (approximation)	20	20

Section 3.8

3.41. Thermal diffusion.

Using the thermal diffusion apparatus of Example 3.22 with two bulbs at 0°C and 123°C, respectively, estimate the mole-fraction difference in H_2 at steady state from a mixture initially consisting of mole fractions 0.1 and 0.9 for D_2 and H_2, respectively.

3.42. Separation in a centrifugal force field.

Estimate the steady-state concentration profile for an aqueous ($\hat{V}_B = 1.0$ cm^3/g) solution of cytochrome C (12×10^3 Da; $x_{A_0} = 1 \times 10^{-6}$; $\hat{V}_A = 0.75$ cm^3/g) subjected to a centrifugal field 50×10^3 times the force of gravity in a rotor held at 4°C.

3.43. Diffusion in ternary mixture.

Two large bulbs, A and B, containing mixtures of H_2, N_2, and CO_2 at 1 atm and 35°C are separated by an 8.6-cm capillary. Determine the quasi-steady-state fluxes of the three species for the following conditions [77]:

	$x_{i,A}$	$x_{i,B}$	D'_{AB}, cm^2/s
H_2	0.0	0.5	$D'_{H_2-N_2} = 0.838$
N_2	0.5	0.5	$D'_{H_2-CO_2} = 0.168$
CO_2	0.5	0.0	$D'_{N_2-CO_2} = 0.681$

Chapter 4

Single Equilibrium Stages and Flash Calculations

§4.0 INSTRUCTIONAL OBJECTIVES

After completing this chapter, you should be able to:

- Explain what an equilibrium stage is and why it may not be sufficient to achieve a desired separation.
- Extend Gibbs phase rule to include extensive variables so that the number of degrees of freedom (number of variables minus the number of independent relations among the variables) can be determined.
- Use T–y–x and y–x diagrams of binary mixtures, with the q-line, to determine equilibrium compositions.
- Understand the difference between minimum- and maximum-boiling azeotropes and how they form.
- Calculate bubble-point, dew-point, and equilibrium-flash conditions.
- Use triangular phase diagrams for ternary systems with component material balances to determine equilibrium compositions of liquid–liquid mixtures.
- Use distribution (partition) coefficients, from activity coefficients, with component material-balance equations to calculate liquid–liquid phase equilibria for multicomponent systems.
- Use equilibrium diagrams with material balances to determine amounts and compositions for solid–fluid systems (leaching, crystallization, sublimation, desublimation, adsorption) and gas absorption in liquids.

The simplest separation process is one in which two phases in contact are brought to physical equilibrium, followed by phase separation. If the separation factor, Eq. (1-4), between two species in the two phases is very large, a single contacting stage may be sufficient to achieve a desired separation between them; if not, multiple stages are required. For example, if a vapor phase is brought to equilibrium with a liquid phase, the separation factor is the relative volatility, α, of a volatile component called the light key, LK, with respect to a less-volatile component called the heavy key, HK, where $\alpha_{\mathrm{LK,\,HK}} = K_{\mathrm{LK}}/K_{\mathrm{HK}}$. If the separation factor is 10,000, a near-perfect separation is achieved in a single equilibrium stage. If the separation factor is only 1.10, an almost perfect separation requires hundreds of equilibrium stages.

In this chapter, only a single equilibrium stage is considered, but a wide spectrum of separation operations is described. In all cases, a calculation is made by combining material balances with phase-equilibrium relations discussed in Chapter 2. When a phase change such as vaporization occurs, or when heat of mixing effects are large, an energy balance must be added to account for a temperature change. The next chapter describes arrangements of multiple equilibrium stages, called cascades, which are used when the desired degree of separation cannot be achieved with a single stage. The specification of both single-stage and multiple-stage separation operations, is not intuitive. For that reason,

this chapter begins with a discussion of Gibbs phase rule and its extension to batch and continuous operations.

Although not always stated, all diagrams and most equations in this chapter are valid only if the phases are at equilibrium. If mass-transfer rates are too slow, or if the time to achieve equilibrium is longer than the contact time, the degree of separation will be less than calculated by the methods in this chapter. In that case, stage efficiencies must be introduced into the equations, as discussed in Chapter 6, or calculations must be based on mass-transfer rates rather than phase equilibrium, as discussed in Chapter 12.

§4.1 GIBBS PHASE RULE AND DEGREES OF FREEDOM

Equilibrium calculations involve *intensive variables*, which are independent of quantity, and *extensive variables*, which depend on quantity. Temperature, pressure, and mole or mass fractions are intensive. Extensive variables include mass or moles and energy for a batch system, and mass or molar flow rates and energy-transfer rates for a flow system.

Phase-equilibrium equations, and mass and energy balances, provide dependencies among the intensive and extensive variables. When a certain number of the variables (called the independent variables) are specified, all other variables (called the dependent variables) become fixed. The number of independent variables is called the *variance*, or the number of *degrees of freedom*, $\mathscr{F}$.

§4.1.1 Gibbs Phase Rule

At physical equilibrium and when only intensive variables are considered, the Gibbs phase rule applies for determining $\mathscr{F}$. The rule states that

$$\mathscr{F} = C - \mathscr{P} + 2 \qquad (4\text{-}1)$$

where C is the number of components and $\mathscr{P}$ is the number of phases. Equation (4-1) is derived by counting the number of intensive variables, $\mathscr{V}$, and the number of independent equations, $\mathscr{E}$, that relate these variables. The number of intensive variables is

$$\mathscr{V} = C\mathscr{P} + 2 \qquad (4\text{-}2)$$

where the 2 refers to temperature and pressure, and $C\mathscr{P}$ is the total number of composition variables (e.g., mole fractions) for components distributed among $\mathscr{P}$ phases. The number of independent equations relating the intensive variables is

$$\mathscr{E} = \mathscr{P} + C(\mathscr{P} - 1) \qquad (4\text{-}3)$$

where the first term, $\mathscr{P}$, refers to the requirement that mole fractions sum to one in each phase, and the second term, $C(\mathscr{P} - 1)$, refers to the number of independent phase-equilibrium equations of the form

$$K_i = \frac{\text{mole fraction of } i \text{ in phase } (1)}{\text{mole fraction of } i \text{ in phase } (2)}$$

where (1) and (2) refer to equilibrium phases. For two phases, there are C independent expressions of this type; for three phases, $2C$; for four phases, $3C$; and so on. For example, for three phases ($V, L^{(1)}, L^{(2)}$), there are $3C$ different K-value equations:

$$K_i^{(1)} = y_i/x_i^{(1)} \qquad i = 1 \text{ to } C$$
$$K_i^{(2)} = y_i/x_i^{(2)} \qquad i = 1 \text{ to } C$$
$$K_{D_i} = x_i^{(1)}/x_i^{(2)} \qquad i = 1 \text{ to } C$$

However, only $2C$ of these equations are independent, because

$$K_{D_i} = K_i^{(2)}/K_i^{(1)}$$

Thus, the number of independent K-value equations is

$$C(\mathscr{P} - 1), \text{ and not } C\mathscr{P}$$

The degrees of freedom for Gibbs phase rule is the number of intensive variables, $\mathscr{V}$, less the number of independent equations, $\mathscr{E}$. Thus, from (4-2) and (4-3), (4-1) is derived:

$$\mathscr{F} = \mathscr{V} - \mathscr{E} = (C\mathscr{P} + 2) - [\mathscr{P} + C(\mathscr{P} - 1)] = C - \mathscr{P} + 2$$

If $\mathscr{F}$ intensive variables are specified, the remaining $\mathscr{P} + C(\mathscr{P} - 1)$ intensive variables are determined from $\mathscr{P} + C(\mathscr{P} - 1)$ equations. In using the Gibbs phase rule, it should be noted that the K-values are not counted as variables because they are thermodynamic functions that depend on the intensive variables.

As an example of the application of the Gibbs phase rule, consider the vapor–liquid equilibrium ($\mathscr{P} = 2$) in Figure

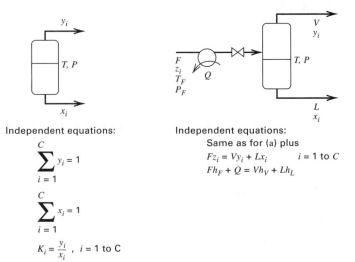

Figure 4.1 Treatments of degrees of freedom for vapor–liquid phase equilibria: (a) Gibbs phase rule (considers intensive variables only); (b) general analysis (considers both intensive and extensive variables).

4.1a, where the intensive variables are labels on the sketch above the list of independent equations relating these variables. Suppose there are $C = 3$ components. From (4-2) there are eight intensive variables: $T, P, x_1, x_2, x_3, y_1, y_2$, and y_3. From (4-1), $\mathscr{F} = 3 - 2 + 2 = 3$.

Suppose these three degrees of freedom are used to specify three variables: T, P, and one mole fraction. From (4-3) there are five independent equations, listed in Figure 4.1a, which are then used to compute the remaining five mole fractions. Similarly, if the number of components were two instead of three, only two variables need be specified.

Irrational specifications must be avoided because they lead to infeasible results. For example, if the components are H_2O, N_2, and O_2, and $T = 100°F$ and $P = 15$ psia are specified, a specification of $x_{N_2} = 0.90$ is not feasible because nitrogen is not this soluble.

§4.1.2 Extension of Gibbs Phase Rule to Extensive Variables

The Gibbs phase rule is limited because it does not deal with the extensive variables of feed, product, and energy streams, whether for a batch or continuous process. However, the rule can be extended for process applications by adding material and energy streams, with their extensive variables (e.g., flow rates or amounts), and additional independent equations. To illustrate this, consider the continuous, single-stage ($\mathscr{P} = 2$) process in Figure 4.1b. By comparison with Figure 4.1a, the additional variables are: z_i, T_F, P_F, F, Q, V, and L, or $C + 6$ additional variables, shown in the diagram of Figure 4.1a. In general, for $\mathscr{P}$ phases, the additional variables number $C + \mathscr{P} + 4$. The additional independent equations, listed below Figure 4.1b, are the C component material balances and the energy balance, for a total of $C + 1$ equations. Note that, like K-values, stream enthalpies are not counted as variables.

For a degrees-of-freedom analysis for phase equilibrium involving one feed phase, $\mathscr{P}$ product phases, and C components, (4-2) and (4-3) are extended by adding the above increments as a number of additional variables and equations:

$$\begin{aligned} \mathscr{V} &= (C\mathscr{P} + 2) + (C + \mathscr{P} + 4) = \mathscr{P} + C\mathscr{P} + C + 6 \\ \mathscr{E} &= [\mathscr{P} + C(\mathscr{P} - 1)] + (C + 1) = \mathscr{P} + C\mathscr{P} + 1 \quad (4\text{-}4) \\ \mathscr{F} &= \mathscr{V} - \mathscr{E} = C + 5 \end{aligned}$$

If the $C + 5$ degrees of freedom are used to specify all z_i and the five variables F, T_F, P_F, T, and P, the remaining variables are found using equations in Figure 4.1.[1] When applying (4-4), determination of the number of phases, $\mathscr{P}$, is implicit in the computational procedure as illustrated later in this chapter.

Next, the Gibbs phase rule, (4-1), and the equation for the degrees of freedom of a flow system, (4-4), are applied to (1) tabular equilibrium data, (2) graphical equilibrium data, and (3) thermodynamic equations for K-values and enthalpies for multiphase systems.

§4.2 BINARY VAPOR–LIQUID SYSTEMS

Experimental vapor–liquid equilibrium data for binary systems are widely available. Sources include *Perry's Handbook* [1], Gmehling and Onken [2], and Hála [3]. Because $y_B = 1 - y_A$ and $x_B = 1 - x_A$, the data are presented in terms of just four intensive variables: T, P, y_A, and x_A. Most commonly T, y_A, and x_A are tabulated at a fixed P for y_A and x_A from 0 to 1, where A is the more-volatile component ($y_A > x_A$). However, if an azeotrope forms, B becomes the more-volatile component on one side of the azeotropic point.

By the Gibbs phase rule, (4-1), $\mathscr{F} = 2 - 2 + 2 = 2$. Thus, with pressure fixed, phase compositions are completely defined if temperature and the relative volatility, (4-5), are fixed.

$$\alpha_{A,B} = \frac{K_A}{K_B} = \frac{(y_A/x_A)}{(y_B/x_B)} = \frac{(y_A/x_A)}{(1 - y_A)/(1 - x_A)} \quad (4\text{-}5)$$

Equilibrium data of the form T–y_A–x_A at 1 atm for three binary systems of importance are given in Table 4.1. Included are values of relative volatility computed from (4-5).

As discussed in Chapter 2, $\alpha_{A,B}$ depends on T, P, and the phase compositions. At 1 atm, where $\alpha_{A,B}$ is approximated well by $\gamma_A P_A^s / \gamma_B P_B^s$, $\alpha_{A,B}$ depends only on T and x_A, since vapor-phase nonidealities are small. Because $\alpha_{A,B}$ depends on x_A, it is not constant. For the three binary systems in Table 4.1, at 1 atm pressure, T–y_A–x_A data are presented. Both phases become richer in the less-volatile component, B, as temperature increases. For $x_A = 1$, the temperature is the boiling point of A at 1 atm; for $x_A = 0$, the temperature is the normal boiling point of B. For the three systems, all other data points are at temperatures between the two boiling points. Except for the pure components ($x_A = 1$ or 0), $y_A > x_A$ and $\alpha_{A,B} > 1$.

Table 4.1 Vapor–Liquid Equilibrium Data for Three Common Binary Systems at 1 atm Pressure

a. Water (A)–Glycerol (B) System
$P = 101.3$ kPa
Data of Chen and Thompson, *J. Chem. Eng. Data*, **15**, 471 (1970)

Temperature, °C	y_A	x_A	$\alpha_{A,B}$
100.0	1.0000	1.0000	
104.6	0.9996	0.8846	333
109.8	0.9991	0.7731	332
128.8	0.9980	0.4742	544
148.2	0.9964	0.3077	627
175.2	0.9898	0.1756	456
207.0	0.9804	0.0945	481
244.5	0.9341	0.0491	275
282.5	0.8308	0.0250	191
290.0	0.0000	0.0000	

b. Methanol (A)–Water (B) System
$P = 101.3$ kPa
Data of J.G. Dunlop, M.S. thesis, Brooklyn Polytechnic Institute (1948)

Temperature, °C	y_A	x_A	$\alpha_{A,B}$
64.5	1.000	1.000	
66.0	0.958	0.900	2.53
69.3	0.870	0.700	2.87
73.1	0.779	0.500	3.52
78.0	0.665	0.300	4.63
84.4	0.517	0.150	6.07
89.3	0.365	0.080	6.61
93.5	0.230	0.040	7.17
100.0	0.000	0.000	

c. *Para*-xylene (A)–*Meta*-xylene (B) System
$P = 101.3$ kPa
Data of Kato, Sato, and Hirata, *J. Chem. Eng. Jpn.*, **4**, 305 (1970)

Temperature, °C	y_A	x_A	$\alpha_{A,B}$
138.335	1.0000	1.0000	
138.491	0.8033	0.8000	1.0041
138.644	0.6049	0.6000	1.0082
138.795	0.4049	0.4000	1.0123
138.943	0.2032	0.2000	1.0160
139.088	0.0000	0.0000	

For the water–glycerol system, the difference in boiling points is 190°C. Therefore, relative volatility values are very high, making it possible to achieve a good separation in a single equilibrium stage. Industrially, the separation is often conducted in an evaporator, which produces a nearly pure water vapor and a glycerol-rich liquid. For example, as seen in the Table 4.1a, at 207°C, a vapor of 98 mol% water is in equilibrium with a liquid containing more than 90 mol% glycerol.

For the methanol–water system, in Table 4.1b, the difference in boiling points is 35.5°C and the relative volatility is an order of magnitude lower than for the water–glycerol

---

[1]Development of (4-4) assumes that the sum of mole fractions in the feed equals one. Alternatively, the equation $\sum_{i=1}^C z_i = 1$ can be added to the number of independent equations (thus forcing the feed mole fractions to sum to one). Then, the degrees of freedom becomes one less or $C + 4$.

system. A sharp separation cannot be made with a single stage. A 30-tray distillation column is required to obtain a 99 mol% methanol distillate and a 98 mol% water bottoms.

For the paraxylene–metaxylene isomer system in Table 4.1c, the boiling-point difference is only 0.8°C and the relative volatility is very close to 1.0, making separation by distillation impractical because about 1,000 trays are required to produce nearly pure products. Instead, crystallization and adsorption, which have much higher separation factors, are used commercially.

Vapor–liquid equilibrium data for methanol–water in Table 4.2 are in the form of $P–y_A–x_A$ for temperatures of 50, 150, and 250°C. The data cover a wide pressure range of 1.789 to 1,234 psia, with temperatures increasing with pressure. At 50°C, α_{AB} averages 4.94. At 150°C, the average α_{AB} is only 3.22; and at 250°C, it is 1.75. Thus, as temperature and pressure increase, α_{AB} decreases. For the data set at 250°C, it is seen that as compositions become richer in methanol, a point is reached near 1,219 psia, at a methanol mole fraction of 0.772, where the relative volatility is 1.0 and distillation is impossible because the vapor and liquid compositions are identical and the two phases become one. This is the critical point for the mixture. It is intermediate between the critical points of methanol and water:

$y_A = x_A$	T_c, °C	P_c, psia
0.000	374.1	3,208
0.772	250	1,219
1.000	240	1,154

Critical conditions exist for each binary composition. In industry, distillation columns operate at pressures well below the critical pressure of the mixture to avoid relative volatilities that approach 1.

The data for the methanol–water system are plotted in three different ways in Figure 4.2: (a) T vs. y_A or x_A at $P = 1$ atm; (b) y_A vs. x_A at $P = 1$ atm; and (c) P vs. x_A at $T = 150°C$. These plots satisfy the requirement of the Gibbs phase rule that when two intensive variables are fixed, all other variables are determined. Of the three diagrams in Figure 4.2, only (a) contains the complete data; (b) does not contain temperatures; and (c) does not contain vapor-phase mole fractions. Mass or mole fractions could be used, but the latter are preferred because vapor–liquid equilibrium relations are always based on molar properties.

Plots like Figure 4.2a are useful for determining phase states, phase-transition temperatures, phase compositions, and phase amounts. Consider the $T–y–x$ plot in Figure 4.3 for the n-hexane (H)–n-octane (O) system at 101.3 kPa.

The upper curve, labeled "Saturated vapor," gives the dependency of vapor mole fraction on the dew-point temperature; the lower curve, labeled "Saturated liquid," shows the bubble-point temperature as a function of liquid-phase mole fraction. The two curves converge at $x_H = 0$, the normal boiling point of n-octane (258.2°F), and at $x_H = 1$, the boiling point of normal hexane (155.7°F). For two phases to exist, a point representing the overall composition of the binary mixture at a given temperature must be located in the two-phase

Table 4.2 Vapor–Liquid Equilibrium Data for the Methanol–Water System at Temperatures of 50, 150, and 250°C

a. Methanol (A)–Water (B) System
  $T = 50°C$
  Data of McGlashan and Williamson, *J. Chem. Eng. Data*, **21**, 196 (1976)

Pressure, psia	y_A	x_A	$\alpha_{A,B}$
1.789	0.0000	0.0000	
2.373	0.2661	0.0453	7.64
3.369	0.5227	0.1387	6.80
4.641	0.7087	0.3137	5.32
5.771	0.8212	0.5411	3.90
6.811	0.9090	0.7598	3.16
7.800	0.9817	0.9514	2.74
8.072	1.0000	0.0000	

b. Methanol (A)–Water (B) System
  $T = 150°C$
  Data of Griswold and Wong, *Chem. Eng. Prog. Symp. Ser.*, **48**(3), 18 (1952)

Pressure, psia	y_A	x_A	$\alpha_{A,B}$
73.3	0.060	0.009	7.03
85.7	0.213	0.044	5.88
93.9	0.286	0.079	4.67
139.7	0.610	0.374	2.62
160.4	0.731	0.578	1.98
193.5	0.929	0.893	1.57
196.5	0.960	0.936	1.64
199.2	0.982	0.969	1.75

c. Methanol (A)–Water (B) System
  $T = 250°C$
  Data of Griswold and Wong, *Chem. Eng. Prog. Symp. Ser.*, **48**(3), 18 (1952)

Pressure, psia	y_A	x_A	$\alpha_{A,B}$
681	0.163	0.066	2.76
818	0.344	0.180	2.39
949	0.487	0.331	1.92
1099	0.643	0.553	1.46
1204	0.756	0.732	1.13
1219	0.772	0.772	1.00
1234	0.797	0.797	1.00

region between the two curves. If the point lies above the saturated-vapor curve, a superheated vapor exists; if the point lies below the saturated-liquid curve, a subcooled liquid exists.

Consider a mixture of 30 mol% H at 150°F. From Figure 4.3, point A is a subcooled liquid with $x_H = 0.3$. When this mixture is heated at 1 atm, it remains liquid until a temperature of 210°F, point B, is reached. This is the *bubble point* where the first bubble of vapor appears. This bubble is a saturated vapor in equilibrium with the liquid at the same

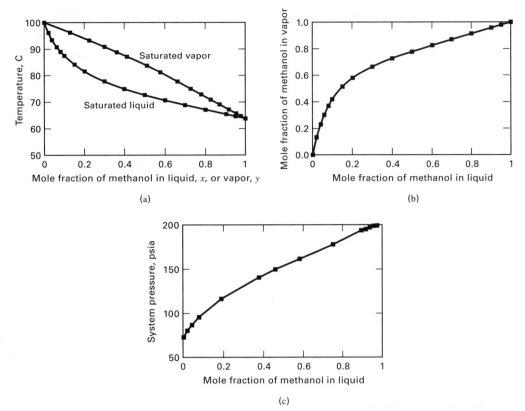

(a)

(b)

(c)

Figure 4.2 Vapor–liquid equilibrium conditions for the methanol–water system: (a) T–y–x diagram for 1 atm pressure; (b) y–x diagram for 1 atm pressure; (c) P–x diagram for 150°C.

temperature. Its composition is determined by following a *tie line*, which in Figure 4.3 is BC, from $x_H = 0.3$ to $y_H = 0.7$. This tie line is horizontal because the phase temperatures are equal. As the temperature of the two-phase mixture is increased to point E, on horizontal tie line DEF at 225°F, the mole fraction of H in the liquid phase decreases to $x_H = 0.17$ (because it is more volatile than O and preferentially vaporizes), and the mole fraction of H in the vapor phase increases

to $y_H = 0.55$. Throughout the two-phase region, the vapor is at its dew point, and the equilibrium liquid is at its bubble point. The overall composition of the two phases remains at a mole fraction of 0.30 for hexane. At point E, the relative phase amounts are determined by the inverse-lever-arm rule using the lengths of line segments DE and EF. Referring to Figures 4.1b and 4.3, $V/L = DE/EF$ or $V/F = DE/DEF$. When the temperature is increased to 245°F, point G, the dew point for $y_H = 0.3$, is reached, where only a differential amount of liquid remains. An increase in temperature to point H at 275°F gives a superheated vapor with $y_H = 0.30$.

Constant-pressure x–y plots like Figure 4.2b are useful because the vapor-and-liquid compositions are points on the equilibrium curve. However, temperatures are not included. Such plots include a 45° reference line, $y = x$. The y–x plot of Figure 4.4 for H–O at 101.3 kPa is convenient for determining compositions as a function of mole-percent vaporization by geometric constructions as follows.

Consider feed mixture F in Figure 4.1b, of overall composition $z_H = 0.6$. To determine the phase compositions if, say, 60 mol% of the feed is vaporized, the dashed-line construction in Figure 4.4 is used. Point A on the 45° line represents z_H. Point B is reached by extending a line, called the *q-line*, upward and to the left toward the equilibrium curve at a slope equal to $[(V/F) - 1]/(V/F)$. Thus, for 60 mol% vaporization, the slope $= (0.6 - 1)/0.6 = -\frac{2}{3}$. Point B at the intersection of line AB with the equilibrium curve is the equilibrium composition $y_H = 0.76$ and $x_H = 0.37$. This computation requires a trial-and-error placement of a horizontal line using

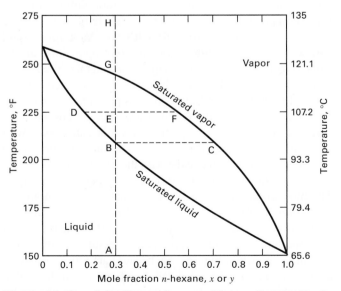

Figure 4.3 Use of the T–y–x phase equilibrium diagram for the n-hexane–n-octane system at 1 atm.

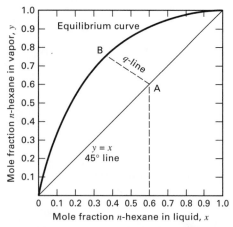

Figure 4.4 Use of the y–x phase-equilibrium diagram for the n-hexane–n-octane system at 1 atm.

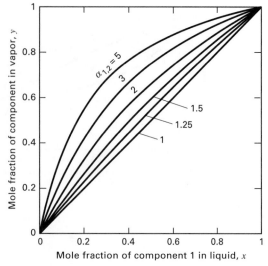

Figure 4.5 Vapor–liquid equilibrium curves for constant values of relative volatility.

Figure 4.3. The derivation of the slope of the q-line in Figure 4.4 follows by combining

$$Fz_H = Vy_H + Lx_H$$

with the total mole balance,

$$F = V + L$$

to eliminate L, giving the q-line equation:

$$y_H = \left[\frac{(V/F) - 1}{(V/F)}\right]x_H + \left[\frac{1}{(V/F)}\right]z_H$$

Thus, the slope of the q-line passing through the equilibrium point (y_H, x_H) is $[(V/F) - 1]/(V/F)$ and the line does pass through the point $z_H = x_H = y_H$.

Figure 4.2c is seldom used, but it illustrates, for a fixed temperature, the extent to which the mixture deviates from Raoult's law, which predicts the total pressure to be

$$\begin{aligned} P &= P_A^s x_A + P_B^s x_B \\ &= P_A^s x_A + P_B^s(1 - x_A) \\ &= P_B^s + x_A(P_A^s - P_B^s) \end{aligned} \quad (4\text{-}6)$$

Thus, in this case, a plot of P versus x_A is a straight line with intersections at the vapor pressure of B for $x_A = 0$ and that of A for $x_B = 0$. The greater the departure from a straight line, the greater the deviation from Raoult's law.

If the vapor phase is as in Figure 4.2c, deviations from Raoult's law are positive, and species liquid-phase activity coefficients are greater than 1; if the curve is concave, deviations are negative and activity coefficients are less than 1. In either case, the total pressure is

$$P = \gamma_A P_A^s x_A + \gamma_B P_B^s x_B \quad (4\text{-}7)$$

For narrow-boiling binary mixtures that exhibit ideal or nearly ideal behavior, the relative volatility, $\alpha_{A,B}$, varies little with pressure. If $\alpha_{A,B}$ is constant over the entire composition range, the y–x phase-equilibrium curve can be determined and plotted from a rearrangement of (4-5):

$$y_A = \frac{\alpha_{A,B} x_A}{1 + x_A(\alpha_{A,B} - 1)} \quad (4\text{-}8)$$

If Raoult's law applies, $\alpha_{A,B}$ can be approximated by

$$\alpha_{A,B} = \frac{K_A}{K_B} = \frac{P_A^s/P}{P_B^s/P} = \frac{P_A^s}{P_B^s} \quad (4\text{-}9)$$

Thus, from a knowledge of just the vapor pressures of the two components at a given temperature, a y–x phase-equilibrium curve can be approximated using only one value of $\alpha_{A,B}$. Families of curves, as shown in Figure 4.5, can be used for preliminary calculations in the absence of detailed experimental data. The use of (4-8) and (4-9) is not recommended for wide-boiling or nonideal mixtures.

§4.3 BINARY AZEOTROPIC SYSTEMS

Departures from Raoult's law commonly manifest themselves in the formation of *azeotropes*; indeed, many close-boiling, nonideal mixtures form azeotropes, particularly those of different chemical types. Azeotropic-forming mixtures exhibit either maximum- or minimum-boiling points at some composition, corresponding, respectively, to negative and positive deviations from Raoult's law. Vapor and liquid compositions are identical for an azeotrope; thus, all K-values are 1, $\alpha_{AB} = 1$, and no separation can take place.

If only one liquid phase exists, it is a *homogeneous* azeotrope; if more than one liquid phase is present, the azeotrope is *heterogeneous*. By the Gibbs phase rule, at constant pressure in a two-component system, the vapor can coexist with no more than two liquid phases; in a ternary mixture, up to three liquid phases can coexist with the vapor, and so on,

Figures 4.6, 4.7, and 4.8 show three types of azeotropes. The most common by far is the minimum-boiling homogeneous azeotrope, e.g., isopropyl ether–isopropyl alcohol, shown in Figure 4.6. At a temperature of 70°C, the maximum total pressure is greater than the vapor pressure of either component, as shown in Figure 4.6a, because activity coefficients are greater than 1. The y–x diagram in Figure 4.6b shows that for a pressure of 1 atm, the azeotropic mixture is at 78 mol% ether. Figure 4.6c is a T–x diagram at 1 atm, where the

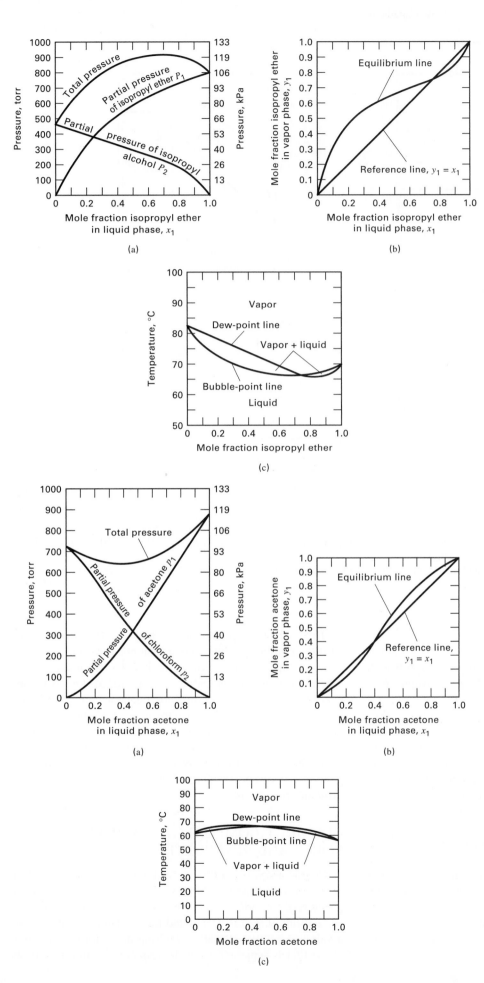

Figure 4.6 Minimum-boiling-point azeotrope, isopropyl ether–isopropyl alcohol system: (a) partial and total pressures at 70°C; (b) vapor–liquid equilibria at 101 kPa; (c) phase diagram at 101 kPa.
[Adapted from O.A. Hougen, K.M. Watson, and R.A. Ragatz, *Chemical Process Principles. Part II*, 2nd ed., John Wiley & Sons, New York (1959).]

Figure 4.7 Maximum-boiling-point azeotrope, acetone–chloroform system: (a) partial and total pressures at 60°C; (b) vapor–liquid equilibria at 101 kPa; (c) phase diagram at 101 kPa pressure.
[Adapted from O.A. Hougen, K.M. Watson, and R.A. Ragatz, *Chemical Process Principles. Part II*, 2nd ed., John Wiley & Sons, New York (1959).]

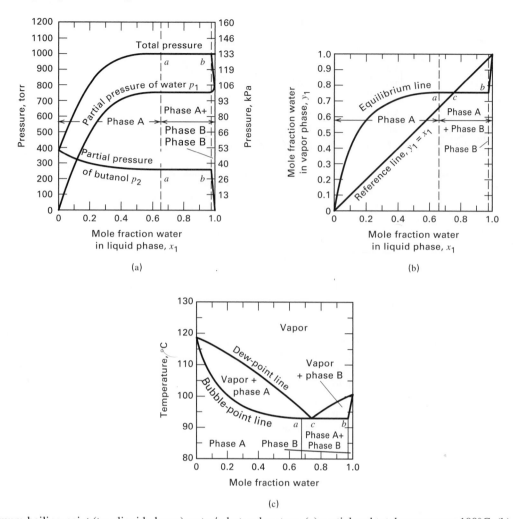

Figure 4.8 Minimum-boiling-point (two liquid phases) water/n-butanol system: (a) partial and total pressures at 100°C; (b) vapor–liquid equilibria at 101 kPa; (c) phase diagram at 101 kPa pressure.
[Adapted from O.A. Hougen, K.M. Watson, and R.A. Ragatz, *Chemical Process Principles. Part II*, 2nd ed., John Wiley & Sons, New York (1959).]

azeotrope is seen to boil at 66°C. In Figure 4.6a, for 70°C, the azeotrope occurs at 123 kPa (923 torr), for 72 mol% ether. Thus, the azeotropic composition and temperature shift with pressure. In distillation, minimum-boiling azeotropic mixtures are approached in the overhead product.

For the maximum-boiling homogeneous azeotropic acetone–chloroform system in Figure 4.7a, the minimum total pressure at 60°C is below the vapor pressures of the pure components because activity coefficients are less than 1. The azeotrope is approached in the bottoms product in a distillation operation. Phase compositions at 1 atm are shown in Figures 4.7b and c.

Heterogeneous azeotropes are minimum-boiling because activity coefficients must be significantly greater than 1 to form two liquid phases. The region a–b in Figure 4.8a for the water–n-butanol system is a two-phase region, where total and partial pressures remain constant as the amounts of the phases change, but phase compositions do not. The y–x diagram in Figure 4.8b shows a horizontal line over the immiscible region, and the phase diagram of Figure 4.8c shows a minimum constant temperature.

To avoid azeotrope limitations, it is sometimes possible to shift the equilibrium by changing the pressure sufficiently to "break" the azeotrope, or move it away from the region where the required separation is to be made. For example, ethyl alcohol and water form a homogeneous minimum-boiling azeotrope of 95.6 wt% alcohol at 78.15°C and 101.3 kPa. However, at vacuums of less than 9.3 kPa, no azeotrope is formed. As discussed in Chapter 11, ternary azeotropes also occur, in which azeotrope formation in general, and heterogeneous azeotropes in particular, are employed to achieve difficult separations.

§4.4 MULTICOMPONENT FLASH, BUBBLE-POINT, AND DEW-POINT CALCULATIONS

A *flash* is a single-equilibrium-stage distillation in which a feed is partially vaporized to give a vapor richer than the feed in the more volatile components. In Figure 4.9a, (1) a pressurized liquid feed is heated and flashed adiabatically across a valve to a lower pressure, resulting in creation of a vapor phase that is separated from the remaining liquid in a flash

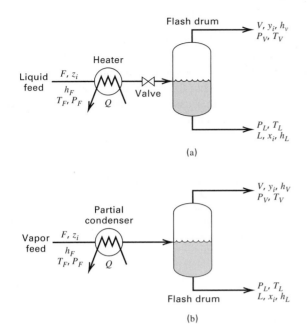

Figure 4.9 Continuous, single-stage equilibrium separations: (a) flash vaporization and (b) partial condensation.

drum, or (2) if the valve is omitted, a liquid can be partially vaporized in a heater and then separated into two phases. Alternatively, a vapor feed can be cooled and partially condensed, as in Figure 4.9b, to give, after phase separation, a liquid richer in the less-volatile components. For properly designed systems, the streams leaving the drum will be in phase equilibrium [4].

Unless the relative volatility, α_{AB}, is very large, flashing (partial vaporization) or partial condensation is not a replacement for distillation, but an auxiliary operation used to prepare streams for further processing.

Single-stage flash calculations are among the most common calculations in chemical engineering. They are used not only for the operations in Figure 4.9, but also to determine the phase condition of mixtures anywhere in a process, e.g. in a pipeline.

For the single-stage operation in Figure 4.9, the $2C + 5$ equations listed in Table 4.3 apply. (In Figure 4.9, T and P are given separately for the vapor and liquid products to emphasize the need to assume mechanical and thermal equilibrium.) The equations relate the $3C + 10$ variables (F, V, L, z_i, y_i, x_i, T_F, T_V, T_L, P_F, P_V, P_L, Q) and leave $C + 5$ degrees of freedom. Assuming that $C + 3$ feed variables F, T_F, P_F, and C values of z_i are known, two additional variables can be specified for a flash calculation. The most common sets of specifications are:

T_V, P_V	Isothermal flash
$V/F = 0, P_L$	Bubble-point temperature
$V/F = 1, P_V$	Dew-point temperature
$T_L, V/F = 0$	Bubble-point pressure
$T_V, V/F = 1$	Dew-point pressure
$Q = 0, P_V$	Adiabatic flash
Q, P_V	Nonadiabatic flash
$V/F, P_V$	Percent vaporization flash

Table 4.3 Equations for Single-Stage Flash Vaporization and Partial Condensation Operations. Feed mole fractions must sum to one.

Equation		Number of Equations
(1) $P_V = P_L$	(mechanical equilibrium)	1
(2) $T_V = T_L$	(thermal equilibrium)	1
(3) $y_i = K_i x_i$	(phase equilibrium)	C
(4) $Fz_i = Vy_i + Lx_i$	(component material balance)	C
(5) $F = V + L$	(total material balance)	1
(6) $h_F F + Q = h_V V + h_L L$	(energy balance)	1
(7) $\sum_i y_i - \sum_i x_i = 0$	(summations)	1
		$\mathscr{E} = 2C + 5$

$K_i = K_i\{T_V, P_V, \mathbf{y}, \mathbf{x}\}$ $\qquad h_F = h_F\{T_F, P_F, \mathbf{z}\}$
$h_V = h_V\{T_V, P_V, \mathbf{y}\}$ $\qquad h_L = h_L\{T_L, P_L, \mathbf{x}\}$

§4.4.1 Isothermal Flash

If the equilibrium temperature T_V (or T_L) and the equilibrium pressure P_V (or P_L) in the drum are specified, values of the remaining $2C + 5$ variables are determined from $2C + 5$ equations as given in Table 4.3.

Isothermal-flash calculations are not straightforward because Eq. (4) in Table 4.3 is a nonlinear equation in the unknowns V, L, y_i, and x_i. The solution procedure of Rachford and Rice [5], widely used in process simulators and described next, is given in Table 4.4.

Equations containing only a single unknown are solved first. Thus, Eqs. (1) and (2) in Table 4.3 are solved, respectively, for P_L and T_L. The unknown Q appears only in (6), so Q is computed after all other equations have been solved. This leaves Eqs. (3), (4), (5), and (7) in Table 4.3 to be solved for V, L, and all values of y and x. These equations can be partitioned to solve for the unknowns in a sequential manner

Table 4.4 Rachford–Rice Procedure for Isothermal-Flash Calculations When K-Values Are Independent of Composition

Specified variables: $F, T_F, P_F, z_1, z_2, \ldots, z_C, T_V, P_V$
Steps
(1) $T_L = T_V$
(2) $P_L = P_V$
(3) Solve

$$f\{\Psi\} = \sum_{i=1}^{C} \frac{z_i(1 - K_i)}{1 + \Psi(K_i - 1)} = 0$$

for $\Psi = V/F$, where $K_i = K_i\{T_V, P_V\}$.
(4) $V = F\Psi$
(5) $x_i = \dfrac{z_i}{1 + \Psi(K_i - 1)}$

(6) $y_i = \dfrac{z_i K_i}{1 + \Psi(K_i - 1)} = x_i K_i$

(7) $L = F - V$
(8) $Q = h_V V + h_L L - h_F F$

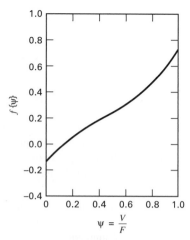

Figure 4.10 Rachford–Rice function for Example 4.1.

by substituting Eq. (5) into Eq. (4) to eliminate L and combining the result with Eq. (3) to obtain Eqs. (5) and (6) in Table 4.4. Here (5) is in x_i but not y_i, and (6) is in y_i but not x_i. Summing these two equations and combining them with $\sum y_i - \sum x_i = 0$ to eliminate y_i and x_i gives Eq. (3) in Table 4.4, a nonlinear equation in V (or $\Psi = V/F$) only. Upon solving this equation numerically in an iterative manner for Ψ and then V, from Eq. (4), the remaining unknowns are obtained directly from Eqs. (5) through (8) in Table 4.4. When T_F and/or P_F are not specified, Eq. (6) of Table 4.3 is not solved for Q.

Equation (3) of Table 4.4 can be solved iteratively by guessing values of Ψ between 0 and 1 until the function $f\{\Psi\} = 0$. A typical function, encountered in Example 4.1, is shown in Figure 4.10. The most widely employed procedure for solving Eq. (3) of Table 4.4 is Newton's method [6]. A value of the Ψ root for iteration $k + 1$ is computed by the recursive relation

$$\Psi^{(k+1)} = \Psi^{(k)} - \frac{f\{\Psi^{(k)}\}}{f'\{\Psi^{(k)}\}} \qquad (4\text{-}10)$$

where the superscript is the iteration index, and the derivative of $f\{\Psi\}$, from Eq. (3) in Table 4.4, with respect to Ψ is

$$f'\{\Psi^{(k)}\} = \sum_{i=1}^{C} \frac{z_i(1 - K_i)^2}{[1 + \Psi^{(k)}(K_i - 1)]^2} \qquad (4\text{-}11)$$

The iteration can be initiated by assuming $\Psi^{(1)} = 0.5$. Sufficient accuracy is achieved by terminating the iterations when $|\Psi^{(k+1)} - \Psi^{(k)}|/\Psi^{(k)} < 0.0001$.

The existence of a valid root ($0 \leq \Psi \leq 1$) must be checked before employing the procedure of Table 4.4, by checking if the equilibrium condition corresponds to subcooled liquid or superheated vapor rather than partial vaporization or condensation. A first estimate of whether a multicomponent feed gives a two-phase mixture is made by inspecting the K-values. If all K-values are > 1, the phase is superheated vapor. If all K-values are < 1, the single phase is a subcooled liquid. If one or more K-values are greater than 1 and one or more K-values are less than 1, the check is made by first

computing $f\{\Psi\}$ from Eq. (3) in Table 4.4 for $\Psi = 0$. If the resulting $f\{0\} > 0$, the mixture is below its bubble point (subcooled liquid). Alternatively, if $f\{1\} < 0$, the mixture is above the dew point (superheated vapor). The Rachford–Rice procedure may fail to converge if K-values are sensitive to composition. In that case, the method of Boston and Britt [19] is employed in some process simulators.

EXAMPLE 4.1   Phase Conditions of a Process Stream.

A 100-kmol/h feed consisting of 10, 20, 30, and 40 mol% of propane (3), n-butane (4), n-pentane (5), and n-hexane (6), respectively, enters a distillation column at 100 psia (689.5 kPa) and 200°F (366.5°K). Assuming equilibrium, what fraction of the feed enters as liquid, and what are the liquid and vapor compositions?

Solution

At flash conditions, from Figure 2.4, $K_3 = 4.2$, $K_4 = 1.75$, $K_5 = 0.74$, $K_6 = 0.34$, independent of compositions. Because some K-values are greater than 1 and some less than 1, it is necessary first to compute values of $f\{0\}$ and $f\{1\}$ for Eq. (3) in Table 4.4 to see if the mixture is between the bubble and dew points.

$$f\{0\} = \frac{0.1(1 - 4.2)}{1} + \frac{0.2(1 - 1.75)}{1}$$
$$+ \frac{0.3(1 - 0.74)}{1} + \frac{0.4(1 - 0.34)}{1} = -0.128$$

Since $f\{0\}$ is not more than zero, the mixture is above the bubble point. Now compute $f\{1\}$:

$$f\{1\} = \frac{0.1(1 - 4.2)}{1 + (4.2 - 1)} + \frac{0.2(1 - 1.75)}{1 + (1.75 - 1)}$$
$$+ \frac{0.3(1 - 0.74)}{1 + (0.74 - 1)} + \frac{0.4(1 - 0.34)}{1 + (0.34 - 1)} = 0.720$$

Since $f\{1\}$ is not less than zero, the mixture is below the dew point. Therefore, the mixture is part vapor. Using the Rachford–Rice procedure and substituting z_i and K_i values into Eq. (3) of Table 4.4 gives

$$0 = \frac{0.1(1 - 4.2)}{1 + \Psi(4.2 - 1)} + \frac{0.2(1 - 1.75)}{1 + \Psi(1.75 - 1)}$$
$$+ \frac{0.3(1 - 0.74)}{1 + \Psi(0.74 - 1)} + \frac{0.4(1 - 0.34)}{1 + \Psi(0.34 - 1)}$$

Solving this equation by Newton's method using an initial guess for Ψ of 0.50 gives the following iteration history:

k	$\Psi^{(k)}$	$f\{\Psi^{(k)}\}$	$f'\{\Psi^{(k)}\}$	$\Psi^{(k+1)}$	$\left\|\dfrac{\Psi^{(k+1)} - \Psi^{(k)}}{\Psi^{(k)}}\right\|$
1	0.5000	0.2515	0.6259	0.0982	0.8037
2	0.0982	−0.0209	0.9111	0.1211	0.2335
3	0.1211	−0.0007	0.8539	0.1219	0.0065
4	0.1219	0.0000	0.8521	0.1219	0.0000

Convergence is rapid, giving $\Psi = V/F = 0.1219$. From Eq. (4) of Table 4.4, the vapor flow rate is $0.1219(100) = 12.19$ kmol/h, and the liquid flow rate from Eq. (7) is $(100 - 12.19) = 87.81$ kmol/h. Liquid and vapor compositions from Eqs. (5) and (6) are

	x	y
Propane	0.0719	0.3021
n-Butane	0.1833	0.3207
n-Pentane	0.3098	0.2293
n-Hexane	0.4350	0.1479
Sum	1.0000	1.0000

A plot of $f\{\Psi\}$ as a function of Ψ is shown in Figure 4.10.

§4.4.2 Bubble and Dew Points

At the *bubble point*, $\Psi = 0$ and $f\{0\} = 0$. By Eq. (3) of Table 4.4

$$f\{0\} = \sum_i z_i(1 - K_i) = \sum z_i - \sum z_i K_i = 0$$

However, $\sum z_i = 1$. Therefore, the bubble-point equation is

$$\sum_i z_i K_i = 1 \qquad (4\text{-}12)$$

At the *dew point*, $\Psi = 1$ and $f\{1\} = 0$. From Eq. (3) of Table 4.4,

$$f\{1\} = \sum_i \frac{z_i(1 - K_i)}{K_i} = \sum \frac{z_i}{K_i} - \sum z_i = 0$$

Therefore, the dew-point equation is

$$\sum_i \frac{z_i}{K_i} = 1 \qquad (4\text{-}13)$$

For a given feed composition, z_i, (4-12) or (4-13) can be used to find T for a specified P or to find P for a specified T.

The bubble- and dew-point equations are nonlinear in temperature, but only moderately nonlinear in pressure, except in the region of the *convergence pressure*, where K-values of very light or very heavy species change drastically with pressure, as in Figure 2.6. Therefore, iterative procedures are required to solve for bubble- and dew-point conditions except if Raoult's law K-values are applicable. Substitution of $K_i = P_i^s/P$ into (4-12) allows direct calculation of bubble-point pressure:

$$P_{\text{bubble}} = \sum_{i=1}^{C} z_i P_i^s \qquad (4\text{-}14)$$

where P_i^s is the temperature-dependent vapor pressure of species i. Similarly, from (4-13), the dew-point pressure is

$$P_{\text{dew}} = \left(\sum_{i=1}^{C} \frac{z_i}{P_i^s}\right)^{-1} \qquad (4\text{-}15)$$

Another exception occurs for mixtures at the bubble point when K-values can be expressed by the modified Raoult's law, $K_i = \gamma_i P_i^s/P$. Substituting into (4-12)

$$P_{\text{bubble}} = \sum_{i=1}^{C} \gamma_i z_i P_i^s \qquad (4\text{-}16)$$

Thus, liquid activity coefficients can be computed at a known bubble-point temperature and composition, since $x_i = z_i$ at the bubble point.

Bubble- and dew-point calculations are used to determine saturation conditions for liquid and vapor streams. Whenever there is vapor–liquid equilibrium, the vapor is at its dew point and the liquid is at its bubble point.

EXAMPLE 4.2 Bubble-Point Temperature.

In Figure 1.9, the nC_4-rich bottoms product from Column C3 has the composition given in Table 1.5. If the pressure at the bottom of the distillation column is 100 psia (689 kPa), estimate the mixture temperature.

Solution

The bottoms product is a liquid at its bubble point with the following composition:

Component	kmol/h	$z_i = x_i$
i-Butane	8.60	0.0319
n-Butane	215.80	0.7992
i-Pentane	28.10	0.1041
n-Pentane	17.50	0.0648
	270.00	1.0000

The bubble-point temperature can be estimated by finding the temperature that will satisfy (4-12), using K-values from Figure 2.4. Because the bottoms product is rich in nC_4, assume the K-value of nC_4 is 1. From Figure 2.4, for 100 psia, $T = 150°F$. For this temperature, using Figure 2.4 to obtain K-values of the other three components and substituting these values and the z-values into (4-12),

$$\sum z_i K_i = 0.0319(1.3) + 0.7992(1.0) + 0.1041(0.47)$$
$$+ 0.0648(0.38)$$
$$= 0.042 + 0.799 + 0.049 + 0.025 = 0.915$$

The sum is not 1.0, so another temperature is assumed and the summation repeated. To increase the sum, the K-values must increase and, thus, the temperature must increase as well. Because the sum is dominated by nC_4, assume its K-value $= 1.09$. This corresponds to a temperature of 160°F, which results in a summation of 1.01. By linear interpolation, $T = 159°F$.

EXAMPLE 4.3 Bubble-Point Pressure.

Cyclopentane is separated from cyclohexane by liquid–liquid extraction with methanol at 25°C. To prevent vaporization, the mixture must be above the bubble-point pressure. Calculate that pressure using the following compositions, activity coefficients, and vapor pressures:

	Methanol	Cyclohexane	Cyclopentane
Vapor pressure, psia	2.45	1.89	6.14
Methanol-rich layer:			
x	0.7615	0.1499	0.0886
γ	1.118	4.773	3.467
Cyclohexane-rich layer:			
x	0.1737	0.5402	0.2861
γ	4.901	1.324	1.074

Solution

Assume the modified Raoult's law in the form of (4-16) applies for either liquid phase. If the methanol-rich-layer data are used:

$$P_{\text{bubble}} = 1.118(0.7615)(2.45) + 4.773(0.1499)(1.89)$$
$$+ 3.467(0.0886)(6.14)$$
$$= 5.32 \text{ psia } (36.7 \text{ kPa})$$

A similar calculation based on the cyclohexane-rich layer gives an identical result because the data are consistent; thus $\gamma_{iL}^{(1)} x_i^{(1)} = \gamma_{iL}^{(2)} x_i^{(2)}$. A pressure higher than 5.32 psia will prevent formation of vapor at this location in the extraction process. Operation at atmospheric pressure is viable.

EXAMPLE 4.4   Distillation column operating pressure.

Propylene (P) is separated from 1-butene (B) by distillation into a vapor distillate containing 90 mol% propylene. Calculate the column pressure if the partial condenser exit temperature is 100°F (37.8°C), the lowest attainable temperature with cooling water. Determine the composition of the liquid reflux. In Figure 4.11, K-values estimated from Eq. (5) of Table 2.3, using the Redlich–Kwong equation of state for vapor fugacity, are plotted and compared to experimental data [7] and Raoult's law K-values.

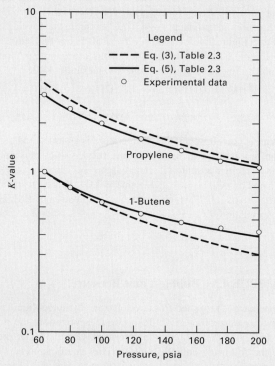

Figure 4.11 K-values for propylene/1-butene system at 100°F.

Solution

The column pressure is at the dew point for the vapor distillate. The reflux composition is that of a liquid in equilibrium with the vapor distillate at its dew point. The method of false position [8] is used to perform the calculations by rewriting (4-13) in the form

$$f\{P\} = \sum_{i=1}^{C} \frac{z_i}{K_i} - 1 \tag{1}$$

The recursion relationship for the method of false position is based on the assumption that $f\{P\}$ is linear in P such that

$$P^{(k+2)} = P^{(k+1)} - f\left\{P^{(k+1)}\right\}\left[\frac{P^{(k+1)} - P^{(k)}}{f\{P^{(k+1)}\} - f\{P^{(k)}\}}\right] \tag{2}$$

This is reasonable because, at low pressures, K-values in (2) are almost inversely proportional to pressure. Two values of P are required to initialize this formula. Choose 100 psia and 190 psia. At $P^{(1)} = 100$ psia, K-values from the solid lines in Figure 4.11, when substituted into Eq. (1) give,

$$f\{P\} = \frac{0.90}{2.0} + \frac{0.10}{0.68} - 1.0 = -0.40$$

Similarly, for $P^{(2)} = 190$ psia, $f\{P\} = +0.02$. Substitution into Eq. (2) gives $P^{(3)} = 186$, and so on. Iterations end when $|P^{(k+2)} - P^{(k+1)}| /P^{(k+1)} < 0.005$. In this example, that occurs when $k = 3$. Thus, the operating pressure at the partial condenser outlet is 186 psia (1,282 kPa). The liquid reflux composition is obtained from $x_i = z_i/K_i$, using K-values at that pressure. The final results are:

Component	Equilibrium Mole Fraction	
	Vapor Distillate	Liquid Reflux
Propylene	0.90	0.76
1-Butene	0.10	0.24
	1.00	1.00

§4.4.3 Adiabatic Flash

When the pressure of a liquid stream is reduced adiabatically across a valve as in Figure 4.9a, an *adiabatic-flash* $(Q = 0)$ calculation is made to determine the resulting phases, temperature, compositions, and flow rates for a specified downstream pressure. The calculation is made by applying the isothermal-flash calculation procedure of §4.4.1 in an iterative manner. First a guess is made of the flash temperature, T_V. Then Ψ, V, x, y, and L are determined, as for an isothermal flash, from steps 3 through 7 in Table 4.4. The guessed value of T_V (equal to T_L) is next checked by an energy balance obtained by combining Eqs. (7) and (8) of Table 4.4 with $Q = 0$ to give

$$f\{T_V\} = \frac{\Psi h_V + (1 - \Psi)h_L - h_F}{1,000} = 0 \tag{4-17}$$

where division by 1,000 makes the terms of the order 1. Enthalpies are computed at $T_V = T_L$. If the computed value of $f\{T_V\}$ is not zero, the entire procedure is repeated. A plot of $f\{T_V\}$ versus T_V is interpolated to determine the correct value of T_V. The procedure is tedious because it involves inner-loop iteration on Ψ and outer-loop iteration on T_V.

Outer-loop iteration on T_V is successful when Eq. (3) of Table 4.4 is not sensitive to the guess of T_V. This is the case for wide-boiling mixtures. For close-boiling mixtures, the algorithm may fail because of sensitivity to the value of T_V. In this case, it is preferable to do the outer-loop iteration on Ψ and solve Eq. (3) of Table 4.4 for T_V in the inner loop, using a guessed value for Ψ to initiate the process, as follows:

$$f\{T_V\} = \sum_{i=1}^{C} \frac{z_i(1 - K_i)}{1 + \Psi(K_i - 1)} = 0 \qquad (4\text{-}18)$$

Then, Eqs. (5) and (6) of Table 4.4 are solved for x and y. Equation (4-17) is then solved directly for Ψ, since

$$f\{\Psi\} = \frac{\Psi h_V + (1 - \Psi)h_L - h_F}{1,000} = 0 \qquad (4\text{-}19)$$

from which

$$\Psi = \frac{h_F - h_L}{h_V - h_L} \qquad (4\text{-}20)$$

If Ψ from (4-20) is not equal to the guessed Ψ, a new Ψ is used to repeat the outer loop, starting with (4-18).

Multicomponent, isothermal-flash, bubble-point, dew-point, and adiabatic-flash calculations are tedious. Especially for nonideal mixtures, required thermodynamic property expressions are complex, and calculations should be made with a process simulator.

EXAMPLE 4.5 Adiabatic Flash of the Feed to a Distillation Column.

Equilibrium liquid from the flash drum at 120°F and 485 psia in Example 2.6 is fed to a so-called "*stabilizer*" distillation tower to remove the remaining hydrogen and methane. Feed-plate pressure of the stabilizer is 165 psia (1,138 kPa). Calculate the percent molar vaporization of the feed and compositions of the vapor and liquid if the pressure is decreased adiabatically from 485 to 165 psia by a valve and pipeline pressure drop.

Solution

This problem, involving a wide-boiling feed, is best solved by using a process simulator. Using the CHEMCAD program with K-values and enthalpies from the P–R equation of state (Table 2.5), the result is:

	kmol/h		
Component	Feed 120°F 485 psia	Vapor 112°F 165 psia	Liquid 112°F 165 psia
Hydrogen	1.0	0.7	0.3
Methane	27.9	15.2	12.7
Benzene	345.1	0.4	344.7
Toluene	113.4	0.04	113.36
Total	487.4	16.34	471.06
Enthalpy, kJ/h	−1,089,000	362,000	−1,451,000

The results show that only a small amount of vapor ($\Psi = 0.0035$), predominantly H_2 and CH_4, is produced. The flash temperature of 112°F is 8°F below the feed temperature. The enthalpy of the feed is equal to the sum of the vapor and liquid product enthalpies for this adiabatic operation.

§4.5 TERNARY LIQUID–LIQUID SYSTEMS

Ternary mixtures that undergo phase splitting to form two separate liquid phases differ as to the extent of solubility of the three components in the two liquid phases. The simplest case is in Figure 4.12a, where only the *solute*, component B, has any appreciable solubility in either the *carrier*, A, or the *solvent*, C, both of which have negligible (although never zero) solubility in each other. Here, equations can be derived for a single equilibrium stage, using the variables F, S, E, and R to refer, respectively, to the flow rates (or amounts) of the feed, solvent, exiting extract (C-rich), and exiting raffinate (A-rich). By definition, the extract is the exiting liquid phase that contains the extracted solute; the raffinate is the exiting liquid phase that contains the portion of the solute, B, that is not extracted. By convention, the extract is shown as leaving from the top of the stage even though it may not have the smaller density. If the entering solvent contains no B, it is convenient to write material-balance and phase-equilibrium equations for the solute, B, in terms of molar or mass flow rates. Often, it is preferable to express compositions as mass or mole ratios instead of fractions, as follows:

Let: F_A = feed rate of carrier A; S = flow rate of solvent C; X_B = ratio *of mass* (or moles) of solute B, to mass (or moles) of the other component in the feed (F), raffinate (R), or extract (E).

Then, the solute material balance is

$$X_B^{(F)} F_A = X_B^{(E)} S + X_B^{(R)} F_A \qquad (4\text{-}21)$$

and the distribution of solute at equilibrium is given by

$$X_B^{(E)} = K'_{D_B} X_B^{(R)} \qquad (4\text{-}22)$$

where K'_{D_B} is the distribution or partition coefficient in terms of mass or mole ratios (instead of mass or mole fractions). Substituting (4-22) into (4-21) to eliminate $X_B^{(E)}$,

$$X_B^{(R)} = \frac{X_B^{(F)} F_A}{F_A + K'_{D_B} S} \qquad (4\text{-}23)$$

A useful parameter is the *extraction factor*, E_B, for the solute B:

$$E_B = K'_{D_B} S / F_A \qquad (4\text{-}24)$$

Large extraction factors result from large distribution coefficients or large ratios of solvent to carrier. Substituting (4-24) into (4-23) gives the fraction of B not extracted as

$$\boxed{X_B^{(R)} / X_B^{(F)} = \frac{1}{1 + E_B}} \qquad (4\text{-}25)$$

Thus, the larger the extraction factor, the smaller the fraction of B not extracted *or* the larger the fraction of B extracted. Alternatively, the fraction of B extracted is 1 minus (4-25) or $E_B/(1 + E_B)$.

Mass (mole) ratios, X, are related to mass (mole) fractions, x, by

$$X_i = x_i/(1 - x_i) \qquad (4\text{-}26)$$

Values of the distribution coefficient, K'_D, in terms of ratios, are related to K_D in terms of fractions as given in (2-20) by

$$K'_{D_i} = \frac{x_i^{(1)}/(1 - x_i^{(1)})}{x_i^{(2)}/(1 - x_i^{(2)})} = K_{D_i}\left(\frac{1 - x_i^{(2)}}{1 - x_i^{(1)}}\right) \qquad (4\text{-}27)$$

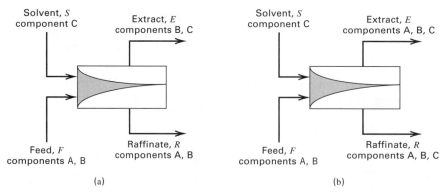

Figure 4.12 Phase splitting of ternary mixtures: (a) components A and C mutually insoluble; (b) components A and C partially soluble.

where (1) and (2) are the equilibrium solvent-rich and solvent-poor liquid phases, respectively. When values of x_i are small, K'_D approaches K_D. As discussed in Chapter 2, the distribution (partition) coefficient, which can be determined from activity coefficients by $K_{D_B} = \gamma_B^{(2)}/\gamma_B^{(1)}$ when mole fractions are used, varies with compositions and temperature. When the raffinate and extract are both dilute, solute activity coefficients can be approximated by values at infinite dilution so that K_{D_B} can be taken as constant at a given temperature. An extensive listing of such K_{D_B} values for various ternary systems is given in *Perry's Chemical Engineers' Handbook* [9]. If values for F_B, $X_B^{(F)}$, S, and K_{D_B} are given, (4-25) can be solved for $X_B^{(R)}$.

EXAMPLE 4.6 Single-Stage Extraction of Acetic Acid.

Methyl isobutyl ketone (C) is used as a solvent to remove acetic acid (B) from a 13,500 kg/h feed of 8 wt% acid in water (A), because distillation would require vaporization of large amounts of water. If the raffinate is to contain 1 wt% acetic acid, estimate the kg/h of solvent for a single equilibrium stage.

Solution

Assume the water and solvent are immiscible. From *Perry's Chemical Engineers' Handbook*, $K_D = 0.657$ in mass-fraction units. For the low concentrations of acetic acid, assume $K'_D = K_D$.

$$F_A = (0.92)(13,500) = 12,420 \text{ kg/h}$$
$$X_B^{(F)} = (13,500 - 12,420)/12,420 = 0.087$$

The raffinate is to contain 1 wt% B. Therefore,

$$X_B^{(R)} = 0.01/(1 - 0.01) = 0.0101$$

From (4-25), solving for E_B,

$$E_B = \frac{X_B^{(F)}}{X_B^{(R)}} - 1 = (0.087/0.0101) - 1 = 7.61$$

From (4-24), the definition of the extraction factor,

$$S = \frac{E_B F_A}{K'_D} = 7.61(12,420/0.657) = 144,000 \text{ kg/h}$$

This solvent/feed flow-rate ratio is very large. The use of multiple stages, as discussed in Chapter 5, could reduce the solvent rate, or a solvent with a larger distribution coefficient could be sought. For example, l-butanol as the solvent, with $K_D = 1.613$, would halve the solvent flow.

In the ternary liquid–liquid system shown in Figure 4.12b, components A and C are partially soluble in each other, and component B distributes between the extract and raffinate phases. This case is the most commonly encountered, and different phase diagrams and computational techniques have been devised for making calculations of equilibrium compositions and phase amounts. Examples of ternary-phase diagrams are shown in Figure 4.13 for the ternary system water (A)–ethylene glycol (B)–furfural (C) at 25°C and a pressure of 101 kPa, which is above the bubble-point pressure. Experimental data are from Conway and Norton [18]. Water–ethylene glycol and furfural–ethylene glycol are completely miscible pairs, while furfural–water is a partially miscible pair. Furfural can be used as a solvent to remove the solute, ethylene glycol, from water, where the furfural-rich phase is the extract, and the water-rich phase is the raffinate.

Figure 4.13a, an equilateral-triangular diagram, is the most common form of display of ternary liquid–liquid equilibrium data. Each apex is a pure component of the mixture. Each edge is a mixture of the two pure components at the terminal apexes of the side. Any point located within the triangle is a ternary mixture. In such a diagram the sum of the lengths of three perpendicular lines drawn from any interior point to the edges equals the altitude of the triangle. Thus, if each of these three lines is scaled from 0 to 100, the percent of, say, furfural, at any point such as M, is simply the length of the line perpendicular to the edge opposite the pure furfural apex. The determination of the composition of an interior point is facilitated by the three sets of parallel lines on the diagram, where each set is in mass-fraction increments of 0.1 (or 10%), and is parallel to an edge of the triangle opposite the apex of the component, whose mass fraction is given. Thus, the point M in Figure 4.13a represents a mixture of feed and solvent (before phase separation) containing 19 wt% water, 20 wt% ethylene glycol, and 61 wt% furfural.

Miscibility limits for the furfural–water binary system are at D and G. The miscibility boundary (saturation or binodal curve) DEPRG for the system is obtained experimentally by a *cloud-point titration*. For example, water is added to a completely miscible (and clear) 50 wt% solution of furfural and glycol, and it is noted that the onset of cloudiness, due to formation of a second phase, occurs when the mixture is 11% water, 44.5% furfural, and 44.5% glycol by weight. Other

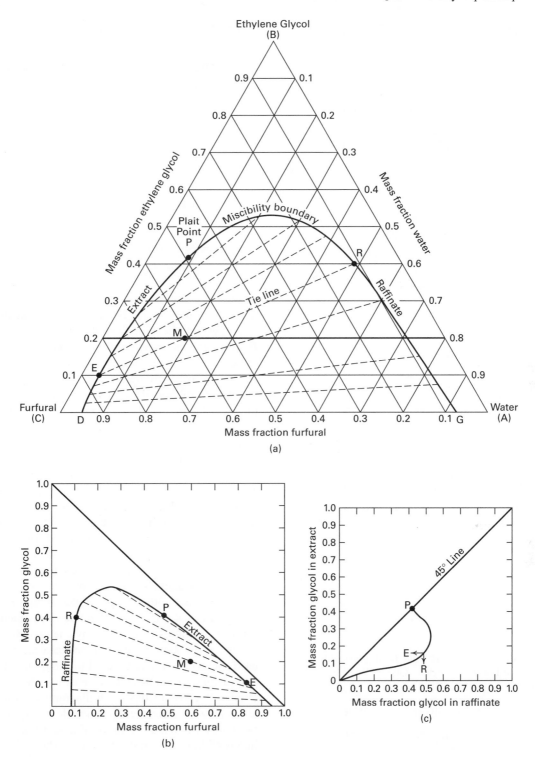

Figure 4.13 Liquid–liquid equilibrium, ethylene glycol–furfural–water, 25°C, 101 kPa: (a) equilateral-triangular diagram; (b) right-triangular diagram; (c) equilibrium solute diagram in mass fractions; (*continues*)

miscibility data are given in Table 4.5, from which the miscibility curve in Figure 4.13a was drawn.

Tie lines, shown as dashed lines below the miscibility boundary, connect equilibrium-phase composition points on the miscibility boundary. To obtain data to construct tie line ER, it is necessary to make a mixture such as M (20% glycol, 19% water, 61% furfural), equilibrate it, and then chemically analyze the resulting equilibrium extract and raffinate phases

E and R (in this case, 10% glycol, 4% water, and 86% furfural; and 40% glycol, 49% water, and 11% furfural, respectively). At point P, the *plait point*, the two liquid phases have identical compositions. Therefore, the tie lines converge to point P and the two phases become one phase. Tie-line data for this system are listed in Table 4.6, in terms of glycol composition.

When there is mutual phase solubility, thermodynamic variables necessary to define the equilibrium system are T, P,

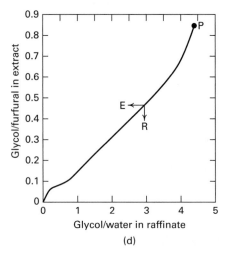

(d)

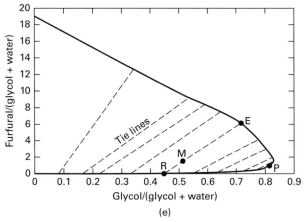

(e)

Figure 4.13 (*continued*) (d) equilibrium solute diagram in mass ratios; (e) Janecke diagram.

Table 4.5 Equilibrium Miscibility Data in Weight Percent for the Furfural–Ethylene Glycol–Water System at 25°C and 101 kPa

Furfural	Ethylene Glycol	Water
95.0	0.0	5.0
90.3	5.2	4.5
86.1	10.0	3.9
75.1	20.0	4.9
66.7	27.5	5.8
49.0	41.5	9.5
34.3	50.5	15.2
27.5	52.5	20.0
13.9	47.5	38.6
11.0	40.0	49.0
9.7	30.0	60.3
8.4	15.0	76.6
7.7	0.0	92.3

Table 4.6 Mutual Equilibrium (Tie-Line) Data for the Furfural–Ethylene Glycol–Water System at 25°C and 101 kPa

Glycol in Water Layer, wt%	Glycol in Furfural Layer, wt%
41.5	41.5
50.5	32.5
52.5	27.5
51.5	20.0
47.5	15.0
40.0	10.0
30.0	7.5
20.0	6.2
15.0	5.2
7.3	2.5

and component concentrations in each phase. According to the phase rule, (4-1), for a three-component, two-liquid-phase system, there are three degrees of freedom. With T and P specified, the concentration of one component in either phase suffices to completely define the equilibrium system. As shown in Figure 4.13a, one value for percent glycol on the miscibility boundary curve fixes the composition and, by means of the tie line, the composition of the other phase.

Figure 4.13b represents the same system on a right-triangular diagram. Here, concentrations in wt% of any two components (normally the solute and solvent) are given. Concentration of the third is obtained by the difference from 100 wt%. Diagrams like this are easier to construct and read than equilateral-triangular diagrams. However, equilateral-triangular diagrams are conveniently constructed with CSpace, which can be downloaded from the web site www.ugr.es/~cspace/Whatis.htm.

Figures 4.13c and 4.13d represent the same ternary system in terms of weight fraction and weight ratios of solute, respectively. Figure 4.13c is simply a plot of equilibrium (tie-line) data of Table 4.6 in terms of solute mass fraction. In Figure 4.13d, mass ratios of solute (ethylene glycol) to furfural and water for the extract and raffinate phases, respectively, are used. Such curves can be used to interpolate tie lines, since only a limited number of tie lines are shown on triangular graphs. Because of this, such diagrams are often referred to as *distribution diagrams*. When mole (rather than mass) fractions are used in a diagram like Figure 4.13c, a nearly straight line is often evident near the origin, whose slope is the distribution coefficient K_D for the solute at infinite dilution.

In 1906, Janecke [10] suggested the data display shown as Figure 4.13e. Here, the mass of solvent per unit mass of solvent-free liquid, furfural/(water + glycol), is plotted as the ordinate versus mass ratio, on a solvent-free basis, of glycol/(water + glycol) as the abscissa. The ordinate and abscissa apply to both phases. Equilibrium conditions are connected by tie lines. Mole ratios can also be used to construct Janecke diagrams.

Any of the diagrams in Figure 4.13 can be used for solving problems involving material balances subject to liquid–liquid equilibrium constraints.

EXAMPLE 4.7 Single-Equilibrium Stage Extraction Using Diagrams.

Determine extract and raffinate compositions when a 45 wt% glycol (B)–55 wt% water (A) solution is contacted with twice its weight of pure furfural solvent (C) at 25°C and 101 kPa. Use each of the five diagrams in Figure 4.13, if possible.

Solution

Assume a basis of 100 g of 45% glycol–water feed. Thus, in Figure 4.12b, the feed (*F*) is 55 g A and 45 g B. The solvent (*S*) is 200 g C. Let *E* denote the extract, and *R* the raffinate.

(a) Using the equilateral-triangular diagram of Figure 4.14:

Step 1. Locate the feed and solvent compositions at points F and S, respectively.

Step 2. Define mixing point M as $M = F + S = E + R$.

Step 3. Apply the inverse-lever-arm rule. Let $w_i^{(1)}$ be the mass fraction of species i in the extract, $w_i^{(2)}$ be the fraction of species i in the raffinate, and $w_i^{(M)}$ be the fraction of species i in the feed-plus-solvent phases.

From a solvent balance, C: $(F + S)w_C^{(M)} = Fw_C^{(F)} + Sw_C^{(S)}$.

$$\frac{F}{S} = \frac{w_C^{(S)} - w_C^{(M)}}{w_C^{(M)} - w_C^{(F)}} \qquad (1)$$

Thus, points S, M, and F lie on a straight line, as they should, and, by the inverse-lever-arm rule,

$$\frac{F}{S} = \frac{SM}{MF} = \frac{1}{2}$$

The composition at point M is 18.3% A, 15.0% B, and 66.7% C.

Step 4. Since M lies in the two-phase region, the mixture must separate along an interpolated dash-dot tie line into an extract phase at point E (8.5% B, 4.5% A, and 87.0% C) and the raffinate at point R (34.0% B, 56.0% A, and 10.0% C).

Step 5. The inverse-lever-arm rule applies to points E, M, and R, so $E = M(\overline{RM}/\overline{ER})$. $M = 100 + 200 = 300$ g. From measurements of line segments, $E = 300(147/200) = 220$ g and $R = M - E = 300 - 220 = 80$ g.

(b) Using the right-triangular diagram of Figure 4.15:

Step 1. Locate the *F* and *S* for the two feed streams.

Step 2. Define the mixing point $M = F + S$.

Step 3. The inverse-lever-arm rule also applies to right-triangular diagrams, so MF/MS = ½.

Step 4. Points R and E are on the ends of the interpolated dash-dot tie line passing through point M.

The numerical results of part (b) are identical to those of part (a).

(c) By the equilibrium solute diagram of Figure 4.13c, a material balance on glycol B,

$$Fw_B^{(F)} + Sw_B^{(S)} = 45 = Ew_B^{(E)} + Rw_B^{(R)} \qquad (2)$$

must be solved simultaneously with a phase-equilibrium relationship. It is not possible to do this graphically using Figure 4.13c in any straightforward manner unless the solvent (C) and carrier (A) are mutually insoluble. The outlet-stream composition can be found, however, by the following iterative procedure.

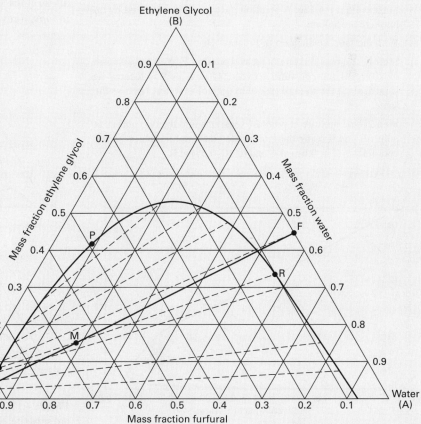

Figure 4.14 Solution to Example 4.7a.

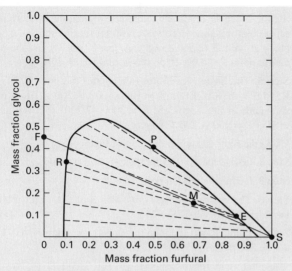

Figure 4.15 Solution to Example 4.7b.

Step 1. Guess a value for $w_B^{(E)}$ and read the equilibrium value, $w_B^{(R)}$, from Figure 4.13c.

Step 2. Substitute these two values into the equation obtained by combining (2) with the overall balance, $E + R = 300$, to eliminate R. Solve for E and then R.

Step 3. Check to see if the furfural (or water) balance is satisfied using the data from Figures 4.13a, 4.13b, or 4.13e. If not, repeat steps 1 to 3 with a new guess for $w_B^{(E)}$. This procedure leads to the results obtained in parts (a) and (b).

(d) Figure 4.13d, a mass-fraction plot, suffers from the same limitations as Figure 4.13c. A solution must again be achieved by an iterative procedure.

(e) With the Janecke diagram of Figure 4.16:

Step 1. The feed mixture is located at point F. With the addition of 200 g of pure furfural solvent, $M = F + S$ is located as shown, since the ratio of glycol to (glycol + water) remains the same.

Step 2. The mixture at point M separates into the two phases at points E and R using the interpolated dash-dot tie line, with the coordinates (7.1, 0.67) at E and (0.10, 0.37) at R.

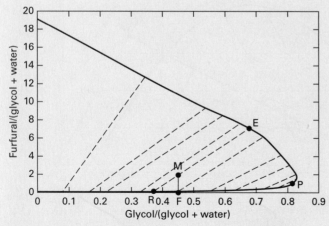

Figure 4.16 Solution to Example 4.7e.

Step 3. Let Z^E and Z^R equal the total mass of components A and B in the extract and raffinate, respectively. Then, the following balances apply:

$$\text{Furfural: } 7.1Z^E + 0.10Z^R = 200$$
$$\text{Glycol: } 0.67Z^E + 0.37Z^R = 45$$

Solving these equations, $Z^E = 27$ g and $Z^R = 73$ g.

Thus, the furfural in the extract $= (7.1)(27 \text{ g}) = 192$ g, the furfural in the raffinate $= 200 - 192 = 8$ g, the glycol in the extract $= (0.67)(27 \text{ g}) = 18$ g, the glycol in the raffinate $= 45 - 18 = 27$ g, the water in the raffinate $= 73 - 27 = 46$ g, and the water in the extract $= 27 - 18 = 9$ g. Total extract is $192 + 27 = 219$ g, which is close to the results of part (a). The raffinate composition and amount can be obtained just as readily.

It should be noted on the Janecke diagram that $\overline{ME}/\overline{MR}$ does not equal R/E; it equals R/E on a solvent-free basis.

In Figure 4.13, two pairs of components are mutually soluble, while one pair is only partially soluble. Ternary systems where two pairs and even all three pairs are only partially soluble also exist. Figure 4.17 shows examples, from Francis [11] and Findlay [12], of four cases where two pairs of components are only partially soluble.

In Figure 4.17a, two two-phase regions are formed, while in Figure 4.17c, in addition to the two-phase regions, a three-phase region, RST, exists. In Figure 4.17b, the two separate two-phase regions merge. For a ternary mixture, as temperature is reduced, phase behavior may progress from Figure 4.17a to 4.17b to 4.17c. In Figures 4.17a, 4.17b, and 4.17c, all tie lines slope in the same direction. In some systems *solutropy*, a reversal of tie-line slopes, occurs.

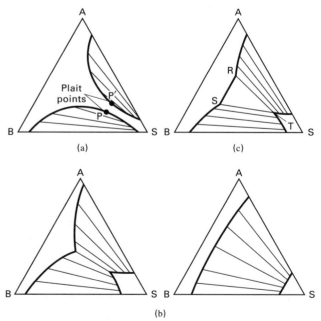

Figure 4.17 Equilibria for 3/2 systems: (a) miscibility boundaries are separate; (b) miscibility boundaries and tie-line equilibria merge; (c) tie lines do not merge and the three-phase region RST is formed.

§4.6 MULTICOMPONENT LIQUID–LIQUID SYSTEMS

Quarternary and higher multicomponent mixtures are encountered in extraction processes, particularly when two solvents are used. Multicomponent liquid–liquid equilibria are complex, and there is no compact, graphical way of representing phase-equilibria data. Accordingly, the computation of equilibrium-phase compositions is best made by process simulators using activity-coefficient equations that account for the effect of composition (e.g., NRTL, UNIQUAC, or UNIFAC). One such method is a modification of the Rachford–Rice algorithm for vapor–liquid equilibrium from Tables 4.3 and 4.4. For extraction, symbol transformations are made and moles are used instead of mass.

Vapor–Liquid Equilibria	Liquid–Liquid Equilibria
Feed, F	Feed, F, + solvent, S
Equilibrium vapor, V	Extract, E ($L^{(1)}$)
Equilibrium liquid, L	Raffinate, R ($L^{(2)}$)
Feed mole fractions, z_i	Mole fractions of combined F and S
Vapor mole fractions, y_i	Extract mole fractions, $x_i^{(1)}$
Liquid mole fractions, x_i	Raffinate mole fractions, $x_i^{(2)}$
K-value, K_i	Distribution coefficient, K_{D_i}
$\Psi = V/F$	$\Psi = E/F$

Industrial extraction processes are commonly adiabatic so, if the feeds are at identical temperatures, the only energy effect is the heat of mixing, which is usually sufficiently small that isothermal assumptions are justified.

The modified Rachford–Rice algorithm is shown in Figure 4.18. This algorithm is applicable for an isothermal vapor–liquid or liquid–liquid stage calculation when K-values depend strongly on phase compositions. The algorithm requires that feed and solvent flow rates and compositions be fixed, and that pressure and temperature be specified. An initial estimate is made of the phase compositions, $x_i^{(1)}$ and $x_i^{(2)}$, and corresponding estimates of the distribution coefficients are made from liquid-phase activity coefficients using (2-30) with, for example, the NRTL or UNIQUAC equations discussed in Chapter 2. Equation (3) of Table 4.4 is then solved iteratively for $\Psi = E/(F + S)$, from which values of $x_i^{(2)}$ and $x_i^{(1)}$ are computed from Eqs. (5) and (6), respectively, of Table 4.4. Resulting values of $x_i^{(1)}$ and $x_i^{(2)}$ will not usually sum to 1 for each phase and are therefore normalized using equations of the form $x'_i = x_i/\Sigma x_j$, where x'_i are the normalized values that force $\sum x'_j$ to equal 1. Normalized values replace the values computed from Eqs. (5) and (6). The iterative procedure is repeated until the compositions $x_i^{(1)}$ and $x_i^{(2)}$ no longer change by more than three or four significant digits from one iteration to the next. Multicomponent liquid–liquid equilibrium calculations are best carried out with a process simulator.

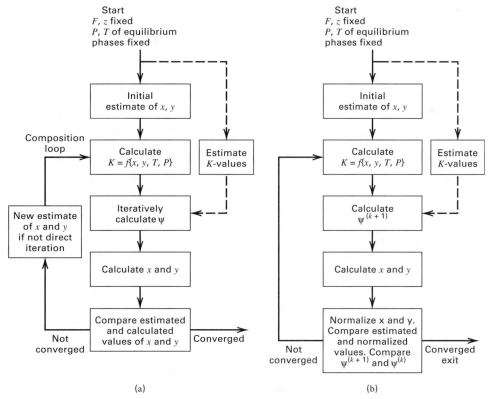

Figure 4.18 Algorithm for isothermal-flash calculation when K-values are composition-dependent: (a) separate nested iterations on Ψ and (x, y); (b) simultaneous iteration on Ψ and (x, y).

EXAMPLE 4.8 Liquid–Liquid Equilibrium for a Four-Component Mixture.

An azeotropic mixture of isopropanol, acetone, and water is dehydrated with ethyl acetate in a system of two distillation columns. Benzene was previously used as the dehydrating agent, but legislation has made benzene undesirable because it is carcinogenic. Ethyl acetate is far less toxic. The overhead vapor from the first column, with the composition below, at 20 psia and 80°C, is condensed and cooled to 35°C, without significant pressure drop, causing the formation of two liquid phases assumed to be in equilibrium. Estimate the amounts of the phases in kg/h and the equilibrium phase compositions in wt%.

Component	kg/h
Isopropanol	4,250
Acetone	850
Water	2,300
Ethyl acetate	43,700

Note that the specification of this problem conforms with the degrees of freedom predicted by (4-4), which for $C = 4$ is 9.

Solution

This example was solved with the CHEMCAD program using the UNIFAC method to estimate liquid-phase activity coefficients. The results are:

	Weight Fraction	
Component	Organic-Rich Phase	Water-Rich Phase
Isopropanol	0.0843	0.0615
Acetone	0.0169	0.0115
Water	0.0019	0.8888
Ethyl acetate	0.8969	0.0382
	1.0000	1.0000
Flow rate, kg/h	48,617	2,483

It is of interest to compare the distribution coefficients from the UNIFAC method to values given in *Perry's Handbook* [1]:

	Distribution Coefficient (wt% Basis)	
Component	UNIFAC	*Perry's Handbook*
Isopropanol	1.37	1.205 (20°C)
Acetone	1.47	1.50 (30°C)
Water	0.0021	—
Ethyl acetate	23.5	—

Results for isopropanol and acetone are in agreement at these dilute conditions, considering the temperature differences.

§4.7 SOLID–LIQUID SYSTEMS

Solid–liquid separations include leaching, crystallization, and adsorption. In leaching (solid–liquid extraction), a multi-component solid mixture is separated by contacting the solid with a solvent that selectively dissolves some of the solid species. Although this operation is quite similar to liquid–liquid extraction, leaching is a much more difficult operation in practice in that diffusion in solids is very slow compared to diffusion in liquids, thus making it difficult to achieve equilibrium. Also, it is impossible to completely separate a solid phase from a liquid phase. A solids-free liquid phase can be obtained, but the solids will always be accompanied by some liquid. In comparison, complete separation of two liquid phases is fairly easy to achieve.

Crystallization or precipitation of a component from a liquid mixture is an operation in which equilibrium can be achieved, but a sharp phase separation is again impossible. A drying step is always needed because crystals occlude liquid.

A third application of solid–liquid systems, adsorption, involves use of a porous solid agent that does not undergo phase or composition change. Instead, it selectively adsorbs liquid species, on its exterior and interior surfaces. Adsorbed species are then desorbed and the solid adsorbing agent is regenerated for repeated use. Variations of adsorption include ion exchange and chromatography. A solid–liquid system is also utilized in membrane-separation operations, where the solid is a membrane that selectively absorbs and transports selected species.

Solid–liquid separation processes, such as leaching and crystallization, almost always involve phase-separation operations such as gravity sedimentation, filtration, and centrifugation.

§4.7.1 Leaching

In Figure 4.19, the solid feed consists of particles of components A and B. The solvent, C, selectively dissolves B. Overflow from the stage is a solids-free solvent C and dissolved B. The underflow is a slurry of liquid and solid A. In an *ideal leaching stage*, all of the solute is dissolved by the solvent, whereas A is not dissolved. Also, the composition of the retained liquid phase in the underflow slurry is identical to the composition of the liquid overflow, and that overflow is free of solids. The mass ratio of solid to liquid in the underflow depends on the properties of the phases and the type of

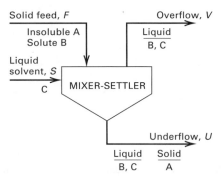

Figure 4.19 Leaching stage.

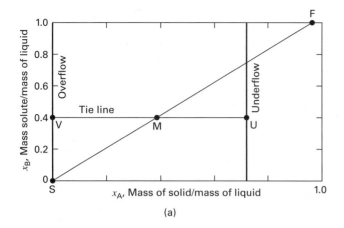

(a)

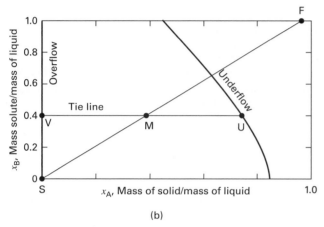

(b)

Figure 4.20 Underflow–overflow conditions for ideal leaching:
(a) constant-solution underflow; (b) variable-solution underflow.

equipment, and is best determined from experience or tests
with prototype equipment. In general, if the viscosity of the
liquid phases increases with increasing solute concentration,
the mass ratio of solid to liquid in the underflow decreases
because the solid retains more liquid.

Ideal leaching calculations can be done algebraically or
with diagrams like Figure 4.20. Let:

F = total mass flow rate of feed to be leached

S = total mass flow rate of entering solvent

U = total mass flow rate of the underflow, including solids

V = total mass flow rate of the overflow

X_A = mass ratio of insoluble solid A to (solute B + solvent
C) in the feed flow, F, or underflow, U

Y_A = mass ratio of insoluble solid A to (solute B + solvent
C) in the entering solvent flow, S, or overflow, V

X_B = mass ratio of solute B to (solute B + solvent C) in the
feed flow, F, or underflow, U

Y_B = mass ratio of solute B to (solute B + solvent C) in the
solvent flow, S, or overflow, V

Figure 4.20a depicts ideal leaching conditions where, in the
underflow, the mass ratio of insoluble solid to liquid, X_A, is a
constant, independent of the concentration, X_B, of solute in
the solids-free liquid. The resulting tie line is vertical. This is

§4.7 Solid–Liquid Systems **159**

constant-solution underflow. Figure 4.20b depicts ideal
leaching conditions when X_A varies with X_B. This is *variable-solution underflow.* In both cases, the assumptions are:
(1) an entering feed, F, free of solvent such that $X_B = 1$; (2) a
solids-free and solute-free solvent, S, such that $Y_A = 0$ and $Y_B
= 0$; and (3) equilibrium between exiting liquid solutions in
underflow, U, and overflow, V, such that $X_B = Y_B$; and (4) a
solids-free overflow, V, such that $Y_A = 0$.

A mixing point, M, can be defined for $(F + S)$, equal to
that for the sum of the products of the leaching stage, $(U +
V)$. Typical mixing points, and inlet and outlet compositions,
are included in Figures 4.20a and b. In both cases, as shown
in the next example, the inverse-lever-arm rule can be applied
to line UMV to obtain flow rates of U and V.

EXAMPLE 4.9 Leaching of Soybeans to Recover Oil.

Soybeans are a predominant oilseed crop, followed by cottonseed,
peanuts, and sunflower seed. While soybeans are not consumed
directly by humans, they can be processed to produce valuable products. Production of soybeans in the United States began after World
War II, increasing in recent years to more than 140 billion lb/yr.
Most soybeans are converted to soy oil and vitamins like niacin and
lecithin for humans, and defatted meal for livestock. Compared to
other vegetable oils, soy oil is more economical and healthier. Typically, 100 pounds of soybeans yields 18 lb of soy oil and 79 lb of
defatted meal.

To recover their oil, soybeans are first cleaned, cracked to loosen
the seeds from the hulls, dehulled, and dried to 10–11% moisture.
Before leaching, the soybeans are flaked to increase the mass-transfer rate of the oil out of the bean. They are leached with hexane to
recover the oil. Following leaching, the hexane overflow is separated
from the soy oil and recovered for recycle by evaporation, while the
underflow is treated to remove residual hexane, and toasted with hot
air to produce defatted meal. Modern soybean extraction plants
crush up to 3,000 tons of soybeans per day.

Oil is to be leached from 100,000 kg/h of soybean flakes, containing 19 wt% oil, in a single equilibrium stage by 100,000 kg/h of
a hexane solvent. Experimental data indicate that the oil content of
the flakes will be reduced to 0.5 wt%. For the type of equipment to
be used, the expected contents of the underflows is as follows:

β, Mass fraction of solids in underflow	0.68	0.67	0.65	0.62	0.58	0.53
Mass ratio of solute in underflow liquid, X_B	0.0	0.2	0.4	0.6	0.8	1.0

Calculate, both graphically and analytically, compositions and
flow rates of the underflow and overflow, assuming an ideal leaching
stage. What % of oil in the feed is recovered?

Solution

The flakes contain $(0.19)(100,000) = 19,000$ kg/h of oil and
$(100,000 − 19,000) = 81,000$ kg/h of insolubles. However, all of
the oil is not leached. For convenience in the calculations, lump the
unleached oil with the insolubles to give an effective A. The flow
rate of unleached oil = $(81,000)(0.5/99.5) = 407$ kg/h. Therefore,
the flow rate of A is taken as $(81,000 + 407) = 81,407$ kg/h and the
oil in the feed is just the amount leached, or $(19,000 − 407) =$

18,593 kg/h of B. Therefore, in the feed, F, $Y_A = (81,407/18,593) = 4.38$, and $X_B = 1.0$.

The sum of the liquid solutions in the underflow and overflow includes 100,000 kg/h of hexane and 18,593 kg/h of leached oil. Therefore, for the underflow and overflow, $X_B = Y_B = [18,593/(100,000 + 18,593)] = 0.157$.

This is a case of variable-solution underflow. Using data in the above table, convert values of β to values of X_A,

$$X_A = \frac{kg/h\ A}{kg/h\ (B + C)} = \frac{\beta U}{(1 - \beta)U} = \frac{\beta}{(1 - \beta)} \qquad (1)$$

Using (1), the following values of X_A are computed from the previous table.

X_A	2.13	2.03	1.86	1.63	1.38	1.13
X_B	0.0	0.2	0.4	0.6	0.8	1.0

Graphical Method

Figure 4.21 is a plot of X_A as a function of X_B. Because no solids leave in the overflow, that line is horizontal at $X_A = 0$. Plotted are the feeds, F, and hexane, S, with a straight line between them. A point for the overflow, V, is plotted at $X_A = 0$ and, from above, $X_B = 0.157$. Since $Y_B = X_B = 0.157$, the value of X_A in the underflow is at the intersection of a vertical line from overflow, V, to the underflow line. This value is $X_A = 2.05$. Lines $\overline{FS}$ and $\overline{UV}$ intersect at point M.

In the overflow, from $X_B = 0.157$, mass fractions of solute B and solvent C are, respectively, 0.157 and $(1 - 0.157) = 0.843$. In the underflow, using $X_A = 2.05$ and $X_B = 0.157$, mass fractions of solids B and C are $[2.05/(1 + 2.05)] = 0.672$, $0.157(1 - 0.672) = 0.0515$, and $(1 - 0.672 - 0.0515) = 0.2765$, respectively.

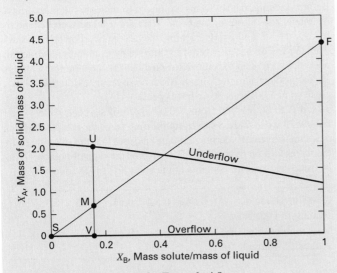

Figure 4.21 Constructions for Example 4.9.

The inverse-lever-arm rule is used to compute the underflow and overflow. The rule applies only to the liquid phases in the two exiting streams because Figure 4.21 is on a solids-free basis. The mass ratio of liquid flow rate in the underflow to liquid flow rate in the overflow is the ratio of line $\overline{MV}$ to line $\overline{MU}$. With M located at $X_A = 0.69$, this ratio = $(0.69 - 0.0)/(2.05 - 0.69) = 0.51$. Thus, the liquid flow rate in the underflow = $(100,000 + 18,593)(0.51)/(1 + 0.51) = 40,054$ kg/h. Adding the flow rates of carrier and unextracted oil gives $U = 40,054 + 81,407 = 121,461$ kg/h or, say, 121,000 kg/h. The overflow rate = $V = 200,000 - 121,000 = 79,000$ kg/h.

Oil flow rate in the feed is 19,000 kg/h. The oil flow rate in the overflow = $Y_B V = 0.157(79,000) = 12,400$ kg/h. Thus, the oil in the feed that is recovered in the overflow = $12,400/19,000 = 0.653$ or 65.3%. Adding washing stages, as described in §5.2, can increase the oil recovery.

Algebraic Method

As with the graphical method, $X_B = 0.157$, giving a value from the previous table of $X_A = 2.05$. Then, since the flow rate of solids in the underflow = 81,407 kg/h, the flow rate of liquid in the underflow = $81,407/2.05 = 39,711$ kg/h. The total flow rate of underflow is $U = 81,407 + 39,711 = 121,118$ kg/h. By mass balance, the flow rate of overflow = $200,000 - 121,118 = 78,882$ kg/h. These values are close to those obtained graphically. The percentage recovery of oil, and the underflow and overflow, are computed as before.

§4.7.2 Crystallization

Crystallization takes place from aqueous or nonaqueous solutions. Consider a binary mixture of two organic chemicals such as naphthalene and benzene, whose solid–liquid equilibrium diagram at 1 atm is shown in Figure 4.22. Points A and B are melting (freezing) points of pure benzene (5.5°C) and pure naphthalene (80.2°C). When benzene is dissolved in liquid naphthalene or vice versa, the freezing point is depressed. Point E is the *eutectic point*, corresponding to a eutectic temperature (−3°C) and composition (80 wt% benzene). "Eutectic" is derived from a Greek word meaning "easily fused," and in Figure 4.22 it represents the binary mixture of naphthalene and benzene with the lowest freezing (melting) point.

Points located above the curve AEB correspond to a homogeneous liquid phase. Curve AE is the solubility curve for benzene in naphthalene. For example, at 0°C solubility is very high, 87 wt% benzene. Curve EB is the solubility curve for naphthalene in benzene. At 25°C, solubility is 41 wt% naphthalene and at 50°C, it is much higher. For most mixtures, solubility increases with temperature.

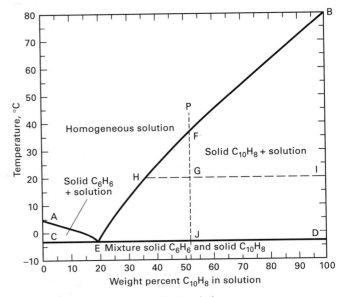

Figure 4.22 Solubility of naphthalene in benzene.

[Adapted from O.A. Hougen, K.M. Watson, and R.A. Ragatz, *Chemical Process Principles. Part I*, 2nd ed., John Wiley & Sons, New York (1954).]

If a liquid solution represented by point P is cooled along the vertical dashed line, it remains liquid until the line intersects the solubility curve at point F.

If the temperature is lowered further, crystals of naphthalene form and the remaining liquid, the *mother liquor*, becomes richer in benzene. When point G is reached, pure naphthalene crystals and a mother liquor, point H on solubility curve EB, coexist, the solution composition being 37 wt% naphthalene. By the Gibbs phase rule, (4-1), with $C = 2$ and $\mathscr{P} = 2$, $\mathscr{F} = 2$. Thus for fixed T and P, compositions are fixed. The fraction of solution crystallized can be determined by the inverse-lever-arm rule. In Figure 4.22, the fraction is kg naphthalene crystals/kg original solution = length of line GH/length of line HI = $(52 - 37)/(100 - 37) = 0.238$.

As the temperature is lowered, line CED, corresponding to the eutectic temperature, is reached at point J, where the two-phase system consists of naphthalene crystals and a mother liquor of eutectic composition E. Any further removal of heat causes the eutectic solution to solidify.

EXAMPLE 4.10 Crystallization of Naphthalene from a Solution with Benzene.

Eight thousand kg/h of a solution of 80 wt% naphthalene and 20 wt % benzene at 70°C is cooled to 30°C to form naphthalene crystals. If equilibrium is achieved, determine the kg of crystals formed and the composition in wt% of the mother liquor.

Solution

From Figure 4.22, at 30°C, the solubility of naphthalene is 45 wt%. By the inverse-lever-arm rule, for an original 80 wt% solution,

$$\frac{\text{kg naphthalene crystals}}{\text{kg original mixture}} = \frac{(80 - 45)}{(100 - 45)} = 0.636$$

The flow rate of crystals = 0.636 (8,000) = 5,090 kg/h.

The remaining 2,910 kg/h of mother liquor is 55 wt% benzene.

Crystallization of a salt from an aqueous solution can be complicated by the formation of water hydrates. These are stable, solid compounds that exist within certain temperature ranges. For example, $MgSO_4$ forms the stable hydrates $MgSO_4 \cdot 12H_2O$, $MgSO_4 \cdot 7H_2O$, $MgSO_4 \cdot 6H_2O$, and $MgSO_4 \cdot H_2O$. The high hydrate exists at low temperatures; the low hydrate exists at higher temperatures.

A simpler example is that of Na_2SO_4 and water. As seen in the phase diagram in Figure 4.23, only one stable hydrate is formed, $Na_2SO_4 \cdot 10H_2O$, known as Glauber's salt. Since the molecular weights are 142.05 for Na_2SO_4 and 18.016 for H_2O, the weight percent Na_2SO_4 in the decahydrate is 44.1, which is the vertical line BFG.

The water freezing point, 0°C, is at A, but the melting point of Na_2SO_4, 884°C, is not on the diagram. The decahydrate melts at 32.4°C, point B, to form solid Na_2SO_4 and a mother liquor, point C, of 32.5 wt% Na_2SO_4. As Na_2SO_4 dissolves in water, the freezing point is depressed slightly along curve AE until the eutectic, point E, is reached. Curves EC and CD represent solubilities of decahydrate crystals and anhydrous sodium sulfate in water. The solubility of Na_2SO_4 decreases slightly with increasing temperature, which is unusual. In the region below GFBHI, a solid solution of anhydrous and decahydrate forms exist. The amounts of coexisting phases can be found by the inverse-lever-arm rule.

EXAMPLE 4.11 Crystallization of Na_2SO_4 from Water.

A 30 wt% aqueous Na_2SO_4 solution of 5,000 lb/h enters a cooling-type crystallizer at 50°C. At what temperature will crystallization begin? Will the crystals be decahydrate or the anhydrous form? At what temperature will the mixture crystallize 50% of the Na_2SO_4?

Solution

From Figure 4.23, the 30 wt% Na_2SO_4 solution at 50°C corresponds to a point in the homogeneous liquid solution region. If a vertical

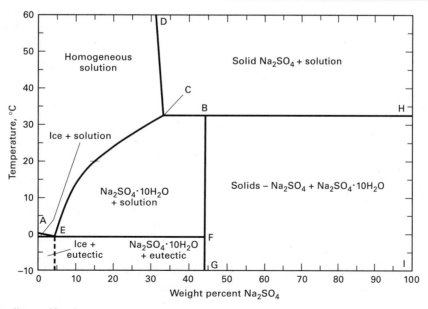

Figure 4.23 Solubility of sodium sulfate in water.

[Adapted from O.A. Hougen, K.M. Watson, and R.A. Ragatz, *Chemical Process Principles. Part I*, 2nd ed., John Wiley & Sons, New York (1954).]

line is dropped from that point, it intersects solubility curve EC at 31°C. Below this temperature, the crystals are the decahydrate.

The feed contains $(0.30)(5,000) = 1,500$ lb/h of Na_2SO_4 and $(5,000 - 1,500) = 3,500$ lb/h of H_2O. Thus, $(0.5)(1,500) = 750$ lb/h are to be crystallized. The decahydrate crystals include water of hydration in an amount given by a ratio of molecular weights or

$$750\left[\frac{(10)(18.016)}{(142.05)}\right] = 950 \text{ lb/h}$$

The total amount of decahydrate is $750 + 950 = 1,700$ lb/h. The water remaining in the mother liquor is $3,500 - 950 = 2,550$ lb/h. The composition of the mother liquor is $750/(2,550 + 750)(100\%)$ $= 22.7$ wt% Na_4SO_4. From Figure 4.23, the temperature corresponding to 22.7 wt% Na_2SO_4 on the solubility curve EC is 26°C.

§4.7.3 Liquid Adsorption

When a liquid contacts a microporous solid, adsorption takes place on the external and internal solid surfaces until equilibrium is reached. The solid *adsorbent* is essentially insoluble in the liquid. The component(s) adsorbed are called *solutes* when in the liquid and *adsorbates* upon adsorption. The higher the concentration of solute, the higher the adsorbate concentration on the adsorbent. Component(s) of the liquid other than the solute(s) are called the *solvent* or *carrier* and are assumed not to adsorb.

No theory for predicting adsorption-equilibrium curves, based on molecular properties of the solute and solid, is universally embraced, so laboratory measurements must be performed to provide data for plotting curves, called *adsorption isotherms*. Figure 4.24, taken from the data of Fritz and Schuluender [13], is an isotherm for the adsorption of phenol from an aqueous solution onto activated carbon at 20°C. Activated carbon is a microcrystalline, nongraphitic form of carbon, whose microporous structure gives it a high internal surface area per unit mass of carbon, and therefore a high capacity for adsorption. Activated carbon preferentially adsorbs organic compounds when contacted with water containing dissolved organics.

As shown in Figure 4.24, as the concentration of phenol in water increases, adsorption increases rapidly at first, then increases slowly. When the concentration of phenol is 1.0 mmol/L (0.001 mol/L of aqueous solution or 0.000001 mol/g of aqueous solution), the concentration of phenol on the activated carbon is somewhat more than 2.16 mmol/g (0.00216

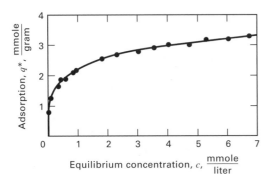

Figure 4.24 Adsorption isotherm for phenol from an aqueous solution in the presence of activated carbon at 20°C.

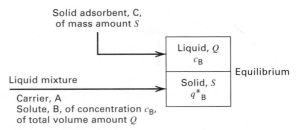

Figure 4.25 Equilibrium stage for liquid adsorption.

mol/g of carbon or 0.203 g phenol/g of carbon). Thus, the affinity of this adsorbent for phenol is high. The extent of adsorption depends on the process used to produce the activated carbon. Adsorption isotherms can be used to determine the amount of adsorbent required to selectively remove a given amount of solute from a liquid.

Consider the ideal, single-stage adsorption process of Figure 4.25, where A is the carrier liquid, B is the solute, and C is the solid adsorbent. Let: $c_B =$ concentration of solute in the carrier liquid, mol/unit volume; $q_B =$ concentration of adsorbate, mol/unit mass of adsorbent; $Q =$ volume of liquid (assumed to remain constant during adsorption); and $S =$ mass of adsorbent (solute-free basis).

A solute material balance, assuming that the entering adsorbent is free of solute and that equilibrium is achieved, as designated by the asterisk superscript on q, gives

$$c_B^{(F)}Q = c_B Q = q_B^* S \qquad (4-28)$$

This equation can be rearranged in the form of a straight line that can be plotted on a graph of the type in Figure 4.24 to obtain a graphical solution for c_B and q_B^*. Solving (4-28) for q_B^*,

$$q_B^* = -\frac{Q}{S}c_B + c_B^{(F)}\frac{Q}{S} \qquad (4-29)$$

The intercept on the c_B axis is $c_B^{(F)}Q/S$, and the slope is $-(Q/S)$. The intersection of (4-29) with the adsorption isotherm is the equilibrium condition c_B and q_B^*.

Alternatively, an algebraic solution can be obtained if the adsorption isotherm for equilibrium-liquid adsorption of a species i can be fitted to an equation. For example, the Freundlich equation discussed in Chapter 15 is of the form

$$q_i^* = Ac_i^{(1/n)} \qquad (4-30)$$

where A and n depend on the solute, carrier, and adsorbent. Constant, n, is greater than 1, and A is a function of temperature. Freundlich developed his equation from data on the adsorption on charcoal of organic solutes from aqueous solutions. Substitution of (4-30) into (4-29) gives

$$Ac_B^{(1/n)} = -\frac{Q}{S}c_B + c_B^{(F)}\frac{Q}{S} \qquad (4-31)$$

which is a nonlinear equation in c_B that is solved numerically by an iterative method, as illustrated in the following example.

EXAMPLE 4.12 Adsorption of Phenol on Activated Carbon.

A 1.0-liter solution of 0.010 mol of phenol in water is brought to equilibrium at 20°C with 5 g of activated carbon having the adsorption isotherm shown in Figure 4.24. Determine the percent adsorption and equilibrium concentration of phenol on carbon by (a) a graphical method, and (b) a numerical algebraic method. For the latter case, the curve of Figure 4.24 is fitted with the Freundlich equation, (4-30), giving

$$q_B^* = 2.16c_B^{(1/4.35)} \tag{1}$$

Solution

From the data, $c_B^{(F)} = 10$ mmol/L, $Q = 1$ L, and $S = 5$ g.

(a) Graphical method.

From (4-29), $q_B^* = -(\frac{1}{5})c_B + 10(\frac{1}{5}) = -0.2c_B + 2$.

Plot this equation, with a slope of -0.2 and an intercept of 2, on Figure 4.24. An intersection with the equilibrium curve will occur at $q_B^* = 1.9$ mmol/g and $c_B = 0.57$ mmol/liter. Thus, the adsorption of phenol is

$$\frac{c_B^{(F)} - c_B}{c_B^{(F)}} = \frac{10 - 0.57}{10} = 0.94 \quad \text{or} \quad 94\%$$

(b) Numerical algebraic method.

Applying Eq. (1) from the problem statement and (4-31),

$$2.16c_B^{0.23} = -0.2c_B + 2 \tag{2}$$

or

$$f\{c_B\} = 2.16c_B^{0.23} + 0.2c_B - 2 = 0 \tag{3}$$

This nonlinear equation for c_B can be solved by an iterative numerical technique. For example, Newton's method [14], applied to Eq. (3), uses the iteration rule:

$$c_B^{(k+1)} = c_B^{(k)} - f^{(k)}\{c_B\}/f'^{(k)}\{c_B\} \tag{4}$$

where k is the iteration index. For this example, $f\{c_B\}$ is given by Eq. (3) and $f'\{c_B\}$ is obtained by differentiating with respect to c_B:

$$f'^{(k)}\{c_B\} = 0.497c_B^{-0.77} + 0.2$$

A convenient initial guess for c_B is 100% adsorption of phenol to give $q_B^* = 2$ mmol/g. Then, from (4-30), $c_B^{(0)} = (q_B^*/A)^n = (2/2.16)^{4.35} = 0.72$ mmol/L, where the (0) superscript designates the starting guess. The Newton iteration rule of Eq. (4-1) is now used, giving the following results:

k	$c_B^{(k)}$	$f^{(k)}\{c_B\}$	$f'^{(k)}\{c_B\}$	$c_B^{(k+1)}$
0	0.72	0.1468	0.8400	0.545
1	0.545	−0.0122	0.9928	0.558
2	0.558	−0.00009	0.9793	0.558

These results indicate convergence to $f\{c_B\} = 0$ for a value of $c_B = 0.558$ after only three iterations. From Eq. (1), $q_B^* = 2.16(0.558)^{(1/4.35)} = 1.89$ mmol/g. Numerical and graphical methods are in agreement.

§4.8 GAS–LIQUID SYSTEMS

Vapor–liquid systems were covered in § 4.2, 4.3, and 4.4, wherein the vapor was mostly condensable. Although the terms *vapor* and *gas* are often used interchangeably, the term *gas* often designates a mixture for which the ambient

temperature is above the critical temperatures of most or all of the species. Thus, in gas–liquid systems, the components of the gas are not easily condensed.

Even when components of a gas mixture are at a temperature above critical, they dissolve in a liquid solvent to an extent that depends on temperature and their partial pressure in the gas mixture. With good mixing, equilibrium between the two phases can be achieved in a short time unless the liquid is very viscous.

No widely accepted theory for gas–liquid solubilities exists. Instead, plots of experimental data, or empirical correlations, are used. Experimental data for 13 pure gases dissolved in water are plotted in Figure 4.26 over a range of temperatures from 0 to 100°C. The ordinate is the gas mole fraction in the liquid when gas pressure is 1 atm. The curves of Figure 4.26 can be used to estimate the solubility in water

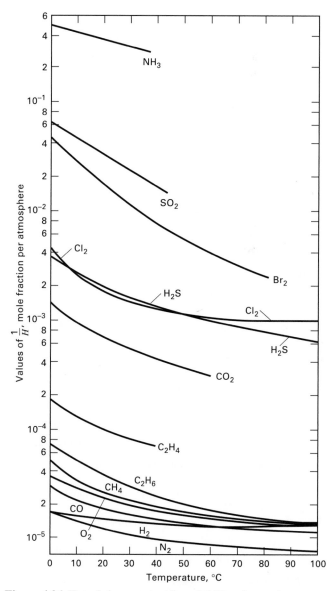

Figure 4.26 Henry's law constant for solubility of gases in water.
[Adapted from O.A. Hougen, K.M. Watson, and R.A. Ragatz, *Chemical Process Principles. Part I*, 2nd ed., John Wiley & Sons, New York (1954).]

at other pressures and for mixtures of gases by applying Henry's law and using the partial pressure of the solute, provided that mole fractions are low and no chemical reactions occur in the gas or water. Henry's law, from Table 2.3, is rewritten for use with Figure 4.26 as

$$x_i = \left(\frac{1}{H_i}\right) y_i P \qquad (4\text{-}32)$$

where H_i = Henry's law constant, atm.

For gases with a high solubility, such as ammonia, Henry's law is not applicable, even at low partial pressures. In that case, experimental data for the actual conditions of pressure and temperature are necessary. Calculations of equilibrium conditions are made, as in previous sections of this chapter, by combining material balances with equilibrium relationships.

EXAMPLE 4.13 Absorption of CO_2 with Water.

An ammonia plant, located at the base of a 300-ft-high mountain, employs a unique absorption system for disposing of byproduct CO_2, in which the CO_2 is absorbed in water at a CO_2 partial pressure of 10 psi above that required to lift water to the top of the mountain. The CO_2 is then vented at the top of the mountain, the water being recirculated as shown in Figure 4.27. At 25°C, calculate the amount of water required to dispose of 1,000 ft³ (STP) of CO_2.

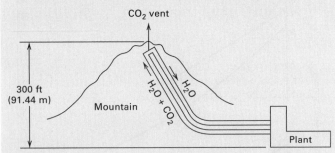

Figure 4.27 Flowsheet for Example 4.13.

Solution

Basis: 1,000 ft³ of CO_2 at 0°C and 1 atm (STP). From Figure 4.26, the reciprocal of the Henry's law constant for CO_2 at 25°C is 6 × 10^{-4} mole fraction/atm CO_2 pressure in the absorber (at the foot of the mountain) is

$$p_{CO_2} = \frac{10}{14.7} + \frac{300 \text{ ft } H_2O}{34 \text{ ft } H_2O/\text{atm}} = 9.50 \text{ atm} = 960 \text{ kPa}$$

At this partial pressure, the concentration of CO_2 in the water is

$$x_{CO_2} = 9.50(6 \times 10^{-4}) = 5.7 \times 10^{-3} \text{ mole fraction } CO_2 \text{ in water}$$

The corresponding ratio of dissolved CO_2 to water is

$$\frac{5.7 \times 10^{-3}}{1 - 5.7 \times 10^{-3}} = 5.73 \times 10^{-3} \text{ mol } CO_2/\text{mol } H_2O$$

The total number of moles of CO_2 to be absorbed is

$$\frac{1,000 \text{ ft}^3}{359 \text{ ft}^3/\text{lbmol (at STP)}} = \frac{1,000}{359} = 2.79 \text{ lbmol}$$

or $(2.79)(44)(0.454) = 55.73$ kg.

Assuming all absorbed CO_2 is vented, the number of moles of water required is $2.79/(5.73 \times 10^{-3}) = 458$ lbmol = 8,730 lb = 3,963 kg.

If one corrects for the fact that that not all the CO_2 is vented, because the pressure on top of the mountain is 101 kPa, 4,446 kg (9,810 lb) of water are required.

EXAMPLE 4.14 Equilibrium Diagram for Air–NH_3–H_2 at 20°C and 1 atm.

The partial pressure of ammonia (A) in air–ammonia mixtures in equilibrium with their aqueous solutions at 20°C is given in Table 4.7. Using these data, and neglecting the vapor pressure of water and the solubility of air in water, construct an equilibrium diagram at 101 kPa using mole ratios Y_A = mol NH_3/mol air and X_A = mol NH_3/mol H_2O as coordinates. Henceforth, the subscript A is dropped. If 10 mol of gas of Y = 0.3 are contacted with 10 mol of solution of X = 0.1, what are the compositions of the resulting phases? The process is assumed to be isothermal at 1 atm.

Table 4.7 Partial Pressure of Ammonia over Ammonia–Water Solutions at 20°C

NH_3 Partial Pressure, kPa	g NH_3/g H_2O
4.23	0.05
9.28	0.10
15.2	0.15
22.1	0.20
30.3	0.25

Solution

Equilibrium data in Table 4.7 are recalculated in terms of mole ratios in Table 4.8 and plotted in Figure 4.28.

Table 4.8 Y–X Data for Ammonia–Water, 20°C

Y, mol NH_3/mol	X, mol NH_3/mol
0.044	0.053
0.101	0.106
0.176	0.159
0.279	0.212
0.426	0.265

Mol NH_3 in entering gas = $10[Y/(1 + Y)] = 10(0.3/1.3) = 2.3$
Mol NH_3 in entering liquid = $10[X/(1 + X)] = 10(0.1/1.1) = 0.91$

A material balance for ammonia about the equilibrium stage is

$$GY_0 + LX_0 = GY_1 + LX_1 \qquad (1)$$

where G = moles of air and L = moles of H_2O. Then $G = 10 - 2.3$ = 7.7 mol and $L = 10 - 0.91 = 9.09$ mol. Solving for Y_1 from (1),

$$Y_1 = -\frac{L}{G}X_1 + \left(\frac{L}{G}X_0 + Y_0\right) \qquad (2)$$

This is an equation of a straight line of slope $(L/G) = -9.09/7.7$ = -1.19, with an intercept of $(L/G)(X_0) + Y_0 = 0.42$. The intersection of this material-balance line with the equilibrium curve, as shown in Figure 4.28, gives the ammonia composition of the gas and liquid leaving the stage as $Y_1 = 0.195$ and $X_1 = 0.19$. This result can be checked by an NH_3 balance, since the amount of NH_3 leaving is

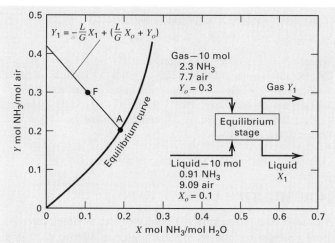

Figure 4.28 Equilibrium for air–NH$_3$–H$_2$O at 20°C, 1 atm, in Example 4.14.

$(0.195)(7.70) + (0.19)(9.09) = 3.21$, which equals the total moles of NH$_3$ entering.

Equation (2), the material-balance line, called an *operating line* and discussed in detail in Chapters 5 to 8, is the locus of all passing stream pairs; thus, X_0, Y_0 (point F) also lies on this operating line.

§4.9 GAS–SOLID SYSTEMS

Gas–solid systems are encountered in sublimation, desublimation, and adsorption separation operations.

§4.9.1 Sublimation and Desublimation

In sublimation, a solid vaporizes into a gas phase without passing through a liquid state. In desublimation, one or more components (solutes) in the gas phase are condensed to a solid phase without passing through a liquid state. At low pressure, both sublimation and desublimation are governed by the solid vapor pressure of the solute. Sublimation of the solid takes place when the partial pressure of the solute in the gas phase is less than the vapor pressure of the solid at the system temperature. When the partial pressure of the solute in the gas phase exceeds the vapor pressure of the solid, desublimation occurs. At equilibrium, the vapor pressure of the species as a solid is equal to the partial pressure of the species as a solute in the gas phase.

EXAMPLE 4.15 Desublimation of Phthalic Anhydride.

Ortho-xylene is completely oxidized in the vapor phase with air to produce phthalic anhydride, PA, in a catalytic reactor at about 370°C and 780 torr. A large excess of air is used to keep the xylene concentration below 1 mol% to avoid an explosive mixture. In a plant, 8,000 lbmol/h of reactor-effluent gas, containing 67 lbmol/h of PA and other amounts of N$_2$, O$_2$, CO, CO$_2$, and water vapor, are cooled to separate the PA by desublimation to a solid at a total pressure of 770 torr. If the gas is cooled to 206°F, where the vapor pressure of solid PA is 1 torr, calculate the number of pounds of PA

condensed per hour as a solid, and the percent recovery of PA from the gas if equilibrium is achieved.

Solution

At these conditions, only PA condenses. The partial pressure of PA is equal to the vapor pressure of solid PA, or 1 torr. Thus, PA in the cooled gas is given by Dalton's law of partial pressures:

$$(n_{PA})_G = \frac{p_{PA}}{P} n_G \tag{1}$$

where

$$n_G = (8,000 - 67) + (n_{PA})_G \tag{2}$$

and n = lbmol/h. Combining Eqs. (1) and (2),

$$
\begin{aligned}
(n_{PA})_G &= \frac{p_{PA}}{P}\left[(8,000 - 67) + (n_{PA})_G\right] \\
&= \frac{1}{770}\left[(8,000 - 67) + (n_{PA})_G\right]
\end{aligned} \tag{3}
$$

Solving this linear equation gives $(n_{PA})_G = 10.3$ lbmol/h of PA. The amount of PA desublimed is $67 - 10.3 = 56.7$ lbmol/h. The percent recovery of PA is $56.7/67 = 0.846$ or 84.6%. The amount of PA remaining in the gas is above EPA standards, so a lower temperature is required. At 140°F the recovery is almost 99%.

§4.9.2 Gas Adsorption

As with liquid mixtures, one or more components of a gas can be adsorbed on the external and internal surfaces of a porous, solid adsorbent. Data for a single solute can be represented by an adsorption isotherm of the type shown in Figure 4.24 or in similar diagrams. However, when two components of a gas mixture are adsorbed and the purpose is to separate them, other methods of representing the data, such as Figure 4.29, are preferred. Figure 4.29 displays the data of Lewis et al. [15] for the adsorption of a propane (P)–propylene (A) gas mixture on silica gel at 25°C and 101 kPa. At 25°C, a pressure of at least 1,000 kPa is required to initiate condensation of a mixture of propylene and propane. However, in the presence of silica gel, significant amounts of gas are adsorbed at 101 kPa.

Figure 4.29a is similar to a binary vapor–liquid plot of the type seen in §4.2. For adsorption, the liquid-phase mole fraction is replaced by the mole fraction in the adsorbate. For propylene–propane mixtures, propylene is adsorbed more strongly. For example, for an equimolar mixture in the gas phase, the adsorbate contains only 27 mol% propane. Figure 4.29b combines data for the mole fractions in the gas and adsorbate with the amount of adsorbate per unit of adsorbent. The mole fractions are obtained by reading the abscissa at the two ends of a tie line. With $y_P = y^* = 0.50$, Figure 4.29b gives $x_P = x^* = 0.27$ and 2.08 mmol of adsorbate/g adsorbent. Therefore, $y_A = 0.50$, and $x_A = 0.73$. The separation factor, analogous to α for distillation, is $(0.50/0.27)/(0.50/0.73) = 2.7$.

This value is much higher than the α for distillation, which, from Figure 2.4 at 25°C and 1,100 kPa, is only 1.13. Accordingly, the separation of propylene and propane by adsorption has received some attention.

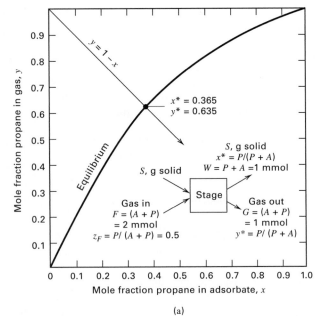

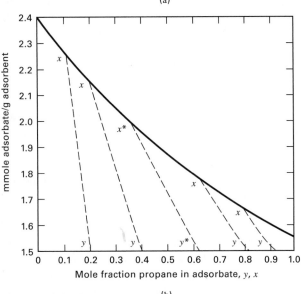

(a)

(b)

Figure 4.29 Adsorption equilibrium at 25°C and 101 kPa of propane and propylene on silica gel.

[Adapted from W.K. Lewis, E.R. Gilliland, B. Chertow, and W. H. Hoffman, *J. Am. Chem. Soc,* **72**, 1153 (1950).]

EXAMPLE 4.16   Separation of Propylene–Propane by Adsorption.

Propylene (A) and propane (P) are separated by preferential adsorption on porous silica gel (S) at 25°C and 101 kPa. Two millimoles of a gas of 50 mol% P and 50 mol% A are equilibrated with silica gel at 25°C and 101 kPa. Measurements show that 1 mmol of gas is adsorbed. If the data of Figure 4.29 apply, what is the mole fraction of propane in the equilibrium gas and in the adsorbate, and how many grams of silica gel are used?

Solution

The process is represented in Figure 4.29a, where W = millimoles of adsorbate, G = millimoles of gas leaving, and z_F = mole fraction

of propane in the feed. The propane mole balance is

$$Fz_F = Wx^* + Gy^* \qquad (1)$$

Because $F = 2$, $z_F = 0.5$, $W = 1$, and $G = F - W = 1$, $1 = x^* + y^*$. The operating (material-balance) line $y^* = 1 - x^*$ in Figure 4-29a is the locus of all solutions of the material-balance equations. It intersects the equilibrium curve at $x^* = 0.365$, $y^* = 0.635$. From Figure 4.29b, at the point x^*, there are 2.0 mmol adsorbate/g adsorbent and $1.0/2 = 0.50$ g of silica gel.

§4.10 MULTIPHASE SYSTEMS

Although two-phase systems predominate, at times three or more co-existing phases are encountered. Figure 4.30 is a schematic of a photograph of a laboratory curiosity taken from Hildebrand [16], which shows seven phases in equilibrium. The phase on top is air, followed by six liquid phases in order of increasing density: hexane-rich, aniline-rich, water-rich, phosphorous, gallium, and mercury. Each phase contains all components in the mixture, but many of the mole fractions are extremely small. For example, the aniline-rich phase contains on the order of 10 mol% n-hexane, 20 mol% water, but much less than 1 mol% each of dissolved air, phosphorous, gallium, and mercury. Note that even though the hexane-rich phase is not in direct contact with the water-rich phase, water (approximately 0.06 mol %) is present in the hexane-rich phase because each phase is in equilibrium with each of the other phases, by the equality of component fugacities:

$$f_i^{(1)} = f_i^{(2)} = f_i^{(3)} = f_i^{(4)} = f_i^{(5)} = f_i^{(6)} = f_i^{(7)}$$

More practical multiphase systems include the vapor–liquid–solid systems present in evaporative crystallization and pervaporation, and the vapor–liquid–liquid systems that occur when distilling certain mixtures of water and hydrocarbons that have a limited solubility in water. Actually, all of the two-phase systems considered in this chapter involve a third phase, the containing vessel. However, as a practical matter, the container is selected on the basis of its chemical inertness and insolubility.

Although calculations of multiphase equilibria are based on the same principles as for two-phase systems (material

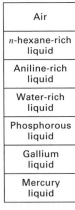

Figure 4.30 Seven phases in equilibrium.

balances, energy balances, and equilibrium), the computations are complex unless assumptions are made, in which case approximate answers result. Rigorous calculations are best made with process simulators.

§4.10.1 Approximate Method for a Vapor–Liquid–Solid System

A simple case of multiphase equilibrium occurs in an evaporative crystallizer involving crystallization of an inorganic compound, B, from its aqueous solution at its bubble point in the presence of water vapor. Assume that only two components are present, B and water, that the liquid is a mixture of water and B, and that the solid is pure B. Then, the solubility of B in the liquid is not influenced by the presence of the vapor, and the system pressure at a given temperature can be approximated by Raoult's law applied to the liquid phase:

$$P = P^s_{H_2O} x_{H_2O} \qquad (4-33)$$

where x_{H_2O} can be obtained from the solubility of B.

EXAMPLE 4.17 Evaporative Crystallizer.

A 5,000-lb batch of 20 wt% aqueous $MgSO_4$ solution is fed to an evaporative crystallizer operating at 160°F. At this temperature, the stable solid phase is the monohydrate, with a $MgSO_4$ solubility of 36 wt%. If 75% of the water is evaporated, calculate: (a) lb of water evaporated; (b) lb of monohydrate crystals, $MgSO_4·H_2O$; and (c) crystallizer pressure.

Solution

(a) The feed solution is $0.20(5,000) = 1,000$ lb $MgSO_4$, and $5,000 - 1,000 = 4,000$ lb H_2O. The amount of water evaporated is $0.75(4,000) = 3,000$ lb H_2O.

(b) Let $W =$ amount of $MgSO_4$ remaining in solution. Then $MgSO_4$ in the crystals $= 1,000 - W$. MW of $H_2O = 18$ and MW of $MgSO_4 = 120.4$. Water of crystallization for the monohydrate $= (1,000 - W)(18/120.4) = 0.15(1,000 - W)$.

Water remaining in solution $= 4,000 - 3,000 - 0.15(1,000 - W) = 850 + 0.15\,W$.

Total amount of solution remaining $= 850 + 0.15\,W + W = 850 + 1.15\,W$. From the solubility of $MgSO_4$,

$$0.36 = \frac{W}{850 + 1.15W}$$

Solving: $W = 522$ pounds of dissolved $MgSO_4$. $MgSO_4$ crystallized $= 1,000 - 522 = 478$ lb. Water of crystallization $= 0.15(1,000 - W) = 0.15(1,000 - 522) = 72$ lb. Total monohydrate crystals $= 478 + 72 = 550$ lb.

(c) Crystallizer pressure is given by (4-33). At 160°F, the vapor pressure of H_2O is 4.74 psia. Then water remaining in solution $= (850 + 0.15W)/18 = 51.6$ lbmol.

$MgSO_4$ remaining in solution $= 522/120.4 = 4.3$ lbmol. Hence,

$$x_{H_2O} = 51.6/(51.6 + 4.3) = 0.923.$$

By Raoult's law, $p_{H_2O} = P = 4.74(0.923) = 4.38$ psia.

§4.10.2 Approximate Method for a Vapor–Liquid–Liquid System

Suitable for an approximate method is the case of a mixture containing water and hydrocarbons (HCs), at conditions such that a vapor and two liquid phases, HC-rich (1) and water-rich (2), coexist. Often the solubilities of water in the liquid HC phase and the HCs in the water phase are less than 0.1 mol% and may be neglected. Then, if the liquid HC phase obeys Raoult's law, system pressure is the sum of pressures of the liquid phases:

$$P = P^s_{H_2O} + \sum_{HCs} P^s_i x^{(1)}_i \qquad (4-34)$$

For more general cases, at low pressures where the vapor phase is ideal but the liquid HC phase may be nonideal,

$$P = P^s_{H_2O} + P \sum_{HCs} K_i x^{(1)}_i \qquad (4-35)$$

which can be rearranged to

$$P = \frac{P^s_{H_2O}}{1 - \sum_{HCs} K_i x^{(1)}_i} \qquad (4-36)$$

Equations (4-34) and (4-36) can be used to estimate the pressure for a given temperature and liquid-phase composition, or iteratively to estimate the temperature for a given pressure. Of importance is the determination of which of six possible phase combinations are present: V, $V–L^{(1)}$, $V–L^{(1)}–L^{(2)}$, $V–L^{(2)}$, $L^{(1)}–L^{(2)}$, and L. Indeed, if a $V–L^{(1)}–L^{(2)}$ solution to a problem exists, $V–L^{(1)}$ and $V–L^{(2)}$ solutions also almost always exist. In that case, the three-phase solution is the correct one. It is important, therefore, to seek the three-phase solution first.

EXAMPLE 4.18 Approximate Vapor–Liquid–Liquid Equilibrium.

A mixture of 1,000 kmol of 75 mol% water and 25 mol% n-octane is cooled under equilibrium conditions at a constant pressure of 133.3 kPa from a temperature of 136°C to a temperature of 25°C. Determine: (a) the initial phase condition, and (b) the temperature, phase amounts, and compositions when each phase change occurs. Assume that water and n-octane are immiscible liquids. The vapor pressure of octane is included in Figure 2.3.

Solution

(a) Initial phase conditions are $T = 136°C = 276.8°F$ and $P = 133.3$ kPa $= 19.34$ psia; vapor pressures are $P^s_{H_2O} = 46.7$ psia and $P^s_{nC_8} = 19.5$ psia. Because the initial pressure is less than the vapor pressure of each component, the initial phase condition is all vapor, with partial pressures

$$p_{H_2O} = y_{H_2O}P = 0.75(19.34) = 14.5 \text{ psia}$$
$$p_{nC_8} = y_{nC_8}P = 0.25(19.34) = 4.8 \text{ psia}$$

(b) As the temperature is decreased, a phase change occurs when either $P^s_{H_2O} = p_{H_2O} = 14.5$ psia or $P^s_{nC_8} = p_{nC_8} = 4.8$ psia. The temperatures where these vapor pressures occur are 211°F for H_2O and 194°F for nC_8. The highest temperature applies.

Therefore, water condenses first when the temperature reaches 211°F. This is the dew-point temperature of the mixture at the system pressure. As the temperature is further reduced, the number of moles of water in the vapor decreases, causing the partial pressure of water to decrease below 14.5 psia and the partial pressure of nC_8 to increase above 4.8 psia. Thus, nC_8 begins to condense, forming a second liquid at a temperature higher than 194°F but lower than 211°F. This temperature, referred to as the *secondary dew point*, must be determined iteratively. The calculation is simplified if the bubble point of the mixture is computed first.

From (4-34),

$$P = 19.34 \text{ psi} = P^s_{H_2O} + P^s_{nC_8} \qquad (1)$$

Thus, a temperature that satisfies (4-17) is sought:

T, °F	$P^s_{H_2O}$, psia	$P^s_{nC_8}$, psia	P, psia
194	10.17	4.8	14.97
202	12.01	5.6	17.61
206	13.03	6.1	19.13
207	13.30	6.2	19.50

By interpolation, $T = 206.7$°F for $P = 19.34$ psia. Below 206.7°F the vapor phase disappears and only two immiscible liquids exist.

To determine the temperature at which one of the liquid phases disappears (the same condition as when the second liquid phase begins to appear, i.e., the secondary dew point), it is noted for this case, with only pure water and a pure HC present, that vaporization starting from the bubble point is at a constant temperature until one of the two liquid phases is completely vaporized. Thus, the secondary dew-point temperature is the same as the bubble-point temperature, or 206.7°F. At the secondary dew point, partial pressures are $p_{H_2O} = 13.20$ psia and $p_{nC_8} = 6.14$ psia, with all of the nC_8 in the vapor. Therefore,

Component	Vapor kmol	y	H_2O-Rich Liquid kmol
H_2O	53.9	0.683	21.1
nC_8	25.0	0.317	0.0
	78.9	1.000	21.1

If desired, additional flash calculations can be made for conditions between the dew point and the secondary dew point. The resulting flash curve is shown in Figure 4.31a. If more than one HC species is present, the liquid HC phase does not evaporate at a constant composition and the secondary dew-point temperature is higher than the bubble-point temperature. Then the flash is described by Figure 4.31b.

§4.10.3 Rigorous Method for a Vapor–Liquid–Liquid System

The rigorous method for treating a vapor–liquid–liquid system at a given temperature and pressure is called a *three-phase isothermal flash*. As first presented by Henley and Rosen [17], it is analogous to the isothermal two-phase flash algorithm in §4.4. The system is shown in Figure 4.32. The usual material balances and phase-equilibrium relations apply for each component:

$$Fz_i = Vy_i + L^{(1)}x_i^{(1)} + L^{(2)}x_i^{(2)} \qquad (4-37)$$

$$K_i^{(1)} = y_i/x_i^{(1)} \qquad (4-38)$$

$$K_i^{(2)} = y_i/x_i^{(2)} \qquad (4-39)$$

A relation that can be substituted for (4-38) or (4-39) is

$$K_{D_i} = x_i^{(1)}/x_i^{(2)} \qquad (4-40)$$

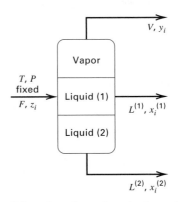

Figure 4.32 Conditions for a three-phase isothermal flash.

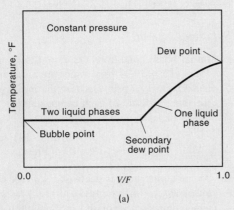

(a)

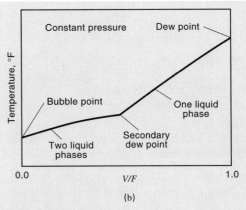

(b)

Figure 4.31 Typical flash curves for immiscible liquid mixtures of water and hydrocarbons at constant pressure: (a) only one hydrocarbon species present; (b) more than one hydrocarbon species present.

These equations are solved by a modification of the Rachford–Rice procedure if we let $\Psi = V/F$ and $\xi = L^{(1)}/(L^{(1)} + L^{(2)})$, where $0 \leq \Psi \leq 1$ and $0 \leq \xi \leq 1$. By combining (4-37), (4-38), and (4-39) with

$$\sum x_i^{(1)} - \sum y_i = 0 \qquad (4\text{-}41)$$

and

$$\sum x_i^{(1)} - \sum x_i^{(2)} = 0 \qquad (4\text{-}42)$$

to eliminate y_i, $x_i^{(1)}$, and $x_i^{(2)}$, two simultaneous equations in Ψ and ξ are obtained:

$$\sum_i \frac{z_i(1 - K_i^{(1)})}{\xi(1 - \Psi) + (1 - \Psi)(1 - \xi)K_i^{(1)}/K_i^{(2)} + \Psi K_i^{(1)}} = 0$$
$$(4\text{-}43)$$

and

$$\sum_i \frac{z_i(1 - K_i^{(1)}/K_i^{(2)})}{\xi(1 - \Psi) + (1 - \Psi)(1 - \xi)K_i^{(1)}/K_i^{(2)} + \Psi K_i^{(1)}} = 0$$
$$(4\text{-}44)$$

Values of Ψ and ξ are computed by solving nonlinear equations (4-43) and (4-44) simultaneously. Then the phase amounts and compositions are determined from

$$V = \Psi F \qquad (4\text{-}45)$$

$$L^{(1)} = \xi(F - V) \qquad (4\text{-}46)$$

$$L^{(2)} = F - V - L^{(1)} \qquad (4\text{-}47)$$

$$y_i = \frac{z_i}{\xi(1 - \Psi)/K_i^{(1)} + (1 - \Psi)(1 - \xi)/K_i^{(2)} + \Psi} \qquad (4\text{-}48)$$

$$x_i^{(1)} = \frac{z_i}{\xi(1 - \Psi) + (1 - \Psi)(1 - \xi)(K_i^{(1)}/K_i^{(2)}) + \Psi K_i^{(1)}}$$
$$(4\text{-}49)$$

$$x_i^{(2)} = \frac{z_i}{\xi(1 - \Psi)(K_i^{(2)}/K_i^{(1)}) + (1 - \Psi)(1 - \xi) + \Psi K_i^{(2)}}$$
$$(4\text{-}50)$$

Calculations for a three-phase flash are difficult because of the strong dependency of K-values on liquid-phase compositions when two immiscible liquids are present. This dependency appears in the liquid-phase activity coefficients (e.g., Eq. (4) in Table 2.3). In addition, it is not obvious how many phases will be present. A typical algorithm for determining phase conditions is shown in Figure 4.33. Calculations are best made with a process simulator, which can also perform adiabatic or nonadiabatic three-phase flashes by iterating on temperature until the enthalpy balance,

$$h_F F + Q = h_V V + h_{L^{(1)}} L^{(1)} + h_{L^{(2)}} L^{(2)} = 0 \qquad (4\text{-}51)$$

is satisfied.

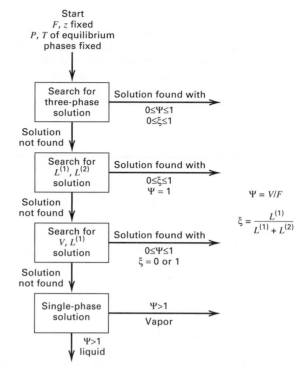

Figure 4.33 Algorithm for an isothermal three-phase flash.

EXAMPLE 4.19 Three-Phase Isothermal Flash.

In a process for producing styrene from toluene and methanol, the gaseous reactor effluent is as follows:

Component	kmol/h
Hydrogen	350
Methanol	107
Water	491
Toluene	107
Ethylbenzene	141
Styrene	350

If this stream is brought to equilibrium at 38°C and 300 kPa. Compute the amounts and compositions of the phases present.

Solution

Because water, hydrocarbons, an alcohol, and a light gas are present, the possibility of a vapor and two liquid phases exists, with methanol distributed among all phases. The isothermal three-phase flash module of the CHEMCAD process simulator was used with Henry's law for H_2 and UNIFAC for activity coefficients for the other components, to obtain:

Component	kmol/h V	$L^{(1)}$	$L^{(2)}$
Hydrogen	349.96	0.02	0.02
Methanol	9.54	14.28	83.18
Water	7.25	8.12	475.63
Toluene	1.50	105.44	0.06
Ethylbenzene	0.76	140.20	0.04
Styrene	1.22	348.64	0.14
Totals	370.23	616.70	559.07

As expected, little H_2 is dissolved in either liquid. The water-rich liquid phase, $L^{(2)}$, contains little of the hydrocarbons, but much methanol. The organic-rich phase, $L^{(1)}$, contains most of the hydrocarbons and small amounts of water and methanol. Additional calculations at 300 kPa indicate that the organic phase condenses first, with a dew point = 143°C and a secondary dew point = 106°C.

SUMMARY

1. The Gibbs phase rule applies to intensive variables at equilibrium. It determines the number of independent variables that can be specified. This rule can be extended to determine the degrees of freedom (number of allowable specifications) for flow systems, including extensive variables. The intensive and extensive variables are related by material- and energy-balance equations and phase-equilibria data.

2. Vapor–liquid equilibrium conditions for binary systems can be represented by T–y–x, y–x, and P–x diagrams. Relative volatility for a binary system tends to 1.0 as the critical point is approached.

3. Minimum- or maximum-boiling azeotropes formed by nonideal liquid mixtures are represented by the same types of diagrams used for nonazeotropic (zeotropic) binary mixtures. Highly nonideal liquid mixtures can form heterogeneous azeotropes having two liquid phases.

4. For multicomponent mixtures, vapor–liquid equilibrium-phase compositions and amounts can be determined by isothermal-flash, adiabatic-flash, and bubble- and dew-point calculations. For non-ideal mixtures, process simulators should be used.

5. Liquid–liquid equilibrium conditions for ternary mixtures are best determined graphically from triangular and other equilibrium diagrams, unless only one of the three components (the solute) is soluble in the two liquid phases. In that case, the conditions can be readily determined algebraically using phase-distribution ratios (partition coefficients) for the solute.

6. Liquid–liquid equilibrium conditions for multicomponent mixtures of four or more components are best determined with process simulators, particularly when the system is not dilute in the solute(s).

7. Solid–liquid equilibrium occurs in leaching, crystallization, and adsorption. In leaching it is common to assume that all solute is dissolved in the solvent and that the remaining solid in the underflow is accompanied by a known fraction of liquid. Crystallization calculations are best made with a phase-equilibrium diagram. For crystallization of salts from an aqueous solution, formation of hydrates must be considered. Adsorption can be represented algebraically or graphically by adsorption isotherms.

8. Solubility of gases that are only sparingly soluble in a liquid are well represented by a Henry's law constant that depends on temperature.

9. Solid vapor pressure can determine equilibrium sublimation and desublimation conditions for gas–solid systems. Adsorption isotherms and y–x diagrams are useful in adsorption-equilibrium calculations for gas mixtures in the presence of solid adsorbent.

10. Calculations of multiphase equilibrium are best made by process simulators. However, manual procedures are available for vapor–liquid–solid systems when no component is found in all phases, and to vapor–liquid–liquid systems when only one component distributes in all phases.

REFERENCES

1. Green, D.W., and R.H. Perry, Eds., *Perry's Chemical Engineers' Handbook*, 8th ed., McGraw-Hill, New York (2008).

2. Gmehling, J., and U. Onken, *Vapor-Liquid Equilibrium Data Collection*, DECHEMA Chemistry Data Series, **1–8**, (1977–1984).

3. Hála, E., *Vapour-Liquid Equilibrium: Data at Normal Pressures* Pergamon Press, New York (1968).

4. Hughes, R.R., H.D. Evans, and C.V. Sternling, *Chem. Eng. Progr*, **49**, 78–87 (1953).

5. Rachford, H.H., Jr., and J.D. Rice, *J. Pet. Tech.*, **4**(10), Section 1, p. 19, and Section 2, p. 3 (Oct.1952).

6. Press, W.H., S.A. Teukolsky, W.T. Vetterling, and B.P. Flannery, *Numerical Recipes in FORTRAN, 2nd ed.*, Cambridge University Press, Cambridge, chap. 9 (1992).

7. Goff, G.H., P.S. Farrington, and B.H. Sage, *Ind. Eng. Chem.*, **42**, 735–743 (1950).

8. Constantinides, A., and N. Mostoufi, *Numerical Methods for Chemical Engineers with MATLAB Applications*, Prentice Hall PTR, Upper Saddle River, NJ (1999).

9. Robbins, L.A., in R.H. Perry, D.H. Green, and J.O. Maloney Eds., *Perry's Chemical Engineers' Handbook*, 7th ed., McGraw-Hill, New York, pp.15-10 to 15-15 (1997).

10. Janecke, E., *Z. Anorg. Allg. Chem.* **51**, 132–157 (1906).

11. Francis, A.W., *Liquid-Liquid Equilibriums*, Interscience, New York (1963).

12. Findlay, A., *Phase Rule*, Dover, New York (1951).

13. Fritz, W., and E.-U. Schuluender, *Chem. Eng. Sci.*, **29**, 1279–1282 (1974).

14. Felder, R.M., and R.W. Rousseau, *Elementary Principles of Chemical Processes*, 3rd ed., John Wiley & Sons, New York, pp. 613–616 (1986).

15. Lewis, W.K., E.R. Gilliland, B. Cherton, and W.H. Hoffman, *J. Am. Chem. Soc.* **72**, 1153–1157 (1950).

16. Hildebrand, J.H., *Principles of Chemistry*, 4th ed., Macmillan, New York (1940).

17. Henley, E.J., and E.M. Rosen, *Material and Energy Balance Computations*, John Wiley & Sons, New York, pp. 351–353 (1969).

18. Conway, J.B., and J.J. Norton, *Ind. Eng. Chem.*, **43**, 1433–1435 (1951).

19. Boston, J., and H. Britt, *Comput. Chem. Engng.*, **2**, 109 (1978).

STUDY QUESTIONS

4.1. What two types of equations are used for single equilibrium stage calculations?

4.2. How do intensive and extensive variables differ?

4.3. What is meant by the number of degrees of freedom?

4.4. What are the limitations of the Gibbs phase rule? How can it be extended?

4.5. When a liquid and a vapor are in physical equilibrium, why is the vapor at its dew point and the liquid at its bubble point?

4.6. What is the difference between a homogeneous and a heterogeneous azeotrope?

4.7. Why do azeotropes limit the degree of separation achievable in a distillation operation?

4.8. What is the difference between an isothermal and an adiabatic flash?

4.9. Why is the isothermal-flash calculation so important?

4.10. When a binary feed is contacted with a solvent to form two equilibrium liquid phases, which is the extract and which the raffinate?

4.11. Why are triangular diagrams useful for ternary liquid–liquid equilibrium calculations? On such a diagram, what are the miscibility boundary, plait point, and tie lines?

4.12. Why is the right-triangular diagram easier to construct and read than an equilateral-triangular diagram? What is, perhaps, the only advantage of the latter diagram?

4.13. What are the conditions for an ideal, equilibrium leaching stage?

4.14. In crystallization, what is a eutectic? What is mother liquor? What are hydrates?

4.15. What is the difference between adsorbent and adsorbate?

4.16. In adsorption, why are adsorbents having a microporous structure desirable?

4.17. Does a solid have a vapor pressure?

4.18. What is the maximum number of phases that can exist at physical equilibrium for a given number of components?

4.19. In a rigorous vapor–liquid–liquid equilibrium calculation (the so-called three-phase flash), is it necessary to consider all possible phase conditions, i.e., all-liquid, all-vapor, vapor–liquid, liquid–liquid, as well as vapor–liquid–liquid?

4.20. What is the secondary dew point? Is there also a secondary bubble point?

EXERCISES

Section 4.1

4.1. **Degrees-of-freedom for a three-phase equilibrium.**

Consider the equilibrium stage shown in Figure 4.34. Conduct a degrees-of-freedom analysis by performing the following steps: (a) list and count the variables; (b) write and count the equations relating the variables; (c) calculate the degrees of freedom; and (d) list a reasonable set of design variables.

4.2. **Uniqueness of three different separation operations.**

Can the following problems be solved uniquely?

(a) The feed streams to an adiabatic equilibrium stage consist of liquid and vapor streams of known composition, flow rate,

temperature, and pressure. Given the stage (outlet) temperature and pressure, calculate the composition and amounts of equilibrium vapor and liquid leaving.

(b) The same as part (a), except that the stage is not adiabatic.

(c) A vapor of known T, P, and composition is partially condensed. The outlet P of the condenser and the inlet cooling water T are fixed. Calculate the cooling water required.

4.3. **Degrees-of-freedom for an adiabatic, two-phase flash.**

Consider an adiabatic equilibrium flash. The variables are all as indicated in Figure 4.10a with $Q = 0$. (a) Determine the number of variables. (b) Write all the independent equations that relate the variables. (c) Determine the number of equations. (d) Determine the number of degrees of freedom. (e) What variables would you prefer to specify in order to solve an adiabatic-flash problem?

4.4. **Degrees of freedom for a nonadiabatic, three-phase flash.**

Determine the number of degrees of freedom for a nonadiabatic equilibrium flash for the liquid feed and three products shown in Figure 4.32.

4.5. **Application of Gibbs phase rule.**

For the seven-phase equilibrium system shown in Figure 4.30, assume air consists of N_2, O_2, and argon. What is the number of degrees of freedom? What variables might be specified?

Section 4.2

4.6. **Partial vaporization of a nonideal binary mixture.**

A liquid mixture containing 25 mol% benzene and 75 mol% ethyl alcohol, in which components are miscible in all proportions, is heated at a constant pressure of 1 atm from 60°C to 90°C. Using the following T–x–y experimental data, determine (a) the temperature where vaporization begins; (b) the composition of the first bubble of vapor; (c) the composition of the residual liquid when 25 mol % has evaporated, assuming that all vapor formed is retained in the apparatus and is in equilibrium with the residual liquid. (d) Repeat

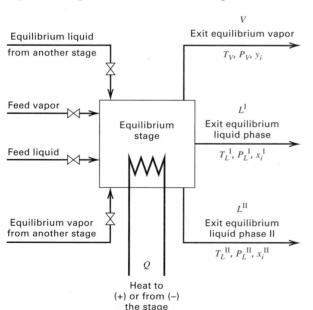

Figure 4.34 Conditions for Exercise 4.1.

part (c) for 90 mol% vaporized. (e) Repeat part (d) if, after 25 mol% is vaporized as in part (c), the vapor formed is removed and an additional 35 mol% is vaporized by the same technique used in part (c). (f) Plot temperature versus mol% vaporized for parts (c) and (e).

T–x–y DATA FOR BENZENE–ETHYL ALCOHOL AT 1 ATM

Temperature, °C:

| 78.4 | 77.5 | 75 | 72.5 | 70 | 68.5 | 67.7 | 68.5 | 72.5 | 75 | 77.5 | 80.1 |

Mole percent benzene in vapor:

| 0 | 7.5 | 28 | 42 | 54 | 60 | 68 | 73 | 82 | 88 | 95 | 100 |

Mole percent benzene in liquid:

| 0 | 1.5 | 5 | 12 | 22 | 31 | 68 | 81 | 91 | 95 | 98 | 100 |

(g) Use the following vapor pressure data with Raoult's and Dalton's laws to construct a T–x–y diagram, and compare it to the answers obtained in parts (a) and (f) with those obtained using the experimental T–x–y data. What are your conclusions?

VAPOR PRESSURE DATA

Vapor pressure, torr:

| 20 | 40 | 60 | 100 | 200 | 400 | 760 |

Ethanol, °C:

| 8 | 19.0 | 26.0 | 34.9 | 48.4 | 63.5 | 78.4 |

Benzene, °C:

| −2.6 | 7.6 | 15.4 | 26.1 | 42.2 | 60.6 | 80.1 |

4.7. Steam distillation of stearic acid.

Stearic acid is steam distilled at 200°C in a direct-fired still. Steam is introduced into the molten acid in small bubbles, and the acid in the vapor leaving the still has a partial pressure equal to 70% of the vapor pressure of pure stearic acid at 200°C. Plot the kg acid distilled per kg steam added as a function of total pressure from 101.3 kPa to 3.3 kPa at 200°C. The vapor pressure of stearic acid at 200°C is 0.40 kPa.

4.8. Equilibrium plots for benzene–toluene.

The relative volatility, α, of benzene to toluene at 1 atm is 2.5. Construct x–y and T–x–y diagrams for this system at 1 atm. Repeat the construction of the x–y diagram using vapor pressure data for benzene from Exercise 4.6 and for toluene from the table below, with Raoult's and Dalton's laws. Use the diagrams for the following: (a) A liquid containing 70 mol% benzene and 30 mol% toluene is heated in a container at 1 atm until 25 mol% of the original liquid is evaporated. Determine the temperature. The phases are then separated mechanically, and the vapors condensed. Determine the composition of the condensed vapor and the liquid residue. (b) Calculate and plot the K-values as a function of temperature at 1 atm.

VAPOR PRESSURE OF TOLUENE

Vapor pressure, torr:

| 20 | 40 | 60 | 100 | 200 | 400 | 760 | 1,520 |

Temperature, °C:

| 18.4 | 31.8 | 40.3 | 51.9 | 69.5 | 89.5 | 110.6 | 136 |

4.9. Vapor–liquid equilibrium for heptane–toluene system.

(a) The vapor pressure of toluene is given in Exercise 4.8, and that of n-heptane is in the table below. Construct the following plots:

(a) an x–y diagram at 1 atm using Raoult's and Dalton's laws; (b) a T–x bubble-point curve at 1 atm; (c) α and K-values versus temperature; and (d) repeat of part (a) using an average value of α. Then, (e) compare your x–y and T–x–y diagrams with the following experimental data of Steinhauser and White [Ind. Eng. Chem., **41**, 2912 (1949)].

VAPOR PRESSURE OF n-HEPTANE

Vapor pressure, torr:

| 20 | 40 | 60 | 100 | 200 | 400 | 760 | 1,520 |

Temperature, °C:

| 9.5 | 22.3 | 30.6 | 41.8 | 58.7 | 78.0 | 98.4 | 124 |

VAPOR–LIQUID EQUILIBRIUM DATA FOR n-HEPTANE/TOLUENE AT 1 ATM

$x_{n\text{-heptane}}$	$y_{n\text{-heptane}}$	T, °C
0.025	0.048	110.75
0.129	0.205	106.80
0.354	0.454	102.95
0.497	0.577	101.35
0.843	0.864	98.90
0.940	0.948	98.50
0.994	0.993	98.35

4.10. Continuous, single-stage distillation.

Saturated-liquid feed of $F = 40$ mol/h, containing 50 mol% A and B, is supplied to the apparatus in Figure 4.35. The condensate is split so that reflux/condensate = 1.

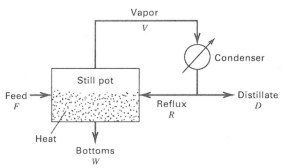

Figure 4.35 Conditions for Exercise 4.10.

(a) If heat is supplied such that $W = 30$ mol/h and $\alpha = 2$, as defined below, what will be the composition of the overhead and the bottoms product?

$$\alpha = \frac{P_A^s}{P_B^s} = \frac{y_A x_B}{y_B x_A}$$

(b) If the operation is changed so that no condensate is returned to the still pot and $W = 3D$, compute the product compositions.

4.11. Partial vaporization of feed to a distillation column.

A fractionation tower operating at 101.3 kPa produces a distillate of 95 mol% acetone (A), 5 mol% water, and a residue containing 1 mol% A. The feed liquid is at 125°C and 687 kPa and contains 57 mol% A. Before entering the tower, the feed passes through an expansion valve and is partially vaporized at 60°C. From the data below, determine the molar ratio of liquid to vapor in the feed. Enthalpy and equilibrium data are: molar latent heat of A = 29,750 kJ/kmol; molar latent heat of H_2O = 42,430 kJ/kmol; molar specific heat of A = 134 kJ/kmol-K; molar specific heat of H_2O = 75.3 kJ/kmol-K; enthalpy of high-pressure, hot feed before adiabatic expansion = 0; enthalpies of feed phases after expansion are h_V = 27,200 kJ/kmol and h_L = −5,270 kJ/kmol. All data except K-values, are temperature-independent.

EQUILIBRIUM DATA FOR ACETONE–H_2O AT 101.3 kPa

			T, °C				
	56.7	57.1	60.0	61.0	63.0	71.7	100
Mol% A in liquid:	100	92.0	50.0	33.0	17.6	6.8	0
Mol% A in vapor:	100	94.4	85.0	83.7	80.5	69.2	0

4.12. Enthalpy-concentration diagram.

Using vapor pressure data from Exercises 4.6 and 4.8 and the enthalpy data provided below: (a) construct an h–x–y diagram for the benzene–toluene system at 1 atm (101.3 kPa) based on Raoult's and Dalton's laws, and (b) calculate the energy required for 50 mol% vaporization of a 30 mol% liquid solution of benzene in toluene at saturation temperature. If the vapor is condensed, what is the heat load on the condenser in kJ/kg of solution if the condensate is saturated, and if it is subcooled by 10°C?

	Saturated Enthalpy, kJ/kg			
	Benzene		Toluene	
T, °C	h_L	h_V	h_L	h_V
60	79	487	77	471
80	116	511	114	495
100	153	537	151	521

Section 4.3

4.13. Azeotrope of chloroform–methanol.

Vapor–liquid equilibrium data at 101.3 kPa are given for the chloroform–methanol system on p. 13-11 of *Perry's Chemical Engineers' Handbook*, 6th ed. From these data, prepare plots like Figures 4.6b and 4.6c. From the plots, determine the azeotropic composition, type of azeotrope, and temperature at 101.3 kPa.

4.14. Azeotrope of water–formic acid.

Vapor–liquid equilibrium data at 101.3 kPa are given for the water–formic acid system on p. 13-14 of *Perry's Chemical Engineers' Handbook*, 6th ed. From these data, prepare plots like Figures 4.7b and 4.7c. From the plots, determine the azeotropic composition, type of azeotrope, and temperature at 101.3 kPa.

4.15. Partial vaporization of water–isopropanol mixture.

Vapor–liquid equilibrium data for mixtures of water and isopropanol at 1 atm are given below. (a) Prepare T–x–y and x–y diagrams. (b) When a solution containing 40 mol% isopropanol is slowly vaporized, what is the composition of the initial vapor? (c) If the mixture in part (b) is heated until 75 mol% is vaporized, what are the compositions of the equilibrium vapor and liquid? (d) Calculate K-values and values of α at 80°C and 89°C. (e) Compare your answers in parts (a), (b), and (c) to those obtained from T–x–y and x–y diagrams based on the following vapor pressure data and Raoult's and Dalton's laws. What do you conclude?

VAPOR–LIQUID EQUILIBRIUM DATA FOR ISOPROPANOL AND WATER AT 1 ATM

	Mol% Isopropanol	
T, °C	Liquid	Vapor
93.00	1.18	21.95
84.02	8.41	46.20
83.85	9.10	47.06
81.64	28.68	53.44
81.25	34.96	55.16
80.32	60.30	64.22
80.16	67.94	68.21
80.21	68.10	68.26
80.28	76.93	74.21
80.66	85.67	82.70
81.51	94.42	91.60

Notes: Composition of the azeotrope: $x = y = 68.54\%$.
Boiling point of azeotrope: 80.22°C.
Boiling point of pure isopropanol: 82.5°C.

Vapor Pressures of Isopropanol and Water

Vapor pressure, torr	200	400	760
Isopropanol, °C	53.0	67.8	82.5
Water, °C	66.5	83	100

Section 4.4

4.16. Vaporization of mixtures of hexane and octane.

Using the y–x and T–y–x diagrams in Figures 4.3 and 4.4, determine the temperature, amounts, and compositions of the vapor and liquid phases at 101 kPa for the following conditions with a 100-kmol mixture of nC_6 (H) and nC_8 (C). (a) $z_H = 0.5$, $\Psi = V/F = 0.2$; (b) $z_H = 0.4$, $y_H = 0.6$; (c) $z_H = 0.6$, $x_C = 0.7$; (d) $z_H = 0.5$, $\Psi = 0$; (e) $z_H = 0.5$, $\Psi = 1.0$; and (f) $z_H = 0.5$, $T = 200°F$

4.17. Derivation of equilibrium-flash equations for a binary mixture.

For a binary mixture of components 1 and 2, show that the phase compositions and amounts can be computed directly from the following reduced forms of Eqs. (5), (6), and (3) of Table 4.4:

$$x_1 = (1 - K_2)/(K_1 - K_2)$$
$$x_2 = 1 - x_1$$
$$y_1 = (K_1K_2 - K_1)/(K_2 - K_1)$$
$$y_2 = 1 - y_1$$
$$\Psi = \frac{V}{F} = \frac{z_1[(K_1 - K_2)/(1 - K_2)] - 1}{K_1 - 1}$$

4.18. Conditions for Rachford–Rice equation to be satisfied.

Consider the Rachford–Rice form of the flash equation,

$$\sum_{i=1}^{C} \frac{z_i(1 - K_i)}{1 + (V/F)(K_i - 1)} = 0$$

Under what conditions can this equation be satisfied?

4.19. Equilibrium flash using a graph.

A liquid containing 60 mol% toluene and 40 mol% benzene is continuously distilled in a single equilibrium stage at 1 atm. What percent of benzene in the feed leaves as vapor if 90% of the toluene entering in the feed leaves as liquid? Assume a relative volatility of 2.3 and obtain the solution graphically.

4.20. Flash vaporization of a benzene–toluene mixture.

Solve Exercise 4.19 by assuming an ideal solution with vapor pressure data from Figure 2.3. Also determine the temperature.

4.21. Equilibrium flash of seven-component mixture.

A seven-component mixture is flashed at a fixed P and T. (a) Using the K-values and feed composition below, make a plot of the Rachford–Rice flash function

$$f\{\Psi\} = \sum_{i=1}^{C} \frac{z_i(1 - K_i)}{1 + \Psi(K_i - 1)}$$

at intervals of Ψ of 0.1, and estimate the correct root of Ψ. (b) An alternative form of the flash function is

$$f\{\Psi\} = \sum_{i=1}^{C} \frac{z_i K_i}{1 + \Psi(K_i - 1)} - 1$$

Make a plot of this equation at intervals of Ψ of 0.1 and explain why the Rachford–Rice function is preferred.

Component	z_i	K_i
1	0.0079	16.2
2	0.1321	5.2
3	0.0849	2.6
4	0.2690	1.98
5	0.0589	0.91
6	0.1321	0.72
7	0.3151	0.28

4.22. Equilibrium flash of a hydrocarbon mixture.

One hundred kmol of a feed composed of 25 mol% n-butane, 40 mol% n-pentane, and 35 mol% n-hexane is flashed. If 80% of the hexane is in the liquid at 240°F, what are the pressure and the liquid and vapor compositions? Obtain K-values from Figure 2.4.

4.23. Equilibrium-flash vaporization of a hydrocarbon mixture.

An equimolar mixture of ethane, propane, n-butane, and n-pentane is subjected to flash vaporization at 150°F and 205 psia. What are the expected amounts and compositions of the products? Is it possible to recover 70% of the ethane in the vapor by a single-stage flash at other conditions without losing more than 5% of nC_4 to the vapor? Obtain K-values from Figure 2.4.

4.24. Cooling of a reactor effluent with recycled liquid.

Figure 4.36 shows a system to cool reactor effluent and separate light gases from hydrocarbons. K-values at 500 psia and 100°F are:

Component	K_i
H_2	80
CH_4	10
Benzene	0.010
Toluene	0.004

(a) Calculate composition and flow rate of vapor leaving the flash drum. (b) Does the liquid-quench flow rate influence the result? Prove your answer analytically.

4.25. Partial condensation of a gas mixture.

The feed in Figure 4.37 is partially condensed. Calculate the amounts and compositions of the equilibrium phases, V and L.

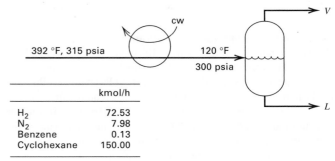

	kmol/h
H_2	72.53
N_2	7.98
Benzene	0.13
Cyclohexane	150.00

Figure 4.37 Conditions for Exercise 4.25.

4.26. Rapid determination of phase condition.

The following stream is at 200 psia and 200°F. Without making a flash calculation, determine if it is a subcooled liquid or a super-heated vapor, or if it is partially vaporized.

Component	lbmol/h	K-value
C_3	125	2.056
nC_4	200	0.925
nC_5	175	0.520

4.27. Determination of reflux-drum pressure.

Figure 4.38 shows the overhead system for a distillation column. The composition of the total distillates is indicated, with 10 mol% being vapor. Determine reflux-drum pressure if the temperature is 100°F. Use the K-values below, assuming that K is inversely proportional to pressure.

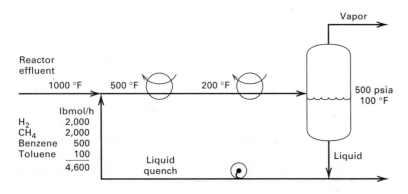

Figure 4.36 Conditions for Exercise 4.24.

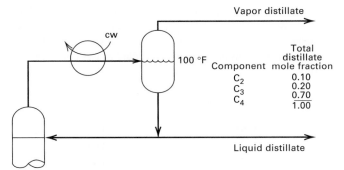

Figure 4.38 Conditions for Exercise 4.27.

Component	K at 100°F, 200 psia
C_2	2.7
C_3	0.95
C_4	0.34

4.28. Flash calculations for different K-value correlations.

Determine the phase condition of a stream having the following composition at 7.2°C and 2,620 kPa.

Component	kmol/h
N_2	1.0
C_1	124.0
C_2	87.6
C_3	161.6
nC_4	176.2
nC_5	58.5
nC_6	33.7

Use a process simulator with the S–R–K and P–R options for K-values, together with one other option, e.g., B–W–R–S. Does the choice of K-value method influence the results significantly?

4.29. Flash calculations at different values of T and P.

A liquid mixture consisting of 100 kmol of 60 mol% benzene, 25 mol% toluene, and 15 mol% o-xylene is flashed at 1 atm and 100°C. Assuming ideal solutions, use vapor pressure data from a process simulator to: (a) Compute kmol amounts and mole-fraction compositions of liquid and vapor products. (b) Repeat the calculation at 100°C and 2 atm. (c) Repeat the calculation at 105°C and 0.1 atm. (d) Repeat the calculation at 150°C and 1 atm.

4.30. Conditions at vapor–liquid equilibrium.

Using equations in Table 4.4, prove that the vapor leaving an equilibrium flash is at its dew point and that the liquid leaving is at its bubble point.

4.31. Bubble-point temperature of feed to a distillation column.

The feed below enters a distillation column as saturated liquid at 1.72 MPa. Calculate the bubble-point temperature using the K-values of Figure 2.4.

Compound	kmol/h
Ethane	1.5
Propane	10.0
n-Butane	18.5
n-Pentane	17.5
n-Hexane	3.5

4.32. Bubble- and dew-point pressures of a binary mixture.

An equimolar solution of benzene and toluene is evaporated at a constant temperature of 90°C. What are the pressures at the beginning and end of the vaporization? Assume an ideal solution and use the vapor pressure curves of Figure 2.3, or use a process simulator.

4.33. Bubble point, dew point, and flash of a water–acetic acid mixture.

The following equations are given by Sebastiani and Lacquaniti [*Chem. Eng. Sci.*, **22**, 1155 (1967)] for the liquid-phase activity coefficients of the water (W)–acetic acid (A) system.

$$\log\gamma_W = x_A^2[A + B(4x_W - 1) + C(x_W - x_A)(6x_W - 1)]$$
$$\log\gamma_A = x_W^2[A + B(4x_W - 3) + C(x_W - x_A)(6x_W - 5)]$$
$$A = 0.1182 + \frac{64.24}{T(\text{K})}$$
$$B = 0.1735 - \frac{43.27}{T(\text{K})}$$
$$C = 0.1081$$

Find the dew point and bubble point of the mixture $x_W = 0.5$, $x_A = 0.5$, at 1 atm. Flash the mixture at a temperature halfway between the dew and bubble points.

4.34. Bubble point, dew point, and flash of a mixture.

Find the bubble point and dew point of a mixture of 0.4 mole fraction toluene (1) and 0.6 mole fraction n-butanol (2) at 101.3 kPa. K-values can be calculated from (2-72) using vapor-pressure data, and γ_1 and γ_2 from the van Laar equation of Table 2.9 with $A_{12} = 0.855$ and $A_{21} = 1.306$. If the same mixture is flashed midway between the bubble and dew points and 101.3 kPa, what fraction is vaporized, and what are the phase compositions?

4.35. Bubble point, dew point, and azeotrope of mixture.

For a solution of a molar composition of ethyl acetate (A) of 80% and ethyl alcohol (E) of 20%: (a) Calculate the bubble-point temperature at 101.3 kPa and the composition of the corresponding vapor using (2-72) with vapor pressure data and the van Laar equation of Table 2.9 with $A_{AE} = 0.855$, $A_{EA} = 0.753$. (b) Find the dew point of the mixture. (c) Does the mixture form an azeotrope? If so, predict the temperature and composition.

4.36. Bubble point, dew point, and azeotrope of a mixture.

A solution at 107°C contains 50 mol% water (W) and 50 mol% formic acid (F). Using (2-72) with vapor pressure data and the van Laar equation with $A_{WF} = -0.2935$ and $A_{FW} = -0.2757$: (a) Compute the bubble-point pressure. (b) Compute the dew-point pressure. (c) Determine if the mixture forms an azeotrope. If so, predict the azeotropic pressure at 107°C and the composition.

4.37. Bubble point, dew point, and equilibrium flash of a ternary mixture.

For a mixture of 45 mol% n-hexane, 25 mol% n-heptane, and 30 mol% n-octane at 1 atm, use a process simulator to: (a) Find the bubble- and dew-point temperatures. (b) Find the flash temperature, compositions, and relative amounts of liquid and vapor products if the mixture is subjected to a flash distillation at 1 atm so that 50 mol % is vaporized. (c) Find how much octane is taken off as vapor if 90% of the hexane is taken off as vapor. (d) Repeat parts (a) and (b) at 5 atm and 0.5 atm.

4.38. Vaporization of column bottoms in a partial reboiler.

In Figure 4.39, 150 kmol/h of a saturated liquid, L_1, at 758 kPa of molar composition propane 10%, n-butane 40%, and n-pentane 50% enters the reboiler from stage 1. Use a process simulator to find the compositions and amounts of V_B and B. What is Q_R, the reboiler duty?

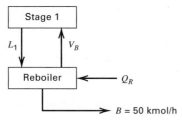

Figure 4.39 Conditions for Exercise 4.38.

4.39. Bubble point and flash temperatures for a ternary mixture.

For a mixture with mole fractions 0.005 methane, 0.595 ethane, and the balance n-butane at 50 psia, and using K-values from Figure 2.4: (a) Find the bubble-point temperature. (b) Find the temperature that results in 25% vaporization at this pressure, and determine the liquid and vapor compositions in mole fractions.

4.40. Heating and expansion of a hydrocarbon mixture.

In Figure 4.40, a mixture is heated and expanded before entering a distillation column. Calculate, using a process simulator, mole percent vapor and vapor and liquid mole fractions at locations indicated by pressure specifications.

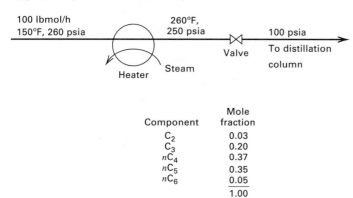

Component	Mole fraction
C_2	0.03
C_3	0.20
nC_4	0.37
nC_5	0.35
nC_6	0.05
	1.00

Figure 4.40 Conditions for Exercise 4.40.

4.41. Equilibrium vapor and liquid leaving a feed stage.

Streams entering stage F of a distillation column are shown in Figure 4.41. Using a process simulator, find the stage temperature and compositions and amounts of streams V_F and L_F if the pressure is 785 kPa.

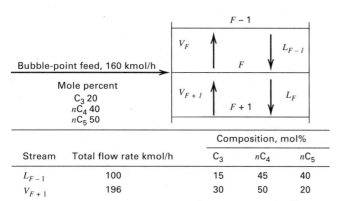

Stream	Total flow rate kmol/h	Composition, mol%		
		C_3	nC_4	nC_5
L_{F-1}	100	15	45	40
V_{F+1}	196	30	50	20

Figure 4.41 Conditions for Exercise 4.41.

4.42. Adiabatic flash across a valve.

The stream below is flashed adiabatically across a valve. Conditions upstream are 250°F and 500 psia, and downstream are 300

psia. Compute using a process simulator the: (a) phase condition upstream of the valve; (b) temperature downstream of the valve; (c) molar fraction vaporized downstream of the valve; and (d) mole-fraction compositions of the vapor and liquid phases downstream of the valve.

Component	z_i
C_2H_4	0.02
C_2H_6	0.03
C_3H_6	0.05
C_3H_8	0.10
iC_4	0.20
nC_4	0.60

4.43. Single-stage equilibrium flash of a clarified broth.

The ABE biochemical process makes acetone, n-butanol, and ethanol by an anaerobic, submerged, batch fermentation at 30°C of corn kernels, using a strain of the bacterium *Clostridia acetobutylicum*. Following fermentation, the broth is separated from the biomass solids by centrifugation. Consider 1,838,600 L/h of clarified broth of S.G. = 0.994, with a titer of 22.93 g/L of ABE in the mass ratio of 3.0:7.5:1.0. A number of continuous bioseparation schemes have been proposed, analyzed, and applied. In particular, the selection of the first separation step needs much study because the broth is so dilute in the bioproducts. Possibilities are single-stage flash, distillation, liquid–liquid extraction, and pervaporation. In this exercise, a single-stage flash is to be investigated. Convert the above data on the clarified broth to component flow rates in kmol/h. Heat the stream to 97°C at 101.3 kPa. Use a process simulator to run a series of equilibrium-flash calculations using the NRTL equation for liquid-phase activity coefficients. Note that n-butanol and ethanol both form an azeotrope with water. Also, n-butanol may not be completely soluble in water for all concentrations. The specifications for each flash calculation are pressure = 101.3 kPa and V/F, the molar vapor-to-feed ratio. A V/F is to be sought that maximizes the ABE in the vapor while minimizing the water in the vapor. Because the boiling point of n-butanol is greater than that of water, and because of possible azeotrope formation and other nonideal solution effects, a suitable V/F may not exist.

4.44. Algorithms for various flash calculations.

Given the isothermal-flash algorithm and Table 4.4, propose algorithms for the following flash calculations, assuming that expressions for K-values and enthalpies are available.

Given	Find
h_F, P	Ψ, T
h_F, T	Ψ, P
h_F, Ψ	T, P
Ψ, T	h_F, P
Ψ, P	h_F, T
T, P	h_F, Ψ

Section 4.5

4.45. Comparison of solvents for single-stage extraction.

A feed of 13,500 kg/h is 8 wt% acetic acid (B) in water (A). Removal of acetic acid is to be by liquid–liquid extraction at 25°C. The raffinate is to contain 1 wt% acetic acid. The following four solvents, with accompanying distribution (partition) coefficients in mass-fraction units, are candidates. Water and each solvent (C) can be considered immiscible. For each solvent, estimate the kg/hr required if one equilibrium stage is used.

Solvent	K_D
Methyl acetate	1.273
Isopropyl ether	0.429
Heptadecanol	0.312
Chloroform	0.178

4.46. Liquid–liquid extraction of ethylene glycol from water by furfural.

Forty-five kg of a solution of 30 wt% ethylene glycol in water is to be extracted with furfural. Using Figures 4.14a and 4.14c, calculate the: (a) minimum kg of solvent; (b) maximum kg of solvent; (c) kg of solvent-free extract and raffinate for 45 kg solvent, and the percentage glycol extracted; and (d) maximum purity of glycol in the extract and the maximum purity of water in the raffinate for one stage.

4.47. Representation of a ternary on a triangular diagram.

Prove that, in a triangular diagram where each vertex represents a pure component, the composition of the system at any point inside the triangle is proportional to the length of the respective perpendicular drawn from the point to the side of the triangle opposite the vertex in question. Note that it is not necessary that the triangle be of a right or equilateral type.

4.48. Liquid–liquid extraction of acetic acid from chloroform by water.

A mixture of chloroform ($CHCl_3$) and acetic acid at 18°C and 1 atm (101.3 kPa) is extracted with water to recover the acid. Forty-five kg of 35 wt% $CHCl_3$ and 65 wt% acid is treated with 22.75 kg of water at 18°C in a one-stage batch extraction. (a) What are the compositions and masses of the raffinate and extract layers? (b) If the raffinate layer from part (a) is extracted again with one-half its weight of water, what are the compositions and weights of the new layers? (c) If all the water is removed from the final raffinate layer of part (b), what will its composition be? Solve this exercise using the following equilibrium data to construct one or more of the types of diagrams in Figure 4.13.

LIQUID–LIQUID EQUILIBRIUM DATA FOR $CHCl_3$–H_2O–CH_3COOH AT 18°C AND 1 ATM

Heavy Phase (wt%)			Light Phase (wt%)		
$CHCl_3$	H_2O	CH_3COOH	$CHCl_3$	H_2O	CH_3COOH
99.01	0.99	0.00	0.84	99.16	0.00
91.85	1.38	6.77	1.21	73.69	25.10
80.00	2.28	17.72	7.30	48.58	44.12
70.13	4.12	25.75	15.11	34.71	50.18
67.15	5.20	27.65	18.33	31.11	50.56
59.99	7.93	32.08	25.20	25.39	49.41
55.81	9.58	34.61	28.85	23.28	47.87

4.49.

Isopropyl ether (E) is used to separate acetic acid (A) from water (W). The liquid–liquid equilibrium data at 25°C and 1 atm are below: (a) One hundred kilograms of a 30 wt% A–W solution is contacted with 120 kg of ether (E). What are the compositions and weights of the resulting extract and raffinate? What would the concentration of acid in the (ether-rich) extract be if all ether were removed? (b) A solution of 52 kg A and 48 kg W is contacted with 40 kg of E. Calculate the extract and raffinate compositions and quantities.

LIQUID–LIQUID EQUILIBRIUM DATA FOR ACETIC ACID (A), WATER (W), AND ISOPROPANOL ETHER (E) AT 25°C AND 1 ATM

Water-Rich Layer			Ether-Rich Layer		
Wt% A	Wt% W	Wt% E	Wt% A	Wt% W	Wt% E
1.41	97.1	1.49	0.37	0.73	98.9
2.89	95.5	1.61	0.79	0.81	98.4
6.42	91.7	1.88	1.93	0.97	97.1
13.30	84.4	2.3	4.82	1.88	93.3
25.50	71.1	3.4	11.4	3.9	84.7
36.70	58.9	4.4	21.6	6.9	71.5
45.30	45.1	9.6	31.1	10.8	58.1
46.40	37.1	16.5	36.2	15.1	48.7

Section 4.6

4.50. Separation of paraffins from aromatics by liquid–liquid extraction.

Diethylene glycol (DEG) is the solvent in the UDEX liquid–liquid extraction process [H.W. Grote, *Chem. Eng. Progr.*, **54**(8), 43 (1958)] to separate paraffins from aromatics. If 280 lbmol/h of 42.86 mol% *n*-hexane, 28.57 mol% *n*-heptane, 17.86 mol% benzene, and 10.71 mol% toluene is contacted with 500 lbmol/h of 90 mol% aqueous DEG at 325°F and 300 psia, calculate, using a process simulator with the UNIFAC L/L method for liquid-phase activity coefficients, the flow rates and molar compositions of the resulting two liquid phases. Is DEG more selective for the paraffins or the aromatics?

4.51. Liquid–liquid extraction of organic acids from water with ethyl acetate.

A feed of 110 lbmol/h includes 5, 3, and 2 lbmol/h, respectively, of formic, acetic, and propionic acids in water. If the acids are extracted in one equilibrium stage with 100 lbmol/h of ethyl acetate (EA), calculate, with a process simulator using the UNIFAC method, the flow rates and compositions of the resulting liquid phases. What is the selectivity of EA for the organic acids?

Section 4.7

4.52. Leaching of oil from soybean flakes by a hexane.

Repeat Example 4.9 for 200,000 kg/h of hexane.

4.53. Leaching of Na_2CO_3 from a solid by water.

Water is used in an equilibrium stage to dissolve 1,350 kg/h of Na_2CO_3 from 3,750 kg/h of a solid, where the balance is an insoluble oxide. If 4,000 kg/h of water is used and the underflow is 40 wt% solvent on a solute-free basis, compute the flow rates and compositions of overflow and underflow.

4.54. Incomplete leaching of Na_2CO_3 from a solid by water.

Repeat Exercise 4.53 if the residence time is sufficient to leach only 80% of the carbonate.

4.55. Crystallization from a mixture of benzene and naphthalene.

A total of 6,000 lb/h of a liquid solution of 40 wt% benzene in naphthalene at 50°C is cooled to 15°C. Use Figure 4.22 to obtain the weight of crystals and the flow rate and composition of mother liquor. Are the crystals benzene or naphthalene?

4.56. Crystallization from a mixture of benzene and naphthalene.

Repeat Example 4.10, except determine the temperature necessary to crystallize 80% of the naphthalene.

4.57. Cooling crystallization for a mixture of benzene and naphthalene.

Ten thousand kg/h of a 10 wt% liquid solution of naphthalene in benzene is cooled from 30°C to 0°C. Determine the amount of crystals and composition and flow rate of the mother liquor. Are the crystals benzene or naphthalene? Use Figure 4.22.

4.58. Cooling crystallization of Na_2SO_4 from an aqueous solution.

Repeat Example 4.11, except let the original solution be 20 wt% Na_2SO_4.

4.59. Neutralization to precipitate the tetrahydrate of calcium citrate.

Although citric acid ($C_6H_8O_7$) can be obtained by solvent extraction from fruits (e.g., lemons and limes) and vegetables, or synthesized from acetone, most commonly it is produced by submerged, batch, aerobic fermentation of carbohydrates (e.g., dextrose, sucrose, glucose, etc.) using the pure culture of a mold such as *Aspergillus niger*. Fermentation is followed by a series of continuous downstream processing steps. First, biomass in the form of suspended or precipitated solids is removed by a rotary vacuum filter, leaving a clarified broth. For a process that produces 1,700,000 kg/yr of anhydrous citric acid crystals, the flow rate of clarified broth is 1,300 kg/h, consisting of 16.94 wt% citric acid, 82.69 wt% water, and 0.37 wt% other solutes. To separate the citric acid from the other solutes, the broth is neutralized at 50°C with the stoichiometric amount of $Ca(OH)_2$ from a 33 wt% aqueous solution, causing calcium citrate to precipitate as the tetrahydrate $[Ca_3(C_6H_5O_7)_2 \cdot 4H_2O]$. The solubility of calcium citrate in water at 50°C is 1.7 g/1,000 g H_2O. (a) Write a chemical equation for the neutralization reaction to produce the precipitate. (b) Complete a component material balance in kg/h, showing in a table the broth, calcium hydroxide solution, citrate precipitate, and mother liquor.

4.60. Dissolving crystals of Na_2SO_4 with water.

At 20°C, for 1,000 kg of a mixture of 50 wt% $Na_2SO_4 \cdot 10H_2O$ and 50 wt% Na_2SO_4 crystals, how many kg of water must be added to completely dissolve the crystals if the temperature is kept at 20°C at equilibrium ? Use Figure 4.23.

4.61. Adsorption of phenol (B) from an aqueous solution.

Repeat Example 4.12, except determine the grams of activated carbon needed to achieve: (a) 75% adsorption of phenol; (b) 90% adsorption of phenol; (c) 98% adsorption of phenol.

4.62. Adsorption of a colored substance from an oil by clay particles.

A colored substance (B) is removed from a mineral oil by adsorption with clay particles at 25°C. The original oil has a color index of 200 units/100 kg oil, while the decolorized oil must have an index of only 20 units/100 kg oil. The following are experimental adsorption equilibrium data measurements:

c_B, color units/100 kg oil	200	100	60	40	10
q_B, color units/100 kg clay	10	7.0	5.4	4.4	2.2

(a) Fit the data to the Freundlich equation. (b) Compute the kg of clay needed to treat 500 kg of oil if one equilibrium contact is used.

Section 4.8

4.63. Absorption of acetone (A) from air by water.

Vapor–liquid equilibrium data in mole fractions for the system acetone–air–water at 1 atm (101.3 kPa) are as follows:

y, acetone in air :   0.004 0.008 0.014 0.017 0.019 0.020
x, acetone in water : 0.002 0.004 0.006 0.008 0.010 0.012

(a) Plot the data as (1) moles acetone per mole air versus moles acetone per mole water, (2) partial pressure of acetone versus g acetone per g water, and (3) y versus x. (b) If 20 moles of gas containing 0.015 mole-fraction acetone is contacted with 15 moles of water, what are the stream compositions? Solve graphically. Neglect water/air partitioning.

4.64. Separation of air into O_2 and N_2 by absorption into water.

It is proposed that oxygen be separated from nitrogen by absorbing and desorbing air in water. Pressures from 101.3 to 10,130 kPa and temperatures between 0 and 100°C are to be used. (a) Devise a scheme for the separation if the air is 79 mol% N_2 and 21 mol% O_2. (b) Henry's law constants for O_2 and N_2 are given in Figure 4.26. How many batch absorption steps would be necessary to make 90 mol% oxygen? What yield of oxygen (based on the oxygen feed) would be obtained?

4.65. Absorption of ammonia from nitrogen into water.

A vapor mixture of equal volumes NH_3 and N_2 is contacted at 20°C and 1 atm (760 torr) with water to absorb some of the NH_3. If 14 m^3 of this mixture is contacted with 10 m^3 of water, calculate the % of ammonia in the gas that is absorbed. Both T and P are maintained constant. The partial pressure of NH_3 over water at 20°C is:

Partial Pressure of NH_3 in Air, torr	Grams of Dissolved NH_3/100 g of H_2O
470	40
298	30
227	25
166	20
114	15
69.6	10
50.0	7.5
31.7	5.0
24.9	4.0
18.2	3.0
15.0	2.5
12.0	2.0

Section 4.9

4.66. Desublimation of phthalic anhydride from a gas.

Repeat Example 4.15 for temperatures corresponding to vapor pressures for PA of: (a) 0.7 torr, (b) 0.4 torr, (c) 0.1 torr. Plot the percentage recovery of PA vs. solid vapor pressure for 0.1 torr to 1.0 torr.

4.67. Desublimation of anthraquinone (A) from nitrogen.

Nitrogen at 760 torr and 300°C containing 10 mol% anthraquinone (A) is cooled to 200°C. Calculate the % desublimation of A. Vapor pressure data for solid A are

T, °C:	190.0	234.2	264.3	285.0
Vapor pressure, torr:	1	10	40	100

These data can be fitted to the Antoine equation (2-39) using the first three constants.

4.68. Separation of a gas mixture of by adsorption.

At 25°C and 101 kPa, 2 mol of a gas containing 35 mol% propylene in propane is equilibrated with 0.1 kg of silica gel adsorbent. Using Figure 4.29, calculate the moles and composition of the gas adsorbed and the gas not adsorbed.

4.69. Separation of a gas mixture by adsorption.

Fifty mol% propylene in propane is separated with silica gel. The products are to be 90 mol% propylene and 75 mol% propane. If 1,000 lb of silica gel/lbmol of feed gas is used, can the desired separation be made in one stage? If not, what separation can be achieved? Use Figure 4.29.

Section 4.10

4.70. Crystallization of $MgSO_4$ from water by evaporation.

Repeat Example 4.17 for 90% evaporation of the water.

4.71. Crystallization of Na_2SO_4 from water by evaporation.

A 5,000-kg/h aqueous solution of 20 wt% Na_2SO_4 is fed to an evaporative crystallizer at 60°C. Equilibrium data are given in Figure 4.23. If 80% of the Na_2SO_4 is crystallized, calculate the: (a) kg of water that must be evaporated per hour; (b) crystallizer pressure in torr.

4.72. Bubble, secondary dew points, and primary dew points.

Calculate the dew-point pressure, secondary dew-point pressure, and bubble-point pressure of the following mixtures at 50°C, assuming that the aromatics and water are insoluble: (a) 50 mol% benzene and 50 mol% water; (b) 50 mol% toluene and 50 mol% water; (c) 40 mol% benzene, 40 mol% toluene, and 20 mol% water.

4.73. Bubble and dew points of benzene, toluene, and water mixtures.

Repeat Exercise 4.71, except compute temperatures for a pressure of 2 atm.

4.74. Bubble point of a mixture of toluene, ethylbenzene, and water.

A liquid of 30 mol% toluene, 40 mol% ethylbenzene, and 30 mol% water is subjected to a continuous flash distillation at 0.5 atm. Assuming that mixtures of ethylbenzene and toluene obey Raoult's law and that the hydrocarbons are immiscible in water and vice versa, calculate the temperature and composition of the vapor phase at the bubble-point temperature.

4.75. Bubble point, dew point, and 50 mol% flash for water–*n*-butanol.

As shown in Figure 4.8, water (W) and *n*-butanol (B) can form a three-phase system at 101 kPa. For a mixture of overall composition of 60 mol% W and 40 mol% B, use a process simulator with the UNIFAC method to estimate: (a) dew-point temperature and composition of the first drop of liquid; (b) bubble-point temperature and composition of the first bubble of vapor; (c) compositions and relative amounts of all three phases for 50 mol% vaporization.

4.76. Isothermal three-phase flash of six-component mixture.

Repeat Example 4.19 for a temperature of 25°C. Are the changes significant?

Chapter 5

Cascades and Hybrid Systems

§5.0 INSTRUCTIONAL OBJECTIVES

After completing this chapter, you should be able to:

- Explain how multi-equilibrium-stage cascades with countercurrent flow can achieve a significantly better separation than a single equilibrium stage.
- Estimate recovery of a key component in countercurrent leaching and washing cascades, and in each of three types of liquid–liquid extraction cascades.
- Define and explain the significance of absorption and stripping factors.
- Estimate the recoveries of all components in a single-section, countercurrent cascade using the Kremser method.
- Explain why a two-section, countercurrent cascade can achieve a sharp separation between two feed components, while a single-section cascade cannot.
- Configure a membrane cascade to improve a membrane separation.
- Explain the merits and give examples of hybrid separation systems.
- Determine degrees of freedom and a set of specifications for a separation process or any element in the process.

One separation stage is rarely sufficient to produce pure commercial products. *Cascades*, which are aggregates of stages, are needed to (1) accomplish separations that cannot be achieved in a single stage, (2) reduce the amounts of mass- or energy-separating agents required, and (3) make efficient use of raw materials.

Figure 5.1 shows the type of *countercurrent cascade* prevalent in *unit operations* such as distillation, absorption, stripping, and liquid–liquid extraction. Two or more process streams of different phase states and compositions are intimately contacted to promote rapid mass and heat transfer so that the separated phases leaving the stage approach physical equilibrium. Although equilibrium conditions may not be achieved in each stage, it is common to design and analyze cascades using equilibrium-stage models. In the case of membrane separations, phase equilibrium is not a consideration and mass-transfer rates through the membrane determine the separation.

In cases where the extent of separation by a single-unit operation is limited or the energy required is excessive, *hybrid systems* of two different separation operations, such as the combination of distillation and pervaporation, can be considered. This chapter introduces both cascades and hybrid systems. To illustrate the benefits of cascades, the calculations are based on simple models. Rigorous models are deferred to Chapters 10–12.

§5.1 CASCADE CONFIGURATIONS

Figure 5.2 shows possible cascade configurations in both horizontal and vertical layouts. Stages are represented by either boxes, as in Figure 5.1, or as horizontal lines, as in Figures 5.2d, e. In Figure 5.2, F is the feed; the mass-separating agent, if used, is S; and products are designated by P_i.

The linear *countercurrent cascade*, shown in Figures 5.1 and 5.2a, is very efficient and widely used. The linear *cross-current cascade*, shown in Figure 5.2b, is not as efficient as the countercurrent cascade, but it is convenient for batchwise configurations in that the solvent is divided into portions fed individually to each stage.

Figure 5.2c depicts a two-dimensional diamond configuration rather than a linear cascade. In batch crystallization, Feed F is separated in stage 1 into crystals, which pass to stage 2, and mother liquor, which passes to stage 4. In other stages, partial crystallization or recrystallization occurs by processing crystals, mother liquor, or combinations thereof with solvent S. Final products are purified crystals and impurity-bearing mother liquors.

Figure 5.1 and the first three cascades in Figure 5.2 consist of *single sections* of stages with streams entering and leaving only from the ends. Such cascades are used to recover components from a feed stream, but are not useful for making a sharp separation between two selected feed components, called *key components*. To do this, a cascade of *two sections* of stages is used, e.g., the countercurrent cascade of Figure 5.2d, which consists of one section above the feed and

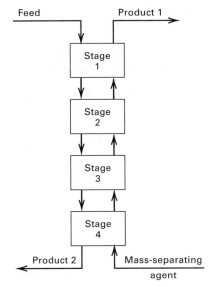

Feed ── Product 1

Stage 1

Stage 2

Stage 3

Stage 4

Product 2 ◄── Mass-separating agent

Figure 5.1 Cascade of contacting stages.

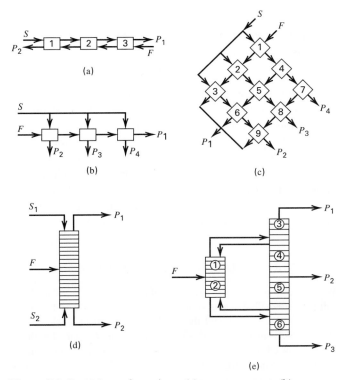

Figure 5.2 Cascade configurations: (a) countercurrent; (b) crosscurrent; (c) two-dimensional, diamond; (d) two-section, countercurrent; (e) interlinked system of countercurrent cascades.

one below. If S_2 is boiling vapor produced by steam or partial vaporization of P_2 by a boiler, and S_1 is liquid reflux produced by partial condensation of P_1, this is a simple distillation column. If two solvents are used, where S_1 selectively dissolves certain components of the feed while S_2 is more selective for the other components, the process is *fractional liquid–liquid extraction.*

Figure 5.2e is an interlinked system of two distillation columns containing six countercurrent cascade sections. Reflux and boilup for the first column are provided by the second column. This system can take a three-component feed, F, and produce three almost pure products, P_1, P_2, and P_3.

In this chapter, a countercurrent, single-section cascade for a leaching or washing process is considered first. Then, cocurrent, crosscurrent, and countercurrent single-section cascades are compared for a liquid–liquid extraction process. After that, a single-section, countercurrent cascade is developed for a vapor–liquid absorption operation. Finally, membrane cascades are described. In the first three cases, a set of linear algebraic equations is reduced to a single relation for estimating the extent of separation as a function of the number of stages, the separation factor, and the ratio of mass- or energy-separating agent to the feed. In later chapters it will be seen that for cascade systems, easily solved equations cannot be obtained from rigorous models, making calculations with a process simulator a necessity.

§5.2 SOLID–LIQUID CASCADES

The N-stage, countercurrent leaching–washing process in Figure 5.3 is an extension of the single-stage systems in §4.7. The solid feed entering stage 1 consists of two components, A and B, of mass flow rates F_A and F_B. Pure solvent, C, which enters stage N at flow rate S, dissolves solute B but not insoluble carrier A. The concentrations of B are expressed in terms of mass ratios of solute-to-solvent, Y. Thus, the liquid *overflow* from each stage, j, contains Y_j mass of soluble material per mass of solute-free solvent. The *underflow* is a slurry consisting of a mass flow F_A of insoluble solids; a constant ratio of mass of solvent-to-mass of insoluble solids, R; and X_j mass of soluble material-to-mass of solute-free solvent. For a given feed, a relationship between the exiting underflow concentration of the soluble component, X_N; the solvent feed rate, S; and the number of stages, N, is derived next.

All soluble material, B, in the feed is leached in stage 1, and all other stages are then washing stages for reducing the amount of soluble material lost in the underflow leaving the last stage, N, thereby increasing the amount of soluble material leaving in the overflow from stage 1. By solvent material balances, for constant R the flow rate of solvent leaving in the

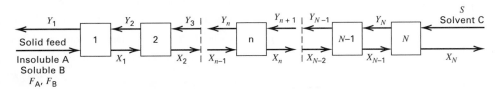

Figure 5.3 Countercurrent leaching or washing system.

overflow from stages 2 to N is S. The flow rate of solvent leaving in the underflow from stages 1 to N is RF_A. Therefore, the flow rate of solvent leaving in the overflow from stage 1 is $S - RF_A$.

A material balance for soluble material B around any interior stage n from $n = 2$ to $N - 1$ is

$$Y_{n+1}S + X_{n-1}RF_A = Y_nS + X_nRF_A \tag{5-1}$$

For terminal stages 1 and N, the material balances on the soluble material are, respectively,

$$Y_2S + F_B = Y_1(S - RF_A) + X_1RF_A \tag{5-2}$$

$$X_{N-1}RF_A = Y_NS + X_NRF_A \tag{5-3}$$

Assuming equilibrium, the concentration of B in each overflow equals the concentration of B in the liquid part of the underflow from the same stage. Thus,

$$X_n = Y_n \tag{5-4}$$

In addition, it is convenient to define a *washing factor*, W, as

$$\boxed{W = \frac{S}{RF_A}} \tag{5-5}$$

If (5-1), (5-2), and (5-3) are each combined with (5-4) to eliminate Y, and the resulting equations are rearranged to allow substitution of (5-5), then,

$$X_1 - X_2 = \left(\frac{F_B}{S}\right) \tag{5-6}$$

$$\left(\frac{1}{W}\right)X_{n-1} - \left(\frac{1+W}{W}\right)X_n + X_{n+1} = 0, \tag{5-7}$$
$$n = 2 \text{ to } N - 1$$

$$\left(\frac{1}{W}\right)X_{N-1} - \left(\frac{1+W}{W}\right)X_N = 0 \tag{5-8}$$

Equations (5-6) to (5-8) are a set of N linear equations in N unknowns, X_n ($n = 1$ to N). The equations form a tridiagonal, sparse matrix. For example, with $N = 5$,

$$\begin{bmatrix} 1 & -1 & 0 & 0 & 0 \\ \left(\frac{1}{W}\right) & -\left(\frac{1+W}{W}\right) & 1 & 0 & 0 \\ 0 & \left(\frac{1}{W}\right) & -\left(\frac{1+W}{W}\right) & 1 & 0 \\ 0 & 0 & \left(\frac{1}{W}\right) & -\left(\frac{1+W}{W}\right) & 1 \\ 0 & 0 & 0 & \left(\frac{1}{W}\right) & -\left(\frac{1+W}{W}\right) \end{bmatrix}$$
$$\times \begin{bmatrix} X_1 \\ X_2 \\ X_3 \\ X_4 \\ X_5 \end{bmatrix} = \begin{bmatrix} \left(\frac{F_B}{S}\right) \\ 0 \\ 0 \\ 0 \\ 0 \end{bmatrix} \tag{5-9}$$

Equations of type (5-9) are solved by Gaussian elimination by eliminating unknowns X_1, X_2, etc., to obtain

$$X_N = \left(\frac{F_B}{S}\right)\left(\frac{1}{W^{N-1}}\right) \tag{5-10}$$

By back-substitution, interstage values of X are given by

$$X_n = \left(\frac{F_B}{S}\right)\left(\frac{\sum_{k=0}^{N-n} W^k}{W^{N-1}}\right) \tag{5-11}$$

For example, with $N = 5$,

$$X_1 = Y_1 = \left(\frac{F_B}{S}\right)\left(\frac{1 + W + W^2 + W^3 + W^4}{W^4}\right)$$

The cascade, for any given S, maximizes Y_1, the amount of B dissolved in the solvent leaving stage 1, and minimizes X_N, the amount of B dissolved in the solvent leaving with A from stage N. Equation (5-10) indicates that this can be achieved for feed rate F_B by specifying a large solvent feed, S, a large number of stages, N, and/or by employing a large washing factor, W, which can be achieved by minimizing the amount of liquid underflow compared to overflow. It should be noted that the minimum amount of solvent required corresponds to zero overflow from stage 1,

$$S_{\min} = RF_A \tag{5-12}$$

For this minimum value, $W = 1$ from (5-5), and all soluble solids leave in the underflow from the last stage, N, regardless of the number of stages; hence it is best to specify a value of S significantly greater than $S_{\min}$. Equations (5-10) and (5-5) show that the value of X_N is reduced exponentially by increasing N. Thus, the countercurrent cascade is very effective. For two or more stages, X_N is also reduced exponentially by increasing S. For three or more stages, the value of X_N is reduced exponentially by decreasing R.

EXAMPLE 5.1   Leaching of Na₂CO₃ with Water.

Water is used to dissolve 1,350 kg/h of Na_2CO_3 from 3,750 kg/h of a solid, where the balance is an insoluble oxide. If 4,000 kg/h of water is used as the solvent and the total underflow from each stage is 40 wt% solvent on a solute-free basis, compute and plot the % recovery of carbonate in the overflow product for one stage and for two to five countercurrent stages, as in Figure 5.3.

Solution

Soluble solids feed rate $= F_B = 1{,}350$ kg/h

Insoluble solids feed rate $= F_A = 3{,}750 - 1{,}350 = 2{,}400$ kg/h

Solvent feed rate $= S = 4{,}000$ kg/h

Underflow ratio $= R = 40/60 = 2/3$

Washing factor $= W = S/RF_A = 4{,}000/[(2/3)(2{,}400)] = 2.50$

Overall fractional recovery of soluble solids $= Y_1(S - RF_A)/F_B$

By overall material balance on soluble solids for N stages,

$$F_B = Y_1(S - RF_A) + X_NRF_A$$

Solving for Y_1 and using (5-5),

$$Y_1 = \frac{(F_B/S) - (1/W)X_N}{(1 - 1/W)}$$

From the given data,

$$Y_1 = \frac{(1.350/4.000) - (1/2.50)X_N}{(1 - 1/2.50)} \quad \text{or} \quad (1)$$

$$Y_1 = 0.5625 - 0.6667X_N$$

where, from (5-10),

$$X_N = \left(\frac{1.350}{4.000}\right)\frac{1}{2.50^{N-1}} = \frac{0.3375}{2.50^{N-1}} \quad (2)$$

The percent recovery of soluble material is

$$Y_1(S - RF_A)/F_B = Y_1[4,000 - (2/3)(2,400)]/1,350 \times 100\%$$
$$= 177.8Y_1 \quad (3)$$

Results for one to five stages, as computed from (1) to (3), are

No. of Stages in Cascade, N	X_N	Y_1	Percent Recovery of Soluble Solids
1	0.3375	0.3375	60.0
2	0.1350	0.4725	84.0
3	0.0540	0.5265	93.6
4	0.0216	0.5481	97.4
5	0.00864	0.5567	99.0

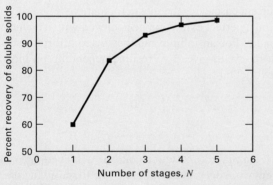

Figure 5.4 Effect of number of stages on percent recovery in Example 5.1.

A plot of the % recovery of B is shown in Figure 5.4. Only a 60% recovery is obtained with one stage, but 99% recovery is achieved for five stages. For 99% recovery with one stage, a water rate of 160,000 kg/h is required, which is 40 times that required for five stages! Thus, use of a countercurrent cascade is more effective than an excess of mass-separating agent in a single stage.

§5.3 SINGLE-SECTION, EXTRACTION CASCADES

Two-stage cocurrent, crosscurrent, and countercurrent, single-section, liquid–liquid extraction cascades are shown in Figure 5.5. The countercurrent arrangement is generally preferred because, as will be shown, this arrangement results in a higher degree of extraction for a given amount of solvent and number of equilibrium stages.

Equation (4-25) (for the fraction of solute, B, that is not extracted) was derived for a single liquid–liquid equilibrium extraction stage, assuming the use of pure solvent and a constant value for the distribution coefficient, K'_{D_B}, of B dissolved in components A and C, which are mutually insoluble. That equation is in terms of X_B, the ratio of mass of solute B to the mass of A, the other component in the feed, and the extraction factor

$$E = K'_{D_B}S/F_A \quad (5-13)$$

where

$$K'_{D_B} = Y_B/X_B \quad (5-14)$$

and Y_B is the ratio of mass of solute B to the mass of solvent C in the solvent-rich phase.

§5.3.1 Cocurrent Cascade

In Figure 5.5a, the fraction of B not extracted in stage 1 is from (4-25),

$$X_B^{(1)}/X_B^{(F)} = \frac{1}{1+E} \quad (5-15)$$

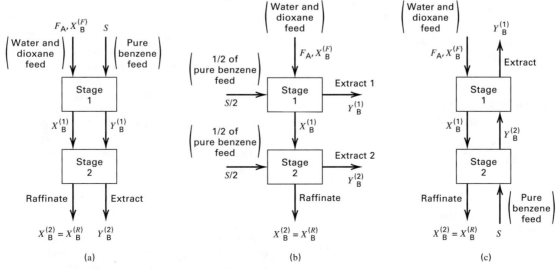

(a)   (b)   (c)

Figure 5.5 Two-stage arrangements: (a) cocurrent cascade; (b) crosscurrent cascade; (c) countercurrent cascade.

Since $Y_B^{(1)}$ is in equilibrium with $X_B^{(1)}$, combining (5-15) with (5-14) to eliminate $X_B^{(1)}$ gives

$$Y_B^{(1)}/X_B^{(F)} = K'_{D_B}/(1 + E) \qquad (5\text{-}16)$$

For the second stage, a material balance for B gives

$$X_B^{(1)}F_A + Y_B^{(1)}S = X_B^{(2)}F_A + Y_B^{(2)}S \qquad (5\text{-}17)$$

with

$$K'_{D_B} = Y_B^{(2)}/X_B^{(2)} \qquad (5\text{-}18)$$

However, no additional extraction can take place after the first stage because the two streams leaving are already at equilibrium when they are recontacted in subsequent stages. Accordingly, a cocurrent cascade has no merit unless required residence times are so long that equilibrium is not achieved in one stage and further capacity is needed. Regardless of the number of cocurrent equilibrium stages, N,

fraction not
extracted $\longrightarrow$
$$\frac{X_B^{(N)}}{X_B^{(F)}} = \frac{1}{1 + E} \qquad (5\text{-}19)$$

§5.3.2 Crosscurrent Cascade

For the crosscurrent cascade in Figure 5.5b, the feed progresses through each stage, starting with stage 1. The solvent flow rate, S, is divided into portions that are sent to each stage. If the portions are equal, the following mass ratios are obtained by application of (4-25), where S is replaced by S/N so that E is replaced by E/N:

$$\begin{aligned}
X_B^{(1)}/X_B^{(F)} &= 1/(1 + E/N) \\
X_B^{(2)}/X_B^{(1)} &= 1/(1 + E/N) \\
&\vdots \\
X_B^{(N)}/X_B^{(N-1)} &= 1/(1 + E/N)
\end{aligned} \qquad (5\text{-}20)$$

Combining equations in (5-20) to eliminate intermediate-stage variables, $X_B^{(n)}$, the final raffinate mass ratio is

$$X_B^{(N)}/X_B^{(F)} = X_B^{(R)}/X_B^{(F)} = \frac{1}{(1 + E/N)^N} \qquad (5\text{-}21)$$

Interstage values of $X_B^{(n)}$ are obtained similarly from

$$X_B^{(n)}/X_B^{(F)} = \frac{1}{(1 + E/N)^n} \qquad (5\text{-}22)$$

The value of X_B decreases in each successive stage. For an infinite number of equilibrium stages, (5-21) becomes

$$X_B^{(\infty)}/X_B^{(F)} = 1/\exp(E) \qquad (5\text{-}23)$$

Thus, even for an infinite number of stages, $X_B^{(R)} = X_B^{(\infty)}$ cannot be reduced to zero to give a perfect extraction.

§5.3.3 Countercurrent Cascade

In the countercurrent arrangement in Figure 5.5c, the feed liquid passes through the cascade countercurrently to the solvent. For a two-stage system, the material-balance and equilibrium equations for solute B for each stage are:

Stage 1: $X_B^{(F)}F_A + Y_B^{(2)}S = X_B^{(1)}F_A + Y_B^{(1)}S \qquad (5\text{-}24)$

$$K'_{D_B} = \frac{Y_B^{(1)}}{X_B^{(1)}} \qquad (5\text{-}25)$$

Stage 2: $X_B^{(1)}F_A = X_B^{(2)}F_A + Y_B^{(2)}S \qquad (5\text{-}26)$

$$K'_{D_B} = \frac{Y_B^{(2)}}{X_B^{(2)}} \qquad (5\text{-}27)$$

Combining (5-24) to (5-27) with (5-14) to eliminate $Y_B^{(1)}$, $Y_B^{(2)}$, and $X_B^{(1)}$ gives

$$X_B^{(2)}/X_B^{(F)} = X_B^{(R)}/X_B^{(F)} = \frac{1}{1 + E + E^2} \qquad (5\text{-}28)$$

Extending (5-28) to N countercurrent stages,

$$\boxed{X_B^{(R)}/X_B^{(F)} = 1 \bigg/ \sum_{n=0}^{N} E^n = \frac{E-1}{E^{N+1}-1}} \qquad (5\text{-}29)$$

Interstage values of $X_B^{(n)}$ are given by

$$X_B^{(n)}/X_B^{(F)} = \sum_{k=0}^{N-n} E^k \bigg/ \sum_{k=0}^{N} E^k \qquad (5\text{-}30)$$

The decrease of X_B for countercurrent flow is greater than that for crosscurrent flow, and the difference increases exponentially with increasing extraction factor, E. Thus, the countercurrent cascade is the most efficient.

Can a perfect extraction be achieved with a countercurrent cascade? For an infinite number of equilibrium stages, the limit of (5-28) gives two results, depending on the value of the extraction factor, E:

$$X_B^{(\infty)}/X_B^{(F)} = 0, \quad 1 \leq E \leq \infty \qquad (5\text{-}31)$$

$$X_B^{(\infty)}/X_B^{(F)} = (1 - E), \quad E \leq 1 \qquad (5\text{-}32)$$

Thus, complete extraction can be achieved in a countercurrent cascade with an infinite N if extraction factor $E > 1$.

EXAMPLE 5.2   Extraction with Different Cascade Arrangements.

Ethylene glycol is catalytically dehydrated to *p*-dioxane (a cyclic diether) by the reaction $2\text{HOCH}_2\text{CH}_2\text{HO} \rightarrow \text{H}_2\text{CCH}_2\text{OCH}_2\text{CH}_2\text{O} + 2\text{H}_2\text{O}$. Water and *p*-dioxane have normal boiling points of 100°C and 101.1°C, respectively, and cannot be separated economically by distillation. However, liquid–liquid extraction at 25°C using benzene as a solvent is possible. A feed of 4,536 kg/h of a 25 wt% solution of *p*-dioxane in water is to be separated continuously with 6,804 kg/h of benzene. Assuming benzene and water are mutually insoluble, determine the effect of the number and arrangement of stages on the percent extraction of *p*-dioxane. The flowsheet is in Figure 5.6.

Solution

Three arrangements of stages are examined: (a) cocurrent, (b) crosscurrent, and (c) countercurrent. Because water and benzene are assumed mutually insoluble, (5-15), (5-21), and (5-29) can be used to estimate $X_B^{(R)}/X_B^{(F)}$, the fraction of dioxane not extracted, as a

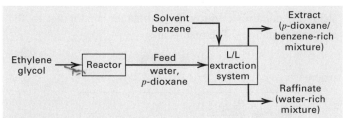

Figure 5.6 Flowsheet for Example 5.2.

function of stages. From the equilibrium data of Berdt and Lynch [1], the distribution coefficient for p-dioxane, $K'_{D_B} = Y_B/X_B$, where Y is the benzene phase and X the water phase, varies from 1.0 to 1.4; assume a value of 1.2. From the given data, $S = 6,804$ kg/h of benzene, $F_A = 4,536(0.75) = 3,402$ kg/h of water, and $X_B^{(F)} = 0.25/0.75 = 1/3$. From (5-13), $E = 1.2(6,804)/3,402 = 2.4$.

Single equilibrium stage:

All arrangements give identical results for one stage. By (5-15),

$$X_B^{(1)}/X_B^{(F)} = 1/(1+2.4) = 0.294$$

The corresponding fractional extraction is

$$1 - X_B^{(1)}/X_B^{(F)} = 1 - 0.294 = 0.706 \quad \text{or} \quad 70.6\%$$

More than one equilibrium stage:

(a) Cocurrent: For any number of stages, extraction is still only 70.6%.

(b) Crosscurrent: For any number of stages, (5-21) applies. For two stages, assuming equal flow of solvent to each stage,

$$X_B^{(2)}/X_B^{(F)} = 1/(1+E/2)^2 = 1/(1+2.4/2)^2 = 0.207$$

and extraction is 79.3%. Results for other N are in Figure 5.7.

(c) Countercurrent: (5-29) applies. For example, for two stages,

$$X_B^{(2)}/X_B^{(F)} = 1/\left(1 + E + E^2\right) = 1/\left(1 + 2.4 + 2.4^2\right) = 0.109$$

and extraction is 89.1%. Results for other N are in Figure 5.7, where a probability-scale ordinate is convenient because for the countercurrent case with $E > 1$, 100% extraction is approached as $N \to \infty$.

For the crosscurrent arrangement, the maximum extraction from (5-23) is 90.9%, while for five stages, the countercurrent cascade achieves 99% extraction.

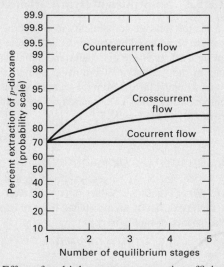

Figure 5.7 Effect of multiple stages on extraction efficiency.

§5.4 MULTICOMPONENT VAPOR–LIQUID CASCADES

Countercurrent cascades are used extensively for vapor–liquid separation operations, including absorption, stripping, and distillation. For absorption and stripping, a *single-section cascade* is used to recover one selected component from the feed. The approximate calculation procedure in this section relates compositions of multicomponent vapor and liquid streams entering and exiting the cascade to the number of equilibrium stages. This procedure is called a *group method* because it provides only an overall treatment of a group of stages in the cascade, without considering detailed changes in temperature, phase compositions, and flows from stage to stage.

§5.4.1 Single-Section Cascades by Group Methods

Kremser [2] originated the group method by deriving an equation for absorption or stripping in a multistage, countercurrent absorber. The treatment here is similar to that of Edmister [3]. An alternative treatment is given by Smith and Brinkley [4].

Consider first a countercurrent absorber of N adiabatic, equilibrium stages in Figure 5.8a, where stages are numbered from top to bottom. The absorbent is pure, and component molar flow rates are v_i and l_i, in the vapor and liquid phases, respectively. In the following derivation, the subscript i is dropped. A material balance around the top, including stages 1 through $N - 1$, for any absorbed species is

$$v_N = v_1 + l_{N-1} \tag{5-33}$$

where

$$v = yV \tag{5-34}$$

$$l = xL \tag{5-35}$$

and $l_0 = 0$. The equilibrium K-value at stage N is

$$y_N = K_N x_N \tag{5-36}$$

Combining (5-34), (5-35), and (5-36), v_N becomes

$$v_N = \frac{l_N}{L_N/(K_N V_N)} \tag{5-37}$$

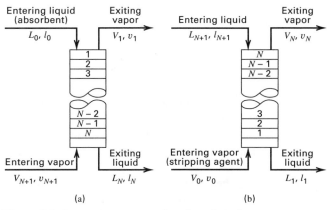

Figure 5.8 Countercurrent cascades of N adiabatic stages: (a) absorber; (b) stripper.

An absorption factor A, analogous to the extraction factor, E, for a given stage and component is defined by

$$A = \frac{L}{KV} \qquad (5\text{-}38)$$

Combining (5-37) and (5-38),

$$v_N = \frac{l_N}{A_N} \qquad (5\text{-}39)$$

Substituting (5-39) into (5-33),

$$l_N = (l_{N-1} + v_1)A_N \qquad (5\text{-}40)$$

The flow rate, l_{N-1}, is eliminated by successive substitution using material balances around successively smaller sections of the top of the cascade. For stages 1 through $N-2$,

$$l_{N-1} = (l_{N-2} + v_1)A_{N-1} \qquad (5\text{-}41)$$

Substituting (5-41) into (5-40),

$$l_N = l_{N-2}A_{N-1}A_N + v_1(A_N + A_{N-1}A_N) \qquad (5\text{-}42)$$

Continuing to the top stage, where $l_1 = v_1 A_1$, converts (5-42) to

$$l_N = v_1(A_1 A_2 A_3 \ldots A_N + A_2 A_3 \ldots A_N + A_3 \ldots A_N + \cdots + A_N) \qquad (5\text{-}43)$$

Combining (5-43) with the overall component balance,

$$l_N = v_{N+1} - v_1 \qquad (5\text{-}44)$$

gives an equation for the exiting vapor in terms of the entering vapor and a recovery fraction, ϕ_A:

$$v_1 = v_{N+1}\phi_A \qquad (5\text{-}45)$$

where, by definition, the recovery fraction is

$$\phi_A = \frac{1}{A_1 A_2 A_3 \ldots A_N + A_2 A_3 \ldots A_N + A_3 \ldots A_N + \cdots + A_N + 1}$$
$$= \text{fraction of species in entering vapor that is not absorbed} \qquad (5\text{-}46)$$

In the group method, an average, effective absorption factor, A_e, replaces the separate absorption factors for each stage, simplifying (5-46) to

$$\phi_A = \frac{1}{A_e^N + A_e^{N-1} + A_e^{N-2} + \cdots + A_e + 1} \qquad (5\text{-}47)$$

When multiplied and divided by $(A_e - 1)$, (5-47) reduces to the Kremser equation:

$$\phi_A = \frac{A_e - 1}{A_e^{N+1} - 1} \qquad (5\text{-}48)$$

Because each component has a different A_e, it has a different ϕ_A. Figure 5.9 from Edmister [3] is a plot of (5-48) with a probability scale for ϕ_A, a logarithmic scale for A_e, and N as a parameter. This plot, in linear coordinates, was first developed by Kremser [2].

Consider next the stripper shown in Figure 5.8b. Assume the components stripped from the liquid are absent in the entering vapor, and ignore absorption of the stripping agent. Stages are numbered from bottom to top. The pertinent

stripping equations follow in a manner analogous to the absorber equations. The results are

$$l_1 = l_{N+1}\phi_S \qquad (5\text{-}49)$$

where

$$\phi_S = \frac{S_e - 1}{S_e^{N+1} - 1}$$
$$= \text{fraction of species in entering liquid that is not stripped} \qquad (5\text{-}50)$$

$$S = \frac{KV}{L} = \frac{1}{A} = \text{stripping factor} \qquad (5\text{-}51)$$

Figure 5.9 also applies to (5-50). As shown in Figure 5.10, absorbers are frequently coupled with strippers or distillation columns to permit regeneration and recycle of absorbent. Since stripping action is not perfect, recycled absorbent contains species present in the vapor feed. Up-flowing vapor strips these as well as absorbed species in the makeup absorbent. A general absorber equation is obtained by combining (5-45) for absorption with a form of (5-49) for stripping species from the entering liquid. For stages numbered from top to bottom, as in Figure 5.8a, (5-49) becomes:

$$l_N = l_0\phi_S \qquad (5\text{-}52)$$

or, since

$$l_0 = v_1 + l_N$$
$$v_1 = l_0(1 - \phi_S) \qquad (5\text{-}53)$$

A balance for a component in both entering vapor and entering liquid is obtained by adding (5-45) and (5-53):

$$v_1 = v_{N+1}\phi_A + l_0(1 - \phi_S) \qquad (5\text{-}54)$$

which applies to each component in the entering vapor. Equation (5-52) is for species appearing only in the entering liquid. The analogous equation for a stripper is

$$l_1 = l_{N+1}\phi_S + v_0(1 - \phi_A) \qquad (5\text{-}55)$$

EXAMPLE 5.3 Absorption of Hydrocarbons by Oil.

In Figure 5.11, the heavier components in a superheated hydrocarbon gas are to be removed by absorption at 400 psia with a high-molecular-weight oil. Estimate exit vapor and liquid flow rates and compositions by the Kremser method, using estimated component absorption and stripping factors from the entering values of L and V, and the component K-values below based on an average entering temperature of $(90 + 105)/2 = 97.5°F$.

Solution

From (5-38) and (5-51), $A_i = L/K_i V = 165/[K_i(800)] = 0.206/K_i$; $S_i = 1/A_i = 4.85K_i$; and $N = 6$ stages. Values of ϕ_A and ϕ_S are from (5-48) and (5-50) or Figure 5.9. Values of $(v_i)_1$ are from (5-54). Values of $(l_i)_6$, in the exit liquid, are computed from an overall component material balance using Figure 5.8a:

$$(l_i)_6 = (l_i)_0 + (v_i)_7 - (v_i)_1 \qquad (1)$$

The computations, made with a spreadsheet, give the following results:

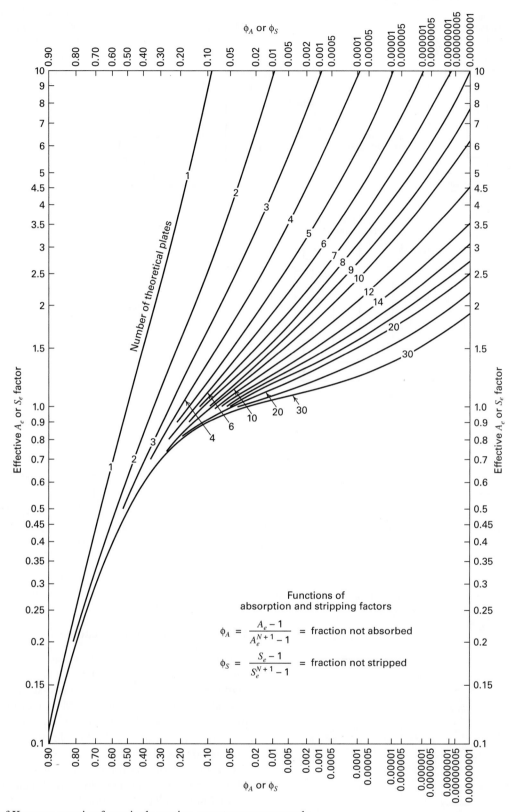

Figure 5.9 Plot of Kremser equation for a single-section, countercurrent cascade.
[From W.C. Edmister, *AIChE J.*, **3**, 165–171 (1957).]

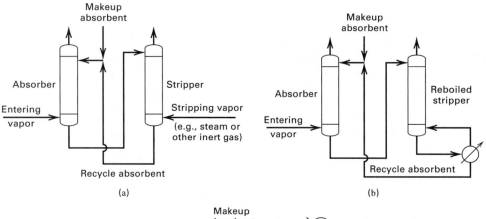

Figure 5.10 Various coupling schemes for absorbent recovery: (a) use of steam or inert gas stripper; (b) use of reboiled stripper; (c) use of distillation.

Component	K@97.5°F, 400 psia	A	S	ϕ_A	ϕ_S	v_1	l_6
C_1	6.65	0.0310	—	0.969	—	155.0	5.0
C_2	1.64	0.126	—	0.874	—	323.5	46.5
C_3	0.584	0.353	—	0.647	—	155.4	84.6
nC_4	0.195	1.06	0.946	0.119	0.168	3.02	22.03
nC_5	0.0713	2.89	0.346	0.00112	0.654	0.28	5.5
Oil	0.0001	—	0.0005	—	0.9995	0.075	164.095
						637.275	327.725

The results indicate that approximately 20% of the gas is absorbed. Less than 0.1% of the absorbent oil is stripped.

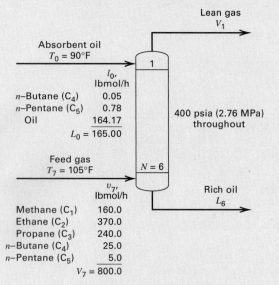

Figure 5.11 Specifications for absorber of Example 5.3.

§5.4.2 Two-Section Cascades

A single-section, countercurrent cascade of the type shown in Figure 5.8 cannot make a sharp separation between two *key components* of a feed. Instead, a two-section, countercurrent cascade, as shown in Figure 5.2d, is required. Consider the distillation of 100 lbmol/h of an equimolar mixture of *n*-hexane and *n*-octane at the bubble point at 1 atm. Let this mixture be heated by a boiler to the dew point. In Figure 5.12a, this vapor mixture is sent to a single-section cascade of three equilibrium stages. Instead of using liquid absorbent as in Figure 5.8a, the vapor leaving the top stage is condensed, with the liquid being divided into a distillate product and a reflux that is returned to the top stage. The reflux selectively absorbs *n*-octane so that the distillate is enriched in the more-volatile *n*-hexane. This set of stages is called a *rectifying section*.

To enrich the liquid leaving stage 1 in *n*-octane, that liquid becomes the feed to a second single-section cascade of three stages, as shown in Figure 5.12b. The purpose of this

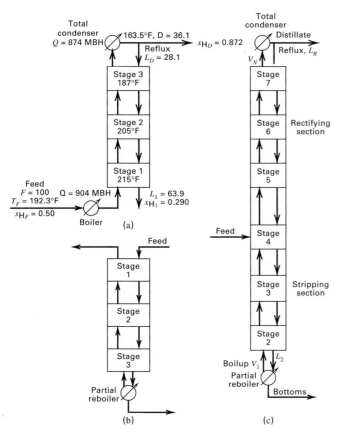

Figure 5.12 Development of a two-section cascade: (a) rectifying section; (b) stripping section; (c) multistage distillation.

§5.5 MEMBRANE CASCADES

Membrane-separation systems often consist of multiple-membrane modules because a single module may not be large enough to handle the required feed rate. Figure 5.13a shows a number of modules of identical size in parallel with retentates and permeates from each module combined. For example, a membrane-separation system for separating hydrogen from methane might require a membrane area of 9,800 ft^2. If the largest membrane module available has 3,300 ft^2 of membrane surface, three modules in parallel are required. The parallel units function as a single stage. If, in addition, a large fraction of the feed is to become permeate, it may be necessary to carry out the membrane separation in two or more stages, as shown in Figure 5.13b for four stages, with the number of modules reduced for each successive stage as the flow rate on the feed-retentate side of the membrane decreases. The combined retentate from each stage becomes the feed for the next stage. The combined permeates for each stage, which differ in composition from stage to stage, are combined to give the final permeate, as shown in Figure 5.13b, where required interstage compressors and/or pumps are not shown.

Single-membrane stages are often limited in the degree of separation and recovery achievable. In some cases, a high purity can be obtained, but only at the expense of a low recovery. In other cases, neither a high purity nor a high recovery can be obtained. The following table gives two examples of the separation obtained for a single stage of gas permeation using a commercial membrane.

Feed Molar Composition	More Permeable Component	Product Molar Composition	Percent Recovery
85% H$_2$ 15% CH$_4$	H$_2$	99% H$_2$, 1% CH$_4$ in the permeate	60% of H$_2$ in the feed
80% CH$_4$ 20% N$_2$	N$_2$	97% CH$_4$, 3% N$_2$ in the retentate	57% of CH$_4$ in the feed

cascade, called a *stripping section*, is similar to that of the stripper shown in Figure 5.8b. However, instead of using a stripping vapor, the liquid leaving the bottom stage enters a partial reboiler that produces the stripping vapor and a bottoms product rich in *n*-octane.

Vapor leaving the top of the bottom section is combined with the vapor feed to the top section, resulting in a distillation column shown in Figure 5.12c. Two-section cascades of this type are the industrial workhorses of the chemical industry because they produce nearly pure liquid and vapor products. The two-section cascade in Figure 5.12c is applied to the distillation of binary mixtures in Chapter 7 and multicomponent mixtures in Chapters 9 and 10.

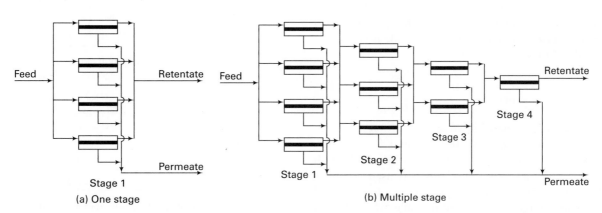

Figure 5.13 Parallel units of membrane separators.

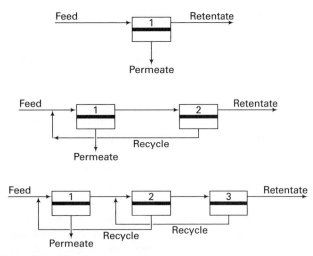

Figure 5.14 Membrane cascades.

In the first example, the permeate purity is quite high, but the recovery is not. In the second example, the purity of the retentate is reasonably high, but the recovery is not. To improve purity and recovery, membrane stages are cascaded with recycle. Shown in Figure 5.14 are three membrane-separation systems, studied by Prasad et al. [5] for the production of pure nitrogen (retentate) from air, using a membrane material that is more permeable to oxygen. The first system is just a single stage. The second system is a cascade of two stages, with recycle of permeate from the second to the first stage. The third system is a cascade of three stages with permeate recycles from stage 3 to stage 2 and stage 2 to stage 1. The two cascades are similar to the single-section, countercurrent stripping cascade shown in Figure 5.8b. Prasad et al. [5] give the following results for the three configurations in Figure 5.14:

Membrane System	Mol% N_2 in Retentate	% Recovery of N_2
Single Stage	98	45
Two Stage	99.5	48
Three Stage	99.9	50

Thus, high purities are obtained with a single-section membrane cascade, but little improvement in the recovery is provided by additional stages. To obtain both high purity and high recovery, a two-section membrane cascade is necessary, as discussed in §14.3.

§5.6 HYBRID SYSTEMS

Hybrid systems, encompassing two or more different separation operations in series, have the potential for reducing energy and raw-material costs and accomplishing difficult separations. Table 5.1 lists hybrid systems used commercially that have received considerable attention. Examples of applications are included. Not listed in Table 5.1 are hybrid systems consisting of distillation combined with extractive

Table 5.1 Hybrid Systems

Hybrid System	Separation Example
Adsorption—gas permeation	Nitrogen—Methane
Simulated moving bed adsorption—distillation	Metaxylene-paraxylene with ethylbenzene eluent
Chromatography—crystallization	—
Crystallization—distillation	—
Crystallization—pervaporation	—
Crystallization—liquid–liquid extraction	Sodium carbonate—water
Distillation—adsorption	Ethanol—water
Distillation—crystallization	—
Distillation—gas permeation	Propylene—propane
Distillation—pervaporation	Ethanol—water
Gas permeation—absorption	Dehydration of natural gas
Reverse osmosis—distillation	Carboxylic acids—water
Reverse osmosis—evaporation	Concentration of wastewater
Stripper—gas permeation	Recovery of ammonia and hydrogen sulfide from sour water

distillation, azeotropic distillation, and/or liquid–liquid extraction, which are considered in Chapter 11.

The first example in Table 5.1 is a hybrid system that combines pressure-swing adsorption (PSA), to preferentially remove methane, with a gas-permeation membrane operation to remove nitrogen. The permeate is recycled to the adsorption step. Figure 5.15 compares this hybrid system to a single-stage gas-permeation membrane and a single-stage pressure-swing adsorption. Only the hybrid system is capable of making a sharp separation between methane and nitrogen. Products obtainable from these three processes are compared in Table 5.2 for 100,000 scfh of feed containing 80% methane and 20% nitrogen. For all processes, the methane-rich

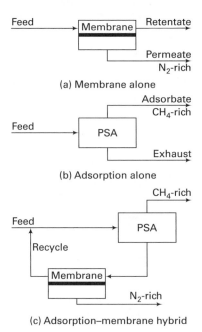

Figure 5.15 Separation of methane from nitrogen.

Table 5.2 Typical Products for Processes in Figure 5.15

	Flow Rate, Mscfh	Mol% CH$_4$	Mol% N$_2$
Feed gas	100	80	20
Membrane only:			
Retentate	47.1	97	3
Permeate	52.9	65	35
PSA only:			
Adsorbate	70.6	97	3
Exhaust	29.4	39	61
Hybrid system:			
CH$_4$-rich	81.0	97	3
N$_2$-rich	19.0	8	92

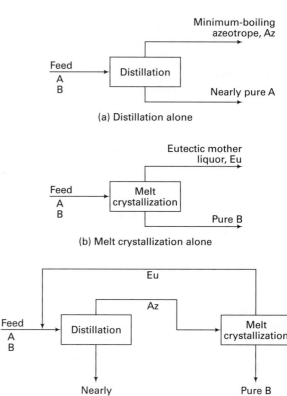

(a) Distillation alone

(b) Melt crystallization alone

(c) Distillation–crystallization hybrid

product contains 97 mol% methane. Only the hybrid system gives a nitrogen-rich product of greater than 90 mol%, and a high recovery of methane (98%). The methane recovery for a membrane alone is only 57%, while the adsorber gives 86%.

No application is shown in Table 5.1 for crystallization and distillation. However, there is much interest in these processes because Berry and Ng [6] show that such systems can overcome limitations of eutectics in crystallization and azeotropes in distillation. Furthermore, although solids are more difficult to process than fluids, crystallization requires just a single stage to obtain high purity. Figure 5.16 includes one of the distillation and crystallization hybrid configurations of Berry and Ng [6]. The feed of A and B, as shown in the phase diagram, forms an azeotrope in the vapor–liquid region, and a eutectic in the liquid–solid region at a lower temperature. The feed composition in Figure 5.16d lies between the eutectic and azeotropic compositions. If distillation alone is used, the distillate composition approaches that of the minimum-boiling azeotrope, Az, and the bottoms approaches pure A. If melt crystallization is used, the products are crystals of pure B and a mother liquor approaching the eutectic, Eu. The hybrid system in Figure 5.16 combines distillation with melt crystallization to produce pure B and nearly pure A. The feed is distilled and the distillate of near-azeotropic composition is sent to the melt crystallizer. Here, the mother liquor of near-eutectic composition is recovered and recycled to the distillation column. The net result is near-pure A obtained as bottoms from distillation and pure B obtained from the crystallizer.

The combination of distillation and membrane pervaporation for separating azeotropic mixtures, particularly ethanol–water, is also receiving considerable attention. Distillation produces a bottoms of nearly pure water and an azeotrope distillate that is sent to the pervaporation step, which produces a nearly pure ethanol retentate and a water-rich permeate that is recycled to the distillation step.

§5.7 DEGREES OF FREEDOM AND SPECIFICATIONS FOR CASCADES

The solution to a multicomponent, multiphase, multistage separation problem involves material-balance, energy-balance, and phase-equilibria equations. This implies that a sufficient

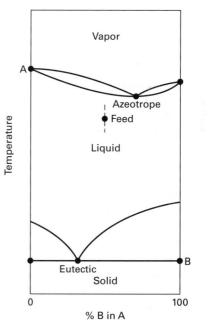

(d) Phase diagram for distillation–crystallization hybrid system.

Figure 5.16 Separation of an azeotropic- and eutectic-forming mixture.

number of design variables should be specified so that the number of remaining unknown variables equals the number of independent equations relating the variables. The degrees-of-freedom analysis discussed in §4.1 for a single equilibrium stage is now extended to one- and multiple-section cascades. Although the extension is for continuous, steady-state processes, similar extensions can be made for batch and semi-continuous processes.

An intuitively simple, but operationally complex, method of finding N_D, the number of independent design variables (*degrees of freedom*, or *variance*), as developed by Kwauk [7], is to enumerate all variables, N_V, and to subtract the total number of independent equations, N_E, that relate the variables:

$$\boxed{N_D = N_V - N_E} \qquad (5\text{-}56)$$

Typically, there are intensive variables such as pressure, composition, and temperature; extensive variables such as flow and heat-transfer rates; and equipment parameters such as number of stages. Physical properties such as enthalpy or K-values are not counted because they are functions of intensive variables. The variables are relatively easy to enumerate, but to achieve an unambiguous count of N_E, it is necessary to find all independent relationships due to mass and energy conservations, phase-equilibria restrictions, process specifications, and equipment configurations.

Separation equipment consists of physically identifiable elements: equilibrium stages, condensers, reboilers, pumps, etc., as well as stream dividers and stream mixers. It is helpful to examine each element separately before considering the complete system.

§5.7.1 Stream Variables

A complete specification of intensive variables for a single-phase stream consists of C mole fractions plus T and P, or $C + 2$ variables. However, only $C - 1$ of the mole fractions are independent, because the other mole fraction must satisfy the mole-fraction constraint:

$$\sum_{i=1}^{c} \text{mole fractions} = 1.0$$

Thus, $C + 1$ intensive stream variables can be specified. This is in agreement with the Gibbs phase rule (4-1), which states that, for a single-phase system, the intensive variables are specified by $C - \mathscr{P} + 2 = C + 1$ variables. To this number can be added the total flow rate, an extensive variable. Although the missing mole fraction is often treated implicitly, it is preferable to include all mole fractions in the list of stream variables and then to include, in the equations, the above mole-fraction constraint, from which the missing mole fraction is calculated. Thus, for each stream there are $C + 3$ variables. For example, for a liquid stream, the variables are liquid mole fractions $x_1, x_2, \ldots, x_C$; total flow rate L; temperature T; and pressure P.

§5.7.2 Adiabatic or Nonadiabatic Equilibrium Stage

For an equilibrium-stage element with two entering and two exiting streams, as in Figure 5.17, the variables are those associated with the four streams plus the heat-transfer rate. Thus,

$$N_V = 4(C + 3) + 1 = 4C + 13$$

The exiting streams V_{OUT} and L_{OUT} are in equilibrium, so there are equilibrium equations as well as component material balances, a total material balance, an energy balance, and mole-fraction constraints. The equations relating these variables and N_E are

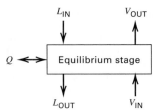

Figure 5.17 Equilibrium stage with heat addition.

Equations	Number of Equations
Pressure equality $\quad P_{V_{\text{OUT}}} = P_{L_{\text{OUT}}}$	1
Temperature equality, $\quad T_{V_{\text{OUT}}} = T_{L_{\text{OUT}}}$	1
Phase-equilibrium relationships, $\quad (y_i)_{V_{\text{OUT}}} = K_i(x_i)_{L_{\text{OUT}}}$	C
Component material balances, $\quad L_{\text{IN}}(x_i)_{L_{\text{IN}}} + V_{\text{IN}}(y_i)_{V_{\text{IN}}}$ $\qquad = L_{\text{OUT}}(x_i)_{L_{\text{OUT}}} + V_{\text{OUT}}(y_i)_{V_{\text{OUT}}}$	$C - 1$
Total material balance, $\quad L_{\text{IN}} + V_{\text{IN}} = L_{\text{OUT}} + V_{\text{OUT}}$	1
Energy balance, $\quad Q + h_{L_{\text{IN}}} L_{\text{IN}} = h_{V_{\text{IN}}} V_{\text{IN}}$ $\qquad = h_{L_{\text{OUT}}} L_{\text{OUT}} + h_{V_{\text{OUT}}} V_{\text{OUT}}$	1
Mole-fraction constraints in entering and exiting streams	4
$\quad$ e.g., $\sum\limits_{i=1}^{C} (x_i)_{L_{\text{IN}}} = 1$	$N_E = 2C + 7$

Alternatively, C, instead of $C - 1$, component material balances can be written. The total material balance is then a dependent equation obtained by summing the component material balances and applying the mole-fraction constraints to eliminate mole fractions. From (5-56),

$$N_D = (4C + 13) - (2C + 7) = 2C + 6$$

Several different sets of design variables can be specified. The following typical set includes complete specification of the two entering streams as well as stage pressure and heat-transfer rate.

Variable Specification	Number of Variables
Component mole fractions, $(x_i)L_{\text{IN}}$	$C - 1$
Total flow rate, L_{IN}	1
Component mole fractions, $(y_i)V_{\text{IN}}$	$C - 1$
Total flow rate, V_{IN}	1
Temperature and pressure of L_{IN}	2
Temperature and pressure of V_{IN}	2
Stage pressure, $(P_{V \text{ OUT}}$ or $P_{L \text{ OUT}})$	1
Heat transfer rate, Q	1
	$N_D = 2C + 6$

Specification of these $(2C + 6)$ variables permits calculation of the unknown variables L_{OUT}, V_{OUT}, $(x_C)_{L_{IN}}$, $(y_C)_{V_{IN}}$, all $(x_i)_{L_{OUT}}$, T_{OUT}, and all $(y_i)_{V_{OUT}}$, where C denotes the missing mole fractions in the two entering streams.

§5.7.3 Single-Section, Countercurrent Cascade

The single-section, countercurrent cascade unit in Figure 5.18 contains N of the adiabatic or nonadiabatic stage elements shown in Figure 5.17. For enumerating variables, equations, and degrees of freedom for combinations of such elements, the number of design variables for the unit is obtained by summing the variables associated with each element and then subtracting from the total variables, the $C + 3$ variables for each of the N_R redundant, interconnecting streams that arise when the output of one element becomes the input to another. Also, if an unspecified number of repetitions, e.g., stages, occurs within the unit, an additional variable is added, one for each group of repetitions, giving a total of N_A additional variables. The number of independent equations for the unit is obtained by summing the values of N_E for the units and then subtracting the N_R redundant mole-fraction constraints. The number of degrees of freedom is obtained as before, from (5-56). Thus,

$$(N_V)_{unit} = \sum_{all\ elements,\ e} (N_V)_e - N_R(C + 3) + N_A \quad (5\text{-}57)$$

$$(N_E)_{unit} = \sum_{all\ elements,\ e} (N_E)_e - N_R \quad (5\text{-}58)$$

Combining (5-56), (5-57), and (5-58),

$$(N_D)_{unit} = \sum_{all\ elements,\ e} (N_D)_e - N_R(C + 2) + N_A \quad (5\text{-}59)$$

or $\qquad (N_D)_{unit} = (N_V)_{unit} - (N_E)_{unit} \quad (5\text{-}60)$

For the N-stage cascade unit of Figure 5.18, with reference to the above degrees-of-freedom analysis for the single adiabatic or nonadiabatic equilibrium-stage element, the total

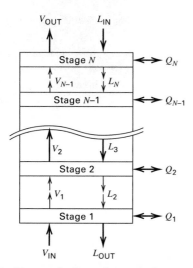

Figure 5.18 An N-stage single-section cascade.

number of variables from (5-57) is

$$(N_V)_{unit} = N(4C + 13) - [2(N - 1)](C + 3) + 1$$
$$= 7N + 2NC + 2C + 7$$

since $2(N - 1)$ interconnecting streams exist. The additional variable is the total number of stages (i.e., $N_A = 1$). The number of independent relationships from (5-58) is

$$(N_E)_{unit} = N(2C + 7) - 2(N - 1) = 5N + 2NC + 2$$

since $2(N - 1)$ redundant mole-fraction constraints exist. The number of degrees of freedom from (5-60) is

$$(N_D)_{unit} = N_V - N_E = 2N + 2C + 5$$

Note, again, that the coefficient of C is 2, the number of streams entering the cascade. For a cascade, the coefficient of N is always 2 (corresponding to P and Q for each stage). One possible set of design variables is:

Variable Specification	Number of Variables
Heat-transfer rate for each stage (or adiabaticity)	N
Stage pressures	N
Stream V_{IN} variables	$C + 2$
Stream L_{IN} variables	$C + 2$
Number of stages	1
	$2N + 2C + 5$

Output variables for this specification include missing mole fractions for V_{IN} and L_{IN}, stage temperatures, and variables associated with V_{OUT}, L_{OUT}, and interstage streams. N-stage cascade units represent absorbers, strippers, and extractors.

§5.7.4 Two-Section, Countercurrent Cascades

Two-section, countercurrent cascades can consist not only of adiabatic or nonadiabatic equilibrium-stage elements, but also elements shown in Table 5.3 for total and partial reboilers; total and partial condensers; stages with a feed, F, or sidestream S; and stream mixers and dividers. These elements can be combined into any of a number of complex cascades by applying to Eqs. (5-57) through (5-60) the values of N_V, N_E, and N_D given in Table 5.3 for the other elements.

The design or simulation of multistage separation operations involves solving relationships for output variables after selecting values of design variables to satisfy the degrees of freedom. Two common cases exist: (1) the *design case*, in which recovery specifications are made and the number of required equilibrium stages is determined; and (2) the *simulation case*, in which the number of stages is specified and product separations are computed. The second case is less complex computationally because the number of stages is specified, thus predetermining the number of equations to be solved. Table 5.4 is a summary of possible variable specifications for each of the two cases for a number of separator types discussed in Chapter 1 and shown in Table 1.1. For all separators in Table 5.4, it is assumed that all inlet streams are completely specified, and that all element and unit pressures and heat-transfer rates (except for condensers and reboilers)

Table 5.3 Degrees of Freedom for Separation Operation Elements and Units

	Schematic	Element or Unit Name	N_V, Total Number of Variables	N_E, Independent Relationships	N_D, Degrees of Freedom
(a)	$L \rightarrow\!\!\!\bigotimes\!\!\!\rightarrow V$ Q	Total boiler (reboiler)	$(2C + 7)$	$(C + 3)$	$(C + 4)$
(b)	$V \rightarrow\!\!\!\bigotimes\!\!\!\rightarrow L$ Q	Total condenser	$(2C + 7)$	$(C + 3)$	$(C + 4)$
(c)	$L_{in} \rightarrow\!\!\!\bigotimes\!\!\!\rightarrow V_{out}, L_{out}$ Q	Partial (equilibrium) boiler (reboiler)	$(3C + 10)$	$(2C + 6)$	$(C + 4)$
(d)	$V_{in} \rightarrow\!\!\!\bigotimes\!\!\!\rightarrow V_{out}, L_{out}$ Q	Partial (equilibrium) condenser	$(3C + 10)$	$(2C + 6)$	$(C + 4)$
(e)	$V_{out}\ L_{in} / V_{in}\ L_{out}$	Adiabatic equilibrium stage	$(4C + 12)$	$(2C + 7)$	$(2C + 5)$
(f)	$V_{out}\ L_{in} / V_{in}\ L_{out}$ $\leftrightarrow Q$	Equilibrium stage with heat transfer	$(4C + 13)$	$(2C + 7)$	$(2C + 6)$
(g)	$F \rightarrow$ $V_{out}\ L_{in} / V_{in}\ L_{out}$ $\leftrightarrow Q$	Equilibrium feed stage with heat transfer and feed	$(5C + 16)$	$(2C + 8)$	$(3C + 8)$
(h)	$a_s \leftarrow$ $V_{out}\ L_{in} / V_{in}\ L_{out}$ $\leftrightarrow Q$	Equilibrium stage with heat transfer and sidestream	$(5C + 16)$	$(3C + 9)$	$(2C + 7)$
(i)	$V_{out}\ L_{in}$ Stage $N \leftrightarrow Q_N$, $\leftrightarrow Q_{N-1}$ ⋯ $\leftrightarrow Q_2$, Stage 1 $\leftrightarrow Q_1$ $V_{in}\ L_{out}$	N-connected equilibrium stages with heat transfer	$(7N + 2NC + 2C + 7)$	$(5N + 2NC + 2)$	$(2N + 2C + 5)$
(j)	$L_1, L_2 \rightarrow\!\!\bigcirc\!\!\rightarrow L_3$ Q	Stream mixer	$(3C + 10)$	$(C + 4)$	$(2C + 6)$
(k)	$b_{L_1} \rightarrow\!\!\bigcirc\!\!\rightarrow L_2, L_3$ Q	Stream divider	$(3C + 10)$	$(2C + 5)$	$(C + 5)$

[a]Sidestream can be vapor or liquid.

[b]Alternatively, all streams can be vapor.

Table 5.4 Typical Variable Specifications for Design Cases

Unit Operation	N_D	Variable Specification[a]	
		Case I, Component Recoveries Specified	Case II, Number of Equilibrium Stages Specified
(a) Absorption (two inlet streams)	$2N + 2C + 5$	1. Recovery of one key component	1. Number of stages
(b) Distillation (one inlet stream, total condenser, partial reboiler)	$2N + C + 9$	1. Condensate at saturation temperature 2. Recovery of light-key component 3. Recovery of heavy-key component 4. Reflux ratio (> minimum) 5. Optimal feed stage[b]	1. Condensate at saturation temperature 2. Number of stages above feed stage 3. Number of stages below feed stage 4. Reflux ratio 5. Distillate flow rate
(c) Distillation (one inlet stream, partial condenser, partial reboiler, vapor distillate only)	$(2N + C + 6)$	1. Recovery of light key component 2. Recovery of heavy key component 3. Reflux ratio (> minimum) 4. Optimal feed stage[b]	1. Number of stages above feed stage 2. Number of stages below feed stage 3. Reflux ratio 4. Distillate flow rate
(d) Liquid–liquid extraction with two solvents (three inlet streams)	$2N + 3C + 8$	1. Recovery of key component 1 2. Recovery of key component 2	1. Number of stages above feed 2. Number of stages below feed
(e) Reboiled absorption (two inlet streams)	$2N + 2C + 6$	1. Recovery of light-key component 2. Recovery of heavy-key component 3. Optimal feed stage[b]	1. Number of stages above feed 2. Number of stages below feed 3. Bottoms flow rate
(f) Reboiled stripping (one inlet stream)	$2N + C + 3$	1. Recovery of one key component 2. Reboiler heat duty[d]	1. Number of stages 2. Bottoms flow rate

(Continued)

Table 5.4 (*Continued*)

Unit Operation	N_D	Case I, Component Recoveries Specified	Case II, Number of Equilibrium Stages Specified
		Variable Specification[a]	
(g) Distillation (one inlet stream, partial condenser, partial reboiler, both liquid and vapor distillates)	$2N + C + 9$	1. Ratio of vapor distillate to liquid distillate 2. Recovery of light-key component 3. Recovery of heavy-key component 4. Reflux ratio ($>$ minimum) 5. Optimal feed stage[b]	1. Ratio of vapor distillate to liquid distillate 2. Number of stages above feed stage 3. Number of stages below feed stage 4. Reflux ratio 5. Liquid distillate flow rate
(h) Extractive distillation (two inlet streams, total condenser, partial reboiler, single-phase condensate)	$2N + 2C + 12$	1. Condensate at saturation temperature 2. Recovery of light-key component 3. Recovery of heavy-key component 4. Reflux ratio ($>$ minimum) 5. Optimal feed stage[b] 6. Optimal MSA stage[b]	1. Condensate at saturation temperature 2. Number of stages above MSA stage 3. Number of stages between MSA and feed stages 4. Number of stages below feed stage 5. Reflux ratio 6. Distillate flow rate
(i) Liquid–liquid extraction (two inlet streams)	$2N + 2C + 5$	1. Recovery of one key component	1. Number of stages
(j) Stripping (two inlet streams)	$2N + 2C + 5$	1. Recovery of one key component	1. Number of stages

[a]Does not include the following variables, which are also assumed specified: all inlet stream variables ($C + 2$ for each stream); all element and unit pressures; all element and unit heat-transfer rates except for condensers and reboilers.

[b]Optimal stage for introduction of inlet stream corresponds to minimization of total stages.

[c]For case I variable specifications, MSA flow rates must be greater than minimum values for specified recoveries.

[d]For case I variable specifications, reboiler heat duty must be greater than minimum value for specified recovery.

are specified. Thus, only variable specifications satisfying the remaining degrees of freedom are listed.

EXAMPLE 5.4 Specifications for a Distillation Column.

Consider the multistage distillation column in Figure 5.19, which has one feed, one sidestream, a total condenser, a partial reboiler, and heat transfer to or from stages. Determine the number of degrees of freedom and a reasonable set of specifications.

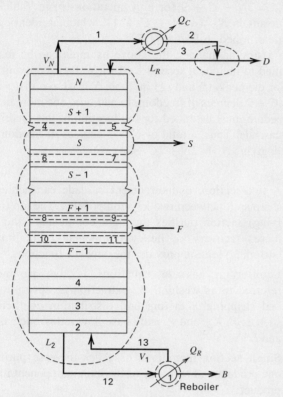

Figure 5.19 Complex distillation unit.

Solution

The separator is assembled from the circled elements and units of Table 5.3. The total variables are determined by summing the variables $(N_V)_e$ for each element from Table 5.3 and subtracting redundant variables due to interconnecting flows. Redundant mole-fraction constraints are subtracted from the sum of independent relationships for each element $(N_E)_e$. This problem was first treated by Gilliland and Reed [8] and more recently by Kwauk [7]. Differences in N_D obtained by various authors are due, in part, to how stages are numbered. Here, the partial reboiler is the first equilibrium stage. From Table 5.3, element variables and relationships are as follows:

Subtracting $(C + 3)$ redundant variables for 13 interconnecting streams, according to (5-57), with $N_A = 0$ (no unspecified repetitions), gives

$$(N_V)_{unit} = \sum (N_V)_e - 13(C + 3) = 7N + 2NC + 5C + 20$$

Subtracting the corresponding 13 redundant mole-fraction constraints, according to (5-58),

$$(N_E)_{unit} = \sum (N_E)_e - 13 = 5N + 2NC + 4C + 9$$

Therefore, from (5-60),

$$N_D = (7N + 2NC + 5C + 20) - (5N + 2NC + 4C + 9)$$
$$= 2N + C + 11$$

Note that the coefficient of C is only 1, because there is only one feed, and, again, the coefficient of N is 2.

A set of feasible design variable specifications is:

Variable Specification	Number of Variables
1. Pressure at each stage (including partial reboiler)	N
2. Pressure at reflux divider outlet	1
3. Pressure at total condenser outlet	1
4. Heat-transfer rate for each stage (excluding partial reboiler)	$(N - 1)$
5. Heat-transfer rate for divider	1
6. Feed mole fractions and total feed rate	C
7. Feed temperature	1
8. Feed pressure	1
9. Condensate temperature (e.g., saturated liquid)	1
10. Total number of stages, N	1
11. Feed stage location	1
12. Sidestream stage location	1
13. Sidestream total flow rate, S	1
14. Total distillate flow rate, D or D/F	1
15. Reflux flow rate, L_R, or reflux ratio, L_R/D	1
	$N_D = (2N + C + 11)$

In most separation operations, variables related to feed conditions, stage heat-transfer rates, and stage pressure are known or set. Remaining specifications have proxies, provided that the variables are mathematically independent. Thus, in the above list, the first nine entries are almost always known or specified. Variables 10 to 15, however, have surrogates. Some of these are: Condenser heat duty, Q_C; reboiler heat duty, Q_R; recovery or mole fraction of one component in bottoms; and recovery or mole fraction of one component in distillate.

Element or Unit	$(N_V)_e$	$(N_E)_e$
Total condenser	$(2C + 7)$	$(C + 3)$
Reflux divider	$(3C + 10)$	$(2C + 5)$
$(N - S)$ stages	$[7(N - S) + 2(N - S)C + 2C + 7]$	$[5(N - S) + 2(N - S)C + 2]$
Sidestream stage	$(5C + 16)$	$(3C + 9)$
$(S - 1) - F$ stages	$[7(S - 1 - F) + 2(S - 1 - F)C + 2C + 7]$	$[5(S - 1 - F) + 2(S - 1 - F)C + 2]$
Feed stage	$(5C + 16)$	$(2C + 8)$
$(F - 1) - 1$ stages	$[7(F - 2) + 2(F - 2)C + 2C + 7]$	$[5(F - 2) + 2(F - 2)C + 2]$
Partial reboiler	$(3C + 10)$	$(2C + 6)$
	$\sum (N_V)_e = 7N + 2NC + 18C + 59$	$\sum (N_E)_e = 5N + 2NC + 4C + 22$

Heat duties Q_C and Q_R are not good design variables because they are difficult to specify. A specified condenser duty, Q_C, might result in a temperature that is not realizable. Similarly, it is much easier to calculate Q_R knowing the total flow rate and enthalpy of the bottom streams than vice versa. Q_R and Q_C are so closely related that both should not be specified. Preferably, Q_C is fixed by distillate rate and reflux ratio, and Q_R is calculated from the overall energy balance.

Other proxies are possible, but the problem of independence of variables requires careful consideration. Distillate product rate, Q_C, and L_R/D, for example, are not independent. It should also be noted that if recoveries of more than two key species are specified, the result can be nonconvergence of the computations because the specified composition may not exist at equilibrium.

As an alternative to the solution to Example 5.4, the degrees of freedom for the unit of Figure 5.19 can be determined quickly by modifying a similar unit in Table 5.4.

The closest unit is (b), which differs from that in Figure 5.19 by only a sidestream. From Table 5.3, an equilibrium stage with heat transfer but without a sidestream (f) has $N_D = (2C + 6)$, while an equilibrium stage with heat transfer and a sidestream (h) has $N_D = (2C + 7)$, or one additional degree of freedom. When this sidestream stage is in a cascade, an additional degree of freedom is added for its location. Thus, two degrees of freedom are added to $N_D = 2N + C + 9$ for unit operation (b) in Table 5.4. The result is $N_D = 2N + C + 11$, which is identical to that determined in Example 5.4.

In a similar manner, the above example can be readily modified to include a second feed stage. By comparing values for elements (f) and (g) in Table 5.3, we see that a feed adds $C + 2$ degrees of freedom. In addition, one more degree of freedom must be added for the location of this feed stage in a cascade. Thus, a total of $C + 3$ degrees of freedom are added, giving $N_D = 2N + 2C + 14$.

SUMMARY

1. A cascade is a collection of stages arranged to: (a) accomplish a separation not achievable in a single stage, and/or (b) reduce the amount of mass- or energy-separating agent.

2. Cascades are single- or multiple-sectioned and configured in cocurrent, crosscurrent, or countercurrent arrays. Cascades are readily computed if equations are linear in component split ratios.

3. Equation (5-10) gives stage requirements for countercurrent solid–liquid leaching and/or washing involving constant underflow and mass transfer of one component.

4. Stages required for single-section, liquid–liquid extraction with constant distribution coefficients and immiscible solvent and carrier are given by (5-19), (5-22), and (5-29) for, respectively, cocurrent, crosscurrent, and (the most efficient) countercurrent flow.

5. Single-section stage requirements for a countercurrent cascade for absorption and stripping can be estimated with the Kremser equations, (5-48), (5-50), (5-54), and (5-55). Such cascades are limited in their ability to achieve high degrees of separation.

6. A two-section, countercurrent cascade can achieve a sharp split between two key components. The top (rectifying) section purifies the light components and increases recovery of heavy components. The bottom (stripping) section provides the opposite functions.

7. Equilibrium cascade equations involve parameters referred to as washing W, extraction E, absorption A, and stripping S factors and distribution coefficients, such as K, K_D, and R, and phase flow ratios, such as S/F and L/V.

8. Single-section membrane cascades increase purity of one product and recovery of the main component in that product.

9. Hybrid systems may reduce energy expenditures and make possible separations that are otherwise difficult, and/or improve the degree of separation.

10. The number of degrees of freedom (number of specifications) for a mathematical model of a cascade is the difference between the number of variables and the number of independent equations relating those equations. For a single-section, countercurrent cascade, the recovery of one component can be specified. For a two-section, countercurrent cascade, two recoveries can be specified.

REFERENCES

1. Berdt, R.J., and C.C. Lynch, *J. Am. Chem. Soc.*, **66**, 282–284 (1944).

2. Kremser, A., *Natl. Petroleum News*, **22**(21), 43–49 (May 21, 1930).

3. Edmister, W.C., *AIChE J.*, **3**, 165–171 (1957).

4. Smith, B.D., and W.K. Brinkley, *AIChE J.*, **6**, 446–450 (1960).

5. Prasad, R., F. Notaro, and D.R. Thompson, *J. Membrane Science*, **94**, Issue 1, 225–248 (1994).

6. Berry, D.A., and K.M. Ng, *AIChE J.*, **43**, 1751–1762 (1997).

7. Kwauk, M., *AIChE J.*, **2**, 240–248 (1956).

8. Gilliland, E.R., and C.E. Reed, *Ind. Eng. Chem.*, **34**, 551–557 (1942).

STUDY QUESTIONS

5.1. What is a separation cascade? What is a hybrid system?

5.2. What is the difference between a countercurrent and a crosscurrent cascade?

5.3. What is the limitation of a single-section cascade? Does a two-section cascade overcome this limitation?

5.4. What is an interlinked system of stages?

5.5. Which is more efficient, a crosscurrent cascade or a countercurrent cascade?

5.6. Under what conditions can a countercurrent cascade achieve complete extraction?

5.7. Why is a two-section cascade used for distillation?

5.8. What is a group method of calculation?

5.9. What is the Kremser method? To what type of separation operations is it applicable? What are the major assumptions of the method?

5.10. What is an absorption factor? What is a stripping factor?

5.11. In distillation, what is meant by reflux, boilup, rectification section, and stripping section?

5.12. Under what conditions is a membrane cascade of multiple stages in series necessary?

5.13. Why are hybrid systems often considered?

5.14. Give an example of a hybrid system that involves recycle.

5.15. Explain how a distillation–crystallization hybrid system works for a binary mixture that exhibits both an azeotrope and a eutectic.

5.16. When solving a separation problem, are the number and kind of specifications obvious? If not, how can the required number of specifications be determined?

5.17. Can the degrees of freedom be determined for a hybrid system? If so, what is the easiest way to do it?

EXERCISES

Section 5.1

5.1. Interlinked cascade arrangement.

Devise an interlinked cascade like Figure 5.2e, but with three columns for separating a four-component feed into four products.

5.2. Batchwise extraction process.

A liquid–liquid extraction process is conducted batchwise as shown in Figure 5.20. The process begins in Vessel 1 (Original), where 100 mg each of solutes A and B are dissolved in 100 mL of water. After adding 100 mL of an organic solvent that is more selective for A than B, the distribution of A and B becomes that shown for Equilibration 1 with Vessel 1. The organic-rich phase is

transferred to Vessel 2 (Transfer), leaving the water-rich phase in Vessel 1 (Transfer). The water and the organic are immiscible. Next, 100 mL of water is added to Vessel 2, resulting in the phase distribution shown for Vessel 2 (Equilibration 2). Also, 100 mL of organic is added to Vessel 1 to give the phase distribution shown for Vessel 1 (Equilibration 2). The batch process is continued by adding Vessel 3 and then 4 to obtain the results shown. (a) Study Figure 5.20 and then draw a corresponding cascade diagram, labeled in a manner similar to Figure 5.2b. (b) Is the process cocurrent, countercurrent, or crosscurrent? (c) Compare the separation with that for a batch equilibrium step. (d) How could the cascade be modified to make it countercurrent? [See O. Post and L.C. Craig, *Anal. Chem.*, **35**, 641 (1963).]

5.3. Two-stage membrane cascade.

Nitrogen is removed from a gas mixture with methane by gas permeation (see Table 1.2) using a glassy polymer membrane that is selective for nitrogen. However, the desired degree of separation cannot be achieved in one stage. Draw sketches of two different two-stage membrane cascades that might be used.

Section 5.2

5.4. Multistage leaching of oil.

In Example 4.9, 83.25% of the oil is leached by benzene using a single stage. Calculate the percent extraction of oil if: (a) two countercurrent equilibrium stages are used to process 5,000 kg/h of soybean meal with 5,000 kg/h of benzene; (b) three countercurrent stages are used with the flows in part (a). (c) Also determine the number of countercurrent stages required to extract 98% of the oil with a solvent rate twice the minimum.

5.5. Multistage leaching of Na$_2$CO$_3$.

For Example 5.1, involving the separation of sodium carbonate from an insoluble oxide, compute the minimum solvent feed rate. What is the ratio of actual solvent rate to the minimum solvent rate? Determine and plot the percent recovery of soluble solids with a cascade of five countercurrent equilibrium stages for solvent flow rates from 1.5 to 7.5 times the minimum value.

5.6. Production of aluminum sulfate.

Aluminum sulfate (alum) is produced as an aqueous solution from bauxite ore by reaction with aqueous sulfuric acid, followed by three-stage, countercurrent washing to separate soluble aluminum sulfate from the insoluble content of the bauxite, which is then followed by evaporation. In a typical process, 40,000 kg/day of solid bauxite containing 50 wt% Al$_2$O$_3$ and 50 wt% inert is crushed and fed with the stoichiometric amount of 50 wt% aqueous sulfuric acid to a reactor, where the Al$_2$O$_3$ is

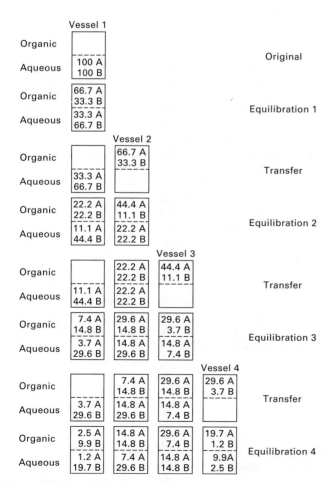

Figure 5.20 Liquid–liquid extraction process for Exercise 5.2.

reacted completely to alum by:

$$Al_2O_3 + 3H_2SO_4 \rightarrow Al_2(SO_4)_3 + 3H_2O$$

The slurry from the reactor (digester), consisting of solid inert material from the ore and an aqueous solution of aluminum sulfate, is then fed to a three-stage, countercurrent washing unit to separate the aqueous aluminum sulfate from the inert material. If the solvent is 240,000 kg/day of water and the underflow from each washing stage is 50 wt% water on a solute-free basis, compute the flows in kg/day of aluminum sulfate, water, and inert solid in the product streams leaving the cascade. What is the recovery of the aluminum sulfate? Would addition of one stage be worthwhile?

5.7. Rinse cycle for washing clothes.

(a) When rinsing clothes, would it be more efficient to divide the water and rinse several times, or should one use all the water in one rinse? Explain. (b) Devise a washing machine that gives the most efficient rinse cycle for a fixed amount of water.

Section 5.3

5.8. Batch extraction of acetic acid.

An aqueous acetic acid solution containing 6.0 mol/L of acid is extracted with chloroform at 25°C to recover the acid (B) from chloroform-insoluble impurities in the water. The water (A) and chloroform (C) are immiscible. If 10 L of solution are to be extracted at 25°C, calculate the % extraction of acid obtained with 10 L of chloroform under the following conditions: (a) the entire quantity of solvent in a single batch extraction; (b) three batch extractions with one-third of the solvent in each batch; (c) three batch extractions with 5 L of solvent in the first, 3 L in the second, and 2 L in the third batch.

Assume the distribution coefficient for the acid $= K''_{D_B} = (c_B)_C/(c_B)_A = 2.8$, where $(c_B)_C =$ concentration of acid in chloroform and $(c_B)_A =$ concentration of acid in water, both in mol/L.

5.9. Extraction of uranyl nitrate.

A 20 wt% solution of uranyl nitrate (UN) in water is to be treated with tributyl phosphate (TBP) to remove 90% of the uranyl nitrate in batchwise equilibrium contacts. Assuming water and TBP are mutually insoluble, how much TBP is required for 100 g of solution if, at equilibrium, (g UN/g TBP) = 5.5(g UN/g H_2O) and: (a) all the TBP is used at once in one stage; (b) half is used in each of two consecutive stages; (c) two countercurrent stages are used; (d) an infinite number of crosscurrent stages is used; and (e) an infinite number of countercurrent stages is used?

5.10. Extraction of uranyl nitrate.

The uranyl nitrate (UN) in 2 kg of a 20 wt% aqueous solution is extracted with 500 g of tributyl phosphate. Using the equilibrium data in Exercise 5.9, calculate and compare the % recoveries for the following alternative procedures: (a) a single-stage batch extraction; (b) three batch extractions with 1/3 of the total solvent used in each batch (solvent is withdrawn after contacting the entire UN phase); (c) a two-stage, cocurrent extraction; (d) a three-stage, countercurrent extraction; (e) an infinite-stage, countercurrent extraction; and (f) an infinite-stage, crosscurrent extraction.

5.11. Extraction of dioxane.

One thousand kg of a 30 wt% dioxane in water solution is to be treated with benzene at 25°C to remove 95% of the dioxane. The benzene is dioxane-free, and the equilibrium data of Example 5.2 applies. Calculate the solvent requirements for: (a) a single batch extraction; (b) two crosscurrent stages using equal amounts of benzene; (c) two countercurrent stages; (d) an infinite number of crosscurrent stages; and (e) an infinite number of countercurrent stages.

5.12. Extraction of benzoic acid.

Chloroform is used to extract benzoic acid from wastewater effluent. The benzoic acid is present at a concentration of 0.05 mol/L in the effluent, which is discharged at 1,000 L/h. The distribution coefficient for benzoic acid is $c^I = K_D^{II} c^{II}$, where $K_D^{II} = 4.2$, $c^I =$ molar concentration of solute in solvent, and $c^{II} =$ molar concentration of solute in water. Chloroform and water may be assumed immiscible. If 500 L/h of chloroform is to be used, compare the fraction benzoic acid removed in: (a) a single equilibrium contact; (b) three crosscurrent contacts with equal portions of chloroform; and (c) three countercurrent contacts.

5.13. Extraction of benzoic acid.

Repeat Example 5.2 with a solvent for $E = 0.90$. Display your results in a plot like Figure 5.7. Does countercurrent flow still have a marked advantage over crosscurrent flow? Is it desirable to choose the solvent and solvent rate so that $E > 1$? Explain.

5.14. Extraction of citric acid from a broth.

A clarified broth from fermentation of sucrose using *Aspergillus niger* consists of 16.94 wt% citric acid, 82.69 wt% water, and 0.37 wt% other solutes. To recover citric acid, the broth would normally be treated first with calcium hydroxide to neutralize the acid and precipitate it as calcium citrate, and then with sulfuric acid to convert calcium citrate back to citric acid. To avoid the need for calcium hydroxide and sulfuric acid, U.S. Patent 4,251,671 describes a solvent-extraction process using N,N-diethyldodecanamide, which is insoluble in water and has a density of 0.847 g/cm³. In a typical experiment at 30°C, 50 g of 20 wt% citric acid and 80 wt% water was contacted with 0.85 g of amide. The resulting organic phase, assumed to be in equilibrium with the aqueous phase, contained 6.39 wt% citric acid and 2.97 wt% water. Determine: (a) the partition (distribution) coefficients for citric acid and water, and (b) the solvent flow rate in kg/h needed to extract 98% of the citric acid in 1,300 kg/h of broth using five countercurrent, equilibrium stages, with the partition coefficients from part (a), but ignoring the solubility of water in the organic phase. In addition, (c) propose a series of subsequent steps to produce near-pure citric acid crystals. In part (b), how serious would it be to ignore the solubility of water in the organic phase?

5.15. Extraction of citric acid from a broth.

A clarified broth of 1,300 kg/h from the fermentation of sucrose using *Aspergillus niger* consists of 16.94 wt% citric acid, 82.69 wt% water, and 0.37 wt% other solutes. To avoid the need for calcium hydroxide and sulfuric acid in recovering citric acid from clarified broths, U.S. Patent 5,426,220 describes a solvent-extraction process using a mixed solvent of 56% tridodecyl lauryl amine, 6% octanol, and 38% aromatics-free kerosene, which is insoluble in water. In one experiment at 50°C, 570 g/min of 17 wt% citric acid in a fermentation liquor from pure carbohydrates was contacted in five countercurrent stages with 740 g/minute of the mixed solvent. The result was 98.4% extraction of citric acid. Determine: (a) the average partition (distribution) coefficient for citric acid from the experimental data, and (b) the solvent flow rate in kg/h needed to extract 98% of the citric acid in the 1,300 kg/h of clarified broth using three countercurrent, equilibrium stages, with the partition coefficient from part (a).

Section 5.4

5.16. Multicomponent, multistage absorption.

(a) Repeat Example 5.3 for $N = 1, 3, 10$, and 30 stages. Plot the % absorption of each of the five hydrocarbons and the total feed gas, as well as % stripping of the oil versus the number of stages, N. Discuss your results. (b) Solve Example 5.3 for an absorbent flow rate of 330 lbmol/h and three theoretical stages. Compare your results to those of Example 5.3. What is the effect of trading stages for absorbent?

5.17. Minimum absorbent flow.

Estimate the minimum absorbent flow rate required for the separation in Example 5.3 assuming the key component is propane, whose exit flow rate in the vapor is to be 155.4 lbmol/hr.

5.18. Isothermal, multistage absorption.

Solve Example 5.3 with the addition of a heat exchanger at each stage so as to maintain isothermal operation of the absorber at: (a) 125°F and (b) 150°F. What is the effect of temperature on absorption in this range?

5.19. Multicomponent, multistage absorption.

One million lbmol/day of a gas of the composition below is absorbed by n-heptane at $-30°F$ and 550 psia in an absorber with 10 theoretical stages so as to absorb 50% of the ethane. Calculate the required flow rate of absorbent and the distribution, in lbmol/h, of all components between the exiting gas and liquid.

Component	Mole Percent in Feed Gas	K-value @ $-30°F$ and 550 psia
C_1	94.9	2.85
C_2	4.2	0.36
C_3	0.7	0.066
nC_4	0.1	0.017
nC_5	0.1	0.004

5.20. Multistage stripper.

A stripper at 50 psia with three equilibrium stages strips 1,000 kmol/h of liquid at 300°F with the following molar composition: 0.03% C_1, 0.22% C_2, 1.82% C_3, 4.47% nC_4, 8.59% nC_5, 84.87% nC_{10}. The stripping agent is 1,000 kmol/h of superheated steam at 300°F and 50 psia. Use the Kremser equation to estimate the compositions and flow rates of the stripped liquid and exiting rich gas. Assume a K-value for C_{10} of 0.20 and that no steam is absorbed. Calculate the dew-point temperature of the exiting gas at 50 psia. If it is above 300°F, what can be done?

Section 5.7

5.21. Degrees of freedom for reboiler and condenser.

Verify the values given in Table 5.3 for N_V, N_E, and N_D for a partial reboiler and a total condenser.

5.22. Degrees of freedom for mixer and divider.

Verify the values given in Table 5.3 for N_V, N_E, and N_D for a stream mixer and a stream divider.

5.23. Specifications for a distillation column.

Maleic anhydride with 10% benzoic acid is a byproduct of the manufacture of phthalic anhydride. The mixture is to be distilled in a column with a total condenser and a partial reboiler at a pressure of 13.2 kPa with a reflux ratio of 1.2 times the minimum value to give a product of 99.5 mol% maleic anhydride and a bottoms of 0.5 mol% anhydride. Is this problem completely specified?

5.24. Degrees of freedom for distillation.

Verify N_D for the following unit operations in Table 5.4: (b), (c), and (g). How would N_D change if two feeds were used?

5.25. Degrees of freedom for absorber and stripper.

Verify N_D for unit operations (e) and (f) in Table 5.4. How would N_D change if a vapor sidestream were pulled off some stage located between the feed stage and the bottom stage?

5.26. Degrees of freedom for extractive distillation.

Verify N_D for unit operation (h) in Table 5.4. How would N_D change if a liquid sidestream was added to a stage that was located between the feed stage and stage 2?

5.27. Design variables for distillation.

The following are not listed as design variables for the distillation operations in Table 5.4: (a) condenser heat duty; (b) stage temperature; (c) intermediate-stage vapor rate; and (d) reboiler heat load. Under what conditions might these become design variables? If so, which variables listed in Table 5.4 could be eliminated?

5.28. Degrees of freedom for condenser change.

For distillation, show that if a total condenser is replaced by a partial condenser, the number of degrees of freedom is reduced by 3, provided the distillate is removed solely as a vapor.

5.29. Replacement of a reboiler with live steam.

Unit operation (b) in Table 5.4 is heated by injecting steam into the bottom plate of the column, instead of by a reboiler, for the separation of ethanol and water. Assuming a fixed feed, an adiabatic operation, 1 atm, and a product alcohol concentration: (a) What is the total number of design variables for the general configuration? (b) How many design variables are needed to complete the design? Which variables do you recommend?

5.30. Degrees-of-freedom for a distillation column.

(a) For the distillation column shown in Figure 5.21, determine the number of independent design variables. (b) It is suggested that a feed of 30% A, 20% B, and 50% C, all in moles, at 37.8°C and 689 kPa, be processed in the unit of Figure 5.21, with 15-plates in a 3-m-diameter column, which operates at vapor velocities of 0.3 m/s and an L/V of 1.2. The pressure drop per plate is 373 Pa, and the condenser is cooled by plant water at 15.6°C.

The product specifications in terms of the concentration of A in the distillate and C in the bottoms have been set by the process department, and the plant manager has asked you to specify a feed rate for the column. Write a memorandum to the plant manager pointing out why you can't do this, and suggest alternatives.

5.31. Degrees of freedom for multistage evaporation.

Calculate the number of degrees of freedom for the mixed-feed, triple-effect evaporator system shown in Figure 5.22. Assume that the steam and all drain streams are at saturated conditions and that the feed is an aqueous solution of a dissolved organic solid. Also, assume all overhead streams are pure steam. If this evaporator system is used to concentrate a feed containing 2 wt% dissolved organic to a product with 25 wt% dissolved organic, using 689-kPa saturated steam, calculate the number of unspecified design variables and suggest likely candidates. Assume perfect insulation against heat loss.

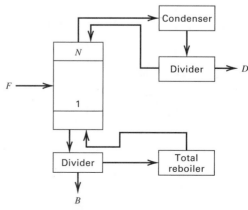

Figure 5.21 Conditions for Exercise 5.30.

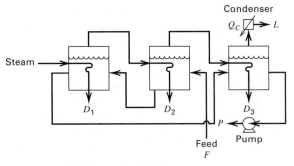

Figure 5.22 Conditions for Exercise 5.31.

5.32. Degrees of freedom for a reboiled stripper.

A reboiled stripper, shown in Figure 5.23, is to be designed. Determine: (a) the number of variables; (b) the number of equations relating the variables; and (c) the number of degrees of freedom. Also indicate (d) which additional variables, if any, need to be specified.

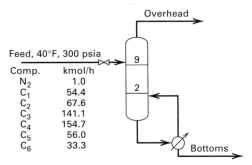

Figure 5.23 Conditions for Exercise 5.32.

5.33. Degrees of freedom of a thermally coupled distillation system.

The thermally coupled distillation system in Figure 5.24 separates a mixture of three components. Determine: (a) the number of

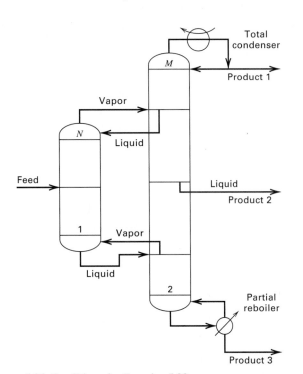

Figure 5.24 Conditions for Exercise 5.33.

variables; (b) the number of equations relating the variables; and (c) the number of degrees of freedom. Also propose (d) a reasonable set of design variables.

5.34. Adding a pasteurization section to distillation column.

When feed to a distillation column contains impurities that are much more volatile than the desired distillate, it is possible to separate the volatile impurities from the distillate by removing the distillate as a liquid sidestream from a stage several stages below the top. As shown in Figure 5.25, this additional section of stages is referred to as a pasteurizing section. (a) Determine the number of degrees of freedom for the unit. (b) Determine a reasonable set of design variables.

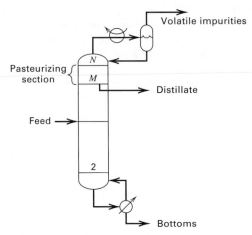

Figure 5.25 Conditions for Exercise 5.34.

5.35. Degrees of freedom for a two-column system.

A system for separating a feed into three products is shown in Figure 5.26. Determine: (a) the number of variables; (b) the number of equations relating the variables; and (c) the number of degrees of freedom. Also propose (d) a reasonable set of design variables.

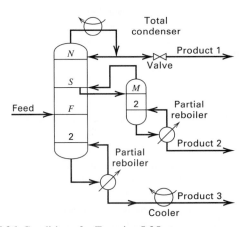

Figure 5.26 Conditions for Exercise 5.35.

5.36. Design variables for an extractive distillation.

A system for separating a binary mixture by extractive distillation, followed by ordinary distillation for recovery and recycle of the solvent, is shown in Figure 5.27. Are the design variables shown sufficient to specify the problem completely? If not, what additional design variables(s) should be selected?

5.37. Design variables for a three-product distillation column.

A single distillation column for separating a three-component mixture into three products is shown in Figure 5.28. Are the design variables shown sufficient to specify the problem completely? If not, what additional design variable(s) would you select?

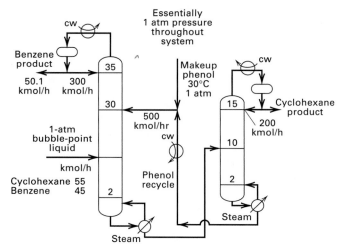

Figure 5.27 Conditions for Exercise 5.36.

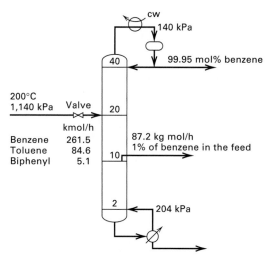

Figure 5.28 Conditions for Exercise 5.37.

Part Two

Separations by Phase Addition or Creation

In Part Two of this book, common industrial chemical separation methods of absorption, stripping, distillation, and liquid–liquid extraction, which involve mass transfer of components from a liquid to a gas, from a gas to a liquid, or from a liquid to another (immiscible) liquid, are described. Separations based on solid–gas or solid–liquid phases are covered in Parts Three and Four.

Second phases are created by thermal energy (energy-separating agent) or addition of mass (mass-separating agent). Design and analysis calculations for counter-current vapor–liquid and liquid–liquid operations are presented in Chapters 6 to 13, where two types of mathematical models are considered: (1) stages that attain thermodynamic phase equilibrium and (2) stages that do not, whose design is governed by rates of mass transfer. Equilibrium-stage models corrected with stage efficiencies are in common use, but wide availability of digital computations is encouraging increased use of more accurate and realistic mass-transfer models.

Absorption and stripping, which are covered in Chapter 6, rely on the addition of a mass-separating agent, but may also use heat transfer to produce a second phase. These operations are conducted in single-section cascades and, therefore, do not make sharp separations but can achieve high recoveries of one key component. The equipment consists of columns containing trays or packing for good turbulent-flow contact of the two phases. Graphical and algebraic methods for computing stages and estimating tray efficiency, column height, and diameter are described.

Distillation of binary mixtures in multiple-stage trayed or packed columns is covered in Chapter 7, with emphasis on the McCabe–Thiele graphical, equilibrium-stage model. To separate non-azeotropic, binary mixtures into pure products, two-section (rectifying and stripping) cascades are required. Energy-use analyses and equipment-sizing methods for absorption and stripping in Chapter 6 generally apply to distillation discussed in Chapter 7.

Liquid–liquid extraction, which is widely used in bioseparations and when distillation is too expensive or the chemicals are heat labile, is presented in Chapter 8. Columns with mechanically assisted agitation are useful when multiple stages are needed. Centrifugal extractors are advantageous in bioseparations because they provide short residence times, avoid emulsions, and can separate liquid phases with small density differences. That chapter emphasizes graphical, equilibrium-stage methods.

Models and calculations for multicomponent mixtures are more complex than those for binary mixtures. Approximate algebraic methods are presented in Chapter 9, while rigorous mathematical methods used in process simulators are developed in Chapter 10.

Chapter 11 considers design methods for *enhanced distillation* of mixtures that are difficult to separate by conventional distillation or liquid–liquid extraction. An important aspect of enhanced distillation is the use of residue-curve maps to determine feasible products. Extractive, azeotropic, and salt distillation use mass-addition as well as thermal energy input. Also included in Chapter 11 is pressure-swing distillation, which involves two columns at different pressures; reactive distillation, which couples a chemical reaction with product separation; and supercritical-fluid extraction, which makes use of favorable properties in the vicinity of the critical point to achieve a separation.

Mass-transfer models for multicomponent separation operations are available in process simulators. These models, described in Chapter 12, are particularly useful when stage efficiency is low or uncertain.

Batch distillation is important in the "specialty product" chemical industry. Calculation methods are presented in Chapter 13, along with an introduction to methods for determining an optimal set of operation steps.

Chapter 6

Absorption and Stripping of Dilute Mixtures

§6.0 INSTRUCTIONAL OBJECTIVES

After completing this chapter, you should be able to:

- Explain the differences among physical absorption, chemical absorption, and stripping.
- Explain why absorbers are best operated at high pressure and low temperature, whereas strippers are best operated at low pressure and high temperature.
- Compare three different types of trays.
- Explain the difference between random and structured packings and cite examples of each.
- Derive the operating-line equation used in graphical methods, starting with a component material balance.
- Calculate the minimum MSA flow rate to achieve a specified key-component recovery.
- Determine—graphically, by stepping off stages, or algebraically—the required number of equilibrium stages in a countercurrent cascade to achieve a specified component recovery, given an MSA flow rate greater than the minimum value.
- Define overall stage efficiency and explain why efficiencies are low for absorbers and moderate for strippers.
- Explain two mechanisms by which a trayed column can flood.
- Enumerate the contributions to pressure drop in a trayed column.
- Estimate column diameter and tray pressure drop for a trayed column.
- Estimate tray efficiency from correlations of mass-transfer coefficients using two-film theory.
- For a packed column, define the height equivalent to a theoretical (equilibrium) stage or plate (HETP or HETS), and explain how it and the number of equilibrium stages differ from *height of a transfer unit*, HTU, and *number of transfer units*, NTU.
- Explain the differences between *loading point* and *flooding point* in a packed column.
- Estimate packed height, packed-column diameter, and pressure drop across the packing.

Absorption is used to separate gas mixtures; remove impurities, contaminants, pollutants, or catalyst poisons from a gas; and recover valuable chemicals. Thus, the species of interest in the gas mixture may be all components, only the component(s) not transferred, or only the component(s) transferred. The species transferred to the liquid absorbent are called *solutes* or *absorbate*.

In *stripping* (*desorption*), a liquid mixture is contacted with a gas to selectively remove components by mass transfer from the liquid to the gas phase. Strippers are frequently coupled with absorbers to permit regeneration and recycle of the absorbent. Because stripping is not perfect, absorbent recycled to the absorber contains species present in the vapor entering the absorber. When water is used as the absorbent, it is common to separate the absorbent from the solute by distillation rather than by stripping.

In bioseparations, stripping of bioproduct(s) from the broth is attractive when the broth contains a small concentration of bioproduct(s) that is (are) more volatile than water.

Industrial Example

For the operation in Figure 6.1, which is accompanied by plant measurements, the feed gas, which contains air, water vapor, and acetone vapor, comes from a dryer in which solid cellulose acetate fibers, wet with water and acetone, are dried. The purpose of the 30-tray (equivalent to 10 equilibrium stages) absorber is to recover the acetone by contacting the gas with a suitable absorbent, water. By using countercurrent gas and liquid flow in a multiple-stage device, a material balance shows an acetone absorption of 99.5%. The exiting gas contains only 143 ppm (parts per million) by weight of acetone, which can be recycled to the dryer (although a small amount of gas must be discharged through a pollution-control device to the atmosphere to prevent argon buildup). Although acetone is the main species absorbed, the material balance indicates that minor amounts of O_2 and N_2 are also absorbed by the water. Because water is present in the feed gas and the absorbent, it can be both absorbed and stripped. As seen in the figure, the net effect is that water is stripped because more water appears in the exit gas than in

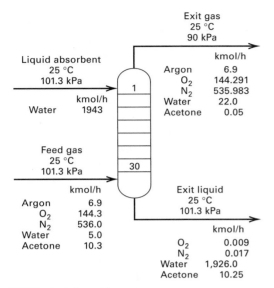

Figure 6.1 Typical absorption process.

the feed gas. The exit gas is almost saturated with water vapor, and the exit liquid is almost saturated with air. The absorbent temperature decreases by 3°C to supply energy of vaporization to strip the water, which in this example is greater than the energy of condensation liberated from the absorption of acetone. In gas absorption, heat effects can be significant.

The amount of each species absorbed depends on the number of equilibrium stages and the component absorption factor, $A = L/(KV)$. For Figure 6.1, K-values and absorption factors based on inlet flow rates are

Component	$A = L/(KV)$	K-value
Water	89.2	0.031
Acetone	1.38	2.0
Oxygen	0.00006	45,000
Nitrogen	0.00003	90,000
Argon	0.00008	35,000

For acetone, the K-value is obtained from Eq. (4) of Table 2.3, the modified Raoult's law, $K = \gamma P^s/P$, with $\gamma = 6.7$ for acetone in water at 25°C and 101.3 kPa. For oxygen and nitrogen, K-values are based on Eq. (6) of Table 2.3, Henry's law, $K = H/P$, using constants from Figure 4.26 at 25°C. For water, the K-value is from Eq. (3) of Table 2.3, Raoult's law, $K = P^s/P$, which applies because the mole fraction of water is close to 1. For argon, the Henry's law constant is from the International Critical Tables [1].

Figure 5.9 shows that if absorption factor $A > 1$, any degree of absorption can be achieved; the larger the A, the fewer the number of stages required. However, very large values of A can correspond to larger-than-necessary absorbent flows. From an economic standpoint, the value of A for the key species should be in the range of 1.25 to 2.0, with 1.4 being a recommended value. Thus, 1.38 for acetone is favorable.

For a given feed and absorbent, factors that influence A are absorbent flow rate, T, and P. Because $A = L/(KV)$, the

larger the absorbent flow rate, the larger the A. The absorbent flow rate can be reduced by lowering the solute K-value. Because K-values for many solutes vary exponentially with T and inversely to P, this reduction can be achieved by reducing T and/or increasing P. Increasing P also reduces the diameter of the equipment for a given gas throughput. However, adjustment of T by refrigeration, and/or adjustment of P by gas compression, increase(s) both capital and operating costs.

For a stripper, the stripping factor, $S = 1/A = KV/L$, is crucial. To reduce the required flow rate of stripping agent, operation of the stripper at high T and/or low P is desirable, with an optimum stripping factor being about 1.4.

For absorption and stripping, design procedures are well developed and commercial processes are common. Table 6.1 lists representative applications. In most cases, the solutes are in gaseous effluents. Strict environmental standards have greatly increased the use of gas absorbers.

When water and hydrocarbon oils are used as absorbents, no significant chemical reactions occur between the absorbent and the solute, and the process is commonly referred to as *physical absorption*. When aqueous NaOH is used as the absorbent for an acid gas, absorption is accompanied by a rapid and irreversible reaction in the liquid. This is *chemical absorption* or *reactive absorption*. More complex examples are processes for absorbing CO_2 and H_2S with aqueous solutions of monoethanolamine (MEA) and diethanolamine (DEA), where there is a more desirable, reversible chemical reaction in the liquid. Chemical reactions can increase the rate of absorption and solvent-absorption capacity and convert a hazardous chemical to an inert compound.

In this chapter, trayed and packed columns for absorption and stripping operations are discussed. Design and analysis of trayed columns are presented in §6.5 and 6.6, while packed columns are covered in §6.7, 6.8, and 6.9. *Equilibrium-based* and *rate-based* (mass-transfer) models, using both graphical and algebraic procedures, for physical absorption and stripping of mainly dilute mixtures are developed. The methods also apply to reactive absorption with irreversible and complete chemical reactions of solutes. Calculations for concentrated mixtures and reactive absorption with reversible chemical reactions are best handled by process simulators, as discussed in Chapters 10 and 11. An introduction to calculations for concentrated mixtures is described in the last section of this chapter.

§6.1 EQUIPMENT FOR VAPOR–LIQUID SEPARATIONS

Methods for designing and analyzing absorption, stripping, and distillation depend on the type of equipment used for contacting vapor and liquid phases. When multiple stages are required, phase contacting is most commonly carried out in cylindrical, vertical columns containing trays or packing of the type descibed next.

Table 6.1 Representative, Commercial Applications of Absorption

Solute	Absorbent	Type of Absorption
Acetone	Water	Physical
Acrylonitrile	Water	Physical
Ammonia	Water	Physical
Ethanol	Water	Physical
Formaldehyde	Water	Physical
Hydrochloric acid	Water	Physical
Hydrofluoric acid	Water	Physical
Sulfur dioxide	Water	Physical
Sulfur trioxide	Water	Physical
Benzene and toluene	Hydrocarbon oil	Physical
Butadiene	Hydrocarbon oil	Physical
Butanes and propane	Hydrocarbon oil	Physical
Naphthalene	Hydrocarbon oil	Physical
Carbon dioxide	Aq. NaOH	Irreversible chemical
Hydrochloric acid	Aq. NaOH	Irreversible chemical
Hydrocyanic acid	Aq. NaOH	Irreversible chemical
Hydrofluoric acid	Aq. NaOH	Irreversible chemical
Hydrogen sulfide	Aq. NaOH	Irreversible chemical
Chlorine	Water	Reversible chemical
Carbon monoxide	Aq. cuprous ammonium salts	Reversible chemical
CO_2 and H_2S	Aq. monoethanolamine (MEA) or diethanolamine (DEA)	Reversible chemical
CO_2 and H_2S	Diethyleneglycol (DEG) or triethyleneglycol (TEG)	Reversible chemical
Nitrogen oxides	Water	Reversible chemical

§6.1.1 Trayed Columns

Absorbers and strippers are mainly trayed towers (plate columns) and packed columns, and less often spray towers, bubble columns, and centrifugal contactors, all shown in Figure 6.2. A trayed tower is a vertical, cylindrical pressure vessel in which vapor and liquid, flowing countercurrently, are contacted on trays or plates that provide intimate contact of liquid with vapor to promote rapid mass transfer. An example of a tray is shown in Figure 6.3. Liquid flows across each tray, over an outlet weir, and into a downcomer, which takes the liquid by gravity to the tray below. Gas flows upward through openings in each tray, bubbling through the liquid on the tray. When the openings are holes, any of the five two-phase-flow regimes shown in Figure 6.4 and analyzed by Lockett [2] may occur. The most common and favored regime is the *froth regime*, in which the liquid phase is continuous and the gas passes through in the form of jets or a series of bubbles. The *spray regime*, in which the gas phase is continuous, occurs for low weir heights (low liquid depths) at high gas rates. For low gas rates, the *bubble regime* can occur, in which the liquid is fairly quiescent and bubbles rise in swarms. At high liquid rates, small gas bubbles may be undesirably emulsified. If bubble coalescence is hindered, an undesirable foam forms. Ideally, the liquid carries no vapor bubbles (*occlusion*) to the tray below, the vapor carries no liquid droplets (*entrainment*) to the tray above, and there is no *weeping* of liquid through the holes in the tray. With good contacting, equilibrium between the exiting vapor and liquid phases is approached on each tray, unless the liquid is very viscous.

Shown in Figure 6.5 are tray openings for vapor passage: (a) perforations, (b) valves, and (c) bubble caps. The simplest is perforations, usually $\frac{1}{8}$ to $\frac{1}{2}$ inch in diameter, used in *sieve (perforated)* trays. A *valve tray* has openings commonly from 1 to 2 inches in diameter. Each hole is fitted with a valve consisting of a cap that overlaps the hole, with legs or a cage to limit vertical rise while maintaining the valve cap in a horizontal orientation. Without vapor flow, each valve covers a hole. As vapor rate increases, the valve rises, providing a larger opening for vapor to flow and to create a froth.

A bubble-cap tray consists of a cap, 3 to 6 inches in diameter, mounted over and above a concentric riser, 2 to 3 inches in diameter. The cap has rectangular or triangular slots cut around its side. The vapor flows up through the tray opening into the riser, turns around, and passes out through the slots and into the liquid, forming a froth. An 11-ft-diameter tray might have 50,000 $\frac{3}{16}$-inch-diameter perforations, or 1,000 2-inch-diameter valve caps, or 500 4-inch-diameter bubble caps.

In Table 6.2, tray types are compared on the basis of cost, pressure drop, mass-transfer efficiency, vapor capacity, and flexibility in terms of turndown ratio (ratio of maximum to minimum vapor flow capacity). At the limiting *flooding* vapor velocity, liquid-droplet entrainment becomes excessive, causing the liquid flow to exceed the downcomer capacity, thus pushing liquid up the column. At too low a vapor rate, liquid weeping through the tray openings or vapor pulsation becomes excessive. Because of their low cost, sieve trays are preferred unless flexibility in throughput is required, in which case valve trays are best. Bubble-cap trays, predominant in pre-1950 installations, are now rarely specified, but may be

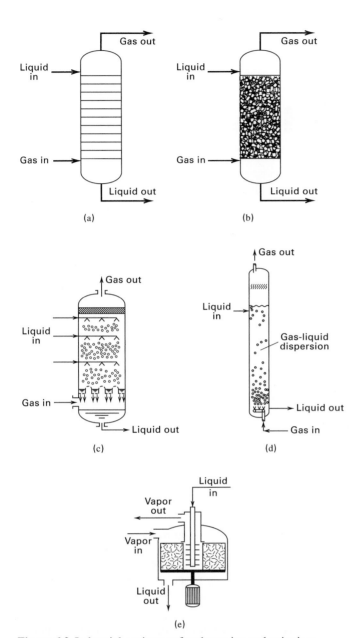

Figure 6.2 Industrial equipment for absorption and stripping: (a) trayed tower; (b) packed column; (c) spray tower; (d) bubble column; (e) centrifugal contactor.

preferred when liquid holdup must be controlled to provide residence time for a chemical reaction or when weeping must be prevented.

§6.1.2 Packed Columns

A *packed column*, shown in Figure 6.6, is a vessel containing one or more sections of packing over whose surface the liquid flows downward as a film or as droplets between packing elements. Vapor flows upward through the wetted packing, contacting the liquid. The packed sections are contained between a gas-injection support plate, which holds the packing, and an upper hold-down plate, which prevents packing movement. A *liquid distributor*, placed above the hold-down plate, ensures uniform distribution of liquid over the cross-sectional area of the column as it enters the packed section.

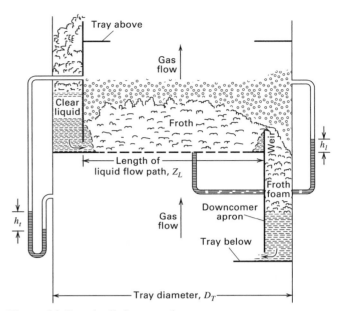

Figure 6.3 Tray details in a trayed tower.

[Adapted from B.F. Smith, *Design of Equilibrium Stage Processes*, McGraw-Hill, New York (1963).]

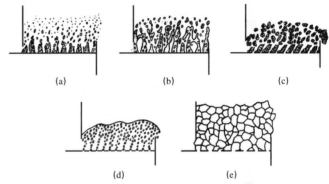

Figure 6.4 Possible vapor–liquid flow regimes for a contacting tray: (a) spray; (b) froth; (c) emulsion; (d) bubble; (e) cellular foam.

[Reproduced by permission from M.J. Lockett, *Distillation Tray Fundamentals*, Cambridge University Press, London (1986).]

If the height of packing is more than about 20 ft, liquid channeling may occur, causing the liquid to flow down near the wall, and gas to flow up the center of the column, thus greatly reducing the extent of vapor–liquid contact. In that case, liquid *redistributors* need to be installed.

Commercial packing materials include *random* (dumped) packings, some of which are shown in Figure 6.7a, and *structured* (arranged, ordered, or stacked) packings, some shown

Table 6.2 Comparison of Types of Trays

	Sieve Trays	Valve Trays	Bubble-Cap Trays
Relative cost	1.0	1.2	2.0
Pressure drop	Lowest	Intermediate	Highest
Efficiency	Lowest	Highest	Highest
Vapor capacity	Highest	Highest	Lowest
Typical turndown ratio	2	4	5

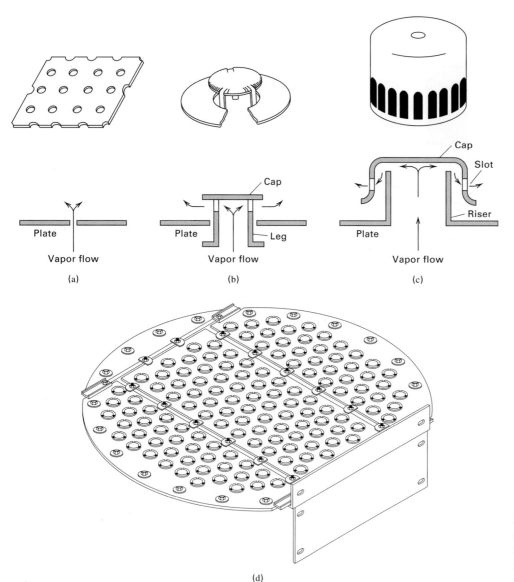

(a) (b) (c)

(d)

Figure 6.5 Three types of tray openings for passage of vapor up into liquid: (a) perforation; (b) valve cap; (c) bubble cap; (d) tray with valve caps.

in Figure 6.7b. Among the random packings, which are poured into the column, are the old (1895–1950) ceramic Raschig rings and Berl saddles. They have been largely replaced by metal and plastic Pall® rings, metal Bialecki® rings, and ceramic Intalox®* saddles, which provide more surface area for mass transfer, a higher flow capacity, and a lower pressure drop. More recently, through-flow packings of a lattice-work design have been developed. These include metal Intalox IMTP®*; metal, plastic, and ceramic Cascade Mini-Rings®*; metal Levapak®; metal, plastic, and ceramic Hiflow® rings; metal Tri-packs®; and plastic Nor-Pac® rings, which exhibit lower pressure drop and higher mass-transfer rates. Accordingly, they are called "high-efficiency" random packings. Most are available in nominal diameters, ranging from 1 to 3.5 inches.

As packing size increases, mass-transfer efficiency and pressure drop decrease. Therefore, an optimal packing size exists. However, to minimize liquid channeling, nominal packing size should be less than one-eighth of the column

diameter. A "fourth generation" of random packings, including VSP® rings, Fleximax®*, and Raschig super-rings, feature an open, undulating geometry that promotes uniform wetting with recurrent turbulence promotion. The result is low pressure drop and a mass-transfer efficiency that does not decrease with increasing column diameter and permits a larger depth of packing before a liquid redistributor is necessary.

Metal packings are usually preferred because of their superior strength and good wettability, but their costs are high. Ceramic packings, which have superior wettability but inferior strength, are used in corrosive environments at elevated temperatures. Plastic packings, usually of polypropylene, are inexpensive and have sufficient strength, but may have poor wettability at low liquid rates.

Structured packings include corrugated sheets of metal gauze, such as Sulzer® BX, Montz® A, Gempak®* 4BG, and Intalox High-Performance Wire Gauze Packing. Newer and less-expensive structured packings, which are fabricated from metal and plastics and may or may not be perforated, embossed, or surface-roughened, include metal and plastic

---

*Trademarks of Koch Glitsch, LP, and/or its affiliates.

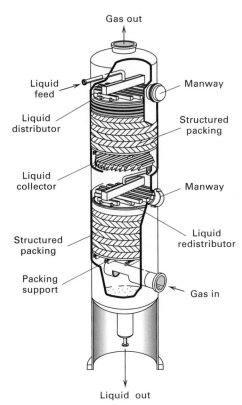

Figure 6.6 Details of internals used in a packed column.

Mellapak® 250Y, metal Flexipac®*, metal and plastic Gempak 4A, metal Montz B1, and metal Intalox High-Performance Structured Packing. These come with different-size openings between adjacent layers and are stacked in the column. Although they are considerably more expensive per unit volume than random packings, structured packings exhibit less pressure drop per theoretical stage and have higher efficiency and capacity.

In Table 6.3, packings are compared using the same factors as for trays. However, the differences between random and structured packings are greater than the differences among the three types of trays in Table 6.2.

§6.1.3 Spray, Bubble, and Centrifugal Contactors

Three other contactors are shown in Figure 6.2. If only one or two stages and very low pressure drop are required, and the solute is very soluble in the liquid, use of the *spray tower* is indicated for absorption. This consists of a vessel through which gas flows countercurrent to a liquid spray.

The *bubble column* for absorption consists of a vertical vessel partially filled with liquid into which vapor is bubbled. Vapor pressure drop is high because of the high head of liquid absorbent, and only one or two theoretical stages can be achieved. This device has a low vapor throughput and is impractical unless the solute has low solubility in the liquid and/or a slow chemical reaction that requires a long residence time.

(a)

Figure 6.7 Typical materials used in a packed column: (a) random packing materials; (*continued*)

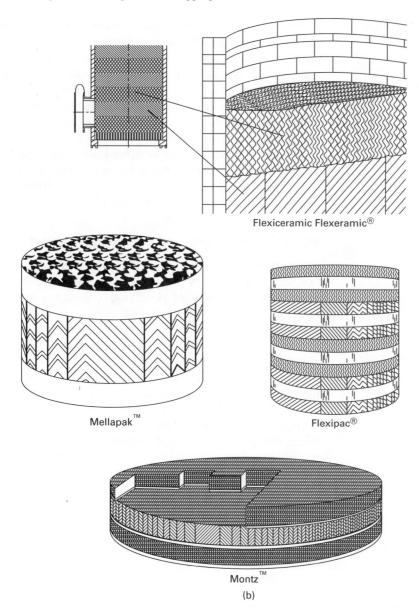

Flexiceramic Flexeramic®

Mellapak™

Flexipac®

Montz™

(b)

Figure 6.7 (*Continued*) (b) structured packing materials.

A novel device is the *centrifugal contactor*, which consists of a stationary, ringed housing, intermeshed with a ringed rotating section. The liquid phase is fed near the center of the packing, from which it is thrown outward. The vapor flows inward. Reportedly, high mass-transfer rates can be achieved. It is possible to obtain the equivalent of several equilibrium stages in a very compact unit. These short-contact-time type of devices are practical only when there are space limitations, in which case they are useful for distillation.

§6.1.4 Choice of Device

The choice of device is most often between a trayed and a packed column. The latter, using dumped packings, is always favored when column diameter is less than 2 ft and the packed height is less than 20 ft. Packed columns also get the nod for corrosive services where ceramic or plastic materials are preferred over metals, particularly welded column internals, and also in services where foaming is too severe for the use of trays, and pressure drop must be low, as in vacuum operations or where low liquid holdup is desirable. Otherwise, trayed towers, which can be designed more reliably, are preferred. Although structured packings are expensive, they are the best choice for installations when pressure drop is a factor or for replacing existing trays (retrofitting) when a higher capacity or degree of separation is required. Trayed towers are preferred when liquid velocities are low, whereas columns with random packings are best for high-liquid

Table 6.3 Comparison of Types of Packing

	Random		
	Raschig Rings and Saddles	"Through Flow"	Structured
Relative cost	Low	Moderate	High
Pressure drop	Moderate	Low	Very low
Efficiency	Moderate	High	Very high
Vapor capacity	Fairly high	High	High
Typical turndown ratio	2	2	2

velocities. Use of structured packing should be avoided at pressures above 200 psia and liquid flow rates above 10 gpm/ft^2 (Kister [33]). Turbulent liquid flow is desirable if mass transfer is limiting in the liquid phase, while a continuous, turbulent gas flow is desirable if mass transfer is limiting in the gas phase. Usually, the (continuous) gas phase is mass-transfer-limiting in packed columns and the (continuous) liquid phase is mass-transfer-limiting in tray columns.

§6.2 GENERAL DESIGN CONSIDERATIONS

Absorber and stripper design or analysis requires consideration of the following factors:

1. Entering gas (liquid) flow rate, composition, T, and P

2. Desired degree of recovery of one or more solutes

3. Choice of absorbent (stripping agent)

4. Operating P and T, and allowable gas pressure drop

5. Minimum absorbent (stripping agent) flow rate and actual absorbent (stripping agent) flow rate

6. Heat effects and need for cooling (heating)

7. Number of equilibrium stages and stage efficiency

8. Type of absorber (stripper) equipment (trays or packing)

9. Need for redistributors if packing is used

10. Height of absorber (stripper)

11. Diameter of absorber (stripper)

The ideal absorbent should have (a) a high solubility for the solute(s); (b) a low volatility to reduce loss; (c) stability and inertness; (d) low corrosiveness; (e) low viscosity and high diffusivity; (f) low foaming proclivities; (g) low toxicity and flammability; (h) availability, if possible, within the process; and (i) a low cost. The most widely used absorbents are water, hydrocarbon oils, and aqueous solutions of acids and bases. The most common stripping agents are steam, air, inert gases, and hydrocarbon gases.

Absorber operating pressure should be high and temperature low to minimize stage requirements and/or absorbent flow rate, and to lower the equipment volume required to accommodate the gas flow. Unfortunately, both compression and refrigeration of a gas are expensive. Therefore, most absorbers are operated at feed-gas pressure, which may be greater than ambient pressure, and at ambient temperature, which can be achieved by cooling the feed gas and absorbent with cooling water, unless one or both streams already exist at a subambient temperature.

Operating pressure should be low and temperature high for a stripper to minimize stage requirements and stripping agent flow rate. However, because maintenance of a vacuum is expensive, and steam jet exhausts are polluting, strippers are commonly operated at a pressure just above ambient. A high temperature can be used, but it should not be so high as to cause vaporization or undesirable chemical reactions. The possibility of phase changes occurring can be checked by bubble-point and dew-point calculations.

For given feed-gas (liquid) flow rate, extent of solute absorption (stripping), operating P and T, and absorbent

(stripping agent) composition, a *minimum absorbent (stripping agent) flow rate* exists that corresponds to an infinite number of countercurrent equilibrium contacts between the gas and liquid phases. In every design problem, a trade-off exists between the number of equilibrium stages and the absorbent (stripping agent) flow rate, a rate that must be greater than the minimum. Graphical and analytical methods for computing the minimum flow rate and this trade-off are developed in the following sections for mixtures that are dilute in solute(s). For this essentially isothermal case, the energy balance can be ignored. As discussed in Chapters 10 and 11, process simulators are best used for concentrated mixtures, where multicomponent phase equilibrium and mass-transfer effects are complex and an energy balance is necessary.

§6.3 GRAPHICAL METHOD FOR TRAYED TOWERS

For the countercurrent-flow, trayed tower for absorption (or stripping) shown in Figure 6.8, stages are numbered from top (where the absorbent enters) to bottom for the absorber; and from bottom (where the stripping agent enters) to top for the stripper. Phase equilibrium is assumed between the vapor and liquid leaving each tray. Assume for an absorber that only solute is transferred from one phase to the other. Let:

L' = molar flow rate of solute-free absorbent

V' = molar flow rate of solute-free gas (carrier gas)

X = mole ratio of solute to solute-free absorbent in the liquid

Y = mole ratio of solute to solute-free gas in the vapor

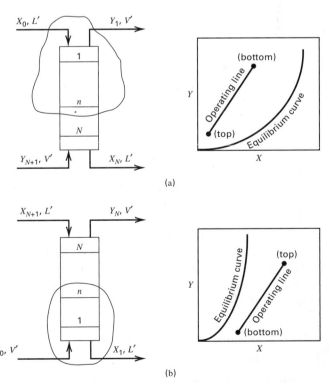

Figure 6.8 Continuous, steady-state operation in a countercurrent cascade with equilibrium stages: (a) absorber; (b) stripper.

Values of L' and V' remain constant through the tower, assuming no vaporization of absorbent into carrier gas or absorption of carrier gas by liquid.

§6.3.1 Equilibrium Curves

For the solute at any stage n, the K-value is

$$K_n = \frac{y_n}{x_n} = \frac{Y_n/(1 + Y_n)}{X_n/(1 + X_n)} \qquad (6\text{-}1)$$

where $Y = y/(1 - y)$ and $X = x/(1 - x)$.

From experimental x–y values, a representative equilibrium curve of Y as a function of X is calculated and plotted, as in Figure 6.8. In general, this curve will not be a straight line, but it will pass through the origin. If the solute undergoes, in the liquid phase, a complete irreversible conversion by chemical reaction to a nonvolatile solute, the equilibrium curve will be a straight line of zero slope passing through the origin.

§6.3.2 Operating Lines (from Material Balances)

At both ends of the towers in Figure 6.8, entering and leaving streams are paired. For the absorber, the pairs at the top are $(X_0, L'$ and $Y_1, V')$ and $(Y_{N+1}, V'$ and $X_N, L')$ at the bottom; for the stripper, $(X_{N+1}, L'$ and $Y_N, V')$ at the top and $(Y_0, V'$ and $X_1, L')$ at the bottom. These terminal pairs can be related to intermediate pairs of passing streams between stages by solute material balances for the envelopes shown in Figure 6.8. The balances are written around one end of the tower and an arbitrary intermediate equilibrium stage, n.

For the absorber,

$$X_0 L' + Y_{n+1} V' = X_n L' + Y_1 V' \qquad (6\text{-}2)$$

or, solving for Y_{n+1},

$$Y_{n+1} = X_n(L'/V') + Y_1 - X_0(L'/V') \qquad (6\text{-}3)$$

For the stripper,

$$X_{n+1} L' + Y_0 V' = X_1 L' + Y_n V' \qquad (6\text{-}4)$$

or, solving for Y_n,

$$Y_n = X_{n+1}(L'/V') + Y_0 - X_1(L'/V') \qquad (6\text{-}5)$$

Equations (6-3) and (6-5) are *operating lines*, plotted in Figure 6.8. The terminal points represent conditions at the top and bottom of the tower. For absorbers, the operating line is above the equilibrium line because, for a given solute concentration in the liquid, the solute concentration in the gas is always greater than the equilibrium value, thus providing a mass-transfer driving force for absorption. For strippers, operating lines lie below equilibrium line, thus enabling desorption. In Figure 6.8, operating lines are straight with a slope of L'/V'.

§6.3.3 Minimum Absorbent Flow Rate (for ∞ Stages)

Operating lines for four absorbent flow rates are shown in Figure 6.9, where each line passes through the terminal point,

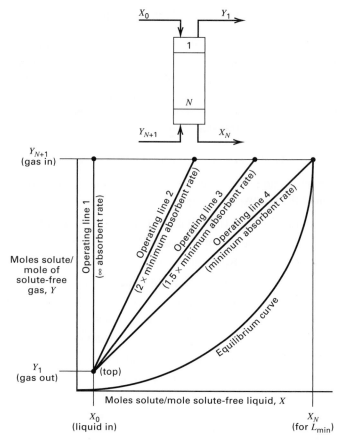

Figure 6.9 Operating lines for an absorber.

(Y_1, X_0), at the top of the column, and corresponds to a different liquid absorbent rate and corresponding operating-line slope, L'/V'. To achieve the desired value of Y_1 for given Y_{N+1}, X_0, and V', the solute-free absorbent flow rate L' must be between an ∞ absorbent flow (line 1) and a minimum absorbent rate (corresponding to ∞ stages), $L'_{\min}$ (line 4). The solute concentration in the outlet liquid, X_N (and, thus, the terminal point at the bottom of the column, Y_{N+1}, X_N), depends on L' by a material balance on the solute for the entire absorber. From (6-2), for $n = N$,

$$X_0 L' + Y_{N+1} V' = X_N L' + Y_1 V' \qquad (6\text{-}6)$$

or

$$L' = \frac{V'(Y_{N+1} - Y_1)}{(X_N - X_0)} \qquad (6\text{-}7)$$

Note that the operating line can terminate at the equilibrium line (line 4), but cannot cross it because that would be a violation of the second law of thermodynamics.

The minimum absorbent flow rate, $L'_{\min}$, corresponds to a value of X_N (leaving the bottom of the tower) in equilibrium with Y_{N+1}, the solute concentration in the feed gas. Note that it takes an infinite number of stages for this equilibrium to be achieved. An expression for $L'_{\min}$ of an absorber can be derived from (6-7) as follows.

For stage N, (6-1) for the minimum absorbent rate is

$$K_N = \frac{Y_{N+1}/(1 + Y_{N+1})}{X_N/(1 + X_N)} \qquad (6\text{-}8)$$

Solving (6-8) for X_N and substituting the result into (6-7),

$$L'_{min} = \frac{V'(Y_{N+1} - Y_1)}{\{Y_{N+1}/[Y_{N+1}(K_N - 1) + K_N]\} - X_0} \quad (6-9)$$

For dilute solutes, where $Y \approx y$ and $X \approx x$, (6-9) approaches

$$L'_{min} = V' \left(\frac{y_{N+1} - y_1}{\frac{y_{N+1}}{K_N} - x_0} \right) \quad (6-10)$$

If, for the entering liquid, $X_0 \approx 0$, (6-10) approaches

$$L'_{min} = V'K_N(\text{fraction of solute absorbed}) \quad (6-11)$$

Equation (6-11) confirms that L'_{min} increases with increasing V', K-value, and fraction of solute absorbed.

In practice, the absorbent flow rate is some multiple of L'_{min}, typically from 1.1 to 2. A value of 1.5 corresponds closely to the value of 1.4 for the optimal absorption factor mentioned earlier. In Figure 6.9, operating lines 2 and 3 correspond to 2.0 and 1.5 times L'_{min}, respectively. As the operating line moves from 1 to 4, the stages required increase from zero to infinity. Thus, a trade-off exists between L' and N, and an optimal L' exists.

A similar derivation of V'_{min}, for the stripper of Figure 6.8, results in an expression analogous to (6-11):

$$V'_{min} = \frac{L'}{K_N}(\text{fraction of solute stripped}) \quad (6-12)$$

§6.3.4 Number of Equilibrium Stages

As shown in Figure 6.10a, the operating line relates the solute concentration in the vapor passing upward between two stages to the solute concentration in the liquid passing downward between the same two stages. Figure 6.10b illustrates that the equilibrium curve relates the solute concentration in the vapor leaving an equilibrium stage to the solute concentration in the liquid leaving the same stage. This suggests starting from the top of the tower (at the bottom of the Y–X diagram) and moving to the tower bottom (at the top of the Y–X diagram) by constructing a staircase alternating between the operating line and the equilibrium curve, as in Figure 6.11a. The stages required for an absorbent flow rate corresponding to the slope of the operating line [which in Figure

6.11a is $(L'/V') = 1.5(L'_{min}/V')$] is stepped off by moving up the staircase, starting from the point (Y_1, X_0) on the operating line and moving horizontally to the right to (Y_1, X_1) on the equilibrium curve. From there, a vertical move is made to (Y_2, X_1) on the operating line. The staircase is climbed until the terminal point (Y_{N+1}, X_N) on the operating line is reached. As shown in Figure 6.11a, the stages are counted at the points on the equilibrium curve. As the slope (L'/V') is increased, fewer equilibrium stages are required. As (L'/V') is decreased, more stages are required until (L'_{min}/V') is reached, at which the operating line and equilibrium curve intersect at a *pinch point*, for which an infinite number of stages is required. Operating line 4 in Figure 6.9 has a pinch point at Y_{N+1}, X_N. If (L'/V') is reduced below (L'_{min}/V'), the specified extent of absorption cannot be achieved.

The stages required for stripping a solute are determined analogously to absorption. An illustration is shown in Figure 6.11b, which refers to Figure 6.8b. For given specifications of Y_0, X_{N+1}, and the extent of stripping of the solute, X_1, V'_{min} is determined from the slope of the operating line that passes through the points (Y_0, X_1) and (Y_N, X_{N+1}). The operating line in Figure 6.11b is for:

$$V' = 1.5V'_{min} \text{ or a slope of } (L'/V') = (L'/V'_{min})/1.5$$

In Figure 6.11, the number of equilibrium stages for the absorber and stripper is exactly three. Ordinarily, the result is some fraction above an integer number and the result is usually rounded to the next highest integer.

EXAMPLE 6.1 Recovery of Alcohol.

In a bioprocess, molasses is fermented to produce a liquor containing ethyl alcohol. A CO_2-rich vapor with a small amount of ethyl alcohol is evolved. The alcohol is recovered by absorption with water in a sieve-tray tower. Determine the number of equilibrium stages required for countercurrent flow of liquid and gas, assuming isothermal, isobaric conditions and that only alcohol is absorbed.

Entering gas is 180 kmol/h; 98% CO_2, 2% ethyl alcohol; 30°C, 110 kPa.

Entering liquid absorbent is 100% water; 30°C, 110 kPa.

Required recovery (absorption) of ethyl alcohol is 97%.

Solution

From §5.7 for a single-section, countercurrent cascade, the number of degrees of freedom is $2N + 2C + 5$. All stages operate adiabatically at a pressure of approximately 1 atm, thus fixing $2N$ design variables. The entering gas is completely specified, taking $C + 2$ variables. The entering liquid flow rate is not specified; thus, only $C + 1$ variables are taken by the entering liquid. The recovery of ethyl alcohol is a specified variable; thus, the total degrees of freedom taken by the specification is $2N + 2C + 4$. This leaves one additional specification to be made, which can be the entering liquid absorbent flow rate at, say, 1.5 times the minimum value, which will have to be determined.

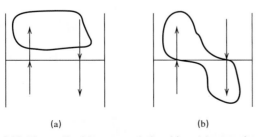

(a)   (b)

Figure 6.10 Vapor–liquid stream relationships: (a) operating line (passing streams); (b) equilibrium curve (leaving streams).

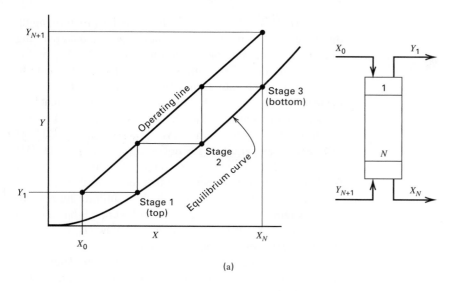

(a)

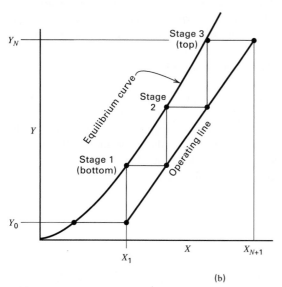

(b)

Figure 6.11 Graphical determination of the number of equilibrium stages for (a) absorber and (b) stripper.

Note that the above degrees-of-freedom analysis assumes an energy balance for each stage. The energy balances are assumed to result in isothermal operation at 30°C.

For dilute ethyl alcohol, the K-value is determined from a modified Raoult's law, $K = \gamma P^S/P$. The vapor pressure at 30°C is 10.5 kPa, and from infinite dilution in water data at 30°C, the liquid-phase activity coefficient of ethyl alcohol is 6. Thus, $K = (6)(10.5)/110 = 0.57$. The minimum solute-free absorbent rate is given by (6-11), where the solute-free gas rate, V', is (0.98)(180) = 176.4 kmol/h. Thus, the minimum absorbent rate for 97.5% recovery is

$$L'_{\min} = (176.4)(0.57)(0.97) = 97.5 \text{ kmol/h}$$

The solute-free absorbent rate at 50% above the minimum is

$$L' = 1.5(97.5) = 146.2 \text{ kmol/h}$$

The alcohol recovery of 97% corresponds to

$$(0.97)(0.02)(180) = 3.49 \text{ kmol/h}$$

The amount of ethyl alcohol remaining in the exiting gas is

$$(1.00 - 0.97)(0.02)(180) = 0.11 \text{ kmol/h}$$

Alcohol mole ratios at both ends of the operating line are:

$$\text{top}\begin{cases} X_0 = 0, & Y_1 = \dfrac{0.11}{176.4} = 0.0006 \end{cases}$$

$$\text{bottom}\begin{cases} Y_{N+1} = \dfrac{0.11 + 3.49}{176.4} = 0.0204, & X_N = \dfrac{3.49}{146.2} = 0.0239 \end{cases}$$

The equation for the operating line from (6-3) with $X_0 = 0$ is

$$Y_{N+1} = \left(\frac{146.2}{176.4}\right)X_N + 0.0006 = 0.829X_N + 0.0006 \quad (1)$$

This is a dilute system. From (6-1), the equilibrium curve, using $K = 0.57$, is

$$0.57 = \frac{Y/(1+Y)}{X/(1+X)}$$

Solving for Y,

$$Y = \frac{0.57X}{1 + 0.43X} \quad (2)$$

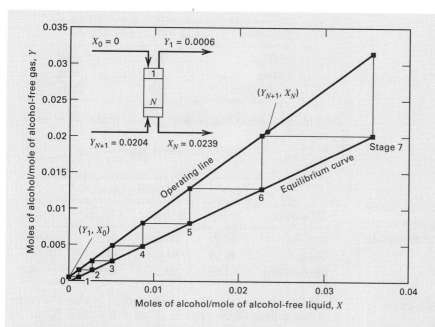

Figure 6.12 Graphical determination of number of equilibrium stages for an absorber.

To cover the entire column, the necessary range of X for a plot of Y versus X is 0 to almost 0.025. From the Y–X equation, (2), the following values are obtained:

Y	X
0.00000	0.000
0.00284	0.005
0.00569	0.010
0.00850	0.015
0.01130	0.020
0.01410	0.025

For this dilute system in ethyl alcohol, the maximum error in Y is 1.0%, if Y is taken simply as $Y = KX = 0.57X$. The equilibrium curve, which is almost straight, and a straight operating line drawn through the terminal points (Y_1, X_0) and (Y_{N+1}, X_N) are shown in Figure 6.12. The theoretical stages are stepped off as shown, starting from the top stage (Y_1, X_0) located near the lower left corner. The required number of stages, N, for 97% absorption of ethyl alcohol is slightly more than six. Accordingly, it is best to provide seven theoretical stages.

§6.4 ALGEBRAIC METHOD FOR DETERMINING N

Graphical methods for determining N have great educational value because graphs provide a visual insight into the phenomena involved. However, these graphical methods become tedious when (1) problem specifications fix the number of stages rather than percent recovery of solute, (2) more than one solute is absorbed or stripped, (3) the best operating conditions of T and P are to be determined so that the equilibrium-curve location is unknown, and (4) very low or very high concentrations of solute force the construction to the corners of the diagram so that multiple Y–X diagrams of varying scales are needed to achieve accuracy. When the

graphical method is unsuitable and commercial software is unavailable, the Kremser method of §5.4 can be applied as follows. Rewrite (5-48) and (5-50) in terms of the fraction of solute absorbed or stripped:

$$\text{Fraction of a solute, } i, \text{ absorbed} = \frac{A_i^{N+1} - A_i}{A_i^{N+1} - 1} \quad (6\text{-}13)$$

$$\text{Fraction of a solute, } i, \text{ stripped} = \frac{S_i^{N+1} - S_i}{S_i^{N+1} - 1} \quad (6\text{-}14)$$

where the solute absorption and stripping factors are

$$A_i = L/(K_i V) \quad (6\text{-}15)$$

$$S_i = K_i V/L \quad (6\text{-}16)$$

L and V in moles per unit time may be taken as entering values. From Chapter 2, K_i depends mainly on T, P, and liquid-phase composition. At near-ambient pressure, for dilute mixtures, some common expressions are

$$K_i = P_i^s/P \quad \text{(Raoult's law)} \quad (6\text{-}17)$$

$$K_i = \gamma_{iL}^\infty P_i^s/P \quad \text{(modified Raoult's law)} \quad (6\text{-}18)$$

$$K_i = H_i/P \quad \text{(Henry's law)} \quad (6\text{-}19)$$

$$K_i = P_i^s/x_i^* P \quad \text{(solubility)} \quad (6\text{-}20)$$

Raoult's law is for ideal solutions involving solutes at subcritical temperatures. The modified Raoult's law is useful for nonideal solutions when activity coefficients are known at infinite dilution. For solutes at supercritical temperatures, use of Henry's law may be preferable. For sparingly soluble solutes at subcritical temperatures, (6-20) is preferred when solubility data x_i^* are available. This expression is derived by considering a three-phase system consisting of an ideal-vapor-containing solute, carrier vapor, and solvent; a pure or near-pure solute as liquid (1); and the solvent liquid (2) with dissolved solute. In that case, for solute i at equilibrium between the two liquid phases,

$$x_i^{(1)} \gamma_{iL}^{(1)} = x_i^{(2)} \gamma_{iL}^{(2)}$$

But, $x_i^{(1)} \approx 1, \quad \gamma_{iL}^{(1)} \approx 1, \quad x_i^{(2)} = x_i^*$

Therefore, $\gamma_{iL}^{(2)} \approx 1/x_i^*$

and from (6-18),

$$K_i^{(2)} = \gamma_{iL}^{(2)} P_i^s / P = P_i^s / \left(x_i^* P \right)$$

The advantage of (6-13) and (6-15) is that they can be solved directly for the percent absorption or stripping of a solute when the number of theoretical stages, N, and the absorption or stripping factor are known.

EXAMPLE 6.2   Stripping of VOCs from Wastewater.

Okoniewski [3] studied the stripping by air of volatile organic compounds (VOCs) from wastewater. At 70°F and 15 psia, 500 gpm of wastewater were stripped with 3,400 scfm of air (60°F and 1 atm) in a 20-plate tower. The wastewater contained three pollutants in the amounts shown in the following table. Included are properties from the 1966 *Technical Data Book—Petroleum Refining* of the American Petroleum Institute. For all compounds, the organic concentrations are less than their solubility values, so only one liquid phase exists.

Organic Compound	Concentration in the Wastewater, mg/L	Solubility in Water at 70°F, mole fraction	Vapor Pressure at 70°F, psia
Benzene	150	0.00040	1.53
Toluene	50	0.00012	0.449
Ethylbenzene	20	0.000035	0.149

99.9% of the total VOCs must be stripped. The plate efficiency of the tower is estimated to be 5% to 20%, so if one theoretical stage is predicted, as many as twenty actual stages may be required. Plot the % stripping of each organic compound for one to four theoretical stages. Under what conditions will the desired degree of stripping be achieved? What should be done with the exiting air?

Solution

Because the wastewater is dilute in the VOCs, the Kremser equation may be applied independently to each organic chemical. The absorption of air by water and the stripping of water by air are ignored. The stripping factor for each compound is $S_i = K_i V / L$, where V and L are taken at entering conditions. The K-value may be computed from a modified Raoult's law, $K_i = \gamma_{iL} P_i^s / P$, where for a compound that is only slightly soluble, take $\gamma_{iL} = 1/x_i^*$ where x_i^* is the solubility in mole fractions. Thus, from (6-22), $K_i = P_i^s / x_i^* P$.

$V = 3,400(60)/(379 \text{ scf/lbmol})$ or 538 lbmol/h

$L = 500(60)(8.33 \text{ lb/gal})/(18.02 \text{ lb/lbmol})$ or $13,870$ lbmol/h

The corresponding K-values and stripping factors are

Component	K at 70°F, 15 psia	S
Benzene	255	9.89
Toluene	249	9.66
Ethylbenzene	284	11.02

From (6-16), a spreadsheet program gives the results below, where

$$\text{Fraction stripped} = \frac{S^{N+1} - S}{S^{N+1} - 1}$$

	Percent Stripped			
Component	1 Stage	2 Stages	3 Stages	4 Stages
Benzene	90.82	99.08	99.91	99.99
Toluene	90.62	99.04	99.90	99.99
Ethylbenzene	91.68	99.25	99.93	99.99

The results are sensitive to the number of stages, as shown in Figure 6.13. To achieve 99.9% removal of the VOCs, three stages are needed, corresponding to the necessity for a 15% stage efficiency in the existing 20-tray tower.

The exiting air must be processed to destroy the VOCs, particularly the carcinogen, benzene [4]. The amount stripped is

$$(500 \text{ gpm})(60 \text{ min/h})(3.785 \text{ liters/gal})(150 \text{ mg/liters})$$
$$= 17,030,000 \text{ mg/h or } 37.5 \text{ lb/h}$$

If benzene is valued at $0.50/lb, the annual value is approximately $150,000. This would not justify a recovery technique such as carbon adsorption. It is thus preferable to destroy the VOCs by incineration. For example, the air can be sent to an on-site utility boiler, a waste-heat boiler, or a catalytic incinerator. Also the amount of air was arbitrarily given as 3,400 scfm. To complete the design procedure, various air rates should be investigated and column-efficiency calculations made, as discussed in the next section.

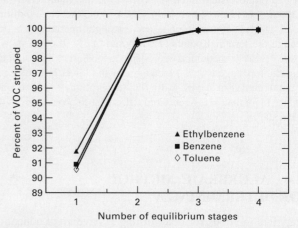

Figure 6.13 Results of Example 6.2 for stripping VOCs from water with air.

§6.5   STAGE EFFICIENCY AND COLUMN HEIGHT FOR TRAYED COLUMNS

Except when temperatures change significantly from stage to stage, the assumption that vapor and liquid phases leaving a stage are at the same temperature is reasonable. However, the assumption of equilibrium at each stage may not be reasonable, and for streams leaving a stage, vapor-phase mole fractions are not related exactly to liquid-phase mole fractions by thermodynamic K-values. To determine the actual number of required plates in a trayed column, the number of equilibrium stages must be adjusted with a *stage efficiency* (*plate efficiency*, *tray efficiency*).

Stage efficiency concepts are applicable when phases are contacted and then separated—that is, when discrete stages and interfaces can be identified. This is not the case for packed columns. For these, the efficiency is embedded into equipment- and system-dependent parameters, like the HETP or HETS (height of packing equivalent to a theoretical stage).

§6.5.1 Overall Stage Efficiency

A simple approach for preliminary design and evaluation of an existing column is to apply an overall stage (or column) efficiency, defined by Lewis [5] as

$$E_o = N_t/N_a \qquad (6\text{-}21)$$

where E_o is the fractional overall stage efficiency, usually less than 1.0; N_t is the calculated number of equilibrium (theoretical) stages; and N_a is the actual number of trays or plates required. Based on the results of extensive research conducted over a period of more than 70 years, the overall stage efficiency has been found to be a complex function of: (a) geometry and design of the contacting trays; (b) flow rates and flow paths of vapor and liquid streams; and (c) compositions and properties of vapor and liquid streams.

For well-designed trays and for flow rates near the column capacity limit (discussed in §6.6), E_o depends mainly on the physical properties of the vapor and liquid streams. Values of E_o can be predicted by four methods: (1) comparison with performance data from industrial columns; (2) use of empirical efficiency equations derived from data on industrial columns; (3) use of semitheoretical models based on mass- and heat-transfer rates; and (4) scale-up from laboratory or pilot-plant columns. These methods are discussed in the following four subsections, and are applied to distillation in the next chapter. Suggested correlations of mass-transfer coefficients for trayed towers are deferred to §6.6, following the discussion of tray capacity. A final subsection presents a method for estimating the column height based on the number of equilibrium stages, stage efficiency, and tray spacing.

§6.5.2 Stage Efficiencies from Column Performance Data

Performance data from trayed industrial absorption and stripping columns generally include gas- and liquid-feed and product flow rates and compositions, pressures and temperatures at the bottom and top of the column, details of the tray design, column diameter, tray spacing, average liquid viscosity, and computed overall tray efficiency with respect to one or more components. From these data, particularly if the system is dilute with respect to the solute(s), the graphical or algebraic

Table 6.4 Effect of Species on Overall Stage Efficiency

Component	Overall Stage Efficiency, %
Ethylene	10.3
Ethane	14.9
Propylene	25.5
Propane	26.8
Butylene	33.8

Source: H.E. O'Connell [8].

methods, described in §6.3 and 6.4, can estimate the number of equilibrium stages, N_t, required. Then, knowing N_a, (6-21) can be applied to determine the overall stage efficiency, E_o, whose values for absorbers and strippers are typically low, especially for absorption, which is often less than 50%.

Drickamer and Bradford [6] computed overall stage efficiencies for five hydrocarbon absorbers and strippers with column diameters ranging from 4 to 5 feet and equipped with bubble-cap trays. Overall stage efficiencies for the key component, n-butane in the absorbers and probably n-heptane in the strippers, varied from 10.4% to 57%, depending primarily on the molar-average liquid viscosity, a key factor for liquid mass-transfer rates.

Individual component overall efficiencies differ because of differences in component physical properties. The data of Jackson and Sherwood [7] for a 9-ft-diameter hydrocarbon absorber equipped with 19 bubble-cap trays on 30-inch tray spacing, and operating at 92 psia and 60°F, is summarized in Table 6.4 from O'Connell [8]. Values of E_o vary from 10.3% for ethylene, the most-volatile species, to 33.8% for butylene, the least-volatile species. Molar-average liquid viscosity was 1.90 cP.

A more dramatic effect of absorbent species solubility on the overall stage efficiency was observed by Walter and Sherwood [9] using laboratory bubble-cap tray columns ranging from 2 to 18 inches in diameter. Stage efficiencies varied over a range from less than 1% to 69%. Comparing data for the water absorption of NH_3 (a very soluble gas) and CO_2 (a slightly soluble gas), they found a lower stage efficiency for CO_2, with its low gas solubility (high K-value); and a high stage efficiency for NH_3, with its high gas solubility (low K-value). Thus, both solubility (or K-value) and liquid-phase viscosity have significant effects on stage efficiency.

§6.5.3 Empirical Correlations for Stage Efficiency

From 20 sets of performance data from industrial absorbers and strippers, Drickamer and Bradford [6] correlated key-component overall stage efficiency with just the molar-average viscosity of the rich oil (liquid leaving an absorber or liquid entering a stripper) over a viscosity range of 0.19 to 1.58 cP at the column temperature. The empirical equation

$$E_o = 19.2 - 57.8 \log \mu_L, \quad 0.2 < \mu_L < 1.6 \, \text{cP} \qquad (6\text{-}22)$$

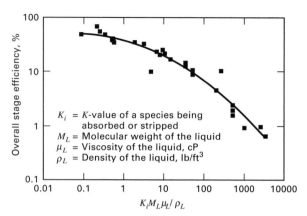

Figure 6.14 O'Connell correlation for plate efficiency of absorbers and strippers.

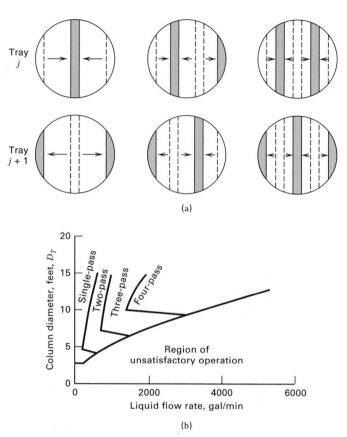

Figure 6.15 Estimation of number of required liquid flow passes. (a) Multipass trays (2, 3, 4 passes). (b) Flow pass selection.

[Derived from *Koch Flexitray Design Manual, Bulletin 960*, Koch Engineering Co., Inc., Wichita, KS (1960).]

where E_o is in % and μ is in cP, fits the data with an average % deviation of 10.3%. Equation (6-24) should not be used for nonhydrocarbon liquids and is restricted to the viscosity range of the data.

Mass-transfer theory indicates that when the solubility or volatility of species being absorbed or stripped covers a wide range, the relative importance of liquid- and gas-phase mass-transfer resistances shifts. O'Connell [8] found that the Drickamer–Bradford correlation, (6-24), was inadequate when species cover a wide solubility or K-value range. O'Connell obtained a more general correlation by using a parameter that includes not only liquid viscosity, but also liquid density and a Henry's law solubility constant. Edmister [10] and Lockhart et al. [11] suggested slight modifications to the O'Connell correlation, shown in Figure 6.14, to permit its use with K-values (instead of Henry's law constants). The correlating parameter suggested by Edmister, $K_i M_L \mu_L / \rho_L$, and appearing in Figure 6.14 with M_L, the liquid molecular weight, has units of $(\text{lb/lbmol})(\text{cP})/(\text{lb/ft}^3)$. The data cover the following range of conditions:

Column diameter:	2 inches to 9 ft
Average pressure:	14.7 to 485 psia
Average temperature:	60 to 138°F
Liquid viscosity:	0.22 to 21.5 cP
Overall stage efficiency:	0.65 to 69%

The following empirical equation by O'Connell predicts most of the data in Figure 6.14 to within about a 15% deviation for water and hydrocarbons:

$$\log E_o = 1.597 - 0.199 \log \left(\frac{K M_L \mu_L}{\rho_L} \right)$$
$$- 0.0896 \left[\log \left(\frac{K M_L \mu_L}{\rho_L} \right) \right]^2 \qquad (6\text{-}23)$$

The data for Figure 6.14 are mostly for columns having a liquid flow path across the tray, shown in Figure 6.3, from 2 to 3 ft. Theory and data show higher efficiencies for longer flow paths. For short liquid flow paths, the liquid flowing across the tray is usually completely mixed. For longer flow

paths, the equivalent of two or more completely mixed, successive liquid zones exists. The result is a greater average driving force for mass transfer and, thus, a higher stage efficiency—even greater than 100%! A column with a 10-ft liquid flow path may have an efficiency 25% greater than that predicted by (6-23). However, at high liquid rates, long liquid-path lengths are undesirable because they lead to excessive liquid (hydraulic) gradients. When the height of a liquid on a tray is appreciably higher on the inflow side than at the overflow weir, vapor may prefer to enter the tray in the latter region, leading to nonuniform bubbling. Multipass trays, as shown in Figure 6.15a, are used to prevent excessive hydraulic gradients. Estimates of the required number of flow paths can be made with Figure 6.15b, where, e.g., a 10-foot-diameter column with a liquid flow rate of 1000 gpm should use a three-pass tray.

EXAMPLE 6.3 Absorber Efficiency.

Performance data, given below for a trayed absorber located in a Texas petroleum refinery, were reported by Drickamer and Bradford [6]. Based on these data, back-calculate the overall stage efficiency for *n*-butane and compare the result with both the Drickamer–Bradford (6-22) and the O'Connell (6-23) correlations. Lean oil and rich gas enter the tower; rich oil and lean gas leave the tower.

Performance Data

Number of plates	16
Plate spacing, inches	24
Tower diameter, ft	4
Tower pressure, psig	79
Lean oil temperature, °F	102
Rich oil temperature, °F	126
Rich gas temperature, °F	108
Lean gas temperature, °F	108
Lean oil rate, lbmol/h	368
Rich oil rate, lbmol/h	525.4
Rich gas rate, lbmol/h	946
Lean gas rate, lbmol/h	786.9
Lean oil molecular weight	250
Lean oil viscosity at 116°F, cP	1.4
Lean oil gravity, °API	21

Stream Compositions, Mol%

Component	Rich Gas	Lean Gas	Rich Oil	Lean Oil
C_1	47.30	55.90	1.33	
C_2	8.80	9.80	1.16	
$C_3^=$	5.20	5.14	1.66	
C_3	22.60	21.65	8.19	
$C_4^=$	3.80	2.34	3.33	
nC_4	7.40	4.45	6.66	
nC_5	3.00	0.72	4.01	
nC_6	1.90		3.42	
Oil absorbent			70.24	100
Totals	100.00	100.00	100.00	100

Solution

First, it is worthwhile to check the consistency of the plant data by examining the overall material balance and the material balance for each component. From the above stream compositions, it is apparent that the compositions have been normalized to total 100%. The overall material balance is

Total flow into tower $= 368 + 946 = 1,314$ lbmol/h
Total flow from tower $= 525.4 + 786.9 = 1,312.3$ lbmol/h

These totals agree to within 0.13%, which is excellent agreement. The component material balance for the oil absorbent is

Total oil in $= 368$ lbmol/h   and
total oil out $= (0.7024)(525.4) = 369$ lbmol/h

These two totals agree to within 0.3%. Again, this is excellent agreement. Component material balances give the following results:

	lbmol/h			
Component	Lean Gas	Rich Oil	Total Out	Total In
C_1	439.9	7.0	446.9	447.5
C_2	77.1	6.1	83.2	83.2
$C_3^=$	40.4	8.7	49.1	49.2
C_3	170.4	43.0	213.4	213.8

$C_4^=$	18.4	17.5	35.9	35.9
nC_4	35.0	35.0	70.0	70.0
nC_5	5.7	21.1	26.8	28.4
nC_6	0.0	18.0	18.0	18.0
	786.9	156.4	943.3	946.0

Again, there is excellent agreement. The largest difference is 6% for pentanes. Plant data are not always so consistent.

For the back-calculation of stage efficiency from performance data, the Kremser equation is used to compute the number of equilibrium stages required for the measured absorption of n-butane.

$$\text{Fraction of } nC_4 \text{ absorbed} = \frac{35}{70} = 0.50$$

From (6-13),
$$0.50 = \frac{A^{N+1} - A}{A^{N+1} - 1} \tag{1}$$

To calculate the number of equilibrium stages, N, using Eq. (1), the absorption factor, A, for n-butane must be estimated from L/KV.

Because L and V vary greatly through the column, let

$$L = \text{average liquid rate} = \frac{368 + 525.4}{2} = 446.7 \text{ lbmol/h}$$

$$V = \text{average vapor rate} = \frac{946 + 786.9}{2} = 866.5 \text{ lbmol/h}$$

The average tower temperature $= (102 + 126 + 108 + 108)/4 = 111°F$. Also assume that the given viscosity of the lean oil at 116°F equals the viscosity of the rich oil at 111°F, i.e., $\mu = 1.4$ cP.

If ambient pressure $= 14.7$ psia, tower pressure $= 79 + 14.7 = 93.7$ psia.

From Figure 2.4, at 93.7 psia and 111°F, $K_{nC_4} = 0.7$. Thus,

$$A = \frac{446.7}{(0.7)(866.5)} = 0.736$$

Therefore, from Eq. (1),

$$0.50 = \frac{0.736^{N+1} - 0.736}{0.736^{N+1} - 1} \tag{2}$$

Solving, $N = N_t = 1.45$

From the performance data, $N_a = 16$.

From (6-21), $E_o = \frac{1.45}{16} = 0.091$ or 9.1%

Equation (6-22) is applicable to n-butane, because it is about 50% absorbed and is one of the key components. Other possible key components are butenes and n-pentane.

From the Drickamer equation (6-22),

$$E_o = 19.2 - 57.8 \log(1.4) = 10.8\%$$

Given that lean oil properties are used for M_L and μ_L, a conservative value for ρ_L is that of the rich oil, which from a process simulator is 44 lb/ft³. From Figure 2.4 for n-butane, K at 93.7 psia and 126°F is 0.77. Therefore,

$$K_i M_L \mu_L / \rho_L = (0.77)(250)(1.4)/44 = 6.1$$

From (6-23)

$$\log E_o = 1.597 - 0.199 \log\left(\frac{K M_L \mu_L}{\rho_L}\right) - 0.0896\left[\log\left(\frac{K M_L \mu_L}{\rho_L}\right)\right]^2 = 1.38$$

$$E_o = 24\%$$

This compares unfavorably to 10.8% for the Drickamer and Bradford efficiency and 9.1% from plant data.

§6.5.4 Semitheoretical Models—Murphree Efficiencies

A third method for predicting efficiency is a semitheoretical model based on mass- and heat-transfer rates. With this model, the fractional approach to equilibrium (the *stage* or *tray efficiency*), is estimated for each component or averaged for the column.

Tray efficiency models, in order of increasing complexity, have been proposed by Holland [12], Murphree [13], Hausen [14], and Standart [15]. All four models assume that vapor and liquid streams entering each tray are of uniform compositions. The *Murphree vapor efficiency*, which is the oldest and most widely used, is derived with the additional assumptions of (1) a uniform liquid composition on the tray equal to that leaving the tray, and (2) plug flow of the vapor passing up through the liquid, as indicated in Figure 6.16 for tray *n*. For species *i*, let:

n = mass-transfer rate for absorption from gas to liquid

K_G = overall gas mass-transfer coefficient based on a partial-pressure driving force

a = vapor–liquid interfacial area per volume of combined gas and liquid holdup (froth or dispersion) on the tray

A_b = active bubbling area of the tray (total cross-sectional area minus liquid downcomer areas)

Z_f = height of combined gas and liquid tray holdup

y_i = mole fraction of *i* in the vapor rising up through the liquid on the tray

y_i^* = vapor mole fraction of *i* in equilibrium with the completely mixed liquid on the tray

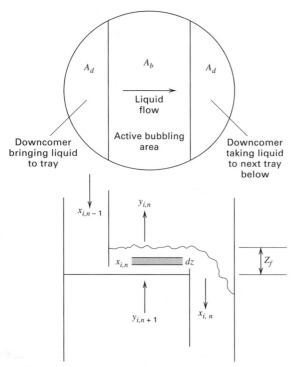

Figure 6.16 Schematic top and side views of tray for derivation of Murphree vapor-tray efficiency.

The differential rate of mass transfer for a differential height of holdup on tray *n*, numbered down from the top, is

$$dn_i = K_G a(y_i - y_i^*) P A_b \, dZ \qquad (6\text{-}24)$$

where, from (3-218), K_G includes both gas- and liquid-phase resistances to mass transfer. By material balance, assuming a negligible change in V across the stage,

$$dn_i = -V \, dy_i \qquad (6\text{-}25)$$

where V = molar gas flow rate up through the tray liquid.

Combining (6-24) and (6-25) to eliminate dn_i, separating variables, and converting to integral form,

$$A_b \int_0^{Z_f} \frac{K_G a P}{V} \, dZ = \int_{y_{i,n+1}}^{y_{i,n}} \frac{dy_i}{y_{i,n}^* - y_i} = N_{OG} \qquad (6\text{-}26)$$

where a second subscript for the tray number, *n*, has been added to the vapor mole fraction. The vapor enters tray *n* as $y_{i,n+1}$ and exits as $y_{i,n}$. This equation defines

N_{OG} = number of overall gas-phase mass-transfer units

Values of K_G, a, P, and V vary somewhat as the gas flows up through the liquid on the tray, but if they, as well as y_i^*, are taken to be constant, (6-30) can be integrated to give

$$N_{OG} = \frac{K_G a P Z_f}{(V/A_b)} = \ln\left(\frac{y_{i,n+1} - y_{i,n}^*}{y_{i,n} - y_{i,n}^*}\right) \qquad (6\text{-}27)$$

Rearrangement of (6-27) in terms of the fractional approach of y_i to equilibrium defines the *Murphree vapor efficiency* as

$$\boxed{E_{MV} = \frac{y_{i,n+1} - y_{i,n}}{y_{i,n+1} - y_{i,n}^*} = 1 - e^{-N_{OG}}} \qquad (6\text{-}28)$$

where
$$N_{OG} = -\ln(1 - E_{MV}) \qquad (6\text{-}29)$$

Suppose that measurements give y_i entering tray $n = y_{i,n+1} = 0.64$, y_i leaving tray $n = y_{i,n} = 0.61$, and, from phase-equilibrium data, y_i^* in equilibrium with x_i on and leaving tray n is 0.60. Then from (6-28),

$$E_{MV} = (0.64 - 0.61)/(0.64 - 0.60) = 0.75$$

or a 75% approach to equilibrium. From (6-29),

$$N_{OG} = -\ln(1 - 0.75) = 1.386$$

The Murphree vapor efficiency does not include the exiting stream temperatures. However, it implies that the completely mixed liquid phase is at its bubble point, so the equilibrium vapor-phase mole fraction, $y_{i,n}^*$, can be obtained.

For multicomponent mixtures, values of E_{MV} are component-dependent and vary from tray to tray; but at each tray the number of independent values of E_{MV} is one less than the number of components. The dependent value of E_{MV} is determined by forcing $\sum y_i = 1$. It is thus possible that a negative value of E_{MV} can result for a component in a multicomponent mixture. Such negative efficiencies are possible because of mass-transfer coupling among concentration gradients in a multicomponent mixture, as discussed in Chapter 12.

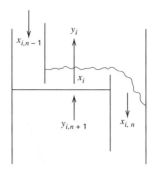

Figure 6.17 Schematic of tray for Murphree vapor-point efficiency.

However, for a binary mixture, values of E_{MV} are always positive and identical.

Only if liquid travel across a tray is short will the tray liquid satisfy the complete-mixing assumption of (6-27). For the more general case of incomplete liquid mixing, a *Murphree vapor-point efficiency* is defined by assuming that liquid composition varies across a tray, but is uniform in the vertical direction. Thus, for species i on tray n, at any horizontal distance from the downcomer, as in Figure 6.17,

$$E_{OV} = \frac{y_{i,n+1} - y_i}{y_{i,n+1} - y_i^*} \qquad (6\text{-}30)$$

Because x_i varies across a tray, y_i^* and y_i also vary. However, the exiting vapor is then assumed to mix to give a uniform $y_{i,n}$ before entering the tray above. Because E_{OV} is a more fundamental quantity than E_{MV}, E_{OV} serves as the basis for semi-theoretical estimates of tray E_{MV} and overall column efficiency.

Lewis [16] integrated E_{OV} over a tray to obtain E_{MV} for several cases. For complete mixing of liquid on the tray (uniform $x_{i,n}$),

$$E_{OV} = E_{MV} \qquad (6\text{-}31)$$

For plug flow of liquid across a tray with no mixing of liquid or diffusion in the horizontal direction, Lewis derived

$$E_{MV} = \frac{1}{\lambda}\left(e^{\lambda E_{OV}} - 1\right) \qquad (6\text{-}32)$$

where

$$\lambda = mV/L \qquad (6\text{-}33)$$

and $m = dy/dx$, the slope of the equilibrium line for a species, using the expression $y = mx + b$. If b is taken as zero, then m is the K-value, and for the key component, k,

$$\lambda = K_k V/L = 1/A_k$$

If A_k, the key-component absorption factor, has a typical value of 1.4, $\lambda = 0.71$.

Suppose the measured or predicted point efficiency is $E_{OV} = 0.25$. From (6-32), $E_{MV} = 1.4\left(e^{0.71(0.25)} - 1\right) = 0.27$, which is only 9% higher than E_{OV}. However, if $E_{OV} = 0.9$, E_{MV} is 1.25, which is more than a theoretical stage. This surprising result is due to the liquid concentration gradient across the length of the tray, which allows the vapor to contact a liquid having an average concentration of species k appreciably lower than that in the liquid leaving the tray.

Equations (6-31) and (6-32) represent extremes between complete mixing and no mixing. A more realistic, but more complex, model that accounts for partial liquid mixing, as developed by Gerster et al. [17], is

$$\frac{E_{MV}}{E_{OV}} = \frac{1 - e^{-(\eta + N_{Pe})}}{(\eta + N_{Pe})\{1 + [(\eta + N_{Pe})/\eta]\}} + \frac{e^{\eta} - 1}{\eta\{1 + [\eta/(\eta + N_{Pe})]\}} \qquad (6\text{-}34)$$

where

$$\eta = \frac{N_{Pe}}{2}\left[\left(1 + \frac{4\lambda E_{OV}}{N_{Pe}}\right)^{1/2} - 1\right] \qquad (6\text{-}35)$$

The dimensionless Peclet number, N_{Pe}, which serves as a partial-mixing parameter, is defined by

$$N_{Pe} = Z_L^2/D_E\theta_L = Z_L u/D_E \qquad (6\text{-}36)$$

where Z_L is liquid flow path length across the tray shown in Figure 6.3, D_E is the eddy diffusivity in the liquid flow direction, θ_L is the average liquid tray residence time, and $u = Z_L/\theta_L$ is the mean liquid velocity across the tray. Equation (6-34) is plotted in Figure 6.18 for ranges of N_{Pe} and λE_{OV}. When $N_{Pe} = 0$, (6-31) holds; when $N_{Pe} = \infty$, (6-32) holds.

Equation (6-36) suggests that the Peclet number is the ratio of the mean liquid velocity to the eddy-diffusion velocity. When N_{Pe} is small, eddy diffusion is important and the liquid is well mixed. When N_{Pe} is large, bulk flow predominates and the liquid approaches plug flow. Measurements of D_E in bubble-cap and sieve-plate columns [18–21] range from 0.02 to 0.20 ft^2/s. Values of u/D_E range from 3 to 15 ft^{-1}. Based on the second form of (6-36), N_{Pe} increases directly with increasing Z_L and, therefore, column diameter. Typically, N_{Pe} is 10 for a 2-ft-diameter column and 30 for a 6-ft-diameter column. For N_{Pe} values of this magnitude, Figure 6.18 shows that values of E_{MV} are larger than E_{OV} for large values of λ.

Lewis [16] showed that for straight, but not necessarily parallel, equilibrium and operating lines, overall stage efficiency is related to the Murphree vapor efficiency by

$$E_o = \frac{\log[1 + E_{MV}(\lambda - 1)]}{\log \lambda} \qquad (6\text{-}37)$$

When the two lines are straight and parallel, $\lambda = 1$, and (6-37) becomes $E_o = E_{MV}$. Also, when $E_{MV} = 1$, then $E_o = 1$ regardless of the value of λ.

§6.5.5 Scale-up of Data with the Oldershaw Column

When vapor–liquid equilibrium data are unavailable, particularly if the system is highly nonideal with possible formation of azeotropes, tray requirements and feasibility of achieving the desired separation must be verified by conducting laboratory tests. A particularly useful apparatus is a small glass or metal sieve-plate column with center-to-side downcomers developed by Oldershaw [22] and shown in Figure 6.19. Oldershaw columns are typically 1 to 2 inches in diameter and can be assembled with almost any number of sieve plates,

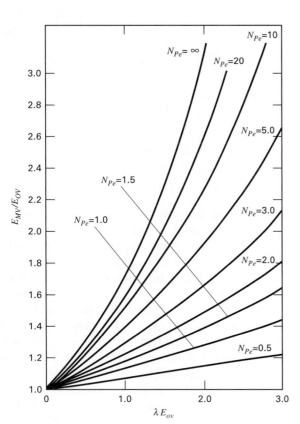

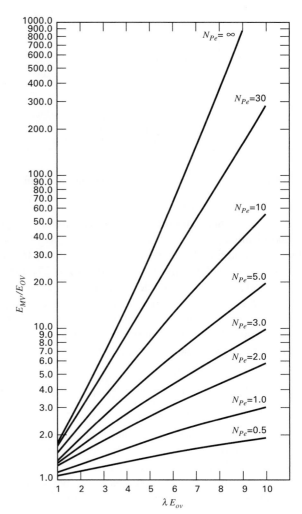

Figure 6.18 (a) Effect of longitudinal mixing on Murphree vapor-tray efficiency. (b) Expanded range for effect of longitudinal mixing on Murphree vapor-tray efficiency.

usually containing 0.035- to 0.043-inch holes with a hole area of approximately 10%. A detailed study by Fair, Null, and Bolles [23] showed that overall plate (stage) efficiencies of Oldershaw columns operated over a pressure range of 3 to 165 psia are in conservative agreement with distillation data obtained from sieve-tray, pilot-plant, and industrial-size columns ranging in size from 1.5 to 4 ft in diameter, when operated in the range of 40% to 90% of flooding (described in

§6.6). It may be assumed that similar agreement might be realized for absorption and stripping.

The small-diameter Oldershaw column achieves essentially complete mixing of liquid on each tray, permitting the measurement of a point efficiency from (6-30). Somewhat larger efficiencies may be observed in much-larger-diameter columns due to incomplete liquid mixing, resulting in a higher Murphree tray efficiency, E_{MV}, and, therefore, higher overall plate efficiency, E_o.

Fair et al. [23] recommend the following scale-up procedure using data from the Oldershaw column: (1) Determine the flooding point, as described in §6.6. (2) Establish operation at about 60% of flooding. (3) Run the system to find a combination of plates and flow rates that gives the desired degree of separation. (4) Assume that the commercial column will require the same number of plates for the same ratio of L to V.

If reliable vapor–liquid equilibrium data are available, they can be used with the Oldershaw data to determine overall column efficiency, E_o. Then (6-37) and (6-34) can be used to estimate the average point efficiency. For commercial-size columns, the Murphree vapor efficiency can be determined from the Oldershaw column point efficiency using (6-34),

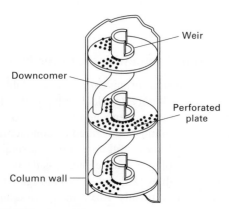

Figure 6.19 Oldershaw column.

which corrects for incomplete liquid mixing. In general, tray efficiency of commercial columns are higher than that for the Oldershaw column at the same percentage of flooding.

EXAMPLE 6.4 Murphree Efficiencies.

Assume that the absorber column diameter of Example 6.1 is 3 ft. If the overall stage efficiency, E_o, is 30% for the absorption of ethyl alcohol, estimate the average Murphree vapor efficiency, E_{MV}, and the possible range of the Murphree vapor-point efficiency, E_{OV}.

Solution

For Example 6.1, the system is dilute in ethyl alcohol, the main component undergoing mass transfer. Therefore, the equilibrium and operating lines are essentially straight, and (6-37) can be applied. From the data of Example 6.1, $\lambda = KV/L = 0.57(180)/151.5 = 0.68$. Solving (6-37) for E_{MV}, using $E_o = 0.30$,

$$E_{MV} = (\lambda^{E_o} - 1)/(\lambda - 1) = (0.68^{0.30} - 1)/(0.68 - 1) = 0.34$$

For a 3-ft-diameter column, the degree of liquid mixing is probably intermediate between complete mixing and plug flow. From (6-31) for the former case, $E_{OV} = E_{MV} = 0.34$. From a rearrangement of (6-36) for the latter case, $E_{OV} = \ln(1 + \lambda E_{MV})/\lambda = \ln[1 + 0.68(0.34)]/0.68 = 0.31$. Therefore, E_{OV} lies in the range of 31% to 34%, a small difference. The differences between E_o, E_{MV}, and E_{OV} for this example are quite small.

---

§6.5.6 Column Height

Based on estimates of N_a and tray spacing, the height of a column between the top tray and the bottom tray is computed. By adding another 4 ft above the top tray for removal of entrained liquid and 10 ft below the bottom tray for bottoms surge capacity, the total column height is estimated. If the height is greater than 250 ft, two or more columns arranged in series may be preferable to a single column. However, perhaps the tallest column in the world, located at the Shell Chemical Company complex in Deer Park, Texas, stands 338 ft tall [*Chem. Eng.*, **84** (26), 84 (1977)].

§6.6 FLOODING, COLUMN DIAMETER, PRESSURE DROP, AND MASS TRANSFER FOR TRAYED COLUMNS

Figure 6.20 shows a column where vapor flows upward, contacting liquid in crossflow. When trays are designed properly: (1) Vapor flows only up through the open regions of the tray between the downcomers. (2) Liquid flows from tray to tray only through downcomers. (3) Liquid neither weeps through the tray perforations nor is carried by the vapor as entrainment to the tray above. (4) Vapor is neither carried (occluded) down by the liquid in the downcomer nor allowed to bubble up through the liquid in the downcomer. Tray design determines tray diameter and division of tray cross-sectional area, A, as shown in Figure 6.20, into active vapor bubbling area, A_a, and liquid downcomer area, A_d. When the tray diameter is fixed, vapor pressure drop and mass-transfer coefficients can be estimated.

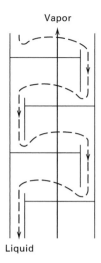

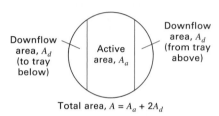

Figure 6.20 Vapor and liquid flow through a trayed tower.

§6.6.1 Flooding and Tray Diameter

At a given liquid flow rate, as shown in Figure 6.21 for a sieve-tray column, a maximum vapor flow rate exists beyond which column flooding occurs because of backup of liquid in the downcomer. Two types of flooding occur: (1) *entrainment flooding* or (2) *downcomer flooding*. For either type, if sustained, liquid is carried out with the vapor leaving the column.

Downcomer flooding takes place when, in the absence of entrainment, liquid backup is caused by downcomers of inadequate cross-sectional area, A_d, to carry the liquid flow. It rarely occurs if downcomer cross-sectional area is at least 10% of total column cross-sectional area and if tray spacing is at least 24 inches. The usual design limit is entrainment flooding, which is caused by excessive carry-up of liquid, at the molar

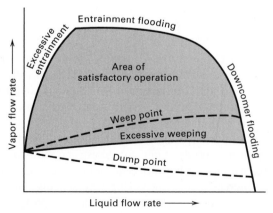

Figure 6.21 Limits of stable operation in a trayed tower.

[Reproduced by permission from H.Z. Kister, *Distillation Design*, McGraw-Hill, New York (1992).]

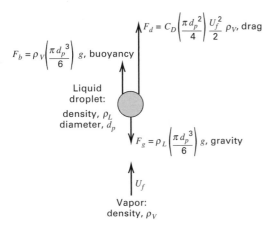

Figure 6.22 Forces acting on a suspended liquid droplet.

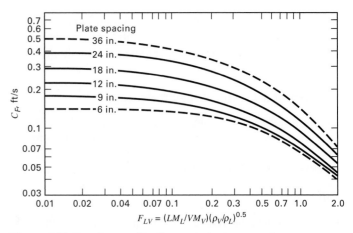

Figure 6.23 Entrainment flooding capacity in a trayed tower.

rate e, by vapor entrainment to the tray above. At incipient flooding, $(e + L) \gg L$ and downcomer cross-sectional area is inadequate for the excessive liquid load $(e + L)$.

To predict entrainment flooding, tray designers assume that entrainment of liquid is due to carry-up of suspended droplets by rising vapor or to throw-up of liquid by vapor jets at tray perforations, valves, or bubble-cap slots. Souders and Brown [24] correlated entrainment flooding data by assuming that carry-up of droplets controls entrainment. At low vapor velocity, a droplet settles out; at high vapor velocity, it is entrained. At incipient entrainment velocity, U_f, the droplet is suspended such that the vector sum of the gravitational, buoyant, and drag forces, as shown in Figure 6.22, is zero. Thus,

$$\sum F = 0 = F_g - F_b - F_d \tag{6-38}$$

In terms of droplet diameter, d_p, terms on the RHS of (6-38) become, respectively,

$$\rho_L \left(\frac{\pi d_p^3}{6}\right) g - \rho_v \left(\frac{\pi d_p^3}{6}\right) g - C_D \left(\frac{\pi d_p^2}{4}\right) \frac{U_f^2}{2} \rho_V = 0 \tag{6-39}$$

where C_D is the drag coefficient. Solving for flooding velocity,

$$U_f = C \left(\frac{\rho_L - \rho_V}{\rho_v}\right)^{1/2} \tag{6-40}$$

where C = capacity parameter of Souders and Brown. According to the above theory,

$$C = \left(\frac{4 d_p g}{3 C_D}\right)^{1/2} \tag{6-41}$$

Parameter C can be calculated from (6-41) if the droplet diameter d_p is known. In practice, d_p is combined with C and determined using experimental data. Souders and Brown obtained a correlation for C from commercial-size columns. Data covered column pressures from 10 mmHg to 465 psia, plate spacings from 12 to 30 inches, and liquid surface tensions from 9 to 60 dyne/cm. In accordance with (6-41), C increases with increasing surface tension, which increases d_p. Also, C increases with increasing tray spacing, since this allows more time for agglomeration of droplets to a larger d_p.

Fair [25] produced the general correlation of Figure 6.23, which is applicable to commercial columns with bubble-cap and sieve trays. Fair utilizes a net vapor flow area equal to the total inside column cross-sectional area minus the area blocked

off by the downcomer, $(A - A_d$ in Figure 6.20). The value of C_F in Figure 6.23 depends on tray spacing and the abscissa ratio $F_{LV} = (LM_L/VM_V)(\rho_V/\rho_L)^{0.5}$, which is a kinetic-energy ratio first used by Sherwood, Shipley, and Holloway [26] to correlate packed-column flooding data. The value of C in (6-41) is obtained from Figure 6.23 by correcting C_F for surface tension, foaming tendency, and ratio of vapor hole area A_h to tray active area A_a, according to the empirical relationship

$$C = F_{ST} F_F F_{HA} C_F \tag{6-42}$$

where

F_{ST} = surface-tension factor = $(\sigma/20)^{0.2}$

F_F = foaming factor

F_{HA} = 1.0 for $A_h/A_a \geq 0.10$ and $5(A_h/A_a) + 0.5$
  for $0.06 \leq A_h/A_a \leq 0.1$

σ = liquid surface tension, dyne/cm

For nonfoaming systems, $F_F = 1.0$; for many absorbers it is 0.75 or less. A_h is the area open to vapor as it penetrates the liquid on a tray. It is total cap slot area for bubble-cap trays and perforated area for sieve trays. Figure 6.23 is conservative for valve trays. This is shown in Figure 6.24, where entrainment-flooding data of Fractionation Research, Inc. (FRI) [27, 28] for a 4-ft diameter column equipped with Glitsch type A-1 and V-1 valve trays on 24-inch spacing are compared to the correlation in Figure 6.23. For valve trays, the slot area A_h is taken as the full valve opening through which vapor enters the tray at a 90° angle with the axis of the column.

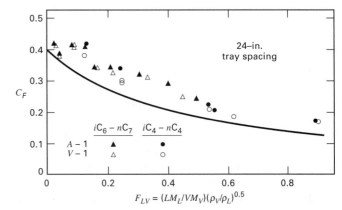

Figure 6.24 Comparison of flooding correlation with data for valve trays.

Column (tower) diameter D_T is based on a fraction, f, of flooding velocity U_f, calculated from (6-40) using C from (6-42) based on C_F from Figure 6.23. By the *continuity equation* (flow rate = velocity × flow area × density), the molar vapor flow rate is related to the flooding velocity by

$$V = (fU_f)(A - A_d)\frac{\rho_V}{M_V} \qquad (6\text{-}43)$$

where A = total column cross-sectional area = $\pi D_T^2/A$. Thus,

$$D_T = \left[\frac{4V M_V}{fU_f\pi(1 - A_d/A)\rho_V}\right]^{0.5} \qquad (6\text{-}44)$$

Typically, the fraction of flooding, f, is taken as 0.80.

Oliver [29] suggests that A_d/A be estimated from F_{LV} in Figure 6.23 by

$$\frac{A_d}{A} = \begin{cases} 0.1, & F_{LV} \leq 0.1 \\ 0.1 + \dfrac{F_{LV} - 0.1}{9}, & 0.1 \leq F_{LV} \leq 1.0 \\ 0.2, & F_{LV} \geq 1.0 \end{cases}$$

Column diameter is calculated at both the top and bottom, with the larger of the two used for the entire column unless the two diameters differ appreciably, in which case the column is swedged. Because of the need for internal access to columns with trays, a packed column, discussed later, is generally used if the diameter from (6-44) is less than 2 ft.

Tray spacing must be selected to use Figure 6.23. As spacing is increased, column height increases, but column diameter is reduced because higher velocities are tolerated. A 24-inch spacing gives ease of maintenance and is common; a smaller spacing is desirable for small-diameter columns with a large number of stages; and larger spacing is used for large columns with few stages.

As shown in Figure 6.21, a minimum vapor rate exists, below which liquid weeps through tray perforations, openings, or risers instead of flowing completely across the active area into the downcomer. Below this minimum, liquid–vapor contact is reduced, causing tray efficiency to decline. The ratio of vapor rate at flooding to the minimum vapor rate is the *turndown ratio*, which is approximately 5 for bubble-cap trays, 4 for valve trays, but only 2 for sieve trays. Thus, valve trays are preferable to sieve trays for operating flexibility.

When vapor and liquid flow rates change from tray to tray, column diameter, tray spacing, or hole area can be varied to reduce column cost and ensure stable operation at high efficiency. Variation of tray spacing is common for sieve trays because of their low turndown ratio.

§6.6.2 High-Capacity Trays

Since the 1990s, high-capacity trays have been retrofitted and newly installed in many industrial columns. By changes to conventional trays, capacity increases of more than 20% of that predicted by Figure 6.23 have been achieved with both perforated trays and valve trays. These changes, which are discussed in detail by Sloley [71], include (1) sloping or stepping the downcomer to make the downcomer area smaller at the bottom than at the top to increase the active flow area; (2) vapor flow through the tray section beneath the downcomer, in addition to the normal flow area; (3) use of staggered, louvered downcomer floor plates to impart horizontal flow to liquid exiting the downcomer, thus enhancing the vapor flow beneath; (4) elimination of vapor impingement from adjacent valves by using bidirectional fixed valves; (5) use of multiple-downcomer trays that provide very long outlet weirs leading to low crest heights and lower froth heights, the downcomers then terminating in the active vapor space of the tray below; and (6) directional slotting of sieve trays to impart a horizontal component to the vapor, enhance plug flow of liquid across the tray, and eliminate dead areas.

Regardless of the tray design, as shown by Stupin and Kister [72], an ultimate capacity, independent of tray spacing, exists for a countercurrent-flow contactor in which vapor velocity exceeds the settling velocity of droplets. Their formula, based on FRI data, uses the following form of (6-40):

$$U_{S,\text{ult}} = C_{S,\text{ult}}\left(\frac{\rho_L - \rho_V}{\rho_V}\right)^{1/2} \qquad (6\text{-}45)$$

where $U_{S,\text{ult}}$ is the superficial vapor velocity in m/s based on the column cross-sectional area. The ultimate capacity parameter, $C_{S,\text{ult}}$ in m/s, is independent of the superficial liquid velocity, L_S in m/s, below a critical value; but above that value, it decreases with increasing L_S. It is given by the smaller of C_1 and C_2, both in m/s, where

$$C_1 = 0.445(1 - F)\left(\frac{\sigma}{\rho_L - \rho_V}\right)^{0.25} - 1.4L_S \qquad (6\text{-}46)$$

$$C_2 = 0.356(1 - F)\left(\frac{\sigma}{\rho_L - \rho_V}\right)^{0.25} \qquad (6\text{-}47)$$

where

$$F = \frac{1}{\left[1 + 1.4\left(\dfrac{\rho_L - \rho_V}{\rho_V}\right)^{1/2}\right]} \qquad (6\text{-}48)$$

and ρ is in kg/m^3, and σ is the surface tension in dynes/cm.

EXAMPLE 6.5 Ultimate Superficial Vapor Velocity.

For the absorber of Example 6.1: (a) Estimate the tray diameter assuming a tray spacing of 24 inches, a foaming factor of $F_F = 0.90$, a fraction flooding of $f = 0.80$, and a surface tension of $\sigma = 70$ dynes/cm. (b) Estimate the ultimate superficial vapor velocity.

Solution

Because tower conditions are almost the same at the top and bottom, column diameter is determined only at the bottom, where gas rate is highest. From Example 6.1,

$$T = 30°C, \quad P = 110\,\text{kPa}$$
$$V = 180\,\text{kmol/h}, \quad L = 151.5 + 3.5 = 155.0\,\text{kmol/h}$$
$$M_V = 0.98(44) + 0.02(46) = 44.0$$
$$M_L = \frac{151.5(18) + 3.5(46)}{155} = 18.6$$
$$\rho_V = \frac{PM}{RT} = \frac{(110)(44)}{(8.314)(303)} = 1.92\,\text{kg/m}^3,$$
$$\rho_L = (0.986)(1,000) = 986\,\text{kg/m}^3$$
$$F_{LV} = \frac{(155)(18.6)}{(180)(44.0)}\left(\frac{1.92}{986}\right)^{0.5} = 0.016$$

(a) Column Diameter

For tray spacing = 24 inches from Figure 6.23, $C_F = 0.39$ ft/s,

$$F_{ST} = \left(\frac{\sigma}{20}\right)^{0.2} = \left(\frac{70}{20}\right)^{0.2} = 1.285, \quad F_F = 0.90$$

Because $F_{LV} < 0.1, A_d/A = 0.1$ and $F_{HA} = 1.0$. From (6-42),

$$C = 1.285(0.90)(1.0)(0.39) = 0.45 \text{ ft/s}$$

From (6-40), $\quad U_f = 0.45\left(\frac{986 - 1.92}{1.92}\right)^{0.5} = 10.2$ ft/s

From (6-44), using SI units and time in seconds, column diameter is

$$D_T = \left[\frac{4(180/3,600)(44.0)}{(0.80)(10.2/3.28)(3.14)(1 - 0.1)(1.92)}\right]^{0.5}$$

$$= 0.81 \text{ m} = 2.65 \text{ ft}$$

(b) Ultimate Superficial Vapor Velocity

From (6-48),

$$F = \frac{1}{\left[1 + 1.4\left(\frac{986 - 1.92}{1.92}\right)^{1/2}\right]} = 0.0306$$

From (6-47),

$$C_2 = 0.356(1 - 0.0306)\left(\frac{70}{986 - 1.92}\right)^{0.25}$$

$$= 0.178 \text{ m/s} = 0.584 \text{ ft/s}$$

If C_2 is the smaller value of C_1 and C_2, then from (6-45),

$$U_{S,\text{ult}} = 0.178\left(\frac{986 - 1.92}{1.92}\right)^{1/2} = 4.03 \text{ m/s} = 13.22 \text{ ft/s}$$

To apply (6-46) to compute C_1, the value of L_S is required. This is related to the value of the superficial vapor velocity, U_S.

$$L_S = U_S\frac{\rho_V LM_L}{\rho_L VM_V} = U_S\frac{(1.92)(155)(18.6)}{(986)(180)(44.0)} = 0.000709 \, U_S$$

With this expression for L_S, (6-46) becomes

$$C_1 = 0.445(1 - 0.0306)\left(\frac{70}{986 - 1.92}\right)^{0.25} - 1.4(0.000709)\, U_S$$

$$= 0.223 - 0.000993 \, U_S, \text{ m/s}$$

If C_1 is the smaller, then, using (6-45),

$$U_{S,\text{ult}} = \left(0.223 - 0.000993 \, U_{S,\text{ult}}\right)\left(\frac{986 - 1.92}{1.92}\right)^{1/2}$$

$$= 5.05 - 0.0225 \, U_{S,\text{ult}}$$

Solving, $U_{S,\text{ult}} = 4.94$ m/s, which gives $C_1 = 0.223 - 0.000993(4.94) = 0.218$ m/s. Thus, C_2 is the smaller value and $U_{S,\text{ult}} = 4.03$ m/s = 13.22 ft/s. This ultimate velocity is 30% higher than the flooding velocity computed in part (a).

§6.6.3 Tray Pressure Drop

As vapor passes up through a column, its pressure decreases. Vapor pressure drop in a tower is from 0.05 to 0.15 psi/tray (0.35 to 1.03 kPa/tray). Referring to Figure 6.3, total pressure

drop (head loss), h_t, for a sieve tray is due to: (1) friction for vapor flow through dry tray perforations, h_d; (2) holdup of equivalent clear liquid on the tray, h_l; and (3) a loss due to surface tension, h_σ:

$$h_t = h_d + h_l + h_\sigma \tag{6-49}$$

where the total and the contributions are commonly expressed in inches of liquid.

Dry sieve-tray pressure drop in inches is given by a modified orifice equation, applied to the tray holes,

$$h_d = 0.186\left(\frac{u_o^2}{C_o^2}\right)\left(\frac{\rho_V}{\rho_L}\right) \tag{6-50}$$

where u_o = hole velocity (ft/s), and C_o depends on the percent hole area and the ratio of tray thickness to hole diameter. For a 0.078-in.-thick tray with $\frac{3}{16}$-inch-diameter holes and a percent hole area (based on the tower cross-sectional area) of 10%, $C_o = 0.73$. Generally, C_o lies between 0.65 and 0.85.

Equivalent height of clear liquid holdup on a tray in inches depends on weir height, liquid and vapor densities and flow rates, and downcomer weir length, as given by an empirical expression developed from experimental data by Bennett, Agrawal, and Cook [30]:

$$h_l = \phi_e\left[h_w + C_l\left(\frac{q_L}{L_w\phi_e}\right)^{2/3}\right] \tag{6-51}$$

where

h_w = weir height, inches

ϕ_e = effective relative froth density (height of clear liquid/froth height) $\tag{6-52}$
   $= \exp(-4.257 K_s^{0.91})$

K_s = capacity parameter, ft/s $= U_a\left(\frac{\rho_V}{\rho_L - \rho_V}\right)^{1/2}$

$\tag{6-53}$

U_a = superficial vapor velocity based on active bubbling area
$A_a = (A - 2A_d)$, of the tray, ft/s
L_w = weir length, inches
q_L = liquid flow rate across tray, gal/min

$$C_l = 0.362 + 0.317\exp(-3.5h_w) \tag{6-54}$$

The second term in (6-51) is related to the Francis weir equation, taking into account the liquid froth flow over the weir. For $A_d/A = 0.1$, $L_w = 73\%$ of the tower diameter.

Pressure drop due to surface tension in inches, which bubbles must overcome, is given by the difference between the pressure inside the bubble and that of the liquid, according to the theoretical relation

$$h_\sigma = \frac{6\sigma}{g\rho_L D_{B(\text{max})}} \tag{6-55}$$

where, except for tray perforations smaller than $\frac{3}{16}$-inch, $D_{B(\text{max})}$, the maximum bubble size, is taken as the perforation diameter, D_H. Consistent dimensions must be used in (6-55), as shown in the example below.

Methods for estimating vapor pressure drop for bubble-cap and valve trays are given by Smith [31] and Klein

[32], respectively, and are discussed by Kister [33] and Lockett [34].

EXAMPLE 6.6 Sieve-Tray Pressure Drop.

Estimate tray pressure drop for the absorber of Example 6.1, assuming sieve trays with a tray diameter of 1 m, a weir height of 2 inches, and a $\frac{3}{16}$-inch hole diameter.

Solution

From Example 6.5, $\rho_V = 1.92$ kg/m^2 and $\rho_L = 986$ kg/m^3. At tower bottom, vapor velocity based on total cross-sectional area is

$$\frac{(180/3,600)(44)}{(1.92)\left[3.14(1)^2/4\right]} = 1.46 \text{ m/s}$$

For a 10% hole area, based on the total tower cross-sectional area,

$$u_o = \frac{1.46}{0.10} = 14.6 \text{ m/s} \quad \text{or} \quad 47.9 \text{ ft/s}$$

Using the above densities, dry tray pressure drop, (6-50), is

$$h_d = 0.186\left(\frac{47.9^2}{0.73^2}\right)\left(\frac{1.92}{986}\right) = 1.56 \text{ in. of liquid}$$

Weir length is 73% of tower diameter, with $A_d/A = 0.10$. Then

$$L_w = 0.73(1) = 0.73 \text{ m or } 28.7 \text{ inches}$$

$$\text{Liquid flow rate in gpm} = \frac{(155/60)(18.6)}{986(0.003785)} = 12.9 \text{ gpm}$$

with $\quad A_d/A = 0.1, \quad A_a/A = (A - 2A_d)/A = 0.8$

Therefore, $\quad U_a = 1.46/0.8 = 1.83$ m/s $= 5.99$ ft/s

From (6-53), $\quad K_s = 5.99[1.92/(986 - 1.92)]^{0.5} = 0.265$ ft/s

From (6-52), $\quad \phi_e = \exp\left[-4.257(0.265)^{0.91}\right] = 0.28$

From (6-54), $\quad C_l = 0.362 + 0.317 \exp[-3.5(2)] = 0.362$

From (6-51), $\quad h_l = 0.28\left[2 + 0.362(12.9/28.7/0.28)^{2/3}\right]$

$$= 0.28(2 + 0.50) = 0.70 \text{ in.}$$

From (6-55), using $D_B(\text{max}) = D_H = \frac{3}{16}$ inch $= 0.00476$ m,

$$\sigma = 70 \text{ dynes/cm} = 0.07 \text{ N/m} = 0.07 \text{ kg/s}^2$$

$$g = 9.8 \text{ m/s}^2, \text{ and } \rho_L = 986 \text{ kg/m}^3$$

$$h_\sigma = \frac{6(0.07)}{9.8(986)(0.00476)} = 0.00913 \text{ m} = 0.36 \text{ in.}$$

From (6-45), total tray head loss is $h_t = 1.56 + 0.70 + 0.36$ $= 2.62$ inches.
For $\rho_L = 986$ kg/m$^3 = 0.0356$ lb/in^3,
tray vapor pressure drop $= h_t \rho_L = 2.62(0.0356) = 0.093$ psi/tray.

§6.6.4 Mass-Transfer Coefficients and Transfer Units for Trayed Columns

After tray specifications are established, the Murphree vapor-point efficiency, (6-30), can be estimated using empirical correlations for mass-transfer coefficients, based on experimental data. For a vertical path of vapor flow up through the froth from a point on the bubbling area of the tray, (6-29) applies. For the Murphree vapor-point efficiency:

$$N_{OG} = -\ln(1 - E_{OV}) \tag{6-56}$$

where

$$N_{OG} = \frac{K_G a P Z_f}{(V/A_b)} \tag{6-57}$$

The overall, volumetric mass-transfer coefficient, $K_G a$, is related to the individual volumetric mass-transfer coefficients by the mass-transfer resistances, which from §3.7 are

$$\frac{1}{K_G a} = \frac{1}{k_G a} + \frac{(K P M_L/\rho_L)}{k_L a} \tag{6-58}$$

where the RHS terms are the gas- and liquid-phase resistances, respectively, and the symbols k_p for the gas and k_c for the liquid used in Chapter 3 have been replaced by k_G and k_L. In terms of individual transfer units, defined by

$$N_G = \frac{k_G a P Z_f}{(V/A_b)} \tag{6-59}$$

and

$$N_L = \frac{k_L a \rho_L Z_f}{M_L(L/A_b)} \tag{6-60}$$

(6-57) and (6-58) give,

$$\frac{1}{N_{OG}} = \frac{1}{N_G} + \frac{(KV/L)}{N_L} \tag{6-61}$$

Important mass-transfer correlations were published by the AIChE [35] for bubble-cap trays, Chan and Fair [36, 37] for sieve trays, and Scheffe and Weiland [38] for one type of valve tray (Glitsch V-1). These correlations were developed in terms of N_L, N_G, k_L, k_G, a, and N_{Sh} for either the gas or the liquid phase. The correlations given in this section are for sieve trays from Chan and Fair [36], who used a correlation for the liquid phase based on the work of Foss and Gerster [39], as reported by the AIChE [40]. Chan and Fair developed a separate correlation for the vapor phase from an experimental data bank of 143 data points for towers of 2.5 to 4.0 ft in diameter, operating at pressures from 100 mmHg to 400 psia.

Experimental data for sieve trays have validated the direct dependence of mass transfer on the interfacial area between the gas and liquid phases, and on the residence times in the froth of the gas and liquid phases. Accordingly, Chan and Fair use modifications of (6-59) and (6-60):

$$N_G = k_G \bar{a} t_G \tag{6-62}$$

$$N_L = k_L \bar{a} t_L \tag{6-63}$$

where $\bar{a}$ is the interfacial area per unit volume of equivalent clear liquid, $\bar{t}_G$ is the average gas residence time in the froth, and $\bar{t}_L$ is the average liquid residence time in the froth.

Average residence times are estimated from the following equations, using (6-51) for the equivalent head of clear liquid on the tray and (6-56) for the effective relative density of the froth:

$$\bar{t}_L = \frac{h_l A_a}{q_L} \tag{6-64}$$

$$\bar{t}_G = \frac{(1 - \phi_e) h_l}{\phi_e U_a} \tag{6-65}$$

where $(1 - \phi_e) h_l / \phi_e$ is the equivalent height of vapor holdup in the froth, with residence time in seconds.

Empirical expressions of Chan and Fair for $k_G \bar{a}$ and $k_L \bar{a}$ in units of s^{-1} are

$$k_G \bar{a} = \frac{1,030 D_V^{0.5} \left(f - 0.842 f^2 \right)}{(h_l)^{0.5}} \tag{6-66}$$

$$k_L \bar{a} = 78.8 D_L^{0.5} (F + 0.425) \tag{6-67}$$

where the variables and their units are: D_V, D_L = diffusion coefficients, cm²/s; h_l = clear liquid height, cm; $f = U_a / U_f$, fractional approach to flooding; and $F = F\text{-factor} = U_a \rho_V^{0.5}$, (kg/m)$^{0.5}$/s.

From (6-66), an important factor influencing $k_G \bar{a}$ is the fractional approach to flooding. This effect is shown in Figure 6.25, where (6-66) is compared to experimental data. At gas rates where the fractional approach to flooding is greater than 0.60, the mass-transfer factor decreases with increasing values of f. This may be due to entrainment, which is discussed in the next subsection. On an entrainment-free basis, the curve in Figure 6.25 should remain at its peak above $f = 0.60$.

From (6-67), the F-factor is important for liquid-phase mass transfer. Supporting data are shown in Figure 6.26, where $k_L \bar{a}$ depends strongly on F, but is almost independent of liquid flow rate and weir height. The Murphree vapor-point efficiency model of (6-56), (6-61), and (6-64) to (6-67) correlates the 143 points of the Chan and Fair [36] data bank with an average absolute deviation of 6.27%. Lockett [34] pointed out that (6-67) implies that $k_L \bar{a}$ depends on tray spacing. However, the data bank included data for spacings from 6 to 24 inches.

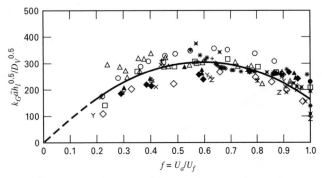

Figure 6.25 Comparison of experimental data to the correlation of Chan and Fair for gas-phase mass transfer.

[From H. Chan and J.R. Fair, *Ind. Eng. Chem. Process Des. Dev.*, **23**, 817 (1984) with permission.]

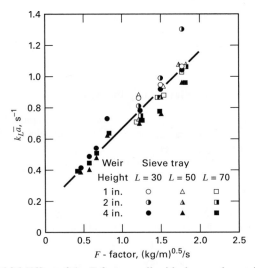

Figure 6.26 Effect of the F-factor on liquid-phase volumetric mass-transfer coefficient for oxygen desorption from water with air at 1 atm and 25°C, where L = gal/(min)/(ft of average flow width).

EXAMPLE 6.7 E_{MV} from Chan and Fair Correlation.

Estimate the Murphree vapor-point efficiency, E_{OV}, for the absorber of Example 6.1 using results from Examples 6.5 and 6.6 for the tray of Example 6.6. In addition, determine the controlling resistance to mass transfer.

Solution

Pertinent data for the two phases are as follows.

	Gas	Liquid
Molar flow rate, kmol/h	180.0	155.0
Molecular weight	44.0	18.6
Density, kg/m³	1.92	986
Ethanol diffusivity, cm²/s	7.86×10^{-2}	1.81×10^{-5}

Pertinent tray dimensions from Example 6.6 are $D_T = 1$ m; $A = 0.785$ m²; $A_a = 0.80 A = 0.628$ m² = 6,280 cm²; $L_w = 28.7$ inches = 0.73 m.

From Example 6.6,

$$\phi_e = 0.28; \quad h_l = 0.70 \text{ in.} = 1.78 \text{ cm};$$
$$U_a = 5.99 \text{ ft/s} = 183 \text{ cm/s} = 1.83 \text{ m/s}$$

From Example 6.5,

$$U_f = 10.2 \text{ ft/s}; \quad f = U_a / U_f = 5.99/10.2 = 0.59$$

$$F = 1.83(1.92)^{0.5} = 2.54 \text{ (kg/m)}^{0.5}/\text{s}$$

$$q_L = \frac{(155.0)(18.6)}{986} \left(\frac{10^6}{3,600} \right) = 812 \text{ cm}^3/\text{s}$$

From (6-64), $\bar{t}_L = 1.78(6,280)/812 = 13.8$ s

From (6-65),

$$\bar{t}_G = (1 - 0.28)(1.78)/[(0.28)(183)] = 0.025 \text{ s}$$

From (6-67),

$$k_L\bar{a} = 78.8(1.81 \times 10^{-5})^{0.5}(2.54 + 0.425) = 0.99 \text{ s}^{-1}$$

From (6-66),

$$k_G\bar{a} = 1.030(7.86 \times 10^{-2})^{0.5}\left[0.59 - 0.842(0.59)^2\right]/(1.78)^{0.5}$$
$$= 64.3 \text{ s}^{-1}$$

From (6-63), $\quad N_L = (0.99)(13.8) = 13.7$

From (6-62), $\quad N_G = (64.3)(0.025) = 1.61$

From Example 6.1, $K = 0.57$. Therefore, $KV/L = (0.57)(180)/155 = 0.662$.

From (6-61),

$$N_{OG} = \frac{1}{(1/1.61) + (0.662/13.7)} = \frac{1}{0.621 + 0.048} = 1.49$$

Thus, mass transfer of ethanol is seen to be controlled by the vapor-phase resistance, which is $0.621/0.048 = 13$ times the liquid-phase resistance. From (6-56), solving for E_{OV},

$$E_{OV} = 1 - \exp(-N_{OG}) = 1 - \exp(-1.49) = 0.77 = 77\%$$

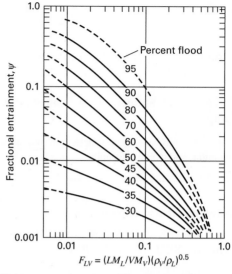

Figure 6.27 Correlation of Fair for fractional entrainment for sieve trays.

[Reproduced by permission from B.D. Smith, *Design of Equilibrium Stage Processes*, McGraw-Hill, New York (1963).]

§6.6.5 Weeping, Entrainment, and Downcomer Backup

For high tray efficiency: (1) weeping of liquid through the tray must be small compared to flow into the downcomer, (2) entrainment of liquid by the gas must not be excessive, and (3) froth height in the downcomer must not approach tray spacing. The tray must operate in the stable region of Figure 6.21. Weeping occurs at the lower limit of gas velocity, while entrainment flooding occurs at the upper limit.

Weeping occurs at low vapor velocities and/or high liquid rates when the clear liquid height on the tray exceeds the sum of dry tray pressure drop and the surface tension effect. Thus, to prevent weeping, it is necessary that

$$h_d + h_\sigma > h_l \tag{6-68}$$

everywhere on the tray active area. If weeping occurs uniformly or mainly near the downcomer, a ratio of weep rate to downcomer liquid rate as high as 0.1 may not cause an unacceptable decrease in tray efficiency. Estimation of weeping rate is discussed by Kister [33].

The prediction of fractional liquid entrainment by the vapor, $\psi = e/(L + e)$, can be made by the correlation of Fair [41] shown in Figure 6.27. Entrainment becomes excessive at high values of fraction of flooding, particularly for small values of the kinetic-energy ratio, F_{LV}. The effect of entrainment on E_{MV} can be estimated by the relation derived by Colburn [42], where E_{MV} is "dry" efficiency and $E_{MV,\text{wet}}$ is the "wet" efficiency:

$$\frac{E_{MV,\text{wet}}}{E_{MV}} = \frac{1}{1 + eE_{MV}/L}$$

$$= \frac{1}{1 + E_{MV}[\psi/(1 - \psi)]} \tag{6-69}$$

Equation (6-69) assumes that $\lambda = KV/L = 1$ and that the tray liquid is well mixed. For a given value of the entrainment ratio, ψ, the larger the value of E_{MV}, the greater the effect of entrainment. For $E_{MV} = 1.0$ and $\psi = 0.10$, the "wet" efficiency is 0.90. An equation similar to (6-69) for the effect of weeping is not available, because this effect depends on the degree of liquid mixing on the tray and the distribution of weeping over the active tray area. If weeping occurs only in the vicinity of the downcomer, no decrease in the value of E_{MV} is observed.

The height of clear liquid in the downcomer, h_{dc}, is always greater than its height on the tray because, by reference to Figure 6.3, the pressure difference across the froth in the downcomer is equal to the total pressure drop across the tray from which liquid enters the downcomer, plus the height of clear liquid on the tray below to which it flows, and plus the head loss for flow under the downcomer apron. Thus, the clear liquid head in the downcomer is

$$h_{dc} = h_t + h_l + h_{da} \tag{6-70}$$

where h_t is given by (6-49) and h_l by (6-51), and the hydraulic gradient is assumed negligible. The head loss for liquid flow under the downcomer, h_{da}, in inches of liquid, can be estimated from an empirical orifice-type equation:

$$h_{da} = 0.03\left(\frac{q_L}{100\,A_{da}}\right)^2 \tag{6-71}$$

where q_L is the liquid flow in gpm and A_{da} is the area in ft^2 for liquid flow under the downcomer apron. If the height of the opening under the apron (typically 0.5 inch less than h_w) is h_a, then $A_{da} = L_w h_a$. The height of the froth in the downcomer is

$$h_{df} = h_{dc}/\phi_{df} \qquad (6\text{-}72)$$

where the froth density, ϕ_{df}, can be taken as 0.5.

EXAMPLE 6.8 Entrainment, Downcomer Backup, and Weeping.

Using Examples 6.5, 6.6, and 6.7, estimate entrainment rate, downcomer froth height, and whether weeping occurs.

Solution

Weeping criterion: From Example 6.6,

$$h_d = 1.56 \text{ in.}, \quad h_\sigma = 0.36 \text{ in.}, \quad h_t = 0.70 \text{ in.}$$

From (6-68), $1.56 + 0.36 > 0.70$

Therefore, if the liquid level is uniform, no weeping occurs.

Entrainment: From Example 6.5, $F_{LV} = 0.016$. From Example 6.7, $f = 0.59$.

From Figure 6.27, $\psi = 0.06$. Therefore, for $L = 155$ kmol/h from Example 6.7, the entrainment rate is $0.06(155) = 9.3$ kmol/h. Assuming (6-69) is reasonably accurate for $\lambda = 0.662$ from Example 6.7, and that $E_{MV} = 0.78$, the effect of ψ on E_{MV} is given by

$$\frac{E_{MV,\text{wet}}}{E_{MV}} = \frac{1}{1 + 0.78(0.06/0.94)} = 0.95$$

or

$$E_{MV} = 0.95(0.78) = 0.74$$

Downcomer backup: From Example 6.6, $h_t = 2.62$ inches; from Example 6.7, $L_w = 28.7$ inches; from Example 6.6, $h_w = 2.0$ inches. Assume that $h_a = 2.0 - 0.5 = 1.5$ inches. Then,

$$A_{da} = L_w h_a = 28.7(1.5) = 43.1 \text{ in.}^2 = 0.299 \text{ ft}^2$$

From Example 6.6, $q_L = 12.9$ gpm

From (6-71), $\quad h_{da} = 0.03 \left[\dfrac{12.9}{(100)(0.299)} \right]^2 = 0.006$ in.

From (6-70), $\quad h_{dc} = 2.62 + 0.70 + 0.006 = 3.33$ inches of clear liquid backup

From (6-72), $\quad h_{df} = \dfrac{3.33}{0.5} = 6.66$ in. of froth in the downcomer

Based on these results, neither weeping nor downcomer backup are problems. A 5% loss in efficiency occurs due to entrainment.

§6.7 RATE-BASED METHOD FOR PACKED COLUMNS

Packed columns are continuous, differential-contacting devices that do not have physically distinguishable, discrete stages. Thus, packed columns are better analyzed by mass-transfer models than by equilibrium-stage concepts. However, in practice, packed-tower performance is often presented on the basis of equivalent equilibrium stages using a packed-height equivalent to a theoretical (equilibrium) plate (stage), called the HETP or HETS and defined by the equation

$$\text{HETP} = \frac{\text{packed height}}{\text{number of equivalent equilibrium stages}} = \frac{l_T}{N_t}$$
$$(6\text{-}73)$$

The HETP concept, however, has no theoretical basis. Accordingly, although HETP values can be related to mass-transfer coefficients, such values are best obtained by back-calculation from (6-73) using experimental data. To illustrate the application of the HETP concept, consider Example 6.1, which involves recovery of bioethyl alcohol from a CO_2-rich vapor by absorption with water. The required N_t is slightly more than 6, say, 6.1. If experience shows that use of 1.5-inch metal Pall rings will give an average HETP of 2.25 ft, then the packed height from (6-73) is $l_T = (\text{HETP})N_t = 2.25(6.1) = 13.7$ ft. If metal Intalox IMTP #40 random packing has an HETP $= 2.0$ ft, then $l_T = 12.3$ ft. With Mellapak 250Y sheet-metal structured packing, the HETP might be 1.2 ft, giving $l_T = 7.3$ ft. Usually, the lower the HETP, the more expensive the packing, because of higher manufacturing cost.

It is preferable to determine packed height from theoretically based methods involving mass-transfer coefficients. Consider the countercurrent-flow packed columns of packed height l_T shown in Figure 6.28. For packed absorbers and strippers, operating-line equations analogous to those of §6.3.2 can be derived. Thus, for the absorber in Figure 6.28a, a material balance around the upper envelope, for the solute, gives

$$x_{\text{in}}L_{\text{in}} + yV_l = xL_l + y_{\text{out}}V_{\text{out}} \qquad (6\text{-}74)$$

or solving for y, assuming dilute solutions such that $V_l = V_{\text{in}} = V_{\text{out}} = V$ and $L_l = L_{\text{in}} = L_{\text{out}} = L$,

$$y = x\left(\frac{L}{V}\right) + y_{\text{out}} - x_{\text{in}}\left(\frac{L}{V}\right) \qquad (6\text{-}75)$$

Similarly for the stripper in Figure 6.28b,

$$y = x\left(\frac{L}{V}\right) + y_{\text{in}} - x_{\text{out}}\left(\frac{L}{V}\right) \qquad (6\text{-}76)$$

In Equations (6-74) to (6-76), mole fractions y and x represent *bulk compositions* of the gas and liquid in contact at any vertical location in the packing. For the case of absorption, with solute mass transfer from the gas to the liquid stream, the two-film theory of §3.7 and illustrated in Figure 6.29 applies. A concentration gradient exists in each film. At the *interface* between the two phases, physical equilibrium exists. Thus, as with trayed towers, an operating line and an

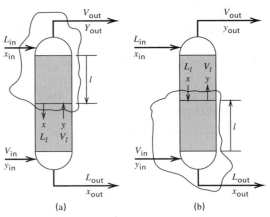

Figure 6.28 Packed columns with countercurrent flow: (a) absorber; (b) stripper.

Two-film theory of mass transfer

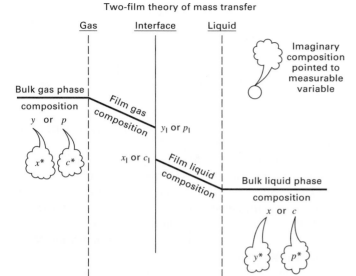

Figure 6.29 Interface properties in terms of bulk properties.

equilibrium line are of great importance for packed towers. For a given problem specification, the location of the two lines is independent of whether the tower is trayed or packed. Thus, the method for determining the minimum absorbent liquid or stripping vapor flow rates in a packed column is identical to that for trayed towers, as presented in §6.3 and illustrated in Figure 6.9.

The rate of mass transfer for absorption or stripping can be expressed in terms of mass-transfer coefficients for each phase. Coefficients, k, based on a unit area for mass transfer could be used, but the area in a packed bed is difficult to determine. Accordingly, as with mass transfer in the froth of a trayed tower, it is common to use *volumetric mass-transfer coefficients*, ka, where the quantity a represents mass transfer area per unit volume of packed bed. At steady state, in the absence of chemical reactions, the rate of solute mass transfer across the gas-phase film must equal the rate across the liquid film. If the system is dilute in solute, unimolecular diffusion (UMD) is approximated by the equations for equimolar counterdiffusion (EMD) discussed in Chapter 3. The mass-transfer rate per unit volume of packed bed, r, is written in terms of mole-fraction driving forces in each phase or in terms of a partial-pressure driving force in the gas phase and a concentration driving force in the liquid, as in Figure 6.29. Using the former, for absorption, with the subscript I to denote the phase interface:

$$r = k_y a(y - y_I) = k_x a(x_I - x) \qquad (6-77)$$

The composition at the interface depends on the ratio $k_x a / k_y a$ because (6-77) can be rearranged to

$$\frac{y - y_I}{x - x_I} = -\frac{k_x a}{k_y a} \qquad (6-78)$$

Thus, as shown in Figure 6.30, a straight line of slope $-k_x a / k_y a$, drawn from the operating line at point (y, x), intersects the equilibrium curve at (y_I, x_I).

The slope $-k_x a / k_y a$ determines the relative resistances of the two phases to mass transfer. In Figure 6.30, the distance AE is the gas-phase driving force $(y - y_I)$, while AF is the

liquid-phase driving force $(x_I - x)$. If the resistance in the gas phase is very low, y_I is approximately equal to y. Then, the resistance resides entirely in the liquid phase. This occurs in the absorption of a slightly soluble solute in the liquid phase (a solute with a high K-value) and is referred to as a liquid-film controlling process. Alternatively, if the resistance in the liquid phase is very low, x_I is nearly equal to x. This occurs in the absorption of a very soluble solute in the liquid phase (a solute with a low K-value) and is referred to as a gas-film controlling process. It is important to know which of the two resistances is controlling so that its rate of mass transfer can be increased by promoting turbulence in and/or increasing the dispersion of the controlling phase.

The composition at the interface between two phases is difficult to measure, so overall volumetric mass-transfer coefficients for either of the phases are defined in terms of overall driving forces. Using mole fractions,

$$r = K_y a(y - y^*) = K_x a(x^* - x) \qquad (6-79)$$

where, as shown in Figure 6.30 and previously discussed in §3.7, y^* is the fictitious vapor mole fraction in equilibrium with the mole fraction, x, in the bulk liquid; and x^* is the fictitious liquid mole fraction in equilibrium with the mole fraction, y, in the bulk vapor. By combining (6-77) to (6-79), overall coefficients can be expressed in terms of separate phase coefficients:

$$\frac{1}{K_y a} = \frac{1}{k_y a} + \frac{1}{k_x a}\left(\frac{y_I - y^*}{x_I - x}\right) \qquad (6-80)$$

and

$$\frac{1}{K_x a} = \frac{1}{k_x a} + \frac{1}{k_y a}\left(\frac{x^* - x_I}{y - y_I}\right) \qquad (6-81)$$

From Figure 6.30, for dilute solutions when the equilibrium curve is a nearly straight line through the origin,

$$\frac{y_I - y^*}{x_I - x} = \frac{\overline{ED}}{\overline{BE}} = K \qquad (6-82)$$

and

$$\frac{x^* - x_I}{y - y_I} = \frac{\overline{CF}}{\overline{FB}} = \frac{1}{K} \qquad (6-83)$$

where K is the K-value for the solute. Combining (6-80) with (6-82), and (6-81) with (6-83),

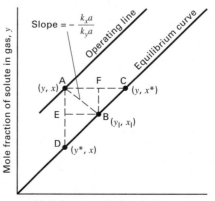

Figure 6.30 Interface composition in terms of the ratio of mass-transfer coefficients.

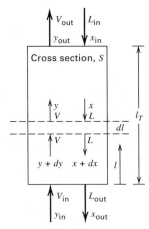

Figure 6.31 Differential contact in a countercurrent-flow, packed absorption column.

$$\frac{1}{K_y a} = \frac{1}{k_y a} + \frac{K}{k_x a} \tag{6-84}$$

and

$$\frac{1}{K_x a} = \frac{1}{k_x a} + \frac{1}{K k_y a} \tag{6-85}$$

Determination of column packed height commonly involves the overall gas-phase coefficient, $K_y a$, because most often the liquid has a strong affinity for the solute, so resistance to mass transfer is mostly in the gas phase. This is analogous to a trayed tower, where the tray-efficiency analysis is commonly based on $K_{OG} a$ or N_{OG}. In the countercurrent-flow absorption column in Figure 6.31 for a dilute system, a differential material balance for a solute being absorbed in a differential height of packing dl gives:

$$-V \, dy = K_y a (y - y^*) S \, dl \tag{6-86}$$

where S is the inside cross-sectional area of the tower. In integral form, with nearly constant terms placed outside the integral, (6-86) becomes

$$\frac{K_y a S}{V} \int_0^{l_T} dl = \frac{K_y a S l_T}{V} = \int_{y_{out}}^{y_{in}} \frac{dy}{y - y^*} \tag{6-87}$$

Solving for the packed height,

$$l_T = \frac{V}{K_y a S} \int_{y_{out}}^{y_{in}} \frac{dy}{y - y^*} \tag{6-88}$$

Chilton and Colburn [43] suggested that the RHS of (6-88) be written as the product of two terms:

$$l_T = H_{OG} N_{OG} \tag{6-89}$$

where

$$H_{OG} = \frac{V}{K_y a S} \tag{6-90}$$

and

$$N_{OG} = \int_{y_{out}}^{y_{in}} \frac{dy}{y - y^*} \tag{6-91}$$

Comparing (6-89) to (6-73), it is seen that H_{OG} is analogous to HETP, as is N_{OG} to N_t.

H_{OG} is the *overall height of a (gas) transfer unit* (HTU). Experimental data show that HTU varies less with V than does $K_y a$. The smaller the HTU, the more efficient the contacting. N_{OG} is the *overall number of (gas) transfer units* (NTU). It represents the overall change in solute mole fraction divided by the average mole-fraction driving force. The larger the NTU, the greater the time or area of contact required.

Equation (6-91) was integrated by Colburn [44], who used a linear equilibrium, $y^* = Kx$, to eliminate y^* and a linear, solute material-balance operating line, (6-75), to eliminate x, giving:

$$\int_{y_{out}}^{y_{in}} \frac{dy}{y - y^*} = \int_{y_{out}}^{y_{in}} \frac{dy}{(1 - KV/L)y + y_{out}(KV/L) - K x_{in}} \tag{6-92}$$

Letting $L/(KV) = A$, and integrating (6-88), gives

$$N_{OG} = \frac{\ln\{[(A-1)/A][(y_{in} - K x_{in})/(y_{out} - K x_{in})] + (1/A)\}}{(A-1)/A} \tag{6-93}$$

Using (6-93) and (6-90), the packed height, l_T, can be determined from (6-89). However, (6-93) is a very sensitive calculation when $A < 0.9$.

NTU (N_{OG}) and HTU (H_{OG}) are not equal to the number of equilibrium stages, N_t, and HETP, respectively, unless the operating and equilibrium lines are straight and parallel. Otherwise, NTU is greater than or less than N_t, as shown in Figure 6.32 for the case of absorption. When the operating and equilibrium lines are straight but not parallel,

$$\text{HETP} = H_{OG} \frac{\ln(1/A)}{(1-A)/A} \tag{6-94}$$

and

$$N_{OG} = N_t \frac{\ln(1/A)}{(1-A)/A} \tag{6-95}$$

Although most applications of HTU and NTU are based on (6-89) to (6-91) and (6-93), alternative groupings have been used, depending on the driving force for mass transfer

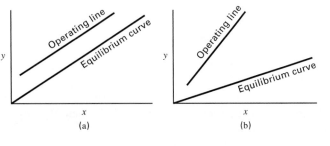

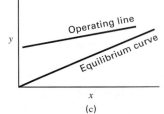

Figure 6.32 Relationship between NTU and the number of theoretical stages N_t: (a) NTU = N_t; (b) NTU > N_t; (c) NTU < N_t.

Table 6.5 Alternative Mass-Transfer Coefficient Groupings

		Height of a Transfer Unit, HTU			Number of a Transfer Unit, NTU	
Driving Force	Symbol	EM Diffusion or Dilute UM Diffusion	UM Diffusion	Symbol	EM Diffusion[a] or Dilute UM Diffusion	UM Diffusion
1. $(y - y^*)$	H_{OG}	$\dfrac{V}{K_y a S}$	$\dfrac{V}{K'_y a (1-y)_{LM} S}$	N_{OG}	$\displaystyle\int \dfrac{dy}{(y-y^*)}$	$\displaystyle\int \dfrac{(1-y)_{LM}dy}{(1-y)(y-y^*)}$
2. $(p - p^*)$	H_{OG}	$\dfrac{V}{K_G a P S}$	$\dfrac{V}{K'_G a (1-y)_{LM} P S}$	N_{OG}	$\displaystyle\int \dfrac{dp}{(p-p^*)}$	$\displaystyle\int \dfrac{(P-p)_{LM}dp}{(P-p)(p-p^*)}$
3. $(Y - Y^*)$	H_{OG}	$\dfrac{V'}{K_Y a S}$	$\dfrac{V'}{K_Y a S}$	N_{OG}	$\displaystyle\int \dfrac{dY}{(Y-Y^*)}$	$\displaystyle\int \dfrac{dY}{(Y-Y^*)}$
4. $(y - y_I)$	H_G	$\dfrac{V}{k_y a S}$	$\dfrac{V}{k'_y a (1-y)_{LM} S}$	N_G	$\displaystyle\int \dfrac{dy}{(y-y_I)}$	$\displaystyle\int \dfrac{(1-y)_{LM}dy}{(1-y)(y-y_I)}$
5. $(p - p_I)$	H_G	$\dfrac{V}{k_p a P S}$	$\dfrac{V}{k'_p a (P-p)_{LM} S}$	N_G	$\displaystyle\int \dfrac{dp}{(p-p_I)}$	$\displaystyle\int \dfrac{(P-p)_{LM}dp}{(P-p)(p-p_I)}$
6. $(x^* - x)$	H_{OL}	$\dfrac{L}{K_x a S}$	$\dfrac{L}{K'_x a (1-x)_{LM} S}$	N_{OL}	$\displaystyle\int \dfrac{dx}{(x^*-x)}$	$\displaystyle\int \dfrac{(1-x)_{LM}dx}{(1-x)(x^*-x)}$
7. $(c^* - c)$	H_{OL}	$\dfrac{L}{K_L a (\rho_L/M_L) S}$	$\dfrac{L}{K'_L a (\rho_L/M_L - c)_{LM} S}$	N_{OL}	$\displaystyle\int \dfrac{dc}{(c^*-c)}$	$\displaystyle\int \dfrac{(\rho_L/M_L - c)_{LM}dx}{(\rho_L/M_L - c)(c^*-c)}$
8. $(X^* - X)$	H_{OL}	$\dfrac{L'}{K_X a S}$	$\dfrac{L'}{K_X a S}$	N_{OL}	$\displaystyle\int \dfrac{dX}{(X^*-X)}$	$\displaystyle\int \dfrac{dX}{(X^*-X)}$
9. $(x_I - x)$	H_L	$\dfrac{L}{k_x a S}$	$\dfrac{L}{k'_x a (1-x)_{LM} S}$	N_L	$\displaystyle\int \dfrac{dx}{(x_I-x)}$	$\displaystyle\int \dfrac{(1-x)_{LM}dx}{(1-x)(x_I-x)}$
10. $(c_I - c)$	H_L	$\dfrac{L}{k_L a (\rho_L/M_L) S}$	$\dfrac{L}{k'_L a (\rho_L/M_L - c)_{LM} S}$	N_L	$\displaystyle\int \dfrac{dc}{(c_I-c)}$	$\displaystyle\int \dfrac{(\rho_L/M_L - c)_{LM}dc}{(\rho_L/M_L - c)(c_I-c)}$

[a]The substitution $K_y = K'_y y_{B_{LM}}$ or its equivalent can be made.

and whether the overall basis is the gas, or the liquid, where H_{OL} and N_{OL} apply. These groupings are summarized in Table 6.5. Included are driving forces based on partial pressures, p; mole ratios, X, Y; and concentrations, c; as well as mole fractions, x, y. Also included in Table 6.5, for later reference, are groupings for unimolecular diffusion (UMD) when solute concentration is not dilute.

It is frequently necessary to convert a mass-transfer coefficient based on one driving force to another coefficient based on a different driving force. Table 3.16 gives the relationships among the different coefficients.

EXAMPLE 6.9 Height of an Absorber.

Repeat Example 6.1 for a tower packed with 1.5-inch metal Pall rings. If $H_{OG} = 2.0$ ft, compute the required packed height.

Solution

From Example 6.1, $V = 180$ kmol/h, $L = 151.5$ kmol/h, $y_{in} = 0.020$, $x_{in} = 0.0$, and $K = 0.57$. For 97% recovery of ethyl alcohol by material balance,

$$y_{out} = \frac{(0.03)(0.02)(180)}{180 - (0.97)(0.02)(180)} = 0.000612$$

$$A = \frac{L}{KV} = \frac{151.5}{(0.57)(180)} = 1.477$$

$$\frac{y_{in}}{y_{out}} = \frac{0.020}{0.000612} = 32.68$$

From (6-93),

$$N_{OG} = \frac{\ln\{[(1.477 - 1)/1.477](32.68) + (1/1.477)\}}{(1.477 - 1)/1.477}$$

$$= 7.5 \text{ transfer units}$$

The packed height, from (6-89), is $l_T = 2.0(7.5) = 15$ ft. N_t was determined in Example 6.1 to be 6.1. The 7.5 for N_{OG} is greater than N_t because the operating-line slope, L/V, is greater than the slope of the equilibrium line, K, so Figure 6.32b applies.

EXAMPLE 6.10 Absorption of SO₂ in a Packed Column.

Air containing 1.6% by volume SO_2 is scrubbed with pure water in a packed column of 1.5 m² in cross-sectional area and 3.5 m in packed height. Entering gas and liquid flow rates are, respectively, 0.062 and 2.2 kmol/s. If the outlet mole fraction of SO_2 in the gas is 0.004 and column temperature is near-ambient with $K_{SO_2} = 40$, calculate (a) N_{OG} for absorption of SO_2, (b) H_{OG} in meters, and (c) $K_y a$ for SO_2 in kmol/m^3-s-(Δy).

Solution

(a) The operating line is straight, so the system is dilute in SO_2.

$$A = \frac{L}{KV} = \frac{2.2}{(40)(0.062)} = 0.89, \quad y_{\text{in}} = 0.016,$$

$$y_{\text{out}} = 0.004, \quad x_{\text{in}} = 0.0$$

From (6-93),

$$N_{OG} = \frac{\ln\{[(0.89 - 1)/0.89](0.016/0.004) + (1/0.89)\}}{(0.89 - 1)/0.89}$$

$$= 3.75$$

(b) $l_T = 3.5$ m. From (6-89), $H_{OG} = l_T/N_{OG} = 3.5/3.75 = 0.93$ m.

(c) $V = 0.062$ kmol/s, $S = 1.5$ m^2. From (6-90), $K_y a = V/H_{OG}S = 0.062/[(0.93)(1.5)] = 0.044$ kmol/m^3-s-(Δy).

EXAMPLE 6.11 Absorption of Ethylene Oxide.

A gaseous reactor effluent of 2 mol% ethylene oxide in an inert gas is scrubbed with water at 30°C and 20 atm. The gas feed rate is 2,500 lbmol/h, and the entering water rate is 3,500 lbmol/h. Column diameter is 4 ft, packed in two 12-ft-high sections with 1.5-in. metal Pall rings. A liquid redistributor is located between the packed sections. At column conditions, the K-value for ethylene oxide is 0.85 and estimated values of $k_y a$ and $k_x a$ are 200 lbmol/h-ft^3-Δy and 165 lbmol/h-ft^3-Δx. Calculate: (a) $K_y a$, and (b) H_{OG}.

Solution

(a) From (6-84),

$$K_y a = \frac{1}{(1/k_y a) + (K/k_x a)} = \frac{1}{(1/200) + (0.85/165)}$$

$$= 98.5 \text{ lbmol/h-ft}^3\text{-}\Delta y$$

(b) $S = 3.14(4)^2/4 = 12.6$ ft^2

From (6-90), $H_{OG} = V/K_y a S = 2,500/[(98.5)(12.6)] = 2.02$ ft. In this example, both gas- and liquid-phase resistances are important. The value of H_{OG} can also be computed from values of H_G and H_L using equations in Table 6.5:

$$H_G = V/k_y a S = 2,500/[(200)(12.6)] = 1.0 \text{ ft}$$

$$H_L = L/k_x a S = 3,500/[(165)(12.6)] = 1.68 \text{ ft}$$

Substituting these two expressions and (6-90) into (6-84) gives the following relationship for H_{OG} in terms of H_G and H_L:

$$H_{OG} = H_G + H_L/A \qquad (6\text{-}96)$$

$$A = L/KV = 3,500/[(0.85)(2,500)] = 1.65$$

$$H_{OG} = 1.0 + 1.68/1.65 = 2.02 \text{ ft}$$

§6.8 PACKED-COLUMN LIQUID HOLDUP, DIAMETER, FLOODING, PRESSURE DROP, AND MASS-TRANSFER EFFICIENCY

Values of volumetric mass-transfer coefficients and HTUs depend on gas and/or liquid velocities, and these, in turn, depend on column diameter. Estimation of column diameter for a given system, packing, and operating conditions requires consideration of liquid holdup, flooding, and pressure drop.

§6.8.1 Liquid Holdup

Data taken from Billet [45] and shown by Stichlmair, Bravo, and Fair [46] for pressure drop in m-water head/m-of-packed height (specific pressure drop), and liquid holdup in m^3/m of packed height (specific liquid holdup) as a function of superficial gas velocity (velocity of gas in the absence of packing) for different values of superficial water velocity, are shown in Figures 6.33 and 6.34 for a 0.15-m-diameter column packed with 1-inch metal Bialecki rings to a height of 1.5 m and operated at 20°C and 1 bar. In Figure 6.33, the lowest curve corresponds to zero liquid flow. Over a 10-fold range of superficial air velocity, pressure drop is proportional to velocity to the 1.86 power. At increasing liquid flows, gas-phase pressure drop for a given velocity increases. Below a limiting gas velocity, the curve for each liquid velocity is a straight line parallel to the dry pressure-drop curve. In this region, the liquid holdup in the packing for a given liquid velocity is constant, as seen in Figure 6.34. For a liquid velocity of 40 m/h, specific liquid holdup is 0.08 m^3/m^3 of packed bed (8% of the packed volume is liquid) until a superficial gas velocity of 1.0 m/s is reached. Instead of a packed-column void fraction, ϵ, of 0.94 (for Bialecki-ring packing) for the gas to flow through, the effective void fraction is reduced by the liquid holdup to $0.94 - 0.08 = 0.86$, occasioning increased pressure drop. The upper gas-velocity limit for a constant liquid holdup is the *loading point*. Below this, the gas phase is the

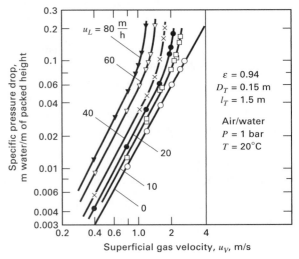

Figure 6.33 Specific pressure drop for dry and irrigated 25-mm metal Bialecki rings.

[From R. Billet, *Packed Column Analysis and Design*, Ruhr-University Bochum (1989) with permission.]

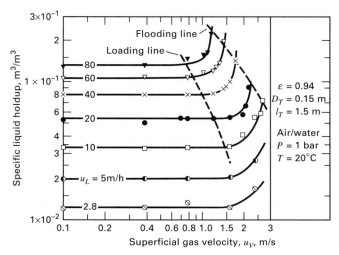

Figure 6.34 Specific liquid holdup for irrigated 25-mm metal Bialecki rings.

[From R. Billet, *Packed Column Analysis and Design*, Ruhr-University Bochum (1989) with permission.]

continuous phase. Above this point, gas begins to hinder the downward flow of liquid, and liquid begins to load the bed, replacing gas and causing a sharp pressure-drop increase. Finally, a gas velocity is reached at which the liquid is continuous across the top of the packing and the column is flooded. At the *flooding point*, the gas drag force is sufficient to entrain the entire liquid. Both loading and flooding lines are included in Figure 6.34.

Between the loading and flooding points is the *loading region*, where liquid entrainment is significant. Here, liquid holdup increases sharply and mass-transfer efficiency decreases with increasing gas velocity. In the loading region, column operation is unstable. Typically, the superficial gas velocity at the loading point is approximately 70% of that at the flooding point. Although a packed column can operate in the loading region, design is best for operation at or below the loading point, in the *preloading region*.

A dimensionless expression for specific liquid holdup, h_L, in the preloading region was developed by Billet and Schultes [47, 69] for a variety of random and structured packings, with dependence on packing characteristics and on the viscosity, density, and superficial velocity of the liquid, u_L:

$$h_L = \left(12 \frac{N_{Fr_L}}{N_{Re_L}}\right)^{1/3} \left(\frac{a_h}{a}\right)^{2/3} \qquad (6\text{-}97)$$

where

$$N_{Re_L} = \text{liquid Reynolds number} = \frac{\text{inertial force}}{\text{viscous force}} \qquad (6\text{-}98)$$
$$= \frac{u_L \rho_L}{a \mu_L} = \frac{u_L}{a \nu_L}$$

and where ν_L is the kinematic viscosity,

$$N_{Fr_L} = \text{liquid Froude number} = \frac{\text{inertial force}}{\text{gravitational force}} \qquad (6\text{-}99)$$
$$= \frac{u_L^2 a}{g}$$

and the ratio of specific hydraulic area of packing, a_h, to specific surface area of packing, a, is given by

$$a_h/a = C_h N_{Re_L}^{0.15} N_{Fr_L}^{0.1} \quad \text{for } N_{Re_L} < 5 \qquad (6\text{-}100)$$

$$a_h/a = 0.85 \, C_h N_{Re_L}^{0.25} N_{Fr_L}^{0.1} \quad \text{for } N_{Re_L} \geq 5 \qquad (6\text{-}101)$$

Values of $a_h/a > 1$ are possible because of droplets and jet flow plus rivulets that cover the packing surface [70].

Values of a and C_h are listed in Table 6.6, together with packing void fraction, ϵ, and other constants for various types and sizes of packing. The liquid holdup is constant in the preloading region, as seen in Figure 6.34, so (6-97) does not involve gas-phase properties or velocity.

At low liquid velocities, liquid holdup is low and it is possible that some of the packing is dry, causing packing efficiency to decrease dramatically, particularly for aqueous systems of high surface tension. For adequate wetting, proven liquid distributors and redistributors must be used, and superficial liquid velocities should exceed the following values:

Type of Packing Material	$u_{L_{min}}$, m/s
Ceramic	0.00015
Oxidized or etched metal	0.0003
Bright metal	0.0009
Plastic	0.0012

EXAMPLE 6.12 Liquid Holdup for Two Packings.

An absorber uses oil absorbent with a kinematic viscosity three times that of water at 20°C. The superficial liquid velocity is 0.01 m/s, which assures good wetting. The superficial gas velocity is in the preloading region. Two packings are considered: (1) randomly packed 50-mm metal Hiflow rings and (2) metal Montz B1-200 structured packing. Estimate the specific liquid holdup for each.

Solution

From Table 6.6,

Packing	a, m²/m³	ϵ	C_h
50-mm metal Hiflow	92.3	0.977	0.876
Montz metal B1-200	200.0	0.979	0.547

At 20°C for water, kinematic viscosity, $\nu_L = \mu/\rho = 1 \times 10^{-6}$ m²/s. Therefore, for the oil, $\mu/\rho = 3 \times 10^{-6}$ m²/s. From (6-98) and (6-99),

Therefore, $\quad N_{Re_L} = \dfrac{0.01}{3 \times 10^{-6} a}, \quad N_{Fr_L} = \dfrac{(0.01)^2 a}{9.8}$

Packing	N_{Re_L}	N_{Fr_L}
Hiflow	36.1	0.00094
Montz	16.67	0.00204

From (6-101), since $N_{Re_L} > 5$ for the Hiflow packing, $a_h/a = (0.85)(0.876)(36.1)^{0.25}(0.000942)^{0.1} = 0.909$. For the Montz packing, $a_h/a = 0.85(0.547)(16.67)^{0.25}(0.00204)^{0.10} = 0.506$.

Table 6.6 Characteristics of Packings

Packing	Material	Size	F_P, ft²/ft³	Characteristics from Billet							
				a, m²/m³	ϵ, m³/m³	C_h	C_p	C_L	C_V	c_s	C_{Fl}
Random Packings											
Berl saddles	Ceramic	25 mm	110	260.0	0.680	0.620	0.719	1.246	0.387	2.916	1.896
Berl saddles	Ceramic	13 mm	240	545.0	0.650	0.833	1.011	1.364	0.232	2.753	1.885
Bialecki rings	Metal	50 mm		121.0	0.966	0.798	0.891	1.721	0.302	2.521	1.856
Bialecki rings	Metal	35 mm		155.0	0.967	0.787	0.378	1.412	0.390	2.970	1.912
Bialecki rings	Metal	25 mm		210.0	0.956	0.692	0.514	1.461	0.331	2.929	1.991
Dinpak® rings	Plastic	70 mm		110.7	0.938	0.991	0.358	1.527	0.326	2.846	1.522
Dinpak rings	Plastic	47 mm		131.2	0.923	1.173	0.338	1.690	0.354	2.987	1.864
EnviPac® rings	Plastic	80 mm, no. 3		60.0	0.955	0.641	0.549	1.603	0.257	2.944	2.012
EnviPac rings	Plastic	60 mm, no. 2		98.4	0.961	0.794	0.851	1.522	0.296	2.694	1.900
EnviPac rings	Plastic	32 mm, no. 1		138.9	0.936	1.039	1.056	1.517	0.459	2.564	1.760
Cascade Mini-Rings	Metal	30 PMK		180.5	0.975	0.930	0.632	1.920	0.450	2.697	1.841
Cascade Mini-Rings	Metal	30 P		164.0	0.959	0.851	0.627	1.577	0.398	2.790	1.870
Cascade Mini-Rings	Metal	1.5"		174.9	0.974	0.935					
Cascade Mini-Rings	Metal	1.5", T		188.0	0.972	0.870					
Cascade Mini-Rings	Metal	1.0"		232.5	0.971	1.040	0.641	2.038	0.495	2.703	1.996
Cascade Mini-Rings	Metal	0.5"		356.0	0.955	1.338	0.882	1.377	0.379	2.644	2.178
Hackettes	Plastic	45 mm		139.5	0.928	0.643	0.399	1.659	0.464	2.832	1.966
Hiflow rings	Ceramic	75 mm	15	54.1	0.868						
Hiflow rings	Ceramic	50 mm	29	89.7	0.809		0.435	1.744	0.465	2.819	1.694
Hiflow rings	Ceramic	38 mm	37	111.8	0.788		0.538	1.168	0.408	2.840	1.930
Hiflow rings	Ceramic	20 mm, 6 stg.		265.8	0.776	0.958	0.621				
Hiflow rings	Ceramic	20 mm, 4 stg.		261.2	0.779	1.167	0.628				
Hiflow rings	Metal	50 mm	16	92.3	0.977	0.876	0.421	1.641	0.402	2.702	1.626
Hiflow rings	Metal	25 mm	42	202.9	0.962	0.799	0.689	1.553	0.369	2.918	2.177
Hiflow rings	Plastic	90 mm	9	69.7	0.968		0.276				
Hiflow rings	Plastic	50 mm, hydr.	20	118.4	0.925		0.311				
Hiflow rings	Plastic	50 mm		117.1	0.924	1.038	0.327	1.487	0.345	2.894	1.871
Hiflow rings	Plastic	25 mm		194.5	0.918		0.741				
Hiflow rings, super	Plastic	50 mm, S		82.0	0.942		0.414	1.577	0.390	2.841	1.989
Hiflow saddles	Plastic	50 mm		86.4	0.938		0.454				
Intalox saddles	Ceramic	50 mm	40	114.6	0.761		0.747				
Intalox saddles	Plastic	50 mm	28	122.1	0.908		0.758	1.219	0.342	2.866	1.702
Nor-Pak® rings	Plastic	50 mm	14	86.8	0.947	0.651	0.350	1.080	0.322	2.959	1.786
Nor-Pak rings	Plastic	35 mm	21	141.8	0.944	0.587	0.371	0.756	0.425	3.179	2.242
Nor-Pak rings	Plastic	25 mm, type B		202.0	0.953	0.601	0.397	0.883	0.366	3.277	2.472
Nor-Pak rings	Plastic	25 mm, 10 stg.		197.9	0.920		0.383	0.976	0.410	2.865	2.083

Nor-Pak rings	Plastic	25 mm	31	180.0	0.927	0.601	0.397	1.278	0.333	3.793	3.024
Nor-Pak rings	Plastic	22 mm		249.0	0.913		0.365	1.192	0.410	2.725	1.580
Nor-Pak rings	Plastic	15 mm		311.4	0.918	0.343	0.233	1.012	0.341	2.629	1.679
Pall rings	Ceramic	50 mm	43	155.2	0.754	1.066	0.763	1.440	0.336	2.627	2.083
Pall rings	Metal	50 mm	27	112.6	0.951	0.784	0.967	1.239	0.368	2.816	1.757
Pall rings	Metal	35 mm	40	139.4	0.965	0.644	0.957	0.856	0.380	2.654	1.742
Pall rings	Metal	25 mm	56	223.5	0.954	0.719	0.990	0.905	0.446	2.696	2.064
Pall rings	Metal	15 mm	70	368.4	0.933	0.590	0.698	1.913	0.370	2.825	2.400
Pall rings	Plastic	50 mm	26	111.1	0.919	0.593	0.927	1.486	0.360	3.612	2.401
Pall rings	Plastic	35 mm	40	151.1	0.906	0.718	0.865	1.270	0.320	3.412	2.174
Pall rings	Plastic	25 mm	55	225.0	0.887	0.528	0.595	1.481	0.341	2.843	1.812
Raflux® rings	Plastic	15 mm		307.9	0.894	0.491	0.485	1.520	0.303	2.843	1.812
Ralu flow	Plastic	1		165	0.940	0.640	0.350	1.320	0.333	2.841	1.989
Ralu flow	Plastic	2		100	0.945	0.640	0.468	1.320	0.333	2.725	1.580
Ralu® rings	Plastic	50 mm, hydr.		94.3	0.939	0.439	0.672	1.192	0.345	2.629	1.679
Ralu rings	Plastic	50 mm		95.2	0.983	0.640	0.800	1.277	0.341	2.627	2.083
Ralu rings	Plastic	38 mm		150	0.930	0.640	0.763	1.440	0.336		
Ralu rings	Plastic	25 mm		190	0.940	0.719	1.003				
Ralu rings	Metal	50 mm		105	0.975	0.784	0.957				
Ralu rings	Metal	38 mm		135	0.965	0.644					
Ralu rings	Metal	25 mm		215	0.960	0.714					
Raschig rings	Carbon	25 mm		202.2	0.720	0.623		1.379	0.471		
Raschig rings	Ceramic	25 mm	179	190.0	0.680	0.577		1.361	0.412		
Raschig rings	Ceramic	15 mm	380	312.0	0.690	0.648		1.276	0.401		
Raschig rings	Ceramic	10 mm	1,000	440.0	0.650	0.791		1.303	0.272		
Raschig rings	Ceramic	6 mm	1,600	771.9	0.620	1.094		1.130			
Raschig rings	Metal	15 mm	170	378.4	0.917	0.455					
Raschig Super-rings	Ceramic	25		190.0	0.680	0.577	1.329	1.361	0.412	2.454	1.899
Raschig Super-rings	Metal	0.3		315	0.960	0.750	0.760	1.500	0.450	3.560	2.340
Raschig Super-rings	Metal	0.5		250	0.975	0.620	0.780	1.450	0.430	3.350	2.200
Raschig Super-rings	Metal	1		160	0.980	0.750	0.500	1.290	0.440	3.491	2.200
Raschig Super-rings	Metal	2		97.6	0.985	0.720	0.464	1.323	0.400	3.326	2.096
Raschig Super-rings	Metal	3		80	0.982	0.620	0.430	0.850	0.300	3.260	2.100
Raschig Super-rings	Plastic	2		100	0.960	0.720	0.377	1.250	0.337	3.326	2.096
Tellerettes	Plastic	25 mm		190.0	0.930	0.588	0.538	0.899		2.913	2.132
Top-Pak rings	Aluminum	50 mm	40	105.5	0.956	0.881	0.604	1.326	0.389	2.528	1.579
VSP rings	Metal	50 mm, no. 2		104.6	0.980	1.135	0.773	1.222	0.420	2.806	1.689
VSP rings	Metal	25 mm, no. 1		199.6	0.975	1.369	0.782	1.376	0.405	2.755	1.970

(Continued)

Table 6.6 (Continued)

Packing	Material	Size	F_P, ft²/ft³	a, m²/m³	ϵ, m³/m³	C_h	C_p	C_L	C_V	c_s	C_{Fl}
						Characteristics from Billet					
				Structured Packings							
Euroform®	Plastic	PN-110		110.0	0.936	0.511	0.250	0.973	0.167	3.075	1.975
Gempak	Metal	A2 T-304		202.0	0.977	0.678	0.344			2.986	2.099
Impulse®	Ceramic	100		91.4	0.838	1.900	0.417	1.317	0.327	2.664	1.655
Impulse	Metal	250		250.0	0.975	0.431	0.262	0.983	0.270	2.610	1.996
Koch-Sulzer	Metal	CY	70								
Koch-Sulzer	Metal	BX	21								
Mellapak	Plastic	250 Y	22	250.0	0.970	0.554	0.292			3.157	2.464
Montz	Metal	B1-100		100.0	0.987	0.626					
Montz	Metal	B1-200	33	200.0	0.979	0.547	0.355	0.971	0.390	3.116	2.339
Montz	Metal	B1-300		300.0	0.930	0.482	0.295	1.165	0.422	3.098	2.464
Montz	Plastic	C1-200		200.0	0.954		0.453	1.006	0.412		
Montz	Plastic	C2-200		200.0	0.900		0.481	0.739		2.653	1.973
Ralu Pak®	Metal	YC-250		250.0	0.945	0.650	0.191	1.334	0.385	3.178	2.558

From (6-97), for the Hiflow packing,

$$h_L = \left[\frac{12(0.000942)}{36.1} \right]^{1/3} (0.909)^{2/3} = 0.0637 \ \text{m}^3/\text{m}^3$$

For the Montz packing,

$$h_L = \left[\frac{12(0.0204)}{16.67} \right]^{1/3} (0.506)^{2/3} = 0.0722 \ \text{m}^3/\text{m}^3$$

Note that for the Hiflow packing, the void fraction available for gas flow is reduced by the liquid flow from $\epsilon = 0.977$ (Table 6.6) to $0.977 - 0.064 = 0.913 \ \text{m}^3/\text{m}^3$. For Montz packing, the reduction is from 0.979 to 0.907 m³/m³.

§6.8.2 Flooding, Column Diameter, and Pressure Drop

Liquid holdup, column diameter, and pressure drop are closely related. The diameter must be such that flooding is avoided and pressure drop is below 1.5 inches of water (equivalent to 0.054 psi)/ft of packed height. General rules for packings also exist. One is that the nominal packing diameter not be greater than 1/8 of the diameter of the column; otherwise, poor distribution of liquid and vapor flows can occur.

Flooding data for packed columns were first correlated by Sherwood et al. [26], who used the liquid-to-gas kinetic-energy ratio, $F_{LV} = (LM_L/VM_V)(\rho_V/\rho_L)^{0.5}$, previously used for trayed towers, and shown in Figures 6.23 and 6.27. The superficial gas velocity, u_V, was embedded in the dimensionless term $u_V^2 a/g\epsilon^3$, arrived at by considering the square of the actual gas velocity, u_V^2/ϵ^2; the hydraulic radius, $r_H = \epsilon/a$, i.e., the flow volume divided by the wetted surface area of the packing; and the gravitational constant, g, to give a dimensionless expression, $u_V^2 a/g\epsilon^3 = u_V^2 F_P/g$. The ratio, a/ϵ^3, is the packing factor, F_P. Values of a, ϵ, and F_P are included in Table 6.6. In some cases, F_P is a modified packing factor from experimental data, chosen to fit a generalized correlation. Additional factors were added to account for liquid density and viscosity and for gas density.

In 1954, Leva [48] used experimental data on ring and saddle packings to extend the Sherwood et al. [26] flooding correlations to include lines of constant pressure drop, the resulting chart becoming known as the generalized pressure-drop correlation (GPDC).

A modern version of the GPDC chart by Leva [49] is shown in Figure 6.35a. The abscissa is F_{LV}; the ordinate is

$$Y = \frac{u_V^2 F_P}{g} \left(\frac{\rho_v}{\rho_{H_2O_{(L)}}} \right) f\{\rho_L\} f\{u_L\} \qquad (6\text{-}102)$$

The functions $f\{\rho_L\}$ and $f\{\mu_L\}$ are corrections for liquid properties given by Figures 6.35b and 6.35c, respectively.

For given flow rates, properties, and packing, the GPDC chart is used to compute $u_{V,f}$, the superficial gas velocity at flooding. Then a fraction of flooding, f, is selected (usually from 0.5 to 0.7), followed by calculation of the tower diameter from an equation similar to (6-44):

$$D_T = \left(\frac{4V M_V}{f u_{V,f} \pi \rho_V} \right)^{0.5} \qquad (6\text{-}103)$$

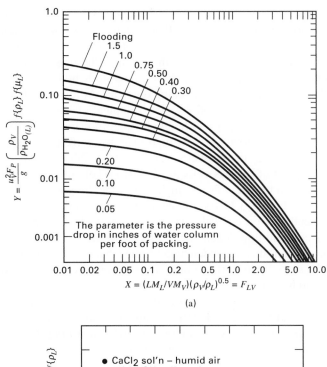

(a)

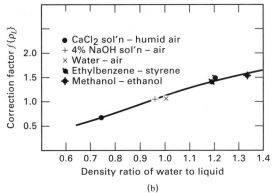

(b)

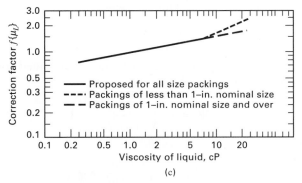

(c)

Figure 6.35 (a) Generalized pressure-drop correlation of Leva for packed columns. (b) Correction factor for liquid density. (c) Correction factor for liquid viscosity.

[From M. Leva, *Chem. Eng. Prog.*, **88** (1), 65–72 (1992) with permission.]

EXAMPLE 6.13 Flooding, Pressure Drop, and Diameter.

Forty lbmol/h of air containing 5 mol% NH_3 enters a packed column at 20°C and 1 atm, where 90% of the ammonia is scrubbed by a countercurrent flow of 3,000 lb/h of water. Use the GPDC chart of Figure 6.35 to estimate the superficial, gas flooding velocity; the column inside diameter for operation at 70% of flooding; and the pressure drop per foot of packing for: (a) 1-inch ceramic Raschig

rings $(F_P = 179 \text{ ft}^2/\text{ft}^3)$ and (b) 1-inch metal IMTP packing $(F_P = 41 \text{ ft}^2/\text{ft}^3)$.

Solution

Calculations are made at the bottom of the column, where the superficial gas velocity is highest.

Inlet gas:

$$M_V = 0.95(29) + 0.05(17) = 28.4, \quad V = 40 \text{ lbmol/h}$$
$$\rho_V = PM_V/RT = (1)(28.4)/[(0.730)(293)(1.8)]$$
$$= 0.0738 \text{ lb/ft}^3$$

Exiting liquid:

NH_3 absorbed $= 0.90(0.05)(40)(17) = 30.6$ lb/h or 1.8 lbmol/h

Water rate (assumed constant) $= 3{,}000$ lb/h or 166.7 lbmol/h

Mole fraction of ammonia $= 1.8/(166.7 + 1.8) = 0.0107$

$$M_L = 0.0107(17) + (0.9893)(18) = 17.9$$
$$L = 1.8 + 166.7 = 168.5 \text{ lbmol/h}$$

Let $\rho_L = 62.4 \text{ lb/ft}^3$ and $\mu_L = 1.0 \text{ cP}$.

$$X = F_{LV}(\text{abscissa in Figure 6.35a})$$
$$= \frac{(168.5)(17.9)}{(40)(28.4)}\left(\frac{0.0738}{62.4}\right)^{0.5} = 0.092$$

From Figure 6.35a, $Y = 0.125$ at flooding. From Figure 6.35b, $f\{\rho_L\} = 1.14$.
From Figure 6.35c, $f\{\mu_L\} = 1.0$. From (6-102),

$$u_V^2 = 0.125\left(\frac{g}{F_P}\right)\frac{62.4}{(0.0738)(1.14)(1.0)} = 92.7 \, g/F_P$$

Using $g = 32.2 \text{ ft/s}^2$,

Packing Material	F_P, ft²/ft³	u_o, ft/s
Raschig rings	179	4.1
IMTP packing	41	8.5

For $f = 0.70$, using (6-103),

Packing Material	$fu_{V,f}$, ft/s	D_T, inches
Raschig rings	2.87	16.5
IMTP packing	5.95	11.5

From Figure 6.35a, for $F_{LV} = 0.092$ and $Y = 0.70^2(0.125) = 0.0613$ at 70% of flooding, the pressure drop is 0.88 inch H_2O/ft for both packings.

The IMTP packing is clearly superior to Raschig rings, by about 50%.

Pressure Drop

Flooding-point data for packing materials are in reasonable agreement with the upper curve of the GPDC chart of Figure 6.35. Unfortunately, this is not always the case for pressure drop. Reasons for this are discussed by Kister [33]. As an example of the disparity, the predicted pressure drop of 0.88 inch H_2O/ft in Example 6.13 for IMTP packing at 70% of flooding is in poor agreement with the 0.63 inch/ft determined from data supplied by the packing manufacturer.

If Figure 6.35a is cross-plotted as pressure drop versus Y for constant F_{LV}, a pressure drop of from 2.5 to 3 inches/ft is predicted at flooding for all packings. However, Kister and Gill [33, 50] for random and structured packings show that pressure drop at flooding is strongly dependent on the packing factor, F_P, by the empirical expression

$$\Delta P_{\text{flood}} = 0.115 \, F_P^{0.7} \qquad (6\text{-}104)$$

where ΔP_{flood} has units of inches H_2O/ft of packed height and F_P is ft^2/ft^3. Table 6.6 shows that F_P is from 10 to 100. Thus, (6-103) predicts pressure drops at flooding from as low as 0.6 to as high as 3 inches H_2O/ft. Kister and Gill also give a procedure for estimating pressure drop, which utilizes data in conjunction with a GPDC-type plot.

Theoretical models for pressure drop have been presented by Stichlmair et al. [46], who used a particle model, and Billet and Schultes [51, 69], who used a channel model. Both extend equations for dry-bed pressure drop to account for the effect of liquid holdup. Billet and Schultes [69] include predictions of superficial vapor velocity at the loading point, $u_{V,l}$, which provides an alternative, perhaps more accurate, method for estimating column diameter. Their model gives

$$u_{V,l} = \left(\frac{g}{\Psi_l}\right)^{1/2} \left[\frac{\epsilon}{a^{1/6}} - a^{1/2}\xi_l^{1/3}\right]\xi_l^{1/6}\left(\frac{\rho_L}{\rho_V}\right)^{1/2} \qquad (6\text{-}105)$$

where $u_{V,l}$ is in m/s, g = gravitational acceleration = 9.807 m/s^2, and

$$\Psi_l = \frac{g}{C^2}\left[F_{LV}\left(\frac{\mu_L}{\mu_V}\right)^{0.4}\right]^{-2n_s} \qquad (6\text{-}106)$$

where ϵ and a are obtained from Table 6.6, F_{LV} = kinetic energy ratio of Figures 6.23 and 6.35a, and

$$\xi_l = \left(12\frac{\mu_L}{g\rho_L}u_{L,l}\right) \qquad (6\text{-}107)$$

where μ_L and μ_V are in kg/m-s, ρ_L and ρ_V are in kg/m^3, and

$u_{L,l}$ = superficial liquid velocity at loading point

$$= u_{V,l}\frac{\rho_V LM_L}{\rho_L VM_V} \text{ in m/s}$$

Values for n_s and C in (6-106) depend on F_{LV} as follows: If $F_{LV} \leq 0.4$, the liquid trickles over the packing as a disperse phase and $n_s = -0.326$, with $C = C_s$ from Table 6.6. If $F_{LV} > 0.4$, the column holdup reaches such a large value that empty spaces within the bed close and liquid flows downward as a continuous phase, while gas rises in the form of bubbles, with $n_s = -0.723$ and

$$C = 0.695\left(\frac{\mu_L}{\mu_V}\right)^{0.1588} C_s \quad \text{(from Table 6.6)} \qquad (6\text{-}108)$$

Billet and Schultes [69] have a model for the vapor velocity at the flooding point, $u_{V,f}$, that involves the flooding constant, C_{Fl}, in Table 6.6; but a more suitable expression is

$$u_{V,f} = \frac{u_{V,l}}{0.7} \qquad (6\text{-}109)$$

For gas flow through packing under conditions of no liquid flow, a correlation from fluid mechanics for the friction factor in terms of a modified Reynolds number can be obtained in a manner similar to that for flow through an empty, straight pipe, as in Figure 6.36 from the study by Ergun [52]. Here, D_P is an effective packing diameter. At low superficial gas velocities (modified $N_{Re} < 10$) typical of laminar flow, the pressure drop per unit height is proportional to the superficial vapor velocity, u_V. At high velocity, corresponding to turbulent flow, the pressure drop per unit height depends on the square of the gas velocity. Industrial columns generally operate in turbulent flow, which is why the dry pressure-drop data in Figure 6.33 for Bialecki rings show an exponential dependency on gas velocity of about 1.86. In Figure 6.33, when liquid flows countercurrently to the gas in the preloading region, this dependency continues, but at a higher pressure drop, because the volume for gas flow decreases due to liquid holdup.

Based on studies using more than 50 different packings, Billet and Schultes [51, 69] developed a correlation for dry-gas pressure drop, ΔP_o, similar in form to that of Figure 6.36. Their dimensionally consistent correlating equation is

$$\frac{\Delta P_o}{l_T} = \Psi_o \frac{a}{\epsilon^3}\frac{u_V^2\rho_V}{2}\frac{1}{K_W} \qquad (6\text{-}110)$$

where l_T = height of packing and K_W = a wall factor.

K_W can be important for columns with an inadequate ratio of packing diameter to inside column diameter ($> \sim 1/8$), and is given by

$$\frac{1}{K_W} = 1 + \frac{2}{3}\left(\frac{1}{1-\epsilon}\right)\frac{D_P}{D_T} \qquad (6\text{-}111)$$

where the effective packing diameter is

$$D_P = 6\left(\frac{1-\epsilon}{a}\right) \qquad (6\text{-}112)$$

The dry-packing resistance coefficient (a modified friction factor), Ψ_o, is given by the empirical expression

$$\Psi_o = C_p\left(\frac{64}{N_{Re_V}} + \frac{1.8}{N_{Re_V}^{0.08}}\right) \qquad (6\text{-}113)$$

where

$$N_{Re_V} = \frac{u_V D_P \rho_V}{(1-\epsilon)\mu_V}K_W \qquad (6\text{-}114)$$

and C_p is tabulated for a number of packings in Table 6.6. In (6-113), the laminar-flow region is characterized by $64/N_{Re_V}$, while the next term characterizes the turbulent-flow regime.

When a packed tower has a downward-flowing liquid, the area for gas flow is reduced by the liquid holdup, and the surface structure exposed to the gas is changed as a result of the coating of the packing with a liquid film. The pressure drop now depends on the holdup and a two-phase flow resistance, which was found by Billet and Schultes [69] to depend on the liquid flow Froude number for flow rates up to the loading point:

$$\frac{\Delta P}{\Delta P_o} = \left(\frac{\epsilon}{\epsilon - h_L}\right)^{3/2}\exp\left[\frac{13{,}300}{a^{3/2}}(N_{Fr_L})^{1/2}\right] \qquad (6\text{-}115)$$

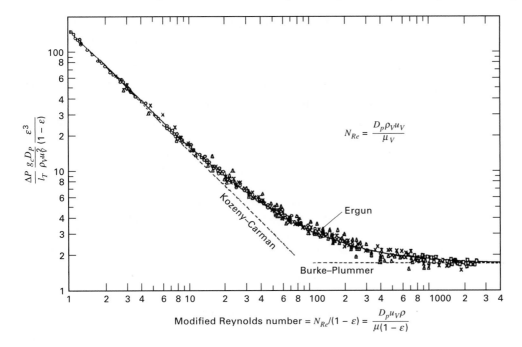

Figure 6.36 Ergun correlation for dry-bed pressure drop.

[From S. Ergun, *Chem. Eng. Prog.*, **48** (2), 89–94 (1952) with permission.]

where h_L is given by (6-97) in m^2/m^3; ϵ and a are given in Table 6.6, where a in (6-115) must be in m^2/m^3; and N_{Fr_L} is given by (6-99).

EXAMPLE 6.14 Holdup, Loading, Flooding, Pressure Drop, and Diameter of a Packed Column.

A column packed with 25-mm metal Bialecki rings is to be designed for the following vapor and liquid conditions:

	Vapor	Liquid
Mass flow rate, kg/h	515	1,361
Density, kg/m³	1.182	1,000
Viscosity, kg/m-s	1.78×10^{-5}	1.00×10^{-3}
Molecular weight	28.4	18.02
Surface tension, kg/s²		2.401×10^{-2}

Using equations of Billet and Schultes, determine the (a) vapor and liquid superficial velocities at loading and flooding points, (b) specific liquid holdup at the loading point, (c) specific pressure drop at the loading point, and (d) column diameter at the loading point.

Solution

(a) From Table 6.6, the following constants apply to Bialecki rings:

$$a = 210 \, m^2/m^3, \quad \epsilon = 0.956, \quad C_h = 0.692,$$
$$C_p = 0.891, \quad \text{and} \quad C_s = 2.521$$

First, compute the superficial vapor velocity at the loading point. From the abscissa label of Figure 6.35a,

$$F_{LV} = \frac{1,361}{515}\left(\frac{1.182}{1,000}\right)^{1/2} = 0.0908$$

Because $F_{LV} < 0.4$, $n_s = -0.326$, and C in (6-106) $= C_s = 2.521$.

From (6-106),

$$\Psi_l = \frac{9.807}{2.521^2}\left[0.0908\left(\frac{0.001}{0.0000178}\right)^{0.4}\right]^{-2(-0.326)} = 0.923$$

$$u_{L,l} = u_{V,l}\frac{\rho_V LM_L}{\rho_L LM_V} = u_{V,l}\frac{(1.182)(1,361)}{(1,000)(515)} = 0.00312\, u_{V,l}$$

From (6-107),

$$\xi_l = \left(12\frac{(0.001)}{(9.807)(1,000)}(0.00312)\, u_{V,l}\right) = 3.82 \times 10^{-9}\, u_{V,l}$$

From (6-105),

$$u_{V,l} = \left(\frac{9.807}{0.923}\right)^{1/2}\left[\frac{0.956}{210^{1/6}} - 210^{1/2}\left(3.82 \times 10^{-9}\, u_{V,l}\right)^{1/3}\right]$$
$$\times \left(3.82 \times 10^{-9}\, u_{V,l}\right)^{1/6}\left(\frac{1,000}{1.182}\right)^{1/2}$$
$$= 3.26\left[0.392 - 0.0227\, u_{V,l}^{1/3}\right]1.15\, u_{V,l}^{1/6}$$
$$= \left(1.47 - 0.0851\, u_{V,l}^{1/3}\right)u_{V,l}^{1/6}$$

Solving this nonlinear equation gives $u_{V,l} =$ superficial vapor velocity at the loading point $= 1.46$ m/s. The corresponding superficial liquid velocity $= u_{L,l} = 0.00312\, u_{V,l} = 0.00312(1.46) = 0.00457$ m/s.

The superficial vapor flooding velocity $= u_{Vf} = \dfrac{u_{V,l}}{0.7} = \dfrac{1.46}{0.7} = 2.09$ m/s.

The corresponding superficial liquid velocity $= u_{Lf} = \dfrac{0.00457}{0.7} = 0.00653$ m/s.

(b) Next, compute the specific liquid holdup at the loading point. From (6-98) and (6-99),

$$N_{Re_L} = \frac{(0.00457)(1,000)}{(210)(0.001)} = 21.8$$

and

$$N_{Fr_L} = \frac{(0.00457)^2(210)}{9.807} = 0.000447$$

Because $N_{Re_L} > 5$, (6-101) applies:

$$a_h/a = 0.85(0.692)(21.8)^{0.25}(0.000447)^{0.1} = 0.588$$

From (6-97), the specific liquid holdup at the loading point is

$$h_L = \left(12\frac{0.000447}{21.8}\right)^{1/3} 0.588^{2/3} = 0.0440 \text{ m}^3/\text{m}^3$$

(c) and (d) Before computing the specific pressure drop at the loading point, compute the column diameter at the loading point. Applying (6-103),

$$D_T = \left[\frac{4(515/3600)}{(1.46)(3.14)(1.182)}\right]^{1/2} = 0.325 \text{ m}$$

From (6-112),

$$D_P = 6\left(\frac{1-0.956}{210}\right) = 0.00126 \text{ m}$$

From (6-111),

$$\frac{1}{K_W} = 1 + \frac{2}{3}\left(\frac{1}{1-0.956}\right)\frac{0.00126}{0.325} = 1.059 \quad \text{and} \quad K_W = 0.944$$

From (6-114),

$$N_{Re_V} = \frac{(1.46)(0.00126)(1.182)}{(1-0.956)(0.0000178)}(0.944) = 2,621$$

From (6-113),

$$\Psi_o = 0.891\left(\frac{64}{2.621} + \frac{1.8}{2,621^{0.08}}\right) = 0.876$$

From (6-110), the specific dry-gas pressure drop is

$$\frac{\Delta P_o}{l_T} = 0.876\frac{(210)(1.46)^2(1.182)}{(0.956)^3(2)}(1.059)$$

$$= 281 \text{ kg/m}^2\text{-s}^2 = \text{Pa/m}$$

From (6-115), the specific pressure drop at the loading point is

$$\frac{\Delta P}{l_T} = 281\left(\frac{0.956}{0.956-0.0440}\right)^{3/2} \exp\left[\frac{13,300}{210^{3/2}}(0.000447)^{1/2}\right]$$

$$= 331 \text{ kg/m}^2\text{-s}^2$$

or 0.406 in. of water/ft.

§6.8.3 Mass-Transfer Efficiency

Packed-column mass-transfer efficiency is included in the HETP, HTUs, and volumetric mass-transfer coefficients. Although the HETP concept lacks a theoretical basis, its simplicity, coupled with the relative ease of making equilibrium-stage calculations, has made it a widely used method for estimating packed height. In the preloading region, with good distribution of vapor and liquid, HETP values depend mainly on packing type and size, liquid viscosity, and surface tension. For preliminary estimates, the following relations, taken from Kister [33], can be used.

1. Pall rings and similar high-efficiency random packings with low-viscosity liquids:

$$\text{HETP, ft} = 1.5 D_P, \text{ in.} \qquad (6\text{-}116)$$

2. Structured packings at low-to-moderate pressure with low-viscosity liquids:

$$\text{HETP, ft} = 100/a, \text{ ft}^2/\text{ft}^3 + 4/12 \qquad (6\text{-}117)$$

3. Absorption with viscous liquid:

$$\text{HETP} = 5 \text{ to } 6 \text{ ft}$$

4. Vacuum service:

$$\text{HETP, ft} = 1.5 D_P, \text{ in.} + 0.5 \qquad (6\text{-}118)$$

5. High-pressure service (> 200 psia):
   HETP for structured packings may be greater than predicted by (6-117).

6. Small-diameter columns, $D_T < 2$ ft:

$$\text{HETP, ft} = D_T, \text{ ft, but not less than 1 ft}$$

In general, lower values of HETP are achieved with smaller-size random packings, particularly in small-diameter columns, and with structured packings, particularly those with large values of a, the packing surface area per packed volume. The experimental data of Figure 6.37 for No. 2 (2-inch-diameter) Nutter rings from Kunesh [53] show that in the preloading region, the HETP is relatively independent of the vapor flow F-factor:

$$F = u_V(\rho_V)^{0.5} \qquad (6\text{-}119)$$

provided that the ratio L/V is maintained constant as the superficial gas velocity, u_V, is increased. Beyond the loading point, and as the flooding point is approached, the HETP can increase dramatically, like the pressure drop and liquid holdup.

Mass-transfer data for packed columns are usually correlated in terms of volumetric mass-transfer coefficients and HTUs, rather than in terms of HETPs. Because the data come from experiments in which either the liquid- or the gas-phase mass-transfer resistance is negligible, the other resistance can be correlated independently. For applications where both resistances are important, they are added according to the two-film theory of Whitman, discussed in §3.7 [54], to describe the overall resistance. This theory assumes

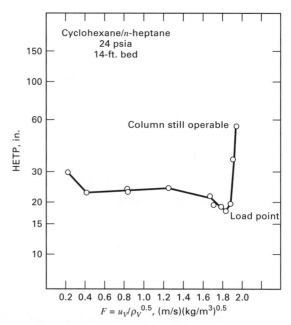

Figure 6.37 Effect of F-factor on HETP.

interface equilibrium, i.e. the absence of mass-transfer resistance at the gas–liquid interface.

Equations (6-84) and (6-77) define the overall gas-phase coefficient in terms of the individual volumetric mass-transfer coefficients and the mass-transfer rates in terms of mole-fraction driving forces and the vapor–liquid K-value:

$$\frac{1}{K_y a} = \frac{1}{k_y a} + \frac{K}{k_x a}$$

$$r = k_y a(y - y_I) = k_x a(x_I - x) = K_y a(y - y^*)$$

Mass-transfer rates can also be expressed in terms of liquid-phase concentrations and gas-phase partial pressure:

$$r = k_p a(p - p_I) = k_L a(c_I - c) = K_G a(p - p^*) \quad (6\text{-}120)$$

If a Henry's law constant is defined at the equilibrium interface between the two phases by

$$p_I = H' c_I \quad (6\text{-}121)$$

and let

$$p^* = H' c \quad (6\text{-}122)$$

then

$$\frac{1}{K_G a} = \frac{1}{k_p a} + \frac{H'}{k_L a} \quad (6\text{-}123)$$

Other formulations for $K_x a$ and $K_L a$ are given in Table 6.5, with the most common units as follows:

	SI Units	American Engineering Units
r	mol/m^3-s	lbmol/ft^3-h
$k_y a, k_x a, K_x a, K_y a$	mol/m^3-s	lbmol/ft^3-h
$k_p a, K_G a$	mol/m^3-s-kPa	lbmol/ft^3-h-atm
$k_L a, k_G a, k_c a$	s^{-1}	h^{-1}
k_L, k_G, k_c	m/s	ft/h

As shown in Table 6.5, mass-transfer coefficients are directly related to HTUs, which have the advantages of: (1) only one dimension (length), (2) variation with column conditions less than mass-transfer coefficients, and (3) being related to an easily understood geometrical quantity, namely, height per theoretical stage. Definitions of individual and overall HTUs are included in Table 6.5 for the dilute solute case. By substituting these into (6-84),

$$H_{OG} = H_G + (KV/L)H_L \quad (6\text{-}124)$$

Alternative expressions for H_{OL} exist. In absorption or stripping of low-solubility gases, the solute K-value or Henry's law constant, H' in (6-112), is large, making the last terms in (6-88), (6-123), and (6-124) large; thus gas-phase resistance is negligible and the rate of mass transfer is liquid-phase-controlled. Such data are used to study the effect of variables on volumetric liquid-phase mass-transfer coefficients and HTUs. Figure 6.38 shows three different Berl-saddle packings for stripping O_2 from water by air, in a 20-inch-I.D. column operating in the preloading region, as reported in a study by Sherwood and Holloway [55]. The effect of liquid velocity on $k_L a$ is pronounced, with $k_L a$ increasing at the 0.75 power of the liquid mass velocity. Gas velocity has no effect on $k_L a$ in the preloading region. Figure 6.38 also contains data

plotted in terms of H_L, where

$$H_L = \frac{M_L L}{\rho_L k_L a S} \quad (6\text{-}125)$$

Clearly, H_L does not depend as strongly as $k_L a$ does on liquid mass velocity, $M_L L/S$.

Another liquid mass-transfer-controlled system is CO_2–air–H_2O. Measurements for a variety of modern packings shown in Figure 6.39 are reported by Billet [45]. The effect of gas velocity on $k_L a$ in terms of the F-factor at a constant liquid rate is shown in Figure 6.40 for the same system, but with 50-mm plastic Pall rings and Hiflow rings. Up to an F-factor of about $1.8 \text{ m}^{-1/2}\text{-s}^{-1}\text{-kg}^{1/2}$, which is in the preloading region, no effect of gas velocity is observed. Above the loading limit, $k_L a$ increases with gas velocity because the larger liquid holdup increases interfacial area for mass transfer. Although not illustrated in Figures 6.38 to 6.40, a major liquid-phase factor is the solute diffusivity. Data in the preloading region can usually be correlated by an empirical expression, which includes only the liquid velocity and diffusivity:

$$k_L a = C_1 D_L^{1/2} u_L^n \quad (6\text{-}126)$$

where n varies from 0.6 to 0.95, with 0.75 being a typical value. The exponent on the diffusivity is consistent with the penetration theory presented in §3.6.

A convenient system for studying gas-phase mass transfer is NH_3–air–H_2O. The low K-value and high solubility of NH_3 in H_2O make the last terms in (6-84), (6-123), and (6-124) negligible, so the gas-phase resistance controls the rate of mass transfer.

Figures 6.41 and 6.42 verify the greater effect of vapor velocity compared to that of liquid velocity, as evidenced by the much greater slope in Figure 6.41, where the coefficient is proportional to the 0.75 power of F. The small effect of liquid velocity in Figure 6.42 is due to increases in holdup and interfacial area.

For a given packing, experimental data on $k_p a$ or $k_G a$ for different systems in the preloading region can usually be correlated satisfactorily with empirical correlations of the form

$$k_p a = C_2 D_G^{0.67} F^{m'} u_L^{n'} \quad (6\text{-}127)$$

where D_G is the gas diffusivity of the solute and m' and n' have been observed by different investigators to vary from 0.65 to 0.85 and from 0.25 to 0.5, respectively, a typical value for m' being 0.8.

Measurement and correlation of gas- and liquid-phase mass-transfer coefficients and HTUs are important aspects of chemical engineering because individual film coefficients are required for the modern, more scientific design methods for separation processes. These theoretical and semitheoretical equations are based mostly on the application of the two-film theory by Fair and co-workers [56–63] and others [64, 65]. In some cases, values of k_G and k_L are reported separately from a; in others, the combinations $k_G a$ and $k_L a$ are used. Important features of some correlations are summarized in Table 6.7. Development of such correlations for packed columns is difficult because, as shown by Billet [66], values of

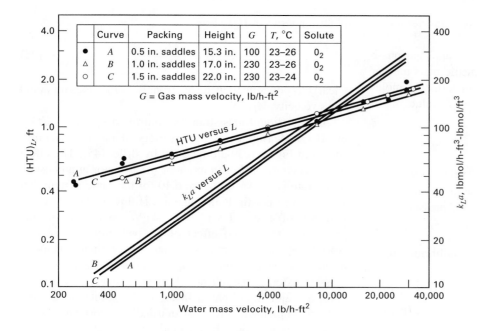

	Curve	Packing	Height	G	T, °C	Solute
●	A	0.5 in. saddles	15.3 in.	100	23–26	O_2
△	B	1.0 in. saddles	17.0 in.	230	23–26	O_2
○	C	1.5 in. saddles	22.0 in.	230	23–24	O_2

G = Gas mass velocity, lb/h-ft²

Figure 6.38 Effect of liquid rate on liquid-phase mass transfer of O_2.

[From T.K. Sherwood and F.A.L. Holloway, *Trans. AIChE.*, **36**, 39–70 (1940) with permission.]

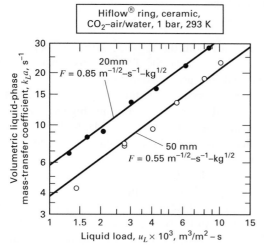

Figure 6.39 Effect of liquid load on liquid-phase mass transfer of CO_2.

[From R. Billet, *Packed Column Analysis and Design*, Ruhr-University Bochum (1989) with permission.]

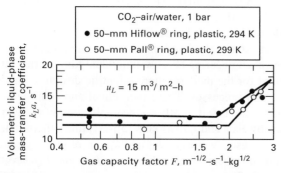

Figure 6.40 Effect of gas rate on liquid-phase mass transfer of CO_2.

[From R. Billet, *Packed Column Analysis and Design*, Ruhr-University Bochum (1989) with permission.]

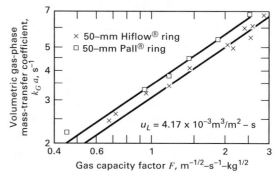

Figure 6.41 Effect of gas rate on gas-phase mass transfer of NH_3.

[From R. Billet, *Packed Column Analysis and Design*, Ruhr-University Bochum (1989) with permission.]

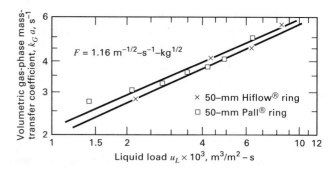

Figure 6.42 Effect of liquid rate on gas-phase mass transfer of NH_3.

[From R. Billet, *Packed Column Analysis and Design*, Ruhr-University Bochum (1989) with permission.]

mass-transfer coefficients are significantly affected by the technique used to pack the column and the number of liquid feed-distribution points across the column, which must be more than 25 points/sq ft.

Billet and Schultes [67] measured and correlated volumetric mass-transfer coefficients and HTUs for 31 different chemical systems, with 67 different types and sizes of packings in columns of diameter 2.4 inches to 4.6 ft, with additional data [69] for Hiflow rings and Raschig Super-rings.

Table 6.7 Generalized Correlations for Mass Transfer in Packed Columns

Investigator	Year	Ref. No.	Type of Correlations	Packings
Shulman et al.	1955	64	k_p, k_L, a	Raschig rings, Berl saddles
Cornell et al.	1960	56, 57	H_G, H_L	Raschig rings, Berl saddles
Onda et al.	1968	65	k_p, k_L, a	Raschig rings, Berl saddles
Bolles and Fair	1979, 1982	58, 59	H_G, H_L	Raschig rings, Berl saddles, Pall rings
Bravo and Fair	1982	60	a	Raschig rings, Berl saddles, Pall rings
Bravo et al.	1985	61	k_G, k_L	Sulzer
Fair and Bravo	1987	62	k_G, k_L, a	Sulzer, Gempak, Mellapak, Montz, Ralu Pak
Fair and Bravo	1991	63	k_G, k_L, a	Flexipac, Gempak, Intalox 2T, Montz, Mellapak, Sulzer
Billet and Schultes	1991	67	$k_G a, k_L a$	14 random packings and 4 structured packings
Billet and Schultes	1999	69	$k_G a, k_L a$	19 random packings and 6 structured packings

The systems include those for which mass-transfer resistance resides mainly in the liquid phase and others where gas-phase resistance is controlling. They assume uniform distribution of gas and liquid and apply the two-film theory of mass transfer (§3.7). For the liquid-phase resistance, they assume that the liquid flows in a thin film through the irregular channels of the packing, with continual remixing of the liquid at points of contact with the packing such that Higbie's penetration theory (§3.6) [68] is applicable. The volumetric mass-transfer coefficient is defined by

$$r = (k_L a_{Ph})(c_{L_i} - c_L) \tag{6-128}$$

From the penetration theory of Higbie, (3-194),

$$k_L = 2(D_L/\pi t_L)^{0.5} \tag{6-129}$$

where t_L = the time of exposure of the liquid film before remixing. Billet and Schultes assume that this time is based on a length of travel equal to the *hydraulic diameter* of the packing:

$$t_L = h_L d_H / u_L \tag{6-130}$$

where d_H, the hydraulic diameter, $= 4r_H$ or $4\epsilon/a$. In terms of height of a liquid transfer unit, (6-129) and (6-130) give

$$H_L = \frac{u_L}{k_L a_{Ph}} = \frac{\sqrt{\pi}}{2}\left(\frac{4h_L\epsilon}{D_L a u_L}\right)^{1/2}\frac{u_L}{a_{Ph}} \tag{6-131}$$

Equation (6-131) was modified to include a constant, C_L, which is back-calculated for each packing to fit the data. The final predictive equation of Billet and Schultes is

$$H_L = \frac{1}{C_L}\left(\frac{1}{12}\right)^{1/6}\left(\frac{4h_L\epsilon}{D_L a u_L}\right)^{1/2}\frac{u_L}{a}\left(\frac{a}{a_{Ph}}\right) \tag{6-132}$$

where values of C_L are included in Table 6.6.

An analogous equation developed by Billet and Schultes for gas-phase resistance, where the time of exposure of the gas between periods of mixing is determined empirically, is

$$H_G = \frac{1}{C_V}(\epsilon - h_L)^{1/2}\left(\frac{4\epsilon}{a^4}\right)^{1/2}(N_{Re_V})^{-3/4}(N_{Sc_V})^{-1/3}\left(\frac{u_V a}{D_G a_{Ph}}\right) \tag{6-133}$$

where C_V is included in Table 6.6 and

$$N_{Re_V} = \frac{u_V \rho_V}{a\mu_V} \tag{6-134}$$

$$N_{Sc_V} = \frac{\mu_V}{\rho_V D_V} \tag{6-135}$$

Equations (6-132) and (6-133) contain an area ratio, a_{Ph}/a, the ratio of the phase-interface area to the packing surface area, which, from Billet and Schultes [69], is not the same as the hydraulic area ratio, a_h/a, given by (6-100) and (6-101). Instead, they give the following correlation:

$$\frac{a_{Ph}}{a} = 1.5(ad_h)^{-1/2}\left(N_{Re_{L,h}}\right)^{-0.2}\left(N_{We_{L,h}}\right)^{0.75}\left(N_{Fr_{L,h}}\right)^{-0.45} \tag{6-136}$$

where d_h = packing hydraulic diameter $= 4\dfrac{\epsilon}{a} \tag{6-137}$

and the following liquid-phase dimensionless groups use the packing hydraulic diameter as the characteristic length:

$$\text{Reynolds number} = N_{Re_{L,h}} = \frac{u_L d_h \rho_L}{\mu_L} \tag{6-138}$$

$$\text{Weber number} = N_{We_{L,h}} = \frac{u_L^2 \rho_L d_h}{\sigma} \tag{6-139}$$

$$\text{Froude number} = N_{Fr_{L,h}} = \frac{u_L^2}{g d_h} \tag{6-140}$$

After calculating H_L and H_G from (6-132) and (6-133), the overall HTU value is obtained from (6-124), the packed height from (6-89), and N_{OG} as described in §6.7.

EXAMPLE 6.15 Packed Height from Mass-Transfer Theory.

For the absorption of ethyl alcohol from CO_2 with water, as considered in Example 6.1, a 2.5-ft-I.D. tower, packed with 1.5-inch metal Pall-like rings, is to be used. It is estimated that the tower will operate in the preloading region with a pressure drop of approximately 1.5 inches H_2O/ft of packed height. From Example 6.9, $N_{OG} = 7.5$. Estimate H_G, H_L, H_{OG}, HETP, and the required packed height in feet using the following estimates of flow conditions and physical properties at the bottom of the packing:

	Vapor	Liquid
Flow rate, lb/h	17,480	6,140
Molecular weight	44.05	18.7
Density, lb/ft^3	0.121	61.5
Viscosity, cP	0.0145	0.63
Surface tension, dynes/cm	—	101
Diffusivity of ethanol, m^2/s	7.75×10^{-6}	1.82×10^{-9}
Kinematic viscosity, m^2/s	0.75×10^{-5}	0.64×10^{-6}

Solution

Cross-sectional area of tower $= (3.14)(2.5)^2/4 = 4.91 \text{ ft}^2$

Volumetric liquid flow rate $= 6,140/61.5 = 99.8 \text{ ft}^3/\text{h}$

$u_L =$ superficial liquid velocity $= 99.8/[(4.91)(3,600)] = 0.0056 \text{ ft/s}$
or 0.0017 m/s

From this section, $u_L > u_{L,\min}$, but the velocity is on the low side.

$\quad u_V =$ superficial gas velocity $= 17,480/[(0.121)(4.91)(3,600)]$
$\quad\quad = 8.17 \text{ ft/s} = 2.49 \text{ m/s}$

Let the packing characteristics for the 1.5-inch metal Pall-like rings be as follows (somewhat different from values for Pall rings in Table 6.6):

$$a = 149.6 \text{ m}^2/\text{m}^3, \quad \epsilon = 0.952$$

$$C_h = \text{approximately } 0.7, \quad C_L = 1.227, \quad C_V = 0.341$$

Estimation of specific liquid holdup, h_L:

From (6-98), $\quad N_{Re_L} = \dfrac{0.0017}{(0.64 \times 10^{-6})(149.6)} = 17.8.$

From (6-99), $\quad N_{Fr_L} = \dfrac{(0.0017)^2(149.6)}{9.8} = 4.41 \times 10^{-5}$

From (6-101),

$$\frac{a_h}{a} = 0.85(0.7)(17.8)^{0.25}\left(4.41 \times 10^{-5}\right)^{0.10} = 0.45$$

$$a_h = 0.45(149.6) = 67.3 \text{ m}^2/\text{m}^3$$

From (6-97),

$$h_L = \left[\frac{12\left(4.41 \times 10^{-5}\right)}{17.8}\right]^{1/3}(0.45)^{2/3} = 0.0182 \text{ m}^3/\text{m}^3$$

Estimation of H_L:

First compute a_{Ph}, the ratio of phase interface area to packing surface area.

From (6-137), $\quad d_h = 4\dfrac{0.952}{149.6} = 0.0255 \text{ m}$

From (6-138), $\quad N_{Re_{L},h} = \dfrac{(0.0017)(0.0255)}{(0.64 \times 10^{-6})} = 67.7$

From (6-139),

$$N_{We_{L},h} = \frac{(0.0017)^2[(61.5)(16.02)](0.0255)}{[(101)(0.001)]} = 0.000719$$

From (6-140),

$$N_{Fr_{L},h} = \frac{(0.0017)^2}{(9.807)(0.0255)} = 1.156 \times 10^{-5}$$

From (6-136),

$$\frac{a_{Ph}}{a} = 1.5(149.6)^{-1/2}(0.0255)^{-1/2}(67.7)^{-0.2}$$
$$\times (0.000719)^{0.75}\left(1.156 \times 10^{-5}\right)^{-0.45} = 0.242$$

From (6-132), using consistent SI units,

$$H_L = \frac{1}{1.227}\left(\frac{1}{12}\right)^{1/6}\left[\frac{(4)(0.0182)(0.952)}{(1.82 \times 10^{-9})(149.6)(0.0017)}\right]^{1/2}$$
$$\times \left(\frac{0.0017}{149.6}\right)\left(\frac{1}{0.242}\right) = 0.31 \text{ m} = 1.01 \text{ ft}$$

Estimation of H_G:

From (6-134),

$$N_{Re_V} = 2.49/\left[(149.6)\left(0.75 \times 10^{-5}\right)\right] = 2,220$$

From (6-135),

$$N_{Sc_V} = 0.75 \times 10^{-5}/7.75 \times 10^{-6} = 0.968$$

From (6-133), using consistent SI units,

$$H_G = \frac{1}{0.341}(0.952 - 0.0128)^{1/2}\left[\frac{(4)(0.952)}{(149.6)^4}\right]^{1/2}$$
$$\times (2220)^{-3/4}(0.968)^{-1/3}\left[\frac{(2.49)}{7.75 \times 10^{-6}(0.242)}\right]$$
$$= 1.03 \text{ m} \quad \text{or} \quad 3.37 \text{ ft}$$

Estimation of H_{OG}:

From Example 6.1, the K-value for ethyl alcohol $= 0.57$,

$$V = 17,480/44.05 = 397 \text{ lbmol/h},$$
$$L = 6,140/18.7 = 328 \text{ lbmol/h},$$

and $\quad\quad 1/A = KV/L = (0.57)(397)/328 = 0.69$

From (6-124),

$$|H_{OG} = 3.37 + 0.69(1.01) = 4.07 \text{ ft}|$$

The mass-transfer resistance in the gas phase is $\gg$ than in the liquid phase.

Estimation of Packed Height:

From (6-89), $\quad l_T = 4.07(7.5) = 30.5 \text{ ft}$

Estimation of HETP:

From (6-94), for straight operating and equilibrium lines, with $A = 1/0.69 = 1.45$,

$$\text{HETP} = 4.07\left[\frac{\ln(0.69)}{(1 - 1.45)/1.45}\right] = 4.86 \text{ ft}$$

§6.9 CONCENTRATED SOLUTIONS IN PACKED COLUMNS

For concentrated solutions, the x–y equilibrium relations and the material balances (operating lines) are curved, so the analytical integrations formerly used for determining N_{OG} and l_T from (6-93) cannot be used. Instead, the following alternative manual methods or the computer methods of Chapters 10 and 11 apply.

For concentrated solutions, the columns in Table 6.5 labeled UM (unimolecular) diffusion apply. To obtain these from the columns labeled EM (equimolar) diffusion, let

$$L' = L(1 - x) \quad \text{and} \quad V' = V(1 - y) \qquad (6\text{-}141\text{a, b})$$

where L' and V' are constant flow rates of the inert (solvent) liquid and (carrier) gas, on a solute-free basis. Then

$$d(Vy) = V'd\left(\frac{y}{1 - y}\right) = V'\frac{dy}{(1 - y)^2} = V\frac{dy}{(1 - y)} \qquad (6\text{-}142)$$

$$d(Lx) = L'd\left(\frac{x}{1 - x}\right) = L'\frac{dx}{(1 - x)^2} = L\frac{dx}{(1 - x)} \qquad (6\text{-}143)$$

Equation (6-88) now becomes

$$l_T = \int_{y_2}^{y_1} \left(\frac{V}{K'_y aS}\right) \frac{dy}{(1 - y)(y - y^*)}$$
$$= \frac{V}{K'_y aS} \int_{y_2}^{y_1} \frac{dy}{(1 - y)(y - y^*)} \qquad (6\text{-}144)$$

where 1 refers to inlet and 2 refers to outlet conditions. Based on the liquid phase,

$$l_T = \int_{x_1}^{x_2} \left(\frac{L}{K'_x aS}\right) \frac{dx}{(1 - x)(x^* - x)}$$
$$= \frac{L}{K'_x aS} \int_{x_1}^{x_2} \frac{dx}{(1 - x)(x^* - x)} \qquad (6\text{-}145)$$

where the overall mass-transfer coefficients are primed to signify UM diffusion.

If the numerators and denominators of (6-144) and (6-145) are multiplied by $(1 - y)_{LM}$ and $(1 - x)_{LM}$, respectively, where $(1 - y)_{LM}$ is the log mean of $(1 - y)$ and $(1 - y^*)$, and $(1 - x)_{LM}$ is the log mean of $(1 - x)$ and $(1 - x^*)$, the expressions in rows 1 and 6 of columns 4 and 7 in Table 6.5 are obtained:

$$l_T = \int_{y_2}^{y_1} \left[\frac{V}{K'_y a(1 - y)_{LM} S}\right] \frac{(1 - y)_{LM} dy}{(1 - y)(y - y^*)}$$
$$= \frac{V}{K'_y a(1 - y)_{LM} S} \int_{y_2}^{y_1} \frac{(1 - y)_{LM} dy}{(1 - y)(y - y^*)} \qquad (6\text{-}146)$$

$$l_T = \int_{x_1}^{x_2} \left[\frac{L}{K'_x a(1 - x)_{LM} S}\right] \frac{(1 - x)_{LM} dx}{(1 - x)(x^* - x)}$$
$$= \frac{L}{K'_x a(1 - x)_{LM} S} \int_{x_1}^{x_2} \frac{(1 - x)_{LM} dx}{(1 - x)(x^* - x)} \qquad (6\text{-}147)$$

The term $K'_y(1 - y)_{LM}$ is equal to the concentration-independent K_y, and $K'_x(1 - x)_{LM}$ is equal to concentration-independent K_x. If there is appreciable absorption, vapor flow decreases from the bottom to the top. However, the values of Ka are also a function of flow rate, so the ratio V/Ka is approximately constant and HTU groupings, $[L/K'_x a(1 - x)_{LM} S]$ and $[V/K'_y a(1 - y)_{LM} S]$, can often be taken out of the integral without incurring errors larger than those inherent in experimental measurements of Ka. Usually, average values of V, L, and $(1 - y)_{LM}$ are used.

A second approach is to leave the terms in (6-146) or (6-147) under the integral sign and evaluate l_T by a stepwise or graphical integration. To obtain the terms $(y - y^*)$ or $(x^* - x)$, equilibrium and operating lines are required. With

reference to Figure 6.28, an overall material balance around the upper part of the absorber gives

$$V + L_{in} = V_{out} + L \qquad (6\text{-}148)$$

A component balance around the upper part of the absorber, assuming a pure-liquid absorbent, gives:

$$Vy = V_{out}y_{out} + Lx \qquad (6\text{-}149)$$

An absorbent balance around the upper part of the absorber is:

$$L_{in} = L(1 - x) \qquad (6\text{-}150)$$

Combining (6-148) to (6-150) to eliminate V and L gives

$$y = \frac{V_{out}y_{out} + [L_{in}x/(1 - x)]}{V_{out} + [L_{in}x/(1 - x)]} \qquad (6\text{-}151)$$

Equation (6-151) allows the y–x operating line to be calculated from a knowledge of terminal conditions only.

A simpler approach to the problem of concentrated gas or liquid mixtures is to linearize the operating line by expressing all concentrations in mole ratios, with the gas and liquid flows on a solute-free basis—that is, $V' = (1 - y)V$, and $L' = (1 - x)L$. Then, in place of (6-146) and (6-147),

$$l_T = \int_{Y_2}^{Y_1} \left(\frac{V'}{K_Y aS}\right) \frac{dY}{(Y - Y^*)} = \frac{V'}{K_Y aS} \int_{Y_2}^{Y_1} \frac{dY}{(Y - Y^*)} \qquad (6\text{-}152)$$

$$l_T = \int_{X_1}^{X_2} \left(\frac{L'}{K_X aS}\right) \frac{dX}{(X^* - X)} = \frac{L'}{K_X aS} \int_{X_1}^{X_2} \frac{dX}{(X^* - X)} \qquad (6\text{-}153)$$

This set of equations is listed in rows 3 and 8 of Table 6.5.

EXAMPLE 6.16 NTU for a Packed Column with Concentrated Solute.

To remove 95% of the ammonia from an air stream containing 40% ammonia by volume, 488 lbmol/h of absorbent per 100 lbmol/h of entering gas are used, which is greater than the minimum requirement.

Equilibrium data are given in Figure 6.43. $P = 1$ atm and $T = 298$ K. Calculate NTU by: (a) Equation (6-146) using a curved operating line from (6-151), and (b) Equation (6-152) using mole ratios.

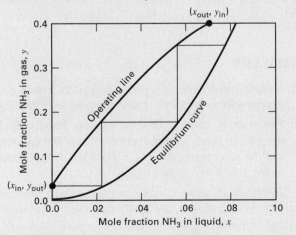

Figure 6.43 Theoretical stages for Example 6.16.

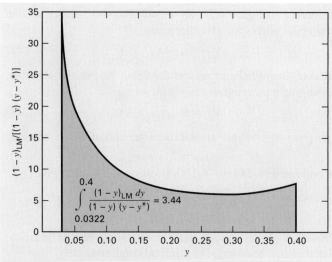

Figure 6.44 Determination of N_{OG} for Example 6.16.

Solution

(a) $L_{in} = 488$ lbmol/h. Then $V_{out} = 100 - (40)(0.95) = 62$ lbmol/h, and $y_{out} = (0.05)(40)/62 = 0.0323$. From (6-151), the curved operating line of Figure 6.43 is constructed. If $x = 0.04$,

$$y = \frac{(62)(0.0323) + [(488)(0.04)/(1 - 0.04)]}{62 + [(488)(0.04)/(1 - 0.04)]} = 0.27$$

Now calculate values of $y, y^*, (1 - y)_{LM} = [(1 - y) - (1 - y^*)]/ \ln[(1 - y)/(1 - y^*)]$, and $(1 - y)_{LM}/[(1 - y)(y - y^*)]$ for use in (6-146). In Figure 6.43, for $x = 0.044$, y (on the operating line) $= 0.30$, and y^* (on the equilibrium curve) $= 0.12$, from which the other four quantities follow.

y	y^*	$(y - y^*)$	$(1 - y)$	$(1 - y)_{LM}$	$\dfrac{(1 - y)_{LM}}{(1 - y)(y - y^*)}$
0.03	0.002	0.028	0.97	0.99	36.47
0.05	0.005	0.045	0.95	0.97	22.68
0.10	0.01	0.09	0.90	0.94	11.60
0.15	0.025	0.125	0.85	0.91	8.56
0.20	0.04	0.16	0.80	0.89	6.95
0.25	0.08	0.17	0.75	0.85	6.66
0.30	0.12	0.18	0.70	0.82	6.51
0.35	0.17	0.18	0.65	0.73	6.24
0.40	0.26	0.14	0.60	0.67	7.97

Note that $(1 - y) \approx (1 - y)_{LM}$, so these two terms frequently cancel out of the NTU equations, particularly when y is small.

Figure 6.44 is a plot of $(1 - y)_{LM}/[(1 - y)(y - y^*)]$ versus y to determine N_{OG}. The integral on the RHS of (6-146), between $y = 0.4$ and $y = 0.0322$, is $3.44 = N_{OG}$. This is approximately 1 more than $N_t = 2.6$, as represented by the steps of Figure 6.43.

(b) It is a simple matter to obtain values for $Y = y/(1 - y)$, $Y^* = y^*/(1 - y^*)$, $(Y - Y^*)$, and $(Y - Y^*)^{-1}$, as given in the following table:

y	Y	y^*	Y^*	$(Y - Y^*)^{-1}$
0.03	0.031	0.002	0.002	34.48
0.05	0.053	0.005	0.005	20.83
0.1	0.111	0.01	0.010	9.9
0.15	0.176	0.025	0.026	6.66
0.20	0.250	0.04	0.042	4.8
0.25	0.333	0.08	0.087	4.06
0.30	0.43	0.12	0.136	3.40
0.35	0.54	0.17	0.205	2.98
0.40	0.67	0.26	0.310	2.78

Graphical integration of the RHS integral of (6-152) determines the area under the curve of Y versus $(Y - Y*)^{-1}$ between $Y = 0.67$ and $Y = 0.033$. The result is $N_{OG} = 3.46$. Alternatively, the integration can be performed on a computer using a spreadsheet.

For concentrated solutions, the assumption of constant temperature may not be valid and could result in a large error. If an overall energy balance predicts a temperature change that alters the equilibrium curve significantly, it is best to use a process simulator that includes the energy balance.

SUMMARY

1. A liquid can selectively absorb components from a gas. A gas can selectively desorb or strip components from a liquid.

2. The fraction of a component that can be absorbed or stripped depends on the number of equilibrium stages and the absorption factor, $A = L/(KV)$, or the stripping factor, $S = KV/L$, respectively.

3. Towers with sieve or valve trays, or with random or structured packings, are most often used for absorption and stripping.

4. Absorbers are most effective at high pressure and low temperature. The reverse is true for strippers. However, high costs of gas compression, refrigeration, and vacuum often preclude operation at the most thermodynamically favorable conditions.

5. For a given gas flow, composition, degree of absorption, choice of absorbent, and operating T and P, there is a minimum absorbent flow rate, given by (6-9) to (6-11), that corresponds to an infinite number of stages. A rate of 1.5 times the minimum typically leads to a reasonable number of stages. A similar criterion, (6-12), holds for a stripper.

6. The equilibrium stages and flow rates for an absorber or stripper can be determined from the equilibrium line, (6-1), and an operating line, (6-3) or (6-5), using graphical, algebraic, or numerical methods. Graphical methods, such as Figure 6.11, offer visual insight into stage-by-stage changes in compositions of the gas and liquid streams and the effects of changes in the variables.

7. Estimates of overall stage efficiency, defined by (6-21), can be made with the correlations of Drickamer and Bradford (6-22), O'Connell (6-23), and Figure 6.14. More accurate procedures involve the use of a laboratory Oldershaw column or semitheoretical equations, e.g., of Chan and Fair, based on mass-transfer considerations, to determine a Murphree vapor-point efficiency, (6-30). The Murphree vapor-tray efficiency is obtained from (6-31) to (6-34), and the overall efficiency from (6-37).

8. Tray diameter is determined from (6-44) based on entrainment flooding considerations shown in Figure 6.23. Vapor pressure drop, weeping, entrainment, and downcomer backup can be estimated from (6-49), (6-68), (6-69), and (6-70), respectively.

9. Packed-column height is determined using HETP, (6-73), or HTU/NTU, (6-89), concepts, with the latter having a more theoretical basis in the two-film theory of mass transfer. For straight equilibrium and operating lines, HETP is related to the HTU by (6-94), and the number of stages to the NTU by (6-95).

10. In the preloading region, liquid holdup in a packed column is independent of vapor velocity. The loading point is typically 70% of the flooding point, and most packed columns are designed to operate in the preloading region from 50% to 70% of flooding. The flooding point from Figure 6.35, the GPDC chart, is used to determine column diameter (6-102) and loading point (6-105).

11. An advantage of a packed column is its low pressure drop, as compared to that in a trayed tower. Packed-column pressure drop is estimated from Figure 6.35, (6-106), or (6-115).

12. Numerous rules of thumb for estimating the HETP of packed columns exist. The preferred approach is to estimate H_{OG} from semitheoretical mass-transfer correlations such as those of (6-132) and (6-133) based on the work of Billet and Schultes.

13. Obtaining theoretical stages for concentrated solutions involves numerical integration because of curved equilibrium and/or operating lines.

REFERENCES

1. Washburn, E.W., Ed.-in-Chief, *International Critical Tables*, McGraw-Hill, New York, Vol. **III**, p. 255 (1928).

2. Lockett, M., *Distillation Tray Fundamentals*, Cambridge University Press, Cambridge, UK, p. 13 (1986).

3. Okoniewski, B.A., *Chem. Eng. Prog.*, **88**(2), 89–93 (1992).

4. Sax, N.I., *Dangerous Properties of Industrial Materials*, 4th ed., Van Nostrand Reinhold, New York, pp. 440–441 (1975).

5. Lewis, W.K., *Ind. Eng. Chem.*, **14**, 492–497 (1922).

6. Drickamer, H.G., and J.R. Bradford, *Trans. AIChE*, **39**, 319–360 (1943).

7. Jackson, R.M., and T.K. Sherwood, *Trans. AIChE*, **37**, 959–978 (1941).

8. O'Connell, H.E., *Trans. AIChE*, **42**, 741–755 (1946).

9. Walter, J.F., and T.K. Sherwood, *Ind. Eng. Chem.*, **33**, 493–501 (1941).

10. Edmister, W.C., *The Petroleum Engineer*, C45–C54 (Jan. 1949).

11. Lockhart, F.J., and C.W. Leggett, in K.A. Kobe and J.J. McKetta, Jr. Eds., *Advances in Petroleum Chemistry and Refining*, Vol. 1, Interscience, New York, Vol. **1**, pp. 323–326 (1958).

12. Holland, C.D., *Multicomponent Distillation*, Prentice-Hall, Englewood Cliffs, NJ (1963).

13. Murphree, E.V, *Ind. Eng. Chem.*, **17**, 747 (1925).

14. Hausen, H., *Chem. Ing. Tech.*, **25**, 595 (1953).

15. Standart, G., *Chem Eng. Sci.*, **20**, 611 (1965).

16. Lewis, W.K., *Ind. Eng. Chem.*, **28**, 399 (1936).

17. Gerster, J.A., A.B. Hill, N.H. Hochgraf, and D.G. Robinson, "Tray Efficiencies in Distillation Columns," Final Report from the University of Delaware, American Institute of Chemical Engineers, New York (1958).

18. *Bubble-Tray Design Manual*, AIChE, New York (1958).

19. Gilbert, T.J., *Chem. Eng. Sci.*, **10**, 243 (1959).

20. Barker, P.E., and M.F. Self, *Chem. Eng. Sci.*, **17**, 541 (1962).

21. Bennett, D.L., and H.J. Grimm, *AIChE J.*, **37**, 589 (1991).

22. Oldershaw, C.F., *Ind. Eng. Chem. Anal. Ed.*, **13**, 265 (1941).

23. Fair, J.R., H.R. Null, and W.L. Bolles, *Ind. Eng. Chem. Process Des. Dev.*, **22**, 53–58 (1983).

24. Souders, M., and G.G. Brown, *Ind. Eng. Chem.*, **26**, 98–103 (1934).

25. Fair, J.R., *Petro/Chem. Eng.*, **33**, 211–218 (Sept. 1961).

26. Sherwood, T.K., G.H. Shipley, and F.A.L. Holloway, *Ind. Eng. Chem.*, **30**, 765–769 (1938).

27. *Glitsch Ballast Tray, Bulletin No. 159*, Fritz W. Glitsch and Sons, Dallas, TX (from FRI report of Sept. 3, 1958).

28. *Glitsch V-1 Ballast Tray, Bulletin No. 160*, Fritz W. Glitsch and Sons, Dallas, TX (from FRI report of Sept. 25, 1959).

29. Oliver, E.D., *Diffusional Separation Processes. Theory, Design, and Evaluation*, John Wiley & Sons, New York, pp. 320–321 (1966).

30. Bennett, D.L., R. Agrawal, and P.J. Cook, *AIChE J.*, **29**, 434–442 (1983).

31. Smith, B.D., *Design of Equilibrium Stage Processes*, McGraw-Hill, New York (1963).

32. Klein, G.F., *Chem. Eng.*, **89**(9), 81–85 (1982).

33. Kister, H.Z., *Distillation Design*, McGraw-Hill, New York (1992).

34. Lockett, M.J., *Distillation Tray Fundamentals*, Cambridge University Press, Cambridge, UK, p. 146 (1986).

35. American Institute of Chemical Engineers (AIChE) *Bubble-Tray Design Manual*, AIChE, New York (1958).

36. Chan, H., and J.R. Fair, *Ind. Eng. Chem. Process Des. Dev.*, **23**, 814–819 (1984).

37. Chan, H., and J.R. Fair, *Ind. Eng. Chem. Process Des. Dev.*, **23**, 820–827 (1984).

38. Scheffe, R.D., and R.H. Weiland, *Ind. Eng. Chem. Res.*, **26**, 228–236 (1987).

39. Foss, A.S., and J.A. Gerster, *Chem. Eng. Prog.*, **52**, 28-J to 34-J (Jan. 1956).

40. Gerster, J.A., A.B. Hill, N.N. Hochgraf, and D.G. Robinson, "Tray Efficiencies in Distillation Columns," Final Report from University of Delaware, American Institute of Chemical Engineers (AIChE), New York (1958).

41. Fair, J.R., *Petro./Chem. Eng.*, **33**(10), 45 (1961).

42. Colburn, A.P, *Ind. Eng. Chem.*, **28**, 526 (1936).

43. Chilton, T.H., and A.P. Colburn, *Ind. Eng. Chem.*, **27**, 255–260, 904 (1935).

44. Colburn, A.P, *Trans. AIChE*, **35**, 211–236, 587–591 (1939).

45. Billet, R., *Packed Column Analysis and Design*, Ruhr-University Bochum (1989).

46. Stichlmair, J., J.L. Bravo, and J.R. Fair, *Gas Separation and Purification*, **3**, 19–28 (1989).

47. Billet, R., and M. Schultes, *Packed Towers in Processing and Environmental Technology*, translated by J.W. Fullarton, VCH Publishers, New York (1995).

48. Leva, M., *Chem. Eng. Prog. Symp. Ser*, **50**(10), 51 (1954).

49. Leva, M., *Chem. Eng. Prog.*, **88**(1), 65–72 (1992).

50. Kister, H.Z., and D.R. Gill, *Chem. Eng. Prog.*, **87**(2), 32–42 (1991).

51. Billet, R., and M. Schultes, *Chem. Eng. Technol*, **14**, 89–95 (1991).

52. Ergun, S., *Chem. Eng. Prog.*, **48**(2), 89–94 (1952).

53. Kunesh, J.G., *Can. J. Chem. Eng.*, **65**, 907–913 (1987).

54. Whitman, W.G., *Chem. and Met. Eng.*, **29**, 146–148 (1923).

55. Sherwood, T.K., and F.A.L. Holloway, *Trans. AIChE.*, **36**, 39–70 (1940).

56. Cornell, D., W.G. Knapp, and J.R. Fair, *Chem. Eng. Prog.*, **56**(7), 68–74 (1960).

57. Cornell, D., W.G. Knapp, and J.R. Fair, *Chem. Eng. Prog.*, **56**(8), 48–53 (1960).

58. Bolles, W.L., and J.R. Fair, *Inst. Chem. Eng. Symp. Ser*, **56**, 3/35 (1979).

59. Bolles, W.L., and J.R. Fair, *Chem. Eng.*, **89**(14), 109–116 (1982).

60. Bravo, J.L., and J.R. Fair, *Ind. Eng. Chem. Process Des. Devel.*, **21**, 162–170 (1982).

61. Bravo, J.L., J.A. Rocha, and J.R. Fair, *Hydrocarbon Processing*, **64**(1), 56–60 (1985).

62. Fair, J.R., and J.L. Bravo, *I. Chem. E. Symp. Ser.*, **104**, A183–A201 (1987).

63. Fair, J.R., and J.L. Bravo, *Chem. Eng. Prob.*, **86**(1), 19–29 (1990).

64. Shulman, H.L., C.F. Ullrich, A.Z. Proulx, and J.O. Zimmerman, *AIChE J.*, **1**, 253–258 (1955).

65. Onda, K., H. Takeuchi, and Y.J. Okumoto, *J. Chem. Eng. Jpn.*, **1**, 56–62 (1968).

66. Billet, R., *Chem. Eng. Prog.*, **63**(9), 53–65 (1967).

67. Billet, R., and M. Schultes, *Beitrage zur Verfahrens-Und Umwelttechnik*, Ruhr-Universitat Bochum, pp. 88–106 (1991).

68. Higbie, R., *Trans. AIChE*, **31**, 365–389 (1935).

69. Billet, R., and M. Schultes, *Chem. Eng. Res. Des., Trans. IChemE*, **77A**, 498–504 (1999).

70. M. Schultes, Private Communication (2004).

71. Sloley, A.W., *Chem. Eng. Prog.*, **95**(1), 23–35 (1999).

72. Stupin, W.J., and H.Z. Kister, *Trans. IChemE.*, **81A**, 136–146 (2003).

STUDY QUESTIONS

6.1. What is the difference between physical absorption and chemical (reactive) absorption?

6.2. What is the difference between an equilibrium-based and a rate-based calculation method?

6.3. What is a trayed tower? What is a packed column?

6.4. What are the three most common types of openings in trays for the passage of vapor? Which of the three is rarely specified for new installations?

6.5. In a trayed tower, what is meant by flooding and weeping? What are the two types of flooding, and which is more common?

6.6. What is the difference between random and structured packings?

6.7. For what conditions is a packed column favored over a trayed tower?

6.8. In general, why should the operating pressure be high and the operating temperature be low for an absorber, and the opposite for a stripper?

6.9. For a given recovery of a key component in an absorber or stripper, does a minimum absorbent or stripping agent flow rate exist for a tower or column with an infinite number of equilibrium stages?

6.10. What is the difference between an operating line and an equilibrium curve?

6.11. What is a reasonable value for the optimal absorption factor when designing an absorber? Does that same value apply to the optimal stripping factor when designing a stripper?

6.12. When stepping off stages on an *Y–X* plot for an absorber or a stripper, does the process start and stop with the operating line or the equilibrium curve?

6.13. Why do longer liquid flow paths across a tray give higher stage efficiencies?

6.14. What is the difference between the Murphree tray and point efficiencies?

6.15. What is meant by turndown ratio? What type of tray has the best turndown ratio? Which tray the worst?

6.16. What are the three contributing factors to the vapor pressure drop across a tray?

6.17. What is the HETP? Does it have a theoretical basis? If not, why is it so widely used?

6.18. Why are there so many different kinds of mass-transfer coefficients? How can they be distinguished?

6.19. What is the difference between the loading point and the flooding point in a packed column?

6.20. When the solute concentration is moderate to high instead of dilute, why are calculations for packed columns much more difficult?

EXERCISES

Section 6.1

6.1.   Stripping in an absorber and absorption in a stripper.

In absorption, the absorbent is stripped to an extent that depends on its K-value. In stripping, the stripping agent is absorbed to an extent depending on its K-value. In Figure 6.1, it is seen that both absorption and stripping occur. Which occurs to the greatest extent in terms of kmol/h? Should the operation be called an absorber or a stripper? Why?

6.2.   Advances in packing.

Prior to 1950, two types of commercial random packings were in common use: Raschig rings and Berl saddles. Since 1950, many new random packings have appeared. What advantages do these newer ones have? By what advances in design and fabrication were achievements made? Why were structured packings introduced?

6.3.   Bubble-cap trays.

Bubble-cap trays were widely used in towers prior to the 1950s. Today sieve and valve trays are favored. However, bubble-cap trays are still specified for operations that require very high turndown ratios or appreciable liquid residence time. What characteristics of bubble-cap trays make it possible for them to operate satisfactorily at low vapor and liquid rates?

Section 6.2

6.4.   Selection of an absorbent.

In Example 6.3, a lean oil of 250 MW is used as the absorbent. Consideration is being given to the selection of a new absorbent. Available streams are:

	Rate, gpm	Density, lb/gal	MW
C_5s	115	5.24	72
Light oil	36	6.0	130
Medium oil	215	6.2	180

Which would you choose? Why? Which are unacceptable?

6.5.   Stripping of VOCs with air.

Volatile organic compounds (VOCs) can be removed from water effluents by stripping with steam or air. Alternatively, the VOCs can be removed by carbon adsorption. The U.S. Environmental Protection Agency (EPA) identified air stripping as the best available technology (BAT). What are the advantages and disadvantages of air compared to steam or carbon adsorption?

6.6.   Best operating conditions for absorbers and strippers.

Prove by equations why, in general, absorbers should be operated at high P and low T, whereas strippers should be operated at low P and high T. Also prove, by equations, why a trade-off exists between number of stages and flow rate of the separating agent.

Section 6.3

6.7.   Absorption of CO_2 from air.

The exit gas from an alcohol fermenter consists of an air–CO_2 mixture containing 10 mol% CO_2 that is to be absorbed in a 5.0-N solution of triethanolamine, containing 0.04 mol CO_2 per mole of amine solution. If the column operates isothermally at 25°C, if the exit liquid contains 78.4% of the CO_2 in the feed gas to the absorber, and if absorption is carried out in a six-theoretical-plate column, use the equilibrium data below to calculate: (a) exit gas composition, (b) moles of amine solution required per mole of feed gas.

Equilibrium Data					
Y	0.003	0.008	0.015	0.023	0.032	0.043
X	0.01	0.02	0.03	0.04	0.05	0.06
Y	0.055	0.068	0.083	0.099	0.12	
X	0.07	0.08	0.09	0.10	0.11	

Y = moles CO_2/mole air; X = moles CO_2/mole amine solution

6.8.   Absorption of acetone from air.

Ninety-five percent of the acetone vapor in an 85 vol% air stream is to be absorbed by countercurrent contact with pure water in a valve-tray column with an expected overall tray efficiency of 50%. The column will operate at 20°C and 101 kPa. Equilibrium data for acetone–water at these conditions are:

Mole-percent acetone in water	3.30	7.20	11.7	17.1
Acetone partial pressure in air, torr	30.00	62.80	85.4	103.0

Calculate: (a) the minimum value of L'/V', the ratio mol H_2O/mol air; (b) N_t, using a value of L'/V' of 1.25 times the minimum; and (c) the concentration of acetone in the exit water.

From Table 5.2 for N connected equilibrium stages, there are $2N + 2C + 5$ degrees of freedom. Specified in this problem are:

Stage pressures (101 kPa)	N
Stage temperatures (20°C)	N
Feed stream composition	$C-1$
Water stream composition	$C-1$
Feed stream T, P	2
Water stream, T, P	2
Acetone recovery	1
L/V	1
	$2N + 2C + 4$

The last specification is the gas feed rate at 100 kmol/h.

6.9.   Absorber-stripper system.

A solvent-recovery plant consists of absorber and stripper plate columns. Ninety percent of the benzene (B) in the inlet gas stream, which contains 0.06 mol B/mol B-free gas, is recovered in the absorber. The oil entering the top of the absorber contains 0.01 mol B/mol pure oil. In the leaving liquid, $X = 0.19$ mol B/mol pure oil. Operating temperature is 77°F (25°C).

Superheated steam is used to strip benzene from the benzene-rich oil at 110°C. Concentrations of benzene in the oil = 0.19 and 0.01, in mole ratios, at inlet and outlet, respectively. Oil (pure)-to-steam (benzene-free) flow rate ratio = 2.0. Vapors are condensed, separated, and removed. Additional data are: MW oil = 200, MW benzene = 78, and MW gas = 32. Benzene equilibrium data are:

Equilibrium Data at Column Pressures

X in Oil	Y in Gas, 25°C	Y in Steam, 110°C
0	0	0
0.04	0.011	0.1
0.08	0.0215	0.21
0.12	0.032	0.33
0.16	0.042	0.47

0.20	0.0515	0.62
0.24	0.060	0.795
0.28	0.068	1.05

Calculate the: (a) molar ratio of B-free oil to B-free gas in the absorber; (b) theoretical plates for the absorber; and (c) minimum steam flow rate required to remove the benzene from 1 mol of oil under given terminal conditions, assuming an infinite-plates column.

6.10.   Steam stripping of benzene from oil.

A straw oil used to absorb benzene (B) from coke-oven gas is to be steam-stripped in a sieve-plate column at 1 atm to recover B. Equilibrium at the operating temperature is approximated by Henry's law. It is known that when the oil phase contains 10 mol% B, its partial pressure is 5.07 kPa. The oil is assumed nonvolatile, and enters containing 8 mol% B, 75% of which is to be recovered. The steam leaving is 3 mol% B. (a) How many theoretical stages are required? (b) How many moles of steam are required per 100 mol of feed? (c) If the benzene recovery is increased to 85% using the same steam rate, how many theoretical stages are required?

Section 6.4

6.11.   Stripping of VOCs from groundwater with air.

Groundwater at a rate of 1,500 gpm, containing three volatile organic compounds (VOCs), is to be stripped in a trayed tower with air to produce drinking water that will meet EPA standards. Relevant data are given below. Determine the minimum air flow rate in scfm (60°F, 1 atm) and the number of equilibrium stages required if an air flow rate of twice the minimum is used and the tower operates at 25°C and 1 atm. Determine the composition in parts per million (ppm) for each VOC in the resulting drinking water.

		Concentration, ppm	
			Max. for
		Ground-	Drinking
Component	K-value	water	water
1,2-Dichloroethane (DCA)	60	85	0.005
Trichloroethylene (TCE)	650	120	0.005
1,1,1-Trichloroethane (TCA)	275	145	0.200

Note: ppm = parts per million by weight.

6.12.   Stripping of SO$_2$ and butadienes with N$_2$.

SO$_2$ and butadienes (B3 and B2) are stripped with nitrogen from the liquid stream shown in Figure 6.45 so that butadiene sulfone (BS) product contains less than 0.05 mol% SO$_2$ and less than 0.5 mol% butadienes. Estimate the flow rate of nitrogen, N$_2$, and the equilibrium stages required. At 70°C, K-values for SO$_2$, B2, B3, and BS are 6.95, 3.01, 4.53, and 0.016, respectively.

6.13.   Trade-off between stages and pressure for absorption.

Determine by the Kremser method the separation that can be achieved for the absorption operation indicated in Figure 6.46 for the following conditions: (a) six equilibrium stages and 75 psia operating pressure, (b) three stages and 150 psia, (c) six stages and 150 psia. At 90°F and 75 psia, the K-value of nC$_{10}$ = 0.0011.

What do you conclude about a trade-off between pressure and stages?

6.14.   Absorption of a hydrocarbon gas.

One thousand kmol/h of rich gas at 70°F with 25% C$_1$, 15% C$_2$, 25% C$_3$, 20% nC$_4$, and 15% nC$_5$ by moles is to be absorbed by 500 kmol/h of nC$_{10}$ at 90°F in an absorber operating at 4 atm. K-values

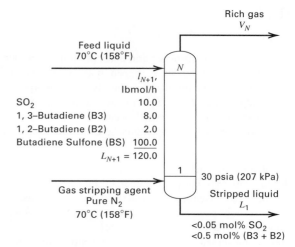

Figure 6.45 Data for Exercise 6.12.

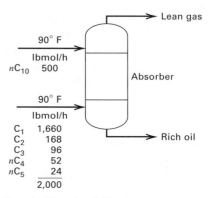

Figure 6.46 Data for Exercise 6.13.

can be obtained from Figure 2.4, except for for nC$_{10}$, which at 80°F and 4 atm is 0.0014. Calculate, by the Kremser method, % absorption of each component for 4, 10, and 30 theoretical stages. What do you conclude ?

Section 6.5

6.15.   Comparison of measured overall stage efficiency with correlations.

Using the data of Example 6.3, back-calculate E_o for propane and compare the result with estimates from the Drickamer–Bradford and O'Connell correlations.

6.16.   Production of 95% H$_2$ by absorption of HCs from a refinery gas.

Fuel cell automotive systems are being considered that will require hydrogen of 95% purity. A refinery stream of 800,000 scfm (at 32°F, 1 atm), containing 72.5% H$_2$, 25% CH$_4$, and 2.5% C$_2$H$_6$, is available. To convert this gas to the required purity, oil absorption, activated charcoal adsorption, and membrane separation are being considered. For oil absorption, an available n-octane stream can be used as absorbent. Because the 95% H$_2$ must be delivered at not less than 375 psia, it is proposed to operate the absorber at 400 psia and 100°F. If at least 80% of the hydrogen fed to the absorber is to leave in the exit gas, determine the: (a) minimum absorbent rate in gpm; (b) absorbent rate if 1.5 times the minimum amount is used; (c) number of theoretical stages; (d) stage efficiency for each of the three species in the feed gas, using the O'Connell correlation; (e) number of trays actually required; (f) exit gas composition, accounting for octane stripping; and (g) if the lost octane in part (f) is not recovered, estimate its value if the process operates 7,900 h/year

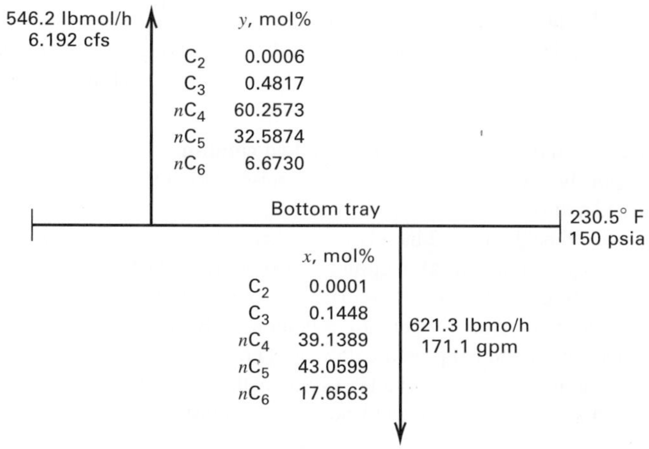

546.2 lbmol/h
6.192 cfs

	y, mol%
C_2	0.0006
C_3	0.4817
nC_4	60.2573
nC_5	32.5874
nC_6	6.6730

Bottom tray

230.5° F
150 psia

	x, mol%
C_2	0.0001
C_3	0.1448
nC_4	39.1389
nC_5	43.0599
nC_6	17.6563

621.3 lbmo/h
171.1 gpm

Figure 6.47 Data for Exercise 6.18.

and the octane is valued at $1.00/gal. Would the octane preclude use of this hydrogen in fuel cells?

6.17. Scale-up of absorber using Oldershaw-column data.

The absorber of Examples 6.1 and 6.4 is being scaled up by a factor of 15, so a column with an 11.5-ft diameter will be needed. Because of the 30% efficiency for the original tray, a new design has been developed and tested in an Oldershaw column. The resulting Murphree vapor-point efficiency, E_{OV}, for the new tray design for this system is 55%. Estimate E_{MV} and E_o. To estimate the length of the liquid flow path, Z_L, use Figure 6.15. Assume that $u/D_E = 6 \text{ ft}^{-1}$.

Section 6.6

6.18. Diameter of a valve-tray column.

Conditions at the bottom tray of a reboiled stripper are in Figure 6.47. If valve trays are used with 24-inch tray spacing, estimate column diameter for 80% of flooding.

6.19. Flooding velocity and diameter of a valve tray.

Determine the flooding velocity and column diameter at the top tray of a valve-tray absorber using the data below:

Pressure	400 psia
Temperature	128°F
Vapor rate	530 lbmol/h
Vapor MW	26.6
Vapor density	1.924 lb/ft^3
Liquid rate	889 lbmol/h
Liquid MW	109
Liquid density	41.1 lb/ft^3
Liquid surface tension	18.4 dynes/cm
Foaming factor	0.75
Tray spacing	24 inches
Fraction flooding	0.85
Valve trays	

6.20. Sieve-tray column design.

For Exercise 6.16, if an octane absorbent flow rate of 40,000 gpm is used in a sieve-tray column with 24-inch tray spacing, a weir height of 2.5 inches, and holes of $\frac{1}{4}$-inch diameter, determine for an 0.80 foaming factor and 0.70 fraction flooding: (a) column diameter based on conditions near the bottom of the column, (b) vapor pressure drop per tray, (c) whether weeping will occur, (d) entrainment rate, (e) fractional decrease in E_{MV} due to entrainment, and (f) froth height in the downcomer.

6.21. Trayed-column absorber design.

Repeat the calculations of Examples 6.5, 6.6, and 6.7 for a column diameter corresponding to 40% of flooding.

6.22. Trayed column for acetone absorption.

For the acetone absorber of Figure 6.1, assuming sieve trays with 10% hole area, $\frac{3}{16}$-inch holes, and 18-inch tray spacing, estimate: (a) column diameter for 0.85 foaming factor and 0.75 fraction flooding; (b) vapor pressure drop per tray; (c) number of transfer units, N_G and N_L, from (6-62) and (6-63), respectively; (d) N_{OG} from (6-61); (e) controlling resistance to mass transfer; and (f) E_{OV} from (6-56). From your results, are 30 actual trays adequate?

6.23. Design of a VOC stripper.

Design a VOC stripper for the flow conditions and separation of Example 6.2, with wastewater and air flow rates twice as high. To develop the design, determine: (a) number of equilibrium stages required, (b) column diameter for sieve trays, (c) vapor pressure drop per tray, (d) Murphree vapor-point efficiencies using the Chan and Fair method, and (e) number of actual trays.

Section 6.7

6.24. Absorption of SO$_2$ in a packed column.

Air containing 1.6 vol% SO$_2$ is scrubbed at 1 atm with pure water in a packed column of 1.5-m^2 cross-sectional area and 3.5m height, packed with No. 2 plastic Super Intalox saddles. Total gas flow rate is 0.062 kmol/s, liquid flow rate is 2.2 kmol/s, and outlet gas SO$_2$ concentration is $y = 0.004$. At the column temperature, the equilibrium relationship is $y^* = 40x$. (a) What is L/L_{min}? (b) Calculate N_{OG} and compare your answer to that for the number of theoretical stages required. (c) Determine H_{OG} and the HETP from the operating data. (d) Calculate K_Ga from the data, based on a partial-pressure driving force as in Item 2 of Table 6.5.

6.25. Absorption of SO$_2$ in packed tower.

An SO$_2$–air mixture is scrubbed with water in a packed tower at 20°C and 1 atm. Solute-free water enters the top at 1,000 lb/h and is well distributed over the packing. The liquor leaving contains 0.6 lb SO$_2$/100 lb of solute-free water. The partial pressure of SO$_2$ in the gas leaving is 23 torr. The mole ratio of water to air is 25. The necessary equilibrium data are given below. (a) What percent of the SO$_2$ in the entering gases is absorbed in the tower? (b) In operation it was found that rate coefficients k_p and k_L remained substantially constant throughout the tower at:

$$k_L = 1.3 \text{ ft/h}$$
$$k_p = 0.195 \text{ lbmol/h-ft}^2\text{-atm}$$

At a point in the tower where the liquid concentration is 0.001 lbmol SO$_2$ per lbmol of water, what is the liquid concentration at the gas–liquid interface in lbmol/ft^3? Solution density is 1 gm/cm^3.

Solubility of SO$_2$ in H$_2$O at 20°C	
lb SO$_2$ / 100 lb H$_2$O	Partial Pressure of SO$_2$ in Air, torr
0.02	0.5
0.05	1.2
0.10	3.2
0.15	5.8
0.20	8.5
0.30	14.1
0.50	26.0
0.70	39.0
1.0	59

6.26. Stripping of benzene from wastewater in a packed column.

Wastewater at 600 gpm, containing 10 ppm (by weight) of benzene, is to be stripped with air in a packed column operating at 25°C and 2 atm to produce water containing 0.005 ppm of benzene. The packing is 2-inch polypropylene Flexirings. The vapor pressure of benzene at 25°C is 95.2 torr. The solubility of benzene in water at 25°C is 0.180 g/100 g. An expert in VOC stripping with air suggests use of 1,000 scfm of air (60°F, 1 atm), At these conditions, for benzene:

$$k_La = 0.067 \text{ s}^{-1} \quad \text{and} \quad k_Ga = 0.80 \text{ s}^{-1}$$

Determine the: (a) minimum air-stripping rate in scfm. (Is it less than the rate suggested by the expert? If not, use 1.4 times your minimum value); (b) stripping factor based on the air rate suggested by the expert; (c) number of transfer units, N_{OG}; (d) overall mass-transfer coefficient, K_Ga, in units of mol/m³-s-kPa and s⁻¹ and which phase controls mass transfer; and (e) volume of packing in m³.

Section 6.8

6.27. Scrubbing of GeCl₄ with caustic in a packed column.

Germanium tetrachloride ($GeCl_4$) and silicon tetrachloride ($SiCl_4$) are used in the production of optical fibers. Both chlorides are oxidized at high temperature and converted to glasslike particles. Because the $GeCl_4$ oxidation is incomplete, the unreacted $GeCl_4$ is scrubbed from its air carrier with a dilute caustic solution in a packed column operating at 25°C and 1 atm. The dissolved $GeCl_4$ has essentially no vapor pressure and mass transfer is controlled by the gas phase. Thus, the equilibrium curve is a straight line of zero slope. Why? The entering gas is 23,850 kg/day of air containing 288 kg/day of $GeCl_4$. The air also contains 540 kg/day of Cl_2, which, when dissolved, also will have no vapor pressure. The two liquid-phase reactions are:

$$GeCl_4 + 5OH^- \rightarrow HGeO_3^- + 4Cl^- + 2H_2O$$
$$Cl_2 + 2OH^- \rightarrow ClO^- + Cl^- + H_2O$$

Ninety-nine % of both $GeCl_4$ and Cl_2 must be absorbed in a 2-ft-diameter column packed to a height of 10 ft with ½-inch ceramic Raschig rings. The column is to operate at 75% of flooding. For the packing: $\epsilon = 0.63$, $F_P = 580 \text{ ft}^{-1}$, and $D_p = 0.01774$ m.

Gas-phase mass-transfer coefficients for $GeCl_4$ and Cl_2 from empirical equations and properties are given by,

$$K_ya = k_ya$$
$$\frac{k_y}{(V/S)} = 1.195\left[\frac{D_PV'}{\mu(1-\epsilon_o)}\right]^{-0.36}(N_{Sc})^{-2/3}$$
$$\epsilon_o = \epsilon - h_L$$
$$h_L = 0.03591(L')^{0.331}$$
$$a = \frac{14.69(808\,V'/\rho^{1/2})^n}{(L')^{0.111}}$$
$$n = 0.01114\,L' + 0.148$$

where S = column cross-sectional area, m²; k_y = kmol/m²-s; V = molar gas rate, kmol/s; D_p = equivalent packing diameter, m; μ = gas viscosity, kg/m-s; ρ = gas density, kg/m³; N_{Sc} = Schmidt number = $\mu/\rho D_i$; D_i = molecular diffusivity of component i in the gas, m²/s; a = interfacial area for mass transfer, m²/m³ of packing; L' = liquid mass velocity, kg/m²-s; and V' = gas mass velocity, kg/m²-s. For the two diffusing species, take

$$D_{GeCl_4} = 0.000006 \text{ m}^2/\text{s}$$
$$D_{Cl_2} = 0.000013 \text{ m}^2/\text{s}$$

Determine the: (a) dilute caustic flow rate in kg/s. (b) required packed height in feet based on the controlling species ($GeCl_4$ or

Cl_2); is the 10 ft of packing adequate? (c) percent absorption of $GeCl_4$ and Cl_2 based on the available 10 ft of packing. If the 10 ft of packing is not sufficient, select an alternative packing that is adequate.

6.28. Stripping of VOCs in a packed column.

For the VOC-stripping task of Exercise 6.26, the expert has suggested a tower diameter of 0.80 m with a pressure drop of 500 N/m²-m of packed height (0.612 inch H₂O/ft). Verify the information from the expert by estimating: (a) fraction of flooding using the GPDC chart of Figure 6.35 with $F_P = 24 \text{ ft}^2/\text{ft}^3$, (b) pressure drop at flooding, (c) pressure drop at the operating conditions of Exercise 6.26 using the GPDC chart, and (d) pressure drop at operating conditions using the correlation of Billet and Schultes, assuming that 2-inch plastic Flexiring packing has the same characteristics as 2-inch plastic Pall rings.

6.29. Mass-transfer coefficients for a packed stripper.

For the VOC stripper of Exercise 6.26, the expert suggested certain mass-transfer coefficients. Check these by using the correlations of Billet and Schultes, assuming that 2-inch plastic Flexiring packing has the characteristics of 2-inch plastic Pall rings.

6.30. Scrubbing of NH₃ with water in a packed column.

A 2 mol% NH₃-in-air mixture at 68°F and 1 atm is scrubbed with water in a tower packed with 1.5-inch ceramic Berl saddles. The inlet water mass velocity is 2,400 lb/h-ft², and the inlet gas mass velocity is 240 lb/h-ft². The gas solubility follows Henry's law, $p = Hx$, where p is the partial pressure of ammonia, x is the mole fraction of ammonia in the liquid, and $H = 2.7$ atm/mole fraction. (a) Calculate the packed height for 90% NH₃ absorption. (b) Calculate the minimum water mass velocity in lb/h-ft² for absorbing 98% of the NH₃. (c) The use of 1.5-inch ceramic Hiflow rings rather than Berl saddles has been suggested. What changes would this cause in K_Ga, pressure drop, maximum liquid rate, K_La, column height, column diameter, H_{OG}, and N_{OG}?

6.31. Absorption of CO₂ into caustic in a packed column.

Your company, for a carbon-credit exchange program, is considering a packed column to absorb CO_2 from air into a dilute-caustic solution. The air contains 3 mol% CO_2, and a 97% recovery of CO_2 is mandated. The air flow rate is 5,000 ft³/minute at 60°F, 1 atm. It may be assumed that the equilibrium curve is $Y^* = 1.75X$, where Y and X are mole ratios of CO_2 to CO_2-free carrier gas and liquid, respectively. A column diameter of 2.5 ft with 2-inch Intalox saddle packing is assumed for the initial design estimates. Assume caustic solution has the properties of water. Calculate the: (a) minimum caustic solution-to-air molar flow rate ratio, (b) maximum possible concentration of CO_2 in the caustic solution, (c) number of theoretical stages at $L/V = 1.4$ times the minimum, (d) caustic solution rate, (e) pressure drop per foot of column height (what does this result suggest?), (f) overall number of gas transfer units, N_{OG}, and (g) height of packing, using a K_Ga of 2.5 lbmol/h-ft³-atm. Is this a reasonable way to get carbon credits?

Section 6.9

6.32. Back-calculation of mass-transfer coefficients.

At a point in an ammonia absorber at 101.3 kPa and 20°C, using water as the absorbent, the bulk gas phase contains 10 vol% NH₃. At the interface, the partial pressure of NH₃ is 2.26 kPa. Ammonia concentration in the body of the liquid is 1 wt%. The rate of ammonia absorption at this point is 0.05 kmol/h-m². (a) Given this information and the equilibrium curve in Figure 6.48, calculate X, Y, Y_I, X_I, X^*, Y^*, K_Y, K_X, k_y, and k_x. (b) What percent of the mass-transfer resistance is in each phase? (c) Verify for these data that $1/K_Y = 1/k_y + H'/k_X$.

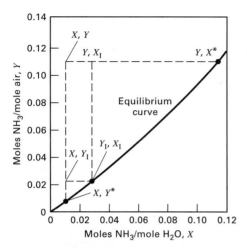

Figure 6.48 Data for Exercise 6.32.

6.33. Desorption of NH₃ from water with air in a packed column.

One thousand ft³/h of a 10 mol% NH₃-in-air mixture is required to produce nitrogen oxides. This mixture is to be obtained by desorbing an aqueous 20 wt% NH₃ solution with air at 20°C. The spent solution should not contain more than 1 wt% NH₃. Calculate the volume of packing required for the desorption column. Vapor–liquid equilibrium data for Exercise 6.32 can be used and $K_G a = 4$ lbmol/h-ft³-atm partial pressure.

6.34. Absorption of NH₃ from air with water in a packed column.

Ammonia, at a partial pressure of 12 torr in a 2,000-lb/hr air stream saturated with water vapor at 68°F and 1 atm, must be removed to the extent of 99.6% by water absorption at the same temperature and pressure. (a) Calculate the minimum amount of water using the equilibrium data for Exercise 6.32 in Figure 6.48. (b) Assuming twice the minimum water flow and one-half the flooding gas velocity, compute the dimensions of a column packed with 38-mm ceramic Berl saddles. (c) Repeat part (b) for 50-mm Pall rings. (d) Which of the two packings would you recommend?

6.35. Absorption of Cl₂ with water in a packed column.

One hundred kmol/h of air containing 20 mol% Cl₂ enters a packed bed, wherein 95% of the Cl₂ is absorbed by water at 20°C and 1 atm. Using the given *x–y* equilibrium data below, obtain: (a) minimum water rate in kg/h, and (b) N_{OG} for twice the minimum water rate. Equilibrium data at 20°C in Cl₂ mole fractions is:

x	0.0001	0.00015	0.0002	0.00025	0.0003
y	0.006	0.012	0.024	0.04	0.06

6.36. Diameter of a packed tower.

Calculate the diameter and height for the column of Example 6.15 if the tower is packed with 1.5-inch metal Pall rings. Assume the absorbing solution has the properties of water, and use conditions at the bottom of the tower, where flow rates are highest.

6.37. Absorption of acetone with water in a packed tower.

You are asked to design a packed column to recover acetone from air by absorption with water at 60°F. The air contains 3 mol% acetone, and a 97% recovery is desired. The gas flow rate is 50 ft³/min at 60°F, 1 atm. The maximum-allowable gas superficial velocity in the column is 2.4 ft/s. In the range of operation, $Y^* = 1.75X$, where Y and X are mole ratios. Calculate the: (a) minimum water-to-air molar flow rate ratio; (b) maximum acetone concentration in the aqueous solution; (c) theoretical stages for flow rate ratio of 1.4 × the minimum; (d) corresponding number of overall gas transfer units; (e) height of packing, assuming $K_Y a = 12.0$ lbmol/h-ft³-molar ratio difference; and (f) height of packing as a function of the molar flow rate ratio, assuming V and HTU remain constant.

6.38. Diameter and height of a packed tower for NH₃ absorption.

Determine the diameter and packed height of a packed tower required to recover 99% of the ammonia from a mixture that contains 6 mol% NH₃ in air. The tower, packed with 1-inch metal Pall rings, must handle 2,000 ft³/min of gas at 68°F and 1 atm. The water-absorbent rate will be twice the minimum, and the gas velocity will be 50% of the flooding velocity. Assume isothermal operation at 68°F and 1 atm. Equilibrium data are in Figure 6.48.

6.39. Absorption of SO₂ with structured packing.

A tower, packed with Montz B1-200 metal structured packing, is to be designed to absorb SO₂ from air by scrubbing with water. The entering gas, at an SO₂-free flow rate of 6.90 lbmol/h-ft² of bed cross section, contains 80 mol% air and 20 mol% SO₂. Water enters at 364 lbmol/h-ft² of bed cross section. The exiting gas is to contain 0.5 mol% SO₂. Assume that neither air nor water is transferred between phases and that the tower operates at 2 atm and 30°C. Equilibrium data in mole fractions for SO₂ solubility in water at 30°C and 2 atm (*Perry's Chemical Engineers' Handbook*, 4th ed., Table 14.31, p. 14-6) have been fitted to the equation:

$$y = 12.697x + 3148.0x^2 - 4.724 \times 10^5 x^3 + 3.001 \times 10^7 x^4 \\ - 6.524 \times 10^8 x^5$$

(a) Derive the following molar material balance operating line for SO₂ mole fractions:

$$x = 0.0189\left(\frac{y}{1-y}\right) - 0.00010$$

(b) Use a spreadsheet to calculate NTU based on the overall gas-phase resistance.

Chapter 7

Distillation of Binary Mixtures

§7.0 INSTRUCTIONAL OBJECTIVES

After completing this chapter, you should be able to:

- Explain the need in distillation for a condenser to produce reflux and a reboiler to produce boilup.
- Determine the five construction lines of the McCabe–Thiele method using material balances and vapor–liquid equilibrium relations.
- Distinguish among five possible phase conditions of the feed.
- Apply the McCabe–Thiele method for determining minimum reflux ratio, minimum equilibrium stages, number of stages for a specified reflux ratio greater than minimum, and optimal feed-stage location, given the required split between the two feed components.
- Use a Murphree vapor-stage efficiency to determine the number of actual stages (plates) from the number of equilibrium stages.
- Extend the McCabe–Thiele method to multiple feeds, sidestreams, and open steam (in place of a reboiler).
- Estimate overall stage efficiency for binary distillation from correlations and laboratory column data.
- Determine the diameter of a trayed tower and the size of the reflux drum.
- Determine packed height and diameter of a packed column.

In *distillation* (*fractionation*), a feed mixture of two or more components is separated into two or more products, including, and often limited to, an overhead distillate and a bottoms product, whose compositions differ from that of the feed. Most often, the feed is a liquid or a vapor–liquid mixture. The bottoms product is almost always a liquid, but the distillate may be a liquid, a vapor, or both. The separation requires that: (1) a second phase be formed so that both liquid and vapor are present and can make contact while flowing countercurrently to each other in a trayed or packed column, (2) components have different volatilities so that they partition between phases to different extents, and (3) the two phases are separable by gravity or mechanical means. Distillation differs from absorption and stripping in that the second fluid phase is usually created by thermal means (vaporization and condensation) rather than by the introduction of a second phase that may contain an additional component or components not present in the feed mixture.

According to Forbes [1], distillation dates back to at least the 1st century A.D. By the 11th century, distillation was used in Italy to produce alcoholic beverages. At that time, distillation was a batch process. The liquid feed was placed into a heated vessel, causing part of the liquid to evaporate. The vapor passed out of the vessel into a water-cooled condenser and dripped into a product receiver. The word *distillation* is derived from the Latin word *destillare*, which means "dripping." By the 16th century, it was known that the extent of separation could be improved by providing multiple vapor–liquid contacts (stages) in a so-called Rectificatorium. The term *rectification* is derived from the Latin words *recte facere*, meaning "to improve." Today, almost pure products are obtained by multistage contacting.

Multistage distillation is the most widely used industrial method for separating chemical mixtures. However, it is a very energy-intensive technique, especially when the relative volatility, α, (2-21), of the key components being separated is low (<1.50). Mix et al. [2] report that the energy consumption for distillation in the United States for 1976 totaled 2×10^{15} Btu (2 quads), which was nearly 3% of the entire national energy consumption. Approximately two-thirds of the distillation energy was consumed by petroleum refineries, where distillation is used to separate crude oil into petroleum fractions, light hydrocarbons (C_2's to C_5's), and other organic chemicals. Distillation is also widely used in the chemical industry, to recover and purify small biomolecules such as ethanol, acetone, and *n*-butanol, and solvents (e.g., organic alcohols, acids, and ketones) in the biochemical industry. However, it is scarcely used in bioseparations involving larger biological metabolites, polymers, or products that are thermolabile.

Industrial Example

The fundamentals of distillation are best understood by the study of *binary distillation*, the separation of a two-component mixture. The more general and mathematically complex case of a multicomponent mixture is covered in Chapters 10 and 11.

A representative binary distillation is shown in Figure 7.1 for the separation of 620 lbmol/h of a mixture of 46 mol% benzene (the more volatile component) from 54 mol% toluene. The purpose of the 80% efficient, 25-sieve-tray column and the partial reboiler that acts as an additional theoretical stage is to separate the feed into a liquid distillate of 99 mol% benzene and a liquid bottoms product of 98 mol% toluene. The column operates at a reflux-drum pressure of 18 psia (124 kPa). For a negligible pressure drop across the condenser and a pressure drop of 0.1 psi/tray (0.69 kPa/tray) for the vapor as it flows up the column, the pressure in the reboiler is $18 + 0.1(25) = 20.5$ psia (141 kPa). In this pressure range, benzene and toluene form near-ideal mixtures with a relative volatility, α, (2-21), of from 2.26 at the bottom tray to 2.52 at the top tray, as determined from Raoult's law, (2-44). The reflux ratio (reflux rate to distillate rate) is 2.215. If an infinite number of stages were used, the required reflux ratio would be 1.708, the minimum value. Thus, the ratio of reflux rate to minimum reflux rate is 1.297.

Most distillation columns are designed for optimal-reflux-to-minimum-reflux ratios of 1.1 to 1.5. If an infinite ratio of reflux to minimum reflux were used, only 10.7 theoretical stages would be required. Thus, the ratio of theoretical stages to minimum theoretical stages for this example is $21/10.7 = 1.96$. This ratio is often about 2 for operating columns. The stage efficiency is 20/25 or 80%. This is close to the average efficiency of trayed distillation columns.

The feed is a saturated (bubble-point) liquid at 55 psia (379 kPa) and 294°F (419 K). When flashed adiabatically across the feed valve to the feed-tray pressure of 19.25 psia (133 kPa), 23.4 mol% of the feed is vaporized, causing the temperature to drop to 220°F (378 K).

A total condenser is used to obtain saturated-liquid reflux and liquid distillate at a bubble-point temperature of 189°F (360 K) at 18 psia (124 kPa). The heat duty of the condenser is 11,820,000 Btu/h (3.46 MW). At the bottom of the column, a partial reboiler is used to produce vapor boilup and a liquid bottoms product. Assuming that the boilup and bottoms are in physical equilibrium, the partial reboiler functions as an additional theoretical stage, for a total of 21 theoretical stages. The bottoms is a saturated (bubble-point) liquid, at 251°F (395 K) and 20.5 psia (141 kPa). The reboiler duty is 10,030,000 Btu/h (2.94 MW), which is within 15% of the condenser duty.

The inside diameter of the column in Figure 7.1 is a constant 5 ft (1.53 m). At the top, this corresponds to 84% of flooding, while at the bottom, 81%. The column is provided with three feed locations. For the design conditions, the optimal feed entry is between trays 12 and 13. Should the feed composition or product specifications change, one of the other feed trays could become optimal.

Columns similar to those in Figure 7.1 have been built for diameters up to at least 30 ft (9.14 m). With a 24-inch (0.61-m) tray spacing, the number of trays is usually no greater than 150. For the sharp separation of a binary mixture with an $\alpha < 1.05$, distillation can require many hundreds of trays, so a more efficient separation technique should be sought. Even when distillation is the most economical separation technique, its second-law efficiency, §2.1, can be less than 10%.

In Figure 1.13, distillation is the most mature of all separation operations. Design and operation procedures are well established (see Kister [3, 4]). Only when vapor–liquid equilibrium, azeotrope formation, or other data are uncertain is a laboratory and/or pilot-plant study necessary prior to design of a commercial unit. Table 7.1, taken partially from Mix et al. [2], lists common commercial binary distillations in decreasing order of difficulty of separation. Included are average values of α, number of trays, typical column operating pressure, and reflux-to-minimum-reflux ratio. Although the data in Table 7.1 refer to trayed towers, distillation is also carried out in packed columns. Frequently, additional distillation capacity is achieved with existing trayed towers by replacing trays with random or structured packing.

Equipment design and operation, as well as equilibrium and rate-based calculational procedures, are covered in this chapter. Trayed and packed distillation columns are mostly identical to absorption and stripping columns discussed previously. Where appropriate, reference is made to Chapter 6 and only important differences are discussed in this chapter.

§7.1 EQUIPMENT AND DESIGN CONSIDERATIONS

Types of trays and packings for distillation are identical to those used in absorption and stripping, as shown in Figures 6.2 to 6.7, and compared in Tables 6.2 and 6.3.

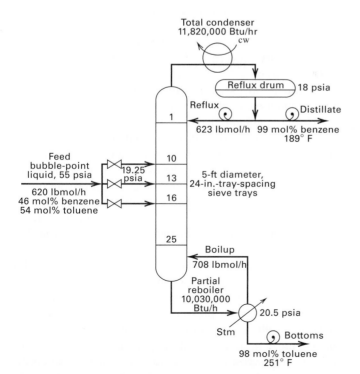

Figure 7.1 Distillation of a binary mixture of benzene and toluene.

Table 7.1 Representative Commercial Binary Distillation Operations [2]

Binary Mixture	Average Relative Volatility	Number of Trays	Typical Operating Pressure, psia	Reflux-to-Minimum-Reflux Ratio
1,3-Butadiene/vinyl acetylene	1.16	130	75	1.70
Vinyl acetate/ethyl acetate	1.16	90	15	1.15
o-Xylene/m-xylene	1.17	130	15	1.12
Isopentane/n-pentane	1.30	120	30	1.20
Isobutane/n-butane	1.35	100	100	1.15
Ethylbenzene/styrene	1.38	34	1	1.71
Propylene/propane	1.40	138	280	1.06
Methanol/ethanol	1.44	75	15	1.20
Water/acetic acid	1.83	40	15	1.35
Ethylene/ethane	1.87	73	230	1.07
Acetic acid/acetic anhydride	2.02	50	15	1.13
Toluene/ethylbenzene	2.15	28	15	1.20
Propane/1,3-butadiene	2.18	40	120	1.13
Ethanol azeotrope/water	2.21	60	15	1.35
Isopropanol/water	2.23	12	15	1.28
Benzene/toluene	3.09	34	15	1.15
Methanol/water	3.27	60	45	1.31
Cumene/phenol	3.76	38	1	1.21
Benzene/ethylbenzene	6.79	20	15	1.14
HCN/water	11.20	15	50	1.36
Ethylene oxide/water	12.68	50	50	1.19
Formaldehyde/methanol	16.70	23	50	1.17
Water/ethylene glycol	81.20	16	4	1.20

§7.1.1 Design and Analysis Factors

Factors that influence the design or analysis of a binary-distillation operation include:

1. Feed flow rate, composition, temperature, pressure, and phase condition

2. Desired degree of component separation

3. Operating pressure (which must be below the critical pressure of the mixture)

4. Pressure drop, particularly for vacuum operation

5. Minimum reflux ratio and actual reflux ratio

6. Minimum number of equilibrium stages and actual number of equilibrium stages (stage efficiency)

7. Type of condenser (total, partial, or mixed)

8. Degrees of liquid reflux subcooling

9. Type of reboiler (partial or total)

10. Type of trays or packing

11. Column height

12. Feed-entry stage

13. Column diameter

14. Column internals, and materials of construction

15. Heat lability and chemical reactivity of feed components

§7.1.2 Some Initial Considerations

Temperature and phase of the feed are determined at the feed-tray pressure by an adiabatic-flash calculation across the feed valve. As the feed vapor fraction increases, the required reflux ratio (L/D) increases, but the boilup ratio (V/B) decreases.

As column operating pressure is increased, temperatures in the column increase, in a manner similar to a vapor pressure plot. The column operating pressure in the reflux drum should correspond to a distillate temperature somewhat higher (e.g., 10 to 50°F or 6 to 28°C) than the supply temperature of the cooling water to the overhead condenser. However, if this pressure approaches the critical pressure of the more volatile component, then a lower pressure must be used and a refrigerant is required as coolant. For example, in Table 7.1, the separation of ethylene/ethane is conducted at 230 psia (1,585 kPa), giving a column top temperature of −40°F (233 K), which requires a refrigerant. Water at 80°F (300 K) cannot be used because the critical temperature of ethylene is 48.6°F (282 K). If the estimated pressure is less than atmospheric, the operating pressure at the top is often set just above atmospheric to avoid vacuum operation, unless the temperature at the bottom of the column is limited by decomposition, polymerization, excessive corrosion, or other chemical reactions. In that case, vacuum operation is necessary.

For example, in Table 7.1, vacuum operation is required for the separation of ethylbenzene from styrene to maintain a temperature low enough to prevent styrene polymerization in the reboiler.

For given (1) feed, (2) desired degree of separation, and (3) operating pressure, a minimum reflux ratio exists that corresponds to an infinite number of theoretical stages; and a minimum number of theoretical stages exists that corresponds to an infinite reflux ratio. The design trade-off is between the number of stages and the reflux ratio. A graphical method for determining the data needed to determine this trade-off and to determine the optimal feed-stage location is developed in the next section.

§7.2 McCABE–THIELE GRAPHICAL METHOD FOR TRAYED TOWERS

Figure 7.2 shows a column containing the equivalent of N theoretical stages; a total condenser in which the overhead vapor leaving the top stage is totally condensed to a bubble-point liquid distillate and a liquid reflux that is returned to the top stage; a partial reboiler in which liquid from the bottom stage is partially vaporized to give a liquid bottoms product and vapor boilup that is returned to the bottom stage; and an intermediate feed stage. By means of multiple, counter-current stages arranged in a two-section cascade with reflux and boilup, as discussed in §5.4, a sharp separation between the two feed components is possible unless an azeotrope exists, in which case one of the two products will approach the azeotropic composition.

The feed, which contains a more volatile (light) component (the *light key*, LK), and a less-volatile (heavy) component (the *heavy key*, HK), enters the column at feed stage f. At feed-stage pressure, the feed of LK mole fraction z_F may be liquid, vapor, or a mixture of the two. The mole fraction of LK is x_D in the distillate and x_B in the bottoms product. Mole fractions of the HK are $1 - z_F$, $1 - x_D$, and $1 - x_B$.

The goal of distillation is to produce a distillate rich in the LK (i.e., x_D approaching 1.0), and a bottoms product rich in the HK (i.e., x_B approaching 0.0). Whether the separation is achievable depends on $\alpha_{1,2}$ of the two components (LK = 1 and HK = 2), where

$$\alpha_{1,2} = K_1/K_2 \tag{7-1}$$

If the two components form ideal solutions and follow the ideal-gas law in the vapor phase, Raoult's law (2-44) applies, giving

$$K_1 = P_1^s/P \quad \text{and} \quad K_2 = P_2^s/P$$

and from (7-1), the relative volatility is given by the ratio of vapor pressures, $\alpha_{1,2} = P_1^s/P_2^s$, and thus is a function only of temperature. As discussed in §4.2, as the temperature (and therefore the pressure) increases, $\alpha_{1,2}$ decreases. At the mixture convergence pressure (e.g. see Figure 2.6), $\alpha_{1,2} = 1.0$, and separation cannot be achieved at this or a higher pressure.

The relative volatility can be expressed in terms of equilibrium vapor and liquid mole fractions from the K-value expressed as $K_i = y_i/x_i$ (2-19). For a binary mixture,

$$\alpha_{1,2} = \frac{y_1/x_1}{y_2/x_2} = \frac{y_1(1 - x_1)}{y_1(1 - y_1)} \tag{7-2}$$

Solving (7-2) for y_1,

$$y_1 = \frac{\alpha_{1,2}x_1}{1 + x_1(\alpha_{1,2} - 1)} \tag{7-3}$$

For components with close boiling points, the temperature change over the column is small and $\alpha_{1,2}$ is almost constant. An equilibrium curve for the benzene–toluene system is shown in Figure 7.3, where the fixed pressure is 1 atm, at which pure benzene and pure toluene boil at 176 and 231°F, respectively. Thus, these two components are not close-

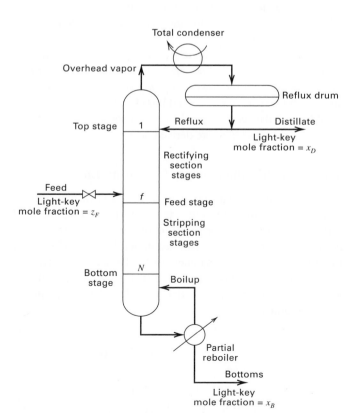

Figure 7.2 Distillation using a total condenser and partial reboiler.

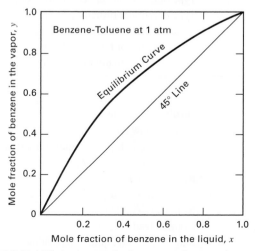

Figure 7.3 Equilibrium curve for benzene–toluene at 1 atm.

boiling. Using (7-3) with experimental $x-y$ data, α varies from 2.6 at the bottom of the column to 2.35 at the top. Equilibrium curves for some average values of α are shown in Figure 4.5. The higher the average value of α, the easier the desired separation. Average values of α in Table 7.1 range from 1.16 to 81.2.

In 1925, McCabe and Thiele [5] published a graphical method for combining the equilibrium curve with material-balance operating lines to obtain, for a binary-feed mixture and selected column pressure, the number of equilibrium stages and reflux required for a desired separation of feed components.

Although computer-aided methods discussed in Chapter 10 are more accurate and easier to apply, the graphical McCabe–Thiele method greatly facilitates visualization of the fundamentals of multistage distillation, and therefore the effort required to learn the method is well justified.

Typical input specifications and results (outputs) from the McCabe–Thiele construction for a single-feed, two-product distillation are summarized in Table 7.2, where it is required that $x_B < z_F < x_D$. The distillate can be a liquid from a total condenser, as shown in Figure 7.2, or a vapor from a partial condenser. The feed-phase condition must be known at column pressure, assumed to be uniform throughout the column. The type of condenser and reboiler must be specified, as well as the ratio of reflux to minimum reflux. From the specification of x_D and x_B for the LK, distillate and bottoms rates, D and B, are fixed by material balance, since

$$Fz_F = x_D D + x_B B$$

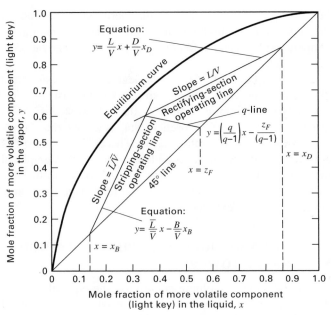

Figure 7.4 Construction lines for McCabe–Thiele method.

But, $B = F - D$ and therefore

$$Fz_F = x_D D + x_B(F - D)$$

or

$$D = F\left(\frac{z_F - x_B}{x_D - x_B}\right)$$

The McCabe–Thiele method determines N, the number of equilibrium stages; N_{min}, the minimum number of equilibrium stages; $R_{min} = L_{min}/D$, the minimum reflux ratio; and the optimal feed-stage location. Lastly, condenser and reboiler heat duties are obtained from energy balances, as discussed in §7.3.5.

Besides the equilibrium curve, the McCabe–Thiele method includes a 45° reference line, operating lines for the upper *rectifying* section and the lower *stripping* section of the column, and a fifth line (the *q-line* or feed line) for the phase or thermal condition of the feed. Typical lines are shown in Figure 7.4. Equations for these lines are derived next.

Table 7.2 Specifications for and Results from the McCabe–Thiele Method for Binary Distillation

Specifications	
F	Total feed rate
z_F	Mole-fraction of LK in the feed
P	Column operating pressure (assumed uniform throughout the column)
	Phase condition of the feed at column pressure
	Vapor–liquid equilibrium curve for the binary mixture at column pressure
	Type of overhead condenser (total or partial)
	Type of reboiler (usually partial)
x_D	Mole-fraction of LK in the distillate
x_B	Mole-fraction of LK in the bottoms
R/R_{min}	Ratio of reflux to minimum reflux

Results	
D	Distillate flow rate
B	Bottoms flow rate
N_{min}	Minimum number of equilibrium stages
R_{min}	Minimum reflux ratio, L_{min}/D
R	Reflux ratio, L/D
V_B	Boilup ratio, $\overline{V}/B$
N	Number of equilibrium stages
	Optimal feed-stage location
	Stage vapor and liquid compositions

§7.2.1 Rectifying-Section Operating Line

Figure 7.2, shows that the *rectifying section* of equilibrium stages extends from the top stage, 1, to just above the feed stage, f. Consider a top portion of the rectifying stages, including the total condenser, as shown by the envelope in Figure 7.5a. A material balance for the LK over the envelope for the total condenser and stages 1 to n is as follows, where y and x refer, respectively, to LK vapor and liquid mole fractions.

$$V_{n+1}y_{n+1} = L_n x_n + D x_D \tag{7-4}$$

Solving (7-4) for y_{n+1} gives the equation for the rectifying-section operating line:

$$y_{n+1} = \frac{L_n}{V_{n+1}}x_n + \frac{D}{V_{n+1}}x_D \tag{7-5}$$

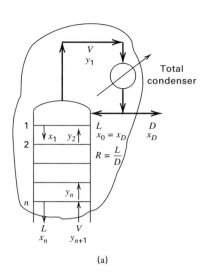

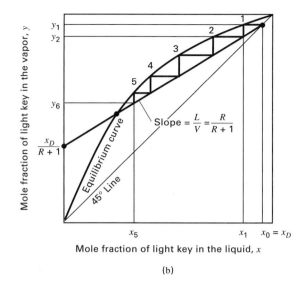

(a)

(b)

Figure 7.5 McCabe–Thiele operating line for the rectifying section.

This equation relates LK compositions y_{n+1} and x_n of *passing streams* V_{n+1} and L_n, respectively. For (7-5) to be plotted as a straight line, $y = mx + b$, which is the locus of compositions of all passing streams in the rectifying section, L and V must not vary from stage to stage in the rectifying section. This is the case if:

1. The two components have equal and constant molar enthalpies of vaporization (latent heats).

2. Component sensible-enthalpy changes ($C_P \Delta T$) and heat of mixing are negligible compared to latent heat changes.

3. The column is insulated, so heat loss is negligible.

4. Column pressure is uniform (thus, no pressure drop).

These are the *McCabe–Thiele assumptions* leading to the condition of *constant molar overflow* in the rectifying section, where the molar flow rates remain constant as the liquid overflows each weir from one stage to the next. Since a total material balance for the rectifying-section envelope in Figure 7.5a gives $V_{n+1} = L_n + D$, if L is constant, then V is also constant for a fixed D. Rewriting (7-5) as in Figure 7.4,

$$y = \frac{L}{V}x + \frac{D}{V}x_D \tag{7-6}$$

Thus, the slope of the operating line in the rectifying section is a constant L/V, with $V > L$ and $L/V < 1$, as in Figure 7.5b.

For constant molar overflow in either the rectifying or the stripping section, only material balances and an equilibrium curve are required. Energy balances are needed only to determine condenser and reboiler duties, as discussed in §7.3.5.

Liquid entering stage 1 at the top is the external reflux rate, L_0, and its ratio to the distillate rate, L_0/D, is reflux ratio R. Because of constant molar overflow, $R = L/D$ is a constant in the rectifying section. Since $V = L + D$, the slope of the operating line is readily related to the reflux ratio:

$$\frac{L}{V} = \frac{L}{L+D} = \frac{L/D}{L/D + D/D} = \frac{R}{R+1} \tag{7-7}$$

Similarly,
$$\frac{D}{V} = \frac{D}{L+D} = \frac{1}{R+1} \tag{7-8}$$

Combining (7-6) to (7-8) produces the most useful form of the operating line for the rectifying section:

$$y = \left(\frac{R}{R+1}\right)x + \left(\frac{1}{R+1}\right)x_D \tag{7-9}$$

If R and x_D are specified, (7-9) plots as a straight line in Figure 7.5b, with an intersection at $y = x_D$ on the 45° line ($y = x$); a slope of $L/V = R/(R+1)$; and an intersection at $y = x_D/(R+1)$ for $x = 0$.

In Figure 7.5b, the stages are stepped off as in §6.3.4 for absorption. Starting from point ($y_1 = x_D, x_0 = x_D$) on the operating line and the 45° line, a horizontal line is drawn to the left until it intersects the equilibrium curve at (y_1, x_1), the compositions of the *equilibrium phases* leaving the top stage. A vertical line is dropped until it intersects the operating line at (y_2, x_1), the compositions of the passing streams between stages 1 and 2. Horizontal- and vertical-line constructions are continued down the rectifying section to give a staircase construction, which is arbitrarily terminated at stage 5. The optimal termination stage is considered in §7.2.3.

§7.2.2 Stripping-Section Operating Line

The stripping section extends from the feed to the bottom stage. In Figure 7.6a, consider a bottom portion of stripping stages, including the partial reboiler and extending up from stage N to stage $m + 1$, below the feed entry. A material balance for the LK over the envelope results in

$$\overline{L}x_m = \overline{V}y_{m+1} + Bx_B \tag{7-10}$$

Solving for y_{m+1}:

$$y_{m+1} = \frac{\overline{L}}{\overline{V}}x_m - \frac{B}{\overline{V}}x_B$$
$$\tag{7-11}$$

or

$$y = \frac{\overline{L}}{\overline{V}}x - \frac{B}{\overline{V}}x_B$$

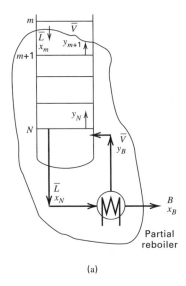

(a)

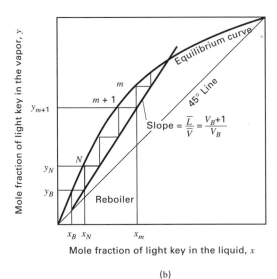

(b)

Figure 7.6 McCabe–Thiele operating line for the stripping section.

where $\overline{L}$ and $\overline{V}$ are total molar flows (which may be different from L and V in the rectifying section because of feed addition), subject to the constant-molar-overflow assumption. The slope of this operating line for the compositions of passing steams in the stripping section is $\overline{L}/\overline{V}$. Because $\overline{L} > \overline{V}, \overline{L}/\overline{V} > 1$, as in Figure 7.6b. This is the inverse of the flow conditions in the rectifying section.

Vapor leaving the partial reboiler is assumed to be in equilibrium with the liquid bottoms product, B, making the partial reboiler an equilibrium stage. The vapor rate leaving it is the *boilup*, $\overline{V}_{N+1}$, and its ratio to the bottoms product rate, $V_B = \overline{V}_{N+1}/B$, is the *boilup ratio*. With the constant-molar-overflow assumption, V_B is constant in the stripping section. Since $\overline{L} = \overline{V} + B$,

$$\frac{\overline{L}}{\overline{V}} = \frac{\overline{V} + B}{\overline{V}} = \frac{V_B + 1}{V_B} \qquad (7\text{-}12)$$

Similarly,
$$\frac{B}{\overline{V}} = \frac{1}{V_B} \qquad (7\text{-}13)$$

Combining (7-11) to (7-13), the stripping-section operating-line equation is:

$$y = \left(\frac{V_B + 1}{V_B}\right)x - \left(\frac{1}{V_B}\right)x_B \qquad (7\text{-}14)$$

If values of V_B and x_B are known, (7-14) can be plotted as a straight line with an intersection at $y = x_B$ on the 45° line and a slope of $\overline{L}/\overline{V} = (V_B + 1)/V_B$, as in Figure 7.6b, which also contains the equilibrium curve and a 45° line. The stages are stepped off, in a manner similar to that described for the rectifying section, starting from $(y = x_B, x = x_B)$ on the operating and 45° lines and moving upward on a vertical line until the equilibrium curve is intersected at $(y = y_B, x = x_B)$, which represents the vapor and liquid leaving the partial reboiler. From that point, the staircase is constructed by drawing horizontal and then vertical lines between the operating line and equilibrium curve, as in Figure 7.6b, where

the staircase is arbitrarily terminated at stage m. Next, the termination of the two operating lines at the feed stage is considered.

§7.2.3 Feed-Stage Considerations—the q-Line

In determining the operating lines for the rectifying and stripping sections, it is noted that although x_D and x_B can be selected independently, R and V_B are not independent of each other, but related by the feed-phase condition.

Consider the five feed conditions in Figure 7.7, where the feed has been flashed adiabatically to the feed-stage pressure. If the feed is a bubble-point liquid, it adds to the reflux, L, from the stage above, to give $\overline{L} = L + F$. If the feed is a dew-point vapor, it adds to the boilup, $\overline{V}$, coming from the stage below, to give $V = \overline{V} + F$. For the partially vaporized feed in Figure 7.7c, $F = L_F + V_F, \overline{L} = L + L_F$, and $V = \overline{V} + V_F$. If the feed is a subcooled liquid, it will cause some of the boilup, $\overline{V}$, to condense, giving $\overline{L} > L + F$ and $V < \overline{V}$. If the feed is a superheated vapor, it will cause a portion of the reflux, L, to vaporize, giving $\overline{L} < L$ and $V > \overline{V} + F$.

For cases (b), (c), and (d) of Figure 7.7, covering feed conditions from a saturated liquid to a saturated vapor, the boilup $\overline{V}$ is related to the reflux L by the material balance

$$\overline{V} = L + D - V_F \qquad (7\text{-}15)$$

and the boilup ratio, $V_B = \overline{V}/B$, is

$$V_B = \frac{L + D - V_F}{B} \qquad (7\text{-}16)$$

Alternatively, the reflux can be obtained from the boilup by

$$L = \overline{V} + B - L_F \qquad (7\text{-}17)$$

Although distillations can be specified by reflux ratio R or boilup ratio V_B, by tradition R or $R/R_{\min}$ is used because the distillate is often the more important product.

For cases (a) and (e) in Figure 7.7, V_B and R cannot be related by simple material balances. An energy balance is

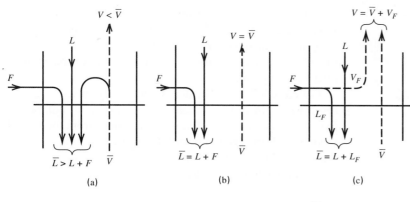

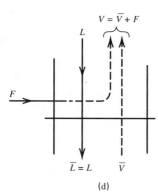

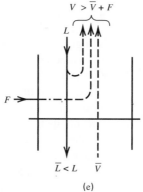

Figure 7.7 Possible feed conditions: (a) subcooled-liquid feed; (b) bubble-point liquid feed; (c) partially vaporized feed; (d) dew-point vapor feed; (e) superheated-vapor feed.

[Adapted from W.L. McCabe, J.C. Smith, and P. Harriott, *Unit Operations of Chemical Engineering*, 5th ed., McGraw-Hill, New York (1993).]

necessary to convert sensible enthalpy of subcooling or superheating into heat of vaporization. This is conveniently done by defining a parameter, q, as the ratio of the increase in molar reflux rate across the feed stage to the molar feed rate,

$$q = \frac{\overline{L} - L}{F} \qquad (7\text{-}18)$$

or by material balance around the feed stage,

$$q = 1 + \frac{\overline{V} - V}{F} \qquad (7\text{-}19)$$

Values of q for the five feed conditions of Figure 7.7 are

Feed Condition	q
Subcooled liquid	>1
Bubble-point liquid	1
Partially vaporized	$L_F/F = 1 -$ molar fraction vaporized
Dew-point vapor	0
Superheated vapor	<0

For subcooled liquids and superheated vapors, a more general definition of q is:

$q =$ enthalpy change to bring the feed to a dew-point vapor divided by enthalpy of vaporization of the feed (dew-point vapor minus bubble-point liquid); that is,

$$q = \frac{(h_F)_{\text{sat'd vapor temperature}} - (h_F)_{\text{feed temperature}}}{(h_F)_{\text{sat'd vapor temperature}} - (h_F)_{\text{sat'd liquid temperature}}} \qquad (7\text{-}20)$$

For a subcooled liquid feed, (7-20) becomes

$$q = \frac{\Delta H^{\text{vap}} + C_{P_L}(T_b - T_F)}{\Delta H^{\text{vap}}} \qquad (7\text{-}21)$$

For a superheated vapor feed, (7-20) becomes

$$q = \frac{C_{P_V}(T_d - T_F)}{\Delta H^{\text{vap}}} \qquad (7\text{-}22)$$

where C_{P_L} and C_{P_V} are molar heat capacities, ΔH^{vap} is the molar enthalpy change from the bubble point to the dew point, and T_F, T_d, and T_b are, respectively, feed, dew-point, and bubble-point temperatures at column operating pressure.

Instead of using (7-14) to locate the stripping operating line on the McCabe–Thiele diagram, it is common to use an alternative method that involves the q-line, shown in Figure 7.4. The q-line, one point of which is the intersection of the rectifying- and stripping-section operating lines, is derived by combining (7-11) with (7-6) to give

$$y(V - \overline{V}) = (L - \overline{L})x + Dx_D + Bx_B \qquad (7\text{-}23)$$

However, overall, $\quad Dx_D + Bx_B = Fz_F \qquad (7\text{-}24)$

and a total material balance around the feed stage gives

$$F + \overline{V} + L = V + \overline{L} \qquad (7\text{-}25)$$

Combining (7-23) to (7-25) with (7-18) gives the q-line equation

$$y = \left(\frac{q}{q-1}\right)x - \left(\frac{z_F}{q-1}\right) \qquad (7\text{-}26)$$

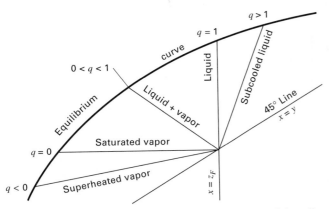

Figure 7.8 Effect of thermal condition of feed on slope of the q-line.

which is located on the McCabe–Thiele diagram of Figure 7.4 by noting that when $x = z_F$, (7-26) reduces to the point $y = z_F = x$, which lies on the 45° line. From (7-26), the q-line slope is $q/(q-1)$. In Figure 7.4, the q-line is constructed for a partially vaporized feed, where $0 < q < 1$ and $-\infty < [q/(q-1)] < 0$. Following placement of the rectifying-section operating line and the *q-line*, the stripping-section operating line is located by drawing a line from the point $(y = x_B, x = x_B)$ on the 45° line to and through the intersection of the q-line and rectifying-section operating line, as in Figure 7.4. The point of intersection must lie somewhere between the equilibrium curve and the 45° line.

As q changes from > 1 (subcooled liquid) to < 0 (superheated vapor), the q-line slope, $q/(q-1)$, changes from positive to negative and back to positive, as shown in Figure 7.8. For a saturated-liquid feed, the q-line is vertical; for a saturated vapor, the q-line is horizontal.

§7.2.4 Number of Equilibrium Stages and Feed-Stage Location

Following construction of the equilibrium curve, the 45° line, the two operating lines, and the q-line, all shown in Figure 7.4, the equilibrium stages required as well as the location of the feed stage are determined by stepping off stages from the top down or from the bottom up until a point of merger is found with the feed stage. An exact integer number of stages is rare; usually fractions of stages arise. Normally the staircase is stepped off from the top and continued to the bottom, starting from the point $(y = x_D, x = x_D)$ on the 45° line, as shown in Figure 7.9a for a partially vaporized feed. In that figure, point P is the intersection of the q-line with the two operating lines. The feed-stage location is the transfer point for stepping off stages between the rectifying-section operating line and the equilibrium curve to stepping off stages between the stripping-section operating line and the equilibrium curve.

The smallest (optimal) number of total equilibrium stages occurs when the transfer is made at the first opportunity after a horizontal line of the staircase passes over point P, as in Figure 7.9a, where the feed stage is stage 3 from the top and a fortuitous total of exactly five stages is required (four plus a partial reboiler).

In Figure 7.9b, the transfer is delayed and the feed stage is stage 5. But now a total of about 6.4 stages is required. The stepping off of stages in the rectifying section could be continued indefinitely, finally approaching, but never reaching, a feed stage at point, K, with total stages $= \infty$.

In Figure 7.9c, the transfer is made early, at feed stage 2, resulting again in more stages than the optimal number of five. If the stepping off of stages had started from the partial reboiler and proceeded upward, the staircase in the stripping section could have been continued indefinitely, approaching, but never reaching, point R.

§7.2.5 Limiting Conditions

For a given set of specifications (Table 7.2), a reflux ratio can be selected anywhere from the minimum, R_{min}, to an infinite value (total reflux), where all of the overhead vapor is condensed and returned to the top stage (thus, no distillate is withdrawn). As shown in Figure 7.10b, the minimum reflux corresponds to the need for ∞ stages, while in Figure 7.10a the infinite reflux ratio corresponds to the minimum number of stages. The McCabe–Thiele method can determine the two limits, N_{min} and R_{min}. Then, for a practical operation, $N_{min} < N < \infty$ and $R_{min} < R < \infty$.

N_{min}, Minimum Number of Equilibrium Stages

As the reflux ratio increases, the rectifying-section operating-line slope given by (7-7) increases from $L/V < 1$ to a limiting value of $L/V = 1$. Correspondingly, as the boilup ratio increases, the stripping-section operating-line slope given by (7-12) decreases from $\overline{L}/\overline{V} > 1$ to a limiting value of $\overline{L}/\overline{V} = 1$. At this limiting condition, shown in Figure 7.11 for a two-stage column, both the rectifying and stripping operating lines coincide with the 45° line, and neither the feed composition, z_F, nor the q-line influences the staircase construction. This is *total reflux* because when $L = V$, $D = B = 0$, and the total condensed overhead is returned as reflux. Also, all liquid leaving the bottom stage is vaporized in the reboiler and returned as boilup.

If both distillate and bottoms flow rates are zero, the feed to the column is zero, which is consistent with the lack of influence of the feed condition. A distillation column can be operated at total reflux to facilitate experimental measurement of tray efficiency because a steady-state operating condition is readily achieved. Figure 7.11 demonstrates that at total reflux, the operating lines are located as far away as possible from the equilibrium curve, resulting in minimum stages.

R_{min}, Minimum Reflux Ratio

As the reflux ratio decreases from the limiting case of total reflux, the intersection of the two operating lines and the q-line moves away from the 45° line and toward the equilibrium curve, thus requiring more equilibrium stages. Finally, a limiting condition is reached when the intersection is on the

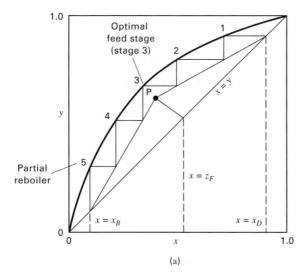

(a)

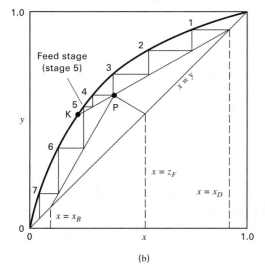

(b)

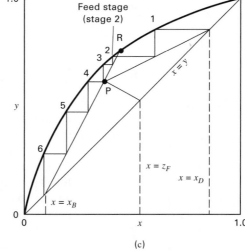

(c)

Figure 7.9 Optimal and non-optimal locations of feed stage: (a) optimal feed-stage location; (b) feed-stage location below optimal stage; (c) feed-stage location above optimal stage.

equilibrium curve, as in Figure 7.12. For the nearly ideal binary system in Figure 7.12a, the intersection, P, is at the feed stage. To reach that stage from the rectifying section or the stripping section, an infinite number of stages is required. P is called a *pinch point* because the two operating lines pinch the equilibrium curve.

For a highly nonideal binary system, the pinch point can occur above or below the feed stage. The former case is illustrated in Figure 7.12b, where the rectifying-section operating line intersects the equilibrium curve at P, before the feed stage is reached, when stepping off stages from the top. The slope of this operating line cannot be reduced further because the line would then cross the equilibrium curve and thereby violate the second law of thermodynamics. This would require spontaneous mass transfer from a region of low concentration to a region of high concentration, which is impossible in a binary system. This is analogous to a second-law violation by a temperature crossover in a heat exchanger. Now, the pinch point occurs entirely in the rectifying section, where an infinite number of stages exists. A column cannot operate at minimum reflux.

The minimum reflux ratio can be determined from the slope of the limiting rectifying-section operating line using (7-7).

$$(L/V)_{\min} = R_{\min}/(R_{\min} + 1)$$

or

$$R_{\min} = (L/V)_{\min}/[1 - (L/V)_{\min}] \qquad (7-27)$$

The limiting condition of infinite stages corresponds to a minimum boilup ratio for $(\overline{L}/\overline{V})_{\max}$. From (7-12),

$$(V_B)_{\min} = 1/[(\overline{L}/\overline{V})_{\max} - 1] \qquad (7-28)$$

Perfect Separation

A third limiting condition is the degree of separation. As a perfect split $(x_D = 1, x_B = 0)$ is approached for $R \geq R_{\min}$, the number of stages required near the top and near the bottom increases rapidly and without limit until pinches are encountered at $x_D = 1$ and $x_B = 0$. If there is no azeotrope, a perfect separation requires ∞ stages in both sections of the column. This is not the case for the reflux ratio. In Figure 7.12a, as x_D is moved from 0.90 toward 1.0, the slope of the

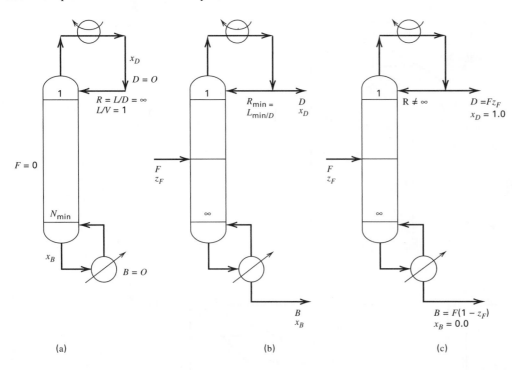

Figure 7.10 Limiting conditions for distillation: (a) total reflux, minimum stages; (b) minimum reflux, infinite stages; (c) perfect separation for nonazeotropic system.

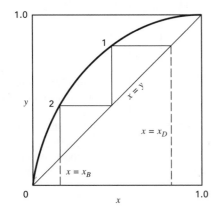

Figure 7.11 Construction for minimum stages at total reflux.

operating line at first increases, but in the range of x_D from 0.99 to 1.0, the slope changes only slightly, so R changes only slightly as it approaches a limiting value. Furthermore, the value of the slope and, therefore, the value of R, is finite for a perfect separation. If the feed is a saturated liquid, application of (7-4) and (7-7) gives an equation for the minimum reflux of a perfect binary separation:

$$R_{\min} = \frac{1}{z_F(\alpha - 1)} \tag{7-29}$$

where relative volatility, α, is at the feed condition.

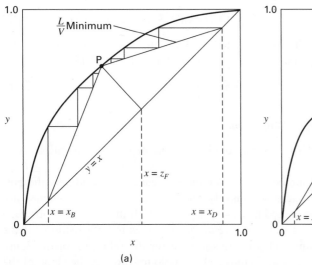

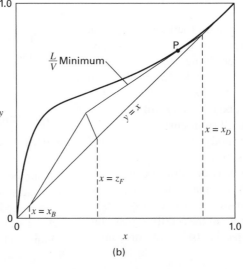

Figure 7.12 Construction for minimum reflux at infinite stages: (a) typical ideal or near-ideal system, pinch point at the feed stage; (b) typical nonideal system, pinch point above the feed stage.

EXAMPLE 7.1 Distillation of a Mixture of Benzene and Toluene.

Four hundred and fifty lbmol/h (204 kmol/h) of a mixture of 60 mol% benzene (LK) and 40 mol% toluene (HK) is to be separated into a liquid distillate and a liquid bottoms product of 95 mol% and 5 mol% benzene, respectively. The feed enters the column with a molar percent vaporization equal to the distillate-to-feed ratio. Use the McCabe–Thiele method to compute, at 1 atm (101.3 kPa): (a) N_{min}, (b) R_{min}, and (c) number of equilibrium stages N, for $R/R_{min} = 1.3$, and the optimal feed-stage location. Also, compare the results with those from a process simulator.

Solution

First calculate D and B. An overall material balance on benzene gives

$$0.60(450) = 0.95D + 0.05B \tag{1}$$

A total balance gives $\quad 450 = D + B \tag{2}$

Combining (1) and (2) and solving, $D = 275$ lbmol/h, $B = 175$ lbmol/h, and $D/F = 0.611$. Thus, the molar vaporization of the feed is 61.1%.

Calculate the slope of the q-line:
$V_F/F = D/F = 0.611$, and q for a partially vaporized feed is

$$\frac{L_F}{F} = \frac{(F - V_F)}{F} = 1 - \frac{V_F}{F} = 0.389$$

From (7-26),

the slope of the q-line is $\dfrac{q}{q-1} = \dfrac{0.389}{0.389 - 1} = 0.637$

(a) In Figure 7.13, where y and x refer to benzene, $x_D = 0.95$ and $x_B = 0.05$, the minimum stages are stepped off between the equilibrium curve and the 45° line, giving $N_{min} = 6.7$.

(b) In Figure 7.14, a q-line is drawn that has a slope of -0.637 and passes through the feed composition ($z_F = 0.60$) on the 45° line. For R_{min}, an operating line for the rectifying section passes through the point $x = x_D = 0.95$ on the 45° line and through the point of intersection of the q-line and the equilibrium curve ($y = 0.684$, $x = 0.465$). The slope of this operating line is 0.55, which from (7-9) equals $R/(R+1)$. Therefore, $R_{min} = 1.22$.

(c) The operating reflux ratio is $1.3 R_{min} = 1.3(1.22) = 1.59$. From (7-9), the stripping-section operating-line slope is

$$\frac{R}{R+1} = \frac{1.59}{1.59 + 1} = 0.614$$

The two operating lines and the q-line are shown in Figure 7.15, where the stripping-section operating line is drawn to pass through the point $x = x_B = 0.05$ on the 45° line and through the intersection of the q-line with the rectifying-section operating line. The equilibrium stages are stepped off, first, between the rectifying-section operating line and the equilibrium curve, and then between the stripping-section operating line and the equilibrium curve, starting from point A (distillate composition) and finishing at point B (bottoms composition). For the optimal feed stage, the transfer from the rectifying-section operating line to the stripping-section operating line takes place at point P, giving $N = 13.2$ equilibrium stages, the feed going into stage 7 from the top, and $N/N_{min} = 13.2/6.7 = 1.97$. The bottom

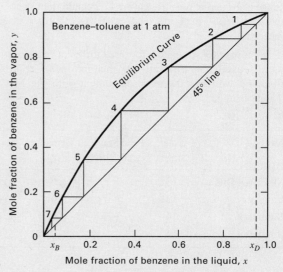

Figure 7.13 Determination of minimum stages for Example 7.1.

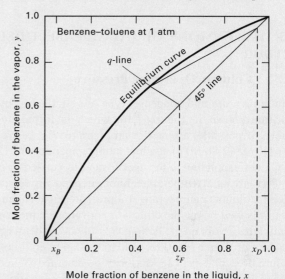

Figure 7.14 Determination of minimum reflux for Example 7.1.

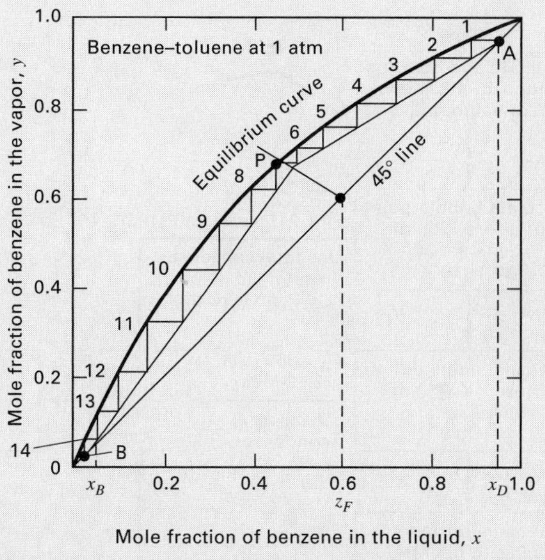

Figure 7.15 Determination of number of equilibrium stages and feed-stage location for Example 7.1.

stage is the partial reboiler, leaving 12.2 equilibrium stages in the column. If the plate efficiency were 0.8, 16 trays would be needed.

(d) Using a process simulator (for three different feed locations), as discussed in Chapter 10, the following results are obtained, which show good agreement with the McCabe–Thiele method.

Method	Total Stages	Feed Stage from Top	Benzene x_D	Benzene x_B
McCabe–Thiele	13.2	7	0.950	0.050
Simulator	13	7	0.947	0.055
Simulator	14	7	0.953	0.045
Simulator	14	8	0.950	0.050

§7.3 EXTENSIONS OF THE McCABE–THIELE METHOD

§7.3.1 Column Operating Pressure

Column pressure and condenser type are established by the algorithm shown in Figure 7.16, which is formulated to achieve, if possible, a reflux-drum pressure, P_D, between 0 and 415 psia (2.86 MPa) at a minimum temperature of 120°F (49°C), corresponding to the use of water as condenser coolant. Pressure and temperature limits depend on economic factors. Columns can operate at a pressure higher than 415 psia if it is well below the critical or convergence pressure of the mixture. To obtain the bottom pressure, a condenser pressure drop of 0 to 2 psi (0 to 14 kPa) and an overall column pressure drop of 5 psi (35 kPa) may be assumed. When the number of trays is known, more refined computations give

approximately 0.1 psi/tray (0.7 kPa/tray) pressure drop for atmospheric and superatmospheric pressure operation, and 0.05 psi/tray (0.35 kPa/tray) pressure drop for vacuum operation. A bottoms bubble-point calculation should be made to ensure that conditions are not near-critical or above product-decomposition temperatures. As the algorithm indicates, if the bottoms temperature is too high, a lower temperature is mandated. The pressure in the reflux drum is then reduced, and the calculation of bottoms pressure and temperature is repeated, until they are acceptable. This may result in vacuum operation and/or the need for a refrigerant, rather than cooling water, for the condenser.

§7.3.2 Condenser Type

Types of condensers are shown in Figure 7.17. A *total condenser* is suitable for reflux-drum pressures to 215 psia (1.48 MPa). A *partial condenser* is appropriate from 215 psia to 365 psia (2.52 MPa) but can be used below 215 psia to give a vapor distillate. A *mixed condenser* can provide both vapor and liquid distillates. A refrigerant is often used as coolant above 365 psia when components are difficult to condense. As illustrated in Example 7.2, a partial condenser provides an additional stage, based on the assumption that liquid reflux leaving the reflux drum is in equilibrium with the vapor distillate.

§7.3.3 Subcooled Reflux

Although most distillation columns are designed so that reflux is a saturated (bubble-point) liquid, such is not always the case. For an operating column with a partial condenser, the reflux is a saturated liquid unless heat losses occur. For a total condenser, however, the reflux is often a subcooled liquid if the condenser is not tightly designed, thus resulting in the distillate bubble-point temperature being much higher than the inlet cooling-water temperature. If the condenser

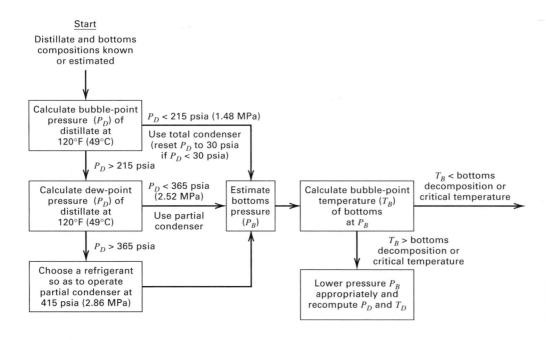

Figure 7.16 Algorithm for establishing distillation-column pressure and condenser type.

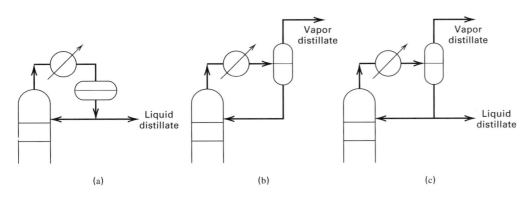

Figure 7.17 Condenser types: (a) total condenser; (b) partial condenser; (c) mixed condenser.

outlet pressure is lower than the top tray pressure, the reflux will be subcooled for all three types of condensers.

When subcooled reflux enters the top tray, it causes some vapor entering the tray to condense. The latent enthalpy of condensation of the vapor provides the sensible enthalpy to heat the subcooled reflux to the bubble point. In that case, the internal reflux ratio within the rectifying section of the column is higher than the external reflux ratio from the reflux drum. The McCabe–Thiele construction should be based on the internal reflux ratio, which can be estimated from the following equation, which is derived from an approximate energy balance around the top tray:

$$R_{\text{internal}} = R \left(1 + \frac{C_{P_L} \Delta T_{\text{subcooling}}}{\Delta H^{\text{vap}}} \right) \qquad (7\text{-}30)$$

where C_{P_L} and ΔH^{vap} are per mole and $\Delta T_{\text{subcooling}}$ is the degrees of subcooling. The internal reflux ratio replaces R, the external reflux ratio in (7-9). If a correction is not made for subcooled reflux, the calculated number of equilibrium stages is somewhat more than required. Thus, subcooled reflux is beneficial.

EXAMPLE 7.2 McCabe–Thiele Method When Using a Partial Condenser.

One thousand kmol/h of 30 mol% n-hexane and 70% n-octane is distilled in a column consisting of a partial reboiler, one equilibrium plate, and a partial condenser, all operating at 1 atm. The feed, a bubble-point liquid, is fed to the reboiler, from which a liquid bottoms is withdrawn. Bubble-point reflux from the partial condenser is returned to the plate. The vapor distillate contains 80 mol% hexane, and the reflux ratio, L/D, is 2. Assume the partial reboiler, plate, and partial condenser are equilibrium stages. (a) Using the McCabe–Thiele method, calculate the bottoms composition and kmol/h of distillate produced. (b) If α is assumed to be 5 (actually, 4.3 at the reboiler and 6.0 at the condenser), calculate the bottoms composition analytically.

Solution

First determine whether the problem is completely specified. From Table 5.4c, $N_D = C + 2N + 6$ degrees of freedom, where N includes the partial reboiler and the stages in the column, but not the partial condenser. With $N = 2$ and $C = 2$, $N_D = 12$. Specified are:

Feed stream variables	4
Plate and reboiler pressures	2
Condenser pressure	1
$Q \, (= 0)$ for plate	1
Number of stages	1
Feed-stage location	1
Reflux ratio, L/D	1
Distillate composition	1
Total	12

Thus, the problem is fully specified and can be solved.

(a) *Graphical solution.* A diagram of the separator is given in Figure 7.18, as is the McCabe–Thiele graphical solution, which is constructed in the following manner:

1. The point $y_D = 0.8$ at the partial condenser is located on the $x = y$ line.

2. Because x_R (reflux composition) is in equilibrium with y_D, the point (x_R, y_D) is located on the equilibrium curve.

3. Since $(L/V) = 1 - 1/[1 + (L/D)] = 2/3$, the operating line with slope $L/V = 2/3$ is drawn through the point $y_D = 0.8$ on the 45° line until it intersects the equilibrium curve. Because the feed is introduced into the partial reboiler, there is no stripping section.

4. Three stages (partial condenser, plate 1, and partial reboiler) are stepped off, and the bottoms composition $x_B = 0.135$ is read.

   The amount of distillate is determined from overall material balances. For hexane, $z_F F = y_D D + x_B B$. Therefore, $(0.3)(1,000) = (0.8)D + (0.135)B$. For the total flow, $B = 1,000 - D$. Solving these two equations simultaneously, $D = 248$ kmol/h.

(b) *Analytical solution.* For $\alpha = 5$, equilibrium liquid compositions are given by a rearrangement of (7-3):

$$x = \frac{y}{y + \alpha(1 - y)} \qquad (1)$$

The steps in the analytical solution are as follows:

1. The liquid reflux at x_R is calculated from (1) for $y = y_D = 0.8$:

$$x_R = \frac{0.8}{0.8 + 5(1 - 0.8)} = 0.44$$

2. y_1 is determined by a material balance about the condenser:

$$V y_1 = D y_D + L x_R \quad \text{with} \quad D/V = 1/3 \quad \text{and} \quad L/V = 2/3$$

$$y_1 = (1/3)(0.8) + (2/3)(0.44) = 0.56$$

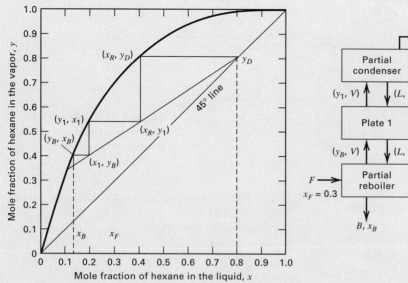

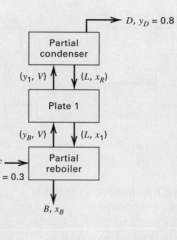

Figure 7.18 Solution to Example 7.2.

3. From (1), for plate 1, $x_1 = \dfrac{0.56}{0.56 + 5(1 - 0.56)} = 0.203$

4. By material balance around plate 1 and the partial condenser,

$$Vy_B = Dy_D + Lx_1$$

and $y_B = (1/3)(0.8) + (2/3)(0.203) = 0.402$

5. From (1), for the partial reboiler,

$$x_B = \frac{0.402}{0.402 + 5(1 - 0.402)} = 0.119$$

By approximating the equilibrium curve with $\alpha = 5$, $x_B = 0.119$ is obtained, rather than the 0.135 obtained in part (a). For a large number of plates, part (b) can be computed with a spreadsheet.

EXAMPLE 7.3   McCabe–Thiele Method for a Column with Only a Feed Plate.

Solve Example 7.2: (a) graphically, assuming that the feed is introduced on plate 1 rather than into the reboiler, as in Figure 7.19; (b) by determining the minimum number of stages required to carry out the separation; (c) by determining the minimum reflux ratio.

Solution

(a) The solution given in Figure 7.19 is obtained as follows:

1. The point x_R, y_D is located on the equilibrium line.

2. The operating line for the rectifying section is drawn through the point $y = x = 0.8$, with a slope of $L/V = 2/3$.

3. Intersection of the q-line, $x_F = 0.3$ (which, for a saturated liquid, is a vertical line), with the enriching-section operating

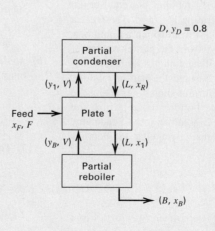

Figure 7.19 Solution to Example 7.3.

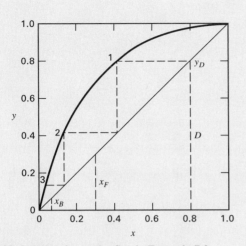

Figure 7.20 Solution for total reflux in Example 7.3.

line is at point P. The stripping-section operating line must also pass through P, but its slope and the point x_B are not known.

4. The slope of the stripping-section operating line is found by trial and error to give three equilibrium contacts in the column, with the middle stage involved in the switch from one operating line to the other. If the middle stage is the optimal feed stage, the result is $x_B = 0.07$, as shown in Figure 7.19. From hexane and total material balances: $(0.3)(1,000) = (0.8D) + 0.07(1,000 - D)$. Solving, $D = 315$ kmol/h.

Comparing this result to that obtained in Example 7.2 shows that the bottoms purity and distillate yield are improved by feeding to plate 1 rather than to the reboiler. This could have been anticipated if the q-line had been constructed in Figure 7.18. The partial reboiler is thus not the optimal feed stage.

(b) The construction corresponding to total reflux ($L/V = 1$, no products, no feed, minimum equilibrium stages) is shown in Figure 7.20. Slightly more than two stages are required for an x_B of 0.07, compared to the three stages previously required.

(c) To determine the minimum reflux ratio, the vertical q-line in Figure 7.19 is extended from point P until the equilibrium curve is intersected, which is the point (0.71, 0.3). The slope, $(L/V)_{min}$ of the operating line for the rectifying section, which connects this point to the point (0.8, 0.8) on the 45° line, is 0.18. Thus $(L/D)_{min} = (L/V_{min})/[1 - (L/V_{min})] = 0.22$. This is considerably less than the $L/D = 2$ specified.

§7.3.4 Reboiler Type

Reboilers for industrial-size distillation columns are usually external heat exchangers of either the *kettle* or the vertical *thermosyphon* type, shown in Figure 7.21. Either can provide the large heat-transfer surface required. In the former case, liquid leaving the sump (reservoir) at the bottom of the column enters the kettle, where it is partially vaporized by transfer of heat from tubes carrying condensing steam or some other heat-transfer fluid. The bottoms product liquid leaving

the reboiler is assumed to be in equilibrium with the vapor returning to the bottom tray. Thus, a kettle reboiler, which is sometimes located in the bottom of a column, is a partial reboiler equivalent to one equilibrium stage.

Vertical thermosyphon reboilers are shown in Figures 7.21b and 7.21c. In the former, bottoms product and reboiler feed are both withdrawn from the column bottom sump. Circulation through the reboiler tubes occurs because of a difference in static heads of the supply liquid and the partially vaporized fluid in the reboiler tubes. The partial vaporization provides enrichment in the exiting vapor. But the exiting liquid is then mixed with liquid leaving the bottom tray, which contains a higher percentage of volatiles. This type of reboiler thus provides only a fraction of a stage and it is best to take no credit for it.

In the more complex and less-common vertical thermosyphon reboiler of Figure 7.21c, the reboiler liquid is withdrawn from the bottom-tray downcomer. Partially vaporized liquid is returned to the column, where the bottoms product from the bottom sump is withdrawn. This type of reboiler functions as an equilibrium stage.

Thermosyphon reboilers are favored when (1) the bottoms product contains thermally sensitive compounds, (2) bottoms pressure is high, (3) only a small ΔT is available for heat transfer, and (4) heavy fouling occurs. Horizontal thermosyphon reboilers may be used in place of vertical types when only small static heads are needed for circulation, when the surface-area requirement is very large, and/or when frequent tube cleaning is anticipated. A pump may be added to a thermosyphon reboiler to improve circulation. Liquid residence time in the column bottom sump should be at least 1 minute and perhaps as much as 5 minutes or more. Large columns may have a 10-foot-high sump.

§7.3.5 Condenser and Reboiler Heat Duties

For saturated-liquid feeds and columns that fulfill the McCabe–Thiele assumptions, reboiler and condenser duties are nearly equal. For all other situations, it is necessary to establish heat duties by an overall energy balance:

$$Fh_F + Q_R = Dh_D + Bh_B + Q_C + Q_{loss} \quad (7\text{-}31)$$

Except for small and/or uninsulated distillation equipment, Q_{loss} can be ignored. With the assumptions of the McCabe–Thiele method, an energy balance for a total condenser is

$$Q_C = D(R + 1)\Delta H^{vap} \quad (7\text{-}32)$$

where ΔH^{vap} = average molar heat of vaporization. For a partial condenser,

$$Q_C = DR\Delta H^{vap} \quad (7\text{-}33)$$

For a partial reboiler,

$$Q_R = BV_B\Delta H^{vap} \quad (7\text{-}34)$$

For a bubble-point feed and total condenser, (7-16) becomes:

$$BV_B = L + D = D(R + 1) \quad (7\text{-}35)$$

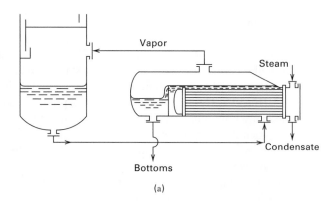

(a)

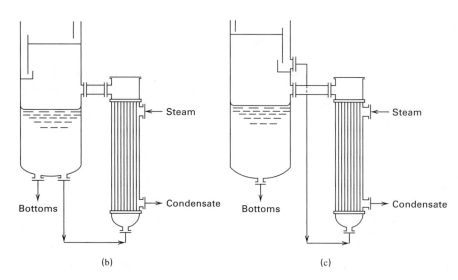

(b)   (c)

Figure 7.21 Reboilers for plant-size distillation columns: (a) kettle-type reboiler; (b) vertical thermosyphon-type reboiler, reboiler liquid withdrawn from bottom sump; (c) vertical thermosyphon-type reboiler, reboiler liquid withdrawn from bottom-tray downcomer.

For partially vaporized feed and a total condenser, the reboiler duty is less than the condenser duty, and is given by

$$Q_R = Q_C \left[1 - \frac{V_F}{D(R+1)} \right] \qquad (7\text{-}36)$$

If saturated steam is the heating medium for the reboiler, the steam rate required is given by the energy balance

$$m_s = \frac{M_s Q_R}{\Delta H_s^{\text{vap}}} \qquad (7\text{-}37)$$

where m_s = mass flow rate of steam, M_S = molecular weight of steam, and ΔH_s^{vap} = molar enthalpy of vaporization of steam.

The cooling water rate for the condenser is

$$m_{\text{cw}} = \frac{Q_C}{C_{P_{\text{H}_2\text{O}}}(T_{\text{out}} - T_{\text{in}})} \qquad (7\text{-}38)$$

where m_{cw} = mass flow rate of cooling water, $C_{P_{\text{H}_2\text{O}}}$ = specific heat of water, and T_{out}, T_{in} = cooling water temperature out of and into the condenser.

Because the cost of reboiler steam is usually an order of magnitude higher than the cost of cooling water, the feed is frequently partially vaporized to reduce Q_R, in comparison to Q_C, as suggested by (7-36).

§7.3.6 Feed Preheat

Feed pressure must be greater than the pressure in the column at the feed tray. Excess feed pressure is dropped across a valve, which may cause the feed to partially vaporize before entering the column. Erosion on feed trays can be a serious problem in column operations.

Second-law thermodynamic efficiency is highest if the feed temperature equals the temperature in the column at the feed tray. It is best to avoid a subcooled liquid or superheated vapor by supplying a partially vaporized feed. This is achieved by preheating the feed with the bottoms product or a process stream that has a suitably high temperature, to ensure a reasonable ΔT driving force for heat transfer and a sufficient available enthalpy.

§7.3.7 Optimal Reflux Ratio

A distillation column operates between the limiting conditions of minimum and total reflux. Table 7.3, which is adapted from Peters and Timmerhaus [6], shows that as R increases, N decreases, column diameter increases, and reboiler steam and condenser cooling-water requirements increase. When the annualized fixed investment costs for the column, condenser, reflux drum, reflux pump, and reboiler are added to the annual cost of steam and cooling water, an

Table 7.3 Effect of Reflux Ratio on Annualized Cost of a Distillation Operation

R/R_{min}	Actual N	Diam., ft	Reboiler Duty, Btu/h	Condenser Duty, Btu/h	Annualized Cost, $/yr			Total Annualized Cost, $/yr
					Equipment	Cooling Water	Steam	
1.00	Infinite	6.7	9,510,160	9,416,000	Infinite	17,340	132,900	Infinite
1.05	29	6.8	9,776,800	9,680,000	44,640	17,820	136,500	198,960
1.14	21	7.0	10,221,200	10,120,000	38,100	18,600	142,500	199,200
1.23	18	7.1	10,665,600	10,560,000	36,480	19,410	148,800	204,690
1.32	16	7.3	11,110,000	11,000,000	35,640	20,220	155,100	210,960
1.49	14	7.7	11,998,800	11,880,000	35,940	21,870	167,100	224,910
1.75	13	8.0	13,332,000	13,200,000	36,870	24,300	185,400	246,570

(Adapted from an example by Peters and Timmerhaus [6].)

optimal reflux ratio of $R/R_{min} = 1.1$ is established, as shown in Figure 7.22 for the conditions of Table 7.3.

Table 7.3 shows that the annual reboiler-steam cost is almost eight times the condenser cooling-water cost. At the optimal reflux ratio, steam cost is 70% of the total. Because the cost of steam is dominant, the optimal reflux ratio is sensitive to the steam cost. At the extreme of zero cost for steam, the optimal R/R_{min} is shifted from 1.1 to 1.32. This example assumes that the heat removed by cooling water in the condenser has no value.

The accepted range of optimal to minimum reflux is from 1.05 to 1.50, with the lower value applying to a difficult separation (e.g., $\alpha = 1.2$). However, because the optimal reflux ratio is broad, for flexibility, columns are often designed for reflux ratios greater than the optimum.

§7.3.8 Large Numbers of Stages

The McCabe–Thiele construction becomes inviable when relative volatility or product purities are such that many stages must be stepped off. In that event, one of the following five techniques can be used:

1. Use separate plots of expanded scales and/or larger dimensions for stepping off stages at the ends of the y–x diagram, e.g., an added plot covering 0.95 to 1.

2. As described by Horvath and Schubert [7] and shown in Figure 7.23, use a plot based on logarithmic coordinates for the bottoms end of the y–x diagram, while for the distillate end, turn the log–log graph upside down and rotate it 90°. As seen in Figure 7.23, the operating lines become curved and must be plotted from (7-9) and (7-14). The 45° line remains straight and the equilibrium curve becomes nearly straight at the two ends.

3. Compute the stages at the two ends algebraically in the manner of part (b) of Example 7.2. This is readily done with a spreadsheet.

4. Use a commercial McCabe–Thiele program.

5. Combine the McCabe–Thiele graphical construction with the Kremser equations of §5.4 for the low and/or high ends, where absorption and stripping factors are almost constant. This preferred technique is illustrated in Example 7.4.

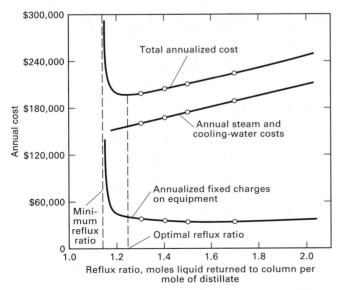

Figure 7.22 Optimal reflux ratio for a representative distillation operation.

EXAMPLE 7.4   McCabe–Thiele Method for a Very Sharp Separation.

Repeat part (c) of Example 7.1 for benzene distillate and bottoms purities of 99.9 and 0.1 mol%, respectively, using a reflux ratio of 1.88, which is about 30% higher than the minimum reflux of 1.44. At the top of the column, $\alpha = 2.52$; at the bottom, $\alpha = 2.26$.

Solution

Figure 7.24 shows the construction for the region $x = 0.028$ to 0.956, with stages stepped off in two directions, starting from the feed stage. In this region, there are seven stages above the feed and eight below, for a total of 16, including the feed stage. The Kremser equation is used to determine the remaining stages.

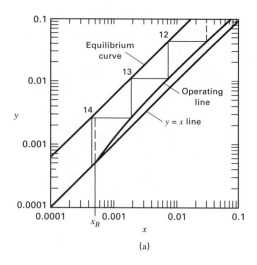

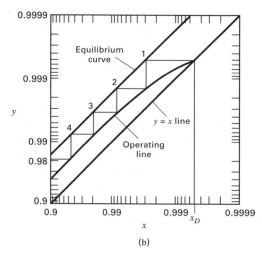

Figure 7.23 Use of log–log coordinates for McCabe–Thiele construction: (a) bottoms end of column; (b) distillate end of column.

Additional stages for the rectifying section: With respect to Figure 5.8a, counting stages from the top down, from Figure 7.24: Using (7-3), for $(x_N)_{benzene} = 0.956$,

$$(y_{N+1})_{benzene} = 0.982 \quad \text{and} \quad (y_{N+1})_{toluene} = 0.018$$

Also, $(x_0)_{benzene} = (y_1)_{benzene} = 0.999$ and $(x_0)_{toluene} = (y_1)_{toluene} = 0.001$.

Combining (5-55), (5-34), (5-35), (5-48), and (5-50):

$$N_R = \frac{\log\left[\frac{1}{A} + \left(1 - \frac{1}{A}\right)\left(\frac{y_{N+1} - x_0 K}{y_1 - x_0 K}\right)\right]}{\log A} \quad (7\text{-}39)$$

where N_R = additional rectifying-section stages. Since this is like an absorption section, it is best to apply (7-39) to toluene, the heavy key. Because $\alpha = 2.52$ at the top of the column, where $K_{benzene}$ is close to 1, take $K_{toluene} = 1/2.52 = 0.397$. Since $R = 1.88$, $L/V = R/(R+1) = 0.653$.

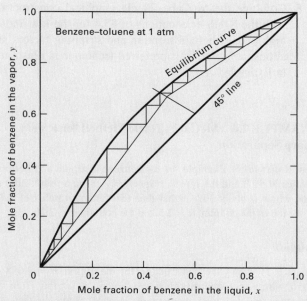

Figure 7.24 McCabe–Thiele construction for Example 7.4 from $x = 0.028$ to $x = 0.956$.

Therefore, $A_{toluene} = L/(V K_{toluene}) = 0.653/0.397 = 1.64$, which is assumed constant in the uppermost part of the rectifying section. Therefore, from (7-39) for toluene,

$$N_R = \frac{\log\left[\frac{1}{1.64} + \left(1 - \frac{1}{1.64}\right)\left(\frac{0.018 - 0.001(0.397)}{0.001 - 0.001(0.397)}\right)\right]}{\log 1.64} = 5.0$$

Additional stages for the stripping section: As in Figure 5.8b, counting stages from the bottom up we have from Figure 7.24: $(x_{N+1})_{benzene} = 0.048$. Also, $(x_1)_{benzene} = (x_B)_{benzene} = 0.001$. Combining the Kremser equations for a stripping section gives

$$N_S = \frac{\log\left[\overline{A} + (1 - \overline{A})\left(\frac{x_{N+1} - x_1/K}{x_1 - x_1/K}\right)\right]}{\log(1/\overline{A})} \quad (7\text{-}40)$$

where, N_S = additional equilibrium stages for the stripping section and $\overline{A}$ = absorption factor in the stripping section = $\overline{L}/K\overline{V}$.

Benzene is stripped in the stripping section, so it is best to apply (7-40) to the benzene. At the bottom, where $K_{toluene}$ is approximately 1.0, $\alpha = 2.26$, and therefore $K_{benzene} = 2.26$. By material balance, with flows in lbmol/h, $D = 270.1$. For $R = 1.88$, $L = 507.8$, and $V = 270.1 + 507.8 = 777.9$. From Example 7.1, $V_F = D = 270.1$ and $L_F = 450 - 270.1 = 179.9$. Therefore, $\overline{L} = L + L_F = 507.8 + 179.9 = 687.7$ lbmol/h and $\overline{V} = V - V_F = 777.9 - 270.1 = 507.8$ lbmol/h.

$$\overline{L}/\overline{V} = 687.7/507.8 = 1.354;$$

$$\overline{A}_{benzene} = \overline{L}/K\overline{V} = 1.354/2.26 = 0.599$$

Substitution into (7-40) gives

$$N_S = \frac{\log\left[0.599 + (1 - 0.599)\left(\frac{0.028 - 0.001/2.26}{0.001 - 0.001/2.26}\right)\right]}{\log(1/0.599)} = 5.9$$

This value includes the partial reboiler. Accordingly, the total number of equilibrium stages starting from the bottom is: partial reboiler + 5.9 + 8 + feed stage + 7 + 5.0 = 26.9.

§7.3.9 Use of Murphree Vapor Efficiency, E_{MV}

In industrial equipment, it is not often practical to provide the combination of residence time and intimacy of contact required to establish equilibrium on each stage. Hence, concentration changes are less than those predicted by equilibrium.

The Murphree vapor efficiency, introduced in §6.5.4, describes tray performance for individual components in either phase and is equal to the change in composition divided by the equilibrium-predicted change. When applied to the vapor phase in a manner similar to (6-28):

$$E_{MV} = \frac{y_n - y_{n+1}}{y_n^* - y_{n+1}} \qquad (7\text{-}41)$$

where E_{MV} is the Murphree vapor efficiency for stage n, $n +$ 1 is the stage below, and y_n^* is the composition in the hypothetical vapor phase in equilibrium with the liquid leaving n. The component subscript in (7-41) is dropped because values of E_{MV} are equal for two binary components.

In stepping off stages, the Murphree vapor efficiency dictates the fraction of the distance taken from the operating line to the equilibrium line. This is shown in Figure 7.25a for the case of Murphree efficiencies based on the vapor phase. In Figure 7.25b, the Murphree tray efficiency is based on the liquid. In effect, the dashed curve for actual exit-phase composition replaces the thermodynamic equilibrium curve for a particular set of operating lines. In Figure 7.25a, $E_{MV} = \overline{EF}/\overline{EG} = 0.7$ for the bottom stage.

§7.3.10 Multiple Feeds, Sidestreams, and Open Steam

The McCabe–Thiele method for a single feed and two products is extended to multiple feed streams and sidestreams by adding one additional operating line for each additional stream. A multiple-feed arrangement is shown in Figure 7.26. In the absence of sidestream L_S, this arrangement has no effect on the material balance in the section above the upper-feed point, F_1. The section of column between the upper-feed point and the lower-feed point, F_2 (in the absence of feed F), is represented by an operating line of slope L'/V', which intersects the rectifying-section operating line. A similar argument holds for the stripping section. Hence it is possible to apply the McCabe–Thiele construction shown in Figure 7.27a, where feed F_1 is a dew-point vapor, while feed F_2 is a bubble-point liquid. Feed F and sidestream L_S of Figure 7.26 are not present. Thus, between the two feed points, the molar vapor flow rate is $V' = V - F_1$ and $\overline{L} = L' + F_2 = L + F_2$. For given $x_B, z_{F_2}, z_{F_1}, x_D,$ and $L/D,$ the three operating lines in Figure 7.27a are constructed.

A sidestream may be withdrawn from the rectifying section, the stripping section, or between multiple feed points, as a saturated vapor or saturated liquid. Within material-balance constraints, L_S and x_S can both be specified. In Figure 7.27b, a saturated-liquid sidestream of LK mole fraction x_S and molar flow rate L_S is withdrawn from the rectifying section above feed F. In the section of stages between the side stream-withdrawal stage and the feed stage, $L' = L - L_S,$ while $V' = V.$ The McCabe–Thiele constructions determine the location of the sidestream stage. However, if it is not located directly above $x_S,$ the reflux ratio must be varied until it does.

For certain distillations, an energy source is introduced directly into the base of the column. Open steam, for example, can be used if one of the components is water or if water can form a second liquid phase, thereby reducing the boiling point, as in the steam distillation of fats, where there is no reboiler and heat is supplied by superheated steam. Commonly, the feed contains water, which is removed as bottoms. In that application, Q_R of Figure 7.26 is replaced by a stream of composition $y = 0$ (pure steam), which, with $x = x_B,$ becomes a point on the operating line, since these passing streams exist at the end of the column. With open steam, the bottoms flow rate

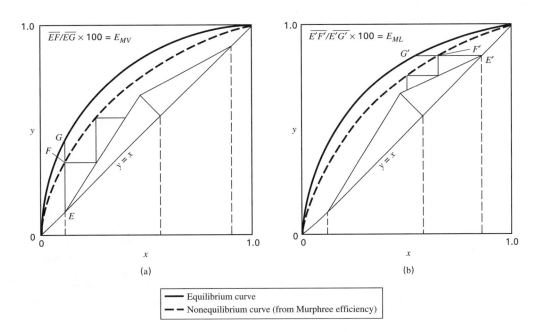

Figure 7.25 Use of Murphree plate efficiencies in McCabe–Thiele construction.

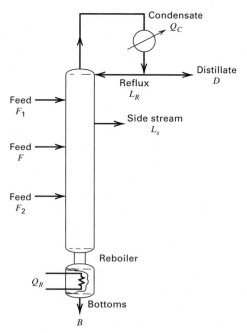

Figure 7.26 Complex distillation column with multiple feeds and sidestream.

is increased by the flow rate of the open steam. The use of open steam rather than a reboiler for the operating condition $F_1 = F_2 = L_S = 0$ is represented graphically in Figure 7.27c.

EXAMPLE 7.5 McCabe–Thiele Method for Column with a Sidestream.

A column equipped with a partial reboiler and total condenser, operating at steady state with a saturated-liquid feed, has a liquid sidestream in the rectifying section. Using the McCabe–Thiele assumptions: (a) derive the two operating lines in the rectifying section; (b) find the point of intersection of the operating lines; (c) find the intersection of the operating line between F and L_S with the diagonal; and (d) show the construction on a y–x diagram.

Solution

(a) By material balance over Section 1 in Figure 7.28, $V_{n-1}y_{n-1} = L_n x_n + D x_D$. For Section 2, $V_{S-2}y_{S-2} = L'_{S-1}x_{S-1} + L_S x_S + D x_D$. The two operating lines for conditions of constant molar overflow become:

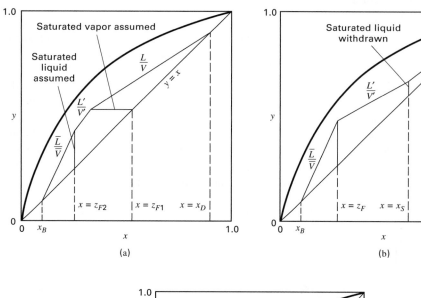

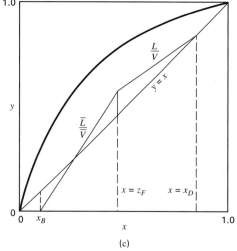

Figure 7.27 McCabe–Thiele construction for complex columns: (a) two feeds (saturated liquid and saturated vapor); (b) one feed, one sidestream (saturated liquid); (c) use of open steam.

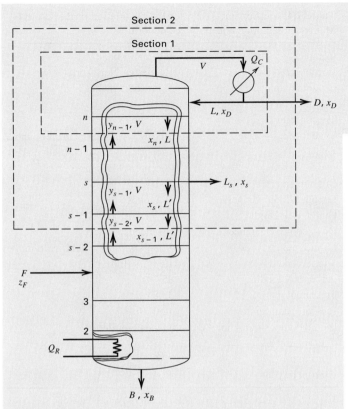

Figure 7.28 Distillation column with sidestream for Example 7.5.

$$y = \frac{L}{V}x + \frac{D}{V}x_D \quad \text{and} \quad y = \frac{L'}{V}x + \frac{L_S x_S + D x_D}{V}$$

(b) Equating the two operating lines, the intersection occurs at $(L - L')x = L_S x_S$ and since $L - L' = L_S$, the point of intersection becomes $x = x_S$.

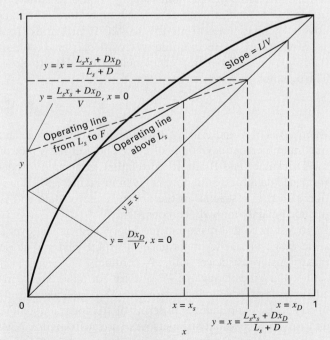

Figure 7.29 McCabe–Thiele diagram for Example 7.5.

(c) The intersection of the lines

$$y = \frac{L'}{V}x + \frac{L_S x_S + D x_D}{V}$$

and $\quad y = x$ occurs at $x = \dfrac{L_S x_S + D x_D}{L_S + D}$

(d) The y–x diagram is shown in Figure 7.29.

§7.4 ESTIMATION OF STAGE EFFICIENCY FOR DISTILLATION

Methods for estimating stage efficiency for binary distillation are analogous to those for absorption and stripping, with one major difference. In absorption and stripping, the liquid phase is often rich in heavy components, and thus liquid viscosity is high and mass-transfer rates are low. This leads to low stage efficiencies, usually less than 50%. For binary distillation, particularly of close-boiling mixtures, both components are near their boiling points and liquid viscosity is low, with the result that stage efficiencies for well-designed trays are often higher than 70% and can even be higher than 100% for large-diameter columns where a crossflow effect is present.

§7.4.1 Performance Data

Techniques for measuring performance of industrial distillation columns are described in *AIChE Equipment Testing Procedure* [8]. Overall column efficiencies are generally measured at conditions of total reflux to eliminate transients due to fluctuations from steady state that are due to feed variations, etc. However, as shown by Williams, Stigger, and Nichols [9], efficiency measured at total reflux ($L/V = 1$) can differ from that at design reflux. A significant factor is how closely to flooding the column is operated. Overall efficiencies are calculated from (6-21) and total reflux data. Individual-tray, Murphree vapor efficiencies are calculated using (6-28). Here, sampling from downcomers leads to variable results. To mitigate this and other aberrations, it is best to work with near-ideal systems. These and other equipment-specific factors are discussed in §6.5.

Table 7.4, from Gerster et al. [10], lists plant data for the distillation at total reflux of a methylene chloride (MC)–ethylene chloride (EC) mixture in a 5.5-ft-diameter column containing 60 bubble-cap trays on 18-inch tray spacing and operating at 85% of flooding at total reflux.

EXAMPLE 7.6 Tray Efficiency from Performance Data.

Using the performance data of Table 7.4, estimate: (a) the overall tray efficiency for the section of trays from 35 to 29 and (b) E_{MV} for tray 32. Assume the following values for $\alpha_{MC,EC}$:

x_{MC}	$\alpha_{MC,EC}$	y_{MC} from (7-3)
0.00	3.55	0.00
0.10	3.61	0.286
0.20	3.70	0.481
0.30	3.76	0.617
0.40	3.83	0.719
0.50	3.91	0.796
0.60	4.00	0.857
0.70	4.03	0.904
0.80	4.09	0.942
0.90	4.17	0.974
1.00	4.25	1.00

Solution

(a) The above x–α–y data are plotted in Figure 7.30. Four equilibrium stages are stepped off from $x_{33} = 0.898$ to $x_{29} = 0.0464$ for total reflux. Since the actual number of stages is also 4, E_o from (6-21) is 100%.

(b) At total reflux conditions, passing vapor and liquid streams have the same composition, so the operating line is the 45° line. Using this, together with the above performance data and the equilibrium curve in Figure 7.30 for methylene chloride, with trays counted from the bottom up:

$$y_{32} = x_{33} = 0.898 \quad \text{and} \quad y_{31} = x_{32} = 0.726$$

From (6-28),

$$(E_{MV})_{32} = \frac{y_{32} - y_{31}}{y_{32}^* - y_{31}}$$

From Figure 7.30, for $x_{32} = 0.726$ and $y_{32}^* = 0.917$,

$$(E_{MV})_{32} = \frac{0.898 - 0.726}{0.917 - 0.726} = 0.90 \quad \text{or} \quad 90\%$$

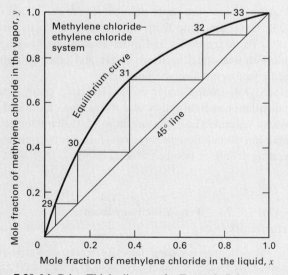

Figure 7.30 McCabe–Thiele diagram for Example 7.6.

Table 7.4 Performance Data for the Distillation of a Mixture of Methylene Chloride and Ethylene Chloride

Company	Eastman Kodak
Location	Rochester, New York
Column diameter	5.5 ft (65.5 inches I.D.)
No. of trays	60
Tray spacing	18 inches
Type tray	10 rows of 3-inch-diameter bubble caps on 4-7/8-inch triangular centers;115 caps/tray
Bubbling area	20 ft^2
Length of liquid travel	49 inches
Outlet-weir height	2.25 inches
Downcomer clearance	1.5 inches
Liquid rate	24.5 gal/min-ft = 1,115.9 lb/min
Vapor F-factor	1.31 ft/s (lb/ft^3)$^{0.5}$
Percent of flooding	85
Pressure, top tray	33.8 psia
Pressure, bottom tray	42.0 psia
Liquid composition, mole % methylene chloride:	
From tray 33	89.8
From tray 32	72.6
From tray 29	4.64

Source: J.A. Gerster, A.B. Hill, N.H. Hochgrof, and D.B. Robinson, *Tray Efficiencies in Distillation Columns, Final Report from the University of Delaware*, AIChE, New York (1958).

§7.4.2 Empirical Correlations of Tray Efficiency

Based on 41 sets of data for bubble-cap-tray and sieve-tray columns distilling hydrocarbons and a few water and miscible organic mixtures, Drickamer and Bradford [11] correlated E_o in terms of the molar-average liquid viscosity, μ, of the tower feed at average tower temperature. The data covered temperatures from 157 to 420°F, pressures from 14.7 to 366 psia, feed liquid viscosities from 0.066 to 0.355 cP, and overall tray efficiencies from 41% to 88%. The equation

$$E_o = 13.3 - 66.8 \log \mu \tag{7-42}$$

where E_o is in % and μ is in cP, fits the data with average and maximum percent deviations of 5.0% and 13.0%. A plot of the Drickamer and Bradford correlation, compared to performance data for distillation, is given in Figure 7.31. Equation (7-42) is restricted to the above range of data and is intended mainly for hydrocarbons.

§6.5 showed that mass-transfer theory predicts that over a wide range of α, the importance of liquid- and gas-phase mass-transfer resistances shifts. Accordingly, O'Connell [12] found that the Drickamer–Bradford formulation is inadequate for feeds having a large α. O'Connell also developed separate correlations in terms of $\mu\alpha$ for fractionators and for absorbers and strippers. As shown in Figure 7.32, Lockhart and Leggett [13] were able to obtain a single correlation using the product of liquid viscosity and an appropriate

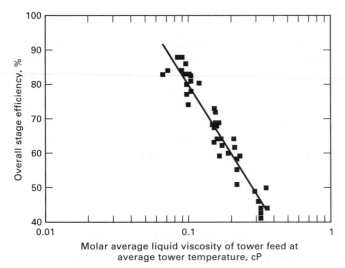

Figure 7.31 Drickamer and Bradford's correlation for plate efficiency of distillation columns.

volatility as the correlating variable. For fractionators, $\alpha_{LK,HK}$ was used; for hydrocarbon absorbers, the volatility was taken as 10 times the K-value of a key component, which must be distributed between top and bottom products. Data used by O'Connell cover a range of α from 1.16 to 20.5. The effect on E_o of the ratio of liquid-to-vapor molar flow rates, L/V, for eight different water and organic binary systems in a 10-inch-diameter column with bubble-cap trays was reported by Williams et al. [9]. While L/V did have an effect, it could not be correlated. For fractionation with L/V nearly equal to 1.0 (i.e., total reflux), their distillation data, which are included in Figure 7.32, are in reasonable agreement with the O'Connell correlation. For the distillation of hydrocarbons in a 0.45-m-diameter column, Zuiderweg, Verburg, and Gilissen [14] found the differences in E_o among bubble-cap, sieve, and valve trays to be insignificant at 85% of flooding. Accordingly, Figure 7.32 is assumed to be applicable to all three tray types, but may be somewhat conservative for well-designed trays. For example, data of Fractionation

Research Incorporated (FRI) for valve trays operating with hydrocarbon systems, also included in Figure 7.32, show efficiencies 10% to 20% higher than the correlation.

For just the distillation data plotted in Figure 7.32, the O'Connell correlation fits the empirical equation

$$E_o = 50.3(\alpha\mu)^{-0.226} \qquad (7-43)$$

where E_o is in %, μ is in cP, and α is at average column conditions.

The data in Figure 7.32 are mostly for columns with liquid flow paths from 2 to 3 ft. Gautreaux and O'Connell [15] showed that higher efficiencies are achieved for longer flow paths because the equivalent of two or more completely mixed, successive liquid zones may be present.

Provided that $\mu\alpha$ lies between 0.1 and 1.0, Lockhart and Leggett [13] recommend adding the increments in Table 7.5 to the value of E_o from Figure 7.32 when the liquid flow path is greater than 3 ft. However, at high liquid flow rates, long liquid-path lengths are undesirable because they lead to excessive liquid gradients and cause maldistribution of vapor

Table 7.5 Correction to Overall Tray Efficiency for Length of Liquid Flow Path ($0.1 \le \mu\alpha \le 1.0$)

Length of Liquid Flow Path, ft	Value to Be Added to E_o from Figure 7.32, %
3	0
4	10
5	15
6	20
8	23
10	25
15	27

Source: F.J. Lockhart and C.W. Leggett, in K.A. Kobe and J.J. McKetta, Jr., Eds., *Advances in Petroleum Chemistry and Refining*, Vol. 1, Interscience, New York, pp. 323–326 (1958).

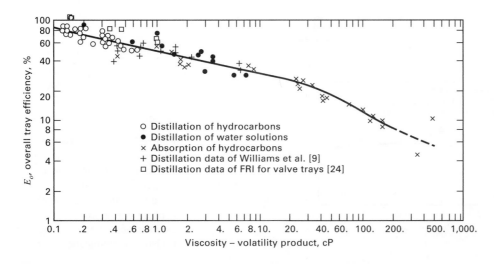

Figure 7.32 Lockhart and Leggett version of the O'Connell correlation for overall tray efficiency of fractionators, absorbers, and strippers.

[Adapted from F.J. Lockhart and C.W. Leggett, in K.A. Kobe and J.J. McKetta, Jr., Eds., *Advances in Petroleum Chemistry and Refining*, Interscience, New York, Vol. 1, pp. 323–326 (1958).]

flow, in which case multipass trays, shown in Figure 6.15 and discussed in §6.5.3, are preferred.

EXAMPLE 7.7 Estimation of Stage Efficiency from Empirical Correlations.

For the benzene–toluene distillation of Figure 7.1, use the Drickamer–Bradford and O'Connell correlations to estimate E_o and the number of actual plates required. Obtain the column height, assuming 24-inch tray spacing with 4 ft above the top tray for removal of entrained liquid and 10 ft below the bottom tray for bottoms surge capacity. The separation requires 20 equilibrium stages plus a partial reboiler that acts as an equilibrium stage.

Solution

The liquid viscosity is determined at the feed-stage condition of 220°F, assuming a liquid composition of 50 mol% benzene; μ of benzene = 0.10 cP; μ of toluene = 0.12 cP; and average μ = 0.11 cP. From Figure 7.3, the average α is

$$\text{Average } \alpha = \frac{\alpha_{\text{top}} + \alpha_{\text{bottom}}}{2} = \frac{2.52 + 2.26}{2} = 2.39$$

From the Drickamer–Bradford correlation (7-42), $E_o = 13.3 - 66.8\log(0.11) = 77\%$. Therefore, $N_a = 20/0.77 = 26$.

Column height $= 4 + 2(26 - 1) + 10 = 64$ ft.

From the O'Connell correlation, (7-43), $E_o = 50.3[(2.39)(0.11)]^{-0.226} = 68\%$.

For a 5-ft-diameter column, the length of the liquid flow path is about 3 ft for a single-pass tray and even less for a two-pass tray. From Table 7.5, the efficiency correction is zero. Therefore, $N_a = 20/0.68 = 29.4$, or round up to 30 trays. Column height $= 4 + 2(30 - 1) + 10 = 72$ ft.

§7.4.3 Semitheoretical Models for Tray Efficiency

In §6.5.4, semitheoretical tray models for absorption and stripping based on E_{MV} (6-28) and E_{OV} (6-30) are developed. These are also valid for distillation. However, because the equilibrium line is curved for distillation, λ must be taken as mV/L (not $KV/L = 1/A$), where $m = $ local slope of the equilibrium curve $= dy/dx$. In §6.6.3, the method of Chan and Fair [16] is used for estimating E_{OV} from mass-transfer considerations. E_{MV} can then be estimated. The Chan and Fair correlation is applicable to binary distillation because it was developed from data for six binary mixtures.

§7.4.4 Scale-up from Laboratory Data

Experimental pilot-plant or laboratory data are rarely necessary prior to the design of columns for ideal or nearly ideal binary mixtures. With nonideal or azeotrope-forming solutions, use of a laboratory Oldershaw column of the type discussed in §6.5.5 should be used to verify the desired degree of separation and to obtain an estimate of E_{OV}. The ability to predict the efficiency of industrial-size sieve-tray columns from measurements with 1-inch glass and 2-inch metal diameter Oldershaw columns is shown in Figure 7.33, from the work of Fair, Null, and Bolles [17]. The measurements are for cyclohexane/n-heptane at vacuum conditions (Figure 7.33a) and near-atmospheric conditions (Figure 7.33b), and for the isobutane/n-butane system at 11.2 atm (Figure 7.33c). The Oldershaw data are correlated by the solid lines. Data for the 4-ft-diameter column with sieve trays of 8.3% and 13.7% open area were obtained, respectively, by Sakata and Yanagi [18] and Yanagi and Sakata [19], of FRI. The Oldershaw column is assumed to measure E_{OV}. The FRI column measured

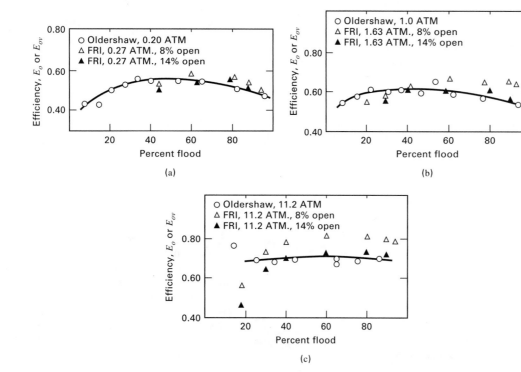

Figure 7.33 Comparison of Oldershaw column efficiency with point efficiency in 4-ft-diameter FRI column with sieve trays: (a) cyclohexane/n-heptane system; (b) cyclohexane/n-heptane system; (c) isobutane/n-butane system.

E_o, but the relations of §6.5.4 were used to convert the FRI data to E_{OV}. The data cover % flooding from 10% to 95%. Data from the Oldershaw column are in agreement with the FRI data for 14% open area, except at the lower part of the flooding range. In Figures 7.33b and 7.33c, FRI data for 8% open area show efficiencies as much as 10% higher.

§7.5 COLUMN AND REFLUX-DRUM DIAMETERS

As with absorbers and strippers, distillation-column diameter is calculated for conditions at the top and bottom trays of the tower, using the method of §6.6.1. If the diameters differ by 1 ft or less, the larger diameter is used for the entire column. If the diameters differ by more than 1 ft, it is often more economical to swage the column, using the different diameters computed for the sections above and below the feed.

§7.5.1 Reflux Drums

As shown in Figure 7.1, vapor flows from the top plate to the condenser and then to a cylindrical reflux drum, usually located near ground level, which necessitates a pump to lift the reflux to the top of the column. If a partial condenser is used, the drum is oriented vertically to facilitate the separation of vapor from liquid—in effect, acting as a flash drum.

Vertical Drums

Vertical reflux and flash drums are sized by calculating a minimum drum diameter, D_T, to prevent liquid carryover by entrainment, using (6-44) in conjunction with the curve for 24-inch tray spacing in Figure 6.23, along with values of $F_{HA} = 1.0$ in (6-42), $f = 0.85$, and $A_d = 0$. To handle process fluctuations and otherwise facilitate control, vessel volume, V_V, is determined on the basis of liquid residence time, t, which should be at least 5 min, with the vessel half full of liquid [20]:

$$V_V = \frac{2LM_L t}{\rho_L} \qquad (7\text{-}44)$$

where L is the molar liquid flow rate leaving the vessel. Assuming a vertical, cylindrical vessel and neglecting head volume, the vessel height H is

$$H = \frac{4V_V}{\pi D_T^2} \qquad (7\text{-}45)$$

However, if $H > 4D_T$, it is generally preferable to increase D_T and decrease H to give $H = 4D$. Then

$$D_T = \frac{H}{4} = \left(\frac{V_V}{\pi}\right)^{1/3} \qquad (7\text{-}46)$$

A height above the liquid level of at least 4 ft is necessary for feed entry and disengagement of liquid droplets from the vapor. Within this space, it is common to install a wire mesh pad, which serves as a mist eliminator.

Horizontal Drums

When vapor is totally condensed, a cylindrical, horizontal reflux drum is commonly employed to receive the condensate. Equations (7-44) and (7-46) permit estimates of the drum diameter, D_T, and length, H, by assuming a near-optimal value for H/D_T of 4 and the liquid residence time suggested for a vertical drum. A horizontal drum is also used following a partial condenser when the liquid flow rate is appreciably greater than the vapor flow rate.

EXAMPLE 7.8 Diameter and Height of a Flash Drum.

Equilibrium vapor and liquid streams leaving a flash drum supplied by a partial condenser are as follows:

Component	Vapor	Liquid
Lbmol/h:		
HCl	49.2	0.8
Benzene	118.5	81.4
Monochlorobenzene	71.5	178.5
Total	239.2	260.7
Lb/h:	19,110	26,480
T, °F	270	270
P, psia	35	35
Density, lb/ft^3	0.371	57.08

Determine the dimensions of the flash drum.

Solution

Using Figure 6.24,

$$F_{LV} = \frac{26,480}{19,110}\left(\frac{0.371}{57.08}\right)^{0.5} = 0.112$$

and C_F at a 24-inch tray spacing is 0.34. Assume, in (6-42), that $C = C_F$.

From (6-40),

$$U_f = 0.34\left(\frac{57.08 - 0.371}{0.371}\right)^{0.5} = 4.2 \text{ ft/s} = 15,120 \text{ ft/h}$$

From (6-44) with $A_d/A = 0$,

$$D_T = \left[\frac{(4)(19,110)}{(0.85)(15,120)(3.14)(1)(0.371)}\right]^{0.5} = 2.26 \text{ ft}$$

From (7-44), with $t = 5$ minutes $= 0.0833$ h,

$$V_V = \frac{(2)(26,480)(0.0833)}{(57.08)} = 77.3 \text{ ft}^3$$

From (7-43),

$$H = \frac{(4)(77.3)}{(3.14)(2.26)^2} = 19.3 \text{ ft}$$

However, $H/D_T = 19.3/2.26 = 8.54 > 4$. Therefore, redimension V_V for $H/D_T = 4$.

From (7-46),

$$D_T = \left(\frac{77.3}{3.14}\right)^{1/3} = 2.91 \text{ ft} \quad \text{and} \quad H = 4D_T = (4)(2.91) = 11.64 \text{ ft}$$

Height above the liquid level is $11.64/2 = 5.82$ ft, which is adequate. Alternatively, with a height of twice the minimum disengagement height, $H = 8$ ft and $D_T = 3.5$ ft.

§7.6 RATE-BASED METHOD FOR PACKED DISTILLATION COLUMNS

Improvements in distributors and fabrication techniques, and more economical and efficient packings, have led to increasing use of packed towers in new distillation processes and the retrofitting of existing trayed towers to increase capacity and reduce pressure drop. Methods in §6.7 and §6.8 for estimating packed-column parameters and packed heights for absorbers are applicable to distillation and are extended here for use in conjunction with the McCabe–Thiele diagram. Both the HETP and HTU methods are covered. Unlike dilute-solution absorption or stripping, where values of HETP and HTU may be constant throughout, values of HETP and HTU can vary, especially across the feed entry, where appreciable changes in vapor and liquid traffic occur. Also, because the equilibrium line for distillation is curved, equations of §6.8 must be modified by replacing $\lambda = KV/L$ with

$$\lambda = \frac{mV}{L} = \frac{\text{slope of equilibrium curve}}{\text{slope of operating line}}$$

where $m = dy/dx$ varies with location. The efficiency and mass-transfer relationships are summarized in Table 7.6.

§7.6.1 HETP Method for Distillation

In the HETP method, equilibrium stages are first stepped off on a McCabe–Thiele diagram, where equimolar counter-diffusion (EMD) applies. At each stage, T, P, phase-flow

ratio, and phase compositions are noted. A suitable packing material is selected, and the column diameter is estimated for operation at, say, 70% of flooding by one of the methods of §6.8. Mass-transfer coefficients for the individual phases are estimated for the stage conditions from correlations in §6.8. From these coefficients, values of H_{OG} and HETP are estimated for each stage and then summed to obtain the packed heights of the rectifying and stripping sections. If experimental values of HETP are available, they are used directly. In computing values of H_{OG} from H_G and H_L, or K_y from k_y and k_x, (6-92) and (6-80) must be modified because for binary distillation, where the mole fraction of the LK may range from almost 0 at the bottom of the column to almost 1 at the top, the ratio $(y_1 - y^*)/(x_1 - x)$ in (6-82) is no longer a constant equal to the K-value, but the ratio is dy/dx equal to the slope, m, of the equilibrium curve. The modified equations are included in Table 7.6.

EXAMPLE 7.9 Packed Height by the HETP Method.

For the benzene–toluene distillation of Example 7.1, determine packed heights of the rectifying and stripping sections based on the following values for the individual HTUs. Included are the L/V values for each section from Example 7.1.

	H_G, ft	H_L, ft	L/V
Rectifying section	1.16	0.48	0.62
Stripping section	0.90	0.53	1.40

Solution

Equilibrium curve slopes, $m = dy/dx$, are from Figure 7.15 and values of λ are from (7-47). H_{OG} for each stage is from (7-52) in Table 7.6 and HETP for each stage is from (7-53). Table 7.7 shows that only 0.2 of stage 13 is needed and that stage 14 is the partial reboiler. From the results in Table 7.7, 10 ft of packing should be used in each section.

Table 7.6 Modified Efficiency and Mass-Transfer Equations for Binary Distillation

$\lambda = mV/L$ (7-47)

$m = dy/dx = $ local slope of equilibrium curve

Efficiency:

Equations (6-31) to (6-37) hold if λ is defined by (7-47)

Mass transfer:

$$\frac{1}{N_{OG}} = \frac{1}{N_G} + \frac{\lambda}{N_L} \tag{7-48}$$

$$\frac{1}{K_{OG}} = \frac{1}{k_G a} + \frac{mPM_L/\rho_L}{k_L a} \tag{7-49}$$

$$\frac{1}{K_y a} = \frac{1}{k_y a} + \frac{m}{k_x a} \tag{7-50}$$

$$\frac{1}{K_x a} = \frac{1}{k_x a} + \frac{1}{mk_y a} \tag{7-51}$$

$$H_{OG} = H_G + \lambda H_L \tag{7-52}$$

$$\text{HETP} = H_{OG} \ln \lambda/(\lambda - 1) \tag{7-53}$$

Table 7.7 Results for Example 7.9

Stage	m	$\lambda = \frac{mV}{L}$ or $m\overline{V}/\overline{L}$	H_{OG}, ft	HETP, ft
1	0.47	0.76	1.52	1.74
2	0.53	0.85	1.56	1.70
3	0.61	0.98	1.62	1.64
4	0.67	1.08	1.68	1.62
5	0.72	1.16	1.71	1.59
6	0.80	1.29	1.77	1.56
Total for rectifying section:				9.85
7	0.90	0.64	1.32	1.64
8	0.98	0.70	1.28	1.52
9	1.15	0.82	1.34	1.47
10	1.40	1.00	1.43	1.43
11	1.70	1.21	1.53	1.40
12	1.90	1.36	1.62	1.38
13	2.20	1.57	1.73	1.37(0.2) = 0.27
Total for stripping section:				9.11
Total packed height:				18.96

§7.6.2 HTU Method for Distillation

In the HTU method, stages are not stepped off on a McCabe–Thiele diagram. Instead, the diagram provides data to perform an integration over the packed height using mass-transfer coefficients or transfer units.

Consider the packed distillation column and its McCabe–Thiele diagram in Figure 7.34. Assume that $V, L, \overline{V},$ and $\overline{L}$ are constant. For equimolar countercurrent diffusion (EMD), the rate of mass transfer of the LK from the liquid to the vapor phase is

$$n = k_x a(x - x_1) = k_y a(y_1 - y) \qquad (7\text{-}54)$$

Rearranging,

$$-\frac{k_x a}{k_y a} = \frac{y_1 - y}{x_1 - x} \qquad (7\text{-}55)$$

In Figure 7.34b, for any point (x, y) on the operating line, the interfacial point (x_I, y_I) on the equilibrium curve is obtained by drawing a line of slope $(-k_x a/k_y a)$ from point (x, y) to the point where it intersects the equilibrium curve.

By material balance over an incremental column height, l, with constant molar overflow,

$$V\,dy = k_y a(y_I - y)S\,dl \qquad (7\text{-}56)$$
$$L\,dx = k_x a(x - x_1)S\,dl \qquad (7\text{-}57)$$

where S is the cross-sectional area of the packed section. Integrating over the rectifying section,

$$(l_T)_R = \int_0^{(l_T)_R} dl = \int_{y_F}^{y_2} \frac{V\,dy}{k_y aS(y_1 - y)} = \int_{x_F}^{x_D} \frac{L\,dx}{k_x aS(x - x_1)} \qquad (7\text{-}58)$$

or

$$(l_T)_R = \int_{y_F}^{y_2} \frac{H_G\,dy}{(y_1 - y)} = \int_{x_F}^{x_D} \frac{H_L\,dx}{(x - x_1)} \qquad (7\text{-}59)$$

Integrating over the stripping section,

$$(l_T)_S = \int_0^{(l_T)_S} dl = \int_{y_1}^{y_F} \frac{V\,dy}{k_y aS(y_1 - y)} = \frac{L\,dx}{k_x aS(x - x_1)} \qquad (7\text{-}60)$$

or

$$(l_T)_S = \int_{y_1}^{y_F} \frac{H_G\,dy}{(y_1 - y)} = \int_{x_1}^{x_F} \frac{H_L\,dx}{(x - x_1)} \qquad (7\text{-}61)$$

Values of k_y and k_x vary over the packed height, causing the slope $(-k_x a/k_y a)$ to vary. If $k_x a > k_y a$, resistance to mass transfer resides mainly in the vapor and, in using (7-61), it is most accurate to evaluate the integrals in y. For $k_y a > k_x a$, the integrals in x are used. Usually, it is sufficient to evaluate k_y and k_x at three points in each section to determine their variation with x. Then by plotting their ratios from (7-55), a locus of points P can be found, from which values of $(y_I - y)$ for any value of y, or $(x - x_I)$ for any value of x, can be read for use with (7-58) to (7-61). These integrals can be evaluated either graphically or numerically.

EXAMPLE 7.10 Packed Height by the HTU Method.

Two hundred and fifty kmol/h of saturated-liquid feed of 40 mol% isopropyl ether in isopropanol is distilled in a packed column operating at 1 atm to obtain a distillate of 75 mol% isopropyl ether and a bottoms of 95 mol% isopropanol. The reflux ratio is 1.5 times the minimum and the column has a total condenser and partial reboiler. The mass-transfer coefficients given below have been estimated from empirical correlations in §6.8. Compute the packed heights of the rectifying and stripping sections.

Solution

From an overall material balance on isopropyl ether,

$$0.40(250) = 0.75D + 0.05(250 - D)$$

Solving,

$$D = 125 \text{ kmol/h and } B = 250 - 125 = 125 \text{ kmol/h}$$

The equilibrium curve at 1 atm is shown in Figure 7.35, where isopropyl ether is the LK and an azeotrope is formed at 78 mol% ether.

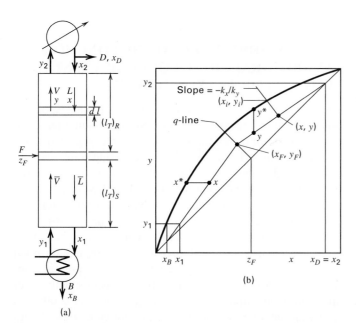

(a)

Figure 7.34 Distillation in a packed column.

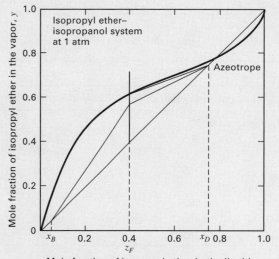

Figure 7.35 Operating lines and minimum reflux line for Example 7.10.

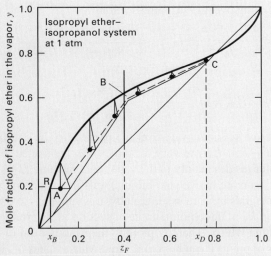

Figure 7.36 Mass-transfer driving forces for Example 7.10.

The distillate composition of 75 mol% is safely below the azeotropic composition. Also shown in Figure 7.35 are the q-line and the rectification-section operating line for minimum reflux. The slope of the latter is measured to be $(L/V)_{min} = 0.39$. From (7-27),

$$R_{min} = 0.39/(1 - 0.39) = 0.64 \quad \text{and} \quad R = 1.5\,R_{min} = 0.96$$

$$L = RD = 0.96(125) = 120\,\text{kmol/h}$$

and

$$V = L + D = 120 + 125 = 245\,\text{kmol/h}$$

$$\bar{L} = L + L_F = 120 + 250 = 370\,\text{kmol/h}$$

$$\bar{V} = V - V_F = 245 - 0 = 245\,\text{kmol/h}$$

Rectification operating-line slope = $L/V = 120/245 = 0.49$

This line and the stripping-section operating line are plotted in Figure 7.35. The partial reboiler, R, is stepped off in Figure 7.36 to give the following end points for determining the packed heights of the two sections, where the symbols refer to Figure 7.34a:

	Stripping Section	Rectifying Section
Top	$(x_F = 0.40,\ y_F = 0.577)$	$(x_2 = 0.75,\ y_2 = 0.75)$
Bottom	$(x_1 = 0.135,\ y_1 = 0.18)$	$(x_F = 0.40,\ y_F = 0.577)$

Mass-transfer coefficients at three values of x are as follows:

x	$k_y a$ kmol/m³-h-(mole fraction)	$k_x a$ kmol/m³-h-(mole fraction)
Stripping section:		
0.15	305	1,680
0.25	300	1,760
0.35	335	1,960
Rectifying section:		
0.45	185	610
0.60	180	670
0.75	165	765

Mass-transfer-coefficient slopes are computed for each point x on the operating line using (7-55), and are drawn from the operating

line to the equilibrium line in Figure 7.36. These are tie lines because they tie the operating line to the equilibrium line. Using the tie lines as hypotenuses, right triangles are constructed, as shown in Figure 7.36. Dashed lines, AB and BC, are then drawn through the points at the 90° triangle corners. Additional tie lines can, as needed, be added to the three plotted lines in each section to give better accuracy. From the tie lines, values of $(y_I - y)$ can be tabulated for operating-line y-values. Column diameter is not given, so the packed volumes are determined from rearrangements of (7-58) and (7-60), with $V = Sl_T$:

$$V_R = \int_{y_F}^{y_2} \frac{V\,dy}{k_y a (y_1 - y)} \tag{7-62}$$

$$V_S = \int_{y_1}^{y_F} \frac{V\,dy}{k_y a (y_1 - y)} \tag{7-63}$$

Values of $k_y a$ are interpolated as necessary. Results are:

y	$(y_I - y)$	$k_y a$	$\dfrac{V\,(\text{or } \bar{V})}{k_y a (y_1 - y)},\text{m}^3$
Stripping section:			
0.18	0.145	307	5.5
0.25	0.150	303	5.4
0.35	0.143	300	5.7
0.45	0.103	320	7.4
0.577	0.030	350	23.3
Rectifying section:			
0.577	0.030	187	43.7
0.60	0.033	185	40.1
0.65	0.027	182	49.9
0.70	0.017	175	82.3
0.75	0.010	165	148.5

By numerical integration, $V_S = 3.6\,\text{m}^3$ and $V_R = 12.3\,\text{m}^3$.

§7.7 INTRODUCTION TO THE PONCHON–SAVARIT GRAPHICAL EQUILIBRIUM-STAGE METHOD FOR TRAYED TOWERS

The McCabe–Thiele method assumes that molar vapor and liquid flow rates are constant. This, plus the assumption of no heat losses, eliminates the need for stage energy balances. When component latent heats are unequal and solutions nonideal, the McCabe–Thiele method is not accurate, but the Ponchon–Savarit graphical method [21, 22], which includes energy balances and utilizes an enthalpy-concentration diagram of the type shown in Figure 7.37 for n-hexane/n-octane, is applicable. This diagram includes curves for enthalpies of saturated-vapor and liquid mixtures, where the terminal points of tie lines connecting these two curves represent equilibrium vapor and liquid compositions, together with vapor and liquid enthalpies, at a given temperature. Isotherms above the saturated-vapor curve represent enthalpies of superheated vapor, while isotherms below the saturated-liquid

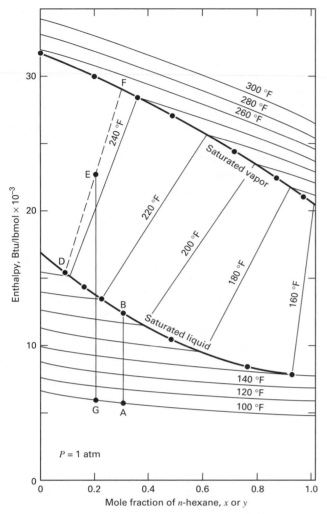

Figure 7.37 Enthalpy-concentration diagram for *n*-hexane/*n*-octane.

where H and h are vapor and liquid molar enthalpies, respectively. The material-balance equations for the light component are

Mixing: $\qquad y_{n-2}V_{n-2} + x_nL_n = z(V_{n-2} + L_n)$ (7-66)

Equilibration: $\qquad z(V_{n-2} + L_n) = y_{n-1}V_{n-1} + x_{n-1}L_{n-1}$

(7-67)

Simultaneous solution of (7-64) and (7-66) gives

$$\frac{H_{n-2} - h_z}{y_{n-2} - z} = \frac{h_z - h_n}{z - x_n}$$ (7-68)

which is the three-point form of a straight line plotted in Figure 7.38. Similarly, solution of (7-65) and (7-67) gives

$$\frac{H_{n-1} - h_z}{y_{n-1} - z} = \frac{h_z - h_{n-1}}{z - x_{n-1}}$$ (7-69)

which is also the equation for a straight line. However, in this case, y_{n-1} and x_{n-1} are in equilibrium and, therefore, the points (H_{n-1}, y_{n-1}) and (H_{n-1}, y_{n-1}) must lie on opposite ends of the tie line that passes through the mixing point (h_z, z), as shown in Figure 7.38.

The Ponchon–Savarit method for binary distillation is an extension of the constructions in Figure 7.38 to countercurrent cascades. A detailed description of the method is not given here because it has been superseded by rigorous, computer-aided calculation procedures in process simulators, which are discussed in Chapter 10. A detailed presentation of the Ponchon–Savarit method for binary distillation is given by Henley and Seader [23].

curve represent subcooled liquid. In Figure 7.37, a mixture of 30 mol% hexane and 70 mol% octane at 100°F (Point A) is a subcooled liquid. By heating it to Point B at 204°F, it becomes a liquid at its bubble point. When a mixture of 20 mol% hexane and 80 mol% octane at 100°F (Point G) is heated to 243°F (Point E), at equilibrium it splits into a vapor phase at Point F and a liquid phase at Point D. The liquid phase contains 7 mol% hexane, while the vapor contains 29 mol% hexane.

Application of the enthalpy-concentration diagram to equilibrium-stage calculations is illustrated by considering a single equilibrium stage, $n - 1$, where vapor from stage $n - 2$ is mixed adiabatically with liquid from stage n to give an overall mixture, denoted by mole fraction z, and then brought to equilibrium. The process is represented in two steps, mixing followed by equilibration, at the top of Figure 7.38. The energy-balance equations for stage $n - 1$ are

Mixing: $\quad V_{n-2}H_{n-2} + L_nh_n = (V_{n-2} + L_n)h_z$ (7-64)

Equilibration: $\quad (V_{n-2} + L_n)h_z = V_{n-1}H_{n-1} + L_{n-1}h_{n-1}$

(7-65)

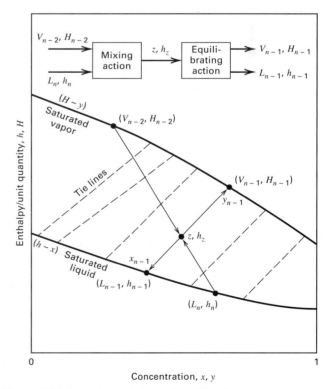

Figure 7.38 Two-phase mixing and equilibration on an enthalpy-concentration diagram.

SUMMARY

1. A binary mixture can be separated economically into two nearly pure products by distillation if $\alpha > \sim 1.05$ and no azeotrope forms.

2. Distillation is the most mature and widely used separation operation, with design and operation practices well established.

3. Product purities depend mainly on the number of equilibrium stages in the rectifying section and the stripping section, and to some extent on the reflux ratio. However, both the number of stages and the reflux ratio must be greater than the minimum values corresponding to total reflux and infinite stages, respectively. The optimal $R/R_{\min}$ is usually in the range of 1.10 to 1.50.

4. Distillation is conducted in trayed towers equipped with sieve or valve trays, or in columns packed with random or structured packings. Many older towers are equipped with bubble-cap trays.

5. Most distillation towers have a condenser, cooled with water, to provide reflux, and a reboiler, heated with steam, for boilup.

6. When the assumption of constant molar overflow is valid, the McCabe–Thiele graphical method is convenient for determining stage and reflux requirements. This method facilitates the visualization of many aspects of distillation and provides a procedure for locating the feed stage.

7. Design of a distillation tower includes selection of operating pressure, type of condenser, degree of reflux subcooling, type of reboiler, and extent of feed preheat.

8. The McCabe–Thiele method can be extended to handle Murphree stage efficiency, multiple feeds, sidestreams, open steam, and use of interreboilers and intercondensers.

9. For trayed columns, estimates of overall stage efficiency, defined by (6-21), can be made with the Drickamer and Bradford, (7-42), or O'Connell, (7-43), correlations. More accurate procedures use data from a laboratory Oldershaw column or the semitheoretical mass-transfer equations of Chan and Fair in Chapter 6.

10. Tray diameter, pressure drop, weeping, entrainment, and downcomer backup can be estimated by procedures in Chapter 6.

11. Reflux and flash drums are sized by a procedure based on vapor entrainment and liquid residence time.

12. Packed-column diameter and pressure drop are determined by procedures presented in Chapter 6.

13. The height of a packed column is established by the HETP method or, preferably, the HTU method. Application to distillations parallels the methods in Chapter 6 for absorbers and strippers, but differs in the manner in which the curved equilibrium line, (7-47), is handled.

14. The Ponchon–Savarit graphical method removes the assumption of constant molar overflow in the McCabe–Thiele method by employing energy balances with an enthalpy-concentration diagram. However, it has been largely supplanted by rigorous programs in process simulators.

REFERENCES

1. Forbes, R.J., *Short History of the Art of Distillation*, E.J. Brill, Leiden (1948).

2. Mix, T.W., J.S. Dweck, M. Weinberg, and R.C. Armstrong, *Chem. Eng. Prog.*, **74**(4), 49–55 (1978).

3. Kister, H.Z., *Distillation Design*, McGraw-Hill, New York (1992).

4. Kister, H.Z., *Distillation Operation*, McGraw-Hill, New York (1990).

5. McCabe, W.L., and E.W. Thiele, *Ind. Eng. Chem.*, **17**, 605–611 (1925).

6. Peters, M.S., and K.D. Timmerhaus, *Plant Design and Economics for Chemical Engineers*, 4th ed., McGraw-Hill, New York (1991).

7. Horvath, P.J., and R.F. Schubert, *Chem. Eng.*, **65**(3), 129–132 (1958).

8. *AIChE Equipment Testing Procedure, Tray Distillation Columns*, 2nd ed., AIChE, New York (1987).

9. Williams, G.C., E.K. Stigger, and J.H. Nichols, *Chem. Eng. Progr.*, **46**(1), 7–16 (1950).

10. Gerster, J.A., A.B. Hill, N.H. Hochgrof, and D.B. Robinson, *Tray Efficiencies in Distillation Columns, Final Report from the University of Delaware*, AIChE, New York (1958).

11. Drickamer, H.G., and J.R. Bradford, *Trans. AIChE*, **39**, 319–360 (1943).

12. O'Connell, H.E., *Trans. AIChE*, **42**, 741–755 (1946).

13. Lockhart, F.J., and C.W. Leggett, in K.A. Kobe and J.J. McKetta, Jr., Eds., *Advances in Petroleum Chemistry and Refining* Vol. 1, Interscience, New York, pp. 323–326 (1958).

14. Zuiderweg, F.J., H. Verburg, and F.A.H. Gilissen, *Proc. International Symposium on Distillation*, Institution of Chem. Eng., London, 202–207 (1960).

15. Gautreaux, M.F., and H.E. O'Connell, *Chem. Eng. Prog.*, **51**(5) 232–237 (1955).

16. Chan, H., and J.R. Fair, *Ind. Eng. Chem. Process Des. Dev.*, **23**, 814–819 (1984).

17. Fair, J.R., H.R. Null, and W.L. Bolles, *Ind. Eng. Chem. Process Des. Dev.*, **22**, 53–58 (1983).

18. Sakata, M., and T. Yanagi, *I. Chem. E. Symp. Ser.*, **56**, 3.2/21 (1979).

19. Yanagi, T., and M. Sakata, *Ind. Eng. Chem. Process Des. Devel.*, **21**, 712 (1982).

20. Younger, A.H., *Chem. Eng.*, **62**(5), 201–202 (1955).

21. Ponchon, M., *Tech. Moderne*, **13**, 20, 55 (1921).

22. Savarit, R., *Arts et Metiers*, pp. 65, 142, 178, 241, 266, 307 (1922).

23. Henley, E.J., and J.D. Seader, *Equilibrium-Stage Separation Operations in Chemical Engineering*, John Wiley & Sons, New York (1981).

24. Glitsch Ballast Tray, Bulletin 159, Fritz W. Glitsch and Sons, Dallas (from FRI report of September 3, 1958).

STUDY QUESTIONS

7.1. What equipment is included in a typical distillation operation?

7.2. What determines the operating pressure of a distillation column?

7.3. Under what conditions does a distillation column have to operate under vacuum?

7.4. Why are distillation columns arranged for countercurrent flow of liquid and vapor?

7.5. Why is the McCabe–Thiele graphical method useful in this era of more rigorous, computer-aided algebraic methods used in process simulators?

7.6. Under what conditions does the McCabe–Thiele assumption of constant molar overflow hold?

7.7. In the McCabe–Thiele method, between which two lines is the staircase constructed?

7.8. What is meant by the reflux ratio? What is meant by the boilup ratio?

7.9. What is the q-line and how is it related to the feed condition?

7.10. What are the five possible feed conditions?

7.11. In the McCabe–Thiele method, are the stages stepped off from the top down or the bottom up? In either case, when is it best, during the stepping, to switch from one operating line to the other? Why?

7.12. Can a column be operated at total reflux? How?

7.13. How many stages are necessary for operation at minimum reflux ratio?

7.14. What is meant by a pinch point? Is it always located at the feed stage?

7.15. What is meant by subcooled reflux? How does it affect the amount of reflux inside the column?

7.16. Is it worthwhile to preheat the feed to a distillation column?

7.17. Why is the stage efficiency in distillation higher than that in absorption?

7.18. What kind of a small laboratory column is useful for obtaining plate efficiency data?

EXERCISES

Note: Unless otherwise stated, the usual simplifying assumptions of saturated-liquid reflux, optimal feed-stage location, no heat losses, steady state, and constant molar liquid and vapor flows apply to each exercise.

Section 7.1

7.1. Differences between absorption, distillation, and stripping.

List as many differences between (1) absorption and distillation and (2) stripping and distillation as you can.

7.2. Popularity of packed columns.

Prior to the 1980s, packed columns were rarely used for distillation unless column diameter was less than 2.5 ft. Explain why, in recent years, some trayed towers are being retrofitted with packing and some new large-diameter columns are being designed for packing rather than trays.

7.3. Use of cooling water in a condenser.

A mixture of methane and ethane is subject to distillation. Why can't water be used as a condenser coolant? What would you use?

7.4. Operating pressure for distillation.

A mixture of ethylene and ethane is to be separated by distillation. What operating pressure would you suggest? Why?

7.5. Laboratory data for distillation design.

Under what circumstances would it be advisable to conduct laboratory or pilot-plant tests of a proposed distillation?

7.6. Economic trade-off in distillation design.

Explain the economic trade-off between trays and reflux.

Section 7.2

7.7. McCabe–Thiele Method.

In the 50 years following the development by Sorel in 1894 of a mathematical model for continuous, steady-state, equilibrium-stage distillation, many noncomputerized methods were proposed for solving the equations graphically or algebraically. Today, the only method from that era that remains in widespread use is the McCabe–Thiele graphical method. What attributes of this method are responsible for its continuing popularity?

7.8. Compositions of countercurrent cascade stages.

For the cascade in Figure 7.39a, calculate (a) compositions of streams V_4 and L_1 by assuming 1 atm pressure, saturated-liquid and -vapor feeds, and the vapor–liquid equilibrium data below, where compositions are in mole %. (b) Given the feed compositions in cascade (a), how many stages are required to produce a V_4 containing 85 mol% alcohol? (c) For the cascade configuration in Figure 7.39b, with $D = 50$ mols, what are the compositions of D and L_1? (d) For the configuration of cascade (b), how many stages are required to produce a D of 50 mol% alcohol?

EQUILIBRIUM DATA, MOLE-FRACTION ALCOHOL

x	0.1	0.3	0.5	0.7	0.9
y	0.2	0.5	0.68	0.82	0.94

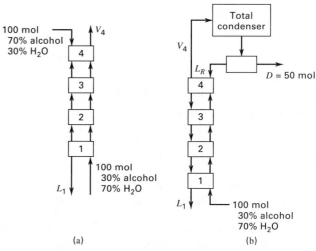

Figure 7.39 Data for Exercise 7.8.

7.9. Stripping of air.

Liquid air is fed to the top of a perforated-tray reboiled stripper operated at 1 atm. Sixty % of the oxygen in the feed is to be drawn off in the bottoms vapor product, which is to contain 0.2 mol% nitrogen. Based on the assumptions and equilibrium data below, calculate: (a) the mole % N_2 in the vapor from the top plate, (b) the vapor generated in the still per 100 moles of feed, and (c) the number of stages required.

Assume constant molar overflow equal to the moles of feed. Liquid air contains 20.9 mol% O_2 and 79.1 mol% N_2. The equilibrium data [*Chem. Met. Eng.*, **35**, 622 (1928)] at 1 atm are:

Temperature, K	Mole-Percent N_2 in Liquid	Mole-Percent N_2 in Vapor
77.35	100.00	100.00
77.98	90.00	97.17
78.73	79.00	93.62
79.44	70.00	90.31
80.33	60.00	85.91
81.35	50.00	80.46
82.54	40.00	73.50
83.94	30.00	64.05
85.62	20.00	50.81
87.67	10.00	31.00
90.17	0.00	0.00

7.10. Using operating data to determine reflux and distillate composition.

A mixture of A (more volatile) and B is separated in a plate distillation column. In two separate tests run with a saturated-liquid feed of 40 mol% A, the following compositions, in mol% A, were obtained for samples of liquid and vapor streams from three consecutive stages between the feed and total condenser at the top:

	Mol% A			
	Test 1		Test 2	
Stage	Vapor	Liquid	Vapor	Liquid
$M + 2$	79.5	68.0	75.0	68.0
$M + 1$	74.0	60.0	68.0	60.5
M	67.9	51.0	60.5	53.0

Determine the reflux ratio and overhead composition in each case, assuming that the column has more than three stages.

7.11. Determining the best distillation procedure.

A saturated-liquid mixture of 70 mol% benzene and 30 mol% toluene, whose relative volatility is 2.5, is to be distilled at 1 atm to produce a distillate of 80 mol% benzene. Five procedures, described below, are under consideration. For each procedure, calculate and tabulate: (a) moles of distillate per 100 moles of feed, (b) moles of total vapor generated per mole of distillate, and (c) mol% benzene in the residue. (d) For each part, construct a y–x diagram. On this, indicate the compositions of the overhead product, the reflux, and the composition of the residue. (e) If the objective is to maximize total benzene recovery, which, if any, of these procedures is preferred?

The procedures are as follows:

1. Continuous distillation followed by partial condensation. The feed is sent to the direct-heated still pot, from which the residue is continuously withdrawn. The vapors enter the top of a helically coiled partial condenser that discharges into a trap. The liquid is returned (refluxed) to the still, while the residual vapor is condensed as a product containing 80 mol% benzene. The molar ratio of reflux to product is 0.5.

2. Continuous distillation in a column containing one equilibrium plate. The feed is sent to the direct-heated still, from which residue is withdrawn continuously. The vapors from the plate enter the top of a helically coiled partial condenser that discharges into a trap. The liquid from the trap is returned to the plate, while the uncondensed vapor is condensed to form a distillate containing 80 mol% benzene. The molar ratio of reflux to product is 0.5.

3. Continuous distillation in a column containing the equivalent of two equilibrium plates. The feed is sent to the direct-heated still, from which residue is withdrawn continuously. The vapors from the top plate enter the top of a helically coiled partial condenser that discharges into a trap. The liquid from the trap is returned to the top plate (refluxed), while the uncondensed vapor is condensed to a distillate containing 80 mol% benzene. The molar ratio of reflux to product is 0.5.

4. The operation is the same as for Procedure 3, except that liquid from the trap is returned to the bottom plate.

5. Continuous distillation in a column with the equivalent of one equilibrium plate. The feed at its boiling point is introduced on the plate. The residue is withdrawn from the direct-heated still pot. The vapors from the plate enter the top of a partial condenser that discharges into a trap. The liquid from the trap is returned to the plate, while the uncondensed vapor is condensed to a distillate of 80 mol% benzene. The molar ratio of reflux to product is 0.5.

7.12. Evaluating distillation procedures.

A saturated-liquid mixture of 50 mol% benzene and toluene is distilled at 101 kPa in an apparatus consisting of a still pot, one theoretical plate, and a total condenser. The still pot is equivalent to an equilibrium stage. The apparatus is to produce a distillate of 75 mol% benzene. For each procedure below, calculate, if possible, the moles of distillate per 100 moles of feed. Assume an α of 2.5.

Procedures: (a) No reflux with feed to the still pot. (b) Feed to the still pot with reflux ratio = 3. (c) Feed to the plate with a reflux ratio of 3. (d) Feed to the plate with a reflux ratio of 3 from a partial condenser. (e) Part (b) using minimum reflux. (f) Part (b) using total reflux.

7.13. Separation of benzene and toluene.

A column at 101 kPa is to separate 30 kg/h of a bubble-point solution of benzene and toluene containing 0.6 mass-fraction toluene into an overhead product of 0.97 mass-fraction benzene and a bottoms product of 0.98 mass-fraction toluene at a reflux ratio of 3.5. The feed is sent to the optimal tray, and the reflux is at saturation temperature. Determine the: (a) top and bottom products and (b) number of stages using the following vapor–liquid equilibrium data.

EQUILIBRIUM DATA IN MOLE- FRACTION BENZENE, 101 kPa

y	0.21	0.37	0.51	0.64	0.72	0.79	0.86	0.91	0.96	0.98
x	0.1	0.2	0.3	0.4	0.5	0.6	0.7	0.8	0.9	0.95

7.14. Calculation of products.

A mixture of 54.5 mol% benzene in chlorobenzene at its bubble point is fed continuously to the bottom plate of a column containing two equilibrium plates, with a partial reboiler and a total condenser. Sufficient heat is supplied to the reboiler to give $\overline{V}/F = 0.855$, and the reflux ratio L/V in the top of the column is constant at 0.50. Under these conditions using the equilibrium data below, what are the compositions of the expected products?

EQUILIBRIUM DATA AT COLUMN PRESSURE, MOLE- FRACTION BENZENE

x	0.100	0.200	0.300	0.400	0.500	0.600	0.700	0.800
y	0.314	0.508	0.640	0.734	0.806	0.862	0.905	0.943

7.15. Loss of trays in a distillation column.

A continuous distillation with a reflux ratio (L/D) of 3.5 yields a distillate containing 97 wt% B (benzene) and a bottoms of 98 wt% T (toluene). Due to weld failures, the 10 stripping plates in the bottom section of the column are ruined, but the 14 upper rectifying plates are intact. It is suggested that the column still be used, with the feed (F) as saturated vapor at the dew point, with $F = 13,600$ kg/h containing 40 wt% B and 60 wt% T. Assuming that the plate efficiency remains unchanged at 50%: (a) Can this column still yield a distillate containing 97 wt% B? (b) How much distillate is there? (c) What is the residue composition in mole %?

For vapor–liquid equilibrium data, see Exercise 7.13.

7.16. Changes to a distillation operation.

A distillation column having eight theoretical stages (seven stages + partial reboiler + total condenser) separates 100 kmol/h of saturated-liquid feed containing 50 mol% A into a product of 90 mol% A. The liquid-to-vapor molar ratio at the top plate is 0.75. The saturated-liquid feed enters plate 5 from the top. Determine: (a) the bottoms composition, (b) the $\overline{L}/\overline{V}$ ratio in the stripping section, and (c) the moles of bottoms per hour.

Unknown to the operators, the bolts holding plates 5, 6, and 7 rust through, and the plates fall into the still pot. What is the new bottoms composition?

It is suggested that, instead of returning reflux to the top plate, an equivalent amount of liquid product from another column be used as reflux. If that product contains 80 mol% A, what is now the composition of (a) the distillate and (b) the bottoms?

EQUILIBRIUM DATA, MOLE FRACTION OF A

y	0.19	0.37	0.5	0.62	0.71	0.78	0.84	0.9	0.96
x	0.1	0.2	0.3	0.4	0.5	0.6	0.7	0.8	0.9

7.17. Effect of different feed conditions.

A distillation unit consists of a partial reboiler, a column with seven equilibrium plates, and a total condenser. The feed is a 50 mol % mixture of benzene in toluene.

It is desired to produce a distillate containing 96 mol% benzene, when operating at 101 kPa.

(a) With saturated-liquid feed fed to the fifth plate from the top, calculate: (1) minimum reflux ratio $(L_R/D)_{min}$; (2) the bottoms composition, using a reflux ratio (L_R/D) of twice the minimum; and (3) moles of product per 100 moles of feed.

(b) Repeat part (a) for a saturated vapor fed to the fifth plate from the top.

(c) With saturated-vapor feed fed to the reboiler and a reflux ratio (L/V) of 0.9, calculate: (1) bottoms composition, and (2) moles of product per 100 moles of feed.

Equilibrium data are in Exercise 7.13.

7.18. Conversion of distillation to stripping.

A valve-tray column containing eight theoretical plates, a partial reboiler, and a total condenser separates a benzene–toluene mixture containing 36 mol% benzene at 101 kPa. The reboiler generates 100 kmol/h of vapor. A request has been made for very pure toluene, and it is proposed to run this column as a stripper, with the saturated-liquid feed to the top plate, employing the same boilup at the still and returning no reflux to the column. Equilibrium data are given in Exercise 7.13. (a) What is the minimum feed rate under the proposed conditions, and what is the corresponding composition of the liquid in the reboiler at the minimum feed? (b) At a feed rate 25% above the minimum, what is the rate of production of toluene, and what are the compositions in mol% of the product and distillate?

7.19. Poor performance of distillation.

Fifty mol% methanol in water at 101 kPa is continuously distilled in a seven-plate, perforated-tray column, with a total condenser and a partial reboiler heated by steam. Normally, 100 kmol/h of feed is introduced on the third plate from the bottom. The overhead product contains 90 mol% methanol, and the bottoms 5 mol%. One mole of reflux is returned for each mole of overhead product. Recently it has been impossible to maintain the product purity in spite of an increase in the reflux ratio. The following test data were obtained:

Stream	kmol/h	mol% alcohol
Feed	100	51
Waste	62	12
Product	53	80
Reflux	94	—

What is the most probable cause of this poor performance? What further tests would you make to establish the reason for the trouble? Could some 90% product be obtained by further increasing the reflux ratio, while keeping the vapor rate constant?

Vapor–liquid equilibrium data at 1 atm [*Chem. Eng. Prog.*, **48**, 192 (1952)] in mole-fraction methanol are

x	0.0321	0.0523	0.075	0.154	0.225	0.349	0.813	0.918
y	0.1900	0.2940	0.352	0.516	0.593	0.703	0.918	0.963

7.20. Effect of feed rate reduction operation.

A fractionating column equipped with a steam-heated partial reboiler and total condenser (Figure 7.40) separates a mixture of 50 mol% A and 50 mol% B into an overhead product containing 90 mol% A and a bottoms of 20 mol% A. The column has three theoretical plates, and the reboiler is equivalent to one theoretical plate. When the system is operated at $L/V = 0.75$ with the feed as a saturated liquid to the bottom plate, the desired products are obtained. The steam to the reboiler is controlled and remains constant. The reflux to the column also remains constant. The feed to the column is normally 100 kmol/h, but it was inadvertently cut back to 25 kmol/h. What will be the composition of the reflux and the vapor leaving the reboiler under these new conditions? Assume

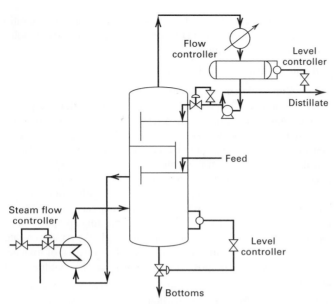

Figure 7.40 Data for Exercise 7.20.

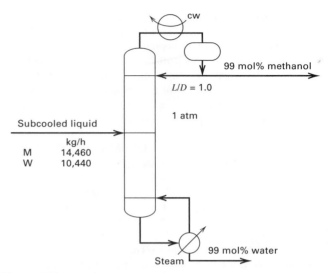

Figure 7.41 Data for Exercise 7.23.

that the vapor leaving the reboiler is not superheated. Relative volatility is 3.0.

7.21. Stages for a binary separation.

A saturated vapor of maleic anhydride and benzoic acid containing 10 mol% acid is a byproduct of the manufacture of phthalic anhydride. It is distilled at 13.3 kPa to give a product of 99.5 mol% maleic anhydride and a bottoms of 0.5 mol%. Calculate the number of theoretical plates using an L/D of 1.6 times the minimum using the data below.

VAPOR PRESSURE, TORR:

Temperature, °C:	10	50	100	200	400
Maleic anhydride	78.7	116.8	135.8	155.9	179.5
Benzoic acid	131.6	167.8	185.0	205.8	227

7.22. Calculation of stages algebraically.

A bubble-point feed of 5 mol% A in B is to be distilled to give a distillate containing 35 mol% A and a bottoms containing 0.2 mol%. The column has a partial reboiler and a partial condenser. If $\alpha = 6$, calculate the following algebraically: (a) the minimum number of equilibrium stages; (b) the minimum boilup ratio $\overline{V}/B$; (c) the actual number of stages for a boilup ratio equal to 1.2 times the minimum.

7.23. Distillation with a subcooled feed.

Methanol (M) is to be separated from water (W) by distillation, as shown in Figure 7.41. The feed is subcooled: $q = 1.12$. Determine the feed-stage location and the number of stages required. Vapor–liquid equilibrium data are given in Exercise 7.19.

7.24. Calculation of distillation graphically and analytically.

A saturated-liquid feed of 69.4 mol% benzene (B) in toluene (T) is to be distilled at 1 atm to produce a distillate of 90 mol% benzene, with a yield of 25 moles of distillate per 100 moles of feed. The feed is sent to a steam-heated reboiler, where bottoms is withdrawn continuously. The vapor from the reboiler goes to a partial condenser and then to a phase separator that returns the liquid reflux to the reboiler. The vapor from the separator, which is in equilibrium with the liquid reflux, is sent to a total condenser to produce distillate. At equilibrium, the mole ratio of B to T in the vapor from the reboiler is 2.5 times the mole ratio of B to T in the bottoms. Calculate

analytically and graphically the total moles of vapor generated in the reboiler per 100 mol of feed.

7.25. Operation at total reflux.

A plant has 100 kmol of a liquid mixture of 20 mol% benzene and 80 mol% chlorobenzene, which is to be distilled at 1 atm to obtain bottoms of 0.1 mol% benzene. Assume $\alpha = 4.13$. The plant has a column containing four theoretical plates, a total condenser, a reboiler, and a reflux drum to collect condensed overhead. A run is to be made at total reflux. While steady state is being approached, a finite amount of distillate is held in a reflux trap. When the steady state is reached, the bottoms contain 0.1 mol% benzene. What yield of bottoms can be obtained? The liquid holdup in the column is negligible compared to that in the reboiler and reflux drum.

Section 7.3

7.26. Trays for a known Murphree efficiency.

A 50 mol% mixture of acetone in isopropanol is to be distilled to produce a distillate of 80 mol% acetone and a bottoms of 25 mol%. The feed is a saturated liquid, the column is operated with a reflux ratio of 0.5, and the Murphree vapor efficiency is 50%. How many trays are required? Assume a total condenser, partial reboiler, saturated-liquid reflux, and optimal feed stage. The vapor–liquid equilibrium data are:

EQUILIBRIUM DATA, MOLE- PERCENT ACETONE

Liquid	0	2.6	5.4	11.7	20.7	29.7	34.1	44.0	52.0
Vapor	0	8.9	17.4	31.5	45.6	55.7	60.1	68.7	74.3

Liquid	63.9	74.6	80.3	86.5	90.2	92.5	95.7	100.0
Vapor	81.5	87.0	89.4	92.3	94.2	95.5	97.4	100.0

7.27. Minimum reflux, boilup, and number of trays for known efficiency.

A mixture of 40 mol% carbon disulfide (CS_2) in carbon tetrachloride (CCl_4) is continuously distilled. The feed is 50% vaporized ($q = 0.5$). The product from a total condenser is 95 mol% CS_2, and the bottoms from a partial reboiler is 5 mol% CS_2. The column operates with a reflux ratio, L/D, of 4 to 1. The Murphree vapor efficiency is 80%. (a) Calculate graphically the minimum reflux, the minimum boilup ratio from the reboiler, $\overline{V}/B$, and the minimum number of stages (including the reboiler). (b) How many trays are required for the actual column at 80% Murphree vapor-tray efficiency by the McCabe–Thiele method? The vapor–liquid equilibrium data at column pressure in terms of CS_2 mole fractions are:

x	0.05	0.1	0.2	0.3	0.4	0.5	0.6	0.7	0.8	0.9
y	0.135	0.245	0.42	0.545	0.64	0.725	0.79	0.85	0.905	0.955

7.28.   Reboiler duty for a distillation.

A distillation unit consists of a partial reboiler, a bubble-cap column, and a total condenser. The overall plate efficiency is 65%. The feed is a bubble-point liquid of 50 mol% benzene in toluene, which is fed to the optimal plate. The column is to produce a distillate containing 95 mol% benzene and a bottoms of 95 mol% toluene. Calculate for an operating pressure of 1 atm the: (a) minimum reflux ratio $(L/D)_{min}$; (b) minimum number of actual plates; (c) number of actual plates needed for a reflux ratio (L/D) of 50% more than the minimum; (d) kg/h of distillate and bottoms, if the feed is 907.3 kg/h; and (e) saturated steam at 273.7 kPa required in kg/h for the reboiler using the enthalpy data below and any assumptions necessary. (f) Make a rigorous enthalpy balance on the reboiler, using the enthalpy data below and assuming ideal solutions. Enthalpies are in Btu/lbmol at reboiler temperature:

	h_L	h_V
benzene	4,900	18,130
toluene	8,080	21,830

Vapor–liquid equilibrium data are given in Exercise 7.13.

7.29.   Distillation of an azeotrope-forming mixture.

A continuous distillation unit, consisting of a perforated-tray column with a partial reboiler and a total condenser, is to be designed to separate ethanol and water at 1 atm. The bubble-point feed contains 20 mol% alcohol. The distillate is to contain 85 mol% alcohol, and the recovery is to be 97%. (a) What is the molar concentration of the bottoms? (b) What is the minimum value of the reflux ratio L/V, the reflux ratio L/D, and the boilup ratio V/B? (c) What is the minimum number of theoretical stages and the number of actual plates, if the overall plate efficiency is 55%? (d) If the L/V is 0.80, how many actual plates will be required?

Vapor–liquid equilibrium for ethanol–water at 1 atm in terms of mole fractions of ethanol are [*Ind. Eng. Chem.*, **24**, 881 (1932)]:

x	y	T, °C	x	y	T, °C
0.0190	0.1700	95.50	0.3273	0.5826	81.50
0.0721	0.3891	89.00	0.3965	0.6122	80.70
0.0966	0.4375	86.70	0.5079	0.6564	79.80
0.1238	0.4704	85.30	0.5198	0.6599	79.70
0.1661	0.5089	84.10	0.5732	0.6841	79.30
0.2337	0.5445	82.70	0.6763	0.7385	78.74
0.2608	0.5580	82.30	0.7472	0.7815	78.41
			0.8943	0.8943	78.15

7.30.   Multiple feeds and open steam.

Solvent A is to be separated from water by distillation to produce a distillate containing 95 mol% A at a 95% recovery. The feed is

available in two saturated-liquid streams, one containing 40 mol% A and the other 60 mol% A. Each stream will provide 50 kmol/h of component A. The α is 3 and since the less-volatile component is water, it is proposed to supply the necessary reboiler heat in the form of open steam. For the preliminary design, the operating reflux ratio, L/D, is 1.33 times the minimum, using a total condenser. The overall plate efficiency is estimated to be 70%. How many plates will be required, and what will be the bottoms composition? Determine analytically the points necessary to locate the operating lines. Each feed should enter the column at its optimal location.

7.31.   Optimal feed plate location.

A saturated-liquid feed of 40 mol% *n*-hexane (H) and 60 mol% *n*-octane is to be separated into a distillate of 95 mol% H and a bottoms of 5 mol% H. The reflux ratio L/D is 0.5, and a cooling coil submerged in the liquid of the second plate from the top removes sufficient heat to condense 50 mol% of the vapor rising from the third plate down from the top. The *x*–*y* data of Figure 7.37 may be used. (a) Derive the equations needed to locate the operating lines. (b) Locate the operating lines and determine the required number of theoretical plates if the optimal feed plate location is used.

7.32.   Open steam for alcohol distillation.

One hundred kmol/h of a saturated-liquid mixture of 12 mol% ethyl alcohol in water is distilled continuously using open steam at 1 atm introduced directly to the bottom plate. The distillate required is 85 mol% alcohol, representing 90% recovery of the alcohol in the feed. The reflux is saturated liquid with $L/D = 3$. Feed is on the optimal stage. Vapor–liquid equilibrium data are given in Exercise 7.29. Calculate: (a) the steam requirement in kmol/h; (b) the number of theoretical stages; (c) the optimal feed stage; and (d) the minimum reflux ratio, $(L/D)_{min}$.

7.33.   Distillation of an azeotrope-forming mixture using open steam.

A 10 mol% isopropanol-in-water mixture at its bubble point is to be distilled at 1 atm to produce a distillate containing 67.5 mol% isopropanol, with 98% recovery. At a reflux ratio L/D of 1.5 times the minimum, how many theoretical stages will be required: (a) if a partial reboiler is used? (b) if no reboiler is used and saturated steam at 101 kPa is introduced below the bottom plate? (c) How many stages are required at total reflux?

Vapor–liquid data in mole-fraction isopropanol at 101 kPa are:

T, °C	93.00	84.02	82.12	81.25	80.62	80.16	80.28	81.51
y	0.2195	0.4620	0.5242	0.5686	0.5926	0.6821	0.7421	0.9160
x	0.0118	0.0841	0.1978	0.3496	0.4525	0.6794	0.7693	0.9442

Notes: Composition of the azeotrope is $x = y = 0.6854$. Boiling point of azeotrope = 80.22°C.

7.34.   Comparison of partial reboiler with live steam.

An aqueous solution of 10 mol% isopropanol at its bubble point is fed to the top of a stripping column, operated at 1 atm, to produce a vapor of 40 mol% isopropanol. Two schemes, both involving the same heat expenditure, with V/F (moles of vapor/mole of feed) = 0.246, are under consideration. Scheme 1 uses a partial reboiler at the bottom of a stripping column, with steam condensing inside a closed coil. In Scheme 2, live steam is injected directly below the bottom plate. Determine the number of stages required in each case. Equilibrium data are given in Exercise 7.33.

7.35.   Optimal feed stages for two feeds.

Determine the optimal-stage location for each feed and the number of theoretical stages required for the distillation separation

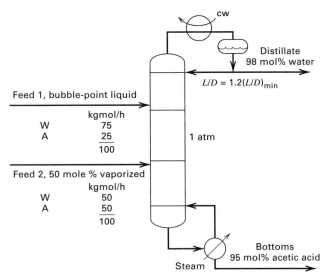

Figure 7.42 Data for Exercise 7.35.

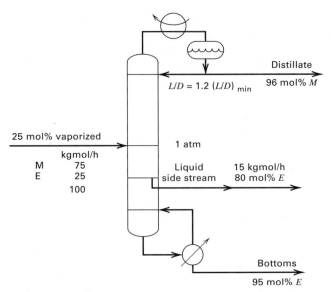

Figure 7.43 Data for Exercise 7.36.

shown in Figure 7.42, using the following vapor–liquid data in mole fractions of water.

WATER (W)/ACETIC ACID (A), 1 ATM

x_W	0.0055	0.053	0.125	0.206	0.297	0.510	0.649	0.803	0.9594
y_W	0.0112	0.133	0.240	0.338	0.437	0.630	0.751	0.866	0.9725

7.36.   Optimal sidestream location.

Determine the number of equilibrium stages and optimal-stage locations for the feed and liquid sidestreams of the distillation process in Figure 7.43, assuming that methanol (M) and ethanol (E) form an ideal solution.

7.37.   Use of an interreboiler.

A mixture of n-heptane (H) and toluene (T) is separated by extractive distillation with phenol (P). Distillation is then used to recover the phenol for recycle, as shown in Figure 7.44a, where the small amount of n-heptane in the feed is ignored. For the conditions shown in Figure 7.44a, determine the number of stages required. Note that heat must be supplied to the reboiler at a high temperature

because of the high boiling point of phenol, which causes second-law inefficiency. Therefore, consider the scheme in Figure 7.44b, where an interreboiler, located midway between the bottom plate and the feed stage, provides 50% of the boilup used in Figure 7.44a. The remainder of the boilup is provided by the reboiler. Determine the stages required for the case with the interreboiler and the temperature of the interreboiler stage. Unsmoothed vapor–liquid equilibrium data at 1 atm [*Trans. AIChE*, **41**, 555 (1945)] are:

x_T	y_T	$T,\,°C$	x_T	y_T	$T,\,°C$
0.0435	0.3410	172.70	0.6512	0.9260	120.00
0.0872	0.5120	159.40	0.7400	0.9463	119.70
0.1248	0.6250	149.40	0.8012	0.9545	115.60
0.2190	0.7850	142.20	0.8840	0.9750	112.70
0.2750	0.8070	133.80	0.9394	0.9861	113.30
0.4080	0.8725	128.30	0.9770	0.9948	111.10
0.4800	0.8901	126.70	0.9910	0.9980	111.10
0.5898	0.9159	122.20	0.9973	0.9993	110.50

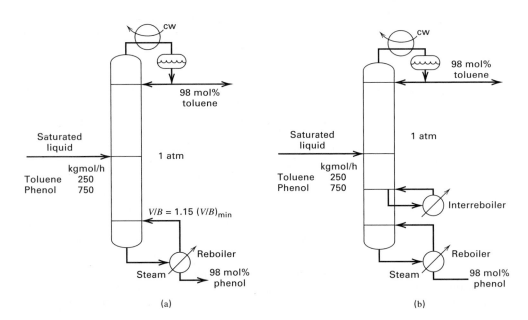

Figure 7.44 Data for Exercise 7.37.

7.38. Addition of intercondenser and interreboiler.

A distillation column to separate *n*-butane from *n*-pentane was recently put on line in a refinery. Apparently, there was a design error because the column did not make the desired separation, as shown below [*Chem. Eng. Prog.*, **61**(8), 79 (1965)].

It is proposed to add an intercondenser in the rectifying section to generate more reflux and an interreboiler in the stripping section to produce additional boilup. Show by use of a McCabe–Thiele diagram how this might improve the operation.

	Design Specification	Actual Operation
Mol% nC_5 in distillate	0.26	13.49
Mol% nC_4 in bottoms	0.16	4.28

7.39. Use of Kremser method to extend McCabe-Thiele method.

Chlorobenzene production by chlorination of benzene produces two close-boiling isomers, *para*-dichlorobenzene (P) and *ortho*-dichlorobenzene (O), which are separated by distillation. The feed to the column is 62 mol% P and 38 mol% O. The pressures at the bottom and top of the column are 20 psia and 15 psia, respectively. The distillate is to be 98 mol% P, and the bottoms 96 mol% O. The feed is slightly vaporized with $q = 0.9$. Calculate the number of equilibrium stages required for an $R/R_{min} = 1.15$. Base your calculations on an average $\alpha = 1.163$ obtained as the arithmetic average between the column top and column bottom using vapor-pressure data and the assumption of Raoult's and Dalton's laws. The McCabe–Thiele construction should be supplemented at the two ends by use of the Kremser equations as in Example 7.4.

7.40. Air separation using a Linde double column.

O_2 and N_2 are obtained by distillation of air using the Linde double column, shown in Figure 7.45, which consists of a lower column at elevated pressure surmounted by an atmospheric-pressure column. The boiler of the upper column is also the reflux condenser for both columns. Gaseous air plus enough liquid to compensate for heat leak into the column (more liquid if liquid-oxygen product is withdrawn) enters the exchanger at the base of the lower column and condenses, giving up heat to the boiling liquid and thus supplying the column vapor flow. The liquid air enters an intermediate point in this column. The rising vapors are partially condensed to form the reflux, and the uncondensed vapor passes to an outer row of tubes and is totally condensed, the liquid nitrogen collecting in an annulus, as shown. By operating this column at 4 to 5 atm, the liquid oxygen boiling at 1 atm is cold enough to condense pure nitrogen. The liquid in the bottom of the lower column contains about 45 mol% O_2 and forms the feed for the upper column. This double column can produce very pure O_2 with high O_2 recovery, and relatively pure N_2. On a single McCabe–Thiele diagram—using equilibrium lines, operating lines, *q*-lines, a 45° line, stepped-off stages, and other illustrative aids—show qualitatively how stage requirements can be computed.

Section 7.4

7.41. Comparison of tray efficiency.

Performance data for a distillation tower separating a 50/50 by weight percent mixture of methanol and water are as follows:

Feed rate = 45,438 lb/h; feed condition = bubble-point liquid at feed-tray pressure;

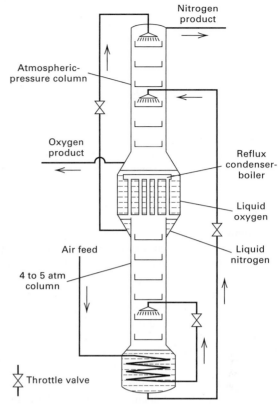

Figure 7.45 Data for Exercise 7.40.

Wt% methanol in distillate = 95.04; Wt% methanol in bottoms = 1.00;

Reflux ratio = 0.947; reflux condition = saturated liquid;

Boilup ratio = 1.138; pressure in reflux drum = 14.7 psia;

Type condenser = total; type reboiler = partial;

Condenser pressure drop = 0.0 psi; tower pressure drop = 0.8 psi;

Trays above feed tray = 5; trays below feed tray = 6;

Total trays = 12; tray diameter = 6 ft; type tray = single-pass sieve tray; flow path length = 50.5 inches;

Weir length = 42.5 inches; hole area = 10%; hole size = 3/16 inch;

Weir height = 2 inches; tray spacing = 24 inches;

Viscosity of feed = 0.34 cP; Surface tension of distillate = 20 dyne/cm;

Surface tension of bottoms = 58 dyne/cm;

Temperature of top tray = 154°F; temperature of bottom tray = 207°F

Vapor–liquid equilibrium data at column pressure in mole fraction of methanol are

y	0.0412	0.156	0.379	0.578	0.675	0.729	0.792	0.915
x	0.00565	0.0246	0.0854	0.205	0.315	0.398	0.518	0.793

Based on the above data: (a) Determine the overall tray efficiency assuming the reboiler is equivalent to a theoretical stage. (b) Estimate the overall tray efficiency from the Drickamer–Bradford correlation. (c) Estimate the overall tray efficiency from the O'Connell correlation, accounting for length of flow path. (d) Estimate the Murphree vapor-tray efficiency by the method of Chan and Fair.

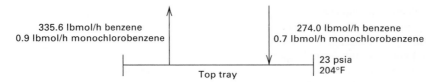

335.6 lbmol/h benzene
0.9 lbmol/h monochlorobenzene

274.0 lbmol/h benzene
0.7 lbmol/h monochlorobenzene

23 psia
204°F

Top tray

Figure 7.46 Data for Exercise 7.43.

7.42.   Oldershaw column efficiency.

For the conditions of Exercise 7.41, a laboratory Oldershaw column measures an average Murphree vapor-point efficiency of 65%. Estimate E_{MV} and E_o.

Section 7.5

7.43.   Column diameter.

Figure 7.46 shows conditions for the top tray of a distillation column. Determine the column diameter at 85% of flooding for a valve tray. Make whatever assumptions necessary.

7.44.   Column sizing.

Figure 7.47 depicts a propylene/propane distillation. Two sieve-tray columns in series are used because a 270-tray column poses structural problems. Determine column diameters, tray efficiency using the O'Connell correlation, number of actual trays, and column heights.

7.45.   Sizing a vertical flash drum.

Determine the height and diameter of a vertical flash drum for the conditions shown in Figure 7.48.

7.46.   Sizing a horizontal flash drum.

Determine the length and diameter of a horizontal reflux drum for the conditions shown in Figure 7.49.

7.47.   Possible swaged column.

Results of design calculations for a methanol–water distillation operation are given in Figure 7.50. (a) Calculate the column diameter at the top and at the bottom, assuming sieve trays. Should the column be swaged? (b) Calculate the length and diameter of the horizontal reflux drum.

7.48.   Tray calculations of flooding, pressure drop, entrainment, and froth height.

For the conditions given in Exercise 7.41, estimate for the top tray and the bottom tray: (a) % of flooding; (b) tray pressure drop in psi; (c) whether weeping will occur; (d) entrainment rate; and (e) froth height in the downcomer.

7.49.   Possible retrofit to packing.

If the feed rate to the tower of Exercise 7.41 is increased by 30%, with conditions—except for tower pressure drop—remaining the same, estimate for the top and bottom trays: (a) % of flooding; (b) tray pressure drop in psi; (c) entrainment rate; and (d) froth height in the downcomer. Will the new operation be acceptable? If not, should you consider a retrofit with packing? If so, should both sections of the column be packed, or could just one section be packed to achieve an acceptable operation?

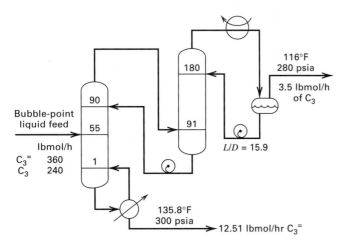

Figure 7.47 Data for Exercise 7.44.

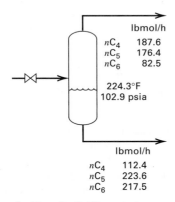

Figure 7.48 Data for Exercise 7.45.

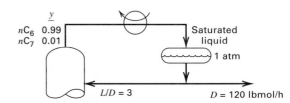

Figure 7.49 Data for Exercise 7.46.

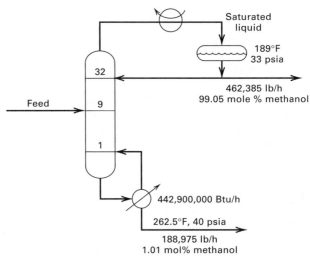

Figure 7.50 Data for Exercise 7.47.

Section 7.6

7.50. HETP calculation.

A mixture of benzene and dichloroethane is used to obtain HETP data for a packed column that contains 10 ft of packing and operates adiabatically at atmospheric pressure. The liquid is charged to the reboiler, and the column is operated at total reflux until equilibrium is established. Liquid samples from the distillate and reboiler give for benzene $x_D = 0.653$, $x_B = 0.298$. Calculate HETP in inches for this packing. What are the limitations of using this calculated value for design?

Data for x–y at 1 atm (in benzene mole fractions) are

x	0.10	0.20	0.30	0.40	0.50	0.60	0.70	0.80	0.90
y	0.11	0.22	0.325	0.426	0.526	0.625	0.720	0.815	0.91

7.51. Plate versus packed column.

Consider a distillation column for separating ethanol from water at 1 atm. The feed is a 10 mol% ethanol bubble-point liquid, the bottoms contains 1 mol% ethanol, and the distillate is 80 mol% ethanol. $R/R_{min} = 1.5$. Phase-equilibrium data are given in Exercise 7.29, and constant molar overflow applies. (a) How many theoretical plates are required above and below the feed if a plate column is used? (b) How many transfer units are required above and below the feed if a packed column is used? (c) Assuming the plate efficiency is 80% and the plate spacing is 18 inches, how high is the plate column? (d) Using an H_{OG} value of 1.2 ft, how high is the packed column? (e) Assuming that HTU data are available only on the benzene–toluene system, how would one go about applying the data to obtain the HTU for the ethanol–water system?

7.52. Design of random and structured packed columns.

Plant capacity for the methanol–water distillation of Exercise 7.41 is to be doubled. Rather than installing a second, identical trayed tower, a packed column is being considered. This would have a feed location, product purities, reflux ratio, operating pressure, and capacity identical to the present trayed tower. Two packings are being considered: (1) 50-mm plastic Nor-Pac rings (a random packing), and (2) Montz metal B1-300 (a structured packing).

For each of these packings, design a column to operate at 70% of flooding by calculating for each section: (a) liquid holdup, (b) column diameter, (c) H_{OG}, (d) packed height, and (e) pressure drop. What are the advantages, if any, of each of the packed-column designs over a second trayed tower? Which packing, if either, is preferable?

7.53. Advantages of a packed column.

For the specifications of Example 7.1, design a packed column using 50-mm metal Hiflow rings and operating at 70% of flooding by calculating for each section: (a) liquid holdup, (b) column diameter, (c) H_{OG}, (d) packed height, and (e) pressure drop. What are the advantages and disadvantages of a packed column as compared to a trayed tower for this service?

Section 7.7

7.54. Use of an enthalpy-concentration diagram.

Figure 7.37 is an enthalpy-concentration diagram for *n*-hexane (H), and *n*-octane (O) at 101 kPa. Use this diagram to determine the: (a) mole-fraction composition of the vapor when a liquid containing 30 mol% H is heated from Point A to the bubble-point temperature at Point B; (b) energy required to vaporize 60 mol% of a

Table 7.8   Methanol–Water Vapor–Liquid Equilibrium and Enthalpy Data for 1 atm (MeOH = Methyl Alcohol)

Mol% MeOH y or x	Enthalpy above 0°C, Btu/lbmol Solution				Vapor Liquid Equilibrium Data		
	Saturated Vapor		Saturated Liquid		Mol% MeOH in		Temperature, °C
	T, °C	h_V	T, °C	h_L	Liquid	Vapor	
0	100	20,720	100	3,240	0	0	100
5	98.9	20,520	92.8	3,070	2.0	13.4	96.4
10	97.7	20,340	87.7	2,950	4.0	23.0	93.5
15	96.2	20,160	84.4	2,850	6.0	30.4	91.2
20	94.8	20,000	81.7	2,760	8.0	36.5	89.3
30	91.6	19,640	78.0	2,620	10.0	41.8	87.7
40	88.2	19,310	75.3	2,540	15.0	51.7	84.4
50	84.9	18,970	73.1	2,470	20.0	57.9	81.7
60	80.9	18,650	71.2	2,410	30.0	66.5	78.0
70	76.6	18,310	69.3	2,370	40.0	72.9	75.3
80	72.2	17,980	67.6	2,330	50.0	77.9	73.1
90	68.1	17,680	66.0	2,290	60.0	82.5	71.2
100	64.5	17,390	64.5	2,250	70.0	87.0	69.3
					80.0	91.5	67.6
					90.0	95.8	66.0
					95.0	97.9	65.0
					100.0	100.0	64.5

Source: J.G. Dunlop, "Vapor–Liquid Equilibrium Data," M.S. thesis, Brooklyn Polytechnic Institute, Brooklyn, NY (1948).

mixture initially at 100°F and containing 20 mol% H (Point G); and (c) compositions of the equilibrium vapor and liquid resulting from part (b).

7.55. Use of an enthalpy-concentration diagram.

Using the enthalpy-concentration diagram of Figure 7.37, determine the following for a mixture of *n*-hexane (H) and *n*-octane (O) at 1 atm: (a) temperature and compositions of equilibrium liquid and vapor resulting from adiabatic mixing of 950 lb/h of a mixture of 30 mol% H in O at 180°F with 1,125 lb/h of a mixture of 80 mol% H in O at 240°F; (b) energy required to partially condense, by cooling, a mixture of 60 mol% H in O from an initial temperature of 260°F to 200°F. What are the compositions and amounts of the resulting vapor and liquid phases per lbmol of original mixture? (c) If the vapor from part (b) is further cooled to 180°F, determine the compositions and relative amounts of the resulting vapor and liquid.

7.56. Plotting an enthalpy-concentration diagram for distillation calculations.

One hundred lbmol/h of 60 mol% methanol in water at 30°C and 1 atm is to be separated by distillation into a liquid distillate containing 98 mol% methanol and a bottoms containing 96 mol% water. Enthalpy and equilibrium data for the mixture at 1 atm are given in Table 7.8. The enthalpy of the feed mixture is 765 Btu/lbmol. (a) Using the given data, plot an enthalpy-concentration diagram. (b) Devise a procedure to determine, from the diagram of part (a), the minimum number of equilibrium stages for the condition of total reflux and the required separation. (c) From the procedure developed in part (b), determine N_{min}. Why is the value independent of the feed condition? (d) What are the temperatures of the distillate and bottoms?

Chapter 8

Liquid–Liquid Extraction with Ternary Systems

§8.0 INSTRUCTIONAL OBJECTIVES

After completing this chapter, you should be able to:

- List situations where liquid–liquid extraction might be preferred to distillation.
- Explain why only certain types of equipment are suitable for extraction in bioprocesses.
- Define the distribution coefficient and show its relationship to activity coefficients and selectivity of a solute between carrier and solvent.
- Make a preliminary selection of a solvent using group-interaction rules.
- Distinguish, for ternary mixtures, between Type I and Type II systems.
- For a specified recovery of a solute, calculate with the Hunter and Nash method, using a triangular diagram, minimum solvent requirement, and equilibrium stages for ternary liquid–liquid extraction in a cascade.
- Design a cascade of mixer-settler units based on mass-transfer considerations.
- Size a multicompartment extraction column, including consideration of the effect of axial dispersion.
- Compare organic-solvent, aqueous two-phase, and supercritical-fluid extraction for recovery of bioproducts.
- Determine effects of pH, temperature, salt, and solute valence on partitioning of bioproducts in organic-solvent and aqueous two-phase extraction.
- Evaluate mass transfer in liquid–liquid extraction using Maxwell–Stefan relations.

In *liquid–liquid extraction* (also called *solvent extraction* or *extraction*), a liquid feed of two or more components is contacted with a second liquid phase, called the *solvent*, which is immiscible or only partly miscible with one or more feed components and completely or partially miscible with one or more of the other feed components. Thus, the solvent partially dissolves certain species of the liquid feed, effecting at least a partial separation of the feed components.

The solvent may be a pure compound or a mixture. If the feed is an aqueous solution, an organic solvent is used; if the feed is organic, the solvent is often water. Important exceptions occur in metallurgy for the separation of metals and in bioseparations for the extraction from aqueous solutions of proteins that are denatured or degraded by organic solvents. In that case, *aqueous two-phase extraction*, described in §8.6.2, can be employed.

Solid–liquid extraction (also called *leaching*) involves recovery of substances from a solid by contact with a liquid solvent, such as the recovery of oil from seeds by an organic solvent, and is covered in Chapter 16.

According to Derry and Williams [1], extraction has been practiced since the time of the Romans, who used molten lead to separate gold and silver from molten copper by extraction. This was followed by the discovery that sulfur could selectively dissolve silver from an alloy with gold. However, it was not until the early 1930s that L. Edeleanu invented the first large-scale extraction process, the removal of aromatic and sulfur compounds from liquid kerosene using liquid sulfur dioxide at 10 to 20°F. This resulted in a cleaner-burning kerosene. Liquid–liquid extraction has grown in importance since then because of the demand for temperature-sensitive products, higher-purity requirements, better equipment, and availability of solvents with higher selectivity, and is an important method in bioseparations.

The first five sections of this chapter cover the simplest liquid–liquid extraction, which involves only a ternary system consisting of two miscible feed components—the *carrier*, *C*, and the *solute*, *A*—plus solvent, *S*, a pure compound. Components *C* and *S* are at most only partially soluble, but solute *A* is completely or partially soluble in *S*. During extraction, mass transfer of *A* from the feed to the solvent occurs, with less transfer of *C* to the solvent, or *S* to the feed. Nearly complete transfer of *A* to the solvent is seldom achieved in just one stage. In practice, a number of stages are used in one- or two-section, countercurrent cascades. This chapter concludes with a section on extraction of bioproducts, which may involve more than three components.

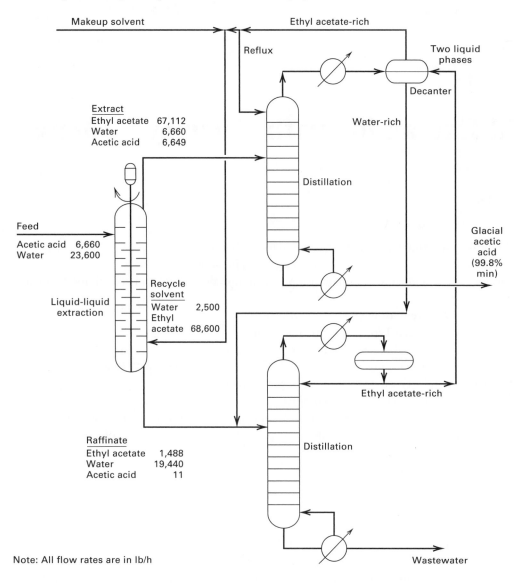

Note: All flow rates are in lb/h

Figure 8.1 Typical liquid–liquid extraction process.

Industrial Example

Acetic acid is produced by methanol carbonylation or oxidation of acetaldehyde, or as a byproduct of cellulose–acetate manufacture. In all cases, a mixture of acetic acid (n. b.p. = 118.1°C) and water (n. b.p. = 100°C) is separated to give glacial acetic acid (99.8 wt% min.). When the mixture contains less than 50% acetic acid, separation by distillation is expensive because of the high heat of vaporization and large amounts of water. So, a solvent-extraction process is attractive. Figure 8.1 shows a typical implementation of extraction, where two distillations are required to recover the solvent for recycle. These additional steps are common to extraction processes. Here, a feed of 30,260 lb/h of 22 wt% acetic acid in water is sent to a single-section extraction column operating at ambient conditions, where it is contacted with 71,100 lb/h of ethyl-acetate solvent (n. b.p. = 77.1°C), saturated with water. The low-density, solvent-rich extract exits from the top of the extractor with 99.8% of the acetic acid in the feed. The high-density, carrier-rich raffinate exiting the extractor bottom contains only 0.05 wt% acetic acid. The extract is sent to a distillation column, where glacial acetic

acid is the bottoms product. The overhead vapor, which is rich in ethyl acetate but also contains appreciable water vapor, splits into two liquid phases when condensed. These are separated in the decanter by gravity. The lighter ethyl-acetate-rich phase is divided into reflux and solvent recycle to the extractor.

The water-rich phase from the decanter is sent, together with the raffinate from the extractor, to a second distillation column, where wastewater is the bottoms product and the ethyl-acetate-rich overhead is recycled to the decanter. Makeup ethyl-acetate solvent is provided for solvent losses to the glacial acetic acid and wastewater.

Six equilibrium stages are required to transfer 99.8% of the acetic acid from feed to extract using a solvent-to-feed weight ratio of 2.35; the recycled solvent is water-saturated. A rotating-disk contactor (RDC), described in §8.1.5, is employed, which disperses the organic-rich phase into droplets by horizontal, rotating disks, while the water-rich phase is continuous throughout the column. Dispersion, subsequent coalescence, and settling take place easily because, at extractor operating conditions, liquid-phase viscosities are less than 1 cP, the phase-density difference is more than 0.08 g/cm^3, and interfacial tension between the two phases is more than 30 dyne/cm.

The column has an inside diameter of 5.5 ft and a total height of 28 ft and is divided into 40 compartments, each 7.5 inches high containing a 40-inch-diameter rotor disk located between a pair of stator (donut) rings of 46-inch inside diameter. Settling zones exist above the top stator ring and below the bottom stator ring. Because the light liquid phase is dispersed, the liquid–liquid interface is maintained near the top of the column. The rotors are mounted on a centrally located shaft driven at 60 rpm by a 5-hp motor equipped with a speed control, the optimal disk speed being determined during plant operation. The HETP is 50 inches, equivalent to 6.67 compartments per equilibrium stage. The HETP would be only 33 inches if no axial (longitudinal) mixing, discussed in §8.5, occurred. Because of the corrosive nature of aqueous acetic acid solutions, the extractor is constructed of stainless steel. Since the 1930s, thousands of similar extraction columns, with diameters ranging up to at least 25 ft, have been built. As discussed in §8.1, a number of other extraction devices are suitable for the process in Figure 8.1.

Liquid–liquid extraction is a reasonably mature operation, although not as mature or as widely applied as distillation, absorption, and stripping. Procedures for determining the stages to achieve a desired solute recovery are well established. However, in the thermodynamics of liquid–liquid extraction, no simple limiting theory, such as that of ideal solutions for vapor–liquid equilibrium, exists. Frequently, experimental data are preferred over predictions based on activity-coefficient correlations. Such data can be correlated and extended by activity-coefficient equations such as NRTL or UNIQUAC, discussed in §2.6. Also, considerable laboratory effort may be required to find an optimal solvent. A variety of industrial equipment is available, making it necessary to consider alternatives before making a final selection. Unfortunately, no generalized capacity and efficiency correlations are available for all equipment types. Often, equipment vendors and pilot-plant tests must be relied upon to determine appropriate equipment size.

The petroleum industry represents the largest-volume application for liquid–liquid extraction. By the late 1960s, more than 100,000 m^3/day of liquid feedstocks were being processed [2]. Extraction processes are well suited to the petroleum industry because of the need to separate heat-sensitive liquid feeds according to chemical type (e.g., aliphatic, aromatic, naphthenic) rather than by molecular weight or vapor pressure. Table 8.1 lists some representative industrial extraction processes. Other applications exist in the biochemical industry, including the separation of antibiotics and recovery of proteins from natural substrates; in the recovery of metals, such as copper from ammoniacal leach liquors; in separations involving rare metals and radioactive isotopes from spent-fuel elements; and in the inorganic chemical industry, where high-boiling constituents such as phosphoric acid, boric acid, and sodium hydroxide need to be recovered from aqueous solutions. In general, extraction is preferred over distillation for:

Table 8.1 Representative Industrial Liquid–Liquid Extraction Processes

Solute	Carrier	Solvent
Acetic acid	Water	Ethyl acetate
Acetic acid	Water	Isopropyl acetate
Aconitic acid	Molasses	Methyl ethyl ketone
Ammonia	Butenes	Water
Aromatics	Paraffins	Diethylene glycol
Aromatics	Paraffins	Furfural
Aromatics	Kerosene	Sulfur dioxide
Aromatics	Paraffins	Sulfur dioxide
Asphaltenes	Hydrocarbon oil	Furfural
Benzoic acid	Water	Benzene
Butadiene	1-Butene	aq. Cuprammonium acetate
Ethylene cyanohydrin	Methyl ethyl ketone	Brine liquor
Fatty acids	Oil	Propane
Formaldehyde	Water	Isopropyl ether
Formic acid	Water	Tetrahydrofuran
Glycerol	Water	High alcohols
Hydrogen peroxide	Anthrahydroquinone	Water
Methyl ethyl ketone	Water	Trichloroethane
Methyl borate	Methanol	Hydrocarbons
Naphthenes	Distillate oil	Nitrobenzene
Naphthenes/ aromatics	Distillate oil	Phenol
Phenol	Water	Benzene
Phenol	Water	Chlorobenzene
Penicillin	Broth	Butyl acetate
Sodium chloride	aq. Sodium hydroxide	Ammonia
Vanilla	Oxidized liquors	Toluene
Vitamin A	Fish-liver oil	Propane
Vitamin E	Vegetable oil	Propane
Water	Methyl ethyl ketone	aq. Calcium chloride

1. Dissolved or complexed inorganic substances in organic or aqueous solutions.

2. Removal of a contaminant present in small concentrations, such as a color former in tallow or hormones in animal oil.

3. A high-boiling component present in relatively small quantities in an aqueous waste stream, as in the recovery of acetic acid from cellulose acetate.

4. Recovery of heat-sensitive materials, where extraction may be less expensive than vacuum distillation.

5. Separation of mixtures according to chemical type rather than relative volatility.

6. Separation of close-melting or close-boiling liquids, where solubility differences can be exploited.

7. Separation of mixtures that form azeotropes.

The key to an effective extraction process is a suitable solvent. In addition to being stable, nontoxic, inexpensive, and easily recoverable, a solvent should be relatively immiscible with feed components(s) other than the solute, and have a different density from the feed to facilitate phase separation by gravity. It must have a high affinity for the solute, from which it should be easily separated by distillation, crystallization, or other means. Ideally, the distribution (partition) coefficient (2-20) for the solute between the liquid phases should be greater than one, or a large solvent-to-feed ratio will be required. When the degree of solute extraction is not particularly high and/or when a large extraction factor (4-24) can be achieved, an extractor will not require many stages. This is fortunate because mass-transfer resistance in liquid–liquid systems is high and stage efficiency is low in contacting devices, even if mechanical agitation is provided.

In this chapter, equipment for liquid–liquid extraction is discussed, with special attention directed to devices for bioseparations. Equilibrium- and rate-based calculation procedures are presented, mainly for extraction in ternary systems. Use of graphical methods is emphasized. Except for systems dilute in solute(s), calculations for multicomponent systems are best conducted using process simulators, as discussed in Chapter 10.

§8.1 EQUIPMENT FOR SOLVENT EXTRACTION

Equipment similar to that used for absorption, stripping, and distillation is sometimes used for extraction, but such devices are inefficient unless liquid viscosities are low and differences in phase density are high. Generally, centrifugal and mechanically agitated devices are preferred. Regardless of the type of equipment, the number of equilibrium stages required is computed first. Then the size of the device is obtained from experimental HETP or mass-transfer-performance-data characteristic of that device. In extraction, some authors use the acronym HETS, height equivalent to a theoretical stage, rather than HETP. Also, the dispersed phase, in the form of droplets, is referred to as the *discontinuous phase*, the other phase being the *continuous phase*.

§8.1.1 Mixer-Settlers

In mixer-settlers, the two liquid phases are first mixed in a vessel (Figure 8.2) by one of several types of impellers (Figure 8.3), and then separated by gravity-induced settling (Figure 8.4). Any number of mixer-settler units may be connected together to form a multistage, countercurrent cascade. During mixing, one of the liquids is dispersed in the form of small droplets into the other liquid. The dispersed phase may be either the heavier or the lighter phase. The mixing is commonly conducted in an agitated vessel with sufficient residence time so that a reasonable approach to equilibrium (e.g., 80% to 90%) is achieved. The vessel may be compartmented as in Figure 8.2. If dispersion is easily achieved and equilibrium rapidly approached, as with liquids of low interfacial tension and viscosity, the mixing step can be achieved by impingement in a jet mixer; by turbulence in a nozzle mixer, orifice mixer, or other in-line mixing device; by shearing action if both phases are fed simultaneously into a

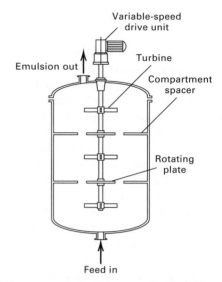

Figure 8.2 Compartmented mixing vessel with turbine agitators.
[Adapted from R.E. Treybal, *Mass Transfer*, 3rd ed., McGraw-Hill, New York (1980).]

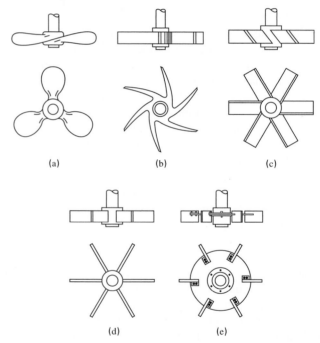

Figure 8.3 Some common types of mixing impellers: (a) marine-type propeller; (b) centrifugal turbine; (c) pitched-blade turbine; (d) flat-blade paddle; (e) flat-blade turbine.
[From R.E. Treybal, *Mass Transfer*, 3rd ed., McGraw-Hill, New York (1980) with permission.]

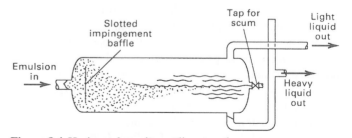

Figure 8.4 Horizontal gravity-settling vessel.
[Adapted from R.E. Treybal, *Liquid Extraction*, 2nd ed., McGraw-Hill, New York (1963) with permission.]

centrifugal pump; or by injectors, wherein the flow of one liquid is induced by another.

The settling step is by gravity in a settler (decanter). In Figure 8.4 a horizontal vessel, with an impingement baffle to prevent the jet of the entering two-phase dispersion (emulsion) from disturbing the gravity-settling process, is used. Vertical and inclined vessels are also common. A major problem in settlers is emulsification in the mixing vessel, which may occur if the agitation is so intense that the dispersed droplet size falls below 1 to 1.5 μm (micrometers). When this happens, coalescers, separator membranes, meshes, electrostatic forces, ultrasound, chemical treatment, or other ploys are required to speed settling. If the phase-density difference is small, the rate of settling can be increased by substituting centrifugal for gravitational force, as discussed in Chapter 19.

Many single- and multistage mixer-settler units are available and described by Bailes, Hanson, and Hughes [3] and Lo, Baird, and Hanson [4]. Worthy of mention is the Lurgi extraction tower [4] for extracting aromatics from hydrocarbon mixtures, where the phases are mixed by centrifugal mixers stacked outside the column and driven from a single shaft. Settling is in the column, with phases flowing interstagewise, guided by a complex baffle design.

§8.1.2 Spray Columns

The simplest and one of the oldest extraction devices is the spray column. Either the heavy phase or the light phase can be dispersed, as seen in Figure 8.5. The droplets of the dispersed phase are generated at the inlet, usually by spray nozzles. Because of lack of column internals, throughputs are large, depending upon phase-density difference and phase viscosities. As in gas absorption, *axial dispersion (backmixing)* in the continuous phase limits these devices to applications where only one or two stages are required. Axial dispersion, discussed in §8.5, is so serious for columns with a large diameter-to-length ratio that the continuous phase is

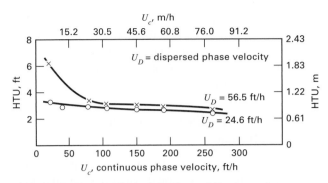

Figure 8.6 Efficiency of 1-inch Intalox saddles in a column 60 inches high with MEK–water–kerosene.

[From R.R. Neumatis, J.S. Eckert, E.H. Foote, and L.R. Rollinson, *Chem. Eng. Progr.*, **67**(1), 60 (1971) with permission.]

completely mixed, and spray columns are thus rarely used, despite their low cost.

§8.1.3 Packed Columns

Axial dispersion in a spray column can be reduced, but not eliminated, by packing the column. This also improves mass transfer by breaking up large drops to increase interfacial area, and promoting mixing in drops by distorting droplet shape. With the exception of Raschig rings [5], the packings used in distillation and absorption are suitable for liquid–liquid extraction, but choice of packing material is more critical.

A material preferentially wetted by the continuous phase is preferred. Figure 8.6 shows performance data, in terms of HTU, for Intalox saddles in an extraction service as a function of continuous, U_C, and discontinuous, U_D, phase superficial velocities. Because of backmixing, the HETP is generally larger than for staged devices; hence packed columns are suitable only when few stages are needed.

§8.1.4 Plate Columns

Sieve plates reduce axial mixing and promote a stagewise type of contact. The dispersed phase may be the light or the heavy phase. For the former, the dispersed phase, analogous to vapor bubbles in distillation, flows up the column, with redispersion at each tray. The heavy phase is continuous, flowing at each stage through a *downcomer*, and then across the tray like a liquid in a distillation tower. If the heavy phase is dispersed, *upcomers* are used for the light phase. Columns have been built with diameters larger than 4.5 m. Holes from 0.64 to 0.32 cm in diameter and 1.25 to 1.91 cm apart are used, and tray spacings are closer than in distillation—10 to 15 cm for low-interfacial-tension liquids. Plates are usually built without outlet weirs on the downspouts. In the Koch Kascade Tower, perforated plates are set in vertical arrays of complex designs.

If operated properly, extraction rates in sieve-plate columns are high because the dispersed-phase droplets coalesce and re-form on each sieve tray. This destroys concentration gradients, which develop if a droplet passes through the

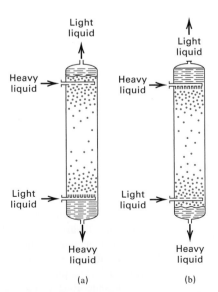

Figure 8.5 Spray columns: (a) light liquid dispersed, heavy liquid continuous; (b) heavy liquid dispersed, light liquid continuous.

entire column undisturbed. Sieve-plate extractors are subject to the same limitations as distillation columns: flooding, entrainment, and, to a lesser extent, weeping. An additional problem is scum formation at phase interfaces due to small amounts of impurities.

§8.1.5 Columns with Mechanically Assisted Agitation

If interfacial tension is high, the density difference between liquid phases is low, and/or liquid viscosities are high, gravitational forces are inadequate for proper phase dispersal and turbulence creation. In that case, mechanical agitation is necessary to increase interfacial area per unit volume, thus decreasing mass-transfer resistance. For packed and plate columns, agitation is provided by an oscillating pulse to the liquid, either by mechanical or pneumatic means. Pulsed, perforated-plate columns found considerable application in the nuclear industry in the 1950s, but their popularity declined because of mechanical problems and the

unreliability of pulse propagation [6]. Now, the most prevalent agitated columns are those that employ rotating agitators, such as those in Figure 8.3, driven by a shaft extending axially through the column. The agitators create shear mixing zones, which alternate with settling zones. Nine of the more popular arrangements are shown in Figure 8.7a-i. Agitation can also be induced in a column by moving plates back and forth in a reciprocating motion (Figure 8.7j) or in a novel horizontal contactor (Figure 8.7k). These devices answer the 1947 plea of Fenske, Carlson, and Quiggle [7] for equipment that can efficiently provide large numbers of stages in a device without large numbers of pumps, motors, and piping. They stated, "Despite . . . advantages of liquid–liquid separational processes, the problems of accumulating twenty or more theoretical stages in a small compact and relatively simple countercurrent operation have not yet been fully solved." In 1946, it was considered impractical to design for more than seven stages, which represented the number of mixer-settler units in the only large-scale, commercial, solvent-extraction process in use.

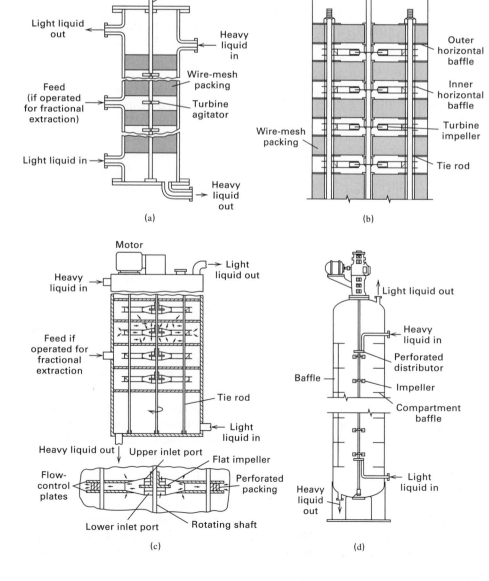

Figure 8.7 Commercial extractors with mechanically assisted agitation:
(a) Scheibel column—first design;
(b) Scheibel column—second design;
(c) Scheibel column—third design;
(d) Oldshue–Rushton (Mixco) column;
(*continued*)

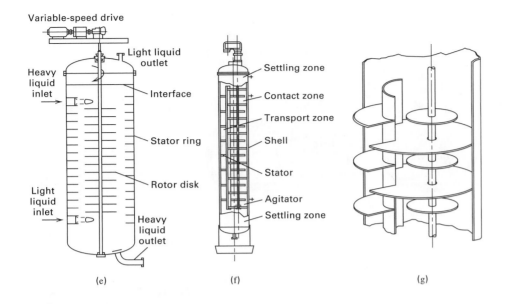

(e)　　　　　　　(f)　　　　　　　(g)

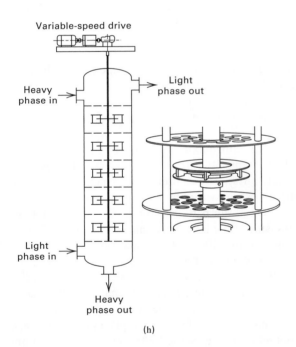

(h)

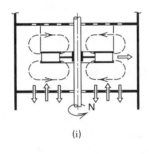

(i)

Figure 8.7 (*Continued*)
(e) rotating-disk-contactor (RDC);
(f) asymmetric rotating-disk contactor
(ARD); (g) section of ARD contactor;
(h) Kuhni column; (i) flow pattern in
Kuhni column.

Perhaps the first mechanically agitated column of importance was the Scheibel column [8] in Figure 8.7a, in which liquid phases are contacted at fixed intervals by unbaffled, flat-bladed, turbine-type agitators (Figure 8.3) mounted on a vertical shaft. In the unbaffled separation zones, located between the mixing zones, knitted wire-mesh packing prevents backmixing between mixing zones, and induces coalescence and settling of drops. The mesh material must be wetted by the dispersed phase. For larger-diameter installations (>1 m), Scheibel [9] added outer and inner horizontal annular baffles (Figure 8.7b) to divert the vertical flow in the mixing zone and promote mixing. For systems with high interfacial tension and viscosities, the wire mesh is removed. The first two Scheibel designs did not permit removal of the agitator shaft for inspection and maintenance. Instead, the entire internal assembly had to be removed. To permit removal of just the agitator assembly shaft, especially for large-diameter columns (e.g., >1.5 m), and allow an access way through the column for inspection, cleaning, and repair, Scheibel [10] offered a third design, shown in Figure 8.7c. Here the agitator assembly shaft can be removed because it has a smaller diameter than the opening in the inner baffle.

The Oldshue–Rushton extractor [11] (Figure 8.7d) consists of a column with a series of compartments separated by annular outer stator-ring baffles, each with four vertical baffles attached to the wall. The centrally mounted vertical shaft drives a flat-bladed turbine impeller in each compartment.

A third type of column with rotating agitators that appeared about the same time as the Scheibel and Oldshue–Rushton columns is the rotating-disk contactor (RDC) [12, 13] (Figure 8.7e), an example of which is described at the beginning of this chapter and shown in Figure 8.1. On a worldwide basis, it is an extensively used device, with hundreds of units in use by 1983 [4]. Horizontal disks, mounted on a centrally located

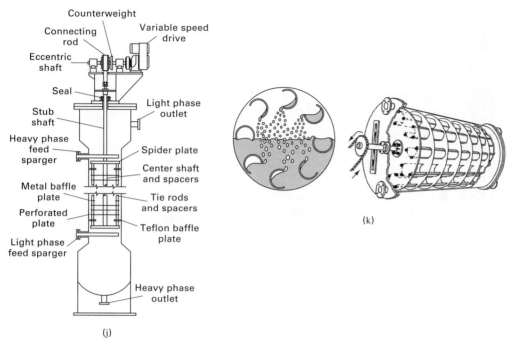

(j)

Figure 8.7 (*Continued*) (j) Karr reciprocating-plate column (RPC); (k) Graesser raining-bucket (RTL) extractor.

rotating shaft, are the agitation elements. The ratio of disk diameter to column diameter is 0.6. The distance, H in m, between disks depends on column diameter, D_C in m, according to $H = 0.13(D_C)^{0.67}$. Mounted at the column wall are annular stator rings with an opening larger than the agitator-disk diameter, typically 0.7 of D_C. Thus, the agitator assembly shaft is easily removed from the column. Because the rotational speed of the rotor controls the drop size, the rotor speed can be continuously varied over a wide range.

A modification of the RDC concept is the asymmetric rotating-disk contactor (ARD) [14], which has been in industrial use since 1965. As shown in Figure 8.7f, the contactor consists of a column, a baffled stator, and an offset multistage agitator fitted with disks. The asymmetric arrangement, shown in more detail in Figure 8.7g, provides contact and transport zones that are separated by a vertical baffle, to which is attached a series of horizontal baffles. This design retains the efficient shearing action of the RDC, but reduces backmixing because of the separate mixing and settling compartments.

Another extractor based on the Scheibel concept is the Kuhni extraction column [15] in Figure 8.7h, where the column is compartmented by a series of stator disks made of perforated plates. The distance, H in m, between stator disks depends on column diameter, D_C in m, according to $0.2 < H < 0.3(D_C)^{0.6}$. A centrally positioned shaft has double-entry, radial-flow, shrouded-turbine mixers, which promote, in each compartment, the circulation action shown in Figure 8.7i. The ratio of turbine diameter to column diameter ranges from 0.33 to 0.6. For columns of diameter greater than 3 m, three turbine-mixer shafts on parallel axes are normally provided to preserve scale-up.

Rather than provide agitation by impellers on a vertical shaft or by pulsing, Karr [16, 17] devised a reciprocating, perforated-plate extractor column in which plates move up and down approximately 2–7 times per second with a 6.5–25 mm stroke, using less energy than for pulsing the entire volume of liquid. Also, the close spacing of the plates (25–50 mm) promotes high turbulence and minimizes axial mixing, thus giving high mass-transfer rates and low HETS. The annular baffle plates in Figure 8.7j are provided periodically in the plate stack to minimize axial mixing. The perforated plates use large holes (typically 9/16-inch diameter) and a high hole area (typically 58%). The central shaft, which supports both sets of plates, is reciprocated by a drive at the top of the column. Karr columns are particularly useful for bioseparations because residence time is reduced, and they can handle systems that tend to emulsify and feeds that contain particulates.

A modification of the Karr column is the vibrating-plate extractor (VPE) of Prochazka et al. [18], which uses perforated plates of smaller hole size and smaller % hole area. The small holes provide passage for the dispersed phase, while one or more large holes on each plate provide passage for the continuous phase. Some VPE columns have uniform motion of all plates; others have two shafts for countermotion of alternate plates.

Another novel device for providing agitation is the Graesser raining-bucket contactor (RTL), developed in the late 1950s [4] primarily for processes involving liquids of small density difference, low interfacial tension, and a tendency to form emulsions. Figure 8.7k shows a series of disks mounted inside a shell on a horizontal, rotating shaft with horizontal, C-shaped buckets fitted between and around the periphery of the disks. An annular gap between the disks and the inside shell periphery allows countercurrent, longitudinal flow of the phases. Dispersing action is very gentle, with each phase cascading through the other in opposite directions toward the two-phase interface, which is close to the center.

High-speed centrifugal extractors have been available since 1944, when the Podbielniak (POD) extractor, shown in Figure 8.7l, with residence times as short as 10 s, was

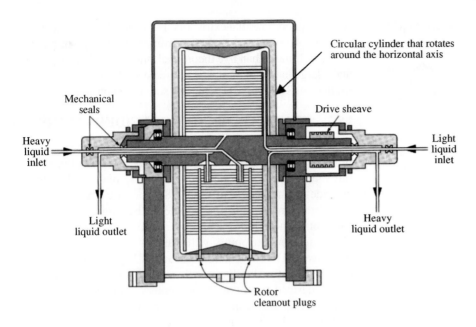

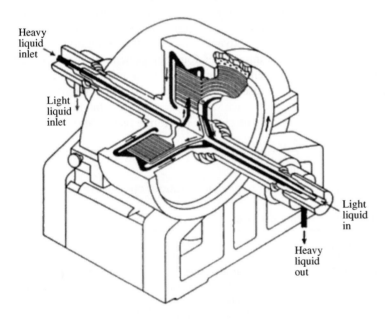

Figure 8.7 (*Continued*) (l) Cross section of a Podbielniak centrifugal extractor (POD).

successfully used in penicillin extraction [19]. Since then, the POD has found wide application in bioseparations because it provides very low holdup, prevents emulsification, and can separate liquid phases of density differences as small as 0.01 g/cm^3. Most fermentation-produced antibiotics are processed in PODs. In the POD, several concentric sieve trays encircle a horizontal axis through which the two liquid phases flow countercurrently. The feed and the solvent enter at opposite ends of the POD. As a result of the centrifugal force and density difference of the liquids, the heavy liquid is forced out to the rim. As it propagates through the perforations, it displaces an equal volume of light liquid flowing toward the shaft. Thus, the two liquids, flowing countercurrently, are forced to pass each other through the perforations on each band, leading to intense contact. Processing time is about one minute, an order of magnitude shorter than that of other devices, which is very important for many of the unstable fermentation products. The

light liquid exits at the end where the heavy liquid enters, and vice versa. The countercurrent series of dispersion and coalescence steps results in multiple stages (from 2 to 7) of extraction. Inlet pressures to 7 atm are required to overcome pressure drop and centrifugal force. The POD is available in the following five sizes, where the smaller total volumetric flows refer to emulsifiable broths. Additional material on centrifugal separators appears in Chapters 14 and 19.

Total Volumetric Flow, m^3/h	Max. Speed, rpm
0.05–0.10	10,000
6–9	3,200
15–34	2,100
30–68	2,100
60–136	1,600

Table 8.2 Maximum Size and Loading for Commercial Liquid–Liquid Extraction Columns

Column Type	Approximate Maximum Liquid Throughout, m^3/m^2-h	Maximum Column Diameter, m
Lurgi tower	30	8.0
Pulsed packed	40	3.0
Pulsed sieve tray	60	3.0
Scheibel	40	3.0
RDC	40	8.0
ARD	25	5.0
Kuhni	50	3.0
Karr	100	1.5
Graesser	<10	7.0

Above data apply to systems of:
1. High interfacial surface tension (30 to 40 dyne/cm).
2. Viscosity of approximately 1 cP.
3. Volumetric phase ratio of 1:1.
4. Phase-density difference of approximately 0.6 g/cm^3.

§8.1.6 Comparison of Industrial Extraction Columns

Maximum loadings and sizes for industrial extraction columns, as given by Reissinger and Schroeter [5, 20] and Lo et al. [4], are listed in Table 8.2. As seen, the Lurgi tower, RDC, and Graesser extractors have been built in very large sizes. Throughputs per unit cross-sectional area are highest for the Karr extractor and lowest for the Graesser extractor.

Table 8.3 lists the advantages and disadvantages of the various types of extractors, and Figure 8.8 shows a selection scheme for commercial extractors. For example, if only a small number of stages is required, a mixer-settler unit might be selected. If more than five theoretical stages, a high throughput, and a large load range (m^3/m^2-h) are needed, and floor space is limited, an RDC or ARD contactor should be considered.

§8.2 GENERAL DESIGN CONSIDERATIONS

Liquid–liquid extractors involve more variables than vapor–liquid operations because liquids have more complex structures than gases. To determine stages, one of the three cascade arrangements in Figure 8.9, or an even more complex arrangement, must be selected. The single-section cascade of Figure 8.9a, which is similar to that used for absorption and stripping, will transfer solute in the feed to the solvent. The two-section cascade of Figure 8.9b is similar to that used for distillation. Solvent enters at one end and reflux, derived from the extract, enters at the other end. The feed enters in between. With two sections, depending on solubilities, it is sometimes possible to achieve a separation between feed components; if not, a dual-solvent arrangement with two sections, as in Figure 8.9c, with or without reflux at the ends, may be advantageous. For this configuration, which involves a minimum of four components (two in the feed and two solvents), calculations by a process simulator are preferred, as discussed in Chapter 10. The configurations in Figure 8.9 are shown with packed sections, but any extraction equipment may be chosen. Operative factors are:

1. Entering feed flow rate, composition, temperature, and pressure
2. Type of stage configuration (one- or two-section)

Table 8.3 Advantages and Disadvantages of Different Extraction Equipment

Class of Equipment	Advantages	Disadvantages
Mixer-settlers	Good contacting Handles wide flow ratio Low headroom High efficiency Many stages available Reliable scale-up	Large holdup High power costs High investment Large floor space Interstage pumping may be required
Continuous, counterflow contactors (no mechanical drive)	Low initial cost Low operating cost Simplest construction	Limited throughput with small density difference Cannot handle high flow ratio High headroom Sometimes low efficiency Difficult scale-up
Continuous, counterflow contactors (mechanical agitation)	Good dispersion Reasonable cost Many stages possible Relatively easy scale-up	Limited throughput with small density difference Cannot handle emulsifying systems Cannot handle high flow ratio
Centrifugal extractors	Handles low-density difference between phases Low holdup volume Short holdup time Low space requirements Small inventory of solvent	High initial costs High operating cost High maintenance cost Limited number of stages (2–7) in single unit

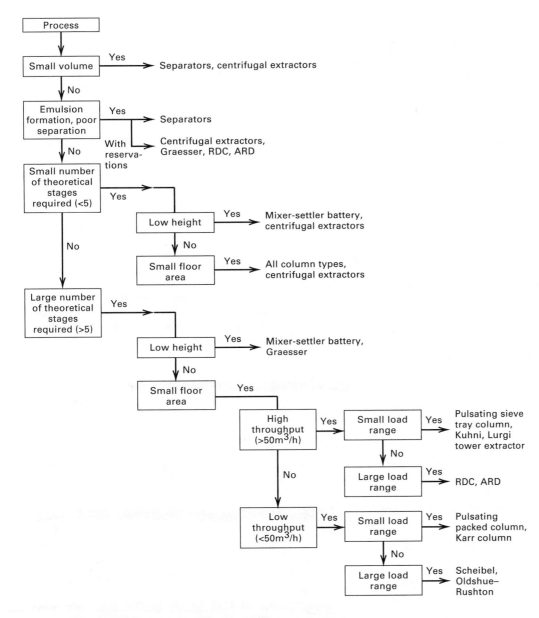

Figure 8.8 Scheme for selecting extractors.

[From K.-H. Reissinger and J. Schroeter, *I. Chem. E. Symp. Ser. No. 54*, 33–48 (1978).]

3. Desired degree of recovery of one or more solutes for one-section cascades

4. Degree of feed separation for two-section cascades

5. Choice of solvent(s)

6. Operating temperature

7. Operating pressure (greater than the bubble point)

8. Minimum-solvent flow rate and actual-solvent flow rate as a multiple of the minimum rate for one-section cascades or reflux rate and minimum reflux ratio for two-section cascades

9. Number of equilibrium stages

10. Emulsification and scum-formation tendency

11. Interfacial tension

12. Phase-density difference

13. Maximum residence time to avoid degradation

14. Type of extractor

15. Extractor cost and horsepower requirement

The ideal solvent has:

1. High selectivity for the solute relative to the carrier to minimize the need to recover carrier from the solvent

2. High capacity for dissolving the solute to minimize solvent-to-feed ratio

3. Minimal solubility in the carrier

4. A volatility sufficiently different from the solute that recovery of the solvent can be achieved by distillation, but not so high that a high extractor pressure is needed, or so low that a high temperature is needed if the solvent is recovered by distillation

5. Stability to maximize the solvent life and minimize the solvent makeup requirement

6. Inertness to permit use of common materials of construction

7. Low viscosity to promote phase separation, minimize pressure drop, and provide a high-solute mass-transfer rate

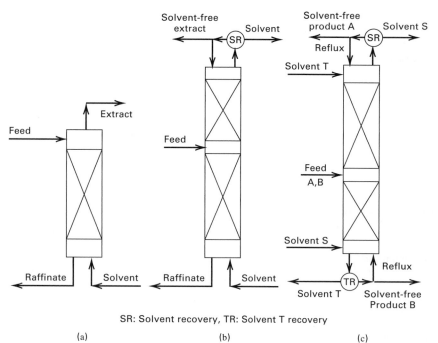

SR: Solvent recovery, TR: Solvent T recovery

(a)　　　　　　　　　　(b)　　　　　　　　(c)

Figure 8.9 Common liquid–liquid extraction cascade configurations: (a) single-section cascade; (b) two-section cascade; (c) dual solvent with two-section cascade.

8. Nontoxic and nonflammable characteristics to facilitate its safe use

9. Availability at a relatively low cost

10. Moderate interfacial tension to balance the ease of dispersion and the promotion of phase separation

11. Large difference in density relative to the carrier to achieve a high capacity in the extractor

12. Compatibility with the solute and carrier to avoid contamination

13. Lack of tendency to form a stable rag or scum layer at the phase interface

14. Desirable wetting characteristics with respect to extractor internals

Solvent selection is a compromise among all the properties listed above. However, initial consideration is usually given to selectivity and environmental concerns, and second to capacity and cost. From (2-20) in Chapter 2, the distribution (partition) coefficient for solute A between solvent S and carrier C is given by:

$$(K_A)_D = (x_A)^{II}/(x_A)^I = (\gamma_A)^I/(\gamma_A)^{II} \quad (8\text{-}1)$$

where II is the extract phase, rich in S, and I is the raffinate phase, rich in C. Similarly, for the carrier and the solvent, respectively,

$$(K_C)_D = (x_C)^{II}/(x_C)^I = (\gamma_C)^I/(\gamma_C)^{II} \quad (8\text{-}2)$$

$$(K_S)_D = (x_S)^{II}/(x_S)^I = (\gamma_S)^I/(\gamma_S)^{II} \quad (8\text{-}3)$$

From (2-22), the relative selectivity of the solute with respect to the carrier is obtained by taking the ratio of (8-1) to (8-2), giving

$$\beta_{AC} = (K_A)_D/(K_C)_D = \frac{(x_A)^{II}/(x_A)^I}{(x_C)^{II}/(x_C)^I} = \frac{(\gamma_A)^I/(\gamma_A)^{II}}{(\gamma_C)^I/(\gamma_C)^{II}} \quad (8\text{-}4)$$

For high selectivity, β_{AC} should be high, so at equilibrium there is a high concentration of A and a low concentration of C in the solvent. A first estimate of β_{AC} is made from predictions of activity coefficients $(\gamma_A)^I$, $(\gamma_A)^{II}$, and $(\gamma_C)^{II}$ at infinite dilution where $(\gamma_C)^I = 1$, or by using equilibrium data for the lowest tie line on a triangular diagram of the type discussed in Chapter 4. If A and C form a nearly ideal solution, the value of $(\gamma_A)^I$ in (8-4) can be taken as one.

For high solvent capacity, $(K_A)_D$ should be high. From (8-2) it is seen that this is difficult to achieve if A and C form nearly ideal solutions such that $(\gamma_A)^I = 1.0$, unless A and S have a great affinity for each other, which would result in a negative deviation from Raoult's law to give $(\gamma_A)^{II} < 1$. Unfortunately, such systems are rare.

For solvent recovery, $(K_S)_D$ should be large and $(K_C)_D$ as small as possible to minimize solvent in the raffinate and carrier in the extract. This will be the case if activity coefficients $(\gamma_S)^I$ and $(\gamma_C)^{II}$ at infinite dilution are large.

If a water-rich feed is to be separated, it is common to select an organic solvent; for an organic-rich feed, an aqueous solvent is often preferred. In either case, the solvent chosen should lower the activity coefficient of the solute. Consideration of molecule group interactions can help narrow the search before activity coefficients are estimated or equilibrium data are sought. Interactions for solvent-screening purposes, as given by Cusack et al. [21], based on the work of Robbins [22], are shown in Table 8.4, where the solute and solvent each belong to any of nine different chemical groups. In this table, a minus (−) sign for a given solute–solvent pair means the solvent will desirably lower the activity coefficient of the solute relative to its value in the feed.

Suppose it is desired to extract acetone from water. In Table 8.4, group 3 applies for this solute, and desirable solvents are given in groups 1 and 6. In particular, trichloroethane, a group 6 compound, is a selective solvent with high

Table 8.4 Group Interactions for Solvent Selection

Group	Solute	Solvent 1	2	3	4	5	6	7	8	9
1	Acid, aromatic OH (phenol)	0	−	−	−	−	0	+	+	+
2	Paraffinic OH (alcohol), water, imide or amide with active H	−	0	+	+	+	+	+	+	+
3	Ketone, aromatic nitrate, tertiary amine, pyridine, sulfone, trialkyl phosphate, or phosphine oxide	−	+	0	+	+	−	0	+	+
4	Ester, aldehyde, carbonate, phosphate, nitrite or nitrate, amide without active H; intramolecular bonding, e.g., o-nitrophenol	−	+	+	0	+	−	+	+	+
5	Ether, oxide, sulfide, sulfoxide, primary and secondary amine or imine	−	+	+	+	0	−	0	+	+
6	Multihaloparaffin with active H	0	+	−	−	−	0	0	+	0
7	Aromatic, halogenated aromatic, olefin	+	+	0	+	0	0	0	0	0
8	Paraffin	+	+	+	+	+	+	0	0	0
9	Monohaloparaffin or olefin	+	+	+	+	+	0	0	+	0

(+) Plus sign means that compounds in the column group tend to raise activity coefficients of compounds in the row group.
(−) Minus sign means a lowering of activity coefficients.
(0) Zero means no effect.
Choose a solvent that lowers the activity coefficient.
Source: R.W. Cusack, P. Fremeaux, and D. Glate, *Chem Eng.*, **98**(2), 66–76 (1991).

capacity for acetone. If the compound is environmentally objectionable, though, it must be rejected.

A sophisticated solvent-selection method, based on the UNIFAC group-contribution method for estimating activity coefficients and utilizing a computer-aided constrained optimization approach, has been developed by Naser and Fournier [23]. A preliminary solvent selection can also be made using tables of partition coefficients and Godfrey Miscibility Numbers in *Perry's Chemical Engineers' Handbook* [106].

§8.2.1 Representation of Ternary Data

Chapter 4 introduced ternary diagrams for representing liquid–liquid equilibrium data at constant temperature. Such diagrams are available for a large number of systems, as discussed by Humphrey et al. [6]. For ternary systems, the most common diagram is Type I, shown in Figure 8.10a; much less

common are Type II systems, in Figure 8.10b, which include: (1) n-heptane/aniline/methylcyclohexane, (2) styrene/ethylbenzene/diethylene glycol, and (3) chlorobenzene/water/methyl ethyl ketone. For Type I, the solute and solvent are miscible in all proportions, while in Type II they are not. For Type I systems, the larger the two-phase region on line $\overline{CS}$, the greater the immiscibility of carrier and solvent. The closer the top of the two-phase region is to apex A, the greater the range of feed composition, along line $\overline{AC}$, that can be separated with solvent S. In Figure 8.11, only feed solutions in the composition range from C to F can be separated because, regardless of how much solvent is added, two liquid phases are not formed in the feed composition range of $\overline{FA}$ (i.e., $\overline{FS}$ does not pass through the two-phase region). Figure 8.11a has a wider range of feed composition than Figure 8.11b. For Type II systems, a high degree of insolubility of S in C and C in S will produce a desirable high relative selectivity, but at

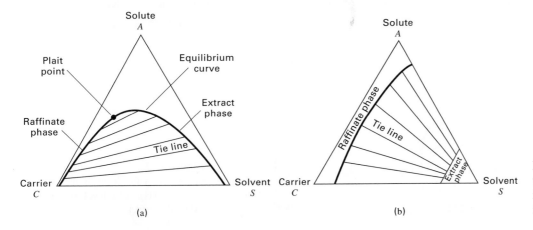

Figure 8.10 Common classes of ternary systems: (a) Type I, one immiscible pair; (b) Type II, two immiscible pairs.

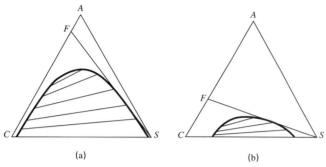

Figure 8.11 Effect of solubility on range of feed composition that can be extracted.

the expense of solvent capacity. Solvents that result in Type I systems are thus more desirable.

Whether a system is of Type I or Type II often depends on temperature. The data of Darwent and Winkler [24] for the ternary system *n*-hexane (H)/methylcyclopentane (M)/aniline (A) for temperatures of 25, 34.5, and 45°C are shown in Figure 8.12. At the lowest temperature, 25°C, the system is Type II because both H and M are only partially miscible in the aniline solvent. As temperature increases, solubility of M in aniline increases more rapidly than the solubility of H in aniline, until 34.5°C, the critical solution temperature for M in aniline, where the system is at the border of Type II and Type I. At 45°C, the system is clearly of Type I, with aniline more selective for M than H. Type I systems have a *plait point* (see Figure 8.10a); Type II systems do not.

Except in the near-critical region, pressure has little effect on liquid–liquid equilibrium because it has little effect on liquid-phase activity coefficients. The operating pressure need only be greater than the bubble-point pressure. Most extractors operate at near-ambient temperature, with the process being nearly isothermal because of the small heat of mixing. An exception is biochemical processes, where a below-ambient temperature may be necessary to prevent bioproduct degradation.

Laboratory or pilot-plant work, using actual plant feed and solvent, is normally necessary to ascertain dispersion and coalescence properties. Although rapid coalescence of drops is desirable, this reduces interfacial area and leads to reduced mass-transfer rates. Thus, compromises are necessary. Coalescence is enhanced when the solvent phase is continuous and mass transfer of solute is from the droplets. This phenomenon, the *Marangoni effect*, is due to a lowering

of interfacial tension by a significant solute presence in the interfacial film. If solvent is the dispersed phase, the interfacial film is depleted of solute, causing an increase in interfacial tension and inhibition of coalescence.

For a given (1) feed liquid, (2) degree of solute extraction, (3) operating pressure and temperature, and (4) solvent for a single-section cascade, there is a minimum solvent-to-feed flow-rate ratio that corresponds to an infinite number of countercurrent, equilibrium contacts. As with absorption and stripping, a trade-off exists between the number of equilibrium stages and the solvent-to-feed ratio. For a two-section cascade, as for distillation, the trade-off involves the reflux ratio and the number of stages. Algebraic methods—similar to those for absorption and stripping—for computing the trade-off are rapid but useful only for very dilute solutions, where the solute activity coefficients are those near infinite dilution. When the carrier and solvent are mutually insoluble, the algebraic method of §5.3–5.4 applies. For more general applications of ternary systems, the graphical methods described in §8.3–8.4 are preferred. For higher-order, multicomponent systems, process simulator methods described in Chapter 10 are necessary.

§8.3 HUNTER–NASH GRAPHICAL EQUILIBRIUM-STAGE METHOD

Stagewise extraction calculations for Type I and Type II systems (Figure 8.10) are most conveniently carried out with equilibrium diagrams [25]. In this section, procedures are developed using mainly triangular diagrams. Other diagrams are covered in §8.4.

Consider a countercurrent-flow, *N*-equilibrium-stage extractor operating isothermally in steady-state, continuous flow, as in Figure 8.13. Stages are numbered from the feed end. Thus, the final extract is E_1 and the final raffinate is R_N. Equilibrium is assumed at each stage, so for any stage *n*, the components in extract E_n and raffinate R_n are in equilibrium. Mass transfer of all species occurs at each stage. The feed, *F*, contains the carrier, *C*, the solute, *A*, and the solvent, *S*, up to the solubility limit. Entering solvent, *S*, can contain *C* and *A*, but preferably contains little of either. Because most liquid–liquid equilibrium data are listed and plotted in mass rather than mole concentrations, let:

F = mass flow rate of feed to the cascade;

S = mass flow rate of solvent to the cascade;

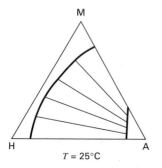
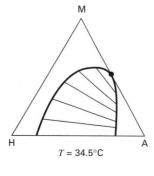
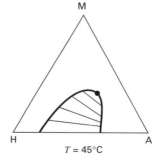

Figure 8.12 Effect of temperature on solubility for the system *n*-hexane (H)/methylcyclopentane (M)/aniline (A).

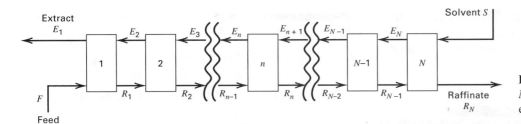

Figure 8.13 Countercurrent-flow, N-equilibrium-stage liquid–liquid extraction cascade.

E_n = mass flow rate of extract leaving stage n;

R_n = mass flow rate of raffinate leaving stage n;

$(y_i)_n$ = mass fraction of species i in extract leaving stage n; and

$(x_i)_n$ = mass fraction of species i in raffinate leaving stage n.

Although Figure 8.13 implies that the extract is the light phase, either phase can be the light phase. Phase equilibrium is represented, as in §4.5, on an equilateral-triangle diagram, as proposed by Hunter and Nash [26], or on a right-triangle diagram, as proposed by Kinney [27]. Assume the ternary system is A (solute), C (carrier), and S (solvent) at a temperature, T, such that liquid–liquid equilibrium data are as shown in Figure 8.14, where the bold line is the equilibrium curve (also called the *binodal curve* because the plait point separates the curve into an extract to the left and a raffinate to the right), and the dashed lines are tie lines connecting compositions of equilibrium phases. Because the tie lines slope upward from the C side toward the S side, at equilibrium, A has a concentration higher in S than in C. Thus, in this example, S is an effective solvent for extracting A. Alternatively, because the tie lines slope downward from the S side toward the C side, C is not an effective solvent for extracting A from S.

Some systems, such as isopropanol–water–benzene, exhibit a phenomenon called *solutropy*, wherein moving from the plait point into the two-phase region of the diagram, the tie lines first slope in one direction; but then the slope diminishes until an intermediate tie line becomes horizontal. Below that tie line, the remaining tie lines slope in the other direction. Sometimes the solutropy phenomenon disappears if mole-fraction coordinates, rather than mass-fraction coordinates, are used.

§8.3.1 Number of Equilibrium Stages

From the degrees-of-freedom discussions in §4.1 and §5.8, the following sets of specifications for the ternary component cascade of Figure 8.13 can be made, where, in addition, all sets include the specification of F, $(x_i)_F$, $(y_i)_S$, and T:

Set 1. S and $(x_i)_{R_N}$	Set 4. N and $(x_i)_{R_N}$
Set 2. S and $(y_i)_{E_1}$	Set 5. N and $(y_i)_{E_1}$
Set 3. $(x_i)_{R_N}$ and $(y_i)_{E_1}$	Set 6. S and N

where $(x_i)_{R_N}$ and $(y_i)_{E_1}$ and all exiting phases lie on the equilibrium curve.

Calculations for sets 1 to 3 involve determination of N, and are made using triangular diagrams. Sets 4 to 6 involve a specified N, and require an iterative procedure. First consider set 1, with the procedures for sets 2 and 3 being minor modifications. The technique, sometimes called the *Hunter–Nash method* [26], involves three kinds of constructions on the triangular diagram, and is more difficult than the McCabe–Thiele staircase method for distillation. Although the procedure is illustrated here only for a Type I system, parallel principles apply to a Type II system. The constructions are shown in Figure 8.14, where A is the solute, C the carrier, and S the solvent. On the equilibrium curve, extract compositions

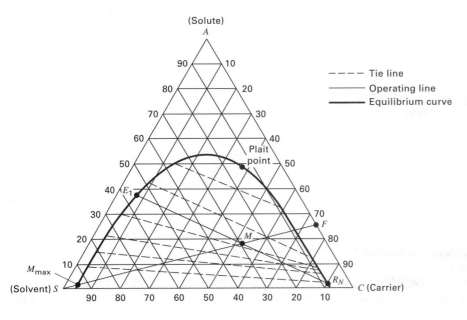

Figure 8.14 Construction 1: Location of product points.

lie to the left of the plait point, while all raffinate compositions lie to the right. Determination of N is as follows, given feed and solvent flow rates and compositions, and raffinate composition:

Construction 1 (Product Composition Points)

First, on Figure 8.14, locate mixing point M, which represents the combined composition of feed F plus entering solvent S. Assume the following feed and solvent specifications, as plotted in Figure 8.14:

Feed	Solvent
$F = 250$ kg	$S = 100$ kg
$(x_A)_F = 0.24$	$(x_A)_S = 0.00$
$(x_C)_F = 0.76$	$(x_C)_S = 0.00$
$(x_S)_F = 0.00$	$(x_S)_S = 1.00$

By material balances, composition M of combined F and S is:

$$M = F + S = 250 + 100 = 350 \text{ kg}$$
$$(x_A)_M M = (x_A)_F F + (x_A)_S S$$
$$= 0.24(250) + 0(100) = 60 \text{ kg}$$
$$(x_A)_M = 60/350 = 0.171$$

$$(x_C)_M M = (x_C)_F F + (x_C)_S S$$
$$= 0.76(250) + 0(100) = 190 \text{ kg}$$
$$(x_C)_M = 190/350 = 0.543$$

$$(x_S)_M M = (x_S)_F F + (x_S)_S S$$
$$= 0(250) + 1(100) = 100 \text{ kg}$$
$$(x_S)_M = 100/350 = 0.286$$

From any two of the $(x_i)_M$ values, point M is located, as shown, in Figure 8.14. Based on properties of the triangular diagram discussed in §4.5, point M must be located on the straight line connecting F and S. Therefore, M can be located knowing just one value of $(x_i)_M$, say, $(x_S)_M$. Also, the ratio S/F is given by the inverse-lever-arm rule as

$$S/F = \overline{MF}/\overline{MS} = 100/250 = 0.400$$

or $\qquad S/M = \overline{MF}/\overline{SF} = 100/350 = 0.286$

Thus, point M can be located by two points or by measurement, employing either of these ratios.

With point M located, the composition of exiting extract E_1 is determined from overall material balances:

$$M = R_N + E_1 = 350 \text{ kg}$$
$$(x_A)_M M = 60 = (x_A)_{R_N} R_N + (x_A)_{E_1} E_1$$
$$(x_C)_M M = 190 = (x_C)_{R_N} R_N + (x_C)_{E_1} E_1$$
$$(x_S)_M M = 100 = (x_S)_{R_N} R_N + (x_S)_{E_1} E_1$$

Specify $(x_A)_{R_N} = 0.025$. Because it lies on the equilibrium curve, R_N can be located and the values of $(x_C)_{R_N}$ and $(x_S)_{R_N}$ can be read from Figure 8.14. A straight line drawn from R_N through M locates E_1 at the equilibrium-curve intersection, from which the composition of E_1 can be read. Values of the flow rates R_N and E_1 can then be determined from the overall

material balances above, or from Figure 8.14 by the inverse-lever-arm rule:

$$E_1/M = \overline{MR_N}/\overline{E_1 R_N}$$
$$R_N/M = \overline{ME_1}/\overline{E_1 R_N}$$

with $M = 350$ kg. By either method the results are:

Raffinate Product	Extract Product
$R_N = 198.6$ kg	$E_1 = 151.4$ kg
$(x_A) = 0.025$	$(x_A) = 0.364$
$(x_C) = 0.90$	$(x_C) = 0.075$
$(x_S) = 0.075$	$(x_S) = 0.561$

Also included in Figure 8.14 is point $M_{\max}$, which lies on the equilibrium curve along the straight line connecting F to S. $M_{\max}$ corresponds to the maximum possible solvent addition if two liquid phases are to exist.

Construction 2 (Operating Point and Operating Lines)

For vapor–liquid cascades, §6.3 and §7.2 describe an operating line that is the locus of passing streams in a cascade. Referring to Figure 8.13, material balances around groups of stages from the feed end are

$$F - E_1 = \cdots = R_{n-1} - E_n = \cdots = R_N - S = P \quad (8\text{-}5)$$

Because the passing streams are differenced, P defines a *difference point*, not a *mixing point*, M. From the same geometric considerations that apply to a mixing point, a difference point also lies on a line through the points involved. However, whereas M lies inside the diagram and between the two end points, P usually lies outside the triangular diagram along an extrapolation of the line through two points such as F and E_1, R_N and S, and so on.

To locate the difference point, two straight lines are drawn through the passing-stream point pairs (E_1, F) and (S, R_N) established by Construction 1 and shown in Figure 8.15. These lines are extrapolated until they intersect at difference point P. These lines and point P are shown in Figure 8.15. From (8-5), straight lines drawn through points for any other pair of passing streams, such as (E_n, R_{n-1}), must also pass through point P. Thus, the difference point becomes an *operating point*, and lines drawn through pairs of points for passing streams and extrapolated to point P are *operating lines*.

The difference point has properties similar to a mixing point. If $F - E_1 = P$ is rewritten as $F = E_1 + P$, F can be interpreted as the mixing point for P and E_1. Therefore, by the inverse-lever-arm rule, the length of line $\overline{E_1 P}$ relative to the length of line $\overline{FP}$ is

$$\frac{\overline{E_1 P}}{\overline{FP}} = \frac{E_1 + P}{E_1} = \frac{F}{E_1} \quad (8\text{-}6)$$

Thus, point P can be located with a ruler using either pair of feed-product passing streams.

The operating point, P, lies on the feed or raffinate side of the diagram in Figure 8.15. Depending on the relative amounts of feed and solvent and the slope of the tie lines,

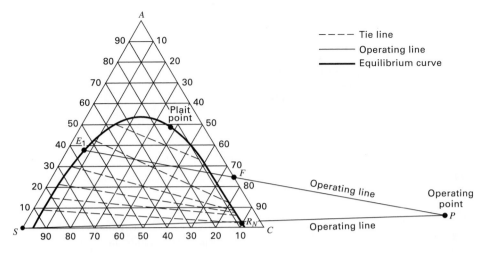

Figure 8.15 Construction 2: Location of operating point.

point P may lie on the solvent or feed side of the diagram, and inside or outside of the diagram.

Construction 3 (Tie Lines and Equilibrium Lines)

The third type of construction involves the *tie lines* that define the equilibrium curve, which is divided into the two sides (raffinate and extract) by the plait point. For Type I diagrams, the plait point is the composition of two equilibrium phases that become one phase. A material balance around stage n for any of the three components is

$$(x_i)_{n-1}R_{n-1} + (y_i)_{n+1}E_{n+1} = (x_i)_n R_n + (y_i)_n E_n \quad (8-7)$$

Because R_n and E_n are in equilibrium, their composition points are at the two ends of a tie line. Typically, a diagram will not contain all tie lines needed; however, they may be added by centering them between existing tie lines, or by using either of two interpolation procedures illustrated in Figure 8.16. In Figure 8.16a, the conjugate line from the plait

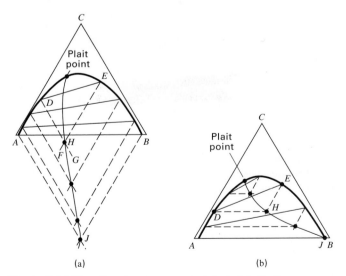

(a)   (b)

Figure 8.16 Use of conjugate curves to interpolate tie lines:
(a) method of *International Critical Tables*, Vol. III, McGraw Hill, New York, p. 393 (1928); (b) method of T.K. Sherwood, *Absorption and Extraction*, McGraw-Hill, New York, p. 242 (1937).

[From R.E. Treybal, *Liquid Extraction*, 2nd ed., McGraw-Hill, New York (1963) with permission.]

point to J is determined from four tie lines and the plait point. From tie line $\overline{DE}$, lines $\overline{DG}$ and $\overline{EF}$ are drawn parallel to triangle sides $\overline{CB}$ and $\overline{AC}$, respectively. The intersection at point H gives a second point on the conjugate curve. Subsequent intersections establish additional points from which the conjugate curve is drawn. Then, using the curve, additional tie lines are drawn by reversing the procedure. If it is desired to keep the conjugate curve inside the two-liquid-phase region of the triangular diagram, the procedure illustrated in Figure 8.16b is used, where lines are drawn parallel to triangle sides $\overline{AB}$ and $\overline{AC}$.

Stepping Off Stages

Equilibrium stages are stepped off by alternate use of tie lines and operating lines, as shown in Figure 8.17, where Constructions 1 and 2 have been employed to locate points F, E, S, R_1, and P. Starting at the feed end from point E_1 and referring to Figure 8.13, it is seen that R_1 is in equilibrium with E_1. Therefore, by Construction 3, R_1 in Figure 8.17 must be at the opposite end of a tie line (shown as a dashed line) connecting to E_1. From Figure 8.13, R_1 passes E_2. Therefore, by Construction 2, E_2 lies at the intersection of the straight operating line drawn through points R_1 and P, and back to the extract side of the equilibrium curve. From E_2, R_2 is located with a tie line by Construction 3; from R_2, E_3 is located by Construction 2. Continuing in this fashion by alternating between equilibrium tie lines and operating lines, the specified point R_N is reached or passed. If the latter, a fraction of the last stage is taken. In Figure 8.17, 2.8 equilibrium stages are required, where equilibrium stages are counted by the number of equilibrium tie lines used.

Procedures for specification sets 2 and 3 are similar to that for set 1. Sets 4 and 5 can be handled by iteration on assumed values for S and following the procedure for set 1. Set 6 can also use the procedure of set 1 by iterating on E_1.

From (8-6), it is seen that if the ratio F/E_1 approaches a value of 1, operating point P will be located at a large distance from the triangular diagram. In that case, by using an arbitrary rectangular-coordinate system superimposed over the triangular diagram, the coordinates of P can be calculated from (8-6) using the equations for the two straight lines

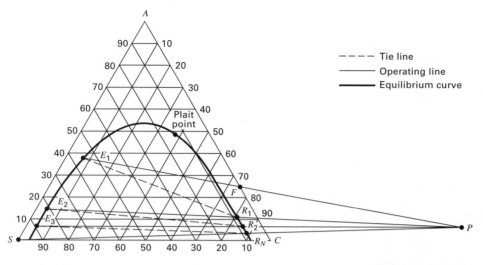

Figure 8.17 Determination of the number of equilibrium stages.

established in Construction 2. Operating lines for intermediate stages can then be located on the triangular diagram so as to pass through P. Details of this procedure are given by Treybal [25].

§8.3.2 Minimum and Maximum Solvent-to-Feed Flow-Rate Ratios

The procedure of §8.3.1 for determining the number of equilibrium stages needed to achieve a desired solute extraction for a given solvent-to-feed ratio presupposes that this ratio is greater than the minimum ratio, which corresponds to an infinite number of stages, and less than the maximum ratio that would prevent the formation of the required second liquid phase. In practice, one usually determines the minimum ratio before solving specification sets 1 or 2. This is done by solving set 4 with $N = \infty$, where, as in distillation, absorption, and stripping, the infinity of stages occurs at an equilibrium-curve and operating-line pinch point. In ternary systems, the pinch point occurs when a tie line coincides with an operating line. The calculation is involved because the pinch point is not always at the feed end of the cascade.

Consider the previous A–C–S system, shown in Figure 8.18. The composition points F, S, and R_N are specified, but E_1 is not because the solvent rate is not known. The operating line $\overline{OL}$ is drawn through the points S and R_N and extended to the left and right. This line is the locus of all possible material balances determined by adding S to R_N. Each tie line is then assumed to be a pinch point by extending each tie line until it intersects the line $\overline{OL}$. In this manner, a sequence of intersections P_1, P_2, etc., is found. If these points lie on the raffinate side of the diagram, as in Figure 8.18, the pinch point corresponds to the point $P_{\min}$ located at the greatest distance from R_N. If the triangular diagram does not have a sufficient number of tie lines to determine that point accurately, additional tie lines are introduced by a method illustrated in Figure 8.16. If it is assumed in Figure 8.18 that no other tie line gives a point P_i that is farther away from R_N than P_1, then $P_1 = P_{\min}$.

With $P_{\min}$ known, an operating line can be drawn through point F and extended to E_1 at an intersection with the extract side of the equilibrium curve. From the compositions of the four points S, R_N, F, and E_1, the mixing point M can be found and material balances can be used to solve for $S_{\min}/F$:

$$F + S_{\min} = R_N + E_1 = M \qquad (8\text{-}8)$$

$$(x_A)_F F + (x_A)_S S_{\min} = (x_A)_M M \qquad (8\text{-}9)$$

from which

$$\frac{S_{\min}}{F} = \frac{(x_A)_F - (x_A)_M}{(x_A)_M - (x_A)_S} \qquad (8\text{-}10)$$

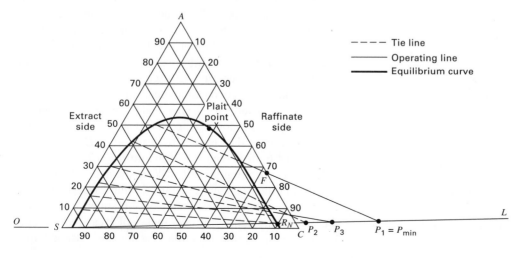

Figure 8.18 Determining minimum solvent-to-feed ratio.

A solvent flow rate greater than S_{min} is required for extraction in a finite number of stages. In Figure 8.18, such a solvent rate results in an operating point P to the right of P_{min}, which is a location farther away from R_N. A reasonable value for S might be 1.5 S_{min}. From Figure 8.18, $(x_A)_M = 0.185$, from which, by (8-10), $S_{min}/F = 0.30$. In Figure 8.17, $S/F = 0.40$, giving $S/S_{min} = 1.33$.

In Figure 8.18 the tie lines slope downward toward the raffinate side. If the tie lines slope downward toward the extract side of the diagram, the above procedure for finding S_{min}/F is modified. The sequence of points P_1, P_2, etc., is now on the other side of the diagram. However, the pinch point now corresponds to the point, P_{min}, that is closest to point S, so an operating point, P, must be chosen between points P_{min} and S. For a system that exhibits solutropy, intersections P_1, P_2, etc., are found on both sides of the diagram. Those on the extract side determine the minimum solvent-to-feed ratio.

In Figure 8.14, mixing point M lies in the two-phase region. As this point moves along the line $\overline{SF}$ toward S, the ratio S/F increases according to the inverse-lever-arm rule. In the limit, a maximum S/F ratio is reached when $M = M_{max}$ arrives at the equilibrium curve on the extract side. Now all of the feed is dissolved in the solvent, no raffinate is obtained, and only one stage is required. To avoid this impractical condition, as well as the other extreme of infinite stages, it is necessary to select a solvent ratio, S/F, such that $(S/F)_{min} < (S/F) < (S/F)_{max}$. In Figure 8.14, the mixing point M_{max} is located as shown, from which $(S/F)_{max}$ is determined to be about 16.

EXAMPLE 8.1 Equilibrium Stages for Extraction.

Acetone is to be extracted from a 30 wt% acetone (A) and 70 wt% ethyl acetate (C) feed at 30°C, using pure water (S) as the solvent, by the cascade shown at the bottom of Figure 8.19, which establishes the nomenclature for this example. The final raffinate is to contain 5 wt% acetone on a water-free basis. Determine the minimum and maximum solvent-to-feed ratios and the equilibrium stages required for two intermediate S/F ratios. The data shown in Figure 8.19 are from Venkataranam and Rao [28] and correspond to a Type I system, but with tie lines sloping downward toward the extract side. Thus, although water is a convenient solvent, it does not have a high capacity, relative to ethyl acetate, for dissolving acetone. Also determine, for the feed, the maximum weight % acetone that can enter the extractor. This example, as well as Example 8.2, are taken largely from an analysis by Sawistowski and Smith [29].

Solution

Point B represents the solvent-free raffinate. By drawing a straight line from B to S, the intersection with the equilibrium curve on the raffinate side, B', is the raffinate composition leaving stage N.

Minimum S/F. Because the tie lines slope downward toward the extract side, the extrapolated tie line that intersects the extrapolated line $\overline{SB}$ closest to S is sought. This tie line, leading to P_{min}, is shown in Figure 8.20. The intersection is not shown because it occurs far to the left of the diagram. Because this tie line is at the feed end of the extractor, location of extract composition D'_{min} is found as shown in Figure 8.20. The mixing point, M_{min}, for $(S/F)_{min}$ is the intersection of lines $\overline{B'D'}_{min}$ and $\overline{SF}$. By the inverse-lever-arm rule, $(S/F)_{min} = \overline{FM}_{min}/\overline{SM}_{min} = 0.60$.

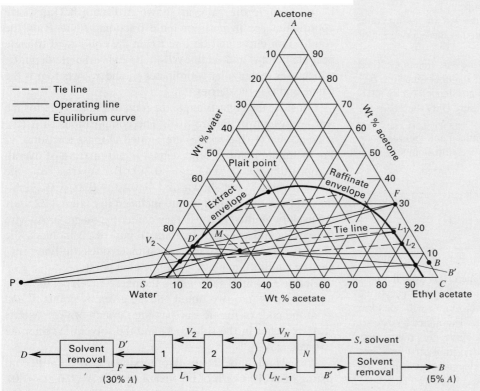

Figure 8.19 Determination of stages for Example 8.1 with $S/F = 1.75$.

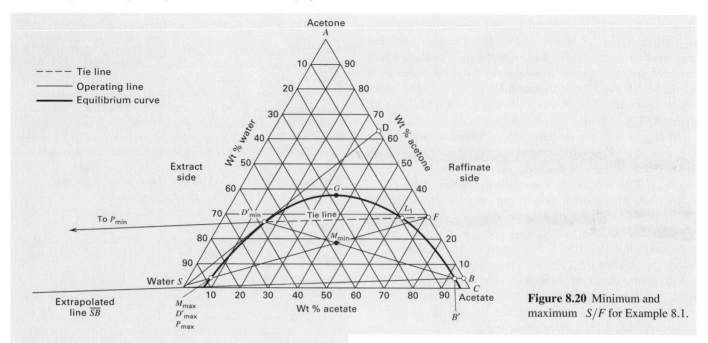

Figure 8.20 Minimum and maximum S/F for Example 8.1.

Maximum S/F. If M in Figure 8.20 is moved along line $\overline{FS}$ toward S, the intersection for $(S/F)_{max}$ occurs at the point shown on the extract side of the binodal curve. By the inverse-lever-arm rule, using line $\overline{FS}$, $(S/F)_{max} = \overline{FM}_{max}/\overline{SM}_{max} = 12$.

Equilibrium stages for other S/F ratios. First consider $S/F = 1.75$. In Figure 8.19, the composition of the saturated extract D' is obtained from an extractor material balance,

$$S + F = D' + B' = M$$

For $S/F = 1.75$, point M is located such that $\overline{FM}/\overline{MS} = 1.75$. A straight line must pass through D', B', and M. Therefore, D' is located by extending $\overline{B'M}$ to the extract envelope.

The difference point P is located to the left of the diagram. Therefore, $P = S - B' = D' - F$. It is located at the intersection of extensions of lines $\overline{FD'}$ and $\overline{B'S}$.

Step off stages, starting at D', by following a tie line to L_1. Then V_2 is located by noting the intersection of the operating line $\overline{L_1P}$ with the phase envelope. Additional stages are stepped off by alternating between the tie lines and operating lines. Only the first stage is shown; four are required.

For $S/F = 5(S/F)_{min} = 3.0$, M is determined and the stages are stepped off in a similar manner to give two equilibrium stages. In summary, for the countercurrent cascade,

S/F (solvent/feed ratio)	0.60	1.75	3	12
N (equilibrium stages)	∞	4	2	1
x_D (wt% acetone, solvent free)	64	62	50	30

If the wt% acetone in the feed mixture is increased from the base value of 30%, a feed composition will be reached that cannot be extracted because two liquid phases in equilibrium will not form (no phase splitting). This feed composition is determined by extending a line from S, tangent to the equilibrium curve, until it intersects $\overline{AC}$ at point D in Figure 8.20. The feed composition is 64 wt% acetone. Feed mixtures with a higher acetone content cannot be extracted with water.

§8.3.3 Use of Right-Triangle Diagrams

Ternary, countercurrent extraction calculations can also be made on a right-triangle diagram, discussed in §4.5 and shown by Kinney [27]; no new principles are involved. The disadvantage is that mass-percent compositions of only two of the components are plotted; the third is determined, when needed, by difference from 100%. The advantage of right-triangle diagrams is that ordinary, rectangular-coordinate graph paper can be used and either one of the coordinates can be expanded, if necessary, to increase construction accuracy.

A right-triangle diagram can be developed from an equilateral-triangle diagram, as shown in Figure 8.21a, where coordinates are in mass or mole fractions. Point P on the equilibrium curve and tie line $\overline{RE}$ in the equilateral triangle become point P and tie line $\overline{RE}$ in the right-triangle diagram, which uses rectangular coordinates x_A and x_C, where A is the solute and C is the carrier.

Consider the right-triangle diagram in Figure 8.22 for the A–C–S system of Figure 8.14. The compositions of S (the solvent) and A (the solute) are plotted in mass fractions, x_i. For example, point M represents a liquid mixture of overall composition $(x_A = 0.43, x_S = 0.28)$. By contrast, x_C, the carrier, which is not shown, is $1 - 0.43 - 0.28 = 0.29$. Although lines of constant x_C are included in Figure 8.22, they are usually omitted because they clutter the diagram. As with the equilateral-triangle diagram, Figure 8.22 includes the binodal curve with extract and raffinate sides, tie lines connecting equilibrium phases, and the plait point.

Because point M falls within the phase envelope, the mixture separates into two liquid phases, given by points A' and A'' at the ends of the tie line passing through M. The extract at A'' is richer in the solute (A) and the solvent (S) than the raffinate at A'.

Point M might be the result of mixing a feed, point F, consisting of 26,100 kg/h of 60 wt% A in C ($x_A = 0.6$, $x_S = 0$), with 10,000 kg/h of pure furfural, point S. The mixture splits

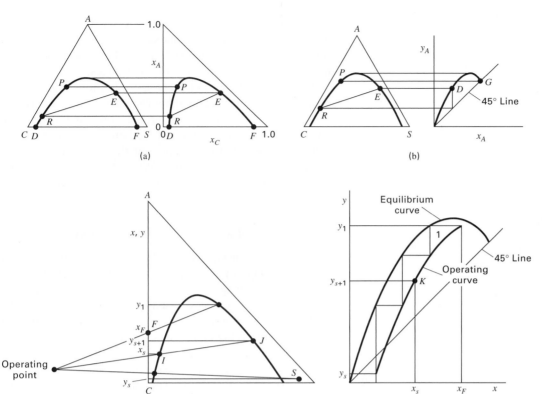

Figure 8.21 Development of other coordinate systems from the equilateral-triangle diagram: (a) to right-triangle diagram; (b) to auxiliary distribution curve. (c) Location of operating point on auxiliary McCabe–Thiele diagram.

[From R.E. Treybal, *Liquid Extraction*, 2nd ed., McGraw-Hill, New York (1963) with permission.]

into phases A' and A''. The location of M and the amounts of extract and raffinate are given by the same mixing rule and inverse-lever-arm rule used for equilateral-triangle diagrams. The mixture separates spontaneously into 11,600 kg/h of raffinate ($x_S = 0.08, x_A = 0.32$) and 24,500 kg/h of extract ($x_S = 0.375, x_A = 0.48$).

Figure 8.23 represents the portion of an *n*-stage, countercurrent-flow cascade, where x and y are weight fractions of

solute in the raffinate and extract, respectively, and L and V are total amounts of raffinate and extract. The feed to stage N is $L_{N+1} = 180$ kg of 35 wt% A in a saturated mixture with C and S ($x_{N+1} = 0.35$), and the solvent to stage 1 is $V_W = 100$ kg of pure S ($y_W = 0.0$). Thus, the solvent-to-feed ratio is $100/180 = 0.556$. These points are shown in Figure 8.24. The mixing point for L_{N+1} and V_W is M_1, as determined by the inverse-lever-arm rule.

Suppose the final raffinate, L_W, leaving stage 1 is to contain $x_W = 0.05$ glycol. By an overall balance,

$$M_1 = V_W + L_{N+1} = V_N + L_W \qquad (8\text{-}11)$$

Because V_W, L_{N+1}, and M_1 lie on a straight line, the mixing rule requires that V_N, L_W, and M_1 also lie on a straight line. Furthermore, because V_N leaves stage N at equilibrium and L_W leaves stage 1 at equilibrium, these streams lie on the extract and raffinate sides, respectively, of the equilibrium

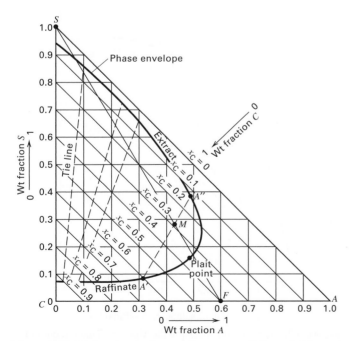

Figure 8.22 Right-triangle diagram for system of Figure 8.14.

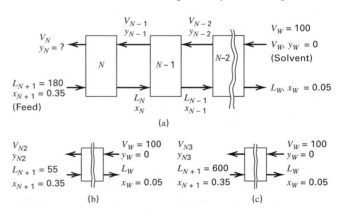

Figure 8.23 Multistage, countercurrent contactors.

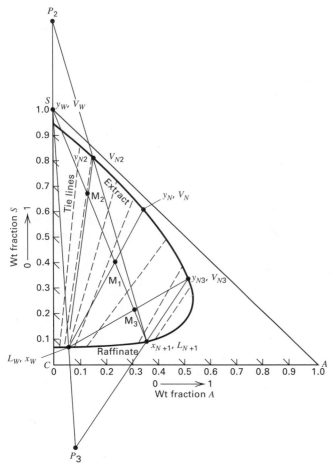

Figure 8.24 Right-triangle diagram constructions for cascade in Figure 8.23.

curve. The resulting points are shown in Figure 8.24, where it is seen that the glycol in the extract is $y_N = 0.34$.

Figures 8.23b, 8.23c, and 8.24 include two additional cases of solvent-to-feed ratio, each with the same compositions for the solvent and feed, and the same value for x_W:

Case	Feed L_{N+1}, kg	Solvent, V_W, kg	Solvent-to Feed Ratio	Extract Designation	Mixing Point
1	180	100	0.556	V_N	M_1
2	55	100	1.818	V_{N2}	M_2
3	600	100	0.167	V_{N3}	M_3

For case 2, a difference point, P_2, is defined in terms of passing streams, as

$$P_2 = V_{N2} - L_{N+1} = V_W - L_W \tag{8-12}$$

In Figure 8.24 P_2 is at the top of the diagram, where $\overline{L_W V_W}$ and $\overline{L_{N+1} V_{N2}}$ intersect. For case 3, difference point P_3 lies at the bottom, where $\overline{L_W V_W}$ and $\overline{L_{N+1} V_{N3}}$ intersect.

Equilibrium stages for Figure 8.24 are stepped off as before by alternating the use of equilibrium tie lines and operating lines that pass through the difference point. Considering case 2, with a solvent-to-feed ratio of 1.818, and stepping off stages from stage N, a tie line from the point y_{N2} gives $x_N = 0.04$. But this is less than the specified $x_W = 0.05$, so only a fraction of a stage is required.

If stages are stepped off for case 3, starting from y_{N3}, the tie line and operating line coincide, giving a pinch point for a solvent-to-feed ratio of 0.167. Thus, this ratio is the minimum corresponding to ∞ equilibrium stages.

For case 1, where the solvent-to-feed ratio is between that of cases 2 and 3, the required number of stages lies between 1 and ∞. The difference point and the steps for this case are not shown in Figure 8.14, but the difference point is located at a very large distance from the triangle because lines $\overline{L_W V_W}$ and $\overline{L_{N+1} V_N}$ are nearly parallel. When the stages are stepped off, using operating lines parallel to $\overline{L_W V_W}$, between one and two stages are required.

§8.3.4 Use of an Auxiliary Distribution Curve

As the number of stages on either one of the two types of triangular diagrams increases, the diagram becomes cluttered, and a triangular diagram in conjunction with the McCabe–Thiele method, devised by Varteressian and Fenske [30], becomes attractive. The y–x diagram discussed in §4.5 and illustrated in Figure 8.21b is simply a plot of the tie-line data in terms of solute mass fractions in the extract (y_A) and equilibrium raffinate (x_A). The curve begins at the origin and terminates at the plait point, where it intersects the 45° line. A tie line, such as $\overline{RE}$ in the triangular diagram, becomes a point in the equilibrium curve of Figure 8.25.

If an operating line is added to the equilibrium curve in Figure 8.21b, the McCabe–Thiele staircase construction determines the equilibrium stages required. However, unlike in distillation, where the operating line is straight, the operating line for liquid–liquid extraction is curved except in the low-solute region. Fortunately, the curved operating line is readily drawn using the technique of Varteressian and Fenske

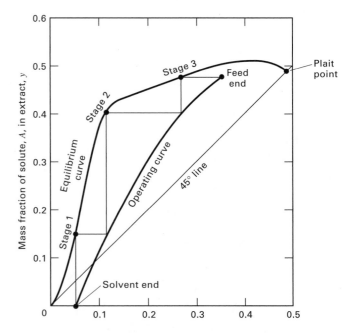

Figure 8.25 Stepping off stages by the McCabe–Thiele method for the A–C–S system.

[30]. In Figure 8.19 for the equilateral-triangle diagram, or in Figure 8.24 for the right-triangle diagram, the intersections of the equilibrium curve with a line drawn through a difference (operating) point represent the compositions of passing streams. Thus, for each such operating line on the triangular diagram, one point on the operating line for the y–x plot is determined. The operating lines passing through the difference point need not correspond to actual passing streams. Usually five or six such fictitious operating-line intersections, covering the range of compositions in the cascade, are sufficient to establish the curved operating line. For example, in Figure 8.21c, the arbitrary operating line that intersects the equilibrium curve at I and J in the right-triangle diagram becomes point K in the y–x diagram operating line. The y–x plot of Figure 8.25 includes an operating line established in this manner, based on the data of Figure 8.24, but with a solvent-to-feed ratio of 0.208—that is, $V_W = 100$, $L_{N+1} = 480$ (25% greater than the minimum ratio of 0.167). The stages are stepped off in the McCabe–Thiele manner starting from the feed end. The result is three equilibrium stages.

§8.3.5 Extract and Raffinate Reflux

A single-section extraction cascade can be refluxed, as in Figure 8.26a, to resemble distillation. In Figure 8.26a, L is used for raffinate flows, V for extract flows, and stages are

numbered from the solvent end of the process. Extract reflux L_R is provided by sending the extract, V_N, to a solvent-recovery step, which removes most of the solvent and gives a solute-rich solution, $L_R + D$, divided into extract reflux L_R, which is returned to stage N, and product D. At the other end of the cascade, a portion, B, of the raffinate, L_1, is withdrawn in a stream divider and added as raffinate reflux, V_B, to fresh solvent, S. The remaining raffinate, B, is sent to a solvent-removal step (not shown) to produce a carrier-rich raffinate product. When using extract reflux, minimum- and total-reflux conditions, corresponding to infinite and minimum stages, bracket the optimal extract reflux ratio. Raffinate reflux is not processed through the solvent-removal unit because fresh solvent is added at this end of the cascade. It is necessary, however, to remove solvent from extract reflux.

The analogy between a two-section liquid–liquid extractor, with feed entering a middle stage, and distillation is considered in some detail by Randall and Longtin [32]. Different aspects of the analogy are listed in Table 8.5. Most important is that the solvent (MSA) in extraction takes the place of heat (ESA) in distillation.

The use of raffinate reflux has been judged to be of little, if any, benefit by Skelland [31], who shows that the amount of raffinate reflux does not affect the number of stages required. Accordingly, only a two-section cascade that includes extract reflux, as shown in Figure 8.26b, is considered here.

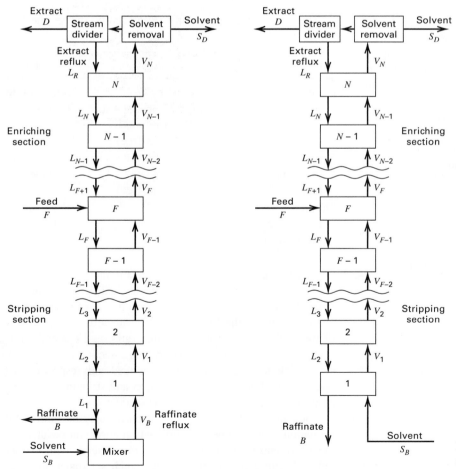

(a)

(b)

Figure 8.26 Liquid–liquid extraction with reflux: (a) with extract and raffinate reflux; (b) with extract reflux only.

Table 8.5 Analogy between Distillation and Extraction

Distillation	Extraction
Addition of heat	Addition of solvent
Reboiler	Solvent mixer
Removal of heat	Removal of solvent
Condenser	Solvent separator
Vapor at the boiling point	Solvent-rich solution saturated with solvent
Superheated vapor	Solvent-rich solution containing more solvent than that required to saturate it
Liquid below the boiling point	Solvent-lean solution, containing less solvent than that required to saturate it
Liquid at the boiling point	Solvent-lean solution saturated with solvent
Mixture of liquid and vapor	Two-phase liquid mixture
Relative volatility	Relative selectivity
Change of pressure	Change of temperature
D = distillate	D = extract product (solute on a solvent-free basis)
B = bottoms	B = raffinate (solvent-free basis)
L = saturated liquid	L = saturated raffinate (solvent-free)
V = saturated vapor	V = saturated extract (solvent-free)
A = more volatile component	A = solute to be recovered
C = less volatile component	C = carrier from which A is extracted
F = feed	F = feed
x = mole fraction A in liquid	X = mole or weight ratio of A (solvent-free), $A/(A + C)$
y = mole fraction A in vapor	$Y = S/(A + C)$

Analysis of a refluxed extractor involves direct extensions of procedures already developed. As will be shown, however, results for a Type I system depend critically on feed composition and the phase diagram, and it is difficult to draw any general conclusions with respect to the effect (or even feasibility) of extract reflux.

For the two-section cascade with extract reflux shown in Figure 8.26b, a degrees-of-freedom analysis can be performed. The result—using as elements two countercurrent cascades, a feed stage, a splitter, and a divider—is $N_D = 2N + 3C + 13$. All but four of the specifications usually are:

Variable Specification	Number of Variables
Pressure at each stage	N
Temperature for each stage	N
Feed stream flow rate, composition, temperature, and pressure	$C + 2$
Solvent composition, temperature, and pressure	$C + 1$
Split of each component in the splitter (solvent-removal step)	C
Temperature and pressure of the two streams leaving the splitter	4
Pressure and temperature of the divider	2
	$2N + 3C + 9$

The four additional specifications can be one of the following sets:

Set 1	Set 2	Set 3
Solvent rate	Reflux ratio	Solvent rate
Solute concentration in extract (solvent-free)	Solute concentration in extract (solvent-free)	Reflux ratio
Solute concentration in raffinate (solvent-free)	Solute concentration in raffinate (solvent-free)	Number of stages
Optimal feed-stage location	Optimal feed-stage location	Feed-stage location

Sets 1 and 2 of §8.3.1 are of interest in the design of a new extractor because two specifications deal with products of designated purities. Set 2 is analogous to the design of a binary distillation using the McCabe–Thiele method, where purities of the distillate and bottoms, reflux ratio, and optimal feed-stage location are specified. For a single-section cascade, it is not feasible to specify the split of the feed with respect to two components. Instead, as in absorption and stripping, recovery of just one component in the feed is specified.

For binary distillations, product purity may be limited by formation of azeotropes. A similar limitation can occur for a Type I system when using a two-section cascade with extract reflux, because of the plait point, which separates the two-liquid region from the homogeneous, single-phase region. This limitation can be determined from a triangular diagram, but it is most readily observed on the Janecke diagram described in Chapter 4 and shown in Figure 4.14e. In Figure 8.27, equilibrium data are repeated for the A–C–S system of Figure 8.14, where A is the solute and S is the solvent. In the triangular representation of Figure 8.27a, the maximum solvent-free solute in the extract, achieved by a countercurrent cascade with extract reflux, is determined by the intersection of line $\overline{SE'}$, drawn tangent to the binodal curve, from pure solvent point S to solvent-free composition line $\overline{AC}$, giving 83 wt% solute. The same value is read from the Janecke diagram in Figure 8.27b as the abscissa for point E' farthest to the right of the binodal curve.

Without extract reflux, the maximum solvent-free solute concentration corresponds to an extract in equilibrium with

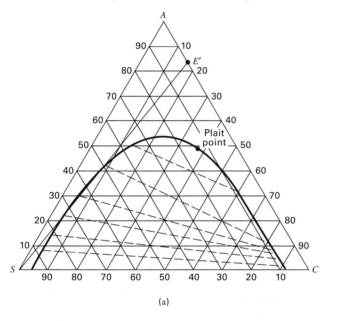

(a)

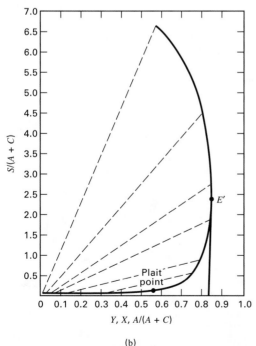

(b)

Figure 8.27 Limitation on product purity: (a) using an equilateral-triangle diagram; (b) using a Janecke diagram.

the feed, when saturated with the extract. If this maximum value is close to the maximum value determined, as in Figure 8.27, then use of extract reflux is of little value. This is often true for Type I systems, as shown in the following example.

EXAMPLE 8.2 Use of Extract Reflux.

In Example 8.1, a feed of 30 wt% acetone and 70 wt% ethyl acetate was extracted with pure water in a single-section, countercurrent cascade to obtain a raffinate of 5 wt% acetone on a water (solvent)-free basis. The maximum solvent-free solute concentration in the extract was found to be 64 wt%, as shown in Figure 8.20 at point D, corresponding to the condition of minimum $S/F = 0.60$ at ∞ stages. For an actual $S/F = 1.75$ with four equilibrium stages, the extract contains 62 wt% acetone on a solvent-free basis. Thus, use of extract reflux for the purpose of producing a more pure (solvent-free) extract is not attractive, given the particular phase-equilibrium diagram and feedstock composition. Nevertheless, to demonstrate the technique, the calculation for extract reflux is illustrated. Also, the minimum number of stages at total reflux and the minimum reflux ratio are determined.

Solution

For the single-section, countercurrent cascade, the extract pinch point is at 57 wt% water, 27 wt% acetone, and 16 wt% acetate, as in Figure 8.20 at $D'_{\min}$. If stages are added above the feed point, as in the two-section, refluxed cascade of Figure 8.26b, it is possible, theoretically, to reduce the water content of the extract to about 28 wt%, as shown by G in Figure 8.20. However, the solvent (water)-free extract would be only 51 wt% acetone, which is determined from the line drawn through points S and G and extended to where it intersects the solvent-free line $\overline{AC}$.

To make this example more interesting, assume an extract containing 50 wt% water is required. Thus, the extraction cascade is that shown in Figure 8.28, rather than that in Figure 8.26b. The difference lies in the location of the solvent-removal step. This saturated-extract product is shown as point D' in Figure 8.28. Assume the ratio S/F to be 1.43, which is more than twice the minimum ratio in Example 8.1. The desired raffinate composition is again 5 wt% acetone (point B in Figure 8.28), which maps to point B' on the raffinate side of the binodal curve on a line connecting points B and S.

As with single-section cascades, the mixing point, M, in Figure 8.28, for streams S and F entering the cascade, is determined from the inverse-lever-arm rule using the S/F ratio, or by computing the composition of M, which in this case is 59 wt% water, 29 wt% acetate, and 12 wt% acetone.

The cascade in Figure 8.28 consists of an enriching section to the left of the feed point, where extract is enriched in solute, and a stripping section to the right, where raffinate is stripped of solute. Difference points P' and P'' are needed, respectively, for the enriching and stripping sections. In the enriching section, referring to the cascade in Figure 8.28, $P' = V_n - L_R = V_{n-1} - L_n$. But, by material balance, $V_n - L_R = D' + S_D$. Therefore, $P' = D' + S_D$, the total flow leaving the extract end of the cascade. Also, by overall material balance, $M = F + S = B' + D' + S_D = B' + P'$. Thus, P' must lie on a line drawn through points B' and M. To locate the position of P' on that line, note that V_n has the same composition as D', and that L_R is simply D' with the solvent removed (point D). However, $P' = V_n - L_R$ or $V_n = P' + L_R$. Thus, point P' must lie on a line drawn through points $V_n(D')$ and $L_R(D)$. Now, point P' is located at the intersection of extended lines $\overline{B'M}$ and $\overline{DD'}$ as in Figure 8.28.

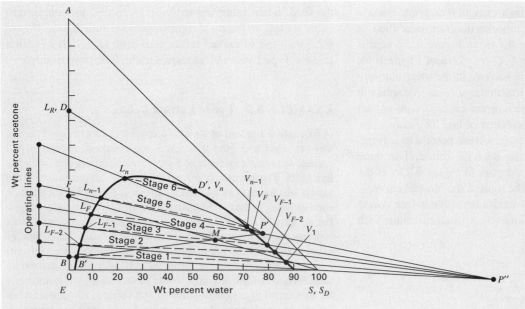

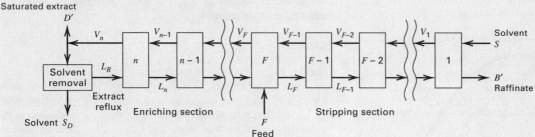

Figure 8.28 Equilibrium stages for Example 8.2.

Difference point P'' for the stripping section is located similarly. It is noted that $P'' = B' - S = F - D' - S_D = F - P'$. Thus, P'' is at the intersection of extended lines $\overline{FP'}$ and $\overline{B'S}$, as shown at the far right of Figure 8.28.

The stages can be stepped off by starting from V_n and using the difference point P' until an operating line crosses feed line $\overline{FP'}$. From there, stages are stepped off using difference point P'' until the raffinate composition is reached. In Figure 8.28, six equilibrium stages are required, with two in the enriching section and four in the stripping section. Feed enters the third stage from the left.

The reflux ratio, defined for this example as $(V_n - D')/D' = (L_R + S_D)/D'$, can also be determined since, from above, $P' = D' + S_D$. Therefore, by the mixing rule, $S_D/D' = \overline{D'P'}/\overline{S_DP'}$. By material balance, $V_n - D' = L_R + S_D$. Therefore,

$$\frac{V_n - D'}{S_D} = \frac{L_R + S_D}{S_D} = \frac{\overline{L_RS_D}}{\overline{L_RD'}} \quad \text{and}$$

$$\frac{V_n - D'}{D'} = \left(\frac{S_D}{D'}\right)\left(\frac{V_n - D'}{S_D}\right) = \left(\frac{\overline{D'P'}}{\overline{S_DP'}}\right)\left(\frac{\overline{L_RS_D}}{\overline{L_RD'}}\right)$$

By measurement from Figure 8.28, $(V_n - D')/D' = (1.2)(2.0) = 2.4$. The reflux ratio is valid only for a solvent-to-feed ratio of 1.43.

Next, consider the case of total reflux for minimum stages. With reference to Figure 8.29, stream compositions are as previously specified or computed. For acetone, there is 30 wt% in F, 4.9 wt% in B', 33 wt% in D', and 62 wt% in L_R. As in the case of the single-section cascade of Figure 8.20, as the solvent-to-feed ratio increases, the mixing point $M = F + S$ moves toward the solvent apex. At maximum solvent addition, M lies at the intersection of the line through F and S with the extract side of the binodal curve. Difference points P' and P'' also move toward S because $P' = D' + S_D$ approaches S_D at total reflux and $P'' = F - P'$ approaches P', recalling that at total reflux,

$F = D' = 0$. As shown in Figure 8.29, the minimum number of stages is three, as stepped off from the S apex.

Lastly, consider the minimum reflux ratio at ∞ stages, which corresponds to the minimum solvent ratio. As the solvent ratio is reduced, point M moves toward feed point F, and point P'' moves away from the binodal curve. Also, point P' moves toward V_n. Ultimately, the S/F value is reached where an operating line in either the enriching section or the stripping section coincides with a tie line, giving a pinch point and ∞ stages. Often, this occurs for the extended tie line that passes through the feed point. Such is the case here, giving a minimum reflux ratio of about 0.6 and a corresponding minimum S/F ratio of 0.75.

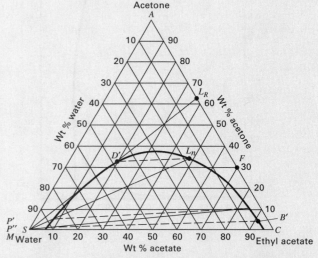

Figure 8.29 Total reflux and minimum stages for Example 8.2.

§8.4 MALONEY–SCHUBERT GRAPHICAL EQUILIBRIUM-STAGE METHOD

For Type II ternary systems, shown in Figure 8.10b, a two-section cascade with extract reflux is particularly desirable. Without a plait point, the two-phase region extends across the entire solute composition. Thus, while solvent-free solute concentrations in the extract are limited for a Type I system, as shown in Figure 8.27, no such limit exists for a Type II system. It is thus possible with extract reflux to achieve as sharp a separation as desired between solute (A) and carrier (C).

When reflux is used, many stages may be required and triangular diagrams are often cumbersome. Instead, a McCabe–Thiele-type diagram—or a Janecke diagram of the type in Figure 4.14e, often in conjunction with a distribution diagram—has proven useful. The Janecke diagram uses rectangular coordinates, with solvent concentration on a solvent-free basis plotted as the ordinate against solute concentration on a solvent-free basis as the abscissa—that is, %S/(%A + %C) against %A/(%A + %C)—in mass or mole percents. The Janecke diagram is analogous to an enthalpy-concentration diagram and is consistent with the distillation–extraction analogy of Table 8.5, where enthalpy is replaced by solvent concentration because an MSA replaces an ESA. Its application to liquid–liquid extraction of a Type II system with the use of reflux is considered in detail by Maloney and Schubert [33], who use an auxiliary distribution diagram of the McCabe–Thiele type, but on a solvent-free basis, to facilitate visualization. This method is also called the *Ponchon–Savarit method for extraction*. Unlike the analogous method for distillation discussed briefly in §7.7, which requires both enthalpy and vapor–liquid equilibrium data, extraction requires only liquid–liquid solubility data, which are more common than combined vapor–liquid enthalpy and equilibrium data. Accordingly, despite the development of computer-aided methods used in process simulators, the Ponchon–Savarit method for extraction has remained useful, while the analogous method for distillation has declined in popularity. Although the Janecke diagram can also be applied to Type I systems, it is difficult to use when the carrier and solvent are highly immiscible, because values of the ordinate can become very large.

§8.4.1 Number of Equilibrium Stages

With the Janecke diagram, construction of tie lines, mixing points, operating points, and operating lines is made in a manner similar to that for a triangular diagram. Consider the case of extraction for a Type II system with extract reflux, shown in Figure 8.26b. A representative Janecke diagram is shown in Figure 8.30, where all concentrations are on a solvent-free mass basis:

$$Y = \frac{\text{mass solvent}}{\text{mass of solvent-free liquid phase}}$$

$$X = \frac{\text{mass solvent}}{\text{mass of solvent-free liquid phase}}$$

Values of the ordinate, Y, especially for the saturated-extract phase, vary over a wide range depending on solute and carrier solubility in the solvent. Values of the abscissa, X, vary

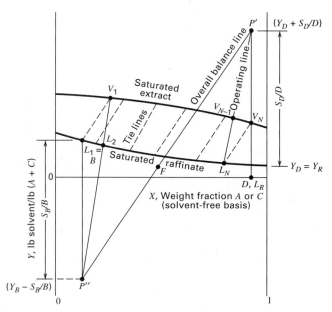

Figure 8.30 Construction of equilibrium stages on a Janecke diagram.

from 0 to 1 (pure carrier to pure solute). Equilibrium tie lines relate concentrations in the saturated extract to concentrations in the saturated raffinate. The Y location of the feed, F, is somewhere between 0 and the saturated-raffinate curve. The extract, V_N, leaving stage N and prior to solvent removal is rich in solute and lies on the saturated-extract curve. Upon solvent removal, extract D and extract reflux L_R, with identical compositions, lie on the $Y = 0$ horizontal line. Solvent removed from the cascade, S_D, and solvent fed to the cascade, S_B, are assumed pure. The raffinate, L_1, leaving stage 1 is rich in the carrier and lies on the saturated-raffinate line.

Points P' and P'' are difference points for the enriching and stripping sections, and are used to draw operating lines. The locations of P' and P'' are derived from solvent-free and solvent material balances around the solvent-recovery step, in terms of passing streams in Figure 8.26b:

$$V_N - L_R = D \tag{8-13}$$

$$Y_{V_N} V_N - Y_D L_R = S_D + Y_D D \tag{8-14}$$

For a solvent-difference balance around a section of top stages down to stage n, located above stage F in Figure 8.26b,

$$Y_{V_n} V_n - Y_{V_{n+1}} L_{n+1} = S_D + Y_D D \tag{8-15}$$

Thus, any solvent flow difference between passing streams in the enriching section above the feed stage is given by $S_D + Y_D D$. If (8-13) and (8-14) are combined to eliminate V_N,

$$\frac{L_R}{D} = \frac{(Y_D + S_D/D) - Y_{V_N}}{Y_{V_N} - Y_D} \tag{8-16}$$

In Figure 8.30, $(Y_{V_N} - Y_D)$ is the vertical distance between V_N and (D, L_R). Difference point P' in Figure 8.30 becomes $(S_D + Y_D D)/D$ or $(Y_D + S_D/D)$. Thus,

$$P' = Y_D + S_D/D \tag{8-17}$$

and

$$L_R/D = \overline{P' V_N} / \overline{V_N L_R} \tag{8-18}$$

In a similar manner for the stripping section,

$$P'' = Y_B - S_B/B \qquad (8\text{-}19)$$

$$D/B = \overline{FP''}/\overline{P'F} \qquad (8\text{-}20)$$

Stages are stepped off as for triangular diagrams, starting from either the extract or the raffinate and alternating between operating lines and tie lines. In Figure 8.30, the process starts from the top at extract D for stages in the enriching section. The solute compositions of D, L_R, and V_N on a solvent-free basis are identical. The operating line for passing streams L_R and V_N is a vertical line passing through the difference point P'. From point V_N on the saturated-extract curve, a tie line is followed down to the equilibrium raffinate phase, L_N. An operating line connecting P' and L_N intersects the extract curve at the passing stream V_{N-1}. Subsequent enriching stages are stepped off until the feed stage is reached. The optimal feed-stage location stage is determined by the intersection of the rectification and stripping operating lines, as in the McCabe–Thiele method. On the Janecke diagram, this intersection is the line $\overline{P'P''}$, which passes through the feed point in Figure 8.30. Thus, the transition from the enriching section (where the difference point P' is used) to the stripping section (where P'' is used) is made when a tie line for a stage crosses the line $\overline{P'P''}$. Following the feed-stage location, the remaining stripping stages are stepped off until the desired raffinate concentration is reached or crossed over.

§8.4.2 Limiting Conditions of Minimum Stages and Minimum Extract Reflux

The Janecke diagram can be used to determine the limiting conditions of total reflux (minimum stages) and minimum reflux (infinite stages). For total reflux, the difference points P' and P'' lie at $Y = +\infty$ and $-\infty$, respectively, because $F = B = D = 0$. Thus, all operating lines become vertical and the minimum number of stages is established in the manner illustrated in Figure 8.31a.

For the condition of minimum extract reflux, a pinch condition is sought at either the feed stage or some other stage location. In Figure 8.31b, where the pinch is assumed to be at the feed stage, an operating line is drawn coincident with a tie line and the feed point, F, to determine P' and P''. To establish if the pinch does occur at the feed stage, tie lines to the right of the feed-stage tie line are extended to intersect with the vertical line through D. If a higher intersection occurs, then that P' may be the correct P'_{min} difference point. In a similar manner, tie lines to the left of the feed-stage tie line are extended to intersect with the vertical line through B. If a lower intersection occurs, then that P'' may determine the

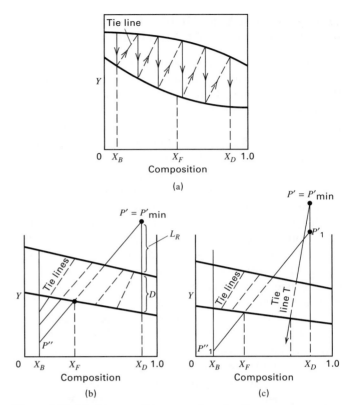

Figure 8.31 Limiting conditions on a Janecke diagram: (a) minimum number of stages at total reflux; (b) minimum reflux determined by a tie line through the feed point; (c) minimum reflux determined by a tie line to the right of the feed point.

minimum reflux. The former case is shown in Figure 8.31c, where P' is higher than P'_1. Thus, $P' = P'_{min}$. In any case, once the controlling P' or P'' is determined, a line through F determines the other difference point, and the minimum reflux ratio is computed from (8-18) using P'_{min}.

EXAMPLE 8.3 Extraction of a Type II System.

The extraction cascade in Figure 8.32, which is equipped with a perfect solvent separator to provide extract reflux, is used to separate methylcyclopentane (A) and n-hexane (C) into a final extract and raffinate containing, on a solvent-free basis, 95 wt% and 5 wt% A, respectively, using aniline (S) as the solvent. The feed is 1,000 kg/h with 55 wt% A, and the mass ratio of S/F is 4.0. The feed contains no aniline, and the fresh and recycle solvents are pure. Equilibrium curves and tie lines are given in Figure 8.33.

(a) Determine the extract reflux ratio and number of stages, with feed entering the optimal stage.

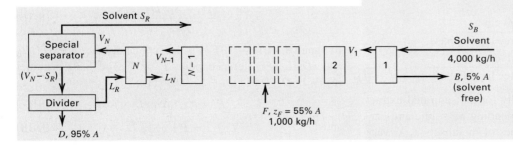

Figure 8.32 Countercurrent extraction cascade with extract reflux for Example 8.3.

(b) Determine the minimum number of stages for the specified solvent-free extract and raffinate compositions.

(c) Determine the minimum extract reflux ratio for the specified feed and product compositions.

Solution

(a) Product rates are determined by material balance. An overall balance is $D + B = 1,000$ kg/h. A solute balance gives $0.95D + 0.05B = (0.55)(1,000) = 550$ kg/hr. Solving these two equations simultaneously, $D = 556$ kg/h and $B = 444$ kg/h. Since $S_B = 4,000$ kg/h, $S_B/B = 9.0$. In Figure 8.30, point P'' is located at a distance S_B/B below the raffinate composition, X_B, at point B. Since Y at point B, from Figure 8.33, is approximately 0.3, point P'' is located at $0.3 - 9.0 = -8.7$.

A line drawn through P'' and F, extended to the intersection with the vertical line through D, gives $P' = 6.7$. By measurement from Figure 8.33, using (8-16), $L_R/D = 3.7$.

Stages are stepped off starting from D. At the third stage $(N - 2)$, the tie line crosses line $\overline{P''FP'}$, so this is the optimal feed stage. Three more stages are required to reach B, for a total of six.

(b) If the construction for minimum stages in Figure 8.31a is used in Figure 8.33, just less than five stages are determined.

(c) If the minimum reflux shown in Figure 8.31b for a pinch at the feed stage is used in Figure 8.33, $P' = 2.90$ is obtained. No other tie line in either section gives a larger value, so $P_{min} = 2.9$. By measurement, using (8-18), $(L_R/D)_{min} = 0.83$. Using Figure 8.30, the corresponding $(S_B/B)_{min}$ is found to be 4.2. Thus, $(S_B)_{min} = 4.2(444) = 1,865$ kg/h or $(S_B/F)_{min} = 1,865/1,000 = 1.865$.

For this example, a high reflux ratio and corresponding solvent-to-feed ratio keep the number of equilibrium stages small. When the number of stages is large, the Janecke diagram becomes cluttered with operating lines and tie lines. In that case, an auxiliary McCabe–Thiele plot of solute mass fraction in the extract layer versus solute mass fraction in the raffinate layer, both on a solvent-free basis, can be drawn, with points on the enriching and stripping operating lines determined, as shown previously, from arbitrary operating lines on the Janecke diagram. Stages are counted in the McCabe–Thiele manner. Figure 8.34 shows an auxiliary McCabe–Thiele diagram [33].

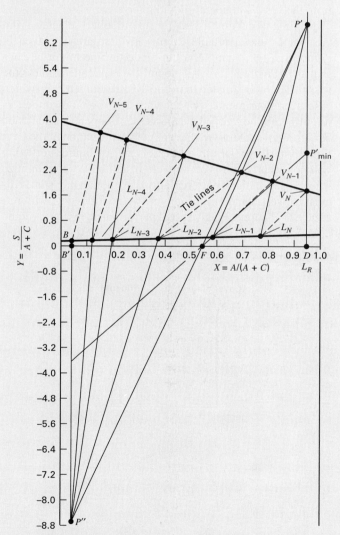

Figure 8.33 Maloney–Schubert constructions on Janecke diagram for Example 8.3.

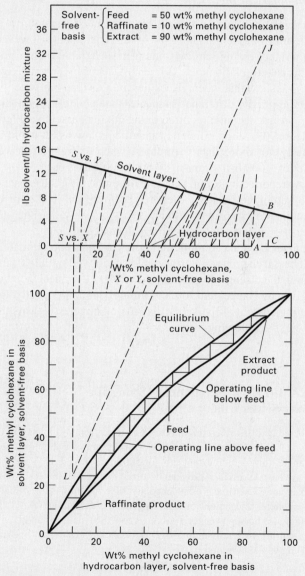

Figure 8.34 Use of Janecke diagram with auxiliary distribution diagram.

[From R.H. Perry and C.H. Chilton, Eds., *Chemical Engineers' Handbook*, 5th ed., McGraw-Hill, New York (1973).]

§8.5 THEORY AND SCALE-UP OF EXTRACTOR PERFORMANCE

Industrial extraction equipment can be selected using the scheme of Figure 8.8. Often in the chemical industry, the choice is between a cascade of mixer-settlers and a multi-compartment, column-type extractor with mechanical agitation, the main considerations being stages required, floor space and headroom available, and capital and operating costs. For biochemical applications, the choice may be between a Karr column and a POD. Methods for estimating size and power requirements of these four extractor types are presented next. Devices with no mechanical agitation are considered briefly.

§8.5.1 Mixer-Settler Units

Preliminary Sizing

Final mixer-settler unit design is done most accurately by scale-up from laboratory or pilot-plant equipment. However, preliminary sizing can be done using available theory and empirical correlations. Experimental data of Flynn and Treybal [34] show that when viscosities are less than 5 cP and the specific-gravity difference is greater than about 0.10, the average residence time required in the mixing vessel to achieve at least 90% stage efficiency may be as low as 30 s and is usually not more than 5 minutes, when an agitator-power input of 1,000 ft-lb$_f$/min-ft^3 (4 hp/1,000 gal) is used.

As reported by Ryan, Daley, and Lowrie [35], the capacity of a settler vessel can be expressed in terms of C gpm of combined extract and raffinate per ft^2 of phase-disengaging area. For a horizontal, cylindrical vessel of length L and diameter D_T, the economic ratio of L to D_T is approximately 4. Thus, if the phase interface is located at the middle of the vessel, the disengaging area is $D_T L$ or $4D_T^2$. A typical value of C given by Happel and Jordan [36] is about 5. Frequently, the settling vessel will be larger than the mixing vessel, as is the case in the following example.

EXAMPLE 8.4 Preliminary Sizing of a Mixer-Settler Unit.

Benzoic acid is to be continuously extracted from a dilute solution in water using toluene in a series of discrete mixer-settler vessels arranged for countercurrent flow of 500 gpm of feed and 750 gpm of solvent. Assuming a residence time, t_{res}, of 2 minutes in each mixer and a settling vessel capacity of 5 gal/min-ft^2, estimate: (a) diameter and height of a mixing vessel, assuming $H/D_T = 1$; (b) agitator horsepower for a mixing vessel; (c) diameter and length of a settling vessel, assuming $L/D_T = 4$; and (d) residence time in a settling vessel in minutes.

Solution

(a) $Q =$ total flow rate $= 500 + 750 = 1,250$ gal/min

$V =$ volume $= Q t_{res} = 1,250(2) = 2,500$ gal or $2,500/7.48 = 334$ ft^3

$$V = \pi D_T^2 H/4, \; H = D_T, \text{ and } V = \pi D_T^3/4$$
$$D_T = (4V/\pi)^{1/3} = [(4)(334)/3.14]^{1/3}$$
$$= 7.52 \text{ ft and } H = 7.52 \text{ ft}$$

(b) Horsepower $= 4(2,500/1,000) = 10$ hp

(c) $D_T L = 1,250/5 = 250$ ft^2; $D_T^2 = 250/4 = 62.5$ ft^2

$$D_T = 7.9 \text{ ft}; \; L = 4D_T = 4(7.9) = 31.6 \text{ ft}$$

(d) Volume of settler $= \pi D_T^2 L/4 = 3.14(7.9)^2(31.6)/4 = 1,548$ ft^3 or $1,548(7.48) = 11,580$ gal

$$t_{res} = V/Q = 11,580/1,250 = 9.3 \text{ min}$$

Power Requirement of a Mixer Unit from a Correlation

Figure 8.35 shows a typical single-compartment mixing tank for liquid–liquid extraction. The vessel is closed, with the two liquid phases entering at the bottom, and the effluent, in the form of a two-phase emulsion, leaving at the top. Rounded heads of the type in Figure 8.2 are preferred, to eliminate stagnant fluid regions. All gases must be evacuated from the vessel, so no gas–liquid interface exists.

Mixing is accomplished by an impeller selected from the many types available, some of which are displayed in Figure 8.3. For example, a flat-blade turbine might be chosen, as in Figure 8.35. A single turbine is adequate unless the vessel height is greater than the vessel diameter, in which case a compartmented vessel with two or more impellers might be employed. When the vessel is open, vertical side baffles are mandatory to prevent vortex formation at the gas–liquid interface. For closed vessels full of liquid, vortexing will not occur, but it is common to install baffles to minimize swirling and improve circulation. Although no standards exist for vessel and turbine geometry, the following, with reference to Figure 8.35, give good performance in liquid–liquid agitation:

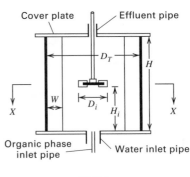

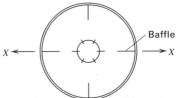

Figure 8.35 Agitated vessel with flat-blade turbine and baffles.

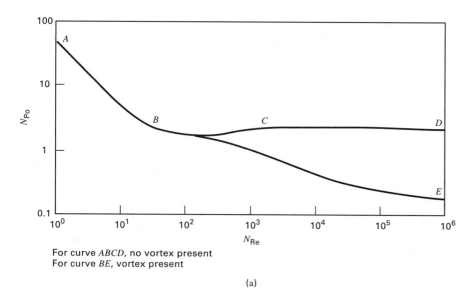

For curve *ABCD*, no vortex present
For curve *BE*, vortex present

(a)

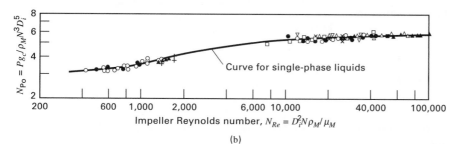

Impeller Reynolds number, $N_{Re} = D_i^2 N \rho_M / \mu_M$

(b)

Figure 8.36 Power consumption of agitated vessels. (a) Typical power characteristics. [From J.H. Rushton and J.Y. Oldshue, *Chem. Eng. Prog.*, **49**, 161–168 (1953).] (b) Power correlation for six-bladed, flat-blade turbines with no vortex. [From D.S. Laity and R.E. Treybal, *AIChE J.*, **3**, 176–180 (1957).]

number of turbine blades = 6;

number of vertical baffles = 4;

$$H/D_T = 1; D_i/D_T = 1/3;$$
$$W/D_T = 1/12, \text{ and } H_i/H = 1/2.$$

To achieve a high extraction efficiency, say, between 90 and 100%, it is necessary to provide vigorous agitation. For a given type of impeller and vessel–impeller geometry, the agitator power, P, can be estimated from an empirical correlation in terms of a power number, N_{Po}, which depends on an impeller Reynolds number, N_{Re}, where

$$N_{Po} = \frac{Pg_c}{N^3 D_i^5 \rho_M} \tag{8-21}$$

$$N_{Re} = \frac{D_i^2 N \rho_M}{\mu_M} \tag{8-22}$$

The impeller Reynolds number is the ratio of the inertial force to the viscous force, with inertial force $\propto (ND_i)^2 \rho_M D_i^2$ and viscous force $\propto \mu_M(ND_i)D_i^2/D_i$, where N = rate of rotation. Thus, the characteristic length in the Reynolds number is the impeller diameter, D_i, and the characteristic velocity is ND_i = impeller peripheral velocity.

The agitator power for a mixer is proportional to the product of volumetric liquid flow produced by the impeller and the applied kinetic energy per unit volume of fluid. The result is: Power, P, is proportional to $(ND_i^3)[\rho_M(ND_i)^2/2g_c]$, which can be rewritten as (8-21), where the constant of proportionality is $2N_{Po}$. Both the impeller Reynolds number and the power number (also called the Newton number) are

dimensionless groups. The power number for an agitated vessel serves the same purpose as the friction factor for flow of fluid through a pipe. This is illustrated, over a wide range of impeller Reynolds numbers, for an impeller in Figure 8.36a, taken from Rushton and Oldshue [37]. The upper curve, ABCD, is for a vessel with baffles, while the lower curve, ABE, pertains to the same tank with no baffles. In the low-Reynolds-number region, AB, viscous forces dominate and the impeller power is proportional to $\mu_M N^2 D_i^3$. Beyond a Reynolds number of about 200, a vortex appears if no baffles are present and the power-number relation is given by BE. In this region, the Froude number, $N_{Fr} = N^2 D_i/g$, the ratio of inertial to gravitational forces, also becomes a factor. With baffles, and $N_{Re} > 1,000$ for region CD, fully developed turbulence exists, inertial forces dominate, and the power is proportional to $\rho_M N^3 D_i^5$. It is clear that baffles greatly increase power requirements.

Experimental data for liquid–liquid mixing in baffled vessels with six-bladed, flat-blade turbines are shown in Figure 8.36b, from Laity and Treybal [38]. The impeller Reynolds number covers only the turbulent–flow region, where there is efficient liquid–liquid mixing. The solid line represents batch mixing of single-phase liquids. The data represent liquid–liquid mixing, where agreement with the single-phase curve is achieved by computing two-phase mixture properties from

$$\rho_M = \rho_C \phi_C + \rho_D \phi_D \tag{8-23}$$

$$\mu_M = \frac{\mu_C}{\phi_C}\left(1 + \frac{1.5\mu_D\phi_D}{\mu_C + \mu_D}\right) \tag{8-24}$$

where ϕ is the volume fraction of tank holdup, with subscripts C for the continuous phase and D the dispersed phase, such that $\phi_D + \phi_C = 1$. For continuous flow from inlets at the bottom of the vessel to a top outlet for the emulsion and with the impeller located at a position above the resting interface, the data correlate with Figure 8.36b.

With fully developed turbulent flow, the volume fraction of a dispersed phase in the vessel closely approximates that in the feed; otherwise, the volume fractions may be different, and the residence times of the two phases will not be the same. At best, spheres of uniform size can pack tightly to give a void fraction of 0.26. Therefore, $\phi_C > 0.26$ and $\phi_D < 0.74$ is quoted, but some experiments have shown a 0.20–0.80 range. At startup, the vessel is filled with the phase to be continuous. Following initiation of agitation, the two-feed liquids are introduced at their desired flow ratio.

Based on the work of Skelland and Ramsey [39] and Skelland and Lee [40], a minimum impeller rotation rate is required for uniform dispersion of one liquid into another. For a flat-blade turbine in a baffled vessel, their equation in terms of dimensionless groups is:

$$\frac{N_{\min}^2 \rho_M D_i}{g \Delta \rho} = 1.03 \left(\frac{D_T}{D_i}\right)^{2.76} \phi_D^{0.106} \left(\frac{\mu_M^2 \sigma}{D_i^5 \rho_M g^2 (\Delta \rho)^2}\right)^{0.084}$$

$$(8\text{-}25)$$

where $\Delta \rho$ is the absolute value of the density difference and σ is the interfacial tension between the liquid phases. The group on the left side of (8-25) is a two-phase Froude number; the group at the far right is a ratio of forces:

$$\frac{(\text{viscous})^2 (\text{interfacial tension})}{(\text{inertial})(\text{gravitational})^2}$$

EXAMPLE 8.5 Design of a Mixer Extraction Unit.

Furfural is extracted from water by toluene at 25°C in an agitated vessel like the one in Figure 8.35. The feed enters at 20,400 lb/h, while the solvent enters at 11,200 lb/h. For a residence time of 2 minutes, estimate, for either phase as the dispersed phase: (a) dimensions of the mixing vessel and diameter of the flat-blade turbine impeller; (b) minimum rate of rotation of the impeller for complete and uniform dispersion; (c) the power requirement of the agitator at the minimum rotation rate.

Solution

Mass flow rate of feed = 20,400 lb/h; feed density = 62.3 lb/ft³;

Volumetric flow rate of feed = Q_F = 20,400/62.3 = 327 ft³/h;

Mass flow rate of solvent = 11,200 lb/h; solvent density = 54.2 lb/ft³;

Volumetric flow rate of solvent = Q_S = 11,200/54.2 = 207 ft³/h

Because the solute in the feed is dilute and there is sufficient agitation to achieve uniform dispersion, assume that fractional volumetric holdups of raffinate and extract in the vessel are equal to the corresponding volume fractions in the combined feed. Thus,

$$\phi_R = 327/(327 + 207) = 0.612; \quad \phi_E = 1 - 0.612 = 0.388$$

(a) Mixer volume = $(Q_F + Q_S)t_{res} = V = (327 + 207)(2/60) = 17.8$ ft³. Assume a cylindrical vessel with $D_T = H$ and neglect the volume of the bottom and top heads and the volume occupied by the agitator and the baffles. Then

$$V = (\pi D_T^2/4)H = \pi D_T^3/4$$

$$D_T = [(4/\pi)V]^{1/3} = [(4/3.14)17.8]^{1/3} = 2.83 \text{ ft}$$

$$H = D_T = 2.83 \text{ ft}$$

Make the vessel 3 ft in diameter by 3 ft high, which gives a volume $V = 21.2$ ft³ = 159 gal. Assume that $D_i/D_T = 1/3$; $D_i = D_T/3 = 3/3 = 1$ ft.

(b) *Case 1—Raffinate phase dispersed:*

$$\phi_D = \phi_R = 0.612; \quad \phi_C = \phi_E = 0.388;$$

$$\rho_D = \rho_R = 62.3 \text{ lb/ft}^3; \quad \rho_C = \rho_E = 54.2 \text{ lb/ft}^3;$$

$$\mu_D = \mu_R = 0.89 \text{ cP} = 2.16 \text{ lb/h-ft};$$

$$\mu_C = \mu_E = 0.59 \text{ cP} = 1.43 \text{ lb/h-ft};$$

$$\Delta \rho = 62.3 - 54.2 = 8.1 \text{ lb/ft}^3;$$

$$\sigma = 25 \text{ dyne/cm} = 719{,}000 \text{ lb/h}^2$$

From (8-23)

$$\rho_M = (54.2)(0.388) + (62.3)(0.612) = 59.2 \text{ lb/ft}^3$$

From (8-24),

$$\mu_M = \frac{1.43}{0.388}\left[1 + \frac{1.5(2.16)(0.612)}{1.43 + 2.16}\right] = 5.72 \text{ lb/h-ft}$$

From (8-25), using AE units, with $g = 4.17 \times 10^8$ ft/h²,

$$\frac{\mu_M^2 \sigma}{D_i^5 \rho_M g^2 (\Delta \rho)^2} = \frac{(5.72)^2 (719.000)}{(1)^5 (59.2)(4.17 \times 10^8)^2 (8.1)^2}$$

$$= 3.47 \times 10^{-14}$$

$$N_{\min}^2 = 1.03 \left(\frac{g\Delta\rho}{\rho_M D_i}\right)\left(\frac{D_T}{D_i}\right)^{2.76} \phi_D^{0.106}(3.47 \times 10^{-14})^{0.084}$$

$$= 1.03 \left[\frac{(4.17 \times 10^8)(8.1)}{(59.2)(1)}\right]\left(\frac{3}{1}\right)^{2.76}(0.612)^{0.106}(0.0740)$$

$$= 8.56 \times 10^7 (\text{rph})^2$$

$$N_{\min} = 9{,}250 \text{ rph} = 155 \text{ rpm}$$

Case 2—Extract phase dispersed: Calculations similar to case 1 result in $N_{\min} = 8{,}820$ rph = 147 rpm.

(c) *Case 1—Raffinate phase dispersed:*

From (8-22), $N_{Re} = \dfrac{(1)^2(9{,}250)(59.2)}{(5.72)} = 9.57 \times 10^4$

From Figure 8.36b, a fully turbulent flow exists, with the power number given by its asymptotic value of $N_{Po} = 5.7$.

From (8-21),

$$P = N_{Po} N^3 D_i^5 \rho_M/g_c$$

$$= (5.7)(9{,}250)^3(1)^5(59.2)/(4.17 \times 10^8)$$

$$= 640{,}000 \text{ ft-lb}_f/\text{h} = 0.323 \text{ hp}$$

$$P/V = 0.323(1000)/159 = 2.0 \text{ hp}/1{,}000 \text{ gal}$$

Case 2—Extract phase dispersed:
Calculations as in case 1 result in $P = 423{,}000$ ft-lb$_f$/h = 0.214 hp.

$$P/V = 0.214(1000)/159 = 1.4 \text{ hp}/1{,}000 \text{ gal}$$

Mass-Transfer Efficiency

When dispersion is complete, both phases in the vessel are perfectly mixed, and the solute concentrations in each phase are uniform and equal to the concentrations in the two-phase emulsion leaving the vessel. This is the CFSTR or CSTR (continuous-flow, stirred-tank reactor) model used in chemical reactor design, sometimes called the completely back-mixed or perfectly mixed model, first discussed by MacMullin and Weber [41]. The Murphree dispersed-phase efficiency for extraction, based on the raffinate as the dispersed phase, is expressed as the fractional approach to equilibrium. In terms of solute,

$$E_{MD} = \frac{c_{D,\text{in}} - c_{D,\text{out}}}{c_{D,\text{in}} - c_D^*} \qquad (8\text{-}26)$$

where c_D^* is the solute concentration in equilibrium with the bulk-solute concentration in the exiting continuous phase, $c_{C,\text{out}}$. The molar rate of solute mass transfer, n, from the dispersed phase to the continuous phase is

$$n = K_{OD}a(c_{D,\text{out}} - c_D^*)V \qquad (8\text{-}27)$$

where the mass-transfer driving force is uniform throughout the vessel and equal to the driving force based on exit concentrations; a is the interfacial area for mass transfer per unit volume of liquid phases; V is the total volume of liquid in the vessel; and K_{OD} is the overall mass-transfer coefficient based on the dispersed phase, given in terms of the resistances of the dispersed and continuous phases by

$$\frac{1}{K_{OD}} = \frac{1}{k_D} + \frac{1}{mk_C} \qquad (8\text{-}28)$$

where equilibrium is assumed at the interface between the phases and $m =$ the slope of the equilibrium curve for the solute, plotted as c_C versus c_D:

$$m = dc_C/dc_D \qquad (8\text{-}29)$$

For dilute solutions, changes in volumetric flow rates of raffinate and extract are small, and the rate of mass transfer based on the change in solute concentration in the dispersed phase is:

$$n = Q_D(c_{D,\text{in}} - c_{D,\text{out}}) \qquad (8\text{-}30)$$

where Q_D is the volumetric flow rate of the dispersed phase. To obtain an expression for E_{MD} in terms of $K_{OD}a$, (8-26), (8-27), and (8-30) are combined. From (8-26),

$$\frac{E_{MD}}{1 - E_{MD}} = \frac{c_{D,\text{in}} - c_{D,\text{out}}}{c_{D,\text{out}} - c_D^*} \qquad (8\text{-}31)$$

Equating (8-27) and (8-30), and noting that the RHS of (8-31) is the number of dispersed-phase transfer units for a perfectly mixed vessel with $c_D = c_{D,\text{out}}$

$$N_{OD} = \int_{c_{D,\text{out}}}^{c_{D,\text{in}}} \frac{dc_D}{c_D - c_D^*} = \frac{c_{D,\text{in}} - c_{D,\text{out}}}{c_{D,\text{out}} - c_D^*} = \frac{K_{OD}aV}{Q_D} \qquad (8\text{-}32)$$

Combining (8-31) and (8-32) and solving for E_{MD},

$$E_{MD} = \frac{K_{OD}aV/Q_D}{1 + K_{OD}aV/Q_D} = \frac{N_{OD}}{1 + N_{OD}} \qquad (8\text{-}33)$$

When $N_{OD} = (K_{OD}aV/Q_D) \gg 1$, $E_{MD} = 1$.

Drop Size and Interfacial Area

Estimates of E_{MD} require experimental data for interfacial area, a, and mass-transfer coefficients k_D and k_C. The droplets in an agitated vessel cover a range of sizes and shapes; hence it is useful to define d_e, the equivalent diameter of a spherical drop, using the method of Lewis, Jones, and Pratt [42],

$$d_e = \left(d_1^2 d_2\right)^{1/3} \qquad (8\text{-}34)$$

where d_1 and d_2 are major and minor axes of an ellipsoidal-drop image. For a spherical drop, d_e is simply the drop diameter. For the drop population, it is necessary to define an average drop diameter as weight-mean, mean-volume, surface-mean, mean-surface, length-mean, or mean-length diameter [43]. For mass-transfer calculations, the surface-mean diameter, d_{vs} (also called the *Sauter mean diameter*), is appropriate because it is the mean drop diameter that gives the same interfacial surface area as the entire population of drops for the same mass of drops. It is determined from drop-size distribution data for N drops by:

$$\frac{\pi d_{vs}^2}{(\pi/6)d_{vs}^3} = \frac{\pi \sum_N d_e^2}{(\pi/6) \sum_N d_e^3}$$

which, when solved for d_{vs}, gives

$$d_{vs} = \frac{\sum_N d_e^3}{\sum_N d_e^2} \qquad (8\text{-}35)$$

With this definition, the interfacial surface area per unit volume of a two-phase mixture is

$$a = \frac{\pi N d_{vs}^2 \phi_D}{\pi N d_{vs}^3/6} = \frac{6\phi_D}{d_{vs}} \qquad (8\text{-}36)$$

Equation (8-36) is used to estimate the interfacial area, a, from a measurement of d_{vs} or vice versa. Early experimental investigations, such as those of Vermeulen, Williams, and Langlois [44], found that d_{vs} depends on a Weber number:

$$N_{\text{We}} = \frac{\text{(inertial force)}}{\text{(interfacial tension force)}} = \frac{D_i^3 N^2 \rho_C}{\sigma} \qquad (8\text{-}37)$$

High Weber numbers give small droplets and high interfacial areas. Gnanasundaram, Degaleesan, and Laddha [45] correlated d_{vs} over a wide range of N_{We}. Below $N_{\text{We}} = 10,000$, d_{vs} depends on dispersed-phase holdup, ϕ_D, because of coalescence effects. For $N_{\text{We}} > 10,000$, inertial forces dominate so that coalescence effects are less prominent and d_{vs} is almost independent of holdup up to $\phi_D = 0.5$. The correlations are

$$\frac{d_{vs}}{D_i} = 0.052(N_{\text{We}})^{-0.6}e^{4\phi_D}, N_{\text{We}} < 10,000 \qquad (8\text{-}38)$$

$$\frac{d_{vs}}{D_i} = 0.39(N_{\text{We}})^{-0.6}, N_{\text{We}} > 10,000 \qquad (8\text{-}39)$$

Typical values of N_{We} for industrial extractors are less than 10,000, so (8-38) applies. Values of d_{vs}/D_i are frequently in the range of 0.0005 to 0.01.

Studies like those of Chen and Middleman [46] and Sprow [47] show that dispersion in an agitated vessel is dynamic. Droplet breakup by turbulent pressure fluctuations dominates near the impeller blades, while for reasonable dispersed-phase holdup, coalescence of drops by collisions dominates away from the impeller. Thus, there is a distribution of drop sizes, with smaller drops in the vicinity of the impeller and larger drops elsewhere. When both drop breakup and coalescence occur, the drop-size distribution is such that $d_{min} \approx d_{vs}/3$ and $d_{max} \approx 3d_{vs}$. Thus, the drop size varies over about a 10-fold range, approximating a Gaussian distribution.

EXAMPLE 8.6 Droplet Size and Interfacial Area.

For the conditions and results of Example 8.5, with the extract phase dispersed, estimate the Sauter mean drop diameter, the range of drop sizes, and the interfacial area.

Solution

$$D_i = 1 \text{ ft}; \quad N = 147 \text{ rpm} = 8,820 \text{ rph};$$
$$\rho_C = 62.3 \text{ lb/ft}^3; \quad \sigma = 718,800 \text{ lb/h}^2$$

From (8-37),
$$N_{We} = (1)^3(8,820)^2(62.3)/718,800 = 6,742; \quad \phi_D = 0.388$$

From (8-38),
$$d_{vs} = (1)(0.052)(6,742)^{-0.6} \exp[4(0.388)] = 0.00124 \text{ ft}$$
or
$$(0.00124)(12)(25.4) = 0.38 \text{ mm}$$

$$d_{min} = d_{vs}/3 = 0.126 \text{ mm}; \quad d_{max} = 3d_{vs} = 1.134 \text{ mm}$$

From (8-36), $a = 6(0.388)/0.00124 = 1,880 \text{ ft}^2/\text{ft}^3$

Mass-Transfer Coefficients

For many reasons, mass transfer in agitated liquid–liquid systems is complex (1) in the dispersed-phase droplets, (2) in the continuous phase, and (3) at the interface. The magnitude of k_D depends on drop diameter, solute diffusivity, and fluid motion within the drop. According to Davies [48], when drop diameter is small (less than 1 mm, interfacial tension is high (> 15 dyne/cm), and trace amounts of surface-active agents are present, droplets are rigid (internally stagnant) and behave like solids. As droplets enlarge, interfacial tension decreases, surface-active agents become ineffective, and internal toroidal fluid circulation patterns, caused by viscous drag of the continuous phase, appear within the drops. For larger-diameter drops, the shape of the drop oscillates between spheroid and ellipsoid or other shapes.

Continuous-phase mass-transfer coefficients, k_C, depend on the motion between the droplets and the continuous phase, and whether the drops are forming, breaking, or coalescing. Interfacial movements or turbulence, called *Marangoni effects*, occur due to interfacial-tension gradients, which induce increases in mass-transfer rates.

A conservative estimate of the overall mass-transfer coefficient, K_{OD}, in (8-28) can be made from estimates of k_D and k_C by assuming rigid drops, the absence of

Marangoni effects, and a stable drop size. For k_D, the asymptotic steady-state solution for mass transfer in a rigid sphere with negligible surrounding resistance is given by Treybal [25] as

$$(N_{Sh})_D = \frac{k_D d_{vs}}{D_D} = \frac{2}{3}\pi^2 = 6.6 \qquad (8\text{-}40)$$

where D_D is the solute diffusivity in the droplet and N_{Sh} is the Sherwood number.

Exercise 3.31 in Chapter 3 for diffusion from the surface of a sphere into an infinite, quiescent fluid gives the continuous-phase Sherwood number as:

$$(N_{Sh})_C = \frac{k_C d_{vs}}{D_C} = 2 \qquad (8\text{-}41)$$

where D_C is the continuous-phase solute diffusivity. However, if other spheres of equal diameter are located near the sphere of interest, $(N_{Sh})_C$ may decrease to a value as low as 1.386, according to Cornish [49]. In an agitated vessel, $(N_{Sh})_C > 1.386$. An estimate can be made with the correlation of Skelland and Moeti [50], which fits data for three different solutes, three different dispersed organic solvents, and water as the continuous phase. Mass transfer was from the dispersed phase to the continuous phase, but only for $\phi_D = 0.01$. They assumed an equation of the form

$$(N_{Sh})_C \propto (N_{Re})_C^y (N_{Sc})_C^x \qquad (8\text{-}42)$$

where
$$(N_{Sh})_C = k_C d_{vs}/D_C \qquad (8\text{-}43)$$

$$(N_{Sc})_C = \mu_C/\rho_C D_C \qquad (8\text{-}44)$$

For the Reynolds number, they assumed the characteristic velocity to be the square root of the mean-square, local fluctuating velocity in the droplet vicinity based on the theory of local isotropic turbulence of Batchelor [51]:

$$\bar{u}^2 \propto \left(\frac{Pg_c}{V}\right)^{2/3} \left(\frac{d_{vs}}{\rho_C}\right)^{2/3} \qquad (8\text{-}45)$$

Thus,
$$(N_{Re})_C = \frac{(\bar{u}^2)^{1/2} d_{vs}\rho_C}{\mu_C} \qquad (8\text{-}46)$$

Combining (8-45) and (8-46) and omitting the proportionality constant, $\propto$:

$$(N_{Re})_C = \frac{d_{vs}^{4/3} \rho_C^{2/3} (Pg_c/V)^{1/3}}{\mu_C} \qquad (8\text{-}47)$$

As discussed previously in conjunction with Figure 8.36, in the turbulent-flow region,

$$Pg_c \propto \rho_M N^3 D_i^5 \text{ or for low } \phi_D, Pg_c/V \propto \rho_C N^3 D_i^5 D_T^3$$

Thus,
$$(N_{Re})_C = \frac{d_{vs}^{4/3} \rho_C N D_i^{5/3}}{\mu_C D_T} \qquad (8\text{-}48)$$

Skelland and Moeti correlated the mass-transfer coefficient data with

$$k_C \propto D_C^{2/3} \mu_C^{-1/3} N^{3/2} d_{vs}^0$$

The exponents in this proportionality determine the y and x exponents in (8-42) as $\frac{2}{3}$ and $\frac{1}{3}$. In addition, based on previous investigations, a droplet Eotvos number,

$$N_{Eo} = \rho_D d_{vs}^2 g / \sigma \qquad (8\text{-}49)$$

where N_{Eo} = (gravitational force)/(surface tension force) and the dispersed-phase holdup, ϕ_D, are incorporated into the final correlation, which predicts 180 data points to an average deviation of 19.71%:

$$(N_{Sh})_C = \frac{k_C d_{vs}}{D_C} = 1.237 \times 10^{-5} \left(\frac{\mu_C}{\rho_C D_C}\right)^{1/3}$$

$$\times \left(\frac{D_i^2 N \rho_C}{\mu_C}\right)^{2/3} \phi_D^{-1/2} \left(\frac{D_i N^2}{g}\right)^{5/12}$$

$$\times \left(\frac{D_i}{d_{vs}}\right)^2 \left(\frac{d_{vs}}{D_T}\right)^{1/2} \left(\frac{\rho_D d_{vs}^2 g}{\sigma}\right)^{5/4} \qquad (8\text{-}50)$$

EXAMPLE 8.7 Mass Transfer in a Mixer.

For the system, conditions, and results of Examples 8.5 and 8.6, with the extract as the dispersed phase, estimate the:

(a) dispersed-phase mass-transfer coefficient, k_D;

(b) continuous-phase mass-transfer coefficient, k_C;

(c) Murphree dispersed-phase efficiency, E_{MD}; and

(d) fractional extraction of furfural.

The molecular diffusivities of furfural in toluene (dispersed) and water (continuous) at dilute conditions are $D_D = 8.32 \times 10^{-5}$ ft²/h and $D_C = 4.47 \times 10^{-5}$ ft²/h. The distribution coefficient for dilute conditions is $m = dc_C/dc_D = 0.0985$.

Solution

(a) From (8-40), $k_D = 6.6(D_D)/d_{vs} = 6.6(8.32 \times 10^{-5})/0.00124 = 0.44$ ft/h.

(b) To apply (8-50) to the estimation of k_C, first compute each dimensionless group:

$N_{Sc} = \mu_C/\rho_C D_C = 2.165/[(62.3)(4.47 \times 10^{-5})] = 777$
$N_{Re} = D_i^2 N \rho_C/\mu_C = (1)^2(8,820)(62.3)/2.165 = 254,000$
$N_{Fr} = D_i N^2/g = (1)(8,820)^2/(4.17 \times 10^8) = 0.187$
$D_i/d_{vs} = 1/0.00124 = 806; \quad d_{vs}/D_T = 0.00124/3 = 0.000413$
$N_{Eo} = \rho_D d_{vs}^2 g/\sigma = (54.2)(0.00124)^2(4.17 \times 10^8)/718,800$
$\quad = 0.0483$

From (8-50),

$N_{Sh} = 1.237 \times 10^{-5}(777)^{1/3}(254,000)^{2/3}(0.388)^{-1/2}(0.187)^{5/12}$

$\quad \times (806)^2(0.000413)^{1/2}(0.0483)^{5/4} = 109$

which is much greater than 2 for a quiescent fluid.

$k_C = N_{Sh} D_C/d_{vs} = (109)(4.47 \times 10^{-5})/0.00124 = 3.93$ ft/h

(c) From (8-28) and the results of Example 8.6,

$$K_{OD}a = \left\{ \frac{1}{1/0.44 + 1/[(0.0985)(3.93)]} \right\} 1,880 = 387\,h^{-1}$$

From (8-32), with $V = \pi D_T^2 H/4 = (3.14)(3)^2(3)/4 = 21.2$ ft²,

$$N_{OD} = K_{OD}aV/Q_D = 387(212)/207 = 39.6$$

From (8-33),

$$E_{MD} = N_{OD}/(1 + N_{OD}) = 39.6/(1 + 39.6) = 0.975 = 97.5\%$$

(d) By material balance,

$$Q_C(c_{C,in} - c_{C,out}) = Q_D c_{D,out} \qquad (1)$$

From (8-26), $\quad E_{MD} = c_{D,out}/c_D^* = m c_{D,out}/c_{C,out} \qquad (2)$

Combining (1) and (2) to eliminate $c_{D,out}$ gives

$$\frac{c_{C,out}}{c_{C,in}} = \frac{1}{1 + Q_D E_{MD}/(Q_C m)} \qquad (3)$$

and

$$f_{Extracted} = \frac{c_{C,in} - c_{C,out}}{c_{C,in}} = 1 - \frac{c_{C,out}}{c_{C,in}} = \frac{Q_D E_{MD}/(Q_C m)}{1 + Q_D E_{MD}/(Q_C m)}$$

$$\frac{Q_D}{Q_C} \frac{E_{MD}}{m} = \frac{(207)(0.975)}{(327)(0.0985)} = 6.27$$

Thus, $\quad f_{Extracted} = \dfrac{6.27}{1 + 6.27} = 0.862 \quad$ or $\quad 86.2\%$

---

§8.5.2 Column Extractors

An extraction column is sized by determining its diameter and height. Column diameter must be large enough to permit the liquid phases to flow through the column counter-currently without flooding. Column height must allow sufficient stages to achieve the desired extraction. A number of scale-up and design procedures have been published, the most detailed being the stagewise model of Kumar and Hartland [105], for the Kuhni, rotating-disk (RDC), pulsed perforated-plate, and Karr reciprocating-plate columns, described in §8.1.5. Their model considers drop-size distribution, droplet breakage and coalescence, drop velocities, dispersed-phase holdup, and backflow in the continuous phase. Only less detailed models are considered here.

For small columns, rough estimates of the diameter and height can be made using the results of Stichlmair [52] with toluene–acetone–water for $Q_D/Q_C = 1.5$. Typical ranges of l/HETS and the sum of the superficial phase velocities for some extractor types are given in Table 8.6.

Table 8.6 Performance of Several Types of Column Extractors

Extractor Type	l/HETS, m⁻¹	$U_D + U_C$, m/h
Packed column	1.5–2.5	12–30
Pulsed packed column	3.5–6	17–23
Sieve-plate column	0.8–1.2	27–60
Pulsed-plate column	0.8–1.2	25–35
Scheibel column	5–9	10–14
RDC	2.5–3.5	15–30
Kuhni column	5–8	8–12
Karr column	3.5–7	30–40
RTL contactor	6–12	1–2

Source: J. Stichlmair, *Chemie-Ingenieur-Technik*, **52**, 253 (1980).

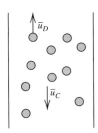

Figure 8.37 Countercurrent flows of dispersed and continuous liquid phases in a column.

Column Diameter. Because of the many variables involved, an accurate assessment of column diameter for extractors is far more uncertain than that for vapor–liquid contactors. Extractor variables include phase-flow rates, phase-density differences, interfacial tension, direction of mass transfer, continuous-phase viscosity and density, rotating or reciprocating speed, and geometry of internals. Column diameter is best determined by scale-up from laboratory or pilot-plant units. The sum of the absolute superficial velocities of the liquid phases in the test unit is assumed to hold for commercial units. This sum is often expressed in gal/h-ft^2 of column cross-sectional area.

In the absence of laboratory data, preliminary estimates of column diameter can be made by a simplification of the theory of Logsdail, Thornton, and Pratt [53], which is compared to other procedures by Landau and Houlihan [54] for rotating-disk contactors. Because of the relative motion between a dispersed-droplet phase and a continuous phase, this theory is based on a concept similar to that in §6.6.1 for liquid droplets dispersed in a vapor.

Consider the case of droplets of the lower-density rising through the denser, downward-flowing, continuous liquid phase, as in Figure 8.37. If the average superficial velocities of the droplet phase and the continuous phase are U_D upward and U_C downward (i.e., both velocities are positive), average velocities relative to the column wall are

$$\bar{u}_D = \frac{U_D}{\phi_D} \tag{8-51}$$

and

$$\bar{u}_C = \frac{U_C}{1 - \phi_D} \tag{8-52}$$

The average droplet-rise velocity relative to the continuous phase is the sum of (8-51) and (8-52):

$$\bar{u}_r = \frac{U_D}{\phi_D} + \frac{U_C}{1 - \phi_D} \tag{8-53}$$

This relative (*slip*) velocity is expressed as a modified form of (6-40), where the continuous-phase density in the buoyancy term is replaced by the mixture density, ρ_M. Thus, noting that drag force, F_d, and gravitational force, F_g, act downward while buoyancy, F_b, acts upward,

$$\bar{u}_r = C \left(\frac{\rho_M - \rho_D}{\rho_C} \right)^{1/2} f\{1 - \phi_D\} \tag{8-54}$$

where C is the parameter in (6-41) and $f\{1 - \phi_D\}$ is a factor that allows for hindered rising due to neighboring droplets. The density ρ_M is a volumetric mean given by

$$\rho_M = \phi_D \rho_D + (1 - \phi_D)\rho_C \tag{8-55}$$

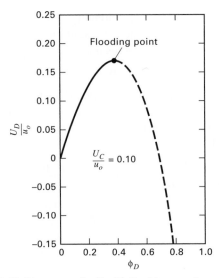

Figure 8.38 Holdup curve for liquid–liquid extraction column.

$$\rho_M - \rho_D = (1 - \phi_D)(\rho_C - \rho_D) \tag{8-56}$$

Substitution of (8-56) into (8-57) yields

$$\bar{u}_r = C \left(\frac{\rho_C - \rho_D}{\rho_C} \right)^{1/2} (1 - \phi_D)^{1/2} f\{1 - \phi_D\} \tag{8-57}$$

Gayler, Roberts, and Pratt [55] expressed the RHS of (8-57) as

$$\bar{u}_r = u_0(1 - \phi_D) \tag{8-58}$$

where u_0 is a characteristic rise velocity for a single droplet, which is a function of all the variables discussed above, except those on the RHS of (8-53). Thus, for a given liquid–liquid system, column design, and operating conditions, combining (8-53) and (8-58) gives

$$\frac{U_D}{\phi_D} + \frac{U_C}{1 - \phi_D} = u_0(1 - \phi_D) \tag{8-59}$$

If u_0 is a constant, (8-59) is cubic in ϕ_D, with a typical solution shown in Figure 8.38 for $U_C/u_0 = 0.1$. Thornton [56] argues that, with U_C fixed, an increase in U_D results in an increased value of the holdup ϕ_D, until flooding is reached, at which $(\partial U_D/\partial \phi_D)_{U_C} = 0$. Thus, in Figure 8.38, only that portion of the curve for $\phi_D = 0$ to $(\phi_D)_f$, the holdup at the flooding point, is observed in practice. Alternatively, with U_D fixed, $(\partial U_C/\partial \phi_D)_{U_D} = 0$ at the flooding point. If these derivatives are applied to (8-59),

$$U_C = u_0[1 - 2(\phi_D)_f][1 - (\phi_D)_f]^2 \tag{8-60}$$

$$U_D = 2u_0[1 - (\phi_D)_f](\phi_D)_f^2 \tag{8-61}$$

where the subscript f denotes flooding. Combining (8-60) and (8-61) to eliminate u_0 gives

$$(\phi_D)_f = \frac{[1 + 8(U_C/U_D)]^{0.5} - 3}{4[(U_C/U_D) - 1]} \tag{8-62}$$

This predicts values of $(\phi_D)_f$ ranging from 0 at $U_D/U_C = 0$ to 0.5 at $U_C/U_D = 0$. At $U_D/U_C = 1$, $(\phi_D)_f = 1/3$. The

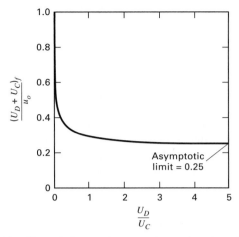

Figure 8.39 Effect of phase ratio on total capacity of a column.

simultaneous solution of (8-59) and (8-62) results in Figure 8.39 for variation of capacity as a function of phase-flow ratio. The largest capacities occur at the smallest ratios of dispersed-phase flow rate to continuous-phase flow rate.

For fixed column geometry and rotor speed, data of Logsdail et al. [53] for a laboratory-scale RDC indicate that the dimensionless group $(u_0 \mu_C \rho_C / \sigma \Delta\rho)$ is approximately constant. For well-designed and operated commercial RDC columns ranging from 8 to 42 inches in diameter, the data of Reman and Olney [57] and Strand, Olney, and Ackerman [58] indicate that this dimensionless group has a value of 0.01 for systems involving water as the continuous or dispersed phase. This value is suitable for preliminary calculations of RDC column diameter, when the sum of the actual superficial phase velocities is taken as 50% of the sum at flooding.

EXAMPLE 8.8 Diameter of an RDC Extractor.

Estimate the diameter of an RDC to extract acetone from a dilute toluene–acetone solution into water at 20°C. The flow rates for the dispersed organic and continuous aqueous phases are 27,000 and 25,000 lb/h, respectively.

Solution

The physical properties are

$$\mu_C = 1.0 \, \text{cP}(0.000021 \, \text{lb}_f\text{-s/ft}^2) \text{and} \, \rho_C = 1.0 \, \text{g/cm}^3$$

$$\Delta\rho = 0.14 \, \text{g/cm}^3 \text{and} \, \sigma = 32 \, \text{dyne/cm}(0.00219 \, \text{lb}_f/\text{ft})$$

$$\frac{U_D}{U_C} = \left(\frac{27,000}{25,000}\right)\left(\frac{\rho_C}{\rho_D}\right) = \left(\frac{27,000}{25,000}\right)\left(\frac{1.0}{0.86}\right) = 1.26$$

From Figure 8.39, $(U_D + U_C)_f/u_0 = 0.29$.
Assume that $u_0 \mu_C \rho_C / \sigma \Delta\rho = 0.01$. Therefore,

$$u_0 = \frac{(0.01)(0.00219)(0.14)}{(0.000021)(1.0)} = 0.146 \, \text{ft/s}$$

$$(U_D + U_C)_f = 0.29(0.146) = 0.0423 \, \text{ft/s}$$

$$(U_D + U_C)_{50\% \text{ of flooding}} = \left(\frac{0.0423}{2}\right)(3,600) = 76.1 \, \text{ft/h}$$

$$\text{Total ft}^3/\text{h} = \frac{27,000}{(0.86)(62.4)} + \frac{25,000}{(1.0)(62.4)} = 904 \, \text{ft}^3/\text{h}$$

$$\text{Column cross-sectional area} = A_c = \frac{904}{76.1} = 11.88 \, \text{ft}^2$$

$$\text{Column diameter} = D_T = \left(\frac{4A_c}{\pi}\right)^{0.5} = \left[\frac{(4)(11.88)}{3.14}\right]^{0.5} = 3.9 \, \text{ft}$$

Note that from Table 8.6, a typical $(U_D + U_C)$ for an RDC is 15 to 30 m/h or 49 to 98.4 ft/h.

Column Height. Despite compartmentalization, mechanically assisted liquid–liquid extraction columns, such as the RDC and Karr columns, operate more like differential devices than staged contactors. Therefore, it is common to consider stage efficiency for such columns in terms of HETS or as some function of mass-transfer parameters, such as HTU. While not theoretically based, HETS is preferred because it can be used to determine column height from the number of equilibrium stages.

The large number of variables that influence efficiency have made general correlations for HETS difficult to develop. However, for well-designed and efficiently operated columns, data indicate that the dominant physical properties influencing HETS are interfacial tension, viscosities, and density difference between the phases. In addition, observations by Reman [59] for RDC units, and by Karr and Lo [60] for Karr columns, show that HETS increases with increasing column diameter because of axial mixing, discussed in §8.5.5.

A prudent procedure for determining column height is to obtain values of HETS from small-scale laboratory experiments and scale these values to commercial-size columns by assuming that HETS varies with column diameter D_T raised to an exponent that may vary from 0.2 to 0.4, depending on the system.

In the absence of data, the crude correlation of Figure 8.40 can be used for preliminary design if phase viscosities are below 1 cP. The data correspond to minimum reported HETS values for RDC and Karr units, with the exponent on the diameter set to $\frac{1}{3}$. The points represent values of HETS that

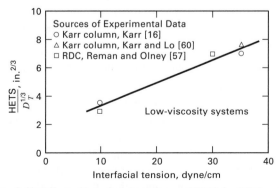

Figure 8.40 Effect of interfacial tension on HETS for RDC and Karr columns.

vary from as low as 6 inches for a 3-inch-diameter column operating with a low-interfacial-tension/low-viscosity system, to as high as 25 inches for a 36-inch-diameter column operating with a high-interfacial-tension/low-viscosity system such as xylenes–acetic acid–water. For systems having one phase of high viscosity, values of HETS can be 24 inches or more, even for a laboratory-size column.

EXAMPLE 8.9   HETS for an RDC Extractor.

Estimate HETS for the conditions of Example 8.8.

Solution

Because toluene has a viscosity of approximately 0.6 cP, this is a low-viscosity system. From Example 8.8, the interfacial tension is 32 dyne/cm. From Figure 8.40, $\text{HETS}/D_T^{1/3} = 6.9$. For $D_T = 3.9$ ft, $\text{HETS} = 6.9[(3.9)(12)]^{1/3} = 24.8$ inches. Note that from Table 8.6, HETS for an RDC varies from 0.29 to 0.40 m or from 11.4 to 15.7 inches for a small column.

More accurate estimates of flooding and HETS are discussed in detail by Lo et al. [4] and by Thornton [61]. Packed-column design is considered by Strigle [62].

§8.5.3   Scale-up of Karr Columns

Reciprocating-plate extraction columns used for bioproduct separations, such as Karr columns, use a stack of closely spaced, vertically reciprocating or vibrating plates to interdisperse liquids that contain suspended solids and easily emulsified mixtures. Amplitude and frequency of reciprocation vary from 3 to 50 mm and up to 1,000 strokes per minute, respectively. Low axial mixing and nearly uniform isotropic turbulence are produced. Scale-up, based on column diameter and rate of reciprocation, maintains plate spacing, solvent-to-feed ratio, stroke length, and throughput per column cross-sectional area (solvent plus feed).

Scale-up methods commonly employed for Karr columns and Podbielniak centrifugal extractors, another bioseparation tool, are based on dilute amounts of bioproduct in the feed, so the partition coefficient can be assumed constant. For the Karr column, the following procedure, based on studies by Karr et al. [107], is useful when data from small columns of 1–3 inches in diameter are scaled up to industrial columns of up to 1.5-m (5-ft) diameter, with a plate-stack height of up to 12.2 m (40 ft).

1. Determine the partition coefficient, K_D, for the solute between the selected solvent and the feed broth.

2. Establish the desired extent of extraction of the solute.

3. Conduct experiments in a small Karr column of known inside diameter, D, and plate-stack height, H, varying the broth and solvent volumetric flow rates, F and S, and the stroke speed per minute, SPM. For each run that achieves or approaches the desired extent of extraction, compute the number of theoretical stages,

N, and the HETS. From those results, select the operating conditions that achieve the highest volumetric efficiency, defined by

Volumetric efficiency

$$= \frac{\text{total volumetric flow rate/column cross-sectional area}}{\text{HETP}}$$

4. Let subscript 1 refer to the optimal operating conditions in step 3 for the small column, and let subscript 2 refer to the scaled-up conditions for the commercial extractor. Then, for a given F_2, use the following scale-up equations to compute the remaining conditions for the commercial unit:

    (a) Use the same solvent-to-feed ratio. Thus, $S_2 = S_1(F_2/F_1)$.

    (b) Use the same total volumetric flow rate/column cross-sectional area. Thus,

    $$D_2 = D_1 \left(\frac{S_2 + F_2}{S_1 + F_1} \right)^{1/2}$$

    (c) Use a conservative estimate of the effect of column diameter on HETS,

    $$\text{HETS}_2 = \text{HETS}_1 \left(\frac{D_2}{D_1} \right)^{0.38}$$

    (d) For the same number of theoretical stages ($N_2 = N_1$), compute the plate-stack height from

    $$H_2 = H_1 \left(\frac{\text{HETS}_2}{\text{HETS}_1} \right) = \text{HETS}_2(N)$$

    (e) Compute the stroke speed from

    $$\text{SPM}_2 = \text{SPM}_1 \left(\frac{D_1}{D_2} \right)^{0.14}$$

EXAMPLE 8.10   Scale-up of a Karr Column for a Bioseparation.

A pharmaceutical company has determined the best operating conditions for a bioreactor. The company produces a whole broth, containing 4 wt% solids with a bioproduct concentration of 1.4 g/L, at volumetric flow rates of up to 200 mL/minute. To separate the whole broth and bioproducts, the company is considering processing the whole broth, without first removing the solids, in a Karr extractor. Process viability has been confirmed by tests in a small glass Karr column, which has an inside diameter of 2.54 cm and a plate-stack height of 3.05 m. The plates, of stainless steel with 8-mm diameter holes and a 62% open area, are on 51-mm plate spacings. The stroke height is 19.1 mm and the SPM can be varied from 220 to 320. The tests were conducted with a suitable solvent, chloroform, which has a partition coefficient for the bioproduct of 2.68, which is constant in the dilute region. The tests sought to extract 99.5% of the bioproduct.

The optimal test conditions for this extent of extraction were found to be $F_1 = 120$ mL/minute and $S_1 = 180$ mL/minute at $\text{SPM}_1 = 250$. Scale these results up to a commercial unit that can process 1,325 L/h of feed.

Solution

For the same S/F ratio, the solvent rate, S_2, for the commercial unit is: $180(1,325/120) = 1,990$ L/h, and $S_2 + F_2 = 1,990 + 1,325 = 3,315$ L/h. Also, $S_1 + F_1 = 180 + 120 = 300$ mL/min $= 18$ L/h.

The diameter for the commercial unit $= D_2 = 2.54(3,315/18)^{0.50} = 34.5$ cm $= 0.345$ m. For this S/F ratio and 99.5% extraction (0.5% unextracted), compute the number of theoretical stages. The extraction factor from (5-14), using volumetric units, is $E = K_D S/F = 2.68(1,990/1,325) = 4.03$.
From (5-29), the unextracted fraction

$$= 0.005 = \frac{E-1}{E^{N+1}-1} = \frac{4.03-1}{4.03^{N+1}-1}.$$

Solving, $N = 3.6$ stages.

For the small unit, $\text{HETS}_1 = H_1/N = 3.05/3.6 = 0.85$ m/stage.

For the commercial unit, $\text{HETS}_2 = 0.85(0.345/0.0254)^{0.38} = 2.3$ m/stage.

Stacked-plate height of the commercial unit $= 2.3(3.6) = 8.3$ m.

Stroke speed of the commercial unit $= \text{SPM}_2 = 250(0.0254/0.345)^{0.14} = 174$.

§8.5.4 Scale-up of Podbielniak Centrifugal Extractors (PODs)

Centrifugal extractors were developed for bioproduct separations to avoid emulsions and rapidly recover unstable antibiotics from liquid phases with small (<0.01 g/cm^3) density differences. Bioseparations remain the primary application for PODs, where heavy and light phases are fed to the extractor near the axis and periphery, respectively, of a stack of perforated, concentric cylinders rotating around a shaft, as shown in Figure 8.7(l). Centrifugal force rapidly drives the heavy (light) phase radially outward (inward) to effect countercurrent contacting, extract the product, and return it to another aqueous phase (e.g., phosphate buffer) within minutes. Attainable dimensionless g-force (G), quantified below, increases from 2,300 (2,100 rpm) at a capacity of 30 m^3/h (sum of feed and solvent rates) to 11,400 (10,000 rpm) at a capacity of 0.05 m^3/h. During scale-up, the equivalent time, Gt, and the ratio of feed volume to solvent volume are kept constant, while the large-scale feed rate is calculated from the contact time.

The following is a method for the scale-up of Podbielniak centrifugal extractors using data from a small unit and a criterion from Harrison et al. [69]:

1. Determine the partition coefficient, K_D, for the solute between the selected solvent and the feed broth.

2. Establish the desired extent of extraction of the solute.

3. Conduct experiments in a small POD of known diameter, D, and length, L, of the rotating cylinder, varying the broth and solvent volumetric flow rates, F and S, and the rotation rate, rpm. For each run that achieves or approaches the desired extent of extraction, compute the number of theoretical stages, N, and a nominal

residence time, t_{nom}, of the combined feed and solvent using the dimensions of the rotating cylinder, thus ignoring the volume of the central shaft and perforated plates. Thus,

$$t_{\text{nom}} = \frac{\pi(D^2/4)L}{F+S}$$

From those results, select the operating conditions that achieve the highest volumetric throughput.

4. Let subscript 1 refer to the optimal operating conditions in step 3 for the small POD, and let subscript 2 refer to the scaled-up conditions for a commercial POD. For each of the commercial POD extractors, compute the volumetric flow rates, F_2 and S_2, and select the POD that will handle the required commercial flow rates using the following equations:

   (a) Use the same solvent-to-feed ratio. Thus, $S_2/F_2 = S_1/F_1$.

   (b) Compute ω, the angular velocity in rad/s for the small POD from

   $$\omega_1 = \frac{2\pi \dfrac{\text{rad}}{\text{rev}}(\text{rpm}_1)}{60 \dfrac{\text{s}}{\text{minute}}}$$

   (c) Compute the dimensionless acceleration, G, for the small POD from the ratio of centrifugal to gravitational acceleration:

   $$G_1 = \frac{\omega_1^2 \left(\dfrac{D_1}{2}\right)}{g}, \text{ where } g = 9.807 \text{ m}^2/\text{s}$$

   (d) For a given commercial unit, compute ω_2 and G_2 from rpm_2 and D_2.

   (e) Assume that the equivalent time, Gt_{nom}, for the small POD is equal to that for the commercial POD. Compute the nominal residence time for the commercial POD from

   $$(t_{\text{nom}})_2 = (t_{\text{nom}})_1 \frac{G_1}{G_2}$$

   (f) For the commercial POD, using $(t_{\text{nom}})_2$, compute F_2 and S_2 and compare to the required values.

EXAMPLE 8.11   Scale-up of a POD for a Bioseparation.

Chloramphenicol is an oral antibiodic sometimes used to treat serious infections such as typhoid fever and cholera. It can be derived from a bacterium or made synthetically. A process is being designed to extract 95% of it from a clarified broth, where the extraction time must be of the order of a few minutes. The solvent is *n*-amyl acetate and the partition coefficient is 16. Optimal data have been obtained from a small POD having a rotating cylinder of 20-cm diameter and 2.5-cm length, operating at 9,000 rpm with a volumetric feed-to-solvent ratio of 4 and a broth feed rate of 920 mL/minute. A commercial-size POD is available with a rotating cylinder of 91-cm diameter and 91-cm length, operating at a maximum rpm of 2,100. Compute the broth feed rate that could be handled by this commercial POD.

Solution

For the optimal run with the small POD, compute the number of theoretical stages, N, for an F/S ratio of 4 and 95% extraction (5% unextracted). The extraction factor from (5-14), using volumetric units, is

$$E = KS/F = 16/4 = 4$$

From (5-29), the fraction unextracted $= 0.5 = \dfrac{E-1}{E^{N+1}-1} = \dfrac{4-1}{4^{N+1}-1}$.

Solving, $N = 1.97$ stages.
Calculate the nominal residence time for the small POD, with solvent rate, S, of $920/4 = 230$ mL/min.

$$t_{\text{nom}} = \frac{3.14\left(20^2/4\right)2.5}{920 + 230} = 0.683 \text{ minute}$$

Compute the angular velocity, ω, and dimensionless acceleration, G, for the small POD:

$$\omega_1 = \frac{2(3.14)(9,000)}{60} = 942 \text{ rad/s and } G_1 = \frac{942^2\left(\dfrac{20/100}{2}\right)}{9.807} = 9,050$$

Compute ω and G for the commercial POD for a 91-cm diameter and 2,100 rpm:

$$\omega_2 = \frac{2(3.14)(2,100)}{60} = 220 \text{ rad/s and } G_2 = \frac{220^2\left(\dfrac{91/100}{2}\right)}{9.807} = 2,240$$

Assuming $G_2(t_{\text{nom}})_2 = G_1(t_{\text{nom}})_1$, calculate $(t_{\text{nom}})_2$:

$$(t_{\text{nom}})_2 = 0.683\left(\frac{9,050}{2,240}\right) = 2.76 \text{ minutes}$$

Compute the volume of the rotating cylinder for the commercial unit:

$$V_2 = \frac{\pi D_2^2 L_2}{4} = \frac{3.14(91)^2 91}{4} = 592,000 \text{ cm}^3$$

Compute the volumetric throughput, $F_2 + S_2$, from V_2 and the nominal residence time:

$$F_2 + S_2 = \frac{V_2}{(t_{\text{nom}})_2} = \frac{592,000}{2.76} = 214,000 \text{ cm}^3/\text{minute}$$

Therefore, $F_2 = (4/5)214,000 = 172,000 \text{ cm}^3/\text{minute or } 10.3 \text{ m}^3/\text{h}$.

More rigorous design methods for centrifuges and cyclones used for particle and phase separations are developed in Chapter 19.

§8.5.5 Axial Dispersion

Because axial concentration gradients in the direction of bulk flow are established in each phase in column extractors, diffusion of a species is superimposed on its bulk flow in that phase. Axial diffusion degrades efficiency of the separation equipment, and in the limit, a multistage separator behaves like a single equilibrium stage. In Figure 8.41, solute concentration profiles for the extract and raffinate phases are shown for plug flow (dashed lines) and for flow with significant axial diffusion in the two phases (solid lines). The continuous phase is the feed/raffinate (x subscript), which enters at the top ($z = 0$). The dispersed phase is the solvent/extract (y subscript), which enters at the bottom ($z = H$). Solute transfer is

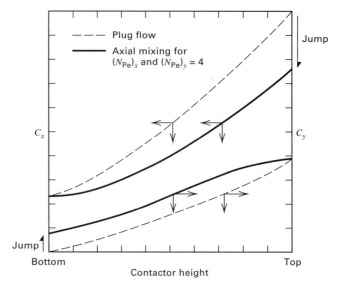

Figure 8.41 Solute concentration profiles for continuous, countercurrent extraction with and without axial mixing.

from the continuous phase to the dispersed phase. Two effects of axial diffusion are seen: (1) concentration curves in the presence of axial diffusion are closer than they are for plug flow and (2) these close proximities are due partially to concentrations at the two ends, which are different from those in the original feed and solvent. These differences are *jumps* due to axial diffusion outside the region in the contactor where the liquid phases are in contact. The top jump is caused by axial diffusion in the feed liquid before it enters the contactor. This causes the concentration of solute in the entering feed to be less than its concentration in the original feed liquid. Similarly, diffusion of solute into the incoming solvent causes the concentration of solute in the entering solvent to be greater than the concentration in the original solvent, which in Figure 8.41 is 0. The overall effect of axial diffusion is a reduction in the average driving force for solute mass transfer between the two phases, necessitating a taller column to accomplish the separation.

The effects shown in Figure 8.41 are due to factors besides diffusion, which are lumped together into an overall effect referred to as *axial dispersion, axial mixing, longitudinal dispersion,* or *backmixing*. These factors include: (1) molecular and turbulent diffusion of the continuous phase along concentration gradients; (2) circulatory motion of the continuous phase due to the droplets of the dispersed phase; (3) transport and shedding of the continuous phase in the wakes attached to the rear of dispersed droplets; (4) circulation of continuous and dispersed phases in mechanically agitated columns; and (5) channeling and nonuniform velocity profiles leading to distributions of residence times in the two phases.

In general, the effect of axial dispersion is most pronounced when (1) a high recovery of solute is desired, (2) the contactor is short, (3) large circulation patterns occur, (4) a range of droplet sizes is present, and/or (5) the feed-to-solvent flow ratio is very small or very large. Axial dispersion is negligible in extractors where phase separation occurs between stages, such as in mixer-settler cascades and sieve-plate columns with downcomers, but it can be significant in

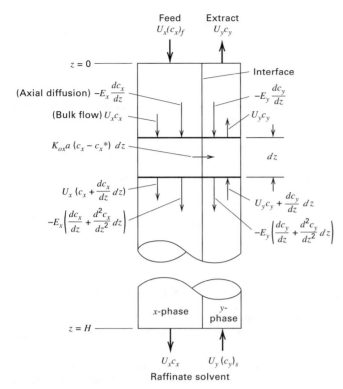

Figure 8.42 Axial dispersion in an extraction column.
[From J.D. Thornton, *Science and Practice of Liquid–Liquid Extraction*, Vol. 1, Clarendon, Oxford (1992) with permission.]

spray columns, packed columns, and RDCs. Although axial dispersion can occur in packed absorbers, packed strippers, and packed distillation columns, it is significant only at high liquid-to-gas ratios.

Two models have been developed for predicting the extent and effect of axial mixing: (1) diffusion models for differential-type contactors, by Sleicher [63] and Miyauchi and Vermeulen [64], and (2) backflow models for staged extractors without complete phase separation between stages, by Sleicher [65] and Miyauchi and Vermeulen [66]. Both types are discussed by Vermeulen et al. [99]. Diffusion models have received the most attention and application.

Consider a differential height, dz, of the differential contactor with two-phase flow, shown in Figure 8.42. Feed enters the top at $z = 0$, while solvent enters at $z = H$. Assume that: (1) axial dispersion is characterized by a constant turbulent-diffusion coefficient, E; (2) phase superficial velocities are uniform over the cross section and constant in the axial direction; (3) the overall volumetric mass-transfer coefficients for the solute are constant; (4) only the solute undergoes mass transfer between the two phases; and (5) the phase-equilibrium ratio for the solute is constant. Then the solute mass-balance equations for the feed/raffinate (x) and solvent/extract (y) phases, are

$$E_x \frac{d^2 c_x}{dz^2} - U_x \frac{dc_x}{dz} - K_{Ox}a(c_x - c_x^*) = 0 \qquad (8\text{-}63)$$

$$E_y \frac{d^2 c_y}{dz^2} - U_y \frac{dc_y}{dz} - K_{Ox}a(c_x - c_x^*) = 0 \qquad (8\text{-}64)$$

where c_x^* is the concentration of solute in the raffinate that is in equilibrium with the solute concentration in the bulk

extract. The boundary conditions, proposed by Danckwerts [100] and elucidated by Wehner and Wilhelm [101], are:

at $z = 0$,
$$U_x c_{x_f} - U_x c_{x_0} = -E_x \frac{dc_x}{dz} \qquad (8\text{-}65)$$

and
$$dc_y/dz = 0 \qquad (8\text{-}66)$$

at $z = H$,
$$U_y c_{y_H} - U_y c_{y_s} = E_y \frac{dc_y}{dz} \qquad (8\text{-}67)$$

and
$$dc_x/dz = 0 \qquad (8\text{-}68)$$

where:

c_{x_f} = concentration of solute in the original feed

c_{x_0} = concentration of solute in the feed at $z = 0$

c_{y_H} = concentration of solute in the solvent at $z = H$

c_{y_s} = concentration of solute in the original solvent

The two terms on the LHSs of (8-65) and (8-67) are the jumps shown in Figure 8.41.

It is customary to convert (8-63) and (8-64) to alternative forms in terms of pertinent dimensionless groups. This is readily done by defining

$$Z = z/H \qquad (8\text{-}69)$$

$$N_{Pe_y} = U_y H/E_y = \text{axial, turbulent Peclet} \\ \text{number for the extract} \qquad (8\text{-}70)$$

$$N_{Pe_x} = U_x H/E_x = \text{axial, turbulent Peclet} \\ \text{number for the raffinate} \qquad (8\text{-}71)$$

$$N_{Ox} = K_{Ox}aH/U_x = K_{Ox}aV/Q_x \qquad (8\text{-}72)$$

Equations (8-63) and (8-64) then become

$$\frac{d^2 c_x}{dZ^2} - N_{Pe_x}\frac{dc_x}{dZ} - N_{Ox}N_{Pe_x}(c_x - c_x^*) = 0 \qquad (8\text{-}73)$$

$$\frac{d^2 c_y}{dZ^2} - N_{Pe_y}\frac{dc_y}{dZ} - \left(\frac{U_x}{U_y}\right)N_{Ox}N_{Pe_y}(c_x - c_x^*) = 0 \qquad (8\text{-}74)$$

The boundary conditions are transformed in a similar way. For a straight equilibrium curve, $c_x^* = mc_y$. Thus, the result is a coupled set of ordinary differential equations, whose solutions for c_y and c_x are functions of c_{y_f}, c_{y_s}, m, N_{Pe_x}, N_{Pe_y}, N_{Ox}, U_x/U_y, and Z.

Further manipulations, involving substitution of dimensionless solute concentrations, reduce the number of variables from 10 to 7. Nevertheless solutions of the axial-dispersion equations as obtained by Sleicher [65] and Miyauchi and Vermeulen [66] are difficult to display in tabular or graphical form. However, the importance of axial dispersion is commonly judged by the magnitudes of the Peclet numbers. A Peclet number of 0 corresponds to complete backmixing, such that the entire column functions like a single mixer stage. A Peclet number of ∞ corresponds to an absence of axial dispersion. Experimental data on several types of liquid–liquid extraction columns indicate that N_{Pe} for the dispersed phase is frequently greater than 50, while N_{Pe} for the continuous phase may be in the range of 5 to 30. As a first approximation, axial dispersion in the dispersed phase can be largely ignored. This effect was observed by Geankoplis and Hixson [102] in a spray extraction column and by Gier and Hougen [103] in spray and packed extraction columns. They reported end-concentration changes at the continuous-phase entrance, but not at the dispersed-phase entrance. Figure 8.41, where axial dispersion is absent in plug flow, shows the importance of the Peclet number.

Approximate solutions to the axial-dispersion equations have been published, e.g., Sleicher [63]. Alternatively, if the entering solvent is free of solute, a rapid and conservative estimate of the effect of axial dispersion can be made by the method of Watson and Cochran [104] from an empirical relation for column efficiency:

$$\frac{H_{\text{plug flow}}}{H_{\text{actual}}} = \frac{(\text{HTU}_{Ox})(\text{NTU}_{Ox})}{H}$$

$$= 1 - \frac{1}{1 + N_{\text{Pe}_x}(\text{HTU}_{Ox}/H) - E + (1/\text{NTU}_{Ox})}$$
$$- \frac{E}{N_{\text{Pe}_y}(\text{HTU}_{Ox}/H) - 1 + E + (1/\text{NTU}_{Ox})}$$

$$(8\text{-}75)$$

where

$H_{\text{actual}} = H$ = height of column taking into account axial dispersion

HTU_{Ox} = height of an overall transfer unit based on the raffinate phase for plug flow

NTU_{Ox} = number of overall transfer units based on the raffinate phase for plug flow

E = extraction factor = mU_x/U_y

$m = dc_x/dc_y$

The product of HTU_{Ox} and NTU_{Ox} is the column height for plug flow, which is $< H$. Thus, the ratio on the left-hand side of (8-75) is a column efficiency. The NTU_{Ox} is approximated by:

$$\text{NTU}_{Ox} = \ln\left(\frac{X}{XE + 1 - E}\right)/(E - 1) \qquad (8\text{-}76)$$

where $\qquad X = (c_x)_{\text{out}}/(c_x)_{\text{in}} \qquad (8\text{-}77)$

The HTU_{Ox} is defined by:

$$\text{HTU}_{Ox} = U_x/K_{Ox}a \qquad (8\text{-}78)$$

For given values of HTU_{Ox}, NTU_{Ox}, E, N_{Pe_x}, and N_{Pe_y}, (8-75) is solved for H. Caution must be exercised in using (8-75) because of its empirical nature. The equation is limited to $\text{NTU}_{Ox} \geq 2$, $E > 0.25$, $N_{\text{Pe}_x}(\text{HTU}_{Ox}/H) > 1.5$, and the calculated value of the column efficiency, $H_{\text{plug flow}}/H_{\text{actual}}$, must be ≥ 0.20. Within these restrictions, an extensive comparison by Watson and Cochran with the exact solution of (8-63) to (8-68) gives conservative efficiency values that deviate by no more than 7%, with the highest accuracy for efficiencies above 50%.

EXAMPLE 8.12 Effect of Axial Dispersion.

Experiments conducted for a dilute system under laboratory conditions approximating plug flow give $\text{HTU}_{Ox} = 3$ ft. If a commercial column is to be designed for $\text{NTU}_{Ox} = 4$, with N_{Pe_x} and N_{Pe_y} estimated to be 19 and 50, respectively, determine the necessary column height if $E = 0.5$.

Solution

For plug flow, column height is $(\text{HTU}_{Ox})(\text{NTU}_{Ox})$. Substitution of the data into (8-75) gives

$$\frac{12}{H} = 1 - \frac{1}{(57/H) + 0.75} - \frac{0.5}{(180/H) - 0.25}$$

This is a nonlinear algebraic equation in H. Solving iteratively,

$$H = 17 \text{ ft}$$

$$\text{Efficiency} = (\text{HTU}_{Ox})(\text{NTU}_{Ox})/H = 12/17 = 0.706 \ (70.6\%)$$

---

§8.6 EXTRACTION OF BIOPRODUCTS

Small biomolecules such as inhibitory metabolites like ethanol and butanol, and antibiotics such as penicillin, erythromycin, tylosin, bacitracin, and cephalosporin, can be extracted directly from fermentation broth into an immiscible organic fluid phase [67–73]. Extraction of an inhibitory metabolite like ethanol reduces its down-regulation of a catalytic enzyme in the metabolic pathway, curtailing synthesis and thus improving ethanol productivity. Penicillin is an example of an antibiotic extracted via methyl isobutyl ketone (MIBK) from aqueous fermentation broth at acid pH values less than the pK_a of the antibiotic [see (2-118)]. Solvent extraction of bioproducts is generally less expensive than membrane (Chapter 14) or chromatographic (Chapter 15) operations, permits continuous operation, and is readily scalable [74]. Its importance as a bioseparation increased in the mid-20th century when it was adopted to recover antibiotics [75]. Larger biopolymers like peptides, proteins, and lipids; cellular particulates such as membrane components, organelles, and cells; and products from solid feeds, whose activity is commonly reduced by organic solvents, can be extracted into separate aqueous or supercritical-fluid phases, as discussed below.

The three extractive solvent systems widely used for bioseparations are aqueous/nonaqueous (organic solvent), aqueous two-phase, and supercritical-fluid extraction. Ideal liquid extractants do not react with solutes; are nondenaturing, nontoxic, inexpensive, and immiscible with the feed; and provide a large distribution coefficient (partition coefficient or partition ratio), K_D, for the product. The density difference between the two phases should be large enough to promote phase separation and allow ready recovery of extracted solutes. Interfacial tension should not be so high as to increase the amount of energy required for sufficient contacting time, nor so low as to encourage the formation of stable emulsions that preclude phase separation. In many cases, centrifugal extractors can overcome problems of density difference and emulsions.

Liquid extraction can significantly reduce process volume, separate bioproducts from cells/debris, and facilitate further product recovery via evaporation and/or crystallization. Loss

Table 8.7 Advantages and Disadvantages of Solvent Systems for Bioproduct Extraction

Extractant	Advantages	Disadvantages
Aqueous/nonaqueous (organic solvent)		
Carboxylic acids, amino acids, alcohols, steroids, antibiotics, small peptides	For organic solvents: $P^s_{\text{solvent}} > P^s_{\text{water}}$ $\Delta H^{\text{vap}}_{\text{solvent}} < \Delta H^{\text{vap}}_{\text{water}}$ Larger, purer crystals formed Inorganic salts remain in water	Biopolymers denatured in organic solvent Expense Flammability Toxicity Waste disposal
Aqueous two-phase		
Proteins, enzymes, virus, organelles, cells	Preserves biopolymer activity Eliminates disadvantages of solvent use	Pure dextran polymer is expensive Small phase density difference
Supercritical-fluid extraction		
Alkaloids, therapeutics, food/beverage processing	Facile solid penetration and solute dissolution CO_2 solvent: innocuous, easily removed	Costly: high-pressure equipment, compression Experimental determination of process conditions needed

of activity (e.g., denaturation) associated with solid membrane barriers or adsorptive phase and bioproduct dilution during elution can be avoided using liquid extraction. On the other hand, water-immiscible organic solvents used to recover small biomolecules like antibiotics or to solubilize lipophilic membrane biomolecules may denature biopolymers and lyse cells. So extraction of proteins and cells often uses two immiscible aqueous phases formed by adding two incompatible, but water-soluble, polymers or an incompatible polymer and concentrated salts. Supercritical-fluid extraction with CO_2, discussed in Chapter 11, is particularly useful with foods and beverages.

Table 8.7 lists advantages and disadvantages of bioproduct extraction by each of the three solvent systems.

§8.6.1 Organic/Aqueous Extraction

Solvents commonly used in biological organic/aqueous extraction are acetone, amyl acetate, butyl acetate, methyl isobutyl ketone, methylene chloride, and hexane. Antibiotics are the chief high-value bioproducts extracted. They include penicillin, clavulanic acid, erythromycin, gramicidin D, cycloheximide, fusidic acid, antimycin A, chloramphenicol, and virginamycin. An abbreviated summary of solvents and corresponding extracted biological solutes is listed in Table 8.8. The organic phase is usually the light, top phase, although methylene chloride, if used (S.G. 1.33 at 20°C), becomes the heavy, bottom phase. Yeast fermentation of ethanol and butanol from renewable agricultural feedstocks

is limited by product inhibition to broth concentrations of ~5 to 10% by volume. Although capital costs of solvent extraction (before considering solvent costs) are 60% higher than distillation [76], *in-situ* extraction offers a low-temperature method for recovering biological alcohols from ongoing fermentation. Identification of a solvent system that provides a suitable partition coefficient while maintaining cell activity remains a primary challenge. As seen in Table 8.8, a number of the partition coefficients, K_D, are very small (< 0.1).

Solvent Selection

Partition coefficients K_{D_i} for organic/aqueous extraction are functions of solubility, temperature, pH, ionic strength, and component concentrations. In §2.9.1, effects of several of these parameters on bioproduct K_a [see (2-119), (2-122), and (2-133)] and pI [see (2-118)] were evaluated. Ionization state

Table 8.8 Solvent Partition Coefficients for Some Bioproducts

Bioproduct	Solvent[a]	T, °C/pH	K_D
Alcohols			
Ethanol	n-Octanol	25/-	0.49
n-Butanol	n-Octanol	25/-	7.6
Ketones			
Acetone	n-Octanol	25/-	0.58
Methyl ethyl ketone	n-Octanol	25/-	1.95
Carboxylic Acids			
Citric acid	n-Butanol	25/-	0.29
	MIBK	25/-	0.009
Shikimic acid	Hexane	25/-	0.01
	Propyl acetate	25/-	0.06
Succinic acid	n-Butanol	25/-	1.2
	n-Octanol	25/-	0.26
Amino Acids			
Glycine	n-Butanol	25/-	0.01
Alanine	n-Butanol	25/-	0.02
Lysine	n-Butanol	25/-	0.2
Glutamic acid	n-Butanol	25/-	0.07
α-aminobutyric acid	n-Butanol	25/-	0.02
α-aminocaproic acid	n-Butanol	25/-	0.3
Antibiotics			
Erythromycin	Amyl acetate	-/6	120
		-/10	0.04
Novobiocin	Butyl acetate	-/7	100
		-/10.5	0.01
Penicillin F	Amyl acetate	-/4	32
		-/6	0.06
Proteins			
Glucose isomerase	PEG 1550/potassium phosphate		3
Fumarase	PEG 1550/potassium phosphate		0.2
Catalase	PEG/crude dextran		3

Source: Garcia et al. [67] and Belter et al. [73].
[a]Bioproduct extracted from H_2O except as indicated for proteins.
PEG = Polyethylene glycol

and local charge of bioproducts significantly impact equilibrium partitioning between immiscible solvents. Values of K_{D_i} for solute i increase exponentially with the difference in chemical potentials of the organic and aqueous phases, in their standard reference states at equilibrium. To illustrate this, (2-7) is used to evaluate the chemical potential of solute i in extract (1) and raffinate (2) phases at equilibrium: $\mu_i^{(1)} = \mu_i^{(2)}$. Writing $\mu_i^{(j)}$ in terms of standard reference-state chemical potential $\mu_i^{(j)o}$ for $j = 1$ and 2 (i.e., $\mu_i^{(j)} = \mu_i^{(j)o} + RT \ln \gamma_i x_i$) and solving for $x_i^{(1)}/x_i^{(2)}$ in (2-20) to obtain K_{D_i} gives

$$K_{D_i} = \frac{x_i^{(1)}}{x_i^{(2)}} = \frac{\gamma_i^{(2)}}{\gamma_i^{(1)}} \exp\left(\frac{\mu_i^{(2)o} - \mu_i^{(1)o}}{RT}\right) \qquad (8\text{-}79)$$

Equation (8-79) shows that K_{D_i} may be increased considerably (making extraction more effective) by reducing the standard-state chemical potential of the extracting phase (1) by changing pH or I, or by using a different extractant. To illustrate practical application of (8-79), Examples 8.13 and 8.14 demonstrate how decreasing pH to a value below the pK_a for a weak acid increases its K_{D_i} in organic solvents. Table 8.8 lists K_{D_i} values that increase with decreasing pH for the antibiotics erythromycin, novobiocin, and penicillin F.

Values of K_{D_i} for solvents that are different from those listed in Table 8.8 or similar sources may be estimated by rewriting (8-79) in terms of the partial molar volume, $\bar{V}$, and solubility parameter, δ, for the extract (1), raffinate (2), and solute (i). The solubility parameter was defined in (2-63) as part of a regular solution model for activity coefficients in the liquid phase. It is a measure of the energy required to remove a unit volume of molecules from neighboring species. This same energy is needed to dissolve a molecule, separate it from like neighbors, and surround it by solvent. Taking the natural log of both sides of (8-79) gives

$$\ln K_{D_i} = \frac{\bar{V}^{(2)}\left(\delta_i - \delta^{(2)}\right)^2 - \bar{V}^{(1)}\left(\delta_i - \delta^{(1)}\right)^2}{RT\bar{V}_i} \qquad (8\text{-}80)$$

Solubility parameters for some common solvents are listed in Table 8.9. Similar values of δ indicate that two components such as polyethylene and diethyl ether will interact strongly with each other. Miscibility, solvation, or swelling will result. Equation (8-80) shows that similar solubility parameters between the biological species to be extracted and the extract (1) increase the partition coefficient. One may insert $\delta = 9.4$ for water along with the δ value for the extracting solvent (from Table 8.9) into (8-80) to back-calculate δ_i for a target solute using a value of K_{D_i} obtained experimentally from that organic/aqueous extraction. This computed solute δ can then be used with (8-80) to estimate K_{D_i} for a different solvent whose δ value is known, in order to identify a solvent with greater extraction potential.

In the absence of data for the partial molar volumes in (8-80), pure-component molar volumes can be substituted. Equations (8-79) and (8-80) indicate that K_{D_i} increases as temperature decreases. This result is analogous to (2-122) after (2-121) has been rewritten in terms of K_{D_i}. The pH influences K_{D_i} by changing solute ionization state. Since un-ionized forms

Table 8.9 Solubility Parameters for Common Solvents

Solvent	δ (cal$^{1/2}$cm$^{-3/2}$)
Amyl acetate	8.0
Benzene	9.2
Butanol	13.6
Butyl acetate	8.5
Carbon tetrachloride	8.6
Chloroform	9.2
Cyclohexane	8.2
Dichloromethane	9.93
Diethyl ether	7.62
Ethyl acetate	9.1
n-Hexane	7.24
Hexanol	10.7
Acetone	7.5
n-Pentane	7.1
Perfluorohexane	5.9
Polyethylene	7.9
2-propanol	11.6
Toluene	8.9
Water	9.4

Compiled from [77] and other sources.

are more soluble in organic solvents, weak biological acids (bases) are extracted from fermentation broths at low (high) pH values. Temperature and pH dependencies explain why aqueous solutions of the weak acid penicillin G in Fig 1.11 ($pK_a = 2.7$–2.75) are buffered to pH $\sim$ 2 to 2.5 and chilled to 0 to 3°C to optimize extraction into n-butyl acetate [78]. Table 8.8 illustrates pH dependence of K_{D_i} for some other bioproduct-pure solvent pairs. Reactive extraction, e.g., extracting an organic acid into a solvent containing an amine or other base, can greatly increase the partition coefficient, as discussed below for extractant/diluent systems.

EXAMPLE 8.13 Dependence of pH on Partition Coefficient on pH.

Derive a general expression that shows pH dependence of the distribution coefficient defined in (2-20) for a weak acid between a fermentation broth and an organic solvent.

Solution

Beginning with the definition in (2-20), the partition coefficient, K_D^o, in an organic (1)–aqueous (2) LLE system, considering only the un-ionized (neutral) species of a weak acid, HA, where the superscript, o, designates un-ionized, may be written as

$$K_D^o \equiv \frac{x_{HA}^{(1)}}{1} \frac{1}{x_{HA}^{(2)}} = \frac{[HA]^{(1)}}{(\rho/M)^{(1)}} \frac{(\rho/M)^{(2)}}{[HA]^{(2)}} \qquad (8\text{-}81)$$

where $(\rho/M)^i$ corresponds to the total moles per unit volume of phase i. The pH dependence of the partition coefficient of the ionized species, K_D, is obtained by multiplying the fractional weak-acid form of the *Henderson–Hasselbach* equation in (2-117) by (8-81). This yields the partition coefficient for the ionized acid in terms of the un-ionized acid:

$$K_D = \frac{x_{HA}^{(1)}}{x_{HA}^{(2)} + x_{A^-}^{(2)}} = K_D^o \frac{[HA]^{(2)}}{[HA]^{(2)} + [A^-]^{(2)}} = \frac{K_D^o}{1 + 10^{(pH - pK_a)}} \quad (8\text{-}82)$$

Equation (8-82) shows that K_D for weak acids decreases in organic solvents, which typically have low dielectric constants, in the range $K_D^o \geq K_D \geq 0$ as pH increases to values that exceed pK_a. The effect of pH on K_D is illustrated in Table 8.8 and in Examples 2.11 and 8.14. Table 8.8 and Example 8.14 illustrate that values of K_D for extracted biological products decrease considerably as pH increases.

For dilute solute concentrations typical of bioproduct extraction, the value of K_D is often constant, independent of solute concentration. Low values of K_D are observed in Table 8.8 for dipolar zwitterions such as amino acids, which exhibit low solubility in polar solvents. See Example 2.11 for an illustration of solubility of zwitterions, weak bio-organic acids, and bases in water.

EXAMPLE 8.14 Effect of pH on the Partition Coefficient.

A monoacidic sugar extracted from water into hexanol has partition coefficients of 4.5 and 0.23 at pH 4.0 and 5.5, respectively, both at the same temperature. Estimate the value of K_D at pH 7.2.

Solution

Using (8-82), with the two given pairs of values for K_D and pH, two nonlinear equations are obtained:

$$4.5 = \frac{K_D^o}{1 + 10^{(4.0 - pK_a)}} \quad \text{and} \quad 0.23 = \frac{K_D^o}{1 + 10^{(5.5 - pK_a)}}$$

A nonlinear solver, e.g. a spreadsheet, gives $pK_a = 3.813$ ($K_a = 1.54 \times 10^{-4}$ mol/L) and $K_D^o = 11.43$. Therefore, applying (8-82) at pH 7.2,

$$K_D = \frac{11.43}{1 + 10^{(7.2 - 3.81)}} = 0.0047$$

Thus, increasing the pH from 4.0 to 7.2 decreases the ability of hexanol to extract the monoacid sugar by a factor of 1,000!

Reactive Extraction-Extractant/Diluent Systems

Because partition coefficients of bioproducts are unfavorably low for many solvents, reactive extraction has received much attention. These systems use a solvent or molecule as an *extractant* to react with or complex the target bioproduct to increase extraction partitioning and specificity, and a *diluent* that controls density and viscosity of the organic phase to ease phase disengagement. The extractant may utilize (1) *hydrogen bonding*, (2) *ion pairing* (e.g., a quaternary amine), or (3) *Lewis acid-base complexation* to interact with the bioproduct at the organic/aqueous interface, as follows:

1. *Hydrogen bonding* can be used to enhance extraction of polar zwitterions such as amino acids by adding organic carriers like trioctylmethylammonium chloride

(TOMAC). Octylmethyl ammonium is a *chaotropic ion* (see §2.9.2) that disrupts water structure and solvates hydrophobic structures like the alkane structures in TOMAC into water. Such carriers form *hydrogen bonds* (see §2.9.3) with biomolecules that displace water dipoles, which surround dissolved biomolecules and *solvate them* (see §2.9.2). This results in desolvation and facilitates extraction into an organic phase. Schügerl [79] summarized amino acid separations using consecutive solvent extraction. Extraction via xylene-containing TOMAC increases partition coefficients to 0.036 for glycine and 0.038 for alanine. Hydrogen bonding between oxygen-donor extractants such as tributyl phosphate (TBP) and trioctylphosphine oxide (TOPO) and bioproduct moieties like carboxylic acid, alcohol, ketone, ester, or ether groups produces weaker bonds ($\sim$2 kJ/mol) than acid-base interactions such as carboxylic acid with amines. Hydrogen bonds may be adjunct or synergistic with acid-base interactions.

2. Organic partitioning of a cation, anion, or zwitterion bioproduct is enhanced, sometimes dramatically, by forming an *ion pair* between the target species and a complementary *ion-pair agent*. Ideal ion-pair agents are water soluble when ionized, hydrophobic when paired to the target, and easily dissociated to return the desired species. Counterions from organic soluble salts (greasy salts) are common ion-pair extractants. Acetate or butyrate, whose organic solubility exceeds that of acetate, ion-pair with cationic biopolymer, while quaternary amines like chaotropic tetrabutylammonium and hexadecyltributylammonium ion-pair favorably with anionic biopolymers at high pH. For example, extracting tetrabutylammonium cation paired with chloride anion into chloroform (1) from water (2) yields a partition coefficient given by

$$K_D = \frac{\left[N\left(C_4H_9\right)_4^+ Cl^-\right]^{(1)}}{\left[N\left(C_4H_9\right)_4^+ Cl^-\right]^{(2)}} = 1.3 \quad (8\text{-}83)$$

whereas adding sodium acetate to the solution yields a coefficient of

$$K_D = \frac{\left[N\left(C_4H_9\right)_4^+ CH_3COO^-\right]^{(1)}}{\left[N\left(C_4H_9\right)_4^+ CH_3COO^-\right]^{(2)}} = 132 \quad (8\text{-}84)$$

Larger counterion extractants are associated with 2° (second-order) effects that decrease effectiveness. Perfluorooctanoate tends to remain ionic after ion-pairing; dodecanoate may form micelles; and linoleate forms liquid crystals in organic solvents.

3. *Acid-base pairing* between a *soft* or moderately hard extractant (see *Ionic interactions* in §2.9.3) and a targeted bioproduct permits removal of competitive water (e.g., hydronium and hydroxyl ions) and mineral salts added for pH adjustment, which are *hard* acids or bases, respectively. Acid-base pairing typically requires 1:1 stoichiometry, and correlation of the partition coefficient of a particular target species with inherent acidity (basicity) of the extractant using *linear*

free-energy relationships. Methanol or acetone precipitation of SO_4^{2-}, PO_4^{3-}, NH_4^+, Cl^-, Na^+, and trace metal ions from the culture medium may be used to eliminate competitive effects of these ions on pH variations that alter partition coefficients of weak Brönsted acid or base bioproducts or extractants.

EXAMPLE 8.15 Dependence of a Weak-Acid Bioproduct on the Partition Coefficient.

Based on chemical equilibrium, obtain a general expression for dependence of the partition coefficient of a weak-acid bioproduct $B^{(2)}$ that interacts with an organic solvent extractant $X^{(i)}$ at the interface of an organic (1)/aqueous (2) system with valence z.

Solution

The law of mass action yields

$$B^{(2)} + zX^{(i)} \leftrightarrow BX_z^{(1)} \qquad (8\text{-}85)$$

for which the chemical equilibrium constant may be written as

$$K_X = \frac{[BX_z]^{(1)}}{[B]^{(2)} \left([X]^{(i)}\right)^z} \frac{(\rho/M)^{(2)}}{(\rho/M)^{(1)}} \qquad (8\text{-}86)$$

where superscript $i = 1$ for an organic-soluble extractant and 2 for a water-soluble ion-pair agent like a quaternary amine. Both sides of (8-86) are multiplied by $([X]^{(i)})^z$ to obtain a form analogous to (8-81). This form is then multiplied by (2-117) to obtain the partition coefficient for the weak-acid bioproduct, HA, as shown in (8-82). This yields

$$K_D = \frac{K_X \left([X]^{(i)}\right)^z}{1 + 10^{(pH - pK_a)}} \qquad (8\text{-}87)$$

The z-dependence of K_D on $[X]^{(i)}$ in (8-87), which increases for $z > 1$, may be evaluated experimentally from the slope of a plot of $\ln K_D$ versus $\ln [X]^{(i)}$, using increasing concentrations of organic extractant at constant pH.

Back Extraction

Extracted bioproducts may be back-extracted to an aqueous solution by *temperature swing, displacement,* or *aqueous-phase reaction.* Increasing the temperature generally decreases K_D, as indicated in (8-79) and (8-80), and decreases acidity or basicity of extractant/diluent systems. Temperature affects dissociation constants of amines like Tris HCl much more than carboxylic acids [see Table 2.13, (2-122), and Example 2.10]. For example, shikimik acid extracted from H_2O into tridodecylamine at 5°C may be back-extracted into water at 80°C via a temperature swing. An extractant may be *displaced* by introducing a compound into the aqueous phase that effectively competes with the solute complex in the organic phase to effect crystallization or back extraction. As an example, oleic acid ($pK_a = 3.33$) readily displaces shikimik acid ($pK_a = 4.25$) from an organic phase. Alternatively, adding a water-soluble base to the aqueous phase yields an *aqueous-phase reaction* with a carboxylic acid, removing the acid from the organic phase via Le Chatelier's principle.

Reverse Micelles

Reverse micelles may be formed by dissolving surfactants in aqueous solutions. A surfactant is an organic molecule that is amphiphilic, meaning it contains both hydrophobic (tail) and hydrophilic (head) groups. Therefore, surfactants are soluble in both organic solvents and water. Above a *critical micelle concentration,* CMC, in water, surfactant molecules aggregate head-to-head and tail-to-tail into spherical *micelles,* in which the heads form an outer layer, with the tails in the interior core. An oil droplet could be encapsulated in the interior, hydrophobic core. In an organic solvent, *reverse micelles* can form, with the heads in the cores. Thus, exposing amphiphilic ionic surfactants such as sodium bis(2-ethylhexyl) sulfosuccinate (AOT), didodecyldimethyl ammonium bromide, or trioctylmethyl ammonium chloride dissolved in an aqueous solution to a nonpolar organic solvent forms *reverse micelles.* Polar head groups of the surfactant molecules turn inward, enveloping a polar environment, which can support ions, biopolymers, and bacterial cells while maintaining a minimum amount of water. Formation of reverse micelles is facilitated by *hydrophobic interactions* and the ability of *chaotropic ions* (see §2.9.2, like di- and octyl-methyl ammonium, to disrupt water structure and solvate hydrophobic structures. The size of reverse micelles decreases as surfactant concentration increases and/or organic-solvent water interfacial tension decreases (depending on surfactant type, hydrophobic chain length, and chemical group). Size is also affected by aqueous salt composition and nonpolar solvent chemistry, which influences surface tension. Bioproducts partition into reverse micelles via electrostatic interactions that depend on pH, salt type and concentration, surfactant concentration, and temperature, as well as factors that influence micelle size. The *Hofmeister series* (§2.9.2) ranks salt ions in terms of their general ability to effect partitioning into micelles and organic phases in general. Characteristics such as isoelectric point, size and shape, hydrophobicity, and charge distribution on a protein surface affect phase transfer. Proteins with net charge opposite to surfactant polar groups are solubilized and extracted into micelles electrostatically at ionic strengths of ~0.1 M for simple salts.

Mass Transfer in Liquid Mixtures

In §3.1–3.7, calculations of mass-transfer rates are based on ideal solutions with concentration driving forces for effective binary mixtures. A more rigorous treatment of mass transfer is presented in §3.8, which considers other driving forces like gradients in T and P, and concentration gradients of other species in systems containing three or more species. For example, performing solvent extraction at hyperbaric pressures and/or cycling between atmospheric and hyperbaric pressures up to 35,000 psi can increase extraction efficiency of biopolymeric proteins, lipids, and polynucleic acids from cells and tissues. High pressure can increase water solubility

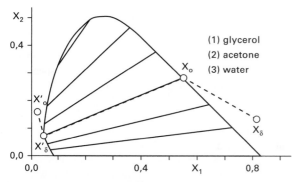

Figure 8.43 Liquid–liquid equilibrium diagram for glycerol–acetone–water system. From [81].

of hydrophobic components like lipid bilayers, and induce conformational changes to dissociate multimeric proteins and disassemble protein–lipid complexes without breaking covalent bonds. High pressure also increases hydrolysis by a variety of catalytic enzymes like trypsin.

In ambient-pressure organic-aqueous extractions, mass transfer is enhanced by concentration gradients of other species in the multicomponent system. As an example, consider mass transfer in the film of the glycerol-rich phase of a glycerol (1)–acetone (2)–water (3) system [80]. This is illustrated in the mole-fraction phase diagram of Figure 8.43 [81]. Tie lines between points at equilibrium on the phase boundary slope upward to the right, indicating water prefers the glycerol phase. Resistance to mass transfer occurs mainly in the viscous glycerol-rich phase. Compositions and activity coefficients (γ) of glycerol and acetone at the interface (o) and bulk (δ) sides of the glycerol-phase film are

$x_{1,o} = 0.5480$	$\gamma_{1,o} = 0.7440$	$x_{2,o} = 0.2838$	$\gamma_{2,o} = 0.2519$
$x_{1,\delta} = 0.7824$	$\gamma_{1,\delta} = 0.8776$	$x_{2,\delta} = 0.1877$	$\gamma_{2,\delta} = 0.1235$

EXAMPLE 8.16 Mass-Transfer Velocities.

Calculate the mass-transfer velocities of the three species in the glycerol-phase film extending from x_o to x_δ. Evaluate the effects of nonideality on the mass-transfer driving forces [81].

Solution

Arithmetic-average film compositions and corresponding arithmetic-average activity coefficients are calculated to be

$$\bar{x}_1 = 0.6652 \quad \bar{\gamma}_1 = 0.8080 \quad \bar{x}_2 = 0.2357 \quad \bar{\gamma}_2 = 0.1764$$

Effects of nonideality are considered by introducing activity coefficients for average concentrations into (3-288), the linearized Maxwell–Stefan difference equation for the ternary system, which relates driving forces to friction using species velocities:

$$\frac{\gamma_{1,\delta}x_{1,\delta} - \gamma_{1,o}x_{1,o}}{\bar{\gamma}_1\bar{x}_1} = \bar{x}_2\frac{\bar{v}_2 - \bar{v}_1}{k_{12}} + \bar{x}_3\frac{\bar{v}_3 - \bar{v}_1}{k_{13}}$$

$$\frac{\gamma_{2,\delta}x_{2,\delta} - \gamma_{2,o}x_{2,o}}{\bar{\gamma}_2\bar{x}_2} = \bar{x}_1\frac{\bar{v}_1 - \bar{v}_2}{k_{12}} + \bar{x}_3\frac{\bar{v}_3 - \bar{v}_2}{k_{23}}$$

(8-88a, b)

The composition of glycerol in the acetone-rich phase is approximately that of the phase boundary at which $x'_1 = 0.04$, so

$$\frac{N_1}{\sum_{i=1}^{3} N_i} \approx 0.04 = \frac{\bar{x}_1\bar{v}_1}{\sum_{i=1}^{3} \bar{x}_i\bar{v}_i}$$

(8-89)

Simultaneous solution of (8-88a, b) and (8-93) yields

$$\bar{v}_1 = 0.000004, \quad \bar{v}_2 = 0.000196, \quad \bar{v}_3 = 0.000199$$

All three velocities are positive. Effects of the nonideality on the driving forces are significant, as seen in comparison with ideal driving forces:

$$\frac{\gamma_{1,\delta}x_{1,\delta} - \gamma_{1,o}x_{1,o}}{\bar{\gamma}_1\bar{x}_1} = 0.519; \quad \frac{x_{1,\delta} - x_{1,o}}{\bar{x}_1} = 0.352$$

$$\frac{\gamma_{2,\delta}x_{2,\delta} - \gamma_{2,o}x_{2,o}}{\bar{\gamma}_2\bar{x}_2} = -1.161; \quad \frac{x_{2,\delta} - x_{2,o}}{\bar{x}_2} = -0.408$$

(8-90a, b)

---

§8.6.2 Aqueous Two-Phase Extraction (ATPE)

Biopolymers can partition between two aqueous phases, each containing 75 to 90% water, formed by dissolving one or two ionic or nonionic polymers, or a polymer and mineral salt (e.g., sodium or potassium phosphate). Polyacrylamide is a common ionic polymer; sodium dodecylsulfate (SDS), an anionic surfactant, is also used. Nonionic (nondissociating) polymers are polyethylene oxide (PEO), polyethylene glycol (PEG), and dextran. These *kosmotropic* polymers (see §2.9.2) order water molecules, sterically stabilize biomolecule structures in solution, and promote hydrophobic interactions. Each polymer in a two-phase system is fully soluble in water, yet incompatible with the other phase in concentration ranges where two aqueous phases are formed. The most common aqueous two-phase systems are PEG–dextran–H_2O and PEG–potassium phosphate–H_2O. An aqueous solution of 5 wt% dextran 500 (avg. MW 500,000) and 3.5% PEG 6000 (avg. MW 6,000) at 20°C partitions into two aqueous phases: a PEG-rich top phase containing 4.9% PEG, 1.8% dextran, and 93.3% H_2O; and a dextran-rich bottom phase containing 2.6% PEG, 7.3% dextran, and 90.1% H_2O.

Biomolecules such as peptides, proteins, nucleic acids, viruses, and cells exhibit different solubilities in the two phases and partition accordingly [82]. Of great importance is that high water activity in each phase preserves biological activity. Therefore, the application and study of aqueous two-phase extraction have grown rapidly in significance. A comprehensive bibliography of aqueous two-phase extraction in biological systems from 1956 to 1985 was compiled by Sutherland and Fisher [83]. Monograms by Albertsson [84] and Zaslavsky [85] contain 50 and 160 aqueous two-phase phase diagrams, respectively. Aqueous two-phase extraction of enzymes was comprehensively examined by Walter and Johansson [86].

Proteins commonly partition into a less-dense PEG-rich phase formed in systems using 10 wt% PEG and 15 wt% dextran, or 15 wt% PEG and 15 wt% potassium phosphate. Values of partition coefficients, K_D, for proteins range from about 0.1 to 10. Contaminating cells, debris, nucleic acids, and polysaccharides are usually removed in the lower phase.

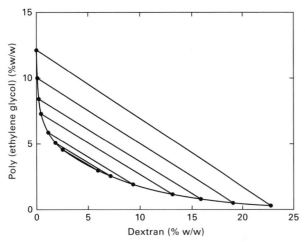

Figure 8.44 Aqueous two-phase diagram for PEG 6000–dextran D48 system at 20°C.

Small ions exhibit K_D of ~1. Aqueous two-phase extraction systems have been developed to recover whole cells and DNA, with K_D ranging from <0.01 to >100. Low interfacial tension between the phases helps maintain biopolymer activity. Aqueous–aqueous phase separation takes place only at compositions in excess of a series of *critical concentrations*—points on a solubility curve that separate the one-phase region (below the curve) from the two-phase region (above the curve), as in Figure 8.44 from [84]. Tie lines connecting points on the top and bottom phases in equilibrium are characterized using the inverse-lever-arm rule.

Optimizing K_D

Many enzymes exhibit partition coefficients from 1 to 3.7 between PEG- and dextran-rich phases, yielding poor single-stage separations. This motivates examining factors to optimize K_D. Aqueous two-phase partitioning is influenced by biopolymer properties of size, charge, surface hydrophobicity/hydrophilicity, composition, and attached affinity ligands—and properties of the respective phases, including size, type and relative concentrations of phase-forming polymers and salts, and tie-line length. Lowering the average PEG MW and increasing the dextran MW tend to increase K_D, particularly for higher MW proteins like catalase (MW 250,000) compared with cytochrome (MW 12,385). The larger the difference between PEG concentrations in the two phases, the more the partition coefficient deviates from unity. Partial hydrolysis of dextran and PEG can increase K_D since lower-MW polymers interact more strongly with proteins. The K_D value of fumarase increases by a factor of 6 when PEG 400 and PEG 4000 are mixed.

EXAMPLE 8.17 Dependence of the Partition Coefficient on Size and Temperature of a Protein.

By analyzing the free-energy change resulting from partitioning in an aqueous two-phase extraction, determine the dependence of the partition coefficient on size and temperature for a globular protein.

Solution

Free energy, the surface energy of the interface between a sphere and a liquid in a phase, changes during partitioning in proportion to the sphere surface area and the relative difference in sphere surface tension, σ_{sl}, for extract (1) and raffinate (2) phases:

$$\Delta G = \pi d^2 \left(\sigma_{sl}^{(1)} - \sigma_{sl}^{(2)} \right) = -RT \ln K_{D_i} \qquad (8\text{-}91)$$

Since surface area of a globular protein, πd^2, is proportional to molecular weight, $\ln K_D$ in (8-91) increases as size decreases. This semilogarithmic dependence was initially identified by Brönsted. It has been observed in a PEG 6000–dextran 500 system when the net charge is zero so that pH = pI for the proteins insulin, lysozyme, papain, trypsin, α-chymotrypsin, ovalbumin, bacterial α-amylase, BSA, human transferrin, and β-galactosidase [87].

Temperature Effect

Phase separation occurs at lower polymer concentrations for lower temperatures in PEG–dextran–H₂O systems, while the opposite is true for PEG–salt–H₂O systems [88]. Lower temperatures usually raise partition coefficients, as indicated in (8-91), and can increase separability of high- and low-molecular-weight proteins.

Salt Effect

Salt strongly affects systems containing polyelectrolytes like diethylaminoethyl (DEAE)–dextran–H₂O or solutes far from their isoelectric point, but only marginally affects phase diagrams of nonionic polymer–polymer–water systems and partition coefficients of uncharged solutes. Higher salt content usually requires less polymer for phase separation. Effects of biopolymer charge on K_D can be modulated by adjusting pH, electrolyte composition, or ionic strength, especially at low salt concentrations (0.1 to 0.2 M), to increase the value of K_D. Increasing KH_2PO_4 concentration from 0.1 to 0.3 M in a 14% PEG 4000/9.5% $(NH_4)_2SO_4$ system, for example, increases K_D more than 10-fold. Anionic proteins tend to exhibit lower K_D values with salt in the order sulfate > fluoride > acetate > chloride > bromide > iodide, and lithium > ammonium > sodium > potassium. This trend essentially follows the *Hofmeister series* (see §2.9.2), which ranks salt ions in terms of their ability to increase solvent surface tension and lower solubility of proteins. Cationic proteins follow the opposite trend. Salts that distribute unevenly between phases establish a Donnan-type electrochemical potential difference that influences partitioning of charged polymers (see §14.5). High ionic strengths in PEG–salt systems may cause protein precipitation at the interface. Adding ion-exchange resins or derivatizing PEG to yield cation- or anion-exchange properties can similarly increase K_D.

Predicting Biopolymer K_{D_i} Values

Each of the three components that forms an ATPE system—solvent (i.e., water), polymer 1 (e.g., PEG), and polymer 2 (e.g., dextran)—partitions between the top (t) and bottom (b) phases. The biopolymer also distributes between the two phases. Partitioning involves forces due to hydrogen, ionic, hydrophobic, and other weak bonds. Effects of solvent (i.e., water for ATPE) and polymer concentration differences, molecular volumes, and biopolymer interactions on K_{D_i} for the biopolymer have been analyzed via the *Flory–Huggins* formalism [89]:

$$\ln K_{D_i} = P_p \sum_{i=1}^{3} \left(\Phi_i^t - \Phi_i^b \right) \left(\frac{1}{P_i} - \chi_{i,p} \right) \qquad (8\text{-}92)$$

where subscripts 1, 2, 3, and p represent solvent, polymer 1, polymer 2, and protein biopolymer, respectively; P_i is molecular volume of component i divided by molecular volume of solvent i; Φ_i is volume fraction; superscripts t and b represent top and bottom phase, respectively; and $\chi_{i,p}$ describes the Flory–Huggins interaction, which accounts for the energy of interdispersing molecules of component i with protein biopolymer p. The value of $\chi_{i,p}$ in (8-92) may be estimated from the respective values of solubility parameter, δ_i, of the species:

$$\chi_{i,p} = \bar{V}_p \left(\delta_i - \delta_p \right)^2 / RT$$

For just the partitioning of the two polymers between the solvent in the absence of the biopolymer, the *Diamond–Hsu* approach [90] simplifies the formalism of Flory–Huggins by replacing volume fractions with weight fractions and consolidating the terms for component molecular volume and Flory–Huggins interaction into a linear coefficient to obtain

$$\ln K_{D_i} = \ln \left(\frac{\omega_i^t}{\omega_i^b} \right) = A_i \left(\omega_1^t - \omega_1^b \right) \qquad (8\text{-}93)$$

where i applies only to the two polymers, the empirical parameter A_i depends on molecular weights of the polymers, and ω_i^p is the weight fraction of the polymer in phase p. Note that the difference term on the RHS of (8-93) is always for polymer 1. From one measurement of phase distribution (i.e., one tie line), the value of A_i can be computed for each polymer, and Equation (8-93) can then predict a series of tie lines, such as those in Figure 8.44. However, to do this requires the additional assumption that all tie lines have the same slope on a plot like Figure 8.44. A refinement of the Diamond–Hsu approach for the partitioning of the two polymers is given by Croll et al., who relax the assumption of constant tie-line slope [111].

EXAMPLE 8.18 Tie Line Calculation for an Aqueous Two-Phase System.

A phase-distribution measurement was made for the PEG 3400–dextran T40–water system at 4°C. The following results give the starting composition and the equilibrium compositions of the two resulting phases, all in wt%.

Component	Start Total	Bottom Phase	Top Phase
PEG (1)	6.50	3.28	8.82
Dextran (2)	8.80	15.83	3.70
Water	84.70	80.89	87.48

Use the Diamond–Hsu method of (8-93) to compute the values of A_i for dextran and PEG. Use those values to produce a tie line with one end at 11.19 wt% PEG in the top phase.

Solution

First, compute the tie-line slope, S_{TL}, for the measured phase compositions.

$$S_{TL} = \frac{(\omega_1^t - \omega_1^b)}{(\omega_2^t - \omega_2^b)} = \frac{(0.0882 - 0.0328)}{(0.0370 - 0.1583)} = -0.457$$

Solving (8-93) for A_i of PEG,

$$A_1 = \ln \left(\frac{\omega_1^t}{\omega_1^b} \right) \bigg/ \left(\omega_1^t - \omega_1^b \right)$$

$$= \ln \left(\frac{0.0882}{0.0328} \right) \bigg/ (0.0882 - 0.0328) = 17.85$$

Solving (8-93) for A_i of dextran,

$$A_2 = \ln \left(\frac{\omega_2^t}{\omega_2^b} \right) \bigg/ \left(\omega_1^t - \omega_1^b \right)$$

$$= \ln \left(\frac{0.0370}{0.1583} \right) \bigg/ (0.0882 - 0.0328) = -26.24$$

Now, compute a tie line for 11.19 wt% PEG in the top phase, using $A_1 = 17.85$. From (1) for PEG,

$$A_1 = 17.85 = \ln \left(\frac{0.1119}{\omega_1^b} \right) \bigg/ \left(0.1119 - \omega_1^b \right)$$

Solving, $\omega_1^b = 0.0228$, compared to an experimental value of 0.0209 from Diamond and Hsu [96].

To solve for ω_2^t and ω_2^b for the new tie line, the following two equations apply:

$$A_2 = -26.24 = \ln \left(\frac{\omega_2^t}{\omega_2^b} \right) \bigg/ (0.1119 - 0.0228)$$

or

$$\ln \left(\frac{\omega_2^t}{\omega_2^b} \right) = -2.338$$

and

$$S_{TL} = -0.457 = \frac{(0.1119 - 0.0228)}{(\omega_2^t - \omega_2^b)}$$

or

$$\left(\omega_2^t - \omega_2^b \right) = -0.195$$

Solving, $\omega_2^t = 0.0208$ and $\omega_2^b = 0.216$. Experimental values are 0.0176 and 0.2121 from Diamond and Hsu [96].

Diamond and Hsu have also applied (8-93) to the prediction of partition coefficients for dipeptids and small proteins (MW < 20,000) in PEG–dextran–water systems [96]. For dipeptides, a knowledge of the K_D in any PEG–dextran–water system enables prediction of K_D in other such systems, regardless of MW. For proteins, knowledge of a K_D on one PEG–dextran–water tie line enables K_D to be predicted for any other tie line for the same phase diagram.

For more accuracy and wider application, King et al. [91] and Haynes et al. [92] developed a virial expansion model for K_D of biomolecules in a solution of polymers PEG (1) and dextran (2), dilute in a protein (p) such as albumin, α-chymotrypsin, or lysozyme, where K_D is defined in terms of protein molalities as mol/1000 g of solvent (water) in the top (t) and bottom (b) phases:

$$\ln K_D = \ln\left(\frac{m_p^t}{m_p^b}\right) = \sum_{i=1}^{2} a_{i,p}\left(m_i^b - m_i^t\right) + \frac{z_p \Im}{RT}\left(\Phi_e^b - \Phi_e^t\right)$$

(8-94)

where $a_{i,p}$ are second virial coefficients for interaction between polymer and biomolecule in the solvent, in L/mol, measured by membrane osmometry or low-angle laser-light scattering; z_F is the net surface charge of the biomolecule; Φ_e are electrical potentials in mV; and $\Im$ is Faraday's constant. In general, K_D is independent of protein concentration up to 30 wt%.

For isoelectric proteins in the absence of salts, (8-94) reduces to

$$\ln K_D = a_{1,p}\left(m_1^b - m_1^t\right) + a_{2,p}\left(m_2^b - m_2^t\right)$$

(8-95)

However, as discussed by Prausnitz et al. [112], the effect of Donnan-type electrochemical potential difference, caused by the unequal partitioning of salts between phases, can be the dominant effect in (8-94). By adjusting it, the partition coefficient can be significantly enhanced.

Effects of pH, ionic strength, and changing salt type and concentration are included in a thermodynamic model by Hartounian et al. [94] for partitioning of biomolecules in PEG–dextran–water systems containing low concentrations of salts. By combining the UNIQUAC (§2.6.8) and extended Debye–Hückel equations, the result is

$$\ln K_D = A(\text{TLL}) + B\left(m_{\text{cations}}^b - m_{\text{cations}}^t\right) + \frac{z_p F}{RT}\left(\Phi_e^b - \Phi_e^t\right)$$

(8-96)

where A and B are constants that depend on the biomolecule, the polymers, and the salt. TLL is the tie-line length as in Figure 8.44. The second term on the RHS, in terms of cation molalities, accounts for the effect of ionic strength. The parameters are obtained by fitting measurements. As an example, selective partitioning of ovalbumin (pI 4.5) in a PEG 6000–dextran 500–water system decreased more than 5-fold from pH 3 to pH 6 in the presence of potassium and sodium chloride, but varied little with corresponding sulfate salts [95].

Equations (8-93) to (8-96), for correlating and predicting partition coefficients of biomolecules in aqueous two-phase systems, assume that K_D is independent of the volumes of the two phases. While this is a good assumption for polymer (1)–polymer (2)–water systems, even with small quantities of salts present, it is not a good assumption to make in the case of polymer–potassium phosphate–water systems, due to the salting out of the biomolecule, e.g., bovine serum albumin, as discussed by Huddleston et al. [93].

Affinity Partition Extraction

Selective partitioning of a particular protein may be increased > 10-fold by coupling a biospecific ligand (§2.9.3) to a polymer in the target phase. In general, a ligand (molecule) binds to an active site on the protein by short-range (20 to 0.1 nm), noncovalent forces. Binding is initiated by electrostatic interactions, which are followed by solvent displacement, steric selection and charge/conformational rearrangement, and finally, rehydration of the stabilized complex. The later steps involve breaking and creating hydrogen, hydrophic, and van der Waals bonds. Ligands commonly used for affinity-partition extraction are reactive dyes like Cibacron blue, Procion red, Procion yellow, or Triazine dye; fatty acids; and NADH (the reduced form of nicotinamide adenine dinucleotide). Ligands are coupled to terminal-free hydroxyls on PEG via reactive intermediates like halide, sulfonate ester, or epoxide. Adding salt, or an effector that competes with the coupled ligand for the bound protein, allows recovery of target protein in the bottom phase. Aspects of affinity-partition extraction that have been reviewed include biomolecule/ligand pairs by Zaslavsky [85] and Diamond and Hsu [90], polymer ligands by Harris and Yalpani [97], and effects of pH, temperature, and competition by Kopperschläger [98]. The costs associated with the ligand itself, and the expense of coupling it to a polymer, limit large-scale application of affinity-partition extraction.

Large-Scale, Aqueous Two-Phase Extraction

Over 30 biomolecules have been purified on a large scale using aqueous two-phase extraction [90]. The cost of purified dextran (hundreds of $/kg) is a limiting factor, and thus lower-cost alternatives such as crude dextran, hydrolyzed crude dextran, and hydroxypropyl starch have been examined. Equipment used for large-scale ATPE is the same used for solvent extraction. Phase separation in PEG–salt–water systems is promoted by relatively large density differences between the two aqueous phases, low viscosity of the salt phase, and large drops generated during mixing. The protein recovered in the target phase is purified by removing salt and polymer via ultrafiltration, or by adding salt to partition the solute to a new salt phase.

EXAMPLE 8.19 Recovery of Lysozyme by Aqueous Two-Phase Extraction.

Lysozyme is an enzyme of the innate immune system, lacking from the diet of children who are fed infant formula, increasing their susceptibility to pathogens like *Salmonella* or *E. coli*. Balasubramaniam et al. [108] discuss genetically engineering tobacco, a ubiquitous gene host, to produce large quantities of recombinant lysozyme. A clarified tobacco broth contains 800 kg/h of water, 5 kg/h of lysozyme (L), and 40 kg/h of other proteins (OP). The lysozyme is to be extracted by ATPE using the PEG 3000–water–Na$_2$SO$_4$ (salt) system at 20°C. A phase diagram for this system, similar to Figure 8.44, has been measured by Hammer et al. [109]. One tie line for this system, in terms of wt% for PEG, salt, and water shows a top-phase composition of 40, 1, and 59, with a bottom-phase composition of 0.6, 18, and 81.4. The partition coefficients based on mass ratios K'_L and K'_{OP} have been measured with the system to be 20 and 2, where K'_i is defined as (kg i in top phase/kg protein-free top phase)/(kg *i* in bottom phase/kg protein-free bottom phase). If 190 kg/h of salt is added to the broth and if equilibrium is achieved in a single extraction stage, determine: (a) the flow rate of the water in the aqueous PEG solvent if it is to contain 406 kg/h of PEG, and if the overall composition of PEG–water–salt for the extraction system is to lie on the aforementioned tie line; (b) the compositions in wt% on a protein-free basis of the equilibrium top phase and bottom phase; (c) the percent of lysozyme and of other proteins extracted to the top phase; and (d) the % purity of the lysozyme in the total extracted protein.

Solution

(a) The flow rate of water needed to form the solvent phase with PEG must place the overall composition of solvent and feed on the tie line. Therefore, develop an equation for the tie line from the data given for its end points. Let y = wt% PEG and x = wt% salt. The two end points for (y, x) are (40, 1) and (0.6, 18). A straight-line fit of these two points gives: $y = -2.31765 + 42.31765x$. The overall composition includes 190 kg/h of salt and 406 kg/h of PEG, giving a ratio, y/x, of 406/190 = 2.13684. Combining this with the equation for the tie line and solving gives an overall composition of $y = 20.30$ wt% PEG and $x = 9.50$ wt% salt. The wt% water = $100 - 20.3 - 9.5 = 70.2$. The ratio of total water to salt = 70.2/9.5 = 7.3895. Therefore, the total water flow rate = 7.3895(190) = 1,404 kg/h. Because the feed contains 800 kg/h of water, the additional water in the solvent solution = $1,404 - 800 = 604$ kg/h. The component flow rates in the feed (after added sulfate) and solvent are included in the material-balance table below.

(b) To calculate the kg/h of the components in the equilibrium top and bottom phases on a protein-free basis, the following material-balance equations apply, using the tie-line compositions. (Note that these equations must be selected carefully to avoid a singular matrix.) Let t_i = flow rate of component i in the top phase, and b_i = flow rate of component i in the bottom phase, with w = water, s = salt, and P = PEG 3000.

$$t_P = 0.40(t_w + t_s + t_P) \tag{1}$$
$$b_P = 0.006(b_w + b_s + b_P) \tag{2}$$
$$t_s = 0.01(t_w + t_s + t_P) \tag{3}$$
$$b_s = 0.18(b_w + b_s + b_P) \tag{4}$$
$$406 = t_P + b_P \tag{5}$$
$$190 = t_s + b_s \tag{6}$$

Solving these six linear equations in six unknowns gives the flow rates in the table below.

(c) From (4-24), the extraction factor for lysozyme (L) on a protein-free basis, noting that the carrier is the protein-free bottom phase and the solvent is the protein-free top phase from the table below, is $E_L = 20(1,000)/(1,000) = 20$, while for the other proteins, $E_{OP} = 2(1,000)/(1,000) = 2$.

From (4-25), the fraction of lysozyme not extracted = $1/(1 + 20) = 0.0476$.

Therefore, % lysozyme extracted = $(1 - 0.0476)100\% = 95.2\%$.

Similarly the % other proteins extracted = 66.7%.

(d) The % purity of lysozyme in extracted protein = $0.952(5)/[0.952(5) + 0.667(40)]$ $100\% = 15.1\%$.

The material balance in kg/h is:

Component	Feed Broth after Added Sulfate	Solvent	Top Phase	Bottom Phase
Water	814	604	590	814
Sodium sulfate	190		10	180
PEG 3000		406	400	6
Lysozyme	5		4.88	0.12
Other proteins	40		33.33	6.67
Total	1,049	1,010	1,038.21	1,006.79

In Example 8.19, the % extraction of lysozyme is high (95.2 %), but the corresponding extraction of the other proteins, whose amount in the broth is much greater than that of lysozyme, is high enough that the lysozyme purity is very low. The purity can be significantly increased by reducing the amount of solvent so as to reduce the extraction factors for lysozyme and the other proteins from 20 and 2 to, say, 1.5 and 0.15, and by using several countercurrent extraction stages instead of just a single stage. The effect of doing this can be seen by studying the plot of the Kremser equation in Figure 5.9, where it is seen that when the stripping factor (by analogy, the extraction factor) is 1.5, the fraction not stripped (or not extracted) is greatly reduced by increasing the number of countercurrent equilibrium stages; but this is not the case when the factor is 0.15. This is the subject of Exercise 8.48.

§8.6.3 Supercritical-Fluid Extraction (SFE)

At conditions beyond critical values of temperature and pressure, substances like CO_2, ethylene, propylene, and nitrous oxide exhibit tunable temperature- and pressure-sensitive density that modulates solubility to extract (back-extract) a bioproduct from (to) an aqueous solution. Adiabatic compression is used to increase density of the supercritical-fluid molecules, which increases solubility of the solute (relative to that in a gas). At the same time, supercritical-fluid viscosities are one to two orders of magnitude lower than that of a liquid, while supercritical-fluid diffusivities are one to two orders of magnitude higher. This allows easy penetration of solid matrices like coffee beans and seeds to extract solutes like caffeine and oils, respectively. Supercritical-fluid extraction (SFE) with inexpensive, apolar CO_2 ($T_c = 31°C$; $P_c = 73$ atm) is used commercially to recover relatively nonpolar, nonionic compounds like therapeutic alkaloids, remove ethanol from an ethanol–water mixture to manufacture alcohol-free beer, and selectively remove α-acids and volatile flavor components from hops. However, CO_2 does not easily extract proteins or carbohydrates. Cosolvents like methanol can increase solubility of more polar, oxygen-containing therapeutics, but conditions for solubility must be determined experimentally. Additional discussion of SFE is presented in Chapter 11.

SFE Operation

Expensive high-pressure equipment and costs for compression currently limit production-scale SFE applications for bioproducts. Feed may be separated from incoming supercritical fluid (SF) by an SF-permeable membrane barrier to efficiently segregate exiting raffinate and extract while maintaining a large interfacial area independent of fluid velocity. A subsequent expansion chamber flashes gaseous CO_2 from the extractant. An ionic surfactant and cosurfactant (e.g., octane) or cosolvent (e.g., isooctane) may be added to

supercritical ethane, ethylene, or methane to form a dispersed phase for reversed-micelle extraction and back extraction of amino acids and proteins. A fluorocarbon surfactant, ammonium carboxylate perfluoropolymer, has been added to CO_2 to lower the critical point of the fluid and extract proteins with reverse micelles [113].

SUMMARY

1. A solvent can be used to selectively extract one or more components from a liquid mixture.

2. Although liquid–liquid extraction is a reasonably mature separation operation, considerable experimental effort is often needed to find a solvent and residence-time requirements or values of HETS, NTU, or mass-transfer coefficients.

3. Mass-transfer rates in extraction are lower than in vapor–liquid systems. Column efficiencies are frequently low.

4. Commercial extractors range from simple columns with no mechanical agitation to centrifugal devices that spin at several thousand revolutions per minute. The selection scheme in Table 8.3 is useful for choosing suitable extractors for a given separation.

5. Solvent selection is facilitated by consideration of a number of chemical and physical factors given in Tables 8.4 and 8.2.

6. For extraction with ternary mixtures, phase equilibrium is conveniently represented on equilateral- or right-triangle diagrams for both Type I (solute and solvent completely miscible) and Type II (solute and solvent not completely miscible) systems.

7. For determining equilibrium-stage requirements of single-section, countercurrent cascades for ternary systems, the graphical methods of Hunter and Nash (equilateral-triangle diagram), Kinney (right-triangle diagram), or Varteressian and Fenske (distribution diagram of McCabe–Thiele type) can be applied. These methods can also determine minimum and maximum solvent requirements.

8. A two-section, countercurrent cascade with extract reflux can be employed with a Type II ternary system to enable a sharp separation of a binary-feed mixture. Obtaining stage requirements of a two-section cascade is conveniently carried out by the graphical method of Maloney and Schubert using a Janecke equilibrium diagram. Addition of raffinate reflux is of little value.

9. When few equilibrium stages are required, mixer-settler cascades are attractive because each mixer can be designed to approach an equilibrium stage. With many ternary systems, the residence-time requirement may be

only a few minutes for a 90% approach to equilibrium using an agitator input of approximately 4 hp/1,000 gal. Adequate phase-disengaging area for the settlers may be estimated from the rule of 5 gal of combined extract and raffinate per minute per square foot of disengaging area.

10. For mixers utilizing a six-flat-bladed turbine in a closed vessel with side vertical baffles, extractor design correlations are available for estimating, for a given extraction, mixing-vessel dimensions, minimum impeller rotation rate for uniform dispersion, impeller horsepower, mean droplet size, range of droplet sizes, interfacial area per unit volume, dispersed- and continuous-phase mass-transfer coefficients, and stage efficiency.

11. For column extractors, with and without mechanical agitation, correlations for determining flooding, and column diameter and height, are suitable only for preliminary sizing. For final extractor selection and design, recommendations of equipment vendors and scale-up procedures based on data from pilot-size equipment are desirable.

12. Sizing of most column extractors must consider axial dispersion, which can reduce mass-transfer driving forces and increase column height. Axial dispersion is most significant in the continuous phase.

13. Small biomolecules (e.g., antibiotics) may be extracted from fermentation broths with common organic solvents. Caffeine, oils, or volatiles may be extracted from solid seeds or beans using supercritical fluids. Labile biopolymers (e.g., proteins) are extracted using aqueous two-phase systems like PEG–dextran–water or PEG–potassium phosphate–water.

14. Partitioning (e.g., K_D values) of bioproducts during organic-solvent or aqueous two-phase extraction is influenced by pH, temperature, salts, and solute valence. Hydrogen bonding, ion pairing and Lewis acid-base complexation also influence partitioning in organic-solvent extraction. Size of solute, and polymer and affinity ligand, affect partitioning in aqueous two-phase extraction. Values of K_D may be predicted from theory using a minimum of experimental data.

REFERENCES

1. Derry, T.K., and T.I. Williams, *A Short History of Technology*, Oxford University Press, New York (1961).

2. Bailes, P.J., and A. Winward, *Trans. Inst. Chem. Eng.*, **50**, 240–258 (1972).

3. Bailes, P.J., C. Hanson, and M.A. Hughes, *Chem. Eng.*, **83**(2), 86–100 (1976).

4. Lo, T.C., M.H.I. Baird, and C. Hanson, Eds., *Handbook of Solvent Extraction*, Wiley-Interscience, New York (1983).

5. Reissinger, K.-H., and J. Schroeter, "Alternatives to Distillation," *I. Chem. E. Symp. Ser. No.* **54**, 33–48 (1978).

6. Humphrey, J.L., J.A. Rocha, and J.R. Fair, *Chem. Eng.*, **91**(19), 76–95 (1984).

7. Fenske, M.R., C.S. Carlson, and D. Quiggle, *Ind. Eng. Chem.*, **39**, 1932 (1947).

8. Scheibel, E.G., *Chem. Eng. Prog.*, **44**, 681 (1948).

9. Scheibel, E.G., *AIChE J.*, **2**, 74 (1956).

10. Scheibel, E.G., U.S. Patent 3,389,970 (June 25, 1968).

11. Oldshue, J., and J. Rushton, *Chem. Eng. Prog.*, **48**(6), 297 (1952).

12. Reman, G.H., *Proceedings of the 3rd World Petroleum Congress*, The Hague, Netherlands, Sec. III, 121 (1951).

13. Reman, G.H., *Chem. Eng. Prog.*, **62**(9), 56 (1966).

14. Misek, T., and J. Marek, *Br. Chem. Eng.*, **15**, 202 (1970).

15. Fischer, A., *Verfahrenstechnik*, **5**, 360 (1971).

16. Karr, A.E., *AIChE J.*, **5**, 446 (1959).

17. Karr, A.E., and T.C. Lo, *Chem. Eng. Prog.*, **72**(11), 68 (1976).

18. Prochazka, J., J. Landau, F. Souhrada, and A. Heyberger, *Br. Chem. Eng.*, **16**, 42 (1971).

19. Barson, N., and G.H. Beyer, *Chem. Eng. Prog.*, **49**(5), 243–252 (1953).

20. Reissinger, K.-H., and J. Schroeter, "Liquid-Liquid Extraction, Equipment Choice," in J.J. McKetta and W.A. Cunningham, Eds., *Encyclopedia of Chemical Processing and Design*, Vol. **21**, Marcel Dekker, New York (1984).

21. Cusack, R.W., P. Fremeaux, and D. Glatz, *Chem. Eng.*, **98**(2), 66–76 (1991).

22. Robbins, L.A., *Chem. Eng. Prog.*, **76**(10), 58–61 (1980).

23. Naser, S.F., and R.L. Fournier, *Comput. Chem. Eng.*, **15**, 397–414 (1991).

24. Darwent, B., and C.A. Winkler, *J. Phys. Chem.*, **47**, 442–454 (1943).

25. Treybal, R.E., *Liquid Extraction*, 2nd ed., McGraw-Hill, New York (1963).

26. Hunter, T.G., and A.W. Nash, *J. Soc. Chem. Ind.*, **53**, 95T–102T (1934).

27. Kinney, G.F., *Ind. Eng. Chem.*, **34**, 1102–1104 (1942).

28. Venkataranam, A., and R.J. Rao, *Chem. Eng. Sci.*, **7**, 102–110 (1957).

29. Sawistowski, H., and W. Smith, *Mass Transfer Process Calculations*, Interscience, New York (1963).

30. Varteressian, K.A., and M.R. Fenske, *Ind. Eng. Chem.*, **28**, 1353–1360 (1936).

31. Skelland, A.H.P., *Ind. Eng. Chem.*, **53**, 799–800 (1961).

32. Randall, M., and B. Longtin, *Ind. Eng. Chem.*, **30**, 1063, 1188, 1311 (1938); **31**, 908, 1295 (1939); **32**, 125 (1940).

33. Maloney, J.O., and A.E. Schubert, *Trans. AIChE*, **36**, 741 (1940).

34. Flynn, A.W., and R.E. Treybal, *AIChE J.*, **1**, 324–328 (1955).

35. Ryon, A.D., F.L. Daley, and R.S. Lowrie, *Chem. Eng. Prog.*, **55**(10), 70–75 (1959).

36. Happel, J., and D.G. Jordan, *Chemical Process Economics*, 2nd ed., Marcel Dekker, New York (1975).

37. Rushton, J.H., and J.Y. Oldshue, *Chem Eng. Prog.*, **49**, 161–168 (1953).

38. Laity, D.S., and R.E. Treybal, *AIChE J.*, **3**, 176–180 (1957).

39. Skelland, A.H.P., and G.G. Ramsey, *Ind. Eng. Chem. Res.*, **26**, 77–81 (1987).

40. Skelland, A.H.P., and J.M. Lee, *Ind. Eng. Chem. Process Des. Dev.*, **17**, 473–478 (1978).

41. MacMullin, R.B., and M. Weber, *Trans. AIChE*, **31**, 409–458 (1935).

42. Lewis, J.B., I. Jones, and H.R.C. Pratt, *Trans. Inst. Chem. Eng.*, **29**, 126 (1951).

43. Coulson, J.M., and J.F. Richardson, *Chemical Engineering*, Vol. 2, 4th ed., Pergamon, Oxford (1991).

44. Vermuelen, T., G.M. Williams, and G.E. Langlois, *Chem. Eng. Prog.*, **51**, 85F (1955).

45. Gnanasundaram, S., T.E. Degaleesan, and G.S. Laddha, *Can. J. Chem. Eng.*, **57**, 141–144 (1979).

46. Chen, H.T., and S. Middleman, *AIChE J.*, **13**, 989–995 (1967).

47. Sprow, F.B., *AIChE J.*, **13**, 995–998 (1967).

48. Davies, J.T., *Turbulence Phenomena*, Academic Press, New York, p. 311 (1978).

49. Cornish, A.R.H., *Trans. Inst. Chem. Eng.*, **43**, T332–T333 (1965).

50. Skelland, A.H.P., and L.T. Moeti, *Ind. Eng. Chem. Res.*, **29**, 2258–2267 (1990).

51. Batchelor, G.K., *Proc. Cambridge Phil. Soc.*, **47**, 359–374 (1951).

52. Stichlmair, J., *Chemie-Ingenieur-Technik*, **52**, 253 (1980).

53. Logsdail, D.H., J.D. Thornton, and H.R.C. Pratt, *Trans. Inst. Chem. Eng.*, **35**, 301–315 (1957).

54. Landau, J., and R. Houlihan, *Can. J. Chem. Eng.*, **52**, 338–344 (1974).

55. Gayler, R., N.W. Roberts, and H.R.C. Pratt, *Trans. Inst. Chem. Eng.*, **31**, 57–68 (1953).

56. Thornton, J.D., *Chem. Eng. Sci.*, **5**, 201–208 (1956).

57. Reman, G.H., and R.B. Olney, *Chem. Eng. Prog.*, **52**(3), 141–146 (1955).

58. Strand, C.P., R.B. Olney, and G.H. Ackerman, *AIChE J.*, **8**, 252–261 (1962).

59. Reman, G.H., *Chem. Eng. Prog.*, **62**(9), 56–61 (1966).

60. Karr, A.E., and T.C. Lo, "Performance of a 36-inch Diameter Reciprocating-Plate Extraction Column," paper presented at the 82nd National Meeting of AIChE, Atlantic City, NJ (Aug. 29–Sept. 1, 1976).

61. Thornton, J.D., *Science and Practice of Liquid-Liquid Extraction*, Vol. 1, Clarendon Press, Oxford (1992).

62. Strigle, R.F., Jr., *Random Packings and Packed Towers*, Gulf Publishing Company, Houston, TX (1987).

63. Sleicher, C.A., Jr., *AIChE J.*, **5**, 145–149 (1959).

64. Miyauchi, T., and T. Vermeulen, *Ind. Eng. Chem. Fund.*, **2**, 113–126 (1963).

65. Sleicher, C.A., Jr., *AIChE J.*, **6**, 529–531 (1960).

66. Miyauchi, T., and T. Vermeulen, *Ind. Eng. Chem. Fund.*, **2**, 304–310 (1963).

67. Garcia, A.A., M.R. Bonen, J. Ramirez-Vick, M. Sadaka, and A. Vuppa, *Bioseparation Process Science*, Blackwell Science, Malden, MA (1999).

68. Ghosh, R. *Principles of Bioseparations Engineering*, World Scientific, Singapore (2006).

69. Harrison, R.G., P. Todd, S.R. Rudge, and D.P. Petrides, *Bioseparations Science and Engineering*, Oxford University Press, New York (2003).

70. Shuler, M.L., and F. Kargi, *Bioprocess Engineering*, 2nd ed., Prentice Hall PTR, Upper Saddle River, NJ (2002).

71. Ward, O.P., *Bioprocessing*, Van Nostrand Reinhold, New York (1991).

72. Gu, T., "Liquid-Liquid Partitioning Methods for Bioseparations," in S. Ahuja, Ed., *Handbook of Bioseparations*, Academic Press, San Diego, CA (2000).

73. Belter, P.A., E.L. Cussler, and W.-S. Hu, *Bioseparations: Downstream Processing for Biotechnology*, John Wiley & Sons, New York (1988).

74. Rydberg, J., "Introduction to Solvent Extraction," in J. Rydberg, C. Musikas, and G.R. Choppin, Eds., *Principles and Practices of Solvent Extraction* pp. 1–17. *Dekker*, New York (1992).

75. Vandamme, E.J., *Biotechnology of Industrial Antibiotics*, Dekker, New York (1984).

76. Essien, D.E., and D.L. Pyle, "Fermentation Ethanol Recovery by Solvent Extraction," in M.S. Verrall and M.J. Hudson, Eds., *Separations for Biotechnology* pp. 320–332, Ellis Horwood, Chichester (1987).

77. Hildebrand, J.H., J.M. Prausnitz, and R.L. Scott, *Regular and Related Solutions*, Van Nostrand, New York (1970) and *Handbook of Chemistry and Physics*, CRC Press, Boca Raton, FL (1986).

78. Likidis, Z., and K. Schugeri, *Biotechnol. Lett.*, **9**(4), 229–232 (1987).

79. Schügerl, K., *Solvent Extraction in Biotechnology: Recovery of Primary and Secondary Metabolites*, Springer-Verlag, Berlin (1994).

80. Krishna, R., C.Y. Low, D.M.T. Newsham, C.G. Olivera-Fuentes, and G.L. Standart, *Chem. Eng. Sci.*, **40**(6), 893–903 (1985).

81. Wesselingh, J.A., and R. Krishna, *Mass Transfer*, Ellis Horwood, Chichester, England (1990).

82. Walter, H., D.E. Brooks, and D. Fisher, Eds., *Partitioning in Aqueous Two-Phase Systems: Theory, Methods, Uses and Applications to Biotechnology*, Academic Press, New York (1985).

83. Sutherland, I.A., and D. Fisher, "Partitioning: A Comprehensive Bibliography," in H. Walter, D.E. Brooks, and D. Fisher, Eds., *Partitioning in Aqueous Two-Phase Systems: Theory, Methods, Uses and Applications to Biotechnology*, pp. 627–676, Academic Press, Orlando, FL (1985).

84. Albertsson, P.A., *Partition of Cell Particles and Macromolecules*, 3rd ed., John Wiley & Sons, New York (1986).

85. Zaslavsky, B.Y., *Aqueous Two-Phase Partitioning: Physical Chemistry and Bioanalytical Applications*, Dekker, New York (1995).

86. Walter, H., and G. Johansson, Eds., *Methods in Enzymology* Vol. **228**, Academic Press, San Diego, CA (1994).

87. Sasakawa, S., and H. Walter, *Biochemistry*, **11**, 2760 (1972).

88. Albertsson, P.-A., and F. Tjerneld, "Phase Diagram," in H. Walter and G. Johansson, Eds., *Methods in Enzymology*, Vol. **228**, pp. 3–13, Academic Press, San Diego, CA (1994).

89. Flory, P.J., *J. Chem. Phys.*, **10**, 51–61 (1942).

90. Diamond, A.D., and J.R. Hsu, "Aqueous Two-Phase Systems for Biomolecule Separation," in C.L. Cooney and A.E. Humphrey, Eds., *Advances in Biochemical Engineering/Biotechnology* Vol. **47**, pp. 89–135, Springer-Verlag, Berlin (1992).

91. King, R.S., H.W. Blanch, and J.M. Prausnitz, *AIChE J.*, **34**, 1585 (1988).

92. Haynes, C.A., H.W. Blanch, and J.M. Prausnitz, *Fluid Phase Equilibrium*, **53**, 463 (1989).

93. Huddleston, J.G., R. Wang, and J.A. Flanagan, *J. Chromatogr. A*, **668**, 3 (1994).

94. Hartounian, H., E.W. Kaler, and S.I. Sandler, *Ind. Eng. Chem. Res.*, **33**, 2294 (1994).

95. Walter, H., S. Sasakawa, and P.-A., Albertsson, *Biochemistry*, **11**, 3880 (1972).

96. Diamond, A.D., and J.T. Hsu, *Biotechnology and Bioengineering*, **34**, 1000–1014 (1989).

97. Harris, J.M., and M. Yalpani, "Polymer Ligands Used in Affinity Partitioning and Their Synthesis," in H. Walter, D.E. Brooks, and D. Fisher, Eds., *Partitioning in Aqueous Two-Phase Systems: Theory, Methods, Uses and Applications to Biotechnology*, pp. 589–626, Academic Press, New York (1985).

98. Kopperschläger, G., "Affinity Extraction with Dye Ligands," in H. Walter and G. Johansson, Eds., *Methods in Enzymology* Vol. **228**, p. 313, Academic Press, San Diego, CA (1994).

99. Vermeulen, T., J.S. Moon, A. Hennico, and T. Miyauchi, *Chem. Eng. Prog.*, **62**(9), 95–101 (1966).

100. Danckwerts, P.V., *Chem. Eng. Sci.*, **2**, 1–13 (1953).

101. Wehner, J.F., and R.H. Wilhelm, *Chem. Eng. Sci.*, **6**, 89–93 (1956).

102. Geankoplis, C.J., and A.N. Hixson, *Ind. Eng. Chem.*, **42**, 1141–1151 (1950).

103. Gier, T.E., and J.O. Hougen, *Ind. Eng. Chem.*, **45**, 1362–1370 (1953).

104. Watson, J.S., and H.D. Cochran, Jr., *Ind. Eng. Chem. Process Des. Dev.*, **10**, 83–85 (1971).

105. Kumar, A., and S. Hartland, *Ind. Eng. Chem. Res.*, **38**, 1040–1056 (1999).

106. Green, D.W., and R.H. Perry, Eds., *Perry's Chemical Engineers' Handbook*, 8th ed., McGraw-Hill, New York (2008).

107. Karr, A.E., W. Gebert, and M. Wang, *Canadian Journal of Chemical Engineering*, **58**, 249–252 (1980).

108. Balasubramaniam, D., C. Wilkinson, K. Van Cott, and C. Zhang, *J. Chromatography A*, **989**, 119–129 (2003).

109. Hammer, S., A. Pfennig, and M. Stumpf, *J. Chem. Eng. Data*, **39**, 409–413 (1994).

110. Diamond, A.D., *AIChE J.*, **36**, 1017–1024 (1990).

111. Croll, T., P.D. Munro, D.J. Winzor, M. Trau, and L.K. Nielsen, *J. Polym. Sci., Part B: Polym Phys.*, **41**, 437–443 (2003).

112. Prausnitz, J.M., R.N. Lichtenthaler, and E.G. de Azevedo, *Molecular Thermodynamics of Fluid-Phase Equilibria*, 3rd ed., Prentice Hall PTR, Upper Saddle River, NJ (1999).

113. Cooper, A.I., and J.M. DeSimone, *Current Opinion in Solid State and Materials Science*, **1**(6) 761–768 (1996).

STUDY QUESTIONS

8.1. When liquid–liquid extraction is used, are other separation operations needed? Why?

8.2. Under what conditions is extraction preferred to distillation?

8.3. What are the important characteristics of a good solvent?

8.4. Can a mixer-settler unit be designed to closely approach phase equilibrium?

8.5. Under what conditions is mechanically assisted agitation necessary in an extraction column?

8.6. What are the advantages and disadvantages of mixer-settler extractors?

8.7. What are the advantages and disadvantages of continuous, counterflow, mechanically assisted extractors?

8.8. What is the difference between a Type I and a Type II ternary system? Can a system transition from one type to the other by changing the temperature? Why?

8.9. What is meant by the mixing point? For a multistage extractor, is the mixing point on a triangular diagram the same for the feeds and the products?

8.10. What happens if more than the maximum solvent rate is used? What happens if less than the minimum solvent rate is used?

8.11. What are extract and raffinate reflux? Which one is of little value? Why?

8.12. What is the typical range of residence time for approaching equilibrium in an agitated mixer when the liquid-phase viscosities are less than 5 cP?

8.13. When continuously bringing together two liquid phases in an agitated vessel, are the residence times of the two phases necessarily the same? If not, are there any conditions where they would be the same?

8.14. Why is liquid–liquid mass transfer so complex in agitated systems?

8.15. What are Marangoni effects? How do they influence mass transfer?

8.16. What is axial dispersion? What causes it, and should it be avoided?

8.17. What are relative advantages and disadvantages of organic-solvent, aqueous two-phase, and supercritical-fluid extraction for recovery of bioproducts?

8.18. How do effects of pH, salt composition, and solute valence on partitioning of bioproducts compare in organic-solvent and aqueous two-phase extraction?

8.19. What information about mass transfer in liquid–liquid extraction does the linearized Maxwell–Stefan relation provide?

8.20. How do polymer and solute size, and affinity ligand, affect partitioning in aqueous two-phase extraction?

EXERCISES

Section 8.1

8.1.   Extraction versus distillation.

Explain why it is preferable to separate a dilute mixture of benzoic acid in water by solvent extraction rather than by distillation.

8.2.   Liquid–liquid extraction versus distillation.

Why is liquid–liquid extraction preferred over distillation for the separation of a mixture of formic acid and water?

8.3.   Selection of extraction equipment.

Based on Table 8.3 and the selection scheme in Figure 8.8, is an RDC appropriate for extraction of acetic acid from water by ethyl acetate in the process in Figure 8.1? What other types of extractors might be considered?

8.4.   Extraction devices.

What is the major advantage of the ARD over the RDC? What is the disadvantage of the ARD compared to the RDC?

8.5.   Selection of extraction devices.

Under what conditions is a cascade of mixer-settler units probably the best choice of extraction equipment?

8.6.   Selection of extraction device.

A petroleum reformate stream of 4,000 bbl/day is to be contacted with diethylene glycol to extract aromatics from paraffins. The ratio of solvent to reformate volume is 5. It is estimated that eight theoretical stages are needed. Using Tables 8.2 and 8.3 and Figure 8.8, which extractors would be suitable?

Section 8.2

8.7.   Selection of extraction solvents.

Using Table 8.4, select possible liquid–liquid extraction solvents for separating the following mixtures: (a) water–ethyl alcohol, (b) water–aniline, and (c) water–acetic acid. For each case, indicate which of the two components should be the solute.

8.8.   Selection of extraction solvents.

Using Table 8.4, select liquid–liquid extraction solvents for removing the solute from the carrier in the following cases:

	Solute	Carrier
(a)	Acetone	Ethylene glycol
(b)	Toluene	*n*-Heptane
(c)	Ethyl alcohol	Glycerine

8.9.   Characteristics of an extraction system.

For extracting acetic acid (A) from a dilute water solution (C) into ethyl acetate (S) at 25°C, estimate or obtain data for $(K_A)_D$, $(K_C)_D$, $(K_S)_D$, and β_{AC}. Does this system exhibit: (a) high selectivity, (b) high solvent capacity, and (c) easy solvent recovery? Can you select a better solvent than ethyl acetate?

8.10.   Estimation of interfacial tension.

Very low values of interfacial tension result in stable emulsions that are difficult to separate, while very high values require large energy inputs to form the dispersed phase. It is best to measure the interfacial tension for the two-phase mixture of interest. However, in the absence of experimental data, propose a method for estimating the interfacial tension of a ternary system using only the compositions of the equilibrium phases and the values of surface tension in air for each of the three components.

Section 8.3

8.11.   Extraction of acetone by trichloroethane.

One thousand kg/h of a 45 wt% acetone-in-water solution is to be extracted at 25°C in a continuous, countercurrent system with pure 1,1,2-trichloroethane to obtain a raffinate containing 10 wt% acetone. Using the following equilibrium data, determine with an equilateral-triangle diagram: (a) the minimum flow rate of solvent; (b) the number of stages required for a solvent rate equal to 1.5 times minimum; (c) the flow rate and composition of each stream leaving each stage.

	Acetone, Weight Fraction	Water, Weight Fraction	Trichloroethane, Weight Fraction
Extract	0.60	0.13	0.27
	0.50	0.04	0.46
	0.40	0.03	0.57
	0.30	0.02	0.68
	0.20	0.015	0.785
	0.10	0.01	0.89
Raffinate	0.55	0.35	0.10
	0.50	0.43	0.07
	0.40	0.57	0.03
	0.30	0.68	0.02
	0.20	0.79	0.01
	0.10	0.895	0.005

The tie-line data are:

Raffinate, Weight Fraction Acetone	Extract, Weight Fraction Acetone
0.44	0.56
0.29	0.40
0.12	0.18

8.12. Using a right-triangle diagram for extraction.

Solve Exercise 8.11 with a right-triangle diagram.

8.13. Extraction of isopropanol with water.

A distillate of 45 wt% isopropyl alcohol, 50 wt% diisopropyl ether, and 5 wt% water is obtained from an isopropyl alcohol finishing unit. The ether is to be recovered by liquid–liquid extraction with water, the solvent, entering the top and the feed entering the bottom, so as to produce an ether containing less than 2.5 wt% alcohol and an extracted alcohol of at least 20 wt%. The unit will operate at 25°C and 1 atm. Using the method of Varteressian and Fenske with a McCabe–Thiele diagram, find the stages required. Is it possible to obtain an extracted alcohol composition of 25 wt%?

Equilibrium data are given below.

PHASE-EQUILIBRIUM (TIE-LINE) DATA AT 25°C, 1 ATM

Ether phase			Water phase		
Wt% Alcohol	Wt% Ether	Wt% Water	Wt% Alcohol	Wt% Ether	Wt% Water
2.4	96.7	0.9	8.1	1.8	90.1
3.2	95.7	1.1	8.6	1.8	89.6
5.0	93.6	1.4	10.2	1.5	88.3
9.3	88.6	2.1	11.7	1.6	86.7
24.9	69.4	5.7	17.5	1.9	80.6
38.0	50.2	11.8	21.7	2.3	76.0
45.2	33.6	21.2	26.8	3.4	69.8

ADDITIONAL POINTS ON PHASE BOUNDARY

Wt% Alcohol	Wt% Ether	Wt% Water
45.37	29.70	24.93
44.55	22.45	33.00
39.57	13.42	47.01
36.23	9.66	54.11
24.74	2.74	72.52
21.33	2.06	76.61
0	0.6	99.4
0	99.5	0.5

8.14. Extraction of trimethylamine from benzene with water.

Benzene and trimethylamine (TMA) are to be separated in a three-equilibrium-stage liquid–liquid extraction column using pure water as the solvent. If the solvent-free extract and raffinate products are to contain, respectively, 70 and 3 wt% TMA, find the original feed composition and the water-to-feed ratio with a right-triangle diagram. There is no reflux. Equilibrium data are as follows:

TRIMETHYLAMINE–WATER–BENZENE COMPOSITIONS ON PHASE BOUNDARY

Extract, wt%			Raffinate, wt%		
TMA	H₂O	Benzene	TMA	H₂O	Benzene
5.0	94.6	0.4	5.0	0.0	95.0
10.0	89.4	0.6	10.0	0.0	90.0
15.0	84.0	1.0	15.0	1.0	84.0
20.0	78.0	2.0	20.0	2.0	78.0
25.0	72.0	3.0	25.0	4.0	71.0
30.0	66.4	3.6	30.0	7.0	63.0
35.0	58.0	7.0	35.0	15.0	50.0
40.0	47.0	13.0	40.0	34.0	26.0

The tie-line data are:

Extract, wt% TMA	Raffinate, wt% TMA
39.5	31.0
21.5	14.5
13.0	9.0
8.3	6.8
4.0	3.5

8.15. Extraction of diphenylhexane from docosane with furfural.

The system docosane–diphenylhexane (DPH)–furfural is representative of complex systems encountered in the solvent refining of lubricating oils. Five hundred kg/h of a 40 wt% mixture of DPH in docosane are to be extracted in a countercurrent system with 500 kg/h of a solvent containing 98 wt% furfural and 2 wt% DPH to produce a raffinate of 5 wt% DPH. Calculate, with a right-triangle diagram, the stages required and the kg/h of DPH in the extract at 45°C and 80°C.

BINODAL CURVES IN DOCOSANE–DIPHENYLHEXANE–FURFURAL SYSTEM [*IND. ENG. CHEM.*, 35, 711 (1943)]

Wt% at 45°C			Wt% at 80°C		
Docosane	DPH	Furfural	Docosane	DPH	Furfural
96.0	0.0	4.0	90.3	0.0	9.7
84.0	11.0	5.0	50.5	29.5	20.0
67.0	26.0	7.0	34.2	35.8	30.0
52.5	37.5	10.0	23.8	36.2	40.0
32.6	47.4	20.0	16.2	33.8	50.0
21.3	48.7	30.0	10.7	29.3	60.0
13.2	46.8	40.0	6.9	23.1	70.0
7.7	42.3	50.0	4.6	15.4	80.0
4.4	35.6	60.0	3.0	7.0	90.0
2.6	27.4	70.0	2.2	0.0	97.8
1.5	18.5	80.0			
1.0	9.0	90.0			
0.7	0.0	99.3			

The tie lines in the docosane–diphenylhexane–furfural system are:

Docosane Phase Composition, wt%			Furfural Phase Composition, wt%		
Docosane	DPH	Furfural	Docosane	DPH	Furfural
Temperature, 45°C:					
85.2	10.0	4.8	1.1	9.8	89.1
69.0	24.5	6.5	2.2	24.2	73.6
43.9	42.6	13.3	6.8	40.9	52.3
Temperature, 80°C:					
86.7	3.0	10.3	2.6	3.3	94.1
73.1	13.9	13.0	4.6	15.8	79.6
50.5	29.5	20.2	9.2	27.4	63.4

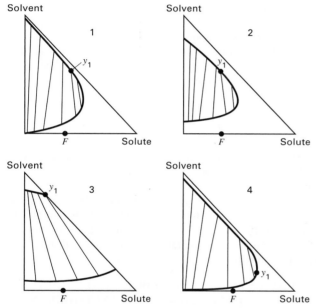

Figure 8.45 Data for Exercise 8.16.

8.16. Selection of extraction method.

For each ternary system in Figure 8.45, indicate whether: (a) simple countercurrent extraction, (b) countercurrent extraction with extract reflux, (c) countercurrent extraction with raffinate reflux, or (d) countercurrent extraction with both extract and raffinate reflux would be the most economical.

8.17. Extraction of acetone from two feeds.

Two feeds—F at 7,500 kg/h containing 50 wt% acetone and 50 wt% water, and F' at 7,500 kg/h containing 25 wt% acetone and 75 wt% water—are to be extracted in a system with 5,000 kg/h of 1,1,2-trichloroethane at 25°C to give a 10 wt% acetone raffinate. Calculate the stages required and the stage to which each feed

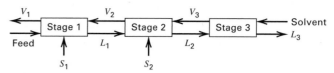

Figure 8.46 Data for Exercise 8.18.

should be introduced using a right-triangle diagram. Equilibrium data are in Exercise 8.11.

8.18. Extraction in a three-stage unit.

The three-stage extractor shown in Figure 8.46 is used to extract the amine from a fluid consisting of 40 wt% benzene (B) and 60 wt% trimethylamine (T). The solvent (water) flow to stage 3 is 5,185 kg/h and the feed flow rate is 10,000 kg/h. On a solvent-free basis, V_1 is to contain 76 wt% T, and L_3 is to contain 3 wt% T. Determine the required solvent flow rates S_1 and S_2 using an equilateral-triangle diagram. Solubility data are in Exercise 8.14.

8.19. Analysis of a multiple-feed, countercurrent extraction cascade.

The extraction process shown Figure 8.47 is conducted without extract or raffinate reflux. Feed F' is composed of solvent and solute, and is an extract-phase feed. Feed F'' is composed of unextracted raffinate and solute and is a raffinate-phase feed. Derive the equations required to establish the three reference points needed to step off the stages in the extraction column. Show the graphical determination of these points on a right-triangle graph.

8.20. Extraction of MCH from heptane with aniline.

Fifty wt% methylcyclohexane (MCH) in n-heptane is fed to a countercurrent, stage-type extractor at 25°C. Aniline is the solvent and reflux is used at both ends of the column. An extract containing 95 wt% MCH and a raffinate containing 5 wt% MCH (both on a solvent-free basis) are required. The minimum extract reflux ratio is 3.49. Using a right-triangle diagram with the data of Exercise 8.22, calculate the: (a) raffinate reflux ratio, (b) amount of aniline that must be removed at the separator "on top" of the column, and (c) amount of solvent added to the solvent mixer at the bottom of the column.

8.21. Extraction of hafnium from zirconium.

Zirconium, which is used in nuclear reactors, is associated with hafnium, which has a high neutron-absorption cross section and must be removed. Refer to Figure 8.48 for a proposed liquid–liquid extraction process wherein tributyl phosphate (TBP) is used as a solvent for the separation. One L/h of 5.10-N HNO_3 containing 127 g of dissolved Hf and Zr oxides per liter is fed to stage 5 of the 14-stage extraction unit. The feed contains 22,000 g Hf per million g of Zr. Fresh TBP enters at stage 14, while scrub water is fed to stage 1. Raffinate is removed at stage 14, while the organic extract phase removed at stage 1 goes to a stripping unit. The stripping operation consists of a single contact between fresh water and the organic phase. (a) Use the data below to complete a material balance for the process. (b) Check the data for consistency. (c) What is the advantage of running the extractor as shown? Would you recommend that all stages be used?

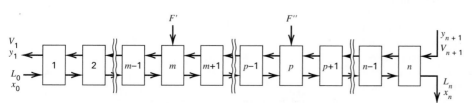

Figure 8.47 Data for Exercise 8.19.

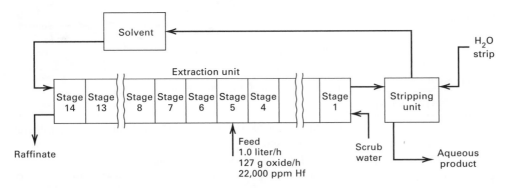

Figure 8.48 Data for Exercise 8.21.

STAGEWISE ANALYSES OF MIXER-SETTLER RUN

Stage	Organic Phase			Aqueous Phase		
	g oxide/ liter	N HNO₃	(Hf/Zr) × (100)	g oxide/ liter	N HNO₃	(Hf/Zn) × (100)
1	22.2	1.95	<0.010	17.5	5.21	<0.010
2	29.3	2.02	<0.010	27.5	5.30	<0.010
3	31.4	2.03	<0.010	33.5	5.46	<0.010
4	31.8	2.03	0.043	34.9	5.46	0.24
5	32.2	2.03	0.11	52.8	5.15	3.6
6	21.1	1.99	0.60	30.8	5.15	6.8
7	13.7	1.93	0.27	19.9	5.05	9.8
8	7.66	1.89	1.9	11.6	4.97	20
9	4.14	1.86	4.8	8.06	4.97	36
10	1.98	1.83	10	5.32	4.75	67
11	1.03	1.77	23	3.71	4.52	110
12	0.66	1.68	32	3.14	4.12	140
13	0.46	1.50	42	2.99	3.49	130
14	0.29	1.18	28	3.54	2.56	72
Stripper		0.65		76.4	3.96	<0.01

[Data from R.P. Cox, H.C. Peterson, and C.H. Beyer, *Ind. Eng. Chem.*, **50** (2), 141 (1958). Exercise adapted from E.J. Henley and H. Bieber, *Chemical Engineering Calculations*, McGraw-Hill, New York, p. 298 (1959).]

8.22. Extraction of diphenylhexane from docosane with furfural.

At 45°C, 5,000 kg/h of a mixture of 65 wt% docosane, 7 wt% furfural, and 28 wt% diphenylhexane is to be extracted with pure furfural to obtain a raffinate with 12 wt% diphenylhexane in a continuous, countercurrent, multistage liquid–liquid extraction system. Phase-equilibrium data are given in Exercise 8.15. Determine the: (a) minimum solvent flow, (b) flow rate and composition of the extract at the minimum solvent flow, and (c) stages if a solvent flow rate of 1.5 times minimum is used.

8.23. Extraction of diphenylhexane from docosane with furfural.

At 45°C, 1,000 kg/h of a mixture of 0.80 mass fraction docosane and 0.20 mass fraction diphenylhexane is extracted with pure furfural to remove some diphenylhexane from the feed. Phase-equilibrium data are given in Exercise 8.15. Determine the: (a) composition and flow rate of the extract and raffinate from a single equilibrium stage for solvent flow rates of 100, 1,000, and 10,000 kg/h; (b) minimum solvent flow rate to form two liquid phases; (c) maximum solvent flow rate to form two liquid phases; (d) composition and flow rate of the extract and raffinate if a solvent flow rate of 2,000 kg/h and two equilibrium stages are used in a countercurrent-flow system.

8.24. Extraction of acetone from water by 1,1,2-trichloroethane.

A liquid of 27 wt% acetone and 73 wt% water is to be separated at 25°C into a raffinate and extract by multistage, countercurrent liquid–liquid extraction with a solvent of pure 1,1,2-trichloroethane. Equilibrium data are given in Exercise 8.11. Determine the: (a) minimum solvent-to-feed ratio to obtain a raffinate essentially free of acetone, (b) composition of extract at the minimum solvent-to-feed ratio, (c) composition of the extract stream leaving stage 2 (see Figure 8.13), if a very large number of equilibrium stages is used with the minimum solvent.

Section 8.4

8.25. Extraction of MCH from heptane by aniline.

A feed mixture of 50 wt% n-heptane (H) and 50 wt% methylcyclohexane (MCH) is to be separated by liquid–liquid extraction into one 92.5 wt% MCH product and another containing 7.5 wt% MCH, both on a solvent-free basis. Aniline is the solvent. Using the equilibrium data below and the graphical method of Maloney and Schubert: (a) What is the minimum number of theoretical stages necessary to effect this separation? (b) What is the minimum extract reflux ratio? (c) If the reflux ratio is 7.0, how many theoretical contacts are required?

LIQUID–LIQUID EQUILIBRIUM DATA FOR THE SYSTEM n-HEPTANE–METHYLCYCLOHEXANE–ANILINE AT 25°C AND AT 1 ATM (101 kPa)

Hydrocarbon Layer		Solvent Layer	
Weight Percent MCH, Solvent-Free Basis	Pounds Aniline/Pound Solvent-Free Mixture	Weight Percent MCH, Solvent-Free Basis	Pounds Aniline/Pound Solvent-Free Mixture
0.0	0.0799	0.0	15.12
9.9	0.0836	11.8	13.72
20.2	0.087	33.8	11.5
23.9	0.0894	37.0	11.34
36.9	0.094	50.6	9.98
44.5	0.0952	60.0	9.0
50.5	0.0989	67.3	8.09
66.0	0.1062	76.7	6.83
74.6	0.1111	84.3	6.45
79.7	0.1135	88.8	6.0
82.1	0.116	90.4	5.9
93.9	0.1272	96.2	5.17
100.0	0.135	100.0	4.92

	Raffinate	Extract
Flow rate, lb/h	21,000	52,000
Density, lb/ft^3	63.5	45.3
Viscosity, cP	3.0	1.0

8.26. Use of extract reflux.

Liquids A and B, which have nearly identical boiling points, are to be separated by liquid–liquid extraction with solvent C. The following data represent the phase equilibrium at 95°C.

EQUILIBRIUM DATA, WT%

Extract Layer			Raffinate Layer		
A, %	B, %	C, %	A, %	B, %	C, %
0	7.0	93.0	0	92.0	8.0
1.0	6.1	92.9	9.0	81.7	9.3
1.8	5.5	92.7	14.9	75.0	10.1
3.7	4.4	91.9	25.3	63.0	11.7
6.2	3.3	90.5	35.0	51.5	13.5
9.2	2.4	88.4	42.0	41.0	17.0
13.0	1.8	85.2	48.1	29.3	22.6
18.3	1.8	79.9	52.0	20.0	28.0
24.5	3.0	72.5	47.1	12.9	40.0
31.2	5.6	63.2		Plait point	

[Adapted from McCabe and Smith, *Unit Operations of Chemical Engineering*, 4th ed., McGraw-Hill, New York, p. 557 (1985).]

Determine the minimum reflux returned from the extract product to produce an extract containing 83% A and 17% B (compositions in this exercise are on a solvent-free basis) and a raffinate product of 10% A and 90% B. The feed contains 35% A and 65% B and is a saturated raffinate. The raffinate is the heavy liquid. Determine the number of theoretical stages on both sides of the feed to produce the same end products from the same feed when the reflux ratio of the extract (pounds of extract reflux per pound of extract product including solvent) is twice the minimum. Calculate the stream masses per 1,000 lb of feed. Use equilateral-triangle coordinates, right-triangle coordinates, and solvent-free coordinates. Which method is best for this exercise?

8.27. Use of extract reflux with the Maloney–Schubert method.

Solve Exercise 8.20 by the Maloney–Schubert method.

Section 8.5

8.28. Design of a mixer-settler unit for extraction.

Acetic acid is extracted from a 3 wt% dilute solution in water with a solvent of isopropyl ether in a mixer-settler unit. The flow rates of the feed and solvent are 12,400 and 24,000 lb/h, respectively. Assuming a residence time of 1.5 minutes in the mixer and a settling vessel capacity of 4 gal/minute-ft^2, estimate: (a) diameter and height of the mixing vessel, assuming $H/D_T = 1$; (b) agitator horsepower for the mixing vessel; (c) diameter, length, and residence time in minutes of the settling vessel, assuming $L/D_T = 4$.

8.29. Extraction in an available unit.

A cascade of six mixer-settler units is available, each unit consisting of a 10-ft-diameter by 10-ft-high mixing vessel equipped with a 20-hp agitator, and a 10-ft-diameter by 40-ft-long settling vessel. If this cascade is used for the acetic acid extraction described in the introduction to this chapter, how many lb/hr of feed can be processed?

8.30. Agitator size for extraction of acetic acid.

Acetic acid is extracted from a dilute aqueous solution with isopropyl ether at 25°C in a countercurrent cascade of mixer-settler units. In one unit, the following conditions apply:

Interfacial tension = 13.5 dyne/cm. If the raffinate is the dispersed phase and the mixer residence time is 2.5 minutes, estimate for the mixer the: (a) dimensions of a closed, baffled vessel; (b) diameter of a flat-bladed impeller; (c) minimum rate of impeller rotation in rpm for uniform dispersion; and (d) agitator power requirement at the minimum rate of rotation.

8.31. Droplet characteristics for extraction of acetic acid.

For Exercise 8.30, estimate: (a) Sauter mean drop size, (b) range of drop sizes, and (c) interfacial area of the two-phase liquid–liquid emulsion.

8.32. Mass-transfer for extraction of acetic acid.

For Exercises 8.30 and 8.31 and the data below, estimate: (a) dispersed-phase mass-transfer coefficient, (b) continuous-phase mass-transfer coefficient, (c) Murphree dispersed-phase efficiency, and (d) fraction of acetic acid extracted.

Diffusivity of acetic acid: in the raffinate is 1.3×10^{-9} m^2/s and in the extract is 2.0×10^{-9} m^2/s. Distribution coefficient for acetic acid: $c_D/c_C = 2.7$.

8.33. Design of a mixer unit.

For the conditions and results of Example 8.4, determine the following when using a six-flat-bladed turbine impeller in a closed vessel with baffles and with the extract phase dispersed, based on the properties given below: (a) minimum rate of rotation of the impeller for complete and uniform dispersion; (b) agitator power requirement at the minimum rotation rate; (c) Sauter mean droplet diameter; (d) interfacial area; (e) overall mass-transfer coefficient, K_{OD}; (f) overall transfer units, N_{OD}; (g) Murphree efficiency, E_{MD}; and (h) fractional benzoic acid extraction.

Interfacial tension = 22 dyne/cm, and distribution coefficient for benzoic acid = $c_D/c_C = 21$.

	Raffinate Phase	Extract Phase
Density, g/cm^3	0.995	0.860
Viscosity, cP	0.95	0.59
Diffusivity of benzoic acid, cm^2/s	2.2×10^{-5}	1.5×10^{-5}

8.34. Diameter of an RDC column.

Estimate the diameter of an RDC column to extract acetic acid from water with isopropyl ether for Exercises 8.28 and 8.30.

8.35. Diameter of a Karr column.

Estimate the diameter of a Karr column to extract benzoic acid from water with toluene for the conditions of Exercise 8.33.

8.36. HETS of an RDC column.

Estimate HETS for an RDC column operating under the conditions of Exercise 8.34.

8.37. HETS of a Karr column.

Estimate HETS for a Karr column operating under the conditions of Exercise 8.35.

8.38. Scale-up of a Karr extraction column.

A Karr column is to be sized to extract 99.95% of the methyl vanillin in water using *o*-xylene. The feed rate is 6 m^3/hr with a vanillin concentration of 40 kg/m^3. The average partition coefficient is

5.6, defined as the concentration in xylene to the concentration in water. Experiments with a Karr column of 2-inch diameter and 1.5-meter plate-stack height give the optimal conditions as 610 mL/minute of feed and 270 mL/minute of xylene, for a stroke height of 19.1 mm and 250 strokes/min. Scale-up this data to determine the diameter and plate-stack height of a commercial Karr column.

8.39. Applicability of an existing Podbielniak centrifugal extractor.

Of the penicillin at a concentration of 0.065 kg/L, 98.5% is to be extracted from 3,000 L/h of a clarified and pH-adjusted broth with *n*-butyl acetate. The partition coefficient is 15. Optimal data from small-scale tests in a POD having a rotating cylinder of 20-cm diameter and 2.5-cm width, operating at 10,000 rpm, are a feed rate of 800 mL/minute and a solvent rate of 130 mL/minute. A commercial POD with a rotating cylinder of 91 cm diameter and 45-cm width, operating at a maximum rpm of 2,100, is available. Determine whether it can handle the required feed rate.

8.40. Solvent flow rate for a POD.

Penicillin in 12 m^3/h of fermentation broth is extracted using iso-amylacetate in a POD. The distribution coefficient at pH = 2 is $K_D = 50$. If the number of countercurrent equilibrium stages in the extractor is 4, determine the flow rate of isoamylacetate required to reduce the broth penicillin concentration from 35 g/L to 0.07 g/L.

8.41. Number of equilibrium stages for a POD.

Penicillin in 12 m^3/h of fermentation broth is extracted using 1.2 m^3/h isoamylacetate in a POD. The distribution coefficient at pH = 2 is $K_D = 50$. Estimate the number of countercurrent equilibrium stages required to reduce the broth penicillin concentration from 35 g/L to 0.035 g/L.

8.42. Number of equilibrium stages for an antibiotic.

Actinomycin D present at 260 mg/L in fermentation broth is to be extracted into butyl acetate. At pH 3.5, the distribution constant is 57. How many stages are required to recover 99% of the antibiotic from a broth flow rate of 450 L/hr using a solvent flow rate of 37 L/hr?

Section 8.6

8.43. Expressions for partition coefficients.

Derive general expressions for (a) the partition coefficient, K_D^o, in an organic (1) aqueous–(2) LLE system for an un-ionized (neutral) form of a weak base given by

$$HB^+ \leftrightarrow H^+ + B$$

and (b) the distribution coefficient between organic and aqueous fluid phases in terms of pH and pK_a.

8.44. Effect of pH on the distribution coefficient.

A monoacidic sugar extracted from water into hexanol has distribution coefficients of 6.0 and 0.47 moles per liter at pH 4.0 and 5.5, respectively. Estimate the value of K_D at pH 7.2.

8.45. Extraction parameters for two solutes.

Compare the required solvent volume, stage number, and purity obtained when penicillin F is extracted from water preferentially to penicillin K into amyl acetate at pH 3.1 relative to the values at pH 3.75. The respective pK_a values for penicillin F and K are 3.51 and 2.77, and the respective K_D^o values are 131 and 215.

8.46. Distribution coefficient with chemical equilibrium.

Based on chemical equilibrium, obtain a general expression for dependence of the distribution coefficient of a weak base bioproduct $B^{(2)}$ described by the equilibrium expression in Exercise 8.43 that interacts with an organic solvent extractant $X^{(1)}$ at the interface of an organic (1)/aqueous (2) system with valence z.

8.47. Extraction of a steroid.

Steroid in water at 6.8 mg/L is extracted using initially pure methylene dichloride in a volume ratio of 1% vol/vol. The distribution constant is 175. Estimate the concentration of steroid in the organic solvent and the fraction of steroid removed.

8.48. ATPE of proteins from tobacco broth.

In Example 8.19, the % extraction of lysozyme is high (95.2%), but the % purity of the lysozyme in the total extracted proteins is low (15.1%). The Kremser plot of Figure 5.9 suggests that by reducing the amount of solvent and increasing the number of stages, the % purity can be increased. To verify this, repeat the calculations of Example 8.19 by reducing the flow rate of PEG 3000 by a factor of 10 to 40.6 kg/h and re-solving Eqs. (1) to (6) to obtain a new, much-lower water flow rate in the solvent, and new protein-free top and bottom phase equilibrium compositions that satisfy the tie line given in Example 8.19. Then, as a function of the number of countercurrent equilibrium stages from 1 to 5, calculate the % extraction of lysozyme and the other proteins, and the % purity of lysozyme in the extracted proteins.

8.49. ATPE of alcohol dehydrogenase from a lysed broth.

The enzyme alcohol dehydrogenase (ADH) catalyzes oxidation of primary and secondary alcohols to aldehydes and ketones. It can be produced from a culture of *Lactobacillus brevis*, and recovered from the lysed broth by countercurrent ATPE with a PEG 8000–dextran T-500–water system at 10°C. Assume the clarified feed to the system is 1,000 kg/h containing 0.5 wt% ADH, 98 wt% water, and 1.5 wt% other solutes. From Diamond [110], a tie line for the PEG–dextran–water system has points in weight fractions on the binodal curve at (0.072, 0.003, 0.925) for the upper phase and (0.01, 0.137, 0.853) for the lower phase. For these conditions, the partition coefficient for ADH is 1.0, defined as the weight fraction of ADH in the upper phase divided by that in the lower phase on a solute-free basis. If 167.6 kg/h of dextran D-500 is added to the lysed solution, an extraction factor of 2.0 is desired, and three countercurrent equilibrium stages are used, calculate the: (a) kg/h of aqueous PEG 8000 solvent required, and the flow rates of PEG and water in the solvent; (b) flow rates in the upper and lower phases; and (c) % extraction of ADH to the upper phase.

Chapter 9

Approximate Methods for Multicomponent, Multistage Separations

§9.0 INSTRUCTIONAL OBJECTIVES

After completing this chapter, you should be able to:

- Select two key components, operating pressure, and type condenser for multicomponent distillation.
- Estimate minimum number of equilibrium stages and distribution of nonkey components by the Fenske equation, minimum reflux ratio by the Underwood method, number of equilibrium stages for a reflux ratio greater than minimum by the Gilliland correlation, and feed-stage location for a specified separation between two key components.
- Estimate stages for multicomponent absorption, stripping, and extraction using the Kremser equation.

Although rigorous methods are available for solving multicomponent separation problems, approximate methods continue to be used for preliminary design, parametric studies to establish optimal design conditions, process synthesis studies to determine optimal separation sequences, and to obtain initial approximations for rigorous, iterative methods.

Approximate methods of Kremser [1] and Edmister [2] for single-section cascades used in absorption were presented in §5.4. This chapter develops an additional approximate method for preliminary design and optimization of simple distillation, called the *Fenske–Underwood–Gilliland* (FUG) method. In addition, the Kremser method is extended to strippers and liquid–liquid extraction. Although all these methods are suitable for manual calculation if physical properties are independent of composition, computer calculations with process simulators are preferred.

§9.1 FENSKE–UNDERWOOD–GILLILAND (FUG) METHOD

Figure 9.1 gives an algorithm for the FUG method, named after the authors of the three steps in the procedure, which will be applied to the distillation column in Figure 9.3. From Table 5.4, the number of degrees of freedom with a total condenser is $2N + C + 9$. For a design case, the variables below are generally specified, with the partial reboiler counted as a theoretical stage:

	Number of Specifications
Feed flow rate	1
Feed mole fractions	$C - 1$
Feed temperature[1]	1
Feed pressure[1]	1
Adiabatic stages (excluding reboiler)	$N - 1$
Stage pressures (including reboiler)	N
Split of LK component	1
Split of HK component	1
Feed-stage location	1
Reflux ratio (as a multiple of R_{min})	1
Reflux temperature	1
Adiabatic reflux divider	1
Pressure of total condenser	1
Pressure at reflux divider	1
Total	$2N + C + 9$

Similar specifications can be written for columns with a partial condenser.

[1]Feed temperature and pressure may correspond to known stream conditions leaving the previous piece of equipment.

§9.1.1 Selection of Two Key Components

For the design case with multicomponent feeds, specification of two key components and their distribution between distillate and bottoms is required. Preliminary guesses of the distribution of nonkey components can be sufficiently difficult to require the iterative procedure indicated in Figure 9.1. However, generally only two and seldom more than three iterations are necessary.

The multicomponent hydrocarbon feed in Figure 9.2 is typical of the feed to the recovery section of an alkylation plant in a petroleum refinery [3]. Components are listed in order of decreasing volatility. For the three products shown, two sequences of distillation columns are possible, both

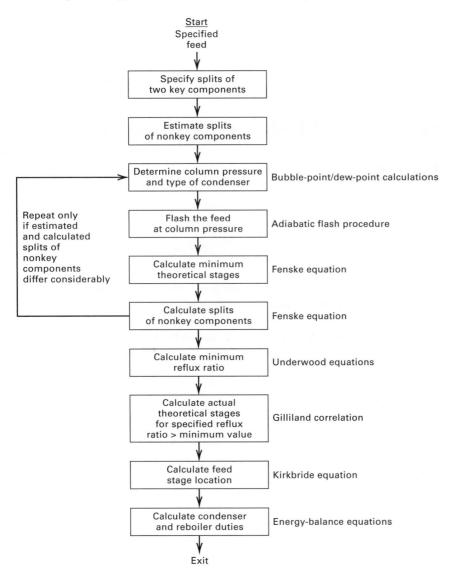

Figure 9.1 Algorithm for multicomponent distillation by FUG method.

including a deisobutanizer and a debutanizer. In Case 1 of Table 9.1, the deisobutanizer is selected as the first column in the sequence. Since the allowable quantities of n-butane in the isobutane recycle, and isobutane in the n-butane product, are specified, isobutane is the LK and n-butane is the HK.

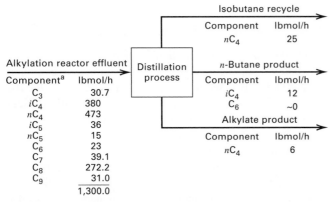

Figure 9.2 Separation specifications for alkylation-reactor effluent.

These two keys are adjacent in volatility. Because a fairly sharp separation between these two keys is indicated and the nonkey components are not close in volatility to the butanes, as a preliminary estimate it is assumed that the nonkey component separation is perfect.

Alternatively, in Case 2, if the debutanizer is placed first in the sequence, specifications in Figure 9.2 require that n-butane be the LK. However, the HK selection is uncertain because no recovery or purity is specified for any component less volatile than n-butane. Possible HK components for the debutanizer are iC_5, nC_5, or C_6. It is simplest to select iC_5 so that the two keys are again adjacent.

For example, suppose that 13 lbmol/h of iC_5 in the feed is allowed to appear in the distillate. Because the split of iC_5 is then not sharp and nC_5 is close in volatility to iC_5, it is probable that the nC_5 in the distillate will not be negligible. An estimate of the distributions of nonkey components for Case 2 is given in Table 9.1. iC_4 may also distribute, but a preliminary estimate of zero is made.

In Case 3, C_6 is selected as the heavy key for the debutanizer at a rate of 0.01 lbmol/h in the distillate, as shown in Table 9.1. Now iC_5 and nC_5 will distribute between the

Table 9.1 Specifications of Key Component Splits and Preliminary Estimation of Nonkey Component Splits for Alkylation Reactor Effluent

Component	Feed, lbmol/h	Case 1, Deisobutanizer Column First, lbmol/h		Case 2, Debutanizer Column First (iC_5 is HK), lbmol/h		Case 3, Debutanizer Column First (C_6 is HK), lbmol/h	
		Distillate	Bottoms	Distillate	Bottoms	Distillate	Bottoms
C_3	30.7	(30.7)	(0)	(30.7)	(0)	(30.7)	(0)
iC_4	380	368[a]	12[b]	(380.0)	(0)	(380.0)	(0)
nC_4	473	25[b]	448[a]	467[a]	6[b]	467[a]	6[b]
iC_5	36	(0)	(36)	13[b]	23[a]	(13)	(23)
nC_5	15	(0)	(15)	(1)	(14)	(1)	(14)
C_6	23	(0)	(23)	(0)	(23)	0.01[b]	22.99[a]
C_7	39.1	(0)	(39.1)	(0)	(39.1)	(0)	(39.1)
C_8	272.2	(0)	(272.2)	(0)	(272.2)	(0)	(272.2)
C_9	31.0	(0)	(31.0)	(0)	(31.0)	(0)	(31.0)
	1,300.0	423.7	876.3	891.7	408.3	891.71	408.29

[a] By material balance.

[b] Specification.

(Preliminary estimate.)

distillate and bottoms; as a preliminary estimate, the distribution in Case 2 is assumed.

The separation in the deisobutanizer is more difficult than the separation in the debutanizer because the deisobutanizer involves the separation of two close-boiling butane isomers. From §1.7.3, Heuristic 2 favors Case 1, while Heuristics 3 and 4 favor Cases 2 and 3. In practice, the deisobutanizer is placed first in the sequence, and the bottoms for Case 1 becomes the debutanizer feed, for which, if nC_4 and iC_5 are selected as the key components, product separations are as

indicated in Figure 9.3, with preliminary estimates for nonkey components shown in parentheses. This separation has been treated by Bachelor [4]. Because nC_4 and C_8 comprise 82.2 mol% of the feed and differ widely in volatility, the temperature difference between distillate and bottoms is likely to be large. Furthermore, the light key split is rather sharp, but the heavy key split is not. As will be shown, this case provides a severe test of the FUG shortcut design procedure.

§9.1.2 Column Operating Pressure

Column pressure is fixed before the design equations are solved by the procedure discussed in §7.2 and shown in Figure 7.16. One reason for this is that there are no fast and easy algorithms for solving the design equations for pressure. If the outcome of the design is unsatisfactory, a new pressure must be assumed, and the calculations become iterative. With column operating pressure established, the column feed can be flashed adiabatically at the feed-tray pressure to determine feed phase condition.

EXAMPLE 9.1 Column Pressure and Condenser Type.

Determine column operating pressure and type of condenser for the debutanizer of Figure 9.3.

Solution

Using the distillate composition in Figure 9.3, the reflux-drum, bubble-point pressure, 79 psia, at 120°F is calculated as in Example 4.2. Thus, a total condenser is indicated. Allowing a 2-psi condenser pressure drop, column top pressure is (79 + 2) = 81 psia; and allowing a 5-psi pressure drop through the column, the bottoms pressure is 86 psia.

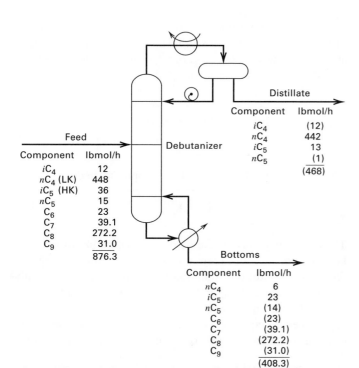

Figure 9.3 Specifications for debutanizer.

§9.1.3 Fenske Equation for Minimum Equilibrium Stages

Minimum equilibrium stages correspond to total reflux, a condition that can be achieved in practice by charging the column with feedstock and bringing it to steady state with no further feed and no withdrawal of products, as shown in Figure 9.4.

To facilitate derivation of the Fenske equation, stages are numbered from the bottom up. All vapor leaving the top at stage N is condensed and returned to stage N as reflux. All liquid leaving stage 1 is vaporized in the reboiler and returned to stage 1 as boilup. For steady-state operation within the column, heat input to the reboiler and heat output from the condenser are made equal (assuming no heat losses). Then, by a material balance, vapor and liquid streams passing between any pair of stages have equal flow rates and compositions, for example, $V_{N-1} = L_N$ and $y_{i,N-1} = x_{i,N}$. However, molar vapor and liquid flow rates will change from stage to stage unless the assumption of constant molar overflow, discussed in §7.2, is valid.

Derivation of an exact equation for the minimum number of equilibrium stages involves only the definition of the K-value and the mole-fraction equality between stages. For component i at stage 1 in Figure 9.4,

$$y_{i,1} = K_{i,1}x_{i,1} \tag{9-1}$$

But for passing streams

$$y_{i,1} = x_{i,2} \tag{9-2}$$

Combining these two equations,

$$x_{i,2} = K_{i,1}x_{i,1} \tag{9-3}$$

Similarly, for stage 2,

$$y_{i,2} = K_{i,2}x_{i,2} \tag{9-4}$$

Combining (9-3) and (9-4),

$$y_{i,2} = K_{i,2}K_{i,1}x_{i,1} \tag{9-5}$$

Equation (9-5) is readily extended in this fashion to give

$$y_{i,N} = K_{i,N}K_{i,N-1} \ldots K_{i,2}K_{i,1}x_{i,1} \tag{9-6}$$

Similarly, for component j,

$$y_{j,N} = K_{j,N}K_{j,N-1} \ldots K_{j,2}K_{j,1}x_{j,1} \tag{9-7}$$

Dividing (9-6) by (9-7),

$$\frac{y_{i,N}}{y_{j,N}} = \alpha_N\alpha_{N-1} \ldots \alpha_2\alpha_1\left(\frac{x_{i,1}}{x_{j,1}}\right) \tag{9-8}$$

or

$$\left(\frac{x_{i,N+1}}{x_{i,1}}\right)\left(\frac{x_{j,1}}{x_{j,N+1}}\right) = \prod_{k=1}^{N_{\min}}\alpha_k \tag{9-9}$$

where $\alpha_k = K_{i,k}/K_{j,k}$, the relative volatility between components i and j. Equation (9-9) relates the relative enrichments of any two components i and j over N theoretical stages to the component relative volatilities. Although (9-9) is exact, it is rarely used in practice because the conditions of each stage must be known to compute the relative volatilities. However, if an average relative volatility is used, (9-9) simplifies to

$$\left(\frac{x_{i,N+1}}{x_{i,1}}\right)\left(\frac{x_{j,1}}{x_{j,N+1}}\right) = \alpha^N \tag{9-10}$$

or

$$N_{\min} = \frac{\log\left\{\left[(x_{i,N+1})/x_{i,1}\right]\left[x_{j,1}/(x_{j,N+1})\right]\right\}}{\log\alpha_{i,j}} \tag{9-11}$$

Equation (9-11) is the very useful *Fenske equation* [5]. When $i =$ the LK and $j =$ the HK, the minimum number of equilibrium stages is influenced by the nonkey components only by their effect (if any) on the relative volatility between the key components.

Equation (9-11) permits a rapid estimation of $N_{\min}$. A more convenient form of (9-11) is obtained by replacing the product of the mole-fraction ratios by the equivalent product of mole-distribution ratios in terms of component distillate and bottoms flow rates d and b, respectively,[2] and by replacing $\alpha_{i,j}$ by a geometric mean of the top- and bottom-stage values. Thus,

$$\boxed{N_{\min} = \frac{\log[(d_i/d_j)(b_j/b_i)]}{\log\alpha_m}} \tag{9-12}$$

where the mean relative volatility is approximated by

$$\boxed{\alpha_m = [(\alpha_{i,j})_N(\alpha_{i,j})_1]^{1/2}} \tag{9-13}$$

Thus, $N_{\min}$ depends on the degree of separation of the two key components and their mean α, but is independent of feed-phase condition. Equation (9-12) in combination with (9-13) is exact for $N_{\min} = 2$. For one stage, it is equivalent to the equilibrium-flash equation.

The Fenske equation is exact only if α does not vary and/or the mixture forms ideal solutions. In general, if the Fenske

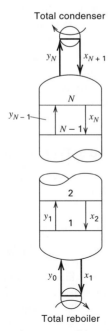

Figure 9.4 Distillation column operation at total reflux.

[2]This substitution is valid even though no distillate or bottoms products are withdrawn at total reflux.

equation is applied with (9-13), it should be with caution, and should be followed by rigorous calculations with a process simulator, as discussed in Chapter 10.

EXAMPLE 9.2 Minimum Stages by the Fenske Equation

For the debutanizer shown in Figure 9.3 and considered in Example 9.1, estimate the $N_{\min}$ by the Fenske equation. Assume a uniform operating pressure of 80 psia (552 kPa) throughout and utilize the ideal K-values given by Bachelor [4] as plotted in Figure 9.5.

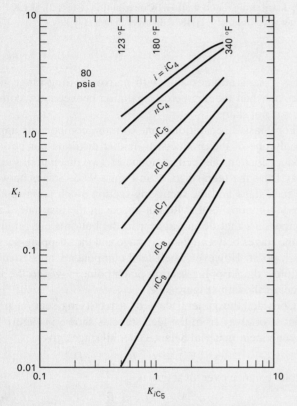

Figure 9.5 Ideal K-values for hydrocarbons at 80 psia.

Solution

The two key components are n-butane and isopentane. Distillate and bottoms conditions based on the estimated product distributions for nonkey components in Figure 9.3 are 123°F at the top and 340°F at the bottom:

Component	$x_{N+1} = x_D$	$x_1 = x_B$
iC_4	0.0256	~0
nC_4 (LK)	0.9445	0.0147
iC_5 (HK)	0.0278	0.0563
nC_5	0.0021	0.0343
nC_6	~0	0.0563
nC_7	~0	0.0958
nC_8	~0	0.6667
nC_9	~0	0.0759
	1.0000	1.0000

From Figure 9.5, at the top-stage,

$$\left(\alpha_{nC_4,iC_5}\right)_N = 1.03/0.495 = 2.08$$

At the bottom,

$$\left(\alpha_{nC_4,iC_5}\right)_1 = 5.20/3.60 = 1.44$$

From (9-13),

$$\alpha_m = [(2.08)(1.44)]^{1/2} = 1.73$$

Noting that $(d_i/d_j) = (x_{D_i}/x_{D_j})$ and $(b_i/b_j) = (x_{B_i}/x_{B_j})$, (9-12) becomes

$$N_{\min} = \frac{\log[(0.9445/0.0278)(0.0563/0.0147)]}{\log 1.73} = 8.88 \text{ stages}$$

§9.1.4 Distribution of Nonkey Components at Total Reflux

The derivation of the Fenske equation in §9.1.3 was not restricted to key components. Therefore, once $N_{\min}$ is known, (9-12) can be used to calculate product molar flow rates d and b for all nonkey components. These values provide an approximation to the product distribution when more stages than $N_{\min}$ are employed.

Let i = a nonkey component and j = the HK or reference component denoted by r. Then (9-12) becomes

$$\left(\frac{d_i}{b_i}\right) = \left(\frac{d_r}{b_r}\right)\left(\alpha_{i,r}\right)_m^{N_{\min}} \tag{9-14}$$

Substituting $f_i = d_i + b_i$ in (9-14) gives

$$b_i = \frac{f_i}{1 + (d_r/b_r)\left(\alpha_{i,r}\right)_m^{N_{\min}}} \tag{9-15}$$

or

$$d_i = \frac{f_i(d_r/b_r)\left(\alpha_{i,r}\right)_m^{N_{\min}}}{1 + (d_r/b_r)\left(\alpha_{i,r}\right)_m^{N_{\min}}} \tag{9-16}$$

Equations (9-15) and (9-16) give the distribution of nonkey components at total reflux as predicted by the Fenske equation.

For accurate calculations, (9-15) and (9-16) are best used to compute the smaller of b_i and d_i. The other quantity is then obtained by an overall material balance.

EXAMPLE 9.3 Split of Nonkey Components

Estimate the product distributions for nonkey components by the Fenske equation for the conditions of Example 9.2.

Solution

All nonkey α values are calculated relative to isopentane using the K-values of Figure 9.5 as follows:

Component	α_{i,iC_5}		
	123°F	340°F	Geometric Mean
iC_4	2.81	1.60	2.12
nC_5	0.737	0.819	0.777
nC_6	0.303	0.500	0.389
nC_7	0.123	0.278	0.185
nC_8	0.0454	0.167	0.0870
nC_9	0.0198	0.108	0.0463

Based on $N_{min} = 8.88$ stages from Example 9.2 and the above geometric-mean values of α, values of $(\alpha_{i,r})_m^{N_{min}}$ are computed relative to isopentane as tabulated below.

From (9-15), using the feed rate specifications in Figure 9.3 for f_i, the distribution of nonkey iC_4 is

$$b_{iC_4} = \frac{12}{1 + (13/23)790} = 0.0268 \text{ lbmol/h}$$

$$d_{iC_4} = f_{iC_4} - b_{iC_4} = 12 - 0.0268 = 11.9732 \text{ lbmol/h}$$

Results of similar calculations for the other nonkey components are

Component	$\left(\alpha_{i,iC_5}\right)_m^{N_{min}}$	d_i	b_i
i_{C_4}	790	11.9732	0.0268
n_{C_4}	130	442.0	6.0
i_{C_5}	1.00	13.0	23.0
n_{C_5}	0.106	0.851	14.149
n_{C_6}	0.000228	0.00297	22.99703
n_{C_7}	3.11×10^{-7}	6.87×10^{-6}	39.1
n_{C_8}	3.83×10^{-10}	5.98×10^{-8}	272.2
n_{C_9}	1.41×10^{-12}	2.48×10^{-11}	31.0
		467.8272	408.4728

§9.1.5 Underwood Equations for Minimum Reflux

The minimum reflux is a useful limiting condition. Unlike minimum stages, there is a feed, and products are withdrawn, but the column is imaginary because it must have ∞ stages.

For distillation of an ideal mixture at minimum reflux, as shown in Figure 7.12a, most of the stages are crowded into a constant-composition zone that bridges the feed stage. In this zone, all vapor and liquid streams have compositions essentially identical to those of the flashed feed. This zone constitutes a single *pinch point* or *point of infinitude* as shown in Figure 9.6a. If nonideal phase conditions are such as to create a point of tangency between the equilibrium curve and the operating line in the rectifying section, as shown in Figure 7.12b, the pinch point occurs within the rectifying section as in Figure 9.6b. Alternatively, the single pinch point can occur in the stripping section.

Shiras, Hanson, and Gibson [6] classified multicomponent systems as having one (Class 1) or two (Class 2) pinch points. For Class 1 separations, all components in the feed distribute to both the distillate and bottoms products. Then the single pinch point bridges the feed stage, as shown in Figure 9.6c.

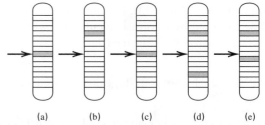

Figure 9.6 Location of pinch-point zones at minimum reflux: (a) binary system; (b) binary system, nonideal conditions giving point of tangency; (c) multicomponent system, all components distributed (Class 1); (d) multicomponent system, not all LLK and HHK distributing (Class 2); (e) multicomponent system, all LLK, if any, distributing, but not all HHK distributing (Class 2). (LLK = lighter than light key; HHK = heavier than heavy key.)

Class 1 separations occur with narrow-boiling-range mixtures or when the degree of separation between key components is not sharp.

For Class 2 separations, one or more components appear in only one of the products. If neither distillate nor bottoms product contains all feed components, two pinch points occur away from the feed stage, as in Figure 9.6d. Stages between the feed stage and the rectifying-section pinch point remove heavy components that do not appear in the distillate. Light components that do not appear in the bottoms are removed by the stages between the feed stage and the stripping-section pinch point. However, if all feed components appear in the bottoms, the stripping-section pinch point moves to the feed stage, as shown in Figure 9.6e.

Consider the general case of a rectifying-section pinch point at or away from the feed stage as shown in Figure 9.7. A component material balance over all stages gives

$$y_{i,\infty}V_\infty = x_{i,\infty}L_\infty + x_{i,D}D \qquad (9\text{-}17)$$

A total balance over all stages is

$$V_\infty = L_\infty + D \qquad (9\text{-}18)$$

Since phase compositions do not change in the pinch zone, the phase-equilibrium relation is

$$y_{i,\infty} = K_{i,\infty}x_{i,\infty} \qquad (9\text{-}19)$$

Combining (9-17) to (9-19) for components i and j to eliminate $y_{i,\infty}$, $y_{j,\infty}$, and V_∞; solving for the internal reflux ratio at

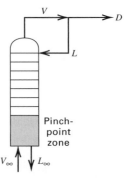

Figure 9.7 Rectifying-section pinch-point zone.

the pinch point; and substituting $(\alpha_{i,j})_\infty = K_{i,\infty}/K_{j,\infty}$,

$$\frac{L_\infty}{D} = \frac{[(x_{i,D}/x_{i,\infty}) - (\alpha_{i,j})_\infty(x_{j,D}/x_{j,\infty})]}{(\alpha_{i,j})_\infty - 1} \quad (9\text{-}20)$$

Class 1

For Class 1 separations, flashed feed- and pinch-zone compositions are identical.[3] Therefore, $x_{i,\infty} = x_{i,F}$ and (9-20) for LK and HK becomes

$$
\frac{(L_\infty)_{min}}{F} =
$$
$$
\frac{(L_F/F)\left[(Dx_{LK,D})/(L_F x_{LK,F}) - (\alpha_{LK,HK})_F(Dx_{HK,D}/L_F x_{HK,F})\right]}{(\alpha_{LK,HK})_F - 1}
$$

$$(9\text{-}21)$$

This equation is attributed to Underwood [7] and can be applied to subcooled-liquid or superheated-vapor feeds by using fictitious values of L_F and $x_{i,F}$ computed by making a flash calculation outside the two-phase region. As with the Fenske equation, (9-21) applies to components other than the key components. Therefore, for a specified split of two key components, the distribution of nonkey components is obtained by combining (9-21) with the analogous equation for component i in place of the light key to give

$$\frac{Dx_{i,D}}{L_F x_{i,F}} = \left[\frac{(\alpha_{i,HK})_F - 1}{(\alpha_{LK,HK})_F - 1}\right]\left(\frac{Dx_{LK,D}}{L_F x_{LK,F}}\right)$$
$$+ \left[\frac{(\alpha_{LK,HK})_F - (\alpha_{i,HK})_F}{(\alpha_{LK,HK})_F - 1}\right]\left(\frac{Dx_{HK,D}}{L_F x_{HK,F}}\right) \quad (9\text{-}22)$$

For a Class 1 separation,

$$0 < \left(\frac{Dx_{i,D}}{Fx_{i,F}}\right) < 1$$

for all nonkey components. If that is so, the external reflux ratio is obtained from the internal reflux by an enthalpy balance around the rectifying section in the form

$$\frac{(L_{min})_{external}}{D} = (R_{min})_{external}$$
$$= \frac{(L_\infty)_{min}(h_{V_\infty} - h_{L_\infty}) + D(h_{V_\infty} - h_V)}{D(h_V - h_L)} \quad (9\text{-}23)$$

where subscripts V and L refer to vapor leaving the top stage and external liquid reflux sent to the top stage, respectively. For conditions of constant molar overflow,

$$(R_{min})_{external} = (L_\infty)_{min}/D$$

Even when (9-21) is invalid, it is useful because, as shown by Gilliland [8], R_{min} computed by assuming a Class 1 separation is $\geq$ the true minimum. This is because the distributing nonkey components in the pinch-point zones increase the separation difficulty, thus increasing the reflux requirement.

---

[3]Assuming the feed is neither subcooled nor superheated.

EXAMPLE 9.4 Minimum Reflux for a Class 1 Separation.

Calculate the minimum internal reflux for Example 9.2 assuming a Class 1 separation. Check the validity of this assumption.

Solution

From Figure 9.5, the α between nC_4 (LK) and iC_5 (HK) at a feed temperature of say 180°F is 1.93. Feed liquid and distillate quantities are given in Figure 9.3 and Example 9.1. From (9-21),

$$(L_\infty)_{min} \frac{759.4[(442/346.5) - 1.93(13/31.4)]}{1.93 - 1} = 389 \text{ lbmol/h}$$

Distribution of nonkey components is determined by (9-22). The most likely nonkey component to distribute is nC_5 because its volatility is close to that of iC_5 (HK), which does not undergo a sharp separation. For nC_5, using data for K-values from Figure 9.5,

$$\frac{Dx_{nC_5,D}}{L_F x_{nC_5,F}} = \left[\frac{0.765 - 1}{1.93 - 1}\right]\left(\frac{442}{346.5}\right) + \left[\frac{1.93 - 0.765}{1.93 - 1}\right]\left(\frac{13}{31.4}\right)$$
$$= 0.1963$$

Therefore, $Dx_{nC_5,D} = 0.1963(13.4) = 2.63$ lbmol/h of nC_5 in the distillate. This is less than the quantity of nC_5 in the total feed. Therefore, nC_5 distributes between the distillate and the bottoms. However, similar calculations for the other nonkey components give negative distillate flow rates for the other heavy components and, in the case of iC_4, a distillate flow rate greater than the feed rate. Thus, the computed reflux rate is not valid. However, it is greater than the true value of 298 lbmol/h found by Bachelor [4].

Class 2

For Class 2 separations, (9-17) to (9-20) still apply. However, (9-20) cannot be used directly to compute the internal minimum reflux ratio since values of $x_{i,\infty}$ are not simply related to feed composition for Class 2 separations. Underwood [9] devised a procedure to overcome this. For the rectifying section, he defined a quantity Φ by

$$\sum \frac{(\alpha_{i,r})_\infty x_{i,D}}{(\alpha_{i,r})_\infty - \Phi} = 1 + (R_\infty)_{min} \quad (9\text{-}24)$$

Similarly, for the stripping section, Underwood defined Φ' by

$$\sum \frac{(\alpha'_{i,r})_\infty x_{i,B}}{(\alpha'_{i,r})_\infty - \Phi'} = 1 - (R'_\infty)_{min} \quad (9\text{-}25)$$

where $R'_\infty = L'_\infty/B$ and the prime refers to the stripping-section pinch-point zone. Underwood assumed that α values are constant in the region between the two pinch-point zones and that $(R_\infty)_{min}$ and $(R'_\infty)_{min}$ are related by the assumption of constant molar overflow between the feed entry and the rectifying-section pinch point and between the feed entry and the stripping-section pinch point. Hence,

$$(L'_\infty)_{min} - (L_\infty)_{min} = qF \quad (9\text{-}26)$$

With these two critical assumptions, Underwood showed that at least one common root θ (where $\theta = \Phi = \Phi'$) exists between (9-24) and (9-25).

Equation (9-24) is analogous to the following equation derived from (9-19) and the relation $\alpha_{i,r} = K_i/K_r$,

$$\sum_i \frac{(\alpha_{i,r})_\infty x_{i,D}}{(\alpha_{i,r})_\infty - L_\infty/[V_\infty (K_r)_\infty]} = 1 + (R_\infty)_{\min} \quad (9\text{-}27)$$

where $L_\infty/[V_\infty (K_r)_\infty]$ is the *absorption factor* for a reference component in the rectifying-section pinch-point zone. Although Φ is analogous to the absorption factor, a different root of Φ is used to solve for $(R_\infty)_{\min}$ (Shiras et al. [6]).

The common root θ is determined by multiplying (9-24) and (9-25) by D and B, respectively, adding the equations, substituting (9-25) to eliminate $(R'_\infty)_{\min}$ and $(R_\infty)_{\min}$, and utilizing the component balance $z_{i,F}F = x_{i,D}D + x_{i,B}B$ to obtain

$$\sum_i \frac{(\alpha_{i,r})_\infty z_{i,F}}{(\alpha_{i,r})_\infty - \theta} = 1 - q \quad (9\text{-}28)$$

where q is the feed thermal condition from (7-20) and r is taken as the HK. When only the two key components distribute, (9-28) is solved iteratively for a root of θ that satisfies $\alpha_{\text{LK,HK}} > \theta > 1$. The following modification of (9-24) is then solved for the internal reflux ratio $(R_\infty)_{\min}$:

$$\sum_i \frac{(\alpha_{i,r})_\infty x_{i,D}}{(\alpha_{i,r})_\infty - \theta} = 1 + (R_\infty)_{\min} \quad (9\text{-}29)$$

If any nonkey components are suspected of distributing, estimated values of $x_{i,D}$ cannot be used directly in (9-29). This is particularly true when nonkey components are intermediate in volatility between the keys. In this case, (9-28) is solved for m roots of θ, where m is one less than the number of distributing components. Each root of θ lies between an adjacent pair of relative volatilities of distributing species. For instance, in Example 9.4, nC_5 distributes at minimum reflux, but nC_6 and heavier do not and iC_4 does not. Thus, two roots of θ are necessary, where

$$\alpha_{nC_4,iC_5} > \theta_1 > 1.0 > \theta_2 > \alpha_{nC_5,iC_5}$$

With these two roots, (9-29) is written twice and solved simultaneously to yield $(R_\infty)_{\min}$ and the unknown value of $x_{nC_5,D}$. The solution must satisfy the condition $\sum x_{i,D} = 1.0$.

With the internal reflux ratio $(R_\infty)_{\min}$ known, the external reflux ratio is computed by enthalpy balance with (9-23). This requires a knowledge of the rectifying-section pinch-point compositions. Underwood [9] shows that

$$x_{i,\infty} = \frac{\theta x_{i,D}}{(R_\infty)_{\min}\left[(\alpha_{i,r})_\infty - \theta\right]} \quad (9\text{-}30)$$

with $y_{i,\infty}$ given by (9-17). The value of θ to be used in (9-30) is the root of (9-29) satisfying the inequality

$$(\alpha_{\text{HNK},r})_\infty > \theta > 0$$

where HNK refers to the heaviest nonkey in the distillate at minimum reflux. This root is equal to $L_\infty/[V_\infty (K_r)_\infty]$ in (9-27). With wide-boiling feeds, the external reflux can be higher than the internal reflux. Bachelor [4] cites a case where the external reflux rate is 55% greater.

For the stripping-section pinch-point composition, Underwood obtains

$$x'_{i,\infty} = \frac{\theta x_{i,B}}{\left[(R'_\infty)_{\min} + 1\right]\left[(\alpha_{i,r})_\infty - \theta\right]} \quad (9\text{-}31)$$

where, here, θ is the root of (9-29) satisfying the inequality $(\alpha_{\text{HNK},r})_\infty > \theta > 0$, where HNK refers to the heaviest non-key in the bottoms product at minimum reflux.

The Underwood minimum reflux equations for Class 2 separations are widely used, but often without examining the possibility of nonkey distribution. In addition, the assumption is frequently made that $(R_\infty)_{\min}$ equals the external reflux ratio. When the assumptions of constant α and constant molar overflow between the two pinch-point zones are not valid, values of the Underwood minimum reflux ratio for Class 2 separations can be appreciably in error because of the sensitivity of (9-28) to the value of q, as will be shown in Example 9.5. When the Underwood assumptions appear valid and a negative minimum reflux ratio is computed, a rectifying section may not be needed for the separation. The Underwood equations show that the minimum reflux depends mainly on feed condition and α and, to a lesser extent, on degree of separation, as is the case with binary distillation as discussed in the sub-section, Perfect Separation, in §7.2.5. As with binary distillation, a minimum reflux ratio exists in a multicomponent system for a perfect separation between the LK and HK.

An extension of the Underwood method for multiple feeds is given by Barnes et al. [10]. Exact methods for determining minimum reflux are also available [11]. For calculations at actual reflux conditions with a process simulator by the computer methods of Chapter 10, knowledge of $R_{\min}$ is not essential, but $N_{\min}$ must be known if the split between two components is to be specified.

EXAMPLE 9.5 Minimum Reflux for a Class 2 Separation.

Repeat Example 9.4 assuming a Class 2 separation and using the Underwood equations. Check the validity of the Underwood assumptions. Also calculate the external reflux ratio.

Solution

From Example 9.4, assume that the only distributing nonkey component is *n*-pentane. Assuming a feed temperature of 180°F for computing relative volatilities in the pinch zone, the following quantities are obtained from Figures 9.3 and 9.5:

Species i	$z_{i,F}$	$(\alpha_{i,\text{HK}})_\infty$
iC_4	0.0137	2.43
nC_4 (LK)	0.5113	1.93
iC_5 (HK)	0.0411	1.00
nC_5	0.0171	0.765
nC_6	0.0262	0.362
nC_7	0.0446	0.164
nC_8	0.3106	0.0720
nC_9	0.0354	0.0362
	1.0000	

The feed q is the mole fraction of liquid in the flashed feed. From a flash of the feed in Example 9.1 at 80 psia and 180°F, the mol% vaporized is 13.34%. Therefore, $q = 1 - 0.1334 = 0.8666$. Applying (9-28),

$$\frac{2.43(0.0137)}{2.43 - \theta} + \frac{1.93(0.5113)}{1.93 - \theta} + \frac{1.00(0.0411)}{1.00 - \theta}$$
$$+ \frac{0.765(0.0171)}{0.765 - \theta} + \frac{0.362(0.0262)}{0.362 - \theta} + \frac{0.164(0.0446)}{0.164 - \theta}$$
$$+ \frac{0.072(0.3106)}{0.072 - \theta} + \frac{0.0362(0.0354)}{0.0362 - \theta} = 1 - 0.8666$$

Solving by a Newton method for two roots of θ that satisfy

$$\alpha_{nC_4,iC_5} > \theta_1 > \alpha_{iC_5,iC_5} > \theta_2 > \alpha_{nC_5,iC_5}$$

or

$$1.93 > \theta_1 > 1.00 > \theta_2 > 0.765$$

$\theta_1 = 1.04504$ and $\theta_2 = 0.78014$. Because distillate rates for nC_4 and iC_5 are specified (442 and 13 lbmol/h, respectively), the following form of (9-29) is preferred:

$$\sum_i \frac{(\alpha_{i,r})_\infty (x_{i,D}D)}{(\alpha_{i,r})_\infty - \theta} = D + (L_\infty)_{\min} \qquad (9\text{-}32)$$

with the restriction that

$$\sum_i (x_{i,D}D) = D \qquad (9\text{-}33)$$

Assuming that $x_{i,D}D$ equals 0.0 for species heavier than nC_5 and 12.0 lbmol/h for iC_4, these relations give three linear equations:

$$D + (L_\infty)_{\min} = \frac{2.43(12)}{2.43 - 1.04504} + \frac{1.93(442)}{1.93 - 1.04504}$$
$$+ \frac{1.00(13)}{1.00 - 1.04504} + \frac{0.765(x_{nC_5,D}D)}{0.765 - 1.04504}$$

$$D + (L_\infty)_{\min} = \frac{2.43(12)}{2.43 - 0.78014} + \frac{1.93(442)}{1.93 - 0.78014}$$
$$+ \frac{1.00(13)}{1.00 - 0.78014} + \frac{0.765(x_{nC_5,D}D)}{0.765 - 0.78014}$$

$$D = 12 + 442 + 13 + (x_{nC_5,D}D)$$

Solving these three equations gives

$$x_{nC_5,D}D = 2.56 \text{ lbmol/h}$$
$$D = 469.56 \text{ lbmol/h}$$
$$(L_\infty)_{\min} = 219.8 \text{ lbmol/h}$$

The distillate rate for nC_5 is very close to the 2.63 in Example 9.4 for a Class 1 separation. The internal minimum reflux rate at the rectifying pinch point is less than the 389 computed in Example 9.4 and is also less than the true internal value of 298 reported by Bachelor [4]. The reason for the discrepancy between 219.8 and the true value of 298 is the invalidity of the constant molar overflow assumption. Bachelor computed the pinch-point region flow rates and temperatures shown in Figure 9.8. The average temperature between the two pinch regions is 152°F (66.7°C), which is appreciably lower than the flashed-feed temperature. The relatively hot feed causes vaporization across the feed zone. The value of q in the region between the pinch points from (7-18) is:

$$q_{\text{eff}} = \frac{L'_\infty - L_\infty}{F} = \frac{896.6 - 296.6}{876.3} = 0.685$$

This is considerably lower than the 0.8666 for q based on the flashed-feed condition. On the other hand, the value of $\alpha_{LK,HK}$ at

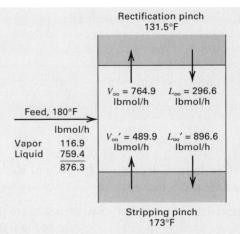

Figure 9.8 Pinch-point region conditions for Example 9.5 from computations by Bachelor.

[From J.B. Bachelor, *Petroleum Refiner*, **36** (6), 161–170 (1957).]

152°F (66.7°C) is not much different from the value at 180°F (82.2°C). If this example is repeated using $q = 0.685$, the resulting value of $(L_\infty)_{\min}$ is 287.3 lbmol/h, which is just 3.6% lower than 298. In practice, this corrected procedure cannot be applied because the true value of q cannot be readily determined.

For the external reflux ratio from (9-23), rectifying pinch-point compositions are calculated from (9-30) and (9-17). The θ root for (9-30) is obtained from (9-29) used above. Thus,

$$\frac{2.43(12)}{2.43 - \theta} + \frac{1.93(442)}{1.93 - \theta} + \frac{1.00(13)}{1.00 - \theta}$$
$$+ \frac{0.765(2.56)}{0.765 - \theta} = 469.56 + 219.8$$

where $0.765 > \theta > 0$. Solving, $\theta = 0.5803$. Liquid pinch-point compositions are obtained from the following form of (9-30):

$$x_{i,\infty} = \frac{\theta(x_{i,D}D)}{(L_\infty)_{\min}\left[(\alpha_{i,r})_\infty - \theta\right]}$$

with $(L_\infty)_{\min} = 219.8$ lbmol/h. For iC_4,

$$x_{iC_4,\infty} = \frac{0.5803(12)}{219.8(2.43 - 0.5803)} = 0.0171$$

From a combination of (9-17) and (9-18),

$$y_{i,\infty} = \frac{x_{i,\infty}L_\infty + x_{i,D}D}{L_\infty + D}$$

For iC_4,

$$y_{iC_4,\infty} = \frac{0.0171(219.8) + 12}{219.8 + 469.56} = 0.0229$$

Similarly, mole fractions of the other distillate components are

Component	$x_{i,\infty}$	$y_{i,\infty}$
iC_4	0.0171	0.0229
nC_4	0.8645	0.9168
iC_5	0.0818	0.0449
nC_5	0.0366	0.0154
	1.0000	1.0000

The rectifying-section pinch-point temperature is obtained from either a bubble-point calculation on $x_{i,\infty}$ or a dew-point calculation on $y_{i,\infty}$. The result is 126°F. Similarly, the liquid-distillate temperature (bubble point) and the vapor temperature leaving the top stage (dew point) are both computed to be 123°F. Because the rectifying-section pinch-point and distillate temperatures are very close, it is expected that $(R_\infty)_{min}$ and $(R_{min})_{external}$ are almost identical. This is confirmed by Bachelor, who obtained an external reflux rate of 292 lbmol/h and an internal reflux rate of 298 lbmol/h.

§9.1.6 Gilliland Correlation for Actual Reflux Ratio and Equilibrium Stages

Capital, and interest-on-capital costs, are related to the number of stages, whereas operating costs are tied to reflux ratio and thus to fuel costs for providing heat to the reboiler. Less reflux is needed if more stages are added, so operating costs decrease as capital costs increase. The Gilliland correlation [13] provides an approximate relationship between number of stages and reflux ratio so that an optimal reflux ratio can be determined with a minimum of calculations.

For a specified separation, the reflux ratio and equilibrium stages must be greater than their minimum values. The actual reflux ratio should be established by economic considerations at some multiple of minimum reflux. The corresponding number of stages is then determined by suitable analytical or graphical methods or, as discussed in this section, by the empirical Gilliland correlation. However, there is no reason why the number of stages could not be specified as a multiple of minimum stages and the corresponding actual reflux computed by the same empirical relationship. As shown in Figure 9.9, from studies by Fair and Bolles [12], an optimal value of R/R_{min} at the time their paper was published was 1.05. However, near-optimal conditions extend over a relatively broad range of mainly larger values of R/R_{min}.

Superfractionators requiring many stages commonly use a value of R/R_{min} of approximately 1.10, while columns requiring a small number of stages are designed for a value of R/R_{min} of approximately 1.50. For intermediate cases, a commonly used rule of thumb is $R/R_{min} = 1.30$.

The number of equilibrium stages required for the separation of a binary mixture assuming constant relative volatility and constant molar overflow depends on $z_{i,F}$, $x_{i,D}$, $x_{i,B}$, q, R, and α. From (9-11), for a binary mixture, N_{min} depends on $x_{i,D}$, $x_{i,B}$, and α, while R_{min} depends on $z_{i,F}$, $x_{i,D}$, q, and α. Accordingly, studies have assumed correlations of the form

$$N = N\{N_{min}\{x_{i,D},\ x_{i,B},\ \alpha\},\quad R_{min}\{z_{i,F},\ x_{i,D},\ q,\ \alpha\},\ R\}$$

Furthermore, they have assumed that such a correlation might exist for nearly ideal multicomponent systems even though additional feed composition variables and values of nonkey α also influence the value of R_{min}.

A successful and simple correlation is that of Gilliland [13] and a later modified version by Robinson and Gilliland [14]. The correlation is shown in Figure 9.10, where three sets of data points, all based on accurate calculations, are the original points from Gilliland [13], and the points of Brown and Martin [15] and Van Winkle and Todd [16]. The 61 data points cover the following ranges:

1. Number of components: 2 to 11	4. α: 1.11 to 4.05
2. q: 0.28 to 1.42	5. R_{min}: 0.53 to 9.09
3. Pressure: vacuum to 600 psig	6. N_{min}: 3.4 to 60.3

The line drawn through the data represents the equation developed by Molokanov et al. [17]:

$$Y = \frac{N - N_{min}}{N + 1} = 1 - \exp\left[\left(\frac{1 + 54.4X}{11 + 117.2X}\right)\left(\frac{X - 1}{X^{0.5}}\right)\right]$$

$$(9\text{-}34)$$

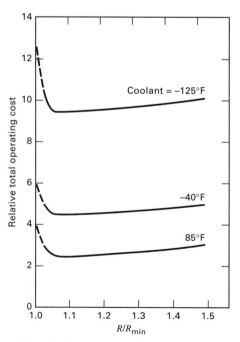

Figure 9.9 Effect of reflux ratio on cost.
[From J.R. Fair and W.L. Bolles, *Chem. Eng.*, **75** (9), 156–178 (1968).]

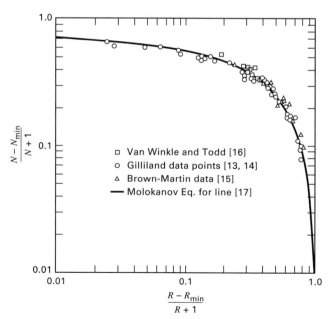

Figure 9.10 Comparison of rigorous calculations with Gilliland correlation.

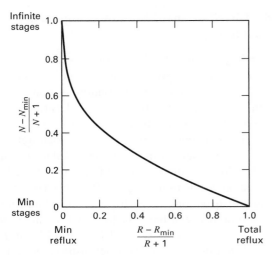

Figure 9.11 Gilliland correlation with linear coordinates.

where

$$X = \frac{R - R_{\min}}{R + 1}$$

This equation satisfies the end points ($Y = 0$, $X = 1$) and ($Y = 1$, $X = 0$). At a value of $R/R_{\min}$ near the optimum of 1.3, Figure 9.10 predicts an optimal ratio for $N/N_{\min}$ of approximately 2. The value of N includes one stage for a partial reboiler and one stage for a partial condenser, if used.

The Gilliland correlation is appropriate only for preliminary exploration of design variables. Although it was never intended for final design, the correlation was widely used before digital computers were invented, making possible accurate stage-by-stage calculations. In Figure 9.11, a replot of the correlation in linear coordinates shows that a small initial increase in R above $R_{\min}$ causes a large decrease in N, but that further changes have a smaller effect on N. The knee in the curve of Figure 9.11 corresponds to the optimal $R/R_{\min}$ in Figure 9.9.

Robinson and Gilliland [14] state that a more accurate correlation should utilize a parameter involving the feed condition q. This effect is shown in Figure 9.12 using data points for the sharp separation of benzene–toluene mixtures from Guerreri [18]. The data, which cover feed conditions ranging from subcooled liquid to superheated vapor ($q = 1.3$ to

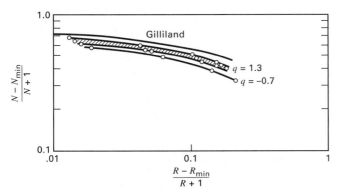

Figure 9.12 Effect of feed condition on Gilliland correlation.
[From G. Guerreri, *Hydrocarbon Processing*, **48** (8), 137–142 (1969).]

−0.7), show a trend toward decreasing stage requirements with increasing feed vaporization. The Gilliland correlation is conservative for feeds having low values of q. Donnell and Cooper [19] state that this effect of q is important only when the $\alpha_{LK,HK}$ is high, or when the feed is low in volatile components.

A problem with the Gilliland correlation can occur when stripping is more important than rectification, because the correlation is based only on reflux and not boilup. For example, Oliver [20] considers a fictitious binary case with specifications of $z_F = 0.05$, $x_D = 0.40$, $x_B = 0.001$, $q = 1$, $\alpha = 5$, $R/R_{\min} = 1.20$, and constant molar overflow. By exact calculations, $N = 15.7$. From the Fenske equation, $N_{\min} = 4.04$. From the Underwood equation, $R_{\min} = 1.21$. From (9-32) for the Gilliland correlation, $N = 10.3$. This is 34% lower than the exact value. This limitation, caused by ignoring boilup, is discussed by Strangio and Treybal [21].

EXAMPLE 9.6   Use of the Gilliland Correlation.

Use the Gilliland correlation to estimate the equilibrium-stage requirement for the debutanizer of Examples 9.1, 9.2, and 9.5 for an external reflux of 379.6 lbmol/h (30% greater than the exact value of the minimum reflux rate from Bachelor).

Solution

From the examples cited, values of $R_{\min}$ and $[(R - R_{\min})/(R + 1)]$ are obtained using a distillate rate from Example 9.5 of 469.56 lbmol/h. Thus, $R = 379.6/469.56 = 0.808$. With $N_{\min} = 8.88$,

$$R_{\min} = 0.479, \quad \text{and} \quad \frac{R - R_{\min}}{R + 1} = X = 0.182$$

From (9-34),

$$\frac{N - N_{\min}}{N + 1} = 1 - \exp\left[\left(\frac{1 + 54.4(0.182)}{11 + 117.2(0.182)}\right)\left(\frac{0.182 - 1}{0.182^{0.5}}\right)\right]$$

$$= 0.476$$

$$N = \frac{8.88 + 0.476}{1 - 0.476} = 17.85$$

$$N - 1 = 16.85$$

where $N - 1$ corresponds to the stages in the tower, allowing one stage for the partial reboiler but no stage for the total condenser.

It should be noted that had the exact value of $R_{\min}$ not been known and a value of R equal to 1.3 times $R_{\min}$ from the Underwood method been used, the value of R would have been 292 lbmol/h. But this, by coincidence, is only the true minimum reflux. Therefore, the desired separation would not have been achieved.

---

§9.1.7   Feed-Stage Location

Implicit in the application of the Gilliland correlation is the specification that stages be distributed optimally between the rectifying and stripping sections. Brown and Martin [15] suggest that the optimal feed stage be located by assuming that the ratio of stages above the feed to stages below is the same as the ratio determined by applying the Fenske equation to the separate sections at total reflux conditions. Thus,

$$
\frac{N_R}{N_S} \simeq \frac{(N_R)_{\min}}{(N_S)_{\min}}
$$

$$
= \frac{\log\left[\left(x_{LK,D}/z_{LK,F}\right)\left(z_{HK,F}/x_{HK,D}\right)\right]\log\left[(\alpha_B\alpha_F)^{1/2}\right]}{\log\left[\left(z_{LK,F}/x_{LK,B}\right)\left(x_{HK,B}/z_{HK,F}\right)\right]\log\left[(\alpha_D\alpha_F)^{1/2}\right]}
$$

$$(9\text{-}35)$$

Unfortunately, (9-35) is not reliable except for fairly symmetrical feeds and separations.

A better approximation of the optimal feed-stage location can be made with the Kirkbride [22] empirical equation:

$$
\frac{N_R}{N_S} = \left[\left(\frac{z_{HK,F}}{z_{LK,F}}\right)\left(\frac{x_{LK,B}}{x_{HK,D}}\right)^2\left(\frac{B}{D}\right)\right]^{0.206}
$$

$$(9\text{-}36)$$

A test of both equations is provided by a fictitious binary-mixture problem of Oliver [20] cited in the previous section. Exact calculations by Oliver and calculations using (9-35) and (9-36) give the following results:

Method	N_R/N_S
Exact	0.08276
Kirkbride (9-34)	0.1971
Fenske ratio (9-33)	0.6408

Although the Kirkbride result is not very satisfactory, the result from the Fenske ratio method is much worse. In practice, distillation columns are provided with several feed entry locations.

EXAMPLE 9.7 Feed-Stage Location.

Use the Kirkbride equation to determine the feed-stage location for the debutanizer of Example 9.1, assuming 18.27 equilibrium stages.

Solution

Assume that the product distribution computed in Example 9.3 for total-reflux conditions is a good approximation to the distillate and bottoms compositions at actual reflux conditions. Then

$$
x_{nC_4,B} = \frac{6.0}{408.5} = 0.0147 \quad x_{iC_5,D} = \frac{13}{467.8} = 0.0278
$$

$$
D = 467.8 \text{ lbmol/h} \quad B = 408.5 \text{ lbmol/h}
$$

From Figure 9.3,

$$
z_{nC_4,F} = 448/876.3 = 0.5112
$$

and

$$
z_{iC_5,F} = 36/876.3 = 0.0411
$$

From (9-36),

$$
\frac{N_R}{N_S} = \left[\left(\frac{0.0411}{0.5112}\right)\left(\frac{0.0147}{0.0278}\right)^2\left(\frac{408.5}{467.8}\right)\right]^{0.206} = 0.445
$$

Therefore, $N_R = (0.445/1.445)(18.27) = 5.63$ stages and $N_S = 18.27 - 5.63 = 12.64$ stages. Rounding the estimated stage requirements leads to one stage as a partial reboiler, 12 stages below the feed, and six stages above the feed.

---

§9.1.8 Distribution of Nonkey Components at Actual Reflux

As shown in §9.1.3 to §9.1.5 for multicomponent mixtures, all components distribute between distillate and bottoms at total reflux; while at minimum reflux conditions, none or only a few of the nonkey components distribute. Distribution ratios for these two limiting conditions are given in Figure 9.13 for the previous debutanizer example. For total reflux, the Fenske equation results from Example 9.3 plot as a straight line on log–log coordinates. For minimum reflux, Underwood equation results from Example 9.5 are a dashed line.

Product-distribution curves for a given reflux might be expected to lie between lines for total and minimum reflux. However, as shown by Stupin and Lockhart [23] in Figure 9.14, this is not the case, and product distributions are complex. Near $R_{\min}$, product distribution (curve 3) lies between the two limits (curves 1 and 4). However, for a high reflux ratio, nonkey distributions (curve 2) may lie outside the limits, thus causing inferior separations.

Stupin and Lockhart provide explanations for Figure 9.14 consistent with the Gilliland correlation. As the reflux ratio is decreased from total reflux while maintaining the key-component splits, stage requirements increase slowly at first, but then rapidly as minimum reflux is approached. Initially, large decreases in reflux cannot be compensated for by increasing stages. This causes inferior nonkey distributions. As $R_{\min}$ is approached, small decreases in reflux are compensated for by large increases in stages; and the separation of nonkey components becomes superior to that at total reflux. It appears reasonable to assume that, at a near-optimal reflux ratio of 1.3, nonkey-component distribution is close to that estimated by the Fenske equation for total-reflux conditions.

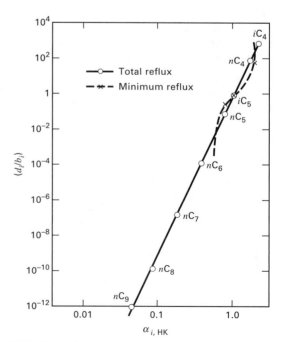

Figure 9.13 Component distribution ratios at extremes of distillation operating conditions.

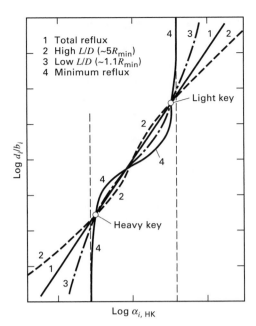

Figure 9.14 Component distribution ratios at various reflux ratios.

§9.2 KREMSER GROUP METHOD

Many separators are cascades where the two contacting phases flow countercurrently. Approximate calculation procedures have been developed to relate compositions of streams entering and exiting to the number of stages required.

These approximate procedures are called *group methods* because they provide only an input–output treatment of the cascade without considering detailed changes in individual stage temperature, flow rates, and composition. This section considers absorption, stripping, and extraction.

Kremser [1] originated the group method. Subsequent articles by Souders and Brown [24], Horton and Franklin [25], and Edmister [26] improved the method. The Kremser equations are derived and applied to absorption in §5.4. These equations are illustrated for strippers and extractors here. An alternative treatment by Smith and Brinkley [27] emphasizes liquid–liquid separations.

§9.2.1 Strippers

The vapor entering a stripper is often steam or an inert gas. When it is pure and not present in the entering liquid, and is not absorbed or condensed in the stripper, the only direction of mass transfer is from the liquid to the gas phase. Then, only values of the effective stripping factor, S_e, as defined by (5-51), are needed to apply the group method via (5-49) and (5-50). The equations for strippers are analogous to those for absorbers.

For optimal stripping, temperatures should be high and pressures low. However, temperatures should not be so high as to cause decomposition, and vacuum should be used only if necessary. The minimum stripping-agent flow rate, for a specified value of ϕ_S for a key component K corresponding to ∞ stages, can be estimated from an equation obtained

from (5-50) with $N = \infty$,

$$(V_0)_{\min} = \frac{L_{N+1}}{K_K}\left(1 - \phi_{S_K}\right) \qquad (9\text{-}37)$$

This equation assumes that $A_K < 1$ and that the fraction of liquid feed stripped is small.

EXAMPLE 9.8 Stripping with Nitrogen.

Sulfur dioxide and butadienes (B3 and B2) are to be stripped with nitrogen from the liquid stream given in Figure 9.15 so that butadiene sulfone (BS) product will contain less than 0.05 mol% SO_2 and less than 0.5 mol% butadienes. Estimate the flow rate of N_2 and the number of equilibrium stages required.

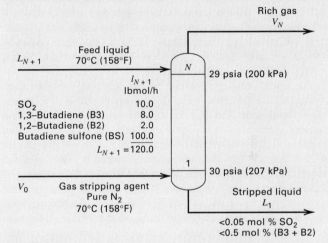

Figure 9.15 Specifications for stripper of Example 9.8.

Solution

Neglecting stripping of BS, the stripped liquid must have the following component flow rates, and corresponding values for ϕ_S:

Species	l_1, lbmol/h	$\phi_S = \dfrac{l_1}{l_{N+1}}$
SO_2	<0.0503	<0.00503
B3 + B2	<0.503	<0.0503
BS	100.0	—

Thermodynamic properties can be computed based on ideal solutions at low pressures. For butadiene sulfone, the vapor pressure is given by

$$P_{BS}^s = \exp\left(17.30 - \frac{11,142}{T + 459.67}\right)$$

where P_{BS}^s is in psia and T is in °F. The liquid enthalpy of BS is

$$(h_L)_{BS} = 50\,T$$

where $(h_L)_{BS}$ is in Btu/lbmol and T is in °F.

The entering flow rate of the stripping agent V_0 is not specified. The minimum rate at infinite stages can be computed from (9-37) if a key component is selected. Select B2, which is the heaviest species to be stripped to a specified extent. At 70°C, the vapor pressure of B2 is 90.4 psia. From Raoult's law at 30 psia,

$$K_{B2} = \frac{90.4}{30} = 3.01$$

From (9-37), using $(\phi_S)_{B2} = 0.0503$,

$$(V_0)_{\min} = \frac{120}{3.01}(1 - 0.0503) = 37.9 \text{ lbmol/h}$$

For this value of $(V_0)_{\min}$, (9-42) can be used to determine ϕ_S for B3 and SO_2. The K-values for these two species are 4.53 and 6.95. From (9-37), at ∞ stages with $V_0 = 37.9$ lbmol/h,

$$(\phi_S)_{B3} = 1 - \frac{4.53(37.9)}{120} = -0.43$$

and

$$(\phi_S)_{SO_2} = 1 - \frac{6.95(37.9)}{120} = -1.19$$

These negative values indicate complete stripping of B3 and SO_2. Therefore, the total butadienes in the stripped liquid would be only $(0.0503)(2.0) = 0.1006$, compared to the specified value of 0.503. A better estimate of $(V_0)_{\min}$ can be obtained by assuming that all butadiene content of the stripped liquid is due to B2. Then $(\phi_S)_{B2} = 0.503/2 = 0.2515$, and $(V_0)_{\min}$ from (9-37) is 29.9 lbmol/h. Values of $(\phi_S)_{B3}$ and $(\phi_S)_{SO_2}$ are still negative.

The actual entering flow rate for the stripping vapor must be greater than the minimum value. To estimate the effect of V_0 on the stage requirements and ϕ_S values for the nonkey components, the Kremser approximation is used with K-values at 70°C and 30 psia, $L = L_{N+1} = 120$ lbmol/h, and $V = V_0$ equal to a series of multiples of 29.9 lbmol/h. The calculations are facilitated if values of N are selected and values of V are determined from (5-51), where S is obtained from Figure 5.9. Because B3 will be found to some extent in the stripped liquid, $(\phi_S)_{B2}$ is held below 0.2515. By making iterative calculations, one can choose $(\phi_S)_{B2}$ so that $(\phi_S)_{B2+B3}$ satisfies the specification of, say, 0.05. For $N = 10$, assuming essentially complete stripping of B3 such that $(\phi_S)_{B2} \approx 0.25$, $S_{B2} \approx 0.76$ from Figure 5.9. From (5-51),

$$V = V_0 = \frac{(120)(0.76)}{3.01} = 30.3 \text{ lbmol/h}$$

For B3, from (5-51),

$$S_{B3} = \frac{(4.53)(30.3)}{120} = 1.143$$

From Figure 5.9, $\qquad (\phi_S)_{B3} = 0.04$

Thus, $\qquad (\phi_S)_{B2+B3} = \frac{0.25(2) + 0.04(8)}{10} = 0.082$

This is above the 0.05 specification; therefore, repeat the calculations with $(\phi_S)_{B2} = 0.09$, and continue to repeat until the specified $(\phi_S)_{B2+B3}$ is achieved. Calculations for various stage numbers were carried out with converged results as follows:

N	V_0, lbmol/h	$V_0/$ $(V_0)_{\min}$	Fraction Not Stripped				
			ϕ_{SO_2}	ϕ_{B3}	ϕ_{B2}	ϕ_{B2+B3}	ϕ_{BS}
∞	29.9	1.00	0.0	0.0	0.2515	0.0503	0.9960
10	33.9	1.134	0.0005	0.017	0.18	0.050	0.9955
5	45.4	1.518	0.0050	0.029	0.117	0.040	0.9940
3	94.3	3.154	0.0050	0.016	0.045	0.022	0.9874

These results show that the SO_2 specification is met for all N.

§9.2.2 Liquid–Liquid Extraction

Figure 9.16 is an extraction cascade, with stages numbered from the top down and solvent V_{N+1} entering at the bottom.[4] The group method can be applied, with the equations written by analogy to absorbers. In place of the K-value, the distribution coefficient (partition ratio) is used:

$$K_{D_i} = \frac{y_i}{x_i} = \frac{v_i/V}{l_i/L} \tag{9-38}$$

Here, y_i is the mole fraction of i in the solvent or extract phase and x_i is the mole fraction in the feed or raffinate phase. Also, in place of the stripping factor, an *extraction factor*, E, is used, where

$$\boxed{E_i = \frac{K_{D_i} V}{L}} \tag{9-39}$$

The reciprocal of E is

$$U_i = \frac{1}{E_i} = \frac{L}{K_{D_i} V} \tag{9-40}$$

The working equations for each component are

$$v_1 = v_{N+1}\phi_U + l_0(1 - \phi_E) \tag{9-41}$$

$$l_N = l_0 + v_{N+1} - v_1 \tag{9-42}$$

where

$$\phi_U = \frac{U_e - 1}{U_e^{N+1} - 1} \tag{9-43}$$

and

$$\phi_E = \frac{E_e - 1}{E_e^{N+1} - 1} \tag{9-44}$$

For the Kremser approximation, values of E_i and U_i in (9-39) and (9-40) are based on the feed and solvent at entering conditions. However, in liquid–liquid extraction, values of V, L, and K_D may change considerably from stage to stage. Therefore, a better approximation is desirable. This is achieved, with reference to Figure 9.16, by the following relations due to Horton

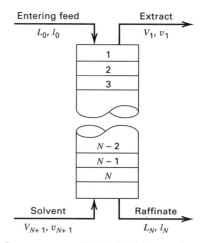

Figure 9.16 Countercurrent, liquid–liquid extraction cascade.

[4]In a vertical extractor, solvent would have to enter at the top if of greater density than the feed.

and Franklin [25] and Edmister [26], which use average values of E_i and U_i based on estimates of values of V, L, and K_D at each end of the cascade. These equations are used in Example 9.9.

$$V_2 = V_1 \left(\frac{V_{N+1}}{V_1} \right)^{1/N} \qquad (9\text{-}45)$$

$$L_1 = L_0 + V_2 - V_1 \qquad (9\text{-}46)$$

$$V_N = V_{N+1} \left(\frac{V_1}{V_{N+1}} \right)^{1/N} \qquad (9\text{-}47)$$

$$E_e = [E_1(E_N + 1) + 0.25]^{1/2} - 0.5 \qquad (9\text{-}48)$$

$$U_e = [U_N(U_1 + 1) + 0.25]^{1/2} - 0.5 \qquad (9\text{-}49)$$

Equations (9-38) through (9-49) can be used with mass or mole units. No enthalpy balances are required because temperature changes in an adiabatic extractor are small unless feed and solvent enter at appreciably different temperatures or the heat of mixing is large. Unfortunately, the group method is not reliable for liquid–liquid extraction cascades when the solute is not dilute. This is because the distribution coefficient is a ratio of activity coefficients, which can vary strongly with solute composition. Accordingly, rigorous methods using a process simulator, as in §10.3, are preferred for nondilute cases.

EXAMPLE 9.9 Extraction of Dimethylformamide (DMF).

As shown in Figure 9.17, extraction with methylene chloride (MC) is used at 25°C to recover DMF from an aqueous stream. Estimate flow rates and compositions of extract and raffinate streams by the group method using mass units. Distribution coefficients for all components except DMF are essentially constant over the expected composition range, and on a mass-fraction basis are:

Component	K_{D_i}
MC	40.2
FA	0.005
DMA	2.2
W	0.003

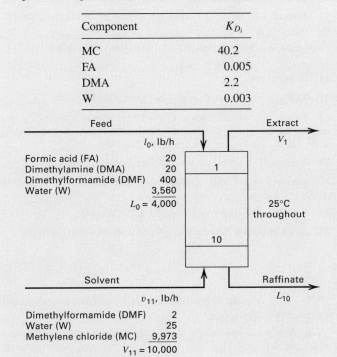

Figure 9.17 Specifications for extractor of Example 9.9.

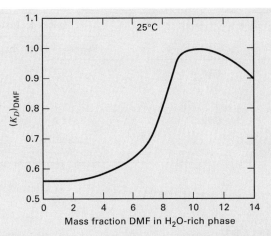

Figure 9.18 Distribution coefficient for DMF between water and MC.

where FA = formamide and W = water. The distribution coefficient for DMF depends on concentration in the water-rich phase as shown in Figure 9.18.

Solution

Although the Kremser approximation could be applied for the first trial calculation, the following values will be assumed from guesses based on the magnitudes of the K_D values.

Component		Pounds per Hour		
	Feed, l_0	Solvent, v_{11}	Raffinate, l_{10}	Extract, v_1
FA	20	0	20	0
DMA	20	0	0	20
DMF	400	2	2	400
W	3,560	25	3,560	25
MC	0	9,973	88	9,885
	4,000	10,000	3,670	10,330

From (9-45) through (9-49), we have

$$V_2 = 10,330 \left(\frac{10,000}{10,330} \right)^{1/10} = 10,297 \text{ lb/h};$$

$$L_1 = 4,000 + 10,297 - 10,330 = 3,967 \text{ lb/h};$$

$$V_{10} = 10,000 \left(\frac{10,330}{10,000} \right)^{1/10} = 10,033 \text{ lb/h}$$

From (9-39), (9-40), (9-48), and (9-49), assuming a mass fraction of 0.09 for DMF in L_1 in order to obtain $(K_D)_{DMF}$ for stage 1,

Component	E_1	E_{10}	U_1	U_{10}	E_e	U_e
FA	0.013	0.014	—	—	0.013	—
DMA	5.73	6.01	—	—	5.86	—
DMF	2.50	1.53	0.400	0.653	2.06	0.579
W	0.0078	0.0082	128	122	0.0078	125
MC	—	—	0.0096	0.0091	—	0.0091

From (9-44), (9-43), (9-41), and (9-42),

Component	ϕ_E	ϕ_U	Pounds per Hour	
			Raffinate, l_{10}	Extract, υ_1
FA	0.9870	—	19.7	0.3
DMA	0.0	—	0.0	20.0
DMF	0.000374	0.422	1.3	400.7
W	0.9922	0.0	3,557.2	37.8
MC	—	0.9909	90.8	9,882.2
			3,669.0	10,331.0

The calculated flow rates L_{10} and V_1 are almost exactly equal to the assumed rates. Therefore, an additional iteration is not necessary. The degree of DMF extraction is very high. More cases with less solvent and/or fewer stages should be calculated.

SUMMARY

1. The Fenske–Underwood–Gilliland (FUG) method for simple distillation of ideal and nearly ideal multicomponent mixtures is useful for preliminary estimates of stage and reflux requirements.

2. Based on a specified split of two key components in the feed mixture, the Fenske equation is used to determine N_{min} at total reflux. The Underwood equations are used to determine R_{min} for an infinite number of stages. The empirical Gilliland correlation relates N_{min} and R_{min} to the actual R and actual N.

3. Distribution of nonkey components and feed-stage location can be estimated with the Fenske and Kirkbride equations, respectively.

4. The Underwood equations are more restrictive than the Fenske equation and must be used with care and caution.

5. The Kremser group method can be applied to strippers and liquid–liquid extractors for dilute solute conditions to make estimates of component recoveries for specified values of entering flow rates and equilibrium stages.

REFERENCES

1. Kremser, A., *Natl. Petroleum News*, **22**(21), 43–49 (1930).

2. Edmister, W.C., *AIChE J.*, **3**, 165–171 (1957).

3. Kobe, K.A., and J.J. McKetta, Jr., Eds, *Advances in Petroleum Chemistry and Refining*, Interscience, New York, Vol. 2, pp. 315–355 (1959).

4. Bachelor, J.B., *Petroleum Refiner.*, **36**(6), 161–170 (1957).

5. Fenske, M.R., *Ind. Eng. Chem.*, **24**, 482–485 (1932).

6. Shiras, R.N., D.N. Hanson, and C.H. Gibson, *Ind. Eng. Chem.*, **42**, 871–876 (1950).

7. Underwood, A.J.V., *Trans. Inst. Chem. Eng.*, **10**, 112–158 (1932).

8. Gilliland, E.R., *Ind. Eng. Chem.*, **32**, 1101–1106 (1940).

9. Underwood, A.J.V., *J. Inst. Petrol.*, **32**, 614–626 (1946).

10. Barnes, F.J., D.N. Hanson, and C.J. King, *Ind. Eng. Chem., Process Des. Dev.*, **11**, 136–140 (1972).

11. Tavana, M., and D.N. Hanson, *Ind. Eng. Chem., Process Des. Dev.*, **18**, 154–156 (1979).

12. Fair, J.R., and W.L. Bolles, *Chem. Eng.*, **75**(9), 156–178 (1968).

13. Gilliland, E.R., *Ind. Eng. Chem.*, **32**, 1220–1223 (1940).

14. Robinson, C.S., and E.R. Gilliland, *Elements of Fractional Distillation*, 4th ed., McGraw-Hill, New York, pp. 347–350 (1950).

15. Brown, G.G., and H.Z. Martin, *Trans. AIChE*, **35**, 679–708 (1939).

16. Van Winkle, M., and W.G. Todd, *Chem. Eng.*, **78**(21), 136–148 (1971).

17. Molokanov, Y.K., T.P. Korablina, N.I. Mazurina, and G.A. Nikiforov, *Int. Chem. Eng.*, **12**(2), 209–212 (1972).

18. Guerreri, G., *Hydrocarbon Processing*, **48**(8), 137–142 (1969).

19. Donnell, J.W., and C.M. Cooper, *Chem. Eng.*, **57**, 121–124 (1950).

20. Oliver, E.D., *Diffusional Separation Processes: Theory, Design, and Evaluation*, John Wiley & Sons, New York, pp. 104–105 (1966).

21. Strangio, V.A., and R.E. Treybal, *Ind. Eng. Chem., Process Des. Dev.*, **13**, 279–285 (1974).

22. Kirkbride, C.G., *Petroleum Refiner.*, **23**(9), 87–102 (1944).

23. Stupin, W.J., and F.J. Lockhart, "The Distribution of Non-Key Components in Multicomponent Distillation," presented at the 61st Annual Meeting of the AIChE, Los Angeles, CA, December 1–5, 1968.

24. Souders, M., and G.G. Brown, *Ind. Eng. Chem.*, **24**, 519–522 (1932).

25. Horton, G., and W.B. Franklin, *Ind. Eng. Chem.*, **32**, 1384–1388 (1940).

26. Edmister, W.C., *Ind. Eng. Chem.*, **35**, 837–839 (1943).

27. Smith, B.D., and W.K. Brinkley, *AIChE J.*, **6**, 446–450 (1960).

STUDY QUESTIONS

9.1. Rigorous, computer-based methods for multicomponent distillation are readily available in process simulators. Why, then, is the FUG method still useful and widely applied for distillation?

9.2. When calculating multicomponent distillation, why is it best to list the components in order of decreasing volatility? In such a list, do the two key components have to be adjacent?

9.3. What does the Fenske equation compute? What assumptions are made in its derivation?

9.4. For what conditions should the Fenske equation be used with caution?

9.5. Is use of the Fenske equation restricted to the two key components? If not, what else can the Fenske equation be used for besides the estimation of the minimum number of equilibrium stages, corresponding to total reflux?

9.6. What is a pinch point or region? For multicomponent distillation, under what conditions is the pinch point located at the feed location? What conditions cause the pinch point to migrate away from the feed location?

9.7. What is the difference between a Class 1 and a Class 2 separation? Why is the Class 1 Underwood equation useful even if the separation is Class 2?

9.8. What is internal reflux? How does it differ from external reflux? Does the Underwood equation compute internal or external reflux? How can one be determined from the other?

9.9. What is the optimal range of values for R/R_{min}?

9.10. What key parameter is missing from the Gilliland correlation?

9.11. When can a serious problem arise with the Gilliland correlation?

9.12. What is the best method for estimating the distribution of nonkey components at the actual (operating) reflux?

9.13. Is the Kremser method a group method? What is meant by a group method?

9.14. Under what conditions can the Kremser method be applied to liquid–liquid extraction?

EXERCISES

Section 9.1

9.1. Type of condenser and operating pressure.

A mixture of propionic and *n*-butyric acids, which can be assumed to form ideal solutions, is to be separated by distillation into a distillate containing 95 mol% propionic acid and a bottoms of 98 mol% *n*-butyric acid. Determine the type of condenser and estimate the distillation operating pressure.

9.2. Type of condenser and operating pressure.

Two distillation columns are used to produce the products indicated in Figure 9.19. Establish the type of condenser and an operating pressure for each column for the: (a) direct sequence (C_2/C_3 separation first) and (b) indirect sequence (C_3/nC_4 separation first). Use K-values from Figures 2.4 and 2.5.

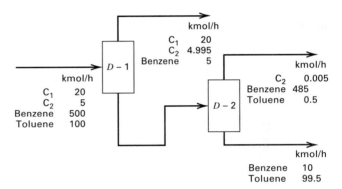

Figure 9.20 Data for Exercise 9.3.

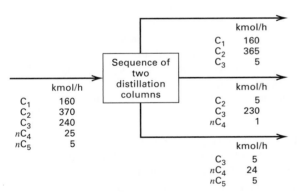

Figure 9.19 Data for Exercise 9.2.

9.3. Type of condenser and operating pressure.

For each of the distillations *D*-1 and *D*-2 indicated in Figure 9.20, establish the type of condenser and an operating pressure.

9.4. Stages for a deethanizer.

For the deethanizer in Figure 9.21, estimate the number of stages, assuming it is equal to 2.5 times N_{min}.

9.5. Fenske equation for a column with a vapor sidestream.

For the complex distillation in Figure 9.22, use the Fenske equation to determine N_{min} between the: (a) distillate and feed, (b) feed and sidestream, and (c) sidestream and bottoms. Use Raoult's law for K-values.

9.6. Comparison of Fenske equation with McCabe–Thiele method.

A 25 mol% mixture of acetone (A) in water (W) is to be separated by distillation at 130 kPa into a distillate containing 95 mol% acetone and a bottoms of 2 mol% acetone. The infinite-dilution activity coefficients are $\gamma_A^\infty = 8.12$ and $\gamma_W^\infty = 4.13$.

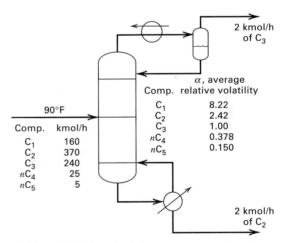

Figure 9.21 Data for Exercise 9.4.

Calculate N_{min} by the Fenske equation. Compare the result to that calculated using the McCabe–Thiele method. Is the Fenske equation reliable for this separation?

9.7. Distribution of nonkeys and minimum stages.

For the distillation in Figure 9.23, calculate N_{min} and the distribution of the nonkey components by the Fenske equation, using Figures 2.4 and 2.5 for K-values.

9.8. Type of condenser, operating pressure, nonkey distribution, and N_{min}.

For the distillation in Figure 9.24, establish the condenser type and operating pressure, calculate N_{min}, and estimate the distribution of the nonkey components. Obtain K-values from Figures 2.4 and 2.5.

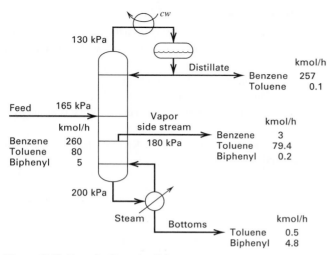

Figure 9.22 Data for Exercise 9.5.

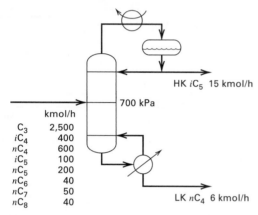

Figure 9.23 Data for Exercise 9.7.

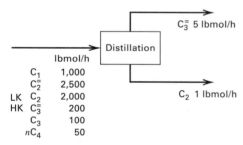

Figure 9.24 Data for Exercise 9.8.

9.9. Effect of distillate rate on key-component recovery.

For $N_{min} = 15$ at 250 psia, calculate and plot the percent recovery of C_3 in the distillate as a function of distillate flow rate for the distillation of 1,000 lbmol/h of a feed containing by moles: 3% C_2, 20% C_3, 37% nC_4, 35% nC_5, and 5% nC_6. Obtain K-values from Figures 2.4 and 2.5.

9.10. Class 1 Underwood equations.

Use the Underwood equations to estimate the minimum external reflux ratio for the separation by distillation of 30 mol% propane in propylene to obtain 99 mol% propylene and 98 mol% propane, if the feed condition at a column pressure of 300 psia is: (a) bubble-point liquid, (b) 50 mol% vaporized, and (c) dew-point vapor. Use K-values from Figures 2.4 and 2.5.

9.11. Class 2 Underwood equations.

For the conditions of Exercise 9.7, with bubble-point liquid feed at column pressure, compute the minimum external reflux and non-key distribution at R_{min} by the Class 2 Underwood equations.

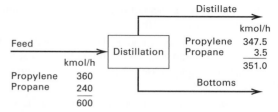

Figure 9.25 Data for Exercise 9.13.

9.12. R_{min} and N_{min} as functions of product purity.

Calculate and plot the external R_{min} and N_{min} against % product purity for the separation by distillation of an equimolar bubble-point liquid feed of isobutane/n-butane at 100 psia. The distillate is to have the same iC_4 purity as the bottoms is to have nC_4 purity. Consider purities from 90% to 99.99%. Discuss the results.

9.13. Reflux ratio by the FUG method.

Use the FUG method to determine the reflux ratio required for the distillation in Figure 9.25 if $N/N_{min} = 2.0$, the average relative volatility = 1.11, and the feed is at the bubble-point temperature at feed-stage pressure. Assume that external reflux equals internal reflux at the upper pinch zone. Assume a total condenser and a partial reboiler.

9.14. Equilibrium stages by the FUG method.

A feed of 62 mol% *para*-dichlorobenzene in *ortho*-dichlorobenzene is separated by distillation at near-atmospheric pressure into a distillate containing 98 mol% para isomer and bottoms of 96 mol% ortho isomer. If a total condenser and partial reboiler are used, $q = 0.9$, average $\alpha = 1.154$, and $R/R_{min} = 1.15$, use the FUG procedure to estimate the theoretical stages required.

9.15. Limitation of the Gilliland correlation.

Explain why the Gilliland correlation can give erroneous results when the ratio of rectifying to stripping stages is small.

9.16. FUG method for HC distillation.

The hydrocarbon feed to a distillation column is a bubble-point liquid at 300 psia with mole fractions: $C_2 = 0.08$, $C_3 = 0.15$, $nC_4 = 0.20$, $nC_5 = 0.27$, $nC_6 = 0.20$, and $nC_7 = 0.10$. Determine: (a) column pressure and type of condenser, if condenser outlet temperature is 120°F, for a sharp separation between nC_4 and nC_5; (b) at total reflux, the separation for eight theoretical stages overall, for 0.01 mole fraction nC_4 in the bottoms product; (c) R_{min} for the separation in part (b); and (d) the number of theoretical stages, at $R/R_{min} = 1.5$, using the Gilliland correlation.

9.17. FUG method for HC distillation.

The following feed mixture is to be separated by ordinary distillation at 120 psia to obtain 92.5 mol% of the nC_4 in the liquid distillate and 82.0 mol% of the iC_5 in the bottoms.

Component	Avg α	lbmol/h
C_3	4.36	5
iC_4	2.36	15
nC_4	1.88	25
iC_5	1.00	20
nC_5	0.84	35
		100

(a) Estimate N_{min} by the Fenske equation. (b) Use the Fenske equation to determine the distribution of nonkey components between distillate and bottoms. (c) Assuming that the feed is at its bubble point, use the Underwood method to estimate R_{min}. (d) Determine N by the Gilliland correlation assuming $R/R_{min} = 1.2$, a partial reboiler, and a total condenser. (e) Estimate feed-stage location.

9.18. FUG method for a chlorination effluent.

Consider the separation by distillation of a chlorination effluent to recover C_2H_5Cl. The feed is a bubble-point liquid at the column pressure of 240 psia with the following composition and K-values.

Component	Mole Fraction	K
C_2H_4	0.05	5.1
HCl	0.05	3.8
C_2H_6	0.10	3.4
C_2H_5Cl	0.80	0.15

Specifications are (x_D/x_B) for $C_2H_5Cl = 0.01$ and (x_D/x_B) for $C_2H_6 = 75$.

Calculate product distribution, N_{min}, R_{min}, and N at $R = 1.5\,R_{min}$; and locate the feed stage. The column is to have a partial condenser and a partial reboiler.

9.19. FUG method for distillation of a ternary mixture.

One hundred kmol/h of a three-component bubble-point mixture to be separated by distillation has the composition:

Component	Mole Fraction	Relative Volatility
A	0.4	5
B	0.2	3
C	0.4	1

(a) For a distillate rate of 60 kmol/h, five stages, and total reflux, calculate the distillate and bottoms compositions by the Fenske equation. (b) Using the separation in part (a) for components B and C, determine R_{min} and minimum boilup ratio by the Class 2 Underwood equations. (c) For $R/R_{min} = 1.2$, determine N and the feed stage.

9.20. Feed-stage location.

For the conditions of Exercise 9.6, determine the ratio of rectifying to stripping equilibrium stages by the: (a) Fenske equation, (b) Kirkbride equation, and (c) McCabe–Thiele diagram. Discuss your results.

Section 9.2

9.21. Effective stripping factor for two stages.

Starting with equations like (5-46) and (5-47), show that for two stages, $S_e = \sqrt{0.25 + S_2(S_1 + 1)} - 0.5$.

9.22. Effect of N and operating pressure on absorption.

Determine, by the Kremser method, the achievable separation for the absorber in Figure 9.26 for the following conditions: (a) six stages and 75-psia pressure, (b) three stages and 150-psia pressure, and (c) six stages and 150-psia pressure. What do you conclude?

9.23. Effect of N on absorption of a paraffin mixture.

One thousand kmol/h of rich gas at 70°F with 25% C_1, 15% C_2, 25% C_3, 20% nC_4, and 15% nC_5 by moles is to be absorbed by 500 kmol/h of nC_{10} at 90°F in an absorber at 4 atm. Calculate by the Kremser method the % absorption of each component for: (a) 4 theoretical stages, (b) 10 theoretical stages, and (c) 30 theoretical stages. Use Figures 2.4 and 2.5 for K-values. What do you conclude?

9.24. Stripping stream flow rate.

For the flashing and stripping operation in Figure 9.27, determine by the Kremser method the kmol/hr of steam if the stripper is operated at 2 atm with five stages.

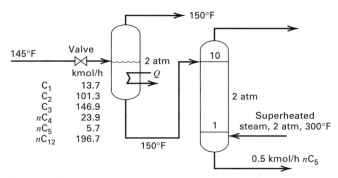

Figure 9.27 Data for Exercise 9.24.

9.25. Stripping with superheated steam.

A stripper at 50 psia with three stages is used to strip 1,000 kmol/h of liquid at 250°F with molar composition 0.03% C_1, 0.22% C_2, 1.82% C_3, 4.47% nC_4, 8.59% nC_5, and 84.87% nC_{10}. The stripping agent is 100 kmol/h of superheated steam at 300°F and 50 psia. Use the Kremser method to estimate compositions and flow rates of the stripped liquid and rich gas.

9.26. Extraction of a HC mixture by DEG.

One hundred kmol/h of an equimolar mixture of benzene (B), toluene (T), n-hexane (C_6), and n-heptane (C_7) is to be extracted at 150°C by 300 kmol/h of diethylene glycol (DEG) in a counter-current, liquid–liquid extractor having five equilibrium stages. Estimate the flow rates and compositions of the extract and raffinate streams by the group method. In mole-fraction units, the distribution coefficients for the hydrocarbon can be assumed constant, with the following values:

Component	$K_{D_i} = y$ (solvent phase)/ x (raffinate phase)
B	0.33
T	0.29
C_6	0.050
C_7	0.043

For DEG, assume $K_D = 30$. [E.D. Oliver, *Diffusional Separation Processes*, John Wiley & Sons, New York, p. 432 (1966).]

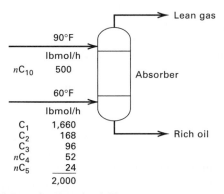

Figure 9.26 Data for Exercise 9.22.

Chapter 10

Equilibrium-Based Methods for Multicomponent Absorption, Stripping, Distillation, and Extraction

§10.0 INSTRUCTIONAL OBJECTIVES

After completing this chapter, you should be able to:

- Write MESH equations for an equilibrium stage in a multicomponent vapor–liquid cascade.
- Explain how equilibrium stages can be combined to form a countercurrent cascade of N equilibrium stages that can be used to model absorption, stripping, distillation, and extraction.
- Discuss different methods to solve the MESH equations and the use of the tridiagonal-matrix algorithm.
- Solve rigorously countercurrent-flow, multi-equilibrium stage, multicomponent separation problems by the bubble-point and sum-rates methods; and, with a process simulator, the Newton–Raphson and inside-out methods. Select the best method to use for a given problem.

Except for simple cases, such as binary distillation, or when physical properties or stage efficiencies are not well known, the design methods described in the previous chapters are suitable only for preliminary-design studies. Final design of multistage, multicomponent separation equipment requires rigorous determination of temperatures, pressures, stream flow rates, stream compositions, and heat-transfer rates at each stage by solving material-balance, energy-balance, and equilibrium relations for each stage. Unfortunately, these relations consist of strongly interacting nonlinear algebraic equations, where solution procedures are difficult and tedious. However, once the procedures are programmed for a high-speed digital computer, solutions are usually achieved rapidly and almost routinely. This chapter discusses the solution methods used in process simulators, with applications to absorption, stripping, distillation, and liquid–liquid extraction. Applications to the more difficult operations of extractive, azeotropic, and reactive distillation are covered in Chapter 11.

This chapter begins in §10.1 with the development of a mathematical model for an equilibrium stage for vapor–liquid contacting. The resulting equations, when collected together for a countercurrent cascade of stages, are called the MESH equations. A number of strategies for solving these equations are summarized in §10.2, with the most important considered in detail. All utilize an algorithm for solving a tridiagonal-matrix equation, described in §10.3. When the feed(s) to the cascade contains components of a narrow boiling-point range, the bubble-point (BP) method is efficient. When the components cover a wide range of volatilities, the sum-rates (SR) method is a better choice. The BP and SR methods are relatively simple, but are restricted to ideal and nearly ideal mixtures, and are limited in allowable specifications. §10.4 and 10.5 present more flexible, complex methods, Newton–Raphson (NR) and inside-out, respectively, which are required for nonideal systems and are widely available in process simulators.

§10.1 THEORETICAL MODEL FOR AN EQUILIBRIUM STAGE

For any stage in a countercurrent cascade, assume (1) phase equilibrium is achieved at each stage, (2) no chemical reactions occur, and (3) entrainment of liquid drops in vapor and occlusion of vapor bubbles in liquid are negligible. Figure 10.1 represents such a stage for the vapor–liquid case, where the stages are numbered down from the top. The same representation applies to liquid–liquid extraction if the higher-density liquid phases are represented by liquid streams and the lower-density liquid phases are represented by vapor streams.

Entering stage j is a single- or two-phase feed of molar flow rate F_j, with overall composition in mole fractions $z_{i,j}$ of component i, temperature T_{F_j}, pressure P_{F_j}, and corresponding overall molar enthalpy h_{F_j}. Feed pressure is equal to or greater than stage pressure P_j. If greater, $\left(P_F - P_j\right)$ is reduced to zero adiabatically across valve F.

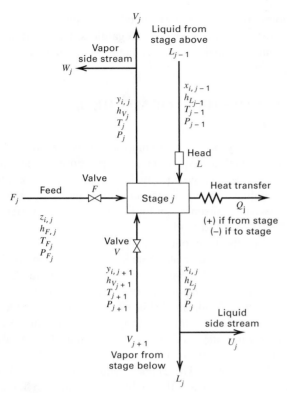

Figure 10.1 General equilibrium stage.

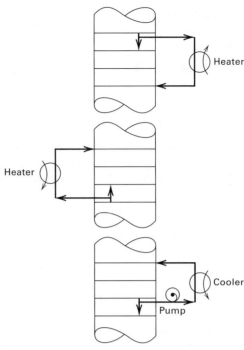

Figure 10.2 Pumparounds.

Also entering stage j is interstage liquid from stage $j - 1$ above, if any, of molar flow rate L_{j-1}, with composition in mole fractions $x_{i,j-1}$, enthalpy $h_{L_{j-1}}$, temperature T_{j-1}, and pressure P_{j-1}, which is equal to or less than the pressure of stage j. Pressure of liquid from stage $j - 1$ is increased adiabatically by hydrostatic head change across head L.

Similarly, from stage $j + 1$ below, interstage vapor of molar flow rate V_{j+1}, with composition in mole fractions $y_{i,j+1}$, enthalpy $h_{V_{j+1}}$, temperature T_{j+1}, and pressure P_{j+1} enters stage j. Any excess pressure $(P_{j+1} - P_j)$ is reduced to zero adiabatically across valve V.

Leaving stage j is vapor of intensive properties $y_{i,j}$, h_{V_j}, T_j, and P_j. This stream can be divided into a vapor sidestream of molar flow rate W_j and an interstage stream of molar flow rate V_j to be sent to stage $j - 1$ or, if $j = 1$, to leave as a product. Also leaving stage j is liquid of intensive properties $x_{i,j}$, h_{L_j}, T_j, and P_j, in equilibrium with vapor $(V_j + W_j)$. This liquid is divided into a sidestream of molar flow rate U_j and an interstage stream of molar flow rate L_j to be sent to stage $j + 1$ or, if $j = N$, to leave as a product.

Heat can be transferred at a rate Q_j from (+) or to (−) stage j to simulate stage intercoolers, interheaters, intercondensers, interreboilers, condensers, or reboilers, as shown in Figure 1.8. The model in Figure 10.1 does not allow for pumparounds of the type shown in Figure 10.2. They are often used in columns having sidestreams, such as crude units in petroleum refineries, in order to conserve energy and balance column vapor loads. Advanced models in process simulators can handle pumparounds.

Associated with each general stage are the following indexed equations expressed in terms of the variable set in Figure 10.1. However, variables other than those shown in Figure 10.1 can be used, e.g. component flow rates can replace mole fractions, and sidestream flow rates can be expressed as fractions of interstage flow rates.

The equations are similar to those of §5.7[1] and are referred to as MESH equations, after Wang and Henke [1].

1. **M equations**—*M*aterial balance for each component (C equations for each stage).

$$M_{i,j} = L_{j-1}x_{i,j-1} + V_{j+1}y_{i,j+1} + F_j z_{i,j}$$
$$- (L_j + U_j)x_{i,j} - (V_j + W_j)y_{i,j} = 0 \qquad (10\text{-}1)$$

2. **E equations**—phase-*E*quilibrium relation for each component (C equations for each stage), from (2-19).

$$E_{i,j} = y_{i,j} - K_{i,j}x_{i,j} = 0 \qquad (10\text{-}2)$$

where $K_{i,j}$ is the phase-equilibrium ratio or *K*-value.

3. **S equations**—mole-fraction *S*ummations (one for each stage),

$$(S_y)_j = \sum_{i=1}^{C} y_{i,j} - 1.0 = 0 \qquad (10\text{-}3)$$

$$(S_x)_j = \sum_{i=1}^{C} x_{i,j} - 1.0 = 0 \qquad (10\text{-}4)$$

---

[1]Unlike the treatment in §5.7, all C component material balances are included here, and the total material balance is omitted. Also, the separate but equal temperature and pressure of the equilibrium phases are replaced by the stage temperature and pressure.

4. *H* equation—energy balance (one for each stage).

$$H_j = L_{j-1}h_{L_{j-1}} + V_{j+1}h_{V_{j+1}} + F_j h_{F_j}$$
$$- (L_j + U_j)h_{L_j} - (V_j + W_j)h_{V_j} - Q_j = 0$$

(10-5)

where kinetic- and potential-energy changes are ignored.

A total material-balance equation can be used in place of (10-3) or (10-4). It is derived by combining these two equations and $\sum_j z_{i,j} = 1.0$ with (10-1) summed over the *C* components and over stages 1 through *j* to give

$$L_j = V_{j+1} + \sum_{m=1}^{j}(F_m - U_m - W_m) - V_1$$

(10-6)

In general, $K_{i,j} = K_{i,j}\{T_j, P_j, \mathbf{x}_j, \mathbf{y}_j\}, h_{V_j} = h_{V_j}\{T_j, P_j, \mathbf{y}_j\},$ and $h_{L_j} = h_{L_j}\{T_j, P_j, \mathbf{x}_j\}$, where $\mathbf{x}_j$ and $\mathbf{y}_j$ (in bold) are vectors of component mole fractions in streams leaving stage *j*. If these property relations are not counted as equations and the three properties are not counted as variables, each equilibrium stage is defined only by the $2C + 3$ MESH equations. A countercurrent cascade of *N* such stages, as shown in Figure 10.3, is represented by $N(2C + 3)$ such equations in $[N(3C + 10) + 1]$ variables. If *N* and all $F_j, z_{i,j}, T_{F_j}, P_{F_j}, P_j, U_j, W_j,$ and Q_j are specified, the model is represented by $N(2C + 3)$ simultaneous algebraic equations in $N(2C + 3)$ unknown (output) variables comprising all $x_{i,j}, y_{i,j}, L_j, V_j,$ and T_j, where the *M*, *E*, and *H* equations are nonlinear. If other

variables are specified, corresponding substitutions are made to the list of output variables. Regardless, the result is a set containing nonlinear equations that must be solved by an iterative technique.

§10.2 STRATEGY OF MATHEMATICAL SOLUTION

A wide variety of iterative solution procedures for solving nonlinear, algebraic equations has appeared in the literature. In general, these procedures make use of equation partitioning in conjunction with equation tearing and/or linearization by Newton–Raphson techniques, which are described in detail by Myers and Seider [2]. Equation tearing was applied in §4.4 for a flash computation.

Early, pre-computer attempts to solve (10-1) to (10-5) or equivalent forms of these equations resulted in the classical *stage-by-stage*, *equation-by-equation* calculational procedures of Lewis–Matheson [3] in 1932 and Thiele–Geddes [4] in 1933 based on equation tearing for solving simple fractionators with one feed and two products. Composition-independent *K*-values and component enthalpies were generally employed. The Thiele–Geddes method was formulated to handle the Case II variable specification in Table 5.4 wherein the number of equilibrium stages above and below the feed, the reflux ratio, and the distillate flow rate are specified, and stage temperatures and interstage vapor (or liquid) flow rates are the iteration (tear) variables. Although widely used for hand calculations in the years following its development, the Thiele–Geddes method was often found to be numerically unstable when attempts were made to program it for a computer. However, Holland [5] developed a Thiele–Geddes procedure called the *theta method*, which was applied with considerable success.

The Lewis–Matheson method was formulated for the Case I variable specification in Table 5.4 to determine stage requirements for specifications of the separation of two key components, a reflux ratio and a feed-stage location criterion. Both outer and inner iterations are required. The outer-loop tear variables are the mole fractions or flow rates of nonkey components in the products. The inner-loop tear variables are the interstage vapor (or liquid) flows. The Lewis–Matheson method was widely used for hand calculations, but was often unstable when implemented on a computer.

Rather than an equation-by-equation solution procedure, Amundson and Pontinen [6], in a significant development in 1958, showed that (10-1), (10-2), and (10-6) of the MESH equations for a Case II specification could be combined and solved component-by-component from simultaneous-linear-equation sets for all *N* stages by an equation-tearing procedure using the same tear variables as the Thiele–Geddes method. Although tedious for hand calculations, the equation sets are easily solved by a digital computer.

In a study in 1964, Friday and Smith [7] systematically analyzed a number of tearing techniques for solving the MESH equations. They considered the choice of output variable for each equation and showed that no one tearing technique could solve all problem types. For separators where the

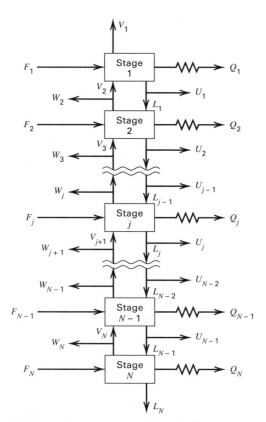

Figure 10.3 General countercurrent cascade of *N* stages.

feed(s) contain(s) only components of similar volatility (narrow-boiling case), a modified Amundson–Pontinen approach, termed the *bubble-point (BP) method*, was recommended. For a feed(s) containing components of widely different volatility (wide-boiling case), the BP method was subject to failure and a so-called sum-rates (SR) method was suggested. For intermediate cases, the equation-tearing technique may fail to converge; in that case, Friday and Smith indicated that either a Newton–Raphson method or a combined tearing and Newton–Raphson technique was necessary.

Boston and Sullivan [8] in 1974 presented an alternative, robust approach to obtaining a solution to the MESH equations. They defined energy and volatility parameters, which are used as the primary successive-approximation variables. A third parameter, which is a combination of the phase flow rates and temperature at each stage, was employed to iterate on the primary variables—thus the name *inside-out method*. Current practice is based mainly on the BP, SR, Newton–Raphson, and inside-out methods, all of which are treated in this chapter. The latter two methods are most widely used in process simulators because they provide flexibility in the choice of specified variables and are capable of solving most problems. However, the first iteration of the BP or SR method is frequently used to provide estimated values to initiate the Newton–Raphson (NR) or inside-out method.

§10.3 EQUATION-TEARING PROCEDURES

The modern equation-tearing procedures are readily programmed, rapid, and require a minimum of computer memory. Although they can be applied to a wider variety of problems than the Thiele–Geddes tearing procedure, they are usually limited to the same choice of specified variables, including most importantly the number of equilibrium stages and feed and product stage locations. Product purities, species recoveries, interstage flow rates, and stage temperatures are not specified.

§10.3.1 Tridiagonal Matrix Algorithm

The key to the BP and SR tearing procedures is the tridiagonal matrix, which results from a modified form of the M equations, (10-1), when they are torn from the other equations by selecting T_j and V_j as the tear variables, leaving the modified M equations linear in the unknown liquid mole fractions. This set of equations, one for each component, is solved by a modified Gaussian–elimination algorithm due to Thomas as applied by Wang and Henke [1]. The modified M equations are obtained by substituting (10-2) into (10-1) to eliminate y and by substituting (10-6) into (10-1) to eliminate L. Thus, equations for calculating y and L are partitioned from the other equations. The result for each component, i, and each stage, j, is as follows, where the i subscripts have been dropped from the B, C, and D terms.

$$A_j x_{i,j-1} + B_j x_{i,j} + C_j x_{i,j+1} = D_j \qquad (10\text{-}7)$$

where

$$A_j = V_j + \sum_{m=1}^{j-1}(F_m - W_m - U_m) - V_1, \quad 2 \le j \le N \quad (10\text{-}8)$$

$$B_j = -\left[V_{j+1} + \sum_{m=1}^{j}(F_m - W_m - U_m) \right. $$
$$\left. - V_1 + U_j + (V_j + W_j)K_{i,j}\right], \quad 1 \le j \le N \qquad (10\text{-}9)$$

$$C_j = V_{j+1}K_{i,j+1}, \quad 1 \le j \le N-1 \qquad (10\text{-}10)$$

$$D_j = -F_j z_{i,j}, \quad 1 \le j \le N \qquad (10\text{-}11)$$

with $x_{i,0} = 0$, $V_{N+1} = 0$, $W_1 = 0$, and $U_N = 0$, as indicated in Figure 10.3. If the modified M equations are grouped by component, they can be partitioned by writing them as a series of separate tridiagonal-matrix equations, one for each component, where the output variable for each matrix equation is x_i over the entire N-stage cascade.

$$
\begin{bmatrix}
B_1 & C_1 & 0 & 0 & 0 & \cdots & & & & & 0 \\
A_2 & B_2 & C_2 & 0 & 0 & \cdots & & & & & 0 \\
0 & A_3 & B_3 & C_3 & 0 & \cdots & & & & & 0 \\
\vdots & & & & & & & & & & \\
& & & & & & & & & & \\
& & & & & & & & & & \\
0 & & & \cdots & 0 & A_{N-2} & B_{N-2} & C_{N-2} & & & 0 \\
0 & & & \cdots & 0 & 0 & A_{N-1} & B_{N-1} & C_{N-1} \\
0 & & & \cdots & 0 & 0 & 0 & A_N & B_N
\end{bmatrix}
\times
\begin{bmatrix}
x_{i,1} \\ x_{i,2} \\ x_{i,3} \\ \cdots \\ \cdots \\ \cdots \\ \cdots \\ \cdots \\ \cdots \\ x_{i,N-2} \\ x_{i,N-1} \\ x_{i,N}
\end{bmatrix}
=
\begin{bmatrix}
D_1 \\ D_2 \\ D_3 \\ \cdots \\ \cdots \\ \cdots \\ \cdots \\ \cdots \\ \cdots \\ D_{N-2} \\ D_{N-1} \\ D_N
\end{bmatrix}
\qquad (10\text{-}12)
$$

Constants B_j and C_j for each component depend only on tear variables T and V if K-values are composition-independent. If not, previous iteration compositions may be used to estimate K-values.

The Thomas algorithm for solving the linearized equation set (10-12) is a Gaussian–elimination procedure involving forward elimination starting from stage 1 and working toward stage N to finally isolate $x_{i,N}$. Other values of $x_{i,j}$ are then obtained, starting with $x_{i,N-1}$ by backward substitution. For five stages, the matrix equations for a given component at the beginning, middle, and end of the procedure are as shown in Figure 10.4.

The details of the procedure are as follows. In the Thomas algorithm, for stage 1, (10-7) is $B_1 x_{i,1} + C_1 x_{i,2} = D_1$, which can be solved for $x_{i,1}$ in terms of unknown $x_{i,2}$ to give

$$x_{i,1} = (D_1 - C_1 x_{i,2})/B_1$$

Let
$$p_1 = \frac{C_1}{B_1} \quad \text{and} \quad q_1 = \frac{D_1}{B_1}$$

$$\begin{bmatrix} B_1 & C_1 & 0 & 0 & 0 \\ A_2 & B_2 & C_2 & 0 & 0 \\ 0 & A_3 & B_3 & C_3 & 0 \\ 0 & 0 & A_4 & B_4 & C_4 \\ 0 & 0 & 0 & A_5 & B_5 \end{bmatrix} \cdot \begin{bmatrix} x_1 \\ x_2 \\ x_3 \\ x_4 \\ x_5 \end{bmatrix} = \begin{bmatrix} D_1 \\ D_2 \\ D_3 \\ D_4 \\ D_5 \end{bmatrix}$$

(a)

$$\begin{bmatrix} 1 & p_1 & 0 & 0 & 0 \\ 0 & 1 & p_2 & 0 & 0 \\ 0 & 0 & 1 & p_3 & 0 \\ 0 & 0 & 0 & 1 & p_4 \\ 0 & 0 & 0 & 0 & 1 \end{bmatrix} \cdot \begin{bmatrix} x_1 \\ x_2 \\ x_3 \\ x_4 \\ x_5 \end{bmatrix} = \begin{bmatrix} q_1 \\ q_2 \\ q_3 \\ q_4 \\ q_5 \end{bmatrix}$$

(b)

$$\begin{bmatrix} 1 & 0 & 0 & 0 & 0 \\ 0 & 1 & 0 & 0 & 0 \\ 0 & 0 & 1 & 0 & 0 \\ 0 & 0 & 0 & 1 & 0 \\ 0 & 0 & 0 & 0 & 1 \end{bmatrix} \cdot \begin{bmatrix} x_1 \\ x_2 \\ x_3 \\ x_4 \\ x_5 \end{bmatrix} = \begin{bmatrix} r_1 \\ r_2 \\ r_3 \\ r_4 \\ r_5 \end{bmatrix}$$

(c)

Figure 10.4 The coefficient matrix for the M equations of a component at various steps in the Thomas algorithm for five stages. The i subscript is deleted from x. (a) Initial matrix. (b) Matrix after forward elimination. (c) Matrix after backward substitution.

Then
$$x_{i,1} = q_1 - p_1 x_{i,2} \tag{10-13}$$

Thus, the coefficients in the matrix become $B_1 \leftarrow 1$, $C_1 \leftarrow p_1$, and $D_1 \leftarrow q_1$, where $\leftarrow$ means "is replaced by." Only values for p_1 and q_1 need be stored in memory.

For stage 2, (10-7) can be combined with (10-13) and solved for $x_{i,2}$ to give

$$x_{i,2} = \frac{D_2 - A_2 q_1}{B_2 - A_2 p_1} - \left(\frac{C_2}{B_2 - A_2 p_1} \right) x_{i,3}$$

Let
$$q_2 = \frac{D_2 - A_2 q_1}{B_2 - A_2 p_1} \quad \text{and} \quad p_2 = \frac{C_2}{B_2 - A_2 p_1}$$

Then
$$x_{i,2} = q_2 - p_2 x_{i,3}$$

Thus, $A_2 \leftarrow 0$, $B_2 \leftarrow 1$, $C_2 \leftarrow p_2$, and $D_2 \leftarrow q_2$. Only values for p_2 and q_2 need be stored in memory.

In general,

$$p_j = \frac{C_j}{B_j - A_j p_{j-1}} \tag{10-14}$$

$$q_j = \frac{D_j - A_j q_{j-1}}{B_j - A_j p_{j-1}} \tag{10-15}$$

Then
$$x_{i,j} = q_j - p_j x_{i,j+1} \tag{10-16}$$

with $A_j \leftarrow 0$, $B_j \leftarrow 1$, $C_j \leftarrow p_j$, and $D_j \leftarrow q_j$. Only p_j and q_j need be stored. Thus, starting with stage 1, values of p_j and q_j are computed recursively in the order p_1, q_1, p_2, q_2, ..., p_{N-1}, q_{N-1}, q_N. For stage N, (10-16) isolates $x_{i,N}$ as

$$x_{i,N} = q_N \tag{10-17}$$

Successive values of x_i are computed recursively by backward substitution from (10-16) in the form

$$x_{i,j-1} = q_{j-1} - p_{j-1} x_{i,j} = r_{j-1} \tag{10-18}$$

Equation (10-18) corresponds to the identity matrix.

The Thomas algorithm avoids buildup of computer truncation errors because none of the steps involves subtraction of nearly equal quantities. Furthermore, computed values of $x_{i,j}$ are almost always positive. The algorithm is superior to alternative matrix-inversion routines. A modified Thomas algorithm for difficult cases is given by Boston and Sullivan [9]. Such cases can occur for columns with large numbers of equilibrium stages and components whose absorption factors, $A = L/KV$, are less than unity in one section and greater than unity in another.

§10.3.2 Bubble-Point (BP) Method for Distillation

Frequently, distillation involves species that cover a relatively narrow range of K-values. A particularly effective procedure for this case was suggested by Friday and Smith [7] and developed in detail by Wang and Henke [1]. It is referred to as the bubble-point (BP) method because a new set of stage temperatures is computed during each iteration from the bubble-point equations in §4.4.2. All equations are partitioned and solved sequentially except for the M equations (10-1), which are solved separately for each component by the tridiagonal-matrix technique.

The Wang–Henke algorithm is shown in Figure 10.5. A FORTRAN program is available [10]. Specifications are conditions and stage location of feeds, stage pressures, flow rates of sidestreams (note that liquid distillate flow rate, if any, is designated as U_1), heat-transfer rates for all stages except stage 1 (condenser) and stage N (reboiler), total stages, bubble-point reflux flow rate, and vapor distillate flow rate. Figure 10.6 shows a sample specification for a complex column with two feeds, two sidestreams, a mixed condenser, and an intercooler.

To initiate the calculations, values of tear variables, V_j and T_j, are assumed. Generally, it is sufficient to establish an initial set of V_j values based on constant-molar interstage flows using the specified reflux, distillate, feed, and sidestream flows. Initial T_j values can be provided by computing the bubble-point temperature of an estimated bottoms product and the dew-point temperature of an assumed distillate product (or computing a bubble-point temperature if distillate is liquid, or a temperature in between the dew point and bubble point if distillate is both vapor and liquid), and then using linear interpolation for the other stage temperatures.

To solve (10-12) for x_i by the Thomas method, $K_{i,j}$ values are required. When they are composition-dependent, initial assumptions for all $x_{i,j}$ and $y_{i,j}$ values are also needed, unless ideal K-values are employed initially. For each iteration, the computed set of $x_{i,j}$ values for each stage are not likely to satisfy the summation constraint given by (10-4). Although not mentioned by Wang and Henke, it is advisable to normalize the set of computed $x_{i,j}$ values by the relation

$$\left(x_{i,j} \right)_{\text{normalized}} = \frac{x_{i,j}}{\displaystyle\sum_{i=1}^{C} x_{i,j}} \tag{10-19}$$

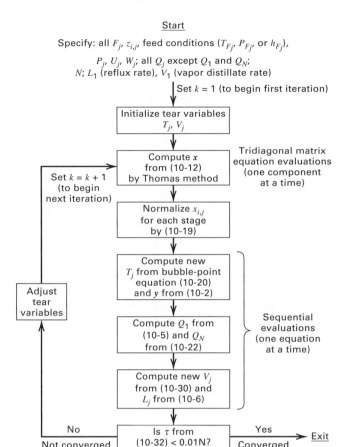

Figure 10.5 Algorithm for Wang–Henke BP method.

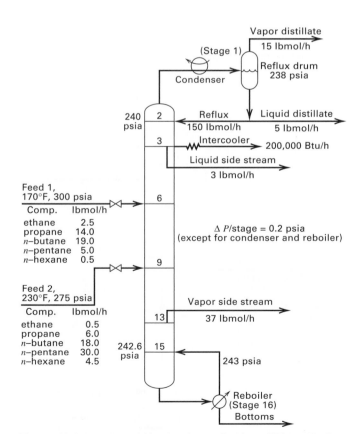

Figure 10.6 Sample specification for application of Wang–Henke BP method to distillation.

New temperatures for the stages are obtained by bubble-point calculations using normalized $x_{i,j}$ values. Friday and Smith [7] showed that bubble-point calculations are effective for mixtures having a narrow range of K-values because temperatures are not sensitive to composition. In the limiting case where all components have identical K-values, the temperature corresponds to $K_{i,j} = 1$ and is not dependent on $x_{i,j}$. At the other extreme, bubble-point calculations can be sensitive to composition for a binary mixture containing one component with a high K-value that changes little with temperature and a second component with a low K-value that changes rapidly with temperature. Such a mixture is methane and n-butane at 400 psia. The effect on bubble-point temperature of small quantities of methane dissolved in liquid n-butane is very large:

Liquid Mole Fraction of Methane	Bubble-Point Temperature, °F
0.000	275
0.018	250
0.054	200
0.093	150

Thus, the BP method is best when components have a relatively narrow range of K-values.

The necessary bubble-point equation is obtained in the manner described in §4.4.2 by combining (10-2) and (10-3) to eliminate $y_{i,j}$, giving

$$\sum_{i=1}^{C} K_{i,j} x_{i,j} - 1.0 = 0 \qquad (10\text{-}20)$$

which is nonlinear in T_j and must be solved iteratively. Wang and Henke prefer to use Muller's iterative method [11] because it is reliable and does not require derivatives. Muller's method requires three initial assumptions of T_j. For each assumption, the value of S_j is computed from

$$S_j = \sum_{i=1}^{C} K_{i,j} x_{i,j} - 1.0 = 0 \qquad (10\text{-}21)$$

The three sets of (T_j, S_j) are fitted to a quadratic equation for S_j in terms of T_j, which is then employed to predict T_j for $S_j = 0$, as required by (10-20). The validity of this value of T_j is checked by using it to compute S_j in (10-21). The quadratic fit and S_j check are repeated with the three best sets of (T_j, S_j) until some convergence tolerance is satisfied, say, $\left| T_j^{(n)} - T_j^{(n-1)} \right| / T_j^{(n)} \leq 0.0001$, with T in °R or K, where n is the iteration number for the temperature loop in the bubble-point calculation; or $S_j \leq 0.0001\, C$ can be used, which is preferred.

Values of $y_{i,j}$ are determined along with the calculation of stage temperatures using the E equations, (10-2). With a consistent set of values for $x_{i,j}$, T_j, and $y_{i,j}$, molar enthalpies are computed for each liquid and vapor stream leaving a stage. Since F_1, V_1, U_1, W_1, and L_1 are specified, V_2 is readily obtained from (10-6), and the condenser duty, a $(+)$ quantity, is obtained from (10-5). Reboiler duty, a $(-)$ quantity, is determined by summing (10-5) for all stages to give

$$Q_N = \sum_{j=1}^{N} \left(F_j h_{F_j} - U_j h_{L_j} - W_j h_{V_j} \right)$$
$$- \sum_{j=1}^{N-1} \left(Q_j - V_1 h_{V_1} - L_N h_{L_N} \right) \tag{10-22}$$

A new set of V_j tear variables is computed by applying a modified energy balance obtained by combining (10-5) and (10-6) twice to eliminate L_{j-1} and L_j. After rearrangement,

$$\alpha_j V_j + \beta_j V_{j+1} = \gamma_j \tag{10-23}$$

where
$$\alpha_j = h_{L_{j-1}} - h_{V_j} \tag{10-24}$$

$$\beta_j = h_{V_{j+1}} - h_{L_j} \tag{10-25}$$

$$\gamma_j = \left[\sum_{m=1}^{j-1} (F_m - W_m - U_m) - V_1 \right] \left(h_{L_j} - h_{L_{j=1}} \right)$$
$$+ F_j \left(h_{L_j} - h_{F_j} \right) + W_j \left(h_{V_j} - h_{L_j} \right) + Q_j \tag{10-26}$$

and enthalpies are evaluated at the stage temperatures last computed rather than at those used to initiate the iteration. Written in didiagonal-matrix form, (10-23) applied over stages 2 to $N - 1$ is:

$$\begin{bmatrix} \beta_2 & 0 & 0 & 0 & .. & & & 0 \\ \alpha_3 & \beta_3 & 0 & 0 & .. & & & 0 \\ 0 & \alpha_4 & \beta_4 & 0 & .. & & & 0 \\ .. & .. & & & & & & .. \\ .. & .. & & & & & & .. \\ 0 & .. & 0 & \alpha_{N-3} & \beta_{N-3} & 0 & & 0 \\ 0 & .. & 0 & 0 & \alpha_{N-2} & \beta_{N-2} & & 0 \\ 0 & .. & 0 & 0 & 0 & \alpha_{N-1} & & \beta_{N-1} \end{bmatrix} \cdot \begin{bmatrix} V_3 \\ V_4 \\ V_5 \\ .. \\ .. \\ V_{N-2} \\ V_{N-1} \\ V_N \end{bmatrix}$$

$$= \begin{bmatrix} \gamma_2 - \alpha_2 V_2 \\ \gamma_3 \\ \gamma_4 \\ .. \\ .. \\ \gamma_{N-3} \\ \gamma_{N-2} \\ \gamma_{N-1} \end{bmatrix} \tag{10-27}$$

Matrix equation (10-27) is solved one equation at a time by starting at the top, where V_2 is known, and working down recursively. Thus,

$$V_3 = \frac{\gamma_2 - \alpha_2 V_2}{\beta_2} \tag{10-28}$$

$$V_4 = \frac{\gamma_3 - \alpha_3 V_3}{\beta_3} \tag{10-29}$$

or, in general

$$V_j = \frac{\gamma_{j-1} - \alpha_{j-1} V_{j-1}}{\beta_{j-1}} \tag{10-30}$$

Corresponding liquid flow rates are obtained from (10-6).

The solution is considered converged when sets of $T_j^{(k)}$ and $V_j^{(k)}$ values are within some prescribed tolerance of corresponding sets of $T_j^{(k-1)}$ and $V_j^{(k-1)}$ values, where k is the iteration index. One convergence criterion is

$$\sum_{j=1}^{N} \left[\frac{T_j^{(k)} - T_j^{(k-1)}}{T_j^{(k)}} \right]^2 + \sum_{j=1}^{N} \left[\frac{V_j^{(k)} - V_j^{(k-1)}}{T_j^{(k)}} \right]^2 \le \epsilon \tag{10-31}$$

where T is an absolute temperature and ϵ is some prescribed tolerance. However, Wang and Henke suggest that the following simpler criterion, which is based on successive sets of T_j values only, is adequate.

$$\tau = \sum_{j=1}^{N} \left[T_j^{(k)} - T_j^{(k-1)} \right]^2 \le 0.01N \tag{10-32}$$

Successive substitution is often employed for iterating the tear variables; that is, values of T_j and V_j generated from (10-20) and (10-30) during an iteration are used directly to initiate the next iteration. It is desirable to inspect, and, if necessary, adjust the generated tear variables prior to beginning the next iteration. Upper and lower bounds should be placed on stage temperatures, and negative values of interstage flow rates should be changed to near-zero positive values. Also, to prevent iteration oscillations, damping can be employed to limit changes in V_j and absolute T_j from one iteration to the next to, say, 10%.

EXAMPLE 10.1 First Iteration of the BP Method.

For the distillation column shown in Figure 10.7, do one iteration of the BP method up to and including the calculation of a new set of T_j values from (10-20). Assume composition-independent K-values.

Solution

By overall total material balance

Liquid distillate $= U_1 = F_3 - L_5 = 100 - 50 = 50 \text{ lbmol/h}$

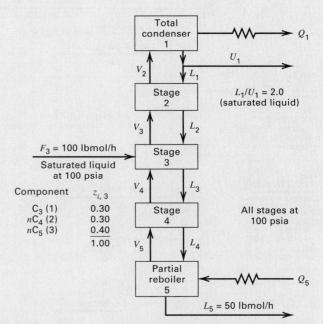

Figure 10.7 Specifications for distillation column of Example 10.1.

Then, $\quad L_1 = (L_1/U_1)U_1 = (2)(50) = 100 \text{ lbmol/h}$

By total material balance around the total condenser,

$$V_2 = L_1 + U_1 = 100 + 50 = 150 \text{ lbmol/h}$$

Initial guesses of tear variables are

Stage j	V_j, lbmol/h	T_j, °F
1	(Fixed at 0 by specifications)	65
2	(Fixed at 150 by specifications)	90
3	150	115
4	150	140
5	150	165

At 100 psia, the estimated K-values at the assumed stage temperatures are

Stage			$K_{i,j}$		
	1	2	3	4	5
$C_3(1)$	1.23	1.63	2.17	2.70	3.33
$nC_4(2)$	0.33	0.50	0.71	0.95	1.25
$nC_5(3)$	0.103	0.166	0.255	0.36	0.49

The matrix equation (10-12) for the first component C_3 is developed as follows. From (10-8) with $V_1 = 0$, $W = 0$,

$$A_j = V_j + \sum_{m=1}^{j-1} (F_m - U_m)$$

Thus, $A_5 = V_5 + F_3 - U_1 = 150 + 100 - 50 = 200$ lbmol/h. Similarly, $A_4 = 200$, $A_3 = 100$, and $A_2 = 100$ in the same units. From (10-9) with $V_1 = 0$, $W = 0$,

$$B_j = -\left[V_{j+1} + \sum_{m=1}^{j} (F_m - U_m) + U_j + V_j K_{i,j} \right]$$

Thus, $B_5 = -[F_3 - U_1 + V_5\,K_{1,5}] = -[100 - 50 + (150)3.33] = -549.5$ lbmol/h. Similarly, $B_4 = -605$, $B_3 = -525.5$, $B_2 = -344.5$, and $B_1 = -150$ in the same units. From (10-10), $C_j = V_{j+1}K_{1,j+1}$. Thus, $C_1 = V_2 K_{1,2} = 150(1.63) = 244.5$ lbmol/h.

Similarly, $C_2 = 325.5$, $C_3 = 405$, and $C_4 = 499.5$ in the same units. From (10-11), $D_j = -F_j z_{1,j}$. Thus, $D_3 = -100(0.30) = -30$ lbmol/h. Similarly, $D_1 = D_2 = D_4 = D_5 = 0$. Substitution of the above values into (10-7) gives

$$\begin{bmatrix} -150 & 244.5 & 0 & 0 & 0 \\ 100 & -344.5 & 325.5 & 0 & 0 \\ 0 & 100 & -525.5 & 405 & 0 \\ 0 & 0 & 200 & -605 & 499.5 \\ 0 & 0 & 0 & 200 & -549.5 \end{bmatrix} \begin{bmatrix} x_{1,1} \\ x_{1,2} \\ x_{1,3} \\ x_{1,4} \\ x_{1,5} \end{bmatrix} = \begin{bmatrix} 0 \\ 0 \\ -30 \\ 0 \\ 0 \end{bmatrix}$$

Using (10-14) and (10-15), the forward step of the Thomas algorithm becomes

$$p_1 = \frac{C_1}{B_1} = 244.5/(-150) = -1.630$$

$$q_1 = \frac{D_1}{B_1} = 0/(-150) = 0$$

$$p_2 = \frac{C_2}{B_2 - A_2 p_1} = \frac{325.5}{-344.5 - 100(-1.630)} = -1.793$$

By similar calculations, the matrix equation after the forward-elimination procedure is

$$\begin{bmatrix} 1 & -1.630 & 0 & 0 & 0 \\ 0 & -1 & -1.793 & 0 & 0 \\ 0 & 0 & 1 & -1.170 & 0 \\ 0 & 0 & 0 & 1 & -1.346 \\ 0 & 0 & 0 & 0 & 1 \end{bmatrix} \begin{bmatrix} x_{1,1} \\ x_{1,2} \\ x_{1,3} \\ x_{1,4} \\ x_{1,5} \end{bmatrix} = \begin{bmatrix} 0 \\ 0 \\ 0.0867 \\ 0.0467 \\ 0.0333 \end{bmatrix}$$

Applying the backward steps of (10-17) and (10-18) gives

$$x_{1,5} = q_5 = 0.0333$$
$$x_{1,4} = q_4 - p_4 x_{1,5} = 0.0467 - (-1.346)(0.0333) = 0.0915$$

Similarly,

$$x_{1,3} = 0.1938, \quad x_{1,2} = 0.3475, \quad x_{1,1} = 0.5664$$

The matrix equations for nC_4 and nC_5 are solved similarly to give

Stage			$x_{i,j}$		
	1	2	3	4	5
C_3	0.5664	0.3475	0.1938	0.0915	0.0333
nC_4	0.1910	0.3820	0.4483	0.4857	0.4090
nC_5	0.0191	0.1149	0.3253	0.4820	0.7806
$\sum_i x_{i,j}$	0.7765	0.8444	0.9674	1.0592	1.2229

After these compositions are normalized, bubble-point temperatures at 100 psia are computed iteratively from (10-20) and compared to the initially assumed values,

Stage	$T^{(2)}$, °F	$T^{(1)}$, °F
1	66	65
2	94	90
3	131	115
4	154	140
5	184	165

The BP convergence rate is unpredictable, and, as shown next in Example 10.2, can depend on the assumed initial set of T_j values. Cases with high reflux ratios can be more difficult to converge than those with low ratios. Orbach and Crowe [12] describe an extrapolation method for accelerating convergence based on periodic adjustment of the tear variables when their values form geometric progressions during at least four successive iterations.

EXAMPLE 10.2 BP Method.

Calculate stage temperatures, interstage vapor and liquid flow rates and compositions, reboiler duty, and condenser duty by the BP method for the column specifications given in Figure 10.8.

Solution

The Wang–Henke computer program of Johansen and Seader [10] was used. In this program, no adjustments to the tear variables are made prior to the start of each iteration, and the convergence criterion is (10-32). K-values and enthalpies are computed from correlations for hydrocarbons. Initial assumptions required are distillate and bottoms temperatures shown below for four cases. The significant effect of initially assumed distillate and bottoms temperatures

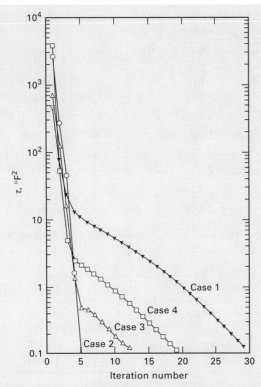

Figure 10.8 Convergence patterns for Example 10.2.

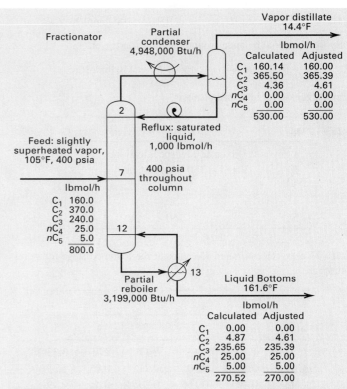

Figure 10.9 Specifications and overall results for Example 10.2.

on the number of iterations required to satisfy (10-32) is indicated by the following results.

Case	Assumed Temperatures, °F		Number of Iterations for Convergence
	Distillate	Bottoms	
1	11.5	164.9	29
2	0.0	200.0	5
3	20.0	180.0	12
4	50.0	150.0	19

The terminal temperatures of Case 1 were within a few degrees of the exact values and were much closer than those of the other three cases. Nevertheless, Case 1 required the largest number of iterations. Figure 10.8 is a plot of τ from (10-32) as a function of the iterations for the four cases. Case 2 converged rapidly to the criterion of $\tau < 0.13$. Cases 1, 3, and 4 converged rapidly for the first three or four iterations, but then moved slowly toward the criterion. This was particularly true of Case 1, for which application of a convergence-acceleration method would be desirable. In none of the cases did tear-value oscillations occur; rather the values approached the converged results monotonically.

The converged calculations, from Case 2, are shown in Figure 10.9. Product component flow rates were not quite in material balance with the feed. Therefore, adjusted values that do satisfy overall material-balance equations were determined by averaging the calculated values and are included in Figure 10.9. A smaller value of τ would have improved the material balance. Figures 10.10 to 10.13 are plots of converged values for stage temperatures, interstage flow rates, and mole-fraction compositions from Case 2. Results from the other three cases were almost identical to those of Case 2. Included in Figure 10.10 is the initial temperature profile.

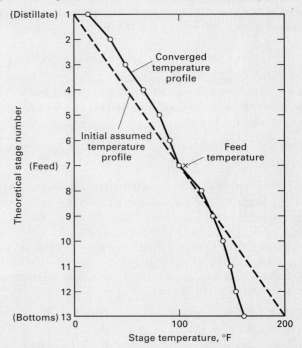

Figure 10.10 Converged temperature profile for Example 10.2.

Except for the bottom stages, the initial temperature profile does not deviate significantly from the converged profile. The jog in the profile at the feed stage is a common occurrence.

Figure 10.11 shows that the assumption of constant interstage molar flow rates does not hold in the rectifying section. Both liquid and vapor flow rates decrease in moving from the top toward the feed stage. Because the feed is vapor near the dew point, the liquid rate changes only slightly across the feed stage. Correspondingly, the vapor rate decreases across the feed stage by an amount equal to the feed rate.

The interstage molar flows are almost constant in the stripping section. However, the assumed vapor flow rate, based on adjusting

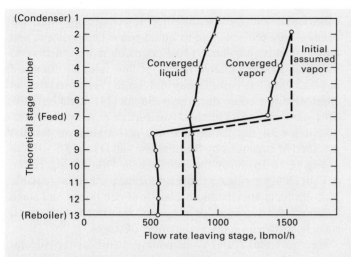

Figure 10.11 Converged interstage flow rates for Example 10.2.

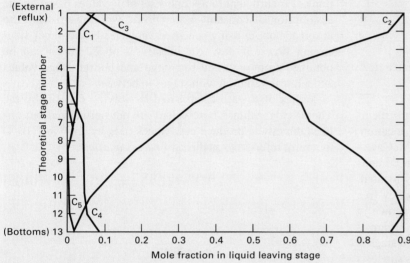

Figure 10.12 Converged liquid composition profiles for Example 10.2.

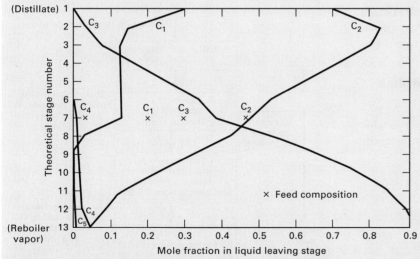

Figure 10.13 Converged vapor composition profiles for Example 10.2.

the rectifying-section rate across the feed zone, is approximately 33% higher than the average converged vapor rate. A better initial vapor rate estimate in the stripping section can be made by computing the reboiler duty from the condenser duty based on the specified reflux rate and then determining the vapor rate from the reboiler.

The separation is between C_2 (LK) and C_3 (HK); thus C_1 is a lighter-than-light key (LLK), and nC_4 and nC_5 are heavier than the heavy key (HHK). Each of these four exhibits a different composition-profile curve, as shown in Figures 10.12 and 10.13. Except at the feed zone and at each end of the column, both liquid and vapor LK mole fractions decrease smoothly and continuously from the top of the column to the bottom. The inverse occurs for C_3 (HK). Mole fractions of methane (LLK) are almost constant over the rectifying section except near the top. Below the feed zone, methane rapidly disappears from both vapor and liquid streams. The inverse is true for the two HHK components. In Figure 10.13, feed composition is somewhat different from the composition of vapor entering the feed stage from below or vapor leaving the feed stage.

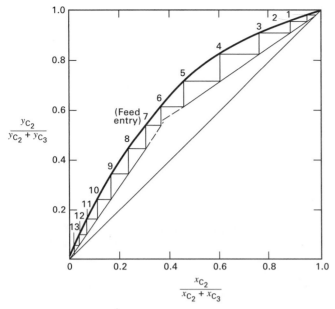

Figure 10.14 Modified McCabe–Thiele diagram for Example 10.2.

For problems where distillate flow rate and N are specified, it is difficult to specify the optimal feed-stage location. However, after a rigorous calculation is made, a McCabe–Thiele plot based on the key components [13] can be constructed to determine if the feed stage is optimally located. For this plot, mole fractions of the LK are on a nonkey-free basis. The resulting diagram for Example 10.2 is shown in Figure 10.14, where the trend toward a pinched-in region is more noticeable in the rectifying section just above stage 7 than in the stripping section just below stage 7. This suggests that a better separation might be made by shifting the feed entry to stage 6. Figure 10.15 shows the effect of feed-stage location on the percent loss of ethane to the bottoms product. As predicted from Figure 10.14, the optimal feed stage is 6.

§10.3.3 Sum-Rates (SR) Method for Absorption and Stripping

The species in most absorbers and strippers cover a wide range of volatility. Hence, the BP method of solving the

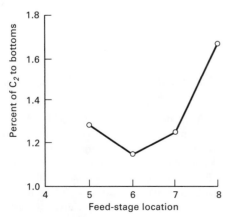

Figure 10.15 Effect of feed-stage location on separation for Example 10.2.

MESH equations fails because bubble-point temperature calculations are too sensitive to liquid-phase composition, and the stage energy balance (10-5) is much more sensitive to stage temperatures than to interstage flow rates, as discussed in §10.3.2. In this case, Friday and Smith [7] showed that an alternative procedure devised by Sujata [14] could be used. This *sum-rates* (SR) *method* was further developed in conjunction with the tridiagonal-matrix formulation for the modified M equations by Burningham and Otto [15].

Figure 10.16 shows the Burningham–Otto SR algorithm. A FORTRAN computer program for the method is available [16]. Problem specifications consist of conditions and stage locations for feeds, stage pressure, sidestream flow rates, stage heat-transfer rates, and number of stages.

Tear variables T_j and V_j are assumed to initiate the calculations. It is sufficient to assume a set of V_j values based on the assumption of constant-molar flows, working up from the absorber bottom using specified vapor feeds and vapor sidestream flows, if any. An initial set of T_j values can be obtained from assumed top-stage and bottom-stage values and a linear variation with stages in between.

Values of $x_{i,j}$ are established by solving (10-12) by the Thomas algorithm. These values are not normalized but utilized directly to produce new values of L_j by applying (10-4) in a form referred to as the *sum-rates equation*:

$$L_j^{(k+1)} = L_j^{(k)} \sum_{i=1}^{C} x_{i,j} \qquad (10\text{-}33)$$

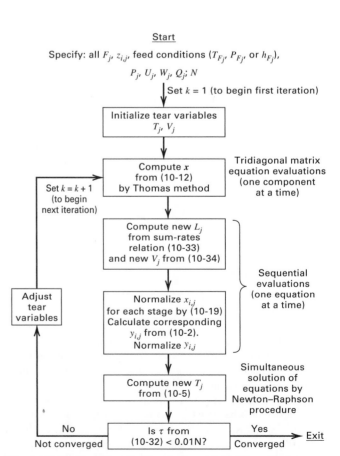

Figure 10.16 Algorithm for Burningham–Otto SR method.

where $L_j^{(k)}$ values are obtained from $V_j^{(k)}$ values by (10-6). Corresponding values of $V_j^{(k+1)}$ are obtained from a total material balance derived by summing (10-1) over the C components, combining the result with (10-3) and (10-4), and summing that result over stages j through N to give

$$V_j = L_{j-1} - L_N + \sum_{m=j}^{N} (F_m - W_m - U_m) \qquad (10\text{-}34)$$

Normalized $x_{i,j}$ values are calculated from (10-19). Corresponding values of $y_{i,j}$ are computed from (10-2).

A new set of T_j is obtained by solving the simultaneous energy-balance relations for the N stages given by (10-5). The temperatures are embedded in the specific enthalpies for the unspecified vapor and liquid flow rates. Typically, these enthalpies are nonlinear in temperature. Therefore, an iterative procedure such as the Newton–Raphson method is required. In that method, the simultaneous equations in terms of variables x_i are written in zero form:

$$f_i\{x_1, x_2, \ldots, x_n\} = 0 \quad i = 1, 2, \ldots, n \qquad (10\text{-}35)$$

Initial guesses, marked by asterisks, are provided for the n variables, and each function is expanded about these guesses in a Taylor's series that is terminated after the first derivatives to give

$$0 = f_i\{x_1, x_2, \ldots, x_n\} \qquad (10\text{-}36)$$

$$\approx f_i\{x_1^*, x_2^*, \ldots, x_n^*\} + \left.\frac{\partial f_i}{\partial x_1}\right|^* \Delta x_1 + \left.\frac{\partial f_i}{\partial x_2}\right|^* \Delta x_2 + \cdots + \left.\frac{\partial f_i}{\partial x_n}\right|^* \Delta x_n$$

$$(10\text{-}37)$$

where $\Delta x_j = x_j - x_j^*$.

Equations (10-36) and (10-37) are linear and can be solved directly for the corrections Δx_i. If all are zero, the guesses are correct and (10-35) has been solved; if not, the corrections are added to the guesses to provide a new set of guesses for (10-36). The procedure is repeated, for r iterations, until all corrections, and thus all the functions, become zero to within some tolerance. In recursion form, (10-36) and (10-37) are

$$\sum_{j=1}^{n} \left[\left(\frac{\partial f_i}{\partial x_i}\right)^{(r)} \Delta x_j^{(r)} \right] = -f_i^{(r)}, \quad i = 1, 2, \ldots, n \qquad (10\text{-}38)$$

$$x_j^{(r+1)} = x_j^{(r)} + \Delta x_j^{(r)}, \quad j = 1, 2, \ldots, n \qquad (10\text{-}39)$$

EXAMPLE 10.3 Newton-Raphson Method.

Solve the simultaneous, nonlinear equations

$$x_1 \ln x_2 + x_2 \exp(x_1) = \exp(1)$$
$$x_2 \ln x_1 + 2x_1 \exp(x_2) = 2\exp(1)$$

for x_1 and x_2 to within ± 0.001, by the Newton–Raphson method.

Solution

In the form of (10-35), the two equations are

$$f_1\{x_1, x_2\} = x_1 \ln x_2 + x_2 \exp(x_1) - \exp(1) = 0$$
$$f_2\{x_1, x_2\} = x_2 \ln x_1 + 2x_1 \exp(x_2) - 2\exp(1) = 0$$

From (10-38), the linearized recursive form of these equations is

$$\left(\frac{\partial f_1}{\partial x_1}\right)^{(r)} \Delta x_1^{(r)} + \left(\frac{\partial f_1}{\partial x_2}\right)^{(r)} \Delta x_2^{(r)} = -f_1^{(r)}$$

$$\left(\frac{\partial f_2}{\partial x_1}\right)^{(r)} \Delta x_1^{(r)} + \left(\frac{\partial f_2}{\partial x_2}\right)^{(r)} \Delta x_2^{(r)} = -f_2^{(r)}$$

The solution of these two equations is readily obtained by the method of determinants to give

$$\Delta x_1^{(r)} = \frac{\left[f_2^{(r)} \left(\frac{\partial f_1}{\partial x_2}\right)^{(r)} - f_1^{(r)} \left(\frac{\partial f_2}{\partial x_2}\right)^{(r)} \right]}{D}$$

and

$$\Delta x_2^{(r)} = \frac{\left[f_1^{(r)} \left(\frac{\partial f_2}{\partial x_1}\right)^{(r)} - f_2^{(r)} \left(\frac{\partial f_1}{\partial x_1}\right)^{(r)} \right]}{D}$$

where

$$D = \left(\frac{\partial f_1}{\partial x_1}\right)^{(r)} \left(\frac{\partial f_2}{\partial x_2}\right)^{(r)} - \left(\frac{\partial f_1}{\partial x_2}\right)^{(r)} \left(\frac{\partial f_2}{\partial x_1}\right)^{(r)}$$

and the derivatives as obtained from the equations are

$$\left(\frac{\partial f_1}{\partial x_1}\right)^{(r)} = \ln\left(x_2^{(r)}\right) + x_2^{(r)} \exp\left(x_1^{(r)}\right),$$

$$\left(\frac{\partial f_2}{\partial x_1}\right)^{(r)} = \frac{x_2^{(r)}}{x_1^{(r)}} + 2\exp\left(x_2^{(r)}\right)$$

$$\left(\frac{\partial f_1}{\partial x_2}\right)^{(r)} = \frac{x_1^{(r)}}{x_2^{(r)}} + \exp\left(x_1^{(r)}\right),$$

$$\left(\frac{\partial f_2}{\partial x_2}\right)^{(r)} = \ln\left(x_1^{(r)}\right) + 2x_1^{(r)} \exp\left(x_2^{(r)}\right)$$

As initial guesses, let $x_1^{(1)} = 2$, $x_2^{(1)} = 2$. Application of the Newton–Raphson procedure gives the following results, where at the sixth iteration, values of $x_1 = 1.0000$ and $x_2 = 1.0000$ correspond closely to the values of 0 for f_1 and f_2.

r	$x_1^{(r)}$	$x_2^{(r)}$	$f_1^{(r)}$	$f_2^{(r)}$	$(\partial f_1/\partial x_1)^{(r)}$	$(\partial f_1/\partial x_2)^{(r)}$	$(\partial f_2/\partial x_1)^{(r)}$	$(\partial f_2/\partial x_2)^{(r)}$	$\Delta x_1^{(r)}$	$\Delta x_2^{(r)}$
1	2.0000	2.0000	13.4461	25.5060	15.4731	8.3891	15.7781	30.2494	−0.5743	−0.5436
2	1.4247	1.4564	3.8772	7.3133	6.4354	5.1395	9.6024	12.5880	−0.3544	−0.3106
3	1.0713	1.1457	0.7720	1.3802	3.4806	3.8541	7.3591	6.8067	−0.0138	−0.1878
4	1.0575	0.9579	−0.0059	0.1290	2.7149	3.9830	6.1183	5.5679	−0.0591	0.0417
5	0.9984	0.9996	−0.0057	−0.0122	2.7126	3.7127	6.4358	5.4244	0.00159	0.000368
6	1.0000	1.0000	5.51×10^{-6}	2.86×10^{-6}	2.7183	3.7183	6.4366	5.4366	12.1×10^{-6}	-3.0×10^{-6}
7	1.0000	1.0000	0.0	-2×10^{-9}	2.7183	3.7183	6.4366	5.4366	—	—

To obtain a new set of T_j from the energy equation (10-5), the Newton–Raphson recursion equation is

$$\left(\frac{\partial H_j}{\partial T_{j-1}}\right)^{(r)} \Delta T_{j-1}^{(r)} + \left(\frac{\partial H_j}{\partial T_j}\right)^{(r)} \Delta T_j^{(r)} + \left(\frac{\partial H_j}{\partial T_{j+1}}\right)^{(r)} \Delta T_{j+1}^{(r)} = -H_j^{(r)}$$

(10-40)

where

$$\Delta T_j^{(r)} = T_j^{(r+1)} - T_j^{(r)} \tag{10-41}$$

$$\frac{\partial H_j}{\partial T_{j-1}} = L_{j-1} \frac{\partial h_{L_{j-1}}}{\partial T_{j-1}} \tag{10-42}$$

$$\frac{\partial H_j}{\partial T_j} = -\left(L_j + U_j\right) \frac{\partial h_{L_j}}{\partial T_j} - \left(V_j + W_j\right) \frac{\partial h_{V_j}}{\partial T_j} \tag{10-43}$$

$$\frac{\partial H_j}{\partial T_{j+1}} = V_{j+1} \frac{\partial h_{V_{j+1}}}{\partial T_{j+1}} \tag{10-44}$$

The partial derivatives depend upon the enthalpy correlations. If composition-independent polynomial equations in temperature are used, then

$$h_{V_j} = \sum_{i=1}^{C} y_{i,j} \left(A_i + B_i T + C_i T^2\right) \tag{10-45}$$

$$h_{L_j} = \sum_{i=1}^{C} x_{i,j} \left(a_i + b_i T + c_i T^2\right) \tag{10-46}$$

and the partial derivatives are

$$\frac{\partial h_{V_j}}{\partial T_j} = \sum_{i=1}^{C} y_{i,j} (B_i + 2C_i T) \tag{10-47}$$

$$\frac{\partial h_{L_j}}{\partial T_j} = \sum_{i=1}^{C} x_{i,j} (b_i + 2c_i T) \tag{10-48}$$

The N relations given by (10-40) form a tridiagonal-matrix equation that is linear in $\Delta T_j^{(r)}$. The form of the matrix equation is identical to (10-12), where, for example, $A_2 = (\partial H_2/\partial T_1)^{(r)}$, $B_2 = (\partial H_2/\partial T_2)^{(r)}$, $C_2 = (\partial H_2/\partial T_3)^{(r)}$, $x_{i,2} \leftarrow \Delta T_2^{(r)}$, and $D_2 = -H_2^{(r)}$. The matrix of partial derivatives is called the *Jacobian correction matrix*. The Thomas algorithm can be employed to solve for the set of corrections $\Delta T_j^{(r)}$. New guesses of T_j are then determined from

$$T_j^{(r+1)} = T_j^{(r)} + t\Delta T_j^{(r)} \tag{10-49}$$

where t is a scalar attenuation factor that is useful when initial guesses and true values are not reasonably close. Generally, as in (10-39), t is taken as 1, but an optimal value can be determined at each iteration to minimize the sum of the squares of the functions,

$$\sum_{j=1}^{N} \left[H_j^{(r+1)}\right]^2$$

When corrections $\Delta T_j^{(r)}$ approach zero, the resulting values of T_j are used with criteria such as (10-31) or (10-32) to determine if convergence has been achieved. If not, before beginning a new k iteration, values of V_j and T_j are adjusted as in Figure 10.16 and as discussed for the BP method. Convergence is rapid for the sum-rates method.

EXAMPLE 10.4 SR Method.

Calculate stage temperatures and interstage vapor and liquid flow rates and compositions by the SR method for the absorber specifications in Figure 5.11.

Solution

The program of Shinohara et al. [16], based on the Burningham–Otto procedure, was used. Initial assumptions for the top-stage and bottom-stage temperatures were 90°F and 105°F. The corresponding number of iterations to satisfy the convergence criterion of (10-32) was seven. Values of τ were as follows.

Iteration Number	τ, $(°F)^2$
1	9,948
2	2,556
3	46.0
4	8.65
5	0.856
6	0.124
7	0.0217

Figure 10.17 presents converged calculations as well as adjusted values of product-component flow rates that satisfy overall material-balance equations. Figures 10.18 to 10.20 are plots of converged stage temperatures, interstage total flow rates, and interstage component vapor flows. Figure 10.18 shows that the initial assumed linear-temperature profile is grossly in error. Because of the substantial absorption and accompanying heat released by absorption, stage temperatures are greater than the two entering stream temperatures. The peak stage temperature is at the midpoint of the column, unless a predominant amount of absorption takes place near the gas inlet.

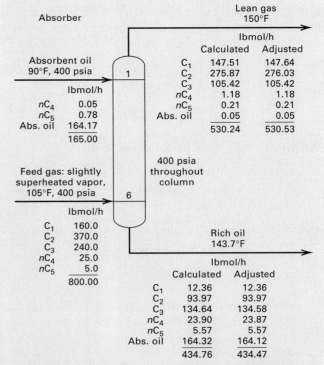

Figure 10.17 Specifications and overall results for Example 10.4.

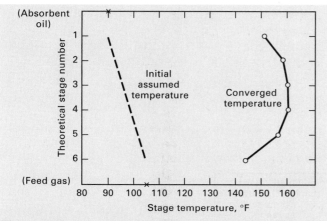

Figure 10.18 Converged temperature profile for Example 10.4.

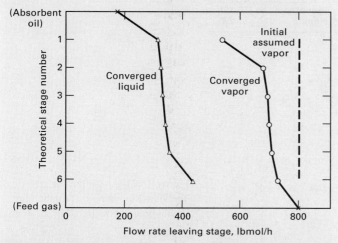

Figure 10.19 Converged interstage flow profiles for Example 10.4.

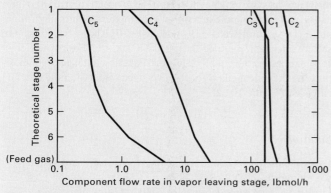

Figure 10.20 Converged vapor flow profiles for Example 10.4.

Figure 10.19 shows that the bulk of the absorption occurs on the two terminal stages. Figure 10.20 shows that absorption of C_1 and C_2 occurs mainly on the top and bottom stages. Absorption of C_3 occurs throughout the column, but mainly at the terminal stages. Absorption of C_4 and C_5 occurs throughout, but for C_5 mainly at the bottom, where vapor first contacts absorption oil.

§10.3.4 Isothermal Sum-Rates (ISR) Method for Liquid–Liquid Extraction

In industry, liquid–liquid extraction is usually adiabatic, and when entering streams are at the same temperature and heat of mixing is negligible, the operation is also isothermal.

If stage temperatures are specified, as discussed by Friday and Smith [7] and Tsuboka and Katayama [17], a simplified isothermal version of the sum-rates method (ISR) is suitable. It is based on the equilibrium-stage model in §10.1.1. With all stage temperatures specified, Q_j can be computed from stage energy balances, which can be partitioned from the other equations and solved in a later, separate step. Particular attention must be paid to the possibility that phase compositions may strongly influence values of partition ratios, $K_{i,j}$.

Figure 10.21 shows the Tsuboka–Katayama ISR algorithm. Liquid-phase and vapor-phase symbols correspond to raffinate and extract, respectively. Problem specifications are flow rates, compositions, and feed-stage locations; stage temperatures (frequently all equal); total flow rates of side-streams; and N. Pressures need not be specified but must be greater than corresponding stage bubble-point pressures to prevent vaporization.

With stage temperatures specified, the only tear variables are V_j (extract flow rates). An initial set is obtained by assuming a perfect separation for feed components and neglecting solvent mass transfer to the raffinate phase. This fixes approximate values for the exiting raffinate and extract flows. Intermediate V_j values are obtained by linear interpolation. Modifications are necessary for side-streams or intermediate feeds. As in Figure 10.21, tear variables are reset in an outer iterative loop.

The effect of compositions on K_D-values can be considerable. It is best to provide initial estimates of $x_{i,j}$ and $y_{i,j}$ from which values of $K_{i,j}$ are computed. Initial values of $x_{i,j}$ are by linear interpolation using the compositions of the known entering and assumed exit streams. Corresponding values of $y_{i,j}$ are obtained by material balance from (10-1). Values of $\gamma_{iL,j}$ and $\gamma_{iV,j}$ are determined using an appropriate correlation from §2.6. Corresponding K_D-values are from an equation equivalent to (2-30).

$$K_{i,j} = \frac{\gamma_{iL,j}}{\gamma_{iV,j}} \tag{10-50}$$

New $x_{i,j}$ values come from solving (10-12) by the Thomas algorithm. These are compared to the assumed values by

$$\tau_1 = \sum_{j=1}^{N} \sum_{i=1}^{C} \left| x_{i,j}^{(r-1)} - x_{i,j}^{(r)} \right| \tag{10-51}$$

where r is an inner-loop index. If $\tau_1 > \epsilon_1$—where, for example, ϵ_1 might be taken as 0.01 NC—the inner loop is used to improve values of $K_{i,j}$ by using normalized values of $x_{i,j}$ and $y_{i,j}$ to compute new values of $\gamma_{iL,j}$ and $\gamma_{iV,j}$.

When the inner loop is converged, values of $x_{i,j}$ are used to get new values of $y_{i,j}$ from (10-2). A new set of tear variables V_j is then obtained from the sum-rates relation

$$V_j^{(k+1)} = V_j^{(k)} \sum_{i=1}^{C} y_{i,j} \tag{10-52}$$

where k is an outer-loop index. Values of $L_j^{(k+1)}$ are obtained from (10-6). The outer loop is converged when

$$\tau_2 = \sum_{j=1}^{N} \left(\frac{V_j^{(k)} - V_j^{(k-1)}}{V_j^{(k)}} \right)^2 \leq \epsilon_2 \tag{10-53}$$

where ϵ_2 may be taken as 0.01 N.

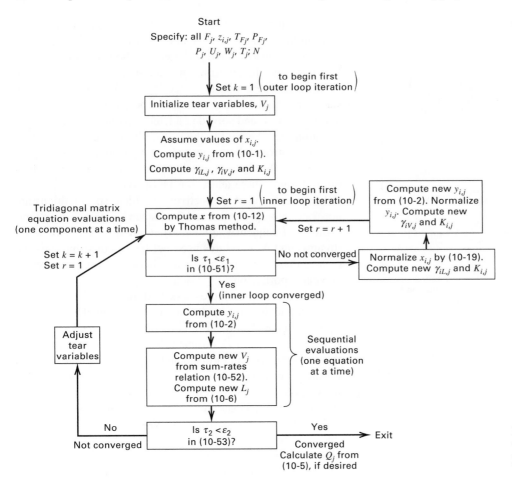

Figure 10.21 Algorithm for Tsuboka–Katayama ISR method for liquid–liquid extraction.

Before beginning a new k iteration, values of V_j are adjusted as previously discussed for the BP method. Convergence of the ISR method is rapid but is subject to the extent to which $K_{i,j}$ depends upon composition.

EXAMPLE 10.5 ISR Method.

The separation of benzene (B) from n-heptane (H) by distillation is difficult, even though the normal boiling points differ by 18.3°C, because of liquid-phase nonideality; the relative volatility decreases to a value less than 1.15 at high benzene concentrations [18]. Alternatively, liquid–liquid extraction with a mixture of dimethylformamide (DMF) and water [19] can be used. The solvent is more selective for benzene than for n-heptane at 20°C. For two different solvent compositions, calculate interstage flow rates and compositions by the ISR method for the five-stage liquid–liquid extraction cascade of Figure 10.22.

Solution

Experimental phase-equilibrium data for the quaternary system [19] were fitted to the NRTL equation by Cohen and Renon [20]. The resulting binary-pair constants in (2-90) to (2-92) are

Binary Pair, ij	τ_{ij}	τ_{ji}	α_{ji}
DMF, H	2.036	1.910	0.25
Water, H	7.038	4.806	0.15
B, H	1.196	−0.355	0.30
Water, DMF	2.506	−2.128	0.253
B, DMF	−0.240	0.676	0.425
B, Water	3.639	5.750	0.203

For Case A, estimates of V_j (the extract phase), $x_{i,j}$, and $y_{i,j}$, based on a perfect separation and linear interpolation by stage, are

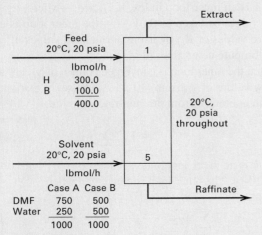

Figure 10.22 Specifications for Example 10.5.

Stage j	V_j	H	B	DMF	Water	H	B	DMF	Water
				$y_{i,j}$				$x_{i,j}$	
1	1100	0.0	0.0909	0.6818	0.2273	0.7895	0.2105	0.0	0.0
2	1080	0.0	0.0741	0.6944	0.2315	0.8333	0.1667	0.0	0.0
3	1060	0.0	0.0566	0.7076	0.2359	0.8824	0.1176	0.0	0.0
4	1040	0.0	0.0385	0.7211	0.2404	0.9375	0.0625	0.0	0.0
5	1020	0.0	0.0196	0.7353	0.2451	1.0000	0.0	0.0	0.0

The converged solution is obtained by the ISR method with the following stage flow rates and compositions:

Stage j	V_j	H	B	DMF	Water	H	B	DMF	Water
				$y_{i,j}$				$x_{i,j}$	
1	1113.1	0.0263	0.0866	0.6626	0.2245	0.7586	0.1628	0.0777	0.0009
2	1104.7	0.0238	0.0545	0.6952	0.2265	0.8326	0.1035	0.0633	0.0006
3	1065.6	0.0213	0.0309	0.7131	0.2347	0.8858	0.0606	0.0532	0.0004
4	1042.1	0.0198	0.0157	0.7246	0.2399	0.9211	0.0315	0.0471	0.0003
5	1028.2	0.0190	0.0062	0.7316	0.2432	0.9438	0.0125	0.0434	0.0003

Computed products for the two cases are:

	Extract, lbmol/h		Raffinate, lbmol/h	
	Case A	Case B	Case A	Case B
H	29.3	5.6	270.7	294.4
B	96.4	43.0	3.6	57.0
DMF	737.5	485.8	12.5	14.2
Water	249.9	499.7	0.1	5.0
	1113.1	1034.1	286.9	365.9

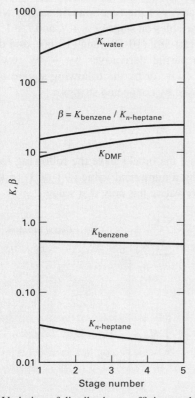

Figure 10.23 Variation of distribution coefficient and relative selectivity for Example 10.5, Case A.

On a percentage extraction basis, the results are:

	Case A	Case B
Percent of benzene feed extracted	96.4	43.0
Percent of n-heptane feed extracted	9.8	1.87
Percent of solvent transferred to raffinate	1.26	1.45

Thus, the solvent with 75% DMF extracts a much larger percentage of the benzene, but the solvent with 50% DMF is more selective between benzene and n-heptane. For Case A, the variations with stage of K-values and the relative selectivity are shown in Figure 10.23, where the relative selectivity is $\beta_{B,H} = K_B/K_H$. The distribution coefficient for n-heptane varies by a factor of almost 1.75 from stage 5 to stage 1, while the coefficient for benzene is almost constant. The relative selectivity varies by a factor of almost 2.

§10.4 NEWTON–RAPHSON (NR) METHOD

BP and SR methods for vapor–liquid systems converge with difficulty or not at all for very nonideal liquid mixtures or for cases where the separator is like an absorber or stripper in one section and a fractionator in another section (e.g., the reboiled absorber in Figure 1.7). Furthermore, BP and SR methods are generally restricted to limited specifications. Universal procedures for solving separation problems are based on the solution of the MESH equations, or combinations thereof, by simultaneous-correction (SC) techniques, which employ the Newton–Raphson (NR) method.

To develop an SC procedure, it is necessary to select and order the unknown variables and corresponding functions (MESH equations). As discussed by Goldstein and Stanfield [21], grouping of functions by type is computationally most efficient for problems involving many components, but few stages. For problems involving many stages but relatively few components, it is most efficient to group the functions according to stage location. The latter grouping, presented here, is described by Naphtali [22] and was implemented by Naphtali and Sandholm [23]. Their procedure utilizes the mathematical techniques presented in §10.3. A computer program for their method is given by Fredenslund et al. [24]. However, that program does not have the flexibility of specifications found in process simulators.

The stage model of Figures 10.1 and 10.3 is again employed. However, rather than solving the $N(2C+3)$ MESH equations simultaneously, (10-3) and (10-4) are combined with the other MESH equations to eliminate $2N$ variables and thus reduce the problem to the simultaneous solution of $N(2C + 1)$ equations. This is done by first multiplying (10-3) and (10-4) by V_j and L_j, respectively to give

$$V_j = \sum_{i=1}^{C} v_{i,j} \qquad (10\text{-}54)$$

$$L_j = \sum_{i=1}^{C} l_{i,j} \qquad (10\text{-}55)$$

where mole-fraction definitions are used:

$$y_{i,j} = \frac{\upsilon_{i,j}}{V_j} \qquad (10\text{-}56)$$

$$x_{i,j} = \frac{l_{i,j}}{L_j} \qquad (10\text{-}57)$$

Equations (10-54) to (10-57) are now substituted into (10-1), (10-2), and (10-5) to eliminate V_j, L_j, $y_{i,j}$, and $x_{i,j}$ and introduce component flow rates $\upsilon_{i,j}$ and $l_{i,j}$. As a result, the following $N(2C+1)$ equations are obtained, where $s_j = U_j/L_j$ and $S_j = W_j/V_j$ are dimensionless sidestream flows.

Material Balance

$$M_{i,j} = l_{i,j}(1 + s_j) + \upsilon_{i,j}(1 + S_j) - l_{i,j-1} - \upsilon_{i,j+1} - f_{i,j} = 0 \qquad (10\text{-}58)$$

Phase Equilibria

$$E_{i,j} = K_{i,j} l_{i,j} \frac{\sum_{k=1}^{C} \upsilon_{k,j}}{\sum_{k=1}^{C} l_{k,j}} - \upsilon_{i,j} = 0 \qquad (10\text{-}59)$$

Energy Balance

$$\begin{aligned}
H_j &= h_{L_j}(1 + s_j)\sum_{i=1}^{C} l_{i,j} + h_{V_j}(1 + S_j)\sum_{i=1}^{C} \upsilon_{i,j} \\
&\quad - h_{L_{j-1}}\sum_{i=1}^{C} l_{i,j-1} - h_{V_{j+1}}\sum_{i=1}^{C} \upsilon_{i,j+1} \\
&\quad - h_{F_j}\sum_{i=1}^{C} f_{i,j} - Q_j = 0
\end{aligned} \qquad (10\text{-}60)$$

where $f_{i,j} = F_j z_{i,j}$.

If N and all $f_{i,j}$, T_{F_j}, P_{F_j}, P_j, s_j, S_j and Q_j are specified, the M, E, and H functions are nonlinear in the $N(2C+1)$ unknown (output) variables $\upsilon_{i,j}$, $l_{i,j}$ and T_j for $i = 1$ to C and $j = 1$ to N. Although other sets of specified and unknown variables are possible, this set is considered first.

Equations (10-58), (10-59), and (10-60) are solved simultaneously by the Newton–Raphson iterative method in which successive sets of the output variables are computed until the values of the M, E, and H functions are driven to within the convergence criteria or zero. During the iterations, nonzero values of the functions are called *discrepancies* or *errors*. Let the functions and output variables be grouped by stage in order from top to bottom. This is done to produce a *block-tridiagonal structure* for the Jacobian matrix of partial derivatives so that a matrix form of the Thomas algorithm can be applied. Let

$$\mathbf{X} = [\mathbf{X}_1, \mathbf{X}_2, \ldots, \mathbf{X}_j, \ldots, \mathbf{X}_N]^T \qquad (10\text{-}61)$$

and

$$\mathbf{F} = [\mathbf{F}_1, \mathbf{F}_2, \ldots, \mathbf{F}_j, \ldots, \mathbf{F}_N]^T \qquad (10\text{-}62)$$

where $\mathbf{X}_j$ is the vector of output variables for stage j arranged in the order

$$\mathbf{X}_j = [\upsilon_{1,j}, \upsilon_{2,j}, \ldots, \upsilon_{i,j}, \ldots, \upsilon_{C,j}, T_j, l_{1,j}, l_{2,j}, \ldots, l_{i,j}, \ldots, l_{C,j}]^T \qquad (10\text{-}63)$$

and $\mathbf{F}_j$ is the vector of functions for stage j arranged in the order

$$\mathbf{F}_j = [H_j, M_{1,j}, M_{2,j}, \ldots, M_{i,j}, \ldots, M_{C,j}, E_{1,j}, E_{2,j}, \ldots, E_{i,j}, \ldots, E_{C,j}]^T \qquad (10\text{-}64)$$

The Newton–Raphson iteration is performed by solving for the corrections $\Delta\mathbf{X}$ to the output variables from (10-38), which in matrix form becomes

$$\Delta\mathbf{X}^{(k)} = -\left[\left(\frac{\partial\mathbf{F}}{\partial\mathbf{X}}\right)^{-1}\right]^{(k)} \mathbf{F}^{(k)} \qquad (10\text{-}65)$$

These corrections are used to compute the next approximation to the set of output variables from

$$\mathbf{X}^{(k+1)} = \mathbf{X}^{(k)} + t\,\Delta\mathbf{X}^{(k)} \qquad (10\text{-}66)$$

where t is a scalar step factor, whose use is described below in the discussion on convergence.

The quantity $(\partial\mathbf{F}/\partial\mathbf{X})$ is the following Jacobian or $(N \times N)$ matrix of $\mathbf{A}$, $\mathbf{B}$, and $\mathbf{C}$ blocks of partial derivatives of all the functions with respect to all the output variables.

$$\frac{\partial\mathbf{F}}{\partial\mathbf{X}} = \begin{bmatrix}
\overline{\mathbf{B}}_1 & \overline{\mathbf{C}}_1 & 0 & 0 & .. & & 0 \\
\overline{\mathbf{A}}_2 & \overline{\mathbf{B}}_2 & \overline{\mathbf{C}}_2 & 0 & .. & & 0 \\
0 & \overline{\mathbf{A}}_3 & \overline{\mathbf{B}}_3 & \overline{\mathbf{C}}_3 & .. & & 0 \\
.. & & & & & & .. \\
.. & & & & & & .. \\
0 & .. & & & & & 0 \\
0 & .. & & & 0 & \overline{\mathbf{A}}_{N-1} & \overline{\mathbf{B}}_{N-1} & \overline{\mathbf{C}}_{N-1} \\
0 & .. & & & 0 & 0 & \overline{\mathbf{A}}_N & \overline{\mathbf{B}}_N
\end{bmatrix} \qquad (10\text{-}67)$$

This Jacobian is of a block-tridiagonal form, like (10-12), because functions for stage j are dependent only on output variables for stages $j-1$, j, and $j+1$. Each $\overline{\mathbf{A}}$, $\overline{\mathbf{B}}$, or $\overline{\mathbf{C}}$ block in (15-67) represents a $(2C+1)$ by $(2C+1)$ submatrix of partial derivatives, where the arrangements of output variables and functions are given by (10-63) and (10-64), respectively. Blocks $\overline{\mathbf{A}}_j$, $\overline{\mathbf{B}}_j$, and $\overline{\mathbf{C}}_j$ correspond to submatrices of partial derivatives of the functions on stage j with respect to the output variables on stages $j-1$, j, and $j+1$, respectively. Thus, using (10-58), (10-59), and (10-60), and denoting only the nonzero partial derivatives by $+$, by row or diagonal strings of $+\cdots+$, or by the following square or rectangular blocks enclosed by connected strings,

$$\begin{matrix}
+ & \cdots & + \\
\vdots & & \vdots \\
+ & \cdots & +
\end{matrix}$$

it is found that the blocks have the following form, where $+$ is replaced by a numerical value (-1 or 1) in the event that the partial derivative has only that value.

$$\tilde{\mathbf{A}}_j = \frac{\partial\overline{\mathbf{F}}_j}{\partial\mathbf{X}_{j-1}} = \text{Functions}\begin{matrix} H_j \\ M_{1,j} \\ . \\ . \\ M_{C,j} \\ E_{1,j} \\ . \\ . \\ E_{C,j} \end{matrix}\overbrace{\begin{bmatrix}
 & + & +\cdots+ \\
 & & -1 \\
 & & \\
 & & \ddots \\
 & & -1 \\
 & & \\
 & & \\
 & & \\
 & &
\end{bmatrix}}^{\upsilon_{1,j-1}\cdots\upsilon_{C,j-1} \quad T_{j-1} \quad l_{1,j-1}\cdots l_{C,j-1}} \qquad (10\text{-}68)$$

$$\bar{\mathbf{B}}_j = \frac{\partial \mathbf{F}_j}{\partial \mathbf{X}_j} = $$

(10-69)

$$\bar{\mathbf{C}}_j = \frac{\partial \mathbf{F}_j}{\partial \mathbf{X}_{j+1}} = $$

(10-70)

Thus, (10-65) consists of a set of $N(2C + 1)$ simultaneous, linear equations in the $N(2C + 1)$ corrections $\Delta \mathbf{X}$. For example, the $2C + 2$ equation in the set is obtained by expanding function H_2 (10-60) into a Taylor's series like (10-36) around the $N(2C + 1)$ output variables. The result is as follows after the usual truncation of terms involving derivatives of order greater than one:

$$0 \left(\Delta v_{1,1} + \cdots + \Delta v_{C,1} \right) - \frac{\partial h_{L_1}}{\partial T_1} \sum_{i=1}^{C} l_{i,1} (\Delta T_1)$$

$$- \left(\frac{\partial h_{L_1}}{\partial l_{1,1}} \sum_{i=1}^{C} l_{i,1} + h_{L_1} \right) \Delta l_{1,1} - \cdots$$

$$- \left(\frac{\partial h_{L_1}}{\partial l_{C,1}} \sum_{i=1}^{C} l_{i,1} + h_{L_1} \right) \Delta l_{C,1} - \cdots$$

$$+ \left[\left(\frac{\partial h_{V_2}}{\partial v_{1,2}} \right)(1 + S_2) \sum_{i=1}^{C} v_{i,2} + h_{V_2}(1 + S_2) \right] \Delta v_{1,2} + \cdots$$

$$+ \left[\left(\frac{\partial h_{V_2}}{\partial v_{C,2}} \right)(1 + S_2) \sum_{i=1}^{C} v_{i,2} + h_{V_2}(1 + S_2) \right] \Delta v_{C,2}$$

$$+ \left[\left(\frac{\partial h_{L_2}}{\partial T_2} \right)(1 + s_2) \sum_{i=1}^{C} l_{i,2} + \left(\frac{\partial h_{V_2}}{\partial T_2} \right)(1 + S_2) \sum_{i=1}^{C} v_{i,2} \right] \Delta T_2$$

$$+ \left[\left(\frac{\partial h_{L_2}}{\partial l_{1,2}} \right)(1 + s_2) \sum_{i=1}^{C} l_{i,2} + h_{L_2}(1 + s_2) \right] \Delta l_{1,2} + \cdots$$

$$+ \left[\left(\frac{\partial h_{L_2}}{\partial l_{C,2}} \right)(1 + s_2) \sum_{i=1}^{C} l_{i,2} + h_{L_2}(1 + s_2) \right] \Delta l_{C,2}$$

$$- \left(\frac{\partial h_{V_3}}{\partial v_{1,3}} \sum_{i=1}^{C} v_{i,3} + h_{V3} \right) \Delta v_{1,3} - \cdots$$

$$- \left(\frac{\partial h_{V_3}}{\partial v_{C,3}} \sum_{i=1}^{C} v_{i,3} + h_{V3} \right) \Delta v_{C,3}$$

$$- \frac{\partial h_{V_3}}{\partial T_3} \sum_{i=1}^{C} v_{i,3} \Delta T_3 + 0 \left(\Delta l_{1,3} + \cdots + \Delta l_{C,3} \right) = -H_2$$

(10-71)

As a further example, the entry in the Jacobian matrix for row $(2C + 2)$ and column $(C + 3)$ is obtained from (10-71) as

$$\frac{\partial H_2}{\partial l_{2,1}} = - \frac{\partial h_{L_1}}{\partial l_{2,1}} \sum_{i=1}^{C} l_{i,1} + h_{L_1} \quad (10\text{-}72)$$

Partial derivatives of enthalpies and K-values depend upon the correlation utilized for these properties and are sometimes simplified by including only the dominant terms—for example, if the Chao–Seader correlation is to be used for K-values. In general,

$$K_{i,j} = K_{i,j} \left\{ P_j, T_j, \frac{l_{i,j}}{\sum_{k=1}^{C} l_{k,j}}, \frac{v_{i,j}}{\sum_{k=1}^{C} v_{k,j}} \right\}$$

In terms of the output variables, the derivatives $\partial K_{i,j} / \partial T_j$; $\partial K_{i,j} / \partial l_{i,j}$; and $\partial K_{i,j} / \partial v_{i,j}$ all exist and can be expressed analytically or evaluated numerically. However, for some problems, the terms that include the first and second of these three groups of derivatives may be the dominant terms, so the third group may be taken as zero.

EXAMPLE 10.6 Derivation of a Derivative.

Derive an expression for $(\partial h_V / \partial T)$ from the Redlich–Kwong equation of state.

Solution

From (2-53),

$$h_V = \sum_{i=1}^{C} \left(y_i h_{iV}^{\circ} \right) + RT \left[Z_V - 1 - \frac{3A}{2B} \ln \left(1 + \frac{B}{Z_V} \right) \right]$$

where h_{iV}°, Z_V, A, and B all depend on T, as determined from (2-36) and (2-46) to (2-50). Thus,

$$\frac{\partial h_V}{\partial T} = \sum_{i=1}^{C} \left[y_i \left(\frac{\partial h_{iV}^{\circ}}{\partial T} \right) \right] + R \left[Z_V - 1 - \frac{3A}{2B} \ln \left(1 + \frac{B}{Z_V} \right) \right]$$

$$+ RT \left\{ \left(\frac{\partial Z_V}{\partial T} \right) - \frac{3}{2} \left(\frac{\partial A/B}{\partial T} \right) \ln \left(1 + \frac{B}{Z_V} \right) \right.$$

$$\left. - \frac{3A}{2B} \left[\frac{1}{Z_V} \left(\frac{\partial B}{\partial T} \right) - \frac{B}{Z_V^2} \left(\frac{\partial Z_V}{\partial T} \right) \right] \right\}$$

From (2-36) and (2-35),

$$\left(\frac{\partial h_{iV}^{\circ}}{\partial T}\right) = \sum_{k=0}^{4} (a_k)_i T^k = \left(C_{P_V}^{\circ}\right)_i$$

From (2-48) and Table 2.5,

$$B = \frac{bP}{RT} \quad \text{and} \quad b = \frac{0.08664RT_c}{P_c}$$

Thus,

$$\frac{\partial B}{\partial T} = -\frac{B}{T}$$

From (2-47) to (2-50) and Table 2.5,

$$\frac{A}{B} = \frac{a}{bRT} \quad \text{and} \quad a = \frac{0.42748R^2 T_c^{2.5}}{P_c T^{0.5}}$$

Thus,

$$\frac{\partial (A/B)}{\partial T} = -1.5 \frac{A}{BT}$$

From (2-46),

$$Z_V^3 - Z_V^2 + (A - B - B^2) Z_V - AB = 0$$

By implicit differentiation,

$$3Z_V^2 \frac{\partial Z_V}{\partial T} - 2Z_V \frac{\partial Z_V}{\partial T} + (A - B - B^2) \frac{\partial Z_V}{\partial T} + \frac{Z_V}{T}$$
$$(-2.5A + B + 2B^2) + \frac{3.5AB}{T} = 0$$

which reduces to

$$\frac{\partial Z_V}{\partial T} = \frac{(Z_V/T)(2.5A - B - 2B^2) - 3.5AB/T}{3Z_V^2 - 2Z_V + (A - B - B^2)}$$

Because the Thomas algorithm can be applied to the block-tridiagonal structure of (10-67), submatrices of partial derivatives are computed only as needed. The solution of (10-65) follows the scheme in §10.3.1, given by (10-13) to (10-18) and represented in Figure 10-4, where matrices and vectors $\bar{\mathbf{A}}_j, \bar{\mathbf{B}}_j, \bar{\mathbf{C}}_j, -\bar{\mathbf{F}}_j$ and $\Delta \mathbf{X}_j$ correspond to variables $A_j, B_j, C_j, D_j,$ and x_j, respectively. However, the multiplication and division operations in §10.3.1 are changed to matrix multiplication and inversion, respectively. The steps are:

Starting at stage 1, $\bar{\mathbf{C}}_1 \leftarrow (\bar{\mathbf{B}}_1)^{-1} \bar{\mathbf{C}}_1$, $\mathbf{F}_1 \leftarrow (\bar{\mathbf{B}}_1)^{-1} \mathbf{F}_1$ and $\bar{\mathbf{B}}_1 \leftarrow \mathbf{I}$ (the identity submatrix). Only $\bar{\mathbf{C}}_1$ and $\mathbf{F}_1$ are saved.

For stages j from 2 to $(N-1)$, $\bar{\mathbf{C}}_j \leftarrow (\bar{\mathbf{B}}_j - \bar{\mathbf{A}}_j \bar{\mathbf{C}}_{j-1})^{-1} \bar{\mathbf{C}}_j$. $\mathbf{F}_j \leftarrow (\bar{\mathbf{B}}_j - \bar{\mathbf{A}}_j \bar{\mathbf{C}}_{j-1})^{-1} (\mathbf{F}_j - \bar{\mathbf{A}}_j \mathbf{F}_{j-1})$. Then $\bar{\mathbf{A}}_j \leftarrow 0$, and $\bar{\mathbf{B}}_j \leftarrow \mathbf{I}$. Save $\bar{\mathbf{C}}_j$ and $\mathbf{F}_j$ for each stage.

For the last stage, $\mathbf{F}_N \leftarrow (\mathbf{B}_N - \mathbf{A}_N \mathbf{C}_{N-1})^{-1} (\mathbf{F}_N - \mathbf{A}_N \mathbf{F}_{N-1})$, $\bar{\mathbf{A}}_N \leftarrow 0$, $\bar{\mathbf{B}}_N \leftarrow \mathbf{I}$, and therefore $\Delta \mathbf{X}_N = -\mathbf{F}_N$. This completes the forward steps. Remaining values of $\Delta \mathbf{X}$ are obtained by successive, backward substitution from $\Delta \mathbf{X}_j = -\mathbf{F}_j \leftarrow -(\mathbf{F}_j - \bar{\mathbf{C}}_j \mathbf{F}_{j+1})$

This procedure is illustrated by the following example.

EXAMPLE 10.7 Block-Tridiagonal-Matrix Equation.

Solve the following block-tridiagonal-matrix equation by the matrix form of the Thomas algorithm.

$$\begin{bmatrix} 1 & 2 & 1 & 2 & 2 & 1 & 0 & 0 & 0 \\ 2 & 1 & 1 & 2 & 1 & 0 & 0 & 0 & 0 \\ 1 & 2 & 2 & 1 & 2 & 0 & 0 & 0 & 0 \\ 0 & 1 & 3 & 1 & 2 & 1 & 1 & 2 & 1 \\ 0 & 0 & 1 & 2 & 2 & 0 & 1 & 2 & 0 \\ 0 & 0 & 2 & 2 & 1 & 1 & 1 & 1 & 0 \\ 0 & 0 & 0 & 0 & 1 & 2 & 2 & 1 & 1 \\ 0 & 0 & 0 & 0 & 0 & 2 & 1 & 1 & 1 \\ 0 & 0 & 0 & 0 & 0 & 1 & 2 & 1 & 2 \end{bmatrix} \cdot \begin{bmatrix} \Delta x_1 \\ \Delta x_2 \\ \Delta x_3 \\ \Delta x_4 \\ \Delta x_5 \\ \Delta x_6 \\ \Delta x_7 \\ \Delta x_8 \\ \Delta x_9 \end{bmatrix} = \begin{bmatrix} 9 \\ 7 \\ 8 \\ 12 \\ 8 \\ 8 \\ 7 \\ 5 \\ 6 \end{bmatrix}$$

Solution

The matrix equation is in the form

$$\begin{bmatrix} \bar{\mathbf{B}}_1 & \bar{\mathbf{C}}_1 & 0 \\ \bar{\mathbf{A}}_2 & \bar{\mathbf{B}}_2 & \bar{\mathbf{C}}_2 \\ 0 & \bar{\mathbf{A}}_3 & \bar{\mathbf{B}}_3 \end{bmatrix} \cdot \begin{bmatrix} \Delta \mathbf{X}_1 \\ \Delta \mathbf{X}_2 \\ \Delta \mathbf{X}_3 \end{bmatrix} = - \begin{bmatrix} \mathbf{F}_1 \\ \mathbf{F}_2 \\ \mathbf{F}_3 \end{bmatrix}$$

Following the procedure just given, starting at the first block row,

$$\bar{\mathbf{B}}_1 = \begin{bmatrix} 1 & 2 & 1 \\ 2 & 1 & 1 \\ 1 & 2 & 2 \end{bmatrix}, \quad \bar{\mathbf{C}}_1 = \begin{bmatrix} 2 & 2 & 1 \\ 2 & 1 & 0 \\ 1 & 2 & 0 \end{bmatrix}, \quad \mathbf{F}_1 = \begin{bmatrix} -9 \\ -7 \\ -8 \end{bmatrix}$$

By standard matrix inversion

$$(\bar{\mathbf{B}}_1)^{-1} = \begin{bmatrix} 0 & 2/3 & -1/3 \\ 1 & -1/3 & -1/3 \\ -1 & 0 & 1 \end{bmatrix}$$

By standard matrix multiplication

$$(\bar{\mathbf{B}}_1)^{-1}(\bar{\mathbf{C}}_1) = \begin{bmatrix} 1 & 0 & 0 \\ 1 & 1 & 1 \\ -1 & 0 & -1 \end{bmatrix}$$

which replaces $\bar{\mathbf{C}}_1$, and

$$(\bar{\mathbf{B}}_1)^{-1}(\mathbf{F}_1) = \begin{bmatrix} -2 \\ -4 \\ 1 \end{bmatrix}$$

which replaces $\mathbf{F}_1$. Also

$$\mathbf{I} = \begin{bmatrix} 1 & 0 & 0 \\ 0 & 1 & 0 \\ 0 & 0 & 1 \end{bmatrix} \quad \text{replaces } \bar{\mathbf{B}}_1$$

For the second block row

$$\bar{\mathbf{A}}_2 = \begin{bmatrix} 0 & 1 & 3 \\ 0 & 0 & 1 \\ 0 & 0 & 2 \end{bmatrix}, \quad \bar{\mathbf{B}}_2 = \begin{bmatrix} 1 & 2 & 1 \\ 2 & 2 & 0 \\ 2 & 1 & 1 \end{bmatrix}.$$

$$\bar{\mathbf{C}}_2 = \begin{bmatrix} 1 & 2 & 1 \\ 1 & 2 & 0 \\ 1 & 1 & 0 \end{bmatrix}, \quad \bar{\mathbf{F}}_2 = \begin{bmatrix} -12 \\ -8 \\ -8 \end{bmatrix}$$

By matrix multiplication and subtraction

$$(\bar{\mathbf{B}}_2 - \bar{\mathbf{A}}_2 \bar{\mathbf{C}}_1) = \begin{bmatrix} 3 & 1 & 3 \\ 3 & 2 & 1 \\ 4 & 1 & 3 \end{bmatrix}$$

which upon inversion becomes

$$(\bar{\mathbf{B}}_2 - \bar{\mathbf{A}}_2 \bar{\mathbf{C}}_1)^{-1} = \begin{bmatrix} -1 & 0 & 3 \\ 1 & 3/5 & -6/5 \\ 1 & -1/5 & -3/5 \end{bmatrix}$$

By multiplication

$$\left(\bar{\mathbf{B}}_2 - \bar{\mathbf{A}}_2\bar{\mathbf{C}}_1\right)^{-1}\bar{\mathbf{C}}_2 = \begin{bmatrix} 0 & -1 & -1 \\ 2/5 & 2 & 1 \\ 1/5 & 1 & 1 \end{bmatrix}$$

which replaces $\bar{\mathbf{C}}_2$. In a similar manner, the remaining steps for this and the third block row are carried out to give

$$\begin{bmatrix} \begin{bmatrix} 1 & 0 & 0 \\ 0 & 1 & 0 \\ 0 & 0 & 1 \end{bmatrix} & \begin{bmatrix} 1 & 0 & 0 \\ 1 & 1 & 1 \\ -1 & 0 & -1 \end{bmatrix} & \begin{bmatrix} 0 & 0 & 0 \\ 0 & 0 & 0 \\ 0 & 0 & 0 \end{bmatrix} \\ \begin{bmatrix} 0 & 0 & 0 \\ 0 & 0 & 0 \\ 0 & 0 & 0 \end{bmatrix} & \begin{bmatrix} 1 & 0 & 0 \\ 0 & 1 & 0 \\ 0 & 0 & 1 \end{bmatrix} & \begin{bmatrix} 0 & -1 & -1 \\ 2/5 & 2 & 1 \\ 1/5 & 1 & 1 \end{bmatrix} \\ \begin{bmatrix} 0 & 0 & 0 \\ 0 & 0 & 0 \\ 0 & 0 & 0 \end{bmatrix} & \begin{bmatrix} 0 & 0 & 0 \\ 0 & 0 & 0 \\ 0 & 0 & 0 \end{bmatrix} & \begin{bmatrix} 1 & 0 & 0 \\ 0 & 1 & 0 \\ 0 & 0 & 1 \end{bmatrix} \end{bmatrix} \cdot \begin{bmatrix} \Delta X_1 \\ \Delta X_2 \\ \Delta X_3 \\ \Delta X_4 \\ \Delta X_5 \\ \Delta X_6 \\ \Delta X_7 \\ \Delta X_8 \\ \Delta X_9 \end{bmatrix} = -\begin{bmatrix} -2 \\ -4 \\ +1 \\ +1 \\ -22/5 \\ -16/5 \\ -1 \\ -1 \\ -1 \end{bmatrix}$$

Thus, $\Delta X_7 = \Delta X_8 = \Delta X_9 = 1$. The remaining backward steps begin with the second block row, where

$$\bar{\mathbf{C}}_2 = \begin{bmatrix} 0 & -1 & -1 \\ 2/5 & 2 & 1 \\ 1/5 & 1 & 1 \end{bmatrix}, \bar{\mathbf{F}}_2 = \begin{bmatrix} 1 \\ -22/5 \\ -16/5 \end{bmatrix}$$

$$\left(\mathbf{F}_2 - \bar{\mathbf{C}}_2\bar{\mathbf{F}}_3\right) = \begin{bmatrix} -1 \\ -1 \\ -1 \end{bmatrix}$$

Thus, $\Delta X_4 = \Delta X_5 = \Delta X_6 = 1$. Similarly, for the first block row,

$$\Delta X_1 = \Delta X_2 = \Delta X_3 = 1$$

It is desirable to specify top- and bottom-stage variables other than condenser and/or reboiler duties, which are so interdependent that specification of both values is not recommended. Specifying other variables is accomplished by removing heat balance functions H_1 and/or H_N from the simultaneous equation set and replacing them with discrepancy functions. Such alternative specifications for a partial condenser are in Table 10.1.

If desired, (10-59) can be modified to accommodate real rather than theoretical stages. Values of the E_{MV} (§6.5.3) must then be specified. These are related to phase compositions by the definition

$$\eta_j = \frac{y_{i,j} - y_{i,j+1}}{K_{i,j}x_{i,j} - y_{i,j+1}} \tag{10-73}$$

In terms of component flow rates, (10-73) becomes the following discrepancy function, which replaces (10-59).

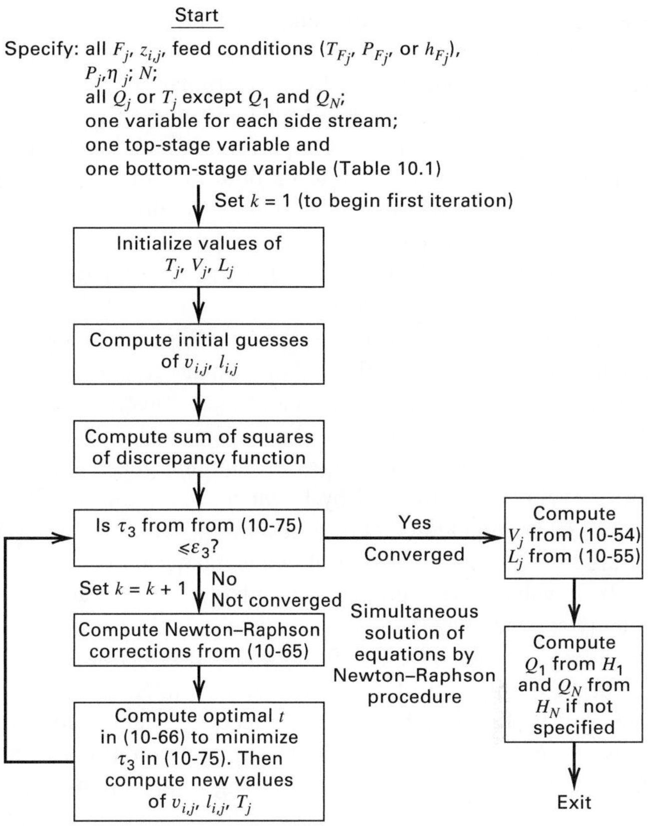

Figure 10.24 Algorithm for the Newton–Raphson method of Naphtali–Sandholm for all vapor–liquid separators.

$$E_{i,j} = \frac{\eta_j K_{i,j} l_{i,j} \sum\limits_{\kappa=1}^{C} \upsilon_{\kappa,j}}{\sum\limits_{\kappa=1}^{C} l_{\kappa,j}} - \upsilon_{i,j} + \frac{\left(1 - \eta_j\right)\upsilon_{i,j+1} \sum\limits_{\kappa=1}^{C} \upsilon_{\kappa,j}}{\sum\limits_{\kappa=1}^{C} \upsilon_{\kappa,j+1}} = 0 \tag{10-74}$$

For a total condenser with subcooling, it is necessary to specify the degrees of subcooling and to replace (10-59) or (10-74) with functions that express identity of reflux and distillate compositions [23].

The algorithm for the Naphtali–Sandholm implementation of the Newton–Raphson method is shown in Figure 10.24. Problem specifications are quite flexible. However, number of stages, and pressure, compositions, flow rates, and stage locations for all feeds are necessary specifications. The thermal condition of each feed can be given in terms of enthalpy, temperature, or fraction vaporized. A two-phase feed can be sent to the same stage or the vapor can be directed to the

Table 10.1 Alternative Functions for H_1 and H_N

Specification	Replacement for H_1	Replacement for H_N
Reflux or reboil (boilup) ratio, (L/D) or (V/B)	$\sum l_{i,1} - (L/D)\sum \upsilon_{i,1} = 0$	$\sum \upsilon_{i,N} - (V/B)\sum l_{i,N} = 0$
Stage temperature, T_D or T_B	$T_1 - T_D = 0$	$T_N - T_B = 0$
Product flow rate, D or B	$\sum \upsilon_{i,1} - D = 0$	$\sum l_{i,N} - B = 0$
Component flow rate in product, d_i or b_i	$\upsilon_{i,1} - d_i = 0$	$l_{i,N} - b_i = 0$
Component mole fraction in product, y_{iD} or x_{iB}	$\upsilon_{i,1} - \left(\sum \upsilon_{i,1}\right)y_{iD} = 0$	$l_{i,N} - \left(\sum l_{i,N}\right)x_{iB} = 0$

stage above. Stage pressures and stage efficiencies can be designated by specifying top- and bottom-stage values with remaining values obtained by linear interpolation. By default, intermediate stages are assumed adiabatic unless Q_j or T_j values are specified. Vapor and/or liquid sidestreams can be designated in terms of total flow or flow rate of a specified component, or by the ratio of the sidestream flow to the flow rate passing to the next stage. The top- and bottom-stage specifications are selected from Q_1 or Q_N, and/or from the other specifications in Table 10.1.

To achieve convergence, the Newton–Raphson procedure requires guesses for the values of all output variables. Rather than provide these *a priori*, they can be generated if T, V, and L are guessed for the bottom and top stages and, perhaps, for one or more intermediate stages. Remaining guessed T_j, V_j, and L_j values are obtained by linear interpolation of the T_j values and computed (V_j/L_j) values. Initial values for $v_{i,j}$ and $l_{i,j}$ are then obtained by either of two techniques.

If K-values are composition-independent or approximately so, one technique is to compute $x_{i,j}$ values and corresponding $y_{i,j}$ values from (10-12) and (10-2), as in the first iteration of the BP or SR method. A cruder estimate is obtained by flashing the combined feeds at average column pressure and a V/L ratio that approximates the ratio of overheads to bottoms products. The resulting compositions of the equilibrium vapor and liquid phases are assumed to hold for each stage. The second technique works surprisingly well, but the first technique is preferred for difficult cases. For either, the initial component flow rates are computed using the $x_{i,j}$ and $y_{i,j}$ values to solve (10-56) and (10-57) for $l_{i,j}$ and $v_{i,j}$ values.

Based on initial guesses for all output variables, the sum of the squares of the discrepancy functions is compared to the convergence criterion

$$\tau_3 = \sum_{j=1}^{N}\left\{\left(H_j\right)^2 + \sum_{i=1}^{C}\left[\left(M_{i,j}\right)^2 + \left(E_{i,j}\right)^2\right]\right\} \leq \epsilon_3 \quad (10\text{-}75)$$

For all discrepancies to be of the same order of magnitude, it is necessary to divide energy-balance functions H_j by a scale factor approximating the latent heat of vaporization (e.g., 1,000 Btu/lbmol). If the convergence criterion is

$$\epsilon_3 = N(2C + 1)\left(\sum_{j=1}^{N}F_j^2\right)10^{-10} \quad (10\text{-}76)$$

resulting converged values will generally be accurate, on the average, to four or more significant figures. When employing (10-76), most problems converge in 10 iterations or fewer.

The convergence criterion is far from satisfied during the first iteration with guessed values for the output variables. For subsequent iterations, Newton–Raphson corrections are computed from (10-65). These can be added directly to the present values of the output variables to obtain a new set of output variables. Alternatively, (10-66) can be employed where t is a nonnegative, scalar step factor. At each iteration, a value of t is applied to all output variables. By permitting t to vary from slightly greater than 0 up to 2, it can dampen or accelerate convergence, as appropriate. For each iteration, a t that minimizes the sum of the squares given by (10-75) is

sought. Generally, optimal values of t proceed from an initial value for the second iteration at between 0 and 1 to a value nearly equal to or slightly greater than 1 when the criterion is almost satisfied. An optimization procedure for finding t at each iteration is the Fibonacci search [25]. If there is no optimal value of t in the designated range, t can be set to 1, or some smaller value, and the sum of squares can be allowed to increase. Generally, after several iterations, the sum of squares decreases for every iteration.

If the application of (10-66) results in a negative component flow rate, Naphtali and Sandholm recommend a mapping equation, which reduces the value of the unknown variable to a near-zero, but nonnegative, quantity:

$$X^{(k+1)} = X^{(k)}\exp\left[\frac{t\Delta X^{(k)}}{X^{(k)}}\right] \quad (10\text{-}77)$$

In addition, it is advisable to limit temperature corrections at each iteration.

The NR method is readily extended to staged separators involving two liquid phases (e.g., extraction) and three coexisting phases (e.g., three-phase distillation), as shown by Block and Hegner [26], and to interlinked separators as shown by Hofeling and Seader [27].

EXAMPLE 10.8 Newton–Raphson Method.

A reboiled absorber is to separate the hydrocarbon vapor feed of Examples 10.2 and 10.4. Absorbent oil of the same composition as that of Example 10.4 enters the top stage. Specifications are given in Figure 10.25. The 770 lbmol/h (349 kmol/h) of bottoms product corresponds to the amount of C_3 and heavier in the two feeds, so the column is designed as a deethanizer. Calculate stage temperatures, interstage vapor and liquid flow rates and compositions, and reboiler duty by the Newton–Raphson method. Assume all stage efficiencies are 100%. Compare the separation to that achieved by ordinary distillation in Example 10.2.

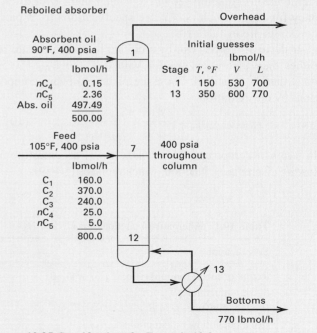

Figure 10.25 Specifications for Example 10.8.

Solution

A digital computer program for the method of Naphtali and Sandholm was used. The K-values and enthalpies were assumed independent of composition and were computed by linear interpolation between tabular values at 100°F increments from 0 to 400°F.

From (10-76), the convergence criterion is

$$\epsilon_3 = 13[2(6) + 1](500 + 800)^2 10^{-10} = 2.856 \times 10^{-2}$$

Figure 10.26 shows the sum-of-squares reduction of the 169 discrepancy functions from iteration to iteration, where t is the scalar step factor in (10-66). Seven iterations were required to satisfy the convergence criterion. The initial iteration was based on values of the unknown variables computed from interpolation of the initial guesses shown in Figure 10.25 together with a flash of the combined feeds at 400 psia and a V/L ratio of 0.688, (530/770). For the first iteration, mole-fraction compositions were computed and assumed to apply to every stage:

Species	y	x
C_1	0.2603	0.0286
C_2	0.4858	0.1462
C_3	0.2358	0.1494
nC_4	0.0153	0.0221
nC_5	0.0025	0.0078
Abs. oil	0.0003	0.6459
	1.0000	1.0000

The corresponding sum of squares of the discrepancy functions, τ_3, of 2.865×10^7, was very large for the first iteration. For iteration 2, the optimal value of t was 0.34, which caused only a moderate reduction in the sum of squares. The optimal value of t was 0.904 for iteration 3, and the sum of squares was reduced by an order of magnitude. For the fourth and subsequent iterations, the effect of t on the sum of squares is included in Figure 10.26. Following iteration 4, the sum of squares was reduced by at least two orders of magnitude

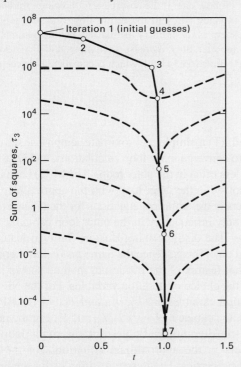

Figure 10.26 Convergence pattern for Example 10.8.

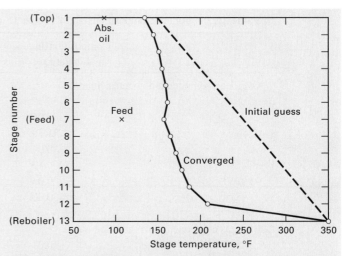

Figure 10.27 Converged temperature profile for Example 10.8.

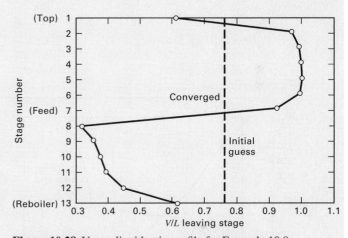

Figure 10.28 Vapor–liquid ratio profile for Example 10.8.

for each iteration. Also, the optimal value of t was rather sharply defined and corresponded closely to 1. An improvement of τ_3 was obtained for every iteration.

In Figures 10.27 and 10.28, converged temperature and V/L profiles are compared to the initial profiles. In Figure 10.27, the converged temperatures are far from linear with respect to stage number. Above the feed stage, the temperature increases from the top down in a gradual, declining manner. The cold feed causes a small temperature drop from stage 6 to 7. Temperature also increases from stage 7 to 13. A dramatic increase occurs in moving from the bottom stage to the reboiler. In Figure 10.28, the converged V/L profile is far from the initial guess.

Figure 10.29 shows component flow-rate profiles for the key components (ethane vapor, propane liquid). The guessed values are in poor agreement with converged values. The propane-liquid profile is regular except at the bottom, where a large decrease occurs because of vaporization in the reboiler. The ethane-vapor profile has changes at the top, where entering oil absorbs ethane, and at the feed stage, where substantial ethane vapor is introduced.

Converged values for the reboiler duty, and overhead and bottoms compositions, are given in Table 10.2. Also included are converged results for solutions using the Chao–Seader (CS) and Soave–Redlich–Kwong (SRK) equations of Chapter 2 for K-values and enthalpies in place of composition-independent tabular properties. With SRK, a somewhat sharper separation between the two key components is predicted, as well as a substantially higher bottoms

Table 10.2 Product Compositions and Reboiler Duty for Example 10.8

	Composition-Independent Tabular Properties	Chao–Seader Correlation	Soave–Redlich–Kwong Equation
Overhead component flow rates, lbmol/h			
C_1	159.99	159.98	159.99
C_2	337.96	333.52	341.57
C_3	31.79	36.08	28.12
nC_4	0.04	0.06	0.04
nC_5	0.17	0.21	0.18
Abs. oil	0.05	0.15	0.10
	530.00	530.00	530.00
Bottoms component flow rates, lbmol/h			
C_1	0.01	0.02	0.01
C_2	32.04	36.4	28.43
C_3	208.21	203.92	211.88
nC_4	25.11	25.09	25.11
nC_5	7.19	7.15	7.18
Abs. oil	497.44	497.34	497.39
	770.00	770.00	770.00
Reboiler duty, Btu/h	11,350,000	10,980,000	15,640,000
Bottoms temperature, °F	346.4	338.5	380.8

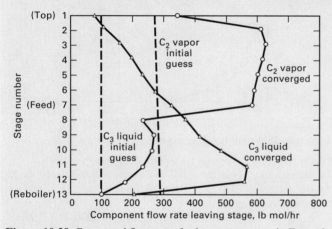

Figure 10.29 Converged flow rates for key components in Example 10.8.

temperature and a much larger reboiler duty. As discussed in Chapter 4, the effect of physical properties on column design can be significant; care must be exercised to choose the most appropriate physical property correlations.

It is interesting to compare the results using the reboiled absorber to the separation achieved by ordinary distillation of the same feed in the same-size column as provided in Example 10.2 and shown in Figure 10.9. The latter results in a sharper separation and a lower bottoms temperature and reboiler duty. However, refrigeration is necessary for the condenser, and the reflux flow rate is twice the absorbent oil flow rate. If the absorbent oil flow rate for the reboiled absorber is made equal to the reflux flow rate, the separation is almost as sharp as for ordinary distillation, but the bottoms temperature and reboiler duty increase to almost 600°F (315.6°C) and 60,000,000 Btu/h (63.3 GJ/h), which is considered unacceptable.

§10.5 INSIDE-OUT METHOD

In the BP, SR, and NR methods, the major computational effort is expended in calculating K-values and enthalpies when rigorous thermodynamic-property models are utilized, because property calculations are made at each iteration. Furthermore, at each iteration, derivatives are required of: (1) all properties with respect to temperature and compositions of both phases for the NR method; (2) K-values with respect to temperature for the BP method, unless Muller's method is used, and (3) vapor and liquid enthalpies with respect to temperature for the SR method.

In 1974, Boston and Sullivan [28] presented an algorithm designed to reduce the time spent computing thermodynamic properties when making column calculations. As shown in Figure 10.30c, two sets of thermodynamic-property models are

employed: (1) a simple, approximate, empirical set used frequently to converge inner-loop calculations, and (2) a rigorous set used less often in the outer loop. The MESH equations are always solved in the inner loop with the approximate set. The parameters in the empirical equations for the inner-loop set are updated only infrequently in the outer loop using the rigorous equations. The distinguishing Boston–Sullivan feature is the inner and outer loops, hence the name *inside-out method*.

Another feature of the inside-out method shown in Figure 10.30 is the choice of iteration variables. For the NR method, the iteration variables are $l_{i,j}$, $\upsilon_{i,j}$, and T_j. For the BP and SR methods, the choice is $x_{i,j}$ $y_{i,j}$, T_j, L_j, and V_j. For the inside-out method, the iteration variables for the outer loop are the parameters in the approximate equations for the thermodynamic properties. The iteration variables for the inner loop are related to stripping factors, $S_{i,j} = K_{i,j}V_j/L_j$.

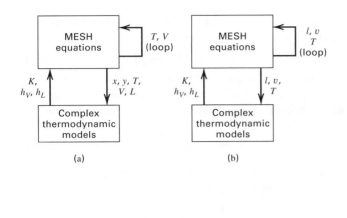
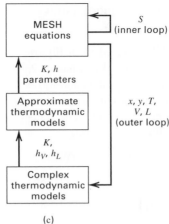

Figure 10.30 Incorporation of thermodynamic-property correlations into interactive loops. (a) BP and SR methods. (b) Newton–Raphson method. (c) Inside-out method.

In the original inside-out method, applications were restricted to moderately nonideal hydrocarbon distillations for the Case II variable specifications in Table 5.4, but with multiple feeds, sidestreams, and intermediate heat exchangers. For these applications, the inside-out method was shown to be rapid and robust. Since 1974, the method has been extended and improved in published articles [29–34] and proprietary process simulation programs. These extensions permit the inside-out method to be applied to almost any type of steady-state, multicomponent, multistage vapor–liquid separation operation. In the extensive implementation of the inside-out method in RADFRAC and MULTIFRAC of ASPEN PLUS, these applications include:

1. Absorption, stripping, reboiled absorption, reboiled stripping, extractive distillation, and azeotropic distillation
2. Three-phase (vapor–liquid–liquid) systems
3. Reactive systems
4. Highly nonideal systems requiring activity-coefficient models
5. Interlinked units, including pumparounds, bypasses, and external heat exchangers
6. Narrow-boiling, wide-boiling, and dumbbell (mostly heavy and light component) feeds
7. Presence of free water
8. Wide variety of specifications other than Case II of Table 5.2 for the reflux ratio and product rates (e.g., product purities)
9. Use of Murphree-stage efficiencies

The inside-out method takes advantage of the following characteristics of the iterative calculations:

1. Component relative volatilities vary much less than component K-values.
2. Enthalpy of vaporization varies less than phase enthalpies.
3. Component stripping factors combine effects of temperature and liquid and vapor flows at each stage.

The inner loop of the inside-out method uses relative volatility, energy, and stripping factors to improve stability and

reduce computing time. A widely used implementation is that of Russell [31], which is described here together with further refinements suggested and tested by Jelinek [33].

§10.5.1 MESH Equations for Inside-Out Method

As with the BP, SR, and NR methods, the equilibrium-stage model of Figures 10.1 and 10.3 is employed. The form of the equations is similar to the NR method in that component flow rates are utilized. However, in addition, the following inner-loop variables are defined:

$$\alpha_{i,j} = K_{i,j}/K_{b,j} \tag{10-78}$$

$$S_{b,j} = K_{b,j}V_j/L_j \tag{10-79}$$

$$R_{L_j} = 1 + U_j/L_j \tag{10-80}$$

$$R_{V_j} = 1 + W_j/V_j \tag{10-81}$$

where K_b is the K-value for a base or hypothetical reference component, $S_{b,j}$ is the stripping factor for the base component, R_{L_j} is a liquid-phase withdrawal factor, and R_{V_j} is a vapor-phase withdrawal factor. For stages without sidestreams, R_{L_j} and R_{V_j} reduce to 1. The defined variables of (10-78) to (10-81), (10-54) to (10-57) still apply, but the MESH equations, (10-58) to (10-60), become as follows, where (10-83) results from the use of (10-80) to (10-82) to eliminate variables in V and the sidestream ratios s and S:

Phase Equilibria:

$$v_{i,j} = \alpha_{i,j}S_{b,j}l_{i,j}, \quad i = 1 \text{ to } C, \quad j = 1 \text{ to } N \tag{10-82}$$

Component Material Balance:

$$l_{i,j-1} - (R_{L_j} + \alpha_{i,j}S_{b,j}R_{V_j})l_{i,j} + (\alpha_{i,j+1}S_{b,j+1})l_{i,j+1}$$
$$= -f_{i,j}, \quad i = 1 \text{ to } C, \quad j = 1 \text{ to } N \tag{10-83}$$

Energy Balance:

$$H_j = h_{L_j}R_{L_j}L_j + h_{V_j}R_{V_j}V_j - h_{L_{j-1}}L_{j-1} - h_{V_{j+1}}V_{j+1}$$
$$- h_{F_j}F_j - Q_j = 0, \quad j = 1 \text{ to } N \tag{10-84}$$

where $S_{i,j} = \alpha_{i,j}S_{b,j}$.

In addition, discrepancy functions as shown in Table 10.1 can be added to the MESH equations to permit any reasonable set of product specifications.

§10.5.2 Rigorous and Complex Thermodynamic-Property Models

The complex thermodynamic models referred to in Figure 10.30 can include any of the models discussed in Chapter 2, including those based on $P–v–T$ equations of state and those based on free-energy models for liquid-phase activity coefficients. These generate parameters in the approximate thermodynamic-property models of the form

$$K_{i,j} = K_{i,j}\{P_j, T_j, \mathbf{x}_j, \mathbf{y}_j\} \tag{10-85}$$

$$h_{V_j} = h_{V_j}\{P_j, T_j, \mathbf{y}_j\} \tag{10-86}$$

$$h_{L_j} = h_{L_j}\{P_j, T_j, \mathbf{x}_j\} \tag{10-87}$$

§10.5.3 Approximate Thermodynamic-Property Models

The approximate models in the inside-out method are designed to facilitate calculation of stage temperatures and stripping factors.

K-values

The approximate K-value model of Russell [31] and Jelinek [33], which differs slightly from that of Boston and Sullivan [28] and originated from a proposal in Robinson and Gilliland [35], is (10-78) combined with

$$K_{b,j} = \exp(A_j - B_j/T_j) \tag{10-88}$$

Either a feed component or a hypothetical reference component can be selected as the base, b, with the latter preferred, and determined from a vapor-composition weighting using the following relations:

$$K_{b,j} = \exp\left(\sum_i w_{i,j} \ln K_{i,j}\right) \tag{10-89}$$

where $w_{i,j}$ are weighting functions given by

$$w_{i,j} = \frac{y_{i,j}[\partial \ln K_{i,j}/\partial(1/T)]}{\sum_i y_{i,j}[\partial \ln K_{i,j}/\partial(1/T)]} \tag{10-90}$$

A unique K_b model and values of $\alpha_{i,j}$ in (10-78) are derived for each stage j from values of $K_{i,j}$ from the rigorous model. At the top stage, the base component will be close to a light component, while at the bottom stage, the base component will be close to a heavy one. The derivatives in (10-90) are obtained numerically or analytically from the rigorous model. To obtain values of A_j and B_j in (10-88), two temperatures must be selected for each stage. For example, the estimated or current temperatures of the two adjacent stages, $j-1$ and $j+1$, might be selected. Calling these T_1 and T_2 and using (10-88) at each stage, b:

$$B = \frac{\ln\left(K_{b_{T_1}}/K_{b_{T_2}}\right)}{\left(\frac{1}{T_2} - \frac{1}{T_1}\right)} \tag{10-91}$$

and

$$A = \ln K_{b_{T_1}} + B/T_1 \tag{10-92}$$

For highly nonideal-liquid solutions, it is advisable to separate the rigorous K-value into two parts, as in (2-27):

$$K_i = \gamma_{iL}(\phi_{iL}/\bar{\phi}_{iV}) \tag{10-93}$$

Then, $(\phi_{iL}/\bar{\phi}_{iV})$ is used to determine K_b and, as proposed by Boston [30], values of γ_{iL} at each stage are fitted at a reference temperature, T^*, to the liquid-phase mole fraction by the linear function

$$\gamma_{iL}^* = a_i + b_i x_i \tag{10-94}$$

to obtain the approximate estimates, γ_{iL}^*. Equation (10-83) is then modified by replacing $\alpha_{i,j}$ with $\alpha_{i,j}\gamma_{iL}^*$, where

$$\alpha_{i,j} = \frac{(\phi_{iL}/\bar{\phi}_{iV})_j}{K_{b,j}} \tag{10-95}$$

rather than the $\alpha_{i,j}$ given by (10-78).

Enthalpies

Boston and Sullivan [28] and Russell [31] employ the same approximate enthalpy models. Jelinek [33] does not use approximate enthalpy models, because the additional complexity involved in the use of two enthalpy models may not always be justified to the extent that approximate and rigorous K-value models are justified.

The basis for the enthalpy calculations is the same as for the rigorous equations discussed in Chapter 2. Thus, for either phase, from Table 2.6,

$$h = h_V^\circ + (h - h_V^\circ) = h_V^\circ + \Delta H \tag{10-96}$$

where h_V° is the ideal-gas mixture enthalpy, as given by the polynomial equations (2-35) and (2-36), based on the vapor-phase composition for h_V and the liquid-phase composition for h_L. The ΔH term is the enthalpy departure, $\Delta H_V = (h_V - h_V^\circ)$, for the vapor phase, which accounts for the effect of pressure, and $\Delta H_L = (h_L - h_V^\circ)$ for the liquid phase, which accounts for the enthalpy of vaporization and the effect of pressure on both liquid and vapor phases, as indicated in (2-57). The enthalpy of vaporization dominates the ΔH_L term. The time-consuming parts of the enthalpy calculations are the two enthalpy-departure terms, which are complex when an equation of state is used. Therefore, in the approximate enthalpy equations, the rigorous enthalpy departures are replaced by the simple linear functions

$$\Delta H_{V_j} = c_j - d_j(T_j - T^*) \tag{10-97}$$

and

$$\Delta H_{L_j} = e_j - f_j(T_j - T^*) \tag{10-98}$$

where the departures are modeled in terms of enthalpy per unit mass instead of per unit mole, and T^* is a reference temperature. The parameters c, d, e, and f are evaluated from the rigorous models at each outer-loop iteration.

§10.5.4 Inside-Out Algorithm

The inside-out algorithm of Russell [31] involves an initialization procedure, inner-loop iterations, and outer-loop iterations.

Initialization Procedure

First, it is necessary to provide reasonably good estimates of all stage values of $x_{i,j}$, $y_{i,j}$, T_j, V_j, and L_j. Boston and Sullivan [28] suggest the following procedure:

1. Specify the number of theoretical stages, conditions of all feeds, feed-stage locations, and pressure profile.

2. Specify stage locations for each product withdrawal (including sidestreams) and for each heat exchanger.

3. Provide an additional specification for each product and each intermediate heat exchanger.

4. If not specified, estimate each product-withdrawal rate, and estimate each value of V_j. Obtain values of L_j from the total material-balance equation, (10-6).

5. Estimate an initial temperature profile, T_j, by combining all feed streams (composite feed) and determining bubble- and dew-point temperatures at average column pressure. The dew-point temperature is the top-stage temperature, T_1, whereas the bubble-point temperature is the bottom-stage temperature, T_N. Intermediate-stage temperatures are estimated by interpolation. Reference temperatures T^* for use with (10-94), (10-97), and (10-98) are set equal to T_j.

6. Flash the composite feed isothermally at the average column pressure and temperature. The resulting vapor and liquid compositions, y_i and x_i, are the estimated stage compositions.

7. With the initial estimates from steps 1 through 6, use the complex thermodynamic-property correlation to determine values of the stagewise outside-loop K and h parameters A_j, B_j, $a_{i,j}$, $b_{i,j}$, c_j, d_j, e_j, f_j, $K_{b,j}$, and $\alpha_{i,j}$ of the approximate models.

8. Compute initial values of $S_{b,j}$, R_{Lj}, and R_{Vj} from (10-79), (10-80), and (10-81).

Inner-Loop Calculation Sequence

An iterative sequence of inner-loop calculations begins with values for the outside-loop parameters listed in step 7, obtained initially from the initialization procedure and later from outer-loop calculations, using results from the inner loop, as shown in Figure 10.30c.

9. Compute component liquid flow rates, $l_{i,j}$, from the set of N equations (10-83) for each of the C components by the tridiagonal-matrix algorithm.

10. Compute component vapor flows, $v_{i,j}$, from (10-82).

11. Compute a revised set of flow rates, V_j and L_j, from the component flow rates by (10-54) and (10-55).

12. To calculate a revised set of stage temperatures, T_j, compute a set of x_i values for each stage from (10-57), then a revised set of $K_{b,j}$ values from a combination of the bubble-point equation, (4-12), with (10-78), which gives

$$K_{b,j} = 1 \bigg/ \sum_{i=1}^{C} (\alpha_{i,j} x_{i,j}) \qquad (10\text{-}99)$$

From this new set of $K_{b,j}$ values, compute a set of stage temperatures from a rearrangement of (10-88):

$$T_j = \frac{B_j}{A_j - \ln K_{b,j}} \qquad (10\text{-}100)$$

At this point in the inner-loop iterative sequence, there is a revised set of $v_{i,j}$, $l_{i,j}$, and T_j, which satisfy the component material-balance and phase-equilibria equations for the estimated properties. However, these values do not satisfy the energy-balance and specification equations unless the estimated base-component stripping factors and product-withdrawal rates are correct.

13. Select inner-loop iteration variables as

$$\ln S_{b,j} = \ln(K_{b,j} V_j / L_j) \qquad (10\text{-}101)$$

together with any other iteration variables. For the simple distillation column in Figure 10.9, no other inner-loop iteration variables would be needed if the condenser and reboiler duties were specified. If the reflux ratio (L/D) and bottoms flow rate (B) are specified rather than the two duties (which is the more common situation), in place of the two (10-84) equations for H_1 and H_N, the two specification equations from Table 10.1 in the form of discrepancy functions D_1 and D_2 are added.

$$D_1 = L_1 - (L/D)V_1 = 0 \qquad (10\text{-}102)$$

$$D_2 = L_N - B = 0 \qquad (10\text{-}103)$$

For each sidestream, a sidestream-withdrawal factor is added as an inner-loop iteration variable, e.g., $\ln(U_j/L_j)$ and $\ln(W_j/V_j)$, together with a specification on purity or some other variable.

14. Compute stream enthalpies from (10-96) to (10-98).

15. Compute normalized discrepancies of H_j, D_1, D_2, etc., from the energy balances (10-84) and (10-102), (10-103), etc., but compute Q_1 from H_1, and Q_N from H_N where appropriate. A typical normalization is discussed in §10.4 for the NR method.

16. Compute the Jacobian of partial derivatives of H_j, D_1, D_2, etc., with respect to the iteration variables of (10-101), etc., by perturbation of each iteration variable and recalculation of the discrepancies through steps 9 to 15, numerically or by differentiation.

17. Compute corrections to the inner-loop iteration variables by a NR iteration of the type discussed for the SR and NR methods in §10.3 and 10.4.

18. Compute new values of the iteration variables from the sum of the previous values and the corrections with (10-66), using damping if necessary to reduce the sum of squares of the normalized discrepancies.

19. Check whether the sum-of-squares is sufficiently small. If so, proceed to the outer-loop calculation procedure given next. If not, repeat steps 15 to 18 using the latest iteration variables. For any subsequent cycles through steps 15 to 18, Russell [31] uses Broyden [36] updates to avoid reestimation of the Jacobian partial derivatives,

whereas Jelinek [33] recommends the standard NR method of recalculating the partial derivatives for each inner-loop iteration.

20. Upon convergence of steps 15 to 19, steps 9 through 12 will have produced an improved set of primitive variables $x_{i,j}$, $v_{i,j}$, $l_{i,j}$, T_j, V_j, and L_j. From (10-56), corresponding values of $y_{i,j}$ can be computed. The values of these variables are not correct until the approximate thermodynamic properties are in agreement with the properties from the rigorous models. The primitive variables are input to the outer-loop calculations to bring the approximate and complex models into successively better agreement.

Outer-Loop Calculation Sequence.

Each outer loop proceeds as follows:

21. Using the values of the primitive variables from step 20, compute relative volatilities and stream enthalpies from the complex thermodynamic models. If they are in close agreement with previous values used to initiate a set of inner-loop iterations, both outer- and inner-loop iterations are converged, and the problem is solved. If not, proceed to step 22.

22. Determine values of the stagewise outside-loop K and h parameters of the approximate models from the complex models, as in initialization step 7.

23. Compute values of $S_{b,j}$, R_{Lj}, and R_{Vj}, as in initialization step 8.

24. Repeat the inner-loop calculation of Steps 9–20.

Although convergence of the inside-out method is not guaranteed, for most problems, the method is robust and rapid. Convergence difficulties arise because of poor initial estimates, which result in negative or zero flow rates at certain locations in the column. To counteract this tendency, all component stripping factors use a scalar multiplier, S_b, called a base stripping factor, to give

$$S_{i,j} = S_b \alpha_{i,j} S_{b,j} \qquad (10\text{-}104)$$

The value of S_b is initially chosen to force the results of the initialization procedure to give a reasonable distribution of component flows throughout the column. Russell [31] recommends that S_b be chosen only once, but Boston and Sullivan [28] compute a new S_b for each new set of $S_{b,j}$ values.

For highly nonideal-liquid mixtures, use of the inside-out method may lead to difficulties, and the NR method should be tried. If the NR method also fails to converge, relaxation or continuation methods, described by Kister [37], are usually successful, but computing time may be an order of magnitude longer than that for similar problems converged successfully with the inside-out method.

EXAMPLE 10.9 Inside-Out Method.

For the conditions of the distillation column shown in Figure 10.7, obtain a converged solution by the inside-out method, using the SRK equation of state for thermodynamic properties.

Solution

A computer solution was obtained with the module TOWR (an inside-out method) of the CHEMCAD process simulator. The only initial assumptions are a condenser outlet temperature of 65°F and a bottoms-product temperature of 165°F. The bubble-point temperature of the feed is computed as 123.5°F. In the initialization procedure, the constants A and B in (10-88), with T in °R, are determined from the SRK equation, with the following results:

Stage	T, °F	A	B	K_b
1	65	6.870	3708	0.8219
2	95	6.962	4031	0.7374
3	118	7.080	4356	0.6341
4	142	7.039	4466	0.6785
5	165	6.998	4576	0.7205

Values of the enthalpy coefficients c, d, e, and f in (10-97) and (10-98) are not tabulated here but are computed for each stage, based on the initial temperature distribution.

In the inner-loop calculation sequence, component flow rates are obtained from (10-83) by the tridiagonal-matrix method of §10.3.1. The resulting bottoms-product flow rate deviates somewhat from the specified value of 50 lbmol/h. By modifying the component stripping factors with a base stripping factor, S_b, in (10-104) of 1.1863, the error in the bottoms flow rate is reduced to 0.73%.

The initial inside-loop error from the solution of the normalized energy-balance equations, (10-84), is found to be only 0.04624. This is reduced to 0.000401 after two iterations through the inner loop.

At this point in the inside-out method, the revised column profiles of temperature and phase compositions are used in the outer loop with the complex SRK thermodynamic models to compute updates of the approximate K and h constants. Only one inner-loop iteration is required to obtain satisfactory convergence of the energy equations. The K and h constants are again updated in the outer loop. After one inner-loop iteration, the approximate K and h values are found to be sufficiently close to the SRK values for overall convergence. Thus, a total of only three outer-loop iterations and four inner-loop iterations are required.

To illustrate the efficiency of the inside-out method, results from each of the three outer-loop iterations are:

Outer-Loop Iteration	Stage Temperatures, °F				
	T_1	T_2	T_3	T_4	T_5
Initial guess	65	—	—	—	165
1	82.36	118.14	146.79	172.66	193.20
2	83.58	119.50	147.98	172.57	192.53
3	83.67	119.54	147.95	172.43	192.43

Outer-Loop Iteration	Total Liquid Flows, lbmol/h				
	L_1	L_2	L_3	L_4	L_5
Specification	100	—	—	—	—
1	100.00	89.68	187.22	189.39	50.00
2	100.03	89.83	188.84	190.59	49.99
3	100.0	89.87	188.96	190.56	50.00

Outer-Loop Iteration	Component Flows in Bottoms Product, lbmol/h			
	C_3	nC_4	nC_5	L_5
1	0.687	12.045	37.268	50.000
2	0.947	12.341	36.697	49.985
3	0.955	12.363	36.683	50.001

It is seen that stage temperatures and total liquid flows are already close to the converged solution after one outer-loop iteration. However, the composition of the bottoms product with respect to the lightest component, C_3, is not close to the converged solution until after two iterations. The inside-out method does not always converge so dramatically but is usually quite efficient, as shown in the following table for four exercises in this chapter.

Problem	Total Number of Inner Loops	Number of Outer-Loop Iterations
Exercise 10.11	7	6
Exercise 10.25	6	3
Exercise 10.37	17	9
Exercise 10.41	16	5

The computing time for each of these four exercises was less than 1 second on a PC with a Pentium 4 processor at 2.4 GHz.

SUMMARY

1. Rigorous methods are readily available for computer-solution of equilibrium-based models for multicomponent, multistage absorption, stripping, distillation, and liquid–liquid extraction.

2. The equilibrium-based model for a countercurrent-flow cascade provides for multiple feeds, vapor sidestreams, liquid sidestreams, and intermediate heat exchangers. Thus, the model can handle almost any type of column configuration.

3. The model equations include component and total material balances, phase-equilibria relations, and energy balances.

4. Some or all of the model equations can usually be grouped to obtain tridiagonal-matrix equations, for which an efficient solution algorithm is available.

5. Widely used for iteratively solving the model equations are the bubble-point (BP) method, the sum-rates (SR) method, the Newton–Raphson (NR) method, and the inside-out method.

6. The BP method is generally restricted to distillation problems involving narrow-boiling feed mixtures.

7. The SR method is generally restricted to absorption and stripping problems involving wide-boiling feed mixtures or, in the ISR form, to extraction problems.

8. The NR and inside-out methods are designed to solve any type of column configuration for any type of feed mixture. Because of its computational efficiency, the inside-out method is often the method of choice; however, it may fail to converge when highly nonideal-liquid mixtures are involved, in which case the slower NR method should be tried. Both permit considerable flexibility in specifications.

9. When the NR and inside-out methods fail, slower relaxation and continuation methods can be resorted to.

REFERENCES

1. Wang, J.C., and G.E. Henke, *Hydrocarbon Processing*, **45**(8), 155–163 (1966).

2. Myers, A.L., and W.D. Seider, *Introduction to Chemical Engineering and Computer Calculations*, Prentice-Hall, Englewood Cliffs, NJ, pp. 484–507 (1976).

3. Lewis, W.K., and G.L. Matheson, *Ind. Eng. Chem.*, **24**, 496–498 (1932).

4. Thiele, E.W., and R.L. Geddes, *Ind. Eng. Chem.*, 25, **290** (1933).

5. Holland, C.D., *Multicomponent Distillation*, Prentice-Hall, Englewood Cliffs, NJ (1963).

6. Amundson, N.R., and A.J. Pontinen, *Ind. Eng. Chem.*, **50**, 730–736 (1958).

7. Friday, J.R., and B.D. Smith, *AIChE J.*, **10**, 698–707 (1964).

8. Boston, J.F., and S.L. Sullivan, Jr., *Can. J. Chem. Eng.*, **52**, 52–63 (1974).

9. Boston, J.F., and S.L. Sullivan, Jr., *Can. J. Chem. Eng.*, **50**, 663–669 (1972).

10. Johanson, P.J., and J.D. Seader, *Stagewise Computations—Computer Programs for Chemical Engineering Education* (ed. by J. Christensen), Aztec Publishing, Austin, TX, pp. 349–389, A-16 (1972).

11. Lapidus, L., *Digital Computation for Chemical Engineers*, McGraw-Hill, New York, pp. 308–309 (1962).

12. Orbach, O., and C.M. Crowe, *Can. J. Chem. Eng.*, **49**, 509–513 (1971).

13. Scheibel, E.G., *Ind. Eng. Chem.*, **38**, 397–399 (1946).

14. Sujata, A.D., *Hydrocarbon Processing*, **40**(12), 137–140 (1961).

15. Burningham, D.W., and F.D. Otto, *Hydrocarbon Processing*, **46**(10), 163–170 (1967).

16. Shinohara, T., P.J. Johansen, and J.D. Seader, *Stagewise Computations—Computer Programs for Chemical Engineering Education* (ed. by J. Christensen), Aztec Publishing, Austin, TX, pp. 390–428, A-17 (1972).

17. Tsuboka, T., and T. Katayama, *J. Chem. Eng. Japan*, **9**, 40–45 (1976).

18. Hála, E., I. Wichterle, J. Polak, and T. Boublik, *Vapor-Liquid Equilibrium Data at Normal Pressures*, Pergamon, Oxford, p. 308 (1968).

19. Steib, V.H., *J. Prakt. Chem.*, **4**, Reihe, Bd. 28, 252–280 (1965).

20. Cohen, G., and H. Renon, *Can. J. Chem. Eng.*, **48**, 291–296 (1970).

21. Goldstein, R.P., and R.B. Stanfield, *Ind. Eng. Chem., Process Des. Develop.*, **9**, 78–84 (1970).

22. Naphtali, L.M., "The Distillation Column as a Large System," paper presented at the AIChE 56th National Meeting, San Francisco, CA, May 16–19, 1965.

23. Naphtali, L.M., and D.P. Sandholm, *AIChE J.*, **17**, 148–153 (1971).

24. Fredenslund, A., J. Gmehling, and P. Rasmussen, *Vapor-Liquid Equilibria Using UNIFAC, A Group Contribution Method*, Elsevier, Amsterdam (1977).

25. Beveridge, G.S.G., and R.S. Schechter, *Optimization: Theory and Practice*, McGraw-Hill, New York, pp. 180–189 (1970).

26. Block, U., and B. Hegner, *AIChE J.*, **22**, 582–589 (1976).

27. Hofeling, B., and J.D. Seader, *AIChE J.*, **24**, 1131–1134 (1978).

28. Boston, J.F., and S.L. Sullivan, Jr., *Can. J. Chem. Engr.*, **52**, 52–63 (1974).

29. Boston, J.F., and H.I. Britt, *Comput. Chem. Engng.*, **2**, 109–122 (1978).

30. Boston, J.F., *ACS Symp. Ser. No.* **124**, 135–151 (1980).

31. Russell, R.A., *Chem. Eng.*, **90**(20), 53–59 (1983).

32. Trevino-Lozano, R.A., T.P. Kisala, and J.F. Boston, *Comput. Chem. Engng.*, **8**, 105–115 (1984).

33. Jelinek, J., *Comput. Chem. Engng.*, **12**, 195–198 (1988).

34. Venkataraman, S., W.K. Chan, and J.F. Boston, *Chem. Eng. Prog.*, **86** (8), 45–54 (1990).

35. Robinson, C.S., and E.R. Gilliland, *Elements of Fractional Distillation*, 4th ed., McGraw-Hill, New York, pp. 232–236 (1950).

36. Broyden, C.G., *Math Comp.*, **19**, 577–593 (1965).

37. Kister, H. Z., *Distillation Design*, McGraw-Hill, Inc., New York (1992).

STUDY QUESTIONS

10.1. Why are rigorous solution procedures difficult and tedious for multicomponent, multistage separation operations?

10.2. In the equilibrium-stage model, can each stage have a feed, a vapor sidestream, and/or a liquid sidestream? How many independent equations apply to each stage for *C* components?

10.3. In the equilibrium-stage model equations, are *K*-values and enthalpies counted as variables? Are the equations used to compute these properties counted as equations?

10.4. For a cascade of *N* countercurrent equilibrium stages, what is the number of variables, number of equations, and number of degrees of freedom? What are typical specifications, and what are the typical computed (output) variables? Why is it necessary to specify the number of equilibrium stages and the locations of all sidestream withdrawals and heat exchangers?

10.5. Early attempts to solve the MESH equations by hand calculations were the Lewis–Matheson and Thiele–Geddes methods. Why are they not favored for computer calculations?

10.6. What are the four methods most widely used to solve the MESH equations?

10.7. How do equation-tearing and Newton–Raphson procedures differ?

10.8. What is a tridiagonal-matrix (TDM) equation? How is it developed from the MESH equations? In the matrix equation, what are the variables and what are the tear variables? What is a tear

variable? Is there one TDM equation for each component? If so, can each equation be solved independently of the others?

10.9. What is meant by normalization of a set of variables?

10.10. In the BP method, which of the MESH equations is used to compute a new set (i.e., update) of total molar vapor flow rates leaving each stage?

10.11. Does the SR method use tridiagonal-matrix equations? How does the SR method differ from the BP method? For what types of problems is the SR method preferred over the BP method? What are the tear variables in the SR method?

10.12. What limitations of the BP and SR methods are overcome by the NR methods? How do the NR methods differ from the BP and SR methods?

10.13. What is the difference between a tridiagonal-matrix (TDM) equation and a block-tridiagonal-matrix (BTDM) equation? How do the algorithms for solving these two types of equations differ?

10.14. What is a Jacobian matrix? How is the Jacobian formulated?

10.15. What types of calculations consume the most time in the BP, SR, and NR methods? How does the inside-out method reduce this time?

10.16. Would it be expected that for a given problem, the NR and inside-out methods converge to the same result?

EXERCISES

Exercises for this chapter are divided into two groups: (1) those that can be solved manually, and (2) those that are best solved with a process simulator. The first group is referenced to chapter section numbers. The second group of problems follows the first group and is referenced to the type of separator.

Section 10.1

10.1. Independency of the MESH equations.

Show mathematically that (10-6) is not independent of (10-1), (10-3), and (10-4).

10.2. Revision of MESH equations.

Revise the MESH equations to account for entrainment, occlusion, and chemical reaction.

Section 10.2

10.3. Revision of MESH equations.

Revise the MESH equations (10-1) to (10-6) to allow for pump-arounds of the type shown in Figure 10.2 and discussed by Bannon

and Marple [*Chem. Eng. Prog.*, **74**(7), 41–45 (1978)] and Huber [*Hydrocarbon Processing*, **56**(8), 121–125 (1977)].

Combine the equations to obtain modified *M* equations similar to (10-7). Can these equations still be partitioned in a series of *C* tridiagonal-matrix equations?

10.4. The Thomas algorithm.

Use the Thomas algorithm to solve the following matrix equation for x_1, x_2, and x_3.

$$\begin{bmatrix} -160 & 200 & 0 \\ 50 & -350 & 180 \\ 0 & 150 & -230 \end{bmatrix} \cdot \begin{bmatrix} x_1 \\ x_2 \\ x_3 \end{bmatrix} = \begin{bmatrix} 0 \\ -50 \\ 0 \end{bmatrix}$$

10.5. The Thomas algorithm.

Use the Thomas algorithm to solve the following tridiagonal-matrix equation for the **x** vector.

$$\begin{bmatrix} -6 & 3 & 0 & 0 & 0 \\ 3 & -4.5 & 3 & 0 & 0 \\ 0 & 1.5 & -7.5 & 3 & 0 \\ 0 & 0 & 4.5 & -7.5 & 3 \\ 0 & 0 & 0 & 4.5 & -4.5 \end{bmatrix} \cdot \begin{bmatrix} x_1 \\ x_2 \\ x_3 \\ x_4 \\ x_5 \end{bmatrix} = \begin{bmatrix} 0 \\ 0 \\ 100 \\ 0 \\ 0 \end{bmatrix}$$

Section 10.3

10.6. Avoiding subtraction errors wth the TDM.

Wang and Henke [1] state that their method of solving the tridiagonal matrix for the liquid-phase mole fractions does not involve subtraction of nearly equal quantities. Prove this.

10.7. Substituting component flow rates for mole fractions.

Derive an equation similar to (10-7), but with $v_{i,j} = y_{i,j} V_j$ as variables instead of liquid mole fractions. Can the equations still be partitioned into a series of C tridiagonal-matrix equations?

10.8. Memory locations for the BP method.

In a computer program for the Wang–Henke bubble-point method, 10,100 memory locations are wastefully set aside for the four indexed coefficients of the tridiagonal-matrix solution of the component material balances for a 100-stage distillation column:

$$A_j x_{i,j-1} + B_j x_{i,j} + C_j x_{i,j+1} - D_j = 0$$

Determine the minimum number of memory locations required if the calculations are conducted in the most efficient manner.

10.9. Newton–Raphson method.

Solve by the Newton–Raphson method the simultaneous, nonlinear equations

$$x_1^2 + x_2^2 = 17$$
$$(8x_1)^{1/3} + x_2^{1/2} = 4$$

for x_1 and x_2 to within ± 0.001. As initial guesses, assume: (a) $x_1 = 2$, $x_2 = 5$; (b) $x_1 = 4$, $x_2 = 5$; (c) $x_1 = 1$, $x_2 = 1$; (d) $x_1 = 8$, $x_2 = 1$.

10.10. Newton–Raphson method.

Solve by the Newton–Raphson method the simultaneous, nonlinear equations

$$\sin(\pi x_1 x_2) - \frac{x_2}{2} - x_1 = 0$$
$$\exp(2x_1)\left[1 - \frac{1}{4\pi}\right] + \exp(1)\left[\frac{1}{4\pi} - 1 - 2x_1 + x_2\right] = 0$$

for x_1 and x_2 to within ± 0.001. As initial guesses, assume(a) $x_1 = 0.4$, $x_2 = 0.9$; (b) $x_1 = 0.6$, $x_2 = 0.9$; (c) $x_1 = 1.0$, $x_2 = 1.0$.

10.11. First iteration of the BP method.

One thousand kmol/h of a saturated-liquid mixture of 60 mol% methanol, 20 mol% ethanol, and 20 mol% n-propanol is fed to the middle stage of a distillation column having three equilibrium stages, a total condenser, a partial reboiler, and an operating pressure of 1 atm. The distillate rate is 600 kmol/h, and the external reflux rate is 2,000 kmol/h of saturated liquid. Assuming ideal solutions with K-values from vapor pressures and constant-molar overflow such that the vapor rate leaving the reboiler and each stage is 2,600 kmol/h, calculate one iteration of the BP method up to and including a new set of T_j values. To initiate the iteration, assume a linear-temperature profile based on a distillate temperature equal to the normal boiling point of methanol and a bottoms temperature equal to the arithmetic average of the normal boiling points of the other two alcohols.

Section 10.4

10.12. Block-tridiagonal-matrix equation.

Solve the nine simultaneous linear equations below, which have a block-tridiagonal-matrix structure, by the Thomas algorithm:

$$x_2 + 2x_3 + 2x_4 + x_6 = 7$$
$$x_1 + x_3 + x_4 + 3x_5 = 6$$
$$x_1 + x_2 + x_3 + x_5 + x_6 = 6$$
$$x_4 + 2x_5 + x_6 + 2x_7 + 2x_8 + x_9 = 11$$
$$x_4 + x_5 + 2x_6 + 3x_7 + x_9 = 8$$
$$x_5 + x_6 + x_7 + 2x_8 + x_9 = 8$$
$$x_1 + 2x_2 + x_3 + x_4 + x_5 + 2x_6 + 3x_7 + x_8 = 13$$
$$x_2 + 2x_3 + 2x_4 + x_5 + x_6 + x_7 + x_8 + 3x_9 = 14$$
$$x_3 + x_4 + 2x_5 + x_6 + 2x_7 + x_8 + x_9 = 10$$

10.13. Matrix structure for equations ordered by type.

Naphtali and Sandholm group the $N(2C + 1)$ equations by stage. Instead, group the equations by type (i.e., enthalpy balances, component balances, and equilibrium relations). Using a three-component, three-stage example, show whether the resulting matrix structure is still block tridiagonal.

10.14. Derivatives for the NR method.

Derivatives of properties are needed in the Naphtali–Sandholm NR method. For the Chao–Seader correlation, determine analytical derivatives for

$$\frac{\partial K_{i,j}}{\partial T_j}, \frac{\partial K_{i,j}}{\partial v_{j,k}}, \frac{\partial K_{i,j}}{\partial l_{i,k}}$$

10.15. A partial NR method.

A rigorous partial NR method for multicomponent, multistage vapor–liquid separations can be devised that is midway between the complexity of the BP/SR methods and the NR methods. The first major step is to solve the modified M equations for the liquid-phase mole fractions by the TDM algorithm. In the second step, new sets of stage temperatures and total vapor flow rates leaving a stage are computed simultaneously by an NR method. These two steps are repeated until a sum-of-squares criterion is satisfied. For this partial NR method:

(a) Write two indexed equations to simultaneously solve for a new set of T_j and V_j.

(b) Write the truncated Taylor series expansions for the two indexed equations in the T_j and V_j unknowns, and derive complete expressions for all partial derivatives, except that derivatives of physical properties with respect to temperature can be left as such. These derivatives depend on the physical property correlations.

(c) Order the resulting linear equations and the new variables ΔT_j and ΔV_j into a Jacobian matrix that has a rapid and efficient solution.

10.16. Thermally coupled distillation.

Revise (10-58) to (10-60) to allow two interlinked columns of the type shown in Figure 10.31 to be solved simultaneously by the NR method. Is the matrix equation that results from the NR procedure still block tridiagonal?

10.17. Ordering of variables and equations in NR method.

In (10-63), why is the variable order selected as v, T, l? What would be the consequence of changing the order to l, v, T? In (10-64), why is the function order selected as H, M, E? What would be the consequence of changing the order to E, M, H?

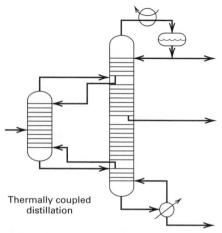

Figure 10.31 Data for Exercise 10.16.

Section 10.5

10.18. Scalar multiplier in the inside-out method.

Suggest in detail a method for determining the scalar multiplier, S_b, in (10-104).

10.19. Error function for the inside-out method.

Suggest in detail an error function, similar to (10-75), that could be used to determine convergence of the inner-loop calculations for the inside-out method.

Distillation Problems

10.20. Rigorous equilibrium-stage calculation for distillation.

Calculate product compositions, stage temperatures, interstage vapor and liquid flow rates and compositions, reboiler duty, and condenser duty for the following column specifications.

Feed (bubble-point liquid at 250 psia and 213.9°F):

Component	lbmol/h
Ethane	3.0
Propane	20.0
n-Butane	37.0
n-Pentane	35.0
n-Hexane	5.0

Column pressure = 250 psia; partial condenser and partial reboiler;

Distillate rate = 23.0 lbmol/h; reflux rate = 150.0 lbmol/hr;

Number of equilibrium stages (exclusive of condenser and reboiler) = 15;

Feed is sent to middle stage.

For this system at 250 psia, K-values and enthalpies may be computed by the Soave–Redlich–Kwong equations.

10.21. Optimal feed-stage location.

Find the optimal feed-stage location for Exercise 10.20.

10.22. Distillation with a vapor sidestream.

Revise Exercise 10.20 so as to withdraw a vapor sidestream at a rate of 37.0 lbmol/h from the fourth stage from the bottom.

10.23. Distillation with intercooler and interheater.

In Exercise 10.20 provide a 200,000 Btu/hr intercooler on the fourth stage from the top and a 300,000 Btu/h interheater on the fourth stage from the bottom.

10.24. Distillation with two feeds.

Using the Peng–Robinson equations for thermodynamic properties, calculate the product compositions, stage temperatures,

interstage vapor and liquid flow rates and compositions, reboiler duty, and condenser duty for the following multiple-feed distillation column, which has 30 equilibrium stages exclusive of a partial condenser and a partial reboiler and operates at 250 psia.

Feeds (both bubble-point liquids at 250 psia):

	Lbmol per Hour	
Component	Feed 1 to stage 15 from the Bottom	Feed 2 to stage 6 from the Bottom
Ethane	1.5	0.5
Propane	24.0	10.0
n-Butane	16.5	22.0
n-Pentane	7.5	14.5
n-Hexane	0.5	3.0

Distillate rate = 36.0 lbmol/hr; Reflux rate = 150.0 lbmol/hr

Determine whether the feed locations are optimal.

10.25. Comparison and distillation calculations.

Use the Chao–Seader or Grayson–Streed correlation for thermodynamic properties to calculate product compositions, stage temperatures, interstage flow rates and compositions, reboiler duty, and condenser duty for the distillation specifications in Figure 10.32. Compare your results with those in the *Chemical Engineers' Handbook*, 8th Edition, pp. 13–35. Why do the two solutions differ?

10.26. Distillation of a light alcohol mixture.

Solve Exercise 10.11 using the UNIFAC method for K-values and obtain the converged solution.

10.27. Distillation with two sidestreams.

Calculate, with the Peng–Robinson equation for properties, the product compositions, stage temperatures, interstage flow rates and compositions, reboiler duty, and condenser duty for the distillation specifications in Figure 10.33, which represents an attempt to obtain four nearly pure products from a single distillation operation. Reflux is a saturated liquid. Why is such a high reflux ratio required?

10.28. Distillation of a hydrocarbon mixture.

Repeat Exercise 10.25, but substitute the following specifications for vapor distillate rate and reflux rate: recovery of nC_4 in distillate = 98% and recovery of iC_5 in bottoms = 98%. If the calculations fail to converge, the number of stages may be less than the minimum value. If so, increase the number of stages, revise the feed location, and repeat until convergence is achieved.

10.29. Distillation with a specified split.

A saturated liquid feed at 125 psia contains 200 lbmol/h of 5 mol % iC_4, 20 mol% nC_4, 35 mol% iC_5, and 40 mol% nC_5. This feed is

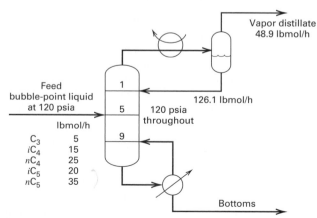

Figure 10.32 Data for Exercise 10.25.

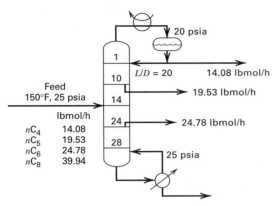

Figure 10.33 Data for Exercise 10.27.

to be distilled at 125 psia in a column equipped with a total condenser and partial reboiler. The distillate is to contain 95% of the nC_4 in the feed, and the bottoms is to contain 95% of the iC_5 in the feed. Use the SRK equation for thermodynamic properties to determine a suitable design. Twice the minimum number of stages, as estimated by the Fenske equation in Chapter 9, should provide a reasonable number of actual equilibrium stages.

10.30. Design of a depropanizer.

A depropanizer distillation column is designed to operate at a feed stage pressure of 315 psia for separating a feed into distillate and bottoms with the following flow rates:

	lbmol/h		
	Feed	Distillate	Bottoms
Methane (C_1)	26	26	
Ethane (C_2)	9	9	
Propane (C_3)	25	24.6	0.4
n-Butane (C_4)	17	0.3	16.7
n-Pentane (C_5)	11		11
n-Hexane (C_6)	12		12
Totals	100	59.9	40.1

The feed is 66 mol% vapor at tower pressure. Steam at 315 psia and cooling water at 65°F are available for the reboiler and condenser. Assume a 2-psi column pressure drop. (a) Should a total condenser be used for this column? (b) What are the feed temperature, K-values, and relative volatilities (with reference to C_3) at the feed temperature and pressure? (c) If the reflux ratio is 1.3 times the minimum reflux, what is the actual reflux ratio? How many theoretical plates are needed in the rectifying and stripping sections? (d) Compute the separation of species. How will the separation differ if a reflux ratio of 1.5, 15 theoretical plates, and feed at the ninth plate are chosen? (e) For part (c), compute the temperature and concentrations on each stage. What is the effect of feed plate location? How will the results differ if a reflux ratio of 1.5 and 15 theoretical plates are used?

10.31. Separation of toluene from biphenyl.

Toluene is to be separated from biphenyl by ordinary distillation. The specifications for the separation are as follows:

	lbmol/h		
	Feed	Distillate	Bottoms
Benzene	3.4		
Toluene	84.6		2.1
Biphenyl	5.1	1.0	

Temperature = 264°F; pressure = 37.1 psia for the feed; reflux ratio = 1.3 times minimum reflux with total condenser; top pressure = 36 psia; bottom pressure = 38.2 psia. (a) Determine the actual reflux ratio and the number of theoretical trays in the rectifying and stripping sections. (b) For a D/F ratio of $(3.4 + 82.5 + 1.0)/93.1$, compute the separation of species. Compare the results to the preceding specifications. (c) If the separation of species computed in part (b) is not sufficiently close to the specified split, adjust the reflux ratio to achieve the specified toluene flow in the bottoms.

10.32. Comparison of two distillation sequences.

A feed at 100°F and 480 psia is to be separated by two ordinary distillation columns into the indicated products.

	lbmol/h			
Species	Feed	Product 1	Product 2	Product 3
H_2	1.5	1.5		
CH_4	19.3	19.2	0.1	
C_6H_6 (benzene)	262.8	1.3	258.1	3.4
C_7H_8 (toluene)	84.7		0.1	84.6
$C_{12}H_{10}$ (biphenyl)	5.1			5.1

Two distillation sequences (see §1.7.3) are to be examined. In the first, CH_4 is the LK in the first column. In the second, toluene is the HK in the first column. Compute the two sequences by estimating the actual reflux ratio and stage requirements for both sequences. Specify a reflux ratio 1.3 times the minimum. Adjust isobaric column pressures to obtain distillate temperatures of about 130°F; however, no column pressure should be less than 20 psia. Specify total condensers, except that a partial condenser should be used when methane is taken overhead.

10.33. Separation of propylene from propane.

A process for the separation of a propylene–propane mixture to produce 99 mol% propylene and 95 mol% propane is shown in Figure 10.34. Because of the high product purities and the low α, 200 stages may be required. A tray efficiency of 100% and tray spacing of 24 inches will necessitate two columns in series, because a single tower would be too tall. Assume a vapor distillate pressure of 280 psia, a pressure drop of 0.1 psi per tray, and a 2-psi drop through the condenser. The stage numbers and reflux ratio shown are only approximate. Determine the necessary reflux ratio for the stage numbers shown. Pay close attention to the determination of the proper feed-stage location so as to avoid pinch or near-pinch conditions wherein several adjacent trays may not be accomplishing any separation.

10.34. Design of stabilizer to remove hydrogen.

So-called stabilizers are distillation columns used in the petroleum industry to perform relatively easy separations between light components and considerably heavier components when one or two single-stage flashes are inadequate. An example of a stabilizer is shown in Figure 10.35 for the separation of H_2, methane, and ethane from benzene, toluene, and xylenes. Such columns can be difficult to calculate because a purity specification for the vapor distillate cannot be readily determined. Instead, it is more likely that the designer will be told to provide a column with 20 to 30 actual trays and a water-cooled partial condenser to provide 100°F reflux at a rate that will provide sufficient boilup at the bottom of the column to meet the purity specification there. It is desired to more accurately design the stabilizer column. The number of theoretical stages shown is just a first approximation and may be varied. A desirable bottoms product has no more than 0.05 mol% methane plus ethane

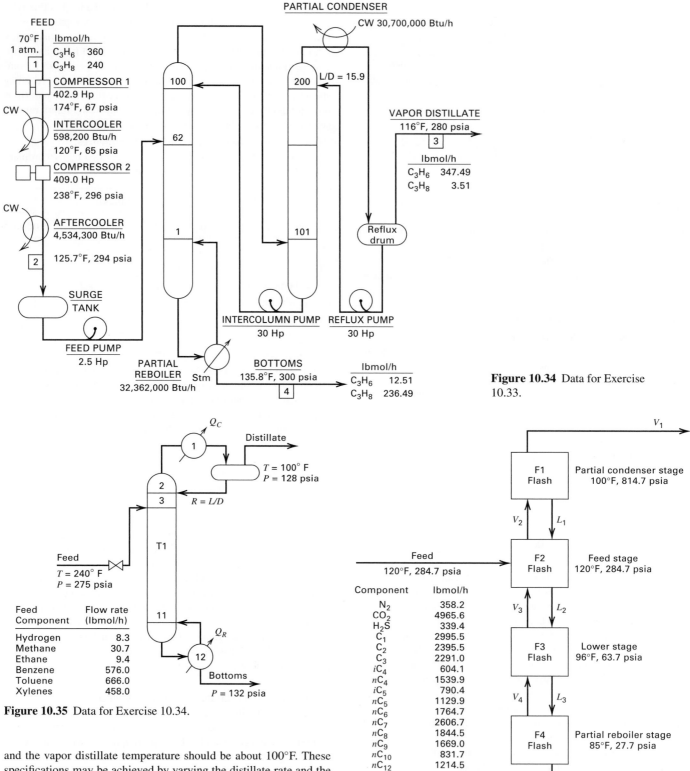

Figure 10.34 Data for Exercise 10.33.

Figure 10.35 Data for Exercise 10.34.

Figure 10.36 Data for Exercise 10.35.

and the vapor distillate temperature should be about 100°F. These specifications may be achieved by varying the distillate rate and the reflux ratio. Reasonable initial estimates for these two quantities are 49.4 lbmol/h and 2, respectively. Assume a tray efficiency of 70%.

10.35. Isothermal distillation.

A multiple recycle-loop problem, formulated by Cavett[2] and shown in Figure 10.36, has been used to test tearing, sequencing, and convergence procedures. The flowsheet is the equivalent of a

four-theoretical-stage, near-isothermal distillation (rather than the conventional near-isobaric type), for which a patent by Gunther[3] exists. The flowsheet does not include necessary mixers, compressors, pumps, valves, or heat exchangers to make it a practical system. For the specifications shown in Figure 10.36, determine the component flow rates for all streams in the process.

[2]R. H. Cavett, *Proc. Am. Petrol. Inst.*, **43**, 57 (1963).

[3]A. Gunther, U.S. Patent 3,575,077 (April 13, 1971).

Absorber and Stripper Problems

10.36. Absorber design.

An absorber is to be designed for a pressure of 75 psia to handle 2,000 lbmol/h of gas at 60°F having the following composition:

Component	Mole Fraction
Methane	0.830
Ethane	0.084
Propane	0.048
n-Butane	0.026
n-Pentane	0.012

The absorbent is an oil, which can be treated as a pure component having a molecular weight of 161. Calculate product rates and compositions, stage temperatures, and interstage vapor and liquid flow rates and compositions for the following conditions:

	Number of Equilibrium Stages	Entering Absorbent Flow Rate lbmol/h	Entering Absorbent Temperature, °F
(a)	6	500	90
(b)	12	500	90
(c)	6	1,000	90
(d)	6	500	60

10.37. Absorption of a hydrocarbon gas.

Calculate product rates and compositions, stage temperatures, and interstage vapor and liquid flow rates and compositions for an absorber having four equilibrium stages with the specifications in Figure 10.37. Assume the oil is nC_{10}.

10.38. An intercooler for an absorber.

In Example 10.4, temperatures of the gas and oil, as they pass through the absorber, increase substantially. This limits the extent of absorption. Repeat the calculations with a heat exchanger that removes 500,000 Btu/h from: (a) stage 2; (b) stage 3; and (c) stage 4. How effective is the intercooler? Which stage is the preferred location for the intercooler? Should the duty of the intercooler be increased or decreased, assuming that the minimum-stage temperature is 100°F using cooling water? The absorber oil is nC_{12}.

10.39. Absorber with two feeds.

Calculate product rates and compositions, stage temperatures, and interstage vapor and liquid flow rates and compositions for the absorber shown in Figure 10.38.

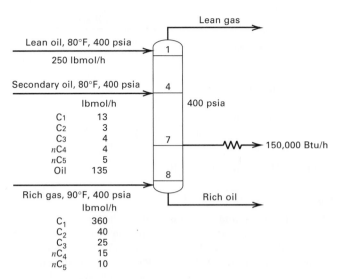

Figure 10.38 Data for Exercise 10.39.

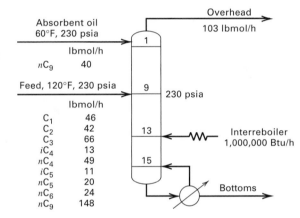

Figure 10.39 Data for Exercise 10.40.

10.40. Reboiled absorber.

Determine product compositions, stage temperatures, interstage flow rates and compositions, and reboiler duty for the reboiled absorber in Figure 10.39. Repeat the calculations without the interreboiler. Is the interreboiler worthwhile? Should an intercooler in the top section of the column be considered?

10.41. Reboiled stripper.

Calculate the product compositions, stage temperatures, interstage flow rates and compositions, and reboiler duty for the reboiled stripper shown in Figure 10.40.

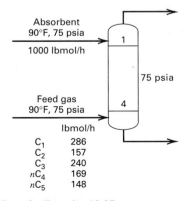

Figure 10.37 Data for Exercise 10.37.

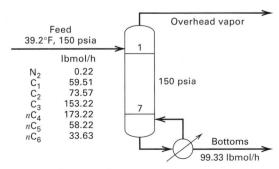

Figure 10.40 Data for Exercise 10.41.

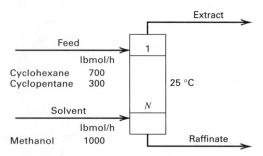

Figure 10.41 Data for Exercise 10.42.

Liquid–Liquid Extraction Problems

10.42.   Liquid–liquid extraction with methanol.

A mixture of cyclohexane and cyclopentane is to be separated by liquid–liquid extraction at 25°C with methanol. Phase equilibria for this system may be predicted by the NRTL or UNIQUAC equations. Calculate product rates and compositions and interstage flow rates and compositions for the conditions in Figure 10.41 with 1, 2, 5, and 10 equilibrium stages.

10.43.   Liquid–liquid extraction of acetic acid with water.

The liquid–liquid extractor in Figure 8.1 operates at 100°F and a nominal pressure of 15 psia. For the feed and solvent flows shown, determine the number of equilibrium stages to extract 99.5% of the acetic acid, using the NRTL equation for activity coefficients. The NRTL constants may be taken as follows, with: 1 = ethyl acetate; 2 = water; and 3 = acetic acid.

I	J	B_{ij}	B_{ji}	α_{ij}
1	2	166.36	1190.1	0.2
1	3	643.30	−702.57	0.2
2	3	−302.63	−1.683	0.2

Compare the computed compositions of the raffinate and extract products to those of Figure 8.1.

Chapter 11

Enhanced Distillation and Supercritical Extraction

§11.0 INSTRUCTIONAL OBJECTIVES

After completing this chapter, you should be able to:

- Explain how enhanced-distillation methods work and how they differ from ordinary distillation.
- Explain how supercritical-fluid extraction differs from liquid–liquid extraction.
- Describe what residue-curve maps and distillation-curve maps represent on triangular diagrams for a ternary.
- Explain how residue-curve maps limit feasible product-composition regions in ordinary and enhanced distillation.
- Calculate, with a simulation program, a separation by extractive distillation.
- Explain how pressure-swing distillation is used to separate a binary azeotropic mixture.
- Calculate, with a simulator and a residue-curve map, a separation by homogeneous azeotropic distillation.
- Calculate, with a process simulator but using a residue-curve map and a bimodal curve, a separation by heterogeneous azeotropic distillation.
- Calculate, with a process simulator, a separation by reactive distillation.
- Explain why enormous changes in properties can occur in the critical region.
- Calculate, with a process simulator, a separation by supercritical-fluid extraction.

When $\alpha < 1.10$, separation by ordinary distillation may be uneconomical, and even impossible if an azeotrope forms. In that event, the following techniques referred to by Stichlmair, Fair, and Bravo [1] as *enhanced distillation*, should be explored:

1. *Extractive Distillation:* Uses large amounts of a relatively high-boiling solvent to alter the liquid-phase activity coefficients (§2.6) so that the α (7-1) of key components becomes more favorable. Solvent enters the column a few trays below the top, and exits from the bottom without forming any azeotropes. If the column feed is an azeotrope, the solvent breaks it. It may also reverse key-component volatilities.

2. *Salt Distillation:* A variation of extractive distillation in which α of the key components is altered by adding to the top reflux a soluble, nonvolatile ionic salt, which stays in the liquid phase as it passes down the column.

3. *Pressure-Swing Distillation:* Separates a mixture that forms a pressure-sensitive azeotrope by utilizing two columns in sequence at different pressures.

4. *Homogeneous Azeotropic Distillation:* A method of separating a mixture by adding an entrainer that forms a homogeneous minimum- or maximum-boiling azeotrope with feed component(s). Where the entrainer is

added depends on whether the azeotrope is removed from the top or the bottom of the column.

5. *Heterogeneous Azeotropic Distillation:* A minimum-boiling heterogeneous azeotrope is formed by the entrainer. The azeotrope splits into two liquid phases in the overhead condenser. One liquid phase is sent back as reflux; the other is sent to another separation step or is a product.

6. *Reactive Distillation:* A chemical that reacts selectively and reversibly with one or more feed constituents is added, and the reaction product is then distilled from the nonreacting components. The reaction is later reversed to recover the separating agent and reacting component. This operation, referred to as *catalytic distillation* if a catalyst is used, is suited to reactions limited by equilibrium constraints, since the product is continuously separated. Reactive distillation also refers to chemical reaction and distillation conducted simultaneously in the same apparatus.

For ordinary multicomponent distillation, determination of feasible distillation sequences, as well as column design and optimization, is relatively straightforward. In contrast, determining and optimizing enhanced-distillation sequences are considerably more difficult. Rigorous calculations frequently fail because of liquid-solution nonidealities and/or the difficulty of specifying feasible separations. To significantly reduce the chances of failure, especially for ternary systems, graphical techniques—described by Partin [2] and developed largely by Doherty and co-workers, and by Stichlmair and co-workers, as referenced later—provide guidance for the feasibility of enhanced-distillation sequences prior to making rigorous column calculations. This chapter presents an introduction to these graphical methods and applies them to enhanced distillation. Doherty and Malone [94], Stichlmair and Fair [95], and Siirola and Barnicki [96] give more detailed treatments.

Also discussed in this chapter is supercritical extraction, which differs considerably from conventional liquid–liquid extraction because of strong nonideal effects, and requires considerable care in the development of an optimal system. The principles and techniques in this chapter are largely restricted to ternary systems; enhanced distillation and supercritical extraction are commonly applied to ternaries because the expense of these operations often requires that a multicomponent mixture first be reduced, by distillation or other means, to a binary or ternary system.

§11.1 USE OF TRIANGULAR GRAPHS

Figure 11.1 shows two isobaric vapor–liquid equilibrium curves for a binary mixture in terms of the mole fractions of the lowest-boiling component (A). All possible equilibrium compositions are located on the diagrams. In Figure 11.1a, compositions of the distillate and bottoms cover the range from pure B to pure A for a *zeotropic* (nonazeotropic) system. Temperatures, although not shown, range from the boiling point of A to the boiling point of B. As the composition changes from pure B to pure A, the temperature decreases.

In Figure 11.1b, a minimum-boiling azeotrope forms at C, dividing the plot into two regions. For Region 1, distillate and bottoms compositions vary from pure B to azeotrope C; in Region 2, they vary only from pure A to azeotrope C. For Region 1, as the composition changes from pure B to azeotrope C, the temperature decreases, as shown in Figure 4.6,

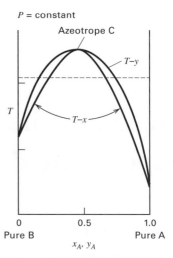

Figure 11.2 Multiple equilibrium solutions for an azeotropic system.

where B is isopropyl alcohol, A is isopropyl ether, and the minimum-boiling azeotrope is 78 mol% isopropyl ether at 66°C and 1 atm. In Region 2, the temperature also decreases as the composition changes from pure A to azeotrope C. A distillation column at 1 atm cannot separate the mixture into two nearly pure products. Depending upon whether the feed composition lies in Region 1 or 2, the column, at best, can produce only a distillate of azeotrope C and a bottoms of either pure B or pure A. However, all equilibrium compositions still lie on the equilibrium curve. From Gibbs phase rule (4-1), with two components and two phases, there are two degrees of freedom. Thus, if the pressure and temperature are fixed, the equilibrium vapor and liquid compositions are fixed. However, as shown in Figure 11.2 for the case of an azeotrope-forming binary, two feasible solutions exist within a certain temperature range. The solution observed depends on the overall composition of the two phases.

In the distillation of a ternary mixture, possible equilibrium compositions do not lie uniquely on a single, isobaric equilibrium curve because the Gibbs phase rule gives an additional degree of freedom. The other compositions are determined only if the temperature, pressure, and composition of one component in one phase are fixed.

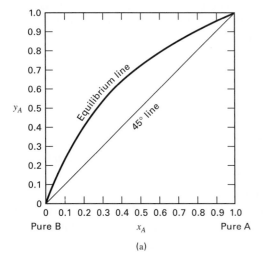

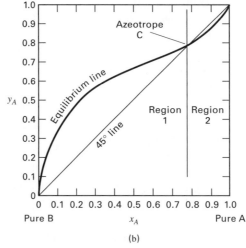

Figure 11.1 Vapor–liquid equilibria for binary systems. (a) Zeotropic system. (b) Azeotropic system.

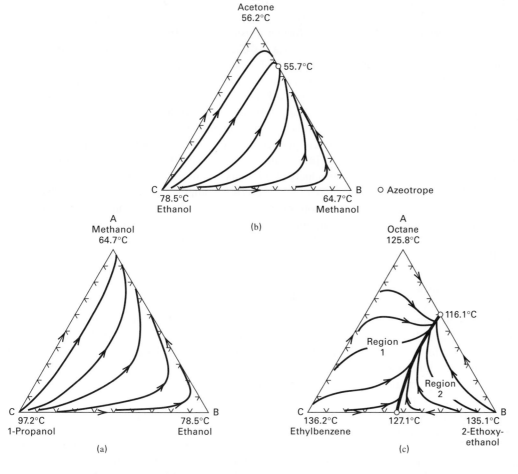

Figure 11.3 Distillation curves for liquid-phase compositions of ternary systems at 1 atm. (a) Mixture not forming an azeotrope. (b) Mixture forming one minimum-boiling azeotrope. (c) Mixture forming two minimum-boiling azeotropes.

§11.1.1   Distillation Regions and Boundaries

From Chapters 4 and 8, the composition of a ternary mixture can be represented on a triangular diagram, either equilateral or right, where the three apexes represent pure components. Although Stichlmair [3] shows that vapor–liquid phase equilibria at a fixed pressure can be plotted by letting the triangular grid represent the liquid phase, with superimposing lines of constant equilibrium-vapor composition for two of the three components, this representation is seldom used. It is more useful, when developing a feasible-separation process for a ternary mixture, to plot only equilibrium-liquid-phase compositions on the triangular diagram. Figure 11.3, where compositions are in mole fractions, shows plots of this type for three different ternary systems. Each curve is the locus of possible equilibrium-liquid-phase compositions during distillation of a mixture, starting from any point on the curve. The boiling points of the three components and their binary and/or ternary azeotropes at 1 atm are included on the diagrams. The zeotropic alcohol system of Figure 11.3a does not form any azeotropes. If a mixture of these three alcohols is distilled, there is only one *distillation region*, similar to the binary system of Figure 11.1a. Accordingly, the distillate can be nearly pure methanol (A), or the bottoms can be nearly pure 1-propanol (C). However, nearly pure ethanol (B), the intermediate-boiling component, cannot be produced as a distillate or bottoms. To separate this ternary mixture into the three components, a sequence of two columns is used, as

shown in Figure 11.4, where the feed, distillate, and bottoms product compositions must lie on a straight, total-material-balance line within the triangular diagram. In the so-called *direct sequence* of Figure 11.4a, the feed, F, is first separated into distillate A and a bottoms of B and C; then B is separated

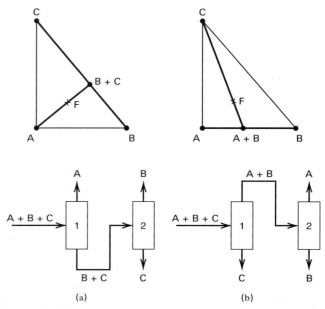

Figure 11.4 Distillation sequences for ternary zeotropic mixtures. (a) Direct sequence. (b) Indirect sequence.

from C in the second column. In the *indirect sequence* of Figure 11.4b, a distillate of A and B and a bottoms of C are produced in the first column, followed by the separation of A from B in the second column.

When a ternary mixture forms an azeotrope, the products from a single distillation column depend on the feed composition, as for a binary mixture. However, unlike the case of the binary mixture, where two distillation regions, shown in Figure 11.1b, are well defined, the determination of distillation regions for azeotrope-forming ternary mixtures is complex. Consider first the example of Figure 11.3b, for a mixture of acetone (A), methanol (B), and ethanol (C), which are in the order of increasing boiling point. The only azeotrope formed at 1 atm is a minimum-boiling binary azeotrope, at 55.7°C, of the two lower-boiling components, acetone and methanol. The azeotrope contains 78.4 mol% acetone. For this type of system, as will be shown later, no distillation boundaries for the ternary mixture exist, even though an azeotrope is present. A feed composition located within the triangular diagram can be separated into two binary products, consistent with the total-material-balance line. Ternary distillate or bottoms products can be avoided if the column split is properly selected. For example, the following five feed compositions can all produce, at a high reflux ratio and a large number of stages, a distillate of the minimum-boiling azeotrope of acetone and methanol, and a bottoms product of methanol and ethanol. That is, little or no ethanol will be in the distillate and little or no acetone will be in the bottoms.

Case	Feed		Distillate		Bottoms	
	$x_{acetone}$	$x_{methanol}$	$x_{acetone}$	$x_{methanol}$	$x_{acetone}$	$x_{methanol}$
1	0.1667	0.1667	0.7842	0.2158	0.0000	0.1534
2	0.1250	0.3750	0.7837	0.2163	0.0000	0.4051
3	0.2500	0.2500	0.7837	0.2163	0.0000	0.2658
4	0.3750	0.1250	0.7837	0.2163	0.0000	0.0412
5	0.3333	0.3333	0.7837	0.2163	0.0000	0.4200

Alternatively, the column split can be a bottoms of nearly pure ethanol and a distillate of acetone and methanol. For either split, the straight, total-material-balance line passing through the feed point can extend to the sides of the triangle.

The more complex case of the ternary mixture of *n*-octane (A), 2-ethoxyethanol (B), and ethylbenzene (C) is presented in Figure 11.3c. A and B form a minimum-boiling binary azeotrope at 116.1°C, and B and C do the same at 127.1°C. A triangular diagram for this system is separated by a *distillation boundary* (shown as a bold curved line) into Regions 1 and 2. A material-balance line connecting the feed to the distillate and bottoms cannot cross this distillation boundary, thus restricting the possible distillation products. For example, a mixture with a feed composition inside Region 2 cannot produce a bottoms of ethylbenzene, the highest-boiling component in the mixture. It can be distilled to produce a distillate of the A–B azeotrope and a bottoms of a mixture of B and C, or a bottoms of B and a distillate of all three components. If the feed lies in Region 1 of Figure 11.3c, it is possible to produce the A–B azeotrope and a bottoms of a mixture

of A and C, or a bottoms of C and a distillate of an A and B mixture. Thus, each region produces unique products.

To further illustrate the restriction in product compositions caused by a distillation boundary, consider a feed mixture of 15 mol% A, 70 mol% B, and 15 mol% C. For this composition in Figure 11.3a or b, a bottoms product of nearly pure C, the highest-boiling component, is obtained with a distillate-to-bottoms ratio of 85/15. If, however, the mixture is that in Figure 11.3c, the same feed split ratio results in a bottoms of nearly pure B, the second-highest-boiling component.

In conclusion, when distillation boundaries are present, products of a ternary mixture cannot be predicted from component and azeotrope compositions and a specified distillate-to-bottoms ratio. These distillation boundaries, as well as the mappings of distillation curves in the ternary plots of Figure 11.3, can be determined by two methods described in §11.1.2 and §11.1.4.

§11.1.2 Residue-Curve Maps

Consider the simple batch distillation (no trays, packing, or reflux) shown schematically in Figure 13.1. For any ternary-mixture component, a material balance for its vaporization from the still, assuming that the liquid is perfectly mixed and at its bubble point, is given by (13-1), which can be written as

$$\frac{dx_i}{dt} = (y_i - x_i)\frac{dW}{Wdt} \tag{11-1}$$

where $x_i =$ mole fraction of component i in W moles of a perfectly mixed liquid residue in the still, and $y_i =$ mole fraction of component i in the vapor leaving the still (instantaneous distillate) in equilibrium with x_i.

Because W decreases with time, t, it is possible to combine W and t into a single variable. Following Doherty and Perkins [4], let this variable be ξ, such that

$$\frac{dx_i}{d\xi} = x_i - y_i \tag{11-2}$$

Combining (11-1) and (11-2) to eliminate $dx_i/(x_i - y_i)$:

$$\frac{d\xi}{dt} = -\frac{1}{W}\frac{dW}{dt} \tag{11-3}$$

Let the initial condition be $\xi = 0$ and $W = W_0$ at $t = 0$. Then the solution to (11-3) for ξ at time t is

$$\xi\{t\} = \ln[W_0/W\{t\}] \tag{11-4}$$

Because $W\{t\}$ decreases monotonically with time, $\xi\{t\}$ must increase monotonically with time and is considered a dimensionless, warped time. Thus, for the ternary mixture, the distillation process can be modeled by the following set of differential-algebraic equations (DAEs), assuming that a second liquid phase does not form:

$$\frac{dx_i}{d\xi} = x_i - y_i, \quad i = 1, 2 \tag{11-5}$$

$$\sum_{i=1}^{3} x_i = 1 \tag{11-6}$$

$$y_i = K_i x_i, \quad i = 1, 2, 3 \tag{11-7}$$

and the bubble-point-temperature equation:

$$\sum_{i=1}^{3} K_i x_i = 1 \qquad (11\text{-}8)$$

where, in the general case, $K_i = K_i\{T, P, \boldsymbol{x}, \boldsymbol{y}\}$.

Thus, the system consists of seven equations in nine variables: $P, T, x_1, x_2, x_3, y_1, y_2, y_3,$ and ξ. With the pressure fixed, the next seven variables can be computed from (11-5) to (11-8) as a function of the ninth variable, ξ, from a specified initial condition. The calculations can proceed in the forward or backward direction of ξ. The results, when plotted on a triangular graph, are *residue curves* because the plot follows, with time, the liquid-residue composition in the still. A collection of residue curves, at a fixed pressure, is a *residue-curve map*. A simple, but inefficient, procedure for calculating a residue curve is illustrated in Example 11.1. Better, but more elaborate, procedures are given by Doherty and Perkins [4] and Bossen, Jørgensen, and Gani [5]. The last procedure is also applicable when two separate liquid phases form, as is a procedure by Pham and Doherty [6].

EXAMPLE 11.1 Residue-Curve Calculation.

Plot a portion of a residue curve for *n*-propanol (1), isopropanol (2), and benzene (3) at 1 atm, starting from a bubble-point liquid with 20 mol% each of 1 and 2, and 60 mol% of component 3. For *K*-values, use Raoult's law (Table 2.3) with regular-solution theory (2-64) for estimating the liquid-phase activity coefficients. The normal boiling points of the three components in °C are 97.3, 82.3, and 80.1, respectively. Minimum-boiling azeotropes are formed at 77.1°C for components 1, 3 and at 71.7°C for 2, 3.

Solution

A bubble-point calculation, using (11-7) and (11-8), gives starting values of $\boldsymbol{y}$ of 0.1437, 0.2154, and 0.6409, respectively, and a value of 79.07°C for the starting temperature, from the ChemSep program of Taylor and Kooijman [7].

For an increment in dimensionless time, ξ, the differential equations (11-5) can be solved for x_1 and x_2 using Euler's method with a spreadsheet. Then x_3 is obtained from (11-6). The corresponding values of $\boldsymbol{y}$ and T are from (11-7) and (11-8). This procedure is repeated for the next increment in ξ. Thus,

from (11-5) for component 1:

$$x_1^{(1)} = x_1^{(0)} + \left(x_1^{(0)} - y_1^{(0)} \right)\Delta\xi$$
$$= 0.2000 + (0.2000 - 0.1437)0.1 = 0.2056$$

where superscripts (0) indicate starting values and superscript (1) indicates the value after the first increment in ξ. The value of 0.1 for $\Delta\xi$ gives reasonable accuracy, since the change in x_1 is only 2.7%. Similarly:

$$x_2^{(1)} = 0.2000 + (0.2000 - 0.2154)0.1 = 0.1985$$

From (11-6):

$$x_3^{(1)} = 1 - x_1^{(1)} - x_2^{(1)} = 1 - 0.2056 - 0.1985 = 0.5959$$

From a bubble-point calculation using (11-7) and (11-8),

$$y^{(1)} = [0.1474, 0.2134, 0.6392]^T \quad \text{and} \quad T^{(1)} = 79.14°C$$

The calculations are continued in the forward direction of ξ only to $\xi = 1.0$, and in the backward direction only to $\xi = -1.0$. The results are in the table below, and that portion of the partial residue curve is plotted in Figure 11.5a. The complete residue-curve map for this system, from Doherty [8], is given on a right-triangle diagram in Figure 11.5b.

ξ	x_1	x_2	y_1	y_2	$T,°C$
−1.0	0.1515	0.2173	0.1112	0.2367	78.67
−0.9	0.1557	0.2154	0.1141	0.2344	78.71
−0.8	0.1600	0.2135	0.1171	0.2322	78.75
−0.7	0.1644	0.2117	0.1201	0.2300	78.79
−0.6	0.1690	0.2099	0.1232	0.2278	78.83
−0.5	0.1737	0.2081	0.1264	0.2256	78.87
−0.4	0.1786	0.2064	0.1297	0.2235	78.91
−0.3	0.1837	0.2047	0.1331	0.2214	78.95
−0.2	0.1889	0.2031	0.1365	0.2194	79.00
−0.1	0.1944	0.2015	0.1401	0.2173	79.05
0.0	0.2000	0.2000	0.1437	0.2154	79.07
0.1	0.2056	0.1985	0.1474	0.2134	79.14
0.2	0.2115	0.1970	0.1512	0.2115	79.19
0.3	0.2175	0.1955	0.1550	0.2095	79.24
0.4	0.2237	0.1941	0.1589	0.2076	79.30
0.5	0.2302	0.1928	0.1629	0.2058	79.34
0.6	0.2369	0.1915	0.1671	0.2041	79.41
0.7	0.2439	0.1902	0.1714	0.2023	79.48
0.8	0.2512	0.1890	0.1758	0.2006	79.54
0.9	0.2587	0.1878	0.1804	0.1989	79.61
1.0	0.2665	0.1867	0.1850	0.1973	79.68

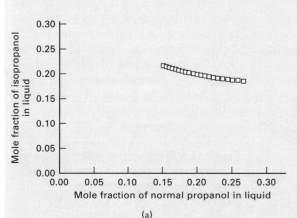

(a)

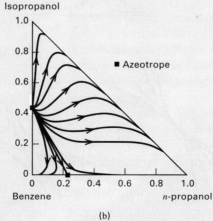

(b)

Figure 11.5 Residue curves for the normal propanol–isopropanol–benzene system at 1 atm for Example 11.1. (a) Calculated partial residue curve. (b) Residue-curve map.

The residue-curve map in Figure 11.5b shows an arrow on each residue curve. The arrows point from a lower-boiling component or azeotrope to a higher-boiling component or azeotrope. In Figure 11.5b, all residue curves originate from the isopropanol–benzene azeotrope (lowest boiling point, 71.7°C). One of the curves terminates at the other azeotrope (*n*-propanol–benzene, which has a higher boiling point, 77.1°C) and is a special residue curve, called a *simple distillation boundary* because it divides the ternary region into two separate regions. All residue curves lying above and to the right of this distillation boundary terminate at the *n*-propanol apex, which has the highest boiling point (97.3°C) for that region. All residue curves lying below and to the left of the distillation boundary are deflected to the benzene apex, whose boiling point of 80.1°C is the highest for this second region.

On a triangular diagram, all pure-component vertices and azeotropic points—whether binary azeotropes on the borders of the triangle, as in Figure 11.5b, or a ternary azeotrope within the triangle—are *singular* or *fixed points* of the residue curves because at these points, $d\boldsymbol{x}/d\xi = 0$. In the vicinity

of these points, the behavior of a residue curve depends on the two eigenvalues of (11-5). At each pure-component vertex, the two eigenvalues are identical. At each azeotropic point, the two eigenvalues are different. Three cases, illustrated by each of three pattern groups in Figure 11.6, are possible:

Case 1: Both eigenvalues are negative. This is the point reached as ξ tends to ∞, and is where all residue curves in a given region terminate. Thus, it is the component or azeotrope with the highest boiling point in the region. This point is a *stable node* because it is like the low point of a valley, in which a rolling ball finds a stable position. In Figure 11.6b, the stable node is pure *n*-propanol.

Case 2: Both eigenvalues are positive. This is the point where all residue curves in a region originate, and is the component or azeotrope with the lowest boiling point in the region. This point is an *unstable node* because it is

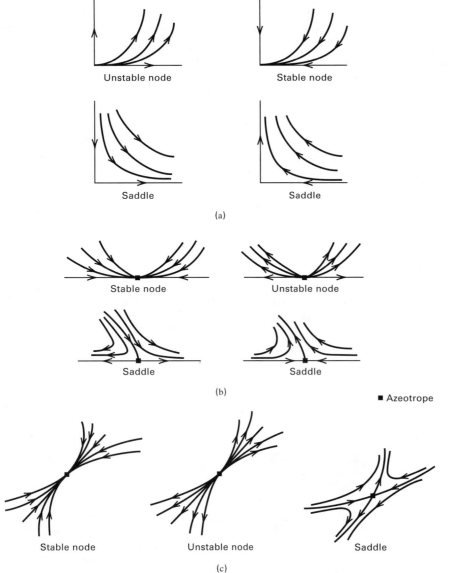

Figure 11.6 Residue-curve patterns: (a) near-pure-component vertices; (b) near-binary azeotropes; (c) near-ternary azeotropes.

[From M.F. Doherty and G.A. Caldarola, *IEC Fundam.*, **24**, 477 (1985) with permission.]

like the top of a mountain from which a ball rolls toward a stable position. In Figure 11.6b, the unstable node is the isopropanol–benzene azeotrope.

Case 3: One eigenvalue is positive and one is negative. Residue curves within the triangle move toward and then away from such *saddle points*. For a given region, all pure components and azeotropes intermediate in boiling point between the stable node and the unstable node are saddles. In Figure 11.5b, the upper region has one saddle at the isopropanol vertex and another saddle at the *n*-propanol–benzene azeotrope.

§11.1.3 Approximate Residue-Curve Maps

From Example 11.1, it is clear that calculation of a residue-curve map requires a considerable effort. However, process simulators such as ASPEN PLUS [9] and CHEMCAD compute residue maps. Alternatively, as developed by Doherty and Perkins [10] and Doherty [8], the classification of singular points as stable nodes, unstable nodes, and saddles provides a rapid method for approximating a residue-curve map, including approximate distillation boundaries, from just the pure-component boiling points and azeotrope boiling points and compositions. Boiling points of pure substances are available in handbooks and databases, and extensive listings of binary azeotropes are found in Horsley [11] and Gmehling et al. [12]. The former lists more than 1,000 binary azeotropes. The latter includes experimental data for more than 20,000 systems involving approximately 2,000 compounds, as well as material on selecting enhanced-distillation systems. The listings of ternary azeotropes are incomplete; however, in lieu of experimental data, a homotopy-continuation method for estimating homogeneous azeotropes of a multi-component mixture from a thermodynamic model (e.g., Wilson, NRTL, UNIQUAC, UNIFAC) has been developed by Fidkowski, Malone, and Doherty [13]. Eckert and Kubicek [97] present an extension for computing heterogeneous azeotropes.

Based on experimental evidence for ternary mixtures, with very few exceptions there are at most three binary azeotropes and one ternary azeotrope. Accordingly, the following set of restrictions applies to a ternary system:

$$N_1 + S_1 = 3 \tag{11-9}$$

$$N_2 + S_2 = B \leq 3 \tag{11-10}$$

$$N_3 + S_3 = 1 \text{ or } 0 \tag{11-11}$$

where N is the number of stable and unstable nodes, S is the number of saddles, B is the number of binary azeotropes, and the subscript is the number of components at the node (stable or unstable) or saddle. Thus, S_2 is the number of binary azeotrope saddles. Doherty and Perkins [10] give a topological relationship among N and S:

$$2N_3 - 2S_3 + 2N_2 - B + N_1 = 2 \tag{11-12}$$

For Figure 11.5b, where there is no ternary azeotrope, $N_1 = 2$, $N_2 = 1$, $N_3 = 0$, $S_1 = 1$, $S_2 = 1$, $S_3 = 0$, and $B = 2$.

Applying (11-12) gives $0 - 0 + 2 - 2 + 2 = 2$. Equation (11-9) gives $2 + 1 = 3$; (11-10) gives $1 + 1 = 2$; and (11-11) gives $0 + 0 = 0$. Thus, all four relations are satisfied.

The topological relationships are useful for rapidly sketching, on a ternary diagram, an approximate residue-curve map, including distillation boundaries, as described in detail by Foucher, Doherty, and Malone [14]. Their procedure involves the following nine steps, which are partly illustrated by an example from their article and are shown in Figure 11.7. The procedure is summarized in Figure 11.8. Approximate maps are usually developed from data at 1 atm.

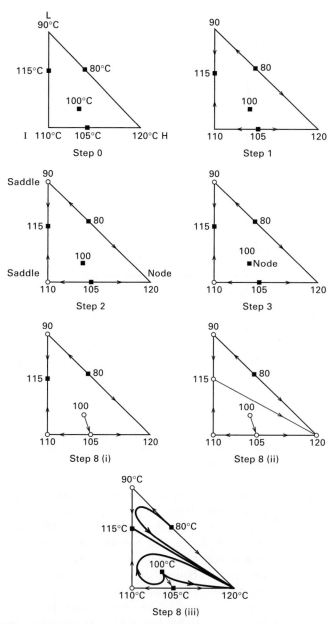

Figure 11.7 Step-by-step development of an approximate residue-curve map for a hypothetical system with two minimum-boiling binary azeotropes, one maximum-boiling binary azeotrope, and one ternary azeotrope.

[From E.R. Foucher, M.F. Doherty, and M.F. Malone, *IEC Res.*, **30**, 764 (1991) with permission.]

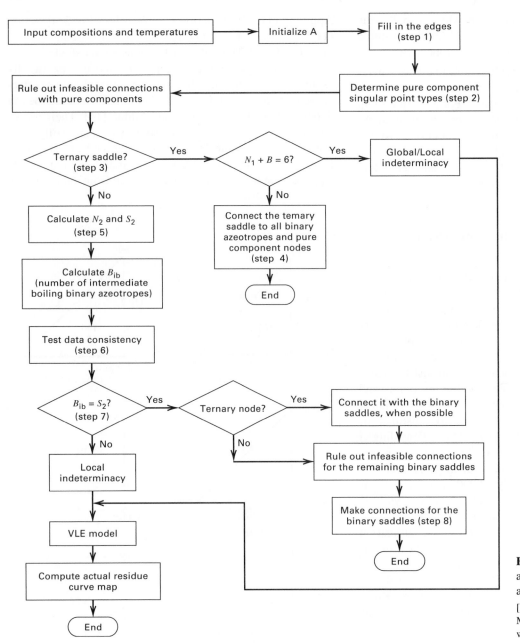

Figure 11.8 Flowchart of algorithm for sketching an approximate residue-curve map.

[From E.R. Foucher, M.F. Doherty, and M.F. Malone, *IEC Res.*, **30**, 763 (1991) with permission.]

In the description, the term *species* refers to both pure components and azeotropes.

Step 0 Label the ternary diagram with the pure-component, normal-boiling-point temperatures. It is preferable to designate the top vertex of the triangle as the low boiler (L), the bottom-right vertex as the high boiler (H), and the bottom-left vertex as the intermediate boiler (I). Plot composition points for the binary and ternary azeotropes and add labels for their normal boiling points. This determines the value of B. See Figure 11.7, Step 0, where two minimum-boiling and one maximum-boiling binary azeotropes and one ternary azeotrope are designated by filled square markers. Thus, $B = 3$.

Step 1 Draw arrows on the edges of the triangle, in the direction of increasing temperature, for each pair of adjacent species. See Figure 11.7, Step 1, where six species are on the edges of the triangle and six arrows have been added.

Step 2 Determine the type of singular point for each pure-component vertex by using Figure 11.6 with the arrows drawn in Step 1 of Figure 11.7. This determines the values for N_1 and S_1. If a ternary azeotrope exists, go to Step 3; if not, go to Step 5. In Figure 11.7, Step 2, L is a saddle because one arrow points toward L and one points away from L; H is a stable node because both arrows point toward H, and I is a saddle. Therefore, $N_1 = 1$ and $S_1 = 2$.

Step 3 **(for a ternary azeotrope):** Determine the type of singular point for the ternary azeotrope, if one exists. The point is a node if (a) $N_1 + B < 4$, and/or (b) excluding the pure-component saddles, the ternary azeotrope has the highest, second-highest, lowest, or second-lowest boiling point of all species. Otherwise, the point is a saddle. This determines the values for N_3 and S_3. If the point is a node, go to Step 5; if a saddle, go to Step 4. In Figure 11.7, Step 3, $N_1 + B = 1 + 3 = 4$.

However, excluding L and I because they are saddles, the ternary azeotrope has the second-lowest boiling point. Therefore, the point is a node, and $N_3 = 1$ and $S_3 = 0$. The type of node, stable or unstable, is still to be determined.

Step 4 **(for a ternary saddle):** Connect the ternary saddle, by straight lines, to all binary azeotropes and to all pure-component nodes (but not to pure-component saddles), and draw arrows on the lines to indicate the direction of increasing temperature. Determine the type of singular point for each binary azeotrope, by using Figure 11.6 with the arrows drawn in this step. This determines the values for N_2 and S_2. These values should be consistent with (11-10) and (11-12). This completes the development of the approximate residue-curve map, with no further steps needed. However, if $N_1 + B = 6$, then special checks must be made, as given in detail by Foucher, Doherty, and Malone [14]. This step does not apply to the example in Figure 11.7, because the ternary azeotrope is not a saddle.

Step 5 **(for a ternary node or no ternary azeotrope):** Determine the number of binary nodes, N_2, and binary saddles, S_2, from (11-10) and (11-12), where (11-12) can be solved for N_2 to give

$$N_2 = (2 - 2N_3 + 2S_3 + B - N_1)/2 \quad (11\text{-}13)$$

For the example of Figure 11.7, $N_2 = (2 - 2 + 0 + 3 - 1)/2 = 1$. From (11-10), $S_2 = 3 - 1 = 2$.

Step 6 Count the binary azeotropes that are intermediate boilers (i.e., that are not the highest- or the lowest-boiling species), and call that number B_{ib}. Make the following two data-consistency checks: (a) The number of binary azeotropes, B, less B_{ib}, must equal N_2, and (b) S_2 must be $\leq B_{ib}$. For the system in Figure 11.7, both checks are satisfied because $B_{ib} = 2$, $B - B_{ib} = 1$, $N_2 = 1$, and $S_2 = 2$. If these two consistency checks are not satisfied, one or more of the boiling points may be in error.

Step 7 If $S_2 \neq B_{ib}$, this procedure cannot determine a unique residue-curve-map structure, which therefore must be computed from (11-5) to (11-8). If $S_2 = B_{ib}$, there is a unique structure, which is completed in Step 8. For the example in Figure 11.7, $S_2 = B_{ib} = 2$; therefore, there is a unique map.

Step 8 In this final step for a ternary node or no ternary azeotrope, the distillation boundaries (connections), if any, are determined and entered on the triangular diagram as straight lines, and, if desired, one or more representative residue curves are sketched as curved lines within each distillation region. This step applies to cases of $S_3 = 0$, $N_3 = 0$ or 1, and $S_2 = B_{ib}$. In all cases, the number of

distillation boundaries equals the number of binary saddles, S_2. Each binary saddle must be connected to a node (pure component, binary, or ternary). A ternary node must be connected to at least one binary saddle. Thus, a pure-component node cannot be connected to a ternary node, and an unstable node cannot be connected to a stable node. The connections are made by determining a connection for each binary saddle such that (a) a minimum-boiling binary saddle connects to an unstable node that boils at a lower temperature and (b) a maximum-boiling binary saddle connects to a stable node that boils at a higher temperature.

It is best to first consider connections with the ternary node and then examine possible connections for the remaining binary saddles. In the example of Figure 11.7, $S_2 = 2$, with these saddles denoted as L-I, a maximum-boiling azeotrope at 115°C, and as I-H, a minimum-boiling azeotrope at 105°C. Therefore, two connections are made to establish two distillation boundaries. The ternary node at 100°C cannot connect to L-I because 100°C is not greater than 115°C. The ternary node can, however, connect, as shown in Step 8 (i), to I-H because 100°C is lower than 105°C. This marks the ternary node as unstable. The connection for L-I can only be to H, as shown in Step 8 (ii), because it is a node (stable), and 120°C is greater than 115°C. This completes the connections. Finally, as shown in Step 8 (iii) of Figure 11.7, three typical, but approximate, residue curves are added to the diagram. These curves originate from unstable nodes and terminate at stable nodes.

Residue-curve maps are used to determine feasible distillation sequences for nonideal ternary systems. Matsuyama and Nishimura [15] showed that the topological constraints just discussed limit the number of possible maps to about 113. However, Siirola and Barnicki [96] show 12 additional maps; all 125 maps are called distillation region diagrams (DRD). Doherty and Caldarola [16] provide sketches of 87 maps that contain at least one minimum-boiling binary azeotrope and also cover industrial applications, since minimum-boiling azeotropes are much more common than maximum-boiling azeotropes.

§11.1.4 Distillation-Curve Maps

A residue curve represents the changes in residue composition with time as the result of a simple, one-stage batch distillation. The curve points in the direction of increasing time, from a lower-boiling state to a higher-boiling state. An alternative representation for distillation on a ternary diagram is a *distillation curve* for continuous, rather than batch, distillation. The curve is most readily obtained for total reflux (see §9.1.3, the Fenske method) at a constant pressure, usually 1 atm. The calculations are made down or up the column, starting from any composition. Consider making the calculations by moving up the column, starting from a stage designated as Stage 1. Between equilibrium stages j and $j+1$, at total reflux, passing vapor and liquid streams have the same composition. Thus,

$$x_{i,j+1} = y_{i,j} \qquad (11\text{-}14)$$

Also, liquid and vapor streams leaving the same stage are in equilibrium.

$$y_{i,j} = K_{i,j} x_{i,j} \qquad (11\text{-}15)$$

To calculate a distillation curve, an initial liquid-phase composition, $x_{i,1}$, is assumed. This liquid is at its bubble-point temperature, which is determined from (11-8), which also gives the equilibrium-vapor composition, $y_{i,1}$ in agreement with (11-15). The composition, $x_{i,2}$, of the passing liquid stream is equal to $y_{i,1}$ by (11-14). The process is then repeated to obtain $x_{i,3}$, then $x_{i,4}$, and so forth. The sequence of liquid-phase compositions, which corresponds to the operating line for the total-reflux condition, is plotted on the triangular diagram. The distillation curve is analogous to the 45° line on a McCabe–Thiele diagram (§7.2). The calculation of a portion of a distillation curve is now illustrated.

EXAMPLE 11.2 Calculation of a Distillation Curve.

Calculate and plot a portion of a distillation curve for the same starting conditions as in Example 11.1.

Solution

The starting values, $x^{(1)}$, are 0.2000, 0.2000, and 0.6000 for components 1, 2, and 3, respectively. From Example 11.1, the bubble-point calculation gives a temperature of 79.07°C and $y^{(1)}$ values of 0.1437, 0.2154, and 0.6409. From (11-14), values of $x^{(2)}$ are 0.1437, 0.2154, and 0.6409. A bubble-point calculation for this composition gives $T^{(2)} = 78.62°C$ and $y^{(2)} = 0.1063$, 0.2360, and 0.6577. Subsequent calculations are summarized in the following table:

Equilibrium Stage	x_1	x_2	y_1	y_2	T, °C
1	0.2000	0.2000	0.1437	0.2154	79.07
2	0.1437	0.2154	0.1063	0.2360	78.62
3	0.1063	0.2360	0.0794	0.2597	78.29
4	0.0794	0.2597	0.0592	0.2846	78.02
5	0.0592	0.2846	0.0437	0.3091	77.80

Figure 11.9 is the resulting distillation curve, where points represent equilibrium stages and are connected by straight lines.

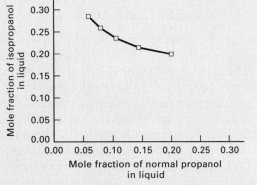

Figure 11.9 Calculated distillation curve for the normal propanol–isopropanol–benzene system at 1 atm for Example 11.2.

Distillation curves can be computed more rapidly than residue curves, and closely approximate them for reasons noted by Fidkowski, Doherty, and Malone [17]. If (11-5) (which must be solved numerically as in Example 11.1) is written in a forward-finite-difference form,

$$(x_{i,j+1} - x_{i,j})/\Delta\xi = x_{i,j} - y_{i,j} \qquad (11\text{-}16)$$

In Example 11.1, $\Delta\xi$ was set to +0.1 for calculations that give increasing values of T and to –0.1 to give decreasing values. If the latter direction is chosen to be consistent with the direction used in Example 11.2 and $\Delta\xi$ is set equal to –1.0, (11-16) becomes identical to (11-14). Thus, residue curves (which are true continuous curves) are equal to distillation curves (which are discrete points through which a smooth curve is drawn), when the residue curves are approximated by a crude forward-finite-difference formulation, using $\Delta\xi = -1.0$.

A collection of distillation curves, including lines for distillation boundaries, is a *distillation-curve map*, an example of which, from Fidkowski et al. [17], is given in Figure 11.10. The Wilson equation was used to compute liquid-phase activity coefficients. The dashed lines are the distillation curves, which approximate the solid-line residue curves. This system has two minimum-boiling binary azeotropes, one maximum-boiling binary azeotrope, and a ternary saddle azeotrope. The map shows four distillation boundaries, designated by A, B, C, and D. These computed boundaries, which define four distillation regions (1 to 4), are all curved lines rather than the approximately straight lines in the sketches of Figure 11.7.

Distillation-curve maps have been used by Stichlmair and associates [1, 3, 18] for the development of feasible-distillation sequences. In their maps, arrows are directed toward the lower-boiling species, rather than toward the higher-boiling species as in residue-curve maps.

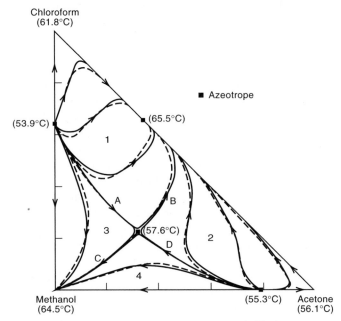

Figure 11.10 Comparison of residue curves to distillation curves.

[From Z.T. Fidkowski, M.F. Malone, and M.F. Doherty, *AIChE J.*, **39**, 1303 (1993) with permission.]

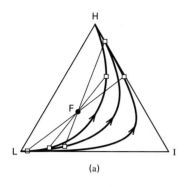

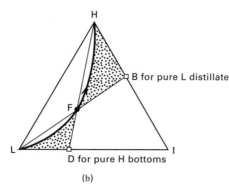

Figure 11.11 Product-composition regions for a zeotropic system. (a) Material-balance lines and distillation curves. (b) Product-composition regions shown shaded.

[From S. Widagdo and W.D. Seider, *AIChE J.*, **42**, 96–130 (1996) with permission.]

§11.1.5   Feasible Product-Composition Regions at Total Reflux (Bow-Tie Regions)

The feasible-distillation regions for azeotrope-forming ternary mixtures are not obvious. Fortunately, residue-curve maps and distillation-curve maps can be used to make preliminary estimates of regions of *feasible-product compositions* for nonideal ternary systems. These regions are determined by superimposing a column material-balance line on either type of curve-map diagram. Consider first the simpler zeotropic ternary system in Figure 11.11a, which shows an isobaric residue-curve map with three residue curves. Assume this map is identical to a corresponding distillation-curve map for total-reflux conditions and to a map for a finite, but very high reflux, ratio. Suppose ternary feed F in Figure 11.11a is continuously distilled isobarically, at a high R, to produce distillate D and bottoms B. A straight line that connects distillate and bottoms compositions must pass through the feed composition at some intermediate point to satisfy material-balance equations. Three *material-balance lines* are included in the figure. For a given line, D and B composition points, designated by open squares, lie on the same distillation curve. This causes the material-balance line to intersect the distillation curve at these two points and be a chord to the distillation curve.

The limiting distillate-composition point for this zeotropic system is pure low-boiling component, L. From the material-balance line passing through F, as shown in Figure 11.11b, the corresponding bottoms composition with the least amount of component L is point B. At the other extreme, the limiting bottoms-composition point is high-boiling component H. A material-balance line from this point, through feed point F, ends at D. These two lines and the distillation curve define the feasible product-composition regions, shown shaded. Note that, because for a given feed both the distillate and bottoms compositions must lie on the same distillation curve, shaded feasible regions lie on the convex side of the distillation curve that passes through the feed point. Because of its appearance, the feasible-product-composition region is called a *bow-tie region*.

For azeotropes, where distillation boundaries are present, a feasible-product-composition region exists for each distillation region. Two examples are shown in Figure 11.12. Figure 11.12a has two distillation regions caused by two minimum-boiling binary azeotropes. A curved distillation boundary connects the minimum-boiling azeotropes. In the lower, right-hand distillation region (1), the lowest-boiling species is the *n*-octane–2-ethoxy-ethanol minimum-boiling azeotrope, while the highest-boiling species is 2-ethoxy-ethanol. Accordingly, for feed *F*1, straight lines are drawn

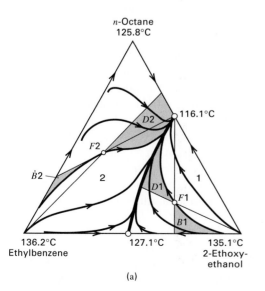

(a)

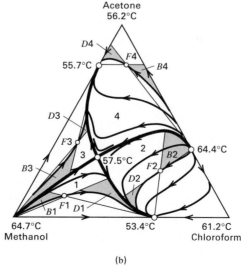

(b)

Figure 11.12 Product-composition regions for given feed compositions. (a) Ternary mixture with two minimum-boiling binary azeotropes at 1 atm. (b) Ternary mixture with three binary and one ternary azeotrope at 1 atm.

from the points for each of these two species, through the point $F1$, and to a boundary (either a distillation boundary or a side of the triangle). Shaded, feasible-product-composition regions are then drawn on the outer side of the distillation curve that passes through the feed point. The result is that distillate compositions are confined to shaded region $D1$ and bottoms compositions are confined to shaded region $B1$. For a given $D1$, $B1$ must lie on a straight line that passes through $D1$ and $F1$. At total reflux, $D1$ and $B1$ must also lie on the same distillation curve.

A more complex distillation-curve map, with four distillation regions, is shown in Figure 11.12b for the acetone–methanol–chloroform system with two minimum-boiling binary azeotropes, one maximum-boiling binary azeotrope, and one ternary azeotrope. One shaded bow-tie region, determined in the same way as for Region 1 in Figure 11.12a, is present for each distillation region. For this system, feasible-product-composition regions are highly restricted.

A complicated situation is observed in distillation Region 1 on the left side of Figure 11.12a, where the lowest-boiling species is the binary azeotrope of octane and 2-ethoxy-ethanol, while the highest-boiling species is ethylbenzene. The complicating factor in Region 1 is that feed $F2$ lies on or close to an inflection point of an S-shaped distillation curve. In this case, as discussed by Wahnschafft et al. [20], feasible-product-composition regions may lie on either side of the distillation curve passing through the feed point. The feasible regions shown are similar to those determined by Stichlmair et al. [1], while other feasible regions are shown for this system by Wahnschafft et al. [20]. As they point out, mass-balance lines of the type drawn in Figure 11.12b do not limit feasible regions.

Hoffmaster and Hauan [98] provide a method for determining extended-product-feasibility regions for S-shaped distillation curves.

In Figures 11.11b, 11.12a, and 11.12b, each bow-tie region is confined to its distillation region, defined by the distillation boundaries. In all cases, the feed, distillate, and bottoms points on the material-balance line lie within a distillation region, with the feed point between the distillate and bottoms points. The material-balance lines do not cross the distillation-boundary lines. Is this always so? The answer is no! Under conditions where the distillation-boundary line is highly curved, it can be crossed by material-balance lines to obtain feasible-product compositions. That is, a feed point can be on one side and the distillate and bottoms points on the other side of the distillation-boundary line.

Consider the example in Figure 11.13, from Widagdo and Seider [19]. The highly curved distillation-boundary line extends from a minimum-boiling azeotrope K of H-I to pure component L. This line divides the triangular diagram into two distillation regions, 1 and 2. Feed F_1 can be separated into products D_1 and B_1, which lie on distillation curve (a). In this case, the material-balance line and the distillation curve are both on the convex side of the distillation-boundary line. However, because feed point F_1 lies close to the highly curved boundary line, F_1 can also be separated into D_2 and B_2 (or B_3), which lie on a distillation curve in Region 2 on the concave side of the boundary. Thus, the material-balance

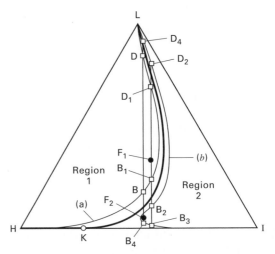

Figure 11.13 Feasible and infeasible crossings of distillation boundaries for an azeotropic system.
[From S. Widagdo and W.D. Seider, *AIChE J.*, **42**, 96–130 (1996) with permission.]

line crosses the boundary from the convex to the concave side. Feed F_2 can be separated into D_4 and B_4, but not into D and B. In the latter case, the material-balance line cannot cross the boundary from the concave to the convex side, because the point F_2 does not lie between D and B on the material-balance line. The determination of the feasible-product-composition regions for Figure 11.13 is left for an exercise at the end of this chapter. A detailed treatment of product-composition regions is given by Wahnschafft et al. [20].

§11.2 EXTRACTIVE DISTILLATION

Extractive distillation is used to separate azeotropes and close-boiling mixtures. If the feed is a minimum-boiling azeotrope, a solvent, with a lower volatility than the key components of the feed mixture, is added to a tray just a few trays below the top of the column so that (1) the solvent is present in the down-flowing liquid, and (2) little solvent is stripped and lost to the overhead vapor. If the feed is a maximum-boiling azeotrope, the solvent enters the column with the feed. The components in the feed must have different solvent affinities so that the solvent causes an increase in α of the key components, to the extent that separation becomes feasible and economical. The solvent should not form an azeotrope with any components in the feed. Usually, a molar ratio of solvent to feed on the order of 1 is required. The bottoms are processed to recover the solvent for recycle and complete the feed separation. The name *extractive distillation* was introduced by Dunn et al. [21] in connection with the commercial separation of toluene from a paraffin–hydrocarbon mixture, using phenol as solvent.

Table 11.1 lists industrial applications of extractive distillation. Consider the case of the acetone–methanol system. At 1 atm, acetone (nbp = 56.2°C) and methanol (nbp = 64.7°C) form a minimum-boiling azeotrope of 80 mol% acetone at a

Table 11.1 Some Industrial Applications of Extractive Distillation

Key Components in Feed Mixture	Solvent
Acetone–methanol	Aniline, ethylene glycol, water
Benzene–cyclohexane	Aniline
Butadienes–butanes	Acetone
Butadiene–butene-1	Furfural
Butanes–butenes	Acetone
Butenes–isoprene	Dimethylformamide
Cumene–phenol	Phosphates
Cyclohexane–heptanes	Aniline, phenol
Cyclohexanone–phenol	Adipic acid diester
Ethanol–water	Glycerine, ethylene glycol
Hydrochloric acid–water	Sulfuric acid
Isobutane–butene-1	Furfural
Isoprene–pentanes	Acetonitrile, furfural
Isoprene–pentenes	Acetone
Methanol–methylene bromide	Ethylene bromide
Nitric acid–water	Sulfuric acid
n-Butane–butene-2s	Furfural
Propane–propylene	Acrylonitrile
Pyridine–water	Bisphenol
Tetrahydrofuran–water	Dimethylformamide, propylene glycol
Toluene–heptanes	Aniline, phenol

temperature of 55.7°C. Using UNIFAC (§2.6.9) to predict vapor–liquid equilibria for this system at 1 atm, the azeotrope was estimated to occur at 55.2°C with 77.1 mol% acetone, reasonably close to measured values. At infinite dilution with respect to methanol, $\alpha_{A,M}$ for acetone (A) with respect to methanol (M), is predicted to be 0.74 by UNIFAC, with a liquid-phase activity coefficient for methanol of 1.88. At infinite dilution with respect to acetone, $\alpha_{A,M}$ is 2.48; by coincidence, the liquid-phase activity coefficient for acetone is also 1.88.

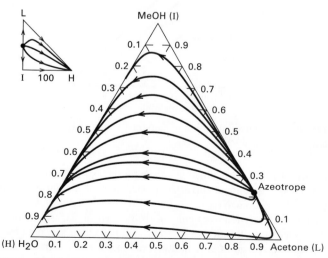

Figure 11.14 Residue-curve map for acetone–methanol–water system at 1 atm.

Water is a possible solvent for the system because at 1 atm: (1) it does not form a binary or ternary azeotrope with acetone and/or methanol, and (2) it boils (100°C) at a higher temperature. The resulting residue-curve map with arrows directed from the azeotrope to pure water, computed by ASPEN PLUS using UNIFAC, is shown in Figure 11.14, where it is seen that no distillation boundaries exist. As discussed by Doherty and Caldarola [16], this is an ideal situation for the selection of an extractive distillation process. Their residue-curve map for this type of system (designated 100) is included as an insert in Figure 11.14.

Ternary mixtures of acetone, methanol, and water at 1 atm give the following separation factors, estimated from the UNIFAC equation, when appreciable solvent is present.

Mol% Water	Relative Volatility, $\alpha_{A,M}$			Liquid-Phase Activity Coefficient at Infinite Dilution	
	Methanol-rich	Acetone-rich	Equimolar	Acetone	Methanol
40	2.48	2.57	2.03	2.12	0.70
50	2.56	2.86	2.29	2.41	0.72

The presence of appreciable water increases the liquid-phase activity coefficient of acetone and decreases that of methanol; thus, over the entire concentration range of acetone and methanol, the α of acetone to methanol is at least 2.0. This makes it possible, with extractive distillation, to obtain a distillate of acetone and a bottoms of methanol and water. The α values of acetone to water and methanol to water average 4.5 and 2.0, thus making it relatively easy to prevent water from reaching the distillate, and, in subsequent operations, to separate methanol from water by distillation.

EXAMPLE 11.3 Extractive Distillation of Acetone and Methanol.

Forty mol/s of a bubble-point mixture of 75 mol% acetone and 25 mol% methanol at 1 atm is separated by extractive distillation, using water as the solvent, to produce an acetone product of not less than 95 mol% acetone, a methanol product of not less than 98 mol% methanol, and a water stream for recycle of at least 99.9 mol% purity. Prepare a preliminary process design using the traditional three-column sequence consisting of ordinary distillation followed by extractive distillation, and then ordinary distillation to recover the solvent, as shown for another system in Figure 11.15.

Solution

In the first column, the feed mixture of acetone and methanol would be partially separated by ordinary distillation, where the distillate composition approaches that of the binary azeotrope. The bottoms would be nearly pure acetone or nearly pure methanol, depending upon whether the feed contains more or less than 80 mol% acetone. In this example, the feed composition is already close to the azeotrope composition; therefore, the first column is not required, and the acetone–methanol feed is sent to the second column, an extractive-distillation column equipped with a total condenser and a partial reboiler, to produce a distillate of at least 95 mol% acetone.

Table 11.2 Material and Energy Balances for Extractive-Distillation Process of Example 11.3

	Material Balances					
Species	Flow Rate, mol/s: Column 2 Feed	Column 2 Solvent	Column 2 Distillate	Column 2 Bottoms	Column 3 Distillate	Column 3 Bottoms
Acetone	30	0	29.86	0.14	0.14	0.0
Methanol	10	0	0.016	9.984	9.926	0.058
Water	0	60	1.35	58.65	0.06	58.59
Total	40	60	31.226	68.774	10.126	58.648

	Energy Balances	
	Column 1	Column 2
Condenser duty, MW	4.71	1.07
Reboiler duty, MW	4.90	1.12

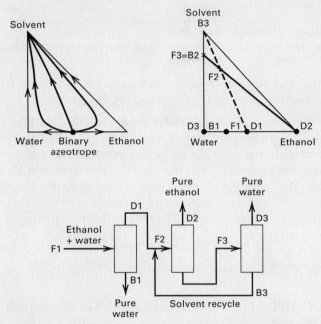

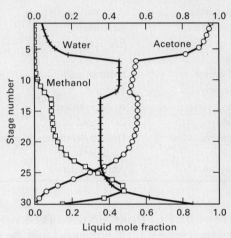

Figure 11.16 Liquid composition profile for extractive-distillation column of Example 11.3.

Figure 11.15 Distillation sequence for extractive distillation.
[From M.F. Doherty and G.A. Caldarola, *IEC Fundam.*, **24**, 479 (1985) with permission.]

Acetone recovery is better than approximately 99% in the extractive distillation step, and the required methanol purity in the recycle column is achieved.

The ChemSep and CHEMCAD programs were used to make the calculations, with the UNIFAC method for activity coefficients. Number of stages, feed-stage location, solvent-entry stage, solvent flow rate, and reflux-ratio requirements were manipulated until a satisfactory design was achieved. The resulting material and energy balances are summarized in Table 11.2.

For the extractive-distillation column, a solvent flow rate of 60 mol/s of water is suitable. Using 28 theoretical trays, a 50°C solvent entry at Tray 6 from the top, a feed entry at Tray 12 from the top, and a reflux ratio of 4, a distillate composition of 95.6 mol% acetone is achieved. The impurity is mainly water. The acetone recovery is 99.5%. A 6-ft-diameter column with 60 sieve trays on 2-ft tray spacing is adequate. A liquid-phase composition profile is shown in Figure 11.16. The mole fraction of water (the solvent) is appreciable, at least 0.35 for all of the stages below the solvent-entry stage.

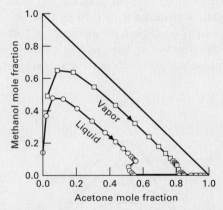

Figure 11.17 Distillation-curve map for Example 11.3. Data points are for theoretical stages.

Figure 11.17 shows distillation curves for the extractive distillation, where vapor and liquid curves are plotted. Arrows are directed from the column bottom to the top.

For the water-recovery column, 16 theoretical stages, a bubble-point feed-stage location of stage 11, and $R = 2$ are adequate for a methanol distillate of 98.1 mol% purity and a water bottoms, suitable for recycle, of 99.9 mol% purity. A McCabe–Thiele diagram in

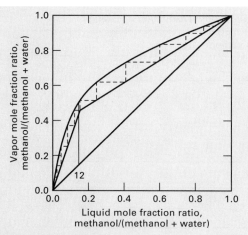

Figure 11.18 McCabe–Thiele diagram for methanol–water distillation in Example 11.3.

Figure 11.18 shows the locations of the theoretical stages. The feed stage is optimally located. Water makeup is less than 1.5 mol/s. A 2.5-ft-diameter column packed with 48 feet of 50-mm metal Pall rings is suitable.

One unfortunate aspect of the extractive-distillation column in Example 11.3 is the relatively low boiling point of water. With a solvent-entry point of Tray 6 from the top, 1.35 mol/s (2.25% of the water solvent) is stripped from the liquid into the distillate. Use of two other, higher-boiling solvents listed in Table 11.1, aniline (nbp = 184°C) or ethylene glycol (nbp = 198°C), results in far less solvent stripping. Other possible solvents include methylethylketone (MEK) and ethanol. MEK behaves in a fashion opposite to that of water, causing the volatility of methanol to be greater than acetone. Thus, methanol is the distillate, leaving acetone to be separated from MEK.

In selecting a solvent for extractive distillation, a number of factors are considered, including availability, cost, corrosivity, vapor pressure, thermal stability, heat of vaporization, reactivity, toxicity, infinite-dilution activity coefficients in the solvent of the components to be separated, and ease of solvent recovery for recycle. In addition, the solvent should not form azeotropes. Initial screening is based on the measurement or prediction of infinite-dilution activity coefficients. Berg [22] discusses selection of separation agents for both extractive and azeotropic distillation. He points out that all successful solvents for extractive distillation are highly hydrogen-bonded liquids (Table 2.7), such as (1) water, amino alcohols, amides, and phenols that form three-dimensional networks of strong hydrogen bonds; and (2) alcohols, acids, phenols, and amines that are composed of molecules containing both active hydrogen atoms and donor atoms (oxygen, nitrogen, and fluorine). It is difficult or impossible to find a solvent to separate components having the same functional groups.

Extractive distillation is also used to separate binary mixtures that form a maximum-boiling azeotrope, as shown in the following example.

EXAMPLE 11.4 Extractive Distillation of Acetone and Chloroform.

Acetone (nbp = 56.16°C) and chloroform (nbp = 61.10°C) form a maximum-boiling homogeneous azeotrope at 1 atm and 64.43°C that contains 37.8 mol% acetone, so they cannot be separated by ordinary distillation at 1 atm. Instead, extractive distillation in a two-column sequence is to be used, shown in Figure 11.19, with benzene (nbp = 80.24°C) as the solvent. Benzene does not form azeotropes with the feed components.

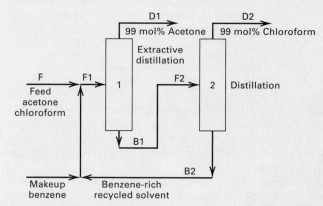

Figure 11.19 Process for the separation of acetone and chloroform in Example 11.4.

In the first column, feed, blended with recycled solvent, produces a distillate of 99 mol% acetone. The bottoms is sent to the second column, where 99 mol% chloroform leaves as distillate, and the bottoms, rich in benzene, is recycled to the first column with makeup benzene. If the fresh feed is 21.89 mol/s of 54.83 mol% acetone, with the balance chloroform, design a feasible two-column system using a ratio of 3.1667 moles of benzene per mole of (acetone + chloroform) in the combined feed to the first column. Both columns operate at 1 atm with total condensers, saturated-liquid reflux, and partial reboilers. Use the UNIFAC method for estimating activity coefficients. The combined feed to the first column is brought to the bubble point before entering the column.

Solution

Figure 11.20 shows the residue-curve map for the ternary system acetone–chloroform–benzene at 1 atm. The only azeotrope is that of

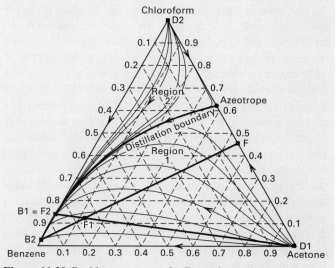

Figure 11.20 Residue-curve map for Example 11.4.

Table 11.3 Material and Energy Balances for Homogeneous Azeotropic Distillation of Example 11.4

Material Balances with Flows in mol/s						
Species	F	F_1	D_1	$B_1 = F_2$	D_2	B_2
Acetone	12.0000	12.0000	11.9948	0.0052	0.0052	0.0000
Chloroform	9.8858	12.0000	0.1046	11.8954	9.7812	2.1142
Benzene	0.0000	76.0000	0.0207	75.9793	0.0934	75.8859

Energy Balances		
Heat duty, kcal/h	Column 1	Column 2
Condenser	950,000	891,600
Reboiler	958,400	1,102,000

acetone and chloroform. A curved distillation boundary extending from that azeotrope to the pure-benzene apex divides the diagram into two distillation regions. The first column, which produces nearly pure acetone, operates in Region 1; the second column operates in Region 2.

This ternary system was studied in detail by Fidkowski, Doherty, and Malone [17]. A design based on their studies that uses the CHEMCAD process simulator is summarized in Table 11.3. The first column contains 65 theoretical stages, with the combined feed entering stage 30 from the top. For $R = 10$, the acetone distillate purity is achieved with an acetone recovery of better than 99.95%. In Column 2, which contains 50 theoretical stages with feed entering at stage 30, an $R = 11.783$, gives the required chloroform purity in the distillate but with a recovery of only 82.23%. This is not serious because the chloroform leaving in the bottoms is recycled with benzene to Column 1, resulting in a 98.9% overall recovery of chloroform. The benzene makeup rate is 0.1141 mol/s. Feed, distillate, and bottoms compositions are designated in Figure 11.20.

§11.3 SALT DISTILLATION

Water, as a solvent in the extractive distillation of acetone and methanol in Example 11.3, has the disadvantages that a large amount is required to adequately alter α and, even though the solvent is introduced into the column several trays below the top, enough water is stripped into the distillate to reduce the acetone purity to 95.6 mol%. The water vapor pressure can be lowered, and thus the purity of acetone distillate increased, by using an aqueous inorganic-salt solution as the solvent. A 1927 patent by Othmer [23] describes use of a concentrated, calcium chloride brine. Not only does calcium chloride, which is highly soluble in water, reduce the volatility of water, but it also has a strong affinity for methanol. Thus, α of acetone with respect to methanol is enhanced. The separation of brine solution from methanol is easily accommodated in the subsequent distillation, with the brine solution recycled to the extractive-distillation column. The vapor pressure of the dissolved salt is so small that it never enters the vapor, provided entrainment is avoided. An even earlier patent by Van Raymbeke [24] describes the extractive distillation of ethanol from water using solutions of calcium chloride, zinc chloride, or potassium carbonate in glycerol.

Salt can be added as a solid or melt into the column by dissolving it in the reflux before it enters the column. This was demonstrated by Cook and Furter [25] in a 4-inch-diameter, 12-tray rectifying column with bubble caps, separating ethanol from water using potassium acetate. At salt concentrations below saturation and between 5 and 10 mol%, an almost pure ethanol distillate was achieved. The salt, which must be soluble in the reflux, is recovered from the aqueous bottoms by evaporation and crystallization.

Salt distillation is accompanied by several problems. First and foremost is corrosion, particularly with aqueous chloride-salt solutions, which may require stainless steel or a more expensive corrosion-resistant material. Feeding and dissolving a salt into the reflux poses problems described by Cook and Furter [25]. The solubility of salt will be low in the reflux because it is rich in the more-volatile component, the salt being most soluble in the less-volatile component. Salt must be metered at a constant rate and the salt-feeding mechanism must avoid bridging and prevent the entry of vapor, which could cause clogging when condensed. The salt must be rapidly dissolved, and the reflux must be maintained near the boiling point to avoid precipitation of already-dissolved salt. In the column, presence of dissolved salt may increase foaming, requiring addition of antifoaming agents and/or column-diameter increase. Concern has been voiced for the possibility of salt crystallization within the column. However, the concentration of the less-volatile component (e.g., water) increases down the column, so the solubility of salt increases down the column while its concentration remains relatively constant. Thus, the possibility of clogging and plugging due to solids formation is unlikely.

In aqueous alcohol solutions, both *salting out* and *salting in* have been observed by Johnson and Furter [26], as shown in the vapor–liquid equilibrium data in Figure 11.21. In (a), sodium nitrate salts out methanol, but in (b), mercuric chloride salts in methanol. Even low concentrations of potassium acetate can eliminate the ethanol–water azeotrope, as shown in Figure 11.21c. Mixed potassium- and sodium-acetate salts were used in Germany and Brazil from 1930 to 1965 for the separation of ethanol and water.

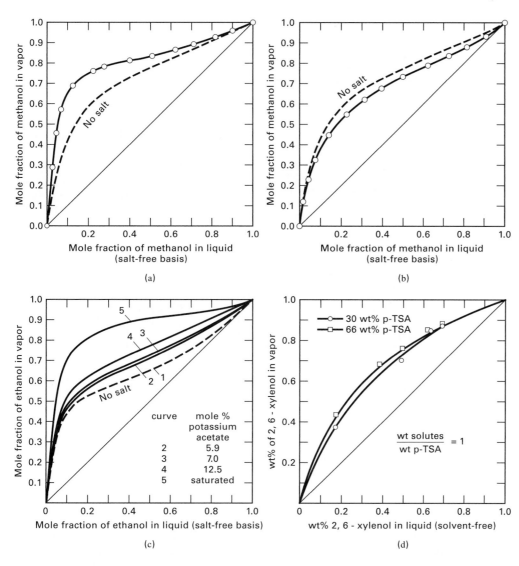

Figure 11.21 Effect of dissolved salts on vapor–liquid equilibria at 1 atm. (a) Salting-out of methanol by saturated aqueous sodium nitrate. (b) Salting-in of methanol by saturated aqueous mercuric chloride. (c) Effect of salt concentration on ethanol–water equilibria. (d) Effect of *p*-toluene-sulfonic acid (*p*-TSA) on phase equilibria of 2,6-xylenol–*p*-cresol.

[From A.I. Johnson and W.F. Furter, *Can. J. Chem. Eng.*, **43**, 356–358 (1965) with permission.]

Surveys of the use of inorganic salts for extractive distillation, including effects on vapor–liquid equilibria, are given by Johnson and Furter [27], Furter and Cook [28], and Furter [29, 30]. A survey of methods for predicting the effect of inorganic salts on vapor–liquid equilibria is given by Kumar [31]. Column-simulation results, using the Newton–Raphson method, are presented by Llano-Restrepo and Aguilar-Arias [99] for the ethanol–water–calcium chloride system and by Fu [100], for the ethanol–water–ethanediol–potassium acetate system, who shows simulation results that compare favorably with those from an industrial column.

Salt distillation can be applied to organic compounds that have little capacity for dissolving inorganic salts by using organic salts called *hydrotropes*. Typical are alkali and alkaline-earth salts of the sulfonates of toluene, xylene, or cymene, and the alkali benzoates, thiocyanates, and salicylates. Mahapatra, Gaikar, and Sharma [32] found that the addition of aqueous solutions of 30 and 66 wt% *p*-toluenesulfonic acid to 2,6-xylenol and *p*-cresol at 1 atm increased the α from approximately 1 to about 3, as shown in Figure 11.21d. Hydrotropes can also enhance liquid–liquid extraction, as shown by Agarwal and Gaikar [33].

§11.4 PRESSURE-SWING DISTILLATION

If a binary azeotrope disappears at some pressure, or changes composition by 5 mol% or more over a moderate range of pressure, consideration should be given to using two ordinary distillation columns operating in series at different pressures. This process is referred to as *pressure-swing distillation*. Knapp and Doherty [34] list 36 pressure-sensitive, binary azeotropes, mainly from the compilation of Horsley [11]. The effect of pressure on temperature and composition of two minimum-boiling azeotropes is shown in Figure 11.22. The mole fraction of ethanol in the ethanol–water azeotrope increases from 0.8943 at 760 torr to more than 0.9835 at 90 torr. Not shown in Figure 11.22b is that the azeotrope disappears at below 70 torr. A more dramatic change in composition with pressure is seen in Figure 11.22b for the ethanol–benzene system, which forms a minimum-boiling azeotrope at 44.8 mol% ethanol and 1 atm. Applications of pressure-swing distillation, first noted by Lewis [35] in a 1928 patent, include separations of the minimum-boiling azeotrope of tetrahydrofuran–water and maximum-boiling azeotropes of hydrochloric acid–water and formic acid–water.

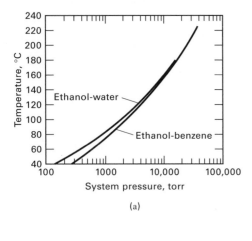

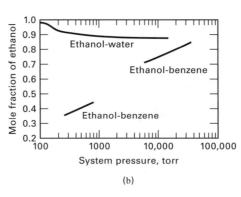

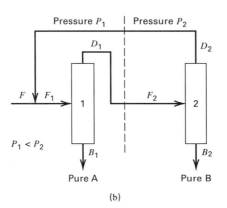

Figure 11.22 Effect of pressure on azeotrope conditions. (a) Temperature of azeotrope. (b) Composition of azeotrope.

[From *Perry's Chemical Engineers' Handbook*, 6th ed., R.H. Perry and D.W. Green, Eds., McGraw-Hill (1984) with permission.]

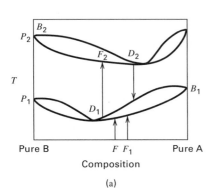

(a)

(b)

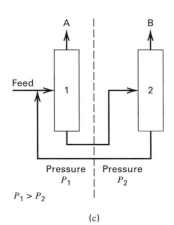

(c)

Figure 11.23 Pressure-swing distillation. (a) T–y–x curves at pressures P_1 and P_2 for minimum-boiling azeotrope. (b) Distillation sequence for minimum-boiling azeotrope. (c) Distillation sequence for maximum-boiling azeotrope.

Van Winkle [36] describes a minimum-boiling azeotrope for A and B, with the T–y–x curves shown in Figure 11.23a. As pressure is decreased from P_2 to P_1, the azeotropic composition moves toward a smaller percentage of A. An operable pressure-swing sequence is shown in Figure 11.23b. The total feed, F_1, to Column 1, operating at lower pressure P_1, is the sum of fresh feed, F, which is richer in A than the azeotrope, and recycled distillate, D_2, whose composition is close to that of the azeotrope at pressure P_2. D_2 and, consequently, F_1 are both richer in A than the azeotrope at P_1. The bottoms, B_1, leaving Column 1 is almost pure A. Distillate, D_1, which is slightly richer in A than the azeotrope but less rich in A than the azeotrope at P_2, is fed to Column 2, where the bottoms, B_2, is almost pure B. Robinson and Gilliland [37] discuss the separation of ethanol and water, where the fresh feed is less rich in ethanol than the azeotrope. For that case, products are still removed as bottoms, but nearly pure B is taken from the first column and A from the second.

Pressure-swing distillation can also be used to separate less-common maximum-boiling binary azeotropes. A sequence is shown in Figure 11.23c, where both products are withdrawn as distillates, rather than as bottoms. In this case, the azeotrope becomes richer in A as pressure is decreased. Fresh feed, richer in A than the azeotrope at the higher pressure, is first distilled in Column 1 at higher pressure P_1, to produce a distillate of nearly pure A and a bottoms slightly

richer in A than the azeotrope at the higher pressure. The bottoms is fed to Column 2, operating at lower pressure P_2, where the azeotrope is richer in A than feed to that column. Accordingly, distillate is nearly pure B, while recycled bottoms from Column 2 is less rich in A than the azeotrope at the lower pressure.

For pressure-swing-distillation sequences, because of the high cost of gas compression, recycle ratio is a key design factor, and depends on the variation in azeotropic composition with column pressure. The next example illustrates the importance of the recycle stream.

EXAMPLE 11.5 Pressure-Swing Distillation.

Ninety mol/s of a mixture of 2/3 by moles ethanol and 1/3 benzene at the bubble point at 101.3 kPa is to be separated into 99 mol% ethanol and 99 mol% benzene. The mixture forms a minimum-boiling azeotrope of 44.8 mol% ethanol at 760 torr and 68°C. If the pressure is reduced to 200 torr, as shown in Figure 11.22b, the azeotrope shifts to 36 mol% ethanol at 35°C. This is a candidate for pressure-swing distillation.

Apply the sequence shown in Figure 11.23b with the first column operating at a top-tray pressure at 30 kPa (225 torr). Because the feed composition is greater than the azeotrope composition at the column pressure, the distillate composition approaches the

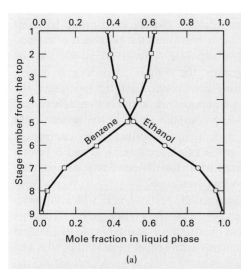

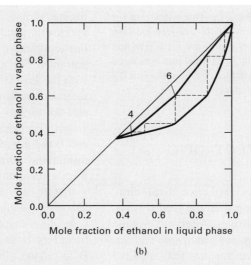

Figure 11.24 Computed results for Column 1 of pressure-swing distillation system in Example 11.5. (a) Liquid composition profiles. (b) McCabe–Thiele diagram.

(a)

(b)

minimum-boiling azeotrope at the top-tray pressure, and the bottoms is 99 mol% ethanol. The distillate is sent to the second column, which has a top-tray pressure of 106 kPa. Feed to this column has an ethanol content less than the azeotrope at the second-column pressure, so the distillate approaches the azeotrope at the top-tray pressure, and the bottoms is 99 mol% benzene. The distillate is recycled to the first column. Design a pressure-swing distillation system for this separation.

Solution

For the first column, which operates under vacuum, reflux-drum and reboiler pressures are set at 26 and 40 kPa, respectively. For the second column, operating slightly above ambient pressure, reflux-drum and reboiler pressures are set at 101.3 and 120 kPa, respectively. Bottoms compositions are specified at the required purities. The distillate for the first column is set at 37 mol% ethanol, slightly greater than the azeotrope composition at 30 kPa. Distillate composition for the second column is 44 mol% ethanol, slightly less than the azeotrope composition at 106 kPa. Material-balance calculations on ethanol and benzene give the following flow rates in mol/s.

Component	F	D_2	F_1	B_1	D_1	B_2
Ethanol	60.0	67.3	127.3	59.7	67.6	0.3
Benzene	30.0	85.6	115.6	0.6	115.0	29.4
Totals:	90.0	152.9	242.9	60.3	182.6	29.7

Equilibrium-stage calculations for the columns were made with the ChemSep program, using total condensers and partial reboilers. For Column 1, runs were made to find optimal feed-tray locations for the fresh feed and the recycle, using a reflux rate that avoided any near-pinch conditions. The selected design uses seven theoretical trays (not counting the partial reboiler), with the recycle stream, at a temperature of 68°C, sent to Tray 3 from the top and the fresh feed to Tray 5 from the top. An $R = 0.5$ is sufficient to meet specifications. The resulting liquid-phase composition profile is shown in Figure 11.24a, where the desired lack of pinch points is observed. The McCabe–Thiele diagram for Column 1 in Figure 11.24b has three operating lines, and the optimal-feed locations are indicated. Because of the azeotrope, the operating lines and equilibrium curve all lie below the 45° line. The condenser duty is 9.88 MW, while the reboiler duty is 8.85 MW. The bottoms temperature is 56°C. This column was sized with the CHEMCAD program for sieve trays on 24-inch tray spacing and a 1-inch weir height to minimize pressure drop. The resulting diameter is 3.2 meters (10.5 ft). A tray efficiency of about 47% is predicted, so 15 trays are required.

A similar procedure was used to establish the optimal-feed tray, total trays, and reflux ratio for Column 2. The selected design turned out to be a refluxed stripper with only three theoretical stages (not counting the partial reboiler). A reflux rate of only 25.5 mol/s achieves the product specifications, with most of the liquid traffic in the stripper coming from the feed. The resulting liquid-phase composition profile is shown in Figure 11.25a, where, again, no

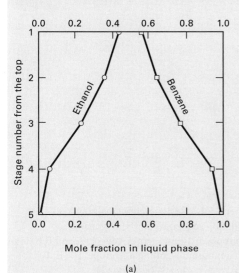

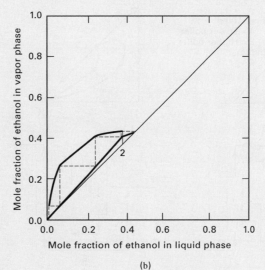

Figure 11.25 Computed results for Column 2 of pressure-swing distillation system in Example 11.5. (a) Liquid composition profiles. (b) McCabe–Thiele diagram.

(a)

(b)

composition pinches are evident. The McCabe–Thiele diagram for Column 2 is given in Figure 11.25b, where an optimal-feed location is indicated. The condenser and reboiler duties are 6.12 MW and 7.07 MW. The bottoms temperature is 84°C. This column was sized like Column 1, resulting in a column diameter of 2.44 meters (8 ft). A tray efficiency of 50% results in 6 actual trays.

§11.5 HOMOGENEOUS AZEOTROPIC DISTILLATION

An azeotrope can be separated by extractive distillation, using a solvent that is higher boiling than the feed components and does not form any azeotropes. Alternatively, the separation can be made by *homogeneous azeotropic distillation*, using an *entrainer* not subject to such restrictions. Like extractive distillation, a sequence of two or three columns is used. Alternatively, the sequence is a hybrid system that includes operations other than distillation, such as solvent extraction.

The conditions that a potential entrainer must satisfy have been studied by Doherty and Caldarola [16]; Stichlmair, Fair, and Bravo [1]; Foucher, Doherty, and Malone [14]; Stichlmair and Herguijuela [18]; Fidkowski, Malone, and Doherty [13]; Wahnschafft and Westerberg [38]; and Laroche, Bekiaris, Andersen, and Morari [39]. If it is assumed that a distillation boundary, if any, of a residue-curve map is straight or cannot be crossed, the conditions of Doherty and Caldarola apply. These are based on the rule that for entrainer E, the two components, A and B, to be separated, or any product azeotrope, must lie in the same distillation region of the residue-curve map. Thus, a distillation boundary cannot be connected to the A–B azeotrope. Furthermore, A or B, but not both, must be a saddle.

The maps suitable for a sequence that includes homogeneous azeotropic distillation together with ordinary distillation are classified into the five groups illustrated in

Figure 11.26a, b, c, d, and e, where each group includes applicable residue-curve maps and the sequence of separation columns used to separate A from B and recycle the entrainer. For all groups, the residue-curve map is drawn, with the lowest-boiling component, L, at the top vertex; the intermediate-boiling component, I, at the bottom-left vertex; and the highest-boiling component, H, at the bottom-right vertex. Component A is the lower-boiling binary component and B the higher. For the first three groups, A and B form a minimum-boiling azeotrope; for the other two groups, they form a maximum-boiling azeotrope.

In Group 1, the intermediate boiler, I, is E, which forms no azeotropes with A and/or B. As shown in Figure 11.26a, this case, like extractive distillation, involves no distillation boundary. Both sequences assume that fresh feed F, of A and B, as fed to Column 1, is close to the azeotropic composition. This feed may be distillate from a previous column used to produce the azeotrope from the original A and B mixture. Either the *direct sequence* or the *indirect sequence* may be used. In the former, Column 2 is fed by the bottoms from Column 1 and the entrainer is recovered as distillate from Column 2 and recycled to Column 1. In the latter, Column 2 is fed by the distillate from Column 1 and the entrainer is recovered as bottoms from Column 2 and recycled to Column 1. Although both sequences show entrainer combined with fresh feed before Column 1, fresh feed and recycled entrainer can be fed to different trays to enhance the separation.

In Group 2, low boiler L is E, which forms a maximum-boiling azeotrope with A. Entrainer E may also form a minimum-boiling azeotrope with B, and/or a minimum-boiling (unstable node) ternary azeotrope. Thus, in Figure 11.26b, any of the five residue-curve maps may apply. In all cases, a distillation boundary exists, which is directed from the maximum-boiling azeotrope of A–E to pure B, the high boiler. A feasible indirect or direct sequence is restricted to the subtriangle bounded by the vertices of pure components A, B,

Residue-curve map arrangement

L = Lowest boiler
I = Intermediate boiler
H = Highest boiler

Applicable residue-curve map

Sequences

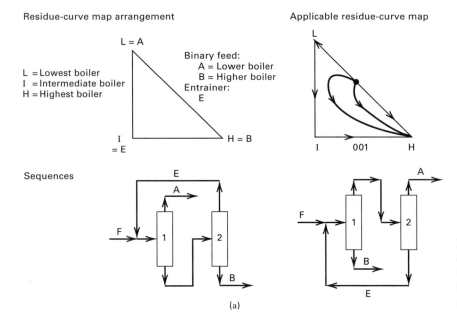

(a)

Figure 11.26 Residue-curve maps and distillation sequences for homogeneous azeotropic distillation. (a) Group 1: A and B form a minimum-boiling azeotrope, I = E, E forms no azeotropes. (*continued*)

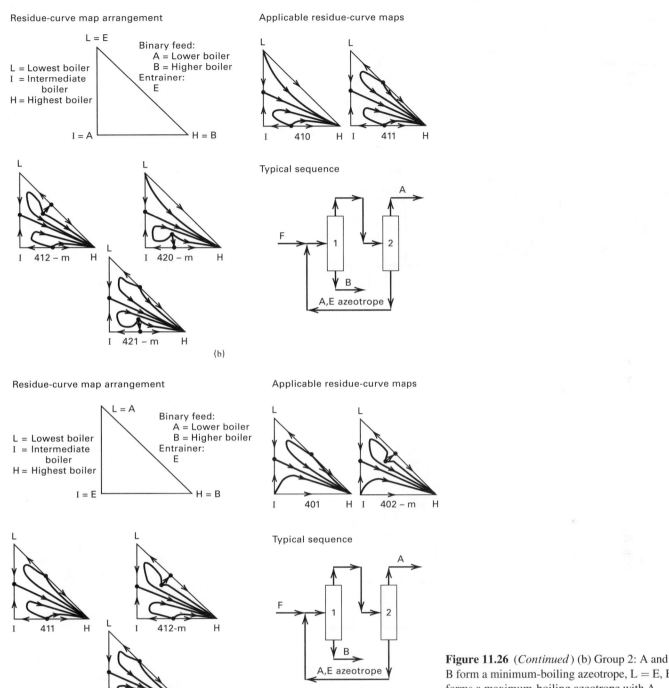

Figure 11.26 (*Continued*) (b) Group 2: A and B form a minimum-boiling azeotrope, L = E, E forms a maximum-boiling azeotrope with A. (c) Group 3: A and B form a minimum-boiling azeotrope, I = E, E forms a maximum-boiling azeotrope with A.

and the binary A–E azeotrope. An example of an indirect sequence is included in Figure 11.26b. Here, the A–E azeotrope is recycled to Column 1 from the bottoms of Column 2. Alternatively, as in Figure 11.26c for Group 3, A and E may be switched to make A the low boiler and E the intermediate boiler, which again forms a maximum-boiling azeotrope with A. All sequences for Group 3 are confined to the same subtriangle as for Group 2.

Groups 4 and 5, in Figures 11.26d and e, are similar to Groups 2 and 3. However, A and B now form a maximum-

boiling azeotrope. In Group 4, the entrainer is the intermediate boiler, which forms a minimum-boiling azeotrope with B. The entrainer may also form a maximum-boiling azeotrope with A, and/or a maximum-boiling (stable node) ternary azeotrope. A feasible sequence is restricted to the subtriangle formed by vertices A, B, and the B–E azeotrope. In the sequence, the distillate from Column 2, which is the minimum-boiling B–E azeotrope, is mixed with fresh feed to Column 1, which produces a distillate of pure A. The bottoms from Column 1 has a composition such that when fed to Column 2, a bottoms of

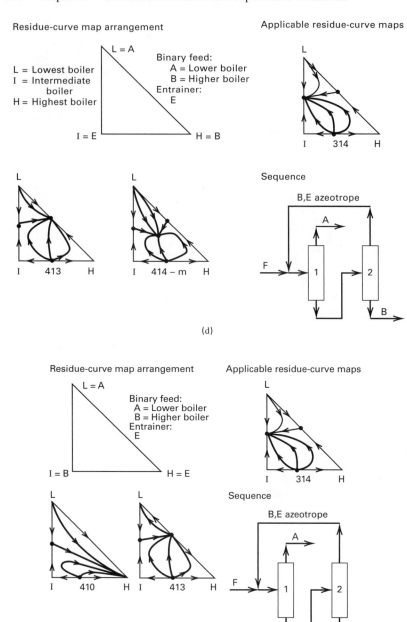

(d)

(e)

Figure 11.26 (*Continued*) (d) Group 4: A and B form a maximum-boiling azeotrope, I = E, E forms a minimum-boiling azeotrope with B. (e) Group 5: A and B form a maximum-boiling azeotrope, H = E, E forms a minimum-boiling azeotrope with B.

pure B can be produced. Although a direct sequence is shown, the indirect sequence can also be used. Alternatively, as shown in Figure 11.26e for Group 5, B and E may be switched to make E the high boiler. In the sequence shown, as in that of Figure 11.26d, the bottoms from Column 1 is such that, when fed to Column 2, a bottoms of pure B can be produced. The other conditions and sequences are the same as for Group 4.

The distillation boundaries for the hypothetical ternary systems in Figure 11.26 are shown as straight lines. When a distillation boundary is curved, it may be crossed, provided that both the distillate and bottoms products lie on the same side of the boundary.

It is often difficult to find an entrainer for a sequence involving homogeneous azeotropic distillation and ordinary distillation. However, azeotropic distillation can also be incorporated into a hybrid sequence involving separation operations other than distillation. In that case, some of the restrictions for the entrainer and resulting residue-curve map may not apply. For example, the separation of the close-boiling and minimum-azeotrope-forming system of benzene and cyclohexane using acetone as the entrainer violates the restrictions for a distillation-only sequence because the ternary system involves only two minimum-boiling binary azeotropes. However, the separation can be made by the sequence shown in Figure 11.27, which involves: (1) homogeneous azeotropic distillation with acetone entrainer to produce a bottoms product of nearly pure benzene and a distillate close in composition to the minimum-boiling binary azeotrope of acetone and cyclohexane; (2) solvent extraction of distillate with water to give a raffinate of cyclohexane and an extract of acetone and water; and (3) ordinary distillation of extract to recover acetone for recycle. In Example 11.6, the azeotropic distillation is subject to product-composition-region restrictions.

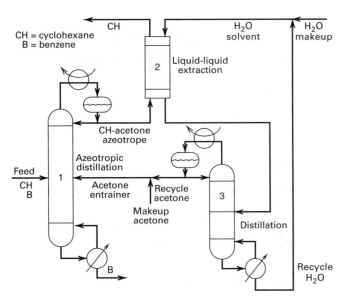

Figure 11.27 Sequence for separating cyclohexane and benzene using homogeneous azeotropic distillation with acetone entrainer.

[From *Perry's Chemical Engineers' Handbook*, 6th ed., R.H. Perry and D.W. Green, Eds., McGraw-Hill, New York (1984) with permission.]

EXAMPLE 11.6   Homogeneous Azeotropic Distillation.

Benzene (nbp = 80.13°C) and cyclohexane (nbp = 80.64°C) form a minimum-boiling homogeneous azeotrope at 1 atm and 77.4°C of 54.2 mol% benzene. Since they cannot be separated by distillation at 1 atm, it is proposed to separate them by homogeneous azeotropic distillation using acetone as entrainer in the sequence shown in Figure 11.27. The azeotropic-column feed consists of 100 kmol/h of 75 mol% benzene and 25 mol% cyclohexane. Determine a feasible acetone-addition rate so that nearly pure benzene can be obtained as bottoms product. Acetone (nbp = 56.14°C) forms a minimum-boiling azeotrope with cyclohexane at 53.4°C and 1 atm at 74.6 mol% acetone. Figure 11.28 is the residue-curve map at 1 atm.

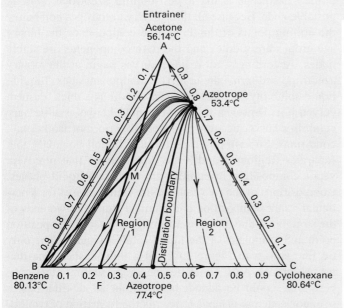

Figure 11.28 Residue-curve map for Example 11.6.

Solution

The residue-curve map shows a slightly curved distillation boundary connecting the two azeotropes and dividing the diagram into distillation regions 1 and 2. The fresh feed is designated in Figure 11.28 by a filled-in box labeled F. If a straight line is drawn from F to the pure acetone apex, A, the mixture of fresh feed and acetone entrainer must lie somewhere on this line in Region 1. Suppose 100 kmol/h of fresh feed is combined with an equal flow rate of entrainer. The mixing point, M, is located at the midpoint of the line connecting F and A. If a line is drawn from the benzene apex, B, through M and to the side that connects the acetone apex to the cyclohexane apex, it does not cross the distillation boundary separating the two regions, but lies completely in Region 1. Thus, the separation into a nearly pure benzene bottoms and a distillate mixture of mainly acetone and cyclohexane is possible. This is confirmed with the ASPEN PLUS process simulator for a column operating at 1 atm with 38 theoretical stages, a total condenser, a partial reboiler, $R = 4$, $B = 75$ kmol/h (equivalent to the benzene flow rate in the column feed), and a bubble-point combined feed to Stage 19 from the top. The product flow rates are listed in Table 11.4 as Case 3, where bottoms of 99.8 mol% benzene is obtained with a benzene recovery of the same value. A higher entrainer flow rate of 125 kmol/h, included as Case 4, also achieves a high benzene-bottoms-product purity and recovery. However, if only 75 kmol/h (Case 2) or 50 kmol/h (Case 1) of entrainer is used, a nearly pure benzene bottoms is not achieved because of the distillation boundary restriction.

Table 11.4 Effect of Acetone-Entrainer Flow Rate on Benzene Purity for Example 11.6

Case	Acetone Flow Rate, kmol/h	Benzene Purity in Bottoms, %
1	50	88.69
2	75	94.21
3	100	99.781
4	125	99.779

§11.6   HETEROGENEOUS AZEOTROPIC DISTILLATION

Required for homogeneous azeotropic distillation is that A and B lie in the same distillation region of the residue-curve map as entrainer E. This is so restrictive that it is difficult, if not impossible, to find a feasible entrainer. The Group 1 map in Figure 11.26a requires that the entrainer not form an azeotrope but yet be the intermediate-boiling component, while the other two components form a minimum-boiling azeotrope. Such systems are rare, because most intermediate-boiling entrainers form an azeotrope with one or both of the other two components. The other four groups in Figure 11.26 require formation of at least one maximum-boiling azeotrope. However, such azeotropes are far less common than minimum-boiling azeotropes. Thus, sequences based on homogeneous azeotropic distillation are rare and a better alternative is needed.

A better, alternative technique that finds wide use is *heterogeneous azeotropic distillation* to separate close-boiling binaries and minimum-boiling binary azeotropes by employing an entrainer that forms a binary and/or ternary heterogeneous (two-phase) azeotrope. As discussed in §4.3, a heterogeneous azeotrope has two or more liquid phases. If it has two, the overall, two-liquid-phase composition is equal to that of the vapor phase. Thus, all three phases have different compositions. The overhead vapor from the column is close to the composition of the heterogeneous azeotrope. When condensed, two liquid phases form in a decanter. After separation, most or all of the entrainer-rich liquid phase is returned to the column as reflux, while most or all of the other liquid phase is sent to the next column for further processing. Because these two liquid phases usually lie in different distillation regions of the residue-curve map, the restriction that dooms homogeneous azeotropic distillation is overcome. Thus, in heterogeneous azeotropic distillation, the components to be separated need not lie in the same distillation region.

Heterogeneous azeotropic distillation has been practiced for a century, first by batch and then by continuous processing. Two of the most widely used applications are (1) the use of benzene or another entrainer to separate the minimum-boiling azeotrope of ethanol and water, and (2) use of ethyl acetate or another entrainer to separate the close-boiling mixture of acetic acid and water. Other applications, cited by Widagdo and Seider [19], include dehydrations of isopropanol with isopropylether, *sec*-butyl-alcohol with *disec*-butylether, chloroform with mesityl oxide, formic acid with toluene, and acetic acid with toluene. Also, tanker-transported feedstocks such as benzene and styrene, which become water-saturated during transport, are dehydrated.

Consider separation of the ethanol–water azeotrope by heterogeneous azeotropic distillation. The two most widely used entrainers are benzene and diethyl ether, but others are feasible—including *n*-pentane, illustrated later, in Example 11.7—and cyclohexane. In 1902, Young [40] discussed the use of benzene as an entrainer for a batch process, in perhaps the first application of heterogeneous azeotropic distillation. In 1928, Keyes obtained a patent [41] on a continuous process, discussed in a 1929 article [42].

A residue-curve map, computed by Bekiaris, Meski, and Morari [43] for the ethanol (E)–water (W)–benzene (B) system at 1 atm, using the UNIQUAC equation (§2.6.8) for liquid-phase activity coefficients (with parameters from ASPEN PLUS), is shown in Figure 11.29. Superimposed on the map is a bold-dashed binodal curve for the two-liquid-phase-region boundary. The normal boiling points of E, W, and B are 78.4, 100, and 80.1°C, respectively. The UNIQUAC equation predicts that homogeneous minimum-boiling azeotropes AZ1 and AZ2 are formed by E and W at 78.2°C and 10.0 mol% W, and by E and B at 67.7°C and 44.6 mol% E, respectively. A heterogeneous minimum-boiling azeotrope AZ3 is predicted for W and B at 69.3°C, with a vapor composition of 29.8 mol% W. The overall two-liquid-phase composition is the same as that of the vapor, but each liquid phase is almost pure. The B-rich liquid phase is predicted to contain

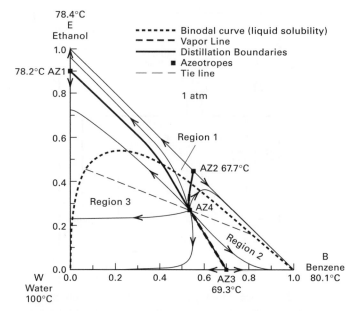

Figure 11.29 Residue-curve map for the ethanol–water–benzene system at 1 atm.

0.55 mol% W, while the W-rich liquid phase contains only 0.061 mol% B. A ternary minimum-boiling heterogeneous azeotrope AZ4 is predicted at 64.1°C, with a vapor composition of 27.5 mol% E, 53.1 mol% B, and 19.4 mol% W. The overall two-liquid-phase composition of the ternary azeotrope equals that of the vapor, but a thin, dashed tie line through AZ4 shows that the benzene-rich liquid phase contains 18.4 mol% E, 79.0 mol% B, and 2.6 mol% W, while the water-rich liquid phase contains 43.9 mol% E, 6.3 mol% B, and 49.8 mol% W.

In Figure 11.29, the map is divided into three distillation regions by three thick, solid-line distillation boundaries that extend from AZ4 to binary azeotropes at AZ1, 2, and 3. Each distillation region contains one pure component. Because the ternary azeotrope is the lowest-boiling azeotrope, it is an unstable node. Because all three binary azeotropes boil below the boiling points of the three pure components, the binary azeotropes are saddles and the pure components are stable nodes. Accordingly, all residue curves begin at the ternary azeotrope and terminate at a pure-component apex. Liquid–liquid solubility (binodal curve) is shown as a thick, dashed, curved line. However, this curve is not like the usual ternary solubility curve, because it is for isobaric, not isothermal, conditions. Superimposed on the distillation boundary that separates Regions 2 and 3 are thick dashes that represent vapor composition in equilibrium with two liquid phases. Compositions of the two equilibrium-liquid phases for a particular vapor composition are obtained from the two ends of the straight tie line that passes through the vapor composition point and terminates at the liquid solubility curve. The only tie line shown in Figure 11.29 is a thin, dashed line that passes through AZ4. Other tie lines, representing other temperatures, could be added; however, in all distillations, a strenuous attempt is made to restrict the formation of two liquid phases to the decanter because when two liquid phases form on a tray, the tray efficiency decreases.

Figure 11.29 clearly shows how a distillation boundary is crossed by the tie line through AZ4 to form two liquid phases in the decanter. This phase split is utilized in a typical operation, where the tower is treated as a column with no condenser, a main feed that enters a few trays below the top of the column, and the reflux of benzene-rich liquid as a second feed. The composition of the combined two feeds lies in Region 1. Thus, from the residue-curve directions, products of the tower can be a bottoms of nearly pure ethanol and an overhead vapor approaching the AZ4 composition. When that vapor is condensed, phase splitting occurs to give a water-rich phase that lies in Region 3 and an entrainer-rich phase in Region 2. If the water-rich phase is sent to a reboiled stripper, the residue curves indicate that a nearly pure-water bottoms can be produced, with the overhead vapor, rich in ethanol, recycled to the decanter. When the entrainer-rich phase in Region 2 is added to the main feed in Region 1, the overall composition lies in Region 1.

To avoid formation of two liquid phases on the top trays of the azeotropic tower, the composition of the vapor leaving the top tray must have an equilibrium liquid that lies outside of the two-phase-liquid region in Figure 11.29. Shown in Figure 11.30, from Prokopakis and Seider [44], are 18 vapor compositions that form two liquid phases when condensed, but are in equilibrium with only one liquid phase on the top tray, as restricted to the very small expanded window. That window is achieved by adding to the entrainer-rich reflux a portion of the water-rich liquid or some condensed vapor prior to separation in the decanter.

Figure 11.31, taken from Ryan and Doherty [45], shows three proposed heterogeneous azeotropic distillation schemes that utilize only distillation for the other column(s). Most common is the three-column sequence in Figure 11.31a, in which an aqueous feed dilute in ethanol is preconcentrated in Column 1 to obtain a nearly pure-water bottoms product and distillate close in composition to the binary azeotrope. The latter is fed to the azeotropic tower, Column 2, where nearly pure ethanol is recovered as bottoms and the tower is

refluxed by most or all of the entrainer-rich liquid from the decanter. The water-rich phase, which contains ethanol and a small amount of entrainer, is sent to the entrainer-recovery column, which is a distillation column or a stripper. Distillate from the recovery column is recycled to the azeotropic column. Alternatively, the distillate from Column 3 could be recycled to the decanter. As shown in all three sequences of Figure 11.31, portions of either liquid phase from the decanter can be returned to the azeotropic tower or to the next column in the sequence to control phase splitting on the top trays of the azeotropic tower.

A four-column sequence is shown in Figure 11.31b. The first column is identical to the first column of the three-column sequence of Figure 11.31a. The second column is the azeotropic column, which is fed by the near-azeotrope distillate of ethanol and water from Column 1 and by a recycle distillate of about the same composition from Column 4. The purpose of Column 3 is to remove, as distillate, entrainer from the water-rich liquid leaving the decanter and recycle it to the decanter. Ideally, the composition of this distillate is identical to that of the vapor distillate from Column 2. The bottoms from Column 3 is separated in Column 4 into a bottoms of nearly pure water, and a distillate that approaches the ethanol–water azeotrope and is therefore recycled to the feed to Column 2. Pham and Doherty [46] found no advantage of the four-column over the three-column sequence.

A novel two-column sequence, described by Ryan and Doherty [45], is shown in Figure 11.31c. The feed is sent to Column 2, a combined pre-concentrator and entrainer-recovery column. The distillate from this column is feed for the azeotropic column. The bottoms from Column 1 is nearly pure ethanol, while Column 2 produces a bottoms of nearly pure water. For feeds dilute in ethanol, Ryan and Doherty found that the two-column sequence has a lower investment cost, but a higher operating cost, than a three-column sequence. For ethanol-rich feeds, the two sequences are economically comparable.

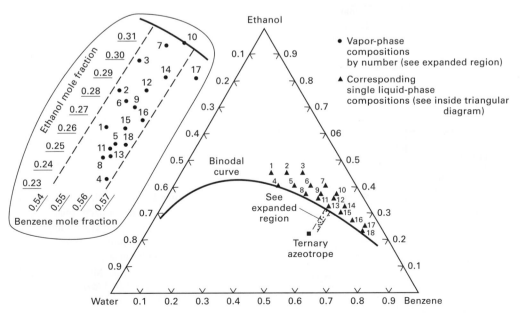

Figure 11.30 Overhead vapor compositions not in equilibrium with two liquid phases.

[From J. Prokopakis and W.D. Seider, *AIChE J.*, **29**, 49–60 (1983) with permission.]

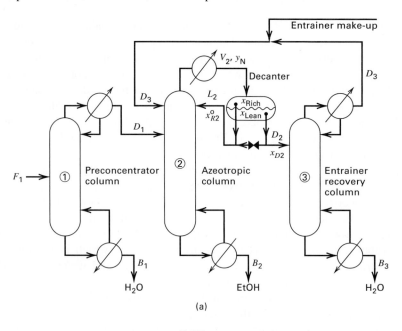

(a)

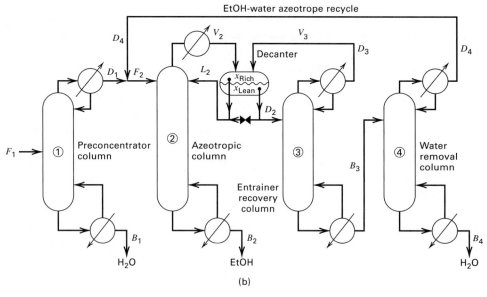

(b)

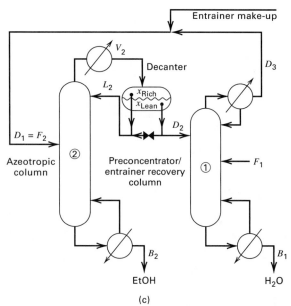

(c)

Figure 11.31 Distillation sequences for heterogeneous azeotropic distillation: (a) Three-column sequence; (b) four-column sequence; (c) two-column sequence.

[From P.J. Ryan and M.F. Doherty, *AIChE J.*, **35**, 1592–1601 (1989) with permission.]

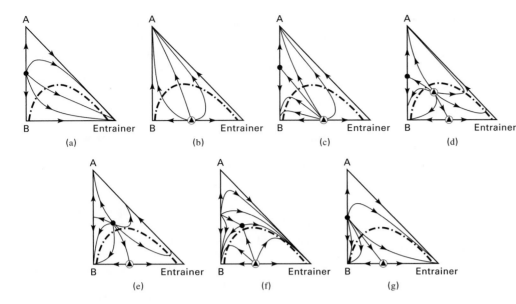

Figure 11.32 Residue-curve maps for heterogeneous azeotropic distillations that lead to feasible distillation sequences.

[H.N. Pham and M.F. Doherty, *Chem. Eng. Sci.*, **45**, 1845–1854 (1990) with permission.]

The ethanol–benzene–water residue-curve map of Figure 11.29 is one of a number of residue-curve maps that can lead to feasible distillation sequences that include heterogeneous azeotropic distillation. Pham and Doherty [46] note that a feasible entrainer is one that causes phase splitting over a portion of the three-component region, but does not cause the two feed components to be placed in different distillation regions. Figure 11.32 shows seven such maps, where the dash-dot lines are liquid–liquid solubility (binodal) curves.

Convergence of computer simulations for heterogeneous azeotropic distillation columns by the methods described in Chapter 10 is difficult, especially when convergence of the entire sequence is attempted. It is preferable to uncouple the columns by using a residue-curve map to establish, by material-balance calculations, flow rates and compositions of feeds and products for each column. This procedure is illustrated for a three-column sequence in Figure 11.33, where the dash-dot lines separate the three distillation regions, the short-dash line is the liquid–liquid solubility curve, and the remaining lines are material-balance lines. Each column in the sequence is computed separately. Even then, the calculations can fail because of nonidealities in the liquid phase and possible phase splitting, making it necessary to use more robust methods such as the boundary-value, tray-by-tray method of Ryan and Doherty [45], the homotopy-continuation method of Kovach and Seider [47], and the collocation method of Swartz and Stewart [48].

§11.6.1 Multiplicity of Solutions

Solutions to nonlinear mathematical models are not always unique. The existence of multiple, steady-state solutions for continuous, stirred-tank reactors has been known since at least 1922 and is described in a number of textbooks on chemical reaction engineering, where typically one or more of the multiple solutions are unstable and, therefore, unoperable. The existence of multiplicity in steady-state separation problems is a relatively new discovery. Gani and Jørgensen

[49] define three types of multiplicity, all of which can occur in distillation simulations:

1. *Output multiplicity*, where all input variables are specified and more than one solution for the output variables, typically sets of product compositions and temperature profiles, are found.

2. *Input multiplicity*, where one or more output variables are specified and multiple solutions are found for the unknown input variables.

3. *Internal-state multiplicity*, where multiple sets of internal conditions or profiles are found for the same values of the input and output variables.

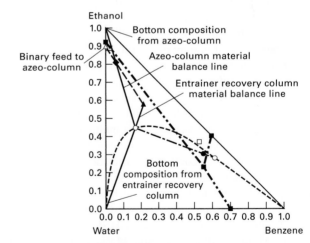

- ● Overall vapor composition from azeo-column
- □ Liquid in equilibrium with overhead vapor composition from azeo-column
- ▲ Distillate composition from entrainer recovery column
- ◆ Overall feed composition to azeo-column
- ■ Azeotrope

Figure 11.33 Material-balance lines for the three-column sequence of Figure 11.31a.

[From P.J. Ryan and M.F. Doherty, *AIChE J.*, **35**, 1592–1601 (1989) with permission.]

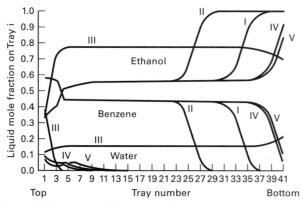

Figure 11.34 Five multiple solutions for a heterogeneous distillation operation.

[From J.W. Kovach III and W.D. Seider, *Computer Chem. Engng.*, **11**, 593 (1987) with permission.]

Output multiplicity for azeotropic distillation was first discovered by Shewchuk [50] in 1974. With different starting guesses, two steady-state solutions for the dehydration of ethanol by heterogeneous azeotropic distillation with benzene were found. In a more detailed study, Magnussen, Michelsen, and Fredenslund [51] found, with difficulty, for a narrow range of ethanol flow rate in the top feed to the column, three steady-state solutions, one of which was unstable. One of the two stable solutions predicts a far purer ethanol bottoms product than the other stable solution. Thus, from a practical standpoint, it is important to obtain all stable solutions when more than one exists. Subsequent studies, some contradictory, show that multiple solutions persist only over a narrow range of D or B, but may exist over a wide range of R, provided there are sufficient stages. Composition profiles of five solutions found by Kovach and Seider [47] for a 40-tray ethanol–water–benzene heterogeneous azeotropic distillation are given in Figure 11.34. The variation in the profiles is extremely large, showing again that it is important to locate all multiple solutions when they exist. Unfortunately, process simulators do not seek multiple solutions, and finding these solutions is difficult because: (1) azeotropic columns are difficult to converge to even one solution, (2) multiple solutions may exist only in a restricted range of input variables, (3) multiple solutions can be found only by changing initial-composition guesses, and (4) choice of an activity-coefficient correlation and interaction parameters can be crucial. The best results are obtained when advanced mathematical techniques such as continuation and bifurcation analysis are employed, as described by Kovach and Seider [47], Widagdo and Seider [19]; Bekiaris, Meski, Radu, and Morari [52]; and Bekiaris, Meski, and Morari [43]. The last two articles provide explanations why multiple solutions occur in azeotropic distillations.

EXAMPLE 11.7 Heterogeneous Azeotropic Distillation.

Studies by Black and Ditsler [53] and Black, Golding, and Ditsler [54] show that *n*-pentane is a superior entrainer for dehydrating ethanol. Like benzene, *n*-pentane forms a minimum-boiling

heterogeneous ternary azeotrope with ethanol and water. Design a system for dehydrating 16.82 kmol/h of 80.94 mol% ethanol and 19.06 mol% water as a liquid at 344.3 K and 333 kPa, using *n*-pentane as an entrainer, to produce 99.5 mol% ethanol, and water with less than 100 ppm (by weight) of combined ethanol and *n*-pentane.

Solution

This ternary system has been studied by Black [55], who proposed the two-column flow diagram in Figure 11.35. Included are an 18-equilibrium-stage heterogeneous azeotropic distillation column (C-1) equipped with a total condenser and a partial reboiler, a decanter (D-1), a 4-equilibrium-stage reboiled stripper (C-2), and a condenser (E-1) to condense the overhead vapor from C-2. Each reboiler adds another equilibrium stage. Column C-1 operates at a bottoms pressure of 344.6 kPa with a column pressure drop of 13.1 kPa. Column C-2 operates at a top pressure of 308.9 kPa, with a column pressure drop of 3.0 kPa. These pressures permit the use of cooling water in the condensers. Purity specifications are placed on the bottoms products. Feed enters C-1 at Stage 3 from the top. The ethanol product is withdrawn from the bottom of C-1. A small *n*-pentane makeup stream, not shown in Figure 11.35, enters Stage 2 from the top. The overhead vapor from C-1 is condensed and sent to D-1, where a pentane-rich liquid phase and a water-rich liquid phase are formed. The pentane-rich phase is returned to C-1 as reflux, while the water-rich phase is sent to C-2, where the water is stripped of residual pentane and ethanol to produce a bottoms of the specified water purity. Twenty % of the condensed vapor from C-2 is returned to D-1. To ensure that two liquid phases form in the decanter, but not on the trays of C-1, the remaining 80% of the condensed vapor from C-2 is combined with the pentane-rich phase from D-1 for use as additional reflux to C-1. The specifications are included on Figure 11.35.

A very important step in the design of a heterogeneous azeotropic distillation column is the selection of a suitable method for predicting liquid-phase activity coefficients and determination of binary interaction parameters. The latter usually involves regression of both vapor–liquid (VLE) and liquid–liquid (LLE) data for all binary pairs. If available, ternary data can also be included in the regression. Unfortunately, for most activity-coefficient prediction methods, it is difficult to simultaneously fit VLE and LLE data. For this reason, different binary interaction parameters are often used for the azeotropic column, where VLE is important, and for the decanter, where LLE is important. This has been found especially desirable for the ethanol–water–benzene system. For this example, however, a single set of binary interaction parameters, with a modification by Black [56] of the Van Laar equation (§2.6.5), was adequate. The binary interaction parameters are listed by Black et al. [54].

The calculations were made with Simulation Sciences, Inc., software using their rigorous distillation routine to model the columns and a three-phase-flash routine (§4.10.3) to model the decanter. Because the entrainer was internal to the system, except for a very small makeup rate, it was necessary to provide initial guesses for the component flow rates in the combined decanter feed. Guessed values in kmol/h were 25.0 for *n*-pentane, 3.0 for ethanol, and 7.5 for water. The converged material balance is given in Table 11.5. Product specifications are met and 22.6 kmol/h of *n*-pentane circulates through the top trays of the azeotropic distillation column. The computed condenser and reboiler duties for Column C-1 are 1,116.5 and 1,135.0 MJ/h, respectively. The reboiler duty for Column C-2 is 486 MJ/h, and the duty for Condenser E-1 is 438 MJ/h.

Because of the large effect of composition on liquid-phase activity coefficients, column profiles for azeotropic columns often show steep fronts. In Figure 11.36a to c, stage temperatures, total vapor

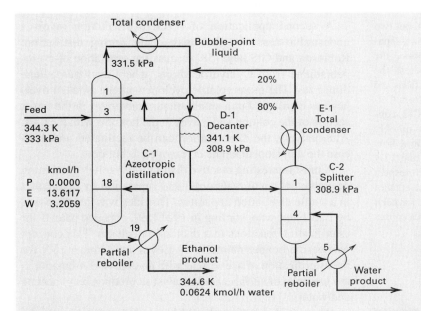

Figure 11.35 Process-flow diagram for Example 11.7. [From *Perry's Chemical Engineers' Handbook*, 6th ed., R.H. Perry and D.W. Green, Eds., McGraw-Hill, New York (1984) with permission.]

and liquid flow rates, and liquid-phase compositions for Column C-1 vary only slightly from the reboiler (Stage 19) up to Stage 13. In this region, the liquid is greater than 99 mol% ethanol, whereas the n-pentane concentration slowly builds up from a negligible concentration in the bottoms to just less than 0.02 mol% at Stage 13. From Stage 13 to Stage 8, the n-pentane mole fraction in the liquid increases rapidly to 53.8 mol%. In the same region, temperature decreases from 385.6 K to 348.4 K. Continuing up the column from Stage 8 to 3, where feed enters, the most significant change is the mole fraction of water in the liquid. Rather drastic changes in all variables occur near Stage 3. Effects of n-C5 concentration on the α of water to ethanol, and of water on n-C5 to ethanol, are shown in Figure 11.36d, where the variation over the column is 10-fold for each pair.

Table 11.5 Converged Material Balance for Example 11.7

Stream	Flowrate, kmol/h			
	n-Pentane	Ethanol	Water	Total
C-1 feed	0.0000	13.6117	3.2059	16.8176
C-1 overhead	22.5565	2.1298	10.7269	35.4132
C-1 bottoms	0.0000	13.6117	0.0624	13.6741
C-1 reflux	22.5565	2.1298	7.5834	32.2697
D-1 nC5-rich	22.5500	1.0637	0.1129	23.7266
D-1 water-rich	0.0081	1.3326	12.4816	13.8223
C-2 overhead	0.0081	1.3326	9.3381	10.6788
C-2 bottoms	0.0000	0.0000	3.1435	3.1435

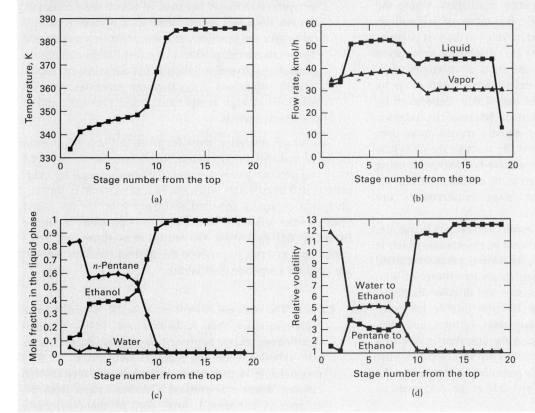

Figure 11.36 Results for azeotropic distillation column of Example 11.7. (a) Temperature profile. (b) Vapor and liquid traffic profiles. (c) Liquid-phase composition profiles. (d) Relative volatility profiles.

No phase splitting occurs on the plates of either column, but two liquid phases of quite different composition are formed and separated in the decanter. The light phase, which is almost twice the quantity of the heavy phase, is 95 mol% *n*-pentane, whereas the heavy phase is 90 mol% water. These different compositions are due to the small amount of ethanol in the overhead vapor from C-1. Because of the high concentration of water in the feed to C-2, concentrations of ethanol and *n*-pentane in the liquid are quickly reduced to ppm. Temperatures, vapor flow rates, and liquid flow rates in C-2 are almost constant at 408 K, 15.6 kmol/h, and 12.4 kmol/h, respectively. Because of the large α of ethanol with respect to water ($\sim$ 9) under the dilute ethanol conditions in C-2, the ethanol mole fraction decreases by almost an order of magnitude for each stage. The very large α of *n*-C5 to water (more than 1,000) causes the *n*-C5 to be entirely stripped in just two stages.

§11.7 REACTIVE DISTILLATION

Reactive distillation denotes simultaneous chemical reaction and distillation. The reaction usually takes place in the liquid phase or at the surface of a solid catalyst in contact with the liquid. One application of reactive distillation, described by Terrill, Sylvestre, and Doherty [57], is the separation of a close-boiling or azeotropic mixture of components A and B, where a chemically reacting entrainer E is introduced into the column. If A is the lower-boiling component, it is preferable that E be higher boiling than B and that it react selectively and reversibly with B to produce reaction product C, which also has a higher boiling point than component A and does not form an azeotrope with A, B, or E. Component A is removed as distillate, and components B and C, together with any excess E, are removed as bottoms. Components B and E are recovered from C in a separate distillation, where the reaction is reversed (C $\rightarrow$ B + E); B is taken off as distillate, and E is taken off as bottoms and recycled to the first column.

Terrill, Sylvestre, and Doherty [57] discuss the application of reactive entrainers to the separation of mixtures of *p*-xylene and *m*-xylene, whose normal boiling points differ by only 0.8°C, resulting in an α of only 1.029. Separation by ordinary distillation is impractical because, to produce 99 mol% pure products from an equimolar feed, more than 500 theoretical stages are required. By reacting the *m*-xylene with a reactive entrainer such as *tert*-butylbenzene using a solid aluminum chloride catalyst, or chelated sodium *m*-xylene dissolved in cumene, stage requirements are drastically reduced.

Closely related to the use of reactive entrainers in distillation is the use of reactive absorbents in absorption, which is widely practiced. For example, sour natural gas is sweetened by removal of H_2S and CO_2 acid gases by absorption into aqueous alkaline solutions of mono- and di-ethanolamines. Fast, reversible reactions occur to form soluble-salt complexes such as carbonates, bicarbonates, sulfides, and mercaptans. The rich solution leaving the absorber is sent to a reboiled stripper, where the reactions are reversed at higher temperatures and often at lower pressures to regenerate the amine solution as the bottoms and deliver the acid gases as overhead vapor.

A second application of reactive distillation involves undesirable reactions that may occur during distillation. Robinson and Gilliland [58] discuss the separation of cyclopentadiene from C_7 hydrocarbons, where cyclopentadiene dimerizes. The more volatile cyclopentadiene is taken overhead as distillate, but a small amount dimerizes in the lower section of the column and leaves in the bottoms with the C_7's. Alternatively, the dimerization can be facilitated, in which case the dicyclopentadiene is removed as bottoms.

A most interesting reactive distillation involves combining desirable chemical reaction(s) and separation by distillation in a single distillation apparatus. This idea was first proposed by Backhaus, who, starting in 1921 [59], obtained patents for esterification reactions in a distillation column. This concept was verified experimentally by Leyes and Othmer [60] for the esterification of acetic acid with an excess of *n*-butanol in the presence of sulfuric acid catalyst to produce butyl acetate and water.

Reactive distillation should be considered whenever the following hold:

1. The chemical reaction occurs in the liquid phase, in the presence or absence of a homogeneous catalyst, or at the interface of a liquid and a solid catalyst.

2. Feasible temperature and pressure for the reaction and distillation are the same. That is, reaction rates and distillation rates are of the same order of magnitude.

3. The reaction is equilibrium-limited so that if one or more of the products can be removed by distillation, the reaction can be driven to completion; thus, a large reactant excess is not necessary for a high conversion. This is particularly advantageous when excess reagent recovery is difficult because of azeotrope formation. For reactions that are irreversible, it is more economical to take the reactions to completion in a reactor and then separate the products in a distillation column. In general, reactive distillation is not attractive for supercritical conditions, for gas-phase reactions, and for reactions at high temperatures and pressures and/or that involve solids.

Careful consideration must be given to the configuration of reactive distillation columns. Important factors are feed entry and product-removal stages, the possible need for intercoolers and interheaters when the heat of reaction is appreciable, and obtaining required residence time for the liquid phase. In the following ideal cases, it is possible, as shown by Belck [61] and others, for several two-, three-, and four-component systems to produce the desired products without the need for a separate distillation.

Case 1: The reaction A $\leftrightarrow$ R or A $\leftrightarrow$ 2R, where R has a higher volatility than A. In this case, only a reboiled rectification section is needed. Feed A is sent to the column reboiler where the reaction takes place. As R is produced, it is vaporized, passing to the rectification column, where it is purified. Overhead vapor from the column is condensed, with part of the condensate returned as reflux. Chemical reaction may also take place in the column. If A and R form a maximum-boiling azeotrope, this configuration is still applicable if the mole fraction of R in the reboiler is greater than the azeotropic composition.

Case 2: The reaction A $\leftrightarrow$ R or 2A $\leftrightarrow$ R, where A has the higher volatility. In this case, only a stripping section is needed. Liquid A is sent to the top of the column, from which it flows down, reacting to produce R. The column is provided with a total condenser and partial reboiler. Product R is withdrawn from the reboiler. This configuration requires close examination because, at a certain location in the column, chemical equilibrium may be achieved, and if the reaction is allowed to proceed beyond that point, the reverse reaction can occur.

Case 3: The reactions 2A $\leftrightarrow$ R + S or A + B $\leftrightarrow$ R + S, where A and B are intermediate in volatility to R and S, and R has the highest volatility. Feed enters an ordinary distillation column somewhere near the middle, with R withdrawn as distillate and S as bottoms. If B is less volatile than A, then B may enter the column separately and at a higher level than A.

Commercial applications of reactive distillation include the following:

1. Esterification of acetic acid with ethanol to produce ethyl acetate and water

2. Reaction of formaldehyde and methanol to produce methylal and water, using a solid acid catalyst [62]

3. Esterification of acetic acid with methanol to produce methyl acetate and water, using a sulfuric acid catalyst, as patented by Agreda and Partin [63], and described by Agreda, Partin, and Heise [64]

4. Reaction of isobutene with methanol to produce methyl-*tert*-butyl ether (MTBE), using a solid, strong-acid ion-exchange resin catalyst, as patented by Smith [65–67] and further developed by DeGarmo, Parulekar, and Pinjala [68]

Consider the esterification of acetic acid (A) with ethanol (B) to produce ethyl acetate (R) and water (S). The respective normal boiling points in °C are 118.1, 78.4, 77.1, and 100. Also, minimum-boiling, binary homogeneous azeotropes are formed by B–S at 78.2°C with 10.57 mol% B, and by B–R at 71.8°C with 46 mol% B. A minimum-boiling, binary heterogeneous azeotrope is formed by R–S at 70.4°C with 24 mol% S, and a ternary, minimum-boiling azeotrope is formed by B–R–S at 70.3°C with 12.4 mol% B and 60.1 mol% R. Thus, this system is exceedingly complex and nonideal. A number of studies, both experimental and computational, have been published, many of which are cited by Chang and Seader [69], who developed a robust computational procedure for reactive distillation based on a homotopy-continuation method. More recently, other algorithms have been reported by Venkataraman, Chan, and Boston [70] and Simandl and Svrcek [71]. Kang, Lee, and Lee [72] obtained binary interaction parameters for the UNIQUAC equation (§2.6.8)

by fitting experimental data simultaneously for vapor–liquid equilibrium and liquid-phase chemical equilibrium.

In all of the computational procedures, a reaction-rate term must be added to the stage material balance. Chang and Seader [69] modified (10-58) of the NR method (§10.4) to include a reaction-rate source term for the liquid phase, assuming the liquid is completely mixed:

$$M_{i,j} = l_{i,j}(1 + s_j) + v_{i,j}(1 + S_j) - l_{i,j-1} - v_{i,j+1} - f_{i,j}$$
$$- (V_{LH})_j \sum_{n=1}^{NRX} v_{i,n} r_{j,n}, \quad i = 1, \ldots C \qquad (11\text{-}17)$$

where

$(V_{LH})_j$ = volumetric liquid holdup on stage j;

$\quad v_{i,n}$ = stoichiometric coefficient for component i and reaction n using positive values for products and negative values for reactants;

$\quad r_{j,n}$ = reaction rate for reaction n on stage j, as the increase in moles of a reference reactant per unit time per unit volume of liquid phase; and

NRX = number of reversible and irreversible reactions.

Typically, each reaction rate is expressed in a power-law form with liquid molar concentrations (where the n subscript is omitted):

$$r_j = \sum_{p=1}^{2} k_p \prod_{q=1}^{NRC} c_{j,q}^m = \sum_{p=1}^{2} A_p \exp\left(-\frac{E_p}{RT_j}\right) \prod_{q=1}^{NRC} c_{j,q}^m \quad (11\text{-}18)$$

where

$\quad c_{j,q}$ = concentration of component q on stage j;

$\quad k_p$ = reaction-rate constant for the pth term, where $p = 1$ indicates the forward reaction and $p = 2$ indicates the reverse reaction, with k_1 positive and k_2 negative;

$\quad m$ = the exponent on the concentration;

NRC = number of components in the power-law expression;

$\quad A_p$ = pre-exponential (frequency) factor; and

$\quad E_p$ = activation energy.

With (11-17) and (11-18), a reaction may be treated as irreversible ($k_2 = 0$), reversible (k_2 negative and not equal to zero), or at equilibrium. The latter can be achieved by using very large values for the volumetric liquid holdup at each stage in the case of a single, reversible reaction, or by multiplying each of the two frequency factors, A_1 and A_2, by the same large number, thus greatly increasing the forward and backward reactions, but maintaining the correct value for the chemical-reaction equilibrium constant. For equilibrium reactions, it is important that the power-law expression for the backward reaction be derived from the power-law expression for the forward reaction and the reaction stoichiometry so as to be consistent with the expression for the chemical-reaction equilibrium constant. The volumetric liquid holdup for a stage, when using a trayed tower, depends on the: (1) active bubbling area, (2) height of the froth as influenced by the weir height, and (3) liquid-volume fraction of the froth. These factors are all considered in §6.6.3. In general, the liquid backup in the downcomer is not included in the estimate of liquid holdup. When large holdups are necessary, bubble-cap trays are preferred because they do not weep. When the chemical reaction is in the reboiler, a large liquid holdup can be provided.

The following example deals with the esterification of acetic acid with ethanol to produce ethyl acetate and water. Here, the single, reversible chemical reaction is assumed to reach chemical equilibrium at each stage. Thus, no estimate of liquid holdup is needed. In a subsequent example, chemical equilibrium is not achieved, so holdup estimates are made, and tower diameter is assumed.

EXAMPLE 11.8   Ethanol by Reactive Distillation.

A reactive-distillation column with 13 theoretical stages and equipped with a total condenser and partial reboiler is used to produce ethyl acetate (R) at 1 atm. A saturated-liquid feed of 90 lbmol/h of acetic acid (A) enters Stage 2 from the top, while 100 lbmol/h of a saturated liquid of 90 mol% ethanol (B) and 10 mol% water (S) (close to the azeotropic composition) enters Stage 9 from the top. Thus, the acetic acid and ethanol are in stoichiometric ratio for esterification. Other specifications are $R = 10$ and $D = 90$ lbmol/h, in the hope that complete conversion to ethyl acetate (the low boiler) will occur. Data for the homogeneous reaction are given by Izarraraz et al. [73], in terms of the rate law:

$$r = k_1 c_A c_B - k_2 c_R c_S$$

with $k_1 = 29{,}000 \exp(-14{,}300/RT)$ and $k_2 = 7{,}380 \exp(-14{,}300/RT)$, both in L/(mol-minute) with T in K. Because the

activation energies for the forward and backward steps are identical, the chemical-equilibrium constant is independent of temperature and equal to $k_1/k_2 = 3.93$. Assume that chemical equilibrium is achieved at each stage. Thus, very large values of liquid holdup are specified for each stage. Binary interaction parameters for all six binary pairs, for predicting liquid-phase activity coefficients from the UNIQUAC equation (§2.6.8), are as follows, from Kang et al. [72]:

Components in Binary Pair, i–j	Binary Parameters	
	$u_{i,j}/R$, K	$u_{j,i}/R$, K
Acetic acid–ethanol	268.54	−225.62
Acetic acid–water	398.51	−255.84
Acetic acid–ethyl acetate	−112.33	219.41
Ethanol–water	−126.91	467.04
Ethanol–ethyl acetate	−173.91	500.68
Water–ethyl acetate	−36.18	638.60

Vapor-phase association of acetic acid and possible formation of two liquid phases on the stages must be considered. Calculate compositions of distillate and bottoms products and determine the liquid-phase-composition and reaction-rate profiles.

Solution

Calculations were made with the NR method of the CHEMCAD process simulation program, where the total condenser is counted as the first stage. The only initial estimates provided were 163 and 198°F for the temperatures of the distillate and the bottoms, respectively. Calculations converged in 17 iterations to the values below:

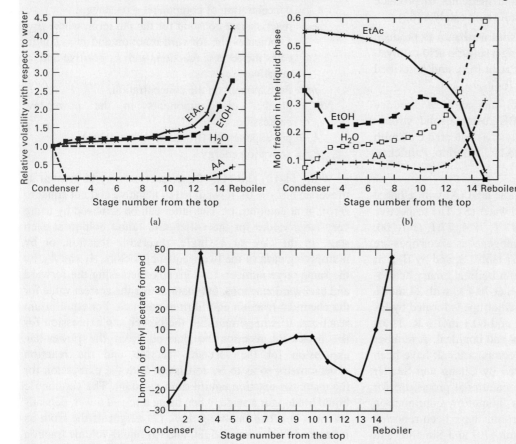

Figure 11.37 Profiles for reactive distillation in Example 11.8. (a) Relative volatility profile. (b) Liquid-phase mole-fraction profiles. (c) Reaction-rate profile.

	Product Flow Rates, lbmol/h	
Component	Distillate	Bottoms
Ethyl acetate	49.52	6.39
Ethanol	31.02	3.07
Water	6.73	59.18
Acetic acid	2.73	31.36
Total	90.00	100.00

All four components appear in both products. The overall conversion to ethyl acetate is only 62.1%, with 88.6% of this going to the distillate. The distillate is 55 mol% acetate; the bottoms is 59.2 mol % water. Only small composition changes occur when feed locations are varied. Two factors in the failure to achieve high conversion and nearly pure products are (1) the highly nonideal nature of the quaternary mixture, accompanied by the large number of azeotropes, and (2) the tendency of the reverse reaction to occur on certain stages. The former is shown in Figure 11.37a, where the α values between ethyl acetate and water and between ethanol and water in the top section are no greater than 1.25, making the separations difficult.

The liquid-phase mole-fraction distribution is shown in Figure 11.37b, where, between feed points, compositions change slowly despite the esterification reaction. In Figure 11.37c, the reaction-rate profile is unusual. Above the upper feed stage (now Stage 3), the reverse reaction is dominant. From that feed point down to the second feed entry (now Stage 10), the forward reaction dominates mainly at the upper feed stage. The reverse reaction is dominant for Stages 11–13, whereas the forward reaction dominates at Stages 14 and 15 (the reboiler). The largest extents of forward reaction occur at Stages 3 and 15. Even with 60 stages, and with the reaction confined to Stages 25 to 35, the distillate contains appreciable ethanol and the bottoms contains a substantial fraction of acetic acid. For this example, development of a reactive-distillation scheme for achieving a high conversion and nearly pure products represents a significant challenge.

EXAMPLE 11.9 Reactive Distillation to Produce MTBE.

Following the ban on addition of tetraethyl lead to gasoline by the U.S. Amendment to the Clean Air Act of 1990, refiners accelerated the addition of methyl-tert butyl ether (MTBE) to gasoline to boost octane number and reduce unburned hydrocarbons and CO_2 in automobile exhaust. More than 50 MTBE plants were constructed, with many using reactive distillation to produce MTBE from methanol. However, by 2002, MTBE also fell into disfavor in the U.S. because when it leaked from an underground tank at a gas station, it dissolved easily and traveled quickly in goundwater, causing contamination in wells and cancer in animals.

Using thermodynamic and kinetic data from Rehfinger and Hoffmann [74] for the formation of methyl-tert-butyl ether (MTBE) from methanol (MeOH) and isobutene (IB), in the presence of n-butene (NB), both Jacobs and Krishna [75] and Nijhuis, Kerkhof, and Mak [76] found drastically different IB conversions when the feed stage for methanol was varied. An explanation for these multiple solutions is given by Hauan, Hertzberg, and Lien [77].

Compute a converged solution, taking into account the reaction kinetics but assuming vapor–liquid equilibrium at each stage. The column has a total condenser, a partial reboiler, 15 stages at V/L equilibrium, and operates at 11 bar. The total condenser is numbered Stage 1 even though it is not an equilibrium stage. The mixed butenes feed, consisting of 195.44 mol/s of IB and 353.56 mol/s of NB, enters Stage 11 as a vapor at 350 K and 11 bar. The methanol, at a flow rate of 215.5 mol/s, enters Stage 10 as a liquid at 320 K and 11 bar. $R = 7$ and $B = 197$ mol/s. The catalyst is provided only for Stages 4 through 11, with 204.1 kg of catalyst per stage. The catalyst is a strong-acid, ion-exchange resin with 4.9 equivalents of acid groups per kilogram of catalyst. Thus, the equivalents per stage are 1,000, or 8,000 for the eight stages. Compute the product compositions and column profiles.

Solution

The RADFRAC model in ASPEN PLUS was used. Because NB is inert, the only chemical reaction considered is

$$IB + MeOH \leftrightarrow MTBE$$

For the forward reaction, the rate law is formulated in terms of mole-fraction concentrations as in Rehfinger and Hoffmann [74]:

$$r_{forward} = 3.67 \times 10^{12} \exp(-92,440/RT) x_{IB}/x_{MeOH} \quad (1)$$

The corresponding backward rate law, consistent with chemical equilibrium, is

$$r_{backward} = 2.67 \times 10^{17} \exp(-134,454/RT) x_{MTBE}/x_{MeOH}^2 \quad (2)$$

where r is in moles per second per equivalent of acid groups, $R = 8.314$ J/mol-K, T is in K, and x_i is liquid mole fraction.

The Redlich–Kwong equation of state (§2.5.1) is used to estimate vapor-phase fugacities with the UNIQUAC equation (§2.6.8) to estimate liquid-phase activity coefficients. The UNIQUAC binary interaction parameters are as follows, where it is important to include the inert NB in the system by assuming it has the same parameters as IB and that the two butenes form an ideal solution. The parameters are defined as follows, with all $a_{ij} = 0$.

$$T_{ij} = \exp\left(-\frac{u_{ij} - u_{jj}}{RT}\right) = \exp\left(a_{ij} + \frac{b_{ij}}{T}\right) \quad (3)$$

	Binary Parameters	
Components in Binary Pair, ij	b_{ij}, K	b_{ji}, K
MeOH–IB	35.38	−706.34
MeOH–MTBE	88.04	−468.76
IB–MTBE	−52.2	24.63
MeOH–NB	35.38	−706.34
NB–MTBE	−52.2	24.63

The only initial guesses provided are temperatures of 350 and 420 K, for Stages 1 and 17, respectively; liquid-phase mole fractions of 0.05 for MeOH and 0.95 for MTBE leaving Stage 17; and vapor-phase mole fractions of 0.125 for MeOH and 0.875 for MTBE leaving Stage 17. The ASPEN PLUS input data are listed in the first and second editions of this book. The converged temperatures for Stages 1 and 17 are 347 and 420 K. Converged products are:

	Flow Rate, mol/s	
Component	Distillate	Bottoms
MeOH	28.32	0.31
IB	7.27	1.31
NB	344.92	8.64
MTBE	0.12	186.74
Total	380.63	197.00

The combined feeds contained a 10.3% mole excess of MeOH over IB. Therefore, IB was the limiting reactant and the preceding product distribution indicates that 95.6% of the IB, or 186.86 mol/s, reacted to form MTBE. The percent purity of the MTBE in the bottoms is 94.8%. Only 2.4% of the inert NB and 1.1% of the unreacted MeOH are in the bottoms. The condenser and reboiler duties are 53.2 and 40.4 MW.

Seven iterations gave a converged solution. Figure 11.38a shows that most of the reaction occurs in a narrow temperature range of 348.6 to 353 K. Figure 11.38b shows that vapor traffic above the two feed entries changes by less than 11% because of small changes in temperature. Below the two feed entries, temperature increases rapidly from 353 to 420 K, causing vapor traffic to decrease by about 20%. In Figure 11.38c, composition profiles show that the liquid is dominated by NB from the top down to Stage 13, thus drastically reducing the reaction driving force. Below Stage 11, liquid becomes richer in MTBE as mole fractions of other components decrease because of increasing temperature. Above the reaction zone, the mole fraction of MTBE quickly decreases as one moves to the top stage. These changes are due mainly to the large differences between the K-values for MTBE and those for the other three components. The α of MTBE with any of the other components ranges from about 0.24 at the top stage to about 0.35 at the bottom. Nonideality in the liquid influences mainly MeOH, whose liquid-phase activity coefficient

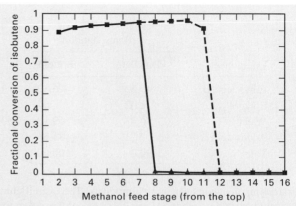

Figure 11.39 Effect of MeOH feed-stage location on conversion of IB to MTBE.

varies from a high of 10 at Stage 5 to a low of 2.6 at Stage 17. This causes the unreacted MeOH to leave mainly with the NB in the distillate rather than with MTBE in the bottoms. The rate-of-reaction profile in Figure 11.38d shows that the forward reaction dominates in the reaction section; however, 56% of the reaction occurs on Stage 10, the MeOH feed stage. The least amount of reaction is on Stage 11.

The literature indicates that conversion of IB to MTBE depends on the MeOH feed stage. In the range of MeOH feed stages from about 8 to 11, both low- and high-conversion solutions exists. This is shown in Figure 11.39, where the high-conversion solutions are in the 90+ % range, while the low-conversion solutions are all less than 10%. However, if component activities rather than mole fractions are used in the rate expressions, the low-conversion solutions are higher because of the large MeOH activity coefficient. The results in Figure 11.39 were obtained starting with the MeOH feed entering Stage 2. The resulting profiles were used as the initial guesses for the run with MeOH entering Stage 3. Subsequent runs were performed in a similar manner, increasing the MeOH feed stage by 1 each time and initializing with the results of the previous run.

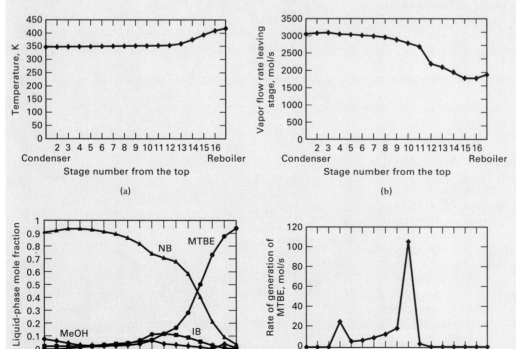

(a)

(b)

(c)

(d)

Figure 11.38 Profiles for reactive distillation in Example 11.9.
(a) Temperature profile. (b) Vapor traffic profile. (c) Liquid-phase mole-fraction profile.
(d) Reaction-rate profile.

High-conversion solutions were obtained for each run, until the MeOH feed stage was lowered to Stage 12, at which point conversion decreased dramatically. Further lowering of the MeOH feed stage to Stage 16 also resulted in a low-conversion solution. However, when the direction of change to the MeOH feed stage was reversed starting from Stage 12, a low conversion was obtained until the feed stage was decreased to Stage 9, at which point the conversion jumped back to the high-conversion result.

Huss et al. [101] present a study of reactive distillation for the acid-catalyzed reaction of acetic acid and methanol to produce methyl acetate and water, including the side reaction of methanol dehydration, using simulation models and experimental measurements. They consider both finite reaction rates and chemical equilibrium, coupled with phase equilibrium. The results include reflux limits and multiple solutions.

§11.8 SUPERCRITICAL-FLUID EXTRACTION

Solute extraction from a liquid or solid mixture is usually accomplished with a liquid solvent at conditions of temperature and pressure substantially below the solvent critical point, as discussed in Chapters 8 and 16, respectively. Following extraction, solvent and dissolved solute are subjected to subsequent separations to recover solvent for recycle and to purify the solute.

In 1879, Hannay and Hogarth [78] reported that solid potassium iodide could be dissolved in ethanol, as a dense gas, at supercritical conditions of $T > T_c = 516$ K and $P > P_c = 65$ atm. The iodide could then be precipitated from the ethanol by reducing the pressure. This process was later called *supercritical-fluid extraction* (SFE), *supercritical-gas extraction*, and, most commonly, *supercritical extraction*. By the 1940s, as chronicled by Williams [79], proposed applications of SFE began to appear in the patent and technical literature. Figure 11.40 shows the supercritical-fluid region for CO_2, which has a critical point of 304.2 K and 73.83 bar.

The solvent power of a compressed gas can undergo an enormous increase in the vicinity of its critical point. Consider the solubility of *p*-iodochlorobenzene (pICB) in ethylene, as shown in Figure 11.41, at 298 K for pressures from 2 to 8

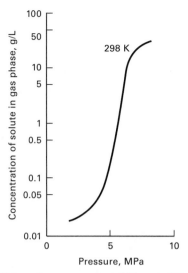

Figure 11.41 Effect of pressure on solubility of pICB in supercritical ethylene.

MPa. This temperature is 1.05 times the critical temperature of ethylene (283 K), and the pressure range straddles the critical pressure of ethylene (5.1 MPa). At 298 K, pICB is a solid (melting point = 330 K) with a vapor pressure of the order of 0.1 torr. At 2 MPa, if pICB formed an ideal-gas solution with ethylene, the concentration of pICB in the gas in equilibrium with pure, solid pICB would be 0.00146 g/L. But the concentration from Figure 11.41 is 0.015 g/L, an order of magnitude higher. If the pressure is increased from 2 MPa to almost the critical pressure at 5 MPa (an increase by a factor of 2.5), the equilibrium concentration of pICB is increased about 10-fold to 0.15 g/L. At 8 MPa, the concentration is 40 g/L, 2,700 times higher than predicted for an ideal-gas solution. It is this dramatic increase in solubility of a solute at near-critical solvent conditions that makes SFE of interest.

Why such a dramatic increase in solvent power? The explanation lies in the change of solvent density while the solute solubility increases. A pressure–enthalpy diagram for ethylene is shown in Figure 11.42, which includes the

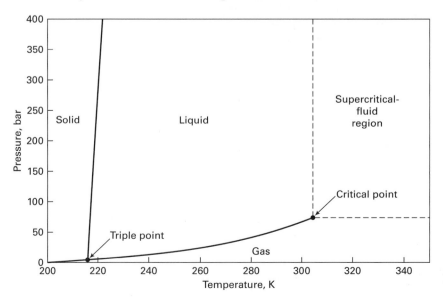

Figure 11.40 Supercritical-fluid region for CO_2.

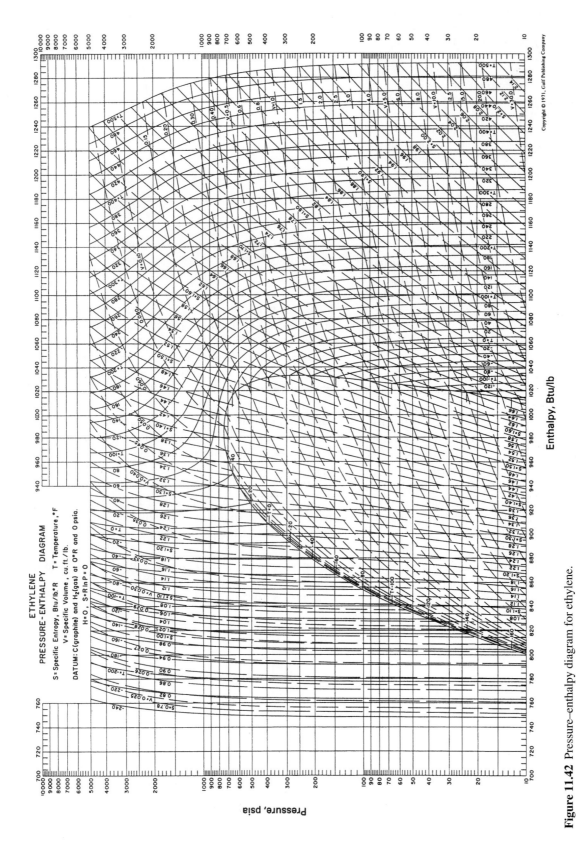

Figure 11.42 Pressure–enthalpy diagram for ethylene.

[From K.E. Starling, *Fluid Thermodynamic Properties for Light Petroleum Systems*, Gulf Publishing, Houston (1973) reprinted with permission.]

specific volume (reciprocal of the density) as a parameter. The range of variables and parameters straddles the critical point of ethylene. The density of ethylene compared to the solubility of pICB is as follows at 298 K:

Pressure, MPa	Ethylene Density, g/L	Solubility of pICB, g/L
2	25.8	0.015
5	95	0.15
8	267	40

Although far from a 1:1 correspondence for the increase of pICB solubility with ethylene density over this range of pressure, there is a meaningful correlation. As the pressure increases, closer packing of the solvent molecules allows them to surround and trap solute molecules. This phenomenon is most useful at reduced temperatures from about 1.01 to 1.12.

Two other effects in the supercritical region are favorable for SFE. Molecular diffusivity of a solute in an ambient-pressure gas is about four orders of magnitude higher than for a liquid. For a near-critical fluid, the diffusivity of solute molecules is usually one to two orders of magnitude higher than in a liquid solvent, thus resulting in a lower mass-transfer resistance in the solvent phase. In addition, the viscosity of the supercritical fluid is about an order of magnitude lower than that of a liquid solvent.

Industrial applications for SFE have been the subject of many studies, patents and venture capital proposals. However, when other techniques are feasible, SFE usually cannot compete because of high solvent-compression costs. SFE is most favorable for extraction of small amounts of large, relatively nonvolatile and expensive solutes, as discussed in §8.6.3 on bioextraction. Applications are also cited by Williams [79] and McHugh and Krukonis [80].

Solvent selection depends on the feed mixture. If only the chemical(s) to be extracted is (are) soluble in a potential solvent, then high solubility is a key factor. If a chemical besides the desired solute is soluble in the potential solvent, then solvent selectivity becomes as important as solubility. A number of gases and low-MW chemicals, including the following, have received attention as solvents.

Solvent	Critical Temperature, K	Critical Pressure, MPa	Critical Density, kg/m^3
Methane	192	4.60	162
Ethylene	283	5.03	218
Carbon dioxide	304	7.38	468
Ethane	305	4.88	203
Propylene	365	4.62	233
Propane	370	4.24	217
Ammonia	406	11.3	235
Water	647	22.0	322

Solvents with $T_c < 373$ K have been well studied. Most promising, particularly for extraction of undesirable, valuable, or heat-sensitive chemicals from natural products, is CO_2, with its moderate P_c, high critical density, low supercritical viscosity, high supercritical molecular diffusivity, and T_c close to ambient. Also, it is nonflammable, noncorrosive, inexpensive, nontoxic in low concentrations, readily available, and safe. Separation of CO_2 from the solute is often possible by simply reducing the extract pressure. According to Williams [79], supercritical CO_2 has been used to extract caffeine from coffee, hops oil from beer, piperine from pepper, capsaicin from chilis, oil from nutmeg, and nicotine from tobacco. However, the use of CO_2 for such applications in the U.S. may be curtailed in the future because of an April, 2009 endangerment finding by the Environmental Protection Agency (EPA) that CO_2 is a pollutant that threatens public health and welfare, and must be regulated.

CO_2 is not always a suitable solvent, however. McHugh and Krukonis [81] cite the energy crisis of the 1970s that led to substantial research on an energy-efficient separation of ethanol and water. The goal, which was to break the ethanol–water azeotrope, was not achieved by SFE with CO_2 because, although supercritical CO_2 has unlimited capacity to dissolve pure ethanol, water is also dissolved in significant amounts. A supercritical-fluid phase diagram for the ethanol–water–CO_2 ternary system at 308.2 K and 10.08 MPa, based on the data of Takishima et al. [82], is given in Figure 11.43. These conditions correspond to $T_r = 1.014$ and $P_r = 1.366$ for CO_2. For the mixture of water and CO_2, two phases exist: a water-rich phase with about 2 mol% CO_2 and a CO_2-rich phase with about 1 mol% water. Ethanol and CO_2 are mutually soluble. Ternary mixtures containing more than 40 mol% ethanol are completely miscible.

If a near-azeotropic mixture of ethanol and water, say, 85 mol% ethanol and 15 mol% water, is extracted by CO_2 at the conditions of Figure 11.43, a mixing line drawn between this composition and a point for pure CO_2 does not cross into the two-phase region, so no separation is possible at these conditions. Alternatively, consider an ethanol–water broth from a fermentation reactor with 10 wt% (4.17 mol%) ethanol. If this

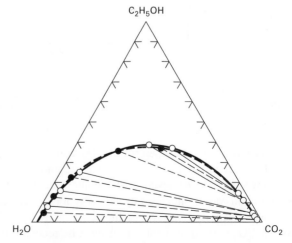

Figure 11.43 Liquid–fluid equilibria for CO_2–C_2H_5OH–H_2O at 308–313.2 K and 10.1–10.34 MPa.

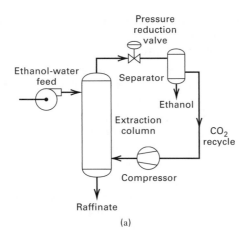

(a)

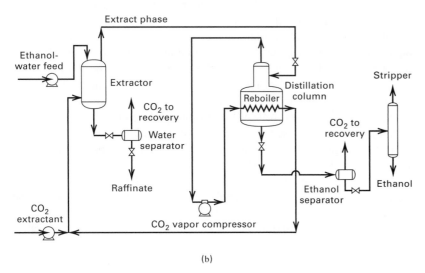

(b)

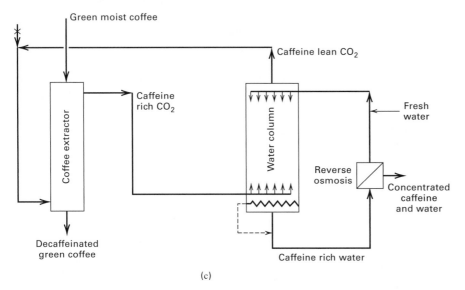

(c)

Figure 11.44 Recovery of CO_2 in supercritical extraction processes. (a) Pressure reduction. (b) High-pressure distillation. (c) High-pressure absorption with water.

mixture is extracted with supercritical CO_2, complete dissolution will not occur and a modest degree of separation of ethanol from water can be achieved, as shown in the next example. The separation can be enhanced by a cosolvent (e.g., glycerol) to improve selectivity, as shown by Inomata et al. [83].

When CO_2 is used as a solvent, it must be recovered and recycled. Three schemes discussed by McHugh and Krukonis [81] are shown in Figure 11.44. In the first scheme for

separation of ethanol and water, the ethanol–water feed is pumped to the pressure of the extraction column, where it is contacted with supercritical CO_2. The raffinate leaving the extractor bottom is enriched with respect to water and is sent to another location for further processing. The top extract stream, containing most of the CO_2, some ethanol, and a smaller amount of water, is expanded across a valve to a lower pressure. In a flash drum downstream of the valve, ethanol–water condensate is collected and the CO_2-rich gas is recycled through a compressor back to the extractor. Unless pressure is greatly reduced across the valve, which results in high compression costs, little ethanol condenses.

A second CO_2 recovery scheme, due to de Filippi and Vivian [84], is given in Figure 11.44b. The flash drum is replaced by high-pressure distillation, at just below the pressure of the extraction column, to produce a CO_2-rich distillate and an ethanol-rich bottoms. The distillate is compressed and recycled through the reboiler back to the extractor. Both raffinate and distillate are flashed to recover CO_2. This scheme, though more complicated than the first, is more versatile.

A third CO_2-recovery scheme, due to Katz et al. [85] for coffee decaffeination, is shown in Figure 11.44c. In the extractor, green, wet coffee beans are mixed with supercritical CO_2 to extract caffeine. The extract is sent to a second column, where the caffeine is recovered with water. The CO_2-rich raffinate is recycled through a compressor (not shown) back to the first column, from which the decaffeinated coffee leaves from the bottom and is sent to a roaster. The caffeine-rich water leaving the second column is sent to a reverse-osmosis unit, where the water is purified and recycled to the water column. All three separation steps operate at high pressure. The concentrated caffeine–water mixture leaving the osmosis unit is sent to a crystallizer.

Multiple equilibrium stages are generally needed to achieve the desired extent of extraction. A major problem in determining the number of stages is the estimation of liquid–supercritical-fluid phase-equilibrium constants. Most commonly, cubic-equation-of-state methods, such as the Soave–Redlich–Kwong (SRK) or Peng–Robinson (PR) equations (§2.5.1), are used, but they have two shortcomings. First, their accuracy diminishes in the critical region. Second, for polar components that form a nonideal-liquid mixture, an appropriate mixing rule that provides a correct bridge between equation-of-state methods and activity-coefficient methods must be employed, e.g., Wong and Sandler [86].

First consider the SRK and PR equations. As discussed in §2.5.1, these equations for pure components contain two parameters, a and b, computed from critical constants. They are extended to mixtures by a *mixing rule* for computing values of a_m and b_m for the mixture from values for pure components. The simplest mixing rule, due to van der Waals, is:

$$a_m = \sum_{i=1}^{C} \sum_{j=1}^{C} x_i x_j a_{ij} \tag{11-19}$$

$$b_m = \sum_{i=1}^{C} \sum_{j=1}^{C} x_i x_j b_{ij} \tag{11-20}$$

where x is a mole fraction in the vapor or liquid. Although these two mixing rules are identical in form, the following *combining rules* for a_{ij} and b_{ij} are different, with the former being a geometric mean and the latter an arithmetic mean:

$$a_{ij} = (a_i a_j)^{1/2} \tag{11-21}$$

$$b_{ij} = (b_i + b_j)/2 \tag{11-22}$$

As stated by Sandler, Orbey, and Lee [87], (11-19) to (11-22) are usually adequate for nonpolar mixtures of hydrocarbons and light gases when critical temperature and/or size differences between molecules are not large.

Molecular-size differences and/or modest degrees of polarity are better handled by the following combining rules:

$$a_{ij} = (a_i a_j)^{1/2}(1 - k_{ij}) \tag{11-23}$$

$$b_{ij} = \left[(b_i + b_j)/2\right](1 - l_{ij}) \tag{11-24}$$

where k_{ij} and l_{ij} are binary interaction parameters back-calculated from vapor–liquid equilibrium and/or density data. Often the latter parameter is set equal to zero. Values of k_{ij} suitable for use with the SRK and PR equations when the mixture contains hydrocarbons with CO_2, H_2S, N_2, and/or CO are given by Knapp et al. [88]. In a study by Shibata and Sandler [89], using experimental phase-equilibria and phase-density data for the nonpolar binary system nitrogen–n-butane at 410.9 K over a pressure range of 30 to 70 bar, good predictions, except in the critical region, were obtained using (11-19) and (11-20), with (11-23) and (11-24), and values of $k_{ij} = -0.164$ and $l_{ij} = -0.233$ in conjunction with the PR equation. Similar good agreement with experimental data was obtained for the systems N_2–cyclohexane, CO_2–n-butane, and CO_2–cyclohexane, and the ternary systems N_2–CO_2–n-butane and N_2–CO_2–cyclohexane.

For high pressures and mixtures with one or more strongly polar components, the preceding rules are inadequate and it is desirable to combine the equation-of-state method with the activity-coefficient method to handle liquid nonidealities. The following theoretically based mixing rule of Wong and Sandler [86] provides the combination. If the PR equation of state and the NRTL activity-coefficient equation are used, the Wong and Sandler mixing rule leads to expressions for computing a_m and b_m for the PR equation:

$$a_m = RTQD/(1 - D) \tag{11-25}$$

$$b_m = Q/(1 - D) \tag{11-26}$$

where

$$Q = \sum_{i=1}^{C} \sum_{j=1}^{C} x_i x_j \left(b - \frac{a}{RT}\right)_{ij} \tag{11-27}$$

$$D = \sum_{i=1}^{C} x_i \frac{a_i}{b_i RT} + \frac{G^{ex}(x_i)}{\sigma RT} \tag{11-28}$$

$$\left(b - \frac{a}{RT}\right)_{ij} = \frac{1}{2}\left[\left(b_i - \frac{a_i}{RT}\right) + \left(b_j - \frac{a_j}{RT}\right)\right](1 - k_{ij}) \tag{11-29}$$

$$\sigma = \frac{1}{\sqrt{2}}\left[\ln(\sqrt{2} - 1)\right] \tag{11-30}$$

$$\frac{G^{ex}}{RT} = \sum_{i=1}^{C} x_i \left(\frac{\sum_{j=1}^{C} x_j \tau_{ji} g_{ji}}{\sum_{k=1}^{C} x_k g_{ki}}\right) \tag{11-31}$$

$$g_{ij} = \exp(-\alpha_{ij} \tau_{ij}) \tag{11-32}$$

with $a_{ij} = a_{ji}$.

Equations (11-25) to (11-32) show that for a binary system using the NRTL equation, there are four adjustable binary interaction parameters (BIPs): k_{ij}, α_{ij}, τ_{ij}, and τ_{ji}. For a temperature and pressure range of interest, these parameters are best obtained by regression of experimental binary-pair data for VLE, LLE, and/or VLLE. The parameters can be used to predict phase equilibria for ternary and higher multi-component mixtures. However, Wong, Orbey, and Sandler [90] show that when values of the latter three parameters are already available, even at just near-ambient temperature and pressure conditions from a data source such as Gmehling and Onken [91], those parameters can be assumed independent of temperature and used to make reasonably accurate predictions of phase equilibria, even at temperatures to at least 200°C and pressures to 200 bar. Regression of experimental data to obtain a value of k_{ij} is also not necessary, because Wong, Orbey, and Sandler show that it can be determined from the other three parameters by choosing its value so that the excess Gibbs free energy from the equation of state matches that from the activity-coefficient model. Thus, application of the Wong–Sandler mixing rule to supercritical extraction is facilitated.

Another phase-equilibrium prediction method applicable to wide ranges of pressure, temperature, molecular size, and polarity is the group-contribution equation of state (GC-EOS) of Skjold-Jørgensen [92]. This method, which combines features of the van der Waals equation of state, the Carnahan–Starling expression for hard spheres, the NRTL activity-coefficient equation, and the group-contribution principle, has been applied to SFE conditions, and is useful when all necessary binary data are not available.

When experimental K-values are available, or when the Wong–Sandler mixing rule or the GC-EOS can be applied,

stage calculations for supercritical extraction can be made using process simulators, as in the next example.

EXAMPLE 11.10 SFE of Ethanol with CO_2.

One mol/s of 10 wt% ethanol in water is extracted by 3 mol/s of carbon dioxide at 305 K and 9.86 MPa in a countercurrent-flow extraction column with five equilibrium stages. Determine the flow rates and compositions of the exiting extract and raffinate.

Solution

This problem, taken from Colussi et al. [93], who used the GC-EOS method, was solved with Tower Plus of the CHEMCAD process simulator, at constant T and P, where composition changes were small enough that K-values were constant and are defined as the extract mole fraction divided by the raffinate mole fraction. They are in good agreement with experimental data:

Component	K-Value
CO_2	34.5
Ethanol	0.115
Water	0.00575

The percent extraction of ethyl alcohol is 33.6%, with an extract of 69 wt% pure ethanol (solvent-free basis) and a raffinate containing 93 wt% water (solvent-free basis). Calculated stagewise flow rates and component mole fractions are listed in Table 11.6, where stages are numbered from the feed end.

Table 11.6 Calculated Flow and Composition Profiles for Example 11.10

Leaving Streams	Stage 1 Extract Mole Fraction	Stage 1 Raffinate Mole Fraction	Stage 2 Extract Mole Fraction	Stage 2 Raffinate Mole Fraction	Stage 3 Extract Mole Fraction	Stage 3 Raffinate Mole Fraction	Stage 4 Extract Mole Fraction	Stage 4 Raffinate Mole Fraction	Stage 5 Extract Mole Fraction	Stage 5 Raffinate Mole Fraction
Carbon dioxide	0.98999	0.02870	0.99002	0.02870	0.99012	0.02870	0.99043	0.02870	0.99138	0.02874
Ethanol	0.00466	0.04053	0.00463	0.04023	0.00452	0.03929	0.00419	0.03645	0.00319	0.02775
Water	0.00535	0.93077	0.00535	0.93107	0.00536	0.93201	0.00538	0.93485	0.00543	0.94351
Total flow, gmol/s	3.0013	1.0298	3.0311	1.0294	3.0308	1.0285	3.0298	1.0255	3.0268	0.9987

SUMMARY

1. Extractive distillation, salt distillation, pressure-swing distillation, homogeneous azeotropic distillation, heterogeneous azeotropic distillation, and reactive distillation are enhanced-distillation techniques to be considered when separation by ordinary distillation is uneconomical or unfeasible. Reactive distillation can be used to conduct, simultaneously and in the same apparatus, a chemical reaction and a separation.

2. For ternary systems, a composition plot on a triangular graph is very useful for finding feasible separations, especially when binary and ternary azeotropes form. With such a diagram, distillation paths, called residue curves or distillation curves, are readily tracked. The curves may be restricted to certain regions of the triangular diagram by distillation boundaries. Feasible-product compositions at total reflux are readily determined.

3. Extractive distillation, using a low-volatility solvent that enters near the top of the column, is widely used to separate azeotropes and very close-boiling mixtures. Preferably, the solvent should not form an azeotrope with any feed component.

4. Certain salts, when added to a solvent, reduce the solvent volatility and increase the relative volatility between the two feed components. In this process, called salt distillation, the salt is dissolved in the solvent or added as a solid or melt to the reflux.

5. Pressure-swing distillation, utilizing two columns operating at different pressures, can be used to separate an azeotropic mixture when the azeotrope can be made to disappear at some pressure. If not, it may still be practical if the azeotropic composition changes by 5 mol% or more over a moderate range of pressure.

6. In homogeneous azeotropic distillation, an entrainer is added to a stage, usually above the feed stage. A minimum- or maximum-boiling azeotrope, formed by the entrainer with one or more feed components, is removed from the top or bottom of the column. Applications of this technique for difficult-to-separate mixtures are not common because of limitations due to distillation boundaries.

7. A more common and useful technique is heterogeneous azeotropic distillation, in which the entrainer forms, with one or more components of the feed, a minimum-boiling heterogeneous azeotrope. When condensed, the overhead vapor splits into organic-rich and water-rich phases. The azeotrope is broken by returning one liquid phase as reflux, with the other sent on as distillate for further processing.

8. A growing application of reactive distillation is to combine chemical reaction and distillation in one column. To be effective, the reaction and distillation must be feasible at the same pressure and range of temperature, with reactants and products favoring different phases so that an equilibrium-limited reaction can go to completion.

9. Liquid–liquid or solid–liquid extraction can be carried out with a supercritical-fluid solvent at temperatures and pressures just above critical because of favorable values for solvent density and viscosity, solute diffusivity, and solute solubility in the solvent. An attractive supercritical solvent is carbon dioxide, particularly for extraction of certain chemicals from natural products.

REFERENCES

1. Stichlmair, J., J.R. Fair, and J.L. Bravo, *Chem. Eng. Progress*, **85**(1), 63–69 (1989).

2. Partin, L.R., *Chem. Eng. Progress*, **89**(1), 43–48 (1993).

3. Stichlmair, J., "Distillation and Rectification," in *Ullmann's Encyclopedia of Industrial Chemistry*, 5th ed., VCH Verlagsgesellschaft Weinheim Vol. B3, pp. 4-1 to 4-94 (1988).

4. Doherty, M.F., and J.D. Perkins, *Chem. Eng. Sci.*, **33**, 281–301 (1978).

5. Bossen, B.S., S.B. Jørgensen, and R. Gani, *Ind. Eng. Chem. Res.*, **32**, 620–633 (1993).

6. Pham, H.N., and M.F. Doherty, *Chem. Eng. Sci.*, **45**, 1837–1843 (1990).

7. Taylor, R., and H.A. Kooijman, *CACHE News*, No. **41**, 13–19 (1995).

8. Doherty, M.F., *Chem. Eng. Sci.*, **40**, 1885–1889 (1985).

9. ASPEN PLUS, "What's New in Release 9," Aspen Technology, Cambridge, MA (1994).

10. Doherty, M.F., and J.D. Perkins, *Chem. Eng. Sci.*, **34**, 1401–1414 (1979).

11. Horsley, L.H., "Azeotropic Data III," in *Advances in Chemistry Series*, American Chemical Society, Washington, D.C., Vol. 116 (1973).

12. Gmehling, J., J. Menke, J. Krafczyk, and K. Fischer, *Azeotropic Data*, 2nd ed. in 3 volumes, Wiley-VCH, Weinheim, Germany (2004).

13. Fidkowski, Z.T., M.F. Malone, and M.F. Doherty, *Computers Chem. Engng.*, **17**, 1141–1155 (1993).

14. Foucher, E.R., M.F. Doherty, and M.F. Malone, *Ind. Eng. Chem. Res.*, **30**, 760–772 (1991) and **30**, 2364 (1991).

15. Matsuyama, H., and H.J. Nishimura, *J. Chem. Eng. Japan*, **10**, 181 (1977).

16. Doherty, M.F., and G.A. Caldarola, *Ind. Eng. Chem. Fundam.*, **24**, 474–485 (1985).

17. Fidkowski, Z.T., M.F. Doherty, and M.F. Malone, *AIChE J.*, **39**, 1303–1321 (1993).

18. Stichlmair, J.G., and J.-R. Herguijuela, *AIChE J.*, **38**, 1523–1535 (1992).

19. Widagdo, S., and W.D. Seider, *AIChE J.*, **42**, 96–130 (1996).

20. Wahnschafft, O.M., J.W. Koehler, E. Blass, and A.W. Westerberg, *Ind. Eng. Chem. Res.*, **31**, 2345–2362 (1992).

21. Dunn, C.L., R.W. Millar, G.J. Pierotti, R.N. Shiras, and M. Souders, Jr., *Trans. AIChE*, **41**, 631–644 (1945).

22. Berg, L., *Chem. Eng. Progress*, **65**(9), 52–57 (1969).

23. Othmer, D.F., *AIChE Symp. Series*, **235**(79), 90–117 (1983).

24. Van Raymbeke, U.S. Patent 1,474,216 (1922).

25. Cook, R.A., and W.F. Furter, *Can. J. Chem. Eng.*, **46**, 119–123 (1968).

26. Johnson, A.I., and W.F. Furter, *Can. J. Chem. Eng.*, **43**, 356–358 (1965).

27. Johnson, A.I., and W.F. Furter, *Can. J. Chem. Eng.*, **38**, 78–87 (1960).

28. Furter, W.F., and R.A. Cook, *Int. J. Heat Mass Transfer*, **10**, 23–36 (1967).

29. Furter, W.F., *Can. J. Chem. Eng.*, **55**, 229–239 (1977).

30. Furter, W.F., *Chem. Eng. Commun.*, **116**, 35 (1992).

31. Kumar, A., *Sep. Sci. and Tech.*, **28**, 1799–1818 (1993).

32. Mahapatra, A., V.G. Gaikar, and M.M. Sharma, *Sep. Sci. and Tech.*, **23**, 429–436 (1988).

33. Agarwal, M., and V.G. Gaikar, *Sep. Technol.*, **2**, 79–84 (1992).

34. Knapp, J.P., and M.F. Doherty, *Ind. Eng. Chem. Res.*, **31**, 346–357 (1992).

35. Lewis, W.K., U.S. Patent 1,676,700 (1928).

36. Van Winkle, M., *Distillation*, McGraw-Hill, New York (1967).

37. Robinson, C.S., and E.R. Gilliland, *Elements of Fractional Distillation*, 4th ed., McGraw-Hill, New York (1950).

38. Wahnschafft, O.M., and A.W. Westerberg, *Ind. Eng. Chem. Res.*, **32**, 1108 (1993).

39. Laroche, L., N. Bekiaris, H.W. Andersen, and M. Morari, *AIChE J.*, **38**, 1309 (1992).

40. Young, S., *J. Chem. Soc. Trans.*, **81**, 707–717 (1902).

41. Keyes, D.B., U.S. Patent 1,676,735 (1928).

42. Keyes, D.B., *Ind. Eng. Chem.*, **21**, 998–1001 (1929).

43. Bekiaris, N., G.A. Meski, and M. Morari, *Ind. Eng. Chem. Res.*, **35**, 207–217 (1996).

44. Prokopakis, G.J., and W.D. Seider, *AIChE J.*, **29**, 49–60 (1983).

45. Ryan, P.J., and M.F. Doherty, *AIChE J.*, **35**, 1592–1601 (1989).

46. Pham, H.N., and M.F. Doherty, *Chem. Eng. Sci.*, **45**, 1845–1854 (1990).

47. Kovach, III, J.W., and W.D. Seider, *Computers and Chem. Engng.*, **11**, 593 (1987).

48. Swartz, C.L.E., and W.E. Stewart, *AIChE J.*, **33**, 1977–1985 (1987).

49. Gani, R., and S.B. Jørgensen, *Computers Chem. Engng.*, **18**, Suppl., S55 (1994).

50. Shewchuk, C.F.,"Computation of Multiple Distillation Towers," Ph.D. Thesis, University of Cambridge (1974).

51. Magnussen, T., M.L. Michelsen, and A. Fredenslund, *Inst. Chem. Eng. Symp. Series No. 56, Third International Symp. on Distillation*, Rugby, England (1979).

52. Bekiaris, N., G.A. Meski, C.M. Radu, and M. Morari, *Ind. Eng. Chem. Res.*, **32**, 2023–2038 (1993).

53. Black, C., and D.E. Ditsler, *Advances in Chemistry Series*, ACS, Washington, D.C., Vol. 115, pp. 1–15 (1972).

54. Black, C., R.A. Golding, and D.E. Ditsler, *Advances in Chemistry Series*, ACS, Washington, D.C., Vol. 115, pp. 64–92 (1972).

55. Black, C., *Chem. Eng. Progress*, **76**(9), 78–85 (1980).

56. Black, C., *Ind. Eng. Chem.*, **50**, 403–412 (1958).

57. Terrill, D.L., L.F. Sylvestre, and M.F. Doherty, *Ind. Eng. Chem. Proc. Des. Develop.*, **24**, 1062–1071 (1985).

58. Robinson, C.S., and E.R. Gilliland, *Elements of Fractional Distillation*, 4th ed., McGraw-Hill, New York (1950).

59. Backhaus, A.A., U.S. Patent 1,400,849 (1921).

60. Leyes, C.E., and D.F. Othmer, *Trans. AIChE*, **41**, 157–196 (1945).

61. Belck, L.H., *AIChE J.*, **1**, 467–470 (1955).

62. Masamoto, J., and K. Matsuzaki, *J. Chem. Eng. Japan*, **27**, 1–5 (1994).

63. Agreda, V.H., and L.R. Partin, U.S. Patent 4,435,595 (March 6, 1984).

64. Agreda, V.H., L.R. Partin, and W.H. Heise, *Chem. Eng. Prog.*, **86**(2), 40–46 (1990).

65. Smith, L.A., U.S. Patent 4,307,254 (Dec. 22, 1981).

66. Smith, L.A., U.S. Patent 4,443,559 (April 17, 1984).

67. Smith, L.A., U.S. Patent 4,978,807 (Dec. 18, 1990).

68. DeGarmo, J.L., V.N. Parulekar, and V. Pinjala, *Chem. Eng. Prog.*, **88**(3), 43–50 (1992).

69. Chang, Y.A., and J.D. Seader, *Computers Chem. Engng.*, **12**, 1243–1255 (1988).

70. Venkataraman, S., W.K. Chan, and J.F. Boston, *Chem. Eng. Progress*, **86**(8), 45–54 (1990).

71. Simandl, J., and W.Y. Svrcek, *Computers Chem. Engng.*, **15**, 337–348 (1991).

72. Kang, Y.W., Y.Y. Lee, and W.K. Lee, *J. Chem. Eng. Japan*, **25**, 649–655 (1992).

73. Izarraraz, A., G.W. Bentzen, R. G. Anthony, and C.D. Holland, *Hydrocarbon Processing*, **59**(6), **195** (1980).

74. Rehfinger, A., and U. Hoffmann, *Chem. Eng. Sci.*, **45**, 1605–1617 (1990).

75. Jacobs, R., and R. Krishna, *Ind. Eng. Chem. Res.*, **32**, 1706–1709 (1993).

76. Nijhuis, S.A., F.P.J.M. Kerkhof, and N.S. Mak, *Ind. Eng. Chem. Res.*, **32**, 2767–2774 (1993).

77. Hauan, S., T. Hertzberg, and K.M. Lien, *Ind. Eng. Chem. Res.*, **34**, 987–991 (1995).

78. Hannay, J.B., and J. Hogarth, *Proc. Roy. Soc. (London) Sec. A*, **29**, 324 (1879).

79. Williams, D.F., *Chem. Eng. Sci.*, **36**, 1769–1788 (1981).

80. McHugh, M., and V. Krukonis, *Supercritical Fluid Extraction— Principles and Practice*, Butterworths, Boston (1986).

81. McHugh, M., and V. Krukonis, *Supercritical Fluid Extraction— Principles and Practice*, 2nd ed., Butterworth-Heinemann, Boston (1994).

82. Takishima, S., A. Saiki, K. Arai, and S. Saito, *J. Chem. Eng. Japan*, **19**, 48–56 (1986).

83. Inomata, H., A. Kondo, K. Arai, and S. Saito, *J. Chem. Eng. Japan*, **23**, 199–207 (1990).

84. de Fillipi, R.P., and J.E. Vivian, U.S. Patent 4,349,415 (1982).

85. Katz, S.N., J.E. Spence, M.J. O'Brian, R.H. Skiff, G.J. Vogel, and R. Prasad, U.S. Patent 4,911,941 (1990).

86. Wong, D.S.H., and S.I. Sandler, *AIChE J.*, **38**, 671–680 (1992).

87. Sandler, S.I., H. Orbey, and B.-I. Lee, in S.I. Sandler, Ed., *Models for Thermodynamic and Phase Equilibria Calculations*, Marcel Dekker, New York, pp. 87–186 (1994).

88. Knapp, H., R. Doring, L. Oellrich, U. Plocker, and J.M. Prausnitz, *Vapor-Liquid Equilibria for Mixtures of Low Boiling Substances*, Chem. Data Ser., Vol. VI, DECHEMA, pp. 771–793 (1982).

89. Shibata, S.K., and S.I. Sandler, *Ind. Eng. Chem. Res.*, **28**, 1893–1898 (1989).

90. Wong, D.S.H., H. Orbey, and S.I. Sandler, *Ind. Eng. Chem. Res.*, **31**, 2033–2039 (1992).

91. Gmehling, J., and U. Onken, *Vapor-Liquid Equilibrium Data Compilation*, DECHEMA Data Series, DECHEMA, Frankfurt (1977).

92. Skjold-Jørgensen, S., *Ind. Eng. Chem. Res.*, **27**, 110–118 (1988).

93. Colussi, I.E., M. Fermeglia, V. Gallo,and I. Kikic, *Computers Chem. Engng.*, **16**, 211–224 (1992).

94. Doherty, M.F., and M.F. Malone, *Conceptual Design of Distillation Systems*, McGraw-Hill, New York (2001).

95. Stichlmair, J.G., and J.R. Fair, *Distillation Principles and Practices*, Wiley-VCH, New York (1998).

96. Siirola, J.J., and S.D. Barnicki,in R.H. Perry and D.W. Green, Eds., *Perry's Chemical Engineers' Handbook*, 7th ed. McGraw-Hill, New York, pp. 13-54 to 13-85 (1997).

97. Eckert, E., and M. Kubicek, *Computers Chem. Eng.*, **21**, 347–350 (1997).

98. Hoffmaster, W.R., and S. Hauan, *AIChE J.*, **48**, 2545–2556 (2002).

99. Llano-Restrepo, M., and J. Aguilar-Arias, *Computers Chem. Engng.*, **27**, 527–549 (2003).

100. Fu, J., *AIChE J.*, **42**, 3364–3372 (1996).

101. Huss, R.S., F. Chen, M.F. Malone, and M.F. Doherty, *Computers Chem. Engng.*, **27**, 1855–1866 (2003).

STUDY QUESTIONS

11.1.   What is meant by enhanced distillation? When should it be considered?

11.2.   What is the difference between extractive distillation and azeotropic distillation?

11.3.   What is the difference between homogeneous and heterogeneous azeotropic distillation?

11.4.   What are the two reasons for conducting reactive distillation?

11.5.   What is a distillation boundary? Why is it important?

11.6.   To what type of a distillation does a residue curve apply? What is a residue-curve map?

11.7.   Is a residue curve computed from an algebraic or a differential equation? Does a residue curve follow the composition of the distillate or the residue?

11.8.   Residue curves involve nodes. What is the difference between a stable and an unstable node? What is a saddle?

11.9.   What is a distillation-curve map? How does it differ from a residue-curve map?

11.10.   What is a region of feasible-product compositions? How is it determined? Why is it important?

11.11.   Under what conditions can a distillation boundary be crossed by a material-balance line?

11.12.   In extractive distillation, why is a large concentration of solvent required in the liquid phase? Why doesn't the solvent enter the column at the top tray?

11.13.   Why is heterogeneous azeotropic distillation a more feasible technique than homogeneous azeotropic distillation?

11.14.   What is meant by multiplicity? What kinds of multiplicity are there? Why is it important to obtain all multiple solutions when they exist?

11.15.   In reactive distillation, does the reaction preferably take place in the vapor or in the liquid phase? Can a homogeneous or solid catalyst be used?

11.16.   What happens to the solvent power of a compressed gas as it passes through the critical region? What happens to physical properties in the critical region?

11.17.   Why is CO_2 frequently a desirable solvent for SFE?

EXERCISES

Section 11.1

11.1.   Approximate residue-curve map.

For the normal hexane–methanol–methyl acetate system at 1 atm, find, in suitable references, all binary and ternary azeotropes, sketch a residue-curve map on a right-triangular diagram, and indicate the distillation boundaries. Determine for each azeotrope and pure component whether it is a stable node, an unstable node, or saddle.

11.2.   Calculation of a residue curve.

For the system in Exercise 11.1, use a process simulator with the UNIFAC equation to calculate a portion of a residue curve at 1 atm starting from a bubble-point liquid with 20 mol% normal hexane, 60 mol% methanol, and 20 mol% methyl acetate.

11.3.   Calculation of a distillation curve.

For the same conditions as in Exercise 11.2, use a process simulator with the UNIFAC equation to calculate a portion of a distillation curve at 1 atm.

11.4.   Distillation boundaries.

For the acetone, benzene, and n-heptane system at 1 atm find, in references, all binary and ternary azeotropes, sketch a distillation-curve map on an equilateral-triangle diagram, and indicate distillation boundaries. Determine for each azeotrope and pure component whether it is a stable node, an unstable node, or saddle.

11.5.   Calculation of a residue curve.

For the same ternary system as in Exercise 11.4, use a simulation program with UNIFAC to calculate a portion of a residue curve at 1 atm starting from a bubble-point liquid composed of 20 mol% acetone, 60 mol% benzene, and 20 mol% n-heptane.

11.6.   Calculation of a distillation curve.

For the same conditions as in Exercise 11.5, use a process simulator with the UNIFAC equation to calculate a portion of a distillation curve at 1 atm.

11.7.   Feasible-product-composition regions.

Develop the feasible-product-composition regions for the system of Figure 11.13, using feed F_1.

11.8.   Feasible-product-composition regions.

Develop the feasible-product-composition regions for the system of Figure 11.10 if the feed is 50 mol% chloroform, 25 mol% methanol, and 25 mol% acetone.

Section 11.2

11.9.   Extraction distillation with ethanol.

Repeat Example 11.3, but with ethanol as the solvent.

11.10.   Extraction distillation with MEK.

Repeat Example 11.3, but with MEK as the solvent.

11.11.   Extraction distillation with toluene.

Repeat Example 11.4, but with toluene as the solvent.

11.12.   Extraction distillation with phenol.

Four hundred lbmol/h of an equimolar mixture of n-heptane and toluene at 200°F and 20 psia is to be separated by extractive distillation at 20 psia, using phenol at 220°F as the solvent, at a flow rate of 1,200 lbmol/h. Design a suitable two-column system to obtain reasonable product purities with minimal solvent loss.

Section 11.4

11.13.   Pressure-swing distillation.

Repeat Example 11.5, but with a feed of 100 mol/s of 55 mol% ethanol and 45 mol% benzene.

11.14.   Pressure-swing distillation.

Determine the feasibility of separating 100 mol/s of a mixture of 20 mol% ethanol and 80 mol% benzene by pressure-swing distillation. If feasible, design such a system.

11.15.   Pressure-swing distillation.

Design a pressure-swing distillation system to produce 99.8 mol% ethanol for 100 mol/s of an aqueous feed containing 30 mol% ethanol.

Section 11.5

11.16.   Homogeneous azeotropic distillation.

In Example 11.6, a mixture of benzene and cyclohexane is separated in a separation sequence that begins with homogeneous azeotropic distillation using acetone as the entrainer. Can the same separation be achieved using methanol as the entrainer? If not, why not? [Ref.: Ratliff, R.A., and W.B. Strobel, *Petro. Refiner.*, **33**(5), 151 (1954).]

11.17. Homogeneous azeotropic distillation.

Devise a separation sequence to separate 100 mol/s of an equimolar mixture of toluene and 2,5-dimethylhexane into nearly pure products. Include in the sequence a homogeneous azeotropic distillation column using methanol as the entrainer and determine a feasible design for that column. [Ref.: Benedict, M., and L.C. Rubin, *Trans. AIChE*, **41**, 353–392 (1945).]

11.18. Homogeneous azeotropic distillation.

A mixture of 16,500 kg/h of 55 wt% methyl acetate and 45 wt% methanol is to be separated into 99.5 wt% methyl acetate and 99 wt% methanol. Use of one homogeneous azeotropic distillation column and one ordinary distillation column has been suggested. Possible entrainers are *n*-hexane, cyclohexane, and toluene. Determine feasibility of the sequence. If feasible, prepare a design. If not, suggest an alternative and prove its feasibility.

Section 11.6

11.19. Heterogeneous azeotropic distillation.

Design a three-column distillation sequence to separate 150 mol/s of an azeotropic mixture of ethanol and water at 1 atm into nearly pure ethanol and nearly pure water using heterogeneous azeotropic distillation with benzene as the entrainer.

11.20. Heterogeneous azeotropic distillation.

Design a three-column distillation sequence to separate 120 mol/s of an azeotropic mixture of isopropanol and water at 1 atm into nearly pure isopropanol and nearly pure water using heterogeneous azeotropic distillation with benzene entrainer. [Ref.: Pham, H.N., P.J. Ryan, and M.F. Doherty, *AIChE J.*, **35**, 1585–1591 (1989).]

11.21. Heterogeneous azeotropic distillation.

Design a two-column distillation sequence to separate 1,000 kmol/h of 20 mol% aqueous acetic acid into nearly pure acetic acid and water. Use heterogeneous azeotropic distillation with *n*-propyl acetate as the entrainer in Column 1.

Section 11.7

11.22. Reactive distillation.

Repeat Example 11.9, with the entire range of methanol feed-stage locations. Compare your results for isobutene conversion with the values shown in Figure 11.39.

11.23. Reactive distillation.

Repeat Exercise 11.22, but with activities, instead of mole fractions, in the reaction-rate expressions. How do the results differ? Explain.

11.24. Reactive distillation.

Repeat Exercise 11.22, but with the assumption of chemical equilibrium on stages where catalyst is employed. How do the results differ from Figure 11.39? Explain.

Section 11.8

11.25. Supercritical-fluid extraction with CO_2.

Repeat Example 11.10, but with 10 equilibrium stages instead of 5. What is the effect of this change?

11.26. Model for SFE of a solute from particles.

An application of supercritical extraction is the removal of solutes from particles of porous natural materials. Such applications include extraction of caffeine from coffee beans and extraction of ginger oil from ginger root. When CO_2 is used as the solvent, the rate of extraction is found to be independent of flow rate of CO_2 past the particles, but dependent upon the particle size. Develop a mathematical model for the rate of extraction consistent with these observations. What model parameter would have to be determined by experiment?

11.27. SFE of β-carotene with CO_2.

Cygnarowicz and Seider [*Biotechnol. Prog.*, **6**, 82–91 (1990)] present a process for supercritical extraction of β-carotene from water with CO_2, using the GC-EOS method of Skjold–Jørgensen to estimate phase equilibria. Repeat the calculations for their design using the Peng–Robinson EOS with the Wong–Sandler mixing rules. How do the designs compare?

11.28. SFE of acetone from water with CO_2.

Cygnarowicz and Seider [*Ind. Eng. Chem. Res.*, **28**, 1497–1503 (1989)] present a design for the supercritical extraction of acetone from water with CO_2 using the GC-EOS method of Skjold–Jørgensen to estimate phase equilibria. Repeat their design using the Peng–Robinson EOS with the Wong–Sandler mixing rules. How do the designs compare?

Chapter 12

Rate-Based Models for Vapor–Liquid Separation Operations

§12.0 INSTRUCTIONAL OBJECTIVES

After completing this chapter, you should be able to:

- Write equations that model a nonequilibrium stage, where equilibrium is assumed only at the interface between phases.
- Explain component-coupling effects in multicomponent mass transfer.
- Explain the bootstrap problem and how it is handled for distillation.
- Cite methods for estimating transport coefficients and interfacial areas required for rate-based calculations.
- Explain differences among ideal vapor–liquid flow patterns that are employed for rate-based calculations.
- Use a process simulator or stand-alone computer program to make a rate-based calculation for a vapor–liquid separation problem.

Equations for equilibrium-based distillation models were first published by Sorel [1] in 1893. They consisted of total and component material balances around top and bottom sections of equilibrium stages, including a total condenser and a reboiler, and corresponding energy balances with provision for heat losses, which are important for small laboratory columns but not for insulated, industrial columns. Sorel used graphs of phase-equilibrium data instead of equations. Because of the complexity of Sorel's model, it was not widely applied until 1921, when it was adapted to graphical-solution techniques for binary systems, first by Ponchon and then by Savarit, who used an enthalpy-concentration diagram. In 1925, a much simpler, but less-rigorous, graphical technique was developed by McCabe and Thiele, who eliminated the energy balances by assuming constant vapor and liquid molar flow rates except across feed or sidestream withdrawal stages. When applicable, the McCabe–Thiele graphical method, presented in Chapter 7, is used even today for binary distillation, because it gives valuable insight into changes in phase compositions from stage to stage.

Because some of Sorel's equations are nonlinear, it is not possible to obtain algebraic solutions, unless simplifying assumptions are made. Smoker [2] did just that in 1938 for the distillation of a binary mixture by assuming not only constant molar overflow, but also constant relative volatility. Smoker's equation is still useful for superfractionators involving close-boiling binary mixtures, where that assumption is valid. Starting in 1932, two iterative, numerical

methods were developed for obtaining a solution to Sorel's model for multicomponent mixtures. The Thiele–Geddes method [3] requires specification of the number of equilibrium stages, feed-stage location, reflux ratio, and distillate flow rate, for which component product distribution is calculated. The Lewis–Matheson method [4] computes the stages required and the feed-stage location for a specified reflux ratio and split between two key components. These two methods were widely used for the simulation and design of single-feed, multicomponent distillation columns prior to the 1960s.

Attempts in the late 1950s and early 1960s to adapt the Thiele–Geddes and Lewis–Matheson methods to digital computers had limited success. The breakthrough in computerization of stage-wise calculations occurred when Amundson and co-workers, starting in 1958, applied techniques of matrix algebra. This led to successful computer-aided methods, based on sparse-matrix algebra, for Sorel's equilibrium-based model. The most useful of these models are presented in Chapter 10. Although the computations sometimes fail to converge, the methods are widely applied and have become flexible and robust.

Methods presented in Chapters 10 and 11 assume that equilibrium is achieved, at each stage, with respect to both heat and mass transfer. Except when temperature changes significantly from stage to stage, the assumption of temperature equality for vapor and liquid phases leaving a stage is usually acceptable. However, for most industrial applications, assuming equilibrium of exiting-phase compositions is not reasonable. In general, exiting vapor-phase mole fractions are not related to exiting liquid-phase mole fractions by thermodynamic *K*-values. To overcome this limitation of equilibrium-based models, Lewis [5], in 1922, proposed an overall stage efficiency for converting theoretical (equilibrium) stages to actual stages. Experimental data show that this efficiency varies, depending on the application, over a range of 5 to 120%, where the highest values are for distillation in large-diameter, single-liquid-pass trays because of a crossflow effect, whereas the lowest values occur in absorption columns with high-viscosity, high-molecular-weight absorbents.

An improved procedure to account for nonequilibrium with respect to mass transfer was introduced by Murphree [6] in 1925. It incorporates the Murphree vapor-phase tray efficiency, $(E_{MV})_{i,j}$, directly into Sorel's model to replace the equilibrium equation based on the *K*-value. Thus,

$$K_{i,j} = y_{i,j}/x_{i,j} \qquad (12\text{-}1)$$

is replaced by

$$(E_{MV})_{i,j} = \left(y_{i,j} - y_{i,j+1}\right)/\left(y_{i,j}^* - y_{i,j+1}\right) \qquad (12\text{-}2)$$

where *i* refers to the component, and *j* the stage, with stages numbered down from the top.

This efficiency is the ratio of the actual change in vapor-phase mole fraction to the change that would occur if equilibrium were achieved. The equilibrium value, $y_{i,j}^*$, is obtained from (12-1), with substitution into (12-2) giving

$$\boxed{(E_{MV})_{i,j} = \left(y_{i,j} - y_{ij+1}\right)/\left(K_{i,j}x_{i,j} - y_{i,j+1}\right)} \qquad (12\text{-}3)$$

Equations (12-2) and (12-3) assume: (1) uniform concentrations in vapor and liquid streams entering and exiting a tray; (2) complete mixing in the liquid flowing across the tray; (3) plug flow of the vapor up through the liquid; and (4) negligible resistance to mass transfer in the liquid.

Application of E_{MV} using empirical correlations has proved adequate for binary and close-boiling, ideal, and near-ideal multicomponent mixtures. However, deficiencies of the Murphree efficiency for multicomponent mixtures have long been recognized. Murphree himself stated clearly these deficiencies for multicomponent mixtures and for cases where the efficiency is low. He even argued that theoretical plates should not be the basis of calculation for multicomponent mixtures.

For binary mixtures, values of E_{MV} are always positive and identical for the two components. However, for multicomponent mixtures, values of E_{MV} differ from component to component and from stage to stage. The independent values of E_{MV} need to force the sum of the mole fractions in the vapor phase to sum to 1, which introduces the possibility of negative values of E_{MV}. When using the Murphree vapor-phase efficiency, the temperatures of the exiting vapor and liquid phases are assumed identical and equal to the exiting-liquid bubble-point temperature. Because the vapor is not in equilibrium with the liquid, the vapor temperature does not correspond to the dew-point temperature. It is even possible, algebraically, for the vapor temperature to correspond to an impossible, below-the-dew-point value.

Values of E_{MV} can be obtained from data or correlations, and are more likely to be Murphree vapor-point (rather than tray) efficiencies, which apply only to a particular location on the tray. To convert them to tray efficiencies, vapor and liquid flow patterns must be assumed after the manner of Lewis [7], as discussed by Seader [8]. If vapor and liquid are completely mixed, point efficiency equals tray efficiency.

Walter and Sherwood [9] found that measured tray efficiencies cover a range of 0.65 to 4.2% for absorption and stripping of carbon dioxide from water and glycerine solutions; 4.7 to 24% for absorption of olefins into oils; and 69 to 92% for absorption of ammonia, humidification of air, and rectification of alcohol.

In 1957, Toor [10] showed that diffusion in a ternary mixture is enormously more complex than that in a binary mixture because of coupling among component concentration gradients, especially when components differ widely in molecular size, shape, and polarity. Toor showed that, in addition to diffusion due to a Fickian concentration driving force, gradient coupling could result in: (1) diffusion against a driving force (reverse diffusion), (2) no diffusion even though a concentration driving force is present (diffusion barrier), and (3) diffusion with zero driving force (osmotic diffusion). Theoretical calculations by Toor and Burchard [11] predicted the possibility of negative values of E_{MV} in multicomponent systems, but E_{MV} for binary systems is restricted to 0–100%.

In 1977, Krishna et al. [12] extended the work of Toor and Burchard and showed that when the vapor mole-fraction driving force of component A is small compared to that of the other components, the transport rate of A is controlled by the other components, with the result that E_{MV} for A is anywhere from minus to plus ∞. They confirmed this prediction by conducting experiments with the ethanol/*tert*-butanol/water system and obtained values of E_{MV} for *tert*-butanol ranging from −2,978% to +527%. In addition, E_{MV} for ethanol and water sometimes differed significantly.

Two other tray efficiencies are defined in the literature: the vaporization efficiency of Holland, which was first touted by McAdams, and the Hausen tray efficiency, which eliminates the assumption in E_{MV} that the exiting liquid is at its bubble point. The former cannot account for the Toor phenomena and can vary widely in a manner not ascribable to the particular component. The latter does appear to be superior to E_{MV}, but is considerably more complicated and difficult to use, and, thus, has not found many adherents.

Because of the difficulties in applying a tray efficiency to an equilibrium-stage model for multicomponent systems, the development of a realistic, nonequilibrium transport- or *rate-based model* has long been a desirable goal. In 1977,

Waggoner and Loud [13] developed a rate-based, mass-transfer model limited to nearly ideal, close-boiling systems. However, an energy-transfer equation was not included (because thermal equilibrium would be closely approximated for a close-boiling mixture), and the coupling of component mass-transfer rates was ignored.

In 1979, Krishna and Standart [14] showed the possibility of applying rigorous, multicomponent mass- and heat-transfer theory to calculations of simultaneous transport. The theory was further developed by Taylor and Krishna [15], which led to the development in 1985 by Krishnamurthy and Taylor [16] of the first rate-based, computer-aided model for trayed and packed distillation columns, and other continuous separation operations. Their model applies the two-film theory of mass transfer discussed in §3.7, with the assumption of phase equilibria at the interface, and provides options for vapor and liquid flow configurations in trayed columns, including plug flow and perfectly mixed flow, on each tray. The model does not require tray efficiencies or values of HETP, but correlations of mass-transfer and heat-transfer coefficients are needed for the particular type of trays or packing employed. The model was extended in 1994 by Taylor, Kooijman, and Hung [17] to include: (1) effect of liquid-droplet entrainment in the vapor and occlusion of vapor bubbles in the liquid, (2) column-pressure profile, (3) interlinking streams, and (4) axial dispersion in packed columns. Unlike the 1985 model, which required the user to specify the column diameter and tray geometry or packing size, the 1994 version includes a design mode that estimates column diameter for a specified fraction of flooding or pressure drop. Rate-based models are available in process simulators, including RATEFRAC [18] of ASPEN PLUS, ChemSep [19], and CHEMCAD. The use of rate-based models is highly recommended for cases of low tray efficiencies (e.g. absorbers) and distillation of highly non-ideal multicomponent systems.

§12.1 RATE-BASED MODEL

A schematic diagram of a nonequilibrium stage, consisting of a tray, a group of trays, or a segment of a packed section, is shown in Figure 12.1. Stages are numbered from the top down.

§12.1.1 Model Variables

Entering stage j, at pressure P_j, are molar flow rates of feed liquid F_j^L and/or feed vapor F_j^V with component i molar flow rates, $f_{i,j}^L$ and $f_{i,j}^V$, and stream molar enthalpies, H_j^{LF} and H_j^{VF}. Also leaving from (+) or entering (−) the liquid and/or vapor phases in the stage are heat-transfer rates Q_j^V and Q_j^L, respectively; and also entering stage j from the stage above is liquid molar flow rate L_{j-1} at temperature T_{j-1}^L and pressure P_{j-1}, with molar enthalpy H_{j-1}^L and component mole fractions $x_{i\,j-1}$. Entering the stage from the stage below is vapor molar flow rate V_{j+1} at temperature T_{j+1}^V and pressure P_{j+1}, with

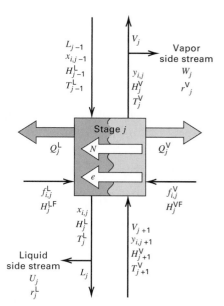

Figure 12.1 Nonequilibrium stage for rate-based method.

molar enthalpy H_{j+1}^V and component mole fractions $y_{i,j+1}$. Within the stage, mass transfer of components occurs across the phase boundary at molar rates $N_{i,j}$ from the vapor phase to the liquid phase (+) or vice versa (−), and heat transfer occurs across the phase boundary at rates e_j from the vapor phase to the liquid phase (+) or vice versa (−). Leaving the stage is liquid at temperature T_j^L and pressure P_j, with molar enthalpy H_j^L; and vapor at temperature T_j^V and pressure P_j, with molar enthalpy H_j^V. A fraction, r_j^L, of the liquid exiting the stage may be withdrawn as a liquid sidestream at molar flow rate U_j, leaving the molar flow rate L_j to enter the stage below or to exit the column. A fraction, r_j^V, of the vapor exiting the stage may be withdrawn as a vapor sidestream at molar flow rate W_j, leaving the molar flow rate V_j to enter the stage above or to exit the column. If desired, entrainment, occlusion, interlink flows, a second immiscible liquid phase, and chemical reaction(s) can be added to the model.

§12.1.2 Model Equations

Recall that the equilibrium-stage model of §10.1 utilizes the $2C + 3$ MESH equations for each stage: C mass balances for components; C phase-equilibria relations; two summations of mole fractions; and one energy balance.

In the rate-based model, the mass and energy balances around each equilibrium stage are replaced by separate balances for each phase around a stage, which can be a tray, a collection of trays, or a segment of a packed section. In residual form, the equations are as follows, where the residuals are on the LHSs and become zero when the computations are converged. When not converged, the residuals are used to determine the proximity to convergence.

Liquid-phase component material balance:

$$M_{i,j}^L \equiv \left(1 + r_j^L\right)L_j x_{i,j} - L_{j-1} x_{i,j-1} - f_{i,j}^L - N_{i,j}^L = 0, \tag{12-4}$$
$$i = 1, 2, \ldots, C$$

Vapor-phase component material balance:

$$M_{i,j}^V \equiv \left(1 + r_j^V\right)V_j y_{i,j} - V_{j+1} y_{i,j+1} - f_{i,j}^V - N_{i,j}^V = 0,$$
$$i = 1, 2, \ldots, C \qquad (12\text{-}5)$$

Liquid-phase energy balance:

$$E_j^L \equiv \left(1 + r_j^L\right)L_j H_j^L - L_{j-1}H_{j-1}^L - \left(\sum_{i=1}^{C} f_{i,j}^L\right)H_j^{LF}$$
$$+ Q_j^L - e_j^L = 0 \qquad (12\text{-}6)$$

Vapor-phase energy balance:

$$E_j^V \equiv \left(1 + r_j^V\right)V_j H_j^V - V_{j+1}H_{j+1}^V - \left(\sum_{i=1}^{C} f_{i,j}^V\right)H_j^{VF}$$
$$+ Q_j^V - e_j^V = 0 \qquad (12\text{-}7)$$

where at the phase interface, I,

$$E_j^I \equiv e_j^V - e_j^L = 0 \qquad (12\text{-}8)$$

Equations (12-4) and (12-5) are coupled by the component mass-transfer rates:

$$R_{i,j}^L \equiv N_{i,j} - N_{i,j}^L = 0, \quad i = 1, 2, \ldots, C-1 \qquad (12\text{-}9)$$

$$R_{i,j}^V \equiv N_{i,j} - N_{i,j}^V = 0, \quad i = 1, 2, \ldots, C-1 \qquad (12\text{-}10)$$

The equations for the mole-fraction summation for each phase are applied at the vapor–liquid interface:

$$S_j^{LI} \equiv \sum_{i=1}^{C} x_{i,j}^I - 1 = 0 \qquad (12\text{-}11)$$

$$S_j^{VI} \equiv \sum_{i=1}^{C} y_{i,j}^I - 1 = 0 \qquad (12\text{-}12)$$

A hydraulic equation for stage pressure drop is given by

$$H_j \equiv P_{j+1} - P_j - \left(\Delta P_j\right) = 0, \quad j = 1, 2, 3, \ldots, N-1 \qquad (12\text{-}13)$$

where the stage is assumed to be at mechanical equilibrium:

$$P_j^L = P_j^V = P_j \qquad (12\text{-}14)$$

and ΔP_j is the gas-phase pressure drop from stage $j+1$ to stage j. Equation (12-13) is optional. It is included only when it is desired to compute one or more stage pressures from hydraulics. Phase equilibrium for each component is assumed to exist only at the phase interface:

$$Q_{i,j}^I \equiv K_{i,j} x_{i,j}^I - y_{i,j}^I = 0, \quad i = 1, 2, \ldots, C \qquad (12\text{-}15)$$

Because only $C-1$ equations are written for the component mass-transfer rates in (12-9) and (12-10), total phase material balances in terms of total mass-transfer rates, $N_{T,j}$, can be added to the system:

$$M_{T,j}^L \equiv \left(1 + r_j^L\right)L_j - L_{j-1} - \sum_{i=1}^{C} f_{i,j}^L - N_{T,j} = 0 \qquad (12\text{-}16)$$

$$M_{T,j}^V \equiv \left(1 + r_j^V\right)V_j - V_{j+1} - \sum_{i=1}^{C} f_{i,j}^V + N_{T,j} = 0 \qquad (12\text{-}17)$$

where $$N_{T,j} = \sum_{i=1}^{C} N_{i,j} \qquad (12\text{-}18)$$

Equations (12-4), (12-5), (12-9), (12-10), (12-16), (12-17), and (12-18) contain terms for component mass-transfer rates, estimated from diffusive and bulk-flow (convective) contributions. The former are based on interfacial area, average mole-fraction driving forces, and mass-transfer coefficients that account for component-coupling effects through binary-pair coefficients. Empirical equations are used for interfacial area and binary mass-transfer coefficients, based on correlations of data for bubble-cap trays, sieve trays, valve trays, random packings, and structured packings. The average mole-fraction driving forces for diffusion depend upon the assumed vapor and liquid flow patterns. The simplest case is perfectly mixed flow for both phases, which simulates small-diameter, trayed columns. Countercurrent plug flow for vapor and liquid simulates a packed column with no axial dispersion.

Equations (12-6) to (12-8) contain heat-transfer rates. These are estimated from convective and enthalpy-flow contributions, where the former are based on interfacial area, average temperature-driving forces, and convective heat-transfer coefficients from the Chilton–Colburn analogy for the vapor phase (§3.5.2), and the penetration theory for the liquid phase (§3.6.2).

K-values in (12-15) are estimated from equation-of-state models (§2.5.1) or activity-coefficient models (§2.6). Tray or packed-segment pressure drops are estimated from suitable correlations of the type discussed in Chapter 6.

§12.1.3 Degrees-of-Freedom Analysis

The total number of independent equations, referred to as the MERSHQ equations, for each nonequilibrium stage, is $5C + 5$, as listed in Table 12.1. These equations apply for N stages—that is, $N_E = N(5C + 5)$ equations—in terms of $7NC + 14N + 1$ variables, listed in Table 12.2. The number of degrees of freedom is $N_D = N_V - N_E = (7NC + 14N + 1) - (5NC + 5N) = 2NC + 9N + 1$.

Table 12.1 Summary of Independent Equations for Rate-Based Model

Equation	No. of Equations
$M_{i,j}^L$	C
$M_{i,j}^V$	C
$M_{T,j}^L$	1
$M_{T,j}^V$	1
E_j^L	1
E_j^V	1
E_j^I	1
$R_{i,j}^L$	$C-1$
$R_{i,j}^V$	$C-1$
S_j^{LI}	1
S_j^{VI}	1
H_j	(optional)
$Q_{i,j}^I$	C
	$5C+5$

Table 12.2 List of Variables for Rate-Based Model

Variable Type No.	Variable	No. of Variables
1	No. of stages, N	1
2	$f_{i,j}^{L}$	NC
3	$f_{i,j}^{V}$	NC
4	T_j^{LF}	N
5	T_j^{VF}	N
6	P_j^{LF}	N
7	P_j^{VF}	N
8	L_j	N
9	$x_{i,j}$	NC
10	r_j^{L}	N
11	T_j^{L}	N
12	V_j	N
13	$y_{i,j}$	NC
14	r_j^{V}	N
15	T_j^{V}	N
16	P_j	N
17	Q_j^{L}	N
18	Q_j^{V}	N
19	$x_{i,j}^{I}$	NC
20	$y_{i,j}^{I}$	NC
21	T_j^{I}	N
22	$N_{i,j}$	NC
		$N_V = 7NC + 14N + 1$

If variable types 1 to 7, 10, 14, and 16 to 18 in Table 12.2 are specified, a total of $2NC + 9N + 1$ variables are assigned values and the degrees of freedom are totally consumed. Then, the remaining $5C + 5$ independent variables in the $5C + 5$ equations are $x_{i,j}$, $y_{i,j}$, $x_{i,j}^{I}$, $y_{i,j}^{I}$, $N_{i,j}$, T_j^{L}, T_j^{V}, T_j^{I}, L_j, and V_j, which are the variables to be computed from the equations. Properties $K_{i,j}^{I}$, H_j^{LF}, H_j^{VF}, H_j^{L}, and H_j^{V} are computed from correlations in terms of the remaining independent variables. Transport rates $N_{i,j}^{L}$, $N_{i,j}^{V}$, e_j^{L}, and e_j^{V} are from transport correlations and certain physical properties, in terms of the remaining independent variables. Stage pressures are computed from pressure drops, ΔP_j, stage geometry, fluid-mechanic equations, and certain physical properties, in terms of the remaining independent variables.

For a distillation column, it is preferable that Q_1^{V} (heat-transfer rate from the vapor in the condenser) and Q_N^{L} (heat-transfer rate to the liquid in the reboiler) are not specified but instead, as in the case of a column with a partial condenser, substitute L_1 (reflux rate) and L_N (bottoms flow rate), which specifications are sometimes referred to as standard specifications for ordinary distillation. For an adiabatic absorber or adiabatic stripper, however, all Q_j^{L} and Q_j^{V} are set equal to 0, with no substitution of specifications.

§12.2 THERMODYNAMIC PROPERTIES AND TRANSPORT-RATE EXPRESSIONS

§12.2.1 Thermodynamic Properties

Rate-based models use the same K-value and enthalpy correlations as equilibrium-based models. However, K-values apply only at the equilibrium interface between vapor and liquid phases on trays or in packing. The K-value correlation, whether based on an equation-of-state or activity-coefficient model, is a function of interface temperature, interface compositions, and tray pressure. Enthalpies are evaluated at conditions of the phases as they exit a tray. For the equilibrium-based model, vapor is at the dew-point temperature and liquid is at the bubble-point temperature, where both temperatures are at the stage temperature. For the rate-based model, liquid is subcooled and vapor is superheated, so they are at different temperatures.

§12.2.2 Transport-Rate Expressions

Accurate enthalpies and, particularly, K-values are crucial to equilibrium-based models. For rate-based models, accurate predictions of heat-transfer rates and, particularly, mass-transfer rates are also required. These depend upon transport coefficients, interfacial area, and driving forces. It is important that mass-transfer rates for multicomponent mixtures account for component-coupling effects through binary-pair coefficients.

The general forms for component mass-transfer rates across the vapor and liquid films, respectively, on a tray or in a packed segment, are as follows, where both diffusive and convective (bulk-flow) contributions are included:

$$N_{i,j}^{V} = a_j^{I} J_{i,j}^{V} + y_{i,j} N_{T,j} \tag{12-19}$$

and

$$N_{i,j}^{L} = a_j^{I} J_{i,j}^{L} + x_{i,j} N_{T,j} \tag{12-20}$$

where a_j^{I} is the total interfacial area for the stage and $J_{i,j}^{P}$ is the molar diffusion flux relative to the molar-average velocity, where P stands for the phase (V or L). For a binary mixture, as discussed in §3.7, these fluxes, in terms of mass-transfer coefficients, are given by

$$J_i^{V} = c_t^{V} k_i^{V} \left(y_i^{V} - y_i^{I} \right)_{avg} \tag{12-21}$$

and

$$J_i^{L} = c_t^{L} k_i^{L} \left(x_i^{I} - y_i^{L} \right)_{avg} \tag{12-22}$$

where c_t^{P} is the total molar concentration, k_i^{P} is the mass-transfer coefficient for a binary mixture based on a mole-fraction driving force, and the last terms in (12-21) and (12-22) are the mean mole-fraction driving forces over the stage. The positive direction of mass transfer is assumed to be from the vapor phase to the liquid phase. From the definition of the molar diffusive flux:

$$\sum_{i=1}^{C} J_i = 0 \tag{12-23}$$

Thus, for the binary system (1, 2), $J_1 = -J_2$.

§12.2.3 Mass-Transfer Coupling

As discussed in detail by Taylor and Krishna [15] and in (§3.8), the general multicomponent case for mass transfer is considerably more complex than the binary case because of component-coupling effects. For example, for the ternary system (1, 2, 3), the fluxes for the first two components are

$$J_1^V = c_t^V \kappa_{11}^V (y_1^V - y_1^I)_{avg} + c_t^V \kappa_{12}^V (y_2 - y_2^I)_{avg} \quad (12\text{-}24)$$

$$J_2^V = c_t^V \kappa_{21}^V (y_1^V - y_1^I)_{avg} + c_t^V \kappa_{22}^V (y_2 - y_2^I)_{avg} \quad (12\text{-}25)$$

The flux for the third component is not independent of the other two, but is obtained from (12-23):

$$J_3^V = -J_1^V - J_2^V \quad (12\text{-}26)$$

In these equations, the binary-pair coefficients, κ^P, are complex functions related to inverse-rate functions described below and called the Maxwell–Stefan mass-transfer coefficients in binary mixtures, which were introduced in §3.8.2.

For the general multicomponent system (1, 2, . . . , C), the independent fluxes for the first $C - 1$ components are given in matrix equation form as

$$\boldsymbol{J}^V = c_t^V [\boldsymbol{\kappa}^V] (\boldsymbol{y}^V - \boldsymbol{y}^I)_{avg} \quad (12\text{-}27)$$

$$\boldsymbol{J}^L = c_t^L [\boldsymbol{\kappa}^L] (\boldsymbol{x}^I - \boldsymbol{x}^L)_{avg} \quad (12\text{-}28)$$

where $\boldsymbol{J}^P$, $(\boldsymbol{y}^V - \boldsymbol{y}^I)_{avg}$, and $(\boldsymbol{x}^I - \boldsymbol{x}^L)_{avg}$ are column vectors of length $C - 1$ and $[\boldsymbol{\kappa}^P]$ is a $(C - 1) \times (C - 1)$ square matrix. The method for determining average mole-fraction driving forces depends, as discussed in the next section, upon the flow patterns of the vapor and liquid phases.

The fundamental theory for multicomponent diffusion is that of Maxwell and Stefan, who, in the period from 1866 to 1871, developed the kinetic theory of gases. Their theory is presented most conveniently in terms of rate coefficients, $\boldsymbol{B}$, which are defined in reciprocal diffusivity terms [15]. Likewise, it is convenient to determine $[\boldsymbol{\kappa}^P]$ from a reciprocal mass-transfer coefficient function, $\boldsymbol{R}$, defined by Krishna and Standart [14]. For an ideal-gas solution:

$$[\boldsymbol{\kappa}^V] = [\boldsymbol{R}^V]^{-1} \quad (12\text{-}29)$$

For a nonideal-liquid solution:

$$[\boldsymbol{\kappa}^L] = [\boldsymbol{R}^L]^{-1} [\boldsymbol{\Gamma}^L] \quad (12\text{-}30)$$

where the elements of $\boldsymbol{R}^P$ in terms of general mole fractions, z_i, are

$$R_{ii}^P = \frac{z_i}{k_{iC}^P} + \sum_{\substack{k=1 \\ k \neq i}}^{C} \frac{z_k}{k_{ik}^P} \quad (12\text{-}31)$$

$$R_{ij}^P = -z_i \left(\frac{1}{k_{ij}^P} - \frac{1}{k_{iC}^P} \right) \quad (12\text{-}32)$$

where j refers to the jth component and not the jth stage, and the values of k are binary-pair mass-transfer coefficients obtained from experimental data.

For a four-component vapor-phase system, the combination of (12-27) and (12-29) gives

$$\begin{bmatrix} J_1^V \\ J_2^V \\ J_3^V \end{bmatrix} = c_t^V \begin{bmatrix} R_{11}^V & R_{12}^V & R_{13}^V \\ R_{21}^V & R_{22}^V & R_{23}^V \\ R_{31}^V & R_{32}^V & R_{33}^V \end{bmatrix}^{-1} \begin{bmatrix} (y_1^V - y_1^I)_{avg} \\ (y_2^V - y_2^I)_{avg} \\ (y_3^V - y_3^I)_{avg} \end{bmatrix} \quad (12\text{-}33)$$

with

$$J_4^V = -(J_1^V + J_2^V + J_3^V) \quad (12\text{-}34)$$

and, for example, from (12-32) and (12-33), respectively:

$$R_{11}^V = \frac{y_1}{k_{14}^V} + \frac{y_2}{k_{12}^V} + \frac{y_3}{k_{13}^V} + \frac{y_4}{k_{14}^V} \quad (12\text{-}35)$$

$$R_{12}^V = -y_1 \left(\frac{1}{k_{12}^V} - \frac{1}{k_{14}^V} \right) \quad (12\text{-}36)$$

The term $[\boldsymbol{\Gamma}^L]$ in (12-30) is a $(C - 1) \times (C - 1)$ matrix of thermodynamic factors that corrects for nonideality, which often is a necessary correction for the liquid phase. When an activity-coefficient model is used:

$$\Gamma_{ij}^L = \delta_{ij} + x_i \left(\frac{\partial \ln \gamma_i}{\partial x_j} \right)_{T, P, x_k, k \neq j = 1, \ldots, C-1} \quad (12\text{-}37)$$

For a nonideal vapor, a $[\boldsymbol{\Gamma}^V]$ term can be included in (12-29), but this is rarely necessary. For either phase, if an equation-of-state model is used, (12-37) can be rewritten by substituting $\bar{\phi}_i$, the mixture fugacity coefficient, for γ_i. The term δ_{ij} is the Kronecker delta, which is 1 if $i = j$ and 0 if not. The thermodynamic factor is required because it is generally accepted that the fundamental driving force for diffusion is the gradient of the chemical potential rather than the mole fraction or concentration gradient.

When mass-transfer fluxes are moderate to high, an additional correction term is needed in (12-29) and (12-30) to correct for distortion of composition profiles. This correction, which can have a serious effect on the results, is discussed in detail by Taylor and Krishna [15]. The calculation of low-mass-transfer flux, according to (12-19) to (12-32), is illustrated by the following example.

EXAMPLE 12.1 Multicomponent Mass-Transfer Rates.

This example is similar to Example 11.5.1 on page 283 of Taylor and Krishna [15]. The following results were obtained for Tray n from a rate-based calculation of a ternary distillation at 14.7 psia, involving acetone (1), methanol (2), and water (3) in a 5.5-ft-diameter column using sieve trays with a 2-inch-high weir. Vapor and liquid phases are assumed to be completely mixed.

Component	y_n	y_{n+1}	y_n^I	K_n^I	x_n
1	0.2971	0.1700	0.3521	2.759	0.1459
2	0.4631	0.4290	0.4677	1.225	0.3865
3	0.2398	0.4010	0.1802	0.3673	0.4676
	1.0000	1.0000	1.0000		1.0000

The computed products of the gas-phase, binary mass-transfer coefficients and interfacial area, using the Chan–Fair correlation of §6.6, are as follows in lbmol/(h-unit mole fraction):

$$k_{12} = k_{21} = 1,955; \; k_{13} = k_{31} = 2,407; \; k_{23} = k_{32} = 2,797$$

(a) Compute the molar diffusion rates. (b) Compute the mass-transfer rates. (c) Calculate the Murphree vapor-tray efficiencies.

Solution

Because rates instead of fluxes are given, the equations developed in this section are used with rates rather than fluxes.

(a) Compute the reciprocal rate functions, **R**, from (12-31) and (12-32), assuming linear mole-fraction gradients such that z_i can be replaced by $(y_i + y_i^I)/2$. Thus:

$$z_1 = (0.2971 + 0.3521)/2 = 0.3246$$

$$z_2 = (0.4631 + 0.4677)/2 = 0.4654$$

$$z_3 = (0.2398 + 0.1802)/2 = 0.2100$$

$$R_{11}^V = \frac{z_1}{k_{13}} + \frac{z_2}{k_{12}} + \frac{z_3}{k_{13}} = \frac{0.3246}{2.407} + \frac{0.4654}{1.955} + \frac{0.2100}{2.407}$$
$$= 0.000460$$

$$R_{22}^V = \frac{z_2}{k_{23}} + \frac{z_1}{k_{21}} + \frac{z_3}{k_{23}} = \frac{0.4654}{2.797} + \frac{0.3246}{1.955} + \frac{0.2100}{2.797}$$
$$= 0.000408$$

$$R_{12}^V = -z_1\left(\frac{1}{k_{12}} - \frac{1}{k_{13}}\right) = -0.3246\left(\frac{1}{1.955} - \frac{1}{2.407}\right)$$
$$= -0.0000312$$

$$R_{21}^V = -z_2\left(\frac{1}{k_{21}} - \frac{1}{k_{23}}\right) = -0.4654\left(\frac{1}{1.955} - \frac{1}{2.797}\right)$$
$$= -0.0000717$$

Thus, in matrix form:

$$[\mathbf{R}^V] = \begin{bmatrix} 0.000460 & -0.0000312 \\ -0.0000717 & 0.000408 \end{bmatrix}$$

From (12-29), by matrix inversion:

$$[\mathbf{\kappa}^V] = [\mathbf{R}^V]^{-1} = \begin{bmatrix} 2,200 & 168.2 \\ 386.6 & 2,480 \end{bmatrix}$$

Because the off-diagonal terms in the preceding 2×2 matrix are much smaller than the diagonal terms, the effect of coupling in this example is small.

From (12-27):

$$\begin{bmatrix} J_1^V \\ J_2^V \end{bmatrix} = \begin{bmatrix} \kappa_{11}^V & \kappa_{12}^V \\ \kappa_{21}^V & \kappa_{22}^V \end{bmatrix} \begin{bmatrix} (y_1 - y_1^I) \\ (y_2 - y_2^I) \end{bmatrix}$$

$$J_1^V = \kappa_{11}^V(y_1 - y_1^I) + \kappa_{12}^V(y_2 - y_2^I)$$
$$= 2,200(0.2971 - 0.3521) + 168.2(0.4631 - 0.4677)$$
$$= -121.8 \; \text{lbmol/h}$$

$$J_2^V = \kappa_{21}^V(y_1 - y_1^I) + \kappa_{22}^V(y_2 - y_2^I)$$
$$= 386.6(0.2971 - 0.3521) + 2,480(0.4631 - 0.4677)$$
$$= -32.7 \; \text{lbmol/h}$$

From (12-23):

$$J_3^V = -J_1^V - J_2^V = 121.8 + 32.7 = 154.5 \; \text{lbmol/h}$$

(b) From (12-19), but with diffusion and mass-transfer rates instead of fluxes:

$$N_1^V = J_1^V + z_1 N_T^V = -121.8 + 0.3246 N_T^V \tag{1}$$

Similarly:

$$N_2^V = -32.7 + 0.4654 N_T^V \tag{2}$$

$$N_3^V = 154.5 + 0.2100 N_T^V \tag{3}$$

To determine component mass-transfer rates, it is necessary to know the total mass-transfer rate for the tray, N_T^V. The problem of determining this quantity when the diffusion rates, J, are known is referred to as the *bootstrap problem* (p. 145 in Taylor and Krishna [15]). In chemical reaction with diffusion, N_T is determined by the stoichiometry. In distillation, N_T is determined by an energy balance, which gives the change in molar vapor rate across a tray. For the assumption of constant molar overflow, $N_T = 0$. In this example, that assumption is not valid, and the change is

$$N_T = V_{n+1} - V_n = -54 \; \text{lbmol/h}$$

From (1), (2), and (3):

$$N_1^V = -121.8 + 0.3246(-54) = -139.4 \; \text{lbmol/h}$$
$$N_2^V = -32.7 + 0.4654(-54) = -57.8 \; \text{lbmol/h}$$
$$N_3^V = -154.5 + 0.2100(-54) = 143.2 \; \text{lbmol/h}$$

(c) Values of the E_{MV} are obtained from (12-3), with K-values at phase-interface conditions:

$$E_{MV_i} = (y_{i,n} - y_{i,n+1})/(K_{i,n}^I x_{i,n} - y_{i,n+1}) \tag{4}$$

From (4):

$$E_{MV_1} = \frac{(0.2971 - 0.1700)}{[(2.759)(0.1459) - 0.1700]} = 0.547$$

$$E_{MV_2} = \frac{(0.4631 - 0.4290)}{[(1.225)(0.3865) - 0.4290]} = 0.767$$

$$E_{MV_3} = \frac{(0.2398 - 0.4010)}{[(0.3673)(0.4676) - 0.4010]} = 0.703$$

General forms for heat-transfer rates across vapor and liquid films of a stage are

$$e_j^V = a_j^I h^V(T^V - T^I) + \sum_{i=1}^{C} N_{i,j}^V \bar{H}_{i,j}^V \tag{12-38}$$

$$e_j^L = a_j^I h^L(T^I - T^L) + \sum_{i=1}^{C} N_{i,j}^L \bar{H}_{i,j}^L \tag{12-39}$$

where $\bar{H}_{i,j}^P$ are the partial molar enthalpies of component i for stage j and h^P are convective heat-transfer coefficients. The second terms on the RHSs of (12-38) and (12-39) account for the transfer of enthalpy by mass transfer. Temperatures T^V and T^L are the temperatures of vapor and liquid exiting the stage.

§12.3 METHODS FOR ESTIMATING TRANSPORT COEFFICIENTS AND INTERFACIAL AREA

Equations (12-31) and (12-32) require binary-pair mass-transfer coefficients for phase contacting devices, which must be estimated from empirical correlations of experimental data for different contacting devices.

As discussed in §6.6 for trayed columns, widely used correlations are the AIChE method [20] for bubble-cap trays, the correlations of Chan and Fair [24] for sieve trays, and correlations of Scheffe and Weiland [36] for Glitsch V-1 valve trays. Other correlations are those of Harris [21] and Hughmark [22] for bubble-cap trays; and Zuiderweg [23], Chen and Chuang [25], Taylor and Krishna [15], and Young and Stewart [37, 38] for sieve trays.

Some mass-transfer correlations are presented in terms of the number of transfer units, N_V and N_L, where, by definition,

$$N_V \equiv k^V a h_f / u_s \qquad (12\text{-}40)$$

$$N_L \equiv k^L a h_f z / (Q_L / W) \qquad (12\text{-}41)$$

where $a =$ interfacial area/volume of froth on the tray, $h_f =$ froth height, $u_s =$ superficial vapor velocity based on tray bubbling area, $z =$ length of liquid-flow path across the bubbling area, $Q_L =$ volumetric liquid flow rate, and $W =$ weir length.

The interfacial area for a tray, a^I, is related to a by

$$a^I = a h_f A_b \qquad (12\text{-}42)$$

where $A_b =$ bubbling area.

Thus, k^P and a^I are from correlations in terms of N_V and N_L.

For random (dumped) packings, empirical correlations for mass-transfer coefficients and interfacial-area density (area/packed volume) have been published by Onda, Takeuchi, and Okumoto [26] and Bravo and Fair [27]. For structured packings, the empirical correlations of Bravo, Rocha, and Fair for gauze packings [28] and for a wide variety of structured packings [29] are available. A semitheoretical correlation by Billet and Schultes [30], based on over 3,500 data points for more than 50 test systems and more than 70 different types of packings, requires five packing parameters and is applicable to both random and structured packings. This correlation is given in §6.8.

Heat-transfer coefficients for the vapor film are usually estimated from the Chilton–Colburn analogy between heat and mass transfer (§3.5.2). Thus,

$$h^V = k^V \rho^V C_P^V (N_{Le})^{2/3} \qquad (12\text{-}43)$$

where

$$N_{Le} = \left(\frac{N_{Sc}}{N_{Pr}} \right) \qquad (12\text{-}44)$$

A penetration model (§3.6.2) is preferred for the liquid-phase film:

$$h^L = k^L \rho^L C_P^L (N_{Le})^{1/2} \qquad (12\text{-}45)$$

A more detailed heat-transfer model, specifically for sieve trays, is given by Spagnolo et al. [39].

§12.4 VAPOR AND LIQUID FLOW PATTERNS

The simplest flow pattern corresponds to the assumption of perfectly mixed vapor and liquid. Under these conditions, mass-transfer driving forces in (12-27) and (12-28) are

$$\left(y^V - y^I \right)_{avg} = \left(y^V - y^I \right) \qquad (12\text{-}46)$$

$$\left(x^I - x^L \right)_{avg} = \left(x^I - x^L \right) \qquad (12\text{-}47)$$

where y^V and x^L are exiting-stage mole fractions. These flow patterns are valid only for trayed towers with a short liquid flow path.

A plug-flow pattern for the vapor and/or liquid assumes that the phase moves through the froth without mixing. This requires that mass-transfer rates be integrated over the froth. An approximation of the integration is provided by Kooijman and Taylor [31], who assume constant mass-transfer coefficients and interface compositions. The resulting expressions for the average mole-fraction driving forces are the same as (12-46) and (12-47), except for a correction factor in terms of N^V or N^L included on the RHS of each equation. Plug-flow patterns are more accurate for trayed towers than perfectly mixed flow patterns and are also applicable to packed towers.

The perfectly mixed flow and plug-flow patterns are the two patterns presented by Lewis [7] to convert E_{OV} to E_{MV}, as discussed in §6.5. They represent the extreme situations. Fair, Null, and Bolles [32] recommend a more realistic partial mixing model that utilizes a turbulent Peclet number, whose value can cover a wide range. This model is a bridge between the two extremes.

For reactive distillation, a rate-based multicell (or mixed-pool) model has proven useful. In this model, the liquid on the tray is assumed to flow horizontally across the tray through a series of perfectly mixed cells (perhaps 4 or 5). In the model of Higler, Krishna, and Taylor [40], which is available in the ChemSep program, the vapor phase is assumed to be perfectly mixed in each cell. If desired, cells for each tray can also be stacked in the vertical direction. Thus, a tray model might consist of a 5×5 cell arrangement, for a total of 25 perfectly mixed cells. It is assumed that the vapor streams leaving the topmost tray cells are collected and mixed before being divided to enter the cells on the next tray. The rate-based multicell model of Pyhalahti and Jakobsson [41] allows one set of cells in the horizontal direction, but vapor streams leaving the tray cells may be mixed or not mixed before entering the cells on the next tray, and the reversal of liquid-flow direction from tray to tray, shown in Figure 6.21 for single-pass trays, is allowed.

§12.5 METHOD OF CALCULATION

As stated in §12.1, the equations to be solved for the single-cell-per-tray, rate-based model of Figure 12.1 is $N(5C + 5)$ when the pressure-drop equations are omitted, as summarized in Table 12.1. The equations contain the variables listed in Table 12.2. Other parameters in the equations are computed from these variables. When the number of equations is subtracted from the number of variables, the degrees of freedom is $2NC + 9N + 1$. If the total number of stages and all column feed conditions, including feed-stage locations ($2NC + 4N + 1$ variables) are specified, the number of remaining degrees of freedom, using the variable designations in Table 12.2, is $5N$. A computer program for the rate-based model would generally require the user to specify these $2NC + 4N + 1$ variables. The degree of flexibility provided to the user in the selection of the remaining $5N$ variables depends on the particular rate-based computer algorithm, three of which are widely available: (1) ChemSep from R. Taylor and H. A. Kooijman, (2) RATEFRAC in ASPEN PLUS, and (3) CHEMCAD. All these algorithms provide a wide variety of correlations for thermodynamic and transport properties and flexibility in the selection of the remaining $5N$ specifications. The basic $5N$ specifications are

$$r_j^L \text{ or } U_j, \ r_j^V \text{ or } W_j, \ P_j, \ Q_j^L, \text{ and } Q_j^V$$

However, substitutions can be made, as discussed next.

§12.5.1 ChemSep Program (www.chemsep.org)

The ChemSep program applies the transport equations to trays or short heights (called segments) of packing. Partial condensers and reboilers are treated as equilibrium stages. The specification options include:

1. r_j^L and r_j^V: From each stage, either a liquid or a vapor sidestream can be specified as (a) a sidestream flow rate or (b) a ratio of the sidestream flow rate to the flow rate of the remaining fluid passing to the next stage:

$$r_j^L = U_j/L_j \quad \text{or} \quad r_j^V = W_j/V_j \text{ in Figure 12.1}$$

2. P_j: Four options are available, all requiring the pressure of the condenser if used:

   (a) Constant pressure for all stages in the tower and reboiler.

   (b) Top tower pressure, and bottom pressure. Pressures of stages intermediate between top and bottom are obtained by linear interpolation.

   (c) Top tower pressure, and specified pressure drop per stage to obtain remaining stage pressures.

   (d) Top tower pressure, with stage pressure drops estimated by ChemSep from hydraulic correlations.

3. Q_j^L and Q_j^V: The heat duty must be specified for all stage heaters and coolers except the condenser and/or reboiler, if present. In addition, a heat loss for the tower that is divided equally over all stages can be specified. When a condenser (total without subcooling, total with subcooling, or partial) is present, one of the following specifications can replace the condenser heat duty: (a) molar reflux ratio; (b) condensate temperature; (c) distillate molar flow rate; (d) reflux molar flow rate; (e) component molar flow rate in distillate; (f) mole fraction of a component in distillate; (g) fractional recovery, from all feeds, of a component in the distillate; (h) molar fraction of all feeds to the distillate; and (i) molar ratio of two distillate components. For distillation, an often-used specification is the molar reflux ratio.

When a reboiler (partial, total with a vapor product, or total with a superheated vapor product) is present, the following list of specification options, similar to those just given for a condenser, can replace the reboiler heat duty: (a) molar boilup ratio; (b) reboiler temperature; (c) bottoms molar flow rate; (d) reboiled-vapor (boilup) molar flow rate; (e) component molar flow rate in bottoms; (f) mole fraction of a component in bottoms; (g) fractional recovery, from all feeds, of a component in the bottoms; (h) molar fraction of all feeds to the bottoms; and (i) molar ratio of two components in the bottoms. For distillation, an often-used specification is the molar bottoms flow rate, which must be estimated if it is not specified.

The preceding number of optional specifications is considerable. In addition, ChemSep also provides "flexible" specifications that can substitute for the condenser and/or reboiler duties. These are advanced options supplied in the form of strings that contain values of certain allowable variables and/or combinations of these variables using the five common arithmetic operators $(+, -, *, /,$ and exponentiation). The variables include stage variables (L, V, x, y, and T) and interface variables (x^I, y^I, and T^I) at any stage. Flow rates can be in mole or mass units.

Certain options and advanced options must be used with care because values might be specified that cannot lead to a converged solution. For example, with a simple distillation column of a fixed number of stages, that N may be less than the N_{min} needed to achieve specified distillate and bottoms purities. As always, it is generally wise to begin a simulation with a standard pair of top and bottom specifications, such as reflux ratio and a bottoms molar flow rate that corresponds to the desired distillate rate. These specifications are almost certain to converge unless interstage liquid or vapor flow rates tend to zero somewhere in the column. A study of the calculated results will provide insight into possible limits in the use of other options.

The equations for the rate-based model, some linear and some nonlinear, are solved by Newton's method in a manner similar to that developed by Naphtali and Sandholm for the equilibrium-based model described in §10.4. Thus, the variables and equations are grouped by stage so that the Jacobian matrix is of block-tridiagonal form. However, the equations to be solved number $5C + 6$ or $5C + 5$ per stage, depending on whether stage pressures are computed or specified, compared to just $2C + 1$ for the equilibrium-based method.

Calculations of transport coefficients and pressure drops require column diameter and dimensions of column internals. These may be specified (simulation mode) or computed (design mode). In the latter case, default dimensions are selected for the internals, with column diameter computed from a specified value for percent of flooding for a trayed or packed column, or a specified pressure drop per unit height for a packed column.

Computing time per iteration for the design mode is only approximately twice that of the simulation mode, which usually requires less than twice the time for the equilibrium-based model. The number of iterations required for the design mode can be two to three times that for the equilibrium-based model. Overall, the total computing time for the design mode is usually less than an order of magnitude greater than that for the equilibrium-based model. With today's fast PCs, computing time for the design mode of the rate-based model is usually less than 1 minute.

Like the Naphtali–Sandholm equilibrium-based method, the rate-based model utilizes mainly analytical partial derivations in the Jacobian matrix, and requires initial estimates of all variables. These estimates are generated automatically by the ChemSep program using a method of Powers et al. [33], in which the usual assumptions of constant molar overflow and a linear temperature profile are employed. The initialization of the stage mole fractions is made by performing several iterations of the BP method (§10.3.2) using ideal K-values for the first iteration and non-ideal K-values thereafter. Initial interface mole fractions are set equal to estimated bulk values, and initial mass-transfer rates are arbitrarily set to values of $\pm 10^{-3}$ kmol/h, with the sign dependent upon the component K-value.

To prevent oscillations and promote convergence of the iterations, corrections to certain variables from iteration to iteration can be limited. Defaults are 10 K for temperature and 50% for flows. When a correction to a mole fraction would result in a value outside the feasible range of 0 to 1, the default correction is one-half of the step that would take the value to a limit. For very difficult problems, homotopy-continuation methods described by Powers et al. [33] can be applied to promote convergence.

Convergence of Newton's method is determined from residuals of the functions, as in the Naphtali–Sandholm method, or from the corrections to the variables. ChemSep applies both criteria and terminates when either of the following are satisfied:

$$\left[\sum_{j=1}^{N} \sum_{k=1}^{N_j} f_{k,j}^2 \right]^{1/2} < \epsilon \qquad (12\text{-}48)$$

$$\sum_{j=1}^{N} \sum_{k=1}^{N_j} |\Delta X_{k,j}|/X_{k,j} < \epsilon \qquad (12\text{-}49)$$

where $f_{k,j}$ = residuals in Table 12.1, N = number of stages, N_j = number of equations for the jth stage. $X_{k,j}$ = unknown variables from Table 12.2, and ϵ = a small number with a default value of 10^{-4}. Unlike the Naphtali–Sandholm method, the residuals are not scaled. Accordingly, the second criterion is usually satisfied first.

From the results of a converged solution, it is highly desirable to back-calculate values of E_{MV}, component by component and tray by tray, from (12-3) for trayed columns, and HETP values for packed towers.

ChemSep can also perform rate-based calculations for liquid–liquid extraction.

EXAMPLE 12.2 Extractive Distillation.

A mixture of n-heptane and toluene cannot be separated at 1 atm by ordinary distillation. Accordingly, an enhanced-distillation scheme with methylethyl ketone as a solvent is used. As part of an initial design study, use the rate-based model of ChemSep with the specifications listed in Table 12.3 to calculate a sieve-tray column.

Solution

The information in Table 12.3 was entered via the ChemSep menu and the program was executed. A converged solution was achieved in 8 iterations in 6 seconds on a PC running Windows 95. Initialization of all variables was done by the program.

Table 12.3 Specifications for Example 12.2

Total condenser delivering saturated liquid
Partial reboiler
Pressure at condenser outlet = 14.7 psia
Pressure at condenser inlet = 15.0 psia
Reflux ratio = 1.5
Bottoms flow rate = 45 lbmol/h
Total number of trays = 20
Feed 1 to Tray 10 from top:
    55 lbmol/h of n-heptane
    45 lbmol/h of toluene
    100 lbmol/h of methylethyl ketone (MEK)
    Saturated liquid at 20 psia
Feed 2 to Tray 15 from top:
    100 lbmol/h of MEK
    Saturated liquid at 20 psia
UNIFAC for liquid-phase activity coefficients
Chan–Fair correlation for mass-transfer coefficients
Plug flow for vapor
Mixed flow for liquid
85% of flooding
Tray spacing = 0.5 m (19.7 inches)
Weir height = 2 inches

The predicted separation is as follows:

Component	Distillate, lbmol/h	Bottoms, lbmol/h
n-Heptane	54.87	0.13
Toluene	0.45	44.55
Methylethyl ketone	199.68	0.32

Predicted column profiles for pressure, liquid-phase temperature, total vapor and liquid flow rates, component vapor and liquid mole fractions, component mass-transfer rates, and values of E_{MV} are shown in Figure 12.2, where stages are numbered from the top down and stages 2 to 21 are sieve trays. Back-calculated Murphree tray efficiencies are summarized as follows:

| Component | Fractional Murphree Efficiencies | |
	Range	Median
n-Heptane	0.52 to 1.10	0.73
Toluene	0.70 to 0.79	0.79
Methylethyl ketone	−3.23 to 1.14	0.76

The median values, based on experience, seem reasonable and give confidence in the rate-based method. The 20 trays are equivalent to approximately 15 equilibrium stages.

For sizing, the column was divided into three sections: 9 trays above the top feed, 5 trays from the top feed to the bottom feed, and 6 trays below the bottom feed. Computed column diameters are, respectively, 1.75 m (5.74 ft), 1.74 m (5.71 ft), and 1.83 m (6.00 ft). Thus, a 1.83-m- (6.00-ft-) diameter column is a reasonable choice. Average predicted pressure drop per tray is 0.06 psi. Computed heat-exchanger duties are: condenser: 2.544 MW (8,680,000 Btu/h); and reboiler: 2.482 MW (8,470,000 Btu/h).

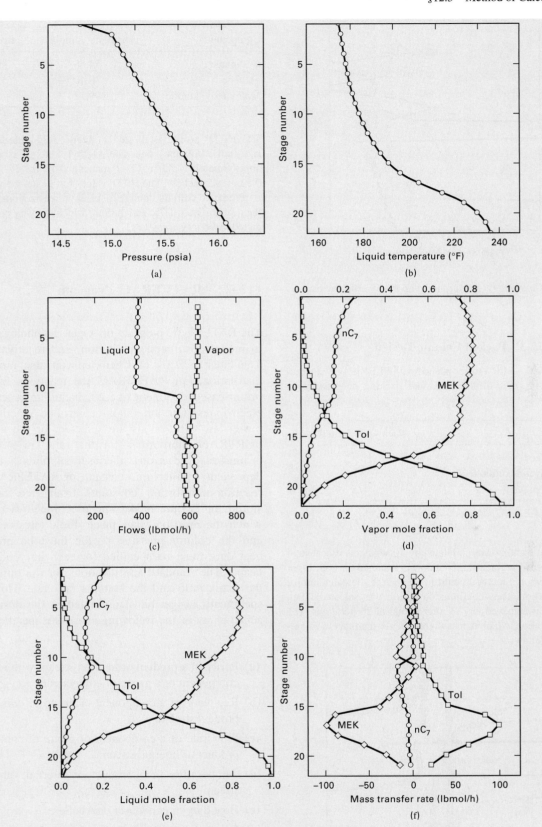

Figure 12.2 Column profiles for Example 12.2: (a) pressure profile; (b) liquid-phase temperature profile;
(c) vapor and liquid flow rate profiles; (d) vapor mole-fraction profiles; (e) liquid mole-fraction profiles;
(f) mass-transfer rate profiles. (*continued*)

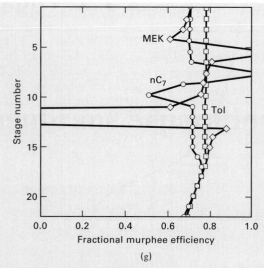

Figure 12.2 (*Continued*) (g) Murphree vapor-tray efficiencies.

EXAMPLE 12.3 Packed Column Design.

Repeat Example 12.2 for a tower packed with FLEXIPAC® 2 structured packing, at 75% of flooding. The packing heights are as follows:

Section	Packing Height, ft
Above top feed	13
Between top and bottom feeds	6.5
Below bottom feed	6.5

Solution

Each 6.5 feet of packing was simulated by 50 segments. Because of the large number of segments, mixed flow is assumed for both vapor and liquid. Newton's method could not converge the calculations. Therefore, the homotopy-continuation option was selected. Then convergence was achieved in 73 s after a total of 26 iterations. The predicted separation, which is just slightly better than that in Example 12.2, is as follows:

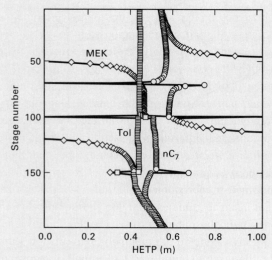

Figure 12.3 Column HETP profiles for Example 12.3.

Component	Distillate, lbmol/h	Bottoms, lbmol/h
n-Heptane	54.88	0.12
Toluene	0.40	44.60
Methylethyl ketone	199.72	0.28

The HETP profile is plotted in Figure 12.3. Median values for *n*-heptane, toluene, and methylethyl ketone, respectively, are approximately 0.55 m (21.7 inches), 0.45 m (17.7 inches), and 0.5 m (19.7 inches). The HETP values for the ketone vary widely.

Predicted column diameters for the three sections, starting from the top, are 1.65, 1.75, and 1.85 m, which are very close to the predicted sieve-tray diameters.

§12.5.2 RATEFRAC Program

The RATEFRAC program of Aspen Technology is designed to model absorbers, distillation, and reactive distillation. The latest version of ChemSep can also model reactive distillation. For RATEFRAC, the reactions can be equilibrium-based or kinetics-based, including reactions among electrolytes. For kinetically controlled reactions, built-in power-law expressions are selected, or the user supplies FORTRAN subroutines for the rate law(s). For equilibrium-based reactions, the user supplies a temperature-dependent equilibrium constant, or RATEFRAC computes reaction-equilibrium constants from free-energy values stored in its data bank. The user specifies the phase in which the reaction takes place. Flow rates of sidestreams and the column-pressure profile must be provided. The heat duty must be specified for each intercooler or interheater. The standard specifications for the rating mode are the reflux ratio and the bottoms flow rate. However, these specifications can be manipulated in the design mode to achieve any of the following substitute specifications:

(a) Purity of a product or internal stream with respect to one component or a group of components.

(b) Recovery of a component or group of components in a product stream.

(c) Flow rate of a component or group of components in a product or internal stream.

(d) Temperature of a product or internal vapor or liquid stream.

(e) Heat duty of condenser or reboiler.

(f) Value of a product or internal stream physical property.

(g) Ratio or difference of any pair of product or internal stream physical properties, where the two streams can be the same or different.

Mass-transfer correlations are built into RATEFRAC for bubble-cap trays, valve trays, sieve trays, and packings. Users may provide their own FORTRAN subroutines for transport coefficients and interfacial area. Newton's method is used to converge the calculations.

EXAMPLE 12.4   Absorber Design.

Use RATEFRAC to predict column profiles for a 3.5-ft-diameter, 20-bubble-cap tray absorber operating at the conditions in Table 12.4.

Table 12.4   Specifications for Example 12.4

Column top pressure = 182 psia
Column bottom pressure = 185 psia
Weir height = 2 inches

Vapor completely mixed on each tray
Liquid completely mixed on each tray
AIChE correlations for binary mass-transfer coefficients and
    interfacial area
Chilton–Colburn analogy for heat transfer
Chao–Seader correlation for K-values

Vapor feed at 123°F and 184 psia:

Component	lbmol/h
Hydrogen	218
Nitrogen	87
Methane	136
Ethane	139
Propane	118
Isobutane	6
n-Butane	2
Isopentane	43
n-Hexane	14
n-Heptane	4
	767

Liquid absorbent feed at 100°F and 182 psia:

Component	lbmol/h
n-Dodecane	165
n-Tridecane	165

Solution

Initial estimates of the variables were provided by RATEFRAC. A total of five iterations were required after an initialization step. Converged results were:

Component	Lean Vapor, lbmol/h	Rich Oil, lbmol/h
Hydrogen	216.6	1.4
Nitrogen	86.1	0.9
Methane	131.4	4.6
Ethane	120.0	19.0
Propane	73.4	44.6
Isobutane	1.1	4.9
n-Butane	0.0	2.0
Isopentane	0.0	43.0
n-Hexane	0.0	14.0
n-Heptane	0.0	4.0
n-Dodecane	0.0	165.0
n-Tridecane	0.0	165.0
	628.6	468.4

The back-calculated fractional Murphree vapor-tray efficiencies are:

Component	E_{MV} Range	E_{MV} Median Value
Hydrogen	−1.25 to 1.21	0.43
Nitrogen	−0.51 to 3.25	0.41
Methane	−0.13 to 0.87	0.41
Ethane	−0.03 to 1.02	0.30
Propane	−1.89 to 2.71	0.33
Isobutane	−6.51 to 1.16	0.42
n-Butane	−0.75 to 5.65	0.50
Isopentane	−3.17 to 1.15	0.57
n-Hexane	0.63 to 1.38	0.63
n-Heptane	0.64 to 0.88	0.64
n-Dodecane	−2.92 to 1.01	0.42
n-Tridecane	−5.04 to 0.97	0.42

It is seen that efficiencies vary from component to component and from tray to tray. For absorber simulation and design, a rate-based model is clearly superior to an equilibrium-based model.

As chemical engineers become more informed of the principles of mass transfer, and improved correlations for mass-transfer and heat-transfer coefficients are developed for trays and packings, use of rate-based models should accelerate. For best results, these models will also benefit from more realistic options for vapor and liquid flow patterns. More comparisons of rate-based models with industrial operating data are needed to gain confidence in the use of such models. Some comparisons are presented by Taylor, Kooijman, and Woodman [34], and Kooijman and Taylor [31]. Comparisons by Ovejero et al. [35], with distillation data obtained in a column packed with spheres and cylinders of known interfacial area, show very good agreement for three binary and two ternary systems assumed at the phase interface.

SUMMARY

1. Rate-based models of multicomponent, multistage, vapor–liquid separation operations became available in the late 1980s. These models are potentially superior to equilibrium-based models for all but near-ideal systems.

2. Rate-based models incorporate rigorous procedures for component-coupling effects in multicomponent mass transfer.

3. The number of equations for a rate-based model is greater than that for an equilibrium-based model because separate balances are needed for each of the two phases. In addition, rate-based models are influenced by the geometry of the column internals. Correlations are used to predict mass-transfer and heat-transfer rates. Tray or packing hydraulics are also incorporated into the rate-based model to enable

prediction of column-pressure profile. Phase equilibrium is assumed only at the phase interface.

4. Computing time for a rate-based model is not generally more than an order of magnitude greater than that for an equilibrium-based model.

5. Both the ChemSep and RATEFRAC rate-based computer programs offer considerable flexibility in user specifications, so much so that inexperienced users can easily specify impossible conditions. Therefore, it is best to begin simulation studies with standard specifications.

REFERENCES

1. Sorel, E., *La rectification de l'alcool*, Paris (1893).

2. Smoker, E.H., *Trans. AIChE*, **34**, 165 (1938).

3. Thiele, E.W., and R.L. Geddes, *Ind. Eng. Chem.*, **25**, 290 (1933).

4. Lewis, W.K., and G.L. Matheson, *Ind. Eng. Chem.*, **24**, 496–498 (1932).

5. Lewis, W.K., *Ind. Eng. Chem.*, **14**, 492 (1922).

6. Murphree, E.V., *Ind. Eng. Chem.*, **17**, 747–750, 960–964 (1925).

7. Lewis, W.K., *Ind. Eng. Chem.*, **28**, 399 (1936).

8. Seader, J.D., *Chem. Eng. Prog.*, **85**(10), 41–49 (1989).

9. Walter, J.F., and T.K. Sherwood, *Ind. Eng. Chem.*, **33**, 493–501 (1941).

10. Toor, H.L., *AIChE J.*, **3**, 198 (1957).

11. Toor, H.L., and J.K. Burchard, *AIChE J.*, **6**, 202 (1960).

12. Krishna, R., H.F. Martinez, R. Sreedhar, and G.L. Standart, *Trans. I. Chem. E.*, **55**, 178 (1977).

13. Waggoner, R.C., and G.D. Loud, *Comput. Chem. Engng.*, **1**, 49 (1977).

14. Krishna, R., and G.L. Standart, *Chem. Eng. Comm.*, **3**, 201 (1979).

15. Taylor, R., and R. Krishna, *Multicomponent Mass Transfer*, John Wiley & Sons, New York (1993).

16. Krishnamurthy, R., and R. Taylor, *AIChE J.*, **31**, 449, 456 (1985).

17. Taylor, R., H.A. Kooijman, and J.-S. Hung, *Comput. Chem. Engng.*, **18**, 205–217 (1994).

18. *ASPEN PLUS Reference Manual—Volume 1*, Aspen Technology, Cambridge, MA (1994).

19. Taylor, R., and H.A. Kooijman, *CACHE News*, No. **41**, 13–19 (1995).

20. AIChE, *Bubble-Tray Design Manual*, New York (1958).

21. Harris, I.J., *British Chem. Engng.*, **10**(6), 377 (1965).

22. Hughmark, G.A., *Chem. Eng. Progress*, **61**(7), 97–100 (1965).

23. Zuiderweg, F.J., *Chem Eng. Sci.*, **37**, 1441 (1982).

24. Chan, H., and J.R. Fair, *Ind. Eng. Chem. Process Des. Dev.*, **23**, 814–827 (1984).

25. Chen, G.X., and K.T. Chuang, *Ind. Eng. Chem. Res.*, **32**, 701–708 (1993).

26. Onda, K., H. Takeuchi, and Y.J. Okumoto, *J. Chem. Eng. Japan*, **1**, 56–62 (1968).

27. Bravo, J.L., and J.R. Fair, *Ind. Eng. Chem. Process Des. Devel.*, **21**, 162–170 (1982).

28. Bravo, J.L., J.A. Rocha, and J.R. Fair, *Hydrocarbon Processing*, **64**(1), 56–60 (1985).

29. Bravo, J.L., J.A. Rocha, and J.R. Fair, *I. Chem. E. Symp. Ser.*, No. 128, A489–A507 (1992).

30. Billet, R., and M. Schultes, *I. Chem. E. Symp. Ser.*, No. 128, B129 (1992).

31. Kooijman, H.A., and R. Taylor, *Chem. Eng. J.*, **57**(2), 177–188 (1995).

32. Fair, J.R., H.R. Null, and W.L. Bolles, *Ind. Eng. Chem. Process Des. Dev.*, **22**, 53–58 (1983).

33. Powers, M.F., D.J. Vickery, A. Arehole, and R. Taylor, *Comput. Chem. Engng.*, **12**, 1229–1241 (1988).

34. Taylor, R., H.A. Kooijman, and M.R. Woodman, *I. Chem. E. Symp. Ser.*, No. 128, A415–A427 (1992).

35. Ovejero, G., R. Van Grieken, L. Rodriguez, and J.L. Valverde, *Sep. Sci. Tech.*, **29**, 1805–1821 (1994).

36. Scheffe, R.D., and R.H. Weiland, *Ind. Eng. Chem. Res.*, **26**, 228–236 (1987).

37. Young, T.C., and W.E. Stewart, *AIChE J.*, **38**, 592–602 with errata on p. 1302 (1993).

38. Young, T.C., and W.E. Stewart, *AIChE J.*, **41**, 1319–1320 (1995).

39. Spagnolo, D.A., E.L. Plaice, H.J. Neuburg, and K.T. Chuang, *Can. J. Chem. Eng.*, **66**, 367–376 (1988).

40. Higler, A., R. Krishna, and R. Taylor, *AIChE J.*, **45**, 2357–2370 (1999).

41. Pyhalahti, A., and K. Jakobsson, *Ind. Eng. Chem. Res.*, **42**, 6188–6195 (2003).

STUDY QUESTIONS

12.1. For binary distillation, what assumption did Smoker add to the McCabe–Thiele assumptions to obtain an algebraic solution?

12.2. What assumptions did Murphree make in the development of his tray efficiency equations?

12.3. For which situations does the Murphree efficiency appear to be adequate? What are its deficiencies?

12.4. What unusual phenomena did Toor find for diffusion in a ternary mixture? Is a theory available to predict these phenomena?

12.5. In the rate-based model, is the assumption of phase equilibrium used anywhere? If so, where? Is it justified?

12.6. The rate-based model requires component mass-transfer coefficients, interfacial areas, and heat-transfer coefficients. How are the latter obtained?

12.7. What are component-coupling effects in mass-transfer-rate equations?

12.8. Does the rate-based model account for the bulk-flow effect in mass transfer?

12.9. Can the rate-based model be applied to packed columns?

12.10. Are tray flow patterns important in the rate-based model? What are the ideal flow-pattern models?

EXERCISES

Section 12.1

12.1. Entrainment and occlusion.
    Modify the rate-based model of (12-4) to (12-18) to include entrainment and occlusion.

12.2. Addition of chemical reaction.
    Modify the rate-based model of (12-4) to (12-18) to include a chemical reaction in the liquid phase under conditions of: (a) chemical equilibrium; (b) kinetic rate law.

12.3. Reducing equations in rate-based models.
    Explain how the number of rate-based modeling equations can be reduced. Would this be worthwhile?

Section 12.2

12.4. Mass-transfer rates and tray efficiencies.
    The following results were obtained at tray n from a rate-based calculation at 14.7 psia, for a ternary mixture of acetone (1), methanol (2), and water (3) in a sieve-tray column assuming that both phases are perfectly mixed.

Component	y_n	y_{n+1}	y_n^1	K_n^1	x_n
1	0.4913	0.4106	0.5291	1.507	0.3683
2	0.4203	0.4389	0.4070	0.900	0.4487
3	0.0884	0.1505	0.0639	0.3247	0.1830

The products of the computed gas-phase, binary mass-transfer coefficients, and interfacial area from the Chan–Fair correlations are as follows in units of lbmol/(h-unit mole fractions).

$$k_{12} = k_{21} = 1,750; \ k_{13} = k_{31} = 2,154; \ k_{23} = k_{32} = 2,503$$

The vapor rates are $V_n = 1,200$ lbmol/h and $V_{n+1} = 1,164$ lbmol/h. Determine: (a) component molar diffusion rates; (b) mass-transfer rates; (c) Murphree vapor-tray efficiencies.

12.5. Reciprocal rate functions.
    Write all the expanded equations (12-31) and (12-32) for $\mathbf{R}^P$ for a five-component system.

12.6. Perfectly mixed tray.
    Repeat the calculations of Example 12.1 but use 1 = methanol, 2 = water, and 3 = acetone. Are the results any different? If not, why not? Prove your conclusion mathematically.

Section 12.3

12.7. Mass-transfer coefficients for trays.
    Compare and discuss the advantages and disadvantages of the available correlations for estimating binary-pair mass-transfer coefficients for trayed columns.

12.8. Mass-transfer coefficients for packings.
    Discuss the advantages and disadvantages of the available correlations for estimating binary-pair mass-transfer coefficients for columns with random (dumped) and structured packings.

Section 12.4

12.9. Modeling flow patterns.
    Discuss how the method of Fair, Null, and Bolles [32] might be used to model the flow patterns in a rate-based model. How would the mole-fraction driving forces be computed?

Section 12.5

12.10. Distillation of a nonideal ternary mixture.
    A bubble-point mixture of 100 kmol/h of methanol, 50 kmol/h of isopropanol, and 100 kmol/h of water at 1 atm is sent to the 25th tray from the top of a 40-sieve-tray column equipped with a total condenser and partial reboiler, operating at a nominal pressure of 1 atm. If the reflux ratio is 5 and the bottoms flow rate is 150 kmol/h, determine the separation achieved if the UNIFAC method is used to estimate K-values and the Chan–Fair correlations are used for mass transfer. Assume that both phases are perfectly mixed on each tray and that operation is at about 80% of flooding.

12.11. Extractive distillation.
    A sieve-tray column, operating at a nominal pressure of 1 atm, is used to separate a mixture of acetone and methanol by extractive distillation using water. The column has 40 trays with a total condenser and partial reboiler. The feed of 50 kmol/h of acetone and 150 kmol/h of methanol at 60°C and 1 atm enters Tray 35 from the top, while 50 kmol/h of water at 65°C and 1 atm enters Tray 5 from the top. Determine the separation for a reflux ratio of 10 and a bottoms flow rate of 200 kmol/h. Use the UNIFAC method for K-values and the AIChE method for mass transfer. Assume a perfectly mixed liquid and a vapor in plug flow on each tray, with operation at 80% of flooding. Determine the equilibrium stages (to the nearest stage) to achieve the same separation.

12.12. Distillation with dumped packing.
    Repeat Exercise 12.10, if a column packed with 2-inch stainless steel Pall rings is used with 25 ft of rings above the feed and 15 ft below. Be sure to use sufficient segments for the calculations.

12.13. Distillation with structured packing.
    Repeat Exercise 12.10, if a column with structured packing is used with 25 ft above the feed and 15 ft below. Be sure to use a sufficient number of segments.

12.14. Effect of % flooding, weir height, and % hole area.
    Solve Exercise 12.10 for combinations of the following values of % flooding, weir height, and % hole area: 40, 60, and 80%; 1, 2, and 3 inches; 6, 10, and 14%.

12.15. Effect of % flooding on E_{MV}.
    The upper column of an air-separation system of the type shown in Exercise 7.40 contains 48 sieve trays and operates at a nominal pressure of 131.7 kPa. A feed at 80 K and 131.7 kPa enters the top plate at 1,349 lbmol/h with a composition of 97.868 mol% nitrogen, 0.365 mol% argon, and 1.767 mol% oxygen. A second feed enters Tray 12 from the top at 83 K and 131.7 kPa at 1,832 lbmol/h with a composition of 59.7 mol% nitrogen, 1.47 mol% argon, and 38.83 mol% oxygen. The column has no condenser, but has a partial reboiler. Vapor distillate leaves the top plate at 2,487 lbmol/h, with remaining

products leaving the reboiler as 50 mol% vapor and 50 mol% liquid. Assume ideal solutions. Determine the effect of flooding on the separation and the median E_{MV} for oxygen, using a rate-based model.

12.16. Extractive distillation in a sieve-tray column.

The following bubble-point, organic-liquid mixture at 1.4 atm is distilled by extractive distillation with the following phenol-rich solvent at 1.4 atm and at the same temperature as the feed:

Component	Feed, kmol/h	Solvent, kmol/h
Methanol	50	0
n-Hexane	20	0
n-Heptane	180	0
Toluene	150	10
Phenol	0	800

The column has 30 sieve trays, with a total condenser and a partial reboiler. The solvent enters the fifth tray and the feed enters Tray 15 from the top. The pressure in the condenser is 1.1 atm; the pressure at the top tray is 1.2 atm; and the pressure at the bottom is 1.4 atm. The reflux ratio is 5 and the bottoms rate is 960 kmol/h. Thermodynamic properties can be estimated with the UNIFAC method for the liquid phase and the SRK equation for the vapor phase. The Antoine equation is suitable for vapor pressure. Use the ChemSep program to estimate the separation. Assume that the vapor and liquid are both well mixed and that the trays operate at 75% of flooding. Specify the Chan–Fair correlation for calculating mass-transfer coefficients. In addition, determine from the tray-by-tray results the average E_{MV} for each component (after discarding values much different than the majority). Try to improve the sharpness of the split by changing the feed and solvent entry tray locations. How can you increase the sharpness of the separation? List as many ideas as you have.

12.17. Equilibrium- and rate-based methods.

A bubble-cap tray absorber is designed to absorb 40% of the propane from a rich gas at 4 atm. The specifications for the entering rich gas and absorbent oil are as follows:

	Absorbent Oil	Rich Gas
Flow rate, kmol/s	11.0	11.0
Temperature, °C	32	62
Pressure, atm	4	4
Mole fraction:		
Methane	0	0.286
Ethane	0	0.157
Propane	0	0.240
n-Butane	0.02	0.169
n-Pentane	0.05	0.148
n-Dodecane	0.93	0

(a) Determine the number of equilibrium stages required and the splits of all components. (b) Determine the actual number of trays required and the splits and Murphree vapor-tray efficiencies of all components. Discuss and compare the equilibrium-based and rate-based results. What do you conclude?

12.18. Equilibrium- and rate-based methods.

A ternary mixture of methanol, ethanol, and water is distilled in a sieve-tray column to obtain a distillate with not more than 0.01 mol% water. The feed to the column is as follows:

Flow rate, kmol/h	142.46
Pressure, atm	1.3
Temperature, K	316
Mole fractions:	
Methanol	0.6536
Ethanol	0.0351
Water	0.3113

For a distillate rate of 93.10 kmol/h, a reflux ratio of 1.2, a condenser outlet pressure of 1.0 atm, and a top-tray pressure of 1.1 atm, determine, using UNIFAC for activity coefficients, the: (a) number of equilibrium stages required and the corresponding split, if the feed enters at the optimal stage; (b) number of actual trays required if the column operates at 85% of flooding and the feed is introduced to the optimal tray. Compare the split to that in part (a). (c) In addition, compute the component Murphree vapor-tray efficiencies. What do you conclude about the two methods of calculation?

12.19. Distillation in a packed bed.

Repeat Exercise 12.18 for a column packed with 2-inch stainless steel Pall rings.

12.20. Sizing a sieve-tray column.

It is required to absorb 96% of the benzene from a gas stream with absorption oil in a sieve-tray column at a nominal pressure of 1 atm. The feed conditions are as follows:

	Vapor	Liquid
Flow rate, kmol/s	0.01487	0.005
Pressure, atm	1.0	1.0
Temperature, K	300	300
Composition, mol fraction:		
Nitrogen	0.7505	0
Oxygen	0.1995	0
Benzene	0.0500	0.005
n-Tridecane (C_{13})	0	0.995

Tray geometry is as follows:

Tray spacing, m	0.5
Weir height, m	0.05
Hole diameter, m	0.003
Sheet thickness, m	0.002

Determine column diameter for 80% of flooding, the number of actual trays required, and the Murphree vapor-tray efficiency profile for benzene for the possible combinations of vapor and liquid flow patterns on a tray. Could the equilibrium-based method be used to obtain a reliable solution to this problem?

Chapter 13

Batch Distillation

§13.0 INSTRUCTIONAL OBJECTIVES

After completing this chapter, you should be able to:

- Derive the Rayleigh equation for a simple batch distillation (differential distillation), and state the necessary assumptions.
- Calculate, by graphical and algebraic means, batch-still temperature, residue composition, and instantaneous and average distillate composition for a binary mixture as a function of time for binary batch distillation.
- Calculate, by modified McCabe–Thiele methods, residue and distillate compositions for binary batch rectification under conditions of equilibrium stages, no liquid holdup, and constant or variable reflux ratio to achieve constant distillate composition.
- Explain the importance of taking into account liquid holdup.
- Calculate, using shortcut and rigorous equilibrium-stage methods with a process simulator, multicomponent, multistage batch rectification that includes a sequence of operating steps to obtain specified products.
- Apply the principles of optimal control to optimize batch distillation.

A familiar example of a batch distillation is the laboratory distillation shown in Figure 13.1, where a liquid mixture is charged to a *still-pot* and heated to boiling. The vapor formed is continuously removed and condensed to produce a distillate. The compositions of the initial charge and distillate change with time; there is no steady state. The still temperature increases and the amount of lower-boiling components in the still pot decreases as distillation proceeds.

Batch operations can be used to advantage when:

1. The capacity of a facility is too small to permit continuous operation at a practical rate.

2. Seasonal or customer demands require distillation in one unit of different feedstocks to produce different products.

3. Several new products are to be produced with one distillation unit for evaluation by potential buyers.

4. Upstream process operations are batchwise and the compositions of feedstocks for distillation vary with time or from batch to batch.

5. The feed contains solids or materials that form solids, tars, or resin that can plug or foul a continuous distillation column.

§13.1 DIFFERENTIAL DISTILLATION

The simple, batch distillation apparatus in Figure 13.1, which functions in an unsteady state manner, was first quantified by Lord Rayleigh [1] and is often referred to as *differential*

distillation. There is no reflux; at any instant, vapor leaving the still pot with composition y_D is assumed to be in equilibrium with liquid (*residue*) in the still, which is assumed to be perfectly mixed. For total condensation, $y_D = x_D$. The still pot is assumed to be the only equilibrium stage because there are no trays above the still pot. This apparatus is useful for separating wide-boiling mixtures.

The following nomenclature is used for variables that are a function of time, t, assuming that all compositions refer to a particular species in the multicomponent feed:

D = instantaneous-distillate rate, mol/h;

$y = y_D = x_D$ = mole fraction in instantaneous distillate leaving the still pot;

W = moles of liquid (residue) left in still;

$x = x_W$ = mole fraction in liquid (residue); and

0 = subscript referring to $t = 0$.

For any component in the mixture: instantaneous rate of output = Dy_D.

$$\left.\begin{array}{c}\text{Instantaneous}\\ \text{rate of depletion}\\ \text{in the still}\end{array}\right\} = -\frac{d}{dt}(Wx_W) = -W\frac{dx_W}{dt} - x_W\frac{dW}{dt}$$

Distillate rate and, therefore, liquid-depletion rate in the still, depend on the heat-input rate, Q, to the still. By component material balance at any instant:

$$\frac{d}{dt}(Wx_W) = W\frac{dx_W}{dt} + x_W\frac{dW}{dt} = -Dy_D \quad (13\text{-}1)$$

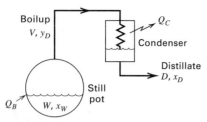

Figure 13.1 Differential (Rayleigh) distillation.

Multiplying (13-1) by dt:

$$W dx_W + x_W dW = y_D(-Ddt) = y_D dW$$

since by total balance, $-Ddt = dW$. Separating variables and integrating from the initial charge condition of W_0 and x_{W_0},

$$\int_{x_{W_0}}^{x_W} \frac{dx_W}{y_D - x_W} = \int_{W_0}^{W} \frac{dW}{W} = \ln\left(\frac{W}{W_0}\right) \quad (13\text{-}2)$$

This is the well-known Rayleigh equation, which was first applied to batch distillation of wide-boiling mixtures such as HCl–H$_2$O, H$_2$SO$_4$–H$_2$O, and NH$_3$–H$_2$O, where one stage is sufficient to get the desired separation. Without reflux, y_D and x_W are assumed to be in equilibrium, and (13-2) simplifies to

$$\int_{x_0}^{x} \frac{dx}{y - x} = \ln\left(\frac{W}{W_0}\right) \quad (13\text{-}3)$$

where x_{W_0} is replaced by x_0.

Equation (13-3) is easily integrated when pressure is constant, temperature change in the still pot is relatively small (close-boiling mixture), and K-values are composition-independent. Then $y = Kx$, where K is approximately constant, and (13-3) becomes

$$\ln\left(\frac{W}{W_0}\right) = \frac{1}{K - 1} \ln\left(\frac{x}{x_0}\right) \quad (13\text{-}4)$$

For a binary mixture, if the relative volatility α, instead of K, is assumed constant, substitution of (4-8) into (13-3), followed by integration and simplification, gives

$$\ln\left(\frac{W_0}{W}\right) = \frac{1}{\alpha - 1}\left[\ln\left(\frac{x_0}{x}\right) + \alpha \ln\left(\frac{1 - x}{1 - x_0}\right)\right] \quad (13\text{-}5)$$

If the equilibrium relationship $y = f\{x\}$ is graphical or tabular, integration of (13-3) can be performed graphically or numerically, as in the following three examples, which illustrate use of the Rayleigh equation for binary mixtures.

EXAMPLE 13.1 Constant Boilup Rate.

A batch still is loaded with 100 kmol of a binary mixture of 50 mol% benzene in toluene. As a function of time, make plots of: (a) still temperature, (b) instantaneous vapor composition, (c) still-pot composition, and (d) average total-distillate composition. Assume a constant boilup rate, and, therefore, constant D of 10 kmol/h, and a constant α of 2.41 at a pressure of 101.3 kPa (1 atm).

Solution

Initially, $W_0 = 100$ kmol and $x_0 = 0.5$. Solving (13.5) for W at values of x from 0.5 in increments of 0.05, and determining corresponding values of time in hours from $t = (W_0 - W)/D$, the following table is generated:

t, h	2.12	3.75	5.04	6.08	6.94	7.66	8.28	8.83	9.35
W, kmol	78.85	62.51	49.59	39.16	30.59	23.38	17.19	11.69	6.52
$x = x_W$	0.45	0.40	0.35	0.30	0.25	0.20	0.15	0.10	0.05

The instantaneous-vapor composition, y, is obtained from (4-8), which is $y = 2.41x/(1 + 1.41x)$, the equilibrium relationship for constant α. The average value of y_D or x_D over the time interval 0 to t is related to x and W at time t by combining overall component and total material balances to give

$$(x_D)_{\text{avg}} = (y_D)_{\text{avg}} = \frac{W_0 x_0 - Wx}{W_0 - W} \quad (13\text{-}6)$$

Equation (13-6) is much easier to apply than an equation that integrates the distillate composition.

To obtain the still temperature, $T–x–y$ data are required for benzene–toluene at 101.3 kPa, as given in Table 13.1. Temperature and compositions as a function of time are shown in Figure 13.2.

Table 13.1 Vapor–Liquid Equilibrium Data for Benzene (B)–Toluene (T) at 101.3 kPa

x_B	y_B	T, °C
0.100	0.208	105.3
0.200	0.372	101.5
0.300	0.507	98.0
0.400	0.612	95.1
0.500	0.713	92.3
0.600	0.791	89.7
0.700	0.857	87.3
0.800	0.912	85.0
0.900	0.959	82.7
0.950	0.980	81.4

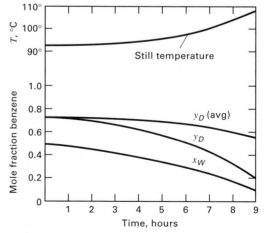

Figure 13.2 Distillation conditions for Example 13.1.

EXAMPLE 13.2   Using Tabular Data.

Repeat Example 13.1, except instead of using $\alpha = 2.41$, use the vapor–liquid equilibrium data in Table 13.1 for benzene–toluene at 101.3 kPa to solve the problem graphically or numerically with (13-3) rather than (13-5).

Solution

Equation (13-3) is solved by graphical integration by plotting $1/(y - x)$ versus x with a lower limit of $x_0 = 0.5$. Using the data of Table 13.1 for y as a function of x, points for the plot in terms of benzene are as follows:

x	0.5	0.4	0.3	0.2	0.1
$1/(y - x)$	4.695	4.717	4.831	5.814	9.259

The area under the curve from $x_0 = 0.5$ to a given value of x is equated to $\ln(W/W_0)$, and W is computed for $W_0 = 100$ kmol. In the region from $x_0 = 0.5$ to 0.3, $1/(y - x)$ changes only slightly; thus a numerical integration by the trapezoidal rule is readily made:

$x_0 = 0.5$ to 0.4:

$$\ln\left(\frac{W}{W_0}\right) = \int_{0.5}^{0.4} \frac{dx}{y - x} \approx \Delta x \left[\frac{1}{y - x}\right]_{avg}$$

$$= (0.4 - 0.5)\left(\frac{4.695 + 4.717}{2}\right) = -0.476$$

$$W/W_0 = 0.625, \quad W = 0.625(100) = 62.5 \text{ kmol}$$

$x = 0.5$ to 0.3:

$$\ln\left(\frac{W}{W_0}\right) = \int_{0.5}^{0.3} \frac{dx}{y - x} \approx \Delta x \left[\frac{1}{y - x}\right]_{avg}$$

$$= (0.3 - 0.5)\left[\frac{4.695 + 4.717 + 4.717 + 4.831}{4}\right] = -0.948$$

$$W/W_0 = 0.388, \quad W = 0.388(100) = 38.8 \text{ kmol}$$

These two values of W are in good agreement with those in Example 13.1. A graphical integration from $x_0 = 0.4$ to $x = 0.1$ gives $W = 10.7$, which is approximately 10% less than the result in Example 13.1, which uses a constant value of the relative volatility.

The Rayleigh equation, (13-1), for differential distillation applies to any two components, i and j, of a multicomponent mixture. Thus, if M_i is the moles of i in the still pot,

$$\frac{dM_i}{dt} = \frac{d}{dt}(Wx_{W_i}) = -Dy_{D_i}$$

Then,

$$dM_i/dM_j = y_{D_i}/y_{D_j} \tag{13-7}$$

For constant $\alpha_{i,j} = y_{D_i}x_{W_j}/y_{D_j}x_{W_i}$, (13-7) becomes

$$dM_i/dM_j = \alpha_{i,j}(x_{W_i}/x_{W_j}) \tag{13-8}$$

Substitution of $M_i = Wx_{W_i}$ for both i and j into (13-8) gives

$$dM_i/M_i = \alpha_{i,j}dM_j/M_j \tag{13-9}$$

Integration from the initial-charge condition gives

$$\ln(M_i/M_{i_0}) = \alpha_{i,j}\ln\left(M_j/M_{j_0}\right) \tag{13-10}$$

The following example shows that (13-10) is useful for determining the effect of relative volatility on the separation achievable by differential distillation.

EXAMPLE 13.3   Effect of Relative Volatility.

The charge to a simple batch still consists of an equimolar, binary mixture of A and B. For values of $\alpha_{A,B}$ of 2, 5, 10, 100, and 1,000 and 50% vaporization of A, determine the mole fraction and % vaporization of B in the total distillate.

Solution

For $\alpha_{A,B} = 2$ and $M_A/M_{A_0} = 1 - 0.5 = 0.5$, (13-10) gives $M_B/M_{B_0} = (M_A/M_{A_0})^{1/\alpha_{A,B}} = (0.5)^{(0.5)} = 0.7071$. % vaporization of B = $(1 - 0.7071)(100) = 29.29\%$.

For 200 moles of charge, the amounts in the distillate are $D_A = (0.5)(0.5)(200) = 50$ mol and $D_B = (0.2929)(0.5)(200) = 29.29$ mol. Mole fraction of B in the total distillate = $\frac{29.29}{50 + 29.29} = 0.3694$.

Similar calculations for other values of $\alpha_{A,B}$ give:

$\alpha_{A,B}$	% Vaporization of B	Mole Fraction of B in Total Distillate
2	29.29	0.3694
5	12.94	0.2057
10	6.70	0.1182
100	0.69	0.0136
1,000	0.07	0.0014

These results show that a sharp separation between A and B for 50% vaporization of A is achieved only if $\alpha_{A,B} \geq 100$. Furthermore, the purity achieved depends on the % vaporization of A. For $\alpha_{A,B} = 100$, if 90% of A is vaporized, the mole fraction of B in the total distillate increases from 0.0136 to 0.0247. For this reason, as discussed §13.6, it is common to conduct a binary, batch-distillation separation of LK and HK in the following manner:

1. Produce a distillate LK cut until the limit of impurity of HK in the total distillate is reached.

2. Continue the batch distillation to produce an intermediate (slop) cut of impure LK until the limit of impurity of LK in the liquid remaining in the still is reached.

3. Empty the HK-rich cut from the still.

4. Recycle the intermediate (slop) cut to the next still charge.

For desired purities of the LK cut and the HK cut, the fraction of intermediate (slop) cut increases as the LK–HK relative volatility decreases.

§13.2   BINARY BATCH RECTIFICATION

For a sharp separation and/or to reduce the intermediate-cut fraction, a trayed or packed column is added between the still and the condenser, and a reflux drum is added after the condenser, as shown in Figure 13.3. In addition, one or more drums are provided to collect distillate cuts. For a column of a given diameter, the molar vapor-boilup rate is usually fixed at a value safely below the column flooding point.

Two modes of operating a batch rectification are cited most frequently because they are the most readily modeled. The first is operation at a constant reflux rate or ratio (same as a constant distillate rate), while the second is operation at a constant distillate composition. With the former, the distillate composition varies with time; with the latter, the reflux ratio or distillate rate varies with time. The first mode is easily implemented because of the availability of rapidly responding flow sensors. For the second mode, a rapidly responding composition sensor is required. In a third mode, referred to as the optimal-control mode, both reflux ratio (or distillate rate) and distillate composition vary with time to maximize the amount of distillate, minimize operation time, or maximize profit. Constant reflux is discussed first, followed by constant composition. Optimal control is deferred to §13.8.

§13.2.1   Constant Reflux Operation

If R or D is fixed, instantaneous-distillate and still-bottoms compositions vary with time. Assume a total condenser, negligible holdup of vapor and liquid in the condenser and column, phase equilibrium at each stage, and constant molar overflow. Then, (13-2) applies with $y_D = x_D$. The analysis is facilitated by a McCabe–Thiele diagram using the method of Smoker and Rose [2].

Initially, the composition of the LK in the liquid in the still is the charge composition, x_{W_0}, which is 0.43 in the McCabe–Thiele diagram of Figure 13.4. If there are two theoretical stages (the still-pot and one equilibrium plate), the initial distillate composition, x_{D_0}, at time 0 can be found by constructing an operating line of slope $L/V = R/(R+1)$, such that exactly two stages are stepped off to the right from x_{W_0} to the $y = x$ line in Figure 13.4. At an arbitrary later time, say, time 1, at still-pot composition $x_W < x_{W_0}$, e.g., $x_W = 0.26$ in Figure 13.4, the instantaneous-distillate

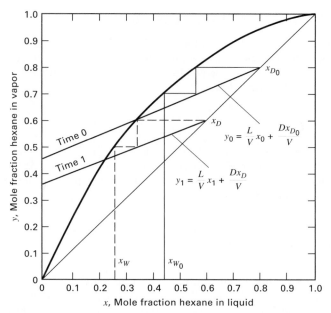

Figure 13.4 Batch binary distillation with fixed L/V and two theoretical stages.

composition is x_D. In this manner, a time-dependent series of points for x_D as a function of x_W is established, with L/V and the number of stages held constant.

Equation (13-2) cannot be integrated analytically because the relationship between $y_D = x_D$ and x_W depends on L/V, the number of theoretical stages, and the phase-equilibrium relationship. However, (13-2) can be integrated graphically with pairs of values for x_W and $y_D = x_D$ obtained from the McCabe–Thiele diagram, as in Figure 13.4, for a series of operating lines of the same slope.

The time t required for this batch rectification can be computed by a total material balance based on a constant boilup rate V, to give an equation due to Block [3]:

$$t = \frac{W_0 - W_t}{V\left(1 - \frac{L}{V}\right)} = \frac{R+1}{V}(W_0 - W_t) \qquad (13\text{-}11)$$

With a constant-reflux policy, instantaneous-distillate purity is above specification at the beginning and below specification at the end of the run. By an overall material balance, the average mole fraction of LK in the accumulated distillate at time t is given by

$$x_{D_{avg}} = \frac{W_0 x_0 - W_t x_{W_t}}{W_0 - W_t} \qquad (13\text{-}12)$$

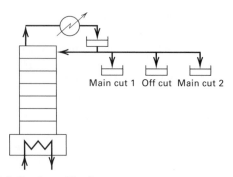

Main cut 1   Off cut   Main cut 2

Figure 13.3 Batch rectification.

EXAMPLE 13.4   Constant Reflux Operation.

A three-stage batch rectifier (first stage is the still-pot) is charged with 100 kmol of a 20 mol% n-hexane in n-octane mixture. At a constant-reflux ratio of 1 $(L/V = 0.5)$, how many moles of charge must be distilled for an average product composition of 70 mol% nC_6? The phase-equilibrium curve at column pressure is given in Figure 13.5. If the boilup rate is 10 kmol/h, calculate distillation time.

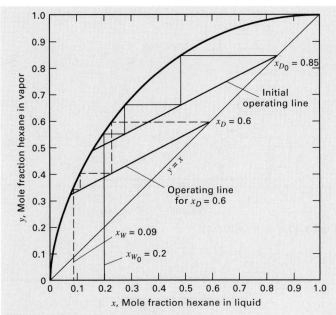

Figure 13.5 Solution to Example 13.4.

Solution

A series of operating lines and values of x_W are located by the trial-and-error procedure described earlier, as shown in Figure 13.5 for $x_{W_0} = 0.20$ and $x_W = 0.09$. Using Figure 13.5, the following table is developed:

$y_D = x_D$	0.85	0.60	0.5	0.35	0.3
x_W	0.2	0.09	0.07	0.05	0.035
$\dfrac{1}{y_D - x_W}$	1.54	1.96	2.33	3.33	3.77

The graphical integration is shown in Figure 13.6. Assuming a final value of $x_W = 0.1$, for instance, integration of (13-2) gives

$$\ln \frac{100}{W} = \int_{0.1}^{0.2} \frac{dx_W}{y_D - x_W} = 0.162$$

Hence, $W = 85$ and $D = 15$.

From (13-12): $(x_D)_{\text{avg}} = \dfrac{100(0.20) - 85(0.1)}{(100 - 85)} = 0.77$

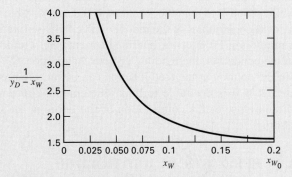

Figure 13.6 Graphical integration for Example 13.4.

The $(x_D)_{\text{avg}}$ is higher than the desired value of 0.70; hence, another x_W must be chosen. By trial, the correct answer is found to be $x_W = 0.06$, with $D = 22$, and $W = 78$, corresponding to a value of 0.25 for the integral. From (13-11), distillation time is $t = \frac{(1+1)}{10}(100 - 78) = 4.4$ h. When differential distillation is used, Figure 13.5 shows that 70 mol% hexane distillate is not achievable because the initial distillate is only 56 mol% hexane.

§13.2.2 Constant Distillate Composition

The constant-reflux-ratio policy described in the previous section is easy to implement. For small batch-rectification systems, it may be the least expensive policy. A more optimal policy is to maintain a constant V but continuously vary R to achieve a constant x_D that meets the specified purity. This requires a more complex control system, including a composition sensor on the distillate, which may be justified only for large batch-rectification systems. Other methods of operating batch columns are described by Ellerbe [5].

Calculations for the policy of constant x_D can also be made with the McCabe–Thiele diagram, as described by Bogart [4] and illustrated in Example 13.5. The Bogart method assumes negligible liquid holdup and constant molar overflow. An overall material balance for the LK, at any time t, is given by a rearrangement of (13-12) at constant x_D, for W as a function of x_W.

$$W = W_0 \left[\frac{x_D - x_{W_0}}{x_D - x_W} \right] \qquad (13\text{-}13)$$

Differentiating (13-13) with respect to t for varying W and x_W gives

$$\frac{dW}{dt} = W_0 \frac{(x_D - x_{W_0})}{(x_D - x_W)^2} \frac{dx_W}{dt} \qquad (13\text{-}14)$$

For constant molar overflow, the distillation rate is given by the rate of loss of charge, W:

$$-\frac{dW}{dt} = (V - L) = \frac{dD}{dt} \qquad (13\text{-}15)$$

where D is now the amount of distillate, not the distillate rate. Substituting (13-15) into (13-14) and integrating:

$$t = \frac{W_0(x_D - x_{W_0})}{V} \int_{x_{W_t}}^{x_0} \frac{dx_W}{(1 - L/V)(x_D - x_W)^2} \qquad (13\text{-}16)$$

For fixed values of W_0, x_{W_0}, x_D, V, and the number of stages, the McCabe–Thiele diagram is used to determine values of L/V for a series of values of still composition between x_{W_0} and the final value of x_W. These values are then used with (13-16) to determine, by graphical or numerical integration, the time for rectification or the time to reach any still composition. The required number of stages can be estimated by assuming total reflux for the final value of x_W. While rectification is proceeding, the instantaneous-distillate rate varies according to (13-15), which can be expressed in terms of L/V as

$$\frac{dD}{dt} = V(1 - L/V) \qquad (13\text{-}17)$$

EXAMPLE 13.5 Constant Distillate Composition.

A three-stage batch still (boiler and the equivalent of two equilibrium plates) is loaded with 100 kmol of a liquid containing a mixture of 50 mol% *n*-hexane in *n*-octane. A liquid distillate of 0.9 mole-fraction hexane is to be maintained by continuously adjusting the reflux ratio, while maintaining a distillate rate of 20 kmol/h. What should the reflux ratio be after 1 h when the accumulated distillate is 20 kmol? Theoretically, when must distillate accumulation be stopped? Assume negligible holdup and constant molar overflow.

Solution

When the accumulated distillate = 20 kmol, $W = 80$ kmol, and the still residue composition with respect to the light-key is given by a rearrangement of (13-13):

$$x_W = \frac{W x_D - W_0(x_D - x_{W_0})}{W} = \frac{0.9(80) - 100(0.9 - 0.5)}{80} = 0.4$$

For $y_D = x_D = 0.9$, a series of operating lines of varying slope, $L/V = R/(R + 1)$, with three stages stepped off is used to determine the corresponding still residue composition, x_W. By trial and error, Line 1 in Figure 13.7 is found for $x_W = 0.4$, corresponding to an $L/V = 0.22$. The reflux ratio $= (L/V)/[1 - (L/V)] = 0.282$. At the highest reflux rate possible, $L/V = 1$ (total reflux), and $x_W = 0.06$, according to the dashed-line construction shown in Figure 13.7. The corresponding time by material balance is $0.06(100 - 20t) = 50 - 20t(0.9)$. Solving, $t = 2.58$ h.

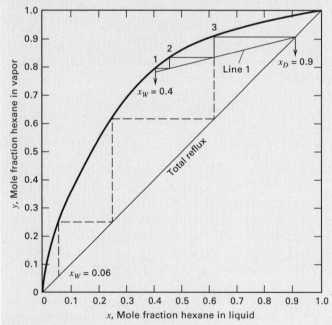

Figure 13.7 Solution to Example 13.5.

§13.3 BATCH STRIPPING AND COMPLEX BATCH DISTILLATION

A batch stripper consisting of a large accumulator (feed tank), a trayed or packed stripping column, and a reboiler is shown in Figure 13.8. The initial charge is placed in the accumulator rather than the reboiler. The mixture in the accumulator is fed to the top of the column and the bottoms cut is

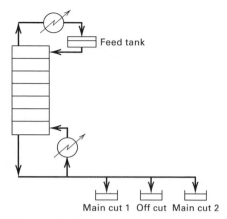

Figure 13.8 Batch stripping.

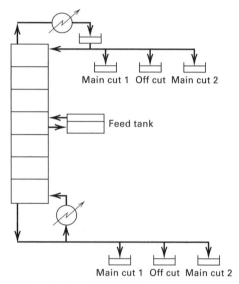

Figure 13.9 Complex batch distillation.

removed from the reboiler. A batch stripper is useful for removing small quantities of volatile impurities. For binary mixtures, the McCabe–Thiele construction applies, and the graphical methods in §13.2 can be modified to follow, with time, the change in composition in the accumulator and the corresponding instantaneous and average composition of the bottoms cut.

Figure 13.9 depicts a complex batch-distillation unit, described by Hasebe et al. [6], Barolo et al. [35], and Phimister and Seider [36], which permits considerable operating flexibility. The charge in the feed tank is fed to a suitable column location. Holdups in the reboiler and condenser are kept to a minimum. Products or intermediate cuts are withdrawn from the condenser, the reboiler, or both. In addition, the liquid in the column at the feed location can be recycled to the feed tank if it is desirable to make the composition in the feed tank close to the composition of the liquid at the feed location.

§13.4 EFFECT OF LIQUID HOLDUP

Except at high pressure, vapor holdup in a rectifying column is negligible in batch distillation because of the small vapor density. However, the effect of liquid holdup on the trays and in the condensing and reflux system can be significant when the molar ratio of holdup to original charge is more than a few percent. This is especially true when a charge contains low concentrations of one or more of the components to be separated. The effect of holdup in a trayed column is greater than in a packed column because of the lower amount of holdup in the latter. For either type of column, liquid holdup can be estimated by methods in §6.6 and 6.8.

A batch rectifier is usually started up under total-reflux conditions for an initial period of time prior to the withdrawal of distillate. During this initial period, liquid holdup in the column increases and approaches a value that is reasonably constant for the remainder of the run. Because of the total-reflux concentration profile, the initial concentration of light components in the remaining charge to the still is less than in the original charge. At high liquid holdups, this causes the initial purity and degree of separation to be reduced from estimates based on methods that ignore liquid holdup. Liquid holdup can reduce the size of product cuts, increase the size of intermediate fractions that are recycled, increase the amount of residue, increase the batch-cycle time, and increase total energy input. Although approximate methods for predicting the effect of liquid holdup are available, the complexity of the holdup effect is such that it is best to use rigorous batch-distillation algorithms described in §13.6 to study the effect on a case-by-case basis.

§13.5 SHORTCUT METHOD FOR BATCH RECTIFICATION

The batch-rectification methods presented in §13.2 are limited to binary mixtures, under the assumptions of constant molar overflow, and negligible vapor and liquid holdup. Shortcut methods for handling multicomponent mixtures under the same assumptions have been developed by Diwekar and Madhaven [7] for the two cases of constant distillate composition and constant reflux, and by Sundaram and Evans [8] for constant reflux. Both methods avoid tedious, stage-by-stage calculations of vapor and liquid compositions by employing the Fenske–Underwood–Gilliland (FUG) shortcut procedure for continuous distillation, described in §9.1, at successive time steps. In essence, they treat batch rectification as a sequence of continuous, steady-state rectifications. As in the FUG method, no estimations of compositions or temperatures are made for intermediate stages.

Sundaram and Evans [8] developed their shortcut method for a column of the type shown in Figure 13.3. An overall mole balance for a constant distillate rate, D, gives

$$-\frac{dW}{dt} = D \qquad (13\text{-}18)$$

Therefore,

$$-\frac{dW}{dt} = \frac{V}{1+R} \qquad (13\text{-}19)$$

For any component i, an instantaneous mole balance around the column gives

$$\frac{d(x_{W_i}W)}{dt} = x_{D_i}\frac{dW}{dt} \qquad (13\text{-}20)$$

Expanding the LHS of (13-20) and solving for dx_{W_i}:

$$dx_{W_i} = (x_{D_i} - x_{W_i})\frac{dW}{W} \qquad (13\text{-}21)$$

In finite-difference form, using Euler's method, (13-19) and (13-21) become, respectively,

$$W^{(k+1)} = W^{(k)} - \left(\frac{V}{1+R}\right)\Delta t \qquad (13\text{-}22)$$

$$x_{W_i}^{(k+1)} = x_{W_i}^{(k)} + \left(x_{D_i}^{(k)} - x_{W_i}^{(k)}\right)\left[\frac{W^{(k+1)} - W^{(k)}}{W^{(k)}}\right] \qquad (13\text{-}23)$$

where k is the time-increment index. For a given Δt time increment, $W^{(k+1)}$ is computed from (13-22), and then $x_{W_i}^{(k+1)}$ is obtained for each component from (13-23), which, however, requires values for $x_{D_i}^{(k)}$.

Calculations are initiated at $k = 0$. The initial charge to the still is $W^{(0)}$. Values of $x_{W_i}^{(0)}$ are equal to the mole fractions in the initial charge. Corresponding values of $x_{D_i}^{(0)}$ depend on the start-up method. If total reflux is the start-up method, as mentioned in §13.4, the Fenske equation of §9.1 can be applied to compute values of $x_{D_i}^{(0)}$ for given $x_{W_i}^{(0)}$ if column and condenser holdups are negligible. For a given number of stages, N, the Fenske equation is

$$N = \frac{\log\left[\left(\dfrac{x_{D_i}}{x_{W_i}}\right)\left(\dfrac{x_{W_r}}{x_{D_r}}\right)\right]}{\log \alpha_{i,r}} \qquad (13\text{-}24)$$

Solving,

$$x_{D_i} = x_{W_i}\left(\frac{x_{D_r}}{x_{W_r}}\right)\alpha_{i,r}^N \qquad (13\text{-}25)$$

where r is an arbitrary reference component, such as the least volatile species. Since

$$\sum_{i=1}^{C} x_{D_i} = 1.0 \qquad (13\text{-}26)$$

substitution of (13-25) into (13-26) gives

$$x_{D_r} = \frac{x_{W_r}}{\displaystyle\sum_{i=1}^{C} x_{W_i}\alpha_{i,r}^N} \qquad (13\text{-}27)$$

The initial distillate composition, $x_{D_r}^{(0)}$, is computed from (13-27). Remaining values of $x_{D_i}^{(0)}$ are computed from (13-25).

Using the initial set of values for $x_{D_i}^{(0)}$, $x_{W_i}^{(1)}$ values are computed from (13-23) following the calculation of $W^{(1)}$ from (13-22). To compute each subsequent set of $x_{W_i}^{(k+1)}$ for $k > 0$, values of $x_{D_i}^{(k)}$ for $k > 0$ are needed. These are obtained by applying the FUG method. Equation (13-24) applies during batch rectification if N is replaced by $N_{min} < N$ with $i = $ LK and $r =$ HK. But N_{min} is related to N by the Gilliland correlation. An approximate equation for that correlation, due to Eduljee [9], is used:

$$\frac{N - N_{min}}{N + 1} = 0.75\left[1 - \left(\frac{R - R_{min}}{R + 1}\right)^{0.5668}\right] \quad (13\text{-}28)$$

An estimate of the minimum reflux ratio, R_{min}, is provided by the Class I Underwood equation of Chapter 9, which assumes that all components in the charge distribute between the two products. Thus:

$$R_{min} = \frac{\left(\dfrac{x_{D_{LK}}}{x_{W_{LK}}}\right) - \alpha_{LK,HK}\left(\dfrac{x_{D_{HK}}}{x_{W_{HK}}}\right)}{\alpha_{LK,HK} - 1} \quad (13\text{-}29)$$

If one or more components fail to distribute, then Class II Underwood equations should be used. Sundaram and Evans use only (13-29) with LK and HK equal to the lightest component, 1, and the heaviest component, C, in the mixture. If (13-25), with $i = 1$, $r = C$, and $N = N_{min}$, and (13-27) with $r = C$ are substituted into (13-29) with LK $= 1$ and HK $= C$, the result is

$$R_{min} = \frac{\alpha_{1,C}^{N_{min}} - \alpha_{1,C}}{(\alpha_{1,C} - 1)\displaystyle\sum_{i=1}^{C} x_{W_i}\alpha_{i,C}^{N_{min}}} \quad (13\text{-}30)$$

For specified values of N and R, (13-28) and (13-30) are solved for R_{min} and N_{min} simultaneously by an iterative method. The value of x_{D_C} is then computed from (13-27) with $N = N_{min}$, followed by the calculation of the other values of x_{D_i} from (13-25). Values of N_{min} and R_{min} change with time.

The procedure of Sundaram and Evans involves an inner loop for the calculation of x_D, and an outer loop for $W^{(k+1)}$ and $x_{W_i}^{(k+1)}$. The inner loop requires iterations because of the nonlinear nature of (13-28) and (13-30). Calculations of the outer loop are direct because (13-22) and (13-23) are linear. Application of the method is illustrated in the following example, where α values are assumed constant.

EXAMPLE 13.6   Approximate Method.

A charge of 100 kmol of a ternary mixture of A, B, and C with composition $x_{W_A}^{(0)} = 0.33$, $x_{W_B}^{(0)} = 0.33$, and $x_{W_C}^{(0)} = 0.34$ is distilled in a batch rectifier with $N = 3$ (including the reboiler), $R = 10$, and $V = 110$ kmol/h. Estimate the variation of the still-pot, instantaneous distillate, and distillate-accumulator compositions as a function of time for 2 h of operation, following an initial start-up period during which a steady-state operation at total reflux is achieved. Use $\alpha_{AC} = 2.0$ and $\alpha_{BC} = 1.5$, and neglect column holdup.

Solution

The method of Sundaram and Evans is applied with $D = V/(1 + R) = 110/(1 + 10) = 10$ kmol/h. Therefore, $100/10 = 10$ h would be required to distill the entire charge.

Start-up Period:

From (13-27), with C as the reference r,

$$x_{D_C}^{(0)} = \frac{0.34}{0.33(2)^3 + 0.33(1.5)^3 + 0.34(1)^3} = 0.0831$$

From (13-25),

$$x_{D_A}^{(0)} = 0.33\left(\frac{0.0831}{0.34}\right)2^3 = 0.6449$$

and

$$x_{D_B}^{(0)} = 0.33\left(\frac{0.0831}{0.34}\right)1.5^3 = 0.2720$$

Take time increments, Δt, of 0.5 h.

At $t = 0.5$ h for outer loop:

From (13-22),

$$W^{(1)} = 100 - \left(\frac{110}{1 + 10}\right)0.5 = 95 \text{ kmol}$$

From (13-23) with $k = 0$,

$$x_{W_A}^{(1)} = 0.33 + (0.6449 - 0.33)\left[\frac{95 - 100}{100}\right] = 0.3143$$

$$x_{W_B}^{(1)} = 0.33 + (0.2720 - 0.33)\left[\frac{95 - 100}{100}\right] = 0.3329$$

$$x_{W_C}^{(1)} = 0.34 + (0.0831 - 0.34)\left[\frac{95 - 100}{100}\right] = 0.3528$$

At $t = 0.5$ h for inner loop:

From (13-28),

$$\frac{3 - N_{min}}{3 + 1} = 0.75\left[1 - \left(\frac{10 - R_{min}}{10 + 1}\right)^{0.5668}\right]$$

Solving for R_{min},

$$R_{min} = 10 - 1.5835\,N_{min}^{1.7643} \quad (1)$$

This equation holds for all values of time t.

From (13-30),

$$R_{min} = \frac{2^{N_{min}} - 2}{(2 - 1)\left[0.3143(2)^{N_{min}} + 0.3329(1.5)^{N_{min}} + 0.3528(1)^{N_{min}}\right]} \quad (2)$$

Equations (1) and (2) are solved simultaneously for R_{min} and N_{min}. This can be done by numerical or graphical methods including successive substitution, Newton's method, or with a spreadsheet by plotting each equation as R_{min} versus N_{min} and determining the intersection. The result is $R_{min} = 1.2829$ and $N_{min} = 2.6294$. From (13-27), with $N = 2.6294$,

$$x_{D_C}^{(1)} = 0.3528/[0.3143(2)^{2.6294} + 0.3329(1.5)^{2.6294} + 0.3528]$$
$$= 0.1081$$

From (13-25):

$$x_{D_A} = 0.3143\left(\frac{0.1081}{0.3528}\right)2^{2.6924} = 0.5959$$

$$x_{D_B} = 0.3329\left(\frac{0.1081}{0.3528}\right)1.5^{2.6294} = 0.2962$$

Subsequent, similar calculations give the results in Table 13.2.

Table 13.2 Results for Example 13.6

Time, h	W, kmol	x_W			N_{min}	R_{min}	x_D			x of Accumulated Distillate		
		A	B	C			A	B	C	A	B	C
0.0	100	0.3300	0.3300	0.3400	—	—	0.6449	0.2720	0.0831	—	—	—
0.5	95	0.3143	0.3329	0.3528	2.6294	1.2829	0.5957	0.2962	0.1081	0.6283	0.2749	0.0968
1.0	90	0.2995	0.3348	0.3657	2.6249	1.3092	0.5803	0.3048	0.1149	0.6045	0.2868	0.1087
1.5	85	0.2839	0.3365	0.3796	2.6199	1.3385	0.5633	0.3142	0.1225	0.5912	0.2932	0.1156
2.0	80	0.2675	0.3378	0.3947	2.6143	1.3709	0.5446	0.3242	0.1312	0.5800	0.2988	0.1212

§13.6 STAGE-BY-STAGE METHODS FOR BATCH RECTIFICATION

Complete stage-by-stage temperature, flow rates, and composition profiles as a function of time are required for final design studies or simulation of multicomponent, batch rectification. Such calculations are tedious, but can be carried out with either of two types of computer-based methods. Both are based on the same differential-algebraic equations for the distillation model, but differ in the way the equations are solved.

§13.6.1 Rigorous Model

Meadows [10] developed the first rigorous, multicomponent batch-distillation model, based on the assumptions of equilibrium stages; perfect mixing of liquid and vapor on each stage; negligible vapor holdup; constant-molar-liquid holdup, M, on a stage and in the condenser system; and adiabatic stages. Distefano [11] extended the model and developed a computer-based method for solving the equations. A more efficient method is presented by Boston et al. [12].

The Distefano model is based on the multicomponent, batch-rectification operation shown in Figure 13.10. The unit consists of a partial reboiler (still-pot), a column with N equilibrium stages or equivalent in packing, and a total condenser with a reflux drum. Also included, but not shown in Figure 13.10, are a number of accumulator or receiver drums equal to the number of overhead product and intermediate cuts. When product purity specifications cannot be made for successive distillate cuts, then intermediate (waste or slop) cuts are necessary. These are usually recycled. To initiate operation, the feed is charged to the reboiler, to which heat is supplied. Vapor leaving Stage 1 at the top of the column is condensed and passes to the reflux drum. At first, a total-reflux condition is established for a steady-state, fixed-overhead vapor flow rate. Depending upon the amount of liquid holdup in the column and in the condenser system, the liquid amount and composition in the reboiler at total reflux differs from the original feed.

Starting at time $t = 0$, distillate is removed from the reflux drum and sent to a receiver (accumulator) at a constant molar flow rate, and a reflux ratio is established. The heat-transfer

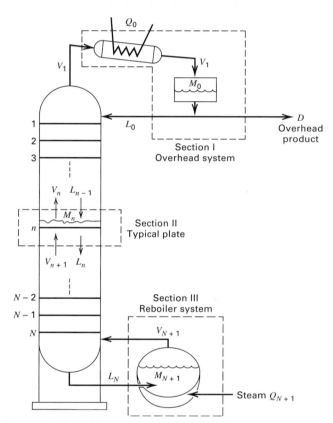

Figure 13.10 Multicomponent, batch-rectification operation.
[From G.P. Distefano, *AIChE J.*, **140**, 190 (1968) with permission.]

rate to the reboiler is adjusted to maintain the overhead-vapor molar flow rate. Model equations are derived for the overhead condensing system, column stages, and reboiler sections, as illustrated in Figure 13.10. For Section I, component material balances, a total material balance, and an energy balance are:

$$V_1 y_{i,1} - L_0 x_{i,0} - D_{x_{i,D}} = \frac{d(M_0 x_{i,0})}{dt} \qquad (13\text{-}31)$$

$$V_1 - L_0 - D = \frac{dM_0}{dt} \qquad (13\text{-}32)$$

$$V_1 h_{V_1} - (L_0 + D)h_{L_0} = Q_0 + \frac{d(M_0 h_{L_0})}{dt} \qquad (13\text{-}33)$$

where the derivative terms are accumulations due to holdup, which is assumed to be perfectly mixed. Also, for phase equilibrium at Stage 1 of the column:

$$y_{i,1} = K_{i,1} x_{i,1} \qquad (13\text{-}34)$$

The working equations are obtained by combining (13-31) and (13-34) to obtain a revised component material balance in terms of liquid-phase compositions, and by combining (13-22) and (13-33) to obtain a revised energy balance that does not include dM_0/dt. Equations for Sections II and III in Figure 13.10 are derived in a similar manner. The resulting working model equations for $t = 0^+$ are as follows, where i refers to the component, j refers to the stage, and M is molar liquid holdup.

1. Component mole balances for the overhead-condensing system, column stages, and reboiler, respectively:

$$\frac{dx_{i,0}}{dt} = -\left[\frac{L_0 + D + \dfrac{dM_0}{dt}}{M_0}\right]x_{i,0} + \left[\frac{V_1 K_{i,1}}{M_0}\right]x_{i,1}, \qquad (13\text{-}35)$$

$$i = 1 \text{ to } C$$

$$\frac{dx_{i,j}}{dt} = \left[\frac{L_{j-1}}{M_j}\right]x_{i,j-1} - \left[\frac{L_j + K_{i,j}V_j + \dfrac{dM_j}{dt}}{M_j}\right]x_{i,j}$$

$$+ \left[\frac{K_{i,j+1}V_{j+1}}{M_j}\right]x_{i,j+1}, \qquad (13\text{-}36)$$

$$i = 1 \text{ to } C, \quad j = 1 \text{ to } N$$

$$\frac{dx_{i,N+1}}{dt} = \left(\frac{L_N}{M_{N+1}}\right)x_{i,N}$$

$$- \left[\frac{V_{N+1}K_{i,N+1} + \dfrac{dM_{N+1}}{dt}}{M_{N+1}}\right]x_{i,N+1}, \qquad (13\text{-}37)$$

$$i = 1 \text{ to } C$$

where $L_0 = RD$.

2. Total mole balances for overhead-condensing system and column stages, respectively:

$$V_1 = D(R+1) + \frac{dM_0}{dt} \qquad (13\text{-}38)$$

$$L_j = V_{j+1} + L_{j-1} - V_j - \frac{dM_j}{dt}, \quad j = 1 \text{ to } N \qquad (13\text{-}39)$$

3. Enthalpy balances around overhead-condensing system, adiabatic column stages, and reboiler, respectively:

$$Q_0 = V_1(h_{V_1} - h_{L_0}) - M_0 \frac{dh_{L_0}}{dt} \qquad (13\text{-}40)$$

$$V_{j+1} = \frac{1}{\left(h_{V_{j+1}} - h_{L_j}\right)}$$

$$\times \left[V_j\left(h_{V_j} - h_{L_j}\right) - L_{j-1}\left(h_{L_{j-1}} - h_{L_j}\right) + M_j \frac{dh_{L_j}}{dt}\right],$$

$$j = 1 \text{ to } N \qquad (13\text{-}41)$$

$$Q_{N+1} = V_{N+1}\left(h_{V_{N+1}} - h_{L_{N+1}}\right) - L_N\left(h_{L_N} - h_{L_{N+1}}\right)$$

$$+ M_{N+1}\left(\frac{dh_{L_{N+1}}}{dt}\right) \qquad (13\text{-}42)$$

4. Phase equilibrium on the stages and in the reboiler:

$$y_{i,j} = K_{i,j}x_{i,j}, \quad i = 1 \text{ to } C, \quad j = 1 \text{ to } N+1 \qquad (13\text{-}43)$$

5. Mole-fraction sums at column stages and in the reboiler:

$$\sum_{i=1}^{C} y_{i,j} = \sum_{i=1}^{C} K_{i,j}x_{i,j} = 1.0, \quad j = 0 \text{ to } N+1 \qquad (13\text{-}44)$$

6. Molar holdups in the condenser system and on the column stages, based on constant-volume holdups, G_j:

$$M_0 = G_0\rho_0 \qquad (13\text{-}45)$$

$$M_j = G_j\rho_j, \quad j = 1 \text{ to } N \qquad (13\text{-}46)$$

where ρ is liquid molar density.

7. Variation of molar holdup in the reboiler, where M_{N+1}^0 is the initial charge to the reboiler:

$$M_{N+1} = M_{N+1}^0 - \sum_{j=0}^{N} M_j - \int_0^t D\,dt \qquad (13\text{-}47)$$

Equations (13-35) through (13-47) constitute an *initial-value problem* for a system of *ordinary differential and algebraic equations* (DAEs). The total number of equations is $(2CN + 3C + 4N + 7)$. If variables N, D, $R = L_0/D$, M_{N+1}^0, and all G_j are specified, and if correlations are available for computing liquid densities, vapor and liquid enthalpies, and K-values, the number of unknown variables, distributed as follows, is equal to the number of equations.

$x_{i,j}$	$CN + 2C$
$y_{i,j}$	$CN + C$
L_j	N
V_j	$N + 1$
T_j	$N + 2$
M_j	$N + 2$
Q_0	1
Q_{N+1}	1
	$2CN + 3C + 4N + 7$

Initial values at $t = 0$ for all these variables are obtained from the steady-state, total-reflux calculation, which depends only on values of N, M_{N+1}^0, x_{N+1}^0, G_j, and V_1.

Equations (13-35) through (13-42) include first derivatives of $x_{i,j}$, M_j, and h_{Lj}. Except for M_{N+1}, derivatives of the latter two variables can be approximated with sufficient accuracy by incremental changes over the previous time step. If the reflux ratio is high, as it often is, the derivative of M_{N+1} can also be approximated in the same manner. This leaves only the $C(N + 2)$ ordinary differential equations (ODEs) for the component material balances to be integrated in terms of the $x_{i,j}$ dependent variables.

§13.6.2 Rigorous Integration Method

The nonlinear equations (13-35) to (13-37) cannot be integrated analytically. Distefano [11] developed a numerical solution method based on an investigation of 11 different numerical integration techniques that step in time. Of particular concern were the problems of *truncation* error and *stability*, which make it difficult to integrate the equations rapidly and accurately. Such systems of ODEs or DAEs constitute so-called *stiff systems* as described later in this section.

Local truncation errors result from using approximations for the functions on the RHS of the ODEs at each time step. These errors may be small, but they can grow through subsequent time steps, resulting in global truncation errors sufficiently large to be unacceptable. As truncation errors become large, the number of significant digits in the computed dependent variables gradually decrease. Truncation errors can be reduced by decreasing the time-step size.

Stability problems are much more serious. When instability occurs, the computed values of the dependent variables become totally inaccurate, with no significant digits at all. Reducing the time step does not eliminate instability until a time-step criterion, which depends on the numerical method, is satisfied. Even then, a further reduction in the time step is required to prevent oscillations of dependent variables.

Problems of stability and truncation error are conveniently illustrated by comparing results obtained by using the explicit- and implicit-Euler methods, both of which are first-order in accuracy, as discussed by Davis [15] and Riggs [16]. Consider the nonlinear, first-order ODE:

$$\frac{dy}{dt} = f\{t, y\} = ay^2 t e^y \quad (13-48)$$

for $y\{t\}$, where initially $y\{t_0\} = y_0$. The explicit- (forward) Euler method approximates (13-48) with a sequence of discretizations of the form

$$\frac{y_{k+1} - y_k}{\Delta t} = ay_k^2 t_k e^{y_k} \quad (13-49)$$

where Δt is the time step and k is the sequence index. The function $f\{t, y\}$ is evaluated at the beginning of the current time step. Solving for y_{k+1} gives the recursion equation:

$$y_{k+1} = y_k + ay_k^2 t_k e^{y_k} \Delta t \quad (13-50)$$

Regardless of the nature of $f\{t, y\}$ in (13-48), the recursion equation can be solved explicitly for y_{k+1} using results from the previous time step. However, as discussed later, this advantage is counterbalanced by a limitation on the magnitude of Δt to avoid instability and oscillations.

The implicit- (backward) Euler method also utilizes a sequence of discretizations of (13-48), but the function $f\{t, y\}$ is evaluated at the end of the current time step. Thus:

$$\frac{y_{k+1} - y_k}{\Delta t} = ay_{k+1}^2 t_{k+1} e^{y_{k+1}} \quad (13-51)$$

Because the function $f\{t, y\}$ is nonlinear in y, (13-51) cannot be solved explicitly for y_{k+1}. This disadvantage is counterbalanced by unconditional stability with respect to selection of Δt. However, too large a value can result in unacceptable truncation errors.

When the explicit-Euler method is applied to (13-35) to (13-47) for batch rectification, as shown in the following example, the maximum value of Δt can be estimated from the maximum, absolute eigenvalue, $|\lambda|_{max}$, of the Jacobian matrix of (13-35) to (13-37). To prevent instability, $\Delta t_{max} \leq 2/|\lambda|_{max}$. To prevent oscillations, $\Delta t_{max} \leq 1/|\lambda|_{max}$. Applications of the explicit- and implicit-Euler methods are compared in the following batch-rectification example.

EXAMPLE 13.7 Selection of Time Step.

One hundred kmol of an equimolar mixture of n-hexane (A) and n-heptane (B) is distilled at 15 psia in a batch rectifier consisting of a total condenser with a constant liquid holdup, M_0, of 0.10 kmol; a single equilibrium stage with a constant liquid holdup, M_1, of 0.01 kmol; and a reboiler. Initially the system is brought to the following total-reflux condition, with saturated liquid leaving the total condenser:

Stage	T, °F	x_A	K_A	K_B	M, kmol
Condenser	162.6	0.85935	—	—	0.1
Plate, 1	168.7	0.70930	1.212	0.4838	0.01
Reboiler, 2	178.6	0.49962	1.420	0.5810	99.89

Distillation begins ($t = 0$) with a reflux rate, L_0, of 10 kmol/h and a distillate rate, D, of 10 kmol/h. Calculate the mole fractions of n-hexane and n-heptane at $t = 0.05$ h (3 min), at each of the three rectifier locations, assuming constant molar overflow and constant K-values for this small elapsed time period. Use explicit- and implicit-Euler methods to determine the influence of the time step, Δt.

Solution

Based on the constant molar overflow assumption, $V_1 = V_2 = 20$ kmol and $L_0 = L_1 = 10$ kmol/h. Using the K-values and liquid holdups given earlier, (13-35) to (13-37), with all $dM_j/dt = 0$, become as follows:

Condenser:

$$\frac{dx_{A,0}}{dt} = -200x_{A,0} + 242.4x_{A,1} \quad (1)$$

$$\frac{dx_{B,0}}{dt} = -200x_{B,0} + 96.76x_{B,1} \quad (2)$$

Plate:

$$\frac{dx_{A,1}}{dt} = 1{,}000x_{A,0} - 3{,}424x_{A,1} + 2{,}840x_{A,2} \qquad (3)$$

$$\frac{dx_{B,1}}{dt} = 1{,}000x_{B,0} - 1{,}967x_{B,1} + 1{,}162x_{B,2} \qquad (4)$$

Reboiler:

$$\frac{dx_{A,2}}{dt} = \left(\frac{10}{M_2}\right)x_{A,1} - \left(\frac{28.40}{M_2}\right)x_{A,2} \qquad (5)$$

$$\frac{dx_{B,2}}{dt} = \left(\frac{10}{M_2}\right)x_{B,1} - \left(\frac{11.62}{M_2}\right)x_{B,2} \qquad (6)$$

where

$$M_2\{t = t\} = M_2\{t = 0\} - (V_2 - V_1)t$$

or

$$M_2 = 99.89 - 10t \qquad (7)$$

Equations (1) through (6) can be grouped by component into the following two matrix equations:

Component A:

$$\begin{bmatrix} -200 & 242.2 & 0 \\ 1{,}000 & -3{,}424 & 2{,}840 \\ 0 & 10/M_2 & -28.40/M_2 \end{bmatrix} \cdot \begin{bmatrix} x_{A,0} \\ x_{A,1} \\ x_{A,2} \end{bmatrix} = \begin{bmatrix} dx_{A,0}/dt \\ dx_{A,1}/dt \\ dx_{A,2}/dt \end{bmatrix} \quad (8)$$

Component B:

$$\begin{bmatrix} -200 & 96.76 & 0 \\ 1{,}000 & -1{,}967 & 1{,}160 \\ 0 & 10/M_2 & -11.62/M_2 \end{bmatrix} \cdot \begin{bmatrix} x_{B,0} \\ x_{B,1} \\ x_{B,2} \end{bmatrix} = \begin{bmatrix} dx_{B,0}/dt \\ dx_{B,1}/dt \\ dx_{B,2}/dt \end{bmatrix} \quad (9)$$

Although (8) and (9) do not appear to be coupled, they are because at each time step, the sums $x_{A,j} + x_{B,j}$ do not equal 1. Accordingly, the mole fractions are normalized at each time step to force them to sum to 1. The initial eigenvalues of the Jacobian matrices, (8) and (9), are computed from any of a number of computer programs, such as MathCad, Mathematica, MATLAB, or Maple, to be as follows, using $M_2 = 99.89$ kmol:

	Component A	Component B
λ_0	−126.54	−146.86
λ_1	−3,497.6	−2,020.2
λ_2	−0.15572	−0.03789

It is seen that $|\lambda|_{max} = 3{,}497.6$. Thus, for the explicit-Euler method, instability and oscillations can be prevented by choosing $\Delta t \leq 1/3{,}497.6 = 0.000286$ h.

If $\Delta t = 0.00025$ h (just slightly smaller than the criterion) is selected, it takes $0.05/0.00025 = 200$ time steps to reach $t = 0.05$ h (3 min). No such restriction applies to the implicit-Euler method, but too large a Δt may result in an unacceptable truncation error.

Explicit-Euler Method

According to Distefano [11], the maximum step size for integration using an explicit method is nearly always limited by stability considerations, and usually the truncation error is small. Assuming this

to be true for this example, the following results were obtained using $\Delta t = 0.00025$ h with a spreadsheet program by converting (8) and (9), together with (7) for M_2, to the form of (13-50). Only the results for every 40 time steps are given.

Time, h	Normalized Mole Fractions in Liquid for n-Hexane			Normalized Mole Fractions in Liquid for n-Heptane		
	Distillate	Plate	Still	Distillate	Plate	Still
0.01	0.8183	0.6271	0.4993	0.1817	0.3729	0.5007
0.02	0.8073	0.6219	0.4991	0.1927	0.3781	0.5009
0.03	0.8044	0.6205	0.4988	0.1956	0.3795	0.5012
0.04	0.8036	0.6199	0.4985	0.1964	0.3801	0.5015
0.05	0.8032	0.6195	0.4982	0.1968	0.3805	0.5018

To show the instability effect, a time step of 0.001 h (four times the previous time step) gives the following unstable results during the first five time steps to an elapsed time of 0.005 h. Also included are values at 0.01 h for comparison to the preceding stable results.

Time, h	Normalized Mole Fractions in Liquid for n-Hexane			Normalized Mole Fractions in Liquid for n-Heptane		
	Distillate	Plate	Still	Distillate	Plate	Still
0.000	0.85935	0.7093	0.49962	0.14065	0.2907	0.50038
0.001	0.859361	0.559074	0.499599	0.140639	0.440926	0.500401
0.002	0.841368	0.75753	0.499563	0.158632	0.24247	0.500437
0.003	0.852426	0.00755	0.499552	0.147574	0.99245	0.500448
0.004	0.809963	0.884925	0.499488	0.190037	0.115075	0.500512
0.005	0.874086	1.154283	0.499546	0.125914	−0.15428	0.500454
0.01	1.006504	0.999254	0.493573	−0.0065	0.000746	0.506427

Much worse results are obtained if the time step is increased 10-fold to 0.01 h, as shown in the following table, where at $t = 0.01$ h, a negative mole fraction has appeared.

Time, h	Normalized Mole Fractions in Liquid for n-Hexane			Normalized Mole Fractions in Liquid for n-Heptane		
	Distillate	Plate	Still	Distillate	Plate	Still
0.00	0.85935	0.7093	0.49962	0.14065	0.2907	0.50038
0.01	0.859456	−0.79651	0.49941	0.140544	1.796512	0.50059
0.02	2.335879	2.144666	0.497691	−1.33588	−1.14467	0.502309
0.03	1.284101	1.450481	0.534454	−0.2841	−0.45048	0.465546
0.04	1.145285	1.212662	8.95373	−0.14529	−0.21266	−7.95373
0.05	1.07721	1.11006	1.191919	−0.07721	−0.11006	−0.19192

Implicit-Euler Method

If (8) and (9) are converted to implicit equations like (13-51), they can be rearranged into a linear, tridiagonal set for each component. For example, the equation for component A on the plate becomes

$$(1{,}000\Delta t)x_{A,0}^{(k+1)} - (1 + 3{,}424\Delta t)x_{A,1}^{(k+1)} + (2{,}840\Delta t)x_{A,2}^{(k+1)} = -x_{A,1}^{(k)}$$

The two tridiagonal equation sets can be solved by the tridiagonal-matrix algorithm of §10.3.1 or with a spreadsheet program using the iterative, circular-reference technique. For the implicit-Euler method, the selection of the time step, Δt, is not restricted by stability considerations. However, too large a Δt can lead to unacceptable truncation errors. Normalized, liquid-mole-fraction results at $t = 0.05$ h for just component A are as follows for a number of choices of Δt, all of which are greater than the 0.00025 h used earlier to obtain stable and oscillation-free results with the explicit-Euler method. Included for comparison is the explicit-Euler result for $\Delta t = 0.00025$ h.

Time = 0.05 h:

Δt, h	Normalized Mole Fractions in Liquid for *n*-Hexane		
	Distillate	Plate	Still
	Explicit-Euler		
0.00025	0.8032	0.6195	0.4982
	Implicit-Euler		
0.0005	0.8042	0.6210	0.4982
0.001	0.8042	0.6210	0.4982
0.005	0.8045	0.6211	0.4982
0.01	0.8049	0.6213	0.4982
0.05	0.8116	0.6248	0.4982

The preceding data show acceptable results with the implicit-Euler method using a time step of 40 times the Δt_{max} for the explicit-Euler method.

Stiffness Problem

Another serious computational problem occurs when integrating the equations; because the liquid holdups on the trays and in the condenser are small, the corresponding liquid mole fractions, x_{ij}, respond quickly to changes. The opposite holds for the reboiler with its large liquid holdup. Hence, the required time step for accuracy is usually small, leading to a very slow response of the overall rectification operation. Systems of ODEs having this characteristic constitute so-called stiff systems. For such a system, as discussed by Carnahan and Wilkes [17], an explicit method of solution must utilize a small time step for the entire period even though values of the dependent variables may all be changing slowly for a large portion of the time period. Accordingly, it is preferred to utilize a special implicit-integration technique developed by Gear [14] and others, as contained in the public-domain software package called ODEPACK. Gear-type methods for stiff systems strive for accuracy, stability, and computational

efficiency by using multistep, variable order, and variable-step-size implicit techniques.

A commonly used measure of the degree of stiffness is the eigenvalue ratio $|\lambda|_{max}/|\lambda|_{min}$, where λ values are the eigenvalues of the Jacobian matrix of the set of ODEs. For the Jacobian matrix of (13-35) through (13-37), the Gerschgorin circle theorem, discussed by Varga [18], can be employed to estimate the eigenvalue ratio. The maximum absolute eigenvalue corresponds to the component with the largest K-value and the tray with the smallest liquid molar holdup. When the Gerschgorin theorem is applied to a row of the Jacobian matrix based on (13-36),

$$|\lambda|_{max} \le \left[\left(\frac{L_{j-1}}{M_j} \right) + \left(\frac{L_j + K_{i,j}V_j}{M_j} \right) + \left(\frac{K_{i,j+1}V_{j+1}}{M_j} \right) \right]$$

$$\approx 2 \left[\frac{L_j + K_{i,j}V_j}{M_j} \right] \qquad (13\text{-}52)$$

where i refers to the most-volatile component and j to the stage with the smallest liquid molar holdup. The minimum absolute eigenvalue almost always corresponds to a row of the Jacobian matrix for the reboiler. Thus, from (13-37):

$$|\lambda|_{min} \le \left[\left(\frac{L_N}{M_{N+1}} \right) + \left(\frac{V_{N+1}K_{i,N+1}}{M_{N+1}} \right) \right]$$

$$\approx \left[\frac{L_N + K_{i,N+1}V_{N+1}}{M_{N+1}} \right] \qquad (13\text{-}53)$$

where i now refers to the least-volatile component and $N+1$ is the reboiler stage. The largest value of the reboiler holdup is M_{N+1}^0. The stiffness ratio, SR, is

$$\text{SR} = \frac{|\lambda|_{max}}{|\lambda|_{min}} \approx 2 \left(\frac{L + K_{lightest}V}{L + K_{heaviest}V} \right) \left(\frac{M_{N+1}^0}{M_{tray}} \right) \qquad (13\text{-}54)$$

From (13-54), the stiffness ratio depends not only on the difference between tray and initial reboiler molar holdups, but also on the difference between K-values of the lightest and heaviest components.

Davis [15] states that SR = 20 is not stiff; SR = 1,000 is stiff; and SR = 1,000,000 is very stiff. For the conditions of Example 13.7, using (13-54),

$$SR \approx 2 \left[\frac{10 + (1.212)(20)}{10 + (0.581)(20)} \right] \left(\frac{100}{0.01} \right) = 31{,}700$$

which meets the criterion of a stiff problem. A modification of the computational procedure of Distefano [11], for solving (13-35) through (13-46), is as follows:

Initialization

1. Establish total-reflux conditions, based on vapor and liquid molar flow rates V_j^0 and L_j^0. V_{N+1}^0 is the desired boilup rate or L_0^0 is based on the desired distillate rate and reflux ratio such that $L_0^0 = D(R + 1)$.

2. At $t = 0$, reduce L_0^0 to begin distillate withdrawal, but maintain the boilup rate established or specified for the total-reflux condition. This involves replacing all L_j^0

with $L_j^0 - D$. Otherwise, the initial values of all variables are those established for total reflux.

Time Step

3. In (13-35) to (13-37), replace liquid-holdup derivatives by total-material-balance equations:

$$\frac{dM_j}{dt} = V_{j+1} + L_{j-1} - V_j - L_j$$

Solve the resulting equations for the liquid mole fractions using an appropriate implicit-integration technique and a suitable time step. Normalize the mole fractions at each stage if they do not sum to 1.

4. Compute a new set of stage temperatures and vapor-phase mole fractions from (13-44) and (13-43), respectively.

5. Compute liquid densities and liquid holdups, and liquid and vapor enthalpies, from (13-45) and (13-46), and then determine derivatives of enthalpies and liquid holdups with respect to time by forward-finite-difference approximations.

6. Compute a new set of liquid and vapor molar flow rates from (13-38), (13-39), and (13-41).

7. Compute the new reboiler molar holdup from (13-47).

8. Compute condenser and reboiler heat-transfer rates from (13-40) and (13-42).

Iteration to Completion of Operation

9. Repeat Steps 3 through 8 for additional time steps until the completion of a specified operation, such as a desired amount of distillate, mole fraction of a component in the distillate, etc.

New Operation

10. Dump the accumulated distillate into a receiver, change operating conditions, and repeat Steps 2 through 9. Terminate calculations following the final operation.

The foregoing procedure is limited to narrow-boiling feeds and the simple configuration shown in Figure 13.10. A more flexible and efficient method, designed to cope with stiffness, is that of Boston et al. [12], which uses a modified inside-out algorithm of the type discussed in §10.5, which can handle feeds ranging from narrow- to wide-boiling for nonideal-liquid solutions. In addition, the method permits multiple feeds, sidestreams, tray heat transfer, vapor distillate, and flexibility in operation specifications.

EXAMPLE 13.8   Rectification by a Rigorous Method.

One hundred kmol of 30 mol% acetone, 30 mol% methanol, and 40 mol% water at 60°C and 1 atm is to be distilled in a batch rectifier consisting of a reboiler, a column with five equilibrium stages, a total condenser, a reflux drum, and three distillate accumulators. The molar liquid holdup of the condenser-reflux drum is 5 kmol,

whereas the molar liquid holdup of each stage is 1 kmol. The pressure is assumed constant at 1 atm throughout the rectifier. The following four events are to occur, each with a reboiler duty of 1 million kcal/h:

Event 1: Establishment of total-reflux conditions.

Event 2: Rectification with a reflux ratio of 3 until the acetone purity of the accumulated distillate in the first accumulator drops to 73 mol%.

Event 3: Rectification with a reflux ratio of 3 and a second accumulator for 21 minutes.

Event 4: Rectification with a reflux ratio of 3 and a third accumulator for 27 minutes.

Determine accumulator and column conditions at the end of each event. Use the Wilson equation from §2.6.6 to compute K-values.

Solution

The following results were obtained with a batch-distillation program. The conditions are as follows:

Event 1: Total Reflux Conditions:

Stage	T, °C	L, kmol/h	Mole Fraction in Liquid		
			Acetone	Methanol	Water
Condenser	55.6	138.9	0.770	0.223	0.007
1	55.6	138.6	0.761	0.227	0.012
2	55.7	138.0	0.747	0.235	0.018
3	55.9	137.0	0.722	0.247	0.031
4	56.2	134.8	0.673	0.269	0.058
5	57.3	128.7	0.560	0.306	0.134
Reboiler	62.2	—	0.252	0.307	0.441

The stiffness ratio, SR, was computed from (13-54) based on total-reflux conditions at the end of Event 1. The charge remaining in the still is $100 - 5 - 5(1) = 90$ kmol. The most-volatile component is acetone, with a K-value at the bottom stage of 1.203, and the least-volatile is water, with $K = 0.428$. The stiffness ratio is

$$\text{SR} \approx 2\left[\frac{128.7 + (1.203)(134.8)}{128.7 + (0.428)(134.8)}\right]\left(\frac{90}{1}\right) \approx 281$$

Thus, this problem is not very stiff. A time step of 0.06 min is used.

Event 2:
The time required to complete Event 2 is computed to be 57.5 minutes. The accumulated distillate in Tank 1 is 32.0 kmol with a composition of 73.0 mol% acetone, 26.0 mol% methanol, and 1.0 mol% water. The 58.0 kmol liquid remaining in the reboiler is 2.8 mol% acetone, 30.0 mol% methanol, and 67.2 mol% water.

Event 3:
The time specified to complete this event is 21 minutes. The accumulated distillate in Tank 2 is 11.3 kmol of 47.2 mol% acetone, 51.8 mol% methanol, and 1.0 mol% water. This intermediate cut is recycled for addition to the next charge.

Event 4:
At the end of the 27-minute specification, the accumulated distillate in Tank 3 is 13.8 kmol of 8.3 mol% acetone, 86.2 mol% methanol, and 5.5 mol% water. The remaining 32.9 kmol in the still is 0.0 mol% acetone, 0.4 mol% methanol, and 99.6 mol% water.

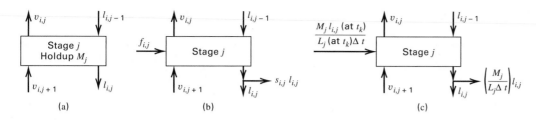

Figure 13.11 Simulation of holdup in a batch rectifier. (a) Stage in a batch rectifier with holdup. (b) Stage in a continuous fractionator. (c) Simulation of batch holdup in a continuous fractionator.

§13.6.3 Rapid-Solution Method

An alternative to integrating the stiff system of differential equations is the quasi-steady-state procedure of Galindez and Fredenslund [13], where the transient conditions are simulated as a succession of a finite number of continuous steady states of short duration, typically 0.05 h (3 minutes). Holdup is taken into account, and the stiffness of the problem is of no consequence. Results compare favorably with those from the rigorous integration method of §13.6.2.

Consider an intermediate theoretical stage, j, with molar holdup, M_j, in the batch rectifier in Figure 13.11a. A material balance for component i, in terms of component flow rates rather than mole fractions, is

$$l_{i,j} + v_{i,j} - l_{i,j-1} - v_{i,j+1} + \frac{d(M_j x_{i,j})}{dt} = 0 \quad (13\text{-}55)$$

Assume constant molar holdup. Also, assume that during a short time period, $dt = \Delta t = t_{k+1} - t_k$, the component flow rates given by the first four terms in (13-55) remain constant at values corresponding to time t_{k+1}. The component holdup term in (13-55) is

$$\frac{d(M_j x_{i,j})}{dt} = M_j \left[\frac{x_{i,j}\{t_{k+1}\} - x_{i,j}\{t_k\}}{\Delta t} \right] \quad (13\text{-}56)$$

But $x_{i,j} = l_{i,j}/L_j$. Therefore, (13-56) can be rewritten as

$$\frac{d(M_j x_{i,j})}{dt} = \frac{M_j l_{i,j}}{L_j \Delta t} - \frac{M_j l_{i,j}\{t_k\}}{L_j\{t_k\}\Delta t} \quad (13\text{-}57)$$

If (13-57) is substituted into (13-55) and terms in the component flow rate $l_{i,j}$ are collected,

$$l_{i,j}\left(1 + \frac{M_j}{L_j\Delta t}\right) + v_{i,j} - l_{i,j-1} - v_{i,j+1} - \frac{M_j l_{i,j}\{t_k\}}{L_j\{t_k\}\Delta t} \quad (13\text{-}58)$$

If (13-58) for unsteady-state (batch) distillation is compared to (10-58) for steady-state (continuous) distillation, it is seen that the term $M_j/(L_j\Delta t)$ in (13-58) corresponds to the liquid sidestream ratio in (10-58), or that $M_j/\Delta t$ corresponds to a liquid sidestream flow rate. Also, the term $M_j l_{i,j}\{t_k\}/(L_j\{t_k\}\Delta t)$ in (13-58) corresponds to a component feed rate in (10-58). The analogy is shown in parts (b) and (c) of Figure 13.11. Thus, the change in component liquid holdup per unit time, $d(M_j x_{i,j})/dt$ in (13-56), is interpreted for a small, finite-time difference as the difference between a component feed rate into the stage and a component flow rate in a liquid sidestream leaving the stage. Similarly, the enthalpy holdup in the stage energy balance is interpreted as the

difference over a small, finite-time interval between a heat input to the stage and an enthalpy output in a liquid side-stream leaving the stage.

The overall result is a system of steady-state equations, identical in form to the equations for the Newton-Raphson and inside-out methods. Either method can be used to solve the system of component-material-balance, phase-equilibrium, and energy-balance equations at each time step. The initial guesses used to initiate each time step are the values at the end of the previous time step. Because the variables generally change by only a small amount for each time step, convergence of either method is achieved in a small number of iterations.

EXAMPLE 13.9 Rectification by the Rapid-Solution Method.

One hundred lbmol of 25 mol% benzene, 50 mol% monochlorobenzene (MCB), and 25 mol% *ortho*-dichlorobenzene (DCB) is distilled in a batch rectifier consisting of a reboiler, 10 equilibrium stages in the column, a reflux drum, and three distillate product accumulators. The condenser-reflux drum holdup is constant at 0.20 ft³, and each stage in the column has a liquid holdup of 0.02 ft³. Pressures are 17.5 psia in the reboiler and 14.7 psia in the reflux drum, with a linear profile in the column from 15.6 psia at the top to 17 psia at the bottom. Following initialization at total reflux, the batch is distilled in three steps, each with a vapor boilup rate of 200 lbmol/h and a reflux ratio of 3. Thus, the distillate rate is 50 lbmol/h. Using the rapid-solution method, determine the amounts and compositions of the accumulated distillate and reboiler holdup, and heat duties at the end of each of the three steps.

Step 1: Terminate when the mole fraction of benzene in the instantaneous distillate drops below 0.100.

Step 2: Terminate when the mole fraction of MCB in the distillate drops below 0.40.

Step 3: Terminate when the mole fraction of DCB in the reboiler rises above 0.98.

Assume ideal solutions and the ideal-gas law.

Solution

This problem has a stiffness ratio of approximately 15,000. The quasi-steady-state procedure of Galindez and Fredenslund [13] in CHEMCAD was used with a time increment of 0.005 h for each of the three steps. Although 0.05 h is normal for the Galindez and Fredenslund method, the high ratio of distillate rate to charge necessitated a smaller Δt. Results are given in Table 13.3, where it is seen that the accumulated distillate cuts from Steps 1 and 3 are quite impure with respect to benzene and DCB, respectively. The cut from

Table 13.3 Results at the End of Each Operation Step for Example 13.9

	Operation Step		
	1	2	3
Operation time, h	0.605	0.805	0.055
No. of time increments	121	161	11
Accumulated distillate:			
Total lbmol	33.65	41.96	2.73
Mole fractions:			
B	0.731	0.009	0.000
MCB	0.269	0.950	0.257
DCB	0.000	0.041	0.743
Reboiler holdup:			
Total lbmol	66.13	24.19	21.46
Mole fractions:			
B	0.006	0.000	0.000
MCB	0.616	0.044	0.018
DCB	0.378	0.956	0.982
Total heat duties, 10^6 Btu:			
Condenser	1.95	2.65	0.19
Reboiler	2.08	2.63	0.18

Table 13.4 Results of Alternative Operating Schedule for Example 13.9

Distillate Cut	Amount, lbmol	Composition, Mole Fractions		
		B	MCB	DCB
Benzene-rich	18	0.993	0.007	0.000
Intermediate 1	18	0.374	0.626	0.000
MCB-rich	34	0.006	0.994	0.000
Intermediate 2	8	0.000	0.536	0.464
DCB-rich residual	22	0.000	0.018	0.982
Total	100			

Changes in mole fractions occur rapidly at certain times during the batch rectification, indicating that relatively pure cuts may be possible. This plot is useful in developing alternative schedules to obtain almost pure cuts. Using Figure 13.12, if relatively rich distillate cuts of B, MCB, and DCB are desired, an initial benzene-rich cut of, say, 18 lbmol might be taken, followed by an intermediate cut for recycle of, say, 18 lbmol. Then, an MCB-rich cut of 34 lbmol, followed by another intermediate cut of 8 lbmol, might be taken, leaving a DCB-rich residual of 22 lbmol. For this series of operation steps, with the same vapor boilup rate of 200 lbmol/h and reflux ratio of 3, the computed results for each distillate accumulation (cut), using a time step of 0.005 h, are given in Table 13.4. As seen, all three product cuts are better than 98 mol% pure. However, $(18 + 8) = 26$ lbmol of intermediate cuts, or about 1/4 of the original charge, would have to be recycled. Further improvements in purities of the cuts or reduction in the amounts of intermediate cuts for recycle can be made by increasing the reflux ratio and/or the number of stages.

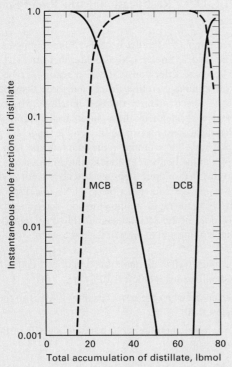

Figure 13.12 Instantaneous-distillate composition profile for Example 13.9.

[*Perry's Chemical Engineers' Handbook*, 6th ed., R.H. Perry and D.W. Green, Eds., McGraw-Hill, New York (1984) with permission.]

Step 2 is 95 mol% pure MCB. The residual left in the reboiler after Step 3 is quite pure in DCB. A plot of the instantaneous-distillate composition as a function of total-distillate accumulation for all steps is shown in Figure 13.12.

§13.7 INTERMEDIATE-CUT STRATEGY

Luyben [19] points out that design of a batch-distillation process is complex because two aspects must be considered: (1) the products to be obtained and (2) the control method to be employed. Basic design parameters are the number of trays, the size of the charge to the still pot, the boilup ratio, and the reflux ratio as a function of time. Even for a binary feed, it may be necessary to take three products: a distillate rich in the most-volatile component, a residue rich in the least-volatile component, and an intermediate cut containing both components. If the feed is a ternary system, more intermediate cuts may be necessary. The next two examples demonstrate intermediate-cut strategies for binary and ternary feeds.

EXAMPLE 13.10 Intermediate Cuts.

One hundred kmol of an equimolar mixture of *n*-hexane (C6) and *n*-heptane (C7) at 1 atm is batch-rectified in a column with a total condenser. It is desired to produce two products, one

Table 13.5 Batch Distillation of a C6–C7 Mixture

	Case 1	Case 2	Case 3	Case 4	Case 5
Reflux ratio	2	3	4	8	9.54
C6 product, kmol	15.1	36.0	42.4	49.2	50.0
C7 product, kmol	34.4	40.7	44.3	49.2	50.0
Intermediate cut, kmol	50.5	23.3	13.3	1.6	0.0
Mole fraction of C6 in intermediate cut	0.67	0.59	0.57	0.54	No intermediate cut
Total operation time, hours	1.97	2.37	2.78	4.57	5.27

with 95 mol% C6 and the other with 95 mol% C7. Neglect holdup and assume a boilup rate of 100 kmol/h. Also assume a constant reflux ratio; thus, the distillate composition will change with time. Determine a reasonable number of equilibrium stages and the effect of reflux ratio on the amount of intermediate cut.

Solution

To determine the number of equilibrium stages, a McCabe–Thiele diagram, based on SRK K-values (§2.5.1), is used in the manner of Figures 13.4 and 13.5. For total reflux ($y = x$, 45° line), the minimum number of stages for a 95 mol% C6 from an initial feed of 50 mol% C6 is 3.1, where one stage is the boiler. For operation at twice R_{min}, 5 stages plus the boiler are required.

For each reflux ratio, the first product is the 95 mol% C6 distillate. At this point, if the residue contains less than 95 mol% C7, then in a second step, a second accumulation of distillate (the intermediate cut) is made until the residue achieves the desired C7 composition. The reflux ratio is held constant throughout. The results, using CHEMCAD, are given in Table 13.5. For no intermediate cut (by material balance), the C6 and C7 products must each be 50 lbmol at 95 mol% purity. From Table 13.5, this is achieved at a constant reflux ratio of 9.54, with an operating time of 5.27 hours.

For lower reflux ratios, an intermediate cut whose amount increases as the reflux ratio decreases is necessary. If the quantity of feed is much larger than the capacity of the still-pot, the feed can be distilled in a sequence of charges. Then the intermediate cut for binary distillation of a batch can be recycled to the next batch. In this manner, each charge consists of fresh feed mixed with recycle intermediate cut. As discussed by Luyben [19], the composition of the intermediate cut is often not very different from the feed. This is confirmed in Table 13.5. If the number of stages is increased from 6, the reflux ratio for eliminating the intermediate cut can be reduced. For example, if 10 equilibrium stages are used, the reflux ratio can be reduced from 9.54 to approximately 6.

Intermediate-cut strategy for batch distillation of a ternary mixture, as discussed by Luyben [19], is considerably more complex, as shown in the following example.

EXAMPLE 13.11 Intermediate-Cut Strategy.

An equimolar ternary mixture of 150 kmol of C6, C7, and normal octane (C8) is to be distilled at 1 atm in a batch-rectification column with a total condenser. It is desired to produce three products: distillates of 95 mol% C6 and 90 mol% C7, and a residue of 95 mol% C8. Neglect holdup and assume a boilup rate of 100 kmol/h. Also assume that column operation is by constant reflux ratio. Thus, the distillate composition will change with time. Further assume that the column will contain 5 equilibrium stages and an equilibrium-stage boiler. Determine the effect of reflux ratio on the intermediate cuts.

Solution

The difficulty in this ternary example lies in determining specifications for termination of the second cut, which, unless R is high enough, is an intermediate cut. Suppose R is held constant at 4 and the intention is to terminate the second cut when the mole fraction of C7 in the instantaneous distillate reaches 90 mol% C7. Unfortunately, computer simulations show that only a value of 88 mol% C7 can be reached. Therefore, R is increased to 8. In addition, the third cut (the C7 product) is terminated when the mole fraction of C8 in that cut rises to 0.09 in the accumulator; and the second intermediate cut is terminated when the mole fraction of C8 in the residue rises to 0.95, the desired purity. Note that no purity specification has been placed on the C7 product. Instead, it has been assumed that the desired purity of 90 mol% C7 will be achieved with impurities of 9 mol% C8 and 1 mol% C6. Acceptable results are almost achieved for the reflux ratio of 8, as shown in Table 13.6 and Figure 13.13, where desired purity of the C7 cut is 89.8 mol%. However, for a reflux ratio of 8, these results may not correspond to the optimal termination specification for the first intermediate cut. With a small adjustment in the reflux ratio, it may be possible to eliminate the second intermediate cut. These two considerations are the subject of Exercise 13.29.

Table 13.6 Batch Distillation of a C6–C7–C8 Mixture

	Case 1	Case 2
Reflux ratio	4	8
C6 product, kmol	35.85	46.70
First intermediate cut:		
Amount, kmol	42.16	16.67
Mole fraction C6	0.373	0.316
Mole fraction C7	0.602	0.672
C7 product:		
Amount, kmol		35.43
Mole fraction C6		0.011
Mole fraction C7	0.877 max	0.898
Second intermediate cut:		
Amount, kmol		4.38
Mole fraction C6		0.000
Mole fraction C7		0.523
C8 product, kmol		46.82
Total operation time, hr		8.48

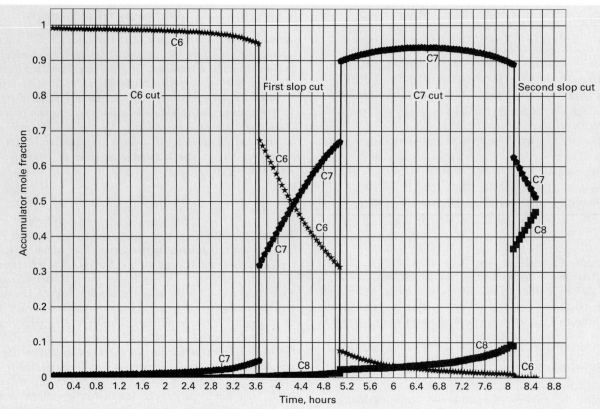

Figure 13.13 Ternary batch distillation with two intermediate (slop) cuts in Example 13.11.

Intermediate cuts and their recycle have been studied by a number of investigators, including Mayur, May, and Jackson [20]; Luyben [19]; Quintero-Marmol and Luyben [21]; Farhat et al. [22]; Mujtaba and Macchietto [23]; Diehl et al. [24]; and Robinson [25].

§13.8 OPTIMAL CONTROL BY VARIATION OF REFLUX RATIO

An operation policy in which the composition of the instantaneous distillate and, therefore, the accumulated distillate, is maintained constant is discussed in §13.2.2. This policy requires a variable reflux ratio and accompanying distillate rate. Although not as simple as the constant-reflux-ratio method of §13.2.1, it can be implemented with a rapidly responding composition (or surrogate) sensor and an associated reflux control system.

Which is the optimal way to control a batch distillation by (1) constant reflux ratio, (2) constant distillate composition, or (3) some other means? With a process simulator, it is fairly straightforward to compare the first two methods. However, the results depend on the objective for the optimization. Diwekar [26] studied the following three objectives when the accumulated-distillate composition and/or the residual composition is specified:

1. Maximize the amount of accumulated distillate in a given time.

2. Minimize the time to obtain a given amount of accumulated distillate.

3. Maximize the profit.

The next example compares the first two control policies with respect to their ability to meet the first two objectives.

EXAMPLE 13.12 Two Control Policies.

Repeat Example 13.10 under conditions of constant distillate composition and compare the results to those of that example for a constant reflux ratio of 4 with respect to both the amount of distillate and time of operation.

Solution

For Example 13.10, from Table 13.5 for a reflux ratio of 4, the amount of accumulated distillate during the first operation step is 42.4 kmol of 95 mol% C6. The time required for this cut, which is not listed in Table 13.5, is 1.98 hours. Using a process simulator, the operation specifications for a constant-composition operation are a boilup rate of 100 kmol/h, as in Example 13.10, with a constant instantaneous-distillate composition of 95 mol% C6. For the maximum distillate objective, the stop time for the first cut is 1.98 hours, as in Example 13.10. The amount of distillate obtained is 43.5 kmol, which is 2.6% higher than for operation at constant reflux ratio. The variation of reflux ratio with time for constant-composition control is shown in Figure 13.14, where the constant reflux ratio of 4 is also shown. The initial reflux ratio, 1.7, rises gradually at first and rapidly at the end. At 1 hour, the reflux ratio is 4, while at 1.98 hours, it is 15.4. For constant composition control, 42.4 kmol of accumulated distillate are obtained in 1.835 hours, compared to 1.98 hours for reflux-ratio control. Constant composition control is more optimal, this time by almost 8%.

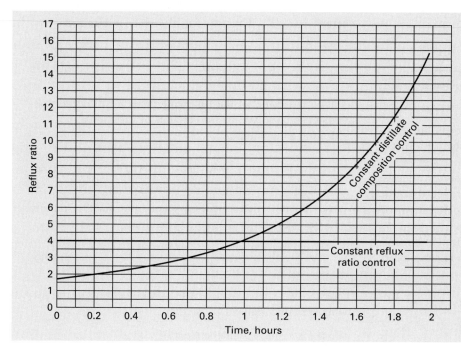

Figure 13.14 Binary batch distillation under distillate-composition control in Example 13.12.

Studies by Converse and Gross [27], Coward [28, 29], and Robinson [25] for binary systems; by Robinson and Coward [30] and Mayur and Jackson [31] for ternary systems; and Diwekar et al. [32] for higher multicomponent systems show that maximization of distillate or minimization of operation time, as well as maximization of profit, can be achieved by using an *optimal-reflux-ratio policy*. Often, this policy is intermediate between the constant-reflux-ratio and constant-composition controls in Figure 13.14 for Example 13.12. Generally, the optimal-reflux curve rises less sharply than that for the constant-distillate-composition control, with the result that savings in distillate, time, or money are highest for the more difficult separations. For relatively easy separations, savings for constant-distillate-composition control or optimal-reflux-ratio control may not be justified over the use of the simpler constant-reflux-ratio control.

Determination of optimal-reflux-ratio policy for complex operations requires a much different approach than that used for simpler optimization problems, which involve finding the optimal discrete values that minimize or maximize some objective with respect to an algebraic function. For example, in §7.3.7, a single value of the optimal reflux ratio for a continuous-distillation operation is found by plotting, as in Figure 7.22, the total annualized cost versus R, and locating the minimum in the curve. Establishing the optimal reflux ratio as a function of time, $R\{t\}$, for a batch distillation, which is modeled with differential or integral equations rather than algebraic equations, requires optimal-control methods that include the calculus of variations, the maximum principle of Pontryagin, dynamic programming of Bellman, and nonlinear programming. Diwekar [33] describes these methods in detail. Their development by mathematicians in Russia and the United States were essential for the success of their respective space programs.

To illustrate one of the approaches to optimal control, consider the classic *Brachistochrone* (Greek for "shortest time") *problem* of Johann Bernoulli, one of the earliest variational problems, whose investigation by famous mathematicians—including Johann and Jakob Bernoulli, Gottfried Leibnitz, Guillaume de L'Hopital, and Isaac Newton—was the starting point for development of the *calculus of variations*, a subject considered in detail by Weinstock [34]. A particle, e.g., a bead, is located in the x–y plane at (x_1, y_1), where the x-axis is horizontal to the right while the y-axis is vertically downward. The problem is to find the frictionless path, $y = f\{x\}$, ending at the point (x_2, y_2), down which the particle will move, subject only to gravity, in the least time. Some possible paths from point 1 to point 2, shown in Figure 13.15,

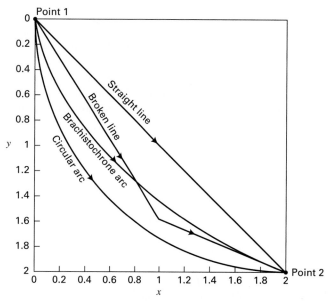

Figure 13.15 Frictionless paths between two points.

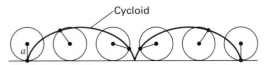

Figure 13.16 Generation of a cycloid from a circle of radius a.

include a straight line, a circular arc, and a broken line consisting of two connected straight lines (one steep followed by one shallow). The shortest distance is the straight line, but Galileo Galilei proposed that the path of shortest time is the circular arc. The other mathematicians proved that the solution is the arc of a cycloid (Brachistochrone arc), which, as included in Figure 13.15, is the locus of a point on the rim of a circle of radius a rolling along a straight line, as generated in Figure 13.16. The cycloid is given in parametric form as

$$x = a(\theta - \sin\theta) \quad \text{and} \quad y = a(1 - \cos\theta) \quad (13\text{-}59)$$

By eliminating θ, the Cartesian equation for the cycloid is

$$x = a\cos^{-1}(1 - y/a) - (2ay - y^2)^{0.5} \quad (13\text{-}60)$$

This optimal solution of the Brachistochrone arc problem is obtained by the calculus of variations as follows.

Let the arc length along the path be s. Then a differential length along the arc is the hypotenuse of a differential triangle, such that

$$ds = \left[(dy)^2 + (dx)^2\right]^{0.5} \quad (13\text{-}61)$$

or

$$ds = dx\sqrt{\left(\frac{dy}{dx}\right)^2 + 1} \quad (13\text{-}62)$$

The time, t_{12}, for the particle to travel from point 1 (P_1) to point 2 (P_2) is given by

$$t_{12} = \int_{P_1}^{P_2} \frac{ds}{v} \quad (13\text{-}63)$$

where v is the speed of the particle. By the conservation of energy, as the particle descends, its kinetic energy will increase as its potential energy decreases. Thus, if m is the mass of the particle and g is the acceleration due to gravity, where the velocity increases as the downward distance increases,

$$\frac{1}{2}mv^2 = mgy \quad (13\text{-}64)$$

Solving for v,

$$v = \sqrt{2gy} \quad (13\text{-}65)$$

Substituting (13-62) and (13-65) into (13-63) gives

$$t_{12} = \int_{P_1}^{P_2} \sqrt{\frac{1 + (y')^2}{2gy}}\,dx \quad (13\text{-}66)$$

where $y' = dy/dx$. Thus, the function to be minimized is

$$f = \sqrt{\frac{1 + (y')^2}{2gy}} \quad (13\text{-}67)$$

Equation (13-67) is of the following general form of a problem that can be solved by the calculus of variations:

Minimize the integral,

$$I = \int_1^2 f(x, y\{x\}, y'\{x\})\,dx \quad (13\text{-}68)$$

Weinstock shows that a necessary condition for the solution of (13-68) is the Euler–Lagrange equation:

$$\frac{\partial f}{\partial y} - \frac{d}{dx}\left(\frac{\partial f}{\partial y'}\right) = 0 \quad (13\text{-}69)$$

There are two special cases of (13-69), resulting in the following simplifications:

1. If f is explicitly independent of y, then

$$\frac{\partial f}{\partial y'} = C_1 \quad (13\text{-}70)$$

2. If f is explicitly independent of x, then

$$y'\left(\frac{\partial f}{\partial y'}\right) - f = C_2 \quad (13\text{-}71)$$

The Brachistochrone arc function of (13-67) is explicitly independent of x, so (13-71) applies, giving

$$(y')^2\left[1 + (y')^2\right]^{-\frac{1}{2}}(2gy)^{-\frac{1}{2}} - \left[1 + (y')^2\right]^{\frac{1}{2}}(2gy)^{-\frac{1}{2}} = C_2 \quad (13\text{-}72)$$

which simplifies to

$$\left[1 + \left(\frac{dy}{dx}\right)^2\right]y = \frac{1}{2gC_2^2} = 2a \quad (13\text{-}73)$$

If point 1 is located at $x = 0$, $y = 0$, the solution of (13-73) is (13-60), which is the arc of a cycloid, shown as the Brachistochrone arc in Figure 13.15.

How much better is the cycloid-arc path compared to the other paths shown in Figure 13.15? If point 2 is taken at $x = 2$ ft and $y = 2$ ft, then with $g = 32.17$ ft/s^2, the calculated travel times, from the application of (13-66) to move from point 1 to point 2, are as follows, where the cycloid arc is just slightly better than a circular arc:

Path in Figure 13.15	Travel time, seconds
Straight line	0.498
Broken line	0.472
Circular arc of radius = 2 ft	0.460
Cycloid arc with a = 1.145836 ft	0.455

Application of the calculus of variations to the determination of the optimal-reflux-control strategy for batch distillation is carried out in a manner similar to that above for the Brachistochrone problem. If it is desired to find the reflux-ratio path that minimizes the time, t, required to obtain an accumulated-distillate amount and composition for a fixed boilup rate, V, using a column with N equilibrium stages, the integral to be minimized is as follows, where the variables for the distillate are replaced by the variables for the residual, W, remaining in the still-pot:

$$t = \int_{x_{W_0}}^{x_W} \frac{W}{V}\left(\frac{R+1}{y_N - x_W}\right) dx_W \qquad (13\text{-}74)$$

SUMMARY

1. The simplest case of batch distillation corresponds to the condensation of a vapor rising from a boiling liquid, called differential or Rayleigh distillation. The vapor leaving the liquid surface is assumed to be in equilibrium with the liquid. The compositions of the liquid and vapor vary as distillation proceeds. The instantaneous vapor and liquid compositions can be computed as a function of time for a given vaporization rate.

2. A batch-rectifier system consists of a reboiler, a column with plates or packing that sits on top of the reboiler, a condenser, a reflux drum, and one or more distillate receivers.

3. For a binary system, a batch rectifier is usually operated at a constant reflux ratio or at a constant distillate composition. For either case, a McCabe–Thiele diagram can be used to follow the process, if assumptions are made of constant molar overflow and negligible liquid holdups in trays (or packing), condenser, and reflux drum.

4. A batch stripper is useful for removing impurities from a charge. For complete flexibility, complex batch distillation involving both rectification and stripping can be employed.

5. Liquid holdup on the trays (or packing) and in the condenser and reflux drum influences the course of batch rectification and the size and composition of distillate cuts. The holdup effect is best determined by rigorous calculations for specific cases.

6. For multicomponent batch rectification, with negligible liquid holdup except in the reboiler, the shortcut method of Sundaram and Evans, based on successive applications of the Fenske–Underwood–Gilliland (FUG) method at a sequence of time intervals, can be used to obtain approximate distillate and charge compositions and amounts as a function of time.

7. For accurate and detailed multicomponent, batch-rectification compositions, the model of Distefano as implemented by Boston et al. should be used. It accounts for liquid holdup and permits a sequence of operation steps to produce multiple distillate cuts. The model consists of algebraic and ordinary differential equations (DAE) that, when stiff, are best solved by Gear-type implicit-integration methods. The Distefano model can also be solved by the method of Galindez and Fredenslund, which simulates the unsteady batch process by a succession of steady states of short duration, which are solved by the NR or inside-out method.

8. Two difficult aspects of batch distillation are (1) determination of the best set of operations for the production of the desired products and (2) determination of the optimal-control method to be used. The first, which involves the possibility that intermediate cuts may be necessary, is solved by computational studies using a simulation program. The second, which requires consideration of the best reflux-ratio policy, is solved by optimal-control techniques.

REFERENCES

1. Rayleigh, J.W.S., *Phil. Mag. and J. Sci.*, Series 6, **4**(23), 521–537 (1902).

2. Smoker, E.H., and A. Rose, *Trans. AIChE*, **36**, 285–293 (1940).

3. Block, B., *Chem. Eng.*, **68**(3), 87–98 (1961).

4. Bogart, M.J.P., *Trans. AIChE*, **33**, 139–152 (1937).

5. Ellerbe, R.W., *Chem. Eng.*, **80**(12), 110–116 (1973).

6. Hasebe, S., B.B. Abdul Aziz, I. Hashimoto, and T. Watanabe, *Proc. IFAC Workshop, London, Sept. 7–8*, 1992, p. 177.

7. Diwekar, U.M., and K.P. Madhaven, *Ind. Eng. Chem. Res.*, **30**, 713–721 (1991).

8. Sundaram, S., and L.B. Evans, *Ind. Eng. Chem. Res.*, **32**, 511–518 (1993).

9. Eduljee, H.E., *Hydrocarbon Processing*, **56**(9), 120–122 (1975).

10. Meadows, E.L., *Chem. Eng. Progr. Symp. Ser. No. 46*, **59**, 48–55 (1963).

11. Distefano, G.P., *AIChE J.*, **14**, 190–199 (1968).

12. Boston, J.F., H.I. Britt, S. Jirapongphan, and V.B. Shah, in R.H.S. Mah and W.D. Seider, Eds., *Foundations of Computer-Aided Chemical Process Design*, AIChE, Vol. II, pp. 203–237 (1981).

13. Galindez, H., and A. Fredenslund, *Comput. Chem. Eng.*, **12**, 281–288 (1988).

14. Gear, C.W., *Numerical Initial Value Problems in Ordinary Differential Equations*, Prentice-Hall, Englewood Cliffs, NJ (1971).

15. Davis, M.E., *Numerical Methods and Modeling for Chemical Engineers*, John Wiley & Sons, New York (1984).

16. Riggs, J.B., *An Introduction to Numerical Methods for Chemical Engineers*, Texas Tech. Univ. Press, Lubbock, TX (1988).

17. Carnahan, B., and J.O. Wilkes, "Numerical Solution of Differential Equations—An Overview," in R.S.H. Mah and W.D. Seider, Eds., *Foundations of Computer-Aided Chemical Process Design,* Engineering Foundation, New York, Vol. I, pp. 225–340 (1981).

18. Varga, R.S., *Matrix Iterative Analysis*, Prentice-Hall, Englewood Cliffs, NJ (1962).

19. Luyben, W.L., *Ind. Eng. Chem. Res.*, **27**, 642–647 (1988).

20. Mayur, D.N., R.A. May, and R. Jackson, *Chem. Eng. Journal*, **1**, 15–21 (1970).

21. Quintero-Marmol, E., and W.L. Luyben, *Ind. Eng. Chem. Res.*, **29**, 1915–1921 (1990).

22. Farhat, S., M. Czernicki, L. Pibouleau, and S. Domenech, *AIChE J.*, **36**, 1349–1360 (1990).

23. Mujtaba, I.M., and S. Macchietto, *Comput. Chem. Eng.*, **16**, S273–S280 (1992).

24. Diehl, M., A. Schafer, H.G. Bock, J.P. Schloder, and D.B. Leineweber, *AIChE J.*, **48**, 2869–2874 (2002).

25. Robinson, E.R., *Chem. Eng. Journal*, **2**, 135–136 (1971).

26. Diwekar, U.M., *Batch Distillation–Simulation, Optimal Design and Control*, Taylor & Francis, Washington, DC. (1995).

27. Converse, A.O., and G.D. Gross, *Ind. Eng. Chem. Fundamentals*, **2**, 217–221 (1963).

28. Coward, I., *Chem. Eng. Science*, **22**, 503–516 (1967).

29. Coward, I., *Chem. Eng. Science*, **22**, 1881–1884 (1967).

30. Robinson, E.R., and I. Coward, *Chem. Eng. Science*, **24**, 1661–1668 (1969).

31. Mayur, D.N., and R. Jackson, *Chem. Eng. Journal*, **2**, 150–163 (1971).

32. Diwekar, U.M., R.K. Malik, and K.P. Madhavan, *Comput. Chem. Eng.*, **11**, 629–637 (1987).

33. Diwekar, U.M., *Introduction to Applied Optimization*, Kluwer Academic Publishers (2003).

34. Weinstock, R., *Calculus of Variations*, McGraw-Hill Book Co., Inc., New York (1952).

35. Barolo, M., G. Guarise, S. Rienzi, and A. Macchietto, *Ind. Eng. Chem. Res.*, **35**, 4612–4618 (1996).

36. Phimister, J.R., and W.D. Seider, *Ind. Eng. Chem. Res.*, **39**, 1840–1849 (2000).

STUDY QUESTIONS

13.1. How does batch distillation differ from continuous distillation?

13.2. When should batch distillation be considered?

13.3. What is differential (Rayleigh) distillation? How does it differ from batch rectification?

13.4. For what kinds of mixtures is differential distillation adequate?

13.5. What is the easiest way to determine the average composition of the distillate from a batch rectifier?

13.6. Which is easiest to implement: (1) the constant-reflux policy, (2) the constant-distillate-composition policy, or (3) the optimal-control policy? Why?

13.7. What is a batch stripper?

13.8. Can a batch rectifier and a batch stripper be combined? If so, what advantage is gained?

13.9. What effects does liquid holdup have on batch rectification?

13.10. What are the assumptions of the rigorous-batch distillation model of Distefano?

13.11. Why is the Distefano model referred to as a differential-algebraic equation (DAE) system?

13.12. What is the difference between truncation error and stability?

13.13. How does the explicit-Euler method differ from the implicit method?

13.14. What is stiffness and how does it arise? What criterion can be used to determine the degree of stiffness, if any?

13.15. In the development of an operating policy (campaign) for batch distillation, what is done with intermediate (slop) cuts?

13.16. What are the common objectives of optimal control of a batch distillation, as cited by Diwekar?

13.17. What is varied to achieve optimal control?

EXERCISES

Section 13.1

13.1. Evaporation from a drum.

A bottle of pure *n*-heptane is accidentally poured into a drum of pure toluene in a laboratory. One of the laboratory assistants suggests that since heptane boils at a lower temperature than toluene, the following purification procedure can be used:

Pour the mixture (2 mol% *n*-heptane) into a simple still pot. Boil the mixture at 1 atm and condense the vapors until all heptane is boiled away. Obtain the pure toluene from the residue.

You, a chemical engineer with knowledge of vapor–liquid equilibrium, immediately realize that such a purification method will not work. (a) Indicate this by a curve showing the composition of the material remaining in the still-pot after various quantities of the liquid have been distilled. What is the composition of the residue after 50 wt% of the original material has been distilled? What is the composition of the cumulative distillate? (b) When one-half of the heptane has been distilled, what is the composition of the cumulative

distillate and the residue? What weight %t of the original material has been distilled?

Equilibrium data at 1 atm [*Ind. Eng. Chem.*, **42**, 2912 (1949)] are:

Mole Fraction *n*-Heptane			
Liquid	Vapor	Liquid	Vapor
0.025	0.048	0.448	0.541
0.062	0.107	0.455	0.540
0.129	0.205	0.497	0.577
0.185	0.275	0.568	0.637
0.235	0.333	0.580	0.647
0.250	0.349	0.692	0.742
0.286	0.396	0.843	0.864
0.354	0.454	0.950	0.948
0.412	0.504	0.975	0.976

13.2. Differential distillation.

A mixture of 40 mol% isopropanol in water is distilled at 1 atm by differential distillation until 70 mol% of the charge has been vaporized (equilibrium data are given in Exercise 7.33). What is the composition of the liquid residue in the still-pot and of the collected distillate?

13.3. Differential distillation.

A 30 mol% feed of benzene in toluene is to be distilled in a batch differential-distillation operation. A product having an average composition of 45 mol% benzene is to be produced. Calculate the amount of residue, assuming $\alpha = 2.5$ and $W_0 = 100$.

13.4. Differential distillation.

A charge of 250 lb of 70 mol% benzene and 30 mol% toluene is subjected to differential distillation at 1 atm. Determine the compositions of the distillate and residue after $1/3$ of the feed has been distilled. Assume Raoult's and Dalton's laws.

13.5. Differential distillation.

A mixture containing 60 mol% benzene and 40 mol% toluene is subjected to differential distillation at 1 atm, under three different conditions:

1. Until the distillate contains 70 mol% benzene
2. Until 40 mol% of the feed is evaporated
3. Until 60 mol% of the original benzene leaves in the vapor

Using $\alpha = 2.43$, determine for each of the three cases: (a) number of moles in the distillate for 100 mol of feed; (b) compositions of distillate and residue.

13.6. Differential distillation.

Fifteen mol% phenol in water is to be differential-batch-distilled at 260 torr. What fraction of the batch is in the still-pot when the total distillate contains 98 mol% water? What is the residue concentration?

Vapor–liquid data at 260 torr [*Ind. Eng. Chem.*, **17**, 199 (1925)]:

x, wt% (H_2O):

   1.54 4.95 6.87 7.73 19.63 28.44 39.73 82.99
   89.95 93.38 95.74

y, wt% (H_2O):

   41.10 79.72 82.79 84.45 89.91 91.05 91.15 91.86
   92.77 94.19 95.64

13.7. Differential distillation with added feed.

A still-pot is charged with 25 mol of benzene and toluene containing 35 mol% benzene. Feed of the same composition is supplied at a rate of 7 mol/h, and the heating rate is adjusted so that the liquid level in the still-pot remains constant. If $\alpha = 2.5$, how long will it be before the distillate composition falls to 0.45 mol% benzene?

13.8. Differential distillation with continuous feed.

A system consisting of a still-pot and a total condenser is used to separate A and B from a trace of nonvolatile material. The still-pot initially contains 20 lbmol of feed of 30 mol% A. Feed of the same composition is supplied to the still-pot at the rate of 10 lbmol/h, and the heat input is adjusted so that the total moles of liquid in the reboiler remain constant at 20. No residue is withdrawn from the still-pot. Calculate the time required for the composition of the overhead product to fall to 40 mol% A. Assume $\alpha = 2.50$.

Section 13.2

13.9. Batch rectification at constant reflux ratio.

Repeat Exercise 13.2 for the case of batch distillation carried out in a two-stage column with $L/V = 0.9$.

13.10. Batch rectification at constant reflux ratio.

Repeat Exercise 13.3 assuming the operation is carried out in a three-stage column with $L/V = 0.6$.

13.11. Batch rectification at constant reflux ratio.

One kmol of an equimolar mixture of benzene and toluene is fed to a batch still containing three equivalent stages (including the boiler). The liquid reflux is at its bubble point, and $L/D = 4$. What is the average composition and amount of product when the instantaneous product composition is 55 mol% benzene? Neglect holdup, and assume $\alpha = 2.5$.

13.12. Differential distillation and batch rectification.

The fermentation of corn produces a mixture of 3.3 mol% ethyl alcohol in water. If this mixture is distilled at 1 atm by a differential distillation, calculate and plot the instantaneous-vapor composition as a function of mol% of batch distilled. If reflux with three theoretical stages (including the boiler) is used, what is the maximum purity of ethyl alcohol that can be produced by batch rectification?

Equilibrium data are given in Exercise 7.29.

13.13. Batch rectification at constant composition.

An acetone–ethanol mixture of 0.5 mole fraction acetone is to be separated by batch distillation at 101 kPa.

Vapor–liquid equilibrium data at 101 kPa are as follows:

	Mole Fraction Acetone									
y	0.16	0.25	0.42	0.51	0.60	0.67	0.72	0.79	0.87	0.93
x	0.05	0.10	0.20	0.30	0.40	0.50	0.60	0.70	0.80	0.90

(a) Assuming an L/D of 1.5 times the minimum, how many stages should this column have if the desired composition of the distillate is 0.90 mole fraction acetone when the residue contains 0.1 mole fraction acetone?

(b) Assume the column has eight stages and the reflux rate is varied continuously so that the top product is maintained constant at 0.9 mole fraction acetone. Make a plot of the reflux ratio versus the still-pot composition and the amount of liquid left in the still-pot.

(c) Assume the same distillation is carried out at constant reflux ratio (and varying product composition). It is desired to have a residue containing 0.1 and an (average) product containing 0.9 mole fraction acetone. Calculate the total vapor generated. Which method of operation is more energy-intensive? Suggest operating policies other than constant reflux ratio and constant distillate compositions that lead to equipment or operating cost savings.

13.14. Batch rectification at constant composition.

Two thousand gallons of 70 wt% ethanol in water having a specific gravity of 0.871 is to be separated at 1 atm in a batch rectifier operating at a constant distillate composition of 85 mol% ethanol with a constant molar vapor boilup rate to obtain a residual wastewater containing 3 wt% ethanol. If the task is to be completed in 24 h, allowing 4 h for charging, start-up, shutdown, and cleaning, determine: (a) the number of theoretical stages; (b) the reflux ratio when the ethanol in the still-pot is 25 mol%; (c) the instantaneous distillate rate in lbmol/h when the concentration of ethanol in the still-pot is 15 mol%; (d) the lbmol of distillate product; and (e) the lbmol of residual wastewater. Vapor–liquid equilibrium data are given in Exercise 7.29.

13.15. Batch rectification at constant composition.

One thousand kmol of 20 mol% ethanol in water is to undergo batch rectification at 101.3 kPa at a vapor boilup rate of 100 kmol/h. If the column has six theoretical stages and the distillate composition is to be maintained at 80 mol% ethanol by varying the reflux ratio, determine the: (a) time in hours for the residue to reach an ethanol mole fraction of 0.05; (b) kmol of distillate obtained when the condition of part (a) is achieved; (c) minimum and maximum reflux ratios during the rectification period; and (d) variation of the distillate rate in kmol/h during the rectification period. Assume constant molar overflow, neglect liquid holdup, and obtain equilibrium data from Exercise 7.29.

13.16. Batch rectification for constant composition.

Five hundred lbmol of 48.8 mol% A and 51.2 mol% B with a relative volatility $\alpha_{A,B}$ of 2.0 is separated in a batch rectifier consisting of a total condenser, a column with seven theoretical stages, and a still-pot. The reflux ratio is varied to maintain the distillate at 95 mol% A. The column operates with a vapor boilup rate of 213.5 lbmol/h. The rectification is stopped when the mole fraction of A in the still is 0.192. Determine the (a) rectification time and (b) total amount of distillate produced.

Section 13.3

13.17. Batch stripping at constant boilup ratio.

Develop a procedure similar to that of §13.3 to calculate a binary batch stripping operation using the equipment arrangement of Figure 13.8.

13.18. Batch stripping at constant boilup ratio.

A three-theoretical-stage batch stripper (one stage is the boiler) is charged to the feed tank (see Figure 13.8) with 100 kmol of 10 mol% n-hexane in n-octane mix. The boilup rate is 30 kmol/h. If a constant boilup ratio (V/L) of 0.5 is used, determine the instantaneous bottoms composition and the composition of the accumulated bottoms product at the end of 2 h of operation.

13.19. Batch distillation with a middle feed vessel.

Develop a procedure similar to that of §13.3 to calculate a complex, binary, batch-distillation operation using the equipment arrangement of Figure 13.9.

Section 13.5

13.20. Effect of holdup on batch rectification.

For a batch rectifier with appreciable column holdup: (a) Why is the charge to the still-pot higher in the light component than at the start of rectification, assuming that total-reflux conditions are established before rectification? (b) Why will separation be more difficult than with zero holdup?

13.21. Effect of holdup on batch rectification.

For a batch rectifier with appreciable column holdup, why do tray compositions change less rapidly than they do for a rectifier with negligible column holdup, and why is the separation improved?

13.22. Effect of holdup on batch rectification.

Based on the statements in Exercises 13.20 and 13.21, why is it difficult to predict the effect of holdup?

Section 13.6

13.23. Batch rectification by shortcut method.

Use the shortcut method of Sundaram and Evans to solve Example 13.7, but with zero condenser and stage holdups.

13.24. Batch rectification by shortcut method.

A charge of 100 kmol of an equimolar mixture of A, B, and C, with $\alpha_{A,B} = 2$ and $\alpha_{A,C} = 4$, is distilled in a batch rectifier containing four theoretical stages, including the still-pot. If holdup can be neglected, use the shortcut method with $R = 5$ and $V = 100$ kmol/h to estimate the variation of the still-pot and instantaneous-distillate compositions as a function of time after total reflux conditions are established.

13.25. Batch rectification by the shortcut method.

A charge of 200 kmol of a mixture of 40 mol% A, 50 mol% B, and 10 mol% C, with $\alpha_{A,C} = 2.0$ and $\alpha_{B,C} = 1.5$, is to be separated in a batch rectifier with three theoretical stages, including the still-pot, and operating at a reflux ratio of 10, with a molar vapor boilup rate of 100 kmol/h. Holdup is negligible. Use the shortcut method to estimate instantaneous-distillate and bottoms compositions as a function of time for the first hour of operation after total reflux conditions are established.

Section 13.7

13.26. Batch rectification by rigorous method.

A charge of 100 lbmol of 35 mol% n-hexane, 35 mol% n-heptane, and 30 mol% n-octane is to be distilled at 1 atm in a batch rectifier, consisting of a still-pot, a column, and a total condenser, at a constant boilup rate of 50 lbmol/h and a constant reflux ratio of 5. Before rectification begins, total-reflux conditions are established. Then, the following three operation steps are carried out to obtain an n-hexane-rich cut, an intermediate cut for recycle, an n-heptane-rich cut, and an n-octane-rich residue:

Step 1: Stop when the accumulated-distillate purity drops below 95 mol% n-hexane.

Step 2: Empty the n-hexane-rich cut produced in Step 1 into a receiver and resume rectification until the instantaneous-distillate composition reaches 80 mol% n-heptane.

Step 3: Empty the intermediate cut produced in Step 2 into a receiver and resume rectification until the accumulated-distillate composition reaches 4 mol% n-octane.

For properties, assume ideal solutions and the ideal-gas law.

Consider conducting the rectification in two different columns, each with the equivalent of 10 theoretical stages, a still-pot, and a total condenser reflux-drum liquid holdup of 1.0 lbmol. For each column, determine with a suitable batch-distillation computer program the compositions and amounts in lbmol of each of the four products.

Column 1: A plate column with a total liquid holdup of 8 lbmol

Column 2: A packed column with a total liquid holdup of 2 lbmol

Discuss the effect of liquid holdup for the two columns. Are the results what you expected?

13.27. Rigorous batch rectification with holdup.

One hundred lbmol of 10 mol% propane, 30 mol% n-butane, 10 mol% n-pentane, and the balance n-hexane is to be separated in a batch rectifier equipped with a still-pot, a total condenser with a liquid holdup of 1.0 ft^3, and a column with the equivalent of eight theoretical stages and a total holdup of 0.80 ft^3. The pressure in the condenser is 50.0 psia and the column pressure drop is 2.0 psi. The rectification campaign, given as follows, is designed to produce cuts of 98 mol% propane and 99.8 mol% n-butane, a residual cut of 99 mol% n-hexane, and two intermediate cuts, one of which may be a relatively rich cut of n-pentane. All five operating steps are conducted at a molar vapor boilup rate of 40 lbmol/h. Use a suitable batch-distillation computer program to determine the amounts and compositions of all cuts.

Step	Reflux Ratio	Stop Criterion
1	5	98% propane in accumulator
2	20	95% *n*-butane in instantaneous distillate
3	25	99.8% *n*-butane in accumulator
4	15	80% *n*-pentane in instantaneous distillate
5	25	99% *n*-hexane in the pot

How might you alter the operation steps to obtain larger amounts of the product cuts and smaller amounts of the intermediate cuts?

13.28. Stability and stiffness.

One hundred lbmol of benzene (B), monochlorobenzene (MCB), and *o*-dichlorobenzene (DCB) is distilled in a batch rectifier that has a total condenser, a column with 10 theoretical stages, and a still-pot. Following establishment of total reflux, the first operation step begins at a boilup rate of 200 lbmol/h and a reflux ratio of 3. At the end of 0.60 h, the following conditions exist for the top three stages in the column:

	Top Stage	Stage 2	Stage 3
Temperature, °F	267.7	271.2	272.5
V, lbmol/h	206.1	209.0	209.5
L, lbmol/h	157.5	158.0	158.1
M, lbmol	0.01092	0.01088	0.01087
Vapor Mole Fractions:			
B	0.0994	0.0449	0.0331
MCB	0.9006	0.9551	0.9669
DCB	0.0000	0.0000	0.0000
Liquid Mole Fractions:			
B	0.0276	0.0121	0.00884
MCB	0.9724	0.9879	0.99104
DCB	0.0000	0.0000	0.00012

In addition, still-pot and condenser holdups at 0.6 h are 66.4 and 0.1113 lbmol, respectively. For benzene, use the preceding data with (13-36) and (13-39) to estimate the liquid-phase mole fraction of benzene leaving Stage 2 at 0.61 h by using the explicit-Euler method with a Δt of 0.01 h. If the result is unreasonable, explain why, with respect to stability and stiffness considerations.

13.29. Batch rectification of a ternary mixture.

One hundred kmol of 30 mol% methanol, 30 mol% ethanol, and 40 mol% *n*-propanol is charged at 120 kPa to a batch rectifier consisting of a still-pot, a column with the equivalent of 10 equilibrium stages, and a total condenser. After establishing a total-reflux condition, the column begins a sequence of two operating steps, each for a duration of 15 h at a distillate flow rate of 2 kmol/h and a reflux ratio of 10. Thus, the two accumulated distillates are equal in moles to the methanol and ethanol in the feed. Neglect the liquid holdup in the condenser and column. The column pressure drop is 8 kPa, with a pressure drop of 2 kPa through the condenser. Using a simulation program with UNIFAC (§2.6.9) for liquid-phase activity coefficients, determine the composition and amount in kmol for the three cuts.

13.30. Batch rectification of a ternary mixture.

Repeat Exercise 13.29 with the following modifications: Add a third operating step. For all three steps, use the same distillate rate and reflux rate as in Exercise 13.29. Use the following durations for the three steps: 13 hours for Step 1, 4 hours for Step 2, and 13 hours for Step 3. The distillate from Step 2 will be an intermediate cut. Determine the mole-fraction composition and amount in kmol of each of the four cuts.

13.31. Batch rectification of a ternary mixture.

One hundred kmol of 45 mol% acetone, 30 mol% chloroform, and 25 mol% benzene is charged at 101.3 kPa to a batch rectifier consisting of a still-pot, a column containing the equivalent of 10 equilibrium stages, and a total condenser. After establishing a total-reflux condition, the column will begin a sequence of two operating steps, each at a distillate flow rate of 2 kmol/h and a reflux ratio of 10. The durations will be 13.3 hours for Step 1 and 24.2 hours for Step 2. Neglect pressure drops and the liquid holdup. Using a process simulator with UNIFAC (§2.6.9) for liquid-phase activity coefficients, determine the mole-fraction composition and amount in kmol of each of the three cuts.

Section 13.8

13.32. Reduction of intermediate cuts.

Using a process simulator, make the following modifications to the C6–C7–C8 example in §13.7: (a) Increase the reflux above 8 to eliminate the second intermediate cut. (b) Change the termination specification on the second step to reduce the amount of the first intermediate cut, without failing to meet all three product specifications.

Part Three

Separations by Barriers and Solid Agents

In recent years, industrial applications of separations using barriers and solid agents have increased because of progress in producing selective membranes and adsorbents. Chapter 14 presents a discussion of mass-transfer rates through membranes, and calculation methods for the more widely used batch and continuous membrane separations for gas and liquid feeds, including bioprocess streams. These include gas permeation, reverse osmosis, dialysis, electrodialysis, pervaporation, ultrafiltration, and microfiltration.

Chapter 15 covers separations by adsorption, ion exchange, and chromatography, which use solid separation agents. Discussions of equilibrium and mass-transfer rates in porous adsorbents are followed by design methods for batch and continuous equipment for liquid and gaseous feeds, including bioprocess streams. These include fixed-bed, pressure-swing, and simulated-moving-bed adsorption. Electrophoresis is also included in Chapter 15.

Chapter 14

Membrane Separations

§14.0 INSTRUCTIONAL OBJECTIVES

After completing this chapter, you should be able to:

- Explain membrane processes in terms of the membrane, feed, sweep, retentate, permeate, and solute-membrane interactions.
- Distinguish effects on membrane mass transfer due to permeability, permeance, solute resistance, selectivity, concentration polarization, fouling, inertial lift, and shear-induced diffusion.
- Explain contributions to mass-transfer coefficients from membrane thickness and tortuosity; solute size, charge, and solubility; and hydrodynamic viscous and shear forces.
- Differentiate between asymmetric and thin-layer composite membranes, and between dense and microporous membranes.
- Distinguish among microfiltration, ultrafiltration, nanofiltration, virus filtration, sterile filtration, filter-aid filtration, and reverse osmosis in terms of average pore size, unique role in a biopurification sequence, and operation in normal flow versus tangential flow.
- Describe four membrane shapes and six membrane modules, and identify the most common types used in bioseparations.
- Distinguish among mass transfer through membranes by bulk flow, molecular diffusion, and solution diffusion.
- Differentiate between predicting flux in normal-flow filtration versus tangential-flow filtration.
- Differentiate among resistances due to cake formation, pore constriction, and pore blockage in both constant-flux and constant-pressure operation, and explain how to distinguish these using data from membrane operation.
- Explain four common idealized flow patterns in membrane modules.
- Differentiate between concentration polarization and membrane fouling, and explain how to minimize effects of each on membrane capacity and throughput.
- Calculate mass-transfer rates for dialysis and electrodialysis, reverse osmosis, gas permeation, pervaporation, normal-flow filtration, and tangential-flow filtration.
- Explain osmosis and how reverse osmosis is achieved.

In a membrane-separation process, a feed consisting of a mixture of two or more components is partially separated by means of a semipermeable barrier (the membrane) through which some species move faster than others. The most general membrane process is shown in Figure 14.1, where the feed mixture is separated into a *retentate* (that part of the feed that does not pass through the membrane) and a *permeate* (that part that does pass through the membrane). Although the feed, retentate, and permeate are usually liquid or gas, in bioprocesses, solid particles may also be present. The barrier is most often a thin, nonporous, polymeric film, but may also be porous polymer, ceramic, or metal material, or even a liquid, gel, or gas. To maintain selectivity, the barrier must not dissolve, disintegrate, or break. The optional sweep, shown in Figure 14.1, is a liquid or gas used to facilitate removal of the permeate. Many of the industrially important membrane-separation operations are listed in Tables 1.2 and Table 14.1.

In membrane separations: (1) the two products are usually miscible, (2) the separating agent is a semipermeable barrier, and (3) a sharp separation is often difficult to achieve. Thus, membrane separations differ in some respects from the more common separation operations of absorption, distillation, and liquid–liquid extraction.

Although membranes as separating agents have been known for more than 100 years [1], large-scale applications have appeared only in the past 60 years. In the 1940s, porous fluorocarbons were used to separate $^{235}UF_6$ from $^{238}UF_6$ [2]. In the mid-1960s, reverse osmosis with cellulose acetate was first used to desalinize seawater to produce potable water (drinkable water with less than 500 ppm by weight of dissolved solids) [3]. Commercial ultrafiltration membranes

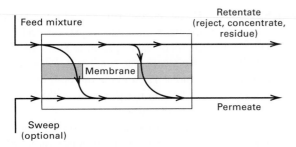

Figure 14.1 General membrane process.

Table 14.1 Industrial Membrane-Separation Processes

1. Reverse osmosis:
    Desalinization of brackish water
    Treatment of wastewater to remove a wide variety of impurities
    Treatment of surface and groundwater
    Concentration of foodstuffs
    Removal of alcohol from beer
2. Dialysis:
    Separation of nickel sulfate from sulfuric acid
    Hemodialysis (removal of waste metabolites and excess body water, and restoration of electrolyte balance in blood)
3. Electrodialysis:
    Production of table salt from seawater
    Concentration of brines from reverse osmosis
    Treatment of wastewaters from electroplating
    Demineralization of cheese whey
    Production of ultra-pure water for the semiconductor industry
4. Microfiltration:
    Sterilization of liquids, gases, and parenteral drugs
    Clarification and biological stabilization of beverages
    Bacterial cell harvest and purification of antibiotics
    Recovery of mammalian cells from cell culture broth
5. Ultrafiltration:
    Preconcentration of milk before making cheese
    Clarification of fruit juice
    Purification of recombinant proteins and DNA, antigens, and antibiotics from clarified cell broths
    Color removal from Kraft black liquor in papermaking
6. Pervaporation:
    Dehydration of ethanol–water azeotrope
    Removal of water from organic solvents
    Removal of organics from water
7. Gas permeation:
    Separation of CO_2 or H_2 from methane
    Separation of uranium isotopes
    Adjustment of the H_2/CO ratio in synthesis gas
    Separation of air into nitrogen- and oxygen-enriched streams
    Recovery of helium
    Recovery of methane from biogas
8. Liquid membranes:
    Recovery of zinc from wastewater in the viscose fiber industry
    Recovery of nickel from electroplating solutions

followed in the 1960s. In 1979, Monsanto Chemical Company introduced a hollow-fiber membrane of polysulfone to separate certain gas mixtures—for example, to enrich hydrogen- and carbon-dioxide-containing streams [4]. Commercialization of alcohol dehydration by pervaporation began in the late 1980s, as did the large-scale application of emulsion liquid membranes for removal of metals and organics from wastewater. Also in the 1980s, the application of membrane separations to bioprocesses began to emerge, particularly ultrafiltration to separate proteins and microfiltration to separate bacteria and yeast. A recent industrial membrane catalog, Filmtec Inc., a subsidiary of the Dow Chemical Company, lists 76 membrane products.

Replacement of more-common separation operations with membrane separations has the potential to save energy and lower costs. It requires the production of high-mass-transfer-flux, defect-free, long-life membranes on a large scale and the fabrication of the membrane into compact, economical modules of high surface area per unit volume. It also requires considerable clean-up of process feeds and careful control of operating conditions to prevent membrane deterioration and to avoid degradation of membrane functionality due to caking, plugging, and fouling.

Industrial Example

A large-scale membrane process, currently uneconomical because its viability depends on the price of toluene compared to that of benzene, is the manufacture of benzene from toluene, which requires the separation of hydrogen from methane. After World War II, during which large amounts of toluene were required to produce TNT (trinitrotoluene) explosives, petroleum refiners sought other markets for toluene. One was the use of toluene for manufacturing benzene, xylenes, and a number of other chemicals, including polyesters. Toluene can be catalytically disproportionated to benzene and xylenes in an adiabatic reactor with the feed entering at 950°F at a pressure above 500 psia. The main reaction is

$$2C_7H_8 \rightarrow C_6H_6 + C_8H_{10} \text{ isomers}$$

To suppress coke formation, which fouls the catalyst, the reactor feed must contain a large fraction of hydrogen at a partial pressure of at least 215 psia. Unfortunately, the hydrogen takes part in a side reaction, the hydrodealkylation of toluene to benzene and methane:

$$C_7H_8 + H_2 \rightarrow C_6H_6 + CH_4$$

Makeup hydrogen is usually not pure, but contains perhaps 15 mol% methane and 5 mol% ethane. Thus, typically, the reactor effluent contains H_2, CH_4, C_2H_6, C_6H_6, unreacted C_7H_8, and C_8H_{10} isomers. As shown in Figure 14.2a for the reactor section of the process, this effluent is cooled and partially condensed to 100°F at a pressure of 465 psia. At these conditions, a good separation between C_2H_6 and C_6H_6 is achieved in the flash drum. The vapor leaving the flash contains most of the H_2, CH_4, and C_2H_6, with the aromatic

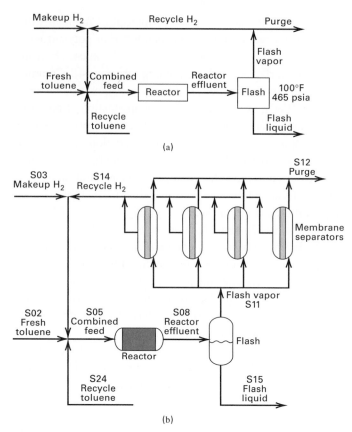

Figure 14.2 Reactor section of process to disproportionate toluene into benzene and xylene isomers. (a) Without a vapor-separation step. (b) With a membrane-separation step. Note: Heat exchangers, compressors, pump not shown.

chemicals leaving in the liquid. The large amount of hydrogen in the flash-drum vapor should be recycled to the reactor, rather than sending it to a flare or using it as a fuel. However, if all of the vapor were recycled, methane and ethane would build up in the recycle loop, since no other exit is provided. Before the development of acceptable membranes for the separation of H_2 from CH_4 by permeation, part of the vapor stream was purged from the process, as shown in Figure 14.2a, to provide an exit for CH_4 and C_2H_6. With the introduction of a suitable membrane in 1979, it became possible to install membrane separators, as shown in Figure 14.2b.

Table 14.2 is the material balance of Figure 14.2b for a plant processing 7,750 barrels (42 gal/bbl) per day of toluene feed. The permeation membranes separate the flash vapor

(stream S11) into an H_2-enriched permeate (S14, the recycled hydrogen), and a methane-enriched retentate (S12, the purge). The feed to the membrane system is 89.74 mol% H_2 and 9.26 mol% CH_4. No sweep gas is necessary. The permeate is enriched to 94.5 mol% H_2, and the retentate is 31.2 mol % CH_4. The recovery of H_2 in the permeate is 90%, leaving only 10% of the H_2 lost to the purge.

Before entering the membrane-separator system, the vapor is heated to at least 200°F (the dew-point temperature of the retentate) at a pressure of 450 psia (heater not shown). Because the hydrogen in the feed is reduced in passing through the separator, the retentate becomes more concentrated in the heavier components and, without the heater, undesirable condensation would occur. The retentate leaves the separator at about the same temperature and pressure as that of heated flash vapor. Permeate leaves at a pressure of 50 psia and a temperature lower than 200°F because of gas expansion.

The membrane is an aromatic polyamide polymer, 0.3-μm thick, with the nonporous layer in contact with the feed, and a much-thicker porous support backing to give the membrane strength to withstand the pressure differential of 450 − 50 = 400 psi. This large pressure difference is needed to force the hydrogen through the membrane, which is in the form of a spiral-wound module made from flat membrane sheets. The average flux of hydrogen through the membrane is 40 scfh (standard ft³/h at 60°F and 1 atm) per ft² of membrane surface area. From the material balance in Table 14.2, the H_2 transported through the membrane is

$$(1,685.1 \text{ lbmol/h})(379 \text{ scf/lbmol}) = 639,000 \text{ scfh}$$

The total membrane surface area required is 639,000/40 = 16,000 ft². The membrane is packaged in pressure-vessel modules of 4,000 ft² each. Thus, four modules in parallel are used. A disadvantage of the membrane process is the need to recompress the recycle hydrogen to the reactor inlet pressure.

Membrane separations are well developed for the applications listed in Table 14.1. Important progress is being made in developing new membrane applications, efficient membrane materials, and the modularization thereof. Applications covering wider ranges of temperature and types of membrane materials are being found. Membrane-separation processes have found wide application in the diverse industries listed in Table 14.1 and Table 1.2. Often, compared to other separation equipment, membrane separators are more compact, less capital intensive, and more easily operated, controlled, and maintained. However, membrane units are modular in

Table 14.2 Material Balance for Toluene Disproportionation Plant; Flow Rates in lbmol/h for Streams in Reactor Section of Figure 14.2b

Component	S02	S03	S24	S14	S05	S08	S15	S11	S12
Hydogen		269.0		1,685.1	1,954.1	1,890.6	18.3	1,872.3	187.2
Methane		50.5		98.8	149.3	212.8	19.7	193.1	94.3
Ethane		16.8			16.8	16.8	5.4	11.4	11.4
Benzene			13.1		13.1	576.6	571.8	4.8	4.8
Toluene	1,069.4		1,333.0		2,402.4	1,338.9	1,334.7	4.2	4.2
p-Xylene			8.0		8.0	508.0	507.4	0.6	0.6
Total	1,069.4	336.3	1,354.1	1,783.9	4,543.7	4,543.7	2,457.4	2,086.3	302.4

construction, with many parallel units required for large-scale applications, in contrast to common separation techniques, where larger pieces of equipment are designed as plant size increases.

A key to an efficient and economical membrane-separation process is the membrane and how it is packaged to withstand large pressure differences. Research and development of membrane processes deals mainly with the discovery of suitably thin, selective membrane materials and their fabrication.

This chapter discusses membrane materials and modules, the theory of transport through membranes, and the scale-up of separators from experimental data. Emphasis is on dialysis, electrodialysis, reverse osmosis, gas permeation, pervaporation, ultrafiltration, and microfiltration. Many of the theoretical principles apply as well to emerging but less-commercialized membrane processes such as membrane distillation, membrane gas absorption, membrane stripping, membrane solvent extraction, perstraction, and facilitated transport, which are not covered here. The status of industrial membrane systems and directions in research to improve existing applications and make possible new applications are considered in detail by Baker et al. [5] and by contributors to a handbook edited by Ho and Sirkar [6], which includes emerging processes. Baker [49] treats theory and technology.

§14.1 MEMBRANE MATERIALS

Originally, membranes were made from processed natural polymers such as cellulose and rubber, but now many are custom-made synthetically, a wide variety of them having been developed and commercialized since 1930. Synthetic polymers are produced by condensation reactions, or from monomers by free-radical or ionic-catalyzed addition (chain) reactions. The resulting polymer is categorized as having (1) a long linear chain, such as linear polyethylene; (2) a branched chain, such as polybutadiene; (3) a three-dimensional, highly cross-linked structure, such as a condensation polymer like phenol–formaldehyde; or (4) a moderately cross-linked structure, such as butyl rubber or a partially cross-linked polyethylene. The linear-chain polymers soften with an increase in temperature, are soluble in organic solvents, and are referred to as *thermoplastic* polymers. At the other extreme, highly cross-linked polymers decompose at high temperature, are not soluble in organic solvents, and are referred to as *thermosetting* polymers. Of more interest in the application of polymers to membranes is a classification based on the arrangement or conformation of the polymer molecules.

Polymers can be classified as *amorphous* or *crystalline*. The former refers to a polymer that is glassy in appearance and lacks crystalline structure, whereas the latter refers to a polymer that is opaque and has a crystalline structure. If the temperature of a glassy polymer is increased, a point called the *glass-transition temperature*, T_g, may be reached where

the polymer becomes *rubbery*. If the temperature of a crystalline polymer is increased, a point called the *melting temperature*, T_m, is reached where the polymer becomes a melt. However, a thermosetting polymer never melts. Most polymers have both amorphous and crystalline regions—that is, a certain degree of crystallinity that varies from 5 to 90%, making it possible for some polymers to have both a T_g and a T_m. Membranes made of glassy polymers can operate below or above T_g; membranes of crystalline polymers must operate below T_m.

Table 14.3 lists *repeat units* and values of T_g and/or T_m for some of the many natural and synthetic polymers from which membranes have been fabricated. Included are crystalline, glassy, and rubbery polymers. Cellulose triacetate is the reaction product of cellulose and acetic anhydride. The repeat unit of cellulose is identical to that shown for cellulose triacetate, except that the acetyl, Ac (CH_3CO), groups are replaced by H. The repeat units (*degree of polymerization*) in cellulose triacetate number ~300. Triacetate is highly crystalline, of uniformly high quality, and hydrophobic.

Polyisoprene (natural rubber) is obtained from at least 200 different plants, with many of the rubber-producing countries located in the Far East. Polyisoprene has a very low glass-transition temperature. Natural rubber has a degree of polymerization of from about 3,000 to 40,000 and is hard and rigid when cold, but soft, easily deformed, and sticky when hot. Depending on the temperature, it slowly crystallizes. To increase strength, elasticity, and stability of rubber, it is vulcanized with sulfur, a process that introduces cross-links.

Aromatic polyamides (also called aramids) are high-melting, crystalline polymers that have better long-term thermal stability and higher resistance to solvents than do aliphatic polyamides such as nylon. Some aromatic polyamides are easily fabricated into fibers, films, and sheets. The polyamide structure shown in Table 14.3 is that of Kevlar, a trade name of DuPont.

Polycarbonates, characterized by the presence of the –OCOO– group in the chain, are mainly amorphous. The polycarbonate shown in Table 14.3 is an aromatic form, but aliphatic forms also exist. Polycarbonates differ from most other amorphous polymers in that they possess ductility and toughness below T_g. Because polycarbonates are thermoplastic, they can be extruded into various shapes, including films and sheets.

Polyimides are characterized by the presence of aromatic rings and heterocyclic rings containing nitrogen and attached oxygen. The structure shown in Table 14.3 is only one of a number available. Polyimides are tough, amorphous polymers with high resistance to heat and excellent wear resistance. They can be fabricated into a wide variety of forms, including fibers, sheets, and films.

Polystyrene is a linear, amorphous, highly pure polymer of about 1,000 units of the structure shown in Table 14.3. Above a low T_g, which depends on molecular weight, polystyrene becomes a viscous liquid that is easily fabricated by extrusion or injection molding. Polystyrene can be annealed

Table 14.3 Common Polymers Used in Membranes

Polymer	Type	Representative Repeat Unit	Glass-Transition Temp., °C	Melting Temp., °C
Cellulose triacetate	Crystalline			300
Polyisoprene (natural rubber)	Rubbery	$\left[CH_2CH=CH_3 \atop CCH_2\right]_n$	−70	
Aromatic polyamide	Crystalline			275
Polycarbonate	Glassy		150	
Polyimide	Glassy		310–365	
Polystyrene	Glassy	$-CH_2CH-$	74–110	
Polysulfone	Glassy		190	
Polytetrafluoroethylene (Teflon)	Crystalline	$-CF_2-CF_2-$		327

(heated and then cooled slowly) to convert it to a crystalline polymer with a melting point of 240°C. Styrene monomer can be copolymerized with a number of other organic monomers, including acrylonitrile and butadiene to form ABS copolymers.

Polysulfones are synthetic polymers first introduced in 1966. The structure in Table 14.3 is just one of many, all of which contain the SO_2 group, which gives the polymers high strength. Polysulfones are easily spun into hollow fibers. Membranes of closely related polyethersulfone have also been commercialized.

Polytetrafluoroethylene is a straight-chain, highly crystalline polymer with a high degree of polymerization of the order of 100,000, giving it considerable strength. It possesses exceptional thermal stability and can be formed into films and tubing, as can polyvinylidenefluoride.

To be effective for separating a mixture of chemical components, a polymer membrane must possess high *permeance* and a high permeance ratio for the two species being separated by the membrane. The permeance for a given species diffusing through a membrane of given thickness is analogous to a mass-transfer coefficient, i.e., the flow rate of that species per unit cross-sectional area of membrane per unit driving force (concentration, partial pressure, etc.) across the membrane thickness. The molar transmembrane flux of species i is

$$N_i = \left(\frac{P_{M_i}}{l_M}\right)(\text{driving force}) = \bar{P}_{M_i}(\text{driving force})$$

(14-1)

where $\bar{P}_{M_i}$ is the permeance, which is defined as the ratio of P_{M_i}, the *permeability*, to l_M, the membrane thickness.

Polymer membranes can be characterized as dense or microporous. For dense, amorphous membranes, pores of microscopic dimensions may be present, but they are generally less than a few Å in diameter, such that most, if not all, diffusing species must dissolve into the polymer and then diffuse through the polymer between the segments of the macromolecular chains. Diffusion can be difficult, but highly selective, for glassy polymers. If the polymer is partly crystalline, diffusion will occur almost exclusively through the amorphous regions, with the crystalline regions decreasing the diffusion area and increasing the diffusion path.

Microporous membranes contain interconnected pores and are categorized by their use in microfiltration (MF), ultrafiltration (UF), and nanofiltration (NF). The MF membranes, which have pore sizes of 200–100,000 Å, are used primarily to filter bacteria and yeast and provide cell-free suspensions. UF membranes have pore sizes of 10–200 Å and are used to separate low-molecular-weight solutes such as enzymes from higher-molecular-weight solutes like viruses. NF membranes have pore sizes from 1 to 10 Å and can retain even smaller molecules. NF membranes are used in osmosis and pervaporation processes to purify liquids. The pores are formed by a variety of proprietary techniques, some of which are described by Baker et al. [5]. Such techniques are valuable for producing symmetric, microporous, crystalline membranes. Permeability for microporous membranes is high, but selectivity is low for small molecules. However, when there are molecules smaller and larger than the pore size, they may be separated almost perfectly by size.

The separation of small molecules presents a dilemma. A high permeability is not compatible with a high separation factor. The beginning of the resolution of this dilemma occurred in 1963 with the fabrication by Loeb and Sourirajan [7] of an asymmetric membrane of cellulose acetate by a novel casting procedure. As shown in Figure 14.3a, the resulting membrane consists of a thin dense skin about 0.1–1.0 μm in. thick, called the *permselective* layer, formed over a much thicker microporous layer that provides support for the skin.

The flux rate of a species is controlled by the permeance of the very thin permselective skin. From (14-1), the permeance of species i can be high because of the very small value of l_M even though the permeability, P_{M_i}, is low because of the absence of pores. When large differences of P_{M_i} exist among molecules, both high permeance and high selectivity can be achieved with asymmetric membranes.

A very thin, asymmetric membrane is subject to formation of minute holes in the permselective skin, which can render the membrane useless. A solution to the defect problem for an asymmetric polysulfone membrane was patented by Henis and Tripodi [8] of the Monsanto Company in 1980. They pulled silicone rubber, from a coating on the skin surface, into the defects by applying a vacuum. The resulting membrane, referred to as a *caulked membrane*, is shown in Figure 14.3b.

Wrasidlo [9] in 1977 introduced the thin-film composite membrane as an alternative to the asymmetric membrane. In the first application, shown in Figure 14.3c, a thin, dense film of polyamide polymer, 250 to 500 Å in thickness, was formed on a thicker microporous polysulfone support. Today, both asymmetric and thin-film composites are fabricated from polymers by a variety of techniques.

Application of polymer membranes is often limited to temperatures below 200°C and to mixtures that are chemically inert. Operation at high temperatures and with chemically active mixtures requires membranes made of inorganic materials. These include mainly microporous ceramics, metals, and carbon; and dense metals, such as palladium, that allow the selective diffusion of small molecules such as hydrogen and helium.

Examples of inorganic membranes are (1) asymmetric, microporous α-alumina tubes with 40–100 Å pores at the inside surface and 100,000 Å pores at the outside; (2) microporous glass tubes, the pores of which may be filled with other oxides or the polymerization–pyrolysis product of trichloromethylsilane; (3) silica hollow fibers with extremely fine pores of 3–5 Å; (4) porous ceramic, glass, or polymer materials coated with a thin, dense film of palladium metal that is just a few μm thick; (5) sintered metal; (6) pyrolyzed carbon; and (7) zirconia on sintered carbon. Extremely fine pores (<10 Å) are necessary to separate gas mixtures. Larger pores (>50 Å) are satisfactory for the separation of large molecules or solid particles from solutions containing small molecules.

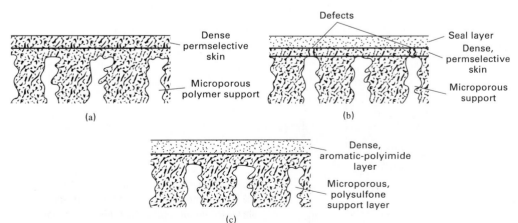

(a)

(b)

(c)

Figure 14.3 Polymer membranes: (a) asymmetric, (b) caulked asymmetric, and (c) typical thin-film composite.

EXAMPLE 14.1 Membrane Flux from Permeability.

A silica-glass membrane, 2-μm thick with pores <10 Å in diameter, has been developed for separating H_2 from CO at a temperature of 500°F. From laboratory data, the membrane permeabilities for H_2 and CO, respectively, are 200,000 and 700 barrer, where the barrer, a common unit for gas permeation, is defined by:

$$1 \text{ barrer} = 10^{-10} \text{ cm}^3 \text{ (STP)-cm/(cm}^2\text{-s-cmHg)}$$

where cm^3 (STP)/(cm^2-s) refers to the volumetric transmembrane flux of the diffusing species in terms of standard conditions of 0°C and 1 atm; cm refers to the membrane thickness; and cmHg refers to the transmembrane partial-pressure driving force for the diffusing species.

The barrer unit is named for R. M. Barrer, who published an early article [10] on diffusion in a membrane, followed by a widely referenced monograph on diffusion in and through solids [11].

If the transmembrane, partial-pressure driving forces for H_2 and CO, respectively, are 240 psi and 80 psi, calculate the transmembrane fluxes in kmol/m^2-s. Compare the H_2 flux to that for H_2 in the industrial application described at the beginning of this chapter.

Solution

At 0°C and 1 atm, 1 kmol of gas occupies 22.42×10^6 cm^3. Also, 2 μm thickness = 2×10^{-4} cm, and 1 cmHg $\Delta P = 0.1934$ psi. Therefore, using (14.1):

$$N_{H_2} = \frac{(200,000)(10^{-10})(240/0.1934)(10^4)}{(22.42 \times 10^6)(2 \times 10^{-4})} = 0.0554 \frac{\text{kmol}}{\text{m}^2\text{-s}}$$

$$N_{CO} = \frac{(700)(10^{-10})(80/0.1934)(10^4)}{(22.42 \times 10^6)(2 \times 10^{-4})} = 0.000065 \frac{\text{kmol}}{\text{m}^2\text{-s}}$$

In the application discussed at the beginning of this chapter, the flux of H_2 for the polymer membrane is

$$\frac{(1685.1)(1/2.205)}{(16,000)(0.3048)^2(3600)} = 0.000143 \frac{\text{kmol}}{\text{m}^2\text{-s}}$$

Thus, the flux of H_2 through the ultra-microporous-glass membrane is more than 100 times higher than the flux through the dense-polymer membrane. Large differences in molar fluxes through different membranes are common.

The following are useful factors for converting barrer to SI and American Engineering units:

Multiply barrer by 3.348×10^{-19} to obtain units of (kmol $\times$ m)/(m^2 $\times$ s $\times$ Pa).

Multiply barrer by 5.584×10^{-12} to obtain units of (lbmol $\times$ ft)/(ft^2 $\times$ h $\times$ psi).

§14.2 MEMBRANE MODULES

The asymmetric and thin-film, composite, polymer-membrane materials in the previous section are available in one or more of the three shapes shown in Figures 14.4a, b, and c. Flat sheets have typical dimensions of 1 m $\times$ 1 m $\times$ 200 μm thick, with a dense skin or thin, dense layer 500 to 5,000 Å in thickness. Tubular membranes are typically 0.5 to 5.0 cm in diameter and up to 6 m long. The thin, dense layer is on the inside, as seen in Figure 14.4b, or on the outside tube surface. The porous tube support is fiberglass, perforated metal, or other suitable material. Very small-diameter hollow fibers, first reported by Mahon [12, 13] in the 1960s, are typically 42 μm i.d. $\times$ 85 μm o.d. $\times$ 1.2 m long with a 0.1- to 1.0-μm-thick dense skin. The hollow fibers shown in Figure 14.4c provide a large membrane surface area per unit volume. A honeycomb, monolithic element for inorganic oxide membranes is included in Figure 14.4d. Elements of hexagonal and circular cross section are available [14]. The circular flow channels are 0.3 to 0.6 cm in diameter, with a 20- to

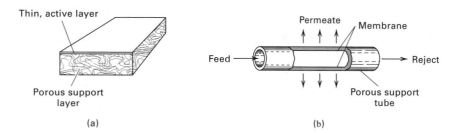

(a)

(b)

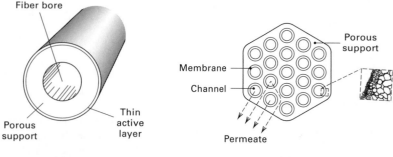

(c)

(d)

Figure 14.4 Common membrane shapes: (a) flat, asymmetric or thin-film composite sheet; (b) tubular; (c) hollow-fiber; (d) monolithic.

40-mm-thick membrane layer. The hexagonal element in Figure 14.4d has 19 channels and is 0.85 m long. Both the bulk support and the thin membrane layer are porous, but the pores of the latter can be as small as 40 Å.

The shapes in Figure 14.4 are incorporated into modules and cartridges, some of which are shown in Figure 14.5. Flat sheets used in plate-and-frame modules are circular, square, or rectangular in cross section. The sheets are separated by support plates that channel the permeate. In Figure 14.5a, a feed of brackish water flows across the surface of each sheet in the stack. Pure water is the permeate, while brine is the retentate.

Flat sheets are also fabricated into spiral-wound modules, as in Figure 14.5b. A laminate, consisting of two membrane sheets separated by spacers for the flow of the feed and permeate, is wound around a central, perforated collection tube

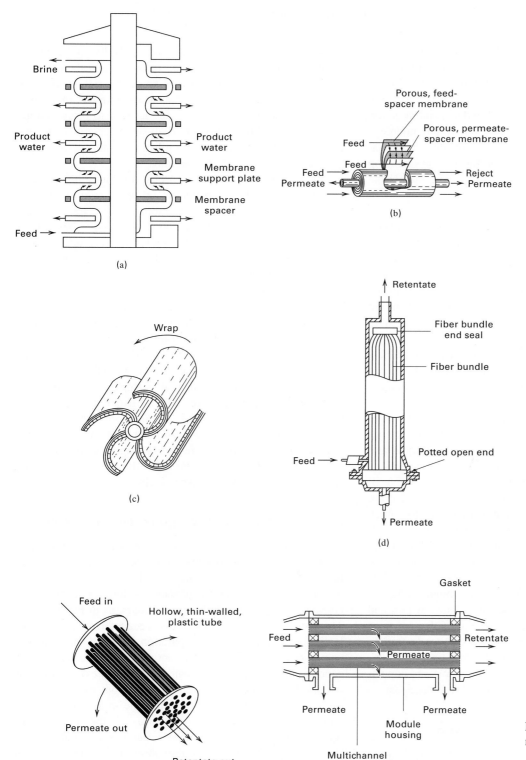

Figure 14.5 Common membrane modules: (a) plate-and-frame, (b) spiral-wound, (c) four-leaf spiral-wound, (d) hollow-fiber, (e) tubular, (f) monolithic.

Table 14.4 Typical Characteristics of Membrane Modules

	Plate-and-Frame	Spiral-Wound	Tubular	Hollow-Fiber
Packing density, m^2/m^3	30 to 500	200 to 800	30 to 200	500 to 9,000
Resistance to fouling	Good	Moderate	Very good	Poor
Ease of cleaning	Good	Fair	Excellent	Poor
Relative cost	High	Low	High	Low
Main applications	D, RO, PV, UF, MF	D, RO, GP, UF, MF	RO, UF	D, RO, GP, UF

Note: D, dialysis; RO, reverse osmosis; GP, gas permeation; PV, pervaporation; UF, ultrafiltration; MF, microfiltration.

to form a module that is inserted into a pressure vessel. Feed flows axially in the channels created between the membranes by porous spacers. Permeate passes through the membrane, traveling inward in a spiral path to the central collection tube. From there, the permeate flows in either axial direction through and out of the tube. A typical spiral-wound module is 0.1–0.3 m in diameter and 3 m long. The four-leaf modification in Figure 14.5c minimizes the permeate pressure drop because the permeate travel is less for the same membrane area.

The hollow-fiber module in Figure 14.5d, for a gas-permeation application, resembles a shell-and-tube heat exchanger. The pressurized feed enters the shell side at one end. While flowing over the fibers toward the other end, permeate passes through the fiber walls into the central fiber channels. Typically, the fibers are sealed at one end and embedded into a tube sheet with epoxy resin at the other end. A module might be 1 m long × 0.1 to 0.25 m in diameter, and contain more than 1 million hollow fibers.

The tubular module in Figure 14.5e also resembles a heat exchanger, but the feed flows through the tubes. Permeate passes through the tube wall into the shell side of the module. Tubular modules contain up to 30 tubes.

The monolithic module in Figure 14.5f contains from 1 to 37 elements in a housing. Feed flows through the circular channels, and permeate passes through the membrane and porous support and into the open region between elements.

Table 14.4 is a comparison of the characteristics of four of the modules shown in Figure 14.5. The packing density is the membrane surface area per unit volume of module, for which hollow-fiber membrane modules are clearly superior.

Although the plate-and-frame module has a high cost and a moderate packing density, it finds use in all membrane applications except gas permeation. It is the only module widely used for pervaporation. The spiral-wound module is very popular for most applications because of its low cost and reasonable resistance to fouling. Tubular modules are used only for low-flow applications or when resistance to fouling and/or ease of cleaning is essential. Hollow-fiber modules, with their very high packing density and low cost, are popular where fouling does not occur and cleaning is not necessary.

§14.3 TRANSPORT IN MEMBRANES

Calculation of membrane surface area for a new application must be based on laboratory data for the selected membrane system. Nevertheless, because both the driving force and the permeability (or permeance) depend markedly on the mechanism of transport, it is important to understand the nature of transport in membranes so that an appropriate membrane process is selected. This section deals with the theoretical aspects of the transport processes that lead to proper choice of membrane. Applications to dialysis, reverse osmosis, gas permeation, pervaporation, ultrafiltration, and microfiltration are presented in subsequent sections.

Membranes can be macroporous, microporous, or dense (nonporous). Only microporous or dense membranes are permselective. Macroporous membranes are used to support thin microporous and dense membranes when significant pressure differences across the membrane are necessary to achieve high flux. The theoretical basis for transport through microporous membranes is more highly developed than that for dense membranes, so porous-membrane transport is discussed first, with respect to bulk flow, liquid diffusion, and then gas diffusion. This is followed by nonporous (dense)-membrane solution-diffusion transport, first for liquid mixtures and second for gas mixtures. External mass-transfer resistances in the fluid films on either side of the membrane are treated where appropriate. It is important to note that, because of the range of pore sizes in membranes, the distinction between porous and nonporous membranes is not always obvious. The distinction can be made based only on the relative permeabilities for diffusion through the pores of the membrane and diffusion through the solid, amorphous regions of the membrane, respectively.

§14.3.1 Transport Through Porous Membranes

Mechanisms for transport of liquid and gas molecules through a porous membrane are depicted in Figures 14.6a, b, and c. If the pore diameter is large compared to the molecular diameter and a pressure difference exists, bulk, convective flow through the pores occurs, as in Figure 14.6a. This flow is generally undesirable because it is not permselective and, therefore, no separation between feed components occurs. If fugacity, activity, chemical-potential, concentration, or partial-pressure differences exist across the membrane for the various components, but the pressure is the same on both sides of the membrane so as not to cause a bulk flow, permselective diffusion of the components through the pores takes place, effecting a separation as shown in Figure 14.6b. If the pores are of the order of molecular size for at least some of the components in the feed mixture, the diffusion of those components will be

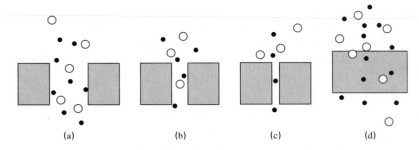

Figure 14.6 Mechanisms of transport in membranes. (Flow is downward.) (a) Bulk flow through pores; (b) diffusion through pores; (c) restricted diffusion through pores; (d) solution diffusion through dense membranes.

restricted (hindered) as shown in Figure 14.6c, resulting in an enhanced separation. Molecules of size larger than the pores will be prevented altogether from diffusing through the pores. This special case is highly desirable and is referred to as *size exclusion* or *sieving*. Another special case exists for gas diffusion in which the pore size and/or pressure (typically a vacuum) is such that the mean free path of the molecules is greater than the pore diameter, resulting in so-called *Knudsen diffusion*, which is dependent on molecular weight.

Bulk Flow

Bulk flow is pressure-driven flow of fluid through a semipermeable barrier (e.g., a membrane used for microfiltration and ultrafiltration), which results in accumulation of retained (sieved) solutes in suspension at the upstream face of the membrane, referred to as a cake. In 1855, Darcy found that the flow rate of water through sand is proportional to pressure drop. Darcy's law, which originally lacked a term for viscosity (as Wakeman and others have observed), has become the basis for describing bulk flow in applications including normal-flow (dead-end) filtration (DEF) modes like continuous rotary filtration using filter aids or sterile filtration; filtration through the cake where Reynolds numbers are <1 (laminar flow) and inertial effects are negligible; and local membrane flux in crossflow (tangential flow, TFF) modes like ultrafiltration. In Darcy's law, the instantaneous rate of filtration, $dV\{t\}/dt$, and the flux of permeate, J (dm³/m²-h or LMH), of viscosity μ through a medium of permeability k and cross-sectional area A_M are proportional to a constant-pressure (or vacuum) differential pressure drop dP through thickness dz:

$$J = \frac{1}{A_M}\frac{dV\{t\}}{dt} = -\frac{k}{\mu}\frac{dP}{dz} \qquad (14\text{-}2)$$

where the flux, J, corresponds to the superficial fluid velocity, $u = J$ (volume flow rate per unit filter area). Bulk flow through a series of media—each with characteristic resistance $R_i = \Delta z_i/k_i$, such as a filter cake (R_c) and filter medium (R_m), which can be as porous as a canvas cloth, is illustrated in Figure 14.7. The series resistances have a common superficial velocity given by

$$u = (P - P_i)/\mu R_c = (P_i - P_o)/\mu R_m \qquad (14\text{-}3)$$

For total pressure drop across cake and medium equal to $\Delta P = P - P_o$,

$$\boxed{J = \frac{1}{A_M}\frac{dV\{t\}}{dt} = \frac{\Delta P}{\mu(R_m + R_c)}} \qquad (14\text{-}4)$$

where bulk flow varies inversely with viscosity; with resistance due to filter medium, R_m; and with accumulated cake, R_c, which have dimensions of reciprocal length. Cake resistance often dominates filter resistance as filtration proceeds and cake accumulates, provided that the cake collects on something like a porous cloth.

Expressions for the resistance terms, R_i, in (14-4), which arise from various hydrodynamic and phenomenological descriptions specific to particular filter media and cakes, allow Darcy's law in (14-4) to be used to select process equipment, identify operating conditions, and troubleshoot filtration processes. Resistance varies inversely with hydraulic membrane permeability, L_p, given by

$$L_p = \frac{J}{\Delta P} \qquad (14\text{-}5)$$

which is a function of pore-size distribution, porosity, and thickness of barrier or cake, as well as solvent properties.

Pore Resistance to Flow

Consider bulk flow of a fluid due to a pressure difference through an idealized straight, cylindrical pore. If the flow is laminar ($N_{Re} = D\upsilon\rho/\mu < 2{,}100$), which is almost always true for small-diameter pores, flow velocity, υ, given by the Hagen–Poiseuille law [15], is directly proportional to the transmembrane pressure drop:

$$\upsilon = \frac{D^2}{32\mu L}(P_0 - P_L) \qquad (14\text{-}6)$$

where D is the pore diameter, large enough to pass all molecules; μ is the fluid viscosity; and L is the length of the pore. This assumes a parabolic velocity profile, a Newtonian fluid, and, if a gas, that the mean free path of the molecules is small compared to the pore diameter. If the membrane contains n such pores per unit cross section

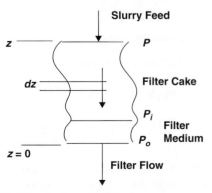

Figure 14.7 Profile of composite filtration.

of membrane surface area normal to flow, the porosity (void fraction) of the membrane is

$$\epsilon = n\pi D^2/4 \tag{14-7}$$

Then the superficial bulk-flow flux (mass velocity), N, through the membrane is

$$N = \upsilon\rho\epsilon = \frac{\epsilon\rho D^2}{32\mu l_M}(P_0 - P_L) = \frac{n\pi\rho D^4}{128\mu l_M}(P_0 - P_L) \tag{14-8}$$

where l_M is the membrane thickness and ρ and μ are fluid properties.

In real porous membranes, pores may not be cylindrical and straight, making it necessary to modify (14-8). Kozeny, in 1927, and Carman, in 1938, replaced cylindrical pores by a bundle of capillary tubes oriented at 45° to the surface. Ergun [16] extended this model by replacing, as a rough approximation, the pore diameter in (14-6) by the hydraulic diameter

$$d_H = 4\left(\frac{\text{Volume available for flow}}{\text{Total pore surface area}}\right)$$

$$= \frac{4\left(\dfrac{\text{Total pore volume}}{\text{Membrane volume}}\right)}{\left(\dfrac{\text{Total pore surface area}}{\text{Membrane volume}}\right)} = \frac{4\epsilon}{a} \tag{14-9}$$

where the membrane volume includes the volume of the pores. The specific surface area, a_υ, which is the total pore surface area per unit volume of just the membrane material (not including the pores), is

$$a_\upsilon = a/(1 - \epsilon) \tag{14-10}$$

Pore length is longer than the membrane thickness and can be represented by $l_M\tau$, where τ is a tortuosity factor >1. Substituting (14-9), (14-10), and the tortuosity factor into (14-8) gives

$$N = \frac{\rho\epsilon^2(P_0 - P_L)}{2(1 - \epsilon)^2\tau a_\upsilon^2\mu l_M} \tag{14-11}$$

In terms of a bulk-flow permeability, (14-11) becomes

$$N = \frac{P_M}{l_M}(P_0 - P_L) \tag{14-12}$$

where

$$P_M = \frac{\rho\epsilon^3}{2(1 - \epsilon)^2\tau a_\upsilon^2\mu} \tag{14-13}$$

Typically, τ is 2.5, whereas a_υ is inversely proportional to the average pore diameter, giving it a wide range of values.

Particulate Resistance to Flow

Equation (14-11) may be compared to the semitheoretical Ergun equation [16], which represents the best fit of data for flow of a fluid through a packed bed:

$$\frac{P_0 - P_L}{l_M} = \frac{150\mu\upsilon_0(1 - \epsilon)^2}{D_P^2\epsilon^3} + \frac{1.75\rho\upsilon_0^2(1 - \epsilon)}{D_P\epsilon^3} \tag{14-14}$$

where D_P is the mean particle diameter, υ_0 is the superficial fluid velocity through the bed, and υ_0/ϵ is the average velocity in the void space outside the particles. The first term on

the RHS of (14-14) applies to the laminar-flow region, and together with the LHS is the Kozeny–Carman equation. It may be included in Darcy's law as shown in (14-17) to estimate cake resistance. The second term applies to the turbulent region. For a spherical particle, the specific surface area is

$$a_\upsilon = \pi D_P^2/\pi\left(\pi D_P^3/6\right) \quad \text{or} \\ D_P = 6/a_\upsilon \tag{14-15}$$

Substitution of (14-15) into (14-14) for laminar flow, and rearrangement into the bulk-flow flux form gives

$$N = \frac{\rho\epsilon^3(P_0 - P_L)}{(150/36)(1 - \epsilon)^2 a_\upsilon^2\mu l_M} \tag{14-16}$$

Comparing (14-16) to (14-11), it is seen that the term $(150/36)$ in (14-16) corresponds to the term 2τ in (14-11), giving $\tau = 2.08$, which seems reasonable. Accordingly, (14-16) can be used as a first approximation to the pressure drop for flow through a porous membrane when the pores are not straight cylinders.

For gas flow, the density is taken as the average of the densities at the two membrane faces. Bulk transport through porous membranes in constant-pressure and constant-flux operation is considered in detail below. In cake filtration, described in Chapter 19, porosity has a slightly different meaning, usually defined being as the volume fraction of the cake occupied by occluded liquid.

EXAMPLE 14.2 Pressure Drop Through a Membrane.

It is desired to pass water at 70°F through a supported polypropylene membrane, with a skin of 0.003-cm thickness and 35% porosity, at the rate of 200 m³/m² membrane surface area/day. The pores can be considered straight cylinders of uniform diameter equal to 0.2 μm. If the pressure on the downstream side of the membrane is 150 kPa, estimate the required pressure on the upstream side of the membrane. The pressure drop through the support is negligible.

Solution

Equation (8) applies, where in SI units:

$N/\rho = 200/(24)(3600) = 0.00232$ m³/m²-s,
$\epsilon = 0.35, D_P = 0.2 \times 10^{-6}$ m, $l_M = 0.00003$ m, $P_L = 150$ kPa $= 150,000$ Pa, $\mu = 0.001$ Pa-s

From a rearrangement of (8),

$$P_0 = P_L + \frac{32\mu l_M(N/\rho)}{\epsilon D_P^2}$$

$$= 150,000 + \frac{(32)(0.001)(0.00003)(0.00232)}{(0.35)(0.2 \times 10^{-6})^2}$$

$$= 309,000 \text{ Pa} \quad \text{or} \quad 309 \text{ kPa}$$

Cake Resistance to Flow

Resistance from a cake, R_c, of incompressible, close-packed, colloidal $(1–10^3$ nm) particulates at low Reynolds numbers (laminar flow) may be described using the first term on

RHS of the Ergun equation (14-14), which is the Kozeny–Carman term:

$$R_c = \frac{150 l_c (1 - \epsilon_c)^2}{D_P^2 \epsilon_c^3} = K_1 \frac{l_c (1 - \epsilon_c)^2}{\epsilon_c^3} \qquad (14\text{-}17)$$

where l_c = cake thickness, which increases with time; ϵ_c = cake porosity; D_P = effective diameter of the matter in the cake; and K_1 = constant for a particular filtration system.

The solid matter in the feed, m_c, retained in the cake as a function of filtration time, t, is

$$m_c\{t\} = c_F V\{t\} = \rho_c (1 - \epsilon_c) A_c l_c\{t\} \qquad (14\text{-}18)$$

where c_F = solid matter per unit volume of liquid in the feed, and A_c = area of the cake when it is very thin relative to the membrane radius of curvature. Equation (14-18) is often rewritten in terms of the mass of dry filter cake per unit filter area, $W = m_c\{t\}/A_c$, and thus the differential change in mass, dW, is related to the differential change in cake thickness, dz, by

$$dW = \rho_c (1 - \epsilon_c) dz \qquad (14\text{-}19)$$

where ϵ_c is the fraction of voids in the filter cake and, thus, is a measure of the volume of flow paths through the medium.

Substituting (14-18) into (14-17) to eliminate l_c shows that cake resistance increases during filtration in proportion to the volume filtered,

$$R_c\{t\} = K_1 \frac{(1 - \epsilon_c)}{\epsilon_c^3} \frac{c_F V\{t\}}{\rho_c A_c} = \alpha \frac{c_F V\{t\}}{A_c} \qquad (14\text{-}20)$$

where α is the average specific resistance of the cake, which can be measured experimentally. Unfortunately, the complex structure of a compressible filter cake and its dependence on pressure, preclude direct calculation of ϵ_c or α from scanning electron micrographs or other means.

As filtration progresses, cake thickness, filtrate volume, and resistance to flow pressure drop increase, but R_m is assumed to remain constant. W, the weight of the dry cake, is related to V and c_F by $W = c_F V$. However, care must be

exercised when applying this formula because the cake is wet, and then dried, so c_F will have different values depending on whether wet or dry cake masses are used.

It is useful to replace R_c by α, which replaces k in the Darcy equation (14-4), where, now, $R_c = \alpha W / A_c = \alpha c_F (V / A_c)$. Thus, if length is in ft, R_c has units of ft^{-1} and α has units of ft/lb$_m$ because ΔP must be multiplied by g_c for dimensional consistency if AE units are used [15, 19]. For SI units, α has dimensions of m/kg. Substituting the definition of α into (14-4) gives an alternative Darcy equation, where $A_c = A_M$:

$$\frac{d(V / A_c)}{dt} = \frac{\Delta P}{\mu [R_m + \alpha c_F (V / A_c)]} \qquad (14\text{-}21)$$

For membranes like capillary or hollow fibers with small inner and outer radii, r_i and r_o, respectively, in which cake thickness is not negligible relative to radius of curvature, the resistance is

$$R_c\{t\} = \alpha r_j \ln \left\{ \left[(r_j + (-1)^a \delta_c\{t\}) / r_j \right]^{(-1)^a} \right\} \qquad (14\text{-}22)$$

where cake thickness is $\delta_c\{t\} = c_F V\{t\}/\rho_c A_c$, with ρ_c being the cake density. For operation in which permeate exits the cylinder, subscript $j = i$ and superscript $a = 1$, while for permeate entering the cylinder, subscript $j = 0$ and $a = 0$.

Summary of Resistance Models

Other resistances neglected in Darcy's law (14-21) arise from matter blocking or constricting the pores and from osmotic pressure. These resistances can be negligible in microfiltration of high-molecular-weight solutes. Table 14.5 from [69] summarizes the four general mechanisms for pore blockage, intermediate pore blockage, pore constriction, and cake formation. Expressions are given for each mechanism for two common regimes of operations: decline of filtrate flow rate, Q, at constant trans-membrane pressure drop, ΔP; and change in pressure drop at constant filtrate rate. Linearized forms of each

Table 14.5   Models for Corresponding Flux and Pressure [69]

Constant Pressure	Flux	Linearized Form
Pore blockage	$\dfrac{Q}{Q_o} = \exp(-\beta t)$	$\ln(Q) = at + b$
Intermediate blockage	$\dfrac{Q}{Q_o} = (1 + \beta t)^{-1}$	$\dfrac{1}{Q} = at + b$
Pore constriction	$\dfrac{Q}{Q_o} = (1 + \beta t)^{-2}$	$\dfrac{t}{V} = at + b$
Cake formation	$\dfrac{Q}{Q_o} = (1 + \beta t)^{-1/2}$	$\dfrac{t}{V} = aV + b$

Constant Flux	Pressure	Linearized Form
Pore blockage	$\dfrac{P}{P_o} = (1 - \beta t)^{-1/2}$	$\dfrac{1}{P^2} = a - bV$
Intermediate blockage	$\dfrac{P}{P_o} = (1 - \beta t)^{-1}$	$\dfrac{1}{P} = a - bV$
Pore constriction	$\dfrac{P}{P_o} = (1 - \beta t)^{-2}$	$\dfrac{1}{P^{1/2}} = a - bV$
Cake formation	$\dfrac{P}{P_o} = 1 + \beta t$	$P = a + bV$

expression (usually in terms of cumulative filtrate volume, V, at filtration time, t) allow model identification, parameter estimation, and data analysis. As described above, fouling due to *cake formation* occurs as foulant deposits on the external membrane surface, producing an additional resistance in series with the membrane resistance. The remaining models assume that parallel arrays of uniform, right-cylindrical pores permeate a membrane barrier. *Pore constriction* occurs as foulant deposits on internal pore surfaces, reducing the effective pore size. Standard and intermediate *pore blockage* occurs when foulant occludes flow through individual pores. Intermediate pore blockage allows for superposition of foulant on the external membrane surface.

A combined pore-blockage, cake-filtration model describes experimental data for fouling by protein aggregates, which exhibits an initial flux decline attributable to pore occlusion leaving the membrane partially permeable to flow, followed by cake formation [70]. It has been extended to account for asymmetric membrane structures [71] and pore interconnectivity [72], which significantly reduce flux-decline rates relative to parallel pore arrays.

Consider application of the cake-formation model in (14-21) and its two forms in Table 14.5 to two regimes of filtration: constant-pressure, and constant-flow rate.

Constant-Pressure Filtration

Operating with constant-pressure drop produces a permeate flow rate and flux that decrease with time as cake thickness increases.

Assuming the filtration area is constant, (14-21) can be integrated from $V = 0$ to $V = V$ for $t = 0$ to $t = t$ to obtain the time-dependent permeate volume $V\{t\}$ and flux $J\{t\}$:

$$\int_0^V \left(R_m + \alpha \frac{c_F V}{A_c} \right) dV = \frac{A_c \Delta P}{\mu} \int_0^t dt \qquad (14\text{-}23)$$

This yields the constant-pressure-drop form of the Ruth equation for sieve (i.e., surface) retention developed in 1933 [20, 68]:

$$V^2\{t\} + 2V\{t\} \frac{R_m A_c}{\alpha c_F} = 2 \frac{A_c^2 \Delta P}{\alpha c_F \mu} t \qquad (14\text{-}24)$$

or

$$V^2\{t\} + 2V\{t\}V_o = Kt \qquad (14\text{-}25)$$

where $V_o = R_m A_c / \alpha c_F$ and $K = 2A_c^2 \Delta P / \alpha c_F \mu$

In Exercise 14.27, the quadratic Ruth equation (14-25) is solved to obtain the positive root for $V\{t\}$. That result is then differentiated with respect to time to yield the permeate flux, $J\{t\}$. The following straight-line form of the Ruth equation shows that filtration time per unit volume, $t/V\{t\}$, increases with filtrate volume in proportion to c_F and α for a given filtration area at constant-pressure drop:

$$\frac{t}{V\{t\}} = \frac{1}{K}(V\{t\} + 2V_o) \qquad (14\text{-}26)$$

Fitting (14-26) to experimental data for $t/V\{t\}$ versus $V\{t\}$ yields slope K^{-1} (to estimate α and determine cake resistance) and intercept $2V_o/K$ (to estimate R_m).

Constant-Flux Cake Filtration

In a plate-and-frame filter or a pressure leaf filter, where centrifugal pumps are used, the early stages of filtration are frequently at a reasonably constant rate; then, as the cake builds up, the ability of the pump to develop pressure becomes the limiting factor and the process continues at a constant pressure and a falling rate. Operating so that permeate flux, J, remains constant corresponds to a constant permeate flow rate, $dV/dt = V/t$, in (14-4) with $V = 0$ at $t = 0$. After the early filtration stages, this requires increasing the pressure differential across the membrane as cake thickness increases. For constant dV/dt, (14-4) becomes

$$J = \frac{V/A_c}{t} = \frac{\Delta P\{t\}}{\mu[R_m + \alpha c_F(V/A_c)]} \qquad (14\text{-}27)$$

The expected value for $\Delta P\{t\}$ is obtained by rearranging (14-27) using an initial pressure drop of $\Delta P_o = J\mu R_m$ to yield

$$\frac{\Delta P\{t\}}{\Delta P_o} = 1 + \alpha \frac{c_F}{R_m} t \qquad (14\text{-}28)$$

The ΔP increase is linear with time, in proportion to specific cake resistance and solid content in the feed. Rewriting (14-28) after substituting $u = V/t$, the superficial velocity of the filtrate through the cake, yields a similar linear equation for the variation of the ΔP with time:

$$\Delta P = au^2 t + bu \qquad (14\text{-}29)$$

where

$$a = \alpha c_F \mu / A_c^2 \qquad (14\text{-}30)$$

$$b = R_m \mu / A_c \qquad (14\text{-}31)$$

Since u must be constant for a constant rate of filtration, (14-29) defines a straight line on a plot of ΔP versus t, as shown below in Example 14.3.

Combined Operation

Yield is improved in a combined operation in which: (1) constant-flux operation is employed in Stage 1 up to a limiting pressure drop, followed by (2) constant-pressure operation in Stage 2 until a minimum flux is reached. Let V_{CF} = volume of permeate obtained during Stage 1 for time period t_{CF}. For constant-permeate flux, with A_c replaced by A_M,

$$V_{CF} = JA_M t_{CF} \qquad (14\text{-}32)$$

Let ΔP_{UL} = the upper limit of the pressure drop across the cake and membrane. Substituting ΔP_{UL} for ΔP into (14-23), using V_{CF} and t_{CF} for the lower limits of volumetric and time integration, respectively; and letting $K_2 = \alpha$, yields a quadratic equation for V. Solving the resulting quadratic equation for the positive root of V,

$$V = -\frac{R_m A_M}{K_2 c_F} + \left[A_M^2 R_m^2 + \frac{2A_M}{K_2 c_F} \right.$$
$$\left. \times \left(R_m V_{CF} + \frac{K_2 c_F V_{CF}^2}{2A_M} + \frac{A_M \Delta P_{UL}(t - t_{CF})}{\mu} \right) \right]^{\frac{1}{2}}$$

$$(14\text{-}33)$$

where $V - V_{CF}$ = volume of permeate obtained during the Stage 2 time period $t - t_{CF}$. During Stage 2, the permeate flux decreases with time, according to an equation obtained by applying the definition of the permeate flux from (14-2) to (14-33).

$$J = \frac{A_M \Delta P_{UL}}{K_2 c_F \mu} \left[\frac{A_M^2 R_m^2}{K_2^2 c_F^2} + \frac{2A_M}{K_2 c_F} \right.$$
$$\left. \times \left(R_m V_{CF} + \frac{K_2 c_F V_{CF}^2}{2A_M} + \frac{A_M \Delta P_{UL}(t - t_{CF})}{\mu} \right) \right]^{-\frac{1}{2}}$$

(14-34)

Application of the above equations for a combined operation is illustrated in the following microfiltration example.

EXAMPLE 14.3 Microfiltration of Skim Milk.

Diluted skim milk with a protein concentration of 4.3 g/L is to undergo DEF microfiltration. Experiments have been performed using a cellulose-acetate membrane with an average pore diameter of 0.45 μm and $A_M = 17.3$ cm^2. For Stage 1 operation at a constant-permeate rate of 15 mL/minute, pressure drop across the cake and membrane increases from 0.3 psi to an upper limit of 20 psi in 400 seconds. The permeate viscosity is 1 cP. If continued in a second stage at constant ΔP at the upper limit until the permeate rate drops to 5 mL/minute, estimate the additional time of operation. Also, prepare plots of permeate volume in mL and permeate flux in mL/cm^2-minute as functions of time.

Solution

In SI units, $c_F = 4.3$ kg/m^3, $A_M = 0.00173$ m^2, volumetric flow rate in Stage 1 $= 0.25 \times 10^{-6}$ m^3/s, ΔP at zero time $= 2,068$ Pa, ΔP at 400 s $= 137,900$ Pa, and $\mu = 0.001$ Pa-s. Based on the experimental results for Stage 1, calculate values of R_m and $K_2 = \alpha$ as follows. From (14-27),

$$R_m = \frac{(\Delta P \text{ at } t = 0)}{J\mu} = \frac{2068}{\left(\frac{0.25 \times 10^{-6}}{0.00173} \right)(0.001)}$$

$$= 1.43 \times 10^{10} \text{ m}^{-1}$$

For $t = 400$ s,

$$K_2 = \frac{\Delta P - J\mu R_m}{J^2 \mu c_F t} = \frac{\Delta P - (\Delta P \text{ at } t = 0)}{J^2 \mu c_F t}$$

$$= \frac{6895(20 - 0.3)}{\left(\frac{0.25 \times 10^{-6}}{0.00173} \right)^2 (0.001)(4.3)(400)} = 3.78 \times 10^{12} \text{ m/kg}$$

Assume that these values of R_m and K_2 are valid during Stage 2. At the end of Stage 2, the permeate flux $= J = (5$ mL/minute)/(17.3 cm^2) $= 0.289$ mL/minute-cm$^2 = 4.82 \times 10^{-5}$ m/s.

Solving (14-34) for t gives 2,025 s. Plots of permeate volume and flux versus time, in Figures 14.8 and 14.9, respectively, are obtained by solving, in time, (14-33) and (14-34) with a spreadsheet.

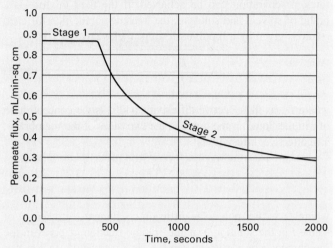

Figure 14.9 Instantaneous permeate flux for Example 14.3.

Compressibility

Filtration of cells or flocculated clays produces cakes that exhibit decreased void volume and increased specific cake resistance, α, as filtration proceeds. This indicates cake compressibility, which may be expressed by [66, 67]

$$\alpha = \alpha'(\Delta P)^s$$

(14-35)

where s is an empirical compressibility factor that ranges from zero for incompressible filter-aid cakes to near unity for highly compressible cakes, and α' is an empirical constant related to the size and shape of cake-forming particles.

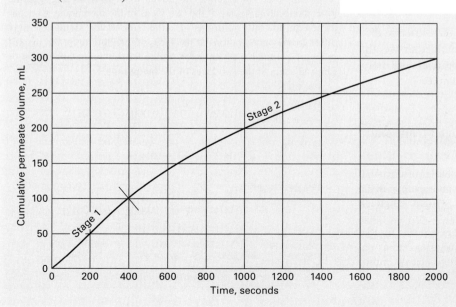

Figure 14.8 Cumulative permeate volume for Example 14.3.

§14.3.2 Liquid Diffusion Through Pores

Consider diffusion through the pores of a membrane from a fluid feed to a sweep fluid when identical total pressures but different component concentrations exist on both sides of the membrane. Bulk flow through the membrane due to a pressure difference does not occur, but if species diffuse at different rates, a separation can be achieved. If the feed mixture is a liquid of solvent and solutes i, the transmembrane flux for each solute is given by a modified form of Fick's law (§3.1.1):

$$N_i = \frac{D_{e_i}}{l_M}(c_{i_0} - c_{i_L}) \qquad (14\text{-}36)$$

where D_{e_i} is the effective diffusivity, and c_i is the concentration of i in the liquid in the pores at the two faces of the membrane. The effective diffusivity is

$$D_{e_i} = \frac{\epsilon D_i}{\tau} K_{r_i} \qquad (14\text{-}37)$$

where D_i is the molecular diffusion coefficient (diffusivity) of solute i in the solution, ϵ is the volume fraction of pores in the membrane, τ is tortuosity, and K_r is a restrictive factor that accounts for pore diameter, d_p, causing interfering collisions of the diffusing solutes with the pore wall, when the ratio of molecular diameter, d_m, to pore diameter exceeds about 0.01. The restrictive factor, according to Beck and Schultz [17], is:

$$K_r = \left[1 - \frac{d_m}{d_p}\right]^4, \quad (d_m/d_p) \leq 1 \qquad (14\text{-}38)$$

From (14-38), when $(d_m/d_p) = 0.01$, $K_r = 0.96$, but when $(d_m/d_p) = 0.3$, $K_r = 0.24$. When $d_m > d_p$, $K_r = 0$, and the solute cannot diffuse through the pore. This is the sieving or size-exclusion effect illustrated in Figure 14.6c. As illustrated in the next example, transmembrane fluxes for liquids through microporous membranes are very small because effective diffusivities are low.

For solute molecules not subject to size exclusion, a useful selectivity ratio is defined as

$$S_{ij} = \frac{D_i K_{r_i}}{D_j K_{r_j}} \qquad (14\text{-}39)$$

This ratio is greatly enhanced by the effect of restrictive diffusion when the solutes differ widely in molecular weight and one or more molecular diameters approach the pore diameter. This is shown in the following example.

EXAMPLE 14.4 Solute Diffusion Through Membrane Pores.

Beck and Schultz [18] measured effective diffusivities of urea and different sugars, in aqueous solutions, through microporous mica membranes especially prepared to give almost straight, elliptical pores of almost uniform size. Based on the following data for a membrane and two solutes, estimate transmembrane fluxes for the two solutes in g/cm²-s at 25°C. Assume the aqueous solutions on either side of the membrane are sufficiently dilute that no multi-component diffusional effects are present.

Membrane:

Material	Microporous mica
Thickness, μm	4.24
Average pore diameter, Angstroms	88.8
Tortuosity, τ	1.1
Porosity, ϵ	0.0233

Solutes (in aqueous solution at 25°C):

Solute	MW	$D_i \times 10^6$ cm²/s	molecular diameter, d_m, Å	c_{i_0} (g/cm³)	c_{i_L} (g/cm³)
1 Urea	60	13.8	5.28	0.0005	0.0001
2 β-Dextrin	1135	3.22	17.96	0.0003	0.00001

Solution

Calculate the restrictive factor and effective diffusivity from (14-38) and (14-37), respectively. For urea (1):

$$K_{r_1} = \left[1 - \left(\frac{5.28}{88.8}\right)\right]^4 = 0.783$$

$$D_{e_1} = \frac{(0.0233)(13.8 \times 10^{-6})(0.783)}{1.1} = 2.29 \times 10^{-7} \text{ cm}^2/\text{s}$$

For β-dextrin (2):

$$K_{r_2} = \left[1 - \left(\frac{17.96}{88.8}\right)\right]^4 = 0.405$$

$$D_{e_2} = \frac{(0.0233)(3.22 \times 10^{-6})(0.405)}{1.1} = 2.78 \times 10^{-8} \text{ cm}^2/\text{s}$$

Because of differences in molecular size, effective diffusivities differ by an order of magnitude. From (14-39), selectivity is

$$S_{1,2} = \frac{(13.8 \times 10^{-6})(0.783)}{(3.22 \times 10^{-6})(0.405)} = 8.3$$

Next, calculate transmembrane fluxes from (36), noting that the given concentrations are at the two faces of the membrane. Concentrations in the bulk solutions on either side of the membrane may differ from concentrations at the faces, depending upon the magnitudes of external mass-transfer resistances in boundary layers or films adjacent to the two faces of the membrane.

For urea:

$$N_1 = \frac{(2.29 \times 10^{-7})(0.0005 - 0.0001)}{4.24 \times 10^{-4}}$$

$$= 2.16 \times 10^{-7} \text{ g/cm}^2\text{-s}$$

For β-dextrin:

$$N_2 = \frac{(2.768 \times 10^{-8})(0.0003 - 0.00001)}{(4.24 \times 10^{-4})}$$

$$= 1.90 \times 10^{-8} \text{ g/cm}^2\text{-s}$$

Note that these fluxes are extremely low.

§14.3.3 Gas Diffusion Through Porous Membranes

When the mixture on either side of a microporous membrane is a gas, rates of diffusion can be expressed in terms of Fick's law (§3.1.1). If pressure and temperature on either side of the membrane are equal and the ideal-gas law holds, (14-36) in terms of a partial-pressure driving force is:

$$N_i = \frac{D_{e_i} c_M}{P l_M}\left(p_{i_0} - p_{i_L}\right) = \frac{D_{e_i}}{RT l_M}\left(p_{i_0} - p_{i_L}\right) \quad (14\text{-}40)$$

where c_M is the total gas-mixture concentration given as P/RT by the ideal-gas law.

For a gas, diffusion through a pore occurs by ordinary diffusion, as with a liquid, and/or in series with Knudsen diffusion when pore diameter is very small and/or total pressure is low. In the Knudsen-flow regime, collisions occur primarily between gas molecules and the pore wall, rather than between gas molecules. In the absence of a bulk-flow effect or restrictive diffusion, (14-14) is modified to account for both mechanisms of diffusion:

$$D_{e_i} = \frac{\epsilon}{\tau}\left[\frac{1}{(1/D_i) + (1/D_{K_i})}\right] \quad (14\text{-}41)$$

where D_{K_i} is the Knudsen diffusivity, which from the kinetic theory of gases as applied to a straight, cylindrical pore of diameter d_p is

$$D_{K_i} = \frac{d_p \bar{v}_i}{3} \quad (14\text{-}42)$$

where $\bar{v}_i$ is the average molecule velocity given by

$$\bar{v}_i = (8RT/\pi M_i)^{1/2} \quad (14\text{-}43)$$

where M is molecular weight. Combining (14-42) and (14-43):

$$D_{K_i} = 4{,}850 d_p (T/M_i)^{1/2} \quad (14\text{-}44)$$

where D_K is cm^2/s, d_p is cm, and T is K. When Knudsen flow predominates, as it often does for micropores, a selectivity based on the permeability ratio for species A and B is given from a combination of (14-1), (14-40), (14-41), and (14-44):

$$\frac{P_{M_A}}{P_{M_B}} = \left(\frac{M_B}{M_A}\right)^{1/2} \quad (14\text{-}45)$$

Except for gaseous species of widely differing molecular weights, the permeability ratio from (14-45) is not large, and the separation of gases by microporous membranes at low to moderate pressures that are equal on both sides of the membrane to minimize bulk flow is almost always impractical, as illustrated in the following example. However, the separation of the two isotopes of UF$_6$ by the U.S. government was accomplished by Knudsen diffusion, with a permeability ratio of only 1.0043, at Oak Ridge, Tennessee, using thousands of stages and acres of membrane surface.

EXAMPLE 14.5 Knudsen Diffusion.

A gas mixture of hydrogen (H) and ethane (E) is to be partially separated with a composite membrane having a 1-μm-thick porous skin with an average pore size of 20 Å and a porosity of 30%. Assume $\tau = 1.5$. The pressure on either side of the membrane is 10 atm and the temperature is 100°C. Estimate permeabilities of the components in barrer.

Solution

From (14-1), (14-40), and (14-41), the permeability can be expressed in mol-cm/cm^2-s-atm:

$$P_{M_i} = \frac{\epsilon}{RT\tau}\left[\frac{1}{(1/D_i) + (1/D_{K_i})}\right]$$

where $\epsilon = 0.30$, $R = 82.06$ cm^3-atm/mol-K, $T = 373$ K, and $\tau = 1.5$.

At 100°C, the ordinary diffusivity is given by $D_H = D_E = D_{H,E} = 0.86/P$ in cm^2/s with total pressure P in atm. Thus, at 10 atm, $D_H = D_E = 0.086$ cm^2/s. Knudsen diffusivities are given by (44), with $d_p = 20 \times 10^{-8}$ cm.

$$D_{K_H} = 4{,}850(20 \times 10^{-8})(373/2.016)^{1/2} = 0.0132 \text{ cm}^2/\text{s}$$

$$D_{K_E} = 4{,}850(20 \times 10^{-8})(373/30.07)^{1/2} = 0.00342 \text{ cm}^2/\text{s}$$

For both components, diffusion is controlled mainly by Knudsen diffusion.

For hydrogen: $\frac{1}{(1/D_H)+(1/D_{K_H})} = 0.0114$ cm^2/s.

For ethane: $\frac{1}{(1/D_E)+(1/D_{K_E})} = 0.00329$ cm^2/s.

$$P_{M_H} = \frac{0.30(0.0114)}{(82.06)(373)(1.5)} = 7.45 \times 10^{-8} \frac{\text{mol-cm}}{\text{cm}^2\text{-s-atm}}$$

$$P_{M_E} = \frac{0.30(0.00329)}{(82.06)(373)(1.5)} = 2.15 \times 10^{-8} \frac{\text{mol-cm}}{\text{cm}^2\text{-s-atm}}$$

To convert to barrer as defined in Example 14.1, note that

$$76 \text{ cmHg} = 1 \text{ atm and } 22{,}400 \text{ cm}^3 \text{ (STP)} = 1 \text{ mol}$$

$$P_{M_H} = \frac{7.45 \times 10^{-8}(22{,}400)}{(10^{-10})(76)} = 220{,}000 \text{ barrer}$$

$$P_{M_E} = \frac{2.15 \times 10^{-8}(22{,}400)}{(10^{-10})(76)} = 63{,}400 \text{ barrer}$$

§14.3.4 Transport Through Nonporous Membranes

Transport through nonporous (dense) solid membranes is the predominant mechanism of membrane separators for reverse osmosis, gas permeation, and pervaporation (liquid and vapor). As indicated in Figure 14.6d, gas or liquid species absorb at the upstream face of the membrane, diffuse through the membrane, and desorb at the downstream face.

Liquid diffusivities are several orders of magnitude less than gas diffusivities, and diffusivities of solutes in solids are a few orders of magnitude less than diffusivities in liquids. Thus, differences between diffusivities in gases and solids are enormous. For example, at 1 atm and 25°C, diffusivities in cm^2/s for water are as follows:

Water vapor in air	0.25
Water in ethanol liquid	1.2×10^{-5}
Dissolved water in cellulose-acetate solid	1×10^{-8}

As might be expected, small molecules fare better than large molecules for diffusivities in solids. From the *Polymer Handbook* [19], diffusivities in cm^2/s for several species in low-density polyethylene at 25°C are

Helium	6.8×10^{-6}
Hydrogen	0.474×10^{-6}
Nitrogen	0.320×10^{-6}
Propane	0.0322×10^{-6}

Regardless of whether a nonporous membrane is used to separate a gas or a liquid mixture, the *solution-diffusion model* of Lonsdale, Merten, and Riley [20] is used with experimental permeability data to design nonporous membrane separators. This model is based on Fick's law for diffusion through solid, nonporous membranes based on the driving force, $c_{i_0} - c_{i_L}$ shown in Figure 14.10b, where concentrations are those for solute dissolved in the membrane.

Concentrations in the membrane are related to the concentrations or partial pressures in the fluid adjacent to the membrane faces by assuming thermodynamic equilibrium for the solute at the fluid–membrane interfaces. This assumption has been validated by Motanedian et al. [21] for permeation of light gases through dense cellulose acetate at up to 90 atm.

Solution-Diffusion for Liquid Mixtures

Figures 14.10a and b show typical solute-concentration profiles for liquid mixtures with porous and nonporous (dense) membranes. Included is the drop in concentration across the membrane, and also possible drops due to resistances in the fluid boundary layers or films on either side of the membrane.

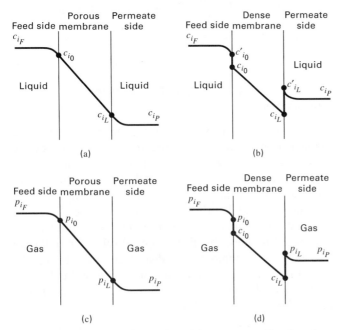

(a)

(b)

(c)

(d)

Figure 14.10 Concentration and partial-pressure profiles for solute transport through membranes. Liquid mixture with (a) a porous and (b) a nonporous membrane; gas mixture with (c) a porous and (d) a nonporous membrane.

For porous membranes considered above, the concentration profile is continuous from the bulk-feed liquid to the bulk-permeate liquid because liquid is present continuously from one side to the other. The concentration c_{i_0} is the same in the liquid feed just adjacent to the membrane surface and in the liquid just within the pore entrance. This is not the case for the nonporous membrane in Figure 14.10b. Solute concentration c'_{i_0} is that in the feed liquid just adjacent to the upstream membrane surface, whereas c_{i_0} is that in the membrane just adjacent to the upstream membrane surface. In general, c_{i_0} is considerably smaller than c'_{i_0}, but the two are related by a thermodynamic equilibrium partition coefficient K_i, defined by

$$K_{i_0} = c_{i_0}/c'_{i_0} \qquad (14\text{-}46)$$

Similarly, at the other face:

$$K_{i_L} = c_{i_L}/c'_{i_L} \qquad (14\text{-}47)$$

Fick's law for the dense membrane of Figure 14.10b is:

$$N_i = \frac{D_i}{l_M}(c_{i_0} - c_{i_L}) \qquad (14\text{-}48)$$

where D_i is the diffusivity of the solute in the membrane. If (14-46) and (14-47) are combined with (14-48), and the partition coefficient is assumed independent of concentration, such that $K_{i_0} = K_{i_L} = K_i$, the flux is

$$N_i = \frac{K_i D_i}{l_M}\left(c'_{i_0} - c'_{i_L}\right) \qquad (14\text{-}49)$$

If the mass-transfer resistances in the two fluid boundary layers or films are negligible:

$$N_i = \frac{K_i D_i}{l_M}(c_{i_F} - c_{i_P}) \qquad (14\text{-}50)$$

In (14-49) and (14-50), $P_{M_i} = K_i D_i$ is the permeability for the solution-diffusion model, where K_i accounts for the solute solubility in the membrane and D_i accounts for diffusion through the membrane. Because D_i is generally very small, it is important that the membrane material have a large value for K_i and/or a small membrane thickness.

D_i and K_i, and therefore P_{M_i}, depend on the solute and the membrane. When solutes dissolve in a polymer membrane, it will swell, causing both D_i and K_i to increase. Other polymer-membrane factors that influence D_i, K_i, and P_{M_i} are listed in Table 14.6. However, the largest single factor is the chemical

Table 14.6 Factors That Influence Permeability of Solutes in Dense Polymers

Factor	Value Favoring High Permeability
Polymer density	low
Degree of crystallinity	low
Degree of cross-linking	low
Degree of vulcanization	low
Amount of plasticizers	high
Amount of fillers	low
Chemical affinity of solute for polymer (solubility)	high

structure of the membrane polymer. Because of the many factors involved, it is important to obtain experimental permeability data for the membrane and feed mixture of interest. The effect of external mass-transfer resistances is considered later in this section.

Solution-Diffusion for Gas Mixtures

Figures 14.10c and d show typical solute profiles for gas mixtures with porous and nonporous membranes, including the effect of the external-fluid boundary layer. For the porous membrane, a continuous partial-pressure profile is shown. For the nonporous membrane, a concentration profile is shown within the membrane, where the solute is dissolved. Fick's law holds for transport through the membrane. Assuming that thermodynamic equilibrium exists at the fluid–membrane interfaces, concentrations in Fick's law are related to partial pressures adjacent to the membrane faces by Henry's law as

$$H_{i_0} = c_{i_0}/p_{i_0} \qquad (14\text{-}51)$$

and

$$H_{i_L} = c_{i_L}/p_{i_L} \qquad (14\text{-}52)$$

If it is assumed that H_i is independent of total pressure and that the temperature is the same at both membrane faces:

$$H_{i_0} = H_{i_L} = H_i \qquad (14\text{-}53)$$

Combining (14-48), (14-51), (14-52), and (14-53), the flux is

$$\boxed{N_i = \frac{H_i D_i}{l_M}\left(p_{i_0} - p_{i_L}\right)} \qquad (14\text{-}54)$$

If the external mass-transfer resistances are neglected, $p_{i_F} = p_{i_0}$ and $p_{i_L} = p_{i_P}$, giving

$$N_i = \frac{H_i D_i}{l_M}\left(p_{i_F} - p_{i_P}\right) = \frac{P_{M_i}}{l_M}\left(p_{i_F} - p_{i_P}\right) \qquad (14\text{-}55)$$

where

$$P_{M_i} = H_i D_i \qquad (14\text{-}56)$$

Thus, permeability depends on both solubility of the gas component in the membrane and its diffusivity when dissolved in the membrane. An acceptable rate of transport can be achieved only by using a very thin membrane and a high pressure on the feed side. The permeability of a gas through a polymer membrane is subject to factors listed in Table 14.6. Light gases do not interact with the polymer or cause it to swell. Thus, a light-gas–permeant–polymer combination is readily characterized experimentally. Often both solubility and diffusivity are measured. An extensive tabulation is given in the *Polymer Handbook* [19]. Representative data at 25°C are given in Table 14.7. In general, diffusivity decreases and solubility increases with increasing molecular weight of the gas species, making it difficult to achieve a high selectivity. The effect of temperature over a modest range of about 50°C can be represented for both solubility and diffusivity by Arrhenius equations. For example,

$$D = D_0 e^{-E_D/RT} \qquad (14\text{-}57)$$

The modest effect of temperature on solubility may act in either direction; however, an increase in temperature can cause an increase in diffusivity, and a corresponding increase in permeability. Typical activation energies of diffusion in polymers, E_D, range from 15 to 60 kJ/mol.

Application of Henry's law to rubbery polymers is well accepted, particularly for low-molecular-weight penetrants, but is less accurate for glassy polymers, for which alternative theories have been proposed. Foremost is the dual-mode model first proposed by Barrer and co-workers [22–24] as the result of a comprehensive study of sorption and diffusion in ethyl cellulose. In this model, sorption of penetrant occurs by ordinary dissolution in polymer chains, as described by Henry's law, and by Langmuir sorption into holes or sites

Table 14.7 Coefficients for Gas Permeation in Polymers

	Gas Species					
	H_2	O_2	N_2	CO	CO_2	CH_4
Low-Density Polyethylene:						
$D \times 10^6$	0.474	0.46	0.32	0.332	0.372	0.193
$H \times 10^6$	1.58	0.472	0.228	0.336	2.54	1.13
$P_M \times 10^{13}$	7.4	2.2	0.73	1.1	9.5	2.2
Polyethylmethacrylate:						
$D \times 10^6$	—	0.106	0.0301	—	0.0336	—
$H \times 10^6$	—	0.839	0.565	—	11.3	—
$P_M \times 10^{13}$	—	0.889	0.170	—	3.79	—
Polyvinylchloride:						
$D \times 10^6$	0.5	0.012	0.0038	—	0.0025	0.0013
$H \times 10^6$	0.26	0.29	0.23	—	4.7	1.7
$P_M \times 10^{13}$	1.3	0.034	0.0089	—	0.12	0.021
Butyl Rubber:						
$D \times 10^6$	1.52	0.081	0.045	—	0.0578	—
$H \times 10^6$	0.355	1.20	0.543	—	6.71	—
$P_M \times 10^{13}$	5.43	0.977	0.243	—	3.89	—

Note: Units: D in cm^2/s; H in cm^3 (STP)/cm^3-Pa; P_M in cm^3 (STP)-cm/cm^2-s-Pa.

between chains of glassy polymers. When downstream pressure is negligible compared to upstream pressure, the permeability for Fick's law is given by

$$P_{M_i} = H_i D_i + \frac{D_{L_i} ab}{1 + bP} \qquad (14\text{-}58)$$

where the second term refers to Langmuir sorption, with D_{L_i} = diffusivity of Langmuir sorbed species, P = penetrant pressure, and ab = Langmuir constants for sorption-site capacity and site affinity, respectively.

Koros and Paul [25] found that the dual-mode theory accurately represents data for CO_2 sorption in polyethylene terephthalate below its glass-transition temperature of 85°C. Above that temperature, the rubbery polymer obeys Henry's law. Mechanisms of diffusion for the Langmuir mode have been suggested by Barrer [26].

The ideal dense-polymer membrane has a high permeance, P_{M_i}/l_M, for the penetrant molecules and a high separation factor between components. The separation factor is defined similarly to relative volatility in distillation:

$$\alpha_{A,B} = \frac{(y_A/x_A)}{(y_B/x_B)} \qquad (14\text{-}59)$$

where y_i is the mole fraction in the permeate leaving the membrane, corresponding to partial pressure p_{i_P} in Figure 14.10d, while x_i is the mole fraction in the retentate on the feed side of the membrane, corresponding to partial pressure p_{i_F} in Figure 14.10d. Unlike distillation, y_i and x_i are not in equilibrium.

For the separation of a binary gas mixture of A and B in the absence of external boundary layer or film mass-transfer resistances, transport fluxes are given by (14-55):

$$N_A = \frac{H_A D_A}{l_M}\left(p_{A_F} - p_{A_P}\right) = \frac{H_A D_A}{l_M}(x_A P_F - y_A P_P)$$
$$(14\text{-}60)$$

$$N_B = \frac{H_B D_B}{l_M}\left(p_{B_F} - p_{B_P}\right) = \frac{H_B D_B}{l_M}(x_B P_F - y_B P_P) \quad (14\text{-}61)$$

When no sweep gas is used, the ratio of N_A to N_B fixes the composition of the permeate so that it is simply the ratio of y_A to y_B in the permeate gas. Thus,

$$\boxed{\frac{N_A}{N_B} = \frac{y_A}{y_B} = \frac{H_A D_A(x_A P_F - y_A P_P)}{H_B D_B(x_B P_F - y_B P_P)}} \qquad (14\text{-}62)$$

If the downstream (permeate) pressure, P_P, is negligible compared to the upstream pressure, P_F, such that $y_A P_P \ll x_A P_F$ and $y_B P_P \ll x_B P_F$, (14-62) can be rearranged and combined with (14-59) to give an *ideal separation factor*:

$$\boxed{\alpha_{A,B}^* = \frac{H_A D_A}{H_B D_B} = \frac{P_{M_A}}{P_{M_B}}} \qquad (14\text{-}63)$$

Thus, a high separation factor results from a high solubility ratio, a high diffusivity ratio, or both. The factor depends on both transport phenomena and thermodynamic equilibria.

When the downstream pressure is not negligible, (14-62) can be rearranged to obtain an expression for $\alpha_{A,B}$ in terms of the pressure ratio, $r = P_P/P_F$, and the mole fraction of A on the retentate side of the membrane. Combining (14-59),

Table 14.8 Ideal Membrane-Separation Factors of Binary Pairs for Two Membrane Materials

	PDMS, Silicon Rubbery Polymer Membrane	PC, Polycarbonate Glassy Polymer Membrane
$P_{M_{He}}$, barrer	561	14
α_{He, CH_4}^*	0.41	50
$\alpha_{He, C_2H_4}^*$	0.15	33.7
$P_{M_{CO_2}}$, barrer	4,550	6.5
α_{CO_2, CH_4}^*	3.37	23.2
$\alpha_{CO_2, C_2H_4}^*$	1.19	14.6
$P_{M_{O_2}}$, barrer	933	1.48
α_{O_2, N_2}^*	2.12	5.12

(14-63), and the definition of r with (14-62):

$$\alpha_{A,B} = \alpha_{A,B}^* \left[\frac{(x_B/y_B) - r\alpha_{A,B}}{(x_B/y_B) - r}\right] \qquad (14\text{-}64)$$

Because $y_A + y_B = 1$, it is possible to substitute into (14-64) for x_B, the identity:

$$x_B = x_B y_A + x_B y_B$$

to give

$$\alpha_{A,B} = \alpha_{A,B}^* \left[\frac{x_B\left(\dfrac{y_A}{y_B} + 1\right) - r\alpha_{A,B}}{x_B\left(\dfrac{y_A}{y_B} + 1\right) - r}\right] \qquad (14\text{-}65)$$

Combining (14-59) and (14-65) and replacing x_B with $1 - x_A$, the separation factor becomes:

$$\alpha_{A,B} = \alpha_{A,B}^* \left[\frac{x_A\left(\alpha_{A,B} - 1\right) + 1 - r\alpha_{A,B}}{x_A\left(\alpha_{A,B} - 1\right) + 1 - r}\right] \qquad (14\text{-}66)$$

Equation (14-66) is an implicit equation for $\alpha_{A,B}$ in terms of the pressure ratio, r, and x_A, which is readily solved for $\alpha_{A,B}$ by rearranging the equation into a quadratic form. In the limit when $r = 0$, (14-66) reduces to (14-63), where $\alpha_{A,B} = \alpha_{A,B}^* = (P_{M_A}/P_{M_B})$. Many investigators report values of $\alpha_{A,B}^*$. Table 14.8, taken from the *Membrane Handbook* [6], gives data at 35°C for various binary pairs with polydimethyl siloxane (PDMS), a rubbery polymer, and bisphenol-A-polycarbonate (PC), a glassy polymer. For the rubbery polymer, permeabilities are high, but separation factors are low. The opposite is true for a glassy polymer. For a given feed composition, the separation factor places a limit on the achievable degree of separation.

EXAMPLE 14.6 Air Separation by Permeation

Air can be separated by gas permeation using different dense-polymer membranes. In all cases, the membrane is more permeable to oxygen. A total of 20,000 scfm of air is compressed, cooled, and treated to remove moisture and compressor oil prior to being sent to a membrane separator at 150 psia and 78°F. Assume the composition of the air is 79 mol% N_2 and 21 mol% O_2. A low-density, thin-film, composite polyethylene membrane with solubilities and diffusivities given in Table 14.7 is being considered.

If the membrane skin is 0.2 μm thick, calculate the material balance and membrane area in ft^2 as a function of the cut, which is defined as

$$\theta = cut = \text{fraction of feed permeated} = \frac{n_P}{n_F} \qquad (14\text{-}67)$$

where n = flow rate in lbmol/h and subscripts F and P refer, respectively, to the feed and permeate. Assume 15 psia on the permeate side with perfect mixing on both sides of the membrane, such that compositions on both sides are uniform and equal to exit compositions. Neglect pressure drop and mass-transfer resistances external to the membrane. Comment on the practicality of the membrane for making a reasonable separation.

Solution

Assume that standard conditions are 0°C and 1 atm (359 ft^3/lbmol).

$$n_F = \text{Feed flow rate} = \frac{20{,}000}{359}(60) = 3{,}343 \text{ lbmol/h}$$

For the low-density polyethylene membrane, from Table 14.7, and applying (14-56), letting A = O_2 and B = N_2:

$$P_{M_B} = H_B D_B = (0.228 \times 10^{-6})(0.32 \times 10^{-6})$$
$$= 0.073 \times 10^{-12} \text{ cm}^3(\text{STP})\text{-cm/cm}^2\text{-s-Pa}$$

or, in AE units,

$$P_{M_B} = \frac{(0.073 \times 10^{-12})(2.54 \times 12)(3600)(101{,}300)}{(22{,}400)(454)(14.7)}$$
$$= 5.43 \times 10^{-12} \frac{\text{lbmol-ft}}{\text{ft}^2\text{-h-psia}}$$

Similarly, for oxygen:

$$P_{M_A} = 16.2 \times 10^{-12} \frac{\text{lbmol-ft}}{\text{ft}^2\text{-h-psia}}$$

Permeance values are based on a 0.2-μm-thick membrane skin (0.66 × 10^{-6} ft).
From (14-1),

$$\bar{P}_{M_i} = P_{M_i}/l_M$$
$$\bar{P}_{M_B} = 5.43 \times 10^{-12}/0.66 \times 10^{-6}$$
$$= 8.23 \times 10^{-6} \text{ lbmol/ft}^2\text{-h-psia}$$
$$\bar{P}_{M_A} = 16.2 \times 10^{-12}/0.66 \times 10^{-6}$$
$$= 24.55 \times 10^{-6} \text{ lbmol/ft}^2\text{-h-psia}$$

Material-balance equations:

For N_2,
$$x_{F_B} n_F = y_{P_B} n_P + x_{R_B} n_R \qquad (1)$$

where n = flow rate in lbmol/h and subscripts F, P, and R refer, respectively, to the feed, permeate, and retentate. Since $\theta = cut = n_P/n_F$, $(1-\theta) = n_R/n_F$.

Note that if all components of the feed have a finite permeability, the cut, θ, can vary from 0 to 1. For a cut of 1, all of the feed becomes permeate and no separation is achieved. Substituting (14-67) into (1) gives

$$x_{R_B} = \frac{x_{F_B} - y_{P_B}\theta}{1 - \theta} = \frac{0.79 - y_{P_B}\theta}{1 - \theta} \qquad (2)$$

Similarly, for O_2,
$$x_{R_A} = \frac{0.21 - y_{P_A}\theta}{1 - \theta} \qquad (3)$$

Separation factor:
From the definition of the separation factor, (14-59), with well-mixed fluids, compositions are those of the retentate and permeate,

$$\alpha_{A,B} = \frac{y_{P_A}/x_{R_A}}{(1 - y_{P_A})/(1 - x_{R_A})} \qquad (4)$$

Transport equations:
The transport of A and B through the membrane of area A_M, with partial pressures at exit conditions, can be written as

$$N_B = y_{P_B} n_P = A_M \bar{P}_{M_B}(x_{R_B} P_R - y_{P_B} P_P) \qquad (5)$$
$$N_A = y_{P_A} n_P = A_M \bar{P}_{M_A}(x_{R_A} P_R - y_{P_A} P_P) \qquad (6)$$

where A_M is the membrane area normal to flow, n_P, through the membrane. The ratio of (6) to (7) is y_{P_A}/y_{P_B}, and subsequent manipulations lead to (14-66),
where

$$r = P_P/P_R = 15/150 = 0.1 \quad \text{and} \quad \alpha^*_{A,B} = \alpha_{O_2,N_2} = \bar{P}_{M_{O_2}}/\bar{P}_{M_{N_2}}$$
$$= (24.55 \times 10^{-6})/(8.23 \times 10^{-6}) = 2.98$$

From (66):

$$\alpha_{A,B} = \alpha = 2.98\left[\frac{x_{R_A}(\alpha - 1) + 1 - 0.1\alpha}{x_{R_A}(\alpha - 1) + 1 - 0.1}\right] \qquad (7)$$

Equations (3), (4), and (7) contain four unknowns: x_{R_A}, y_{P_A}, θ, and $\alpha_{A,B} = \alpha$. The variable θ is bounded between 0 and 1, so values of θ are selected in that range. The other three variables are computed in the following manner. Combine (3), (4), and (7) to eliminate α and x_{R_A}. Solve the resulting nonlinear equation for y_{P_A}. Then solve (3) for x_{R_A} and (4) for α. Solve (6) for the membrane area, A_M. The following results are obtained:

θ	x_{R_A}	y_{P_A}	$\alpha_{A,B}$	A_M, ft^2
0.01	0.208	0.406	2.602	22,000
0.2	0.174	0.353	2.587	462,000
0.4	0.146	0.306	2.574	961,000
0.6	0.124	0.267	2.563	1,488,000
0.8	0.108	0.236	2.555	2,035,000
0.99	0.095	0.211	2.548	2,567,000

Note that the separation factor remains almost constant, varying by only 2% with a value of about 86% of the ideal. The maximum permeate O_2 content (40.6 mol%) occurs with the smallest amount of permeate ($\theta = 0.01$). The maximum N_2 retentate content (90.5 mol%) occurs with the largest amount of permeate ($\theta = 0.99$). With a retentate equal to 60 mol% of the feed ($\theta = 0.4$), the N_2 retentate content has increased only from 79 to 85.4 mol%. Furthermore, the membrane area requirements are very large. The low-density polyethylene membrane is thus not a practical membrane for this separation. To achieve a reasonable separation, say, with $\theta = 0.6$ and a retentate of 95 mol% N_2, it is necessary to use a membrane with an ideal separation factor of 5, in a membrane module that approximates crossflow or countercurrent flow of permeate and retentate with no mixing and a higher O_2 permeance. For higher purities, a cascade of two or more stages is needed. These alternatives are developed in the next two subsections.

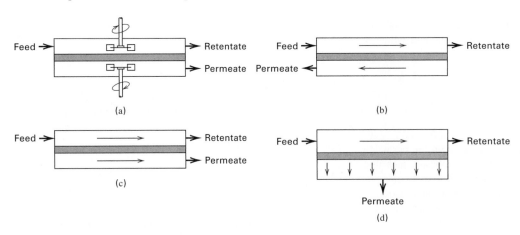

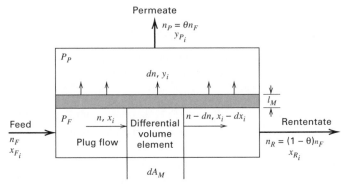

Figure 14.11 Idealized flow patterns in membrane modules: (a) perfect mixing; countercurrent flow; (b) cocurrent flow; (c) crossflow.

§14.3.5 Module Flow Patterns

In Example 14.6, perfect mixing, as shown in Figure 14.11a, was assumed. The three other idealized flow patterns shown, which have no mixing, have received considerable attention and are comparable to the idealized flow patterns used to design heat exchangers. These patterns are (b) countercurrent flow; (c) cocurrent flow; and (d) crossflow. For a given θ (14-67), the flow pattern can significantly affect the degree of separation and the membrane area. For flow patterns (b) to (d), fluid on the feed or retentate side of the membrane flows along and parallel to the upstream side of the membrane. For countercurrent and cocurrent flow, permeate fluid at a given location on the downstream side of the membrane consists of fluid that has just passed through the membrane at that location plus the permeate fluid flowing to that location. For the crossflow case, there is no flow of permeate fluid along the membrane surface. The permeate fluid that has just passed through the membrane at a given location is the only fluid there.

For a given module geometry, it is not obvious which idealized flow pattern to assume. This is particularly true for the spiral-wound module of Figure 14.5b. If the permeation rate is high, the fluid issuing from the downstream side of the membrane may continue to flow perpendicularly to the membrane surface until it finally mixes with the bulk permeate fluid flowing past the surface. In that case, the idealized crossflow pattern might be appropriate. Hollow-fiber modules are designed to approximate idealized countercurrent, cocurrent, or crossflow patterns. The hollow-fiber module in Figure 14.5d is approximated by a countercurrent-flow pattern.

Walawender and Stern [27] present methods for solving all four flow patterns of Figure 14.11, under assumptions of a binary feed with constant-pressure ratio, r, and constant ideal separation factor, $\alpha^*_{A,B}$. Exact analytical solutions are possible for perfect mixing (as in Example 14.6) and for crossflow, but numerical solutions are necessary for countercurrent and cocurrent flow. A reasonably simple, but approximate, analytical solution for the crossflow case, derived by Naylor and Backer [28], is presented here.

Consider a module with the crossflow pattern shown in Figure 14.12. Feed passes across the upstream membrane surface in plug flow with no longitudinal mixing. The

pressure ratio, $r = P_P/P_F$, and the ideal separation factor, $\alpha^*_{A,B}$, are assumed constant. Boundary-layer (or film) mass-transfer resistances external to the membrane are assumed negligible. At the differential element, local mole fractions in the retentate and permeate are x_i and y_i, and the penetrant molar flux is dn/dA_M. Also, the local separation factor is given by (14-66) in terms of the local x_A, r, and $\alpha^*_{A,B}$. An alternative expression for the local permeate composition in terms of y_A, x_A, and r is obtained by combining (14-59) and (14-64):

$$\frac{y_A}{1 - y_A} = \alpha^*_{A,B}\left[\frac{x_A - ry_A}{(1 - x_A) - r(1 - y_A)}\right] \quad (14\text{-}68)$$

A material balance for A around the differential-volume element gives

$$y_A\,dn = d(nx_A) = x_A\,dn + n\,dx_A \quad \text{or} \quad \frac{dn}{n} = \frac{dx_A}{y_A - x_A} \quad (14\text{-}69)$$

which is identical in form to the Rayleigh equation (13-2) for batch differential distillation. If (14-59) is combined with (14-69) to eliminate y_A,

$$\frac{dn}{n} = \left[\frac{1 + (\alpha - 1)x_A}{x_A(\alpha - 1)(1 - x_A)}\right]dx_A \quad (14\text{-}70)$$

where $\alpha = \alpha_{A,B}$.

In the solution to Example 14.6, it was noted that $\alpha = \alpha_{A,B}$ is relatively constant over the entire range of cut, θ. Such is generally the case when the pressure ratio, r, is small. If the assumption of constant $\alpha = \alpha_{A,B}$ is made in (14-70) and integration is carried out from the intermediate location of the

Figure 14.12 Crossflow model for membrane module.

differential element to the final retentate, that is, from n to n_R and from x_A to x_{R_A}, the result is

$$n = n_R\left[\left(\frac{x_A}{x_{R_A}}\right)^{\left(\frac{1}{\alpha-1}\right)}\left(\frac{1-x_{R_A}}{1-x_A}\right)^{\left(\frac{1}{\alpha-1}\right)}\right] \quad (14\text{-}71)$$

The mole fraction of A in the final permeate and the total membrane surface area are obtained by integrating the values obtained from solving (14-68) to (14-70):

$$y_{P_A} = \int_{x_{F_A}}^{x_{R_A}} y_A\,dn/\theta n_F \quad (14\text{-}72)$$

By combining (14-72) with (14-70), (14-71), and the definition of α, the integral in n can be transformed to an integral in x_A, which when integrated gives

$$y_{P_A} = x_{R_A}^{\left(\frac{1}{1-\alpha}\right)}\left(\frac{1-\theta}{\theta}\right)$$
$$\times\left[(1-x_{R_A})^{\left(\frac{\alpha}{\alpha-1}\right)}\left(\frac{x_{F_A}}{1-x_{F_A}}\right)^{\left(\frac{\alpha}{\alpha-1}\right)} - x_{R_A}^{\left(\frac{\alpha}{1-\alpha}\right)}\right] \quad (14\text{-}73)$$

where $\alpha = \alpha_{A,B}$ can be estimated from (14-70) by using $x_A = x_{F_A}$.

The differential rate of mass transfer of A across the membrane is given by

$$y_A\,dn = \frac{P_{M_A}\,dA_M}{l_M}[x_A P_F - y_A P_P] \quad (14\text{-}74)$$

from which the total membrane surface area can be obtained by integration:

$$A_M = \int_{x_{R_A}}^{x_{F_A}} \frac{l_M y_A\,dn}{P_{M_A}(x_A P_F - y_A P_P)} \quad (14\text{-}75)$$

The crossflow model is illustrated in the next example.

EXAMPLE 14.7 Gas Permeation in a Crossflow Module.

For the conditions of Example 14.6, compute exit compositions for a spiral-wound module that approximates crossflow.

Solution

From Example 14.6: $\alpha_{A,B}^* = 2.98$; $r = 0.1$; $x_{F_A} = 0.21$

From (14-66), using $x_A = x_{F_A}$: $\alpha_{A,B} = 2.60$

An overall module material balance for O_2 (A) gives

$$x_{F_A}n_F = x_{R_A}(1-\theta)n_F + y_{P_A}\theta n_F \quad \text{or} \quad x_{R_A} = \frac{(x_{F_A} - y_{P_A}\theta)}{(1-\theta)} \quad (1)$$

Solving (1) and (14-73) simultaneously with a program such as Mathcad, Matlab, or Polymath gives the following results:

θ	x_{R_A}	x_{P_A}	Stage α_S
0.01	0.208	0.407	2.61
0.2	0.168	0.378	3.01
0.4	0.122	0.342	3.74
0.6	0.0733	0.301	5.44
0.8	0.0274	0.256	12.2
0.99	0.000241	0.212	1,120.

Comparing these results to those of Example 14.6, it is seen that for crossflow, the permeate is richer in O_2 and the retentate is richer in N_2. Thus, for a given cut, θ, crossflow is more efficient than perfect mixing, as might be expected.

Also included in the preceding table is the calculated degree of separation for the stage, α_S, defined on the basis of the mole fractions in the permeate and retentate exiting the stage by

$$(\alpha_{A,B})_S = \alpha_S = \frac{(y_{P_A}/x_{R_A})}{(1-y_{P_A})/(1-x_{R_A})} \quad (2)$$

The ideal separation factor, $\alpha_{A,B}^*$, is 2.98. Also, if (2) is applied to the perfect mixing case of Example 14.6, α_S is 2.603 for $\theta = 0.01$ and decreases slowly with increasing θ until at $\theta = 0.99$, $\alpha_S = 2.548$. Thus, for perfect mixing, $\alpha_S < \alpha^*$ for all θ. Such is not the case for crossflow. In the above table, $\alpha_S < \alpha^*$ for $\theta > 0.2$, and α_S increases with increasing θ. For $\theta = 0.6$, α_S is almost twice α^*.

Calculating the degree of separation of a binary mixture in a membrane module utilizing cocurrent- or countercurrent-flow patterns involves numerical solution of ODEs. These and computer codes for their solution are given by Walawender and Stern [27]. A representative solution is shown in Figure 14.13 for the separation of air (20.9 mol% O_2) for conditions of $\alpha^* = 5$ and $r = 0.2$. For a given cut, θ, it is seen that the best separation is with countercurrent flow. The curve for cocurrent flow lies between crossflow and perfect mixing. The computed crossflow case is considered to be a conservative estimate of membrane module performance. The perfect mixing case for binary mixtures is extended to multicomponent mixtures by Stern et al. [29]. As with crossflow, countercurrent flow also offers the possibility of a separation

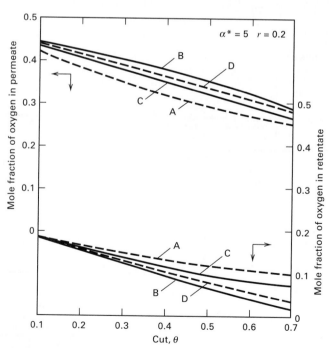

Figure 14.13 Effect of membrane module flow pattern on degree of separation of air. A, perfect mixing; B, countercurrent flow; C, cocurrent flow; D, crossflow.

factor for the stage, α_S, defined by (2) in Example 14.6, that is considerably greater than α^*.

§14.3.6 Cascades

A single membrane module or a number of such modules arranged in parallel or in series without recycle constitutes a single-stage membrane-separation process. The extent to which a feed mixture can be separated in a single stage is limited and determined by the separation factor, α. This factor depends, in turn, on module flow patterns; the permeability ratio (ideal separation factor); the cut, θ; and the driving force for membrane mass transfer. To achieve a higher degree of separation than is possible with a single stage, a counter-current cascade of membrane stages—such as used in distillation, absorption, stripping, and liquid–liquid extraction—or a hybrid process that couples a membrane separator with another type of separator can be devised. Membrane cascades were presented briefly in §5.5. They are now discussed in detail and illustrated with an example.

A countercurrent recycle cascade of membrane separators, similar to a distillation column, is depicted in Figure 14.14a. The feed enters at stage F, somewhere near the middle of the column. Permeate is enriched in components of high permeability in an enriching section, while the retentate is enriched in components of low permeability in a stripping section. The final permeate is withdrawn from stage 1, while the final retentate is withdrawn from stage N. For a cascade, additional factors that affect the degree of separation are the number of stages and the recycle ratio (permeate recycle rate/permeate product rate). As discussed by Hwang and Kammermeyer [30], it is best to manipulate the cut and reflux rate at each stage so as to force compositions of the two streams entering each stage to be identical. For example, the composition of retentate leaving stage 1 and entering stage 2 would be identical to the composition of permeate flowing from stage 3 to stage 2. This corresponds to the least amount of entropy production for the cascade and, thus, the highest second-law efficiency. Such a cascade is referred to as "ideal".

Calculation methods for cascades are discussed by Hwang and Kammermeyer [30] and utilize single-stage methods that depend upon the module flow pattern, as discussed in the previous section. The calculations are best carried out on a computer, but results for a binary mixture can be conveniently displayed on a McCabe–Thiele-type diagram (§7.2) in terms of the mole fraction in the permeate leaving each stage, y_i, versus the mole fraction in the retentate leaving each stage, x_i. For a cascade, the equilibrium curve becomes the selectivity curve in terms of the separation factor for the stage, a_S.

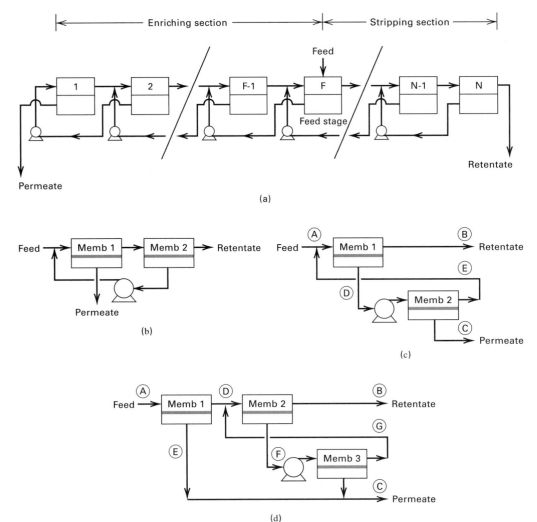

Figure 14.14 Countercurrent recycle cascades of membrane separators. (a) Multiple-stage unit. (b) Two-stage stripping cascade. (c) Two-stage enriching cascade. (d) Two-stage enriching cascade with additional premembrane stage.

In Figure 14.14, it is assumed that pressure drop on the feed or upstream side of the membrane is negligible. Thus, only the permeate must be pumped to the next stage if a liquid, or compressed if a gas. In the case of gas, compression costs are high. Thus, membrane cascades for gas permeation are often limited to just two or three stages, with the most common configurations shown in Figures 14.14b, c, and d.

Compared to one stage, the two-stage stripping cascade is designed to obtain a purer retentate, whereas a purer permeate is the goal of the two-stage enriching cascade. Addition of a premembrane stage, shown in Figure 14.14d, may be attractive when feed concentration is low in the component to be passed preferentially through the membrane, desired permeate purity is high, separation factor is low, and/or a high recovery of the more permeable component is desired. An example of the application of enrichment cascades is given by Spillman [31] for the removal of carbon dioxide from natural gas (simulated by methane) using cellulose-acetate membranes in spiral-wound modules that approximate crossflow. The ideal separation factor, α_{CO_2,CH_4}^*, is 21. Results of calculations are given in Table 14.9 for a single stage (not shown in Figure 14.14), a two-stage enriching

cascade (Figure 14.14c), and a two-stage enriching cascade with an additional premembrane stage (Figure 14.14d). Carbon dioxide flows through the membrane faster than methane. In all three cases, the feed is 20 million (MM) scfd of 7 mol% CO_2 in methane at 850 psig (865 psia) and the retentate is 98 mol% in methane. For each stage, the downstream (permeate-side) membrane pressure is 10 psig (25 psia). In Table 14.9, for all three cases, stream A is the feed, stream B is the final retentate, and stream C is the final permeate. Case 1 achieves a 90.2% recovery of methane. Case 2 increases that recovery to 98.7%. Case 3 achieves an intermediate recovery of 94.6%. The following degrees of separation are computed from data given in Table 14.9:

	α_s for Membrane Stage		
Case	1	2	3
1	28	—	—
2	28	57	—
3	20	19	44

Table 14.9   Separation of CO_2 and CH_4 with Membrane Cascades

Case 1: Single Membrane Stage:

	Stream		
	A Feed	B Retentate	C Permeate
Composition (mole%)			
CH_4	93.0	98.0	63.4
CO_2	7.0	2.0	36.6
Flow rate (MM SCFD)	20.0	17.11	2.89
Pressure (psig)	850	835	10

Case 2: Two-Stage Enriching Cascade (Figure 14.14c):

	Stream				
	A	B	C	D	E
Composition (mole%)					
CH_4	93.0	98.0	18.9	63.4	93.0
CO_2	7.0	2.0	81.1	36.6	7.0
Flow rate (MM SCFD)	20.00	18.74	1.26	3.16	1.90
Pressure (psig)	850	835	10	10	850

Case 3: Two-Stage Enriching Cascade with Premembrane Stage (Figure 14.14d):

	Stream						
	A	B	C	D	E	F	G
Composition (mole%)							
CH_4	93.0	98.0	49.2	96.1	56.1	72.1	93.0
CO_2	7.0	2.0	50.8	3.9	43.9	27.9	7.0
Flow rate (MM SCFD)	20.00	17.95	2.05	19.39	1.62	1.44	1.01
Pressure (psig)	850	835	10	840	10	10	850

Note: MM = million.

It is also possible to compute overall degrees of separation for the cascades, α_C, for cases 2 and 3, giving values of 210 and 51, respectively.

§14.3.7 External Mass-Transfer Resistances

Thus far, resistance to mass transfer has been associated only with the membrane. Thus, concentrations in the fluid at the upstream and downstream faces of the membrane have been assumed equal to the respective bulk-fluid concentrations. When mass-transfer resistances external to the membrane are not negligible, gradients exist in the boundary layers (or films) adjacent to the membrane surfaces, as is illustrated for four cases in Figure 14.10. For given bulk-fluid concentrations, the presence of these resistances reduces the driving force for mass transfer across the membrane and, therefore, the flux of penetrant.

Gas permeation by solution-diffusion (14-54) is slow compared to diffusion in gas boundary layers or films, so external mass-transfer resistances are negligible and $P_{i_F} = P_{i_0}$ and $P_{i_P} = P_{i_L}$ in Figure 14.10d. Because diffusion in liquid boundary layers and films is slow, concentration polarization, which is the accumulation of non-permeable species on the upstream surface of the membrane, cannot be neglected in membrane processes that involve liquids, such as dialysis, reverse osmosis, and pervaporation. The need to consider the effect of concentration polarization is of particular importance in reverse osmosis, where the effect can reduce the water flux and increase the salt flux, making it more difficult to obtain potable water.

Consider a membrane process of the type in Figure 14.10a, involving liquids with a porous membrane. At steady state, the rate of mass transfer of a penetrating species, i, through the three resistances is as follows, assuming no change in area for mass transfer across the membrane:

$$N_i = k_{i_F}(c_{i_F} - c_{i_0}) = \frac{D_{e_i}}{l_M}(c_{i_0} - c_{i_L}) = k_{i_P}(c_{i_L} - c_{i_P})$$

where D_{e_i} is given by (14-38). If these three equations are combined to eliminate the intermediate concentrations, c_{i_0} and c_{i_L},

$$N_i = \frac{c_{i_F} - c_{i_P}}{\dfrac{1}{k_{i_F}} + \dfrac{l_M}{D_{e_i}} + \dfrac{1}{k_{i_P}}} \tag{14-76}$$

Now consider the membrane process in Figure 14.10b, involving liquids with a nonporous membrane, for which the solution-diffusion mechanism, (14-49), applies for mass transfer through the membrane. At steady state, for constant mass-transfer area, the rate of mass transfer through the three resistances is:

$$N_i = k_{i_F}\left(c_{i_F} - c'_{i_0}\right) = \frac{K_i D_i}{l_M}\left(c'_{i_0} - c'_{i_L}\right) = k_{i_P}\left(c'_{i_L} - c_{i_P}\right)$$

If these three equations are combined to eliminate the intermediate concentrations, c'_{i_0} and c'_{i_L},

$$N_i = \frac{c_{i_F} - c_{i_P}}{\dfrac{1}{k_{i_F}} + \dfrac{l_M}{K_i D_i} + \dfrac{1}{k_{i_P}}} \tag{14-77}$$

where in (14-76) and (14-77), k_{i_F} and k_{i_P} are mass-transfer coefficients for the feed-side and permeate-side boundary layers (or films). The three terms in the RHS denominator are the resistances to the mass flux. Mass-transfer coefficients depend on fluid properties, flow-channel geometry, and flow regime. In the laminar-flow regime, a long entry region may exist where the mass-transfer coefficient changes with distance, L, from the entry of the membrane channel. Estimation of coefficients is complicated by fluid velocities that change because of mass exchange between the two fluids. In (14-76) and (14-77), the membrane resistances, l_M/D_{e_i} and $l_M/K_i D_i$, respectively, can be replaced by l_M/P_{M_i} or $\bar{P}_{M_i}$.

Mass-transfer coefficients for channel flow can be obtained from the general empirical film-model correlation [32]:

$$N_{Sh} = k_i d_H/D_i = a N_{Re}^b N_{Sc}^{0.33}(d_H/L)^d \tag{14-78}$$

where $N_{Re} = d_H \upsilon \rho/\mu$, $N_{Sc} = \mu/\rho D_i$, d_H = hydraulic diameter, and υ = velocity.

Values for constants a, b, and d are as follows:

Flow Regime	Flow Channel Geometry	d_H	a	b	d
Turbulent, ($N_{Re} > 10{,}000$)	Circular tube	D	0.023	0.8	0
	Rectangular channel	$2hw/(h+w)$	0.023	0.8	0
Laminar, ($N_{Re} < 2{,}100$)	Circular tube	D	1.86	0.33	0.33
	Rectangular channel	$2hw/(h+w)$	1.62	0.33	0.33

where w = width of channel, h = height of channel, and L = length of channel.

EXAMPLE 14.8 Solute Flux Through a Membrane.

A dilute solution of solute A in solvent B is passed through a tubular-membrane separator, where the feed flows through the tubes. At a certain location, solute concentrations on the feed and permeate sides are 5.0×10^{-2} kmol/m^3 and 1.5×10^{-2} kmol/m^3, respectively. The permeance of the membrane for solute A is given by the membrane vendor as 7.3×10^{-5} m/s. If the tube-side Reynolds number is 15,000, the feed-side solute Schmidt number is 500, the diffusivity of the feed-side solute is 6.5×10^{-5} cm^2/s, and the inside diameter of the tube is 0.5 cm, estimate the solute flux through the membrane if the mass-transfer resistance on the permeate side of the membrane is negligible.

Solution

Flux of the solute is from the permeance form of (14-76) or (14-77):

$$N_A = \frac{c_{A_F} - c_{A_P}}{\dfrac{1}{k_{A_F}} + \dfrac{1}{\bar{P}_{M_A}} + 0}$$

$$c_{A_F} - c_{A_P} = 5 \times 10^{-2} - 1.5 \times 10^{-2} = 3.5 \times 10^{-2} \text{ kmol/m}^3 \tag{1}$$

$$\bar{P}_{M_A} = 7.3 \times 10^{-5} \text{ m/s}$$

From (14-78), for turbulent flow in a tube, since $N_{Re} > 10,000$:

$$k_{A_F} = 0.023 \frac{D_A}{D} N_{Re}^{0.8} N_{Sc}^{0.33}$$

$$= 0.023 \left(\frac{6.5 \times 10^{-5}}{0.5} \right) (15,000)^{0.8} (500)^{0.33}$$

$$= 0.051 \text{ cm/s or } 5.1 \times 10^{-4} \text{ m/s}$$

From (1),

$$N_A = \frac{3.5 \times 10^{-2}}{\dfrac{1}{5.1 \times 10^{-4}} + \dfrac{1}{7.3 \times 10^{-5}}} = 2.24 \times 10^{-6} \text{ kmol/s-m}^2$$

The fraction of the total resistance due to the membrane is

$$\frac{\dfrac{1}{7.3 \times 10^{-5}}}{\dfrac{1}{5.1 \times 10^{-4}} + \dfrac{1}{7.3 \times 10^{-5}}} = 0.875 \text{ or } 87.5\%$$

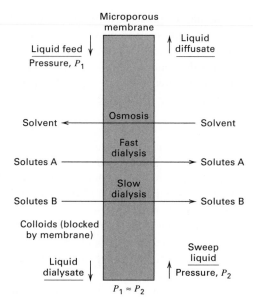

Figure 14.15 Dialysis.

§14.3.8 Concentration Polarization and Fouling

When gases are produced during electrolysis, they accumulate on and around the electrodes of the electrolytic cell, reducing the flow of electric current. This is referred to as polarization. A similar phenomenon, *concentration polarization*, occurs in membrane separators when the membrane is permeable to A, but relatively impermeable to B. Thus, molecules of B are carried by bulk flow to the upstream surface of the membrane, where they accumulate, causing their concentration at the surface of the membrane to increase in a "polarization layer." The equilibrium concentration of B in this layer is reached when its back-diffusion to the bulk fluid on the feed-retentate side equals its bulk flow toward the membrane.

Concentration polarization is most common in pressure-driven membrane separations involving liquids, such as reverse osmosis and ultrafiltration, where it reduces the flux of A. The polarization effect can be serious if the concentration of B reaches its solubility limit on the membrane surface. Then, a precipitate of gel may form, the result being fouling on the membrane surface or within membrane pores, with a further reduction in the flux of A. Concentration polarization and fouling are most severe at high values of the flux of A. Theory and examples of concentration polarization and fouling are given in §14.6 and §14.8 on reverse osmosis and ultrafiltration.

§14.4 DIALYSIS

In the dialysis membrane-separation process, shown in Figure 14.15, the feed is liquid at pressure P_1 and contains solvent, solutes of type A, and solutes of type B and insoluble, but dispersed, colloidal matter. A sweep liquid or wash of the same solvent is fed at pressure P_2 to the other side of the membrane. The membrane is thin, with micropores of a size such that solutes of type A can pass through by a concentration-driving force. Solutes of type B are larger

in molecular size than those of type A and pass through the membrane only with difficulty or not at all. This transport of solutes through the membrane is called dialysis. Colloids do not pass through the membrane. With pressure $P_1 = P_2$, the solvent may also pass through the membrane, but by a concentration-driving force acting in the opposite direction. The transport of the solvent is called osmosis. By elevating P_1 above P_2, solvent osmosis can be reduced or eliminated if the difference is higher than the osmotic pressure. The products of a dialysis unit (dialyzer) are a liquid *diffusate* (permeate) containing solvent, solutes of type A, and little or none of type B solutes; and a *dialysate* (retentate) of solvent, type B solutes, remaining type A solutes, and colloidal matter. Ideally, the dialysis unit would enable a perfect separation between solutes of type A and solutes of type B and any colloidal matter. However, at best only a fraction of solutes of type A are recovered in the diffusate, even when solutes of type B do not pass through the membrane.

For example, when dialysis is used to recover sulfuric acid (type A solute) from an aqueous stream containing sulfate salts (type B solutes), the following results are obtained, as reported by Chamberlin and Vromen [33]:

	Streams in		Streams out	
	Feed	Wash	Dialysate	Diffusate
Flow rate, gph	400	400	420	380
H_2SO_4, g/L	350	0	125	235
$CuSO_4$, g/L as Cu	30	0	26	2
$NiSO_4$, g/L as Ni	45	0	43	0

Thus, about 64% of the H_2SO_4 is recovered in the diffusate, accompanied by only 6% of the $CuSO_4$, and no $NiSO_4$.

Dialysis is closely related to other membrane processes that use other driving forces for separating liquid mixtures, including (1) reverse osmosis, which depends upon a transmembrane pressure difference for solute and/or solvent transport; (2) electrodialysis and electro-osmosis, which depend upon a transmembrane electrical-potential difference for solute and solvent transport, respectively; and (3) thermal osmosis, which depends upon a transmembrane temperature difference for solute and solvent transport.

Dialysis is attractive when concentration differences for the main diffusing solutes are large and permeability differences between those solutes and the other solute(s) and/or colloids are large. Although dialysis has been known since the work of Graham in 1861 [34], commercial applications of dialysis do not rival reverse osmosis and gas permeation. Nevertheless, dialysis has been used in separations, including: (1) recovery of sodium hydroxide from a 17–20 wt% caustic viscose liquor contaminated with hemicellulose to produce a diffusate of 9–10 wt% caustic; (2) recovery of chromic, hydrochloric, and hydrofluoric acids from contaminating metal ions; (3) recovery of sulfuric acid from aqueous solutions containing nickel sulfate; (4) removal of alcohol from beer to produce a low-alcohol beer; (5) recovery of nitric and hydrofluoric acids from spent stainless steel pickle liquor; (6) removal of mineral acids from organic compounds; (7) removal of low-molecular-weight contaminants from polymers; and (8) purification of pharmaceuticals. Also of great importance is hemodialysis, in which urea, creatine, uric acid, phosphates, and chlorides are removed from blood without removing essential higher-molecular-weight compounds and blood cells in a device called an artificial kidney. Dialysis centers servicing those suffering from incipient kidney failure are common in shopping centers.

Typical microporous-membrane materials used in dialysis are hydrophilic, including cellulose, cellulose acetate, various acid-resistant polyvinyl copolymers, polysulfones, and polymethylmethacrylate, typically less than 50 μm thick and with pore diameters of 15 to 100 Å. The most common membrane modules are plate-and-frame and hollow-fiber. Compact hollow-fiber hemodialyzers, such as the one shown in Figure 14.16, which are widely used, typically contain several thousand 200-μm-diameter fibers with a wall thickness of 20–30 μm and a length of 10–30 cm. Dialysis membranes can be thin because pressures on either side of the membrane are essentially equal. The differential rate of solute mass transfer across the membrane is

$$dn_i = K_i(c_{i_F} - c_{i_P})\, dA_M \qquad (14\text{-}79)$$

where K_i is the overall mass-transfer coefficient, in terms of the three coefficients from the permeability form of (14-76):

$$\frac{1}{K_i} = \frac{1}{k_{i_F}} + \frac{l_M}{P_{M_i}} + \frac{1}{k_{i_P}} \qquad (14\text{-}80)$$

Membrane area is determined by integrating (14-79), taking into account module flow patterns, bulk-concentration gradients, and individual mass-transfer coefficients in (14-80).

One of the oldest membrane materials used with aqueous solutions is porous cellophane, for which solute permeability is given by (14-37) with $P_{M_i} = D_{e_i}$ and $\bar{P}_{M_i} l_M$. If immersed, cellophane swells to about twice its dry thickness. The wet thickness should be used for l_M. Typical values of parameters in (14-36) to (14-38) for commercial cellophane are as follows: Wet thickness $= l_M = 0.004$ to 0.008 cm; porosity $= \epsilon = 0.45$ to 0.60; tortuosity $= \tau = 3$ to 5; pore diameter $= D = 30$ to 50 Å.

If a solute does not interact with the membrane material, diffusivity, D_{e_i}, in (14-37) is the ordinary molecular-diffusion coefficient, which depends only on solute and solvent properties. In practice, the membrane may have a profound effect on solute diffusivity if membrane–solute interactions such as covalent, ionic, and hydrogen bonding; physical adsorption and chemisorption; and increases in membrane polymer flexibility occur. Thus, it is best to measure $\bar{P}_{M_i}$ experimentally using process fluids.

Although transport of solvents such as water, usually in a direction opposite to the solute, can be described in terms of Fick's law, it is common to measure the solvent flux and report a so-called *water-transport number*, which is the ratio of the water flux to the solute flux, with a negative value indicating transport of solvent in the solute direction. The membrane can also interact with solvent and curtail solvent transport. Ideally, the water-transport number should be a small value, less than +1.0. Design parameters for dialyzers are best measured in the laboratory using a batch cell with a variable-speed stirring mechanism on both sides of the membrane so that external mass-transfer resistances, $1/k_{i_F}$ and $1/k_{i_P}$ in (14-80), are made negligible. Stirrer speeds >2,000 rpm may be required.

A common dialyzer is the plate-and-frame type of Figure 14.5a. For dialysis, the frames are vertical and a unit might contain 100 square frames, each 0.75 m $\times$ 0.75 m on 0.6-cm spacing, equivalent to 56 m^2 of membrane surface. A typical dialysis rate for sulfuric acid is 5 lb/day-ft^2. Recent dialysis units utilize hollow fibers of 200-μm inside diameter, 16-μm

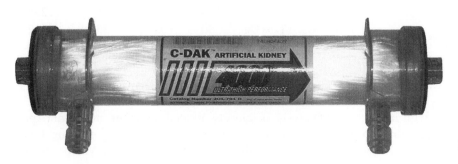

Figure 14.16 Artificial kidney.

wall thickness, and 28-cm length, packed into a heat-exchanger-like module to give 22.5 m² of membrane area in a volume that might be one-tenth of the volume of an equivalent plate-and-frame unit.

In a plate-and-frame dialyzer, the flow pattern is nearly countercurrent. Because total flow rates change little and solute concentrations are small, it is common to estimate solute transport rate by assuming a constant overall mass-transfer coefficient with a log-mean concentration-driving force. Thus, from (14-79):

$$n_i = K_i A_M (\Delta c_i)_{LM} \qquad (14\text{-}81)$$

where K_i is from (14-80). This design method is used in the following example.

EXAMPLE 14.9   Recovery of H₂SO₄ by Dialysis.

A countercurrent-flow, plate-and-frame dialyzer is to be sized to process 0.78 m³/h of an aqueous solution containing 300 kg/m³ of H₂SO₄ and smaller amounts of copper and nickel sulfates, using a wash water sweep of 1.0 m³/h. It is desired to recover 30% of the acid at 25°C. From batch experiments with an acid-resistant vinyl membrane, in the absence of external mass-transfer resistances, a permeance of 0.025 cm/min for the acid and a water-transport number of +1.5 are measured. Membrane transport of copper and nickel sulfates is negligible. Experience with plate-and-frame dialyzers indicates that flow will be laminar and the combined external liquid-film mass-transfer coefficients will be 0.020 cm/min. Determine the membrane area required in m².

Solution

$m_{H_2SO_4}$ in feed $= 0.78(300) = 234\,\text{kg/h}$;

$m_{H_2SO_4}$ transferred $= 0.3(234) = 70\,\text{kg/h}$;

m_{H_2O} transferred to dialysate $= 1.5(70) = 105\,\text{kg/h}$;

m_{H_2O} in entering wash $= 1.0(1,000) = 1,000\,\text{kg/h}$;

m_P leaving $= 1,000 - 105 + 70 = 965\,\text{kg/h}$

For mixture densities, assume aqueous sulfuric acid solutions and use the appropriate table in *Perry's Chemical Engineers' Handbook*:

$\rho_F = 1,175\,\text{kg/m}^3$; $\rho_R = 1,114\,\text{kg/m}^3$; $\rho_P = 1,045\,\text{kg/m}^3$;

$m_F = 0.78(1,175) = 917\,\text{kg/h}$; m_R leaving $= 917 + 105 - 70$

$= 952\,\text{kg/h}$

Sulfuric acid concentrations:

$$c_F = 300\,\text{kg/m}^3; \quad c_{wash} = 0\,\text{kg/m}^3$$

$$c_R = \frac{(234 - 70)}{952}(1,114) = 192\,\text{kg/m}^3$$

$$c_P = \frac{70}{965}(1,045) = 76\,\text{kg/m}^3$$

The log-mean driving force for H₂SO₄ with countercurrent flow of feed and wash:

$$(\Delta c)_{LM} = \frac{(c_F - c_P) - (c_R - c_{wash})}{\ln\left(\dfrac{c_F - c_P}{c_R - c_{wash}}\right)} = \frac{(300 - 76) - (192 - 0)}{\ln\left(\dfrac{300 - 76}{192 - 0}\right)}$$

$$= 208\,\text{kg/m}^3$$

The driving force is almost constant in the membrane module, varying only from 224 to 192 kg/m³.

From (14-80),

$$K_{H_2SO_4} = \cfrac{1}{\cfrac{1}{\overline{P}_M} + \left(\cfrac{1}{k}\right)_{combined}} = \cfrac{1}{\cfrac{1}{0.025} + \cfrac{1}{0.020}}$$

$$= (0.0111\,\text{cm/min or } 0.0067\,\text{m/h})$$

From (14-81), using mass units instead of molar units:

$$A_M = \frac{m_{H_2SO_4}}{K_{H_2SO_4}(\Delta c_{H_2SO_4})_{LM}} = \frac{70}{0.0067(208)} = 50\,\text{m}^2$$

§14.5   ELECTRODIALYSIS

Electrodialysis dates back to the early 1900s, when electrodes and a direct current were used to increase the rate of dialysis. Since the 1940s, electrodialysis has become a process that differs from dialysis in many ways. Today, electrodialysis refers to an electrolytic process for separating an aqueous, electrolyte feed into concentrate and dilute or desalted water diluate by an electric field and ion-selective membranes. An electrodialysis process is shown in Figure 14.17, where the four ion-selective membranes are of two types arranged in an alternating-series pattern. The cation-selective membranes (C) carry a negative charge, and thus attract and pass positively charged ions (cations), while retarding negative ions (anions). The anion-selective membranes (A) carry a positive charge that attracts and permits passage of anions. Both types of membranes are impervious to water. The net result is that both anions and cations are concentrated in compartments 2 and 4, from which concentrate is withdrawn, and ions are depleted in compartment 3, from which the diluate is withdrawn. Compartment pressures are essentially equal. Compartments 1 and 5 contain the anode and cathode, respectively. A direct-current voltage causes current to flow through the cell by ionic conduction from the cathode to the anode. Both electrodes are chemically neutral metals, with the anode being typically stainless steel and the cathode platinum-coated tantalum, niobium, or titanium. Thus, the electrodes are neither oxidized nor reduced.

The most easily oxidized species is oxidized at the anode and the most easily reduced species is reduced at the cathode. With inert electrodes, the result at the cathode is the reduction of water by the half reaction

$$2H_2O + 2e^- \rightarrow 2OH^- + H_{2(g)}, E^0 = -0.828\,\text{V}$$

The oxidation half reaction at the anode is

$$H_2O \rightarrow 2e^- + \tfrac{1}{2}O_{2(g)} + 2H^+, E^0 = -1.23\,\text{V}$$

or, if chloride ions are present:

$$2Cl^- \rightarrow 2e^- + Cl_{2(g)}, E^0 = -1.360\,\text{V}$$

where the electrode potentials are the standard values at 25°C for 1-M solution of ions, and partial pressures of 1 atmosphere for the gaseous products. Values of E^0 can be corrected for nonstandard conditions by the Nernst equation

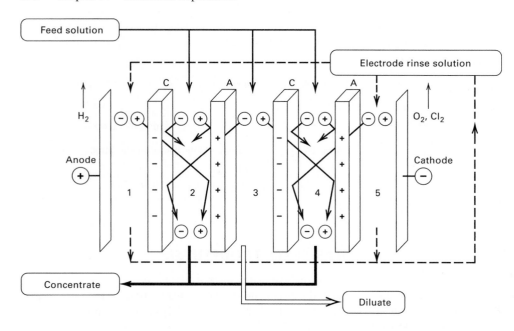

Figure 14.17 Schematic diagram of the electrodialysis process. C, cation-transfer membrane; A, anion-transfer membrane.

[Adapted from W.S.W. Ho and K.K. Sirkar, Eds., *Membrane Handbook*, Van Nostrand Reinhold, New York (1992).]

[92]. The corresponding overall cell reactions are:

$$3H_2O \rightarrow H_{2_{(g)}} + \tfrac{1}{2}O_{2_{(g)}} + 2H^+ + 2OH^-$$

or

$$2H_2O + 2Cl^- \rightarrow 2OH^- + H_{2_{(g)}} + Cl_{2_{(g)}}, E^0_{cell} = -2.058 \text{ V}$$

The net reaction for the first case is

$$H_2O \rightarrow H_{2_{(g)}} + \frac{1}{2}O_{2_{(g)}}, E^0_{cell} = -2.188 \text{ V}$$

The electrode rinse solution that circulates through compartments 1 and 5 is typically acidic to neutralize the OH ions formed in compartment 1 and prevent precipitation of compounds such as $CaCO_3$ and $Mg(OH)_2$.

The most widely used ion-exchange membranes for electrodialysis, first reported by Juda and McRae [35] in 1950, are: (1) cation-selective membranes containing negatively charged groups fixed to a polymer matrix, and (2) anion-selective membranes containing positively charged groups fixed to a polymer matrix. The former, shown schematically in Figure 14.18, includes fixed anions, mobile cations (called counterions), and mobile anions (called co-ions). The latter are almost completely excluded from the polymer matrix by

electrical repulsion, called the Donnan effect. For perfect exclusion, only cations are transferred through the membrane. In practice, the exclusion is better than 90%.

A cation-selective membrane may be made of polystyrene cross-linked with divinylbenzene and sulfonated to produce fixed sulfonate, $-SO_3^-$, anion groups. An anion-selective membrane of the same polymer contains quaternary ammonium groups such as $-NH_3^+$. Membranes are 0.2–0.5 mm thick and are reinforced for mechanical stability. The membranes are flat sheets, containing 30 to 50% water and have a network of pores too small to permit water transport.

A cell pair or unit cell contains one cation-selective membrane and one anion-selective membrane. A commercial electrodialysis system consists of a large stack of membranes in a plate-and-frame configuration, which, according to Applegate [2] and the *Membrane Handbook* [6], contains 100 to 600 cell pairs. In a stack, membranes of 0.4 to 1.5 m^2 surface area are separated by 0.5 to 2 mm with spacer gaskets. The total voltage or electrical potential applied across the cell includes: (1) the electrode potentials, (2) overvoltages due to gas formation at the two electrodes, (3) the voltage required to overcome the ohmic resistance of the electrolyte in each compartment, (4) the voltage required to overcome the resistance in each membrane, and (5) the voltage required to overcome concentration-polarization effects in the electrolyte solutions adjacent to the membrane surface. For large stacks, the latter three voltage increments predominate and depend upon the current density (amps flowing through the stack per unit surface area of membranes). A typical voltage drop across a cell pair is 0.5–1.5 V. Current densities are in the range of 5–50 mA/cm^2. Thus, a stack of 400 membranes (200 unit cells) of 1 m^2 surface area each might require 200 V at 100 A. Typically 50 to 90% of brackish water is converted to water, depending on concentrate recycle. As the current density is increased for a given membrane surface area, the concentration-polarization effect increases. Figure 14.19 is a schematic of this effect for a cation-selective membrane, where c_m refers to cation concentration in the membrane, c_b refers to bulk electrolyte cation concentration, and

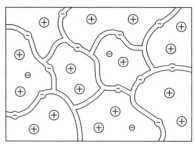

≈ Matrix with fixed charges
⊕ Counterion
⊖ Co-ion

Figure 14.18 Cation-exchange membrane.

[From H. Strathmann, *Sep. and Purif. Methods*, **14** (1), 41–66 (1985) with permission.]

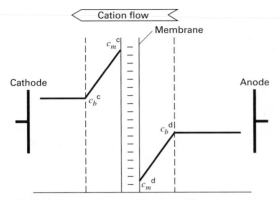

Figure 14.19 Concentration-polarization effects for a cation-exchange membrane.

[From H. Strathmann, *Sep. and Purif. Methods*, **14** (1), 41–66 (1985) with permission.]

superscripts c and d refer to concentrate side and dilute side. The maximum or limiting current density occurs when c_m^d reaches zero. Typically, an electrodialysis cell is operated at 80% of the limiting current density, which is determined by experiment, as is the corresponding cell voltage or resistance.

The gases formed at the electrodes at the ends of the stack are governed by *Faraday's law of electrolysis*. One Faraday (96,520 coulombs) of electricity reduces at the cathode and oxidizes at the anode an equivalent of oxidizing and reducing agent corresponding to the transfer of 6.023×10^{23} (Avogadro's number) electrons through wiring from the anode to the cathode. In general, it takes a large quantity of electricity to form appreciable quantities of gases in an electrodialysis process.

Of importance in design of an electrodialysis process are the membrane area and electrical-energy requirements, as discussed by Applegate [2] and Strathmann [36]. The membrane area is estimated from the current density, rather than from permeability and mass-transfer resistances, by applying Faraday's law:

$$A_M = \frac{FQ\Delta c}{i\xi} \qquad (14\text{-}82)$$

where A_M = total area of all cell pairs, m²; F = Faraday's constant (96,520 amp-s/equivalent); Q = volumetric flow rate of the diluate (potable water), m³/s; Δc = difference between feed and diluate ion concentration in equivalents/m³; i = current density, amps/m² of a cell pair, usually about 80% of i_{max}; and ξ = current efficiency < 1.00.

The efficiency accounts for the fact that not all of the current is effective in transporting ions through the membranes. Inefficiencies are caused by a Donnan exclusion of less than 100%, some transfer of water through the membranes, current leakage through manifolds, etc. Power consumption is given by

$$P = IE \qquad (14\text{-}83)$$

where P = power, W; I = electric current flow through the stack; and E = voltage across the stack. Electrical current is given by a rearrangement of (14-82),

$$I = \frac{FQ\Delta c}{n\xi} \qquad (14\text{-}84)$$

where n is the number of cell pairs.

The main application of electrodialysis is to the desalinization of brackish water in the salt-concentration range of 500 to 5,000 ppm (mg/L). Below this range, ion exchange is more economical, whereas above this range, to 50,000 ppm, reverse osmosis is preferred. However, electrodialysis cannot produce water with a very low dissolved-solids content because of the high electrical resistance of dilute solutions. Other applications include recovery of nickel and copper from electroplating rinse water; deionization of cheese whey, fruit juices, wine, milk, and sugar molasses; separation of salts, acids, and bases from organic compounds; and recovery of organic compounds from their salts. Bipolar membranes, prepared by laminating a cation-selective membrane and an anion-selective membrane back-to-back, are used to produce H_2SO_4 and NaOH from a Na_2SO_4 solution.

EXAMPLE 14.10 Electrodialysis of Brackish Water.

Estimate membrane area and electrical-energy requirements for an electrodialysis process to reduce the salt (NaCl) content of 24,000 m³/day of brackish water from 1,500 mg/L to 300 mg/L with a 50% conversion. Assume each membrane has a surface area of 0.5 m² and each stack contains 300 cell pairs. A reasonable current density is 5 mA/cm², and the current efficiency is 0.8 (80%).

Solution

Use (14-82) to estimate membrane area:

$$F = 96{,}520 \text{ A/equiv}$$

$$Q = (24{,}000)(0.5)/(24)(3{,}600) = 0.139 \text{ m}^3/\text{s}$$

$$\text{MW}_{\text{NaCl}} = 58.5, \; i = 5 \text{ mA/cm}^2 = 50 \text{ A/m}^2$$

$$\Delta c = (1{,}500 - 300)/58.5 = 20.5 \text{ mmol/L or } 20.5 \text{ mol/m}^3$$

$$= 20.5 \text{ equiv/m}^3$$

$$A_M = \frac{(1)(96{,}520)(0.139)(20.5)}{(50)(0.8)} = 6{,}876 \text{ m}^2$$

Each stack contains 300 cell pairs with a total area of 0.5(300) = 150 m². Therefore, the number of stacks = 6,876/150 = 46 in parallel. From (14-84), electrical current flow is given by

$$I = \frac{(96{,}500)(0.139)(20.5)}{(300)(0.8)}$$

$$= 1{,}146 \text{ A or } I/\text{stack} = 1{,}146/46 = 25 \text{ A/stack}$$

To obtain the electrical power, the average voltage drop across each cell pair is needed. Assume a value of 1 V. From (93) for 300 cell pairs:

$$P = (1{,}146)(1)(300) = 344{,}000 \text{ W} = 344 \text{ kW}$$

Additional energy is required to pump feed, recycle concentrate, and electrode rinse.

It is instructive to estimate the amount of feed that would be electrolyzed (as water to hydrogen and oxygen gases) at the electrodes. From the half-cell reactions presented earlier, half a molecule of H_2O is electrolyzed for each electron, or 0.5 mol H_2O is electrolyzed for each faraday of electricity.

$$1{,}146 \text{ amps} = 1{,}146 \text{ coulombs/s}$$

$$\text{or } (1{,}146)(3{,}600)(24) = 99{,}010{,}000 \text{ coulombs/day}$$

$$\text{or } 99{,}010{,}000/96{,}520 = 1{,}026 \text{ faradays/day}$$

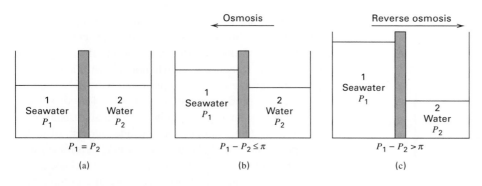

Figure 14.20 Osmosis and reverse-osmosis phenomena. (a) Initial condition. (b) At equilibrium after osmosis. (c) Reverse osmosis.

This electrolyzes $(0.5)(1,026) = 513$ mol/day of water. The feed rate is 12,000 m³/day, or

$$\frac{(12,000)(10^6)}{18} = 6.7 \times 10^8 \text{ mol/day}$$

Therefore, the amount of water electrolyzed is negligible.

§14.6 REVERSE OSMOSIS

Osmosis, from the Greek word for "push," refers to passage of a solvent, such as water, through a membrane that is much more permeable to solvent (A) than to solute(s) (B) (e.g., inorganic ions). The first recorded account of osmosis was in 1748 by Nollet, whose experiments were conducted with water, an alcohol, and an animal-bladder membrane. Osmosis is illustrated in Figure 14.20, where all solutions are at 25°C. In the initial condition (a), seawater of approximately 3.5 wt% dissolved salts and at 101.3 kPa is in cell 1, while pure water at the same pressure is in cell 2. The dense membrane is permeable to water, but not to dissolved salts. By osmosis, water passes from cell 2 to the seawater in cell 1, causing dilution of the dissolved salts. At equilibrium, the condition of Figure 14.20b is reached, wherein some pure water still resides in cell 2 and seawater, less concentrated in salt, resides in cell 1. Pressure P_1, in cell 1, is now greater than pressure P_2, in cell 2, with the difference, π, referred to as the *osmotic pressure.*

Osmosis is not a useful separation process because the solvent is transferred in the wrong direction, resulting in mixing rather than separation. However, the direction of transport of solvent through the membrane can be reversed, as shown in Figure 14.20c, by applying a pressure, P_1, in cell 1, that is higher than the sum of the osmotic pressure and the pressure, P_2, in cell 2: that is, $P_1 - P_2 > \pi$. Now water in the seawater is transferred to the pure water, and the seawater becomes more concentrated in dissolved salts. This phenomenon, called *reverse osmosis*, is used to partially remove solvent from a solute–solvent mixture. An important factor in developing a reverse-osmosis separation process is the osmotic pressure, π, of the feed mixture, which is proportional to the solute concentration. For pure water, $\pi = 0$.

In reverse osmosis (RO), as shown in Figure 14.21, feed is a liquid at high pressure, P_1, containing solvent (e.g., water) and solubles (e.g., inorganic salts and, perhaps, colloidal matter). No sweep liquid is used, but the other side of the membrane is maintained at a much lower pressure, P_2. A dense membrane such as an acetate or aromatic polyamide, permselective for the solvent, is used. To withstand the large ΔP, the membrane must be thick. Accordingly, asymmetric or thin-wall composite membranes, having a thin, dense skin or layer on a thick, porous support, are needed. The products of reverse osmosis are a permeate of almost pure solvent and a retentate of solvent-depleted feed. A perfect separation between solvent and solute is not achieved, since only a fraction of the solvent is transferred to the permeate.

Reverse osmosis is used to desalinate and purify seawater, brackish water, and wastewater. Prior to 1980, multistage, flash distillation was the primary desalination process, but by 1990 this situation was dramatically reversed, making RO the dominant process for new construction. The dramatic shift from a thermally driven process to a more economical, pressure-driven process was made possible by Loeb and Sourirajan's [7] development of an asymmetric membrane that allows pressurized water to pass through at a high rate, while almost preventing transmembrane flows of dissolved salts, organic compounds, colloids, and microorganisms. Today more than 1,000 RO desalting plants are producing more than 750,000,000 gal/day of potable water.

According to Baker et al. [5], use of RO to desalinize water is accomplished mainly with spiral-wound and

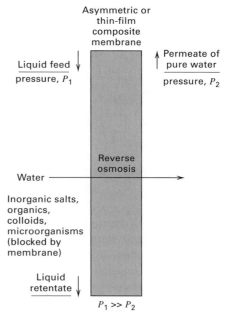

Figure 14.21 Reverse osmosis.

hollow-fiber membrane modules utilizing cellulose triacetate, cellulose diacetate, and aromatic polyamide membrane materials. Cellulose acetates are susceptible to biological attack, and to acidic or basic hydrolysis back to cellulose, making it necessary to chlorinate the feed water and control the pH to within 4.5–7.5. Polyamides are not susceptible to biological attack, and resist hydrolysis in the pH range of 4–11, but are attacked by chlorine.

The preferred membrane for the desalinization of seawater, which contains about 3.5 wt% dissolved salts and has an osmotic pressure of 350 psia, is a spiral-wound, multileaf module of polyamide, thin-film composite operating at a feed pressure of 800 to 1,000 psia. With a transmembrane water flux of 9 gal/ft^2-day (0.365 m^3/m^2-day), this module can recover 45% of the water at a purity of about 99.95 wt%. A typical module is 8 inches in diameter by 40 inches long, containing 365 ft^2 (33.9 m^2) of membrane surface. Such modules resist fouling by colloidal and particulate matter, but seawater must be treated with sodium bisulfate to remove oxygen and/or chlorine.

For desalinization of brackish water containing less than 0.5 wt% dissolved salts, hollow-fiber modules of high packing density, containing fibers of cellulose acetates or aromatic polyamides, are used if fouling is not serious. Because the osmotic pressure is much lower (<50 psi), feed pressures can be <250 psia and transmembrane fluxes may be as high as 20 gal/ft^2-day.

Other uses of reverse osmosis, usually on a smaller scale than the desalinization of water, include: (1) treatment of industrial wastewater to remove heavy-metal ions, nonbiodegradable substances, and other components of possible commercial value; (2) treatment of rinse water from electroplating processes to obtain a metal-ion concentrate and a permeate that can be reused as a rinse; (3) separation of sulfites and bisulfites from effluents in pulp and paper processes; (4) treatment of wastewater in dyeing processes; (5) recovery of constituents having food value from wastewaters in food-processing plants (e.g., lactose, lactic acid, sugars, and starches); (6) treatment of municipal water to remove inorganic salts, low-molecular-weight organic compounds, viruses, and bacteria; (7) dewatering of certain food products such as coffee, soups, tea, milk, orange juice, and tomato juice; and (8) concentration of amino acids and alkaloids. In such applications, membranes must have chemical, mechanical, and thermal stability to be competitive with other processes.

As with all membrane processes where feed is a liquid, three resistances to mass transfer must be considered: the membrane resistance and the two fluid-film or boundary-layer resistances on either side of the membrane. If the permeate is pure solvent, then there is no film resistance on that side of the membrane.

Although the driving force for water transport is the concentration or activity difference in and across the membrane, common practice is to use a driving force based on osmotic pressure. Consider the reverse-osmosis process of Figure 14.20c. At thermodynamic equilibrium, solvent chemical potentials or fugacities on the two sides of the membrane must be equal. Thus,

$$f_A^{(1)} = f_A^{(2)} \tag{14-85}$$

From definitions in Table 2.2, rewrite (14-85) in terms of activities:

$$a_A^{(1)} f_A^0\{T, P_1\} = a_A^{(2)} f_A^0\{T, P_2\} \tag{14-86}$$

For pure solvent, A, $a_A^{(2)} = 1$. For seawater, $a_A^{(1)} = x_A^{(1)} \gamma_A^{(1)}$. Substitution into (14-86) gives

$$f_A^0\{T, P_2\} = x_A^{(1)} \gamma_A^{(1)} f_A^0\{T, P_1\} \tag{14-87}$$

Standard-state, pure-component fugacities f^0 increase with increasing pressure. Thus, if $x_A^{(1)} \gamma_A^{(1)} < 1$, then from (14-87), $P_1 > P_2$. The pressure difference $P_1 - P_2$ is shown as a hydrostatic head in Figure 14.20b. It can be observed experimentally, and is defined as the osmotic pressure, π.

To relate π to solvent or solute concentration, the Poynting correction of (2-28) is applied. For an incompressible liquid of specific volume, v_A,

$$f_A^0\{T, P_2\} = f_A^0\{T, P_1\} \exp\left[\frac{v_{A_L}(P_2 - P_1)}{RT}\right] \tag{14-88}$$

Substitution of (14-87) into (14-88) gives

$$\boxed{\pi = P_1 - P_2 = -\frac{RT}{v_{A_L}} \ln\left(x_A^{(1)} \gamma_A^{(1)}\right)} \tag{14-89}$$

Thus, osmotic pressure replaces activity as a thermodynamic variable.

For a mixture on the feed or retentate side of the membrane that is dilute in the solute, $\gamma_A^{(1)} = 1$. Also, $x_A^{(1)} = 1 - x_B^{(1)}$ and $\ln\left(1 - x_B^{(1)}\right) \approx -x_B^{(1)}$. Substitution into (14-89) gives

$$\pi = P_1 - P_2 = RT x_B^{(1)} / v_{A_L} \tag{14-90}$$

Finally, since $x_B^{(1)} \approx n_B/n_A$, $n_A v_{A_L} = V$, and $n_B/V = c_B$, (14-100) becomes

$$\boxed{\pi \approx RT c_B} \tag{14-91}$$

which was used in Exercise 1.8. For seawater, Applegate [2] suggests the approximate expression

$$\pi = 1.12T \sum \bar{m}_i \tag{14-92}$$

where π is in psia, T is in K, and $\sum \bar{m}_i$ is the summation of molarities of dissolved ions and nonionic species in the solution in mol/L. More exact expressions for π are those of Stoughton and Lietzke [38].

In the general case, when there are solutes on each side of the membrane, at equilibrium $(P_1 - \pi_1) = (P_2 - \pi_2)$. Accordingly, as discussed by Merten [37], the driving force for solvent transport through the membrane is $\Delta P - \Delta \pi$, and the rate of mass transport is

$$\boxed{N_{H_2O} = \frac{P_{M_{H_2O}}}{l_M}(\Delta P - \Delta \pi)} \tag{14-93}$$

where ΔP = hydraulic pressure difference across the membrane = $P_{feed} - P_{permeate}$, and $\Delta \pi$ = osmotic pressure difference across the membrane = $\pi_{feed} - \pi_{permeate}$.

If the permeate is almost pure solvent, $\pi_{permeate} \approx 0$.

The flux of solute (e.g., salt) is given by (14-49) in terms of membrane concentrations, and is independent of ΔP across the membrane. Accordingly, the higher the ΔP, the

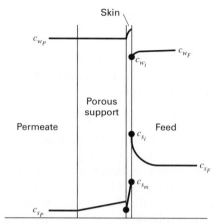

Figure 14.22 Concentration-polarization effects in reverse osmosis.

purer the permeate water. Alternatively, the flux of salt may be conveniently expressed in terms of *salt passage, SP,*

$$SP = (c_{\text{salt}})_{\text{permeate}} / (c_{\text{salt}})_{\text{feed}} \qquad (14\text{-}94)$$

Values of *SP* decrease with increasing ΔP. *Salt rejection* is given by $SR = 1 - SP$.

For brackish water of 1,500 mg/L NaCl, at 25°C, (14-92) predicts $\pi = 17.1$ psia. For seawater of 35,000 mg/L NaCl, at 25°C, (14-92) predicts $\pi = 385$ psia, while Stoughton and Lietzke [38] give 368 psia. From (14-93), ΔP must be $> \Delta\pi$ for reverse osmosis to occur. For desalination of brackish water by RO, ΔP is typically 400–600 psi, while for seawater, it is 800–1,000 psi.

The feed water to an RO unit contains potential foulants, which must be removed prior to passage through the membrane unit; otherwise, performance and membrane life are reduced. Suspended solids and particulate matter are removed by screening and filtration. Colloids are flocculated and filtered. Scale-forming salts require acidification or water softening, and biological materials require chlorination or ozonation. Other organic foulants are removed by adsorption or oxidation.

Concentration polarization is important on the feed side of RO membranes and is illustrated in Figure 14.22, where concentrations are shown for water, c_w, and salt, c_s. Because of the high pressure, activity of water on the feed side is somewhat higher than that of near-pure water on the permeate

side, thus providing the driving force for water transport through the membrane. The flux of water to the membrane carries with it salt by bulk flow, but because the salt cannot readily penetrate the membrane, salt concentration adjacent to the surface of the membrane, c_{s_i}, is $> c_{s_F}$. This difference causes mass transfer of salt by diffusion from the membrane surface back to the bulk feed. The back rate of salt diffusion depends on the mass-transfer coefficient for the boundary layer (or film) on the feed side. The lower the mass-transfer coefficient, the higher the value of c_{s_i}. The value of c_{s_i} is important because it fixes the osmotic pressure, and influences the driving force for water transport according to (14-93).

Consider steady-state transport of water with back-diffusion of salt. A salt balance at the upstream membrane surface gives

$$N_{\text{H}_2\text{O}} c_{s_F}(SR) = k_s(c_{s_i} - c_{s_F})$$

Solving for c_{s_i} gives

$$c_{s_i} = c_{s_F}\left(1 + \frac{N_{\text{H}_2\text{O}}(SR)}{k_s}\right) \qquad (14\text{-}95)$$

Values of k_s are estimated from (14-78). The concentration-polarization effect is seen to be most significant for high water fluxes and low mass-transfer coefficients.

A quantitative estimate of the importance of concentration polarization is derived by defining the concentration-polarization factor, Γ, by a rearrangement of the previous equation:

$$\Gamma \equiv \frac{c_{s_i} - c_{s_F}}{c_{s_F}} = \frac{N_{\text{H}_2\text{O}}(SR)}{k_s} \qquad (14\text{-}96)$$

Values of *SR* are in the range of 0.97–0.995. If $\Gamma >$, say, 0.2, concentration polarization may be significant, indicating a need for design changes to reduce Γ.

Feed-side pressure drop is also important because it causes a reduction in the driving force for water transport. Because of the complex geometries used for both spiral-wound and hollow-fiber modules, it is best to estimate pressure drops from experimental data. Feed-side pressure drops for spiral-wound modules and hollow-fiber modules range from 43 to 85 and 1.4 to 4.3 psi, respectively [6].

A schematic diagram of a reverse-osmosis process for desalination of water is shown in Figure 14.23. The source of feed water may be a well or surface water, which is pumped

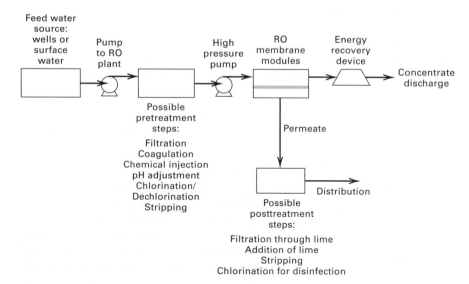

Figure 14.23 Reverse-osmosis process.

through a series of pretreatment steps to ensure a long membrane life. Of particular importance is pH adjustment. The pretreated water is fed by a high-pressure discharge pump to a parallel-and-series network of reverse-osmosis modules. The concentrate, which leaves the membrane system at a high pressure that is 10–15% lower than the inlet pressure, is then routed through a power-recovery turbine, which reduces the net power consumption by 25–40%. The permeate, which may be 99.95 wt% pure water and about 50% of the feed water, is sent to a series of post-treatment steps to make it drinkable.

EXAMPLE 14.11 Polarization Factor in Reverse Osmosis.

At a certain point in a spiral-wound membrane, the bulk conditions on the feed side are 1.8 wt% NaCl, 25°C, and 1,000 psia, while bulk conditions on the permeate side are 0.05 wt% NaCl, 25°C, and 50 psia. For this membrane the permeance values are 1.1×10^{-5} g/cm^2-s-atm for H$_2$O and 16×10^{-6} cm/s for the salt. If mass-transfer resistances are negligible, calculate the flux of water in gal/ft^2-day and the flux of salt in g/ft^2-day. If $k_s = 0.005$ cm/s, estimate the polarization factor.

Solution

Bulk salt concentrations are

$$c_{S_F} = \frac{1.8(1,000)}{58.5(98.2)} = 0.313 \text{ mol/L on feed side}$$

$$c_{S_P} = \frac{0.05(1,000)}{58.5(99.95)} = 0.00855 \text{ mol/L on permeate side}$$

For water transport, using (14-92) for osmotic pressure and noting that dissolved NaCl gives 2 ions per molecule:

$$\Delta P = (1,000 - 50)/14.7 = 64.6 \text{ atm}$$

$$\pi_{\text{feedside}} = 1.12(298)(2)(0.313) = 209 \text{ psia} = 14.2 \text{ atm}$$

$$\pi_{\text{permeate side}} = 1.12(298)(2)(0.00855) = 5.7 \text{ psia} = 0.4 \text{ atm}$$

$$\Delta P - \Delta \pi = 64.6 - (14.2 - 0.4) = 50.8 \text{ atm}$$

$$P_{M_{\text{H}_2\text{O}}}/l_M = 1.1 \times 10^{-5} \text{ g/cm}^2\text{-s-atm}$$

From (14-93),

$$N_{\text{H}_2\text{O}} = (1.1 \times 10^{-5})(50.8) = 0.000559 \text{ g/cm}^2\text{-s} \quad \text{or}$$

$$\frac{(0.000559)(3,600)(24)}{(454)(8.33)(1.076 \times 10^{-3})} = 11.9 \text{ gal/ft}^2\text{-day}$$

For salt transport:

$$\Delta c = 0.313 - 0.00855 = 0.304 \text{ mol/L} \quad \text{or} \quad 0.000304 \text{ mol/cm}^3$$

$$P_{M_{\text{NaCl}}}/l_M = 16 \times 10^{-6} \text{ cm/s}$$

From (14-49):

$$\text{or} \quad N_{\text{NaCl}} = 16 \times 10^{-6}(0.000304) = 4.86 \times 10^{-9} \text{ mol/cm}^2\text{-s}$$

$$\frac{(4.86 \times 10^{-9})(3,600)(24)(58.5)}{1.076 \times 10^{-3}} = 22.8 \text{ g/ft}^2\text{-day}$$

The flux of salt is much smaller than the flux of water.

To estimate the concentration-polarization factor, first convert the water flux through the membrane into the same units as the salt mass-transfer coefficient, k_s, i.e., cm/s:

$$N_{\text{H}_2\text{O}} = \frac{0.000559}{1.00} = 0.000559 \text{ cm/s}$$

From (14-94), the salt passage is

$$SP = 0.00855/0.313 = 0.027$$

Therefore, the salt rejection = $SR = 1 - 0.027 = 0.973$.

From (14-96), the concentration-polarization factor is

$$\Gamma = \frac{0.000559(0.972)}{0.005} = 0.11$$

Here polarization is not particularly significant.

§14.7 GAS PERMEATION

Figure 14.24 shows gas permeation (GP) through a thin film, where feed gas, at high pressure P_1, contains some low-molecular-weight species (MW < 50) to be separated from small amounts of higher-molecular-weight species. Usually a sweep gas is not needed, but the other side of the membrane is maintained at a much lower pressure, P_2, often near-ambient to provide an adequate driving force. The membrane, often dense but sometimes microporous, is permselective for the low-molecular-weight species A. If the membrane is dense, these species are absorbed at the surface and then transported through the membrane by one or more mechanisms. Then, permselectivity depends on both membrane absorption and transport rate. Mechanisms are formulated in terms of a partial-pressure or fugacity driving force using the solution-diffusion model of (14-55). The products are a permeate enriched in A and a retentate enriched in B. A near-perfect separation is generally not achievable. If the membrane is microporous, pore size is extremely important because it is necessary to block the passage of species B. Otherwise, unless molecular weights of A and B differ appreciably, only a very modest separation is achievable, as was discussed in connection with Knudsen diffusion, (14-45).

Since the early 1980s, applications of GP with dense polymeric membranes have increased dramatically. Major applications include: (1) separation of hydrogen from methane; (2) adjustment of H$_2$-to-CO ratio in synthesis gas; (3) O$_2$ enrichment of air; (4) N$_2$ enrichment of air; (5) removal of

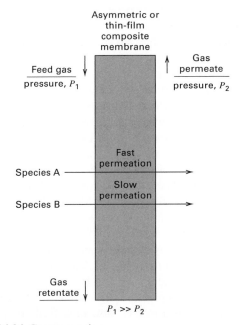

Figure 14.24 Gas permeation.

CO_2; (6) drying of natural gas and air; (7) removal of helium; and (8) removal of organic solvents from air.

Gas permeation competes with absorption, pressure-swing adsorption, and cryogenic distillation. Advantages of gas permeation, as cited by Spillman and Sherwin [39], are low capital investment, ease of installation, ease of operation, absence of rotating parts, high process flexibility, low weight and space requirements, and low environmental impact. In addition, if the feed gas is already high pressure, a gas compressor is not needed, and thus no utilities are required.

Since 1986, the most rapidly developing application for GP has been air separation, for which available membranes have separation factors for O_2 with respect to N_2 of 3 to 7. However, product purities are economically limited to a retentate of 95–99.9% N_2 and a permeate of 30–45% O_2.

Gas permeation also competes favorably for H_2 recovery because of high separation factors. The rate of permeation of H_2 through a dense polymer membrane is more than 30 times that for N_2. GP can achieve a 95% recovery of 90% pure H_2 from a feed gas containing 60% H_2.

Early applications of GP used nonporous membranes of cellulose acetates and polysulfones, which are still predominant, although polyimides, polyamides, polycarbonates, polyetherimides, sulfonated polysulfones, Teflon, polystyrene, and silicone rubber are also finding applications for temperatures to at least 70°C.

Although plate-and-frame and tubular modules can be used for gas permeation, almost all large-scale applications use spiral-wound or hollow-fiber modules because of their higher packing density. Commercial membrane modules for gas permeation are available from many suppliers. Feed-side pressure is typically 300 to 500 psia, but can be as high as 1,650 psia. Typical refinery applications involve feed-gas flow rates of 20 million scfd, but flow rates as large as 300 million scfd have been reported [40]. When the feed contains condensables, it may be necessary to preheat the feed gas to prevent condensation as the retentate becomes richer in the high-molecular-weight species. For high-temperature applications where polymers cannot be used, membranes of glass, carbon, and inorganic oxides are available, but are limited in their selectivity.

For dense membranes, external mass-transfer resistances or concentration-polarization effects are generally negligible, and (14-55) with a partial-pressure driving force can be used to compute the rate of membrane transport. As discussed in §14.3.5 on module flow patterns, the appropriate partial-pressure driving force depends on the flow pattern. Cascades are used to increase degree of separation.

Progress is being made in the prediction of permeability of gases in glassy and rubbery homopolymers, random copolymers, and block copolymers. Teplyakov and Meares [41] present correlations at 25°C for the diffusion coefficient, D, and solubility, S, applied to 23 different gases for 30 different polymers. Predicted values for glassy polyvinyltrimethylsilane (PVTMS) and rubbery polyisoprene are listed in Table 14.10. D and S values agree with data to within ±20% and ±30%, respectively.

Gas-permeation separators are claimed to be relatively insensitive to changes in feed flow rate, feed composition, and loss of membrane surface area [42]. This claim is tested in the following example.

Table 14.10 Predicted Values of Diffusivity and Solubility of Light Gases in a Glassy and a Rubbery Polymer

Permeant	$D \times 10^{11}$, m²/s	$S \times 10^4$, gmol/m³-Pa	P_M, barrer
\multicolumn{4}{c}{Polyvinyltrimethylsilane (Glassy Polymer)}			
He	470	0.18	250
Ne	87	0.26	66
Ar	5.1	1.95	30
Kr	1.5	6.22	29
Xe	0.29	20.6	18
Rn	0.07	69.6	15
H_2	160	0.54	250
O_2	7.6	1.58	37
N_2	3.8	0.84	9
CO_2	4.0	13.6	160
CO	3.7	1.28	14
CH_4	1.9	3.93	22
C_2H_6	0.12	30.2	10
C_3H_8	0.01	98.1	2.8
C_4H_{10}	0.001	347	1.2
C_2H_4	0.23	17.8	12
C_3H_6	0.038	77.6	9
C_4H_8	0.0052	293	4.5
C_2H_2	0.58	16.8	32
C_3H_4 (m)	0.17	138.1	70
C_4H_6 (e)	0.053	318.5	50
C_3H_4 (a)	0.15	186.5	83
C_4H_6 (b)	0.03	226.1	20
\multicolumn{4}{c}{Polyisoprene (Rubber-like Polymer)}			
He	213	0.06	35
Ne	77.4	0.08	18
Ar	14.6	0.58	25
Kr	7.2	1.78	25
Xe	2.7	5.68	45
Rn	1.2	18.7	64
H_2	109	0.17	54
O_2	18.4	0.47	26
N_2	12.2	0.26	10
CO_2	12.6	3.80	140
CO	12.1	0.38	14
CH_4	8.0	1.14	27
C_2H_6	3.3	8.13	79
C_3H_8	1.6	25.4	123
C_4H_{10}	1.5	86.4	390
C_2H_4	4.3	4.84	62
C_3H_6	2.7	20.3	163
C_4H_8	1.5	73.3	333
C_2H_2	5.7	4.64	80
C_3H_4 (m)	4.1	35.3	433
C_4H_6 (e)	2.9	79.6	690
C_3H_4 (a)	4.5	47.4	640
C_4H_6 (b)	3.4	40.0	410

Note: m, methylacetylene; e, ethylacetylene; a, allene; b, butadiene.

EXAMPLE 14.12 Recovery of H₂ Permeation.

The feed to a membrane separator consists of 500 lbmol/h of a mixture of 90% H_2 (H) and 10% CH_4 (M) at 500 psia. Permeance values based on a partial-pressure driving force are

$$\bar{P}_{M_H} = 3.43 \times 10^{-4} \text{ lbmol/h-ft}^2\text{-psi}$$

$$\bar{P}_{M_M} = 5.55 \times 10^{-5} \text{ lbmol/h-ft}^2\text{-psi}$$

The flow patterns in the separator are such that the permeate side is well mixed and the feed side is in plug flow. The pressure on the permeate side is constant at 20 psia, and there is no feed-retentate pressure drop. (a) Compute the membrane area and permeate purity if 90% of the hydrogen is transferred to the permeate. (b) For the membrane area determined in part (a), calculate the permeate purity and hydrogen recovery if: (1) the feed rate is increased by 10%, (2) the feed composition is reduced to 85% H_2, and (3) 25% of the membrane area becomes inoperative.

Solution

The following independent equations apply to all parts of this example. Component material balances:

$$n_{i_F} = n_{i_R} + n_{i_P}, \quad i = \text{H, M} \qquad (1, 2)$$

Dalton's law of partial pressures:

$$P_k = p_{H_k} + p_{M_k}, \quad k = F, R, P \qquad (3, 4, 5)$$

Partial-pressure–mole relations:

$$p_{H_k} = P_k n_{H_k}/(n_{H_k} + n_{M_k}), \quad k = F, R, P \qquad (6, 7, 8)$$

Solution-diffusion transport rates are obtained using (14-55), assuming a log-mean partial-pressure driving force based on the exiting permeate partial pressures on the downstream side of the membrane because of the assumption of perfect mixing.

$$n_{i_P} = \bar{P}_{M_i} A_M \left[\frac{p_{i_F} - p_{i_R}}{\ln\left(\dfrac{p_{i_F} - p_{i_P}}{p_{i_R} - p_{i_P}}\right)} \right], \quad i = \text{H, M} \qquad (9, 10)$$

Thus, a system of 10 equations has the following 18 variables:

$$A_M \quad n_{H_F} \quad n_{M_F} \quad P_F \quad P_R \quad P_P$$
$$\bar{P}_{M_H} \quad n_{H_R} \quad n_{M_R} \quad p_{H_F} \quad p_{H_R} \quad p_{H_P}$$
$$\bar{P}_{M_M} \quad n_{H_P} \quad n_{M_P} \quad p_{M_F} \quad p_{M_R} \quad p_{M_P}$$

To solve the equations, eight variables must be fixed. For all parts of this example, the following five variables are fixed:

$$\bar{P}_{M_H} \text{ and } P_{M_M} \text{ given above}$$

$$P_F = 500 \text{ psia} \quad P_R = 500 \text{ psia} \quad P_P = 20 \text{ psia}$$

For each part, three additional variables must be fixed.

(a)
$$n_{H_F} = 0.9(500) = 450 \text{ lbmol/h}$$
$$n_{M_F} = 0.1(500) = 50 \text{ lbmol/h}$$
$$n_{H_P} = 0.9(450) = 405 \text{ lbmol/h}$$

Solving Equations (1)–(10) above, using a program such as MathCad, Matlab, or Polymath,

$$A_M = 3,370 \text{ ft}^2$$
$$n_{M_P} = 20.0 \text{ lbmol/h} \quad n_{H_R} = 45.0 \text{ lbmol/h} \quad n_{M_R} = 30.0 \text{ lbmol/h}$$
$$p_{H_F} = 450 \text{ psia} \quad p_{M_F} = 50 \text{ psia} \quad p_{H_R} = 300 \text{ psia}$$
$$p_{M_R} = 200 \text{ psia} \quad p_{H_P} = 19.06 \text{ psia} \quad p_{M_P} = 0.94 \text{ psia}$$

(b) Calculations are made in a similar manner using Equations (1)–(10). Results for parts (1), (2), and (3) are:

	Part		
	(1)	(2)	(3)
Fixed:			
n_{H_F}, lbmol/h	495	425	450
n_{M_F}, lbmol/h	55	75	50
A_M, ft²	3,370	3,370	2,528
Calculated, in lbmol/h:			
n_{H_P}	424.2	369.6	338.4
n_{M_P}	18.2	25.9	11.5
n_{H_R}	70.8	55.4	111.6
n_{M_R}	36.8	49.1	38.5
Calculated, in psia:			
p_{H_F}	450	425	450
p_{M_F}	50	75	50
p_{H_R}	329	265	372
p_{M_R}	171	235	128
p_{H_P}	19.18	18.69	19.34
p_{M_P}	0.82	1.31	0.66

From the above results:

	Part			
	(a)	(b1)	(b2)	(b3)
Mol% H_2 in permeate	95.3	95.9	93.5	96.7
% H_2 recovery in permeate	90	85.7	87.0	75.2

It is seen that when the feed rate is increased by 10% (Part b1), the H_2 recovery drops about 5%, but the permeate purity is maintained. When the feed composition is reduced from 90% to 85% H_2 (Part b2), H_2 recovery decreases by about 3% and permeate purity decreases by about 2%. With 25% of the membrane area inoperative (Part b3), H_2 recovery decreases by about 15%, but the permeate purity is about 1% higher. Overall, percentage changes in H_2 recovery and purity are less than the percentage changes in feed flow rate, feed composition, and membrane area, thus confirming the insensitivity of gas-permeation separators to changes in operating conditions.

§14.8 PERVAPORATION

Figure 14.25 depicts pervaporation (PV), which differs from dialysis, reverse osmosis, and gas permeation in that the phase on one side of the pervaporation membrane is different from that on the other. Feed to the membrane module is a liquid mixture at pressure P_1, which is high enough to maintain a liquid phase as the feed is depleted of species A and B to produce liquid retentate. A composite membrane is used that is selective for species A, but with some finite permeability for species B. The dense, thin-film side of the membrane is in contact with the liquid side. The retentate is enriched in species B. Generally, a sweep fluid is not used on the other

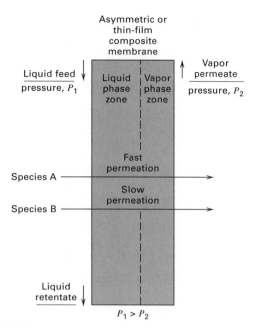

Figure 14.25 Pervaporation.

side of the membrane, but a pressure P_2, which may be a vacuum, is held at or below the dew point of the permeate, making it vapor. Vaporization may occur near the downstream face such that the membrane operates with two zones, a liquid-phase zone and a vapor-phase zone, as shown in Figure 14.25. Alternatively, the vapor phase may exist only on the permeate side of the membrane. The vapor permeate is enriched in species A. Overall permeabilities of species A and B depend on solubilities and diffusion rates. Generally, solubilities cause the membrane to swell.

The term *pervaporation* is a combination of the words "*perm*selective" and "e*vaporation*." It was first reported in 1917 by Kober [43], who studied several experimental techniques for removing water from albumin–toluene solutions. The economic potential of PV was shown by Binning et al. [44] in 1961, but commercial applications were delayed until the mid-1970s, when adequate membrane materials became available. Major commercial applications now include: (1) dehydration of ethanol; (2) dehydration of other organic alcohols, ketones, and esters; and (3) removal of organics from water. The separation of close-boiling organic mixtures like benzene–cyclohexane is receiving much attention.

Pervaporation is favored when the feed solution is dilute in the main permeant because sensible heat of the feed mixture provides the permeant enthalpy of vaporization. If the feed is rich in the main permeant, a number of membrane stages may be needed, with a small amount of permeant produced per stage and reheating of the retentate between stages. Even when only one membrane stage is sufficient, the feed liquid may be preheated.

Many pervaporation schemes have been proposed [6], with three important ones shown in Figure 14.26. A hybrid process for integrating distillation with pervaporation to produce 99.5 wt% ethanol from a feed of 60 wt% ethanol is shown in Figure 14.26a. Feed is sent to a distillation column operating at near-ambient pressure, where a bottoms product

of nearly pure water and an ethanol-rich distillate of 95 wt% is produced. The distillate purity is limited by the 95.6 wt% ethanol–water azeotrope. The distillate is sent to a pervaporation unit, where a permeate of 25 wt% alcohol and a retentate of 99.5 wt% ethanol is produced. The permeate vapor is condensed under vacuum and recycled to the distillation column, the vacuum being sustained with a vacuum pump. The dramatic difference in separability by pervaporation as compared to vapor–liquid equilibrium for distillation is shown in Figure 14.27 from Wesslein et al. [45], with a 45° line for reference. For pervaporation, compositions refer to a liquid feed (abscissa) and a vapor permeate (ordinate) at 60°C for a polyvinylalcohol (PVA) membrane and a vacuum of 15 torr. There is no limitation on ethanol purity, and the separation index is high for feeds of > 90 wt% ethanol.

A pervaporation process for dehydrating dichloroethylene (DCE) is shown in Figure 14.26b. The liquid feed, which is DCE saturated with water (0.2 wt%), is preheated to 90°C at 0.7 atm and sent to a PVA membrane system, which produces a retentate of almost pure DCE (<10 ppm H_2O) and a permeate vapor of 50 wt% DCE. Following condensation, the two resulting liquid phases are separated, with the DCE-rich phase recycled and the water-rich phase sent to an air stripper, steam stripper, adsorption unit, or hydrophobic, pervaporation membrane system for residual DCE removal.

For removal of VOCs (e.g., toluene and trichloroethylene) from wastewater, pervaporation with hollow-fiber modules of silicone rubber can be used, as shown in Figure 14.23c. The retentate is almost pure water (<5 ppb of VOCs) and the permeate, after condensation, is (1) a water-rich phase that is recycled to the membrane system and (2) a nearly pure VOC phase.

A pervaporation module may operate with heat transfer or adiabatically, with the enthalpy of vaporization supplied by feed enthalpy. Consider the adiabatic pervaporation of a binary liquid mixture of A and B. Ignore heat of mixing. For an enthalpy datum temperature of T_0, an enthalpy balance—in terms of mass flow rates m, liquid sensible heats C_P, and heats of vaporization ΔH^{vap}—gives

$$(m_{A_F} C_{P_A} + m_{B_F} C_{P_B})(T_F - T_0)$$
$$= [(m_{A_F} - m_{A_P})C_{P_A} + (m_{B_F} - m_{B_P})C_{P_B}](T_R - T_0)$$
$$+ (m_{A_P}C_{P_A} + m_{B_P}C_{P_B})(T_P - T_0) + m_{A_P}\Delta H_A^{vap}$$
$$+ m_{B_P}\Delta H_B^{vap}$$

$$(14\text{-}97)$$

where enthalpies of vaporization are evaluated at T_P. After collection of terms, (14-98) reduces to

$$(m_{A_F}C_{P_A} + m_{B_F}C_{P_B})(T_F - T_R)$$
$$= (m_{A_P}C_{P_A} + m_{B_P}C_{P_B}) \times (T_P - T_R) \qquad (14\text{-}98)$$
$$+ (m_{A_P}\Delta H_A^{vap} + m_{B_P}\Delta H_B^{vap})$$

Permeate temperature, T_P, is the dew point at the permeate vacuum upstream of the condenser. The retentate temperature is computed from (14-98).

Membrane selection is critical in the commercial application of PV. For water permeation, hydrophilic membrane materials are preferred. For example, a three-layer composite

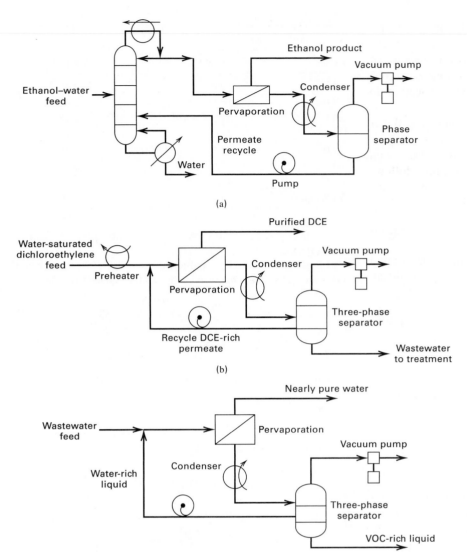

(a)

(b)

(c)

Figure 14.26 Pervaporation processes.
(a) Hybrid process for removal of water from ethanol. (b) Dehydration of dichloroethylene. (c) Removal of volatile organic compounds (VOCs) from wastewater.

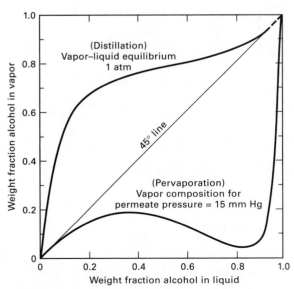

Figure 14.27 Comparison of ethanol–water separabilities.
[From M. Wesslein et al., *J. Membrane Sci.*, **51**, 169 (1990).]

membrane is used for the dehydration of ethanol, with water being the main permeating species. The support layer is porous polyester, which is cast on a microporous polyacrylonitrile or polysulfone membrane. The final layer, which provides the separation, is dense PVA of 0.1 μm in thickness. This composite combines chemical and thermal stability with adequate permeability. Hydrophobic membranes, such as silicone rubber and Teflon, are preferred when organics are the permeating species.

Commercial membrane modules for PV are almost exclusively of the plate-and-frame type because of the ease of using gasketing materials that are resistant to organic solvents and the ease of providing heat exchange for vaporization and high-temperature operation. Hollow-fiber modules are used for removal of VOCs from wastewater. Because feeds are generally clean and operation is at low pressure, membrane fouling and damage is minimal, resulting in a useful membrane life of 2–4 years.

Models for transport of permeant through a membrane by pervaporation have been proposed, based on solution-diffusion (§14.3.4). They assume equilibrium between the upstream liquid and the upstream membrane surface, and between the downstream vapor and its membrane side. Membrane transport follows Fick's law, with a permeant concentration gradient as the driving force. However, because of phase change and nonideal-solution feed, simple equations like (14-79) for dialysis and (14-55) for gas permeation do not apply.

A convenient PV model is that of Wijmans and Baker [46], who express the driving force for permeation in terms of a partial-vapor-pressure difference. Because pressures on both sides of the membrane are low, the gas phase follows the ideal-gas law. Therefore, at the upstream membrane surface (1), permeant activity for component i is

$$a_i^{(1)} = f_i^{(1)}/f_i^{(0)} = p_i^{(1)}/P_i^{s(1)} \tag{14-99}$$

where P_i^s is the vapor pressure at the feed temperature. Liquid on the upstream side of the membrane is generally nonideal. Thus, from Table 2.2:

$$a_i^{(1)} = \gamma_i^{(1)} x_i^{(1)} \tag{14-100}$$

Combining (14-99) and (14-100):

$$p_i^{(1)} = \gamma_i^{(1)} x_i^{(1)} P_i^{s(1)} \tag{14-101}$$

On the vapor side of the membrane (2), partial pressure is

$$p_i^{(2)} = y_i^{(2)} P_P^{(2)} \tag{14-102}$$

Thus, the driving force can be expressed as

$$\left(\gamma_i^{(1)} x_i^{(1)} P_i^{s(1)} - y_i^{(2)} P_P^{(2)} \right)$$

The corresponding permeant flux, after dropping unnecessary superscripts, is

$$N_i = \frac{P_{M_i}}{l_M} \left(\gamma_i x_i P_i^s - y_i P_P \right) \tag{14-103}$$

or

$$N_i = \bar{P}_{M_i} \left(\gamma_i x_i P_i^s - y_i P_P \right) \tag{14-104}$$

where γ_i and x_i refer to feed-side liquid, P_i^s is the vapor pressure at the feed-side temperature, y_i is the mole fraction in the permeant vapor, and P_P is total permeant pressure. Unlike gas permeation, where P_{M_i} depends mainly on permeant, polymer, and temperature, the permeability for pervaporation depends also on the concentrations of permeants in the polymer, which can be large enough to cause swelling and cross-diffusion. It is thus best to back-calculate and correlate permeant flux with feed composition at a given feed temperature and permeate pressure. Because of nonideal effects, selectivity can be a strong function of feed concentration and permeate pressure, causing inversion of selectivity, as illustrated next.

EXAMPLE 14.13 Pervaporation.

Wesslein et al. [45] present the following experimental data for the pervaporation of liquid mixtures of ethanol (1) and water (2) at a feed temperature of 60°C for a permeate pressure of 76 mmHg, using a commercial polyvinylalcohol membrane:

| wt% Ethanol | | Total Permeation Flux |
Feed	Permeate	kg/m²-h
8.8	10.0	2.48
17.0	16.5	2.43
26.8	21.5	2.18
36.4	23.0	1.73
49.0	22.5	1.46
60.2	17.5	0.92
68.8	13.0	0.58
75.8	9.0	0.40

At 60°C, vapor pressures are 352 and 149 mmHg for ethanol and water, respectively.

Liquid-phase activity coefficients at 60°C for the ethanol (1)–water (2) system are given by the van Laar equations (§2.6.5):

$$\ln \gamma_1 = 1.6276 \left[\frac{0.9232 x_2}{1.6276 x_1 + 0.9232 x_2} \right]^2$$

$$\ln \gamma_2 = 0.9232 \left[\frac{1.6276 x_1}{1.6276 x_1 + 0.9232 x_2} \right]^2$$

Calculate permeance for water and ethanol from (14-104).

Solution

For the first row of data, mole fractions in the feed (x_i) and permeate (y_i), with $MW_1 = 46.07$ and $MW_2 = 18.02$, are

$$x_1 = \frac{0.088/46.07}{\dfrac{0.088}{46.07} + \dfrac{(1.0 - 0.088)}{18.02}} = 0.0364$$

$$x_2 = 1.0 - 0.0364 = 0.9636$$

$$y_1 = \frac{0.10/46.07}{\dfrac{0.10}{46.07} + \dfrac{0.90}{18.02}} = 0.0416$$

$$y_2 = 1.0 - 0.0416 = 0.9584$$

Activity coefficients for the feed mixture are

$$\gamma_1 = \exp\left\{ 1.6276 \left[\frac{0.9232(0.9636)}{1.6276(0.0364) + 0.9232(0.9636)} \right]^2 \right\} = 4.182$$

$$\gamma_2 = \exp\left\{ 0.9232 \left[\frac{1.6276(0.0364)}{1.6276(0.0364) + 0.9232(0.9636)} \right]^2 \right\} = 1.004$$

From the total mass flux, component molar fluxes are

$$N_1 = \frac{(2.48)(0.10)}{46.07} = 0.00538 \frac{kmol}{h - m^2}$$

$$N_2 = \frac{(2.48)(0.90)}{18.02} = 0.1239 \frac{kmol}{h - m^2}$$

From (14-104), permeance values are

$$\bar{P}_{M_1} = \frac{0.00538}{(4.182)(0.0364)(352) - (0.0416)(76)}$$

$$= 0.000107 \frac{kmol}{h - m^2 - mmHg}$$

$$\bar{P}_{M_2} = \frac{0.1239}{(2.004)(1.0 - 0.0364)(149) - (1.0 - 0.0416)(76)}$$

$$= 0.001739 \frac{kmol}{h - m^2 - mmHg}$$

Results for other feed conditions are computed in a similar manner:

wt% Ethanol		Activity Coefficient in Feed		Permeance, kmol/h-m^2-mmHg	
Feed	Permeate	Ethanol	Water	Ethanol	Water
8.8	10.0	4.182	1.004	1.07×10^{-4}	1.74×10^{-3}
17.0	16.5	3.489	1.014	1.02×10^{-4}	1.62×10^{-3}
26.8	21.5	2.823	1.038	8.69×10^{-5}	1.43×10^{-3}
36.4	23.0	2.309	1.077	6.14×10^{-5}	1.17×10^{-3}
49.0	22.5	1.802	1.158	4.31×10^{-5}	1.10×10^{-3}
60.2	17.5	1.477	1.272	1.87×10^{-5}	8.61×10^{-4}
68.8	13.0	1.292	1.399	7.93×10^{-6}	6.98×10^{-4}
75.8	9.0	1.177	1.539	3.47×10^{-6}	6.75×10^{-4}

The PVA membrane is hydrophilic. As concentration of ethanol in the feed liquid increases, sorption of feed liquid by the membrane decreases, reducing polymer swelling. As swelling is reduced, the permeance of ethanol decreases more rapidly than that of water, thus increasing selectivity for water. For example, selectivity for water can be defined as

$$\alpha_{2,1} = \frac{(100 - w_1)_P / (w_1)_P}{(100 - w_1)_F / (w_1)_F}$$

where w_1 = weight fraction of ethanol. For cases of 8.8 and 75.8 wt% ethanol in the feed, the selectivities for water are, respectively, 0.868 (more selective for ethanol) and 31.7 (more selective for water).

§14.9 MEMBRANES IN BIOPROCESSING

Semipermeable membranes are widely used to selectively retain and/or permeate biological species based on relative size or solubility in a membrane phase. Membrane bioseparations are the subject of many reviews [50–54] and textbooks [55, 56]. Operations for bioproducts include reverse osmosis (RO), electrodialysis (ED), pervaporation (PV), microfiltration (MF), ultrafiltration (UF), nanofiltration (NF), and virus filtration (VF). Selection of a well-suited operation is guided by considering: (1) physical features of the biological species to be separated; (2) attributes of its matrix; (3) pore-size distribution and surface properties of the membrane; and (4) transport features of a particular operation. These determine species-specific *selectivity* of a membrane. Table 14.11 illustrates nominal membrane pore sizes and species diameters of some membrane bioprocesses and solutions.

Membrane installation and operating costs relate directly to selectivity, membrane *capacity* (volume of feed processed per unit of *surface area* before regeneration or replacement), and *permeate flux* (volumetric flow rate per unit area). Capacity and flux are optimized by reducing buildup of dissolved solids near the membrane wall (*concentration polarization*, CP) and preventing deposition of dissolved or suspended solutes on or in the membrane (*fouling*) [69].

Surface charges on suspended species, filter surfaces, and filter cake, as well as solution state (pH, ionic strength, temperature), all affect retention and resolution of targeted species. Operation is usually at or near physiological conditions to maintain biological activity. Solutions are buffered to pH 7.2 (37°C). Pressure-driving forces are $5 \leq \Delta P \leq 50$ psi. Temperatures are in the range $4 \leq T \leq 30$°C, with 4°C often selected to minimize protease or nuclease activity and growth of contaminant microorganisms.

Factors affecting selection of a membrane for bioseparation include selectivity, biocompatibility, chemical inertness, mechanical stability, and economics.

1. **Selectivity** (e.g., relative solute rejection). The membrane must retain active species and pass contaminants at targeted specifications. Table 14.12 lists sizes of some common solutes in biological streams.

2. **Biocompatibility.** The membrane should resist inactivation, plugging, and fouling by biological species or solution components. *Hydrophilic* membranes resist protein inactivation and fouling better than hydrophobic membranes. Examples of hydrophilic membranes include cellulose (cellulose acetate, regenerated

Table 14.11 Nominal Size in meters of Membrane Pores and Filtered Species in Membrane Bioprocesses

Size, m	10^{-10} (Å) RO	10^{-9} (nm) UF, NF	10^{-8} UF, VF	10^{-7} MF	10^{-6} (μm) MF	10^{-5} MF
Process water	water, salts, ions: monovalent	ions: divalent	colloids		bacteria	
Whole milk	lactose, salts		proteins	fats		casein micelles
Whole blood	salts	amino acids	peptides, proteins	platelets		erythrocytes
Cell culture broth	salts, glucose, vitamins, antibiotics	amino acids, lipids			bacteria	mammalian cells
Cell lysate			virus, proteins, organic macro molecules	virus	cell debris	

Adapted from [51].

Table 14.12   Sizes of Biological Solutes

	MW (dalton)		Diameter (nm)	
	lower	upper	lower	upper
yeast, fungi	-	-	1000.0	10000.0
human red blood cell	-	-	7000.0	8000.0
bacteria	-	-	300.0	1000.0
virus	-	2×10^8	20.0	300.0
protein, polysaccharide	10000	1×10^6	2.0	12.0
antibiotic	300	1000	0.6	1.2
mono-disaccharide	200	400	0.8	1.0
organic acid	100	500	0.4	0.8
inorganic acid	10	100	0.2	0.4
water	18	-	0.2	-

Adapted from [50].

cellulose, cellulose nitrate), polyamide, polyethersulfones (PES), borosilicate glass, or poly(vinylidene difluoride) (i.e., PVDF or Kynar®). *Hydrophobic* membranes are readily fouled by fatty acids, surfactants, and antifoams. They include polyethylene, polypropylene, polycarbonate, and polysulfones, which adsorb and denature proteins; and nonwetted poly(tetrafluoroethylene) (i.e., PTFE or Teflon®).

3. **Chemical inertness.** The membrane, filter housing, and associated glues, resins, potting agents, and adhesives must tolerate use and validatable cleaning and/or sterilization cycles without causing side reactions (e.g., denaturation) or producing significant leaching or *extractables*. Levels of leachables and extractables during operation and cleaning may be validated to be low ($<$1–10 µg/mL) by comparing NMR spectra of final bulks to profiles from model process streams. Filters for parenteral solutions (for injection or infusion) must comply with United States Pharmacopeia (USP) limits for extractables.

4. **Mechanical stability.** The membrane, its housing, and affiliated components must be able to withstand pressure and temperature ranges employed during use and validatable cleaning and/or sterilization cycles (including backflushing). Membrane housings typically consist of Type 304, 316, or 316L stainless steel that has

been passivated and electropolished to ensure a measured roughness average (RA) $<$ 20.

5. **Economics.** The filter operation and membrane consumables must reliably provide adequate filtrate flux (minimal concentration polarization) and capacity (minimal surface-area requirement) at reasonable operating conditions (minimal pressure drop to decrease energy consumption) over its use period when projected capital and consumable costs are compared with acceptable alternatives. This includes validatable cleanability and/or sterilizability and re-use, easy module replacement, rapid implementation and scale-up, and long-term vendor support.

§14.9.1   Membrane Operations in Bioprocess Purification Trains

Economic and reliable recovery of active biological product often employs a series of membrane filter operations to harvest cells, clarify debris, concentrate product, remove impurities, reduce bioburden to provide sterility, and exchange buffers. At each filtration step, purified bioproduct must be recovered and its activity preserved while satisfying current good manufacturing practice (cGMP) requirements. Table 14.13 gives a typical sequence of purification steps with corresponding membrane operations and comparable alternatives. Membrane biofiltration technology closely parallels filtration technology used by the chemical industry, as described in Chapter 19. The differences are that in the chemical industry, particles are much larger, woven cloths or metal screens rather than membranes are used as filters to retain the particles, and the hydrodynamics may be different (tangential flow, for example, is not used).

Initial *harvest* of cells can use MF, UF, or centrifugation to recover and concentrate cells from fermentation broth or culture media. This immediately reduces processing volume and associated costs. After harvest, intracellular bioproducts can be recovered from cells by *disruption* using shear, pressure, temperature, or chemical means. *Clarification* using MF alone, or in conjunction with a filter aid (e.g., diatomaceous earth), retains spent cells, fragments, and protein debris while permeating soluble bioproducts.

Purification by using UF can be used with, or instead of, precipitation or chromatography to concentrate product and remove soluble cellular impurities such as nucleic acids and

Table 14.13   Filtration Steps in Biopurification Trains

Step: Purpose	Filter Type	Alternatives
Harvest: Concentrate cells from broth	MF, UF	Centrifuge
Disrupt: Lyse cells		
Debris Clarification: Remove cell debris	MF, filter aid	Centrifuge; expanded-bed chromatography
Purification: Concentrate product; remove impurity	UF	Crystallization; precipitation; chromatography
Polish	MF, UF	Crystallization; ultracentrifuge; chromatography
Sterile Filtration: Reduce bioburden	0.22 µm MF	
Buffer Exchange	MF, UF	

proteins. *Polishing* with direct-flow (dead-end) MF, described in §14.9.3, may be used in lieu of ultracentrifugation to remove residual insoluble particulate and precipitated impurities. *Sterile filtration* in pharmaceutical operations uses a validatable sterilizing-grade 0.22 μm MF to reduce bioburden in preparation for subsequent formulation. One or more sterile *buffer exchanges* often follow sterile filtration to incorporate excipients or adjuvants into the final bulk product prior to filling vials. Compared to centrifugation, membrane filtration of biological products is energy-efficient and less capital intensive, with less product shear and less severe operating conditions.

§14.9.2 Biofiltration Operating Modes

MF, NF, UF, and VF of bioproducts may be conducted by flowing feed normal to a dead-end membrane surface (referred to as direct flow, normal flow, in-line or *dead-end filtration*, DEF) or tangentially across the surface (called *crossflow* or *tangential-flow filtration*, TFF). Figure 14.28 compares normal- and tangential-flow modes. In DEF, a batch of feed solution is forced under pressure through the membrane, causing retained material to accumulate on and within the membrane. The pressure required to maintain a desired flow rate must increase, or permeate flux will decrease. A combined operation, as described in §14.3.1 and illustrated in Example 14.3, in which constant-flux operation is employed up to a limiting pressure, followed by constant-pressure operation until a minimum flux is reached, is superior to either constant-pressure or constant-flux operation. DEF has lower capital cost, lower complexity, and higher operating cost relative to TFF. DEF is better suited for dilute solutions, while TFF can be employed for concentrated solutions.

In TFF, which is more suitable for large-scale, continuous filtration, feed flows along the surface, with only a fraction of the solvent passing through the membrane, while retained matter is carried out with the retentate fluid. Retentate is usually recycled through the filter at tangential-flow velocities parallel to the membrane surface in the 3–25-ft/s range. TFF gives up to 10-fold-higher flux values than DEF [57].

The tangential-flow mode is also used, almost exclusively, for RO, as discussed in §14.6, and for UF. Improvements in product yield and throughput in TFF have been demonstrated by operating to maintain flux rather than *transmembrane pressure drop* (TMP). Concentration factors up to 100-fold in single-stage UF systems have been demonstrated using high membrane-packing density and reduced holdup volumes. Maintaining constant retained protein concentration at the membrane surface (c_{wall}) has been shown to enhance product yield and minimize membrane area for large variations in feed quality and membrane properties [58]. Flux data at successively higher TMP values taken at multiple concentrations is fit to stagnant film and osmotic pressure models to estimate values of mass-transfer coefficients, osmotic virial coefficients, and fouled membrane resistance to guide operation to maintain constant c_{wall}, a variable that is not known *a priori*.

High-performance TFF (HPTFF) uses optimal values of buffer pH, ionic strength, and membrane charge to maximize differences in hydrodynamic volume between product and impurity to enhance mass throughput and selectivity as a function of local, pressure-dependent flux [59, 60]. Cocurrent flow on the membrane filtrate side maintains uniform TMP at or below the point at which filtrate flux becomes pressure independent. HPTFF can separate equally sized proteins based on charge differences, monomers from dimers, and single-amino-acid variants in real, dilute feeds, significantly improving yield and purification factors. Scalable UF devices are available that permit 1000-fold volumetric increases with consistent protein yield and permeate flux by increasing channel number in hollow-fiber cassettes or by decreasing channel width in flat-sheet cassettes while maintaining pressure, fluid flow, concentration profile, and channel length [61].

Membrane Geometries for Bioseparations

The most common membrane geometries used in bioprocessing are flat plate, spiral wound, tubular (internal diameter [i.d.] > 0.635 cm), capillary (0.1 < i.d. < 0.635 cm), and hollow fibers (0.025 < i.d. < 0.1 cm), which need clarified feed to avoid clogging. Flat-plate membranes are commonly used in plate-and-frame, filter-leaf, Nutsch, and rotating filter configurations. Plate-and-frame and filter-leaf (pleated) cartridges are typically used for MF. In the latter, the membrane is pleated and then folded around a permeate core. Many module types are inexpensive and disposable. A typical disposable cartridge is 2.5 inches in diameter by 10 inches long, with 3 ft^2 of membrane area. The cartridge may include a prefilter to extend filter life by removing large particles, leaving the microporous membrane to make the required separation. For UF, newer composite-regenerated cellulose membranes that are mechanically strong, easily cleaned, and foul less than synthetic polymers provide better permeability and retention [62, 63]. Covalent surface modification with quaternary amine or sulfonic-acid groups improves membrane selectivity, particularly for HPTFF applications.

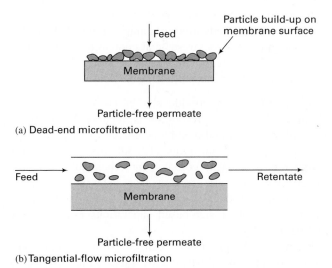

(a) Dead-end microfiltration

(b) Tangential-flow microfiltration

Figure 14.28 Common modes of microfiltration.

Membrane Casting

Polymer membranes used widely in MF, UF, and RO of bio-products are typically prepared by casting a polymer that has been dissolved in a mixture of solvent and high-boiling non-solvent as a film of precise thickness on a conveyer in an environmentally controlled chamber [64]. The casting process produces membranes in which pores result from inter-connected openings between polyhedral cells formed by progressive evaporation of solvent that causes phase separation. The nonsolvent coalesces into droplets surrounded by a shell of polymer, which gels out of solution and concentrates at phase interfaces. Further solvent evaporation deposits additional polymer that thickens swelling polymer shells, which come into mutual contact as solvent disappears. Area-minimizing forces consolidate the shells into clusters that are distorted into polyhedral cells filled with nonsolvent. Cell edges accumulate polymer, thinning the walls, which rupture and create interconnections between adjacent cells. Metering pores of the membrane consist of the interconnected openings between the polyhedral cells. The concentration of polymer in solution determines intersegmental separation of flexible chain segments that coil and overlap, as opposing electrical attractive and repulsive forces maintain separation of long polymer molecules, increasing pore size at greater dilution.

Membrane Requirements for Biotechnology

Process filters to prepare biopharmaceutical agents described in §1.9—like recombinant proteins or DNA, vaccine antigens, or viral vectors for gene therapy—have the following unique requirements when compared with bioprocess filters used to prepare food and beverages, or to purify other non-pharmacological bioproducts:

1. **Preserve biological activity.** Denaturation, proteolytic cleavage, or misforming of protein projects must be avoided. Immunogenicity of a targeted vaccine antigen, for example, must be maintained.

2. **Satisfy cGMP requirements.** Depending on the application, these may include biocompatibility, sterilizability, and flushout of extractables.

3. **Accommodate modest scales of operation.** Dose sizes of mg or less may be required for vaccine antigens or recombinant proteins. Milligrams to grams of active agents may be recovered from just 10 to 1,000 liters of broth, so process scales are relatively small, particularly for orphan drugs that treat rare diseases.

4. **Include batch operation.** A defined volume of pharmaceutical product undergoes a battery of assays to verify activity, purity, sterility, and other mandates in the Code of Federal Regulations (CFR) to be approveable by the Food and Drug Administration (FDA).

Batch bioprocess volumes are a consequence of the volume of fermentation or cell culture required to produce sufficient active bioproduct to economically satisfy market demands. This batch volume is processed discretely from inception to final release to eliminate carryover contamination that may compromise multiple batches. The batch nature and release criteria of biopharmaceutical operations distinguish them from large-scale, continuous bioprocesses.

Challenges Unique to Filtration in Biotechnology

There are also the following unique challenges to implementing filtration in vaccine bioprocesses in the pharmaceutical industry:

1. **Integrated process.** The process may "define the product," particularly when complete physicochemical characterization of a biological antigen to satisfy FDA regulatory requirements is not possible. Therefore, filtration cannot be implemented or optimized in isolation, but must be approached as an integral part of the entire series of fermentation, purification, and formulation steps.

2. **Compressed development.** Pressing market need for biotechnology products to prevent or treat public health problems drives accelerated timelines for development. Consequently, as little as weeks to months may be available to select and optimize filters in the lab.

3. **Limited raw materials.** Only mL to L of fermentation or cell culture broth may be initially available for filter selection, characterization, and optimization.

4. **Variable fermentation or cell culture.** Membrane filter operations must accommodate wide variations in cell culture and fermentation composition and productivity while providing consistent yield and purity. Such variability often occurs during scale-up and in campaigns to produce actives for clinical trials.

5. **Operability.** Filter operations that maximize the robustness of process operations must be selected to provide consistent purity and yield, resulting in an economical, validatable process.

6. **Virus removal.** Endogenous virus-like particles in mammalian cells used to manufacture rDNA products and adventitious viruses that contaminate cell cultures (e.g., 20-nm parvovirus) must be reduced to a level of less than one virus particle per 10^6 doses.

Membrane filters provide size-based virus removal in which maximum virus resolution is obtained by optimizing pH, ionic strength, and membrane charge to distinguish proteins (4–12 nm) from virus (12–300 nm) by exploiting charge repulsion. This complements chemical inactivation (chaotropes, low pH, solvents, or detergents), physical inactivation (heat or UV), adsorption (ion-exchange chromatography), or other size-based separations (size-exclusion chromatography).

Membrane bioprocessing can contribute unique benefits to society, as illustrated by membrane filtration of vaccine antigens [65]. Vaccines have virtually eliminated

Table 14.14 Typical Specifications for Nuclepore™ Track-etch Microfiltration Membranes

Specified Pore Size, μm	Pore-size Range, μm	Nominal Pore Density, Pores/cm^2	Nominal Membrane Thickness, μm	Typical Flow Rates at 10 lb/in^2 ΔP, 70°F Water, gal/(min)(ft^2)
8.0	6.9–8.0	1×10^5	8.0	144.0
5.0	4.3–5.0	4×10^5	8.6	148.0
3.0	2.5–3.0	2×10^6	11.0	121.0
1.0	0.8–1.0	2×10^7	11.5	67.5
0.8	0.64–0.80	3×10^7	11.6	48.3
0.6	0.48–0.60	3×10^7	11.6	16.3
0.4	0.32–0.40	1×10^8	11.6	17.0
0.2	0.16–0.20	3×10^8	12.0	3.1
0.1	0.08–0.10	3×10^8	5.3	1.9
0.08	0.064–0.080	3×10^8	5.4	0.37
0.05	0.040–0.050	6×10^8	5.4	1.12
0.03	0.024–0.030	6×10^8	5.4	0.006

incidence and associated effects of diseases such as measles, mumps, rubella, polio, diphtheria, and invasive *H. flu*, which once were widespread. Prevention of viral infections that cause Hepatitis A, Hepatitis B, or chicken pox, and bacteria infections from *S. pneumonia* and *B. burgdorferi* (Lyme disease) is now possible. Prophylactic candidates against viral Hepatitis C, HIV, and other agents that cause significant public health issues are under active investigation. Successful clinical investigation of candidate vaccines and manufacture of marketable vaccine products relies on developing consistent, scalable processes to purify bulk vaccine antigen from complex bacterial fermentation and cell culture broths.

§14.9.3 Dead-End Membrane Biofiltration

Dead-end filtration (DEF) is used to recover, concentrate, clarify, and sterilize biological species by depth and surface mechanisms. Microfilters are most commonly employed in the DEF mode. Microfilters have pore sizes ranging from 0.05 to 10 μm, with $\sim 10^{12}$ pores/m^2. Recovery of ~ 0.1–20 μm particles like animal cells, yeasts, bacteria, or fungi from large volumes of culture broth by MF provides rapid volume reduction to improve process economics. Clarification of gases, media, intermediate process streams, virus-containing solutions, and beverages like wine, juice, and beer by MF removes insoluble particulate solids. Microfilters rated with 0.1- or 0.2-μm pore sizes are used to sterilize water for injection (WFI), parenteral solutions (for injection or infusion), antibiotics, buffer solutions, and culture media. MF is performed at low TMP, typically less than 50 psi (3.4 bar; 0.35 MPa), to yield high permeate fluxes ranging from 10^{-4} to 10^{-2} m^3 permeate/m^2 membrane areas for unfouled membranes.

Membrane structures for MF include screen filters that collect retained matter on the surface and depth filters that trap particles at constrictions within the membrane. As Porter

discusses in Schweitzer [73], depth filters include: (1) thick, high-porosity (80–85%) cast–cellulose–ester membranes having an open, tortuous, sponge-like structure; and (2) thin, low-porosity (nominal 10%) polyester or polycarbonate track-etch membranes of a sieve-like structure with narrow distribution of straight-through, cylindrical pores. The latter have a much sharper cut-off, resulting in enhanced separation factors. For example, a Nuclepore™ Type 2 membrane can separate a male-determining sperm from a female-determining sperm. As shown in Table 14.14, Nuclepore™ membranes come in pore sizes from 0.03 to 8.0 μm, with water permeate fluxes at 70°F and a TMP of 10 psi, ranging from 0.006–144 gal/min-ft^2.

This section discusses scale-up of DEF and its application to several steps in bioproduct purification: harvesting cell lysates using filter aids, virus filtration, sterile filtration of bioproduct solutions, culture media, room air, and nanofiltration.

Scale-Up of DEF

Filter area required for manufacturing-scale DEF may be quantitatively estimated from lab-scale biofiltration data by determining the filter capacity using a model for filter resistance that is consistent with the data. The V_{max} method [74] uses a pore-constriction model to determine filter capacity faster, and with less feed volume, than by measuring the cumulative filtrate volume that reduces Q to $\sim 10\%$ of Q_o (flow-decay method). Pore constriction assumes that the membrane removes sub-pore-sized particles that interact with membrane surfaces inside pore cavities, where they are retained by adsorption [64]. An increase in species retention at lower TMP and/or at a lower challenge level suggests pore constriction is occurring due to adsorptive sequestration rather than sieving of the species by a filter rated for absolute retention at a given particle size.

EXAMPLE 14.14 Normal-Flow (DEF) Microfiltration.

Consider the time (t) versus volume collected (V) data for a 0.18 m^2 normal-flow MF in the following table. Estimate the capacity, V_{max}, of the filter and the initial flow rate, Q_o. Confirm whether the data fit a pore-constriction model for flux decline at constant TMP. Determine the flow rate at any point during filtration.

Normal-Flow Filtration Data

t (min)	V (L)
0	0
0.25	2.3
0.5	4.3
0.75	6.1
1	7.7
1.25	9.1
1.5	10.3
1.75	11.4
2	12.4

Solution

The linearized form for pore constriction at constant pressure in Table 14.5 is

$$\frac{t}{V} = \frac{t}{V_o} + \frac{1}{Q_o} \tag{1}$$

with $a = 1/V_o$ and $b = 1/Q_o$, where V_o corresponds to the initial (maximum) filter volumetric capacity and Q_o is the initial flow rate. Fitting the data to a plot of t/V versus t yields $V_o = 33$ L and $Q_o = 10$ L/minute. Then $V_{max} = 33$ L/0.18 m^2 = 183 L/m^2.

A pore-constriction mechanism may be confirmed by fitting Q/Q_o versus V/V_o data to the form

$$\frac{Q}{Q_o} = \left(1 - \frac{V}{V_o}\right)^2 \tag{14-105}$$

Equation (14-105) shows that an initial flow rate Q_o will decay to zero when the feed volume V entirely saturates the initial volumetric capacity, V_o, of a filter. Figure 14.29 shows that the data fit the form of (14-105). The instantaneous flux, Q/A, where A equals filter area, can be predicted from (14-105) for any value of fractional capacity V/V_o.

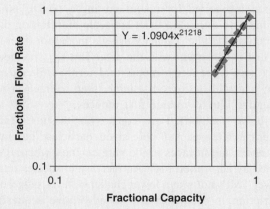

Figure 14.29 Effect of fractional capacity on fractional flow rate to confirm V_{max} filter sizing in Example 14.14.

The V_{max} filter-sizing method is faster and requires less feed volume than the flow-decay method, but may significantly underestimate capacity when pores are interconnected or fouling is not due to pore constriction. Scaling is more reliable by identifying capacity at constant flux in Table 14.5, using the linearized form for intermediate blockage for which $a = P_o^{-1}$ and $b = (P_o V_{max})^{-1}$.

Filter Aids

Adding filter aids (e.g., diatomaceous earth, DE) to clarify lysed bacterial or cultured animal cell feeds by MF can substantially improve throughput, capacity, and clarity. *Diatoms* are million-year-old skeletons of aquatic creatures with diameters of 1–200 μm and pores between 0.1 and 30 μm, resulting in permeabilities of 0.06–30 darcy (1 darcy = 0.9869 μm^2, where 0.9869 is the conversion factor from bar to atm). Calcining adheres diatoms and results in solubilities of approximately 0.1 wt% in dilute acid. Diatomite is 90 wt% silica, with remaining nonsilica elements bound as silicates. Diatomaceous earth clarification of bacterial lysates is attractive relative to centrifugal or membrane clarification in terms of capital cost, fouling, shear, aerosol generation, and scalability.

A typical DE filtration process consists of first layering a 1/16-inch precoat of DE onto a filtration matrix by recirculating a slurry of DE. Then lysate and *body feed* (continuous addition of small amounts of filter aid) are combined at a predetermined rate and pumped onto the precoated matrix at constant flow rate. Solids accumulate with DE on the precoated matrix until back-pressure reaches a preselected target, usually between 30 and 50 psig, at which point pumping is halted, the cake is removed, and the process repeated. Body feed, driving force, and settling times affect permeability, mass recovery, and protein recovery.

Relative to centrifugation or unaided MF, DE filter products have lower turbidities, <20 nephelometric turbidity units (NTU), and at a lower cost ($0.10–$0.30 per lb DE, 0.2–10 lb DE per lb feed) without problems associated with centrifugal shear or limited membrane capacity, although dust is a potential safety issue during handling of bulk DE. There is a $2 billion world market for DE products, including applications in blood fractionation; clarification of beverages including juice, beer, and wine; and processing of oils, pharmaceuticals, chemicals, waste liquids, and sludges [75].

Clarity and Productivity

Use of DE filtration seeks to maximize both clarity and productivity, often leading to competing outcomes that must be balanced. Clarity is optimized by selecting a grade of DE that has an average pore size near the mean particle size of the feed to maximize particle retention. For example, *E. coli* batch fermentations typically yield about 4% by weight with particles > 0.1 μm. Antibiotic fermentation broth is 6 wt% solids, consisting of 1–2 μm particles. To optimize productivity, a ratio of body feed-to-solids content is selected to maximize Darcy permeability. Body feed-to-solids ratios (BF:S) range from 0.25 for rigid particulates to 5 for gelatinous debris. This results in fluxes from 0.1 to 1.0 gal/ft^2-minute, or flow rates from 30 gpm (sludge) to 10,000 gpm (water). However, a high-permeability DE grade yields high flow rates, whereas a low-permeability DE grade gives high clarity. DE filtration is useful because it can accommodate a range of flow rates and back-pressures, although its performance can be subject to vibration.

Filter-Aided Harvest of Antibiotic Fermentation

Large-scale harvest of mycelial protein from fermentation of antibiotics like penicillin or streptomycin commonly employs continuous rotary filters or rotary vacuum precoat filters [76, 77, 84] described in Chapter 19. Rotation of the drum at a constant rotational velocity, n (rps), exposes a fraction, φ, of the drum-surface area to a reservoir in which mycelia coat the drum. Accumulated mycelia are then washed, dewatered, and finally removed by a string (penicillin) or knife-blade scraper (streptomycin). *Streptomyces* mycelia are more difficult to process than *Penicillium* mycelia and require addition of a filter aid like diatomaceous earth (silica skeletons of algae-like diatoms insoluble in strong acids and alkalis) or perlite (porous aluminum silicate used for rough filtrations). To use diatomaceous earth, a 1/16-inch precoat of filter aid is slurried onto the filter using a vacuum maintained in the drum. Broth mixed with 1% to 5% filter aid or another coagulating agent is suspended in the slurry reservoir. A mixed layer of cells and filter aid adheres to the rotating drum and thickens as it moves toward the wash, dewatering, and final discharge from the drum by a knife blade.

EXAMPLE 14.15 Use of Filter Aid.

Identify a general expression to characterize bulk flow in a filter-aided harvest of bacterial cells for antibiotic recovery. Identify strategies specific for improving filter-aid MF rates for streptomycin and penicillin.

Solution

A variable broth volume of V'/n is filtered with a drum during a time of φ/n corresponding to one revolution of the drum, where $V' =$ filtrate volume per unit time. These variables are substituted into the Ruth equation (14-25) for constant-pressure operation:

$$\left(\frac{V'}{n}\right)^2 + 2\frac{V'}{n}V_o = K\frac{\varphi}{n} \qquad (14\text{-}106)$$

This equation may be rearranged to its straight-line form analogous to (14-26) to estimate average specific cake resistance, from which the compressibility factor may be determined using (14-35).

Filtration rate may be increased by decreasing specific resistance of the cake by five approaches (typical harvest values for particular antibiotics are given in parentheses):

1. Increase filter aid in the slurry (2%–3% filter aid for streptomycin).
2. Decrease pH (pH 3.6 for streptomycin).

3. Extend fermentation time (180 to 200 h for penicillin).
4. Coagulate mycelial protein by heat pretreatment before filtering (30–60 min at $T = 80$–$90°C$ for streptomycin).
5. Minimize cake compression, e.g., by lowering ΔP or raising filter-aid content.

§14.9.4 Sterile Filtration

Solutions

Microporous membranes with pore sizes nominally rated to 0.22 or 0.1 μm are used to sterilize water, nutrient media, buffer formulations, or pharmacological actives during bioprocessing; or to perform sterile fill operations.

Validation of particle removal and retention capacity of 0.22 μm sterile filters is performed using a challenge suspension containing 10^7 cells/cm^2 of *Brevundimonas diminuta* ATCC 12146 bacteria ($\sim$0.3 o.d. × 1.0 μm long). The mycoplasma *Acholeplasma laidlawii* is used to validate rated 0.1-μm filters. Base materials for sterile filters include polyether sulphone (PES), polyvinylidene fluoride (PVDF), nylon, polypropylene (PP), and cellulose esters. Asymmetric membranes with a graded pore-size distribution that varies with membrane depth are used to minimize TMP requirements in the standard dead-end configuration. Turbid protein aggregates that result from air–water interfaces during microcavitation in multiple passes through pumps and valves form insoluble particulates that can blind sterile filters and reduce capacity. Particulates may be removed via sieving, depth filtration (e.g., inertial and Brownian impaction), or adsorptive sequestration by adding a depth filter either upstream or as a layer atop the sterile filter (i.e., multilayer) or by casting the sterile filter onto a membrane substrate that has a pore size and/or surface chemistry that removes particulates. Sterile filters are capsulized in self-contained cartridges that are presterilized by gamma irradiation or assembled into a cartridge housing like a Code 7 design, which is available in 10-, 20-, 30-, and 40-inch configurations. Further discussion of cartridge filters is found in Chapter 19.

Integrity Testing

The sterile filter assembly is tested "in-place" using a "bubble-point" or "pressure-hold" (gas-diffusion) method to ensure integrity of the filter and its assembly within the housing. Bubble-point testing measures gas flow through a fully wetted membrane at successively higher pressures [78]. A hydrophilic sterile filter, wetted with water or an aqueous alcohol (isopropanol) solution, is pressurized on its feed side by sterile-filtered, compressed air or N_2. At the bubble point, feed gas overcomes surface tension, σ, of the largest membrane pore and passes through the membrane to appear in the permeate as a stream of bubbles and cause an inflection in a plot of gas flow rate versus pressure. The bubble-point pressure, P_{BP}, required to just displace a liquid from a wetted membrane pore of diameter d_p is

$$P_{BP} = \frac{4\phi\sigma\cos\theta}{d_p} \qquad (14\text{-}107)$$

where ϕ is the pore-shape correction factor and θ is the contact angle for the wetting fluid in contact with the membrane material. Membrane pore sizes of 0.65–0.1 μm typically yield bubble points of 0.2×10^5 to 8×10^5 Pa (0.2–8 bar). In a pressure hold or "diffusion test," a wetted sterile filter is pressurized at ~80% of its bubble point. Total diffusive flux of sterile-filtered gas through the membrane is measured using an inverted graduated cylinder or flowmeter to meet manufacturer's specification. All connections to and from the assembled filter cartridge must be validated to be sterile.

Scale-Up

The required sterile filtration area depends on the maximum allowable load of organisms possible in the feed solution, dose volume, membrane capacity, and targeted *sterility assurance limit* (SAL). The maximum allowable load is the bioburden or viral load specification in the feed to the sterile filtration. The SAL is the calculated probability of a single unit of product containing a single microorganism (expected to be $<10^{-3}$ for aseptic processes, and designed for at least an extra order of magnitude). Filtration area is then increased, if necessary, to complete sterile filtration within one 8-hour shift to preclude "grow-through" (i.e., retained microbes that colonize the filter and grow through to the other side). The sterile filter and any associated depth filters must be compatible with the feed to prevent adsorption of a biologically active ingredient or any excipient, such as a preservative. Validation of pre-use removal of extractables (membrane monomers, storage solutions such as glycerol or ethanol) from the sterile filter assembly prior to use is required.

§14.9.5 Virus Filtration

Removal of endogenous or adventitious contaminating virus to sterilize solutions uses virus filtration membranes intermediate between MF and UF that have 20–70 nm pore sizes. Virus filtration uses composite membranes made from hydrophilic PES (e.g., Millipore's Viresolve® NFR), PVDF (e.g., Millipore's Viresolve® 70, 180, and NFP; Pall's Ultipor® DV50 and DV20), or regenerated cellulose materials to remove $\geq$ 4-logs of virus per filtration step, which is reported as log reduction value (LRV),

$$\text{LRV} = -\log_{10}\left(\frac{c_{i,P}}{c_{i,b}}\right) \qquad (14\text{-}108)$$

where $c_{i,b}$ and $c_{i,P}$ are concentrations in moles of solute i per volume, e.g., dm^3, of liquid in bulk, b, and permeate, P, solutions, respectively. A *bulk* solution is a reservoir of solute at a uniform concentration that is typically found adjacent to the region of interest (e.g., membrane surface or adsorptive particle surface) and often has a volume that is large relative to that of others (e.g., boundary layers or pore volumes) in a system. A bulk solution can correspond to the *feed* to a membrane system in some circumstances, but it often is *not* the feed, as illustrated in Figures 14.30 and 14.31, where the bulk consists of the *retentate* and the *feed + recycled retentate*, respectively. Using subscript b, rather than F for feed, for the general case clarifies and preserves this distinction. The subscript F may be substituted for b in the appropriate specific cases, but b is more accurate globally and is consistent with widely used references in the field of bioseparations. Removal of both enveloped and non-enveloped viruses is validated by spiking with high titers of model viruses such as animal parvovirus (~22 μm; e.g., minute virus of mice, MVM), poliovirus, SV40, sindbis virus, or reovirus. Parvovirus filters (20 nm pores) like Pall Ultipor® DV20 are designed to remove viruses as small as 20 nm, while retrovirus filters (50–70 nm pores) like Pall Ultipor® DV50 remove viruses $\geq$ 50 nm. Bacteriophage, more readily obtained at high purity and titer, can be used for initial evaluation of LRV values.

Room air is prepared using high-efficiency particulate air (HEPA) filters, which are large, high-throughput ventilation, depth-type filters made of compacted fibrous glass wool onto which microbes or other airborne particulates are impacted. Such filters reduce particulate load in a room to class 10,000 (airborne particles per m^3) "acceptable for biotechnology processing" required for antechamber to sterile work areas, and class 100 (particles per m^3) "clean or Aseptic" levels required for sterile filling. Filters are sized according to anticipated flow rate using the pressure differential between one room and an adjacent room or corridor (0.2- to 0.6-inch water) concerning air changes/hour or linear flow rate, usually specified in laminar-flow hoods (5 to 20 ft/s) or aseptic areas. Filters are integrity-tested for 10^3 reduction in aerosol spray of diisoctyl phthalate (DOP). DOP aerosol generators produce "most penetrating particle" droplets ~0.3 μm in size, which are less likely to deposit by inertial impaction than larger particles whose trajectory remains constant as fluid veers due to small diffusivities, or than smaller particles that easily traverse adjacent streamlines via Brownian motion due to higher diffusivities [64].

Equipment gases (typically air or N_2) are sterilized by hydrophobic, asymmetric-membrane vent filters rated to 0.2 μm that are installed on all vessels (fermentors, holding tanks, filter canisters) that must be filled or drained to prevent aerosol contamination of, or by, pathogenic batch contents. These filters are sized for area based on maximum anticipated flow rate to allow flow in both directions. Methods for testing them are subject to government mandates, discussed in Chapter 19.

§14.9.6 Nanofiltration

Nanofiltration employs membranes in which nm-sized cylindrical through-pores penetrate the membrane barrier at right angles to its surface [79]. Nanofilter membranes are made primarily from polymer thin films (e.g., porous polycarbonate, polyethylene terephthalate, or polyimide or metal [aluminum] [80]). Pores in thin-film polymer membranes are formed by bombarding (or "tracking") the film with high-energy particles, which creates damage tracks that are chemically developed (or "etched"). Pore dimensions are controlled by pH, temperature, and time during development, with pore densities ranging from 1 to $\sim 10^6$ pores per cm^2. Track-etch membranes are often thicker and less porous than asymmetric UF membranes.

Alumina membranes are made by electrochemically growing a thin, porous layer of aluminum oxide from aluminum metal in acidic media. Pores of 5–200 nm are arranged in hexagonally packed arrays with densities as high as 10^{11} pores/cm^2. Nanofilters can "soften" water by retaining scale-forming, hydrated divalent ions (e.g., Ca^{2+}, Mg^{2+}) while passing smaller hydrated monovalent ions without adding extra Na$^+$ ions used in ion exchangers [64].

Proteins, nucleic acids, and enantiomers of drugs have been separated in nanotube membranes. The selectivity and flux of species that selectively translocate nanometer-scale pores via free or electrophoresis-assisted diffusion can be controlled by pore characteristics and by incorporating molecular-recognition chemistries (e.g., antibodies, nucleic acids) in nanotube walls. Membranes electrolessly plated with gold and coated with polyethylene glycol (PEG) to give 20- and 45-nm pores selectively pass lysozyme (Lys, MW = 14 kDa) from a solution of Lys and bovine serum albumin (BSA, MW = 67 kDa) based on relative size and diffusivity. Electrophoretic transport due to electrophoretic mobility of proteins based on the Nernst–Planck equation in a transmembrane potential applied using electrodes on feed and permeate sides separates Lys (pI = 11), BSA (pI = 4.9), and hemoglobin (Hb, pI = 7.0, MW = 65 kDa). Coating alumina membranes with antibody Fab fragments distinguishes RS and SR forms of enantiomeric drug 4-[3-(4-fluorophenyl)-2-hydroxy-1-[1,2,4]triazol-1-yl-propyl]-benzonitrile, an inhibitor of aromatase enzyme activity, with selectivities from 2 to 4.5. Selective passage of 18-bp deoxyribonucleic acid (DNA) molecules containing 0 mismatches (perfect complement) versus 1-, and 7-base mismatch DNA gives selectivity coefficients of 3 and 7, respectively, after coating the pores with 30-base DNA hairpin with an 18-base loop, and 1 and 5 after coating the pores with 18-bp linear DNA. Nanotube membrane pores with ligand, voltage, or electromechanical gating can function as ion-channel mimics.

§14.9.7 Tangential-Flow Membrane Biofiltration

Tangential-flow (crossflow) *filtration* (TFF) sweeps a membrane surface with parallel feed flow to enhance flux values relative to direct-flow (dead-end) filtration by reducing cake formation and concentration polarization. *Ultrafiltration* (UF), a term used to identify separations that employ membranes with pore sizes between 0.001 and 0.02 μm, is performed almost exclusively in TFF mode. *Microfiltration* (MF), a term used to identify separations that employ membranes with pore sizes ranging from 0.02 to 10 μm, is also often employed in TFF mode.

UF can selectively retain bioproducts with a molecular-weight range of 300 to 500,000 [81]. Macromolecules like proteins, starches, or DNA and larger species like plasmid DNA [82] and virus-like particles [83] are retained, while smaller solutes like salts, simple sugars, amino acids, and surfactants or replacing buffers are permeated. Ultrafilters have up to $\sim 10^{12}$ pores/cm^2. A UF filter with a molecular weight cut-off (MWCO) of 50,000 retains 90% of globular protein with the corresponding MW. Molecular weights of some widely studied globular proteins are summarized in Table 14.15. Hydrodynamic diameter of a protein is governed by its folding and solution conditions like pH and ionic strength. Therefore, a UF filter is typically selected that has a MWCO value that is 50% of the MW of the retained protein target.

Applications of UF

Cell-concentration factors of 15–50 are reported for harvesting *E. coli*, mycoplasma (for veterinary vaccines), and influenza virus (whole virus vaccine) by UF with a 100,000 MWCO [77]. In addition to cell harvesting, UF is used to process blood and plasma, remove fever-producing (pyrogenic) mucopolysaccharides from medical-grade water, concentrate virus from surface water for assay detection, fractionate immunocomplexes from residual haptens (small molecules that elicit an immune response only when attached to a large carrier such as protein), and concentrate and fractionate other biological species. UF is a large-scale analog of osmotically driven batch dialysis in which unwanted, low-molecular-weight solutes are removed or buffers are exchanged in protein or DNA solutions [85].

Hollow-fiber membranes and flat-sheet membranes configured in plate-and-frame systems are most common for TFF in bioprocessing. Hollow-fiber membranes offer the highest surface area per unit volume and validatable cleanability, whereas plate-and-frame systems incur higher initial capital costs. Recent developments that make UF a mainstay for protein concentration and buffer exchange are: (1) a composite UF membrane consisting of defect-free, low-protein-binding, regenerated cellulose filtration layer bonded atop a mechanically robust polyethylene microporous substrate; (2) simple, effective sanitizing (peroxyacetic acid) and storage (0.1–N NaOH) solutions; (3) linearly scalable module designs.

Process Considerations for TFF

Dead legs (peripheral piping that results in unmixed holdup volumes) should be eliminated. Holdup volume should be minimized during design and fabrication of TFF skids to ensure complete buffer exchange and to minimize holdup losses during recovery of product using: (1) cone-bottomed tanks to minimize final concentrated volume and (2) the return of retentate through the cone bottom using a tee-outlet to aid mixing. To minimize deactivation of proteins during long recirculation times required for TFF: (1) operate at 4°C; (2) eliminate air–water interfaces at which proteins denature, e.g., submerge the retentate return line below the liquid level in the feed tank; and (3) use a large-lobe sanitary lobe pump for recirculation to minimize degradation due to pump shear.

UF or MF operated in TFF mode is often used to harvest *E. coli* or yeast cell suspensions. Harvesting separates a concentrate of intact cells from cell-free supernatant. MF is more frequently used to sieve species such as bacterial or mammalian cells ranging from 0.1 to 10 μm. MF pore sizes range from 0.02 to 10 μm, though MF is typically categorized by a nominal removal rating that may be unrelated to pore size.

Membrane Selectivity

Species *selectivity* in TFF is related to the solute sieving coefficient, S,

$$S_i = \frac{c_{i,P}}{c_{i,b}} \tag{14-109}$$

where $c_{i,b}$ and $c_{i,P}$ are concentrations (mol/cm^3 of liquid) of solute i in bulk feed, b, and permeate, P, solutions, respectively. Solute passage or rejection by a semipermeable-membrane filter is measured using a rejection coefficient, σ_i, for solute i (also, solute reflection coefficient) from thermodynamics of irreversible processes [81]:

$$\sigma_i = \frac{c_{i,b} - c_{i,P}}{c_{i,b}} = 1 - \frac{c_{i,P}}{c_{i,b}} = 1 - S_i \tag{14-110}$$

Unrestricted passage of solute i through the membrane with the solvent corresponds to $\sigma_i = 0$, while little to no passage of solute i retained by a membrane typically yields $\sigma_i \sim 0.95$–0.98.

The size of a globular macromolecule is denoted by its molecular weight, MW:

$$MW = \rho N_A \frac{4}{3} \pi a^3 \tag{14-111}$$

where N_A is Avogadro's number, a is macromolecular radius, and ρ is the globular density. Units for MW are usually reported in kDa (1 dalton = 1 g/mol), as illustrated in Table 14.15. There is reasonable agreement between (14-111) and empirical data written in the form $MW = \alpha a^n$, where values for coefficient α (6.1×10^{22}; 1.46×10^{21}; 3.1×10^{25}) and exponent n (2.17; 2; 2.72) have been obtained experimentally for dextran, polyethylene glycol (PEG), and proteins [54].

The MWCO of a UF membrane represents the MW of a globular protein that exhibits $\sigma_i = 0.9$. Retention, MWCO, and MW may be related [54] by

$$\sigma_i = \left\{ 1 - \left[1 - \left(\frac{MW}{MWCO} \right)^{1/3} \right]^2 \right\}^2 \tag{14-112}$$

In practice, the value of σ_i is influenced by membrane characteristics (porosity, chemistry) as well as by external influences (TMP, solute concentration(s) in the feed, temperature, pH, ionic strength) and thus may vary over the course of an operation. Negative zeta potentials (measured across the membrane) are typical for materials such as cellulose acetate or sulphonated polysulphone for pH > 3, as well as for chemically neutral materials such as PVDF or PES due to strong adsorption of anions from the buffer or electrolyte solution. It is common for σ_i to vary from 0 to 1 over a range of solute MW between about 10 to 10^2-fold. Thus, complete separation by UF of biological species that differ in MW by less than about 10-fold is rare, and partial retention to some degree of similar-sized compounds occurs most frequently. For UF membranes, retention-cut-off curves are established experimentally. Figure 14.30 shows two generic curves. A sharp cut-off is desirable, but more typical is the diffuse cut-off curve, because of the difficulty in producing a membrane with a narrow pore-size distribution.

Electrostatic Effects

Decreasing salt concentration from 100 to 1 mM decreases the protein-sieving coefficient $\geq$ 100-fold [86], an effect attributed to electrostatic and electrokinetic effects.

Table 14.15 Physical Parameters of Some Widely Studied Proteins

Protein	Molecular Weight (kDa)	Stokes Radius (nm)	pI
Urease	480		5.0
Collagen (gelatin)	345		
γ Globulin	170		6.6
β-galactosidase (β-gal) (*Escherichia coli*)	116		
Human tissue plasminogen activator (tPA)	70		
Bovine serum albumin (BSA)	66.2	3.6	4.9
Hemoglobin	68		6.8
Bovine hemoglobin (Hb)	65		7.0
Chicken ovalbumin (OA)	45		
Horseradish peroxidase (*Amoricia rusticana*)	44		
Protein A (*Staphylococcus aureus*)	42		
Egg albumin	33.8–40.5		4.6
Pepsin	34.5		1
Chymotrypsinogen	25		9.5
β-lactoglobulin	18.3		5.2
Human calmodulin	18.2		4.46
Myoglobin	16.7		7.0
Hen egg white lysozyme (HEW)	14.4	2	11
Cytochrome C (Cyt C)	12.4		10.6

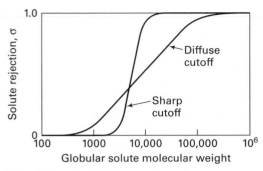

Figure 14.30 Molecular-weight cut-off curves.

Electrostatic interactions between a solute and counterions and co-ions in solution produce a diffuse ion cloud or electrical double layer that increases its effective size. Distortion of this double layer adjacent to pore walls and electrostatic interactions between charged solute and charged membrane surface also affect sieving. Buffer conductivities < 50 mS (millisiemen)/cm (1 siemen = 1 ohm^{-1}) enhance permeability and anionic protein selectivity of negatively charged, composite, regenerated cellulose membranes functionalized with sulfonic acid.

Concentration Polarization

Partially or completely retained dissolved or suspended solutes increase in concentration from bulk solution toward the membrane surface during membrane filtration, creating reversible *concentration polarization* (CP), a major factor limiting TFF, MF, and UF. Concentration polarization can reduce solute flux and change solute-rejection characteristics via increasing osmotic resistance to pressure-driving force or inducing solute–solute or solute–surface coagulative interactions that result in aggregation, cake formation, or pore plugging [81]. At steady state, partially rejected solute in a static film of thickness δ adjacent to the membrane surface is transported away from the surface by pore convection and diffusion at a rate equal to its bulk convective flux toward the surface:

$$Jc_{i,P} + D_e \frac{dc_i}{dz} = Jc_i \qquad (14\text{-}113)$$

where D_e is the effective solute diffusivity in the liquid film (cm^2/s), J is the z-directed volumetric filtration flux of solvent (cm^3/cm^2-s) normal to the surface from (14-4), c_i is the concentration of solute i (mol/cm^3 of liquid), subscript P indicates permeate concentration in an adjacent membrane pore, and subscript b indicates bulk (e.g., feed) concentration. Separating variables in (14-113) and integrating across the mass boundary-layer thickness with boundary conditions $c_i\{z = 0\} = c_{i,b}$ and $c_i\{z = \delta\} = c_{i,w}$ at the membrane wall yields the classical stagnant-film model [87]:

$$J = \frac{D_e}{\delta} \ln \frac{c_{i,w} - c_{i,P}}{c_{i,b} - c_{i,P}} \qquad (14\text{-}114)$$

Substituting (14-110) for solute rejection into (14-114) gives

$$\frac{c_{i,w}}{c_{i,b}} = \frac{\exp(J\delta/D_e)}{\sigma_i + (1 - \sigma_i)\exp(J\delta/D_e)} \qquad (14\text{-}115)$$

For a completely rejected species, $\sigma = 1$ and $c_{i,P} = 0$, and (14-114) reduces to

$$J = \frac{D_e}{\delta} \ln \frac{c_{i,w}}{c_{i,b}} \qquad (14\text{-}116)$$

which shows that permeate flux is proportional to $\ln(c_{i,b})^{-1}$, causing permeate flux to slow as UF concentration of a desired biological product proceeds, until flux attains a maximum pressure-independent value. Maximum $c_{i,w}$ for solid particles is $\sim 74\%$, corresponding to hexagonal close packing, whereas for deformable particles [e.g., red blood cells (RBC)], it may increase up to 95%. It also shows that species concentration at the wall relative to its bulk value—i.e., the polarization modulus, $c_{i,w}/c_{i,b}$, a measure of the extent of CP—increases in exponential proportion to a dimensionless ratio of bulk convective transport to Brownian diffusive transport in the film given by

$$\frac{J\delta}{D_e} = J/k_c \qquad (14\text{-}117)$$

where k_c, in dm^3 m^{-2} h^{-1} (LMH) or in m/s, is a diffusive mass-transfer coefficient. High membrane permeability (high J) and/or high MW solutes (small D_e) may produce severe CP with $c_{i,w}/c_{i,b} > 10$, which drives solute–membrane (e.g., adsorptive) or solute–solute (e.g., precipitation) interactions. Eventually, the solubility limit for solute i may be reached, maximizing $c_{i,w}$ and eliminating increases in permeate flux J in (14-116), even with a larger ΔP driving force, which is negated by gel-layer formation. Solvent flux as a function of solute concentration is shown in Figure 14.31 for two protein solutes. Instead of correlating flux with the logarithm of solute concentration, it may be correlated with concentration factor, CF, which is defined in terms of volumetric flow rates of feed and retentate:

$$CF = \frac{Q_F}{Q_R} \qquad (14\text{-}118)$$

Higher transmembrane velocities and local vortices or eddies induced by obstructions to local flow, macroscopic turbulent flow, or rotation of the sieving surface decrease film thickness, δ, and resistive polarization effects. Concentration polarization may be reduced by:

1. **Module design.** Introducing features like tangential flow, mixing, or turbulence promoters to disrupt CP; membrane protrusions such as dimples, or corrugation.

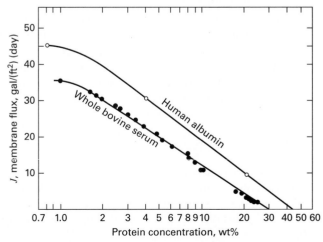

Figure 14.31 Effect of solute concentration on membrane flux.

2. **Hydrodynamic flow management.** Increasing tangential-flow rate, decreasing convective flux below the critical flux for fouling to reduce CP buildup rate, pulsing or reversing feed flow to disrupt CP.

3. **Treatment of feed.** Adding dynamic, solid particles to feed to scour the CP.

4. **Membrane treatment.** Modifying electrostatics to minimize coagulative solute–membrane and solute–solute interactions.

5. **Periodic membrane cleaning.** Including in-line procedures such as back-pulsing, intermittently spiking filtrate pressure to induce temporary back-flow of filtrate into the static film and disrupt the CP layer.

Fouling

Solute interactions in biofiltration may result in membrane fouling, the primary factor limiting microfiltration permeate flux. Fouling results from kinetic adsorption or flow-induced deposition onto, or intrusion into, the membrane by macromolecules (e.g., proteins), colloids, and particles with low diffusion coefficients. Aggregates generated by microcavitation at the pump or by shear at the membrane surface in the case of proteins, or by velocity gradients imparted by an impeller, pump, or flow in a duct in the case of colloids, contribute to fouling. Fouling phenomena constrict, and may eventually blind (block), membrane pores. A sequence of four periods usually produces fouling in MF [88]:

1. **Macromolecular sorption.** Dissolved macromolecules introduced in the feed adsorb rapidly to membrane surfaces, decreasing permeate rate in proportion to coverage, until a pseudo–steady state is reached.

2. **Particle deposition.** The first sublayer builds as colloids, like suspended cells, slowly deposit, decreasing permeate flux as monolayer coverage is approached.

3. **Sublayer rearrangement.** Additional sublayers build, reducing the cross section for axial flow, which increases the wall shear rate and axial pressure gradient. More shear increases Brownian and shear-induced back-diffusion of solids and inertial lift while higher TMP grows and compresses the sublayers, reducing the flux.

4. **Non-Newtonian viscous effects.** Densification of sublayers increases bulk concentration until it increases rapidly and bulk viscosity becomes sharply non-Newtonian, precipitously dropping permeate flux. Concentrated particle suspensions exhibit a concentration-dependent effective shear viscosity that can be correlated by Euler's equation [89]:

$$\mu\{\phi\} = \mu_o\left(1 + \frac{\frac{1}{2}[\mu]\phi}{1 - \phi/\phi_{max}}\right) \qquad (14\text{-}119)$$

where fitted parameters $[\mu]$ and ϕ_{max} are intrinsic velocity and maximum particle volume fraction with best-fit values of 3.0 and 0.58, respectively, at low shear rates. Equation (14-119) shows that $\mu/\mu_o = 1.5, 2.0, 2.9,$ and 5.4 for $\phi = 0.2, 0.3, 0.4,$ and 0.5, respectively.

Fouling lowers permeate flux and alters membrane selectivity. Protein fouling may be minimized by various chemical, physical, and hydrodynamic means. Chemically, selecting sorption-minimizing hydrophilic membrane materials (PVDF, for example), and reducing nominal pore size of the skin below the MW of suspended proteins, decreases CP and fouling. As an example, bovine serum albumin (BSA, 69,000 Da), which forms full and partial monolayers in the ultra-thin skin of asymmetric membranes of sizes $\geq$ 300,000 and 100,000 MWCO, respectively, exhibits no measurable skin adsorption on 50,000 MWCO membranes [90]. Physically, the membrane may be back-flushed using pressure or electrical driving forces, or seed particles added to drag macromolecules away from the membrane. Hydrodynamically, fluid shear rate at the membrane surface may be increased via turbulence, inserts, or rotating disks. Fouling or CP may be disrupted by inducing flow instabilities via surface roughness, pulsation, or flow reversal, or by creating secondary flows using vortices [91]. Taylor vortices are created via Couette flow in annuli of cylindrical devices. Dean vortices arise from flow in a helically coiled channel. Simultaneous application of more than one hydrodynamic method yields flux improvements ranging from 2.5- to 9-fold [88], which are offset by increased energy requirement, equipment complexity, and difficulty in membrane replacement.

Cleaning

It is good practice to measure the clean water flux, J_{H_2O}, of a TFF membrane prior to its initial use at anticipated operating conditions (T, ΔP, J, pH). Between batches, when J_{H_2O} has decreased to an unacceptably low value, say, 50% of J_{H_2O}, the membrane may be cleaned using hydraulic or chemical methods to restore J_{H_2O}. Hydraulic cleaning generally uses 45 L of clean, 40–50°C water/m^2 of membrane area at a crossflow velocity $\geq$ 1.5 m/s to dislodge and wash away gross soil. Pulsed reversal of permeate flow (backflushing) at a TMP that is a fraction of that in forward flow may assist in disrupting plugs or cake. One or more chemical cleaning agents such as alkalis (0.1 to 1.0-M NaOH, pH of 10–14); enzymes (proteases, amylases, 0.2% Terg-A-zyme®, pH 10); disinfectants (300 ppm sodium hypochlorite, pH 10; hydrogen peroxide); or nonionic alkaline detergents [0.1% sodium dodecyl sulfate (SDS) or Tween 80, pH 5–8] to remove organic deposits, or agents like acids (0.1-N H$_3$PO$_4$, pH 1) or complexing agents [e.g., EDTA] to remove inorganics are often alternated between clean-water flushes to remove soils. Manufacturers will typically recommend cleaning agents at concentrations compatible with a particular membrane. Membranes are usually stored in 0.1-M NaOH to prevent microbial growth between batches. An irreversible decline in J_{H_2O} usually results from a series of periodic use and cleaning cycles and ultimately requires membrane replacement.

Predicting Permeate Flux from Boundary-Layer Mass Transport

The film mass-transfer coefficient, $k_c = D_e/\delta$ in (14-117), may be obtained from Sherwood-number correlations for laminar flow and turbulent flow using $N_{Sh} = k_c d_H / D_e$, where d_H is the hydraulic diameter

$$d_H = \frac{4(\text{cross-sectional area for flow})}{\text{wetted perimeter of flow channel}} \qquad (14\text{-}120)$$

In §3.4, N_{Sh} was evaluated in laminar flow using the general Graetz solution for fully developed flow in a straight circular tube of diameter $D = d_H$ between limiting values of $N_{Sh} = 3.656$ for large distances, x, down the flow channel, and

$$N_{Sh} = \alpha \left(\frac{N_{Re} N_{Sc}}{x/D} \right)^{1/3} \qquad (14\text{-}121)$$

for small x, where $\alpha = 1.077$ for $100 < N_{Re} N_{Sc} d_H / L < 5000$, as derived by Leveque [93]. In the latter regime, permeate flux, J, from TFF increases in proportion to the 1/3 power of average axial velocity. Values of N_{Sh} for a range of geometries and flow conditions may be obtained using a generalized form of (14-121) [32],

$$N_{Sh} = \alpha \left(\frac{d_H}{L} \right)^a \left(\frac{\rho d_H u_L}{\mu} \right)^b \left(\frac{\mu}{\rho D_e} \right)^c \qquad (14\text{-}122)$$

where L is the length of the flow channel, u_L is the average axial velocity of the feed, and a, b, and c are coefficients. Values of b are often obtained empirically, whereas values of a and c are usually derived theoretically. Table 14.16 lists values of these coefficients for UF and MF in laminar and turbulent flow for tubes and channels. For example, in turbulent flow, flux (theoretically) increases in proportion to the 0.875 power of average axial velocity.

For UF in which laminar flow is fully developed, the Porter equation relates $k_c = D_e/\delta$ to a geometry-dependent shear rate, γ [93]:

$$k_c = 0.816 \left(\frac{\gamma}{L} D^2 \right)^{0.33} \qquad (14\text{-}123)$$

where $\gamma = 8u_L / d_H$ for tubes and $\gamma = 6u_L / d_H$ for rectangular channels of height h.

Thus, flow may be increased, albeit to the power of 1/3, by raising channel velocity or decreasing channel height. Experimental data confirm application of (14-113)–(14-123) for a large number of macromolecular solutions including proteins as well as suspensions of colloidal particles such as latex beads [81].

Table 14.16 Coefficients of Sherwood Number for Mass Transport in TFF

	α	a	b	c
Laminar:				
UF, empirical	[93]	0	0.5	1/3
MF of cells, empirical	[93]	0	0.8	1/3
Tube, theoretical [93]	1.62	1/3	1/3	1/3
Channel, theoretical	1.86	1/3	1/3	1/3
Turbulent:				
UF	0.023	0	0.083 to 1.0	1/3
MF of cells	0.023	0	1.3	1/3
Theoretical	0.023	0	0.875	1/4

Shear-Induced Diffusion

TFF of particles above ~ 1 μm in size yields experimental flux values 1 to 2 logs higher than (14-117) evaluated with (14-121) [88]. This flux enhancement has been attributed to *shear-induced diffusion* of particles with diameters from about 1–30 μm and inertial lift of larger particles. Interactions ("collisions") between particles concentrated on neighboring streamlines in shear flow cause transient displacements perpendicular to streamlines, which increase in proportion to shear rate and to the square of particle size. Each particle rotates in shear flow, producing rotational flow in nearby fluid that exerts drag forces on neighboring particles.

Effects of shear-induced diffusion may be examined by replacing Brownian diffusivity in (14-121) with a random-walk type of shear-induced hydrodynamic diffusivity D_s given by [94]:

$$D_s = \alpha \gamma_w a^2 \qquad (14\text{-}124)$$

for small particles of radius a that constitute volume fraction $0.2 < \phi_b < 0.45$ in the bulk, where γ_w is the fluid shear at the membrane surface. The empirical α is reported to be ~ 0.025 for 1.6-mm disks and spheres. A value of 0.03 was applied to analyze shear-induced diffusion in UF [95].

EXAMPLE 14.16 Shear-Induced Diffusivity.

Compare hydrodynamic and shear-induced diffusivity values for a 1-μm particle at a shear rate of 1,000 s⁻¹ [88].

Solution

From (14-124), shear-induced diffusivity is 3×10^{-7} cm²/s. From (3-38), hydrodynamic diffusivity is 2×10^{-9} cm²/s. Shear-induced diffusivity is about 150 times larger.

Substituting (14-124) and (14-123) into (14-121) shows that shear-induced diffusion enhances mass transport of 1–40-μm particles by a factor of $2.4 u_L a^2 / D_e D$. Using (14-116) gives a steady length-averaged transmembrane flux [53],

$$\langle J \rangle = \alpha \gamma_w \left(\frac{a^4}{L} \right)^{1/3} \ln \left(\frac{c_{i,w}}{c_{i,b}} \right) \qquad (14\text{-}125)$$

where the coefficient α ranges from 0.126 for constant-viscosity fluids to 0.072 for fluids with a concentration-dependent viscosity [96]. The term $\ln(c_w/c_b)$ may be replaced by $(\phi_w/\phi_b)^{1/3}$ [88].

EXAMPLE 14.17 Membrane Flux for Fluid Shear.

Estimate the flux for a membrane module with $L = 30$ cm, in which particles of radius $a = 0.5$ μm at a relative concentration of $c_w/c_b = 10^3$ are separated using a fluid velocity that produces a typical shear rate of $\gamma_w = 4000$ s⁻¹.

Table 14.17 Predicted MF Flux Dependence on Brownian and Shear-Induced Diffusion and Inertial-Lift Transport Mechanisms [51]

Dominant Mechanism	Exponent	Brownian Diffusion	Shear-Induced Diffusion	Inertial Lift
Shear rate		Low	Intermediate	High
Particle size, a (μm)		<1	0.5–~40	>~30
Shear rate, γ_w	n	0.33	1	2
Particle size, a	m	−0.67	1.33	3
Volume fraction, ϕ_b	p	−0.33	−0.33	0
Filter length, L	q	−0.33	−0.33	0
Suspension viscosity, μ	r	−1	−0.33	−1

Solution

Using (14-125),

$$\langle J \rangle = 0.126 \left(4 \times 10^3\right) \left[\frac{\left(5 \times 10^{-7}\right)^4}{0.3} \right]^{1/3} \ln\left(10^3\right)(3600)(1000)$$

$$= \text{a stable operating flux of 74 L/m}^2\text{-h}.$$

Inertial Lift

Nonlinear hydrodynamic interactions arising from streamline distortions in the gap between particles greater than about 20 μm in diameter and the flow boundary of the surrounding flow field result in inertial lift, which carries dilute suspensions away from membrane walls with thin fouling layers. Inertial lift results in steady transmembrane flux in fast laminar flow of [97]:

$$J = \frac{0.036\, \rho a^3 \gamma_w^2}{\mu} \tag{14-126}$$

for permeate with viscosity μ and density ρ [88].

A general form for steady permeate flux shown in (14-125) and (14-126) may be written as

$$J = c\gamma_w^n a^m \phi_b^p L^q \mu^r \tag{14-127}$$

where c is a model-specific constant and theoretical values of exponents n, m, p, q, and r are summarized in Table 14.17 for the three mechanisms. In each case, flux increases with γ_w and decreases with μ to varying degrees. Higher values of volume fraction and filter length decrease flux in the diffusion models but have no anticipated effect on inertial lift.

Table 14.18 Experimental Permeate Flux Dependence on Shear Rate for Laminar Flow [51]

Suspension	Shear-Rate Dependence, n
Styrene–butadiene latex polymers (5–50% solids by weight) [93]	0.5–0.85
Whole plasma [93]	0.33
Whole blood [93]	0.6
Bacteria (1% solids by weight) [98]	0.5–0.8
Colloidal impurities (5–10 μm) [99]	0.49–0.86
Yeast [100, 101]	0.4–0.7, 1.1
Bovine blood [102]	0.9

Flux decreases with particle size in Brownian diffusion, but increases with size in shear-induced diffusion and inertial lift. Experimental values for n from intermediate-sized particles in Table 14.18 are in a range consistent with the shear-induced diffusion model.

A larger net charge on a protein increases its diffusivity so that mass-transfer coefficients increase as $|\text{pH} - \text{pI}|$ increases and as buffer conductivity decreases. For example, mass-transfer coefficients for a monoclonal antibody increase from 49 to 73 L/m^2-h as conductivity decreases from 20 to 1 mS/cm [53].

EXAMPLE 14.18 Membrane Flux Mechanisms.

Compare the steady-state flux in cm/s for typical conditions of $\gamma_w = 10^3$ s^{-1}, $T = 293$ K, $\mu = 0.01$ g/cm-s, $\rho = 1.0$ g/cm^3, $\phi_w = 0.6$, $\phi_b = 0.01$, $L = 10$ cm, and $h = 0.1$ cm, for 1- and 50-μm particles using models for Brownian diffusion, shear-induced diffusion, and inertial lift [51].

Solution

Values for steady-state flux for each of the particles are summarized in Table 14.19. Shear-induced diffusion provides the largest flux for both particles. Inertial lift provides flux comparable to shear-induced diffusion for the 50-μm particles. For the 1-μm particles, diffusion provides a flux that is ~fourfold lower.

Table 14.19 Predicted Flux from Different Transport Mechanisms

Predicted flux (cm/s)	Brownian Diffusion	Shear-Induced Diffusion	Inertial Lift
$a = 1$ μm	6.3×10^{-5}	2.4×10^{-4}	4.5×10^{-7}
$a = 50$ μm	4.6×10^{-6}	4.4×10^{-2}	5.6×10^{-2}

Permeate Flux

An average value of permeate flux for a TFF system in (14-116) may be obtained from (14-4) by calculating an average value of TMP, ΔP_M, the driving force for TFF across a tangential-flow filter,

$$\Delta P_M = \frac{P_i + P_o}{2} - P_f \tag{14-128}$$

Table 14.20 Typical Values of TMP in TFF

	TMP (psi)
MF	<15
UF	15–150
RO	450–1200

where subscript *i* represents inlet or feed, *o* represents outlet or retentate, and *f* is filtrate, whose pressure is often atmospheric pressure. Typical TMP values for TFF are summarized in Table 14.20.

Increasing ΔP_M eventually raises $c_{i,w}$ in (14-114)–(14-116) to a limiting solubility point at which accumulated solute forms a semisolid gel [103]. Further increases in ΔP_M beyond gel formation increase the thickness of the gel layer and decrease solvent flux. From (14-4) and (14-128), the flux corresponds to

$$J = \frac{\Delta P_M - R_i \Delta \pi}{\mu\left(R_g + R_M\right)} \qquad (14\text{-}129)$$

where $\Delta \pi$ is the osmotic pressure of the polarized solute, defined in (14-90) to (14-93), which resists the superimposed ΔP_M driving force; R_g is the resistance due to the gel formed by CP, which varies with solute composition, concentration, and tangential velocity across the membrane; and R_M is the membrane resistance. Osmotic pressure of a 20°C solution of 50 kg/m³ sucrose (MW 342) is 0.356 MPa (51.6 psi), whereas it is 1.22×10^{-5} MPa (1.77×10^{-3} psi) for a large colloid (~20 nm; MW = 10^7) at the same concentration and temperature. Permeate flux is maximized in practice by identifying optimal values of transmembrane velocity and ΔP_M, below which filtration rate increases linearly with TMP and above which filtration rate decreases due to reduced velocity.

Economics

The cost of installing a fully automated sanitary-filter system, complete with pumps, piping, tanks, and membrane, is summarized in Table 14.21 for five different membrane systems [105]. Also shown is the cost for periodic replacement of membrane modules. Permeate flux varies with the chosen configuration, which impacts the total installed and consumables cost per unit volume of feed.

To estimate cost of energy consumed and process fluid heating, the pump power, P_o, in kW required to achieve a tangential-flow rate Q in m³/h for a given pressure increase across the pump ΔP_{pump} in psi may be estimated as

$$P_o = \frac{Q \Delta P_{\text{pump}}}{522\eta} \qquad (14\text{-}130)$$

where the pump and motor fractional efficiency, η, has typical values ranging from 0.85 for positive displacement pumps to 0.65 for centrifugal pumps.

Process Configurations

Four configurations or combinations thereof are used for TFF: (1) batch TFF, (2) continuous bleed-and-feed TFF, (3) batch diafiltration, and (4) continuous bleed-and-feed diafiltration. Each of these configurations is next discussed in detail and illustrated by examples.

Batch TFF

Figure 14.32 shows a batch configuration where the membrane unit consists of cartridges in parallel. Initially, the feed tank is filled with a batch, V_F, of feed with solute concentration c_F. The solution is pumped through the cartridges, where permeate is continuously removed but retentate is recycled, usually at a high volumetric flow rate, Q, to minimize fouling. As solvent selectively passes through the membrane, the retained volume of solution in the system, $V\{t\}$, decreases and its retentate solute concentration, $c_R\{t\}$, increases. Operation is terminated when the desired solute retentate concentration, c_R, is reached. At that point, the feed tank and associated equipment contain the final retentate, V_R, which can be drained to another tank. After cleaning, another batch is processed. The required time for batch processing depends on the membrane area, A, and permeate flux J, which decreases with time due to increasing solute concentration on the upstream side, as evidenced in Figure 14.31.

Assume the feed contains completely rejected solutes and only partially rejected solutes, and that the flux is a linear function of the logarithm of the concentration factor, CF,

Table 14.21 Costs of Sanitary-Filter System Installation and Membrane Replacement [105]

Membrane System	Cost ($/m²)	
	System Installation Capital	Consumable Replacement
Spiral-wound	150–600	30–80
Tubular	1000–1500	100–200
Hollow-fiber	1500–2000	110–160
Plate-and-frame	1500–5000	300–700
Ceramic	5000–15,000	2000–2500

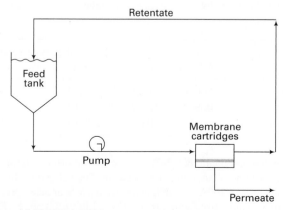

Figure 14.32 Batch TFF.

where for the batch process of Figure 14.32, CF as a function of time is

$$CF = \frac{V_F}{\text{retained } V\{t\}} \qquad (14\text{-}131)$$

Then, it can shown that the average flux, J_{avg}, for the batch process is approximately

$$J_{\text{avg}} = J\{c_F\} - 0.33[J\{c_F\} - J\{c_R\}] \qquad (14\text{-}132)$$

where values of $J\{c_F\}$ and $J\{c_R\}$ are from experimental data like that in Figure 14.31. The required membrane area as a function of batch processing time, t, becomes

$$A = \frac{V_P}{t\,J_{\text{avg}}} = \frac{V_F - V_R}{t\,J_{\text{avg}}} \qquad (14\text{-}133)$$

To obtain a solute material balance, note that solute concentration in the retained volume is a function of both the reduction in retained volume and the amount of solute that passes through the membrane. A solute, i, material balance for a differential volume passing through the membrane is, by analogy to (14-69) for gas permeation,

$$\frac{dV}{V} = \frac{dc_{i_R}}{c_{i_P} - c_{i_R}} \qquad (14\text{-}134)$$

Combining with (14-117) for the definition of rejection, σ_i,

$$\frac{dV}{V} = \frac{dc_{i_R}}{\sigma_i c_{i_R}} \qquad (14\text{-}135)$$

Integrating this equation from initial feed to final retentate gives an equation for retentate solute concentration as a function of retained volume, where if the retained volume is the final volume, its solute concentration is c_{i_F}.

$$c_{i_R} = c_{i_F} \left(\frac{V_F}{V_R} \right)^{\sigma_i} = c_{i_F}(CF)^{\sigma_i} \qquad (14\text{-}136)$$

The yield, Y_i, of solute i, defined as the amount of feed solute that is retained in the retentate, is obtained from (14-136):

$$Y_i = \frac{c_{i_R} V_R}{c_{i_F} V_F} = \left(\frac{V_F}{V_R} \right)^{\sigma_i} \left(\frac{V_R}{V_F} \right) = \left(\frac{V_F}{V_R} \right)^{\sigma_i - 1} = CF^{\sigma_i - 1} \qquad (14\text{-}137)$$

Application of (14-131) to (14-137) is illustrated in the following analysis of ultrafiltration using batch TFF.

EXAMPLE 14.19 Batch UF of an Aqueous Feed.

An aqueous feed of 1,000 L is to undergo batch UF with a polysulfone membrane. Solute concentrations and their measured rejection values are:

Solute	Type Molecule	MW	Concentration, c, g/L	Rejection, σ
Albumin	Globular	67,000	10	1.00
Cytochrome C	Globular	13,000	10	0.70
Polydextran	Linear	100,000	10	0.05

Polydextran has the highest MW, but the lowest rejection because it is a linear rather than a globular molecule. The volume of the final retentate is to be 200 L, which is achieved in a 4-hour batch-processing time. Thus, from (14-118), $CF = 1,000/200 = 5$. From

experimental measurements, the flux values are 30 L/m²-h at $CF = 1$ and 10 L/m²-h at $CF = 5$. Calculate the solute concentration in the final retentate, the yield of each solute, and the membrane area. Neglect changes in solution density.

Solution

From (14-132), the average flux $= 30 - 0.33(30 - 10) = 23.4$ L/m²-h. The total permeate volume $= 1,000 - 200 = 800$ L. From (14–133), for $t = 4$ h,

$$A = \frac{800}{4(23.4)} = 8.55 \text{ m}^2$$

Using (14-137) and (14-136), the yield and concentration of each solute in the final retentate are

Solute	Concentration in Final Retentate, g/L	% Yield
Albumin	50.0	100.0
Cytochrome C	30.9	61.7
Polydextran	10.8	21.7

Note that although polydextran has a very low rejection, value neither the final concentration in the retentate nor the % yield approaches zero.

Batch TFF may damage proteins or cells due to retentate recycle, or allow bacterial growth if residence times are too long. In such circumstances, continuous UF, which is widely used for large-scale processes, is preferred.

Continuous Feed-and-Bleed TFF

Although, as shown in Figure 14.20, continuous reverse osmosis usually operates in a single-pass mode, continuous TFF operates in a multipass mode, called single-stage *feed-and-bleed*, as shown in Figure 14.33. This is achieved by recycling, at steady state, a large fraction of the retentate. In effect, membrane feed is the sum of fresh feed and recycle retentate. The bleed is that portion of the retentate not recycled, but withdrawn as product retentate.

At startup the entire retentate is recycled until the desired retentate concentration is achieved, at which time bleed is initiated. The advantages and disadvantages of feed-and-bleed operation are considered by Cheryan [106] and Zeman and Zydney [56]. The single-pass mode is usually unsuitable

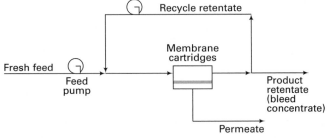

Figure 14.33 Single-stage continuous feed-and-bleed ultrafiltration.

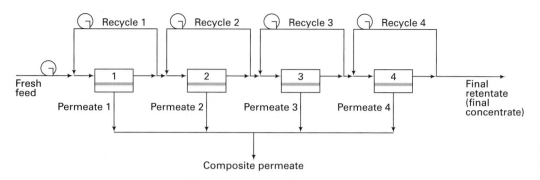

Figure 14.34 Multistage, continuous feed-and-bleed TFF.

for TFF because the main product is the concentrate rather than permeate (as in reverse osmosis), and high yields of permeate are required to adequately concentrate retentate solutes. Typically, the concentration factor, CF, defined by (14-118), has a value of 10. As a result, a single-pass TFF requires a long membrane path or a very large area. A disadvantage of the feed-and-bleed mode, however, is that with the high recycle ratio, the concentration of solutes on the retentate side is highest, resulting, as shown in Figure 14.31, in the lowest flux, with a resulting membrane area larger than that for the batch mode. To counter this, the feed-and-bleed mode is often conducted with four stages, as in Figure 14.34, where the retentate (bleed) from each stage is sent to the next stage, while the permeates from the stages are collected into a final composite permeate. Solute concentrations increase incrementally as the retentates pass through the system. The final and highest concentration is present only in the final stage. As a result, retentate concentrations are lower and fluxes higher than for a single-stage, bleed-and-feed system, for all but the final stage; thus, the total membrane area is smaller. In practice, three to four stages are optimal.

For a single-stage, continuous bleed-and-feed TFF, the material-balance equations in terms of volumetric flow rates and concentrations are:

Total balance:
$$Q_F = Q_R + Q_P \qquad (14\text{-}138)$$

Solute total balance:
$$c_{i_F} Q_F = c_{i_R} Q_R + c_{i_P} Q_P \qquad (14\text{-}139)$$

If the recycle rate is sufficiently high, concentration of the stream flowing on the upstream side of the membrane will be the retentate. Then, if (14-138) and (14-139) are combined with (14-110) and (14-118), rejection in the stage and CF are constant and based on retentate, such that the equation for computing the solute retentate concentration becomes:

$$c_{i_R} = c_{i_F} \left[\frac{CF}{CF(1 - \sigma_i) + \sigma_i} \right] \qquad (14\text{-}140)$$

Area is given by (14-133) in continuous-process form as

$$A = \frac{Q_P}{J\{\text{at } CF\}} \qquad (14\text{-}141)$$

Solute yield in continuous-process form is given by combining (14-137) with (14-118) and (14-140):

$$Y_i = \frac{c_{i_R} Q_R}{c_{i_F} Q_F} = \frac{1}{CF(1 - \sigma_i) + \sigma_i} \qquad (14\text{-}142)$$

For the four-stage, continuous feed-and-bleed TFF system in Figure 14.34, (14-138) to (14-142) are applied to each stage.

It is assumed that the most desirable multistage system is one in which all stages have the same membrane area, to reduce cost of maintenance. The calculations, as described below in Example 14.20, are iterative in nature, using an outer loop in which membrane area per stage is assumed, and an inner loop in which an overall concentration parameter is assumed.

Diafiltration

As seen in Figure 14.31, when a high degree of separation is desired, the flux drops to a low value. To overcome this when it is necessary to continue removing permeable solutes from solutes of little or no permeability, diafiltration, which involves the addition of solvent (usually water) to the retentate, followed by filtration, can be employed. Additional solvent dilutes the retentate to increase the flux in order to achieve a defined solute concentration. The final retentate is not as concentrated in retained solutes, but contains a smaller fraction of permeable solutes.

Diafiltration is conducted in the same modes as UF, i.e., batch or continuous feed-and-bleed, including multistage systems. The added amount of solvent is a variable whose value, for preliminary calculations, may be set equal to the amount of permeate.

Consider a batch diafiltration in which retentate from the previous step is added to the feed tank and recycled, without permeate withdrawal from the membrane unit during startup. Dilution solvent is then added at a continuous rate to the feed tank, under perfect-mixing conditions, with permeate withdrawal at a rate equal to the solvent-addition rate. This operation is sometimes referred to as *fed-batch* or semicontinuous. If the recycle rate is very high, the concentrations of solutes in the membrane unit will be uniform on each side of the membrane; thus rejection in the membrane at any instant is given by (14-110), where both concentrations change with time. Let c_i = the instantaneous solute concentration in the recycle retentate. Initially, before solvent is added, its value is that of the feed, c_{i_F}. If c_{i_P} = the instantaneous permeate solute concentration leaving the membrane unit, (14-110) becomes

$$\sigma_i = 1 - \frac{c_{i_P}}{c_i} \qquad (14\text{-}143)$$

With a constant volume, V_F, in the feed tank before solvent is added, an instantaneous solute material balance equates the decrease in the amount of solute in the feed tank to the amount of solute appearing in the permeate. But permeate flow rate, Q_P, is equal to the solvent addition rate, $Q_S = dV_S/dt$, giving for a solute material balance

$$-V_F \frac{dc_i}{dt} = c_{i_P} Q_P = c_{i_P} \frac{dV_S}{dt} \qquad (14\text{-}144)$$

Combining (14-143) and (14-144) to eliminate c_{i_P} gives, in integral form over time for diafiltration to the final retentate concentration,

$$\int_{c_{i_F}}^{c_{i_R}} \frac{dc_i}{c_i} = -\frac{(1-\sigma_i)}{V_F} \int_0^{V_{S_{\text{total}}}} dV_S \qquad (14\text{-}145)$$

Integration gives an equation for computing the final retentate concentration:

$$c_{i_R} = c_{i_F} \exp\left[-\frac{V_{S_{\text{total}}}}{V_F}(1-\sigma_i)\right] \qquad (14\text{-}146)$$

Continuous diafiltration design is similar to continuous ultrafiltration design, as will be illustrated in Example 14.20.

A major industrial application of UF is in processes for manufacturing protein concentrates from skim milk. The milk is coagulated to render two products: (1) a thick precipitate called curd, rich in a phosphoprotein called casein, which is used to make cheese, plastics, paints, and adhesives; and (2) whey (or cheese whey), a watery, residual liquid. One hundred pounds of skim milk yields approximately 10 pounds of curd and 90 pounds of whey. Typically, whey consists, on a mass basis, of 93.35% water; 0.6% true protein (TP) of molecular weight ranging from 10,000 to 200,000; 0.3% nonprotein nitrogen compounds (NPN); 4.9% lactose (a sugar of empirical formula $C_{12}H_{22}O_{11}$ and MW of 342, which has an ambient solubility in water of about 10 wt%); 0.2% lactic acid ($C_3H_6O_3$) of molecular weight 90, which is very soluble in water; 0.6% ash (salts of calcium, sodium, phosphorus, and potassium) of MW from 20 to 100; and 0.05% butter fat.

Proteins are macromolecules consisting of sequences of amino acids, which contain both amino and carboxylic-acid functional groups. When digested, proteins become sources of amino acids, which are classified as nutritionally essential or nonessential. The *nonessential amino acids* are synthesized by a healthy body from metabolized food. *Essential amino acids* cannot be synthesized by the body, but must be ingested. Amino acids are building blocks for health that repair body cells, build and repair muscles and bones, regulate metabolic processes, and provide energy.

Proteins in whey, on a mass basis, are betalactoglobulin (50–55%), alpha-lactalbumin (20–25%), immunoglobulins (10–15%), bovine serum albumin (5–10%), and smaller amounts of glycomacropeptide, lactoferrin, lactoperoxidase, and lysozyme. The first five of these eight proteins provide an excellent source of all eight essential amino acids: isoleucine, leucine, lysine, methionine, phenylalanine, threonine, tryptophan, and valine. Approximately 35 wt% of proteins in whey provide amino acids. The nonprotein nitrogen compounds include ammonia, creatine, creatinine, urea, and uric acid, with MW of 17–168.

To obtain dry protein concentrate from whey requires a number of processes. Most involve UF, separating by size exclusion based on MW and shape. For separation purposes, the compounds in whey consist of five groups: (1) true protein and butter fat, (2) nonprotein nitrogen, (3) lactose, (4) lactic acid and ash, and (5) water. A typical process is shown in Figure 14.35. Whey is pumped to UF Section I, where the exiting retentate (concentrate) contains all of the protein. The other whey-feed components leave in the exiting permeate. The retentate is further concentrated in an evaporator and then spray-dried to produce a *whey-protein concentrate*. Permeate is pumped to UF Section II, where all remaining lactose is retained and sent to a second spray dryer to produce *lactose-rich concentrate*, while the permeate is sent to wastewater treatment. Whey-protein concentrate produced by this process contains too high a lactose content for the millions of individuals who are "lactose intolerant" because of susceptibility to digestive disorders. To produce so-called *whey-protein isolate* of 90–97 wt% protein and almost no lactose or fat, the process of Figure 14.35 is modified by additional ultrafiltration. The following example, based on information in the 2001 AIChE National Student Design Competition, involves an ultrafiltration section for producing a protein concentrate.

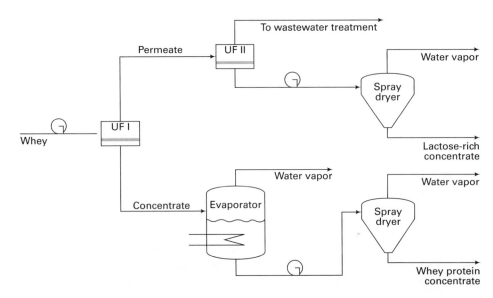

Figure 14.35 Whey process to produce protein and lactose concentrates.

EXAMPLE 14.20 Ultrafiltration Process for Whey.

A cheese plant produces a byproduct stream of 1,000,000 lb/day of whey, to be further processed to obtain a dry powder containing 85 wt% combined TP (true protein) and NPN (nonprotein nitrogen compounds). The process includes three sections: (1) four stages of continuous bleed-and-feed ultrafiltration to reach 55 wt% (dry basis), followed by (2) four stages of continuous diafiltration to reach 75 wt% (dry basis), followed by (3) one stage of batch diafiltration to reach the final 85 wt% (dry basis), with a batch-time limit of 4 hours. Diafiltration must be used above 55 wt% because the retentate from UF becomes too viscous. Each section will use PM 10 ultrafiltration hollow-fiber membrane cartridges from Koch Membrane Systems, which are 3 inches in diameter by 40 inches long, with 26.5 ft^2 of membrane area, at a cost of $200.00 each. For each cartridge, the recirculation (recycle) rate is 23 gpm. The number of cartridges is to be the same in each stage for Sections 1 and 2. The inlet pressure to each is 30 psig, with a crossflow pressure drop of 15 psi and a permeate pressure of 5 psig. For these conditions, the membrane flux has been measured for the whey and correlated as a function of the concentration factor, CF, by:

$$\text{membrane flux, gal/ft}^2\text{-day} = 27.9 - 5.3\ln(CF) \qquad (1)$$

where CF, for any stage n, is defined by reference to the fresh feed to Section 1, as

$$CF_n = F_{\text{Section 1}}/R_n$$

Whey composition and membrane–solute rejections, σ, are:

Component	Wt% in Whey	Flow Rate in Whey, lb/day	Solute Rejection, σ
Water	93.35	933,500	—
True protein, TP	0.6	6,000	0.970
Nonprotein nitrogen, NPN	0.3	3,000	0.320
Lactose	4.9	49,000	0.085
Ash	0.8	8,000	0.115
Butter fat	0.05	500	1.000

Based on σ values, the membrane increases the concentration of TP while selectively removing the low-MW solutes of lactose and ash. Whey and all retentate and permeate streams have a density of 8.5 lb/gal. The continuous sections of the process will operate 20 hr/day, leaving 4 hr/day to remove accumulated membrane foulants and sterilize equipment. For each section, calculate: (1) component material balances in lb/day of operation, including dilution water for diafiltration; (2) percent recovery from whey of TP and NPN in the intermediate and final 85 wt% concentrate; and (3) number of membrane cartridges. Also, for Section 1, make calculations for a single continuous stage, compare results to those for four stages, and discuss advantages and disadvantages of four stages versus one stage.

Solution

The flow diagram for the ultrafiltration–diafiltration process is shown in Figure 14.36.

Single Continuous UF Stage for Section 1 to Reach 55 wt% TP + NPN

First compute results for Section 1 using the single-stage continuous bleed-and-feed UF shown in Figure 14.33. Assume $CF = 10$ and

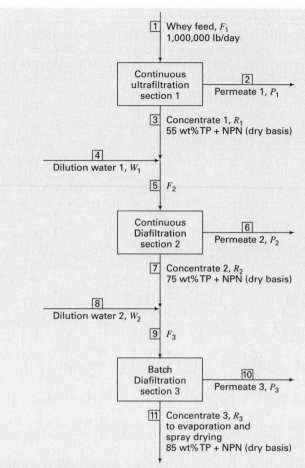

Figure 14.36 Process for Example 14.20.

compute by material balance from the whey-feed rate—$F_1 = 1,000,000$ lb/day—flow rates of retentate (concentrate) R_1, and permeate P_1. By definition of CF for this type of ultrafiltration, $R_1 = F_1/CF = 1,000,000/10 = 100,000$. Therefore, $P_1 = F_1 - R_1 = 1,000,000 - 100,000 = 900,000$ lb/day. Next, use a mass flow rate form of the yield equation, (14-142), to compute each solute flow rate in the concentrate. For TP,

$$(m_{TP})_{R_1} = \frac{(m_{TP})_{F_1}}{[CF(1-\sigma)+\sigma]}$$

$$= \frac{6000}{[10(1-0.97)+0.97]} = 4724 \text{ lb/day}$$

Similarly, flow rates of other solutes in the concentrate R_1 are computed as follows, where the water rate is by difference:

Concentrate for a Single Stage of Continuous Ultrafiltration in Section 1, for an Assumed $CF = 10$

Component	Wt% in Concentrate, C_1	Flow Rate in Concentrate, C_1, lb/day
Water	88.157	88,157
True protein, TP	4.724	4,724
Nonprotein nitrogen, NPN	0.421	421
Lactose	5.306	5,306
Ash	0.892	892
Butter fat	0.500	500
Total	100.000	100,000

From this table, the wt% TP + NPN in the concentrate on a dry basis is $(4,724 + 421)/(100,000 - 88,157) = 0.4344$ or 43.44 wt% (dry basis), which is less than the 55-wt% target. Therefore, the assumed $CF = 10$ is too low. Using a spreadsheet, the Solver function finds $CF = 24.955$, which meets the 55-wt% target. This result gives the following concentrate:

Concentrate for a Single Stage of Continuous Ultrafiltration in Section 1, for the Correct $CF = 24.955$

Component	Wt% in Concentrate, C_1, lb/day	Flow Rate in Concentrate, C_1, lb/day
Water	83.372	33,409
True protein, TP	8.713	3,491
Nonprotein nitrogen, NPN	0.433	174
Lactose	5.335	2,138
Ash	0.899	360
Butter fat	1.248	500
Total	100.000	40,072

The wt% TP + NPN in the concentrate is now $(3,491 + 174)/(40,072 - 33,409) = 0.5500$ or 55.00%, the specified value.

It is also of interest to compute the % yield of TP + NPN:

$$(3491 + 174)/(6000 + 3000) = 0.4072 \text{ or } 40.72\%$$

Membrane area for this single stage is computed from (14-141). The permeate rate is $P_1 = F_1 - R_1 = 1,000,000 - 40,072 = 959,928$ lb/day or $959,928/8.5 = 112,933$ gal/day. For 20-h/day operation, the volumetric permeate rate $= 112,933/20 = 5,647$ gal/h. From (1), for the computed CF,

$$\text{membrane flux} = 27.9 - 5.3\ln(CF) = 27.9 - 5.3\ln(24.955)$$
$$= 10.85 \text{ gal/ft-day or } 10.85/24 = 0.452 \text{ gal/ft}^2\text{-h}$$

Therefore, from (14-141), the membrane area $= 5,647/0.452 = 12,490$ ft^2. Each cartridge has an area of 26.5 ft^2; therefore, $12,490/26.5 = 471$ parallel cartridges are needed. Total fresh feed rate based on 20 hr of operation is

$$\frac{1.000,000}{8.5(20)} = 5,882 \text{ gal/h}$$

Fresh feed rate to each cartridge is

$$\frac{5,882}{60(471)} = 0.208 \text{ gal/min/cartridge}$$

The combined flow rate (fresh plus recycle) to each cartridge is $0.208 + 23 = 23.208$ gal/min, which is a desirable recycle ratio.

To increase the % yield and decrease the number of cartridges, a multistage section is needed. In the problem statement, four stages in series are specified. These are computed next.

Four Continuous UF Stages for Section 1 to Reach 55 wt% TP + NPN

Calculations are based on an equal membrane area for the four stages in Figure 14.34. They are made by a double "trial-and-error" (nested-iteration) procedure, which is best carried out using a spreadsheet with a Solver function. Assume a membrane area per stage. Because the single-stage calculation resulted in an area of 12,500 ft^2, and there are four stages, the total area for four stages will be smaller. First, assume a total area of 8,000 ft^2 or 2,000 ft^2 per stage $= A$. Next, find, by iteration, the overall concentration

factor, $\overline{CF}$, that gives the fresh feed rate to the first stage, as calculated above, of 5,882 gal/h. This is done with a spreadsheet starting from Stage 4 and working backward to Stage 1, using the following equations, where J_n = hourly membrane flux = $(1)/24$ of Eq. (1), based on a CF using F_1 and R_n. For Stage 4, $CF_4 =$ the assumed $\overline{CF}$ and $R_4 = F_1/CF_4$. Then, for the calculations back to Stage 1:

$$P_n = AJ_n; \quad R_{n-1} = P_n + R_n; \quad CF_{n-1} = F_1/R_{n-1}$$

When Stage 1 is reached by calculation of P_1, the fresh feed rate is computed from $F_1 = P_1 + R_1$. If F_1 is not 5,882 gal/h, new values of $\overline{CF}$ are assumed until the correct value of F_1 is obtained. This iteration can be done with the spreadsheet Solver function. If $\overline{CF} = 20$ and $A = 2,000$ ft^2 are assumed, the results are as follows, where all flows are in gal/h:

$\overline{CF}$	R_4	P_4	R_3	P_3	R_2	P_2	R_1	P_1	F_1
20	294	1002	1296	1657	2953	2021	4974	2250	7224
61.2	96.2	508	604	1320	1924	1831	3755	2127	5882

The tabulation shows $F_1 = 7,224$ gal/h, which is too high. Using the Solver function, $\overline{CF} = 61.2$ gives the correct F_1, with the corresponding computed values of CF_n and J_n:

$\overline{CF}$	CF_4	J_4	CF_3	J_3	CF_2	J_2	CF_1	J_1
61.2	61.2	0.254	9.735	0.660	3.057	0.916	1.566	1.063

However, the assumed membrane area per stage may not be correct. To check this, calculations similar to those above are carried out with a spreadsheet, starting with Stage 1 and proceeding stage-by-stage to Stage 4. Pertinent results for $A = 2,000$ ft^2 per stage and $\overline{CF} = 61.2$ are:

Stage	1	2	3	4
Wt% TP + NPN in retentate from stage (dry basis)	17.47	25.46	44.96	74.16

Because TP + NPN in the retentate from Stage 4 is 74.16 wt% (dry basis), which is higher than the specified 55 wt%, calculations must be repeated for other values of membrane area per stage. For each assumed membrane area, a new value of $\overline{CF}$ that gives the correct fresh feed rate must be found. The following spreadsheet results are obtained when iterating on A and $\overline{CF}$:

A, Membrane Area per Stage, ft^2	$\overline{CF}$ for Correct Fresh Feed Rate	Wt% TP + NPN in Final Retentate (dry basis)
2,000	61.2	74.16
1,750	26.9	63.22
1,700	22.5	60.26
1,650	18.8	57.14
1,600	15.7	53.95
1,617	16.65	55.00

For a continuous, four-stage UF system, with equal membrane area per stage, the desired value of 55 wt% (dry basis) for TP + NPN in the retentate (concentrate) from Stage 4 corresponds to $A = 1,617$ ft^2 per stage and an overall concentration factor, $\overline{CF}$, of 16.65. From these results, the material balance—which includes the combined permeate from four UF stages and the computed retentate (concentrate) of 55 wt% that leaves Stage 4 and becomes feed to the continuous diafiltration section to increase TP + NPN to 75 wt% (dry basis)—is as presented in the following table:

Concentrate and Combined Permeate from a Four-Stage Continuous UF in Section 1, for a $\overline{CF}$ of 16.65 and $A = 1,617$ ft^2/Stage, which Meets the 55 wt% Specification

Component	Flow Rate of Whey, lb/day	Flow Rate of Concentrate, lb/day	Flow Rate of Combined Permeate, lb/day
Water	933,500	49,897	883,603
True protein, TP	6,000	5,245	755
Nonprotein nitrogen, NPN	3,000	353	2,647
Lactose	49,000	3,476	45,524
Ash	8,000	603	7,397
Butter fat	500	500	0
Total	1,000,000	60,074	939,926

For Section 1, the number of ultrafiltration cartridges required is $1,617/26.5 = 61$ cartridges per stage or a total of 244 cartridges for four stages. The % yield of TP + NPN in the concentrate is $(5,245 + 353)/(6,000 + 3,000) \times 100\% = 62.20\%$. These results compare to 471 cartridges and a % yield of 40.72% for a single stage in Section 1. On both counts, the four-stage system is preferred. However, a definitive economic analysis would include additional piping and instrumentation costs for a four-module system.

Four Continuous Diafiltration Stages for Section 2 to Reach 75 wt% TP + NPN

The following procedure is based on equal membrane areas for the four stages, based on a flow diagram similar to Figure 14.34, differing only in the addition of water to the feed to each stage. Calculations for a continuous, multistage diafiltration system require iteration on a single variable, the added water rate, to achieve the specified 75 wt% TP + NPN. This is done with a spreadsheet using the Solver function.

For each diafiltration stage, the added water rate, W_n, is the same and is equal to W. The permeate rate, P_n, for each stage is set equal to the added water rate. Therefore, the feed rates to the stages, F_1 for the first stage and R_{n-1} for the succeeding three stages (i.e., before the added water and the recycle), are all equal to the retentate (concentrate) rate, R_n, sent to the next stage, and all retentate rates are the same, i.e., $F_1 = R_n$. These simplifications result in the same concentration factor, $\overline{CF}$, for every stage:

$$\overline{CF} = \frac{W + F_1}{F_1} \qquad (2)$$

For a continuous diafiltration system of n stages, with solute i, flow rates in the concentrate from the final stage are given by an equation, obtained by applying (14-142) successively to each stage, that sets the solute flow rate in the feed to Stages 2, 3, and 4 equal to the flow rate in the retentate from the preceding stage:

$$(m_i)_{R_n} = (m_i)_{F_1} \left[\frac{1}{\overline{CF}(1 - \sigma_i) + \sigma_i} \right]^n \qquad (3)$$

Using a spreadsheet, (2) and (3) are solved, where values of $(m_i)_{F_1}$ are solute component flow rates in the concentrate leaving Section 1, as given in the table above. Solute rejections, σ_i, are given in the problem statement. A value is assumed for the added water rate to each stage, W, and $\overline{CF}$ is computed from (2). From (3), values of $(m_i)_{R_n}$ are computed for each solute. The wt% TP + NPN (dry basis) is then calculated, and if it is not the specified 75 wt%, a new value of W is chosen. Assume W at half the feed rate F_1, or $60,074/2 = 30,037$ lb/day. From (1), $\overline{CF} = (30,037 + 60,074)/60,074 = 1.50$. For TP, from (3),

$$(m_{TP})_{R_4} = 5,245 \left[\frac{1}{1.50(1 - 0.97) + 0.97} \right]^4 = 4,942 \text{ lb/day}$$

The calculations are repeated for other components, and the water rate in the concentrate from Stage 4 of Section 2 is determined so that the total concentrate flow rate equals that of the feed, 60,074 lb/day. The wt% TP + NPN in the concentrate is then calculated, with a result of 78.2 wt%, which is higher than the specified 75 wt%. Using the Solver function, the correct water rate for each stage is found to be 23,332 lb/day or a total of 93,328 lb/day for the four stages, with a corresponding $\overline{CF} = 1.388$. The resulting material balance—which includes the combined permeate and the computed retentate (concentrate) of 75 wt% that leaves Stage 4 to become the feed to the batch diafiltration section to increase the wt% TP + NPN to 85% (dry basis)—is as presented in the next table.

Concentrate and Combined Permeate from a Four-Stage Continuous Diafiltration in Section 2, and an Added Water Rate of 23,332 lb/day per Stage, which Meets the 75 wt% Specification

Component	Flow Rate in Feed to Section 2, lb/day	Flow Rate in Concentrate, lb/day	Flow Rate in Combined Permeate, lb/day
Water	49,897	53,214	90,011
True protein, TP	5,245	5,007	238
Nonprotein nitrogen, NPN	353	138	215
Lactose	3,476	1,030	2,446
Ash	603	185	418
Butter fat	500	500	0
Total	60,074	60,074	93,328

From these results, the yield of TP + NPN from diafiltration is $(5,007 + 138)/(5,245 + 353) \times 100\% = 91.91\%$ for an overall yield, to this point, of $(0.9191)(0.6220) \times 100\% = 57.17\%$. The membrane flux for each stage is obtained from (1). However, the CF used in that equation is the ratio of the whey feed for the process to the retentate rate from the stage, which for the four stages of diafiltration is the same as that for the last stage of the ultrafiltration in Section 1. A value of $CF = 1,000,000/60,074 = 16.65$ applies, which results in a membrane flux of 0.5415 gal/h-ft^2. The volumetric permeate flow rate per stage $= 93,328/[(20)(4)(8.5)] = 137$ gal/h. The membrane area required per stage $= 137/0.5415 = 253$ ft^2. The number of cartridges per stage $= 253/26.5 = 9.5 \approx 10$ cartridges per diafiltration stage, for a total of 40 diafiltration cartridges.

These results for four stages of diafiltration may be compared to the results obtained with just a single continuous diafiltration stage, which gives an added water rate of 167,200 gal/day (compared to 93,328 for four stages) and an overall TP + NPN yield of 55% (compared to 57% for four stages).

A Single Batch Diafiltration Stage for Section 3 to Reach 85 wt% TP + NPN

The final membrane section is a single batch diafiltration with maximum batch time of 4 hr. A feed tank is filled with concentrate from Section 2, and water is added continuously over the 4-hr period to maintain the liquid level in the tank. With a high recycle ratio, the concentration of solutes in the retentate is maintained constant, with solute i, and flow rates in the concentrate feed are given by (14-146) in mass flow form:

$$(m_i)_R = (m_i)_F \exp\left[-\frac{W}{F}(1 - \sigma_i)\right] \quad (4)$$

where W and F are amounts of additional water and feed processed during the 4-hr period. To reach 85 wt%, it is important to remove the lactose from the feed. For example, suppose the daily amount of added water is equal to the daily amount of feed from Section 2. Then, $W/F = 1$. From the preceding table, that feed contains 1,030 lb/day of lactose, which has the rejection $\sigma = 0.085$. Substitution into (71) gives

$$(m_{\text{lactose}})_R = 1,030 \exp[-1(1 - 0.085)] = 413 \text{ lb/day}$$

Flows of other solutes in the concentrate from Section 3 are computed similarly and the wt% TP + NPN (dry basis) is obtained. The spreadsheet Solver function determined that the added water needed to achieve 85 wt% is 80,520 lb/day.

The following material balance includes permeate from the batch diafiltration stage and computed retentate (concentrate) of 85 wt% (dry basis) from Section 3.

Concentrate and Permeate from a Single-Stage Batch Diafiltration in Section 3, for an Added Water Rate of 80,520 lb/day, to Meet the 85 wt% Specification

Component	Flow Rate of Feed to Section 3, lb/day	Flow Rate of Concentrate, lb/day	Flow Rate of Permeate, lb/day
Water	53,214	54,351	79,383
True protein, TP	5,007	4,810	197
Nonprotein nitrogen, NPN	138	55	83
Lactose	1,030	302	728
Ash	185	56	129
Butter fat	500	500	0
Total	60,074	60,074	80,520

The % yield of TP + NPN in Section 3 is $(4,810 + 55)/(5,007 + 138) \times 100\% = 94.56\%$. The overall yield of TP + NPN from the whey feed is $(4,810 + 55)/(9,000) \times 100\% = 54.06\%$.

The membrane flux is 0.5415 gal/h-ft^2, as in Section 2. The volumetric permeate flow rate over a 4-hour batch operation = $80,520/[(4)(8.5)] = 2,368$ gal/h. Therefore, the membrane area required = $2,368/0.5415 = 4,373$ ft^2. The number of cartridges needed in Section 3 is $4,373/26.5 = 165$.

SUMMARY

1. The separation of liquid and gas mixtures with membranes is an emerging separation operation. Applications began accelerating in the 1980s. The products of separation are retentate and permeate.

2. The key to an efficient and economical membrane-separation process is the membrane. It must have good permeability, high selectivity, solute compatibility, high capacity, stability, freedom from fouling, and a long life.

3. Commercialized membrane-separation processes include dialysis, electrodialysis, reverse osmosis, gas permeation, pervaporation, ultrafiltration, and microfiltration.

4. Most membranes for commercial separation processes are natural or synthetic, or glassy or rubbery polymers cast as a film from a solvent mixture. However, for high-temperature (>200°C) operations with chemically reactive mixtures, ceramics, metals, and carbon find applications.

5. To achieve high permeability and selectivity, dense, nonporous membranes are preferred. For mechanical integrity, membranes 0.1–1.0 mm thick are incorporated as a surface layer or film onto or as part of a thicker asymmetric or composite membrane.

6. To achieve a high surface area per unit volume, membranes are fabricated into spiral-wound or hollow-fiber modules. Less surface is available in plate-and-frame, tubular, and monolithic modules.

7. Permeation through a membrane occurs by many mechanisms. For a microporous membrane, mechanisms include bulk flow (no selectivity), liquid and gas diffusion, Knudsen diffusion, restrictive diffusion, sieving, and surface diffusion. For a nonporous membrane, a solution-diffusion mechanism applies.

8. Flow patterns in membrane modules have a profound effect on overall permeation rates. Idealized flow patterns for which theory has been developed include perfect mixing, countercurrent flow, cocurrent flow, and crossflow. To overcome separation limits of a single membrane module stage, modules can be arranged in series and/or parallel cascades.

9. In gas permeation, boundary-layer or film mass-transfer resistances on either side of the membrane are usually negligible compared to the membrane resistance. For separation of liquid mixtures, external mass-transfer effects and concentration polarization can be significant.

10. For most membrane separators, the component mass-transfer fluxes through the membrane can be formulated as the product of two terms: concentration, partial pressure, fugacity, or activity-driving force; and a permeance $\bar{P}_{M_i}$, which is the ratio of the permeability, P_{M_i}, to the membrane thickness, l_M.

11. In the dialysis of a liquid mixture, small solutes of type A are separated from the solvent and larger solutes of type B with a microporous membrane. The driving force is the concentration difference across the membrane. Transport of solvent can be minimized by adjusting pressure differences across the membrane to equal osmotic pressure.

12. In electrodialysis, a series of alternating cation- and anion-selective membranes are used with a direct-current voltage across an outer anode and an outer cathode to concentrate an electrolyte.

13. In reverse osmosis, the solvent of a liquid mixture is selectively transported through a dense membrane. By this means, seawater can be desalinized. The driving force for solvent transport is fugacity difference, which is commonly expressed in terms of $\Delta P - \Delta \pi$, where π is the osmotic pressure.

14. In gas permeation, mixtures of gases are separated by differences in permeation rates through dense membranes. The driving force for each component is its partial pressure difference, Δp_i, across the membrane. Both permeance and permeability depend on membrane absorptivity for the particular gas species and species diffusivity. Thus, $P_{M_i} = H_i D_i$.

15. In pervaporation, a liquid mixture is separated with a dense membrane by pulling a vacuum on the permeate side of the membrane so as to evaporate the permeate. The driving force may be approximated as a fugacity difference expressed by $\left(\gamma_i x_i P_i^s - y_i P_P\right)$. Permeability can vary with concentration because of membrane swelling.

16. Polymer membranes cast from solvent mixtures are used for harvest, clarification, purification, polishing, sterile filtration, and buffer exchange in bioprocessing. Selection of a membrane for biofiltration is guided by its solute selectivity, capacity, and flux, which are impacted by concentration polarization and fouling.

17. Microfiltration, ultrafiltration, and virus filtration are pressure-driven operations that selectively retain species in aqueous solutions based on their size, charge, and composition. They may be operated in normal-flow (dead-end) or tangential-flow modes. To be used in bioseparations, these membranes must preserve biological activity, satisfy cGMP requirements, and allow batch processing.

18. Bulk transport in biofiltration is modeled by Darcy's law using resistances for membrane and solute that is retained as a cake to restrict or completely block pores. Constant-flux and constant-pressure equations allow identification of the appropriate model for resistance.

19. Normal-flow operation is used primarily to clarify debris or remove infectious agents from solutions, gases, and parenteral drug suspensions. Filter aids increase capacity of debris removal. Capacity, throughput, and scale-up of normal-flow filters can be characterized using $V_{\max}$ models.

20. Tangential-flow operation is used primarily for cell harvest, species concentration, and purification, or buffer exchange. Selectivity in tangential-flow filtration is determined by solute size, charge, and composition.

21. Permeate flux in tangential-flow filtration is predicted using mass-transfer models for concentration polarization, boundary-layer mass transport, shear-induced diffusion, and inertial lift. Mass-transport coefficients in tangential-flow filtration are functions of species size and charge, solution composition, system geometry, and hydrodynamic shear and viscosity.

22. Tangential-flow filtration may be configured in four ways: batch, feed-and-bleed, batch diafiltration, and continuous feed-and-bleed diafiltration.

REFERENCES

1. Lonsdale, H.K., *J. Membrane Sci.*, **10**, 81 (1982).

2. Applegate, L.E., *Chem. Eng.*, **91**(12), 64–89 (1984).

3. Havens, G.G., and D.B. Guy, *Chem. Eng. Progress Symp. Series*, **64** (90), 299 (1968).

4. Bollinger, W.A., D.L. MacLean, and R.S. Narayan, *Chem. Eng. Progress*, **78**(10), 27–32 (1982).

5. Baker, R.W., E.L. Cussler, W. Eykamp, W.J. Koros, R.L. Riley, and H. Strathmann, *Membrane Separation Systems—A Research and Development Needs Assessment*, Report DE 90-011770, Department of Commerce, NTIS, Springfield, VA (1990).

6. Ho, W.S.W., and K.K. Sirkar, Eds., *Membrane Handbook*, Van Nostrand Reinhold, New York (1992).

7. Loeb, S., and S. Sourirajan, *Advances in Chemistry Series*, Vol. **38**, *Saline Water Conversion II* (1963).

8. Henis, J.M.S., and M.K. Tripodi, U.S. Patent 4,230,463 (1980).

9. Wrasidlo, W.J., U.S. Patent 3,951,815 (1977).

10. Barrer, R.M., *J. Chem. Soc.*, 378–386 (1934).

11. Barrer, R.M., *Diffusion in and through Solids*, Cambridge Press, London (1951).

12. Mahon, H.I., U.S. Patent 3,228,876 (1966).

13. Mahon, H.I., U.S. Patent 3,228,877 (1966).

14. Hsieh, H.P., R.R. Bhave, and H.L. Fleming, *J. Membrane Sci.*, **39**, 221–241 (1988).

15. Bird, R.B., W.E. Stewart, and E.N. Lightfoot, *Transport Phenomena*, John Wiley & Sons, New York, pp. 42–47 (1960).

16. Ergun, S., *Chem. Eng. Progress*, **48**, 89–94 (1952).

17. Beck, R.E., and J.S. Schultz, *Science*, **170**, 1302–1305 (1970).

18. Beck, R.E., and J.S. Schultz, *Biochim. Biophys. Acta*, **255**, 273 (1972).

19. Brandrup, J., and E.H. Immergut, Eds., *Polymer Handbook*, 3rd ed., John Wiley & Sons, New York (1989).

20. Lonsdale, H.K., U. Merten, and R.L. Riley, *J. Applied Polym. Sci.*, **9**, 1341–1362 (1965).

21. Motamedian, S., W. Pusch, G. Sendelbach, T.-M. Tak, and T. Tanioka, *Proceedings of the 1990 International Congress on Membranes and Membrane Processes*, Chicago, Vol. **II**, pp. 841–843.

22. Barrer, R.M., J.A. Barrie, and J. Slater, *J. Polym. Sci.*, **23**, 315–329 (1957).

23. Barrer, R.M., and J.A. Barrie, *J. Polym. Sci.*, **23**, 331–344 (1957).

24. Barrer, R.M., J.A. Barrie, and J. Slater, *J. Polym. Sci.*, **27**, 177–197 (1958).

25. Koros, W.J., and D.R. Paul, *J. Polym. Sci., Polym. Physics Edition*, **16**, 1947–1963 (1978).

26. Barrer, R.M., *J. Membrane Sci.*, **18**, 25–35 (1984).

27. Walawender, W.P., and S.A. Stern, *Separation Sci.*, **7**, 553–584 (1972).

28. Naylor, R.W., and P.O. Backer, *AIChE J.*, **1**, 95–99 (1955).

29. Stern, S.A., T.F. Sinclair, P.J. Gareis, N.P. Vahldieck, and P.H. Mohr, *Ind. Eng. Chem.*, **57**(2), 49–60 (1965).

30. Hwang, S.-T., and K.L. Kammermeyer, *Membranes in Separations*, Wiley-Interscience, New York, pp. 324–338 (1975).

31. Spillman, R.W., *Chem. Eng. Progress*, **85**(1), 41–62 (1989).

32. Strathmann, H., "Membrane and Membrane Separation Processes," in *Ullmann's Encyclopedia of Industrial Chemistry*, VCH, FRG, Vol. **A16**, p. 237 (1990).

33. Chamberlin, N.S., and B.H. Vromen, *Chem. Engr.*, **66**(9), 117–122 (1959).

34. Graham, T., *Phil. Trans. Roy. Soc. London*, **151**, 183–224 (1861).

35. Juda, W., and W.A. McRae, *J. Amer. Chem. Soc.*, **72**, 1044 (1950).

36. Strathmann, H., *Sep. and Purif. Methods*, **14**(1), 41–66 (1985).

37. Merten, U., *Ind. Eng. Chem. Fundamentals*, **2**, 229–232 (1963).

38. Stoughton, R.W., and M.H. Lietzke, *J. Chem. Eng. Data*, **10**, 254–260 (1965).

39. Spillman, R.W., and M.B. Sherwin, *Chemtech*, 378–384 (June 1990).

40. Schell, W.J., and C.D. Houston, *Chem. Eng. Progress*, **78**(10), 33–37 (1982).

41. Teplyakov, V., and P. Meares, *Gas Sep. and Purif.*, **4**, 66–74 (1990).

42. Rosenzweig, M.D., *Chem. Eng.*, **88**(24), 62–66 (1981).

43. Kober, P.A., *J. Am. Chem. Soc.*, **39**, 944–948 (1917).

44. Binning, R.C., R.J. Lee, J.F. Jennings, and E.C. Martin, *Ind. Eng. Chem.*, **53**, 45–50 (1961).

45. Wesslein, M., A. Heintz, and R.N. Lichtenthaler, *J. Membrane Sci.*, **51**, 169 (1990).

46. Wijmans, J.G., and R.W. Baker, *J. Membrane Sci.*, **79**, 101–113 (1993).

47. Rautenbach, R., and R. Albrecht, *Membrane Processes*, John Wiley & Sons, New York (1989).

48. Rao, M.B., and S. Sircar, *J. Membrane Sci.*, **85**, 253–264 (1993).

49. Baker, R., *Membrane Technology and Applications*, 2nd ed., John Wiley & Sons, New York (2004).

50. Chisti, Y., "Principles of Membrane Separation Processes," in G. Subramanian, Ed., *Bioseparation and Bioprocessing*, Wiley-VCH Verlag GmbH & Co. KGaA, Weinheim (2007).

51. Belfort, G., R.H. Davis, and A.J. Zydney, *J. Membrane Science.*, **96**, 1–58 (1994).

52. Van Reis, R., and A.L. Zydney, *Curr. Opinion Biotechnol.*, **12**, 208–211 (2001).

53. Van Reis, R., and A.L. Zydney, *J. Membr. Sci.*, **297**(1), 16–50 (2007).

54. Aimar, P., *Membranes in Bioprocessing*, 113–139 (1993).

55. McGregor, W.C., Ed., *Membrane Separations in Biotechnology*, Marcel Dekker, Inc., New York (1986).

56. Zeman, L.J., and A.L. Zydney, *Microfiltration and Ultrafiltration: Principles and Applications*, Marcel Dekker, New York (1996).

57. Porter, M.C., *Ind. Eng. Chem. Prod. Res. Dev.*, **11**, 234 (1972).

58. Van Reis, R., E.M. Goodrich, C.L. Yson, L.N. Frautschy, R. Whitely, and A.L. Zydney, *J Membr. Sci.*, **130**, 123–140 (1997).

59. Van Reis, R., J.M. Brake, J. Charkoudian, D.B. Burns, and A.L. Zydney, *J Membr. Sci.*, **159**, 133–143 (1999).

60. Zeman, L.J., and A.L. Zydney, *Microfiltration and Ultrafiltration, Principles and Applications*, Marcel Dekker, Inc., New York (1996).

61. Van Reis, R., E.M. Goodrich, C.L. Yson, L.M. Frautschy, S. Dzengeleski, and H. Lutz, *Biotechnol Bioeng*, **55**, 737–746 (1997).

62. Tucceli, R., and P.V. McGrath, Cellulosic ultrafiltration membrane. U.S. Patent 5,736,051 (1996).

63. Van Reis, R., and A.L. Zydney, "Protein Ultrafiltration," in M.C. Flickinger and S.W. Drew, Eds., *Encyclopedia of Bioprocess Technology: Fermentation, Biocatalysis and Bioseparation*, John Wiley & Sons, New York, pp. 2197–2214 (1999).

64. Meltzer, T.H., "Modus of Filtration," in *Adv, Biochem. Engin./Biotechnol.*, Springer-Verlag, Heidelberg, Vol. **9**, pp. 27–71 (2006).

65. Roper, D.K., A. Johnson, A. Lee, J. Taylor, C. Trimor, and E. Wen, *First International Conference on Membrane and Filtration Technology in Biopurification*, Cambridge, U.K., April 7–9, 1999.

66. Harrison, R.G., P. Todd, S.R. Rudge, and D.P. Petrides, *Bioseparations Science and Engineering*. Oxford University Press, New York (2003).

67. Grace, H.P., *Chem. Eng. Progr.*, **49**, 303 (1953).

68. Ruth, B.F., G.H. Montillon, and R.E. Montanna, *Ind. Eng. Chem.* **25**, 76–82 (1933).

69. Hermia, J., *Trans. Inst. Chem. Eng. –Lond.* **60**, 183 (1982).

70. Ho C.-C. and A.L. Zydney, *J Colloid Interface Sci.*, **232**, 389 (2000).

71. Ho C.-C. and A.L. Zydney, *Ind. Eng. Chem. Res.*, **40**, 1412 (2001).

72. Zydney, A.L. and C.-C. Ho, *Biotech. Bioeng.*, **83**, 537 (2001).

73. Schweitzer, P. A., *Handbook of Separation Techniques for Chemical Engineers*, 2nd ed., Section 2.1 by M.C. Porter, McGraw-Hill Book Co., New York (1988).

74. Badmington, G., M. Payne, R. Wilkins, and E. Honig, *Pharmaceut. Tech.*, **19**, 64 (1995).

75. Meltzer, T.H., and M.W. Jornitz, *Filtration in the Biopharmaceutical Industry*, Marcel Dekker, Inc., New York (1998).

76. Shuler, M.L., and F. Kargi, *Bioprocess Engineering*, 2nd ed., Prentice Hall PTR, Upper Saddle River, NJ (2002).

77. Bailey, J.E., and D.F. Ollis, *Biochemical Engineering Fundamentals*, 2nd ed., McGraw-Hill, New York (1986).

78. Emory, S., *Pharm. Technol.*, **13**, 68 (1980).

79. Baker, L.A., and C.R. Martin, *Nanotechnology in Biology and Medicine: Methods, Devices and Applications*, Vol. 9, pp. 1–24 (2007).

80. Baker, L.A., T. Choi, and C.R. Martin, *Current Nanoscience*, **2**(3), 243–255 (2006).

81. Belfort, G., "Membrane Separation Technology: An Overview," in H.R. Bungay and G. Belfort, Eds., *Advanced Biochemical Engineering*, John Wiley & Sons, New York, p. 253 (1987).

82. Kahn, D.W., M.D. Butler, G.M. Cohen, J.W. Kahn, and M.E. Winkler, *Biotechnol. Bioeng.*, **69**, 101–106 (2000).

83. Cruz, P.E., C.C. Peixoto, K. Devos, J.L. Moreira, E. Saman, and M.J. T. Carrondo, *Enzyme Microb. Technol.*, **26**, 61–70 (2000).

84. Ladisch, M.R., *Biseparations Engineering*, Wiley-Interscience, New York (2001).

85. Scopes, R.K., *Protein Purification*, 2nd ed., Springer-Verlag, New York (1987).

86. Pujar, N.S., and A.L. Zydney, *Ind. Eng. Chem. Res.*, **22**, 2473 (1994).

87. Blatt, W.F., A. Dravid, A.S. Michaels, and L. Nelson, "Solute Polarization and Cake Formation in Membrane Ultrafiltration: Causes Consequences, and Control Techniques," in J.E. Flinn, Ed., *Membrane Science and Technology*, Plenum Press, New York, pp. 47–97 (1970).

88. Belfort, G., R.H. Davis, and A.J. Zydney, *J. Membrane Sci.*, **96**, 1–58 (1994).

89. Leighton, D.T., and A. Acrivos, *J. Fluid Mech.*, **181**, 415–439 (1987).

90. Robertson, B.C., and A.L. Zydney, *J. Colloid Interface Sci.*, **134**, 563 (1990).

91. Winzeler, H.B., and G. Belfort, *J Membrane Sci.*, **80**, 157–185 (1993).

92. Sandler, S.I., *Chemical, Biochemical, and Engineering Thermodynamics*, 4th ed., John Wiley & Sons, New York (2006).

93. Porter, M.C., *Ind. Eng. Chem. Prod. Res. Dev.*, **11**, 234 (1972).

94. Eckstein, E.C., P.G. Bailey, and A.H. Shapiro, *J. Fluid Mech.*, **7**, 191 (1977).

95. Zydney, A.L., and C.K. Colton, *Chem. Eng. Commun.*, **47**, 1 (1986).

96. Davis, R.H., and J.D. Sherwood, *Chem. Eng. Sci.*, **45**, 3204–3209 (1990).

97. Drew, D.A., J.A. Schonber, and G. Belfort, *Chem. Eng. Sci.*, **46**, 3219–3224 (1991).

98. Henry, J.D., "Cross Flow Filtration," in N.N. Li, Ed., *Recent Developments in Separation Science*, CRC Press, Cleveland, OH, Vol. **2**, pp. 205–225 (1972).

99. Belfort, G., P. Chin, and D.M. Dziewulski, "A New Gel-Polarization Model Incorporating Lateral Migration for Membrane Fouling," *Proc. World Filtration Congress III*, Philadelphia, PA, p. 91 (1982).

100. Taddei, C., P. Aimar, J.A. Howell, and J.A. Scott, *J. Chem. Technol. Biotechnol.*, **47**, 365–376 (1990).

101. Le, M.S., *J. Chem. Technol. Biotechnol.*, **37**, 59–66 (1987).

102. Sakai, K., K. Ozawa, K. Ohashi, R. Yoshida, and H. Sakarai, *Ind. Eng. Chem. Res.*, **28**, 57–64 (1989).

103. Geankoplis, C.J., *Transport Processes and Separation Process Principles*, 4th ed., Prentice Hall PTR, New Jersey (2003).

104. Nielsen, W.K., Ed., *Membrane Filtration and Related Molecular Separation Technologies*, International Dairy Books, Aarhus, Denmark (2000).

105. Cheryan, M., *Ultrafiltration Handbook*, Technomic Publishing Co., Lancaster, PA (1986).

STUDY QUESTIONS

14.1. What are the two products from a membrane separation called? What is a sweep?

14.2. What kinds of materials are membranes made from? Can a membrane be porous or nonporous? What forms pores in polymer membranes?

14.3. What is the basic equation for computing the rate of mass transfer through a membrane? Explain each of the four factors in the equation and how they can be exploited to obtain high rates of mass transfer.

14.4. What is the difference between permeability and permeance? How are they analogous to diffusivity and the mass-transfer coefficient?

14.5. For a membrane separation, is it usually possible to achieve both a high permeability and a large separation factor?

14.6. What are the three mechanisms for mass transfer through a porous membrane? Which are the best mechanisms for making a separation? Why?

14.7. What is the mechanism for mass transfer through a dense (nonporous) membrane? Why is it called solution-diffusion? Does this mechanism work if the polymer is completely crystalline? Explain.

14.8. How do the solution-diffusion equations differ for liquid transport and gas transport? How is Henry's law used for solution-diffusion for gas transport? Why are the film resistances to mass transfer on either side of the membrane for gas permeation often negligible?

14.9. What are the four idealized flow patterns in membrane modules? Which is the most effective? Which is the most difficult to calculate?

14.10. What is osmosis? Can it be used to separate a liquid mixture? How does it differ from reverse osmosis? For what type of mixtures is it well suited?

14.11. Can a near-perfect separation be made with gas permeation? If not, why not?

14.12. What is pervaporation?

14.13. How do microfiltration and ultrafiltration differ from reverse osmosis with respect to pore size, pressure drop, and the nature of the permeate?

14.14. What is the evidence that concentration polarization and fouling are occurring during biofiltrations, and what steps are taken to minimize these effects?

14.15. What are the four common configurations for ultrafiltration?

14.16. What is continuous feed-and-bleed ultrafiltration? What are its limitations?

14.17. What is diafiltration? How does it differ from continuous feed-and-bleed ultrafiltration? Under what conditions is diafiltration used in conjunction with continuous feed-and-bleed ultrafiltration?

14.18. In microfiltration, why is an operation that combines constant-flux and constant-pressure operations used?

EXERCISES

Section 14.1

14.1. Differences between membrane separations and other separations.

Explain, as completely as you can, how membrane separations differ from: (a) absorption and stripping; (b) distillation; (c) liquid–liquid extraction; (d) extractive distillation.

14.2. Barrer units for permeabilities.

For the commercial application of membrane separators discussed at the beginning of this chapter, calculate the permeabilities of hydrogen and methane in barrer units.

14.3. Membrane separation of N_2 from CH_4.

A new asymmetric, polyimide polymer membrane has been developed for the separation of N_2 from CH_4. At 30°C, permeance values are 50,000 and 10,000 barrer/cm for N_2 and CH_4, respectively. If this new membrane is used to perform the separation in Figure 14.37, determine the membrane surface area in m^2, and the kmol/h of CH_4 in the permeate. Base the driving force for diffusion on the arithmetic average of the partial pressures of the entering feed and the exiting retentate, with the permeate-side partial pressures at the exit condition.

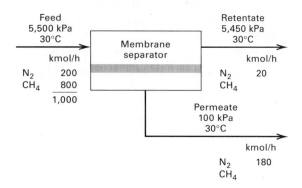

Figure 14.37 Data for Exercise 14.3.

Section 14.2

14.4. Characteristics of a hollow-fiber module.

A hollow-fiber module has 4,000 ft^2 of membrane surface area based on the size of the fibers, which are 42 μm i.d. × 85 μm o.d. × 1.2 m long each. Determine the: (a) number of hollow fibers in the module; (b) diameter of the module, assuming the fibers are on a square spacing of 120 μm center-to-center; and (c) membrane surface area per unit volume of module (packing density) m^2/m^3. Compare your result with that in Table 14.4.

14.5. Geometry of a membrane module.

A spiral-wound module made from a flat sheet of membrane material is 0.3 m in diameter and 3 m long. If the packing density (membrane surface area/unit module volume) is 500 m^2/m^3, what is the center-to-center spacing of the membrane in the spiral, assuming a collection tube 1 cm in diameter?

14.6. Characteristics of a monolithic element.

A monolithic membrane element of the type shown in Figure 14.4d contains 19 flow channels of 0.5 cm in inside diameter by 0.85 m long. If 9 of these elements are in a cylindrical module of the type in Figure 14.5, determine values for: (a) module volume in m^3; and (b) packing density in m^2/m^3. Compare your value with values for other membrane modules given in Table 14.4.

Section 14.3

14.7. Porous membrane with pressure differential.

Water at 70°C is passed through a polyethylene membrane of 25% porosity with an average pore diameter of 0.3 μm and an average tortuosity of 1.3. The pressures on the downstream and upstream sides of the membrane are 125 and 500 kPa, respectively. Estimate the flux of water in m^3/m^2-day.

14.8. Knudsen flow in a membrane.

A porous-glass membrane, with an average pore diameter of 40 Å, is used to separate light gases at 25°C when Knudsen flow may be dominant. The pressures are 15 psia downstream and not > 120 psia upstream. The membrane has been calibrated with pure helium

gas, giving a constant permeability of 117,000 barrer. Experiments with pure CO_2 give a permeability of 68,000 barrer. Assuming that helium is in Knudsen flow, predict the permeability of CO_2. Is it in agreement with the experimental value? If not, suggest an explanation. Reference: Kammermeyer, K., and L.O. Rutz, *C.E.P. Symp. Ser.*, **55** (24), 163–169 (1959).

14.9. Partial condensation and surface diffusion.

Two mechanisms for the transport of gas through a porous membrane not discussed in §14.3 or illustrated in Figure 14.6 are (1) partial condensation in the pores by some components of the gas mixture to the exclusion of other components, and subsequent transport of the condensed molecules through the pore, and (2) selective adsorption on pore surfaces of some components and subsequent surface diffusion across the pores. In particular, Rao and Sircar [48] have found that the latter mechanism provides a potentially attractive means for separating hydrocarbons from hydrogen for low-pressure gas streams. In porous-carbon membranes with continuous pores 4–15 Å in diameter, little pore void space is available for Knudsen diffusion of hydrogen when the hydrocarbons are selectively adsorbed.

Typically, the membranes are not more than 5 μm in thickness. Measurements at 295.1 K of permeabilities for five pure components and a mixture of the five components are as follows:

Component	Permeability, barrer		mol% in the Mixture
	As a Pure Gas	In the Mixture	
H_2	130	1.2	41.0
CH_4	660	1.3	20.2
C_2H_6	850	7.7	9.5
C_3H_8	290	25.4	9.4
nC_4H_{10}	155	112.3	19.9
			100.0

A refinery waste gas mixture of the preceding composition is to be processed through such a porous-carbon membrane. If the pressure of the gas is 1.2 atm and an inert sweep gas is used on the permeate side such that partial pressures of feed-gas components on that side are close to zero, determine the permeate composition on a sweep-gas-free basis when the composition on the upstream pressure side of the membrane is that of the feed gas. Explain why the component permeabilities differ so much between pure gas and the gas mixture.

14.10. Module flow pattern and membrane area.

A mixture of 60 mol% propylene and 40 mol% propane at a flow rate of 100 lbmol/h and at 25°C and 300 psia is to be separated with a polyvinyltrimethylsilane polymer (see Table 14.10 for permeabilities). The membrane skin is 0.1 μm thick, and spiral-wound modules are used with a pressure of 15 psia on the permeate side. Calculate the material balance and membrane area in m^2 as a function of the cut (fraction of feed permeated) for: (a) perfect-mixing flow pattern and (b) crossflow pattern.

14.11. Membrane area for gas permeation.

Repeat part (a) of Exercise 14.10 for a two-stage stripping cascade and a two-stage enriching cascade, as shown in Figure 14.14. However, select just one set of reasonable cuts for the two stages of each case so as to produce 40 lbmol/h of final retentate.

14.12. Dead-end microfiltration of skim milk.

Using the membrane and feed conditions of and values for R_m and K_2 determined in Example 14.3 for DE microfiltration, compute and plot the permeate flux and cumulative permeate volume as a function of time. Assume a combined operation with Stage 1 at a constant permeate rate of 10 mL/minute to an upper-limit pressure drop of 25 psi, followed by Stage 2 at this pressure drop until the permeate rate drops to a lower limit of 5 mL/minute.

14.13.   Concentration polarization in dialysis.

Repeat Example 14.8 with the following changes: tube-side Reynolds number = 25,000; tube inside diameter = 0.4 cm; permeate-side mass-transfer coefficient = 0.06 cm/s. How important is concentration polarization?

Section 14.4

14.14.   Dialysis to separate Na_2SO_4.

An aqueous process stream of 100 gal/h at 20°C contains 8 wt% Na_2SO_4 and 6 wt% of a high-molecular-weight substance (A). This stream is processed in a continuous countercurrent-flow dialyzer using a pure water sweep of the same flow rate. The membrane is a microporous cellophane with pore volume = 50%, wet thickness = 0.0051 cm, tortuosity = 4.1, and pore diameter = 31Å. The molecules to be separated have the following properties:

	Na_2SO_4	A
Molecular weight	142	1,000
Molecular diameter, Å	5.5	15.0
Diffusivity, cm²/s × 10⁵	0.77	0.25

Calculate the membrane area in m² for only a 10% transfer of A through the membrane, assuming no transfer of water. What is the % recovery of the Na_2SO_4 in the diffusate? Use log-mean concentration-driving forces and assume the mass-transfer resistances on each side of the membrane are each 25% of the total mass-transfer resistances for Na_2SO_4 and A.

14.15.   Removal of HCl by dialysis.

A dialyzer is to be used to separate 300 L/h of an aqueous solution containing 0.1-M NaCl and 0.2-M HCl. Laboratory experiments with the microporous membrane to be used give the following values for the overall mass-transfer coefficient K_i in (14-79) for a log-mean concentration-driving force:

	K_i, cm/min
Water	0.0025
NaCl	0.021
HCl	0.055

Determine the membrane area in m² for 90, 95, and 98% transfer of HCl to the diffusate. For each case, determine the complete material balance in kmol/h for a sweep of 300 L/h.

Section 14.5

14.16.   Desalinization by electrodialysis.

A total of 86,000 gal/day of an aqueous solution of 3,000 ppm of NaCl is to be desalinized to 400 ppm by electrodialysis, with a 40% conversion. The process will be conducted in four stages, with three stacks of 150 cell pairs in each stage. The fractional desalinization will be the same in each stage and the expected current efficiency is 90%. The applied voltage for the first stage is 220 V. Each cell pair has an area of 1,160 cm². Calculate the current density in mA/cm², the current in A, and the power in kW for the first stage. Reference: Mason, E.A., and T.A. Kirkham, *C.E.P. Symp. Ser.*, **55** (24), 173–189 (1959).

Section 14.5

14.17.   Reverse osmosis of seawater.

A reverse-osmosis plant is used to treat 30,000,000 gal/day of seawater at 20°C containing 3.5 wt% dissolved solids to produce 10,000,000 gal/day of potable water, with 500 ppm of dissolved solids and the balance as brine containing 5.25 wt% dissolved solids. The feed-side pressure is 2,000 psia, while the permeate pressure is 50 psia. A single stage of spiral-wound membranes is used that approximates crossflow. If the total membrane area is 2,000,000 ft², estimate the permeance for water and the salt passage.

14.18.   Reverse osmosis with multiple stages.

A reverse-osmosis process is to be designed to handle a feed flow rate of 100 gpm. Three designs have been proposed, differing in the % recovery of potable water from the feed:

Design 1: A single stage consisting of four units in parallel to obtain a 50% recovery

Design 2: Two stages in series with respect to the retentate (four units in parallel followed by two units in parallel)

Design 3: Three stages in series with respect to the retentate (four units in parallel followed by two units in parallel followed by a single unit)

Draw the three designs and determine the percent recovery of potable water for Designs 2 and 3.

14.19.   Concentration of Kraft black liquor by two-stage reverse osmosis.

Production of paper requires a pulping step to break down wood chips into cellulose and lignin. In the Kraft process, an aqueous solution known as white liquor and consisting of dissolved inorganic chemicals such as Na_2S and NaOH is used. Following removal of the pulp (primarily cellulose), a solution known as weak Kraft black liquor (KBL) is left, which is regenerated to recover white liquor for recycle. In this process, a 15 wt% (dissolved solids) KBL is concentrated to 45 to 70 wt % by multieffect evaporation. It has been suggested that reverse osmosis be used to perform an initial concentration to perhaps 25 wt%. Higher concentrations may not be feasible because of the high osmotic pressure, which at 180°F and 25 wt% solids is 1,700 psia. Osmotic pressure for other conditions can be scaled with (14-102) using wt% instead of molality.

A two-stage RO process, shown in Figure 14.38, has been proposed to carry out this initial concentration for a feed rate of 1,000 lb/h at 180°F. A feed pressure of 1,756 psia is to be used for the first stage to yield a permeate of 0.4 wt% solids. The feed pressure to the second stage is 518 psia to produce water of 300 ppm dissolved solids and a retentate of 2.6 wt% solids. Permeate-side pressure for both stages is 15 psia. Equation (14-93) can be used to estimate membrane area, where the permeance for water can be taken as 0.0134 lb/ft²-hr-psi in conjunction with an arithmetic mean osmotic pressure for plug flow on the feed side. Complete the material balance for the process and estimate the required membrane areas for each stage. Reference: Gottschlich, D.E., and D.L. Roberts. Final Report DE91004710, SRI International, Menlo Park, CA, Sept. 28, 1990.

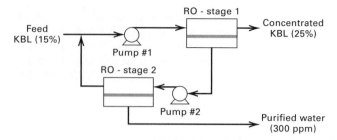

Figure 14.38 Data for Exercise 14.19.

Section 14.7

14.20. Recovery of VOCs by gas permeation.

Gas permeation can be used to recover VOCs (volatile organic compounds) from air at low pressures using a highly selective membrane. In a typical application, 1,500 scfm (0°C, 1 atm) of air containing 0.5 mol% acetone (A) is fed to a spiral-wound membrane module at 40°C and 1.2 atm. A liquid-ring vacuum pump on the permeate side establishes a pressure of 4 cmHg. A silicone-rubber, thin-composite membrane with a 2-μm-thick skin gives permeabilities of 4 barrer for air and 20,000 barrer for acetone.

If the retentate is to contain 0.05 mol% acetone and the permeate is to contain 5 mol% acetone, determine the membrane area required in m^2, assuming crossflow. References: (1) Peinemann, K.-V., J.M. Mohr, and R.W. Baker, *C.E.P. Symp. Series*, **82** (250), 19–26 (1986); (2) Baker, R.W., N. Yoshioka, J.M. Mohr, and A.J. Khan, *J. Membrane Sci.*, **31**, 259–271 (1987).

14.21. Separation of air by gas permeation.

Separation of air into N$_2$ and O$_2$ is widely practiced. Cryogenic distillation is most economical for processing 100 to 5,000 tons of air per day, while pressure-swing adsorption is favorable for 20 to 50 tons/day. For small-volume users requiring less than 10 tons/day, gas permeation finds applications where for a single stage, either an oxygen-enriched air (40 mol% O$_2$) or 98 mol% N$_2$ can be produced. It is desired to produce a permeate of 5 tons/day (2,000 lb/ton) of 40 mol% oxygen and a retentate of nitrogen, ideally of 90 mol% purity, by gas permeation. Assume pressures of 500 psia (feed side) and 20 psia (permeate). Two companies who can supply the membrane modules have provided the following data:

	Company A	Company B
Module type	Hollow-fiber	Spiral-wound
$\bar{P}_M$ for O$_2$, barrer/μm	15	35
$\bar{P}_{M_{O_2}}/\bar{P}_{M_{N_2}}$	3.5	1.9

Determine the required membrane area in m^2 for each company. Assume that both module types approximate crossflow.

14.22. Removal of CO$_2$ and H$_2$S by permeation.

A joint venture has been underway for several years to develop a membrane process to separate CO$_2$ and H$_2$S from high-pressure, sour natural gas. Typical feed and product conditions are:

	Feed Gas	Pipeline Gas
Pressure, psia	1,000	980
Composition, mol%:		
CH$_4$	70	97.96
H$_2$S	10	0.04
CO$_2$	20	2.00

To meet these conditions, the following hollow-fiber membrane material targets have been established:

	Selectivity
CO$_2$–CH$_4$	50
H$_2$S–CH$_4$	50

where selectivity is the ratio of permeabilities. $P_{M_{CO_2}} = 13.3$ barrer, and membrane skin thickness is expected to be 0.5 μm. Make calculations to show whether the targets can realistically meet the pipeline-gas conditions in a single stage with a reasonable membrane area. Assume a feed-gas flow rate of 10×10^3 scfm (0°C, 1 atm) with crossflow. Reference: Stam, H., in L. Cecille and J.-C. Toussaint, Eds., *Future Industrial Prospects of Membrane Processes*, Elsevier Applied Science, London, pp. 135–152 (1989).

Section 14.8

14.23. Separation by pervaporation.

Pervaporation is to be used to separate ethyl acetate (EA) from water. The feed rate is 100,000 gal/day of water containing 2.0 wt% EA at 30°C and 20 psia. The membrane is dense polydimethylsiloxane with a 1-μm-thick skin in a spiral-wound module that approximates crossflow. The permeate pressure is 3 cmHg. The total measured membrane flux at these conditions is 1.0 L/m^2-h with a separation factor given by (14-59) of 100 for EA with respect to water. A retentate of 0.2 wt% EA is desired for a permeate of 45.7 wt% EA. Determine the required membrane area in m^2 and the feed temperature drop. Reference: Blume, I., J.G. Wijans, and R.W. Baker, *J. Membrane Sci.*, **49**, 253–286 (1990).

14.24. Permeances for pervaporation.

For a temperature of 60°C and a permeate pressure of 15.2 mmHg, Wesslein et al. [45] measured a total permeation flux of 1.6 kg/m^2-h for a 17.0 wt% ethanol-in-water feed, giving a permeate of 12 wt% ethanol. Otherwise, conditions were those of Example 14.13. Calculate the permeances of ethyl alcohol and water for these conditions. Also, calculate the selectivity for water.

14.25. Second stage of a pervaporation process.

The separation of benzene (B) from cyclohexane (C) by distillation at 1 atm is impossible because of a minimum-boiling-point azeotrope at 54.5 mol% benzene. However, extractive distillation with furfural is feasible. For an equimolar feed, cyclohexane and benzene products of 98 and 99 mol%, respectively, can be produced. Alternatively, the use of a three-stage pervaporation process, with selectivity for benzene using a polyethylene membrane, has received attention, as discussed by Rautenbach and Albrecht [47]. Consider the second stage of this process, where the feed is 9,905 kg/h of 57.5 wt% B at 75°C. The retentate is 16.4 wt% benzene at 67.5°C and the permeate is 88.2 wt% benzene at 27.5°C. The total permeate mass flux is 1.43 kg/m^2-h and selectivity for benzene is 8. Calculate flow rates of retentate and permeate in kg/h and membrane surface area in m^2.

Section 14.9

14.26. Permeability of a nanofiltration membrane.

Obtain general expressions for hydraulic membrane permeability, L_p, and membrane resistance, R_m, for laminar flow through a nanofiltration membrane of thickness L that is permeated by right-cylindrical pores of radius r in terms of surface porosity σ, the total area of pore mouths per m^2.

14.27. Constant-pressure cake filtration.

Beginning with the Ruth equation (14-24), obtain general expressions for time-dependent permeate volume, $V\{t\}$, and time-dependent flux, $J\{t\}$, in terms of operating parameters and characteristics of the cake for constant-pressure cake filtration.

14.28. Pore-constriction model.

Derive a general expression for the total filtration time necessary to filter a given feed volume V using the pore-constriction model. From this expression, predict the average volumetric flux during a filtration and the volumetric capacity necessary to achieve a given filtration time, based on laboratory-scale results.

14.29. Minimum filter area for sterile filtration.

Derive a general expression for the minimum filter area requirement per a sterility assurance limit (SAL) in terms of (a) concentration of microorganisms in the feed; (b) volume per unit parenteral dose; (c) sterility assurance limit; and (d) filter capacity.

14.30. Cheese whey ultrafiltration process.

Based on the problem statement of Example 14.20, calculate for just Section 1 the component material balance in pounds per day of operation, the percent recovery (yield) from the whey of the TP and NPN in the final concentrate, and the number of cartridges required if two stages are used instead of four.

14.31. Four-stage diafiltration section.

Based on the problem statement of Example 14.20, design a four-stage diafiltration section to take the 55 wt% concentrate from Section 1 and achieve the desired 85 wt% concentrate, thus eliminating Section 3.

Chapter 15

Adsorption, Ion Exchange, Chromatography, and Electrophoresis

§15.0 INSTRUCTIONAL OBJECTIVES

After completing this chapter, you should be able to:

- Explain why a few grams of porous adsorbent can have an adsorption area as large as a football field.
- Differentiate between chemisorption and physical adsorption.
- Explain how ion-exchange resins work.
- Compare three major expressions (so-called isotherms) used for correlating adsorption-equilibria data.
- List steps involved in adsorption of a solute, and which steps may control the rate of adsorption.
- Describe major modes for contacting the adsorbent with a fluid containing solute(s) to be adsorbed.
- Describe major methods for regenerating adsorbent.
- Calculate vessel size or residence time for any of the major modes of slurry adsorption.
- List and explain assumptions for ideal fixed-bed adsorption and explain the concept of width of mass-transfer zone. Explain the concept of breakthrough in fixed-bed adsorption.
- Calculate bed height, bed diameter, and cycle time for fixed-bed adsorption.
- Compute separations for a simulated-moving-bed operation.
- Calculate rectangular and Gaussian-distribution pulses in chromatography.
- Describe electrophoresis of biomolecules, including factors that affect mobility as well as effects of electro-osmosis and convective Joule heating
- Distinguish different electrophoretic modes (native gel electrophoresis, SDS-PAGE, isoelectric focusing, isotachophoresis, 2-D gel electrophoresis, and pulsed field gel electrophoresis) in terms of denaturants used, pH and electrolyte content, and application of electric-field gradients.

Adsorption, ion exchange, and chromatography are *sorption* operations in which components of a fluid phase (*solutes*) are selectively transferred to insoluble, rigid particles suspended in a vessel or packed in a column. Sorption, a general term introduced by J.W. McBain [*Phil. Mag.*, **18**, 916–935 (1909)], includes selective transfer to the surface and/or into the bulk of a solid or liquid. Thus, absorption of gas species into a liquid and penetration of fluid species into a nonporous membrane are sorption operations. In a sorption process, the sorbed solutes are referred to as *sorbate*, and the sorbing agent is the *sorbent*.

In an *adsorption* process, molecules, as in Figure 15.1a, or atoms or ions, in a gas or liquid, diffuse to the surface of a solid, where they bond with the solid surface or are held by weak intermolecular forces. Adsorbed solutes are referred to as *adsorbate*, whereas the solid material is the *adsorbent*. To achieve a large surface area for adsorption per unit volume, porous solid particles with small-diameter, interconnected

pores are used, with adsorption occurring on the surface of the pores.

In an *ion-exchange* process, as in Figure 15.1b, ions of positive charge (*cations*) or negative charge (*anions*) in a liquid solution, usually aqueous, replace dissimilar and displaceable ions, called *counterions*, of the same charge contained in a solid *ion exchanger*, which also contains immobile, insoluble, and permanently bound *co-ions* of the opposite charge. Thus, ion exchange can be cation or anion exchange. Water softening by ion exchange involves a cation exchanger, in which a reaction replaces calcium ions with sodium ions:

$$Ca^{2+}_{(aq)} + 2NaR_{(s)} \leftrightarrow CaR_{2(s)} + 2Na^{+}_{(aq)}$$

where R is the ion exchanger. The exchange of ions is reversible and does not cause any permanent change to the solid ion-exchanger structure. Thus, it can be used and reused unless fouled by organic compounds in the liquid feed that attach to exchange sites on and within the ion exchange resin.

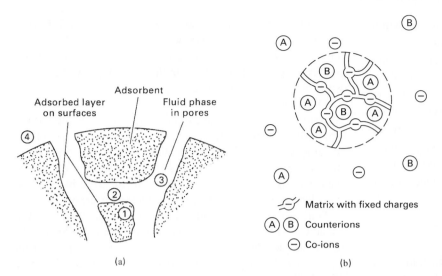

Matrix with fixed charges

(A) (B) Counterions

(−) Co-ions

Figure 15.1 Sorption operations with solid-particle sorbents. (a) Adsorption. (b) Ion exchange.

(a)

(b)

The ion-exchange concept can be extended to the removal of essentially all inorganic salts from water by a two-step *demineralization* process or *deionization.* In step 1, a cation resin exchanges hydrogen ions for cations such as calcium, magnesium, and sodium. In step 2, an anion resin exchanges hydroxyl ions for strongly and weakly ionized anions such as sulfate, nitrate, chloride, and bicarbonate. The hydrogen and hydroxyl ions combine to form water. Regeneration of the cation and anion resins is usually accomplished with sulfuric acid and sodium hydroxide.

In *chromatography*, the sorbent may be a solid adsorbent; an insoluble, nonvolatile liquid absorbent contained in the pores of a granular solid support; or an ion exchanger. In any case, the solutes to be separated move through the chromatographic separator, with an inert, eluting fluid, at different rates because of different sortion affinities during repeated sorption, desorption cycles.

During adsorption and ion exchange, the solid separating agent becomes saturated or nearly saturated with the molecules, atoms, or ions transferred from the fluid phase. To recover the sorbed substances and allow the sorbent to be reused, the asorbent is regenerated by desorbing the sorbed substances. Accordingly, these two separation operations are carried out in a cyclic manner. In chromatography, regeneration occurs continuously, but at changing locations in the separator.

Adsorption processes may be classified as *purification* or *bulk separation*, depending on the concentration in the feed of the components to be adsorbed. Although there is no sharp dividing concentration, Keller [1] has suggested 10 wt%. Early applications of adsorption involved only purification. Adsorption with charred wood to improve the taste of water has been known for centuries. Decolorization of liquids by adsorption with bone char and other materials has been practiced for at least five centuries. Adsorption of gases by a solid (charcoal) was first described by C.W. Scheele in 1773.

Commercial applications of bulk separation by gas adsorption began in the early 1920s, but did not escalate until

the 1960s, following inventions by Milton [2] of synthetic molecular-sieve zeolites, which provide high adsorptive selectivity, and by Skarstrom [3] of the *pressure-swing cycle*, which made possible a fixed-bed, cyclic gas-adsorption process. The commercial separation of liquid mixtures also began in the 1960s, following the invention by Broughton and Gerhold [4] of the simulated moving bed for adsorption.

Uses of ion exchange date back at least to the time of Moses, who, while leading his followers out of Egypt, sweetened the bitter waters of Marah with a tree [Exodus 15:23–26]. In ancient Greece, Aristotle observed that the salt content of water is reduced when it percolates through certain sands. Studies of ion exchange were published in 1850 by both Thompson and Way, who experimented with cation exchange in soils before the discovery of ions.

The first major application of ion exchange occurred over 100 years ago for water treatment to remove calcium and other ions responsible for water hardness. Initially, the ion exchanger was a porous, natural, mineral zeolite containing silica. In 1935, synthetic, insoluble, polymeric-resin ion exchangers were introduced. Today they are dominant for water-softening and deionizing applications, but natural and synthetic zeolites still find some use.

Since the 1903 invention of chromatography by M. S. Tswett [5], a Russian botanist, it has found widespread use as an analytical, preparative, and industrial technique. Tswett separated a mixture of structurally similar yellow and green chloroplast pigments in leaf extracts by dissolving the extracts in carbon disulfide and passing the solution through a column packed with chalk particles. The pigments were separated by color; hence, the name chromatography, which was coined by Tswett in 1906 from the Greek words *chroma*, meaning "color," and *graphe*, meaning "writing." Chromatography has revolutionized laboratory chemical analysis of liquid and gas mixtures. Large-scale, commercial applications described by Bonmati et al. [6] and Bernard et al. [7] began in the 1980s.

Also included in this chapter is *electrophoresis*, which involves the size- and charge-based separation of charged solutes that move in response to an electric field applied across an electrophoretic medium. Positively charged solutes migrate to the negative electrode; negatively charged solutes migrate toward the positive electrode. Typical media include agarose, polyacrylamide, and starch, which form gels with a high H_2O content that allows passage of large solutes through their porous structures. Electrophoresis is widely used to separate and purify biomolecules, including proteins and nucleic acids.

Industrial Example

Pressure-swing gas adsorption is used for air dehydration and for separation of air into nitrogen and oxygen. A small unit for the dehydration of compressed air is described by White and Barkley [8] and shown in Figure 15.2. The unit consists of two fixed-bed adsorbers, each 12.06 cm in diameter and packed with 11.15 kg of 3.3-mm-diameter Alcoa F-200 activated-alumina beads to a height of 1.27 m. The external porosity (void fraction) of the bed is 0.442 and the alumina-bead bulk density is 769 kg/m^3.

The unit operates on a 10-minute cycle, with 5 minutes for adsorption of water vapor and 5 minutes for regeneration, which consists of depressurization, purging of the water vapor, and a 30-s repressurization. While one bed is adsorbing, the other bed is being regenerated. The adsorption (drying) step takes place with air entering at 21°C and 653.3 kPa (6.45 atm) with a flow rate of 1.327 kg/minute, passing through the bed with a pressure drop of 2.386 kPa. The dew-point temperature of the air at system pressure is reduced from 11.2 to −61°C by the adsorption process. During the 270-s purge period, about one-third of the dry air leaving one bed is directed to the other bed as a downward-flowing purge to regenerate the adsorbent. The purge is exhausted at a pressure of 141.3 kPa. By conducting the purge flow countercurrently to the entering air flow, the highest degree of water-vapor desorption is achieved.

Other equipment shown in Figure 15.2 includes an air compressor, an aftercooler, piping and valving to switch the beds from one step in the cycle to the other, a coalescing filter to remove aerosols from the entering air, and a particulate filter to remove adsorbent fines from the exiting dry air. If the dry air is needed at a lower pressure, an air turbine can be installed to recover energy while reducing air pressure.

During the 5-minute adsorption period of the cycle, the capacity of the adsorbent for water must not be exceeded. In this example, the water content of the air is reduced from 1.27×10^{-3} kg H_2O/kg air to the very low value of 9.95×10^{-7} kg H_2O/kg air. To achieve this exiting water-vapor content, only a small fraction of the adsorbent capacity is utilized during the adsorption step, with most of the adsorption occurring in the first 0.2 m of the 1.27-m bed height.

The bulk separation of gas and liquid mixtures by adsorption is an emerging separation operation. Important progress is being made in the development of more-selective adsorbents and more-efficient operation cycles. In addition, attention is being paid to hybrid systems that include membrane and other separation steps. The three sorption operations addressed in this chapter have found many applications, as given in Table 15.1, compiled from listings in Rousseau [9]. These cover a wide range of solute molecular weights.

This chapter discusses: (1) sorbents, including their equilibrium, sieving, transport, and kinetic properties with respect to solutes removed from solutions; (2) techniques for conducting cyclic operations; and (3) equipment configuration and design. Both equilibrium-stage and rate-based models are developed. Although emphasis is on adsorption, basic principles of ion exchange, chromatography, and electrophoresis are also presented. Further descriptions of sorption operations are given by Rousseau [9] and Ruthven [10].

§15.1 SORBENTS

To be suitable for commercial use, a sorbent should have: (1) high selectivity to enable sharp separations; (2) high capacity to minimize amount of sorbent; (3) favorable kinetic and transport properties for rapid sorption; (4) chemical and thermal stability, including extremely low solubility in the contacting fluid, to preserve the amount of sorbent and its properties; (5) hardness and mechanical strength to prevent crushing and erosion; (6) a free-flowing tendency for ease of filling or emptying vessels; (7) high resistance to fouling for long life; (8) no tendency to promote undesirable chemical reactions; (9) capability of being regenerated when used with commercial feedstocks containing trace quantities of high-MW species that are strongly sorbed and difficult to desorb; and (10) low cost.

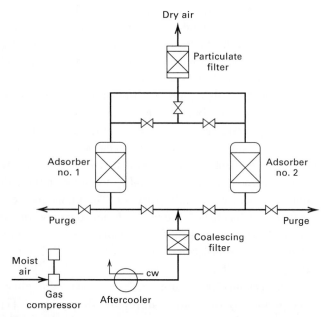

Figure 15.2 Pressure-swing adsorption for the dehydration of air.

Table 15.1 Industrial Applications of Sorption Operations

1. Adsorption

Gas purifications:

    Removal of organics from vent streams

    Removal of SO_2 from vent streams

    Removal of sulfur compounds from gas streams

    Removal of water vapor from air and other gas streams

    Removal of solvents and odors from air

    Removal of NO_x from N_2

    Removal of CO_2 from natural gas

Gas bulk separations:

    N_2/O_2

    H_2O/ethanol

    Acetone/vent streams

    C_2H_4/vent streams

    Normal paraffins/isoparaffins, aromatics

    CO, CH_4, CO_2, N_2, A, NH_3, H_2

Liquid purifications:

    Removal of H_2O from organic solutions

    Removal of organics from H_2O

    Removal of sulfur compounds from organic solutions

    Decolorization of solutions

Liquid bulk separations:

    Normal paraffins/isoparaffins

    Normal paraffins/olefins

    p-xylene/other C_8 aromatics

    p- or m-cymene/other cymene isomers

    p- or m-cresol/other cresol isomers

    Fructose/dextrose, polysaccharides

2. Ion Exchange

    Water softening

    Water demineralization

    Water dealkalization

    Decolorization of sugar solutions

    Recovery of uranium from acid leach solutions

    Recovery of antibiotics from fermentation broths

    Recovery of vitamins from fermentation broths

3. Chromatography

    Separation of sugars

    Separation of perfume ingredients

    Separation of C_4–C_{10} normal and isoparaffins

§15.1.1 Adsorbents

Most solids adsorb species from gases and liquids, but few have a sufficient selectivity and capacity to qualify as serious candidates for commercial adsorbents. Of importance is a large specific surface area (area per unit volume), which is achieved by manufacturing techniques that result in solids with a microporous structure. Pore sizes are usually given in angstroms, Å; nanometers, nm; or micrometers (microns), μm, which are related to meters, m, and millimeters, mm, by:

$$1 \text{ m} = 10^2 \text{ cm} = 10^3 \text{ mm} = 10^6 \text{ μm} = 10^9 \text{ nm} = 10^{10} \text{ Å}$$

Hydrogen and helium atoms are approximately 1 Å in size. By the International Union of Pure and Applied Chemistry (IUPAC) definitions, a *micropore* is <20 Å, a *mesopore* is 20–500 Å, and a *macropore* is >500 Å. Typical commercial adsorbents, which may be granules, spheres, cylindrical pellets, flakes, and/or powders of diameter ranging from 50 μm to 1.2 cm, have specific surface areas from 300 to 1,200 m^2/g. Thus, a few grams of adsorbent can have a surface area equal to that of a football field (120 × 53.3 yards or 5,350 m^2)! This large area is made possible by a particle porosity from 30 to 85 vol% with pore diameters from 10 to 200 Å. To quantify this, consider a cylindrical pore of diameter d_p and length L. The surface area-to-volume ratio is

$$S/V = \pi d_p L \Big/ \left(\pi d_p^2 L/4 \right) = 4/d_p \qquad (15\text{-}1)$$

If the fractional particle porosity is ϵ_p and the particle density is ρ_p, the specific surface area, S_g, in area per unit mass of adsorbent is

$$S_g = 4\epsilon_p/\rho_p d_p \qquad (15\text{-}2)$$

Thus, if ϵ_p is 0.5, ρ_p is 1 $g/cm^3 = 1 \times 10^6$ g/m^3, and d_p is 20 Å (20 × 10^{-10} m), use of (15-2) gives $S_g = 1,000$ m^2/g.

Depending upon the forces between fluid molecules and solid molecules, adsorption may be *physical adsorption* (van der Waals adsorption) or *chemisorption* (activated adsorption). Physical adsorption from a gas occurs when intermolecular attractive forces between solid and gas molecules are greater than those between gas molecules. In effect, the resulting adsorption is like condensation, which is exothermic and accompanied by a release of heat. The magnitude of the heat of adsorption can be > or < than heat of vaporization, and changes with amount of adsorption.

Physical adsorption occurs rapidly, and may be a monomolecular (unimolecular) layer, or two or more layers thick (multimolecular). If unimolecular, it is reversible; if multimolecular, such that capillary pores are filled, hysteresis may occur. The adsorbate density is of the order of magnitude of the liquid rather than the vapor. As physical adsorption takes place, it begins as a monolayer, becomes multilayered, and then, if the pores are close to the size of the molecules, capillary condensation occurs, and pores fill with adsorbate. Accordingly, maximum capacity of a porous adsorbent is related more to pore volume than to surface area. However, for gases at temperatures above their critical temperature, adsorption is confined to a monolayer.

Chemisorption involves formation of chemical bonds between adsorbent and adsorbate in a monolayer, often with a release of heat larger than the heat of vaporization. Chemisorption from a gas generally takes place only at temperatures greater than 200°C and may be slow and irreversible. Commercial adsorbents rely on physical adsorption to achieve separations; solid catalysts rely on chemisorption to catalyze chemical reactions.

Adsorption from liquids is difficult to measure or describe. When the fluid is a gas, the amount of gas adsorbed in a confined space is determined from the measured decrease in total pressure. For a liquid, no simple procedure for determining the extent of adsorption from a pure liquid exists; consequently, experiments are conducted using liquid mixtures. When porous particles of adsorbent are immersed in a liquid

Table 15.2   Representative Properties of Commercial Porous Adsorbents

Adsorbent	Nature	Pore Diameter d_p, Å	Particle Porosity, ϵ_p	Particle Density ρ_p, g/cm³	Surface Area S_g, m²/g	Capacity for H₂O Vapor at 25°C and 4.6 mmHg, wt% (Dry Basis)
Activated alumina	Hydrophilic, amorphous	10–75	0.50	1.25	320	7
Silica gel:	Hydrophilic/hydrophobic, amorphous					
Small pore		22–26	0.47	1.09	750–850	11
Large pore		100–150	0.71	0.62	300–350	—
Activated carbon:	Hydrophobic, amorphous					
Small pore		10–25	0.4–0.6	0.5–0.9	400–1200	1
Large pore		>30	—	0.6–0.8	200–600	—
Molecular-sieve carbon	Hydrophobic	2–10	—	0.98	400	—
Molecular-sieve zeolites	Polar-hydrophilic, crystalline	3–10	0.2–0.5	1.4	600–700	20–25
Polymeric adsorbents	—	40–25	0.4–0.55	—	80–700	—

mixture, the pores, if sufficiently larger in diameter than the liquid molecules, fill with liquid. At equilibrium, because of differences in the extent of physical adsorption among liquid molecules, composition of the liquid in pores differs from that of bulk liquid surrounding adsorbent particles. The observed exothermic heat effect is referred to as the *heat of wetting*, which is much smaller than the heat of adsorption for a gas. As with gases, the extent of equilibrium adsorption of a given solute increases with concentration and decreases with temperature. Chemisorption can also occur with liquids.

Table 15.2 lists, for six major types of solid adsorbents: the nature of the adsorbent and representative values of the mean pore diameter, d_p; particle porosity (internal void fraction), ϵ_p; particle density, ρ_p; and specific surface area, S_g. In addition, for some adsorbents, the capacity for adsorbing water vapor at a partial pressure of 4.6 mmHg in air at 25°C is listed, as taken from Rousseau [9]. Not included is specific pore volume, V_p, which is given by

$$V_p = \epsilon_p / \rho_p \qquad (15\text{-}3)$$

Also not included in Table 15.2, but of interest when the adsorbent is used in fixed beds, are bulk density, ρ_b, and bed porosity (external void fraction), ϵ_b, which are related by

$$\epsilon_b = 1 - \frac{\rho_b}{\rho_p} \qquad (15\text{-}4)$$

In addition, the true solid particle density (also called the crystalline density), ρ_s, can be computed from a similar expression:

$$\epsilon_p = 1 - \frac{\rho_p}{\rho_s} \qquad (15\text{-}5)$$

Surface Area and the BET Equation

Specific surface area of an adsorbent, S_g, is measured by adsorbing gaseous nitrogen, using the well-accepted *BET*

method (Brunauer, Emmett, and Teller [11]). Typically, the BET apparatus operates at the normal boiling point of N₂ (−195.8°C) by measuring the equilibrium volume of pure N₂ physically adsorbed on several grams of the adsorbent at a number of different values of the total pressure in a vacuum of 5 to at least 250 mmHg. Brunauer, Emmett, and Teller derived an equation to model adsorption by allowing for formation of multimolecular layers. They assumed that the heat of adsorption during monolayer formation (ΔH_{ads}) is constant and that the heat effect associated with subsequent layers is equal to the heat of condensation (ΔH_{cond}). The BET equation is

$$\frac{P}{v(P_0 - P)} = \frac{1}{v_m c} + \frac{(c-1)}{v_m c}\left(\frac{P}{P_0}\right) \qquad (15\text{-}6)$$

where P = total pressure, P_0 = vapor pressure of adsorbate at test temperature, v = volume of gas adsorbed at STP (0°C, 760 mmHg), v_m = volume of monomolecular layer of gas adsorbed at STP, and c = a constant related to the heat of adsorption $\approx \exp[(\Delta H_{cond} - \Delta H_{ads})/RT]$.

Experimental data for v as a function of P are plotted, according to (15-6), as $P/[v(P_0 - P)]$ versus P/P_0, from which v_m and c are determined from the slope and intercept of the best straight-line fit of the data. The value of S_g is then computed from

$$S_g = \frac{\alpha v_m N_A}{V} \qquad (15\text{-}7)$$

where N_A = Avogadro's number = 6.023×10^{23} molecules/mol, V = volume of gas per mole at STP conditions (0°C, 1 atm) = 22,400 cm³/mol, and α is surface area per adsorbed molecule. If spherical molecules arranged in close two-dimensional packing are assumed, the projected surface area is:

$$\alpha = 1.091\left(\frac{M}{N_A \rho_L}\right)^{2/3} \qquad (15\text{-}8)$$

where M = molecular weight of the adsorbate, and ρ_L = density of the adsorbate in g/cm³, taken as the liquid at the test temperature.

Although the BET surface area may not always represent the surface area available for adsorption of a particular molecule, the BET test is reproducible and widely used to characterize adsorbents.

Pore Volume and Distribution

Specific pore volume, typically cm^3 of pore volume/g of adsorbent, is determined for a small mass of adsorbent, m_p, by measuring the volumes of helium, V_{He}, and mercury, V_{Hg}, displaced by the adsorbent. Helium is not adsorbed, but fills the pores. At ambient pressure, mercury cannot enter the pores because of unfavorable interfacial tension and contact angle. Specific pore volume, V_p, is then determined from

$$V_p = (V_{Hg} - V_{He})/m_p \qquad (15\text{-}9)$$

Particle density is

$$\rho_p = \frac{m_p}{V_{Hg}} \qquad (15\text{-}10)$$

and true solid density is

$$\rho_s = \frac{m_p}{V_{He}} \qquad (15\text{-}11)$$

Particle porosity is then obtained from (15-3) or (15-5).

Distribution of pore volumes over the range of pore size is of great importance in adsorption. It is measured by mercury porosimetry for large-diameter pores (>100 Å); by gaseous-nitrogen desorption for pores of 15–250 Å in diameter; and by molecular sieving, using molecules of different diameter, for pores <15 Å in diameter. In mercury porosimetry, the extent of mercury penetration into the pores is measured as a function of applied hydrostatic pressure. A force balance along the axis of a straight pore of circular cross section for the pressure and interfacial tension between mercury and the adsorbent surface gives the following equation, which is identical to (14-107) for the bubble test of a sterile filter:

$$d_p = -\frac{4\sigma_I \cos\theta}{P} \qquad (15\text{-}12)$$

where for mercury, σ_I = interfacial tension = 0.48 N/m and θ = contact angle = 140°. With these values, (15-12) becomes

$$d_p(\text{Å}) = \frac{21.6 \times 10^5}{P(\text{psia})} \qquad (15\text{-}13)$$

Thus, forcing mercury into a 100-Å-diameter pore requires a very high pressure of 21,600 psia.

The nitrogen desorption method for determining pore-size distribution in the 15–250-Å-diameter range is an extension of the BET method for measuring specific surface area. By increasing nitrogen pressure above 600 mmHg, multilayer adsorbed films reach the point where they bridge the pore, resulting in capillary condensation. At $P/P_0 = 1$, the entire pore volume is filled with nitrogen. Then, by reducing the pressure in steps, nitrogen is desorbed selectively, starting with larger pores. This selectivity occurs because of the effect of pore diameter on vapor pressure of the condensed phase in the pore, as given by the Kelvin equation:

$$P_p^s = P^s \exp\left(-\frac{4\sigma v_L \cos\theta}{RT d_p}\right) \qquad (15\text{-}14)$$

where P_p^s = vapor pressure of liquid in pore, P^s = the normal vapor pressure of liquid on a flat surface, σ = surface tension of liquid in pore, and v_L = molar volume of liquid in pore.

Vapor pressure of the condensed phase in the pores is less than its normal vapor pressure for a flat surface. The effect of d_p on P_p^s can be significant. For example, for liquid nitrogen at $-195.8°C$, $P^s = 760$ torr, $\sigma = 0.00827$ N/m, $\theta = 0$, and $v_L = 34.7$ cm^3/mol. Equation (15-14) then becomes

$$d_p(\text{Å}) = 17.9/\ln(P^s/P_p^s) \qquad (15\text{-}15)$$

From (15-15) for $d_p = 30$ Å, $P_p^s = 418$ torr, a reduction in vapor pressure of almost 50%. At 200 Å, the reduction is only about 10%. At 418 torr pressure, only pores less than 30 Å in diameter remain filled with liquid nitrogen. For greater accuracy in applying the Kelvin equation, a correction is needed for the thickness of the adsorbed layer. This correction is discussed in detail by Satterfield [12]. For a monolayer, this thickness for nitrogen is about 0.354 nm, corresponding to a P/P_0 in (15-6) of between 0.05 and 0.10. At $P/P_0 = 0.60$ and 0.90, the adsorbed thicknesses are 0.75 and 1.22 nm, respectively. The correction is applied by subtracting twice the adsorbed thickness from d_p in (15-14) and (15-15).

EXAMPLE 15.1 Particle Porosity.

Using data from Table 15.2, determine the volume fraction of pores in silica gel (small-pore type) filled with adsorbed water vapor when its partial pressure is 4.6 mmHg and the temperature is 25°C. At these conditions, the partial pressure is considerably below the vapor pressure of 23.75 mmHg. In addition, determine whether the amount of water adsorbed is equivalent to more than a monolayer, if the area of an adsorbed water molecule is given by (15-8) and the specific surface area of the silica gel is 830 m^2/g.

Solution

Take 1 g of silica gel particles as a basis. From (15-3) and data in Table 15.2, $V_p = 0.47/1.09 = 0.431$ cm^3/g. Thus, for 1 g, pore volume is 0.431 cm^3. From the capacity value in Table 15.2, amount of adsorbed water = $0.11/(1 + 0.11) = 0.0991$ g. Assume density of adsorbed water is 1 g/cm^3, volume of adsorbed water = 0.0991 cm^3, fraction of pores filled with water = $0.0991/0.431 = 0.230$, and surface area of 1 g = 830 m^2. From (15-8):

$$\alpha = 1.091\left[\frac{18.02}{(6.023 \times 10^{23})(1.0)}\right]^{2/3} = 10.51 \times 10^{-16} \text{ cm}^2/\text{molecule}$$

Number of H$_2$O molecules adsorbed $= \dfrac{(0.0991)(6.023 \times 10^{23})}{18.02}$
$= 3.31 \times 10^{21}$

Number of H$_2$O molecules in a monolayer for 830 m^2

$$= \frac{830(100)^2}{10.51 \times 10^{-16}} = 7.90 \times 10^{21}$$

Therefore, only 3.31/7.90 or 42% of one monolayer is adsorbed.

Activated Alumina

The four most widely used adsorbents in decreasing order of commercial usage are carbon (activated and molecular-sieve), molecular-sieve zeolites, silica gel, and activated alumina. In Table 15.2, activated alumina, Al_2O_3, which includes activated bauxite, is made by removing water from hydrated colloidal alumina. It has a moderately high S_g, with a capacity for adsorption of water sufficient to dry gases to less than 1 ppm moisture. Because of this, activated alumina is widely used for removal of water from gases and liquids.

Silica Gel

SiO_2, made from colloidal silica, has a high S_g and high affinity for water and other polar compounds. Related silicate adsorbents include magnesium silicate, calcium silicate, various clays, Fuller's earth, and diatomaceous earth. Silica gel is also desirable for water removal. Small-pore and large-pore types are available.

Activated Carbon

Partial oxidation of materials like coconut shells, fruit nuts, wood, coal, lignite, peat, petroleum residues, and bones produces activated carbon. Macropores within the carbon particles help transfer molecules to the micropores. Two commercial grades are available, one with large pores for processing liquids and one with small pores for gas adsorption. As shown in Table 15.2, activated carbon is relatively hydrophobic and has a large surface area. Accordingly, it is widely used for purification and separation of gas and liquid mixtures containing nonpolar and weakly polar organic compounds, which adsorb much more strongly than water. In addition, the bonding strength of adsorption on activated carbon is low, resulting in a low heat of adsorption and ease of regeneration.

Molecular-Sieve Carbon

Unlike activated carbon, which typically has pore diameters starting from 10 Å, molecular-sieve carbon (MSC) has pores ranging from 2 to 10 Å, making it possible to separate N_2 from air. In one process, small pores are made by depositing coke in the pore mouths of activated carbon.

Molecular-Sieve Zeolites

Most adsorbents have a range of pore sizes, as shown in Figure 15.3, where the cumulative pore volume is plotted against pore diameter. Exceptions are molecular-sieve zeolites, which are crystalline, inorganic polymers of aluminosilicates

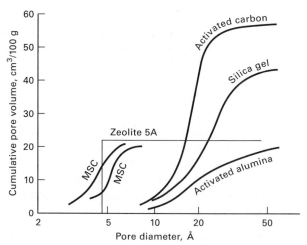

Figure 15.3 Representative cumulative pore-size distributions of adsorbents.

and alkali or alkali-earth elements, such as Na, K, and Ca, with the stoichiometric, unit-cell formula $M_{x/m}[(AlO_2)_x (SiO_2)_y]z\,H_2O$, where M is the cation with valence m, z is the number of water molecules in each unit cell, and x and y are integers such that $y/x \geq 1$. The cations balance the charge of the AlO_2 groups, each having a net charge of -1. To activate the zeolite, the water molecules are removed by raising the temperature or pulling a vacuum. This leaves the remaining atoms spatially intact in interconnected, cagelike structures with six identical window apertures each of from 3.8 to about 10 Å, depending on the cation and crystal structure. These apertures act as sieves, which permit small molecules to enter the crystal cage, but exclude large molecules. Thus, compared to other types of adsorbents, molecular-sieve zeolites are highly selective because all apertures have the same size.

The properties and applications of five of the most commonly used molecular-sieve zeolites are given in Table 15.3, from Ruthven [13]. Zeolites separate not only by molecular size and shape, but also by polarity, so they can also separate molecules of similar size. Zeolites have circular or elliptical apertures. Adsorption in zeolites is a selective and reversible filling of crystal cages, so cage volume is a pertinent factor. Although natural zeolite minerals have been known for more than 200 years, molecular-sieve zeolites were first synthesized by Milton [2], using reactive materials at temperatures of $25-100°C$.

A type A zeolite is shown in Figure 15.4a as a three-dimensional structure of silica and alumina tetrahedra, each formed by four oxygen atoms surrounding a silicon or aluminum atom. Oxygen and silicon atoms have two negative and

Table 15.3 Properties and Applications of Molecular-Sieve Zeolites

Designation	Cation	Unit-Cell Formula	Aperture Size, Å	Typical Applications
3A	K^+	$K_{12}[(AlO_2)_{12}(SiO_2)_{12}]$	2.9	Drying of reactive gases
4A	Na^+	$Na_{12}[(AlO_2)_{12}(SiO_2)_{12}]$	3.8	H_2O, CO_2 removal; air separation
5A	Ca^{2+}	$Ca_5Na_2[(AlO_2)_{12}(SiO_2)_{12}]$	4.4	Separation of air; separation of linear paraffins
10X	Ca^{2+}	$Ca_{43}[(AlO_2)_{86}(SiO_2)_{106}]$	8.0 ⎫	Separation of air;
13X	Na^+	$Na_{86}[(AlO_2)_{86}(SiO_2)_{86}]$	8.4 ⎭	removal of mercaptans

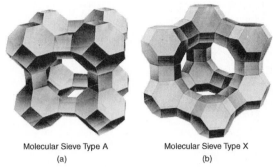

Figure 15.4 Structures of molecular-sieve zeolites: (a) Type A unit cell. (b) Type X unit cell.

four positive charges, respectively, causing the tetrahedra to build uniformly in four directions. Aluminum, with a valence of 3, causes the alumina tetrahedron to be negatively charged. The added cation provides the balance. In Figure 15.4a, an octahedron of tetrahedra is evident with six faces, with one near-circular window aperture at each face. A type X zeolite is shown in Figure 15.4b. This unit-cell structure results in a larger window aperture. Zeolites are treated in monographs by Barrer [14] and Breck [15].

Polymeric Adsorbents

Of lesser commercial importance are polymeric adsorbents. Typically, they are spherical beads, 0.5 mm in diameter, made from microspheres about 10^{-4} mm in diameter. They are produced by polymerizing styrene and divinylbenzene for use in adsorbing nonpolar organics from aqueous solutions, and by polymerizing acrylic esters for adsorbing polar solutes. They are regenerated by leaching with organic solvents.

§15.1.2 Ion Exchangers

The first ion exchangers were naturally occurring inorganic aluminosilicates (zeolites) used in experiments in the 1850s

to exchange ammonium ions in fertilizers with calcium ions in soils. Industrial water softeners using zeolites were introduced about 1910, but they were unstable in the presence of mineral acids. The instability problem was solved by Adams and Holmes [16] in 1935, when they synthesized the first organic-polymer, ion-exchange resins by the polycondensation of phenol and aldehydes. Depending upon the phenolic group, the resin contains either sulfonic $\left(-SO_3^-\right)$ or amine $\left(-NH_3^+\right)$ groups for the reversible exchange of cations or anions. Today, the most widely used ion exchangers are synthetic, organic-polymer resins based on styrene- or acrylic-acid-type monomers, as described by D'Alelio in U.S. Patent 2,366,007 (Dec. 26, 1944).

Ion-exchange resins are generally solid gels in spherical or granular form, which consist of (1) a three-dimensional polymeric network, (2) ionic functional groups attached to the network, (3) counterions, and (4) a solvent. Strong-acid, cation-exchange resins and strong-base, anion-exchange resins that are fully ionized over the entire pH range are based on the copolymerization of styrene and a cross-linking agent, divinylbenzene, to produce the three-dimensional, cross-linked structure shown in Figure 15.5a. Degree of cross-linking is governed by the ratio of divinylbenzene to styrene. Weakly acid, cation exchangers are sometimes based on the copolymerization of acrylic acid and methacrylic acid, as shown in Figure 15.5b. These two cross-linked copolymers swell in the presence of organic solvents and have no ion-exchange properties.

To convert the copolymers to water-swellable gels with ion-exchange properties, ionic functional groups are added to the polymeric network by reacting copolymers with various chemicals. For example, if the styrene–divinylbenzene copolymer is sulfonated, as shown in Figure 15.6a, the cation-exchange resin, shown in Figure 15.6b, is obtained with $\left(-SO_3^-\right)$ groups permanently attached to the polymeric network to give a negatively charged matrix and exchangeable, mobile, positive hydrogen ions (cations). The hydrogen ion can be exchanged on an equivalent basis with other cation

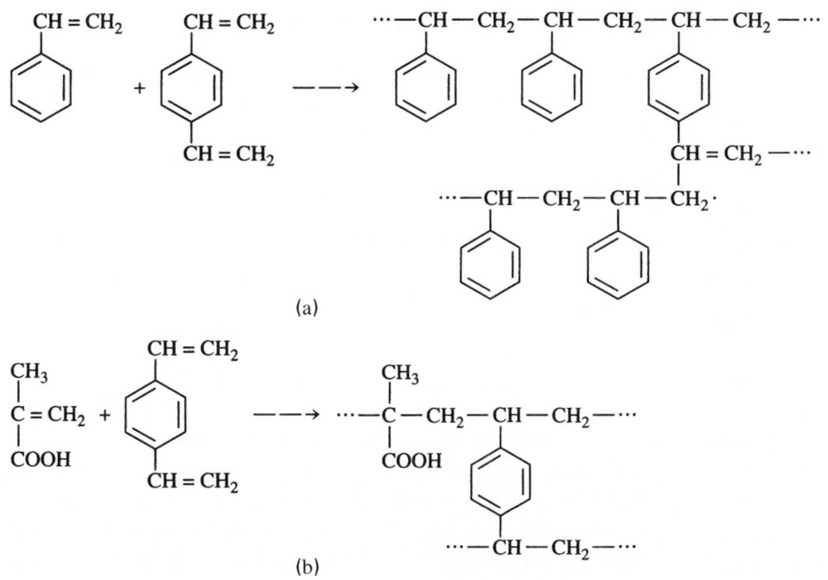

Figure 15.5 Ion-exchange resins: (a) Resin from styrene and divinylbenzene; (b) Resin from acrylic and methacrylic acid.

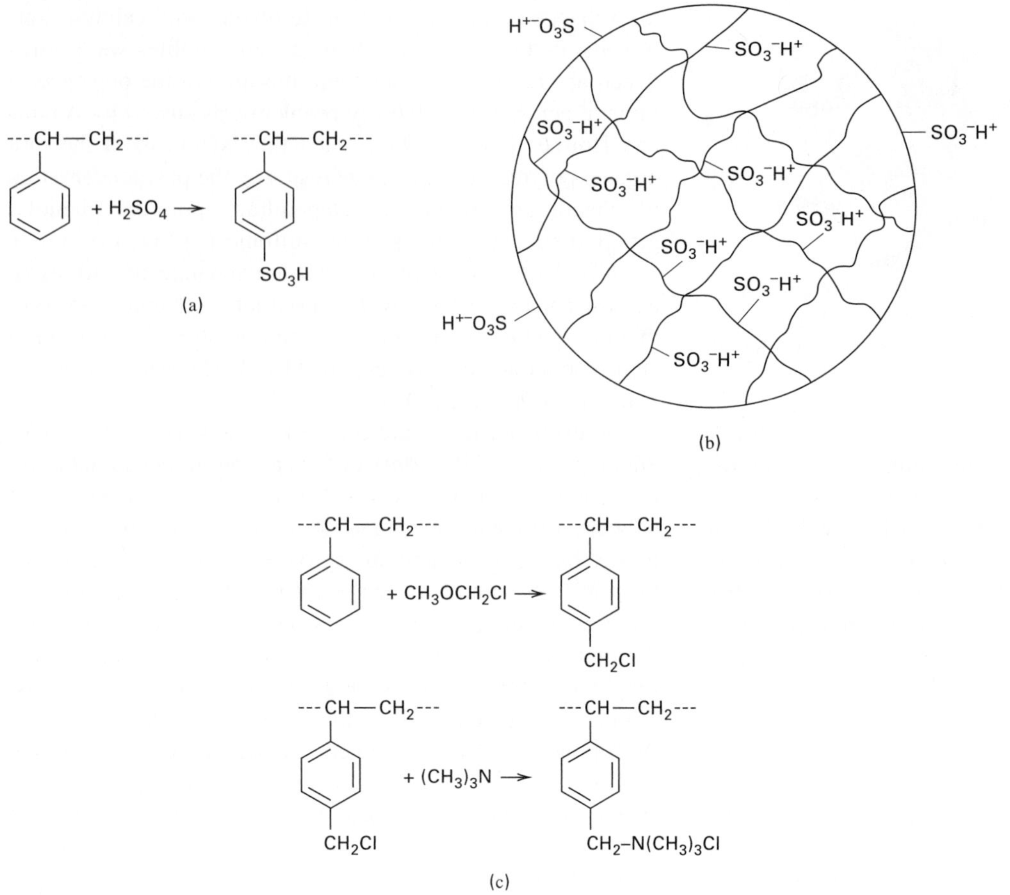

---CH—CH$_2$---   ---CH—CH$_2$---

+ H$_2$SO$_4$ →

SO$_3$H

(a)

H$^+{}^-$O$_3$S   SO$_3^-$H$^+$ SO$_3^-$H$^+$ SO$_3^-$H$^+$ SO$_3^-$H$^+$ SO$_3^-$H$^+$ SO$_3^-$H$^+$ SO$_3^-$H$^+$ SO$_3^-$H$^+$ H$^+{}^-$O$_3$S   SO$_3^-$H$^+$

(b)

---CH—CH$_2$---   ---CH—CH$_2$---

+ CH$_3$OCH$_2$Cl →

CH$_2$Cl

---CH—CH$_2$---   ---CH—CH$_2$---

+ (CH$_3$)$_3$N →

CH$_2$Cl   CH$_2$–N(CH$_3$)$_3$Cl

(c)

Figure 15.6 Introducing ionic functional groups into resins. (a) Sulfonation to a cation exchanger. (b) Fixed and mobile ions in a cation exchanger. (c) Chloromethylation and amination to an anion exchanger.

counterions, such as Na$^+$, Ca^{2+}, K$^+$, or Mg^{2+}, to maintain charge neutrality of the polymer. For example, two H$^+$ ions are exchanged for one Ca^{2+} ion. The liquid whose ions are being exchanged also contains other ions of unlike charge, such as Cl$^-$ for a solution of NaCl, where Na$^+$ is exchanged. These other ions are called *co-ions*. Often the liquid treated is H$_2$O, which dissolves to some extent in the resin and causes it to swell. Other solvents, such as methanol, are also soluble in the resin. If the styrene–divinylbenzene copolymer is chloromethylated and aminated, a strong-base, anion-exchange resin is formed, as shown in Figure 15.6c, which can exchange Cl$^-$ ions for other anions, such as OH$^-$, HCO$_3^-$, SO$_4^{2-}$, and NO$_3^-$.

Commercial ion exchangers in the H, Na, and Cl form are available under the trade names of Amberlite^{TM}, Duolite^{TM}, Dowex^{TM}, Ionac$^®$, and Purolite$^®$, typically in the form of spherical beads from 40 μm to 1.2 mm in diameter. When saturated with water, the beads have typical moisture contents from 40 to 65 wt%. When water-swollen, $\rho_p = 1.1$–1.5 g/cm^3. When packed into a vessel, $\rho_b = 0.56$–0.96 g/cm^3 with ϵ_b of 0.35–0.40.

Before water is demineralized by ion exchange, potential organic foulants must be removed. As discussed by McWilliams [17], this can be accomplished by coagulation, clarification, prechlorination, and use of ion-exchanger traps that exchange inorganic anions for anionic organic molecules.

The maximum ion-exchange capacity of a strong-acid cation or strong-base anion exchanger is stoichiometric, based on the number of equivalents of mobile charge in the resin. Thus, 1 mol H$^+$ is one equivalent, whereas 1 mol Ca^{2+} is two equivalents. Exchanger capacity is usually quoted as eq/kg of dry resin or eq/L of wet resin. Wet capacity depends on resin water content and degree of swelling, whereas dry capacity is fixed. For copolymers of styrene and divinylbenzene, maximum capacity is based on the assumption that each benzene ring in the resin contains one sulfonic-acid group.

EXAMPLE 15.2   Ion-Exchange Capacity.

A commercial ion-exchange resin is made from 88 wt% styrene and 12 wt% divinylbenzene. Estimate the maximum ion-exchange capacity in eq/kg resin (same as meq/g resin).

Solution

Basis: 100 g of resin before sulfonation.

	M	g	gmol
Styrene	104.14	88	0.845
Divinylbenzene	130.18	12	0.092
		100	0.937

Sulfonation at one location on each benzene ring requires 0.937 mol of H$_2$SO$_4$ to attach a sulfonic acid group ($M = 81.07$) and split out one water molecule. This is 0.937 equivalent, with a weight addition of 0.937(81.07) = 76 g. Total dry weight of sulfonated resin $= 100 + 76 = 176$ g maximum ion-exchange capacity, or

$$\frac{0.937}{(176/1{,}000)} = 5.3 \, \text{eq/kg(dry)}$$

Depending on the extent of cross-linking, resins from copolymers of styrene and divinylbenzene are listed as having actual capacities of from 3.9 (high degree of cross-linking) to 5.5 (low degree of cross-linking). Although a low degree of cross-linking favors dry capacity, almost every other ion-exchanger property, including wet capacity and selectivity, is improved by cross-linking, as discussed by Dorfner [18].

§15.1.3 Sorbents for Chromatography

Sorbents (called *stationary phases*) for chromatographic separations come in many forms and chemical compositions because of the diverse ways that chromatography is applied. Figure 15.7 shows a classification of analytical chromatographic systems, taken from Sewell and Clarke [19]. The mixture to be separated, after injection into the carrier fluid to form the *mobile phase*, may be a liquid (*liquid chromatography*) or a gas (*gas chromatography*). Often, the mixture is initially a liquid, but is vaporized by the carrier gas, giving a gas mixture as the mobile phase. Gas carriers are inert and do not interact with the sorbent or feed. Liquid carriers (solvents) can interact and must be selected carefully.

The stationary sorbent phase is a solid, a liquid supported on or bonded to a solid, or a gel. With a porous-solid adsorbent, the mechanism of separation is adsorption. If an ion-exchange mechanism is desired, a synthetic, polymer ion exchanger is used. With a polymer gel or a microporous solid, a separation based on sieving, called *exclusion*, can be operative. Unique to chromatography are liquid-supported or -bonded solids, where the mechanism is absorption into the liquid, also referred to as a *partition mode* of separation or *partition chromatography*. With mobile liquid phases, the stationary liquid phase may be stripped or dissolved. Accordingly, methods of chemically bonding the stationary liquid phase to the solid support have been developed.

In packed columns >1 mm inside diameter, the sorbents are in the form of particles. In capillary columns <0.5 mm inside diameter, the sorbent is the inside wall or a coating on that wall. If coated, the capillary column is referred to as a wall-coated, open-tubular (WCOT) column. If the coating is a layer of fine particulate support material to which a liquid adsorbent is added, the column is a support-coated, open-tubular (SCOT) column. If the wall is coated with a porous adsorbent only, the column is a porous-layer, open-tubular (PLOT) column.

Each type of sorbent can be applied to sheets of glass, plastic, or aluminum for use in *thin-layer* (or planar) *chromatography* or to a sheet of cellulose material for use in *paper chromatography*. If a pump, rather than gravity, is used to pass a liquid mobile phase through a packed column, the name *high-performance liquid chromatography* (HPLC) is used.

The two most common adsorbents used in chromatography are porous alumina and porous silica gel. Of lesser importance are carbon, magnesium oxide, and carbonates.

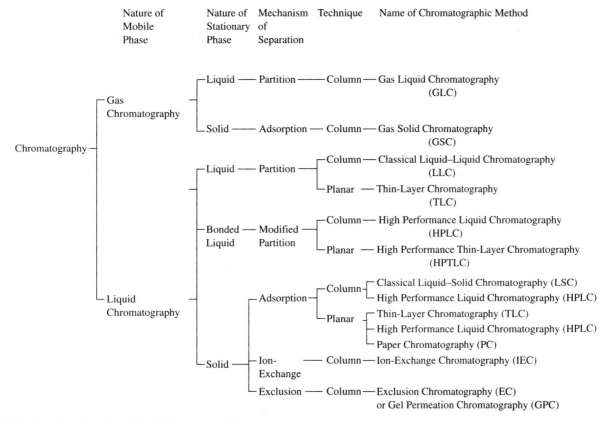

Figure 15.7 Classification of analytical chromatographic systems.

[From P.A. Sewell and B. Clarke, *Chromatographic Separations*, John Wiley & Sons, New York (1987) with permission.]

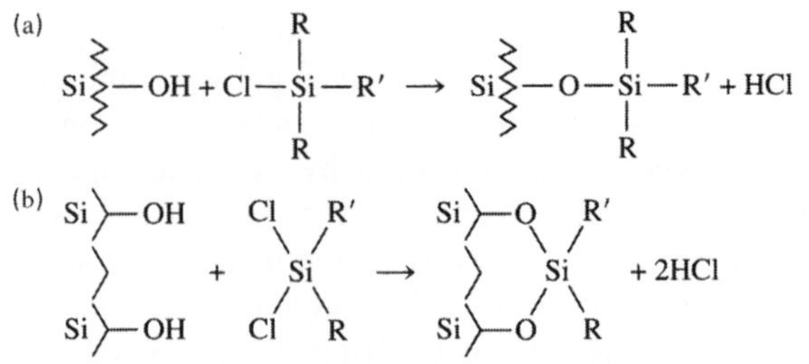

(a)

(b)

Figure 15.8 Bonded phases from the reaction of surface silanol groups with (a) monofunctional and (b) bifunctional chlorosilanes.

Alumina is a polar adsorbent and is preferred for the separation of components that are weakly or moderately polar, with more polar compounds retained more selectively by the adsorbents and therefore eluted from the column last. Alumina is a basic adsorbent, preferentially retaining acidic compounds. Silica gel is less polar than alumina and is an acidic adsorbent, preferentially retaining basic compounds, such as amines. Carbon is a nonpolar (apolar) stationary phase with the highest attraction for nonpolar molecules.

Adsorbent-type sorbents are better suited for separation of a mixture on the basis of chemical type (e.g., olefins, esters, acids, aldehydes, alcohols) than for separation of individual members of a homologous series. For the latter, *partition chromatography*—wherein an inert-solid support, often silica gel, is coated with a liquid phase—is preferred. For *gas chromatography*, that liquid must be nonvolatile. For *liquid chromatography*, the stationary liquid phase must be insoluble in the mobile phase, but since this is difficult to achieve, the stationary liquid phase is usually bonded to the solid support. An example of a bonded phase is the result of reacting silica with a chlorosilane. Both monofunctional and bifunctional silanes are used, as shown in Figure 15.8, where R is a methyl (CH_3) group and R′ is a hydrocarbon chain (C_6, C_8, or C_{18}) where the terminal CH_3 group is replaced with a polar group, such as —CN or —NH_2. If the resulting stationary phase is more polar than the mobile phase, it is *normal-phase chromatography*; otherwise, it is *reverse-phase chromatography*.

In liquid chromatography, the order of elution of solutes in the mobile phase can be influenced by the solvent carrier of the mobile phase by matching the solvent polarity with the solutes and using more-polar adsorbents for less-polar solutes and less-polar adsorbents for more-polar solutes.

EXAMPLE 15.3 Chromatography Mode.

For the separation of each of the following mixtures, select an appropriate mode of chromatography from Figure 15.7: (a) gas mixture of O_2, CO, CO_2, and SO_2; (b) vaporized mixture of anthracene, phenanthrene, pyrene, and chrysene; and (c) aqueous solution containing Ca^{2+} and Ba^{2+}.

Solution

(a) Use gas–solid chromatography, that is, with a gas mobile phase and a solid-adsorbent stationary phase.

(b) Use partition or gas–liquid chromatography, that is, with a gas mobile phase and a bonded liquid coating on a solid for the stationary phase.

(c) Use ion-exchange chromatography, that is, with a liquid as the mobile phase and polymer resin beads as the stationary phase.

---

§15.2 EQUILIBRIUM CONSIDERATIONS

In adsorption, a dynamic equilibrium is established for solute distribution between the fluid and solid surface. This is expressed in terms of: (1) concentration (if the fluid is a liquid) or partial pressure (if the fluid is a gas) of the adsorbate in the fluid and (2) solute *loading* on the adsorbent, expressed as mass, moles, or volume of adsorbate per unit mass or per unit BET adsorbent surface area. Unlike vapor–liquid and liquid–liquid equilibria, where theory is often applied to estimate phase distribution in the form of *K*-values (§2.2.2), no acceptable theory has been developed to estimate fluid–solid adsorption equilibria. It is thus necessary to obtain equilibrium data for a particular solute, or mixture of solutes and/or solvent, and the solid-adsorbent material of interest. If the data are taken over a range of fluid concentrations at a constant temperature, a plot of adsorbent solute loading versus solute concentration or partial pressure in the fluid, called an *adsorption isotherm*, is made. This equilibrium isotherm places a limit on the extent to which a solute is adsorbed from a specific fluid mixture on a given adsorbent for one set of conditions. The rate at which solute is adsorbed is discussed in §15.3.

§15.2.1 Pure-Gas Adsorption

The five experimental physical-adsorption isotherms for pure gases shown in Figure 15.9 are due to Brunauer, as described in [20, 21]. The simplest isotherm is Type I, which corresponds to unimolecular adsorption, as characterized by a maximum limit in the amount adsorbed. This type describes gases at temperatures above their critical temperature. The more complex Type II isotherm is associated with multimolecular BET adsorption and is observed for gases at temperatures below their critical temperature and for pressures below, but approaching, the saturation pressure. The heat of adsorption for the first adsorbed layer is greater than that for the succeeding layers, each of which is assumed to have a heat of adsorption equal to the heat of condensation. Both Types I and II are desirable isotherms, exhibiting strong adsorption.

The Type III isotherm in Figure 15.9 is convex and undesirable because the extent of adsorption is low except at high pressures. According to BET theory, it corresponds to multimolecular adsorption where heat of adsorption of the first layer is less than that of succeeding layers. Fortunately, this type of isotherm is rare, an example being adsorption of iodine vapor on silica gel. In the limit, as heat of adsorption of the first layer approaches zero, adsorption is delayed until the saturation pressure is approached.

Derivation of the BET equation (15-6) assumes that an infinite number of molecular layers can be adsorbed, thus

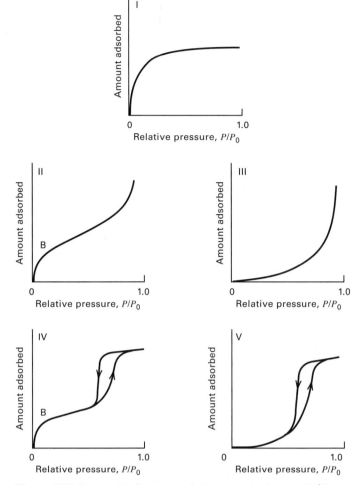

Figure 15.9 Brunauer's five types of adsorption isotherms. ($P/P_0 =$ total pressure/vapor pressure.)

Figure 15.10. The five adsorption isotherms of Figure 15.10a cover pressures from vacuum to almost 800 mmHg and temperatures from -23.5 to $151.5°C$. For ammonia, the normal boiling point is $-33.3°C$ and the critical temperature is $132.4°C$. For the lowest-temperature isotherm, up to 160 cm^3 (STP) of ammonia per gram of charcoal is adsorbed, which is equivalent to 0.12 g NH_3/g charcoal. All five isotherms are of Type I. When the amount adsorbed is low (<25 cm^3/g), isotherms are almost linear and a form of Henry's law, called the *linear isotherm*, is obeyed:

$$q = kp \qquad (15\text{-}16)$$

where q is equilibrium *loading* or amount adsorbed of a given species/unit mass of adsorbent; k is an empirical, temperature-dependent constant for the component; and p is the partial pressure of the species. As temperature increases, the amount adsorbed decreases because of Le Chatelier's principle for an exothermic process. This is shown clearly in the crossplot of *adsorption isobars* in Figure 15.10b, where absolute temperature is employed. Another crossplot of the data yields *adsorption isosteres* in Figure 15.10c. These curves for constant amounts adsorbed, resemble vapor-pressure plots, for which the adsorption form of the Clausius–Clapeyron equation,

$$\frac{d\ln p}{dT} = \frac{-\Delta H_{ads}}{RT^2} \qquad (15\text{-}17)$$

or

$$\frac{d\log p}{d(1/T)} = \frac{-\Delta H_{ads}}{2.303RT} \qquad (15\text{-}18)$$

is used to determine the exothermic heat of adsorption, shown in Figure 15.10d, where $-\Delta H_{ads}$ is initially 7,300 cal/mol, but decreases as the amount adsorbed increases, reaching 6,100 cal/mol at 100 cm^3/g. These values can be compared to the heat of vaporization of NH_3, which at $30°C$ is 4,600 cal/mol.

Adsorption-isotherm data for 18 different pure gases and a variety of solid adsorbents are analyzed by Valenzuela and Myers [23]. The data show that adsorption isotherms for a given pure gas at fixed temperature vary with the adsorbent. For propane vapor at $25–30°C$, as shown in Figure 15.11, for pressures up to 101.3 kPa, the highest specific adsorption is with Columbia G-grade activated carbon, while the lowest is with Norton Z-900H, a zeolite molecular sieve. Columbia G-grade activated carbon has about twice the adsorbate capacity of Cabot Black Pearls activated carbon.

Literature data compiled by Valenzuela and Myers [23] also show that for a given adsorbent, loading depends strongly on the gas. This is illustrated in Table 15.4 for a temperature of $38°C$ and a pressure range of 97.9 to 100 kPa from the data of Ray and Box [24] for Columbia L activated carbon. Included in the table are normal boiling points and critical temperatures. As might be expected, the species are adsorbed in approximately the inverse order of volatility.

Correlation of experimental gas adsorption isotherms is the subject of numerous articles and books. As summarized by Yang [25], approaches have ranged from empirical to theoretical. In practice, the classic equations of Freundlich and

precluding the possibility of capillary condensation. In a development by Brunauer et al. [20] before the development of the BET equation, the number of layers is restricted by pore size, and capillary condensation is assumed to occur at a reduced vapor pressure in accordance with the Kelvin equation (15-14). The resulting equation is complex, but also predicts adsorption isotherms of Types IV and V in Figure 15.9, where the maximum extent of adsorption occurs before the saturation pressure is reached. Types IV and V are the capillary-condensation versions of Types II and III, respectively.

As shown in Figure 15.9, hysteresis occurs in multimolecular adsorption regions for isotherms of Types IV and V. The upward adsorption branch of the hysteresis loop is due to simultaneous, multimolecular adsorption and capillary condensation. Only the latter occurs during the downward desorption branch. Hysteresis can also occur in any isotherm when strongly adsorbed impurities are present. Measurements of pure-gas adsorption require adsorbents with clean pore surfaces, which is achieved by pre-evacuation.

Linear Isotherm

Physical-adsorption data of Titoff [22] for ammonia gas on charcoal, as discussed by Brunauer [21], are shown in

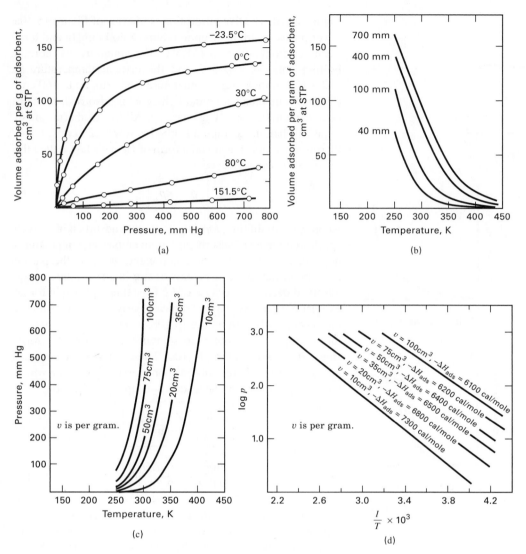

Figure 15.10 Different displays of adsorption-equilibrium data for NH_3 on charcoal.
(a) Adsorption isotherms.
(b) Adsorption isobars.
(c) Adsorption isosteres.
(d) Isosteric heats of adsorption.

[From S. Brunauer, *The Adsorption of Gases and Vapors*, Vol. I, Princeton University Press (1943) with permission.]

Langmuir, discussed next, are dominant because of their simplicity and ability to correlate Type I isotherms.

Freundlich Isotherm

The equation attributed to Freundlich [26], but which, according to Mantell [27], was devised earlier by Boedecker and van Bemmelen, is empirical and nonlinear in pressure:

$$q = kp^{1/n} \qquad (15\text{-}19)$$

where k and n are temperature-dependent constants for a particular component and adsorbent. The constant, n, lies in the range of 1 to 5, and for $n = 1$, (15-19) reduces to the linear

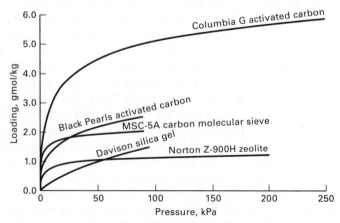

Figure 15.11 Adsorption isotherms for pure propane vapor at 298–303 K.

Table 15.4 Comparison of Equilibrium Adsorption of Pure Gases on 20–40 mesh Columbia L Activated Carbon Particles ($S_g = 1,152$ m^2/g) at 38°C and ∼1 atm

Pure gas	q, mol/kg	T_b, °F	T_c, °F
H_2	0.0241	−423.0	−399.8
N_2	0.292	−320.4	−232.4
CO	0.374	−313.6	−220.0
CH_4	0.870	−258.7	−116.6
CO_2	1.64	−109.3	87.9
C_2H_2	2.67	−119	95.3
C_2H_4	2.88	−154.6	48.6
C_2H_6	3.41	−127.5	90.1
C_3H_6	4.54	−53.9	196.9
C_3H_8	4.34	−43.7	216.0

isotherm (15-16). Experimental q–p isothermal data can be fitted to (15-19) by a nonlinear curve fit or by converting (15-19) to the following linear form, and using a graphical method or a linear-regression program to obtain k and n;

$$\log q = \log k + (1/n)\log p \qquad (15\text{-}20)$$

In the graphical method, data are plotted as $\log q$ versus $\log p$; the best straight line through the data has a slope of $(1/n)$ and an intercept of $\log k$. In general, k decreases, while n increases with increasing temperature, approaching a value of 1 at high temperatures. Although (15-19) is empirical, it can be derived by assuming a heterogeneous surface with a nonuniform distribution of heat of adsorption (Brunauer [21]).

Langmuir Isotherm

The Langmuir equation [28] is restricted to Type I isotherms. It is derived from mass-action kinetics, assuming that chemisorption is the reaction. Let the surface of the pores of the adsorbent be homogeneous ($\Delta H_{ads} = $ constant), with negligible interaction forces between adsorbed molecules. If θ is the fraction of surface covered by adsorbed molecules, $(1 - \theta)$ is the fraction of bare surface, and the net rate of adsorption is the difference between the rate of adsorption on the bare surface and the rate of desorption on the covered surface:

$$dq/dt = k_a p(1 - \theta) - k_d \theta \qquad (15\text{-}21)$$

where k_a and k_d are the adsorption and desorption kinetic constants. At equilibrium, $dq/dt = 0$ and (15-21) reduces to

$$\theta = \frac{Kp}{1 + Kp} \qquad (15\text{-}22)$$

where $K = k_a/k_d$ is the adsorption-equilibrium constant and

$$\theta = q/q_m \qquad (15\text{-}23)$$

where q_m is the maximum loading corresponding to complete surface coverage. Thus, the Langmuir adsorption isotherm is restricted to a monomolecular layer. Combining (15-23) with (15-22) results in the *Langmuir isotherm*:

$$\boxed{q = \frac{Kq_m p}{1 + Kp}} \qquad (15\text{-}24)$$

At low pressures, if $Kp \ll 1$, (15-24) reduces to the linear isotherm, (15-16), while at high pressures where $Kp \gg 1$, $q = q_m$. At intermediate pressures, (15-24) is nonlinear in pressure. Although originally devised by Langmuir for chemisorption, (15-24) is widely applied to physical-adsorption data.

In (15-24), K and q_m are treated as constants obtained by fitting the nonlinear equation to experimental data or by employing the following linearized form, numerically or graphically:

$$\frac{p}{q} = \frac{1}{q_m K} + \frac{p}{q_m} \qquad (15\text{-}25)$$

Using (15-25), the best straight line is drawn through a plot of points p/q versus p, giving a slope of $(1/q_m)$ and an intercept of $1/(q_m K)$. Theoretically, K should change rapidly with temperature but q_m should not, because it is related through v_m by (15-7) to S_g. The Langmuir isotherm predicts an asymptotic limit for q at high pressure, whereas the Freundlich isotherm does not, as shown e.g. by the curve for Columbia G activated carbon in Figure 15.11.

Other Adsorption Isotherms

Valenzuela and Myers [23] fit isothermal, pure-gas adsorption data to the three-parameter isotherms of Toth:

$$q = \frac{mp}{(b + p^t)^{1/t}} \qquad (15\text{-}26)$$

where m, b, and t are constants for a given adsorbate–adsorbent system and temperature.

Honig and Reyerson devisied the three-constant UNILAN equation:

$$q = \frac{n}{2s}\ln\left[\frac{c + pe^s}{c + pe^{-s}}\right] \qquad (15\text{-}27)$$

where n, s, and c are the constants for a given system and temperature. The Toth and UNILAN isotherms reduce to the Langmuir isotherm for $t = 1$ and $s = 0$, respectively.

EXAMPLE 15.4 Freundlich and Langmuir Isotherms.

Data for the equilibrium adsorption of pure methane gas on activated carbon (PCB from Calgon Corp.) at 296 K were obtained by Ritter and Yang [*Ind. Eng. Chem. Res.*, **26**, 1679–1686 (1987)]:

q, cm^3 (STP) of CH$_4$/g carbon	45.5	91.5	113	121	125	126	126
$P = p$, psia	40	165	350	545	760	910	970

Fit the data to: (a) the Freundlich isotherm, and (b) the Langmuir isotherm. Which isotherm provides a better fit? Do the data give a reasonable fit to the linear isotherm?

Solution

Using the linearized forms of the isotherm equations, a spreadsheet or other program can be used to do a linear regression:

(a) Using (15-20), $\log k = 1.213$, $k = 16.34$, $1/n = 0.3101$, and $n = 3.225$. Thus, the Freundlich equation is:

$$q = 16.34 p^{0.3101}$$

(b) Using (15-25), $1/q_m = 0.007301$, $q_m = 137.0$, $1/(q_m K) = 0.5682$, and $K = 0.01285$. Thus, the Langmuir equation is

$$q = \frac{1.760p}{1 + 0.01285p}$$

The predicted values of q from the two isotherms are:

	q, cm^3 (STP) of CH$_4$/g carbon		
p, psia	Experimental	Freundlich	Langmuir
40	45.5	51.3	46.5
165	91.5	79.6	93.1
350	113	101	112
545	121	115	120
760	125	128	124
910	126	135	126
970	126	138	127

The Langmuir isotherm fits the data significantly better than the Freundlich. Average percent deviations, in q, are 1.01% and 8.64%, respectively. One reason for the better Langmuir fit is the trend to an asymptotic value for q at the highest pressures. Clearly, the data do not fit a linear isotherm well at all.

§15.2.2 Gas Mixtures and Extended Isotherms

Commercial applications of physical adsorption involve mixtures rather than pure gases. If adsorption of all components except one (A) is negligible, then adsorption of A is estimated from its pure-gas-adsorption isotherm using the partial pressure of A in the mixture. If adsorption of two or more components in the mixture is significant, the situation is complicated. Experimental data show that one component can increase, decrease, or have no influence on adsorption of another, depending on interactions of adsorbed molecules. A simple theoretical treatment is the extension of the Langmuir equation by Markham and Benton [29], who neglect interactions and assume that the only effect is the reduction of the vacant surface area for the adsorption of A because of adsorption of other components. For a binary gas mixture of A and B, let θ_A = fraction of surface covered by A and θ_B = fraction of surface covered by B. Then, $(1 - \theta_A - \theta_B)$ = fraction of vacant surface. At equilibrium:

$$(k_A)_a p_A (1 - \theta_A - \theta_B) = (k_A)_d \theta_A \quad (15\text{-}28)$$

$$(k_B)_a p_B (1 - \theta_A - \theta_B) = (k_B)_d \theta_B \quad (15\text{-}29)$$

Solving these equations simultaneously, and combining results with (15-23), gives

$$q_A = \frac{(q_A)_m K_A p_A}{1 + K_A p_A + K_B p_B} \quad (15\text{-}30)$$

$$q_B = \frac{(q_B)_m K_B p_B}{1 + K_A p_A + K_B p_B} \quad (15\text{-}31)$$

where $(q_i)_m$ is the maximum amount of adsorption of species i for coverage of the entire surface. Equations (15-30) and (15-31) are readily extended to a mixture of j components:

$$q_i = \frac{(q_i)_m K_i p_i}{1 + \sum_j K_j p_j} \quad (15\text{-}32)$$

In a similar fashion, as shown by Yon and Turnock [30], the Freundlich and Langmuir equations can be combined to give the following extended relation for gas mixtures:

$$q_i = \frac{(q_i)_0 k_i p_i^{1/n_i}}{1 + \sum_j k_j p_j^{1/n_j}} \quad (15\text{-}33)$$

where $(q_i)_0$ is the maximum loading, which may differ from $(q_i)_m$ for a monolayer. Equation (15-33) represents data for nonpolar, multicomponent mixtures in molecular sieves reasonably well. Broughton [31] has shown that both the extended-Langmuir and Langmuir–Freundlich equations lack thermodynamic consistency. Therefore, (15-32) and (15-33) are frequently referred to as nonstoichiometric isotherms. Nevertheless, for practical purposes, their simplicity often makes them the isotherms of choice.

Both (15-32) and (15-33) are also referred to as *constant-selectivity equilibrium* equations because they predict a separation factor (selectivity), $\alpha_{i,j}$, for each pair of components, i, j, in a mixture that is constant for a given temperature and independent of mixture composition. For example, (15-32) gives

$$\alpha_{i,j} = \frac{q_i/q_j}{p_i/p_j} = \frac{(q_i)_m K_i}{(q_j)_m K_j}$$

As with vapor–liquid and liquid–liquid phase equilibria for three or more components, data for binary and multicomponent gas–solid adsorbent equilibria are scarce and less accurate than corresponding pure-gas data. Valenzuela and Myers [23] include experimental data on adsorption of gas mixtures from 9 published studies on 29 binary systems, for which pure-gas-adsorption isotherms were also obtained. They also describe procedures for applying the Toth and UNILAN equations to multicomponent mixtures based on the ideal-adsorbed-solution (IAS) theory of Myers and Prausnitz [32]. Unlike the extended-Langmuir equation (15-32), which is explicit in the amount adsorbed, the IAS theory, though more accurate, is not explicit and requires an iterative solution procedure. Additional experimental data for higher-order (ternary and/or higher) gas mixtures are given by Miller, Knaebel, and Ikels [33] for 5A molecular sieves and by Ritter and Yang [34] for activated carbon. Yang [25] presents a discussion of mixture adsorption theories, together with comparisons of these theories with mixture data for activated carbon and zeolites. The data on zeolites are the most difficult to correlate, with the statistical thermodynamic model (SSTM) of Ruthven and Wong [35] giving the best results.

EXAMPLE 15.5 Extended-Langmuir Isotherm.

The experimental work of Ritter and Yang, cited in Example 15.4, also includes adsorption isotherms for pure CO and CH$_4$, and a binary mixture of CH$_4$ (A) and CO (B). Ritter and Yang give the following Langmuir constants for pure A and B at 294 K:

	q_m, cm³(STP)/g	K, psi⁻¹
CH₄	133.4	0.01370
CO	126.1	0.00624

Use these constants with the extended-Langmuir equation to predict the specific adsorption volumes (STP) of CH₄ and CO for a vapor mixture of 69.6 mol% CH₄ and 30.4 mol% CO at 294 K and a total pressure of 364.3 psia. Compare the results with the following experimental data of Ritter and Yang:

Total volume adsorbed, cm³/(STP)/g	114.1
Mole fractions in adsorbate:	
CH₄	0.867
CO	0.133

Solution

$$p_A = y_A P = 0.696(364.3) = 253.5 \text{ psia}$$
$$p_B = y_B P = 0.304(364.3) = 110.8 \text{ psia}$$

From (15-30):

$$q_A = \frac{133.4(0.0137)(253.5)}{1 + (0.0137)(253.5) + (0.00624)(110.8)} = 89.7 \text{ cm}^3(\text{STP})/\text{g}$$

$$q_B = \frac{126.1(0.00624)(110.8)}{1 + (0.0137)(253.5) + (0.00624)(110.8)} = 16.9 \text{ cm}^3(\text{STP})/\text{g}$$

The total amount adsorbed $= q = q_A + q_B = 89.7 + 16.9 = 106.6$ cm³ (STP)/g, which is 6.6% lower than the experimental value. Mole fractions in the adsorbate are $x_A = q_A/q = 89.7/106.6 = 0.841$ and $x_B = 1 - 0.841 = 0.159$. These adsorbate mole fractions deviate from the experimental values by 0.026. For this example, the extended-Langmuir isotherm gives reasonable results.

§15.2.3 Liquid Adsorption

When porous adsorbent particles are immersed in a confined pure gas, the pores fill with gas, and the amount of adsorbed gas is determined by the decrease in total pressure. With a liquid, the pressure does not change, and no simple experimental procedure has been devised for determining the extent of adsorption of a pure liquid. If the liquid is a homogeneous binary mixture, it is customary to designate one component the solute (1) and the other the solvent (2). The assumption is then made that the change in composition of the bulk liquid in contact with the porous solid is due entirely to adsorption of the solute; solvent adsorption is thus tacitly assumed not to occur. If the liquid mixture is dilute in solute, the consequences are not serious. If, however, experimental data are obtained over the entire concentration range, the distinction between solute and solvent is arbitrary and the resulting adsorption isotherms, as discussed by Kipling [36], can exhibit curious shapes, unlike those obtained for pure gases or gas mixtures. To illustrate this, let $n^0 = $ total moles of binary liquid contacting the adsorbent, $m = $ mass of adsorbent, $x_1^0 = $ mole fraction of solute before contact with adsorbent, $x_1 = $ mole fraction of solute in the bulk solution after adsorption equilibrium is achieved, and $q_1^e = $ apparent moles of solute adsorbed per unit mass of adsorbent.

A solute material balance, assuming no adsorption of solvent and a negligible change in the total moles of liquid mixture, gives

$$q_1^e = \frac{n^0\left(x_1^0 - x_1\right)}{m} \tag{15-34}$$

If isothermal data are obtained over the entire concentration range, processed with (15-34), and plotted as adsorption isotherms, the resulting curves are not as shown in Figure 15.12a. Instead, curves of the type in Figures 15.12b and c are obtained, where negative adsorption appears to occur in Figure 15.12c. Such isotherms are best referred to as *composite isotherms* or *isotherms of concentration change*, as suggested by Kipling [36]. Likewise, adsorption loading, q_1^e, of (15-34) is more correctly referred to as *surface excess*.

Under what conditions are composite isotherms of the shapes in Figures 15.12b and c obtained? This is shown by examples in Figure 15.13, where various combinations of hypothetical adsorption isotherms for solute (A) and solvent (B) are shown together with resulting composite isotherms. When the solvent is not adsorbed, as seen in Figure 15.13a, a composite curve without negative adsorption is obtained. In all other cases of Figure 15.13, negative values of surface excess appear.

Valenzuela and Myers [23] tabulate literature values for equilibrium adsorption of 25 different binary-liquid mixtures. With one exception, all 25 mixtures give composite isotherms of the shapes shown in Figures 15.12b and c. The exception is a mixture of cyclohexane and *n*-heptane with silica gel, for which surface excess is almost negligible $(0 \pm 0.05 \text{ mmol/g})$ from $x_1 = 0.041$ to 0.911. They also include literature references to 354 sets of binary, 25 sets of ternary, and 3 sets of data for higher-order liquid mixtures.

When data for the binary mixture are available only in the dilute region, solvent adsorption, if any, may be constant, and all changes in total amount adsorbed are due to the solute. In that case, the adsorption isotherms are of the form of Figure 15.12a, which resembles the shape obtained with pure gases.

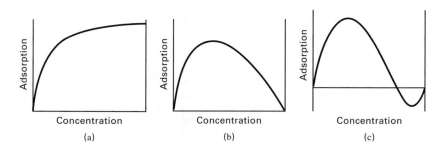

Figure 15.12 Representative isotherms of concentration change for liquid adsorption.

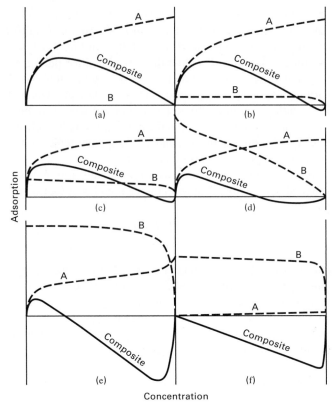

Figure 15.13 Origin of various types of composite isotherms for binary-liquid adsorption.

[From J.J. Kipling, *Adsorption from Solutions of Non-electrolytes*, Academic Press, London (1965) with permission.]

It is then common to fit the data with concentration forms of the Freundlich (15-19) or Langmuir (15-24) equations:

$$q = kc^{1/n} \qquad (15\text{-}35)$$

$$q = \frac{Kq_mc}{1 + Kc} \qquad (15\text{-}36)$$

Candidate systems for this case are small amounts of organics dissolved in water and small amounts of water dissolved in hydrocarbons. For liquid mixtures dilute in two or more solutes, multicomponent adsorption may be estimated from a concentration form of the extended-Langmuir equation (15-32) based on constants, q_m and K, obtained from experiments on single solutes. However, when solute–solute interactions are suspected, it may be necessary to determine constants from multicomponent data. As with gas mixtures, the concentration form of (15-32) also predicts constant selectivity for each pair of components in a mixture.

EXAMPLE 15.6 Adsorption of VOCs.

Small amounts of VOCs in water can be removed by adsorption. Generally, two or more VOCs are present. An aqueous stream containing small amounts of acetone (1) and propionitrile (2) is to be treated with activated carbon. Single-solute equilibrium data from

Radke and Prausnitz [37] have been fitted to the Freundlich and Langmuir isotherms, (15-35) and (15-36), with the average deviations indicated, for solute concentrations up to 50 mmol/L:

Acetone in Water (25°C):		Absolute Average Deviation of q, %
$q_1 = 0.141c_1^{0.597}$	**(1)**	14.2
$q_1 = \dfrac{0.190c_1}{1 + 0.146c_1}$	**(2)**	27.3
Propionitrile in water (25°C):		
$q_2 = 0.138c_2^{0.658}$	**(3)**	10.2
$q_2 = \dfrac{0.173c_2}{1 + 0.0961c_2}$	**(4)**	26.2

where q_i = amount of solute adsorbed, mmol/g, and c_i = solute concentration in aqueous solution, mmol/L.

Use these single-solute results with an extended Langmuir-type isotherm to predict the equilibrium adsorption in a binary-solute, aqueous system containing 40 and 34.4 mmol/L, respectively, of acetone and propionitrile at 25°C with the same adsorbent. Compare the results with the following experimental values from Radke and Prausnitz [37]:

$q_1 = 0.715$ mmol/g, $q_2 = 0.822$ mmol/g, and $q_{\text{total}} = 1.537$ mmol/g

Solution

From (15-32), the extended liquid-phase Langmuir isotherm is

$$q_i = \frac{(q_i)_m K_i c_i}{1 + \sum_j K_j c_j} \qquad (5)$$

From (2), $(q_1)_m = 0.190/0.146 = 1.301$ mmol/g.

From (4), $(q_2)_m = 0.173/0.0961 = 1.800$ mmol/g.

From (5):

$$q_1 = \frac{1.301(0.146)(40)}{1 + (0.146)(40) + (0.0961)(34.4)} = 0.749 \text{ mmol/g}$$

$$q_2 = \frac{1.800(0.0961)(34.4)}{1 + (0.146)(40) + (0.0961)(34.4)} = 0.587 \text{ mmol/g}$$

Summing, $q_{\text{total}} = 1.336$ mmol/g.

Compared to experimental data, the percent deviations for q_1, q_2, and q_{total}, respectively, are 4.8%, −28.6%, and −13.1%. Better agreement is obtained by Radke and Prausnitz using an IAS theory. It is expected that a concentration form of (15-33) would also give better agreement, but that requires that the single-solute data be refitted for each solute to a Langmuir–Freundlich isotherm of the form

$$q = \frac{q_0 kc^{1/n}}{1 + kc^{1/n}} \qquad (6)$$

§15.2.4 Ion-Exchange Equilibria

Ion exchange differs from adsorption in that one sorbate (a counterion) is exchanged for a solute ion, the process being governed by a reversible, stoichiometric, chemical-reaction equation. Thus, selectivity of the ion exchanger for one counterion over another may be just as important as the ion-exchanger capacity. Accordingly, the law of mass action is used to obtain an equilibrium ratio rather than to fit data to a sorption isotherm such as the Langmuir or Freundlich equation.

As discussed by Anderson [38], two cases are important. In the first, the counterion initially in the ion exchanger is exchanged with a counterion from an acid or base solution, e.g.,

$$Na^+_{(aq)} + OH^-_{(aq)} + HR_{(s)} \leftrightarrow NaR_{(s)} + H_2O_{(l)}$$

Note that hydrogen ions leaving the exchanger immediately react with hydroxyl ions to form water, leaving no counterion on the right-hand side of the reaction. Accordingly, ion exchange continues until the aqueous solution is depleted of sodium ions or the exchanger is depleted of hydrogen ions.

In the second, more-common, case, the counterion being transferred from exchanger to fluid remains as an ion. For example, exchange of counterions A and B is expressed by:

$$A^{n\pm}_{(l)} + nBR_{(s)} \leftrightarrow AR_{n_{(s)}} + nB^{\pm}_{(l)} \qquad (15\text{-}37)$$

where A and B must be either cations (positive charge) or anions (negative charge). For this case, at equilibrium, a chemical-equilibrium constant based on the law of mass action can be defined:

$$K_{A,B} = \frac{q_{AR_n} c^n_{B^\pm}}{q^n_{BR} c_{A^{n\pm}}} \qquad (15\text{-}38)$$

where molar concentrations c_i and q_i refer to the liquid and ion-exchanger phases, respectively. The constant, $K_{A,B}$ is not a rigorous equilibrium constant because (15-38) is in terms of concentrations instead of activities. Although it could be corrected by including activity coefficients, it is used in the form shown, with $K_{A,B}$ referred to as a *molar selectivity coefficient* for A displacing B. For the resin phase, concentrations are in equivalents per unit mass or unit bed volume of ion exchanger. For the liquid, concentrations are in equivalents per unit volume of solution. For dilute solutions, $K_{A,B}$ is constant for a given pair of counterions and a given resin.

When exchange is between two counterions of equal charge, (15-38) reduces to a simple equation in terms of equilibrium concentrations of A in the liquid solution and in the ion-exchange resin. Because of (15-37), the total concentrations, C and Q, in equivalents of counterions in the solution and the resin, remain constant during the exchange. Accordingly:

$$c_i = Cx_i/z_i \qquad (15\text{-}39)$$

$$q_i = Qy_i/z_i \qquad (15\text{-}40)$$

where x_i and y_i are equivalent fractions, rather than mole fractions, of A and B, such that

$$x_A + x_B = 1 \qquad (15\text{-}41)$$

$$y_A + y_B = 1 \qquad (15\text{-}42)$$

and $z_i =$ valence of counterion i. Combining (15-38) with (15-42) gives for counterions A and B of equal charge,

$$K_{A,B} = \frac{y_A(1 - x_A)}{x_A(1 - y_A)} \qquad (15\text{-}43)$$

Table 15.5 Relative Molar Selectivities, K, for Cations with 8% Cross-Linked Strong-Acid Resin

Li^+	1.0	Zn^{2+}	3.5
H^+	1.3	Co^{2+}	3.7
Na^+	2.0	Cu^{2+}	3.8
NH_4^+	2.6	Cd^{2+}	3.9
K^+	2.9	Be^{2+}	4.0
Rb^+	3.2	Mn^{2+}	4.1
Cs^+	3.3	Ni^+	3.9
Ag^+	8.5	Ca^{2+}	5.2
UO_2^{2+}	2.5	Sr^{2+}	6.5
Mg^{2+}	3.3	Pb^{2+}	9.9
		Ba^{2+}	11.5

At equilibrium, x_A and y_A are independent of total equivalent concentrations C and Q. Such is not the case when the two counterions are of unequal charge, as in the exchange of Ca^{2+} and Na^+. A derivation for this general case gives

$$K_{A,B} = \left(\frac{C}{Q}\right)^{n-1} \frac{y_A(1 - x_A)^n}{x_A(1 - y_A)^n} \qquad (15\text{-}44)$$

Thus, for unequal counterion charges, $K_{A,B}$ depends on the ratio C/Q and on the ratio of charges, n, as defined in (15-37).

When $K_{A,B}$ data for a system of counterions with a particular ion exchanger are not available, the method of Bonner and Smith [39], as modified by Anderson [38], is used for screening purposes or preliminary calculations. In this method, $K_{A,B}$ is

$$K_{ij} = K_i/K_j \qquad (15\text{-}45)$$

where values for relative molar selectivities K_i and K_j are given in Table 15.5 for cations with an 8% cross-linked, strong-acid resin, and in Table 15.6 for anions with strong-base resins. For values of K in these tables, the units of C and Q are, respectively, eq/L of solution and eq/L of bulk bed volume of water-swelled resin.

A typical cation-exchange resin of the sulfonated styrene–divinylbenzene type, such as Dowex 50, as described by Bauman and Eichhorn [40] and Bauman, Skidmore, and Osmun [41], has an exchangeable ion capacity of 5 ± 0.1 meq/g of dry resin. As shipped, water-wet resin might contain 41.4 wt% water. Thus, wet capacity is $5(58.6/100) = 2.9$ meq/g of wet resin. If bulk density of a drained bed of wet resin is 0.83 g/cm^3, bed capacity is 2.4 eq/L of resin bed.

Table 15.6 Approximate Relative Molar Selectivities, K, for Anions with Strong-Base Resins

I^-	8	OH^- (Type II)	0.65
NO_3^-	4	HCO_3^-	0.4
Br^-	3	CH_3COO^-	0.2
HSO_4^-	1.6	F^-	0.1
NO_2^-	1.3	OH^- (Type I)	0.05–0.07
CN^-	1.3	SO_4^{2-}	0.15
Cl^-	1.0	CO_3^{2-}	0.03
BrO_3^-	1.0	HPO_4^{2-}	0.01

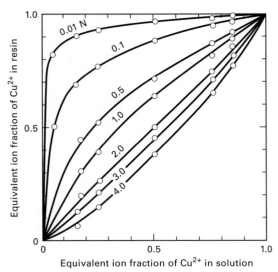

Figure 15.14 Isotherms for ion exchange of Cu^{2+} and Na^+ on Dowex 50-X8 as a function of total normality in the bulk solution.

[From A.L. Myers and S. Byington, *Ion Exchange Science and Technology*, M. Nijhoff, Boston (1986) with permission.]

As with other separation processes, a separation factor, $SP = S_{A,B}$ (§1.8), which ignores the valence of the exchanging ions, can be defined for an equilibrium stage. For binaries in terms of equivalent ionic fractions:

$$S_{A,B} = \frac{y_A(1 - x_A)}{x_A(1 - y_A)} \quad (15\text{-}46)$$

which is identical to (15-43). Data for an exchange between Cu^{2+} (A) and Na^+ (B) (counterions of unequal charge) with Dowex 50 cation resin over a wide range of total-solution normality at ambient temperature are shown in terms of y_A and x_A in Figure 15.14, from Subba Rao and David [42]. At low total-solution concentration, the resin is highly selective for copper ion, whereas at high total-solution concentration, the selectivity is reversed to slightly favor sodium ions. A

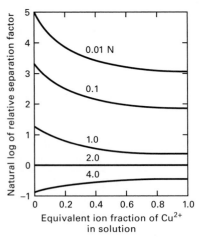

Figure 15.15 Relative separation factor of Cu^{2+} and Na^+ for ion exchange on Dowex 50-X8 as a function of total normality in the bulk solution.

[From A.L. Myers and S. Byington, *Ion Exchange Science and Technology*, M. Nijhoff, Boston (1986) with permission.]

similar trend was observed by Selke and Bliss [43, 44] for exchange between Ca^{2+} and H^+ using a similar resin, Amberlite IR-120. Selectivity sensitivity is shown dramatically in Figure 15.15, from Myers and Byington [45], where the natural logarithm of the separation factor, $S_{Cu^{2+},\,Na^+}$, as computed from data of Figure 15.14 with (15-46), is plotted as a function of equivalent ionic fraction, $x_{Cu^{2+}}$. For dilute solutions of Cu^{2+}, $S_{Cu^{2+},\,Na^+}$ ranges from 0.5 at a total concentration of 4 N to 60 at 0.01 N. In terms of $K_{Cu^{2+},\,Na^+}$ of (15-44), with $n = 2$, the corresponding variation is only from 0.6 to 2.2.

EXAMPLE 15.7 Ion-Exchange Equilibrium.

An Amberlite IR-120 ion-exchange resin similar to that of Example 15.2, but with a maximum ion-exchange capacity of 4.90 meq/g of dry resin, is used to remove cupric ion from a waste stream containing 0.00975-M $CuSO_4$ (19.5 meq Cu^{2+}/L solution). The spherical resin particles range in diameter from 0.2 to just over 1.2 mm. The equilibrium ion-exchange reaction is of the divalent–monovalent type:

$$Cu^{2+}_{(aq)} + 2HR_{(s)} \leftrightarrow CuR_2, +2H^+_{(aq)}$$

As ion exchange takes place, the meq of cations in the aqueous solution and in the resin remain constant.

Experimental measurements by Selke and Bliss [43, 44] show an equilibrium curve like Figure 15.14 at ambient temperature that is markedly dependent on total equivalent concentration of the aqueous solution, with the following equilibrium data for cupric ions with a 19.5 meq/L solution:

c, meq Cu^{2+}/L Solution	0.022	0.786	4.49	10.3
q, meq Cu^{2+}/g Resin	0.66	3.26	4.55	4.65

These data follow a highly nonlinear isotherm.

(a) From the data, compute the molar selectivity coefficient, K, at each value of c for Cu^{2+} and compare it to the value estimated from (15-45) using Table 15.5. (b) Predict the milliequivalents of Cu^{2+} exchanged at equilibrium from 10 L of 20 meq Cu^{2+}/L, using 50 g of dry resin with 4.9 meq of H^+/g.

Solution

(a) Selke and Bliss do not give a value for the resin capacity, Q, in eq/L of bed volume. Assume a value of 2.3. From (15-44):

$$K_{Cu^{2+},\,H^+} = \left(\frac{C}{Q}\right)\frac{y_{Cu^{2+}}(1 - x_{Cu^{2+}})^2}{x_{Cu^{2+}}(1 - y_{Cu^{2+}})^2}$$

where $(C/Q) = 0.0195/2.3 = 0.0085$.

$$x_{Cu^{2+}} = c_{Cu^{2+}}/19.5 \quad \text{and} \quad y_{Cu^{2+}} = q_{Cu^{2+}}/4.9$$

Using the above values of c and q from Selke and Bliss:

q, meq Cu^{2+}/g	$x_{Cu^{2+}}$	$y_{Cu^{2+}}$	$K_{Cu^{2+},\,H^+}$
0.66	0.00113	0.135	1.35
3.26	0.0403	0.665	1.15
4.55	0.230	0.929	4.04
4.65	0.528	0.949	1.30

The average value of K is 2.0. Values in Table 15.5 when substituted into (15-45) predict $K_{Cu^{2+},\,H^+} = 3.8/1.3 = 2.9$, which is somewhat higher.

(b) Assume a value of 2.0 for $K_{Cu^{2+},\,H^+}$ with $Q = 2.3$ eq/L. The total-solution concentration, C, is 0.02 eq/L. Equation (15-44) becomes

$$2.0 = \left(\frac{0.02}{2.3}\right)\frac{y_{Cu^{2+}}(1-x_{Cu^{2+}})^2}{x_{Cu^{2+}}(1-y_{Cu^{2+}})^2} \qquad (1)$$

Initially, the solution contains $(0.02)(10) = 0.2$ equivalent of cupric ion with $x_{Cu^{2+}} = 1.0$. Let $a =$ equivalents of Cu exchanged. Then, at equilibrium, by material balance:

$$x_{Cu^{2+}} = \frac{0.02 - (a/10)}{0.02} \qquad (2)$$

$$y_{Cu^{2+}} = \frac{(a/50)}{0.0049} \qquad (3)$$

Substitution of (2) and (3) into (1) gives

$$2.0 = 0.0087\,\frac{\left[\frac{(a/50)}{0.0049}\right]\left[1 - \frac{0.02-(a/10)}{0.02}\right]^2}{\left[\frac{0.02-(a/10)}{0.02}\right]\left[1 - \frac{(a/50)}{0.0049}\right]^2} \qquad (4)$$

Solving (4), a nonlinear equation, for a gives 0.1887 eq of Cu exchanged. Thus, $0.1887/[(0.020)(10)] = 0.944$ or 94.4% of the cupric ion is exchanged.

Equilibria in Chromatography

As discussed in §15.1.3, separation by chromatography involves many sorption mechanisms, including adsorption on porous solids, absorption or extraction (partitioning) in liquid-supported or bonded solids, and ion transfer in ion-exchange in resins. Thus, at equilibrium, depending upon the sorption mechanism, equations such as (15-19), (15-24), (15-32), and (15-33) for gas adsorption; (15-35) and (15-36) for liquid adsorption; (6-17) to (6-20) for gas absorption; (8-1) for liquid extraction; and (15-38), (15-43), and (15-44) for ion exchange apply.

At equilibrium, the distribution (partition) constant for solute, i, is

$$K_i = q_i/c_i \qquad (15\text{-}47)$$

where q is concentration in the stationary phase and c is concentration in the mobile phase. Solutes with the highest equilibrium constants will elute from the chromatographic column at a slower rate than solutes with smaller constants.

§15.3 KINETIC AND TRANSPORT CONSIDERATIONS

Multicomponent mixtures may often be separated efficiently in a single unit by selective partitioning of targeted specie(s) between a flowing fluid phase and stationary sorptive phase. The latter is typically porous or resin particles packed into a cylindrical vessel. Equipment configurations and operating procedures to accomplish adsorptive separations take various forms, including slurry adsorption, temperature- and

Table 15.7 Recombinant Pharmaceuticals Purified by Adsorption

Product	Annual Sales, Billions of $
Insulin	1.0
Growth hormone	0.9
Interferons	0.5
Erythropoietin	0.4
Tissue plasminogen activator	0.2

pressure-swing adsorption, ion exchange, chromatography, and simulated-moving-bed systems.

Examples of industrial-scale, packed-bed adsorption for purification and bulk separation of mixtures of small molecules are included in Table 15.1. Large molecules produced in biochemical processes are also separated by adsorption. Important examples are found in biopharmaceutical separations that include resolution of synthetic chiral mixtures to produce pure enantiomeric drugs such as the antidepressant fluoxetine (trade names: Prozac®, Sarafem®); the separation of sweeteners fructose and glucose by high-throughput simulated moving beds; and anion-exchange adsorption of infectious, nonreplicative adenovirus type 5 for purification as a viral vector for gene therapy. Some biopharmaceuticals produced by recombinant methods that rely on adsorptive separations are listed in Table 15.7.

To accomplish separation by adsorption in a packed bed, solutes dispersed in a mobile fluid solvent are transported by bulk flow (i.e., convection) through interstices between packed particles. Solutes diffuse between the moving fluid phase and a stationary fluid phase within pores of the solid adsorbent. Instantaneous equilibrium partitioning of solute between fluid and adsorptive surfaces in the stationary phase, due to a solute-specific thermodynamic driving force, is resisted by consecutive mass-transfer processes illustrated in Figure 15.16, which is adapted from Athalye et al. [46]:

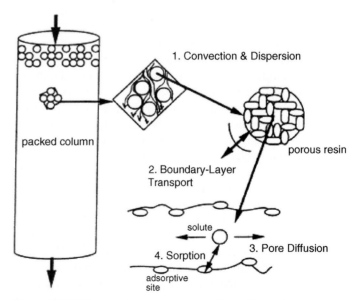

Figure 15.16 Transport-rate processes in adsorption.

1. Solute transport by bulk flow (convection) and dispersion through interstices of the bed of adsorptive particles

2. External (interphase) solute transport from the bulk flow to the outer perimeter of the adsorbent particle through a thin film or boundary layer

3. Internal (intraphase) solute transport by diffusion in quiescent fluid-filled pores in the adsorbent particle

4. Surface diffusion along the internal porous surface of the adsorbent particle.

At the adsorptive surface, kinetic interaction occurs between solute and sorptive sites within the adsorbent particle.

External transport includes convective dispersion of the solute within the bulk fluid, and diffusion through a boundary layer surrounding adsorbent particles. Axial dispersion of individual solute molecules in the bulk fluid occurs primarily by microscale, fluid-phase phenomena such as mixing via solid obstructions to flow, eddies, and recirculation from regional pressure gradients. Boundary-layer transport, internal transport, and surface diffusion occur via random Brownian motion. Kinetic interaction, which depends on solute orientation and frequency of surface collisions, is virtually instantaneous for physical adsorption of gases and small solutes, but can become controlling in bioprocesses for orientation-specific sorption such as antibody interaction with fixed antigen, or chemisorption (bond formation).

During regeneration of the adsorbent, the reverse of the four steps occurs, following desorption. Adsorption and desorption may be accompanied by heat transfer arising from exothermic heat of adsorption and endothermic heat of desorption. Although external mass transfer is limited to convection, external heat transfer from the particle outer surface occurs not only by convection through a thermal boundary layer surrounding each particle, but also by thermal radiation between particles when the fluid is a gas, and by conduction at points of contact by adjacent particles. Conduction and radiation mechanisms for heat transfer also exist within particles.

In a fixed bed of adsorbent particles, solute concentration and temperature change continuously with time and location during adsorption and desorption processes, when significant thermal effects due to sorption occur. For a given particle at a particular time, profiles illustrating temperature and solute concentration in the fluid are as shown in Figures 15.17a and b for adsorption and desorption, respectively, where subscripts b and s refer to bulk fluid and particle outer surface, respectively. The fluid concentration gradient is usually steepest within the particle, whereas the temperature gradient is usually steepest in the boundary layer surrounding the particle. Thus, although resistance to heat transfer is mainly external to the adsorbent particles, the major resistance to mass transfer resides in the adsorbent particle. However, dilute aqueous solute solutions typical of bioseparations rarely result in temperature gradients external to, or within, the particles.

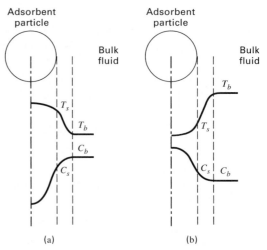

Figure 15.17 Solute concentration and temperature profiles for a porous adsorbent particle surrounded by a fluid: (a) adsorption, (b) desorption.

Costs associated with adsorbent usage and handling and consumption of solvent and regenerant [47, 48] often dominate manufacturing expenses [49]. This is especially true in purification of high-value-added pharmaceuticals [50] or biotechnology products such as antibiotics or recombinant DNA [51]. Thus, understanding how transport resistances affect separation is important to improve packed-bed adsorption efficiency and enhance process economics. Applicable descriptions of mass transport in chromatographic separations are also valuable to predict scale-up and to control operating performance [52]. Therefore, a general model that describes all steps in the adsorption process is presented first to illustrate the important features of the process. This is followed by correlations and methods for predicting the individual resistances in the general model and by specialized cases of the general model.

§15.3.1   Convection-Dispersion Model

Solute convected by dispersed, plug flow of fluid at average interstitial (actual) velocity u in the axial direction z, within a uniform bed of particles, has a position- and time-dependent, bulk, fluid-phase concentration $c_f\{z, t\}$ given by Ruthven [10] and others as

$$\frac{\partial c_f}{\partial t} + u\frac{\partial c_f}{\partial z} - E\frac{\partial^2 c_f}{\partial z^2} = -\frac{1 - \epsilon_b}{\epsilon_b}\frac{\partial \bar{c}_b}{\partial t} \qquad (15\text{-}48)$$

where ϵ_b is the fractional, interstitial void volume (bed porosity); E is a Fickian, convective, axial-dispersion coefficient that accounts for both axial molecular diffusion and the interaction of lateral molecular diffusion with local nonuniformities in fluid velocity caused by complicated flow paths around the stationary particles; and $\bar{c}_b$ is the volume-averaged stationary-phase concentration per unit mass, which for a spherical particle of radius R_p is given by

$$\bar{c}_b = \frac{3}{R_p^3}\int_0^{R_p} r^2 c_b \, dr \qquad (15\text{-}49)$$

The superficial velocity of the fluid through the bed is $u_s = \epsilon_b u$. At a particular time t and axial location in the bed z, the first term on the LHS of (15-48) accounts for the change in solute concentration in the bulk fluid with time. The second term on the LHS accounts for the change in c_f with axial location arising from bulk convection. The third term accounts for axial dispersion, and the term on the RHS accounts for the change in solute loading on the stationary phase.

The equilibrium *loading*, q, measures the mass of solute adsorbed on *surfaces* of the stationary phase per unit mass of adsorbent, whereas $\bar{c}_b$ includes both adsorbed solute *and* unadsorbed solute diffusing in the pore volume of the stationary phase. For example, at equilibrium in nonadsorbing gel-filtration chromatography, $q = 0$, while $\bar{c}_b = \epsilon_p^* c_f$. Occasionally q and $\bar{c}_b$ are interchanged in the literature. The change in $\bar{c}_b$ with time for a rapidly equilibrating particle may be related to flux at its external surface using a linear driving force (LDF) approximation introduced by Glueckauf [53],

$$\frac{4}{3}\pi R_p^3 \frac{\partial \bar{c}_b}{\partial t} = -k_{c,tot} 4\pi R_p^2 (\bar{c}_b \alpha - c_f) \qquad (15\text{-}50)$$

that uses an overall mass-transfer coefficient, $k_{c,tot}$, defined below in (15-58), which consists of a series of transport-rate resistance terms for concentration-independent equilibrium and transport properties illustrated in Figure 15.16. At equilibrium, solute is partitioned between the bulk fluid and average stationary-phase loading according to

$$\alpha = \frac{c_f}{\bar{c}_b} = \frac{1}{\epsilon_p^*(1 + K_d)} \qquad (15\text{-}51)$$

with $K_d = k_a/k_d = c_s/c_p$ being the constant equilibrium distribution coefficient given by the kinetic rate constant for adsorption relative to desorption, with subscripts s and p referring to surface and pore volume of the porous stationary phase, respectively. An inclusion porosity, ϵ_p^*, defines the effective stationary-phase volume fraction accessible to a specific solute. Particle porosity ϵ_p is the internal void fraction of a particle, whereas ϵ_p^* includes only voids penetrable by a particular solute due to size or steric hindrance. For example, Source 15Q^{TM} anion-exchange resin has $\epsilon_p \sim 0.4$, whereas for binding of adenovirus type 5 (MW 165 MDa), its $\epsilon_p^* = 0.0$.

Modes of Time-Dependent Sorption

The mobile phase consists of fluid solvent (also called *carrier* or *eluent*) and solutes, which are components that may be adsorbed and desorbed, dissolved, or suspended in eluent. A solution of eluent and desorbed solute(s) is often referred to as *eluate*. Time-dependent (de)sorption described in (15-48) to (15-51) may be carried out in any of the following three major adsorption operating modes [54] to separate a mixture of solutes:

1. **Frontal.** This mode is widely used for purification of small molecules and biomolecules in fixed-bed adsorbers. The mobile phase (gas or liquid) is fed continuously to the stationary phase until the bed approaches saturation with the solute. The solute is then desorbed by an appropriate mechanism to obtain a relatively pure product. Desorption mechanisms include: increasing bed temperature or reducing bed pressure in the case of a gas, or introducing an isocratic or gradient change in solvent composition (e.g., I, pH, hydrophobicity) to reduce sorbate equilibrium partitioning to the solid phase. The adsorption-desorption cycle may be repeated, sometimes after an intermediate, validated cleaning step, to increase stationary-phase utilization. During the adsorption part of the cycle, solute concentration and loading fronts move with time through the bed. This method can be used to partially separate one of a mixture of solutes, such as oxygen or nitrogen from air, as illustrated in §15.3.5.

2. **Displacement.** This mode is widely used (e.g., displacement chromatography) to separate and concentrate target protein(s) from a mixture. Relative to frontal mode, mobile-phase feeds in displacement mode nearly always contain two or more solutes to be separated. Following near-saturation of the bed with solute(s), desorption (elution) in consecutive zones of pure substances is effected by substitution with a more strongly adsorbed solute (displacer) that is fed into the bed. The cycle may be repeated after removing the displacer from the bed.

3. **Differential.** This mode is used to recover a variety of bioproducts. A small pulse of solute dissolved in the mobile phase is loaded onto the bed. Rather than saturating the bed as in frontal or displacement modes, the solute pulse is carried through the bed (*eluted*) at a rate lower than pure solvent. This reduced rate is due to solute-specific interactions with the stationary phase that are modulated by solvent composition (e.g., I, pH, hydrophobicity), which may be unchanged (*isocratic elution*) or changed in a linear or nonlinear fashion (*gradient elution*). Solute product(s) emerge(s) from the column diluted by the solvent in the form of exponentially modified Gaussian concentration peaks.

Mass transfer in each of the three modes determines efficiency of separation, utilitization of adsorbent and solvent, and recovery of solute.

Solute Concentration Distribution

To identify transport-rate effects on a separation, consider an asymptotic solution to (15-48)–(15-51) obtained by Reis et al. [55] for differential, isocratic elution of a sharp-pulse input of

$$c\{z,0\} = \frac{m_o \delta\{z\}}{A\epsilon_b} \qquad (15\text{-}52)$$

in a long, packed bed, where m_o is solute mass, A is bed cross-sectional area, and $\delta\{z\}$ is the *Dirac delta function*, which is zero everywhere except at $z = 0$, where its magnitude is infinity. The resulting fluid-phase solute concentration illustrated in Figure 15.18 is distributed normally around the mean solute position, $z_o = \omega u t$, along the axis, z, when

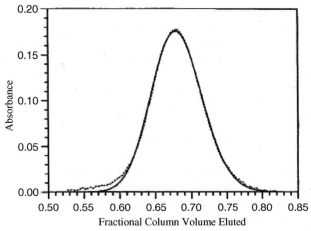

Figure 15.18 Elution of 0.1 mL injection of 10 gm/mL hemoglobin eluted at 0.6 mL/min from an 11.5×2.5 cm i.d. glass column packed with Toyopearl HW65C (- - -). The corresponding Gaussian expression (-) yields $\epsilon_p = 0.498$ and $H = 0.312$ mm [46].

time, t, is treated as a parameter,

$$c_f\{z, t\} = \frac{m_o \omega}{A \epsilon_b \sqrt{2\pi H z_o}} \exp\left[\frac{-(z - z_o)^2}{2 H z_o}\right] \quad (15\text{-}53)$$

where ω is the fraction of solute in the moving fluid phase at equilibrium, given by

$$\omega = \frac{1}{1 + \dfrac{1 - \epsilon_b}{\epsilon_b} \dfrac{1}{\alpha}} \quad (15\text{-}54)$$

and H is the height of a theoretical chromatographic plate, and N is the number of theoretical chromatographic plates as defined by Glueckauf [56]. Equation (15-53) is a Gaussian expression with solute mean residence time

$$\bar{t} = \frac{NH}{\omega u} \quad (15\text{-}55)$$

and variance s, where

$$s^2 = \frac{\bar{t} H}{\omega u} \quad (15\text{-}56)$$

which reflects random solute partitioning in an adsorptive bed. These features are illustrated in Figure 15.19. This solution is comparable to those of Giddings [58], Aris and Amundson [59], Ruthven [10], and Guiochon [60], and may be extended to describe gradient elution [61] and membrane chromatography [62] or to evaluate countercurrent adsorption [63].

Separation Efficiency (Plate Height or Bandwidth)

Each transport-rate resistance to equilibrium partitioning broadens the adsorptive band in (15-53), as shown in Figures 15.18 to 15.20, and lowers separation efficiency, which is proportional to H^{-1}. This is measured by noting that peak width measured at 60.6% of maximum peak height corresponds to the square root of the variance of the solute distribution, s, in (15-56), as shown by van Deemter et al. [64] and Karol [65], and by examining the relation between plate

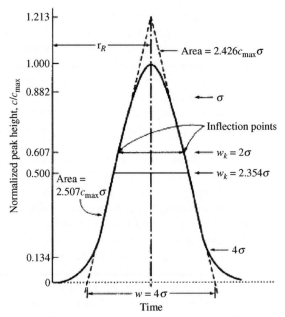

Figure 15.19 Gaussian peak with the properties: maximum peak height c_{max}; standard deviation σ; width at inflection points, w_i; width at half height, w_h; width at base intercept, w; and average retention time, t_R.

[From Horvath and Melander [57] with permission.]

height, H, and the overall mass-transfer coefficient in linear differential chromatography (LDC), LDC $k_{c,tot}$:

$$H = 2\left(\frac{E}{u} + \frac{\omega(1 - \omega)R_b u}{3\alpha k_{c,tot}}\right) \quad (15\text{-}57)$$

$$\frac{1}{k_{c,tot}} = \frac{1}{k_c} + \frac{R_p}{5\epsilon_p^* D_e} + \frac{3}{R_p k_a \epsilon_p^*}\left(\frac{K_d}{1 + K_d}\right)^2 \quad (15\text{-}58)$$

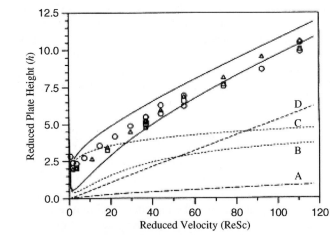

Figure 15.20 van Deemter plot comparing reduced plate height, $h = H/2R_p$, of 0.1-mL bovine hemoglobin at 10 (○), 50 (△), and 100 (□) mg/mL eluted from 11.5×2.5 cm i.d. glass column packed with Toyopearl HW65C. Contributions to h (dotted lines) are due to (A) boundary-layer transport; (B) axial dispersion by [174, 175]; (C) axial dispersion by [70]; (D) intraparticle diffusion. Total h values (lines) arise from A + B + D (lower) and A + C + D (upper). $N_{Re} N_{Sc} = 100$ at $u = 1.5$ cm/minute.

[From Athalye et al. [46] with permission.]

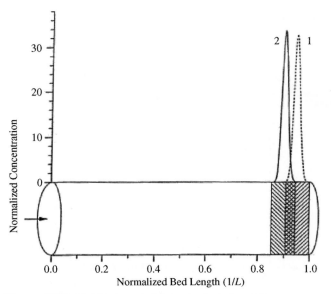

Figure 15.21 Fluid concentrations c_1 and c_2 plotted for chromatographic separation of 1 and 2 where $t = 0.95t_1$, $\delta_{12} = 0.05$, and $R_{12} = 1$ at $z/L = 1$.

where k_c is the fluid-phase mass-transfer coefficient, and D_e represents effective solute diffusivity in the pore liquid. The first term on the RHS of (15-57) represents transport-rate resistance due to convective dispersion. The three terms on the RHS of (15-58) represent series resistances associated with the particle, arising from external boundary-layer transport, internal intraparticle diffusion, and kinetic sorption rate, respectively. Contributions of these individual transport-rate resistances to H are illustrated in van Deemter plots [64], like that in Figure 15.20 for a size-exclusion resin.

Resolving Solute Mixtures

Resolution, R, of similar components 1 and 2, illustrated in Figure 15.21, occurs due to solute-specific partitioning between moving fluid and stationary bulk phases and is defined by the ratio of peak separation relative to the average peak width:

$$R \equiv \frac{|\bar{t}_1 - \bar{t}_2|}{2(s_1 + s_2)} = \frac{\delta\sqrt{N}}{4} \tag{15-59}$$

where δ is the fractional difference in migration velocities in the moving fluid phase of species 1 and 2 or, in general, i and j:

$$\delta_{i,j} = 2\frac{|\omega_i - \omega_j|}{\omega_i + \omega_j} \tag{15-60}$$

The differential fluid-phase/stationary-phase partitioning factor, $\delta_{i,j}$, equals twice the product of relative selectivity and retention factors [58], and contains only measurable geometric and thermodynamic parameters peculiar to any species solid-phase pair i and j.

§15.3.2 Correlations for Transport-Rate Coefficients

Equation (15-59) shows that decreasing s [i.e., decreasing H in (15-56)] improves resolution, R, in a separation. Equation (15-57) shows that H is decreased by reducing axial convective dispersion (i.e., coefficient E). Equation (15-58) shows that H is also decreased by increasing fluid-phase mass transfer, represented by k_c; effective pore diffusivity, D_e; and/or kinetic-adsorption rate, k_a. Identifying correlations for these four transport-rate coefficients allows the user to select geometric and operating parameters that maximize separation efficiency, R.

Convective Dispersion

For low values of the diffusion Peclet number for axial convection dispersion, $N_{Pe} = N_{Re}\,N_{Sc} = 2R_p u\epsilon_b/D_i \ll 1$, the convective dispersion coefficient is given by $E = D_i/\tau_f$, where D_i is the molecular diffusivity of the solute and τ_f is the interstitial tortuosity factor, ~ 1.4 for a packed bed of spheres [66, 67]. For very high $N_{Re}N_{Sc}$ values, the dispersion Peclet number, $N_{Pe_E} = 2R_p u\epsilon_b/E$, asymptotically approaches the limit of pure hydrodynamic dispersion [67–70]:

$$N_{Pe_E} = \frac{2p}{1 - p} \tag{15-61}$$

where $p = 0.17 + 0.33\exp(-24/N_{Re})$. For intermediate values of $N_{Re}N_{Sc}$, convective axial-dispersion coefficients can be predicted reasonably well over a wide range of conditions for both gases and liquids by [174, 175]:

$$\frac{1}{N_{Pe_E}} = \frac{1 - p}{p}\left[Y + Y^2\left(\exp(-Y^{-1}) - 1\right)\right] + \frac{\epsilon_b}{\tau N_{Re}N_{Sc}} \tag{15-62}$$

where $Y = p(1 - p)N_{Re}N_{Sc}/[23.16(1 - \epsilon_b)]$. These correlations may be used to select values of u and/or R_p for a particular set of D_i and ϵ_b to minimize E and convective-dispersion band broadening.

Boundary-Layer (External) Transport

Wakao and Funazkri [71] proposed a correlation for mass transport of species i from the bulk fluid flowing through the bed to the outer surface of the bed particles of diameter $D_p = 2R_p$ through the boundary layer or film around the particles. It is given by

$$N_{Sh_i} = \frac{k_{c,i}D_p}{D_i} = 2 + 1.1\left(\frac{\rho u\epsilon_b D_p}{\mu}\right)^{0.6}\left(\frac{\mu}{\rho D_i}\right)^{1/3} \tag{15-63}$$

and was the result of reanalyzing 37 sets of previously published mass-transfer data for particles packed in a bed, with Sherwood-number corrections for axial dispersion. This correlation is compared to 12 sets of gas-phase and 11 sets of liquid-phase data in Figure 15.22. The data cover a Schmidt number range of 0.6 to 70,600, a Reynolds number range of 3

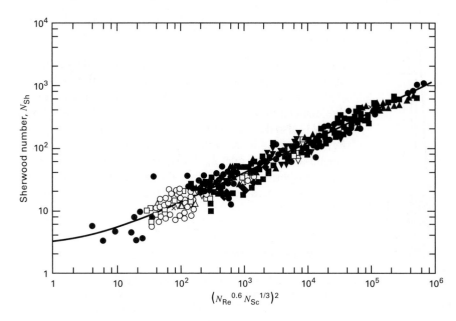

Figure 15.22 Correlation of experimental data for Sherwood number in a packed bed.
[From N. Wakao and T. Funazkri, *Chem. Eng. Sci.*, **33**, 1375 (1978) with permission [71].]

to 10,000, and a particle diameter from 0.6 to 17.1 mm. Particle shapes include spheres, short cylinders, flakes, and granules.

The value of 2 for the first term on the RHS of (15-63) corresponds to $N_{Sh} = k_c D_p/D_i = 2$, the Sherwood (Nusselt) number value for steady-state mass (heat) transport to a spherical particle surrounded by an infinite, quiescent fluid, which is obtained by solving (3-74) using boundary conditions for constant concentrations (temperatures) at the particle surface and far away, respectively. (See Exercise 3.31.) A fluid flowing with momentum-to-mass (heat) diffusivity ratio $N_{Sc} = \mu/\rho D_i \left(N_{Pr} = C_p\mu/k\right)$ past the particle at a dimensionless rate $N_{Re} = D_p G/\mu$, where G is the mass velocity equal to $\rho u \epsilon_b$ in (15-63), adds the second term on the RHS of (15-63), which raises the Sherwood (Nusselt) number to values as high as 160 (30), as shown by Ranz and Marshall [72, 73], who proposed an earlier correlation. Other correlations for packed beds have the form of Chilton and Colburn [74] *j*-factors, as first proposed by Gamson et al. [75]:

$$j_D = (N_{St_M})(N_{Sc})^{2/3} = f\{N_{Re}\} \tag{15-64}$$

$$j_H = (N_{St})(N_{Pr})^{2/3} = f\{N_{Re}\} \tag{15-65}$$

with $N_{Re} = D_p G/\mu,$

and $N_{St_M} = k_c \rho/G$ and $N_{St} = h/C_P G$

Some Chilton–Colburn-type correlations that use $N_{Re} = D_p G/\epsilon_b \mu$ to account for bed void fraction, ϵ_b, are reported by Sen Gupta and Thodos [76], Petrovic and Thodos [77], and Dwivedi and Upadhay [78].

When (15-63) for mass transfer and its analog for heat transfer, given in the following example, are used with beds packed with nonspherical particles, D_p is the equivalent diameter of a spherical particle. The following suggestions have been proposed for computing the equivalent diameter from geometric properties of the particle. These suggestions

may be compared by considering a short cylinder with diameter, D, equal to the length, L.

1. D_p = diameter of a sphere with the same external surface area:

$$\pi D_p^2 = \pi DL + \pi D^2/2$$

and $D_p = \left(DL + D^2/2\right)^{0.5} = 1.225D$

2. D_p = diameter of a sphere with the same volume:

$$\pi D_p^3/6 = \pi D^2 L/4 \text{ and } D_p = \left(3D^2 L/2\right)^{1/3} = 1.145D$$

3. D_p = 4 times the hydraulic radius, r_H, where for a packed bed, $4r_H = 6/a_v$, where a_v = external particle surface area/volume of the particle.

Thus,

$$a_v = \frac{\pi DL + \pi D^2/2}{\pi D^2 L/4} = \frac{6}{D} \text{ and } D_p = 4r_H = \frac{6D}{6} = 1.0D$$

The hydraulic radius concept is equivalent to replacing D_p in the Reynolds number by $\psi D_p'$, where Ψ is sphericity, given by Suggestion 2. The sphericity is defined by:

$$\psi = \frac{\text{Surface area of a sphere of same volume as particle}}{\text{Surface area of particle}}$$

For a cylinder of $D = L$,

$$\psi = \frac{\pi D_p^2}{\pi DL + 2\left(\dfrac{\pi D^2}{4}\right)} = \frac{\pi(1.145D)^2}{\dfrac{3}{2}\pi D^2} = 0.874$$

and $\psi D_p' = (0.874)(1.145D) = D$, the diameter of the cylinder.

Suggestions 2 and 3 are widely used. Suggestion 3 is conveniently applied to crushed particles of irregular surface, but with no obvious longer or shorter dimension, and isotropic in shape. In that case, D_p' is taken as the size of the particle and the sphericity is approximately 0.65, as discussed by Kunii and Levenspiel [79].

EXAMPLE 15.8 Transport Coefficients.

Acetone vapor in a nitrogen stream is removed by adsorption in a fixed bed of activated carbon. At a location in the bed where the pressure is 136 kPa, the bulk gas temperature is 297 K, and the bulk mole fraction of acetone is 0.05, estimate the external gas-to-particle mass-transfer coefficient for acetone and the external particle-to-gas heat-transfer coefficient. Additional data are as follows: Average particle diameter = 0.0040 m, and gas superficial molar velocity = 0.00352 kmol/m²-s.

Solution

Because the temperature and composition are known only for the bulk gas and not at the particle external surface, use gas properties at bulk gas conditions. Relevant fluid properties for use in (15-63) and its heat-transfer analog,

$$N_{Nu} = \frac{hD_p}{k} = 2 + 1.1 \left(\frac{D_p G}{\mu}\right)^{0.6} \left(\frac{C_p \mu}{k}\right)^{1/3} \quad (15\text{-}66)$$

are as follows: $\mu = 0.0000165$ Pa-s (kg/m-s); $\rho = 1.627$ kg/m³; $k = 0.0240$ W/m-K $= 0.024 \times 10^{-3}$ kJ/m-K-s; heat capacity at constant pressure = 31.45 kJ/kmol-K; molecular weight = $M = 29.52$; Thus, specific heat $C_P = 31.45/29.52 = 1.065$ kJ/kg-K. Also, gas mass velocity $G = 0.00352(29.52) = 0.1039$ kg/m²-s.

Assume $\psi = 0.65$; therefore, $D_p = 0.65(0.004) = 0.0026$ m. The diffusivity, D_i, of acetone in nitrogen at 297 K and 136 kPa is independent of composition and is 0.085×10^{-4} m²/s.

$$N_{Re} = D_p G/\mu = 0.0026(0.1039)/(0.0000165) = 16.4$$

$$N_{Sc} = \mu/\rho D_i = 0.0000165/(1.627)(0.0000085) = 1.19$$

$$N_{Pr} = C_P \mu/k = (1.065)(0.0000165)/(0.000024) = 0.73$$

From (15-63):
$$N_{Sh} = 2 + 1.1(16.4)^{0.6}(1.19)^{1/3} = 8.24$$

which from Figure 15.22 is well within the data range of the correlation. Thus, the mass-transfer coefficient for acetone is

$$k_{c_i} = N_{Sh}(D_i/D_p) = 8.24(0.0000085/0.0026)$$
$$= 0.027 \text{ m/s} = 0.088 \text{ ft/s}$$

From (15-66):

$$N_{Nu} = 2 + 1.1(16.4)^{0.6}(0.73)^{1/3} = 7.31;$$

$$h = N_{Nu}(k/D_p) = 7.31(0.0240/0.0026)$$

$$= 67.5 \text{ W/m}^2\text{-K} \quad \text{or } 11.9 \text{ Btu/h-ft}^2\text{-}°\text{F}$$

Internal Transport and Effective Pore Diffusivity

The largest transport-rate resistance to equilibrium solute partitioning arises from diffusion within tortuous, fluid-containing pores inside adsorbent particles or surface diffusion along surfaces lining the pores. Effective values of liquid diffusivity (14-14) or gas diffusivity (14-18) introduced for pore diffusion, D_p, in membranes may be used directly in (15-58). However, they neglect the possibility of an additional internal transport mechanism, surface diffusion. Schneider and Smith [80] suggested a modification to account for surface diffusion, $D_{i,surf}$, arising from the radial concentration gradient inside an adsorbent particle characterized by linear adsorption:

$$D_{e_i} = \frac{\epsilon_p^*}{\tau}\left[\left(D_i^{-1} + D_K^{-1}\right)^{-1} + D_{i,surf}\frac{\rho_p K_d}{\epsilon_p^*}\right] \quad (15\text{-}67)$$

Because intraparticle tortuosity, τ, is not necessarily the same for pore-volume diffusion as for surface diffusion, (15-67) must be used with caution, as discussed by Riekert [81]. Sladek, Gilliland, and Baddour [82] reported surface diffusivity values of physical adsorption of light gases in the range 5×10^{-3} to 10^{-6} cm²/s, with larger values associated with low differential heat of adsorption. They proposed correlating surface diffusivity for nonpolar adsorbates in cm²/s using

$$D_{i,surf} = 1.6 \times 10^{-2}\exp[-0.45(-\Delta H_{ads})/mRT] \quad (15\text{-}68)$$

where $m = 1$ for insulating adsorbents and $m = 2$ for conducting adsorbents (e.g., carbon). A detailed review of surface diffusion is given by Kapoor, Yang, and Wong [83].

EXAMPLE 15.9 Effective Diffusivity in Porous Silica Gel.

Porous silica gel of 1.0 mm particle diameter, with a particle density of 1.13 g/cm³, an inclusion porosity of 0.486, an average pore radius of 11 Å, and a tortuosity of 3.35 is to be used to adsorb propane from helium. At 373 K, diffusion in the pores is controlled by Knudsen and surface diffusion. Estimate the effective diffusivity. The differential heat of adsorption is −5,900 cal/mol. At 100°C, the adsorption-equilibrium constant, K_d (for a linear isotherm), is 19 cm³/g.

Solution

Pore diameter, $d_p, = 22$ Å $= 22 \times 10^{-10}$ m $= 22 \times 10^{-8}$ cm.

Molecular weight of propane, $M_i, = 44.06$.

From (14-21), propane Knudsen diffusivity is $D_K = 4,850$ $(22 \times 10^{-8})(373/44.06)^{1/2} = 3.7 \times 10^{-3}$ cm²/s. From (15-68), using $m = 1$, $D_s = 1.6 \times 10^{-2} \exp\{(-0.45)(5,900)/[(1)(1.987)(373)]\}$ $= 4.45 \times 10^{-4}$ cm²/s. Equation (15-67) reduces to $D_e = (\epsilon_p^*/\tau)D_K + (\rho_p K_d/\epsilon_p^*)D_{i,surf} = (0.486/3.35)(3.17 \times 10^{-3}) + (1.13)(19)(4.45 \times 10^{-4}/3.35 = 0.46 \times 10^{-3} + 2.85 \times 10^{-3} = 3.31 \times 10^{-3}$ cm²/s.

Experiments by Schneider and Smith [80] give a value of 1.22×10^{-3} cm²/s for D_e, with a value of 0.88×10^{-3} for the contribution of surface diffusion. Thus, the estimated contribution from surface diffusion is high, by a factor of about 3. In either case, the fractional contribution due to surface diffusion is large.

Measurement and analysis of D_e in adsorbent media is often undertaken to examine ways to reduce transport-rate resistance due to pore diffusivity. For example, Helfferich [84] reports that pore diffusion is controlling in ion-exchange beds at ion concentrations above 1.0 N, whereas resistance external to ion-exchange resins dominates below 0.01 N. Figure 15.23 shows concentration-independent self-diffusion coefficients of Na^+, Zn^{2+}, and Y^{3+} in a sulfonated styrene–divinylbenzene cation-exchange resin measured using isotope ions at 0.2 and 25°C. Values of self-diffusivity increase

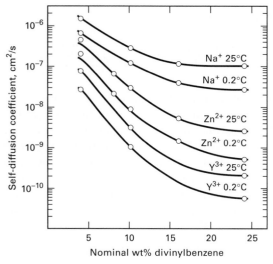

Figure 15.23 Self-diffusion coefficients for cations in a resin as a function of cross-linking with divinylbenzene.

[From B.A. Soldano, *Ann. NY Acad. Sci.*, **57**, 116 (1953) with permission [85].]

for smaller, monovalent cations at higher temperature in resins with lower-percentage cross-linking.

Kinetic Adsorption Rate

As shown by (15-58), chemisorption (bond formation) or orientation-controlled affinity interactions may be so slow as to contribute significantly to band broadening. Biological affinity processes, e.g., antibody–antigen complexation, often become the rate-limiting mass-transport process [86]. On the other hand, transport-rate resistance due to kinetic solute–adsorbent interaction is typically neglected for physical adsorption of gases and small solutes. Adsorption onto ion-exchange and hydrophobic-interaction media, for example, is considered instantaneous [87]. For fast adsorption kinetics, i.e., $k_a \gg k_c$, (15-58) reduces to

$$\frac{1}{k_{c,tot}} = \frac{1}{k_c} + \frac{R_p}{5\epsilon_p^* D_e} \qquad (15\text{-}69)$$

EXAMPLE 15.10 Resolving Power.

In nonadsorbing, size-exclusion chromatography (SEC), solutes are retained in a packed bed to the degree in which they penetrate porous stationary-phase media. Retention time varies approximately inversely with log(MW). SEC is often used to desalt, or exchange buffers, of protein solutions. A 10 mg/mL aqueous solution of bovine hemoglobin is to be purified by size-exclusion chromatography using a 46.5 (L) × 2.5 (D) cm i.d. glass column packed with 75-μm-diameter Toyopearl HW65C gel matrix. Intraparticle effective diffusivity, D_e, of hemoglobin is 3.0×10^{-7} cm²/s. Estimate the resolving power per unit thermodynamic driving force, R/δ, for this protein in this column at $u = 0.5$, 1.0, and 1.5 cm/min. What resolution, R, is expected from a protein of comparable diffusivity whose inclusion porosity, ϵ_p^*, is 111.4% of bovine hemoglobin? What size column would be required to obtain $R = 1$, as in Figure 15.21?

Solution

The resolving power, R, for a unit value of δ may be estimated from $N = LH$ using (15-57) to obtain the height equivalent to a theoretical plate, H. Calculation of H requires estimating coefficients associated with each transport-rate process, i.e., convective dispersion, E; boundary-layer transport, k_c; intraparticle effective diffusivity, D_e; and kinetic sorption, k_a, K_d. The necessary parameters, their sources, and corresponding values are tabulated below. Superscripts in the table correspond to appended notes that provide detailed descriptions and illustrative calculations for each step of the solution.

Parameter	Source	Parameter Value
Inclusion porosity of hemoglobin, ϵ_p^*	Data (Fig. 15.18)	0.498
Partition coefficient, α	(15-51)[1]	2.01
Void volume, ϵ_b	Data[2]	0.38
Fractional migration, velocity, ω	(15-54)[3]	0.552
Convective-dispersion coefficient, E	(15-61)[4]	$f\{N_{Re}N_{Sc}\}$[5]
Boundary-layer coefficient, k_c	(15-67)	$f\{N_{Re}N_{Sc}\}$[6]
Height equivalent to a theoretical plate, H	(15-57)	$f\{N_{Re}N_{Sc}\}$[7]
Resolution, R	(15-59)	[8]

Notes:

[1] Hemoglobin that does not adsorb in SEC has $K_d \approx 0$. So the partition coefficient may be calculated using (15-51) as

$$\alpha = \frac{1}{0.498} = 2.01$$

[2] A well-packed column has a void volume $\epsilon_b \approx 0.38$.

[3] The fractional migration velocity, ω, may be calculated using (15-54) as

$$\omega = \left(1 + \frac{1 - 0.38}{0.38} 0.498\right)^{-1} = 0.552$$

[4] The value of the convective-dispersion coefficient, $E = 2R_b u / N_{Pe_E}$, may be calculated using (15-61) to obtain N_{Pe}.

[5] For the conditions of this problem, values of E/uR_p that correspond to $u = 0.5$ cm/minute (at $N_{Re}N_{Sc} = 33.33$), $u = 1.0$ cm/minute ($N_{Re}N_{Sc} = 66.67$), and $u = 1.5$ cm/minute ($N_{Re}N_{Sc} = 100$) have been calculated and plotted as line B in Figure 15.20. It is not necessary to graphically obtain these values to find H using (15-57), since $H/2R_p$ is also shown in Figure 15.20 for the requested values of u.

[6] k_c for spherical, stationary-phase particles in the range $10^{-2} < N_{Re}N_{Sc} < 10^3$ may be estimated using [46]

$$N_{Sh} = \frac{2R_p k_c}{D_i} = \left[\left(\frac{1.09}{\epsilon_b}\right)^3 N_{Re}N_{Sc} + 4^3\right]^{1/3} \qquad (1)$$

Generally, a value of k_c obtained from (1) would be substituted into (15-68) to obtain $k_{c,tot}$, which would then be substituted into (15-57) to obtain H. However, it is not necessary to calculate k_c directly for this problem, since $H/2R_p$ is also shown in Figure 15.20 for requested values of u.

[7] The magnitude of $H/2R_p$ varies with $N_{Re}N_{Sc}$, as illustrated in Figure 15.20 (lower solid line). Reduced values of $h = H/2R_p$ from Figure 15.20 for $u = 0.5$ cm/minute (at $N_{Re}N_{Sc} = 33.33$), $u = 1.0$ cm/minute ($N_{Re}N_{Sc} = 66.67$), and $u = 1.5$ cm/minute ($N_{Re}N_{Sc} = 100$) are 4.3, 7.1, and 9.6, respectively. Therefore the corresponding values of

H (in μm) are 322.5, 532.5 and 720, respectively. From these values of H, using the definition, $N = L/H$, values of N at 0.5, 1.0, and 1.5 cm/minute are calculated, using $L = 46.5$ cm, to be 1440, 870, and 645, respectively. As an example:

$$N = \frac{L}{(2R_p)h} = \frac{465(1000)}{75(4.3)} = 1442 \quad (2)$$

[8] Using (15-59), resolution per unit driving force, R/δ, at 0.5, 1.0, and 1.5 cm/minute is then calculated to be 9.5, 7.4, and 6.4, respectively. As an example,

$$\frac{R}{\delta} = \frac{\sqrt{N}}{4} = \frac{\sqrt{1442}}{4} = 9.5 \quad (3)$$

Resolving hemoglobin from a protein with an inclusion porosity $\epsilon_p^* = 1.114(0.498) = 0.555$ gives a driving force $\delta = 0.05$. Using (15-59), this yields $R = 0.47$, 0.37, and 0.32 at 0.5, 1.0, and 1.5 cm/min, respectively. As an example,

$$R = \frac{\delta\sqrt{N}}{4} = 9.5(0.05) = 0.47 \quad (4)$$

From (15-59), to obtain $R = 1$ with $\delta = 0.05$, the number of theoretical stages would be $N = 6,400$. For $H = 320$ μm at 0.5 cm/minute, a column 206 cm long would be required.

§15.3.3 Biochromatography Adsorbents

Rapid advances in biochromatography for recovery of biopolymers rely on increased understanding of biomolecules and intermolecular forces to select or formulate suitable adsorbent stationary phases. Of major importance are resin particles of *silica* or *polymer* that are conjugated with chemistries that allow separation of bioproducts using ion-exchange and hydrophobic-interaction, affinity, reversed-phase, or size-exclusion chromatographies [88, 89]. Table 15.8 summarizes characteristics of these five classes of adsorptive phases. The thermodynamic basis for the physicochemical interaction associated with each adsorbent phase is provided in §2.9.2. A range of purification factors is achieved using four stationary phases common to biopharmaceutical fluid–solid separations. Table 15.9 shows examples of these factors.

With few exceptions (e.g., capillary columns), adsorbents used in protein chromatography are porous. Use of nonporous pellicular adsorbents in biochromatography is rare. Each of the five types of chromatographic adsorbents summarized in Table 15.8 is widely used as a resin in a packed bed, as illustrated in Figures 15.16 and 15.21.

Silica Resin

Silica resins ($\sim$1–25 μm) are incompressible at typical high-pressure liquid chromatography (HPLC) pressures, and have high internal surface areas for adsorption (100 to 1,500 m^2/g). But silica may irreversibly bind or denature some proteins and is unstable in common basic regenerants like 1.0-M NaOH (pH 14). Selective, reversible adsorption of hydrophilic species to uncoated silica is called *normal-phase* chromatography. More commonly, long-chain alkanes are bound to silica to selectively adsorb hydrophobic species (e.g., small molecules, peptides, proteins, and DNA) in *reversed-phase* chromatography. The bonded alkanes can form monolayers and be polymerized or end-capped. End-capping adds an organic layer after the bonding step to coat all bare silica surfaces. Excellent resolution makes reversed-phase chromatography attractive for species that do not lose activity upon interaction with bonded silica.

Polymer Resin

Low-cost, pH-stable polymer resins ($\sim$10–100 μm) are made by adding a cross-linking agent (e.g., bis-acrylamide for polyacrylamide resins) to an emulsion of polymer in an immiscible solvent. Styrene divinylbenzene (STDVB) forms a rigid, pH-stable, mildly hydrophobic backbone, primarily derivatized for ion exchange by reaction. Polyacrylamide (PA) forms a hydrogel useful in *size-exclusion* chromatography. Natural, hydrophilic, hydrogel-forming polymers like agarose, large-pore dextran, and microcrystalline cellulose can separate enzymes, antibodies, and virus by size exclusion as well as by interaction with derivatized phenol, antibodies, dyes, heavy metals, nonspecific ion-exchange groups, and biospecific epitopes. Substituted cross-linked agarose gels like Sepharose resist shrinking with changes in pH or ionic strength, I.

Ion-Exchange Chromatography

Table 15.10 identifies the type, group, approximate pK_a, and formula at physiological pH (§2.9.1) of the most common ion-exchange groups used to electrostatically bind small

Table 15.9 Biochromatographic Purification Factors

Type	Purification factor[1]	Examples
Biospecific affinity	50–10,000	Protein immunoglobulins
Dye affinity	10–100	Blue dextran/protein
Cation exchange	2–40	Cytochrome C
Size exclusion	2–20	Hemoglobin

[1]Purification factor was defined in §1.9.4

Table 15.8 Large-Scale Protein Chromatography

Type	Basis for Separation	Resolution/Speed/Capacity	Application
Ion exchange	Charge	High/high/high	Protein, whole virus
Hydrophobic interaction	Surface hydrophobicity	Good/good/high	Polypeptide
Affinity	Bioaffinity	Excels/high/high	Antibody–antigen, dye–ligand
Reversed phase	Surface hydrophobicity	Excels/high/high	DNA, plasmid
Size exclusion	Size	Moderate/good/low	Protein, plasmid, DNA

Table 15.10 Ion-Exchange Resins

Na$^+$ Type	Group	Formula	pK_a
Strong acid	Sulfopropyl, SP	SO_3^-	2
Weak acid	Carboxymethyl, CM	COO^-	4–4.5
Weak base	Diethylamino-ethyl, DEAE	$2C_2H_5N^+HC_2H_5$	9–9.5
Strong base	Quaternary ethyl amine, QEA	$4C_2H_5N^+$	12

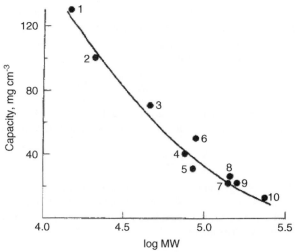

Figure 15.24 Capacity of CM-cellulose at $\alpha = 0$ for (1) hen egg white lysozyme; (2) AMP kinase; (3) phosphoglycerate kinase; (4) phosphoglycerate mutase; (5) creatine kinase; (6) enolase; (7) lactate dehydrogenase; (8) glyceraldehydes phosphate dehydrogenase; (9) aldolase; (10) pyruvate kinase.
[From Scopes [90] with permission.]

molecules (e.g., peptides, antibiotics), biopolymers (proteins, nucleic acids), and particles (virus). Derivatized ion-exchange groups exhibit typical *charge capacities* of 0.5 mmol/cm^3 (0.5 M). These charges are balanced by counterions that are displaced by a binding species. Adjacent DEAE substituents electrostatically repel protons. This lowers the local pK_a, which allows substantial titration around pH 6–8, and increases the pH in the micro-environment of the matrix relative to the surrounding buffer, which affects species solubility (see Donnan effects in §2.9.1). The effect occurs in the opposite direction for CM-cellulose. Protein stability, typically higher slightly above physiological pH than below, suggests using anion, rather than cation, exchangers for protein recovery. Dissolved nucleic acids may also interfere with cation exchange, unless first removed by polycation (protamine) precipitation.

To minimize interactions between buffer and ion-exchange groups, aqueous buffers used with an ion-exchange resin should possess the same sign in their charged form and contain only simple counterions (e.g., Na$^+$, K$^+$, Cl$^-$, CH$_3$COO$^-$). Anion exchange DEAE-cellulose, for example, could be used with Tris HCl buffer at pH 8.0 (see Table 2.13) since it contains HTris$^+$, Cl$^-$ counterion, and Tris (neutral) species. A dissolved, anionic protein neutralized by HTris$^+$ would displace Cl$^-$ associated with DEAE while discharging HTris-Cl, the acid salt of Tris. Every 1 mg/mL protein adsorbed releases 1 mM (millimolar) of buffer salt. From (2-115) and (2-120), *buffer salt discharge* can decrease pH and increase I of the mobile phase, reducing anticipated adsorption. Effects of buffer salt discharge are mitigated by using $\geq$ 10 mM of a suitable buffer [88] within 0.3 unit of its pK_a and at most 5 mg/mL of protein adsorbate.

Ion-exchange *adsorptive capacity* varies inversely with log(MW) until biological species are excluded from resin pores, as shown in Figure 15.24 for α (given by (15-51) = 0. Exclusion occurs at MW $\sim 10^6$ for cellulose and $\sim 10^7$ for Trisacryl. Increasing ionic strength, typically Na$^+$ or K$^+$ and Cl$^-$, up to $I \sim 0.5$ to 1.0 M, is usually used to elute adsorbed species by shielding electrostatic interactions between target and resin (2-135). Adjustment to pH is rarely used due to high matrix buffering capacities. However, continuously adding the acid form of a high-buffering ampholyte buffer, which has low ionic strength, to a column containing polyethyleneimine–agarose ion exchange, which allows continuous titration, yields a very steady pH gradient in which proteins emerge at or above their isolectric point (§2.9.1), a

technique called chromatofocusing [91]. Regeneration of ion-exchange resins by 1.0-M NaOH (pH 14) is common.

Reversed-Phase Chromatography

Hydrophobicity in reversed-phase change varies with the chain length and density of alkanes that are typically bonded to silica: octyl (C$_8$), octadecyl (C$_{18}$), phenyl (C$_6$H$_6$), and methyl (C$_1$). Steric effects yield exposed silica after bonding that can be covered by polymerizing alkyl chains at their attachment point or end-capping the exposed silica with methyl or ethyl groups. Anionic, strong-acid counterions like trifluoroacetate, acetate, or chloride that selectively partition with the targeted co-ion and alter its hydrophobicity are typically used to separate species in a biological mixture. Targeted nonpolar species are dissolved in aqueous mobile phase with minimal organic content added to promote phase interaction, and introduced to reversed-phase resins where they partition based on hydrophobic content, or mixed-mode interactions. The organic (e.g., acetonitrile, methanol, isopropanol) content is gradually increased in the mobile phase to selectively elute adsorbed species.

Hydrophobic-Interaction Chromatography

Proteins with low water solubility—like globulins, membrane-associated proteins, or enzymes that precipitate at 20–40% $(NH_4)_2SO_4$—adhere to polymer resins derivatized with C$_4$, C$_6$, C$_8$, and C$_{10}$; linear aliphatic chains; or phenyl moieties. Other candidates may bind hydrophobic resins at pH values close to their pK_a and/or at high I. In contrast to reversed-phase-type two-phase partitioning in hydrobic interactions, lyophilic salts strip proteins of solvated water, precipitating aggregation or nucleation of the protein onto the surface (see §2.9). Reducing salt content redissolves the adsorbed proteins. Reproducibility is sensitive to temperature, buffer type, salt used, and pH.

Affinity Chromatography

Biospecific interactions introduced in §2.9.3, like enzyme–ligand, enzyme–cofactor, receptor–agonist (antagonist), or antibody–antigen, are the basis for affinity chromatography. One member of the interacting pair is conjugated to a polymer resin to selectively bind the other from a biological mixture. Affinity ligands include starch for binding amylases and glycogen-metabolizing enzymes, cellulose for binding cellulases, and phosphocellulose for binding nucleic-acid binding proteins. *Pseudo-affinity* dye adsorbents like Cibacron Blue F3GA, an analog of ADP-ribose that binds purine-nucleotide-binding enzymes, or Procion Red H-E3B, which binds NADP-binding proteins, are also used. To avoid steric hindrance of affinity interaction, spacer arms and nonobstructing attachment methods are used. Interaction energies > 35 kJ/mol required for affinity binding typically require supplemental nonspecific hydrophobic interactions (see §2.9), usually provided by a hydrophobic spacer arm of hexamethylene, or equivalent. Elution by displacement with a compound that shows higher avidity to the binding ligand is superior to elution via nonspecific changes in I or pH. Column regeneration by changing I or pH is common. Though selectivity of affinity chromatography is excellent, the expense of procuring and derivatizing the conjugated epitope restricts its use to analytical applications such as affinity recognition of cloned epitopes and high-throughput screening, or to preparation of high-value-added bioproducts.

Affinity chromatography is often portrayed as a simple "lock-and-key" mechanism, e.g., between a receptor, -⟨, and a complementary target, ◊. However, the actual mechanistic interaction is much more complex, consisting of several steps that include electrostatic interactions, solvent displacement, steric selection, and charge and conformational rearrangement, as described in §2.9.3.

Immobilized Metal Affinity Chromatography

Electron-donor amino acid residues in proteins like histidine, tryptophan, and cysteine that are surface-accessible form metal coordination complexes with divalent transition metal ions like Ni^{2+}, Cu^{2+}, and Zn^{2+}. This complexation is the basis for immobilized metal affinity chromatography (IMAC) (see Example 2.14 and Table 2.18). Metalloproteins, which require metal centers for activity, are another target of metal ions immobilized with spacer arms to the resin. Complexation is typically enabled by conjugating iminodiacetic acid (IDA) or tris (carboxymethyl) ethylene diamine (TED) to a polymer gel via a spacer arm. The chelating IDA or TED is charged with a small, concentrated (50 mM) pulse of metal salt up to half the length of the column, to allow for metal-ion migration. The mobile phase contains ∼1-M salt to minimize nonspecific ion-exchange interactions and high pH to de-protonate donor groups on targeted proteins. Elution typically employs a stronger complexing agent such as imidazole

or glycine buffer at pH 9. Figure 2.23 shows the fundamental effect of separation distance on electrostatic interactions between adjacent particles like biomolecule and adsorbent.

Size-Exclusion Chromatography

Large molecules (MW ∼ 1–2 × 10⁶) that are excluded from the largest pores of underivatized polymer gels (like hydrophilic agarose and cross-linked dextran or hydrophobic polyacrylamide) elute from the column in the void volume. This volume, V_o, is 30 to 35% of total column volume, V_t. Smaller molecules, down to MW ∼ 1 × 10⁴, exhibit size- and shape-dependent permeability and elute in order of decreasing apparent size. Resins are available with pores that provide 90% exclusion of molecules whose volume is 5–6 times larger than those excluded from 10% of the bead volume. Size-exclusion chromatography, also called *gel permeation* or *molecular sieving*, is limited in capacity by lack of binding. Elution is typically isocratic, unless mixed-mode adsorption requires increasing I. Figures 15.18 and 15.20 illustrate isocratic elution of protein in size exclusion.

§15.3.4   Reducing Transport-Rate Resistances in the Bed: Scale-Up and Process Alternatives

Individual contributions from transport-rate resistances to theoretical plate height, H, illustrated in Figures 15.16 and 15.20, show that H in packed beds generally increases as operating velocity, u, (and throughput) rise. Examination of (15-57) and (15-58) reveals that smaller-diameter adsorbent particles decrease H and increase separation efficiency, primarily by reducing the resistance due to pore-volume diffusion. However, ∼2.5 μm is generally regarded as a practical lower limit for adsorbent radius, R_p, in high-pressure liquid chromatography (HPLC) systems, since pressure drop rises (1) in inverse proportion to decreases in R_p for highly turbulent packed-bed flow; and (2) in inverse proportion to decreases in R_p^2 for laminar packed-bed flow, as found in Ergun's equation (14-10).

Scale-Up

Sopher and Nystrom [92], Janson and Hedman [93], and Pharmacia [94] recommend scaling up chromatographic separations by maintaining H, u, and L while increasing volumetric throughput, mass loading, and gradient slope in proportion to an increase in column cross-sectional area. However, frictional forces from the column wall that support packed particles disappear below a length-to-diameter ratio of roughly 2.5 [95]. This causes settling and cracking in scaled-up packed beds. These phenomena were linked by Janson and Hedman [93, 96] and Love [97] to channeling and backmixing. Deterioration in large-scale packed beds may be counteracted using dynamic compression to increase packing homogeneity and long-term bed stability, while gradually increasing bed density, as reported by Guiochon and co-workers [60]. Alternatively, Grushka [98] and Wankat [99] suggest increasing the length-to-diameter aspect ratio during scale-up, which would require increasing R_p to lessen the rise in pressure resulting from lengthening the packed bed.

Process Alternatives to Chromatography

Transport-rate limitations in process-scale adsorption motivate examining alternatives to fluid–solid partitioning in packed beds that increase separation efficiency as well as throughput, particularly for preparative adsorption of high-value, high-molecular-weight biomolecules from liquid solutions. Growing demand for chromatography as a preparative tool in biotechnology [100–102] and pharmaceutical [103] applications has motivated development of perfusive [104, 105], "hyperdiffusive" [106], chromarod [107, 108] and adsorptive-membrane [109] medias to reduce or virtually eliminate mass-transfer resistance by intraparticle diffusion. Operating strategies have advanced to include counterflow, recycle [60], and displacement to more efficiently utilize chromatographic columns. Adsorptive-membrane separation and countercurrent contacting of bulk liquid and adsorptive-solid phases are two promising alternatives that reduce costs associated with adsorbent, regenerant, and solvent, and increase throughput. Adsorptive stacked membranes essentially eliminate transport-rate resistance due to intraparticle diffusion by derivatizing adsorptive sites on the surfaces of micron-scale, flow-through pores. Countercurrent contacting, which increases the local average thermodynamic driving force for equilibrium partitioning, is nearly achieved by timed-valve delivery to a modest number of packed-bed sections in simulated-moving-bed (SMB) operations, described in §15.4.

Adsorptive Membranes

Membrane adsorption typically utilizes a rigid cylindrical column, illustrated in Figure 15.25. Microporous, $\sim$200-μm-thick, hydrophilic, polymeric membrane rounds derivatized with interactive moieties are layered one on top of another and compression-gasketed at the periphery to prevent bypassing. The adsorptive-membrane cross section perpendicular to the flow direction is considerably longer than the flow path,

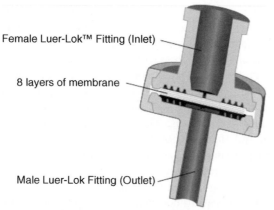

Female Luer-Lok™ Fitting (Inlet)

8 layers of membrane

Male Luer-Lok Fitting (Outlet)

Figure 15.25 Cut-away view of a ChromaSorb™ 0.08 mL Small-Scale Screening and Development Membrane Adsorber.

yielding up to 100-fold smaller back-pressures [110–113], 10-fold smaller process volumes, and shorter response times relative to packed beds [114]. This can increase throughput 10-fold or more [115, 116], and allow as much as 10-fold decreases in processing times, solvent and tankage requirements, and solute residence times [117]. Spiral-wound, hollow-fiber, cross-linked polymer rod, single-sheet, and radial-flow configurations are also used, with similar benefits.

Of the organic and inorganic membrane materials introduced in Chapter 14, adsorptive membranes are often composed of hydrophilic native or regenerated cellulose, reduced with borohydride to neutralize ion-exchange activity of residual carboxylic and aldehyde side groups, or with acrylic copolymers synthesized by free-radical polymerization of a mixture of monovinyl monomer, such as styrene or methacrylate, and divinyl monomer, such as divinylbenzene, in a heated mold [107, 108, 111, 112, 115]. Macroporous poly (glycidyl methacrylate-coethylene dimethacrylate) (GMA-EDMA) is a commonly used copolymer. Epoxy groups are modified to furnish functional-group sites for hydrophobic-interaction (HIC), ion-exchange (IEC), or affinity membrane adsorption [113].

In adsorptive membranes, the length scale for solute diffusion to an adsorptive site [R_p in (15-57) and (15-58)] is reduced to much less than the size of flow-through membrane pores ($\sim$1 μm) [118]. This allows adsorptive-membrane capacity to be maintained at substantially higher throughputs. Eliminating diffusional resistance reduces the expression for theoretical plate height in (15-57) to $H = 2E/u$, which is evaluated for $N_{Re}N_{Sc} > 1$ using an N_{Pe_E} correlation obtained by analysis of creeping flow in high-void-fraction, random configurations of fixed spheres [119]:

$$\frac{1}{N_{Pe}} = \frac{3}{8}\epsilon_b + \frac{\pi^2}{12}\epsilon_b(1 - \epsilon_b)\ln\left(\frac{N_{Re}N_{Sc}}{2}\right) + \frac{\epsilon_b}{N_{Re}N_{Sc}} \quad (15\text{-}70)$$

For use in (15-70), an equivalent mean particle diameter, D_p, for the membrane bed is estimated from its average pore size, d_p, and bulk porosity, ϵ_b, using $D_p = 3d_p(1 - \epsilon_b)/3\epsilon_b$.

Measured values of adsorptive-membrane plate heights and capacities are shown in Tables 15.11 and 15.12, respectively. van Deemter equations like (15-57) and van Deemter plots like Figure 15.20 show that values of plate heights first decrease and then increase as velocity increases. Variations in measured plate height from different sources, or from

Table 15.11 Experimental Adsorptive-Membrane Theoretical Plate Heights (H); from [118]

H Range, micron	Velocity, u, Range, cm/min
0.59–3.3	0.52–3.8
3–7	0.07–0.22
25	0.1–2
50–110	1.5–4
80–160	1–45
400	0.04–1
250–800	0.035–6.5

Table 15.12 Reported Capacity Values of Adsorptive Membranes for Several Biological Macromolecules: Monoclonal Antibody (MAb), Malate Dehydrogenase (Md), Human Serum Albumin (HSA), Ribonuclease (Rib), Lysozyme (Lys), Ovalbumin (Ova), Bovine Serum Albumin (BSA), Gamma-Globulin (G-G), Immunglobulin G (IgG), and a Mixture of IgG and IgA (BGG); from [118]

Membrane	Static Capacity (mg/mL)	Dynamic Capacity (mg/mL)
C-4	200–400	50
Cation exchange		50 (MAb)
Dye affinity	50.8 (Md)	45.7 (Md)
Anion exchange		5.8 (HSA)
Copolymer	20 (Rib), 26 (Lys), 47 (Ova)	5 (Rib), 0 (Lys), 5 (Ova)
Copolymer	40 (Ova)	
L-Phe affinity hollow fiber		50 mg BGG/g fiber
Anion exchange		30–40 g BSA/g membrane
Anion exchange	20	
Dye affinity	8.6 (Lys), 5.6 (BSA)	7.8 (Lys), 7.6 (BSA)
Protein-A/IgG affinity	4.74 (IgG-rabbit), 0.51 (protein A)	
Protein-A affinity	3.3 (G-G)	2.9 (G-G)
Ion exchange		8 IgG/cartridge

theoretical prediction, have been quantitatively shown to arise from differences in the parameters in (15-57), as well as from *external* contributions to plate height such as band broadening due to mixing in extra-column peripheral volumes such as injectors, detectors, tubing, and valves and non-uniform flow in adsorptive-membrane beds. Capacity may be measured experimentally and predicted using (15-51). Example 15.11 illustrates prediction of capacity for virus adsorption. Static capacity, measured under nonflowing (e.g., batch) conditions, typically exceeds dynamic capacity, measured under flowing conditions, since slow, diffusive, mass-transfer limits complete utilitization of surface area of the stationary phase.

EXAMPLE 15.11   Capacity of an Anion-Exchange Resin.

Adenovirus type 5 is a candidate viral vector for gene therapy. Estimate the static capacity of 15-μm-diameter SourceQ^{TM} anion-exchange resin for binding 120-nm-diameter adenovirus type 5. Compare the estimate with an experimental static capacity of $\sim 5 \times 10^{11}$ virus/mL reported for adenovirus on this resin, and with protein capacities of anion-exchange resins and membrane monoliths.

Solution

Using a packing factor of 0.547 for random sequential adsorption, an estimate for static capacity of 120-nm virions with a projected surface area of A_v adsorbed on total outer surface, A_p, of 15 μm SourceQ^{TM} beads packed to a void volume of 0.38 is given by

$$\frac{\text{virions}}{\text{mL resin}} = \frac{A_p}{\dfrac{1}{0.547}A_v}\frac{1-\epsilon_b}{V_p} = \frac{3(0.62)(0.547)}{R_v^2 R_p}$$

where A, V, and R are surface area, volume, and radius; and subscripts p and v represent resin particle and virus.

Substituting $R_v = 0.06 \times 10^{-4}$ cm and $R_p = 7.5 \times 10^{-4}$ into the above equation gives a static capacity of 1.2×10^{13} virions per mL. This value is 1/24th of the reported capacity value, suggesting negligible effective virus penetration into pores of the resin. Adenovirus is comprised of 87% protein and 13% nucleic acid with a total viral mass of 1.65×10^8 Da. Estimated static virus capacity is therefore

$$\frac{1.2 \times 10^{13} \text{ virions}}{\text{mL resin}}\left[\frac{1.65 \times 10^8 \text{ grams}}{6.02 \times 10^{23} \text{ virions}}\right] = 3.3 \text{ mg/mL}$$

This value is comparable to capacity of protein A-affinity interaction in Table 15.12. The experimentally reported capacity is 0.137 mg virus/mL resin, which is about an order of magnitude smaller than measured values for protein chromatography.

Static capacities reported for protein adsorption on several membrane monoliths measured 3.3 to $\sim$50 mg/mL, whereas typical chromatographic protein capacities are 25–60 mg/mL for Poros^{TM} and MonoQ^{TM} media, 110–115 mg/mL for HyperD$^{®}$ resin, and 300 mg/mL for soft Sephadex$^{®}$.

Counterflow

Reductions in counterflow solvent and adsorbent usage relative to packed-bed adsorption at comparable purities may be evaluated using an equilibrium-stage description of steady-state counterflow introduced by Kremser [120], as described in §9.2, and Souders and Brown [121]. Klinkenberg et al. [123] specialized this description for continuous counter-current adsorption of solute 1 and 2 from small feed mass flow rate, F, relative to pure fluid (U) and solid-phase (S) mass flow rates into an M-stage enricher and an N-stage stripper, respectively, separated by a feed stage, e, in a model column, shown in Figure 15.26. The fluid-phase mass fraction of solute i exiting the stripper, $y_{N,i}$, and the solid-phase mass fraction exiting the enricher, $x_{1,i}$, are related recursively to

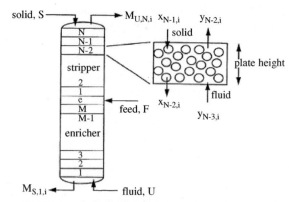

Figure 15.26 Equilibrium-stage representation of ideal steady counterflow. The enricher and stripper consist of M and N stages, respectively. Feed enters stage e. Pure solid (S) and fluid (U) streams enter the stripper and enricher, respectively. Mass flow rates of component i exiting the column from the stripper, $M_{U,N,i}$, and the enricher, $M_{S,N,i}$ are shown. In the column, component i enters and exits a representative stage $j = N - 1$ of height H_{SC} with mass fraction values in the lower solid phase, $x_{j,i}$ and upper fluid phase, $y_{j,i}$, as shown.

the mass fraction of i at a central feed stage, $y_{e,i}$, by

$$y_{N,i} = \frac{y_{e,i}}{\sum_{j=0}^{N} \Gamma_i^{-j}} \qquad (15\text{-}71)$$

$$x_{1,i} = \frac{y_{e,i}}{\alpha_i \sum_{j=0}^{M} \Gamma_i^{j}} \qquad (15\text{-}72)$$

where the extraction ratio, $\Gamma_i = (\alpha_i \rho_s / \rho_u) |U/S|$, summarizes the phase-partitioning of solute between steady mass flow rates of fluid eluent, $U = (1 - r)\epsilon_b VT$, and solid adsorbent, $S = -r(1 - \epsilon_b)VT$. Equation (15-51) defines the solute partition coefficient, α_i. Each stage, j, has volume V, which is transferred at a rate T, and r is the fractional relative motion of the solid phase, i.e., the modulus of the ratio of solid-phase to fluid-phase velocities. Fluid mass flow rate U is related to the countercurrent interstitial liquid velocity, u_{SC}, by $U = A_{SC}\epsilon_b u_{SC}\rho_u$, where A_{SC} is the counterflow-column cross-sectional area. Equilibrated fluid- and solid-phase mass fractions exiting the jth stage are related by $y_{j,i} = \alpha_i x_{j,i}\rho_s/\rho_u$. The total number of counterflow stages, $N_{SC,tot} = N + M + 1$, required to attain a fractional purity, $P_{U,N,i}$, of species i exiting the column at stage N in fluid-phase U in an optimal binary split is given by [63].

$$N_{SC,tot} = \frac{4}{\delta} \ln\left(\frac{P_{U,N,i}}{1 - P_{U,N,i}}\right) - 1 \qquad (15\text{-}73)$$

where δ is given by (15-60) and, in general, $P_{p,j,i} = M_{p,j,i}/(M_{p,j,i} + M_{p,j,not\_i})$, for mass flow rate $M_{p,j,i}$ of solute i leaving stage j in phase p, is the product of solute mass fraction and its corresponding phase flow rate i.e., $M_{U,j,i} = Uy_{j,i}$. Fractional purity, P, obtained in differential adsorption may be evaluated in terms of resolution, R, in (15-59) as $P = \Phi\{2R\}$, where Φ is the cumulative distribution function,

which may be obtained from statistical tables, e.g., Hogg and Ledolter [124]. Comparing (15-73) and (15-59) reveals that the number of counterflow stages required for a selected separation increases proportionally to δ^{-1}, whereas the number of chromatographic stages required in differential operation increases proportionally to δ^{-2}. For close separations, i.e., $\delta \ll 1$, the higher local driving force in counterflow results in (requires) significantly fewer theoretical stages.

Relative Solvent and Adsorbent Usage

Using (15-52)–(15-56), (15-59)–(15-60), and (15-71)–(15-73), the relative solid phase required for steady counterflow (SC) relative to differential chromatography to achieve comparable purities can be evaluated using [63]:

$$\frac{N_{SC,tot}A_{SC}H_{SC}}{NAH} = \frac{3}{32}\frac{U}{F}\frac{\delta^3}{R^4}\ln\left[\frac{\Phi\{2R\}}{\Phi\{-2R\}}\right] \qquad (15\text{-}74)$$

Solid-phase savings using counterflow increases for high resolution of closely related solutes, e.g., racemic mixtures. The physical basis for this decreased adsorbent usage appears when Figure 15.21 is compared with Figure 15.27: 5% of the chromatographic bed actively separates 1 from 2, whereas 100% of the counterflow column actively separated 1 from 2. In Figure 15.27, the profiles for the easier and the more difficult separation appear consistent because (1) unity resolution is specified in both cases and (2) *scaled axes* are used to represent the data. Equation (15-73) shows that to achieve the same final purity, a 100-fold increase of N_{SC} is required for the more difficult ($\delta = 0.00209$) separation.

The counterflow solvent volume, $V_{solv,SC}$, required relative to differential chromatography (DC) may be estimated as

$$\frac{V_{solv,SC}}{V_{solv,DC}} = \frac{2U}{F\sqrt{N}} \qquad (15\text{-}75)$$

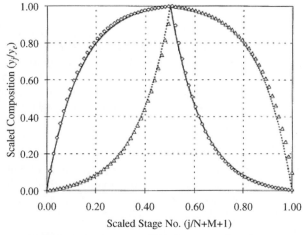

Figure 15.27 Relative concentrations in counterflow for two separations. In the easier ($\delta = 0.209$) separation, solute 1 (Δ) partitions more to the upper fluid phase while 2 ($\Diamond$) tends to the lower solid phase. In the more difficult ($\delta = 0.00209$) separation, solute 1 (---) and solute 2 (-) partition likewise. Stage numbers $M + N$ are selected to resolve 1 and 2 to unity ($P_{U,1} = 0.9772$).

Solvent savings also increases for high purity separations, but less dramatically than adsorbent savings. While adsorbent- and solvent-usage requirements are lower in steady counterflow for high-resolution purification of closely related solutes, separation is limited to binary separations. Like distillation, counterflow separations effect a binary split between key components. Additional components in a multi-component feed that partition more strongly to the solid or liquid phases are separated more efficiently, but may present additional complexity in terms of solid- or fluid-phase regeneration in real systems. In packed-bed adsorption, the number of species that can be resolved in batch or semi-batch operation is limited only by the magnitude of the thermodynamic driving force that distinguishes partitioning of the respective solutes and the separation efficiency of the system.

EXAMPLE 15.12 Steady Counterflow Separation of Albumins.

Bovine serum albumin ($\epsilon_p^* = 0.30$) and ovalbumin ($\epsilon_p^* = 0.34$) in an equimolar mixture are to be purified by size exclusion to a resolution of $R = 1.0$ in a packed bed of Toyopearl® (TSK gel) HW55F, which has a bed porosity of $\epsilon_b = 0.34$. The differential migration velocity, δ, for these two solutes is 0.048. Assume a volumetric feed stream/fluid-phase dilution in counterflow of 10. Determine the relative number of equilibrium stages, the relative solvent requirement, and the relative solid-phase requirement for steady counterflow separation relative to differential chromatography.

Solution

Using tables for a cumulative normal distribution, $P = \Phi\{2R\}$ for $R = 1$ corresponds to mutual fractional purity, $P_{S,1,OA} = P_{U,N,BSA} = 0.9772$. Substituting $R = 1$ and $P_{U,N,BSA}$ into (15-59) and (15-73), respectively, and using $\delta = 0.048$ yields $N_{SC,tot}/N = 313/6,975$. Using (15-74) with $U/F = 10$ and $N = 6,975$ stripping stages shows that only 24% as much solvent is needed for steady counterflow relative to differential chromatography. Substituting values for U/F, $P_{U,N,BSA}$, and δ into (15-73) indicates that 0.04% as much adsorbent might be expected for steady counterflow relative to differential chromatography.

§15.3.5 Mitigating Transport-Rate Resistances: Frontal Loading

Selective partitioning of a solute from mobile fluid to the stationary adsorbent phases may be used to saturate a packed bed by continuous addition of a mobile feed at a volumetric flow rate Q_F containing dilute solute at concentration c_F in *frontal loading mode*, also referred to as *percolation* or simply *fixed-bed adsorption*. Frontal loading concentrates dilute solute, since $c_s > c_b \gg c_f \leq c_F$ usually characterizes loading, and reduces dilution caused by transport-rate resistances, which spread an initially sharp solute pulse $c_f\{z,0\}$ in (15-52) across a bed volume of ~ 3.3 s $= 3.3 N^{1/2} AH/f_s u$ given by (15-56). Subsequent application of a fluid eluent at a thermodynamic state (temperature, pressure, composition) that favors desorption can selectively recover concentrated solute at $c_f > c_F$.

Ideal Fixed-Bed Adsorption

Ideal (local-equilibrium) fixed-bed adsorption represents the limiting case of: (1) neglible external and internal transport-rate resistances; (2) ideal plug flow; and (3) adsorption isotherm beginning at the origin. Local equilibrium between fluid and adsorbent is thus achieved instantaneously, resulting in a shock-like *stoichiometric front*, shown in Figure 15.28, that moves at a constant velocity throughout the bed.

The bed is divided into two zones or sections: (1) Upstream of the stoichiometric front, fluid-phase solute concentration, c_f, equals the feed concentration, c_F, and *spent* adsorbent is saturated with adsorbate at a loading c_b^* in equilibrium with c_F. The length (height) and weight of this section are LES and WES, respectively, where ES refers to the equilibrium section, called the equilibrium zone. (2) Downstream of the stoichiometric front and in the exit fluid, $c_f = 0$, the adsorbent is adsorbate-free. The length and weight of this section are LUB and WUB, respectively, where UB refers to unused bed.

After a stoichiometric time t_s, the stoichiometric wave front reaches the end of the bed; the value of $c_{f,\text{out}}$ abruptly rises to the inlet value, c_F; no further adsorption is possible; and the adsorption step is terminated. This point is referred to as the breakpoint and the stoichiometric wave front becomes the ideal breakthrough curve. For ideal adsorption in a packed bed of length L_B, the location of the concentration wave front $L_{\text{ideal}} \leq L_B$ in Figure 15.28, as a function of time, is obtained by equating the solute entering in the feed to that in the adsorbate:

$$Q_F c_F t_{\text{ideal}} = \bar{c}_b^* A (1 - \epsilon_b) L_{\text{ideal}} \qquad (15\text{-}76)$$

where c_b^* is the loading in equilibrium with c_F, and A is the bed cross-sectional area. Defining the total mass of adsorbent

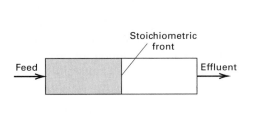

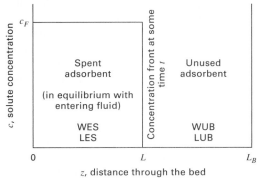

Figure 15.28 Stoichiometric (equilibrium) concentration front for ideal fixed-bed adsorption.

in the bed by $S = A(1 - \epsilon_b)L_B$ and rearranging (15-76) gives, for ideal fixed-bed adsorption corresponding to Figure 5.28,

$$L_{ideal} = LES = \frac{Q_F c_F t_{ideal}}{\bar{c}_b^* S} L_B \qquad (15\text{-}77)$$

$$LUB = L_B - LES \qquad (15\text{-}78)$$

$$WES = S \frac{LES}{L_B} \qquad (15\text{-}79)$$

$$WUB = S - WES \qquad (15\text{-}80)$$

Solute Concentration Distributions in Frontal Loading

Actual solute concentration distributions during frontal loading are not ideal, but may be obtained from (15-48)–(15-51) by superimposing solutions of the form (15-53) using Green's functions [125]. This produces *concentration profiles* for c_f, illustrated in Figure 15.29a, that are broadened by transport-rate resistances summarized in (15-57) and (15-58).

The c_f profiles in 15.29a are normalized relative to feed concentration, c_F, and plotted as a function of axial distance, z, within the column at successive times t_1, t_2, and t_b after loading begins. A corresponding S-shaped *breakthrough curve* for c_f/c_F, shown in Figure 15.29b, is plotted as a function of time, t, at the column outlet, $z = L_B$. In Figure 15.29a,

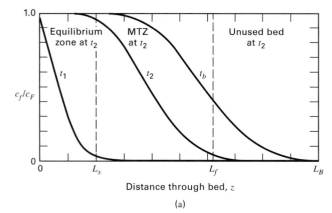

(a)

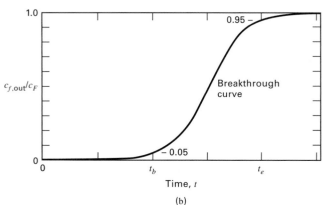

(b)

Figure 15.29 Solute wave fronts in a fixed-bed adsorber with mass-transfer effects. (a) Concentration-distance profiles. (b) Breakthrough curve.

at t_1, no part of the bed is saturated. At t_2, the bed is almost saturated for a distance L_s. At L_f, the bed is almost clean. Beyond L_f, little mass transfer occurs at t_2 and the adsorbent is still unused. The region between L_s and L_f is called the mass-transfer zone, MTZ, at t_2, where adsorption takes place. Because it is difficult to determine where the MTZ zone begins and ends, L_f can be taken where $c_f/c_F = 0.05$, with L_s at $c_f/c_F = 0.95$. From time t_2 to time t_b, the S-shaped front moves through the bed.

At the breakthrough point, t_b, the leading point of the MTZ just reaches the end of the bed. Feeding is discontinued at t_b to prevent loss of unadsorbed, dilute solute, whose outlet concentration begins to rise rapidly. Rather than using $c_f/c_F = 0.05$, the breakthrough concentration can instead be taken as the minimum detectable or maximum allowable solute concentration in the effluent fluid. When feeding inadvertently continues after t_b, the time to reach $c_{f,out}/c_F = 0.95$ is designated t_e.

Analytical Solution

For a single solute and an initially clean bed free of solute adsorbate, Anzelius [126] obtained an analytical solution from (15-48)–(15-51) and (15-69) for frontal loading, neglecting axial dispersion, that is summarized by Ruthven [10] and discussed by Klinkenberg [127], who provided this useful approximation for solute concentration distribution with respect to axial distance and time [126]:

$$\frac{c_f}{c_F} \approx \frac{1}{2}\left[1 + \text{erf}\left(\sqrt{\tau} - \sqrt{\xi} + \frac{1}{8\sqrt{\tau}} + \frac{1}{8\sqrt{\xi}}\right)\right] \qquad (15\text{-}81)$$

where $\text{erf}\{x\}$ is the error function defined in (3-76) and ξ and τ are dimensionless distance and displacement-corrected time coordinates, respectively, given by

$$\xi = \frac{3k_{c,tot}z}{R_p u}\left(\frac{1 - \epsilon_b}{\epsilon_b}\right) \qquad (15\text{-}82)$$

$$\tau = \frac{3\alpha k_{c,tot}}{R_p}\left(t - \frac{z}{u}\right) \qquad (15\text{-}83)$$

where $3/R_p = a_v$ is the surface area per unit volume for a sphere, and resistances due to external transport, pore diffusivity, and kinetics (if present) are included in $k_{c,tot}$, as shown in (15-58). For gas separations at low loadings, $a_v k_{c,tot}$ may be represented by the experimentally obtained product kK, where K is an equilibrium constant defined by (15-16) equivalent to α^{-1} in (15-51), and k is an overall mass-transfer coefficient obtained from experimental data. This approximation is accurate to < 0.6% error for $\xi > 2.0$. Klinkenberg [128] also provided an approximate solution for profiles of solute concentration in equilibrium with the average sorbent loading:

$$\frac{c_f^*}{c_F} = \frac{\bar{c}_b}{\bar{c}_b^*} \approx \frac{1}{2}\left[1 + \text{erf}\left(\sqrt{\tau} - \sqrt{\xi} - \frac{1}{8\sqrt{\tau}} - \frac{1}{8\sqrt{\xi}}\right)\right] \qquad (15\text{-}84)$$

where $c_f^* = \bar{c}_b \alpha$ and $\bar{c}_b^*$ is the loading in equilibrium with c_F.

EXAMPLE 15.13   Breakthrough Curves Using the Klinkenberg Equations.

Air at 70°F and 1 atm, containing 0.9 mol% benzene, enters a fixed-bed adsorption tower at 23.6 lb/minute. The tower is 2 ft in inside diameter and packed to a height of 6 ft with 735 lb of 4 × 6 mesh silica gel (SG) particles with a 0.26-cm effective diameter and an external void fraction of 0.5. The adsorption isotherm for benzene has been determined to be linear for the conditions of interest:

$$q = Kc^* = 5{,}120c^* \qquad (1)$$

where q = lb benzene adsorbed per ft^3 of silica gel particles, and c^* = equilibrium concentration of benzene in the gas, in lb benzene per ft^3 of gas.

Mass-transfer experiments simulating conditions in the 2-ft-diameter bed have been fitted to a linear-driving-force (LDF) model:

$$\frac{\partial \bar{q}}{\partial t} = 0.206K(c - c^*) \qquad (2)$$

where time is in minutes and 0.206 is the constant k in minute^{-1}, which includes resistances both in the gas film and in the adsorbent pores, with the latter resistance dominant.

Using the approximate concentration-profile equations of Klinkenberg [127], compute a set of breakthrough curves and the time when the benzene concentration in the exiting air rises to 5% of the inlet. Assume isothermal, isobaric operation. Compare breakthrough time with time predicted by the equilibrium model.

Solution

For the equilibrium model, the breakthrough curve is vertical, and the bed becomes completely saturated with benzene at c_F.

MW of entering gas = 0.009(78) + 0.991(29) = 29.44.

Density of entering gas = (1)(29.44)/(0.730)(530) = 0.076/lb/ft^3.

Gas flow rate = 23.6/0.0761 = 310 ft^3/minute.

Benzene flow rate in entering gas = $\dfrac{(23.6)}{29.44}(0.009)(78)$

$\qquad\qquad = 0.562$ lb/minute and

$$c_F = \frac{0.562}{310} = 0.00181 \text{ lb benzene/ft}^3 \text{ of gas}$$

From (1),

$$q = 5{,}120(0.00181) = 9.27 \frac{\text{lb benzene}}{\text{ft}^3 \text{ SG}}$$

The total adsorption of benzene at equilibrium

$$= \frac{9.27(3.14)(2)^2(6)(0.5)}{4} = 87.3 \text{ lb}$$

Time of operation $= \dfrac{87.3}{0.562} = 155$ minutes

For the actual operation, taking into account external and internal mass-transfer resistances, and replacing $a_v k_{c,tot}$ in (15-82) and (15-83) with kK obtained from experimental data,

$$\xi = \frac{(0.206)(5{,}120)z}{u}\left(\frac{1 - 0.5}{0.5}\right) = 1{,}055\, z/u$$

$$u = \text{interstitial velocity} = \frac{310}{0.5\left(\dfrac{3.14 \times 2^2}{4}\right)} = 197 \text{ ft/min} \qquad (3)$$

$$\xi = \frac{1{,}055}{197}z = 5.36z, \text{ where } z \text{ is in ft.}$$

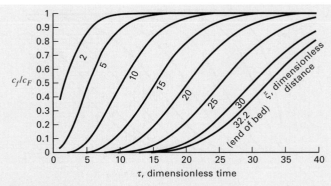

Figure 15.30 Gas concentration breakthrough curves for Example 15.13.

When z = bed height = 6 ft, ξ = 32.2 and

$$\tau = 0.206\left(t - \frac{z}{197}\right) \qquad (4)$$

For t = 155 minutes (the ideal time), and z = 6 ft, using (4), τ = 32.

Thus, breakthrough curves should be computed from (15-81) for values of τ and ξ no greater than about 32. For example, when ξ = 32.2 (exit end of the bed) and τ = 30, which corresponds to a time t = 145.7 minutes, the concentration of benzene in the exiting gas, from (15-81), is

$$\frac{c}{c_F} = \frac{1}{2}\left[1 + \text{erf}\left(30^{0.5} - 32.2^{0.5} + \frac{1}{8(30)^{0.5}} + \frac{1}{8(32.2)^{0.5}}\right)\right]$$

$$= \frac{1}{2}[1 + \text{erf}(-0.1524)] = \frac{1}{2}\text{erfc}(0.1524)$$

$$= 0.4147 \text{ or } 41.47\%$$

This far exceeds the specification of c/c_F = 0.05, or 5%, at the exit. Thus, the time of operation of the bed is considerably less than the ideal time of 155 minutes. Figure 15.30 shows breakthrough curves computed from (15-84) over a range of the dimensionless time, τ, for values of the dimensionless distance, ξ, of 2, 5, 10, 15, 20, 25, 30, and 32.2, where the last value corresponds to the bed exit. For c/c_F = 0.05 and ξ = 32.2, τ is seen to be nearly 20.

From (4), with z = 6 ft, the time to breakthrough is $t = \frac{20}{0.206} + \frac{6}{197} = 97.1$ minutes, which is 62.3% of the ideal time.

Figure 15.29 or (15-84) can be used to compute the bulk concentration of benzene at various locations in the bed for τ = 20. The results are as follows:

ξ	z, ft	c/c_F
2	0.373	1.00000
5	0.932	0.99948
10	1.863	0.97428
15	2.795	0.82446
20	3.727	0.53151
25	4.658	0.25091
30	5.590	0.08857
32.2	6.000	0.05158

At τ = 20, the adsorbent loading, at various positions in the bed, can be computed from (15-84), using q = 5,120c. The maximum loading corresponds to c_F. Thus, q_{max} = 9.28 lb benzene/ft^3 of SG. Breakthrough curves for the solid loading are plotted in Figure 15.31. As expected, those curves are displaced to the right from the curves of Figure 15.30. At τ = 20:

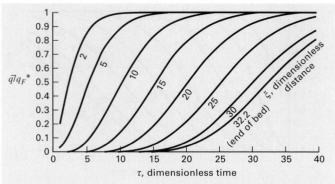

Figure 15.31 Adsorbent loading breakthrough curves for Example 15.13.

ξ	z, ft	$\dfrac{c^*}{c_F} = \dfrac{\bar{q}}{q_F^*}$	$\bar{q}, \dfrac{\text{lb benzene}}{\text{ft}^3\text{SG}}$
2	0.373	0.99998	9.28
5	0.932	0.99883	9.27
10	1.863	0.96054	8.91
15	2.795	0.77702	7.21
20	3.727	0.46849	4.35
25	4.658	0.20571	1.909
30	5.590	0.06769	0.628
32.2	6.000	0.03827	0.355

Values of $\bar{q}$ are plotted in Figure 15.32 and integrated over the bed length to obtain the average bed loading:

$$\bar{q}_{\text{avg}} = \int_0^6 \bar{q}\,dz / 6$$

The result is 5.72 lb benzene/ft^3 of SG, which is 61.6% of the maximum loading based on inlet benzene concentration.

If the bed height were increased by a factor of 5, to 30 ft, $\xi = 161$. The ideal time of operation would be 780 minutes or 13 h. With mass-transfer effects taken into account as before, the dimensionless operating time to breakthrough is computed to be $\tau = 132$, or breakthrough time is

$$t = \frac{132}{0.206} + \frac{30}{197} = 641 \text{ minutes}$$

which is 82.2% of the ideal time. This represents a substantial increase in bed utilization.

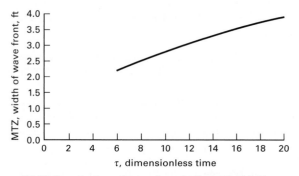

Figure 15.33 Broadening of wave front in Example 15.13.

Favorable Adsorption Isotherms Sharpen Breakthrough

Broadening of the wave front in Example 15.13 due to transport-rate resistance is summarized in Figure 15.33 by plotting the MTZ width for $0.95 \geq c_f/c_F \geq 0.05$ versus dimensionless time τ up to a value of 20, where the front breaks through the 6-ft-long bed. MTZ broadening increases from ~2 feet at $\tau = 6$ to ~4 feet at $\tau = 20$. The rate of broadening slows as τ increases; however, broadening in a deeper bed persisted even at $\tau = 100$. This is typical of frontal loading performed with a linear adsorption isotherm (curve A in Figure 15.34a) or with an unfavorable Type III isotherm (curve C in Figure 15.34a). On the other hand, a favorable Type I Langmuir or Freundlich isotherm (curve B in Figure 15.34a) rapidly diminishes wave-front broadening to produce a "self-sharpening" wave front, as illustrated in Figure 15.29. This has been evaluated by DeVault [129] and others. Solute velocity at the concentration wave front, u_c, within a packed bed of significant capacity, i.e., $K_d \gg 1$, is obtained by substituting (15-51) into (15-54) and multiplying by u:

$$u_c = \frac{u}{1 + \dfrac{1 - \epsilon_b}{\epsilon_b}\epsilon_p^* K_d} \qquad (15\text{-}85)$$

Solute velocity in (15-85) is relatively small at lower values of c_f (i.e., higher K_d) in Figure 15.34a, curve B, but increases as c_f increases. Thus, lagging wave-front regions at higher solute concentration move faster than leading wave-front regions at lower solute concentrations, as shown in Figure 15.34b. Self-sharpening of breakthrough curves via nonlinear Type I adsorption isotherms mitigates broadening due to consecutive transport-rate resistances and thus allows more adsorptive bed capacity to be efficiently utilized.

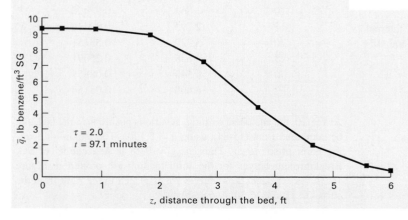

Figure 15.32 Adsorbent loading profile for Example 15.13.

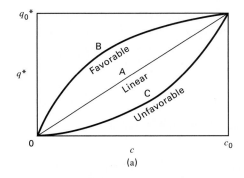

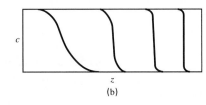

Figure 15.34 Effect of shape of isotherm on sharpness of concentration wave front. (a) Isotherm shapes. (b) Self-sharpening wave front caused by a favorable adsorption isotherm.

Scale-Up Using Constant-Pattern Front

Persistent transport-rate resistance eventually limits self-sharpening, and an asymptotic or *constant-pattern front* (CPF) is developed. For such a front, MTZ becomes constant and curves of c_f/c_F and $\bar{c}_b/\bar{c}_b^*$ become coincident. The bed depth at which CPF is approached depends upon the non-linearity of the adsorption isotherm and the importance of adsorption kinetics. Cooney and Lightfoot [130] proved the existence of an asymptotic wave-front solution, including effects of axial dispersion. Initially, the wave front broadens because of mass-transfer resistance and/or axial dispersion. Analytical solutions for CPF from Sircar and Kumar [131] and a rapid approximate method based on Freundlich and Langmuir isotherms from Cooney [132] are available to estimate CPF concentration profiles and breakthrough curves using mass-transfer and equilibrium parameters.

When the constant-pattern-front assumption is valid, it can be used to determine the length of a full-scale adsorbent bed from breakthrough curves obtained in small-scale laboratory experiments. This widely used technique is described by Collins [133] for purification applications.

Total bed length is taken to be

$$L_B = \text{LES} + \text{LUB} \tag{15-86}$$

the sum of the length of an ideal, equilibrium-adsorption section, LES, unaffected by mass-transfer resistance

$$\text{LES} = \frac{c_F Q_F t_b}{q_F \rho_b A} \tag{15-87}$$

where Q_F is the volumetric feed flow rate, plus a length of unused bed, LUB, determined by the MTZ width and the c_f/c_F profile within that zone, using

$$\text{LUB} = \frac{L_e}{t_s}(t_s - t_b) \tag{15-88}$$

where L_e/t_s = the ideal wave-front velocity. The stoichiometric time t_s divides the MTZ (e.g., CPF zone) into equal areas A and B as shown in Figure 15.35, and L_e/t_s corresponds to the ideal wave-front velocity. Alternatively, t_s, may be determined using

$$t_s = \int_0^{t_e} \left(1 - \frac{c_f}{c_F}\right) dt \tag{15-89}$$

For example, if t_s equalizes areas A and B when it is equidistant between t_b and t_e, then LUB = MTZ/2. A conservative estimate of MTZ = 4 ft may be used in the absence of experimental data.

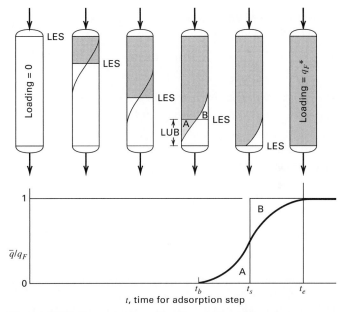

Figure 15.35 Determination of bed length from laboratory measurements.

EXAMPLE 15.14 Scale-Up for Fixed-Bed Adsorption.

Collins [133] reports the experimental data below for water-vapor adsorption from nitrogen in a fixed bed of 4A molecular sieves for bed depth = 0.88 ft, temperature = 83°F, pressure = 86 psia, G = entering gas molar velocity = 29.6 lbmol/h-ft^2, entering water content = 1,440 ppm (by volume), initial adsorbent loading = 1 lb/100 lb sieves, and bulk density of bed = 44.5 lb/ft^3. For the entering gas moisture content, c_F, the equilibrium loading, q_F, = 0.186 lb H$_2$O/lb solid.

c_{exit}, ppm (by volume)	Time, h	c_{exit}, ppm (by volume)	Time, h
<1	0–9.0	650	10.8
1	9.0	808	11.0
4	9.2	980	11.25
9	9.4	1,115	11.5
33	9.6	1,235	11.75
80	9.8	1,330	12.0
142	10.0	1,410	12.5
238	10.2	1,440	12.8
365	10.4	1,440	13.0
498	10.6		

Determine the bed height required for a commercial unit to be operated at the same temperature, pressure, and entering gas mass velocity and water content to obtain an exiting gas with no more than 9 ppm (by volume) of water vapor with a breakthrough time of 20 h.

Solution

$$c_F = \frac{1,440(18)}{106} = 0.02592 \text{ lb } H_2O/\text{lbmol } N_2$$

$$G = \frac{Q_F}{\pi D^2/4} = 29.6 \text{ lbmol } N_2/\text{h-ft}^2 \text{ of bed cross-section}$$

Initial moisture content of bed = 0.01 lb H_2O/lb solid

From (15-87), revised for a gas flow rate based on the lbmol of N_2 instead of the volume in ft^3 of N_2,

$$\text{LES} = \frac{(0.02592)(29.6)(20)}{(0.186 - 0.01)(44.5)} = 1.96 \text{ ft}$$

Use the integration method to obtain LUB. From the data:

Take $t_e = 12.8$ h (1,440 ppm) and $t_b = 9.4$ h (9 ppm).

By numerical integration of breakthrough-curve data, using (15-88): $t_s = 10.93$ h.

From (15-88),

$$\text{LUB} = \left(\frac{10.93 - 9.40}{10.93}\right)(0.88) = 0.12 \text{ ft}$$

From (15-86), $L_B = 1.96 + 0.12 = 2.08$ ft, or a bed utilization of $\frac{1.96}{2.08} \times 100\% = 94.2\%$.

Alternatively, an approximate calculation can be made. Let t_b, the beginning of breakthrough, be 5% of the final ppm, or 0.05 (1,440) = 72 ppm. Using the experimental data, this corresponds to $t_b = 9.76$ h. Let t_e, the end of breakthrough, be 95% of the final ppm, or 0.95(1,440) = 1,370 ppm, corresponding to $t_e = 12.25$ h.

Let t_s = the midpoint or (9.76 + 12.25)/2 = 11 h. The ideal wavefront velocity = L_e/t_s = 0.88/11 = 0.08 ft/h. From (15-87), LUB = 0.08(11 − 9.76) = 0.1 ft. MTZ = 0.2 ft and $L_B = 1.96 + 0.1 = 2.06$ ft.

§15.3.6 Multicomponent Differential Chromatography

Most separation systems discussed thus far perform a binary split between key components (e.g., LK and HK in distillation). Chromatography can separate multicomponent mixtures into 2+ products, whose elution time, $t = z_L/f_s u$, increases proportionally to relative thermodynamic partitioning of each species from moving fluid to stationary adsorptive phase (15-51). Transport-rate processes (e.g., 15-57 and 15-58) dilute the product concentration and mix adjacent solutes. Figure 15.36 illustrates differential chromatography, also called batch or elution chromatography, in which a feed mixture insufficient to load the sorbent is pulsed into the feed end of the column.

An eluant carrier gas or solvent moving at a constant plug-flow interstitial velocity u that has little or no affinity for the sorbent moves the fraction of solute desorbed into the fluid phase (15-54) along the length of the column at solute wave (migration) velocity $u_c = \omega u$, given in (15-85), as solute readsorbs and desorbs in succession via mass action. Transport-rate processes broaden each solute peak as it proceeds along the column. The solute migration velocity, given by (15-85), is smaller for solutes with higher affinity, i.e., smaller α corresponding to a larger K_d in (15-51). Initially, overlapping solute peaks are gradually separated by the fractional difference in migration velocities of adjacent, noninteracting

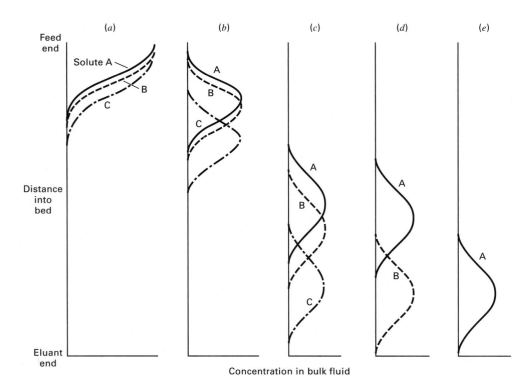

Figure 15.36 Movement of concentration waves during separation in a chromatographic column.

species (15-60) to ultimately achieve a resolution, R, given in (15-59), degraded only by transport-rate resistances that broaden each peak, as summarized in (15-56)–(15-58). The area under each solute peak is proportional to the mass of the component represented by the peak.

Equilibrium Wave Pulse Theory for Linear Isotherms

Solutes pulsed into a plug-flow column with negligible external and internal transport-rate resistances and linear isotherms partition instantaneously between moving fluid and stationary adsorbent phases. This yields square waves for concentration distributions, whose leading and trailing edges move as a stoichiometric front at a constant interstitial fluid velocity given by (15-85), as shown in Figure 15.37a–c, which is used extensively by Wankat [134]. Weakly adsorbed solutes, e.g., A in Figure 15.37a, which have small $K_d \ll 1$ values, progress quickly toward the column outlet at $u_c \sim u$. Figure 15.37b shows that the time for each pulse to move through the column is $z_L/\omega u$. Figure 15.37c shows the product concentration ratios c_f/c_F for A and B, respectively, at the end of the bed as a function of time. Estimates of separation achievable in pulse chromatography are available from this simple equilibrium wave pulse theory. Unfortunately, the estimates are not conservative when computing the necessary column length, because the waves broaden, as shown after the following example.

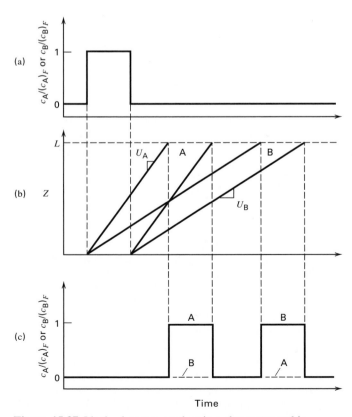

Figure 15.37 Ideal solute wave pulses in a chromatographic column.

EXAMPLE 15.15   Separation by Pulse Chromatography.

An aqueous solution of 3 g/cm³ each of glucose (G), sucrose (S), and fructose (F) is to be separated in a chromatographic column packed with an ion-exchange resin of the calcium form. For the expected solute concentrations, sorption isotherms are independent with $q_i = K_i c_i$, where q_i is in grams sorbent per 100 cm³ resin and c_i is in grams solute per 100 cm³ solution. From experiment:

Solute	K
Glucose	0.26
Sucrose	0.40
Fructose	0.66

The superficial solution velocity, $u_s = \epsilon_b u$, is 0.031 cm/s and bed void fraction is 0.39. If a 500-second feed pulse, t_P, is followed by elution with pure water, what length of column packing is needed to separate the three solutes if sorption equilibrium is assumed? How soon after the first pulse begins can a second 500-second pulse begin?

Solution

Interstitial solution velocity $= u_s/\epsilon_b = 0.031/0.39 = 0.0795$ cm/s. Wave (migration) velocity for glucose from (15-85):

$$u_G = \frac{0.0795}{1 + \left(\dfrac{1 - 0.39}{0.39}\right)(0.26)} = 0.0565 \text{ cm/s}$$

Similarly, $u_S = 0.0489$ cm/s and $u_F = 0.0391$ cm/s.

The smallest difference in wave velocities is between glucose and sucrose, so the separation between these two waves determines the column length. The minimum column length, assuming equilibrium, corresponds to the time at which the trailing edge of the glucose wave pulse, together with the leading edge of the sucrose wave pulse, leaves the column. Thus, if t_P is the duration of the first pulse and L is the length of the packing:

$$t_P + \frac{L}{u_G} = \frac{L}{u_S} \tag{1}$$

Thus,

$$500 + \frac{L}{0.0565} = \frac{L}{0.0489}$$

Solving, $L = 182$ cm. The glucose just leaves the column at $500 + \dfrac{182}{0.0565} = 3{,}718$ s.

The locations of the three wave fronts in the column at 3,718 s are shown in Figure 15.38.

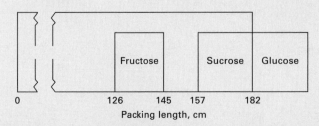

Figure 15.38 Locations of solute waves of first pulse for Example 15.15 at 3,718 s.

The time at which the second pulse begins is determined so that the trailing edge of the first fructose wave pulse just leaves the column as the second pulse of glucose begins to leave the column. This time, based on the fructose, is $500 + 182/0.0391 = 5,155$ s. It takes the leading edge of a glucose wave $182/0.0565 = 3,220$ s to pass through the column. Therefore, the second pulse can begin at $5,155 - 3,220 = 1,935$ s. This establishes the following ideal cycle: pulse: 500 s, elute: 1,435 s, pulse: 500 s, elute: 1,435 s, etc. In the next example, a more realistic case is given, where account is taken of mass-transfer resistance. The column will be longer.

Differential Chromatographic Pulses

A plot of solute concentrations in the elutant as a function of time is a *chromatogram*. When mass-transfer resistances, axial dispersion, and/or other nonideal phenomena are not negligible (almost always the case), the solute concentrations in a chromatogram will not appear as square waves, but will exhibit the wave shapes in Figure 15.36. Carta [135] developed an analytical solution for chromatographic response to periodic, rectangular feed pulses of duration t_F separated by an elution period t_E neglecting axial dispersion and slow adsorption using the LDF approximation and linear solute isotherms. For each solute in the feed, he obtained an analytical solution to (15-48) by neglecting axial dispersion, and using periodic rectangular feed pulses of duration t_F, followed by an elution period of duration t_E. Accordingly, his boundary conditions are:

Initial condition: At $t = 0$, $c_i\{z, 0\} = 0$

Feed pulse: At $z = 0$, $c_i\{0, t\} = (c_i)_F$ for $(j - 1)$ $(t_F + t_E) < t < [j(t_F + t_E) - t_E]$

Elution period: At $z = 0$, $c_i\{0, t\} = 0$ for $j(t_F + t_E) - t_E < t < j(t_F + t_E)$

where $j = 1, 2, 3, \ldots$ is an index that accounts for the periodic nature of the feed and elution pulses. Thus, with $j = 1$, the feed pulse takes place from $t = 0$ to $t = t_F$ and the elution pulse is from t_F to $t_F + t_E$.

Carta solved the linear system of equations for $X = c_f/c_F$ for each solute by the Laplace transform method to obtain the following series solution, in terms of dimensionless parameters.

$$X = \frac{r_F}{2r} + \frac{2}{\pi} \sum_{m=1}^{\infty} \left[\frac{1}{m} \exp\left(-\frac{m^2 n_f}{m^2 + r^2}\right) \sin\left(\frac{m\pi r_F}{2r}\right) \right.$$
$$\left. \times \cos\left(\frac{m\theta_f}{r} - \frac{m\pi r_F}{2r} - \frac{m\beta n_f}{r} - \frac{mr n_f}{m^2 + r^2}\right) \right] \tag{15-90}$$

where,

$$r = \frac{k\alpha}{2\pi}(t_F + t_E) \tag{15-91}$$

$$r_F = \frac{k\alpha}{\pi} t_F \tag{15-92}$$

$$n_f = \frac{(1 - \epsilon_b)kz}{\epsilon_b u} \tag{15-93}$$

$$\theta_f = \alpha k t \tag{15-94}$$

$$\beta = \frac{\alpha \epsilon_b}{1 - \epsilon_b} \tag{15-95}$$

$$k = \frac{3k_{c,tot}}{R_p} \tag{15-96}$$

This solution is not applicable for nonlinear isotherms like those of Freundlich (15-35) and Langmuir (15-36). However, the method of lines, using 5-point, biased, upwind, finite-difference approximations, as described in the section on thermal-swing adsorption below, can be applied to obtain numerical solutions. In (15-96), internal resistance (pore diffusivity) is accounted for in $k_{c,tot}$, as shown by its definition in (15-58).

EXAMPLE 15.16 Calculation of a Chromatogram.

Use Carta's equation with the following properties to compute the chromatogram for the conditions of Example 15.21 with a packing length of 182 cm, $\epsilon_b = 0.39$, $R_p = 0.0025$ cm, $u = 0.0795$ cm/s, $z = 182$ cm, $t_E = 2,000$ s, and $t_F = 500$ s. Does a significant overlap of peaks result?

Property	Glucose	Sucrose	Fructose
$1/\alpha$	0.26	0.40	0.66
D_e, cm^2/s	1.1×10^{-8}	1.8×10^{-8}	2.8×10^{-8}
k_c, cm/s	5.0×10^{-3}	5.0×10^{-3}	5.0×10^{-3}

Solution

Values of k and the computed dimensionless parameters from (15-90) to (15-95) are as follows:

	Glucose	Sucrose	Fructose
r	40.22	42.66	40.06
r_F	16.09	17.07	16.03
n_f	94.13	153.6	238.0
θ_f	$0.1011\,t$	$0.1072\,t$	$0.1007\,t$
β	2.459	1.598	0.9687
k, s^{-1}	0.0263	0.0429	0.0665

where t is in seconds.

Values of $X = c/c_F$ are computed with these parameters using (15-90) for values of time, t, in the neighborhood of times for the equilibrium-based waves. The resulting chromatogram for glucose is shown in Figure 15.39a, compared to the equilibrium rectangular wave (shown as a dashed line) determined in Example 15.15 using (15-85). The areas under the two curves should be identical. The equilibrium-based wave appears to be centered in time within the mass-transfer-based wave.

In Figure 15.39b, the complete computed chromatogram is plotted for the three carbohydrates. It is seen that the effect of mass transfer is to cause the peaks to overlap significantly. To obtain a sharp separation, it is necessary to lengthen the column or reduce the feed pulse time, t_F.

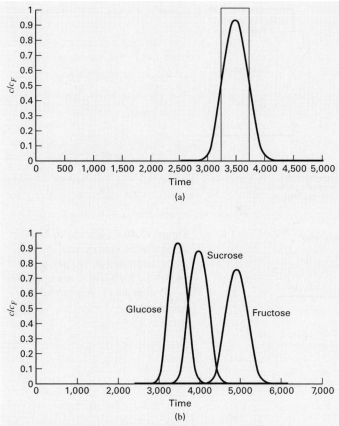

Figure 15.39 Computed chromatograms for Example 15.16.
(a) Comparison of ideal to nonideal wave for fructose.
(b) Computed chromatogram for nonideal eluant.

§15.4 EQUIPMENT FOR SORPTION OPERATIONS

A variety of configurations and operating procedures are employed for commercial, sorption-separation equipment due mainly to the wide range of sorbent particle sizes and the need, in most applications, to regenerate the solid sorbent.

§15.4.1 Adsorption

For adsorption, widely used equipment and operations are listed in Table 15.13. For analysis purposes, the listed devices

are classified into the three operating modes in Figure 15.40. In 15.40a, a powdered adsorbent such as activated carbon, of $d_p < 1$ mm, is added with water, to an agitated tank to form a slurry. The internal resistance to mass transfer within the pores of small particles is small. However, even with good stirring, the external resistance to mass transfer from bulk liquid to the external surface of the adsorbent particles may not be small because small particles tend to move with the liquid. Thus, the rate of adsorption may be controlled by external mass transfer. The main application of this operation is removal of small amounts of large, dissolved molecules, such as coloring agents, from water. Spent adsorbent, which is removed from the slurry by sedimentation or filtration, is discarded because it is difficult to desorb large molecules. The slurry system, also called *contact filtration*, can be operated continuously.

The fixed-bed, cyclic-batch operating mode, shown in Figure 15.40b, is widely used with both liquid and gas feeds. Adsorbent particle sizes range from 0.05 to 1.2 cm. Bed pressure drop decreases with increasing particle size, but solute transport rate increases with decreasing particle size. The optimal particle size is determined mainly from these two considerations. To avoid jiggling, fluidizing the bed, or blowing out fines during adsorption, the feed flow is often downward. For removal of small amounts of dissolved hydrocarbons from water, spent adsorbent is removed from the vessel and reactivated thermally at high temperature or discarded. Fixed-bed adsorption, also called *percolation*, is used for removal of dissolved organic compounds from water. For purification or bulk separation of gases, the adsorbent is almost always regenerated in-place by one of the five methods listed in Table 15.13 and considered next.

In *thermal (temperature)-swing-adsorption* (TSA), the adsorbent is regenerated by desorption at a temperature higher than used during adsorption, as shown in Figure 15.41. Bed temperature is increased by (1) heat transfer from heating coils located in the bed, followed by pulling a moderate vacuum or (2) more commonly, by heat transfer from an inert, nonadsorbing, hot purge gas, such as steam. Following desorption, the bed is cooled before adsorption is resumed. Because heating and cooling of the bed requires hours, a typical cycle time for TSA is hours to days. Therefore, if the quantity of adsorbent in the bed is to be reasonable, TSA is practical only for purification involving small adsorption rates.

Table 15.13 Common Commercial Methods for Adsorption Separations

Phase Condition of Feed	Contacting Device	Adsorbent Regeneration Method	Main Application
Liquid	Slurry in an agitated vessel	Adsorbent discarded	Purification
Liquid	Fixed bed	Thermal reactivation	Purification
Liquid	Simulated moving bed	Displacement purge	Bulk separation
Gas	Fixed bed	Thermal swing (TSA)	Purification
Gas	Combined fluidized bed–moving bed	Thermal swing (TSA)	Purification
Gas	Fixed bed	Inert-purge swing	Purification
Gas	Fixed bed	Pressure swing (PSA)	Bulk separation
Gas	Fixed bed	Vacuum swing (VSA)	Bulk separation
Gas	Fixed bed	Displacement purge	Bulk separation

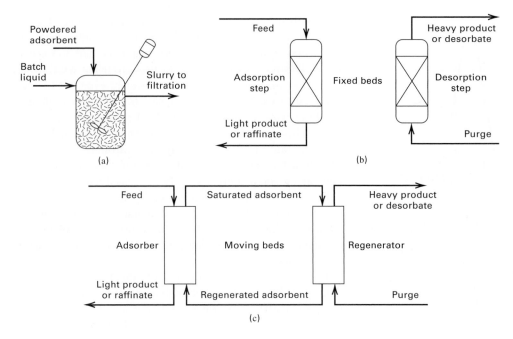

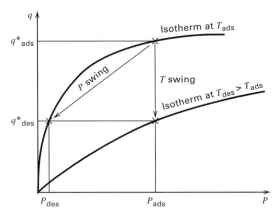

Figure 15.40 Contacting modes for adsorption and ion exchange
(a) Stirred-tank, slurry operation.
(b) Cyclic fixed-bed, batch operation.
(c) Continuous countercurrent operation.

Figure 15.41 Schematic representation of pressure-swing and thermal-swing adsorption.

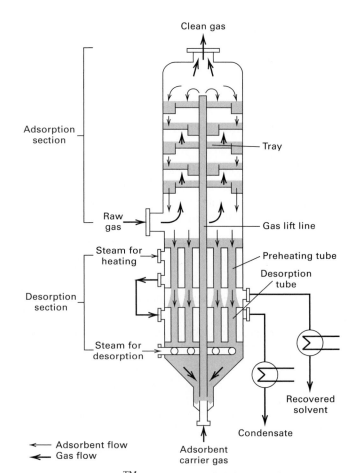

A fluidized bed can be used instead of a fixed bed for adsorption and a moving bed for desorption, as shown in Figure 15.42, provided that particles are attrition-resistant. In the adsorption section, sieve trays are used with raw gas passing up through the perforations and fluidizing the adsorbent. The fluidized particles flow like a liquid across the tray, into the downcomer, and onto the tray below. In the food industry, this type of tray is rotated. From the adsorption section, the solids pass to the desorption section, where, as moving beds, they first flow down through preheating tubes and then through desorption tubes. Steam is used for indirect heating in both sets of tubes and for stripping in the desorption tubes. Moving beds, rather than fluidized beds on trays, are used in desorption because the stripping-steam flow rate is insufficient for fluidizing the solids. At the bottom of the unit, the regenerated solids are picked up by a carrier gas, which flows up through a gas-lift line to the top, where the solids settle out on the top tray to repeat the adsorption cycle. Keller [136] reports that this configuration, which was announced in 1977, is used in more than 50 units worldwide to remove small amounts of solvents from air. Other applications of

Figure 15.42 Purasiv™ process with a fluidized bed for adsorption and moving bed for desorption.

[From G.E. Keller, "Separations: New Directions for an Old Field," *AIChE Monograph Series*, **83** (17) (1987) with permission.]

TSA include removal of moisture, CO_2, and pollutants from gas streams.

In an *inert-purge-swing* regeneration, desorption is at the same temperature and pressure as the adsorption step, because the gas used for purging is nonadsorbing (inert) or only weakly adsorbing. This method is used only when the solute is weakly adsorbed, easily desorbed, and of little or no value. The purge gas must be inexpensive so that it does not have to be purified before recycle.

In *pressure-swing adsorption* (PSA), adsorption takes place at an elevated pressure, whereas desorption occurs at near-ambient pressure, as shown in Figure 15.41. PSA is used for bulk separations because the bed can be depressurized and repressurized rapidly, making it possible to operate at cycle times of seconds to minutes. Because of these short times, the beds need not be large even when a substantial fraction of the feed gas is adsorbed. If adsorption takes place at near-ambient pressure and desorption under vacuum, the cycle is referred to as *vacuum-swing adsorption* (VSA). PSA and VSA are widely used for air separation. If a zeolite adsorbent is used, equilibrium is rapidly established and nitrogen is preferentially adsorbed. Nonadsorbed, high-pressure product gas is a mixture of oxygen and argon with a small amount of nitrogen. Alternatively, if a carbon molecular-sieve adsorbent is used, the particle diffusivity of oxygen is about 25 times that of nitrogen. As a result, the selectivity of adsorption is controlled by mass transfer, and oxygen is preferentially adsorbed. The resulting high-pressure product is nearly pure nitrogen. In both cases, the adsorbed gas, which is desorbed at low pressure, is quite impure. For the separation of air, large plants use VSA because it is more energy-efficient than PSA. Small plants often use PSA because that cycle is simpler.

In *displacement-purge* (*displacement-desorption*) *cycles*, a strongly adsorbed purge gas is used in desorption to displace adsorbed species. Another step is then required to recover the purge gas. Displacement-purge cycles are viable only where TSA, PSA, and VSA cannot be used because of pressure or temperature limitations. One application is separation of medium-MW linear paraffins (C_{10}–C_{18}) from branched-chain and cyclic hydrocarbons by adsorption on 5A zeolite. Ammonia, which is separated from the paraffins by flash vaporization, is used as purge.

Most commercial applications of adsorption involve fixed beds that cycle between adsorption and desorption. Thus, compositions, temperature, and/or pressure at a given bed location vary with time. Alternatively, a continuous, countercurrent operation, where such variations do not occur, can be envisaged, as shown in Figure 15.40c and discussed by Ruthven and Ching [137]. A difficulty with this scheme is the need to circulate solid adsorbent in a moving bed to achieve steady-state operation. The first commercial application of countercurrent adsorption and desorption was the moving-bed Hypersorption process for recovery, by adsorption on activated carbon, of light hydrocarbons from various gas streams in petroleum refineries, as discussed by Berg [138]. However, only a few units were installed because of problems with adsorbent attrition, difficulties in regenerating the adsorbent when heavier hydrocarbons in the feed gas were adsorbed, and unfavorable economics compared to distillation. Newer adsorbents with a much higher resistance to attrition and possible applications to more difficult separations are reviving interest in moving-bed units.

A successful countercurrent system for commercial separation of liquid mixtures is the simulated-moving-bed system, shown as a hybrid system with two added distillation columns in Figure 15.43, and known as the UOP Sorbex process. As described by Broughton [139], the bed is held stationary in one column, which is equipped with a number (perhaps 12) of liquid feed entry and discharge locations. By shifting, with a rotary valve (RV), the locations of feed entry, desorbent entry, extract (adsorbed) removal, and raffinate (nonadsorbed) removal, countercurrent movement of solids is simulated by a downward movement of liquid. For the valve positions shown in Figure 15.43, Lines 2 (entering desorbent), 5 (exiting extract), 9 (entering feed), and 12 (exiting raffinate) are operational, with all other numbered lines closed. Liquid is circulated down through and, externally, back up to the top of the column by a pump. Ideally, an infinite number of entry and exit locations exist and the valve would continuously change the four operational locations. Since this is impractical, a finite number of locations are

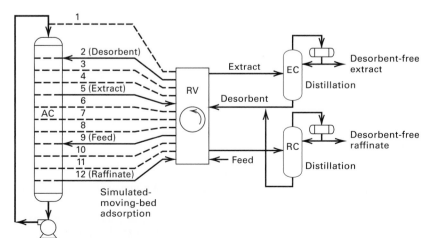

Figure 15.43 Sorbex hybrid simulated-moving-bed process for bulk separation. AC, adsorbent chamber; RV, rotary valve; EC, extract column; RC, raffinate column.

[From D.B. Broughton, *Chem. Eng., Progress,* **64** (8), 60–65 (1968) with permission.]

used and valve changes are made periodically. In Figure 15.43, when the valve is moved to the next position, Lines 3, 6, 10, and 1 become operational. Thus, raffinate removal is relocated from the bottom of the bed to the top of the bed. Thus, the bed has no top or bottom. Gembicki et al. [140] state that 78 Sorbex-type commercial units were installed during 1962–1989 for the bulk separation of *p*-xylene from C8 aromatics; *n*-paraffins from branched and cyclic hydrocarbons; olefins from paraffins; *p*- or *m*-cymene (or cresol) from cymene (or cresol) isomers; and fructose from dextrose and polysaccharides. Humphrey and Keller [141] cite 100 commercial Sorbex installations and more than 50 different demonstrated separations.

§15.4.2 Ion Exchange

Ion exchange, shown in Figure 15.40, employs the same modes of operation as adsorption. Although use of fixed beds in a cyclic operation is most common, stirred tanks are used for batch contacting, with an attached strainer or filter to separate resin beads from the solution after equilibrium is approached. Agitation is mild to avoid resin attrition, but sufficient to achieve suspension of resin particles.

To increase resin utilization and achieve high efficiency, efforts have been made to develop continuous, countercurrent contactors, two of which are shown in Figure 15.44. The Higgins contactor [142] operates as a moving, packed bed by using intermittent hydraulic pulses to move incremental portions of the bed from the ion-exchange section up, around, and down to the backwash region, down to the regenerating section, and back up through the rinse section to the ion-exchange section to repeat the cycle. Liquid and resin move countercurrently. The Himsley contactor [143] has a series of trays on which the resin beads are fluidized by upward flow of liquid. Periodically the flow is reversed to move incremental amounts of resin from one stage to the stage below. The batch of resin at the bottom is lifted to the wash column, then

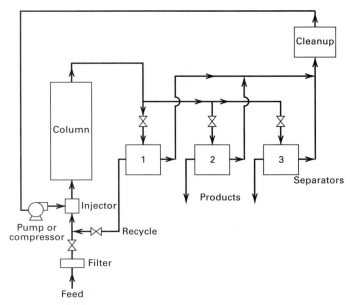

Figure 15.45 Large-scale, batch elution chromatography process.

to the regeneration column, and then back to the top of the ion-exchange column for reuse.

§15.4.3 Chromatography

Operation modes for industrial-scale chromatography are of two major types, as discussed by Ganetsos and Barker [144]. The first, and most common, is a transient mode that is a scaled-up version of an analytical chromatograph, referred to as large-scale, batch (or elution) chromatography. Packed columns of diameter up to 4.6 m and packed heights to 12 m have been reported. As shown in Figure 15.45 and discussed by Wankat in Chapter 14 of a handbook edited by Rousseau [9], a recycled solvent or carrier gas is fed continuously into a sorbent-packed column. The feed mixture and recycle is

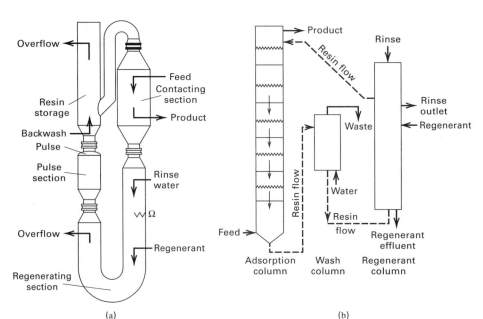

Figure 15.44 Continuous countercurrent ion-exchange contactors.
(a) Higgins moving packed-bed process.
(b) Himsley fluidized-bed process.

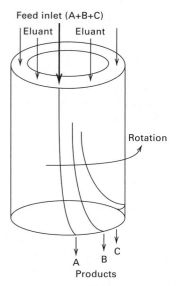

Figure 15.46 Rotating, crosscurrent, annular chromatograph.

pulsed into the column by an injector. A timer or detector (not shown) splits the column effluent by residence time, sending it to different separators (condensers, evaporators, distillation columns, etc.). Each separator is designed to remove a particular feed component from the carrier fluid. An additional cleanup step is required to purify the carrier fluid before it is recycled to the column. Separator 1 produces no product because it handles an effluent pulse containing carrier fluid and two or more feed components, which are recovered and recycled to the column. Thus, the batch chromatograph operates somewhat like a batch-distillation column, producing a nearly pure cut for each component in the feed and slop cuts for recycle. The system shown in Figure 15.45 is designed to separate a binary system. If, say, three more separators are added, the system can separate a five-component feed into five products.

The second major large-scale chromatograph is a countercurrent-flow or simulated-moving-bed mode unit used for adsorption. This mode is more efficient, but more complicated, and can separate a mixture into only two products. A third mode is the continuous, crosscurrent (or rotating) chromatograph, first conceived by Martin [145] and shown in Figure 15.46. The packed annular bed rotates slowly about its axis, past the feed-inlet point. *Eluant* (solvent or carrier gas) enters the top of the bed uniformly over the entire cross-sectional area. Both feed and eluant are fed continuously and are carried downward and around by bed rotation. Because of different selectivities of sorbent for feed components, each traces a different helical path since each spends a different amount of time in contact with sorbent. Thus, each component is eluted from the bottom of the packed annulus at a different location. In principle, a multicomponent feed can be separated continuously into nearly pure components following separation of the carrier fluid from each eluted fraction. Units of up to 12 inches in diameter have successfully separated sugars, proteins, and metallic elements.

§15.5 SLURRY AND FIXED-BED ADSORPTION SYSTEMS

Design procedures are presented and illustrated by example in this section for most of the common sorption operations, including three modes of slurry adsorption; thermal-swing adsorption; pressure-swing adsorption; continuous, countercurrent adsorption; simulated-moving-bed systems; and an ion-exchange cycle.

§15.5.1 Slurry Adsorption (Contact Filtration)

Three modes of adsorption from a liquid in an agitated vessel are of interest. First is the *batch mode*, in which a batch of liquid is contacted with a batch of adsorbent for a period of time, followed by discharge of the slurry from the vessel, and filtration to separate solids from liquid. The second is a *continuous mode*, in which liquid and adsorbent are continuously added to and removed from the agitated vessel. In the third, *semi-batch* or *semicontinuous mode*, liquid is continuously fed, then removed from the vessel, where it is contacted with adsorbent, which is retained in a contacting zone of the vessel until it is nearly spent.

Design models for batch, continuous, and semicontinuous modes are developed next, followed by an example of their applications. In all models, the slurry is assumed to be perfectly mixed in the turbulent regime to produce a fluidized-like bed of sorbent. Perfect mixing is approached by using a liquid depth of from one to two vessel diameters, four vertical wall baffles, and one or two marine propellers or pitched-blade turbines on a vertical shaft. With proper impeller speed, axial flow achieves complete suspension. For semicontinuous operation, a clear liquid region is maintained above the suspension for liquid withdrawal.

Because small particles are used in slurry adsorption and the relative velocity between particles and liquid in an agitated slurry is low (small particles tend to move with the liquid), the rate of adsorption is assumed to be controlled by external, rather than internal, mass transfer.

Batch Mode

The rate of solute adsorption, as controlled by external mass transfer, is

$$-\frac{dc}{dt} = k_L a(c - c^*) \qquad (15\text{-}97)$$

where c is the concentration of solute in the bulk liquid; c^* is the concentration in equilibrium with the adsorbent loading, q; k_L is an external liquid-phase mass-transfer coefficient; and a is external surface area of adsorbent per unit volume of liquid. Starting from feed concentration, c_F, the instantaneous bulk concentration, c, at time t, is related to the instantaneous adsorbent loading, q, by material balance

$$c_F Q = cQ + qS \qquad (15\text{-}98)$$

where the adsorbent is assumed to be initially free of adsorbate, Q is the liquid volume (assumed to remain constant for dilute feeds), and S is the mass of adsorbent. Equilibrium concentration, c^*, is given by an appropriate adsorption isotherm: a linear isotherm, Langmuir isotherm (15-36), or Freundlich isotherm (15-35). For example, a rearrangement of the latter gives

$$c^* = (q/k)^n \tag{15-99}$$

To solve the equations for c and q as a function of time, starting from c_F at $t = 0$, (15-98) is combined with an equilibrium isotherm, for example, (15-99), to eliminate q. The resulting equation is combined with (15-97) to eliminate c^* to give an ODE for c in t, which is integrated analytically or numerically. Values of q are obtained from (15-98).

If the equilibrium is represented by a linear isotherm,

$$c^* = q/k \tag{15-100}$$

an analytical integration gives

$$c = \frac{c_F}{\beta}[\exp(-k_L a \beta t) + \alpha] \tag{15-101}$$

where

$$\beta = 1 + \frac{Q}{Sk} \tag{15-102}$$

$$\alpha = \frac{Q}{Sk} \tag{15-103}$$

As contact time approaches ∞, adsorption equilibrium is approached and for the linear isotherm, from (15-101) or combining (15-98), with $c = c^*$, and (15-100):

$$c\{t = \infty\} = c_F \alpha / \beta \tag{15-104}$$

Continuous Mode

When both liquid and solids flow continuously through a perfectly mixed vessel, (15-97) converts to an algebraic equation because, as in a perfectly mixed reaction vessel (CSTR), concentration c throughout the vessel is equal to the outlet concentration, c_{out}. In terms of vessel residence time, t_{res}:

$$\frac{c_F - c_{out}}{t_{res}} = k_L a(c_{out} - c^*) \tag{15-105}$$

or, rearranging,

$$c_{out} = \frac{c_F + k_L a t_{res} c^*}{1 + k_L a t_{res}} \tag{15-106}$$

Equation (15-78) becomes

$$c_F Q = c_{out} Q + q_{out} S \tag{15-107}$$

where Q and S are now flow rates. An appropriate adsorption isotherm relates c^* to q_{out}. For a linear isotherm, (15-100) becomes $c^* = q_{out}/k$, which when combined with (15-107) and (15-106) to eliminate c^* and q_{out}, gives

$$c_{out} = c_F \left(\frac{1 + \gamma \alpha}{1 + \gamma + \gamma \alpha} \right) \tag{15-108}$$

where α is given by (15-103) and

$$\gamma = k_L a t_{res} \tag{15-109}$$

The corresponding q_{out} is given by rearranging (15-107):

$$q_{out} = \frac{Q(c_F - c_{out})}{S} \tag{15-110}$$

For a nonlinear adsorption isotherm, such as (15-35) or (15-36), (15-105) and (15-107) are combined with the isotherm equation, but it may not be possible to express the result explicitly in q_{out}. Then a numerical solution is required, as illustrated in Example 15.17.

Semicontinuous Mode

The most difficult mode to model is the semicontinuous mode, where adsorbent is retained in the vessel, but feed liquid enters and exits the vessel at a fixed, continuous flow rate. Both concentration, c, and loading, q, vary with time. With perfect mixing, the outlet concentration is given by (15-106), where t_{res} is the liquid residence time in the suspension, and c^* is related to q by an appropriate adsorption isotherm. Variation of q in the batch of solids is given by (15-97), rewritten in terms of the change in q, rather than c:

$$S\frac{dq}{dt} = k_L a(c_{out} - c^*)t_{res} Q \tag{15-111}$$

where, for this mode, S is the batch mass of adsorbent in suspension and Q is the steady, volumetric-liquid flow rate.

Both (15-111) and (15-106) involve c^*, which can be replaced by a function of instantaneous q by selecting an appropriate isotherm. The resulting two equations are combined to eliminate c_{out}, and the resulting ODE is then integrated analytically or numerically to obtain q as a function of time, from which c_{out} as a function of time can be determined from (15-106) and the isotherm. The time-average value of c_{out} is then obtained by integration of c_{out} with respect to time. These steps are illustrated in Example 15.17. For a linear isotherm, the derivation is left as an exercise.

EXAMPLE 15.17 Three Modes of Slurry Adsorption.

An aqueous solution containing 0.010 mol phenol/L is to be treated at 20°C with activated carbon to reduce the concentration of phenol to 0.00057 mol/L. From Example 4.12, the adsorption-equilibrium data are well fitted to the Freundlich equation:

$$q = 2.16c^{1/4.35} \tag{1}$$

or

$$c^* = (q/2.16)^{4.35} \tag{2}$$

where q and c are in mmol/g and mmol/L, respectively. In terms of kmol/kg and kmol/m^3, (2) becomes

$$c^* = (q/0.01057)^{4.35} \tag{3}$$

All three modes of slurry adsorption are to be considered. From Example 4.12, the minimum amount of adsorbent is 5 g/L of solution. Laboratory experiments with adsorbent particles 1.5 mm in diameter in a well-agitated vessel have confirmed that the rate of adsorption is controlled by external mass transfer with $k_L = k_c = 5 \times 10^{-5}$ m/s. Particle surface area is 5 m^2/kg of particles.

(a) Using twice the minimum amount of adsorbent in an agitated vessel operated in the batch mode, determine the time in seconds to reduce phenol content to the desired value. (b) For operation in the continuous mode with twice the minimum amount of adsorbent, determine the required residence time in seconds. Compare it to the batch time of part (a). (c) For semicontinuous operation with 1,000 kg of activated carbon, a liquid feed rate of 10 m³/h, and a liquid residence time equal to 1.5 times the value computed in part (b), determine the run time to obtain a composite liquid product with the desired phenol concentration. Are the results reasonable?

Solution

(a) Batch mode:

$$S/Q = 2(5) = 10 \text{ g/L} = 10 \text{ kg/m}^3; \; k_L a = 5 \times 10^{-5}(5)(10)$$
$$= 2.5 \times 10^{-3} \text{ s}^{-1}; \; c_F = 0.010 \text{ mol/L} = 0.010 \text{ kmol/m}^3$$

From (15-98),

$$q = \frac{c_F - c}{S/Q} = \frac{0.010 - c}{10} \tag{4}$$

Substituting (4) into (3),

$$c^* = \left(\frac{0.10 - c}{0.1057}\right)^{4.35} \tag{5}$$

Substituting (5) into (15-97),

$$-\frac{dc}{dt} = 2.5 \times 10^{-3} \left[c - \left(\frac{0.010 - c}{0.1057}\right)^{4.35}\right] \tag{6}$$

where $c = c_F = 0.010 \text{ kmol/m}^3$ at $t = 0$, and t for $c = 0.00057$ kmol/m³ is wanted. By numerical integration, $t = 1,140$ s.

(b) Continuous mode:
Equation (15-105) applies, where all quantities are the same as those determined in part (a) and $c_{out} = 0.00057$ kmol/m³. Thus,

$$t_{res} = \frac{c_F - c_{out}}{k_L a(c_{out} - c^*)}$$

where c^* is given by with $q = q_{out}$, and q_{out} is obtained from (15-107). Thus,

$$t_{res} = \frac{0.010 - 0.00057}{2.5 \times 10^{-3}\left[0.00057 - \left(\frac{0.010 - 0.00057}{0.1057}\right)^{4.35}\right]} = 6,950 \text{ s.}$$

This is appreciably longer than the batch residence time of 1,140 s. In the batch mode, the concentration-driving force for external mass transfer is initially $(c - c^*) = c_F = 0.010$ kmol/m³ and gradually declines to a final value, at 1,140 s, of

$$(c - c^*) = c_{final} - \left(\frac{0.010 - c_{final}}{0.1057}\right)^{4.35} = 0.000543 \text{ kmol/m}^3$$

For the continuous mode the concentration-driving force for external mass transfer is always at the final batch value of 0.000543 kmol/m³, which here is very small.

(c) Semicontinuous mode:
Equation (15-111) applies with: $S = 1,000$ kg; $c_F = 0.010$ kmol/m³; $Q = 10$ m³/h; $t_{res} = 10,425$ s; $k_L a = 2.5 \times 10^{-3}$ s⁻¹; c^* given in terms of q by (3) and c_{out} given by (15-106).
Combining (15-111), (3), and (15-106) to eliminate c^* and c_{out} gives, after simplification,

$$\frac{dq}{dt} = \left(\frac{\gamma}{1+\gamma}\right)\frac{Q}{S}\left[c_F - \left(\frac{q}{0.01057}\right)^{4.35}\right] \tag{7}$$

Table 15.14 Results for Part (c), Semicontinuous Mode, of Example 15.17

Time t, h	q, kmol/kg	c_{out} kmol/m³	c_{cum} kmol/m³
0.0	0.0	0.000370	0.000370
5.0	0.000481	0.000371	0.000370
10.0	0.000962	0.000398	0.000375
15.0	0.001440	0.000535	0.000401
15.7	0.001506	0.000570	0.000407
20.0	0.001905	0.000928	0.000476
21.0	0.001995	0.001052	0.000501
22.0	0.002084	0.001195	0.000529
23.0	0.002172	0.001356	0.000561
23.2	0.002189	0.001390	0.000568
23.3	0.002197	0.001407	0.000572

where γ is given by (15-109) and time, t, is the time the adsorbent remains in the vessel. For values of γ, Q/S, and c_F equal, respectively, to 26.06, 0.01 m³/h-kg, and 0.010 kmol/m³, (7) reduces to

$$\frac{dq}{dt} = 0.00963\left[0.010 - \left(\frac{q}{0.01057}\right)^{4.35}\right] \tag{8}$$

where t is in hours and q is in kmol. By numerical integration of (8), starting from $q = 0$ at $t = 0$, q as a function of t becomes as given in Table 15.14. Included are corresponding values of c_{out} computed from (15-106) combined with (3) to eliminate c^*:

$$c_{out} = \frac{c_F + \gamma(q/0.01057)^{4.35}}{1+\gamma} = \frac{0.010 + 26.06(q/0.01057)^{4.35}}{27.06}$$

Also included in Table 15.14 are the cumulative values of c for all of the liquid effluent that exits the vessel during the period from $t = 0$ to $t = t$, as obtained by integrating c_{out} with respect to time: $c_{cum} = \int_0^t c_{out} dt/t$.

From the results in Table 15.14, it is seen that the loading, q, increases almost linearly during the first 10 h, while the instantaneous phenol concentration, c_{out}, in the exiting liquid remains almost constant. At 15.7 h, c_{out} has increased to the specified value of 0.00057 kmol/m³, but c_{cum} is only 0.000407 kmol/m³. Therefore, the operation can continue. Finally, at between 23.2 and 23.3 h, c_{cum} reaches 0.00057 kmol/m³ and the operation must be terminated. During operation, the vessel contains 1,000 kg or 2 m³ of adsorbent particles. With a liquid residence time of almost 3 h, the vessel must contain 10(3) = 30 m³. Thus, the vol% solids in the vessel is 6.7. This is reasonable. If adsorbent in the vessel is 2,000 kg, giving almost 12 vol% solids, the time of operation is doubled to 46.5 h.

§15.5.2 Thermal-Swing Adsorption

Thermal (temperature)-swing adsorption (TSA) is carried out with two fixed beds in parallel, operating cyclically, as in Figure 15.40b. While one bed is adsorbing solute at near-ambient temperature, $T_1 = T_{ads}$, the other bed is regenerated by desorption at a higher temperature, $T_2 = T_{des}$, as illustrated in Figure 15.47. Although desorption might be accomplished in the

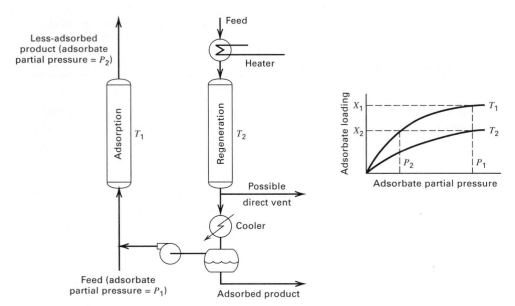

Figure 15.47 Temperature-swing adsorption cycle.

absence of a purge fluid by simply vaporizing the adsorbate, some readsorption of solute vapor would occur upon cooling; thus, it is best to remove desorbed adsorbate with a purge. The desorption temperature is high, but not so high as to cause deterioration of the adsorbent. TSA is best applied to removal of contaminants present at low concentrations in the feed so that nearly isothermal adsorption and desorption is achieved.

An ideal cycle involves four steps: (1) adsorption at T_1 to breakthrough, (2) heating of the bed to T_2, (3) desorption at T_2 to a low adsorbate loading, and (4) cooling of the bed to T_1. Practical cycles do not operate with isothermal steps. Instead, Steps 2 and 3 are combined, with the bed being simultaneously heated and desorbed with preheated purge gas until effluent temperature approaches that of the inlet purge. Steps 1 and 4 may also be combined because, as discussed by Ruthven [10], the thermal wave precedes the MTZ front. Thus, adsorption occurs at feed-fluid temperature.

The heating and cooling steps cannot be accomplished instantaneously because of the low bed thermal conductivity. Although heat transfer can be done indirectly from jackets surrounding the beds or from coils within the beds, temperature changes are more readily achieved by preheating or precooling a purge fluid, as shown in Figure 15.47. The purge fluid can be a portion of the feed or effluent, or some other fluid, and can also be used in the desorption step. When the adsorbate is valuable and easily condensed, the purge fluid might be a noncondensable gas. When the adsorbate is valuable but not easily condensed, and is essentially insoluble in water, steam may be used as the purge fluid, followed by condensation of the steam to separate it from the desorbed adsorbate. When the adsorbate is not valuable, fuel and/or air can be used as the purge fluid, followed by incineration. Often the amount of purge in the regeneration step is much less than the feed in the adsorption step. In Figure 15.47, the feed fluid is a gas and the spent bed is heated and regenerated with preheated feed gas, which is cooled to condense desorbed adsorbate.

Because of the time to heat and cool a fixed bed, cycle times for TSA are long, usually hours or days. Longer cycle times require longer bed lengths, which result in a greater percent bed utilization during adsorption. However, for a given cycle time, when the MTZ width is an appreciable fraction of bed length such that bed capacity is poorly utilized, a *lead-trim-bed* arrangement of two absorbing beds in series should be considered. When the lead bed is spent, it is switched to regeneration. At this time, the trim bed has an MTZ occupying a considerable portion of the bed, and that bed becomes the lead bed, with a regenerated bed becoming the trim bed. In this manner, only a fully spent bed is switched to regeneration and three beds are used. If the feed flow rate is very high, beds in parallel may be required.

Adsorption is usually conducted with the feed fluid flowing downward. Desorption can be either downward or upward, but the upward, countercurrent direction is preferred because it is more efficient. Consider the loading fronts shown in Figure 15.48 for regeneration countercurrent to adsorption. Although the bed is shown horizontal, it must be positioned vertically. The feed fluid flows down, entering at the left and leaving at the right. At time $t = 0$, breakthrough has occurred, with a loading profile as shown at the top, where the MTZ is about 25% of the bed. If the purge fluid for regeneration also flows downward (entering at the left), the adsorbate will move through the unused portion of the bed, and some desorbed adsorbate will be readsorbed in the unused section and then desorbed a second time. If countercurrent regeneration is used, the unused portion of the bed is never in contact with desorbed adsorbate.

During a countercurrent regeneration step, the loading profile changes progressively with time, as shown in Figure 15.48. The right-side end of the bed, where purge enters, is desorbed first. After regeneration, residual loading may be uniformly zero or, more likely, finite and nonuniform, as shown at the bottom of Figure 15.48. If the latter, then the useful cyclic capacity, called the *delta loading*, is as shown in Figure 15.49.

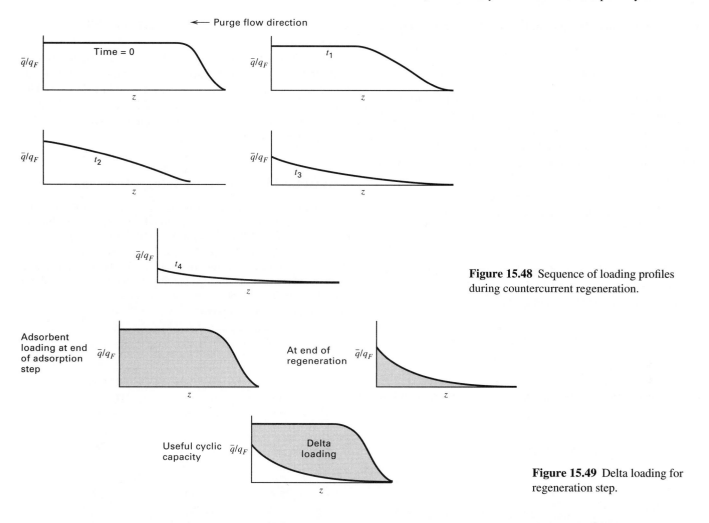

Figure 15.48 Sequence of loading profiles during countercurrent regeneration.

Figure 15.49 Delta loading for regeneration step.

Calculations of the concentration and loading profiles during desorption are only approximated by (15-81) and (15-84) because the loading is not uniform at the beginning of desorption. A numerical solution for the desorption step can be obtained using a procedure discussed by Wong and Niedzwiecki [146]. Although their method was developed for adsorption, it is readily applied to desorption. In the absence of axial dispersion and for constant fluid velocity, (15-48) and (15-50) are rewritten as

$$u\frac{\partial \phi}{\partial z} + \frac{\partial \phi}{\partial t} + \left(\frac{1 - \epsilon_b}{\epsilon_b}\right)kK(\phi - \psi) = 0 \qquad (15\text{-}112)$$

$$\frac{\partial \psi}{\partial t} = k(\phi - \psi) \qquad (15\text{-}113)$$

where

$$\phi = c/c_F \qquad (15\text{-}114)$$

$$\psi = \bar{q}/q_F^* \qquad (15\text{-}115)$$

and c_F and q_F^* are taken at the beginning of the adsorption step. Boundary conditions are:

At $t = 0$: $\phi = \phi\{z\}$ at the end of the adsorption step and $\psi = \psi\{z\}$ at the end of the adsorption step where, for countercurrent desorption, it is best to let z start from the bed bottom (called z') and increase in the direction of purge-gas flow. Thus, u in (15-112) is positive.

At $z' = 0$: $\phi = 0$ (no solute in the entering purge gas) and $\psi = 0$

Partial differential equations (15-112) and (15-113) in independent variables z and t can be converted to a set of ordinary differential equations (ODEs) in independent variable t by the method of lines (MOL), which was first applied to parabolic PDEs by Rothe in 1930, as discussed by Liskovets [147], and subsequently to elliptic and hyperbolic PDEs. The MOL is developed by Schiesser [148]. The lines refer to the z'-locations of the ODEs. To obtain the set of ODEs, the z'-coordinate is divided into N increments or $N + 1$ grid points that are usually evenly spaced, with 20 increments sufficient. Letting i be the index for each grid point in z', starting from the end where the purge gas enters, and discretizing $\partial \phi / \partial z'$, (15-112) and (15-113) become

$$\frac{d\phi_i}{dt} = -u\left(\frac{\Delta \phi}{\Delta z'}\right)_i - \left(\frac{1 - \epsilon_b}{\epsilon_b}\right)kK(\phi_i - \psi_i), \quad i = 1, N+1$$

$$(15\text{-}116)$$

$$\frac{d\psi_i}{dt} = k(\phi_i - \psi_i), \quad i = 1, N+1 \qquad (15\text{-}117)$$

where initial conditions ($t = 0$) for ϕ_i and ψ_i are as given above. Before (15-116) and (15-117) can be integrated, a suitable approximation for $(\Delta \phi / \Delta z')$ must be provided. In general, for a moving-front problem of the hyperbolic type as in adsorption and desorption, the simple central difference

$$\left(\frac{\Delta\phi}{\Delta z'}\right)_i \approx \frac{\phi_{i+1} - \phi_{i-1}}{2\Delta z'}$$

is not adequate. Wong and Niedzwiecki [146] found that a five-point, biased, upwind, finite-difference approximation, used by Schiesser [148], is very effective. This approximation, which is derived from a Taylor's series analysis, places emphasis on conditions upwind of the moving front. At an interior grid point:

$$\left(\frac{\Delta\phi}{\Delta z'}\right)_i \approx \frac{1}{12\Delta z'}\left[-\phi_{i-3} + 6\phi_{i-2} - 18\phi_{i-1} + 10\phi_i + 3\phi_{i+1}\right]$$

$$(15\text{-}118)$$

Note that the coefficients of the ϕ-factors, inside the square brackets, sum to 0. At the last grid point, $N + 1$, where the purge gas exits, (15-118) is replaced by

$$\left(\frac{\Delta\phi}{\Delta z'}\right)_{N+1} \approx \frac{1}{12\Delta z'}\left[3\phi_{N-3} - 16\phi_{N-2} + 36\phi_{N-1}\right.$$
$$\left. - 48\phi_N + 25\phi_{N+1}\right] \quad (15\text{-}119)$$

For the first three node points, the following approximations replace (15-118):

$$\left(\frac{\Delta\phi}{\Delta z'}\right)_1 \approx \frac{1}{12\Delta z'}\left[-25\phi_1 + 48\phi_2 - 36\phi_3 + 16\phi_4 - 3\phi_5\right]$$

$$(15\text{-}120)$$

$$\left(\frac{\Delta\phi}{\Delta z'}\right)_2 \approx \frac{1}{12\Delta z'}\left[-3\phi_1 - 10\phi_2 + 18\phi_3 - 6\phi_4 + \phi_5\right]$$

$$(15\text{-}121)$$

$$\left(\frac{\Delta\phi}{\Delta z'}\right)_3 \approx \frac{1}{12\Delta z'}\left[\phi_1 - 8\phi_2 + 0\phi_3 + 8\phi_4 - \phi_5\right] \quad (15\text{-}122)$$

Because values of ϕ_1 (at $z' = 1$) are given as a boundary condition, (15-120) is not needed.

Equations (15-116) to (15-122) with boundary conditions for ϕ_1 and ψ_1, constitute a set of $2N$ ODEs as an initial-value problem, with t as the independent variable. However, values of ϕ_i and ψ_i at the different axial locations can change with t at vastly different rates. For example, in Figure 15.46 for desorption fronts, if bed length, L, is divided into 20 equal-length increments, starting from the right-hand side where the purge gas enters, it is seen that initially ψ_{21}, where the purge gas exits, is not changing at all, while ψ_5 is changing rapidly. Near the end of the desorption step, ψ_{21} is changing rapidly, while ψ_5 is not. Identical observations hold for ϕ_i. This type of response is referred to as *stiffness*, as described by Schiesser [149] and in *Numerical Recipes* by Press et al. [150]. If attempts are made to integrate the ODEs with simple Euler or Runge–Kutta methods, not only are truncation errors encountered, but, with time, values of ϕ_i and ψ_i go through enormous instability, characterized by wild swings between large and impossible positive and negative values. Even if the length is divided into more than 20 increments

and very small time steps are used, instability is often encountered.

Integration of a stiff set of ODEs is most efficiently carried out by variable-order/variable-step-size implicit methods first developed by Gear [151]. These are included in a widely available software package called ODEPACK, described by Byrne and Hindmarsh [152]. The subject of stiffness is also discussed in §13.6.2.

EXAMPLE 15.18 Thermal-Swing Adsorption.

In Example 15.13, benzene is adsorbed from air at 70°F and 1 atm onto silica gel in a 6-ft-long fixed-bed adsorber. Breakthrough occurs at close to 97.1 minutes for $\phi = 0.05$. At that time, values of $\phi = c/c_F$ and $\psi = \bar{q}/q_F^*$ in the bed are distributed as follows, where z' is measured backward from the bed exit for the adsorption step. These results were obtained by applying the numerical method just described, to the adsorption step, and are in close agreement with the Klinkenberg solution given in Example 15.13.

z', ft	$\phi = c/c_F$	$\psi = \bar{q}/q_F^*$
0	0.05227	0.03891
0.3	0.07785	0.05913
0.6	0.11314	0.08776
0.9	0.16008	0.12690
1.2	0.22017	0.17850
1.5	0.29394	0.24387
1.8	0.38042	0.32310
2.1	0.47678	0.41459
2.4	0.57825	0.51469
2.7	0.67861	0.61786
3.0	0.77108	0.71728
3.3	0.84969	0.80603
3.6	0.91057	0.87858
3.9	0.95281	0.93207
4.2	0.97848	0.96690
4.5	0.99172	0.98636
4.8	0.99731	0.99531
5.1	0.99921	0.99857
5.4	0.99987	0.99960
5.7	1.00000	1.00000
6.0	1.00000	1.00000

If the bed is regenerated isothermally with pure air at 1 atm and 145°F, and benzene desorption during the heat-up period is neglected, determine the loading, $\bar{q}$, profile at times of 15, 30, and 60 minutes for air stripping at interstitial velocities of: (a) 197 ft/minute, and (b) 98.5 ft/minute. At 145°F and 1 atm, the adsorption isotherm, in the same units as in Example 15.13, is

$$q = 1{,}000c^* \quad (1)$$

giving an equilibrium loading of about 20% of that at 70°F. Assume that k is unchanged from the value of 0.206 in Example 15.13.

Solution

This problem was solved by the MOL with 20 increments in z', using the subroutine LSODE in ODEPACK to integrate the set of ODEs. A FORTRAN MAIN program and a subroutine FEX, for the

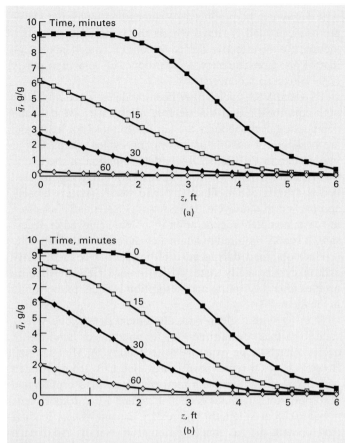

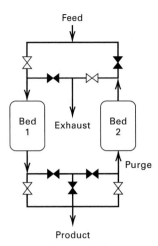

Feed

Figure 15.51 Pressure-swing-adsorption cycle.

Figure 15.50 Regeneration loading profiles for Example 15.18.
(a) Regeneration air interstitial velocity = 197 ft/min.
(b) Regeneration air interstitial velocity = 98.5 ft/min.

derivative functions given by (15-116) to (15-119) and (15-121) to (15-122), were written for both adsorption and desorption steps, with desorption conditions of 30 minutes at 197 ft/minute. The code may be found in Table 15.9 of Chapter 15 of the second edition of Seader and Henley [153]. This example can also be solved with Aspen Adsorption.

The computed loading profiles are plotted in Figures 15.50a and b, for desorption interstitial velocities of 197 and 98.5 ft/minute, where z is distance from the feed gas inlet end for adsorption. The curves are similar to those in Figure 15.29. For the 197-ft/minute case, desorption is almost complete at 60 minutes with less than 1% of the bed still loaded with benzene. If this velocity were used, this would allow $97.1 - 60 = 37.1$ minutes for heating and cooling the bed before and after desorption. For 98.5 ft/minute at 60 minutes, about 5% of the bed is still loaded with benzene. This may be acceptable, but the resulting adsorption step would take a little longer because initially, the bed would not be clean. Several cycles are required to establish a cyclic steady state, whose development is considered in the next section, on pressure-swing adsorption.

§15.5.3 Pressure-Swing Adsorption

Pressure-swing adsorption (PSA) and vacuum-swing adsorption (VSA), in their simplest configurations, are carried out with two fixed beds in parallel, operating in a cycle, as in Figure 15.51. Unlike TSA, where thermal means are used to effect the separation, PSA and VSA use mechanical work to

increase pressure or create a vacuum. While one bed is adsorbing at one pressure, the other bed is desorbing at a lower pressure, as illustrated in Figure 15.41. Unlike TSA, which can be used to purify gases or liquids, PSA and VSA process only gases, because a change in pressure has little or no effect on equilibrium loading for liquid adsorption.

PSA was originally used only for purification, as in the removal of moisture from air by the "heatless drier," which was invented by C. W. Skarstrom in 1960 to compete with TSA. However, by the 1970s, PSA was being applied to bulk separations, such as partial separation of air to produce either nitrogen or oxygen, and to removal of impurities and pollutants from other gas streams. PSA can also be used for vapor recovery, as discussed and illustrated by Ritter and Yang [154].

Steps in the Skarstrom cycle, operating with two beds, are shown in Figure 15.52. Each bed operates alternately in two half-cycles of equal duration: (1) pressurization followed by adsorption, and (2) depressurization (blowdown) followed by a purge. Feed gas is used for pressurization, while a portion of the effluent product gas is used for purge. Thus, in Figure 15.52, while adsorption is taking place in Bed 1, part of the

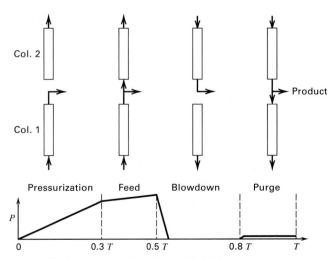

Figure 15.52 Sequence of cycle steps in PSA.

gas leaving Bed 1 is routed to Bed 2 to purge that bed in a direction countercurrent to the direction of flow of feed gas during the adsorption step. When moisture is to be removed from air, the dry-air product is produced during the adsorption step in each of the two beds. In Figure 15.52, the adsorption and purge steps represent less than 50% of the total cycle time. In many commercial applications of PSA, these two steps consume a much greater fraction of the cycle time because pressurization and blowdown can be completed rapidly. Therefore, cycle times for PSA and VSA are short, typically seconds to minutes, and small beds have relatively large throughputs.

With the valving shown in Figure 15.51, the cyclic sequence can be programmed to operate automatically. With some valves open and others closed, as in Figure 15.51, adsorption takes place in Bed 1 and purge in Bed 2. During the second half of the cycle, valve openings and beds are switched.

Improvements have been made to the Skarstrom cycle to increase product purity, product recovery, adsorbent productivity, and energy efficiency, as discussed by Yang [25] and by Ruthven, Farooq, and Knaebel [155]. Among these modifications are use of (1) three, four, or more beds; (2) a pressure-equalization step in which both beds are equalized in pressure following purge of one bed and adsorption in the other; (3) pretreatment or guard beds to remove strongly adsorbed components that might interfere with separation of other components; (4) purge with a strongly adsorbing gas; and (5) use of an extremely short cycle time to approach isothermal operation, if a longer cycle causes an undesirable increase in temperature during adsorption and an undesirable decrease in temperature during desorption.

Separations by PSA and VSA are controlled by adsorption equilibrium or adsorption kinetics, where the latter refers to mass transfer external and/or internal to adsorbent particle. Both types of control are important commercially. For the separation of air with zeolites, adsorption equilibrium is the controlling factor, with N_2 more strongly adsorbed than O_2 and argon. For air with 21% O_2 and 1% argon, O_2 of about 96% purity can be produced. When carbon molecular sieves are used, O_2 and N_2 have almost the same adsorption isotherms, but the effective diffusivity of O_2 is much larger than that of N_2. Consequently, a N_2 product of very high purity (>99%) can be produced.

PSA and VSA cycles have been modeled successfully for both equilibrium and kinetic-controlled cases. Models and computational procedures are similar to those for TSA, and are particularly useful for optimizing cycles. Of particular importance in PSA and TSA is determination of the cyclic steady state. In TSA, following desorption, the regenerated bed is usually clean. Thus, a cyclic steady state is closely approached in one cycle. In PSA and VSA, this is not often the case; complete regeneration is seldom achieved or necessary. It is only required to attain a cyclic steady state whereby product obtained during adsorption has the desired purity and, at cyclic steady state, the difference between loading profiles after adsorption and desorption is equal to the solute in the feed.

Starting with a clean bed, attainment of a cyclic steady state for a fixed cycle time may require tens or hundreds of cycles. Consider an example from a study by Mutasim and Bowen [156] on removal of ethane and CO_2 from nitrogen with 5A zeolite, at ambient temperature with adsorption and desorption for 3 minutes each at 4 bar and 1 bar, respectively, in beds 0.25 m in length. Figures 15.53a and b show loading development and gas concentration profiles at the end of each adsorption step for ethane, starting from a clean bed. After the first cycle, the bed is still clean beyond about 0.11 m. By the end of the 10th cycle, a cyclic steady state has almost been attained, with the bed being clean only near the very end. Experimental data points for ethane loading at the end of 10 cycles agree with the computed profile.

PSA and VSA cycle models are constructed with the same equations as for TSA, but the assumptions of negligible axial diffusion and isothermal operation may be relaxed. For each cycle, the pressurization and blowdown steps are often ignored and initial conditions for adsorption and desorption become the final conditions for desorption and adsorption of

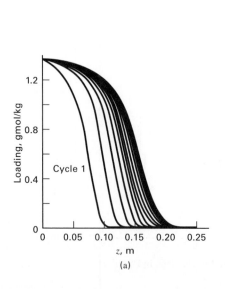

(a)

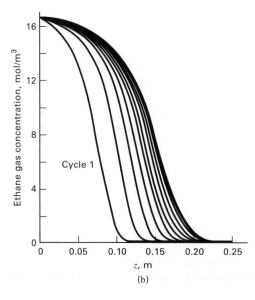

(b)

Figure 15.53 Development of cyclic steady-state profiles. (a) Loading profiles for first 11 cycles. (b) Ethane gas concentration profiles for first 16 cycles.

the previous cycle, as is illustrated in the following example. Calculations can also be made with Aspen Adsorption.

EXAMPLE 15.19   Pressure-Swing Adsorption.

Ritter and Yang [154] conducted a study of PSA to recover dimethyl methylphosphonate (DMMP) vapor from air. For their data and operating conditions, starting with a clean bed, use MOL with a stiff integrator, or Aspen Adsorption, to estimate the bed concentration and loading profiles, % of feed gas recovered as essentially pure air, and average mole fraction of DMMP in the effluent gas leaving the desorption step during the third cycle.

Feed-Gas Conditions: 236 ppm by volume of DMMP in dry air at 294 K and 3.06 atm

Adsorbent: BPL activated carbon, 5.25 g in each bed, 0.07 cm average particle diameter, and 0.43 bed porosity

Bed dimensions: 1.1 cm i.d. by 12.8 cm each

Langmuir adsorption isotherm: $q = \dfrac{48,360 p_{DMMP}}{1 + 98,700 p_{DMMP}}$

where q is in g/g and p is in atm

Overall mass-transfer coefficient: $k = 5 \times 10^{-3} \text{ s}^{-1}$

Cycle conditions (all at 294 K):

1. Pressurization with pure air from p_L to p_H in negligible time.

2. Adsorption at $p_H = 3.06$ atm with feed gas for 20 minutes. $u =$ interstitial velocity = 10.465 cm/s.

3. Blowdown from p_H to p_L with no loss of DMMP from the adsorbent or gas in the bed voids in negligible time.

4. Desorption at $p_L = 1.07$ atm with product gas (pure air) for 20 minutes. Interstitial velocity, u, corresponding to use of 41.6% of product gas leaving adsorption.

Solution

This example is solved using equations and numerical techniques employed in Example 15.18, but noting that units of q are different and that a Langmuir isotherm replaces a linear isotherm. If the bed is not clean following the first desorption step, results for the second and third cycles will differ from the first. The results are not presented here, but the calculations are required for Exercise 15.35.

§15.6 CONTINUOUS, COUNTERCURRENT ADSORPTION SYSTEMS

In previous subsections, slurry and fixed-bed adsorption modes, shown in Figures 15.40a and b, were considered. A third mode of operation, shown in Figure 15.40c, is continuous, countercurrent operation, which has an important advantage because, as in a heat exchanger, an adsorber, and other separation cascades, countercurrent flow maximizes the average driving force for transport. In adsorption, this increases adsorbent use efficiency.

In Figure 15.40c, both liquid or gas mixtures undergoing separation and solid adsorbent particles move through the system. However, as discussed in detail by Ruthven and Ching [157] and Wankat [134], the advantage of countercurrent operation can also be achieved by a simulated-moving-bed (SMB) operation, with one widely used implementation shown in Figure 15.43, where adsorbent particles remain fixed in a bed. In §15.6.1 and 15.6.2, the continuous, countercurrent system shown in Figure 15.40c is considered, while §15.6.3 covers the SMB. Both types of operation can be used for purification or bulk separation.

§15.6.1 McCabe–Thiele and Kremser Methods for Purification

Consider a binary mixture, dilute in a solute to be removed by adsorption in the continuous, countercurrent system shown in Figure 15.54a. Only the solute is adsorbed and feed F, with solute concentration c_F, enters the adsorption section, ADS, at Plane P_1, from which adsorbent S leaves with a solute loading q_F. Purified feed called raffinate, with solute concentration c_R, leaves the adsorption section at Plane P_2, countercurrent to adsorbent of loading q_R, which enters at the top. At Plane P_3, a purge called the desorbent, D, with solute concentration c_D, enters the bottom of the desorption section, DES, from which the adsorbent leaves to enter the adsorption section. It is assumed that desorbent does not adsorb but exits from DES as extract E, with solute concentration c_E, at Plane 4, where recycled adsorbent enters the desorption bed to complete the cycle.

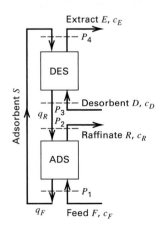

(a)

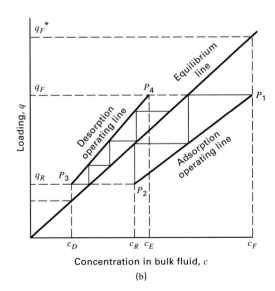

(b)

Figure 15.54 Continuous, countercurrent adsorption–desorption system.
(a) System sections and flow conditions.
(b) McCabe–Thiele diagram.

If the system is dilute in solute, and solute adsorption isotherms for feed solvent and purge fluid are identical, and if the system operates at constant temperature and pressure, a McCabe–Thiele diagram for the solute resembles that shown in Figure 15.54b, where the operating and equilibrium lines are straight because of the dilute condition. Note that proper directions for mass transfer require that adsorption operating lines lie below and desorption operating lines lie above the equilibrium line. These three lines are represented by the following equations:

Adsorption Operating Line:

$$q = \frac{F}{S}(c - c_F) + q_F \qquad (15\text{-}123)$$

Desorption Operating Line:

$$q = \frac{D}{S}(c - c_D) + q_R \qquad (15\text{-}124)$$

Equilibrium Line:

$$q = Kc \qquad (15\text{-}125)$$

where F, S, and D are solute-free mass flow rates, and all solute concentrations are per solute-free carrier.

In Figure 15.54b, as solute concentration in the entering desorbent (purge), c_D, approaches zero, and solute concentration in the exiting raffinate, c_R, approaches zero, it is necessary, in order to avoid a large number of stages, to select adsorbent and desorbent flow rates so that

$$\frac{F}{S} < K < \frac{D}{S}$$

Because more purge, D, than feed, F, is required, this system is economical only when purge fluid is inexpensive. From the equilibrium and operating lines in Figure 15.54b, 2 and 3.3 equilibrium stages are determined for the adsorption and desorption sections by stepping off stages in the McCabe–Thiele diagram. When the equilibrium and operating lines are straight, as in Figure 15.54b, the Kremser method, rather than the graphical McCabe–Thiele method, can be employed.

The Kremser equation, discussed in §6.4, is written in the end-point form for adsorption or desorption:

$$N_t = \frac{\ln\left[\dfrac{c_1 - q_1/K}{c_2 - q_2/K}\right]}{\ln\left[\dfrac{c_1 - c_2}{q_1/K - q_2/K}\right]} \qquad (15\text{-}126)$$

where 1 and 2 refer to opposite ends, such as Planes 1 and 2 in Figure 15.55a, which are chosen so $q_1 > q_2$.

If the temperature or pressure for the two sections can be altered to place the equilibrium line for desorption below that for adsorption, it becomes possible to use a portion of the raffinate for desorption. This situation, shown in Figure 15.55, is achieved by desorbing at elevated temperature or, in the case of gas adsorption, at reduced pressure. Now, as shown in Figure 15.55, F/S can be greater than D/S. With a portion of raffinate used in Bed 2 (DES), the net raffinate product is $F - D$. In this case, the two operating lines must intersect at the point (q_R, c_R). By adjusting D/F, this point can be moved closer to the origin to increase raffinate purity, c_R, but at the expense of more stages and deeper beds.

For a number of theoretical stages, N_t, in the adsorption or desorption sections, bed height L can be determined from

$$L = N_t(\text{HETP}) \qquad (15\text{-}127)$$

Values of HETP, which depend on mass-transfer resistances and axial dispersion, must be established from experimental measurements. For large-diameter beds, values of HETP are in the range of 0.5–1.5 ft [158, 176].

§15.6.2 McCabe–Thiele Method for Bulk Separation

Figure 15.56 illustrates a continuous, countercurrent adsorption–desorption process for bulk separation of a binary mixture. Feed consists of component A, which is more strongly adsorbed than B. The process consists of four sections (zones), numbered from the bottom up. Adsorbent S is circulated through the system, passing downward through the four sections, adsorbing A and leaving B to preferentially pass

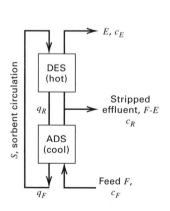

(a)

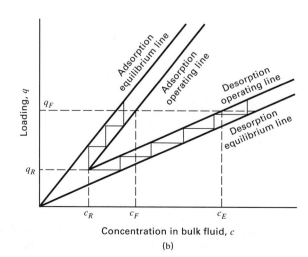

(b)

Figure 15.55 Continuous, countercurrent system with a temperature swing.
(a) System sections and flow conditions.
(b) McCabe–Thiele diagram.

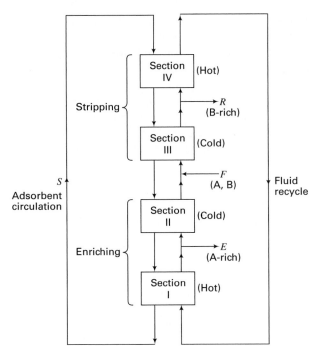

Figure 15.56 Continuous, countercurrent system for bulk separation.

upward. To provide flexibility, a thermal swing is used, with Sections II and III operating at low or ambient temperature, while Sections I and IV operate at elevated temperature. The feed, F, enters between Sections II and III, passing up through Section III, where A is preferentially adsorbed at a relatively cold temperature. Product R, rich in B, is removed between Sections III and IV. At the higher temperature in Section IV, residual A and B are desorbed, with the fluid leaving from Section IV recycled to Section I. Adsorbent with mainly adsorbed A passes downward from Section III to Section II and then to Section I, where component A is desorbed to produce product E, rich in A, which is removed between Sections I and II.

The system in Figure 15.56 resembles an inverted distillation column, with the top two sections (III and IV) providing a stripping action to produce a product rich in the less strongly adsorbed component, while two bottom sections provide an enriching action to produce a product rich in component A. An arrangement similar to that in Figure 15.56 was used in the Hypersorption moving-bed process [159] for separating hydrogen and methane from ethane and heavier hydrocarbons, except that Section IV was a cooler, Section I was a steam stripper, and gas leaving Section IV was used to lift adsorbent from Section I to Section IV. Additional flexibility can be achieved for the system in Figure 15.56 by separate adsorbent-circulation loops for the top two and bottom two sections.

EXAMPLE 15.20 Countercurrent Moving-Bed System.

One hundred lb/minute (dry basis) of air at 80°F and 1 atm with 65% relative humidity is dehumidified isothermally and isobarically to 10% relative humidity in a continuous, countercurrent, moving-

bed adsorption unit. The adsorbent is dry silica gel (SG) having a particle-diameter range of 1.42 to 2.0 mm. The adsorption isotherm is given by Eagleton and Bliss [160] as

$$q_{H_2O} = 29c_{H_2O} \tag{1}$$

with concentration in lb H_2O/lb dry air and loading in lb H_2O/lb dry SG. If 1.5 times the minimum flow rate of silica gel is used, determine the number of equilibrium stages required.

Solution

For relative humidities of 65% and 10%, the corresponding moisture contents, from a humidity chart, are 0.0143 and 0.0022 lb H_2O/lb dry air, respectively.

In this case, Figure 15.54b describes the adsorption section. Using the nomenclature in that figure: $F = 100$ lb/minute, $c_F = 0.0143$ lb H_2O/lb dry air, $c_R = 0.0022$ lb H_2O/lb dry air, and $q_R = 0$.

The value of q_F depends on adsorbent flow rate, S, which is 1.5 times the minimum value. At minimum-adsorbent rate, exiting adsorbent is in equilibrium with the entering gas. Therefore, from (1): $q_F^* = 29(0.0143) = 0.415$ lb H_2O/lb dry SG. The amount of water vapor adsorbed is $F(c_F - c_R) = 100(0.0143 - 0.0022) = 1.21$ lb/minute. Therefore, $S_{min} = \dfrac{1.21}{0.415} = 2.92$ lb dry SG/minute. If 1.5 times the minimum amount of silica gel is used: $S = 1.5 S_{min} = 1.5(2.92) = 4.38$ lb dry SG/minute. By material balance: $q = \dfrac{1.21}{4.38} = 0.276$ lb H_2O/lb dry SG. From (15-126), with $K = 29$ from (15-1) and letting F be at Plane 1 and R at Plane 2:

$$N_t = \frac{\ln\left[\dfrac{0.0143 - 0.276/29}{0.0022 - 0}\right]}{\ln\left[\dfrac{0.0143 - 0.0022}{0.276/29 - 0}\right]} = 3.2 \text{ stages}$$

§15.6.3 Simulated-Moving-Bed Systems

Continuous, countercurrent moving-bed systems, referred to as *true-moving-bed* (TMB) systems, encounter operating difficulties including adsorbent abrasion, failure to achieve particle plug flow, and fluid channeling. Alternatively, as shown in the implementation in Figure 15.43, a continuous, countercurrent operation can be simulated by using a column containing a series of fixed beds and periodically moving the locations at which streams enter and leave the column. *Simulated-moving-bed* (SMB) systems have found widespread commercial application for liquid separations in the petrochemical, food, biochemical, pharmaceutical, and fine chemical industries when employing a circulating desorbent D (diluent or eluent) to aid in the separation. In some cases the properties of D are such that it can be adsorbed so it can displace solutes A and/or B from the sorbent pores, while A and/or B can displace D. In that case, a hybrid process of SMB adsorption and distillation, as shown in Figure 15.43, is advantageous, where following the SMB, a D-free extract of A and D-free raffinate of B are obtained by distillation, with recovered D recycled to the SMB. In other cases, D is a feed component and is not adsorbed, but simply acts as a carrier and stripping agent for separation of A from B. For example, an aqueous solution of glucose and fructose is separated by an SMB into an extract of aqueous glucose and a raffinate of aqueous fructose. In the literature, SMBs are often referred to as chromatographic, rather than adsorptive, separations. An SMB can be treated as a countercurrent cascade of sections (or zones) rather than stages, where stream entry or withdrawal points bound the sections.

Zang and Wankat [161] review two-, three-, and four-section systems for producing two products, and a nine-section system for three products, with the four-section system of Figure 15.57a being the most common commercial design. More recently, Kim and Wankat [162] proposed SMB designs with from 12 to 32 sections for separation of quaternary mixtures.

An SMB is best understood by studying the two representations of a four-section system and accompanying fluid composition profile in Figure 15.57. The schematic in Figure 15.57a shows a TMB, with circulation of solid adsorbent S

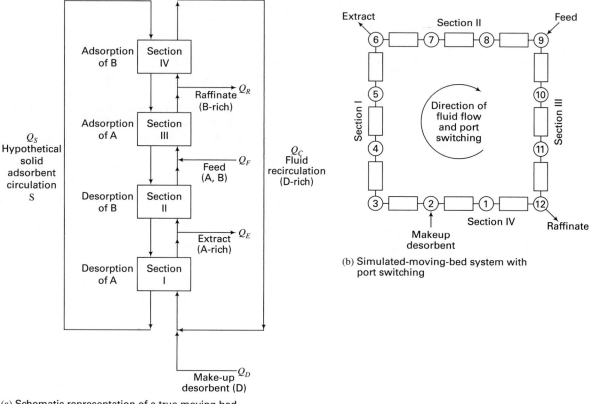

(a) Schematic representation of a true moving bed.

(b) Simulated-moving-bed system with port switching

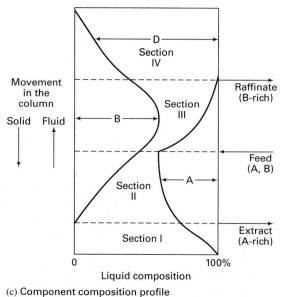

(c) Component composition profile

Figure 15.57 Four-section system.

down through four dense-bed sections in a closed cycle, while Figure 15.57b represents the equivalent SMB system comprised of four sections divided into 12 fixed-bed subsections, shown as rectangles, with periodic movement of fluid inlet and outlet ports, shown as circles. The sections in Figure 15.57a are sometimes referred to as zones, with the fixed-bed subsections referred to as beds and sometimes as columns. In the TMB case of Figure 15.57a, fluid of changing composition with respect to feed components A and B, and desorbent D, flows upward through the downward-flowing adsorbent beds. From the top of Section IV, fluid rich in D is recirculated to Section I. Fluid feed is shown as a binary mixture of A and B entering between Sections II and III. Component A is more strongly adsorbed than D, which is more strongly adsorbed than B. As a result, A is almost completely separated from B, but appreciable amounts of D may appear in both the raffinate and extract, so makeup D must be added to the recirculated fluid.

The four sections in Figure 15.57a perform the primary functions listed in the figure. More detail follows for the case where D, as well as A and B, are adsorbed. A typical component-composition profile is shown in Figure 15.57c.

Section I: Desorption of A. Entering S contains adsorbed A and D. Ideally, entering fluid is nearly pure D. Exiting S contains adsorbed D. Exiting fluid is A and D, part of which is withdrawn as A-rich extract.

Section II: Desorption of B. Entering S contains adsorbed A, B, and D. Entering fluid is A and D. Exiting S contains adsorbed A and D. Exiting fluid is A, B, and D.

Section III: Adsorption of A. Entering S contains adsorbed B and D. Entering fluid is A, B, and D from Section II and fresh feed of A and B. Exiting S contains adsorbed A, B, and D. Exiting fluid is B and D, part of which is withdrawn as B-rich raffinate.

Section IV: Adsorption of B. Entering S contains adsorbed D. Entering fluid is B and D. Exiting S contains adsorbed B and D. Ideally, exiting fluid is nearly pure D.

The steady-state separation achieved by the TMB in Figure 15.57a can be a close approximation to that achieved by the SMB, shown for a commercial Sorbex system in Figure 15.43 and by a simpler representation in Figure 15.57b. In both figures, four sections are provided, with a total of 12 ports for fluid feeds to enter or products to exit. In Figure 15.57b, ports divide each section into subsections, four for Section I, three for Section II, three for Section III, and two for Section IV. As each section is divided into more subsections (thereby adding more ports), the SMB system more closely approaches the separation achieved in a corresponding TMB. In Figure 15.57b, only ports 2, 6, 9, and 12 are open. After an increment of time (called the switching time or port-switching interval, t^*), those ports are closed and 3, 7, 10, and 1 are opened. In this manner, the ports are closed and opened in sequence in the direction shown. By periodically shifting feed and product positions by one port position in the direction of fluid flow, movement of solid adsorbent in the opposite direction is simulated. Because of stream

additions and withdrawals between sections, flow rates in the four sections are different. Figure 15.43 shows a pump for controlling fluid flow rate at the bottom of the SMB. Although sections are switched, the pump is not; therefore, the pump must be programmed for four different flow rates, each depending on the section to which the pump is currently connected.

A number of models are available for designing and analyzing SMBs. These include: (1) TMB equilibrium-stage models using a McCabe–Thiele-type analysis, (2) TMB local-adsorption-equilibrium models, (3) TMB rate-based models, and (4) SMB rate-based models. The first three assume steady-state conditions with continuous, countercurrent flows of fluid and solid adsorbent, approximating SMB operation with a TMB. The SMB rate-based model applies to transient operation for startup, approach to cyclic steady state, and shutdown.

The simplest of the four approaches is the TMB equilibrium-stage model even though it is difficult to apply to systems with nonlinear adsorption-equilibrium isotherms. The TMB local-adsorption-equilibrium model, although ignoring effects of axial dispersion and fluid-particle mass transfer, has proven useful for establishing reasonable operating flow rates in multiple sections of an SMB because, for many applications, behavior of an SMB is determined largely by adsorption equilibria. For a linear adsorption isotherm, Wankat [163] has successfully applied this method to dilute-feed SMBs with up to 32 sections. Methods for solving the TMB local-adsorption-equilibrium model for multicomponent systems, including concentrated mixtures with nonlinear adsorption isotherms, have been presented by a number of investigators including Storti et al. [164], who extended the pioneering work of Rhee, Aris, and Amundson [165] for a single section to the commonly used four-section unit, and Mazzotti et al. [166] for multicomponent systems.

For a final design, rate-based models are preferred. These account for axial dispersion in the bed, particle-fluid mass-transfer resistances, and nonlinear adsorption isotherms, and are available in Aspen Chromatography for both TMB steady-state operation and SMB dynamic operation. Adsorption-equilibrium and rate-based models are described next, followed by three examples, two solved with Aspen Chromatography. Equations are presented for four-section units, but are readily extended to more sections.

Steady-State Local-Adsorption-Equilibrium TMB Model

TMB models describe continuous, steady-state, multicomponent adsorption with countercurrent flow of fluid and solid adsorbent, as shown in Figure 15.58 for a single section of height Z of a multisection system, subject to the assumptions of: (1) one-dimensional plug flow of both phases with no channeling; (2) constant volumetric flow rates, of Q for the liquid and Q_S for the solid; (3) constant external void fraction, ϵ_b, of the solids bed; (4) negligible axial dispersion and particle-fluid mass-transfer resistances; (5) local adsorption equilibrium between solute concentrations, c_i, in the bulk liquid and adsorption loading, q_i, on the solid; (6) isothermal and isochoric conditions.

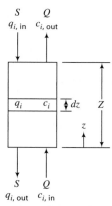

Figure 15.58 TMB local-adsorption-equilibrium model for a single section.

For a differential-bed thickness, dz, where i undergoes mass transfer between two phases, the mass balance is:

$$Q\frac{dc_i}{dz} - S\frac{dq_i}{dz} = 0 \qquad (15\text{-}128)$$

Boundary conditions are

$$z = 0, c_i = c_{i,\text{in}} \text{ and } z = Z, q_i = q_{i,\text{in}}$$

The solution to (15-128) depends on the choice of equilibrium adsorption isotherm (§15.2). Typically, when the fluid is a liquid dilute in solutes, a linear (Henry's law) isotherm, $q_i = K_i c_i$, is used, where q_i is on a particle volume basis so that K_i is dimensionless. For bulk separation of liquid mixtures, where concentrations of feed components and desorbent are not small, a nonlinear, extended-Langmuir equilibrium-adsorption isotherm of the constant-selectivity form from Example 15.6, is appropriate:

$$q_i = \frac{(q_i)_m K_i c_i}{1 + \sum_j K_j c_j} \qquad (15\text{-}129)$$

In either case, the solution of Rhee, Aris, and Amundson [165], when extended to multiple (e.g., four) sections, as by Storti et al. [164], predicts constant concentrations in each

section, but with discontinuities at either or both section boundaries. Typical profiles are shown in Figure 15.59 for a four-solute system (1, 2, 3, and 4), where a set of stationary rectangular (shock-like) waves of constant concentration exists in the fluid phase in each section. The concentration profile for the desorbent (component 5) is not shown. For the local-equilibrium assumption, only desorbent is present in Sections I and IV.

Usefulness of local-equilibrium theory lies in approximate determinations of the amount of solid adsorbent and fluid flow rates, in each TMB section, to achieve a perfect separation of two solutes. The description of the method, first developed by Ruthven and Ching [167] and extended by Zhong and Guiochon [168], is facilitated by applying local-adsorption-equilibrium theory to the case of a feed dilute in binary solutes A and B, which are to be completely separated. Assume that diluent D does not adsorb and that adsorption equilibrium is linear, with $K_A > K_B$ (i.e., A is more strongly adsorbed). First, a set of flow rate ratios, m_j, are defined, one for each section, j, as

$$m_j = \frac{Q_j}{Q_s} = \frac{\text{volumetric fluid phase flow rate}}{\text{volumetric solid particle phase flow rate}}$$

$$(15\text{-}130)$$

For local adsorption equilibrium, the necessary and sufficient conditions at each section for complete separation are:

$$K_A < m_I < \infty \qquad (15\text{-}131)$$

$$K_B < m_{II} < K_A \qquad (15\text{-}132)$$

$$K_B < m_{III} < K_A \qquad (15\text{-}133)$$

$$0 < m_{IV} < K_B \qquad (15\text{-}134)$$

Constraint (15-132) ensures that net flow rates of components A and B will be positive (upward) in Section I. Constraint (15-134) ensures that the net flow rates of components A and B will be negative (downward) in Section IV. Constraints (15-132) and (15-133) are most important because they ensure sharpness of the separation by causing net flow rates of A and B to be negative (downward) and positive (upward),

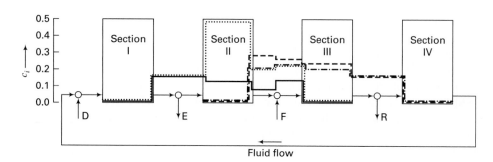

Component		Relative adsorption selectivity
1	—·—·—·	1.00
2	– – – –	1.12
3	··············	2.86
4	————	5.71
5 (not shown)		1.90

Figure 15.59 Typical solute-concentration profiles for local adsorption equilibrium in a four-section unit.

respectively, in the two central Sections II and III. Inequality constraints (15-131) to (15-134) may be converted to equality constraints, where β, the safety margin, is discussed shortly.

$$Q_I/Q_S = K_A\beta \qquad (15\text{-}135)$$

$$(Q_I - Q_E)/Q_S = K_B\beta \qquad (15\text{-}136)$$

$$(Q_I - Q_E + Q_F)/Q_S = K_A/\beta \qquad (15\text{-}137)$$

$$(Q_I - Q_E + Q_F - Q_R)/Q_S = K_B/\beta \qquad (15\text{-}138)$$

Solving (15-135) to (15-138) by eliminating Q_1 gives

$$Q_S = \frac{Q_F}{K_A/\beta - K_B\beta} \qquad (15\text{-}139)$$

$$Q_E = Q_S(K_A - K_B)\beta \qquad (15\text{-}140)$$

$$Q_R = Q_S(K_A - K_B)/\beta \qquad (15\text{-}141)$$

Then, using (15-135),

$$Q_1 = Q_C + Q_D = Q_S Q_A \beta \qquad (15\text{-}142)$$

Therefore,
$$Q_C = Q_S K_A \beta - Q_D \qquad (15\text{-}143)$$

where Q_C = fluid recirculation rate before adding makeup desorbent. By an overall material balance,

$$Q_D = Q_E + Q_R - Q_F \qquad (15\text{-}144)$$

Restrictions on flow rate ratios, m_{II} and m_{III} in inequality constraints (15-132) and (15-133), are conveniently represented by the *triangle method* of Storti et al. [169], as shown in Figure 15.60. If values of m_{II} and m_{III} within the triangular region are selected, a perfect separation is possible. However, if $m_{II} < K_B$, some B will appear in the extract; if $m_{III} > K_A$, some A will appear in the raffinate. If $m_{II} < K_B$ and $m_{III} > K_A$, extract will contain some B and raffinate will contain some A.

The permissible range for safety margin, β, in (15-135) to (15-141) is determined from inequality constraints (15-132) and (15-133). Let

$$\gamma_{i,j} = \frac{m_j}{K_i} = \frac{Q_j}{Q_S K_i} \qquad (15\text{-}145)$$

In Section II, it is required that $\gamma_{A,II} > 1$ and $\gamma_{B,II} < 1$. In terms of safety margin, β, (15-145) can be used to give corresponding equalities $Q_{II}/Q_S = K_A/\beta$ and $Q_{II}/Q_S = K_B\beta$, assuming equal β in all four sections. Equating these two equalities for the same safety margin gives $\beta = \sqrt{K_A/K_B}$, which is the maximum value of β for a perfect separation, the minimum value being 1.0. Above a maximum β, some sections will encounter negative fluid flow rates, and below a β of 1.0, perfect separations will not be achieved.

As the value of β increases from minimum to maximum, fluid flow rates in the sections increase, often exponentially. Thus, estimation of operating flow rates is generally carried out using a value of β close to, but above, 1.0, e.g., 1.05 (unless it exceeds the maximum value of β). Note that as separation factor K_A/K_B approaches 1.0, not only does separation become more difficult, but the permissible range of β also becomes smaller. In the triangle method illustrated in Figure 15.60, the triangle's upper left corner corresponds to

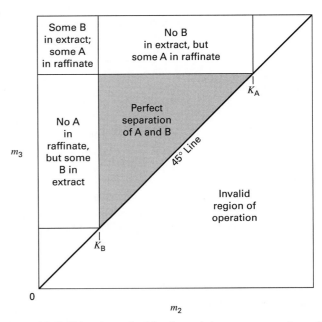

Figure 15.60 Triangle method for determining necessary values of flow rate ratios.

β = 1, while the maximum β occurs when $m_{II} = m_{III}$, which falls on the 45° line between the values K_A and K_B. Extensions of the binary procedures for estimating operating flow rates to cases of both constant-selectivity Langmuir adsorption isotherms (§15.2) and complex nonlinear and multicomponent isotherms are given by Mazzotti et al. [166, 170]. With nonlinear adsorption isotherms, the right triangle of Figure 15.60 is distorted to a shape with one or more curved sides.

EXAMPLE 15.21 Operation with a TMB.

Fructose (A) is separated from glucose (B) in a four-section SMB unit. The aqueous feed of 1.667 mL/minute contains 0.467 g/minute of A, 0.583 g/minute of B, and 0.994 g/minute of water. For the adsorbent and expected concentrations and temperature, Henry's law holds, with constants $K_A = 0.610$ and $K_B = 0.351$ for fluid concentrations in g/mL and loadings in g/mL of adsorbent particles. Water is assumed not to adsorb. Estimate operating flow rates in mL/minute to achieve a perfect separation of fructose from glucose using a TMB. Note that extract will contain fructose and raffinate will contain glucose. Conversion of the results to SMB operation will be made in Example 15.22.

Solution

Equations (15-139) to (15-144) apply. The minimum value of β is 1.0, while the maximum value is $\sqrt{K_A/K_B} = \sqrt{0.610/0.351} = 1.32$. Calculations are most conveniently carried out with a spreadsheet. With reference to Figure 15.57 for the case of a TMB, the results for values of β = 1.0, 1.05, 1.20 are:

	Volumetric Flow Rates, mL/min		
	β = 1.0	β = 1.05	β = 1.20
Feed, Q_F	1.667	1.667	1.667
Solid particles, Q_S	6.436	7.848	19.132
Extract, Q_E	1.667	2.134	5.946

	Volumetric Flow Rates, mL/min		
	$\beta = 1.0$	$\beta = 1.05$	$\beta = 1.20$
Raffinate, Q_R	1.667	1.936	4.129
Recirculation, Q_C	2.259	2.624	5.596
Makeup desorbent, Q_D	1.667	2.403	8.408
Q_I	3.926	5.027	14.004
Q_{II}	2.259	2.893	8.058
Q_{III}	3.926	4.560	9.725
Q_{IV}	2.259	2.624	5.596

The lowest section fluid flow rates, $Q_I - Q_{IV}$, correspond to $\beta = 1.0$. At $\beta = 1.20$, section fluid flow rates, as well as the adsorbent particles flow rate, become significantly higher. The most concentrated products (extract and raffinate) and the smallest flow rate of makeup desorbent are also achieved with the lowest β value.

Steady-State TMB Model

This model—which assumes isothermal, isobaric, plug-flow and constant-fluid-velocity conditions in each section j ($j = 1$ to 4)—requires for each component i ($i = 1$ to C) the following equations, where each section begins at $z = 0$ (where fluid enters) and ends at $z = L_j$. Unlike the previous local-adsorption-equilibrium model, axial dispersion and fluid-particle mass transfer are considered.

(1) Mass-balance equation for the bulk fluid phase, f, [similar to (15-48)]:

$$-D_{L_j}\frac{d^2 c_{i,j}}{dz^2} + u_{f_j}\frac{dc_{i,j}}{dz} + \frac{(1-\epsilon_b)}{\epsilon_b}J_{i,j} = 0 \quad (15\text{-}146)$$

where the first term accounts for axial dispersion, J_i is the mass-transfer flux between the bulk fluid phase and the sorbate in the pores, and u_f is the interstitial fluid velocity, where for an adsorbent bed of cross-sectional area, A_b,

$$u_{f_j} = \frac{Q_j}{\epsilon_b A_b} \quad (15\text{-}147)$$

(2) Mass-balance for sorbate, s, on the solid phase:

$$u_s\frac{d\bar{q}_{i,j}}{dz} - J_{i,j} = 0 \quad (15\text{-}148)$$

where u_s is the true moving-solid velocity:

$$u_s = \frac{Q_S}{(1-\epsilon_b)A_b} \quad (15\text{-}149)$$

(3) Fluid-to-solid mass transfer:

$$J_{i,j} = k_{i,j}\left(q_{i,j}^* - \bar{q}_{i,j}\right) \quad (15\text{-}150)$$

(4) Adsorption isotherm [e.g., the multicomponent, extended-Langmuir equation of (15-129)]:

$$q_{i,j}^* = f\{\text{all } c_{i,j}\} \quad (15\text{-}151)$$

This system of $4C$ second-order ODEs and $4C$ first-order ODEs, together with algebraic equations for mass transfer and adsorption equilibria, requires the following $12C$ boundary conditions, i.e., $3C$ for each section.

At the section entrance, $z = 0$, a boundary condition that accounts for axial dispersion is required. This is discussed by Danckwerts [171]. Most often used is

$$u_{f_j}\left(c_{i,j,0} - c_{i,j}\right) = -\epsilon_b D_{L_j}\frac{dc_{i,j}}{dz} \quad (15\text{-}152)$$

where $c_{i,j,0}$ is the concentration of component i entering ($z = 0$) section j.

For continuity of bulk fluid concentrations and sorbate loadings in moving from one section to another, the following apply at boundaries of adjacent sections:

At Sections I and II where extract is withdrawn:

$$c_{i,I,z=L_j} = c_{i,II,z=0} \quad (15\text{-}153)$$

$$q_{i,I,z=L_j} = q_{i,II,z=0} \quad (15\text{-}154)$$

At Sections III and IV where raffinate is withdrawn:

$$c_{i,III,z=L_j} = c_{i,IV,z=0} \quad (15\text{-}155)$$

$$q_{i,III,z=L_j} = q_{i,IV,z=0} \quad (15\text{-}156)$$

At Sections II and III where feed enters:

$$c_{i,III,z=0} = \frac{Q_{II}}{Q_{III}}c_{i,II,z=L_{II}} + \frac{Q_F}{Q_{III}}c_{i,F} \quad (15\text{-}157)$$

$$q_{i,II,z=L_j} = q_{i,III,z=0} \quad (15\text{-}158)$$

At Sections IV and I where make-up desorbent enters:

$$c_{i,I,z=0} = \frac{Q_{IV}}{Q_I}c_{i,IV,z=L_{II}} + \frac{Q_D}{Q_I}c_{i,D} \quad (15\text{-}159)$$

$$q_{i,IV,z=L_j} = q_{i,I,\,z=0} \quad (15\text{-}160)$$

where the volumetric fluid flow rates, which change from section to section, are subject to

$$Q_I = Q_{IV} + Q_D \quad (15\text{-}161)$$

$$Q_{II} = Q_I - Q_E \quad (15\text{-}162)$$

$$Q_{III} = Q_{II} + Q_F \quad (15\text{-}163)$$

$$Q_{IV} = Q_{III} - Q_E \quad (15\text{-}164)$$

For an SMB, solid particles do not flow down, but are retained in stationary beds in each section. To obtain the same true velocity difference between fluid and solid particles, upward fluid velocity in the SMB must be the sum of the absolute true velocities in the upward-moving fluid and the downward-moving solid particles in the TMB. Using (15-147) and (15-149),

$$(Q_j)_{\text{SMB}} = (Q_j)_{\text{TMB}} + \left(\frac{\epsilon_b}{1-\epsilon_b}\right)(Q_S)_{\text{TMB}} \quad (15\text{-}165)$$

TMB models can be solved by techniques reviewed by Constantinides and Mostoufi [172], the Newton shooting method being preferred. A steady-state TMB model example is solved after the next subsection on the dynamic SMB model.

Dynamic SMB Model

The equations for this model are subject to the same assumptions as steady-state TMB models. Changes in the equations permit taking into account time of operation, t, and using a fluid velocity relative to the stationary solid particles. In addition, equations must be written for each bed subsection (also referred to as a column), k, between adjacent ports, as shown in Figure 15.55b. The revised equations are

(1) Mass-balance equation for bulk fluid, f [similar to (15-48)]:

$$\frac{\partial c_{i,k}}{\partial t} - D_{L_j} \frac{\partial^2 c_{i,k}}{\partial z^2} + u_{f_k} \frac{\partial c_{i,k}}{\partial z} + \frac{(1-\epsilon_b)}{\epsilon_b} J_{i,k} = 0 \quad (15\text{-}166)$$

(2) Mass-balance equation for sorbate on the solid phase:

$$\frac{\partial \bar{q}_{i,k}}{\partial t} - J_{i,k} = 0 \quad (15\text{-}167)$$

where the interstitial fluid velocity for SMB operation is related to that for TMB operation at a particular location by

$$(u_f)_{\text{SMB}} = (u_f)_{\text{TMB}} + |(u_s)|_{\text{TMB}} \quad (15\text{-}168)$$

SMB and TMB models are further connected by an equation that relates solid velocity in the TMB model to a port-switching time, t^*, and bed height between adjacent ports, L_k, for use in this SMB-model:

$$u_s = \frac{L_k}{t^*} \quad (15\text{-}169)$$

Boundary conditions for TMB models apply to SMB models. In addition, initial conditions are needed for fluid concentrations, $c_{i,j}$, and sorbate loadings, $\bar{q}_{i,j}$, throughout the adsorbent beds—e.g., at $t = 0$, $c_{i,k} = 0$, and $\bar{q}_{i,k} = 0$.

The SMB model, which involves PDEs rather than ODEs, is more difficult to solve than the steady-state TMB model because it involves moving concentration fronts. In Aspen Chromatography, dynamic SMB equations are solved by discretizing the first- and second-order PDE spatial terms to obtain a large set of ODEs and algebraic equations, which constitute a DAE (differential-algebraic equation system) for which discretization or differencing methods are provided (see §13.6). Each complete SMB-model cycle provides a different result, which ultimately approaches a cyclic steady state. If the number of bed subsections per section is at least four and there are 10 or more cycles, the steady-state TMB result closely approximates the SMB result. Therefore, if only steady-state results are of interest, the simpler TMB model is best.

All four models apply to gas or liquid mixtures, with the latter being the most widely used in industrial separations. Regardless of the model used for SMB design (dynamic SMB or steady-state TMB), the information required is: (1) flow rate and feed composition; (2) adsorbent, S, and desorbent, D; (3) nominal bed operating temperature and pressure; (4) adsorption isotherm for all components, with known constants at bed operating conditions; (5) desired separation, which may be purity (on a desorbent-free basis) and desired recovery of the most strongly adsorbed component in the extract.

Not known initially but required before calculations can start are: (6) total bed height and inside diameter of adsorption column; (7) amount of adsorbent in the column; (8) desorbent recirculation rate; (9) flow rates for extract and raffinate; (10) overall mass-transfer coefficients for transport of solutes between bulk fluid and sorbate layer; (11) eddy diffusivity for axial dispersion; (12) spacing of inlet and outlet ports.

Guidance on values for items 6, 10, and 11 is sometimes found in patents for similar separations. For example, for the separation of xylene mixtures using para diethylbenzene as desorbent, Minceva and Rodrigues [173] suggest: (1) molecular-sieve zeolite adsorbent with a spherical particle diameter, d_p, between 0.25 and 1.00 mm and a particle density, ρ_p, of 1.39 g/cm^3; (2) operating temperature between 140°C and 185°C, with a pressure sufficient to maintain a liquid phase; (3) liquid interstitial velocity, u_f, between 0.4 and 1.2 cm/s; and (4) four sections with 8 to 24 subsections (beds).

For a commercial-size unit, the following are suggested: (5) bed height, L_k, in each subsection from 40 to 120 cm; (6) Equation (15-58) for estimating overall mass-transfer coefficient, $k_{i,j}$, for solute transport between bulk fluid and sorbate layer on the adsorbent; and (7) an axial diffusivity, D_{L_j}, defined in terms of a Peclet number, where

$$N_{\text{Pe}} = \frac{u_f(\text{characteristic length})}{D_L} \quad (15\text{-}170)$$

Characteristic lengths equal to bed depth or particle diameter have been used. Most common for TMB and SMB is bed depth, with Peclet numbers in the 1,000–2,000 range.

EXAMPLE 15.22 Operation with an SMB.

Use the fructose–glucose separation of Example 15.21, for $\beta = 1.05$, with the steady-state TMB model of Aspen Chromatography to estimate product compositions obtained with the following laboratory-size SMB unit: number of sections = 4; number of subsections (beds) in each section (column) = 2; all bed diameters = 2.54 cm; all bed heights = 10 cm; bed void fraction = 0.40; particle diameter = 500 microns (0.5 mm); overall mass-transfer coefficient for A and B = 10 min^{-1}; Peclet number high enough that axial dispersion is negligible.

Solution

To use Aspen Chromatography, the recirculating liquid flow rate for a TMB must be converted to an SMB using (15-165), and solid particle flow rate must be converted to a port-switching time given by (15-169). From (15-165), using the results for $\beta = 1.05$ in Example 15.21,

$$(Q_C)_{\text{SMB}} = 2.624 + \left(\frac{0.40}{1-0.40}\right) 7.848 = 7.856 \text{ mL/min}$$

The total liquid rate in SMB Section I is

$$(Q_I)_{\text{SMB}} = (Q_C)_{\text{SMB}} + Q_D = 7.856 + 2.403 = 10.259 \text{ mL/min}$$

This is the maximum volumetric flow rate in the SMB, and it is of interest to calculate a corresponding interstitial fluid velocity from (15-147),

$$(u_{f_1})_{\text{SMB}} = \frac{(Q_I)_{\text{SMB}}}{\epsilon_b A_b} = \frac{10.259}{0.40 \left[\frac{3.14(2.54)^2}{4}\right]}$$

$$= 5.06 \text{ cm/min} = 0.0844 \text{ cm/s}$$

This fluid velocity is low, but it corresponds to a desirable (bed-diameter)/(particle-diameter) ratio of $2.54/0.05 = 49$. To increase fluid velocity to, say, 0.4 cm/s, the bed diameter would be decreased to 1.17 cm, giving a (bed-diameter)/(particle-diameter) ratio of 23, which would still be acceptable.

From (15-149), the true velocity of the solid particles in each bed is

$$u_s = \frac{Q_S}{(1 - \epsilon_b)A_b} = \frac{7.848}{(1 - 0.40)\left[\dfrac{3.14(2.54)^2}{4}\right]} = 2.58 \text{ cm/min}$$

From (15-169), port-switching time for subsection bed height, L, of 10 cm is

$$t^* = \frac{L}{u_s} = \frac{10}{2.58} = 3.88 \text{ min}$$

The following results were obtained from Aspen Chromatography for a steady-state TMB:

	Feed	Desorbent	Extract	Raffinate
Flow rate, mL/min	1.667	2.403	2.134	1.936
Concentrations, g/L:				
Fructose	280.0	0.0	211.6	12.7
Glucose	350.0	0.0	8.4	295.3
Water	596.0	996.0	861.7	795.8
Mass fraction on water-free basis:				
Fructose	0.444		0.962	0.040
Glucose	0.556		0.038	0.960

A reasonably sharp separation between fructose and glucose is achieved. In Exercise 15.44, modifications to the input data are studied in an attempt to improve separation sharpness.

EXAMPLE 15.23 Recovery of Paraxylene in an SMB.

Minceva and Rodrigues [173] consider the industrial-scale separation of paraxylene from a liquid mixture of C_8 aromatics in a four-section SMB. Feed is 1,450 L/minute with the component data given below. The adsorbent is a molecular-sieve zeolite with a particle density of 1.39 g/cm^3 and a diameter of 0.092 cm that packs to a bed with an external void fraction of 0.39. The desorbent is paradiethylbenzene (PDEB). With reference to Figure 15.57, the numbers of subsections are 6, 9, 6, and 3, respectively, in Sections I to IV. The height of each is 1.135 m, with a bed diameter of 4.117 m. The operation is at 180°C and a pressure above 12 bar to prevent vaporization. An extended Langmuir adsorption isotherm (15-129) correlates adsorption equilibrium, yielding the following constants. This is a constant-selectivity isotherm; therefore, the selectivity relative to paradiethylbenzene is tabulated.

Component	q_m, mg/g	K, cm^3/mg	Selectivity
Paraxylene	130.3	1.0658	0.9969
Paradiethylbenzene	107.7	1.2935	1.0000
Ethylbenzene	130.3	0.3067	0.2689
Metaxylene	130.3	0.2299	0.2150
Orthoxylene	130.3	0.1884	0.1762

The desorbent does not have the most desirable equilibrium adsorption property because its selectivity does not lie between that of paraxylene and the C_8 components. The overall mass-transfer coefficient between sorbate and bulk fluid in (15-150) is 2 minutes^{-1} for each component. For axial dispersion, assume a Peclet number of 700 in (15-170) with bed height as characteristic length.

Using Aspen Chromatography with the TMB model as an SMB approximation, determine steady-state flow rates and compositions of extract and raffinate, with composition profiles in the four sections for the following operating conditions: extract flow rate = 1,650 L/minute; raffinate flow rate = 2,690 L/minute; circulation flow rate, $(Q_C)_{SMB}$, before adding makeup DPEB = 5,395 L/minute; and port-switching interval, $t^* = 1.15$ minute.

Solution

By an overall material balance, the DPEB makeup flow rate is

$$Q_D = Q_E + Q_R - Q_F = 1,650 + 2,690 - 1,450 = 2,890 \text{ L/minute}$$

From the switching time, using (15-169), with a 1.135-m bed height,

$$u_s = 1.135/1.15 = 0.987 \text{ m/minute} = 98.7 \text{ cm/minute}$$

The bed cross-sectional area, $A_b = 3.14(4.117)^2/4 = 13.31 \text{ m}^2$. From (15-149), the solid particle volumetric flow rate in a TMB is

$$Q_S = u_s(1 - \epsilon)A_b = 0.987(1 - 0.39)(13.31) = 8.014 \text{ m}^3/\text{minute} = 8,014 \text{ L/minute}$$

Liquid flow rates in the four sections are as follows, where both $(Q_j)_{SMB}$ and $(Q_j)_{TMB}$ are included, the former from material balances and the latter from (15-165). For example,

$$(Q_I)_{SMB} = (Q_C)_{SMB} + Q_D = 5,395 + 2,890 = 8,285 \text{ L/min}$$
$$(Q_I)_{TMB} = (Q_I)_{SMB} - [0.39/(1 - 0.39)]Q_S$$
$$= 8,285 - 0.639(8,014) = 3,164 \text{ L/min}$$

Section in Figure 15.57	$(Q_j)_{SMB}$, L/minute	$(Q_j)_{TMB}$, L/minute
I	8,285	3,164
II	6,635	1,514
III	8,085	2,964
IV	5,395	274

Aspen Chromatography results for the steady-state TMB model, but on an SMB basis, are:

Wt% of component	Feed	Desorbent	Extract	Raffinate
Ethylbenzene	14.0	0.0	0.00	7.63
Metaxylene	49.7	0.0	0.00	27.09
Orthoxylene	12.7	0.0	0.00	6.92
PDEB	0.0	100.0	80.79	57.85
Paraxylene	23.6	0.0	19.21	0.51

There is an excellent separation between paraxylene and the other feed components. However, both extract and raffinate contain a substantial fraction of desorbent, PDEB. The desorbent in both products is recovered by the hybrid SMB-distillation process in Figure 15.43. Concentration profiles in the four sections, as computed by Aspen Chromatography, are shown in Figure 15.61. In Sections I and III particularly, they differ considerably from the flat profile predictions of the simple, local-equilibrium TMB model. The circulating desorbent is predicted to be essentially pure PDEB.

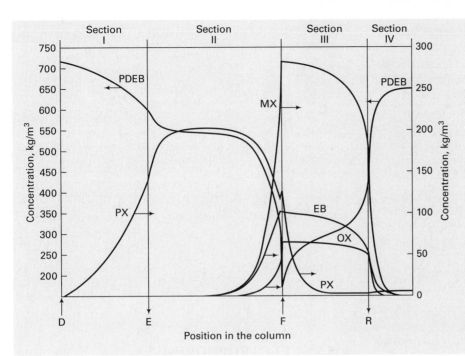

Figure 15.61 Concentration profiles in the liquid for SMB of Example 15.23.

§15.7 ION-EXCHANGE CYCLE

Although ion exchange has a wide range of applications, water softening with gel resins in a fixed bed continues to be its dominant use. It operates in a four-step cycle: (1) loading, (2) displacement, (3) regeneration, and (4) washing. The solute ions removed from water in the loading step are mainly Ca^{2+} and Mg^{2+}, which are absorbed by resin while an equivalent amount of Na^+ is transferred from resin to water. If mass transfer is rapid, solution and resin are at equilibrium throughout the bed. With a divalent ion (e.g., Ca^{2+}) replacing a monovalent ion (e.g., Na^+), the equilibrium expression is (15-44), where A is the divalent ion. If $(Q/C)^{n-1} K_{A,B} \gg 1$, equilibrium for the divalent ion is very favorable (see Figure 15.34a) and the type of self-sharpening front in Figure 15.34b develops. In that case, which is common, ion exchange is well approximated using simple stoichiometric or shock-wave front, plug-flow theory for adsorption. As the front moves down through the bed, the resin behind the front is in equilibrium with the feed composition, while ahead of the front, water is free of the divalent ion(s). Breakthrough occurs when the front reaches the end of the bed.

Suppose the only cations in the feed are Na^+ and Ca^{2+}. Then, from (15-44), with $n = 2$,

$$K_{Ca^{2+}, Na^+}\left(\frac{Q}{C}\right) = \frac{y_{Ca^{2+}}\left(1 - x_{Ca^{2+}}\right)^2}{x_{Ca^{2+}}\left(1 - y_{Ca^{2+}}\right)^2} \qquad (15\text{-}171)$$

where Q is total concentration of the two cations in the resin, in eq/L of wet resin, and C is total concentration of the two ions in the solution, in eq/L of solution. One mole of Na^+ is one equivalent, while 1 mole of Ca^{2+} is two equivalents, and y_i and x_i are equivalent (rather than mole) fractions. From Table 15.5, using (15-45), the molar selectivity factor is

$$K_{Ca^{2+}, Na^+} = 5.2/2.0 = 2.6$$

For a given loading step during water softening, values of Q and C remain constant. Thus, for a given equivalent fraction $x_{Ca^{2+}}$ in the feed, (15-171) is solved for the equilibrium $y_{Ca^{2+}}$. By material balance, for a given bed volume, the time t_L for the loading step is computed. The loading wave-front velocity is $u_L = L/t_L$, where L is the height of the bed. Equivalent fractions ahead of and behind the loading front are in Figure 15.62a. Typically, feed-solution superficial mass velocities are about 15 gal/h-ft^2, but can be much higher at the expense of larger pressure drops.

At the end of the loading step, bed voids are filled with feed solution, which must be displaced. This is best done with a regeneration solution, which is usually a concentrated salt solution that flows upward through the bed. Thus, the displacement and regeneration steps are combined. Following displacement, mass transfer of Ca^{2+} from the resin beads to the regenerating solution takes place while an equivalent amount of Na^+ is transferred from solution to resin. In order for equilibrium to be favorable for regeneration with Na^+, it

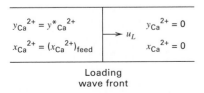

Loading
wave front

(a)

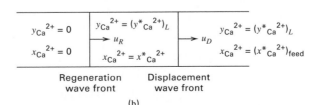

Regeneration   Displacement
wave front   wave front

(b)

Figure 15.62 Ion exchange in a cyclic operation with a fixed bed. (a) Loading step. (b) Displacement and regeneration steps.

is necessary for $(Q/C)K_{Ca^{2+}, Na^+} \ll 1$. In that case, which is the opposite for loading, the wave front during regeneration sharpens quickly into a shock-like wave. This criterion is satisfied by using saturated salt solution to give a large value for C.

During displacement and regeneration, two concentration waves, shown in Figure 15.62b, move through the bed. The first is the displacement front; the second, the regeneration front. For plug flow and negligible mass-transfer resistance, resin and solution are in equilibrium at all bed locations, and (15-171) is used to solve for the equilibrium equivalent fractions shown for the displacement and regeneration steps in Figure 15.62b. The displacement time, t_D, is determined from the interstitial velocity, u_D, of the fluid during displacement:

$$t_D = L/u_D \qquad (15\text{-}172)$$

Regeneration time, t_R, is determined by material balance, from which the regeneration wave-front velocity, which is generally less than the feed velocity, is $u_R = L/t_R$. The cycle, illustrated in the following example, is completed by water displacement of salt solution in the bed voids.

EXAMPLE 15.24 Ion-Exchange Cycle.

Hard water containing 500 ppm (by weight) $MgCO_3$ and 50 ppm NaCl is to be softened at 25°C in an existing fixed, gel resin bed with a cation capacity of 2.3 eq/L of bed volume. The bed is 8.5 ft in diameter and packed to a height of 10 ft, with a wetted-resin void fraction of 0.38. During loading, the recommended throughput is 15 gal/minute-ft². During displacement, regeneration, and washing, flow rate is reduced to 1.5 gal/minute-ft². The displacement and regeneration solutions are water-saturated with NaCl (26 wt%). Determine: (a) feed flow rate, L/minute; (b) loading time to breakthrough, h; (c) loading wave-front velocity, cm/minute; (d) regeneration solution flow rate, L/minute; and (e) displacement time, h.

Solution

Molecular weight, M, of $MgCO_3 = 83.43$

Concentration of $MgCO_3$ in feed $= \dfrac{500(1,000)}{83.43(1,000,000)} =$ 0.006 mol/L or 0.012 eq/L

M of $NaCl = 58.45$

Concentration of NaCl in feed $= \dfrac{50(1,000)}{58.45(1,000,000)} =$ 0.000855 mol/L or eq/L

(a) Bed cross-sectional area $= 3.14(8.5)^2/4 = 56.7 \text{ ft}^2$.
 Feed-solution flow rate $= 15(56.7) = 851$ gpm or 3,219 L/minute

(b) Behind the loading wave front:

$$x_{Mg^{2+}} = \frac{0.012}{0.012 + 0.000855} = 0.9335$$

Since no NaCl in the feed is exchanged, $C = 0.012$ eq/L and $Q = 2.3$ eq/L.
From Table 15.5,

$$K_{Mg^{2+}, Na^+} = 3.3/2 = 1.65$$

From (15-171), for Mg^{2+} instead of Ca^{2+} as the exchanging ion, with $x_{Mg^{2+}} =$ that of the feed from Figure 15.62a:

$$1.65\left(\frac{2.3}{0.012}\right) = \frac{y_{Mg^{2+}}(1 - 0.9335)^2}{0.9335(1 - y_{Mg^{2+}})^2}$$

Solving: $y^*_{Mg^{2+}} = 0.7733$; bed volume $= (56.7)(10) = 567 \text{ ft}^3$ or 16,060 L; total bed capacity $= 2.3(16,060) = 36,940$ eq; Mg^{2+} absorbed by resin $= 0.9961(36,940) = 36,796$ eq; Mg^{2+} entering bed in feed solution $= 0.012(3,219) = 38.63$ eq/minute

$$t_L = \frac{36,796}{38.63} = 953 \text{ minutes or } 15.9 \text{ h}$$

(c) $u_L = L/t_L = 10/953 = 0.0105$ ft/minute or 0.32 cm/minute.

(d) Regeneration solution flow $= (1.5/15)(3,219) = 321.9$ L/minute.

(e) Displacement time for 321.9 L/minutes to displace liquid in voids.

---

§15.8 ELECTROPHORESIS

Electrophoresis separates charged molecules (e.g., nucleic acids and proteins) according to their size, shape, and charge in an electric field. Table 15.15 lists several different modes of electrophoresis that are described in this section.

Analytical applications include DNA sequencing and fingerprinting, and identifying factors for diseases like cystic fibrosis and leukemia. Figure 15.63 illustrates electrophoretic separation of bands (black horizontal bars) of linear DNA fragments, whose concentration (and observed intensity) increases from left to right in series C1 to C3, and X1 to X3. Each band consists of daughter strands amplified from an initial template DNA strand using polymerase chain reaction (PCR). Bands are compared side-by-side by adding a microliter sample from the PCR product to a series of wells located at the bottom of the gel. Application of an electric field in the vertical direction moves DNA anions in each band toward the top of the gel in individual columns or 'lanes'. Lane B on the LHS contains a control sample of DNA with about 10 sets of bands consisting of linear strands with successively larger numbers of base pairs, bp, which are indicated on the LHS of the gel (50, 100, 150 bp).

In general, a particle of diameter d_p and electrostatic charge q moves at constant terminal velocity u_t in an electric field gradient, E in volts/cm, when drag force F_d given by

Stokes' law $\qquad F_d = 3\pi\mu d_p u_t \qquad (15\text{-}173)$

Table 15.15 Electrophoresis Modes

Electrophoresis Mode	Basis for Separation
Native gel	Native biopolymer size
Denaturing gel	Protein MW
Isoelectric focusing	Isoelectric point (pI)
Isotachophoresis	Mobility in an electric field
2-D gel	pI and MW
Pulsed field gel	Biopolymer strand length

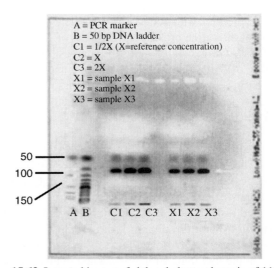

Figure 15.63 Inverted image of slab gel electrophoresis of 110-bp β-globin DNA fragments stained with methylene blue. Lane B contains a 50-bp DNA ladder. Lanes C1 to C3 show PCR amplicons from increasing base concentrations of the template. Lanes X1 to X3 are experimental data. The 110-bp β-globin DNA fragment is associated with anemic phenotype in β-Thalassemia.

due to viscosity, μ, is just balanced (i.e., at $\Sigma F = 0$) by the electrostatic force, $F_{el} = qE$, to give

$$qE = 3\pi\mu d_p u_t \qquad (15\text{-}174)$$

The isoelectric point, pI, of each protein relative to solution pH yields a net electrostatic charge that, along with its size and solution viscosity, determines its characteristic electrophoretic mobility, U_{el}, in an electric field:

$$U_{el} = \frac{u_t}{E} = \frac{q}{3\pi\mu d_p} \qquad (15\text{-}175)$$

Values of U_{el} can be measured using Schlieren optics to track moving boundaries that correspond to voltage-induced, free-solution migration of a species toward an electrode placed in either arm of a U-tube. This method was described by Arne Tiselius [177], who was awarded the Nobel Prize in Chemistry in 1948. One arm of the U-tube is filled with ~10 mg of protein in solution and the other with buffer. A semi-empirical expression for free solution mobility specific to peptides with $n = 3$ to 39 amino acids and $0.33 \leq q \leq 14$, with q in coulombs, is given [178] by

$$U_{el} = D\frac{\ln(q + 1)}{n^{0.43}} \qquad (15\text{-}176)$$

where slope D is fit to experimental data to determine the electrophoretic mobility in (15-174).

The apparent electrophoretic mobility, U, whose direction parallel or antiparallel to the applied electric field is determined by the sign of q, is given by

$$U = U_{el} + U_o \qquad (15\text{-}177)$$

which may include electro-osmotic flow, U_o, the movement of ion-associated carrier liquid (i.e., H_2O) in an open channel

as a result of an applied electric field. U_o is proportional to the zeta potential (electrical double layer in §2.9). Electro-osmotic flow of ion-associated carrer liquid (e.g. H_2O) balances cathode-directed flow of dissolved *cations* that are attracted to fixed charges like hydroxyl groups on paper or silanol groups on glass capillary walls, which are ionized to SiO^- above pH 3. Electro-osmotic flow thus counteracts field-induced electrophoresis of *anions* like DNA. Increasing ionic strength decreases zeta potential and reduces electro-osmosis, but increases conductivity and Joule heating.

Performing electrophoresis in a polymer *gel* matrix, i.e., *gel electrophoresis*, minimizes U_o in (15-177). Typical gels used are agarose, an uncharged polymer extracted from red seaweed used to analyze DNA, or polyacrylamide, a synthetic, cross-linked polymer introduced by Ornstein [179] to electrophorese proteins or nucleic acids. Paper can replace gels, but paper has fixed hydroxyls that produce electro-osmotic effects. Cellulose-acetate membranes, in which acetate esters replace two of every three hydroxyls, often replace gel as the matrix.

To perform gel electrophoresis, dissolved biomolecules are pipetted into a well formed at the distal end of the gel, and move down a lane toward the anode, separating into defined zones based on size, shape, and charge, as shown in Figure 15.63. Biomolecule interactions are reduced using bulky organic buffering species with charge type $n = \pm 1$ (§2.9) at total ionic strengths of 0.05 to 0.15, rather than mobile, current-carrying ions (e.g., metal ions Na^+, K^+, Mg^{2+}, and simple anions F^-, Cl^-, Br^-, SO_4^{2-}, HPO_4^{2-}) that increase Joule heating. Electrophoresed biomolecules are chemically stained or radioactively labeled for analytical detection, excised for further evaluation, or eluted from the end of the gel for preparative recovery.

§15.8.1   Resistive Heating

When compared with free-solution electrophoresis, electrophoresis performed in slab gels reduces effects of *resistive Joule heating*, which can denature protein products and produce thermal convection (which distorts resolved bands) or generate electrolysis gases. Temperature increases in an electrophoretic medium of density ρ, heat capacity C_p, and characteristic length l (e.g. length of voltage drop) in proportion to the applied power, $P = I^2R$, *viz.*,

$$\rho C_P \frac{dT}{dt} = \frac{I^2R}{l^3} = \frac{VI}{l^3} \qquad (15\text{-}178)$$

where I is current, R is electrical resistance, and V is applied voltage. Increasing media resistance, i.e. by lowering ionic strength or including polymer in the matrix, can decrease current density in (15-178) normal to electrophoretic cross section I/l^2 via Ohm's law ($I = V/R$) and lower resistive heating. The applied voltage per unit length in (15-178), V/l, is limited by heat dissipation to ~300 V/cm in capillaries. Heat dissipation increases with the ratio of surface area to volume, in which the field is applied. Therefore, 0.5- to 1.5-mm-thick gel slabs sandwiched between glass plates or ~0.1-mm-i.d. fused-silica capillaries are widely used for analytical electrophoresis. Operating electrophoresis at 4°C minimizes temperature increases in (15-178) by maximizing ρ_{H_2O}, since water density is highest at this temperature. Resistive heating can produce thermal gradients that decrease μ in (15-174), causing a distribution of electrophoretic mobilities, U_{el}, and thermal convection, both of which result in zone spreading. Operating at low values of the Rayleigh number (N_{Ra}), the ratio of free to forced convection, reduces thermal convection, where N_{Ra} is given by

$$N_{Ra} \equiv \frac{\beta g h^3 \Delta T}{\alpha \mu} \qquad (15\text{-}179)$$

and where β is the thermal expansion coefficient, h is the characteristic free vertical dimension (e.g. length in the direction of buoyant heat rise) or gap distance (e.g., capillary diameter or diameter of porous gel), α is the thermal diffusivity, and g is the gravitational constant. Employing small capillaries, porous gels, or microgravity conditions (albeit infeasible) reduces h and limits thermal convection. Although gels provide a porous barrier to thermal convection, scale-up of electrophoretic throughput by increasing cross sectional area is limited by resistive heating as the surface area-to-volume ratio decreases in preparative systems.

§15.8.2 Electrophoretic Modes

Electrophoresis is conducted in several different operating modes, as follows. **Native gel electrophoresis** is performed under non-denaturing conditions at a constant pH of ~8–9, at which most proteins are anionic (acidic) and migrate toward the anode. Proteins retain their native (non-denatured) configuration. Using a low-conductivity tris-borate buffer at pH 8–9 minimizes protein–protein interactions and lowers buffer mobility since it is partially uncharged: borate$^-$ $\leftrightarrow$ H-borate. Planar *slab gels* typically use 5 to 25 μg of protein loaded in a 0.5–5 μL sample volume into a well at the outer end of the gel. More sensitive detection (e.g., using silver stain) allows even smaller quantities of dilute protein sample to be used.

Capillary electrophoresis (CE), performed in gel-filled or open capillaries, requires less than ~1% of these loading values (~5–50 nL) and can tolerate $E \geq 300$ V/cm due to efficient heat dissipation. Adequate buffering (pH buffers in §2.9) of electrophoresis is needed to mitigate electrolysis of H_2O at the electrodes, which produces $^1\!/_2 H_2 + OH^-$ at the reductive cathode and $^1\!/_2 O_2 + e^- + 2H^+$ at the oxidative anode.

Denaturing gel electrophoresis is performed by interacting hydrophobic side chains of proteins with sodium dodecyl sulfate (SDS), an ionic surfactant (chaotropes in §2.9). This occurs at a mass ratio of 0.9 to 1.0 g SDS/g protein. The interaction disrupts protein folding, solubilizes the side chains, denatures the protein, and yields a relatively constant ratio of charge to MW for MW of ~180 to 40,000 daltons [180]. SDS-denatured proteins migrate in polyacrylamide gel electrophoresis (SDS-PAGE), with mobilities inversely proportional to log(MW). Comparing denatured protein *bands* in sample *lanes* on a slab gel with a standard *ladder* containing

a series of proteins with decreasing molecular weights in the direction of migration provides an estimate of sample protein MW. SDS-PAGE generally yields the sharpest overall resolution ($\geq 1\%$ of U_{el}) and cleanest zones of any method.

Comparing denaturing SDS-PAGE results with native gel electrophoresis results can identify the number of subunits formed per protein molecule. Proteins in which internal disulfide bonds have been reduced (disrupted) by a thiol reagent like β-mercaptoethanol can bind ~1.4 g SDS/g protein. The charge-to-mass ratio in the resulting rod-like polypeptides is essentially equal. The number of internal disulfide bonds can be determined by comparing reduced and nonreduced protein samples. Disruption of hydrogen bonds in a protein by first dissolving it in 6- to 8-M urea may be used to characterize hydrogen bonding and/or to solubilize proteins in the low ionic strengths needed in electrophoresis. Binding of ionic species to the biomolecule in SDS-PAGE increases zeta potential, which can increase U_o in (15-177).

In **isoelectric focusing** (IEF), a pH gradient is developed and maintained along the direction of migration to stop a protein zwitterion at the point where local pH equals its isoelectric point (pI in §2.9) and its net charge becomes zero. In practice, pI may be buffer dependent: phosphate and citrate ions commonly bind to enzymes and lower the pI. U_{el} is often taken to increase in proportion to distance x from the focal point, i.e.,

$$U_{el} = -\frac{dU_{el}}{dx}x = -\frac{dU_{el}}{d(\text{pH})}\frac{d(\text{pH})}{dx}x \qquad (15\text{-}180)$$

since (15-175) shows U_{el} decreases as q approaches zero. Formation and maintenance of a constant pH gradient using carrier molecules called ampholytes is described in Example 15.25. The chain rule has been used on the RHS of (15-180). IEF reaches steady state after 3 to 30 h in slab gels, or ~30 min in capillaries, where E may be increased due to lower resistive heating. Current, I, decreases to just μa (microamperes) as IEF nears completion since R increases sharply when all carrier and protein ampholytes converge on their respective values of pI. IEF offers a useful free-flow preparative method capable of cell separation [181].

EXAMPLE 15.25 Isoelectric Focusing.

Quantitatively describe the unique focusing capability of IEF and identify factors that increase resolution of proteins in IEF.

Solution

IEF concentrates dilute proteins and mitigates zone spreading because at steady state, particle diffusion, D, away from the focal point is just balanced by field-gradient-induced flow toward it:

$$D\frac{dC}{dx} = U_{el}CE \qquad (15\text{-}181)$$

where, C is concentration. Substituting (15-180) into (15-181), separating variables, and integrating gives the distribution

$$C = C_{max}\exp\left(\frac{-x^2}{2\sigma_{IEF}^2}\right) \qquad (15\text{-}182)$$

where diffusive band width generated in isoelectric focusing, σ_{IEF}, is defined by

$$\sigma_{IEF}^2 = \frac{D}{E \dfrac{dU_{el}}{d(\text{pH})}\dfrac{d(\text{pH})}{dx}} \qquad (15\text{-}183)$$

Equation (15-183) shows that IEF bandwidth decreases (i.e., resolution increases) for bioproducts with low diffusivity, D, whose mobility changes substantially with respect to pH. Proteins that differ in pI by ≥ 0.02 may be concentrated and separated from dilute samples by IEF. High field values and a sharp pH gradient also increase resolution in (15-183). A sharp pH gradient is developed by adding carrier *ampholyte*—an N- (amino) and C- (carboxyl) terminated polymethylene, $-(CH_2)_{-n}$, whose ionization and isoelectric point depends on n, the number of repeat units. Self-buffering ampholytes that bracket a selected pH range migrate until pH equals ampholyte pI, creating a spatially distributed pH gradient. Ampholyte cost and protein instability and insolubility near pI values (see pH Affects Solubility in §2.9.1) limit preparative applications of IEF.

In **isotachophoresis**, bands in a protein sample are sandwiched between a leading electrolyte with high mobility (e.g., Tris-chloride buffer, pH 6.7) and a trailing electrolyte with lower mobility (e.g., Tris-glycine buffer, pH 8.3) in a two-buffer system [182]. Components are separated on the basis of mobility per unit electric field: two ions possessing like charges, but different mobilities, segregate until all faster ions lead the slower ones. Counterions traveling in the opposite direction (e.g., $HTris^+$) must be balanced at each location, which prevents formation of a void between adjacent ions, maintains identical velocities of each component, and forms a sharp boundary (i.e., Kohlrausch discontinuity) between adjacent ions. Because conductivity is inversely proportional to voltage, field strength self-adjusts, becoming lowest for the high-mobility leading band, highest for the trailing band, and constant across each band. The position-dependent electric field experienced by solutes due to changes in buffer pH and the high electrolyte concentration creates self-sharpening zones in the field gradient between the leading and trailing electrolytes.

In preparative isotachophoresis of proteins, where low solute conductivity can produce high, degradative potentials, ampholytes with mobilities intermediate between adjacent protein components are added to increase conductivity, stabilize boundaries, and act as spacer ions to further increase resolution and provide band separation. In stacking or disc gel electrophoresis [183], illustrated in Figure 15.64, protein bands are distinguished and concentrated via isotachophoresis in a dilute stacking region ($\sim$2.5% acrylamide at pH 6.7), and then separated via zone gel electrophoresis in a subsequent separating region ($\sim$15% acrylamide at pH 8.9).

Two-dimensional gel electrophoresis first separates proteins in a polyacrylamide gel according to pI by isoelectric focusing. Proteins are then coated with an anionic surfactant and separated by molecular weight in an orthogonal direction using another polyacrylamide gel. Up to 1,000 individual protein components may be identified from crude cell extracts. Detection via chemical staining by autoradiography or other methods discussed below reveals spots of proteins with unique charge and size on a 2-D array that typically displays isoelectric points vertically and MW horizontally.

The pattern on the 2-D gel provides a characteristic fingerprint of protein expression in a particular biological system that exhibits 100-fold better resolution than other protein-analysis techniques. Protein spots cut out of the gel may be further analyzed by microchemical techniques, to determine amino acid sequence, for example. Two-dimensional electrophoresis is also used in *crossed immunoelectrophoresis*, in which a longitudinal strip cut out from an electrophoresed slab is placed on an antibody-containing gel into which proteins are electrophoresed sideways to form an antigen–antibody precipitin pattern.

In **pulsed field gel electrophoresis** (PFGE), an electric field, E, is applied to an agarose gel for duration t_{pulse} in a

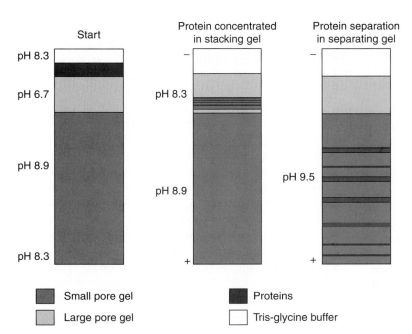

Figure 15.64 Protein concentration and separation in stacking gel electrophoresis.

direction relative to a primary x-axis given by an angle $90° < \theta < 270°$. Changing θ intermittently causes long, linear DNA strands to kink and become entangled in the agarose [184]. Longer strands move in slow, zigzag motion relative to shorter, rod-like strands that quickly reorient to the new direction and move more readily through the agarose maze. Quantitatively, the distance d_x traveled by a strand with respect to the x-axis is

$$d_x = u_t t_{pulse} \cos\theta \qquad (15\text{-}184)$$

The distance, J_x, between the front and rear ends of a strand of length, l, relative to the x-axis is

$$J_x = (u_t t_{pulse} - l)\cos\theta \qquad (15\text{-}185)$$

After one pulse, strands i and j are separated by a distance

$$\Delta J_x = -(l_i - l_j)\cos\theta \qquad (15\text{-}186)$$

The strand separation after n pulses of equal duration in each of m directions is

$$\Delta S_x = -n(l_i - l_j)\sum_{k=1}^{m} \cos\theta_k \qquad (15\text{-}187)$$

where $n = $ (total time)/t_{pulse}. Strand separation thus increases proportionally with the number of pulses and inversely to the pulse duration, as confirmed experimentally. Using $\theta = 180°$ requires forward and reverse pulses to have unequal duration. Resolution of long DNA strands—i.e., 20 kilo-basepairs (kbp) to mega-basepairs (Mbp)—where $l < u_t t_{pulse}$, in PFGE is higher than in a continuous, unidirectional electric field, in which resolution decreases in inverse proportion to log(MW).

EXAMPLE 15.26   Pulsed Field Gel Electrophoresis.

How many pulses of equal duration are required to separate DNA strands of length 100 kbp and 200 kbp, respectively, by 1 mm, using PFGE and alternating between angles of 135° and 225°?

Solution

DNA measures 0.34 nm (i.e. 3.4×10^{-7} mm linear length) per base pair. To obtain the solution, the following variable values are obtained from the problem statement: $\Delta S_x = 1$ mm; $\cos\theta_k = 0.7071$ for $\theta = 135°$ at $k = 1$ and $\theta = 225°$ at $k = 2$; and $l_i - l_j = -100,000$ base pairs $\times 3.4 \times 10^{-7}$ mm per base pair. Next, (15-187), is rearranged to solve for n. Substituting the values into the rearranged equation gives

$$n = \Delta S_x \Big/ \left[-(l_i - l_j)\sum_{k=1}^{m}\cos\theta_k \right]$$
$$= 1 \Big/ - \left[(-1 \times 10^5)(3.4 \times 10^{-7})2(0.7071) \right]$$
$$= 20.8$$

So the number of equal-duration pulses to separate the two strands is $n = 21$, and 42 total pulses at alternating angles would be required. Upon measuring the electrophoretic mobility for this system, (15-175) could be used to identify the terminal velocity, from which the time required per pulse could be estimated, and the total time required could be determined.

§15.8.3   Electrophoretic Gels, Geometries, and Detection

Porous *polyacrylamide gel*, the most frequently used matrix, is synthesized in-place by mixing a neurotoxic acrylamide monomer such as $CH_2{=}CHCONH_2$, with a cross-linker such as bis-acrylamide, ethlyene-diacrylate, or N,N'-di-allyl-tartar diamide and catalysts. The catalysts—free-radical former ammonium persulfate (*ca.* 1.5–2 mM), together with free-radical scavenger tetramethylenediamine (TEMED; *ca.* 0.05–0.1% vol/vol)—initiate polymerization. Sieving properties in *polyacrylamide gel electrophoresis* (PAGE) are determined by effective pore size. Pore size is controlled by varying the mass % of acrylamide + co-monomer added per volume of solution, defined as %T, from 3 to 30%, and/or by adjusting the mass ratio of co-monomer per 100 g of acrylamide, defined as %C, from 0.1 to 1%. High %C increases opacity of gels, which diminishes subsequent visualization. Gelation occurs within 30 minutes at room temperature. Large proteins up to 10^6 daltons require 3–4% T and 0.1% C, whereas small proteins of 10^4 daltons may use 15% T and/or up to 1% C. Polyacrylamide yields pore sizes down to <80 nm and lacks fixed charges. These features minimize electro-osmosis and nonspecific convection due to Joule heating in PAGE.

Agarose gels are a nontoxic, less-expensive alternative to polyacrylamide. Agarose sets to a firm gel at only 2% wt/vol and maintains higher porosity, an advantage in some situations. For example, large pores diminish the viscous barrier to mobility and allow faster equilibration to the focal point in IEF. Agarose is dissolved into a buffer with attendant heating, then poured into the casting apparatus.

Gels for *slab electrophoresis* are formed between two vertical, parallel glass plates sealed on the bottom and sides, using a *comb* along the top to form indented *wells* into which samples will be loaded. Plates are separated by as little as <1 mm to minimize heating in the electrophoretic cross section, I/l^2 in (15-178), and to increase diffusion of dye into the slab during staining and destaining. Horizontal-plate orientation is used for fragile gels. *Capillary electrophoresis* (CE) is performed in gel-filled or open 10–300 μm i.d. fused silica capillaries coated externally with polyimide (to provide flexibility) and chemically modified internally with a hydrophilic coating to prevent interactions with surface silanols and reduce electro-osmosis. Carrier fluids in CE are aqueous buffers that contain no hazardous or expensive solvents, an advantage relative to analytical reversed-phase chromatography, which uses such solvents.

Detection of electrophoresed biological species uses chemical stains, fluorescence, radioactivity, immunological probes, or spectroscopy/spectrometry. Most dyes associate with positively charged lysines, arginines, and histidines on proteins. Therefore, more basic polypeptides exhibit relatively higher densitometric intensity. Detection methods commonly "fix" (denature) bands in-place on slab gels using a 1-h soak in a fixative. Typical fixatives include dilute acetic acid, methanol/acetic acid/water at 3:1:6, trichloroacetic acid (a protein precipitant), or 5–10% perchloric acid. SDS and dye-binding ampholytes must be washed out during fixing to allow protein to bind dye. After fixing and washout, electrophoresed bands can be visualized directly in the optical spectrum by *chemical staining* with Co-omassie blue or silver.

Co-omassie brilliant blue (e.g., G250, or especially R250 for proteins) is compatible with either agarose or polyacrylamide matrices and provides high sensitivity. Silver staining can yield up to 100-fold greater sensitivity relative to that of Co-omassie blue, but since it binds nonstoichiometrically, subsequent scanning by densitometry may not produce intensities in proportion to sample mass. Excess dye or stain is washed out to provide contrast before visualization.

EXAMPLE 15.27 Banding Pattern.

Describe the banding pattern expected after fixing, washout, and staining of an SDS-PAGE of protein with subunit structure $\alpha_2\beta_2$, where $MW_\alpha = 50$ kDa and $MW_\beta = 30$ kDa, using Co-omassie brilliant blue G250.

Solution

Ideally, densitometry should reveal two bands: a distal band at 50 kDa and a proximal band at 30 kDa. The 30-kDa band should exhibit 60% of the intensity of the 50-kDa band, in proportion to its relative mass in the gel.

Fluorescent labeling, which commonly uses fluorescamine or ethidium bromide, can improve sensitivity, albeit with increased cost and instrumentation relative to dyes. Fluorescamine yields a fluorescent product by reacting with primary amines on proteins at basic pH and room temperature. It can be applied either before or after electrophoresis. Ethidium bromide intercalates into DNA and enhances its quantum efficiency, allowing visualization under a UV lamp. *Radioactive* samples exposed to an adjacent X-ray film produce a mirror image of the banding pattern called an *autoradiograph*.

Blotting allows identification and detection via specific hybridization or immunological interactions using reagents whose access to electrophoresed bands may be restricted by the gel matrix itself. Transfer of protein (by electrophoresis) or nucleic acid (by capillary action) to an adjacent nitrocellulose paper, nylon, or other high-affinity membrane allows access by the specific hybridizing or immunological probe. Table 15.16 lists common blotting techniques used with electrophoresis. Detection of each technique may occur by autoradiography, fluorescence, chemiluminescence, or colorimetry.

Bioproducts resolved by CE are detected *in-line* using spectroscopic (i.e., absorptive or fluorescent) or electrochemical detectors, or *post-column* using mass spectrometry (MS). The small capillary i.d. confines the sample to a small volume and reduces the distance required to reach an electrode surface, which increases electrochemical-conversion efficiency, but its low optical path length reduces absorbance relative to conventional spectroscopic flow cells. *Electropherograms* display results of CE detected by spectroscopic or electrochemical detectors. Interfacing CE with MS by on-line electrospray ionization is made difficult by the

Table 15.16 Common Electrophoretic Blotting Methods

Method	Analyte	Membrane	Probe
Southern blot	DNA	Nitro-cellulose	Radiolabeled or nonradiolabeled complementary RNA or DNA
Northern blot	RNA	Nitro-cellulose	Radiolabeled or nonradiolabeled complementary RNA or DNA
Western blot	Proteins	Nitro-cellulose or nylon	Whole antisera or antibodies

Adapted from Harrison et al. [89]

incompatibility of the interface with common electrophoresis buffer salts and solvents.

Band dispersion in zone electrophoresis occurs in a manner similar to convective dispersion of a Heaviside-function pulse in linear differential chromatography. Ideally, solutes move due to an applied field gradient and partition completely to the carrier fluid phase, thus eliminating boundary-layer and intraparticle resistances to transport.

EXAMPLE 15.28 Analogy Between Electrophoresis and Pulse Chromatography.

Use analogies in mass transport between a band moving via a field gradient in electrophoresis and a sharp pulse moving via bulk convection in linear differential chromatography to develop general expressions for electrophoretic plate height, residence time, variance, and average time required to separate adjacent bands via electrophoresis.

Solution

As in pulse chromatography, a band of electrophoresed protein disperses about its center of mass as it migrates toward the anode due to Fickian convective axial dispersion (§15.3), resulting in a Gaussian distribution given by (15-53). In ideal electrophoresis, the fraction of solute in the moving fluid phase, ω in (15-54), is 1. Substituting $\omega = 1$ and (15-54) into (15-57) eliminates the second term in parentheses on the RHS of (15-57). Rearranging (15-175) provides the terminal velocity, $u_t = U_{el}V$, which corresponds to the interstitial velocity, u, in (15-57). Substituting u_t for u in the first term in parentheses on the RHS of (15-57) yields, for the electrophoretic plate height,

$$H = 2\frac{E}{U_{el}E} \tag{15-188}$$

where values of the electrophoretic convective dispersion coefficient, E, may be estimated using (15-70). Comparing (15-55) and (15-184) for $\theta = 180°$ shows that electrophoretic distance traveled, d_x, during application of electric field t_{pulse} is analogous to mean distance traveled by a pulse, z_o, during average residence time $\bar{t}$ in linear differential chromatography. Combining (15-55), (15-184) and (15-175) gives

$$\bar{t} = \frac{NH}{U_{el}E} \tag{15-189}$$

for the average electrophoretic residence time. For large numbers of electrophoretic plates N, and comparable solutes—i.e.,

$(NH)_1 \sim (NH)_2$—(15-189) may be substituted into (15-59) and evaluated at $R = 1.5$ for complete band separation. Rearranging to find the pulse application time required to achieve complete band separation yields

$$t_{pulse} = \frac{3(s_1 + s_2)}{\bar{U}_{el}\left(\dfrac{1}{U_{el,1}} - \dfrac{1}{U_{el,2}}\right)} \qquad (15\text{-}190)$$

where $\bar{U}_{el}$ is the mean electrophoretic mobility of the solutes and the electrophoretic temporal variance, s, is

$$s^2 = \frac{\bar{t}H}{U_{el}E} = \frac{LH}{(U_{el}E)^2} \qquad (15\text{-}191)$$

using the identity $L = NH$.

SUMMARY

1. Sorption is a generic term for the selective transfer of a solute from the bulk of a liquid or gas to the surface and/or into the bulk of a solid or liquid. Thus, sorption includes adsorption and absorption. The sorbed solute is the *sorbate*.

2. For commercial applications, a sorbent should have high selectivity, high capacity, rapid solute transport rates, stability, strength, and ability to be regenerated. An adsorbent should have small pores to give a large surface area per unit volume.

3. Physical adsorption of pure gases and gas mixtures is easily measured; adsorption of pure liquids and liquid mixtures is not.

4. Widely used adsorbents are carbon (activated and molecular-sieve), molecular-sieve zeolites, silica gel, and activated alumina.

5. The most widely used ion exchangers are water-swellable, solid gel resins based on the copolymerization of styrene and a cross-linking agent such as divinylbenzene. They can be cation or anion exchangers. Ions are exchanged stoichiometrically on an equivalent basis. Thus, Ca^{2+} is exchanged for $2\,Na^+$.

6. Sorbents for chromatographic separations are typically solid, liquid absorbents supported on or bonded to an inert solid, or gel.

7. The most commonly used adsorption isotherms for gases and liquids are linear (Henry's law), Freundlich's, and Langmuir's. The latter asymptotically approaches the linear isotherm at low concentrations, and at high concentrations gives an asymptotic value representing maximum surface coverage. For mixtures, extended versions of the isotherms are available.

8. Ion-exchange equilibrium is represented by an equilibrium constant based on the law of mass action. For dilute conditions in chromatography, a linear equilibrium isotherm can be employed.

9. Resolution of species via adsorption is reduced by mass-transfer resistances due to convective dispersion, boundary-layer transfer, intraparticle diffusion, and kinetic adsorption. Correlations available for each mass-transfer resistance allow prediction, design, operation, and scale-up of efficient adsorptive separations. Analytical solutions to a model of adsorption for single- and multicomponent differential chromatography, frontal loading, and constant-pattern front permit characterization and analysis.

10. When the physical adsorption rate is almost instantaneous after solutes reach the sorbing surface, only external and internal mass-transfer resistances need be considered. External mass-transfer coefficients may be obtained from correlations of the Chilton–Colburn *j*-factor type. Internal mass transfer may be analyzed using a modified Fick's first law by an effective diffusivity that depends on particle porosity, pore tortuosity, bulk molecular diffusivity, and surface diffusivity. Diffusivities in ion-exchange resin gels depend strongly on the degree of cross-linking.

11. Silica or polymer resins are conjugated with chemistries that allow separation of bioproducts using reversed-phase, size-exclusion, ion-exchange, hydrophobic-interaction, or affinity interactions. Resolution, speed, and capacity determine the utility of these adsorbents at analytical, preparation, and production scales.

12. Alternatives to fixed-bed adsorption minimize transport-rate limits of conventional operations to reduce costs for solvent, sorbent, and regenerant, and to speed processing. Adsorptive-membrane separations eliminate intraparticle diffusion and reduce back-pressure, increasing throughput. Countercurrent contacting increases the local average thermodynamic force for adsorptive partitioning, thereby reducing media costs and improving resolution per unit stage. Simulated-moving-bed operations achieve countercurrent contacting without moving the solid phase.

13. A wide variety of sorption systems are used, including slurries in various modes of operation, and fixed-bed and simulated continuous, countercurrent units. When sorbent regeneration is necessary, the system is operated on a cycle. For fixed beds, the common cycles are temperature-swing adsorption (TSA) and pressure-swing adsorption (PSA). Ion exchange includes a regeneration step using a displacement fluid. In chromatography, adsorption and regeneration take place in the same column.

14. For the design and operation of sorption systems, the adsorption isotherm is of great importance because it relates, at equilibrium, the concentration of the solute in the fluid to its loading as a sorbate in and/or on the

sorbent. Most commonly, the overall rate of adsorption is expressed in the form of a linear-driving-force (LDF) model, where driving force is the difference between bulk concentration and concentration in equilibrium with the loading. The coefficient in the LDF equation combines an overall mass-transfer coefficient and an area for sorption.

15. In ideal, fixed-bed operation, solute–sorbate equilibrium between the flowing fluid and the static bed is assumed everywhere. For plug flow and negligible axial dispersion, the result is a sharp concentration front that moves like a shock wave (stoichiometric front) through the bed. Upstream of the front, the sorbent is spent and in equilibrium with the feed mixture. Downstream, the sorbent is clean of sorbate. The stoichiometric front travels through the bed at a much slower velocity than the interstitial feed velocity. The time for the front to reach the end of the bed is the breakthrough time.

16. When mass-transfer effects are included, the concentration front broadens into an S-shaped curve such that at breakthrough, only a portion of the sorbent is fully loaded. When mass-transfer coefficients and sorption isotherms are known, these curves can be computed using Klinkenberg's equations. When shapes of experimental concentration fronts exhibit a constant pattern, because of favorable adsorption equilibrium, commercial-size beds can be scaled-up from laboratory breakthrough data by the method of Collins.

17. Thermal-swing adsorption (TSA) is used to remove small concentrations of solutes from gas and liquid mixtures. Adsorption is carried out at ambient temperature and desorption at an elevated temperature. Because bed heating and cooling between adsorption and desorption are not instantaneous, TSA cycles are long, typically hours or days. The desorption step, starting with a partially loaded bed, can be computed by the method of lines, using a stiff integrator.

18. Pressure-swing adsorption (PSA) is used to separate air and enrich hydrogen-containing streams. Adsorption is carried out at an elevated or ambient pressure, whereas desorption occurs at a lower pressure or vacuum; the latter is called vacuum-swing adsorption (VSA). Because pressure swings can be made rapidly, PSA cycles are short, typically seconds or minutes. It is not necessary to regenerate the bed completely, but if not, a number of cycles are needed to approach a cyclic steady-state operation.

19. Although continuous, countercurrent adsorption with a moving bed is difficult to achieve successfully in practice, a simulated-moving-bed (SMB) system is popular, particularly for separation of solutes in dilute aqueous solutions and for bulk-liquid separations. Design procedures for SMB systems, which require solution of differential-algebraic equations (DAEs), are highly developed.

20. Design calculations for ion-exchange operations are based on an equilibrium assumption for the loading and regeneration steps.

21. In the basic mode of chromatography, feed is periodically pulsed into a column packed with sorbent. Between feed pulses, an elutant is passed through the column, causing the less strongly sorbed solutes to move through the column more rapidly than slower solutes. If the column is long enough, a multicomponent feed can be completely separated, with solutes eluted one by one from the column. In the absence of mass-transfer resistances, a rectangular feed pulse is separated into individual-solute rectangular pulses, whose position–time curves are readily established. When mass-transfer effects are important, rectangular pulses take on a Gaussian distribution predicted by analytical solutions that use a linear adsorption isotherm.

22. Electrophoresis separates solutes based on relative mass and charge driven by an electric-field gradient. It commonly occurs in a gel matrix of synthetic or natural polymer, developed in-place between parallel glass plates or inside a silica capillary to minimize electro-osmosis and resistive Joule heating.

23. Several electrophoretic modes are widely used to isolate and concentrate biomolecules. They are distinguished by denaturants, matrix pH, and electrolyte content relative to the direction of the applied field gradient, all of which influence the electrophoretic mobility of the biomolecule.

24. Chemical stains, fluorescence, immunological probes, and spectroscopy/spectrometry are used to visualize and recover biomolecules distinguished by electrophoresis.

REFERENCES

1. Keller, G.E., II, in T.E. Whyte, Jr., C.M. Yon, and E.H. Wagner, Eds., *Industrial Gas Separations*, ACS Symposium Series No. 223, American Chemical Society, Washington, D.C., p. 145 (1983).

2. Milton, R.M., U.S. Patents 2,882,243 and 2,882,244 (1959).

3. Skarstrom, C.W., U.S. Patent 2,944,627 (1960).

4. Broughton, D.B., and C.G. Gerhold, U.S. Patent 2,985,589 (May 23, 1961).

5. Ettre, L.S., and A. Zlatkis, Eds., *75 Years of Chromatography—A Historical Dialog*, Elsevier, Amsterdam (1979).

6. Bonmati, R.G., G. Chapelet-Letourneux, and J.R. Margulis, *Chem. Engr.*, **87**(6), 70–72 (1980).

7. Bernard, J.R., J.P. Gourlia, and M.J. Guttierrez, *Chem. Engr.*, **88**(10), 92–95 (1981).

8. White, D.H., Jr., and P.G. Barkley, *Chem. Eng. Progress*, **85**(1) 25–33 (1989).

9. Rousseau, R.W., Ed., *Handbook of Separation Process Technology*, Wiley-Interscience, New York (1987).

10. Ruthven, D.M., *Principles of Adsorption and Adsorption Processes*, John Wiley & Sons, New York (1984).

11.   Brunauer, S., P.H. Emmett, and E. Teller, *J. Am. Chem. Soc.*, **60**, 309 (1938).

12.   Satterfield, C.N., *Heterogeneous Catalysis in Practice*, McGraw-Hill, New York (1980).

13.   Ruthven, D.M., in *Kirk-Othmer Encyclopedia of Chemical Technology*, Vol. **1**, 4th ed., Wiley-Interscience, New York (1991).

14.   Barrer, R.M., *Zeolites and Clay Minerals as Sorbents and Molecular Sieves*, Academic Press, New York (1978).

15.   Breck, D.W., *Zeolite Molecular Sieves*, John Wiley & Sons, New York (1974).

16.   Adams, B.A., and E.L. Holmes, *J. Soc. Chem. Ind.*, **54**, 1–6T (1935).

17.   McWilliams, J.D., *Chem. Engr.*, **85**(12), 80–84 (1978).

18.   Dorfner, K., *Ion Exchangers, Properties and Applications*, 3rd ed., Ann Arbor Science, Ann Arbor, MI (1971).

19.   Sewell, P.A., and B. Clarke, *Chromatographic Separations*, John Wiley & Sons, New York (1987).

20.   Brunauer, S., L.S. Deming, W.E. Deming, and E. Teller, *J. Am. Chem. Soc.*, **62**, 1723–1732 (1940).

21.   Brunauer, S., *The Adsorption of Gases and Vapors*, Vol. I, *Physical Adsorption*, Princeton University Press (1943).

22.   Titoff, A., *Z. Phys. Chem.*, **74**, 641 (1910).

23.   Valenzuela, D.P., and A.L. Myers, *Adsorption Equilibrium Data Handbook*, Prentice-Hall, Englewood Cliffs, NJ (1989).

24.   Ray, G.C., and E.O. Box, Jr., *Ind. Eng. Chem.*, **42**, 1315–1318 (1950).

25.   Yang, R.T., *Gas Separation by Adsorption Processes*, Butterworths, Boston (1987).

26.   Freundlich, H., *Z. Phys. Chem.*, **73**, 385–423 (1910).

27.   Mantell, C.L., *Adsorption, 2nd ed.*, McGraw-Hill, New York, p. 25 (1951).

28.   Langmuir, J., *J. Am. Chem. Soc.*, **37**, 1139–1167 (1915).

29.   Markham, E.C., and A.F. Benton, *J. Am. Chem. Soc.*, **53**, 497–507 (1931).

30.   Yon, C.M., and P.H. Turnock, *AIChE Symp. Series*, **67**(117), 75–83 (1971).

31.   Broughton, D.B., *Ind. Eng. Chem.*, **40**, 1506–1508 (1948).

32.   Myers, A.L., and J.M. Prausnitz, *AIChE J.*, **11**, 121–127 (1965).

33.   Miller, G.W., K.S. Knaebel, and K.G. Ikels, *AIChE J.*, **33**, 194–201 (1987).

34.   Ritter, J.A., and R.T. Yang, *Ind. Eng. Chem. Res.*, **26**, 1679–1686 (1987).

35.   Ruthven, D.M., and F. Wong, *Ind. Eng. Chem. Fundam.*, **24**, 27–32 (1985).

36.   Kipling, J.J., *Adsorption from Solutions of Nonelectrolytes*, Academic Press, London (1965).

37.   Radke, C.J., and J.M. Prausnitz, *AIChE J.*, **18**, 761–768 (1972).

38.   Anderson, R.E., "Ion-Exchange Separations," in P.A. Schweitzer, Ed., *Handbook of Separation Techniques for Chemical Engineers*, 2nd ed., McGraw-Hill, New York (1988).

39.   Bonner, O.D., and L.L. Smith, *J. Phys. Chem.*, **61**, 326–329 (1957).

40.   Bauman, W.C., and J. Eichhorn, *J. Am. Chem. Soc.*, **69**, 2830–2836 (1947).

41.   Bauman, W.C., J.R. Skidmore, and R.H. Osmun, *Ind. Eng. Chem.*, **40**, 1350–1355 (1948).

42.   Subba Rao, H.C., and M.M. David, *AIChE Journal*, **3**, 187–190 (1957).

43.   Selke, W.A., and H. Bliss, *Chem. Eng. Prog.*, **46**, 509–516 (1950).

44.   Selke, W.A., and H. Bliss, *Chem. Eng. Prog.*, **47**, 529–533 (1951).

45.   Myers, A.L., and S. Byington, in A.E. Rodrigues, Ed., *Ion Exchange: Science and Technology*, Martinus Nijhoff, Boston, pp. 119–145 (1986).

46.   Athalye, A.M., S.J. Gibbs, and E.N. Lightfoot, *J. Chromatogr.*, **589**, 71–85 (1992).

47.   Rossiter, G.J., and C.A. Tolbert.Paper presented at the *AIChE Annual Meeting*, Los Angeles, CA., November 17–22, 1991.

48.   Asenjo, J.A., J. Parrado, and B.A. Andrews, *Ann. N. Y. Acad. Sci.*, **646**, 334–356 (1991).

49.   Carson, K.L., *GEN*, **14**(6), 12 (1994).

50.   Schadle, P. Paper presented at *PrepTech94, Industrial Separations Technology Conference*, Secaucus, NJ, March 22–24, 1994.

51.   Dean, R.C., *Equipment Technology Overview*, Univ. of Virginia (1988).

52.   Paul, E.L., and C. B. Rosas, *Chem. Eng. Progr.*, **12**, 17–25 (1990).

53.   Glueckauf, E., *Trans. Farad. Soc.*, **51**, 34–44 (1955).

54.   Tiselius, A., *Kolloid Zeitschrift*, **105**, 101–109 (1943).

55.   Reis, J.F.C., E.N. Lightfoot, P.T. Noble, and A.S. Chiang, *Sep. Sci. Technol.*, **14**(5), 367–394 (1979).

56.   Glueckauf, E., *Proc. R. Soc. London, Ser. A*, **186**, 35 (1946).

57.   Horvath, C., and W.R. Melander, "Theory of Chromatography," in E. Heftmann, Ed., *Fundamentals and Applications of Chromatographic and Electrophoretic Methods*, Part A, Elsevier Scientific, Amsterdam, Netherlands, p. A41 (1983).

58.   Giddings, J.C., *Unified Separation Science*, John Wiley & Sons, Inc., New York (1991).

59.   Aris, R., and N.R. Amundson, *Mathematical Methods in Chemical Engineering*, Prentice Hall, Englewood Cliffs, NJ (1973).

60.   Sarker, M., and G. Guiochon, *LC-GC*, **12**(4), 300–306 (1994).

61.   Gibbs, S.J., and E.N. Lightfoot, *Ind. Eng. Chem. Fundam.*, **25**, 490–498 (1986).

62.   Roper, D.K., and E.N. Lightfoot, *J. Chromatogr. A.*, **702**, 69–80 (1995).

63.   Roper, D.K., and E.N. Lightfoot, *J. Chromatogr. A.*, **654**, 1–16 (1993).

64.   van Deemter, J.J., F.J. Zuiderweg, and A. Klinkenberg, *Chem. Eng. Sci.*, **5**, 271–298 (1956).

65.   Karol, P.J., *Anal. Chem.*, **61**, 1937–1941 (1989).

66.   Edwards, M.F., and J.F. Richardson, *Chem. Eng. Sci.*, **23**, 109–123 (1968).

67.   Gunn, D.J., and C. Pryce, *Trans. Instn. Chem. Engrs.*, **47**, T341–T350 (1969).

68.   Ebach, E.A., and R.R. White, *AIChE J.*, **4**(2), 161–169 (1958).

69.   Jacques, G.L., A. Hennico, J.S. Moon, and T. Vermeulen, *Lawrence Radiation Lab. Rpt. UCRL-1069*, Part II (1964).

70.   Miller, S.F., and C.J. King, *AIChE J.*, **12**(4), 767–773 (1966).

71.   Wakao, N., and T. Funazkri, *Chem. Eng. Sci.*, **33**, 1375–1384 (1978).

72.   Ranz, W.E., and W.R. Marshall, Jr., *Chem. Eng. Prog.*, **48**, 173–180 (1952).

73.   Ranz, W.E., *Chem. Eng. Prog.*, **48**, 247–253 (1952).

74.   Chilton, T.H., and A.P. Colburn, *Ind. Eng. Chem.*, **26**, 1183–1187 (1934).

75.   Gamson, B.W., G. Thodos, and O.A. Hougen, *Trans. AIChE*, **39**, 1–35 (1943).

76.   Sen Gupta, A., and G. Thodos, *AIChE J.*, **9**, 751–754 (1963).

77.   Petrovic, L.J., and G. Thodos, *Ind. Eng. Chem. Fundamentals*, **7**, 274–280 (1968).

78. Dwivedi, P.N., and S.N. Upadhay, *Ind. Eng. Chem. Process Des. Dev.*, **16**, 157–165 (1977).

79. Kunii, D., and O. Levenspiel, *Fluidization Engineering*, 2nd ed., Butterworth-Heinemann, Boston, Chap. 3 (1991).

80. Schneider, P., and J.M. Smith, *AIChE J.*, **14**, 886–895 (1968).

81. Riekert, L., *AIChE J.*, **31**, 863–864 (1985).

82. Sladek, K.J., E.R. Gilliland, and R.F. Baddour, *Ind. Eng. Chem. Fundam.*, **13**, 100–105 (1974).

83. Kapoor, A., R.T. Yang, and C. Wong, "Surface Diffusion," *Catalyst Reviews*, **31**, 129–214 (1989).

84. Helfferich, F., *Ion Exchange*, McGraw-Hill, New York (1962).

85. Soldano, B.A., *Ann. NY Acad. Sci.*, **57**, 116–124 (1953).

86. Arnold, F.H., H.W. Blanch, and C.R. Wilke, *Chem. Eng. J.*, **30**, B9–B23 (1985).

87. Unarska, M. P.S. Davies, M.P. Esnout, and B.J. Bellhouse, *J. Chromatogr.*, **519**, 53–67 (1990).

88. Scopes, R.K., *Protein Purification. Principles and Practice*, Springer-Verlag, New York (1987).

89. Harrison, R.G., P. Todd, S.R. Rudge, and D.P. Petrides, *Bioseparations Science and Engineering*, Oxford University Press, New York (2003).

90. Scopes, R.K., *Anal. Biochem.*, **114**(1), 8–18 (1981).

91. Sluyterman, L.A.A., and O. Elgersma, *J. Chromatogr.*, **150**, 17–44 (1978).

92. Sopher, G.K., and L.E. Nystrom, *Process Chromatography*, Academic Press, San Diego, CA (1989).

93. Janson, J.C., and P. Hedman, *Adv. Biochem. Eng.*, **43**, 43–99 (1982).

94. Pharmacia Fine Chemicals, *Scale Up to Process Chromatography*, Uppsala, Sweden (1982).

95. Cox, G.B., *LC-GC*, **8**(9), 690–694 (1990).

96. Janson, J.C., and P. Hedman, *Biotechnol. Progr.*, **3**, 9–13 (1987).

97. Love, R. et al., "Large-Scale Production of Recombinant Protein A and Its Use in Purifying Monoclonal Antibodies," in R. Burgess, Ed., *Protein Purification: Micro to Macro*, 1st ed., Alan R. Liss, Inc., New York (1987).

98. Grushka, E., *Preparative-Scale Chromatography*, Marcel Dekker Inc., New York (1989).

99. Wankat, P., *Ind. Eng. Chem. Res.*, **26**, 9–13 (1987).

100. Horvath, C., *GEN*, **14**(9), 4 (1994).

101. Majors, R.E., *LC-GC*, **12**(4), 278 (1994).

102. Majors, R.E., and D. Hardy, *LC-GC*, **10**(5), 356–362 (1992).

103. Lawlis, V.B., and H. Heinsohn, *LG-GC*, **11**(10), 720–729, 1993.

104. Afeyan, N.B., N.F. Gordon, L. Mazsarof, L. Varady, S.P. Fulton, Y.B. Yang, and F.E. Regnier, *J. Chromatogr.*, **519**, 1–29 (1990).

105. Fulton, S., M. Meys, J. Protentis, N.B. Afeyan, J. Carlton, and J. Haycock, *BioTechniques*, **12**(5), 742–747 (1992).

106. DePalma, A., *GEN*, **13**(1), 1993.

107. Wang, Q.C., F. Svec, and J.M.J. Frechet, *Anal Chem.*, **65**, 2243–2248 (1993).

108. Wang, Q.C., F. Svec, and J.M.J. Frechet, *J. Chromatogr. A*, **669**, 230–235 (1994).

109. Le, M.S., and J. A. Sanderson. Device for Liquid Chromatography or Immobilized Enzyme Reaction. US. Patent 4,895,806 (1990).

110. Lightfoot, E.N., A.M. Athalye, J.L. Coffman, D.K. Roper, and T.W. Root, *J. Chromatogr. A*, **707**, 45–55 (1995).

111. Tennikova, T.B., and F. Svec, *J. Chromatogr.*, **646**, 279–288 (1993).

112. Josic, D., J. Reusch, K. Loster, O. Baum, and W. Reutter, *Chromatogr.*, **590**, 59–76 (1992).

113. Kim, M., K. Saito, S. Furusaki, T. Suga, and I. Ishigaki, *J. Chromatogr.*, **586**, 27–33 (1991).

114. Briefs, K.-G., and M.-R. Kula, *Chem. Eng. Sci.*, **47**, 141–149 (1992).

115. Luska, J., V. Menart, S. Milicic, B. Kus, V. Gaberck-Porekar, and D. Josic, *J. Chromatogr. A*, **661**, 161–168 (1994).

116. Lutkemeyer, D., M. Brethscneider, H. Buntemeyer, and J. Lehmann, *J. Chromatogr.*, **639**, 57–66 (1993).

117. Iwata, H., K. Saito, S. Furusaki, T. Sugao, and I. Ishigaki, *J. Chromatogr.*, **17**, 412–418 (1991).

118. Roper, D.K., and E.N. Lightfoot, *J. Chromatogr. A.*, **702**, 3–26 (1995).

119. Koch, D.L., and J.F. Brady, *J. Fluid Mech.*, **154**, 399–427 (1985).

120. Kremser, A., *Natl. Pet. News*, **22**(21), 42 (1930).

121. Souders, M., and G.G. Brown, *Ind. Eng. Chem.*, **24**, 519–522 (1932).

122. Klinkenberg, A, *Chem. Eng. Sci.*, **1**(2), 86–92 (1951).

123. Klinkenberg, A., H.A. Lauwerier, and G.H. Reman, *Chem. Eng. Sci.*, **1**(2), 93–99 (1951).

124. Hogg, R.V., and J. Ledolter, *Engineering Statistics*, 1st ed., Macmillan, New York (1987).

125. Stakgold, I., *Greens Functions and Boundary Value Problems*, John Wiley & Sons, New York (1979).

126. Anzelius, A.Z., *Angew. Math u. Mech.*, **6**, 291–294 (1926).

127. Klinkenberg, A., *Ind. Eng. Chem.*, **46**, 2285–2289 (1954).

128. Klinkenberg, A., *Ind. Eng. Chem.*, **40**, 1992–1994 (1948).

129. DeVault, D., *J. Am. Chem Soc.*, **65**, 532–540 (1943).

130. Cooney, D.O., and E.N. Lightfoot, *IEC Fundamentals*, **4**, 233–236 (1965).

131. Sircar, S., and K. Kumar, *Ind. Eng. Chem. Process Des. Dev.*, **22**, 271–280 (1983).

132. Cooney, D.O., *Chem. Eng. Comm.*, **91**, 1–9 (1990).

133. Collins, J.J., *Chem. Eng. Prog. Symp. Ser.*, **63**(74), 31–35 (1967).

134. Wankat, P.C., *Rate-Controlled Separations*, Elsevier Applied Science, New York (1990).

135. Carta, G., *Chem. Eng. Sci.*, **43**, 2877–2883 (1988).

136. Keller, G.E., "Separations: New Directions for an Old Field," *AIChE Monograph Series*, **83**(17) (1987).

137. Ruthven, D.M., and C.B. Ching, *Chem. Eng. Sci.*, **44**, 1011–1038 (1989).

138. Berg, C., *Trans. AIChE*, **42**, 665–680 (1946).

139. Broughton, D.B., *Chem. Eng. Progress*, **64**(8), 60–65 (1968).

140. Gembicki, S.A., A.R. Oroskar, and J.A. Johnson, in *Encyclopedia of Chemical Technology*, 4th ed., John Wiley & Sons, New York, Vol. **1**, pp. 573–600 (1991).

141. Humphrey, J.L., and G.E. Keller, II, *Separation Process Technology*, McGraw-Hill, New York (1997).

142. Higgins, I.R., and J.T. Roberts, *Chem. Engr. Prog. Symp. Ser.*, **50**(14), 87–92 (1954).

143. Himsley, A., Canadian Patent 980,467 (Dec. 23, 1975).

144. Ganetsos, G., and P.E. Barker,Eds., *Preparative and Production Scale Chromatography*, Marcel Dekker, New York (1993).

145. Martin, A.J.P., *Disc. Faraday Soc.*, **7**, 332 (1949).

146. Wong, Y.W., and J.L. Niedzwiecki, *AIChE Symposium Series*, **78**(219), 120–127 (1982).

147. Liskovets, O.A., *Differential Equations* (a translation of *Differentsial'nye Uravneniya*) **1**, 1308–1323 (1965).

148. Schiesser, W.E., *The Numerical Method of Lines Integration of Partial Differential Equations*, Academic Press, San Diego (1991).

149. Schiesser, W.E., *Computational Mathematics in Engineering and Applied Science*, CRC Press, Boca Raton, FL (1994).

150. Press, W.H., S.A. Teukolsky, W.T. Vetterling, and B.P. Flannery, *Numerical Recipes in FORTRAN*, 2nd ed., Cambridge University Press, Cambridge (1992).

151. Gear, C.W., *Numerical Initial Value Problems in Ordinary Differential Equations*, Prentice-Hall, Englewood Cliffs, NJ (1971).

152. Byrne, G.D., and A.C. Hindmarsh, *J. Comput. Phys.*, **70**, 1–62 (1987).

153. Seader, J.D., and E.J. Henley, *Separation Process Principles*, 2nd ed., John Wiley & Sons, Inc., New York (2006).

154. Ritter, J.A., and R.T. Yang, *Ind. Eng. Chem. Res.*, **30**, 1023–1032 (1991).

155. Ruthven, D.M., S. Farooq, and K.S. Knaebel, *Pressure-Swing Adsorption*, VCH, New York (1994).

156. Mutasim, Z.Z., and J.H. Bowen, *Trans. I. Chem. E.*, **69**, Part A, 108–118 (March 1991).

157. Ruthven, D.M., and C.B. Ching, *Chem. Eng. Sci.*, **44**, 1011–1038 (1989).

158. Wankat, P.C., *Large-Scale Adsorption and Chromatography*, Vols. I and II, CRC Press, Inc., Boca Raton (1986).

159. Berg, C., *Trans. AIChE*, **42**, 665–680 (1946).

160. Eagleton, L.C., and H. Bliss, *Chem. Eng. Progress*, **49**, 543–548 (1953).

161. Zang, Y., and P.C. Wankat, *Ind. Eng. Chem. Res.*, **41**, 5283–5289 (2002).

162. Kim, J.K., and P.C. Wankat, *Ind. Eng. Chem. Res.*, **43**, 1071–1080 (2004).

163. Wankat, P.C., *Ind. Eng. Chem. Res.*, **40**, 6185–6193 (2001).

164. Storti, G., M. Masi, S. Carra, and M. Morbidelli, *Chem. Eng. Sci.*, **44**, 1329 (1989).

165. Rhee, H.-K., R. Aris, and N.R. Amundson, *Phil. Trans. Royal Soc. London, Series A*, **269** (No. 1194), 187–215 (Feb. 5, 1971).

166. Mazzotti, M., G. Storti, and M. Morbidelli, *AIChE Journal*, **40**, 1825–1842 (1994).

167. Ruthven, D.M., and C.B. Ching, *Chem. Eng. Sci.*, **44**, 1011–1038 (1989).

168. Zhong, G., and G. Guiochon, *Chem. Eng. Sci.*, **51**, 4307–4319 (1996).

169. Storti, G., M. Mazzotti, M. Morbidelli, and S. Carra, *AIChE Journal*, **39**, 471–492 (1993).

170. Mazzotti, M., G. Storti, and M. Morbidelli, *J. Chromatography A*, **769**, 3–24 (1997).

171. Danckwerts, P.V., *Chem. Eng. Sci.*, **2**, 1 (1953).

172. Constantinides, A., and N. Mostoufi, *Numerical Methods for Chemical Engineers with MATLAB Applications*, Prentice Hall PTR, Upper Saddle River, NJ (1999).

173. Minceva, M., and A.E. Rodrigues, *Ind. Eng. Chem. Res.*, **41**, 3454–3461 (2002).

174. Gunn, D.J., *Trans. Instn. Chem. Engrs.*, **47**, T351–T359 (1969).

175. Gunn, D.J. *Chem. Eng. Sci.*, **42**(2), 363–373 (1987).

176. Broughton, D.B., R.W. Neuzil, J.M. Pharis, and C.S. Brearby, *Chem. Eng. Prog.*, **66**(9), 70–75 (1970).

177. Tiselius, A., *Trans. Faraday Society*, **33**, 524 (1937).

178. Grossman, P.D., "Free-Solution Capillary Electrophoresis," in P.D. Grossman and J.C. Colburn, Eds., *Capillary Electrophoresis: Theory and Practice*, Academic Press: San Diego, CA (1992).

179. Ornstein, L., *Ann. N. Y. Acad. Sci.*, **121**, 321–349 (1964).

180. Scopes, R.K., *Protein Purification*, Springer-Verlag, New York (1982).

181. Hjerten, S., "Isoelectric Focusing in Capillaries," in P.D. Grossman and J.C. Colburn, Eds., *Capillary Electrophoresis: Theory and Practice*, Academic Press, San Diego, CA (1992).

182. Demarest, C.W., E.A. Monnot-Chase, J. Jiu, and R. Weinberger, "Separation of Small Molecules by High Performance Capillary Electrophoresis," in P.D. Grossman and J.C. Colburn, Eds., *Capillary Electrophoresis: Theory and Practice*, Academic Press, San Diego, CA (1992).

183. Garcia, A.A., M.R. Bonen, J. Ramirez-Vick, M. Sadaka, and A. Vuppu, *Bioseparation Process Science*, Blackwell Science, Malden, MA (1999).

184. Southern, E.M., and J.K. Elder, in A.P. Monaco, Ed., *Pulsed Field Gel Electrophoresis: A Practical Approach*, Oxford University Press, Oxford, UK (1995).

STUDY QUESTIONS

15.1. How is a large surface area achieved for adsorption?

15.2. What is meant by ion exchange? How does ion exchange differ from deionization?

15.3. In adsorption processes, what distinguishes purification from bulk separation?

15.4. What is meant by regeneration?

15.5. Why is it easy to measure the amount of adsorption of a pure gas, but difficult to measure adsorption of a pure liquid?

15.6. What is the BET equation used for? Does it assume physical or chemical adsorption? Does it assume monomolecular- or multimolecular-layer adsorption?

15.7. How is it possible to use a liquid sorbent in chromatography?

15.8. What is meant by loading in adsorption?

15.9. What is an adsorption isotherm? How can the heat of adsorption be determined from a series of isotherms?

15.10. What are the four steps that occur during the adsorption of a solute from a gas or liquid mixture? How do they affect adsorptive bandwidth, resolution, and throughput? Which step may be almost instantaneous such that equilibrium at the fluid-sorbent interface can be assumed?

15.11. Within a porous particle, why are mass and heat transfer not analogous?

15.12. For mass transfer outside a single spherical particle that is not close to a wall or other particles, what is the smallest value of the Sherwood number? What is the basis for this value?

15.13. What is the difference between slurry adsorption (contact filtration) and fixed-bed adsorption (percolation)? When should each be considered and not considered?

15.14. How do pressure-swing and thermal-swing adsorption differ? What are inert-purge swing and displacement purge?

15.15. What is ideal fixed-bed adsorption? What assumptions are necessary for it to apply? What is meant by the stoichiometric front? What is meant by breakthrough?

15.16. What is a mass-transfer zone (MTZ) and what causes it? Is it desirable? If not, why not?

15.17. For fixed-bed adsorption, what is the difference between superficial velocity, interstitial velocity, and concentration wave velocity? Which is the largest and which is the smallest?

15.18. What is the difference between a true-moving-bed (TMB) system and a simulated-moving-bed (SMB) system?

15.19. How is port switching used in an SMB system?

15.20. In chromatography, what causes different solutes to pass through the chromatograph with different residence times?

15.21. What are the differences among native gel electrophoresis, SDS-PAGE, isoelectric focusing, isotachophoresis, 2-D gel electrophoresis, and pulsed-field gel electrophoresis in terms of denaturants used, pH and electrolyte content, and application of electric field gradient?

15.22. In electrophoresis, describe factors that affect mobility, electro-osmosis, and convective Joule heating, and their effects on electrophoretic resolution.

15.23. What are dispersion, residence time, variance, and plate height in zone electrophoresis relative to pulse linear differential chromatography?

EXERCISES

Section 15.1

15.1. Adsorbent characteristics.
Porous particles of activated alumina have a BET surface area of 310 m^2/g, $\epsilon_p = 0.48$, and $\rho_p = 1.30$ g/cm^3. Determine: (a) V_p in cm^3/g; (b) ρ_s, in g/cm^3; and (c) d_p in Å.

15.2. Surface area of molecular sieves.
Carbon molecular sieves are available in two forms from a Japanese manufacturer:

	Form A	Form B
Pore volume, cm^3/g	0.18	0.38
Average pore diameter	5 Å	2.0 μm

Estimate S_g of each form in m^2/g.

15.3. Characteristics of silica gel.
Representative properties of small-pore silica gel are: $d_p = 24$ Å, $\epsilon_p = 0.47$, $\rho_p = 1.09$ g/cm^3, and $S_g = 800$ m^2/g. (a) Are these values reasonably consistent? (b) If the adsorption capacity for water vapor at 25°C and 6 mmHg partial pressure is 18% by weight, what fraction of a monolayer is adsorbed?

15.4. Specific surface area from BET data.
The following data were obtained in a BET apparatus for adsorption equilibrium of N$_2$ on silica gel (SG) at −195.8°C. Estimate S_g in m^2/g of silica gel. How does your value compare with that in Table 15.2?

N$_2$ Partial Pressure, torr	Volume of N$_2$ Adsorbed in cm^3 (0°C, 1 atm) per gram SG
6.0	6.1
24.8	12.7
140.3	17.0
230.3	19.7
285.1	21.5
320.3	23.0
430	27.7
505	33.5

15.5. Maximum ion-exchange capacity.
Estimate the maximum ion-exchange capacity in meq/g resin for an ion-exchange resin made from 8 wt% divinylbenzene and 92 wt% styrene.

Section 15.2

15.6. Adsorption isotherms and heat of adsorption.
Shen and Smith [*Ind. Eng. Chem. Fundam.*, **7**, 100–105 (1968)] measured equilibrium-adsorption isotherms at four temperatures for benzene vapor on silica gel having the following properties: $S_g = 832$ m^2/g, $V_p = 0.43$ cm^3/g, $\rho_p = 1.13$ g/cm^3, and average $d_p = 22$ Å. The adsorption data are:

Partial Pressure of Benzene, atm	Moles Adsorbed/g Gel × 10^5			
	70°C	90°C	110°C	130°C
5.0×10^{-4}	14.0	6.7	2.6	1.13
1.0×10^{-3}	22.0	11.2	4.5	2.0
2.0×10^{-3}	34.0	18.0	7.8	3.9
5.0×10^{-3}	68.0	33.0	17.0	8.6
1.0×10^{-2}	88.0	51.0	27.0	16.0
2.0×10^{-2}	—	78.0	42.0	26.0

(a) For each temperature, obtain a best fit of the data to (1) linear, (2) Freundlich, and (3) Langmuir isotherms. Which isotherm(s), if any, fit(s) the data reasonably well? (b) Do the data represent less than a monolayer of adsorption? (c) From the data, estimate the heat of adsorption. How does it compare to the heat of vaporization (condensation) of benzene?

15.7. Adsorption-equilibria data.
The separation of propane and propylene is accomplished by distillation, but at the expense of more than 100 trays and a reflux ratio greater than 10. Consequently, the use of adsorption has been investigated in a number of studies. Jarvelin and Fair [*Ind. Eng. Chem. Research*, **32**, 2201–2207 (1993)] measured adsorption-equilibrium data at 25°C for three different zeolite molecular sieves (ZMSs) and activated carbon. The data were fitted to the Langmuir isotherm with the following results:

Adsorbent	Sorbate	q_m	K
ZMS 4A	C$_3$	0.226	9.770
	C$_3^=$	2.092	95.096
ZMS 5A	C$_3$	1.919	100.223
	C$_3^=$	2.436	147.260
ZMS 13X	C$_3$	2.130	55.412
	C$_3^=$	2.680	100.000
Activated carbon	C$_3$	4.239	58.458
	C$_3^=$	4.889	34.915

where q and q_m are in mmol/g and p is in bar. (a) Which component is most strongly adsorbed by each adsorbent? (b) Which adsorbent has the greatest capacity? (c) Which adsorbent has the greatest selectivity? (d) Based on equilibrium considerations, which adsorbent is best?

15.8. Fitting zeolite data to isotherms.

Ruthven and Kaul [*Ind. Eng. Chem. Res.*, **32**, 2047–2052 (1993)] measured adsorption isotherms for a series of gaseous aromatic hydrocarbons on well-defined crystals of NaX zeolite over ranges of temperature and pressure. For 1,2,3,5-tetramethylbenzene at 547 K, the following equilibrium data were obtained with a vacuum microbalance:

q, wt%	7.0	9.1	10.3	10.8	11.1	11.5
p, torr	0.012	0.027	0.043	0.070	0.094	0.147

Obtain a best fit of the data to the linear, Freundlich, and Langmuir isotherms, with q in mol/g and pressure in atm. Which isotherm gives the best fit?

15.9. Fitting adsorption data to isotherms.

Lewis, Gilliland, Chertow, and Hoffman [*J. Am. Chem. Soc.*, **72**, 1153–1157 (1950)] measured adsorption equilibria for propane, propylene, and binary mixtures thereof on activated carbon and silica gel. Adsorbate capacity was high on carbon, but selectivity was poor. Selectivity was high on silica gel, but capacity was low. For silica gel (751 m²/g), the pure-component data below were obtained at 25°C:

Propane		Propylene	
P, torr	q, mmol/g	P, torr	q, mmol/g
11.1	0.0564	34.2	0.3738
25.0	0.1252	71.4	0.7227
43.5	0.1980	91.6	0.7472
71.4	0.2986	194.3	1.129
100.0	0.3850	198.3	1.168
158.9	0.5441	271.5	1.401
227.5	0.7020	353.2	1.562
304.2	0.843	550.7	1.918
387.0	1.010	555.2	1.928
468.0	1.138	760.6	2.184
569.0	1.288		
677.8	1.434		
775.0	1.562		

The following mixture data were measured with silica gel at 25°C, over a pressure range of 752–773 torr:

Total Pressure, torr	Millimoles of Mixture Adsorbed/g	y_{C_3}, Mole Fraction in Gas Phase	x_{C_3}, Mole Fraction in Adsorbate
769.2	2.197	0.2445	0.1078
760.9	2.013	0.299	0.2576
767.8	2.052	0.4040	0.2956
761.0	2.041	0.530	0.2816
753.6	1.963	0.5333	0.3655
766.3	1.967	0.5356	0.3120
754.0	1.974	0.6140	0.3591
753.6	1.851	0.6220	0.5550
754.0	1.701	0.6252	0.7007
760.0	1.686	0.7480	0.723
—	2.180	0.671	0.096
760.0	1.993	0.8964	0.253
760.0	1.426	0.921	0.401

(a) Fit the pure-component data to Freundlich and Langmuir isotherms. Which gives the best fit? Which component is most strongly adsorbed? (b) Use the results of the Langmuir fits in part (a) to predict binary mixture adsorption using the extended-Langmuir equation, (15-32). Are the predictions adequate? (c) Ignoring the pure-component data, fit the binary-mixture data to the extended-Langmuir equation, (15-32). Is the fit better than that obtained in part (b)? (d) Ignoring pure-component data, fit the binary mixture data to the extended-Langmuir–Freundlich equation, (15-33). Is the fit adequate? Is it better than that in part (c)? (e) For the binary-mixture data, compute the relative selectivity, $\alpha_{C_3, C_3^=} = y_{C_3}(1 - x_{C_3})/\left[x_{C_3}(1 - y_{C_3})\right]$, for each condition. Does α vary widely or is the assumption of constant α reasonable?

15.10. Extended-Langmuir-Freundlich isotherm.

In Example 15.6, pure-component, liquid-phase adsorption data are used with the extended-Langmuir isotherm to predict a binary-solute data point. Use the mixture data below to obtain the best fit to an extended-Langmuir–Freundlich isotherm of the form

$$q_i = \frac{(q_0)_i k_i c_i^{1/n_i}}{1 + \sum_j k_j c_j^{1/n_j}} \tag{1}$$

Data for binary-mixture adsorption on activated carbon (1,000 m²/g) at 25°C for acetone (1) and propionitrile (2) are:

Solution Concentration, mol/L		Loading, mmol/g	
c_1	c_2	q_1	q_2
5.52E – 5	7.46E – 5	0.0192	0.0199
6.14E – 5	7.71E – 5	0.0191	0.0198
1.06E – 4	1.35E – 4	0.0308	0.0320
1.12E – 4	1.46E – 4	0.0307	0.0319
3.03E – 4	2.32E – 3	0.0378	0.263
3.17E – 4	2.34E – 3	0.0378	0.264
3.25E – 4	3.89E – 4	0.0644	0.0672
1.42E – 3	1.58E – 3	0.161	0.169
1.42E – 3	1.61E – 3	0.161	0.169
1.43E – 3	1.60E – 3	0.161	0.169
2.09E – 3	3.84E – 4	0.250	0.0390
2.17E – 3	3.85E – 4	0.251	0.0392
4.99E – 3	5.24E – 3	0.291	0.307
5.06E – 3	5.31E – 3	0.288	0.305
7.41E – 3	2.42E – 2	0.237	0.900
7.52E – 3	2.47E – 2	0.236	0.896
2.79E – 2	7.59E – 3	0.802	0.251
4.00E – 2	3.44E – 2	0.715	0.822
4.02E – 2	3.42E – 2	0.717	0.834

15.11. Liquid mixture adsorption data.

Sircar and Myers [*J. Phys. Chem.*, **74**, 2828–2835 (1970)] measured liquid-phase adsorption at 30°C for a binary mixture of cyclohexane (1) and ethyl alcohol (2) on activated carbon. Assuming no adsorption of ethyl alcohol, (15-34) gave these results:

x_1	q_1^e, mmol/g	x_1	q_1^e, mmol/g
0.042	0.295	0.440	0.065
0.051	0.485	0.470	0.000
0.072	0.517	0.521	−0.129
0.148	0.586	0.537	−0.362
0.160	0.669	0.610	−0.643
0.213	0.661	0.756	−1.230
0.216	0.583	0.848	−1.310
0.249	0.595	0.893	−1.180
0.286	0.532	0.920	−1.230
0.341	0.383	0.953	−0.996
0.391	0.192	0.974	−0.470

(a) Plot the data as q_1^e against x_1. Explain the shape of the curve. (b) In what concentration region does the Freundlich isotherm fit the data? Make the fits.

15.12. Fitting liquid adsorption-equilibria data.

Both the adsorptive removal of small amounts of toluene from water and small amounts of water from toluene are important in the process industries. Activated carbon is particularly effective for removing soluble organic compounds (SOCs) from water. Activated alumina is effective for removing soluble water from toluene. Fit each of the following two sets of equilibrium data for 25°C to the Langmuir and Freundlich isotherms. For each case, which isotherm provides a better fit? Can a linear isotherm be used?

Toluene (in Water) Activated Carbon		Water (in Toluene) Activated Alumina	
c, mg/L	q, mg/g	c, ppm (by Weight)	q, g/100g
0.01	12.5	25	1.9
0.02	17.1	50	3.1
0.05	23.5	75	4.2
0.1	30.3	100	5.1
0.2	39.2	150	6.5
0.5	54.5	200	8.2
1	90.1	250	9.5
2	70.2	300	10.9
5	125.5	350	12.1
10	165	400	13.3

15.13. Ion exchange and regeneration.

Derive (15-44) and use it to solve the following problem. Sulfate ion is to be removed from 60 L of water by exchanging it with chloride ion on 1 L of a strong-base resin with relative molar selectivities as listed in Table 15.6 and an ion-exchange capacity of 1.2 eq/L of resin. The water to be treated has a sulfate-ion concentration of 0.018 eq/L and a chloride-ion concentration of 0.002 eq/L. Following the attainment of equilibrium ion exchange, the treated water will be removed and the resin will be regenerated with 30 L of

10 wt% aqueous NaCl. (a) Write the ion-exchange reaction. (b) Determine the value of $K_{SO_4^{2-}, Cl^-}$ (c) Calculate equilibrium concentrations $c_{SO_4^{2-}}$, c_{Cl^-}, $q_{SO_4^{2-}}$, and q_{Cl^-} in eq/L for the initial ion-exchange step. (d) Calculate the concentration of Cl^- in eq/L for the regenerating solution. (e) Calculate $c_{SO_4^{2-}}$, c_{Cl^-}, $q_{SO_4^{2-}}$, and q_{Cl^-} upon reaching equilibrium in the regeneration step. (f) Are the separations sufficiently selective?

15.14. Equivalent fractions for ion exchange.

Silver ion in methanol was exchanged with sodium ion using Dowex 50 cross-linked with 8% divinylbenzene by Gable and Stroebel [*J. Phys. Chem.*, **60**, 513–517 (1956)]. The molar selectivity coefficient was found to vary somewhat with the equivalent fraction of Na^+ in the resin as follows:

x_{Na^+}	0.1	0.3	0.5	0.7	0.9
K_{Ag^+, Na^+}	11.2	11.9	12.3	14.1	17.0

If the wet capacity of the resin is 2.5 eq/L and the resin is initially saturated with Na^+, calculate the equilibrium equivalent fractions if 50 L of 0.05-M Ag^+ in methanol is treated with 1 L of wet resin.

15.15. Recovery of glycerol by ion exclusion.

Ion exclusion is a process that uses ion-exchange resins to separate nonionic organic compounds from ionic species contained in a polar solvent, usually water. The resin is presaturated with the same ions as in the solution, thus eliminating ion exchange. However, in the presence of the polar solvent, resins undergo considerable swelling by absorbing the solvent. Experiments have shown that a nonionic solute will distribute between the bulk solution and solution within the resin, while ions only exchange.

A feed solution of 1,000 kg contains 6 wt% NaCl, 35 wt% glycerol, and 47 wt% water. It is to be treated with Dowex-50 ion-exchange resin in the sodium form, after prewetting with water, to recover 75% of the glycerol. The data for the glycerol distribution coefficient,

$$K_d = \frac{\text{mass fraction in solution inside resin}}{\text{mass fraction in solution outside resin}}$$

are from Asher and Simpson [*J. Phys. Chem.*, **60**, 518–521 (1956)]:

Mass Fraction Glycerol in Solution Outside Resin	K_d	
	6 wt% NaCl	12 wt% NaCl
0.10	0.75	0.91
0.20	0.80	0.93
0.30	0.83	0.95
0.40	0.85	0.97

If the prewetted resin contains 40 wt% water, determine the kg of resin (dry basis) required.

Section 15.3

15.16. External transport coefficients in a fixed-bed.

Benzene vapor in an air stream is adsorbed in a fixed bed of 4 × 6 mesh silica gel packed to an external void fraction of 0.5. The bed is 2 ft in inside diameter and air flow is 25 lb/minute (benzene-free basis). At a bed location where pressure is 1 atm, temperature is 70°F, and the benzene mole fraction is 0.005, estimate the external, gas-to-particle mass-transfer and heat-transfer coefficients.

15.17.   External transport coefficients in a fixed-bed.

Water vapor in an air stream is to be adsorbed in a 12.06-cm-i.d. column packed with 3.3-mm-diameter Alcoa F-200 activated-alumina beads of external porosity of 0.442. At a location in the bed where the pressure is 653.3 kPa, temperature is 21°C, gas flow rate is 1.327 kg/minute, and dew-point temperature is 11.2°C, estimate the external, gas-to-particle mass-transfer and heat-transfer coefficients.

15.18.   Effective diffusivity of acetone.

For the conditions of Example 15.8, estimate the effective diffusivity of acetone vapor in the pores of activated carbon whose properties are: $\rho_p = 0.85$ g/cm^3, $\epsilon_p = 0.48$, average $d_p = 25$ Å, and tortuosity $= 2.75$. Consider bulk and Knudsen diffusion, but ignore surface diffusion.

15.19.   Effective diffusivity of benzene vapor.

For the conditions of Exercise 15.16, estimate the effective diffusivity of benzene vapor in the pores of silica gel with the following properties: $\rho_p = 1.15$ g/cm^3, $\epsilon_p = 0.48$, average $d_p = 30$ Å, and tortuosity $= 3.2$. Consider all mechanisms of diffusion. The adsorption-equilibrium constant is given in Example 15.13, and the differential heat of adsorption is $-11,000$ cal/mol.

15.20.   Effective diffusivity of water vapor.

Using data in Exercise 15.17, estimate the effective diffusivity of water vapor in the pores of activated alumina with the following properties: $\rho_p = 1.38$ g/cm^3, $\epsilon_p = 0.52$, average $d_p = 60$ Å, and tortuosity $= 2.3$. Consider all mechanisms of diffusion except surface diffusion.

Section 16.2

15.21.   Intraparticle diffusion resistance.

The volume-average adsorbate loading of a spherical particle is given by (15-49). Show that a parabolic adsorbate loading profile proposed by Liaw, C.H. et al. [*AIChE J.*, **25**, 376–381 (1979)], $q = a_0 + a_1r + a_2r^2$, may be used to obtain a transport-rate resistance term for intraparticle diffusion for a thermodynamic driving force given by the difference between the solute concentration on the particle at its outer surface ($r = R_p$) and the volume-average adsorbate loading, by equating the rate of accumulation of adsorbate within the particle to Fick's first law for diffusion into the particle evaluated at the particle outer surface.

15.22.   Purification by size exclusion.

Cytochrome C ($\epsilon_p^* = 0.38$) and ovalbumin ($\epsilon_p^* = 0.34$) in an equimolar mixture are to be purified by size exclusion to a resolution of $R = 1.0$ in a packed bed of Toyopearl® (TSK gel) HW55F, which has a bed porosity of $\epsilon_b = 0.34$. The differential migration velocity, δ, for these two solutes is 0.05. Assume a volumetric feed stream/fluid-phase dilution in counterflow of 10. Determine the relative number of equilibrium stages, the relative solvent requirement, and the relative solid-phase requirement for steady counterflow separation relative to differential chromatography.

15.23.   Purification of proteins.

Compare the advantages and disadvantages of chromatographic purification of proteins bovine serum albumin, cytochrome C, and ovalbumin by size-exclusion, reversed-phase, hydrophobic-interaction, ion-exchange, affinity chromatography, countercurrent contacting, and stacked-membrane chromatography. Consider resolution, speed, capacity, throughput, recovery, activity, and final purity.

15.24.   Size-exclusion chromatography.

A 50 mg/mL aqueous solution of bovine hemoglobin is to be purified by size-exclusion chromatography using a 75×2.5 cm i.d. glass column packed with 75 μm (diameter) Toyopearl HW65C. Intraparticle diffusivity of hemoglobin is 3.0×10^{-7} cm^2/s.

Estimate the resolving power per unit thermodynamic driving force for this protein in this column at 0.4, 0.8, and 1.2 cm/minute. What resolution is expected from a protein of comparable diffusivity whose inclusion porosity is 118% of hemoglobin? What size column would be required to obtain $R = 1$, as in Figure 15.21?

15.25.   Static binding capacity.

Adeno-associated virus is a candidate viral vector for gene therapy. Estimate the static binding capacity of 20-nm adeno-associated virus on 15 μm Resource™ ion-exchange resin. Compare the estimate with protein capacities of anion-exchange resins and membrane monoliths.

Section 15.5

15.26.   Comparison of slurry adsorption modes.

Adsorption with activated carbon, made from bituminous coal of soluble organic compounds (SOCs), to purify surface and groundwater is a proven technology, as discussed by Stenzel [*Chem. Eng. Prog.*, **89** (4), 36–43 (1993)]. The less-soluble organic compounds, such as chlorinated organic solvents and aromatic solvents, are the more strongly adsorbed. Water containing 3.3 mg/L of trichloroethylene (TCE) is to be treated with activated carbon to obtain an effluent with only 0.01 mg TCE/L. At 25°C, adsorption-equilibrium data for TCE on activated carbon are correlated with the Freundlich equation:

$$q = 67\, c^{0.564} \tag{1}$$

where $q =$ mg TCE/g carbon and $c =$ mg TCE/L solution.

TCE is to be removed by slurry adsorption using powdered activated carbon with an average d_p of 1.5 mm. In the absence of any laboratory data on mass-transfer rates, assume the small-particle absorption rate is controlled by external mass transfer with a Sherwood number of 30. Particle surface area is 5 m^2/kg. The molecular diffusivity of TCE in low concentrations in water at 25°C may be obtained from the Wilke–Chang equation. (a) Determine the minimum amount of adsorbent needed. (b) For operation in the batch mode with twice the minimum amount of adsorbent, determine the time to reduce TCE content to the desired value. (c) For operation in the continuous mode using twice the minimum amount of adsorbent, obtain the required residence time. (d) For operation in the semicontinuous mode at a feed rate of 50 gpm and for a liquid residence time equal to 1.5 times that computed in part (c), determine the amount of activated carbon to give a reasonable vol% solids in the tank and a run time of not less than 10 times the liquid residence time.

15.27.   Comparison of slurry adsorption modes.

Repeat Exercise 15.26 for water containing 0.324 mg/L of benzene (B) and 0.630 mg/L of *m*-xylene (X). Adsorption isotherms at 25°C for these low concentrations are independent and given by

$$q_B = 32\, c_B^{0.428} \tag{1}$$

$$q_X = 125\, c_X^{0.333} \tag{2}$$

The SOC concentrations are to be reduced to 0.002 mg/L each.

15.28.   Comparison of slurry adsorption modes.

Repeat Exercise 15.26 for water containing 0.223 mg/L chloroform whose concentration is to be reduced to 0.010 mg/L. The adsorption isotherm at 25°C is given by

$$q = 10\, c^{0.564} \tag{1}$$

15.29.   MTZ concept.

Three fixed-bed adsorbers containing 10,000 lb of granules of activated carbon ($\rho_b = 30$ lb/ft^3) each are to be used to treat

250 gpm of water containing 4.6 mg/L of 1,2-dichloroethane (D) to reduce the concentration to less than 0.001 mg/L. Each carbon bed has a height equal to twice its diameter. Two beds are to be placed in series so that when Bed 1 (the lead bed) becomes saturated with D at the feed concentration, that bed is removed. Bed 2 (the trailing bed), which is partially saturated at this point, depending upon the width of the MTZ, becomes the lead bed, and previously idle Bed 3 takes the place of Bed 2. While Bed 1 is off-line, its spent carbon is removed and replaced with fresh carbon. The spent carbon is incinerated. The equilibrium adsorption isotherm for D is given by $q = 8 c^{0.57}$, where q is in mg/g and c is in mg/L. Once the cycle is established, how often must the carbon bed be replaced? What maximum MTZ width allows saturated loading of the lead bed?

15.30. MTZ concept.

The fixed-bed adsorber arrangement of Exercise 15.29 is to be used to treat 250 gpm of water containing 0.185 mg/L of benzene (B) and 0.583 mg/L of m-xylene (X). However, because the two solutes may have considerably different breakthrough times, more than two operating beds in series may be needed. Adsorption isotherms are given in Exercise 15.27, where q is in mg/g and c in mg/L. From laboratory measurements, widths of the mass-transfer zones are estimated to be $MTZ_B = 2.5$ ft and $MTZ_X = 4.8$ ft. Once the cycle is established, how often must the carbon be replaced?

15.31. Comparison of models.

Air at 80°F, 1 atm, 80% relative humidity, and a superficial velocity of 100 ft/minute passes through a 5-ft-high bed of 2.8-mm-diameter spherical particles of silica gel ($\rho_b = 39$ lb/ft^3). The adsorption-equilibrium isotherm at 80°F is given by

$$q_{H_2O} = 15.9 p_{H_2O} \tag{1}$$

where q is in lb H_2O/lb gel and p is in atm. The overall mass-transfer coefficient can be estimated from (15-106), using an effective diffusivity of 0.05 cm^2/s, and with k_c estimated from (15-65). Using the approximate concentration-profile equations of Klinkenberg, compute a set of breakthrough curves and determine the time when the exiting-air humidity reaches 0.0009 lb H_2O/lb dry air. Assume isothermal and isobaric operation. Compare the breakthrough time with the time for the equilibrium model. At breakthrough, what is the approximate width of the mass-transfer zone? What is the average loading of the bed at breakthrough?

15.32. Treatment of wastewater by fixed-bed adsorption.

A train of four 55-gallon cannisters of activated carbon is to be used to reduce the nitroglycerine (NG) content of 400 gph of wastewater from 2,000 ppm by weight to less than 1 ppm. Each cannister has a diameter of 2 ft and holds 200 lb activated carbon ($\rho_b = 32$ lb/ft^3). Each cannister is also equipped with a liquid-flow distributor to promote plug flow through the bed. The effluent from the first cannister is monitored so that when a 1-ppm threshold of NG is reached, that cannister is removed from the train and a fresh cannister is added to the end of the train. The spent carbon is mixed with coal for use as a fuel in a coal-fired power plant at the process site. Using the following pilot-plant data, estimate how many $700 cannisters are needed each month and the monthly cost.

Pilot-plant data:

Tests with the 55-gallon cannister used in the commercial process: water rate = 10 gpm; NG content in feed = 1,020 ppm by weight.

Breakthrough correlation:

$t_B = 3.90 L - 2.05$, where t_B = time, h, at breakthrough of the 1-ppm threshold and L = bed depth of carbon in ft.

15.33. Sizing of fixed-bed adsorption unit.

Air at a flow rate of 12,000 scfm (60°F, 1 atm) containing 0.5 mol% ethyl acetate (EA) and no water vapor is to be treated with activated carbon (C) ($\rho_b = 30$ lb/ft^3) with an equivalent particle diameter of 0.011 ft in a fixed-bed adsorber to remove EA, which will be subsequently stripped from C by steam at 230°F. Based on the following data, determine the diameter and height of the carbon bed, assuming adsorption at 100°F and 1 atm and a time to breakthrough of 8 h with a superficial gas velocity of 60 ft/minute. If the bed height-to-diameter ratio is unreasonable, what change in design basis would you suggest?

Adsorption isotherm data (100°F) for EA:

p^{EA}, atm	q, lb EA/lb C	p^{EA}, atm	q, lb EA/lb C
0.0002	0.125	0.0020	0.227
0.0005	0.164	0.0050	0.270
0.0010	0.195	0.0100	0.304

Breakthrough data at 100°F and 1 atm for EA in air at a gas superficial velocity of 60 ft/minute in a 2-ft dry bed:

Mole Fraction EA in Effluent	Time, Minutes	Mole Fraction EA in Effluent	Time, Minutes
0.00005	60	0.00100	95
0.00010	66	0.00250	120
0.00025	75	0.00475	160
0.00050	84		

15.34. Desorption of benzene.

In Examples 15.13 and 15.18, benzene is adsorbed from air at 70°F in a 6-ft-high bed of silica gel and then stripped with air at 145°F. If the bed height is changed to 30 ft, the following data are obtained for breakthrough at 641 minutes for the adsorption step:

z, ft	$\phi = c/c_F$	$\psi = \bar{q}/q_F^*$	z, ft	$\phi = c/c_F$	$\psi = \bar{q}/q_F^*$
16	0.999	0.999	24	0.599	0.575
17	0.997	0.997	25	0.468	0.444
18	0.992	0.990	26	0.343	0.321
19	0.978	0.975	27	0.235	0.217
20	0.951	0.944	28	0.150	0.137
21	0.901	0.890	29	0.090	0.081
22	0.825	0.808	30	0.050	0.044
23	0.722	0.701			

If the bed is regenerated isothermally with pure air at 1 atm and 145°F, and the desorption of benzene during the heat-up period is neglected, determine the loading, $\bar{q}$, profile at a time sufficient to remove 90% of the benzene from the 30-ft bed if an interstitial pure air velocity of 98.5 ft/minute is used. Values of k and K at 145°F are given in Example 15.18.

15.35. Cyclic steady state for PSA.

Use the method of lines with a five-point, biased, upwind, finite-difference approximation and a stiff integrator to perform PSA cycle calculations that approach the cyclic steady state for the data and design basis in Example 15.19, starting from: (a) a clean bed and (b) a saturated bed. Are the two cyclic steady states essentially the same?

15.36. Vacuum-swing adsorption.

Solve Example 15.19 for $P_L = 0.12$ atm and an interstitial velocity during desorption that corresponds to the use of 44.5% of the product gas from the adsorption step.

15.37. Separation of air by PSA.

For the separation of air by PSA, adsorption of both O_2 and N_2 must be considered. Develop a model for this case, taking into account the two species' mass balances, overall mass balance, the two species' mass-transfer rates, and two extended-Langmuir isotherms. Each of the two main steps can be isothermal and isobaric. Can your PDE equations still be solved by the method of lines with a stiff integrator? If so, outline a procedure for doing it.

15.38. Parametric pumping and cycling-zone adsorption.

Two adsorption-based separation processes not considered in this book because of lack of significant commercial applications are (1) parametric pumping, first conceived by R. H. Wilhelm in the early 1960s, and (2) cycling-zone adsorption, invented by R. L. Pigford and co-workers in the late 1960s. The status of and future for these two processes was assessed by Sweed in 1984 [*AIChE Symp. Series*, **80** (233), 44–53 (1984)]. Describe each of these processes. Can either be used for both gas-phase and liquid-phase adsorption?

Section 15.6

15.39. Separation of propylene from propane.

A 55 mol% propane and 45 mol% propylene gas mixture is to be separated into products containing 10 and 90 mol% propane by adsorption in a continuous, countercurrent system operating at 25°C and 1 atm. The adsorbent is silica gel, for which equilibrium data are given in Exercise 15.9. Determine by the McCabe–Thiele method the: (a) adsorbent flow rate per 1,000 m³ of feed gas at 25°C and 1 atm for 1.2 times the minimum, and (b) number of stages required.

15.40. Softening of hard water.

Repeat Example 15.24, except for a feed containing 400 ppm (by weight) of $CaCl_2$ and 50 ppm of NaCl.

15.41. Equilibrium model for a chromatographic column.

An aqueous solution, buffered to a pH of 3.4 by sodium citrate and containing 20 mol/m³ each of glutamic acid, glycine, and valine, is separated in a chromatographic column packed with Dowex 50W-X8 in the sodium form to a depth of 470 mm. The resin is 0.07 mm in diameter and packs to a bed void fraction of 0.374. Equilibrium data follow a linear law, as in Example 15.15, with the following dimensionless constants determined by Takahashi and Goto [*J. Chem. Eng. Japan*, **24**, 121–123 (1991)]:

Solute	K
Glutamic acid	1.18
Glycine	1.74
Valine	2.64

The superficial solution velocity is 0.025 cm/s. Using equilibrium theory, what pulse duration achieves complete separation? What is the time interval between elution and the second pulse?

15.42. Mass-transfer resistances in a chromatographic column.

Repeat Exercise 15.41 using Carta's equation to account for mass transfer with the following effective diffusivities:

Solute	D_e, cm²/s
Glutamic acid	1.94×10^{-7}
Glycine	4.07×10^{-7}
Valine	3.58×10^{-7}

Assume $k_c = 1.5 \times 10^{-3}$ cm/s. Establish a cycle of feed pulses and elution periods that will give the desired separation.

15.43. Simulated moving bed.

A dilute feed of 3-phenyl-1-propanol (A) and 2-phenyl ethanol (B) in a 60/40 wt% ratio methanol–water mixture is to be fed to a four-section laboratory SMB to separate A from B. The feed rate is 0.16 mL/minute, with 0.091 g/L of A and 0.115 g/L of B. The desorbent is a 60/40 wt% methanol–water mixture. For this dilute mixture of A and B and the adsorbent to be used, a linear adsorption isotherm applies, with $K_A = 2.36$ and $K_B = 1.40$. Assume that neither methanol nor water adsorb. The external void fraction, ϵ_b, of the adsorbent beds is 0.572. A switching time of 10 minutes is to be used. Using the steady-state, local-composition TMB model for a perfect separation of A from B with $\beta = 1.15$, estimate initial values for the volumetric flow rates of the extract, raffinate, desorbent, and solid particles. Convert the value of the recirculation rate for the TMB to that for the SMB and compute the resulting volumetric-liquid flow rates in each of the four sections.

15.44. Steady-state TMB.

In the steady-state TMB run of Example 15.22, results for Example 15.21, from application of the steady-state, local-equilibrium TMB model, were used with a β of 1.05 to establish flow rates of the raffinate, extract, makeup desorbent, recirculation, and solid particles so as to approach a perfect separation between fructose and glucose. While the separation was not perfect, it was reasonably good. What results are achieved in Example 15.22 if the overall mass-transfer coefficients approach infinity?

Section 15.8

15.45. Separation of polypeptides.

Determine the field gradient required to separate two polypeptides ($D = 5 \times 10^{-6}$ cm²/s), with respective pI values of 4.2 and 4.8, using IEF with a linear field gradient of 0.3/cm. Use a standard criteria for IEF that peak-to-peak separation must be three times the average peak width.

15.46. Separation of DNA strands.

Determine the number of pulses of equal duration that would be required to separate DNA strands of length 125 kbp and 250 kbp, respectively, by 1 mm, using PFGE. The electric field gradient in PFGE is to alternate between angles of 150° and 210°.

15.47. Resolution of bovine serum albumin.

Estimate the %T, %C, and voltage gradient required to completely resolve protein BSA ($U_{el} = 1.04 \times 10^{-5}$ cm²/V-s) from antibody IgG ($U_{el} = 1.02 \times 10^{-5}$ cm²/V-s) on a 20-cm polyacrylamide slab if their diffusivities are approximately equal ($D = 1.0 \times 10^{-7}$ cm²/s).

Part Four

Separations that Involve a Solid Phase

Chapters 16, 17 and 18 describe separations in which one or more components in a solid phase undergo mass transfer to or from a fluid phase. Chapter 16 covers selective leaching from a solid to a liquid solvent. This operation is widely used in the food industry. Crystallization from a liquid and desublimation from a vapor are discussed in Chapter 17, where evaporation, which often precedes crystallization, is included. Both solution crystallization to produce inorganic crystals and melt crystallization to produce organic crystals are considered. Chapter 18 is devoted to drying of solids and the myriad types of equipment used industrially. Drying is important in the pharmaceutical industry, where many products are prepared in solution and sold as dry powders in tablet form. A section on psychrometry is also included.

Chapter 16

Leaching and Washing

§16.0 INSTRUCTIONAL OBJECTIVES

After completing this chapter, you should be able to:

- Describe equipment used for batch and continuous leaching.
- Explain differences between leaching, washing, and expression.
- List assumptions for an ideal, equilibrium leaching or washing stage.
- Calculate recovery of a solute for a continuous, countercurrent leaching and washing system for a constant ratio of liquid to solids in the underflow or for a variable-underflow ratio.
- Apply rate-based leaching to food-processing applications.
- Develop and apply the rate-based, shrinking-core model to reactive mineral processing.

Leaching, sometimes called solid–liquid (or liquid–solid) extraction, involves the removal of a soluble fraction (the solute or *leachant*) of a solid material by a liquid solvent. The solute diffuses from inside the solid into the surrounding solvent. Either the extracted solid fraction or the insoluble solids, or both, may be valuable products. Leaching is widely used in the metallurgical, natural product, and food industries. In metallurgy, leaching may involve oxidation or reduction reactions of the solid with the solvent. Equipment is available to conduct leaching under batch, semicontinuous, or continuous operating conditions. Effluents from a leaching stage are essentially solids-free liquid, called the *overflow*, and wet solids, the *underflow*. To reduce the concentration of solute in the liquid portion of the underflow, leaching is often accompanied by countercurrent-flow washing stages. The combined process produces a final overflow, referred to as *extract*, which contains some of the solvent and most of the solute; and a final underflow, the *extracted* or *leached solids*, which are wet with almost pure solvent. Ideally, the soluble solids are perfectly separated from the insoluble solids, but solvent is distributed to both products. Therefore, additional processing of the extract and the leached solids is necessary to recover solvent for recycle.

Industrial applications of leaching include: (1) removal of copper from ore using sulfuric acid, (2) recovery of gold from ore using sodium-cyanide solution, (3) extraction of sugar from sugar beets using hot water, (4) extraction of tannin from tree bark using water, (5) removal of caffeine from green coffee beans using supercritical CO_2, and (6) recovery of proteins and other natural products from bacterial cells.

Industrial Example

Extraction of vegetable oil from soybeans with a commercial hexane solvent in a pilot-plant-size, countercurrent-flow, multistage leaching unit, as described by Othmer and Agarwal [1], is an example of a commercial leaching process. Edible oils can be extracted from a number of different field and tree crops, including coconuts, cottonseeds, palms, peanuts, grape seeds, soybeans, and sunflower seeds, but the highest percentage of edible oil is from soybeans. In 2004–2005, world production of soybean oil was 28 million tons. Oil from soybeans is high in polyunsaturated fats and thus less threatening from a cholesterol standpoint. When oil content of seeds and beans is high, some of the oil can be removed by compressing the solids in a process known as *expression*, as discussed in Green and Perry [2]. For soybeans, whose oil content is typically less than 0.30 lb per lb of dry and oil-free solids, leaching is more desirable than expression because a higher yield of oil is achieved using the former.

The soybeans are dehulled, cleaned, cracked, and flaked before being fed to the extractor. Typically, they contain 8 wt% moisture and 20 wt% oil. Dry, oil-free soybeans have a particle density of 1.425 g/cm^3. The oil has a density of 0.907 g/cm^3 and a viscosity at 25°C of 50 cP. Approximately 50% of the flake volume is oil, moisture, and air. Whole beans, rather than flakes, could be fed to the extractor, with leaching taking place by molecular diffusion of solvent into the seed, followed by oil diffusion through the solvent and out of the seed. If so, the mass-transfer process within the seed could be modeled with Fick's second law. However, experiments by King, Katz, and Brier [3] found that although extraction of oil with solvent in uniformly porous inorganic solids, like porous clay plates, obeys Fick's law of molecular diffusion, extraction of oil from soybeans does not, presumably because of the complex internal structure of soybeans. Furthermore, Othmer and Agarwal [1], using whole and cut-in-half soybeans, found that diffusion is extremely slow. After 168 hours in contact with hexane, less than 0.08% of the oil in the whole beans and less than 0.19% of the oil in the half beans was extracted. Such a slow diffusion rate for particles that are about 5 mm in diameter is due to the location of the oil within insoluble cell walls, requiring that oil pass through the walls, driven by low osmotic pressure differences.

The extent and rate of oil extraction is greatly enhanced by flaking the soybeans to thicknesses of 0.005–0.02-inch. Flaking ruptures the cell walls, greatly facilitating contact of oil with solvent. Using trichloroethylene [3] or *n*-hexane [1] as the solvent, with flakes of diameters from 0.04 to 0.24 inch, approximately 90% of the oil can be extracted in 100 minutes. The ideal solvent for commercial leaching of soybeans should have high oil solubility to minimize the amount of solvent, a high volatility to facilitate recovery of solvent from oil by evaporation or distillation, nonflammability to eliminate fires and explosions, low cost, ready availability, chemical stability, low toxicity, and compatibility with inexpensive materials of construction. In many respects, especially nonflammability, trichloroethylene is an ideal solvent, but it is classified as a hazardous, toxic chemical. The favored solvent is thus commercial hexane (mostly *n*-hexane), which presents a fire hazard but has a low toxicity.

The pilot-plant leaching unit used by Othmer and Agarwal, known as the Kennedy extractor, is shown in Figure 16.1. Soybeans enter continuously at the low end and are leached in a countercurrent cascade of tubs by hexane solvent, which enters at the upper end. The flakes and solvent are agitated, and underflows are pushed uphill from one tub to the next, by slowly rotating paddles and scrapers, while overflows move downhill from tub to tub. The paddles are perforated to drain the solids when they are lifted above the liquid level in the tub by the paddle. Othmer and Agarwal used 15 tubs.

Soybean flakes of 0.012-inch average thickness, 10.67 wt% moisture, and 0.2675 g oil/g dry, oil-free flakes were fed to the Kennedy extractor at a rate of 6.375 lb/h. Solvent flow was 10.844 lb/h. Leaching took place at ambient conditions, and after 11 hours of operation at steady state, an extract, the *miscella*, of 7.313 lb/hr, containing 15.35 wt% oil, was produced. The leached solids contained 0.0151 g oil/g dry, oil-free flakes; thus, 94.4% of the oil was extracted. Residence time in each tub was 3 minutes, giving a total residence time of 45 minutes. From these data, a mass-balance check can be made for oil and solvent, and the liquid-to-solids ratio in the leached solids can be estimated. These calculations are left as an exercise.

§16.1 EQUIPMENT FOR LEACHING

Leachable solids generally undergo pretreatment before being fed to leaching equipment so that reasonable leaching times are obtained. For example, seeds and beans are dehulled, cracked, and flaked, as described above for soybeans. When vegetable and animal material cannot be flaked, it may be possible to cut it into thin slices, as is done with sugar beets prior to leaching of the sugar with water. In this case, the cell walls are left largely intact to minimize the leaching of undesirable material, such as colloids and albumens. Metallurgical ores are crushed and ground to small particles because small regions of leachable material may be surrounded by relatively impermeable insoluble material. When that material is quartzite, leaching may be extremely slow. Van Arsdale [4] cites the very important effect of particle size on the time required for effective leaching of a copper ore by aqueous sulfuric acid. The times for particle diameters of 150 mm, 6 mm, and less than 0.25 mm are approximately 5 years, 5 days, and 5 hours, respectively. When leachable solids contain a high % of solute, pretreatment may not be necessary because disintegration of the remaining skeleton of insoluble material takes place at the surface of the particles as leaching progresses. When the entire solid is soluble, leaching may be rapid, such that only one stage of extraction is required as *dissolution* takes place.

Industrial equipment for solid–liquid extraction is designed for batchwise or continuous processing. The method of contacting solids with solvent is either by *percolation* of solvent through a bed of solids or by *immersion* of the solid in the solvent followed by agitation of the mixture. When immersion is used, countercurrent, multistage operation is common. With percolation, either a stagewise or a differential contacting device is appropriate. An extractor must be efficient to minimize the need for solvent because of the high cost of solvent recovery.

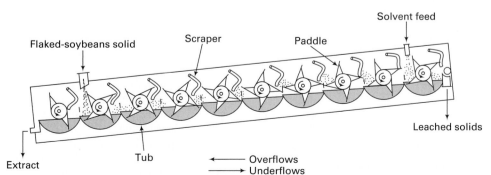

Figure 16.1 Kennedy extractor for leaching of oil from soybeans.

§16.1.1 Batch Extractors

When the solids to be leached are in the form of fine particles, perhaps smaller than 0.1 mm in diameter, batch leaching is conveniently conducted in an agitated vessel. A simple configuration is the Pachuca tank [5], depicted in Figure 16.2a and used extensively in the metallurgical industry. The tank is a tall, cylindrical vessel constructed of wood, concrete, or metal that can be lined with an inert, noncorrosive, nontoxic material. Solvent and solids are placed in the tank and agitation is achieved by an air lift, whereby air bubbles entering at the bottom of a circular tube, concentric with the tank, cause upward flow and subsequent circulation of the solid–liquid suspension. During agitation, air continuously enters and leaves the vessel. When the desired degree of leaching is accomplished, agitation stops and solids are allowed to settle into a sludge at the bottom, where it is removed with the assistance of air. The supernatant extract is removed by siphoning from the top of the tank. Agitation can also be achieved by a paddle stirrer or by the use of a propeller mounted in a draft tube to provide upward flow and circulation of the solid–liquid suspension, much like that in the Pachuca tank.

When the solids are too coarse to be easily suspended by immersion in a stirred solvent, percolation techniques can be used. Again a tall, cylindrical vessel is employed. Solids to be leached are dumped into the vessel, followed by percolation of solvent down through the bed of solids, much like in fixed-bed adsorption. To achieve a high concentration of solute in the solvent, a series of vessels is arranged in a multibatch, countercurrent-leaching technique developed in 1841 by James Shanks and called a Shanks extraction battery. This technique can be used for such applications as batch removal of tannin from wood or bark, sugar from sugar beets, and water-soluble substances from coffee, tea, and spices. A typical vessel arrangement is shown in Figure 16.2b, where Vessel 1 is off-line for emptying and refilling of solids. Solvent enters and percolates down through the solids in Vessel 2, and then percolates through Vessels 3 and 4, leaving as final extract from Vessel 4. The extraction of solids in Vessel 2 is completed first. When that occurs, Vessel 2 is taken off-line for emptying and refilling of solids and Vessel 1 is placed on-line. Fresh solvent first enters Vessel 3, followed by Vessels 4 and 1. In this manner, fresh solvent always contacts solids that have been leached for the longest time, thus realizing the benefits of countercurrent contacting. Heat exchangers are

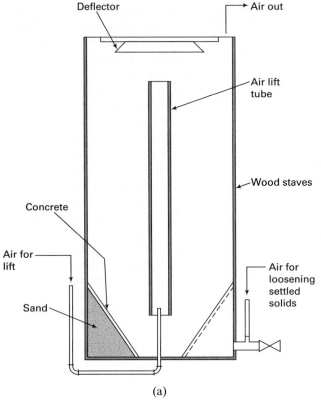

(a)

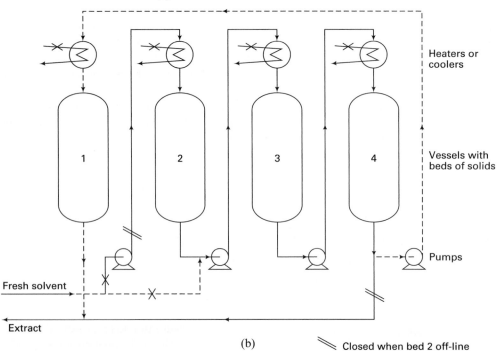

(b)

Closed when bed 2 off-line

Figure 16.2 (a) Pachuca tank for batch leaching of small particles.

[From *Handbook of Separation Techniques for Chemical Engineers*, 2nd ed., P.A. Schweitzer, Editor-in-chief, McGraw-Hill, New York (1988) with permission.]

(b) Shanks countercurrent, multibatch battery system for leaching of large particles by percolation.

[From C.J. King, *Separation Processes*, 2nd ed., McGraw-Hill, New York (1980) with permission.]

provided between vessels to adjust the liquid temperature, and pumps can be used to move liquid from vessel to vessel. Any number of vessels can be included in a battery. Note that although the system is batchwise with respect to solids, it is continuous with respect to solvent and extract.

§16.1.2 Espresso Machine

A universally used batch-leaching machine is the espresso coffee maker. Although a beverage from coffee beans was first made about 1100 B.C. on the Arabian peninsula, it was not until many centuries later that a method and a machine were devised to produce a high-quality coffee, called espresso (a term that connotes a cup of coffee expressly for you, made quickly and individually, and intended to be drunk immediately). The prototype of the espresso machine was created in France in 1822, and the first commercial espresso machine was manufactured in Italy in 1905. By the 1990s, espresso machines were in many countries, producing billions of cups annually. Today, coffee, and, particularly, espresso, is a world commodity second only to oil.

A photograph of a consumer espresso machine is shown in Figure 16.3a; a simpler diagram to help understand its operation is presented in Figure 16.3b. In the machine, 7 to 9 grams of coffee beans are ground to a powder of particle size 250–750 microns by a burr grinder that minimizes temperature increase. The bed of powder, contained as a thin layer in a filter housing, is tamped to increase its uniformity. Water is pumped to a pressure of 9–15 atmospheres and heated rapidly to 88–92°C. The high pressure is required for pressure infusion of hot water into the particles so that extraction can proceed rapidly. During a period of 20 to 30 seconds, hot water is percolated through the bed of coffee powder to produce a 45-mL shot, which has a viscosity of warm honey, and is topped by a thick, dark, golden-cream foam ("crema"). A typical machine produces two shots, which can be added to water, milk, or other liquids or blends to produce various beverages, including Americano, Breve, Cappuccino, Latte, Lungo, Macchiato, Mocha, or Ristretto.

The brief time interval between grinding and leaching, the short residence time of the leaching, the small coffee particle size, and the controlled water temperature and pressure during leaching combine to maximize extraction of the flavor- and-aroma chemicals and minimize extraction of the chemicals associated with bitterness, such as quinine and caffeine. In addition, crema traps the aroma, preventing its escape into the surrounding air. At the pull of a lever or push of a button, espresso coffee is produced that is concentrated, full in body, and rich in flavor and aroma. Studies of espresso machines are provided by Andueza et al. [16–18].

§16.1.3 Continuous Extractors

When leaching is carried out on a large scale, it is preferable to use an extraction device that operates with continuous flow of both solids and liquid. Many such patented devices are available, especially for the food industry, as discussed by Schwartzberg [6]. Some of the widely discussed extractors are shown schematically in Figure 16.4. These differ mainly with respect to the manner in which solids are transported and the degree to which agitation of solid–liquid mixtures is provided.

According to Schwartzberg, several extractors described in the literature are now either obsolete or infrequently used because of various limitations, including ineffective contacting of solid and liquid phases, bypassing, and fines entrainment. These include Hildebrand, Detrex, Anderson, Allis Chalmers, and Bonotto extractors. The Kennedy extractor for oil extraction from soybeans may have a low efficiency in some applications, but it is still used and still available.

The Bollman vertical, moving-basket, conveyor extractor in Figure 16.4a is used to extract oil from flaked seeds and beans. Baskets with perforated bottoms are moved around a vertical loop by a motor-driven chain drive. Solvent percolates down, from basket to basket, through the solids. When a basket reaches the top of the extractor, it is inverted to dump extracted solids and then filled with fresh solids. Flow of liquid is countercurrent to solids in ascending baskets and cocurrent in descending baskets. Fresh solvent enters near the top of the ascending leg and collects as "half miscella" in the left-hand part of the bottom sump. From there, half miscella is pumped to the top of the descending leg, from which it flows down to the right-hand part of the sump and is withdrawn as final extract, "full miscella." A typical Bollman extractor is 14 m high, with each basket filled with solids to a depth of about 0.5 m. According to Coulson et al. [7], baskets

(a)

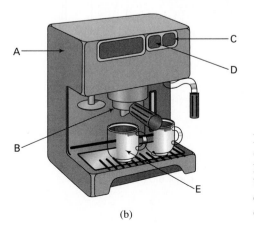

(b)

Figure 16.3 (a) Consumer espresso machine. (b) Operation: (A) pressure vessel; (B) portafilter holding ground coffee; (C) on/off switch, with built-in pressure indicator; (D) solenoid valve for espresso coffee; (E) shot cup holding leached, espresso coffee.

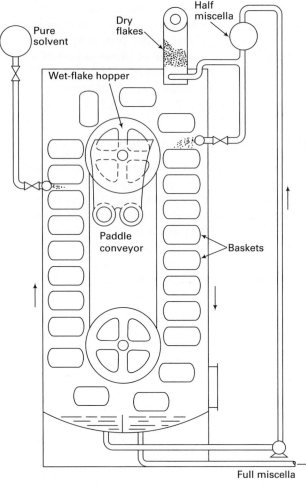

(a) Bollman vertical, moving-basket, conveyor extractor

are rotated slowly, at about 1 rph, to give solids residence times of 60 minutes. Each basket contains approximately 350 kg of flaked solids. For the 23 baskets shown in Figure 16.4a, almost 200,000 kg of solids can be extracted per day. About equal mass flows of solids and solvent are fed to the extractor, and full miscella is essentially solids-free, with about 25 wt% oil.

Another widely used continuous extractor for flaked seeds and beans is the Rotocel extractor in Figure 16.4b. In this device, which resembles a carousel and simulates a Shanks system, walled, annular sectors, called cells, on a horizontal plane, are slowly rotated by a motor. The cells, which hold solids and are perforated for solvent drainage, successively pass a solids-feed area, a series of solvent sprays, a final spray and drainage area, and a solids-discharge area. Fresh solvent is supplied to the cell located just below the final spray and drainage area, from where drained liquid is collected and pumped to the preceding cell location. The drainage from that cell is collected and pumped to the cell preceding that cell and so on. In this manner, a countercurrent flow of solids and liquid is achieved. The extracted solids may contain 25–30 wt% liquid. Rotocel extractors are typically 3.4–11.3 m in diameter, 6.4–7.3 m in height, and with bed depths of 1.8–3.0 m. They process up to 3 million kg/day of flaked soybeans. The number of cells can be varied and residence time controlled by the rate of rotation. A popular variation of Rotocel extractors is the French stationary-basket extractor in Figure 16.4c, which has about the size and capacity of a Rotocel extractor. Instead of the sectored cells moving, the solids-feed spout and solids-discharge zone rotate, with periodic switching of solvent feed and discharge connections. Thus, the weight of moving parts is reduced.

Continuous, perforated-belt extractors, as shown in Figure 16.4d, are used to process sugar cane, sugar beets, oil seeds,

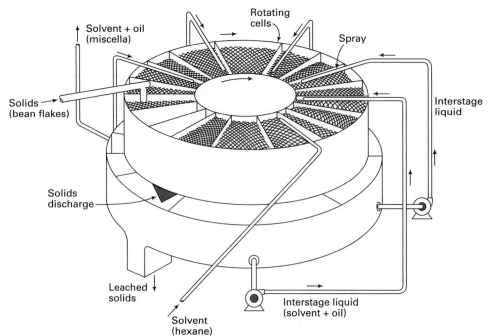

(b) Rotocel extractor

Figure 16.4 Equipment for continuous leaching.

[From *Handbook of Separation Techniques for Chemical Engineers*, 2nd ed., P.A. Schweitzer, Editor-in-chief, McGraw-Hill, New York (1988) with permission.]

[From R.E. Treybal, *Mass-Transfer Operations*, 3rd ed., McGraw-Hill, New York (1980) with permission.] (*Continued*)

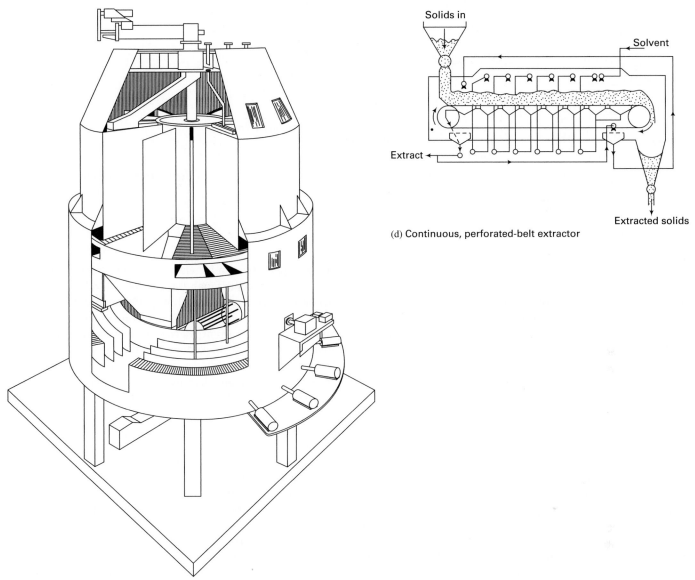

(c) French stationary-basket extractor

(d) Continuous, perforated-belt extractor

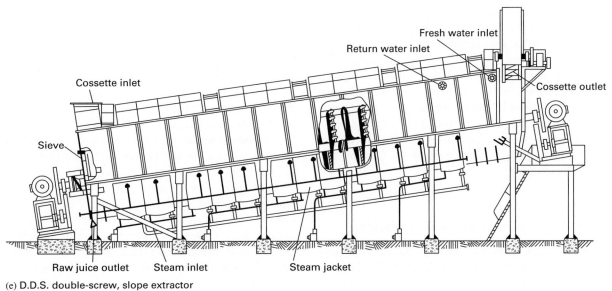

(e) D.D.S. double-screw, slope extractor

Figure 16.4 (*Continued*)

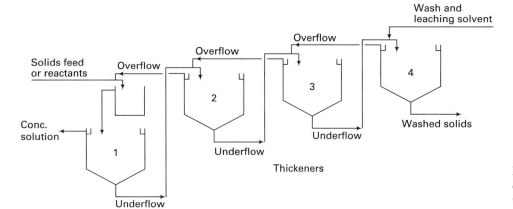

Figure 16.5 Continuous, countercurrent washing system using thickeners.

and apples (for apple juice). The feed solids are fed from a hopper to a slow-moving, continuous, and nonpartitioned perforated belt driven by motorized sprockets at either end. The height of solids on the belt can be controlled by a damper at the feed hopper outlet. Belt speed is automatically adjusted to maintain the desired depth of solids. Extracted solids are discharged into an outlet hopper at the end of the belt by a scraper, and side walls prevent solids from falling off the sides of the belt. Below the belt are compartments for collecting solvent. Fresh solvent is sprayed over solids and above compartments in a countercurrent fashion, starting from the discharge end of the belt, in as many as 17 passes. Bed depths range from 0.8 to 2.6 m and units from 7 to 37 m long with belts from 0.5 to 9.5 m wide have processed as much as 7,000,000 kg/day of sugar cane or sugar beets.

The D.D.S. (De Danske Sukker–fabriker) double-screw, slope extractor in Figure 16.4e is a very versatile unit. Although used mainly for extraction of sugar beets, the device has been applied successfully to a range of other feed materials, including sugar cane, flaked seeds and beans, apples, pears, grapes, cherries, ginger, licorice, red beets, carrots, fishmeal, coffee, and tea. The opposite-turning screws of the metal ribbons are pitched so that both screws move the solids uphill in parallel, cylindrical troughs. Extract flows through the screw surface downhill to achieve a differential, countercurrent flow with the solids. A novel feature is the ability to turn one screw slightly faster and then slightly slower than the other screw, causing the solids to be periodically squeezed. Units range in size from 2–3.7 m in diameter and 21–27 m in length and have been used to process as much as 3,000,000 kg/day of sugar beets in the form of cossettes (long, thin strips).

§16.1.4 Continuous, Countercurrent Washing

When leaching is very rapid, as with small particles containing very soluble solutes or when leaching has already been completed or when solids are formed by chemical reactions in a solution, it is common to countercurrently wash the solids to reduce the solute concentration in the liquid adhering

to the solids. This can be accomplished in a series of *gravity thickeners* or centrifugal thickeners, called *hydroclones*, arranged for countercurrent flow of the underflows and overflows as shown in Figure 16.5, and sometimes called a continuous, countercurrent decantation system. A typical continuous gravity thickener is shown in detail in Figure 16.6a. Combined feed to the thickener consists of feed solids or underflow from an adjacent thickener, together with fresh solvent or overflow from an adjacent thickener. The thickener must first thoroughly mix liquid and solids to obtain a uniform concentration of solute in the liquid, and it must then produce an overflow free of solids and an underflow with as high a fraction of solids as possible.

A thickener consists of a large-diameter, shallow tank with a flat or slightly conical bottom. The combined feed enters the tank near the center by means of a feed launder that discharges into a feed well. Settling and sedimentation of solid particles occur by gravity due to a solid-particle density that is greater than the liquid density. In essence, solids flow downward and liquid flows upward. Around the upper, inner periphery of the tank is an overflow launder or weir for continuously removing clarified liquid. Solids settling to the tank bottom are moved inward toward a thick sludge discharge by a slowly rotating motor-driven rake. Thickeners as large as 100 m in diameter and 3.5 m high have been constructed. In large thickeners, rakes revolve at about 2 rpm.

Residence times of solids and liquids in a gravity thickener are often large (minutes or hours) and, as such, are sufficient to provide adequate residence time for mass transfer and mixing when small particles are involved. When long residence times are not needed and the overflow need not be perfectly clear of solids, the hydroclone, shown in Figure 16.6b, may be appropriate. Here, pressurized feed slurry enters tangentially to create, by centrifugal force, a downward-spiraling motion. Higher-density, suspended solids are, by preference, driven to the wall, which becomes conical as it extends downward, and discharged as a thickened slurry at the hydroclone bottom. The liquid, which is forced to move inward and upward as a spiraling vortex, exits from a vortex-finder pipe extending downward from the closed hydroclone top to a location just below feed entry.

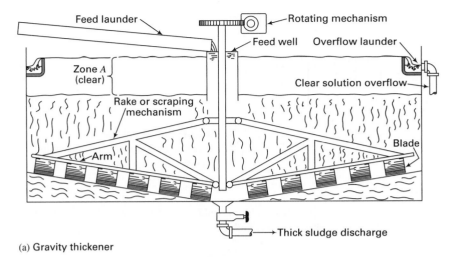

(a) Gravity thickener

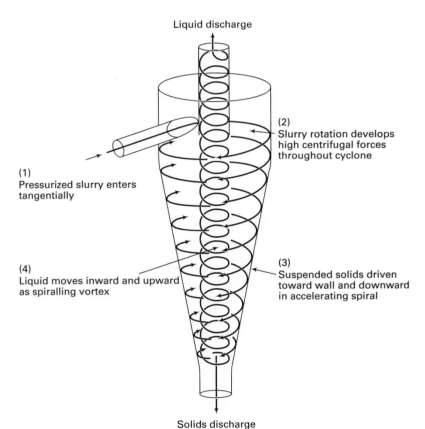

(b) Hydroclone, centrifugal thickener

Figure 16.6 Thickeners.

[From *Handbook of Separation Techniques for Chemical Engineers*, 2nd ed., P.A. Schweitzer, Editor-in-chief, McGraw-Hill, New York (1988) with permission.]

§16.2 EQUILIBRIUM-STAGE MODEL FOR LEACHING AND WASHING

The simplest model for a continuous, countercurrent leaching and washing system, as shown in Figure 16.7, is similar to the model developed in §5.2. It assumes that the solid feed consists of a solute that is completely soluble in the solvent and an inert substance or carrier that is not soluble. Leaching is assumed to be rapid such that it is completed in a single leaching stage, which is followed by a series of one or more washing stages to reduce concentration of solute in the liquid adhering to the solids in the underflow. All overflow streams are assumed to be free of solids. In Figure 16.7: S = mass flow rate of inert solids, which is constant from stage to stage; V = mass flow rate of entering solvent or overflow liquid (solvent plus solute), which varies from stage to stage; L = mass flow rate of underflow liquid (solvent plus solute), which varies from stage to stage; y = mass fraction of solute in the overflow liquid; and x = mass fraction of solute in the underflow liquid.

Alternatively, V and L can refer to mass flow rates of solvent on a solute-free basis and the symbols Y and X can be used to refer to mass ratios of solute to solvent in the overflow liquid and underflow liquid, respectively. Mole or volume flow rates can also be used.

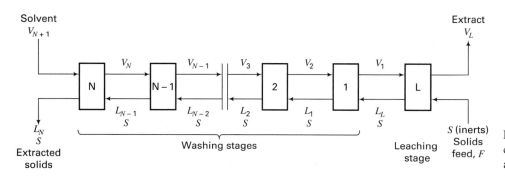

Figure 16.7 Continuous, countercurrent, ideal-stage leaching and washing system.

An ideal leaching or washing stage is defined by Baker [8] as one where: (1) Any entering solid solute is completely dissolved into the liquid in the stage (assuming the liquid contains sufficient solvent). (2) The composition of liquid in the stage is uniform throughout, including any liquid within pores of the inert solid. (3) Solute is not adsorbed on the surfaces of the inert solid. (4) Inert solids leaving in the underflow from each stage are wet with liquid, such that mass ratio of solvent in that liquid (or the total liquid) to inert solids is constant from stage to stage. (5) Because of item 2, concentration of solute in the overflow is equal to that in the liquid portion of the underflow. This is equivalent to an equilibrium assumption. (6) Overflows contain no solids. (7) Solvent is not vaporized, adsorbed, or crystallized.

For the continuous, countercurrent system of ideal leaching stages in Figure 16.7, solute and total-liquid material balances can be used to solve problems, including: (1) determination of ideal stages required to achieve a specified degree of washing; and (2) determination of the effect of washing for a specified degree of washing with a certain number of ideal stages. Depending on the problem, either an algebraic or a graphical method may be preferred. For most problems, it is best to consider the leaching stage separately from the washing stages, as illustrated in the following examples.

EXAMPLE 16.1   Leaching and Washing Cascade.

A finely divided solids feed, F, of 150 kg/h, containing 1/3 water-soluble Na_2CO_3 and 2/3 insoluble ash is to be leached and washed at 30°C in a two-stage, countercurrent system with 400 kg/h of water. The leaching stage consists of an agitated vessel that discharges slurry into a thickener. The washing stage consists of a second thickener. Experiments show that the sludge underflow from each thickener will contain 2 kg of liquid (water and carbonate) per kg of insoluble ash. Assuming ideal stages: (a) Calculate the % recovery of carbonate in the final extract. (b) If a third stage is added, calculate the additional carbonate recovered.

Solution

At 30°C, solubility of carbonate in water is 38.8 kg/100 kg of water. Therefore, there is sufficient water to dissolve all the carbonate. Referring to Figure 16.7:

(a) $N = 1$ (L is the leaching stage); $S = \frac{2}{3}(150) = 100$ kg/h of insoluble ash;

Na_2CO_3 in entering solids $= \frac{1}{3}(150) = 50$ kg/h; $V_2 =$ entering solvent $= 400$ kg/h;

$$L_L = L_1 = 2S = 200 \text{ kg/h}.$$

By total-liquid material balances on Stage 1 and Stage L,

$$V_1 = V_2 + L_L - L_1 = 400 + 200 - 200 = 400 \text{ kg/h}$$
$$V_L = V_1 + 50 - L_L = 400 + 50 - 200 = 250 \text{ kg/h}$$

Na_2CO_3 material balance around washing Stage 1:

$$x_L L_L = y_1 V_1 + x_1 L_1 \tag{1}$$
$$200 x_L = 400 y_1 + 200 x_1 \tag{2}$$

But $y_1 = x_1$ from item 5 above for an ideal stage. Combining (1) and (2),

$$x_L = 3 x_1 \tag{3}$$

Na_2CO_3 material balance around leaching Stage L:

$$y_1 V_1 + 50 = x_L L_L + y_L V_L \tag{4}$$
$$400 y_1 + 50 = 200 x_L + 250 y_L \tag{5}$$

But, again, $y_1 = x_1$ and $y_L = x_L$ \hfill (6)

Combining (4), (5), and (6),

$$x_1 = 1.125 x_L - 0.125 \tag{7}$$

Combining (3) and (7) and solving, $x_L = 0.158$

Therefore, $y_L = 0.158$.
From (7), $x_1 = 0.0526$

$$\text{Recovery of } Na_2CO_3 = \frac{y_L V_L}{50} = \frac{(0.158)(250)}{50} = 0.79 \text{ or } 79\%$$

(b) $N = 2$ washing stages; $V_3 = 400$ kg/h; $L_L = L_1 = L_2 = 2S = 200$ kg/h; $V_2 = V_1 = 400$ kg/h; $V_L = 250$ kg/h.

Na_2CO_3 material balance around Stage 2:

$$x_1 L_1 = y_2 V_2 + x_2 L_2$$
$$200 x_1 = 400 y_2 + 200 x_2 \tag{8}$$

But, $y_2 = x_2$ \hfill (9)

Combining (8) and (9), $x_1 = 3 x_2$ \hfill (10)

Na_2CO_3 material balance around Stage 1:

$$y_2 V_2 + x_L L_L = x_1 L_1 + y_1 V_1$$
$$400 y_2 + 200 x_L = 200 x_1 + 400 y_1 \tag{11}$$

But, $$y_2 = x_2 \qquad (12)$$

$$y_1 = x_1 \qquad (13)$$

Combining (11), (12), and (13),

$$x_L = 3x_1 - 2x_2 \qquad (14)$$

Na_2CO_3 material balance around Stage L:

This is the same as (7) in part (a).

Solving (10), (14), and (7), which are all linear,

$$x_L = 0.1795; y_L = 0.1795; x_1 = 0.0769; x_2 = 0.0256$$

$$\text{Recovery of } Na_2CO_3 = \frac{y_L V_L}{50} = \frac{(0.1795)(250)}{50} = 0.898 \text{ or } 89.8\%$$

From part (a), for two stages, recovery of Na_2CO_3 is 0.158 (250) = 39.5 kg/h.

For three stages, recovery is 0.1795(250) = 44.9 kg/h.

Recover 44.9 − 39.5 = 5.4 kg/h more Na_2CO_3 with three stages.

For this example, it is difficult to use a graphical McCabe–Thiele-type method because only the slope, and not either end point, of the operating line is known.

EXAMPLE 16.2   Leaching of Wax with Kerosene.

Baker [8] presented the following problem, for which a McCabe–Thiele graphical solution is appropriate. Two tons per day of waxed paper containing 25 wt% soluble wax and 75 wt% insoluble pulp are to be dewaxed by leaching with kerosene in the continuous, counter-current system shown in Figure 16.7. The wax is completely dissolved by kerosene in the leaching stage, L. Subsequent washing stages reduce the wax content in the liquid adhering to the pulp leaving the last stage, N, to 0.2 lb wax/100 lb pulp. Kerosene entering the system is recycled from a solvent-recovery system and contains 0.05 lb wax/100 lb kerosene. The final extract is to contain 5 lb wax/100 lb kerosene. Experiments show that underflow from each stage contains 2 lb kerosene/lb insoluble pulp. Determine the washing stages required.

Solution

Referring to Figure 16.7 and using the nomenclature defined at the beginning of this section for the case of concentrations on a mass-ratio basis and flow rates on a solute-free basis, the following material-balance equations apply:

Overall mass balance on solute (wax):

$$0.25(4,000) + \frac{0.05}{100} V_{N+1} = \frac{5}{100} V_L + \frac{0.2}{100}(0.75)(4,000)$$

or $$0.05 V_L - 0.0005 V_{N+1} = 994 \qquad (1)$$

Overall mass balance on solvent (kerosene):

$$V_{N+1} = V_L + 2(0.75)(4,000)$$

or $$V_{N+1} = V_L + 6,000 \qquad (2)$$

Solving (1) and (2), V_{N+1} = kerosene in entering solvent = 26,140 lb/day and V_L = kerosene in exiting extract = 20,140 lb/day.

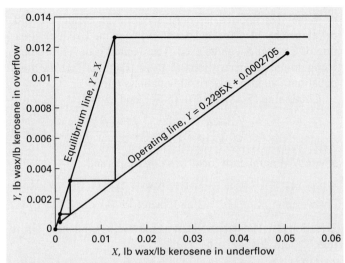

Figure 16.8 McCabe–Thiele diagram for Example 16.2.

Thus, the final underflow contains 26,140 − 20,140 = 6,000 lb/day of kerosene. Also,

$$X_N = \left[\frac{0.2}{100}(0.75)(4,000) \right] \Big/ 6,000$$

$$= 0.001 \text{ lb wax/lb kerosene in the final underflow}$$

Material balances can now be made around the leaching stage.

Mass balance on kerosene:

$$V_1 = V_L + 2(0.75)(4,000)$$

Thus, $$V_1 = 20,140 + 6,000 = 26,140 \text{ lb/day}$$

Mass balance on wax:

$$26,140 Y_1 + (0.25)(4,000) = \frac{5}{100}(20,140) \\ + X_L(2)(0.75)(4,000) \qquad (3)$$

But, $$X_L = Y_L = 0.05 \text{ lb wax/lb kerosene} \qquad (4)$$

Substituting (4) into (3) and solving, $Y_1 = 0.01174$ lb wax/lb kerosene.

The end points (solute compositions) at the two ends of the washing cascade are now established. Referring to Figure 16.7,

$$Y_{N+1} = 0.0005 \quad \text{and} \quad X_N = 0.001 \text{ in lb wax/lb kerosene}$$
$$Y_1 = 0.01174 \quad \text{and} \quad X_L = 0.05 \text{ in lb wax/lb kerosene}$$

For the washing section, the mass ratio of kerosene in the underflow to kerosene in the overflow is constant at a value of

$$\frac{L_{n-1}}{V_n} = \frac{2(0.75)(4,000)}{26,140} = 0.2295$$

The operating line on a McCabe–Thiele plot of Y versus X will be a straight line through the two end points, with a slope of 0.2295: $Y = 0.2295X + 0.0002705$. The equilibrium line is simply $Y = X$, which is plotted, along with the operating line, in Figure 16.8. Ideal washing stages, which can be stepped off starting from either end of the operating line, are stepped off here from stage N. Slightly less than three ideal washing stages are needed.

§16.2.1   McCabe–Smith Algebraic Method

It is seen in Figure 16.8 that it can be difficult to accurately step off washing stages. When the above ideal-stage model of Baker [8] applies, a more accurate algebraic method, developed by McCabe and Smith [9] from the Kremser equation of Chapter 5, can be used. The method is developed here using solute concentrations in mass fractions, but the final equations can also be derived with mass ratios.

Combining (5-48), (5-50), (5-51), and (5-54),

$$y_{N+1} = y_1\left(\frac{1 - A^{N+1}}{1 - A}\right) - y_0^*A\left(\frac{1 - A^N}{1 - A}\right) \qquad (16\text{-}1)$$

where N = number of ideal washing stages and

$$y_0^* = Kx_0 = Kx_L = y_L \qquad (16\text{-}2)$$

For washing, $K = y/x = 1$ and, therefore, $A = L/KV = L/V$.

Equation (16-1) can be written as follows by collecting terms in A^{N+1} and A:

$$A^{N+1}(y_1 - y_L) = A(y_{N+1} - y_L) + (y_1 - y_{N+1}) \qquad (16\text{-}3)$$

and simplified by writing an overall solute balance around all washing stages:

$$y_{N+1}V_{N+1} + x_LL_L = y_1V_1 + x_NL_N \qquad (16\text{-}4)$$

But for ideal washing stages, since

$$V_{N+1} = V_1, \quad L_L = L_N, \quad \text{and } A = L/V,$$

(16-4) simplifies to

$$y_{N+1} = y_1 + Ax_N - Ax_L \qquad (16\text{-}5)$$

But $y_L = x_L$. Therefore, (16-5) becomes

$$(y_1 - y_{N+1}) = A(y_L - x_N) \qquad (16\text{-}6)$$

Combining (16-3) and (16-6) and rearranging gives

$$A^N = \left(\frac{y_{N+1} - x_N}{y_1 - y_L}\right) \qquad (16\text{-}7)$$

Solving (16-7) for N with $A = L/V$ gives

$$N = \frac{\log\left(\dfrac{x_N - y_{N+1}}{y_L - y_1}\right)}{\log(L/V)} \qquad (16\text{-}8)$$

The argument of the log term of the denominator can be written in terms of end points to give

$$\frac{L}{V} = \left(\frac{y_1 - y_{N+1}}{y_L - x_N}\right) = \left(\frac{y_1 - y_{N+1}}{y_L - x_N}\right) \qquad (16\text{-}9)$$

Combining (16-8) and (16-9),

$$N = \frac{\log\left(\dfrac{x_N - y_{N+1}}{y_L - y_1}\right)}{\log\left(\dfrac{y_1 - y_{N+1}}{y_L - x_N}\right)} \qquad (16\text{-}10)$$

When the constant underflow liquid is given in terms of total solvent plus solute, (16-8) or (16-9) is used directly, where L and V are total liquid flows. When the underflow liquid is given in terms of just solvent, y and x solute mass fractions in (16-8) and (16-9) are replaced by Y and X solute mass ratios, and V and L are liquid flows of solute-free solvent.

EXAMPLE 16.3 McCabe–Smith Method.

Solve for the number of ideal, countercurrent washing stages for the conditions of Example 16.2 using the McCabe–Smith equations.

Solution

From the problem statement and the overall material-balance and leaching-stage calculations for Example 16.2,

$$Y_{N+1} = 0.0005,$$
$$X_N = 0.001,$$
$$Y_2 = 0.01174,$$
$$Y_L = X_L = 0.05, \quad \text{and}$$
$$L/V = 0.2295$$

From the mass-ratio form of (16-8), the number of ideal leaching stages is:

$$N = \frac{\log\left(\dfrac{0.001 - 0.0005}{0.05 - 0.01174}\right)}{\log(0.2295)} = 2.95$$

The same result is obtained if (16-10) is used.

When the leaching rate is slow, several countercurrent stages may be required, and the effect of washing will be diminished. This is illustrated in the next example.

EXAMPLE 16.4 Two Leaching Stages Needed.

In Example 16.1, part (b), leaching was assumed to be completed in one stage, with two additional stages provided for washing. The recovery of Na_2CO_3 in the extract was 89.8%. Recalculate that example assuming that 50% of the carbonate is leached in the first stage and the remaining 50% in the second stage, leaving only the last stage as a true washing stage. Number the stages as in Figure 16.9, which includes given and easily computed flows.

Solution

Na_2CO_3 material balance around Stage 3:

$$x_2L_2 = y_3V_3 + x_3L_3$$
$$200x_2 = 400y_3 + 200x_3 \qquad (1)$$

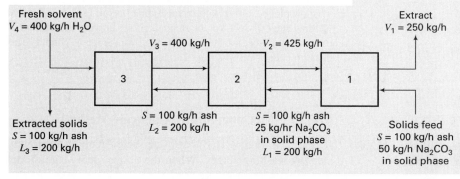

Fresh solvent
$V_4 = 400$ kg/h H_2O

$V_3 = 400$ kg/h $V_2 = 425$ kg/h

Extract
$V_1 = 250$ kg/h

| 3 | 2 | 1 |

Extracted solids
$S = 100$ kg/h ash
$L_3 = 200$ kg/h

$S = 100$ kg/h ash
$L_2 = 200$ kg/h

$S = 100$ kg/h ash
25 kg/hr Na_2CO_3
in solid phase
$L_1 = 200$ kg/h

Solids feed
$S = 100$ kg/h ash
50 kg/h Na_2CO_3
in solid phase

Figure 16.9 Leaching and washing system for Example 16.4.

But $$y_3 = x_3 \qquad (2)$$

Combining (1) and (2), $$x_2 = 3x_3 \qquad (3)$$

Na₂CO₃ material balance around Stage 2:

$$y_3 V_3 + x_1 L_1 = x_2 L_2 + y_2 V_2$$
$$400 y_3 + 200 x_1 + 25 = 200 x_2 + 425 y_2 \qquad (4)$$

But, $$y_2 = x_2 \qquad (5)$$

Combining (2), (4), and (5),

$$x_1 = 3.125 x_2 - 2 x_3 - 0.125 \qquad (6)$$

Na₂CO₃ material balance around Stage 1:

$$y_2 V_2 + 50 = x_1 L_1 + y_1 V_1 + 25$$
$$425 y_2 + 50 = 200 x_1 + 250 y_1 + 25 \qquad (7)$$

But, $$y_1 = x_1 \qquad (8)$$

Combining (5), (7), and (8),

$$x_2 = 1.059 x_1 - 0.0588 \qquad (9)$$

Solving (3), (6), and (9), $x_1 = y_1 = 0.1681$, $x_2 = 0.1192$, and $x_3 = 0.0397$.

Recovery of

$$Na_2CO_3 = \frac{y_1 V_1}{50} = \frac{(0.1681)(250)}{50} = 0.841 \text{ or } 84.1\%$$

which is almost 6% less than in Part (b) of Example 16.1, where leaching was completed in one stage.

§16.2.2 Variable Underflow

In Examples 16.1–16.4, ratios of liquid to solids in the underflow from stage to stage were assumed constant. Experiments by Ravenscroft [10] on extraction of oil from granulated halibut livers by diethylether showed that ratios of liquid to solids in the underflow increased significantly with increasing concentration of oil in the liquid in the range of 0.04–0.64 gal oil/gal liquid. In the leaching experiments, equilibrium was achieved in 2–3 minutes of agitation, but 10 minutes was used. Thirty minutes was allowed for settling of extracted livers, after which the free solution (overflow) was decanted, leaving the underflow. Ravenscroft ascribed variable-underflow effects to an increase in liquid viscosity and density with oil concentration. When neither density nor viscosity varies significantly, experimental data of Othmer and Agarwal [1] show that the main variable affecting the liquid-to-solid ratio in the underflow is solid surface area-to-volume ratios. When underflow varies, operating lines on a McCabe–Thiele diagram are curved and the curvature must be established by computing intermediate points, as illustrated in the next example.

EXAMPLE 16.5 Leaching with Variable Underflow.

Oil is to be extracted from 10,000 lb/h of granulated halibut livers, based on oil-free livers, which contain 0.043 gal of extractable oil per lb of oil-free livers. It is desired to extract 95% of the oil in a countercurrent extraction system using oil-free diethylether as

solvent. The final extract is to contain 0.65 gal oil per gal of extract. Assume volumes of oil and ether are additive and that leaching will be completed in one stage. Ravenscroft's underflow data show scatter, so use the following smoothed data to predict underflow rates:

Gal Oil per Gal Liquid	Gal Liquid Retained per lb Oil-Free Livers
0.00	0.035
0.10	0.042
0.20	0.049
0.30	0.058
0.40	0.069
0.50	0.083
0.60	0.100
0.70	0.132

Determine the number of ideal stages required.

Solution

First, determine the flow rate of solvent.

Oil in liver feed = 10,000(0.043) = 430 gal liquid/h

Oil in extract = 430(0.95) = 408.5 gal/h

Oil in underflow of extracted livers = 430 − 408.5 = 21.5 gal/h

By an iterative procedure, determine the gal/h of ether in the final underflow of extracted livers.
Assuming 0.10 gal oil/gal liquid:

From the above data, there are 0.042 gal liquid/lb oil-free livers. Therefore, 0.042(10,000) = 420 gal liquid/h and 0.10(420) = 42 gal/h oil, which is higher than the required value of 21.5 gal/h from above.

Assuming 0.05 gal oil/gal liquid:

Linear interpolation of the above data table gives 0.0385 gal liquid/lb oil-free livers.

Now, 0.0385(10,000) = 385 gal liquid/h and 0.05(385) = 19.3 gal oil/h, which is too low. By interpolation, there are 0.055 gal oil/gal liquid and 0.0389 gal liquid/lb oil-free livers. Therefore, final underflow contains 0.0389(10,000) − 21.5 = 367.5 gal ether/h. Final extract contains 408.5 gal oil/h, with a 65 vol % oil. Therefore, the final extract contains

$$\frac{0.35}{0.65}(408.5) = 220 \text{ gal ether/h}$$

Ether in solvent feed = 220 + 367.5 = 587.5 gal/h. Overall material balance is

	Solvent Feed	Livers Feed	Final Extract	Final Underflow
Oil-free livers, lb/h	0	10,000	0	10,000
Ether, gal/h	587.5	0	220	367.5
Oil, gal/h	0	430	408.5	21.5

Ideal Leaching Stage:

Referring to Figure 16.7, underflow leaving leaching stage L will have the same concentration of oil as the final extract. That concentration is the specification of 0.65 gal oil/gal liquid. From the above

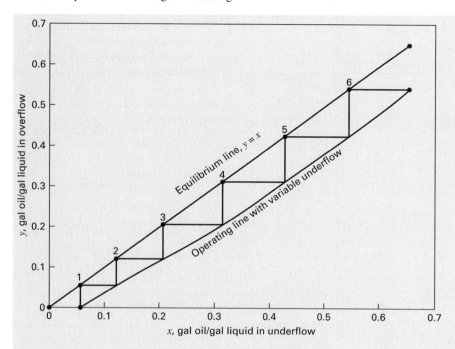

Figure 16.10 McCabe–Thiele diagram for Example 16.5.

data table, by linear interpolation, gal liquid retained/lb oil-free livers = 0.116. Therefore, letting x_j and y_j = volume fractions, $L_L = 0.116(10,000) = 1,160$ gal liquid and $x_L = 0.65$ gal oil/gal liquid.

Oil material balance around leaching stage:

$$y_1 V_1 + 430 = x_L L_L + 408.5$$

or $\quad y_1 V_1 = 0.65(1,160) + 408.5 - 430 = 732.5$ gal oil

Ether material balance around leaching stage:

$$(1 - y_1)V_1 = 0.35(1,160) + 220 = 626 \text{ gal ether/h}$$

$$y_1 = \frac{732.5}{732.5 + 626} = 0.539 \text{ gal oil/gal liquid}$$

End points of the operating line for the washing stages are:

Stage N	Stage 1
$y_{N+1} = 0$	$y_1 = 0.539$
$x_N = 0.055$	$x_L = 0.65$

These two pairs of points, together with the straight equilibrium line, $y = x$, are plotted on the McCabe–Thiele diagram of Figure 16.10. Because the underflow is variable, a straight line does not connect the operating-line end points.

Calculation of Intermediate Points on the Operating Line:
Determine, by oil and total-liquid material balances, the values of y_{n+1} for $x_n = 0.30$ and 0.50. The material balance can be made from the solvent feed end to an arbitrary stage n. For $x_n = 0.3$, with the above data table giving 0.058 gal liquid/lb oil-free livers,

Oil mass balance:

$$0.3(10,000)(0.058) = y_{n+1}V_{n+1} + 21.5 \qquad (1)$$

Total liquid mass balance:

$$10,000(0.058) + 587.5 = V_{n+1} + (367.5 + 21.5) \qquad (2)$$

Solving (1) and (2), $V_{n+1} = 778.5$ gal/h; $y_{n+1} = 0.196$ gal oil/gal liquid.

For $x_n = 0.50$, with 0.083 gal liquid/lb oil-free livers,

Oil mass balance:

$$0.5(10,000)(0.083) = y_{n+1}V_{n+1} + 21.5 \qquad (3)$$

Total liquid mass balance:

$$10,000(0.083) + 587.5 = V_{n+1} + (367.5 + 21.5) \qquad (4)$$

Solving (3) and (4), $V_{n+1} = 1,028.5$ gal/h; $y_{n+1} = 0.383$ gal oil/gal liquid.

When these two sets of x_n, y_{n+1} points are plotted, the curved operating line in Figure 16.10 is obtained. Almost exactly six washing stages can be stepped off. Adding the leaching stage gives a total of seven ideal stages.

§16.3 RATE-BASED MODEL FOR LEACHING

Leaching involves transfer of solute from the interior of a solid into the bulk of a liquid solvent. The process can be modeled by considering two steps (in series): (1) molecular diffusion of solute through the solid, and (2) convection and eddy diffusion of solute through the exterior solvent or extract. Practical rates of molecular diffusion through the solid are achieved only after solvent penetrates the solid to become occluded liquid, unless the solvent is initially present in the solid. The solute then dissolves into that liquid and diffuses at a reasonable rate to the surface of the solid, leaving behind insoluble solids and any sparingly soluble materials in the form of a framework. If relative motion exists between the solids and the exterior solvent solution, resistance to mass transfer in the fluid phase is negligible compared to that in the solid, and leaching can be modeled by diffusion through the solid only.

§16.3.1 Food Processing

Schwartzberg and Chao [11] present a summary of published experimental and theoretical studies of solute diffusion in the leaching of food in the form of slices, near-cylinders, and nearly spherical particles, including a compilation of effective diffusivities. As with diffusion of liquids and gases in porous-solid adsorbents, diffusivity can be expressed as a true molecular diffusivity in the occluded fluid phase, or as an effective diffusivity through the entire solid, including the insoluble-solid framework, called the *marc*, and occluded liquid. When an effective diffusivity, D_e, and Fick's laws are used, the concentration-driving force is concentration of solute X_i in mass per unit volume of solid particle. Thus, if r is the direction of diffusion, Fick's first law for solute i, from §3.1.1, is

$$n_i = -D_e A \left(\frac{\partial X_i}{\partial r} \right) \tag{16-11}$$

Fick's second law, from §3.3.1, is

$$\frac{\partial X_i}{\partial t} = \frac{D_e}{r^{v-1}} \frac{\partial}{\partial r} \left[r^{v-1} D_e \frac{\partial X_i}{\partial r} \right] \tag{16-12}$$

where by comparison to (3-69), (3-70), and (3-71), $v = 1, 2,$ and 3 for rectangular, cylindrical, and spherical coordinates, respectively.

The rate of mass transfer through the solvent external to the solid can be written in terms of a mass-transfer coefficient and the concentration of solute in solvent or extract Y_i, as in (3-105), where subscripts s and b refer to the solid–liquid interface and bulk liquid:

$$n_i = k_c A \left[(Y_i)_s - (Y_i)_b \right] \tag{16-13}$$

At the solid–liquid interface at the exterior solid surface, (16-11) and (16-13) can be equated:

$$-D_e \left(\frac{\partial X_i}{\partial r} \right)_s = k_c \left[(Y_i)_s - (Y_i)_b \right] \tag{16-14}$$

Let $m = Y_i / X_i$, and

$$\begin{gathered} a = \text{characteristic dimension of the solid,} \\ \text{e.g., radius of a cylinder or spherical} \\ \text{particle, or the half-thickness of a slice} \end{gathered} \tag{16-15}$$

Combining (16-14) and (16-15) and expressing results in the form of dimensionless groups,

$$-\left[\frac{\partial \left(\frac{Y_i}{(Y_i)_s - (Y)_b} \right)}{\partial (r/a)} \right]_s = \left(\frac{mk_c a}{D_e} \right) \tag{16-16}$$

The dimensionless group on the RHS of (16-16) is the Biot number for mass transfer:

$$(N_{\text{Bi}})_M = \frac{mk_c a}{D_e} \tag{16-17}$$

which is analogous to the more common Biot number for heat transfer,

$$(N_{\text{Bi}}) = \frac{ha}{k} \tag{16-18}$$

Biot numbers are quantitative measures of the ratio of internal (solid) resistance to external (fluid) resistance to transport. In §3.3, transient solutions are given to (16-12) for different geometries for an initial uniform concentration, $X_o = c_o$, of solute in the solid. At time $t = 0$, solute concentration in the solid phase at the solid–fluid interface is suddenly brought to and then held at $X_s = c_s$. The solutions given are

Geometry	Concentration Profile	Rate of Mass Transfer at Interface	Average Concentration in Solid
Semi-infinite	(3-75)	(3-78)	—
Slab of finite thickness	(3-80), (3-81), and Figure 3.8	(3-82)	(3-85) and Figure 3.9
Infinite cylinder	Figure 3.10	—	Figure 3.9
Sphere	Figure 3.11	—	Figure 3.9

These solutions apply to the case in which the Biot number for mass transfer is infinite, such that the resistance in the fluid phase is negligible and $(Y_i)_b = (Y_i)_s$.

For an infinite Biot number, Figures 3.8 and 3.9 give the solute concentration profile and average solute concentration in the solid as a function of time. In these plots, a dimensionless time, the Fourier number for mass transfer, is used:

$$(N_{\text{Fo}})_M = \frac{D_e t}{a^2} \tag{16-19}$$

where a = flake or slice half-thickness. When internal resistance to mass transfer is negligible, which is almost never the case in food leaching, the solution for uniform concentration of solute in the solid is given by (16-20), whose derivation is left as an exercise:

$$\frac{X_i - \frac{(Y_i)_b}{m}}{(X_i)_o - \frac{(Y_i)_b}{m}} = \exp \left(\frac{-k_c t m}{a} \right) \tag{16-20}$$

When $(N_{\text{Bi}})_M > 200$, the external (fluid) mass-transfer resistance is negligible and Figures 3.8 and 3.9 apply. When $(N_{\text{Bi}})_M < 0.001$, internal (solid) mass-transfer resistance is negligible and (16-20) applies. When $(N_{\text{Bi}})_M$ lies between these extremes, both resistances must be taken into account and solutions given by Schwartzberg and Chao [11] are operative.

Effective diffusivities for solutes in solids depend on molecular forces, solubility, cell structure, volume fraction, concentration, temperature, tortuosity, solvent concentration, and extent of adsorption of solute by the marc. Solute effective diffusivities in a variety of foods, with water as solvent, are tabulated by Schwartzberg and Chao [11]. Values for sucrose, when the cell walls are hard (e.g., sugar cane and coffee), range from 0.5 to 1.0×10^{-6} cm^2/s. When cell walls are soft (e.g., sugar beets, potatoes, apples, celery, and onions), values are higher, ranging from 1.5 to 4.5×10^{-6} cm^2/s. When the solute is a salt (e.g., NaCl and KCl), effective diffusivities are about four times higher. As previously discussed, diffusion of oil from flaked oil seeds does not follow Fick's law. If, nevertheless, Fick's law is applied to determine effective diffusivity, values are found to decrease significantly with time. For example, data of Karnofsky [12], who leached oil from soybeans, cottonseeds, and flaxseeds with hexane, give effective diffusivities that decrease over the course of extraction by about one order of magnitude. Other foods exhibit the same trend under certain conditions. Frequently, diffusivity is not a constant, but varies with flake or slice thickness and solute concentration. Schwartzberg [11] discusses possible reasons for this.

A thin slice or flake of solid can be treated as a slab of finite thickness, with mass transfer from the thin edges ignored. For this case, (3-85) or Figure 3.9 can be used to determine the effective diffusivity from experimental leaching data or to predict rates of leaching. In (3-85), $E_{avg_{slab}}$ is the fractional unaccomplished approach to equilibrium for extraction, which decreases with time. As seen in Figure 3.9 and as can be demonstrated with (3-85), when $(N_{Fo})_M > 0.10$, the series solution is converged to less than a 2% error, with only one infinite series term,

$$E_{ave_{slab}} = \frac{8}{\pi^2} \exp\left(-\frac{(N_{Fo})_M \pi^2}{4}\right)$$

or

$$\ln E_{avg_{slab}} = \ln(8/\pi^2) - \frac{\pi^2}{4}(N_{Fo})_M \quad (16\text{-}21)$$

If Fick's law holds and diffusivity is constant, a plot of data as log $E_{avg_{slab}}$ against time should yield a straight line with a negative slope from which effective diffusivities can be determined, as illustrated in the following example.

EXAMPLE 16.6 Leaching of Sugar Beets.

In the commercial extraction of sugar (sucrose) from sugar beets with water, the process is controlled by diffusion through the sugar beet. Yang and Brier [13] conducted diffusion experiments with beets sliced into cossettes 0.0383 inch thick by 0.25 inch wide and 0.5–1.0 inches long. Typically, the cossettes contained 16 wt% sucrose, 74 wt% water, and 10 wt% insoluble fiber. Experiments were conducted at temperatures from 65 to 80°C, with solvent water rates from 1.0 to 1.2 lb/lb fresh cossettes. For a temperature of 80°C and a water rate of 1.2 lb/lb fresh cossettes, the following smoothed data were obtained:

E_{avg}	t, min.
1.0	0
0.39	10
0.19	20
0.10	30
0.050	40
0.025	50
0.0135	60

These data are plotted in Figure 16.11, where a straight line can be passed through the data in the time range of 10–60 minutes. From the slope of this line, using (16-19) and (16-21),

$$\frac{\pi^2}{4}\left(\frac{D_e}{a^2}\right) = 0.00113 \text{ sec}^{-1}$$

Since $a = $ half thickness $= \dfrac{0.0383}{2}(2.54) = 4.86 \times 10^{-2}$ cm

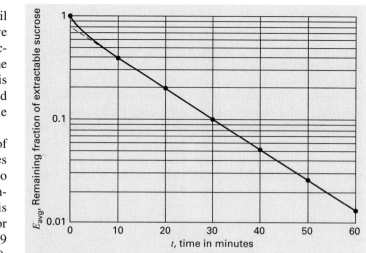

Figure 16.11 Experimental data for leaching of sucrose from sugar beets with water for Example 16.6.

Therefore,

$$D_e = \frac{0.00113(4.86 \times 10^{-2})^2(4)}{(3.14)^2} = 1.1 \times 10^{-6} \text{ cm}^2/\text{s}$$

For a continuous, countercurrent extractor, (16-21) can be used to determine the approximate time for leaching the solids. Time is given in terms of $E = E_{avg}$ by

$$t = \int_{E_{in}}^{E_{out}} \frac{dE}{\left(\dfrac{dE}{dt}\right)} \quad (16\text{-}22)$$

If the solute diffusivity is constant, (dE/dt), except for small values of time, can be obtained by differentiating (16-21) and combining the result to eliminate time, t, to give

$$\frac{dE}{dt} = -\frac{\pi^2 D_e E}{4a^2} \quad (16\text{-}23)$$

Substitution of (16-23) into (16-22), followed by integration, gives

$$t = \frac{4a^2}{\pi^2 D_e} \ln\left(\frac{E_{in}}{E_{out}}\right) \quad (16\text{-}24)$$

When solute diffusivity is not constant, which is more common, data plots of E as a function of time can be used directly to obtain values of (dE/dt) for use in (16-22), which can be graphically or numerically integrated, as shown by Yang and Brier [13].

EXAMPLE 16.7 Extraction of Sucrose.

The sucrose in 10,000 lb/h of sugar beets containing 16 wt% sucrose, 74 wt% water, and 10 wt% insoluble fiber is extracted in a countercurrent extractor at 80°C with 12,000 lb/h of water. If 98% of the sucrose is extracted and no net mass transfer of water occurs, determine the residence time in minutes for the beets. Assume the beets are sliced to 1 mm in thickness and that the effective sucrose diffusivity is that computed in Example 16.6.

Solution

By material balance, extracted beets contain $0.02(0.16)(10,000) = 32$ lb/h sucrose; $0.74(10,000) = 7,400$ lb/h water; and $0.10(10,000) = 1,000$ lb/h insoluble fiber, for a total of $8,432$ lb/h.

Thus, $X_{out} = 32/8,432 = 0.0038$ lb/lb and $X_{in} = 1,600/10,000 = 0.160$ lb/lb, where X is expressed on a weight fraction basis.

At the beet inlet end,

$$E_{in} = \frac{0.16 - (Y/m)_{extract\ out}}{0.16 - (Y/m)_{extract\ out}} = 1.0$$

At the beet outlet end, $(Y/m)_{solvent\ in} = 0$. Therefore,

$$E_{out} = \frac{0.0038}{1.0} = 0.0038$$

From (16-24),

$$t = \frac{4\left(\frac{0.1}{2}\right)^2}{(3.14)^2\left(1.1 \times 10^{-6}\right)} \ln\left(\frac{1.0}{0.0038}\right) = 5,140\ s = 85.6\ min$$

§16.3.2 Mineral Processing

Leaching is used to recover valuable metals from low-grade ores, which is accomplished by reacting part of the ore with a constituent of the leach liquor, to produce metal ions soluble in the liquid. In general, the reaction is

$$A_{(l)} + bB_{(s)} \rightarrow Products \qquad (16\text{-}25)$$

Removal of reactant B from the ore leaves pores for reactant A to diffuse through to reach reactant B in the interior of the particle.

Figure 16.12 shows a spherical mineral particle undergoing leaching. As the process proceeds, an outer porous-leached shell develops, leaving an unleached core. The steps involved are: (1) mass transfer of reactant A from the bulk liquid to the outer surface of the particle; (2) pore diffusion of reactant A through the leached shell; (3) chemical reaction at the interface between the leached shell and the unleached core; (4) pore diffusion of the reaction products back through the leached shell; (5) mass transfer of the reaction products back into the bulk liquid surrounding the particle.

Because the diameter of the unleached core shrinks with time, a mathematical model for the process, first conceived for application to gas–solid combustion reactions by Yagi and Kunii [14] in 1955 and then extended to liquid–solid leaching by Roman, Benner, and Becker [15] in 1974, is referred to as the *shrinking-core model*. Although any one of the above five steps can control the process, the rate of leaching is often controlled by Step 2. Although general models have been developed, the leaching model presented here is derived on the assumption that Step 2 is controlling.

Referring to Figure 16.12, assume that dr_c/dt, the rate of reaction-interface movement at r_c, is small with respect to the diffusion velocity of reactant A in (16-25) through the porous, leached layer. This is the so-called *pseudo-steady-state* assumption. Although it is valid for gas–solid cases, it is less satisfactory for the liquid–solid case here. The importance of this assumption is that it permits neglecting the accumulation

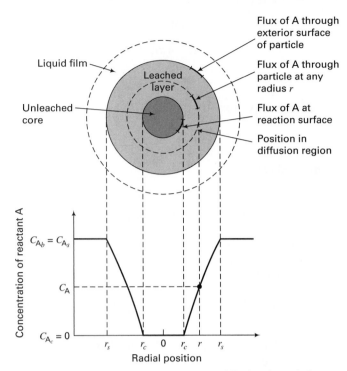

Figure 16.12 Shrinking-core model when diffusion through the leached shell is controlling.

of reactant A as a function of time in the leached layer as that layer increases in thickness, with the result that the model can be formulated as an ODE rather than as a PDE. Thus, the rate of diffusion of reactant A through the porous, leached layer is given by Fick's second law, (3-71), ignoring the LHS term and replacing the molecular diffusivity with an effective diffusivity:

$$\frac{D_e}{r^2}\frac{d}{dr}\left(r^2\frac{dc_A}{dr}\right) = 0 \qquad (16\text{-}26)$$

with boundary conditions

$$c_A = c_{A_s} = c_{A_b} \quad at \quad r = r_s$$
$$c_A = 0 \quad at \quad r = r_c$$

These boundary conditions hold because mass-transfer resistance in the liquid film or boundary layer is assumed negligible and the interface reaction is assumed to be instantaneous and complete.

If (16-26) is integrated twice and boundary conditions are applied, the result after simplification is

$$c_A = c_{A_b}\left[\frac{1 - \dfrac{r_c}{r}}{1 - \dfrac{r_c}{r_s}}\right] \qquad (16\text{-}27)$$

To obtain a relationship between r_c and time t, differentiate (16-27) with respect to r and evaluate the differential at $r = r_c$:

$$\left.\frac{dc_A}{dr}\right|_{r=r_c} = \frac{c_{A_b}}{r_c\left(1 - \dfrac{r_c}{r_s}\right)} \qquad (16\text{-}28)$$

The rate of diffusion at $r = r_c$ is given by Fick's first law:

$$n_A = \frac{d\mathcal{N}_A}{dt} = 4\pi r_c^2 D_e\left(\frac{dc_A}{dr}\right)_{r=r_s} \qquad (16\text{-}29)$$

where $\mathcal{N}_A$ = moles of A. Combining (16-28) and (16-29),

$$-\frac{d\mathcal{N}_A}{dt} = \frac{4\pi r_c D_e c_{A_b}}{\left(1 - \dfrac{r_c}{r_s}\right)} \quad (16\text{-}30)$$

By stoichiometry, from (16-25),

$$\frac{d\mathcal{N}_A}{dt} = \frac{1}{b}\frac{d\mathcal{N}_B}{dt} \quad (16\text{-}31)$$

By material balance,

$$\frac{d\mathcal{N}_B}{dt} = \frac{\rho_B}{M_B}\frac{d}{dt}\left(\frac{4}{3}\pi r_c^3\right) = \frac{4\pi r_c^2 \rho_B}{M_B}\frac{dr_c}{dt} \quad (16\text{-}32)$$

where ρ_B = initial mass of reactant B per unit volume of solid particle and M_B = molecular weight of B. Combining (16-30) with (16-32),

$$\frac{-\rho_B}{M_B}\left(\frac{1}{r_c} - \frac{1}{r_s}\right)r_c^2 dr_c = b D_e c_{A_b}\,dt \quad (16\text{-}33)$$

Integrating (16-33) and applying the boundary condition $r_c = r_s$ at $t = 0$ gives

$$t = \frac{\rho_B r_s^2}{6 D_e b M_B c_{A_b}}\left[1 - 3\left(\frac{r_c}{r_s}\right)^2 + 2\left(\frac{r_c}{r_s}\right)^3\right] \quad (16\text{-}34)$$

For complete leaching, $r_s = 0$, and (16-34) becomes

$$t = \frac{\rho_B r_s^2}{6 D_e b M_B c_{A_b}} \quad (16\text{-}35)$$

EXAMPLE 16.8 Shrinking-Core Model.

A copper ore containing 2 wt% CuO is to be leached with 0.5-M H_2SO_4. The reaction is

$$\frac{1}{2}CuO_{(s)} + H^+_{(aq)} \rightarrow \frac{1}{2}Cu^{2+}_{(aq)} + \frac{1}{2}H_2O_{(aq)} \quad (1)$$

The leaching process is controlled by diffusion of hydrogen ions through the leached layer. The effective diffusivity, D_e, of the hydrogen ion by laboratory tests is 0.6×10^{-6} cm²/s. The SG of the ore is 2.7. For ore particles of 10-mm diameter, estimate the time required

to leach 98% of the copper, assuming that CuO is uniformly distributed throughout the particles. Also, check the validity of the pseudo-steady-state assumption by comparing the amount of hydrogen ions held up in the liquid in the pores with the amount reacted with CuO.

Solution

If 98% of the cupric oxide is leached, then r_c corresponds to 2% of the particle volume. Thus,

$$\frac{4}{3}\pi r_c^3 = (0.02)\frac{4}{3}\pi r_s^3$$

or $r_c = (0.02)^{1/3} r_s = (0.02)^{1/3}(0.5) = 0.136$ cm.
Density of CuO in the ore = $\rho_B = 0.02(2.7) = 0.054$ g/cm³.
Molecular weight of CuO = $M_B = 79.6$.
From (16-25) and reaction (1), $b = 0.5$.

For 0.5-M H_2SO_4, $c_{H_b}^+ = \dfrac{2(0.5)}{1000} = 0.001$ mol/cm³.

From (16-34), with $r_c/r_s = 0.136/0.500 = 0.272$.

$$t = \frac{(0.054)(0.5)^2}{6(0.6 \times 10^{-6})(0.5)(79.6)(0.001)}\left[1 - 3(0.272)^2 + 2(0.272)^3\right]$$

$$= 77,000 \text{ sec} = 21.4 \text{ h}$$

Now, check the pseudo-steady-state assumption:

SG of CuO is 6.4 g/cm³; 100 g of ore occupies $100/2.7 = 37.0$ cm³.
The CuO in this amount of ore occupies $0.02(100)/6.4 = 0.313$ cm³ or 0.845% of the particle volume. Volume of one particle = $\frac{4}{3}\pi r_s^3 = \frac{4}{3}(3.14)(0.5)^3 = 0.523$ cm³.

Volume of CuO as pores in one particle = $0.00845(0.523) = 0.0044$ cm³.

Mols of H^+ in pores, based on the bulk concentration (to be conservative) = $0.001(0.0044) = 4.4 \times 10^{-6}$ mol.

98% of CuO leached in a particle, in mol units, =

$$\frac{0.98(0.02)(2.7)(0.523)}{79.6} = 3.5 \times 10^{-4} \text{ mol}$$

which requires 7.0×10^{-4} mol H^+ for reaction.

Because this value is approximately two orders of magnitude larger than the conservative estimate of H^+ in the pores, the pseudo-steady-state assumption is valid.

SUMMARY

1. Leaching is similar to liquid–liquid extraction, except that solutes initially reside in a solid. Leaching is widely used to remove solutes from foods, minerals, and living cells.

2. When leaching is rapid, it can be accomplished in one stage. However, the leached solid will retain surface liquid that contains solute. To recover solute in the extract, it is desirable to add one or more washing stages in a countercurrent arrangement.

3. Leaching of large solids can be very slow because of small solid diffusivities. Therefore, it is common to reduce size of the solids by crushing, grinding, flaking, slicing, etc.

4. Industrial leaching equipment is available for batch or continuous processing. Solids are contacted with solvent by either percolation or immersion. Large, continuous, countercurrent extractors can process up to 7,000,000 kg/day of food solids.

5. Washing large flows of leached solids is commonly carried out in thickeners that can be designed to produce a clear liquid overflow and a concentrated solids underflow. When a clear overflow is not critical, hydroclones can replace thickeners.

6. An equilibrium-stage model is widely used for continuous, countercurrent systems when leaching is rapid and washing is needed for high solute recovery. The

model assumes that concentration of solute in the over-flow leaving a stage equals that in the underflow liquid retained on the solid leaving the stage.

7. When the ratio of liquid to solids in the underflow is constant from stage to stage, the equilibrium-stage model can be applied algebraically by a modified Kremser method or graphically by a modified McCabe–Thiele method. If the underflow is variable, the graphical method with a curved operating line is appropriate.

8. When leaching is slow, as with food solids or low-grade ores, leaching calculations must be done on a rate basis. In some cases, the diffusion of solutes in food solids does not obey Fick's law, because of complex membrane and fiber structures.

9. Leaching of low-grade ores by reactive leaching is conveniently carried out with a shrinking-core diffusion model, using a pseudo-steady-state assumption.

REFERENCES

1. Othmer, D.F., and J.C. Agarwal, *Chem. Eng. Progress*, **51**, 372–373 (1955).

2. D.W. Green, and R.H. Perry, Eds, *Perry's Chemical Engineers' Handbook*, 8th ed. McGraw-Hill, New York, Section 18 (2008).

3. King, C.O., D.J. Katz, and J.C. Brier, *Trans. AIChE*, **40**, 533–537 (1944).

4. Van Arsdale, G.D., *Hydrometallurgy of Base Metals*, McGraw-Hill, New York (1953).

5. Lamont, A.G.W., *Can. J. Chem. Eng.*, **36**, 153 (1958).

6. Schwartzberg, H.G., *Chem. Eng., Progress*, **76**(4), 67–85 (1980).

7. Coulson, J.M., J.F. Richardson, J.R. Backhurst, and J.H. Harker, *Chemical Engineering*, 4th ed. Pergamon Press, Oxford, Vol. **2** (1991).

8. Baker, E.M., *Trans. AIChE.*, **32**, 62–72 (1936).

9. McCabe, W.L., and J.C. Smith, *Unit Operations of Chemical Engineering*, McGraw-Hill, New York, pp. 604–608 (1956).

10. Ravenscroft, E.A., *Ind. Eng. Chem.*, **28**, 851–855 (1936).

11. Schwartzberg, H.G., and R.Y. Chao, *Food Tech.*, **36**(2), 73–86 (1982).

12. Karnofsky, G., *J. Am. Oil Chem. Soc.*, **26**, 564–569 (1949).

13. Yang, H.H., and J.C. Brier, *AIChE J.*, **4**, 453–459 (1958).

14. Yagi, S., and D. Kunii, *Fifth Symposium (International) on Combustion*, Reinhold, New York, pp. 231–244 (1955).

15. Roman, R.J., B.R. Benner, and G.W. Becker, *Trans. Soc. Mining Engineering of AIME*, **256**, 247–256 (1974).

16. Andueza, S., L. Maeztu, B. Dean, M.P. de Pena, J. Pello, and C. Cid, *J. Agric. Food Chem.*, **50**, 7426–7431 (2002).

17. Andueza, S., L. Maeztu, L. Pascual, C. Ibanez, M.P. de Pena, and C. Cid, *J. Sci. Food Agric.*, **83**, 240–248 (2003).

18. Andueza, S., M.P. de Pena, and C. Cid, *J. Agric. Food Chem.*, **51**, 7034–7039 (2003).

STUDY QUESTIONS

16.1. Is leaching synonymous with solid–liquid and/or liquid–solid extraction?

16.2. In a leaching operation, what is the leachant, the overflow, and the underflow?

16.3. Why does the underflow consist of both leached solids and liquid containing leached material?

16.4. Why is pretreatment of the solids to be leached often necessary?

16.5. Under what conditions would leaching be expected to be very slow?

16.6. What is dissolution?

16.7. What is the difference between suspension leaching and percolation leaching? For what conditions is each method used?

16.8. What are the advantages of the espresso machine over the drip method?

16.9. Why do many leaching processes include multistage, countercurrent washing after the leaching stage?

16.10. What are the assumptions for an ideal leaching or washing stage?

16.11. What is meant by variable underflow and what causes it?

16.12. How does the shrinking-core model used for mineral leaching differ from the simpler model used for leaching of food materials?

16.13. Why is an effective diffusivity that is obtained by experiment preferred for estimating the rate of leaching of food materials?

16.14. What is the pseudo-steady-state assumption used in the shrinking-core leaching model?

EXERCISES

Section 16.1

16.1. Mass-balance check on leaching data.

Using experimental data from pilot-plant tests of soybean extraction by Othmer and Agarwal summarized in the Industrial Example at the beginning of this chapter, check mass balances for oil and hexane around the extractor, assuming the moisture is retained in the flakes, and compute the mass ratio of liquid oil to flakes in leached solids leaving the extractor.

Section 16.2

16.2. Manufacture of barium carbonate.

$BaCO_3$, which is water insoluble, is to be made by precipitation from a solution containing 120,000 kg/day of water and 40,000 kg/day of BaS, with a stoichiometric amount of solid Na_2CO_3. The reaction, which produces a byproduct of water-soluble Na_2S, will be carried out in a continuous, countercurrent system of five thickeners. Complete reaction will take place in the first thickener, which will be fed solid Na_2CO_3, an aqueous solution of BaS, and overflow from the second thickener. Sufficient fresh water will enter the last thickener so that overflow from the first thickener will be 10 wt% Na_2S, assuming the underflow from each thickener contains two parts of water per one part of $BaCO_3$ by weight. (a) Draw a process flow diagram and label it with all given information. (b) Determine the kg/day of Na_2CO_3 required and kg/day of $BaCO_3$ and Na_2S produced. (c) Determine the kg/day of fresh water needed, wt% of Na_2S in the liquid portion of the underflow that leaves each thickener, and kg/day of Na_2S that will remain with the $BaCO_3$ product after it is dried.

16.3. Leaching of calcium carbonate.

$CaCO_3$ precipitate can be produced by reaction of an aqueous solution of Na_2CO_3 and CaO, the byproduct being NaOH. Following decantation, slurry leaving the precipitation tank is 5 wt% $CaCO_3$, 0.1 wt% NaOH, and the balance water. One hundred thousand lb/h of slurry is fed to a two-stage, continuous, countercurrent washing system to be washed with 20,000 lb/h of fresh water. Underflow from each thickener will contain 20 wt% solids. Determine % recovery of NaOH in the extract and wt% NaOH in the dried $CaCO_3$ product. Is it worthwhile to add a third stage?

16.4. Recovery of zinc from a ZnS ore.

Zn is to be recovered from an ore containing ZnS. The ore is first roasted with oxygen to produce ZnO, which is leached with aqueous H_2SO_4 to produce water-soluble $ZnSO_4$ and an insoluble, worthless residue called gangue. The decanted sludge of 20,000 kg/h contains 5 wt% water, 10 wt% $ZnSO_4$, and the balance as gangue. This sludge is to be washed with water in a continuous, countercurrent washing system to produce an extract, called a strong solution, of 10 wt% $ZnSO_4$ in water, with a 98% recovery of $ZnSO_4$. Assume that underflow from each washing stage contains, by weight, two parts of water (sulfate-free basis) per part of gangue. Determine the stages required.

16.5. Leaching of oil from flaked soybeans.

Fifty thousand kg/h of flaked soybeans, containing 20 wt% oil, is leached of oil with the same flow rate of n-hexane in a countercurrent-flow system consisting of an ideal leaching stage and three ideal washing stages. Experiments show the underflow from each stage contains 0.8 kg liquid/kg soybeans (oil-free basis). (a) Determine % recovery of oil in the final extract. (b) If leaching requires three of the four stages such that one-third of the leaching occurs in each stage, followed by one washing stage, determine the % recovery of oil in the final extract.

16.6. Recovery of sodium carbonate.

One hundred tons per hour of a feed containing 20 wt% Na_2CO_3 and the balance insoluble solids is to be leached and washed with water in a continuous, countercurrent system. Assume leaching will be completed in one ideal stage. It is desired to obtain a final extract containing 15 wt% solute, with a 98% recovery of solute. The underflow from each stage will contain 0.5 lb solution/lb insoluble solids. Determine the number of ideal washing stages.

16.7. Production of titanium dioxide.

Titanium dioxide, the most common white pigment in paint, can be produced from the titanium mineral rutile by chlorination to $TiCl_4$, followed by oxidation to TiO_2. To purify insoluble TiO_2, it is washed free of soluble impurities in a continuous, countercurrent system of thickeners with water. Two hundred thousand kg/h of 99.9 wt% TiO_2 pigment is to be produced by washing, followed by filtering and drying. The feed contains 50 wt% TiO_2, 20 wt%

soluble salts, and 30 wt% water. Wash liquid is pure water at a flow rate equal to the feed on a mass-flow basis. (a) Determine the number of washing stages required if the underflow from each stage is 0.4 kg solution/kg TiO_2. (b) Determine the number of washing stages required if the underflow varies as follows:

Concentration of Solute, kg solute/kg solution	Retention of Solution, kg solution/kg TiO_2
0.0	0.30
0.2	0.34
0.4	0.38
0.6	0.42

Section 16.3

16.8. Rate of leaching from a flake.

Derive (16-20), assuming that $(Y_i)_b$, k_c, m, and a are constants and that $(X_i)_o$ is uniform through the solid.

16.9. Rate of leaching from a cossette.

Derive (16-24).

16.10. Effective diffusivity from experimental data.

Data of Othmer and Agarwal [1] for the batch extraction of oil from soybeans by oil-free n-hexane at 80°F are as follows:

Time, min	Oil Content of Soybeans, g/g Dry, Oil-Free Soybeans
0	0.203
0.5	0.1559
1	0.1359
2	0.1190
4	0.0981
7	0.0775
12	0.0591
20	0.04197
35	0.03055
60	0.02388
120	0.02107

Determine whether these data are consistent with a constant effective diffusivity of oil in soybeans.

16.11. Diffusivity of sucrose in water.

Estimate the molecular diffusivity of sucrose in water at infinite dilution at 80°C, noting that the value is 0.54×10^{-5} cm^2/s at 25°C. Give reasons for the difference between the value you obtain and the value for effective diffusivity in Example 16.6.

16.12. Leaching of sucrose from coffee particles.

Sucrose in ground coffee particles of an average diameter of 2 mm is to be extracted with water in a continuous, countercurrent extractor at 25°C. Diffusivity of sucrose in the particles has been determined to be about 1.0×10^{-6} cm^2/s. Estimate the time in minutes to leach 95% of the sucrose. For a sphere with $N_{Fo_M} > 0.10$,

$$E_{avg} = \frac{6}{\pi^2} \exp\left(\frac{-\pi^2 D_e t}{a^2}\right)$$

16.13. Leaching of CuO from ore.

For the conditions of Example 16.8, determine the effect on leaching time of particle sizes from 0.5 mm to 50 mm.

16.14. Shrinking-core model.

For the conditions of Example 16.8, determine the effect of % recovery of copper over the range of 50–100%.

16.15. Shrinking-core model.

Repeat Example 16.8 using ore that contains 3 wt% Cu_2O.

16.16. Shrinking-core model.

For the shrinking-core model, if the rate of leaching is controlled by an interface chemical reaction that is first order in the concentration of reactant A, derive the expression

$$t = \frac{\rho_B r_s}{b M_B k C_{A_b}} \left(1 - \frac{r_c}{r_s}\right)$$

where $k =$ first-order rate constant.

Chapter 17

Crystallization, Desublimation, and Evaporation

§17.0 INSTRUCTIONAL OBJECTIVES

After completing this chapter, you should be able to:

- Explain how crystals grow.
- Explain how crystal-size distribution can be measured, tabulated, and plotted.
- Explain the importance of supersaturation in crystallization.
- Use mass-transfer theory to determine rate of crystal growth.
- Apply the MSMPR model to design of a continuous, vacuum, evaporating crystallizer of the draft-tube baffled (DTB) type.
- Understand precipitation.
- Apply mass-transfer theory to a falling-film melt crystallizer.
- Differentiate between crystallization and desublimation.
- Describe evaporation equipment.
- Derive and apply the ideal evaporator model.
- Design multiple-effect evaporation systems.
- Describe advantages and challenges of bioproduct crystallization relative to inorganic crystallization.
- Use expressions from moment analysis of crystal-size distribution in dilution batch crystallizers to relate crystal size to reaction time and cooling rate.
- Obtain batch cooling curves for seeded/unseeded and monosized/distributed size particles.
- Describe effects of mixing on supersaturation, mass transfer, growth, and scale-up of crystallization.

Crystallization is a solid–fluid separation in which crystalline particles are formed from a homogeneous fluid phase. Ideally, crystals are pure chemicals, are obtained in a high yield with a desirable shape, have a reasonably uniform and desirable size, and, if food or pharmaceutical products, are without loss of taste, aroma, and physiologic activity. Crystallization is one of the oldest known separation operations, recovery of sodium chloride as salt crystals from water by evaporation dating back to antiquity. Even today, many processes involve crystallization from aqueous solution of inorganic salts, a short list of which is in Table 17.1, where all examples are *solution crystallization* because the inorganic salt is clearly the solute crystallized, and water is the solvent remaining as a liquid. The phase diagram for systems suitable for solution crystallization is a solubility curve like Figure 17.1.

For formation of organic crystals, organic solvents such as acetic acid, ethyl acetate, methanol, ethanol, acetone, ethyl ether, chlorinated hydrocarbons, benzene, and petroleum fractions may be preferred choices, but they must be used with care when they are toxic or flammable and have low flash points and wide explosive limits.

For aqueous or organic solutions, crystallization is effected by cooling a solution, evaporating the solvent, or a combination of the two. In some cases, a mixture of two or more solvents may be best, examples of which include water with the lower alcohols, and normal paraffins with chlorinated solvents. Addition of a second solvent is sometimes used to reduce solute solubility. When water is the additional solvent, the process is called *watering-out*; when an organic solvent is added to an aqueous salt solution, the process is *salting-out*. For both cases of solvent addition, fast crystallization called *precipitation* can occur, resulting in large numbers of very small crystals. Precipitation also occurs when one product of two reacting solutes is a solid with low solubility. For example, when aqueous solutions of silver nitrate and sodium chloride are mixed, insoluble silver chloride is precipitated, leaving a solution of mainly soluble sodium nitrate.

When both components of a homogeneous, binary solution have melting (freezing) points not far removed from each other, the solution is referred to as a *melt*. If, as in Figure 17.1b, the phase diagram for the melt exhibits a eutectic point, it is possible to obtain in one step, called *melt crystallization*, pure

Table 17.1 Inorganic Salts Recovered from Aqueous Solutions

Chemical Name	Formula	Common Name	Crystal System
Ammonium chloride	NH_4Cl	sal-ammoniac	cubic
Ammonium sulfate	$(NH_4)_2SO_4$	mascagnite	orthorhombic
Barium chloride	$BaCl_2 \cdot 2H_2O$		monoclinic
Calcium carbonate	$CaCO_3$	calcite	rhombohedral
Copper sulfate	$CuSO_4 \cdot 5H_2O$	blue vitriol	triclinic
Magnesium sulfate	$MgSO_4 \cdot 7H_2O$	Epsom salt	orthorhombic
Magnesium chloride	$MgCl_2 \cdot 6H_2O$	bischofite	monoclinic
Nickel sulfate	$NiSO_4 \cdot 6H_2O$	single nickel salt	tetragonal
Potassium chloride	KCl	muriate of potash	cubic
Potassium nitrate	KNO_3	nitre	hexagonal
Potassium sulfate	K_2SO_4	arcanite	orthorhombic
Silver nitrate	$AgNO_3$	lunar caustic	orthorhombic
Sodium chlorate	$NaClO_3$		cubic
Sodium chloride	$NaCl$	salt, halite	cubic
Sodium nitrate	$NaNO_3$	chile salt petre	rhombohedral
Sodium sulfate	$Na_2SO_4 \cdot 10H_2O$	Glauber's salt	monoclinic
Sodium thiosulfate	$Na_2S_2O_3 \cdot 5H_2O$	hypo	monoclinic
Zinc sulfate	$ZnSO_4 \cdot 7H_2O$	white vitriol	orthorhombic

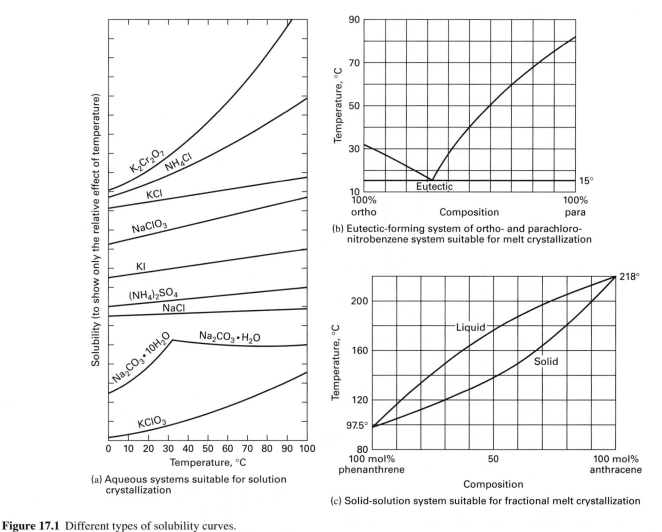

(a) Aqueous systems suitable for solution crystallization

(b) Eutectic-forming system of ortho- and parachloro-nitrobenzene system suitable for melt crystallization

(c) Solid-solution system suitable for fractional melt crystallization

Figure 17.1 Different types of solubility curves.

[From *Handbook of Separation Techniques for Chemical Engineers*, 2nd ed., P.A. Schweitzer, Editor-in-chief, McGraw-Hill, New York (1988) with permission.]

crystals of one component or the other, depending on whether the melt composition is to the left or right of the eutectic composition of two solid phases. If, however, solid solutions form, as shown in Figure 17.1c, a process of repeated melting and freezing steps, called *fractional melt crystallization*, is required to obtain nearly pure crystalline products. A higher degree of purity can be achieved by a technique called *zone melting* or *refining*. Examples of binary organic systems that form eutectics include metaxylene–paraxylene and benzene–naphthalene. Binary systems of naphthalene–beta naphthol and naphthalene–β naphthylamine form less-common solid solutions.

Crystallization can occur from a vapor mixture by *desublimation*. Compounds, including phthalic anhydride and benzoic acid, are produced in this manner. When two or more compounds tend to desublime, a fractional desublimation process can obtain near-pure products.

Crystallization of a compound from a dilute, aqueous solution is often preceded by *evaporation* in one or more vessels, called *effects*, to concentrate the solution, followed by partial separation and washing of the crystals from the resulting slurry, called the *magma*, by centrifugation or filtration. The process is completed by drying the crystals to a specified moisture content.

Pharmaceuticals and powdered food products, which are predominately large organic molecules, are normally precipitated from aqueous solutions and dried rapidly at low temperatures to preserve aroma, taste, and biological activity. A change in solubility due primarily to cooling or dilution with a water-miscible solvent induces bioproduct crystal growth in *batch* volumes. *Cooling curves* that maintain constant *supersaturation* are generated to determine

temperature change as a function of dimensionless time to produce uniform, large crystals and prevent scaling on heat-transfer surfaces. The complexity of these curves increases from a batch crystallizer seeded at uniform particle size and neglecting nucleation, to a batch crystallizer that considers crystal-size distribution as well as nucleation. Effective *micromixing*, which can be provided by *impingement*, is widely used to maintain uniform supersaturation, number density, and rapid mass transfer throughout the batch crystallizer volume. Many biological products form *flocs* rather than crystals. Flocculation and precipitation are treated in §19.7.2.

Industrial Example

$MgSO_4 \cdot 7H_2O$ (Epsom salt) is crystallized industrially from an aqueous solution. A solid–liquid phase diagram for $MgSO_4 \cdot H_2O$ at 1 atm is shown in Figure 17.2, where, depending on the temperature, four different hydrated forms of $MgSO_4$ exist: $MgSO_4 \cdot H_2O$, $MgSO_4 \cdot 6H_2O$, $MgSO_4 \cdot 7H_2O$, and $MgSO_4 \cdot 12H_2O$. Furthermore, a eutectic of the latter hydrate with ice is possible. To obtain the preferred heptahydrate, crystallization must occur in the temperature range of 36–118°F (Point b to Point c), where $MgSO_4$ solubility (anhydrous or hydrate-free basis) increases almost linearly from 21 to 33 wt%.

A representative commercial process for producing 4,205 lb/hr (dry basis) of $MgSO_4 \cdot 7H_2O$ crystals from a 10 wt% aqueous solution at 1 atm and 70°F is shown in Figure 17.3. This feed is first concentrated in a double-effect evaporation system with forward feed, and then mixed with recycled mother liquors from the hydroclone and centrifuge. The combined feed of 14,326 lb/h containing 31.0 wt% $MgSO_4$ at 120°F and 1 atm enters an evaporative, vacuum crystallizer

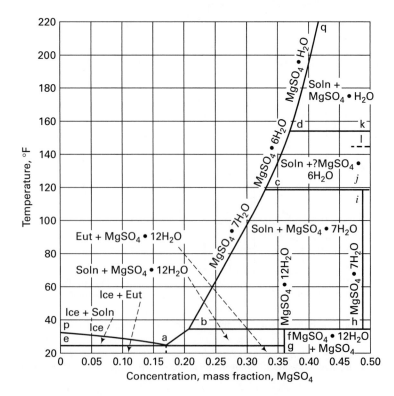

Figure 17.2 Solid–liquid phase diagram for the $MgSO_4 \cdot H_2O$ system at 1 atm.

[From W.L. McCabe, J.C. Smith, and P. Harriott, *Unit Operations of Chemical Engineering*, 5th ed., McGraw-Hill, New York (1993) with permission.]

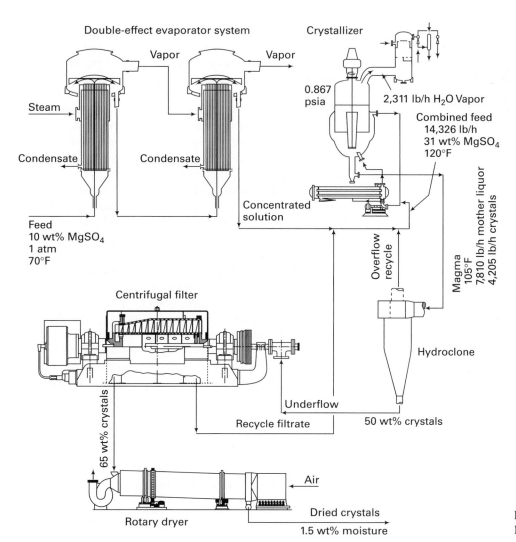

Figure 17.3 Process for production of $MgSO_4 \cdot 7H_2O$.

constructed of 316 stainless steel, shown in more detail in Figure 17.4.

The crystallizer utilizes internal circulation of 6,000 gpm of magma up through a draft tube equipped with a 3-hp marine-propeller agitator to obtain near-perfect mixing of the magma. Mother liquor, separated from crystals during upward flow outside of the skirt baffle, is circulated externally at 625 gpm by a 10-hp stainless-steel pump, through a 300-ft² stainless steel, plate-and-frame heat exchanger, where 2,052,000 Btu/hr of heat is transferred to the solution from 2,185 lb/h of condensing 20 psig steam to provide supersaturation and energy to evaporate 2,311 lb/h of water.

Vapor leaving the top of the crystallizer is condensed by direct contact with cooling water in a barometric condenser, attached to which are ejectors to pull a vacuum of 0.867 psia in the crystallizer. The product magma, at 105°F, consists of 7,810 lb/h of mother liquor saturated with 30.6 wt% $MgSO_4$ and 4,205 lb/h of heptahydrate crystals. This magma contains 35% crystals by weight or 30.2% crystals by volume, based on a crystal density of 1.68 g/cm³ and a mother liquor density of 1.35 g/cm³. The boiling-point elevation of the saturated mother liquor at 105°F is 8°F. Thus, vapor leaving the crystallizer is superheated by 8°F. The magma residence time in the crystallizer is 4 hours, which is sufficient to produce the following crystal-size

distribution: 35 wt% on 20 mesh U.S. screen, 80 wt% on 40 mesh U.S. screen, and 99 wt% on 100 mesh U.S. screen.

The crystallizer is 30 ft high, with a vapor-space diameter of 5-1/2 ft and a magma-space diameter of 10 ft. The magma is thickened to 50 wt% crystals in a hydroclone (§19.2.5), from which the mother-liquor overflow is recycled to the crystallizer and the underflow slurry is sent to a continuous centrifuge (§19.2.5), where it is thickened to 65 wt% crystals and washed. Filtrate mother liquor from the centrifuge is recycled to the crystallizer. Centrifuge cake goes to a continuous direct-heat rotary dryer (§18.1.2) to reduce crystal moisture content to 1.5 wt%.

§17.1 CRYSTAL GEOMETRY

In a solid, the motion of molecules, atoms, or ions is restricted largely to oscillations about fixed positions. In amorphous solids, these positions are not arranged in a regular or lattice pattern, whereas in crystalline solids, they are. Amorphous solids are isotropic, i.e., physical properties are independent of the direction of measurement; crystalline solids are anisotropic, unless the crystals are cubic in structure.

When crystals grow, unhindered by other surfaces such as container walls and other crystals, they form polyhedrons with flat sides and sharp corners; they are never spherical in

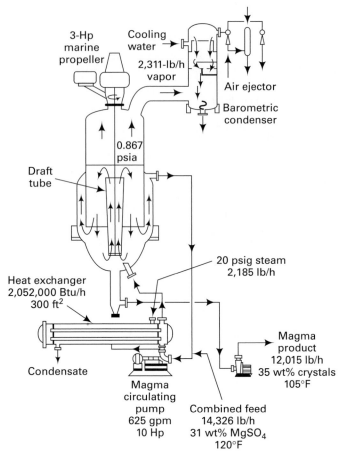

Figure 17.4 Crystallizer for production of $MgSO_4 \cdot 7H_2O$ crystals.

As discussed by Mullin [1], early investigators found that crystals consist of many units, each shaped like the larger crystal. This led to the concept of a space lattice as a regular arrangement of points (molecules, atoms, or ions) such that if a line is drawn between any two points and then extended in both directions, the line will pass through other lattice points with an identical spacing. In 1848, Bravais showed that only the 14 space lattices shown in Figure 17.5 are possible. Based on the symmetry of the three mutually perpendicular axes with respect to their relative lengths (a, b, c) and the angles (α, β, γ) between the axes, the 14 lattices can be classified into the seven crystal systems listed in Table 17.2. For example, the cubic (regular) system includes the simple cubic lattice, the body-centered cubic lattice, and the face-centered lattice. Examples of the seven crystal systems are included in Table 17.1. The five sodium salts in that table form three of the seven crystal systems.

Crystals of a given substance and a given system exhibit markedly different appearances when the faces grow at different rates, particularly when these rates vary greatly, from stunted growth in one direction to give plates, to exaggerated growth in another direction to give needles. For example, potassium sulfate, which belongs to the orthorhombic system, can take on any of the crystal habits shown in Figure 17.6, including plates, needles, and prisms. When product crystals of a particular habit are desired, research is required to find the necessary processing conditions. Modifications of crystal habit are most often accomplished by addition of impurities.

§17.1.1 Crystal-Size Distributions

Crystallizer magmas contain a distribution of crystal sizes and shapes. It is highly desirable to characterize a batch of crystals (or particles in general) by an average crystal size and a crystal-size distribution, by defining a characteristic crystal dimension. However, as shown in Figure 17.6, some crystal shapes require two characteristic dimensions. One solution to this problem, which assists in the correlation of transport rates involving particles, is to relate the irregular-shaped particle to

shape. Although two crystals of a given chemical may appear quite different in size and shape, they obey the *Law of Constant Interfacial Angles* proposed by Hauy in 1784. This states that the angles between corresponding faces of all crystals of a given substance are constant even if the crystals vary in size and in the development of the various faces (the *crystal habit*). The interfacial angles and lattice dimensions can be measured by X-ray crystallography.

Table 17.2 The Seven Crystal Systems

Crystal System	Space Lattices	Length of Axes	Angles Between Axes
Cubic (regular)	Simple cubic Body-centered cubic Face-centered cubic	$a = b = c$	$\alpha = \beta = \gamma = 90°$
Tetragonal	Square prism Body-centered square prism	$a = b < c$	$\alpha = \beta = \gamma = 90°$
Orthorhombic	Simple orthorhombic Body-centered orthorhombic Base-centered orthorhombic Face-centered orthorhombic	$a \neq b \neq c$	$\alpha = \beta = \gamma = 90°$
Monoclinic	Simple monoclinic Base-centered monoclinic	$a \neq b \neq c$	$\alpha = \beta = 90°$ $\gamma \neq 90°$
Rhombohedral (trigonal)	Rhombohedral	$a = b = c$	$\alpha = \beta = \gamma \neq 90°$
Hexagonal	Hexagonal	$a = b \neq c$	$\alpha = \beta = 90°$ $\gamma \neq 120°$
Triclinic	Triclinic	$a \neq b \neq c$	$\alpha \neq \beta \neq \gamma \neq 90°$

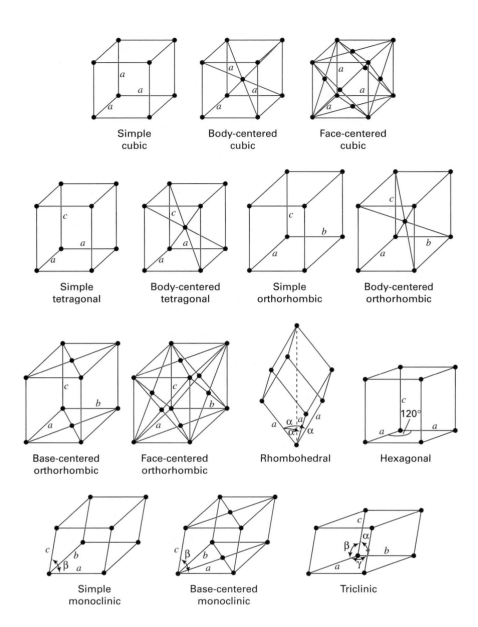

Figure 17.5 The 14 space lattices of Bravais.

a sphere by the sphericity, ψ, defined as

$$\psi = \frac{\text{surface area of a sphere with the same volume as the particle}}{\text{surface area of the particle}}$$

(17-1)

For a sphere, $\psi = 1$, while for all other particles, $\psi < 1$. For a spherical particle of diameter D_p, the surface area (s_p)-to-volume (υ_p) ratio is

$$\left(s_p/\upsilon_p\right)_{\text{sphere}} = \left(\pi D_p^2\right)/\left(\pi D_p^3/6\right) = 6/D_p$$

Therefore, (17-1) becomes

$$\psi = \frac{6}{D_p}\left(\frac{\upsilon_p}{s_p}\right)_{\text{particle}}$$

(17-2)

EXAMPLE 17.1   Sphericity, ψ, of a Cube.

Estimate the sphericity of a cube of dimension a on each side.

Solution

$$\upsilon_{\text{cube}} = a^3 \quad \text{and} \quad s_{\text{cube}} = 6a^2$$

From (17-2),

$$\psi = \frac{6}{D_p}\left(\frac{a^3}{6a^2}\right) = \frac{a}{D_p}$$

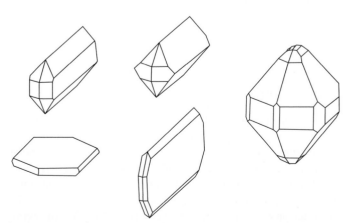

Figure 17.6 Some crystal habits of orthorhombic potassium-sulfate crystals.

Because the volumes of the sphere and the cube must be equal,

$$\pi D_p^3 / 6 = a^3$$

Solving, $D_p = 1.241\, a$.

Then,
$$\psi = a/(1.241\, a) = 0.806$$

Table 17.3 Methods of Measuring Particle Size

Method	Size Range, Microns
Woven-wire screen	32–5600
Coulter electrical sensor	1–200
Gravity sedimentation	1–50
Optical microscopy	0.5–150
Laser-light scattering	0.04–2000
Centrifugal sedimentation	0.01–5
Electron microscopy	0.001–5

Table 17.4 U.S. Standard Screens ASTM EII

Mesh Number	Opening of Square Aperture		
	inch	mm	μm
3-1/2	0.220	5.60	5600
4	0.187	4.75	4750
5	0.157	4.00	4000
6	0.132	3.35	3350
7	0.110	2.80	2800
8	0.0929	2.36	2360
10	0.0787	2.00	2000
12	0.0669	1.70	1700
14	0.0551	1.40	1400
16	0.0465	1.18	1180
18	0.0394	1.000	1000
20	0.0335	0.850	850
25	0.0280	0.710	710
30	0.0236	0.600	600
35	0.0197	0.500	500
40	0.0167	0.425	425
45	0.0140	0.355	355
50	0.0118	0.300	300
60	0.00984	0.250	250
70	0.00835	0.212	212
80	0.00709	0.180	180
100	0.00591	0.150	150
120	0.00492	0.125	125
140	0.00417	0.106	106
170	0.00354	0.090	90
200	0.00295	0.075	75
230	0.00248	0.063	63
270	0.00209	0.053	53
325	0.00177	0.045	45
400	0.00150	0.038	38
450	0.00126	0.032	32

The most common methods of measuring particle size are listed in Table 17.3, together with their particle-size ranges. Because of the irregular shapes of crystals, it should not be surprising that the different methods can give results that differ by as much as 50%. Crystal-size distributions are most often determined with U.S. (or British) standard wire-mesh screens [ASTM Ell (1989)] derived from the earlier Tyler standard screens. The U.S. standard is based on a 1-mm (1000-μm)-square *aperture-opening* screen called Mesh No. 18 because there are 18 apertures per inch. The standard Mesh numbers are listed in Table 17.4, where each successively smaller aperture differs from the preceding aperture by a factor of $2^{1/4}$. Mechanical shaking of a stack of ordered screens is used in sieving operations.

When wire-mesh screens are used to determine size distribution, crystal size is taken to be the screen aperture through which the crystal just passes. Because of particle-shape irregularity, this is considered a nominal value, particularly for plates and needles, as shown in Figure 17.7.

Particle-size-distribution data, called a *screen analysis*, are displayed in the form of a table, from which *differential* and *cumulative* plots can be made, usually on a mass-fraction basis. Consider the following laboratory screen-analysis data of Graber and Taboada [2] for crystals of $Na_2SO_4 \cdot 10H_2O$ (Glauber's salt) grown at 18°C during a residence time of 37.2 minutes in a well-mixed laboratory cooling crystallizer. The smallest screen was 140 mesh, with particles passing through it being retained on a pan.

Mesh Number	Aperture, D_p, mm	Mass Retained on Screen, Grams	% Mass Retained
14	1.400	0.00	0.00
16	1.180	9.12	1.86
18	1.000	32.12	6.54
20	0.850	39.82	8.11
30	0.600	235.42	47.95
40	0.425	89.14	18.15
50	0.300	54.42	11.08
70	0.212	22.02	4.48
100	0.150	7.22	1.47
140	0.106	1.22	0.25
Pan	—	0.50	0.11
		491.00	100.00

§17.1.2 Differential Screen Analysis

A *differential screen analysis* is made by determining the arithmetic-average aperture for each mass fraction that passes through one screen but not the next. From the above table, a mass fraction of 0.0186 passes through a screen of 1.400-mm aperture, but not through a screen of 1.180-mm. The average of these apertures is 1.290 mm, which is the nominal particle size for that mass fraction. The following differential analysis is computed in this manner, where the

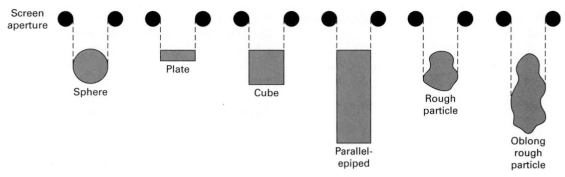

Figure 17.7 Different particle shapes that just pass through the same screen.

designation $-14+16$ refers to particles passing through a 14-mesh screen and retained on a 16-mesh screen.

Mesh Range	$\bar{D}_p$, Average Particle Size, mm	Mass Fraction, x_i
$-14 + 16$	1.290	0.0186
$-16 + 18$	1.090	0.0654
$-18 + 20$	0.925	0.0811
$-20 + 30$	0.725	0.4795
$-30 + 40$	0.513	0.1815
$-40 + 50$	0.363	0.1108
$-50 + 70$	0.256	0.0448
$-70 + 100$	0.181	0.0147
$-100 + 140$	0.128	0.0025
$-140 + (170)$	0.098	0.0011
		1.0000

A plot of the differential screen analysis is shown in Figure 17.8 both as (a) an $x-y$ plot and as (b) a histogram. If a wide range of screen aperture is covered, it is best to use a log scale for aperture.

§17.1.3 Cumulative Screen Analysis

Screen analysis data can also be plotted as cumulative-weight-percent oversize or (which is more common) undersize as a function of screen aperture. For the above data of

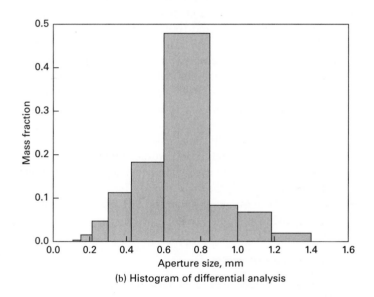

(b) Histogram of differential analysis

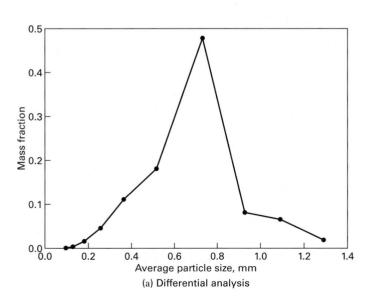

(a) Differential analysis

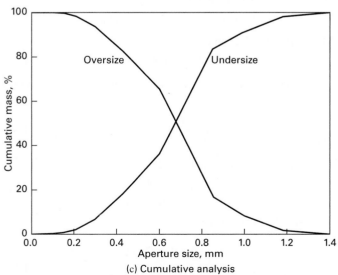

(c) Cumulative analysis

Figure 17.8 Screen analyses for data of Graber and Taboada [2].

Graber and Taboada [2], the two types of *cumulative screen analysis* are as follows:

Aperture, Dp, mm	Cumulative wt% Undersize	Cumulative wt% Oversize
1.400	100.00	0.00
1.180	98.14	1.86
1.000	91.60	8.40
0.850	83.49	16.51
0.600	35.54	64.46
0.425	17.39	82.61
0.300	6.31	93.69
0.212	1.83	98.17
0.150	0.36	99.64
0.106	0.11	99.89

Because 0.11 wt% passed through a 0.106-mm aperture but was retained on a pan with no indication of how small the retained particles were, cumulative wt% undersize and oversize cannot be taken to 0 and 100%, respectively.

The cumulative screen analyses are plotted in Figure 17.8c. The curves, which are mirror images of each other, cross at a median size where 50 wt% is larger in size and 50 wt% is smaller. As with differential plots, a log scale is preferred if a large range of screen aperture is covered. A log scale for the cumulative wt% may also be preferred if an appreciable fraction of the data lie below 10%.

A number of different mean particle sizes that are derived from screen analysis are used in practice, depending upon the application. Of these, the most useful are: (1) surface-mean diameter, (2) mass-mean diameter, (3) arithmetic-mean diameter, and (4) volume-mean diameter.

Surface-Mean Diameter, $\bar{D}_S$

The specific surface area (area/mass) of a particle of spherical or other shape is defined by

$$A_w = s_p/m_p = s_p/\upsilon_p\rho_p \tag{17-3}$$

Combining (17-2) and (17-3),

$$A_w = 6/\psi\rho_p D_p \tag{17-4}$$

For n mass fractions x_i, each of average aperture $\bar{D}_{p_i}$, from a screen analysis, the overall specific surface area is given by

$$A_w = \sum_{i=1}^{n} \frac{6x_i}{\psi\rho_p\bar{D}_{p_i}} = \frac{6}{\psi\rho_p}\sum \frac{x_i}{\bar{D}_{p_i}} \tag{17-5}$$

The surface-mean diameter is defined by

$$A_w = \frac{6}{\psi\rho_p\bar{D}_S} \tag{17-6}$$

Combining (17-5) and (17-6),

$$\boxed{\bar{D}_S = \frac{1}{\sum_{i=1}^{n} \dfrac{x_i}{\bar{D}_{p_i}}}} \tag{17-7}$$

which can be used to determine the surface-mean diameter, $\bar{D}_S$, from a screen analysis. This mean diameter is sometimes referred to as the Sauter mean diameter and as the volume-surface-mean diameter. It is often used for skin friction, heat-transfer, and mass-transfer calculations involving particles.

Weight or Mass-Mean Diameter, $\bar{D}_W$

The mass-mean diameter is defined by

$$\bar{D}_W = \sum_{i=1}^{n} x_i\bar{D}_{p_i} \tag{17-8}$$

Arithmetic-Mean Diameter, $\bar{D}_N$

The arithmetic-mean diameter is defined in terms of the number of particles, N_i, in each size range:

$$\bar{D}_N = \frac{\sum_{i=1}^{n} N_i\bar{D}_{p_i}}{\sum N_i} \tag{17-9}$$

The number of particles is related to the mass fraction of particles by

$$x_i = \frac{\text{mass of particles of average size } \bar{D}_{p_i}}{\text{total mass}}$$
$$= \frac{N_i f_\upsilon\left(\bar{D}_{p_i}\right)^3 \rho_p}{M_t} \tag{17-10}$$

where

M_t = total mass and
f_υ = volume shape factor defined by $\upsilon_p = f_\upsilon\bar{D}_{p_i}^3$ (17-11)

For spherical particles, $f_\upsilon = \pi/6$.

If (17-10) is solved for N_i, substituted into (17-9), and simplified,

$$\bar{D}_N = \frac{\sum_{i=1}^{n} \left(\dfrac{x_i}{\bar{D}_{p_i}^2}\right)}{\sum_{i=1}^{n} \left(\dfrac{x_i}{\bar{D}_{p_i}^3}\right)} \tag{17-12}$$

Volume-Mean Diameter, $\bar{D}_V$

The volume-mean diameter, $\bar{D}_V$, is defined by

$$\left(f_\upsilon\bar{D}_V^3\right)\sum_{i=1}^{n} N_i = \sum_{i=1}^{n} \left(f_\upsilon\bar{D}_{p_i}^3\right)N_i \tag{17-13}$$

Solving (17-13) for $\bar{D}_V$ for a constant value of f_υ gives

$$\bar{D}_V = \left(\frac{\sum_{i=1}^{n} N_i\bar{D}_{p_i}^3}{\sum_{i=1}^{n} N_i}\right) \tag{17-14}$$

The corresponding relation in terms of x_i rather than N_i is obtained by combining (17-14) with (17-10), giving

$$\bar{D}_V = \left(\frac{1}{\sum \dfrac{x_i}{\bar{D}_{p_i}^3}}\right)^{1/3} \tag{17-15}$$

EXAMPLE 17.2 Mean Particle Diameters.

Using the screen analysis data of Graber and Taboada given above, compute all four mean diameters.

Solution

Since the data are given in weight (mass) fractions, use (17-6), (17-8), (17-12), and (17-15).

$\bar{D}_p$ mm	x	$x/\bar{D}_p$	$x\bar{D}_p$	$x/\bar{D}_p^2$	$x/\bar{D}_p^3$
1.290	0.0186	0.0144	0.0240	0.0112	0.0087
1.090	0.0654	0.0600	0.0713	0.0550	0.0505
0.925	0.0811	0.0877	0.0750	0.0948	0.1025
0.725	0.4795	0.6614	0.3476	0.9122	1.2583
0.513	0.1815	0.3538	0.0931	0.6897	1.3444
0.363	0.1108	0.3052	0.0402	0.8409	2.3164
0.256	0.0448	0.1750	0.0115	0.6836	2.6703
0.181	0.0147	0.0812	0.0027	0.4487	2.4790
0.128	0.0025	0.0195	0.0003	0.1526	1.1921
0.098	0.0011	0.0112	0.0001	0.1145	1.1687
	1.0000	1.7695	0.6658	4.0032	12.5909

From (17-7), $$\bar{D}_S = \frac{1}{1.7695} = 0.565 \text{ mm}$$

From (17-8), $$\bar{D}_W = 0.666 \text{ mm}$$

From (17-12), $$\bar{D}_N = \frac{4.0032}{12.5909} = 0.318 \text{ mm}$$

From (17-15), $$\bar{D}_V = \left(\frac{1}{12.5909}\right)^{1/3} = 0.430 \text{ mm}$$

Mean diameters vary significantly!

§17.2 THERMODYNAMIC CONSIDERATIONS

Important thermodynamic properties for crystallization include melting point, heat of fusion, solubility, heat of crystallization, heat of solution, heat of transition, specific heat, and supersaturation.

§17.2.1 Solubility and Mass Balances

For systems of water and soluble inorganic and organic chemicals, Mullin [1] presents extensive tables of solubility as a function of temperature, and heat of solution at infinite dilution and room temperature (18–25°C). Data for the inorganic salts of Table 17.1 in water are listed in Table 17.5, where solubilities are given on a hydrate-free basis. Solubilities are seen to vary widely from as low as 4.8 g/100 g of

Table 17.5 Solubility and Heat of Solution at Infinite Dilution of Some Inorganic Compounds in Water (a Positive Heat of Solution Is Endothermic)

Compound	Heat of Solution of Stable Hydrate (at Room Temperature) kcal/mole Compound	Solubility (Hydrate-free Basis) g/100 g H_2O at T, °C								Stable Hydrate at Room Temperature
		0	10	20	30	40	60	80	100	
NH_4Cl	+3.8	29.7	33.4	37.2	41.4	45.8	55.2	65.6	77.3	0
$(NH_4)_2SO_4$	+1.5	71.0	73.0	75.4	78.0	81.0	88.0	95.3	103.3	0
$BaCl_2$	+4.5	31.6	33.2	35.7	38.2	40.7	46.4	52.4	58.3	2
$CuSO_4$	+2.86	14.3	17.4	20.7	25.0	28.5	40.0	55.0	75.4	5
$MgSO_4$	+3.18	22.3	27.8	33.5	39.6	44.8	55.3	56.0	50.0	7
$MgCl_2$	−3.1	52.8	53.5	54.5	56.0	57.5	61.0	66.0	73.0	6
$NiSO_4$	+4.2	26	32	37	43	47	55	63	—	7
KCl	+4.4	27.6	31.0	34.0	37.0	40.0	45.5	51.1	56.7	0
KNO_3	+8.6	13.3	20.9	31.6	45.8	63.9	110	169	247	0
K_2SO_4	+6.3	7.4	9.3	11.1	13.1	14.9	18.3	21.4	24.2	0
$AgNO_3$	+5.4	122	170	222	300	376	525	669	952	0
$NaClO_3$	+5.4	80	89	101	113	126	155	189	233	0
$NaCl$	+0.93	35.6	35.7	35.8	36.1	36.4	37.1	38.1	39.8	0
$NaNO_3$	+5.0	72	78	85	92	98	—	133	163	0
Na_2SO_4	+18.7	4.8	9.0	19.4	40.8	48.8	45.3	43.7	42.5	10
$Na_2S_2O_3$	+11.4	52	61	70	84	103	207	250	266	5
Na_3PO_4	+15.0	1.5	4	11	20	31	55	81	108	12

water for Na_2SO_4 (as the decahydrate) at $0°C$ to 952 g/100 g of water for $AgNO_3$ at $100°C$. For KNO_3, solubility increases by a factor of 18.6 for the same temperature increase.

Solubility of an inorganic compound can be much lower than that shown for Na_2SO_4. Such compounds are considered to be just slightly or sparingly soluble or almost insoluble, and their solubility is expressed as a solubility product, K_c, in terms of ion concentration. $Al(OH)_3$, which is sparingly soluble with a solubility product, from Table 17.6, of $K_c = 1.1 \times 10^{-15}$ at $18°C$, dissolves according to

$$Al(OH)_{3(s)} \Leftrightarrow Al^{3+}_{(aq)} + 3OH^-_{(aq)}$$

By the *law of mass action*, the equilibrium constant, called the *solubility product* for dissolution, is given by

$$K_c = \frac{(c_{Al^{3+}})(c_{OH^-})^3}{a_{Al(OH)_3}} = (c_{Al^{3+}})(c_{OH^-})^3 = 1.1 \times 10^{-15}$$

where concentrations are used for the ions, and the activity of solid $Al(OH)_3$ is taken as 1.0. By stoichiometry,

$$(c_{OH^-}) = 3(c_{Al^{3+}}) \quad \text{and} \quad K_c = (c_{Al^{3+}})^4(3)^3 = 1.1 \times 10^{-15}$$

Then, $\quad (c_{Al^{3+}}) = c_{\text{dissolved Al(OH)}_3} = 8 \times 10^{-5}$ gmoles/L

which is a very small concentration.

For less sparingly soluble compounds, the equilibrium constant, K_a, is the more rigorous form, expressed in terms of ionic activities or activity coefficients:

$$K_a = \frac{(a_{Al^{3+}})(a_{OH^{-1}})^3}{a_{Al(OH)_3}} = (\gamma_{Al^{3+}})(c_{Al^{3+}})(\gamma_{OH^-})^3(c_{OH^-})^3$$

In general, $\gamma \approx 1.0$ for $c < 1 \times 10^{-3}$ mol/L. As c rises above 1×10^{-3} mol/L, γ decreases, but may pass through a minimum and then increase. Mullin [1] presents activity-coefficient data at $25°C$ for soluble inorganic compounds over a wide range of concentration.

Solubility of most inorganic compounds increases with temperature, but a few common compounds exhibit a negative or inverted solubility in certain ranges of temperature, where solubility decreases with increasing temperature. These compounds are the "hard salts," which include anhydrous Na_2SO_4 and $CaSO_4$.

A change in the solubility curve can occur when a phase transition from one stable hydrate to another takes place. For example, in Table 17.5, $Na_2SO_4 \cdot 10H_2O$ is the stable form from $0°C$ to $32.4°C$. In that temperature range, the solubility increases rapidly from 4.8 to 49.5 g (hydrate-free basis)/100 g H_2O. From $32.4°C$ to $100°C$, the stable form is Na_2SO_4, whose solubility decreases slowly from 49.5 to 42.5 g/100 g H_2O. In the phase diagram of Figure 17.2 for the $MgSO_4$–water system, solubility-temperature curves of the four hydrated forms have distinctive slopes.

The solubility curve is the most important property for determining the best method for causing crystallization and the ease or difficulty of growing crystals. Crystallization by cooling is attractive only for compounds having a solubility that decreases rapidly with decreasing temperature. Such is not the case for most of the compounds in Table 17.5. For

Table 17.6 Concentration Solubility Products of Some Sparingly Soluble Inorganic Compounds

Compound	T, $°C$	K_c
Ag_2CO_3	25	6.15×10^{-12}
$AgCl$	25	1.56×10^{-10}
$Al(OH)_3$	15	4×10^{-13}
$Al(OH)_3$	18	1.1×10^{-15}
$BaSO_4$	18	0.87×10^{-10}
$CaCO_3$	15	0.99×10^{-8}
$CaSO_4$	10	1.95×10^{-4}
$CuSO_4$	16–18	2×10^{-47}
$Fe(OH)_3$	18	1.1×10^{-36}
$MgCO_3$	12	2.6×10^{-5}
ZnS	18	1.2×10^{-23}

$NaCl$, crystallization by cooling would be undesirable because solubility decreases only by about 10% when the temperature decreases from 100 to $0°C$. For most inorganic compounds, cooling by evaporation is the preferred technique.

Solid compounds with low solubility can be produced by reacting two soluble compounds. For example, in Table 17.6, solid $Al(OH)_3$ can be formed by the reaction

$$AlCl_{3(aq)} + 3NaOH_{(aq)} \Leftrightarrow Al(OH)_{3(ppt)} + 3NaCl_{(aq)}$$

The reaction is so fast that only very fine crystals, called a precipitate, are produced, with no simple method to cause them to grow to large crystals.

EXAMPLE 17.3 Mass Balance on a Crystallization System.

Concentrate from an evaporation system is 4,466 lb/h of 37.75 wt% $MgSO_4$ at $170°F$ and 20 psia. It is mixed with 9,860 lb/h of a saturated aqueous recycle filtrate of $MgSO_4$ at $85°F$ and 20 psia and sent to a vacuum crystallizer, operating at $85°F$ and 0.58 psia, to produce water vapor and a magma of 20.8 wt% crystals and 79.2 wt% saturated solution. The magma is sent to a filter, from which filtrate is recycled. Determine the lb/h of water evaporated and the maximum crystal production rate in tons/day on a dry basis.

Solution

For saturated filtrate at $85°F$, the wt% of $MgSO_4$, from Figure 17.2, is 28. Therefore, $MgSO_4$ in the recycle filtrate is $9,860(0.28) = 2,760$ lb/h. By mass balance around the mixing step,

	lb/h		
Component	Feed	Recycle Filtrate	Crystallizer Feed
$MgSO_4$	1,686	2,760	4,446
H_2O	2,780	7,100	9,880
	4,466	9,860	14,326

The crystallizer mass balance is made on $MgSO_4$. At $85°F$, from Figure 17.2, magma is 20.8 wt% $MgSO_4 \cdot 7H_2O$ crystals and 79.2 wt% of 28 wt% aqueous $MgSO_4$ liquid. Because the MW of $MgSO_4$

and $MgSO_4 \cdot 7H_2O$ are 120.4 and 246.4, respectively, the crystals are $120.4/246.4 = 0.4886$ mass fraction $MgSO_4$.

By a $MgSO_4$ balance, $\qquad 4{,}446 = 0.28\,L + 0.4886\,S \qquad$ (1)

where $L =$ lb/h of liquid and $S =$ lb/h of crystals,

Also, $\qquad\qquad S = 0.208(S + L) \qquad\qquad$ (2)

Solving (1) and (2) simultaneously,

$$L = 10{,}876 \text{ lb/h} \quad \text{and} \quad S = 2{,}856 \text{ lb/h}$$

By a total material balance around the crystallizer, $F = V + L + S$, where $F =$ total feed rate and $V =$ evaporation rate. Therefore,

$$14{,}326 = V + 10{,}876 + 2{,}856 \qquad (3)$$

Solving, $V = 594$ lb/h. The results in tabular form are:

Component	lb/h for crystallizer			
	Feed	Vapor	Liquid	Crystals
$MgSO_{4(aq)}$	4,446	0	3,045	0
$MgSO_4 \cdot 7H_2O_{(s)}$	0	0	0	2,856
H_2O	9,880	594	7,831	0
	14,326	594	10,876	2,856

The maximum production rate of crystals is

$$\frac{2{,}856}{2{,}000}(24) = 34.3 \text{ tons/day}$$

A number of organic compounds, particularly organic acids with relatively moderate melting points (125–225°C), are soluble in water. Some data are given in Table 17.7, where it is seen that solubility often increases significantly with increasing temperature. For example, the solubility of o-phthalic acid increases from 0.56 to 18.0 g/100 g H_2O when temperature rises from 20 to 100°C.

EXAMPLE 17.4 Crystallization by Cooling.

Oxalic acid is to be crystallized from a saturated aqueous solution initially at 100°C. To what temperature does the solution have to be cooled to crystallize 95% of the acid as the dihydrate?

Solution

Assume a basis of 100 g of water. From Table 17.7, the dissolved oxalic acid at 100°C is 84.4 g. The amount to be crystallized $= 0.95(84.4) = 80.2$ g. The oxalic acid left in the solution $= 84.4 - 80.2 = 4.2$ g. MW of oxalic acid $= 90.0$. MW of water $= 18.0$.

Water of hydration for $2H_2O = \dfrac{2(18.0)}{90.0} = 0.4\,\dfrac{\text{g } H_2O}{\text{g oxalic acid}}$.

Water of crystallization $= 0.4(80.2) = 32.1$ g H_2O.

Liquid water remaining $= 100 - 32.1 = 67.9$ g.

Final solubility $= \dfrac{4.2}{67.9} \times 100 = 6.19\,\dfrac{\text{g}}{100 \text{ g } H_2O}$.

From Table 17.7, by linear interpolation, temperature $= 10.6$°C.

§17.2.2 Energy Balances

When an anhydrous solid compound, whose solubility increases with increasing temperature, dissolves isothermally in a solvent, heat is absorbed by the solution. This amount of heat per mole of compound in an infinite amount of solvent, which varies with temperature, is the *heat of solution at infinite dilution*, $(\Delta H_{sol}^{\infty})$. In Table 17.5, the solubility of anhydrous NaCl is seen to increase slowly with increasing temperature from 10 to 100°C. Correspondingly, the heat of solution at infinite dilution in Table 17.5 is modestly endothermic (+) at room temperature. In contrast, the solubility of anhydrous KNO_3 increases more rapidly with increasing temperature, resulting in a higher endothermic heat of solution. For compounds that form hydrates, heat of solution at infinite dilution may be exothermic (−) for the anhydrous form, but becomes less negative and often positive as higher hydrates are formed by

$$A \cdot nH_2O_{(s)} \rightarrow A_{(aq)} + nH_2O$$

Table 17.7 Solubility and Melting Point of Some Organic Compounds in Water

Compound	Melting Point, °C	Solubility, g/100 g H_2O at T, °C							
		0	10	20	30	40	60	80	100
Adipic acid	153	0.8	1.0	1.9	3.0	5.0	18	70	160
Benzoic acid	122	0.17	0.20	0.29	0.40	0.56	1.16	2.72	5.88
Fumaric acid (trans)	287	0.23	0.35	0.50	0.72	1.1	2.3	5.2	9.8
Maleic acid	130	39.3	50	70	90	115	178	283	—
Oxalic acid	189	3.5	6.0	9.5	14.5	21.6	44.3	84.4	—
o-phthalic acid	208	0.23	0.36	0.56	0.8	1.2	2.8	6.3	18.0
Succinic acid	183	2.8	4.4	6.9	10.5	16.2	35.8	70.8	127
Sucrose	d	179	190	204	219	238	287	362	487
Urea	133	67	85	105	135	165	250	400	730
Uric acid	d	0.002	0.004	0.006	0.009	0.012	0.023	0.039	0.062

d: decomposes

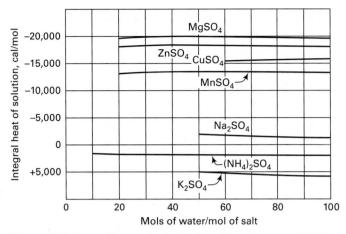

Figure 17.9 Integral heats of solution for sulfates in water at 25°C.

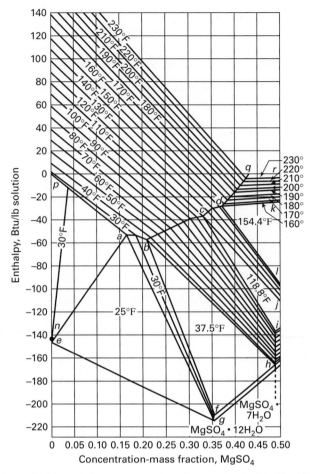

Figure 17.10 Enthalpy-concentration diagram for the $MgSO_4$–H_2O system at 1 atm.

The heats of solution at infinite dilution in kJ/mol of a compound at 18°C for hydrates of $MgSO_4$ clearly show this:

$MgSO_4$	-88.3
$MgSO_4 \cdot H_2O$	-58.6
$MgSO_4 \cdot 6H_2O$	-2.3
$MgSO_4 \cdot 7H_2O$	$+13.3$

Heats of a solution for a number of hydrated and anhydrous compounds are listed in Table 17.5.

As crystals continue to dissolve in a solvent, the heat of solution, which is now referred to as an integral heat of solution, varies, as shown in Figure 17.9 as a function of concentration. The integral heat of solution at saturation is numerically equal—but opposite in sign—to the heat of crystallization. The difference between the integral heat of solution at saturation and the heat of solution at infinite dilution is the heat of dilution:

$$\Delta H_{sol}^{sat} - \Delta H_{sol}^{\infty} = \Delta H_{dil}$$

with

$$\Delta H_{sol}^{sat} = -\Delta H_{crys}$$

As indicated in Figure 17.9, heats of dilution are relatively small; therefore, it is common to use

$$\Delta H_{crys} \approx -\Delta H_{sol}^{\infty}$$

Energy balances around a crystallizer are complex because they involve not only integral heats of solution and/or heat of crystallization, but also solute- and solvent-specific heats and solvent heat of vaporization. However, calculations are readily made if an enthalpy-mass fraction diagram (§7.6) is available, including solubility and phase-equilibria data. Mullin [1] lists 11 aqueous binary systems for which such diagrams have been constructed. For $MgSO_4$–H_2O, a diagram is given in Figure 17.10. Enthalpy datums are pure liquid water at 32°F (consistent with the steam tables in American Engineering Units) at Point p and solid $MgSO_4$ at 32°F (not shown in Figure 17.10).

Points a to l, n, p, and q in an enthalpy-mass fraction plot, Figure 17.10, correspond to the same points in Figure 17.2. In Figure 17.10, the isotherms in the region above Curve $pabcdq$ pertain to enthalpies of unsaturated solutions of $MgSO_4$. Straightness of these isotherms indicates that the heat of dilution is almost negligible. In this solids-free region, a 30 wt% aqueous solution of $MgSO_4$ at 110°F has a specific enthalpy of -31 Btu/lb solution.

In the region below the solubility curve $pabcdq$, in both Figures 17.2 and 17.10, the following phases exist:

Region	Temperature Range, °F	Phases
pae	25–32	ice and aqueous solution of $MgSO_4$
ea	25	ice and eutectic mixture
ag	25	eutectic and $MgSO_4 \cdot 12H_2O$
$abfg$	25–37.5	saturated solution and $MgSO_4 \cdot 12H_2O$
$bcih$	37.5–118.8	saturated solution and $MgSO_4 \cdot 7H_2O$
$cdlj$	118.8–154.4	saturated solution and $MgSO_4 \cdot 6H_2O$
$dqrk$	154.4–	saturated solution and $MgSO_4 \cdot H_2O$

Pure ice exists at Point e, where in Figure 17.10, the specific enthalpy is -147 Btu/lb, which is the heat of crystallization of water at 32°F. If a 30 wt% aqueous solution of $MgSO_4$

is cooled from 110°F to 70°F, equilibrium magma consists of a saturated solution of 26 wt% $MgSO_4$ and crystals of $MgSO_4 \cdot 7H_2O$ (49 wt% $MgSO_4$), as determined from the ends of the 70°F tie line that extends from solubility curve bc to $MgSO_4 \cdot 7H_2O$ solid line ih. The relative amounts of the two equilibrium phases can be computed from a $MgSO_4$ balance. For a basis of 100 lb of mixture, let S be pounds of crystals and A pounds of saturated aqueous solution. The $MgSO_4$ balance is thus

$$0.30(100) = 0.49\,S + 0.26\,A$$

where

$$100 = S + A$$

Solving, $S = 17.4$ lb and $A = 82.6$ lb. Mixture enthalpy at 70°F is −65 Btu/lb, which is equivalent to enthalpies of −46 and −155 Btu/lb, respectively, for the solution and crystals.

EXAMPLE 17.5 Use of an Enthalpy-Concentration Diagram.

For Example 17.3, calculate the Btu/h of heat addition.

Solution

An overall energy balance around the crystallizer gives

$$m_{feed}H_{feed} + Q_{in} = m_{vapor}H_{vapor} + m_{liquid}H_{liquid} + m_{crystals}H_{crystals}$$
$$\tag{1}$$

where liquid refers to the saturated-liquid portion of the magma.
From Example 17.3, feed consists of two streams:

$$m_{feed1} = 4,466 \text{ lb/hr of } 37.75 \text{ wt\% } MgSO_4 \text{ at } 170°F$$
$$m_{feed2} = 9,860 \text{ lb/hr of } 28.0 \text{ wt\% } MgSO_4 \text{ at } 85°F$$

From Figure 17.10, $H_{feed1} = -20$ Btu/lb and $H_{feed2} = -43$ Btu/lb.
Therefore, $m_{feed}H_{feed} = 4,466(-20) + 9,860(-43) = -513,000$ Btu/h and $m_{vapor} = 594$ lb/h at 85°F and 0.58 psia.

Vapor enthalpy does not appear in Figure 17.10, but enthalpy tables for steam can be used since they are based on the same datum, i.e., liquid water at 32°F.

Therefore, $H_{vapor} = 1099$ Btu/lb from steam tables and $m_{vapor}H_{vapor} = 594(1099) = 653,000$ Btu/h.

Liquid plus crystals can be treated together as magma. From the solution to Example 17.3,

$$m_{liquid} + m_{crystals} = m_{magma} = 10,876 + 2,856$$
$$= 13,732 \text{ lb/h of } 32.4 \text{ wt\% } MgSO_4 \text{ at } 85°F$$

From Figure 17.10,

$$H_{magma} = -67 \text{ Btu/lb and } m_{magma}H_{magma}$$
$$= 13,732(-67) = -920,000 \text{ Btu/lb}$$

From (17-1),

$$Q_{in} = 653,000 - 920,000 - (-513,000) = 246,000 \text{ Btu/h}$$

In the absence of an enthalpy-mass fraction diagram, a reasonable energy balance can be made if data for heat of crystallization and specific heats of solutions are available or can be estimated and if heat of dilution is neglected, as shown in the next example.

EXAMPLE 17.6 Heat Removal for a Crystallizer.

Feed to a cooling crystallizer is 1,000 lb/h of 32.5 wt% $MgSO_4$ in water at 120°F. This solution is cooled to 70°F to form crystals of heptahydrate. Estimate the heat removal rate in Btu/h.

Solution

Material balance

From Figure 17.2, feed at 120°F contains no crystals, but magma at 70°F consists of heptahydrate crystals and mother liquor of 26 wt% $MgSO_4$. By material balance in the manner of Example 17.3, the following results are obtained:

	lb/hr		
	Feed	Mother Liquor	Crystals
H_2O	675	530	0
$MgSO_4$	325	186	0
$Mg_2SO_4 \cdot 7H_2O$	0	0	284
	1,000	716	284

Take a thermodynamic path consisting of cooling the feed from 120°F to 70°F followed by crystallization at 70°F. From Hougen et al. [3], the specific heat of the feed is constant over the temperature range at 0.72 Btu/lb-°F. Heat removed to cool the feed to 70°F is $1,000(0.72)(120 - 70) = 36,000$ Btu/h.

The heat of crystallization can be taken as the negative of the heat of solution at infinite dilution: −13.3 kJ/mole of heptahydrate or −23.2 Btu/lb of heptahydrate.

Therefore, heat removed during heptahydrate crystallization is $284(23.2) = 6,600$ Btu/h. Total heat removal is $36,000 + 6,600 = 42,600$ Btu/h.

If this example is solved using Figure 17.10 in the manner of Example 17.5, the result is 44,900 Btu/h, which is 5.4% higher.

§17.3 KINETICS AND MASS TRANSFER

Crystallization is a complex phenomenon that involves three steps: (1) *nucleation*, (2) mass transfer of solute to the crystal surface, and (3) incorporation of solute into the crystal lattice. Collectively, these three are referred to as *crystallization kinetics*. Experiments have shown that the driving force for all three steps is *supersaturation*.

§17.3.1 Supersaturation

Solubility properties discussed in the previous section refer to relatively large crystals that can be seen by the naked eye, i.e., larger than ∼20 μm in diameter. As crystal size decreases, solubility noticeably increases, making it possible to supersaturate a solution if it is cooled slowly without agitation. This phenomenon, based on the work of Miers and Isaac [4] in 1907, is represented in Figure 17.11, where the normal solubility curve c_s is represented as a solid line. The solubility of very small crystals can fall in the metastable region, which is shown to have a metastable limiting solubility, c_m, given by the dashed line.

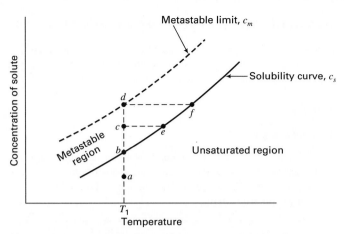

Figure 17.11 Representative solubility–supersolubility diagram.

Consider a solution at temperature T_1, the vertical dashed line in Figure 17.11. If the concentration is given by Point a, the solution is undersaturated and crystals of all sizes dissolve. At Point b, equilibrium exists between a saturated solution and crystals that can be seen by the naked eye. In the metastable region at Point c, crystals can grow but cannot nucleate. If no crystals are present, none can form. Clouds at high altitude are a good example of this. They contain tiny droplets of water, which, in the absence of seed crystals, do not form ice despite being at a temperature below the freezing point. For the concentration at Point c, the difference between the temperature at Point e on the solubility curve and Point c in the metastable region is the supersaturation temperature difference, which may be about 2°F. The *supersaturation*, $\Delta c = c - c_s$, is the difference in concentration between Points c and b.

At Point d, *spontaneous nucleation* of very small crystals, those that are invisible to the naked eye, occurs. The difference in temperature between Points f and d is the limit of the supersaturation temperature difference. The *limiting supersaturation* is $\Delta c_{\text{limit}} = c_m - c_s$.

The relationship between solubility and crystal size is given quantitatively by the Kelvin equation:

$$\ln\left(\frac{c}{c_s}\right) = \left(\frac{4v_s\sigma_{s,L}}{vRTD_p}\right) \qquad (17\text{-}16)$$

where v_s = molar volume of the crystals; $\sigma_{s,L}$ = interfacial tension; v = number of ions/molecule of solute; and c/c_s = *supersaturation ratio* = S.

Measured values of interfacial tension (also called surface energy) range from as low as 0.001 J/m² for very soluble compounds to 0.170 for compounds of low solubility. Equation (17-16), in a more general form, also describes the effect of droplet diameter on vapor pressure and solubility in another liquid phase, based on studies by Gibbs, Thompson, and Ostwald.

It is common to define a *relative supersaturation*, s, by

$$s = \frac{c - c_s}{c_s} = \frac{c}{c_s} - 1 = S - 1 \qquad (17\text{-}17)$$

In practice, s is usually less than 0.02 or 2%. For such small values, $\ln(c/c_s)$ in (17-16) can be approximated by s with no more than a 1% error.

EXAMPLE 17.7   Effect of Crystal Size on Solubility.

Determine the effect of crystal diameter on the solubility of KCl in water at 25°C.

Solution

From Table 17.5, by interpolation, $c_s = 35.5$ g/100 g H_2O.
Because KCl dissociates into K^+ and Cl^-, $v = 2$.
MW of KCl = 74.6; density of KCl crystals = 1980 kg/m³.

$$v_s = \frac{74.6}{1980} = 0.037 \text{ m}^3/\text{kmol}$$

$$T = 298 \text{ K}$$

$$R = 8314 \text{ J/kmol-k}$$

From Mullin [1], page 200, $\sigma_{s,L} = 0.028$ J/m².
From (17-16),

$$c = c_s \exp\left(\frac{4v_s\sigma_{s,L}}{vRT\,D_p}\right)$$

$$= 35.5 \exp\left[\frac{4(0.0376)(0.028)}{2(8315)(298)D_p}\right] \text{ for } D_p \text{ in m, or} \qquad (1)$$

$$= 35.5 \exp(0.00085/D_p, \mu\text{m})$$

The following results of calculations with (1) show only a small effect of crystal diameter above 0.10 μm.

D_p, μm	c/c_s	c, g KCl/100 g H_2O
0.01	1.0887	38.65
0.10	1.0085	35.80
1.00	1.00085	35.53
10.00	1.000085	35.50
100.00	1.0000085	35.50

§17.3.2   Nucleation

To determine volume or residence time for magma in a crystallizer, the rate of *nucleation* (birth) of crystals and their rate of growth must be established. Relative rates of nucleation and growth are important because they determine crystal size and size distribution. Nucleation may be *primary* or *secondary*, depending on whether supersaturated solutions are free of crystalline surfaces or contain crystals, respectively. Primary nucleation requires high supersaturation and is the principal mechanism in precipitation. The theory is fully developed and also applies to condensation of liquid droplets from supersaturated vapor and formation of droplets of a second liquid from an initial liquid phase. However, secondary nucleation is the key in commercial crystallizers, where crystalline surfaces are present and large crystals are desired.

Primary Nucleation

Primary nucleation can be homogeneous or heterogeneous. The former occurs with supersaturated solutions in the absence of foreign matter, such as dust. Molecules in the solution first associate to form a *cluster*, which may dissociate or grow. If a cluster gets large enough to take on the appearance of a lattice structure, it becomes an *embryo*. Further growth can result in a stable crystalline *nucleus* whose size exceeds that given by (17-16) for the prevailing degree of supersaturation.

The rate of homogeneous nucleation is given by classical chemical kinetics in conjunction with (17-16), as discussed by Nielsen [5]. The resulting expression is

$$B^o = A \exp\left[\frac{-16\pi v_s^2 \sigma_{s,L}^3 N_a}{3v^2(RT)^3 \left[\ln\left(\dfrac{c}{c_s}\right)\right]^2}\right] \qquad (17\text{-}18)$$

where B^o = rate of homogeneous primary nucleation, number of nuclei/cm^3-s; A = frequency factor; and N_a = Avogadro's number = 6.022×10^{26} molecules/kmol.

Theoretically, $A = 10^{30}$ nuclei/cm^3-s; however, observed values are different due to the unavoidable presence of foreign matter. Thus, (17-18) can also be applied to heterogeneous primary nucleation, where A is determined experimentally and may be orders of magnitude different from the theoretical value, with a value of 10^{25} often quoted. The rate of primary nucleation is sensitive to the relative supersaturation ratio, S, as illustrated in the next example.

EXAMPLE 17.8   Effect of Supersaturation on Nucleation.

Using the data in Example 17.7, estimate the effects of the supersaturation ratio, S, on the primary homogeneous nucleation of KCl from an aqueous solution at 25°C for values of $S = c/c_s$ of 2.0, 1.5, and 1.1.

Solution

For $c/c_s = 2.0$, $\ln(c/c_s) = 0.693$. From (17-18), using data in Example 17.7,

$$B^o = 10^{30} \exp\left[\frac{-16(3.14)(0.0376)^2(0.028)^3(6.022 \times 10^{26})}{3(2)^2(8315)^3(298)^3(0.693)^2}\right]$$
$$= 2.23 \times 10^{25} \text{ nuclei/cm}^3\text{-s}$$

Calculations for the other values of c/c_s are obtained in the same manner, with the following results:

$S = c/c_s$	B^o, nuclei/cm^3-s
2.0	2.23×10^{25}
1.5	2.60×10^{16}
1.1	0

Since large values of S ($c/c_s > 1.02$) are essentially impossible for crystallization of solutes of moderate to high solubility (e.g., solutes listed in Tables 17.5 and 17.7), primary nucleation for these solutes never occurs. However, for relatively insoluble solutes (e.g., solutes listed in Table 17.6), large values of c/c_s can be generated rapidly from ionic reactions that cause rapid precipitation of very fine particles. If $A = 10^{25}$ is used, B^o is divided by 10^5.

Secondary Nucleation

Nucleation in industrial crystallizers occurs mainly by secondary nucleation caused by existing crystals in the supersaturated solution. It is initiated by: (1) fluid shear past crystal surfaces that sweeps away nuclei, (2) collisions of crystals with each other, and (3) collisions of crystals with metal surfaces such as the crystallizer vessel wall or agitator blades. The latter two mechanisms, which are referred to as *contact nucleation*, are most common since they happen at the low values of relative supersaturation, s, typical of industrial applications.

In the absence of a theory for the complex phenomena of secondary nucleation, an empirical power-law function, which correlates much of the experimental data, is used:

$$B = k_N s^b M_T^j N^r \qquad (17\text{-}19)$$

where B = rate of secondary nucleation, M_T = mass of crystals per volume of magma, and N = agitation rate (e.g., rpm of an impeller). The constants k_N, b, j, and r are determined from experiments on the system of interest, and will be discussed in §17.5.1 in the context of a crystallizer model.

§17.3.3   Crystal Growth

In 1897, Noyes and Whitney [6] presented a mass-transfer theory of crystal growth based on equilibrium at the crystal solution interface. They postulated

$$dm/dt = k_c A(c - c_s) \qquad (17\text{-}20)$$

where dm/dt = rate of mass deposited on the crystal surface, A = surface area of the crystal, k_c = mass-transfer coefficient, c = mass solute concentration in the bulk supersaturated solution, and c_s = solute mass concentration in the solution at saturation. Nernst [7] proposed the existence of a thin, stagnant film of solution adjacent to the crystal face through which solute molecular diffusion takes place. Thus, $k_c = D/\delta$, where D = diffusivity and δ = film thickness, the latter assumed to depend on velocity of the solution past the crystal as determined by the degree of agitation.

The theory of Noyes and Whitney was challenged by Miers [8], who showed experimentally that an aqueous solution in contact with crystals of sodium chlorate is not saturated at the crystal–solution interface but is supersaturated. This finding led to a two-step theory of crystal growth, referred to as the diffusion–reaction theory, as described by Valeton [9]. Mass transfer of solute from the bulk of the solution to the crystal–solution interface occurs in the first step, as given by a modification of (17-20):

$$dm/dt = k_c A(c - c_i) \qquad (17\text{-}21)$$

where c_i is the supersaturated concentration at the interface. In the second step, a first-order reaction is assumed to occur at the crystal–solution interface, in which solute molecules are integrated into the crystal-lattice structure. Thus, for this kinetic step,

$$dm/dt = k_i A(c_i - c_s) \qquad (17\text{-}22)$$

If (17-21) and (17-22) are combined to eliminate c_i,

$$\boxed{dm/dt = \frac{A(c - c_s)}{1/k_c + 1/k_i}} \qquad (17\text{-}23a)$$

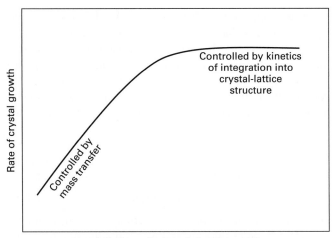

Figure 17.12 Effect of solution velocity past crystal on the rate of crystal growth.

Typically, k_c depends on the velocity of the solution relative to the crystal, as shown in Figure 17.12. At low velocities, growth rate may be controlled by the first step (mass transfer). The second step (kinetics) is important when solution velocity past the crystal surface is high, with the result that k_c is large compared to k_i. In adsorption, the kinetic step is rarely important. It is also unimportant in dissolution, the reverse of crystallization.

The mass-transfer coefficient, k_c, for the first step is independent of the crystallization process and can be estimated from fluid–solid particle mass-transfer-coefficient correlations described in §3.5.2 and §15.3.1. The kinetic coefficient, k_i, is unique to the crystallization process. Theories advanced to model the kinetic step are discussed in Myerson [10]. The theory of Burton, Cabrera, and Frank [11], based on a growth spiral starting from a screw dislocation, shown in Figure 17.13, has been verified in experimental studies using scanning-electron microscopy. A dislocation is an imperfection in the crystal structure. The screw-dislocation theory predicts a growth rate proportional to $(c_i - c_s)^2$ at low supersaturation and to $(c_i - c_s)$ at high supersaturation. Unfortunately, the theory does not predict k_i. Accordingly, (17-23a) is used with k_c estimated from correlations and k_i back-calculated from experimental data.

Although crystals do not grow as spheres, an equation can be derived for the diameter increase of a spherical crystal.

Rewriting (17-23a) with an overall coefficient,

$$dm/dt = K_c A(c - c_s) \qquad (17\text{-}23b)$$

Since $A = \pi D_p^2$ and $m = \dfrac{\pi D_p^3}{6}\rho$, (17-23b) becomes

$$\frac{dD_p}{dt} = \frac{2K_c(c - c_s)}{\rho} = \frac{2K_c(\Delta c)}{\rho} \qquad (17\text{-}24)$$

If growth rate is controlled by k_i, which is assumed to be independent of D_p, then

$$\frac{\Delta D_p}{\Delta t} = \frac{2k_i \Delta c}{\rho} \qquad (17\text{-}25)$$

and the crystal-size increase is linear in time for a constant supersaturation. If growth rate is controlled by k_c at a low velocity, then, from (15-60),

$$K_c = k_c = 2\mathcal{D}/D_p \qquad (17\text{-}26)$$

where $\mathcal{D}$ is solute diffusivity. Substitution of (17-26) into (17-24) gives

$$\frac{dD_p}{dt} = \frac{4\mathcal{D}(\Delta c)}{D_p \rho} \qquad (17\text{-}27)$$

Integrating from D_{p_0} to D_p,

$$\frac{D_p^2 - D_{p_0}^2}{2} = \frac{4\mathcal{D}(\Delta c)}{\rho} t \qquad (17\text{-}28)$$

If $D_{p_0} \ll D_p$, (17-28) reduces to

$$D_p = \left(\frac{8\mathcal{D}(\Delta c)t}{\rho}\right)^{1/2} \qquad (17\text{-}29)$$

In this case, the increase in crystal diameter slows with time.

At higher solution velocities where k_c still controls, use of (15-62) results in

$$K_c = k_c = C_1/D_p^{1/2} \qquad (17\text{-}30)$$

For this case, the increase in crystal diameter also slows with time, but not as rapidly as predicted by (17-29). It is common to assume that crystal growth rates are controlled by k_i. Hence, they are not dependent on crystal size and are invariant with time. This assumption is made in §17.5.1, where a useful crystallizer model is constructed.

EXAMPLE 17.9 Use of Seed Crystals.

The heptahydrate of $MgSO_4$ is to be crystallized batchwise from a seeded aqueous solution. Low supersaturation is to be used to

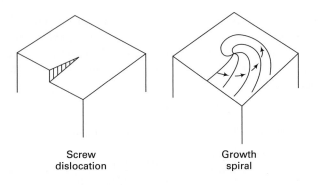

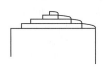

Screw dislocation

Growth spiral

Resulting crystal growth

Figure 17.13 Screw-dislocation mechanism of crystal growth.

avoid primary nucleation and mild agitation is to be used to minimize secondary nucleation. Temperature will be maintained at 35°C, at which the solubility of $MgSO_4$ in water is 30 wt%. The crystallizer will be charged with 3,000 lb of a saturated solution at 35°C. To this solution will be added 2 lb of heptahydrate seed crystals of 50 μm in diameter. A supersaturation of 0.01 gm of heptahydrate per gm of solution at 35°C will be maintained during crystallization by operating the crystallizer at vacuum and using heat exchange and the heat of crystallization to gradually evaporate water. Based on the assumptions and data listed below, determine the following, if the final crystal size is to be 400 μm: (a) lb of heptahydrate crystals, (b) number of crystals, (c) lb of water evaporated, (d) product magma density in lb of crystals per lb of solution, (e) crystallizer volume in gallons if the volume occupied by the magma during operation is at most 50% of the crystallizer volume, (f) crystallizer pressure in psia and the boiling-point elevation in °F, (g) time in minutes to grow the crystals to the final size, (h) amount of heat transfer in Btu, and (i) whether crystal growth is controlled by mass transfer, by surface reaction (incorporation into the lattice), or both.

Assumptions and Data:

1. No primary or secondary nucleation.

2. Properties of aqueous 30 wt% $MgSO_4$ at 35°C: density = 1.34 g/cm^3; viscosity = 8 cP; and diffusivity of $MgSO_4$ = 1.10×10^{-5} cm^2/s.

3. Density of the crystals = 1.68 g/cm^3.

4. Crystal shape can be approximated as a sphere.

5. Average crystal-face growth rate, including effects of both mass transfer and surface reaction, = 0.005 mm/minute.

6. Solution velocity past crystal face = 5 cm/s.

Solution

Molecular weights: $MgSO_4$ 120.5; $MgSO_4 \cdot 7H_2O$ 246.6; H_2O 18.02

Initial charge: 900 lb aq $MgSO_4$ + 2,100 lb H_2O = 3,000 lb and 2 lb $MgSO_4 \cdot 7H_2O$ crystal seeds

(a) Crystals grow from 50 μm to 400 μm in diameter:

$$\text{yield} = 2 \left(\frac{400}{50} \right)^3 = 1,024 \text{ lb } MgSO_4 \cdot 7H_2O$$

(b) Number of crystals, based on the crystal seeds,

$$
= \frac{\text{mass of seeds}}{\text{mass/seed}}
$$

$$
= \frac{2(454)}{1.68 \left(\frac{3.14}{6} \right) (50 \times 10^{-4})^3}
$$

$$
= 8.26 \times 10^9, \text{where mass/seed} = \rho_p V_p = \rho \frac{\pi D_p^3}{6}
$$

(c) Pounds of heptahydrate crystallized = 1,024 − 2 = 1,022 lb

$$H_2O \text{ of hydration} = 1,022 \left(\frac{246.6 - 120.5}{246.6} \right) = 523 \text{ lb}$$

$MgSO_4$ in crystals = 1,022 − 523 = 499 lb

$MgSO_4$ left in solution = 900 − 499 = 401

$$H_2O \text{ left in solution} = 401 \left(\frac{70}{30} \right) = 936 \text{ lb}$$

H_2O evaporated = 2,100 − 523 − 936 = 641 lb

(d) Final mother liquor = 936 + 401 = 1,337 lb

$$\text{Magma density} = \left(\frac{1,024}{1,337} \right) = 0.766 \left(\frac{\text{lb crystals}}{\text{lb mother liquor}} \right)$$

(e) Initially, using the factor of 8.33 lb/gal for 1.0 g/cm^3,

$$\text{Solution volume} = \frac{3,000}{8.33(1.34)} = 269 \text{ gal}$$

Finally,

$$\text{Solution volume} = \frac{1,337}{8.33(1.34)} = 120 \text{ gal}$$

$$\text{Crystal volume} = \frac{1,024}{8.33(1.68)} = 73 \text{ gal}$$

$$\text{Total volume} = 193 \text{ gal}$$

Therefore, initial conditions before crystallization will control the volume.

$$\text{Make crystallizer volume} = \frac{269}{0.5} = 538 \text{ gal}$$

(f) Calculate H_2O partial pressure by Raoult's law applied to mother liquor:

$$\text{lbmol } MgSO_4 = \frac{401}{120.5} = 3.33$$

$$\text{lbmol } H_2O = \frac{936}{18.02} = 51.95$$

$$\text{Total} = 55.28 \text{ lbmol}$$

$$x_{H_2O} = \frac{51.95}{55.28} = 0.94$$

At 35°C = 95°F, $P_{H_2O}^s = 0.8153$ psia.

Therefore, $p_{H_2O} = x_{H_2O} P_{H_2O}^s = 0.94(0.8153) = 0.766$ psia. This corresponds to a saturation temperature of 93°F. Therefore, boiling-point elevation = 2°F.

(g) Growth rate = 0.005 mm/minute. Therefore, diameter grows at 0.01 mm/minute. Must grow from 50 μm = 0.05-mm to 0.40-mm diameter.

$$\text{Time} = \frac{0.40 - 0.05}{0.01} = 35 \text{ min}$$

(h) Enthalpy balance:

$$FH_F + Q = VH_V + MH_M \tag{1}$$

Assume charge is at 35°C = 95°F; then, from Figure 17.10, $H_F = -40$ Btu/lb.

From the steam tables, $H_V = 1,102.2 + 0.9$ (for the boiling-point elevation) = 1,103.1 Btu/lb.

$$\text{Wt\% } MgSO_4 \text{ in magma} = \frac{(499 + 401 + 1)}{(3,000 + 2 - 641)} = \frac{901}{2,361} = 0.38$$

From Figure 17.10, $H_M = -90$ Btu/lb. From (17-1),

$$Q = 641(1,103.1) + 2,361(-90) - 3,002(-40)$$

$$= 615,000 \text{ Btu} = \text{heat transferred to crystallizer}$$

(i) Assume mass-transfer-controlled, using molar amount of crystals, n, and molar concentrations, c. The molar form of (17-21) is

$$\frac{dn}{dt} = 4\pi r^2 k_c (\Delta c), \tag{2}$$

where r is crystal radius,

$$n = \frac{4}{3} \pi r^3 \rho_M, \text{ using a molar density.}$$

Therefore,
$$\frac{dn}{dt} = 4\pi r^2 \rho_M \frac{dr}{dt} \qquad (3)$$

Equating (2) and (3),

$$4\pi r^2 k_c(\Delta c) = 4\pi r^2 \rho_M \frac{dr}{dt}$$

$$\frac{dr}{dt} = \frac{k_c \Delta c}{\rho_M} \qquad (4)$$

From (15-62),

$$N_{Sh} = \frac{k_c D_p}{\mathcal{D}} = \frac{2k_c r}{\mathcal{D}} = 2 + 0.6(N_{Re})^{1/2}(N_{Sc})^{1/3}$$

For $r = \dfrac{50}{2} = 25\ \mu m$,

$$N_{Re} = \frac{D_p \upsilon \rho}{\mu} = \frac{(50 \times 10^{-6})(100)(5)(1.34)}{0.08} = 0.42$$

$$N_{Sc} = \frac{\mu}{\rho \mathcal{D}} = \frac{0.08}{1.34(1.10 \times 10^{-5})} = 5,430$$

$$N_{Sh} = 2 + 0.6(0.42)^{1/2}(5,430)^{1/3}$$

$$= 2 + 6.8 = 8.8 = \frac{2k_c r}{\mathcal{D}}$$

$$k_c = \frac{(8.8)(1.1 \times 10^{-5})}{2(25 \times 10^{-4})} = 0.019\ \text{cm/s}$$

For $r = \dfrac{400}{2} = 200\ \mu m$,

$$N_{Re} = 0.42\left(\frac{400}{50}\right) = 3.36$$

$$N_{Sh} = 2 + 0.6(3.36)^{1/2}(5,430)^{1/3}$$

$$= 2 + 19.3 = 21.3$$

$$k_c = \frac{(21.3)(1.1 \times 10^{-5})}{2(200 \times 10^{-4})} = 0.006\ \text{cm/s}$$

Thus, k_c changes by a factor of 3 as crystal size changes from 25 to 200 μm.

Assume a Δc based on the given supersaturation of 0.01 g crystal per g solution.

$$\Delta c = \frac{0.01(1.34)}{246.6} = 54.3 \times 10^{-6}\ \frac{\text{mol}}{\text{cm}^3}$$

$$\rho_M \text{ of crystals} = \frac{1.68}{246.6} = 6.81 \times 10^{-3}\ \frac{\text{mol}}{\text{cm}^3}$$

From (4),
$$\frac{dr}{dt} = k_c \frac{54.3 \times 10^{-6}}{6.81 \times 10^{-3}} = 0.008\ k_c$$

For the lowest value of $k_c = 0.006$ cm/s, $\dfrac{dr}{dt} = 0.008(0.006) = 48 \times 10^{-6}$ cm/s, or $480 \times 10^{-6}(60) = 0.029$ mm/minute.

This is six times faster than the growth rate; therefore, crystal growth is largely controlled by surface reaction.

§17.4 EQUIPMENT FOR SOLUTION CRYSTALLIZATION

Before developing a mathematical crystallizer model in §17.5, it is useful to consider the equipment for solution crystallization listed in Table 17.8 and classified by operating mode, method of achieving supersaturation, and features. Although equipment for both batch and continuous operation is listed, the latter is generally preferred. The choice of method for achieving supersaturation depends on the effect of temperature on solubility. From Table 17.5, it is seen that

Table 17.8 Classification of Equipment for Solution Crystallization

Operation Modes	Methods for Achieving Supersaturation	Crystallizer Features for Achieving Desired Crystal Growth
Batch Continuous	Cooling Evaporation	Agitated or nonagitated Baffled or unbaffled Circulating liquor or circulating magma Classifying or nonclassifying Controlled or uncontrolled Cooling jacket or cooling coils

for many inorganic compounds in the near-ambient temperature range, e.g., 10–40°C, the change in solubility is small and insufficient to utilize the cooling method. This is certainly the case for $MgCl_2$ and $NaCl$. For KNO_3 and Na_2SO_4, crystallization by cooling may be feasible. The majority of industrial crystallizers use the evaporation method or a combination of cooling and evaporation. Direct-contact cooling with agitation and evaporation can be achieved by bubbling air through the magma.

To produce crystals of a desired size distribution, attention must be paid to selection of the design features in Table 17.8. Use of mechanical agitation can result in smaller and more uniformly sized crystals of a higher purity that are produced in less time. The use of vertical baffles promotes more uniform mixing. Supersaturation and uniformity can be controlled by circulation between a crystallizing zone and a supersaturation zone. In a circulating-liquor design, only the mother liquor is circulated, while in the circulating-magma design, the mother liquor and crystals are circulated together. Circulation may be limited to the crystallizer vessel or may include pumping through an external heat exchanger. In a classifying crystallizer, the smaller crystals are separated from the larger and retained in the crystallizing zone for further growth or are removed from the zone and redissolved. In a controlled design, one or more techniques are used to control the degree of supersaturation to avoid undesirable nucleation. Cooling crystallizers may use a vessel jacket or internal coils, with the former preferred because of the ease of wiping the crystals off the cooling surface.

Many of the patented commercial crystallizer designs are described in Myerson [10] and Mullin [1]. Only four solution crystallizers are described here because these designs suffice to illustrate features found in other designs.

§17.4.1 Circulating-Batch Crystallizers

Although batch crystallizers can be operated without agitation or circulation by simply charging a hot solution to an open vessel and allowing the solution to stand as it cools by natural convection, the resulting crystals may be undesirably large, interlocked, impure because of entrapment of mother liquor, and difficult to remove from the vessel. It is thus

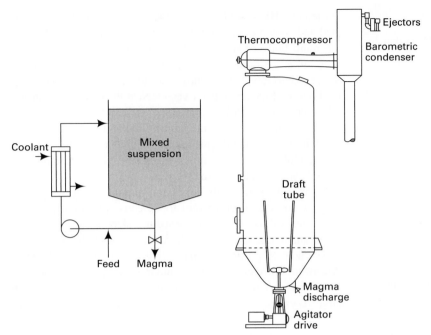

(a) Circulation of magma through an external, cooling heat exchanger

(b) Internal circulation with a draft tube

Figure 17.14 Circulating, batch-cooling crystallizers.

preferable to use a more elaborate configuration similar to either of the two batch crystallizers shown in Figure 17.14. In the design with external circulation, a high magma velocity is used through the tubes of the heat exchanger to obtain a reasonable heat-transfer rate with a small temperature-driving force and minimal crystal formation on the tubes. This design is satisfactory for continuous crystallization, when the solubility-temperature curve dictates cooling crystallization.

In Figure 17.14b, crystallization is accomplished by evaporation under a vacuum pulled by a two-stage, steam-jet-ejector system through a water-cooled condenser or by a mechanical vacuum system. The magma is circulated internally through a draft tube by a propeller. Energy for evaporation is supplied by the hot feed. A typical cycle, including charging the feed, crystallization, and removal of the magma, is 2 to 8 h.

§17.4.2 Continuous Cooling Crystallizers

Figure 17.15 is a schematic diagram of a typical scraped-surface crystallizer known as the Swenson–Walker continuous cooling crystallizer, described in detail by Seavoy and Caldwell [12]. The feed flows through a semicylindrical trough, typically 1 m wide × 3–12 m long. The trough has a water-cooled jacket and is provided with a low-speed (3–10 rpm), helical agitator-conveyor that scrapes the wall, prevents growth of crystals on the trough wall, and promotes crystal growth by gentle agitation. Standard-size units can be linked together. The crystallization process is controlled by the rate of heat transfer, with the major resistance due to the magma on the inside. Overall heat-transfer coefficients of only 10–25 Btu/h-ft²-°F (57–142 W/m²-K) are typically observed, based on a ΔT_{LM} between the magma and the coolant. Production rates of up to 20 tons per day of crystals of $Na_3PO_4 \cdot 12\,H_2O$ and $Na_2SO_4 \cdot 10\,H_2O$ of moderate size and uniformity have

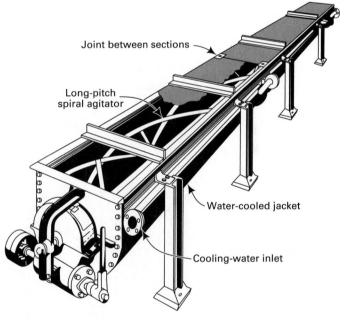

Figure 17.15 Swenson–Walker continuous cooling crystallizer.

been achieved. Both salts show a significant decrease in solubility with temperature, making the cooling crystallizer viable.

EXAMPLE 17.10 Sizing a Cooling Crystallizer.

The cooling crystallizer of Example 17.6 is to be a scraped-surface unit with 3 ft² of cooling surface per foot of running length of crystallizer. Cooling will be provided by a countercurrent flow of chilled water entering the cooling jacket at 60°F and leaving at 85°F. The overall heat-transfer coefficient, U, is expected to be 20 Btu/hr-ft²-°F. What length of crystallizer is needed?

Solution

From Example 17.6, using Figure 17.10, the required heat-transfer rate is 44,900 Btu/h. The log mean temperature driving force is

$$\Delta T_{LM} = \frac{(120 - 85) - (70 - 60)}{\ln\left(\dfrac{120 - 85}{70 - 60}\right)} = 20°F$$

The area for heat transfer is

$$A = \frac{Q}{U \Delta T_{LM}} = \frac{44,900}{20(20)} = 112 \text{ ft}^2$$

The crystallizer length = 112/3 = 37 ft (11.3 m).

§17.4.3 Continuous Vacuum Evaporating Crystallizers

A particularly successful and widely used design for continuous vacuum evaporating crystallizers is the Swenson draft-tube, baffled (DTB) crystallizer, described by Newman and Bennett [13] and shown in one of several variations in Figure 17.16. In the main body of the crystallizer, evaporation occurs, under vacuum, at the boiling surface, which is located several inches above a draft tube that extends down to within several inches of the bottom of the crystallizer vessel. Near the bottom and inside of the draft tube is a low-rpm propeller that directs the magma upward through the draft tube toward the boiling surface under conditions of a small degree of supercooling and in the absence of any violent flashing action. Thus, nucleation and buildup of crystals on the walls are minimized.

Surrounding the draft tube is an annular space where the magma flows back downward for re-entry into the draft tube. The outer wall of the annular space is a skirt baffle, surrounded by an annular settling zone, whose outer wall is the wall of the crystallizer. A portion (perhaps 10%) of the magma flowing downward through the first annular space turns around and flows outward and upward through the settling zone where larger crystals can settle, leaving a mother liquor containing

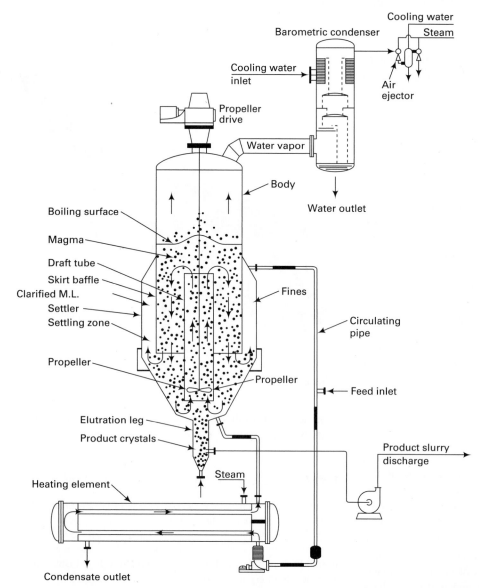

Figure 17.16 Swenson draft-tube, baffled (DTB) crystallizer.

fine crystals that flows to a circulating pipe, where it is joined by the feed and flows upward through a pump and then a heat exchanger. The circulating solution is heated several degrees to provide energy for feed preheat and subsequent evaporation, and to dissolve finer crystals.

Circulating magma re-enters the main body of the crystallizer just below the bottom of the draft tube. Further classification of crystals by size can be accomplished by providing an elutriation leg, as shown in Figure 17.16 at the bottom of the main body of the crystallizer. In that case, product magma is withdrawn through a pipe from a nozzle located near the bottom of the elutriation leg, where the largest crystals are present. Otherwise, the product magma may be withdrawn from the lower part of the annular region surrounding the draft tube.

§17.5 THE MSMPR CRYSTALLIZATION MODEL

Because of the popularity of the DTB crystallizer, a mathematical model, due to Randolph [14] for its design and analysis, is useful and is found in process simulators. It is referred to as the *Mixed-Suspension, Mixed-Product-Removal* (MSMPR) model and is based on the following assumptions: (1) continuous, steady-flow, steady-state operation; (2) perfect mixing of the magma; (3) no classification of crystals; (4) uniform degree of supersaturation of the magma; (5) crystal growth rate independent of crystal size; (6) no crystals in the feed, but seeds are added initially; (7) no crystal breakage; (8) uniform temperature; (9) mother liquor in product magma in equilibrium with the crystals; (10) nucleation rate is constant, uniform, and due to secondary nucleation by crystal contact; (11) crystal-size distribution (CSD) is uniform in the crystallizer and equal to that in the magma; and (12) all crystals have the same shape.

Modifications to the model to account for classification of crystals due to settling, elutriation, and dissolving of fines, and variable growth rate are discussed by Randolph and Larson [15]. The core of the MSMPR model is the estimation, by a crystal-population balance, of the crystal-size distribution (CSD), which is determined by the rpm of the draft-tube propeller and external circulation rate. It is relatively easy to conduct experiments in a laboratory crystallizer that approximates the MSMPR model and can provide crystal nucleation rate and growth-rate data to design an industrial crystallizer.

§17.5.1 Crystal-Population Balance

The crystal-population balance accounts for all crystals in the magma and, with the mass balance, makes possible the determination of the CSD. Let L = characteristic crystal size (e.g., from a screen analysis), N = cumulative number of crystals of size L and smaller in the magma in the crystallizer, and V_{ML} = volume of the mother liquor in the crystallizer magma. A cumulative-numbers undersize plot based on these variables is shown in Figure 17.17, where the slope of the curve, n, at a given value of L is the number of crystals per unit size per unit volume,

$$n = \frac{d(N/V_{ML})}{dL} = \frac{1}{V_{ML}}\frac{dN}{dL} \qquad (17\text{-}31)$$

The limits of n, as shown in Figure 17.17, vary from n^o at $L = 0$ to 0 at $L = L_T$, the largest crystal size. In the MSMPR model, the cumulative plot of Figure 17.17 is independent of time and location in the magma. The plot is, in fact, the numbers-cumulative CSD for product-magma crystals.

For a constant, crystal-size growth rate independent of crystal size, let $G = dL/dt$, or

$$\boxed{\Delta L = G\Delta t} \qquad (17\text{-}32)$$

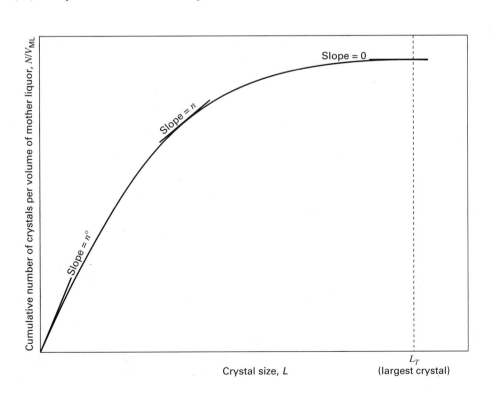

Figure 17.17 Typical cumulative-numbers undersize distribution.

and
$$L = Gt_L \qquad (17\text{-}33)$$

where t_L = residence time in the magma in the crystallizer for crystals of size L.

Equation (17-32) is the ΔL *law of McCabe* [16], who found that it correlated experimental data on the growth of crystals of KCl and $CuSO_4 \cdot 5H_2O$ surprisingly well. Although McCabe's experiments were conducted in a laboratory crystallizer under ideal conditions, his resulting ΔL law is applied to commercial crystallization even though conditions may not be ideal.

From (17-31), $dN = nV_{ML}dL$ = number of crystals in the size range dL. Now, $(\Delta n \, dL)$ crystals per unit volume in time increment Δt are withdrawn. Because of the perfect-mixing assumption for the magma,

$$\frac{\text{number of crystals withdrawn}}{\text{mother-liquor volume withdrawn}}$$
$$= \frac{\text{number of crystals}}{\text{mother-liquor volume in the crystallizer}}$$

Therefore,

$$\frac{\text{number of crystals withdrawn}}{\text{number of crystals in crystallizer}}$$
$$= \frac{\text{mother-liquor volume withdrawn}}{\text{mother-liquor volume in the crystallizer}}$$

or
$$\frac{\Delta n \, dL}{n \, dL} = -\frac{\Delta n}{n} = \frac{Q_{ML} \Delta t}{V_{ML}} \qquad (17\text{-}34)$$

where Q_{ML} = volumetric flow rate of mother liquor in the withdrawn product magma and V_{ML} = volume of mother liquor in the crystallizer. Combining (17-32) and (17-34) and taking the limit,

$$-\frac{dn}{dL} = \frac{Q_{ML}\, n}{GV_{ML}} \qquad (17\text{-}35)$$

which is a simplified version of the following more-general, transient-population balance equation that allows for crystals in the feed and a nonuniform growth rate, but assumes a constant volume of mother liquor in the crystallizer:

$$\frac{\partial n}{\partial t} + \frac{\partial (nG)}{\partial L} + \frac{(Q_{ML})_{out}\, n}{V_{ML}} - \frac{(Q_{ML})_{in}\, n_{in}}{V_{ML}} = 0 \qquad (17\text{-}36)$$

The retention time of mother liquor in the crystallizer is $\tau = V_{ML}/Q_{ML}$. Therefore, (17-35) can be rewritten as

$$-\frac{dn}{n} = \frac{dL}{G\tau} \qquad (17\text{-}37)$$

If (17-37) is integrated for a constant growth rate and residence time,

$$n = n^\circ \exp(-L/G\tau) \qquad (17\text{-}38)$$

This equation is the starting point for determining distribution curves for crystal population, crystal size or length, crystal surface area, and crystal volume or mass.

For example, to obtain crystal population, the number of crystals per unit volume of mother liquor below size L is

$$N/V_{ML} = \int_0^L n\, dL \qquad (17\text{-}39)$$

The number of crystals per unit volume of mother liquor is

$$N_T/V_{ML} = \int_0^\infty n\, dL \qquad (17\text{-}40)$$

Combining (17-38) to (17-40), the cumulative number of crystals of size smaller than L, as a fraction of the total, is

$$x_n = \frac{\int_0^L n^\circ e^{-L/G\tau}dL}{\int_0^\infty n^\circ e^{-L/G\tau}dL} = 1 - \exp(-L/G\tau) \qquad (17\text{-}41)$$

Or, if a dimensionless crystal size is defined as

$$z = L/G\tau \qquad (17\text{-}42)$$

then,
$$x_n = 1 - e^{-z} \qquad (17\text{-}43)$$

The plot of (17-43) in Figure 17.18a is referred to as the cumulative distribution or cumulative crystal population. For a given value of z, x_n is the fraction of crystals having a smaller z-value. A corresponding differential plot of dx_n/dz is given in Figure 17.18b, where from (17-43),

$$\frac{dx_n}{dz} = e^{-z} \qquad (17\text{-}44)$$

The differential plot gives the fraction of crystals in a given interval of z. At small values of z, the fraction is seen to be large, while at large values of z, the fraction is small. Figures 17.18a and b each show four different distribution plots. From statistics, (17-41) is one of a number of *moment equations*, which for a relation $n = f(z)$, are given by

$$x_k = \frac{\int_0^z nz^k dz}{\int_0^\infty nz^k dz} \qquad (17\text{-}45)$$

where k is the order of the moment. Thus, (17-43) is obtained by setting $k = 0$, which is the zeroth moment of the distribution. Results for this moment and the corresponding first (length or size), second (area), and third (volume or mass) moments are summarized in Table 17.9. Corresponding cumulative and differential plots are included in Figure 17.18.

Of particular interest in design and operation of a crystallizer is the predominant crystal size, L_{pd}, in terms of the volume or mass distribution. This size corresponds to the peak of the differential-mass distribution and is derived as follows. From Table 17.9,

$$dx_m/dz = (z^3/6)e^{-z} \qquad (17\text{-}46)$$

At the peak,
$$\frac{d\left(\dfrac{dx_m}{dz}\right)}{dz} = 0 = \frac{3z^2 e^{-z}}{6} - \frac{z^3 e^{-z}}{6} \qquad (17\text{-}47)$$

Solving (17-47) for z,
$$z = 3 = \frac{L}{G\tau} \qquad (17\text{-}48)$$

Therefore,
$$\boxed{L_{pd} = 3G\tau} \qquad (17\text{-}49)$$

Similar developments using the differential expressions in Table 17.9 show that the most populous sizes in terms of number of crystals, size of crystals, and surface area are 0, $G\tau$, and $2\,G\tau$, respectively. If L_{pd} is selected, (17-41) and

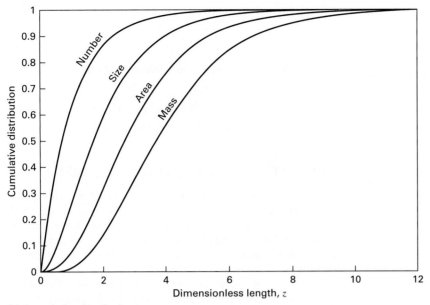

(a) Cumulative distributions

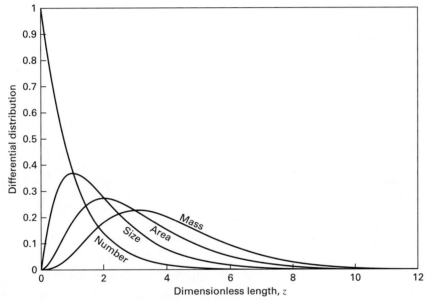

(b) Differential distributions

Figure 17.18 Crystal-size distributions predicted by the MSMPR model.

Table 17.9 Cumulative and Differential Plots for Moments of Crystal Distribution for Constant Growth Rate

Moment	Distribution Basis	Cumulative	Differential
Zeroth	Number	$x_n = 1 - e^{-z}$	$dx_n/dz = e^{-z}$
First	Size or length	$x_L = 1 - (1 + z)e^{-z}$	$dx_L/dz = ze^{-z}$
Second	Area	$x_a = 1 - \left(1 + z + \frac{z^2}{2}\right)e^{-z}$	$dx_a/dz = \frac{z^2}{2}e^{-z}$
Third	Volume or mass	$x_m = 1 - \left(1 + z + \frac{z^2}{2} + \frac{z^3}{6}\right)e^{-z}$	$dx_m/dz = \frac{z^3}{6}e^{-z}$

$z = L/G\tau$

(17-43) can be used to estimate cumulative and differential screen analyses.

To utilize distributions in Table 17.9, values of G and τ are needed. Growth rate depends on the supersaturation and degree of agitation, and residence time depends on crystallizer design and operation. It is also useful to know B^o and n^o, which are related as follows:

$$\frac{1}{V_{ML}}\frac{dN}{dt} = \frac{1}{V_{ML}}\frac{dN}{dL}\left(\frac{dL}{dt}\right)$$

$$\lim_{L\to 0}\frac{1}{V_{ML}}\frac{dN}{dt} = B^\circ$$

$$\frac{dL}{dt} = G \qquad (17\text{-}50)$$

$$\lim_{L\to 0}\frac{1}{V_{ML}}\frac{dN}{dL} = n^\circ$$

Therefore, $\qquad\qquad B^\circ = Gn^\circ \qquad (17\text{-}51)$

Combining (17-51) with (17-38),

$$n = \frac{B^\circ}{G}\exp(-L/G\tau) \qquad (17\text{-}52)$$

This equation is used with experimental data in the following example to obtain nucleation and growth rates for a set of operating conditions when the assumptions of the MSMPR model hold. The effect of operating conditions on B° are expressed as a power-law function of the form of (17-19). However, since the growth rate can be proportional to the relative supersaturation, s, raised to an exponent, (17-19) can be rewritten as

$$B^\circ = k_N' G^i M_T^j N^r \qquad (17\text{-}53)$$

Unfortunately, k_N' can be sensitive to the size of the equipment and is, therefore, best determined from data for a commercial crystallizer, as discussed by Zumstein and Rousseau [17]. The exponents i, j, and r can be determined from small-scale experiments.

The necessary nucleation rate for a crystallization operation is related to the predominant crystal size by the MSMPR model. From (17-40) and (17-38),

$$n_c = N_T/V_{ML} = \int_0^\infty n\,dL = n^\circ\tau G\int_0^\infty e^{-z}dz = n^\circ\tau G \qquad (17\text{-}54)$$

The mass of crystals per unit volume of mother liquor is

$$m_c = \int_v^\infty m_p n\,dL \qquad (17\text{-}55)$$

where m_p = mass of a particle, given by

$$m_p = f_v L^3 \rho_p \qquad (17\text{-}56)$$

where f_v is defined in (17-11).

Combining (17-55), (17-56), and (17-38), followed by integration, gives

$$m_c = 6f_v\rho_p n^\circ(G\tau)^4 \qquad (17\text{-}57)$$

Combining (17-54) and (17-57), the number of crystals per unit mass of crystals is

$$\frac{n_c}{m_c} = \frac{1}{6f_v\rho_p(G\tau)^3} \qquad (17\text{-}58)$$

Combining (17-48) with (17-58),

$$\frac{n_c}{m_c} = \frac{9}{2f_v\rho_p L_{pd}^3} \qquad (17\text{-}59)$$

or the corresponding required nucleation rate is

$$B^\circ = \frac{n_c C}{m_c V_{ML}} = \frac{9C}{2f_v\rho_p V_{ML} L_{pd}^3} \qquad (17\text{-}60)$$

where C = mass rate of production of crystals.

EXAMPLE 17.11 Analysis of a DTB Crystallizer.

A continuous vacuum evaporating crystallizer of the DTB type is used to produce 2,000 lb/hr of $Al_2(SO_4)_3\cdot 18\,H_2O$ ($\rho_p = 105\,lb/ft^3$). The magma contains 0.15 ft^3 crystals/ft^3 magma and the magma residence time in the crystallizer is 2 h. The desired L_{pd} on a mass basis is 0.417 mm. Estimate: (a) required crystal growth rate in ft/h, (b) necessary nucleation rate in nuclei/h-ft^3 of mother liquor, (c) number of crystals produced per hour, (d) tables and plots of estimated cumulative and differential screen analyses of the product crystals on a mass or volume basis. Also explain how the required growth and nucleation rates for the operating crystallizer might be achieved.

Solution

(a) From (17-48),

$$G = \frac{L_{pd}}{3\tau} = \frac{(0.417/304.8)}{3(2)} = 2.28\times 10^{-4}\text{ ft/h}$$

(b) Need volume of mother liquor in the crystallizer:

Volume of crystals produced $= \dfrac{2,000}{105} = 19.1\text{ ft}^3/\text{h}.$

Volume of crystals in crystallizer $= 19.1(2) = 38.2\text{ ft}^3.$

Volume of magma in crystallizer $= \dfrac{38.2}{0.15} = 255\text{ ft}^3.$

Volume of mother liquor in crystallizer $= V_{ML} = 255 - 38.2 = 217\text{ ft}^3.$

From (17-60), assuming $f_v = 0.5$,

$$B^\circ = \frac{9(2,000)}{2(0.5)(105)(217)(0.417/304.8)^3} = 3.1\times 10^8\,\frac{\text{nuclei}}{\text{h-ft}^3}$$

(c) Number of crystals produced $= 3.1\times 10^8\,(217) = 6.7\times 10^{10}/\text{h}.$

(d) $z = \dfrac{L}{G\tau} = \dfrac{L}{(2.28\times 10^{-4})(2)}$

$\qquad = 2.2\times 10^3 L$, ft or $7.19L$, mm

From Table 17.9,

$$x_m = 1 - \left(1 + z + \frac{z^2}{2} + \frac{z^3}{6}\right)e^{-z} \qquad (1)$$

$$\frac{dx_m}{dz} = \frac{z^3}{6}e^{-z} \qquad (2)$$

Using (1) and (2) with a spreadsheet for the older Tyler mesh sizes, the following results are obtained:

Tyler Mesh	Opening, mm	Dimensionless Length, z	Cumulative Screen Analysis, %	Differential Screen Analysis, %
8	2.357	16.96	100.00	0.00
10	1.667	11.99	99.77	0.18
14	1.179	8.48	96.95	2.11

20	0.833	5.99	84.82	8.95
28	0.589	4.24	61.16	18.31
35	0.417	3.00	35.20	22.40
48	0.295	2.12	16.50	19.05
65	0.208	1.50	6.54	12.53
100	0.147	1.06	2.29	6.87
150	0.104	0.75	0.73	3.31
200	0.074	0.53	0.22	1.46

A plot of the two screen analyses is shown in Figure 17.19.

Both growth and nucleation rates depend on supersaturation. Growth may also depend on the relative velocity between the crystals and the mother liquor. The nucleation rate depends upon the degree of agitation. With a DTB crystallizer, the agitator rpm and the magma circulation rate through the external heat exchanger can be adjusted to achieve the required growth and nucleation rates.

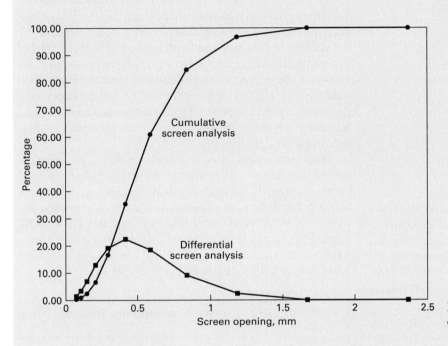

Figure 17.19 Predicted cumulative and differential screen analyses for Example 17.11.

§17.6 PRECIPITATION

In solution crystallization, as discussed in §17.4, a liquid containing a solute is cooled or partially evaporated, causing the solute to exceed its solubility sufficiently to partially crystallize. The degree of supersaturation is small, primary nucleation is avoided, and crystal growth is slow. As shown in Example 17.11, if the process is carried out under controlled conditions, crystals of a desirable size can be grown. In many respects, *precipitation* is the opposite of solution crystallization. As discussed by Nielsen [18], precipitation involves solutes that are only sparingly soluble. In a process known as *reactive crystallization*, the precipitate is formed by changing pH, solvent concentration, or solution temperature, or by adding a reagent that reacts with the solute to produce an insoluble chemical. The degree of supersaturation produced by the reaction is large, causing a high degree of primary nucleation. Although some growth occurs as the supersaturation is depleted, precipitates generally consist of very small particles that form quickly and may be crystalline in nature, but are frequently *aggregates* and *agglomerates*. Aggregates are masses of crystallites that are weakly bonded together. Agglomeration can follow, cementing aggregates together. Chemicals produced by precipitation,

sometimes called *fast crystallization*, include the sparingly soluble compounds listed in Table 17.6.

In precipitation, particle size is related to solubility by (17-16). For precipitates formed from ionic reactions in solution, the supersaturation ratio, $S = c/c_s$, is replaced by $(\pi/K_c)^{1/\upsilon}$, where π = the ionic concentration product for the reaction, K_c = (equilibrium) solubility product, and υ = sum of the cations and anions that form the precipitated compound. Thus, for aluminum hydroxide at 15°C, $\pi = (c_{Al^{+3}})(c_{OH^-})^3$, with K_c given in Table 17.6 and $\upsilon = 1 + 3 = 4$. The Kelvin relation, (17-16), is important in precipitation because nucleation is due to homogeneous primary nucleation.

Extent of supersaturation is also important in precipitation. When a reagent is added to a solution to form a sparingly soluble compound, a high supersaturation, which depends on the ionic concentrations in solution, develops. For example, consider the formation of a $BaSO_4$ precipitate from an aqueous solution containing Ba^{++} ions and an aqueous solution of H_2SO_4. From Table 17.6, the solubility product of $BaSO_4$ at 18°C is 0.87×10^{-10}. Figure 17.20a, taken from Nielsen [19], is a plot of sulfate ion concentration versus barium ion concentration, both in the solution just after mixing and before precipitation, with contours of constant

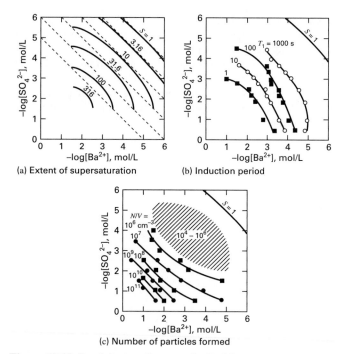

(a) Extent of supersaturation

(b) Induction period

(c) Number of particles formed

Figure 17.20 Precipitation diagrams for $BaSO_4$.

supersaturation ratio, S. The dashed lines refer to an ideal situation in which activity coefficients are 1.0 (concentrations = activities) and no complexes of Ba^{++} and $SO_4^=$ are formed. The solid, curved contours take these deviations from ideality into account. Note the large supersaturation ratio of 100 when the $SO_4^=$ concentration in solution is 0.01 mol/L and the Ba^{++} concentration is 4.7×10^{-4} mol/L.

Precipitation does not take place immediately after development of a large supersaturation because of the slow growth of very small particles. However, after a period of time (the *induction period*, T_1), visible precipitation begins. As shown in Figure 17.20b, this period depends on initial ion concentrations and, by superposition of Figure 17.20a, on the supersaturation ratio. For example, at a ratio of 300, T_1 is only one second, while at a ratio of 10, T_1 is more than one minute.

As shown in Figure 17.20c, the number of particles formed per unit volume of solution, N/V, also depends on initial concentrations of anion and cation and, therefore, on the supersaturation at high ion concentrations. A minimum number of particles per unit volume or a corresponding maximum particle size is frequently observed. For $BaSO_4$, this maximum size occurs for a supersaturation ratio of 300. The number of particles formed also depends on particle growth rate, which may be controlled by mass transfer of ions to a particle surface and/or integration of ions into the particle crystalline lattice (surface reaction). Particle growth rates during precipitation follow laws that exhibit different dependencies on relative supersaturation, $s = S - 1$:

Linear: $\qquad\qquad G = k_1 s \qquad\qquad$ (17-61)

Parabolic: $\qquad\quad G = k_2 s^2 \qquad\qquad$ (17-62)

Exponential: $\qquad\; G = k_3 f\{s\} \qquad\quad$ (17-63)

The linear rate law, which often applies for $G > 10$ nm/s, indicates mass-transfer control or surface-adsorption control,

where the latter depends more strongly on temperature. The parabolic rate law, for which G may be < 10 nm/s, applies to screw-dislocation-controlled growth. The exponential law, where $f\{s\}$ can involve a complex exponential, log, and/or power-law dependency, corresponds to growth control by surface nucleation. The latter two mechanisms may occur in parallel. Rate constants for many electrolytes for the three rate laws are given by Nielsen [20]. When growth is rapid, co-precipitation of soluble electrolytes may occur by entrapment, making it difficult to obtain a pure precipitate.

Because a precipitate is formed at considerable supersaturation, resulting particle shapes may be far from what corresponds to a minimum Gibbs energy, which depends on the particle surface area and interfacial tension. If precipitate and mother liquor are allowed to age, then precipitate particle sizes and shapes tend toward equilibrium values by: (1) flocculation and sintering of fine particles, (2) ion transport over the surface, and (3) ripening by dissolution and redeposition. Ripening can result in the release of co-precipitates, thus increasing precipitate purity.

Small particles produced in abundance during precipitation have a tendency to cluster together by interparticle collisions, variously referred to as agglomeration, aggregation, and flocculation. Such clusters, or agglomerates, are common when the number of particles/cm^3 of solution exceeds 10^7. For $BaSO_4$, Figures 17.20a and c show that agglomeration requires a very large initial supersaturation ratio. Agglomeration is important when particle sizes are between 1 μm and 50 μm.

EXAMPLE 17.12 Crystal-Size and Impeller rpm.

Fitchett and Tarbell [21] studied the effect of impeller rpm on crystal-size distributions in the continuous precipitation of barium sulfate when mixing solutions of sodium sulfate and barium chloride. Contents of the 1.8-L crystallizer were assumed to be perfectly mixed. In their Run 15, they used feeds with dissolved salts in stoichiometric ratio to give a sodium chloride concentration of 0.15 mol/L and an average residence time of 38 s.

Results for two different impeller speeds were as follows:

	In n, Number Density of Crystals, in ($\mu m^{-1} L^{-1}$)	
Size, μm	950 rpm, 0.361 J/s · kg Feeds	400 rpm, 0.028 J/s · kg Feeds
7	22.07	23.45
9	21.66	22.75
11	21.35	22.08
13	20.97	21.36
15	20.77	20.75
17	20.41	20.14
19	20.04	19.57
21	19.77	18.94
23	19.48	18.43
25	19.09	17.75
27	18.85	
29	18.49	
31	18.11	
33	17.87	

Using the MSMPR model, determine from the two sets of data: (a) n°, number density of nuclei, nuclei/μm-L; (b) G, linear growth rate, μm/s; (c) B°, nucleation rate, nuclei/L-s; (d) mean crystal length, μm; (e) n_c, number of crystals/volume of mother liquor, crystals/m^3.

Solution

(a) From (17-38), if each set of data is fitted to

$$\ln n = \ln n^\circ - L/G\tau \qquad (1)$$

the best straight line yields an intercept of $\ln n^\circ$ and a slope of $-1/G\tau$, from which n° and G can be determined for $\tau = 38$ s.

Using a spreadsheet, the results for 950 rpm are intercept = 23.13, slope = −0.1601. Therefore, $n^\circ = \exp(23.13) = 1.11 \times 10^{10}$ crystals/μm·L.

(b) $G = 1/[(38)(0.1601)] = 0.164$ μm/s

(c) The nucleation rate is given by (17-51):

$$B^\circ = Gn^\circ = 0.164(1.11 \times 10^{10}) = 1.82 \times 10^9 \text{ nuclei/L} \cdot \text{s}$$

(d) Mean particle length is obtained from the value of z for the maximum value of dx_L/dz, which, from Table 17.9, is ze^{-z}.

$$d\frac{(ze^{-z})}{dz} = e^{-z} - ze^{-z} = 0$$

Therefore, $z = 1 = L_{mean}/G\tau$ from (17-42) and $L_{mean} = G\tau = 0.164(38) = 6.23$μm.

(e) The number of crystals per unit volume of mother liquor is given by (17-54):

$$n_c = n^\circ\tau G = B^\circ\tau = 1.82 \times 10^9 (38)$$
$$= 6.92 \times 10^{10} \text{ crystals/L} = 6.92 \times 10^7 \text{ crystals/m}^3$$

(f) In a similar manner, the following results are obtained for 400 rpm:

Intercept = 25.53; slope = −0.313; $n^\circ = 1.22 \times 10^{11}$ crystals/μm-L; $G = 0.0841$ μm/s; $B^\circ = 1.03 \times 10^{10}$ nuclei/L-s; $L_{mean} = 3.20$ μm; $n_c = 3.91 \times 10^8$ crystals/m^3

Comparing the two sets of results, it is seen that a higher agitator speed gives a larger mean crystal size, a larger growth rate, and a lower nucleation rate.

§17.7 MELT CRYSTALLIZATION

Solution crystallization is commonly conducted with aqueous solutions of dissolved inorganic salts. The phase-equilibrium diagram for a water–salt system, e.g., Figure 17.2, includes temperatures ranging from the eutectic temperature below 0°C to a temperature exceeding the melting point of ice, but not greater than 220°C. Since melting points of inorganic salts exceed 220°C, the salt melting point is not included.

For mixtures of organic compounds, the situation is quite different. An analysis by Matsuoka et al. [22] found that more than 70% of the common organic compounds had melting points between 0 and 200°C. For binary mixtures of such compounds, phase-equilibrium diagrams include melting points of both compounds. In one example, Figure 17.1b, crystals of ortho-chloronitrobenzene can form if the feed composition is less than the eutectic with para-chloronitrobenzene; otherwise, pure para-chloronitrobenzene can form. The exception is the eutectic composition. *Eutectic-forming* systems consist of compounds that cannot substitute for each other in the crystal lattice, so the eutectic mixture consists of two different solid phases. The two solubility curves separating a liquid-phase region from the two solid–liquid regions are freezing-point curves, and mixtures at conditions in the liquid-phase region are referred to as *melts*.

Much less common are *solid-solution-forming* systems of the type shown in Figure 17.1c for phenanthrene–anthracene. These consist of compounds so nearly alike in structure that they can substitute for each other in the crystal lattice to form a single crystalline phase over a wide range of composition. The liquid–solid phase diagram resembles that for vapor–liquid equilibrium, §4.2, where freezing-point and melting-point curves replace dew-point and bubble-point curves. Mixtures in the liquid-phase region above the freezing-point curve are also referred to as melts. A mixture in the region between the two curves separates into a liquid phase and a crystalline phase, neither of which is pure.

Crystallization of melts from eutectic-forming or solid-solution-forming mixtures is called *melt crystallization*. Theoretically, melt crystallization of eutectic-forming systems, like solution crystallization of such systems, can produce pure crystals. However, the product from commercial, single-stage crystallizers may not meet purity specifications, for reasons discussed by Wilcox [23]. In that case, repeated stages of melting and crystallization may be necessary to produce high-purity crystals.

Separation of organic mixtures is most commonly achieved by distillation. However, if distillation: (1) requires more than 100 theoretical stages, (2) cannot produce products that meet specifications (e.g., purity and color), (3) causes decomposition of feed components, or (4) requires extreme conditions of temperature or pressure (e.g., vacuum), other separation methods should be considered. According to Wynn [24], if the compounds to be separated: (1) are disubstituted benzenes, diphenyl alkyls, phenones, secondary or tertiary aromatic or aliphatic amines, isocyanates, fused-ring compounds, heterocyclic compounds, or carboxylic acids of MW < 150; (2) have a melting-temperature range from 0 to 160°C; (3) are required to be high-purity products; or if (4) a laboratory test produces a clearly defined solid phase from which the liquid phase drains freely, then melt crystallization should be considered as an alternative or supplement to distillation.

§17.7.1 Equipment for Melt Crystallization

As with solution crystallization, a large number of crystallizer designs have been proposed for melt crystallization. Only widely used commercial units are discussed here; Myerson [10] provides a wider panorama. In all cases, crystallization is caused by cooling the mixture.

Two major methods used in melt crystallizers are *suspension crystallization* and *layer crystallization* by progressive

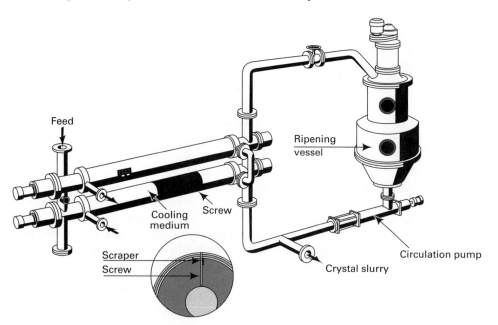

Figure 17.21 Two-stage, scraped-wall melt crystallizer.

freezing. In the former, crystals of a desired size distribution are grown slowly in a suspension by subcooling a seeded-feed melt. In the latter, crystals of uncontrolled size are grown rapidly on a cooled surface, wherein subcooling is supplied through the crystallized layer. In suspension crystallization, the remaining melt must be separated from the crystals by centrifugation, filtration, and/or settling. In layer crystallization, the remaining melt or residual liquid is drained from the solid layer, followed by melting of the solid.

Figure 17.21 shows a two-stage, scraped-wall-crystallizer system used for suspension crystallization. A cooling medium is used to control the surface temperatures of the two scraped-wall units, causing crystals to grow, which are subsequently scraped off by screws. The melt mixture is circulated through a ripening vessel. The two scraped-wall units are typically 3.6 m long with 3.85 m^2 of heat-transfer area. The screws are driven by a 10-kW motor.

Of greater commercial importance is the falling-film crystallizer in Figure 17.22, developed by Sulzer Brothers Ltd. This equipment produces high-purity crystals (>99.9%) at high capacity (>10,000 tons/yr). A large pair of units, each 4 m in diameter and containing 1,100 12-m-high tubes, can produce 100,000 tons/yr of very pure crystals, with typical layer growth rates of 1 inch/h. The feed melt flows as a film down the inside of the tubes over a crystal layer that forms and grows by progressive freezing because the wall of the tube is cooled from the outside. When a predetermined crystal-layer thickness, typically 5–20 mm, is reached, the feed is stopped and the tubes are warmed to cause partial melting, called sweating, to remove impurities that may be bonded to the crystal layer. This is followed by complete melting of the remaining layer, which is of high purity. During the initial crystallization phase, melt is circulated at a high rate, compared to the crystallization rate, so that a uniform temperature and melt composition are approached down the length of the tube. The coolant also flows as a film down along the outside surface of the tubes.

Consider the freezing step in a falling-film crystallizer for which a temperature profile is shown in Figure 17.23. Melt enters at the top of the tube and flows as a film down the inside wall. A coolant, at a temperature below the freezing point of the melt, also enters at the top and flows as a film down the

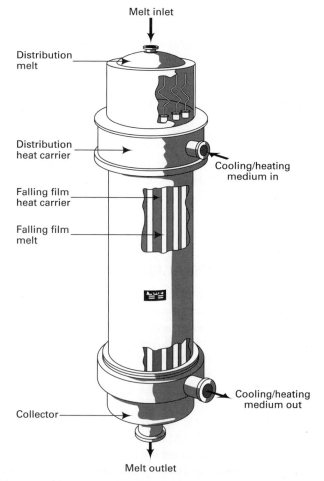

Figure 17.22 Sulzer falling-film melt crystallizer.

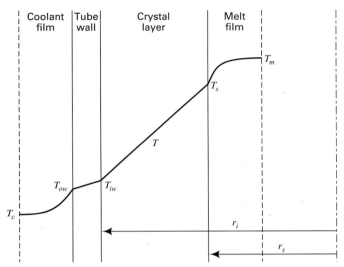

Figure 17.23 Temperature profile for melt crystallization from a falling film at a vertical location along the tube.

outside wall. Heat is transferred from the melt to the coolant, causing the melt to form a crystal layer at the inside tube wall. As melt and coolant flow down the tube, their temperatures decrease and increase, respectively. The thickness of the crystal layer increases with time, with the latent heat of fusion transferred from the crystal-melt interface to the coolant.

For a eutectic-forming, binary melt, only one component crystallizes, although small amounts of the other component may be trapped in the crystal layer, particularly if the rate of crystal formation is too rapid. Temperature at the interface of the crystal layer and the melt is assumed to be the equilibrium temperature corresponding to the saturated melt composition on the solubility curve. If mass transfer of the crystallizing component from the melt film to the phase interface is rapid, the interface temperature will correspond to the melt-film composition at that vertical location. Also, if thermal resistances of coolant film, tube wall, and melt film are negligible compared to that of the crystal layer, and if the heat capacities of the crystal layer and metal wall are negligible, then a simple model for rate of increase of crystal-layer thickness with time can be constructed, as follows.

At a particular vertical location, $T_s \approx T_m$, the melting point, and $T_{iw} \approx T_c$, the coolant bulk temperature. The rate of heat released by freezing, ΔH_f, is equal to the rate of heat conduction through the crystal layer. Thus, for a planar wall, referring to Figure 17.23, the heat evolved is equated to the rate of heat conduction through the crystal layer to give

$$\Delta H_f \frac{dm}{dt} = -A\rho_c(\Delta H_f)\frac{dr_s}{dt} = \frac{k_c A(T_m - T_c)}{r_i - r_s} \quad (17\text{-}64)$$

with an initial condition $r_s = r_i$ at $t = 0$. Integration of (17-64) gives

$$\frac{(r_i - r_s)^2}{2} = \frac{k_c(T_m - T_c)t}{\rho_c \Delta H_f} \quad (17\text{-}65)$$

or the crystal-layer thickness is

$$(r_i - r_s) = \sqrt{\frac{2k_c(T_m - T_c)t}{\rho_c \Delta H_f}} \quad (17\text{-}66)$$

A similar derivation for a cylindrical-tube wall, where $r_i =$ inside radius of the tube, gives

$$\frac{1}{4}(r_i^2 - r_s^2) - \frac{r_s^2}{2}\ln\left(\frac{r_i}{r_s}\right) = \frac{k_c(T_m - T_c)t}{\rho_c \Delta H_f} \quad (17\text{-}67)$$

The value of the LHS of (17-67) approaches the value of the LHS of (17-65) as the value of r_s approaches the value of r_i. For the planar wall, (17-66) shows that the crystal layer grows as the square root of the time. Thus, the growth during the first half of the time period is $\left(1/\sqrt{2}\right) = 70.7\%$ of the total growth. For the cylindrical-tube wall, if the growth is from the wall to half of the radius, 67.9% of that growth occurs during the first half of the time period, to produce 75.2% of the crystal layer. The time required for this thickness of growth for the cylindrical tube is only 80% of the time for the planar wall. Thus, a conservative result is obtained by using the simpler planar wall of (17-66).

During operation of the falling-film crystallizer, temperature of the melt film decreases as it flows down the tube because of heat transfer from melt to the colder crystal layer. Based on the earlier assumptions, the melt temperature at any elevation will be the temperature corresponding to the solubility curve for the melt composition at that elevation. If it is assumed that: (1) the sensible heat from cooling of the melt layer is negligible compared to the latent heat of fusion; (2) any sensible-heat storage in the crystal layer and tube wall can be neglected; (3) vertical conduction in the tube wall, crystal layer, and melt layer can be neglected; and (4) the coolant temperature is constant, then (17-66) can be coupled with a material balance for depletion of the crystallizing component as it flows down the inside tube wall. The result, which is left as an exercise, gives the crystal-layer thickness as a function of time and vertical location.

EXAMPLE 17.13 Falling-Film Crystallizer.

A melt of 80 wt% naphthalene and 20 wt% benzene at the saturation temperature is fed to a falling-film crystallizer, where coolant enters at 15°C. Estimate the time required for the crystal-layer thickness near the top of the 8-cm i.d. tubes to reach 2 cm.

Solution

By extrapolation from Figure 17.24, the saturation temperature of the melt is 62°C. Therefore, naphthalene will crystallize, with

$$T_m - T_c = (62 - 15)1.8 = 84.6°F$$

$$r_i = 8/2 = 4\,\text{cm} = 0.131\,\text{ft}$$

$$r_s = 4 - 2 = 2\,\text{cm} = 0.0655\,\text{ft}$$

The estimate can be made with (17-67), based on the following properties for naphthalene: $\rho_c = 71.4$ lb/ft^3; $k_c = 0.17$ Btu/h-ft-°F; and $\Delta H_f = 63.9$ Btu/lb.

From (17-67), solving for time, t,

$$t = \frac{(71.4)(63.9)}{(0.17)(84.6)}$$

$$\times \left[\frac{1}{4}(0.131^2 - 0.0655^2) - \frac{0.0655^2}{2}\ln\left(\frac{0.131}{0.0655}\right)\right] = 0.549\,\text{h}$$

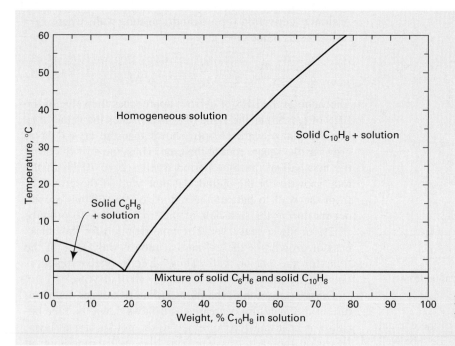

Figure 17.24 Solid–liquid phase diagram for the naphthalene–benzene system.

§17.8 ZONE MELTING

When melt consists of two components that form a solid solution, as seen in Figure 17.1c, liquid and solid phases at equilibrium contain both components, as in vapor–liquid equilibrium. Accordingly, multiple stages of crystallization are required to obtain products of high purity. A useful technique for doing this is *zone melting*, as developed by Pfann [25, 26] and discussed by Zief and Wilcox [27].

Zone melting, as carried out batchwise, is illustrated in Figure 17.25. Starting with an impure crystal slab, a melt zone, whose length is a small fraction of slab length, is passed slowly (typically at 1 cm/h for organic mixtures) along the slab from one end to the other by fixing the slab and using a moving heat source or, less commonly, by moving the slab through a fixed heat source. Radio frequency (RF) induction heating, §18.1.2, is particularly convenient and creates mixing in the melt. The slab can be arranged horizontally or vertically, with the latter

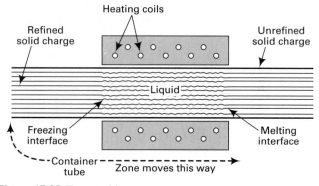

Figure 17.25 Zone melting.

[From *Perry's Chemical Engineers' Handbook*, 6th ed., R.H. Perry, D.W. Green, and J.O. Maloney, Eds., McGraw-Hill, New York (1984) with permission.]

preferred because it takes advantage of the density difference between crystals and melt. Zone melting can be applied to the composition region near either end of the phase-equilibrium diagram. For the phenanthrene (P)–anthracene (A) mixture in Figure 17.1c, zone melting can remove small amounts of A from P, or P from A.

A solid–liquid equilibrium distribution coefficient, K, can be defined as in liquid–liquid equilibrium:

$$K_i = \frac{\text{concentration of impurity,} i, \text{in the solid phase}}{\text{concentration of impurity,} i, \text{in the melt phase}} \quad (17\text{-}68)$$

If concentration is expressed in mole fractions and the impurity is anthracene, as in Figure 17.1c, then at 120°C, $K_i = 0.30/0.12 = 2.50$. At 200°C, with phenanthrene as the impurity, $K_i = 0.11/0.28 = 0.393$. Thus, K_i can be greater or less than 1. When > 1, impurities raise melting points and concentrate in the solid phase; when < 1, impurities lower melting points and concentrate in the melt. However, when $K_i \rightarrow 1$, purification by zone melting becomes very difficult.

Near either end of the composition range, an equilibrium curve for the solid and liquid phases approaches a straight line, and the value of K_i becomes constant. In these ranges, an equation is readily developed to predict impurity concentration in the solid phase upstream of a moving melt zone as a function of distance along the crystal layer in the direction of melt-zone movement. Figure 17.26 shows the position of the melt zone during zone melting, where the zone moves a distance dz. Assume that: (1) the melt zone of width ℓ is perfectly mixed, with impurity weight fraction w_L; (2) no diffusion of impurity in the solid phases occurs; (3) initial concentration of the impurity is uniform at w_o; and (4) impurity concentration in the melt zone is in equilibrium with that in the solid phase upstream of the melt zone. A mass balance on the impurity for a dz melt-zone

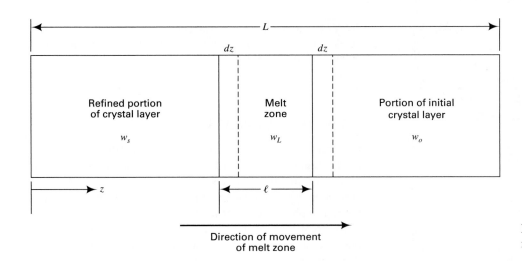

Figure 17.26 Ideal zone-melting model.

movement is given by:

(Mass of impurity added to melt zone)
− (mass of impurity removed from melt zone)
= (increase in mass of impurity in the melt zone)

Thus,

$$w_o \rho_c A_c dz - w_s \rho_c A_c dz = \rho_L A_c \ell dw_L \qquad (17\text{-}69)$$

From (17-68), let

$$K = w_s/w_L \qquad (17\text{-}70)$$

Therefore,

$$dw_L = dw_s/K \qquad (17\text{-}71)$$

Combining (17-69) to (17-71) to eliminate w_L and assuming $\rho_L = \rho_c$,

$$\frac{K}{\ell} \int_0^z dz = \int_{K_{w_o}}^{w_s} \frac{dw_s}{w_o - w_s} \qquad (17\text{-}72)$$

Integration gives

$$\frac{w_s}{w_o} = 1 - (1 - K)\exp(-Kz/\ell) \qquad (17\text{-}73)$$

for

$$z/\ell = 0 \quad \text{to} \quad z/\ell = \frac{L}{\ell} - 1$$

Solving (1) for values of $z/\ell = 0$ to 9 gives

z/ℓ	w_s
0	0.0036
1	0.0055
2	0.0069
3	0.0078
4	0.0085
5	0.0089
6	0.0093
7	0.0095
8	0.0096
9	0.0097

In the melt zone, $w_L = w_s/K = 0.0097/0.36 = 0.0269$. The predicted profile is shown in Figure 17.27. To further refine anthracene, additional zone-melting passes can be made to move more impurity into the melt zone. However, for each pass after the first, (17-73) is not valid because at the beginning of each additional pass, w_o is not

EXAMPLE 17.14 Zone Melting.

A crystal layer of 1 wt% phenanthrene and 99 wt% anthracene is subjected to zone melting with a melt-zone width equal to 0.1 of the length of the crystal layer ($\ell/L = 0.1$). The distribution coefficient for phenanthrene in the dilute composition region is 0.36. Determine the phenanthrene concentration profile in weight fractions at the conclusion of zone melting when the melt reaches the last 10% of the crystal-layer length.

Solution

From Figure 17.1c and the value of K, the phenanthrene favors distribution to the melt phase. Therefore, as the melt zone moves along the crystal layer, the phenanthrene will migrate from the upstream portion of the crystal layer into the melt zone. When the leading edge of the melt zone reaches the end of the crystal layer, all of the migrated phenanthrene will be in the melt layer at a uniform concentration equal to $w_s\{z/\ell = 0.9\}/K$. Assume instant freezing of this melt zone so as to maintain uniform composition. From (17-73), with $K = 0.36$ and $w_o = 0.01$,

$$w_s = 0.01[1 - 0.64\exp(-0.36\,z/l)] \qquad (1)$$

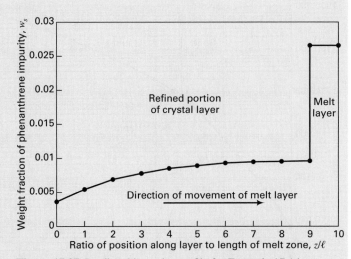

Figure 17.27 Predicted impurity profile for Example 17.14.

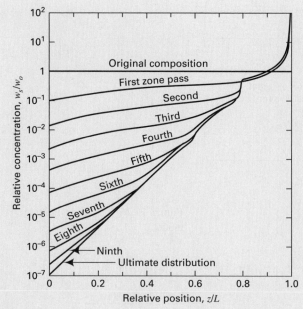

Figure 17.28 Concentration profiles for multiple zone-melting passes.

uniform. For example, at the beginning of the second pass, w_o becomes w_s in Figure 17.27. For the additional passes, it is necessary to numerically solve the following ODE form of (17-72).

$$\frac{dw_s}{dz} = \frac{K}{\ell}(w_o - w_s) \qquad (17\text{-}74)$$

where $w_o = w_o\{z = z\}$ from the results for w_s from the previous pass, and $w_s\{z = 0\} = Kw_o\{z = 0\}$.

Typical impurity-concentration profiles for multiple melt-zone passes are shown in Figure 17.28 from calculations by Burris, Stockman, and Dillon [28] for the case $K = 0.1$ and a melt-zone width of 20% of the crystal-layer length ($\ell/L = 0.2$). In their calculations, the melt zone was allowed to diminish to zero as the heater moved away from the crystal layer, resulting in a steep gradient from $z/L = 0.8$ to 1.0. It is seen that an ultimate impurity distribution is approached after 9 passes. The number of passes required to approach the ultimate distribution is given approximately by

$2(L/\ell) + 1$, as observed by Herington [29]. In this example, a highly pure crystalline product can be achieved if a portion of the purer end of the final crystal layer is taken.

§17.9 DESUBLIMATION

Crystallization from a vapor, rather than from a liquid or melt, is called *desublimation*. The reverse, i.e., vaporization of a solid directly to a vapor, is *sublimation*. To understand how such a phase change can occur without going through the liquid phase, consider the phase diagram for naphthalene in Figure 17.29. As with most chemicals that have a vapor pressure of much less than 1 atm at the melting temperature, that temperature coincides, within a fraction of 1°C, with the triple point. Below the triple-point temperature of 80.2°C, which corresponds to a vapor pressure of 7.8 torr, naphthalene cannot exist as a liquid, regardless of the pressure, but can exist as a solid, provided the pressure is greater than the vapor pressure of naphthalene at the prevailing temperature. However, if the pressure falls below the vapor pressure, solid naphthalene sublimes directly to a vapor.

If naphthalene is present in an ideal-gas mixture with a noncondensable inert gas, sometimes called an entrainer, at a temperature below the triple point of 80.2°C, desublimation to a crystalline solid occurs if the partial pressure of naphthalene gas is increased to a value that exceeds its vapor pressure. Consider a vapor mixture of 5 mol% naphthalene in nitrogen at 70°C and a total pressure of 40 torr. The partial pressure of naphthalene is 0.05(40) = 2 torr. From Figure 17.29, this partial pressure is less than the vapor pressure of 3.8 torr. Desublimation begins when the total pressure is increased to 3.8/0.05 = 76 torr. Alternatively, if the pressure is maintained at 40 torr but the temperature is reduced, desublimation begins at 61°C, where the vapor pressure of naphthalene is 2 torr. Unless nitrogen is entrapped, naphthalene crystals will be pure.

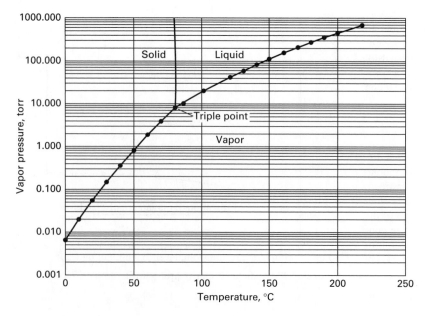

Figure 17.29 Vapor pressure of naphthalene.

Table 17.10 Chemicals Amenable to Purification by Desublimation

Aluminum chloride	Molybdenum Trioxide
Anthracene	Naphthalene
Anthranilic acid	β-Naphthol
Anthraquinone	Phthalic anhydride
Benzanthrone	o-Phthalimide
Benzoic acid	Pyrogallol
Calcium	Salicylic acid
Camphor	Sulfur
Chromium chloride	Terephthalic acid
Cyanuric chloride	Titanium tetrachloride
Ferric chloride	Thymol
Hafnium tetrachloride	Uranium hexafluoride
Iodine	Zirconium tetrachloride
Magnesium	

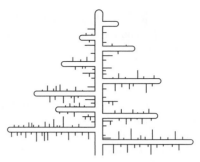

Figure 17.30 Dendritic growth of a crystal.

The most common occurrence of desublimation is the formation of snow crystals by moisture condensation from air at temperatures below 0°C. At temperatures above −40°C, snow crystals form by heterogeneous nucleation on fine mineral particles of 10^{-5} to 10^{-2} mm in diameter. At temperatures below −40°C, homogeneous nucleation can take place. As snow crystals fall through the atmosphere, they cluster together into snowflakes. The difference between the melting point of ice and the triple point of water is less than 0.01°C, with a vapor pressure of only 4.6 torr.

As discussed by Nord [30] and Kudela and Sampson [31], a number of chemicals are amenable to purification by desublimation, preceded, perhaps, by sublimation. Table 17.10 provides a list of chemicals that are solids at ambient conditions and have solid vapor pressures greater than 5 μm Hg at moderate operating temperatures. Applications of desublimation to obtain near-pure solid chemicals or pure-chemical solid films on substrates include crystallization of water-insoluble organic chemicals from mixtures with inert, non-volatile gases and vapor-deposition of metals.

Desublimation is almost always effected by cooling gas mixtures at constant pressure. Holden and Bryant [32] describe four cooling techniques: (1) heat transfer from the gas through a solid surface, on which the crystals form; (2) quenching by addition of a vaporizable liquid; (3) quenching by addition of a cold, noncondensable gas; (4) expansion of the gas mixture through a nozzle.

Technique 1, which is widely used, has the disadvantage that crystals must be removed by scraping or melting from the heat-transfer surface. Technique 2 is less common, and needs a liquid in which the sublimate is not soluble. If excess water is used, this technique will produce a slurry of the crystals and nonvaporized liquid. Technique 3, which is also common, produces dry crystals, called a snow. Technique 4 is not widely used because the degree of cooling necessary requires either a high pressure upstream of the nozzle or a vacuum downstream.

The number, size, and shape of the crystals produced in the snow of the third technique depend on the relative rates of nucleation, crystal growth, and crystal agglomeration. Frequently, as commonly observed with snow crystals, dendritic growth occurs, producing undesirable crystals of the shape shown in Figure 17.30, in which the main stem grows rapidly, followed by slower rates of growth of primary branches, and much slower growth of secondary branches. However, if dendritic crystals can be suspended in the vapor long enough, spaces between the branches can fill in to produce a more-desirable, dense shape.

§17.9.1 Desublimation in a Heat Exchanger

Desublimation is used in industry to recover organic chemicals, including anthracene, maleic anhydride, naphthalene, phthalic anhydride, and salicylic acid from gas streams. Crystals are deposited on the outside of tubes while a coolant flows through the inside. If the rate of desublimation is controlled by the rate of conduction of heat through deposited crystal layers, then a relationship for the time needed to deposit a crystal layer of given thickness is derived in a manner similar to that for the falling-film melt crystallizer in §17.7. The result—which is similar to (17-67), except that crystal layers grow outward from the outside radius of cylindrical tube r_o instead of inward from inside radius r_i as in (17-67)—is

$$t = \frac{\rho_c \Delta H_s}{k_c (T_g - T_c)} \left[\frac{r_s^2}{2} \ln\left(\frac{r_s}{r_o}\right) - \frac{1}{4}(r_s^2 - r_o^2) \right] \quad (17\text{-}75)$$

where r_s is the radius to the crystal layer–gas interface. Equation (17-75) ignores any sensible heat associated with cooling of the gas. If this is not negligible, then it must be added in (17-75) to the heat of sublimation, ΔH_s. If the heat-transfer resistance in the gas phase is negligible, the temperature, T_g in (17-75), corresponds to the temperature where the partial pressure of the desubliming solute equals its vapor pressure. An extended derivation of (17-75) that includes the sensible-heat correction is given by Singh and Tawney [33].

EXAMPLE 17.15 Desublimation in a Heat Exchanger.

A desublimation heat exchanger is to be sized for recovery of 100 kg/h of naphthalene (N) from a gas stream, where the other components are noncondensable. The heat-exchanger tubes are 1 m long with an outside diameter of 2.5 cm. Tube spacing is such that N can build up to a maximum thickness of 1.25 cm. The gas enters the unit at 800 torr and 80°C with a mole fraction for N of 0.0095. The water coolant flows through the inside of the tubes countercurrently to the gas. Cooling water enters at 25°C and exits at 45°C. Determine the number of tubes needed and the time required to reach the maximum thickness if the gas leaves at 60°C.

Solution

The properties are: C_P of gas = 0.26 cal/g-°C; MW of naphthalene = 128.2; k_c of solid naphthalene = 1.5 cal/h-cm-°C; ΔH_s of naphthalene = 115 cal/g; ρ_c of solid naphthalene = 1.025 g/cm³.

When the maximum thickness of N is achieved, the amount of sublimate per tube, if uniform, is

$$m = \pi(r_s^2 - r_o^2)L\rho_c$$
$$= 3.14\left[(1.25 + 1.25)^2 - 1.25^2\right]100(1.025)$$
$$= 1,510\,g = 1.51\,kg$$

Entering gas has a partial pressure for N of 0.0095(800) = 7.6 torr. From Figure 17.29, the saturation temperature for this partial pressure is 79.7°C. This is just slightly below the entering-gas temperature of 80°C, which is less than the triple-point temperature. Therefore, N will condense as a solid. At the exit-gas temperature, 60°C, the vapor pressure of N is 1.8 torr. Assuming saturation at the exit and no pressure drop, the mole fraction of N in the exit gas is 1.8/800 = 0.00225. Thus, per mole of entering gas, 0.0073 mole of N will be condensed. If the gas is assumed to be nitrogen and N, then M_r, the mass ratio of the gas mixture to the condensed N, is

$$M_r = \frac{28(0.9905) + 128.2(0.0095)}{(128.2)(0.0073)} = 30.94$$

The sensible heat plus the heat of fusion is

$$M_r C_{P_g}(T_{in} - T_{out})_g + \Delta H_s = 30.94(0.26)(80 - 60) + 115$$
$$= 276\,cal/g$$

For this example, with a very low mole fraction undergoing desublimation, the sensible-heat effect is large. From (17-75), the time to reach the maximum thickness of N, using the temperature-driving force across the solid N of 35°C, is

$$t = \frac{(1.025)(276)}{(1.5)(35)}\left[\frac{(2.5)^2}{2}\ln\left(\frac{2.5}{1.25}\right) - \frac{1}{4}(2.5^2 - 1.25^2)\right]$$
$$= 5.37\,h$$

The number of tubes required is

$$\frac{(100)(5.37)}{1.51} = 356$$

§17.10 EVAPORATION

Before crystallizing an inorganic solute from an aqueous solution, it is customary to bring the solute concentration close to saturation. This is accomplished by evaporating water in an evaporator, which is also used to concentrate solutions even when the solute is not subsequently crystallized, e.g., solutions of sodium hydroxide. When the vapor formed is essentially pure, there is no mass-transfer resistance in the vapor. When the liquid is agitated, mass transfer is sufficiently rapid that the rate of solvent evaporation can be determined by the rate of heat transfer from the indirect heating medium, usually condensing steam, to the solution.

Evaporators differ in configuration and degree of liquid agitation. The five most widely used continuous-flow

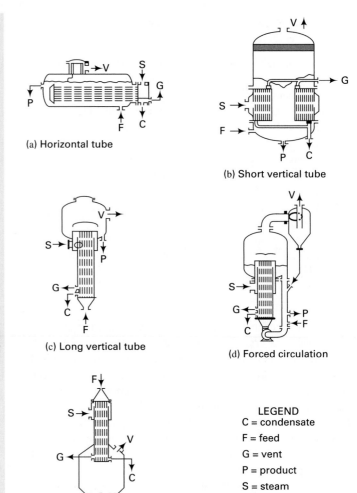

(a) Horizontal tube

(b) Short vertical tube

(c) Long vertical tube

(d) Forced circulation

(e) Falling film

LEGEND
C = condensate
F = feed
G = vent
P = product
S = steam
V = vapor

Figure 17.31 Common types of evaporators.

[From *Perry's Chemical Engineers' Handbook*, 6th ed., R.H. Perry, D.W. Green, and J.O. Maloney, Eds., McGraw-Hill, New York (1984) with permission.]

evaporators are shown schematically in Figure 17.31. Their main characteristics are as follows:

(a) Horizontal-tube evaporator. This unit, shown in Figure 17.31a, consists of a horizontal cylindrical vessel equipped in the lower section with a horizontal bundle of tubes, inside of which steam condenses and outside of which the solution to be concentrated boils. Agitation is provided only by the movement of the bubbles leaving the evaporator as vapor. This type of unit is suitable only for low-viscosity solutions that do not deposit scale on the heat-transfer surfaces.

(b) Short-vertical-tube evaporator. This unit, shown in Figure 17.31b, differs significantly from the horizontal-tube evaporator. The tube bundle is arranged vertically, with the solution inside the tubes and steam condensing outside. Boiling inside the tubes causes the solution to circulate, providing additional agitation and higher heat-transfer coefficients. This type of evaporator is not suitable for very viscous solutions.

(c) **Long-vertical-tube evaporator.** By lengthening the vertical tubes and providing a separate vapor–liquid disengagement chamber, as shown in Figure 17.31c, a higher liquid velocity can be achieved and, thus, an even higher heat-transfer coefficient.

(d) **Forced-circulation evaporator.** To handle very viscous solutions, a pump is used to force the solution upward through relatively short tubes, as shown in Figure 17.31d.

(e) **Falling-film evaporator.** This unit, shown in Figure 17.31e, is popular for concentrating heat-sensitive solutions such as fruit juices. The solution enters at the top and flows as a film down the inside walls of the tubes. Concentrate and vapor produced are separated at the bottom.

§17.10.1 Boiling-Point Elevation

For a given pressure in the vapor space of an evaporator, the boiling temperature of an aqueous solution will be equal to that of pure water if the solute is not dissolved but consists of small, insoluble, colloidal material. If the solute is soluble, the boiling temperature will be greater than that of pure water

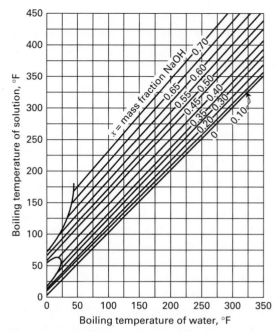

Figure 17.32 Dühring chart for aqueous solutions of sodium hydroxide.

[From W.L. McCabe, J.C. Smith, and P. Harriott, *Unit Operations of Chemical Engineering*, 5th ed., McGraw-Hill, New York (1993) with permission.]

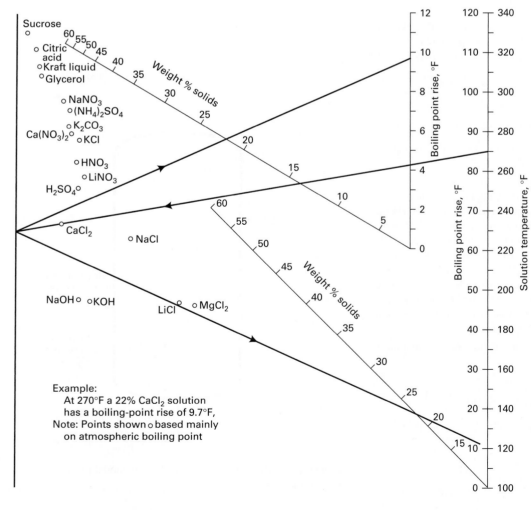

Figure 17.33 Nomograph for boiling-point elevation of aqueous solutions.

[From *Perry's Chemical Engineers' Handbook*, 6th ed., R.H. Perry, D.W. Green, and J.O. Maloney, Eds., McGraw-Hill, New York (1984) with permission.]

by an amount known as the *boiling-point elevation*. If, as is usually the case, the solute has little or no vapor pressure, the evaporator pressure is the partial pressure of water in the solution, so, by Raoult's law (§2.3),

$$P = p_{H_2O} = \gamma_{H_2O} x_{H_2O} P^s_{H_2O} \qquad (17\text{-}76)$$

For a given P and x_{H_2O}, the temperature to give the required $P^s_{H_2O}$ can be determined, provided γ_{H_2O} can be estimated. For solutions dilute in the solute, $\gamma_{H_2O} \rightarrow 1.0$. For concentrated solutions, γ_{H_2O} can be estimated from correlations for electrolyte solutions as discussed by Poling, Prausnitz, and O'Connell [34].

Alternatively, boiling temperatures for solutions can be estimated by using a *Dühring-line chart*, if available for a particular solute. Such a chart is shown in Figure 17.32 for sodium hydroxide–water solutions. The straight lines on this chart for different mass fractions of NaOH obey *Dühring's rule*, which states that as the pressure is increased, the boiling temperature of the solution increases linearly with boiling temperature of the pure solvent. A nomograph for other solutes in water is given in Figure 17.33. The use of Figures 17.32 and 17.33, and (17-76) is illustrated in the next example.

EXAMPLE 17.16 Boiling-Point Elevation.

An aqueous NaOH solution is evaporated at 6 psia. If the solution is 35-wt% NaOH, determine the: (a) boiling temperature of the solution, (b) boiling-point elevation, (c) activity coefficient for water from (17-76).

Solution

From steam tables, water has a vapor pressure of 6 psia at 170°F.

(a) From the Dühring-line chart of Figure 17.32, the boiling temperature of the solution is 207°F.

(b) The boiling-point elevation is: $207 - 170 = 37°F$.

Alternatively, the nomograph of Figure 17.33 may be used by drawing a straight line through the point for NaOH and a solution (boiling-point) temperature of 207°F. That line is extended to the left to the intersection with the leftmost vertical line. A straight line is then drawn from that intersection point through the lower, inclined line labeled "Weight % solids" at 35 (wt% NaOH). This second line is extended so as to intersect the right, vertical line. The value of the boiling-point elevation at this point of interaction is read as 36°F, which is close to the value of 37°F from the Dühring chart.

(c) The mole fraction of water in the solution for complete ionization of NaOH is:

$$x_{H_2O} = \frac{0.65/18}{0.65/18 + 2(0.35)/40} = 0.674$$

Vapor pressure of water at 207°F = 13.3 psia. From (17-76),

$$\gamma_{H_2O} = \frac{6}{0.674(13.3)} = 0.67$$

---

§17.10.2 Evaporator Model

Figure 17.34 is a schematic diagram for the model used to make mass-balance, energy-balance, and heat-transfer calculations to size evaporators operating under continuous-flow, steady-state conditions. A so-called thin liquor at temperature T_f, with weight-fraction solute w_f, is fed to the evaporator at mass flow rate m_f. A heating medium, e.g., saturated steam, is fed to the heat-exchanger tubes at T_s, P_s, and mass flow rate m_s. Saturated condensate leaves the heat exchanger at the same temperature and pressure. Heat-transfer rate Q to the solution in the evaporator at temperature T_e causes the solution to partially evaporate to produce vapor at temperature T_v, with flow rate m_v. The thick-liquor concentrate leaves at temperature T_p, with weight-fraction solute w_p at mass flow rate m_p. The heat exchanger has a heat-transfer area A and overall heat-transfer coefficient U.

Key assumptions in formulating the model are:

1. The feed has only one volatile component, e.g., water.

2. Only the latent heat of vaporization of steam at T_s is available for heating and vaporizing the solution.

3. Boiling action on the heat-exchanger surfaces agitates the solution in the evaporator sufficiently to achieve perfect mixing. Therefore, the solution temperature equals the exiting temperature of the thick-liquor concentrate. Thus, $T_e = T_p$ and $T_v = T_p$.

4. Because of Assumptions 2 and 3, the overall temperature-driving force for heat transfer $= \Delta T = T_s - T_p$.

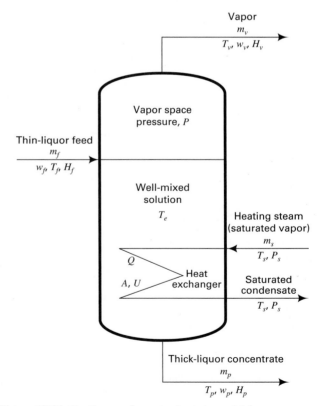

Figure 17.34 Continuous-flow, steady-state model for an evaporator.

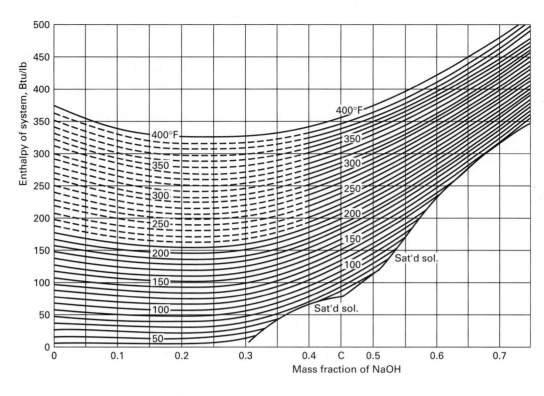

Figure 17.35 Enthalpy-concentration diagram for sodium hydroxide–water system.

[From W.L. McCabe, J.C. Smith, and P. Harriott, *Unit Operations of Chemical Engineering*, 5th ed., McGraw Hill, New York (1993) with permission.]

(y-axis) Enthalpy of system, Btu/lb

(x-axis) Mass fraction of NaOH

5. The ΔT is high enough to achieve *nucleate boiling* but not so high as to cause undesirable *film boiling*.

6. The exiting vapor temperature, T_v, corresponds to evaporator vapor-space pressure, P, taking into account boiling-point elevation of the solution, unless the solute is small, insoluble particles such as colloidal matter.

7. No heat is lost from the evaporator.

Based on the above, the evaporator model is:

Total mass balance:
$$m_f = m_p + m_v \qquad (17\text{-}77)$$

Mass balance on the solute:
$$w_f m_f = w_p m_p \qquad (17\text{-}78)$$

Energy (enthalpy) balance on the solution:
$$Q = m_v H_v + m_p H_p - m_f H_f \qquad (17\text{-}79)$$

where H = enthalpy per unit mass.

Energy (enthalpy) balance on the heating steam:
$$Q = m_s \Delta H^{\text{vap}} \qquad (17\text{-}80)$$

Heat-transfer rate:
$$Q = UA(T_s - T_p) \qquad (17\text{-}81)$$

The procedure used to solve these five equations depends on the problem specifications, the following example being typical. The solution of the energy-balance equations is greatly facilitated if an enthalpy-concentration diagram for the solute–solvent system is available. A diagram for the NaOH–water system is given in Figure 17.35, where the enthalpy datum for water is the pure liquid at 32°F, the same datum as the steam tables. For NaOH, the datum is NaOH at infinite dilution in water at 20°C (68°F). Figure 17.35, and the Dühring chart of Figure 17.32, were first prepared by McCabe [35].

EXAMPLE 17.17 Concentration of Aqueous NaOH.

An existing forced-circulation evaporator with a heat-transfer area of 232 m² is to be used to concentrate 44 wt% NaOH to 65 wt% NaOH using steam at 3 atm gage (barometer reads 1 atm). The feed temperature will be 40°C and pressure in the vapor space of the evaporator will be 2.0 psia. The overall heat-transfer coefficient for the given conditions is estimated to be 2,000 W/m²-°C. The density of the feed solution is 1,450 kg/m³. Neglecting heat losses from the evaporator, determine the: (a) temperature of the solution in the evaporator in °C; (b) heating steam in kg/h; (c) m³/h of feed; (d) kg/h of concentrated NaOH solution; (e) rate of evaporation in kg/h.

Solution

(a) At 2.0 psia, the boiling temperature of water is 126°F. From Figure 17.32, for 65 wt% NaOH, the solution boiling point is 240°F or 116°C. This is a considerable boiling-point elevation of 114°F.

(b) In American Engineering Units, $A = 232/0.0929 = 2{,}497$ ft²; $U = 2{,}000/5.674 = 352.5$ Btu/h-ft²-°F; $T_s = 291$°F for 4 atm saturated steam. The driving force for heat transfer is $\Delta T = T_s - T_p = 291 - 240 = 51$°F.

From (17-81),
$$Q = UA\,\Delta T = 352.5(2{,}497)(51) = 44{,}900{,}000 \text{ Btu/h}.$$

The heat of vaporization of steam at 291°F is 917 Btu/lb. From (17-80),
$$m_s = Q/\Delta H^{\text{vap}} = 44{,}900{,}000/917 = 48{,}950 \text{ lb/h}.$$

or   48,950(0.4536) = 22,200 kg/h.

The heat flux is 44,900,000/2,497 = 18,000 Btu/h-ft². This heat flux is safely in the nucleate-boiling region.

(c) From (17-79), using Figure 17.35 for NaOH solutions and the steam tables for water vapor, the energy balance on the solution is as follows, where the evaporated water is superheated steam at 240°F and 2.0 psia, and $m_v = m_f - m_p$:

$$44,900,000 = (m_f - m_p)(1168) + m_p(340) - m_f(115) \quad (1)$$

From (17-78),

$$0.44m_f = 0.65m_p \quad (2)$$

Substituting (2) into (1) to eliminate m_p,

$$m_f = \frac{44,900,000}{\left(1 - \dfrac{0.44}{0.65}\right)(1,168) + \left[\dfrac{0.44}{0.65}(340) - 115\right]}$$

$$= 91,170 \text{ lb/h}$$

or $m_f = (91,170)(0.4536) = 41,350 \text{ kg/h}.$

(d) From (2), $m_p = \dfrac{0.44}{0.65}(41,350) = 27,990 \text{ kg/h}$

(e) $m_v = m_f - m_p = 41,350 - 27,990 = 13,360 \text{ kg/h}$

§17.10.3 Multiple-Effect-Evaporator Systems

When condensing steam is used to evaporate water from an aqueous solution, the heat of condensation of the higher-temperature condensing steam is less than the heat of vaporization of the lower-temperature boiling water, so less than 1 kg of vapor is produced per kg of steam. This ratio is called the *economy*. In Example 17.17, the economy is 13,360/22,200 = 0.602. To reduce steam consumption and, thereby, increase economy, a series of evaporators, called *effects*, can be used, as diagrammed in Figure 17.36. The increased economy is achieved by operating effects at different pressures, and thus at different boiling points, so vapor from one effect can supply heat for another.

In Figure 17.36a, a *forward-feed, triple-effect-evaporator* system, approximately 1/3 of the total evaporation occurs in each effect. Fresh feed and steam both enter the first effect, which operates at P_1. Concentrate from the first effect is sent to the second effect. The vapor produced in the first effect is sent to the steam chest of the second effect, where it condenses, giving up its heat of condensation to cause additional evaporation. To achieve a temperature-driving force for heat transfer, $P_2 < P_1$. This procedure is repeated in the third effect. For three effects, the flow rate of steam entering the first is only about 1/3 of the amount of steam required if only one effect were used. However, the temperature-driving force in each of the three effects is only about 1/3 of that in a single effect. Therefore, the heat-transfer area of each of the three evaporators in a triple-effect system is approximately

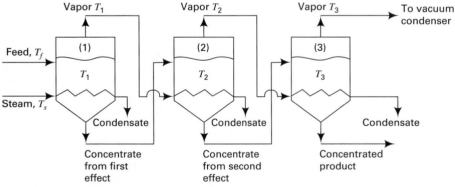

(a) Forward-feed, triple-effect

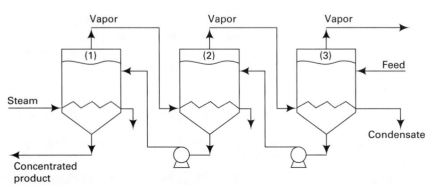

(b) Backward-feed, triple-effect

Figure 17.36 Multiple-effect evaporator systems.

the same as that for the one evaporator in a single-effect unit; thus, savings in steam cost must offset the additional equipment capital cost.

When the fresh-feed temperature is significantly below its saturation temperature, the *backward-feed* operation in Figure 17.36b is desirable. Cold fresh feed is sent to the third effect, which operates at the lowest P and, therefore, the lowest T. Unlike the forward-feed system, pumps are required to move concentrate from one effect to the next because $P_1 > P_2 > P_3$. However, unlike gas compressors, liquid pumps are not high-cost items.

Calculations for multieffect evaporators involve mass-balance, energy-balance, and heat-transfer equations that parallel those for single effects. These equations are usually solved by an iterative method, especially when boiling-point elevations occur. The particular iterative procedure depends on the problem, as demonstrated in the next example.

EXAMPLE 17.18 Triple-Effect-Evaporator System.

A feed of 44,090 lb/h of an aqueous solution containing 8 wt% colloids is to be concentrated to 45 wt% colloids in a triple-effect-evaporator system using forward feed. The feed enters the first effect at 125°F, and the third effect operates at 1.94 psia in the vapor space. Fresh saturated steam at 29.3 psia is used for heating the first effect. Specific heat of the colloids is 0.48 Btu/lb-°F. Overall heat-transfer coefficients are estimated to be:

Effect	U, Btu/h-ft^2-°F
1	350
2	420
3	490

If the heat-transfer areas of each of the three effects are to be equal, determine: (a) evaporation temperatures, T_1 and T_2, in the first two effects; (b) heating steam flow rate, m_s; and (c) solution flow rates, m_1, m_2, and m_3, leaving the three effects.

Solution

The unknowns, which number 7, are $A(=A_1 = A_2 = A_3)$, T_1, T_2, m_s, m_1, m_2, and m_3. Therefore, 7 independent equations are needed. Because the solute is colloids (insolubles), boiling-point elevations do not occur.

It is convenient to add 3 additional unknowns Q_1, Q_2, and Q_3, and, therefore, 3 additional equations, making a total of 10 equations. The 10 equations, which are similar to (17-78) to (17-81), are:

Overall colloid mass balance:

$$w_f m_f = w_3 m_3 \tag{1}$$

Energy balances on the solutions:

$$Q_1 = (m_f - m_1)H_{v_1} + m_1 H_1 - m_f H_f \tag{2}$$

$$Q_2 = (m_1 - m_2)H_{v_2} + m_2 H_2 - m_1 H_1 \tag{3}$$

$$Q_3 = (m_2 m_3)H_{v_3} + m_3 H_3 - m_2 H_2 \tag{4}$$

Energy balances on steam and water vapors:

$$Q_1 = m_s \Delta H_s^{\text{vap}} \tag{5}$$

$$Q_2 = (m_f - m_1)\Delta H_2^{\text{vap}} \tag{6}$$

$$Q_3 = (m_1 - m_2)\Delta H_3^{\text{vap}} \tag{7}$$

Heat-transfer rates:

$$Q_1 = U_1 A_1 (T_s - T_1) = U_1 A_1 \Delta T_1 \tag{8}$$

$$Q_2 = U_2 A_2 (T_1 - T_2) = U_2 A_2 \Delta T_2 \tag{9}$$

$$Q_3 = U_3 A_3 (T_2 - T_3) = U_3 A_3 \Delta T_3 \tag{10}$$

From (1),

$$m_3 = (w_f/w_3)m_f = (0.08/0.45)(44,090) = 7,838 \text{ lb/h}$$

Also, the flow rate of colloids in the feed is (0.08) $(44,090) = 3,527$ lb/h.

Initial estimates of solution temperature in each effect:

With no boiling-point elevation, the temperature of the solution in the third effect is the saturation temperature of water at the specified pressure of 1.94 psia or 125°F. The temperature of the heating steam entering the first effect is the saturation temperature of 249°F at 29.3 psia. Thus, if only one effect were used, the temperature-driving force for heat transfer, ΔT, would be $249 - 125 = 124$°F. With no boiling-point elevations because the colloids are not in solution, the ΔT_s for the three effects must sum to the value for one effect. Thus,

$$\Delta T_1 + \Delta T_2 + \Delta T_3 = 124°\text{F} \tag{11}$$

As a first approximation, assume that the ΔTs for the three effects are, using (8)–(10), inversely proportional to the given values of U_1, U_2, and U_3. Thus,

$$\Delta T_1 = \frac{U_3}{U_1} \Delta T_3 \tag{12}$$

$$\Delta T_2 = \frac{U_3}{U_2} \Delta T_3 \tag{13}$$

Solving (11), (12), and (13), $\Delta T_1 = 48.6°$F; $\Delta T_2 = 40.6°$F; $\Delta T_3 = 34.8°$F; and

$$T_1 = T_s - \Delta T_1 = 249 - 48.6 = 200.4°\text{F}$$
$$T_2 = T_1 - \Delta T_2 = 200.4 - 40.6 = 159.8°\text{F}$$
$$T_3 = T_2 - \Delta T_3 = 159.8 - 34.8 = 125°\text{F}$$

Initial Estimates of m_1 and m_2:

The total evaporation rate for the three effects is $m_f - m_3 = 44,090 - 7,838 = 36,252$ lb/h. Assume, as a first approximation, that equal amounts of vapor are produced in each effect. Then,

$$m_f - m_1 = 36,252/3 = 12,084 \text{ lb/h}$$
$$m_1 = 44,090 - 12,084 = 32,006 \text{ lb/h}$$
$$m_2 = 32,006 - 12,084 = 19,922 \text{ lb/h}$$
$$m_3 = 19,922 - 12,084 = 7,838 \text{ lb/h}$$

Corresponding estimates of the mass fractions of colloids are

$$w_1 = 3{,}527/32{,}006 = 0.110$$

$$w_2 = 3{,}527/19{,}922 = 0.177$$

$$w_3 = 3{,}527/7{,}838 = 0.450 \text{ (given)}$$

The remaining calculations are iterative in nature to obtain corrected values of T_1, T_2, m_1, and m_2, as well as values of A, m_s, Q_1, Q_2, and Q_3. These calculations are best carried out on a spreadsheet. Each iteration consists of the following steps:

Step 1

Combine (2) through (7) to eliminate Q_1, Q_2, and Q_3. Using the approximations for T_1, T_2, T_3, w_1, and w_2, the specific enthalpies for the resulting equations are calculated and the equations are solved for new approximations of m_s, m_1, and m_2. Corresponding approximations for w_1 and w_2 are computed.

For the first iteration, the enthalpy values are

$$\Delta H_s^{vap} = 946.2 \text{ Btu/lb}$$

$$\Delta H_2^{vap} = 977.6 \text{ Btu/lb}$$

$$\Delta H_3^{vap} = 1{,}002.6 \text{ Btu/lb}$$

$$H_{v_1} = 1{,}146 \text{ Btu/lb}$$

$$H_{v_2} = 1{,}130 \text{ Btu/lb}$$

$$H_{v_3} = 1{,}116 \text{ Btu/lb}$$

$$H_f = 0.92(92.9) + 0.08(0.48)(125 - 32) = 89.0 \text{ Btu/lb}$$

$$H_1 = 0.89(168.4) + 0.110(0.48)(200.4 - 32) = 158.8 \text{ Btu/lb}$$

$$H_2 = 0.823(127.7) + 0.177(0.48)(159.8 - 32) = 116.0 \text{ Btu/lb}$$

$$H_3 = 0.55(92.9) + 0.45(0.48)(125 - 32) = 71.2 \text{ Btu/lb}$$

When these enthalpy values are substituted into the combined energy balances, the following equations are obtained:

$$m_s = 49{,}250 - 1.043\, m_1 \tag{14}$$

$$44{,}090 = 1.994\, m_1 - 1.037\, m_2 \tag{15}$$

$$8{,}168 = 1.997\, m_2 - m_1 \tag{16}$$

Solving (14), (15), and (16), $m_s = 15{,}070$ lb/h; $m_1 = 32{,}770$ lb/h, and $m_2 = 20{,}500$ lb/h.

Corresponding values of colloid mass fractions are: $w_1 = 0.108$ and $w_2 = 0.172$.

It may be noted that these values of m_1, m_2, w_1, and w_2 are close to the first approximations. This is often the case.

Step 2

Using the values computed in Step 1, values of Q are determined from (5), (6), and (7); values of A are determined from (8), (9), and (10).

$$Q_1 = 15{,}070(946.2) = 14{,}260{,}000 \text{ Btu/h}$$

$$Q_2 = (44{,}090 - 32{,}770)(977.6) = 11{,}070{,}000 \text{ Btu/h}$$

$$Q_3 = (32{,}770 - 20{,}500)(1{,}002.6) = 12{,}400{,}000 \text{ Btu/h}$$

$$A_1 = \frac{14{,}260{,}000}{(350)(48.6)} = 838 \text{ ft}^2$$

$$A_2 = \frac{11{,}070{,}000}{(420)(40.6)} = 649 \text{ ft}^2$$

$$A_3 = \frac{12{,}400{,}000}{(490)(34.8)} = 727 \text{ ft}^2$$

Step 3

Because the three areas are not equal, calculate the arithmetic average, heat-transfer area, and a new set of ΔT driving forces from (8), (9), and (10). Normalize these values so that they sum to the overall ΔT (124°F in this example). From the ΔT values, compute T_1 and T_2.

$$A_{avg} = \frac{838 + 649 + 727}{3} = 738 \text{ ft}^2$$

$$\Delta T_1 = \frac{14{,}260{,}000}{(350)(738)} = 55.2°\text{F}$$

$$\Delta T_2 = \frac{11{,}070{,}000}{(420)(738)} = 35.7°\text{F}$$

$$\Delta T_3 = \frac{12{,}400{,}000}{(490)(738)} = 34.3°\text{F}$$

These values sum to 125.2°F. Therefore, they are normalized to

$$\Delta T_1 = 55.2(124/125.2) = 54.7°\text{F}$$

$$\Delta T_2 = 35.7(124/125.2) = 35.3°\text{F}$$

$$\Delta T_3 = 34.3(124/125.2) = 34.0°\text{F}$$

$$T_1 = 249 - 54.7 = 194.3°\text{F}$$

$$T_2 = 194.3 - 35.3 = 159.0°\text{F}$$

Steps 1 through 3 are now repeated using new values of T_1 and T_2 from Step 3 and new values of w_1 and w_2 from Step 1. The iterations are continued until values of the unknowns no longer change significantly and $A_1 = A_2 = A_3$. The subsequent iterations for this example are left as an exercise. Based on the results of the first iteration, the economy of the three-effect system is

$$\frac{44{,}090 - 7{,}838}{15{,}070} = 2.41 \quad \text{or} \quad 241\%$$

§17.10.4 Overall Heat-Transfer Coefficients in Evaporators

In an evaporator, the overall heat-transfer coefficient, U, depends mainly on the steam-side condensing coefficient, the solution-side forced-convection or boiling coefficient, and a scale or fouling resistance on the solution side. An additional wall resistance is present in glass-lined evaporators. The conduction resistance of the metal wall of the heat-exchanger tubes is usually negligible. Steam condensation is generally film, rather than dropwise. When boiling occurs on the surfaces of the heat-exchanger tubes, it is nucleate-boiling rather than film-boiling. In the absence of boiling on the tube surfaces, heat transfer is by forced convection to the solution. Local coefficients for film condensation, nucleate boiling, and forced convection of aqueous solutions are all relatively large, of the order of 1,000 Btu/h-ft^2-°F (5,700 W/m^2-K). Thus, the overall coefficient for clean tubes would be about 50% of this. However, when fouling occurs, the overall coefficient can be substantially less. Table 17.11, taken from Geankoplis [36], lists ranges of overall heat-transfer

Table 17.11 Typical Heat-Transfer Coefficients in Evaporators

	U	
Type Evaporator	Btu/h-ft^2-°F	W/m^2-K
Horizontal-tube	200–500	1,100–2,800
Short-tube-vertical	200–500	1,100–2,800
Long-tube-vertical	200–700	1,100–3,900
Forced-circulation	400–2,000	2,300–11,300

coefficients for different types of evaporators. Higher coefficients in forced-circulation evaporators are mainly a consequence of greatly reduced fouling due to high liquid velocity in the tubes.

§17.11 BIOPRODUCT CRYSTALLIZATION

High-purity (99.9%) crystals of bioproducts can be produced by evaporating, cooling, or diluting a homogeneous, aqueous solution with an organic solvent such as alcohol to supersaturate the dissolved species. Crystallization simultaneously purifies the targeted species and reduces process volume without requiring a costly sorbent or membrane. It also facilitates subsequent filtering and drying, and prepares bioproducts in an attractive final form that is convenient for therapeutic dose administration. For these reasons, *production-scale* crystallization is frequently the final purification or *polishing* step for antibiotics, enzyme inhibitors, and some proteins. Commodity bioproducts like sucrose and glucose are crystallized at quantities exceeding 100 million ton/yr. *Small-scale* crystallization can yield 0.2- to 0.9-mm crystals of proteins, whose three-dimensional structures can be characterized by X-ray diffraction. Figure 17.37 shows a crystal of the protein lysozyme. Pure crystals of bioproducts exhibit prolonged *stability* (§1.9) at low temperatures in the presence of stabilizing agents like $(NH_4)_2SO_4$, glycerol, and sucrose, which are added to formulate bioproduct crystals for use as pharmaceutical actives.

Bioproducts are typically crystallized batchwise by adjusting pH to the isoelectric point to reduce solubility [37], and/or by lowering the temperature, and/or by slowly adding salts, nonionic polymers, or organic solvents. Thermodynamic and kinetic considerations dictate cooling rates, purity, and number, shape, and size distribution of the crystals produced [38]. Examples include the cholesterol-synthesis inhibitor lovastatin, which is crystallized by

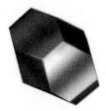

Figure 17.37 A single crystal of lysozyme protein, commonly found in egg whites.

cooling a vacuum-distilled, butyl–acetate extract from *Aspergillus terreus* fermentation broth to below 40°C [39]; alcohol oxidase protein, which is crystallized by dialysis or diafiltration to remove salt due to its low solubility at low ionic strength [40]; and ovalbumin protein, which is crystallized by seeding a mixture containing conalbumin and lysozyme impurities with small ovalbumin crystals [41].

§17.11.1 Comparing Inorganic and Biological Crystals

Like inorganic salts, crystallization of bioproducts involves nucleation, solute mass transfer to the surface, and incorporation of solute into the lattice structure. This is typically a multiphase, multicomponent, thermodynamically unstable process in which heat and mass transfer occur simultaneously. Unlike many inorganic salts, phase diagrams or kinetic data for bioproducts are usually unavailable. So, empirical determination of conditions for crystallization of proteins or antibiotics is usually required. Crystallization at low temperature and high concentration (10–100 g/L of protein) minimizes degradation of heat-sensitive materials, reduces unit cost, and maximizes purity and yield, as defined in §1.9.3.

As with inorganic compounds, crystals of antibiotics and proteins are composed of planar faces joined at an angle, characteristic of a particular substance, to form a polyhedron solid. The solid, three-dimensional space lattices formed by joining planar faces at characteristic angles are grouped into the seven crystal *systems* listed in Table 17.2. The relative size of adjoining faces in a crystal system may change due to impurities, solvent, or other conditions, thus forming various crystal shapes, or habits. For example, hexagonal crystals, as shown in Figure 17.38, may form tabular, prismatic, or needle-shaped, acicular habits, depending on the growth rate in the vertical direction. Unlike inorganic crystals, protein crystals typically contain significant solvent content capable of absorbing small molecules; are only weakly birefringent under cross polarizers; and powder, crumble, or break easily. Recrystallization in fresh solvent may be necessary, as with lovastatin (the first drug used to lower cholesterol), to reduce impurities that are substituted at lattice sites.

§17.11.2 Nucleation of Bioproducts

Bioproducts that are moved from a stable, unsaturated zone across a solubility curve, c_s, into a metastable zone (see Figure 17.11) can form new nuclei or grow existing crystals. High surface energies of small crystals form a thermodynamic barrier that perpetuates stability of a supersaturated solution in the metastable region. Above the metastable limit, c_m, in a labile zone, new nuclei form spontaneously from clear solution. Upon nucleation, a bioproduct transitions thermodynamically from a species dissolved in a fluid phase to a lattice element in a condensed solid phase. Changes in several physical properties distinguish this transition [42]. Crystallization of bioproducts involves formation of particles, too

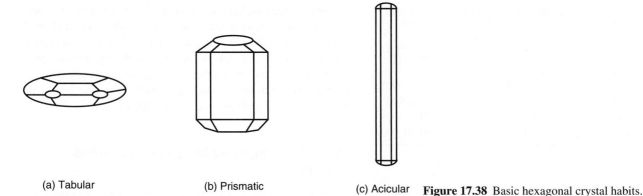

(a) Tabular   (b) Prismatic   (c) Acicular   **Figure 17.38** Basic hexagonal crystal habits.

small to be seen by an ordinary microscope, by primary and secondary nucleation and by attrition.

Primary nucleation of bioproducts occurs in the absence of crystals, either with (heterogeneous) or without (homogeneous) foreign particles present. In homogenous nucleation, clusters formed by diffusing solutes combine at high supersaturation into embryonic nuclei, which grow into macroscopic crystals, whose rate of homogeneous nucleation is given by (17-18). This requires a perfectly clean vessel with no rough surfaces, which is rare in practice. Therefore, formation of crystals of bioproducts via heterogeneous nucleation or secondary nucleation is modeled using power-law expressions similar to (17-19):

$$B = \frac{dN}{dt} = k_n M_T^j (c - c_s)^i \qquad (17\text{-}82)$$

where B is the number of nuclei formed per unit volume per unit time; N is the number of nuclei per unit volume; k_n is typically a function of T and impeller tip speed, N_i; M_T is the suspension density or mass of crystals per volume of suspension; c is the instantaneous solute concentration; and c_s is the saturation concentration at which the solute is in equilibrium with the crystal. The difference between instantaneous and saturation concentrations in (17-82) is supersaturation, Δc, represented by

$$\Delta c = c - c_s \qquad (17\text{-}83)$$

Related definitions for the supersaturation ratio, S, and the relative supersaturation, s, for biological crystallization are defined as in (17-16) and (17-17), respectively. For primary heterogeneous nucleation, exponent j in (17-82) is 0 and exponent i can range up to 10, with 3 to 4 being most common.

Secondary nucleation of bioproducts occurs due to existing crystal particles, which can be introduced by *seeding*. Mechanisms for secondary nucleation include fluid shear on growing crystal faces, which causes *shear nucleation*; and collision of crystals with each other, the impeller, or vessel internal surfaces, which causes *contact nucleation*. For secondary nucleation, exponent j in (17-82) can range up to 1.5, with 1 being typical, and exponent i ranges up to 5, with 2 being most probable. *Attrition* involves breakup of existing crystals into new particles

§17.11.3   Growth and Kinetics

Bioproduct crystals grow after nucleation as dissolved molecules move to their surfaces and are incorporated into crystal lattices, as given by (17-23a). To specialize this expression for a particular geometry, consider a cubic crystal with side length s, mass $m = \rho s^3$, crystal density, ρ, and surface area $A = 6s^2$. A characteristic crystal length L is given by [42],

$$L = \frac{6m}{\rho A} \qquad (17\text{-}84)$$

For a cube, $L = s$. Otherwise, L is generally close to $\bar{D}_p$ in §17.1. These metrics may be generalized using shape-specific geometric factors Φ_i for surface area (A) and volume (V), to define

$$m = \rho \Phi_V L^3 \qquad (17\text{-}85)$$

$$A = 6 \Phi_A L^2 \qquad (17\text{-}86)$$

where $\Phi_A = \Phi_V = 1$ for a cube. Substituting (17-85) and (17-86) into (17-23a) and rearranging yields a linear growth rate, G, given by

$$G = \frac{dL}{dt} = \frac{2 \Phi_A (c - c_s)}{\Phi_V \rho (1/k_c + 1/k_i)} \approx k_g (c - c_s)^i \qquad (17\text{-}87)$$

where k_c and k_i are rate constants for diffusive mass transfer of dissolved solutes to the crystal surface, and subsequent integration into the crystal lattice, respectively. The semi-empirical expression on the RHS of (17-87) contains k_g, a shape-specific growth rate constant that may include terms for effectiveness factor, η_r, and species molecular weight [44]. It also has an exponent, i, that is usually between 0 and 2.5, with 1 being the most common. Equation (17-87) indicates geometrically similar bioproduct crystals of the same material grown at a linear rate, G, referred to in §17.5.1 as the ΔL law. When linear growth rate is *not* independent of crystal size, the empirical Abegg–Stevens–Larson equation may be used [45]:

$$G = G^o (1 + aL)^b \qquad (17\text{-}88)$$

with empirical constants a and b and nuclei growth rate, G^o.

EXAMPLE 17.19   Seeded Batch Crystallization.

Antibiotics like tetracycline may be crystallized by controlled cooling of an aqueous solution in a stirred tank to grow crystal seeds [43]. Develop an expression for a *cooling curve* for a seeded batch crystallization conducted at constant supersaturation (neglecting nucleation). The expression is to give reactor temperature, T, as a function of time, t, in terms of the following system parameters: $T_o =$ initial temperature, $T_p =$ final product temperature, $m_s =$ mass of seed crystal, $m_p =$ mass of product crystal less the seed mass, $G =$ linear crystal growth rate, $L_s =$ seed crystal dimension, and $L_p =$ product crystal dimension.

In these circumstances, the change in solubility with time is balanced by crystal growth given in (17-23a). Assume a single crystal size, L_s, and a constant, linear growth rate, G, as given in (17-87). Neglect variation of solubility with temperature (dc_s/dT) over small temperature ranges.

Solution

At constant supersaturation, neglecting nucleation, the change in solubility with time balances crystal growth according to

$$V\frac{dc_s}{dt} = \frac{dm}{dt} = -\frac{A_{tot}}{1/k_c + 1/k_i}(c - c_s) \quad (1)$$

where V is crystal volume. For a single crystal size, expressions for seed crystal length, L_s, and constant linear growth rate, G, in (17-87) may be substituted into L in (17-86) to describe the changing area of a single crystal:

$$A = 6\Phi_A(L_s + Gt)^2 \quad (2)$$

The number of seed crystals, NV, is the ratio of the total seed mass, m_s, to the mass per crystal in (17-85), given by

$$NV = \frac{m_s}{\rho\Phi_V L_s^3} \quad (3)$$

where N is the number of crystals per total solution volume, V. The total crystal area in the solution, $A_{tot} =$ (crystal number) × (area per crystal) in (1), is given by multiplying (2) by (3),

$$A_{tot} = \left(\frac{m_s}{\rho\Phi_V L_s^3}\right)\left[6\Phi_A(L_s + Gt)^2\right] \quad (4)$$

where ρ, m_s, and L_s are the density and initial mass and size, respectively, of the seed crystals, and G is given by (17-87). Applying the chain rule to the RHS of (1) gives the saturation concentration as a function of temperature:

$$\frac{dc_s}{dT}\frac{dT}{dt} = \frac{dc_s}{dt} \quad (5)$$

Substituting (4), (5), and (17-87) into (1) and rearranging gives the time change in temperature needed to sustain constant linear crystal growth,

$$\frac{dT}{dt} = -\left(\frac{m_s/V}{dc_s/dt}\right)\frac{3G}{L_s^3}(L_s + Gt)^2 \quad (6)$$

Variation of solubility with temperature can be neglected over small temperature ranges. This allows integration of (6) from initial

temperature, T_0, to yield

$$T = T_0 - \left(\frac{m_s/V}{dc_s/dt}\right)\frac{3Gt}{L_s}\left[1 + \frac{Gt}{L_s} + \frac{1}{3}\left(\frac{Gt}{L_s}\right)^2\right] \quad (7)$$

Equation (7) is often written in dimensionless terms. The seed mass, m_s, is scaled by m_p, the maximum possible mass of crystalline product less the seed mass, m_p, given by

$$m_p = (T_0 - T_p)V\frac{dc_s}{dT} \quad (8)$$

where T_p is the final product temperature. The seed size is written in terms of η, the fractional increase in product size, L_p, relative to seed size, L_s, is given by

$$\eta = \frac{L_p - L_s}{L_s} = Gt/L_p - L_s \quad (9)$$

The time is normalized by the total time to produce product t_p, to give

$$\tau = \frac{t}{t_p} = Gt/L_p - L_s \quad (10)$$

Substituting (8), (9), and (10) into (7) yields a tractable quadratic expression,

$$\frac{T - T_0}{T_P - T_0} = \frac{m_s}{m_p}3\eta\tau\left[1 + \eta\tau + \frac{1}{3}(\eta\tau)^2\right] \quad (11)$$

EXAMPLE 17.20   Cooling Curve Plot.

Use the results of Example 17.19 to construct a cooling curve—i.e., a plot of scaled temperature, T/T_0, versus dimensionless time, τ—to crystallize tetracycline antibiotic by reducing the temperature 20°C from an initial ambient value of 20°C. Initial seed concentration and length are 27 ppm and 0.01 cm, respectively. The maximum possible mass of 0.095-cm crystalline product less the seed mass is 57 ppm.

Solution

The *cooling curve* given by (11) is plotted in Figure 17.39 (dotted line) as T/T_o versus τ, after some rearrangement of (11). Parameter

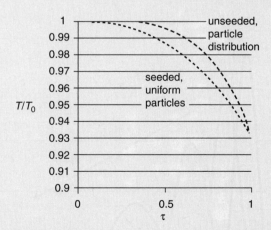

Figure 17.39 Cooling curves for batch crystallization. Growth of uniform particles in seeded vessel (dotted; Eq. 11 of Example 17.19) versus distribution of particles in an unseeded vessel (dashed; Eq. 4 of Example 17.23). Parameter values are $T_0 = 293$ K, $\eta = 8.5$, $m_s = 27$ ppm, $T_p = 273$ K, and $m_p = 57$ ppm.

values used to generate the curve are $T_0 = 293$ K, $L_p = 0.095$ cm, $L_s = 0.01$ cm($\eta = 8.5$), $m_s = 27$ ppm, $T_p = 273$ K, and $m_p = 57$ ppm. Temperature is reduced according to this expression in order to minimize excessive early nucleation, which produces small, nonuniform product. Following the cooling curve in (11) also prevents formation and buildup of condensed-product *scale* on heat-transfer surfaces. Exercise 17.43 illustrates the use of this method. A cooling approach for unseeded crystallization that accounts for crystal-size distribution (dashed line) is developed in Example 17.23.

§17.11.4 Crystal-Size Distributions

Distributions of the number, length, area, and volume or mass of crystals for an average overall linear growth rate can be calculated using the method of moments, as illustrated in Figure 17.18 and Table 17.9. Exponent values in (17-82) and (17-87) show that as supersaturation increases, nucleation rate typically increases faster than growth rate. Thus, low supersaturation (aided by seeding) favors large crystals, while supersaturation near the metastable limit favors more nuclei and smaller crystals. Reducing supersaturation can decrease impurities contained in inclusions of mother liquor within individual crystals.

Three *mechanisms* for growth of biological crystals that result from different relative contributions from nucleation and mass-transfer yield different *crystal-size distributions* (CSD) over time, called *chronomals* [42]. *Diffusion-limited* growth yields CSDs that sharpen and increase in magnitude with time. Weak crystal surfaces with essentially one nucleation site per crystal result in nucleation-limited *mononuclear* growth, characterized by CSDs that broaden and decay with time, as shown in Figure 17.40. Multiple nucleation sites, together with weak crystal surfaces, balance nucleation and diffusion and produce *polynuclear* growth that exhibits CSDs that remain relatively constant over time. These mechanisms and their corresponding chronomals are modeled by different driving forces, geometric parameters, and rate constants.

§17.11.5 Biological Crystallization vs. Precipitation

Crystallization and precipitation both yield solid particles from a solution. Bioproducts of crystallization are generally soluble, with large, well-defined, regular morphologies. Crystals nucleate at low rates via controllable secondary nucleation at low supersaturation, where product concentration is in a labile (i.e., open to change) region. In contrast, precipitated bioproducts are sparingly soluble, with small, ill-defined, irregular morphologies. Precipitates nucleate at high rates via poorly controlled primary nucleation at high supersaturation, where product concentration exceeds a labile region. Precipitation of bioproducts is often used in the early steps of the bioseparation process to reduce process volume, and is treated in §19.7.1.

§17.11.6 Dilution Batch Crystallization

Crystallization of antibiotics from aqueous solution typically occurs in *batch* operation (see §1.9) by adding a miscible organic *diluent*, such as ethanol, to reduce solubility of the target species [49]. Solubility may also be reduced by adding a salt-containing diluent (*salting-out*) or by causing a chemical reaction of the species by adding/removing a proton or coupling two molecules. For dilution batch crystallization, solute solubility may be written in terms of diluent concentration as

$$c_s = c_{s,0}\exp(-k_d c_d) \tag{17-89}$$

where c_d and k_d are diluent volume per total solvent volume, and proportionality constant, respectively.

EXAMPLE 17.21 Dilution Batch Crystallization.

Dilution batch crystallization is often performed at a constant rate of diluent addition, k_m, given by

$$\frac{dc_d}{dt} = k_m \tag{1}$$

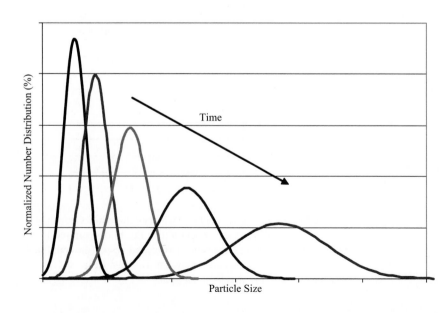

Figure 17.40 Growth chronomals for mononuclear crystal-growth mechanism.

In terms of k_m, the batch solution volume, $V\{t\}$, which includes solvent plus diluent volumes, varies according to

$$V\{t\} = V_0 + V\{t\}k_m t \qquad (2)$$

where the initial batch volume, V_0, is free of diluent.

General expressions for both the time rate of change of target species solubility, c_s, in (17-89) and of batch solution volume, $V\{t\}$, in (2) at a constant rate of diluent addition are useful to predict outcomes of dilution batch crystallization. One of these outcomes is the time required to obtain a target crystal size, which is examined in Example 17.22. It is necessary to differentiate (17-89) and (2) to obtain general expressions for the time rate of change of target species solubility and of batch solution volume. A typical rate for (1) is 10^{-6} L of diluent per L of solvent per second. A value of 10 L of solvent plus diluent per L of diluent for k_d in (17-89) is common. What are the magnitudes of the general expressions for these typical values of diluent addition rate and solubility proportionality constant, if the initial solvent volume is 1 L?

Solution

To obtain an expression for time rate of change of batch solution volume, (2) is rearranged to give

$$V\{t\} = \frac{V_0}{1 - k_m t} \qquad (3)$$

Differentiating (3) with respect to time gives the change in batch solution volume:

$$\frac{dV\{t\}}{dt} = \frac{V_0 k_m}{(1 - k_m t)^2} = \frac{V^2}{V_0} k_m \qquad (4)$$

For typical conditions and a 1-L solvent volume, this gives

$$\frac{dV\{t\}}{dt} = 10^{-6} V^2 \qquad (5)$$

To obtain an expression for time rate of change of target species solubility, (17-89) is differentiated with respect to time,

$$\frac{dc_s}{dt} = -c_{s,0} k_d \frac{dc_d}{dt} \exp(-k_d c_d) = -k_d k_m c_s \qquad (6)$$

For typical diluent addition rate and solubility proportionality constant,

$$\frac{dc_s}{dt} = -\frac{10^{-5}}{\text{sec}} c_s$$

Both of these rates of change are relatively small.

§17.11.7 Moment Equations for Population Balance

As illustrated in §17.5 for *continuous* crystallizers, moments analysis provides estimates of nucleation and growth rates in systems where a distribution of crystal sizes is present. In dilution *batch* crystallization used to recover bioproducts, the working volume varies with time, as illustrated in Example 17.21. To account for this time variation, refined definitions for the population density distribution function, $n\{L\}$; for the number of crystals per unit size per unit volume in (17-31); and for nuclei formation rate, B, in (17-82) are

introduced using the relations

$$\hat{n}\{L\} = n\{L\}V\{t\} \qquad (17\text{-}90)$$

$$\hat{B} = BV\{t\} \qquad (17\text{-}91)$$

where the hat symbol (^) indicates quantities that depend on total working volume in the crystallizer. Since no crystals flow into or out of a batch reactor, the population balance for a perfectly mixed dilution batch crystallizer with negligible attrition and agglomeration reduces from (17-36) to

$$\frac{\partial \hat{n}}{\partial t} + \frac{\partial \hat{n}G}{\partial L} = 0 \qquad (17\text{-}92)$$

where only the leftmost terms for accumulation and growth remain. Solution of this equation is simplified by a variable transformation that combines variables for time, t, and size-independent growth rate, $G\{t\}$, into a new crystal dimension, y, that nucleated at time $t = 0$, by the relation

$$dy = G\{t\}dt \qquad (17\text{-}93)$$

Substituting (17-90) and (17-93) into (17-92) gives a population balance for a crystal system where growth rate, $G\{t\}$, is size-independent

$$\frac{\partial \hat{n}}{\partial y} + \frac{\partial \hat{n}}{\partial L} = 0 \qquad (17\text{-}94)$$

which has boundary conditions ($y = 0$) for the nuclei population density given by

$$\hat{n}\{t,0\} = \hat{n}_0 = \frac{\hat{B}}{G} \qquad (17\text{-}95)$$

Now the definition for a moment equation in (17-45) may be applied to transform (17-94) in order to average the distribution with respect to internal coordinate properties. This leads to the following set of moment equations [46]:

$$\frac{d\hat{N}}{dy} = \hat{n}\{0,y\} \qquad (17\text{-}96)$$

$$\frac{d\hat{L}}{dy} = \hat{N} \qquad (17\text{-}97)$$

$$\frac{d\hat{A}}{dy} = 2\Phi_A \hat{L} \qquad (17\text{-}98)$$

$$\frac{d\hat{m}}{dy} = \frac{3\Phi_V \rho}{\Phi_A} \hat{A} \qquad (17\text{-}99)$$

which are dilution-batch analogs of the differential crystal-size distributions for continuous crystallizers summarized in Table 17.9 and illustrated in Figure 17.18b. In these new relations, $\hat{N}$, $\hat{L}$, $\hat{A}$, and $\hat{m}$ represent the number, characteristic length, area, and mass of crystals based on total reactor volume V, respectively (i.e., $\hat{N} = NV\{t\}$, etc.); ρ is crystal density; and Φ_i are shape-specific geometric factors for surface area, A, and volume, V, defined in (17-85) and (17-86).

Tavare et al. [46] summarized solutions of (17-96) to (17-99) for the batch population density function for seeding with crystals of uniform size and negligible nucleation, and for generalized sets of initial conditions with both size-independent and size-dependent growth rates. Their application is illustrated in Example 17.22 using a constant-nuclei population density.

EXAMPLE 17.22 Batch Crystallization of Tetracycline.

Based on the crystal-size distribution, determine the time required to obtain cubic, 0.06-cm tetracycline crystals ($\rho = 1.06$ g/cm^3) from a liter of batch volume [53]. At 0.01 g/cm^3 supersaturation, the nucleation rate is 100 per second and diffusion-limited growth has a mass-transfer coefficient of 6.5×10^{-5} cm/s. Use a constant-diluent addition rate, k_m, of 10^{-6} L of diluent per L of solvent per second and a solubility proportionality constant, k_d, of 10 L of solvent plus diluent per L of diluent. Solubility of tetracycline acid in aqueous solution is 100 g/L.

Begin by using (17-95) to (17-99) to obtain a general expression for the time t required to obtain crystals of a desired size y for an unseeded dilution batch crystallizer at a constant rate of diluent addition, k_m, given by (1) in Example 17.21, and a constant-nuclei population density, $\hat{n}\{0,0\}$, based on the initial supersaturation level. In this expression, the time required will be a function of y, k_m, $\hat{n}\{0,0\}$, ρ, V, Φ_V, k_d, and $c_{s,0}$ (solubility).

Solution

From (17-95) the nuclei population density remains constant at its initial ($t = 0$) value of

$$\hat{n}\{0,0\} = \hat{n}_0^0 = \frac{\hat{B}}{G} \tag{1}$$

In the crystallizer, change in solute concentration, $\hat{c} = cV\{t\}$, balances crystal formation, $\hat{m} = mV\{t\}$:

$$d\hat{c} + d\hat{m} = 0 \tag{2}$$

The definition of supersaturation in (17-83) can be rewritten to include time-varying working volume $V\{t\}$ as in (17-90) to obtain an expression for time-varying supersaturation, $\Delta\hat{c}$. Substituting this volume-dependent supersaturation into (2) yields

$$\frac{d\Delta\hat{c}}{dy} + \frac{d(Vc_s)}{dy} + \frac{d\hat{m}}{dy} = 0 \tag{3}$$

The middle term in (3) may be expanded via the chain rule. Then substituting (4) and (5) from Example 17.21 into the resulting expansion gives

$$\frac{d(Vc_s)}{dy} = \frac{d(Vc_s)}{dt}\frac{dt}{dy} = \left(V\frac{dc_s}{dt} + c_s\frac{dV}{dt}\right)\frac{dt}{dy} = c_s k_m V\left(\frac{V}{V_0} - k_d\right)\frac{dt}{dy} \tag{4}$$

Introducing $\hat{n}_0^0$ into (17-96) to (17-98) and integrating, yields for (17-99):

$$\frac{d\hat{m}}{dy} = \Phi_V \rho \hat{n}_0^0 y^3 \tag{5}$$

Note in (5) that the Φ_A term in (17-98) has canceled the corresponding term in (17-99). Then, for small changes in volume ($V \approx V_0$) and solubility ($c_s \approx c_{s,0}$), substituting (4) and (5) into (3), followed by rearrangement and integration, yields

$$t = \frac{\Phi_V \rho \hat{n}_0^0}{4c_{s,0}k_m V_0(k_d - 1)} y^4 \tag{6}$$

Equation (6) reveals that the time required in a dilution batch crystallizer is proportional to the fourth power of the desired crystal dimension when the CSD is taken into account.

To find the growth rate, G, values of $\rho = 1.06$ g/cm^3, $\Delta c = 0.01$ g/cm^3, $k_c = 6.5 \times 10^{-5}$ cm/s, and $\Phi_A \approx \Phi_V \approx 1$ for

cubic crystals are inserted into (17-87), giving

$$G = (2)6.5 \times 10^{-5}\left(\frac{1}{1(1.06)}\right)0.01 = 1.2 \times 10^{-6} \text{cm/s}$$

Substituting parameter values into (6):

$$t = \frac{(1)(1.06)\left(\dfrac{10^2}{1.2 x 10^{-6}}\right)}{4(100)(10^{-6})(1)(10-1)}0.06^4 = 3.1 \times 10^5\text{s} = 86 \text{ h}$$

§17.11.8 Constant Supersaturation

High cooling rates at the initiation of batch crystallization cause excessive nucleation, yielding small, nonuniform product as well as resistive scale on heat-transfer surfaces. One strategy to produce uniform crystals is to maintain constant supersaturation, Δc, during crystallization [45]. This can be achieved by maintaining the change in temperature, T, during cooling in constant proportion, k_T^{-1}, to the change in solute concentration:

$$\frac{dT}{dt} = -\frac{1}{k_T}\frac{dc}{dt}. \tag{17-100}$$

In like manner, the uniformity of crystals *salted-out* by addition of an ionogenic agent or water-soluble polar organic solvent may be improved by controlling additive concentration, c^S. To describe this effect, c^S is substituted for c_d in (17-83), and an experimental constant is added. After taking the natural log of both sides, an expression for solute solubility in terms of additive concentration results.

$$\ln\frac{c}{c_0} = a - bc^S \tag{17-101}$$

where a and b are experimentally determined constants.

EXAMPLE 17.23 Constant Supersaturation.

Use (17-100) and the solute mass balance in (2) in Example 17.22 to derive a relationship between dimensionless temperature and dimensionless time that may be used to control temperature so as to maintain constant supersaturation in a cooled batch crystallizer with no seeding. Apply this relationship to generate a cooling curve for crystallizing cubic tetracycline antibiotic ($\rho = 1.06$ g/cm^3) by reducing the temperature 20°C from an initial ambient value of 20°C. At 0.01 g/cm^3 supersaturation, the nucleation rate is 100 per L-s and diffusion-limited growth has a mass-transfer coefficient of 6.5×10^{-5} cm/s. Use a proportionality constant of 7.8×10^{-14} crystals/°C-cm^3.

Solution

Writing (2) of Example 17.22 in terms of time, at constant volume, V, and growth rate, G, yields

$$\frac{dm}{dt} = -\frac{dc}{dt} \tag{1}$$

Writing (5) of Example 17.22 in terms of time, at constant volume and growth rate using (17-92) to replace dy with dt yields

$$\frac{dm}{dt} = \Phi_V \rho n_0^0 G^3 t^3 \tag{2}$$

Substituting (17-100) and (2) into (1) and integrating with the temperature at which crystal formation begins as $T\{t = 0\} = T_0$ yields

$$T_0 - T = \frac{\Phi_V \rho n_0^0 G^3 t^4}{4k_T} \qquad (3)$$

Writing the temperature difference and time relative to final values of T, T_f, and time t_f yields

$$\frac{T_0 - T}{T_0 - T_f} = \left(\frac{t}{t_f}\right)^4 \qquad (4)$$

with

$$t_f = \left[\frac{4k_T(T_0 - T_f)}{\Phi_V \rho n_0^0 G^3}\right]^{1/4} \qquad (5)$$

Controlling the *unaccomplished temperature* given by (4) is intended to maintain supersaturation and produce more uniform crystals. From (17-87), $G = 1.2 \times 10^{-6}$ cm/s. From (5), the value for t_f is

$$t_f = \left[\frac{4(7.8 \times 10^{-14})(293 - 273)}{(\Phi_V)(1.06)\left(\frac{100}{1.2 \times 10^{-6}}\right)\left(\frac{1}{1000}\right)(1.2 \times 10^{-6})^3}\right]^{1/4} = 2.5$$

The *cooling curve* given by (4) is plotted in Figure 17.39 (dashed line) as T/T_o versus $t/t_f \equiv \tau$, after some rearrangement of (4). Initial and final temperatures of $T_o = 293$ K and $T_f = 273$ K are used to generate the curve. The slope of the cooling curve for constant supersaturation that accounts for CSD begins more gradually, but ends more steeply than the slope of the cooling curve for constant supersaturation from Example 17.20, which neglects nucleation. Exercise 17.44 illustrates use of a similar approach to control salting-out of the solute using (17-101).

§17.11.9   Micromixing

Uniform supersaturation, number density, and rapid mass transfer throughout the process volume are provided by effective mixing at the microscopic scale, i.e., *micromixing* [47]. In a homogeneous, isotropic model for turbulent micromixing developed by Kolmogoroff, mass is transferred instantaneously between small eddies that are formed, and shed, via *eddy dispersion*. The mean eddy length, l_{eddy}, within which mass transfer is diffusion-limited, is

$$l_{\text{eddy}} = \left(\frac{\rho \nu^3}{P/V}\right)^{\frac{1}{4}} \qquad (17\text{-}102)$$

where ρ and ν are density and kinematic viscosity of the liquid, P is power dissipated, and V is volume of liquid. The mixing time, t_m, required for molecules to diffuse across a spherical eddy of diameter l_{eddy} is

$$t_m = \frac{l_{\text{eddy}}^2}{8D} \qquad (17\text{-}103)$$

where D is diffusivity in the medium. The mixing time in (17-103) corresponds to the time required for the system to equilibrate after a change of composition and, therefore,

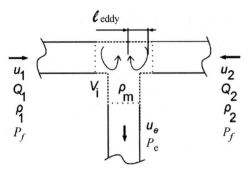

Figure 17.41 Tee-mixing schematic.

(1) limits rates of addition or cooling that affect crystallization and (2) controls the rate of mass transfer in crystal growth.

Impingement Mixing

Two opposing streams intersecting on a common axis can produce a jet-impingement plane that maximizes power dissipation per unit volume, P/V. This minimizes l_{eddy} in (17-102) and t_m in (17-103), and optimizes micromixing and mass transfer. Impingement mixing is used at production scale to crystallize statin inhibitors of cholesterol-forming enzymes. Intersecting opposing streams in a tee (i.e., *tee-mixing*) approaches the power dissipation of jet impingement. Key parameters for tee-mixing are summarized in Figure 17.41, from [48]. Inelastic collision of two opposing streams, identified by subscript $k = 1$ or 2, dissipates power in proportion to the kinetic energy in each stream,

$$P = \frac{1}{2}\left(\rho_1 Q_1 u_1^2 + \rho_2 Q_2 u_2^2\right) \qquad (17\text{-}104)$$

where Q_k is the volumetric flow rate and u_k is the linear stream velocity. Power dissipation occurs over a mass of fluid, m, given by

$$m = \rho_m V \qquad (17\text{-}105)$$

where ρ_m is the mixture stream density and V is the impingement volume, proportional to the mixing-tee volume. In Exercise 17.45, dimensional analysis of tee-mixing is performed to identify the parametric dependence of eddy length and mixing time, in order to design and operate tee-mixers for crystallization.

§17.11.10   Scale-Up

Ideally, scale-up of crystallization maintains local *mixing* conditions that provide *uniform* supersaturation, number density, and mass transport that are consistent with values from laboratory and pilot-plant studies. In mixed vessels, this is done by maintaining geometric similarity,

$$V \propto D_i^3 \qquad (17\text{-}106)$$

where D_i is impeller diameter, at constant power per unit volume, given by (8-21), in turbulent flow determined by the impeller Reynolds number, given by (8-22).

EXAMPLE 17.24 Batch Crystallizer Scale-Up.

Identify conditions listed in the following table for a 50-fold scale-up of batch crystallization of an antibiotic in a 1-L vessel that provides well-mixed results at 250 rpm.

	Laboratory (1)	Plant (2)
Volume	1 L	50 L
Rotation rate (rpm)	250	$N_{i,2}$
Impeller D_i	10.8 cm	$D_{i,2}$

Solution

From conditions identified in §8.5 for ideal, cylindrical mixing vessels (vessel height = vessel diameter and impeller diameter = 1/3 of the vessel diameter), the diameter, $D_{T,1}$, of the 1-L vessel in cm is

$$D_{T,1} = \left(\frac{4V}{\pi}\right)^{1/3} = \left[\frac{4(1000)}{\pi}\right]^{1/3} = 10.8 \text{ cm}$$

and its impeller diameter is $D_T/3 = 3.5$ cm. From (17-106), the large-scale impeller diameter is

$$D_{i,2} = D_{i,1}\left(\frac{V_2}{V_1}\right)^{1/3} = 3.5\left(\frac{50}{1}\right)^{1/3} = 12.9 \text{ cm}$$

and the large-scale vessel diameter and height are $H_{T,2} = D_{T,2} = 3D_{i,2} = 38.7$ cm. From (8-21), constant power per volume requires

$$N_{i,2}^3 D_{i,2}^2 = N_{i,1}^3 D_{i,1}^2$$

so the large-scale rotation rate is

$$N_{i,2} = N_{i,1}\left(\frac{D_{i,1}}{D_{i,2}}\right)^{2/3} = 250\left(\frac{3.5}{12.9}\right)^{2/3} = 105 \text{ rpm}$$

To use (8-21), the flow is confirmed to be turbulent at the 1-L scale (and therefore at the 50-L scale) with (8-22). Using CGS units and assuming properties of water,

$$N_{Re} = \frac{D_i^2 N \rho}{\mu} = \frac{(3.5)^2 (250/60)(1)}{0.01} = 5,400$$

§17.11.11 Crystal Recovery

Crystallization and recovery are interrelated because the nature and size of the crystals affect centrifugation and washing rates [52]. Batch Nutsche-type filters are often used to recover high-purity crystals of bioproducts like antibiotics. Further information about filters and centrifuges suitable for recovery of bioproduct crystals are discussed in §19.4 and 19.5. Acceptable dryers are discussed in §18.6.

SUMMARY

1. Crystallization involves formation of solid crystalline particles from a homogeneous fluid phase. If crystals are formed directly from a gas, the process is desublimation.

2. In crystalline solids, as opposed to amorphous solids, molecules, atoms, and/or ions are arranged in a regular lattice pattern. When crystals grow unhindered, they form polyhedrons with flat sides and sharp corners. Although faces of a crystal may grow at different rates, referred to as crystal habit, the Law of Constant Interfacial Angles restricts the angles between corresponding faces to be constant. Crystals can form only 7 different crystal systems, which include 14 different space lattices. Because of crystal habit, a given crystal system can take on different shapes, e.g., plates, needles, and prisms, but not spheres.

3. Crystal-size distributions can be determined or formulated in terms of differential or cumulative analyses, which are convertible, one from the other. A number of different mean-particle sizes can be derived from crystal-size distribution data.

4. Important thermodynamic properties for crystallization calculations are melting point, solubility, heat of fusion, heat of crystallization, specific heat, heat of solution, heat of transition, and supersaturation. Solubilities of inorganic salts in water can vary widely, from a negligible value to concentrations of greater than 50 wt%. Many salts crystallize in hydrated forms, with the number of waters of crystallization of a stable hydrate depending

upon temperature. The solubility of sparingly soluble compounds is usually expressed in terms of a solubility product. When available, phase diagrams and enthalpy-concentration diagrams are useful for mass- and energy-balance calculations.

5. Crystals smaller in size than can be seen by the naked eye ($< \sim 20$ mm) are more soluble than the normally listed solubility. Supersaturation ratio for a given crystal size is the ratio of the actual solubility of a small-size crystal to the solubility of larger crystals that can be seen by the naked eye. The driving force for nucleation and growth of crystals is supersaturation.

6. Primary nucleation, which requires a high degree of supersaturation, occurs in systems free of crystalline surfaces, and can be homogeneous or heterogeneous. Secondary nucleation occurs when crystalline surfaces are present. Crystal growth involves mass transfer of solute to the crystal surface followed by incorporation of the solute into the crystal-lattice structure.

7. Equipment for solution crystallization can be classified according to operation mode (batch or continuous), method for achieving supersaturation (cooling or evaporation), and features for achieving desired crystal growth (e.g., agitation, baffles, circulation, and classification). Of primary importance is the effect of temperature on solubility. Three of the most widely used types of equipment for solution crystallization are: (1) batch

crystallizer with external or internal circulation; (2) continuous, cooling crystallizer; and (3) continuous, vacuum, evaporating crystallizer.

8. The MSMPR crystallization model is widely used to simulate the continuous, vacuum, evaporating draft-tube, baffled crystallizer. Some of the assumptions are perfect mixing of the magma, no classification of crystals, uniform degree of supersaturation throughout the magma, crystal growth rate independent of crystal size, no crystals in the feed, no crystal breakage, uniform temperature, equilibrium in product magma between mother liquor and crystals, constant and uniform nucleation rate due to secondary nucleation by crystal contact, uniform crystal-size distribution, and uniform crystal shape.

9. For a specified crystallizer feed, magma density, magma residence time, and predominant crystal size, the MSMPR model can predict the required nucleation rate and crystal-growth rate, number of crystals produced per unit time, and size distribution.

10. Precipitation, leading to very small crystals, occurs with solutes that are only sparingly soluble. The precipitate is often produced by reactive crystallization from the addition of two soluble salt solutions, producing one soluble and one insoluble salt. Unlike solution crystallization, which takes place at a low degree of supersaturation, precipitation occurs at a high supersaturation that results in very small crystals.

11. When both components of a mixture can be melted at reasonable temperatures (e.g., certain mixtures of organic compounds), melt crystallization can be used to separate the components. If the components form a eutectic mixture, pure crystals of one of the components can be formed. However, if components form a solid solution, repeated stages of melting and crystallization are required for high purity.

12. Many crystallizer designs have been proposed for melt crystallization. Two major methods are suspension crystallization and layer crystallization. Of particular importance is the falling-film crystallizer, which can be designed for high production rates when the components form eutectic mixtures. For components that form solid solutions, the zone-melting technique developed by Pfann can be employed to produce nearly pure compounds.

13. A number of chemicals are amenable to purification by desublimation, preceded perhaps by sublimation. Desublimation is almost always achieved by cooling a gas mixture at constant pressure by heat transfer, quenching with a vaporizable liquid, or quenching with a cold, non-condensable gas.

14. Evaporation is used to concentrate a solute prior to solution crystallization. Common evaporators include the horizontal-tube unit, short-vertical-tube unit, long-vertical-tube unit, forced-circulation unit, and falling-film unit. For a given evaporation pressure, the presence of a solute can cause a boiling-point elevation.

15. The most widely used evaporator model assumes the liquor being evaporated is well-mixed, so the temperature and solute concentration are uniform and at exiting conditions.

16. Economy of an evaporator is defined as the mass ratio of water evaporated to heating steam required. It can be increased by using multiple evaporator effects that operate at different pressures such that vapor produced in one effect can be used as heating steam in a subsequent effect. The solution being evaporated can progress through the effects in forward, backward, or mixed directions.

17. Evaporators typically operate so that solutions are in the nucleate-boiling regime. Overall, heat-transfer coefficients are generally high because boiling occurs on one side and condensation on the other side of the tubes.

18. Crystallization of bioproducts takes advantage of decreased solubility upon cooling, pH adjustment or slow addition of salts, and use of nonionic polymers or organic solvents to produce stable, high-purity crystals in an attractive final form. Sugars, antibiotics, enzyme inhibitors, and proteins are common examples of bioproducts requiring crystallization.

19. Primary and secondary nucleation of bioproduct crystals, followed by growth via diffusion and lattice incorporation, is characterized by semi-empirical power-law expressions. Together with solubility data, these expressions are useful to obtain operating curves for cooling or solvent addition to maintain supersaturation and produce large, uniform bioproduct crystals.

20. Expressions for crystal-size distributions due to diffusion- or kinetic-controlled growth and dilution crystallization are useful for characterizing final bioproducts and selecting operating parameters to achieve targeted size distributions.

21. Uniform mixing is key to maintaining constant conditions in order to achieve targeted, predicted crystal-size distributions for bioproducts, and maintain performance during scale-up. Jet impingement and tee-mixing provide high-energy dissipation that minimizes eddy length and mixing times in crystallization.

REFERENCES

1. Mullin, J.W., *Crystallization*, 3rd ed., Butterworth-Heinemann, Boston (1993).

2. Graber, T.A., and M.E. Taboada, *Chem. Eng. Ed.*, **25**, 102–105 (1991).

3. Hougen, O.A., K.M. Watson, and R.H., Ragatz, *Chemical Process Principles, Part I, Material and Energy Balances*, 2nd ed., John Wiley & Sons (1954).

4. Miers, H.A., and F. Isaac, *Proc. Roy. Soc.*, **A79**, 322–351 (1907).

5. Nielsen, A.E., *Kinetics of Precipitation*, Pergamon Press, New York (1964).

6. Noyes, A.A., and W.R. Whitney, *J. Am. Chem. Soc.*, **19**, 930–934 (1897).

7. Nernst, W., *Zeit. für Physik. Chem.*, **47**, 52–55 (1904).

8. Miers, H.A., *Phil. Trans.*, **A202**, 492–515 (1904).

9. Valeton, J.J.P., *Zeit. für Kristallographie*, 59, 483 (1924).

10. Myerson, A.S., Ed., *Handbook of Industrial Crystallization*, Butterworth-Heinemann, Boston (1993).

11. Burton, W.K., N. Cabrera, and F.C. Frank, *Phil. Trans.*, **A243**, 299–358 (1951).

12. Seavoy, G.E., and H.B. Caldwell, *Ind. Eng. Chem.*, **32**, 627–636 (1940).

13. Newman, H.H., and R.C. Bennett, *Chem. Eng. Prog.*, **55**(3), 65–70 (1959).

14. Randolph, A.D., *AIChE Journal*, **11**, 424–430 (1965).

15. Randolph, A.D., and M.A. Larson, *Theory of Particulate Processes*, 2nd ed., Academic Press, New York (1988).

16. McCabe, W.L., *Ind. Eng. Chem.*, **21**, 30–33 and 112–119 (1929).

17. Zumstein, R.C., and R.W. Rousseau, *AIChE Symp. Ser.*, **83**(253), 130 (1987).

18. Nielsen, A.E., *Kinetics of Precipitation*, Pergamon Press, Oxford, England (1964).

19. Nielsen, A.E., Chapter 27, in I.M. Kolthoff and P.J. Elving, Eds., *Treatise on Analytical Chemistry, Part 1*, Volume 3, 2nd ed., John Wiley & Sons, New York (1983).

20. Nielsen, A.E., *J. Crys. Gr.*, **67**, 289–310 (1984).

21. Fitchett, D.E., and J.M. Tarbell, *AIChE J.*, **36**, 511–522 (1990).

22. Matsuoka, M., M. Ohishi, A. Sumitani, and K. Ohori, World Congress III of Chemical Engineers, Tokyo, Sept. 21, 1986, pp. 980–983.

23. Wilcox, W.R., *Ind. Eng. Chem.*, **60**(3), 13–23 (1968).

24. Wynn, N., *Chemical Engineering*, **98**(7), 149–154 (1991).

25. Pfann, W.G., *Trans. AIME.*, **194**, 747 (1952).

26. Pfann, W.G., *Zone Melting*, 2nd ed., John Wiley & Sons, New York (1966).

27. Zief, M., and W.R. Wilcox, *Fractional Solidification*, Marcel Dekker, New York (1967).

28. Burris, Jr., L., C.H. Stockman, and I.G. Dillon, *Trans. AIME*, **203**, 1017 (1955).

29. Herington, E.F.G., *Zone Melting of Organic Compounds*, John Wiley & Sons, New York (1963).

30. Nord, M., *Chem. Eng.*, **58**(9), 157–166 (1951).

31. Kudela, L., and M.J. Sampson, *Chem. Eng.*, **93**(12), 93–98 (1986).

32. Holden, C.A., and H.S. Bryant, *Sep. Sci.*, **4**(1), 1–13 (1969).

33. Singh, N.M., and R.K. Tawney, *Ind. J. Tech.*, **9**, 445–447 (1971).

34. Poling, B.E., J.M. Prausnitz, and J.P. O'Connell, *The Properties of Gases and Liquids*, 5th ed., McGraw-Hill Book Co., New York, p. 8.191 (2001).

35. McCabe, W.L., *Trans. AIChE*, **31**, 129–164 (1935).

36. Geankoplis, C.J., *Transport Processes and Unit Operations*, 3rd ed., Prentice Hall, Englewood Cliffs, NJ (1993).

37. Jacobsen, C., J. Garside, and M. Hoare, *Biotechnol. Bioeng.*, **57**, 666 (1998).

38. McCabe, W.L., J.C. Smith, and P. Harriott, *Unit Operations in Chemical Engineering*, 6th ed., McGraw Hill, New York (2000).

39. Hajko, P., T. Vesel, I. Radez, and M. Pokorny, U.S. Patent 5,712,130 (1998).

40. Harrison, R.G., and L.P. Nelles, U.S. Patent 4,956,290 (1990).

41. Judge, R.A., M.R. Johns, and E.T. White, *Biotechnol. Bioeng.*, **48**, 316 (1995).

42. Ring, T.A., *Fundamentals of Ceramic Powder Processing and Synthesis*, Academic Press, San Diego (1996).

43. Belter, P.A., E.L. Cussler, and W.-S. Hu, *Bioseparations: Downstream Processing for Biotechnology*, John Wiley & Sons, New York (1988).

44. Nallet, V., D. Mangin, and J.P. Klein, *Mixing and Crystallization*, Kluwer Academic Publishers, Boston (1998).

45. Garcia, A.A., M.R. Bonen, J. Ramirez-Vick, M. Sadaka, and A. Vuppu, *Bioseparation Process Science*, Blackwell Science, Malden, MA (1999).

46. Tavare, N.S., J. Garside, and M.R. Chivate, *Ind. Eng. Chem. Process Des. Dev.*, **19**, 653–665 (1980).

47. Tosun, G., *Ind. Eng. Chem. Res.*, **26**, 1184–1193 (1987).

48. Kamerath, N.R., *M. S. Thesis*, University of Utah (2008).

49. Harrison, R.G., P. Todd, S.R. Rudge, and D.P. Petrides, *Bioseparations Science & Engineering*, Oxford University Press, New York (2003).

50. de Nevers, N., *Fluid Mechanics for Chemical Engineers*, 3rd ed., McGraw Hill, New York (2004).

51. Baird, C.T., *Guide to Petroleum Product Blending*, HPI Consultants, Inc. (1989).

52. Shuler, M.L., and F. Kargi, *Bioprocess Engineering—Basic Concepts*, 2nd ed., Prentice Hall PTR, Upper Saddle River, NJ (2002).

53. Tavare, N.S., and M.R. Chivate, *J. Chem. Eng. Jpn.*, **13**, 371 (1980).

STUDY QUESTIONS

17.1. How does solution crystallization differ from melt crystallization?

17.2. Under what conditions does precipitation occur?

17.3. What are the two main methods used to cause crystallization from an aqueous solution? Which is more common and why?

17.4. What is the difference between crystallization and desublimation?

17.5. What is the difference between mother liquor and magma?

17.6. Why are crystals never spherical in shape?

17.7. What is meant by crystal habit?

17.8. What is the difference between differential screen analysis and cumulative screen analysis?

17.9. Does the solubility of most inorganic compounds in water increase or decrease with temperature?

17.10. Can an inorganic compound have more than one form of hydrate?

17.11. Does the commonly reported solubility of an inorganic compound in water pertain to large crystals or small crystals? Why?

17.12. What is supersaturation? Under what conditions is it possible to supersaturate a solution? What is the metastable region?

17.13. In physical adsorption, the resistance to the rate of adsorption at the solid–fluid interface is negligible. Is that also true for the incorporation of the solute into the crystal-lattice structure for solution crystallization? If not, why?

17.14. Why is the draft-tube, baffled (DTB) crystallizer popular? What are its main features? What is a draft tube?

17.15. What is a eutectic? What is the difference between a eutectic-forming system and a solid-solution-forming system?

17.16. Why do some evaporators operate under vacuum? How is the vacuum produced?

17.17. Why do aqueous inorganic solutions exhibit a boiling-point rise over the boiling point of pure water? Is it possible to have a boiling-point decrease?

17.18. What are the key assumptions in the evaporator model?

17.19. What are the advantages of crystallization relative to filtration or adsorption for bioproduct purification?

17.20. Why is maintaining constant supersaturation important during bioproduct crystallization?

EXERCISES

Section 17.1

17.1. Estimation of crystal sphericities.

Estimate sphericities of the following simple particle shapes: (a) a cylindrical needle with a height, H, equal to 5 times the diameter, D; (b) a rectangular prism of sides a, $2a$, and $3a$.

17.2. Sphericity of a thin circular plate.

A certain circular plate of diameter D and thickness t has a sphericity of 0.594. What is the ratio of t to D?

17.3. Differential and cumulative screen analysis plots.

A laboratory screen analysis for a batch of crystals of hypo (sodium thiosulfate) is as follows. Prepare both differential and cumulative-undersize plots of the data using a spreadsheet.

U.S. Screen	Mass Retained, gm
6	0.0
8	8.8
12	21.3
16	138.2
20	211.6
30	161.7
40	81.6
50	44.1
70	28.7
100	13.2
140	9.6
170	8.8
230	7.4
	735.0

In preparing your plots, determine whether arithmetic, semilog, or log-log plots are preferred.

17.4. Mean diameters.

Derive expressions for the surface-mean and mass-mean diameter from a particle-size analysis based on counting, rather than weighing, particles in given size ranges, letting N_i be the number of particles in a given size range of average diameter $\bar{D}_{p_i}$.

17.5. Particle diameters from screen analyses.

Using the screen analysis of Exercise 17.3, calculate, with a spreadsheet, the surface-mean, mass-mean, arithmetic-mean, and volume-mean crystal diameters, assuming that all particles have the same sphericity and volume shape factor.

17.6. Particle-size analysis from particle numbers.

A precipitation process for producing perfect spheres of silica has been developed. The individual particles are so small that most cannot be discerned by the naked eye. Using optical microscopy, the particle-size distribution has been measured, with results given in the table below. (a) Using these data with a spreadsheet program,

produce plots of the differential and cumulative particle-size analyses. (b) Determine: (1) surface-mean diameter, (2) arithmetic-mean diameter, (3) mass-mean diameter, and (4) volume-mean diameter.

PSD of Silica Spheres	
Particle-Size Interval, μm	Number of Particles
1.0–1.4	2
1.4–2.0	5
2.0–2.8	14
2.8–4.0	60
4.0–5.6	100
5.6–8.0	190
8.0–12.0	250
12.0–16.0	160
16.0–22.0	110
22.0–30.0	70
30.0–42.0	28
42.0–60.0	10
60.0–84.0	1
Total	1,000

17.7. Screen analysis and average diameters.

A screen analysis for a sample of Glauber's salt from a commercial crystallizer is as follows, where the crystals can be assumed to have a uniform sphericity and volume shape factor.

U.S. Screen	Mass Retained, gm
14	0.0
16	0.9
18	25.4
20	111.2
25	113.9
30	225.9
35	171.7
40	116.5
45	55.1
50	31.5
60	8.7
70	10.5
80	4.4
	875.7

Use a spreadsheet to determine in microns: (a) a plot of the differential analysis, (b) a plot of the cumulative oversize analysis,

(c) a plot of the cumulative undersize analysis, (d) the surface-mean diameter, (e) the mass-mean diameter, (f) the arithmetic-mean diameter, and (g) the volume-mean diameter.

Section 17.2

17.8.   Dissolution of sparingly soluble solids.

One thousand grams of water is mixed with 50 g of Ag_2CO_3 and 100 g of AgCl. At equilibrium at 25°C, calculate the concentrations in mol/L of Ag^+, Cl^-, and CO_3^{2-} ions and g of Ag_2CO_3 and AgCl in the solid phases.

17.9.   Crystallization by cooling and evaporation.

Five thousand lb/h of a saturated aqueous solution of $(NH_4)_2SO_4$ at 80°C is cooled to 30°C. At equilibrium, what is the amount of crystals formed in lb/h? If during the cooling process, 50% of the water is evaporated, what amount of crystals is formed in lb/h?

17.10.   Crystallization by cooling.

7,500 lb/h of a 50 wt% aqueous solution of $FeCl_3$ at 100°C is cooled to 20°C. At 100°C, the solubility of the $FeCl_3$ is 540 g/100 g of water. At 20°C, the solubility is 91.8 g/100 g water, and crystals of $FeCl_3$ are the hexahydrate. At equilibrium at 20°C, determine the lb/h of crystals formed.

17.11.   Continuous vacuum crystallization.

A concentrate from an evaporation system is 5,870 lb/h of 35 wt% $MgSO_4$ at 180°F and 25 psia. It is mixed with 10,500 lb/h of saturated aqueous recycle filtrate of $MgSO_4$ at 80°F and 25 psia. The mixture is sent to a vacuum crystallizer, operating at 85° F and 0.58 psia, to produce steam and a magma of 25 wt% crystals and 75 wt% saturated solution. Determine the lb/h of water evaporated and the maximum production of crystals in tons/day (dry basis).

17.12.   Crystallization by cooling and evaporation.

Urea is to be crystallized from an aqueous solution that is 90% saturated at 100°C. If 90% of the urea is to be crystallized in the anhydrous form and the final solution temperature is to be 30°C, what fraction of the water must be evaporated?

17.13.   Heat addition to a crystallizer feed.

In Examples 17.3 and 17.5, crystallizer heat addition is by an external heat exchanger through which magma circulates, as in Figure 17.16. If instead the heat is added to the feed, determine the new feed temperature. Which is the preferred way to add the heat?

17.14.   Heat addition to a crystallizer.

For Exercise 17.11, determine the rate at which heat must be added to the system.

17.15.   Heat removal from a cooling crystallizer.

For Example 17.4, calculate the amount of heat in calories/100 grams of water that must be removed to cool the solution from 100 to 10.6°C.

Section 17.3

17.16.   Effect of crystal size on solubility.

Based on the following data, compare the effect of crystal size on solubility in water at 25°C for (1) KCl (see Example 17.7), a soluble inorganic salt, with that for (2) $BaSO_4$, an almost insoluble inorganic salt, and (3) sucrose, a very soluble organic compound.

$\sigma_{s,L}$ for barium sulfate = 0.13 J/m^2 and $\sigma_{s,L}$ for sucrose = 0.01 J/m^2

What conclusions can you draw from the results?

17.17.   Required supersaturation ratio.

Determine the supersaturation ratio, S, required to permit 0.5-μm-diameter crystals of sucrose (MW = 342 and ρ_c = 1,590 kg/m^3) to grow if $\sigma_{s,L}$ = 0.01 J/m^2.

17.18.   Maximum crystal solubility.

The Kelvin equation, (17-16), predicts that solubility increases to ∞ as the crystal diameter decreases to 0. However, measurements by L. Harbury [*J. Phys. Chem.*, **50**, 190–199 (1946)] for several inorganic salts in water show a maximum in the solubility curve and a solubility that approaches 0 as crystal size is reduced to 0. Harbury's explanation is that the surface energy of the crystals depends not only on interfacial tension, but also on surface electrical charge, given by

$$2q^2 v_s / \pi \kappa R T D_p^4$$

where q = electrical charge on the crystal, and κ = dielectric constant.

Modify (17-16) to take into account electrical charge. Make sure your equation predicts a maximum.

17.19.   Primary homogeneous nucleation.

Using the following data, compare the effect of supersaturation ratio over the range of 1.005 to 1.02 on the primary homogeneous nucleation of $AgNO_3$, $NaNO_3$, and KNO_3 from aqueous solutions at 25°C:

	$AgNO_3$	$NaNO_3$	KNO_3
Crystal density, g/cm^3	4.35	2.26	2.11
Interfacial tension, J/m^2	0.0025	0.0015	0.0030

17.20.   Primary, homogeneous nucleation.

Estimate the effect of relative supersaturation on the primary, homogeneous nucleation of $BaSO_4$ from an aqueous solution at 25°C, if crystal density = 4.50 g/cm^3 and interfacial tension = 0.12 J/m^2.

17.21.   Controlling resistance in crystal growth.

Repeat parts (g) and (i) of Example 17.9 if the solution velocity past the crystal face is reduced from 5 cm/s to 1 cm/s.

Section 17.4

17.22.   Heat-transfer area of a cooling crystallizer.

Feed to a cooling crystallizer is 2,000 kg/h of 30 wt% Na_2SO_4 in water at 40°C. This solution is to be cooled to a temperature at which 50% of the solute crystallizes as decahydrate. Estimate the required heat-transfer area in m^2 if an overall heat-transfer coefficient of 15 Btu/h-ft^2-°F can be achieved. Assume a constant specific heat for the solution of 0.80 cal/g-°C. Cooling water at 10°C flows countercurrently to the solution, and exits at a temperature that gives a log mean driving force of at least 10°C.

17.23.   Number of cooling crystallizer units.

Two ton/h of the dodecahydrate of sodium phosphate $(Na_3PO_4 \cdot 12H_2O)$ is to be crystallized by cooling, in a cooling crystallizer, an aqueous solution that enters saturated at 40°C and leaves at 20°C. Chilled cooling water flows countercurrently, entering at 10°C and exiting at 25°C. The overall heat-transfer coefficient is 20 Btu/h-ft^2-°F. Solution average specific heat is 0.80 cal/g-°C. Estimate the: (a) tons (2,000 lb) per hour of feed solution; (b) heat-transfer area in ft^2; (c) number of crystallizer units required if each 10-ft-long unit contains 30 ft^2 of heat-transfer surface.

Section 17.5

17.24.   Application of MSMPR model.

An aqueous feed of 10,000 kg/h, saturated with $BaCl_2$ at 100°C, enters a crystallizer that can be simulated with the MSMPR model.

However, crystallization is achieved with negligible evaporation. The magma leaves the crystallizer at 20°C with crystals of the dihydrate. The crystallizer has a volume (vapor-space-free basis) of 2.0 m³. From laboratory experiments, the crystal growth rate is essentially constant at 4.0×10^{-7} m/s. Density of the dihydrate crystals = 3.097 g/cm³. Density of an aqueous, saturated solution of barium chloride at 20°C = 1.29 g/cm³.

Determine the: (a) kg/h of crystals in the magma product; (b) predominant crystal size in mm; and (c) mass fraction of crystals in the size range from U.S. Standard 20 to 25 mesh.

17.25.   Operation of an MSMPR crystallizer.

The feed to a continuous crystallizer that can be simulated with the MSMPR model is 5,000 kg/h of 40 wt% sodium acetate in water. Monoclinic crystals of the trihydrate are formed. The pressure in the crystallizer and the heat-transfer rate in the associated heat exchanger are such that 20% of the water in the feed is evaporated at a crystallizer temperature of 40°C. The crystal growth rate, G, is 0.0002 m/h and a predominant crystal size, L_{pd}, of 20 mesh is desired.

Determine the: (a) kg/h of crystals in the exiting magma; (b) kg/h of mother liquor in the exiting magma; and (c) volume in m³ of magma in the crystallizer if density of the crystals = 1.45 g/cm³ and density of the mother liquor = 1.20 g/cm³.

Solubility data:

T, °C	Solubility, g sodium acetate/100 g H_2O
30	54.5
40	65.5
60	139

17.26.   Design of an MSMRP crystallizer.

An MSMPR-type crystallizer is to be designed to produce 2,000 lb/h of crystals of the heptahydrate of magnesium sulfate with a predominant crystal size of 35 mesh. The magma will be 15 vol% crystals. The temperature in the crystallizer will be 50°C and the residence time 2 h. The densities of the crystals and mother liquor are 1.68 and 1.32 g/cm³, respectively. Determine: (a) the exiting flow rates in ft³/h of crystals, mother liquor, and magma; (b) the crystallizer volume in gallons, if the vapor space equals the magma space; (c) the approximate dimensions in feet of the crystallizer, if the body is cylindrical with a height equal to twice the diameter; (d) the required crystal growth rate in ft/h; (e) the necessary nucleation rate in nuclei/h-ft³ of mother liquor in the crystallizer; (f) the number of crystals produced per hour; (g) a screen analysis table covering a U.S. mesh range of 3-1/2 to 200, giving the predicted % cumulative and % differential screen analyses of the product crystals; (h) plots of the screen analyses predicted in part (g).

Section 17.6

17.27.   Precipitation using the MSMPR model.

Refer to Example 17.12. In Run 15, Fitchett and Tarbell also made measurements of the number density of crystals at 200 rpm, for which the data can be fitted by the equation

$$\ln n = 26.3 - 0.407\,L \qquad (1)$$

where n = number density of crystals and L = crystal size, μm.

Using the MSMPR model, determine in the same units as for Example 17.12:

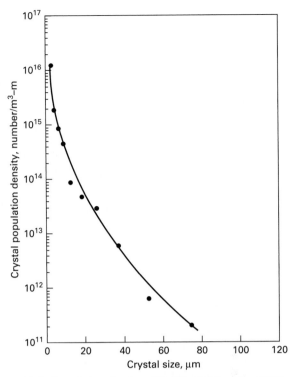

Figure 17.42 Population density of $CaCO_3$ for Exercise 17.28.

(a) n°, (b) G, (c) B°, (d) mean crystal length, and (e) n_c.

(f) Are your results consistent with the trends found in Example 17.12? (g) Using your results and those in Example 17.12, predict the growth rate and mean crystal length if no agitation is used.

17.28.   Precipitation of $CaCO_3$.

Tai and Chen [*AIChE J.*, **41**, 68–77 (1995)] studied the precipitation of calcium carbonate by mixing aqueous solutions of sodium carbonate and calcium chloride in an MSMPR crystallizer with pH control, such that the form of $CaCO_3$ was calcite rather than aragonite or vaterite. In Run S-2, which was conducted at 30°C, a pH of 8.65, and 800 rpm, with a residence time of 100 minutes, the population density data were as shown in Figure 17.42.

The data do not plot as a straight line, so they do not fit (17-38). (a) Develop an empirical equation that will fit the data and determine, by regression, the constants. (b) Can nucleation rate and growth rate be determined from the data? If so, how?

17.29.   Precipitation of $Mg(OH)_2$.

Tsuge and Matsuo ["Crystallization as a Separation Process," *ACS Symposium Series 438*, edited by Myerson and Toyokura, ACS, Washington, DC (1990), pp. 344–354] studied precipitation of $Mg(OH)_2$ by reacting aqueous solutions of $MgCl_2$ and $Ca(OH)_2$ in a 1-L MSMPR crystallizer operating at 450 rpm and 25°C. Crystal sizes were measured by a scanning-electron microscope (SEM) and analyzed by a digitizer. Crystal size was taken to be the maximum length. A plot of the crystal-size distribution is given in Figure 17.43 for an assumed residence time of 5 minutes. Assuming that the number of crystals is proportional to $\exp(-L/G\tau)$, as in (17-38), determine: (a) growth rate, (b) nucleation rate, and (c) predominant crystal size.

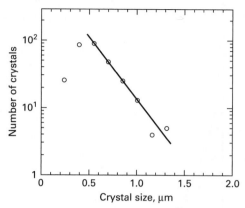

Figure 17.43 Crystal-size distribution of Mg(OH)$_2$ for Exercise 17.29.

Section 17.7

17.30. Falling-film crystallizer.

Feed to the top of a falling-film crystallizer is a melt of 60 wt% naphthalene and 40 wt% benzene at saturation conditions. If the coolant enters the top at 10°C, determine the crystal-layer thickness for up to 2 cm as a function of time. Necessary physical-property data are given in Example 17.13.

17.31. Falling-film crystallizer.

Paradichlorobenzene melts at 53°C, while orthodichlorobenzene melts at −17.6°C. They form a eutectic of 87.5 wt% of the ortho isomer at −23°C. The normal boiling points of these two isomers differ by about 5°C. A mixture of 80 wt% of the para isomer at the saturation temperature of 43°C is fed to the top of a falling-film crystallizer, where coolant enters at 15°C. If 8-cm i.d. tubes are used, determine the time for the crystal-layer thickness at the top of the tube to reach 2 cm. Which isomer will crystallize? From *Perry's Chemical Engineers' Handbook*, $\rho_c = 1.458$ g/cm^2 and $\Delta H_f = 29.67$ cal/g. Crystal thermal conductivity is 0.15 Btu/h-ft-°F for either isomer.

17.32. Melt crystallization in a cylindrical tube.

Derive (17-67).

Section 17.8

17.33. Zone melting with a single or partial pass.

Derive the following expression for the average impurity concentration over a particular length of crystal layer, $z_2 - z_1$, after one pass or partial pass of zone melting.

$$w_{\text{avg}} = w_0 \left\{ \frac{\ell(1 - K)}{K(z_2 - z_1)} [\exp(-z_2 K/\ell) - \exp(-z_1 K/\ell)] + 1 \right\} \quad (1)$$

Using the results of Example 17.14, calculate w_{avg} for $z_1 = 0$ and $z_2/\ell = 9$.

17.34. Zone melting with a single or partial pass.

In Example 17.14, let the last 20% of the crystal layer be removed, following the first pass, to $z/\ell = 9$. Calculate from (1), in Exercise 17.33, the average impurity concentration in the remaining crystal layer.

17.35. Zone melting with a single pass.

A bar of 98 wt% Al with 2 wt% of Fe impurity is subjected to one pass of zone refining. The solid–liquid distribution coefficient for the impurity is 0.29. If $z/\ell = 10$ and the resulting bar is cut off at $z_2 = 0.75z$, calculate the concentration profile for Fe and the average concentration from (1) in Exercise 17.33.

Section 17.9

17.36. Desublimation in a heat-exchanger.

A desublimation unit of the heat-exchanger type is to be sized for the recovery of 200 kg/h of benzoic acid (BA) from a gas stream containing 0.8 mol% BA and 99.2 mol% N$_2$. The gas enters the unit at 780 torr at 130°C and leaves without pressure drop at 80°C. The coolant is pressurized cooling water that enters at 40°C and leaves at 90°C, in countercurrent flow to the gas. The heat-exchanger tubes are of the type in Example 17.15. In addition to benzoic-acid properties in Exercise 17.37: k_c of solid benzoic acid = 1.4 cal/h-cm-°C and ρ_c of solid benzoic acid = 1.316 g/cm^3. Determine the number of tubes needed and the time required to reach the maximum thickness of benzoic acid of 1.25 cm.

17.37. Bulk-phase desublimation.

Benzoic acid is to be crystallized by bulk-phase desublimation from N$_2$ using a novel method described by Vitovec, Smolik, and Kugler [*Coll. Czech. Chem. Commun.*, **42**, 1108–1117 (1977)]. The gas, containing 6.4 mol% benzoic acid and the balance N$_2$, flows at 3 m^3/h at 1 atm and a temperature of 10°C above the dew point. The gas is directly cooled by the vaporization of 150 cm^3/h of a water spray at 25°C. The gas is further cooled in two steps by nitrogen quench gas at 1 atm as follows:

Step	Quench Gas Flow Rate, m^3/h	Quench Gas Temp., °C
1	1.5	105
2	2.0	25

The quench gases enter through porous walls of the vessel to prevent crystallization on the vessel wall. Based on the following data for benzoic acid, determine the final gas temperature and fractional yield of benzoic-acid crystals, assuming equilibrium in the exiting gas. Properties are: melting point = 122.4°C, specific heat of solid and vapor = 0.32 cal/g-°C, and heat of sublimation = 134 cal/g.

Vapor-pressure data:

T, °C	Vapor Pressure, torr
96	1
105	1.7
119.5	5
132.1	10
146.7	20
162.6	40
172.8	60

The vapor-pressure data can be extrapolated to lower temperatures by the Antoine equation.

17.38. Desublimation on the outside of a tube.

Derive (17-75).

Section 17.10

17.39. Single-effect evaporator.

Fifty thousand lb/h of a 20 wt% aqueous solution of NaOH at 120°F is to be fed to an evaporator operating at 3.7 psia, where the solution is concentrated to 40 wt% NaOH. The heating medium is saturated steam at a temperature 40°F higher than the exiting temperature of the caustic solution. Determine the: (a) boiling-point elevation of the solution; (b) saturated-heating-steam temperature and pressure; (c) evaporation rate; (d) heat-transfer rate; (e) heating-steam flow rate; (f) economy; and (g) heat-transfer area if $U = 300$ Btu/h-ft^2-°F.

17.40.   Single- and double-effect evaporation.

A 10 wt% aqueous solution of NaOH at 100°F and a flow rate of 30,000 lb/h is to be concentrated to 50 wt% by evaporation using saturated steam at 115 psia. (a) If a single-effect evaporator is used with $U = 400$ Btu/h-ft^2-°F and a vapor-space pressure of 4 inches Hg, determine the heat-transfer area and the economy. (b) If a double-effect evaporator system is used with forward feed and $U_1 = 450$ Btu/h-ft^2-°F, $U_2 = 350$ Btu/h-ft^2-°F, and a vapor-space pressure of 4 inches Hg in the second effect, determine the heat-transfer area of each effect, assuming equal areas, and the overall economy.

17.41.   Double-effect evaporation with forward feed.

A 10 wt% aqueous solution of MgSO$_4$ at 14.7 psia and 70°F is sent to a double-effect evaporator system with forward feed at a flow rate of 16,860 lb/h, to be concentrated to 30 wt% MgSO$_4$. The pressure in the second effect is 2.20 psia. The heating medium is saturated steam at 230°F. Estimated heat-transfer coefficients in Btu/h-ft^2-°F are 400 for the first effect and 350 for the second. If the heat-transfer areas of the two effects are the same, and boiling-point elevations are neglected, determine: (a) pressure in the first effect, (b) percent of the total evaporation occurring in the first effect, (c) heat-transfer area of each effect, and (d) economy.

Section 17.11

17.42.   Cooled batch crystallization in mixed solvents.

Belter et al. [43] report that cooled batch crystallization of tetracycline in mixed solvents exhibits linear changes in solubility with temperature: 1.14×10^{-3} g antibiotic/cm^3 solvent/K. Suppose 0.01-cm seeds ($\rho = 1.06$ g/cm^3) are added at 27 μg/cm^3 to obtain 0.095-cm crystals at supersaturation of about 0.077 g/cm^3. Calculate the cooling curve for this process, assuming constant, diffusion-controlled growth of cubic crystals with a mass-transfer coefficient of 6.5×10^{-5} cm/sec. Estimate the process time and fractional increase in product size per seed.

17.43.   Control of solvent addition in a batch crystallizer.

Derive a relationship between solvent concentration, c^S, and time to control solvent addition rate, in order to maintain constant supersaturation in a cooled, batch crystallizer with no seeding. Use (17-101) for solubility as a function of solvent content and apply the solute mass balance in (2) of Example 17.21.

17.44.   Mixing time in a tee-mixer.

Determine the parametric dependence of eddy length in (17-102) and mixing time in (17-103) for the tee-mixer in Figure 17.41, summarized in the following general expressions:

$$l_{eddy} = f\{\nu_m, \rho_m, \rho_k, L, P_f, d\}$$
$$t_m = f\{\nu_m, \rho_m, \rho_k, D, L, P_f, d\}$$

The pressures and velocities in each jet will be matched, with velocity being uniquely determined by parameters in the general expression. Begin with Bernoulli's equation:

$$\left(\frac{P_f - P_e}{\rho_m} - \frac{u_e^2}{2}\right) = -\Im$$

where P represents the pressure in Pascals, with subscripts f and e for feed and exit conditions, respectively; u_e is the exit velocity of the mixed stream; and d is the pipe diameter in m. The friction loss in the system, $\Im$ in m^2/s^2, is given [50] by:

$$\Im = 4f \frac{\Delta x}{d} \frac{u_e^2}{2}$$

where Δx is the length, L, over which the pressure drop occurs plus equivalent lengths, L_e/d, that account for pressure drop through fittings,

$$\Delta x = L + d \sum \frac{L_e}{d}$$

and friction factor, f, is given by

$$f = 0.001375\left[1 + \left(20000\frac{\epsilon_R}{d} + \frac{10^6}{N_{Re}}\right)^{1/3}\right]$$

where ϵ_R is the pipe roughness in m, which ranges from 10^{-2} to 10^{-6} m. This equation is applicable for $N_{Re} > 4,000$. Values of equivalent length range from $L_e/d = 2$ for a union up to 350 for a globe valve. Mixture kinematic viscosity may be estimated using the Refutas equation [51]:

$$\nu = \exp\left[\exp\left(\frac{VBN - 10.975}{14.534}\right)\right] - 0.8$$

where VBN is the viscosity blending number of the mixture, given by:

$$VBN = [x_A VBN_A] + [x_B VBN_B] + \ldots + [x_n VBN_n]$$

with x_n being the mass fraction of each of the components. Component VBN values are found by inverting the Refutas equation.

Chapter **18**

Drying of Solids

§18.0 INSTRUCTIONAL OBJECTIVES

After completing this chapter, you should be able to:

- Describe two common modes of drying.
- Discuss industrial drying equipment.
- Use a psychrometric chart to determine drying temperature.
- Differentiate between the adiabatic-saturation and wet-bulb temperatures.
- Explain equilibrium-moisture content of solids.
- Explain types of moisture content used in making dryer calculations.
- Describe the four different periods in direct-heat drying.
- Calculate drying rates for different periods.
- Apply models for a few common types of dryers.

Drying is the removal of moisture (either water or other volatile compounds) from solids, solutions, slurries, and pastes to give solid products. In the feed to a dryer, moisture may be: embedded in a wet solid, a liquid on a solid surface, or a solution in which a solid is dissolved. The term drying also describes a gas mixture in which a condensable vapor is removed from a non-condensable gas by cooling, as discussed in Chapter 4, and the removal of moisture from a liquid or gas by sorption, as discussed in Chapters 6 and 16. This chapter deals only with drying operations that produce solid products.

Drying is widely used to remove moisture from: (1) crystalline particles of inorganic salts and organic compounds to produce a free-flowing product; (2) biological materials, including foods, to prevent spoilage and decay from microorganisms that cannot live without water; (3) pharmaceuticals; (4) detergents; (5) lumber, paper, and fiber products; (6) dyestuffs; (7) solid catalysts; (8) milk; and (9) films and coatings, and (10) products where high water content entails excessive transportation and distribution costs. Not all drying processes have been successful; the beer industry, for decades, has been trying to market dehydrated beer with no success whatsoever.

Drying can be expensive, especially when large amounts of water, with its high heat of vaporization, must be evaporated. Water and energy conservation measures, and advances in equipment design, have broadened the use of pre-feed dewatering operations by mechanical means such as expression; gravity, vacuum, or pressure filtration; settling; and centrifugation, which also diminish the length of drying cycles.

Because drying involves vaporization of moisture, heat must be transferred to the material being dried. The common modes of heat transfer are: (1) convection from a hot gas in contact with the material; (2) conduction from a hot, solid surface in contact with the material; (3) radiation from a hot gas or surface; and (4) heat generation within the material by dielectric, radio frequency, or microwave heating. These different modes can sometimes be used symbiotically, depending on whether the moisture to be removed is on the surface or inside the solid.

Of importance is the temperature at which the moisture evaporates. When convection from a hot gas is employed and the moisture is on the surface or rapidly migrates to the surface from the interior of the solid, the rate of evaporation is independent of the properties of the solid and is governed by the rate of convective heat transfer from the gas to the surface. Then, the evaporating surface is at the wet-bulb temperature of the gas if the dryer operates adiabatically.

If the convective heat transfer is supplemented by radiation, the temperature of the evaporating surface is higher than the wet-bulb temperature. In the absence of contact with a convective-heating gas, as in the latter three modes, and when a sweep gas is not present, such that the dryer operates nonadiabatically, the evaporating moisture is at its boiling-point temperature at the pressure in the dryer. If the moisture contains dissolved, nonvolatile substances, the boiling-point temperature will be elevated.

Industrial Example

The continuous production of 69,530 lb/day of $MgSO_4 \cdot 7H_2O$ crystalline solids containing 0.015 lb H_2O/lb dry solid is an example of an industrial drying operation. The feed to the dryer in Figure 18.1 consists of a filter cake from a rotary-

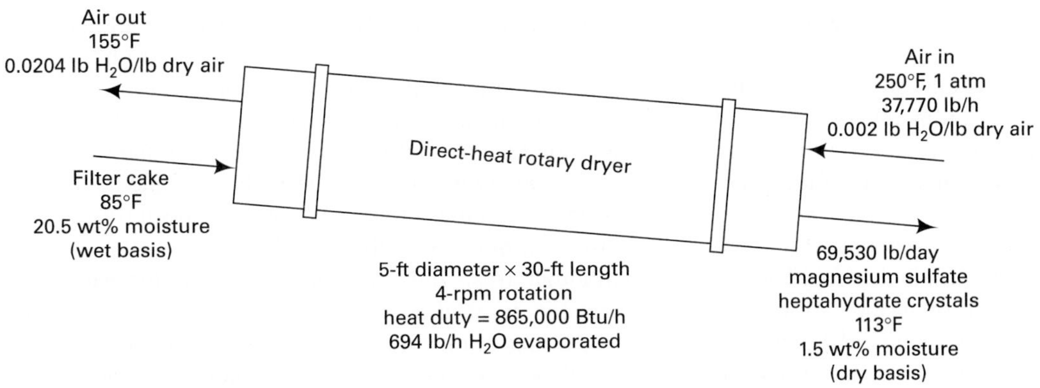

Figure 18.1 Process for drying magnesium–sulfate–heptahydrate filter cake.

Air out
155°F
0.0204 lb H₂O/lb dry air

Filter cake
85°F
20.5 wt% moisture
(wet basis)

Direct-heat rotary dryer

5-ft diameter × 30-ft length
4-rpm rotation
heat duty = 865,000 Btu/h
694 lb/h H₂O evaporated

Air in
250°F, 1 atm
37,770 lb/h
0.002 lb H₂O/lb dry air

69,530 lb/day
magnesium sulfate
heptahydrate crystals
113°F
1.5 wt% moisture
(dry basis)

drum vacuum filter (see §19.2.7). The cake is at 85°F and contains 20.5 wt% moisture on a wet basis. Because the feed crystals are relatively coarse, free flowing, and nonsticking, a direct-heat rotary dryer consisting of a slightly inclined, rotating, cylindrical shell is used. The filter-cake feed enters the high end of the dryer from an inclined, vibrated chute. Heated air at 250°F and atmospheric pressure, with an absolute humidity of 0.002 lb H₂O/lb dry air, enters the other end at a flow rate of 37,770 lb/h. To obtain good contact between the wet crystals and hot air, the dryer is provided with internal, longitudinal flights that extend the entire shell length. As the shell rotates, the flights lift the solids until they reach their angle of repose and then shower down through the hot air in countercurrent flow to the direction of net movement of the solids. The dry solids discharge at 113°F through a rotating valve into a screw conveyor. The air, which has been cooled to 155°F and humidified to 0.0204 lb H₂O/lb dry air by contact with the wet solids, exits at the other end through a fan to pollution-control units.

The hot air causes evaporation of 694 lb/h of water, mostly at a temperature of 94.5°F, which is the average of the entering and exiting gas wet-bulb temperatures of 95.5 and 93.5°F, respectively. In addition, the hot air must heat the solids from 85°F to 113°F and the evaporated moisture to 155°F. The total rate of convective heat transfer, Q, from the gas to the solids is 865,000 Btu/h. This ignores heat loss from the dryer shell to the surroundings and thermal radiation to the solids from the hot gas or the inside shell surface.

Of the total heat load, approximately 83% is required to evaporate moisture, with the balance supplying sensible heat. Therefore, a reasonably accurate log mean temperature-driving force is based on the assumption of a constant temperature at the gas–wet solids interface equal to the average air wet-bulb temperature of 94.5°F:

$$\Delta T_{LM} = \frac{(250 - 94.5) - (155 - 94.5)}{\ln\left(\dfrac{250 - 94.5}{155 - 94.5}\right)} = 100.6°F$$

For a direct-heat, rotary dryer, convective heat transfer is characterized by an overall volumetric heat-transfer coefficient, Ua, which for this example is 14.6 Btu/h-ft³ of dryer volume-°F. The required cylindrical shell volume, V, from $Q = UaV\Delta T_{LM}$, is 590 ft³ and the dryer diameter is 5 ft, which gives an entering superficial hot-air velocity of

9.56 ft/s, which is sufficiently low to prevent entrainment of solid particles in the air. The cylindrical shell is 30 ft long and rotates at 4 rpm. While moving through the dryer, the bulk solids, with a bulk density of 62 lb dry solids/ft³, occupy 8 vol% of the dryer, and have a residence time of one hour.

§18.1 DRYING EQUIPMENT

Material sent to drying equipment includes granular solids, pastes, slabs, films, slurries, fabrics, and liquids. Accordingly, different types of feed- and product-specific dryers have been developed.

§18.1.1 Classification of Dryers

Dryers can be classified in a number of ways; perhaps most importantly is the mode of operation, batch or continuous. Batch operation is generally indicated when the production rate is less than 500 lb/h of dried solid, while continuous operation is preferred for a production rate of more than 2,000 lb/h. In the example above, the production rate is 2,900 lb/h and continuous drying was selected.

A second classification method is the mode of heat transfer to evaporate moisture. As mentioned, *direct-heat* (also called convective or adiabatic) dryers contact material with a hot gas, which not only provides the energy to heat the material and evaporate the moisture, but also sweeps away the moisture. When the continuous mode of operation is used, the hot gas can flow countercurrently, cocurrently, or in crossflow to the material being dried. Countercurrent flow is the most efficient, but cocurrent flow may be required if the material being dried is temperature-sensitive.

Indirect-heat (also called nonadiabatic) dryers provide heat to the material by conduction and/or radiation from a hot surface. Energy may also be generated within the material by dielectric, radio frequency, or microwave heating. Indirect-heat dryers may be operated under vacuum to reduce the temperature at which the moisture is evaporated. A sweep gas is not necessary, but can be provided to help remove moisture. Capital costs for direct-heat dyers are higher, but indirect-heat dryers are more expensive to operate and are used only when the material is either temperature-sensitive or subject to crystal breakage and dust or fines formation.

A third method for classifying dryers is the degree to which the material is agitated. In some dryers, the feed is stationary while being processed. At the opposite extreme is the fluidized-bed dryer, in which agitation increases the rate of heat transfer but, if too severe, can cause crystal breakage and dust formation. Agitation may be necessary if the material is sticky.

The more widely used commercial dryers are described here. A more complete coverage is given in the *Handbook of Industrial Drying* [1]. Extensive performance data for many types of dryers are given in *Perry's Chemical Engineers' Handbook* [2] and by Walas [3]. Batch dryers are discussed first, followed by continuous dryers, and then other dryers that use special means for evaporating moisture.

§18.1.2 Batch Dryers

Equipment for drying batches includes: (1) tray (also called cabinet, compartment, or shelf) dryers; and (2) agitated dryers. Together, these two types cover many of the modes of heat transfer and agitation discussed above.

Tray Dryers

The oldest and simplest batch dryer is the tray dryer, which is shown schematically in Figure 18.2 and is useful when low production rates of multiple products are involved and when drying times vary from hours to days. The material to be dried is loaded to a depth of typically 0.5–4 inches in removable trays that may measure $30 \times 30 \times 3$ inches and are

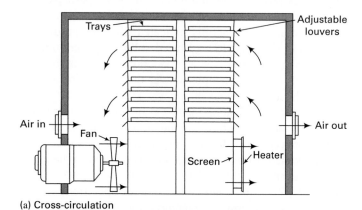

(a) Cross-circulation

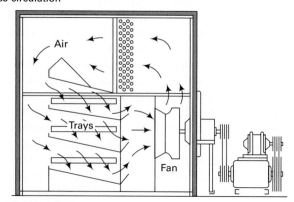

(b) Through-circulation

Figure 18.2 Tray dryers.

stacked by a forklift on shelves about 3 inches apart in a cabinet. If the wet solids are granular or shaped into briquettes, noodles, or pellets with appreciable voids, the tray bottom can be perforated so that heating gas can be passed down through the material (*through-circulation*) as shown in Figure 18.2b. Otherwise, the tray bottom is solid and the hot gas is passed at velocities of 3–30 ft/s over the open tray surface (*cross-circulation*), as in Figure 18.2a.

Although fresh hot gas might be used for each pass through the dryer, it is more economical to recirculate the gas, providing venting and makeup gas at rates of 5–50% of the circulation rate to maintain the humidity at an acceptable level. Gas is heated with an annular, finned-tube heat exchanger by steam condensing inside the tubes. If the moisture being evaporated is water, steam requirements can range from 1.5 to 7.5 lb steam/lb water evaporated. It is important to baffle tray dryers to promote uniform distribution of hot gas to achieve uniform drying.

Tray dryers are available for vacuum operation and with indirect heating. In one configuration, the trays are placed on hollow shelves that carry condensing steam and act as heat exchangers. Heat is transferred by conduction to a tray from the top of the shelf supporting it and by radiation from the bottom of the shelf directly above the tray. Typical performance data for direct-heat, crossflow-circulation tray dryers are given in Table 18.1.

Agitated Dryers

As discussed by van't Land [4] and Uhl and Root [5], indirect heat with agitation and, perhaps, under vacuum, is desirable for batch drying when any of the following conditions exist: (1) material oxidizes or becomes explosive or dusty during drying; (2) moisture is valuable, toxic, flammable, or explosive; (3) material tends to agglomerate or set up if not agitated; and (4) maximum product temperature is less than about 30°C. Heat-transfer rates are controlled mostly by contact resistance at the inner wall of the jacketed vessel and by conduction into the material being dried. A wide variety of heating fluids can be used, including hot liquids, steam, Dowtherm, hot air, combustion gases, and molten salt.

When only Condition 3 applies, the atmospheric, agitated-pan dryer shown in Figure 18.3a is useful, particularly when

Table 18.1 Performance Data for Direct-Heat, Crossflow-Circulation Tray Dryers

Material	Aspirin-Base Granules	Chalk	Filter Cake
Number of trays	20	72	80
Area/tray, ft^2	3.5	15.7	3.5
Total loading, lb wet	56	1,800	2,800
Depth of loading, inches	0.5	2.0	1.0
% Initial moisture	15	46	70
% Final moisture	0.5	2	1
Maximum air temp., °F	122	180	200
Drying time, h	14	4.5	45

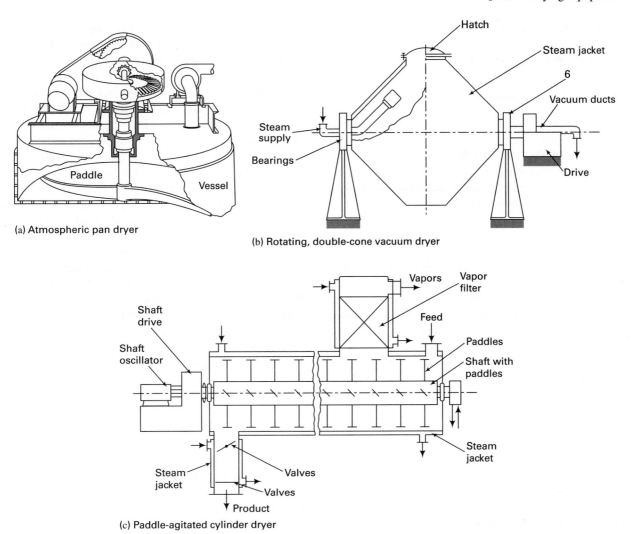

Figure 18.3 Agitated dryers.
[From *Perry's Chemical Engineers' Handbook*, 6th ed., R.H. Perry, D.W. Green, and J.O. Maloney, Eds., McGraw-Hill, New York (1984) with permission.]

the feed is a liquid, slurry, or paste. This dryer consists of a shallow (2–3-ft high), jacketed, flat-bottomed vessel, equipped with a paddle agitator that rotates at 2–20 rpm and scrapes the inner wall to help prevent cake buildup. Units range in size from 3 to 10 ft in diameter, with a capacity of up to 1,000 gallons and from 15 to 300 ft² of heat-transfer surface. When using steam in the jacket, overall heat-transfer coefficients vary from 5 to 75 Btu/h-ft²-°F. The material to be dried occupies about 2/3 of the vessel volume. The degree of agitation can be varied during the drying cycle. With a thin-liquid feed, agitation may vary from very low initially to very high if a sticky paste forms, followed by moderate agitation when the granular solid product begins to form. Typically, several hours are required for drying. Vacuum units are also available.

When any or all of the above four conditions apply, but only mild agitation is required, the jacketed, rotating, double-cone (also called tumbler) vacuum dryer, shown schematically in Figure 18.3b, can be used. V-shaped tumblers are also available. The conical shape facilitates discharge of dried product, but, except for the tumbling, no means is

provided to prevent cake buildup on the inner walls. Double-cone volumes range from 0.13 to 16 m³, with heat-transfer surface areas of 1 to 56 m². Additional heat-transfer surface can be provided by internal tubes or plates. Up to 70% of the volume can be occupied by feed. A typical evaporation rate when operating at 10 torr, with heating steam at 2 atm, is 1 lb/h-ft² of heat-transfer surface.

A more widely used agitated dryer, applicable when any or all of the above four conditions are relevant, is the ribbon- or paddle-agitated, horizontal-cylinder dryer, shown in the paddle form in Figure 18.3c. The cylinder is jacketed and stationary. The ribbons or paddles provide agitation and scrape the inner walls to prevent solids buildup. As discussed by Uhl and Root [5], cylinder dimensions range up to diameters of 6 ft and lengths up to 40 ft. The agitator can be rotated from 4 to 140 rpm, resulting in overall heat-transfer coefficients of 5 to 35 Btu/h-ft²-°F. Typically from 20 to 70% of the cylinder volume is filled with feed, and drying times vary from 4 to 16 hours. In more advanced versions, discussed by McCormick [6], one or two parallel rotating shafts can be provided that intermesh with stationary, lump-breaking bars to increase the range of application. The paddles can also be hollow to provide additional heat-transfer surface. This type of dryer can also be operated in a continuous mode.

§18.1.3 Continuous Dryers

A wide variety of industrial drying equipment for continuous operation is available. The following descriptions cover most types, organized by the nature of the wet feed: (1) granular, crystalline, and fibrous solids, cakes, extrusions, and pastes; (2) liquids and slurries; and (3) sheets and films. In addition, infrared, microwave, and freeze-drying are described.

Tunnel Dryers

The simplest, most widely applicable, and perhaps oldest continuous dryers are the tunnel dryers, which are suitable for any material that can be placed into trays and is not subject to dust formation. The trays are stacked onto wheeled trucks, which are conveyed progressively in series through a tunnel where the material in the trays is contacted by cross-circulation of hot gases. As shown in Figure 18.4, the hot gases can flow countercurrently, cocurrently, or in more complex flow configurations to the movement of the trucks. As a truck of dried material is removed from the discharge end of the tunnel, a truck of wet material enters at the feed end. The overall drying operation is not truly continuous because wet material must be loaded into the trays and dried material removed from the trays outside the tunnel, often with dump truck devices. Tray spacings and dimensions, as well as hot-gas velocities, are the same as for batch tray dryers. A typical tunnel might be 100 ft long and house 15 trucks.

Belt or Band Dryers

A more continuous operation can be achieved by carrying the solids as a layer on a belt conveyor, with hot gases passing over the material. The endless belt is constructed of hinged, slotted-metal plates, or, preferably a thin metal band, which is ideal for slurries, pastes, and sticky materials. The bands are up to 1.5 m wide × 1 mm thick.

More common are screen or perforated-belt or band-conveyor dryers, which, as shown in Figure 18.5a, use circulation of heated gases upward and/or downward through a moving, permeable, layered bed of wet material from 1 to 6 inches deep. As shown in Figure 18.5b, multiple sections, each with a fan and set of gas-heating coils, can be arranged in series to provide a dryer, with a single belt as long as 150 ft with a 6-ft width, giving drying times up to 2 h, with a belt speed of about 1 ft/minute. To be permeable, the wet material must be granular. If it is not, the material can be preformed by scoring, granulation, extrusion, pelletization, flaking, or briquetting. Particle sizes usually range from 30 mesh to 2 inches. Hot-gas superficial velocities through the bed range from 0.5 to 1.5 m/s, with maximum bed pressure drops of 50-mm of water. Heating gases are provided by heat transfer from condensing steam in finned-tube heat exchangers at 50–180°C, but temperatures up to 325°C are feasible. Continuous, through-circulation conveyor dryers are used to remove moisture from a variety of materials, some of which are listed in Table 18.2, which includes, in parentheses, the method of preforming, if necessary.

A perforated-band-conveyor dryer 50-ft long × 75-inches wide can produce 1,800 lb/h of calcium carbonate with a moisture content of 0.005 lb H_2O/lb carbonate, in a residence time of 40 minutes, from 6-mm-diameter carbonate extrusions with a moisture content of 1.5 lb H_2O/lb carbonate, using air heated to 320°F by 160 psig steam and passing through the bed of extrusions at a superficial velocity of 2.7 ft/s. Steam consumption is 1.75 lb/lb H_2O evaporated.

Turbo-Tray Tower Dryers

When floor space is limited but headroom is available, the turbo-tray or rotating-shelf dryer, shown in Figure 18.6, is a good choice for rapid drying of free-flowing, nondusting

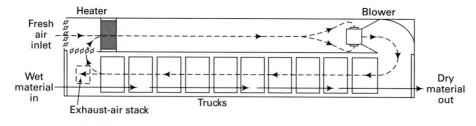

(a) Countercurrent flow

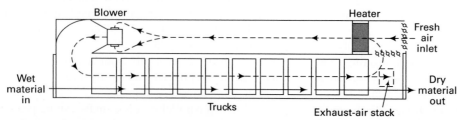

(b) Cocurrent flow

Figure 18.4 Tunnel dryer.

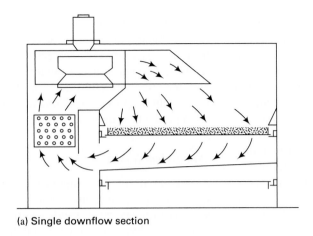

(a) Single downflow section

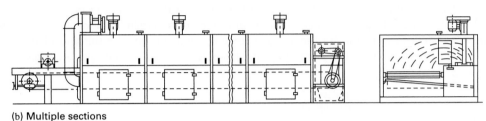

(b) Multiple sections

Figure 18.5 Perforated-belt or band-conveyor dryer.

Table 18.2 Materials Dried in Through-Circulation Conveyor Dryers

Aluminum hydrate (scored on filter)
Aluminum stearate (extruded)
Asbestos fiber
Breakfast food
Calcium carbonate (extruded)
Cellulose acetate (granulated)
Charcoal (briquetted)
Cornstarch
Cotton linters
Cryolite (granulated)
Dye intermediates (granulated)
Fluorspar
Gelatin (extruded)
Kaolin (granulated)
Lead arsenate (granulated)
Lithopone (extruded)
Magnesium carbonate (extruded)
Mercuric oxide (extruded)
Nickel hydroxide (extruded)
Polyacrylic nitrile (extruded)
Rayon staple and waste
Sawdust
Scoured wool
Silica gel
Soap flakes
Soda ash
Starch (scored on filter)
Sulfur (extruded)
Synthetic rubber (briquetted)
Tapioca
Titanium dioxide (extruded)
Zinc stearate (extruded)

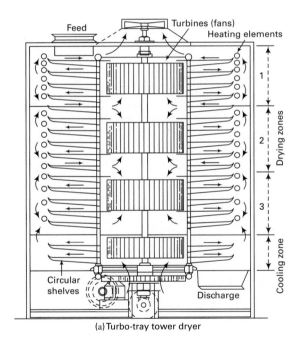

(a) Turbo-tray tower dryer

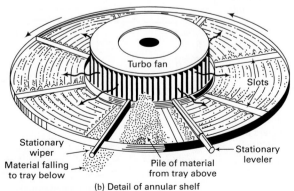

(b) Detail of annular shelf

Figure 18.6 Rotating-shelf dryer.

granular solids. Annular shelves, mounted one above the other, are slowly rotated at up to 1 rpm by a central shaft. Wet feed enters through the roof onto the top shelf as it rotates under the feed opening. At the end of one revolution, a stationary wiper causes the material to fall through a radial slot onto the shelf below, where it is spread into a pile of uniform thickness by a stationary leveler. This action is repeated on each shelf until the dried material is discharged from the bottom of the unit. Also mounted on the central shaft are fans that provide cross-circulation of hot gases at velocities of 2 to 8 ft/s across the shelves, and heating elements located at the unit's outer periphery. The bottom shelves can be used as a solids-cooling zone. Because solids are showered through the hot gases and redistributed from shelf to shelf, drying time is less than for cross-circulation, stationary-tray dryers. Typical turbo-tray dryers are from 2 to 20 m in height and 2 to 11 m in diameter, with shelf areas to 1,675 m^2. Overall heat-transfer coefficients (based on shelf area) of 30–120

J/m^2-s-K have been observed, giving moisture-evaporation rates comparable to those of through-circulation, belt-, or band-conveyor dryers. Materials successfully handled in turbo-tray dryers include calcium hypochlorite, urea, calcium chloride, sodium chloride, antibiotics, antioxidants, and water-soluble polymers. Capacities of up to 24,000 lb/h of dried product are quoted.

Direct-Heat Rotary Dryers

A popular dryer for evaporating water from free-flowing granular, crystalline, and flaked solids of relatively small size, when breakage of solids can be tolerated, is the direct-heat rotary dryer. As shown in Figure 18.7a, it consists of a rotating, cylindrical shell that is slightly inclined from the horizontal with a slope of less than 8 cm/m. Wet solids enter through a chute at the high end and dry solids discharge from the low end. Hot gases (heated air, flue gas, or superheated

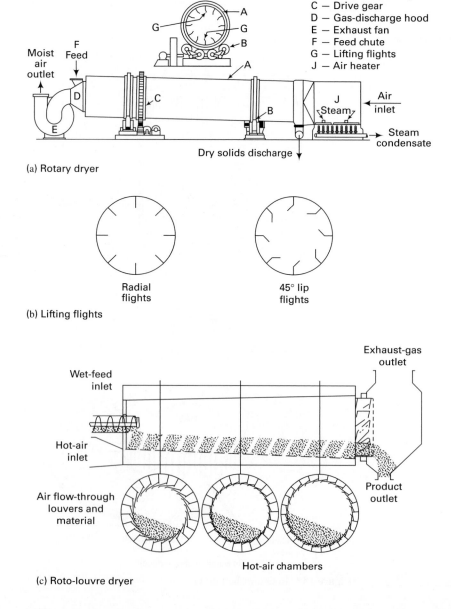

A — Dryer shell
B — Shell-supporting rolls
C — Drive gear
D — Gas-discharge hood
E — Exhaust fan
F — Feed chute
G — Lifting flights
J — Air heater

(a) Rotary dryer

(b) Lifting flights

Radial flights

45° lip flights

(c) Roto-louvre dryer

Figure 18.7 Direct-heat rotary dryer.

[From W.L. McCabe, J.C. Smith, and P. Harriott, *Unit Operations of Chemical Engineering*, 5th ed., McGraw-Hill, New York (1993) with permission.] [From *Perry's Chemical Engineers' Handbook*, 6th ed., R.H. Perry, D.W. Green, and J.O. Maloney, Eds., McGraw-Hill, New York (1984) with permission.]

Table 18.3 Materials Dried in Direct-Heat Rotary Dryers

Ammonium nitrate prills	Sand
Ammonium sulfate	Sodium chloride
Blast furnace slag	Sodium sulfate
Calcium carbonate	Stone
Cast-iron borings	Polystyrene
Cellulose acetate	Sugar beet pulp
Copper	Urea crystals
Fluorspar	Urea prills
Illmenite ore	Vinyl resins
Oxalic acid	Zinc concentrate

steam) flow countercurrently to the solids, but cocurrent flow can be employed for temperature-sensitive solids. With cocurrent flow, the cylinder may not need to be inclined because the gas will help move the solids. To enhance the gas-to-solids heat transfer, longitudinal lifting flights—available in several different designs, two of which are shown in Figure 18.7b—are mounted on the inside of the rotating shell, causing the solids to be lifted, then showered through the hot gas during each cylinder revolution. Typically the bulk solids occupy 8–18% of the cylinder volume, with residence times from 5 minutes to 2 h. Resulting water-evaporation rates are 5–50 kg/h-m^3 of dryer volume. The gas blower can be located to push or pull the gas through the dryer, with the latter favored if the material tends to form dust. Knockers, on the outside shell wall, can be used to prevent solids from sticking to the inside shell wall. Rotary dryers are available from 1 to 20 ft in diameter and 4–150 ft long. Superficial-gas velocities, which may be limited by dust entrainment, are 0.5–10 ft/s. The peripheral shell velocity is typically 1 ft/s. A variety of materials, some of which are listed in Table 18.3, are dried in direct-heat rotary

dryers. The detailed mechanical designs of rotary dryers are industry specific in the sense that the standard designs are modified to accommodate starch, sugar, salt, cement and other products, each of which has unique surface and bulk properties.

Roto-Louvre Dryers

A further improvement in the rate of heat transfer from hot gas to solids in a rotating cylinder is the through-circulation action achieved in the Roto-Louvre dryer in Figure 18.7c. A double wall provides an annular passage for hot gas, which passes through louvers and then through the rotating bed of solids. Because gas pressure drop through the bed may be significant, both inlet and outlet gas blowers are often provided to maintain an internal pressure close to atmospheric. These dryers range from 3 to 12 ft in diameter and 9–36 ft long, with water-evaporation rates reported as high as 12,300 lb/hr. They are useful for processing coarse, free-flowing, dust-free solids.

Indirect-Heat, Steam-Tube Rotary Dryers

When materials are: (1) free flowing and granular, crystalline, or flaked; (2) wet with water or organic solvents; and/or (3) subject to undesirable breakage, dust formation, or contamination by air or flue gases, an indirect-heat, steam-tube rotary dryer is often selected. A version of this dryer, shown in Figure 18.8, consists of a rotating cylinder that houses two concentric rows of longitudinal finned or unfinned tubes that carry condensing steam and rotate with the cylinder. Wet solids are fed into one end of the cylinder through a chute or by a screw conveyor. A gentle solids-lifting action is provided by the tubes. Dried product discharges from the other end

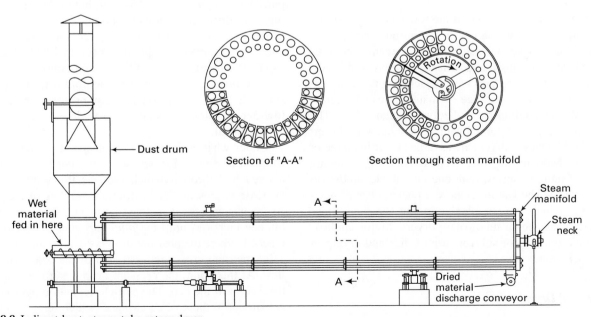

Figure 18.8 Indirect-heat, steam-tube rotary dryer.
[From *Perry's Chemical Engineers' Handbook*, 6th ed., R.H. Perry, D.W. Green, and J.O. Maloney, Eds., McGraw-Hill, New York (1984) with permission.]

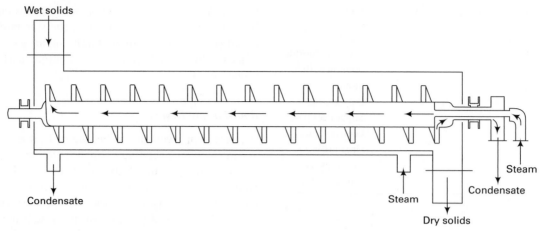

Figure 18.9 Screw-conveyor dryer.

after suitable contact with the hot-tube surfaces. The moisture (water or solvent) evaporates at about the boiling temperature, but can be swept out by a purge of inert gas. Steam enters the tubes through a central revolving inlet manifold. Condensate is discharged into a collection ring. With unfinned tubes, overall heat-transfer coefficients based on the surface area of the tubes range from 30 to 85 J/m²-s-K, when solids occupy 10–20% of the dryer volume. Steam-tube rotary dryers range in size from 3 to 8 ft in diameter by 15–80 ft long, with one or two rows of 14–90 tubes, 2.5–4.5 inches in diameter. The largest-size dryers contain a single row of 90 tubes. Rotation rates are from 3 to 6 rpm. Materials successfully dried include inorganic crystals, silica, mica, flotation concentrates, pigment filter cakes, precipitated calcium carbonate, distillers' grains, brewers' grains, citrus pulp, cellulose acetate, starch, and high-moisture organic compounds.

Screw-Conveyor Dryers

Less popular than rotary dryers is the screw-conveyor dryer, shown in Figure 18.9, which consists of a trough or cylinder that carries a hollow screw, inside of which steam condenses to provide heat for drying the material being conveyed. Additional heat transfer can be provided by jacketing the trough or cylindrical shell. A wide range of materials can be dried, including slurries, solutions, and solvent-laden solids. The boiling moisture can be purged with a small amount of inert gas. Standard conveyor dryers are as large as 3 ft in diameter by 20 ft long. More drying time can be provided by arranging a number of units in series, with one unit above another to save floor space. The last unit can be a cooler. Overall heat-transfer coefficients are comparable to, but less than, those for indirect-heat, steam-tube rotary dryers. Major applications include removal of solvents from solids and drying of fine and sticky materials.

Fluidized-Bed Dryers

Free-flowing, moist particles can be dried continuously with a residence time of a few minutes by contact with hot gases

in a fluidized-bed dryer, such as that shown in Figure 18.10a. This dryer consists of a cylindrical or rectangular fluidizing chamber to which wet particles are fed from a bin through a star valve or by a screw conveyor, and fluidized by hot gases blown through a heater and into a plenum chamber below the bed, from where the particles pass into the fluidizing chamber through a distributor plate, which must have a pressure drop of from 15 to 50% of the static bed head. The hot gases pass up through the bed, transferring heat for evaporation of the moisture, and pass out the top of the fluidizing chamber and through demisters and cyclones for dust removal. The solids are circulated by the action of the hot gases in the bed and by baffles, and sometimes mixers, but eventually pass out of the chamber through an overflow duct, which also serves to establish the height of the fluidized bed.

Fluidized-bed dryers have become very popular in recent years because they: (1) have no moving parts; (2) provide rapid heat and mass transfer between gas and particles; (3) provide intensive mixing of the particles, leading to uniform conditions throughout the bed; (4) provide ease of control; (5) can be designed for hazardous solids and a wide range of temperatures (up to 1200°C), pressures (up to 100 psig), residence times, and atmospheres; (6) can operate on electricity, natural gas, fuel oil, thermal fluids, steam, hot air, or hot water; (7) can process very fine and/or low-density particles; and (8) provide very efficient emissions control.

Under what conditions will the solid particles be fluidized? At low gas velocities, solids are not fluidized but form a fixed bed through which the gas flows upward with a decrease in pressure due to friction and drag of the particles. As the gas velocity is increased, the gas pressure drop across the bed increases until the *minimum fluidization velocity* is reached, where the pressure drop is equal to the weight of the solids per unit cross-sectional area of the bed normal to gas flow. At this point, the pressure drop is sufficient to support the weight of the bed. The particle-levitation hydrodynamics are similar to those of an airplane, which remains suspended because the pressure below the wings is higher than that above the wings. Further increases in gas velocity cause the

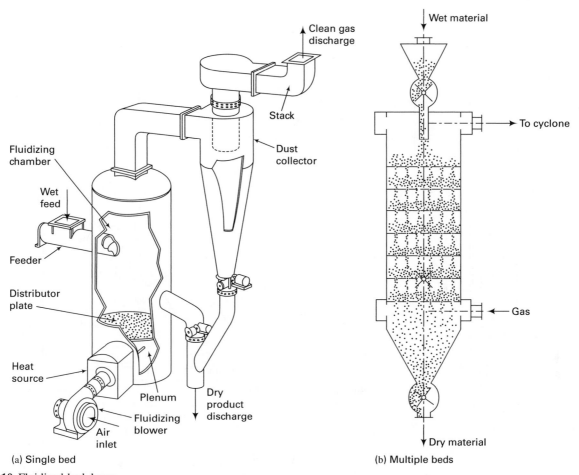

Figure 18.10 Fluidized-bed dryers.

[From W.L. McCabe, J.C. Smith, and P. Harriott, *Unit Operations of Chemical Engineering*, 5th ed., McGraw-Hill, New York (1993) with permission.]

bed to expand with little or no increase in gas pressure drop. Typically, fluidized-bed dryers are designed for gas velocities of no more than twice the minimum required for fluidization. That value depends on particle size and density, and gas density and viscosity. Superficial-gas velocities in fluidized-bed dryers are from 0.5 to 5.0 ft/s, which provide stable, *bubbling fluidization*. Higher velocities can lead to undesirable slugging of large gas bubbles through the bed.

The capital and operating cost of a blower to provide sufficient gas pressure for the pressure drops across the distributor plate and the bed is substantial. Therefore, required solids-residence time for drying is achieved by a shallow bed height and a large chamber cross-sectional area. Fluidized-bed heights can range from 0.5 to 5.0 ft or more, with chamber diameters from 3 to 10 ft. However, chamber heights are much greater than fluidized-bed heights, because it is desirable to provide at least 6 ft of free-board height above the top surface of the fluidized bed, unless demisters are installed, so that the larger dust particles can settle back into the bed rather than be carried by the gas into the cyclone. Because of intense mixing, temperatures of the gas and solids in a fluidized bed are equal and uniform at the temperature of the discharged gas and solids.

There is a substantial residence-time distribution for the particles in the bed, which can be mitigated by baffles, multi-staging, and mechanical agitators. Otherwise, a fraction of the particles short-circuit from the feed inlet to the discharge duct with little residence time and opportunity to dry. Another fraction of the particles spend much more than the necessary residence time for complete drying. Thus, the nonuniform moisture content of the product solids may not meet specifications. When the final moisture content is critical, it may be advisable to smooth out the residence-time distribution by using a more elaborate, multistage fluidized-bed dryer such as the one shown in Figure 18.10b. Alternatively, the stages can be arranged side by side horizontally. Starch dryers have been fabricated with 20 such stages. Materials that are successfully dried in fluidized-bed dryers include coal, sand, limestone, iron ore, clay granules, granular fertilizer, granular desiccant, sodium perborate, polyvinylchloride (PVC), starch, sugar, coffee, sunflower seeds, and salt. Large fluidized-bed dryers for coal and iron ore produce more than 500,000 lb/h of dried material. For metallurgical applications and catalyst regeneration, fluidized beds are frequently heated electrically and carry price tags of from three to six million dollars depending on the temperature and metallurgy requirements.

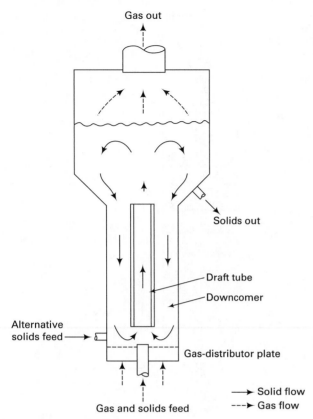

Figure 18.11 Spouted-bed dryer.

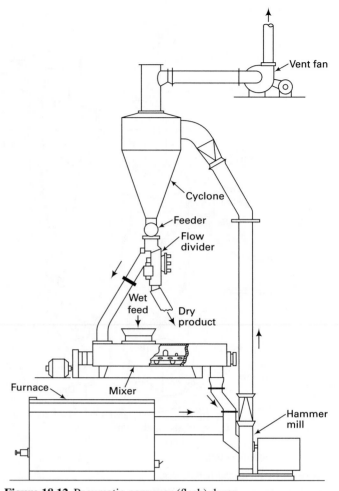

Figure 18.12 Pneumatic-conveyor (flash) dryer.

[From W.L. McCabe, J.C. Smith, and P. Harriott, *Unit Operations of Chemical Engineering*, 5th ed., McGraw-Hill, New York (1993) with permission.]

Spouted-Bed Dryers

When wet, free-flowing particles are larger than 1 mm in diameter but uniform in size and of low density, as in the case of various grains, a spouted-bed dryer, shown in Figure 18.11, is a good choice, particularly when the required drying time is more than just a few minutes. A high-velocity, hot gas enters the bottom of the drying chamber, entrains particles, and flows upward through a draft tube, above which the cross-sectional area for gas flow is significantly increased in a conical section, causing the gas velocity to decrease such that the particles are released to an annular-downcomer region. A fraction of the circulated solids are discharged from a duct in the conical section. The gas exits at the top of the vessel from a free-board region above the bed. For the drying of grains, entrainment of dust in the exiting gas is minimal.

Pneumatic-Conveyor (Flash) Dryers

When only surface moisture must be evaporated from materials that can be reduced to particles by a pulverizer, disintegration mill, or other deagglomeration device, a pneumatic-conveyor (gas-lift or flash) dryer is particularly desirable when the material is temperature-sensitive, oxidizable, explosive, and/or flammable. A flash dryer configuration is shown in Figure 18.12. Wet solids are fed into a paddle-conveyor mixer and dropped into a hammer mill, where solids are disintegrated. Air is pulled, by an exhaust fan, through a furnace into the hammer mill, where the disintegrated solids are

picked up and further deagglomerated into discrete particles while being pneumatically conveyed upward at high velocity in a duct where much of the drying takes place. The particle-gas mixture is separated in a cyclone separator, from which the solids are discharged. Because the particles travel upward in the drying duct at a velocity almost equal to that of the gas, a residence time of less than 5 s is provided. If additional time is needed, up to 30 s can be achieved by partial recycle of the solids leaving the cyclone separator. Recycle of solids is also useful for the disintegration of materials that are sticky or pasty. The deagglomerated particle sizes are from −30 to −300 mesh. If the particles are crystalline or friable, they may be subject to excessive breakage. Pneumatic conveying velocities range from 10 to 30 m/s, usually about 3 m/s greater than the terminal (free-fall) velocity of the largest particle to be conveyed out of the disintegrator. The distribution of remaining moisture in the product particles can be wide because of the distribution of particle-residence times. However, surface drying is rapid because inlet gas temperatures as high as 1,500°F can be employed. Nevertheless, because: (1) the particles flow cocurrently with the gas, (2) the gas temperature decreases significantly, and (3) the particle-residence time is short and evaporation of moisture is incomplete, the particles do not attain temperatures greater than about 200°F.

Large flash dryers are provided with pneumatic-conveying dryer ducts 1 m in diameter and 12 m high, with water-evaporation capacities up to 36,000 lb/h. Compared to many other dryers, they have small floor-area requirements and are used for drying filter cakes, centrifuge cakes and slurries, yeast cakes, whey, starch, sewage sludge, gypsum, fruit pulp, copper sulfate, clay, coal, chicken droppings, adipic acid, polystyrene beads, ammonium sulfate, and hexamethylene tetramine.

Spray Dryers

When solutions, slurries, or pumpable pastes—containing more than 50 wt% moisture, at rates greater than 1,000 lb/h—are to be dried, a spray dryer should be considered. In the configuration in Figure 18.13a the drying chamber has a conical-shaped bottom section with a top diameter that may be nearly equal to the chamber height. Feed is pumped to the top center of the chamber, where it is dispersed into droplets or particles from 2 to 2,000 μm by any of three types of atomizers: (1) single-fluid pressure nozzles, (2) pneumatic nozzles, and (3) centrifugal disks or spray wheels. Hot gas enters the chamber, causing moisture in the atomized feed to rapidly evaporate. Gas flows cocurrently to the solids, and dried solids and gas are either partially separated in the chamber, followed by removal of dust from the gas by a cyclone separator, or, as shown in Figure 18.13a, are sent together to a cyclone separator, bag filter, or other gas–solid separator. The hot gas can be moved by a fan.

In many respects, spray dryers are similar in operating conditions to a pneumatic-conveyor dryer because particles are small, entering gas temperature can be high, residence time of the particles is short, mainly surface moisture is removed, and temperature-sensitive materials can be handled. However, a unique feature of spray dryers is their ability, with some materials such as dyes, foods, and detergents, to produce, from a solution, rounded porous particles of fairly uniform size that can be rapidly dissolved or reacted in subsequent applications.

Although residence times are less than 5 s if only surface moisture is removed, residence times of up to 30 s can be provided for evaporating internal moisture. Spray drying is also unique in that it combines, into one compact piece of equipment, evaporation, crystallization or precipitation, filtration or centrifugation, size reduction, classification, and drying.

A critical part of a spray dryer is the atomizer. Each of the three atomizer types has advantages and disadvantages. Pneumatic (two-fluid) nozzles impinge the gas on the feed at relatively low pressures of up to 100 psig, but are not efficient at high capacities. Consequently, they find applications only in pilot plants and low-capacity commercial plants. Exceptions are the dispersion of stringy and fibrous materials, thick pastes, certain filter cakes, and polymer solutions because with high atomizing gas-to-feed ratios, small particles are produced.

Pressure (single-fluid) nozzles, with orifice diameters of 0.012–0.15 inch, require solution inlet pressures of 300–4,000 psig to achieve breakup of the feed stream. These nozzles can deliver the narrowest range of droplet sizes, but the droplets are the largest delivered by the three types of atomizers, and multiple nozzles are required in large-diameter spray dryers. Also, orifice wear and plugging can be problems with some feeds. Because the spray is largely downward, chambers are relatively slender and tall, with height-to-diameter ratios of 4–5.

The centrifugal disks (spray wheels) shown in Figure 18.13b handle solutions or slurries, delivering thin sheets of feed that break up into small droplets in a nearly radial direction at high capacities. Disks have the largest-diameter spray pattern and therefore require the largest-diameter drying chambers to prevent particles from striking the chamber wall while in a sticky state.

Centrifugal disks range in diameter from a few inches up to 32 inches in large units. Disks spin at 3,000–50,000 rpm, and can operate over a range of feed and rotation rates without significantly affecting the particle-size distribution that occurs with variation of the feed pressure or by enlargement of or other damage to the orifice of a pressure nozzle.

Industrial spray-dryer diameters are large, with 8–30 ft being common. Evaporation rates of up to 2,600 lb/h have been achieved in an 18-ft-diameter by 18-ft-high spray dryer equipped with a centrifugal-disk atomizer and supplied with an aqueous feed solution of 7 wt% dissolved solids and 11,000 cfm of air at 600°F. Larger spray dryers can evaporate up to 15,000 lb/h. Solutions, slurries, and pastes of the following materials are spray-dried: detergents, blood, milk, eggs, starch, yeast, zinc sulfate, lignin, aluminum hydroxide, silica gel, magnesium chloride, manganese sulfate, aluminum

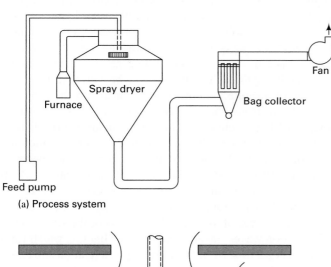

(a) Process system

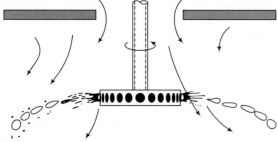

(b) Centrifugal disk atomizer

Figure 18.13 Spray dryer.

sulfate, urea resin, sodium sulfide, coffee extract, tanning extract, color pigments, tea, tomato juice, polymer resins, and ceramics.

Drum Dryers

Approximately 100 years ago and well before the 1920s, when spray dryers were introduced for drying milk and detergent solutions, drum dryers were in use to process solutions, slurries, and pastes with indirect heat. The first such dryer was the double-drum dryer, shown in Figure 18.14a, which is still the most versatile and widely used drum dryer. It consists of two metal, cylindrical drums of identical size (1–5 ft diameter by 1.5–12 ft long), mounted side by side. One drum is movable horizontally so that the distance between the two drums (the nip) can be adjusted. The drums are heated on the inside by condensing steam at pressures as high as 12 atm. The feed enters at the top from a perforated pipe that runs the length of the two drums or from a pipe that swings like a pendulum from end-to-end of the drums. The drums are rotated toward each other at the top, as shown, causing the feed to form, on the hot surface of the drums, a clinging coating, whose thickness is controlled by the nip. As the drums rotate, heat is transferred to the coating, causing it to dry. If the moisture is water, it exits as steam through a vapor hood. If the moisture is a solvent or if the solid is dustable, a dryer enclosure can be provided. When the coating has made about 3/4 of a complete rotation, it is scraped off the drum surfaces by doctor blades that run the length of the drums. Dried material falls off—as surface powder, chips, or, more commonly, as flakes, which are 1–3 mm thick—into conveyors. By adjusting (1) drum-rotation rate from 1 to 30 rpm; (2) drum-surface temperature, usually just a few degrees below the inside, condensing-steam temperature; (3) feed temperature; and (4) coating thickness, the moisture content of the dried material can be controlled. Drying times are 3–20 s. Performance data given by Walas [3] show capacities of double-drum dryers to be in the range of 1–60 kg of dried product/h-m^2 of drum surface for feed moisture contents of 10–90 wt%.

For drum dryers to be effective, the coating must adhere to the drum surface, which is often chrome-plated. When necessary, other drum and feeding arrangements can be employed. The twin-drum dryer with top feed shown in Figure 18.14b is not influenced by drum spacing because the drums rotate away from each other at the top. Thicker coatings can then be formed, and materials like inorganic crystals that might score or cause damage to closely spaced drums can be processed. To improve the likelihood of the feed adhering to the drums, feed may be splashed onto the surface of the drum, as in Figure 18.14c.

For very viscous solutions or pastes that might cause undue pressure to be put on the surfaces of a double-drum dryer, a single-drum dryer can be used. The coating can be applied by using a top feed with applicator rolls, as shown in Figure 18.14d. If a porous product (e.g., malted milk) is desired or if the material is temperature-sensitive, a single-drum or double-drum dryer, as shown in Figure 18.14e, can be enclosed so that a vacuum can be pulled to reduce the boiling point.

Drum dryers are used to dry a wide variety of materials, including brewer's yeast, potatoes, skim milk, malted milk, coffee, tanning extract, and vegetable glue; slurries of $Mg(OH)_2$, $Fe(OH)_2$, and $CaCO_3$; and solutions of sodium acetate, Na_2SO_4, Na_2HPO_4, $CrSO_4$, and various organic compounds. Drum dryers belong to a class of hot-cylinder dryers. Units with large numbers of cylinders in series and parallel are used to dry continuous sheets of woven fabrics and paper pulp at evaporation rates of about 10 kg/h-m^2.

§18.1.4 Other Dryers

A number of other dryers have been developed for special situations. These use infrared radiant energy, generation of heat within the solid by dielectric drying using radio or microwave frequencies, and freeze-drying by sublimation of frozen moisture.

Infrared Drying

In direct-heat dryers, the transfer of heat by convection from hot gases to the wet material is often inadvertently supplemented by thermal radiation from hot surfaces that surround the material. This radiant-heat contribution is usually minor, and ignored. For the drying of certain films, sheets, and coatings, however, use of thermal radiation as the major source of heat is a proven technology.

Radiant energy is released from matter as a result of oscillations and transitions of its electrons. For gases and transparent solids and liquids, radiation can be emitted from throughout the volume of the matter. For opaque solids and liquids, this internal radiation is quickly absorbed by adjoining molecules so that the net transfer of energy by radiation is only from the surface. Of great importance in radiation heat transfer for drying is the transfer of radiation from a hot, opaque surface through a nonabsorbing gas or vacuum to the material being dried. This transfer can be viewed as the propagation of discrete photons (quanta) and/or as the propagation of electromagnetic waves, consisting, as shown in Figure 18.15a, of electric and magnetic fields that oscillate at right angles to each other and to the direction in which the radiation travels. As shown in the electromagnetic spectrum of Figure 18.15b, the wavelength, λ, of the radiation, which depends on the manner in which it is generated, covers an exceedingly wide range, from gamma rays of 10^{-8} μm to long radio waves of 10^{10} μm. Regardless of the wavelength, all radiation waves travel at the speed of light, c, which for a vacuum is 2.998×10^8 m/s. Accordingly, a relationship exists between frequency of the wave, υ, and its wavelength, λ:

$$\upsilon = c/\lambda \qquad (18\text{-}1)$$

The frequency is usually expressed in Hz, which is one cycle/s. The energy transmitted by the wave depends on its frequency and is expressed in terms of the energy, E, of a photon by

$$E = h\upsilon \qquad (18\text{-}2)$$

where h = Planck's constant = 6.62608×10^{-34} J-s.

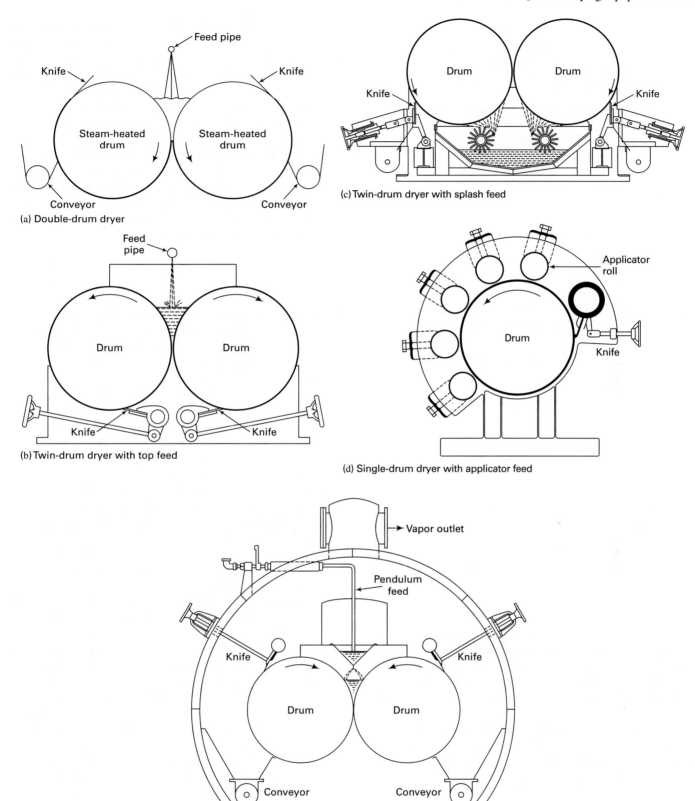

(a) Double-drum dryer

(b) Twin-drum dryer with top feed

(c) Twin-drum dryer with splash feed

(d) Single-drum dryer with applicator feed

(e) Vacuum double-drum dryer

Figure 18.14 Drum dryers.

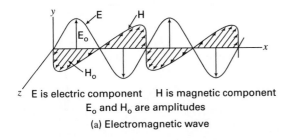

z   E is electric component   H is magnetic component
E_o and H_o are amplitudes

(a) Electromagnetic wave

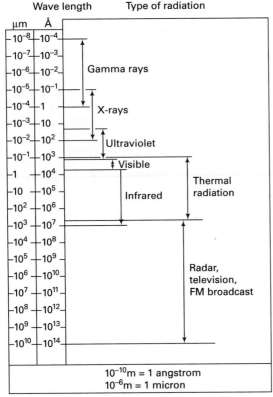

10^{-10} m = 1 angstrom
10^{-6} m = 1 micron

(b) Electromagnetic spectrum

Figure 18.15 Radiation.

A solid, opaque surface can emit infrared radiation by virtue of its temperature. This type of radiation is invisible and has a wavelength, as shown in Figure 18.15b, in the range of 0.75–300 μm. If the surface emitting the radiation is a so-called blackbody, b, such that the maximum amount of radiation will be emitted, that amount will be distributed over a range of wavelengths, depending on the temperature, as governed by the Planck distribution, which, in terms of radiant heat leaving diffusely from a unit area of surface, is

$$E_{\lambda,b} = C_1 / \{\lambda^5 [\exp(C_2/\lambda T) - 1]\} \qquad (18\text{-}3)$$

where the units of $E_{\lambda,b}$ are W/m²-μm, $C_1 = 3.742 \times 10^8$, $C_2 = 1.439 \times 10^4$, T is in K, and λ is in μm. When (18-3) is integrated over the entire range of wavelengths, the result is the Stefan–Boltzmann equation,

$$E_b = \sigma T^4 \qquad (18\text{-}4)$$

where $\sigma = 5.67051 \times 10^{-8}$ W/m²-K⁴ and T is in K. Thus, as the temperature of the infrared heat source is increased, the rate of heat transfer increases exponentially.

Sources of infrared radiant heat at surface temperatures in the range of 600–2,500 K are electrically heated metal-sheath rods, quartz tubes, and quartz lamps; and ceramic-enclosed gas burners. When the radiant energy reaches the material being dried, it is absorbed at the surface, from which it is transferred into the interior by conduction. In this respect, infrared radiant heat transfer to the surface is much like convective heat transfer. However, if the effective thermal conductivity of the material is low, the surface temperature may rise to an undesirable value, particularly if high-temperature, infrared-radiation sources are used. Applications of growing interest include drying of paper, paints, enamels, inks, glue-on flaps, and textiles. Continuous infrared dryers are more common than batch infrared dryers.

Dielectric Drying

In contrast to infrared drying, dielectric drying involves the low-frequency, long-wavelength end of the electromagnetic spectrum of Figure 18.15b, where radio waves and microwaves reside. With nonelectrically conducting materials, heat is not absorbed at the surface but is generated throughout the material, reducing the importance of heat conduction within the material and, thus, making this type of drying unique and making it possible to control the rate of energy dissipation in the material over a wide range. Other advantages over more conventional drying methods include: (1) efficiency of energy usage because the energy dissipation occurs mainly in the moisture rather than in the solid material; (2) operation at low temperatures, thus avoiding high material surface temperatures; and (3) more rapid drying. Dielectric drying is particularly useful for preheating materials and for removing the final traces of internal moisture. *Radio frequency (RF) drying* is confined to frequencies, υ, between 1 and 150 MHz ($\lambda = 3 \times 10^8$ to 2×10^6 μm or 300 to 2 m), while microwave drying utilizes frequencies from 300 MHz to 300 GHz ($\lambda = 10^6$ to 10^3 μm or 1 m to 1 mm). By international agreement, *microwave drying* is done at only 915 and 2,450 MHz, as discussed by Mujumdar [1]. For RF drying, the U.S. Federal Communications Commission (FCC) has reserved frequencies of 13.56 MHz ± 0.05%, 27.12 MHz ± .60%, and 40.68 MHz ± 0.05%. Equipment for dielectric drying consists of an energy generator and an applicator. A generator is used to boost 50–60 Hz line voltage to the much higher values quoted above. A negative-grid triode tube is used with dielectric systems, while magnetron or klystron tubes are used with microwave systems. Dielectric energy is usually applied by electrodes of various types and shapes, between which is placed the material to be dried. Microwave systems often use hollow, rectangular, metallic waveguides.

RF systems are used to dry bulky materials such as lumber, ceramic monoliths, foam rubber, breakfast cereals, dog biscuits, crackers, biscuits, and cookies, as well as films, coatings, and materials such as paper, inks, adhesives, textiles, and penicillin, where high surface temperatures must be avoided. Ceramic catalytic-converter extrusions are dried

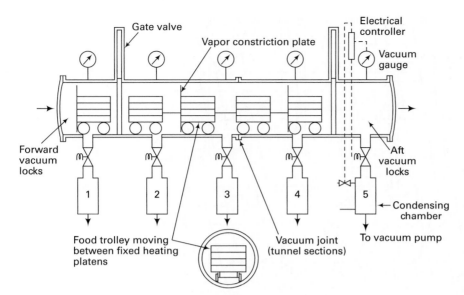

Figure 18.16 Tunnel freeze-dryer.

quickly and uniformly by RF drying. Microwave systems are used to dry pasta, onions, seaweed, baseball bats, potato chips, pharmaceuticals, ceramic filters, and sand casting molds.

Freeze-Drying

In freeze-drying (lyophilization), moisture in the feed is first frozen, by cooling, and then sublimed by conductive, convective, and/or radiant heating. Because the structure and properties of solid material to be dried are hardly altered by freeze-drying, it has been adapted widely to the drying of biological materials, pharmaceuticals, and foodstuffs. Products of freeze-drying are porous and nonshrunken. When foodstuffs are dehydrated by freeze-drying and then stored under a dry, inert gas, they evade deterioration almost indefinitely and can be rehydrated almost perfectly to their original state for later consumption. The first major application of freeze-drying was for the preservation of blood plasma during World War II.

When the moisture is water, the material must be cooled to at least 0°C to freeze the water if it is free, and even lower if the water is dissolved in the material. Most freeze-drying is conducted at −10°C or lower. At this temperature, ice has a vapor pressure of only 2 torr; therefore, freeze-drying must be conducted under a high vacuum. Heat for sublimation is transferred from the heat source to the material under controlled conditions so that the moisture does not reach the melting point. In some cases, an even lower temperature, called the scorch point, must not be exceeded, or degradation of the material will occur. During the drying period, which may take 20 hours, resistance to heat transfer increases because an interface develops between the porous freeze-dried layer and the frozen material, which gradually recedes into the material.

For small quantities of biological and pharmaceutical materials, freeze-drying is conducted batchwise on trays in vacuum cabinets, where the drying step follows the freezing. The sublimed ice is desublimed on a cold metal plate that resides either inside the cabinet and adjacent to the trays or in a separate, adjoining vessel. During sublimation, heat transfer is usually by conduction from the bottom- and side-tray surfaces, which contain coils or passages through which a heating fluid, e.g., vacuum steam, passes. For large quantities of foodstuffs, continuous freeze-drying can be employed, as shown in Figure 18.16, using trays of prefrozen materials transported through a tunnel past fixed heating platens, with vacuum locks at either end. With granular materials, drying times of less than 1 h can be achieved. Continuous freeze-drying of small particles can also be accomplished rapidly in a fluidized bed, where heat transfer for sublimation is by convection and radiation. However, bed turbulence may result in particle breakage and dusting. In some cases, freeze-drying can utilize infrared and microwave heating. Freeze-drying is used for the sublimation of moisture from seafood, meat, vegetables, fruits, coffee, concentrated beverages, pharmaceuticals, blood plasma, and biological materials.

§18.2 PSYCHROMETRY

If moisture is to evaporate from a wet solid, it must be heated to a temperature at which its vapor pressure exceeds the partial pressure of the moisture in the gas in contact with the wet solid. In an indirect-heat dryer, where little or no gas is used to carry away the moisture as vapor, the partial pressure of the moisture approaches the total pressure, and the temperature of evaporation approaches the boiling point of the moisture at the prevailing pressure, as long as the moisture is free liquid at the surface of the solid. If the moisture interface recedes into the solid, a temperature above the boiling point is necessary at the solid–gas interface to transfer the heat for evaporation to the liquid–gas interface. If the moisture level drops to a point where it is entirely sorbed, its vapor pressure is less than the pure vapor pressure and an even higher temperature is required to evaporate it. In a direct-heat dryer, similar situations occur, except that the temperature at which moisture evaporates depends on the gas-moisture content.

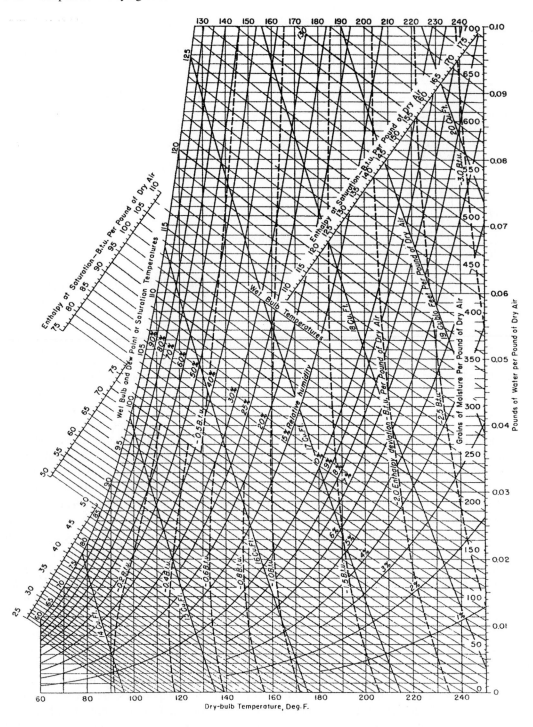

Figure 18.17 Psychrometric (humidity) chart for air–water at 1 atm.

§18.2.1 Psychrometric Chart

Calculations involving the properties of moisture–gas mixtures for application to drying are most conveniently carried out with psychrometric charts. A typical chart, given in Figure 18.17, is that for air–water vapor mixtures at 1-atm total pressure. Included in this chart are properties that are listed and defined in Table 18.4, which applies to general moisture–gas mixtures that obey the ideal-gas law. The definitions given by (18-5)–(18-9) for humidity and by (18-10)–(18-12) for humid volume, humid heat, and enthalpy are in terms of a unit mass or mole of moisture-free gas. For a water vapor–air mixture, the term "dry air" is substituted for "moisture-free gas." The following example, which makes use of Figure 18.17 and Table 18.4, illustrates the use of the dry-air or "dry" basis.

EXAMPLE 18.1 Use of the Psychrometric Chart.

Air at 131°F and 1 atm enters a direct-heat dryer with a humidity, $\mathcal{H}$, of 0.03 lb H_2O (A)/lb H_2O-free air (B). Determine by the psychrometric chart of Figure 18.17 and the relationships of Table 18.4: (a) molal humidity, $\mathcal{H}_m$; (b) saturation humidity, $\mathcal{H}_s$; (c) relative humidity, $\mathcal{H}_R$; (d) percentage humidity, $\mathcal{H}_P$; (e) humid volume, v_H; (f) humid (specific) heat, C_s; and (g) enthalpy, H.

Table 18.4 Definitions of Quantities Useful in Psychrometry: A = moisture; B = moisture-free gas, ideal-gas conditions

Quantity	Definition	Relationship	
Absolute, mass humidity	Moisture content of a gas by mass	$\mathcal{H} = \dfrac{M_A p_A}{M_B(P - p_A)}$	(18-5)
Molal humidity	Moisture content of a gas by mols	$\mathcal{H}_m = \dfrac{p_A}{P - p_A}$	(18-6)
Saturation humidity	Humidity at saturation	$\mathcal{H}_s = \dfrac{M_A P_A^s}{M_B(P - P_A^s)}$	(18-7)
Relative humidity (relative saturation as a percent)	Ratio of partial pressure of moisture to partial pressure of moisture at saturation	$\mathcal{H}_R = 100\% \times \dfrac{p_A}{P_A^s}$	(18-8)
Percentage humidity (percent saturation)	Ratio of humidity to humidity at saturation	$\mathcal{H}_P = 100\% \times \dfrac{\mathcal{H}}{\mathcal{H}_s}$	(18-9)
Humid volume	Volume of moisture–gas mixture per unit mass of moisture-free gas	$\upsilon_H = \dfrac{RT}{P}\left(\dfrac{1}{M_B} + \dfrac{\mathcal{H}}{M_A}\right)$	(18-10)
Humid heat	Specific heat of moisture–gas mixture per unit mass of moisture-free gas	$C_s = (C_P)_B + (C_P)_A \mathcal{H}$	(18-11)
Total enthalpy	Enthalpy of moisture–gas mixture per unit mass of moisture-free gas referred to temperature, T_o	$H = C_s(T - T_o) + \Delta H_o^{\text{vap}}\mathcal{H}$	(18-12)
Dew-point temperature	Temperature at which moisture begins to condense when mixture is cooled	T_{dew}	
Dry-bulb temperature	Temperature of mixture	T_d	
Wet-bulb temperature	Steady-state temperature attained by a wet-bulb thermometer	T_w	
Adiabatic-saturation temperature	Temperature attained when a gas is saturated with moisture in an adiabatic process	T_s	

Solution

At 131°F, the vapor pressure of water is 118 torr = 0.155 atm.

(a) Combining (18-5) and (18-6),

$$\mathcal{H}_m = \frac{M_B}{M_A}\mathcal{H} = \frac{M_{\text{air}}}{M_{H_2O}}\mathcal{H} = \frac{28.97}{18.02}(0.03) = 0.048\ \frac{\text{lb mol } H_2O}{\text{lb mol dry air}}$$

(b) From (18-7),

$$\mathcal{H}_s = \frac{18.02}{28.97}\left(\frac{0.155}{1 - 0.155}\right) = 0.114\ \frac{\text{lb } H_2O}{\text{lb dry air}}$$

(c) From a rearrangement of (18-6),

$$p_{H_2O} = \frac{P\mathcal{H}_m}{1 + \mathcal{H}_m} = \frac{(1)(0.048)}{1 + 0.048} = 0.0458\ \text{atm}$$

From (18-8), $\quad \mathcal{H}_R = 100\left(\dfrac{0.0458}{0.155}\right) = 29.5\%$

The same result is obtained from Figure 18.17.

(d) From (18-9), $\quad \mathcal{H}_P = 100\left(\dfrac{0.03}{0.114}\right) = 26.3\%$

(e) From (18-10), for $R = 0.730\ \dfrac{\text{atm-ft}^3}{\text{lbmol-°R}}$, $T = 131 - 460 = 591°\text{R}$,

$$\upsilon_H = \frac{0.730(591)}{1}\left(\frac{1}{28.97} + \frac{0.03}{18.02}\right) = 15.6\ \text{ft}^3/\text{lb dry air}$$

which agrees with Figure 18.17.

(f) From (18-11), using $(C_P)_{\text{air}} = 0.24$ Btu/lb-°F and $(C_P)_{\text{steam}} = 0.45$ Btu/lb-°F,

$$C_s = 0.24 + (0.45)(0.03) = 0.254\ \frac{\text{Btu}}{\text{lb dry air}}$$

(g) Equation (18-12) assumes that the enthalpy datum refers to air as a gas and water as a liquid. Taking $T_o = 32°\text{F}$ and $\Delta H_o^{\text{vap}} = 1075$ Btu/lb, (18-12) gives

$$H = 0.254(131 - 32) + 1,075(0.03) = 57.4\ \text{Btu/lb dry air}.$$

EXAMPLE 18.2 Humidity for Benzene as the Moisture.

In a dryer where benzene (A) is evaporated from a solid, nitrogen gas (B) at 50°F and 1.2 atm has a relative humidity for benzene of 35%. Determine: (a) benzene partial pressure if its vapor pressure at

$50°F = 45.6$ torr, (b) humidity of the nitrogen–benzene mixture, (c) saturation humidity of the mixture, and (d) percentage humidity of the mixture.

Solution

$$\mathcal{H}_R = 35\%, \; P = 1.2 \, \text{atm} = 912 \, \text{torr},$$

$$M_A = 78.1, \; M_B = 28, \; P^s_{\text{benzene}} = 45.6 \, \text{torr}$$

(a) From (18-8),

$$p_{\text{benzene}} = \frac{P^s_{\text{benzene}}\mathcal{H}_R}{100} = \frac{(45.6)(35)}{100} = 16 \, \text{torr}$$

(b) From (18-5),

$$\mathcal{H} = \left(\frac{78.1}{28}\right)\left(\frac{16}{912-16}\right) = 0.050 \frac{\text{lb benzene}}{\text{lb dry nitrogen}}$$

(c) From (18-7),

$$\mathcal{H}_s = \left(\frac{78.1}{28}\right)\left(\frac{45.6}{912-45.6}\right) = 0.147 \frac{\text{lb benzene}}{\text{lb dry nitrogen}}$$

(d) From (99), $\quad \mathcal{H}_P = 100\left(\frac{0.050}{0.147}\right) = 34\%$

§18.2.2 Wet-Bulb Temperature

The temperature at which moisture evaporates in a direct-heat dryer is difficult to determine and varies from the dryer inlet to the dryer outlet. When the dryer operates isobarically and adiabatically, with all energy for moisture evaporation supplied from the hot gas by convective heat transfer, with no energy required for heating the wet solid to the evaporation temperature, use of simplified heat- and mass-transfer equations leads to an expression for the temperature of evaporation at a particular location in a dryer operating under continuous, steady-state conditions, or at a particular time in a batch dryer cycle.

If it is further assumed that the moisture being evaporated is free liquid exerting its full vapor pressure at the surface of the solid, this temperature of evaporation is called the *wet-bulb temperature*, T_w, because it can be measured by covering a thermometer bulb with a wick saturated with the liquid being evaporated and passing a partially saturated gas past the wick, as indicated in Figure 18.18a.

In Figure 18.18b, where the wetted wick is replaced by an incremental amount of wet solid, assume the heat-transfer area = mass-transfer area = A. At steady state, the rate of convective heat transfer from the gas to the wet solid is given by Newton's law of cooling:

$$Q = h(T - T_w)A \tag{18-13}$$

The molar rate of mass transfer of evaporated moisture from the wet surface of the solid, A, is

$$n_A = \frac{k_y(y_{A_w} - y_A)A}{(1-y_A)_{LM}} \tag{18-14}$$

An enthalpy balance on the moisture evaporated and heated to the gas temperature couples the heat- and mass-transfer equations to give

$$Q = n_A M_A \left[\Delta H_w^{\text{vap}} + (C_P)_A(T-T_w)\right] \tag{18-15}$$

To obtain a simplified relationship for the coupling in terms of T and $\mathcal{H}$, assume that the mole fraction of moisture in the bulk gas and at the wet solid–gas interface is small. Then, the bulk-flow effect in (18-14) becomes $(1-y_A)_{LM} \approx 1.0$. Also, from (18-5), replacing p_A with $y_A P$,

$$y_A = \frac{\mathcal{H}M_A}{\dfrac{1}{M_B}+\dfrac{\mathcal{H}}{M_A}} \approx \frac{\mathcal{H}M_B}{M_A} \tag{18-16}$$

If the latent heat in (18-15) is much greater than the sensible heat,

$$\Delta H_w^{\text{vap}} + (C_P)_A(T-T_w) \approx \Delta H_w^{\text{vap}} \tag{18-17}$$

Simplifying, and combining (18-13) to (18-16),

$$T_w = T - \frac{k_y M_B \Delta H_w^{\text{vap}}}{h}(\mathcal{H}_w - \mathcal{H}) \tag{18-18}$$

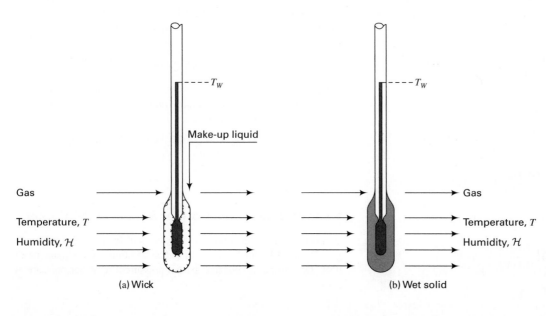

(a) Wick   (b) Wet solid

Figure 18.18 Wet-bulb temperature.
[From W.L. McCabe, J.C. Smith, and P. Harriott, *Unit Operations of Chemical Engineering*, 5th ed., McGraw-Hill, New York (1993) with permission.]

Table 18.5 Lewis Number for Liquids Evaporating into Air at 25°C

Liquid	N_{Le}	$(N_{Le})^{2/3} =$ Psychrometric Ratio
Benzene	2.44	1.812
Carbon tetrachloride	2.67	1.923
Chloroform	3.08	2.114
Ethyl acetate	2.58	1.880
Ethylene tetrachloride	3.05	2.101
Metaxylene	3.18	2.165
Methanol	1.37	1.233
Propanol	1.85	1.506
Toluene	2.64	1.908
Water	0.855	0.901

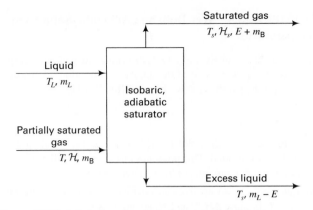

Figure 18.19 Adiabatic saturation of a gas with a liquid.

If the Chilton–Colburn analogy for mass and heat transfer applies, then from (3-165) and (3-228),

$$\frac{k_y M}{G} N_{Sc}^{2/3} = \frac{h}{C_P G} N_{Pr}^{2/3} \quad (18\text{-}19)$$

If $M \approx M_B$, (18-19) becomes

$$\frac{k_y M_B}{h} = \frac{1}{(C_P)_B} \left(\frac{1}{N_{Le}}\right)^{2/3} \quad (18\text{-}20)$$

where $\quad N_{Le} = \text{Lewis number} = N_{Sc}/N_{Pr} \quad (18\text{-}21)$

The reciprocal of the Lewis number is referred to as the Luikov number, N_{Lu}. Substituting (18-20) into (18-18),

$$T_w = T - \frac{\Delta H_w^{vap}}{C_P} \left(\frac{1}{N_{Le}}\right)^{2/3} (\mathcal{H}_w - \mathcal{H}) \quad (18\text{-}22)$$

In Equation (18-18), which defines the wet-bulb temperature, or in (18-22), it is important to note that $\mathcal{H}_w$ is the saturation humidity at temperature T_w.

By coincidence, $(N_{Le})^{2/3}$ for air–water vapor at 25°C is close to 1.0 (actually, 0.901). For the air–organic-vapor systems listed in Table 18.5, which is taken from Keey [7], $(N_{Le})^{2/3}$ is much less than 1.0. As shown in the next section, $(N_{Le})^{2/3}$ has an impact on the variation with location or time of the temperature of moisture evaporation in a direct-heat dryer.

§18.2.3 Adiabatic-Saturation Temperature

To determine the change in the wet-bulb temperature of a wet solid, as it dries in the dryer, it is necessary to consider changes in the temperature and humidity of the gas as it cools, because of heat transfer to the wet solid and transfer of moisture from the wet solid. A simplified relationship between gas temperature and humidity can be derived by considering the adiabatic saturation of a gas with an excess of liquid. Referring to Figure 18.19, partially saturated gas at temperature T, humidity $\mathcal{H}$, and mass flow rate m_B (dry basis), together with liquid at temperature T_L and mass flow rate m_L, enters an isobaric and adiabatic chamber where a fraction of the liquid, (E/m_l), is vaporized to saturate the gas. The gas and excess liquid leave the chamber in equilibrium at temperature T_s. An enthalpy balance using (18-1) for enthalpy of the gas phase, but with the reference temperature $T_o = T_s$ (so as to simplify the balance), gives

$$m_L(C_{P_A})_L(T_L - T_s) + m_B\left[C_{s_{in}}(T - T_s) + \Delta H_s^{vap}\mathcal{H}\right]$$
$$= m_B\left[C_{s_{out}}(T_s - T_s) + \Delta H_s^{vap}\mathcal{H}_s\right] \quad (18\text{-}23)$$
$$+ (m_L - E) + (C_{P_A})_L(T_s - T_s)$$

Assume that the sensible heat required to heat the liquid from T_L to T_s is negligible. Then (18-23) simplifies to an equation for the adiabatic-saturation temperature:

$$T_s = T - \frac{\Delta H_s^{vap}}{(C_s)_{in}}(\mathcal{H}_s - \mathcal{H}) \quad (18\text{-}24)$$

Equation (18-24) can be used to determine gas temperatures and humidities between T and T_s, and $\mathcal{H}$ and $\mathcal{H}_s$, if sensible heat for the liquid is ignored. If (18-24) is compared to (18-18), it is seen that wet-bulb and adiabatic-saturation temperatures are equal if

$$\text{psychrometric ratio} = \left(\frac{h}{k_y M_B}\right)\bigg/ C_{s_{in}} \approx (N_{Le})^{2/3} = 1$$

For the air–water system, as shown by Lewis [8] and referred to as the *Lewis relation*, this is almost the case, with

$$\frac{h}{k_y M_B} \approx 0.216 \text{ Btu/lb-}°\text{F}$$

compared to $C_s \approx 0.24$ Btu/lb-°F. Normally, a small amount of thermal radiation heat transfer to the wet solid supplements the convective heat transfer, so if h in the psychrometric ratio is replaced by $(h_c + h_r)$, h_r is an effective heat-transfer coefficient for thermal radiation, and the corrected ratio is almost identical to the humid heat at low humidities. Accordingly, for the air–water system, the wet-bulb temperature is set equal to the adiabatic-saturation temperature and only one family of lines is shown on the psychrometric chart. From (18-24), these lines have a negative slope of $\Delta H_s^{vap}/(C_s)_m$. The following example illustrates use of the chart in Figure 18.17 to determine the relationship between gas (dry-bulb) temperature, wet-bulb temperature, and humidity. The example also illustrates use of (18-24).

EXAMPLE 18.3 Wet-Bulb and Adiabatic-Saturation Temperatures.

For the conditions of Example 18.1, determine the wet-bulb temperature, assuming it is equal to the adiabatic-saturation temperature, by using (a) the psychrometric chart and (b) (18-24).

Solution

(a) In Figure 18.17, the point $T = T_d = 131°F$ and $\mathcal{H} = 0.03$ lb H$_2$O/lb dry air is plotted. This point lies just above the adiabatic-saturation-temperature line of 95°F at about 96.5°F.

(b) Equation (18-24) involves an iterative calculation to determine $T_w = T_s$ because ΔH_s^{vap} and $\mathcal{H}_s$ are unknown.

Assume: $T_s = 95°F$, $\Delta H_s^{vap} = 1039.8$ Btu/lb from steam tables, $(C_s)_{in} = 0.254$ Btu/lb from Example 18.1, and $P_A^s = 42.2$ torr = vapor pressure of water from Example 18.1.

From (18-7), $\mathcal{H}_s = \dfrac{18.02}{28.97}\left(\dfrac{42.2}{760 - 42.2}\right) = 0.0366\,\dfrac{\text{lb H}_2\text{O}}{\text{lb dry air}}$

From (18-24), $\quad T_s = 131 - \dfrac{1039.8}{0.254}(0.0366 - 0.03) = 104°F$

(So, try again.)
Assume:

$\quad T_s = 97°F$, $\Delta H_s^{vap} = 1038.7$ Btu/lb, and $P_A^s = 45$ torr, then

$$\mathcal{H}_s = \dfrac{18.02}{28.97}\left(\dfrac{45}{760 - 45}\right) = 0.0391\,\dfrac{\text{lb H}_2\text{O}}{\text{lb dry air}}$$

$$T_s = 131 - \dfrac{1038.7}{0.254}(0.0391 - 0.03) = 94°F$$

As can be seen, this iterative calculation is very sensitive. By interpolation, $T_s = 95.5°C$, which is close to the value of 96.5°F read from Figure 18.17.

For systems other than air–water, the Lewis relation does not hold, and it is common to employ the psychrometric ratio, $h/(k_y C_s M_B)$. Calculated values of this ratio, using the Chilton–Colburn analogy, are included in Table 18.5. Comparing (18-18) for wet-bulb temperatures to (18-24) for adiabatic-saturation temperatures, it can be noted that since the inverse of the psychrometric ratio for all air–organic vapor systems in Table 18.5 is less than 1, these systems will have slopes of adiabatic-saturation lines that are less than slopes of wet-bulb lines. An example is the psychrometric chart for air–toluene at 1 atm shown in Figure 18.20. Use of this chart is illustrated in the next example.

EXAMPLE 18.4 Psychrometric Chart for Air-Toluene.

Air is used to dry a solid wet with toluene at 1 atm. At a location where the air has a temperature of 140°F and a relative humidity of

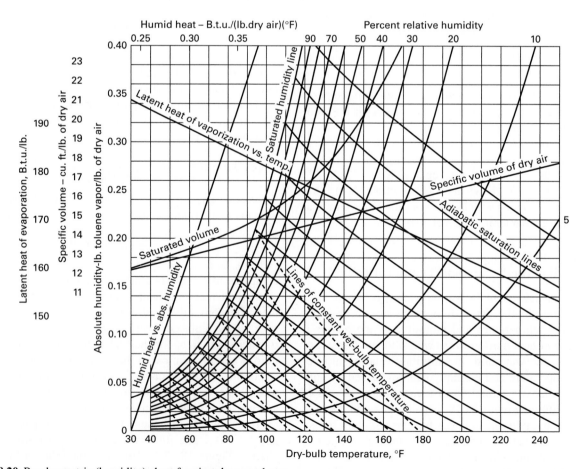

Figure 18.20 Psychrometric (humidity) chart for air–toluene at 1 atm.

[From *Perry's Chemical Engineers' Handbook*, 6th ed., R.H. Perry, D.W. Green, and J.O. Maloney, Eds., McGraw-Hill, New York (1984) with permission.]

10%, determine the humidity, the adiabatic-saturation temperature, and the wet-bulb temperature.

Solution

From Figure 18.20, the humidity is 0.062 lb toluene/lb dry air at the intersection point of a vertical temperature line and a curved percent relative-humidity line. By following an interpolated adiabatic-saturation line from that intersection point to the saturated-humidity line, $T_s = 83°F$. By following an interpolated wet-bulb line from the intersection point to the saturated-humidity line, $T_w = 92°F$. Thus, the wet-bulb temperature is higher than the adiabatic-saturation temperature. In the next section, this causes a reduction in the driving force available for heat transfer during the drying of solids wet with organic moisture.

§18.2.4 Moisture-Evaporation Temperature

The adiabatic-saturation temperature combined with the wet-bulb temperature can be used to track gas temperature, T_g, and moisture-evaporation (solid) temperature, T_v, with respect to location or time when removing surface moisture from a wet solid with direct heat. The accuracy of the tracking is subject to the validity of the assumptions made. If the moisture is water, T_v will be constant and equal to the constant, T_w, of the gas. If the moisture is an organic compound with properties similar to those in Table 18.5, T_v will still be equal to T_w, but, as shown next, will not be constant but will decrease as gas temperature decreases.

Let T_o and $\mathcal{H}$ be hypothetical entering conditions for a gas used to dry a wet solid in an adiabatic, direct-heat dryer. At any point in the dryer, the conditions of the gas are given from (18-24) as

$$T_o - T_g = \frac{\Delta H_s^{\text{vap}}}{(C_s)_o}\left(\mathcal{H}_g - \mathcal{H}_o\right) \qquad (18\text{-}25)$$

Assuming a quasi-steady-state transport condition at any point, the gas–liquid moisture interface conditions of T_v and $\mathcal{H}_v$ are related by the following form of (18-18):

$$T_v - T_g = \frac{k_y M_{\text{B}} \Delta H_s^{\text{vap}}}{h}\left(\mathcal{H}_g - \mathcal{H}_v\right) \qquad (18\text{-}26)$$

For a continuous dryer, (18-26) holds regardless of whether gas and wet solid flows are countercurrent or cocurrent. Equations (18-25) and (18-26) can be combined to give

$$\frac{T_v - T_g}{T_o - T_g} = \left(\frac{k_y M_{\text{B}} (C_s)_o}{h}\right)\left(\frac{\mathcal{H}_g - \mathcal{H}_v}{\mathcal{H}_g - \mathcal{H}_o}\right) \qquad (18\text{-}27)$$

The coefficient of the RHS of (18-27) is the inverse of the psychrometric ratio. For air–water, it is ≈ 1.0, giving

$$\frac{T_v - T_g}{T_o - T_g} = \frac{\mathcal{H}_g - \mathcal{H}_v}{\mathcal{H}_g - \mathcal{H}_o} \qquad (18\text{-}28)$$

Assume that the wet solid in contact with the initial or entering gas is at the gas wet-bulb temperature, T_w. Then,

$$\frac{T_w - T_g}{T_o - T_g} = \frac{\mathcal{H}_g - \mathcal{H}_w}{\mathcal{H}_g - \mathcal{H}_o} \qquad (18\text{-}29)$$

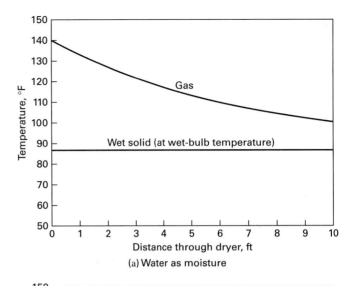

(a) Water as moisture

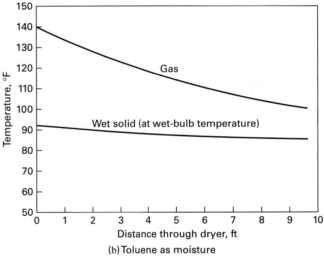

(b) Toluene as moisture

Figure 18.21 Temperature profiles for dryer of Example 18.5.

Since T_o is fixed, T_v must remain at the value T_w regardless of the value of T_g. The result, for Example 18.5 below, is shown in Figure 18.21 for air–water and air–toluene.

For moisture other than water, (18-27) is appropriate. Assume that the wet solid is dried batchwise, or flows cocurrently to the gas in a continuous dryer, with an initial temperature equal to the wet-bulb temperature of the gas. Because the inverse of the psychrometric ratio is less than 1.0, T_w will decrease as T_g decreases, as in Example 18.5.

EXAMPLE 18.5 Temperature Variation in a Dryer.

Air enters a continuous, adiabatic direct-heat dryer at 140°F and 1 atm with the relative humidity below, and exits at 100°F. The wet solid enters in cocurrent flow at the wet-bulb temperature of the entering gas. Plot the variation of the moisture-evaporation temperature, T_v, as a function of the distance through the dryer, z, for an exponential decrease of T_g according to the relation

$$T_g - T_s = (T_o - T_s)\exp(-0.1377z) \qquad (1)$$

where z is in feet and temperatures are °F, for: (a) water moisture with entering air of 12.5% $\mathcal{H}_R$, and (b) toluene moisture with entering air of 10% $\mathcal{H}_R$.

Solution

$$T_o = 140°F, \ T_g \text{ in} = 140°F, \text{ and } T_g \text{ out} = 100°F$$

(a) From Figure 18.17, $\mathcal{H}_o = 0.015$ lb H_2O/lb dry air. $T_w = 86.5°F = T_s$. As the gas cools, its humidity follows the adiabatic-saturation line. Using (1) and Figure 18.17,

z, ft	T_g, °F	$\mathcal{H}_g$, $\dfrac{\text{lb } H_2O}{\text{lb dry air}}$
0	140.0	0.015
2	127.1	0.018
4	117.3	0.020
6	109.9	0.022
8	104.3	0.0235
10	100.0	0.0245

Equation (18-28) is satisfied only for $T_v = T_w = 86.5°F$ and $\mathcal{H}_v = \mathcal{H}_w = 0.0275$ lb H_2O/lb dry air

For example, take $T_g = 109.9°F$, $\mathcal{H}_g = 0.022$ lb H_2O/lb dry air. Using (18-28), take values of $T_v = 80°F$ and $90°F$, and compute the temperature and humidity ratios:

T_v, °F	$\mathcal{H}_v$, $\dfrac{\text{lb } H_2O}{\text{lb dry air}}$	$\dfrac{T_v - T_g}{T_o - T_g}$	$\dfrac{\mathcal{H}_g - \mathcal{H}_v}{\mathcal{H}_g - \mathcal{H}_o}$
86.5	0.0275	−0.777	−0.786
80.0	0.0223	−0.993	−0.043
90.0	0.0310	−0.661	−1.300

(b) From Figure 18.20 for $T_o = 140°F$, $\mathcal{H}_R = 10\%$: $\mathcal{H}_o = 0.062$ lb toluene/lb dry air, T_w for entering air $= 92°F$, and $T_s = 83°F$.

From Table 18.4, 1/(psychrometric ratio) $= 1/1.908 = 0.524$. From (18-27),

$$\frac{T_v - T_g}{140 - T_g} = 0.524\left(\frac{\mathcal{H}_g - \mathcal{H}_v}{\mathcal{H}_g - 0.062}\right) \quad (2)$$

From (1), $T_g = T_v + (140 - T_v)\exp(-0.1377z)$ (3)

Equation (2) is solved iteratively for T_v for each value of T_g, where $\mathcal{H}_v$ is the saturation humidity at T_v, as determined from Figure 18.20. For example, for $T_g = 115.9°F$, following the adiabatic-saturation line, $\mathcal{H}_v = 0.095$, and (2) becomes:

$$\frac{T_v - 115.9}{140 - 115.9} = 0.524\left(\frac{0.095 - \mathcal{H}_v}{0.095 - 0.062}\right)$$

or $\quad T_v = 115.9 + 382.7(0.095 - \mathcal{H}_v)$ (4)

This equation is solved by assuming T_v, determining $\mathcal{H}_v$, and then computing T_v.

Assumed T_v, °F	$\mathcal{H}_v$ from Fig. 18.20, lb toluene/lb dry air	T_v, °F from (4)
90	0.180	83.4
88	0.173	86.0
87	0.165	89.0

By interpolation, $T_v = 87.5°F$.

This result can also be obtained graphically from Figure 18.20 using the construction shown in Figure 18.22, with an essentially identical result. Calculations for the other values of T_g give

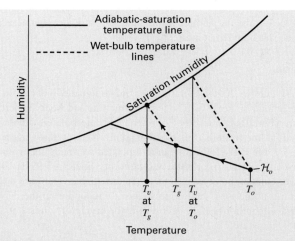

Figure 18.22 Adiabatic drying path for general vapor–moisture mixtures.

T_g, °F	T_v, °F
140.0	92.0
126.3	90.0
115.9	87.5
107.9	86.5
100.0	85.5

From (23), the following values of z are computed:

T_g, °F	z, ft
140.0	0.0
126.3	2.3
115.9	4.5
107.9	6.7
100.0	9.6

Thus, the moisture-evaporation temperature, plotted in Figure 18.21b, decreases with decreasing gas temperature.

§18.3 EQUILIBRIUM-MOISTURE CONTENT OF SOLIDS

Faust et al. [9] group wet solids into two categories according to their drying behavior:

1. *Granular or crystalline solids that hold moisture in open pores between particles.* These are mainly inorganic materials, examples of which are crushed rocks, sand, catalysts, titanium dioxide, zinc sulfate, and sodium phosphates. During drying, the solid is unaffected by moisture removal, so selection of drying conditions and drying rate is not critical to the properties and appearance of the dried product. Materials in this category can be dried rapidly to very low moisture contents.

2. *Fibrous, amorphous, and gel-like materials that dissolve moisture or trap moisture in fibers or very fine pores.* These are mainly organic solids, including tree,

plant, vegetable, and animal materials such as wood, leather, soap, eggs, glues, cereals, starch, cotton, and wool. These materials are affected by moisture removal, often shrinking when dried and swelling when wetted. With these materials, drying in the later stages can be slow. If the surface is dried too rapidly, moisture and temperature gradients can cause checking, warping, case hardening, and/or cracking. Therefore, selection of drying conditions is a critical factor. Drying to low moisture contents is possible only when using a gas of low humidity.

In a direct-heat drying process, the extent to which moisture can be removed from a solid is limited, particularly for the second category, by the *equilibrium-moisture content* of the solid, which depends on factors that include temperature, pressure, and moisture content of the gas. Even if the drying conditions produce a completely dry solid, subsequent exposure of the solid to a different humidity can result in an increase in moisture content.

Terms used to describe equilibrium-moisture content are shown in Figure 18.23 with reference to a hypothetical equilibrium isotherm. Moisture content, X, is expressed as mass of moisture per 100 mass units of bone-dry solid. This is the most common way to express moisture content and is equivalent to wt% moisture on a dry-solid basis. This is analogous

to expressions for humidity and is most convenient in drying calculations where the mass of bone-dry solid and dry gas remain constant while moisture is transferred from solid to gas. Less common is wt% moisture on a wet-solid basis, W. The two moisture contents are related by the expression

$$X = \frac{100W}{100 - W} \qquad (18\text{-}30)$$

or

$$W = \frac{100X}{100 + X} \qquad (18\text{-}31)$$

Rarely used is moisture content on a volume basis because wet solids of the second category shrink during drying. Also, moisture content is never expressed on a mole basis because the molecular weight of the dry solid may not be known.

In Figure 18.23, equilibrium-moisture content, X^*, is plotted for a second-category solid for a given temperature and pressure, against relative humidity, $\mathcal{H}_R$. In some cases, humidity, $\mathcal{H}$, is used with a limit of the saturation humidity, $\mathcal{H}_S$. At $\mathcal{H}_R = 100\%$, equilibrium-moisture content is called *bound moisture*, X_B. If the wet solid has a total moisture content, $X_T > X_B$, the excess, $X_T - X_B$, is *unbound moisture*. At a relative humidity $< 100\%$, the excess of X_T over the equilibrium-moisture content, i.e., $X_T - X^*$, is the *free-moisture* content.

In the presence of a saturated gas, only unbound moisture can be removed during drying. For a partially saturated gas, only free moisture can be removed. But if $\mathcal{H}_R = 0$, all solids, given enough time, may be dried to a bone-dry state. Solid materials that can contain bound moisture are *hygroscopic*. Bound moisture exhibits a vapor pressure less than the normal vapor pressure. The bound-moisture content of cellular materials such as wood is referred to as the fiber-saturation point.

Experimental equilibrium-moisture isotherms at 25°C and 1 atm are shown in Figure 18.24 for second-category materials. At low values of $\mathcal{H}_R$, e.g., $<10\%$, moisture is bound to the solid on its surfaces as an adsorbed monomolecular layer. Such bound moisture can also be present on solids of the first category. At intermediate values of $\mathcal{H}_R$, e.g., 20–60%, multimolecular layers may build up on the monolayer. At large values of $\mathcal{H}_R$, e.g., $>60\%$, moisture is held in micropores so small (e.g., <1 μm in radius) that vapor-pressure lowering occurs, as predicted by the Kelvin equation (15-14). In cellular materials such as plant and tree matter, some moisture is held osmotically in fibers behind semipermeable membranes of cell walls.

Temperature has a significant effect on equilibrium-moisture content, an example of which is shown in Figure 18.25 for cotton at 96–302°F. At an $\mathcal{H}_R$ of 20%, equilibrium-moisture content decreases from 0.037 to 0.012 lb H_2O/lb dry cotton. Experimental determination of equilibrium-moisture isotherms is complicated by a hysteresis effect, shown in Figure 18.26 for sulfite pulp. Sorption and desorption curves were obtained, respectively, by wetting and drying the solid, and it is seen that equilibrium-moisture content in drying is always somewhat higher, particularly in the relative-humidity

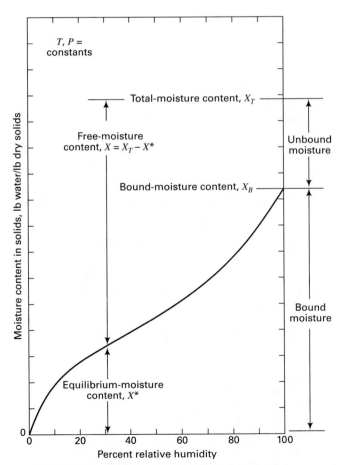

Figure 18.23 Typical isotherm for equilibrium-moisture content of a solid.

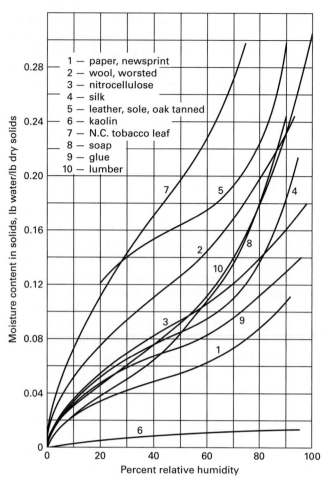

Figure 18.24 Equilibrium-moisture content at 25°C and 1 atm.

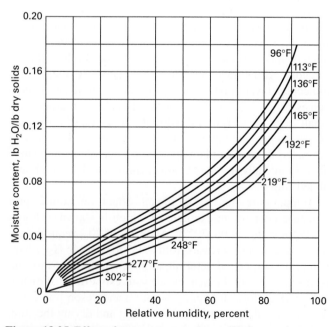

Figure 18.25 Effect of temperature on the equilibrium-moisture content of raw cotton at 1 atm.

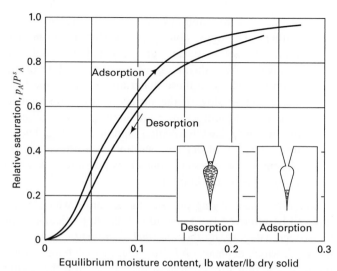

Figure 18.26 Effect of hysteresis on equilibrium-moisture content of sulfite pulp.

range of 30% to 80%. According to Luikov [10], the hysteresis effect may be due to either: (1) failure to achieve true equilibrium or (2) irreversibility of evaporation and condensation in capillaries. For the latter, a possible explanation is based on representations of moisture in necked capillaries, as shown in Figure 18.26. For drying (desorption), the capillary contains more moisture than for wetting (sorption). Thus, for a given relative humidity, the equilibrium-moisture content for drying is less than that for wetting.

Bound moisture can also be defined as moisture held chemically, as, for example, water of hydration of inorganic crystals. This is one example of bound moisture dissolved in a solid where the vapor pressure is lowered significantly below the true vapor pressure. For inorganic salts that form one or more hydrates, the hydrated form depends not only on temperature, but also on relative humidity of the gas in contact with the crystals. The effect of the latter for $CuSO_4$, in terms of the partial pressure of water, is shown in Figure 18.27. At 25°C, the stable hydrate is $CuSO_4 \cdot 5H_2O$. However, if the partial pressure of H_2O is 5.6–7.8 torr, the

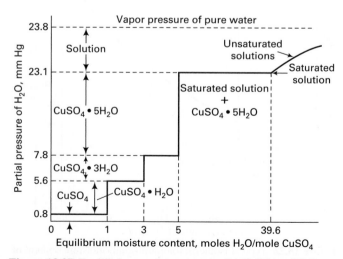

Figure 18.27 Equilibrium-moisture content for $CuSO_4$ at 25°C.

trihydrate forms; at 0.8–5.6 torr, the monohydrate is favored. Below 0.8 torr, $CuSO_4$ crystals are free of water.

EXAMPLE 18.6 Effect of Equilibrium-Moisture Content.

One-kg blocks of wet Borax laundry soap with an initial H_2O content of 20.2 wt% on a dry basis are dried with air in a tunnel dryer at 1 atm. In the limit, if the soap were brought to equilibrium with the air at 25°C and a relative humidity of 20%, determine the kg of moisture evaporated from each block.

Solution

The initial moisture content of the soap on a wet basis is obtained from a rearrangement of (18-30):

$$W = \frac{100X}{100+X} = \frac{100(20.2)}{100+20.2} = 16.8 \text{ wt\%}$$

Initial weight of moisture $= 0.168(1.0) = 0.168$ kg H_2O

Initial weight of dry soap $= 1 - 0.168 = 0.832$ kg dry soap

From Figure 18.24, for soap at $\mathcal{H}_R = 0.20$, $X^* = 0.037$

Final weight of moisture $= 0.037(0.832) = 0.031$ kg

Moisture evaporated $= 0.168 - 0.031 = 0.137$ kg H_2O/kg soap

§18.4 DRYING PERIODS

The decrease in average moisture content, X, as a function of time, t, for drying either category of solids in a direct-heat dryer was observed experimentally by Sherwood [11,12] to exhibit the type of relationship shown in Figure 18.28a, provided the exposed surface of the solid is initially covered with observable moisture. If that curve is differentiated with respect to time and multiplied by the ratio of the mass of dry solid to the interfacial area between the mass of wet solid and the gas, a plot can be made of drying-rate flux, R,

$$R = \frac{dm_v}{A\,dt} = -\frac{m_s}{A}\frac{dX}{dt} \tag{18-32}$$

where m_v = mass of moisture evaporated and m_s = mass of bone-dry solid as a function of moisture content, as shown in Figure 18.28b. In Figure 18.28a, the final equilibrium-moisture content is X^*. Although both plots exhibit four drying periods, the periods are more distinct in the drying-rate curve. For some wet materials and/or some hot-gas conditions, fewer than four drying periods are observed. From A to B, the wet solid is being preheated to an exposed-surface temperature equal to the wet-bulb gas temperature, while moisture is evaporated at an increasing rate. At the end of the preheat period, if the wet solid is of the granular, first category, a cross section has the appearance of Figure 18.29a, where the exposed surface is still covered by a film of moisture. A wet solid of the second category is covered on the exposed surface by free moisture. The drying rate now

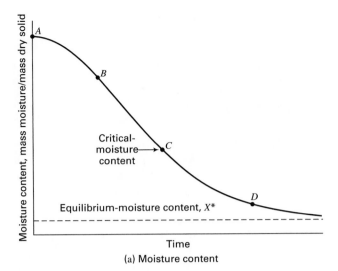

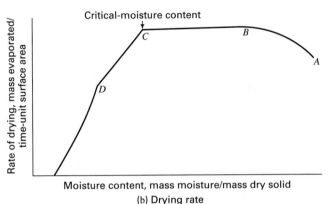

Figure 18.28 Drying curves for constant drying conditions.

becomes constant during the period from B to C, which prevails as long as free moisture covers the exposed surface.

This surface moisture may be part of the original moisture that covered the surface, or it may be moisture brought to the surface by *capillary action* in the case of wet solids of the first category or by *liquid diffusion* in the case of wet solids of the second category. In either case, the rate of drying is controlled by external mass and heat transfer between the exposed surface of the wet solid and the bulk gas. Migration of moisture from the interior of the wet solid to the exposed surface is not a rate-affecting factor. This period, the *constant-rate drying period*, terminates at point C, the *critical moisture content*. When drying wet solids of the first category under agitated conditions—as in a direct-heat rotary dryer, fluidized-bed dryer, flash dryer, or agitated batch dryer—such that all particle surfaces are in direct contact with the gas, the constant-rate drying period may extend all the way to X^*.

At C, the moisture just barely covers the exposed surface; and then until point D is reached, as shown in Figure 18.29b, the surface tends to a dry state because the rate of liquid travel by diffusion or capillary action to the exposed surface is not sufficiently fast. In this period, the exposed-surface temperature remains at the wet-bulb temperature if heat conduction is adequate, but the wetted exposed area for mass transfer decreases. Consequently, the rate of drying decreases

Drying-gas flow

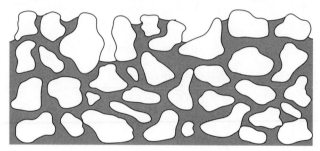

(a) Constant-rate period

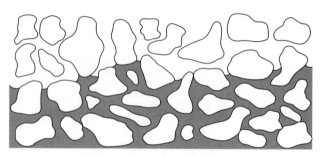

(b) First falling-rate period

(c) Second falling-rate period

Figure 18.29 Drying stages for granular solids.

linearly with decreasing average moisture content. This is the *first falling-rate drying period*. It is not always observed with wet solids of the second category.

During the period from C to D, the liquid in the pores of wet solids of the first category begins to recede from the exposed surface. In the final period from D to E, as shown in Figure 18.29c, evaporation occurs from liquid surfaces in the pores, where the wet-bulb temperature prevails. However, the temperature of the exposed surface of the solid rises to approach the dry-bulb temperature of the gas. During this period, the *second falling-rate drying period*, the rate of drying may be controlled by vapor diffusion for wet solids of the first category and by liquid diffusion for wet solids of the second category. The rate falls exponentially with decreasing moisture content.

§18.4.1 Constant-Rate Drying Period

In direct-heat equipment, drying involves transfer of heat from the gas to the surface and interior of the wet solid, and

mass transfer of moisture from the interior and surface of the solid to the gas. During the constant-rate period, the rate of mass transfer is determined by gas-phase boundary-layer or film resistance at the wet surface of the solid. The wet solid is assumed to be at a uniform temperature, so the only resistance to convective heat transfer is in the gas phase. The rate of moisture evaporation can then be based on convective heat transfer or mass transfer, according to conventional, but simplified, transport relationships in which thermal radiation, the bulk-flow effect for mass transfer, and the sensible-heat effect for the evaporated moisture are often ignored:

$$\frac{dm_v}{dt} = \frac{h(T_g - T_i)A}{\Delta H_i^{\text{vap}}} = M_A k_y (y_i - y_g)A \qquad (18\text{-}33)$$

where subscript i refers to the gas–solid interface.

As discussed previously, the interface at these conditions is at the wet-bulb temperature, T_w. Although drying-rate calculations could be based on mass transfer using (18-33), it is more common to use the heat-transfer relation of (18-33) when air is the gas and water is the moisture because of the wide availability of the psychrometric chart for that system, the equality of wet-bulb and adiabatic-saturation temperatures, and a wider availability of correlations for convective heat transfer than for mass transfer, although analogies can be used to derive one from the other. Combining (18-32) and (18-33), the drying-rate flux for constant-rate drying period R_c, in terms of heat transfer, becomes

$$R_c = \frac{h(T_g - T_w)}{\Delta H_w^{\text{vap}}} \qquad (18\text{-}34)$$

while an equivalent, but less-useful, mass-transfer form is obtained by combining (18-16), (18-32), and (18-33):

$$R_c = M_B k_y (\mathcal{H}_w - \mathcal{H}_d) \qquad (18\text{-}35)$$

where subscripts d and w refer to gas dry-bulb and wet-bulb conditions.

For some dryers, it is preferable to use a volumetric heat-transfer coefficient, (ha), defined by

$$\frac{dm_v}{dt} = \frac{(ha)(T_g - T_i)V}{\Delta H_w^{\text{vap}}} \qquad (18\text{-}36)$$

where a = external surface area of wet solids per unit volume of dryer, and V = volume of dryer. Then, the drying rate per unit dryer volume during the constant-rate drying period is

$$(R_c)_V = \frac{(ha)(T_d - T_w)}{\Delta H_w^{\text{vap}}} \qquad (18\text{-}37)$$

Interphase heat-transfer coefficients were discussed for several geometries in §3.5. Empirical equations useful for drying-rate calculations are summarized in Mujumdar [1], and representative equations, when the gas is air, are listed in Table 18.6, where in (1), G is the mass velocity of air in the flow channel that passes over the wet surface. In (2), G is the mass velocity of the air impinging on the wet surface. In (3) to (8), d_p is the particle diameter and G is the superficial mass velocity.

Table 18.6 Empirical Equations for Interphase Heat-Transfer Coefficients for Application to Dryers (h in W/m²-K, G in kg/hr-m², d_p in m)

Geometry	Equation	
Flat-plate, parallel flow	$h = 0.0204G^{0.8}$ $(T_d = 45-150°C,$ $G = 2,450-29,300)$	(1)
Flat-plate, perpendicular, impingement flow	$h = 1.17G^{0.37}$ $(G = 3,900-19,500)$	(2)
Packed beds, through-circulation	$h = 0.151G^{0.59}/d_p^{0.41},$ $(N_{Re} > 350)$	(3)
	$h = 0.214G^{0.49}/d_p^{0.51},$ $(N_{Re} < 350)$	(4)
Fluidized beds	$N_{Nu} = 0.0133N_{Re}^{1.6}$ $(0 < N_{Re} < 80)$	(5)
Pneumatic conveyors	$N_{Nu} = 0.316N_{Re}^{0.8}$ $(8 < N_{Re} < 500)$	(6)
Droplets in spray dryers	$N_{Nu} = 2 + 1.05N_{Re}^{0.5}N_{Pr}^{1/3}N_{Gu}^{0.175}$ $(N_{Re} < 1000)$	(7)
Spouted beds	$N_{Nu} = 0.0005N_{Re_s}^{1.46}(u/u_s)^{1/3}$	(8)

$N_{Re} = d_pG/\mu, N_{Nu} = hd_p/k, N_{Pr} = C_P\mu/k, N_{Re_s} = d_pG_s/\mu$

G_s = mass velocity for incipient spouting
u = velocity, u_s = incipient spouting velocity
$N_{Gu} = (T_d - T_w)/T_d$ in absolute temperature
d_p = particle size, C_P = specific heat of gas
μ = viscosity of gas, k = thermal conductivity of gas

The dramatic effect of exposed surface area of wet solids in drying was shown by Marshall and Hougen [13] and is illustrated in the next two examples, which deal with batch drying. In Example 18.7, cross-circulation, batch tray drying is used to dry slabs of filter cake. In Example 18.8, the filter cake is extruded and then dried by through-circulation. The difference in the two drying times for the constant-rate drying period is found to be very significant.

EXAMPLE 18.7 Batch Drying with Cross-Circulation.

$CaCO_3$ filter cake in a tray is to be dried by cross-circulation from the top surface. Each tray is 2.5 cm high, with an area of 1.5 m², and is filled with 73 kg of wet filter cake having a water content of 30% on the dry basis. The heating medium is air at 1 atm and 170°F with a relative humidity of 10%. The velocity of air passing across the wet solid is 4 m/s. Estimate time in hours needed to reach the experimentally determined, critical moisture content (end of the constant-rate period) of 10% on the dry basis, if the preheat period is neglected.

Solution

$$H_2O \text{ in wet cake} = \left(\frac{30}{130}\right)(73) = 16.8 \text{ kg},$$

$$H_2O \text{ in cake at } X_c = 0.10(73 - 16.8) = 5.6 \text{ kg},$$
$$m_v = H_2O \text{ evaporated} = 16.8 - 5.6 = 11.2 \text{ kg}$$

For the constant-rate drying period, the heat-transfer form of (18-33) applies, which upon integration gives

$$t_c = \frac{m_v\Delta H_w^{vap}}{h(T_d - T_w)A}$$

where t_c is the time to reach the critical moisture content. From the humidity chart of Figure 18.17,

$$T_w = 100°F \text{ and } \mathcal{H} = 0.026\frac{\text{lb } H_2O}{\text{lb dry air}}$$

$$T_d = 170°F = 76.7°C \text{ and } T_d - T_w = 170 - 100 = 70°F$$
$$= 38.9 \text{ K}$$

At $T_w = 100°F, \Delta H_w^{vap} = 1037.2 \text{ Btu/lb} = 2,413 \text{ kJ/kg}$

From (18-10), Table 18.4,

$$v_H = 0.730(170 + 460)\left(\frac{1}{28.97} + \frac{0.026}{18.02}\right) = 16.5 \text{ ft}^3/\text{lb dry air}$$

or,

$$16.5/(1 + 0.026) = 16.1 \text{ ft}^3/\text{lb moist air} = 1.004 \text{ m}^3/\text{kg moist air}$$
$$= 1/\rho$$
$$G = u_{avg}\rho = u_{avg}/v = 4/1.004 = 3.98 \text{ kg/m}^2\text{-s} = 14,300 \text{ kg/m}^2\text{-h}$$

and $A = 1.5 \text{ m}^2$

From Table 18.6, (1) applies for turbulent flow with T_d and G within the allowable range.

$$h = 0.0204(14,300)^{0.8} = 43 \text{ W/m}^2\text{-K} = 43 \text{ J/s-m}^2\text{-K}$$

From (1), using SI units

$$t_c = \frac{(11.2)[(2,413)(1,000)]}{(43)(38.9)(1.5)} = 10,800 \text{ s} = 2.99 \text{ h}$$

EXAMPLE 18.8 Batch Drying with Through-Circulation.

The filter cake of Example 18.7 is extruded into cylindrical-shaped pieces of 1/4-inch diameter and 1/2-inch length to form a bed 1.5 m² in cross-sectional area and 5 cm high, with an external porosity of 50%. Air at 170°F and 10% relative humidity passes through the bed at a superficial velocity of 2 m/s (average interstitial velocity of 4 m/s). Estimate the time in hours needed to reach the critical-moisture content, if the preheat period is neglected. Compare this time to that estimated in Example 18.7.

Solution

Compared to the tray of Example 18.7, the bed is twice as high with the same cross-sectional area. Therefore, for a porosity of 50%, the bed contains the same amount of wet solids. Thus, as in Example 18.7,

$m_{wet cake} = 73 \text{ kg}, m_v = 11.2 \text{ kg } H_2O \text{ evaporated}, \Delta H_w^{vap} = 2,413 \text{ kJ/kg},$
and $T_d - T_w = 38.9 \text{ K}$

Assume the extrusion density equals filter-cake density.

$$\rho_{\text{filter cake}} = \frac{73}{1.5\left(\frac{2.5}{100}\right)} = 1,950 \text{ kg/m}^3$$

$$\text{Volume of one extrusion} = \pi D^2 L/4 = \frac{3.14(0.25)^2(0.5)}{4}$$

$$= 0.0245 \text{ in}^3 = 4.01 \times 10^{-7} \text{ m}^3$$

$$\text{Number of extrusions} = 1.5\left(\frac{2.5}{100}\right)(0.5)/4.01 \times 10^{-7}$$

$$= 46,800$$

$$\text{Surface area/extrusion} = \pi DL + \pi D^2/2$$

$$= 3.14\left[(0.25)(0.50) + \frac{(0.25)^2}{2}\right]$$

$$= 0.49 \text{ in.}^2 = 0.000316 \text{ m}^2$$

$$A = 46,800(0.000316) = 14.8 \text{ m}^2$$

Thus, the transport area is $14.8/1.5 = 9.9$ times that for Example 18.7. From Table 18.6, (3) or (4) applies for estimating h, depending on N_{Re}. From Example 18.7, but with a superficial bed velocity of 50% of the crossflow velocity, $G = 3.98/2 = 1.96$ kg/m²-s.

Equations (3) and (4) refer to the work of Gamson, Thodos, and Hougen [14] for $N_{Re} > 350$ and Wilke and Hougen [15] for $N_{Re} < 350$, respectively. For both correlations, d_p is taken as the diameter of a sphere of the same surface area as the particle. For the extrusions of this example with $L = 2D$,

$$\pi d_p^2 = \frac{\pi D^2}{2} + 2\pi D^2 = 2.5\pi D^2$$

Solving (1),

$$d_p = D\sqrt{2.5} = 0.25\sqrt{2.5} = 0.395 \text{ in.} = 0.010 \text{ m}$$

$$\mu \approx 0.02 \text{ cP} = 2 \times 10^{-5} \text{ kg/m-s}$$

$$N_{Re} = \frac{d_p G}{\mu} = \frac{(0.010)(1.96)}{2 \times 10^{-5}} = 980$$

Therefore, (3) applies and

$$h = 0.151(14,300/2)^{0.59}/(0.010)^{0.41} = 188 \frac{J}{\text{s-m}^2\text{-K}}$$

The h is $188/43 = 4.4$ times greater than in Example 18.7. From (1) in that example,

$$t_c = \frac{(11.2)[(2,413)(1,000)]}{(188)(38.9)(14.8)} = 250 \text{ s} = 4.16 \text{ min}$$

This, and the preceding, example show that cross-circulation drying takes hours, whereas through-circulation drying may require only minutes.

§18.4.2 Falling-Rate Drying Period

When the drying rate in the constant-rate period is high and/or the distance that interior moisture must travel to reach the surface is large, moisture may fail to reach the surface fast enough to maintain a constant drying rate, and a transition to the falling-rate period occurs. In Examples 18.7 and 18.8, the constant drying rates from (18-34) are

$$R_c = \frac{43(38.9)(3,600)}{(2,413)(1,000)} = 2.50 \text{ kg/h-m}^2$$

and
$$R_c = \frac{(188)(38.9)(3,600)}{(2,413)(1,000)} = 10.9 \text{ kg/h-m}^2$$

However, in Example 18.7, the moisture may have to travel from as far away as 25 mm to reach the exposed surface, while in Example 18.8, the distance is only 3.2 mm. Therefore, as a first approximation, it might be expected that the critical moisture contents for the two examples might not be the same. The value of 10% on the dry basis was taken from through-circulation drying experiments.

When moisture travels from the interior of a wet solid to the surface, a moisture profile develops in the wet solid. The profile's shape depends on the nature of the moisture movement, as discussed by Hougen, McCauley, and Marshall [16]. If the wet solid is of the first category, where the moisture is not held in solution or in fibers but is free moisture in the interstices of particles like soil and sand, or is moisture above the fiber-saturation point in paper and wood, then moisture movement occurs by capillary action. For wet solids of the second category, the internal moisture is bound moisture, as in the last stages of drying of paper and wood, or soluble moisture, as in soap and gelatin. This type of moisture migrates to the surface by liquid diffusion. Moisture can also migrate by gravity, external pressure, and by vaporization–condensation sequences in the presence of temperature gradients. In addition, vapor diffusion through solids can occur in indirect-heat dryers when heating and vaporization occur at opposed surfaces.

A moisture profile for capillary flow is shown in Figure 18.30a. It is concave upward near the exposed surface, concave downward near the opposed surface, with a point of inflection in between. For flow of moisture by diffusion, as in Figure 18.30b, the profile is concave downward throughout. If the diffusivity is independent of moisture content, the solid curve applies. If, as is often the case, the diffusivity decreases with moisture content, due mainly to shrinkage, the dashed profile applies.

During the falling-rate period, idealized theories for capillary flow and diffusion can be used to estimate drying rates. Alternatively, estimates could be made by a strictly empirical approach that ignores the mechanism of moisture movement, but relies on experimental determination of drying rate as a function of average moisture content for a particular set of conditions.

Empirical Approach

The empirical approach relies on experimental data in the form of Figure 18.31a (Case 1), and Figures 18.31b (Case 2) and 18.31c (Case 3), where, for all cases, the preheat period is ignored. In these plots, the abscissa is the free-moisture content, $X = X_T - X$, shown in Figure 18.23, which allows all three plots to be extended to the origin, if all free moisture is removed.

From (18-32),

$$\int dt = -\frac{m_s}{A}\int \frac{dX}{R} \tag{18-38}$$

Ignoring preheat, for the constant-rate period, $R = R_c =$ constant. Starting from an initial free-moisture content of X_o at time $t = 0$, the time to reach the critical free-moisture

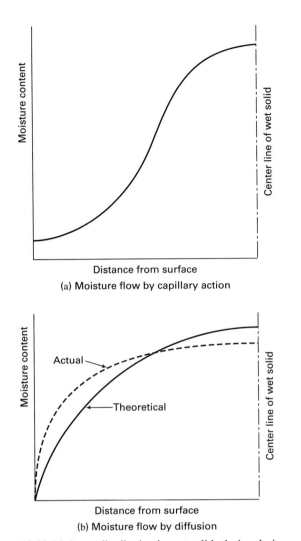

(a) Moisture flow by capillary action

(b) Moisture flow by diffusion

Figure 18.30 Moisture distribution in wet solids during drying.
[From W.L. McCabe, J.C. Smith, and P. Harriott, *Unit Operations of Chemical Engineering*, 5th ed., McGraw-Hill, New York (1993) with permission.]

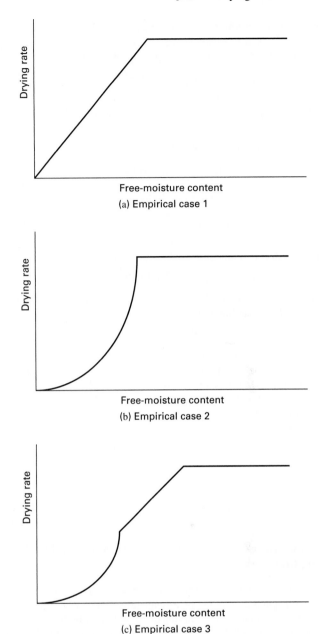

(a) Empirical case 1

(b) Empirical case 2

(c) Empirical case 3

Figure 18.31 Drying-rate curves.

content, X_c, at time $t = t_c$ is obtained by integrating (18-38):

$$t_c = \frac{m_s(X_o - X_c)}{AR_c} \qquad (18\text{-}39)$$

For Case 1 (Figure 18.31a) of the falling-rate period, the rate of drying is linear with X and terminates at the origin, according to

$$R = R_c X / X_c \qquad (18\text{-}40)$$

Substituting (18-40) into (18-39) and integrating t from t_c to $t > 0$ and X from X_c to $X > 0$ gives the following expression for the drying time in the falling-rate period, t_f:

$$t_f = t - t_c = \frac{m_s X_c}{AR_c} \ln\left(\frac{X_c}{X}\right) = \frac{m_s X_c}{AR_c} \ln\left(\frac{R_c}{R}\right) \qquad (18\text{-}41)$$

The total drying time, t_T, is the sum of (18-39) and (18-41):

$$t_T = t_c + t_f = \frac{m_s}{AR_c}\left[(X_o - X_c) + X_c \ln\left(\frac{X_c}{X}\right)\right] \qquad (18\text{-}42)$$

EXAMPLE 18.9   Constant- and Falling-Rate Periods.

Marshall and Hougen [13] present experimental data for the through-circulation drying of 5/16-inch extrusions of ZnO in a bed of 1 ft^2 cross section by 1-inch high, using air of $T_d = 158°F$ and $T_w = 100°F$ at a flow rate of 340 ft^3/min. The data show a constant-rate period from $X_o = 33\%$ to $X_c = 13\%$, with a drying rate of 1.42 lb H$_2$O/h-lb bone-dry solid, followed by a falling-rate period that approximates Case 1 in Figure 18.31a. Calculate the drying time for the constant-rate period and the additional time in the falling-rate period to reach a free-moisture content, X, of 1%.

Solution

In (18-38), the drying rate, R, corresponds to mass of moisture evaporated per unit time per unit of exposed area of wet material. In this example, drying rate is not given per unit area, but per mass of bone-dry solid, with some associated exposed area. Equations (18-38) and (18-42) are rewritten in terms of $R' = RA/m_s$ as

$$\int dt = -\int \frac{dX}{R'} \tag{1}$$

and

$$t_T = t_c + t_f = \frac{1}{R'_c}\left[(X_o - X_c) + X_c \ln\left(\frac{X_c}{X}\right)\right] \tag{2}$$

From (2), for just the constant-rate period,

$$t_c = \frac{1}{1.42}[0.33 - 0.13] = 0.141\,\text{h} = 8.45\,\text{min}$$

From (2), for just the falling-rate period,

$$t_f = \frac{1}{1.42}\left[0.13\,\ln\left(\frac{0.13}{0.01}\right)\right] = 0.235\,\text{h} = 14.09\,\text{min}$$

The total drying time, ignoring the preheat period, is

$$t_T = t_c + t_f = 8.45 + 14.09 = 22.5\,\text{min}$$

For Case 2 of Figure 18.31b, R in the falling-rate period can be expressed as a parabolic function.

$$R = aX + bX^2 \tag{18-43}$$

The values of the parameters a and b are obtained by fitting (18-43) to the experimental drying-rate plot, subject to the constraint that $R = R_c$ at $X = X_c$. If (18-43) is substituted into (18-38) and the result is integrated for the falling-rate period from $X = X_c$ to some final value of X_f and corresponding R_f, the time for the falling-rate period is found to be

$$t_f = t - t_c = \frac{m_s}{aA}\ln\left[\frac{X_c(a + bX_f)}{X_f(a + bX_c)}\right] = \frac{m_s}{aA}\ln\left[\frac{X_c^2 R_f}{X_f^2 R_c}\right] \tag{18-44}$$

EXAMPLE 18.10 Falling-Rate Period by Empirical Equation.

Experimental data for through-circulation drying of 1/4-inch-diameter spherical pellets of a nonhygroscopic carburizing compound exhibit constant-rate drying of 1.9 lb H_2O/h-lb dry solid from $X_o = 30\%$ to $X_c = 21\%$, followed by a falling-rate period to $X_f = 4\%$ that fits (18-43) with $a = 3.23$ and $b = 27.7$ (both in lb H_2O/h-lb dry solid) for X as a fraction and R replaced by R' in lb H_2O/h-lb dry solid. Calculate the time for drying in the falling-rate period. Note that the values of a and b satisfy the constraint of $R'_c = 1.9$ at $X_c = 0.21$.

Solution

For R' in the given units, (18-44) becomes

$$t_f = \frac{1}{a}\ln\left[\frac{X_c(a + bX_f)}{X_f(a + bX_c)}\right] \tag{1}$$

Thus,

$$t_f = \frac{1}{3.23}\ln\left\{\frac{0.21}{0.04}\left[\frac{3.23 + 27.7(0.04)}{3.23 + 27.7(0.21)}\right]\right\} = 0.286\,\text{h} = 17.1\,\text{min}$$

For Case 3 of Figure 18.31c, the falling-rate period consists of two subregions. In the first subregion, which is linear,

$$R = \alpha X + \beta \tag{18-45}$$

with the constraints that $R = R_{c_1}$ at X_{c_1}, and that $R = R_{c_2}$ at X_{c_2}. In the second subregion, (18-44) applies, but with the constraint that $R = R_{c_2}$ at X_{c_2}.

EXAMPLE 18.11 Complex Falling-Rate Period.

Experimental data of Sherwood [12] for the surface drying of a 3.18-cm-thick $\times$ 6.6-cm^2 cross-sectional area slab of a thick paste of $CaCO_3$ (whiting) from both sides by air at $T_d = 39.8°C$ and $T_w = 23.5°C$ and a cross-circulation velocity of 1 m/s exhibit the complex type of drying-rate curve shown in Figure 18.31c, with the following constants:

Constant-rate period:

$$X_o = 10.8\%, X_{c_1} = 8.3\%, \text{ and } R_{c_1} = 0.053\,\text{g } H_2O/\text{h-cm}^2$$

First falling-rate period:

$$X_{c_2} = 3.7\% \text{ and } R_{c_2} = 0.038\,\text{g } H_2O/\text{h-cm}^2$$

Second falling-rate period to $X = 2.2\%$:

$$R = 29.03\,X^2 - 0.048\,X \tag{1}$$

Determine the time to dry a slab of the same dimensions at the same drying conditions, but from $X_o = 0.14$ to $X = 0.01$, ignoring the preheat period. Assume an initial weight of 46.4 g.

Solution

Constant-rate period:

$$X_o = 0.14, X_{c_1} = 0.083, \text{ and } R_{c_1} = 0.053\,\text{g/h-cm}^2$$

$$m_s = 46.4\left(\frac{1}{1.14}\right) = 40.7\,\text{g of moisture-free solid}$$

$$A = 2(6.6) = 13.2\,\text{cm}^2\,(\text{drying is from both sides})$$

From (18-39), $\quad t_c = \dfrac{40.7(0.14 - 0.083)}{13.2(0.053)} = 3.32\,\text{h}$

First falling-rate period:

In this period, R is linear with end points $(R_{c_1} = 0.053, X_{c_1} = 0.083)$ and $(R_{c_2} = 0.038, X_{c_2} = 0.037)$. This gives for (18-45),

$$R = 0.0259 + 0.326X \tag{2}$$

Substituting (2) into (18-38) and integrating,

$$
\begin{aligned}
t_{f_1} &= -\frac{m_s}{A}\int_{X_{c_1}}^{X_{c_2}} \frac{dX}{0.0259 + 0.326X} \\
&= \frac{m_s}{A}\frac{1}{0.326}\ln\left(\frac{0.0259 + 0.326X_{c_1}}{0.0259 + 0.326X_{c_2}}\right) \\
&= \frac{m_s}{A}\frac{1}{0.326}\ln\left(\frac{R_{c_1}}{R_{c_2}}\right) \\
&= \frac{40.7}{13.2(0.326)}\ln\left(\frac{0.053}{0.038}\right) = 3.15\,\text{h}
\end{aligned}
$$

Second falling-rate period:

This period extends from $X_{c_2} = 0.037$ to $X = 0.022$, with R given by (1) for (18-43), with $a = -0.048$ and $b = 29.03$. From (18-44),

$$t_{f_2} = \frac{40.7}{(-0.048)(13.2)}\ln\left\{\frac{0.037[-0.048 + 29.03(0.022)]}{0.022[-0.048 + 29.03(0.037)]}\right\} = 2.08\,\text{h}$$

and the total drying time is $t_T = 3.32 + 3.15 + 2.08 = 8.6\,\text{h}$

For drying-rate curves of shapes other than those of Figure 18.31, time for drying from any X_o to any X can be determined by numerical or graphical integration of (18-38) or (1) in Example 18.9, as illustrated in the following example.

EXAMPLE 18.12 Drying Time from Data.

Marshall and Hougen [13] present the following experimental data for the through-circulation drying of rayon waste. Determine the drying time if $X_0 = 100\%$ and the final X is 10%. Assume that all moisture is free moisture.

X, lb H_2O/lb dry solid	R', $\dfrac{\text{lb } H_2O}{\text{h-lb dry solid}}$
1.40	24
1.00	24
0.75	24
0.73	21
0.70	18
0.65	15.3
0.55	13
0.475	12.3
0.44	12.2
0.40	11
0.20	5.5
0	0

Solution

The data are plotted in Figure 18.32, where three distinct drying-rate periods are seen, but the two falling-rate periods are in the reverse order of Figure 18.31c. By numerical integration of Equation (1) in Example 18.9 with a spreadsheet, the following drying times are obtained, noting that $R' = 2.75$ lb H_2O/h-lb dry solid at $X = 0.10$, $R_{c_1} = 24$ at $X_{c_1} = 0.75$, and $R_{c_2} = 12.2$ at $X_{c_2} = 0.44$.

$t_c = 0.027$ h $= 1.63$ minutes, $t_{f_1} = 1.28$ minutes, and $t_{f_2} = 3.21$ minutes

$t_T = 1.63 + 1.28 + 3.21 = 6.12$ minutes

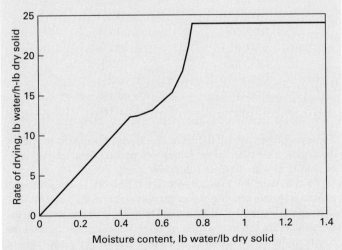

Figure 18.32 Data for through-circulation drying of rayon waste.

Liquid-Diffusion Theory

The empirical approach for determining drying time in the falling-rate period is limited to the conditions for which the experimental drying-rate curve is established. A more general approach, particularly for nonporous wet solids of the second category, is the use of Fick's laws of diffusion. Once the diffusion coefficient is established from experimental data for a wet solid, Fick's laws can be used to predict drying rates and moisture profiles for wet solids of other sizes and shapes and drying conditions during the falling-rate period.

Mathematical formulations of liquid diffusion in solids are readily obtained by analogy to the solutions available for transient heat conduction in solids, as summarized, for example, by Carslaw and Jaeger [17] and discussed in §3.3. Two solutions are of particular interest for drying of slabs in the falling-rate period, where the area of the edges is small compared to the area of the two faces, or the edges are sealed to prevent escape of moisture. As in heat-conduction calculations, the equations also apply when one of the two faces is sealed. Two general moisture-distribution cases are considered.

Case 1: Initially uniform moisture profile in the wet solid with negligible resistance to mass transfer in the gas phase.

Case 2: Initially parabolic moisture profile in the wet solid with negligible resistance to mass transfer in the gas phase.

Although the equations for these two cases are developed here only for a slab with sealed edges, other solutions are available in Carslaw and Jaeger [17]. When edges of slabs and cylinders are not sealed, Newman's method [18], as discussed in §3.3, is suitable.

Case 1. This case models slow-drying materials for which the rate of drying is controlled by internal diffusion of moisture to the exposed surface. This occurs if, initially, the wet solid has no surface liquid film and external resistance to mass transfer is negligible, thus eliminating the constant-rate drying period. Alternatively, the wet solid can have a surface liquid film, but during the evaporation of that film in a constant-rate drying period controlled by gas-phase mass transfer, no moisture diffuses to the surface, and after completion of evaporation of that film, resistance to mass transfer is due to internal diffusion in a falling-rate period.

The slab, of thickness $2a$, is pictured in Figure 3-7a, where the edges at $x = \pm c$ and $y = \pm b$ are sealed to mass transfer. Internal diffusion of moisture is in the z-direction only toward exposed faces at $z = \pm a$. Alternatively, the slab may be of thickness a with the face at $z = 0$ sealed to mass transfer. Initially, the moisture content throughout the slab, not counting any surface liquid film, is assumed uniform at X_o. At the beginning of the falling-rate period, $t = 0$, the exposed face(s) is(are) brought to the equilibrium-moisture content, X^*. For constant moisture diffusivity, D_{AB}, Fick's second law, as discussed in §3.3, applies:

$$\frac{\partial X}{\partial t} = D_{AB} \frac{\partial^2 X}{\partial z^2} \quad (18\text{-}46)$$

for $t \geq 0$ in the region $-a \leq z \leq a$, where the boundary conditions are $X = X_o$ at $t = 0$ for $-a < z < a$ and $X = X^*$ at $z = \pm a$ for $t \geq 0$.

The solution to (18-46) for the moisture profile as a function of time under these boundary conditions, as discussed in §3.3 and first proposed for drying applications by Sherwood [11], is in terms of the unaccomplished free-moisture change, and a modification of (3-80) applies:

$$E = \frac{X - X^*}{X_o - X^*} = \frac{4}{\pi} \sum_{n=0}^{\infty} \frac{(-1)^n}{(2n+1)}$$

$$\times \exp\left[-\frac{\pi^2(2n+1)^2}{4}\left(\frac{D_{AB}t}{a^2}\right)\right] \cos\left[\frac{\pi(2n+1)}{2}\left(\frac{z}{a}\right)\right]$$

$$(18\text{-}47)$$

Thus, E is a function of two dimensionless groups, the Fourier number for diffusion, $N_{Fo_M} = D_{AB}t/a^2$, and the position ratio, z/a. This solution is plotted as $(1 - E)$ in terms of these two groups in Figure 3.8.

The rate of mass transfer from one face is given by (3-82), which in terms of R, the drying rate in mass of moisture evaporated per unit time per unit area, is

$$R = \frac{2D_{AB}(X_o - X^*)\rho_s}{a}$$

$$\times \sum_{n=0}^{\infty} \exp\left[-\frac{\pi^2(2n+1)^2}{4}\left(\frac{D_{AB}t}{a^2}\right)\right]$$

$$(18\text{-}48)$$

where ρ_s = mass of dry solid/volume of slab.

Also of interest is the average moisture content of the slab during drying. From (3-85),

$$E_{avg} = \frac{X_{avg} - X^*}{X_o - X^*}$$

$$= \frac{8}{\pi^2} \sum_{n=0}^{\infty} \frac{1}{(2n+1)^2} \exp\left[-\frac{\pi^2(2n+1)^2}{4}\left(\frac{D_{AB}t}{a^2}\right)\right]$$

$$(18\text{-}49)$$

Equations (18-47)–(18-49) can be used to determine the moisture diffusivity, D_{AB}, from experimental data, and then that value can be used to estimate drying rates for other conditions, as illustrated in the next example. However, such calculations must be made with caution because often the diffusivity is not constant, as shown by Sherwood [11] for drying of slabs of soap, but decreases with decreasing moisture content due to of shrinkage and/or case hardening. In that case, numerical solutions are necessary.

EXAMPLE 18.13 Drying by Liquid Diffusion.

A piece of poplar wood 15.2 cm long × 15.2 cm wide × 1.9 cm thick, with the edges sealed with a waterproofing cement, was dried from both faces in a tunnel dryer using cross-circulation of air at 1 m/s. Initial moisture content was 39.7% on the dry basis, initial weight of the wet piece was 264 g, and no shrinkage occurred during drying. The direction of diffusion was perpendicular to the grain. The equilibrium-moisture content was 5% on the dry basis. Data were obtained for the moisture content as a function of time.

Included are values of E_{avg}, computed from its definition in (18-49), and values of t/a^2, where $a = 0.5 (1.90) = 0.95$ cm.

t, h	X_{avg}, g H_2O/g dry wood	t/a^2, h/cm^2	E_{avg}
0.36	0.362	0.40	0.900
0.90	0.328	1.00	0.800
1.53	0.303	1.70	0.730
1.94	0.291	2.15	0.694
2.89	0.267	3.20	0.626
3.47	0.255	3.85	0.591
4.02	0.245	4.45	0.562
4.92	0.230	5.45	0.520
5.82	0.218	6.45	0.483
6.95	0.204	7.70	0.443
8.03	0.192	8.90	0.409
8.98	0.183	9.95	0.382

Using the data, determine the average value of the diffusivity by nonlinear regression of (18-49), and use that value to determine the drying time from $X_o = 45\%$ to $X = 10\%$ with $X^* = 6\%$ for a piece of poplar measuring 72 inches long × 12 inches wide × 1 inch thick, neglecting mass transfer from the edges and assuming only a falling-rate period, with negligible resistance in the gas phase.

Solution

$$m_s = 264\left(\frac{1}{1 + 0.397}\right) = 189 \text{ g dry wood}$$

A for two faces $= 2(15.2)^2 = 462$ cm^2

At any instant, from (18-38),

$$R = -\frac{m_s}{A}\frac{dX_{avg}}{dt} \qquad (1)$$

From a plot of the data, approximate values of R as a function of X_{avg} are computed to be

R, g H_2O/h-cm^2	X_{avg}, g H_2O/g dry solid
0.02622	0.345
0.01573	0.315
0.01258	0.297
0.01019	0.279
0.00847	0.261
0.00760	0.250
0.00661	0.238
0.00582	0.224
0.00503	0.211
0.00446	0.198
0.00404	0.187

These results are plotted in Figure 18.33, where it appears that all of the drying takes place in the falling-rate period. Thus, the data may be consistent with the Case 1 diffusion theory.

To determine the average moisture diffusivity, a spreadsheet is used to prepare a semilog plot of the data points as E_{avg} against t/a^2, as shown in Figure 18.34. Equation (18-49) is then evaluated on the spreadsheet for different values of the moisture diffusivity until the best fit of the data is obtained, based on minimizing the error sum of squares (ESS) of the differences between E_{avg} of the data points and the corresponding E_{avg} values calculated from (18-49).

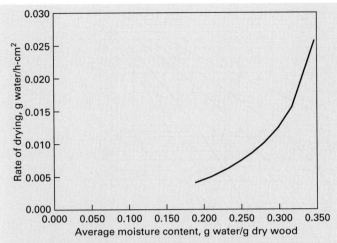

Figure 18.33 Experimental data for drying poplar wood.

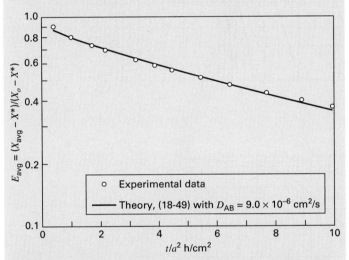

Figure 18.34 Best fit by diffusion theory of experimental data for drying poplar wood.

The best fit is for $D_{AB} = 9.0 \times 10^{-6}$ cm^2/s, with an ESS $= 0.001669$. The best fit of (18-49) is included as a line in Figure 18.34. For values of $N_{Fo_M} > 0.1$, only the first term in the infinite series of (18-49) is significant, and therefore (18-49) approaches a straight line on a semilog plot, as can be observed for the theoretical line in Figure 18.34, when $t/a^2 > 3.2$ h/cm^2.

To determine the drying time for the 72-inch $\times$ 12-inch $\times$ 1-inch poplar, assume that all drying takes place in the diffusion-controlled, falling-rate period, with mass transfer from the edges negligible and a drying time long enough that $N_{Fo_M} > 0.1$. Then (18-49) reduces to

$$\ln\left(\frac{X_{avg} - X^*}{X_o - X^*}\right) = \ln\left(\frac{8}{\pi^2}\right) - \frac{\pi^2}{4}\left(\frac{D_{AB}t}{a^2}\right) \quad (2)$$

Solving (2) for $(D_{AB}t/a^2)$,

$$N_{Fo_M} = \frac{D_{AB}t}{a^2} = \frac{4}{\pi^2}\ln\left[\frac{8}{\pi^2}\left(\frac{X_o - X^*}{X_{avg} - X^*}\right)\right] \quad (3)$$

$$X_{avg} = 0.10, \quad X_o = 0.45, \quad \text{and} \quad X^* = 0.06$$

From (3),

$$N_{Fo_M} = \frac{4}{(3.14)^2}\ln\left[\frac{8}{(3.14)^2}\left(\frac{0.45 - 0.06}{0.10 - 0.06}\right)\right] = 0.839$$

Since $N_{Fo_M} > 0.1$, (2) and (3) are valid, and

$$a = 0.5 \text{ in.} = 1.27 \text{ cm}$$

$D_{AB} = 9.0 \times 10^{-6}$ cm^2/s from the above experiments

$$t = \frac{a^2 N_{Fo_M}}{D_{AB}} = \frac{(1.27)^2(0.839)}{(9.0 \times 10^{-6})(3600)} = 41.8 \text{ h}$$

Case 2. When a liquid-diffusion-controlled, falling-rate drying period is preceded by a constant-rate period, that rate of drying is determined by external mass transfer in the gas phase, as discussed earlier, but diffusional resistance to the flow of moisture in the solid causes a parabolic moisture profile to be established in the solid, as discussed by Sherwood [19] and Gilliland and Sherwood [20].

For the slab of Figure 3.7a, Fick's second law, as given by (18-46), still applies, with $X = X_o$ at $t = 0$ for $-a < z < a$. However, during the constant-rate drying period, the slab–gas interface boundary conditions are changed from those of Case 1 to the conditions $\partial X/\partial z = 0$ at $z = 0$ for $t \geq 0$ and $R_c = -D_{AB}\rho_s(\partial X/\partial z)$ at $z = \pm a$ for $t \geq 0$. This latter boundary condition is more conveniently expressed in the form

$$\frac{\partial X}{\partial z} = -\frac{R_c}{\rho_s D_{AB}} \quad (18\text{-}50)$$

where the term on the RHS is a constant during the constant-rate period. This is analogous to a constant-heat-flux boundary condition in heat transfer. The solution for the moisture profile as a function of time during the constant-rate drying period is given by Walker et al. [21] as:

$$X = X_o - \frac{R_c a}{D_{AB}\rho_s}\left\{\frac{1}{2}\left(\frac{z}{a}\right)^2 - \frac{1}{6} + \frac{D_{AB}t}{a^2}\right.$$
$$\left. - \frac{2}{\pi^2}\sum_{m=1}^{\infty}\frac{(-1)^m}{m^2}\exp\left[-m^2\pi^2\left(\frac{D_{AB}t}{a^2}\right)\right]\cos\left(\frac{\pi m z}{a}\right)\right\}$$
$$(18\text{-}51)$$

where for small values of $D_{AB}t/z^2$, the infinite series term is significant and converges very slowly.

The average moisture content in the slab at any time during the constant-rate period is defined by

$$X_{avg} = \frac{1}{a}\int_0^a X\,dz \quad (18\text{-}52)$$

If (18-52) is integrated after substitution of X from (18-51),

$$(X_o - X_{avg})\frac{D_{AB}\rho_s}{R_c a} = \frac{D_{AB}t}{a^2} = N_{Fo_M} \quad (18\text{-}53)$$

From (18-51), it is seen that the generalized moisture profile during the constant-rate drying period, $(X_o - X)$ $D_{AB}\rho_s/(R_c a)$, is a function of the dimensionless position ratio, z/a, and N_{Fo_M}, where the latter is equal to the generalized, average moisture content given by (18-53). A plot of (18-51) for six position ratios, is given in Figure 18.35a.

Equation (18-51) is based on the assumption that during the constant-rate drying period, moisture will be supplied to the surface by liquid diffusion at a rate sufficient to maintain a constant moisture-evaporation rate. As discussed above, the

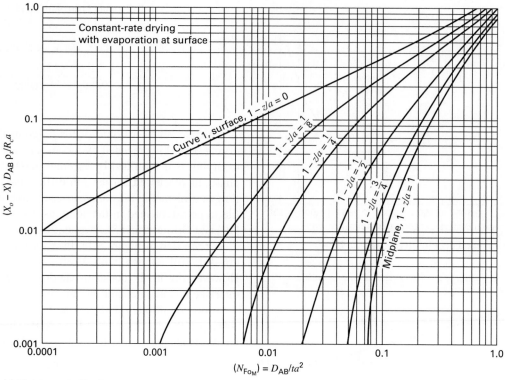

(a) Moisture profile change

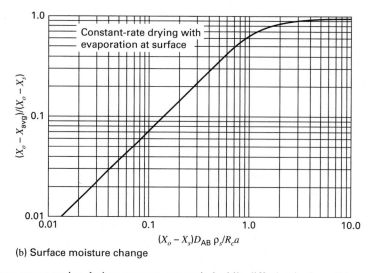

(b) Surface moisture change

Figure 18.35 Changes in moisture concentration during constant-rate period while diffusion in the solid occurs.

[From W.H. Walker, W.K. Lewis, W.H. McAdams, and E.R. Gilliland, *Principles of Chemical Engineering*, 3rd ed., McGraw-Hill, New York (1937) with permission.]

average moisture content at which the constant-rate period ends and the falling-rate period begins is called the critical moisture content, X_c. In the empirical approach to the falling-rate period, X_c must be known from experiment for the particular conditions because X_c is not a constant for a given material but depends on a number of factors, including moisture diffusivity, slab thickness, initial- and equilibrium-moisture contents, and all factors that influence moisture evaporation in the constant-rate drying period. A useful aspect of (18-51) is that it can be used to predict X_c. The basis

for the prediction is the assumption that the falling-rate period will begin when the moisture content at the surface reaches the equilibrium-moisture content corresponding to the conditions of the surrounding gas. This prediction is facilitated, as described by Walker et al. [21], by replotting an extension of Curve 1 in Figure 18.35a for the moisture content at the surface, X_s, in the form shown in Figure 18.35b. Use of Figure 18.35b and the predicted influence of several variables on the value of X_c is illustrated in the following example.

EXAMPLE 18.14 Critical Moisture Content.

Experiments by Gilliland and Sherwood [20] with brick clay mix show that for certain drying conditions, moisture profiles conform reasonably well to the Case 2 diffusion theory. Use Figure 18.35b to predict the critical moisture content for the drying of clay slabs from the two faces only under three different sets of conditions. For all three sets, $X_o = 0.30$, $X^* = 0.05$, $\rho_s = 1.6$ g/cm^3, and $D_{AB} = 0.3$ cm^2/h. The other conditions are

	Set 1	Set 2	Set 3
a, slab half-thickness, cm	0.5	0.5	1.0
R_c, drying rate in constant-rate drying period, g/cm^2-h	0.2	0.4	0.2

Solution

For Set 1, using $X_s = X^* = 0.05$,

$$(X_o - X_s)\frac{D_{AB}\rho_s}{R_c a} = (0.30 - 0.05)\frac{(0.3)(1.6)}{(0.2)(0.5)} = 1.20$$

From Figure 18.35b, $\dfrac{X_o - X_{avg}}{X_o - X_s} = 0.7$

Solving, $X_{avg} = X_c = 0.25 - 0.7(0.25 - 0.05) = 0.11$.

In a similar manner, X_c for Set 2 = 0.16 and X_c for Set 3 = 0.16. These results show that doubling the rate of drying in the constant-rate period or doubling the slab thickness substantially increases X_c.

For sufficiently large values of time, corresponding to $N_{FoM} = \frac{D_{AB}t}{a^2} > 0.5$, the term for the infinite series in (18-51) approaches a value of 0, and, at all locations in the slab, X becomes a parabolic function of z.

A simple equation for the parabolic distribution can be formulated as follows from (18-51) in terms of the moisture contents at the surface and midplane of the slab. At the surface $z = \pm a$, the long-time form is

$$X_o - X_s = \frac{R_c a}{D_{AB}\rho_s}\left[\frac{1}{3} + N_{FoM}\right] \tag{18-54}$$

Similarly, at the midplane, $z = 0$, where $X = X_m$,

$$X_o - X_m = \frac{R_c a}{D_{AB}\rho_s}\left[-\frac{1}{6} + N_{FoM}\right] \tag{18-55}$$

Combining (18-51), (18-54), and (18-55), the dimensionless moisture-content profile becomes

$$\frac{X_m - X}{X_m - X_s} = \left(\frac{z}{a}\right)^2 \tag{18-56}$$

EXAMPLE 18.15 Parabolic Moisture-Profile.

For Example 18.14, determine the drying time for the constant-rate drying period and whether the parabolic moisture-content profile is closely approached by the end of that period.

Solution

From (18-39), $$t_c = \frac{m_s(X_o - X_c)}{AR_c} \tag{1}$$

For a half-slab of thickness a, $\quad m_s = \rho_s a A \tag{2}$

Combining (1) and (2), $\quad t_c = \dfrac{\rho_s a}{R_c}(X_o - X_c) \tag{3}$

For Set 1 of Example 18.14,

$$t_c = \frac{(1.6)(0.5)}{(0.2)}(0.30 - 0.11) = 0.76\,\text{h}$$

$$N_{FoM} = \frac{D_{AB}t_c}{a^2} = \frac{(0.3)(0.76)}{(0.5)^2} = 0.91$$

Because $N_{FoM} > 0.5$, a parabolic profile is closely approached. Similarly, the following results are obtained for Sets 2 and 3:

	Set 2	Set 3
t_c, h	0.28	1.12
N_{FoM}	0.34	0.34
Parabolic profile closely approached?	no	no

For Sets 2 and 3, the parabolic moisture-content profiles are not closely approached. However, the absolute errors in $X_o - X$ at the surface and midplane are determined from (18-51) to be only 1.1% and 4.3%, respectively.

An approximate theoretical estimate of the additional drying time required for the falling-rate period is derived as follows from the development by Walker et al. [21]. At the end of the constant-rate period, the rate of flow of moisture by Fickian diffusion to the surface of the slab, where it is then evaporated, may be equated to the reduction in average moisture content of the slab. Thus,

$$R = -\frac{\rho_s a A}{A}\frac{dX_{avg}}{dt} = -D_{AB}\rho_s\frac{dX}{dz}s \tag{18-57}$$

From the parabolic moisture profile of (18-56), at the surface $z = +a$,

$$-\frac{dX}{dz}\bigg|_{z=+a} = \frac{2}{a}(X_m - X_s) \tag{18-58}$$

However, it is more desirable to convert this expression from one in terms of X_m to one in terms of X_{avg}. To do this, (18-56) can be substituted into (18-52) for the definition of X_{avg}, followed by integration to give

$$X_{avg} = \frac{2}{3}X_m + \frac{1}{3}X_s \tag{18-59}$$

which can be rewritten as

$$X_m - X_s = \frac{3}{2}(X_{avg} - X_s) \tag{18-60}$$

Substitution of (18-60) into (18-58), followed by substitution of the result into (18-57), gives

$$R = -a\rho_s\frac{dX_{avg}}{dt} = \frac{3D_{AB}\rho_s}{a}(X_{avg} - X_s) \tag{18-61}$$

The falling-rate period is assumed to begin with $X_s = X^*$. If the parabolic moisture profile exists during the falling-rate period and if $X_s = X^*$ remains constant, then (18-61) applies during that period and the straight-line, falling-rate period shown in Figure 18.31a is obtained. Integrating (18-61) from the start of the falling-rate period when $X_{avg} = X_c$,

$$t_f = \frac{a^2}{3D_{AB}} \ln \left[\frac{X_c - X^*}{X_{avg} - X^*} \right] \qquad (18\text{-}62)$$

Thus, the falling-rate-period duration is predicted to be directly proportional to the square of the slab half-thickness and inversely proportional to the moisture liquid diffusivity. Equation (18-62) gives reasonable predictions for nonporous slabs of materials such as wood, clay, and soap when the slabs are thick and D_{AB} is low. However, serious deviations can occur when D_{AB} depends strongly on X and/or temperature. In that case, an average D_{AB} can be used to obtain an approximate result. A summary of experimental average moisture liquid diffusivities for a wide range of water-wet solids is tabulated in Chapter 4 of Mujumdar [1].

EXAMPLE 18.16 Falling-Rate Period in Drying.

Gilliland and Sherwood [20] obtained data of the drying of water-wet $7 \times 7 \times 2.54$-cm slabs of 193.9 g (bone-dry) brick clay mix for direct-heat convective air drying from the two faces in both the constant- and falling-rate periods. For $X_o = 0.273$, $X^* = 0.03$, the rate of drying in the constant-rate period to $X_c = 0.165$ was 0.157 g/h-cm^2. The air velocity past the two faces was 15.2 m/s, with $T_d = 25°C$ and $T_w = 17°C$. During the falling-rate period, experimental average slab moisture contents were as follows:

Time from Start of the Constant-Drying Rate Period, minutes	X_{avg}
67	0.165 (critical value)
87	0.145
102	0.134
119	0.124
138	0.114
162	0.106
183	0.099
205	0.095
216	0.090

At the end of the constant-drying-rate period, the moisture profile is assumed parabolic. Other experiments give $D_{AB} = 0.72 \times 10^{-4}$ cm^2/s.

Use (18-62) to predict values of X_{avg} during the falling-rate period and compare predicted values to experimental values.

Solution

Solving (18-62),

$$X_{avg} = X^* + (X_c - X^*)\exp(-3D_{AB}t_f/a^2) \qquad (1)$$

where t_f is the time from the start of the falling-rate period.
For $t_f = 87 - 67 = 20$ min, from (1),

$X_{avg} = 0.03 + (0.165 - 0.03)$

$\times \exp[-3(0.72 \times 10^{-4})(20)(60)/(1.27)^2] = 0.145$ cm^2/s

Calculations for other values of time give the following results:

t_f, Time from Start of Falling-Rate Period, minutes	Experimental X_{avg}	Predicted X_{avg}
0	0.165	0.165
20	0.145	0.145
35	0.134	0.132
52	0.124	0.119
71	0.114	0.106
95	0.106	0.093
116	0.099	0.083
138	0.095	0.075
149	0.090	0.071

Comparing predicted values of X_{avg} with experimental values, the deviation increases with increasing time. If the value of D_{AB} is reduced to 0.53×10^{-4} cm^2/s, much better agreement is obtained with the ESS decreasing from 0.0013 to 0.000154 cm^4/s^2.

Capillary-Flow Theory

For wet solids of the first category, as discussed in §18.3, moisture is held as free moisture in the interstices of the particles. Movement of moisture from the interior to the surface can occur by capillary action in the interstices, but may be opposed by gravity.

Cohesive forces hold liquid molecules together. Also, liquid molecules may be attracted to a solid surface by adhesive forces. Thus, water in a glass tube will creep up the side of the tube until adhesive forces are balanced by the weight of the liquid. For an ideal case of a capillary tube of small diameter partially immersed vertically in a liquid, the liquid rises in the tube to a height above the surface of the liquid in the reservoir. At equilibrium, the height, h, will be

$$h = 2\sigma/\rho_L gr \qquad (18\text{-}63)$$

where σ is the surface tension of the liquid and r is the radius of the capillary. The smaller the radius of the capillary, the larger the capillary effect. Unlike mass transfer by diffusion, which causes moisture to move from a region of high to low concentration, liquid in interstices flows because of capillary effects, regardless of concentration.

For capillary flow in granular beds of wet solids, the variable size and shape of the particles make it extremely difficult to develop a usable theory for predicting the rate of drying in the falling-rate period in terms of permeability and capillarity. Interesting discussions and idealized theories are presented by Keey [7, 23] and Ceaglske and Kiesling [22], but for practical calculations, it appears that, despite pleas to the contrary, it is common to apply diffusion theory with effective diffusivities determined from experiment. In general, these diffusivities are lower than those for true diffusion of moisture in nonporous materials. Some values are included in a tabulation in Chapter 4 of Mujumdar [1]. For example, effective diffusivities of water in beds of sand particles cover a range of 1.0×10^{-2} to 8.0×10^{-4} cm^2/s.

§18.5 DRYER MODELS

Previous sections developed general mathematical models for estimating drying rates and moisture profiles for batch tray dryers of the cross-circulation and through-circulation types. More specific models for continuous dryers have been developed over the years, and this section presents three of them: (1) belt dryer with through-circulation, (2) direct-heat rotary dryer, and (3) fluidized-bed dryer, all of which are categorized as direct-heat dryers. Other models are considered by Mujumdar [1] and in a special issue of *Drying Technology*, edited by Genskow [24].

§18.5.1 Material and Energy Balances for Direct-Heat Dryers

Consider the continuous, steady-state, direct-heat dryer shown in Figure 18.36. Although countercurrent flow is shown, the following development applies equally well to cocurrent flow and crossflow. Assume that the dryer is perfectly insulated so that the operation is adiabatic. As the solid is dried, moisture is transferred to the gas. No solid is entrained in the gas, and changes in kinetic energy and potential energy are negligible. The flow rates of dry solid, m_s, and dry gas, m_g, do not change as drying proceeds. Therefore, a material balance on the moisture is

$$X_{ws}m_s + \mathcal{H}_{gi}m_g = X_{ds}m_s + \mathcal{H}_{go}m_g \qquad (18\text{-}64)$$

The rate of moisture evaporation, m_v, is given by a rearrangement of (18-64):

$$m_v = m_s(X_{ws} - X_{ds}) = m_g(H_{go} - H_{gi}) \qquad (18\text{-}65)$$

where the subscripts are *ws* (wet solid), *ds* (dry solid), *gi* (gas in), and *go* (gas out).

An energy balance can be written in terms of enthalpies or in terms of specific heat and heat of vaporization. In either case, it is convenient to treat the dry gas, dry solid, and moisture (liquid and vapor) separately, and assume ideal mixtures. In terms of enthalpies, the energy balance is as follows, where *s*, *g*, and *m* refer, respectively, to dry solid, dry gas, and moisture:

$$m_s(H_s)_{ws} + X_{ws}m_s(H_m)_{ws} + m_g(H_g)_{gi} + \mathcal{H}_{gi}m_g(H_m)_{gi}$$
$$= m_s(H_s)_{ds} + X_{ds}m_s(H_m)_{ds} \qquad (18\text{-}66)$$
$$+ m_g(H_g)_{go} + \mathcal{H}_{go}m_g(H_m)_{go}$$

A factored rearrangement of (18-66) is

$$m_s\big[(H_s)_{ds} - (H_s)_{ws} + X_{ds}(H_m)_{ds} - X_{ws}(H_m)_{ws}\big]$$
$$= m_g\big[(H_g)_{gi} - (H_g)_{go} + \mathcal{H}_{gi}(H_m)_{gi} \qquad (18\text{-}67)$$
$$- \mathcal{H}_{go}(H_m)_{go}\big]$$

where any convenient reference temperatures can be used to determine the enthalpies.

When the system is air, water, and a solid, a more convenient form of (18-67) can be obtained by evaluating the enthalpies of the solid and the air from specific heats, and obtaining moisture enthalpies from the steam tables. Often, the specific heat of the solid is almost constant over the temperature range of interest, and in the range from 25°C (78°F) to 400°C (752°F), the specific heat of dry air increases by less than 3%, so the use of a constant value of 0.242 Btu/lb-°F introduces little error. If the enthalpy reference temperature of the water is taken as T_o (usually 0°C (32°F) for liquid water when using the steam tables), (18-67) can be rewritten as

$$m_s\big[(C_P)_s(T_{ds} - T_{ws}) + X_{ds}(H_{H_2O})_{ds} - X_{ws}(H_{H_2O})_{ws}\big]$$
$$= m_g\big[(C_P)_{\text{air}}(T_{gi} - T_{go}) + \mathcal{H}_{gi}(H_{H_2O})_{gi}\big] \qquad (18\text{-}68)$$
$$- \mathcal{H}_{go}(H_{H_2O})_{go}\big]$$

A further simplification in the energy balance for the air–water–solid system can be made by replacing enthalpies for water by their equivalents in terms of specific heats for liquid water and steam and the heat of vaporization. In the range from 25°C (78°F) to 100°C (212°F), the specific heat of liquid water and steam are almost constant at 1 Btu/lb-°F and 0.447 Btu/lb-°F, respectively. The heat of vaporization of water over this same range decreases from 1,049.8 to 970.3 Btu/lb, a change of almost 8%. Combining (18-65) with (18-68) and taking a thermodynamic path of water evaporation at the moisture-evaporation temperature, denoted T_v, the simplified energy balance is

$$Q = m_s\Big\{(C_P)_s(T_{ds} - T_{ws}) + X_{ws}(C_P)_{H_2O(\ell)}(T_v - T_{ws})$$
$$+ X_{ds}(C_P)_{H_2O(\ell)}(T_{ds} - T_v)$$
$$+ (X_{ws} - X_{ds})\big[\Delta H_v^{\text{vap}} + (C_P)_{H_2O(g)}(T_{go} - T_v)\big]\Big\}$$
$$= m_g\Big\{\big[(C_P)_{\text{air}} + \mathcal{H}_{gi}(C_P)_{H_2O(v)}\big](T_{gi} - T_{go})\Big\} \qquad (18\text{-}69)$$

Equations (18-64)–(18-69) are useful for determining the required gas flow rate for drying a given flow rate of wet solids, as illustrated in the next example. Also of interest for sizing the dryer is the required heat-transfer rate, Q, from the gas to the solid. For the air–water–solid system, this rate is equal to either the LHS or the RHS of (18-69), as indicated. In the general case,

$$Q = m_g\Big\{(H_g)_{gi} - (H_g)_{go} + \mathcal{H}_{gi}\big[(H_m)_{gi} - (H_m)_{go}\big]\Big\}$$
$$= m_s\Big\{(H_s)_{ds} - (H_s)_{ws} + X_{ds}(H_m)_{ds} - X_{ws}(H_m)_{ws}$$
$$+ m_g[(H_m)_{go}(\mathcal{H}_{go} - \mathcal{H}_{gi})]\Big\} \qquad (18\text{-}70)$$

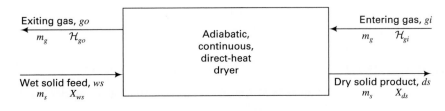

Figure 18.36 General configuration for a continuous, direct-heat dryer.

EXAMPLE 18.17   Balance for a Direct-Heat Dryer.

A continuous, cocurrent-flow direct-heat dryer is to be used to dry crystals of Epsom salt (magnesium sulfate heptahydrate). The feed to the dryer, a filter cake from a rotary, vacuum filter, is 2,854 lb/h of crystals (dry basis) with a moisture content of 25.8 wt% (dry basis) at 85°F and 14.7 psia. Air enters at 14.7 psia, with dry-bulb and wet-bulb temperatures of 250°F and 117°F. The final moisture content of the dried crystals is to be 1.5 wt% (dry basis), at no more than 118°F to prevent decomposition of the heptahydrate (see Figure 17.2). Determine: (a) rate of moisture evaporation, (b) outlet temperature of the air, (c) rate of heat transfer, and (d) entering air flow rate. The average specific heat of Epsom salt is 0.361 Btu/lb-°F.

Solution

$m_s = 2,854$ lb/h, $X_{ws} = 0.258, X_{ds} = 0.015, T_{ws} = 85°F,$
$T_{ds} = 118°F, T_{gi} = 250°F,$ and $T_v = 117°F$

From Figure 18.17 for $T_{db} = 250°F$ and $T_{wb} = 117°F, \mathcal{H}_{gi} = 0.0405.$

(a) From (18-65), $m_v = 2,854(0.258 - 0.015) = 694$ lb/h.

(b) Because the dryer operates cocurrently, the outlet temperature of the gas must be greater than the outlet temperature of the dry solid, which is taken as 118°F. The best value for T_{go} is obtained by optimizing the cost of the drying operation. A reasonable value for T_{go} can be estimated by using the concept of the number of heat-transfer units, which is analogous to the number of transfer units for mass transfer, as developed in §6.7. For heat transfer in a dryer, where the solids temperature throughout most of the dryer will be at T_v, the number of heat-transfer units is

$$N_T = \ln\left[\frac{T_{gi} - T_v}{T_{go} - T_v}\right] \qquad (1)$$

where economical values of N_T are usually in the range of 1.0–2.5. Assume a value of 2.0. From (1),

$$2 = \ln\left[\frac{250 - 117}{T_{go} - 117}\right]$$

from which $T_{go} = 135°F.$

(c) The rate of heat transfer is obtained from (18-69) using the conditions for the solid flow.

$$\begin{aligned} Q = {} & 2,854\{0.361(118 - 85) + 0.258(1)(117 - 85) \\ & + 0.015(1)(118 - 117) + (0.258 - 0.015) \\ & \times [1,027.5 + (0.447)(135 - 117)]\} \\ = {} & 2,854[11.9 + 8.3 + 0.02 + 249.7 + 2.0] \\ = {} & 2,854(271.9) = 776,000 \text{ Btu/h} \end{aligned}$$

It should be noted that the heat required to vaporize the 694 lb/h of moisture at 117°F is $(249.7/271.9) \times 100\% = 91.8\%$ of the total heat load.

(d) The entering air flow rate is obtained from (18-69) using the far RHS of that equation with the above value of Q.

$$m_g = \frac{776,000}{[(0.242) + (0.0405)(0.447)](250 - 135)} = 25,940 \text{ lb/h}$$

The total entering air, including the humidity, is 25,940 $(1 + 0.0405) = 27,000$ lb/h.

§18.5.2   Belt Dryer with Through-Circulation

Consider the continuous, two-zone through-circulation belt dryer in Figure 18.37a. A bed of wet-solid particles is conveyed continuously into Zone 1, where contact is made with hot gas passing upward through the bed. Because the temperature of the gas decreases as it passes through the bed, the temperature-driving force decreases so that the moisture content of solids near the bottom of the moving bed decreases more rapidly than for solids near the top. To obtain a dried solid of more uniform moisture content, the gas flow direction through the bed is reversed in Zone 2.

Based on the work of Thygeson and Grossmann [25], a mathematical model for Zone 1 can be developed using the coordinate system shown in Figure 18.37b, based on six assumptions:

1. Wet solids enter Zone 1 with a uniform moisture content of X_o on the dry basis.

2. Gas passes up through the moving bed in plug flow with no mass transfer in the vertical direction (i.e., no axial dispersion).

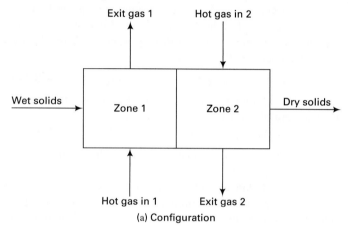

(a) Configuration

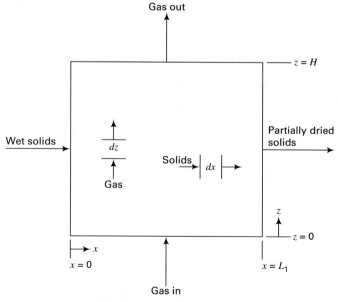

(b) Coordinate system for zone 1

Figure 18.37 Continuous, two-zone through-circulation belt dryer.

3. Drying takes place in the constant-rate period, controlled by the rate of heat transfer from the gas to the surfaces of the solid particles, where the temperature is the adiabatic-saturation temperature.

4. Sensible-heat effects are negligible compared to latent-heat effects.

5. The void fraction of the bed is uniform and constant, and no mixing of solid particles occurs.

6. The solids are conveyed at a uniform linear speed, S.

Based on these assumptions, the gas temperature decreases with increasing distance z from the bottom of the bed, and is independent of the distance, x, in the direction in which the solids are conveyed, i.e., $T = T\{z\}$. The moisture content of the solids varies in both z- and x-directions, decreasing more rapidly near the bottom of the bed, where the gas temperature is higher, i.e., $X = X\{z, x\}$.

Zone 1

With no mixing of solids, a material balance on the moisture in the solids at any vertical location, z, is given by

$$\frac{dX_1}{dt} = S\frac{dX_1}{dx} = -\frac{(ha)(T_1 - T_v)}{\Delta H_v^{vap}(\rho_b)_{ds}} \quad (18\text{-}71)$$

where $a =$ surface area of solid particles per unit volume of bed; $T_1 =$ bulk temperature of the gas in Zone 1, which depends on z; and $(\rho_b)_{ds}$ is the bulk density of solids when dry. The initial condition is $X_1 = X_o$ for $x = 0$.

An energy balance for the gas phase at any location x is

$$\rho_g(C_P)_g u_s \frac{dT_1}{dz} = -(ha)(T_1 - T_v) \quad (18\text{-}72)$$

where $\rho_g =$ gas density and $u_s =$ superficial velocity of gas through the bed. The initial condition is $T_1 = T_{gi}$ for $z = 0$.

Equation (18-71) is coupled to (18-72), which is independent of (18-71). It is possible to separate variables and integrate (18-72) to obtain

$$T_1 = T_v + (T_{gi} - T_v)\exp\left(-\frac{h\,az}{\rho_g(C_P)_g u_s}\right) \quad (18\text{-}73)$$

At $z = H$ at the top of the bed,

$$T_{go_1} = T_v + (T_{gi} - T_v)\exp\left(-\frac{h\,aH}{\rho_g(C_P)_g u_s}\right) \quad (18\text{-}74)$$

Equation (18-71) can now be solved by combining it with (18-73) to eliminate T_1, followed by separation of variables and integration. The result is

$$\frac{X_1}{X_o} = 1 - \frac{x\,ha(T_{gi} - T_v)\exp\left(-\dfrac{h\,az}{\rho_g(C_P)_g u_s}\right)}{S\Delta H_v^{vap}(\rho_b)_{ds}} \quad (18\text{-}75)$$

The moisture content $(X_1)_{L_1}$ at $x = L_1$ is obtained by replacing x with L_1. If desired, X_{avg} at $x = L_1$ can be determined from

$$X_{avg} = \int_0^H (H_1)_{L_1} dz \quad (18\text{-}76)$$

Zone 2

In Zone 2, (18-71) still applies, with X_1 and T_1 replaced by X_2 and T_2, but the initial condition for X_o is $(X_1)_{L_1}$ from (18-75) for $x = L_1$, which depends on z. Equation (18-72) also applies, with T_1 replaced by T_2, but the initial condition is $T_2 = T_{gi}$ at $z = H$. The integrated result is

$$T_2 = T_v + (T_{gi} - T_v)\exp\left[-\frac{h\,a(H - z)}{\rho_g(C_P)_g u_s}\right] \quad (18\text{-}77)$$

With T_{go_2} given by (18-74), where T_{go_1} is replaced by T_{go_2}, and

$$\frac{X_2}{(X_1)_{L_1}} = 1 - \frac{x\,h\,a(T_{gi} - T_v)\exp\left[-\dfrac{h\,a(H - z)}{\rho_g(C_P)_g u_s}\right]}{S\Delta H_v^{vap}(\rho_b)_{ds}} \quad (18\text{-}78)$$

where $(X_1)_{L_1}$ is the value from (18-75) for $x = L_1$ at the value of z in (18-78). The value of x in (18-78) is the distance from the start of Zone 2. Values of $(X_2)_{L_2}$ at any z are obtained from (18-76) for $x = L_2$. The average moisture content over the height of the moving bed leaving Zone 2 is then obtained from (18-76), with $(X_1)_{L_1}$ replaced by $(X_2)_{L_2}$. The above relationships are illustrated in the next example.

EXAMPLE 18.18 Through-Circulation Drying.

The filter cake of $CaCO_3$ in Example 18.8 is to be dried continuously on a belt dryer using through-circulation. The dryer is 6 ft wide, has a belt speed of 1 ft/minute and consists of two drying zones, each 12 ft long. Air at 170°F and 10% relative humidity enters both zones, passing upward through the bed in the first zone, and downward in the second, at a superficial velocity of 2 m/s. Bed height on the belt is 2 inches. Predict the moisture-content distribution with height at the end of each zone, and the average moisture content at the end of Zone 2. Assume all drying is in the constant-rate period and neglect preheat.

Solution

From data in Examples 18.7 and 18.8,

$$X_o = 0.30$$
$$(\rho_b)_{ds} = \frac{1.00}{1.30}(1,950) = 1,500 \text{ kg/m}^3$$
$$\epsilon_b = 0.50$$
$$T_v = 37.8°C = 311 \text{ K}, T_{gi} = 76.7°C = 350 \text{ K}$$
$$\Delta H_v^{vap} = 2,413 \text{ kJ/kg}$$

From extrusion area and volume in Example 18.8,

$$a = \frac{(3.16 \times 10^{-4})(0.5)}{(4.01 \times 10^{-7})} = 395 \text{ m}^2/\text{m}^3 \text{ bed}$$

For $u_s = 2$ m/s, $h = 0.188(\text{kJ/s-m}^2\text{-K}^2)$ from Example 18.8.

$$(C_P)_g = 1.09 \text{ kJ/kg-K}, \rho_g \text{ at 1 atm} = 0.942 \text{ kg/m}^3,$$
$$S = 1 \text{ ft/min} = 0.00508 \text{ m/s}$$

The cross-sectional area of the moving bed normal to the conveying direction is $6(2/12) = 1 \text{ ft}^2 = 0.0929 \text{ m}^2$. For a belt speed of 1 ft/min = 0.305 m/minute, the volumetric flow of solids is $(0.0929)(0.305)$

$= 0.0283$ m³/minute. The mass rate of flow is $0.0283(1,500) = 42.5$ kg/min (dry basis).

Zone 1

$$H = 0.167 \text{ ft} = 0.0508 \text{ m and } L_1 = 12 \text{ ft} = 3.66 \text{ m}$$

From (18-74), the gas temperature leaving the bed is

$$T_{go_1} = 37.8 + (76.7 - 37.8)\exp\left[-\frac{(0.188)(395)(0.0508)}{(0.942)(1.09)(2)}\right]$$

$$= 44°C = 317 \text{ K}$$

The moisture-content distribution at $x = L_1$ is obtained from (18-75). For $z = 0$,

$$X_1 = 0.30\left[1 - \frac{(3.66)(0.188)(395)(76.7 - 37.8)}{(0.00508)(2,413)(1,500)}\right] = 0.127$$

For other values of z, a spreadsheet gives:

z, m	$(X_1)_{L_1}$
0	0.127
0.0127	0.191
0.0254	0.231
0.0381	0.257
0.0508	0.273

Because of the decrease in gas temperature as it passes through the bed, moisture content varies considerably over the bed depth.

Zone 2

The flow of air is reversed to further the drying and smooth out the moisture-content distribution. The value of X_o is replaced by the above values of $(X_1)_{L_1}$ for corresponding values of z. Using (18-78) with a spreadsheet, the following distribution is obtained at $L_2 = 3.66$ m for a total length of both zones $= 24$ ft $= 7.32$ m:

z, m	$(X_2)_{L_2}$
0	0.116
0.0127	0.163
0.0254	0.178
0.0381	0.163
0.0508	0.116

A much more uniform moisture distribution is achieved. From (18-76) for Zone 2, using numerical integration with a spreadsheet, $(X_2)_{avg} = 0.155$.

§18.5.3 Direct-Heat Rotary Dryer

As discussed by Kelly in Mujumdar [1], design of a direct-heat rotary dryer, of the type shown in Figure 18.7, for drying solid particles at a specified feed rate, initial moisture content X_{ws}, and final moisture content X_{ds}, involves determination of heating-gas inlet and outlet conditions, heating-gas velocity and flow direction, dryer-cylinder diameter and length, dryer-cylinder slope and rotation rate, number and type of lifting flights, solids holdup as a % of dryer-cylinder volume, and solids-residence time.

A commercial-size direct-heat rotary dryer should be scaled up from pilot-plant data. However, if a representative sample of the wet solid is not available, the following procedure and model, based on test results with several materials in both pilot-plant-size and commercial-size dryers, is useful for a preliminary design.

The hot gas can flow countercurrently or cocurrently to the flow of the solids. Cocurrent flow is used for very wet, sticky solids with high inlet-gas temperatures, and for non-hygroscopic solids. Countercurrent flow is preferred for low-to-moderate inlet-gas temperatures, where thermal efficiency becomes a factor. When solids are not subject to thermal degradation, melting, or sublimation, an inlet-gas temperature up to ~1,000°F can be used. The exit-gas temperature is determined from economics, as discussed in Example 18.17, where Equation (1) can be used with N_T in the range of 1.5–2.5. Generally, more gas flow and higher gas temperatures increase operating costs, but decrease capital costs, because the larger temperature-driving force increases the heat-transfer rate. Allowable gas velocities are determined from the dusting characteristics of the particles, and can vary widely with particle-size distribution and particle density. Some typical values for allowable gas velocity are as follows:

Material	Particle Density, ρ_p, lb/ft³	Average Particle Size, d_p, μm	Allowable Gas Velocity, u_{all}, ft/s
Plastic granules	69	920	3.5
Ammonium nitrate	104	900	4.5
Sand	164	110	1.0
Sand	164	215	2.0
Sand	164	510	5.0
Sawdust	27.5	640	1.0

Using an appropriate allowable gas velocity, u_{all}, with mass flow rate and density of the gas at the gas-discharge end, $(m_g)_{exit}$ and $(\rho_g)_{exit}$, the dryer diameter, D, can be estimated by the continuity equation

$$D = \left[\frac{4(m_g)_{exit}}{\pi u_{all}(\rho_g)_{exit}}\right]^{0.5} \quad (18\text{-}79)$$

Residence time of the solids in the dryer, θ, is related to the fractional volume holdup of solids, V_H, by

$$\theta = \frac{LV_H}{F_V} \quad (18\text{-}80)$$

where L = length of dryer cylinder and F_V = solids volumetric velocity in volume/unit cross-sectional area-unit time. A conservative estimate of the holdup, including the effect of gas velocity, is obtained by combining (18-80) with a relation in [2]:

$$V_H = \frac{0.23F_V}{SN^{0.9}D} \pm 0.6\frac{G(5/d_p^{0.5})}{\rho_p} \quad (18\text{-}81)$$

where F_V = ft³ solids/(ft² cross section)-h; S = dryer-cylinder slope, ft/ft; N = dryer-cylinder rate of rotation, rpm; D =

dryer diameter, ft; G = gas superficial mass velocity, lb/h-ft^2; and d_p = mass-average particle size, μm. The plus (+) sign on the second term corresponds to countercurrent flow that tends to increase the holdup, while the minus (−) sign denotes cocurrent flow. Equation (18-81) holds for dryers having lifting flights with lips, but is limited to gas velocities less than 3.5 ft/s. A more complex model by Matchett and Sheikh [26] is valid for gas velocities up to 10 ft/s. Optimal solids holdup is 10–18% of dryer volume so that flights run full and all or most of the solids are showered during each revolution.

When drying is in the constant-rate period such that the rate can be determined from the rate of heat transfer from the gas to the wet surface of the solids at the wet-bulb temperature, a volumetric heat-transfer coefficient, ha, can be used, which is defined by

$$Q = (ha)V\Delta T_{LM} \tag{18-82}$$

where V = volume of dryer cylinder = $\pi D^2 L/4$;

$$\Delta T_{LM} = \frac{(T_g)_{in} - (T_g)_{out}}{\ln\left[\dfrac{(T_g)_{in} - T_v}{(T_g)_{out} - T_v}\right]} \tag{18-83}$$

and ha = volumetric heat-transfer coefficient based on dryer-cylinder volume as given by the empirical correlation of McCormick [27], when the heating gas is air:

$$ha = KG^{0.67}/D \tag{18-84}$$

where ha is in Btu/h-ft^3-°F, G is in lb/h-ft^2, and D is in ft.

$K = 0.5$ is recommended in [2] for dryers operating at a peripheral cylinder speed of 1.0–1.25 ft/s and with a flight count of 2.4D to 3.0D per circle. When K is available from pilot-plant data, (18-84) can be used for scale-up to a larger diameter and a different value of G.

It might be expected that a correlation for the volumetric heat-transfer coefficient, ha, would take into account the particle diameter because the solids are lifted and showered through the gas. However, the solids shower as curtains of some thickness, with the gas passing between the curtains. Thus, particles inside the curtains do not receive significant exposure to the gas, and the effective heat-transfer area is more likely determined by the areas of the curtains. Nevertheless, (18-84) accounts for only two of the many possible variables, and the inverse relation with dryer diameter is not well supported by experimental data. A complex model for heat transfer that treats h and a separately is that of Schofield and Glikin [28], as modified by Langrish, Bahu, and Reay [29].

EXAMPLE 18.19 Direct-Heat Rotary Dryer.

Ammonium nitrate, at 70°F with a moisture content of 15 wt% (dry basis), is fed into a direct-heat rotary dryer at a feed rate of 700 lb/minute (dry basis). Air at 250°F and 1 atm, with a humidity of 0.02 lb H$_2$O/lb dry air, enters the dryer and passes cocurrently with the solid. The final solid moisture content is to be 1 wt% (dry basis) and all drying will take place in the constant-rate period. Make a preliminary estimate of the dryer diameter and length, assuming that such dryers are available in: (1) diameters from 1 to 5 ft in increments of 0.5 ft and from 5 to 20 ft in increments of 1.0 ft, and (2) lengths from 5 to 150 ft in increments of 5 ft.

Solution

From the psychrometric chart (Figure 18.17), $T_{wb} = 107°F$. Assume that all drying is at this temperature for the solid. A reasonable outlet temperature for the air can be estimated from (1) of Example 18.17, assuming $N_T = 1.5$. From that equation,

$$1.5 = \ln\left[\frac{250 - 107}{T_{go} - 107}\right]$$

Solving, $T_{go} = 140°F$. Assume solids outlet temperature = T_{ds} = 135°F.

Heat-transfer rate:

$$m_s = 700(60) = 42,000 \text{ lb/h of solids (dry basis),}$$
$$(C_P)_s = 0.4 \text{ Btu/lb-°F}, T_{ws} = 70°F, T_v = T_{wb} = 107°F,$$
$$X_{ws} = 0.15, X_{ds} = 0.01, \text{ and } \Delta H_v^{vap} = 1,033 \text{ Btu/lb}$$

From (18-65), $m_v = 42,000(0.15 - 0.01) = 5,880$ lb/h H$_2$O evaporated. From (18-69),

$$Q = 42,000\{(0.4)(135 - 70) + (0.15)(1)(107 - 70)$$
$$+ (0.01)(1)(135 - 70) + (0.15 - 0.01)[1,033 + (0.447)(140 - 107)]\}$$
$$= 7,510,000 \text{ Btu/h}$$

Air flow rate:

$$m_g = \frac{7,510,000}{[(0.242) + (0.02)(0.447)](250 - 135)}$$
$$= 260,000 \text{ lb/h entering dry air}$$

Dryer diameter:

Assume an allowable gas velocity at the dryer exit of 4.5 ft/s.

$$(m_g)_{exit} = 260,000(1 + 0.02) + 5,880 = 271,000 \text{ lb/h total gas}$$

$$\left(\rho_g\right)_{exit} = \frac{PM}{RT_{go}}$$

$$M = \frac{271,000}{\dfrac{260,000}{29} + \dfrac{11,000}{18}} = 28.3$$

$$\left(\rho_g\right)_{exit} = \frac{(1)(28.3)}{(0.730)(600)} = 0.0646 \text{ lb/ft}^3$$

From (18-79), $\quad D = \left[\dfrac{4(271,000)}{(3.14)(4.5)(3,600)(0.0646)}\right]^{0.5} = 18 \text{ ft}$

Dryer length:

$$G = G_{exit} = \frac{(271,000)(4)}{(3.14)(18)^2} = 1,070 \text{ lb/h-ft}^2$$

From (18-84), $ha = 0.5(1,070)^{0.67}/18 = 3$ Btu/h-ft^3-°F.

From (18-83), neglecting the periods of wet solids heating up to 107°F and the dry solids heating up to 135°F, because the heat transferred is a small % of the total,

$$\Delta T_{LM} = \frac{250 - 140}{\ln\left[\dfrac{250 - 107}{140 - 107}\right]} = 75°F$$

From (18-82), $\quad V = \dfrac{7,510,000}{(3)(75)} = 33,400 \text{ ft}^3$

Cross-sectional area = $(3.14)(18)^2/4 = 254$ ft^2.

$$L = \frac{33,400}{254} = 130 \text{ ft}$$

§18.5.4   Fluidized-Bed Dryer

The behavior of a bed of solid particles when a gas is passed up through the bed is shown in Figure 18.38. At a very low gas velocity, the bed remains fixed. At a high gas velocity, the bed disappears; the particles are pneumatically transported by the gas when its local velocity exceeds the particle terminal settling velocity. and the system becomes a "gas lift." At an intermediate gas velocity, the bed is expanded, but particles are not carried out by the gas. Such a bed is said to be fluidized, because the bed of solids takes on the properties of a fluid. *Fluidization* is initiated when the gas velocity reaches the point where all the particles are suspended by the gas. As the gas velocity is increased further, the bed expands and bubbles of gas are observed to pass up through the bed. This regime of fluidization is referred to as *bubbling fluidization* and is the most desirable regime for most fluidized-bed operations, including drying. If the gas velocity is increased further, a transition to *slugging fluidization* eventually occurs, where bubbles coalesce and spread to a size that approximates the diameter of the vessel. To some extent, this behavior can be modified by placing baffles and low-speed agitators in the bed.

Before fluidization occurs, when the bed of solids is fixed, the pressure drop across the bed for gas flow, ΔP_b, is predicted by the Ergun [30] equation, discussed in §6.8.2:

$$\frac{\Delta P_b}{L_b} = 150 \frac{(1 - \epsilon_b)^2}{\epsilon_b^3} \frac{\mu u_s}{(\phi_s d_p)^2} + 1.75 \frac{(1 - \epsilon_b)}{\epsilon_b^3} \frac{\rho_g u_s^2}{\phi_s d_p}$$

$$(18\text{-}85)$$

where L_b = bed height, u_s = superficial-gas velocity, and ϕ_s = particle sphericity. The first term on the RHS is dominant at low-particle Reynolds numbers where streamline flow exists, and the second term dominates at high-particle Reynolds numbers where turbulent flow exists.

The onset of fluidization occurs when the drag force on the particles by the upward-flowing gas becomes equal to the weight of the particles (accounting for displaced gas):

$$\left(\begin{array}{c} \Delta P \\ \text{across} \\ \text{bed} \end{array} \right) \left(\begin{array}{c} \text{Cross-sectional} \\ \text{area} \\ \text{of bed} \end{array} \right) = \left(\begin{array}{c} \text{Volume} \\ \text{of} \\ \text{bed} \end{array} \right)$$

$$\times \left(\begin{array}{c} \text{Volume} \\ \text{fraction} \\ \text{of solid} \\ \text{particles} \end{array} \right) \left[\left(\begin{array}{c} \text{Density} \\ \text{of} \\ \text{solid} \\ \text{particles} \end{array} \right) - \left(\begin{array}{c} \text{Density} \\ \text{of} \\ \text{displaced} \\ \text{gas} \end{array} \right) \right]$$

Thus, $\Delta P_b A_b = A_b L_b (1 - \epsilon_b)[(\rho_p - \rho_g)g]$ (18-86)

The minimum gas-fluidization superficial velocity, u_{mf}, is obtained by solving (18-85) and (18-86) simultaneously for $u = u_{mf}$. For $N_{Re,p} = d_p u_{mf} \rho_g / \mu < 20$, the turbulent-flow contribution to (18-85) is negligible and the result is

$$u_{mf} = \frac{d_p^2 (\rho_p - \rho_g) g}{150 \mu} \left(\frac{\epsilon_b^3 \phi_s^2}{1 - \epsilon_b} \right)$$

$$(18\text{-}87)$$

For operation in the bubbling fluidization regime, a superficial-gas velocity of $u_s = 2u_{mf}$ is a reasonable choice. At this velocity, the bed will be expanded by about 10%, with no further increase in pressure drop across the bed. In this regime, the solid particles are well mixed and the bed temperature is so uniform that fluidized beds are used industrially to calibrate thermocouples and thermometers. If the fluidized bed is operated continuously at steady-state conditions rather than batchwise with respect to the particles, the particles will have a residence-time distribution like that of a fluid in a continuous-stirred-tank reactor (CSTR). Some particles will be in the dryer for only a very short period of time and will experience almost no decrease in moisture content. Other particles will be in the dryer for a long time and may come to equilibrium before that time has elapsed. Thus, the dried solids will have a distribution of moisture content. This is in contrast to a batch-fluidization process, where all particles have the same residence time and, therefore, a uniform final moisture content. This is an important distinction because continuous, fluidized-bed dryers are usually scaled up from data obtained in small, batch fluidized-bed dryers. Therefore, it is important to have a relationship between batch drying time and continuous drying time.

The distribution of residence times for effluent from a perfectly mixed vessel operating at continuous, steady-state conditions is given by Fogler [31] as

$$E\{t\} = \exp(-t/\tau)/\tau$$

$$(18\text{-}88)$$

where τ is the average residence time and $E\{t\}$ is defined such that $E\{t\}dt$ = the fraction of effluent with a residence time between t and $t + dt$. Thus, $\int_0^{t_1} E\{t\}dt$ = fraction of the effluent with a residence time less than t_1. For example, if the average particle-residence time is 10 min, 63.2% of the particles will have a residence time of less than 10 minutes. If the particles are small and nonporous such that all drying takes place in the constant-rate period, and θ is the time for complete drying, then

$$\frac{X}{X_o} = 1 - \frac{t}{\theta}, \quad t \leq \theta$$

$$(18\text{-}89)$$

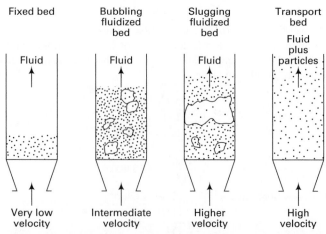

Figure 18.38 Regimes of fluidization of a bed of particles by a gas.

The average moisture content of the dried solids leaving the fluidized-bed is obtained by integrating the expression below from 0 to only θ because $X = 0$ for $t > \theta$.

$$\bar{X}_{ds} = \int_0^\theta XE\{t\}dt = \int_0^\theta X_0\left(1 - \frac{t}{\theta}\right)E\{t\}dt \qquad (18\text{-}90)$$

Combining (18-88) and (18-90) and integrating gives

$$\bar{X}_{ds} = X_o\left[1 - \frac{1 - \exp(-\theta/\tau)}{\theta/\tau}\right] \qquad (18\text{-}91)$$

If the particles are porous and without surface moisture such that all drying takes place in the falling-rate period, diffusion theory may apply such that the following empirical exponential expression may be used for the moisture content as a function of time:

$$\frac{X}{X_o} = \exp(-Bt) \qquad (18\text{-}92)$$

In this case, the combination of (18-92) with (18-90), followed by integration from $t = 0$ to $t = \infty$, gives

$$\bar{X}_{ds} = 1/(1 + B\tau) \qquad (18\text{-}93)$$

Values of θ and B are determined from experiments with laboratory batch fluidized-bed dryers for scale-up to large dryers operating under the same conditions.

EXAMPLE 18.20 Fluidized-Bed Dryer.

Ten thousand lb/h of wet sand at 70°F with a moisture content of 20% (dry basis) is to be dried to a moisture content of 5% (dry basis) in a continuous, fluidized-bed dryer operating at a pressure of 1 atm in the free-board region above the bed. The sand has a narrow size range, with an average particle size of 500 μm; a sphericity, ϕ_s, of 0.67; and a particle density of 2.6 g/cm³. When the sand bed is dry, its void fraction, ϵ_b, is 0.55. Fluidizing air will enter the bed at a temperature of 1,000°F with a humidity of 0.01 lb H$_2$O/lb dry air. The adiabatic-saturation temperature is estimated to be 145°F. Batch pilot-plant tests with a fluidization velocity of twice the minimum show that drying takes place in the constant-rate period and that all moisture can be removed in 8 minutes using air at the same conditions and with a bed temperature of 145°F. Determine the bed height and diameter for the large, continuous unit.

Solution

$$d_p = 500 \ \mu m = 0.0500 \ \text{cm}$$

Heat-transfer rate:

$$(C_P)_s = 0.20 \ \text{Btu/lb-°F}, \ T_v = 145°F = T_{go} = T_{ds},$$
$$m_s = 10,000/(1 + 0.2) = 8,330 \ \text{lb/h dry sand, and}$$
$$\Delta H_v^{\text{vap}} = 1,011 \ \text{Btu/lb}, \ T_{ws} = 70°F.$$

From (18-69),

$$Q = 8,330\{0.20(145 - 70) + 0.20(1)(145 - 70)$$
$$+ (0.20 - 0.05)(1,011)\} = 1,510,000 \ \text{Btu/h}$$

Air rate:

$$m_g = \frac{1,510,000}{[(0.242) + (0.01)(0.447)](1,000 - 145)}$$
$$= 7,170 \ \text{lb/h dry air}$$

From (18-65), $m_v = 8,330(0.20 - 0.05) = 1,250$ lb/h evaporated moisture.

$$\mathcal{H}_{go} = \frac{(7,170)(0.01) + 1,250}{7,170} = 0.184 \ \text{lb H}_2\text{O/lb dry air}$$

Total exiting gas flow rate $= 7,170(1 + 0.184) = 8,500$ lb/h

Minimum fluidization velocity:

$$\text{M of existing gas} = \frac{8,500}{\dfrac{7,170}{29} + \dfrac{1,330}{18}} = 26.5$$

$$(\rho_g)_{\text{exit}} = \frac{PM}{RT_g} = \frac{(1)(26.5)}{(0.730)(605)} = 0.060 \ \text{lb/ft}^3$$

$$= 0.00096 \ \text{g/cm}^3$$

$$\mu = 0.048 \ \text{lb/ft-h} = 0.00020 \ \text{g/cm-s}$$

For small particles, assume streamline flow at u_{mf} so that (18-87) applies, but check to see if $N_{\text{Re},p} < 20$. Using cgs units,

$$u_{mf} = \frac{(0.0500)^2(2.6 - 0.00096)(980)(0.55)^3(0.67)^2}{150(0.00020)(1 - 0.55)}$$

$$= 35.3 \ \text{cm/s}$$

$$N_{\text{Re},p} = \frac{d_p u_{mf} \rho_g}{\mu} = \frac{(0.0500)(35.3)(0.00096)}{0.00020} = 8.5$$

Since $N_{\text{Re},p} < 20$, (18-87) does apply.

Use a superficial-gas velocity of twice $u_{mf} = 2(35.3) = 70.6$ cm/s $= 8,340$ ft/h.

Bed diameter:

Equation (18-79) applies:

$$D = \left[\frac{4(8,500)}{3.14(8,340)(0.060)}\right]^{0.5} = 4.7 \ \text{ft}$$

Bed density:

Fixed-bed density $= \rho_s(1 - \epsilon_b) = 2.6(1 - 0.55)(62.4) = 73.0$ lb/ft³. Assume the bed expands by 10% upon fluidization to $u = 2u_{mf}$:

$$\rho_b = \frac{73.0}{1.10} = 66 \ \text{lb/ft}^3 \text{(dry basis)}$$

Average particle-residence time:

For constant-rate drying in a batch dryer, all particles have the same residence time. From pilot-plant data, $\theta = 8$ min for drying.

For the large, continuous operation, (18-91) applies, with $\bar{X}_{ds} = 0.05$ and $X_o = 0.20$. Thus,

$$0.05 = 0.20\left[1 - \frac{1 - \exp\left(-\dfrac{8}{\tau}\right)}{(8/\tau)}\right]$$

Solving this nonlinear equation, $\tau = 13.2$ minutes average residence time for particles. Only

$$\frac{(0.20 - 0.05)}{(0.20 - 0.0)}(8) = 6 \ \text{min}$$

residence time would be required in a batch dryer to dry to 5% moisture. Therefore, more than double the residence time is needed in the continuous unit.

Bed height:

To achieve the average residence time of 13.2 minutes $= 0.22$ h, the expanded-bed volume, and corresponding bed height, must be

$$V_b = \frac{m_s \tau}{\rho_b} = \frac{8,330(0.22)}{66} = 27.8 \text{ ft}^3$$

$$H_b = \frac{V_b}{\pi D^2/4} = \frac{27.8(4)}{3.14(4.7)^2} = 1.6 \text{ ft}$$

§18.6 DRYING OF BIOPRODUCTS

The selection of a dryer is often a critical step in the design of a process for the manufacture of a bioproduct. As discussed in several chapters of the *Handbook of Industrial Drying* [32], drying may be needed to preserve required properties and maintain activity of bioproducts. If a proper drying method is not selected or adequately designed, the bioproduct may degrade during dewatering or exposure to elevated temperatures. For example, the bioproduct may be subject to oxidation and thus require drying in a vacuum or in the presence of an inert gas. It may degrade or be contaminated in the presence of metallic particles, requiring a dryer constructed of polished stainless steel. Enzymes may require pH control during drying to prevent destabilization. Some bioproducts may require gentle handling during the drying process.

Of major concern is the fact that many bioproducts are thermolabile, in that they are subject to destruction, decomposition, or great change by moderate heating. Table 18.7 lists several examples of bioproduct degradation that can occur during drying at elevated temperatures. As shown, the

Table 18.7 Examples of Degradation of Bioproducts at Elevated Temperatures

Product	Type of Reaction	Degradation Processes	Result
Live microorganisms	Microbiological changes	Destruction of cell membranes	Denaturation of protein Death of cells
Lipids	Enzymatic reactions	Peroxidation of lipids (discoloration of the product)	Reaction with other components (including proteins and vitamins)
Proteins	Enzymatic and chemical reactions	Total destruction of amino acids	Denaturation of proteins and enzymes
		Derivation of some individual amino acids	Partial denaturation, loss of nutritive value
		Cross-linking reaction between amino acids	Change of protein functionality Enzyme reaction
Polymer carbohydrates	Chemical reactions	Gelatination of starch	Improved digestibility and energy utilization
		Hydrolysis	Fragmentation of molecule
Vitamins	Chemical reactions	Derivation of some amino acids	Partial inactivation
Simple sugars	Physical changes	Caramelization (Maillard-Browning reaction)	Loss of color and flavor
		Melting	

Source: *Handbook of Industrial Drying* [32]

Table 18.8 Selection of Dryer for Representative Bioprocesses

Bioproduct	Dryer Type	Comments
Citric acid	Fluidized-bed dryer	Feed is wet cake from a rotary vacuum filter
Pyruvic acid	Fluidized-bed dryer	Feed is wet cake from a rotary vacuum filter
L-Lysine (amino acid)	Spray dryer	Feed is solution from an evaporator
Riboflavin (Vitamin B2)	Spray dryer	Feed is solution from a decanter
α-Cyclodextrin (polysaccharide)	Fluidized-bed dryer	Feed is wet cake from a rotary vacuum filter
Penicillin V (acid)	Fluidized-bed dryer	Feed is a wet cake from a basket centrifuge
Recombinant human serum albumin (protein)	Freeze-dryer	Feed is from sterile filtration
Recombinant human insulin (protein)	Freeze-dryer	Feed is wet cake from a basket centrifuge
Monoclonal antibody (cell)	No dryer	Product is a phosphate-buffered saline (PBS) solution
α-1-Antitrypsin (protein)	No dryer	Product is a PBS solution
Plasmid DNA (parasitic DNA)	No dryer	Product is a PBS solution

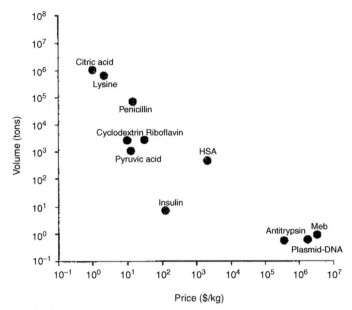

Figure 18.39 Price and production volume of representative bioproducts [35].

result of such exposure is serious and unacceptable. To avoid such degradation, many bioproducts are dried at near-ambient or cryogenic temperatures. The most widely used dryers for sensitive bioproducts, particularly solutions of enzymes and other proteins, are spray dryers and freeze-dryers (i.e., lyophilizers) [33, 34].

Heinzle et al. [35] consider dryer selection for 11 different bioprocesses, as listed in Table 18.8. The bioproducts cover more than a seven-fold range of product value and more than a six-fold range of annual production rate, as shown in Figure 18.39. It is interesting to note that the three most expensive

bioproducts are not dried, but produced as phosphate-buffered saline solutions. The least-expensive and highest-volume bioproducts use either fluidized-bed or spray dryers. The fluidized-bed dryers are used with relatively stable biomolecules, and operate at near-ambient temperatures. The two bioproducts at intermediate levels of price and volume use freeze-dryers.

Intermittent Drying of Bioproducts

As discussed in §13.8, batch-distillation operations can be improved by controlling the reflux ratio. Similarly, batch-drying operations can be improved, particularly for heat-sensitive bioproducts, by varying conditions during the drying operation. This technique is referred to as *intermittent drying*. Although the concept has been known for decades, it is only in recent years that it has received wide attention, as discussed by Chua et al. [36]. The intermittent supply of heat is beneficial for materials that begin drying in a constant-rate period, but dry primarily in the falling-rate period, where the rate of drying is controlled by internal heat and mass transfer. In traditional drying, the external conditions are constant and the surface temperature of the material being dried can rise to unacceptable levels. In intermittent drying, the external conditions are altered so that the surface temperature does not exceed a limiting value. In the simplest case, the heat input to the material is reduced to zero during a so-called tempering phase, while interior moisture moves to the surface so that a constant-rate period can be resumed. The benefits of intermittent drying have been demonstrated for a number of products, including grains, potatoes, guavas, bananas, carrots, rice, corn, clay, cranberries, apples, peanuts, pineapples, sugar, beans, ascorbic acid, and β-carotene.

SUMMARY

1. Drying is the removal of moisture (water or another volatile liquid) from wet solids, solutions, slurries, and pastes.

2. The two most common modes of drying are direct, by heat transfer from a hot gas, and indirect, by heat transfer from a hot wall. The hot gas is frequently air, but can be combustion gas, steam, nitrogen, or any other non-reactive gas.

3. Industrial drying equipment can be classified by operation (batch or continuous), mode (direct or indirect), or the degree to which the material being dried is agitated. Batch dryers include tray dryers and agitated dryers. Continuous dryers include: tunnel, belt or band, turbo-tray tower, rotary, screw-conveyor, fluidized-bed, spouted-bed, pneumatic-conveyor, spray, and drum. Drying can also be accomplished with electric heaters, infrared radiation, radio frequency and microwave radiation, and also from the frozen state by freeze-drying.

4. Psychrometry, which deals with the properties of air–water mixtures and other gas–moisture systems, is useful for making drying calculations. Psychrometric

(humidity) charts are used for obtaining the temperature at which surface moisture evaporates.

5. For the air–water system, the adiabatic-saturation temperature and the wet-bulb temperature are, by coincidence, almost identical. Thus, surface moisture is evaporated at the wet-bulb temperature. This greatly simplifies drying calculations.

6. Most wet solids can be grouped into one of two categories. Granular or crystalline solids that hold moisture in open pores between particles can be dried to very low moisture contents. Fibrous, amorphous, and gel-like materials that dissolve moisture or trap it in fibers or very fine pores can be dried to low moisture contents only with a gas of low humidity. The second category of materials can exhibit a significant equilibrium-moisture content that depends on temperature, pressure, and humidity of the gas.

7. For drying calculations, moisture content of a solid and a gas is usually based on the bone-dry solid and bone-dry gas. The bound-moisture content of a material in contact with a gas is the equilibrium-moisture content when the gas is saturated with the moisture. The excess-moisture

content is the unbound-moisture content. When a gas is not saturated, excess moisture above the equilibrium-moisture content is the free-moisture content. Solid materials that can contain bound moisture are hygroscopic. Bound moisture can be held chemically as water of hydration.

8. Drying by direct heat often takes place in four periods. The first is a preheat period accompanied by a rise in temperature but with little moisture removal. This is followed by a constant-rate period, during which surface moisture is evaporated at the wet-bulb temperature. This moisture may be originally on the surface or moisture brought rapidly to the surface by diffusion or capillary action. The third period is a falling-rate period, during which the rate of drying decreases linearly with time with little change in temperature. A fourth period may occur when the rate of drying falls off exponentially with time and the temperature rises.

9. Drying rate in the constant-rate period is governed by the rate of heat transfer from the gas to the surface of the solid. Empirical expressions for the heat-transfer coefficient are available for different types of direct-heat dryers.

10. The drying rate in the falling-rate period can be determined by using empirical expressions with experimental data. Diffusion theory can be applied in some cases when moisture diffusivity is available or can be measured.

11. For direct-heat dryer models, material and energy balances are used to determine rates of heat transfer from the gas to the wet solid, and the gas flow rate.

12. A useful model for a two-zone belt dryer with through-circulation describes the changes in solids-moisture content both vertically through the bed and in the direction of belt travel.

13. A model for preliminary sizing of a direct-heat rotary dryer is based on the use of a volumetric heat-transfer coefficient, assuming that the gas flows through curtains of cascading solids.

14. A model for sizing a large fluidized-bed dryer is based on the assumption of perfect solids mixing in the dryer when operating in the bubbling-fluidization regime. The procedure involves taking drying-time data from batch operation of a laboratory fluidized-bed dryer and correcting it for the expected solid-particle-residence-time distribution in the large dryer.

15. Many bioproducts are thermolabile and thus require careful selection of a suitable dryer. Most popular are fluidized-bed dryers, spray dryers, and freeze-dryers.

REFERENCES

1. *Handbook of Industrial Drying*, 2nd ed., A.S. Mujumdar, Ed., Marcel Dekker, New York (1995).

2. *Perry's Chemical Engineers' Handbook*, 8th ed., D.W. Green and R.H. Perry, Eds., McGraw-Hill, New York (2008).

3. Walas, S.M., *Chemical Process Equipment*, Butterworths, Boston (1988).

4. van't Land, C.M., *Industrial Drying Equipment*, Marcel Dekker, New York (1991).

5. Uhl, V.W., and W.L. Root, *Chem. Eng. Progress*, **58**, 37–44 (1962).

6. McCormick, P.Y., in *Encyclopedia of Chemical Technology*, 4th ed., John Wiley & Sons, New York, Vol. 8, pp. 475–519 (1993).

7. Keey, R.B., *Introduction to Industrial Drying Operations*, Pergamon Press, Oxford (1978).

8. Lewis, W.K., *Mech. Eng.*, **44**, 445–446 (1922).

9. Faust, A.S., L.A. Wenzel, C.W. Clump, L. Maus, and L.B. Anderson, *Principles of Unit Operations*, John Wiley & Sons, New York (1960).

10. Luikov, A.V., *Heat and Mass Transfer in Capillary-Porous Bodies*, Pergamon Press, London (1966).

11. Sherwood, T.K., *Ind. Eng. Chem.*, **21**, 12–16 (1929).

12. Sherwood, T.K., *Ind. Eng. Chem.*, **21**, 976–980 (1929).

13. Marshall, W.R., Jr., and O.A. Hougen, *Trans. AIChE*, **38**, 91–121 (1942).

14. Gamson, B.W., G. Thodos, and O.A. Hougen, *Trans. AIChE*, **39**, 1–35 (1943).

15. Wilke, C.R., and O.A. Hougen, *Trans. AIChE*, **41**, 445–451 (1945).

16. Hougen, O.A., H.J. McCauley, and W.R. Marshall, Jr., *Trans. AIChE*, **36**, 183–209 (1940).

17. Carslaw, H.S., and J.C. Jaeger, *Heat Conduction in Solids*, 2nd ed., Oxford University Press, London (1959).

18. Newman, A.B., *Trans. AIChE*, **27**, 310–333 (1931).

19. Sherwood, T.K., *Ind. Eng. Chem.*, **24**, 307–310 (1932).

20. Gilliland, E.R., and T.K. Sherwood, *Ind. Eng. Chem.*, **25**, 1134–1136 (1933).

21. Walker, W.H., W.K. Lewis, W.H. McAdams, and E.R. Gilliland, *Principles of Chemical Engineering*, 3rd ed., McGraw-Hill, New York (1937).

22. Ceaglske, N.H., and F.C. Kiesling, *Trans. AIChE*, **36**, 211–225 (1940).

23. Keey, R.B., *Drying Principles and Practice*, Pergamon Press, Oxford (1972).

24. Genskow, L.R., Ed., *Scale-Up of Dryers, in Drying Technology*, **12**(1, 2), 1–416 (1994).

25. Thygeson, J.R., Jr., and E.D. Grossmann, *AIChE Journal*, **16**, 749–754 (1970).

26. Matchett, A.J., and M.S. Sheikh, *Trans. Inst. Chem. Engrs.*, **68**, Part A, 139–148 (1990).

27. McCormick, P.Y., *Chem. Eng. Progress*, **58**(6), 57–61 (1962).

28. Schofield, F.R., and P.G. Glikin, *Trans. Inst. Chem. Engrs.*, **40**, 183–190 (1962).

29. Langrish, T.A.G., R.E. Bahu, and D. Reay, *Trans. Inst. Chem. Engrs.*, **69**, Part A, 417–424 (1991).

30. Ergun, S., *Chem. Eng. Progr.*, **48**, (2), 89–94 (1952).

31. Fogler, H.S., *Elements of Chemical Reaction Engineering*, 3rd ed. Prentice-Hall, Upper Saddle River, NJ (1999).

32. *Handbook of Industrial Drying*, 3rd ed., A.S. Mujumdar, Ed., Taylor and Francis, Boca Raton, FL (2007).

33. Afdull-Fattah, A.M., D.S. Kalonia, and M.J. Pikal, *J. of Pharmaceutical Sciences*, **96**(8) 1886–1916 (2007).

34. Tang, X., and M.J. Pikal, *Pharmaceutical Research*, **21**(2) 191–200 (2004).

35. Heinzle, E., A.P. Biwer, and C.L. Cooney, *Development of Sustainable Bioprocesses*, John Wiley & Sons, England (2006).

36. Chua, K.J., A.S. Mujumdar, and S.K. Chou, *Bioresource Technology*, **90**, 285–295 (2003).

STUDY QUESTIONS

18.1. What are the most commonly employed modes of heat transfer for drying? Does the temperature of the solid during drying depend on the mode?

18.2. Why is there such a large variety of drying equipment?

18.3. What is the difference between a direct-heat dryer and an indirect-heat dryer?

18.4. For what types of wet solids can fluidized-bed, spouted-bed, and pneumatic-conveyor dryers be used?

18.5. What is freeze-drying and when is it a good choice?

18.6. What is psychrometry?

18.7. What are the differences among absolute humidity, relative humidity, and percentage humidity?

18.8. What is the wet-bulb temperature? How is it measured? How does it differ from the dry-bulb temperature?

18.9. What is the adiabatic-saturation temperature? Why is it almost identical to the wet-bulb temperature for the air–water system, but not for other systems?

18.10. Under what drying conditions is the moisture-evaporation temperature equal to the wet-bulb temperature?

18.11. Distinguish among total-moisture content, free-moisture content, equilibrium-moisture content, unbound moisture, and bound moisture.

18.12. What are the different periods that may occur during a drying operation and under what conditions do they occur?

18.13. What is the critical moisture content?

18.14. What are the two most applied theories to the falling-rate drying period?

18.15. In the dryer models for a belt dryer with through-circulation and a direct-heat rotary dryer, is the rate of drying based on heat transfer or mass transfer? Why?

18.16. What are the regimes of fluidization of a bed of particles by a gas? What regime of operation is preferred for drying?

18.17. When selecting a dryer type, why do bioproducts require special considerations?

EXERCISES

Section 18.1 (Use of the Internet is encouraged for the exercises of this section.)

18.1. **Continuous dryer selection.**

The surface moisture of 0.5-mm average particle size NaCl crystals is to be removed in a continuous, direct-heat dryer without a significant change in particle size. What types of dryers would be suitable? How high could the gas feed temperature be?

18.2. **Batch-dryer selection.**

A batch dryer is to be selected to dry 100 kg/h of a toxic, temperature-sensitive material (maximum of 50°C) of an average particle size of 350 μm. What dryers are suitable?

18.3. **Dryer selection for a milky liquid.**

A thin, milk-like liquid is to be dried to produce a fine powder. What types of continuous, direct-heat dryers would be suitable? The material should not be heated above 200°C.

18.4. **Dryer selection for different feeds.**

The selection of a batch or continuous dryer is determined largely by feed condition, temperature-sensitivity of the material, and the form of the dried product. Select types of batch and continuous dryers that would be suitable for the following cases: (a) A temperature-insensitive paste that must be maintained in slab form. (b) A temperature-insensitive paste that can be extruded. (c) A temperature-insensitive slurry. (d) A thin liquid from which flakes are to be produced. (e) Pieces of lumber. (f) Pieces of pottery. (g) Temperature-insensitive inorganic crystals for which particle size is to be maintained and only surface moisture is to be removed. (h) Orange juice to produce a powder.

18.5. **Solar drying for organic materials.**

Solar drying has been used for centuries to dry, and thus preserve, fish, fruit, meat, plants, spices, seeds, and wood. What are the advantages and disadvantages of this type of drying? What other types of dryers can be used to dry such materials? What type of dryer would you select to continuously dry beans?

18.6. **Advantages of fluidized-bed dryers.**

Fluidized-bed dryers are used to dry a variety of vegetables, including potato granules, peas, diced carrots, and onion flakes. What are the advantages of this type of dryer for these types of materials?

18.7. **Production of powdered milk.**

Powdered milk can be produced from liquid milk in a three-stage process: (1) vacuum evaporation in a falling-film evaporator to a high-viscosity liquid of less than 50 wt% water; (2) spray drying to 7 wt% moisture; and (3) fluidized-bed drying to 3.6 wt% moisture. Give reasons why this three-stage process is preferable to a single-stage process involving just spray drying.

18.8. **Drying pharmaceutical products.**

Deterioration must be strictly avoided when drying pharmaceutical products. Furthermore, such products are often produced from a nonaqueous solvent such as ethanol, methanol, acetone, etc. Explain why a closed-cycle spray dryer using nitrogen is frequently a good choice of dryer.

18.9. **Drying of paper.**

Paper is made from a suspension of fibers in water. The process begins by draining the fibers to a water-to-fiber ratio of 6:1, followed by pressing to a 2:1 ratio. What type of dryer could then be used to dry a continuous sheet to an equilibrium-moisture content of 8 wt% (dry basis)?

18.10. **Importance of drying green wood.**

Green wood contains from 40 to 110 wt% moisture (dry basis) and must be dried before use to just under its equilibrium-moisture content when in the final environment. This moisture content is usually in the range from 6 to 15 wt% (dry basis). Why is it important to dry the wood, and what is the best way to do it so as to avoid distortion, cracks, splits, and checks?

18.11. Drying of wet coal.

Wet coal is usually dried to a moisture content of less than 20 wt% (dry basis) before being transported, briquetted, coked, gasified, carbonized, or burned. What types of direct-heat dryers are suitable for drying coal? Can a spouted-bed dryer be used? If air is used as the heating medium, is there a fire and explosion hazard? Could superheated steam be used as the heating medium?

18.12. Drying of coated paper, films, tapes, and sheets.

Drying is widely used to remove solvents from coated webs, which include coated paper and cardboard, coated plastic films and tapes (e.g., photographic films and magnetic tapes), and coated metallic sheets. The coatings may be water-based or other-solvent-based. Solid coatings are also used. Typical coatings are 0.1 mm in wet thickness. Much of the drying time is spent in the falling-rate period, where the rate of drying decreases in an exponentially decaying fashion with time. What types of dryers can be used with coated webs? Are infrared dryers a possibility? Why or why not?

18.13. Drying of polymer beads.

A number of polymers, including polyvinylchloride, polystyrene, and polymethylmethacrylate, are made by suspension or emulsion polymerization, in which the product is finely divided, solvent- or water-wet beads. For large production rates, direct-heat dryers are commonly used with air, inert gas, or superheated steam as the heating medium. Why are rotary dryers, fluidized-bed dryers, and spouted-bed dryers popular choices for drying polymer beads?

18.14. Air and superheated steam as heating media.

What are the advantages and disadvantages of superheated steam compared to air as the heating medium? Why might superheated steam be superior to air for the drying of lumber?

Section 18.2

18.15. Psychrometric chart and equations.

A direct-heat dryer is to operate with air entering at 250°F and 1 atm with a wet-bulb temperature of 105°F. Determine from the psychrometric chart and/or relationships of Table 18.3: (a) humidity; (b) molal humidity; (c) percentage humidity; (d) relative humidity; (e) saturation humidity; (f) humid volume; (g) humid heat; (h) enthalpy; (i) adiabatic-saturation temperature; and (j) mole fraction of water in the air.

18.16. Psychrometric chart and equations.

Air at 1 atm, 200°F, and a relative humidity of 15% enters a direct-heat dryer. Determine the following from the psychrometric chart and/or relationships of Table 18.3: (a) wet-bulb temperature; (b) adiabatic-saturation temperature; (c) humidity; (d) percentage humidity; (e) saturation humidity; (f) humid volume; (g) humid heat; (h) enthalpy; and (i) partial pressure of water in the air.

18.17. Psychrometric chart and equations.

Repeat Example 18.1 with the air at 1.5 atm instead of 1.0 atm.

18.18. Humidity for *n*-hexane/N₂.

n-hexane is being evaporated from a solid with nitrogen gas. At a point in the dryer where the gas is at 70°F and 1.1 atm, with a relative humidity for hexane of 25%, determine: (a) partial pressure of hexane at that point; (b) humidity of the nitrogen–hexane mixture; (c) percentage humidity of the nitrogen–hexane mixture; and (d) mole fraction of hexane in the gas.

18.19. Humidity for toluene and air.

At a location in a dryer for evaporating toluene from a solid with air, the air is at 180°F, 1 atm, and a relative humidity of 15%. Determine humidity, adiabatic-saturation temperature, wet-bulb temperature, and the psychrometric ratio.

18.20. Wet-solid temperature profile.

Repeat Example 18.5 for water only, with air entering at 180°F and 1 atm, with a relative humidity of 15%, for an exit temperature of 120°F. Plot temperatures through the dryer.

18.21. Wet-bulb temperature of high-temperature air.

Air enters a dryer at 1,000°F with a humidity of 0.01 kg H_2O/kg dry air. Determine the wet-bulb temperature if the air pressure is: (a) 1 atm, (b) 0.8 atm, (c) 1.2 atm.

18.22. Drying of paper with two dryers.

Paper is being dried with recirculating air in a two-stage drying system at 1 atm. Air enters the first dryer at 180°F, where it is adiabatically saturated with moisture. The air is then reheated in a heat exchanger to 174°F before entering the second dryer, where it is adiabatically humidified to 80% relative humidity. The air is then cooled to 60°F in a second heat exchanger, causing some moisture to be condensed. This is followed by a third heater to heat the air to 180°F before it returns to the first dryer. (a) Draw a process-flow diagram of the system and enter all of the given data. (b) Determine the lb H_2O evaporated in each dryer per lb of dry air being circulated. (c) Determine the lb H_2O condensed in the second heat exchanger per lb of dry air circulated.

18.23. Dehumidification of air.

Before being recirculated to a dryer, air at 96°F, 1 atm, and 70% relative humidity is to be dehumidified to 10% relative humidity. Cooling water is available at 50°F. Determine a method for carrying out the dehumidification, draw a labeled flow diagram of your process, and calculate the cooling-water requirement in lb H_2O per lb of dry air circulated.

Section 18.3

18.24. Drying of nitrocellulose fibers.

Nitrocellulose fibers with an initial total water content of 40 wt% (dry basis) are dried in trays in a tunnel dryer operating at 1 atm. If the fibers are brought to equilibrium with air at 25°C and a relative humidity of 30%, determine the kg of moisture evaporated per kg of bone-dry fibers. The equilibrium-moisture content of the fibers is in Figure 18.24.

18.25. Slow drying of lumber.

Wet lumber of the type in Figure 18.24 is slowly dried from an initial total moisture content of 50 wt% to a moisture content in equilibrium with atmospheric air at 25°C and 40% relative humidity. Determine: (a) unbound moisture in the wet lumber before drying in lb water/lb bone-dry lumber; (b) bound moisture in the wet lumber before drying in lb water/lb bone-dry lumber; (c) free moisture in the wet lumber before drying, referred to as the final dried lumber, in lb water/lb bone-dry lumber; (d) lb of moisture evaporated per lb of bone-dry wood.

18.26. Drying of cotton cloth.

Fifty pounds of cotton cloth containing 20% total moisture content (dry basis) is hung in a closed room containing 4,000 ft³ of air at 1 atm. Initially, the air is at 100°F at a wet-bulb temperature of 69°F. If the air is kept at 100°F, without admitting new air or venting the air, and the air is brought to equilibrium with the cotton cloth, determine the moisture content of the cotton cloth and the relative humidity of the air. Assume the equilibrium-moisture content for cotton cloth at 100°F is the same as that of glue at 25°C, as shown in Figure 18.24. Neglect the effect of the increase in air pressure, but calculate the final air pressure.

Section 18.4

18.27. Batch drying of raw cotton.

Raw cotton having an initial total moisture content of 95% (dry basis) and a dry density of 43.7 lb/ft^3 is to be dried batchwise to a final moisture content of 10% (dry basis) in a cross-circulation tray dryer. The trays, which are insulated on the bottom, are each 3 cm high, with an area of 1.5 m^2, and are completely filled. The heating medium, which is air at 160°F and 1 atm with a relative humidity of 30%, flows across the top surface of the tray at a mass velocity of 500 lb/h-ft^2. Equilibrium-moisture content isotherms for the cotton are given in Figure 18.25. Experiments have shown that under the given conditions, critical moisture content will be 0.4 lb water/lb bone-dry cotton, and the falling-rate drying period will be like that of Figure 18.31a, based on free-moisture content. Determine: (a) amount of raw cotton in pounds (wet basis) that can be dried in one batch if the dryer contains 16 trays; (b) drying time for the constant-rate period; (c) drying time for the falling-rate period; (d) total drying time if the preheat period is 1 h.

18.28. Batchwise drying of filter cake.

Slabs of filter cake with a bone-dry density of 1,600 kg/m^3 are to be dried from an initial free-moisture content of 110% (dry basis) to a final free-moisture content of 5% (dry basis) batchwise in trays that are 1 m long by 0.5 m wide, with a depth of 3 cm. Drying will take place from the top surface only. The drying air conditions are 1 atm, 160°C, and a 60°C wet-bulb temperature. The air velocity across the trays is 3.5 m/s. Experiments under these drying conditions show a critical free-moisture content of 70% (dry basis), with a falling-rate period like that of Figure 18.31a, based on free-moisture content. Determine: (a) drying time for the constant-rate period; (b) drying time for the falling-rate period.

18.29. Batchwise drying of extrusions.

The filter cake of Exercise 18.28 is extruded into cylindrical-shaped pieces measuring 1/4 inch in diameter by 3/8 inch long that are placed in trays that are 6 cm high × 1 m long × 0.5 m wide and through which the air passes. The external porosity is 50%. If the superficial air velocity, at the same conditions as in Exercise 18.28, is 1.75 m/s, determine: (a) drying time for the constant-rate period; (b) time for the falling-rate period.

18.30. Tray drying.

It takes 5 h to dry a wet solid, contained in a tray, from 36 to 8% moisture content, using air at constant conditions. Additional experiments give critical- and equilibrium-moisture contents of 15% and 5%, respectively. If the length of the preheat period is negligible and the falling-rate period is like that of Figure 18.31a, determine, for the same conditions, drying time if the initial moisture content is 40% and a final moisture content of 7% is desired. All moisture contents are on the dry basis.

18.31. Tunnel drying.

A tunnel dryer is to be designed to dry, by crossflow with air, a wet solid that will be placed in trays measuring 1.5 m long × 1.2 m wide × 25 cm deep. Drying will be from both sides. The initial total moisture content is 116% (dry basis) and the desired final average moisture content is 10% (dry basis). Air is at 90°F and 1 atm with a relative humidity of 15%. The laboratory data below were obtained under the same conditions:

Determine the time needed to dry the solid from 110% to 10% moisture content (dry basis) if the air conditions are changed to 125°F and 20% relative humidity. Assume the critical moisture content will not change and that the drying rate is proportional to the difference between dry- and wet-bulb air temperatures.

Equilibrium-Moisture Content								
% relative humidity	10	20	30	40	50	60	70	90
Moisture content, % (dry basis)	3.0	3.2	4.1	4.8	5.4	6.1	7.2	10.7

Drying Test		Drying Test	
Time, min	Moisture content, % (dry basis)	Time, min	Moisture content, % (dry basis)
0	116	362	31.4
36	106	415	28.6
125	81	465	24.8
194	61.8	506	22.8
211	57.4	601	15.4
242	49.6	635	13.5
277	42.8	785	11.4
313	37.1	822	10.2

18.32. Drying time for wood.

A piece of hemlock wood measuring 15.15 × 14.8 × 0.75 cm is to be dried from the two large faces from an initial total moisture content of 90% to a final average total moisture content of 10% (both dry basis), for drying taking place in the falling-rate period with liquid-diffusion controlling. The moisture diffusivity has been experimentally determined as 1.7×10^{-6} cm^2/s. Estimate the drying time if bone-dry air is used.

18.33. Drying mode in the falling-rate period.

Gilliland and Sherwood [20] obtained data for the drying of a water-wet piece of hemlock wood measuring 15.15 × 14.8 × 0.75 cm, where only the two largest faces were exposed to drying air, which was at 25°C and passed over the faces at 3.7 m/s. The wet-bulb temperature of the air was 17°C and pressure was 1 atm. The data below were obtained for average moisture content (dry basis) of the wood as a function of time. From these data, determine whether Case 1 or Case 2 for the diffusion of moisture in solids applies. If Case 1 is chosen, determine the effective diffusivity; if Case 2, determine: (a) drying rate in g/h-cm^2 for the constant-rate period, assuming a wood density of 0.5 g/cm^3 (dry basis) and no shrinkage upon drying; (b) critical moisture content; (c) predicted parabolic moisture-content profile at the beginning of the falling-rate period; (d) effective diffusivity during the falling-rate period. In addition, for either case, describe what else could be determined from the data and explain how it could be verified.

	Avg. Moisture		Avg. Moisture
Time, h	Content, % (dry basis)	Time, h	Content, % (dry basis)
0	127	9	41.8
1	112	10	38.5
2	96.8	12	30.8
3	83.5	14	26.4
4	73.6	16	20.9
5	64.9	18	16.5
6	57.2	20	14.3
7	51.7	22	12.1
8	46.1	∞	6.6

18.34. Falling-rate period equations.

When Case 1 of liquid diffusion is controlling during the falling-rate period, the time for drying can be determined from (3) in Example 18.13. Using that equation, derive an equation for the rate of drying to show that it varies inversely with the square of the solid thickness. If capillary movement controls the falling-rate period, an equation for the rate of drying can be derived by assuming laminar flow of moisture from the solid's interior to the surface, so the rate of drying varies linearly with the average free-moisture content. If this is the case, derive equations for the rate of drying and the time for drying in the falling-rate period to show that the rate of drying is inversely proportional to just the thickness of the solid. Outline an experimental procedure to determine whether diffusion or capillary flow govern in a given material.

18.35. Cross-circulation tray drying.

In a cross-circulation tray dryer, equations for the constant-rate period neglect radiation and assume the bottoms of the trays are insulated, so heat transfer takes place only by convection from the gas to the surface of the solid. Under these conditions, surface evaporation occurs at the wet-bulb temperature of the gas (when the moisture is water). In actual tray dryers, the bottoms of the trays are not insulated and heat transfer can take place by convection from the gas to the tray bottom and thence by conduction through the tray and the wet solid. Derive an equation similar to (18-34) where heat transfer by convection and conduction from the bottom is taken into account. Assume that the tray-bottom conduction resistance can be neglected. Show by combining your equation with the mass-transfer equation (18-35) that evaporation now takes place at a temperature higher than the gas wet-bulb temperature. What effect would heat transfer by radiation from the bottom surface of a tray to the tray below have on the evaporation temperature?

Section 18.5

18.36. Tunnel drying of raw cotton.

A tunnel dryer is to be used to dry 30 lb/h of raw cotton (dry basis) with a countercurrent flow of 1,800 lb/h of air (dry basis). Cotton enters at 70°F with a moisture content of 100% (dry basis) and exits at 150°F with 10% moisture (dry basis), and air enters at 200°F and 1 atm with a relative humidity of 10%. The specific heat of dry cotton can be taken constant at 0.35 Btu/lb-°F. Calculate: (a) rate of evaporation of moisture; (b) outlet temperature of the air; (c) rate of heat transfer.

18.37. Spray drying of an aqueous coffee.

A 25 wt% solution of coffee in water at 70°F is spray-dried to a moisture content of 5% (dry basis) with air entering at 450°F and 1 atm with a humidity of 0.01 lb/lb (dry basis) and exiting at 200°F. Assuming the specific heat of coffee = 0.3 Btu/lb-°F, calculate: (a) air rate in lb dry air/lb coffee solution; (b) temperature of evaporation; (c) heat-transfer rate in Btu/lb coffee solution.

18.38. Flash drying of clay particles.

Seven thousand lb/h of wet, pulverized clay particles with 27% moisture (dry basis) at 15°C and 1 atm enter a flash dryer, where they are dried to a moisture content of 5% (dry basis) with a cocurrent flow of air that enters at 525°C. The dried solids exit at the air wet-bulb temperature of 50°C, while the air exits at 75°C. Assuming

the specific heat of clay = 0.3 Btu/lb-°F, calculate: (a) flow rate of air in lb/h (dry basis); (b) rate of evaporation of moisture; (c) heat-transfer rate in Btu/h.

18.39. Drying of isophthalic acid crystals.

Five thousand lb/h of wet isophthalic acid crystals with 30 wt% moisture (wet basis) at 30°C and 1 atm enter an indirect-heat steam-tube rotary dryer, where they are dried to a moisture content of 2 wt% (wet basis) by 25 psig steam (14 psia barometer) condensing inside the tubes. Evaporation takes place at 100°C, which is also the exit temperature of the crystals. The specific heat of isophthalic acid is 0.2 cal/g-°C. Determine: (a) rate of evaporation of moisture; (b) rate of heat transfer; (c) quantity of steam required in lb/h.

18.40. Through-circulation belt dryer with three zones.

The extruded filter cake of Examples 18.8 and 18.18 is to be dried under the same conditions as in Example 18.18, except that three drying zones 8 ft long each will be used, with flow upward in the first and third zones and downward in the second. Predict the moisture-content distribution with height at the end of each zone and the final average moisture content.

18.41. Through-circulation belt dryer with two zones.

Repeat the calculations of Example 18.18 if the extrusions are 3/8 inch in diameter × 1/2 inch long. Compare your results with those of Example 18.18 and comment.

18.42. Countercurrent-flow rotary dryer.

A direct-heat, countercurrent-flow rotary dryer with a 6-ft diameter and 60-ft length is available to dry titanium dioxide particles at 70°F and 1 atm with a moisture content of 30% (dry basis) to a moisture content of 2% (dry basis). Hot air is available at 400°F with a humidity of 0.015 lb/lb dry air. Experiments show that an air-mass velocity of 500 lb/h-ft^2 will not cause serious dusting. The specific heat of solid titanium dioxide is 0.165 Btu/lb-°F, and its true density is 240 lb/ft^3. Determine: (a) a reasonable production rate in lb/h of dry titanium dioxide (dry basis); (b) the heat-transfer rate in Btu/h; (c) a reasonable air rate in lb/h (dry basis); (d) reasonable exit-air and exit-solids temperatures.

18.43. Fluidized-bed dryer.

A fluidized-bed dryer is to be sized to dry 5,000 kg/h (dry basis) of spherical polymer beads that are uniformly 1 mm in diameter. The beads will enter the dryer at 25°C with a moisture content of 80% (dry basis). The drying medium is superheated steam at 250°C. The pressure in the vapor space above the bed will be 1 atm. A fluidization velocity of twice the minimum will be used to obtain bubbling fluidization. The bed, exit-solids, and exit-vapor temperatures can all be assumed to be 100°C. The beads are to be dried to a moisture content of 10% (dry basis). The bed void fraction before fluidization is 0.47, the specific heat of dry polymer is 1.15 kJ/kg-K, and the density is 1,500 kg/m^3. Batch fluidization experiments show that drying takes place in the falling-rate period, as governed by diffusion according to (18-92), where 50% of the moisture is evaporated in 150 s. Bed expansion is expected to be about 20%. Determine the dryer diameter, average bead residence time, and expanded bed height. Is the dryer size reasonable? If not, what changes in operation could be made to make the size reasonable? In addition, calculate the entering superheated-steam flow rate and the necessary heat-transfer rate.

Part Five

Mechanical Separation of Phases

Previous chapters of this book deal with separation of chemical species in a mixture by phase creation (distillation, drying), phase addition (absorption, extraction), transport through a barrier (membrane), addition of a solid agent (adsorption), and the imposition of a force field or gradient (electrophoresis). In previous chapters, descriptions of these processes focused on the movement of species, and heat and momentum transfer from one phase to another to achieve a processing goal. However, for many separations, transfer of species, heat, and momentum from one phase to another does not complete the process because the phases must then be disengaged. This is done using mechanical, phase separation devices such as filters, precipitators, settlers, and centrifuges, whose function and design is the subject of Chapter 19. Exceptions occur in distillation, absorption, stripping, and extraction columns, where phase separation takes place in the column.

Chapter 19

Mechanical Phase Separations

§19.0 INSTRUCTIONAL OBJECTIVES

After completing this chapter, you should be able to:

- Have a broad understanding of the entities that are responsible for air pollution.
- Understand the spectrum of particle sizes and identify the devices that could be used to separate them from the fluids in which they are suspended.
- Understand the force balances on which settling equations are based.
- Know the sizes of the particles whose behavior is described by Newton's law, Stokes' law, or Brownian motion.
- Use settling equations modified by hindered settling parameters and adjusted for centrifugal forces to design particle-fluid or microorganism-fluid separators.
- Describe a knock-out drum, coalescer, vane filter, cartridge filter, demister, baghouse, electrostatic precipitator, settling pond, plate-and-frame filter, cyclone, vacuum drum filter, and settling tank.
- Analyze filtration data to ascertain if the operation is at constant or variable rate, or constant pressure, and if the filter cake is compressible.
- Use mathematical models based on Darcy's law or flow through packed beds to predict required filter areas for drum and plate-and-frame filters.
- Understand the nature of filter aids and the utility of washing and pressing cycles.
- Know how centrifuges are designed and what they are used for.
- Describe how extracellular and intracellular bioproducts from fermentors are recovered (prior to purification).
- Understand precipitation, flocculation, and agglomeration processes, and the need for cell disruption in bioprocesses.

This chapter does not deal with separations where one or more chemicals are removed from a feed mixture; instead, it describes mechanical devices used to separate one bulk phase from another. Mundane, household examples of such devices include air conditioning and heat-pump filters to prevent dust and solid particles from clogging heat-exchange surfaces, paper filters in drip coffee makers to prevent coffee grounds from entering the brew, water filters attached to home water-supply units in locations where water quality is suspect, and electric precipitators used in homes where an occupant has serious pollen and dust allergies.

The word "filter" was derived from the Latin "filtrum," which in turn may be traced to the word "feltrum," which describes felt or compressed wool, and which in turn is further related to the Greek word for wool or hair. An Egyptian papyrus dating from the third century A.D. and known as the "Stockholm Papyrus" describes the process of producing caustic soda and the use of a filter for clarifying it, including the application of clay as a filter aid. The first patent for a filtering device was granted to Joseph Amy in 1789 by the French government. Thereafter, for the next 50 years, most filter patents pertained to treating water or sewage. The modern rotary-drum vacuum filter and pressure leaf filter were developed in the late 1800s by mining engineers, including W. J. Hart, E. Sweetland, E. L. Oliver, and J. V. N. Dorr for use in the cyanide process for recovering gold [1].

To be described in this chapter are means of:

- Removing airborne liquids, solid particles, microorganisms, and vapors from air streams when a clean, sterile-air supply is required to prevent contamination or infection of a product, process stream, or the environment.
- Separating entrained liquids from vapor streams as in a flash distillation chamber, or partial condenser.
- Designing an optimal air-purification system comprised of multiple particle-capture devices.
- Condensing vapors from air streams when downstream conditions favor an undesirable condensation.
- Eliminating pollutant particles, mists, and fogs from gases that are vented to the atmosphere from manufacturing plants.

- Removing droplets of one liquid suspended in another as in hydrocarbon-water decanters.
- Recovering, as a cake, solid particles suspended in liquids, by means of plate-and-frame, drum, leaf, and other filters; and determining cake wash cycles.
- Operating filters at constant pressure and variable rates, using pump characteristic curves.
- Designing and analyzing cyclones and centrifuges.
- Applying mechanical separations to bioprocesses: cell disruption, precipitation, and flocculation (preceded by coagulation).

This is only a sample of a plethora of applications. In a sense, the membrane processes described in Chapter 14 could also be classified as mechanical separations and included in this chapter. However, there is a major difference between the two, insofar as the design methods for most membrane devices typically involve molecular diffusion. Pressure drops are high and mass transfer is slow. The devices described in this chapter are bulk-flow units operating mostly at relatively low pressure drops, and the design equations are based on hydrodynamics involving settling velocities of macroparticles rather than molecular diffusion of individual species. Though both are "mechanical separations," there are major differences between design methods for calculating the membrane area required for a diffusing molecular flux, and the demister area needed to retain dust particles or droplets, or filtration areas for pressure- or vacuum-driven devices for solid–liquid filtration. In solid–liquid filtration the screen is not the filter; the particles form a "cake" and this, rather than the screen or fabric, is the filter. This situation is quite different from that when a membrane is the filter, as in Chapter 14.

An important aspect of filtration is that particles suspended in the liquid or gas are retained, as in a "drip grind" coffee maker, where the grounds are retained on the filter medium. Periodic "blowback," scraping, or other particle-removal methods are required; otherwise, both the filter and the retained particles must be disposed of or processed. If the concentration of particulate matter in the gas or solution is high, large amounts of solids must be removed. For large quantities of inexpensive industrial liquids and wastes, use of filters is prohibitively expensive, so the liquids are placed in retention (holding) tanks or ponds, where the particles are allowed to settle, often with the help of coagulants, settling agents, and mild, directed stirring to speed the settling. If the particle is an industrial product, it is not processed in a retention pond or settler. What is used instead is a filter press that can handle a high concentration of particles in the 10–50 micron range inexpensively and in large volumes. Here, pressure or vacuum drive a solution through fabrics or screens, frequently precoated with "filter aids." As the retained particles accumulate on the screen, they form a "cake," which then becomes the filter.

This chapter has its own vocabulary. To be encountered in the upcoming pages are settlers, decanters, coalescers, vanes, centrifuges, demisters, knock-out drums, electrostatic precipitators, mesh pads, cyclones, impingement separators, bag filters, and drum, plate-and-frame, and vacuum filters. Design methods and applications for these devices are unique in the sense that each is designed for a specific purpose and a specific range of particle sizes. A device whose design was based on inertial impingement would be inadequate for an application involving particles whose hydrodynamic behavior is characterized by Brownian motion. As a rule, preliminary selection of a specific mechanical separation device is based on particle size and phase. After the device is selected, laboratory and/or pilot-plant data are analyzed to establish values of empirical constants in the design equations used to size the plant unit. The design variables include particle size and density; fluid velocity, density, and viscosity; the external force field; and device parameters. Except for vacuum and drum filters and centrifugal units, the design equations for the devices are largely based on settling velocities predicted by Newton's gravitational law or Stokes' law.

Industrial Example

The problem of producing enormous quantities of sterile air for aerobic fermentation exists only in biochemical engineering. In the 1960s, cotton plugs were satisfactory for use in test tubes or shake flasks. For pilot-plant fermentors, small fibrous filters were deemed satisfactory. For plant applications, cotton fibers and activated carbon were standard, but glass fibers were recognized as being a better filter medium, and 6-foot-deep beds containing glass fibers 5–19 microns (μm) in diameter came to be in use. The sterilization process was aided and abetted by inefficient air compressors, which heated the air to about 150°C (which is not close enough to the 220°C at 5 minutes required to kill bacterial spores). The pressure drop through the various filters, plus the spargers and 20-or-more-feet-high fermentors, is well over 1 atm, so today, expensive compressors or blowers, rather than fans, are used. These must be protected from damage by solid particles and vapor droplets in the inlet air by suitable means, which include knock-out drums and coalescent filters. A serious source of contamination is the oil mist and oil/water emulsions emitted by compressors.

As discussed by Shuler and Kargi [2], one 50,000-liter aerobic fermentor, during a 5-day fermentation, requires about 2×10^8 liters of absolutely sterile air. Because banks of 10 or more 100,000-liter fermentors may be housed in one building, sterilization processes such as UV radiation, steam, ozone, or scrubbers are not economical. Minimization of pressure drop is critical, as is dependable protection of the high-value product in the fermentors. Not only must the air entering be sterile; if the fermentation involves pathogens or recombinant DNA, the concentration of microorganisms in the exiting gas, which is far higher than that in the inlet gas, must also be reduced to zero. Catalytic combustion has been used to accomplish this, but membrane products, which are much less expensive, have made serious inroads into the market.

Particulate concentrations in air streams vary widely. Populations of microorganisms, which vary from 0.5 to 1 μm in size, have been measured in all parts of the world and range from 1 to 10 per liter, which is $10^3–10^4/m^3$. Because of their small size, this translates into only about 10^{-8} mg/m^3, which is relatively insignificant compared to the dust, vapor, and oil loadings. Reasonable numbers for concentrations of particles in city air are 35 mg/m^3, which is about $500 × 10^6$ particles/m^3, 80% of which cannot be seen by the human eye [3]. The haze that one sees when flying over any large city in the world is not usually visible from the street. Additional problems arise from the concentration of water vapor in the air, which is highly variable, as are the fumes emitted by automobiles.

Air compressors are housed in a building, but provisions must be made for ambient fog, rain, snow, and possible airborne construction-site debris. The sterile supply of air is critical, and although air-purification units upstream of the compressors can be serviced, no downstream maintenance is feasible because of the possibility of contamination. Air cleanup after compression is required because of the oil mist and water/oil emulsions emitted by compressors, which can be as much a source of contamination as the entering air stream. Air-filtering devices downstream of the compressors should be cleanable by blowback, and there must be a certain amount of redundancy. An ever-present danger is wetting of the filters, which increases pressure drop and provides paths for pollutant short circuits.

Historically, *depth filters* with glass–wool fibers akin to building insulation were used, but these have been almost totally replaced with membrane or cartridge surface filters of the type described in Chapter 14 and later in this chapter. Possible mechanisms for capture of bacteria, about 1 μm in size, are direct interception, electrostatic effects, and inertial effects. In depth filters, as the gas flows in streamlines around the fibers, particles with sufficient mass will, because of an inertial effect, maintain a straight-line trajectory and be captured by the fiber. Brownian motion is unimportant for bacteria, but for viruses, which are smaller than bacteria, it is important. In the case of surface filters used for sterilizing air, sieving effects play a major role.

Figure 19.1 shows a cyclone and compressor air-intake filters. The cyclone removes particles above 10 μm with high efficiency and will prolong the life of the suction-line filter, which may be a simple panel or cartridge filter but more

likely is a more modern coalescing filter, possibly preceded by a vane impingement device. Most viruses, bacteria, vapors, odors, and submicron particles pass these filters. Water vapor will generally pass, and will not condense in the compressor because of the adiabatic temperature rise, but it will condense when the compressed gas is cooled.

Downstream of the compressor is an aftercooler, a prefilter to drain condensate, an adsorption dryer, and a high-efficiency HEPA filter to reduce microorganisms to a level of 100 particles/m^3, which is standard for a sterile work area. The prefilter will generally be a two-stage, self-draining, coalescing glass-fiber device specifically designed to reduce oil carry over to 0.001 mg/m^3. The dryer can be activated alumina or zeolite, and if oil vapor is a potential problem, it may be backed up with an activated carbon adsorber to remove hydrocarbon gases. All of these filtration units are subject to governmental and industry standards and performance tests.

§19.1 SEPARATION-DEVICE SELECTION

Separation-device selection is based largely on the size of the particles carried by the fluid. Other considerations such as density, viscosity, particle concentration, and flow rate also enter into the selection process, as do particle and fluid dollar value and the device particle-capture efficacy and cost, but they are secondary. Furthermore, as will be seen shortly, mathematical models for many filtration devices are based on particle settling velocities, and these are based on hydrodynamic equations in which particle size is a key variable. The lists in Tables 19.1 and 19.2 were compiled from manufacturers' product bulletins and various other sources, many of which differ considerably, and must be viewed as a preliminary guide to the selection process.

Helpful guidelines to the interpretation of the particle-size data in Table 19.1 are the entries regarding the limit of visibility and the size of a human hair. It is often surprising to see the amount of condensate dripping out of a coalescer when the air entering the coalescer seemed clear of mist. The range of particle sizes given in Table 19.1 is indicative not only of the fact that the sizes of the particles in atmospheric fog vary from day to day and location to location, but also that within a given fog, the particles have different sizes. This is important because the particle-capture efficiency for the devices listed is a function of particle size. American Filter Co. Inc., for example, distributes product literature that states that their high-efficiency cyclone has a 50% efficiency for

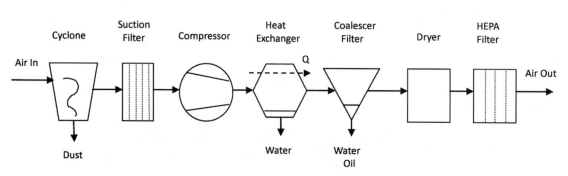

Figure 19.1 Air-purification system.

Table 19.1 Typical Particle Sizes

Particle	Size, μm
Large molecules	0.001–0.004
Smoke	0.005–1
Fume	0.01–0.1
Tobacco smoke	0.01–0.12
Smog	0.01–1
Virus	0.03–0.1
Mist	0.1–10
Fog	0.1–30
Spores	0.5–1.80
Bacteria	0.5–10
Prokaryotic cells	1–10
Dust	1–100
Limit of visibility	10–40
Liquid slurries	10–50
Eukaryotic cells	10–100
Drizzle	10–400
Spray	10–1000
Pollen	20–80
Mist	50–100
Human hair	50–200
Rain	100–1,400
Heavy industrial dust	100–5,000

Table 19.2 Particle-Size Ranges for Particle-Capture Devices

Particle-Capture Device	Size Range, μm
Membranes	0.00001–0.0001
Ultracentrifuges	0.001–1
Electrical precipitators	0.002–20
Centrifuge	0.05–5
Cloth collectors	0.05–500
Fiber panels and candles	0.10–10,000
Elutriation	1–100
Air filters	2–50
Centrifugal separators	2–1,000
Impingement separators	5–2,000
Vane arrays	5–10,000
Cyclones (high efficiency)	6–35
Filter presses	10–50
Cyclones (low efficiency)	15–250
Cloth and fibers	20–1000
Gravity sedimentation	45–10,000
Screens and strainers	50–1,000
Sieving screens	50–20,000

capturing a 6-μm particle, and a 99% efficiency for capturing a 25-μm particle.

The nomenclature in this field is far from standardized. The term *aerosol*, for example, is used to describe suspended liquid or solid particles that are slow to settle, be they sub-micron or 50 μm in size. Mists are generally described as particles upward of 0.1 μm in size that arise because of vapor condensation. Sprays are the result of intentional or unintentional atomization processes.

In developing a flowsheet for a particle-collection system, it is well to remember the strongest of the process design heuristics: "Cheapest first." In terms of the devices listed in Table 19.2, this means removing large particles by inexpensive settling chambers, vane arrays, or impingement devices, and then removing the small amount of remaining particles with the higher-capital-cost units like membranes, centrifuges, or electric precipitators.

§19.2 INDUSTRIAL PARTICLE-SEPARATOR DEVICES

The operative mechanisms for the particle separators to be described are: (1) gravity settling, where the force field is elevation; (2) inertial (including centrifugal) impaction, where the force field is a velocity gradient; (3) flow-line (direct) interception or impingement, where the particle is assumed to have size, but no mass, and follows a streamline; (4) diffusional (Brownian) deposition, where the force field is a concentration gradient; (5) electrostatic attraction due to an electric-field gradient; (6) agglomeration by particle–particle collisions; and (7) sieving, where the flow pathway is smaller than the particle. Mechanisms 2–4 are depicted in Figure 19.2. Note that in interception the particle follows the streamline, while in impaction it follows a direct path. Generally, devices that operate by a combination of mechanisms 2 and 3 combine impaction and interception in one empirical design equation. In many devices, synergistic mechanisms are used. In cyclones, for example, gravity settling is abetted by centrifugal force. A generic consideration in collection devices is the problem of re-entrainment. The inertial forces that deposit a particle on a fiber can also blow the particle off the fiber. Cyclones, for example, are more efficient for liquid droplets than for solid particles because droplets are more likely to coalesce and agglomerate at the bottom than are solid particles.

§19.2.1 Gravity Settlers

If the velocity of the carrier fluid is sufficiently low, all particles whose density is above that of the carrier will eventually settle. Terminal velocities of droplets and solid particles are such that the required size of the settling chamber usually

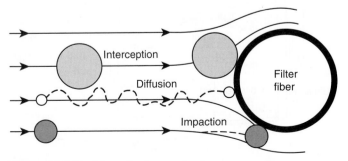

Figure 19.2 Particle-collection mechanisms.

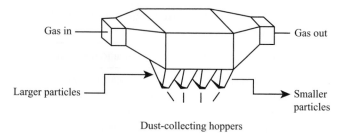

Figure 19.3 Horizontal settling chamber.

becomes excessive for droplets smaller than 50 μm and for solid dusts smaller than 40 μm. For solid particles, air velocities greater than 10 ft/s lead to re-entrainment of all but the heaviest particles.

In the horizontal settling chamber of Figure 19.3, the gas velocity, upon entering the chamber, is greatly reduced. The key design variable, the particle-residence time, computed as the length of the chamber divided by the gas velocity, determines whether or not the chamber is long enough to allow the particle to fall to the bottom. The width of the chamber must be such that the gas velocity is below the "pick-up" velocity that will cause re-entrainment. For low-density materials such as starch, this is 5.8 ft/s. For gas–solid systems, settling chambers have advantages of minimal cost and maintenance, rapid and simple construction, low pressure drop, and dry disposal of solids.

A crude *classification* of solids takes place in the sense that the first of the dust-collecting hoppers contains larger particles than the ones that follow, but little use is made of that because particle sizes overlap. Many variations of the simple enclosure in Figure 19.3 exist. The height a particle has to fall can be decreased by banks of trays set within the chamber, as in the Howard multitray settling chamber. Baffles can be used to direct the gas flow downward to add a momentum effect to the gravitational force. Baffles and tortuous paths also aid particle capture by inertial mechanisms, but the cost in terms of pressure drop is high.

For solid–liquid systems, devices based on gravity are called *sedimenting* separators, *clarifiers*, *thickeners*, *flocculators*, and *coagulators*. Coagulation is the precipitation of colloids, by floc formation, caused by addition of simple electrolytic salts, which modify electrostatic forces between the particles and fluid. The

term *flocculation* is generally used to describe the action of water-soluble, organic, polymeric molecules that may or may not carry a charge, such as polyacrylamide, which promotes settling. Figure 19.4 depicts a liquid-settling device of the type widely used for wastewater treatment, which is equipped with a slowly moving *rake* that revolves at about 2 rph and moves the sludge downward to promote particle agglomeration. The volume of clear liquid produced depends primarily on the cross-sectional area and is almost independent of the tank depth.

Liquid–liquid gravity separators are important in the oil industry, where mixtures of water and oil are commonplace, and in the chemical industry, where extractive distillations and liquid–liquid extractions are carried out extensively. In liquid–liquid separators, called *decanters*, there is often a continuous phase with a discontinuous phase of dispersed droplets. The two phases must be held for a sufficient time for the droplets to settle if heavy, or rise if light, so that the two phases disengage cleanly. A completely clean disengagement is a rarity because, unless the liquids are unusually pure, dirt and impurities concentrate at the interface to form a scum, or worse yet, an emulsion that must be drained off. Figure 19.5 shows a continuous-flow gravity decanter designed to separate an oil layer from a water layer that contains oil droplets. It does not show the perforated underflow and interface baffles, outlet nozzles, or inlet flow distributors.

The unit does not run full, and the design involves balancing the liquid heights due to the density difference of the phases, and determining the settling velocities of droplets moving up or down from the dispersed to the continuous phase. Needless to say, rules of thumb and years of experience are required to design units that work well. Some design methods are based on the time it takes particles to move through a semihypothetical interface between the heavy and light fluids. Example 19.6 shows how the dimensions for a continuous-flow decanter are obtained. Methods for designing a vertical decanter are given in Exercise 19.8.

§19.2.2 Impaction and Interception Separators

Inertial impaction and interception mechanisms, shown above in Figure 19.2, consist of a particle colliding with a

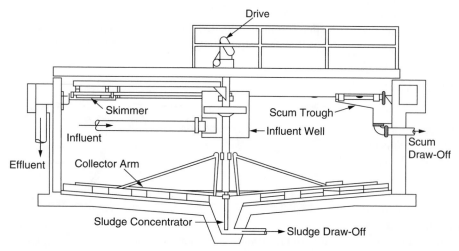

Figure 19.4 Liquid sedimentation and flocculation.

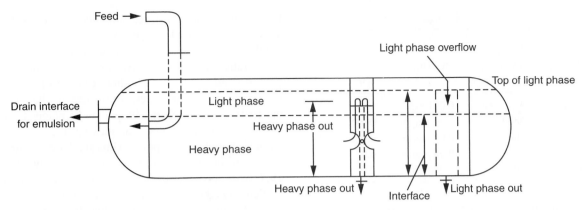

Figure 19.5 Gravity-flow decanter.

target that can be anything from a screen, a bed of fibers, staggered channels, or louvers. Inertial forces accelerate large particles less than small particles, and this, coupled with re-entrainment and variable drag coefficients due to shape, make theoretical prediction of capture efficiency and velocity distributions within a cloth or mesh filter virtually impossible. Instead, impingement separators are designed on the basis of system-specific constants provided by device manufacturers and used in conjunction with the Souders–Brown equation, (6-40), developed in §6.6.1 to describe droplet behavior in distillation columns [3]. Also provided by the manufacturer are recommendations on allowable gas or liquid velocities and pressure drops. For particle-capture devices, performance parameters cannot be calculated from physical properties; if the velocity is lower than what is recommended, impingement of small particles may not take place, and if it is too high, re-entrainment will occur. In addition, use is frequently made of generalized or device-specific information regarding collection efficiency as a function of Reynolds number or particle size. When impingement

devices are used to capture liquid droplets, they coalesce and the liquid must be drained from the collector device. Often, modern coalescence devices combine vane and channel impingements with waffled filters.

An endless array of governmental and industry standards and regulations apply to products manufactured for the purpose of removing particles and contaminants from air streams. Not only do public health laws, with respect to the quality of the air emitted, exist, but there are also industry standards for how devices that impact the environment are to be tested. Based on these tests, products are graded and categorized. This is typical for industrial products intended for a specific use such as filtering air for hospital operating rooms or removing oil mists generated by air compressors. The Eurovent standards for flat-panel ventilation filters shown in Table 19.3 were set by the quasi-governmental agency the European Committee of Air Handling and Refrigeration Equipment Manufacturers, and apply to both glass-fiber media and synthetic organic fibers. Parallel specifications have been set by American manufacturers and trade organizations

Table 19.3 Cen/Eurovent Filter Classification

Type Class	Eurovent Designation	Efficiency, %	Measured by
Coarse dust filter	EU1	<65	Synthetic dust
	EU2	65–80	
	EU3	80–90	
	EU4	>90	
Fine dust filter	EU5	40–60	Atmospheric
	EU6	60–80	Dust spot
	EU7	80–90	Efficiency
	EU8	90–95	
	EU9	>95	
High-efficiency	EU10	85	Sodium chloride
particulate air	EU11	95	or liquid
filter (HEPA)	EU12	99.5	aerosol
	EU13	99.95	
	EU14	99.995	
Ultra low	EU15	99.9995	Liquid aerosol
penetration air	EU16	99.99995	
filter (ULPA)	EU17	99.999995	

Table 19.4 Coalescing-Filter Media Grades

Grade Code	Color Efficiency (%)	Coalescing Carryover	Maximum Oil	Pressure (bar) Dry	Pressure (bar) Wet
2	green	99.999+	0.001 mg/m^3	0.1	0.34
4	yellow	99.995	0.004 mg/m^3	0.085	0.24
6	white	99.97	0.01 mg/m^3	0.068	0.17
8	blue	98.5	0.25 mg/m^3	0.034	0.19
10	orange	95	1.0 mg/m^3	0.034	0.05

such as the American Petroleum Institute (API). Not shown in this table are specifications regarding particle size, but they do exist [4].

Table 19.4 shows the internationally accepted grading system for coalescing filter media used to capture liquid oil, oil/water emulsions, and oil aerosols emitted by oil-lubricated compressors. These are glass microfibers in the 0.5–0.75 μm range, which will trap up to 99.99999% of oil/water aerosols and dirt particles in compressed air, down to a size of 0.01 μm. The mechanical sandwich construction of the two-stage

filter element, held between stainless steel support sleeves, is shown in Figure 19.6. Because of the coalescing filter medium, the condensate is drained, and the elements are self-regenerative as far as removal of liquid is concerned. However, it is advisable that prefilters capable of removing particles down to 5 μm or less be placed in the line ahead of the coalescing filter or it will quickly be plugged. In this table the coalescing efficiency was measured using 0.30–0.6 μm particles based on 50 ppm maximum inlet concentration.

A well-designed filtration system, as shown in Figure 19.7, will have elements such as an inexpensive coarse-particle pre-filter collector like a screen filter or cyclone, followed by an extended-surface filter that is effective down to the micron level, and then a submicron filter where the velocity is lower and the particle capture is principally by Brownian motion and/or sieving.

§19.2.3 Fabric Collectors

A very common industrial filtration device is a fabric dust collector. In industry, multiple collectors are housed in enclosures called *baghouses*. These are relatively inexpensive installations

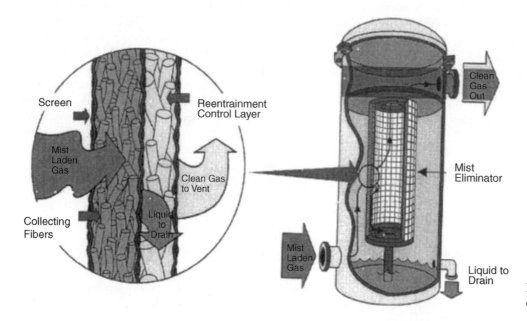

Figure 19.6 Brink fiber-bed mist collector. (Courtesy of MECS, Inc.)

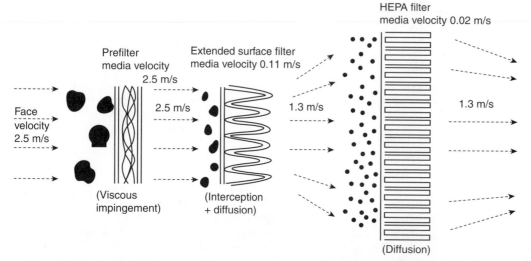

Figure 19.7 Multistage filter system.

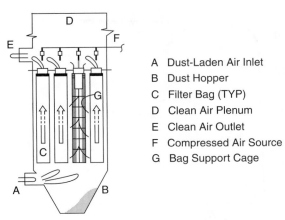

Figure 19.8 Tubular bag filter with pulse jet cleaning.

A   Dust-Laden Air Inlet
B   Dust Hopper
C   Filter Bag (TYP)
D   Clean Air Plenum
E   Clean Air Outlet
F   Compressed Air Source
G   Bag Support Cage

capable of capturing particles down to 0.05 μm. As shown in Figure 19.8, particles are collected on the outside of a fabric-encased, porous, cylindrical *candle*. The device has a vibratory or compressed-air blowback system to remove the particles trapped on the outside of the filter element. Liquids as well as solids are processed in units of this type. For both liquids and gases, as the particles on the cloth build up, they form a *cake* that acts as a filter, and often is a more effective filter than the fabric or screen. This makes screen and fabric collectors system specific; there is no way to predict performance other than to take laboratory data because the filtering action of the cake cannot be predicted analytically.

§19.2.4   Vanes and Louvers

Another device that falls in the aerodynamic-impingement category is the vane or louvered particle collector. Here, the carrier fluid is forced through a maze, changing direction frequently. This type of device is most effective for collecting droplets or mists and fogs that coalesce and can then be drained from the system. Most often, if pressure drop allows, vane units are used as prefilters for mesh filters, particularly for very small droplets that coalesce upon impingement.

§19.2.5   Cyclones and Centrifuges

For a centrifuge or cyclone, centrifugal acceleration is substituted for gravitational acceleration in the appropriate fluid-dynamics equations. The complicating factors are that centrifugal force depends on the distance from the axis of rotation, which depends on the complex geometry and flow patterns in the device, and that the concentration of particles may be so high that hindered (by neighboring particles)-settling equations are necessary. A typical design method, applied to a Podbielniak centrifugal extractor, was demonstrated in Example 8.11. This design strategy consists of finding the optimal conditions for the centrifuge from test runs using a small laboratory unit, and then using a set of scientifically deduced, semi-empirical rules for scale-up to a large industrial unit. This methodology, as will be seen, is also used to design cyclones.

Because cyclones are inexpensive and durable, with a decent collection efficiency for particles larger than about

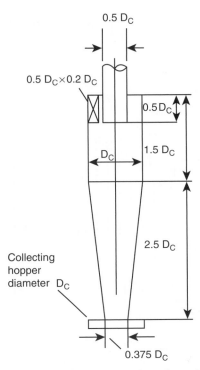

Figure 19.9 Standard high-efficiency cyclone dimensions.

5 μm, they are the most widely used device for industrial dust collection. If the efficiency is not high enough, multiple units can be placed in series. The dust-laden stream enters the top section of the cylindrical device tangentially, which imparts a spinning motion. Centrifugal force sends the particles to the wall, where they agglomerate and fall to the bottom. The spinning gas also travels toward the wall, but it reverses direction and leaves the device from a sleeve at the top, whose bottom extends to below the inlet, as shown in Figure 19.9, which includes standard-dimension relations. The path is usually axial, there being an inner up-flow vortex inside the downward vortex. In liquid cyclones (hydroclones), the upward flow is separated from the downward flow by an outer jacket wherein the liquid flows up. Separation depends on settling velocities, particle properties, and geometry of the device. By directing the inlet flow tangent to the top of the cyclone, centrifugal force can be utilized to greatly enhance particle collection. Well-designed cyclones can separate liquid droplets as small as 10 μm from an air stream. Small cyclones are more efficient than large ones and can generate forces 2,500 times that of gravity. For solids, re-entrainment problems can be reduced by water sprays and vortex baffles at the outlet.

§19.2.6   Electrostatic Precipitators

Electrostatic precipitators are best suited for the collection of fine mists and submicron particles. The first practical application was fashioned by Cottrell in 1907 for abating sulfuric-acid mists. A particle suspended in an ionized gas stream within an electrostatic field will become charged and migrate to a collecting surface. Care must be taken that the particles do not re-entrain, but are removed from the device. Two types of devices are available, one in which ionization and

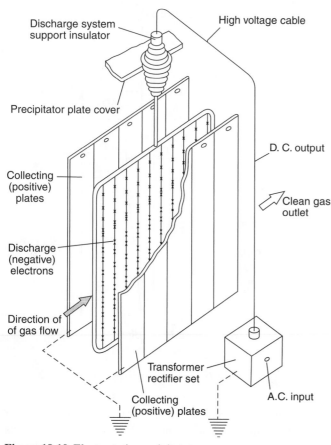

Figure 19.10 Electrostatic precipitator.

collection are combined, and one in which they are separated. In Figure 19.10, the chambers are combined. To obtain ionization, the voltage must be high enough to initiate a corona discharge, but not so high as to cause sparking.

Recent innovations to electrostatic precipitators include adding water sprays and two-stage ionizing wet scrubbers. Of course, any device that adds water to a dry powder complicates the waste-disposal problem and increases power consumption. Waste disposal of particle-laden water streams is a general problem with such devices as spray chambers, wet scrubbers, packed absorption columns, and plate scrubbers, which are not elaborated on in this brief introduction to mechanical separations.

Example 19.1, adapted from Nonhebel [6], illustrates the role of electrostatic precipitators in an industrial environment, and the importance of particle-size distribution in assessing the effectiveness of a pollution-control system.

EXAMPLE 19.1 Cyclones in Series with an Electrostatic Precipitator.

The first two columns in the table below give the dust content of a feed to a bank of precleaning, parallel cyclones in series with an electrostatic precipitator having a contact time of 3 s. Given the efficiency versus particle-size performance data for each of the two devices in Figures 19.11 and 19.12, obtain emission-collection efficiencies for each of the devices and for the overall system if the gas flow rate is 240,000 m^3/hr, gas density is 10 g/m^3, and particle loading is 10/m^3, all at the same standard conditions.

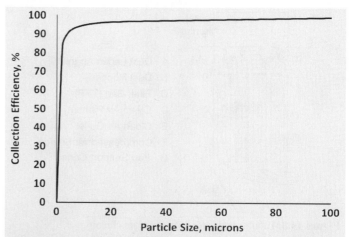

Figure 19.11 Collection efficiency for cyclone.

Solution

Feed		Performance of Parallel Cyclones			
Particle Size, μm	% of Total Particles in Feed	% Collection Efficiency	% of Particles Collected	% of Particles Not Collected	% of Total Particles Not Collected
104–150	3	100.0	3.0	—	—
75–104	7	99.1	6.9	0.1	0.6
60–75	10	98.5	9.9	0.1	0.6
40–60	15	97.3	14.6	0.4	2.5
30–40	10	96.0	9.6	0.4	2.5
20–30	10	94.3	9.4	0.6	3.8
15–20	7	92.0	6.4	0.6	3.8
10–15	8	89.3	7.1	0.9	5.7
7.5–10	4	84.2	3.4	0.6	3.8
5–7.5	6	76.7	4.6	1.4	8.8
2.5–5	8	64.5	5.2	2.8	17.6
0–2.5	12	33.5	4.0	8.0	50.3
	100%		84.1%	15.9%	100%

The numbers in Column 3 (C3) were obtained from Figure 19.11; numbers in C4 are computed from (C2 × C3/100); numbers in C5 are from (C2 – C4); and numbers in C6 are from (C5 × 100/ sum of C5). Total collection efficiency for the cyclones is 84.1%, but particle capture for particle sizes below 5 μm is low, so the

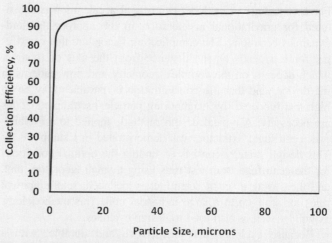

Figure 19.12 Collection efficiency for electrostatic precipitator.

effluent from the cyclones is sent to an electrostatic precipitator. The feed to the precipitator has a particle loading of $(10)(100 - 84.10)/100 = 1.59$ g/m^3.

The performance of the electrostatic precipitator, set forth in the table below, is obtained in exactly the same manner as for the cyclones, using Figure 19.12 for the collection efficiency of the precipitator. From the results, it operates at an efficiency of 85.4%, with an exit dust loading of $1.59(100 - 85.4)/100 = 0.23$ g/m^3.

Feed from Cyclones		Performance of Electrostatic Precipitator			
Particle Size, μm	% of Total Particles in Feed	% Collection Efficiency	% of Particles Collected	% of Particles not Collected	% of Total Particles not Collected
75–104	0.6	99.2	0.6	—	trace
60–75	0.6	98.7	0.6	—	trace
40–60	2.5	97.7	2.4	0.1	0.7
30–40	2.5	96.8	2.4	0.1	0.7
20–30	3.8	96.5	3.7	0.1	0.7
15–20	3.8	96.0	3.7	0.1	0.7
10–15	5.7	95.5	5.4	0.3	2.1
7.5–10	3.8	94.7	3.6	0.2	1.4
5–7.5	8.8	94.0	8.3	0.5	3.4
2.5–5	17.6	90.5	16.0	1.6	11.0
0–2.5	50.3	77.0	38.7	11.6	79.3
	100%		85.4%	14.6%	100%

The overall collection efficiency for the system of cyclones and precipitator is 97.7%. It is interesting to note that although these engineering calculations, made in the 1960s, are the same as those made today, the allowable air-pollution standards have been tightened.

§19.2.7 Filter-Cake Filtration Devices

In the above description of bag filters, it was pointed out that particles collected on the outside of the cloth form a cake, which also acts as a particle collector. For solid–liquid systems, there is a class of equipment where the slurry is pumped or vacuum driven through a cloth filter, with the cake acting as a filter medium. Not all slurries can be treated this way, but if a cake-based filter is an option, the first step is to make laboratory tests to ascertain under what conditions the solid will form a suitable cake. Usually, as discussed in §14.8, the proper concentration and type of filter aid that needs to be added to the slurry must be researched, and, because the pressure may not be constant through the entire filter cycle, the effect of pressure on cake permeability must be studied. Generally, the slurry should contain less than 35 vol % solids and the particle size should be above a few microns. P^c-SELECT, an expert system to guide the laboratory study and device selection, has been formulated by Wakeman and Tarleton [7]. The use of a laboratory test leaf filter, usually of 0.1 ft^2 filter area, is highly recommended and is described in detail in [11].

Shown in Figure 19.13 is a rotary-drum vacuum filter, which operates continuously and consists of a hollow,

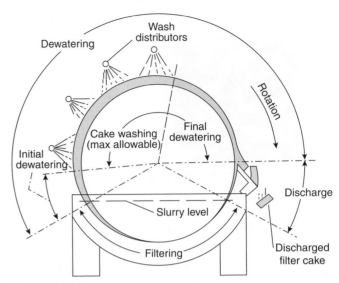

Figure 19.13 Rotary-drum vacuum filter.

rotating drum over which a fabric sleeve is stretched. The volume inside the drum is divided into zones. One zone of the drum, which rotates at from 0.1 to 10 rpm, is in contact with an agitated trough containing the slurry, which is drawn onto the filter cloth by a vacuum of about 500 torr inside the drum. As the drum, along with the newly formed cake, rotates out of the trough, the cake enters a washing zone where water-soluble impurities are washed out of the cake. The wash water may be added to the filtrate. In the next zone, the cake may be dewatered by vacuum, mechanical pressing, and/or an inflatable diaphragm. After that, the cake is removed from the cloth by blowback pressure for high rotation rates, a knife blade or scraper for low rotation rates, or by other means. Following this, there may be a brush to clean the filter cloth; sometimes a precoat of diatomaceous earth (silica) or perlite fiber is applied. There are many possible drum variations, including pressurized drums.

In recent years, vacuum belt filters have taken a share of the market for large, continuous industrial filters. A schematic of a vacuum belt filter is shown in Figure 19.14. It is sectioned in the same way as a rotary-drum vacuum filter.

Since the early days of the chemical industry, the venerable plate-and-frame filter press has been an industry warhorse. Many variations of this press, which is suitable only for batch operations, are in use. The design permits

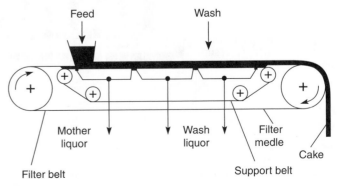

Figure 19.14 Belt filter.

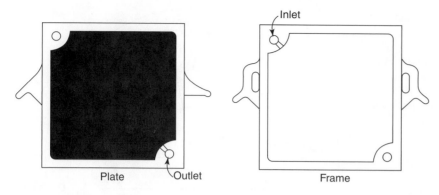

Plate

Outlet

Frame

Inlet

Figure 19.15 Plate-and-frame pair.

delivery of the slurry to the filter cloth, which is backed by a metal plate; discharge of the filtrate and retention of the cake; and addition of wash water, with, in some models, the impurities leaving through a different port. The device can have from two to four separate ports, and some presses embody features such as inflatable diaphragms that enable cake dewatering by compression, a process called *expression*. After the cycle, the press is disassembled, and the cakes are collected manually.

Figures 19.15 and 19.16 show the most common, simplest, two-port configuration, which consists of alternate plates and frames hung on a rack and pressed together with a closing (and opening) screw device. The filter cloths, which have holes to align with the inlet and outlet ports, are hung over the plates and act as gaskets when the press is closed. A very large plate-and-frame filter press may have as many as 100 plates and frames, and up to 300 square meters of filter area. Slurry feed enters from the bottom, and feeds the cavities in parallel. The filtrate flows through the cloth, channels in the plate, and out the top, while the cake builds up in the frame. The frame is full when the cakes, which build up on both sides of the frame, meet. Other versions of the plate-

and-frame filter press have three- and four-port systems, which facilitate washing, when required, because if the slurry fills the frame, the wash water may be blocked if it enters through the slurry feed lines.

A type of filter press that competes with plate-and-frame devices in batch-production processes is the *pressure leaf filter*, which has the advantage of not having to be disassembled completely after each cycle. Most leaf filters resemble the baghouse device shown in Figure 19.8. Horizontal and vertical versions of pressure leaf and plate-and-frame filters are available. Choice of filter equipment is governed mostly by economic factors, which include relative cost of labor, capital, energy, and product loss, but attention must be paid [8] to: (1) fluid viscosity, density, and chemical reactivity; (2) solid particle size, size distribution, shape, flocculation tendency, and compressibility; (3) feed slurry concentration; (4) throughput; (5) value of the product; (6) waste-disposal costs and environmental problems; and (7) completeness of separation and material yields. Experimental data are required to establish these parameters, and pilot-plant testing is a necessity. Proper choice and concentration of filter aid, and the choice and pretreatment of the

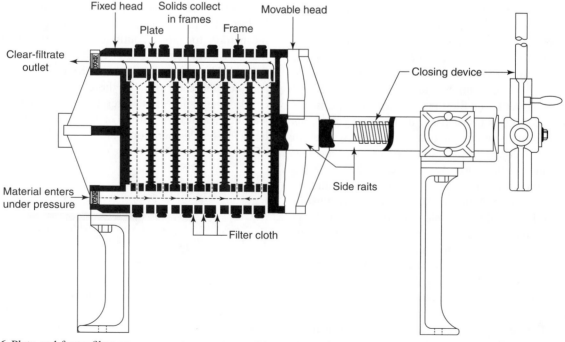

Fixed head

Solids collect in frames

Movable head

Plate

Frame

Clear-filtrate outlet

Closing device

Material enters under pressure

Side raits

Filter cloth

Figure 19.16 Plate-and-frame filter press.

filter cloth, are also critical. As in all engineering ventures, careful attention to details is a necessity.

§19.3   DESIGN OF PARTICLE SEPARATORS

§19.3.1   Empirical Design Equations

What is called an "empirical design equation" here is an equation that, although it may be scientifically based, has constants that are device- and substance-specific. Engineering handbooks, for example, describe generalized graphs and equations that allow the design of piping systems for all common fluids, and are valid for all sizes and configurations; no experimental data are required. This is not the case for particle-collection devices; their performance has to be calibrated. Its collection efficiency for talc particles may be entirely different from its efficiency for collecting gypsum particles even though the particle-size distributions are the same. Particle compressibility, electrical charge, aerodynamic shape, and agglomeration tendency are some of the variables for which no general science-based equation exists. Design equations are empirical, with constants that are use-specific. Frequently, the equations are not dimensionless, but must be used only with specified dimensions.

§19.3.2   Mesh Filters

The most frequently encountered equation in the specification of mesh and vane filters is the Souders–Brown equation (6-40), which was developed in 1924 to calculate settling velocities for the purpose of modeling entrainment in distillation columns. In the version used here for mist-eliminator design, the empirical constant K_L is known as a system load factor, or simply load factor. Its value is such that the velocity it predicts, the bulk impingement velocity, u, is normally considerably higher than the particle settling velocity widely used in mechanical separations and calculated by the following equation, which looks exactly like (6-40), but with K_L in place of C, which was obtained in §6.6.1 from a rigorous force balance.

$$u = K_L \left(\frac{\rho_p - \rho_f}{\rho_f} \right)^{1/2} \qquad (19\text{-}1)$$

General industrial practice is for a manufacturer to provide K_L values that are appropriate for its devices and a specific use. Amistco, in Alvin, Texas, a manufacturer of particle-removal equipment, has the following entries in its product literature [9]:

Table 19.5   Recommended Design Values for K_L in (19-1)

Typical Wire-Mesh Pad, (No-co-knit yarn)
Vertical Flow .... $K_L = 0.35$ ft/s
Horizontal Flow .... $K_L = 0.42$ ft/s

Typical Vane Unit
Vertical Flow .... $K_L = 0.50$ ft/s
Horizontal Flow .... $K_L = 0.65$ ft/s

K_L values for horizontal units are higher than those for vertical units because the filter is designed to remove liquid droplets from gas streams, where the droplets are coalesced and drain from the unit. Drainage of liquid is facilitated so that the gas velocity can be higher, and re-entrainment minimized.

For mesh pads and vanes, as with all particle-fluid separators, pressure drop is a significant operating-cost factor. As will be seen in the next example, this information can also be correlated using K_L factors.

EXAMPLE 19.2   Removal of Droplets with a Mesh Filter.

Determine the diameter, D, of a vertical cylindrical vessel, wherein a 6-inch-thick TM-1109 Amistco pad is used to separate water droplets from air at 70°F. Also, obtain the pressure drop through the mesh. Pertinent data, which apply to water droplets in air, are: Q, volumetric vapor flow = 200 ft³/s; ρ_p, droplet (water) density = 62.3 ft³/lb; ρ_f, fluid (air) density = 0.0749 ft³/lb; and manufacturer's suggested $K_L = 0.35$ ft/s.

Solution

Substituting the above values into (19-1),

$$u = 0.35 \left(\frac{62.3 - 0.0749}{0.0749} \right)^{1/2} = 10.09 \text{ ft/s}$$

Flow area = $(200)/(10.09) = 19.8$ ft²; vessel cross-sectional area = $(3.14)(D)^2/4 = 19.8$. Solving, vessel diameter = $D = 5$ ft.

Knowing the vessel diameter is not enough to complete the detailed design. The particle-size distribution in the incoming gas feed and the collection efficiency of the mesh for each particle size, as in Example 19.1, are needed. Also, it must be ascertained that the calculated velocity is high enough for inertial-capture mechanisms, but not so high as to initiate re-entrainment. With respect to the pressure drop, this can be obtained from the manufacturer. A typical plot is Figure 19.17, where the pressure drop is shown as a function of superficial pressure drop, with liquid loading as a parameter. The pressure drop for Example 19.2 is from approximately 0.2 inch of water at this low liquid loading.

Another example of an empirical approach is the determination of the depth of mesh filters required to reduce particle concentrations to a desired level. Here it is assumed that every differential layer of mesh removes the same fraction of particles. If N is the number of particles per unit volume and x the filter depth, it follows that

$$dN/dx = -KN$$

Integration from $N = N_o$, $x = 0$ to $N = N$, $x = x$ gives

$$\ln(N_o/N) = Kx \qquad (19\text{-}2)$$

The constant K is evaluated experimentally using a mesh whose depth is known. This equation must be used carefully because it is valid only for a very narrow particle-size distribution. An alternative is to use a different K for each size range in the distribution, if data are available.

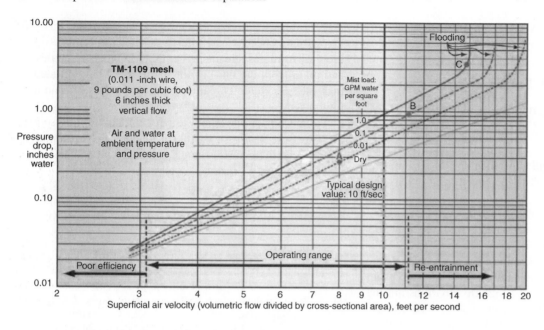

Figure 19.17 Pressure drop for Amistco filter.

[From Amistco Separation Products, Inc., Alvin, TX, with permission.]

EXAMPLE 19.3 Removal of Microorganisms with a Filter.

A 15-cm-thick mesh filter is being used to filter uniformly sized microorganisms from a 10 m³/minute air stream, effectively reducing the concentration from 200 microorganisms/m³ to 10/m³ over a 100-hour fermentation period. New regulations mandate a reduction to 1/m³. If the air velocity and ambient conditions do not change, what depth of the same type of filter is required?

Solution

First, K is evaluated for the current 15-cm filter using (19-2) and the performance data. The original contamination of 200 microorganisms/m³ was reduced to 10. Thus, $\ln[(200/10)] = K(15)$. Solving, $K = 0.2$ cm^{-1}. For the new mesh, x is obtained from $\ln[(200/1)] = (0.2)x$. Solving, $x = 26.5$ cm. The filter required to meet the new regulations will cost almost twice as much as the old filter. However, these specifications are not attainable by any present-day filters.

§19.3.3 Cyclone Design

For cyclones, the effect of feed and device parameters is complex, and interdependencies are to be expected. Larger particles go to the wall quickly, but the smaller ones are separated from the gas near the bottom vortex where the gas reverses direction. Design methods, first developed by Stairmand [10], are based on obtaining particle-collection efficiency data for a cyclone of diameter D and establishing geometric ratios that permit scaling up or down. Design methods for solid–liquid cyclone separators are similar to those for solid–gas or liquid–gas units. Stairmand's design procedure, as presented by Towler and Sinott [5], who show detailed calculations, is as follows:

1. Obtain a collection-efficiency versus particle-size curve for a feed mixture from the literature for the

standard or a prototype cyclone, as in Figure 19.18.

2. Get an estimate of the particle-size distribution in the feed stream to be treated.

3. Estimate the number of cyclones in parallel required.

4. Calculate the cyclone diameter for an inlet velocity of 15 m/s. Scale the other cyclone dimensions from Figure 19.9.

5. Calculate d_2 using d_1, the mean-particle diameter, from Figure 19.18 and

$$d_2 = d_1 \left[(D_{c2}/D_{c1})^3 (Q_1/Q_2)(\Delta\rho_1)/(\Delta\rho_2)(\mu_2/\mu_1) \right]^{1/2}$$

(19-3)

where d_2 = mean diameter of the particles separated in the proposed design, at the same separation efficiency; D_{c1} = diameter of the standard cyclone = 8 in. (203 mm); D_{c2} = diameter of the proposed cyclone, mm;

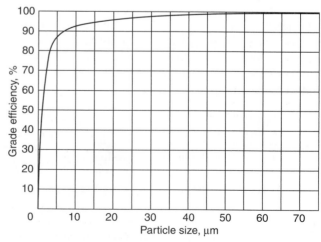

Figure 19.18 High-efficiency cyclone performance curve.

Q_1 = standard fluid flow rate, 223 m³/h; Q_2 = proposed fluid flow rate, m³/h; $\Delta\rho_1$ = particle-fluid density difference, standard cyclone, 2,000 kg/m³; $\Delta\rho_2$ = density difference, proposed design, kg/m³; μ_1 = standard fluid viscosity (air at STP) = 0.018 mN-s/m²; and μ_2 = viscosity, proposed fluid, mN-s/m².

6. Calculate the cyclone performance and recovery of particles (efficiency). If the results are unsatisfactory, try a smaller diameter.

7. Calculate the pressure drop using (19-4) and select a blower

$$\Delta P = \rho_f/203\left\{u_1^2 + \left(u_1^2\right)2\varphi^2[(2r_t/r_e) - 1] + 2u_2^2\right\} \quad (19\text{-}4)$$

where ΔP = cyclone pressure drop, mbar; ρ_f = fluid density, kg/m³; u_1 = inlet duct velocity, m/s; u_2 = exit velocity, m/s; r_t = radius of circle to which the center-line of the inlet is tangential, m; r_e = radius of exit pipe, m; φ = factor given in [5]; ψ = parameter in [5] given by $\psi = f_c(A_s/A_t)$; f_c = friction factor, 0.005 for gases; A_1 = area of inlet duct, m²; and A_s = surface area of the cyclone exposed to the spinning fluid, where length equals total height times cross-sectional area of a cylinder with the same diameter, m².

§19.3.4 Hydrodynamic-Based Equations

Mathematical models used to describe the behavior of particles that separate from fluids, primarily because of gravitational forces, invariably are based on the *terminal velocity* of the particle, u_t, which is defined as the fluid velocity that renders a particle, subject to gravitational force, motionless when suspended unhindered in an upward-flowing fluid stream. At that condition, the drag force on the particle plus the buoyant force balance the force of gravity. The terminal-velocity concept was previously encountered in §6.6.1, where

it was used to derive (6-40) to model distillation-column and reflux-drum diameters. This equation, a combination of (6-40) and (6-41), is

$$u_t = \left[\frac{4d_p g\left(\rho_p - \rho_f\right)}{3C_D\rho_f}\right]^{1/2} \quad (19\text{-}5)$$

From (6-39), the drag coefficient C_D in (19-5) is related to the drag force F_d on the projected area, A_p, of a spherical particle by

$$F_d = C_D\left(\frac{\pi d_p^2}{4}\right)\left(\frac{u^2}{2}\right)\rho_f \quad (19\text{-}6)$$

where: d_p is particle diameter; ρ_p is particle density; ρ_f is fluid density; u_t is terminal velocity (or settling velocity in a quiescent fluid); g is acceleration due to gravity; and C_D is the dimensionless drag coefficient. If AE units are used in (19-6), the denominator must include g_c (e.g., 32.174 lb$_m$-ft/lb$_f$-s²), to convert mass to force.

§19.3.5 Drag Coefficient

Essential to the use of Equation (19-5) are numerical values for the drag coefficient. Fortunately, measurements of drag coefficients and their theoretical interpretations have been the subject of extensive research, and correlations such as Figure 19.19, which is a plot of C_D versus particle Reynolds number, $N_{Re} = d_p u\rho/\mu$, are available. Use of this plot in conjunction with (19-5) frequently leads to trial-and-error calculations because the Reynolds number, which contains particle velocity as a variable, must be known before C_D values can be obtained. Drag coefficients may also be a function of variables not displayed in the plot, which leads to additional correlations and equations. These include: (1) particle velocity history, (2) particle shape, (3) the effect of walls and collisions with other particles, and (4) random Brownian movement, if the particles are very small.

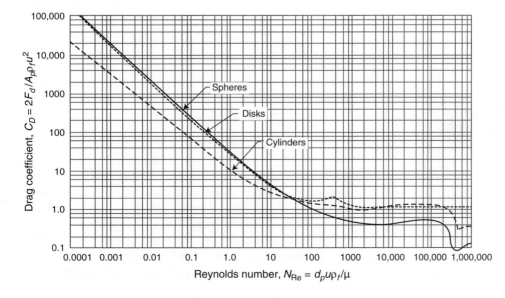

Figure 19.19 Effect of particle Reynolds number on drag coefficients.

[From Lapple and Shepherd, *Ind. Eng. Chem.*, **32**, 605 (1930).]

A particle's history, which includes movement at any velocity other than the terminal velocity, is universally neglected in evaluating C_D, although methods of correcting for past accelerations have been researched. The particle-shape factor poses more of a problem, as can be inferred from the fact that there are separate curves for disks, cylinders, and spheres in Figure 19.19. Falling objects may rotate as they fall, which makes the calculation difficult, with the net effect of increasing drag on the particle; and circulation can occur in droplets. The drag coefficient is discussed in depth in advanced fluid-mechanics textbooks. Other impediments to falling are collisions with walls and other particles. This leads to the development of equations for *hindered settling* (in contrast to *free settling*). The last of the factors to be considered here, Brownian motion, is random particle motion occasioned by the collision of very small particles with surrounding gas molecules or atoms. This will be treated as a separate subject but applied, if necessary, as a correction factor to the settling velocity equations.

§19.3.6   Settling Equations

To avoid having to make trial-and-error calculations for C_D, and to facilitate calculations for settling velocities, it is convenient to divide Figure 19.1 for spherical particles into low, high, and intermediate regions, wherein equations can be written to relate u to C_D.

Low Reynolds Number Region, Stokes' Law

Stokes' law, $F_d = 3\pi\mu u d_p$, which applies at $N_{Re} < \approx 2$, gives $C_D = 24/N_{Re}$. Substitution into (19-5) gives the settling velocity for a spherical particle of diameter d_p as

$$u_t = \frac{g d_p^2 \left(\rho_p - \rho_f\right)}{18\mu} \qquad (19\text{-}7)$$

This equation can be used for Reynolds numbers from 0.001 to 2, with an error for C_D, at the highest N_{Re}, of about 10%. This translates into a usually negligible 5% error in particle velocity. Note that (19-7) and the ensuing equations in this section are limited to spheres falling in a gas or liquid of low viscosity. For a listing of recent literature on particles falling through Bingham plastics and other non-Newtonian fluids, and corrections for nonspherical particles, see [11]. Small variations in the numerical constants in the above equations as well as in other equations in this section appear in the literature. The values here stem from [12].

High Reynolds Number Region, Newton's Law

For Reynolds numbers between 500 and 200,000, the drag coefficient is almost independent of the Reynolds number, and the corresponding settling velocity and drag force for a spherical particle are, respectively, $C_D \cong 0.44$ and $F_d = (0.055)$

$\pi d_p^2 u^2 \rho_f$, resulting in

$$u_t = 1.74 \left[\frac{d_p g \left(\rho_p - \rho_f\right)}{\rho_f}\right]^{1/2} \qquad (19\text{-}8)$$

Intermediate Reynolds Number Region

Between the Stokes and the Reynolds regions, where N_{Re} lies between 2 and 500, $C_D \cong 18.5 N_{Re}^{-0.6}$, resulting in

$$u_t = 0.153 g^{0.71} d_p^{1.14} \left(\rho_p - \rho_f\right)^{0.71} \Big/ \rho_f^{0.29}\mu^{0.43} \qquad (19\text{-}9)$$

Cunningham Correction to Stokes' Law

A correction to Stokes' law is important for particles under 3 μm in diameter for settling in gases and under 0.01 μm for settling in liquids. In gases, small particles can slip between the gas molecules with less drag, resulting in a terminal velocity higher than that predicted by Stokes' law, (19-7). This occurs when the mean free path of the gas, λ (0.0065 μm for ambient air), is comparable to the particle diameter. The increase in terminal velocity can be predicted with the Cunningham slip correction factor, K_m, which is a multiplier to the Stokes settling velocity, u_t, given by

$$K_m = 1 + \left(\frac{\lambda}{d_p}\right)\left[1.644 + 0.552 \exp\left(-\frac{0.656 d_p}{\lambda}\right)\right] \qquad (19\text{-}10)$$

For a 0.01-μm particle falling in ambient air, $K_m = 2.2$. Thus, the particle falls more than twice as fast as predicted by Stokes' law.

Brownian Motion

Oscillatory, zigzag motion of particles whose size falls in the 0.1–0.001-μm range was first observed in 1826 by the British botanist Robert Brown. It was the first visual confirmation of the correctness of the kinetic theory of matter, which predicted that this motion is due to unbalanced impacts of molecules or atoms on particles. Einstein [13] was the first to obtain the following theoretical expression for the average distance Δx moved through by a particle of radius r in a liquid of viscosity μ during time t, where N_A is Avogadro's number and T is absolute temperature. There is a corresponding equation for the rotary movement.

$$(\Delta x)^2 = RTt/3\pi\mu N_A r \qquad (19\text{-}11)$$

Particles larger than 2–3 μm are collected by inertial impaction and direct interception, but for smaller particles, Brownian movement becomes important. The displacements due to Brownian movement for water droplets in air have been measured by Brink [14] and are given in Table 19.6. Because of this motion, submicron particles, given enough time, will coagulate. In general, in fiber and other types of fine-particle mist collectors, where the gas

Table 19.6 Brownian Displacement

Particle Diameter, μm	Brownian Displacement, μm/s
0.1	29.4
0.25	14.2
0.5	8.92
1.0	5.91
2.5	3.58
5.0	2.49
10.0	1.75

velocity is less than 0.075–0.20 m/s, Brownian movement is the controlling mechanism for particle collection; however, the design techniques combine all hydrodynamic mechanisms into some type of an empirical correlation based on experimental data.

Brownian motion is superimposed on the particle velocity, and it lowers the efficiency of capture devices that are based on collisions of particles with fibers because small particles, at very low velocities, will follow the streamlines around the fiber collectors.

Criteria for Settling Equations

Although the range of applicability of the above settling equations has been stated, a commonly used concept is to use a single criterion based on the highest Reynolds number for which the equation applies, which for Stokes' law is 2. The criteria is based on a general procedure used by McCabe et al. [15] and Carpenter [16] that eliminates the terminal-velocity factor, u_t, from the Reynolds number by substituting one of the settling equations. This results in an empirical equation,

$$K_c = 34.81 d_p \left[(\rho_f)(\rho_p - \rho_f)/\mu_f^2 \right]^{1/3} \quad (19\text{-}12)$$

where d_p is in inches and μ_f is in cP, with densities in lb/ft^3. The constant K_c, which is listed in Table 19.7 along with the range of applicability of the settling equations, can be used to determine if the equation is suitable for the particle size in question. Having obtained K_c, it is then convenient to calculate the settling velocity from a general settling-velocity equation,

$$u_t = \left[4g\left(d_p\right)^{(1+n)}\left(\rho_p - \rho_f\right)/3b\mu_f^n \rho_f^{(1-n)} \right]^{1/(2-n)} \quad (19\text{-}13)$$

where the constants are:

Law	b	n
Stokes	24	1.0
Intermediate	18.5	0.6
Newton	0.44	0

Hindered Settling

The settling-velocity equations apply to single particles and predict higher settling velocities than are observed when the concentrations of particles are high enough that settling is hindered by particle–particle collisions. Various approaches for correcting terminal velocities for hindered settling appear in the literature; the one used here is due to Carpenter [16]. For a spherical particle of uniform size,

$$u_{sh} = u_t\left(1 - \phi_{particles}\right)^a \quad (19\text{-}14)$$

where u_{sh} = hindered settling velocity, $\phi_{particles}$ = the volume fraction of particles, and a = an exponent whose value is given in Table 19.8.

Table 19.8 Hindered Settling

N_{Re}	a
≤ 0.5	4.65
$0.5 \leq N_{Re} \leq 1300$	$4.374(N_{Re})^{-0.0875}$
$N_{Re} \geq 1300$	2.33

EXAMPLE 19.4 Settling of Particles.

For a particle 0.01 inch in diameter, determine (a) the proper equation to use for the settling velocity, (b) the terminal (unhindered) velocity, (c) the hindered settling velocity, and (d) the velocity in a centrifugal separator where the acceleration is 30 g. The pertinent data are: $\rho_f = 0.08$ lb$_m$/ft^3, $\mu_f = 13.44 \times 10^{-6}$ lb$_m$/ft-s $= 0.02$ cP, ρ_p 500 lb$_m$/ft^3, and $\phi_{particles} = 0.1$.

Solution

(a) and (b) Solving (19-12),

$$K_c = 34.81(0.01)\left[0.08(500 - 0.08)/(0.02)^2\right]^{1/3} = 16.16$$

Table 19.7 Ranges of Settling Equations

	Newton's Law	Intermediate Law	Stokes' Law	Stokes–Cunningham Law	Brownian Movement
d_p	100,000–1,500	1,500–100	100–3	3–0.1	0.1–0.001
N_{Re}	200,000–500	500–2	2–0.0001		
K_c	≥ 43.6	$3.3 \leq K_c \leq 43.6$	3.3		

with d_p in microns

For $K_c = 16.16$ in Table 19.7, the settling velocity is in the intermediate range, so $b = 18.5$ and $n = 0.6$. Using (19-13) for the settling velocity,

$$u_t = [4(32.2)(0.01/12)^{1.6}(500 - 0.08)/$$
$$3(18.5)(13.44 \times 10^{-6})^{0.6}(0.08)^{0.4}]^{1/1.4} = 11.78 \text{ ft/s}$$

It is possible to check if the selection of the intermediate range is correct by calculating the particle Reynolds number. $N_{Re} = d_p u_t \rho_f / \mu_f = [(0.01/12)(11.78)(0.08)/(13.44 \times 10^{-6})] = 58.4$. Checking Table 19.7, it is seen that the correct choice of region was made.

(c) For $N_{Re} = 58.4$, by Table 19.5, $a = 4.374(N_{Re})^{-0.0875} = 3.064$.

Thus, for a 0.1 volume fraction of solids, the hindered settling velocity from (19-14) is $u_{sh} = 11.78(1 - 0.1)^{3.064} = 8.53$ ft/s. This is a 28% reduction in velocity.

(d) To find the velocity at 30 g acceleration, it is necessary to multiply g by 30 in (19-13), so u_t in the centrifugal field is the velocity from part (c) multiplied by 30 to the $1/(2 - n)$ power, where $n = 0.6$. Thus, $u_t = 8.53(30)^{0.714} = 96.7$ ft/s.

§19.3.7   Particle Classification

Separation of particles in accordance with their size is called *classification*. If two particles have different terminal velocities in air, it is possible to adjust the air velocity so that one particle remains suspended while the particle having the higher terminal velocity falls. Likewise, as is illustrated in Figure 19.20, if a group of particles is injected into a moving body of water, the particle with the lowest terminal velocity will be found farthest downstream. For two groups of different-density particles, 1 and 2, with a range of sizes, and with 1 denser than 2, complete separation is unlikely because the size range overlaps. This overlap occurs when particles in the two groups have equal terminal velocities. If $N_{Re} < \approx 2$ for all particles, from (19-7),

$$u_t = g(d_1)^2(\rho_1 - \rho_f)/18\mu = g(d_2)^2(\rho_2 - \rho_f)/18\mu$$
(19-15)

Dividing the two RHS equalities,

$$(d_1)^2/(d_2)^2 = (\rho_2 - \rho_f)/(\rho_1 - \rho_f)$$
(19-16)

For all N_{Re}, the general result from (19-5) is

$$(d_1)/(d_2) > [(\rho_2 - \rho_f)/(\rho_1 - \rho_f)]^n$$
(19-17)

where $n = \frac{1}{2}$ for laminar flow; $n = 1$ for turbulent flow; $n = 0.625$ for intermediate flow.

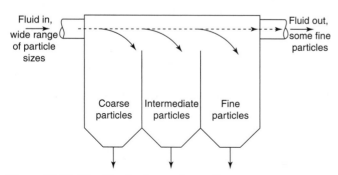

Figure 19.20 Classification by gravity settling.

Examination of (19-17) shows that the separation can be sharpened by choosing a fluid that has a density close to that of one of the particles. A liquid-phase density can be altered by adding a thickener or very fine particles that do not settle. Density adjustment is also the basis for separating enzymes and other biological systems by aqueous two-phase extraction. For this application, centrifuges are used [17].

EXAMPLE 19.5   Separation of Particles by Classification.

A mixture of particles A and B is to be separated by classification using water. The size range for both A and B is between 7 and 70 μm, with $\rho_A = 8$ g/cm^3 and $\rho_B = 2.75$ g/cm^3. Assume unhindered settling and a water viscosity of 1.0 cP (0.01 poise). (a) What velocity will give a pure A product? (b) What is the size of the largest A particle swept out with the B particles?

Solution

(a) Since A will settle faster than B, the water velocity must be larger than the settling velocity of the largest B particle. Assuming that u_t of the largest B particle is in the Stokes' settling domain, (19-7), with CGS units, gives

$$u_t = 980.7(0.007)^2(2.75 - 1)/18(0.01) = 0.468 \text{ cm/s}$$

The Reynolds number, $N_{Re} = (0.007)(0.468)(1)/0.01 = 0.33$, which is in the Stokes' law region, as assumed.

(b) It is necessary to calculate the size of the heavier A particle that settles at 0.468 cm/s. From a rearrangement of (19-7),

$$d_A = \left[\frac{18u_t\mu}{g(\rho_A - \rho_f)}\right]^{1/2} = \left[\frac{18(0.468)(0.01)}{980.7(8 - 1)}\right]^{1/2} = 0.0035 \text{ cm}$$
$$= 35 \text{ μm}$$

Thus, any A particle smaller than 35 μm will be swept out along with all the B particles.

§19.3.8   Gravity Decanter

The design method suggested here for a liquid-liquid separation is due to Schweitzer [18], as used by Coker [12].

EXAMPLE 19.6   Separation of Oil from Water by Settling.

A decanter to separate oil from water is needed. The oil flow is 8,500 lb/hr, and the water rate is 42,000 lb/hr. It is anticipated that there will be oil droplets in the water layer. Obtain the dimensions of the horizontal decanter. The tank will have an L/D (length-to-diameter) ratio of 5. Ignore the hindered settling effect. The following nomenclature and data apply: $u_t =$ oil-droplet terminal velocity, ft/s; $g =$ acceleration of gravity, 32.2 ft/s^2; $d_p =$ oil-droplet diameter, assumed uniform at 150 μm (0.00049 ft); $\rho_p =$ oil density, 56 lb/ft^3; $\rho_f =$ fluid (water) density, 62.4 lb/ft^3; $\mu_f =$ fluid (water) viscosity, 0.71 cP or 4.77×10^{-4} lb$_m$/ft-s; $\mu_{oil} =$ oil viscosity, 9.5 cP or 63.84×10^{-4} lb$_m$/ft-s; $Q_{oil} = 8,500/(56)(3,600) = 0.0422$ ft^3/s; and $Q_{water} = 42,000/(62.3)(3,600) = 0.1873$ ft^3/s

Solution

Although $Q_{\text{oil}} \ll Q_{\text{water}}$, it is best to make sure that oil is the dispersed phase. The criteria suggested by Schweitzer [18] are based on a parameter, θ, in terms of light, l, and heavy, h, phases:

$$\Theta = (Q_l/Q_h)[\rho_l\mu_h/\rho_h\mu_l]^{0.3} \qquad (1)$$

where

Θ	RESULT
<0.3	Light phase always dispersed
0.3–0.5	Light phase probably dispersed
0.5–0.2	Phase inversion probable
2.0–3.3	Heavy phase probably dispersed
>3.3	Heavy phase dispersed

Applying (1),

$$\Theta = (0.0422/0.1873)\left[(56)(4.77 \times 10^{-4})/(62.3)(63.84 \times 10^{-4})\right]^{0.3}$$
$$= 0.10$$

Clearly, oil will be the dispersed phase. Next it is necessary to decide which settling law applies, so K_c is calculated using (19-12), with AE units, and substituting the absolute density difference for $(\rho_p - \rho_f)$:

$$K_c = 34.8(0.00049)\left[(62.4)(6.4)/(0.71)^2\right]^{1/3} = 0.128$$

With reference to Table 19.7, it is seen that K_c is in the Stokes' law range.

Since turbulence in a gravity settler is undesirable, it is also necessary to check the Reynolds number of the fluid after the vessel dimensions are established. By the criteria in Table 19.7, N_{Re} should be < 2.

By Stokes' law, (19-7), the settling velocity (in this case, the rise velocity of the oil droplets) is

$$u_t = 32.2(0.00049)^2(56 - 62.4)/\left[18(4.77 \times 10^{-4})\right] = -0.0058 \text{ ft/s}$$

The negative value arises because the oil droplets rise rather than settle. It is now possible to obtain the vessel dimensions. Assuming that the length of the vessel is five times the diameter, $L/D = 5$, and that the width of the interface, which is not at the top, is $0.8D$, the phase-interface area, A, is $(0.8D)(5D) = 4D^2$, and since $Q_{\text{water}}/A \leq u_t$, $D \geq 0.5(0.1873/0.0058)^{1/2} = 2.84$ ft, and $L = (5)(2.84) = 14.2$ ft. For additional design calculations to establish Reynolds numbers and specifications for this separator, see [12].

§19.4 DESIGN OF SOLID–LIQUID CAKE-FILTRATION DEVICES BASED ON PRESSURE GRADIENTS

Pressure-filtration devices consist of a cloth or mesh barrier (the medium) that retains suspended solids (the cake), while allowing the fluid in which the solids are suspended (the filtrate) to pass through. As shown in Figure 19.21, it is customary to treat the pressure drops through the cake and the medium as separate entities. Filtration models in this chapter and (14-81) are based on Darcy's law, developed in 1855 to describe the flow of water through sand beds. In his

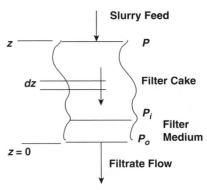

Figure 19.21 Filtration profile.

experiments he found the flow rate to be proportional to the pressure drop, indicating laminar flow, which is also the case for cake filtration when Reynolds numbers are below one and inertial effects are negligible. As has been pointed out by Wakeman and others [7], Darcy did not include viscosity in his original equation, which, with viscosity for application to filtration, is

$$J = -\frac{k}{\mu}\frac{dP}{dz} \qquad (19\text{-}18)$$

where dP is the pressure drop through thickness dz of a medium of permeability k and $J = u$ is the fluid velocity (volume flow/unit filter area). For cake filtration, it is customary to replace the permeability, k, with α, the specific cake resistance, and dP/dz with dP/dW, where W is the mass of the dry filter cake per unit filter area and dW and dz are related by

$$dW = \rho_c(1 - \epsilon)dz \qquad (19\text{-}19)$$

where ϵ is the fraction of voids in the filter cake and, thus, is a measure of the volume of flow paths through the medium. Unfortunately, for filter cakes, the complexity of their structure, and their dependence on pressure, preclude direct calculation by either ϵ or α from scanning-electron micrographs or other means. Nevertheless, Kozeny, in 1927, and Carman, in 1938, developed a theoretical flow model based on pores being replaced by a bundle of capillary tubes whose orientation is at $45°$ to the surface. Their equation is based on the Poiseuille equation for laminar flow through a straight capillary, but with the straight capillary being replaced by more complex geometric constructs. §14.3 and 6.8 contain expositions on this subject as it applies to flows of liquids through membranes and packed beds. The Kozeny–Carman equations have limited applicability to cake filtration because ϵ and α must be evaluated experimentally. A more empirical and widely used approach to finding a relationship between process variables in filtration is to consider the two pressure drops in Figure 19.21, one through the medium and the other through the cake. Denoting the medium resistance to flow by R_m and the cake resistance to flow as R_c, Darcy's law, (19-18), can be applied to each resistance,

$$u = (P - P_i)/\mu R_c = (P_i - P_o)/\mu R_m \qquad (19\text{-}20)$$

Letting the total pressure across the cake and medium $\Delta P = (P - P_o)$, V = volume of filtrate, A_c = cake area, and

$u = (dV/dt)(1/A_c)$.Combining with (19-20) gives

$$dV/dt = \Delta P A_c / \mu (R_m + R_c) \qquad (19\text{-}21)$$

which states that dV/dt is directly proportional to A_c and ΔP, and inversely proportional to the filtrate viscosity, μ, and the sum of the resistances of the cake and medium. Usually, R_m is much smaller than R_c after the cake begins to build up. When pressure is in AE units, the RHS must be multiplied by the gravitational constant, g_c; R_c and R_m have dimensions of reciprocal length.

As filtration progresses, cake thickness, filtrate volume, and resistance to flow increase, but R_m is assumed to remain constant. W, the weight of the dry cake, is related to V, filtrate volume, and c_F, the dry cake mass per unit volume of filtrate, by $W = c_F V$. However, care must be exercised when applying this formula because the cake is wet, and then dried, so c_F will have different values depending on whether wet or dry cake masses are used. It now becomes useful to replace R_c by α, the *specific cake resistance*, which replaces k in the original Darcy equation, (19-18), where now $R_c = \alpha W/A_c = \alpha c_F V/A_c$. Thus, if length is in ft, R_c has units of 1/ft and α has units of hr²/ft because the compressibility has been multiplied by g_c for dimensional consistency [15, 19]. For SI units, α has dimensions of m/kg. Substituting the definition of α into (19-21),

$$\frac{d(V/A_c)}{dt} = \frac{\Delta P}{\mu[R_m + \alpha c_F(V/A_c)]} \qquad (19\text{-}22)$$

It is common to consider the application of (19-22) to two regimes of filtration: constant-pressure and constant-flow rate.

§19.4.1 Constant-Pressure Filtration

Assuming the filtration area is constant and the pressure drop is constant, (19-22) can be integrated from $V = 0$ to $V = V$ for $t = 0$ to $t = t$ to yield the constant-pressure form of what was termed the Ruth equation in §14.3.1, and which was developed in 1933 [20]:

$$V^2 + 2VV_o = K_t \quad \text{or} \quad t/V = (V + 2V_o)/K \qquad (19\text{-}23)$$

where $V_o = R_m A_c / \alpha c_F$ and $K = 2A_c^2 \Delta P / \alpha c_F \mu$

It is well to consider what kind of a pump can deliver a slurry to a filtration unit at constant pressure. Such a pump is certainly not a positive displacement pump, because it operates at constant flow rate. Centrifugal pumps are sold with performance charts (characteristic curves) that display their flow rate as a function of pump pressure or head, so it is possible to devise a pump control system that forces a centrifugal pump, up to a point, to maintain a constant pressure even though the flow diminishes. If the pump is not controlled, it will maintain an output flow and pressure that follow the characteristic curve supplied by the manufacturer.

In practice, regardless of what type of filtration device is used, the cake will be wet, and if it is the final product, it must be dried before being sold. Often before drying, which is energy-intensive, the cake is subjected to *expression* to wring out excess moisture. Predrying devices include machines that are vice-like presses, centrifugal separators like a laundry dryer without a heater, or inflatable

diaphragms inserted between frames of a plate-and-frame filter. Frequently, the filtrate imbibed in the filter cake contains water-soluble impurities, which must be washed out of the cake prior to expression and final drying. The optimal economics is to use the filtration apparatus to also conduct the water wash and the expression so that, if necessary, a "wash cycle" and time for expression are appended to the filtration cycle, and the throughput rate for a filtration system is obtained by dividing the total throughput by the sum of the required wash, expression, and filtration times. Once the V_o and K constants are obtained, the Ruth equation can be used to obtain either wash or filtration rates and filter areas.

When the constants V_o and K are evaluated from constant-pressure laboratory or pilot-plant data, (19-22) can be used to model large-scale units operating with the same feed, concentration of filter aids (if any), and pressure drops. For example, (19-23) can be modified to model a continuously rotating drum filter operating at constant speed of n rpm, where θ is the fraction of the drum immersed in the slurry tank, V'/n represents the filtrate volume filtered in one revolution, and A' is the drum area. The modified equation is

$$(V'/n)^2 + 2(V'/n)(V_o) = K(\theta/n) \qquad (19\text{-}24)$$

If the resistance of the medium is negligible in comparison to the resistance of the cake, $V_o = 0$ and (19-1) becomes

$$\text{Volume of filtrate per unit time} = V'n = A' \left[\frac{2n\theta\Delta P}{\alpha c_F \mu} \right]^{1/2} \qquad (19\text{-}25)$$

As discussed in §14.8, in many applications, α, the specific resistance of the cake, is a function of the pressure drop across the cake because the filter cake is compressible. This is particularly true for bacteria and other "soft" cakes, where an increase in pressure does not, as predicted by (19-22), produce a directly proportional increase in the volume of filtrate and cake. In that case, an adjustment to all of the above equations that contain α should be made by relating the pressure difference to the cake compressibility by an empirical equation. Table 19.9 [21, 22] lists cake compressibility factors for several inorganic filter cakes for the equation

$$\alpha = \alpha'(\Delta P)^s \qquad (19\text{-}26)$$

where ΔP is in psi. Note that in the table, α' varies by a factor of 10,000 and s varies from 0.2664 to 1.01.

The constant s, which is zero for an incompressible cake, must be evaluated from experiments in which the filtration pressure is varied. All filtration equations should be modified

Table 19.9 Filter-Cake Compressibility

Substance	$\alpha' \times 10^{-10}$, m/kg	Exponent s
Calcium carbonate	1.604	0.2664
Kaolin, Hong Kong	101	0.33
Solkofloc	0.0024	1.01
Talc	8.66	0.51
Titanium dioxide	32	0.32
Zinc sulfide	14	0.69

by substituting (19-26) for α when the filter cake is compressible. Modified equations and examples can be found in [21]. The wide range in values for s and α' underscores the need for experimental data prior to undertaking a new design. Reliable predictive equations are not available.

Tables of cake compressibility in the literature also often contain values for ϵ, the cake void fraction. If sample calculations are not provided, it is best to consult the original article to determine how the α and ϵ were calculated. If separately, then α can be used with (19-23) and similar equations that do not require void fractions. The values of ϵ can then be used to obtain cake properties such as particle surface area and to calculate an accurate cake thickness, as will be shown in Example 19.8. Sometimes, however, the ϵ and α are calculated simultaneously by curve fits, so they are interrelated, and care must be taken in applying the values. In either case, it is always worthwhile to consult the original journal references unless sample calculations are offered. Another consideration is that these "constants" are pressure dependent and represent average values over a range of pressure, so unless this range is reported, nothing is known with certainty because extrapolations should not be made.

In Examples 19.7 and 19.8, it will be seen that α and ϵ are both functions of pressure. According to the laws of Darcy and Kozeny, the hydraulic pressure gradient should be linear, and when it is not, it is because of substantial variations in α and ϵ through the cake. The more compressible the cake, the larger the changes. It has been inferred from experiments that, for cubic packed, hard spheres, there is no change in void fraction through the cake, but for highly compressible latex slurries, the flow area is reduced from 5% to 50% from the top of the cake to the surface of the medium. In that case, the solid actually flows through the medium as it replaces the fluid in the voids at a velocity that can be 19–50% of the liquid velocity.

EXAMPLE 19.7   Design of a Filtration System.

A process transmittal from R&D to the engineering department requests that a filtration system be designed based on information from three laboratory filtrations conducted at constant pressure as follows:

Filtrate Volume (L)	Test 1, $\Delta P = 5.5$ psi		Test 2, $\Delta P = 18$ psi		Test 3, $\Delta P = 53$ psi	
	t/V, s/L	t, s	t/V, s/L	t, s	t/V, s/L	t, s
0 (extrapolated)	27	—	20	—	5	—
0.5	38	19	25	12.5	8	4
1	38	38	28	28	9	9
2	48	96	36	72	14	28
3	59	177	42	126	17	51
4	70	280	51	204	21	84
5	—	—	58	370	26	130
6	—	—	66	396	30	180

It is reported that the filter area was 0.75 ft^2, $\mu = 6 \times 10^{-4}$ lb$_m$/ft-s, cake density is 200 lb/ft^3, and $c_F = 1.5$ lb/ft^3. (a) Obtain values for constants V_o and K in (19-23) and specific cake resistance α, and R_m

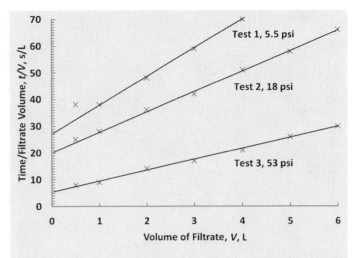

for each run. (b) Use the data to obtain the cake compressibility factor, α', and s in (19-1). (c) The data will be used to size a production unit that will process 300 ft^3 of filtrate, with a filtration time of one hour for each cycle. If a plate-and-frame filter press is used, what filter area will be required if the anticipated pressure drop is 5.5 psi? (d) A rotary-drum vacuum filter is available at the plant. The fraction of the drum area submerged is 0.30, and the rotation speed is 10 rph. The drum is 6 ft in diameter and 10 ft wide, and the system is expected to run at $\Delta P = 5.5$ psi. Is this device suitable for the application?

Solution

Inspection of the data reveals that, as expected for constant-pressure runs, the rate of filtration decreases with time because the cake thickness increases. That the rate is linear can be deduced from the fact that the data for each of the three test runs, after a short time, can be plotted as straight lines, as shown in Figure 19.22. This verifies the mathematical model, (19-23), which has a slope of $1/K$ and an intercept of $2V_o/K$ on a plot of t/V versus V. The data points below 1 L were neglected and the intercept was obtained by extrapolation because often there is a brief, higher *constant-rate* period before the cake starts to build up. If the specific cake resistance is a function of pressure, this can be ascertained by calculating α for each of the three runs, which were at different pressures.

Once V_o and K are known, R_m and α are obtained from their defining equations below (19-23). Another way of solving the problem is to use any two of the data points from a given run, substitute them into (19-23), and solve the two equations simultaneously for V_o and K. However, this method is not as reliable as making a plot and fitting the best line through all five data points because the two points chosen may be unrepresentative. Using AE units, calculations are illustrated for Test 1, using the best line through the data in Figure 19.22.

(a) For Test 1: $K = 1/\text{slope} = (4 - 1)/(70 - 38) = 0.0938$ L^2/s $= 1.17 \times 10^{-4}$ ft^6/s

Intercept $= 27$ s/L $= 764$ s/ft^3

Using the equations just below (19-23) to obtain V_o and R_m, Intercept $= 2V_o/K$. Therefore,

$$V_o = (764)(1.17 \times 10^{-4})/2 = 0.045 \text{ ft}^3$$

$$R_m = (A\Delta P/\mu)(\text{Intercept}) = (0.75)[(5.5)(144)](764)(32.2)/$$
$$6 \times 10^{-4} = 2.44 \times 10^{10}/\text{ft}$$

$$\alpha = R_m A/V_o c_F = 2.44 \times 10^{10}(0.75)/[(0.045)(1.5)] = 2.71$$
$$\times 10^{11} \text{ ft/lb}_m$$

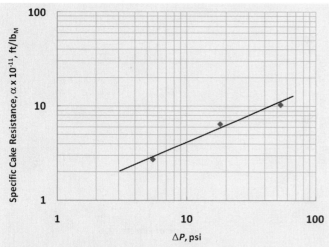

Figure 19.23 Cake compressibility factors for Example 19.7.

Similarly, the values for Test 2 and Test 3 were calculated and are listed in the following summary:

	Test 1	Test 2	Test 3
V_o, ft^3	0.045	0.048	0.021
K, ft^6/s	1.17×10^{-4}	1.62×10^{-4}	4.61×10^{-4}
α, ft/lb$_m$	2.71×10^{11}	6.45×10^{11}	10.35×10^{11}
R_m, 1/ft	2.44×10^{10}	5.90×10^{10}	4.35×10^{10}

(b) All α values listed above are plotted against pressure drop in Figure 19.23. A least-squares fit of the data using Equation (19-1), $\alpha = \alpha'(\Delta P)^s$, gives $\alpha' = 1.04 \times 10^{11}$ and $s = 0.59$.

(c) From part (a), for $\Delta P = 5.5$ psi, $R_m = 2.44 \times 10^{10}$ ft^{-1}, and $\alpha = 2.71 \times 10^{11}$ ft/lb$_m$.

Using (19-23), $(300)(300) + (2)(300)V_o = Kt$.

$$V_o = R_m A_c / \alpha c_F = 2.44 \times 10^{10} A_c / [(2.71 \times 10^{11})(1/5)]$$

$$= 0.0600 A_c \text{ ft}^3 \text{ with } A_c \text{ in ft}^2$$

$$K = 2A_c^2 \Delta P g_c / \alpha c_F \mu = 2A_c^2(5.5)(144)32.2/$$

$$[(2.71 \times 10^{11})(1.5)(6 \times 10^{-4})] = 2.1 \times 10^{-4} A_c^2 \text{ ft}^3/\text{s}$$

with A in ft^2

Time, t, $= 1$ hour $= 3,600$ s

Therefore, (19-23) becomes $90,000 + 600(0.0600)A_c = 2.1 \times 10^{-4} A_c^2 (3,600)$.

Solving for the positive root, $A_c = 370$ ft^2.

(d) Assume the same type of filter cloth is used on the large filter.

Area of drum $=$ (perimeter)(width) $= (3.14)(6)(10) = 188.4$ ft^2

Available area for filtration per revolution $= (0.3)(188.4)$

$= 56.5$ ft^2

Time per rotation $= 1/10 = 0.1$ h $= 360$ s

Time for filtration per rotation $= 0.3(360) = 108$ s

By interpolation of the test at $\Delta P = 5.5$ psi for 108 s, get 2.18 L of filtrate for 0.75 ft^2.

Therefore, for 56.5 ft^2 of the drum, get $2.18(56.5/0.75) = 164$ L of filtrate/rotation, or $164(0.0353) = 5.8$ ft^3 of filtrate/revolution.

At 10 rph, the drum filter can process $(10)(5.8) = 58$ ft^3 of filtrate/h, which is much less than 300 ft^3/h. So the existing rotary-drum filter is inadequate.

EXAMPLE 19.8 Selection of a Filter from Lab Data.

The characteristics of an aqueous slurry at 68°F are being investigated in a laboratory apparatus to determine what class of filter equipment would be suitable. The mass fraction of solids in the slurry, x_s, is 0.01, and their specific gravity is 2.67. Laboratory runs were made at constant pressure drops of 10, 20, 35, and 50 psi, with the data at 10 psi shown in the table below, and data for the other runs plotted in Figure 19.24, where V/A is the filtrate volume flow rate divided by the filter area. The filter area is 0.01 ft^2, and x_c, the weight fractions of moisture in the cakes, were 0.403, 0.431, 0.455, and 0.470, respectively, as measured after each run. Determine cake thicknesses at 30 minutes and average cake porosities.

		Experimental Data for 10 psi Run $A = 0.01$ ft^2, $x_c = 0.403$		
t, min	V, ml	V/A, ft^3/ft^2	$t/(V/A) \times 10^{-3}$, s/ft
1	18	0.063	0.94
3	42.5	0.150	1.20
5	59.0	0.208	1.44
10	96.0	0.338	1.78
15	120.0	0.424	2.12
20	143.0	0.505	2.37
25	165.0	0.583	2.58
30	181.0	0.639	2.82

Solution

The experimental data above and the plot in Figure 19.24 were, as will be shown, used to calculate the values in the table below, where c_F is the lb dry cake/ft^3 of filtrate, α is the cake compressibility, and R_m is the medium resistance.

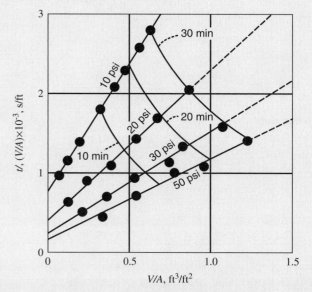

Figure 19.24 Experimental filtration data for Example 19.8.

ΔP, psi	c_F, lb/ft^3	ϵ, void fraction	L, cake thickness, inch	Slope $\times 10^{-3}$ Fig. 19.24	Intercept $\times 10^{-3}$ Fig. 19.24	$\alpha \times 10^{-11}$, ft/lb	$R_m \times 10^{-10}$, ft^{-1}
10	0.640	0.798	0.144	3.252	0.72	6.99	4.97
20	0.639	0.779	0.180	1.946	0.370	8.60	5.10
35	0.638	0.762	0.210	1.314	0.200	10.17	4.82
50	0.638	0.750	0.231	1.025	0.140	11.32	4.83

Calculated Values for Constant-Pressure Filtration Tests of Figure 19.24

The slope and intercept were used to obtain values for α and R_m as in Example 19.7. To obtain ϵ, the void fraction (porosity) of the cake and the thickness of the cake, the following equations apply, where W/A = mass of dry cake/unit filter area, x_c = mass-fraction solids in the cake, V/A = volume filtrate/unit filter area, ρ_f = filtrate density, ρ_s = the true density of the solids in the cake, L = cake thickness, and c_F = mass of dry cake/volume of filtrate.

$$(W/A) = c_F(V/A) \tag{1}$$

$$(W/A) = \rho_s(1 - \epsilon_{\text{avg}})L \tag{2}$$

For a unit volume of cake, x_c is given by

$$x_c = \rho_s(1 - \epsilon_{\text{avg}}) / \left[\rho_s(1 - \epsilon_{\text{avg}}) + \rho_f\epsilon_{\text{avg}}\right] \tag{3}$$

Equation (3) can be solved for the porosity, using $x_c = 0.403$, $\rho_s = 2.67(62.4)$ lb/ft^3, and $\rho_f = 62.4$ lb/ft^3:

$$0.403 = (62.4)(2.67)(1 - \epsilon_{\text{avg}}) / \left[(62.4)(2.67)(1 - \epsilon_{\text{avg}}) + 62.4\epsilon_{\text{avg}}\right]$$

which yields $\epsilon_{\text{avg}} = 0.798$. For a filtering time of 30 min, $V/A = 0.639$ ft. From (1), the corresponding $W/A = 0.640(0.639) = 0.409$ lb/ft^2. Solving (2) gives

$$L = \frac{W/A}{\rho_s(1 - \epsilon_{\text{avg}})} = \frac{0.409}{62.4(2.67)(1 - 0.798)} = 0.012 \text{ ft} = 0.146 \text{ in.}$$

From the above table of calculated values for all runs, there is a fairly strong variation of α with pressure, and some dependence of ϵ.

Using the method of Example 19.7 with (19-26), $\alpha = 3.50 \times 10^{11}$ $(\Delta P)^{0.3}$ and, similarly, $(1 - \epsilon) = (0.15)(\Delta P)^{0.13}$.

There is no numerical or theoretical relationship that links ϵ to α for compressible cakes. For incompressible cakes, where the Kozeny–Carman formulation is valid and when the cake particles are spherical of diameter d_p:

$$\alpha = 150\frac{(1 - \epsilon)}{\rho_s d_p^2 \epsilon^3}$$

The calculated cake thicknesses are small for a filtration time of 30 minutes. For a rotary-drum vacuum filter with 30% submergence, the rate of rotation would be only 0.01 rpm, which is too low. Either centrifugation or pressure filtration is required. Another point to note is that the analytical equations for compressibility and void fractions as a function of pressure should not be extrapolated.

§19.4.2   Constant-Rate Filtration

In a plate-and-frame filter or a pressure leaf filter, where centrifugal pumps are used, the early stages of filtration are frequently at a reasonably constant rate; then, as the cake builds up, the ability of the pump to develop pressure becomes the limiting factor and the process continues at constant pressure and a falling rate. For constant dV/dt, (19-23) becomes

$$\frac{V/A_c}{t} = \frac{\Delta P\{t\}}{\mu[R_m + \alpha c_F(V/A_c)]} \tag{19-27}$$

Rearranging (19-27) after substituting $u = V/t$, the superficial velocity of the filtrate through the cake, the variation of the pressure drop with time is

$$\Delta P = au^2t + bu \tag{19-28}$$

where

$$a = \alpha c_F\mu/A_c^2 \tag{19-29}$$

$$b = R_m\mu/A_c \tag{19-30}$$

Since for a constant rate of filtration, u must be constant, (19-28) defines a straight line on a plot of ΔP versus t, as was shown in Example 14.8.

§19.4.3   Variable-Rate Filtration

The most realistic, and in many respects the simplest, filtration scenario is when the flow and pressure both vary, and the filtration rate varies in accordance with the pump characteristic curve provided by the pump manufacturer. Figure 19.25 shows such a curve. The following example, adapted from Svarovsky [23], demonstrates the procedure.

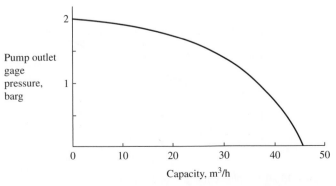

Figure 19.25 Pump characteristic curve for Example 19.9.

EXAMPLE 19.9   Filtration Time in a Plate-and-Frame Filter.

Tests conducted on a laboratory plate-and-frame filter produced the data given below. Determine the time required to process 50 m^3 of the same filtrate in a filter press with an area of 50 m^2, using the same cloth and filter aid as the laboratory unit.

Data: $\rho_c = 2{,}710$ kg/m^3; medium resistance, $R_m = 6.462 \times 10^{10}$/m; filtrate viscosity, $\mu = 2.78 \times 10^{-7}$ N-h/m^2; $c_F = 10.037$ kg/m^3; $\alpha = 1.069 \times 10^{11}$ m/kg.

Solution

The pump characteristic curve, Figure 19.25, shows the pump discharge pressure, as a function of volumetric flow rate through the pump, Q. Assume the flow rate is the filtrate, where $Q = dV/dt$. The time required for a volume of filtrate, V, is

$$t = \int_0^V \frac{dV}{Q} \tag{1}$$

which can be numerically integrated as follows. Assume the ΔP across the filter medium and cake in bar = the discharge pressure of the pump in barg. Rearrange the Darcy equation, (19-22), so that V is a function of Q:

$$Q = A_c \Delta P / \mu [R_m + a\mu c_F (V/A_c)] \tag{2}$$

Solving for V,

$$V = \frac{A_c}{\mu \alpha c_F} \left[\frac{A_c \Delta P}{Q} - \mu R_m \right] \tag{3}$$

Thus, using SI units,

$$V = \frac{50}{2.78 \times 10^{-7} (1.069 \times 10^{11})(10.037)}$$
$$\times \left[\frac{50(\Delta P)}{Q} - 2.78 \times 10^{-7} (6.462 \times 10^{10}) \right] \tag{4}$$
$$= 0.00838[\Delta P/Q - 359]$$

Using the pump characteristic curve of Figure 19.25, tabulate Q, ΔP, V from (4), and $1/Q$ starting from $Q = 45$ m^3/h and marching down in increments of 5 m^3/h. Then plot $1/Q$ versus V until it just exceeds 50 m^3/h. From (1), by graphical integration, the area under the curve from $V = 0$ to 50 m^3/h, as shown in Figure 19.26, is equal to the filtration time.

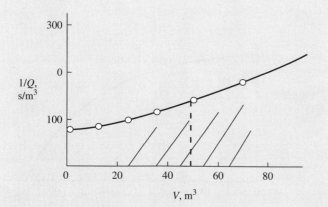

Figure 19.26 Graphical integration for Example 19.9.

Q, m^3/h	$\Delta P \times 10^{-5}$, N/m^2	V, m^3	$1/Q$, s/m^3
45	0.2	0.72	80
40	0.75	12.71	90
35	1.15	24.53	103
30	1.4	36.10	120
25	1.6	50.62	144
20	1.75	70.31	180
15	1.8	97.55	240

Graphical integration gives 1.5 h.

§19.5   CENTRIFUGE DEVICES FOR SOLID–LIQUID SEPARATIONS

Centrifuge devices can greatly increase the rate of sedimentation or filtration, particularly when particles are very small (<10 mm), the liquid is very viscous, and/or the density difference between particles and liquid is very small. They are also favored over cake-filtration devices when the liquid is the main product. Two general types are: (1) sedimentation centrifuges and (2) filtering centrifuges. Three common production centrifuges, shown schematically in Figure 19.27, are (a) the tubular-bowl centrifuge, (b) the disc-stack centrifuge with or without a nozzle discharge, and (c) the basket-filtering centrifuge.

The tubular-bowl centrifuge, whose bowl is suspended from an upper bearing and motor drive through a flexible-drive spindle, is widely used in the biochemical, food, and pharmaceutical industries, particularly for the separation of cells and viruses from broths. The slurry feed is introduced from the bottom through an orifice, followed by a distributor and baffle to accelerate the feed to the very high circumferential speeds. The clarified liquid (centrate) exits from the top by overflowing a ring weir. Solids sediment moves to the bowl wall. When its buildup begins to reduce the clarity of the centrate, the process is stopped and solids are removed. Bowl diameters, D, range from 1.75–5 inches. Corresponding rotation rates, $\omega/2\pi$, vary from 50,000–15,000 rpm, resulting in centrifugal forces of up to 62,100 times the gravitational force, as computed from $\omega^2 D/2g$. Typically, the solids are less than 1% of the feed, and the centrifuge can handle no more than 10 pounds of solids. Liquid throughputs are as low as 0.05 gpm and as high as 20 gpm.

The disc-stack centrifuge is also vertically mounted. The feed enters at the top near the vertical axis, flows to the bottom, and is then accelerated by a radial vane assembly. It then passes through a stack of 50–150 closely spaced (0.4–3 mm) conical disks at an angle with the vertical of 35–50°. Solids settle against the underside of the disks, from where they move to the bowl wall. Liquid flows upward and exits through overflow ports. If the fraction of solids in the feed is small, they remain in the centrifuge until the process is stopped so that they can be manually removed. For larger feed concentrations, the centrifuge can be fitted with an outer

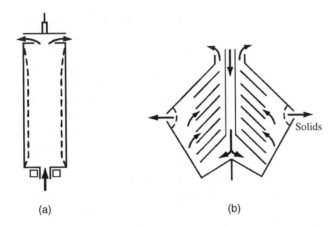

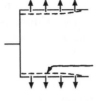

Figure 19.27 Common solid–liquid centrifuges: (a) tubular-bowl, (b) disc-stack, (c) basket-filtering.

housing for periodic solids removal or with nozzles for continuous removal. For the latter type, bowl diameters range from 10–30 inches with corresponding liquid throughputs of 10–400 gpm and bowl speeds of 10,000–3,300 rpm. Corresponding solids throughputs are 0.1–11 tons/h.

When the solids are the main product and high product purity, low cake moisture content, and/or high solids recovery is desired, the basket-filtering centrifuge, operating batchwise in a cyclic operation that can include washing, may be preferred. These devices have a large solids-holding capacity, but are not suitable for soft biological solids. As discussed in great detail in [11], many designs are available, including both vertical (top or bottom unloading) and horizontal units.

Centrifugal separators are widely used in bioseparations where density differences between particles and fluids are so small that gravity settlers are not effective. A typical application is separation of microorganisms from fermentation broths. A common device is the tubular-bowl centrifuge, where the solids move to the wall and form a sludge. Key variables for this centrifuge, shown in Figure 19.28, are rotational speed, ω in radians per unit time; R_1, the distance from the tube center to the surface of the sludge; and R_2, the

distance from the tube center to the inner wall of the tube. The feed enters at the bottom, passing up a length, L, and exiting at the top. Large-scale centrifuges of this type are sized from laboratory data using a sigma scale-up theory, developed next.

For unhindered settling by centrifugal force that obeys Stokes' law, the terminal radial velocity is given by a modification of (19-7), where gravitational acceleration, g, is replaced by centrifugal acceleration, $R\omega^2$:

$$u_t = \frac{dR}{dt} = \frac{R\omega^2 d_p^2 \left(\rho_p - \rho_f \right)}{18\mu} \tag{19-31}$$

The velocity of the liquid in the upward direction, z, is

$$\frac{dz}{dt} = \frac{Q}{A} = \frac{Q}{\pi \left(R_0^2 - R_1^2 \right)} \tag{19-32}$$

Dividing (19-31) by (19-32) gives an equation for the particle trajectory, dR/dz. Integrating it between $R = R_1$ to R_0 and $z = 0$ to L, and multiplying and dividing by g, gives

$$Q = \left[\frac{gd_p^2 \left(\rho_p - \rho_f \right)}{18\mu} \right] \left[\frac{\pi L \omega^2 \left(R_0^2 - R_1^2 \right)}{g \ln \left(\dfrac{R_0}{R_1} \right)} \right] = (u_t)_g \sum \tag{19-33}$$

where $(u_t)_g$ is given by (19-7), Stokes' law for gravity settling, and $\sum_A$ with units of length^2 is a *centrifuge sigma factor* for centrifuge A. For scale-up from a small laboratory centrifuge (A) to a large production centrifuge (B), with different dimensions and for operation at a different rotation rate, assuming that $(u_t)_g$ remains the same, application of (19-33) gives

$$Q_B = Q_A \left[\frac{\sum_B}{\sum_A} \right] \tag{19-34}$$

The assumptions in (19-34) and sigma theory are: (1) The particles are evenly distributed in the continuous liquid and the concentrations are low, so settling is not hindered. (2) Streamline flow at a Reynolds number below 0.2, with the liquid rotating at the same velocity as the bowl. (3) No re-entrainment, displacement of the flow pattern by the deposited material, or nonuniform liquid feed.

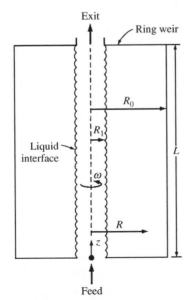

Figure 19.28 Tubular-bowl centrifuge.

Despite these limitations, the theory has been shown to work well for scale-up calculations for centrifuges of the same type [1, 20].

Use of the sigma scale-up theory for a tubular-bowl centrifuge is illustrated in the following example, where the capacity of a laboratory centrifuge is estimated and compared to the measured value. In Exercise 19.19, a plant centrifuge is selected. Application of the sigma scale-up theory to a disc-stack centrifuge is considered in Exercise 19.20.

EXAMPLE 19.10 Feed Capacity of a Tubular-Bowl Centrifuge.

A laboratory tubular-bowl centrifuge has the following dimensions, with respect to Figure 19.28, and operating conditions: bowl speed 800 rps, $R_0 = 0.875$ inch, $R_1 = 0.65$ in., and bowl length $= L = 4.5$ inches. When used to remove *E. coli* cells from the following fermentation broth, a satisfactory volumetric feed capacity of the centrifuge, Q, of 0.11 gpm is achieved.

Broth: $\rho_f = 1.01$ g/cm^3 and $\mu = 1.02 \times 10^{-3}$ kg/m-s

E. coli: smallest diameter $= d_{p_{min}} = 0.7$ μm and $\rho_p = 1.04$ g/cm^3

Assuming the applicability of Stokes' law, estimate the feed capacity of the centrifuge.

Solution

Compute the sigma factor for the laboratory centrifuge from (19-33) using the given dimensions and operating conditions. The rotation rate, ω, in radians/s $= 2(3.14)(800) = 5,030$ s^{-1}. Using AE units,

$$\Sigma = \left[\frac{\pi L \omega^2 \left(R_0^2 - R_1^2 \right)}{g \ln \left(\dfrac{R_0}{R_1} \right)} \right]$$

$$= \left[\frac{3.14 \left(\dfrac{4.5}{12} \right) 5030^2 \left(\left(\dfrac{0.875}{12} \right)^2 - \left(\dfrac{0.65}{12} \right)^2 \right)}{32.2 \ln \left(\dfrac{0.875}{0.65} \right)} \right] = 7,400 \, \text{ft}^2$$

From (19-7), using CGS units with $d_p = 0.7 \times 10^{-4}$ cm and $\mu = 1.02 \times 10^{-2}$ g/cm-s,

$$u_t = \frac{981 \left(0.7 \times 10^{-4} \right)^2 (1.04 - 1.01)}{18 \left(1.02 \times 10^{-2} \right)} = 7.85 \times 10^{-7} \, \text{cm/s}$$

From (19-33), using SI units with $\Sigma = 690$ m^2 and $u_t = 7.85 \times 10^{-9}$ m/s, $Q = 7.85 \times 10^{-9} (690) = 5.42 \times 10^{-6}$ m^3/s $= 0.0860$ gpm, which is 78% of the measured rate.

§19.6 WASH CYCLES

Many filtrations and centrifugations are followed by a wash cycle whose purpose is to recover solutes, filter aids and additives, and otherwise cleanse the cake, if the cake is the product. Or the purpose may be to wash out the retained liquid in the cake and add it to the filtrate, if the filtrate is the product. The symbol S will be used to denote the instantaneous concentration in the effluent washed out of the cake, based

usually on the solutes. The symbol S_o will represent the initial concentration of solubles in the cake before washing.

When wash liquid enters a homogeneous cake, it first displaces solubles and liquid having the same composition as the initial filtrate retained in the cake after filtration. If the wash cycle is "ideal," the minimum volume of wash water required is equal to the exact volume of liquid lodged in the voids of the cake. The wash liquid simply replaces a volume equal to its own volume. Unfortunately, ideal, plug-flow wash cycles do not exist because some washable material diffuses out of the cake very slowly and, more importantly, the wash liquid may carve channels in the cake, allowing pure wash liquid to break through the cake without displacing initial filtrate. At that point, the instantaneous concentration of exiting wash is different from the average concentration remaining in the cake. Thus, three wash phases exist [1]: (1) Liquid is displaced in the pores by wash liquid. Normally, at least 50% of the material is removed. (2) A mixture of breakthrough wash liquid appears along with the displaced liquid. This phase ends when roughly 70–95% of the washables have been removed from the cake, at which point the wash effluent equals about twice the cake void volume. (3) The displacement has ended; solutes (or liquor) can be removed only by diffusion. At this point, it is probably advantageous to repulp the cake and wash in additional countercurrent or concurrent cascades, as described in §16.2.

To establish a wash cycle, it is preferable to conduct laboratory experiments to measure the instantaneous concentration of solutes in the wash effluent in terms of S/S_o, as a function of the wash ratio W, defined as the volume of wash liquid used/initial volume of filtrate in the cake. A typical plot is shown as Curve 2 in Figure 19.29. Included is Curve 1 for the ideal, plug-flow wash cycle case. Note that S/S_o refers to washable entities, which must be carefully defined and are usually, but not always, just the solubles.

It is useful to convert the experimental instantaneous effluent concentration plot to a plot of R, the fraction of initial solutes still retained in the cake, as a function of W. By a solute material balance, the total amount of solutes minus the

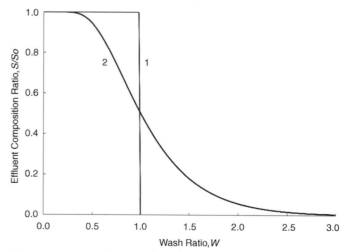

Figure 19.29 Instantaneous effluent concentration as a function of wash ratio.

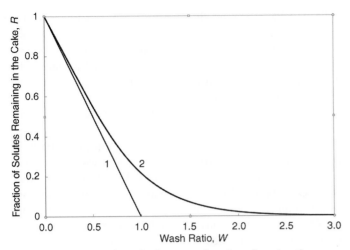

Figure 19.30 Retention of solutes as a function of wash ratio.

W	S/S_o	$\int_0^W \left(\dfrac{S}{S_o}\right) dW$	R
0.00	1.000	0.000	1.000
0.10	1.000	0.100	0.902
0.20	1.000	0.200	0.803
0.30	1.000	0.300	0.705
0.40	1.000	0.400	0.606
0.50	1.000	0.500	0.508
0.60	0.998	0.600	0.409
0.70	0.979	0.699	0.312
0.80	0.897	0.793	0.220
0.90	0.725	0.874	0.140
1.00	0.500	0.935	0.080
1.10	0.295	0.975	0.040
1.20	0.151	0.997	0.018
1.30	0.068	1.008	0.008
1.40	0.028	1.013	0.003
1.50	0.010	1.015	0.001
1.60	0.004	1.015	0.000
1.70	0.001	1.015	0.000
1.80	0.000	1.015	0.000
1.90	0.000	1.015	0.000
2.00	0.000	1.015	0.000

amount of solutes in the effluent equals the solutes still retained in the cake. Thus,

$$\int_0^\infty \left(\frac{S}{S_o}\right) dW - \int_0^W \left(\frac{S}{S_o}\right) dW = R \int_0^\infty \left(\frac{S}{S_o}\right) dW \tag{19-35}$$

Rearranging (19-35), R as a function of W can be computed by integrating, e.g., Curve 2 in Figure 19.29. The result of that integration for Curves 1 and 2 is shown in Figure 19.30.

$$R\{W\} = 1 - \frac{\int_0^W \left(\dfrac{S}{S_o}\right) dW}{\int_0^\infty \left(\dfrac{S}{S_o}\right) dW} \tag{19-36}$$

There are two approaches to designing wash cycles. The first is to establish, by experiment, tables of data or plots like Curve 2 in Figure 19.29, and then manipulate the data to find the wash cycle that satisfies design objectives, which might be reducing the concentration of dissolved solutes to a certain level, displacing a certain amount of mother liquor with wash liquid, or, if the wash takes place on a rotary vacuum filter, determining the amount of wash liquid required to remove a stipulated amount of material in a given amount of time. Example 19.11 illustrates typical data manipulations. An alternative is to devise general mathematical models for washing in much the same way as leaching and drying operations are modeled [7].

EXAMPLE 19.11 Washing a Biomass Cake.

Following fermentation, biomass is separated from the broth, often by centrifugation. Then the wet biomass is washed to recover the occluded broth, if the desired product is in the broth, e.g., if the bioproduct is extracellular. A 1,000-kg biomass sample has been freed of broth and found to weigh 450 kg. (a) What is the average porosity of the biomass if the density of the broth is 1,050 kg/m^3, the biomass density is 1,150 kg/m^3, and the wash liquid is water at 1,000 kg/m^3? (b) The following experimental wash data were obtained for S/S_o as a function of W:

What fractions of broth are recovered for wash ratios of 0.5, 1.0, 1.5, and 2.0? (c) What mass of wash water is required to recover 98% of the broth?

Solution

(a) One thousand kg of wet cake consists of 550 kg broth and 450 kg dry biomass. Broth volume = 550 kg/m^3/1,050 kg/m^3 = 0.524 m^3 and biomass volume = 450 kg/1,150 kg/m^3 = 0.391 m^3. Thus, the total volume is 0.915 m^3 and volume fractions are biomass 0.427 and broth = 0.573. The wash water replaces the broth, so the void volume ϵ_{avg} = 0.573. The volume of water for W = 1 (0.573)(0.915 m^3) = 0.524 m^3, or a wash water mass of 0.524 (1,000) = 524 kg.

(b) From (19-36), values of R can be computed by graphical integration for values of the wash ratio, W. Because the experimental data are closely spaced, reasonable accuracy is achieved using the trapezoidal rule with a spreadsheet. For each interval in W, the arithmetic-average value of S/S_o is determined and multiplied by ΔW. For example, in the interval of W from 0.80 to 0.90, the average value of S/S_o is (0.897 + 0.725)/2 = 0.811, which, multiplied by $\Delta W = 0.1$, is 0.0811. In the above table, the change in the integral for this interval in W is (0.874 − 0.793) = 0.081. Values of the integral and R for a sequence of values of W are included in the above table. Equation (19-36) requires the value of the integral for W = ∞. In the above table, it is seen that by W = 1.60, the integral is no longer changing; so that value of 1.015 can be used.

(c) For recovery of 98% of the broth, the above table shows that a wash ratio of 1.2 is sufficient, or a mass of wash water = 1.2 (524) = 629 kg. In this example, washing is very efficient. Often, it is not.

§19.7 MECHANICAL SEPARATIONS IN BIOTECHNOLOGY

Figure 19.31 is a schematic of the processing steps necessary to separate bioproducts obtained from plants and fermentation of bacteria, molds, and fungi from mammalian cells or by recombinant methods, which include insertion of DNA into appropriate hosts. An introduction to these methods was given in §1.9. When the bioproduct is produced *extracellularly*, the biomass is separated from the broth by vacuum or pressure filtration, centrifugation, or by membranes (microfiltration or ultrafiltration). *Expression*, the de-liquoring of the biomass by compression, may be done if it is economically viable. The filtrate is then subject to an initial purification, which will include precipitation from solution or methods described in previous chapters of this book. The subsequent candidate purification and concentration operations are all described in previous chapters.

If the product resides *intracellularly*, the cells must first be *harvested* (separated from the broth). Then they are subject to *cell disruption*, a *homogenization* process wherein the cell walls are breached so the product can be extracted. Intracellular products include recombinant insulin and growth factors. A number of recombinant products form relatively insoluble inclusion bodies; others, such as porcine insulin, need to be removed from pig pancreas. Different types of

cells can be disrupted differently. Gram-positive bacteria have a cell wall about 0.3 μm thick composed of peptidoglycan, teichoic acid, and polysaccharides, which is followed by a fragile membrane made of proteins and phospholipids. The cell wall of gram-positive bacteria is susceptible to lysis by the enzyme lysozyme, which degrades peptidoglycan.

Gram-negative bacteria are enveloped by multilayer membranes significantly thinner than the walls of gram-positive bacteria, and cannot be lysed. Osmotic shock (simply immersing a cell in distilled water) can be used to recover periplasmic proteins if the cell wall is breached or nonexisting. Yeast and mold cells have walls 0.1–0.2 μm thick, but mammalian cells do not have walls and are relatively fragile. In general, the fragile plasma membranes are readily destabilized by acids, alkali, detergents, or solvents.

Cell debris are removed by centrifugation, microfiltration, or filtration under vacuum or pressure. The broth, which characteristically contains very low concentrations of the target species, then undergoes an initial purification to increase the product concentration, to reduce the cost of subsequent purification steps, and to prevent fouling of ion exchangers, adsorbents, chromatography columns, etc. Precipitation or extraction are possible venues. Both the range of products and the media in which they are produced are enormous, so generalizations are difficult. Special attention must be paid to

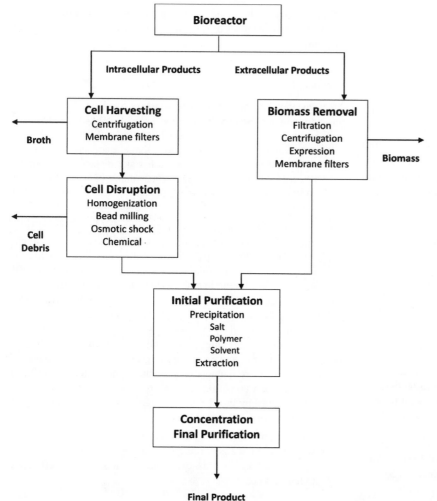

Figure 19.31 Sequencing of bioseparations.

maintaining delicate protein structures. Damage by shear, temperature, pH changes, contamination, and deactivation by endogenous proteases are primary considerations, particularly for products (and byproducts) that are biologically active.

Of the technologies listed in Figure 19.31, only cell disruption has not, as yet, been described in this book, and precipitation has received only brief mention. Interestingly enough, of the six most commonly employed bioproduct separation operations, these two are the most commonly employed. They appear on 80% and 40%, respectively, of bioprocess flowsheets [11].

§19.7.1 Precipitation

The objective of the initial purification is to recover and concentrate the product, which can be in solution or in the precipitate. Precipitation can be induced by temperature, pH adjustment, or addition of salts, solvents, polymers, or biospecific agents. The precipitates are generally amorphous because of occlusions consisting of salts, solvents, or impurities, so accurate phase diagrams of the type seen in Chapter 17 for crystals are not obtainable. Crystallization is a special type of precipitation in which the product is crystalline and is produced slowly under very controlled conditions. Nevertheless, in precipitation, the physical chemistry principles described in §17.11 and Chapter 1 apply.

Temperature

Temperature change, one way of precipitating crystals, is not useful, by itself, for bioproducts. However, cooling, used in conjunction with solvents and salts to maintain system stability, is widespread. Temperatures below 0°C are not uncommon.

pH

The solubility of proteins depends on pH, with the minimum generally at the isoelectric point. In theory, this can be the basis of a protein-separation process; however, the differences in the isoelectric points of proteins are not large enough for commercialization.

Salts

The effectiveness of salts in precipitating proteins follows the Hofmeister series with ammonium sulfate, which is antichaotropic, being the most commonly used. The protein–salt aggregates formed are shear-sensitive, so diaphragm pumps or gravity feed are required. Downstream complications include salt removal and disposal. The salting-out mechanism is complex: it occurs partly because the salt removes water by associating with water molecules, leaving fewer for the proteins, and partly by shielding the electrostatic protein charges that account for protein–protein repulsion. An equation due to Cohn can be used to predict the results of salt-induced protein separations [30]:

$$\ln(S) = B - K_s C_s \qquad (19\text{-}37)$$

where B = the natural log of the solubility of the protein in water, which depends on pH and temperature; K_s = salting-out

constant in m³/kmol; C_s = salt concentration in kmol/m³; and S = protein remaining in solution in kmol/m³. Exercise 19.23 demonstrates use of this equation.

Organic Solvents

The most commonly used organic solvents, acetone and ethanol, can denature protein products, so low temperatures are employed to mitigate protein denaturation. Precautions need to be taken in mixing the solvents with aqueous solutions to prevent regions of high, local solvent concentrations. The precipitates formed with solvent addition are frequently very fine powders, which are difficult to grow by aging, compared to powders formed by salting-out and polyelectrolytes. Nevertheless, organic solvents are widely used for RNA, DNA, and plasma–protein precipitation. They lower the dielectric constant of the solution, thus disrupting the electrolyte stability. The general equation that models solvent-based precipitation of proteins [29] is:

$$\ln(S/S_w) = (A/RT)[(1/e_w) - (1/e)] \qquad (19\text{-}38)$$

where S = solubility of protein in the medium (kmol/m³), S_w = solubility of protein in water (kmol/m³), A = a constant, e = dielectric constant of the medium, and e_w = dielectric constant of water.

EXAMPLE 19.12 Precipitation of Ovalbumin.

The solubility of ovalbumin (OA) in water at 20°C is 42 mg/mL. Thirty percent of the ovalbumin precipitates when 25 mL of ethanol is added to 100 mL of a 40 mg/mL aqueous solution of ovalbumin. How much ethanol would have to be added to the original aqueous solution to precipitate 90% of the ovalbumin at the same temperature? Assume no volume of mixing, and that the dielectric constant is linear with volumetric composition. At 25°C, the dielectric constants for water and ethanol are 78.4 and 24.4, respectively.

Solution

The amount of OA in the aqueous solution = 42(100) = 4,200 mg. For the addition of 25 mL of ethanol, 30% of the OA is precipitated, leaving 70%, or (0.70)(4,200) = 2,940 mg of OA left in a solution of 25 + 100 = 125 mg. Thus, the solubility of OA in this solution is 2,940/125 = 23.5 mg/mL. The dielectric constant of this solution of 25/125, or 20%, ethanol is 24.4 + (1 − 0.20)(78.4 − 24.4) = 67.6. Substituting these values into (19-38) gives

$$\ln(23.5/42) = (A/RT)[(1/78.4) - (1/67.6)]$$

Solving, $A/RT = 285$. Let x = the volume in mL of added ethanol needed to precipitate 90% of the OA. The amount of OA that would be left in the solution = (1 − 0.90)(4,200) = 420 mg. The solubility of OA in the resulting solution would be 420/(100 + x). The dielectric constant of the resulting solution would be 24.4 + [x/(100 + x)](78.4 − 24.4). Substituting these values into (19-38) gives

$$\ln\left(\frac{420}{(100+x)} \bigg/ 42\right) =$$
$$285[(1/78.4) - (1/\{24.4 + [1 - x/(100 + x)](78.4 - 24.4)\})]$$

Solving, the volume of ethanol required = x = 257 mL. This is a very large amount, and thus this method may not be practical to precipitate the OA.

Water-soluble powders of the type used in flocculation, which were introduced in §19.2, can be used to precipitate proteins. In the research stage are affinity precipitants, where a conformal ligand attached to the polymer can couple with a target protein to further enhance aggregation. Here, as with the other precipitation processes, pH is important since proteins exhibit their lowest solubility at the isoelectric point.

§19.7.2 Coagulation, Flocculation, Clarification, and Sedimentation

A precise lexicographic definition of these processes is not possible because they may be proceeding simultaneously and be viewed functionally. *Sedimentation*, in *Perry's Chemical Engineers' Handbook* [11], is defined as "the partial separation or concentration of suspended solid particles from a liquid by gravity settling. This process may be divided into the functional operations of *thickening* and *clarification*. The purpose of thickening is to increase the concentration of suspended solids while that of clarification is to produce a clear effluent." In all aspects but one, clarifiers and thickeners are identical. The one difference is that clarifiers are usually lighter in construction because the average density and viscosity are lower, because the suspended solid concentration is lower. This makes the definition function specific.

Small particles dispersed in a suspension are stabilized by forces due to the surface charges of the particles, which is why they do not agglomerate spontaneously due to Brownian motion. Bacterial cells and most solids suspended in water possess negative charges at neutral pH. The source of the surface charges is the surface groups, which are capable of ionization. A second source of surface charge is the preferential adsorption of ions in the solution.

The physical process of sedimentation is enhanced by *coagulation* and *flocculation*, which may occur sequentially, as in Figure 19.32, but often occur simultaneously, as do precipitation and agglomeration of proteins if a polyelectrolyte is present when the temperature of a saturated solution of proteins is lowered. Flocculation is thus defined as the further *agglomeration* of the small, slowly settling floc formed during coagulation to form a larger aggregated floc particle. The relative sizes of suspended particles encountered in biological systems are shown in Table 19.10.

Table 19.10 Relative Sizes of Suspended Particles

Class	Diameter, mm
Colloids	0.0000001–0.001
Dispersed	0.001–0.1
Coagulated	0.1–1.0
Flocculated	1.0–10

Organic particles below the size visible to the human eye, approximately 0.04 mm (40 microns), generally have settling times that are unreasonably long, and thus coagulation and flocculation, as well as mild agitation, are required to achieve economically sized equipment.

Table 19.11 provides a list of inorganic and organic coagulants as well as some coagulant/flocculant aids, which are used, in part, because they shorten settling times by increasing the density of the suspended microorganisms. These are hydrophyllic and associated with both internal and surface-bound water, so their density is very close to that of the broth.

Inorganic coagulants are water-soluble inorganic acids, bases, or salts that, when dissolved, produce cations or hydrolyzed cations. Increasing the concentration of salt compresses the electrical double layer surrounding a suspended particle and decreases the repulsive interaction between particles, thus destabilizing them. In flocculation there is further agglomeration by an organic polyelectrolyte. One end of a flocculant molecule attaches itself to the surface of one particle at one or more adsorption sites, and the other extended, unadsorbed end of the same molecule bridges and adsorbs to one or more additional particles, thus forming a larger aggregate of floc particles.

The coagulant/flocculant aids in Table 19.11 are insoluble particulates generally used to enhance solid–liquid separations, where slime and glue-like interactions are troublesome. It is known, for example, that broth cultures of actinomycetes, such as *Streptomyces greisius*, are difficult to filter or settle and require the addition of about 2–3% diatomaceous filter aid to form a satisfactory cake. Usually, large quantities of these filter aids are required, and this raises the need of recovery or waste-disposal processes. In general, filtration of biosystems is difficult and centrifugation is preferred.

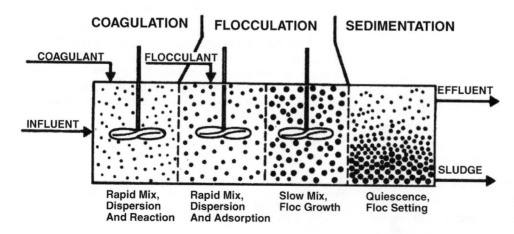

Figure 19.32 Coagulation, flocculation, sedimentation sequence.

Table 19.11 Coagulants/Flocculants

Inorganic Coagulants	Polyelectrolyte Coagulants				Coagulant/Flocculant Aids
	Type	Charge	Examples		
Acids					
Hydrochloric	Anionic	Negative	Polyacrylamide		Activated carbon
Sulfuric			Polyacrylic acid		Activated silica
Bases			Polyacrylate		Bentonite
Calcium hydroxide			Polystyrene sulfonate		Clay
Sodium hydroxide	Cationic	Positive	Polyalkylene polyamine		Metal oxides
Salts			Polyepichlorohydrin		Paper pulp
Aluminum chloride			Polyethylenimine		
Aluminum sulfate			Polyaminoethyl		
Ammonium sulfate			polyacrylamide		
Calcium chloride			Polyvinylbenzyl trimethyl		
Ferrous chloride			ammonium chloride		
Ferric chloride			Polydimethyl diallyl		
Ferric sulfate			ammonium chloride		
Ferrous sulfate	Nonionic	Neutral	Polyacrylamide (1–5%) hydrolyzed		
			Polyethyleneoxide		
	Miscellaneous		Albinic acid, dextran, guar gum, starch		

Destabilization of suspended biological particles is relatively difficult compared to the problems seen in chemical plants and wastewater-treatment facilities. Very frequently, pretreatment is required prior to centrifugation or filtration. The salt content, pH, and temperature of the system affect the surface charges of the suspended solids. The signs, magnitudes, and distribution of the surface charges influence the type and quantity of coagulant/flocculant used. Negatively charged moieties are flocculated by cationic flocculants. Negatively charged solids may also be coagulated by inorganic ions and flocculated by anionic flocculants. "Whatever it takes" is an appropriate cliché. Natural systems are complex and laboratory studies are a necessity.

§19.7.3 Cell Disruption

If the target protein resides intracellularly, cell walls need to be breached, and the microorganism homogenized prior to extracting the product. Bacteria, molds, and yeast are only about 1 μm in diameter. Proteins may be present as soluble or folded, or as insoluble, misfolded *inclusion products*, where all cysteine amino acid residues are fully reduced, making homogenization a considerable challenge. Inclusion aggregates need to be recovered (usually by centrifugation because they are denser than the cell debris), dissolved, and then re-suspended, and the proteins folded to restore biological activity. The substrate is shear-, temperature-, and pH-sensitive, and mild conditions and buffering are mandated.

The two prevalent approaches to cell disruption are mechanical and chemical. The chemical methods are costly, and removal of additives is always an issue. Enzymes such as lysozyme, which is effective for lysing gram-positive bacteria, is less effective for gram-negative bacteria. Other options are detergents like Triton™ X-100, the chaotropes urea and guanidine hydrochloride, or the chelating agent ethylenediaminetetraacetic acid (EDTA), which extracts divalent ions from the cell wall. All of these chemical methods are expensive, and are accompanied by protein denaturation. Other laboratory methods include heat-shocking, freeze-drying, and changes in osmotic pressure induced by adding salts or immersing cells in distilled water. Osmotic shock is effective for mammalian cells, or gram-negative bacterial or fungal cells, whose walls have been weakened by methods such as immersion in an isotonic fluid.

Mechanical means are more attractive from an industrial standpoint. Ultrasound generators, producing pressure waves at about 20 kcycles/s by piezoelectric, titanium transducers, are effective in small vessels. In larger vessels, field inhomogeneity, heat dissipation, cell fragmentation, and enzyme denaturation make the method less suitable for large-scale application.

Industrial-sized equipment includes ball mills wherein cells are mixed with as many as 80%, by volume, of 20–50 mesh beads and then passed through discs rotating at high speed. The cell walls are broken by shear and impact. A unit of this type, called the Dyno-Mill, is capable of processing more than 4,000 lb/h. Even higher capacities, up to 40,000 lb/h, are achieved by pressure-based homogenizers, where cells suspended in an aqueous medium are forced to flow through narrow, adjustable gaps at high speed and pressures up to 50,000 psig. Rotor-stator devices are also on the market.

Large amounts of heat and energy are involved in cell disruption, which is analogous to particle-size reduction, a widely practiced art in chemical manufacture. Prediction of power consumption for these so-called crushing and grinding operations is correlated by Kick's law, which is based on particle-radius-size reduction, or Rittinger's law of grinding, where the energy is correlated to the change in surface area. Ghosh [29] suggests that both of these laws are suited for tissue grinding, but are not useful for modeling cell disruption, and instead suggests an equation of the form

$$C/C_{max} = [1 - \exp(-t/\theta)]^n \qquad (19\text{-}39)$$

where C = concentration of released product (kg/m^3), C_{max} maximum concentration of released material (kg/m^3), t = time (s), θ = time constant (s), and n = number of passes through the device.

EXAMPLE 19.13 Ultrasonification of a Cell Suspension.

A cell suspension has an intracellular antibiotic concentration of 10 mg/mL. If, at the end of 5 minutes of ultrasonification, 20% of the antibiotic is released, how long will it take to release 95% of the antibiotic?

Solution

First compute the time constant using the given data. Using (19-39) with $n = 1$,

$$\frac{0.20(10)}{10} = [1 - \exp(-300/\theta)]^1$$

Solving, $\theta = 1,345$ s. Then, for 95% release,

$$\frac{0.95(10)}{10} = [1 - \exp(-t/1345)]^1$$

Solving, $t = 4,028$ s $= 67$ minutes.

Product quality and yield are key considerations for all cell-disruption schemes. If there are six processing steps in the product-recovery and purification scenario, and the yields in each step are 90%, then 47% of the valuable product is lost. In cell disruption, losses from the release of proteases can lead to enzymatic degradation, and loss of product from heat or oxygen degradation needs to be avoided. Therefore, reducing and chelating agents as well as buffers are often used, particularly if residence times are long. Some laboratory devices disrupt the cells by freezing, or apply mechanical shear to partially frozen paste to avoid loss of biological activity.

SUMMARY

1. Particle size can be used as a criterion for an initial appraisal of what type of mechanical separation device is suitable.

2. Except for devices employing force fields other than gravity, equations for particle settling velocities are the basis for many of the mathematical models used to design industrial separators.

3. The major settling laws, including Newton's and Stokes', can be modified to include centrifugal forces and hindered particle settling. Equations used to design centrifuges, cyclones, and filtering devices generally include particle terminal velocity as a variable.

4. Particle-capture devices can be designed on the basis of efficiency and pressure-drop data provided by manufacturers. Many standards and test procedures are regulated by trade organizations and government agencies.

5. Pressure filtration using plate-and-frame, leaf, or rotary vacuum drum filtration devices is suitable for separating solids approximately 10–50 μm in size and in concentrations of about 1–35 vol% from liquids. At much higher concentrations and particle sizes, settling devices may be more economical; at lower concentrations, particle-size sieving should be considered.

6. The model used for design of plate-and-frame or vacuum solid–liquid filtration devices assumes the filtration rate is proportional to the pressure gradient and inversely proportional to filtrate viscosity. The model can be used for constant-pressure and variable- or constant-rate filtration, depending on pump characteristic curves.

7. An alternative model for filtration is based on the Hagen–Poiseuille formulation for pressure drop through a packed bed. Here, void fractions and cake flow paths must be identified.

8. The Ruth equation, used to model solid–liquid filtration, has three constants and predicts a straight-line relationship between filtrate volume and time, divided by filtrate volume for a constant-pressure filtration.

9. Two of the three constants in the Ruth equation can be eliminated if the filter cake is incompressible and if the filter medium pressure drop is insignificant compared to the cake pressure drop.

10. The diverse and unpredictable nature of filter-cake compressibility has hindered development of generalized correlations and mathematical models.

11. Centrifuge devices can greatly increase sedimentation and filtering rates. A commonly used design scale-up method is the sigma method.

12. Squeezing liquid out of a filter cake is called expression.

13. Washing is done to remove undesirable solutes or to recover occluded product.

14. Precipitation can be accomplished by addition of salts, solvents, polymers, and surface-active agents, aided by temperature and pH changes.

15. When the bioproduct of a fermentation resides intracellularly, the cell must be disrupted so that the bioproduct can be recovered.

REFERENCES

1. Tiller, F.M., *Theory and Practice of Solid-Liquid Separation*, Chemical Engineering Department, University of Houston (1978).

2. Shuler, M.L., and F. Kargi, *Bioprocess Engineering*, Prentice Hall PTR, Upper Saddle River, NJ (2002).

3. Souders, M., and G.G. Brown, *Ind. Eng. Chem.*, **26**(1), 96 (1934).

4. Sutherland, K., *Filters and Filtration Handbook*, 5th ed. Buttersworth-Heinemann, Burlington, MA (2008).

5. Towler, G., and R. Sinott, *Chemical Engineering Design*, Elsevier, Burlington, MA (2008).

6. Nonhebel, G., *Processes for Air Pollution Control*, Butterworth & Co., Cleveland, OH (1972).

7. Wakeman, R.J., and E.S. Tarleton, *Filtration*, Elsevier Science, New York (1999).

8. Foust, A.S., L.A. Wenzel, C.W. Clump, L. Maus, and L.B. Anderson, *Principles of Unit Operations*, J. Wiley & Sons, New York (1960).

9. Amistco Corporation, Alvin, Texas.

10. Stairmand, C.J., *Trans. Inst. Chem. Eng.*, **29**, 356 (1951).

11. *Perry's Chemical Engineers' Handbook*, 8th ed., D.W. Green and R.H. Perry, Eds., McGraw-Hill, New York (2008)

12. Coker, A.K., Chapter 6, "Mechanical Separations," in *Ludwig's Applied Process Design for Chemical and Petroleum Plants*, 4th ed. Vol. 1, Elsevier Publishing, New York (2007).

13. Einstein, A., *Ann Physik*, **17**(4), 549 (1905).

14. Brink, J., *Can. J. Chem. Eng.*, **41**, 134 (1963).

15. McCabe, W.L., J.C. Smith, and P. Harriott, *Unit Operations of Chemical Engineering*, 4th ed. McGraw-Hill Book Co., New York (1985).

16. Carpenter, C.R., *Chem. Eng.*, **90**(23) 227–231 (1983).

17. Kula, M.R., K.H. Kroner, and H. Hustedt, *Advances in Biochemical Engineering*, **24**, 73 (1984).

18. Schweitzer, P.A., *Handbook of Separation Techniques for Chemical Engineers*, McGraw-Hill Book Co., New York (1979).

19. Peters, S.M., K.D. Timmerhaus, and R.E. West, *Plant Design and Economics for Chemical Engineers*, 5th ed. McGraw-Hill, New York (2003).

20. Ruth, B.F., G.H. Montillion, and R.E. Montonna, *Ind. Eng. Chem.*, **25**, 76, 153 (1933).

21. Tiller, F.M., *Chem. Eng.*, **73**, (13) 151 (1966).

22. Silla, H., *Chemical Process Engineering*, Marcel Dekker Inc., New York (2003).

23. Svarovsky, L., *Solid-Liquid Filtration*, 3rd ed. Butterworths, London (1990).

24. Chopey, N., *Handbook of Chemical Engineering Calculations*, 3rd ed. McGraw-Hill Book Co., New York (2003).

25. Aiba, S., A.E. Humphrey, and N.F. Mills, *Biochemical Engineering*, Academic Press, New York (1965).

26. Blasewitz, A.G., and B.F. Judson, "Filtration of Radioactive Aerosols by Glass Fibers," *Chem. Eng. Progress*, **51**(1), 6 (1955).

27. Stairmand, C.J., *Trans. Inst. of Chem. Engrs.*, **28**, 131 (1950).

28. Walas, S.M., *Chemical Process Equipment*, Butterworths, Boston (1988).

29. Ghosh, R., *Principles of Bioseparations Engineering*, World Scientific Publishing Co., Hackensack, NJ (2006).

STUDY QUESTIONS

19.1. Why is particle size the main parameter used in selecting a mechanical phase-separation device?

19.2. At the particle settling velocity, what force balances the drag force plus the buoyant force?

19.3. Into what four regions are settling equations for particles divided?

19.4. What form of the Souders–Brown equation is used to correlate empirical settling data?

19.5. What criteria have been developed for deciding which settling equation is applicable for a given particle diameter?

19.6. How are settling velocity equations modified to take into account particle–particle collisions and particle-shape differences?

19.7. What empirical equations, with constants obtained from experimental data, are frequently used to design many particle-fluid separation devices?

19.8. Why do governmental regulatory agencies and trade organizations set many design and performance specifications for particle emissions?

19.9. Why is centrifugal force frequently applied to speed up and facilitate particle-fluid separation?

19.10. Why have theoretical analyses that treat voids in filter cakes as flow channels not been applied industrially?

19.11. In a filtration cycle, why does constant-pressure filtration usually occur near the end of the cycle and constant-rate filtration at the beginning?

19.12. For what particle-size and particle-concentration ranges are vacuum rotary-drum, leaf, and plate-and-frame filters generally used?

19.13. For what assumptions do filtration data plot as a straight line for V versus t/V coordinates?

19.14. Why are precoat and filter aids generally used in solid–liquid plate-and-frame or vacuum rotary-drum filtrations?

19.15. Why are wash periods followed by expression often part of the filtering cycle?

19.16. What are the assumptions in the Ruth equation for filtration?

19.17. How are empirical constants in filtration models determined?

19.18. Why are pump characteristic curves important in pressure filtration?

19.19. What is the sigma theory and how is it applied?

19.20. How do processes for separating extracellular and intracellular bioproducts differ?

19.21. What steps can be taken to speed coagulation of particulates from bioreactors?

19.22. How are washing cycles determined?

19.23. Name five methods for cell disruption.

EXERCISES

Section 10.3

19.1.   Particle settling velocity.

For a solid, spherical particle of 0.8 mm in diameter and a density of 2,600 kg/m^3 that is immersed in a fluid of density 1,200 kg/m^3 and a viscosity of 1.0 cP, calculate (a) the unhindered terminal velocity in m/s, and (b) the hindered terminal velocity if the volume fraction of such particles is 0.05.

19.2.   Particle velocity prior to terminal velocity.

Consider a spherical particle 4 microns in diameter with a density of 3,000 kg/cm^3 falling through water of 0.001 N-s/m^2 viscosity. At 10^{-5} s, what is the particle velocity? Confirm that the particle is in the Stokes' law range. Assume the particle is at its terminal velocity. *Hint:* An inertial term must be added to (6-40).

19.3.   Settling velocity of microorganisms.

For separations by settling and centrifugation of bacteria, yeast, fungi, and mixed-culture activated sludge from fermentations and sewage systems, values of cell density, equivalent diameter, settling velocity, and volume fraction of cells in suspension must be estimated. The data below and procedures used to obtain these values are described in [23].

Inlet dust particle-size range, μm	d_p, average diameter, μm	Grains/ std ft^3	Weight % of particles
0–20	10	0.0062	2.7
20–30	25	0.0159	6.9
30–40	35	0.0216	9.4
40–50	45	0.0242	10.5
50–60	55	0.0242	10.5
60–70	65	0.0218	9.5
70–80	75	0.0161	7.0
80–94	87	0.0218	9.5
94+	94+	0.0782	34.0

Adapted from [24].

Species	Volume Fraction	u_t, cm/s, $\times 10^{-4}$	ρ_p, g/cm^3	ρ_f, g/cm^3	μ, g/cm-s, $\times 10^{-3}$	gZ, cm/s^2	d_p, microns
Yeast	0.041	0.83	1.09	1.00	11.3	981	5.5
Bacteria	0.31	10.13	1.03	1.00	8.3	981 × 10^3	1.0
Fungi	0.353	1.92	1.003	1.002	9.7	981	143
Sludge	0.279	4.09	1.013	1.00	12.5	981	(71)

In the above table, gZ is the gravitational constant times the centrifugal field strength. The bacteria-settling study was conducted in centrifuges, while the others were done in simple gravity settlers. The particle size was observed microscopically, except for the sludge, which was calculated from the settling velocity using Stokes' law. The authors claim good correlations between microorganism diameters calculated from settling velocities and those determined microscopically. Verify their claim by calculating the diameters from the settling velocities, including a verification of their calculated settling velocity for the sludge. Discuss your results in terms of possible hindered settling, aggregation, and particle-shape properties.

19.4.   Efficiency of a settling chamber.

A settling chamber at a coal-burning installation has the following chamber dimensions, operating conditions, and inlet particle-size distribution, where 1 pound = 7,000 grains. Determine the collection efficiency for each particle-size range, and the overall particle-collection efficiency. Assume the settling velocity, u_t, is one-half of that computed by Stokes' law.

Chamber width = 3.29 m, chamber height = 0.75 m, and chamber length = 4.57 m.

Volumetric gas feed rate = 70.6 ft^3/s at std. conditions of 32°F and 1 atm.

Actual gas temperature and pressure is 446°F and 1 atm.

Actual gas viscosity = 2.60 × 10^{-5} N-s/m^2, and gas density is negligible compared to particle density, which is 2.65 g/cm^3.

19.5.   Diameter of demister pad.

A 6-inch demister pad is to be used to separate liquid entrainment from a gas in a horizontal flash drum. Calculate the demister diameter using the following data:

Vapor flow rate = 465 cfm at 110°F and 50 psia; vapor density 0.30 lb/ft^3; and liquid density = 33 lb/ft^3, $K = 0.35$.

19.6.   Aerosol filtration by glass fiber mats.

An ambient air stream at a superficial velocity of 9.84 ft/min, containing an aerosol of 1-μm-diameter particles at an estimated concentration of 10^4 particles/m^3, is to be filtered through a bed of glass fibers, 19 μm in diameter, with a bed void fraction, ϵ_b, of 0.0033. An empirical equation for the particle-collection efficiency as a function of bed thickness for glass fibers of this type is given by [26]:

$$\text{Fractional efficiency} = 1 - 10^{-0.075L^{0.9}\rho_b u_s^{-0.4}} \qquad (1)$$

where L = bed depth in inches, ρ_b = bed bulk density = 5.15 lb/ft^3, and u_s = superficial air velocity in ft/min. The pressure drop across the filter per unit bed depth is given approximately in terms of a modified drag coefficient as

$$\frac{\Delta P}{L} = C_D \frac{2\rho u_s^2}{\pi g_c d_p} \qquad (2)$$

where ρ = air density = 1.2 × 10^{-3} g/cm^3, d_p = glass fiber diameter 19 × 10^{-4} cm, and the drag coefficient is given by $C_D = 50/N_{Re}$, where $N_{Re} = d_p \rho u_s / \mu$ and μ = air viscosity = 1.8 × 10^{-4} g/cm-s.

Calculate: (a) bed depths in cm for a series of particle-collection efficiencies over a range of 90 to 99.99%; and (b) pressure drop per unit bed depth in kg/m^2-m. Adapted from [25].

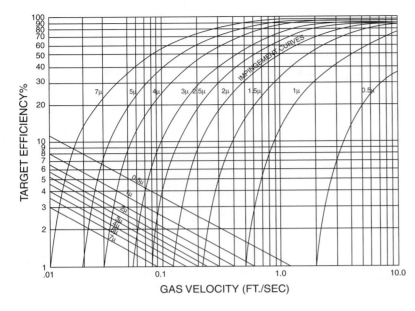

Figure 19.33 Diffusion and impingement target efficiencies for Exercise 19.9. Adapted from Stairmand [27]. In the graph, (TS, symbol for Greek mu), is microns.

19.7. Particle settling.

Aerosol particles 0.001 inch in diameter with specific gravity 0.8 are to be settled from an air stream. If the settling chamber is 2 ft deep, what should the residence time be? At the process conditions, $\rho_f = 0.08$ lb/ft^3, $\rho_p = (0.8)62.4 = 49.92$ lb/ft^3, and μ (air) $= 0.02$ cP.

19.8. Design a vertical decanter.

Use the design method suggested by Towler and Sinott [5] to design a vertical decanter to separate oil droplets suspended in water (the continuous phase). The design method suggests that an oil-droplet particle size, d_p, of 150 μm be assumed; that the height of the decanter be increased by 10% to allow for a dispersion band at the oil/water interface (where the feed enters); that the velocity of the continuous phase, u_c, be less than the settling velocity of the oil droplets, u_s; and that the decanter length be twice the diameter. Check the dimensions of your decanter to make certain that the residence time is reasonable (2–5 minutes), and for the maximum size of water particles that may be carried out by the oil phase.

For the oil, flow rate $= 1,000$ kg/h, density $= 900$ kg/m^3, and viscosity $= 3$ cP.

For the water, flow rate $= 5,000$ kg/h, density $= 1,000$ kg/m^3, and viscosity $= 1$ cP.

19.9. Collection efficiency of a bag filter.

Calculate, using single-fiber collection efficiencies, the overall collection efficiency of a bag filter comprised of glass fibers of mean diameter 10 μm in the form of a cloth 1-mm thick, taking into account both diffusion (Brownian motion) and impingement. The gas contains 1 g/m^3 of dust of specific gravity 2 g/cm^3 and has a velocity of 0.1 ft/s. The particle-size distribution is as follows:

Dust particle size, μm	3–5	2–3	1–2	<1
Mass fraction of total dust	0.45	0.20	0.20	0.15

Under these conditions, impaction does not occur and reentrainment can be neglected. Figure 19.33 gives the single-layer target collection efficiencies, η_t, as a function of gas velocity for diffusion (straight lines) and impingement (curved lines) of dust particles ranging in size from 0.5 to 7 μm for the fiber diameter

and dust density of this exercise. If each particle collision results in capture, the overall % efficiency, η_o, for each particle size is in terms of the fractional single-layer target efficiency of Figure 19.33, given by

$$\eta_o = 100[1 - \exp(-\eta_t S_o)]$$

where S_o is the number of layers or stages of filter. Assume the 1-mm-thick filter cloth has 50 stages.

19.10. Cyclone rating.

A cyclone of diameter D_c of 2 ft, whose dimension ratios are as given in Figure 19.9, is being considered to remove dust from a cement kiln. The gas feed, at inlet velocity u_i of 20 ft^3/s, has 0.5 grain/ft^3 of particles of average diameter d_p of 7 μm and density ρ_p of 175 lb/ft^3, where 1 lb $= 7,000$ grains. At the operating conditions, air viscosity is 1.21×10^{-5} lb/ft-s, and the air density is negligible compared to the particle density. The local air-pollution authority requires an effluent of less than 0.1 grain/ft^3. Can this be achieved with the present cyclone? This can be determined by computing the fractional collection efficiency for the 7 μm particles from

$$E = \left[1 + \left(D_{\text{pcrit}}/d_p\right)^2\right]^{-1} \qquad (1)$$

where $D_{\text{pcrit}} = $ the diameter of the smallest particle that is theoretically separated from the gas stream at a collection efficiency of 50%. From [28],

$$D_{\text{pcrit}} = \left[\frac{9\mu D_c}{4\pi N_t u_i \left(\rho_p - \rho_f\right)}\right]^{\frac{1}{2}} \qquad (2)$$

where N_t, the number of rotations made by the gas stream in the cyclone, is given by

$$N_t = u_i\left[0.1079 - 0.00077u_i + 1.924\left(10^{-6}\right)u_i^2\right] \qquad (3)$$

with u_i in ft/s.

19.11. Scale-up of test leaf filter data.

A test leaf filter is used to determine the filtration rate for an unclarified broth from a fermentor. At a pressure drop of 50 kPa, 150 mL of filtrate are collected in 30 minutes. If the resistance of the filter cloth is negligible, will 300 mL of filtrate be collected in 30 minutes if the pressure drop is doubled? If not, how many mL of filtrate will be collected?

Section 19.4

19.12.   Constant-pressure filtration.

A slurry is being filtered in a plate-and-frame filter press at a constant pressure drop of 10 psi under conditions where the resistance of the medium is negligible. The fluid viscosity is 2.0 lb/hr-ft; 5 lb of dry cake are collected for every ft^3 of filtrate; and the cake is incompressible, with a specific cake resistance, α, of 8.34×10^{10} ft/lb$_m$. What filter area is required to process 50 ft^3 in one hour?

19.13.   Scale-up for a rotary vacuum filter.

A pilot-plant rotary-drum vacuum filter, with 10 ft^2 of filter area in tests involving a new product, delivers 9 ft^3/minute (cfm) of filtrate at a total pressure differential of 15 psi. The drum rotates at 2.5 rpm and 25% of the drum is submerged in the feed reservoir. The pilot-plant data are to be used to design a large plant unit, made by the same manufacturer, to deliver 100 cfm of filtrate. The new unit is also expected to run at a ΔP of 15 psi, but with an rpm of 2.0 and a submergence of 30%. The cake compressibility, filtrate viscosity, as well as the medium and cake resistance in the pilot and plant units are expected to be very much alike. (a) What should be the drum area for the plant unit? (b) Your boss has asked you to recommend desirable and future actions to be taken if production rates have to be increased by 50–400%.

19.14.   Scale-up of filtration data.

A filtration system for a fermentation effluent is being designed on the basis of the data taken using a pilot-plant plate-and-frame filter press. The filter area is 0.3 m^2, $c_F = 2{,}000$ kg/m^3, $\Delta P = 2.5 \times 10^{-4}$ N/m^2, and filtrate $\mu = 3 \times 10^{-3}$ kg/m-s. The liters of filtrate, V, collected during 120 minutes were:

V, L	120	360	700	840	1,100
t, min	5	20	45	75	120

(a) Obtain values for V_o and K in (19-23). (b) Calculate α, the specific cake resistance, in m/kg. (c) If the anticipated plant plate-and-frame filter unit operates at the same pressure, what area is required to handle 5,000 L/h of feed? (d) A rotary-drum vacuum filter is being considered for the same application. The fraction of the drum area submerged in the slurry is 0.35, the pressure drop maintained by the vacuum pump is the same as for the plate-and-frame filter press, and $n = 1$ rpm. What drum area is needed?

19.15.   Filter resistances from test data.

Filtration tests were carried out with a plate-and-frame filter press at 20°C under the following conditions: $\rho_s = 2{,}710$ kg/m^3, filter area 0.37 m^2, filtrate $\mu = 0.001$ N-s/m^2, and $c_F = 10.037$ kg/m^3.

Filtration data from Svarovsky [23]

$\Delta P \times 10^{-5}$, N/m^2	t, s	V, m^3	$(t - t_s)/(V - V_s)$, s/m^3
0.4	447	0.04	
0.5	851	0.07	
0.7	1,262	0.10	
0.8	1,516	0.13	
1.1	1,886	0.16	
1.3	2,167	0.19	
1.3	2,552	0.22	
1.3	2,909	0.25	
1.5	3,381	0.28	
1.5	3,686	0.30	
1.5	4,043	0.32	17,850
1.5	4,398	0.34	17,800
1.5	4,793	0.36	18,450
1.5	5,190	0.38	18,800
1.5	5,652	0.40	19,660
1.5	6,117	0.42	20,260
1.5	6,610	0.44	20,890
1.5	7,100	0.46	21,340
1.5	7,608	0.48	21,790
1.5	8,136	0.50	22,250
1.5	8,680	0.52	22,700
1.5	9,256	0.54	23,210

The filter pump was controlled manually until the pressure became constant at 150,000 N/m^2, which occurred at approximately $V = 0.3$ m^3, $t = 3{,}686$ s. Make a plot of the data, t/V versus V, and determine the specific cake resistance, α, and the medium resistance, R_m.

19.16.   Filtration at constant rate, followed by constant pressure.

A slurry is passed through a filter cloth 0.02 m^2 in area at a constant rate, with 4×10^{-5} m^3/s of filtrate being collected. After 100 s, the pressure is 4×10^4 N/m^2; after 500 s, it is 1.2×10^5 N/m^2.

The same filter-cloth material will be used in a plate-and-frame filter press having an area of 0.5 m^2, which can hold a cake of thickness, L, of 0.04 m, and which will process the same slurry. Assume the filtration will be at a constant rate, and then at a constant pressure after the pressure reaches 8×10^4 N/m^2. If the volume of cake per volume of filtrate is 0.02, calculate the time required to fill the filter frame.

19.17.   Area of a rotary-drum vacuum filter.

Determine the surface area of a rotary-drum filter that processes 20 m^3/hr of a calcium carbonate slurry at a pressure drop of 0.679 bar with 37.5% submergence, and a drum rotation rate of 0.2 rpm. Pertinent data are: $T = 20°C$, mass-fraction carbonate in the feed 0.15, mass fraction of water in the filter cake $= 0.40$, filtrate density 998.3 kg/m^3, filtrate viscosity $= 1$ cP, and dry carbonate density 2,709 kg/m^3. Assume the media resistance is negligible. Use Table 19.8 for cake compressibility.

19.18.   Constant-rate filtration of compressible talc.

Talc is to be filtered at a constant filtration rate under conditions listed below. The pressure drop is expected to rise from 2.0 psi to 100 psi. Generate a table of pressure versus time at 10, 20, 40, 60, 80, and 100 psi. Assume R_m in (19-27) is negligible, but justify your assumption. Obtain α_{avg} by integrating (19-26), which gives $\alpha_{avg} = (1 - s)\alpha'(\Delta P)^s$ [23]. The conditions are: percent solids in slurry 1%, filtrate density $= 62.4$ lb/ft^3, viscosity $= 1.49$ cP, talc density 167 lb/ft^3, filtration rate $= 0.000362$ ft^3/s-ft^2, cake compressibility factor $= \alpha = 8.66 \times 10^{10}(\Delta P, \text{psi})^{0.506}$ m/kg, and cake porosity $= \epsilon = 0.86(\Delta P, \text{psi})^{-0.045}$.

Section 19.5

19.19.   Sigma factor for a centrifuge.

In Example 19.10, a laboratory test was conducted with a small, tubular-bowl centrifuge on a fermentation broth. At the operating conditions, a sigma value of 7,400 ft^2 was computed, with a measured volumetric flow rate of 0.11 gpm. For the commercial plant that will process the same broth, the largest tubular-bowl centrifuge available has the following characteristics: bowl speed = 15,000

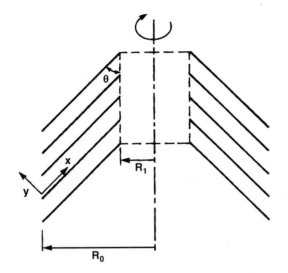

Figure 19.34 Disc-stack centrifuge.

rpm, $R_0 = 5$ cm, $R_1 = 2$ cm, and bowl length $= L = 10$ cm. Using the sigma scale-up theory, calculate how many gpm this unit can process.

19.20. Disc-stack, centrifuge.

A dilute aqueous slurry of viscosity 0.001 N-s/m² and density 1,000 kg/m³ is to be separated from the solid particles, which have a density of 3,000 kg/m³, by centrifugal sedimentation. Pilot-plant runs at a feed rate of 10×10^{-6} m³/s using a disc-stack centrifuge, shown in Figure 19.34, with $n = 20$ disks; a disk angle, θ, of 45° from the vertical; an outer radius $R_0 = 0.075$ m and an internal radius R_1 of 0.025 m; and operating at a bowl speed of 20,000 rpm, give satisfactory results. The same separation is to be carried out in a plant using a disc centrifuge with $R_0 = 0.25$ m and $R_1 = 0.075$ m, 100 disks, a disk angle of 50°, and rotating at 14,000 rpm. The sigma scale-up theory used in Example 19.10 with (19-33) applies, except that for a disc centrifuge [24],

$$\sum = \frac{2\pi n \omega^2 \left(R_0^3 - R_1^3\right)\cot\theta}{3g} \quad (1)$$

where ω is the rotation rate in radians/unit time. What is the expected production rate in m³/s of liquid?

Section 19.6

19.21. Filter-cake washing.

The experimental data given below are the instantaneous concentration of solute S, which is to be washed out of a filter cake as a function of W, the wash ratio (volume of wash liquid/volume of liquid in the cake), and S_o, the initial solute concentration prior to washing. What is the fractional recovery of solute for $W = 1$ and 2?

Section 19.7

19.22. Cell disruption.

The release of protein from a suspension of *Saccharomyces crevisiae* by disruption in an industrial homogenizer was investigated by Hetherington. [*Trans. Instn. Chem. Engrs.*, **49**, 142 (1971)]. The homogenizer pressure was varied, and it was operated on a recycle basis. The results were correlated by the equation

$$\log[R_m/(R_m - R)] = KNP^c$$

where R is the amount of protein released (mg/g yeast), R_m is the maximum amount of protein available for release, K is a temperature-dependent dimensional constant, P is pressure in kgf/cm², N is the number of passes through the homogenizer, and c is a constant. Make a plot of $\log[R_m/(R_m - R)]$ versus N with P as a parameter, using the data below, and determine the value of the constants K and c.

$\log[R_m/(R_m - R)]$	$P(N)$				
1	500(4.7)	400(7.2)			
0.8	500(3.7)	400(5.7)			
0.6	500(2.8)	400(4.2)	360(8.3)		
0.4	500(1.9)	400(2.9)	360(5.5)	270(9.3)	
0.2	500(0.9)	400(1.5)	360(3.0)	270(4.8)	
0.1	500(0.4)	400(0.7)	360(1.5)	270(2.8)	200(7)
0	500(0)	400(0)	360(0)	270(0)	200(0)

19.23. Precipitation of monoclonal antibody with ammonium sulfate.

In a laboratory experiment using 10 L of cell culture, ammonium sulfate powder is added to the solution to precipitate a monoclonal antibody. Eighty percent of the antibody, whose initial concentration was 0.8 mg/mL, with MW of 150,000, precipitates in the 1.0-M solution. If the solubility of the antibody in pure water is 350 kg/m³, what concentration of ammonium sulfate is needed to precipitate 98% of it?

19.24. Rotor-stator mechanical cell disrupter.

A rotor-stator cell disruptor consists of a tapered cavity (stator) and a cone-shaped rotating rotor. The cell suspension is pumped through the very small gap between the rotor and stator, thus being disrupted by the high shear rate. It was found that at a residence time of 10 s, an intracellular antibiotic release was 2.65 mg/mL, and at a residence time of 30 s, it was 4.77 mg/mL. If the residence time is increased to 60 s (by slowing the feed rate), what concentration of antibiotic is predicted? What is the percent extraction?

S/S_o	1.0	1.0	0.9	0.80	0.7	0.55	0.43	0.23	0.15	0.1	0.04	0.02	0.1	0.04	0.02
W	0	0.2	0.4	0.6	0.8	0.9	1.0	1.2	1.4	1.6	1.8	2.0	1.6	1.8	2.0

Answers to Selected Exercises

Chapter 1
1.8 2,750 kPa
1.14(b) 80.8%
1.15(b) 96.08%
1.17(b) 98.8%

Chapter 2
2.1 2,060 kJ/h
2.2 924,000 Btu/h
2.3(e) 3.05%
2.4(b) 4,883,000 Btu/h
2.7(b) 0.00257
2.13 427 kg/m^3

Chapter 3
3.2 991 lb/day
3.4(a) 2,260 h
3.8(a) 1.91×10^{-7} mol/s-cm^2
3.14 0.218 cm
3.17 2.1×10^{-5} cm^2/s
3.20(c) 169,500 h
3.23 54 h
3.24(a) 3.4×10^{-5} cm^2/s
3.27 4.4 cm
3.29(a) 2.54×10^{-4} kmol/s-m^2
3.30 3.73 m
3.31(d) 380 s
3.33 3.44×10^{-5} kmol/s-m^2
3.36(b) 4.08 s
3.38(b) 1.24×10^{-4} mol/s-cm^2-atm

Chapter 4
4.1(c) $4C + 10$
4.3(d) $C + 3$
4.11 4
4.12(b) 196 kJ/kg of feed
4.27 126 psia
4.31 211°F
4.37(c) 65% of nC_8
4.39(a) −61°F
4.46(a) 4.4 kg, **(b)** 504 kg
4.55 2,500 lb/h of crystals
4.57 3,333 kg/h of crystals
4.60 3,800 kg added water
4.65 99.96%
4.67 97.73%
4.71(a) 3,565 kg/h

Chapter 5
5.4(b) 98.8%
5.8(c) 85.5%
5.9(e) 13.1 g
5.10(b) 74.3%, **(e)** 100%
5.11(d) 1,748 kg
5.14(b) 7,169 kg/h
5.15(b) 2,766 kg/h
5.17 165 lbmol/h
5.23 Need 5 more specs
5.29 Need 5 more specs
5.31 $N_D = 19$
5.33(c) $2(N + M) + C + 16$
5.35(c) $2(N + M) + C + 15$

Chapter 6
6.7(a) 1.74
6.9(b) 9–10 stages
6.11 0.005 ppm DCA
6.16(b) 375,000 gpm
6.18 3.56 ft
6.20(b) 0.23 psi/tray
6.21(c) 76%
6.23(d) 6.4%
6.27(b) 2.53 m
6.33 1.61 ft^3
6.35(b) 3.57
6.37(a) 1.70, **(b)** 0.01736,
(c) 7.5, **(d)** 8.7, **(e)** 16 ft

Chapter 7
7.9(a) 90.4%, **(c)** 8
7.13(b) 10 + reboiler
7.14 0.90 and 0.28 for benzene
7.15(b) 43.16 kmol/h
7.21 8 + reboiler
7.23 20 stages + reboiler, feed at 17 from top
7.25 63.49 kmol
7.27 9–10 trays + reboiler
7.29(d) 26 plates + reboiler
7.31(b) 4 + reboiler
7.33(c) 8
7.35 32 + reboiler, feeds at 17 and 27 from the top
7.39 102 + reboiler
7.41(d) 77.4%
7.47(a) 14.7 and 21.2 ft
7.51(c) 42.5 ft, **(d)** 32 ft
7.53(a) 3.8 ft, **(e)** 0.11 psia

Chapter 8
8.11(a) 233 kg/h, **(b)** 5
8.13 2.5 stages
8.15 5 stages

8.17 5 stages
8.23(b) 64 kg/h, **(c)** 142,000 kg/h
8.27(a) 1.66
8.29 139,000 lb/h
8.31(a) 0.32 mm, **(c)** 1,234 ft^2/ft^3
8.33(a) 81 rpm, **(b)** 5.6 Hp,
(c) 0.53 mm, **(d)** 2,080 ft^2/ft^3,
(e) 0.202 ft/h, **(f)** 26.3, **(h)** 96.8%

Chapter 9
9.4 28.3 stages
9.5(a) 7.1, **(b)** 5.3, **(c)** 2.3
9.7 8.4 stages
9.11 272.8 kmol/h reflux rate
9.17(c) 1.175, **(e)** 6 or 7 from top
9.24 65 kmol/h

Chapter 10
10.4 0.3674, 0.2939, 0.1917
10.5 $-16.67, -33.333, -33.333, -33.333, -33.333$
10.9(a) 1.0, 4.0
10.10(a) 0.2994, 0.9030
10.12 $x_1 = -2.62161, x_2 = 3.72971$
10.21 Stage 8 or 9
10.23 Reb. duty = 1,014,000 Btu/h
10.25 Cond. duty = 1,002,000 Btu/h
10.27 Reb. duty = 3,495,000 Btu/h
10.33 14.00
10.41 Reb. duty = 4,470,000 Btu/h
10.43 5 to 6 stages

Chapter 11
11.9 57 stages, 70 mol/s solvent, 7 ft diam. for extractive colm.
11.16 100 stages, 115 kmol/h methanol, $R = 10$

Chapter 12
12.17(a) 15, **(b)** 20 trays
12.18(a) 24, feed to stage 20 from the condenser
12.19 23 m above, 4 m below

Chapter 13
13.1(b) 29.9 wt% distilled
13.3 57.9 moles
13.7 2.14 h
13.9 0, 0.571 isopropanol
13.10 35 lbmol
13.11 0.549 kmol distilled
13.13(a) 9 stages
13.15(a) 7.28 h
13.16(a) 4.60 h
13.23 0.695 and 0.498 for A
13.27(a) 22.8 h, **(b)** 50.6 lbmol
13.30 26, 8, 26, 40 kmol
13.31 26.6, 48.4, 25 kmol

Chapter 14
14.3 48,800 m^2, 281 kmol/h
14.5 2 mm

14.7 41,700 m^3/m^2-day
14.11 Case 1, 195 m^2/stage, permeate = 60 lbmol/h
14.14 545 m^2, 65% recovery
14.16 73 amp/m^2, 25.6 amp
14.18 75% recovery for Des. 2
14.20 548,000 ft^2
14.22 197,000 ft^2 for crossflow
14.24 EtOH permeance = 4.62×10^{-5} kmol/h-m^2-mm Hg
14.30 141 cartridges/stage

Chapter 15
15.1(a) 0.369 cm^3/g, **(b)** 2.5 g/cm^3, 47.6 angstroms
15.3(b) 0.87
15.4 52.1 m^2/g
15.5 5.35 meq/g
15.13(b) 0.15
15.15 4,170 kg dry resin
15.16 0.0741 m/s and 170.1 J/m^2-s-K
15.19 0.0054 cm^2/s
15.22 0.045, 0.24, 2×10^{-5}
15.26(a) 0.66 g/L, **(b)** 12.2 h
15.27(c) 1,845 h
15.29 13.8 days
15.30 279 days
15.31 679 min ideal, 451 min actual, 2.6 ft
15.32 84 cannisters
15.39(a) 21,850 kg/h, **(b)** 10
15.41 660 s, 1,732 s

Chapter 16
16.2(c) 130,600 kg/d water
16.3 90.44% recovery
16.4 10 stages total
16.5(a) 88.54%, **(b)** 76.6%
16.6 2 washing stages
16.12 42.2 minutes
16.15 17.9 h

Chapter 17
17.1(a) 0.696, **(b)** 0.727
17.5 0.579, 0.911, 0.132, 0.272 mm
17.9 1,442 lb/h
17.10 1,320 lb/h
17.11 1,490 lb/h, 44.6 tons/day
17.12 0.512
17.13 257°F
17.17 1.007
17.22 87 m^2
17.23(a) 4.84 tons/h, **(c)** 43 units
17.24(a) 1,777 kg/h, **(b)** 1.22 mm
17.25(c) 4.96 m^3
17.26(f) 3.9×10^{10} crystals/h
17.27(b) 0.0647 micron/s
17.29(c) 0.705 micron
17.34 0.0079
17.36 1,350 tubes, 14.2 h
17.37 101°C, 93%
17.39(a) 44°F, **(e)** 29,100 lb/h
17.41(c) 330 ft^2 each

Chapter 18
18.15(a) 0.0158 lb/lb, **(d)** 1.22%
18.16(a) 127.5°F, **(c)** 3.69%,
(e) 2.262 lb/lb, **(i)** 1.73 psia
18.17(a) 0.048 mol/mol, **(c)** 44.3%,
(d) 41.7%, **(e)** 10.4 ft^3/lb
18.18(a) 0.041 atm, **(c)** 22.2% **(d)** 0.037
18.19 0.21 lb/lb, 112°F, 118°F, 1.908
18.21(a) 151°F, **(d)** 158°F
18.23 2.48 lb/lb
18.25(a) 0.194 lb/lb, **(c)** 0.425 lb/lb
18.27(d) 70.9 h
18.29(b) 0.154 h
18.31 530 minutes
18.33(d) 1.8×10^{-6} cm^2/s
18.36(c) 30,200 Btu/h
18.37(a) 13.6, **(c)** 832

18.39(c) 1,782 lb/h
18.41 18.1% average
18.43 11.5 ft diameter

Chapter 19
19.1(b) 0.0839 m/s
19.3 5.5 μm for yeast
19.5 4.4 ft
19.7 43 s
19.9 88.7% of particles
19.12 42 ft^2
19.13(a) 114 ft^2
19.16 26.9 minutes
19.17 27.2 m^2
19.19 0.29 gpm
19.21 83.7% for $W = 1$
19.23 1.30 M

Index

Physical Constants

Universal (ideal) gas law constant, R
    1.987 cal/mol · K or Btu/lbmol · °F
    8315 J/kmol · K or Pa-m^3/kmol · K
    8.315 kPa · m^3/kmol · K
    0.08315 bar · L/mol · K
    82.06 atm · cm^3/mol · K
    0.08206 atm · L/mol · K
    0.7302 atm · ft^3/lbmol · °R
    10.73 psia · ft^3/lbmol · °R
    1544 ft · lbf/lbmol · °R
    62.36 mmHg · L/mol · K
    21.9 in. Hg · ft^3/lbmol · °R

Atmospheric pressure (sea level)
    101.3 kPa = 101,300 Pa = 1.013 bar
    760 torr = 29.92 in. Hg
    1 atm = 14.696 psia

Avogadro's number
    6.022×10^{23} molecules/mol

Boltzmann constant
    1.381×10^{-23} J/K · molecule

Faraday's constant
    96490 charge/g-equivalent

Gravitational acceleration (sea level)
    9.807 m/s^2 = 32.174 ft/s^2

Joule's constant (mechanical equivalent of heat)
    4.184 J/cal
    778.2 ft · lb$_f$/Btu

Planck's constant
    6.626×10^{-34} J · s/molecule

Speed of light in vacuum
    2.998×10^8 m/s

Stefan-Boltzmann constant
    5.671×10^{-8} W/m^2 · K^4
    0.1712×10^{-8} Btu/h · ft^2 · °R^4